C000008579

1 MONTH OF

FREE

READING

at

www.ForgottenBooks.com

By purchasing this book you are eligible for one month membership to ForgottenBooks.com, giving you unlimited access to our entire collection of over 1,000,000 titles via our web site and mobile apps.

To claim your free month visit:

www.forgottenbooks.com/free952207

* Offer is valid for 45 days from date of purchase. Terms and conditions apply.

ISBN 978-0-260-49853-3
PIBN 10952207

This book is a reproduction of an important historical work. Forgotten Books uses state-of-the-art technology to digitally reconstruct the work, preserving the original format whilst repairing imperfections present in the aged copy. In rare cases, an imperfection in the original, such as a blemish or missing page, may be replicated in our edition. We do, however, repair the vast majority of imperfections successfully; any imperfections that remain are intentionally left to preserve the state of such historical works.

Forgotten Books is a registered trademark of FB &c Ltd.
Copyright © 2018 FB &c Ltd.
FB &c Ltd, Dalton House, 60 Windsor Avenue, London, SW19 2RR.
Company number 08720141. Registered in England and Wales.

For support please visit www.forgottenbooks.com

THE LIBRARY OF THE

JUL 23 1973

UNIVERSITY OF ILLINOIS
AT URBANA-CHAMPAIGN

SELECTED WATER SOURCES BSTRACTS

VOLUME 6, NUMBER 13
JULY 1, 1973

-07801 -- W73-08450

SELECTED WATER RESOURCES ABSTRACTS is published semimonthly for the Water Resources Scientific Information Center (WRSIC) by the National Technical Information Service (NTIS), U.S. Department of Commerce. NTIS was established September 2, 1970, as a new primary operating unit under the Assistant Secretary of Commerce for Science and Technology to improve public access to the many products and services of the Department. Information services for Federal scientific and technical report literature previously provided by the Clearinghouse for Federal Scientific and Technical Information are now provided by NTIS.

SELECTED WATER RESOURCES ABSTRACTS is available to Federal agencies, contractors, or grantees in water resources upon request to: Manager, Water Resources Scientific Information Center, Office of Water Resources Research, U.S. Department of the Interior, Washington, D. C. 20240.

SELECTED WATER RESOURCES ABSTRACTS is also available on subscription from the National Technical Information Service. Annual subscription is $45 (domestic), $56.25 (foreign), single copy price $3. Certain documents abstracted in this journal can be purchased from the NTIS at prices indicated in the entry. Prepayment is required.

SELECTED
WATER RESOURCES ABSTRACTS

A Semimonthly Publication of the Water Resources Scientific Information Center,
Office of Water Resources Research, U.S. Department of the Interior

VOLUME 6, NUMBER 13
JULY 1, 1973

W73-07801 -- W73-08450

ıe Secretary of the U. S. Department of the Interior has determined that the publication of this periodical · necessary in the transaction of the public business required by law of this Department. Use of funds ır printing this periodical has been approved by the Director of the Office of Management and Budget ırough August 31, 1973.

As the Nation's principal conservation agency, the Department of the Interior has basic responsibilities for water, fish, wildlife, mineral, land, park, and recreational resources. Indian and Territorial affairs are other major concerns of America's "Department of Natural Resources."

The Department works to assure the wisest choice in managing all our resources so each will make its full contribution to a better United States—now and in the future.

FOREWORD

Selected Water Resources Abstracts, a semimonthly journal, includes abstracts of current and earlier pertinent monographs, journal articles, reports, and other publication formats. The contents of these documents cover the water-related aspects of the life, physical, and social sciences as well as related engineering and legal aspects of the characteristics, conservation, control, use, or management of water. Each abstract includes a full bibliographical citation and a set of descriptors or identifiers which are listed in the Water Resources Thesaurus. Each abstract entry is classified into ten fields and sixty groups similar to the water resources research categories established by the Committee on Water Resources Research of the Federal Council for Science and Technology.

WRSIC IS NOT PRESENTLY IN A POSITION TO PROVIDE COPIES OF DOCUMENTS ABSTRACTED IN THIS JOURNAL. Sufficient bibliographic information is given to enable readers to order the desired documents from local libraries or other sources.

Selected Water Resources Abstracts is designed to serve the scientific and technical information needs of scientists, engineers, and managers as one of several planned services of the Water Resources Scientific Information Center (WRSIC). The Center was established by the Secretary of the Interior and has been designated by the Federal Council for Science and Technology to serve the water resources community by improving the communication of water-related research results. The Center is pursuing this objective by coordinating and supplementing the existing scientific and technical information activities associated with active research and investigation program in water resources.

To provide WRSIC with input, selected organizations with active water resources research programs are supported as "centers of competence" responsible for selecting, abstracting, and indexing from the current and earlier pertinent literature in specified subject areas.

Additional "centers of competence" have been established in cooperation with the Environmental Protection Agency. A directory of the Centers appears on inside back cover.

Supplementary documentation is being secured from established discipline-oriented abstracting and indexing services. Currently an arrangement is in effect whereby the BioScience Information Service of Biological Abstracts supplies WRSIC with relevant references from the several subject areas of interest to our users. In addition to Biological Abstracts, references are acquired from Bioresearch Index which are without abstracts and therefore also appear abstractless in SWRA. Similar arrangements with other producers of abstracts are contemplated as planned augmentation of the information base.

The input from these Centers, and from the 51 Water Resources Research Institutes administered under the Water Resources Research Act of 1964, as well as input from the grantees and contractors of the Office of Water Resources Research and other Federal water resources agencies with which the

Center has agreements becomes the information base from which this journal is, and other information services will be, derived; these services include bibliographies, specialized indexes, literature searches, and state-of-the-art reviews.

Comments and suggestions concerning the contents and arrangements of this bulletin are welcome.

Water Resources Scientific Information Center
Office of Water Resources Research
U.S. Department of the Interior
Washington, D. C. 20240

CONTENTS

SELECTED WATER RESOURCES ABSTRACTS

01. NATURE OF WATER

1A. Properties

AN EXPERIMENTAL STUDY OF THE STRUCTURE, THERMODYNAMICS AND KINETIC BEHAVIOR OF WATER,
Midwest Research Inst., Kansas City, Mo.
F. T. Greene, J. Beachey, and T. A. Milne.
Available from the National Technical Information Service a PB-214 974, $3.00 in paper copy, $1.45 in microfiche. Office of Saline Water Research and Development Progress Report No. 789, July 1972. 46 p, 26 fig, 3 tab, 9 ref. 14-01-0001-1479.

Descriptors: *Thermodynamics, *Water structure, *Water properties, *Thermodynamic behavior, *Molecular structure, *Hydrogen bonding, Kinetics, Polymers.
Identifiers: Mass spectrometers, Molecular beam technique, Equilibrium, Water dimers, Water trimers.

A combination of mass spectrometric and molecular beam techniques has been applied to the study of the equilibrium thermodynamics and the rates of formation of water clusters. Reliable equilibrium thermochemical data have been obtained for water dimer while preliminary data have been gathered for water trimer and several larger polymers. These results support a hydrogen bond strength in water of 5 kcal per mole. Extensive measurements were also made of the concentrations of several water clusters following the free-jet expansion of water vapor from higher pressures. These data contain unique kinetic information. A kinetic model for the formation of dimers in free-jet expansions has been developed and temperature dependent rate constants for the formation of argon dimer deduced. (OSW)
W73-08188

1B. Aqueous Solutions and Suspensions

THERMOCHEMICAL INVESTIGATION OF DIMETHYLMERCURY IN AQUEOUS AND NONAQUEOUS SOLUTIONS,
Missouri Univ., Rolla. Dept. of Chemistry.
For primary bibliographic entry see Field 05A.
W73-07806

02. WATER CYCLE

2A. General

INTRODUCTION OF TIME VARIANCE TO LINEAR CONCEPTUAL CATCHMENT MODELS,
Institute of Hydrology, Wallingford (England).
A. N. Mandeville, and T. O'Donnell.
Water Resources Research, Vol 9, No 2, p 298-310, April 1973. 7 fig, 7 ref.

Descriptors: *Rainfall-runoff relationships, *Mathematical models, *Unit hydrographs, *Variability, Time series analysis, Runoff forecasting, Mathematical studies.

Time-variant versions of the usual linear conceptual models relating rainfall and runoff based on the linear channel and the linear reservoir are defined. Expressions are obtained for their impulse responses. The convolution integral is used both to study different combinations of the two basic time-variant components and to obtain the response of any time-variant linear system to a given input. The theoretical background of more advanced models is discussed and it is suggested that time variance may be introduced into any linear model founded on these two conceptual components. (Knapp-USGS)
W73-07885

DETERMINATION OF OPTIMAL KERNELS FOR SECOND-ORDER STATIONARY SURFACE RUNOFF SYSTEMS,
Technion - Israel Inst. of Tech., Haifa. Dept. of Civil Engineering.
M. H. Diskin, and A. Boneh.
Water Resources Research, Vol 9, No 2, p 311-325, April 1973. 6 fig, 3 tab, 24 ref.

Descriptors: *Rainfall-runoff relationships, *Mathematical models, Systems analysis, Optimization, Streamflow forecasting, Runoff forecasting.

Discretization of input, output, and kernel functions of a second-order Volterra series representation of the surface runoff system leads to a set of linear equations for the unknown ordinates of the two-kernel functions. The equations may be solved by an iterative descent optimization procedure. The objective function for the optimization procedure is the sum of squared deviations between observed and predicted output ordinates for all storms included in the record. The constraining equations are included by an iterative scheme based on the penalty function method. The algorithm converges to the optimal solution in a finite number of iterations, and its rate of convergence is independent of initial values adopted for the unknowns. A numerical example based on an eight-storm record is included. (Knapp-USGS)
W73-07886

STORM FLOW FROM HARDWOOD-FORESTED AND CLEARED WATERSHEDS IN NEW HAMPSHIRE,
Forest Service (USDA), Durham, N.H. Northeastern Forest Experiment Station.
For primary bibliographic entry see Field 04C.
W73-07889

MEASURE OF THREE-DIMENSIONAL DRAINAGE BASIN FORM,
Cambridge Univ. (England). Dept. of Geography.
M. G. Anderson.
Water Resources Research, Vol 9, No 2, p 378-383, April 1973. 3 fig, 1 tab, 19 ref.

Descriptors: *Terrain analysis, *Maps, *Mapping, *Topography, *Watersheds (Basins), Geomorphology, Topographic mapping, Valleys, Slopes, Small watersheds, River basins, Channel morphology.
Identifiers: *Map analysis.

The form of a small basin can be generated from a minimum of six coefficients derived from map analysis. The coefficients are: A, area; l, length of the basin; G, gradient of the mainstream; and 3, polynomial coefficients. The model reproduces correctly the important characteristics of surface roughness and spatial autocorrelation. (Knapp-USGS)
W73-07892

MODELING INFILTRATION DURING A STEADY RAIN,
Minnesota Univ., St. Paul. Dept. of Agricultural Engineering.
For primary bibliographic entry see Field 02G.
W73-07893

MULTISITE DAILY FLOW GENERATOR,
Department of the Environment, Ottawa (Ontario). Water Management Service.
R. L. Pentland, and D. R. Cuthbert.
Water Resources Research, Vol 9, No 2, p 470-473, April 1973. 2 fig, 7 ref.

Descriptors: *Simulation analysis, *Statistical methods, *Streamflow forecasting, Canada, Regression analysis, Hydrograph analysis, Flood forecasting, Synthetic hydrology.
Identifiers: *Fraser River (Canada).

A daily flow generation model was developed to synthesize correlated flood hydrographs on the main stem and tributaries of a large river system. Statistical parameters on which the synthetic data are based are derived from historical data. The historical data are first subjected to a logarithmic transform, and regression equations are established between the monthly totals and the standard deviations of the flow logarithms for each calendar month of interest. These relationships are used later in the procedure for destandardizing the generated data so that flow variability will be related to the monthly mean to the extent that it is related in nature. Before serial correlation relationships are derived, the historical data are also standardized (by subtracting the long-term mean and dividing by the standard deviation) and normalized by using a Pearson type 3 transform. (Knapp-USGS)
W73-07899

STUDIES ON THE FLUVIAL ENVIRONMENT, ARCTIC COASTAL PLAIN PROVINCE, NORTHERN ALASKA, VOLUMES I AND II,
For primary bibliographic entry see Field 02C.
W73-07906

WATER RESOURCES OF LAND AREAS (FORMIROVANIYE RESURSOV VOD SUSHI).

Izdatel'stvo 'Nauka', Moscow, G. P. Kalinin, editor, 1972. 136 p.

Descriptors: *Hydrology, *Water resources, *Land, Water types, Water levels, Runoff, Streamflow, Surface-groundwater relationships, Water storage, Water balance, Evaporation, Precipitation (Atmospheric), Air temperature, Heat balance, Soil types, Infiltration, Water quality, Model studies, Variability, Maps.
Identifiers: *USSR, Continental hydrology.

This collection, published by the USSR Academy of Sciences' Institute of Water Problems, contains 9 papers devoted to investigations of the relationships between oceanic water and water of continental land masses, surface waters and groundwater, and heat and water balances. The subjects of the individual papers are: (1) variability of annual atmospheric precipitation and air temperature in Europe and North America; (2) variations in secular sea levels; (3) use of space-time characteristics of runoff variability for extrapolation; (4) space-time distribution of surface-groundwater relationships; (5) infiltration of melt water into soil; (6) application of a statistical model of infiltration to water-quality analysis; (7) relation between heat and water balances in European Russia in connection with redistribution of streamflow; (8) water-balance investigations of irrigated soils between periods of irrigation; and (9) minimum discharge in rivers of Iraq. (Josefson-USGS)
W73-07910

WATER TEMPERATURE VARIATIONS IN THE NORTH ATLANTIC IN 1948-1968,
Leningrad Higher School of Marine Engineering (USSR).
A. A. Zverev.
Oceanology, Vol 12, No 2, p 182-186, 1972. 3 fig, 2 tab, 4 ref. Translated from Okeanologiya (USSR), Vol 12, No 2, 1972.

Descriptors: *Oceanography, *Atlantic Ocean, *Water temperature, *Variability, Fluctuations, Meteorology, Equations.
Identifiers: *USSR, Ocean weather stations.

Long-period variations in water temperature anomalies in the North Atlantic were analyzed from weather ship data collected in 1948-68. Between 1948 and 1953, average water temperature for all ocean weather stations in the area increased from minus 0.42 deg C to plus 0.36 deg C, or by 0.78 deg C. Between 1953 and 1963, the water temperature decreased by 0.48 deg C (from plus 0.36 deg C to minus 0.12 deg C), followed by a slight increase in 1964-66. Changes in water temperature reveal long-period variations on which sharp short-period fluctuations are superimposed. The change in sea temperatures in the North Atlantic exhibits an opposite (mirror image) pattern in western and northern regions. (Josefson-USGS)
W73-07914

INSTANTANEOUS UNIT HYDROGRAPHS, PEAK DISCHARGES AND TIME LAGS IN URBANIZING WATERSHEDS,
Purdue Univ., Lafayette, Ind. School of Civil Engineering.
R. A. Rao, and J. W. Delleur.
Research report, (1972). 7 fig, 3 tab, 28 equ, 20 ref. OWRR-B-002-IND (3).

Descriptors: *Urban hydrology, *Watersheds (Basins), *Unit hydrographs, *Runoff, Analytical techniques, Estimating, Time lag, Peak discharge, Floods, Hydrologic data, Reservoirs, Equations, Mathematical models, Regression analysis, Rainfall-runoff relationships.
Identifiers: Physiographic characteristics, Storm characteristics, Fourier transform method, Single linear reservoir method.

Aspects of investigating the effects of urbanization on runoff by using the Instantaneous Unit Hydrograph (IUH), and aspects of analysis of magnitudes and times to peak of annual maximum floods are discussed. The effects of urbanization of a watershed on the runoff has been investigated in the past by using linear conceptual models in which the time lag appears as an important parameter. However, in this approach the effects of noise in the data, of sampling rate, and of errors due to the lack of synchronization between the effective rainfall and runoff on the instantaneous unit hydrograph do not become readily apparent. A case in which the cumulative effect of these factors is predominant is presented as an example of the possible difficulties which might be encountered in the analysis of urban hydrologic data by the unit hydrograph methods. The disadvantages of relating the peak discharge, the time to peak discharge and the time lag to only the physiographic characteristics are discussed. Alternative regression relationships which involve storm characteristics a long with the physiographic characteristics to estimate the peak discharge, time to peak discharge, and time lag are presented. (Bell-Cornell)
W73-07917

WATER-RESOURCES RECONNAISSANCE OF THE OZARK PLATEAUS PROVINCE, NORTHERN ARKANSAS,
Geological Survey, Washington, D.C.
For primary bibliographic entry see Field 07C.
W73-08069

SEPARATION OF THE SURFACE AND SUBSURFACE FLOW FROM SNOWMELT BY THE USE OF RADIOACTIVE TRACERS,
Department of the Environment (Alberta). Water Resources Div.
For primary bibliographic entry see Field 02C.
W73-08146

INFLUENCE OF LATE SEASON PRECIPITATION ON RUNOFF OF THE KINGS RIVER,
Sierra Hydrotechnology, Placerville, Calif.
J. F. Hannaford.
In: Proceedings of the 40th Annual Meeting of the Western Snow Conference, April 18-20, 1972. Phoenix, Ariz: Printed by Colorado State University, Fort Collins, p 67-74, 1972. 9 fig, 3 ref.

Descriptors: *Mathematical models, *Water yield, *Hydrograph analysis, *Snowmelt, *Rainfall-runoff relationships, Base flow, Data collections, Meteorological data, Climatic data, *California, Storms, Precipitation (Atmospheric).
Identifiers: *Kings River (Calif), *Sierra Nevada (Calif).

In the Kings River basin of the southern Sierra Nevadas, late-season, high-elevation storms are important factors in the hydrology of the area. Numerous occurrences producing from a few hundred to perhaps 10,000 acrefeet or more may be expected to occur almost every year. In aggregate they may represent from five to twenty percent of the total runoff in the May-September period. While the supplemental water produced by late-season precipitation cannot be accurately forecast, accurate knowledge of the quantity and timing of such supplemental flow would be valuable to the water manager. Resent studies involving weather modification during the May-September period suggest that perhaps substantial augmentation of runoff from late season storms may be a possibility. (See also W73-08138) (Knapp-USGS)
W73-08151

A PLAN FOR STUDY OF WATER RESOURCES IN THE PLATTE RIVER BASIN, NEBRASKA--WITH SPECIAL EMPHASIS ON THE STREAM-AQUIFER RELATIONS,
Geological Survey, Lincoln, Nebr. Water Resources Div.
C. F. Keech, J. E. Moore, and P. A. Emery.
Geological Survey open-file report, January 1973. 36 p, 3 fig, 57 ref.

Descriptors: *Surface-groundwater relationships, *Mathematical models, *Nebraska, Hydrogeology, Water resources development.
Identifiers: *Platte River (Nebr).

A study is being made of the Platte River basin in Nebraska. The study is a federal and state interagency effort to formulate a comprehensive plan for the conservation, development, and management of the water and related land resources of the Platte River basin. A quantitative description of the operation of the hydrologic system emphasizes the relation of groundwater to surface water. A digital model will simulate the physical character and operation of the stream-aquifer system. Some of the uses of the model are evaluation of effects of proposed conjunctive use projects, effects of changes in water use, and study of future water problems such as water-logged areas, groundwater mining, and streamflow depletion. (Knapp-USGS)
W73-08357

2B. Precipitation

ATMOSPHERIC ENERGETICS AS RELATED TO CYCLOGENESIS OVER THE EASTERN UNITED STATES,
Texas A and M Univ., College Station.
P. W. West.
National Aeronautics and Space Administration Contractor Report CR-2189, January 1973. 103 p, 31 fig, 4 tab, 40 ref. NAS Grant 8-26751.

Descriptors: *Atmospheric physics, *Energy budget, *Equations, *Cyclones, Forecasting, Data collections, Mathematical studies, Kinetics, Cycles, Precipitation (Atmospheric).
Identifiers: *Cyclogenesis, Atmospheric energetics, Eastern United States.

A method is presented for investigating the atmospheric energy budget as related to cyclogenesis. Energy budget equations represent basic physical processes which produce changes in atmospheric energy and provide a means to study the interaction of the cyclone with the larger scales of motion. An extension of previous studies is presented. Computations are carried out over a limited atmospheric volume which encompasses the cyclone, and boundary fluxes of energy that were ignored in most previous studies are evaluated. Two examples of cyclogenesis over the eastern United States were chosen for study. Results indicate that diabatic processes can be a significant factor in the development of a cyclone; the downward transport of kinetic energy from the jet-stream level can be an important source of energy for a developing cyclone; and there is considerable interaction between a cyclone and its environment. (Woodard-USGS)
W73-08083

TRANSPOSITION OF STORMS FOR ESTIMATING FLOOD PROBABILITY DISTRIBUTIONS,
Colorado State Univ., Fort Collins. Dept. of Civil Engineering.
For primary bibliographic entry see Field 02E.
W73-08085

PROCEEDINGS OF THE WESTERN SNOW CONFERENCE,
For primary bibliographic entry see Field 02C.
W73-08138

AIR TEMPERATURE OBSERVATIONS AND FORECASTS--THEIR RELATIONSHIP TO THE PREDICTION OF SPRING SNOWMELT IN THE EAGEL RIVER BASIN, COLORADO,
National Weather Service, Salt Lake City, Utah. River Forecast Center.
For primary bibliographic entry see Field 02C.
W73-08144

THE LABOR DAY STORM OF 1970 IN ARIZONA,
Arizona Univ., Tucson. Dept. of Watershed Management.
D. B. Thorud, and P. F. Ffolliott.
In: Proceedings of the 40th Annual Meeting of the Western Snow Conference, Arpil 18-20, 1972. Phoenix, Ariz: Printed by Colorado State University, Fort Collins, p 37-42, 1972. 2 fig, 9 ref.

Descriptors: *Floods, *Storms, *Arizona, *Tropical cyclones, Meteorology, Climatology, Disasters, Erosion, Storm structure, Weather patterns, Flood forecasting.

The 1970 Labor Day storm caused more loss of life than any other storm in Arizona's recent history. The storm began on September 2nd, when moist air, associated with tropical storm Norma, flowed into Arizona from the Pacific Ocean and the Gulf of California. During the next two days the air mass reached sufficient depth to allow the formation of thunderstorms. A convergent flow of air in the lower atmosphere over southern Arizona on the 4th caused heavy rainfall. This rainfall ended late on the 4th. On the morning of the 5th, a cold front extended from southwestern Utah into southern Nevada, and an associated deep upper trough was located over Nevada and southern California. In advance of the cold front, a surface trough extended from Las Vegas, Nevada, to Palm Springs, California. Strong, southerly winds developed early on the 5th. Orographic rainfall increased sharply over the mountains of central Arizona as the troughs approached. A combination of the advancing trough and normal daytime heating generated lines of thunderstorms by midafternoon. The heavy rainfall brought flooding throughout central and northeastern Arizona, southeastern Utah and southwestern Colorado. (See also W73-08138) (Knapp-USGS)
W73-08145

WATERSHED HYDROMETEOROLOGICAL DATA REQUIRED FOR WEATHER MODIFICATION,
North American Weather Consultants, Santa Barbara, Calif.
For primary bibliographic entry see Field 03B.
W73-08150

THE BOREAL BIOCLIMATES,
Toronto Univ., (Ontario). Dept. of Geography.
F. K. Hare, and J. C. Ritchie.
Geogr Rev. Vol 62, No 3, p 335-365. 1972. Illus.
Identifiers: Albedo, *Bio climates, *Boreal, Climates, Energy, Forests, Tundra, Woodland, *Remote sensing.

The long-established zonal divisions of the boreal forest (forest-tundra, open woodland and closed forest)are examined in the light of new information about energy income and of satellite photographs of the divisions themselves. The North American divisions are found to lie fairly consistently between certain values of mean annual net radiation, but these values are much influenced by the vegetation structure because of the importance of snow cover in determining spring albedo. The phytomass and net production data from the Bazilevish-Rodin synthesis are examined in the light of energy income. Efficiency of energy conversion (all-wave) varies from below 0.5% in the tundra to 1.2% or above in the closed forest.--Copyright 1973, Biological Abstracts, Inc.
W73-08437

2C. Snow, Ice, and Frost

HEAT AND MASS TRANSFER OF FREEZING WATER-SOIL SYSTEM,
British Columbia Univ., Vancouver. Dept. of Chemical Engineering.
G. F. Kennedy, and J. Lielmezs.
Water Resources Research, Vol 9, No 2, p 395-400, April 1973. 7 fig, 9 ref.

Descriptors: *Freezing, *Frozen soils, *Frost, *Mass transfer, *Heat transfer, Water chemistry, Soil water, Soil water movement, Flow, Convection, Diffusion, Thermodynamics.

Equations are given to describe fluxes of heat and mass, and transfer potential distributions for temperature and moisture content for a freezing soil system. Within a given set of physical and mathematical restrictions the derived equations will account for transient conditions, multiphase systems, phase changes, and multiextensive property transfers. The development of these transfer equations involves defining the system and stating the physical conditions, developing conservation equations, developing flux equations subject to the proper use of the principles of thermodynamics of irreversible processes (linearized Onsager flux equations), and combining conservation and flux equations to yield the transfer equation. (Knapp-USGS)
W73-07894

SPRING DISCHARGE OF AN ARCTIC RIVER DETERMINED FROM SALINITY MEASUREMENTS BENEATH SEA ICE,
Louisiana State Univ., Baton Rouge. Coastal Studies Inst.
H. J. Walker.
Water Resources Research, Vol 9, No 2, p 474-480, April 1973. 4 fig, 3 tab, 15 ref. ONR Contract N00014-69-A-0211-0003.

Descriptors: *Discharge measurement, *Streamflow, *Alaska, *Ice breakup, *Sea ice, Mixing, Salinity, Stream gages, Discharge (Water), Permafrost, Arctic.
Identifiers: *Colville River (Alaska).

Salinity measurements under sea ice seaward of the Colville delta in Alaska made possible the calculation of the river's discharge during breakup in 1971. Between May 27 and June 15 the discharge was 5.70 billion cu m, which is about 58% of the total for 1971. The entire drainage basin of the Colville River is confined to the zone of continuous permafrost. In winter both surface water and groundwater freeze, and the river ceases to flow. This cessation of flow allows seawater to occupy completely the delta front and to replace river water in the lower reaches of the river. After flushing the saltwater from the river channels, the floodwater intrudes between sea ice and seawater as it flows into the ocean. (Knapp-USGS)
W73-07900

GLAZE AND RIME (GOLOLED I IZMOROZ'),
Ye. P. Dranevich.
Gidrometeoizdat, Leningrad, 1971. 228 p.

Descriptors: *Sleet, *Rime, *Meteorology, *Atmospheric physics, *Ice, Freezing, Dew, Humidity, Water vapor, Wind velocity, Snow, Rain, Fog, Physical properties, Topography, Maps, Synoptic analysis, Weather data, Isotherms, Forecasting.
Identifiers: *USSR, *Glaze (Ice), Hoarfrost.

Meteorological and synoptic conditions of formation and space-time distribution of glaze (sleet) and rime were investigated in the Leningrad, Novgorod, and Pskov Oblasts in Northwest USSR. The effects of topography on frequency, duration, and extent of glaze and rime phenomena are examined, and a detailed analysis is made of cases of especially dangerous ice deposits. Techniques for forecasting ice accretion on ground surfaces and formation of glaze zones associated with moving warm fronts are accompanied by a map showing the areal distribution of glaze and freezing rain. (Josefson-USGS)
W73-07902

STUDIES ON THE FLUVIAL ENVIRONMENT, ARCTIC COASTAL PLAIN PROVINCE, NORTHERN ALASKA, VOLUMES I AND II,
R. I. Lewellen.
Available from NTIS, Springfield, Va 22151 Vol I-AD-74 9150 Price $6.00 printed copy; Vol II-AD-74 9151 Price $3.00 printed copy, $1.45 microfiche each. Robert I. Lewellen, Littleton, Colorado, 1972. 282 p, 208 fig, 90 tab, 245 ref, 6 append.

Descriptors: *Geomorphology, *Coastal plains, *Alaska, *Artic, Fluvial sediments, Geologic formations, Surveys, Hydrologic data, Data collections, Ice, Streamflow, Stream gages, Sediment transport, Geology, Climatic data, Seasonal, Thawing, Soils, Frozen ground, Melting, Runoff, Tidal effects, Shores, Erosion, Meteorological data, Water chemistry, Vegetation, Hydrographs, Maps, Photography, Data processing.
Identifiers: *Artic Coastal Plain Province (Northern Alaska).

Studies on the fluvial environment of the Arctic Coastal Plain Province, Northern Alaska, include research which ranges in magnitude from small polygon troughs to the Inaru River Basin. The 208 figures include stereograms, ground and aerial photographs, graphs, curves, and maps. Ninety tables appear in the publication. Complete hydrographs for two tundra streams are included. Discussions include soil subsidence, soil consolidation, details of specific study sites and the findings, and a section on the Late Pleistocene and Recent. Instrumentation, methods, and the uses of aerial photography are included. Over 240 bibliographic entries provide references and a foundation for future research. Chronologies of physical events which occurred in the drainage basins are listed. Basic statistics and reproductions of computer printouts of the microclimatic data are presented in the Appendices. The information (contained in 2 volumes) provides a guide to management and developers and can be utilized as engineering criteria. (Woodard-USGS)
W73-07906

THE WEAR OF SANDSTONE BY COLD, SLIDING ICE,
Newcastle-upon-Tyne Univ. (England). Dept. of Geography.
For primary bibliographic entry see Field 02J.
W73-08074

TORS, ROCK WEATHERING AND CLIMATE IN SOUTHERN VICTORIA LAND, ANTARCTICA,
Keele Univ. (England). Dept. of Geography.
For primary bibliographic entry see Field 02J.
W73-08075

VALLEY ASYMMETRY AND SLOPE FORMS OF A PERMAFROST AREA IN THE NORTHWEST TERRITORIES, CANADA,
Cambridge Univ. (England). Dept. of Geography.
B. A. Kennedy, and M. A. Melton.
In: Polar Geomorphology; Institute of British Geographers Special Publication No 4, p 107-121, June 1972. 6 fig, 4 tab, 24 ref. US Army Grant DA-ARO-D-31-124-G939.

Descriptors: *Geomorphology, *Arctic, *Erosion, *Mass wasting, *Valleys, *Permafrost, Profiles, Slopes, Topography, Degradation (Slope), Snow cover, Creep, Solifluction, Freezing, Thawing.

Valley asymmetry and slope forms in a small area of sedimentary rocks adjacent to the Mackenzie River delta are related to the major variations in climate, available relief, and geomorphic processes. The area is greatly varied geomorphologically and it does not fit the classic periglacial pattern. Asymmetry in maximum slope angles reverses between (a) areas of more severe climate and low available relief--where north-facing slopes are significantly steeper than those facing south, and (b) the zone of milder climate and deeper valleys, where south-facing slopes are the steeper. Slopes that are directly under the control of fluvial erosion show less response to differences in the degree of basal corrasion than to variations in aspect. This is unlike nonpermafrost areas. (Knapp-USGS)
W73-08076

THE NATURE OF THE ICE-FOOT ON THE BEACHES OF RADSTOCK BAY, SOUTH-WEST DEVON ISLAND, N.W.T., CANADA IN THE SPRING AND SUMMER OF 1970,
McMaster Univ., Hamilton (Ontario). Dept. of Geography.
S. B. McCann, and R. J. Carlisle.
In: Polar Geomorphology; Institute of British Geographers Special Publication No 4, p 175-186, June 1972. 6 fig, 14 ref.

Descriptors: *Sea ice, *Beaches, *Arctic, Melting, Ice breakup, Canada, Geomorphology, Beach erosion, Ice, Waves (Water), Surf, Ablation.
Identifiers: *Ice foot (Beaches), *Devon Island (Canada).

The ice foot is that part of the sea ice which is frozen to the shore and is therefore unaffected by tidal movements. The ice foot present along the coast of southwest Devon Island, Canada, in the early summer of 1970 is described and series of profiles across the feature, surveyed at intervals during the period of breakup, are presented. The form of the ice foot along Radstock Bay in the spring and summer of 1970 was the most pronounced observed in three seasons of field work on the coast of southwest Devon Island. The size and extent of the ice foot in any year clearly depend on the sea conditions existing at the time of freezing in the previous autumn. The 1970 ice

foot originated in conditions of medium-sized waves for the area. As there were only a few floes included in the ice foot and little evidence of buckling, the bay was relatively ice-free at the time of ice-foot formation, probably in mid- or late October 1969. The control of beach slope on ice-foot width and thickness, as seen in the Radstock Bay measurements, is such that as slope increases so the width and thickness of the ice foot decrease. Erosion of channels by water draining seaward from the melting snow in the backshore zone, early in the season, is important in breaching the ice foot. The sea ice had moved out of the outer part of Radstock Bay, adjacent to the study beach, by August 9, and the ice foot had been ablated and eroded away by August 15. (Knapp-USGS)
W73-08080

PROCESS AND SEDIMENT SIZE ARRANGEMENT ON HIGH ARCTIC TALUS, SOUTHWEST DEVON ISLAND, N.W.T., CANADA.
Alberta Univ., Edmonton. Dept. of Geography.
For primary bibliographic entry see Field 02J.
W73-08090

NATURE AND RATE OF BASAL TILL DEPOSITION IN A STAGNATING ICE MASS, BURROUGHS GLACIER, ALASKA,
Ohio State Univ. Research Foundation, Columbus. Inst. of Polar Studies.
For primary bibliographic entry see Field 02J.
W73-08091

PROCEEDINGS OF THE WESTERN SNOW CONFERENCE,

Proceedings of the 40th Annual Meeting of the Western Snow Conference, April 18-20, 1972, Phoenix, Ariz: Printed by Colorado State University, Fort Collins, 1972. 88 p.

Descriptors: *Snow, *Water resources, *Conferences, *Water resources development, Weather modification, Urbanization, Conservation, Water supply.

Recent feeling about the environment as it concerns the snow resource is changing with the spirit of the times. This is the keynote of the Seminar on 'Environment and the Snow Resource,' given in the 40th annual Western Snow Conference. The basic water and snow resource is not changing, but man's influence on this resource is producing change. In the case of urbanization, the effects of man on natural patterns of water supply may be largely inadvertent and often harmful, but man's intentional activities may often improve the environment to yield either more direct benefit to water users or greater aesthetic and recreational advantages. (See W73-08139 thru W73-08154) (Knapp-USGS)
W73-08138

COPING WITH POPULATION GROWTH AND A LIMITED RESOURCE,
Arizona Water Commission, Phoenix.
For primary bibliographic entry see Field 06A.
W73-08140

RENOVATING SEWAGE EFFLUENT BY GROUNDWATER RECHARGE,
Agricultural Research Service, Phoenix, Ariz. Water Conservation Lab.
For primary bibliographic entry see Field 05D.
W73-08141

THE APPLICATION OF SNOWMELT FORECASTING TO COMBAT COLUMBIA RIVER NITROGEN SUPERSATURATION PROBLEMS,
Corps of Engineers, Portland, Oreg. North Pacific Div.
D. D. Speers.
In: Proceedings of the 40th Annual Meeting of the Western Snow Conference, April 18-20, 1972, Phoenix, Ariz: Printed by Colorado State University, Fort Collins, p 17-22, 1972. 6 fig, 5 ref.

Descriptors: *Reservoir operation, *Columbia River, *Nitrogen, *Snowmelt, *Streamflow forecasting, Simulation analysis, Fish management, Fisheries, Reservoir releases, Water quality, Water quality control.
Identifiers: Columbia River basin.

Nitrogen supersaturation on the Columbia River causes fisheries losses. The cause of nitrogen supersaturation and its effect on fish are reviewed. An effective means of reducing the problem is regulation of upstream reservoirs, and snowmelt forecasting plays an important role in this regulation. Two types of forecasts are utilized; volumetric forecasts determined by multiple regression procedures, and daily simulation of runoff using the SSARR computer model. These procedures are briefly described and examples of their application in combating the nitrogen supersaturation problem are given. (See also W73-08138) (Knapp-USGS)
W73-08142

FORECAST STUDY FOR PRAIRIE PROVINCES WATER BOARD,
Water Survey of Canada, Calgary (Alberta). Alberta and Northwest Territories District Office.
For primary bibliographic entry see Field 04A.
W73-08143

AIR TEMPERATURE OBSERVATIONS AND FORECASTS--THEIR RELATIONSHIP TO THE PREDICTION OF SPRING SNOWMELT IN THE EAGEL RIVER BASIN, COLORADO,
National Weather Service, Salt Lake City, Utah. River Forecast Center.
A. L. Zimmerman.
In: Proceedings of the 40th Annual Meeting of the Western Snow Conference, April 18-20, 1972, Phoenix, Ariz: Printed by Colorado State University, Fort Collins, p 30-36, 1972. 4 fig, 7 ref.

Descriptors: *Snowmelt, *Streamflow forecasting, *Weather data, Snowpacks, Temperature, Data collections, Meteorology, *Colorado.
Identifiers: *Eagle River basin (Colorado).

A simple temperature index to snowmelt based on daily maximum temperature performs quite well in the Eagle River basin in Colorado in the spring snowmelt season during clear-weather melt conditions. During weather situations involving a significant change of air mass, air temperature is likely to be a poorer index to snowmelt than during relatively stable, dry periods. Therefore, additional air temperature data for a given river basin might not significantly improve snowmelt prediction during such periods. The predictability of air temperatures in a time frame of several days improves as the general circulation prognoses improve. (See also W73-08138) (Knapp-USGS)
W73-08144

THE LABOR DAY STORM OF 1970 IN ARIZONA,
Arizona Univ., Tucson. Dept. of Watershed Management.
For primary bibliographic entry see Field 02B.
W73-08145

SEPARATION OF THE SURFACE AND SUBSURFACE FLOW FROM SNOWMELT BY THE USE OF RADIOACTIVE TRACERS,
Department of the Environment (Alberta). Water Resources Div.
G. R. Holecek, and A. A. Noujaim.
In: Proceedings of the 40th Annual Meeting of the Western Snow Conference, April 18-20, 1972, Phoenix, Ariz: Printed by Colorado State University, Fort Collins, p 43-48, 1972. 3 fig, 2 tab, 5 ref.

Descriptors: *Snowmelt, *Surface runoff, *Subsurface runoff, *Tracers, *Tracking techniques, Water yield, Radioisotopes, Radioactivity techniques, Piezometers, Surface-groundwater relationships, Soil water movement, Recharge, Water balance, Hydrograph analysis, Recession curves.

The separation of surface and subsurface flow resulting from snowmelt from a 15 acre plot in North-Central Alberta was accomplished by means of a radioisotope tracer. The snow cover was sprayed with 50 to 150 millicuries of Iodine-125. Surface flow retained its radioactive characteristics while the subsurface flows lost its contamination because of filtering action by the soil mantle. This method can be used to study the impact of land utilization on the hydrologic cycle. (See also W73-08138) (Kanpp-USGS)
W73-08146

SNOW, RELATED ACTIVITIES AND CONCERNS ON THE GRAND MESA, WESTERN COLORADO,
Forest Service (USDA), Delta, Colo. Grand Mesa-Uncompahgre National Forests.
J. J. Christner.
In: Proceedings of the 40th Annual Meeting of the Western Snow Conference, April 18-20, 1972, Phoenix, Ariz: Printed by Colorado State University, Fort Collins, p 49-52, 1972. 4 ref.

Descriptors: *Snowpacks, *Snow cover, *Winter sports, Skiing, Recreation, Environmental effects, Snowmelt, Water yield, Water conservation, *Colorado.
Identifiers: *Grand Mesa (Colo), *Snowmobiling.

Melt water from accumulated snow on the Grand Mesa plays a very important role in the agriculture of western Colorado. The recent large increase in people engaging in wintertime on-the-snow activities has resulted in the altering of snow redistribution patterns and the changing of snow density. The effects of these changes on forest and range ecosystems are largely unknow. Further studies are required to evaluate the changes that are occurring. (See also W73-08138) (Knapp-USGS)
W73-08147

A WILDERNESS SNOW COURSE,
Forest Service (USDA), Kalispell, Mont. Flathead National Forest.
R. Delk.
In: Proceedings of the 40th Annual Meeting of the Western Snow Conference, April 18-20, 1972, Phoenix, Ariz: Printed by Colorado State University, Fort Collins, p 53-57, 1972. 1 fig, 1 tab, 5 ref.

Descriptors: *Snow surveys, *Montana, *Runoff forecasting, Snowpacks, Forests, Forest watersheds, Land use.
Identifiers: *Wilderness areas, *Bob Marshall Wilderness Area (Mont).

The Holbrook Snow Course, located in the Bob Marshall Wilderness area, Flathead National Forest, Montana, has become the focal point of controversy concerning the use of motorized vehicles in wilderness areas. Travel to Holbrook was by fixed-wing aircraft and helicopter from 1951 through 1969, and beginning on March 1, 1970, by Forest Service personnel traveling to the area on skis. The Wilderness Act of 1964 (PL-88-577),

which requires use of 'primitive' means of travel, and its relation to wilderness snow courses are discussed. Problems and hazards encountered during the ski trips are discussed. The importance of the snow course and the value of the wilderness resource are evaluated. A plan to move the course to another site in an attempt to consolidate administrative uses in the Wilderness area is also presented. This plan has resulted in a problem of data correlation which has not yet been solved. (See also W73-08138) (Knapp-USGS)
W73-08148

WILDERNESS IN THE NATIONAL PARKS,
National Park Service, Denver, Colo. Denver Service Center.
For primary bibliographic entry see Field 04C.
W73-08149

WATERSHED HYDROMETEOROLOGICAL DATA REQUIRED FOR WEATHER MODIFICATION,
North American Weather Consultants, Santa Barbara, Calif.
For primary bibliographic entry see Field 03B.
W73-08150

INFLUENCE OF LATE SEASON PRECIPITATION ON RUNOFF OF THE KINGS RIVER,
Sierra Hydrotechnology, Placerville, Calif.
For primary bibliographic entry see Field 02A.
W73-08151

AVANLANCHE AWARENESS AND SAFETY FOR SNOW SCIENTISTS IN THE FIELD,
Geological Survey, Sacramento, Calif.
W. R. Hotchkiss.
In: Proceedings of the 40th Annual Meeting of the Western Snow Conference, April 18-20, 1972, Phoenix, Ariz: Printed by Colorado State University, Fort Collins, p 75-80, 1972. 1 fig.

Descriptors: *Avalanches, *Snow surveys, *Safety, Hazards, Snow management, Geomorphology.

During the winter of 1970-71, avalanches in the United States caught 58 people, buried 46, and killed 12. Snow scientists should periodically update their knowledge of how to classify avalanche events and to evaluate avalanche hazards. A practical, three-part classification of avalanches is proposed. Part I includes static observations made following the event; Part II includes dynamic observations made at the time of the event; and Part III includes genetic deductions based on a description of the meteorological background. Evaluation of hazard is possible through careful observation of the definitive factors which cause avalanches: accumulation of newly fallen and wind-transported snow, free water percolating through the snowpack, and progressive weakening of internal layers of the snowpack. Knowledge of avalanche classification and hazard evaluation together with the use of sound judgment should promote avalanche awareness and safety. (See also W73-08138) (Knapp-USGS)
W73-08152

SECTION 22, 'SNOW SURVEY AND WATER SUPPLY FORECASTING,' OF THE SCS NATIONAL ENGINEERING HANDBOOK,
Soil Conservation Service, Portland, Oreg. Water Supply Forecast Unit.
M. Barton, and G. L. Pearson.
In: Proceedings of the 40th Annual Meeting of the Western Snow Conference, April 18-20, 1972, Phoenix, Ariz: Printed by Colorado State University, Fort Collins, p 81-82, 1972.

Descriptors: *Snow surveys, *Water yield, *Runoff forecasting, Water supply, Precipitation gages, Telemetry, Data collections, Data processing.

The Soil Conservation Service is issuing Section 22 entitled 'Snow Survey and Water Supply Forecasting' of the SCS National Engineering Handbook. Section 22 provides detailed information on many aspects of snow surveys and water supply forecasting. It will be useful not only for specialists in this field but also for scholars and students because it is a compilation of information not previously available in a single publication. It is available from U. S. Government Printing Office. Section 22 is divided into nine chapters. Major topics are: Data Collection for Water Supply Forecasting, Telemetry in Data Collection, Travel to Collect Data, Data Processing, Water Supply Forecasting, and Maintenance of Installations and Equipment. Section 22 is the first comprehensive text on snow surveys and water supply forecasting since Dr. J. E. Church introduced modern snow survey techniques at Mt. Rose, Nev., in 1904 to 1910. (See also W73-08138) (Knapp-USGS)
W73-08153

SOUTH DAKOTA WEATHER MODIFICATION PROGRAM,
South Dakota Weather Control Commission, Pierre.
For primary bibliographic entry see Field 03B.
W73-08154

AVALANCHES ON SAKHALIN AND THE KURIL ISLANDS (LAVINY SAKHALINA I KURIL'SKIKH OSTROVOV).

Sakhalinskoye Upravleniye Gidrometeorologicheskoy Sluzhby; Gidrometeoizdat, K. F. Voytkovskiy, and V. Ye. Barabash, editors, Leningrad, 1971. 180 p.

Descriptors: *Avalanches, *Snow, *Snow cover, *Snowfall, *Islands, Snow surveys, Meteorology, Climatology, Topography, Orography, Mountains, Winds, Storms, Disasters, Safety, Engineering structures, Control structures, Transporation, Communication, Seasonal.
Identifiers: *USSR, *Sakhalin Island, *Kuril Islands, Avalanche hazard, Avalanche control, Snow density, Snow loads.

Evolution and occurrence of avalanches on Sakhalin and the Kuril Islands were investigated for avalanche-hazard evaluation and avalanche prevention and control in this collection of 14 papers prepared on the basis of research conducted by a number of organizations in the USSR. These include the Sakhalin Administration of the Hydrometeorological Service, the Department of Geography of Moscow State University, the Institute 'Sakhalingrazbdanproyekt,' the Novosibirsk Institute of Railroad Transportation Engineers, and the Novosibirsk Scientific Research Institute of Power Engineering. Among the topics discussed are: (1) avalanche activity on Sakhalin Island; (2) avalanche hazard on the Kuril Islands; (3) construction and testing of avalanche-control installations; (4) mass avalanching on Southern Sakhalin in the winter of 1969-70; (5) snow avalanches on sections of railway lines in Southern Sakhalin; (6) determination of snow loads on avalanche-control structures; and (7) distribution of snow in mountainous areas of Sakhalin. (See W73-08166 thru W73-08171) (Josefson-USGS)
W73-08165

A GENERAL REVIEW OF AVALANCHE ACTIVITY ON SAKHALIN ISLAND (OBSHCHIY OBZOR LAVINNOGO REZHIMA O. SAKHALIN),
A. V. Ivanov.
In: Laviny Sakhalina i Kuril'skikh ostrovov. Sakhalinskoye Upravleniye Gidrometeorologicheskoy Sluzhby; Gidrometeoizdat, Leningrad, p 4-25, 1971. 4 fig, 2 tab, 14 ref.

Descriptors: *Avalanches, *Snow, *Snow cover, Snowfall, Storms, Winds, Precipitation (Atmospheric), Air temperature, Rime, Meteorology, Climatology, Topography, Orography, Vegetation, Forests, Earthquakes, Disasters, Mapping.
Identifiers: *USSR, *Sakhalin Island, Avalanche hazard, Avalanche classification, Avalanche cones, Snow stability.

Investigations of snow avalanches on Sakhalin Island were based on data collected in 1965-70 by the Sakhalin Administration of the Hydrometeorological Service. The major factors responsible for avalanche formation on the island are topography, climate, and mountain vegetation. About 50% of all avalanches observed had a volume of less than 2,000 cu m, while 15% or fewer had a volume of more than 10,000 cu m. Despite their relatively small size, the avalanches pose a serious threat to railroad and highway construction and have caused much damage to forested areas. A regionalization of the island by degree of avalanche hazard is based on frequency of avalanche occurrence, snow conditions, forest cover, and character of the relief and vegetation. (See also W73-08165) (Josefson-USGS)
W73-08166

A DESCRIPTION OF AVALANCHE HAZARD ON THE KURIL ISLANDS (KHARAKTERISTIKA LAVINNOY OPASNOSTI KURIL'SKIKH OSTROVOV),
Moscow State Univ. (USSR). Problemnaya Laboratoriya Nezhnykh Lavin i Selei.
N. A. Volodicheva.
In: Laviny Sakhalina i Kuril'skikh ostrovov. Sakhalinskoye Upravleniye Gidrometeorologicheskoy Sluzhby; Gidrometeoizdat, Leningrad, p 26-39, 1971. 4 fig, 15 ref.

Descriptors: *Islands, *Avalanches, *Snow, *Snow cover, Snowfall, Storms, Winds, Precipitation (Atmospheric), Meteorology, Climatology, Topography, Orography, Vegetation, Volcanoes, Craters, Aerial photography, Maps.
Identifiers: *USSR, *Kuril Islands, *Avalanche hazard, Snowslides, Geobotany.

Preparation of a map of snow avalanche-hazard regions in the Kuril Islands was based on geographical analysis of topography, climate, and vegetation; field surveys of mountainous areas; aerial photographic interpretation; and aircraft observations. The islands are grouped according to avalanche danger into four classes as (1) very high (northern islands of Paramushir, Atlasova, Onekotan, Shiashkotan, and other smaller islands southward to Ketoy); (2) slight (Simushir, Urup and the group of small islands between them); (3) potential (Kunashir and Iturup); and (4) none (islands of the Lesser Kuril chain). (See also W73-08165) (Josefson-USGS)
W73-08167

MASS AVALANCHING ON SOUTHERN SAKHALIN IN THE WINTER OF 1969-70 (MASSOVYY SKHOD LAVIN NA YUZHNOM SAKHALINE ZIMOY 1969-70 G.),
A. V. Ivanov.
In: Laviny Sakhalina i Kuril'skikh ostrovov. Sakhalinskoye Upravleniye Gidrometeorologicheskoy Sluzhby; Gidrometeoizdat, Leningrad, p 74-87, 1971. 5 fig, 3 tab, 5 ref.

Descriptors: *Avalanches, *Snow, *Snowfall, *Diseasters, *Meteorology, Precipitation intensity, Snow cornice, Cyclones, Air temperature, Winds, Storms, Topography, Slopes, Mountains, Seasonal, Winter.
Identifiers: *USSR, *Sakhalin Island, Snow density, Geobotany.

Excessive precipitation, strong winds, and prolonged and intense snowfalls were the principal factors responsible for mass movement of avalanches in mountains of Southern Sakhalin in the winter of 1969-70. Precipitation intensity during a 24-hour avalanche period for different parts of Southern Sakhalin varied between 0.6 and 2.8 mm/hr. The rate of increase of snow depth during snowfall averaged 0.5-1.0 cm/hr, reaching maximum values of 4-6 cm/hr in individual 3-hour periods. Dry-powder snow avalanches caused considerable damage to the national economy. Road and rail traffic was disrupted between Yuzhno-Sakhalinsk and Kholmsk and between Shebunino and Il'inskiy for long periods of time; forested areas were devastated; and numerous buildings and structures were demolished. Except for isolated cases of avalanching on stretches of highways and railway lines, wet-snow avalanches created no serious problems. On the basis of 1965-70 observation data, reports of local inhabitants, and geobotanical investigations, the greatest volumes and distances of flowage of avalanches in the area in the last 30-50 years occurred in the winter of 1969-70. (See also W73-08165) (Josefson-USGS)
W73-08168

SNOW AVALANCHES ON SECTIONS OF RAILWAY LINES IN SOUTHERN SAKHALIN (O SNEZHNYKH LAVINAKH NA UCHASTKAKH ZHELEZNYKH DOROG YUZHNOGO SAKHALINA),
Institut Inzhenerov Zheleznodorozhnogo Transporta, Novosibirsk (USSR).
E. P. Isayenko, and Yu. A. Marin.
In: Laviny Sakhalina i Kuril'skikh ostrovov. Sakhalinskoye Upravleniye Gidrometeorologicheskoy Sluzhby; Gidrometeoizdat, Leningrad, p 102-108, 1971. 2 fig, 4 tab, 4 ref.

Descriptors: *Avalanches, *Snow, *Snow cover, *Snowfall, *Railroads, Slopes, Mountains, Storms, Winds, Forecasting.
Identifiers: *USSR, *Sakhalin Island, Avalanche hazard, Avalanche control, Snowslides.

Snow avalanches on railway lines in mountainous areas of Southern Sakhalin were investigated for avalanche prevention and control. Data are presented on exposure of avalanche-hazard slopes on the rail line between Yuzhno-Sakhalinsk and Kholmsk, and conditions conducive to snow-avalanche formation are examined for avalanche-hazard forecast. (See also W73-08165) (Josefson-USGS)
W73-08169

FORMATION AND PHYSICO-MECHANICAL PROPERTIES OF SNOW COVER ON AVALANCHE-HAZARD SLOPES ALONG THE RAILWAY LINE BETWEEN YUZHNO-SAKHALINSK AND KHOLMSK (OSOBENNOSTI FORMIROVANIYA I FIZIKO-MEKHANICHESKIYE SVOYSTVA SNEZHNOGO POKROVA NA LAVINOOPASNKH SKLONAKH VDOL' ZHELEZNOY DOROGI YUZHNO-SAKHALINSK--KHOLMSK),
Institut Inzhenerov Zheleznodorozhnogo Transporta, Novosibirsk (USSR).
E. P. Isayenko, Yu. A. Marin, and V. I. Yadroshnikov.
In: Laviny Sakhalina i Kuril'skikh ostrovov. Sakhalinskoye Upravleniye Gidrometeorologicheskoy Sluzhby; Gidrometeoizdat, Leningrad, p 109-123, 1971. 7 fig, 7 tab, 3 ref.

Descriptors: *Avalanches, *Snow cover, *Slopes, *Railroads, *Properties, Mechanical properties, Physical properties, Temperature, Thermocline, Meteorology, Cyclones, Winds, Storms, Mountains, Snow surveys, Seasonal.
Identifiers: *USSR, *Sakhalin Island, *Avalanche hazard, Avalanche control, Snow stability, Snow density, Snow classification.

Snow deposition, stratigraphy, temperature regime, and physico-mechanical properties were investigated on avalanche-hazard slopes along a 25-km line of railroad track between Yuzhno-Sakhalinsk and Kholmsk in the southern half of Sakhalin Island. Investigations were carried out in 1967-70 by the Novosibirsk Institute of Railroad Transportation Engineers for protection of railroads from snow avalanches. The data obtained provide information on the formation and development of snowpacks in the vicinity of the railway line and can be used for planning avalanche control. (See also W73-08165) (Josefson-USGS)
W73-08170

DISTRIBUTION OF SNOW IN MOUNTAINOUS AREAS OF SAKHALIN (RASPREDELENIYE SNEZHNOGO POKROVA V GORNYKH RAYONAKH SAKHALINA),
I. F. Monastyrskiy.
In: Laviny Sakhalina i Kuril'skikh ostrovov. Sakhalinskoye Upravleniye Gidrometeorologicheskoy Sluzhby; Gidrometeoizdat, Leningrad, p 140-144, 1971. 2 fig, 1 tab, 3 ref.

Descriptors: *Snow, *Snow cover, *Mountains, Distribution, Depth, Snowfall, Snowmelt, Water equivalent, Altitude, Snow surveys, Seasonal.
Identifiers: *USSR, *Sakhalin Island, Snow density.

Annual duration of snow cover on plains in the central part of Sakhalin Island is 5-6 months and in the southern part of the island about 5 months. Because of slower melting in spring, snow is retained 1-2 months longer in the mountains than on the plains. Snow depth in the mountains usually reaches maximum values during the second half of March and averages 60-100 cm in the central part and 100-150 cm in the southern part. At lower altitudes (0-300 m), snow disappears completely in the south by the end of April and in the north by the middle of March. In the central part of the island, snow on the east coast lasts 20-30 days longer than snow on the west coast. Provided depth of snow cover is less than 1 m, a linear relation exists between snow depth and density from mid-December to mid-March. On the basis of snow-survey data, snow density in March varies between 0.18 and 0.35 g/cu cm and in April between 0.23 and 0.40 g/cu cm. For 1961-70, the vertical gradient of maximum snow-water equivalent was assumed to equal 40 mm/100m for central regions of the island and 70 mm/100 m for regions in the south. (See also W73-08165) (Josefson-USGS)
W73-08171

AN ANALYSIS OF YEARLY DIFFERENCES IN SNOWPACK INVENTORY-PREDICTION RELATIONSHIPS,
Arizona Univ., Tucson. Dept. of Watershed Management.
P. F. Ffolliott, D. B. Thorud, and R. W. Enz.
In: Hydrology and Water Resources in Arizona and the Southwest, Proceedings of the 1972 meetings of the Arizona Section--American Water Resources Assn. and the Hydrology Section--Arizona Academy of Science, May 5-6, 1972, Prescott, Arizona, Vol 2, (1972), p 31-42. 3 fig, 2 tab, 7 ref. OWRR A-014-ARIZ (8). 14-31-0001-3203.

Descriptors: *Runoff forecasting, *Snowpacks, *Spatial distribution, *Forest management, *Vegetation effects, Forecasting, Census, Forests, Watershed management, Snow surveys, Water equivalent, Water yield, Snow management, Canopy, *Arizona.
Identifiers: *Inventory-prediction, Ponderosa pine.

Inventory-prediction relationships between snowpack conditions and forest attributes may be useful in estimating water yields derived from snow, but such relationships are developed usually from source data collected over a short time period. Analyses of long-term data suggest inventory-prediction relationships developed from limited data may have more general application, however. Available records from 18 snow courses in the ponderosa pine type in Arizona provided source data in this study, which was designed to empirically analyze inventory-prediction relationships developed from long-term Snow Survey records. The primary hypothesis tested and evaluated by statistically analyzing the family of regression equations representing a snow course, was that, given a precipitation input, the distribution of snowpack water equivalent at peak seasonal accumulation is determined by the spatial arrangement of the forest cover, e.g. basal area. Generally 12 of the 18 snow courses evaluated appeared to support the hypothesis, three courses did not, and three courses were considered inconclusive. (White-Arizona)
W73-08301

REVERSING BARCHAN DUNES IN LOWER VICTORIA VALLEY, ANTARCTICA,
Lunar Science Inst., Houston, Tex.
For primary bibliographic entry see Field 02J.
W73-08373

VENTIFACT EVOLUTION IN WRIGHT VALLEY, ANTARCTICA,
Lunar Science Inst., Houston, Tex.
For primary bibliographic entry see Field 02J.
W73-08374

PLANT SUCCESSION ON TUNDRA MUDFLOWS: PRELIMINARY OBSERVATIONS,
Carleton Univ., Ottawa (Ontario). Dept. of Biology.
For primary bibliographic entry see Field 02G.
W73-08409

2D. Evaporation and Transpiration

COMPARATIVE STUDIES ON THE TRANSPIRATION IN PEDUNCULATE OAK-HORNBEAM COPSE AND ON AGRICULTURAL FIELDS IN PETROVINA (TUROPOLJE NEAR ZAGREB), (IN SERBO-CROATIAN),
M. Gracanin, L. Ilijanic, V. Gazi, and N. Hulina.
Acta Bot Croat. 30 p 57-84. 1971. Illus.
Identifiers: Agriculture, Copse, Erigeron canadensis D, Hornbeam D, Light, Moisture, *Oak D, Opening, Pedunculate, *Petrovina (Yugoslavia), Soils, Stomate, Temperature, *Transpiration, Wheat M.

Studies were made in humid, moderately warm area (yearly precipitation 930 mm, average temperature 10 degrees C) with moderately podzolized loam-clay soils. Studies were made on various tree species, 3 varieties of wheat and species of weeds. Measurements of transpiration were made from April 29 to Sept. 4, 1969 every 2 hr daily from 8 a.m. to 6 p.m. Relative humidity, temperature of the air and soil, light intensity and soil dampness were also measured. The condition of the stomates (size of opening) were also measured with the infiltration method (alcohol and xylene). The tree species were found to have a significantly lower transpiration rate than the wheat or weed species. The daily sum of transpiration was from 2.59 to 7.22 g/g fresh weight for the forest species and 9.65 to 11.73 g/g for the wheat varieties. The weed species showed a high transpiration capacity. Erigeron canadensis had a maximum daily sum of transpiration of 19.68 g/g. Wheat and weed species appear to affect the soil water balance much more than the forest species. The agricultural soil was much dryer than the forest soil. External conditions (light intensity, temperature, relative humidity) can effect transpiration intensity. The greater

heating of the leaves on agricultural areas and the subsequent increased vapor pressure in intercellular areas can significantly increase transpiration. The size of the stomate opening was much larger in wheat than in the trees.--Copyright 1972, Biological Abstracts, Inc.
W73-08057

THE REGRESSION OF NET RADIATION UPON SOLAR RADIATION,
Oregon State Univ., Corvallis. Water Resources Research Inst.
L. W. Gay.
Arch. Met. Geoph. Biokl., Ser. B, 19, p 1-14, 1971, also published as Paper 710, Forest Research Lab, Oregon State University. 3 fig, 14 p. OWRR A-001-ORE (10).

Descriptors: *Regression analysis, *Radiation, Net radiation, Solar radiation, Evapotranspiration, Statistical model, Model studies.

A number of studies have sought to relate net radiation over natural surfaces to incoming global radiation. Several deficiencies are noted in interpretation of the simple regression models used for this purpose in the past. A modification proposed for correction of these deficiencies introduces a new longwave exchange coefficient, lambda, that relates the change in net longwave radiation to the change in net shortwave radiation. This coefficient is a conceptual improvement of beta, the surface heating coefficient that has gained some acceptance. The definition of a unique lambda for various surfaces is shown to be the same problem faced by past studies that have sought to develop a unique regression equation over each surface. The coefficient lambda may serve as an index of the surface properties that govern the dissipation of absorbed global radiation. This index is also affected by environmental characteristics. Interpretation of the index is demonstrated for three types of surfaces under cloudless skies: a desert, a forest, and an irrigated crop. Further consideration of the coefficient, lambda reveals that the apparent success of regression relations linking net and global radiation may often be attributed to negligible variation of longwave exchange, rather than to the soundness of a particular regression model.
W73-08175

OBSERVATIONS ON THE TRANSPIRATION OF TWO SPECIES OF AMARANTHUS, (IN PORTUGUESE),
Sao Paulo Univ. (Brazil). Escola Superior de Agricultura Luiz de Queiroz.
H. Reyes-Zumeta, and P. N. Camargo.
Cienc Cult. Vol 23, No 3, P 351-361. 1971, Illus. (English summary).
Identifiers: *Amaranthus, Amaranthus hybridus D, Amaranthus viridis D, Brazil, Coffee D, Plantation, Species, *Transpiration.

Observations were made on the variation of transpiration in Amaranthus hybridus and A. viridis in Piracicaba, Brazil, during the rainy season. There was no apparent restriction in the total transpiration during the day. The relative transpiration (T/E%) showed that A. hybridus had no restriction, while A. viridis showed a strong restriction between 10:25 a.m. and 12:05 p.m. This behavior may be due to the habitats of the plants, A. viridis grew inside the coffee plantation, and A. hybridus grew 3 m from the crop border. The different transpirational behavior of the 2 spp. may be attributed to the stronger competition for water by the coffee plants with A. viridis than with A. hybridus. --Copyright 1972, Biological Abstracts, Inc.
W73-08240

IDENTIFICATION OF A DAY OF MOISTURE STRESS IN MAIZE AT CEDARA,
Agricultural Research Inst., Cedara (South Africa).
J. B. Mallett, and J. M. DeJager.
Agroplantae. Vol 3, No 3, p 45-50. 1971. Illus.
Identifiers: *Cedara (So. Africa), *Evaporation, *Maize, *Moisture stress, Soils, Transpiration.

As atmospheric evaporative demand increases towards 10 mm per day the available soil moisture required to prevent stress from occurring steadily increases up to 75%. Above 75% plant available moisture, wilting is seldom induced. Evaporative demand was measured using a USA Weather Bureau Class A evaporation pan, while soil moisture and evapotranspiration were determined by means of mass-measuring lysimeters.--Copyright 1972, Biological Abstracts, Inc.
W73-08337

THE BIONIC RELATIONSHIP BETWEEN TRANSPIRATION IN VASCULAR PLANTS AND THE HEAT PIPE,
Carnegie-Mellon Univ., Pittsburgh, Pa. Dept. of Mechanical Engineering.
J. F. Osterle, and J. G. McGowan.
Math Biosci, Vol 14, No 3/4, p 317-323, 1972, Illus.
Identifiers: *Bionics, *Heat pipe, Plants, Relationship, *Transpiration, *Vascular plants, Model studies.

The relationship between the transpiration system in vascular plants and the heat pipe was viewed from the standpoint of bionics--the science of systems whose function is based on living systems. The 2 systems were identical as revealed by a mathematical model based on equilibrium and nonequilibrium thermodynamics. Both energy conversion devices were found to be thermal pumps operating in a special mode.--Copyright 1973, Biological Abstracts, Inc.
W73-08348

EFFECTS OF ABSCISIC ACID AND ITS ESTERS ON STOMATAL APERTURE AND THE TRANSPIRATION RATIO,
King's Coll., London (England). Dept of Botany.
For primary bibliographic entry see Field 02I.
W73-08420

2E. Streamflow and Runoff

AN INEXPENSIVE, RECORDING TIDE GAUGE,
Brookhaven National Lab., Upton, N.Y. Dept. of Biology.
For primary bibliographic entry see Field 07B.
W73-07852

HYDROGRAPH ANALYSIS AND SOME RELATED GEOMORPHIC VARIABLES,
Nebraska Univ., Omaha. Dept. of Civil Engineering.
For primary bibliographic entry see Field 02J.
W73-07878

PREDICTING TIME-OF-TRAVEL IN STREAM SYSTEMS,
Illinois State Water Survey, Urbana.
For primary bibliographic entry see Field 02J.
W73-07879

HYDROGEOMORPHOLOGY OF SUSQUEHANNA AND DELAWARE BASINS,
State Univ., of New York, Binghamton. Dept. of Geology.
For primary bibliographic entry see Field 02J.
W73-07880

EUREKA--IT FITS A PEARSON TYPE 3 DISTRIBUTION,
Geological Survey, Washington, D.C.
For primary bibliographic entry see Field 04A.
W73-07883

APPLICATION OF NONLINEAR SYSTEM IDENTIFICATION TO THE LOWER MEKONG RIVER, SOUTHEAST ASIA,
Geological Survey, Menlo Park, Calif. Water Resources Div.
For primary bibliographic entry see Field 04A.
W73-07884

INTRODUCTION OF TIME VARIANCE TO LINEAR CONCEPTUAL CATCHMENT MODELS,
Institute of Hydrology, Wallingford (England).
For primary bibliographic entry see Field 02A.
W73-07885

DETERMINATION OF OPTIMAL KERNELS FOR SECOND-ORDER STATIONARY SURFACE RUNOFF SYSTEMS,
Technion - Israel Inst. of Tech., Haifa. Dept. of Civil Engineering.
For primary bibliographic entry see Field 02A.
W73-07886

IDENTIFICATION OF MULTIPLE REACH CHANNEL PARAMETERS,
California Univ., Los Angeles. Dept. of Engineering Systems.
L. Becker, and W. W-G. Yeh.
Water Resources Research, Vol 9, No 2, p 326-335, April 1973. 3 fig, 3 tab, 8 ref, 2 append.

Descriptors: *Mathematical models, *Open channel flow, *Numerical analysis, *Routing, Finite element analysis, Computer programs, Streamflow forecasting, Hydrograph analysis.

The influence coefficient algorithm for the solution of the parameter identification problem is extended to multiple-reach channel flows. The stage hydrograph, velocity data, and observation station location requirements for an effective determination of the parameters are discussed. Application involves identification of the individual reach friction parameters, which are not physically measureable and which have to be determined from the mathematical model by the use of concurrent input and output measurements. The influence coefficient algorithm is incorporated with numerical solutions of the unsteady equations of flow through explicit finite difference formulations of these equations. The explicit formulations simplify the computer program but necessitate special precautions to avoid computational instabilities. The stability criteria which must be observed are indicated. (Knapp-USGS)
W73-07887

STORM FLOW FROM HARDWOOD-FORESTED AND CLEARED WATERSHEDS IN NEW HAMPSHIRE,
Forest Service (USDA), Durham, N.H. Northeastern Forest Experiment Station.
For primary bibliographic entry see Field 04C.
W73-07889

MEASURE OF THREE-DIMENSIONAL DRAINAGE BASIN FORM,
Cambridge Univ. (England). Dept. of Geography.
For primary bibliographic entry see Field 02A.
W73-07892

SPRING DISCHARGE OF AN ARCTIC RIVER DETERMINED FROM SALINITY MEASUREMENTS BENEATH SEA ICE,
Louisiana State Univ., Baton Rouge. Coastal Studies Inst.

For primary bibliographic entry see Field 02C.
W73-07900

BOTTOM CURRENTS IN THE HUDSON CANYON,
National Oceanic and Atmospheric Administration, Miami, Fla. Atlantic Oceanographic and Meteorological Labs.
G. H. Keller, D. Lambert, G. Rowe, and N. Staresinic.
Science, Vol 180, No 4082, p 181-183, April 13, 1973. 3 fig, 14 ref.

Descriptors: *Sediment transport, *Submarine canyons, *Hudson River, *Currents (Water), Streamflow, Tides, Stream gages, Current meters, Data collections, Flow measurement, Path of pollutants.
Identifiers: *Hudson submarine canyon.

In the Hudson Canyon the current regime is characterized by a pronounced reversal of flow up and down the canyon. Velocities are commonly of the order of 8 to 15 cm/sec, reaching 27 cm/sec on occasion in the upper and central portion of the canyon. Although a 2.5-day recording of currents showed a net transport upcanyon, a combination of 66 current measurements from the submersible Alvin, analysis of sediment texture and organic carbon, and determination of the benthic fauna-nutrient relationship indicate that over the long term there is a net transport of fine material through the canyon to the outer continental rise. (Knapp-USGS)
W73-07903

STUDIES ON THE FLUVIAL ENVIRONMENT, ARCTIC COASTAL PLAIN PROVINCE, NORTHERN ALASKA, VOLUMES I AND II,
For primary bibliographic entry see Field 02C.
W73-07906

REGULATION OF STREAMFLOW (REGULIROVANIYE RECHNOGO STOKA),
For primary bibliographic entry see Field 04A.
W73-07909

ROTARY CURRENTS, ACCORDING TO MEASUREMENTS IN THE OCEAN,
Akademiya Nauk SSSR, Gelendzhik. Institut Okeanologii.
V. B. Titov.
Oceanology, Vol 12, No 2, p 177-181, 1972. 2 fig, 1 tab, 1 ref. Translated from Okeanologiya (USSR), Vol 12, No 2, 1972.

Descriptors: *Oceanography, *Currents (Water), *Rotational flow, *Ocean currents, Rotations, Measurement, Instrumentation, Buoys, Depth, Velocity.
Identifiers: *USSR, *Rotary currents.

Five instances of unusual, clearly defined, periodic, cyclonic rotary ocean currents with an average period of about 1.5 hours have been observed in recent years on expeditions conducted by the Institute of Oceanology, USSR Academy of Sciences. A possible cause of formation of these rotary currents is the local transverse irregularity of the flow velocity field, which generates vortices with vertical axes of rotation. (Josefson-USGS)
W73-07915

INSTANTANEOUS UNIT HYDROGRAPHS, PEAK DISCHARGES AND TIME LAGS IN URBANIZING WATERSHEDS,
Purdue Univ., Lafayette, Ind. School of Civil Engineering.
For primary bibliographic entry see Field 02A.
W73-07917

TRANSPOSITION OF STORMS FOR ESTIMATING FLOOD PROBABILITY DISTRIBUTIONS,
Colorado State Univ., Fort Collins. Dept. of Civil Engineering.
V. K. Gupta.
Colorado State University Hydrology Papers, No 59, November 1972. 35 p, 30 fig, 7 tab, 18 ref, 1 append. NSF Grant No. 11444.

Descriptors: *Storms, *Floods, *Flood forecasting, *Flood frequency, Methodology, Theoretical analysis, River basins, Stochastic processes, Hydrologic data, Historic floods, Flood data, Model studies, Rainfall-runoff relationships, Regional flood, Geomorphology, Streamflow, Flow characteristics.

Contemporary literature in hydrology usually contains the concepts of maximum probable precipitation and maximum probable flood along with methods used to arrive at these limits. These limits signify some physical upper limits for precipitation and flood; however, it is difficult to find physical justification for existence of these limits and more so the methods used to compute them. Also, the use of the word 'probable' is incorrect because these 'probable limits' are not assigned any probabilities. In view of the misconceptions, practical methodology is described with a theoretical framework for estimating the probability of occurrence of floods in a unit time interval, based on the random characteristics of storms. In general, many random characteristics can be defined for a storm, but as a first step only a three-dimensional random vector has been defined for the random characteristics of storms. The random vector is comprised of the coordinates of storm center location and storm orientation. The developed estimation methodology uses all information on historic storms observed in a region that contains the river basin. (Woodard-USGS)
W73-08085

INFLUENCE OF LATE SEASON PRECIPITATION ON RUNOFF OF THE KINGS RIVER,
Sierra Hydrotechnology, Placerville, Calif.
For primary bibliographic entry see Field 02A.
W73-08151

DRAINAGE AREAS, HARTFORD NORTH QUADRANGLE, CONNECTICUT,
Geological Survey, Washington, D.C.
For primary bibliographic entry see Field 07C.
W73-08173

EFFECTIVE HYDRAULIC ROUGHNESS FOR CHANNELS HAVING BED ROUGHNESS DIFFERENT FROM BANK ROUGHNESS,
Army Engineer Waterways Experiment Station, Vicksburg, Miss.
For primary bibliographic entry see Field 08B.
W73-08350

A PLAN FOR STUDY OF WATER RESOURCES IN THE PLATTE RIVER BASIN, NEBRASKA—WITH SPECIAL EMPHASIS ON THE STREAM-AQUIFER RELATIONS,
Geological Survey, Lincoln, Nebr. Water Resources Div.
For primary bibliographic entry see Field 02A.
W73-08357

APPLICATIONS OF REMOTE SENSING TO STREAM DISCHARGE PREDICTION,
National Aeronautics and Space Administration, Huntsville, Ala. George C. Marshall Space Flight Center.
For primary bibliographic entry see Field 07B.
W73-08359

A new method of studying alluvial fans on the piedmont plains of Central Asia, USSR, uses a combination of geophysical and landscape methods to replace a great amount of drilling. Geophysical methods include direct-current electrical surveying and induced potentials. The proposed combination of methods of investigation may solve the following geological-hydrogeological problems: (1) the grouping of alluvial fans by types according to morphology, source of material, and possible degree of flooding; (2) determination of thickness and lithology of the unconsolidated deposits in fans of each type; (3) study of the hydrogeological conditions of these deposits (determination of water table, discrimination of zones of different mineralization, qualitative evaluation of reservoir properties of aquifers). On the basis of the data obtained, a hydrogeological map of the alluvial fans may be prepared. (Knapp-USGS)
W73-07844

LANDSCAPE-INDICATOR INVESTIGATIONS OF KARST,
Moskovskoe Obshchestvo Ispytatelei Prirody (USSR). Geographic Div.
A. G. Chikishev.
In: Landscape Indicators--New Techniques in Geology and Geography: Consultants Bureau, Div of Plenum Press, London and New York, p 48-63, 1973. 5 fig, 54 ref. (Translated from Russian. Proceedings of Conference of Moscow Society of Naturalists, May 21-22, 1968, Moscow, Nauka Press).

Descriptors: *Terrain analysis, *Karst, *Hydrogeology, *Geomorphology, Karst hydrology, Mapping, Aerial photography, Remote sensing, Vegetation effects, Data collections, Groundwater, Topography.
Identifiers: *USSR.

In mutiple investigations of karst by landscape indicators, aerial photography and direct aerial observation are of fundamental importance. The use of air photos and preliminary aerial flights over the region make it possible to obtain the most complete information concerning the extent of karst development in the region, the morphological aspects of karst forms, and the hydrological conditions of karst formations without laborious surface work. In mountainous regions karst in limestones that crop out at the surface is reliably recognized by a characteristic variegated-porous picture of the photo image and pitted microrelief, emphasized by the darker tone of vegetation associated with the low damper segments. Along with relief and hydrography, vegetation reacts sensitively to the physical and chemical properties of the rocks. The depth to which plants may indicate bedrock covered by unconsolidated deposits varies for different natural zones. The lithology may be indicated by plants in tundra regions to a depth of 1-2 m, in forest zones to a depth of 10 m, and in deserts to a depth of 20 m. (In relation to the principal types of karst formation, plants are divided into calciphytes (confined to carbonate rocks), gypsophytes (on gypseous rocks), and halophytes (on halide rocks)). The contacts between carbonate rocks and other rocks varieties may be drawn with great reliability. Stands of trees on carbonate rocks, because of the extreme dryness of the underlying rock and the harmful effect of calcium in excess, are normally thinner than in neighboring districts underlain by other rocks. (Knapp-USGS)
W73-07845

INTERFACE REFRACTION AT THE BOUNDARY BETWEEN TWO POROUS MEDIA,
Technion - Israel Inst. of Tech., Haifa. Dept. of Civil Engineering.
For primary bibliographic entry see Field 02L.
W73-07896

STEADY SEEPAGE FLOW TO SINK PAIRS SYMMETRICALLY SITUATED ABOVE AND BELOW A HORIZONTAL DIFFUSING INTERFACE: 1. PARALLEL LINE SINKS,
Johns Hopkins Univ., Baltimore, Md. Dept. of Environmental Engineering.
E. J. Wolanski, and R. A. Wooding.
Water Resources Research, Vol 9, No 2, p 415-425, April 1973. 7 fig, 16 ref.

Descriptors: *Saline water-freshwater interfaces, *Diffusion, *Groundwater movement, *Saline water intrusion, Disperison, Encroachment, Withdrawal, Steady flow, Saturated flow, Safe yield, Boundary layers.

The flow regime close to a pair of wells situated on opposite sides of a diffusing interface between freshwater overlying saline water is illustrated using an idealized symmetrical system. The interface is horizontal, and gravitational effects are neglected. A uniform flow is parallel to the interface, and the fresh and saline fluids are separated by a thin impermeable layer at a finite distance upstream from the sinks. The edge of the impermeable layer provides a definite starting point for diffusive mixing at the interface. As part of a large-scale gravity system, flow in the upper fluid should bear a qualitative resemblance to real situations where freshwater is moving slowly over nearly stationary saltwater. Boundary layer theory is used to treat the growth of the diffusive mixing layer, modified by the presence of the sinks. The flux of salt to the upper sink is calculated as a function of the spacing and strength of the uniform flow. A periodic system of sink pairs without an applied flow is also considered. Exact solutions of the boundary layer equations are readily obtained for the two-dimensional case. (Knapp-USGS)
W73-07897

UNSTEADY FLOW TO A PARTIALLY PENETRATING, FINITE RADIUS WELL IN AN UNCONFINED AQUIFER,
Washington Univ., Seattle.
K. L. Kipp, Jr.
Water Resources Research, Vol 9, No 2, p 448-462, April 1973. 8 fig, 18 ref.

Descriptors: *Unsteady flow, *Groundwater movement, *Equations, *Water yield, *Drawdown, Saturated flow, Mathematical studies, Numerical analysis, Water wells, Hydraulic conductivity, Aquifer testing, Hydrogeology.
Identifiers: *Partially penetrating wells.

Unsteady flow to a single, partially penetrating well of finite radius in an unconfined aquifer is solved theoretically. The aquifer is homogeneous, isotropic, and infinite both in thickness and lateral extent. Perturbation expansion techniques linearize the free surface boundary conditions, so that the solution satisfies the boundary conditions through first order provided that the drawdowns remain small, and that a time limit is imposed. The basic potential field is created by distributing dipole moments over the surface of the well bore and solving the resulting integral equation numerically. The solution can be used to model pumped well behavior for the initial period after the start of pumping. This solution is not restricted to the constant flow rate or constant head modes of simulation. The assumption of constant discharge operation in earlier, more approximate solutions to this problem is more realistic than the assumption of constant head operation. (Knapp-USGS)
W73-07898

THE TRANSIENT FLOW PROBLEM - A ONE-DIMENSIONAL DIGITAL MODEL,
Wyoming Univ., Laramie. Dept. of Civil and Architectural Engineering.
R. Herrmann.
MSc thesis, January, 1972. 58 p, 16 fig, 41 ref, 2 append. OWRR-A-001-WYO (53).

Descriptors: *Groundwater basins, *Groundwater movement, *Numerical analysis, *Simulation analysis, Digital computers, *Flow, Equations, Water table, Porous media, Rivers, Drains, Wells, Saturated flow, Unsaturated flow, Continuity equation, Management, Dupuit-Forchheimer theory, Darcy's law, Laplaces equation, Steady flow, Unsteady flow, Surface waters, Mathematical models, Systems analysis.
Identifiers: Finite difference techniques, Boussinesq equation, Glover equation.

Analysis of groundwater basins can be improved by utilizing numerical techniques and digital models to simulate underground flow situations. A solution of the combined flow problem is developed from existing theories of groundwater motion. The flow equations are developed for saturated flow below and unsaturated flow above the water table; a review of the development of these equations leads to a one dimensional equation set which solves a two dimensional flow problem by an implicit iterative procedure. A digital model is developed that solves the boundary value problem for time dependent flow through porous media with a free water table. The model is postulated so that extensive knowledge of flow theory and the physics of flow in porous media are not necessary for understanding the model equation development. The method of solution is based on a finite difference approximation of theoretical time dependent flow equations. The practical applications are for flow to a river, drain, or well. The assumptions necessary to the development of the practical solution of a field or hypothetical flow problem are discussed and justified. (Bell-Cornell)
W73-07916

EVENT-RELATED HYDROLOGY AND RADIONUCLIDE TRANSPORT AT THE CANNIKIN SITE, AMCHITKA ISLAND, ALASKA,
Nevada Univ., Reno. Desert Research Inst.
For primary bibliographic entry see Field 05B.
W73-07961

HYDROGEOLOGIC CHARACTERISTICS OF THE VALLEY-FILL AQUIFER IN THE ARKANSAS RIVER VALLEY, BENT COUNTY, COLORADO,
Geological Survey, Washington, D.C.
For primary bibliographic entry see Field 07C.
W73-08068

HYDRAULIC TESTS IN HOLE UAE-3, AMCHITKA ISLAND, ALASKA,
Geological Survey, Lakewood, Colo.
For primary bibliographic entry see Field 04B.
W73-08071

GEOHYDROLOGY AND ARTIFICIAL-RECHARGE POTENTIAL OF THE IRVINE AREA, ORANGE COUNTY, CALIFORNIA,
Geological Survey, Menlo Park, Calif. Water Resources Div.
For primary bibliographic entry see Field 04B.
W73-08072

HYDROGEOLOGY AND ENGINEERING GEOLOGY (GIDROGEOLOGIYA I INZHENERNAYA GEOLOGIYA).
Gornyi Institut, Leningrad (USSR).

Zapiski Leningradskogo Gornogo Instituta im. G. V. Plekhanova, Vol 62, No 2, Leningrad, Tolstikhin, N. I., and Kiryukhin, V. A., editors, 1971. 136 p.

Descriptors: *Hydrogeology, *Engineering geology, *Investigations, Groundwater, Groundwater movement, Aquifers, Reservoirs, Intakes, Permafrost, Properties, Rocks, Clays, Salts, Mineralogy, Mining, Quarries, Water types, Water

pollution, Analytical techniques, Maps, Dispersion.
Identifiers: *USSR, Mineral deposits, Icings.

A wide range of hydrogeological and geologic-engineering problems is examined in this collection of 19 papers published by the Leningrad Mining Institute. The topics discussed include groundwater classification and distribution; groundwater in regions of perennially frozen ground; subsurface flow in southern regions of the Soviet Far East; dispersion halos in suprapermafrost waters; salt migration in humid regions; chemical and physico-mechanical properties of clay rocks; and icings in river basins of East Siberia. Techniques are presented for compiling geologic-engineering maps, and investigations are made of mine-water pollution in quarries and of changes in geologic-engineering conditions of mineral deposits after open-pit mining. (Josefson-USGS)
W73-08163

A PLAN FOR STUDY OF WATER RESOURCES IN THE PLATTE RIVER BASIN, NEBRASKA--WITH SPECIAL EMPHASIS ON THE STREAM-AQUIFER RELATIONS,
Geological Survey, Lincoln, Nebr. Water Resources Div.
For primary bibliographic entry see Field 02A.
W73-08357

HYDROLOGICAL EFFECTS OF THE CANNIKIN EVENT,
Geological Survey, Denver, Colo.
For primary bibliographic entry see Field 04B.
W73-08367

MAP SHOWING APPROXIMATE GROUNDWATER CONDITIONS IN THE PARKER QUADRANGLE, ARAPAHOE AND DOUGLAS COUNTIES, COLORADO,
Geological Survey, Washington, D.C.
For primary bibliographic entry see Field 07C.
W73-08369

AVAILABILITY OF GROUNDWATER, HARTFORD NORTH QUADRANGLE, CONNECTICUT,
Geological Survey, Washington, D.C.
For primary bibliographic entry see Field 07C.
W73-08370

FUMAROLE WITH PERIODIC WATER FOUNTAINING, VOLCAN ALCEDO, GALAPAGOS ISLANDS,
Arizona Univ., Tucson. Dept. of Geosciences.
For primary bibliographic entry see Field 02K.
W73-08377

NATURAL AND ARTIFICIAL GROUND-WATER RECHARGE, WET WALNUT CREEK, CENTRAL KANSAS,
Geological Survey, Lawrence, Kans.
For primary bibliographic entry see Field 04B.
W73-08379

2G. Water in Soils

SPLASH EROSION RELATED TO SOIL ERODIBILITY INDEXES AND OTHER FOREST SOIL PROPERTIES IN HAWAII,
Forest Service (USDA), Berkeley, Calif. Pacific Southwest Forest and Range Experiment Station.
For primary bibliographic entry see Field 02J.
W73-07888

SEEPAGE STEPS IN THE NEW FOREST, HAMPSHIRE, ENGLAND,
For primary bibliographic entry see Field 02J.
W73-07891

MODELING INFILTRATION DURING A STEADY RAIN,
Minnesota Univ., St. Paul. Dept. of Agricultural Engineering.
R. G. Mein, and C. L. Larson.
Water Resources Research, Vol 9, No 2, p 384-394, April 1973. 10 fig, 2 tab, 23 ref.

Descriptors: *Infiltration, *Rainfall-runoff relationships, *Mathematical models, Wetting, Soil water movement, Rainfall intensity, Hydraulic conductivity.

A simple two-stage model describes infiltration under a constant intensity rainfall into a homogeneous soil with uniform initial moisture content. The first stage predicts the volume of infiltration to the moment at which surface ponding begins. The second stage, which is the Green-Ampt model modified for the infiltration prior to surface saturation, describes the subsequent infiltration behavior. A method is given for estimating the mean suction of the wetting front. Comparison of the model predictions with experimental data and numerical solutions of the Richards equation for several soil types shows excellent agreement. (Knapp-USGS)
W73-07893

HEAT AND MASS TRANSFER OF FREEZING WATER-SOIL SYSTEM,
British Columbia Univ., Vancouver. Dept. of Chemical Engineering.
For primary bibliographic entry see Field 02C.
W73-07894

APPROXIMATE SOLUTIONS FOR NON-STEADY COLUMN DRAINAGE,
Asian Inst. of Tech., Bangkok (Thailand). Div. of Water Science and Engineering.
A. Arbhabhirama, and Z. U. Ahmed.
Water Resources Research, Vol 9, No 2, p 401-408, April 1973. 5 fig, 11 ref.

Descriptors: *Infiltration, *Percolation, *Drainage, *Soil water movement, Unsaturated flow, Capillary action, Equations, Porous media, Porosity, Permeameters.

Approximate solutions for nonsteady column drainage are obtained by application of pore size distribution in a capillary analogy. The analysis is based on the assumption that the drainable pore space of porous media can be represented by a set of capillary tubes of various sizes having a size distribution that provides the same capillary pressure-saturation relationship as the real media. The solutions obtained approximately describe the cumulative outflow and the saturated front as functions of time and measurable soil parameters. Theoretical solutions are verified by using experimental and numerical results. (Knapp-USGS)
W73-07895

A TECHNIQUE USING POROUS CUPS FOR WATER SAMPLING AT ANY DEPTH IN THE UNSATURATED ZONE,
Geological Survey, Lubbock, Tex.
W. W. Wood.
Water Resources Research, Vol 9, No 2, p 486-488, April 1973. 2 fig, 2 ref.

Descriptors: *Sampling, *Soil water, *Soil moisture, *Lysimeters, *Zone of aeration, Instrumentation, Unsaturated flow.
Identifiers: *Suction lysimeters, *Soil water sampling, Porous cup lysimeters.

Porous cups or suction lysimeters provide a simple and direct method for collecting water samples in the unsaturated zone. A new procedure is described in which a check valve is placed in the sample collection assembly. This construction permits complete collection at any depth without the loss of samples. A detailed description of construction and operation illustrates the advancement over previous designs. (Knapp-USGS)
W73-07901

MOVEMENT OF CHEMICALS IN SOILS BY WATER,
Illinois Univ., Urbana. Dept. of Agronomy.
L. T. Kurtz, and S. W. Melsted.
Soil Science, Vol 115, No 3, p 231-239, March 1973. 3 tab, 42 ref.

Descriptors: *Water chemistry, *Leaching, *Weathering, *Soil water movement, Ion transport, Translocation, Reviews, Porous media, Tracers, Tracking techniques, Diffusion, Adsorption, Ion exchange.

The movement of water through soils and the transport of chemicals in soil are reviewed. Sometimes the path of water permits deductions about the direction and extent that solutes have been leached and weathering products have been transported. Conversely the best way to trace the movement of soil water is often to dissolve something in it that can be easily traced. Dyes that can be detected visually or by fluorescence, radioactive ions or compounds, stable isotopes, and various salts and ions have been used as tracers. A major problem is that the solvent and solute rarely move at the same rate. Equations have been derived to describe leaching through soils. Although these have not been applied very extensively to weathering, horizon formation, and the other processes of soil development, curves showing the concentration of clay down through different profiles often resemble curves showing the concentration of a salt being leached down through different profiles. In the temperate zone, water is the primary agent of weathering and the agent of transport of the weathering products. The composition of drainage waters can give at least a partial indication of the composition of the soil and the soil moisture. Likewise the composition of stream waters at low flow partially reflects the composition of water and the degree of weathering in the soils of the region. (Knapp-USGS)
W73-07904

MODELING THE MOVEMENT OF CHEMICALS IN SOILS BY WATER,
Illinois Univ., Urbana. Dept. of Agronomy.
C. W. Boast.
Soil Science, Vol 115, No 3, p 224-230, March 1973. 2 tab, 31 ref.

Descriptors: *Leaching, *Ion transport, *Soil water movement, Model studies, Reviews, Porous media, Dispersion, Diffusion, Adsorption, Ion exchange, Translocation.

The movement of chemicals through soil may be described by applying modeling techniques to the soil-water system. Some of the macroscopic continuum theories are reviewed and are presented in table form. The emphasis of most studies of the movement of chemicals has been on the influence of steady water flow. Recently approximations have been made to give better solutions for nonsteady flows. In principle, flow models can be used to describe the processes of soil formation; however, the complexity of such mathematical models may be prohibitive. (Knapp-USGS)
W73-07905

SYSTEMS ANALYSIS IN IRRIGATION AND DRAINAGE,
California Univ., Riverside. Dry-Lands Research Inst.
For primary bibliographic entry see Field 03F.
W73-07923

RADIONUCLIDE DISTRIBUTION IN SOIL MANTLE OF THE LITHOSPHERE AS A CONSEQUENCE OF WASTE DISPOSAL AT THE NATIONAL REACTOR TESTING STATION.
Idaho Operations Office (AEC), Idaho Falls.
For primary bibliographic entry see Field 05B.
W73-07958

CONTRIBUTION TO THE STUDY OF THE MIGRATION OF RU106 IN SOILS,
Commissariat a l'Energie Atomique, Cadarache (France). Centre d'Etudes Nucleaires.
For primary bibliographic entry see Field 05B.
W73-07963

STATE-OF-THE-ART REVIEW OF THE DEFINITION OF QUALITY OF A SOIL SAMPLE AND MODERN TECHNIQUES OF SOIL SAMPLING,
Technische Universitaet, Berlin (West Germany).
For primary bibliographic entry see Field 08D.
W73-08063

THE ADAPTATION OF CRITICAL STATE SOIL MECHANICS THEORY FOR USE IN FINITE ELEMENTS,
For primary bibliographic entry see Field 08D.
W73-08066

INFLUENCE OF 'BOUND' WATER ON THE CALIBRATION OF A NEUTRON MOISTURE METER,
Ibadan Univ., (Nigeria). Dept. of Agronomy.
O. Babalola.
Soil Science, Vol 114, No 4, p 323-324, October 1972. 1 fig, 1 tab, 3 ref.

Descriptors: *Nuclear moisture meters, *Calibrations, *Soil moisture meters, *Water of hydration, Sorption, Clays, Clay minerals, Water of crystallization, Adsorption.
Identifiers: Bound water.

The calibration of a neutron moisture meter relates the count ratio to the volumetric content of water in soil dried at 105 deg C, the usual oven temperature. The count ratio that corresponds to a particular moisture content is influenced by the presence of hydrogen atoms as one of the components of clay lattice and organic matter present in the soil, the soil bulk density and certain elements in the soil which have a high capacity for capture of thermal neutrons. Heating Mashi soil to 600 deg C revealed appreciable quantities of bound water in quantities varying with depth. The neutron moisture meter was equally sensitive to these as to the water content as usually defined. The amount of bound water seems to correlate with the clay content. Although the measurement of an absolute soil water content, which does not indicate the amount of plant available water, is not essential in crop water consumption studies, it may more accurately define the calibration relationship when a profile is nonuniform and generally high in clay content and make the calibration for individual soil horizons unnecessary. (Knapp-USGS)
W73-08087

APPLICATION OF RHEOLOGICAL MEASUREMENTS TO DETERMINE LIQUID LIMIT OF SOILS,
Central Building Research Inst., Roorkee (India).
R. B. Hajela, and J. M. Bhatnagar.
Soil Science, Vol 114, No 2, p 122-130, August 1972. 7 fig, 2 tab, 9 ref.

Descriptors: *Rheology, *Liquid limits, *Soil properties, *Soil tests, Hysteresis, Laboratory tests, Shear, Shear strength, Viscosity, Stress, Clays, Soils.

The flow properties of clay paste from red, black and alluvial soils containing different clay minerals at small shearing stresses were studied using a viscometer. The Bingham yield stress obtained by extrapolation of the hysteresis curve of clay water system when plotted against moisture content in the soil gives an inflection at a moisture content corresponding to liquid limit of the soil. The results obtained by this method are comparable within 1% of those obtained by the Cup method. The Bingham yield stress at liquid limit was also found to be in conformity with the reported values. (Knapp-USGS)
W73-08089

GLOSSARY OF TERMS IN SOIL SCIENCE.
Canada Department of Agriculture Publication 1459, 1972. 66 p, 1 fig.

Descriptors: *Data collections, *Soil science, Translations, Canada.
Identifiers: *Glossaries.

The present glossary of technical terms used in soil science in Canada updates a preliminary edition of the glossary, based larely upon the glossary of the Soil Science Society of America (Soil Sci. Soc. Amer. Proc. 29:330-351, 1965) and printed in English as Part II of the Proceedings of the Canadian Society of Soil Science, 1967. The glossary is published by the Research Branch of the Canada Department of Agriculture in English and French. In the English edition the French equivalents follow the English terms. In the French edition the English equivalents follow the French terms. (Josefson-USGS)
W73-08161

EFFECT OF IRRIGATION TREATMENTS FOR APPLE TREES ON WATER UPTAKE FROM DIFFERENT SOIL LAYERS,
Volcani Inst. of Agricultural Research, Bet-Dagan (Israel).
For primary bibliographic entry see Field 03F.
W73-08327

AN IMPROVED VARIABLE-INTENSITY SPRINKLING INFILTROMETER,
Hebrew Univ., Jerusalem (Israel). Dept. of Soil Science.
For primary bibliographic entry see Field 07B.
W73-08340

THEORY OF WATER MOVEMENT IN SOILS: 8. ONE-DIMENSIONAL INFILTRATION WITH CONSTANT FLUX AT THE SURFACE,
Connecticut Agricultural Experiment Station, New Haven.
J-Y. Parlange.
Soil Sci, Vol 114, No 1, p 1-4, 1972, Illus.
Identifiers: *Soil water movement, Absorption, Constant, One-Dimensional, Equation, Flux, *Infiltration, Integration, Soils, Surface, Theory.

The 1-dimensional movement of water in a porous medium, when the flux of water is imposed at the surface, is solved analytically. The result agrees with the direct numerical integration of the infiltration equation and with results based on similarity arguments in the case of absorption. (See also W72-02728, W73-04106, W73-04225, and W73-05097 thru-05099)--Copyright 1972, Biological Abstracts, Inc.
W73-08341

ON TOTAL, MATRIC AND OSMOTIC SUCTION,
Saskatchewan Univ., Saskatoon.
J. Krahn, and D. G. Fredlund.
Soil Science, Vol 114, No 5, p 339-348, November 1972. 10 fig, 1 tab, 19 ref.

Descriptors: *Moisture tension, *Osmotic pressure, *Soil moisture, *Vapor pressure, Moisture stress, Osmosis, Water chemistry, Pore water, Laboratory tests, Moisture content, Pore pressure.
Identifiers: *Total suction (Soil moisture), *Osmotic suction (Soil moisture), *Matric suction (Soil moisture).

Independent measurements were made of matric, osmotic, and total suction where dry density and water content are used as the basis for comparison of all soil suction components. The values of osmotic suction determined on a saturation extract differ significantly from values obtained by using pore water obtained by squeezing. Applying a linear dilution factor to the saturation extract values also produces values that are substantially different from those obtained by the squeezer technique. The values obtained on the squeezed pore fluid were in much closer agreement with the difference between matric and total suction than were the values obtained using the saturation extract techniques. The squeezer technique appears to be a satisfactory way of obtaining pore fluid for the determination of the osmotic suction. The sum of independent measurements of matric and osmotic suction is equal to the measured total suction. Therefore, the generally accepted subdivision of total suction is experimentally verified. For remolded, compacted soils, the matric and total suctions are dependent on the molding water content but essentially independent of the dry density. (Knapp-USGS)
W73-08349

SULFIDE DETERMINATION IN SUBMERGED SOILS WITH AN ION-SELECTIVE ELECTRODE,
Ministry of Agriculture, Cairo (Egypt).
For primary bibliographic entry see Field 02K.
W73-08351

SOIL STRUCTURE AND ITS EFFECTS ON HYDRAULIC CONDUCTIVITY,
Punjab Agricultural Univ., Ludhiana (India). Dept. of Civil Engineering.
P. Basak.
Soil Science, Vol 114, No 6, p 417-422, December 1972. 6 fig, 1 tab, 9 ref.

Descriptors: *Hydraulic conductivity, *Particle shape, *Clay minerals, *Anisotropy, *Soil water movement, Permeameters, Permeability, Soil structure, Porosity.

Soil structure influences coefficients of radial and vertical conductivity. The flaky shape of clay particles and their orientation give more hydraulic conductivity in the horizontal than in the vertical direction. The hydraulic conductivity ratio increases as the void ratio or porosity decreases. (Knapp-USGS)
W73-08352

ESTIMATION OF THE SOIL MOISTURE CHARACTERISTIC AND HYDRAULIC CONDUCTIVITY: COMPARISON OF MODELS,
Agricultural Research Service, Beltsville, Md.
A. S. Rogowski.
Soil Science, Vol 114, No 6, p 423-429, December 1972. 4 fig, 12 ref.

Descriptors: *Soil water movement, *Mathematical models, *Hydraulic conductivity, Porosity, Soil moisture, Saturation, Unsaturated flow, Saturated flow, Reviews.

Three ways of modeling the moisture characteristic and two ways of modeling hydraulic conductivity of soils are compared. A modified Brooks and Corey conductivity model and the moisture characteristic associated with it are quite similar in form to a modified (ASR) pore-size interaction model of Green and Corey. The results from both compare well with experimental values

at higher moisture contents. However, the ASR model approximates the experimental results better at lower values of water content and over a wider moisture range. The moisture content and pressure at air entry are significant parameters of the soil water system. A linear model of the soil moisture characteristic underestimates experimental results when used as input into the modified pore-size interaction model of hydraulic conductivity. (Knapp-USGS)
W73-08353

PROCEDURE AND TEST OF AN INTERNAL DRAINAGE METHOD FOR MEASURING SOIL HYDRAULIC CHARACTERISTICS IN SITU,
Hebrew Univ., Rehovot (Israel). Dept. of Agriculture.
D. Hillel, V. D. Krentos, and Y. Stylianou.
Soil Science, Vol 114, No 5, p 395-400, November 1972. 5 fig, 2 tab, 17 ref.

Descriptors: *Soil water movement, *Soil moisture, *On-site tests, Hydraulic conductivity, Moisture tension, Percolation, Unsaturated flow, Moisture content.

A simplified procedure is given for determining the intrinsic hydraulic properties of a complete soil profile in situ. The need for determining the hydraulic properties of soil profiles in the field and available methods are reviewed. The vertical transient-state internal drainage process of the soil is analyzed. An instantaneous profile method for determining soil hydraulic properties is based on simultaneously monitoring the changing wetness and matric suction profiles during internal drainage. The experimental procedure and a technique for handling the data are described and illustrated. The method is not applicable where lateral movement of soil moisture is appreciable, but it is otherwise not limited to homogeneous profiles and can serve for layered profiles. The method is simple and practical enough for routine use. The relation of hydraulic conductivity to wetness can be applied to the analysis of drainage and evapotranspiration in actual field management. (Knapp-USGS)
W73-08354

WIND ERODIBILITY AS INFLUENCED BY RAINFALL AND SOIL SALINITY,
Agricultural Research Service, Manhattan, Kans. Soil and Water Conservation Research Div.
For primary bibliographic entry see Field 02J.
W73-08355

EQUATION FOR DESCRIBING THE FREE-SWELLING OF MONTMORILLONITE IN WATER,
Agricultural Research Service, Phoenix, Ariz. Water Conservation Lab.
D. H. Fink, and F. S. Nakayama.
Soil Science, Vol 114, No 5, p 355-358, November 1972. 1 fig, 1 tab, 6 ref.

Descriptors: *Expansive clays, *Hydration, *Mineralogy, *Montmorillonite, Osmosis, Adsorption, Expansive soils, Hysteresis, Soil physics.

The hydration and swelling properties of a soil depend in large part on the type (as well as the total amount) of colloidal material present. A theoretical equation was developed to describe the linear expansion of free-swelling Na- and Li-saturated montmorillonite clays in water. The equation permits a quantitative differentiation of the water of clay-swelling into three distinct parts: (1) water associated with free-swelling internal surfaces; (2) water associated with limited-swelling internal surfaces and with high-energy adsorption sites; and (3) external water associated with extrapacket surfaces and voids. This technique also should be applicable for studying the hydration properties of the component clay minerals in clay type mixtures which contain montmorillonite. (Knapp-USGS)
W73-08356

PLANT SUCCESSION ON TUNDRA MUDFLOWS: PRELIMINARY OBSERVATIONS,
Carleton Univ., Ottawa (Ontario). Dept. of Biology.
J. D. H. Lambert.
Arct J Arct Inst North Am. Vol 25, No 2, p 99-106, 1972. Illus.
Identifiers: Grasses, Herbs, *Mudflows, *Plant succession, Sedges, Soil, Succession, *Tundra, Soil types, *Canada.

Tundra mudflows are one of the characteristic features of arctic slopes with unstable soils. They generally occur during the early part of the thaw period, but may occur after a heavy rainfall. Only 2 relatively short-lived vegetation elements were evident and both are characteristic of disturbed sites. Islands of vegetation and soil of the type that dominate the slope before the mudflow are left scattered within the flow lines. Once a turf of grasses, sedges and herbs has formed the island vegetation is able to colonize the turf mat. Areas where previous mudflows have occurred are clearly recognizable both by a long depression parallel to the direction of the slope and terminal fan of debris. Detailed studies on such naturally-occurring phenomena would be of great value in view of increased use of heavy vehicular equipment by the oil and mining companies in the Canadian North.--Copyright 1973, Biological Abstracts, Inc.
W73-08409

ANNUAL CYCLES OF SOIL MOISTURE AND TEMPERATURE AS RELATED TO GRASS DEVELOPMENT IN THE STEPPE OF EASTERN WASHINGTON,
Washington State Univ., Pullman. Dept. of Botany.
R. Daubenmire.
Ecology. Vol 53, No 3, p 419-424. 1972. Illus.
Identifiers: Grass, *Soil moisture, Phenologies, Soil profiles, Soils, *Steppes, Soil temperature, *Washington.

The annual cycle of soil moisture use and recharge was followed in 8 climax steppe communities by making gravimetric analyses to a depth of 1 m. Soil temperatures were measured at depths of 50 and 100 cm. The data were related to the phenologies of the dominant grasses. Differences in soil moisture and temperature seem to contribute more toward explaining the distributions of these steppe communities than do chemical content of the soils or their profile characters.--Copyright 1973, Biological Abstracts, Inc.
W73-08436

AVAILABLE SOIL MOISTURE AND PLANT COMMUNITY DIFFERENTIATION IN DAVIES ISLAND, MIDDLE TENNESSEE,
Tennessee Technological Univ., Cookeville.
S. M. Stubblefield, and S. K. Ballal.
J Tenn Acad Sci. Vol 47, No 3, p 112-117. 1972. Illus.
Identifiers: *Soil moisture, Beech, Cedar, *Davies Island (Tenn), Differentiation, Hickory, Islands, Maple, Oak, Plant communities, Soils, *Tennessee, Vegetation, Soil types.

One of the dominating factors determining community differentiation is available soil moisture, which is intimately coupled with soil types. The moisture contents of various soil types found at Davies Island located in the Center Hill Reservoir are determined and correlated with the vegetational types supported by such soils. Species of plants found in 4 sites, namely a beech-maple complex, an oak-history complex, old fields, and cedar woods are listed, and their occurrence is discussed in relation to the available soil moisture.--Copyright 1973, Biological Abstracts, Inc.
W73-08440

2H. Lakes

ORGANIC WASTES AS A MEANS OF ACCELERATING RECOVERY OF ACID STRIP-MINE LAKES,
Missouri Univ., Columbia. Dept. of Civil Engineering.
For primary bibliographic entry see Field 05G.
W73-07808

WATER QUALITY CHANGES IN AN IMPOUNDMENT AS A CONSEQUENCE OF ARTIFICIAL DESTRATIFICATION,
North Carolina Univ., Chapel Hill. School of Public Health.
For primary bibliographic entry see Field 05C.
W73-07818

EXTRAPOLATION OF INDICATOR SCHEMES WITHIN SALT MARSHES,
Moskovskoe Obshchestvo Ispytatelei Prirody (USSR). Geographic Div.
M. A. Monakhov.
In: Lanscape Indicators--New Techniques in Geology and Geography: Consultants Bureau, Div of Plenum Press, London and New York, p 141-147, 1973. 52 ref. (Translated from Russian. Proceedings of Conference of Moscow Society of Naturalists, May 21-22, 1968, Moscow, Nauka Press).

Descriptors: *Terrain analysis, *Vegetation effects, *Salt marshes, Salt tolerance, Vegetation, Aerial photography, Remote sensing, Wetlands, Salinity, Chlorides, Sulfates, Sodium, Calcium.
Identifiers: *USSR.

Landscape indicator schemes may be used in some types of salt flats. Wet, incrusted or armored layers are barren of vegetation. In wet salt marsh with glasswort and annual species of Suaeda, the salt is chloride, mostly sodium chloride, and salinization is sometimes merely surficial, possibly diminishing appreciably with depth. In hummocky salt marsh with Halocnemum the salt is chloride; but with considerable sulfate content it is chiefly sodium chloride with appreciable calcium. In hilly or hummocky salt marshes, with hummocks about clumps of tamarisk, niter bush, and boxthorn, chlorides and sulfides are almost equally represented in the salts. This type of salt marsh is commonly combined with hummocky marshes with Halocnemum. In salt-marsh meadows with species of Puccinellia and Aeluropus, salt is chiefly sulfate. (Knapp-USGS)
W73-07848

TRANSIENT AND STEADY-STATE SALT TRANSPORT BETWEEN SEDIMENTS AND BRINE IN CLOSED LAKES,
Northwestern Univ., Evanston, Ill. Dept. of Geological Sciences.
A. Lerman, and B. F. Jones.
Limnology and Oceanography, Vol 18, No 1, p 72-85, January 1973. 5 fig, 1 tab, 11 ref.

Descriptors: *Salinity, *Saline lakes, *Ion transport, *Bottom sediments, Sedimentation, Leaching, Brines, Connate water, Ion exchange, Limnology, Diffusion, *Oregon, Saline water, Lakes, Salts.
Identifiers: *Lake Abert (Oregon).

A diffusional transport model predicts the rates of salt transport from pore fluids into lake waters. In a lake without outflow, dissolved salts may migrate across the sediment-water interface in response to a concentration difference between

lake and interstitial brine. Transport of salt upward is transient; its direction can be reversed by external input of salt or by depletion of salts stored in the sediments, and a steady-state concentration in lake water is not attainable. Downward transport can be a stationary process if the sedimentation rate is rapid compared with molecular diffusion of salt in interstitial brine, but characteristic rates are too slow to lead to steady-state concentrations within the lifetime of a closed lake. In Lake Abert, Oregon, diffusional flux upward was much more important than input of salt from other sources; 45% of the salt of lake brine in 1963-1964 was added from the sediment pore space during the preceding 25 years, only 0.1% from external inflow. The sediment source will dominate input during high water level. (Knapp-USGS)
W73-07850

THE ST. CLAIR RIVER DELTA: SEDIMENTARY CHARACTERISTICS AND DEPOSITIONAL ENVIRONMENTS,
Wisconsin Univ., Green Bay. Coll. of Environmental Sciences.
For primary bibliographic entry see Field 02J.
W73-07858

LAKES OF BELORUSSIA (BELORUSSKOYE POOZER'YE),
Belorussian State Univ., Minsk (USSR). Laboratoriya Ozerovedeniya.
O. F. Yakushko.
Izdatel'stvo 'Vysheyshaya Shkola', Minsk, 1971. 336 p.

Descriptors: *Limnology, *Paleolimnology, *Lakes, *Lake morphometry, *Lake morphology, Lake stages, Lake basins, Lake sediments, Sediment distribution, Particle size, Thermal properties, Water levels, Water properties, Water chemistry, Biology, Aquatic life, Geomorphology, Hydraulics, Hydrology, Geologic time.
Identifiers: *USSR, *Belorussia, Lake classifications, Macrophytes, Isopleths.

Present conditions of lakes in northern Belorussia are considered against the background of their origin and history in the Upper Pleistocene and Holocene. Morphometric, geomorphological, hydrological, hydrochemical, and biological indices are used to classify the lakes on the basis of nutrient content. The regional classification of lakes, based on a complex of geographical and limnological indices, reflects the interrelations of lake-basin morphometry and dynamics of water mass to the hydrological, hydrochemical, and biological features of lakes, and to the character of sediment accumulation. (Josefson-USGS)
W73-07907

PRODUCTIVITY OF FRESH WATERS,
G. Marlier.
Nat Belg. Vol 52, No 6, p 281-295, 1971. Illus.
Identifiers: *Daphnia, Insect, Larvae, Phytoplankton, *Productivity, Rotifers, *Zooplankton.

Phytoplankton, Cladocera (Daphnia), zooplankton, rotifers, insect larvae and others were studied in Mirwart pond to understand the paths which energy takes through the biocoenosis before becoming dissipated as heat, and of the routes which materials follow through the pond life before returning to their original state.—Copyright (c) 1972, Biological Abstracts, Inc.
W73-07930

INTERACTION OF YELLOW ORGANIC ACIDS WITH CALCIUM CARBONATE IN FRESHWATER,
Michigan State Univ., Hickory Corners. W. K. Kellogg Biological Station.
For primary bibliographic entry see Field 05B.
W73-07931

CHALK RIVER NUCLEAR LABORATORIES PROGRESS REPORT APRIL 1 TO JUNE 30, 1972, BIOLOGY AND HEALTH PHYSICS DIVISION, ENVIRONMENTAL RESEARCH BRANCH AND HEALTH PHYSICS BRANCH,
Atomic Energy of Canada Ltd., Chalk River (Ontario). Chalk River Nuclear Labs.
For primary bibliographic entry see Field 05C.
W73-07955

BACTERIAL AND ALGAL CHLOROPHYLL IN TWO SALT LAKES IN VICTORIA, AUSTRALIA,
For primary bibliographic entry see Field 05A.
W73-08018

REEDS CONTROL EUTROPHICATION OF BALATON LAKE,
Research Inst. for Water Resources Development, Budapest (Hungary).
For primary bibliographic entry see Field 05G.
W73-08025

THE LITTORAL MICROPHYTIC VEGETATION OF LAKE OHRID (IN SERBO-CROATIAN),
Hidrobioloski Zavod, Ochrida (Yugoslavia).
For primary bibliographic entry see Field 05C.
W73-08061

REPRODUCTION OF LAKE SEVAN KHRAMULYA (IN RUSSIAN),
E. M. Malkhin.
Tr Molodykh Uch Vses Nauchno-Issled Inst Morsk Rybn Khoz Okeanogr. 3. p 147-157, 1970, English summary.
Identifiers: Lakes, *Reproduction, *USSR, *Lake Sevan Khramulya.

Analysis of the structure of the spawning and fattening stages in the period preceding the start of spawning established the absence of spawning gaps in sexually mature khramulya. There were no strictly river spawning stages in khramulya, but another stage evidently exists, part of which involves spawning in rivers while another part involves spawning near the mouths of lakes.—Copyright 1972, Biological Abstracts, Inc.
W73-08104

ECOLOGICAL NOTES ON A LAKE-DWELLING POPULATION OF LONGNOSE DACE (RHINICHTHYS CATARACTAE),
Manitoba Univ., Winnipeg. Dept. of Zoology.
J. H. Gee, and K. Machniak.
J Fish Res Board Can. Vol 29 No 3, p 330-332. 1972. Illus.
Identifiers: *Dace, *Fish physiology, Ecology, Lakes, Fish population, Rhinichthys cataractae, Spawning.

Longnose dace were found from early July to Sept. in onshore waters on the east side of Hecla island in the southern basin of Lake Winnipeg. They were most abundant on rocky substrates, where they spawned in early July. It is suggested that at other times of the year these fish occupy deep channels between islands where current is present. While in onshore waters fish less than 5 g adjusted their buoyancy by altering swimbladder volume in response to presence or adsence of wave action. Swimbladder volume was similar to that of river-dwelling populations of this species.—Copyright 1972, Biological Abstracts, Inc.
W73-08130

THE FORMATION OF IRON AND MANGANESE-RICH LAYERS IN THE HOLOCENE SEDIMENTS OF THUNDER BAY, LAKE SUPERIOR,
Lakehead Univ., Thunder Bay (Ontario).
For primary bibliographic entry see Field 02J.
W73-08135

THE RESULTS OF INTRODUCING MONODACNA COLORATA EICHW. INTO THE ZAPOROZHYE RESERVOIR, (IN RUSSIAN),
Dnepropetrovskii Gosudarstvennyi Universitet (USSR). Institut Gidrobiologii.
I. P. Lubyanov, V. L. Bulakhov, and V. I. Zolotareva.
Gidrobiol Zh. Vol 8 No 1, p 103-105. 1972.
Identifiers: Fish food, *Monodacna-colorata, Reservoirs, USSR, *Zaporozhye Reservoir.

One promising method of enriching the natural food base of fish is the acclimatization of estaurine Caspian invertebrates. Data on the location, living conditions, density, morphological characteristics, growth rate and age composition suggest that M. colorata has not only survived in the reservoir, but is in one of the final stages of acclimatization and naturalization.—Copyright 1972, Biological Abstracts, Inc.
W73-08136

LAKE SURVEY CENTER 1972 RESEARCH PROGRAM.
National Ocean Survey, Detroit, Mich. Lake Survey Center.

Available NTIS, Springfield, Va. 22151 as COM 72-10677 Price $3.00 printed copy; $1.45 microfiche. Technical Memorandum NOS LSCD 2, January 1972. 59 p, 2 fig.

Descriptors: *Limnology, *Great Lakes, *Hydrology, *Data collections, Lake morphology, Currents (Water), Water supply, Water quality, Water pollution effects, Environmental effects, Ice-water interfaces, Snow, Streamflow, Inflow, Shores, Shore protection, Sediment transport, Sedimentology, Water level fluctuations, Waves (Water), Precipitation (Atmospheric), Evaporation, Mathematical models.

The Great Lakes research program of the Limnology Division of NOAA consists of data collection, investigations, and studies in five specific fields: water motion within the lakes and connecting rivers; lake water characteristics and their relationships to natural and manmade processes; limnogeology, including interaction of water masses with shoreline and lake bottom; lake hydrology and factors affecting water supply to lakes including effects of ice cover and snow accumulation; and limnologic systems studies with the aim of establishing complex interrelationships that exist between individual processes within the Great Lakes and their immediate environment. All research investigations are closely coordinated with other agencies and educational institutions engaged in related work. For example, cooperative programs of data collection have been established with the Corps of Engineers, EPA, AEC, the University of Michigan, Canadian Department of Transport, the University of Toronto, and others. The techniques, facilities, and equipment used in the data collection investigations are described. (Woodard-USGS)
W73-08160

CHARACTERISTICS OF TWO DISCRETE POPULATIONS OF ARCTIC CHAR (SALVELINUS ALPINUS L.) IN A NORTH SWEDISH LAKE,
N. A. Nilsson, and O. Filipsson.
Ref Inst Freshwater Res Drottningholm. 51 p 90-108. 1971. Illus.
Identifiers: Arctic, *Char, Daphnia, Discrete, Scandinavia fish diets, Gammarus, Fish growth, Lakes, *Fish populations, Salvelinus alpinus.

Several subpopulations of arctic char have long been recognized in Scandinavia by both local fishermen and biologists. A description is given of 2 populations in a lake in southern Swedish Lapland, referred to as 'ordinary char' and 'blattjen.' The 2 forms could be distinguished from the shape of their otoliths, 1 type being characterized as 'ar-

row-shaped' (ordinary char), the other as 'drop-shaped' (blattjen). A slight difference in gillraker counts of the 2 populations could be discerned, but none for pyloric caeca. Coloration of the 2 populations at spawning time is different, as is flesh coloration in autumn. Most ordinary char were caught with floated gill nets in the pelagic; most blattjen with sinking nets in the littoral region of the lake. In autumn ordinary char fed on planktonic Crustacea, mainly Daphnia, while blattjen fed on bottom animals, mainly Gammarus. Even fish that were caught at the same stations in the littoral region displayed a similar food segregration. There was a highly significant difference in growth rate between the 2 forms, the ordinary char being more fast-growing than the blattjen. Ordinary char caught with sinking gill nets as well as blattjen display a greater variability in growth rate than do char caught with floated nets. A greater fraction of the stomachs of the ordinary char had a high degree of filling as compared with blattjen. It is suggested that differences in stomach fullness as well as in growth rates result from a more severe competition between brown trout and blattjen. There was an evident difference in the average size of the spawning fish belonging to the different populations, the blattjen being smaller at maturity. They were on average older than the ordinary char at maturity. The 2 populations are regarded as separate species. Although there are many parallels with the char populations of other lakes in Scandinavia, further analysis is needed before detailed generalization as to the whole complex can be made.--Copyright 1972, Biological Abstracts, Inc.
W73-08244

EUTROPHICATION AND LOUGH NEAGH,
New Univ. of Ulster, Coleraine (Northern Ireland).
For primary bibliographic entry see Field 05C.
W73-08252

STIMULATION OF PHYTOPLANKTON GROWTH BY MIXTURES OF PHOSPHATE, NITRATE, AND ORGANIC CHELATORS,
Virginia Inst. of Marine Science, Gloucester Point.
For primary bibliographic entry see Field 05C.
W73-08253

ROLE OF PHOSPHORUS IN EUTROPHICATION AND DIFFUSE SOURCE CONTROL,
Wisconsin Univ., Madison. Water Chemistry Program.
For primary bibliographic entry see Field 05C.
W73-08255

ECOLOGICAL TOPOGRAPHY OF FISH POPULATIONS IN RESERVOIRS, (IN RUSSIAN),
A. G. Poddubnyi.
Nauka: Leningrad. 1971. 312p. Illus.
Identifiers: *Abramis-Brama, Book, *Ecological studies, *Fish populations, Reservoirs, Topography.

The reservoirs in the Volga system and some other reservoirs, lakes and rivers were studied. Factural data were obtained by studying the Abramis brama population of the Rybinskii, Gorkii and Kuibyshev reservoirs.--Copyright 1972, Biological Abstracts, Inc.
W73-08270

SPAWNING AND SPAWNING GROUNDS OF THE BURBOT LOTA LOTA (L.), (IN RUSSIAN),
V. N. Sorokin.
Vopr Ikhtiol. Vol 11, No 6, p 1032-1041, 1971. Illus.
Identifiers: *Burbot, Lakes, Lota-Lota, *Spawning grounds, USSR, *Lakes Baikal.

Spawning of the burbot (Lota lota) in the Lake Baikal system was investigated. The spawning grounds of the burbot in the Bugul'deika and Kichera rivers are described and information is given on the density of the burbot eggs at the spawning grounds and the characteristics of their development.--Copyright 1972, Biological Abstracts, Inc.
W73-08277

STRUCTURE OF SPAWNING SCHOOLS, GROWTH AND SEXUAL MATURATION OF THE BREAM ABRAMIS BRAMA (L.) OF THE KREMENCHUG RESERVOIR, (IN RUSSIAN).
Akademiya Nauk URSR, Kiev. Instytut Hidrobiologii.
V. P. Bruenko, and I. E. Dyachuk.
Vopr Ikhtiol. Vol 11, No 6, p 955-968, 1971. Illus.
Identifiers: Abramis-brama, *Bream, Growth, *Kremenchug Reservoir (USSR), Reservoir, *Sexual maturation, Spawning, USSR.

The dynamics of the sex, size-weight, and age composition of spawning schools of A. brama from the time of formation of the Kremenchug reservoir (1962) to the present was investigated. The average age of spawners has increased in recent years in comparison with the early years of existence of the reservoir, but the number of age groups in the spawning population has decreased. The growth rate of the bream has decreased regularly as the reservoir has formed and commercial fish fauna has become established in it. The rate of sexual development is slowing down.--Copyright 1972, Biological Abstracts, Inc.
W73-08278

A SURVEY ON CLONORCHIS SINENSIS AND SOME OTHER SPECIES OF METACERCARIAE IN FRESHWATER FISHES FROM LAKE IZU-NUMA, MIYAGI PREFECTURE,
N. Suzuki.
Res Bull Meguro Parasitol Mus. 5, p 19-20, 1971.
Identifiers: *Clonorchis-sinensis, Fishes, Lake Izu-Numa, *Japan, Lake, *Metacercariae, Miyagi, Prefecture, Species, Survey.

Scales, fins, opercula and muscles of 5 species of fishes collected from the lake were examined for metacercaria under a dissecting microscope. Four were all found infected with the metacercariae. Clonorchiasis is still present around the lake.--Copyright 1972, Biological Abstracts, Inc.
W73-08283

GAMETOGENESIS OF SOCKEYE FRY IN LAKE DALNEM, (IN RUSSIAN),
M. Ya. Ievleva.
Izv Tikhookean Nauchno-Issled Inst Rybn Khoz Okeanogr. 78, p 81-104, 1970. English summary.
Identifiers: Fecundity, Fishery, Forms, Fry, *Gametogenesis, Lakes, *Sockeye fry, USSR, *Lake Dalem (USSR).

According to spermato- and oogenesis of sockeye juveniles a stage of gonadal development was exposed for the differentiation of future dwarfs which mature in the lake and juveniles of the migrating portion of the school. The earliest signs indicating transition of a male juvenile to a dwarf is intensive reproduction of spermatogonia beginning less than a year before spawning of dwarfs. In the predownstream migrant period oocytes in the early phases of trophoplasmatic growth are found in both future dwarf females and 3 and 4 yr old downstream migrants. During this period dwarf females are more clearly distinguished from downstream juveniles by their low fecundity.--Copyright 1972, Biological Abstracts, Inc.
W73-08285

THE BIOLOGY OF LAMPREYS, VOL 1.

Academic Press: London, England; New York, N. Y., 1971. M. W. Hardisty, I. C. Potter, Ed. p 423, Illus. Maps. Pr. $23.50.
Identifiers: Biology, *Distribution, Ecology, *Embryology, *Lampreys, Phylogeny, Systematics, *Great Lakes, Aquatic life.

Information on the systematics, distribution, life histories and ecology of lampreys. A chapter on the distribution, phylogeny and taxonomy of the lampreys is followed by a discussion on lampreys, in the fossil record, including descriptions of fossils, calcifications in the head skeleton of modern lampreys and systematics. Chapter 3 presents a detailed discussion on the behavior, ecology and growth of larval lampreys, followed by a chapter on the general biology of adult lampreys, including downstream migration and parasitic feeding, the spawning phase and the size and sex composition of adult populations. Next, a chapter is presented on sea lampreys in the Great Lakes of North America, including the following aspects; sea lamprey invasion of the Great Lakes; distribution of lampreys in the Great Lakes before effective control; lamprey control; effects of chemical control; biological changes in the sea lamprey; and the future of sea lamprey control in the Great Lakes. Chapter 6, paired species, covers the distribution of species, the basis of brook lamprey evolution and speciation, followed by a discussion in chapter 7 of lamprey chromosomes, including techniques, chromosome numbers, size and form of chromosomes, karyotype evolution and taxonomic considerations. Chapter 8, gonadogenesis, sex differentiation and gametogenesis, covers gonadogenesis and sex differentiation, quantitative aspects of gonadogenesis, gonad development in the post-metamorphic period and the somatic tissues of the gonad. The final chapter, embryology, discusses stages in embryonic development, features of lamprey embryology and experimental investigations.--Copyright 1973, Biological Abstracts, Inc.
W73-08326

MACROFAUNA COMMUNITIES OF THE LITTORAL OF MIKOLAJSKIE LAKE,
Polish Academy of Sciences, Warsaw (Poland). Inst. of Experimental Biology.
K. W. Opaliuski.
Pol Arch Hydrobiol. Vol 18, No 3, p 275-285. 1971. Illus.
Identifiers: Chironomidae, Communities, Lakes, *Littoral, *Macro fauna, Mikolajskie Lake (Poland), Habitats.

Discussed are 4 types of macrofauna communities connected with various substrata accessible for colonization in the shallow lake littoral. The tips of last year's reeds, broken under the water surface, are newly discovered habitat. They are colonized by a rich and almost monospecific community of Chironomidae larvae. Only 3 Chironomidae spp. are common for all the 4 communities.--Copyright 1972, Biological Abstracts, Inc.
W73-08336

THE CHEMICAL NATURE OF THE SUBSTANCES THAT INDUCE THE REDDENING OF LAKE TOVEL: III. STUDY OF A REMARKABLE PHENOMENON OF PRE-REDDENING (IN ITALIAN),
For primary bibliographic entry see Field 05C.
W73-08344

SEASONAL VARIATION OF A LIMNIC BEACH,
Catholic Univ. of America, Washington, D.C. Dept. of Geography.
For primary bibliographic entry see Field 02J.
W73-08371

FEEDING PATTERNS OF POTENTIAL MOSQUITO VECTORS,
Walter Reed Army Inst. of Research, Washington, D.C.
For primary bibliographic entry see Field 05B.
W73-08447

2I. Water in Plants

CONDITION OF THE FOOD BASE OF PLANKTOPHAGIC FISH IN ARAKUMSK BODIES OF WATER, (IN RUSSIAN),
O. N. Nasukhov.
Tr Vses Nauchno-Issled Inst Prudovogo Rybn Khoz. 18 p, 149-154. 1971. (English summary).
Identifiers: *Algae, *Arakumsk, Blue-green algae, Cladocera, Copepoda, Diatoms, Fish, Food, *Planktophagic, Fish, Rotifer, USSR, Deltas.

The Arakumsk bodies of water are located in the Terek River delta and have an area equal to 16,700 ha. Species diversity was pronounced in stations located in reed thickets. The maximum index of biomass during the period of observation was observed in the lower body of water. Juvenile fish in these waters fed mainly on Cladocera and Copepoda. Rotifers, blue-green and diatomous algae were also found in the stomachs of young fish.--Copyright 1972, Biological Abstracts, Inc.
W73-07832

THE INFLUENCE OF WEATHER UPON THE ACTIVITY OF SLUGS,
Birmingham Univ. (England). Dept. of Genetics.
T. J. Crawford-Sidebotham.
Oecologia (Berl). Vol 9, No 2, p 141-154. 1972.
Identifiers: Arion hortensis, Arion subfuscus, Milax budapestensis, *Slugs, *Weather.

The activity of slugs was measured by a catch per unit effort sampling system based upon night searching, and was related to the microclimatological conditions in the habitat by regression analysis. The activity of Arion hortensis Fer., A. subfuscus (Drap.), A. lusitanicus Mab., Milax budapstensis (Hazay), and of all slugs irrespective of species, was best related to temperature and vapor pressure deficit. The relevance of these results to the application of effective methods for the control of slugs is discussed. --Copyright 1972, Biological Abstracts, Inc.
W73-08012

COMPARATIVE BEHAVIOR AND HABITAT UTILIZATION OF BROOK TROUT (SALVELINUS FONTINALIS) AND CUTTHROAT TROUT (SALMO CLARKI) IN SMALL STREAMS IN NORTHERN IDAHO,
J. S. Griffith, Jr.
J Fish Res Board Can. Vol 29, No 3, p 265-273. 1972.
Identifiers: *Fish behavior, Brook trout, Cutthroat trout, Habitats, *Idaho, Salmo clarki, Salvelinus fontinalis, Streams, *Trout.

Individual S. fontinalis and S. clarki trout communicated with similar behavioral signals, both in laboratory stream-channels and in northern Idaho streams. Underyearling brook trout were less active socially than equal-sized cutthroat trout in laboratory observations. In study streams, brook trout maintained a 20-mm size advantage over cutthroat of the same age-groups throughout their lives, as they emerged from the gravel before cutthroat. Because of this size advantage, underyearling brook trout of sizes found in study streams in Sept. consistently dominated in experiments the underyearling cutthroat with which they normally lived. But in study streams underyearlings of the 2 spp. utilized different microhabitats, particularly with respect to water depth, and so minimized chances for interaction. Yearlings and older brook trout initiated 40% fewer aggressive encounters under laboratory conditions than did equal-sized cutthroat, and did not displace the cutthroat. In study streams with sympatric populations, cutthroat trout of these age-groups occupied territories with focal points of higher water velocities (averaging 10.2-10.3 cm/sec) than those occupied by brook trout (averaging 7.6-9.6 cm/sec). Considerable interspecific overlap in other habitat characteristics occurred for trout of age-groups I and II. The oldest members of the 2 spp. segregated more distinctly, as the brook trout lived closer to overhead cover. Copyright 1972, Biological Abstracts, Inc.
W73-08158

THE VALUE OF SOME QUALITY INDEX NUMBERS OF SOME MARSH PLANTS OF THE DANUBE DELTA,
Societatea de Stiinte Biologice din R.S.R., Bucharest.
G. Tarita.
Commun Bot. 12: p 539-544. 1971. Illus. English summary.
Identifiers: *Nutrients, Carex pseudocyperus M, *Danube Delta, *Marsh plants, Phragmites communis M, Plants, Protein, Romania, Scirpus lacustris M, Typha aungustifolia M, Index numbers.

Nutritional components of Phragmites communis, Typha angustifolia, Scirpus lacustris and Carex pseudocyperus were studied. Young P. communis was richest in nutrients: Typha and Scirpus were medium and Carex was very low. The protein index number changes in accordance with the vegetation period.--Copyright 1972, Biological Abstracts, Inc.
W73-08242

THE FISH FAUNA OF NERA RIVER AND ITS PROTECTION, (IN RUMANIAN)
Academia R.S.R., Bucharest. Institutul de Biologie.
For primary bibliographic entry see Field 05G.
W73-08269

SEASONAL AND AGE-RELATED CHANGES IN THE DIET OF SILVER SALMON IN THE PLOTNIKOV AND AVECHE RIVERS, (IN RUSSIAN),
Z. K. Zorbidi.
Izv Tikhookean Nauchno-Issled Inst Khoz Okeanogr, Vol 78, p 129-150, 1970. English summary.
Identifiers: *Age, Aveche River, Plotnikov River, Salmon, *Seasonal, *Silver salmon, USSR, *Fish diets.

In the spring-summer period silver salmon feeding in the river is characterized by high indices of gastric accumulation. The species composition of the consumed organisms and dimensions of the same species change according to the season. The differences exist in the diet of silver salmon young of different age groups.--Copyright 1972, Biological Abstracts, Inc.
W73-08272

BIOLOGY OF PACIFIC LAMPREY LARVAE, (IN RUSSIAN),
V. G. Svirskii.
Uch Zap Dal'nevost Univ., Vol 15, No 3, p 114-119, 1971.
Identifiers: Biology, *Khabarovsk (USSR), *Lamprey larvae, *Pacific lamprey, USSR, Wintering.

Wintering larvae of the Pacific lamprey were found in the area of Khabarovsk on a branch of the Talga. In particular many ammocoetes were observed in the littoral area at ground water outlets. Movement to these areas was associated with a decreased O2 content in the winter period of 0.7-0.2 mg/l, which was the main reason for the death of lampreys in water bodies associated with the Amur. Lamprey wintered in the ground; they

reached 2960 specimen/m2 in some areas. A small portion of the head protruded from the ground with the oral funnel. In contrast to larvae, young lamprey which had completed metamorphosis never buried themselves in the sand. Metamorphosis of Pacific lamprey larvae begain in the first days of Dec. The sexual ration in larvae was about 1:1. Determination of sex began at a length of 80-90 mm.--Copyright 1972, Biological Abstracts, Inc.
W73-08273

CONTRIBUTIONS TO THE KNOWLEDGE OF THE GREEN FROG PARASITE-FAUNA (RANA RIDIBUNDA AND RANA ESCULENTA) FROM THE DANUBE FLOODED LAND AND DELTA, (IN RUMANIAN),
Ion Radulescu, E. Cristea, A. Cristea, and B. Demetriuc.
Bul Cercet Piscic. Vol 29, No 3, p 92-96, 1970. English summary.
Identifiers: Acanthocephala, *Danube, Delta, Fauna, *Frogs (Green), Land, Nematodes, Parasites, Protozoa, *Rana-Esculenta, *Rana-Ridibunda, Trematodes.

A study made in 1968 on 67 adult frogs (R. ridibunda and R. esculenta) captured in 2 different biotopes showed infestation (86.16%) with protozoans, trematodes, nematodes and acanthocephala. The principal ways of infesting the frog with parasites are presented, as well as the possibilities of controlling them. The best frogs for eating and marketing are considered those captured during cold seasons, when infestation is reduced.--Copyright 1972, Biological Abstracts, Inc.
W73-08280

HELMINTH FAUNA OF BIRDS OF TWO POND SYSTEMS OF THE MILICZ PONDS RESERVE,
Szkola Glowna Gospodarstwa Wiejskiego, Warsaw (Poland).
T. Sulgostowska, and W. Korpaczewska.
Acta Parasitol Pol. Vol 20, No 1-11, p 75-94, 1972. Illus.
Identifiers: Acanthocephalan, Birds, Cestode, *Dubininolepis-podicipina, Fauna, *Helminth bauna, Milicz, Nematodes, *Poland, Ponds, Trematode.

Examination of 281 birds of 8 families revealed 27 trematode, 40 cestode, a nematode and an acanthocephalan species. A trematode and 2 cestode species are new records for Poland. New hosts were recorded for 8 trematode and 7 cestode species in Poland, 5 trematode and 10 cestode species are reported from new hosts. Dubininolepis podicipina is regarded a valid species. The cestode fauna was much more numerous than that of trematodes, both as regards incidence and intensity of infection.--Copyright 1972, Biological Abstracts, Inc.
W73-08281

SCOLYTIDAE (COLEOPTERA) FROM THE SITE OF THE FUTURE DAM LAKE OF THE IRON GATES OF THE DANUBE, (IN RUMANIAN),
S. Negru.
Trav Mus Hist Nat Grigore Antipa. 11, p 175-189, 1971. English summary.
Identifiers: *Coleoptera, Dam, *Danube, Host, Iron-Gates, Lake, Plant, *Scolytidae, Site.

Data are presented on 37 spp. (and 4 forms) of Scolytidae that were found for first time within the Iron Gates area and on 2 spp. already mentioned in the literature and now found again in the same zone. Some of these species are of great economic interest as harmful insects. Other data are given on the host-plants on which were found the mentioned Scolytidae. A table of the host-plants with the respective species of Scolytidae is given.--Copyright 1972, Biological Abstracts, Inc.
W73-08284

REVIEW OF STUDIES ON FEEDING OF AQUATIC INVERTEBRATES CONDUCTED AT THE INSTITUTE OF BIOLOGY OF INLAND WATERS, ACADEMY OF SCIENCE, USSR,
Akademiya Nauk SSSR, Moscow. Institut Biologii Vnutrennykh Vod.
A. V. Monakov.
J Fish Res Board Can. Vol 29, No 4, p 363-383. 1972. Illus.
Identifiers: *Aquatic animals, *Food habits, Biology, Chironomidae, Cladocera, Copepoda, Feeding, *Invertebrates, Molluscs, Oligochaeta, Rotifera, *Reviews.

The main results obtained at the Borok Institute during the last decade are reviewed. Food and methods of feeding by various aquatic invertebrates (Rotatoria, Oligochaeta, Mollusca, Cladocera, Copepoda, Chironomidae larvae), daily food consumption, and assimilation of food were investigated. Most invertebrates are omnivores although some species live on only one type of food. Daily food consumption changes with food concentration, temperature, and size of consumer. At 15-22 C and a concentration of food close to optimum, mean daily rations of most invertebrates studied usually range from 25-100% of body weight. Only in pulmonate Gastropoda and silt-eating Tubificidae does it greatly exceed body weight. In rare cases, at very high concentration of food unusual in nature, the so-called 'extra feeding' may take place under experimental conditions. For most invertebrates feeding on natural food at optimum concentrations, index of assimilation varies widely, but rarely exceeds 50%. The assimilability of plant food was 45-55% in the majority of investigated species and appears to be considerably higher when animal food is used.--Copyright 1973, Biological Abstract, Inc.
W73-08325

SILVER-FOIL PSYCHROMETER FOR MEASURING LEAF WATER POTENTIAL IN SITU,
Agricultural Research Service, Riverside, Calif. Salinity Lab.
G. J. Hoffman, and S. L. Rawlins.
Science (Wash). Vol 177, No 4051, p 802-804. 1972. Illus.
Identifiers: *Hygrometry, *Instrumentation, Leaves, Measurement, Water potential, Psychrometers (Silver-foil), Temperature.

The water potential of leaves in situ was measured without temperature control with a miniature, single-junction psychrometer constructed from silver foil and attached to the leaf with a silver-impregnated, conductive coating. The temperature of the psychrometer stayed within 0.025C of the temperature of a simulated leaf when the latter temperature was changing at a rate of 1C/min. Leaf water potentials were measured with a precision of plus or minus 1 bar.--Copyright 1973, Biological Abstracts, Inc.
W73-08332

EFFECT OF DRY SEASON DROUGHT ON UPTAKE OF RADIOACTIVE PHOSPHORUS BY SURFACE ROOTS OF THE OIL PALM (ELAEIS GUINEENSIS JACQ.),
University of the West Indies, St. Augustine (Trinidad).
S. C. M. Forde.
Agron J. Vol 64, No 5, p 622-623. 1972.
Identifiers: Dieback, *Droughts, Elaeis guineensis, Moisture, *Oil Palm, *Phosphorus, Radioactive, Roots, Seasons, Soils, Surface, *Nigeria.

The hypothesis that the feeding roots of the oil palm die back because of the effects of pronounced dry season drought in Nigeria was tested using radioactive 32p as a tracer to study P uptake as influenced by different levels of soil moisture. Two trials were carried out in 1964 and 1965, respectively, with 3 treatments: (A) no irrigation during the dry season and 32p applied in solution to the soil; (B) 1 irrigation of 50.8 mm of water shortly before application of 32P; and (C) i rigation at the rate of 50.8 mm of water/palm/w throughout the dry season and 32P applied to th soil. Leaf samples were taken and the activity c 32P was determined. In both trials the uptake c 32P in treatment C was significantly higher tha either treatment A or B and supported th hypothesis that the lower activity was caused b the dieback of the absorbing roots during the dr season drought.--Copyright 1973, Biological Ab stracts, Inc.
W73-08334

STUDIES ON THE FISHERY AND BIOLOG OF A FRESHWATER TELEOST, RITA RITA: MATURITY AND SPAWNING,
K. N. Government Degree Coll., Gyanpur (India Dept. of Zoology.
Maya Shanker Lal.
Indian J Zootomy, Vol 11, No 1, p 41-52, 1970, I lus.
Identifiers: Biology, Fishery, *Maturity (Fish Rita-Rita, *Spawning, *Teleost.

Seven maturity stages were differentiated on th basis of the macroscopic observations and th range of the ova diameters. The seasonal progre sion of size frequency distribution of ova indicate that the fish spawns once a year only and th spawning season is short and restricted, extendin from June to Aug. The lengths of the testis an ovary were plotted against the total length. straight line relationship was found.--Copyrigh 1972, Biological Abstracts, Inc.
W73-08343

POSSIBILITY OF UTILIZING ATMOSPHERI CONDENSATE IN THE HIGHER PLAN SYSTEM,
N. T. Nilovskaya, S. V. Chizhov, Y. E. Sinyak, N N. Shekhovtsova, and M. I. Egorova.
Kosm Biol Med, Vol 5, No 6, p 82-84, 1971, Illus.,
Identifiers: *Water conservation, *Atmospher condensate, *Chinese cabbage, Plants.

Regenerating the moisture from atmosphere in closed air system and utilization of this moisture a higher plant (chinese cabbage) system a discussed. An absorption method was used, an the study showed that the inhibiting effect of u purified atmosphere is thus removed.--Copyrigh 1973, Biological Abstracts, Inc.
W73-08345

AN ECOLOGICAL STUDY ON THE PROCES OF PLANT COMMUNITY FORMATION I TIDAL LAND, (IN KOREAN), C. S. KIM.
Mokpo Teachers' Coll. (South Korea).
For primary bibliographic entry see Field 02L.
W73-08403

OBSERVATIONS ON THE ECOLOGY OF A TRAGALUS TENNESSEENIS,
Kentucky Univ., Lexington. Dept. of Botany.
C. C. Baskin, J. M. Baskin, and E. Quarterman.
Am Midland Nat. Vol 88, No 1, p 167-182, 1972.
Identifiers: *Astragalus-Tennesseensis, Ced glades, Ecology, Glades, Habitat, Morpholog *Legumes, *Tennessee.

A. tennesseensis Gray, a decumbent, perenni legume, is endemic to the cedar glades of north Alabama and central Tennessee. The prese center of distribution is in central Tennessee, a only one population is known in northe Alabama. Its geographical range once extended north-central Illinois and west-central Indiana, b within the past few decades it apparently has di appeared from these area. Within the cedar glad of central Tennessee, the species grows mostly the transition zone between the open glades a glade thickets or woods. Environmental conditio in this zone with respect to light, temperature, a

A psychrometric technique was developed for measuring the water potential of attached growing roots of Zea mays. A measure of the root-water potential as well as of the potential in the adjacent soil reflects the ability of the plants to extract water from soils of varying potentials and provides information regarding the limits of this ability. This should be useful in determining the ability of plants to extract water from soils of varying potentials, and the magnitude of the root resistance to water flow.–Copyright 1973, Biological Abstracts, Inc.
W73-08418

EFFECTS OF ABSCISIC ACID AND ITS ESTERS ON STOMATAL APERTURE AND THE TRANSPIRATION RATIO,
King's Coll., London (England). Dept of Botany.
R. J. Jones, and T. A. Mansfield.
Physiol Plant. Vol 26, No 3, p 321-327, 1972. Illus.
Identifiers: *Abscisic-acid, Carbon, Esters, Oxides, *Stomatal aperture, *Transpiration ratio, Xanthium-strumarium.

A single surface application of abscisic acid or its methyl and phenyl esters suppressed stomatal opening on leaves of Xanthium strumarium. The effect was restricted to the treated parts of the leaf blades, there being no detectable translocation to untreated parts. There were no increases in CO2 compensation to which stomatal closure could be attributed. Abscisic acid and its esters acted successfully as antitranspirants when applied one to leaf surfaces of young barley plants. Over a 9-day period there was a reduction of about 50% in the amount of H2O transpired without any detectable reduction in the rate of dry weight increase. The treatments reduced transpiration relatively more than dry matter accumulation, and hence there was an increase in the water use efficiency. The effect of the treatments became progressively less over 9 days, but even at the end of the experiment (day 9) both the esters reduced transpiration by 20-25%. The esters were slightly more effective than abscisic acid itself. Field trails of the anti-transpirant properties of these compounds were recommended.–Copyright 1973, Biological Abstracts, Inc.
W73-08420

SOME WATER RELATIONS OF CERCOSPORELLA HERPOTRICHOIDES,
Washington State Univ., Pullman. Dept. of Plant Pathology.
For primary bibliographic entry see Field 03F.
W73-08421

ECOLOGICAL SIGNIFICANCE OF VEGETATION TO NORTHERN PIKE, ESOX LUCIUS, SPAWNING,
Nebraska Game and Parks Commission, Lincoln.
D. B. McCarraher, and R. E. Thomas.
Trans Am Fish Soc. Vol 101, No 3, p 560-563. 1972.
Identifiers: Ecological studies, Esox-Lucius, *Pike, *Spawning, Vegetation.

The importance of aquatic vegetation on the growth, survival and reproduction of pike is discussed.–Copyright 1973, Biological Abstracts, Inc.
W73-08435

THREATENED FRESHWATER FISHES OF THE UNITED STATES,
Michigan Univ., Ann Arbor. Museum of Zoology.
R. R. Miller.
Trans Am Fish Soc. Vol 101, No 2, p 239-252. 1972.
Identifiers: *Fishes, Freshwater fish, Legislation, Protective regulations, Threatened fish species.

Threatened, native freshwater fishes are listed for 49 of the 50 states, the first such compilation. Over 300 kinds are included in a formal classification, cross-indexed to states, followed by state lists and the status of each fish, whether rare, endangered, depleted, or undetermined. The concern for native fishes and the important factors responsible for threats to their existence are briefly outlined. Although the lists vary from those based on extensive recent state surveys to others in which current information is sparse, publication is expected to enhance the chances for survival through protective legislation (already enacted by a number of states) and stronger concern for such natural resources.–Copyright 1973, Biological Abstracts, Inc.
W73-08439

TISSUE WATER POTENTIAL, PHOTOSYNTHESIS, 14C-LABELED PHOTOSYNTHATE UTILIZATION, AND GROWTH IN THE DESERT SHRUB LARREA DIVARICATA CAV,
McGill Univ., Montreal (Quebec). Dept. of Biology.
W. C. Oechel, B. R. Strain, and W. R. Odening.
Ecol Monogr. Vol 42, No 2, p 127-141. 1972. Illus.
Identifiers: Adaptations, Carbon-14, Deserts, Growth, Larrea-Divaricata, *Photosynthate utilization, *Photosynthesis, Shrubs, *Water potential.

Tissue water potential is the most important factor throughout the seasons controlling phenological events, photosynthesis, and productivity of L. divaricata growing in Deep Canyon near Palm Desert, California. Growth of reproduction structures was initiated at the time of highest tissue water potential and ceased as water potential decreased. Percentage foliation correlated strongly with drawn water potential (r ± 0.89). The elongation rate of stems and the rate of node production were both dependent on tissue water potential. Leaf growth and node growth proceeded at varying rates throughout the year, providing a continuous sink for photosynthates. Photosynthesis rates ranged from 9.02 mg CO2 incorporated/day/g dry weight of leaf tissue in Sept. to an estimated 74.7 mg CO2 in early Feb. Net photosynthesis and relative productivity correlated very strongly with dawn water potential (r ± 0.93 and r ± 0.97, respectively). Larrea plants were labeled at 1-to 2-mo intervals with photosynthetically incorporated 14CO2 to determine the utilization in growth and storage of photosynthate fractions produced at various times throughout the year. Tissue was subsampled at similar intervals, and the activity in various metabolic compounds (sugar, starch, lipid, organic acid, amino acid, protein, cellulose, and cell-water materials) was analyzed. The utilization of photosynthates in the various fractions was similar in all seasons. No appreciable mobilization into and out of storage materials was apparent. Never dormant, Larrea remains metabolically active and forms new tissue throughout the year. This growth pattern may be an important adaptation allowing Larrea to exist in a wide range of geographical and climatic areas, and perhaps owing to the species' tropical affinities, it might have been a preadaptation to the desert environment.–Copyright 1973, Biological Abstracts, Inc.
W73-08444

2J. Erosion and Sedimentation

INDICATION OF ROCKS BY DRAINAGE PATTERNS,
Moskovskoe Obshchestvo Ispytatelei Prirody (USSR). Geographic Div.
N. A. Gvozdetskii, and I. P. Chalaya.
In: Landscape Indicators–New Techniques in Geology and Geography: Consultants Bureau, Div of Plenum Press, London and New York, p 106-112, 1973. 5 fig, 1 ref. (Translated from Russian, Proceedings of Conference of Moscow Society of

Naturalists, May 21-22, 1968, Moscow, Nauka Press).

Descriptors: *Terrain analysis, *Geomorphology, *Drainage patterns (Geologic), Erosion, Topography, Valleys, Karst, Aerial photography, Remote sensing.
Identifiers: *USSR.

Peculiarities of a valley network in limestones and dolomites, together with general character of dissection, nature of steep slopes, and karst forms may represent a reliable interpretive feature even where exposures are poor and forests are common (as on the Nakeral'skii karst-modified plateau in the western transcaucasian region of the USSR). Interpretation of air photos of many parts of the Tien Shan shows that different rock complexes, differing in origin and in physicochemical and petrographic properties, exhibit correspondence to specific drainage patterns and valley networks. The following groups of rocks may be distinguished: (1) Proterozoic and Paleozoic magmatic rocks, (2) Proterozoic and Paleozoic noncarbonate sedimentary and metamorphic rocks, with some inclusion of volcanic sequences, (3) Proterozoic and Paleozoic carbonate sedimentary and metamorphic rocks, (4) Mesozoic-Cenozoic noncarbonate sedimentary rocks, (5) Mesozoic-Cenozoic carbonate sedimentary rocks. Usually, regions composed of rocks belonging to any one of these groups exhibit distinctive relief with specific forms of erosional dissection. From the drainage pattern and structural features of the relief, it is possible to determine the distribution of particular rock complexes. (Knapp-USGS)
W73-07846

THE USE OF DRAINAGE PATTERNS FOR INTERPRETING THE MORPHOSTRUCTURES OF THE MOSCOW DISTRICT,
Moskovskoe Obshchestvo Ispytatelei Prirody (USSR). Geographic Div.
N. P. Matveev.
In: Landscape Indicators—New Techniques in Geology and Geography: Consultants Bureau, Div of Plenum Press, London and New York, p 113-117, 1973. 5 ref. (Translated from Russian. Proceedings of Conference of Moscow Society of Naturalists, May 21-22, 1968, Moscow, Nauka Press).

Descriptors: *Terrain analysis, *Structural geology, *Geomorphology, *Profiles, *Degradation (Stream), Stream erosion, Erosion, Sedimentation, Geologic control, Mapping, Hydrogeology, Drainage patterns (Geologic).
Identifiers: *USSR, *Moscow.

The drainage pattern is very sensitive to different physicogeographic factors, especially to tectonic deformation and to lithology of a region. Tectonics and lithology affect not only the drainage pattern but also the longitudinal profile of the streams. The relations in the Moscow region are used to show the relationship between longitudinal profiles of streams and structure and to consider the possible interpretation of morphostructures from such longitudinal profiles. Recent tectonic deformation embracing a small area apparently does not substantially affect the stream profile. When the profile is near equilibrium downcutting is absent, and tectonic deformation may change the profile. (Knapp-USGS)
W73-07847

TRANSIENT AND STEADY-STATE SALT TRANSPORT BETWEEN SEDIMENTS AND BRINE IN CLOSED LAKES,
Northwestern Univ., Evanston, Ill. Dept. of Geological Sciences.
For primary bibliographic entry see Field 02H.
W73-07850

PH BUFFERING OF PORE WATER OF RECENT ANOXIC MARINE SEDIMENTS,
California Univ., Los Angeles. Dept. of Geology.
For primary bibliographic entry see Field 05B.
W73-07851

EXAMINATION OF TEXTURES AND STRUCTURES MUD IN LAYERED SEDIMENTS AT THE ENTRANCE OF A GEORGIA TIDAL INLET,
Skidaway Inst. of Oceanography, Savannah, Ga.
G. F. Oertel.
Journal of Sedimentary Petrology, Vol 43, No 1, p 33-41, March 1973. 9 fig, 19 ref.

Descriptors: *Deltas, *Sedimentary structures, *Beds (Stratigraphic), *Deposition (Sediments), Sedimentation, Ripple marks, Stratigraphy, Clays, Silts, Sands, Bed load, Suspended load, *Georgia.
Identifiers: Doboy Sound (Ga).

Mud layers at the entrance of a Georgia tidal inlet are of three texturally and structurally different types. Type I mud layers are composed of laminations of clay-size detritus that occur as flaser, wavy and lenticular bedding. Type II mud layers are composed of foresets and sand-size fecal and organic detritus that also occur as flaser, wavy and lenticular bedding. Type III mud layers are composed of foresets of mud pebbles and occur was wavy lenticular bedding. The types of layers form in response to hydraulically different depositional processes combined with local variation of availability of mud-grain sizes. Grains which constitute mud layers are transported and deposited from the bed load as well as the suspension load. In some areas, deposition from the bedload is considerably more pronounced than deposition from suspension load. Type II and III mud layers result from bedload deposition which does not necessitate a period of slack-water in the tidal cycle or a specific set of conditions of wave activity, suspended-matter concentration, and current velocity. (Knapp-USGS)
W73-07855

ASPECTS OF SEDIMENTATION AND DEVELOPMENT OF A CARBONATE BANK IN THE BARRACUDA KEYS, SOUTH FLORIDA,
Georgia Univ., Athens. Dept. of Geology.
P. B. Basan.
Journal of Sedimentary Petrology, Vol 43, No 1, p 42-53, March 1973. 5 fig, 30 ref.

Descriptors: *Sedimentation, *Carbonate rocks, *Florida, *Tidal waters, Geologic control, Reefs, Sediment transport, Deposition (Sediments), Currents (Water), Tides, Tidal streams, Tidal marshes, Intertidal areas, Waves (Water).
Identifiers: *Carbonate banks, *Barracuda Keys (Fla).

A carbonate bank in the Barracuda Keys, Florida, was studied to find the factors influencing its growth and configuration. Hydrodynamically or biologically controlled sedimentary subenvironments were distinguished: tidal channels, unstable banks, stable banks (including bare-sand, Thalassia, and mangrove island) and silty lagoons. The bank is a closed system wherein local biological production of sediment is in equilibrium with physical dispersal of sediment. Sediment is generally of uniform size, and responds to current flow more as unit 'sheets' than as individual particles, thereby permitting a maximum amount of sediment transport. The major constructional process is the flood tide current, which transports sediment by traction, saltation, and to a lesser extent, suspension and flotation. Development of this bank may be summarized as follows: preferential accumulation of fine sediment in sink holes, forming coalescing silty banks; contemporaneous colonization of these banks by calcareous algae and marine grasses; entrapment and accumulation of coarse sediment by these marine plants, forming a single, contiguous sand bank; and continued growth by accretion of sediment over avalanche slopes. The bank is probably extending itself into the adjoining lagoon by a process of differential growth. This process is dependent upon stabilization of one part of the bank while growth continues in another. (Knapp-USGS)
W73-07856

CARBON ISOTOPE COMPOSITION OF DIAGENETIC CARBONATE NODULES FROM FRESHWATER SWAMP SEDIMENTS,
Louisiana State Univ., Baton Rouge. Coastal Studies Inst.
T. Whelan, III., and H. H. Roberts.
Journal of Sedimentary Petrology, Vol 43, No 1, p 54-58, March 1973. 3 fig, 11 ref. NR 388 002, ONR Contract N00014-69-A-0211-0003.

Descriptors: *Diagenesis, *Carbonates, *Stable isotopes, *Swamps, Water chemistry, Carbonate rocks, Limestones, Carbon, Sedimentation, Bicarbonates, Water circulation, Geochemistry, Sedimentation rates, Louisiana, Mississippi River basin.
Identifiers: Carbon isotopes.

The carbon isotope composition of freshwater diagenetic carbonate nodules ranges from -0.99% (vs. NBS 20) in well-drained environments to -1.91% in poorly-drained environments of a freshwater swamp. The C-13 values of carbonate nodules are not a function of depth of burial. Depositional environment determines the isotopic composition. Data from two cores taken in different sedimentary sequences suggest that in poorly drained sections transformations of isotopically light organic matter yield carbon, which is incorporated into the diagenetic carbonate. In well drained sections, renewal of freshwater, in which carbon species are dissolved, provides the source of carbon for diagenetic carbonate nodules. (Knapp-USGS)
W73-07857

THE ST. CLAIR RIVER DELTA: SEDIMENTARY CHARACTERISTICS AND DEPOSITIONAL ENVIRONMENTS,
Wisconsin Univ., Green Bay. Coll. of Environmental Sciences.
J. M. Pezzetta.
Journal of Sedimentary Petrology, Vol 43, No 1, p 168-187, March 1973. 8 fig, 3 tab, 61 ref.

Descriptors: *Sedimentation, *Lakes, *Deltas, Deposition (Sediments), *Michigan, Stratigraphy, Statistics, Particle size, Sediment transport, Distribution patterns, Waves (Water), Currents (Water), Silts, Sands, Clays.
Identifiers: *Lake St. Clair (Mich).

Progradation of the northeastern shoreline of Lake St. Clair by the St. Clair River has created a modern, freshwater delta with a birdfoot configuration. The recent sediments are moderately sorted very fine sand to very coarse silt with a mean grain size of 3.35 phi. Grain-size frequency distributions are slightly asymmetric and moderately leptokurtic. Coarser textured and somewhat poorly sorted sediments are associated with the occurrence of higher current and wind speeds. Winnowing action by currents may not be as dominant as the introduction of coarser fractions into finer grained sedimentary environments. Primary distributive processes are wind-driven currents and wave action along the delta front where the finer components are dispersed to the west and south. The depositional environments of the lake are: the open lake and interdistributary troughs where high energy processes are dominant; the interdistributary bay margins including the levees, backswamps and marshes; shallow areas of sparse aquatic vegetation and medium energy sedimentary processes; low energy environments with thick overgrowths of aquatic

maximum capacity to adsorb metals. However, under field conditions, there was no indication that the maximum capacity was reached. (Warman-Alabama)
W73-07860

A NOTE ON ASYMMETRICAL STRUCTURES CAUSED BY DIFFERENTIAL WIND EROSION OF A DAMP, SANDY FOREBEACH,
California State Univ., San Diego. Dept. of Geology.
R. W. Berry.
Journal of Sedimentary Petrology, Vol 43, No 1, p 205-206, March 1973. 1 fig, 2 ref.

Descriptors: *Sedimentary structures, *Beaches, *Wind erosion, Sands, Beach erosion, Winds, *Mexico.
Identifiers: Differential erosion, *Baja California.

Asymmetrical structures 5 to 10 mm high, 10 to 20 mm long,and 5 to 10 mm wide on a damp, sandy beach at Bahia de los Frailes, Territory of Baja California, Mexico were formed as a result of differential erosion by wind. They were elongated parallel to the wind with the steep slopes of each structure facing into the wind. Nonuniform compaction of sand grains along the beach is suggested as a fundamental reason for the development of the structures. (Knapp-USGS)
W73-07861

ZONE OF INFLUENCE – INNER CONTINENTAL SHELF OF GEORGIA,
Georgia Inst. of Tech., Atlanta.
G. N. Bigham.
Journal of Sedimentary Petrology, Vol 43, No 1, p 207-214, March 1973, 5 fig, 16 ref.

Descriptors: *Sediment transport, *Clays, *Continental shelf, *Ocean currents, *Clay minerals, *Georgia, Littoral drift, Bottom sediments, Settling velocity, Distribution patterns, Mineralogy, Suspended load, Sedimentation, Coastal plains, Atlantic Coastal Plain.

Sediment transport was studied on the inner continental shelf of Georgia. Ninety-three suspended-matter samples were collected from 52 stations during the summer of 1970. Bottom sediment, near-surface samples, and near-bottom suspended-matter samples were taken. Salinity, temperature, and current direction and velocity measurements were also made to determine the nature of shelf-sediment transport processes. Sediments and suspended matter were analyzed by x-ray diffraction. Georgia rivers contribute kaolinite, smectite, and illite to the coastal region, while a kaolinite-illite clay mineral suite is transported southward by longshore drift so that net transport direction may be inferred by suspended-matter clay mineralogy patterns. The shelf-water circulation pattern during the summer months is a complex system of tidal-current and wind-generated eddies superimposed on a predominantly southward drift. Differential settling characteristics explain the suspended clay mineral distribution and establish a zone of influence which extends 3 to 10 miles offshore. This zone is the maximum seaward extent of present-day river-derived suspended detritus. Particulate and dissolved pollutants are probably restricted to the zone of influence, and are not contributed to the Florida current. (Knapp-USGS)
W73-07862

SURFACE SEDIMENTS OF THE GULF OF PANAMA,
Woods Hole Oceanographic Institution, Mass.
J. C. MacIlvaine, and D. A. Ross.
Journal of Sedimentary Petrology, Vol 43, No 1, p 215-223, March 1973. 8 fig, 2 tab, 12 ref. ONR Contract CO-N-00014-66-0241.

Descriptors: *Bottom sediments, *Continental shelf, *Pacific Ocean, *Provenance, *Distribution patterns, Sediment transport, Sedimentation, Ocean currents, Ocean circulation, Clays, Mineralogy, Silts, Sands, Sedimentology.
Identifiers: *Gulf of Panama.

The surface sediments of the Gulf of Panama consist of a relict sand covering the central and outer portion of the shelf, and nearshore recent fine-grained sediments with local accumulations of coarser sediments. Clays, with occasional patches of rocks and pebbles, are typical of the continental slope. Deposition of sand in the central portion of the Gulf was controlled primarily by the topography of the exposed shelf during the last low stand of sea level. During this time streams distributed sediment from the east and west towards the center of the Gulf. During part of the Pleistocene, the main drainage entering the Gulf through San Miguel Bay passed north of the Archipelago de Las Perlas and across the shelf in the prominent central submarine valley. Presently fine-grained sediments are being transported and deposited in a counter-clockwise direction around the inner part of the Gulf, producing a deposit which is nearly homogeneous in mineralogy. Fine-grained sediment is forming a prism moving out from the shore towards the center of the Gulf; some has bypassed the shelf and is being deposited on the slope. Heavy mineral determinations were made by conventional optical and x-ray techniques. Distribution patterns derived using x-ray peak heights and a Q-mode factor analysis on data from both techniques were compared and showed very good similarity. (Knapp-USGS)
W73-07863

CONTINENTAL-SHELF SEDIMENTS OFF NEW JERSEY,
Rensselaer Polytechnic Inst., Troy, N.Y. Dept. of Geology.
G. M. Friedman.
Journal of Sedimentary Petrology, Vol 43, No 1, p 224-237, March 1973. 9 fig, 1 tab, 59 ref.

Descriptors: *Atlantic Coastal Plain, *Continental shelf, *New Jersey, *Bottom sediments, *Mineralogy, Provenance, Sedimentation, Sedimentology, Sediment transport, Quaternary period, Stratigraphy.

Surface sediment samples from a study area 20 miles wide across the New Jersey continental shelf north of Atlantic City were studied. The sediment consists mostly of moderately well-sorted, medium-grained sand with a remarkable absence of particles finer than 125 microns in size. This absence of fine particles is related to the reworking during the Holocene transgression. Mapping of statistical parameters suggests that during the Holocene submergence stillstands occurred at about 40 and 20 fathoms. Morphologic terraces at about 80 and 68 fathoms may be distinguished. The abundance of hornblende and garnet and the presence of magnetite in the shelf samples contrasts with the absence of these minerals in Cretaceous, Miocene or Quaternary rocks on the adjacent coastal plain. It is suggested that the continental-shelf sediments may have been derived from an ancestral Hudson River. (Knapp-USGS)
W73-07864

TEXTURAL CHANGES AS AN INDICATOR OF SEDIMENT DISPERSION IN THE NORTHERN CHANNEL ISLAND PASSAGES, CALIFORNIA,
University of Southern California, Los Angeles. Dept. of Geological Sciences.
J. S. Booth.
Journal of Sedimentary Petrology, Vol 43, No 1, p 238-250, March 1973. 16 fig, 3 tab, 17 ref. ONR Contract N00014-67-A-0269-0009C.

Descriptors: *Sediment transport, *Dispersion, *Distribution patterns, *California, Tidal waters,

Currents (water), Statistics, Bottom sediments, Winds, Littoral drift, Tracers, Tracking techniques, Continental shelf.
Identifiers: *Channel Island (Calif).

A technique for determining sediment dispersion in shallow, high energy marine environments is based on changes in textural properties with changes in energy. In order to test this technique, samples were collected from the Northern Channel Island Passages off the southern California coast. The four textural moments were calculated for each sample and polynomial trend surface contour maps were constructed for mean and sorting for each passage. San Miguel Passage is characterized by sediment dispersion to the southeast. The energy level is highest in the center of the passage and there is a gradual decrease in energy toward the perimeter. Santa Cruz Channel is more complex. A lobe of coarse sediment, part of which appears to be relict, in the northern section of the channel shows dispersion to the east. However, finer sediment is moving into the southern part of the area from the east, where it is intercepted by the head of Santa Cruz Canyon. Because there is movement to the east and west, this channel may represent a shear zone between two currents. Anacapa Passage shows dominant westward dispersion along both the southern and northern margins, with the southern one being the most significant. Relict sediment appears to exist in this passage as well. Wind-driven currents are more important than either tidal currents or wave action in accfnting for sediment distribution. (Knapp-USGS)
W73-07865

ARGENTINE BASIN SEDIMENT SOURCES AS INDICATED BY QUARTZ SURFACE TEXTURES,
Queens Coll., Flushing, N.Y. Dept. of Geology.
D. Krinsley, P. E. Biscaye, and K. K. Turekian.
Journal of Sedimentary Petrology, Vol 43, No 1, p 251-257, March 1973. 4 fig, 1 tab, 16 ref.

Descriptors: *Provenance, *Bottom sediments, *Particle shape, *Sands, *Quartz, Stratigraphy, Sedimentation, Sedimentology, Mineralogy, Abrasion, Atlantic Ocean, Antarctic Ocean.
Identifiers: *Argentine Basin (Pacific Ocean).

Distinctive quartz surface textures, as determined by scanning electron microscopy, can be sued to identify sources of sediment in the Argentine Basin. Glacial quartz from the south, fluvial-littoral quartz from the continental margin, and windblown hot-dune sand can be identified and their distributions tentatively assessed. (Knapp-USGS)
W73-07866

ALKALINITY DETERMINATION IN INTERESTITIAL WATERS OF MARINE SEDIMENTS,
Scripps Institution of Oceanography, La Jolla, Calif.
J. M. Gieskes, and W. C. Rogers.
Journal of Sedimentary Petrology, Vol 43, No 1, p 272-277, March 1973. 1 fig, 2 tab, 16 ref.

Descriptors: *Alkalinity, *Pore water, *Connate water, *Bottom sediments, Carbonates, Sulfides, Ammonia, Ion exchange, Calcium, Magnesium, Sulfates, Phosphates, Water chemistry, Hydrogen ion concentration, Diagenesis.

The titration alkalinity of interstitial water is defined in terms of the ionic species that contribute to the alkalinity. The significance and the magnitude of these contributions may be estimated from available analytical data. A method for the potentiometric titration of the alkalinity is described with special reference to the various techniques available for the evaluation of the equivalence point. The possible effect of the sulfate content of the samples on the mathematical evaluation of the end point is discussed. The accuracy of the method is estimated to be 0.5%. (Knapp-USGS)
W73-07867

CHANGES IN CLAY MINERAL ASSEMBLAGES BY SAMPLER TYPE,
George Washington Univ., Washington, D.C. Dept. of Geology.
F. R. Siegel, and J. W. Pierce.
Journal of Sedimentary Petrology, Vol 43, No 1, p 287-290, March 1973. 3 tab, 9 ref. NSF Grant GA-16499.

Descriptors: *Sampling, *Bottom sediments, *Clay minerals, *Bottom sampling, Cores, Dredging, Heterogeneity, Reliability, Variability.

Clay mineral contents of samples of Recent marine sediments from the Golfo San Matias, Argentina, taken as piston cores, gravity cores, and grab samples, show values which differ according to the type of sampler used. Averages for the components of the clay mineral assemblages are similar for the piston core and grab sample groups, whereas those for the Phleger samples are significantly different. Comparisons of clay mineral distributions in Recent marine environments can be made with confidence only if the same type sampling device is used to extract all the bottom sediments. The data from grab and piston core samples may, however, be treated together. (Knapp-USGS)
W73-07868

QUANTITATIVE GEOMORPHOLOGY—SOME ASPECTS AND APPLICATIONS.
Proceedings of 2nd Annual Geomorphology Symposia Series held at Binghamton, N.Y., Oct 15-16, 1971: Binghamton, State Univ of New York Publications in Geomorphology, M., Morisawa, editor, 1972. 315 p, 17 ref.

Descriptors: *Geomorphology, *Mathematical studies, *Statistical methods, *Statistics, Topography, Terrain analysis, Land forming, Erosion, Sedimentation, Hydrology, Sediment transport, Reviews.
Identifiers: *Quantitative geomorphology.

One of the most important trends in geomorphology in the past 25 years has been the orientation of the field toward more objective, mathematical methods of studying landforms and the processes which create them. This second Annual Geomorphology Symposium volume presents papers exemplifying the history, use and some problems of quantitative techniques in geomorphology. Papers in Part I deal with abstract concerns in geomorphic quantification. Part II illustrates the wide variety of geomorphic fields into which quantitative investigations have been carried since their inception. Geomorphic phenomena such as mass movements, glaciation, and shoreline changes have become areas for mathematical and statistical methodology. Part III includes four papers illustrating the practical applications of quantitative description and analysis of fluvial systems. Cumulative frequency curves of slope and relief were used in evaluating an area for suitability as a site for testing military trafficability, mobility and visibility. (Knapp-USGS)
W73-07869

THREADS OF INQUIRY IN QUANTITATIVE GEOMORPHOLOGY,
Iowa Univ., Iowa City. Dept. of Geography.
N. E. Salisbury.
In: Quantitative Geomorphology—Some Aspects and Applications, Proc of 2nd Annual Geomorphology Symposia Series, p 9-60, 1972. 3 tab, 398 ref.

Descriptors: *Geomorphology, *Mathematic studies, *Statistical methods, *Statistics, Topography, Terrain analysis, Land forming, Erosio Sedimentation, Hydrology, Sediment transpo Reviews.
Identifiers: *Quantitative geomorphology.

Quantitative geomorphology involves descripti of land surface form, including empirical expla tions of interrelationships between form elemen These descriptions include geometry of chann and valleys, stream profiles, development drainage nets, topology, explanations of strea flow and systems of landscape explanation, cluding slope development. Much of the effort geology can be traced to the impetus provided Horton's 1945 paper on 'Erosional development streams and their drainage basins.' Through tim geographers became more involved with fluvi processes, and by the late 1960's there was little distinguish their work from that of geologists, a somewhat greater attention to spatial variables generalizations. Quantitative techniques that we originally introduced largely in fluv geomorphology with some opposition, are n widely accepted in all phases of geomorpholog (See also W73-07869) (Knapp-USGS)
W73-07870

PROBLEMS OF INTERPRETATION OF SIM LATION MODELS OF GEOLOG PROCESSES,
Virginia Univ., Charlottesville. Dept. of Envir mental Sciences.
A. D. Howard.
In: Quantitative Geomorphology—Some Aspe and Applications, Proc of 2nd Ann Geomorphology Symposia Series, p 62-82, 1972 fig, 1 tab, 40 ref.

Descriptors: *Mathematical mode *Geomorphology, *Simulation analysis, Mo studies, Numerical analysis, Stochastic process, Variability, Reviews.
Identifiers: *Quantitative geomorphology.

Mathematical models of geologic systems co monly can only give numerical predictions using the technique of simulation. Simulation troduces additional assumptions, whose effe are minimized in the case of solution of d ferential equations by small spatio-temporal inc ments and in the case of probabilistic models sufficient repetitions of the simulation to give curate values of statistical parameters. In so cases models with very different assumptio about the nature of the processes involved g similar predictions due to the dominance in det mining results by a seemingly less important sumption common to all the models. Relative i portance of the various theoretical constructs, the model can generally be determined only noting effects of modification of the hypothe upon the outcome. Nearly identical predictions sometimes made by competing models which ha no apparent common theoretical constru Models differ in degree of generality of the vario assumptions incorporated in the model; mode give more satisfactory explanations of nature, the degree that they incorporate general laws su as those of mechanics instead of experiment empirical, or ad hoc assumptions. Inherent r domness can never be proven or disproven; ho ever, the scientist may show by constructing mo accurate deterministic models the maximum lim of any proposed inherent randomness. (See a W73-07869) (Knapp-USGS)
W73-07871

THE TWO DIMENSIONAL SPATIAL O GANIZATION OF CLEAR CREEK AND O MAN CREEK, IOWA,
Harvard Univ., Cambridge, Mass. Lab. for Co puter Graphics and Spatial Analysis.
M. Woldenberg.

In: Quantitative Geomorphology–Some Aspects and Applications, Proc of 2nd Annual Geomorphology Symposia Series, p 121-179, 1972. 29 fig, 4 tab, 17 ref.

Descriptors: *Mass wasting, *Wisconsin, *Geomorphology, *Degradation (slope), *Instrumentation, *On-site data collections, On-site investigations, Measurement, Creep, Loess, Slopes.
Identifiers: *Quantitative geomorphology.

Techniques and equipment for study of mass-movement processes were evaluated on slopes near Madison, Wisconsin. Special bench marks consisted of solid iron rods cemented vertically into bedrock and separated from the soil mantle by much larger diameter capped casings. Within fenced enclosures, point-gage measurements of small stones proved best for determining both horizontal and vertical movements. The point gages were used from permanently mounted steel and aluminum platforms cemented to bedrock. Dial indicators on special mounts on the same platforms measured displacements of small stones, and linear-motion transducers recorded movements continuously even under this snow. However, difficulties with the mounts from wind effects and friction from various causes in the mechanical linkages made those two systems unreliable without frequent attention. Strain gages in the soil failed before a year cycle was completed. Eighty blocks of dolomite 20 cm to 5 m across in a pasture with cattle were drilled and plugged with rods on which fine crosses were scratched. One 75-m line of 21 blocks on a slope of 4 deg to 25 deg lengthened 97.2 mm, with lengthening mostly confined to the steeper slope. At one fenced site, vertical displacement of five small stones on and in loessial soil on nearly level ground was up to 18 mm. Wetting and drying often were more important than freezing and thawing. Downslope motion of surface soil and small stones on steeper nonwooded slopes in the study area averages perhaps 5 mm each year. This equivalent to a vertical lowering of the surface of about 2mm per year, or 200 cm per 1000 years. Gentle or wooded slopes and larger blocks of rock move much less. (See also W73-07869) (Knapp-USGS)
W73-07874

THE SURFICIAL FABRIC OF ROCKFALL TALUS,
Ohio State Univ. Research Foundation, Columbus. Inst. of Polar Studies.
E. R. McSaveney.
In: Quantitative Geomorphology–Some Aspects and Applications, Proc of 2nd Annual Geomorphology Symposia Series, p 181-197, 1972. 7 fig, 4 tab, 10 ref.

Descriptors: *Talus, *Geomorphology, *Sedimentary structures, *Petrofabrics, Degradation (slope), Mass wasting, Particle shape, Particle size, Numerical analysis, Mathematical models.
Identifiers: *Quantitative geomorphology, *Rockfall talus.

Rockfall talus in the Front Range, Colorado, has a distinct fabric. Three-dimensional analysis shows a subhorizontal girdle on which are superimposed several modes. The principal mode parallels slope direction but plunges less than the surface slope, indicating imbrication in an upslope direction. Transverse modes are at angles of 90 deg or less to the downslope mode. The girdle reflects the overall geometry of the talus surface, while the modes are produced by the motion of particles over the surface, with rolling and sliding playing equal roles. Fabric strength is a function of surface roughness, and decreases with increasing particle size downslope. Treatment of talus orientation data only by purely numerical methods, including two- and three-dimensional vector analyses and their associated tests of significance, obscures more than it clarifies. Use of these methods as the sole determinant of the presence or absence of fabric should be avoided. (See also W73-07869) (Knapp-USGS)
W73-07875

COMPUTATIONAL METHODS FOR ANALYSIS OF BEACH AND WAVE DYNAMICS,
Massachusetts Univ., Amherst. Dept. of Civil Engineering.
J. M. Colonell, and V. Goldsmith.
In: Quantitative Geomorphology–Some Aspects and Applications, Proc of 2nd Annual Geomorphology Symposia Series, p 198-222, 1972. 15 fig, 15 ref. USCE Contract DACW 72-67-C-0004.

Descriptors: *Geomorphology, *Data processing, *Beaches, *Beach erosion, *Massachusetts, *Surf, Littoral drift, Waves (water), Computer programs, Sedimentation, Sediment transport, Refraction (water waves), Profiles, Dunes, Sands, Numerical analysis.
Identifiers: *Quantitative geomorphology.

Computer programs (including graphic desplays of results) are available for storm wave forecasting, wave refraction, volumetric changes of beach profiles, and statistical analyses of grain size and internal dune geometry. The primary goal in these analyses is to relate observed changes in the intertidal portions of the beaches to processes active in the adjacent offshore areas. At twelve permanent beach profile locations on Monomoy Island and four on Nauset Spit, Cape Cod, Massachusetts, there are large variations in the amount of erosion and accretion occurring along the length of this 16-mile barrier island coastline. Analysis of the wave behavior in the adjacent offshore area with the aid of storm wave forecasting techniques and over 200 wave refraction diagrams yields a correlation between zones of wave energy concentration and zones of increased erosion. Beach sediments are essentially in equilibrium with the wave energy conditions as there is no decrease in grain size 'downdrift'. Instead, zones of wave energy concentrations are suggested by small areas of coarser-grained sediment observed at several locations along the beach. The coastal sand dunes have a distinctive internal geometry characterized by low dip angles and azimuths that correlate with prevailing wind directions rather than with storm winds. These conclusions are statistically valid at any dune elevation. (See also W73-07869) (Knapp-USGS)
W73-07876

QUANTITATIVE ANALYSIS OF THE FACTORS CONTROLLING THE DISTRIBUTION OF CORRIE GLACIERS IN OKOA BAY, EAST BAFFIN ISLAND: (WITH PARTICULAR REFERENCE TO GLOBAL RADIATION),
Colorado Univ., Boulder. Inst. of Arctic and Alpine Research.
J. T. Andrews.
In: Quantitative Geomorphology–Some Aspects and Applications, Proc of 2nd Annual Geomorphology Symposia Series, p 223-241, 1972. 8 fig, 2 tab, 19 ref.

Descriptors: *Glaciers, *Arctic, *Cirques, *Geomorphology, *Glaciation, Canada, Climates, Statistical methods, Correlation analysis, Solar radiation, Pleistocene epoch, Weather.
Identifiers: *Quantitative geomorphology.

Corrie glaciers near Okoa Bay, east Baffin Island, N.W.T., Canada were studied using quantitative methods. Indirect glaciological climatological measures (such as elevation, geometry and size) were studied using multiple stepwise discriminant analysis. Of five variables selected the most important are residuals from two trend surfaces on corrie lip and mountain summit elevations. Ice-filled and ice-free corries differ on the average by only 200 m elevation. The importance of global radiation on present glacier distribution was analyzed. Dis-

criminant analysis shows the degree of difference in terms of global radiation receipts between presently glacierized valleys (north- and east-facing) and two empty (south-facing) valleys. Average daily totals for August 21st were 311 and 398 cal per sq cm, respectively. The discriminant equation was used to find the effects of changes in global radiation due to variations in the obliquity of the ecliptic and elevation on the state of glacierization. These factors can account for up to a 25% shift in the state of glacierization. Other factors, included in any Quaternary climatic model. (See also W73-07869) (Knapp-USGS)
W73-07877

HYDROGRAPH ANALYSIS AND SOME RELATED GEOMORPHIC VARIABLES,
Nebraska Univ., Omaha. Dept. of Civil Engineering.
W. F. Rogers.
In: Quantitative Geomorphology – Some Aspects and Applications, Proc of 2nd Annual Geomorphology Symposia Series, p 245-257, 1972. 9 fig, 6 ref.

Descriptors: *Hydrograph analysis, *Geomorphology, *Drainage patterns (geologic), Hortons law, Runoff, Rainfall-runoff relationships, Hydrographs, Infiltration, Drainage density.
Identifiers: *Quantitative geomorphology.

A comparison of the frequency histogram of first order channel distances for drainage basins in Pennsylvania and their hydrographs of runoff from general storms show marked similarity. This close correspondence indicates the shape of the surface runoff hydrograph and is largely controlled by the distribution of first order channel distances. First order drainage channels originate when the tractive force exerted by flowing water is sufficient to move surface sediment. The amount of runoff available to move sediment is a function of the geology and climate. Soils derived from fine grained rocks have lower infiltration rates and higher runoff volume than soils derived from coarser grained rocks in a semiarid climate. Root density and penetration increases in a more humid climate and increases infiltration rates. The number of first order channels is inversely proportional to the infiltration capacity of the soil. Each first order channel acts as a source area for surface runoff. The distribution of first order channel distances from the gage determines the timing of the delivery of water to the gage. (See also W73-07869) (Knapp-USGS)
W73-07878

PREDICTING TIME-OF-TRAVEL IN STREAM SYSTEMS,
Illinois State Water Survey, Urbana.
J. B. Stall.
In: Quantitative Geomorphology – Some Aspects and Applications, Proc of 2nd Annual Geomorphology Symposia Series, p 259-271, 1972. 7 fig, 1 tab, 7 ref.

Descriptors: *Travel time, *Streamflow, *Dye releases, *Illinois, Path of pollutants, Hydraulics, Channel morphology, Geomorphology.
Identifiers: *Quantitative geomorphology.

Stream discharge, cross-sectional area, velocity, width and depth are consistently related to drainage area and flow frequency in the 5250-square-mile basin of the Kaskaskia River basin in Illinois. A set of five hydraulic geometry equations represent quantitatively these interrelationships. Mean velocities at various locations in this stream system are averaged for various stream reaches to provide average time-of-travel. Computed time-of-travel for a 65-mile reach of the Kaskaskia River was compared to actual time-of-travel of an injection of dye. Computed values were good at high flows, and fair at medium flows. At low flow the computed value is a minimum expected time-of-travel. Actual time-of-travel may be much longer due to trapping of the dye in pools in the streambed upstream from riffles. Computed time-of-travel curves are presented for this 65-mile reach of the Kaskaskia River. (See also W73-07869) (Knapp-USGS)
W73-07879

HYDROGEOMORPHOLOGY OF SUSQUEHANNA AND DELAWARE BASINS,
State Univ., of New York, Binghamton. Dept. of Geology.
D. R. Coates.
In: Quantitative Geomorphology – Some Aspects and Applications, Proc of 2nd Annual Geomorphology Symposia Series, p 273-306, 1972. 12 fig, 8 tab, 18 ref.

Descriptors: *Geomorphology, *River basins, *Streamflow, *Delaware River, Hydrogeology, Water yield, Hydrology, Surface waters, River systems, Valleys, Topography, Alluvial channels, Alluvium, Low flow, Rainfall-runoff relationships.
Identifiers: *Quantitative geomorphology, *Susquehanna River Basin.

A unified study of hydrology, geology, and geomorphology of river basins in the glaciated Appalachian Plateau of New York and Pennsylvania was made to evaluate the U.S. Geological Survey stream gaging network and to develop parameters of low flow in ungaged streams. New indices developed include a sandstone index, a rock massiveness index, a stream relief number, a valley fill ratio, and a morphologic classification system. The 13 Delaware basins in the rugged Catskill Mountains with terrigenous sandstones are contrasted with 12 Susquehanna basins of more subdued topograph with marine shales. The Catskill region has greater relief, smaller drainage density, steeper topographic slopes and stream gradients, higher sandstone and massiveness numbers, less valley fill, and more precipitation and streamflow. Delaware basins have more uniform characteristics than Susquehanna basins. More than 90 percent of the low streamflow variance in Delaware basins is explained by precipitation and discharge parameters, whereas hydrologic indicators are unimportant predictors of low streamflow in Susquehanna basins. Susquehanna low streamflow characteristics are related to valley fill materials. Glaciation has greatly influenced both the physical nature of Susquehanna basins and their hydrology response system, but is less important in Delaware basins. (See also W73-07869) (Knapp-USGS)
W73-07880

QUANTITATIVE GEOMORPHOLOGY FOR TEST FACILITY EVALUATION AT YUMA PROVING GROUND,
Army Topographic Command, Washington, D.C.
J. A. Millett.
In: Quantitative Geomorphology–Some Aspects and Applications, Proc of 2nd Annual Geomorphology Symposia Series, p 307-313, 1972. 2 fig.

Descriptors: *Geomorphology, *Mapping, *Deserts, *Topography, *Military aspects, *Arizona, Terrain analysis, On-site investigations, Slopes, Surveys.
Identifiers: *Quantitative geomorphology, *Military geomorphology, *Yuma (Ariz).

A simplified method for quantifying desert terrain at Yuma Proving Ground uses field measurements of slope and relief, and mapping of unconsolidated materials. The field data, presented primarily as cumulative frequency curves, are useful in engineering evaluation of test-area suitability for traffic mobility and visibility. The data are further applicable to an overall description of terrain, to an estimate of the accuracy of maps, including seven identified mapping units, and to establish base areas for environmental testing. (See also W73-07869) (Knapp-USGS)
W73-07881

SPLASH EROSION RELATED TO SOIL ERODIBILITY INDEXES AND OTHER FOREST SOIL PROPERTIES IN HAWAII,
Forest Service (USDA), Berkeley, Calif. Pacific Southwest Forest and Range Experiment Station.
T. Yamamoto, and H. W. Anderson.
Water Resources Research, Vol 9, No 2, p 33[illegible]-345, April 1973. 2 fig, 6 tab, 18 ref.

Descriptors: *Soil erosion, *Impact (Rainfall), *Hawaii, *Forest soils, Erosion, Simulated rainfall, Regression analysis, Moisture content, Organic matter, Infiltration.
Identifiers: *Splash erosion.

Losses of Hawaiian forest soils under simulated rainfall were used to test indexes of erodibility based on soil aggregate size and Middleton's suspension percent. Soil samples were collected on the Koolau and Waianae ranges on Oahu, Hawaii. Soil losses were related by regression to principal components to eight factors: soil erodibility index, bulk density, saturation soil moisture content, precipitation excess, organic matter content, geologic type, vegetation type, and climate zone at the sampling site. Equations that included the percent of water stable aggregates 0.25-... mm in size produced the highest explained variation: 81% in gross splash erosion and 66% in maximum splash rate. Gross splash was related to soil erodibility index, bulk density, and infiltration and saturation moisture content; in contrast, maximum splash erosion variation was related to organic matter content as well as to an erodibility index and the bulk density of the soil. Ash and basalt colluvium soils required more careful management than basalt soils because of their higher maximum splash rates. (Knapp-USGS)
W73-07888

SEEPAGE STEPS IN THE NEW FOREST, HAMPSHIRE, ENGLAND,
C. G. Tuckfield.
Water Resources Research, Vol 9, No 2, p 3[illegible]-377, April 1973. 9 fig, 1 tab, 4 ref.

Descriptors: *Geomorphology, *Erosion, *Topography, *Seepage, *Springs, Valleys, Groundwater, Soil water, Slopes, Degradation (Slope).
Identifiers: *New Forest (England), *Seepage steps.

Seepage steps are erosional scarps produced on hillslopes by the removal of material on the downslope side by concentrated seepage. In plan the scarps form accurate lines approximately following the contour. In profile, four elements can be distinguished: a convex slope above the step, an almost vertical face, a convex debris slope below the face, and often a concave slope beyond the debris slope. In the New Forest, Hampshire, England, the seepage is caused by concentration of water above the junction between a permeable sandstone and an underlying impermeable clay. The seepage face is composed partly of superficial material and partly of the permeable rock, the contact between the two strata being obscured by the debris and found in one case to be 8 meters below the level of the base of the step. The water table was always less than 1 m below the level of the base of the step. The position of the step is determined directly by the level of the water table and indirectly by the clay-sandstone junction. The height of the step varies considerably: the mean maximum height for 54 valley sides is 2.16 m; the absolute maximum height recorded is 6 m. Erosion of the face is active at some points; at other points the feature appears to be relict. It is usually most active near the heads of valleys. (Knapp-USGS)
W73-07891

VISCOELASTIC PROPERTIES OF MARINE SEDIMENTS,
Texas A and M Univ., College Station. Dept. of Civil Engineering.
S. H. Carpenter, L. J. Thompson, and W. R. Bryant.
Available from NTIS, Springfield, Va., 22151 as AD-748 647, Price $3.00 printed copy; $1.45 microfiche. Texas A and M University Department of Oceanography Technical Report 72-8-T, September 1972. 56 p, 16 fig, 2 tab, 18 ref, 2 append. ONR Contract N00014-68-0308 (0002).

Descriptors: *Sediments, *Oceans, *Bottom sediments, *Viscosity, *Viscometers, Stability, Shear strength, Moisture content, Specific gravity, Testing procedures, Curves, Soil types, Soil properties.
Identifiers: Anchor resistance, Submarine bearing capacity.

Because of their high moisture content, marine sediments have characteristics of both solids and liquids. Being neither liquid nor solid it is difficult to evaluate material constants using conventional soil testing equipment or conventional liquid viscometers. However, these properties must be evaluated if problems such as anchor resistance to motion, submarine bearing capacity, and slope instability problems are to be solved. Before properties can be evaluated a theory must exist. Nonlinear viscoelastic theory was used and a simple shear device called the shear viscometer was developed. From cores taken in the Gulf of Mexico the shear resistance of marine sediment was the sum of an exponential function of shear deformation and a function of the one-third power of the rate of shear deformation. The shear resistance can also be empirically related to more conventional tests such as vane shear, moisture content, liquid limit, specific gravity, and the depth from which the sample was taken. (Woodard-USGS)
W73-08070

THE WEAR OF SANDSTONE BY COLD, SLIDING ICE,
Newcastle-upon-Tyne Univ. (England). Dept. of Geography.
R. Whitehouse.
In: Polar Geomorphology; Institute of British Geographers Special Publication No 4, p 21-31, June 1972. 6 fig, 11 ref.

Descriptors: *Erosion, *Glaciation, *Scour, *Sandstones, *Ice, Glaciers, Friction, Laboratory tests, Abrasion, Glaciology, Geomorphology, Mechanical properties.

Erosion of rock by sliding ice is a function of friction. Friction increases with load and with decrease in temperature, but decreases with speed of sliding and duration of sliding in laboratory tests. This is caused by the formation, at the ice sliding surface, of a thin layer of slightly deformed ice crystals with their c axes normal to the surface; ice so oriented has a very low coefficient of friction. Increase in load or in speed of sliding causes a jerky motion. This stick-slip motion frequently leaves ice adhering to the rock, producing an ice-to-ice sliding surface. The sliding ice shows layers of rock particles analogous to dirt bands in glaciers. Production of rock debris by sliding ice is small, but pebbles in the rock shatter after only seven cycles of freeze-thaw. (Knapp-USGS)
W73-08074

TORS, ROCK WEATHERING AND CLIMATE IN SOUTHERN VICTORIA LAND, ANTARCTICA,
Keele Univ. (England). Dept. of Geography.
E. Derbyshire.
In: Polar Geomorphology; Institute of British Geographers Special Publication No 4, p 93-95, June 1972. 8 fig, 35 ref.

Descriptors: *Weathering, *Antarctic, *Erosion, *Recent epoch, Glaciation, Paleoclimatology, Climatology, Clays, Pleistocene epoch, Snow cover, Deserts, Geomorphology.
Identifiers: *Wright Valley (Antarctica), *Tors.

Morphological and weathering characteristics of a group of tors are described at Sandy Glacier (Wright Valley) in the McMurdo oasis of southern Victoria Land, Antarctica. Clay minerals at and beneath the surface of the rounded corestones indicate chemical weathering of the dolerite. The juxtaposition of rounded and angular tors on the summit of the arete is caused by local variations in microclimate, especially as it affects snow cover. Both types of tors are the product of the prevailing polar desert conditions. Such conditions have probably prevailed throughout the Pleistocene with only brief episodes of slightly more maritime climate. (Knapp-USGS)
W73-08075

VALLEY ASYMMETRY AND SLOPE FORMS OF A PERMAFROST AREA IN THE NORTHWEST TERRITORIES, CANADA,
Cambridge Univ. (England). Dept. of Geography.
For primary bibliographic entry see Field 02C.
W73-08076

MODIFICATION OF LEVEE MORPHOLOGY BY EROSION IN THE MACKENZIE RIVER DELTA, NORTHWEST TERRITORIES, CANADA,
Alberta Univ., Edmonton. Dept. of Geography.
D. Gill.
In: Polar Geomorphology; Institute of British Geographers Special Publication No 4, p 123-138, June 1972. 13 fig, 19 ref. Alberta Univ. Grant (55-32254).

Descriptors: *Geomorphology, *Deltas, *Arctic, *Permafrost, *Erosion, Sedimentation, Levees, Alluvial channels, Degradation (Stream), Waves (Water), Surf, Beach erosion, Ice cover, Sea ice, Channel morphology, Stream erosion.
Identifiers: *Levees (Natural).

Both erosion and unequal deposition of sediments control the gross morphology of arctic deltas. In addition to normal fluvial erosion, levee morphology in an arctic delta may be significantly modified by wave action, ice abrasion, slumping, solifluction, and thermal degradation of perenially frozen sediment. These processes are analyzed and their effect on levee retreat and rates of aggradation and progradation of slip-off and point bars along shifting channels in the Mackenzie River delta is described. (Knapp-USGS)
W73-08077

RELATIONSHIPS BETWEEN PROCESS AND GEOMETRICAL FORM ON HIGH ARCTIC DEBRIS SLOPES, SOUTH-WEST DEVON ISLAND, CANADA,
McMaster Univ., Hamilton (Ontario). Dept. of Geography.
P. J. Howarth, and J. G. Bones.
In: Polar Geomorphology; Institute of British Geographers Special Publication No 4, p 139-153, June 1972. 4 fig, 2 tab, 16 ref.

Descriptors: *Degradation (Slope), *Geomorphology, *Arctic, Canada, Mass wasting, Topography, Stream erosion, Erosion, Slopes, Surf, Beach erosion, Bank erosion.
Identifiers: *Devon Island (Canada).

Debris slope profiles were measured at four different locations on southwest Devon Island, N.W.T., Canada. The nature and effects of basal erosion, rockfall and of processes dominated by melt water are described. There are significant differences in angle between slopes affected by different processes. The steepest slopes, tending to

be convex in form, are produced by basal erosion, which results from both ice push and storm waves. In noncoastal locations, slopes dominated by rockfall are significantly steeper than those affected by processes involving water, although both types of slope tend to be concave in form. The effects of these tow processes are also seen in the significant differences between talus sheets (rockfall dominant) and talus cones (melt-water dominant). In coastal areas, however, rockfall slopes and avalanche slopes are both steep because debris is rafted away by sea ice to prevent basal accumulation. There appears to be little difference in geometrical form between High Arctic and mid-latitude slopes if angles and profiles are compared on slopes where similar processes operate in the two environments. (Knapp-USGS)
W73-08078

PROCESSES OF SOIL MOVEMENT IN TURF-BANKED SOLIFLUCTION LOBES, OKSTINDAN, NORTHERN NORWAY,
University Coll. of Swansea (Wales). Dept. of Geography.
C. Harris.
In: Polar Geomorphology; Institute of British Geographers Special Publication No 4, p 155-174, June 1972. 9 fig, 7 tab, 13 ref.

Descriptors: *Creep, *Solifluction, *Arctic, *Degradation (Slope), *Mass wasting, *Geomorphology, Permafrost, Soil erosion, Freezing, Thawing, Frost action, Snow cover.

The total annual rates of soil movement on an east-facing arctic slope in Norway were measured, together with soil-moisture conditions, winter frost heave, soil temperatures, the vertical profile of soil movement with depth, and the mechanical properties of the soils. Frost heave is greatest when penetration of the freezing plane through the soil is slow. Soil creep results from both frost heave and expansion and contraction of the soil owing to varying moisture content. Saturated conditions occur for a short period during the spring thaw as a result of drainage being impeded by a frozen subsoil layer. Maximum annual soil movement occurs where saturated conditions persist longest, not where the slope is greatest. Movement takes place by a combined process of creep and flow. (Knapp-USGS)
W73-08079

THE NATURE OF THE ICE-FOOT ON THE BEACHES OF RADSTOCK BAY, SOUTH-WEST DEVON ISLAND, N.W.T., CANADA IN THE SPRING AND SUMMER OF 1970,
McMaster Univ., Hamilton (Ontario). Dept. of Geography.
For primary bibliographic entry see Field 02C.
W73-08080

THE SOLUTION OF LIMESTONE IN AN ARCTIC ENVIRONMENT,
Bristol Univ. (England). Dept. of Geography.
For primary bibliographic entry see Field 02K.
W73-08081

PROCESSES OF SOLUTION IN AN ARCTIC LIMESTONE TERRAIN,
McMaster Univ., Hamilton (Ontario). Dept. of Geography.
For primary bibliographic entry see Field 02K.
W73-08082

APPLICATION OF RHEOLOGICAL MEASUREMENTS TO DETERMINE LIQUID LIMIT OF SOILS,
Central Building Research Inst., Roorkee (India).
For primary bibliographic entry see Field 02G.
W73-08089

PROCESS AND SEDIMENT SIZE ARRANGEMENT ON HIGH ARCTIC TALUS, SOUTHWEST DEVON ISLAND, N.W.T., CANADA.
Alberta Univ., Edmonton. Dept. of Geography.
J. G. Bones.
Arctic and Alpine Research, Vol 5, No 1, p 29-40, 1973. 4 fig, 3 tab, 19 ref.

Descriptors: *Talus, *Sediment sorting, *Arctic, *Mass wasting, Rockslides, Variability, Statistics, Correlation analysis, Degradation (Slope).
Identifiers: *Devon Island (Canada).

The three dominant processes operating on talus of southwest Devon Island, Canada, produce characteristic arrangements in size of surface fragments. On 25 of the 27 surfaces measured, zonal (upslope-downslope) variance accounts for a much higher proportion of size variation than does lateral (cross-slope) variation. This characteristic supports the hypothesis of fall-sorting and reverse fall-sorting of rock fragments as the fundamental mode of talus formation. At present, 66% of the rockfall taluses have statistically significant zonal size arrangements, compared to 50% on talus with basal erosion and only 40% on alluvial talus. Basal erosion and meltwater activity may either reinforce or obscure the original downslope arrangement, depending upon the form of the process, its magnitude and frequency. Comparison of similar studies by way of statistical power analysis reveals considerable support for these findings. (Knapp-USGS)
W73-08090

NATURE AND RATE OF BASAL TILL DEPOSITION IN A STAGNATING ICE MASS, BURROUGHS GLACIER, ALASKA,
Ohio State Univ. Research Foundation, Columbus. Inst. of Polar Studies.
D. M. Mickelson.
Arctic and Alpine Research, Vol 5, No 1, p 17-27, 1973. 5 fig, 2 tab, 17 ref. NSF Grant GA-12300.

Descriptors: *Till, *Glaciers, *Alaska, *Sedimentation rates, Glaciation, Glacial sediments, Glacial drift, Arctic.
Identifiers: *Burroughs Glacier (Alaska).

Rates of basal till deposition ranging from 0.5 to 2.5 cm per year were observed in the Burroughs Glacier in southeast Alaska. Because of the emergence of hills during deglaciation, a change in ice flow direction of up to 90 deg has occurred near the southeast terminus. Because this change is recorded by maps and photographs dating to 1892, a rate of change of ice flow direction could be estimated. Till fabric measurements and till composition at two or three depths in the till at seven localities reflect this change. Estimates of the rate of till deposition were obtained by assuming that the fabric azimuth represents the ice flow direction at the time the till was deposited. Most till deposition took place during late stages of deglaciation. At two locations fabric of till just above bedrock or a paleosol records a post-1892 flow direction. (Knapp-USGS)
W73-08091

CHARACTERISTICS AND GEOMORPHIC EFFECTS OF EARTHQUAKE-INITIATED LANDSLIDES IN THE ADELBERT RANGE, PAPUA NEW GUINEA,
Australian National Univ., Canberra. Dept. of Biogeography and Geomorphology.
C. F. Pain.
Engineering Geology, Vol 6, No 4, p 261-274, December 1972. 8 fig, 1 tab, 13 ref.

Descriptors: *Landslides, *Earthquakes, *Mudflows, *Tropical regions, *Debris avalanches, Mass wasting, Vegetation effects, Geomorphology, Slopes, Forests.
Identifiers: *Papua New Guinea.

On November 1, 1970, an earthquake of magnitude 7.0 occurred 32 km north of Madang on the north coast of Papua New Guinea, and on the fringes of the Adelbert Range. Dense landsliding occurred over an area of 240 sq km. Debris avalanches removed shallow soil and forest vegetation from slopes of 45 deg. Earthflows occurred on deeper soils and lower angled slopes. The nature of the landslides and disposition of the vegetation debris suggest that falling trees triggered the landslides during the earthquake. Logs in the deposits were an important influence on the movement of landslide debris in the channel systems. (Knapp-USGS)
W73-08092

POSSIBLE ACCUMULATION OF AUTHIGENIC, EXPANDABLE-TYPE CLAY MINERALS IN THE SUBSTRUCTURE OF TUTTLE CREEK DAM, KANSAS, U.S.A.,
New Mexico Univ., Albuquerque. Dept. of Geology.
For primary bibliographic entry see Field 02K.
W73-08093

THE FORMATION OF IRON AND MANGANESE-RICH LAYERS IN THE HOLOCENE SEDIMENTS OF THUNDER BAY, LAKE SUPERIOR,
Lakehead Univ., Thunder Bay (Ontario).
J. S. Mothersill, and R. J. Shegelski.
Canadian Journal of Earch Sciences, Vol 10, No 4, p 571-576, April 1973. 3 fig, 1 tab, 19 ref.

Descriptors: *Bottom sediments, *Lake Superior, *Iron, *Manganese, Connate water, Chemical precipitation, Ion transport, Mass transfer.

Thin iron- and manganese-rich layers occur within or just below the Holocene sediments of Thunder Bay, Lake Superior. Most of the iron and manganese was solubilized by connate waters under the reducing conditions of the early burial stage, subsequently migrated upward with the connate waters along vertical fractures during compaction and then eventually deposited as oxides along favorable Eh horizons. (Knapp-USGS)
W73-08135

ATLANTIC CONTINENTAL SHELF AND SLOPE OF THE UNITED STATES—SAND-SIZE FRACTION OF BOTTOM SEDIMENTS, NEW JERSEY TO NOVA SCOTIA,
Geological Survey, Washington, D.C.
For primary bibliographic entry see Field 02L.
W73-08137

QUALITY OF SURFACE WATERS OF THE UNITED STATES, 1968: PART I. NORTH ATLANTIC SLOPE BASINS.
Geological Survey, Washington, D.C.
For primary bibliographic entry see Field 07C.
W73-08155

SURFICIAL SEDIMENTS OF BARKLEY SOUND AND THE ADJACENT CONTINENTAL SHELF, WEST COAST VANCOUVER ISLAND,
British Columbia Univ., Vancouver. Dept. of Geology.
For primary bibliographic entry see Field 02L.
W73-08156

MAP OF DEPOSITS ESPECIALLY SUSCEPTIBLE TO COMPACTION OR SUBSIDENCE, PARKER QUADRANGLE, ARAPAHOE AND DOUGLAS COUNTIES, COLORADO,
Geological Survey, Washington, D.C.
For primary bibliographic entry see Field 07C.
W73-08174

FROSTED BEACH-SAND GRAINS ON THE NEWFOUNDLAND CONTINENTAL SHELF,
Memorial Univ. of Newfoundland, St. John's. Dept. of Geology.
R. M. Slatt.
Geological Society of America Bulletin, Vol 84, No 5, p 1807-1812, May 1973. 4 fig, 24 ref. Canada Grants 29-71, A-8395, and E-2164.

Descriptors: *Sands, *Beaches, *Continental shelf, *Particle shape, *Particle size, Distribution patterns, Sedimentology, Electron microscopy, Frequency analysis, Sediments, Stratigraphy.
Identifiers: Newfoundland.

Frosted, relict beach sands are found in 90 m of water on part of the Newfoundland continental shelf. In this case, the common interpretation that frosted sands on the Atlantic continental shelf are indicative of aeolian deposition during periods of Pleistocene eustatic lowering of sea level would be incorrect. Combined grain-size frequency distribution analysis and grain-surface texture analysis with a scanning electron microscope are probably necessary to properly evaluate the depositional environment of frosted grains. (Knapp-USGS)
W73-08372

REVERSING BARCHAN DUNES IN LOWER VICTORIA VALLEY, ANTARCTICA,
Lunar Science Inst., Houston, Tex.
J. F. Lindsay.
Geological Society of America Bulletin, Vol 84, No 5, p 1799-1806, May 1973. 5 fig, 13 ref. NSR 09-051-001.

Descriptors: *Dunes, *Antarctic, Sediment transport, Dune sands, Sands, Particle size.
Identifiers: *Victoria Valley (Antarctica).

Approximately 30 reversing barchan dunes occur along the northern side of lower Victoria Valley, Antarctica. Laminae which dip steeply upwind suggest that much of the internal structure of the dunes develops during periods when the wind reverses itself—possibly during winter. Although the dune cores are frozen and the migration of the dunes are similar in texture to some barchans of more temperate climates. (Knapp-USGS)
W73-08373

VENTIFACT EVOLUTION IN WRIGHT VALLEY, ANTARCTICA,
Lunar Science Inst., Houston, Tex.
J. F. Lindsay.
Geological Society of America Bulletin, Vol 84, No 5, p 1791-1798, May 1973. 10 fig, 12 ref. NSR 09-051-001.

Descriptors: *Erosion, *Antarctic, *Wind erosion, *Particle shape, Sampling, Weathering, Particle size.
Identifiers: *Ventifacts, *Wright Valley (Antarctica).

Ventifacts occurring on extensive wind-deflated surfaces throughout ice-free Wright Valley, Antarctica, are the product of complex evolutionary processes. Initially, the wind removes the -2.0 to 1.5 phi fraction of valley soils to produce a lag gravel. The lag gravel then continues to evolve at a reduced rate as coarser granule and gravel fractions are removed. The distribution of wind-polished faces is determined largely by the shape of the original unpolished rock fragments. Once ventifaction is initiated, the number of faces per clast declines rapidly as minor faces are removed by the polishing. As ventifaction proceeds, the trend is reversed as the ventifacts are reoriented by the wind and new faces form. Salt weathering is also a major factor in determining the morphology of the Wright Valley ventifacts, proceeding more slowly than ventifaction. (Knapp-USGS)
W73-08374

SIMULATION MODEL FOR STORM CYCLES AND BEACH EROSION ON LAKE MICHIGAN,
Williams Coll., Williamstown, Mass.
W. T. Fox, and R. A. Davis, Jr.
Geological Society of America Bulletin, Vol 84, No 5, p 1769-1790, May 1973. 15 fig, 1 tab, 38 ref. ONR Contract NR 388-092.

Descriptors: *Beach erosion, *Lake Michigan, *Simulation analysis, *Mathematical models, Sand bars, Surf, Waves (Water), Profiles, Littoral drift, Currents (Water), Storms, Mapping, Geomorphology.

A mathematical simulation model relates storm cycles, beach erosion, and nearshore bar migration in Lake Michigan. The model is based on Fourier analysis of weather and wave data collected during the summers of 1969 and 1970. Barometric pressure is used as the independent variable with longshore current velocity computed as the first derivative and breaker height as a filtered version of the second derivative of barometric pressure. The simulated curves are used to compute wave and longshore current energy for each storm cycle and poststorm recovery. A gently sloping linear plus quadratic surface is used to represent the barless topography, with bars and troughs generated by normal curves. Bar distance is computed as a function of wave energy and bottom slope. Position of the bar and trough along the shore is determined by wave and longshore current energy. Simulated maps are produced for each storm cycle and recovery. (Knapp-USGS)
W73-08375

EFFECTS OF HURRICANE GINGER ON THE BARRIER ISLANDS OF NORTH CAROLINA,
Virginia Univ., Charlottesville. Dept. of Environmental Sciences.
R. Dolan, and P. Godfrey.
Geological Society of America Bulletin, Vol 84, No 4, p 1329-1333, April 1973. 4 fig, 8 ref.

Descriptors: *Sediment transport, *Beach erosion, *Barrier islands, *Surf, *Geomorphology, Beaches, North Carolina, Hurricanes, Waves (Water), Sand bars, Seashores.
Identifiers: *Cape Hatteras (NC).

The two barrier-island systems of North Carolina responded to the storm waves and surges of Hurricane Ginger in strikingly different manners. In the northern sector, which has been stabilized by artificial barrier dunes, erosion and dune recession were extensive. In the southern sector, as yet relatively unmodified, overwash and associated deposition were the dominant processes. Overwash is the major means by which the low barrier islands of the mid-Atlantic retreat before the rising sea, and it is the manner in which coarse sand and shell are transported inland from the beach. The barrier-island systems are constructed mostly of overwash-type sediments from the littoral zone. If sea level continues to rise, the resources required to maintain artifical barrier dunes may exceed the economic and psychologic value attached to their existence. The barrier islands, in their natural condition, are able to survive severe perturbations of hurricanes and extratropical storms by the wide runup and overwash profile they present to surges. These unmodified islands are not being washed away, as some engineers and developers have suggested in the past, but rather they are moving back by natural processes fundamental to their origin. (Knapp-USGS)
W73-08378

BEACH PROTECTION SYSTEM,
For primary bibliographic entry see Field 08A.
W73-08393

LAKE SEDIMENTS IN NORTHERN SCOTLAND,
Leicester Univ. (England).

W. Pennington, E. Y. Haworth, A. P. Bonny, and J. P. Lishman.
Philos Trans R Soc Lond Ser B Biol Sci. Vol 264, No 861, p 191-294. 1972. Illus.
Identifiers: Diatoms, *Lake sediments, Pollen, *Scotland, *Sediments, Zones.

A survey of deep-water sediments in 11 lakes in northern Scotland showed that only under certain conditions does a complete and conformable series of deposits accumulate. In lochs exposed to strong winds there may be no permanent settling of organic sediments in water depths of up to 50 m. Three lake cores (representative of 3 regions of northern Scotland), which proved to be complete and conformable profiles, were analyzed in detail for pollen and certain chemical elements; one was also analyzed for diatoms. A series of 14C dates was obtained for 2 of these profiles. Changes in pollen content were found to be very consistently related to changes in sediment composition. Pollen zones were defined in terms of characteristic taxa, and variance in sediment composition was expressed as the first component of a Principal Components Analysis; changes in this first component invariably coincided with pollen zone boundaries based on changes in pollen spectra. This close relationship is explained as the consequence of the derivation of these lake sediments from soils on the catchments. Significance of the pollen spectra and analysis of the core samples are discussed.--Copyright 1973, Biological Abstracts, Inc.
W73-08425

2K. Chemical Processes

THERMOCHEMICAL INVESTIGATION OF DIMETHYLMERCURY IN AQUEOUS AND NONAQUEOUS SOLUTIONS,
Missouri Univ., Rolla. Dept. of Chemistry.
For primary bibliographic entry see Field 05A.
W73-07806

PH BUFFERING OF PORE WATER OF RECENT ANOXIC MARINE SEDIMENTS,
California Univ., Los Angeles. Dept. of Geology.
For primary bibliographic entry see Field 05B.
W73-07851

MORE ON NOBLE GASES IN YELLOWSTONE NATIONAL PARK HOT WATERS,
Israel Atomic Energy Commission, Rehovoth; and Weizmann Inst. of Science, Rehovoth (Israel).
E. Mazor, and R. O. Fournier.
Geochimica et Cosmochimica Acta, Vol 37, No 3, p 515-525, March 1973. 5 fig, 2 tab, 14 ref.

Descriptors: *Gases, *Thermal water, *Argon, *Helium, Hydrothermal studies, Krypton radioisotopes, Water circulation, Groundwater movement, Water sources, Geysers.
Identifiers: *Yellowstone National Park, *Noble gases.

Water and gas samples from research wells in hydrothermal areas of Yellowstone National Park were analyzed for their rare gas contents and isotopic composition. The rare gases originate from infiltrating runoff water, saturated with air at 10 to 20 deg C. The atmospheric rare gas retention values found for the water varied between 3% and 87%. The fine structure of the Ar, Kr and Xe abundance pattern in the water reveals fractionational enrichment of the heavier gases due to partial outgassing of the waters. Radiogenic He and Ar were also detected. No positive evidence was found for magmatic water contribution. If present, the proportion of magmatic water is significantly less than 13% to 36%. (Knapp-USGS)
W73-07853

ALKALINITY DETERMINATION IN INTERSTITIAL WATERS OF MARINE SEDIMENTS,
Scripps Institution of Oceanography, La Jolla, Calif.
For primary bibliographic entry see Field 02J.
W73-07867

MOVEMENT OF CHEMICALS IN SOILS BY WATER,
Illinois Univ., Urbana. Dept. of Agronomy.
For primary bibliographic entry see Field 02G.
W73-07904

MODELING THE MOVEMENT OF CHEMICALS IN SOILS BY WATER,
Illinois Univ., Urbana. Dept. of Agronomy.
For primary bibliographic entry see Field 02G.
W73-07905

SR IN WATER OF THE CASPIAN AND AZOV SEAS,
For primary bibliographic entry see Field 05B.
W73-07970

WATER QUALITY MONITORING IN DISTRIBUTION SYSTEMS: A PROGRESS REPORT,
National Sanitation Foundation, Ann Arbor, Mich.
For primary bibliographic entry see Field 05A.
W73-08027

SEPARATION AND QUANTITATIVE DETERMINATION OF THE YTTRIUM GROUP LANTHANIDES BY GAS-LIQUID CHROMATOGRAPHY,
Iowa State Univ., Ames. Inst. for Atomic Research; and Iowa State Univ., Ames. Dept. of Chemistry.
For primary bibliographic entry see Field 05A.
W73-08033

SUPPORT-BONDED POLYAROMATIC COPOLYMER STATIONARY PHASES FOR USE IN GAS CHROMATOGRAPHY,
Applied Automation, Inc., Bartlesville, Okla. Systems Research Dept.
For primary bibliographic entry see Field 07B.
W73-08034

ELECTROCHEMICAL CELL AS A GAS CHROMATOGRAPH-MASS SPECTROMETER INTERFACE,
Northgate Lab., Hamden, Conn.
W. D. Deneker, D. R. Rushneck, and G. R. Shoemake.
Analytical Chemistry, Vol 44, No 11, p 1753-1758, September, 1972. 8 fig, 12 ref.

Descriptors: *Gas chromatography, *Mass spectrometry, *Instrumentation, Design, Performance, Methodology, Construction, Operations, Laboratory equipment.
Identifiers: *Palladium diffusion electrodes, *Electrochemical cell, *Mechanical interface, Detection limits.

An electrochemical cell employing palladium alloy diffusion electrodes has been developed to remove the hydrogen carrier gas exiting a gas chromatograph. Electrochemical pumping removes greater than 99.9996 percent of the hydrogen, resulting in mass spectrometer inlet pressures less than .01 micro Torr. Construction details, electrode activation procedures, and performance characteristics of the cell are described. (Long-Battelle)
W73-08035

COLORIMETRIC DETERMINATION OF CALCIUM USING REAGENTS OF THE GLYOXAL BIS (2-HYDROXYANIL) CLASS,
Clemson Univ., S.C. Dept. of Chemistry and Geology.
For primary bibliographic entry see Field 05A.
W73-08040

ELECTROCHEMICAL CHARACTERISTICS OF THE GOLD MICROMESH ELECTRODE,
Wisconsin Univ., Madison. Dept. of Chemistry.
W. J. Blaedel, and S. L. Boyer.
Analytical Chemistry, Vol. 45, No. 2, p258-263, February 1973. 9 fig, 2 tab, 12 ref.

Descriptors: *Design, *Construction, Electrochemistry, Flow rates, Zeta potential, Physical properties, Electric currents.
Identifiers: *Electrochemical properties, *Gold micromesh electrode, Ion selective electrodes.

The design and construction of a flow-through gold micromesh electrode are described. Current-voltage curves are reported for various flow rates. Measured limiting currents are shown to be directly proportional to the number of screens (N) in the electrode, to the concentration of electroactive material (C), and to the cube root of the volume flow rate (Vf) of the solution through the electrode. Various mesh sizes are examined. Application is made to the measurement of submicromolar concentrations. (Holoman-Battelle)
W73-08044

OSCILLOMETRIC TITRATION OF INORGANIC HALIDES WITH MERCURY (II) ACETATE,
Orszagos Gyogyszereszeti Intezet, Budapest (Hungary).
For primary bibliographic entry see Field 05A.
W73-08050

THE SOLUTION OF LIMESTONE IN AN ARCTIC ENVIRONMENT,
Bristol Univ. (England). Dept. of Geography.
D. I. Smith.
In: Polar Geomorphology; Institute of British Geographers Special Publication No 4, p 187-200, June 1972. 3 fig, 2 tab, 33 ref.

Descriptors: *Weathering, *Limestones, *Arctic, *Snow cover, Carbon dioxide, Vegetation effects, Solubility, Water chemistry, Melt water, Temperature, Weather, Canada, Geomorphology.

The weathering of limestone was studied in a high latitude Arctic environment. The study area is in a limestone region of northwestern Somerset Island at latitude 74 deg N in arctic Canada. Some 200 water samples were analyzed in the field for their calcium and magnesium content and pH; additional analyses were made of bedrock samples. The total hardness values were generally less than 95 ppm; taking precipitation and evapotranspiration figures into account, this suggests a weathering rate equivalent to about 2 mm/1000 years. Both the concentration of solutes and the weathering rate are considerably less than those found in lower latitudes. The lack of soil cover is thought to be the significant factor. There is evidence to suggest that solution of limestone is concentrated at the snow-rock interface. The concentration of carbon dioxide is no greater than that found in the free atmosphere. The area has been ice free for about the last 10,000 years and the rates of solution indicate that no major development of subnival hollows by solutional erosion could have occurred during this period. (Knapp-USGS)
W73-08081

PROCESSES OF SOLUTION IN AN ARCTIC LIMESTONE TERRAIN,
McMaster Univ., Hamilton (Ontario). Dept. of Geography.

W. B. Clarke, and G. Kugler.
Economic Geology, Vol 68, No 2, p 243-251, March-April 1973. 6 fig, 1 tab, 15 ref.

Descriptors: *Helium, *Uranium radioisotopes, *Tritium, *Groundwater, *Isotopes studies, Radioactive dating, Radiochemical analysis, Water chemistry, *Canada.
Identifiers: *Thorium.

Measurement of dissolved helium in groundwater should indicate the presence of uranium (or thorium) mineralization; such measurements were made near two known deposits, near Elliot Lake, Ontario, and near Inda Lake, Labrador. Helium contents of up to 600 times normal solubility were found, while neon contents were all within 20% of solubility equilibrium with the atmosphere. The ratio of He-4 to He-3 is highly correlated with He-4 content, although He-3 is enriched above solubility (up to a factor of 6) in some cases. Tritium content of the Labrador samples indicates that the water residence time is less than 20 years. This rules out the possibility that the high helium values are due to rocks of 'normal' uranium and thorium content. (Knapp-USGS)
W73-08157

RECORDS OF WELLS AND TEST BORINGS IN THE SUSQUEHANNA RIVER BASIN, NEW YORK,
Geological Survey, Albany, N.Y.
For primary bibliographic entry see Field 07C.
W73-08159

MAP SHOWING GENERAL CHEMICAL QUALITY OF GROUNDWATER IN THE SALINAQUADRANGLE, UTAH,
Geological Survey, Washington, D.C.
For primary bibliographic entry see Field 07C.
W73-08172

AN EXPERIMENTAL STUDY OF THE STRUCTURE, THERMODYNAMICS AND KINETIC BEHAVIOR OF WATER,
Midwest Research Inst., Kansas City, Mo.
For primary bibliographic entry see Field 01A.
W73-08188

AUTOMATED DATA HANDLING USING A DIGITAL LOGGER,
Virginia Polytechnic Inst. and State Univ., Blacksburg. Dept. of Chemistry.
For primary bibliographic entry see Field 07C.
W73-08257

DEMOUNTABLE RING-DISK ELECTRODE,
Illinois Univ., Urbana. School of Chemical Sciences.
For primary bibliographic entry see Field 07B.
W73-08260

LINEAR AND NONLINEAR SYSTEM CHARACTERISTICS OF CONTROLLED-POTENTIAL ELECTROLYSIS CELLS,
California Univ., Livermore. Lawrence Livermore Lab.
J. E. Harrar, and C. L. Pomernacki.
Analytical Chemistry, Vol 45, No 1, p 57-78, January 1973. 24 fig, 2 tab, 98 ref.

Descriptors: *Electrochemistry, Instrumentation, Control systems, *Electrolysis, Design, Model studies.
Identifiers: *Ion selective electrodes, *Transfer functions, Mercury electrodes, Platinum electrodes, Electrolytes.

A detailed study was made of the characteristics of three-electrode controlled-potential electrolysis cells as components of control systems. In the absence of significant faradaic current, these cells can be represented as linear, bridge-T type networks. Important parameters that determine cell response are the reference electrode resistance and a parasitic capacitance that couples the cell input to the output. Cells whose electrodes are arranged for optimum dc potential distribution were also found to have the minimum phase shift for a given attenuation, and on the basis of the circuit model the phase shift will not exceed 90 degrees. Cells with poor geometry exhibit excessive phase shift in their transfer functions. In the presence of significant faradaic current, the fundamental frequency transfer function is altered considerably and at applied potentials near the current-potential waves the cells are nonlinear. Negative-admittance reactions can cause the cell phase shift to be more negatice than -90 degrees, but most faradaic reactions cause the cell to exhibit less phase shift than the background solution value. Sufficient conditions for system stability, taking into account the time-varying, nonlinear, and other complicating characteristics of the cells, can be rigorously obtained using the circle criterion. Several aspects of electrochemical system design and measurement are discussed. (Little-Battelle)
W73-08261

THE EFFECT OF SODIUM ALKYLBENZENE-SULPHONATE ON THE DRAINAGE OF WATER THROUGH SAND,
Westfield Coll., London (England). Dept. of Zoology.
J. E. Webb, and C. M. Earle.
Environ Pollut, Vol 3, No 2, p 157-169, 1972, Illus.
Identifiers: *Alkylbenzenesulfonate, Cations, *Drainage, Pores, Rates, *Sand, Size, Sodium, Surfactant, Soil-water movement.

Sodium alkylbenzene sulfonate at 125-700 ppm caused an 80% reduction in the drainage rate of water through Leighton Buzzard sand in which the size of the pores formed by 3 grains in contact did not exceed 0.10 mm. This effect was not due to high viscosity and disappeared in coarser sands. It also disappeared in sand with the O2 atoms at the quartz surface replaced by trimethylsiloxy groups, suggesting an interaction between the sodium alkylbenzene sulfonate ion-pairs and the quartz surface at O2 sites. Molar equivalents of K and Mg cations, introduced as the sulfate to compete for the O2 sites, eliminated the reduction in drainage rate within 15 min. But, at 5% of the molar equivalent, the full effect was delayed for 4 hr, indicating that the ion's progress in water was impeded. Al cations at 1% molar equivalent eliminated the effect in 15 min, possibly due to the production of H ions. It appeared that the low rate of drainage of the solution could have been due to an O2-Na-sulfonate linkage at the quartz/water interface and the formation of a Na alkylbenzene sulfonate-water lattice in the bulk of the solution. At the concentration used Na alkylbenzene sulfonate ion-pairs were present in the ratio of 1:38 molecules of water in a linear series, so that the formation of a lattice would seem to imply structuring of the water. Such a lattice might be destroyed by H ions. Evidence of this came first from the addition of low concentrations of H ions (2 ppm H2SO4) which eliminated the drainage effect. No effect was found in the absence of the alkyl chain, when sodium benzene sulfonate was tested. The alkyl chain is thought to act either as an intermeshing fiber or as a template to align the water molecules and facilitate the formation of bonds between them. The disappearance of the drainage rate reduction at 700 ppm Na alkylbenzene sulfonate seems to coincide with the formation of micelles.–Copyright 1972, Biological Abstracts, Inc.
W73-08342

SALT TOLERANCE OF ORNAMENTAL SHRUBS AND GROUND COVERS,
Agricultural Research Service, Riverside, Calif. Salinity Lab.
For primary bibliographic entry see Field 03C.

W73-08347

SULFIDE DETERMINATION IN SUBMERGED SOILS WITH AN ION-SELECTIVE ELECTRODE,

Ministry of Agriculture, Cairo (Egypt).
A. I. Allam, G. Pitts, and J. P. Hollis.
Soil Science, Vol 114, No 6, p 456-467, December 1972. 3 fig, 6 tab, 27 ref. NSF Grant GB-8653.

Descriptors: *Aquatic soils, *Hydrogen sulfide, *Electrodes, *Instrumentation, *Chemical potential, Oxidation-reduction potential, Calibrations, On-site tests, Hydrogen ion concentration, Sulfides, Electrochemistry, *Rice, *Louisiana.
Identifiers: *Submerged soils.

The specific silver-sulfide membrane electrode (sulfide electrode) was evaluated for measurement of the soluble sulfide levels in submerged soils 'in situ' for measuring sorption of sulfide by clay in response to sulfide concentrations, and for the testing of these observed sulfide levels against theoretical predictions. A silver sulfide electrode was used for measurement of soil sulfide levels in Louisiana rice fields. The observed potential (E) obeyed the Nernst equation as a function of sulfide ion activity or concentration. Sulfide ion concentrations could be determined by direct potentiometry or potentiometric titration. The response rate of such electrodes suggests that continuous monitoring of some changing systems is feasible. Hydrogen sulfide levels ranged from 0.00005 to 0.64128 ppm in Louisiana rice fields during the tillering and ripening stages of rice plant development. A peak of H2S accumulation coincided with the highly reduced conditions occurring at the heading-flowering stage of the rice plant. H2S levels prevalent in rice fields during the heading-flowering stage were toxic to rice plants in vitro and significantly higher than those predicted from chemical equilibrium theory. The two most important factors regulating H2S accumulation in Louisiana rice fields were soil pH and oxidizable carbon. H2S is removed from the soil solution by the soil clay fraction. (Knapp-USGS)
W73-08351

FUMAROLE WITH PERIODIC WATER FOUNTAINING, VOLCAN ALCEDO, GALAPAGOS ISLANDS,

Arizona Univ., Tucson. Dept. of Geosciences.
B. E. Nordlie, and W. E. Colony.
Geological Society of America Bulletin, Vol 84, No 5, p 1709-1719, May 1973. 8 fig, 1 tab, 11 ref.

Descriptors: *Springs, *Volcanoes, *Sedimentation, *Geysers, Gases, Sands, Hydrogeology, Water chemistry.
Identifiers: *Galapagos Islands, *Fumaroles.

A water fountain and three associated pools are perched on the wall of the Volcan Alcedo caldera, in the Galapagos Islands, at the end of a line of active fumaroles. In July 1970, the lower basin was dry and the overflow pool was partially filled; the fountain pool was nearly full and was boiling. Gases forcefully expelled from a vent caused continuous fountaining. Opal shoreline deposits mark former maximum water levels, during which the overflow pool level is controlled by a spillway to the lower basin. The pools, full in August 1968, became completely dry by October 1970 and partially refilled in 1971. Throughout the cycle, sulfurous steam flowed from the vent. The water of the overflow pool is not in equilibrium with its deposits but is equilibrated with the atmosphere. The fountain pool shows the opposite conditions. Silica content and the location of geyserite deposits indicate that opal saturation occurs as the water cools and circulates through the overflow pool. Strata in the levee show cycles of mud, opal, and sand deposition. A sand cone is associated with the high-pressure fumarole. A turbulent gas flow velocity of about 6 m per sec would transport the sand grains. The source of water in the pools is largely meteoric. (Knapp-USGS)
W73-08377

CHEMICAL ANALYSES OF SELECTED PUBLIC DRINKING WATER SUPPLIES (INCLUDING TRACE METALS),

Wisconsin Dept. of Natural Resources, Madison.
For primary bibliographic entry see Field 05A.
W73-08424

ACETONE CONTAMINATION OF WATER SAMPLES DURING QUICKFREEZING,

Maryland Univ., Solomons. Natural Resources Inst.
For primary bibliographic entry see Field 05A.
W73-08442

2L. Estuaries

SYSTEMS ENGINEERING OF OYSTER PRODUCTION,

Delaware Univ., Newark. Dept. of Mechanical and Aerospace Engineering.
For primary bibliographic entry see Field 06C.
W73-07824

AN ECONOMIC APPROACH TO LAND AND WATER RESOURCE MANAGEMENT: A REPORT ON THE PUGET SOUND STUDY,

Washington Univ., Seattle. Dept. of Economics.
For primary bibliographic entry see Field 06B.
W73-07826

EXAMINATION OF TEXTURES AND STRUCTURES MUD IN LAYERED SEDIMENTS AT THE ENTRANCE OF A GEORGIA TIDAL INLET,

Skidaway Inst. of Oceanography, Savannah, Ga.
For primary bibliographic entry see Field 02J.
W73-07855

INTERFACE REFRACTION AT THE BOUNDARY BETWEEN TWO POROUS MEDIA,

Technion - Israel Inst. of Tech., Haifa. Dept. of Civil Engineering.
Y. Mualem.
Water Resources Research, Vol 9, No 2, p 409-414, April 1973. 5 fig, 2 tab, 4 ref.

Descriptors: *Interfaces, *Saturated flow, *Hydraulic conductivity, Oil-water interfaces, Saline water-freshwater interfaces, Boundary processes, Hydrogeology, Saline water intrusion, Permeability.
Identifiers: *Refraction (Interfaces).

In steady flow, refraction of the interface between fluids takes place at the boundary between porous layers of different hydraulic conductivities. An analytical and graphical solution of this problem is presented. The steady interface, which is a streamline, is refracted so that the ratio between the tangents of the angles that the interface makes with the normal to the surface separating the two layers equals the ratio between the corresponding hydraulic conductivities. In addition to the condition of refraction of a streamline, the angles of refraction of the interface can take only definite values depending on the slope of the boundary between the layers and the ratio of their permeabilities. Laboratory experiments, carried out on a Hele-Shaw analog, verified the analytical solution. (Knapp-USGS)
W73-07896

STEADY SEEPAGE FLOW TO SINK PAIRS SYMMETRICALLY SITUATED ABOVE AND BELOW A HORIZONTAL DIFFUSING INTERFACE: 1. PARALLEL LINE SINKS,

Johns Hopkins Univ., Baltimore, Md. Dept. of Environmental Engineering.
For primary bibliographic entry see Field 02F.
W73-07897

SHELF SEDIMENTS IN THE BAY OF BISCAY,

Akademiya Nauk SSR, Kaliningrad. Institut Okeanologii.
A. I. Blazhchishin.
Oceanology, Vol 12, No 2, p 235-246, 1972. 4 fig, 2 tab, 20 ref. Translated from Okeanologiya (USSR), Vol 12, No 2, 1972.

Descriptors: *Oceanography, *Marine geology, *Continental shelf, *Sedimentation, *Sediments, Bottom sediments, Provenance, Sediment distribution, Sediment transport, Particle size, Geomorphology, Mineralogy, Petrography, Bedrock, Littoral, Analysis.
Identifiers: *USSR, *Bay of Biscay, Relict sediments, Oozes, Isobaths.

Investigations of specific lithological differences between recent and relict deposits on the continental shelf of the Bay of Biscay were based on particle-size, mineralogical, petrographic, and chemical analyses of samples of bottom sediments and bedrock collected by the 1964-68 expedition of the Atlantic Division of the Scientific Research Institute of Sea Fisheries and Oceanography. The bottom deposits identified include terrigenous (<30% CaCO3), and biogenic-calcareous (>30% CaCO3), mixed terrigenous-biogenic sediments. Relict deposits occupy about 70% of the shelf area and are buried in different ways over wide and narrow shelves by recent sediments. The assumption that all depressions in the shelf have already been filled with sediments and that burial of adjacent relict sands is underway is confirmed by particle-size and mineral analyses. Given the low volume of incoming clastic material and the prohibitive conditions for its deposition in the northern part of the shelf, the role of biogenic calcareous sedimentation increases. (Josefson-USGS)
W73-07912

SOME CHARACTERISTICS OF TURBIDITY CURRENTS,

Sakhalin Kompleksnyi Nauchno-Issledovatelskii Institut, Yuzhno-Sakhalinsk (USSR).
N. L. Leonidova.
Oceanology, Vol 12, No 2, p 223-226, 1972. 1 tab, 25 ref. Translated from Okeanologiya (USSR), Vol 12, No 2, 1972.

Descriptors: *Marine geology, *Sedimentation, *Sediments, *Turbidity currents, *Flow characteristics, Submarine canyons, Continental slope, Waves (Water), Tsunamis, Earthquakes, Energy, Kinetics, Density, Velocity, Equations.
Identifiers: *USSR, *Marine sediments, *Slump (Mass movement).

Downslope flowage of unconsolidated marine sediments at the head of submarine canyons was investigated for development of turbidity currents. An estimate of the energy of these currents shows that it is comparable to that of tsunami waves observed during earthquakes. The comparable values may be further proof of the existence of dense high-velocity turbidity flow. (Josefson-USGS)
W73-07913

THE ECOLOGY OF THE PLANKTON OF THE CHESAPEAKE BAY ESTUARY, PROGRESS REPORT DEC 1970-AUG 1972,

Johns Hopkins Univ., Baltimore, Md.
For primary bibliographic entry see Field 05C.
W73-07932

A WATER-QUALITY SIMULATION MODEL FOR WELL MIXED ESTUARIES AND COASTAL SEAS: VOL. II, COMPUTATION PROCEDURES,

New York City-Rand Inst., N.Y.
For primary bibliographic entry see Field 05B.
W73-07935

ATLANTIC CONTINENTAL SHELF AND SLOPE OF THE UNITED STATES—SAND-SIZE FRACTION OF BOTTOM SEDIMENTS, NEW JERSEY TO NOVA SCOTIA,
Geological Survey, Washington, D.C.
J. V. A. Trumbull.
Available from GPO, Washington, D.C. 20402 - Price 55 cents (paper cover). Geological Survey Professional Paper 529-K, 1972. 45 p, 16 fig, 116 ref.

Descriptors: *Sedimentology, *Continental shelf, *Atlantic Ocean, *Northeast U.S., *Bottom sediments, Sediment distribution, Particle size, Sands, Physical properties, Geology, Sediment transport, Erosion, Data collections, Mapping, Coasts.
Identifiers: Sand-size fraction.

The sand-size fraction of surface sediments divides the continental shelf off the Northeastern United States into three distinctive areas: the glaciated Gulf of Maine and Nova Scotia shelf, the shallow high-energy Georges Bank-Nantucket Shoals area, and the more normal continental shelf south of New England and Long Island and east of New Jersey. The area of continental shelf under consideration is about 230,000 sq km, or about two-thirds of the area of the entire continental shelf along the east coast of the United States. Over all the continental shelf the primary components of the sand-size fraction are quartz and feldspar, with a strong admixture of rock fragments and dark minerals in the galciated Gulf of Maine and Nova Scotian shelf, and with a variable admixture of foraminiferan tests and shell fragments. Locally, very high concentrations of glauconite are found in the bight between New Jersey and Long Island and South of Long Island. (Woodard-USGS)
W73-08137

SURFICIAL SEDIMENTS OF BARKLEY SOUND AND THE ADJACENT CONTINENTAL SHELF, WEST COAST VANCOUVER ISLAND,
British Columbia Univ., Vancouver. Dept. of Geology.
L. Carter.
Canadian Journal of Earth Sciences, Vol 10, No 4, p 441-459, April 1973. 13 fig, 5 tab, 36 ref.

Descriptors: *Bottom sediments, *Coastal plains, *Estuaries, *Canada, Sediment transport, Sedimentation, Pacific Ocean, Provenance, Particle size, Distribution patterns, Sands, Gravels, Muds.
Identifiers: *Barkley Sound (Canada).

The bathymetry and sediment distribution of Barkley Sound and the adjacent continental shelf off the west coast of Vancouver Island have been markedly affected by the late Pleistocene glaciation and modern sedimentary processes. Several fjords widen and coalesce to form the sound, which is continuous with glacially eroded basins on the inner continental shelf. These basins are flanked by flat-topped banks, the larger of which merge with the outer shelf. Modern sediments are restricted mainly to Barkley Sound where the glaciated basin and sill topography and an estuarine circulatory system prevent the detrictus from reaching the continental shelf. Relict sands and gravels cover most of the shelf except within basins and drowned river valleys where muds prevail. This relict cover was initially dispersed by glaciers and meltwater streams, then later inundated during the Holocene Transgression, and is now being partly reworked by the present hydraulic regime. Near the shelf-break relict sediments are sparse; authigenic sands (glaucontized mudstone pellets) predominate together with residual sediments derived from submarine exposures of Tertiary mudstone. (Knapp-USGS)
W73-08156

OCEANOGRAPHIC COMMISSION.
Washington Natural Resources and Recreation Agencies, Olympia.
For primary bibliographic entry see Field 06E.
W73-08199

TIDELANDS—URGENT QUESTIONS IN SOUTH CAROLINA WATER RESOURCES LAWS,
South Carolina Univ., Columbia. School of Law.
For primary bibliographic entry see Field 06E.
W73-08201

MARINE WASTE DISPOSAL - A COMPREHENSIVE ENVIRONMENTAL APPROACH TO PLANNING,
For primary bibliographic entry see Field 05B.
W73-08247

MIREX AND DDT RESIDUES IN WILDLIFE AND MISCELLANEOUS SAMPLES IN MISSISSIPPI - 1970,
Mississippi State Univ., State College. Dept. of Biochemistry.
For primary bibliographic entry see Field 05A.
W73-08267

EFFECTS OF HURRICANE GINGER ON THE BARRIER ISLANDS OF NORTH CAROLINA,
Virginia Univ., Charlottesville. Dept. of Environmental Sciences.
For primary bibliographic entry see Field 02J.
W73-08378

AN ECOLOGICAL STUDY ON THE PROCESS OF PLANT COMMUNITY FORMATION IN TIDAL LAND, (IN KOREAN), C. S. KIM.
Mokpo Teachers' Coll. (South Korea).

Korean J Bot. Vol 14, No 4, p 27-33, 1971. Illus. English summary.
Identifiers: Aster-Subulata, Aster-Tripolium, Atriplex-Gmelini, Chloride, Community, Cyperus-Iria, Diplachne-Fusca, Echinochloa-Hispidula, Ecological studies, Formation, Limonium-Tetragonium, *Plant communities, Salicornia-Herbacea, Salt, Scirpus-Maritimus, Setaria-Lutescens, Soils, Suaeda-Maritima, Succession, *Tidal land, *Korea.

An attempt was made to investigate the plant community structure and the process of its formation in the tidal area surrounding Makpo City (Korea); the examined area included a stand in Sam-Hak Do where sands had infiltrated the community, and a stand in Kat-Ba-Woo which had been left as tidal soil land. Two hundred stands were sampled. Frequency, cover, density, standing (g/m2), contained Cl in the soil, and pH were obtained. The rank of dominant species is Salicornia herbacea L., Suaeda maritima Dum., Diplachne fusca L., Echinochloa hispidula Nak., Cyperus iria L., Setaria lutescens Hubb. in Sam-Hak Do, and Suaeda maritima, E. hispidula, Aster tripolium L.? Scirpus maritimus L., Salicornia herbacea, D. fusca, in Kat-Ba-Woo. Among them are 5 kinds of halophytes Salicornia herbacea, Suaeda maritima, Atriplex Gmelini C. A. Mey., Aster tripolium and Limonium tetragonum Bull., and 2 kinds of naturalized plants D. fusca and Aster subulata M. In the stands from Sam-Hak Do there was evidence of secondary succession in the presence of Cyperaceae such as C. liria, Juncus decipiens Nak., and Fimbristylis longispica Steu., which could not be found in the stands from Kat-Ba-Woo. The further inland from the floodgate, the higher the number of species; that is, the lower the content of Cl, the higher the number of species. On the distribution of the vegetation; comparing DFD (Density-Frequency-Dominance) index and Cl content, the main plants are Salicornia herbacea L., Suaeda maritima Dumorties, Atriplex Gmelini, D. fusca, E. hispidula where the Cl content of soil is more than 13.2%. Salicornia herbacea which has high resistance of salt, was half the total standing crop, with production of 1090/m2 while Suaeda maritima was 1/4, D. fusca L. 1/8, and Echinochloa the less than 1/8. The main factor in plant

PROPERTY OF U. OF F. LIBRARY

community formation in tidal land is tolerance for Cl; a plan for utilizing the halophytes which are abundant in tidal land should be devised.–Copyright 1973, Biological Abstracts, Inc.
W73-08403

IDENTIFICATION OF AMAZON RIVER WATER AT BARBARDOS, WEST INDIES, BY SALINITY AND SILICATE MEASUREMENTS,
McGill Univ., Montreal (Quebec). Marine Sciences Centre.
For primary bibliographic entry see Field 05B.
W73-08430

COMMUNITY STRUCTURE OF THE BENTHOS IN SCOTTISH SEALOCHS: I. INTRODUCTION AND SPECIES DIVERSITY,
Dunstaffnage Marine Research Lab., Oban (Scotland).
For primary bibliographic entry see Field 05C.
W73-08431

INFECTION OF BROWN SHRIMP, PENAEUS AZTECUS IVES BY PROCHRISTIANELLA PENAEI KRUSE (CRESTODA: TRYPANORHYNCHA) IN SOUTHEASTERN LOUISIANA BAYS,
Nicholls State Univ., Thibodaux, La.
For primary bibliographic entry see Field 05C.
W73-08433

DEVELOPMENT OF A NEW ENGLAND SALT MARSH,
A. C. Redfield.
Ecol Monogr. Vol 42, No 2, p 201-237. 1972. Illus.
Identifiers: Carbon, Geomorphology, Halophytes, *Marshes, Peat, Regime, *Salt marshes, Tidal marshes, *Massachusetts.

The salt marsh at Barnstable, Massachusetts, occupies an embayment into which it has spread during the past 4000 yr. It exhibits all stages of development from the seedling of bare sand flats through the development of intertidal marsh, to the formation of mature high marsh underlain by peat deposits more than 20 ft deep. Observations and measurements of the stages of its formation are presented. The geomorphology of the marsh is considered in relation to the factors which have influenced its development, that is, the ability of halophytes to grow at limited tide levels, the tidal regime, the processes of sedimentation, and the contemporary rise in sea level. The rates at which the early stage of development takes place have been determined by observations during a period of 12 yr and the time sequence of earlier stages by radiocarbon analyses.–Copyright 1973, Biological Abstracts, Inc.
W73-08443

SPECIES DIVERSITY OF MARINE MACROBENTHOS IN THE VIRGINIA AREA,
Queensland Univ., Brisbane (Australia). Dept. of Zoology.
D. F. Boesch.
Chesapeake Sci. Vol 13, No 3, p 206-211. 1972. Illus.
Identifiers: *Benthos, Diversity, Estuaries, Marine macrobenthos, Pollution, Salinity, Species, *Virginia.

Species diversity of benthic macro-organisms as measured byShannon's formular was highest on the outer continental shelf. Benthic diversity was higher in polyhaline zones of estuaries than on the shallow shelf and decreased sharply into the mesohaline zone, declining to the lowest in oligohaline zones. In addition to environmental stability and salinity regime, sediment grain size and pollution also affect species diversity. Analysis of the components of informational diversity, species richness and equitability, indicates that the richness component accounts for most of the observed pattern, although both components are important to within-habitat differences.–Copyright 1973, Biological Abstracts, Inc.
W73-08445

03. WATER SUPPLY AUGMENTATION AND CONSERVATION

3A. Saline Water Conversion

BRACKISH WATER DESALINATED BY THE 'SIROTHERM' PROCESS,
H. A. J. Battaerd, N. V. Blesing, B. A. Bolto, A. F. G. Cope, and G. K. Stephens.
Australian Chemical Processing and Engineering, Vol 25, No 8, p 19-21, August, 1972. 4 fig, 1 tab, 4 ref.

Descriptors: *Desalination, *Resins, *Ion exchange, *Pilot plants, Economics, Treatment facilities, Separation techniques, Water treatment, Saline water, Salinity, Water quality, Water quality control, Water supply, *Australia.
Identifiers: *Sirotherm process, Resin regeneration.

The 'sirotherm' process is an Australian invention designed for the reduction of salinity of brackish waters containing up to 3000 parts per million total dissolved solids. It utilizes a mixed bed of ion exchange resins which is simply regenerated with hot water rather than with costly acid and alkali used in conventional ion exchange. A flow sheet for the operation of the process in a fixed bed is included. Although pilot plant trials of the ion exchange process have been in progress since January 1970, it will not be possible to assess the economics of the process in its various forms other than on an individual case study basis until more information becomes available on resin life pretreatment needs, and the cost of manufacturing and marketing the 'sirotherm' resin. (Smith-Texas)
W73-07840

DESALINIZATION PLANTS, VIRGIN ISLANDS (FINAL ENVIRONMENTAL IMPACT STATEMENT).
Department of Housing and Urban Development, San Juan, Puerto Rico. Region II.

Available from the National Technical Information Service as EIS-VI-72-5071-F, $6.00 in paper copy, $0.95 in microfiche. August 11, 1972. 71 p, 4 fig, 3 map.

Descriptors: *Virgin Islands, *Environmental effects, *Desalination, *Desalination plants, Water treatment, Water quality, Treatment facilities, Water supply, Water demand, Municipal water, Potable water, Industrial water, Desalination apparatus, Social aspects, Area redevelopment, Desalination wastes.
Identifiers: *Environmental Impact Statements, *St. Thomas and St. Croix (Virgin Islands).

This action consists of the proposed construction of deslainization plants in St. Thomas and St. Croix, Virgin Islands. Two 2,250,000 gallon per day sea water desalinization plants are proposed: one on St. Thomas, adjacent to the existing generating facilities at Krum Bay and another on St. Croix, adjacent to that island's existing facilities. Neither plant will have any impact on wildlife habitat, nor influence the scenic values or land use patterns. Alternatives include the use of other locations, barging water from Puerto Rico, and no project. The only irretrievable commitment of resources involved will be the additional fuel burned by the boilers and the materials used in construction of the projects. The plants will contribute to the living standards of the population by supplying desperately needed potable water to low-income families, residential users and commercial facilities; sustain and enhance employment opportunities for Virgin Islanders; and maintain the quality of life. The discharge of desalination wastes will have a very insignificant impact on the environment. (Mockler-Florida)
W73-07977

HYDROCASTING REVERSE OSMOSIS MEMBRANES,
Hydronautics, Inc., Laurel, Md.
A. Gollan, M. P. Tulin, and C. Elata.
Available from the National Technical Information Service as PB-215 035, $3.00 in paper copy, $1.45 in microfiche. Research and Development Progress Report No. 806, June 1972. Office of Saline Water. 204 p, 68 fig, 17 tab, 36 ref, 2 append. 14-30-2528.

Descriptors: *Desalination, *Reverse osmosis, *Membranes, *Cellulose acetate, *Porosity, Nylon, *Permeability.
Identifiers: *Hydrocasting, Gelling solvents, Salt gelling, Porous tubes, Drain casting, Polymer properties, In-situ casting, Water flux, Salt rejection.

The development of the hydrocasting method for the formation of tubular skinned reverse osmosis membranes of about 1 mm diameter (tubules) was continued. Hydrocasting in single non-porous tubes utilizing modified Loeb-Sourirajan casting solutions has been further perfected and extended to casting in multiple non-porous tubes. Various means were found effective in suppressing the undesirable macropores formed during membrane tubule fabrication by hydrocasting. These improvements in membrane morphology and strength were due to: hydrocasting relatively thin membranes by casting in glass tubes of 1.5 mm I.D. or smaller; utilization of longer gas bubbles; gelling with water of lower chemical activity; and control of gelling water velocity. The way was thus opened for relatively high pressure testing (600-800 psi) of short (approx 8 in) supported sections as well as low pressure characterization (up to 220 psi) of unsupported membranes of almost full produced length (approx 30 in). The tests results demonstrated that Cellulose Acetate membrane tubules of high intrinsic strength and good desalting characteristics can be produced by hydrocasting. Work on hydrocasting has been accompanied by the development of a process for the fabrication of porous nylon tubes of high water permeability and strength. This important development opens up the way for hydrocasting inside porous tubes of strength sufficient to allow relatively high internal pressure operation. The development of a drain casting method which utilizes gravity draining to cast membranes in relative large diameter (more than 1/4 in) tubes in conjunction with the development of a Multiple Tube Module, suitable for sea-water operation, has also been performed. (OSW)
W73-08185

DEVELOPMENT OF REVERSE OSMOSIS MODULES FOR SEAWATER DESALINATION,
Gulf General Atomic Co., San Diego, Calif.
R. L. Riley, G. R. Hightower, H. K. Lonsdale, J. F. Loos, and C. R. Lyons.
Available from the National Technical Information Service as PB-214 973, $3.00 in paper copy, $1.45 in microfiche. Office of Saline Water Research and Development Progress Report No. 799, July 1972. 41 p, 11 fig, 7 tab, 15 ref. 14-30-2685.

Descriptors: *Reverse osmosis, *Membranes, Seawater, *Desalination, *Ion transport.
Identifiers: *Modules, *Ultrathin membranes, Casting device, Composit membranes, Abrasion resistance, Flexibility, Fold resistance, Crosslinking, Porous support, Prototype unit.

ments, that the structuring of water in or near membranes appears to play an important and, at times dominating role in determining functional properties of membranes, regardless of the level of morphological complexity or the operational intricacies of the membranes. (OSW)
W73-08190

TWO-STAGE FLUID TREATMENT SYSTEM,
Puredesal, Inc., Levittown, Pa. (assignee).
W. E. Bradley.
U. S. Patent No. 3,707,231, 7 p, 1 fig, 7 ref; Official Gazette of the United States Patent Office, Vol 905, No 4, p 748, December 26, 1972.

Descriptors: *Patents, Osmosis, *Semi-permeable membranes, *Reverse osmosis, Water treatment, Water purification, Potable water, *Desalination, Demineralization, Saline water, Brackish water.

The two-stage fluid treatment system involves the use of two semi-permeable membranes and an intermediate fluid circulated between them. The untreated liquid is introduced to the first semi-permeable membrane which operates as a direct absorption membrane. Compaction is avoided by operating the first membrane at near zero pressure differential. The intermediate fluid is circulated rapidly on the opposite side of the first membrane and is so selected as to facilitate osmosis of the liquid to be treated. A high pressure pump transfers the intermediate fluid together with the fluid passing through the first membrane to the second membrane which is operated under reverse osmosis conditions. The second membrane is impermeable to the intermediate fluid, passes the treated fluid, while retaining the intermediate fluid which is eventually recirculated to the first membrane. (Sinha-OEIS)
W73-08308

SUPPORT MODULE FOR REVERSE OSMOSIS MEMBRANE,
Westinghouse Electric Corp., Pittsburgh, Pa. (assignee).
N. A. Salemi.
U. S. Patent No. 3,707,234, 3 p, 3 fig, 2 ref; Official Gazette of the United States Patent Office, Vol 905, No 4, p 749, December 26, 1972.

Descriptors: *Patents, *Reverse osmosis, Membranes, *Semipermeable membranes, Water treatment, *Desalination, Demineralization, Water purification.

This reverse osmosis module comprises semipermeable tubular osmotic membranes. There is an influent conduit for supplying pressurized influent liquid to the inside of the tubular osmotic membranes. It has a conduit for draining purified liquid from the outside of the tubular osmotic membranes and a support structure for supporting the tubular osmotic membranes to prevent rupture from internal pressure. The support structure comprises plates fastened together to form a rigid stack. Each plate has holes which register with holes in adjacent plates, and the tubular osmotic membranes are disposed in these holes. Adjacent plates have sufficient spacing between them to pass liquid from the tubular osmotic membranes to the drain conduit. (Sinha-OEIS)
W73-08310

MULTISTAGED FLASH EVAPORATOR AND A METHOD OF OPERATING THE SAME WITH SPONGE BALL DESCALING TREATMENT,
Hitachi, Ltd., Tokyo (Japan). (assignee).
S. Takahashi, K. Otake, and T. Horiuchi.
U. S. Patent No. 3,707,442, 5 p, 14 fig, 11 ref; Official Gazette of the United States Patent Office, Vol 905, No 4, p 796, December 26, 1972.

Descriptors: *Patents, *Flash distillation, *Evaporators, Brine, Potable water, Distillation, Water treatment, *Desalination, *Descaling.

A method of operating a multistage flash evaporator has flashing chambers of successively lower pressures and temperatures. The condenser part in each chamber is composed of at least one brine tube and a flashing part. Brine is passed successively through the condensers to preheat the brine and condense steam evaporated in each flashing chamber. Brine is heated to a temperature beyond a scale eduction temperature occurring at 80C. Sponge balls are introduced to effect continuous removal of scale attached on the inner walls of the brine tubes and brine heaters. (Sinha-OEIS)
W73-08312

REVERSE OSMOSIS MODULE,
Westinghouse Electric Corp., Pittsburgh, Pa. (assignee)
For primary bibliographic entry see Field 05D.
W73-08391

APPARATUS FOR PERFORMING THE IMMISCIBLE REFRIGERANT FREEZE PROCESS FOR PURIFYING WATER,
United Kingdom Atomic Energy Authority, London (England). (assignee)
M. J. S. Smith, J. H. Wilson, and B. R. Parr.
U. S. Patent No 3,712,075, 5 p, 5 fig, 8 ref; Official Gazette of the United States Patent Office, Vol 906, No 4, p 1214, January 23, 1973.

Descriptors: *Patents, *Desalination, *Water treatment, Sea water, *Freezing, *Crystallization, Separation techniques, Water purification, Brine.
Identifiers: *Ice crystals.

The immiscible refrigerant freeze process comprises boiling the immiscible refrigerant in the impure water (sea water) to be treated. The ice crystals so produced are melted to produce purified water. The apparatus has a crystallizer section with two sub-sections. The first forms a refrigerant injection zone in which ice crystal slurry is produced. The second forms a disengagement zone in which entrained refrigerant is removed from the slurry and ice crystals are allowed to grow before transferring the slurry to the brine separating and washing section. (Sinha-OEIS)
W73-08392

REVERSE OSMOSIS MEMBRANE MODULE AND APPARATUS USING THE SAME,
Aqua-Chem, Inc., Waukesha, Wis. (assignee)
G. B. Clark.
U. S. Patent No 3,708,069, 8 p, 6 fig, 1 ref; Official Gazette of the United States Patent Office, Vol 906, No 1, p 182-183, January 2, 1973.

Descriptors: *Patents, *Desalination apparatus, *Reverse osmosis, Separation techniques, *Water treatment, Water quality control, Treatment facilities, *Membranes.

A support is provided for a membrane in reverse osmosis equipment that need not be capable of withstanding the high pressures encountered in reverse osmosis operations. The module uses disposable tubular membranes which are arranged in concjunction with a tubular casing having substantial hoop strength. This balances out the forces due to high pressure within the casing between adjacent support tubes so that substantially all pressure is borne by the exterior casing. Each of the tubes has a discontinuous outer surface, in the shape of a hexagon, and all are held so that each tube is in contact with another tube or the interior of the casing about its entire periepheral extent. A motor driven pump within the casing receives a liquid mixture and raises its pressure to reverse osmosis operating status. The pump also drives an impeller and baffles which operate to recirculate the liquid mixture through the membrane cell. (Sinha-OEIS)
W73-08397

3B. Water Yield Improvement

WATERSHED HYDROMETEOROLOGICAL DATA REQUIRED FOR WEATHER MODIFICATION,
North American Weather Consultants, Santa Barbara, Calif.
R. D. Elliott, and J. Hannaford.
In: Proceedings of the 40th Annual Meeting of the Western Snow Conference, April 18-20, 1972. Phoenix, Ariz: Printed by Colorado State University, Fort Collins, p 61-66, 1972. 2 fig, 2 ref.

Descriptors: *Weather modification, *Data collections, *Meteorological data, *Snowpacks, *Mountains, Meteorology, Snow surveys, Precipitation gages, Snow cover, Snowmelt, Cloud seeding.

Hydrometeorological data in five western mountain watersheds were reviewed and deficiencies in existing data required for planning of large scale research weather modification projects were noted. Ideally the additional data needed for each major watershed includes a base station for project control, several climate stations to record areal coverage of precipitation and standard meteorological parameters, and one mountain observatory unit for observation of special meteorological parameters and snow quality data. Existing data reveal little or no tendency for snowpack to level off or to decrease at higher elevations. A 10% increase in precipitation during the October through Arpil seeding periods could increase the average annual runoff by 2.5 million acrefeet within the five large watersheds and an additional 1 million acrefeet around the periphery of these watersheds. Extending the seeding period into May and June could increase the runoff from the five study basins by 440,000 acrefeet but caution is needed in seeding during this period because of undesirable flood potentials. (See also W73-08138) (Knapp-USGS)
W73-08150

SOUTH DAKOTA WEATHER MODIFICATION PROGRAM,
South Dakota Weather Control Commission, Pierre.
M. Williams.
In: Proceedings of the 40th Annual Meeting of the Western Snow Conference, April 18-20, 1972, Phoenix, Ariz: Printed by Colorado State University, Fort Collins, p 83-88, 1972. 2 fig.

Descriptors: *Weather modification, *South Dakota, Cloud seeding, Research and development, Legislation, Environmental effects, Social aspects, Legal aspects.

Research efforts directed toward the development of a weather modification capability in South Dakota have been underway for approximately 10 years. An economic evaluation of the effect on crop production is under way. This involves development of a model in the form of a statistical regression to predict and detect changes in crop yield as the program progresses. Evaluation of sociological effects is a continuing part of the program. Environmental effects are also monitored. (See also W73-08138) (Knapp-USGS)
W73-08154

WATER AND LAND RESOURCE ACCOMPLISHMENTS, 1971.
Bureau of Reclamation, Washington, D.C.
For primary bibliographic entry see Field 08A.
W73-08191

POSSIBILITY OF UTILIZING ATMOSPHERIC CONDENSATE IN THE HIGHER PLANT SYSTEM,
For primary bibliographic entry see Field 02I.
W73-08345

THE MANAGEMENT OF MOUNTAIN CATCHMENTS BY FORESTRY,
Stellenbosch Univ. (South Africa). Dept. of Silviculture.
C. L. Wicht.
S Afr For J. Vol 77, p 6-12, 1971. Illus.
Identifiers: *Catchments, Forestry, Management, Mountain catchment, *South Africa, *Water conservation.

The management of mountain catchments in South Africa to maintain and improve water resources is still inadequate. The water yields from mountain areas can be strongly influenced by the control of vegetation and this is at present the only practical manner of generally increasing the basic water resources of the Republic, which are derived from precipitation.--Copyright 1973, Biological Abstracts, Inc.
W73-08401

WATER STRESS IN SALT CEDAR,
Georgia Experiment Station, Experiment.
R. E. Wilkinson.
Bot Gaz. Vol 133, No 1, p 73-77. 1972. Illus.
Identifiers: Humidity, *Salt-cedar, *Solar radiation, Temperature, *Water stress, Winds, Evaporation.

Salt cedar (Tamarix pentandra Pall.) cladophylls developed water potentials of -5 bars by early July and -20 bars by late Sept. while the plants grew in a deep sand with a high water table. Plant water utilization closely paralleled pan evaporation and solar radiation. Relative water content was correlated with season, solar radiation, air temperature, wind velocity, relative humidity, and prior growing conditions. Relative water content of trees growing on a 3-ft water table was not significantly different from the relative water content of trees growing on a water table deeper than 10ft. Relative water content decreased throughout the summer but increased in Sept. and reached 100% in mid-Oct.--Copyright 1973, Biological Abstracts, Inc.
W73-08422

3C. Use of Water of Impaired Quality

EXTRAPOLATION OF INDICATOR SCHEMES WITHIN SALT MARSHES,
Moskovskoe Obshchestvo Ispytatelei Prirody (USSR). Geographic Div.
For primary bibliographic entry see Field 02H.
W73-07848

SALT TOLERANCE OF ORNAMENTAL SHRUBS AND GROUND COVERS,
Agricultural Research Service, Riverside, Calif. Salinity Lab.
L. Bernstein, L. E. Francois, and R. A. Clark.
J Am Soc Hortic Sci, Vol 97, No 4, p 550-556, 1972, Illus.
Identifiers: Bougainvillea, Ground covers, Guava, Holly, Injury, Leaves, Loss, Natal-Plum, *Ornamental shrubs, Pittosporum, Rose, Rosemary, *Salt tolerance, Shrubs, Star-Jasmine.

The salt tolerance of 25 shrub and ground-cover species was determined in plots artificially salinized with NaCl plus $CaCl_2$. Chloride and Na injury was observed in sand cultures of the same species with 4 different salt treatments. Tolerant species, like bougainvillea, Natal plum, and rosemary were affected little, if at all, by soil salinities of 8 mmho/cm (electrical conductivity of the saturation extract: ECe), whereas sensitive species like star jasmine, guava, holly, and rose were severely damaged or killed at ECe's of 4 mmho/cm. Salt tolerance was not well correlated with injury by Cl or Na, although many species exhibited leaf burn; nor was survival under highly saline conditions necessarily a good index of salt tolerance. Five-gallon specimens of sensitive, slow-growing species such as pittosporum were more tolerant than 1-gal specimens and delaying salination of such species increased salt tolerance somewhat. Leaves with symptoms like those of Cl or of Na injury but containing very little of these ions were frequently observed in landscape plantings of a number of shrub species. The injury was attributed to inadequate watering. It is suggested that Cl or Na accumulation in leaves of shrubs may cause injury by interfering with normal stomatal closure, causing excessive water loss and leaf injury symptoms like those of drought.--Copyright 1973, Biological Abstracts, Inc.
W73-08347

3D. Conservation in Domestic and Municipal Use

A MANUAL ON COLLECTION OF HYDROLOGIC DATA FOR URBAN DRAINAGE DESIGN,
Hydrocomp, Inc., Palo Alto, Calif.
For primary bibliographic entry see Field 07C.
W73-07801

REGIONAL WATER RESOURCE PLANNING FOR URBAN NEEDS: PART 1,
North Carolina Univ., Chapel Hill. Dept. of City and Regional Planning.
For primary bibliographic entry see Field 06B.
W73-07819

MODELS OF INDUSTRIAL POLLUTION CONTROL IN URBAN PLANNING,
RAND Corp., Santa Monica, Calif.
For primary bibliographic entry see Field 05G.
W73-07831

INSTANTANEOUS UNIT HYDROGRAPHS, PEAK DISCHARGES AND TIME LAGS IN URBANIZING WATERSHEDS,
Purdue Univ., Lafayette, Ind. School of Civil Engineering.
For primary bibliographic entry see Field 02A.
W73-07917

THE FINANCIAL FEASIBILITY OF REGIONALIZATION,
Arkansas Univ., Fayetteville. Dept. of Agricultural Economics and Rural Sociology.
N. C. Williams, and J. M. Redfern.
Journal of the American Water Works Association, Vol 65, No 3, p 159-168, March, 1973. 1 fig, 5 tab, 12 ref.

Descriptors: *Financial feasibility, *Regional development, *Water supply, *Project planning, *Cost analysis, Methodology, Water distribution (Applied), Water demand, Water quantity, Water users, Estimating, Return (Monetary), Rates, Amortization, Capital costs, Domestic water, Industrial water, Municipal water, Rural areas, Transmission lines, Wells, Operating costs, Governments, Human population, *Arkansas.
Identifiers: *Regional water systems, *Economic analysis, Incremental investment, Economies of size, Annual saving, Commercial water, Benton County, Washington County, Beaver Reservoir.

Municipal official concerned with assuring a community an adequate water supply at least cost should consider a regional water system. The feasibility of providing water on a regional basis, rather than having each municipal system augment its own supply is examined. A financially feasible project is defined as one that guarantees revenues that suffice to cover all costs, including interest on funds borrowed to finance the project. Herein, the

WATER SUPPLY PLANNING STUDY OF THE CAPITOL REGION OF CONNECTICUT,
Connecticut Univ., Storrs. Dept. of Civil Engineering.
For primary bibliographic entry see Field 06B.
W73-08302

3E. Conservation in Industry

SYSTEMS APPROACH TO POWER PLANNING,
Bechtel Corp., San Francisco, Calif. Hydro and Community Facilities Div.
For primary bibliographic entry see Field 08C.
W73-07925

AGRI-INDUSTRIAL PARK WATER IMPROVEMENTS, SEWAGE TREATMENT FACILITIES, BUSINESS DEVELOPMENT LOAN (FINAL ENVIRONMENTAL IMPACT STATEMENT).
Economic Development Administration, Austin, Tex. Southwestern Region.
For primary bibliographic entry see Field 05D.
W73-07978

THE ENERGY NEEDS OF THE NATION AND THE COST IN TERMS OF POLLUTION,
Atomic Energy Commission, Washington, D.C.
For primary bibliographic entry see Field 06G.
W73-07995

ENTROPY AS A MEASURE OF THE AREAL CONCENTRATION OF WATER-ORIENTED INDUSTRY,
Tennessee Univ., Knoxville. Coll. of Business Administration.
A. S. Paulson, and C. B. Garrison.
Water Resources Research, Vol 9, No 2, p 263-269, April, 1973. 2 tab, 5 equ, 13 ref.

Descriptors: *Water, *Industries, *Areal, *Employment, *Entropy, *Numerical analysis, *Tennessee, Regional analysis, Data, Water use, Equations.
Identifiers: *Numbers equivalent, Water-oriented industry, Data analysis.

The concepts of entropy and numbers equivalent are applied to the configuration of water intensive and non-water intensive employment data by country in the Tennessee Valley region to provide an overall measure of the areal concentration of employment and population. If the underlying distribution of employees is assumed to be governed by the multinomial distribution, it is shown that there have been shifts in microlocational characteristics, since the role chance can play is minimal. Non-water intensive manufacturing employment has become much more areally dispersed over the study period of 1959-1968, but it appears that there has been neither a decrease nor an increase in concentration in water intensive employment. The two types of employment are concentrated to markedly different degrees. Application of the concepts of entropy and numbers equivalent shows that the areal concentration of water intensive manufacturing employment in the Tennessee Valley region is substantially greater than the corresponding all-manufacturing employment. (Bell-Cornell)
W73-08131

3F. Conservation in Agriculture

ECONOMIC EVALUATION OF ALTERNATIVE FARM WATER SOURCES IN THE CLAYPAN AREA OF ILLINOIS,
Illinois Univ., Urbana. Dept. of Agricultural Economics.
For primary bibliographic entry see Field 05G.
W73-07804

SYSTEMS ANALYSIS IN IRRIGATION AND DRAINAGE,
California Univ., Riverside. Dry-Lands Research Inst.
W. A. Hall.
Journal of the Hydraulics Division American Society of Civil Engineers, Vol 99, No HY4, Proceedings paper 9659, p 567-571, April 1973. 5 p.

Descriptors: *Irrigation, *drainage, *Systems analysis, *soil moisture, Water resources, Decision making, Foods, Crops, Hydraulics, Agriculture, Plants, Nutrients, Optimization, Streamflow, Risks, Alternative planning, Mathematical models, Water rights, Water demand, Water costs, Water shortage, Salinity, Droughts, Irrigation water.
Identifiers: Stochastic water supplies, Stochastic unregulated streamflow.

It is imperative that water engineers and planners utilize systems analysis in order that costly irreversible water decisions may be based upon sound judgments. Growing population, problems of inadequate nutrition, and rising standards of living all demand increases in food production. The rate of increasing agricultural productivity cannot be maintained by continued use of fertilizer, pest control, and genetics; only irrigation and its correlative drainage can be counted on for certain to increase food and fibre supplies. But water shortages and resulting high costs pose great difficulties. Because of such food production and water cost squeezes, and a serious political squeeze wherein potential water for agriculture is given to cities, irrigation and drainage must be brought to a point of maximum efficiency. Needed is a new technology for agricultural water use based on systems analysis, for: (1) precision control of soil moisture related factors of production; (2) optimization of the use of unregulated stochastic streamflows; (3) optimization of risk and return from water use under uncertainty; and (4) optimal salinity and drought strategies. This step must be taken, regardless of important analytical limitations and considerable additional fundamental research requirements. (Bell-Cornell)
W73-07923

RESEARCH NEEDS FOR IRRIGATION RETURN FLOW QUALITY CONTROL,
Colorado State Univ., Fort Collins. Dept. of Agricultural Engineering.
For primary bibliographic entry see Field 05G.
W73-07965

POLECAT BENCH AREA OF THE SHOSHONE EXTENSIONS UNIT, WYOMING.

Hearing--Subcomm. of Water and Power Resources--Comm. on Interior and Insular Affairs, United States Senate, 92nd Cong, 2d Sess, September 19, 1972. 53 p, 1 tab.

Descriptors: *Irrigation, *Wyoming, *Irrigation programs, *Water supply development, Mississippi River Basin, Missouri River, Environmental effects, Crop production, Public benefits, Water resources development, Project planning, River basin development, Dam construction, Impoundments, Construction, Reservoirs, Wildlife conservation, Water distribution, Water management (Applied).
Identifiers: *Congressional hearings, Federal Water Project Recreation Act.

Testimony is reported on a bill to reauthorize the Secretary of the Interior to construct, operate and maintain the Polecat Bench area of the Shoshone extensions unit, Missouri River Basin Project, Wyoming. The development would provide irrigation water to 19,200 acres of irrigable lands and provide outdoor recreation and fish and wildlife conservation. The response from the Secretary of the Interior indicates that cost considerations are prohibitive. The State of Wyoming endorsed the

plan of development. By resolution the Board of Commissioners of the Polecat Bench Irrigation District, which encompasses the lands proposed for irrigation development, expressed the district's support for the development. The Wyoming Recreation Commission has requested that recreation be included as a proposed development and has indicated Wyoming's intent to comply with the provisions of the Federal Water Project Recreation Act. The project itself consists of a dam, reservoir and distribution and drainage systems. (Smith-Adam-Florida)
W73-07985

HYDRAULIC LABORATORY STUDIES OF A 4-FOOT-WIDE WEIR BOX TURNOUT STRUCTURE FOR IRRIGATION USE,
Bureau of Reclamation, Denver, Colo. Engineering and Research Center.
For primary bibliographic entry see Field 08B.
W73-08086

FOREST METEOROLOGICAL, SOIL CLIMATOLOGICAL AND GERMINATION INVESTIGATION, (IN NORWEGIAN),
Norske Skogforsoksvesen, Vollebekk.
K. Bjor.
Medd Nor Skogforsoksves. Vol 28, No 8, p 429-526, 1971, Illus. (English summary).
Identifiers: Climatology, Ecology, *Seedling establishment, *Forests, *Germination, Light, Meteorology, Moisture, Soils, Temperature.

The most important aim was to reveal the ecological conditions for germination and seedling establishment. The experimental area was situated near Elverum (Norway) about 200 m above sea level. Measurements were made of groundwater level, temperature, relative air humidity, evaporation, wind speed, sunshine and soil moisture.—Copyright 1972, Biological Abstracts, Inc.
W73-08099

SOUTH DAKOTA WEATHER MODIFICATION PROGRAM,
South Dakota Weather Control Commission, Pierre.
For primary bibliographic entry see Field 03B.
W73-08154

WATER OF SIBERIAN RIVERS FOR ARID LANDS OF THE SOUTH (VODU SIBIRSKIKH REK--ZASUSHLIVYM ZEMLYAM YUGA),
For primary bibliographic entry see Field 04A.
W73-08162

WATER AND LAND RESOURCE ACCOMPLISHMENTS, 1971.
Bureau of Reclamation, Washington, D.C.
For primary bibliographic entry see Field 08A.
W73-08191

SAN LUIS UNIT, CENTRAL VALLEY PROJECT, CALIFORNIA (FINAL ENVIRONMENTAL IMPACT STATEMENT),
Bureau of Reclamation, Sacramento, Calif. Mid-Pacific Regional Office.
For primary bibliographic entry see Field 08A.
W73-08223

EFFECT OF WATER COMPOSITION AND FEEDING METHOD ON SOIL NUTRIENT LEVELS AND ON TOMATO FRUIT YIELD AND COMPOSITION,
Foras Taluntais, Dublin (Ireland).
T. R. Gormley.
Ir J Agric Res. Vol 11, No 1, p 101-115. 1972. Illus.
Identifiers: *Irrigation systems, *Crop productions, Blossoms, Conductivity, Feeding, Fruit, Methods, Nutrients, Rot, Soils, Tomato, Hydrogen ion concentration, *Water properties.

Trickle feeding and irrigation of spring and autumn crop tomatoes grown in peat gave a lower soil pH and higher soil specific conductivity (SC) and K content than did feeding by hose or low-level sprayline methods. The use of hard water for making up feed and for irrigating gave a higher soil pH and SC than did moderately soft water. The trickle system gave the tallest plants in the autumn crop. In the spring crop plants were taller initially with the trickle system but the sprayline system gave the tallest plants later on. Hard water decreased height in both crops. Plants fed and irrigated with hard water yielded more marketable fruit in the spring crop than those treated with moderately soft water. The trickle system gave highest yields in both crops, and reduced the incidence of blossom-end rot in the spring crop. Values for fruit acidity, percentage soluble solids and K were lower in tricklefed tomatoes, but water type had little effect on fruit composition.—Copyright Biological Abstracts, Inc.
W73-08324

EFFECT OF IRRIGATION TREATMENTS FOR APPLE TREES ON WATER UPTAKE FROM DIFFERENT SOIL LAYERS,
Volcani Inst. of Agricultural Research, Bet-Dagan (Israel).
I. Levin, R. Assaf, and B. Bravdo.
J Am Soc Hortic Sci. Vol 97, No 4, p 521-526. 1972. Illus.
Identifiers: *Irrigation effects, *Crop response, Apples, Climate, Evaporation, Fruit, Irrigation, Layers, Size, Soils, Treatments, Trees, Moisture uptake.

Six irrigation treatments consisting of replenishing the water extracted from the 0-60 cm or 0-120 cm layer, were applied to a 10-yr-old apple orchard. The highest yield and fruit size were obtained by irrigating to 60 cm depth when soil moisture to this depth dropped to 40% available water during the 2 mo. of intensive fruit growth. During the rest of the season this treatment was irrigated to 60 cm whenever the 0-60 cm layer dropped to wilting point and to 120 cm whenever the 60-120 cm layer dropped to 60% available water. The relative water extraction from the 60-90 cm layer was the highest in this treatment. Increasing water uptake from layers below 120 cm by watering them was not effective. Climatic conditions favoring high rates of evaporation increased the relative contribution of layers deeper than 90 cm in all plots. The proportion of water loss from the 0-30 cm layer increased with the number of irrigations.—Copyright 1973, Biological Abstracts, Inc.
W73-08327

EFFECT OF THE SOIL AND PLANT WATER POTENTIALS ON THE DRY MATTER PRODUCTION OF SNAP BEANS,
Dep. Soils, Univ. Concepcion, Chillan, Chile Concepcion Univ. (Chile). Dept. of Soils.
A. A. Millar, and W. R. Gardner.
Agron J. Vol 64, No 5, p 559-562. 1972.
Identifiers: *Soil-water-plant relationships, *Beans, Dry, Growth, Phaseolus vulgaris, Plants, Potentials, Production, Rates, Resistance, Soils, Stomatal, Transpiration.

The dry matter production rate of snap beans (Phaseolus vulgaris L., cv. 'Bush Blue Lake') growing under field conditions on a sandy soil is analyzed during a drying period. Measurements of plant- and soil-water potentials, dry matter accumulation, and stomatal resistance were made as soil-water was depleted, while the transpiration rate were obtained by a model for a loosely structured canopy. The transpiration and dry matter production rates decreased curvilinearly with soil-water potential. When the soil-water potential decreased from -0.28 to -0.40 bar, there was 47% reduction in the dry matter production rate. This is related to the turgor pressure-operated stomatal mechanism. The adaxial stomatal resistances increased at leaf-water potentials lower than -8 bars, which coincided with a rapid decrease in the dry matter production rate. Stomatal closure due to water stress resulted in a greater reduction of growth rate than in transpiration.—Copyright 1973, Biological Abstracts, Inc.
W73-08328

EFFECTS OF POLYETHYLENE MULCH ON THE CHANGE OF TOPGROWTH AND THE YIELD OF TUBERS IN RELATION TO THE RATE OF NITROGEN ABSORPTION IN SWEET POTATO (IN JAPAN),
Niigata Univ., Nagaoka (Japan).
T. Morita.
Mem Fac Educ Niigata Univ. 13 p, 72-79. 1971. Illus. English summary.
Identifiers: *Crop production, Absorption, Polyethylene, Plant growth, Moisture, *Mulching, *Nitrogen, Rates, Soils, *Sweet potato, Temperature, Tubers, Weeds.

The soil beneath the polyethlene film had a lower moisture level throughout the test period in both sandy soil and clayey soil, while the soil temperature to a depth of 10 cm during the day was higher in mulched plots. In clayey soil polyethylene mulches influenced N absorption more than in unmulched plots. Furthermore, whereas little influence was observed in sandy soil, polyethylene mulch in clayey soil was beneficial in the hastening vine elongation. As a result, in clayey soil tuber root formation was favored in mulched plots due to the limited top growth during tuber formation. Polyethylene mulches in sandy soil had no influence on tuber formation or thickening. Weediness in mulched plots was markedly higher, especially in clayey soil.—Copyright 1973, Biological Abstract, Inc.
W73-08329

IDENTIFICATION OF A DAY OF MOISTURE STRESS IN MAIZE AT CEDARA,
Agricultural Research Inst., Cedara (South Africa).
For primary bibliographic entry see Field 02D.
W73-08337

EFFECT OF A MOISTURE STRESS DAY UPON MAIZE PERFORMANCE,
Agricultural Research Inst., Cedara (South Africa).
J. B. Mallett, and J. M. DeJager.
Agroplantae. Vol 3, No 2, p 15-19. 1971. Illus.
Identifiers: Grain, *Maize, *Moisture stress, Crop production, Crop response.

Moisture stress applied 3 wk before silking caused grain yields to be reduced by 3.2% leaf areas by 3.1% and plant height by 2.7% per day of stress. Stress applied after pollination caused yields to be reduced by 4.2% per stress day.—Copyright 1972, Biological Abstracts, Inc.
W73-08338

EFFECT OF ORGANIC AND MINERAL FERTILIZER ON THE HYDROCHEMICAL SYSTEM OF RICE PADDIES STOCKED WITH FISH WHICH WERE EXPOSED UNDER WATER VAPOR (IN RUSSIAN),
For primary bibliographic entry see Field 05C.
W73-08339

EFFECTS OF FLOODING AND GASEOUS COMPOSITION OF THE ROOT ENVIRONMENT ON GROWTH OF CORN,
Illinois Univ., Urbana. Dept. of Agronomy.
A. C. Purvis, and R. E. Williamson.
Agron J., Vol 64, No 5, p 674-678, 1972.
Identifiers: *Plant growth, *Soil gases, Aeration, Carbon dioxide, *Corn, Deficiency, Drainage, Environment, Flooding, *Gaseous composition, Injury, Oxygen, Roots.

ing use to precipitation.--Copyright 1973, Biological Abstracts, Inc.
W73-08400

IN SITU MEASUREMENT OF ROOT-WATER POTENTIAL,
Duke Univ., Durham, N.C. Dept. of Botany.
For primary bibliographic entry see Field 02I.
W73-08418

SOME WATER RELATIONS OF CERCOSPORELLA HERPOTRICHOIDES,
Washington State Univ., Pullman. Dept. of Plant Pathology.
G. W. Bruehl, and J. Manandhar.
Plant Dis Rep. Vol 56, No 7, p 594-596. 1972. Illus.
Identifiers: *Cercosporella-Herpotrichoides, Foot rot, Rot, Water temperature, Wheat, *Winterwheat, *Washington.

Straw breaker foot rot of winter wheat was widespread in much of eastern Washington in 1970-71, even in portions of Adams and Lincoln counties that average only 8-10 in. (20-25 cm) annual precipitation. The in vitro response of isolates of Cercosporella herpotrichoides from dryland and from wetter areas to water potential gave no evidence of dryland ecotypes. C. herpotrichoides grew on agar media amended with salts from the highest (wettest) water potential tested (-1 bar) to about -90 to -100 bars. Growth was stimulated by an osmotic water potential of about -4 to -10 bars between 1 and 25 degrees C and by -8 to -22 bars at 29 degrees. Growth was reduced to 50% of maximum at near -28 bars between 5 and 25 degrees C and at near -38 bars or lesser water potentials at 29degrees. The slight shift in optimum water potential from wetter to drier as the temperature rises may coincide with conditions within the host.--Copyright 1973, Biological Abstracts, Inc.
W73-08421

THE RELATIONSHIP BETWEEN ENVIRONMENTAL FACTORS AND BEHAVIOUR OF STOMATA IN THE RICE PLANT: 2. ON THE DIURNAL MOVEMENT OF THE STOMATA (IN JAPANESE),
Tokyo Univ. of Agriculture and Technology (Japan). Faculty of Agriculture.
K. Ishihara, Y. Ishida, and T. Ogura.
Proc Crop Sci Soc Jap. Vol 40, No 4, p 497-504. 1971. Illus. English summary.
Identifiers: *Diurnal, *Rice, *Stomata, Sun.

Rice grown in a submerged paddy field was much affected by weather conditions from day to day. On sunny days the aperture reached the maximum at about 8.30-9.00 a.m. and then decreased very quickly to only 1/2 or less of the maximum in the afternoon. On cloudy days the aperture increased slowly in the morning to reach the maximum at about noon and in the afternoon the aperture was wide for some time. From the tillering stage to the heading stage the maximum of the aperture per day was practically the same irrespective of weather conditions except under very low light intensity. After the heading stage the maximum per day lessened. After reaching the maximum the aperture decreased more quickly compared with before heading. The aperture decreases in the afternoon of sunny days or after heading due to the water unbalance in the leaves.--Copyright 1973, Biological Abstracts, Inc.
W73-08449

04. WATER QUANTITY MANAGEMENT AND CONTROL

4A. Control of Water on the Surface

A MANUAL ON COLLECTION OF HYDROLOGIC DATA FOR URBAN DRAINAGE DESIGN,
Hydrocomp, Inc., Palo Alto, Calif.
For primary bibliographic entry see Field 07C.
W73-07801

INITIAL RESULTS FROM THE UPPER WABASH SIMULATION MODEL,
Purdue Univ., Lafayette, Ind. Water Resources Research Center.
T. P. Chang, and G. H. Toebes.
Available from the National Technical Information Service as PB-219 478, $3.00 in paper copy, $1.45 in microfiche. Report No. 33, 1973. 89 p, 26 fig, 23 tab, append. OWRR A-016-IND (3) and A-012-IND (5).

Descriptors: *Reservoir operation, *Reservoir storage, *Simulation analysis, *Multiple-purpose reservoirs, *Flood control, Model studies, Drainage systems, Low-flow augmentation, Recreation, Runoff, Water supply, *Indiana.
Identifiers: *Wabash River (Ind), Drainage-area ratio, Storage-volume ratio.

A recently built simulation model for the Upper Wabash reservoir-river system in Indiana was used to study how best to operate that system. The construction of the model and of the three daily operating policies for it (that presently employed by the Corps of Engineers, the Drainage-Area Ratio and the Storage-Volume Ratio) were outlined in two preceding reports. This report discusses results that were obtained with each of the three policies applied to various reservoir configurations having up to five reservoirs, using a variety of runoff input, and for several alternative values of official flood-stage flows. The DAR and SVR policies were both superior to that used by the Corps when the runoff was less than 10 inches. Results obtained for the addition of a small water supply demand at one reservoir indicated that small changes in the mix of project purposes require careful alteration of the operating policies throughout the system. The major conclusion was that this practical model and its operating policies can be a useful aid to design, planning and regulatory agencies.
W73-07815

THE MISSISSIPPI RIVER--A WATER SOURCE FOR TEXAS. (EVALUATION OF A PROPOSED WATER DIVERSION),
Louisiana State Univ., Baton Rouge. Dept. of Civil Engineering.
For primary bibliographic entry see Field 06B.
W73-07816

AN ECONOMIC APPROACH TO LAND AND WATER RESOURCE MANAGEMENT: A REPORT ON THE PUGET SOUND STUDY,
Washington Univ., Seattle. Dept. of Economics.
For primary bibliographic entry see Field 06B.
W73-07826

NATIONAL PROGRAM FOR MANAGING FLOOD LOSSES, GUIDELINES FOR PREPARATION, TRANSMITTAL, AND DISTRIBUTION OF FLOOD-PRONE AREA MAPS AND PAMPHLETS,
Geological Survey, Washington, D.C.

For primary bibliographic entry see Field 07C.
W73-07849

EUREKA--IT FITS A PEARSON TYPE 3 DISTRIBUTION,
Geological Survey, Washington, D.C.
N. C. Matalas, and J. R. Wallis.
Water Resources Research, Vol 9, No 2, p 281-289, April 1973. 8 tab, 11 ref.

Descriptors: *Statistical methods, *Probability, *Streamflow forecasting, Time series analysis, Variability.
Identifiers: *Pearson distribution.

Under the assumption that a random variable is distributed as Pearson type 3, a comparison was made between moment and maximum likelihood estimates of the parameter values of the distribution and the variate values at specified probability levels. For the region where maximum likelihood solutions may be obtained, maximum likelihood estimates yield solutions that are less biased and less variable than the comparable moment estimates. When these results are extended to quite small samples, they become quite pronounced as the probability becomes greater than N/ (N + 1). (Knapp-USGS)
W73-07883

APPLICATION OF NONLINEAR SYSTEM IDENTIFICATION TO THE LOWER MEKONG RIVER, SOUTHEAST ASIA,
Geological Survey, Menlo Park, Calif. Water Resources Div.
S. M. Zand, and J. A. Harder.
Water Resources Research, Vol 9, No 2, p 290-297, April 1973. 4 fig, 14 ref.

Descriptors: *Systems analysis, *Simulation analysis, *Mathematical models, *Streamflow forecasting, Synthetic hydrology, Input-output analysis, Time series analysis, Variability, Hydrograph analysis.
Identifiers: *Mekong River (Cambodia).

The generalized functional series representation of systems may be solved numerically by using a transformation of input that considerably reduces the computational difficulties, with a multiple-regression analysis that establishes an optimum model. Computer programs for constructing such models were applied in the construction of two models for the lower Mekong River. Both models have single outputs, the daily gage height of the river at Chaudoc, South Vietnam. One model is of the single-input type, in which the input is the net average daily discharge below Phnom Penh, Cambodia. The second model is of the double-input type; it has as a second input, the daily rainfall at Takeo, Cambodia. A total of 1,497 daily measurements from January 1, 1964 to February 5, 1968, were used in the analysis. The predicting capability of the technique was tested by using the model constructed for 1,104 days as a predictor for the remaining 393 days. (Knapp-USGS)
W73-07884

INTRODUCTION OF TIME VARIANCE TO LINEAR CONCEPTUAL CATCHMENT MODELS,
Institute of Hydrology, Wallingford (England).
For primary bibliographic entry see Field 02A.
W73-07885

DETERMINATION OF OPTIMAL KERNELS FOR SECOND-ORDER STATIONARY SURFACE RUNOFF SYSTEMS,
Technion - Israel Inst. of Tech., Haifa. Dept. of Civil Engineering.
For primary bibliographic entry see Field 02A.
W73-07886

IDENTIFICATION OF MULTIPLE REACH CHANNEL PARAMETERS,
California Univ., Los Angeles. Dept. of Engineering Systems.
For primary bibliographic entry see Field 02E.
W73-07887

APPROXIMATE SOLUTIONS FOR NON-STEADY COLUMN DRAINAGE,
Asian Inst. of Tech., Bangkok (Thailand). Div. of Water Science and Engineering.
For primary bibliographic entry see Field 02G.
W73-07895

MULTISITE DAILY FLOW GENERATOR,
Department of the Environment, Ottawa (Ontario). Water Management Service.
For primary bibliographic entry see Field 02A.
W73-07899

SPRING DISCHARGE OF AN ARCTIC RIVER DETERMINED FROM SALINITY MEASUREMENTS BENEATH SEA ICE,
Louisiana State Univ., Baton Rouge. Coastal Studies Inst.
For primary bibliographic entry see Field 02C.
W73-07900

REGULATION OF STREAMFLOW (REGULIROVANIYE RECHNOGO STOKA),
Ya. F. Pleshkov.
Gidrometeoizdat, Leningrad, 1972. 508 p.

Descriptors: *Streamflow, *Regulated flow, *Regulation, *Reservoirs, *Reservoir operation, Reservoir storage, Reservoir releases, Reservoir yield, Flood routing, Hydrographs, Water resources, Water management (Applied), Water supply, Water utilization, Water consumption (Except consumptive use), Water loss, Water yield, Water quality control, Electric powerplants, Project planning.
Identifiers: *USSR, Mineralization, Nomograms.

Principles and techniques of project planning and streamflow regulation for water supply, irrigation, hydroelectric and thermal power, flood control, and other beneficial water uses are discussed. Schedules and guides for reservoir operation are developed to determine the most effective use of reservoir storage and release of stored water for conservation purposes, and changes in water-quality characteristics of reservoirs are examined from the standpoint of the protection and improvement of community water supplies. (Josefson-USGS)
W73-07909

WATER RESOURCES OF LAND AREAS (FORMIROVANIYE RESURSOV VOD SUSHI).
For primary bibliographic entry see Field 02A.
W73-07910

OPTIMIZATION OF DEAD END WATER DISTRIBUTION SYSTEMS,
Roorkee Univ. (India). Dept. of Civil Engineering.
P. K. Swamee, V. Kumar, and P. Khanna.
Journal of the Environmental Engineering Division, American Society of Civil Engineers, Vol 99, No EE2, Proceedings paper 9650, p 123-134, April, 1973. 4 fig, 37 equ, 6 ref.

Descriptors: *Water distribution (Applied), *Environmental engineering, *Optimization, *Economics, *Rural areas, *Water supply, Withdrawal, Networks, Pipelines, Mathematical models, Systems analysis.
Identifiers: Savings, Loops, Pump heads.

A single dead end system with multiple withdrawals has been synthesized. The distribution network of any water supply system involves a major portion of the total cost of the system; this portion increases with decreasing population. Dead end water distribution systems are encountered frequently in rural supply systems. Their design problem consists essentially of optimizing a nonlinear objective function subject to nonlinear constraints which are themselves functions of flow direction and therefore not uniquely definable. The solution is presented in a form directly usable by a design engineer, providing optimal pipe diameters, pumping head, hydraulic gradient line, and the minimal cost. The solution has been generalized for a continuous withdrawal of discharge. The case of two withdrawals is depicted in graphical form and provides a clear insight into the variation of the various parameters. Substantial saving can be achieved by designing the water distribution facilities at minimal costs. (Bell-Cornell)
W73-07920

WALKER BRANCH WATERSHED: A STUDY OF TERRESTRIAL AND AQUATIC SYSTEM INTERACTION,
Oak Ridge National Lab., Tenn.
For primary bibliographic entry see Field 04D.
W73-07947

RECTIFICATION OF DEFICIENCIES IN COMPLETED LOCAL PROTECTION PROJECT, WELLSVILLE, NEW YORK (FINAL ENVIRONMENTAL IMPACT STATEMENT).
Army Engineer District, Buffalo, N.Y.
For primary bibliographic entry see Field 08A.
W73-07975

BRANTLEY PROJECT, NEW MEXICO (FINAL ENVIRONMENTAL IMPACT STATEMENT).
Bureau of Reclamation, Denver, Colo.
For primary bibliographic entry see Field 08D.
W73-07976

COPAN LAKE, LITTLE CANEY RIVER, OKLAHOMA (FINAL ENVIRONMENTAL IMPACT STATEMENT).
Army Engineer District, Tulsa, Okla.
For primary bibliographic entry see Field 08A.
W73-07979

LOST CREEK LAKE PROJECT, ROGUE RIVER, OREGON (SUPPLEMENT TO FINAL ENVIRONMENTAL IMPACT STATEMENT).
Army Engineer District, Portland, Oreg.
For primary bibliographic entry see Field 08A.
W73-07980

SNAGGING AND CLEARING PROJECT ON MILL CREEK AT RIPLEY, WEST VIRGINIA (FINAL ENVIRONMENTAL IMPACT STATEMENT).
Army Engineer District, Huntington, W. Va.

Available from the National Technical Information Service as EIS-WV-72-5281-F, $3.00 in paper copy, $1.45 in microfiche. May 15, 1972. 12 p, 1 plate, 1 map, 1 illus.

Descriptors: *West Virginia, *Environmental effects, *Flood control, *Channel improvement, *Flow control, Flood protection, Overflow, Surface runoff, Flood stages, Flow rates, Non-structural alternatives, Fish kill, Water pollution effects, Hydraulic structures, Stream improvement, Fisheries, Fish passages, Open channels, Banks, Channel flow, Clogging, Streamflow.
Identifiers: *Environmental Impact Statements, *Mill Creek (W. Va.).

The proposed project consists of snagging and clearing of the Mill Creek channel for a distance of

pproximately 2.5 miles downstream from Ripley, Vest Virginia. There is dense growth on the banks f the creek and considerable debris and fallen rees in the stream, all of which contribute to poor tream flow and chronic flooding. The city of ipley maintains a sewage lagoon adjacent to the tream which is subject to flooding. The project is esigned to reduce flood stages and flood damage t Ripley, and alleviate stream pollution by reducng the frequency of inundation of the existing ewage lagoon. Adverse environmental effects inlude a temporary increase in the sediment in the tream and a loss of some fish and wildlife cover long the banks of the stream. Alternatives conidered included channel widening and the contruction of levees. However, these measures vere considered more expensive and more disruptive to the environment than the proposed project. Adams-Florida)
V73-07981

OLECAT BENCH AREA OF THE SHOSHONE XTENSIONS UNIT, WYOMING.
For primary bibliographic entry see Field 03F.
V73-07985

(ANNUAL REPORT OF THE DELAWARE RIVER BASIN COMMISSION, 1972).
Delaware River Basin Commission, Trenton, N.J.
For primary bibliographic entry see Field 06E.
V73-07990

IMITING FEDERAL RESERVED WATER RIGHTS THROUGH THE STATE COURTS,
For primary bibliographic entry see Field 06E.
V73-07991

VATER LAW--PRIMARY JURISDICTION OF THE BOARD OF CONTROL OVER QUESTIONS OF WATER RIGHTS.
For primary bibliographic entry see Field 06E.
V73-07992

REGULATION OF RIVERS, LAKES, AND STREAMS.
For primary bibliographic entry see Field 06E.
V73-08002

REEDS CONTROL EUTROPHICATION OF BALATON LAKE,
Research Inst. for Water Resources Development, Budapest (Hungary).
For primary bibliographic entry see Field 05G.
V73-08025

THE PHYSIOGRAPHY AND CHARACTER OF THE SUBSTRATUM OF THE DRAINAGE AREAS OF STREAMS OF THE POLISH HIGH TATRA MOUNTAINS,
Polish Academy of Sciences, Krakow. Zaklad Biologii Wod.
K. Pasternak.
Acta Hydrobiol, Vol 13, No 4, p 363-378, 1971. Illus.
Identifiers: Benthic animals, *Drainage, Mountains, *Physiography, Streams, Substratum, Tatra Mountains.

Important abiotic factors (relief, structure, and physico-chemical properties of the substratum) of the aqueous medium of the principal streams of the Polish High Tatra Mts. were studied. The zonal differentiation of hydrological conditions, the chemical composition of water, and the substratum of the bottom of these streams chiefly depend on the quality of the substratum of the drainage area and on climatic conditions. The shift (glacial and contemporaneous) down valleys and beds of streams of fragments of higher-lying crystalline rocks increased the range of their influence on the quality of the water and the settlement of benthic animals.--Copyright 1972, Biological Abstracts, Inc.
W73-08039

BOUNDARY EFFECTS ON STABILITY OF STRUCTURES,
Uttar Pradesh Irrigation Research Inst., Roorkee (India).
For primary bibliographic entry see Field 08B.
W73-08058

TRANSPOSITION OF STORMS FOR ESTIMATING FLOOD PROBABILITY DISTRIBUTIONS,
Colorado State Univ., Fort Collins. Dept. of Civil Engineering.
For primary bibliographic entry see Field 02E.
W73-08085

THE APPLICATION OF SNOWMELT FORECASTING TO COMBAT COLUMBIA RIVER NITROGEN SUPERSATURATION PROBLEMS,
Corps of Engineers, Portland, Oreg. North Pacific Div.
For primary bibliographic entry see Field 02C.
W73-08142

FORECAST STUDY FOR PRAIRIE PROVINCES WATER BOARD,
Water Survey of Canada, Calgary (Alberta). Alberta and Northwest Territories District Office.
W. Nemanishen.
In: Proceedings of the 40th Annual Meeting of the Western Snow Conference, April 18-20, 1972, Phoenix, Ariz: Printed by Colorado State University, Fort Collins, p 23-29, 1972. 3 fig, 3 tab, 8 ref.

Descriptors: *Water yield, *Streamflow forecasting, *Snowmelt, Data collections, Hydrologic data, Climatic data, Meteorological data, Regional analysis, Weather data, *Canada, Snowpacks, Snow surveys.
Identifiers: *South Saskatchewan River.

The Prairie Provinces Water Board of western Canada initiated a study to improve and coordinate the water supply forecast for Alberta, Saskatchewan and Manitoba. Initial forecasts for the April to October period are based on winter storage precipitation data, summer rainfall, and winter base flows. The use of correction factors based on data for three integrating loss basins significantly improved the accuracy of forecast for the South Saskatchewan River. Extreme residuals were reduced from 23% to 8%. (See also W73-08138) (Knapp-USGS)
W73-08143

AIR TEMPERATURE OBSERVATIONS AND FORECASTS--THEIR RELATIONSHIP TO THE PREDICTION OF SPRING SNOWMELT IN THE EAGEL RIVER BASIN, COLORADO,
National Weather Service, Salt Lake City, Utah. River Forecast Center.
For primary bibliographic entry see Field 02C.
W73-08144

WATER OF SIBERIAN RIVERS FOR ARID LANDS OF THE SOUTH (VODU SIBIRSKIKH REK-ZASUSHLIVYM ZEMLYAM YUGA),
I. Geradi.
Gidrotekhnika i Melioratsiya, No 12, p 13-23, December 1972. 1 fig, 2 tab.

Descriptors: *Rivers, *Arid lands, *Diversion, *Alteration of flow, *Water resources development, Water utilization, Water management (Applied), Water distribution (Applied), Water control, Water supply, Water requirements, Crop production, Irrigable land, Land reclamation, Irrigation, Flooding, Projects, Planning, Feasibility studies, Forecasting.
Identifiers: *USSR, *Siberia, Soviet Central Asia, Kazakhstan.

Specific measures for diversion of Siberian water to arid zones of Soviet Central Asia and Kazakhstan are outlined in a decree of the Central Committee of the Soviet Communist Party and the USSR Council of Ministers 'On Further Reclamation of Lands and Their Agricultural Use in 1971-75.' The areas involved are in the territory known as the Midland Region of the USSR, whose boundaries extend from the Urals and the Caspian Sea on the west to the Yenisey River on the east. In scale, the diversion of part of the discharge of Siberian rivers to the Midland Region would surpass anything thus far attempted anywhere in the world. Major problems that need to be resolved are the choice of places where water would be tapped from the Siberian rivers, the alignment of the diversion routes, and the economically desirable volume of water diversion. A schematic map of proposed diversion projects in the region shows existing and proposed canals, pumping stations, existing reservoirs, section of the Irtysh to be augmented by Ob' waters, and the direction of diversion. (Josefson-USGS)
W73-08162

DRAINAGE AREAS, HARTFORD NORTH QUADRANGLE, CONNECTICUT,
Geological Survey, Washington, D.C.
For primary bibliographic entry see Field 07C.
W73-08173

WATER, SEWER AND STORM DRAINAGE PLAN FOR THE CUMBERLAND PLATEAU PLANNING DISTRICT.
Thompson and Litton, Inc., Wise, Va.
For primary bibliographic entry see Field 06D.
W73-08179

URBAN STORM DRAINAGE AND FLOOD CONTROL IN THE DENVER REGION--FINAL REPORT.
Denver Regional Council of Governments, Colo.

Final report, August 1972. 219 p, 40 tab, 25 fig.

Descriptors: *Planning, *Urban drainage, *Drainage programs, *Urbanization, *Flood control, *Colorado, Flood protection, Regional analysis, Drainage engineering, Storm runoff, Local governments, Administration.
Identifiers: *Denver (Colorado).

Realizing that urbanization affects society and the quality of life, both positively and negatively, Project REUSE (Renewing the Environment through Urban Systems Engineering) was concerned with two aspects of the urban environment in the Denver Region--storm drainage and flood control, and solid waste management. This report includes a 20-year regional program for major drainage in the study area. Included are discussions of major drainage systems, management responsibilities, criteria, assumptions and uncertainties, four alternative concepts or programs for consideration, and an evaluation of these programs. Basically, the four plans are: (1) the current 1970-1974 program, (2) master planning with initial emphasis on preventive master planning followed by design master planning, (3) master planning with construction to be implemented on a county basis as soon as masterplanning is accomplished, and (4) the same plan as item number three except that all preventive master planning would be completed by 1975. The latter is evaluated to be the best plan and costs are estimated to be more than $8,500,000 for all planning and construction. (Poertner)
W73-08180

WATER AND LAND RESOURCE ACCOMPLISHMENTS, 1971.
Bureau of Reclamation, Washington, D.C.
For primary bibliographic entry see Field 08A.
W73-08191

GREAT DISMAL SWAMP AND DISMAL SWAMP CANAL.
For primary bibliographic entry see Field 06E.
W73-08193

PRESERVATION AND ENHANCEMENT OF THE AMERICAN FALLS AT NIAGARA.
For primary bibliographic entry see Field 06G.
W73-08194

A SURVEY OF STATE REGULATION OF DREDGE AND FILL OPERATIONS IN NON-NAVIGABLE WATERS,
Florida Univ., Gainesville. School of Law.
R. C. Ausness.
Land and Water Law Review, Vol 8, No 1, p 65-91, 1973. 132 ref.

Descriptors: *Legislation, *Non-navigable waters, *Dredging, *Constitutional law, Eminent domain, Water law, Water policy, Legal aspects, Judicial decisions, Navigable waters, Riparian rights, Wetlands, Conservation, Protection, State jurisdiction, Maine, Massachusetts, California, Zoning, Land use, Land development, Regulation.
Identifiers: Public trust doctrine, Nuisance (Legal aspects).

This note examines decisions arising under recently enacted state legislation regulating dredge and fill operations in certain classes of nonnavigable waters to determine the nature and extent of constitutional limitations on such regulation. Most states have used the public trust doctrine to sustain regulation over dredging and filling in navigable waters; but until lately control over such operations in non-navigable waters has been left to private remedies based upon nuisance and riparian right theories. Since dredge and fill operations frequently cause ecological harm, regulation is desirable regardless of whether or not the waters involved are navigable. Most legislation to date has been limited to coastal wetland areas, but similar legislation in landlocked areas might promote protection of water quality, conservation of fish and wildlife and flood control. Based on cases discussing the Massachusetts, Maine and California laws, protection of wetlands by regulation of dredge and fill operations has met with judicial approval, and no challenge based on a denial of due process has been successful. (Glickman-Florida)
W73-08197

THE WATER RESOURCES COUNCIL,
National Water Commission, Arlington, Va.
For primary bibliographic entry see Field 06E.
W73-08198

TIDELANDS--URGENT QUESTIONS IN SOUTH CAROLINA WATER RESOURCES LAWS,
South Carolina Univ., Columbia. School of Law.
For primary bibliographic entry see Field 06E.
W73-08201

LEGAL ASPECTS OF COASTAL ZONE MANAGEMENT IN ESCAMBIA AND SAN ROSA COUNTIES, FLORIDA (ESCAROSA),
For primary bibliographic entry see Field 06E.
W73-08204

ARTIFICIAL ADDITIONS TO RIPARIAN LAND: EXTENDING THE DOCTRINE OF ACCRETION,
For primary bibliographic entry see Field 06E.
W73-08206

SMITHVILLE LAKE, LITTLE PLATTE RIVER, MISSOURI (FINAL ENVIRONMENTAL IMPACT STATEMENT).
Army Engineer District, Kansas City, Mo.
For primary bibliographic entry see Field 08A.
W73-08210

WALKER DAM IMPOUNDMENT, AQUATIC PLANT CONTROL PROJECT, NEW KENT COUNTY, VIRGINIA (FINAL ENVIRONMENTAL IMPACT STATEMENT).
Army Engineer District, Norfolk, Va.
For primary bibliographic entry see Field 05G.
W73-08212

COW CREEK WATERSHED, STEPHENS AND JEFFERSON COUNTIES, OKLAHOMA (FINAL ENVIRONMENTAL IMPACT STATEMENT).
Soil Conservation Service, Washington, D.C.
For primary bibliographic entry see Field 08A.
W73-08214

WILLOW ISLAND LOCKS AND DAM OHIO RIVER, OHIO AND WEST VIRGINIA (FINAL ENVIRONMENTAL IMPACT STATEMENT).
Army Engineer District, Huntington, W. Va.
For primary bibliographic entry see Field 08A.
W73-08216

DETAILED PROJECT REPORT, INVESTIGATION FOR FLOOD PROTECTION, MUNDAY, TEXAS, BRAZOS RIVER BASIN, TEXAS (FINAL ENVIRONMENTAL IMPACT STATEMENT).
Army Engineer District, Fort Worth, Tex.
For primary bibliographic entry see Field 08A.
W73-08217

KAHULUI HARBOR WEST BREAKWATER REPAIR, MAUI, HAWAII (FINAL ENVIRONMENTAL IMPACT STATEMENT).
Army Corps of Engineers, Honolulu, Hawaii. Pacific Ocean Div.
For primary bibliographic entry see Field 08A.
W73-08218

GILA RIVER BASIN, NEW RIVER AND PHOENIX CITY STREAMS, ARIZONA, DREAMY DRAW DAM, MARICOPA COUNTY, ARIZONA (FINAL ENVIRONMENTAL IMPACT STATEMENT).
Army Engineer District, Los Angeles, Calif.
For primary bibliographic entry see Field 08A.
W73-08219

T OR C WILLIAMSBURG ARROYOS WATERSHED, SIERRA COUNTY, NEW MEXICO (FINAL ENVIRONMENTAL IMPACT STATEMENT).
Soil Conservation Service, Washington, D.C.
For primary bibliographic entry see Field 08A.
W73-08220

PEARL RIVER BASIN, EDINBURG DAM AND LAKE, MISSISSIPPI AND LOUISIANA (FINAL ENVIRONMENTAL IMPACT STATEMENT),
Army Engineer District, Mobile, Ala.
For primary bibliographic entry see Field 08A.
W73-08221

KANAWHA RIVER COMPREHENSIVE BASIN STUDY, NORTH CAROLINA, VIRGINIA, AND WEST VIRGINIA, (FINAL ENVIRONMENTAL IMPACT STATEMENT).
Ohio River Basin Commission, Cincinnati.

scrubby growth wherever there is sufficient soil to support it. It is suggested that water relations are more important than the soil nutrient status for the establishment of plants. A variety of herbaceous communities appear, but none is entirely distinct; there are intermediate communities in some cases. The community categories are: bog, field layer beneath open shrub canopy, community on loose rocks of screes and landslides, rock crevice communities on open slopes, community on shallow sandy soils subject to drought and flooding, and the communities on felled sites. A table shows the herbaceous species found in 14 sites in a variety of summit zone habitats of altitudes exceeding 11,000 ft.--Copyright 1973, Biological Abstracts, Inc.
W73-08333

THE EFFECT OF SODIUM ALKYLBENZENE-SULPHONATE ON THE DRAINAGE OF WATER THROUGH SAND,
Westfield Coll., London (England). Dept. of Zoology.
For primary bibliographic entry see Field 02K.
W73-08342

APPLICATION OF REMOTE SENSING TO SOLUTION OF ECOLOGICAL PROBLEMS,
IBM Federal Systems Div., Bethesda, Md.
For primary bibliographic entry see Field 07B.
W73-08358

APPLICATIONS OF REMOTE SENSING TO STREAM DISCHARGE PREDICTION,
National Aeronautics and Space Administration, Huntsville, Ala. George C. Marshall Space Flight Center.
For primary bibliographic entry see Field 07B.
W73-08359

SATELLITE OBSERVATIONS OF TEMPORAL TERRESTRIAL FEATURES,
Allied Research Associates, Inc., Concord, Mass.
For primary bibliographic entry see Field 07B.
W73-08362

INTERDISCIPLINARY APPLICATIONS AND INTERPRETATIONS OF REMOTELY SENSED DATA,
Pennsylvania State Univ., University Park.
For primary bibliographic entry see Field 07B.
W73-08363

RIVERBED FORMATION,
Colorado State Univ., Fort Collins. Dept. of Civil Engineering.
For primary bibliographic entry see Field 07B.
W73-08365

GEOLOGICAL AND GEOHYDROLOGICAL STUDIES FOR ANGOSTURA DAM, CHIAPAS, MEXICO,
Comision Federal de Electricidad, Mexico City.
For primary bibliographic entry see Field 08A.
W73-08376

A PROBABILISTIC MODEL FOR STRUCTURING A DRAINAGE NETWORK,
Army Project Mobile Army Sensor Systems Test Evaluation and Review Activity, Fort Hood, Tex.
R. T. Robinson, and A. J. Swartz.
Available from NTIS, Springfield, Va 22151 as AD-750 371 Price $3.00 printed copy; $1.45 microfiche. 1972. 14 p, 3 fig, 2 tab, 5 ref.

Descriptors: *Geomorphology, *Terrain analysis, *Hortons law, *Statistical models, *Drainage patterns (Geologic), Stochastic processes, Probability.

A stochastic model describes the behavior of a mature drainage network in terms of four network parameters. The principal parameters are stream length ratio and bifurcation ratio. The model is amenable to computer solution and may be used to estimate the number, sizes and interfluvial distances of streams to be crossed when traversing a drainage basin, or succession of basins, with a path of varying width. Reliability of the model is materially enhanced by quantifying all four parameters within the geographical area of intended model use. However, the model reliability is little reduced if the basin shape and drainage density parameters are assumed to have equilibrium values. (Knapp-USGS)
W73-08380

SURFACE WATER SUPPLY OF THE UNITED STATES, 1966-1970: PART 6-MISSOURI RIVER BASIN, VOLUME 4-MISSOURI RIVER BASIN BELOW NEBRASKA CITY, NEBRASKA.
Geological Survey, Washington, D.C. Water Resources Div.
For primary bibliographic entry see Field 07C.
W73-08381

THE ECONOMICS OF WATER TRANSFER,
Hawaii Univ., Honolulu. Dept. of Agricultural and Resource Economics.
For primary bibliographic entry see Field 06B.
W73-08387

GREAT BASIN STATION: SIXTY YEARS OF PROGRESS IN RANGE AND WATERSHED RESEARCH,
W. M. Keck.
For Serv Res Pap Int. 118. p 1-48. 1972. Illus.
Identifiers: *Great Basin, History, *Range research, Utah, *Watershed research.

A brief history is given of the Great Basin Experimental Range from its establishment in 1912 as the Utah Experiment Station. Key problems in management of watershed and rangelands and the experiments devised to solve them are described and applications of this research are indicated.--Copyright 1973, Biological Abstracts, Inc.
W73-08398

AGRICULTURE, LAND USE, AND SMALLHOLDER FARMING PROBLEMS IN THE SIGATOKA VALLEY,
Department of Agriculture, Suva (Fiji). Research Div.
For primary bibliographic entry see Field 03F.
W73-08399

THE MANAGEMENT OF MOUNTAIN CATCHMENTS BY FORESTRY,
Stellenbosch Univ. (South Africa). Dept. of Silviculture.
For primary bibliographic entry see Field 03B.
W73-08401

FORESTS AND FLOODS IN THE EASTERN UNITED STATES,
Forest Service (USDA), Upper Darby, Pa. Northeastern Forest Experiment Station.
H. W. Lull, and K. G. Reinhart.
US For Serv Res Pap Ne. Vol 226, p 1-94, 1972.
Identifiers: Control, Cover, Erosion, *Floods, Flow, Forestation, *Forests, Hydrologic studies, Runoff, *Eastern US.

A historical background is presented as a backdrop for discussion of the hydrologic processes affecting flood flows and erosion, the impact of various land uses, and the potentials for management. The forest is the best of all possible natural cover for minimizing overland flow, runoff, and erosion. The flood-reduction potential of the forest can be realized through continued fire protection and careful logging; reforestation of abandoned land can provide additional benefits.--Copyright 1973, Biological Abstracts, Inc.
W73-08419

ANNUAL CYCLES OF SOIL MOISTURE AND TEMPERATURE AS RELATED TO GRASS DEVELOPMENT IN THE STEPPE OF EASTERN WASHINGTON,
Washington State Univ., Pullman. Dept. of Botany.
For primary bibliographic entry see Field 02G.
W73-08436

AVAILABLE SOIL MOISTURE AND PLANT COMMUNITY DIFFERENTIATION IN DAVIES ISLAND, MIDDLE TENNESSEE,
Tennessee Technological Univ., Cookeville.
For primary bibliographic entry see Field 02G.
W73-08440

4B. Groundwater Management

FATE OF TRACE-METALS (IMPURITIES) IN SUBSOILS AS RELATED TO THE QUALITY OF GROUND WATER,
Tuskegee Inst., Ala. Carver Research Foundation.
For primary bibliographic entry see Field 05B.
W73-07802

THE MISSISSIPPI RIVER--A WATER SOURCE FOR TEXAS. (EVALUATION OF A PROPOSED WATER DIVERSION),
Louisiana State Univ., Baton Rouge. Dept. of Civil Engineering.
For primary bibliographic entry see Field 06B.
W73-07816

UNSTEADY FLOW TO A PARTIALLY PENETRATING, FINITE RADIUS WELL IN AN UNCONFINED AQUIFER,
Washington Univ., Seattle.
For primary bibliographic entry see Field 02F.
W73-07898

THE TRANSIENT FLOW PROBLEM - A ONE-DIMENSIONAL DIGITAL MODEL,
Wyoming Univ., Laramie. Dept. of Civil and Architectural Engineering.
For primary bibliographic entry see Field 02F.
W73-07916

HYDROLOGIC INVESTIGATIONS AND THE SIGNIFICANCE OF U-234/U-238 DISEQUILIBRIUM IN THE GROUND WATERS OF CENTRAL TEXAS,
Rice Univ., Houston, Tex.
For primary bibliographic entry see Field 05B.
W73-07949

AGRI-INDUSTRIAL PARK WATER IMPROVEMENTS, SEWAGE TREATMENT FACILITIES, BUSINESS DEVELOPMENT LOAN (FINAL ENVIRONMENTAL IMPACT STATEMENT).
Economic Development Administration, Austin, Tex. Southwestern Region.
For primary bibliographic entry see Field 05D.
W73-07978

HYDROGEOLOGIC CHARACTERISTICS OF THE VALLEY-FILL AQUIFER IN THE ARKANSAS RIVER VALLEY, BENT COUNTY, COLORADO,
Geological Survey, Washington, D.C.
For primary bibliographic entry see Field 07C.
W73-08068

WATER-RESOURCES RECONNAISSANCE OF THE OZARK PLATEAUS PROVINCE, NORTHERN ARKANSAS,
Geological Survey, Washington, D.C.
For primary bibliographic entry see Field 07C.
W73-08069

HYDRAULIC TESTS IN HOLE UAE-3, AMCHITKA ISLAND, ALASKA,
Geological Survey, Lakewood, Colo.
W. C. Ballance.
Available from NTIS, Springfield, Va. 22151, Price $4.00 printed copy; $1.45 in microfiche. Geological Survey Report USGS-474-26. Revision-1 (Amchitka-17, Rev-1), April 1973. 30 p, 17 fig, 1 tab, 3 ref. AT (29-2)-474.

Descriptors: *Aquifer testing, *Specific capacity, Groundwater movement, Drawdown, Transmissivity, Water yield, Aquifer characteristics.
Identifiers: *Amchitka Island (Alaska).

During August through November 1967, the U.S. Geological Survey hydraulically tested hole UAe-3 on Amchitka Island, Alaska. Inflatable straddle packers were used to isolate and test selected intervals. Packer seats were poor in the uncased part of the hole because of unstable wall conditions, and leakage around packers occurred during some tests. However, leakage generally was slight and had little effect on the tests. The static water levels in the intervals tested ranged from 31.6 meters below land surface in the upper interval tested to about 115 meters below land surface in the lower interval tested, indicating decreasing head with depth. The relative specific capacities of isolated zones ranged from less than 0.001 cubic meter per day per meter to 0.898 cubic meter per day per meter of drawdown. (Knapp-USGS)
W73-08071

GEOHYDROLOGY AND ARTIFICIAL-RECHARGE POTENTIAL OF THE IRVINE AREA, ORANGE COUNTY, CALIFORNIA,
Geological Survey, Menlo Park, Calif. Water Resources Div.
J. A. Singer.
Geological Survey Water Resources Division open-file report, January 8, 1973. 41 p, 16 fig, 2 tab, 22 ref.

Descriptors: *Hydrogeology, *Artificial recharge, *Coastal plains, *Water spreading, *California, Aquifer characteristics, Aquitards, Transmissivity, Water levels, Withdrawal, Water yield, Groundwater resources, Drawdown.
Identifiers: *Orange County (Calif), *Irvine (Calif).

The Irvine area is in hydraulic continuity with the rest of the coastal plain in Orange County, California. Rapid facies change and the large percentage of silt and clay in the section locally result in confining conditions. The aquifer, most of which is included in the Fernando Formation, is as much as 1,300 feet thick beneath parts of the plain. The alluvium overlying the Fernando Formation averages about 200 feet in thickness and also contains significant amounts of silt and clay. Transmissivities range from 25,000 to 100,000 gallons per day per foot in the Irvine area, values which are much lower than those in the rest of the coastal plain in Orange County. Water levels have recovered as much as 60 feet from the low levels of the early 1950's. In the winter nonpumping season water tends to move toward upper Newport Bay and the rest of the coastal plain. During the summer pumping season a cone of depression develops, reversing the winter gradient. The average dissolved-solids content of the groundwater is about 800 milligrams per liter. The most prevalent cations are sodium and calcium; the most prevalent anions are bicarbonate and sulfate. No long-term degradation of water quality has occurred, with the exception of a slight increase in dissolved solids. No areas in the Irvine area are suitable for the large-scale spreading of water for artificial recharge. Clay and silt predominate in the section beneath the Tustin plain, and in the foothill areas either bedrock is close to the surface or the alluvium is fine grained. (Knapp-USGS)
W73-08072

POLLUTION OF SUBSURFACE WATER BY SANITARY LANDFILLS, VOLUME 1,
Drexel Univ., Philadelphia, Pa.
For primary bibliographic entry see Field 05B.
W73-08073

POSSIBLE ACCUMULATION OF AUTHIGENIC, EXPANDABLE-TYPE CLAY MINERALS IN THE SUBSTRUCTURE OF TUTTLE CREEK DAM, KANSAS, U.S.A.,
New Mexico Univ., Albuquerque. Dept. of Geology.
For primary bibliographic entry see Field 02K.
W73-08093

HEAVY METALS REMOVAL IN WASTE WATER TREATMENT PROCESSES: PART 1,
Orange County Water District, Santa Ana, Calif.
For primary bibliographic entry see Field 05D.
W73-08117

EFFECT OF INCLUDING WATER PRICE ON THE CONJUNCTIVE OPERATION OF A SURFACE WATER AND GROUNDWATER SYSTEM,
Plan Organization, Tehran (Iran).
F. Mobasheri, and S. Grant.
Water Resources Research, Vol 9, No 2, p 463-469, April, 1973. 2 fig, 17 equ, 15 ref.

Descriptors: *Optimization, *Water demand, *Surface waters, *Groundwater, *Prices, Cost-benefit analysis, *Operating costs, *Water supply, Management, Equilibrium prices, Constraints, Algorithms, Mathematical models, Operations research, Computer programs, Water distribution (Applied).
Identifiers: *Nonlinear programming, *Net benefits, *Conjunctive operation, Penalty cost, Residential water.

A mathematical model is developed to study the impact of water price on residential water demand and on the conjunctive operation policy of a surface water and groundwater system. The economic objective is to maximize the present worth of net benefits from operation of the supply system. The objective function and constraints are nonlinear. A penalty cost is introduced into the objective function to take into consideration the cost of operating the system beyond the planning time horizon; the penalty cost is a function of the final groundwater state. Nonlinear programming is used to find the optimum operation strategy for the conjunctive management of the supply sources. Calculated simultaneously are the optimum demand, the distribution of supply, and the equilibrium price for each time period in the planning horizon. The model is applied to a hypothetical residential area to test the efficiency of the computer program; the planning time horizon is 25 years. (Bell-Cornell)
W73-08132

RENOVATING SEWAGE EFFLUENT BY GROUNDWATER RECHARGE,
Agricultural Research Service, Phoenix, Ariz. Water Conservation Lab.
For primary bibliographic entry see Field 05D.
W73-08141

ATURAL AND ARTIFICIAL GROUND-VATER RECHARGE, WET WALNUT CREEK, ENTRAL KANSAS,
eological Survey, Lawrence, Kans.
B. Gillespie, and S. E. Slagle.
ansas Water Resources Board Bulletin No 17, 972. 94 p, 55 fig, 7 tab, 19 ref.

escriptors: *Groundwater resources, *Irrigation, Kansas, *Groundwater recharge, Aquifers, quifer characteristics, Streams, Natural charge, Artificial recharge, Water quality, hemical analysis, Streamflow, Water wells, Vater yield, Water level fluctuations.

he withdrawal of groundwater for irrigation is accelerating rapidly in western and central Kansas. atural recharge or artificial recharge to the quifers is needed to ensure an adequate supply of ater for the future. Wet Walnut Valley in central ansas is a narrow alluvial valley in which an interrupted meandering stream dominates the ydrologic system. Wet Walnut Creek is generally losing stream in most of the reach through the udy area. The average annual streamflow leaving e area is 50,530 acre-feet, most of which occurs storm runoff. Recharge to the aquifer from the sing reach of the creek during low-flow periods only about 0.05 cubic foot per second per mile of hannel because of a relatively impermeable layer f fine-grained material with an average thickness f about 2 feet that lies between the bottom of the hannel and the sand and gravel of the lower alluium. The maximum amount of water available for rtificial recharge is streamflow, which averages bout 50,000 acre-feet per year. The average mount probably will be less than the potential for echarge after floodwater retarding structures and rtificial recharge systems have been placed in service. (Woodard-USGS)
W73-08379

C. Effects on Water of Man's Non-Water Activities

TORM FLOW FROM HARDWOOD-FORESTED AND CLEARED WATERSHEDS IN NEW HAMPSHIRE,
orest Service (USDA), Durham, N.H. Northeastrn Forest Experiment Station.
W. Hornbeck.
Vater Resources Research, Vol 9, No 2, p 346-54, April 1973. 3 fig, 3 tab, 10 ref.

escriptors: *Storm runoff, *Forest watersheds, Clear-cutting, *New Hampshire, *Rainfall-runoff elationships, Forest management, Water yield, loods, Forests, Hydrographs, Hydrograph analyis, Demonstration watersheds.

hanges in storm flow as a result of forest clearing were determined for a small mountainous watershed in New Hampshire by using a paired watershed as a control. Reduction of transpiration nd interception losses produced wetter soils with ss opportunity for storing rainfall. Consequently, uick flow volumes and instantaneous peaks increased during the growing season. The absence of e hardwood forest canopy also caused earlier nd more rapid snowmelt and affected most spring tormflow events involving snow water. In contrast, storm events occurring after soil moisture echarge in the fall and before the start of spring nowmelt were unaffected by forest clearing. Although changes in the spring and summer stormlow were readily detectable, their magnitude was ot great. The maximum increase in individual uick flow was 30 mm for the summer streamflow eason and just over 50 mm during spring snowmelt. Changes in mean quick flow were much ower. The relatively small amount of forest clearng currently taking place in New England headwaters should not increase downstream flood otential. (Knapp-USGS)

W73-07889

EFFECTS OF LAND USE ON WATER RESOURCES,
Federal Water Pollution Control Administration, Washington, D.C.
W. E. Bullard.
Journal of Water Pollution Control Federation, Vol 38, No 4, p 645-659, April 1966. 23 ref.

Descriptors: *Watershed management, *Water quality control, *Surface runoff, *Urban runoff, Sediment control, Erosion control, Nutrients, Toxins, Environmental effects, Management, Urbanization.

The basic operations of a watershed, i.e. infiltration and runoff, and the effect of various land uses on these operations, particularly in relation to water quality, are discussed. Three broad areas are identified that affect water quality: erosion and sedimentation, toxins and nutrients, and wastes. Under each of these categories and principal sources are identified, the water quality effects are discussed, and means to minimize the adverse effects are offered. In general, the problems can be broken into a non-urban versus urban dichotomy. The non-urban land uses such as agriculture, forestry, mining, and recreation can usually be handled best through improved management practices. The urban runoff problem, however, with its dirt, chemicals, oil, and nutrients from urban activities may require major treatment and storage facilities in addition to management attempts. (Elfers-North Carolina)
W73-08054

THE CHANGING WATER RESOURCE,
Atmospherics, Inc., Fresno, Calif.
T. J. Henderson.
In: Proceedings of the 40th Annual Meeting of the Western Snow Conference, April 18-20, 1972, Phoenix, Ariz: Printed by Colorado State University, Fort Collins, p 2-5, 1972. 2 tab.

Descriptors: *Snow, *Water resources, *Conferences, *Water resources development, Weather modification, Urbanization, Conservation, Water supply.

Water resources are not changing, but man's influence on this resource and his need for water are rapidly changing. The effects of man on natural patterns of water supply are usually largely inadvertent. But man's intentional activities may often improve the supply of water for either more direct benefit or improve the landscape for greater aesthetic advantage. Man has built dams, diverted large rivers, drained swamps, reclaimed deltas, developed enormous irrigation schemes, started to use saltwater conversion, and has modified the weather. (See also W73-08138) (Knapp-USGS)
W73-08139

COPING WITH POPULATION GROWTH AND A LIMITED RESOURCE,
Arizona Water Commission, Phoenix.
For primary bibliographic entry see Field 06A.
W73-08140

SNOW, RELATED ACTIVITIES AND CONCERNS ON THE GRAND MESA, WESTERN COLORADO,
Forest Service (USDA), Delta, Colo. Grand Mesa-Uncompahgre National Forests.
For primary bibliographic entry see Field 02C.
W73-08147

A WILDERNESS SNOW COURSE,
Forest Service (USDA), Kalispell, Mont. Flathead National Forest.
For primary bibliographic entry see Field 02C.
W73-08148

WILDERNESS IN THE NATIONAL PARKS,
National Park Service, Denver, Colo. Denver Service Center.
J. Henneberger.
In: Proceedings of the 40th Annual Meeting of the Western Snow Conference, April 18-20, 1972. Phoenix, Ariz: Printed by Colorado State University, Fort Collins, p 58-60, 1972.

Descriptors: *Recreation, *National parks, *Recreation facilities, *Surveys, Conservation, Planning.
Identifiers: *Wilderness areas.

A wilderness studies program has been underway in the National Park Service since 1964 when the Wilderness Act directed the Service to study all roadless areas of 5,000 acres or more in the parks and other areas of the System that existed in 1964. This would include a total of 65 areas. Public hearings have been held on 43 and up through fiscal year 1972 reports on 40 of these areas have been completed. Substantial progress on the balance has been made and completion by 1974 is anticipated. To date, Congress has created wilderness units in the Petrified Forest National Park and Craters of the Moon National Monument. (See also W73-08138) (Knapp-USGS)
W73-08149

URBAN HYDROLOGY--A SELECTED BIBLIOGRAPHY WITH ABSTRACTS,
Geological Survey, Washington, D.C.
G. L. Knapp, and J. P. Glasby.
Available from the National Technical Information Service as PB-219 105, $6.75 in paper copy, $1.45 in microfiche. Geological Survey Water-Resources Investigations 3-72, 1972. 211 p, 651 ref.

Descriptors: *Bibliographies, *Abstracts, *Urban hydrology, *Urban runoff, Rainfall-runoff relationships, Storm runoff, Water pollution sources, Groundwater, Climatology, Urbanization, Cities, Urban drainage, Land use, Storm drains, Mathematical models, Surburban areas.

This bibliography of 651 selected references on urban hydrology is intended as a source document for scientific and water-management needs. It was stimulated by increasing interest in the problems of runoff and water quality caused by increasing urbanization. The bibliography brings together abstracts with citations that pertain to the rainfall-runoff process, urban groundwater problems, urban water pollution sources, urban climatic changes, and urban runoff modeling. Emphasis is given to technical advances of the past ten years as well as to needs for new research. The bibliography is arranged alphabetically by author and has separate geographic and subject indexes. Each abstract is followed by several added key words to relate it to other similar references. (USGS)
W73-08164

GILA RIVER BASIN, NEW RIVER AND PHOENIX CITY STREAMS, ARIZONA, DREAMY DRAW DAM, MARICOPA COUNTY, ARIZONA (FINAL ENVIRONMENTAL IMPACT STATEMENT).
Army Engineer District, Los Angeles, Calif.
For primary bibliographic entry see Field 08A.
W73-08219

HYDROLOGICAL EFFECTS OF THE CANNIKIN EVENT,
Geological Survey, Denver, Colo.
For primary bibliographic entry see Field 04B.
W73-08367

4D. Watershed Protection

WALKER BRANCH WATERSHED: A STUDY OF TERRESTRIAL AND AQUATIC SYSTEM INTERACTION,
Oak Ridge National Lab., Tenn.
G. S. Henderson, J. W. Elwood, W. F. Harris, H. H. Shugart, and R. I. Van Hook.
Available from NTIS, Springfield, Va., as ORNL-4848; $3.00 paper copy, $1.45 microfiche. In: Environmental Sciences Division Annual Progress Report for Period Ending September 30, 1972, Report No. ORNL-4848, Feb. 1973, p 9-19, 7 fig, 3 tab.

Descriptors: *Water resources, *Forest watersheds, *Biochemistry, *Watersheds (Basins), *Water yield, *Ecosystems, Biological communities, Hydrology, Minerology, Cycles, Water quality, Aquatic populations, Comparative benefits, *Tennessee.
Identifiers: *Walker Branch Watershed.

The primary objective of the Walker Branch Watershed project is quantification of biogeochemical cycles in a forested landscape. To accomplish this objective, both terrestrial and aquatic ecosystems of the watershed are being studied to (1) establish quantitative relationships between the hydrologic and mineral cycles, (2) relate water quality and aquatic productivity to characteristics of the adjacent terrestrial system, (3) provide information on natural terrestrial and aquatic system interactions for comparisons with those modified by cultural treatments, and (4) apply the knowledge gained from this small, controlled drainage basin study to broader landscale units to evaluate the impact of man's activities on the total ecosystem. (Houser-ORNL)
W73-07947

RECTIFICATION OF DEFICIENCIES IN COMPLETED LOCAL PROTECTION PROJECT, WELLSVILLE, NEW YORK (FINAL ENVIRONMENTAL IMPACT STATEMENT).
Army Engineer District, Buffalo, N.Y.
For primary bibliographic entry see Field 08A.
W73-07975

ILLINOIS AND DESPLAINES RIVERS; KASKASKIA RIVER WATERSHED.
For primary bibliographic entry see Field 06E.
W73-08003

EFFECTS OF LAND USE ON WATER RESOURCES,
Federal Water Pollution Control Administration, Washington, D.C.
For primary bibliographic entry see Field 04C.
W73-08054

BACON CREEK WATERSHED, PLYMOUTH AND WOODBURY COUNTIES, IOWA, (FINAL ENVIRONMENTAL IMPACT STATEMENT).
Soil Conservation Service, Washington, D.C.
For primary bibliographic entry see Field 08A.
W73-08211

VIRGINIA BEACH, VIRGINIA--BEACH EROSION CONTROL AND HURRICANE PROTECTION (FINAL ENVIRONMENTAL IMPACT STATEMENT).
Army Engineer District, Norfolk, Va.
For primary bibliographic entry see Field 08A.
W73-08213

COW CREEK WATERSHED, STEPHENS AND JEFFERSON COUNTIES, OKLAHOMA (FINAL ENVIRONMENTAL IMPACT STATEMENT).
Soil Conservation Service, Washington, D.C.
For primary bibliographic entry see Field 08A.
W73-08214

T OR C WILLIAMSBURG ARROYOS WATERSHED, SIERRA COUNTY, NEW MEXICO (FINAL ENVIRONMENTAL IMPACT STATEMENT).
Soil Conservation Service, Washington, D.C.
For primary bibliographic entry see Field 08A.
W73-08220

GREAT BASIN STATION: SIXTY YEARS OF PROGRESS IN RANGE AND WATERSHED RESEARCH,
For primary bibliographic entry see Field 04A.
W73-08398

FORESTS AND FLOODS IN THE EASTERN UNITED STATES,
Forest Service (USDA), Upper Darby, Pa. Northeastern Forest Experiment Station.
For primary bibliographic entry see Field 04A.
W73-08419

05. WATER QUALITY MANAGEMENT AND PROTECTION

5A. Identification of Pollutants

THERMOCHEMICAL INVESTIGATION OF DIMETHYLMERCURY IN AQUEOUS AND NONAQUEOUS SOLUTIONS,
Missouri Univ., Rolla. Dept. of Chemistry.
G. L. Bertrand.
Available from the National Technical Information Service as PB-219 262, $3.00 in paper copy, $1.45 in microfiche. Missouri Water Resources Research Center, Columbia Completion Report, 1973. 13 p, 1 fig, 4 tab, 4 ref. OWRR A-045-MO (1). 14-31-0001-3525.

Descriptors: Physicochemical properties, *Solubility, *Aqueous solutions, *Mercury, Pollutant identification, Analytical techniques, Water analysis, Ions.
Identifiers: *Dimethylmercury, *Thermochemical studies.

Calorimetric investigations were made of the interactions of dimethylmercury (DMM) with various organic solvents and compounds in solution. Heats of solution of DMM in inert solvents cyclohexane and carbon tetrachloride and in active solvents benzene, pyridine, and p-dioxane were measured. The effects of nitrobenzene, thiophenol, and thiourea were also investigated. All heats of solution were endothermic and less than 0.6 kcal/mole. No indications of specific interactions were observed for any of these compounds. The apparent solubility of DMM in water and in aqueous electrolyte solutions was determined using flameless atomic absorption spectrophotometry. The term 'apparent solubility' is used to specify the total concentration of all forms of mercury in an aqueous phase in prolonged contact with an excess of DMM. The solubility of DMM in pure deaerated water is 3.0 micro g Hg/g water, and over the range 5-45 degrees C, Solubility (micro g Hg/g water) ± 127.7 exp (-1109/T) plus or minus 0.05. The presence of halide ions was found to greatly increase the apparent solubility of DMM, with more than a tenfold increase in 1 M NaCl. It is suspected that this increase is due to a reaction forming the more soluble methylmercury halide.
W73-07806

RADIOACTIVITY OF WASTE WATERS IN NUCLEAR RESEARCH INSTITUTE, REZ, AND ENVIRONMENT, (VLIV RADIOAKTIVITY ODPADNICH VOD VYZKUMNEHO JADERNEHO CENTRA V REZI NA OKOLI),
Ceskoslovenska Akademie Ved, Rez. Ustav Jaderneho Vyzkumu.
For primary bibliographic entry see Field 05B.
W73-07926

RESULTS FROM MULTI-TRACE-ELEMENT NEUTRON ACTIVATION ANALYSES OF MARINE BIOLOGICAL SPECIMENS,
California Univ., Irvine. Dept. of Chemistry.
For primary bibliographic entry see Field 05C.
W73-07927

MONITORING OF RADIOACTIVE ISOTOPES IN ENVIRONMENTAL MATERIALS,
Atomic Energy of Canada Ltd., Chalk River (Ontario). Chalk River Nuclear Labs.
W. E. Grummitt.
In: International Symposium of Identification and Measurement of Environmental Pollutants, June 14-17, 1971, Ottawa, Ontario, National Research Council of Canada, Ottawa, p 399-403. 4 fig, 3 ref.

Descriptors: *Nuclear wastes, *Monitoring, *Radioisotopes, *Gamma rays, Neutron activation analysis, Effluents, Spectroscopy, Canada, Rivers, Cobalt radioisotopes, Instrumentation, Pollutant identification, Water pollution control, Path of pollutants.

Several gamma-emitting nuclides may be identified simultaneously using a Ge (Li) detector. The background in the Co60 photopeak channel is very low. It is seldom that anything but fallout radionuclides are seen downstream in the Ottawa River; however, Co and Mn could be detected at 0.02 picoCuries/liter. To locate sources of contamination, effluents are monitored before dilution in the river. Contamination from a failed fuel rod would result in an order of magnitude higher level of neutron capture products (Mb, Mn, Sc, Fe, Zn, and Co) as compared with fission products (I131, Ba140, La140, Ru103). In purging air from water used for reactor shielding, identification was made of Ar41, Kr85, Kr87, Kr88, Xe135, Xe138, and Rb and Cs daughters. Iodine and Br radionuclides were absent. (Bopp-ORNL)
W73-07929

ENVIRONMENTAL MONITOR NG REPORT FOR THE NEVADA TEST SITE JANUARY--DECEMBER 1971.
National Environmental Research Center, Las Vegas, Nev.
For primary bibliographic entry see Field 05B.
W73-07936

DETERMINATION OF PLUTONIUM IN ENVIRONMENTAL SAMPLES,
Kanazawa Univ. (Japan).
M. Sakanoue, M. Nakaura, and T. Imai.
In: Proceedings of an International Symposium, Rapid Methods for Measuring Radioactivity in the Environment, July 5-9, 1971, Neuherberg (Germany), p 171-181. 6 fig, 2 tab, 23 ref.

Descriptors: *Nuclear wastes, *Radiochemical analysis, *Analytical techniques, *Environmental effects, Water analysis, Sea water, Soil contamination, Sediments, Asia, Pacific Ocean, Separation techniques, Solvent extractions, Radioactivity techniques, Path of pollutants, Pollutant identification, Monitoring.
Identifiers: Plutonium radioisotopes.

Pu239 and Sr90 in bottom sediment of Nagasaki water reservoirs are highest in those nearest the blast (100-200 picoCuries Pu239/kg). In recent coral samples from the Pacific coast, Pu239 concentration is about 2 picoCuries/kg; Pu238, about 0.8. In coastal sea water, Pu239 concentration is 0.6-0.8 picoCurie/kiloliter; Pu238, 0.1-0.5. The

ty in natural gas for industrial and domestic consumption. (Houser-ORNL)
W73-07941

WATER AND WASTE WATER STUDIES, 1971 (WASSER- UND ABWASSERCHEMISCHE UNTERSUCHUNGEN),
Gesellschaft fuer Kernforschung m.b.H., Karlsruhe (West Germany).
H. Guesten, W. Kluger, W. Koelle, H. Rohde, and H. Ruf.
Available from NTIS, Springfield, Va., as KFK 1690 UF; $3.00 paper copy, $1.45 microfiche. Report KFK 1690 UF, Oct. 1972. 68 p, 19 fig, 11 tab, 24 ref.

Descriptors: *Water pollution, *Chlorinated hydrocarbon pesticides, *Mercury, *Activated carbon, Analytical techniques, Chromatography, Gas chromatography, Adsorption, Instrumentation, Pollutant identification, Organic wastes, Water quality, Water quality control, Organic loading, Solvent extractions, Halogenated pesticides, Organic pesticides, Ethers, Aromatic compounds, Rivers, Water analysis, Path of pollutants, Water treatment, Public health, Fish, Algae.

Five papers concern analytical methods for organic pollutants. Data (1970-1971) concerning the organic content of German rivers are reviewed critically. Methods described include column chromatography for various organic compounds, thin-layer chromatography for chloro-organics, and neutron-activation analysis for Hg in algae and fish. Activated carbon filters from waterworks were analyzed to give the following order of decreasing quantity for organics which had resisted biologic decomposition: chloro-organics (including polychlorobiphenylenes and hexachlorocyclohexane but no weed-killing agents), aromatic nitro compounds, aromatic compounds with tertiary butyl groups, esters, and ethers. A sixth paper concerns study of activated carbon in waterworks technology by its sorption-desorption characteristics for dimethylformamide. (Bopp-ORNL)
W73-07944

CONTINUOUS MEASUREMENT OF ALPHA AND BETA RADIOACTIVITY OF WATER IN VIEW OF MAXIMUM PERMISSIBLE CONCENTRATIONS FOR THE PUBLIC,
Commissariat a l'Energie Atomique, Saclay (France).
J. Matutano, P. Hory, and L. Girvaud.
Available from NTIS, Springfield, Va., as CEA-CONF-2007; $3.00 per copy, $1.45 microfiche. Report CEA-CONF-2007, March 1972. 11 p, 7 fig, 1 tab.

Descriptors: *Radioactivity, *Measurement, *Nuclear powerplants, *Effluents, *Assay, *Radioisotopes, Lakes, Fish, Regulation, Potable water, Monitoring, Evaporation, Equipment, Aerosols, Filtration, Analytical techniques, Efficiencies.
Identifiers: *France.

Some liquid effluents from the CNEN-Saclay were put into the Saclay lakes which contain large quantities of fish. The radioactivity level was measured to be less than or equal to the maximum permissible concentrations of drinking water. A detector consisting of two flow counter which operates as a proportional counter was developed for the continuous measurement of the activity (10 to the minus 6th power to 10 to the minus 8th power) of liquid effluents. The equipment used in the measurements concentrates the radioactivity of a volume of water by evaporation. The evaporation is effected by the atomization of the water in a hot air flow; the dry aerosol is recovered using air filtration. The efficiency and performance of the apparatus are given. (Houser-ORNL)
W73-07945

FALLOUT PROGRAM QUARTERLY SUMMARY REPORT, SEPTEMBER 1, 1972 - DECEMBER 1, 1972,
New York Operations Office (AEC), N.Y. Health and Safety Lab.
E. P. Hardy, Jr.
Available from NTIS, Springfield, Va., as HASL-268. $3.00 per copy, $1.45 microfiche. Report No. HASL-268, Jan. 1, 1973. 179 p, 8 fig, 25 tab, 19 ref, 6 append.

Descriptors: *Fallout, *Radioactivity, *Monitoring, *Measurement, *Asssay, Sampling, Analytical techniques, Radiochemical analysis, Data collections, Atlantic Ocean, Strontium, Tritium, Food chains, Public health, Bibliographies, Publications.

Current data are presented from the HASL Fallout Program; the National Radiation Laboratory in New Zealand, and the EURATOM Joint Nuclear Research Centre at Ispra, Italy. Interpretive reports are presented on strontium-90 fallout over the Atlantic, fallout tritium and dose commitment, and quality control analyses of surface air, fallout, diet, and bone analyses during 1971. Tabulations of radionuclide levels in fallout, surface air, stratospheric air, milk, and tap water are included. A bibliography of recent publications related to radionuclide studies, is also presented. (See also W73-07951) (Houser-ORNL)
W73-07950

APPENDIX TO QUARTERLY SUMMARY REPORT, SEPT. 1, 1972, THROUGH DEC. 1, 1972. HEALTH AND SAFETY LABORATORY, FALLOUT PROGRAM,
New York Operations Office (AEC), N.Y. Health and Safety Lab.
E. P. Hardy, Jr.
Available from NTIS, Springfield, Va., as HASL-268, Appendix. $3.00 per copy, $1.45 microfiche. Report No. HASL-268, Appendix, Jan. 1, 1973. 448 p, 6 Append.

Descriptors: *Fallout, *Data collections, *Sampling, *Assay, *Strontium, *Food chains, *Population, Public health, Air pollution, Water pollution, Atlantic Ocean, Lead, Milk, Domestic water.

Presents fallout data in the following six appendices: (A) Strontium in Monthly Deposition at World Land Sites; (B) Radiostrontium Deposition at Atlantic Ocean Weather Stations; (C) Radionuclides and Lead in Surface Air; (D) Radiostrontium in Milk and Tap Water, (E) Table of Conversion Factors, and (F) Table of Radionuclides. (See also W73-07950) (Houser-ORNL)
W73-07951

ENVIRONMENTAL RADIOACTIVITY IN GREENLAND IN 1971,
Danish Atomic Energy Commission, Risoe. Health Physics Dept.
For primary bibliographic entry see Field 05B.
W73-07953

DETERMINATION OF TRACE METALS AND FLUORIDE IN MINERALOGICAL AND BIOLOGICAL SAMPLES FROM THE MARINE ENVIRONMENT,
Naval Research Lab., Washington, D.C.
D. J. Bressan, R. A. Carr, P. J. Hannan, and P. E. Wilkniss.
Available from NTIS, Springfield, Va., as CONF-721010-8; $3.00 paper copy, $1.45 microfiche. Report CONF-721010-8, 1972. 10 p, 4 tab, 10 ref. From International Conference on Modern Trends in Activation Analysis, Oct. 2, 1972, Saclay, France.

Descriptors: *Neutron activation analysis, *Mercury, *Marine algae, *Air pollution, Dusts,

Oceans, Water pollution sources, Gamma rays, Spectroscopy, Path of pollutants, Fluorides, Absorption, Trace elements, Tracers, Analytical techniques, Food chains.

Hg or its volatile compounds must condense on atmospheric particles or react with them, since neutron-activation analysis shows that Hg is about two orders of magnitude higher in over-ocean dust than the average in the earth's crust. That the atmospheric dust is of continental origin is suggested by higher F (330-875 ppm) than for sea salt (37 ppm) as determined by neutron-activation analysis and other methods. The neutron activation technique for analysis of Hg in marine algae was checked by atomic absorption spectroscopy. Carrier-free Hg197, produced by irradiation of gold targets, was used in determining Hg uptake by marine algae at varying Hg concentrations in the water. (Bopp-ORNL)
W73-07959

BACTERIAL AND ALGAL CHLOROPHYLL IN TWO SALT LAKES IN VICTORIA, AUSTRALIA,
S. U. Hussainy.
Water Research, Vol 6, No 11, p 1361-1365, November 1972. 3 fig, 1 tab, 15 ref.

Descriptors: *Water analysis, *Chlorophyll, *Algae, *Bacteria, *Dissolved solids, Distribution patterns, Australia, Optical properties, Separation techniques, Saline lakes.
Identifiers: *Optical density, *Grammetric analysis, Lake Keilambete, Lake Gnotuk.

Water samples were collected at depths of 1, 2, 3, 5, 10, and 20 m from Lake Gnotuk and at 10 m from Lake Keilambete (Australia) and analyzed for total dissolved solids, algal chlorophyll, and bacterial chlorophyll. Solids were estimated grammetrically; algal chlorophyll was determined by filtering the sample, extracting with acetone, and recording the optical density; bacterial chlorophyll was estimated by applying a provisional equation derived by Takahashi and Ichimura to the optical density data. Total dissolved solids concentrations were 59.3 and 60.2 g per l for Lakes Keilambete and Gnotuk, respectively. The standing crops of autotrophic sulphur bacteria in terms of bacterial chlorophyll were found to be in larger quantities than algal chlorophyll a. Both at the surface and at the bottom, bacterial chlorophyll was about ten times as much as algal chlorphyll a. Their population in smaller quantities at other depths may be due to the heavy grazing by the zooplankton. It is suggested that the bacteria may be a good food source for the zooplankton. It is also suggested that the deep orange pigmentation in the Copepoda may be due to their grazing on the pink bacteria. (Little-Battelle)
W73-08018

THE USE OF AGAR DIP-SLIDES FOR ESTIMATES OF BACTERIAL NUMBER IN POLLUTED WATERS,
Nairobi Univ. (Kenya). Dept. of Civil Engineering.
D. D. Mara.
Water Research, Vol 6, No 12, p 1605-1607, December 1972. 2 fig, 1 tab, 3 ref.

Descriptors: *Sewage bacteria, *Aquatic bacteria, *Monitoring, *Estimating, Water pollution, Sewage, Methodology, Pollutant identification.
Identifiers: *Agar dip-slides, Culture media, Agars, Plate counts.

Oxoid and 'Uricult' agar dip-slides were used to estimate to the nearest order of magnitude total and coliform counts in sewage and polluted rivers. Samples of sewage and polluted river water were collected in sterile 8-oz (228-ml) bottles. Standard plate counts were obtained in Oxoid yeast extract agar after 24 h incubation at 35 degrees C. Total coliform and faecal coliform pour-plate counts were obtained in lactose teepol agar (Jameson and Emberley, 1956) after 4 h incubation at 30 degrees C followed 18-20 h at either 35 degrees or 44 degrees C (Mara, in preparation). These counts were used to judge the accuracy of the dip-slide counts. Two Oxoid dip-slides were immersed in the sewage or river flow for ca. 5 s and incubated, one at 35 degrees C and the other 44 degrees C for 20-24 h. At some sampling stations 'Uricult' dip-slides (Orion Pharmaceutical, Helsinki) were also used. These estimates agreed closely with the corresponding pour-plate counts in yeast extract agar and lactose teepol agar. The dip-slide technique is simple and suitable for routine monitoring of effluent quality and river pollution. (Holoman-Battelle)
W73-08021

A NEW SIMPLE WATER FLOW SYSTEM FOR ACCURATE CONTINOUS FLOW TESTS,
Kristinebergs Zoologiska Station, Fiskebackskil (Sweden).
For primary bibliographic entry see Field 07B.
W73-08022

A SIMPLE METHOD FOR CONCENTRATING AND DETECTING VIRUSES IN WASTEWATER,
Central Public Health Engineering Research Inst., Nagpur (India).
V. C. Rao, U. Chandorkar, N. U. Rao, P. Kurmaran, and S. B. Lakhe.
Water Research, Vol 6, No 12, p 1565-1576, December, 1972. 1 fig, 13 tab, 11 ref.

Descriptors: *Waste water (Pollution), *Isolation, *Methodology, *Viruses, Sewage, Sewage effluents, Hydrogen ion concentration, Laboratory equipment, Biochemical oxygen demand, Pollutant identification, Efficiencies, Sampling, Cultures.
Identifiers: *Sample preparation, Recovery, *Membrane filters, Poliovirus type I, Coxsackie virus B3, Coxsackie virus B5, Echo virus 9, Tissue culture, Elution, Enteroviruses.

A simple method has been developed for routine analysis of sewage effluents for detecting viruses using adsorption at pH 3 on a 0.45 -micron 47-mm diameter membrane filter and elution at pH 8. It was tested on viruses added to autoclaved sewage. Homogenizing the sample for 4 min in a Waring blender and clarification by centrifugation at 1800 g and later at 9230 g facilitated easy filtration without any loss of virus. Retention of the eluant for 30 min on the millipore membrane and then elution in situ under suction provided a sterile eluate with 100 per cent recovery of viruses. Viruses added to fecal suspensions with 600 mg/l BOD were completely recovered when the sample pH was adjusted to 3 and its salt concentration increased by adding 1200 mg/l of Mg (2+) as the chloride. This procedure eliminated the need for passing the samples through ion exchange resins for removing membrane coating components. In a 1 yr program of monitoring of raw sewage from a middle income group community in Nagpur, a maximum of 3150 PFU/l during monsoon and 11575 PFU/l during winter was obtained. High efficiency and reproducibility of the method allowed the use of sample volumes of 40 ml of raw sewage and 320 ml of treated effluent for the detection of viruses. (Holoman-Battelle)
W73-08023

WATER QUALITY MONITORING IN DISTRIBUTION SYSTEMS: A PROGRESS REPORT,
National Sanitation Foundation, Ann Arbor, Mich.
N. I. McClelland, and K. H. Mancy.
Journal of the American Water Works Association, Vol 64, No 12, p 795-803, December, 1972. 21 fig, 4 tab, 3 ref.

Descriptors: *Potable water, *Monitoring, *Instrumentation, *Water temperature, *Dissolved oxygen, *Hydrogen ion concentration, *Conductivity, *Hardness, *Chlorides, *Flourides, *Turbidity, *Alkalinity, *Nutrients, Nitrates, Chlorine, Copper, Cadmium, Lead, Corrosion, Scaling, Water analysis, Electrical equipment, Research equipment, Sampling, Calibrations, Calcium carbonate, Water quality.
Identifiers: *Heavy metals, Ion selective electrodes, Nephelometers, Galvanic cells, Anodic stripping voltammetry, Membrane electrodes, Sample preparation, Residual chlorine, Glass electrodes, Differential anodic stripping voltammetry.

The progress of the Water Quality Monitoring Project which is being conducted by the National Sanitation Foundation is reviewed. A prototype potable-water quality monitor is on-stream and can measure: temperatures using thermistors; DO with a voltammetric membrane electrode; pH with a glass electrode; conductivity with an a-c conductivity cell; hardness, nitrates, chlorides, and fluorides with ion-selective electrodes; turbidity with a nephelometer; free residual chlorine with a galvanic cell; alkalinity with a glass electrode; Cu by anodic stripping voltammetry; Cd and Pb by differential anodic stripping voltammetry; and corrosion by a polarization admittance technique. Sample analyses for hardness, nitrates, chlorides, fluorides, and heavy metals are included. A laboratory test for measuring scaling potential using a rotating ring disc electrode has also been developed. (Little-Battelle)
W73-08027

MIREX RESIDUES IN WILD POPULATIONS OF THE EDIBLE RED CRAWFISH (PROCAMBARUS CLARKI),
Animal and Plant Health Inspection Service, Gulfport, Miss. Plant Protection and Quarantine Programs.
G. P. Markin, J. H. Ford, and J. C. Hawthorne.
Bulletin of Environmental Contamination and Toxicology, Vol 8, No 6, p 369-374, December, 1972. 1 tab, 10 ref.

Descriptors: *Crawfish, *Pesticide residues, Crustaceans, Invertebrates, Insecticides, DDT, Gas chromatography, Chlorinated hydrocarbon pesticides, Polychlorinated biphenyls, Solvent extractions, Chemical analysis, Pesticide toxicity, Water pollution effects, TDE, DDE.
Identifiers: *Mirex, *Procambarus clarki, Aroclor 1260, Sample preparation, Chlordane, Toxaphene, Detection limits, Arthropods, Macroinvertebrates, Decapods, Electron capture gas chromatography, Metabolites.

Red crawfish from southcentral Louisiana were analyzed in order to determine if mirex residues in specimens from treated areas are beginning to reach those levels found to affect crawfish under laboratory conditions. Crawfish samples were washed to remove adhering materials, ground and mixed in a blender, a 50-gram subsample extracted in organic solvents, and analyzed by electron capture gas chromatography. Samples were also analyzed for DDT and its metabolited (TDE, DDE), chlordane, Toxaphene, and Aroclor 1260. The level of detection was 0.01 ppm. DDT residues ranged from 0.02-0.44 ppm while mirex residues were barely detectable. PCB residues were detected in few instances and 0.11 ppm chlordane was detected in one sample. No results are given for TDE, DDE, toxaphene or Aroclor 1260. There is no real evidence that the insecticide mirex has any significant effect on crawfish populations. (Holoman-Battelle)
W73-08030

PCB RESIDUES IN ATLANTIC ZOOPLANKTON,
California Univ., Berkeley. Inst. of Marine Resources.
R. W. Risebrough, V. Vreeland, G. R. Harvey, H. P. Miklas, and G. M. Carmignani.

with 97.1 recovery with a relative mean deriation of plus or minus 2.3 pph and a relative standard percent deviation of plus or minus 3.1 pph. Detection limits were determined for all of the rare earths. The detection limit was taken to be that amount of mixed ligand-complex necessary to give a chromatographic peak response equal to or greater than twice the background response. The detection limit observed was 20 micrograms metal for all the rare earth. The response to the flame ionization detector varied from metal to metal. When the electron capture detector was employed, it was extremely sensitive to the complexes; however, concentrations of 20 micrograms metal were sufficiently high to overload the detector even at purge flow rates of 400 ml/min. When the concentration was lowered, no response was observed. (Long-Battelle)
W73-08033

SUPPORT-BONDED POLYAROMATIC COPOLYMER STATIONARY PHASES FOR USE IN GAS CHROMATOGRAPHY,
Applied Automation, Inc., Bartlesville, Okla. Systems Research Dept.
For primary bibliographic entry see Field 07B.
W73-08034

ELECTROCHEMICAL CELL AS A GAS CHROMATOGRAPH-MASS SPECTROMETER INTERFACE,
Northgate Lab., Hamden, Conn.
For primary bibliographic entry see Field 02K.
W73-08035

PLASMA CHROMATOGRAPHY OF THE MONO-HALOGENATED BENZENES,
Waterloo Univ. (Ontario). Dept. of Chemistry.
F. W. Karasek, and O. S. Tatoue.
Analytical Chemistry, Vol 44, No 11, p 1758-1763, September, 1972. 7 fig, 2 tab, 14 ref.

Descriptors: *Chemical analysis, Instrumentation, Aromatic compounds.
Identifiers: *Plasma chromatography, *Monohalogenated benzenes, Chromatography peaks, Chlorobenzene, Bromobenzene, Iodobenzene, Fluorobenzene, Thermal electrons, Sample preparation.

Using thermal electrons and positive reactant ions from nitrogen gas, both positive and negative plasmagram patterns have been obtained for fluorobenzene, chlorobenzene, bromobenzene, and iodobenzene. The plasmagrams give characteristic qualitative data. Positive plasmagrams show protonated molecular ions containing one and two molecules; the negative plasmagrams, except for the fluorobenzene, show only a strong halogen ion peak, which provides experimental evidence for dissociative electron capture by thermal electrons. (Long-Battelle)
W73-08037

PIEZOELECTRIC DETECTORS FOR ORGANOPHOSPHORUS COMPOUNDS AND PESTICIDES,
Louisiana State Univ., New Orleans. Dept. of Chemistry.
E. P. Scheide, and G. G. Guilbault.
Analytical Chemistry, Vol 44, No 11, p 1764-1768, September, 1972. 6 fig, 2 tab, 25 ref.

Descriptors: *Pollutant identification, *Organophosphorus pesticides, *Chemical analysis, Organic compounds, Instrumentation, Selectivity, Methodology.
Identifiers: *Piezoelectric detector, Paraoxon, O O-diethyl-o-p-nitrophenyl phosphate, Chemical interference, Sensitivity, Sample preparation, Detection limits.

A quartz piezoelectric crystal coated with a substrate has been used for the detection of small mass changes caused by the selective adsorption of organophosphorus compounds and pesticides. Incorporation of the crystal into a variable oscillator circuit and measurement of the change in frequency of the crystal due to the increase in mass allows a highly sensitive indication of the amount of organophosphorus compound present in the atmosphere down to the part per million level. Instrumentation is relatively inexpensive and can be easily used in the field. Analysis is nondestructive and requires very little time. AT cut quartz crystals with fundamental frequencies of 9.0 MHz were coated with various inorganic substrates and these were evaluated as to slectivity and sensitivity with respect to organophosphorus pollutants. Other parameters that affect the efficiency of the detector were also studied and evaluated. The detector has potential use as both an air pollution sensor and a specific gas chromatography detector. (Long-Battelle)
W73-08038

COLORIMETRIC DETERMINATION OF CALCIUM USING REAGENTS OF THE GLYOXAL BIS (2-HYDROXYANIL) CLASS,
Clemson Univ., S.C. Dept. of Chemistry and Geology.
C. W. Milligan, and F. Lindstrom.
Analytical Chemistry, Vol 44, No 11, p 1822-1829, September, 1972. 10 fig, 5 tab, 16 ref.

Descriptors: *Clorimetry, *Chemical analysis, *Calcium, *Chelation, Chemical reactions, Alkali metals, Cations, Anions, Color reaction, Hydrogen ion concentration, Sodium, Potassium, Magnesium, Strontium, Aluminum, Iron, Cobalt, Nickel, Zinc, Cadmium, Copper, Lead, Nitrates, Phosphates, Chlorides, Silicates, Sulfates.
Identifiers: *Glyoxal bis (2-hydroxyanil), *Reagents, Chemical interference, Metal chelates, Reproducibility, Mixtures, Absorption spectra, Lithium, Barium, Uranyl, Tin, Chlorates.

The determination of calcium in solution from 0.1 to 15 micrograms per milliliter is easy, accurate, and reproducible with reagents of the glyoxal bis (2-hydroxyanil) class. A few precautions are needed: obviously all chemicals must be calcium-free and the sequence of adding the reagents is critical. However, no extraction is needed and a simple, inexpensive colorimeter is all the instrumentation necessary. The reagents chelate calcium yielding red colors at a high pH so interferences are limited. The various reagents have been evaluated and the chelate combining ratios and the apparent formation constants measured. (Long-Battelle)
W73-08040

DETERMINATION OF NANOGRAM QUANTITIES OF SIMPLE AND COMPLEX CYANIDES IN WATER,
Department of the Environment, Ottawa (Ontario). Water Quality Div.
P. D. Goulden, B. K. Afghan, and P. Brooksbank.
Analytical Chemistry, Vol 44, No 11, p 1845-1849, September 1972. 7 fig, 3 tab, 6 ref.

Descriptors: *Methodology, *Distillation, *Colorimetry, *Water analysis, Automation, Industrial wastes, Pollutant identification, Iron, Copper, Chemical analysis, Separation techniques, Laboratory equipment, Chlorides, Sulfates, Zinc, Carbonates, Nitrates.
Identifiers: *Cyanides, Detection limits, Sample preparation, Ottawa River, Synthetic water, Chemical interference, Standard methods, Bisulfites, Chemical recovery, Sulfides, Thiocyanates.

Two methods for analysis of cyanide in water are described. In the first, modifications to the normal distillation procedure given in 'Standard Methods'

are made to lower the limits of detection of this method to 5 micrograms/liter CN. An automated method is described that enables 10 samples per hour to be analyzed with a limit of detection of 1 microgram/liter CN. With a sample size of 7 ml, this corresponds to a detection limit of 7 nanograms CN. Distinction is made between simple and complex cyanides by irradiation with ultraviolet light. This irradiation breaks down complex cyanides, including those of cobalt and iron. (Long-Battelle)
W73-08041

IMPROVED UREA ELECTRODE,
Louisiana State Univ., New Orleans. Dept. of Chemistry.
G. G. Guilbault, and G. Nagy.
Analytical Chemistry, Vol 45, No 2, p 417-419, February 1973. 4 fig, 7 ref.

Descriptors: *Aqueous solutions, *Pollutant identification, *Ureas, Selectivity, Potassium, Ions, Enzymes, Electrical stability, Zeta potential, Construction, Methodology, Electrochemistry.
Identifiers: *Urea electrode, *Ion selective electrodes, Biological fluids, Response time, Urease, Ammonium electrode.

An improved urea electrode was made by covering the active surface of a solid type emmonium electrode with a physically immobilized urease (enzyme) reaction layer. This layer was made according to the procedure of Guilbault and Montalvo (1970). A nylon net was placed over the sensor surface and fixed with rubber rings. A solution of 175 mg urease in 0.9 ml monomer solution was dropped onto the netting and polymerized by light for 60 min. The ammonium selective electrode was made using the antibiotic Nonactin as the active ingredient embedded in a silicone rubber matrix. Studies of the electrode show it to have good stability and response characteristics. It allows a convenient determination of urea in water solution, an estimation of urea content of biological fluids of unknown potassium ion concentration, and an accurate measurement of the urea concentration in biological fluids of approximately known potassium concentration. Some theoretical aspects of the enzyme electrode are discussed. (Holoman-Battelle)
W73-08042

ACTIVITY MEASUREMENTS AT HIGH IONIC STRENGTHS USING HALIDE-SELECTIVE MEMBRANE ELECTRODES,
State Univ. of New York, Buffalo. Dept. of Chemistry.
J. Bagg, and G. A. Rechnitz.
Analytical Chemistry, Vol 45, No 2, p 271-276, February 1973. 5 tab, 22 ref.

Descriptors: *Halides, *Measurements, Bromides, Chlorides, Fluorides, Electrochemistry, Iodides, Sodium chloride, Electrolytes, Anions.
Identifiers: *Ion selective electrodes, *Membrane electrodes, *Halide selective electrodes, *Ionic strength, *Ionic activity, Potassium chloride, Potassium bromide, Potassium fluoride, Lithium chloride.

The newer halide-selective electrodes have been examined in a cell with a homoionic liquid-junction which does not require a reversible cation-electrode or the use of extra-thermodynamic assumptious in the calculation of theoretical emf. These electrodes have been shown to have nearly theoretical potentiometric response in chloride, bromide, and fluoride solutions up to 4-5 molal. The iodide-selective electrode is restricted by deterioration of the electrode surface to solutions less than 0.5 molal. The single-ion activity convention, based upon hydration considerations, proposed by Bates, Staples, and Robinson, combined with the Henderson equation for residual liquid-junction potentials, fitted the data for a cell with heteroionic junction up to 6m NaCl, 4m KCl, 4m KBr, 3m KF, and 1m LiCl. The results are consistent with previously proposed mechanistic models for the operation of crystal membrane electrodes. (Holoman-Battelle)
W73-08043

ELECTROCHEMICAL CHARACTERISTICS OF THE GOLD MICROMESH ELECTRODE,
Wisconsin Univ., Madison. Dept. of Chemistry.
For primary bibliographic entry see Field 02K.
W73-08044

ORGANOCHLORINE PESTICIDE RESIDUES IN WATER, SEDIMENT, ALGAE, AND FISH, HAWAII - 1970-71,
Hawaii Univ., Honolulu. Dept. of Agricultural Biochemistry.
A. Bevenue, J. W. Hylin, Y. Kawano, and T. W. Kelley.
Pesticides Monitoring Journal, Vol 6, No 1, p 56-64, June 1972. 3 fig, 8 tab, 27 ref.

Descriptors: *Chlorinated hydrocarbon pesticides, *Pesticides residues, *Sediments, *Algae, *Fish, Chemical analysis, *Sewage effluents, Water analysis, Potable water, Rain water, *Hawaii, DDT, Dieldrin, DDE, DDD, Gas chromatography, Methodology, Heptachlor, Aldrin, Solvent extractions, Separation techniques, Plankton, Aroclors.
Identifiers: Lindane, Heptachlor epoxide, Chlordane, Sample preparation, Detection limits, Electron capture gas chromatography, Guppy, Molly, p p' DDD, p p' DDE, p p' DDT, Ala Wai Canal, Kapalamca Canal, Kalihi Stream, Sand Island Outfall.

Rainwater, drinking water, and nonpotable waters in Hawaii were sampled and found to contain chlorinated insecticide residues in the low parts-per-trillion range. A portion of each sample was tested for chloride ion concentration and pH and the remainder subjected to hexame extraction, evaporation and gas chromatography. Sewage water samples were prepared for (1) pentachlorophenol identification by pH adjustment, hexane extraction, diazomethane treatment, hexane dissolution, and gas chromatography, and for (2) chlorinated pesticides other than PCP by the combined Florisil and silicic acid procedure of Armour and Burke. Dredged sediment samples were subjected to sodium sulfate, hexane extraction and the Mills' Florisil cleanup procedure, and then gas chromatographed. The cleanup procedure used for algae and fish samples prior to gas chromatography was a modification of the method of Kadoum. Dieldrin, p,p'-DDT, and lindane were the pesticides most prevalent; pentachlorophenol was present in samples from a sewage fallout. The ratio of chlorinated pesticide residues in canal waters to residues in algae, sediment, and fish from the same canals was 1:4,000:9,000:32,000, respectively. According to proposed water quality standards, results of this study indicated that pollution of Hawaii's water by organochlorine pesticides does not occur to any significant degree. (Mackan-Battelle)
W73-08047

ORGANOCHLORINE PESTICIDE RESIDUES IN COMMERCIALLY CAUGHT FISH IN CANADA - 1970,
Fisheries Research Board of Canada, Winnipeg (Manitoba). Freshwater Inst.
J. Reinke, J. F. Uthe, and D. Jamieson.
Pesticides Monitoring Journal, Vol 6, No 1, p 43-49, June 1972. 1 fig, 1 tab, 7 ref.

Descriptors: *Chlorinated hydrocarbon pesticides, *Polychlorinated biphenyls, *Commercial fish, *Canada, *Pesticide residues, DDT, DDE, DDD, Dieldrin, Heptachlor, Aldrin, Separation techniques, Water analysis, Freshwater fish, Saline water fish, Chemical analysis, Gas chromatography, Salmon, Yellow perch, Mu
Rainbow trout, Catfishes, Carp, Lake trout,
heads, Walleye, Lake Erie, Lake Ontario, l
Michigan, Lake Superior, St. Lawrence R
Lake Huron, Solvent extractions, Pikes, Su
Sauger.
Identifiers: *Gas liquid chromatography, *
layer chromatography, Heptachlor epoxide,
dane, Chlordane, Metabolites, Isomers, Preci
Chemical recovery, Detection limits, S
preparation, Burbot, Mukturk, Coho
Whitefish, Kokanee, Yellow pickerel, Stur
Sheepshead, Alewife, Crappies, Tullibee, I
Winnipeg, Sturgeon River, Ottawa River, Lak
Clair, p p' DDT, p p' DDD, p p' DDE, p p' I
Lake Nipigan, Lake St. Paul, Cold River,
River.

A modified Mills extraction method was cou
with a thin-layer chromatographic confirm
and a gas chromatographic quantification o
ganochlorine pesticide residues in commerc
caught fish from 78 locations in 68 central
an lakes and rivers. Only a few of thes
yielded fish with appreciable concentrat
DDT and its analogs (greater than 1 ppm)
only a few cases did the concentrations ex
maximum permissible level of 5 ppm. Of t
organochlorine pesticides commonly used,
lindane, aldrin, heptachlor, heptachlor
endrin, dieldrin, and chlordane, only diel
found at significant levels in a number of
but these amounts were still below the
permissible level. Trace amounts of lind
found in some samples. The pres
polychlorinated biphenyls (PCB's) was
samples from the Great Lakes and the s
of Lake Winnipeg. PCB's were separat
DDE on aluminum oxide G (type E) plate
triethylamine-hexane solvent system. (Ma
Battelle)
W73-08048

CHEMICAL RESIDUES IN LAKE ERIE 1970-71,
Food and Drug Administration, Cincinnati, O
R. L. Carr, C. E. Finsterwalder, and M. J. Sch
Pesticides Monitoring Journal, Vol 6, No 1,
26, June 1972. 1 fig, 2 tab, 6 ref.

Descriptors: *Lake Erie, *Pesticide resi
*Mercury, *Freshwater fish, DDE, DDT,
drin, Polychlorinated biphenyls, Yellow pe
Channel catfish, Carp, Drums, White bass, P
tant identification, Chlorinated hydrocarbon
cides, Great Lakes, Ohio.
Identifiers: TDE, Gas liquid chromatogra
Thin layer chromatography, Coho salmon, Bi
ical magnification, Accumulation,
flavescens, Oncorhynchus kisutch, Ro
chrysops, Aplodinotus grunniens, Ictalurus
tatus, Cyprinus carpio.

Yellow perch, coho salmon, carp, channel cat
freshwater drum, and white bass from the
shore of Lake Erie were analyzed during 197
for residues of chlorinated pesticides (DDE,
DDT, and dieldrin), polychlorinated biphe
(PCB's), and mercury. The method employe
extraction and cleanup of samples to deter
DDT residues, dieldrin, and PCB's was
described by Porter, Young, and Burke.
layer chromatography and gas-liquid chrom
raphy were used for confirmation and quantit
analysis of the residues. All but 1 of the 80 sam
analyzed contained DDT and/or its metabol
PCB's were found in all samples. Fifty-three o
80 samples were analyzed for mercury, an
were found positive. Average levels of resi
for the species sampled ranged from 0.06 to
ppm for DDE; 0.07 to 0.52 ppm, TDE; 0.03 to
ppm, DDT; 0.18 to 0.90 ppm, total DDT; 0.0
0.07 ppm, dieldrin; 0.08 to 4.4 ppm, PCB's;
0.12 to 0.64 ppm, mercury. The highest ave
residue levels of total DDT were in coho
and channel catfish. Average levels of PCB's

Copy available from GPO Sup Doc as EP1.23/2:72-079, $0.55; microfiche from NTIS as PB-219 669, $1.45. Environmental Protection Technology Series Report No EPA-R2-72-079, October 1972. 15 p, 3 fig, 3 tab, 6 ref. EPA Project No 16020 EWC.

Descriptors: *Carbamate pesticides, *Pollutant identification, Organic pesticides, Chromatography, Analytical techniques.
Identifiers: *Liquid chromatography, *Retention time, Baygon, Furadan, Matacil, Mobam, Sevin, UC 10854, UC 8454, Carbanolate, RE 5305, Mesurol, Zectran, Ultraviolet detectors, Carbaryl, Aprocarb, Carbofuran, Aminocarb, Terbutol, Dazomet, Methonyl, Sirmate (2-3-isomer), Sirmate (3-4-isomer), Bux, Azak, Benomyl, IPC, CIPC, Dimetilan, Temik, Swep, Barban, Thiram, Mylone, Lannate.

Standard solutions of 23 carbamate pesticides prepared in isopropanol were analyzed using a Du-Pont Model 820 liquid chromatograph equipped with an ultraviolet photometric detector. Stainless steel columns (1 m x 2 mm i.d.) packed with either Permaphase ODS (octadecyl silane) or Permaphase ETH (ether) were used with the following mobile phases: 6 and 30 percent MeOH in water, hexane, one percent isopropanol/hexane, and 4 percent isopropanol/hexane. The retention times of the pesticides are given. The UV detector required 20-1500 ng for the pesticides studied to give a 25 percent full-scale recorder response. (Holoman-Battelle)
W73-08129

QUALITY OF SURFACE WATERS OF THE UNITED STATES, 1968: PART I. NORTH ATLANTIC SLOPE BASINS.
Geological Survey, Washington, D.C.
For primary bibliographic entry see Field 07C.
W73-08155

DISSOLVED HELIUM IN GROUNDWATER: A POSSIBLE METHOD FOR URANIUM AND THORIUM PROSPECTING,
McMaster Univ., Hamilton (Ontario). Dept. of Physics.
For primary bibliographic entry see Field 02K.
W73-08157

TECHNIQUE FOR MEASURING METALLIC SALT EFFECTS UPON THE INDIGENOUS HETEROTROPHIC MICROFLORA OF A NATURAL WATER,
Simon Fraser Univ. Burnaby (British Columbia). Dept. of Biological Sciences.
L. J. Albright, J. W. Wentworth, and E. M. Wilson.
Water Research, Vol 6, No 12, p 1589-1596, December 1972. 2 fig, 3 tab, 12 ref.

Descriptors: *Metals, *Salts, Water pollution effects, *Aquatic microorganisms, *Methodology, Sodium chloride, Radioactivity techniques, Bioassay, Heavy metals, Cations, Chlorides, Cadmium, Sodium, Alkali metals, Chromium, Mercury, Copper, Nickel, Lead, Zinc, Spectrometers, Aquatic bacteria.
Identifiers: *Heterotrophic bacteria, *Viability, Substrate utilization, Silver sulfate, Barium chloride, Cadmium chloride, Chromium chloride, Mercury chloride, Copper chloride, Nickel chloride, Lead chloride, Zinc chloride, Glucose, Silver, Sodium metaarsenate, Barium, Mercuric chloride, Cupric chloride, Scintillation counting.

A heterotrophic activity assay method is described which may be used to determine the effects of several metallic salts, Ag2SO4, NaAsO3, BaCl2.2H2O, CdCl2.2.5 H2O, CrCl2.6H2O, CuCl2, HgCl2, NaCl, NiCl2.6H2O, PbCl2 and ZnCl2, at very low concentrations, upon the net activity of the native heterotrophic microflora of an aquatic ecosystem, without unduly disturbing in situ conditions. The method is based upon the uptake and mineralization of a radioactivity labeled metabolite (C-14-glucose) by native heterotrophic microflora and the analysis of data by Michaelis-Menten enzyme kinetics equations. A 100-ml water sample was collected, temperature determined, and divided into 25-ml portions, each of which was treated with 0.0538, 0.2690, 0.5375 or 0.8075 microgram amounts of C-14-glucose. Five milliliters were removed from each portion and placed in separate 25-ml erlenmeyer flasks which were sealed immediately. The reaction and control were incubated with reciprocal shaking (60 strokes/min) in the dark at the same temperature of the water when sampled. After 0.5 h incubation the reaction flasks were acidified to about pH 1.0 to stop microbial activity and to drive off C-14 02 which was trapped on phenethylamine-impregnated Whatman paper. After further incubation for an hour, the Watman papers were removed, placed in scintillation vials containing a toluene-based cocktail of 2,5-diphenyloxazole (0.4 percent) and 1,4-bis- (5-phenyloxazoly 1-2)-benzene (0.01 percent). The aquatic contents of each flask were filtered, washed, dried at 60C for 15 min and added to vials contianing the same cocktail. Counting was done with a Beckman LS-250 scintillation spectrometer. Salt effects upon bacterial viability were determined using nutrient agar plate counts immediately before and 0.5 h after treatment of water samples. Petri plates were incubated at 15 degrees C for 96 h before counting. Concentrations of metallic salts which resulted in bacterial death also caused erratic uptake and mineralization rates of C-14-glucose whereas sublethal concentrations, as determined by nutrient agar plate counts, caused a non-competitive inhibition of maximum heterotrophic activity and markedly increased the turnover time of the glucose substrate. (Holoman-Battelle)
W73-08236

UTILIZING METAL CONCENTRATION RELATIONSHIPS IN THE EASTERN OYSTER (CRASSOSTREA VIRGINICA) TO DETECT HEAVY METAL POLLUTION,
Virginia Inst. of Marine Science, Gloucester Point.
R. J. Huggett, M. E. Bender, and H. D. Slone.
Water Research, Vol 7, No 3, p 451-460, March 1973. 9 fig, 4 ref.

Descriptors: *Oysters, *Pollutant identification, *Heavy metals, *Water pollution sources, Mollusks, Invertebrates, Salinity, Sampling, Cadmium, Copper, Zinc, Least squares methods, *Virginia, Industrial wastes, Chesapeake Bay, Sediments, Estuaries, Saline water-freshwater interfaces, Statistical methods.
Identifiers: *Crassostrea virginica, Macroinvertebrates, Atomic absorption spectrophotometry, James River, Hampton Roads, Elizabeth River, Data interpretation.

A total of 495 oysters were collected during February-May, 1971, at various sites in the Chesapeake Bay area, removed from the shell without puncturing, digested in concentrated HNO3, and analyzed for Cd, Cu, and Zn using atomic absorption spectrophotometry. Examination of the data showed that oysters from the same sampling location often differed in metal concentration as much as 100 percent and occasionally 300 percent. These variable concentrations are assumed to be normally distributed around some population mean, therefore the sample mean from each location should approximate the population mean. Means were used only to ascertain the areal distribution of metals in the various river systems. The means showed that a concentration gradient existed in all systems and that each metal increased in concentration as fresh water was approached. Several assumptious were made. (1) The metals (Cu, Cd, and Zn) available to oysters in non-industrialized areas are from the natural weathering of rocks. (2) The ratio of copper to zinc in the weathering rocks is relatively constant within a drainage basin. (3)

Oysters accumulate a constant percentage of each element available to them. Statistical analysis of the metal concentration data showed that a linear relationship exists between Cu and Zn, and Cd levels in contaminated and uncontaminated oysters. No single concentration for an action level can be set for cadmium, copper or zinc in oysters which will definitely indicate pollution sources. However, the approach described has been proven valid in the Chesapeake Bay and may be of use elsewhere. (Holoman-Battelle)
W73-08237

NUTRIENT STUDIES IN TEXAS IMPOUNDMENTS,
Union Carbide Corp., Tonawanda, N.Y. Linde Div.
For primary bibliographic entry see Field 05C.
W73-08241

DEEP-SEA BENTHIC COMMUNITY RESPIRATION: AN IN SITU STUDY AT 1850 METERS,
Woods Hole Oceanographic Institution, Mass.
K. L. Smith, Jr., and J. M. Teal.
Science, Vol 179, No 4070, p 282-283, January 19, 1973. 1 tab, 13 ref.

Descriptors: *Biochemical oxygen demand, *Respiration, *Metabolism, *Benthos, Research equipment, Biological communities, Continental slope, Chemical oxygen demand, Oceans, Sea water.
Identifiers: *Respirometers, Total oxygen demand, Ion selective electrodes, Recorders.

In situ measurements of oxygen uptake, as a measure of metabolic activity, were made on undisturbed deep-sea benthic communities by placing respirometers (bell jars) at a depth of 1850 meters on the continental slope south of New England. The respirometers consisted of two capped Plexiglas cylinders which enclosed 48 sq cm of sediment. A polarographic oxygen electrode in each chamber fed a signal to a Rustrak recorder housed in a glass sphere atop the unit. Each chamber was stirred by a magnetically driven stirrer. Uptake measurements were made over periods of 48-72 hours. Formalin injection was used to poison the biological oxygen demand. Additional measurements of total oxygen uptake and chemical oxygen demand were made from a research vessel using a drill rig equipped with a 2000-m drill pipe, a television camera pod, and a hook apparatus. Oxygen uptake under the bell jars ranged from 0.39-0.55 mg/sq m/hr. Values obtained from the research vehicle were 0.62 ml/sq m/hr. These values were two orders of magnitude lower than values from shallow depths. After treatment of the sediments with formalin there was no measurable chemical oxygen uptake which shows that the total uptake is biological (community respiration). It is concluded that metabolic activity of deep-sea benthic communities is low. (Little-Battelle)
W73-08245

LOWER PH LIMIT FOR THE EXISTENCE OF BLUE-GREEN ALGAE: EVOLUTIONARY AND ECOLOGICAL IMPLICATIONS,
Wisconsin Univ., Madison. Dept. of Bacteriology.
For primary bibliographic entry see Field 05C.
W73-08248

THE MICROBIOLOGY OF AN ACTIVATED SLUDGE WASTE-WATER TREATMENT PLANT CHEMICALLY TREATED FOR PHOSPHORUS REMOVAL,
Pennsylvania State Univ., University Park. Dept. of Microbiology.
For primary bibliographic entry see Field 05D.
W73-08250

HIGH RESOLUTION FIELD IONIZATION MASS SPECTROMETRY OF BACTERIAL PYROLYSIS PRODUCTS,
Bonn Univ. (West Germany). Institut fuer Physikalische Chemie.
H. R. Schulten, H. D. Beckey, H. L. C. Meuzelaar, and A. J. H. Boerboom.
Analytical Chemistry, Vol 45, No 1, p 191-195, January 1973. 1 fig, 1 tab, 20 ref.

Descriptors: *Pseudomonas, *Organic compounds, *Pollutant identification, Mass spectrometry.
Identifiers: *Field ionization-mass spectrometry, Sample preparation, Pseudomonas putida, Mass spectra, Pyrolysis.

The purpose was to explore the potentials of high resolution field ionization-mass spectrometry (FI-MS) for the analysis of extremely complex multicomponent mixutres and to perform a general survey of the chemical nature of bacterial pyrolysis products. The spectra were obtained with a double-focusing mass spectrometer equipped with an FI-ion source and specially designed emitter-adjusting manipulator. The sample was 5 mg of Pseudomonas putida bacteria, freeze-dried and pyrolyzed in a vacuum at 500C. Over 200 lines were revealed on the developed photoplate. Density measurements were made on about 180 lines. Accurate mass measurements are listed for 119 of the strongest lines, and proposed names are included for some compounds. The results show that the range of compounds that can be anlyzed by FI-MS is greater than that of GLC-MS. The usefulness of FI-MS is limited by its inability to separate and identify isomers without additional information. Consequently, the two methods may be used to supplement each other. Differentiation of bacterial strains may be possible in this way. (Little-Battelle)
W73-08256

AUTOMATED DATA HANDLING USING A DIGITAL LOGGER,
Virginia Polytechnic Inst. and State Univ., Blacksburg. Dept. of Chemistry.
For primary bibliographic entry see Field 07C.
W73-08257

DEMOUNTABLE RING-DISK ELECTRODE,
Illinois Univ., Urbana. School of Chemical Sciences.
For primary bibliographic entry see Field 07B.
W73-08260

LINEAR AND NONLINEAR SYSTEM CHARACTERISTICS OF CONTROLLED-POTENTIAL ELECTROLYSIS CELLS,
California Univ., Livermore. Lawrence Livermore Lab.
For primary bibliographic entry see Field 02K.
W73-08261

A PORTABLE VIRUS CONCENTRATOR FOR TESTING WATER IN THE FIELD,
Baylor Coll. of Medicine, Houston, Tex. Dept. of Virology and Epidemiology.
C. Wallis, A. Homma, and J. L. Melnick.
Water Research, Vol 6, No 10, p 1249-1256, October, 1972. 3 fig, 12 ref.

Descriptors: *Viruses, *On-site tests, *Equipment, *Pollutant identification, *Methodology, On-site investigations, Adsorption, Potable water, Filtration, Monitoring.
Identifiers: *Portable virus concentrator, Adsorbents, Virus assays, Poliovirus, Metal complexes, Particulate matter, Plaque forming units, Detection limits.

A system is described for concentrating viruses from large volumes of water. The system consists of a water pump, an electric generator, a series (clarifiers, a virus adsorbent, a virus reconcentr tor, a 5- and a 1-gal pressure vessel with a sma tank of nitrogen as a source of positive pressur and ancillary equipment, all mounted on 2-whe carts for easy portability. Standardization of th system was achieved by use of minute amounts (poliovirus. The virus was added to dechlorinate city tap water so that it could not be detected w less the virus was first concentrated. In th system, raw tap water containing virus is seriall passed through clarifying filters of porosities of ! 1 microns to remove particulate matter, and the through a 1-micron cotton textile filter to electr statically remove submicron ferric and othe heavy metallic complexes. These filters do not de tectably remove virus. Salts are then added to th running tap water to enhance the adsorption c firus to a fiberglass or cellulose acetate filter. Ra water could be processed at the rate of 300 gallon per hour, with total virus removal from the wate and with 80 percent elution of the virus from th adsorbent. (Holoman-Battelle)
W73-08262

IDENTITIES OF POLYCHLORINATE BIPHENYL ISOMERS IN AROCLORS,
Environmental Protection Agency, Athens, Ga Southeast Water Lab.
R. G. Webb, and A. C. McCall.
Journal of the Association of Official Analytica Chemists, Vol 55, No 4, p 746-752, July, 1972. : fig, 2 tab, 15 ref.

Descriptors: *Polychlorinated biphenyls *Aroclors, *Pollutant identification, DDE, Ga chromatography, Water pollution sources, Mas spectrometry, Spectrometers, Chemical reactions Chemical analysis, Organic compounds.
Identifiers: *Infrared spectroscopy, *Flame ionization gas chromatography, *Sample preparation, Gas liquid chromatography, Chlorinated hydrocarbons, Isomers, 2 3 5-trichloroaniline, 2 4 5-trichloriodobenzene, Gomberg reaction, Ullmann reaction, Aroclor 1221, Aroclor 1232, Aroclor 1242, Aroclor 1248, Aroclor 1254.

Twenty-seven polychlorinated biphenyls (PCB) in Aroclors 1221, 1242, and 1254 were separated and identified by matching both their gas-liquid chromatographic (GLC) retention times and infrared spectra with known compounds prepared by the Gomberg or Ullmann reactions. Each Aroclor was mixed with p,p'-DDE prior to chromatographic analysis in order to relate observed PCB retention times to those for p,p'-DDE. The compounds identified by these methods are tabulated as well as those compounds prepared in this study that failed either the GLC or IR tests and were concluded to be absent from the Aroclors. (Long-Battelle)
W73-08263

LIQUID CHROMATOGRAPHY OF POLYCYCLIC AROMATIC HYDROCARBONS,
Food and Drug Administration, Washington, D.C. Div. of Chemistry and Physics.
N. F. Ives, and L. Giuffrida.
Journal of the Association of Official Analytical Chemists, Vol 55, No 4, p 757-761, July, 1972. 2 fig, 1 tab, 22 ref.

Descriptors: *Liquid chromatography, *Polycyclic compounds, *Aromatic hydrocarbons, *Ultraviolet spectroscopy, *Detection systems, Food additives, Preservatives, Duropak OPN, Cellulose acetate, Scanning spectrometer, Sample preparation.

Two column materials (Durapak-OPN and 40 percent cellulose acetate), used for liquid chromatographic analysis, were compared using a continuous monitoring UV detection system for the orders of elution and resolution of 18 polycyclic aromatic hydrocarbons. The detection system con-

MIREX AND DDT RESIDUES IN WILDLIFE AND MISCELLANEOUS SAMPLES IN MISSISSIPPI - 1970,
Mississippi State Univ., State College. Dept. of Biochemistry.
K. P. Baetcke, J. D. Cain, and W. E. Poe.
Pesticides Monitoring Journal, Vol 6, No 1, p 14-22, June 1972. 4 tab, 11 ref.

Descriptors: *Pesticide residues, *Wildlife, *DDT, *Estuaries, Persistence, Food chains, Bird eggs, *Mississippi, Pollutant identification, Path of pollutants, Solvent extractions, Organic pesticides, DDD, DDE, Catfishes, Sunfishes, Channel catfish, Game birds, Non-game birds, Song birds, Cattle, Milk, Annelids, Silage, Insects, Deer, Isopids, Oligochaetes, Insect control, Grasses, Freshwater fish.
Identifiers: *Mirex, Electron capture gas chromatography, Thin layer chromatography, Infrared spectroscopy, Biological samples, Sample preparation, Detection limits, Chemical recovery, Metabolites, Adipose tissue, Brain, Liver, Arthropods, Crickets, Spiders, Blue heron, Cattle egret, Ictalurus punctatus, Lepomis cyanellus, Solenopsis saevissima, Fire ant, Bettles, Grasshoppers, Arachnids, Gas liquid chromatography, Festuca, Odocoileus virginianus, Gallus gallus, Colinus virginianus, Toxostoma rufum, Cyanocitta cristata, Sturnella magna, Meleagris gallopavo, Tyrannus tyrannus, Turdus migratorius, Strix varia, Bubulcus ibis, Florida caerulea, Lumbricus terrestris.

Samples of wildlife and a few miscellaneous samples, such as beef, were collected in Mississippi in 1970 and analyzed for the presence of mirex and DDT and its analogs. Analytical procedures chosen for extraction and cleanup for residue analysis involved methods described in the Pesticide Analytical Manual, Volumes I and III, and other previously described methods. Primary identification and quantification were accomplished by electron capture gas-liquid chromatography in the determinative step. Thin-layer chromatography and infrared spectroscopy were utilized for confirmation in selected samples. Levels of mirex residues were found to range from O ppm to a high of about 104 ppm; residues of DDT and its metabolites (DDTR equals DDT plus DDE plusDDD) were found to range from less than 0.001 ppm to 126 ppm. Comparisons of the amount of the two pesticides found in individual samples showed that mirex residues often exceeded DDT residues. The high levels of mirex residues found in some of these samples 1 year after treatment indicate that mirex can be considered a persistent pesticide. (Mackan-Battelle)
W73-08267

OIL SPILL SOURCE IDENTIFICATION,
Esso Research and Engineering Co., Florham Park, N.J. Government Research Div.
M. Lieberman.
Copy available from GPO Sup Doc as EP1.23/2:73-102, $2.35; microfiche from NTIS as PB-219 822, $1.45. Environmental Protection Agency, Technology Series Report EPA-R2-73-102, February 1973. 175 p, 28 fig, 19 tab, 7 ref, 6 append. EPA Project 15080 HDL, 68-01-0058.

Descriptors: *Oil spills, *Pollutant identification, *Chemical analysis, *Correlation analysis, *Weathering, Gas chromatography, Mass spectrometry, Tagging, Water pollution sources.
Identifiers: Passive tagging, n-Paraffins, Polynuclear aromatics, Naphthenes, Nickel, Vanadium, Nitrogen, Sulfur.

Five different crude oils, two residual fuel oils (No. 4 and No. 5) and one distillate fuel oil (No. 2) were subjected to simulated weathering in the laboratory. Samples were weathered for 10 and 21 days at 55 and 80F, under high and low salt water washing rates. 'Weathered' and 'unweathered' oil samples were analyzed by low voltage mass spectroscopy (polynuclear aromatics), high voltage mass spectroscopy (naphthenes), gas chromatograph (n-paraffins), emission spectroscopy (nickel/vandium), X-ray total sulfur and Kjeldahl total nitrogen techniques. Several compound indices were adequately stable toward simulated weathering to discriminate between like and unlike pairs of oils. Discriminant function analysis was used to select the best compound indices for the oils used. Using these indices, weathered and unweathered samples were correctly paired with high statistical confidence. (EPA)
W73-08289

THE APPEARANCE AND VISIBILITY OF THIN OIL FILMS ON WATER,
Edison Water Quality Research Lab., N.J.
B. Hornstein.
Copy available from GPO Sup Doc as EP1.23/2:72-039, $2.50; microfiche from NTIS as PB-219 825, $1.45. Environmental Protection Agency, Technology Series Report EPA-R2-72-039, August 1972. 95 p, 31 fig, 5 tab, 8 ref.

Descriptors: *Oil-water interfaces, *Thin films, *Oil wastes, *Color, Theoretical analysis, Laboratory tests, On-site investigations, Water pollution, Oil pollution, Water pollution sources, *Pollutant identification.
Identifiers: *Visibility, Iridescent films, Optical interference, Reflectivity.

Oil films of controlled thickness up to 3000 nanometers, upon water surfaces in the laboratory, confirm an inherent and orderly thickness-appearance relationship which is independent of oil type and water type. These laboratory studies also investigated the effects of viewing conditions upon the ease of visibility of these thin films. Out-of-doors observations were made; these and the observations reported by other sources were found to correspond with the laboratory results. The visibility of a thin oil film depends not only upon its thickness-dependent inherent appearance, but also upon conditions external to the film. These include nature of illumination and sky conditions, sun angle, color and depth of water, color of bottom, and viewing angle. Color photographs are included for illustration of the points discussed. (EPA)
W73-08295

MICROFLORA OF THE NEUSTON AND ITS ROLE IN WATER BODIES,
Polskie Towarzystwo Przyrodnikow im. Kopernika, Warsaw.
For primary bibliographic entry see Field 05C.
W73-08335

SULFIDE DETERMINATION IN SUBMERGED SOILS WITH AN ION-SELECTIVE ELECTRODE,
Ministry of Agriculture, Cairo (Egypt).
For primary bibliographic entry see Field 02K.
W73-08351

APPLICATION OF REMOTE SENSING TO SOLUTION OF ECOLOGICAL PROBLEMS,
IBM Federal Systems Div., Bethesda, Md.
For primary bibliographic entry see Field 07B.
W73-08358

INTERDISCIPLINARY APPLICATIONS AND INTERPRETATIONS OF REMOTELY SENSED DATA,
Pennsylvania State Univ., University Park.
For primary bibliographic entry see Field 07B.
W73-08363

USING LIGHT TO COLLECT AND SEPARATE ZOOPLANKTON,
Michigan State Univ., East Lansing. Dept. of Fisheries and Wildlife.

For primary bibliographic entry see Field 07B.
W73-08406

CHEMICAL ANALYSES OF SELECTED PUBLIC DRINKING WATER SUPPLIES (INCLUDING TRACE METALS),
Wisconsin Dept. of Natural Resources, Madison.
R. Baumeister.
Wis Dep Nat Resour Tech Bull. 53 p 1-16. 1972. Illus.
Identifiers: *Chemical analysis, Copper, Lead, Metals, *Trace elements, Water supply, Zinc, *Potable water.

Drinking water supplies utilizing ground and surface water sources were sampled for trace elements in addition to the standard chemical analysis. None of the raw water samples exceeded the Public Health Service Drinking Water Standards for chemical quality, and 1 sample from a distribution system exceeded the standards. The 1 parameter exceeded was Pb (.06 mg/l reported, .01 mg/l higher than the standard) which leached from a service line because of corrosive water in the distribution system. Corrosive water in other systems also caused increased concentrations of Cu and Zn.--Copyright 1973, Biological Abstracts, Inc.
W73-08424

THE EXPENDABLE BATHYOXYMETER,
Oregon State Univ., Corvallis. Dept. of Oceanography.
For primary bibliographic entry see Field 07B.
W73-08438

ACETONE CONTAMINATION OF WATER SAMPLES DURING QUICKFREEZING,
Maryland Univ., Solomons. Natural Resources Inst.
C. W. Keefe, D. H. Hamilton, and D. A. Flemer.
Chesapeake Sci. Vol 13, No 3, p 226-229. 1972.
Identifiers: *Acetone contamination, *Freezing, Quick-freezing, Samples, *Pollutant identification.

Water samples for nutrient concentration analysis 'quick-frozen' in an acetone-dry ice bath may become contaminated with acetone, even though tightly sealed in screw-cap polyethylene bottles. Acetone interferes with oxidation steps in ammonia and total P analyses, as well as the oxidative determination of dissolved organic C. Samples taken for such purposes should not be quick-frozen in an acetone-dry ice bath. Alternatives are discussed.--Copyright 1973, Biological Abstracts, Inc.
W73-08442

5B. Sources of Pollution

FATE OF TRACE-METALS (IMPURITIES) IN SUBSOILS AS RELATED TO THE QUALITY OF GROUND WATER,
Tuskegee Inst., Ala. Carver Research Foundation.
W. E. Nelson.
Available from the National Technical Information Service as PB-219 401, $3.00 in paper copy, $1.45 in microfiche. Tuskegee Institute School of Applied Sciences report, September 1972. 163 p, 77 fig, 66 tab, 48 ref, 1 append. OWRR B-028-ALA (3). 14-01-0001-3053.

Descriptors: *Heavy metals, *Path of pollutants, Soil contamination, Animal wastes, Cation exchange, Chelation, *Soil contamination effects, Chemical precipitation, Diffusion, Groundwater, Adsorption, Manganese, Runoff, Organic matter, Calcium, Potassium, Cobalt, Sodium, Chromium, Lead, Strontium, Magnesium, Water quality.

Laboratory and field studies were conducted to evaluate metal adsorptive capacities of six soils. These and stability constants of soil-metal complexes were used to understand the ability of the soils to complex trace metals, and to explain the leachability or retention of toxic metals by certain soils. Maximum adsorption capacities, relative amounts of metal complexed with one mole of soil, and stability constants (log K values) were calculated from adsorption data of cations retained by soils. The soil-metal adsorption patterns obtained were similar to the Langmuir-Fruendlich adsorption isotherms. Adsorption studies indicated that soil complexes metals and renders them insoluble regardless of genesis, organie matter content, and their physicochemical properties. Readiness to complex, however, depends on these properties and the nature of the metal rendered insoluble. Multivalent and divalent and the more electronegative cations make relatively more stable complexes, particularly with soils high in organic matter. Mechanisms mainly responsible for rendering the metals insoluble were considered to be chelation, surface adsorption, precipitation, and diffusion and physical entrapment. Spectral studies were used to investigate the structure of metal ligands and whether complex formation had occurred nondestructively. The laboratory models indicated that the soil has maximum capacity to adsorb metals. However, under field conditions, there was no indication that the maximum capacity was reached. (Warman-Alabama)
W73-07802

ORGANIC WASTES AS A MEANS OF ACCELERATING RECOVERY OF ACID STRIP-MINE LAKES,
Missouri Univ., Columbia. Dept. of Civil Engineering.
For primary bibliographic entry see Field 05G.
W73-07808

INDUSTRIAL WASTELINE STUDY - A SYSTEM FOR CONTROLLED OCEAN DISPOSAL,
Franklin Inst., Philadelphia, Pa. Labs. for Research and Development.
For primary bibliographic entry see Field 05E.
W73-07812

THE CONTEMPORARY STATE OF DECONTAMINATING WASTE WATER,
For primary bibliographic entry see Field 05D.
W73-07835

PH BUFFERING OF PORE WATER OF RECENT ANOXIC MARINE SEDIMENTS,
California Univ., Los Angeles. Dept. of Geology.
S. Ben-Yaakov.
Limnology and Oceanography, Vol 18, No 1, p 86-94, January 1973. 3 fig, 2 tab, 20 ref.

Descriptors: *Hydrogen ion concentration, *Pore water, *Connate water, *Bottom sediments, *Oxidation-reduction potential, Water chemistry, Biodegradation, Sulfates, Sulfur bacteria, Sulfides, Carbonates, Sea water.
Identifiers: *Buffering (pH).

A model is proposed to explain the relative pH stability in pore water of recent anoxic marine sediments. The model assumes that the pH of the pore waters is controlled by the byproducts of organic decomposition, sulfate reduction, and precipitation of sulfide and carbonate. The model predicts that the pH of pore waters should remain in the range 6.9 to 8.3, which is in agreement with measured values. (Knapp-USGS)
W73-07851

PREDICTING TIME-OF-TRAVEL IN STREAM SYSTEMS,
Illinois State Water Survey, Urbana.
For primary bibliographic entry see Field 02J.
W73-07879

EFFECT OF LONGITUDINAL DISPERSION ON DYNAMIC WATER QUALITY RESPONSE OF STREAMS AND RIVERS,
Manhattan Coll., Bronx, N.Y. Environmental Engineering and Science Program.
R. V. Thomann.
Water Resources Research, Vol 9, No 2, p 355-366, April 1973. 11 fig, 4 tab, 8 ref.

Descriptors: *Dispersion, *Path of pollutants, Mixing, Advection, Mathematical models, Numerical analysis, Kinetics, Water pollution, Chemical degradation.

The amplitude and phase characteristics of a river subjected to a time variable waste input are computed for two cases: zero dispersion with no mixing and dispersion levels representative of streams and large rivers. The frequency response depends on a set of dimensionless numbers that characterize the reactive and dispersive nature of the river. For upland streams and rivers small amounts of dispersion may be important when the input varies with periods of about 7 days or less. For large deep rivers the effect of dispersion can generally not be neglected when the input is time variable. Longitudinal dispersion in water quality response in streams and rivers is analyzed based on frequency response. The analysis provides some guidelines for deciding whether a no-mixing plug flow model is suitable when the problem context involves a time-varying waste input. (Knapp-USGS)
W73-07890

BOTTOM CURRENTS IN THE HUDSON CANYON,
National Oceanic and Atmospheric Administration, Miami, Fla. Atlantic Oceanographic and Meteorological Labs.
For primary bibliographic entry see Field 02E.
W73-07903

MOVEMENT OF CHEMICALS IN SOILS BY WATER,
Illinois Univ., Urbana. Dept. of Agronomy.
For primary bibliographic entry see Field 02G.
W73-07904

MODELING THE MOVEMENT OF CHEMICALS IN SOILS BY WATER,
Illinois Univ., Urbana. Dept. of Agronomy.
For primary bibliographic entry see Field 02G.
W73-07905

A REGIONAL PLANNING MODEL FOR WATER QUALITY CONTROL,
Virginia Polytechnic Inst. and State Univ., Blacksburg.
D. E. Pingry, and A. B. Whinston.
(1972). 46 p, 2 fig, 7 tab, 71 equ, 1 map, 22 ref. OWRR-B-020-IND (15). Supported by Army Research Office.

Descriptors: Water quality control, *Water temperature, *Dissolved oxygen, *Water pollution treatment, *Costs, *Simulation analysis, *Optimization, Resource allocation, Comprehensive planning, River basins, Regional analysis, Effluents, Economic efficiency, Alternative water use, Biochemical oxygen demand, Thermal pollution, Waste water (Pollution), Cooling towers, Flow augmentation, Constraints, Computer programs, *Indiana.
Identifiers: *West Fork White River (Ind.), External effects, Regional planning, Opportunity costs, By-pass piping, Treatment plants, Nonlinear programming.

The information problem when dealing with external effects of water pollution is immense. Usually the external effect (cost or benefit) is transferred to the affected party through a complex series of physical, chemical or biological processes; an in-

Identifiers: *Czechoslovakia.

A system of drainage of liquid wastes and a method of liquidation of radioactive wastes in the Nuclear Research Institute are briefly described. The radioactivity of waste waters discharged into the river is controlled discontinuously. The gross beta-gamma activity is determined in samples. The control also includes a continuous measurement of the activity of chemical wastes discharged separately. Both methods are adequate in view of the isotopes that might be present in a given mixture. In the period of 1963 to 1971, more than 2000 samples were measured. The values obtained were very low, which shows a good function of the liquidation station. No effect on the activity of the river Vltava, into which the wastes were discharged, was found. Therefore, a measurable increase in the exposition by ionizing radiation could not occur in the inhabitants using the Vltava water. The continuous measurement showed several times an accidental uncontrolled discharge of the activity into the chemical drainage and into the river. However, it was found that the maximum permissible concentrations for drinking water could not be exceeded in the river Vltava, which was also proved. (Houser-ORNL)
W73-07926

INTERACTION OF YELLOW ORGANIC ACIDS WITH CALCIUM CARBONATE IN FRESHWATER,
Michigan State Univ., Hickory Corners. W. K. Kellogg Biological Station.
A. Otsuki, and R. G. Wetzel.
Available from NTIS, Springfield, Va., as COO-1599-63; $3.00 in paper copy, $1.45 in microfiche. Report COO-1599-63, 1972. 8 p, 2 fig, 11 ref. AEC-AT (11-1)-1599.

Descriptors: *Path of pollutants, *Humic acids, *Sorption, *Calcium carbonate, Chemical precipitation, Trace metals, Iron, Carbon cycle, Lakes, Hardness (Water), Carbon radioisotopes, Tracers, Radioactivity techniques, Hydrogen ion concentration, Balance of nature.

In connection with determination of the organic carbon budget of a small hardwater lake, experiments were conducted to determine yellow-acid removal during calcium carbonate precipitation. (Yellow acids are a major part of the soluble humic substances in natural waters.) The results indicated substantial removal of yellow acids and probably of complexed trace metals such as iron by this pathway. It appeared likely that yellow acids were incorporated into crystals both by surface absorption and into crystal nuclei during rapid growth. (Bopp-ORNL)
W73-07931

ENVIRONMENTAL RADIOACTIVITY, ISPRA, 1970,
European Atomic Energy Community, Ispra (Italy). Joint Nuclear Research Center.
M. de Bortoli, and P. Gaglione.
Available from NTIS, Springfield, Va., as EUR-4805e; $3.00 in paper copy, $1.45 in microfiche. Report EUR-4805e, 1972. 46 p, 8 fig, 23 tab, 2 ref.

Descriptors: *Fallout, *Nuclear wastes, *Europe, *Food chains, Public health, Path of pollutants, Rivers, Sediments, Lakes, Milk, Freshwater fish, Potable water, Air pollution, Rainfall, Strontium radioisotopes, Absorption.
Identifiers: Cesium radioisotopes.

The abnormally low rainfall for 1970 as compared with 1969 may account for lower deposition by fallout although the radioactivity of the air was higher. In accord with the greater discharge from the Ispra center, Cs137 increased in downstream sediments. In the water of four lakes, Cs137 increased, but Sr90 decreased. In lake fish and in offsite vegetation, Sr90 and Cs137 were the same as in 1969, but were about 10% lower in milk. (Bopp-ORNL)
W73-07933

A WATER-QUALITY SIMULATION MODEL FOR WELL MIXED ESTUARIES AND COASTAL SEAS: VOL. II, COMPUTATION PROCEDURES,
New York City-Rand Inst., N.Y.
J. J. Leendertse, and E. C. Gritton.
Report R-708-NYC, July 1971. 53 p, 13 fig, 8 ref.

Descriptors: *Mathematical models, *Water quality, *Estuarine environment, *New York, Evaluation, Assessment, Discharge (Water), Effluents, Water quality control, Systems analysis, Research and development, Biological properties, Biochemistry, Water properties, Computer programs, Mixing, Shallow water.
Identifiers: *Jamaica Bay (Long Island).

This and companion volumes describe methods used and difficulties encountered in designing a simulation model to study the effects of combined sewage overflows and other discharges on Jamaica Bay, Long Island. Basics are reported in Volume 1; preliminary results in Volume 3. This report (Volume 2) describes the design philosophy used in further development, including the addition of a biochemical and biological reaction model. Tidal flow and dispersion, and transport of constituents, are approximated by a system of partial differential equations. The model simulates changes in boundaries that occur in shallow areas of estuaries as a result of the changing tide level. Procedures used to make time-dependent boundary changes have been revised and are outlined. Examples are given of input and output from a computer program. (Bopp-ORNL)
W73-07935

ENVIRONMENTAL MONITOR NG REPORT FOR THE NEVADA TEST SITE JANUARY-DECEMBER 1971.
National Environmental Research Center, Las Vegas, Nev.

Available from NTIS, Springfield, Va., as NERC-LV-539-1, $3.00 per copy, $1.45 microfiche. Report No. NERC-LV-539-1, Sept. 1972. 91 p, 7 fig, 7 tab.

Descriptors: *Monitoring, *Surveys, *Radioactivity, *Testing, *Assay, *Evaluation, *Air pollution, *Water pollution, Soil contamination, Water pollution sources, Path of pollutants, Public health, Milk, Data collections, Analytical techniques, Sampling, Regulation.
Identifiers: *Nevada Test Site.

Surveillance of the Nevada Test Site environs during 1971 showed that the concentrations of radioactivity and levels of radiation in the environment were within the Radiation Protection Standards of the Atomic Energy Commission. The surveillance data show that most of the environmental radioactivity in the NTS environs was due to naturally occurring radionuclides and world-wide fallout. Increases in gross beta concentrations in air and increases in Sr89, Sr90, and Cs137 in milk during the late spring and early summer were attributed to the seasonal trend of world-wide fallout. Increases in the gross beta concentrations and measurements of fresh fission products in the air during November at many of the Air Surveillance Network Stations were attributed to the nuclear detonation on November 18, 1971, by the People's Republic of China. (Houser-ORNL)
W73-07936

STUDIES ON THE RADIOACTIVE CONTAMINATION OF THE SEA, ANNUAL REPORT, 1971.
Comitato Nazionale per l'Energia Nucleare, La Spezia (Italy). Laboratorio per lo Studio della Contaminazione Radioattiva de Mare.

Available from NTIS, Springfield, Va., as EUR-4865e; $3.00 per copy; $1.45 microfiche. Report EUR-4856e, Aug. 1972. 162 p, 47 fig, 28 tab, 59 ref, 3 annex.

Descriptors: *Environment, *Fallout, *Water pollution, *Water pollution sources, *Ecology, *Ecosystems, Radioecology, Fish, Microorganism, Food chains, Gastropods, Sea water, Algae, Sampling, Analytical techniques, Radioisotopes, Strontium, Cesium, Tritium, Iodine, Assay, Public health.
Identifiers: *Mediterranean Sea.

The eighth Annual Report of the CNEN-EURATOM Contract of Association is presented. The program laid down in this contract calls for studies of the factors which influence the uptake, accumulation and loss of radioisotopes by different inorganic and organic constituents of the marine environment. The program is divided into two parts: (a) the investigation of relevant radioecological and ecological factors in nature and under laboratory conditions; (b) the investigation of the outfall area off-shore of the CNEN-TRISAIA Centre in the Gulf of Taranto (fuel reprocessing plant). The task of carrying out this program has been divided into six groups: Chemistry, Botany, Zooplankton, Fisheries Biology, Microbiology and Special Developments. Results obtained in 1971 are presented. (Houser-ORNL)
W73-07942

RESEARCH AND DEVELOPMENT IN PROGRESS, BIOMEDICAL AND ENVIRONMENTAL RESEARCH PROGRAM.
Division of Biomedical and Environmental Research (AEC), Washington, D.C.

Available from NTIS, Springfield, Va., as TID-4060, Second Edition; $6.00 paper copy, $1.45 microfiche. Report TID-4060, Second Edition, December 1972. 364 p.

Descriptors: *Research and development, *Projects, *Radioecology, *Radioactivity, Radioactivity techniques, Radioactivity effects, Nuclear wastes, Radioisotopes, Absorption, Marine biology, Estuarine environment, Freshwater, Soil-water-plant relationships, Analytical techniques, Sedimentation, Oceanography, Environmental effeets, Path of pollutants.

Summaries of 1473 projects are presented in the form furnished by the investigators to the Science Information Exchange of the Smithsonian Institution. In addition to listing by category, indexing is by contractor, investigator, and subject. The categories dealing with radionuclides in the environment include (numbers of summaries in parentheses): terrestrial systems (70), soils and soil-plant relations (56), freshwater systems (45), transport in soil, food and man (17), analytical procedures (8), radionuclide uptake in marine systems (23), sedimentation and chemical interactions (17), circulation and mixing (11), and other oceanographic studies (8). (Bopp-ORNL)
W73-07943

CONTINUOUS MEASUREMENT OF ALPHA AND BETA RADIOACTIVITY OF WATER IN VIEW OF MAXIMUM PERMISSIBLE CONCENTRATIONS FOR THE PUBLIC,
Commissariat a l'Energie Atomique, Saclay (France).
For primary bibliographic entry see Field 05A.
W73-07945

RADIONUCLIDE CYCLING IN TERRESTRIAL ENVIRONMENTS,
Oak Ridge National Lab., Tenn.
R. C. Dahlman, E. A. Bondietti, and F. S. Brinkley.
Available from NTIS, Springfield, Va., as ORNL-4848; $3.00 paper copy; $1.45 microfiche. In: Environmental Sciences Division Annual Progress Report for Period Ending September 30, 1972, ORNL-4848, February 1973, p 1-8, 3 fig, 1 tab, 4 ref.

Descriptors: *Nuclear wastes, *Radioecology, *Soil-water-plant relationships, *Soil microorganisms, Soil fungi, Soil bacteria, Analytical techniques, Cadmium, Fallout, Zinc, Leaching, Litter, Sulfur, Soil contamination, Radiochemical analysis, Plant physiology, Root zone, Leaves, Foliar application, Food chains, Path of pollutants, Insects, Birds, Forests, Trace elements, Cycling nutrients, X-ray diffraction, Waste water (Pollution), Algae, Cobalt radioisotopes.

Co60 uptake by millet plants was several orders of magnitude less from soil contaminated for 25 years as compared to soil contaminated about 6 months. By density-gradient-centrifugation analysis, a large part of the Co60 (as well as Ru106 and Sb125) was shown to be in roots, 33% through rainout and leaf drop. Soil microbial activity decreased mobility of Zn, Cd and S; but not that of N. By microbial action, leaching of cadmium nitrate from soil was increased; leaching from sand was decreased. Algae and detritus absorbed cadmium oxide more readily than cadmium nitrate. Radioecological studies were made of Cd turnover in arthropods and in birds. Effects of sodium on clay aggregation were studied by x-ray diffraction. Pu was determined in soil by a rapid amine-chelation method. The adenosine triphosphate in fungal mass was correlated with other growth-rate parameters. (Bopp-ORNL)
W73-07946

APPLIED AQUATIC STUDIES,
Oak Ridge National Lab., Tenn.
B. G. Blaylock, C. P. Allen, and M. Frank.
Available from NTIS, Springfield, Va., as ORNL-4848; $3.00 paper copy, $1.45 microfiche. In: Environmental Sciences Division Annual Progress Report for Period Ending September 30, 1972, ORNL-4848, February 1973, p 79-85, 2 fig, 2 tab, 11 ref.

Descriptors: *Path of pollutants, *Radioactivity, *Radioisotopes, *Nuclear wastes, Sediments, Absorption, Freshwater fish, Tritium, Radioactivity effects, Water pollution effects, Fish eggs, Chromium, Toxicity, Hatching, Chlorination, Water treatment, Chromosomes, Larvae, Radioecology, Water pollution sources, Sewage treatment, Waste water (Pollution), Soil contamination, Cobalt radioisotopes, Catfishes, Potable water.
Identifiers: Cesium radioisotopes.

The maximum gamma activity (62-86% Cs137) is now found at a depth of 16-34 cm in the sediment of a holding pond as a result of a decrease in the radioactivity of waste-water discharges. Biological half-lives for elimination of tritium from fish kept in the pond for 36 days were: for tritium in body water, 0.2 hr and 0.9 hr; for tritium in tissue, 8.7 days. Natural populations of insect larvae (C. tentans) in the pond were unchanged in chromosomal ploymorphism (1960-1973). After exposure to 1500 rads of Co60 irradiation, some catfish survived for 98 days at 15 C, for 32 days at 25 C; none for 8 days at 30 C. Contamination of water with 1-10 ppm Cr greatly increased the time to hatching of fish eggs. An attempt is being made to study the production of chlorinated organics during purification of natural waters by a Cl-36 tracer-gas chromatographic method which was successful in a study of the chlorination of primary sewage effluent. (Bopp-ORNL)
W73-07948

HYDROLOGIC INVESTIGATIONS AND THE SIGNIFICANCE OF U-234/U-238 DIS-

itation (Atmospheric), Vegetation, Potable Environment, Potable water, Food chains, Toxicity, Public health, Human pathology,

fiers: *Greenland.

irments of fall-out radioactivity in Greenn 1971 are reported. Sr-90 (and Cs-137 in instances) was determined in samples of itation, sea water, vegetation, animals, and ng water. Estimates of the mean contents of and Cs-137 in the human diet in Greenland in re given. Also reported is the mean level of ium in human bone in Greenland. (Houser-L.)
7953

K RIVER NUCLEAR LABORATORIES RESS REPORT APRIL 1 TO JUNE 30, BIOLOGY AND HEALTH PHYSICS DIVI- ENVIRONMENTAL RESEARCH CH AND HEALTH PHYSICS BRANCH,
ic Energy of Canada Ltd., Chalk River (On- Chalk River Nuclear Labs.
imary bibliographic entry see Field 05C.
7955

MBIA RIVER EFFECTS IN THE HEAST PACIFIC: PHYSICAL STUDIES. RT OF PROGRESS, JUNE 1971 UGH JUNE 1972.
ington Univ., Seattle. Dept. of Oceanog-
imary bibliographic entry see Field 02J.
7956

ONUCLIDE DISTRIBUTION IN SOIL TLE OF THE LITHOSPHERE AS A CON-ENCE OF WASTE DISPOSAL AT THE ONAL REACTOR TESTING STATION.
Operations Office (AEC), Idaho Falls.

able from NTIS, Springfield, Va., as IDO-, $3.00 paper copy, $1.45 microfiche. Report DO-10049, Oct. 1972. 80 p, 18 fig, 27 tab, 34 append.

iptors: *Radioactive waste disposal, *Water ion, *Water pollution sources, *Soil conation, *Assay, Cobalt, Strontium, Plutoni-Lake bed, Ponds, Leaching, Migration, ment, Distribution patterns.

istribution of radionuclides in the soil resultom disposal of liquid and solid waste at the nal Reactor Testing Station (NRTS) was stu-The situation involving liquid waste is apd by: (a) mathematical models using parameletermined in the laboratory; (b) physical s using soil samples and simulated waste on; and (c) sampling of soil and water where is being discharged. In case of strontium-90, ta indicate that the sorption capacity of the al deposits between the disposal pond botand the basalt bedrock has been reached. The for cesium were erratic but the most pesic interpretation indicates that the capacity alluvium has not been reached. The amount balt-60 estimated to be in the alluvium is r than that reported to have been discharged. ischarge estimates do not take into account ore frequent usage of unsheathed cobalt for neutron flux measurements during the history of the reactors. Cobalt-60 retention in il is attributed to reactions other than ionnge. A hypothetical evaluation is included of vironmental problems remaining after active tion of a disposal site has been discontinued. oncluded that disposal sites can be used for purposes with certain restrictions. (Houser-L)
7958

DETERMINATION OF TRACE METALS AND FLUORIDE IN MINERALOGICAL AND BIOLOGICAL SAMPLES FROM THE MARINE ENVIRONMENT,
Naval Research Lab., Washington, D.C.
For primary bibliographic entry see Field 05A.
W73-07959

RADIOACTIVITY AND WATER SUPPLIES,
Interuniversitair Reactor Institutt, Delft (Netherlands).
For primary bibliographic entry see Field 05D.
W73-07960

EVENT-RELATED HYDROLOGY AND RADIONUCLIDE TRANSPORT AT THE CANNIKIN SITE, AMCHITKA ISLAND, ALASKA,
Nevada Univ., Reno. Desert Research Inst.
P. R. Fenske.
Available from NTIS, Springfield, Va., as NVO-1253-1; $3.00 paper copy, $1.45 microfiche. Report NVO-1253-1, 1972. 37 p, 17 fig, 21 ref.

Descriptors: *Nuclear wastes, *Nuclear explosions, *Tritium, *Alaska, Path of pollutants, Fracture permeability, Permeability, Oceans, Saline water-freshwater interfaces, Water properties, Retention, Retardance, Conductivity, Fluctuations, Radioisotopes, Absorption, Groundwater movement, Darcys Law, Ion uptake, Soil contamination effects, Analog models, Aquifer characteristics.

Evaluation of possible ground-water contamination from the Cannikin site involved consideration of thermal effects in the rubble chimney, groundwater transit time along probable flow paths, and the rate of dilution of fresh-water seepage with sea water. Measurements of temperature and pressure in test holes and water analyses were used to locate the freshwater-saltwater interface. The hydraulic conductivity was estimated from the response in test holes to barometric and tidal fluctuations. Electrical analogue simulation of the hydraulic conductivity of fracture networks gave a residence time of 14,000 years for the underlying saline aquifer. The transit time through the freshwater lens to the sea is about 3000-4000 years - such that tritium is eliminated by radioactive delay. (Bopp-ORNL)
W73-07961

TERRESTRIAL AND FRESHWATER RADIOECOLOGY, A SELECTED BIBLIOGRAPHY, SUPPLEMENT 8,
Washington State Univ., Pullman. Dept. of Zoology.
For primary bibliographic entry see Field 05C.
W73-07962

CONTRIBUTION TO THE STUDY OF THE MIGRATION OF RU106 IN SOILS,
Commissariat a l'Energie Atomique, Cadarache (France). Centre d'Etudes Nucleaires.
J. P. Amy.
Available from NTIS, Springfield, Va., as RFP-Trans-113; $3.00 in paper copy, $1.45 cents in microfiche. Report RFP-Trans-113, 1972. 4 p.

Descriptors: *Nuclear wastes, *Irrigation water, *Soil contamination, *Radioisotopes, Sorption, Permeability, Soil types, Soil chemical properties, Hydrogen ion concentration, Physicochemical properties, Hydrolysis, Ion transport, Ion exchange, Leaching, Movement, Absorption, Path of pollutants.

Ru106 mobility was studied in agricultural soils irrigated with Rhine water at one time contaminated by nuclear wastes. Its many chemical forms make the behavior of Ru complex. Nitrodinitrato complexes which exhibit mainly cationic characteristics resemble Cs, but an anionic component which increases with increasing acidity is less readily adsorbed. Minimum mobility is found for soil that is calcareous, clayey, of low permeability and rich in organic matter. Artifical augmentation of the alkalinity of an acid soil augments sorption. (Bopp-ORNL)
W73-07963

DISTRIBUTION OF RADIONUCLIDES IN ORGANISMS OF MARINE FAUNA. APPLICATION OF CRITICAL CONCENTRATION FACTORS,
A. Ya. Zesenko.
Available from NTIS, Springfield, Va., as Part of AEC-tr-7299, $6.00 in paper copy, $1.45 microfiche. In: Marine Radioecology, p 105-146, Trans. from Morskaya Radioekologiya, 1970.

Descriptors: *Bioindicators, *Fallout, *Nuclear wastes, *Marine animals, Path of pollutants, Estuarine environment, Animal metabolism, Radioecology, Salt water, Oceans, Mollusks, Marine fish, Crabs, Radioisotopes, Absorption, Crustaceans, Trace elements, Tracers.

A series of studies is reviewed on uptake of radionuclides (Ag110, Zn65, Y91, Ce144, Zr95, Nb95, P32, S31, Ru106, Cu106) by organs of marine organisms (fish, mussels, grass crabs). Gills, byssus, shell, and chitinous skin (crustaceans) are indicators of uptake of certain radionuclides (Ag110, Cs137, Zn65, Sr90, Y91, Ce144, Zr95, Nb95, Ru106). Filtration by marine mollusks is of slight effectiveness for uptake of Y91, Ce144, Zr95, Nb95, or Ru106. (Bopp-ORNL)
W73-07966

ACCUMULATION OF RADIONUCLIDES BY ROE AND LARVAE OF BLACK SEA FISH,
V. N. Ivanov.
Available from NTIS, Springfield, Va., as Part of AEC-tr-7299; $6.00 in paper copy, $1.45 in microfiche. In: Marine Radioecology, Translation from Marskaya Radioekologiya, 1970, p 147-157, 3 fig, 1 tab, 10 ref.

Descriptors: *Radioisotopes, *Absorption, *Larval growth stage, *Marine fish, Food chains, Public health, Path of pollutants, Laboratory tests, Nuclear wastes, Fallout, Hatching, Water pollution effects, Strontium radioisotopes.
Identifiers: Zirconium radioisotopes, Cesium radioisotopes, Yttrium radioisotopes, Ruthenium radioisotopes.

Radionuclide concentration factor ranges for several species (1A) for newly hatched larvae and (b) for eggs before hatching are - Zr: a, 34-43; b, 14-35. Cs: a, 10; b, 9. Sr: a, 1.3-1.7; b, 0.9-4.4. Y: a, 0.5-11; b, 57-233. Ce: a, 1-4; b, 22-495. Ru: a, 0.5-3.6; b, 12-21. More-limited data are given for C, P, S, Mn, Fe, Co, and W. Concentration factors for older larvae were several-fold higher than for newly hatched larvae with Zr, Cs, Sr, Ce, Y, and Ru. Absorption on the exterior of the egg membrane is believed to account for the relatively high uptake of Y and Ce. (Bopp-ORNL)
W73-07967

EFFECT OF INCORPORATED RADIONUCLIDES ON CHROMOSOME APPARATUS OF OCENA FISH,
V. G. Tsytsugina.
Available from NTIS, Springfield, Va., as AEC-tr-7299; $6.00 in paper copy, $1.45 microfiche. In: Marine Radioecology, p 157-165. Translation from Morskaya Radioekologiya, 1970, 1 fig, 4 tab, 18 ref.

Descriptors: *Nuclear wastes, *Marine fish, *Larval growth stage, *Chromosomes, Cytological studies, Water pollution effects, Radioactivity effects, Fallout, Hatching, Radioisotopes, Absorption, Strontium radioisotopes, Carbon radioisotopes.
Identifiers: Yttrium radioisotopes.

Ruff and flounder embryo from eggs reared in sea water containing above 10 nanoCuries/ liter Sr89 and 180 nanoCuries/liter of C14 showed increased chromosome breaks relative to controls. That increasing the radioactivity several orders of magnitude produced little or no additional effect is explained by the relatively high sensitivity of the initial growth stage, by effects from nonviable aberrations, and perhaps by acceleration of mutation processes in exposed cells. Increased chromosome breaks in 1-day-old flounder larvae resulted from hatching in seawater containing 0.3 nanoCurie/liter Y91. (Bopp-ORNL)
W73-07968

RADIOECOLOGICAL STUDIES ON THE DANUBE RIVER AND ADJOINING PART OF THE BLACK SEA,
V. I. Timoshchuk, and I. A. Sokolova.
Available from NTIS, Springfield, Va., as AEC-tr-7299; $6.00 paper copy, $1.45 microfiche. In: Marine Radioecology, p 174-185, Translation from Morskaya Radioekologiya, 1970, 5 fig, 1 tab, 16 ref.

Descriptors: *Estuarine environment, *Marine fish, *Strontium radioisotopes, *Absorption, Path of pollutants, Fallout, Nuclear wastes, Radioecology, Oceans, Freshwater fish, Mussels, Fish migration, Geomorphology, Climatology, Europe, On-site data collections, Food chains, Sea water, Water analysis.

The geomorphology and climatology are reviewed of the Danube River, its coastal regions, and the northwestern part of the Black Sea. Sr90 concentration (in units of picoCuries/liter) was measured in April 1966, 1967, at the delta head (2.40, 1.75 units) and 50 miles offshore, where the value (0.6-0.8 unit) is typical of the Black Sea. Concentration factors for Sr90 uptake were: Danube fish (carp, rudd), 43-102; semimigratory fish (pike-perch, common carp), 12-26; migratory fish and herring, 7; mussels, including the shell, 50-617. (Bopp-ORNL)
W73-07969

SR IN WATER OF THE CASPIAN AND AZOV SEAS,
V. I. Timoshchuk.
Available from NTIS, Springfield, Va., as Part of AEC-tr-7299; $3.00 in paper copy; $1.45 in microfiche. In: Marine Radioecology, Translation of Morskaya Radioekologiya, 1970, p 185-196. 8 fig, 2 tab, 10 ref.

Descriptors: *Strontium radioisotopes, *Absorption, *Marine fish, *Nuclear wastes, Fallout, Path of pollutants, Food chains, Strontium, Estuarine environment, Migration, Fish behavior, Public health, Saline water fish, Salinity, Leaching, Salt balance, Water pollution effects, Sedimentary rocks, Chemical precipitation, Evaporation, Inflow, Mixing.
Identifiers: USSR, Sea of Azov, Black Sea, Caspian Sea.

Leaching of shoreline rocks gives high Sr concentrations (21-25 mg/liter) in certain regions of the Caspian Sea relative to regions diluted by river discharges (0.5-2.5 mg/liter). Sr in the Sea of Azov (3.1-6.1 mg/liter) is affected by exchange with the Sivash Bay where evaporation gives high Sr concentrations (26 mg/liter). Concentration factors for Sr90 measured in 1966 were higher for semimigratory fish (roaeh, 40; carp, 33) than for migratory fish (kutum, 2.5). (Bopp-ORNL)
W73-07970

SR90 IN AQUATIC ORGANISMS OF THE BLACK SEA,
V. P. Parchevskii, L. G. Kulebakina, I. A. Sokolova, and A. A. Bachurin.
Available from NTIS, Springfield, Va., as Part of AEC-tr-7299; $6.00 paper copy, $1.45 microfiche. In: Marine Radioecology, p 196-221, Translation of Morskaya Radioekologiya, 1970, 21 tab, 30 ref.

Descriptors: *Strontium radioisotopes, *Radioecology, *Estuarine environment, *Marine algae, Fallout, Nuclear wastes, Path of pollutants, Absorption, Marine plants, Marine fish, Marine animals, Mollusks, Crustaceans, Europe, Plant physiology, Crabs, Radioactivity techniques, Marine biology, Foreign research, Oceans, Water analysis, Basic data collections, On-site data collections, Seasonal, Food chains, Sea water, Water analysis.

Studies in 1965-1966 showed low variability in Sr90 uptake within the same species. Of the total Sr90 in plant communities and in water in the vicinity, the percentage absorbed by the plants varied according to species (Cystoseira brown alga, 30%; Carallia red alga, 23%; Padera brown alga, 17%; Zostera flowering plant, 11%; other algae, 1%; Cystoseira biocoenosis that includes animals and other plants, 40%). For Cystoseira samples in Sept. 1964 and Dec. 1965, Sr90 concentration was about twice as high in the stem as in the branches; but in July 1965 there was no difference (a seasonal effect). Sr/Ca atomic ratios were nearly the same in organisms from related taxonomic groups and agreed with values obtained by others. Isotopic exchange between Sr90 and Sr was nearly complete. (Bopp-ORNL)
W73-07971

MN, CU AND ZN IN WATER AND ORGANISMS OF THE SEA OF AZOV,
L. I. Rozhanskaya.
Available from NTIS, Springfield, Va., as Part of AEC-tr-7299; $6.00 paper copy, $1.45 microfiche. In: Marine Radioecology, p 222-255, Translation of Morskaya Radioekologiya, 1970. 9 fig, 9 tab, 50 ref.

Descriptors: *Trace elements, *Absorption, *Radioecology, *Marine algae, Marine fish, Phytoplankton, Zooplankton, Strontium radioisotopes, Copper, Zinc, Path of pollutants, Marine plants, Reduction (Chemical), Oxygen demand, Seasonal, Trace elements, Manganese, Estuarine environment, Fallout, Biomass, On-site data collections, Europe.

Results are compared with similar studies in other seas. Mn in near-bottom water varied with season and location and was inversely related to the oxygen content. The average Cu/Zn ratio was 0.39. Phytoplankton, zooplankton, and benthic organisms had generally higher concent ation factors for trace elements (Mn, 200-10,000; Cu, 250-3,000; Zn, 500-7,000) than fish (Mn, 30-400; Cu, 150-700; Zn, 700-1,800). Migration of Sr90 in coastal zones is affected by absorption into the macrophyte biomass. (Bopp-ORNL)
W73-07972

RADIOECOLOGY OF CENTRAL AMERICAN SEAS,
G. G. Polikarpov, Yu. P. Zaitsev, V. P. Parchevskii, A. A. Bachurin, and I. A. Sokolova.
Available from NTIS, Springfield, Va., as Part of AEC-tr-7299; $3.00 in paper copy, $1.45 in microfiche. In: Marine Radioecology, p 256-288, Translation of Morskaya Radioekologiya, 1970. 20 fig, 14 tab, 6 ref.

Descriptors: *Radioisotopes, *Absorption, *Marine biology, *Radioecology, Nuclear wastes, Path of pollutants, Strontium radioisotopes, Atlantic Ocean, Marine algae, Kelp, Plankton nets, Benthic flora, Zooplankton, Vertical migration, Rhodophyta, Phaeophyta, Chlorophyta, Fallout.
Identifiers: Caribbean Sea, Cerium radioisotopes, Ruthenium radioisotopes, Cesium radioisotopes, Sargassum.

the waste water of hospitals needs no treatment before disposal into city sewers. The disposal of primary treated hospital effluents into streams or their use for irrigation should, however, be considered with care. (Holoman-Battelle)
W73-08024

CHEMICAL COMPOSITION OF WATER OF STREAMS OF POLISH HIGH TATRA MOUNTAINS, PARTICULARLY WITH REGARD TO THE SUCHA WODA STREAM,
Polish Academy of Sciences, Krakow. Zaklad Biologii Wod.

Acta Hydrobio, Vol 13, No 4, p 379-391, 1971. Illus.
Identifiers: *Chemical composition, Climate, *Mountain streams, Pollution, Streams, Sucha, *Tatra mountains, Woda.

The chemical composition of water depends on the geological structure of the drainage area, hypsometric differences, and the climate related with them. Owing to this, trophically varying types of waters develop within very short basins. Attention is also drawn to the local pollution of pure high-mountain streams in the region of shelter-houses and settlements.–Copyright 1972, Biological Abstracts, Inc.
W73-08026

WATER QUALITY MONITORING IN DISTRIBUTION SYSTEMS: A PROGRESS REPORT,
National Sanitation Foundation, Ann Arbor, Mich.
For primary bibliographic entry see Field 05A.
W73-08027

HETERÓTROPHIC NITRIFICATION IN SAMPLES OF NATURAL ECOSYSTEMS,
Cornell Univ., Ithaca, N.Y. Lab. of Soil Microbiology.
W. Verstraete, and M. Alexander.
Environmental Science and Technology, Vol 7, No 1, p 39-42, January 1973. 1 fig, 6 tab, 14 fig.

Descriptors: *Nitrification, *Ecosystems, *Aquatic bacteria, *Aquatic environment, Nitrogen compounds, Sewage sludge, Freshwater, Nitrites, Rivers, Aquatic soils, Cultures.
Identifiers: Enrichment, Arthrobacter, Heterotrophic bacteria, Hydroxylamine, 1-Nitrosoenthanol, Hydroxamic acid.

Since studies of axenic cultures of a nitrifying stain of Arthrobacter revealed that hydroxylamine, hydroxaic acid, 1-nitrosoethanol, nitrate, and nitrite were excreted, an investigation was begun to determine whether this or physiologically related species could bring about the same type of nitrification in natural environments. Products of the reaction and their persistence in samples taken from several different ecosystems were also determined. Samples (200 ml) of different aquatic environments (sewage, river water, lake water, soil) were enriched with sodium acetate and ammonium sulfate to concentrations of 3.0 mg Cml and 1.0 N/ml and incubated on a rotary shaker (120 rpm). Hydroxylamine, 1-nitrosethanol, nitrite, and nitrate were formed in samples of sewage, river water, lake water, and soils amended with ammonium and acetate. A carbon source was needed for the occurrence of this pattern of nitrification, which is apparently heterotrophic. Of the carbon sources tested, only acetate and succinate supported this newly described kind of nitrification. The data suggest that the active microorganisms nitrify at neutral pH values under conditions which do not promote abundant growth of other heterotrophs, but in environments that allowed luxuriant microbial proliferation, these microorganisms competed successfully and nitrified only at alkaline pH values. Hydroxylamine was rapidly inactivated in sewage and in soil, whereas 1-nitrosoethanol was quite persistent in aqueous solutions but disappeared rapidly from soil. (Holoman-Battelle)
W73-08029

MIREX RESIDUES IN WILD POPULATIONS OF THE EDIBLE RED CRAWFISH (PROCAMBARUS CLARKI),
Animal and Plant Health Inspection Service, Gulfport, Miss. Plant Protection and Quarantine Programs.
For primary bibliographic entry see Field 05A.
W73-08030

METABOLISM OF DDT BY FRESH WATER DIATOMS,
Manitoba Univ., Winnipeg. Dept. of Entomology.
S. Miyazaki, and A. J. Thorsteinson.
Bulletin of Environmental Contamination and Toxicology, Vol 8, No 2, p 81-83, August, 1972. 1 tab, 13 ref.

Descriptors: *Diatoms, *Metabolism, *Absorption, *DDT, *DDE, Path of pollutants, Pesticide residues, Pesticides, Chlorinated hydrocarbon pesticides, Chromatography, Cultures, Radioactivity techniques, Biodegradation.
Identifiers: Biotransformation, Thin layer chromatography, Bioaccumulation, Chlorinated hydrocarbons, Nitzschia, Culture media, Metabolites, C-14, Recovery, Biological samples.

Ten diatom cultures were isolated from a sample of ditch water by adding 5 ml of sample to Warner's agar. After incubation and additional culturing, pure cultures were exposed for 2 weeks to 0.71 ppm C-14 labeled DDT dissolved in benzene. After the 2-week period glacial acetic acid was added to the flasks and the mixture immediately extracted 3 times. The medium was then filtered and the cells were extracted 3 times with acetone. Thin-layer chromatography was used to separate DDT from metabolites. On the basis of these tests, Nitzschia sp, and an unidentified diatom were selected for further studies. The procedure was repeated in 4 replicates of each culture and a control without diatom inoculation. The results showed DDE to be the only metabolite produced by either culture. The unidentified diatom culture degraded more DDT to DDE than the Nitzschia species but most of the DDT added to the media remained unchanged in both cultures. The fact that the total radioactivity recovered from the diatom culture media was less than from the control suggests that some of the DDT or its metabolite (s) were bound intracellularly, or were otherwise not extractable by the solvent system used. Evaporation of DDT from the chromatogram may also account for some loss of radioactivity. The results suggest that some species of freshwater diatoms may be significant in the degradation of DDT to the non-insecticidal metabolite, DDE in nature. (Little-Battelle)
W73-08036

RAPID BIODEGRADATION OF NTA BY A NOVEL BACTERIAL MUTANT,
Department of the Environment, Burlington (Ontario). Centre for Inland Waters.
P. T. S. Wong, D. Liu, and B. J. Dutka.
Water Research, Vol 6, No 12, p 1577-1584, December 1972. 3 fig, 3 tab, 8 ref.

Descriptors: *Nitrilotriacetic acid, *Isolation, *Microbial degradation, *Sewage bacteria, Hydrogen ion concentration, Temperature, Environmental effects, Growth rates, Organic acids, Biodegradation, Spectrophotometry, Ultraviolet radiation, Microorganisms.
Identifiers: *Mutants, *Substrate utilization, NTA-metal complexes, Iminodiacetic acid, Glycine, Degradation rates, Acclimatization, Mutagenization.

A bacterial mutant was isolated from sewage after ultraviolet mutagenization and penicillin selection during a study which was initiated to explore the possibility of isolating a potent bacterium which could degrade NTA rapidly. Bacterial flora from sewage after u.v. mutagenization and penicillin selection were plated onto NTA-agar plates. Colonies which developed after 5 days incubation at 20 degrees C, were picked off and purified by repeated transfers onto fresh agar medium. One culture of Gram-negative short rod bacteria was found to grow most rapidly in 0.5 percent NTA broth. The following conclusions were reached concerning the mutant bacterium: (1) This mutant was able to grow without acclimatization in NTA concentrations as high as 2.5 percent as sole carbon, nitrogen and energy source. (2) The mutant could degrade NTA at a wide range of temperatures from 4 degrees to 37 degrees C with the optimal temperature at 20 degrees C. (3) The optimal pH of NTA degradation was pH 7. (4) The mutant could grow on NTA as well as its intermediate products (glycine, iminodiacetic acid). (5) The bacteria were capable of utilizing NTA present in lake water and sewage. (6) The rate of NTA degradation was very rapid. Almost all the NTA was degraded after 4 days incubation at an initial concentration of 0.2 percent NTA. (7) The NTA-metal complexes had no obvious effect on the bacterial degradation of NTA. (Holoman-Battelle)
W73-08046

ORGANOCHLORINE PESTICIDE RESIDUES IN WATER, SEDIMENT, ALGAE, AND FISH, HAWAII - 1970-71,
Hawaii Univ., Honolulu. Dept. of Agricultural Biochemistry.
For primary bibliographic entry see Field 05A.
W73-08047

ORGANOCHLORINE PESTICIDE RESIDUES IN COMMERCIALLY CAUGHT FISH IN CANADA - 1970,
Fisheries Research Board of Canada, Winnipeg (Manitoba). Freshwater Inst.
For primary bibliographic entry see Field 05A.
W73-08048

CHEMICAL RESIDUES IN LAKE ERIE FISH - 1970-71,
Food and Drug Administration, Cincinnati, Ohio.
For primary bibliographic entry see Field 05A.
W73-08049

PROCEEDING 1971 TECHNICAL CONFERENCE ON ESTUARIES OF THE PACIFIC NORTHWEST.
Oregon State Univ., Corvallis.

Available from the National Technical Information Service as COM-71-01115, $6.00 in paper copy, $1.45 in microfiche. OSU Sea Grant Circular No 42, 1971. Nath, J.N.; Slotta, L.S. (editors) 343 p, 104 fig, 11 tab, 119 ref.

Descriptors: *Estuaries, *Water quality, *Model studies, Water pollution sources, Water management (Administrative), Water resources development, Conference, Bays, Estuarine environment, Estuarine fisheries, Aquatic environment, Mathematical models, Numerical analysis, Physical models, Hydrological models, Benthos, Data collections.
Identifiers: *Pacific Northwest, Hydro-ecology, Infrared imagery, Bellingham Harbor, Umpqua estuary, Grays Harbor, Tillamook estuary, San Diego Bay, San Francisco Bay, Tidal currents, New York Bay, Dredging, Dye dispersion, Saline intrusion, Schrewsbury River, Navesink River, Waste loadings, Yaquina River, Coos Bay, Stommel's model, Salinity profile, Upwelling, Thomann's Model, Oxygen deficiency, Flushing rate, Tidal Prism Models.

The theme of the 1971 Technical Conference on Estuaries of the Pacific Northwest was management and planning for water quality in the Pacific Northwest estuaries. Some of the topics presented included theme-related model studies: 'The Potential of Physical Models to Investigate Estuarine Water Quality Problems', 'Applications of Some Numerical Models to Pacific Northwest Estuaries', and 'Mathematical Modeling of Estuarine Benthal Systems'. Other papers dealt with data acquisition techniques or results of experimentation done on waste disposal in the estuaries. Included among these were: 'Remote Sensing Acquisition of Tracer Dye and Infrared Imagery Information and Interpretation for Industrial Discharge Management', 'Studies of Sediment Transport in the Columbia River', and 'A Study of Sediments from Bellingham Harbor as Related to Marine Disposal'. Topics concerned with some of the ecological, legal, and historical aspects of the estuaries and their management include: 'Legal Protection of the Pacific Northwest Estuaries', 'Hydro-Ecological Problems of Marinas on Puget Sound', 'Historical Changes in Estuarine Topography with Questions on Future Management Policies', 'Effects of Institutional Constraints and Resources Planning on Growth in and Near Estuaries', and 'Recent Federal Policies Affecting Marine Science and Engineering Developments'. (Mackan-Battelle)
W73-08051

POLLUTION OF SUBSURFACE WATER BY SANITARY LANDFILLS, VOLUME 1,
Drexel Univ., Philadelphia, Pa.
A. A. Fungaroli.
Available from Sup Doc, GPO, Washington, D.C. 20402, Price $1.50. Environmental Protection Agency Interim report SW-12 rg, 1971. 186 p, 61 fig, 11 tab, 15 ref, 5 append. EPA Grant 000162.

Descriptors: *Landfills, *Garbage dumps, *Model studies, *Leachate, Laboratory tests, On-site tests, Water pollution sources, Path of pollutants, Computer programs, Mathematical models, Lysimeters.

The behavior of sanitary landfills in southeastern Pennsylvania and a large portion of the region between Washington, D.C., and Boston was studied using a lysimeter and a field test. The laboratory facility was operated under controlled conditions, while the field facility was operated under natural conditions. The lysimeter functioned as a closed system representative of the center of a large sanitary landfill, the depth of which was small in comparison to its areal extent. Temperatures, gases, and leachate quality were collected on a routine basis. The laboratory landfill behavior pattern is representative of young low-compaction density refuse. Within ten days of its initiation, refuse temperatures reached 150 deg F at the center, and stabilized at approximately 80 deg F after 60 days. The refuse was initially in the aerobic state, and after 60 days an anaerobic condition became dominant. The lysimeter began to produce leachate almost immediately, even though the refuse had a very low moisture content. At field capacity, net infiltration and leachate quantities were approximately equal. A computer program is given for model studies. First appearance of leachate is dependent on site conditions, including surface grading, vegetation, and soil. The early leachate is highly polluted, acidic pH of 5.5, and carries many dissolved and suspended solids. (See also W72-06103) (Knapp-USGS)
W73-08073

PREDICTED EFFECTS OF PROPOSED NAVIGATION IMPROVEMENTS ON RESIDENCE TIME AND DISSOLVED OXYGEN OF THE SALT WEDGE IN THE DUWAMISH RIVER ESTUARY, KING COUNTY, WASHINGTON,
Geological Survey, Tacoma, Wash.
W. L. Haushild, and J. D. Stoner.

run in half-filled and in completely filled bottles. The oily water was extracted with CCl4, and infrared analyses and the (1958) API 733 method were used to determine hydrocarbon content; oxygen determination was made using the Winkler method; and bacterial counts were made after filtration on a Sartorius 14005 membrane containing a nutritive substrate. It can be concluded from the results that 2 mechanisms are primarily involved in forecasting natural hydrocarbon pollution phenomena: evaporation and biodegradation. Evaporation is very important in perfectly still water and may be a primary factor when mixing takes place. Biodegradation is always involved in the natural elimination of mineral oil. Monod's model gives a good approximation for the forecasting of its course, provided it is recognized that certain oil components, which are in fact oxidized much more slowly, must be treated as nonbiodegradable. It seems safe to assert that anaerobic conditions will arise when initial hydrocarbons levels are around 20 ppm, except in cases where the water is already markedly oxygen-deficient and all forms of aeration are excluded. (Holoman-Battelle)
W73-08238

CELL REPLICATION AND BIOMASS IN THE ACTIVATED SLUDGE PROCESS,
Illinois Univ., Urbana. Dept. of Sanitary Engineering.
R. E. Speece, R. S. Engelbrecht, and D. R. Aukamp.
Water Research, Vol 7, No 3, p 361-374, March 1973. 11 fig, 6 tab, 8 ref.

Descriptors: *Activated sludge, *Biomass, *Microorganisms, Suspended solids, Growth kinetics, Methodology, Organic loading, Organic matter, Sewage bacteria, Bioindicators, Waste water treatment.
Identifiers: *Substrate utilization, *Cell replication, Deoxyribonucleic acid, Substrates, Optical density, Nucleic acids.

Because of the inherent difficulty in enumerating cell numbers in a flocculent suspension, deoxyribonucleic acid (DNA) was chosen as an indicator of cell numbers in the activated sludge process. A direct relationship between the concentration of DNA in the sludge and plate count was found with a dispersed growth of mixed culture microorganisms. Therefore, it was assumed that a direct relationship existed between cell numbers and DNA concentration in the sludge in a flocculent suspension of mixed culture microorganisms as found in activated sludge. An increase in DNA was therefore assumed to be an indication of cell replication. An increase in biomass before an increase in DNA indicated a storage of substrate in some form and not replication of organisms. The average increase in the weight per cell was determined by dividing the weight of biomass just prior to an increase in DNA by the initial weight of organisms present. Storage was a function of the loading rate to which the organisms were acclimated. A sludge acclimated to a loading rate of 2.0 per day increased 270 percent in biomass before replication. Sludges acclimated to lower loading rates showed an extended time lag before cell replication occurred, while higher loading rates maintained the sludges in a more active state having shorter time lags before replication occurred. Increased frequency of feeding also resulted in less time before replication occurred. Replication commenced as long as 4 h after the external substrate was exhausted and the maximum biomass was reached in the F/M equal 0.4 system. Thus, in the contact stabilization process, cell replication would be expected in the stabilization tank with only substrate storage taking place in the contact tank, due to low loading rates used. (Holoman-Battelle)
W73-08243

NITROGEN FIXATION BY A BLUE-GREEN EPIPHYTE ON PELAGIC SARGASSUM,
Woods Hole Oceanographic Institution, Mass.
For primary bibliographic entry see Field 05C.
W73-08246

MARINE WASTE DISPOSAL - A COMPREHENSIVE ENVIRONMENTAL APPROACH TO PLANNING,
D. P. Norris, L. E. Birke, Jr., R. T. Cockburn, and D. S. Parker.
Journal Water Pollution Control Federation, Vol 45, No 1, p 52-70, January 1973. 12 fig, 4 tab, 12 ref.

Descriptors: *Bioassay, *Toxicity, *Pacific Ocean, Worms, Sticklebacks, *Sewage, *Sculpins, Sea basses, Crabs, Shrimp, Snails, Clams, Mussels, *Estuaries, Dispersion, On-site studies, Plankton, Benthos, Currents (Water), Mixing, Tracers, Water temperature, Salinity, Dissolved oxygen, Instrumentation, Waste disposal, Waste water (Pollution), Dyes, Aquatic drift, Tides, Sediments, Suspended solids, Water pollution sources, Water pollution effects, Outfall sewers, Aerial photography, Chlorides, Population, Sewage disposal.
Identifiers: *Species diversity, Polychaetes, Tubifex, Three-spined stickleback, Cymatogaster aggregata, Hyperprosodon argenteum, Citharichthys sordidus, Ophiodon elongatus, Scorpaena guttata, Sebastodes, Gasterosteus aculeatus, Pagurus samuelis, Hemigrapsis oregonensis, Emerita analoga, Crago, Callianassa californiensis, Tegula funebralis.

The city of San Francisco undertook a comprehensive study of the marine environment (San Francisco Bay and the Pacific Ocean) to determine where and in what quantities it is feasible to dispose of the city's dry- and wet-weather wastewater effluents. The first phase of the study defined oceanographic characteristics of potential discharge sites and the fate of the discharges. This included determination of mass water movement, drift of particulate matter on the water surface, dispersion characteristics, and water characteristics by aerial photography, shipboard instrumentation, in situ equipment, and tracer studies. The second phase of the study was an ecological study which included plankton studies, benthic studies, diving studies of near-shore areas, intertidal studies, in situ bioassays with fish, static and continuous-flow bioassay with fish and macroinvertebrates, microcosm studies, stickleback blood studies, and biostimulation studies. The results of the two phases indicated that marine disposal is feasible and that the marine environment can be adequately protected by discharging chlorinated primary effluent through one or more submarine outfalls with properly designed diffuser systems. The study indicated that additional treatment is not necessary to protect the marine ecosystem. (Little-Battelle)
W73-08247

EUTROPHICATION AND LOUGH NEAGH,
New Univ. of Ulster, Coleraine (Northern Ireland).
For primary bibliographic entry see Field 05C.
W73-08252

ROLE OF PHOSPHORUS IN EUTROPHICATION AND DIFFUSE SOURCE CONTROL,
Wisconsin Univ., Madison. Water Chemistry Program.
For primary bibliographic entry see Field 05C.
W73-08255

MIREX AND DDT RESIDUES IN WILDLIFE AND MISCELLANEOUS SAMPLES IN MISSISSIPPI - 1970,
Mississippi State Univ., State College. Dept. of Biochemistry.

For primary bibliographic entry see Field 05A.
W73-08267

CAUSES OF MASSIVE INFECTION OF PINK SALMON WITH SAPROLEGNIA IN THE RIVERS OF ITURUP (ETOROFU) ISLAND, (IN RUSSIAN),
V. N. Ivankov.
Uch Zap Dal'nevost Univ., Vol 15, No 3, p 124-126, 1971.
Identifiers: Crustaceans, Infection, Islands, *Iturup (Etorofu) Island (USSR), Massive, *Pink salmon, Salmon, *Saprolegnia, USSR.

Infection of pink salmon with Saprolegnia was especially strong in 1963. Saprolegnia grew in areas where parasitic crustaceans were attached. In the rivers most of the fish were free from the parasites except those in infected areas. Some of the fish recovered in marine water.--Copyright 1972, Biological Abstracts, Inc.
W73-08271

ROTENONE AND ITS USE IN ERADICATION OF UNDESIRABLE FISH FROM PONDS,
Freshwater Fisheries Research Station, Chandpur (Bangladesh).
K. A. Haque.
Pak J Sci Ind Res. Vol 14, No 4/5, p 385-387, 1971.
Identifiers: Crustaceans, Fish, Frogs, *Ponds, *Rotenone, Snakes, *Bangladesh.

Rotenone, its origin and use are described. Results of application of the chemical to 3 ponds of the Fish Seed Multiplication Farm at Jamalpur, Mymensingh, for eradication of undesirable species are given. Rotenone takes time to reach the deep bottom of ponds in absence of any effective agitation of water. Snakes, frogs and crustaceans are not readily affected by this plant derivative as they can escape the action through terrestrial respiration. A list of fishes and other aquatic organisms in order of their susceptibility to rotenone is presented. A concentration of 1 ppm at summer temperature, around 30 deg C in this region, was adequate to kill fishes.--Copyright 1972, Biological Abstracts, Inc.
W73-08279

OIL SPILLS CONTROL MANUAL FOR FIRE DEPARTMENTS,
Alpine Geophysical Associates, Inc., Norwood, N.J.
For primary bibliographic entry see Field 05G.
W73-08288

THE INFLUENCE OF LOG HANDLING ON WATER QUALITY,
Oregon State Univ., Corvallis. Dept. of Civil Engineering.
F. D. Schaumburg.
Copy available from GPO Sup Doc as EP1.23/2:73-085, $1.25; microfiche from NTIS as PB-219 824, $1.45. Environmental Protection Agency, Technology Series Report EPA-R2-73-085, February 1973. 105 p, 33 fig, 20 tab, 39 ref, 4 append. EPA Project 12100 EBG.

Descriptors: *Bark, *Leachate, Toxicity, Water pollution, Oxygen demand, Biochemical oxygen demand, *Lumbering, Northwest US, Organic matter, Leaching, *Water storage.
Identifiers: Forest industry pollution, Logging wastes, Bark sinkage, Bark deposits, *Log storage, *Benthic deposits.

The water storage of logs is widely practiced in the Pacific Northwest. An investigation has been made to determine the effect of this practice on water quality. Soluble organic matter and some inorganics leach from logs floating in water and from logs held in sprinkled land decks. The character and quantity of leachate from Douglas fir, ponderosa pine and hemlock logs have been examined. Measurements including BOD, COD (1.0-4.2gm/ft2 per week), PBI, solids and toxicity (no kill to 20% TLm 96) have shown that in most situations the contribution of soluble leachates to holding water is not a significant water pollution problem. The most significant problem associated with water storage appears to be the loss of bark from logs during dumping, raft transport and raft storage. Bark losses from 6.2% to 21.7% were measured during logging and raft transport. Dislodged bark can float until it becomes water logged and sinks, forming benthic deposits. Floating bark is aesthetically displeasing and could interfere with other beneficial uses of a lake, stream or estuary. Benthic deposits exert a small, but measurable oxygen demand and may influence the biology of the benthic zone. Implementation of corrective measures by the timber industry to reduce bark losses could make the water storage of logs a practice which is compatible with a high quality environment. (EPA)
W73-08294

MIGRATION AND METABOLISM IN A TEMPERATE STREAM ECOSYSTEM,
North Carolina Univ., Chapel Hill. Dept. of Zoology.
A. S. Hall.
Ecology, Vol 53, No 4, p 585-604, Summer 1972. 9 fig, 3 tab, 99 ref. OWRR B-007-NC (5).

Descriptors: Ecological distribution, *Energy budget, *Respiration, *Fish migration, Fish behavior, Metabolism, *Cycling nutrients, *Phosphorus, *North Carolina, Water pollution sources, Water pollution effects.
Identifiers: *New Hope Creek (NC), Morgan Creek (NC), Orange County (NC), Durham County (NC), *Cape Fear River (NC).

Fish migration, total stream metabolism, and phosphorus were studied in New Hope Creek, N.C., from April 1968 to June 1970. Upstream and downstream movement of fish was monitored using weirs with traps. Most of the 27 species had a consistent pattern of larger fish moving upstream and smaller fish moving downstream. Diurnal oxygen series were run to measure the metabolism of the aquatic community. Gross photosynthesis ranged from 0.21 to almost 9 g O2 m-2 day-1, and community respiration from 0.4 to 13 g O2 m-2 day-1 (mean of 290 and 479 g O2 m-2 yr-1). Both were highest in the spring. Production per volume and respiration per volume were always much larger near the headwaters than farther downstream, apparently due to the dilution effect of the deeper water downstream. Migration may maintain young fish in areas of high productivity. Other effects of migration may include: prey control, recolonization of defaunated regions, genetic exchange, and mineral distribution. An energy diagram was drawn comparing energies of isolation, leaf inputs, currents, total community respiration, fish populations, and migrations. About 1% of the total respiration of the stream was from fish populations, and over 1 year about 0.04% of the total energy used by the ecosystem was used for the process of migration. Each Calorie invested by a fish population in migration returns at least 3 Calories. Analysis of phosphorus entering and leaving the watershed indicated that flows were small relative to storages and that this generally undisturbed ecosystem is in approximate phosphorus balance. Upstream migrating fish were important in maintaining phosphorus reserves in the headwaters. (McJunkin-North Carolina)
W73-08303

INFECTION BY ICHTHYOPHTHIRIUS MULTIFILIIS OF FISH IN UGANDA,
Makerere Univ., Kampala (Uganda). Dept. of Zoology.
For primary bibliographic entry see Field 05C.
W73-08330

ON THE ECOLOGY OF AN ADULT DIGENETIC TREMATODE PROCTOECES SUBTENUIS FROM A LAMELLIBRANCH HOST SCROBICULARIA PLANA,
For primary bibliographic entry see Field 05C.
W73-08331

THE EFFECT OF SODIUM ALKYLBENZENE-SULPHONATE ON THE DRAINAGE OF WATER THROUGH SAND,
Westfield Coll., London (England). Dept. of Zoology.
For primary bibliographic entry see Field 02K.
W73-08342

TIME-OF-TRAVEL STUDY, BATTEN KILL FROM 0.6 MILE EAST OF VERMONT-NEW YORK BORDER TO CLARKS MILLS, NEW YORK,
Geological Survey, Albany, N.Y.
H. L. Shindel.
New York Department of Environmental Conservation Report of Investigation RI-12, 1973. 18 p, 14 fig, 4 tab, 7 ref.

Descriptors: *Travel time, *Time lag, *Streamflow, *New York, *Path of pollutants, Dye releases, Tracers, Fluorescent dye, Dispersion.
Identifiers: *Batten Kill (NY).

Time of travel was determined for the 30.8-mile reach of Batten Kill between BM 543 bridge, 0.6 mile east of the New York-Vermont border, and Clarks Mills, New York, using Rhodamine B and Rhodamine WT dyes. Cumulative peak time of travel for the peak concentration flow of about 46% duration was 47 hours and for a flow of approximately 89% duration was 101 hours. Relationships between peak, centroid, leading, and trailing-edge times of travel and discharge through the subreaches are shown graphically. Time-of-travel data for each subreach as well as cumulative time graphs for the entire reach for different discharges are given. Dye dispersion and peak-analysis information are also given. (Knapp-USGS)
W73-08368

AN INVENTORY AND SENSITIVITY ANALYSIS OF DATA REQUIREMENTS FOR AN OXYGEN MANAGEMENT MODEL OF THE CARSON RIVER,
Nevada Univ., Reno. Dept. of Civil Engineering.
R. G. Orcutt, and J. G. Gonzales.
Cooperative Report Series Publication No EN-1, Civil Engineering Department, College of Engineering, University of Nevada Reno, and Engineering Report No 47 in cooperation with the Center for Water Resources Research, Desert Research Institute, Reno, September, 1972. 53 p, 18 fig, 3 tab, 36 ref, 4 append.

Descriptors: Water resources development, Water quality, Management, Rivers, Wastes, *Oxygen, *Simulation analysis, *Low flow augmentation, *Dissolved oxygen, *Waste assimilative capacity, Streamflow, Waste dilution, Computer programs, Benthos, Sewage, Organic matter, Water pollution control, Mathematical models, Systems analysis, *Nevada.
Identifiers: *Carson River Basin (Nev.), *Sensitivity analysis, Residual waste loads, Data requirements.

The rapid expansion of population in Nevada has compounded the problem of increased waste discharge. Extensive water use has resulted in a decrease in streamflows, causing an increase in the ratio of wastes to dilution water and reducing the assimilative capacity of streams. It is important to know the assimilative characteristics of a river so that a rational plan for wastewater reclamation and utilization can be developed. Using simulation and sensitivity analysis, an assessment has been made of the need for developing more

almost entirely overcome by fusicoccin. Attempts were made to measure the solute potential of the guard cells under the various treatments. Abscisic acid clearly increased their solute potential, but no absolute measurements could be made in the presence of fusicoccin owing to a failure of plasmolysis even with mannitol solutions of solute potential as low as -35 bars. Experiments using isotopically labelled mannitol indicated a massive uptake into the epidermis in the presence of fusicoccin. The effectiveness of this toxin under natural conditions may depend on its ability to counteract effects of abscisic acid, the stress hormone that induces stomatal closure.—Copyright 1973, Biological Abstracts, Inc.
W73-08408

PRELIMINARY RESEARCH ON THE PSYCHROPHILIC LIPOLYTIC BACTERIA OF THE SOIL AND WATER, (IN FRENCH),
Ottawa Univ. (Ontario). Dept. of Biology.
C. Breuil, and A. M. Gounot.
Can J Microbiol. Vol 18, No 9, p 1445-1451. 1972. (English summary).
Identifiers: *Bacteria, *Lipolytic populations, *Pseudomonas, Psychrophilic microorganisms, Soils, Winter.

Different soil extract media and other media were assayed for a comparative evaluation of both total and lipolytic populations of bacteria taken from soil and water samples during winter. By incubating at a low temperature, psychrophilic microorganisms were counted and isolated. Gram-negative bacteria, especially Pseudomonas species, were the most numerous.—Copyright 1973, Biological Abstracts, Inc.
W73-08423

IDENTIFICATION OF AMAZON RIVER WATER AT BARBARDOS, WEST INDIES, BY SALINITY AND SILICATE MEASUREMENTS,
McGill Univ., Montreal (Quebec). Marine Sciences Centre.
D. M. Steven, and A. L. Brooks.
Mar Biol (Berl). Vol 14, No 4, p 345-348. 1972. Illus.
Identifiers: Amazon River, *Barbados (West Indies), Identification, Measurements, Rivers, *Salinity, *Silicates, West-Indies.

Salinity and silicate concentrations were studied at about fortnightly intervals for 21 mo. at a station near Barbados, W. Indies; latitude 13 deg 15 min N, longitude 59 deg 42 min W. A sensitive inverse correlation was found to exist at 5 and 25 m, but not at greater depths. Salinity near the surface varied between 33.5 and 36.0%, and silicate between a little less than 1 and 4 micrograms at/l. Low salinity water, rich in silicate, was found from Feb. to July; salinity increased and silicate decreased from Sept. to Dec. It is argued that the low salinity water at Barbados can be identified with the areas of reduced salinity found by Ryther et al. (1967) about latitude 8 deg to 10 deg N, longitude 50 deg to 55 deg W, and that this water originates from the Amazon River. Local precipitation does not seem to be a significant factor.—Copyright 1973, Biological Abstracts, Inc.
W73-08430

WATER POLLUTION IN SUEZ BAY,
Red Sea Inst. of Oceanography and Fisheries, Al Ghurdaqah (Egypt).
For primary bibliographic entry see Field 05G.
W73-08432

SALICYLANILIDE I, AN EFFECTIVE NON-PERSISTENT CANDIDATE PISCICIDE,
Bureau of Sport Fisheries and Wildlife, La Crosse, Wis. Fish Control Lab.
For primary bibliographic entry see Field 05C.
W73-08434

DEVELOPMENT OF A NEW ENGLAND SALT MARSH,
For primary bibliographic entry see Field 02L.
W73-08443

ECOLOGY OF ARBOVIRUSES IN A MARYLAND FRESHWATER SWAMP: II. BLOOD FEEDING PATTERNS OF POTENTIAL MOSQUITO VECTORS,
Walter Reed Army Inst. of Research, Washington, D.C.
J. W. Leduc, W. Suyemoto, B. F. Eldridge, and E. S. Saugstad.
Am J Epidemiol. Vol 96, No 2, p 123-128. 1972.
Identifiers: Aedes-Atlanticus, Aedes-Canadensis, *Arboviruses, Birds, Blood, Culex-Salinarius, Culiseta-Melanura, Ecology, Mammals, *Maryland, *Mosquito vectors, Psorophora-Ferox, *Swamps, Vectors, Water pollution effects.

Blood engorged mosquito specimens from the Pocomoke Cypress Swamp, Maryland, were collected from May through Nov., 1969. Five mosquito species (Aedes atlanticus, A. canadensis, Culex salinarius, Culiseta melanura and Psorophora ferox) were examined by capillary type precipitin tests of engorged material and by comparisons of human biting collections and collections of mosquitoes attracted to caged animals. A. atlanticus and P. ferox had similar feeding patterns, both most frequently feeding on sylvatic mammals, while C. salinarius appeared to feed mostly on domestic mammals adjacent to the swamp. A. canadensis was an omnivorus feeder, while C. melanura fed almost exclusively on birds. The potential of these mosquitoes as vectors of arboviruses known to be present in the swamp is discussed. (See also W73-08446)—Copyright 1973, Biological Abstracts, Inc.
W73-08447

5C. Effects of Pollution

WATER QUALITY CHANGES IN AN IMPOUNDMENT AS A CONSEQUENCE OF ARTIFICIAL DESTRATIFICATION,
North Carolina Univ., Chapel Hill. School of Public Health.
C. M. Weiss, and B. W. Breedlove.
Available from the National Technical Information Service as PB-219 390, \$6.75 in paper copy, \$1.45 in microfiche. North Carolina Water Resources Research Institute, Report No. 80, January 1973. 216 p, 108 fig, 28 tab, 22 ref. OWRR B-007-NC (6). 14-01-0001-1933.

Descriptors: *Water quality control, *Destratification, *North Carolina, Impoundments, Reservoir operation, Lakes, Epilimnion, *Hypolimnion, *Diptera, Oxygen requirements, Aeration, Distribution patterns, Water temperature.
Identifiers: *University Lake (N.C.).

Destratification of a water supply impoundment was studied over a three year period. Prior to destratification, baseline information on physical, chemical and biological parameters was established. Destratification was accomplished by the use of the 'Air-aqua' system which creates vertical circulation in a body of water by the release of small bubbles from hoses laid on the lake bottom. Effectiveness of destratification distinctly evident in temperature distribution in the epilimnion and hypolimnion. Deep waters of the hypolimnion did not lose oxygen to the point of becoming anaerobic, as they had under stratified conditions. However, quantity of oxygen present showed a limited degree of aeration from the transport of surface water downward, and showed rapid rate of deoxygenation characteristic of the hypolimnion of University Lake. Striking changes in numbers and population characteristics of the phytoplanktonic organisms clearly evident in each of the two years of destratification. Benthic forms, particularly the Chironomidae, shifted from spe-

cies tolerant of micro-aerophylic conditions to those requiring higher levels of oxygen. Total period of observation was too short to establish clear changes in characteristics of fish populations, but it was evident that due to the behavior of fish relative to the air bubbles, the number of fish caught per man hour increased. Overall distribution of organic material occurred in the lake as the body of water became more homogeneous. During the summer months, this resulted in an increase in chlorine demand at the water treatment plant which uses University Lake as a water supply.
W73-07818

LAKES OF BELORUSSIA (BELORUSSKOYE POOZER'YE),
Belorussian State Univ., Minsk (USSR). Laboratoriya Ozerovedeniya.
For primary bibliographic entry see Field 02H.
W73-07907

RESULTS FROM MULTI-TRACE-ELEMENT NEUTRON ACTIVATION ANALYSES OF MARINE BIOLOGICAL SPECIMENS,
California Univ., Irvine. Dept. of Chemistry.
V. P. Guinn, and R. Kishore.
Available from NTIS, Springfield, Va., as CONF-721010-10; $3.00 paper copy, $1.45 microfiche. Report CONF-721010-10, Oct 1972. 5 p, 6 ref.

Descriptors: *Marine biology, *Trace elements, *Neutron activation analysis, *Analytical techniques, Gamma rays, Spectroscopy, Mercury, Zinc, Marine animals, Mammals, Marine fish, Kelps, Pacific Ocean, Public health, Food chains, Fish physiology, Animal physiology.
Identifiers: Selenium.

In Pacific Ocean Marine mammals and large fish, much higher levels of Se and Zn were found in liver than in muscle tissue. With the mammals the same was the case for Hg; however, with the fish the level of Hg was about the same in both liver and muscle tissue. There was appreciable variation between different specimens of the same species and size, and even within a given tissue of a single specimen - such that it is important that large numbers of samples are analyzed. Levels as low as 0.01 ppm Hg, 0.02 ppm Se, and 0.2 ppm Zn could be detected using Ge (Li) gamma spectrometry. (Bopp-ORNL)
W73-07927

NUCLEAR POWER,
Princeton Univ., N. J. Center of International Studies.
G. Garvey.
In: Energy, Ecology, Economy, by Gerald Garvey, W. W. Norton and Company, Inc., New York. 1972. p 135-155. 2 fig, 1 tab, 11 ref.

Descriptors: *Nuclear powerplants, *Radioactivity effects, *Water pollution effects, *Food chains, Feasibility studies, Cost analysis, Hazards, Safety, Waste disposal, Nuclear wastes, Public health, Toxicity, Economic prediction, Future planning (Projected), Risks, Evaluation, Path of pollutants, Insurance, Costs, Compensation.

Nuclear-power hazards are reviewed including those from mining and processing U, and from concentration of fission products in food chains. Using the conservative linear extrapolation of effects (increased susceptibility to disease and genetic effects) to a low-dose level, the cost in medical care per person per year in the 1980's will be about $2.40, with lethal effects resulting in a very small minority of cases. It is considered that although insurance cannot be obtained for all future contingencies, the cost for insurance against the trivially small hazard of disastrous accidents should be relatiely small. (Bopp-ORNL
W73-07928

THE ECOLOGY OF THE PLANKTON OF THE CHESAPEAKE BAY ESTUARY, PROGRESS REPORT DEC 1970-AUG 1972,
Johns Hopkins Univ., Baltimore, Md.
W. R. Taylor, V. Grant, J. J. McCarthy, G. MacKiernan, and S. S. Storms.
Available from NTIS, Springfield, Va., as COO-3279-3; $3.00 in paper copy, $1.45 in microfiche. Report COO-3279-3, August 1972. 95 p, 12 fig, 7 tab, 82 ref. AEC-AT (11-1)-3279.

Descriptors: *Primary productivity, *Estuarine environment, *Water pollution effects, *Copepods, Organic matter, Carbon, Silts, Reviews, Nutrient removal, Photosynthesis, Light, Temperature, Sampling, Seasonal, Vertical migration, Ecological distribution, Dominant organisms, Limiting factors, Environmental effects, *Chesapeake Bay, Phytoplankton, Zooplankton, Nutrient requirements, Marine algae.

The present hypothesis is that primary productivity is controlled by light and temperature, and that nutrient concentration is not limiting during most of the year. The upper Chesapeake Bay is influenced by the Susquehanna-River silt load. The midbay showed a midsummer peak. The lower bay showed spring and late summer peaks. Dissolved organic carbon was calculated from oxidations at 150 and 950 degrees centigrade. Samples of two copepods were collected over the entire bay, and concurrently, environmental parameters were measured. Also the distribution and some aspects of the behavior responses were studied of a cladoceran which is of ecological importance. Feeding and assimilation rates of the two copepods are being correlated with environmental parameters. (Bopp-ORNL)
W73-07932

STUDIES ON THE RADIOACTIVE CONTAMINATION OF THE SEA, ANNUAL REPORT, 1971.
Comitato Nazionale per l'Energia Nucleare, La Spezia (Italy). Laboratorio per lo Studio della Contaminazione Radioattiva de Mare.
For primary bibliographic entry see Field 05B.
W73-07942

APPLIED AQUATIC STUDIES,
Oak Ridge National Lab., Tenn.
For primary bibliographic entry see Field 05B.
W73-07948

APPENDIX TO QUARTERLY SUMMARY REPORT, SEPT. 1, 1972, THROUGH DEC. 1, 1972. HEALTH AND SAFETY LABORATORY, FALLOUT PROGRAM,
New York Operations Office (AEC), N.Y. Health and Safety Lab.
For primary bibliographic entry see Field 05A.
W73-07951

CHALK RIVER NUCLEAR LABORATORIES PROGRESS REPORT APRIL 1 TO JUNE 30, 1972, BIOLOGY AND HEALTH PHYSICS DIVISION, ENVIRONMENTAL RESEARCH BRANCH AND HEALTH PHYSICS BRANCH,
Atomic Energy of Canada Ltd., Chalk River (Ontario). Chalk River Nuclear Labs.
C. A. Mawson, and G. Cowper.
Available from NTIS, Springfield, Va., as AECL-4272; $3.00 paper copy, $1.45 microfiche. Report AECL-4272, September 1972. p 33-65. 3 fig.

Descriptors: *Nuclear wastes, *Path of pollutants, *Canada, *Tritium, Lakes, Lake sediments, Paleoclimatology, Paleohydrology, Glaciers, Water pollution effects, Absorption, Carbon radioisotopes, Tracers, Aquatic insects, Iron, Cobalt, Iodine radioisotopes, Strontium radioisotopes, Strontium, Phosphates, Primary productivity, Monitoring, Radioactivity techniques.

EFFECT OF INDUSTRIAL WASTES ON OXIDATION POND PERFORMANCE,

M. Moshe, N. Betzer, and Y. Kott.

Water Research, Vol 6, No 10, p 1165-1171, October 1972. 1 fig, 4 tab, 12 ref.

Descriptors: *Bioassay, *Heavy metals, *Oxidation lagoons, *Toxicity, *Industrial wastes, Cadmium, Copper, Nickel, Zinc, Chromium, Aquatic algae, Laboratory tests, Water pollution effects, Inhibition, Growth rates, Cations, Dissolved oxygen, Chlorophyta, Biochemical oxygen demand, Hydrogen ion concentration, Domestic wastes, Sewage, Sewage bacteria, Coliforms.
Identifiers: Chlorella sorokiniana, Pollutant effects, Most probable number test, Algal counts.

Cadmium, copper, nickel, zinc, and hexavalent chromium ions were tested in a bench-bioassay experiment for toxicity limits and possible application to experimental oxidation ponds. Domestic sewage was placed into test tubes where predetermined concentrations of metal ions were added together with known initial concentrations of Chlorella sorokiniana. The test tubes were incubated under controlled illumination (1500 1x) at 29 C. Before and after incubation coliform counts (MPN) were carried out according to Standard Methods (1965). Algal counts were performed using a haemocytometer. Experimental ponds of 50-70 1 volume were fed with diluted domestic sewage (BOD equal 200 mg/l). Predetermined quantities of metal salts had been previously added to give the desired concentration of metal ions in the inflowing sewage. At the final stage of the study, an aquarium of 80 1 capacity was operated as experimental pond. To this pond a mixture of metal ions (Cr, Cd, Cu, Ni, and Zn) was introduced, beginning with 3 mg/l and increasing to 12 mg/l of each ion. Samples taken from the ponds were subjected to the following tests: pH, dissolved oxygen, BOD, MPN, algal count and determination of metal ion concentration. The samples were taken from the influent, effluent and bottom sludge. It was found that the metal ions are toxic, inhibiting Chlorella growth. However, when added at concentrations of 0.5-1.5 mg/l to influent of oxidation ponds, the ponds continued to operate normally. Higher concentrations of 3 and 6 mg/l did not effect adversely pond performance - not even a concentration of 6 mg/l of each ion (a total metal ion concentration of 30 mg/l). A mixture of 60 mg/l metal ions brought about a decrease in algal numbers and caused a sharp drop in dissolved oxygen concentration. It is believed that since high pH causes metal ions to precipitate, oxidation ponds operating normally above pH 8.0 will tolerate metal ions in sewage containing industrial wastes for a long time before sludge accumulation will affect pond performance. (Holoman-Battelle)
W73-08015

BACTERIAL CHEMORECEPTION: AN IMPORTANT ECOLOGICAL PHENOMENON INHIBITED BY HYDROCARBONS,

Harvard Univ., Cambridge, Mass. Lab. of Applied Microbiology.
R. Mitchell, S. Fogel, and I. Chet.

Water Research, Vol 6, No 10, p 1137-1140, October 1972. 4 tab, 7 ref.

Descriptors: *Marine bacteria, *Marine algae, Predation, *Organic compounds, *Oil pollution, Water pollution effects, *Microbial degradation, Diatoms, Inhibition, Phenols, Chrysophyta, Biodegradation, Phytoplankton, Enteric bacteria, Carbohydrates, Amino acids, Attractants, Artificial substrates.
Identifiers: *Chemoreception, Substrates, Chemotaxis, Toluene, Crude oil, Skeletonema costatum, Glucose, Ribose, Proline, Serine, Adenine.

Motile marine bacteria have been shown to display chemoreception, with each microorganism exhibiting a highly specific response, and are attracted to a wide range of organic compounds. Chemoreception is also involved in the biodegradation of phytoplankton and enteric bacteria by bacterial predators. Marine bacteria were isolated from seawater samples on seawater nutrient agar. The predators were isolated by enrichment culture on an artificial seawater medium containing the microbial prey as sole C source. Bacterial chemotaxis was detected using this method: A 5-micron capillary tube sealed at one end and containing the test chemical was placed in a suspension of test bacteria placed in seawater on a microscope slide. Bacterial attraction was observed microscopically. Quantitative data were obtained by plating the contents of the capillary tube on seawater nutrient agar. Nutrients, in very low concentrations, were detected very rapidly. Most carbohydrates and amino acids stimulated chemotaxis at concentrations as low as 0.01 microM. The isolated bacterial predators were capable of degrading the diatom, Skeletonema costatum. The addition of specific aromatic hydrocarbons (phenol, toluene, crude oil) to seawater totally inhibited the chemotactic response of all bacteria without immobilization. The ecological implications of this type of sublethal effect on the self-purifying capacity of the sea and on the behavior of marine animals are discussed. (Holoman-Battelle)
W73-08016

INACTIVATION ASSAYS OF ENTEROVIRUSES AND SALMONELLA IN FRESH AND DIGESTED WASTE WATER SLUDGES BY PASTEURIZATION,

J. M. Foliguet, and F. Doncoeur.

Water Research, Vol 6, No 11, p 1399-1407, November 1972. 5 fig, 4 tab, 4 ref.

Descriptors: *Sludge treatment, *Disinfection, Enteric bacteria, Salmonella, Sewage sludge, Sewage bacteria, Cultures, Waste water treatment.
Identifiers: *Poliovirus, *Coxsackievirus, *Salmonella paratyphi, *Pasteurization, Inactivation.

To determine the effectiveness of a pasteurization process for inactivating enteroviruses and Salmonella, three types of fresh sludge, three types of digested sludge, and pure cultures of poliovirus, Coxsackievirus, Salmonella paratyphi B were subjected to the treatment. The pasteurization procedures consisted of homogenization for 30 minutes, heating from 6-15 C to 80 C in less than 10 minutes, and maintenance at 80 C for 10 minutes. After treatment the samples were rapidly cooled to minus 70 C and stored at minus 25 C until use. Culturing procedures are described by which the samples were assayed. The results showed that the treatment provides relatively thorough inactivation of the pathogenic germs, thereby reducing the infection risk of the sludges. The samples cannot, however, be considered to be entirely sterile since they contain sporulated germs. (Little-Battelle)
W73-08017

BACTERIAL AND ALGAL CHLOROPHYLL IN TWO SALT LAKES IN VICTORIA, AUSTRALIA,

For primary bibliographic entry see Field 05A.
W73-08018

THE EFFECTS OF FLUORIDE ON ESTUARINE ORGANISMS,

National Inst. for Water Research, Pretoria (South Africa).
J. Hemens, and R. J. Warwick.

Water Research, Vol 6, No 11, p 1301-1308, November 1972. 2 tab, 6 ref.

Descriptors: Bioassay, *Toxicity, *Fluorides, Industrial wastes, *Crabs, *Mullets, *Mussels, Algae, Water pollution effects, Shrimp, Grasses, Diatoms.

Identifiers: *Eel grass, *Prawns, Bioaccumulation, South Africa, Mhalatuzi River, *Richards Bay (So Afr), Aluminum smelters, Mugil cephalus, Ambassis safgha, Therapon jarbua, Penaeus monodon, Perna perna, Palaemon pacificus, Tylodiplax blephariskios, Mud crabs, Zostera.

Experiments to determine the possible effects of fluoride discharged in the effluent from an aluminum smelter on the fauna and flora of the receiving estuary in Zululand, South Africa showed no toxic effects on three species of fish (juvenile mullet, Ambassis safgha, and Therapon jarbua) and two species of penaeid prawns (Penaeus indicus and Penaeus monodon) during 96 h exposure at concentrations up to 100 mg F per liter. The brown mussel Perna perna showed evidence of toxic effects after 5 days exposure at a concentration of 7.2 mg per liter. Long-term (72 days) exposure in recirculated outdoor laboratory estuary models without external food supply and with 2.0 percent salinity and 52 mg F per liter showed physical deterioration and increased mortality in the mullet Mugil cephalus and the crab Tylodiplax blephariskios and the reproductive processes of the shrimp Palaemon pacificus appeared to be adversely affected. Eel grass and algae grown in the models showed no evidence of fluoride accumulation but all the introduced animals accumulated fluoride, the highest concentration of 7743 microns F per g ash being reached in the mullet compared to 148.1 in the control system. It was concluded that fluoride was accumulated mainly from the water and not via the food materials. (Little-Battelle)
W73-08019

PRIMARY PRODUCTIVITY OF A SHALLOW EUTROPHIC WATER DURING AN APHANIZOMENON FLOSAQUAE-BLOOM,
Rijksfaculteit der Landbouwwetenschappen, Ghent (Belgium).
J. De Maeseneer.
Meded Fac Landbouwwet Rijksuniv Gent. Vol 36, No 4, p 1441-1448. 1971. Illus. English summary.
Identifiers: Aphanizomenon-Flos-Aquae, *Belgium, Blooms, Eutrophic waters, Gent, *Primary productivity, Shallow water.

Measurements of the primary productivity were made following the Gran-method in the 'Nationale Watersportbaan Georges Nachez,' a shallow eutrophic water at Gent during an A. flos-aquae water-bloom period in the autumn of 1971. The results obtained show that the gross primary productivity is very high and must approach or surpass the maximum productivity stated by Steemann-Nielsen. A linear relationship on logarithmic (primary production)-linear (depth) paper was obtained in only 1 case. By extrapolation it can be concluded that the primary production in the superficial layer is about 8 mg C/m3/day. --Copyright 1972, Biological Abstracts, Inc.
W73-08020

THE LOAD OF INFECTIOUS MICRO-ORGANISMS IN THE WASTE WATER OF TWO SOUTH AFRICAN HOSPITALS,
National Inst. for Water Research, Pretoria (South Africa).
For primary bibliographic entry see Field 05B.
W73-08024

GROWTH RATE DETERMINATIONS OF THE MACROPHYTE ULVA IN CONTINUOUS CULTURE,
Harvard Univ., Cambridge, Mass. Lab. of Applied Microbiology.
T. D. Waite, L. A. Spielman, and R. Mitchell.
Environmental Science and Technology, Vol 6, No 13, p 1096-1100, December 1972. 6 fig, 1 tab, 12 ref.

Descriptors: *Marine algae, *Chlorophyta, *Growth rates, Cultures, Plant growth, Photosynthesis, Biomass.
Identifiers: *Growth kinetics, *Ulva lactuca, Continuous cultures, Macrophytes.

Continuous culture experiments were run with the benthic macrophyte Ulva lactuca. Using oxygen evolution as a monitor of photosynthesis and dry weight determination for biomass synthesis, growth rates and stoichiometric growth constants were evaluated. The data showed that the ratio of oxygen production to algal mass synthesis is relatively independent of nutrient concentration and growth rate, but is affected by light intensity. The data also showed that the amount of oxygen evolved per unit of algal material was almost a factor of 10 higher than is predicted from carbohydrate synthesis. It appears that Ulva is capable of synthesizing compounds with carbon oxidation states of plus 1 or plus 2, thus estimates of biomass synthesis may be in error when the average algal material is assumed to be carbohydrate. (Holoman-Battelle)
W73-08028

INSECTICIDE TOLERANCES OF TWO CRAYFISH POPULATIONS (PROCAMBARUS ACUTUS) IN SOUTH-CENTRAL TEXAS,
Texas A and M Univ., College Station. Dept. of Wildlife and Fisheries Sciences.
D. W. Albaugh.
Bulletin of Environmental Contamination and Toxicology, Vol 8, No 6, p 334-338, December, 1972. 1 tab, 12 ref.

Descriptors: *Bioassay, *Crayfish, *DDT, Resistance, *Pesticide toxicity, Water pollution effects, Crustaceans, Invertebrates, Insecticides, Chlorinated hydrocarbon pesticides, Organophosphorus pesticides, Phosphothioate pesticides, Aquatic animals, Texas.
Identifiers: *Toxaphene, *Methyl parathion, *Procambarus acutus, Macroinvertebrates, Decapods, Arthropods.

Five consecutive bioassays were carried out on Procambarus acutus to determine levels of tolerance to DDT, toxaphene, and methyl parathion, and to compare the tolerance of specimens from an area of intensive insecticide use with that of specimens from an area where use was minimal. Equal numbers of male and female specimens were subjected to each insecticide treatment and crayfish from both areas were tested at 3-5 concentrations of each pesticide in each bioassay. The 48-hr LC50 values for DDT, methyl parathion, and toxaphene were 2.4, 1.4, and 1.5 times greater, respectively, for animals from the area of high use than for those from the area of low insecticide use. For crayfish from the clean area, DDT and methyl parathion had similar toxicity, but the LC50 for toxaphene was more than 20 times greater. (Holoman-Battelle)
W73-08032

METABOLISM OF DDT BY FRESH WATER DIATOMS,
Manitoba Univ., Winnipeg. Dept. of Entomology.
For primary bibliographic entry see Field 05B.
W73-08036

EFFECTS OF OXATHIIN SYSTEMIC FUNGICIDES ON VARIOUS BIOLOGICAL SYSTEMS,
Montana State Univ., Bozeman. Dept. of Botany and Microbiology.
D. E. Mathre.
Bulletin of Environmental Contamination and Toxicology, Vol 8, No 5, p 311-316, November 1972. 3 tab, 18 ref.

Descriptors: *Toxicity, *Enteric bacteria, Fungicides, Chlorella, Growth rates, Metabolism, *Molds, Photosynthesis, Inhibition, Slime.
Identifiers: *Carboxin, *Oxycarboxin, Proteus vulgaris, Bacillus cereus, Pseudomonas aeruginosa, Nocardia rubra, Lactobacillus casei, Azotobacter chroococcum, Streptomyces, Sarcina lutea, Mycobacterium phlei.

Several bacteria, a slime mold, and Chlorella pyrenoidosa were exposed to carboxin (5,6-dihydro-2-methyl-1,4-oaxathiin-3-carboxanilide) and its oxidized products to determine their toxic effects. In the presence of .0001 M carboxin, the following bacteria were inhibited in growth from 0-10 percent: Proteus vulgaris, Bacillus cereus, Pseudomonas aeruginosa, Nocardia rubra, Lactobacillus casei, and Azotobacter chroococcum. Streptomyces sp., Sarcina lutea, and Mycobacterium phlei was inhibited from 10-20 percent. The metabolism of C-14-acetate was somewhat more sensitive to .0001 M carboxin in that the release of C-1402 was inhibited by 34 percent in P. vulgaris and 37 percent in S. leutea. The development of sporangia by the slime mold D. discoideum was not affected by .0001M carboxin, F831, or oxycarboxin. Metabolism of C-14-acetate by Chlorella cells was not inhibited by .0001 M carboxin, F831, or oxycarboxin. However, photosynthesis was inhibited by 52 percent with .0001 M carboxin but not with .0001 M F831 or oxycarboxin. (Little-Battelle)
W73-08045

RAPID BIODEGRADATION OF NTA BY A NOVEL BACTERIAL MUTANT,
Department of the Environment, Burlington (Ontario). Centre for Inland Waters.
For primary bibliographic entry see Field 05B.
W73-08046

THE LITTORAL MICROPHYTIC VEGETATION OF LAKE OHRID (IN SERBO-CROATIAN),
Hidrobioloski Zavod, Ochrida (Yugoslavia).
I. Cado.
Acta Bot Croat. 30: p 85-94. 1971. Illus. English summary.
Identifiers: Bangia, Cladophora, Gloeocapsa, *Lake Ohrid, Littoral, Microcoleus, Microphytic vegetation, Rivularia, Schizothrix, Scytonema, *Vegetation, Yugoslavia, Zonation.

Under the modified Mediterranean climatic conditions, the predominant calcareous structure of the massifs of the Lake basin and the features of the shore with its numerous karstic, surface and sublacustrine sources, a classic littoral zone has been differentiated in Lake Ohrid with its more characteristic facies: residue facies, sandy-shore facies, pebble-stone and stone facies, and rock facies. The most numerous within the lithophytic association are the representatives of the following groups; Cyanophyta, Bacillariophyceae, Chlorophyta, Rhodophyta, Bryophyta, Lichenes and bacteria. Forms of a wide ecological range which can endure frequent thermal excesses, long spells of dry weather and intensive insolation predominate. The lithophytic vegetation in the littoral zone is stratified. Supralittoral, eulittoral and infralittoral zones were distinguished. The supralittoral could be defined as the Gloeocapsa-Scytonema zone, the eulittoral as the Microcoleus-Schizothrix-Rivularia zone on the one hand and as the Bangia-Cladophora zone on the other. The infralittoral is characterized by an increased number of forms and by an increased quotient in relation to Cyanophyta and other groups.--Copyright 1972, Biological Abstracts, Inc.
W73-08061

TOXICITY OF CHEMICALS IN PAPER FACTORY EFFLUENTS,
Danmarks Fiskeri- og Havundersogelser, Charlottenlund (Denmark).
B. Norup.
Water Research, Vol 6, No 12, p 1585-1588, December 1972. 2 fig, 20 ref.

Descriptors: *Effluents, *Pulp and paper industry, Water pollution effects, Industrial wastes, *Toxicity, Fish, Freshwater fish, Resistance, Bioassay, Laboratory tests, Mercury, Fish physiology.
Identifiers: *Pentachlorophenol, *Lebistes reticulatus, Sodium pentachlorophenolate, Guppy, Slimicides, Phenols, Chlorinated hydrocarbons, Sodium pentachlorophenate, Mercury compounds, Median survival time.

Pentachlorophenol (PCP), a common toxic substance discharged from pulp and paper factories, was compared with mercuric compounds, and its effect on fish resistance was investigated at sublethal PCP-levels. Female guppies (Lebistes retuculatus), acclimated for at least 5 days at 24 plus or minus 0.5C, were placed in aerated glass tanks in groups of 5-10 per mg Na-PCP. The resistance of the guppy to the sodium salt, Na-PCP, has been shown to increase after acclimation to sublethal levels (1 ppm). The mean survival time of the guppy placed in 5 ppm Na-PCP after acclimation changed significantly from 65 min to 104 min. Such resistance may lead to increased tolerance of accumulated PCP in the organism where severe metabolic distortions, delayed sexual maturity and increased mortality may result. The guppy has been shown to have the fastest reaction and the greatest tolerance among fish. It has been demonstrated by this research that PCP is as toxic to fish as the dangerous, previously used slimicides containing mercury, and less efficient as a controllant of slime organisms and that the use of PCP should be restricted in a manner similar to mercuric compounds to ensure the survival of fish life downstream from paper manufacturing processes. (Holoman-Battelle)
W73-08235

TECHNIQUE FOR MEASURING METALLIC SALT EFFECTS UPON THE INDIGENOUS HETEROTROPHIC MICROFLORA OF A NATURAL WATER,
Simon Fraser Univ. Burnaby (British Columbia). Dept. of Biological Sciences.
For primary bibliographic entry see Field 05A.
W73-08236

CHANGES IN THE MICROBIAL POPULATIONS OF A RESERVOIR TREATED WITH THE HERBICIDE PARAQUAT,
University of Wales Inst. of Science and Tech., Cardiff.
J. C. Fry, M. P. Brooker, and P. L. Thomas.
Water Research, Vol 7, No 3, p 395-407, March 1973. 7 fig, 2 tab, 33 ref.

Descriptors: *Microbial degradation, *Paraquat, Aquatic microorganisms, Water pollution effects, *Pesticide toxicity, *Viability, *Resistance, Chemical analysis, Mud, Aquatic algae, Enzymes, Herbicides, Pondweeds, Chlorophyta, Water sampling, Aquatic weed control, Soil analysis, Aquatic soils, Standing crops, Hydrogen ion concentration.
Identifiers: Angiosperms, Amylase, Protease, Cellulase, Chara globularis, Myriophyllum spicatum, Potamogeton pectinatus, Heterotrophic bacteria, Culture media, Organic carbon, Macrophytes, Esgram.

A freshwater fishing reservoir was treated with paraquat for the control of weeds and the response of microbial populations studied. Estimates of the standing crop of macrophytes (P. pectinatus, M. spicatum, and C. globularis) were made periodically by determining the organic carbon content. A paraquat formulation (Esgram was sprayed evenly over the reservoir. Water, weed and mud were sampled frequently after each spraying and analyzed. Water samples taken in sterile bottles and mud samples taken from the surface of the mud were used in the microbial determinations. Counts of (1) viable heterotrophic microorganisms, (2) amlyase producers, and (3) viable protease and cellulase producers were made on CPS medium using different method of development. Viable paraquat resistant microorganisms were enumerated with the addition of 50 micrograms/ml paraquat, as Esgram, to the complete CPS medium. Submerged angiosperms were completely eradicated by the application of 1.0 mg/l paraquat, but the subsequent growth of the macrophytic alga, Chara sp., was resistant to a second application of the herbicide. Some changes in the microbial populations of the reservoir over the period of study were consistent with the movement of paraquat within the system and others with the death of the plants. Numbers of arbitrarily classified 'paraquat resistant' micro-organisms increased in the water and mud immediately after both herbicide applications, and after the first application a reduction in total viable heterotroph counts was observed. Accompanying the death of the angiosperms were increases in the counts of viable heterotrophs and some exoenzyme producers in the mud and water but after the second application of paraquat, when there was no plant death, these micro-organisms showed little response. (Holoman-Battelle)
W73-08239

NUTRIENT STUDIES IN TEXAS IMPOUNDMENTS,
Union Carbide Corp., Tonawanda, N.Y. Linde Div.
V. H. Huang, J. R. Mase, and E. G. Fruh.
Journal Water Pollution Control Federation, Vol 45, No 1, p 105-118, January 1973. 10 fig, 11 tab, 6 ref.

Descriptors: *Limiting factors, *Nutrients, *Photosynthesis, *Bioassay, *On-site tests, *Cyanophyta, *Nitrogen fixation, *Dominant organisms, *Chlorophyta, Carbon, Nitrogen, Phosphorus, Iron, Growth rates, *Texas, Colorado River, Chlorella, Diatoms, Succession, Anabaena.
Identifiers: Acetylene reduction, Lake Livingston, Lake Travis, Trinity River, Chlorella pyrenoidosa, Oscillatoria phomidium.

The objective was to determine the limiting nutrients in two distinctly different reservoirs in Texas, Lake Livingston on the Trinity River and Lake Travis on the Colorado River. The former is laden with relatively high organic and inorganic nutrient concentrations; the other has a low nutrient loading. Phytoplankton and water quality samples were collected and returned to the laboratory for nutrient enrichment tests as well as C-14 and nitrogen fixation tests. The latter two tests were also conducted in situ. Enrichment tests were conducted with natural populations and with inoculations of Chlorella pyrenoidosa. Growth rates were determined every 2 days by optical density measurements. C-14 tests were made with scintillation counts of laboratory and in situ samples with and without added nutrients. Nitrogen fixation was determined by the acetylene reduction method. The results showed that in the high-nutrient system, nitrogen was the limiting nutrient in summer; blue-green nitrogen-fixing algae became dominant in the late summer and from laboratory tests seem to be regulated by the available phosphorus. In the low-nutrient system, nitrogen, phosphorus, or iron could limit phytoplankton growth at different times of the year. With phosphorus enrichment of this system, algae with nitrogen-fixing capabilities could-develop. (Little-Battelle)
W73-08241

NITROGEN FIXATION BY A BLUE-GREEN EPIPHYTE ON PELAGIC SARGASSUM,
Woods Hole Oceanographic Institution, Mass.
E. J. Carpenter.
Science, Vol 178, No 4066, p 1207-1209, December 15, 1972. 2 tab, 16 ref.

Descriptors: *Nitrogen fixation, *Nitrogen cycle, Cyanophyta, Sea water.
Identifiers: *Dichothrix fucicola, *Sargasso Sea, Acetylene reduction, Enrichment.

Nitrogen fixation by Dichotrix fucicola, an epiphyte on pelagic Sargassum, was measured by acetylene reduction in May and June, 1972, in the western Sargasso Sea and the Gulf Stream. This is the first report of nitrogen fixation by a heterocyst-bearing blue-green alga in the open ocean, and also the first observation of nitrogen fixation in the genus Dichothrix. Cellular carbon-nitrogen ratios suggested that the Dichothrix was nitrogen-starved. In dense aggregations of Sargassum, such as rafts or windrows, the enrichment of surface seawater with combined nitrogen from nitrogen fixation may be pronounced. (Little-Battelle)
W73-08246

LOWER PH LIMIT FOR THE EXISTENCE OF BLUE-GREEN ALGAE: EVOLUTIONARY AND ECOLOGICAL IMPLICATIONS,
Wisconsin Univ., Madison. Dept. of Bacteriology.
T. D. Brock.
Science, Vol 179, No 4072, p 480-483, February 2, 1973. 1 tab, 25 ref.

Descriptors: *Hydrogen ion concentration, *Cyanophyta, *Limiting factors, *Plant growth, Cultures, Lakes, Algal control.

Observations on a wide variety of acidic environments, both natural and man-made, reveal that blue-green algae (Cyanophyta) are completely absent from habitats in which the pH is less than 4 or 5, whereas eukaryotic algae flourish. By using enrichment culture with inocula from habitats of various pH values, the absence of blue-green algae at low pH was confirmed. The ecological implications of the conclusions are clear. Blue-green algal blooms should never occur in acid lakes, and the pollution of lakes and streams with acid mine drainage should eliminate blue-green algae from these waters. Since even in mildly acidic waters (pH 5 to 6) blue-green algae are uncommon, mild acidification of lakes may control or eliminate blue-green algal blooms. (Little-Battelle)
W73-08248

BLUE-GREEN ALGAE: WHY THEY BECOME DOMINANT,
Minnesota Univ., Minneapolis. Limnological Research Center.
J. Shapiro.
Science, Vol 179, No 4071, p 382-384, January 26, 1973. 1 fig, 1 tab, 5 ref.

Descriptors: *Chlorophyta, *Cyanophyta, Limiting factors, *Carbon dioxide, *Hydrogen ion concentration, *Nitrogen, *Phosphorus, *Dominant organisms, Bioassay, Nitrates, Phosphates, Competition, Succession, On-site investigations, Lakes, *Minnesota.
Identifiers: *Lake Emily (Minn).

Mixed populations of algae were subjected to a variety of treatments including high concentrations of nitrogen plus phosphorus, high concentrations of CO_2 and high or low pH (5-6) to test the hypothesis that blue-green algae become dominant because they are more efficient at obtaining CO_2 from low concentrations than green algae. The algae were put in plastic bags suspended from a raft in Lake Emily, Minnesota. Although initially the populations consisted of blue-green algae, samples subjected to CO_2 plus nutrients were dominated by green algae. Somewhat similar results occurred with lowered pH and nutrients. It appears that the addition of free CO_2 or lowering the pH make more CO_2 available to green algae and allow them to become dominant. Since green algae are more desirable than blue-greens, injection of CO_2 in lakes may be a way of controlling blue-greens where nutrient sources cannot be controlled. (Little-Battelle)
W73-08249

EXPERIENCE WITH ALGAL BLOOMS AND THE REMOVAL OF PHOSPHORUS FROM SEWAGE,
M. A. Simmonds.
Water Research, Vol 7, Nos 1/2, p 255-264, January/February 1973. 6 tab, 3 ref.

Descriptors: *Algae, *Hydrogen ion concentration, *Alkalinity, *Carbon dioxide, Limiting factors, Growth rates, Waste water treatment, Absorption, Nutrients, Phosphorus, Phosphates, *Australia.
Identifiers: *Phosphorus removal.

Based upon observations of algal blooms in water treatment plants during the period 1930-1940 when phosphate occurred primarily from natural sources, the conclusion is made that the mechanism which triggers algal blooms may be neither nutrient concentration nor the concentration of organic matter. Instead the pH, alkalinity, carbon dioxide equilibrium condition is a major factor, not only in promoting, but also in maintaining algal blooms. The mechanism involved is the conversion of bicarbonates to carbonates at high pH and the consequent release of carbon dioxide which is utilized by algae. Use of algae for removing phosphates from sewage sludge is discussed. Algae were capable of removing large amounts of phosphate, but were themselves difficult to remove from the sewage. (Little-Battelle)
W73-08251

EUTROPHICATION AND LOUGH NEAGH,
New Univ. of Ulster, Coleraine (Northern Ireland).
R. B. Wood, and C. E. Gibson.
Water Research, Vol 7, Nos 1/2, p 173-187, January/February 1973. 5 fig, 5 tab, 27 ref.

Descriptors: *Eutrophication, Water quality, *Phosphorus, *Limiting factors, *Diatoms, Primary productivity, Industrial wastes, Municipal wastes, Cyanophyta, Lakes, Nitrates, Phosphates, Nitrogen, Nutrients, Chlorophyll, Phytoplankton, Zooplankton, Algae, Midges, Invertebrates, Chrysophyta, Isopods.
Identifiers: *Lough Neagh, Chlorophyll a, Macroinvertebrates, *Ireland, Oscillatoria, Cyclops spp., Diaptomus spp, Diaptomus, Asellus, Glyptotendipes paripes, Glyptotendipes, Chironomus anthracinus, Procladius spp, Mysis relicta, Cyclotella comensis, Cyclotella ocellata, Tabellaria flocculosa, Melosira italica, Stephanodiscus astraea, Stephanodiscus hantzschii.

Comparison of biological and chemical characteristics of Lough Neagh (Ireland) with those of other lakes shows that Lough Neagh is among the most eutrophic of the world's major lakes. Phosphorus appears to be the key factor which limits the growth of algae. It is estimated that Lough Neagh receives 300 tons of P per year, 70-80 percent of which is probably from urban and industrial sewage. It is concluded that reduction of P content of effluents from sewage works could have a beneficial effect on the eutrophic condition of the lake. (Little-Battelle)
W73-08252

STIMULATION OF PHYTOPLANKTON GROWTH BY MIXTURES OF PHOSPHATE, NITRATE, AND ORGANIC CHELATORS,
Virginia Inst. of Marine Science, Gloucester Point.
R. A. Jordan, and M. E. Bender.
Water Research, Vol 7, Nos 1/2, p 189-195, January/February 1973. 2 fig, 3 tab, 8 ref.

Descriptors: *Nitrates, *Phosphates, *Growth rates, *Bioassay, *Primary productivity, *Algae, Nutrients, Cultures, *Michigan, Lakes, Phytoplankton, Phosphorus, Nitrogen, Diatoms, Chrysophyta, Cyanophyta, Chlorophyta, Aquatic algae.
Identifiers: *Crystal Lake (Mich), Synedra nana, Fragilaria crotonensis, Synedra radians, Achnanthes, Synechocystis aquatilus, Rhodomonas minuta, Cyclotella ocellata, Cryptomonas ovata, Cyclotella stelligera, Pediastrum boryanum, EDTA, Nitzschia.

An in situ nutrient enrichment experiment was conducted in which mixed treatments of nitrate, phosphate, and EDTA were applied to natural lake phytoplankton communities. Changes in community productivity and species composition in response to the treatments revealed strong interactions among the components of the treatment mixture. On the community level, phosphate exerted a stimulatory effect that was reduced by EDTA, enhanced by nitrate, and enhanced even more by nitrate and EDTA together. Examination of 15 individual species revealed that the treatment effects were highly variable from species to species. Seven of the 15 species were stimulated by the nutrient treatments, and the growth patterns of 5 of these accounted for essentially all of the features of the productivity response patterns. The eight other species either failed to respond to any treatment or declined in response to containment or treatments. Phosphate was the key substance in all of the positive treatment effects, and its omission from the treatment mixture essentially eliminated all growth responses. (Little-Battelle)
W73-08253

THE ROLE OF PHOSPHORUS IN THE GROWTH OF CLADOPHORA,
Aston Univ., Birmingham (England). Dept. of Biological Sciences.
C. E. R. Pitcairn, and H. A. Hawkes.
Water Research, Vol 7, Nos 1/2, p 159-171, January/Febrauary 1973. 7 fig, 8 tab, 13 ref.

Descriptors: *Phosphorus, *Growth rates, *Bioassay, *Eutrophication, *Limiting factors, *Cladophora, Water pollution effects, Nitrogen, Standing crops, Chlorophyta, Algae, Nutrients, Phosphates, Sewage, Cultures, Statistical methods, Rivers, Aquatic algae.
Identifiers: *Culture media, *England.

An examination of river survey data showed standing crops of Cladophora to be correlated with phosphorus concentration. In general, river water containing less than 1.0 mg/l total inorganic P produced only modest growths of Cladophora. Culture experiments with supplemented river water confirmed the importance of phosphorus by showing that growth of Cladophora in waters upstream of sewage discharges could be increased to downstream levels by addition of phosphorus. Growth experiments in synthetic media containing levels of phosphorus from 1 to 7 mg/l indicated no significant growth increase above 1 mg P/l but a significant reduction below 1 mg P/l. In natural water, the maximum level of phosphorus for growth was found to vary, being 2.5 mg P/l at 3.2 mg N/l NO3 and 0.95 mg P/l at 5.25 mg N/l NO3. A 3 x 4 factorial experiment utilizing synthetic media, confirmed an interaction between nitrogen and phosphorus. The highest level of NO3 (7.7 mg N/l) enhanced growth at the lowest phosphorus level (0.5 mg P/l) but at higher levels of phosphorus, growth was reduced. The importance of such interactions is discussed briefly in connection with entrophication and nutrient stripping. (Little-Battelle)
W73-08254

ROLE OF PHOSPHORUS IN EUTROPHICATION AND DIFFUSE SOURCE CONTROL,
Wisconsin Univ., Madison. Water Chemistry Program.
G. F. Lee.
Water Research, Vol 7, Nos 1/2, p 111-128, January/February 1973. 1 tab, 22 ref.

Descriptors: *Eutrophication, *Phosphat *Lakes, *Limiting factors, *Water pollution co trol, Bioassay, Cycling nutrients, Detergents, A gae, Arsenic, Sediments, Phosphorus, Nutrien Absorption, Analytical techniques, Domes wastes, Water pollution sources.
Identifiers: Orthophosphates, Mobilization.

Many lakes and some streams and estuaries a showing signs of excessive fertilization due to t input of aquatic plant nutrients from man-a sociated sources. The key element often fou limiting aquatic plant populations is phosphoru The attempt to control phosphorus input to natur waters as the overall approach for controlling e cessive fertilization is technically sound a economically feasible for many natural water However, a much better understanding of the rel tionship between the phosphorus input to a lak and the excessive growths of aquatic plants wi the lake must be developed. This development require a combined biological and chemical approach toward assessing the role of phosphorus in eutrophication for a specific water body. The biological approach will use tissue content, enzymatic and kinetic uptake analysis of phosphorus limitations as well as bioassays of phosphorus availability in order to determine the limiting nutrient for a body of water. The chemical approach will utilize amounts of each of the forms of phosphorus present in the lake and the rates of interchange of phosphorus between these various forms. There will be some waters where control of phosphorus from treatment of domestic waste water input and removal of phosphorus from detergents will not result in significant improvement in water quality. This is because these waters derive their phosphorus from diffuse sources, such as urban and rural stormwater drainage, the atmosphere and ground waters. In these instances, it may be necessary to initiate in-lake control of phosphorus by the addition of alum or iron salts. (Little-Battelle)
W73-08255

PHOSPHORUS IN PRIMARY AQUATIC PLANTS,
University Coll. of North Wales, Menai Bridge. Marine Science Labs.
G. E. Fogg.
Water Research, Vol 7, Nos 1/2, p 77-91, January/February 1973. 65 ref.

Descriptors: *Algae, *Phosphates, *Limiting factors, *Growth rates, *Absorption, Nutrients, Cyanophyta, Metabolism, Chlorophyta, Chrysophyta, Rhodophyta, Diatoms, Phosphorus, Nutrient requirements, Aquatic plants.
Identifiers: Orthophosphates, Biotransformation.

A review of the relationships between algae and phosphorus shows that many species can absorb orthophosphate from solutions containing less than 1 ppm P and, when phosphorus-deficient, most species are capable of producing powerful surface or extracellular phosphatases which enable them to obtain phosphate from a great variety of inorganic and organic phosphorus compounds, including synthetic detergents. In the presence of sufficient phosphate algal cells are able to accumulate a store of polyphosphate which suffices for several cycles of cell division in the absence of a further supply. As a result of excretion of phosphates at certain stages of the life cycle and extracellular phosphatase activity there is rapid recycling of phosphorus so that algal activity may be high even when the concentration of free phosphate in the water is low. Behavioral patterns may impose further complexity. There is evidence that planktonic blue-green algae possess a buoyance control mechanism operating via their gas vacuoles that may enable them to descend at night to phosphate-rich water at the bottom of the photic zone and rise nearer the surface in the morning. Because of these complications no clear relationships between the amount of algal growth

M. Koryak, M. A. Shapiro, and J. L. Sykora.
Water Research, Vol 6, No 10, p 1239-1247, October, 1972. 4 fig, 16 ref.

Descriptors: *Benthic fauna, *Mine drainage, *Acid streams, *Invertebrates, *Aquatic animals, *Coals, Water pollution effects, Amphipoda, Oligochaetes, Biomass, Bioindicators, Aquatic insects, Larvae, Bottom sampling, Chemical analysis, Water analysis, Water velocity, Crustaceans, Annelids, Surface waters, Dissolved oxygen, Depth, Iron, Midges, Diptera, Water beetles, Mayflies, Stoneflies, Hydrogen ion concentration, Biochemical oxygen demand, Acidity.
Identifiers: *Receiving waters, Macroinvertebrates, Turtle Creek, Haymaker Creek, Trafford Road Run, Beyers Creek, Lyons Run, Atomic absorption spectrophotometry, Surber sampler, Riffles, Tendipes gr. riparius, Scuds, Ulothrix tenerrima, Psychoda, Antocha.

The bottom fauna of a stream polluted by acid mine drainage was studied using the standard methods of sample collecting. In localities immediately influenced by mine drainage where very low pH values and high acidities prevail, the effect of acid mine wastes on the ecology and composition of the benthic fauna is, in general, similar to the effect of organic pollution. In these areas high numbers of individuals comprised of a few species were found. In the zones of ctive neutralization where iron hydroxides are deposited, species diversity slightly increases but the biomass is very low. The most numerous invertebrates in the stream section exhibiting high acidity and low pH are midge larvae, especially Tendipes gr. riparius. The number of insect groups present increases steadily with progressive neutralization until Crustaces (Amphipoda) and Oligochaeta appear, indicating considerable improvement in water quality. The supply of desirable benthic fish food (Tendipes ssp.) is very high in the parts of the stream where low pH, high acidity, and high ferrous iron concentrations prevail. Unfortunately, fish cannot survice under these conditions to utilize this abundant food supply. On the other hand, in the less drastically diminishes the total biomass of benthic organisms and therefore severely limits fish populations, (Holoman-Battelle)
W73-08268

CAUSES OF MASSIVE INFECTION OF PINK SALMON WITH SAPROLEGNIA IN THE RIVERS OF ITURUP (ETOROFU) ISLAND, (IN RUSSIAN),
For primary bibliographic entry see Field 05B.
W73-08271

SOME CHARACTERISTICS OF GENERAL METABOLISM IN CARP YEARLINGS DURING RETAINER RAISING IN WARM WATER, (IN RUSSIAN),
L. A. Korneeva, and A. N. Korneev.
Tr Vses Nauchno-Issled Inst Prudovogo Rybn Khoz. Vol 17, p 220-224, 1971. English summary.
Identifiers: *Carp, *Metabolism, Retainer, Warm water, Yearlings, *Respiration.

The intensity of respiration in yearling carp raised in net retainers without forced washing through in cooling waters of a thermal electric power station was studied. The level of metabolism in carp during rearing in retainers was lower than that usually observed in pond fish. Conditions of rearing, such as compactness of the retainer and daily diet, affected the level of respiration in the fish. The O2 consumption was directly related to the compactness of the retainer and inversely related to the amount of daily ration.—Copyright 1972, Biological Abstracts, Inc.
W73-08276

THE INFLUENCE OF LOG HANDLING ON WATER QUALITY,
Oregon State Univ., Corvallis. Dept. of Civil Engineering.
For primary bibliographic entry see Field 05B.
W73-08294

MIGRATION AND METABOLISM IN A TEMPERATE STREAM ECOSYSTEM,
North Carolina Univ., Chapel Hill. Dept. of Zoology.
For primary bibliographic entry see Field 05B.
W73-08303

INFECTION BY ICHTHYOPHTHIRIUS MULTIFILIIS OF FISH IN UGANDA,
Makerere Univ., Kampala (Uganda). Dept. of Zoology.
I. Paperna.
Prog Fish-Cult. Vol 34, No 3, p 162-164. 1972. Illus.
Identifiers: Carp, *Fish diseases, *Ichthyophthirius multifiliis, Infection, *Uganda.

Infection by I. multifiliis of fish in Uganda is reported for the first time from tropical Africa. Infection developed in fish from local fish ponds 7 days after being introduced into aquariums and 1 small outdoor pool. Many species of native fish were susceptible to infection while carp were fairly refractory. The possible sources of this infection are discussed.—Copyright 1973, Biological Abstracts, Inc.
W73-08330

ON THE ECOLOGY OF AN ADULT DIGENETIC TREMATODE PROCTOECES SUBTENUIS FROM A LAMELLIBRANCH HOST SCROBICULARIA PLANA,
I. C. White.
J Mar Biol Assoc U K. Vol 52, No 2, p 457-467. 1972. Illus.
Identifiers: Digenetic, Ecology, Host, *Lamellibranch, Proctoeces subtenuis, Scrobicularia plana, *Trematodes, *Thames estuary.

P. subtenuis (L inton), an adult digenetic trematode parasitic within the kidney region of the lamellibranch S. plana (da Costa) was found only in specimens of the host collected from localities along the north coast of the Thames estuary, although the lamellibranch was common in neighboring areas. An investigation of the S. plana from 8 locations along the north coast of the Thames estuary revealed that the abundance of the parasite was far from uniform with S. plana collected from certain localities being heavily infected whereas those collected from short distances away (one mile or less) were often only rarely infected. This pattern was repeated in each of the 3 yr of study. The investigation of a heavily infected population of S. plana over the period of study demonstrated that the parasite was very successful. From a level of infection of 2-3 P. subtenuis per host in 1967 an increase occurred to a level of infection in 1969/70 at which over 95% of all S. plana collected were infected and with an average of 4-5 P. subtenuis per host. As many as 14 P. subtenuis were recovered from a single host and it was demonstrated that the number of P. subtenuis per S. plana increased pari passu with the size of the host. As well as living parasities, dead but preserved P. subtenuis were found in the kidney region of some hosts but their significance is obscure.—Copyright 1973, Biological Abstracts, Inc.
W73-08331

EFFECT OF DRY SEASON DROUGHT ON UPTAKE OF RADIOACTIVE PHOSPHORUS BY SURFACE ROOTS OF THE OIL PALM (ELAEIS GUINEENSIS JACQ.),
University of the West Indies, St. Augustine (Trinidad).
For primary bibliographic entry see Field 02I.
W73-08334

MICROFLORA OF THE NEUSTON AND ITS ROLE IN WATER BODIES,
Polskie Towarzystwo Przyrodnikow im. Kopernika, Warsaw.
S. Niewolak.
Wszechswiat. 4 p 91-93. 1971.
Identifiers: *Microflora, Algae, Bacteria, *Neuston, Protozoa.

A significant role is played in the life of aquatic organisms by the surface water film of inland water bodies. This surface biocenosis is called neuston and is characterized by the presence of numerous bacteria, algae and protozoa. The surface microflora species found in the Ilawa Lakes are described, as also a number of microflora species found by other authors in the Black Sea. Results of chemical analyses of organic matter found in the surface water layer are given. The importance of the water-air phase in the development of surface water aquatic organisms is described. Biology of the surface water film is also discussed.–Copyright 1972, Biological Abstracts, Inc.
W73-08335

EFFECT OF ORGANIC AND MINERAL FERTILIZER ON THE HYDROCHEMICAL SYSTEM OF RICE PADDIES STOCKED WITH FISH WHICH WERE EXPOSED UNDER WATER VAPOR (IN RUSSIAN),
Z. F. Sorokina.
Tr Vses Nauchno-Issled Inst Prudovogo Rybn Khoz. 20: p 3-17. 1971. English summary.
Identifiers: *Fertilizers, *Fish farming, Humus, Paddies, *Rice M, Vapor, Weeds.

Rice paddies were fertilized with different doses of mineral and organic fertilizers. The best results in fish and rice productivity were obtained with superphosphate in a dose of 180 kg active substance/ha. High doses of N fertilizer (180 kg of active substance/ha) had a negative effect. After raising fish for 2 yr on 'resting' (fallow) rice paddies the soil fertility increased, the content of organic matter increased by 30% and the humus content increased 2 times. The weediness of the rice fields decreased to 3-4% of the original amount.–Copyright 1972, Biological Abstracts, Inc.
W73-08339

THE CHEMICAL NATURE OF THE SUBSTANCES THAT INDUCE THE REDDENING OF LAKE TOVEL: III. STUDY OF A REMARKABLE PHENOMENON OF PRE-REDDENING (IN ITALIAN),
V. Gerosa.
Studi Trentini Sci Nat Sez B Biol, Vol 47, No 2, p 107-132, 1970, Illus, English summary.
Identifiers: *Water pollution effects, Chemical properties, *Glenodinium sanguineum, Lakes, *Reddening, *Lake Tovel, *Cartenogenesis.

Initial carotenogenesis in the presence of large quantities of Glenodinium sanguineum and its relationship to the pre-reddening of Lake Tovel during the summer months is discussed.–Copyright 1972, Biological Abstracts, Inc.
W73-08344

EFFECTS OF FLOODING AND GASEOUS COMPOSITION OF THE ROOT ENVIRONMENT ON GROWTH OF CORN,
Illinois Univ., Urbana. Dept. of Agronomy.
For primary bibliographic entry see Field 03F.
W73-08346

STUDIES ON THE PRODUCTIVE STRUCTURE IN SOME LAKES IN KOREA, (IN KOREAN),
Seoul National Univ. (Republic of Korea). Dept. of Botany.
K. B. Uhm.
Korean J Bot. Vol 14, No 1, p 15-23, 1971. Illus. English summary.
Identifiers: Changja, *Chlorophyll, Distribution, Hwajinpo, *Korea, Lakes, Seasonality, Trophic type, Vertical stratification, Yongrang, *Phytoplankton, *Productivity.

The productivity of summer phytoplankton communities in Lake Hwajinpo, Lake Yongrang and Lake Changja was studies by measuring vertical variation of chlorophyll-a. Lakes were classified on the basis of the amount of chlorpbyll in the water. In Lake Changja, the seasonal changes of stratification of chlorophyll were studies. In Lake Hwajinpo, the productive structure of the phytoplankton community in summer was found to be L-shaped and of the mesotrophic type. In Lake Yongrang, the productive structure of the phytoplankton community in summer was also L-shaped and of the mesotrophic type. And maximum chlorophyll layer was near the lake bottom below the compensation depth. In Lake Changja, the structure of phytoplankton community in summer was reversed L-shaped and of the eutrophic type, with the maximum chlorophyll layer just below the surface. The vertical distribution of chlorophyll amounts as a measure of the productive structure almost always formed a stratified distribution except in Sept. and sometimes in May, in Lake Changja. In Sept. homogeneous distribution was observed.–Copyright 1973, Biological Abstracts, Inc.
W73-08402

SEASONAL CHANGES IN PHYTOPLANKTON AND WATER CHEMISTRY OF MOUNTAIN LAKE, VIRGINIA,
Chana Univ., Legon. Volta Basin Research Project.
For primary bibliographic entry see Field 02H.
W73-08404

THE EFFECT OF KRAFT PULP MILL EFFLUENTS ON THE GROWTH OF ZALERION MARITIUM,
Simon Fraser Univ., Burnaby (British Columbia). Dept. of Biological Sciences.
L. M. Chruchland, and M. McClaren.
Can J Bot. Vol 50, No 6, p 1269-1273. 1972.
Identifiers: Effluents, Growth, Kraft mill effluents, Nutrients, *Pulp wastes, *Zalerion-Maritimum, *Aquatic fungi.

Growth of the marine fungus Z. maritimum was measured in Kraft pulp mill effluents. The effluents used were caustic effluent, acidic effluent, and acidic effluent adjusted to pH 8. The effluent and the seawater control flasks were supplemented, in some instances, with basal nutrients. Two concentrations of effluent, 50% (50% effluent, 50% seawater) and 100%, were used. When basal nutrients were added, dry weight was significantly greater (95% probability level) in 100% caustic effluent than in 50% caustic effluent or seawater alone. Without basal nutrients, growth was lower in 100% caustic effluent than in seawater or 50% effluent. This suggests that growth of Z. maritimum would not be stimulated by Kraft effluent under field conditions. Unlike the caustic effluent, the acidic effluent inhibited growth of Z. maritimum. With added nutrients, growth was lower in 50% and 100% acidic effluent than in seawater. When the acidic effluent solutions were adjusted to pH 8, growth was lower than in the seawater medium. –Copyright 1973, Biological Abstracts, Inc.
W73-08426

A YEARS' STUDY OF THE DRIFTING ORGANISMS IN A BROWNWATER STREAM OF ALBERTA, CANADA,
Alberta Univ., Edmonton. Dept. of Zoology.
H. F. Clifford.
Can J Zool. Vol 50, No 7, p 975-983. 1972. Illus.
Identifiers: Alberta, *Brown water streams, Canada, Cladocerans, Cyclopoids, *Drifting stream insects, Organisms, Ostracods, Streams, *Entomostracans.

Ten 24-hr drift samples were taken from a brownwater stream of Alberta, Canada over a 1-yr period with drift nets having a mesh size of 320 micro. Cladocerans, cyclopoids, and ostracods, collectively called entomostracans, made up a large part of the drift by numbers and contributed substantially to the total biomass of the drift. Drift densities of entomostracans tended to increase as the ice-free season progressed, but drift densities of immature insects remained relatively constant throughout the ice-free season. Total daily drift of both the entomostracans and non-entomostracan fractions tended to decrease as the ice-free season progressed, being dependent on water volume. Drift densities, total daily drift, and number of taxa in the drift were very low in winter. Most of the species exhibited nighttime behavioral drift. At the sampling site, the entomostracans and immature aquatic insects were found to be essentially evenly distributed throughout the water column. For part of the study period, drift densities of taxa caught in the 320-micro net were compared with drift densities of the same taxa caught in a 720-micro net. The 720-micro net caught a much smaller fraction of the aquatic insects than did the 320-micro net, and almost all the entomostracans passed through the 720-micro net. When compared with other regional drift studies, the large fraction of entomostracans in the brown-water stream seems to be a unique feature; there is evidence that most of the drifting entomostracans originate in the marshy area drained by the main stream. –Copyright 1973, Biological Abstracts, Inc.
W73-08427

DRIFT OF INVERTEBRATES IN AN INTERMITTENT STREAM DRAINING MARSHY TERRAIN OF WEST-CENTRAL ALBERTA,
Alberta Univ., Edmonton. Dept. of Zoology.
H. F. Clifford.
Can J Zool. Vol 50, No 7, p 985-991. 1972. Illus.
Identifiers: *Alberta, Canada, Chironomids, Cyclopoid, Draining, Drift samples, Harpacticoids, Intermittent streams, *Invertebrates, Marshy land, Naupli, Nematodes, Rotifers, *Plankton.

Seven 24-hr drift samples were taken with a plankton net (pore size: 76 micro) over 1-yr period from an intermittent stream that drains marshy, muskeg-type terrain of west-central Alberta, Canada. The drift was mainly composed of planktonic and benthic animals originating in the marsh. The only abundant lotic taxon in the drift was simuliid larvae. Rotifers and cyclopoid nauplii were numerically the most important taxa. Drift densities for the fauna as a whole tended to decrease as the ice-free season progressed, but there was no consistent correlation between drift densities and flow. However total daily drift across a point varied directly with flow. All the abundant taxa drifted more during the day than at night, and nematodes, harpacticoids, simuliid larvae, chironomid larvae, chydorids, and rotifers were found in significantly ($P<0.05$) greater numbers in the daytime drift. Drift rates of taxa caught in the plankton net were compared with drift rates of the same taxa caught in a 320-micro drift net. Rotifers, entomostracans (especially the immature stages), and even small simuliid and chironomid larvae would have been seriously underestimated using only the 320-micro net. The marshy areas via drift through the intermittent tributaries contribute a very large number of small organisms to the main stream. Draining the wetlands might have a pronounced detrimental effect on the main stream's ecosystem.–Copyright 1973, Biological Abstracts, Inc.
W73-08428

STUDIES ON BIOLOGICAL METABOLISM IN A MEROMICTIC LAKE SUIGETSU,
Nagoya Univ. (Japan). Water Research Lab.

but in Lake Borgne it rose suddenly from 38% to 88% and leveled off. In the delta complex west of the Mississippi River, it increased at a slower but more regular rate and reached a maximum of 63% in the last sampling week. Prochristianella penaei was found in 75% of 150 P. setiferus obtained from Lakes Pontchartrain and Borgne on 2 successive wk (July 24-Aug. 3) following a sharp decline in the availability of P. aztecus. Infection patterns of both species are discussed relative to the ecology of sampled areas and habits of hosts involved in the life cycle of P. penaei. Shrimp drawn concomitantly from different parts of a given estuary showed marked differences in infection. It appears that coarser substrata found at some stations may have been limiting to at least 1 host, possibly shrimp. No definite relationship was found between infection and length of either host in any of the estuaries sampled. Results are discussed relative to the potential utility of P. penaei as a living shrimp 'tag.'—Copyright 1973, Biological Abstracts, Inc.
W73-08433

SALICYLANILIDE I, AN EFFECTIVE NON-PERSISTENT CANDIDATE PISCICIDE,
Bureau of Sport Fisheries and Wildlife, La Crosse, Wis. Fish Control Lab.
L. L. Marking.
Trans Am Fish Soc. Vol 101, No 3, p 526-533. 1972.
Identifiers: Cyprinids, Fishery, Ictalurids, Nonpersistent, *Piscicides, *Salicylanilide, Salmonids, *Toxicity.

Salicylanilide I (2', 5-dichloro-3-tert-butyl-6-methyl-4'-nitrosalicylanilide) was tested for its toxicity to 20 spp. of freshwater fish in laboratory bioassays and to 15 spp. in outdoor pool bioassays. It is extremely toxic to all species of fish tested, and the 96-hr LC50's range from 0.3 to 8.6 ppb in standard tests. Ictalurids are about as sensitive as salmonids to Salicylanilide I. Cyprinids are equally or more sensitive to the chemical than some sentrarchids. Salicylanilide I effectively kills fish at similar concentrations in soft and hard water, in cold or warm water, and in acid or alkaline water. In outdoor pools, Salicylanilide I killed all fish of 15 spp. at concentrations of 40 and 60 ppb and all fish of 12 spp. at a concentration of 20 ppb. The chemical detoxifies in water, but detoxification appears to be inhibited by colder temperatures. The broad spectrum piscicidal activity of Salicylanilide I in waters of various qualities and temperatures offers advantages over presently used fish toxicants.--Copyright 1973, Biological Abstracts, Inc.
W73-08434

PHYSIO-MORPHOLOGICAL EFFECTS OF ABRUPT THERMAL STRESS ON DIATOMS,
Smithsonian Institution, Washington, D.C. Office of Environmental Sciences.
G. R. Lanza, and J. Cairns, Jr.
Trans Am Microsc Soc. Vol 91, No 3, p 276-298. 1972. Illus.
Identifiers: *Diatoms, Ecology, Effluents, Electric plants (Steam), Microscopical surveys, *Morphological studies, Stress, Surveys, *Thermal pollution.

The physio-morphological effects of several categories of defined abrupt temperature increases on diatoms were evaluated. Temperature increases which could result from entrainment through cooling lines of steam electric generating facilities and downstream additions of thermal effluents were simulated. In addition to established criteria such as standard microscopical surveys on cell morphology and effects on population growth, a new approach involving total cellular fluorescent patterns was developed and initially tested in the measurement of cellular alterations following stress. The changes in cellular fluorescent patterns prior to and following severe internal cellular destruction of diatoms are discussed, and ecological and physiological implications are indicated.--Copyright 1973, Biological Abstracts, Inc.
W73-08441

SPECIES DIVERSITY OF MARINE MACROBENTHOS IN THE VIRGINIA AREA,
Queensland Univ., Brisbane (Australia). Dept. of Zoology.
For primary bibliographic entry see Field 02L.
W73-08445

ECOLOGY OF ARBOVIRUSES IN A MARYLAND FRESHWATER SWAMP: I. POPULATION DYNAMICS AND HABITAT DISTRIBUTION OF POTENTIAL MOSQUITO VECTORS,
Walter Reed Army Inst. of Research, Washington, D.C.
E. S. Saugstad, J. M. Dalrymple, and B. F. Eldridge.
Am J Epidemiol. Vol 96, No 2, p 114-122. 1972. Illus.
Identifiers: Aedes-Canadensis, Aedes-Cantator, *Arboviruses, Culex-Salinarius, Culiseta-Melanura, Ecology, *Maryland, *Mosquito vectors, Vectors, Water pollution effects.

Entomological aspects of arbovirus ecology were studied in the Pocomoke Cypress Swamp in eastern Maryland. During 1969 nearly 350,000 adult and 10,000 larval mosquitoes were collected in the swamp and surrounding areas. Aedes canadensis, Culiseta melanura, Culex salinarius and A. cantator accounted for 89% of the total adult catch. Analyses of variance of the capture rates of adults of these species demonstrated highly significant differences in capture rates between 5 habitats (based on vegetative differences) sampled for 3 out of the 4 species, but no significant differences between collection sites within the same habitats. In some instances, interhabitat differences in adult density were related to differences in suitable larval breeding sites; in other cases differences appeared related to the availability of suitable hosts for blood feeding. Population peaks of several species of mosquitoes coincided with the peak of virus activity in the swamp, but C. melanura was the only species from which a group A virus was isolated. Fourteen isolates of Western Equine Encephalitis and 5 of Eastern were recovered from July 15 to Sept. 8, an average of 1 isolate for every 3,881 females. The significance of population dynamics and habitat distribution of the dominant mosquito species of the swamp to virus transmission is discussed. (See also W73-08447)—Copyright 1973, Biological Abstracts, Inc.
W73-08446

RECURRENT ST. LOUIS ENCEPHALITIS INFECTION IN RESIDENTS OF A FLOOD PLAIN OF THE TRINITY RIVER, ROOSEVELT HEIGHTS (DALLAS, TEXAS),
Texas Univ., Dallas. Southwestern Medical School.
J. P. Luby, and R. W. Haley.
Am J Epidemiol. Vol 96, No 2, p 107-113. 1972. Illus.
Identifiers: Culex-Quinquefasciatus, Culex-Tarsalis, Cycles, *Dallas, *Encephalitis (St. Louis), Equine, Floodplain, Infection, River, Texas, Trinity River (Tex).

A serologic survey was conducted among nonwhite persons residing in a circumscribed community (Roosevelt Heights) in Dallas, Texas, situated on a flood plain of the Trinity River. In total, 214 sera were collected. St. Louis encephalitis (SLE) neutralizing antibody (Ab) was found in 13.6% of the sample and rates revealed a statistically significant trend to increase by length of residence, suggesting that this community had experienced recurrent SLE infection. Western equine encephalitis (WEE) Ab was found in 1.9% of the survey population. The finding of a commu-

nity with recurrent SLE infection approximately 100 miles west of the towns that were previously investigated was thought basic to the epidemiologic proof that an interaction did exist between the 2 established transmission cycles for SLE virus in Texas (SLE, WEE-Culex tarsalis mosquitoes and SLE-C. quinquefasciatus mosquitoes).--Copyright 1973, Biological Abstracts, Inc.
W73-08448

5D. Waste Treatment Processes

A METHODOLOGY FOR PLANNING OPTIMAL REGIONAL WASTEWATER MANAGEMENT SYSTEMS,
Massachusetts Univ., Amherst. Water Resources Research Center.
D. D. Adrian, B. B. Berger, R. J. Giglio, F. C. Kaminsky, and R. F. Rikkers.
Available from the National Technical Information Service as PB-219 388, $6.75 in paper copy, $1.45 in microfiche. Publication No. 26, (1972). 243 p, 19 fig, 10 tab, 81 ref, 7 append. OWRR B-011-MASS (10).

Descriptors: Waste water (Pollution), *Regional economics, *Planning, Methodology, Human population, Combined sewers, Interceptors, *Cost allocation, *Massachusetts, *Decision making, Regional analysis, *Treatment facilities, *Construction costs, Optimal development plans, *Regional development.
Identifiers: Springfield (Mass), *Capacity expansion.

The research described is directed towards the development of a methodology and mathematical/computer models which can aid planning agencies to make decisions concerning the development of regional wastewater management plans. The type of region under consideration is one in which a set of communities, commercial establishments, and industries discharge their treated effluent to a common stream. The regional plan produced by the methodology does not yield detailed engineering plans for each treatment plant to be constructed in the region. It does yield a long-range construction program for the region which has been determined to be best in such a way that a quality environment results with a reasonable expenditure of funds. The report is divided into the following sections: (a) Summary of results, (b) Needs for further work, (c) Planning methodology, (d) Using the methodology, (e) Appendices. The seven appendices describe: (a) Population dynamics, (b) Optimal interceptor networks, (c) Optimal facility location-service model, (d) Optimal timing of capacity expansions, (e) Spacing effluent discharges, (f) Combined sewer problems, (g) Apportioning costs among participants in regional systems.
W73-07805

A KINETIC APPROACH TO BIOLOGICAL WASTEWATER TREATMENT DESIGN AND OPERATION,
Cornell Univ., Ithaca, N.Y. Water Resources and Marine Sciences Center.
A. W. Lawrence, and P. L. McCarty.
Available from the National Technical Information Service as PB-219 402, $5.25 in paper copy, $1.45 in microfiche. Technical Report No. 23, 1969. 56 p, 7 fig, 6 tab, 59 ref, 2 append. OWRR A-016-NY (4). 14-01-000-1-1400, 14-01-0001-1852, 14-31-0001-3032, 14-31-0001-3232.

Descriptors: *Biological treatment, Sanitary engineering, Activated sludge, Anaerobic digestion, *Design, *Waste water treatment, Kinetics, Water quality control, Models studies.

A unified basis for design and operation of biological waste treatment systems employing suspensions of microorganisms is developed from microbial kinetic concepts and continuous culture of microorganisms theory. Biological Solids Retention Time, average time period a unit of biological mass is retained in the system, is suggested as an independent parameter for process design and control. Biological Solids Retention Time is functionally related to process performance and is a readily controlled operational parameter. Steady-state kinetic models are presented for three process configurations, i.e., completely mixed reactor without solids recycle, completely mixed reactor with solids recycle, and plug flow reactor with solids recycle. Reported values of kinetic coefficients are tabularized for: (1) aerobic treatment of carbonaceous wastes, (2) aerobic biological nitrification, and (3) anaerobic methanogenic fermentation of carbonaceous wastes. These coefficient values are substituted into the models to determine lower limits, i.e., minimum values of Biological Solids Retention Time, for each process. Minimum values of Biological Solids Rentention Time are compared with Biological Solids Retention Time values for actual treatment systems to identify the safety factors implicit in current design practice.
W73-07809

DISCHARGES OF INDUSTRIAL WASTE TO MUNICIPAL SEWER SYSTEMS,
Cornell Univ., Ithaca, N.Y. Water Resources and Marine Sciences Center.
V. C. Behn.
Available from the National Technical Information Service as PB-219 361, $3.00 in paper copy, $1.45 in microfiche. Technical Report No. 60, February 1973. 18 p, 3 fig, 4 tab. OWRR A-017-NY (1). 14-01-0001-1400.

Descriptors: *Industrial wastes, Industries, *Waste water treatment, Cities, Legislation, *Municipal wastes, New York, Rates, *Combined sewers.

Studies by the Federal Government have shown that approximately one-fourth of industrial wastes are treated jointly with municipal wastes. A survey was made of ordinances and rate structures currently in use in New York State. A close examination shows that the ordinances are very similar in nature being mainly for the purpose of protecting the sewer system. Charges and surcharges are extremely variable. The basic charges from city to city can be as much as 10 to 1, while surcharges can vary by about 4 to 1. Thus, there is not much that can be accomplished within the existing framework insofar as joint treatment is concerned. However, the flow equalization step does lend itself to making the combined treatment more practical. By making flow equalization of industrial wastes mandatory through the mechanism of the sewer ordinance it is felt that municipalities will be in a much better position regarding accepting the industrial waste into their system. At the same time, attention could be paid to surcharges, with some relief given to firms who provide a uniform rate to the municipal sewer system.
W73-07810

DEVELOPMENT OF MULTIPURPOSE WATER PURIFICATION UNIT FOR ARMY FIELD USE.
Army Mobility Equipment Research and Development Center, Fort Belvoir, Va.
For primary bibliographic entry see Field 05F.
W73-07833

AMMONIATED MIXED BEDS GIVE MIXED RESULTS,
E. Salem.
Power Engineering, Vol 75, p 52-55, March, 1971. 2 fig, 3 tab.

Descriptors: *Waste water treatment, *Water treatment, *Separation techniques, *Ion exchange, Resins, Anions, Cations, Economics, Ammonium, Design criteria, Treatment facilities, Flotation, Water quality control, Water pollution control.
Identifiers: *Seprex, *Sodium throw, *Sulfate throw.

Many proposals have been made because of economic considerations that ammonium form cation exchange resin be substituted for the hydrogen form cation exchange resin in the condensate polisher mixed bed. Tests conducted and the results show that the performance of mixed beds containing ammonium form cation resins varies greatly with pH levels. Of all the resins tested none had a volumetric throughput in excess of 14 months of full power operation under today's normal design criteria. To cope with this problem, a new process called Seprex was developed which eliminates sodium and sulfate throw. This process involves separating the entrained cation from the anion resin by flotation. (Smith-Texas)
W73-07834

THE CONTEMPORARY STATE OF DECONTAMINATING WASTE WATER,
V. YE. Privalov, S. N. Lazorin, and V. M. Korniyenko.
Available from the National Technical Information Service as AD-747 514, $3.00 in paper copy, $1.45 in microfiche. Air Force Systems Command Foreign Technology Division Translation F70-MT-24-1460-71, April, 1972. 15 p, 2 tab, 20 ref. (Trans. of Koks i Khimiya, No 5, p 33-38, 1969).

Descriptors: *Pilot plants, *Water pollution, *Phenols, *Wastewater treatment, Chemical oxygen demand, Biological treatment, Oil wastes, Oil pollution, Water pollution treatment, Industrial wastes, Water supply, Lime, Decontamination.
Identifiers: *Russia, *Coke plants.

On the basis of a study conducted by the World Health Organization, a review is given of water pollution that is caused by the coke chemical industry. By using active silt at a pilot plant, an investigation was conducted to check on the purification of concentrated phenol wastewaters. Performance data from this pilot plant are compiled in two tables. If the inflow of limed water into biological installations is eliminated, the average detention time of water in biological basins could be reduced 3 to 4 times, the rate of destruction of phenols rose 5 to 6 times, and the chemical oxygen demand was lowered 2-3 times. (Smith-Texas)
W73-07835

ADVANCED STUDIES OF THE SUBMERGED ANAEROBIC FILTER FOR BREWERY PRESS LIQUOR STABILIZATION,
Kentucky Univ., Lexington.
E. G. Foree.
Available from the National Technical Information Service as PB-210 976, $3.00 in paper copy, $1.45 in microfiche. Technical Report UKY48-72-CE14, 41 p, May, 1972. 9 fig, 2 tab, 14 ref.

Descriptors: *Anaerobic treatment, *Anaerobic bacteria, *Filtration, *Wastewater treatment, Industrial wastes, Chemical oxygen demand, Organic loadings, Water quality control, Water pollution control, Water pollution sources, Water quality.
Identifiers: *Brewery press liquor waste, *Anaerobic filter units.

The anaerobic filter process evaluated consisted of passing liquid brewery press waste upward through a submerged rock packed column. The rocks provide the microorganisms that are necessary for stabilization with suitable media upon which to attach themselves, which in turn insures a long solids retention time necessary for optimum treatment. The COD removal efficiencies stabilized above 90% for all loadings except for dropping below 30% when the filter was loaded at

WATER SUPPLY AND WASTE DISPOSAL SERIES, VOL 6, OXIDATION DITCH SEWAGE WASTE TREATMENT PROCESS,
Federal Highway Administration, Washington, D.C.
H. W. Parker.
For sale by the Superintendent of Documents, U.S. Government Printing Office, Washington, D.C. 20402 Price $0.60. Staff Report, April, 1972. 52 p, 31 fig, 6 tab, 17 ref.

Descriptors: *Oxidation lagoons, *Oxidation, *Waste water treatment, *Waste treatment, Domestic sewage, Sewage treatment, Pilot plant, Water pollution control, Aeration, Sludge, Mixing, Research and development, Activated sludge, Biological treatment, Municipal wastes.
Identifiers: *Roadside rest areas, *Oxidation ditch.

Theory, design, specifications, construction, operation and testing of the oxidation ditch waste treatment process are described. The oxidation ditch waste treatment process was chosen for use at roadside rest areas because of its simplicity of operation, reliablility of performance, ease of maintenance and cost advantage. It is also applicable to small domestic sewage plant use. It was found that the process had very few operational problems although some problems have occurred during intermittent operations. Therefore, the system should be activated and left in continuous operation. (Smith-Texas)
W73-07839

BRACKISH WATER DESALINATED BY THE 'SIROTHERM' PROCESS,
For primary bibliographic entry see Field 03A.
W73-07840

PRACTICAL EXPERIENCE IN THE USE OF POLYELECTROLYTES,
Newcastle and Gateshead Water Co., Newcastle-upon-Tyne (England).
F. Bell.
Water Treatment and Examination, Vol 20, No 3, p 179-181, 1971. 6 ref.

Descriptors: *Polyelectrolytes, *Filtration, *Alum, Water treatment, *Waste water treatment, Head loss, Color, Water quality control, Water pollution control, Economics, Coagulation, Chlorination.
Identifiers: *Filter aids.

The use of polyelectrolytes particularly as filter aids at the Henderson Filter Plant is described. Before the use of polyelectrolytes filter, runs had to be ended before maximum head loss was reached because of excessive penetration and breakthrough of color and residual alum in the water. These problems are solved with the use of polyelectrolytes as filter aids. In addition, they are more economical because of the slight reduction in the alum dose. (Smith-Texas)
W73-07841

SLUDGE FILTER PRESSING AND INCINERATION AT SHEFFIELD,
Sheffield Corp., Wincobank (England).
For primary bibliographic entry see Field 05E.
W73-07842

A REGIONAL PLANNING MODEL FOR WATER QUALITY CONTROL,
Virginia Polytechnic Inst. and State Univ., Blacksburg.
For primary bibliographic entry see Field 05B.
W73-07918

DEACTIVATION OF RADIOACTIVE SEWAGE BY THE METHOD OF TWO-STEP COAGULATION OF FERRIC HYDROXIDE,
Ural Polytechnic Inst., Sverdlovsk (USSR).
V. L. Zolotavin, A. A. Konstantinovich, V. N. Sanatina, V. V. Pushkarev, and V. S. Petrov.
Soviet Radiochemistry, Vol 13, p 167-169, 1971. 3 tab, 7 ref. (Trans. from Radiokhimiya, Vol 13, No 1, p 164-166, Jan-Feb 1971).

Descriptors: *Iron compounds, *Radioactive waste disposal, *Waste water treatment, *Flocculation, Filtration, Settling velocity, Sewage treatment, Filters, Chemical precipitation, Water pollution treatment.

Two-stage flocculation with iron sulfate proved more efficient than a single stage. The alpha radioactivity of the effluent was reduced from the initial value of 0.3-0.9 microCurie/liter to 0.1-9 nanoCuries/liter after stage 1, and to 24-60 picoCuries/liter after stage 2. The residual beta radioactivity was an order of magnitude higher. Comparable results were obtained in stage 1 by decantation after 24 hours settling, or by filtration through sand or filter paper after 2 hours aging. (Bopp-ORNL)
W73-07937

WATER AND WASTE WATER STUDIES, 1971 (WASSER- UND ABWASSERCHEMISCHE UNTERSUCHUNGEN),
Gesellschaft fuer Kernforschung m.b.H., Karlsrube (West Germany).
For primary bibliographic entry see Field 05A.
W73-07944

RADIOACTIVE EFFLUENTS FROM NUCLEAR POWERPLANTS (BETRIEBLICHE ABLEITUNGEN RADIOAKTIVER STOFFE AUS KERNTECHNISCHEN ANLAGEN).
Technischer Ueberwachungs-Verein e. V., Cologne (West Germany). Institut fuer Reaktorsicherheit.

Available from NTIS, Springfield, Va., as IRS-T-23; $3.00 paper copy, $1.45 microfiche. Report IRS-T-23, July 1972. Proceedings of the Seventh Conference, Nov. 8-9, 1971, Koln, West Germany. 158 p.

Descriptors: *Radioactivity effects, *Nuclear wastes, *Nuclear powerplants, *Effluents, Europe, Conferences, Standards, Regulation, Water pollution control, *Waste treatment, Monitoring, Forecasting, Environmental effects.

German nuclear powerplants under construction detain Xe 40-60 days and Kr 2.5 days in charcoal absorption systems, which also remove iodine aerosols. The radioactivity of liquid effluents is reduced to less than 50 microCuries/cubic meter (by treatment by chemical precipitation, evaporation, deposition-filtration, and ion exchange) before discharge with the cooling water. The average dose to individuals living in the vicinity is about 1 mr/year. The adequacy of the 30-mr/year limit is discussed. English abstracts and the discussion following the papers are included. (Bopp-ORNL)
W73-07952

RADIOACTIVITY AND WATER SUPPLIES,
Interuniversitair Reactor Institutt, Delft (Netherlands).
G. Lettinga.
Available from NTIS, Springfield, Va., as EUR-4866e. $3.00 per copy, $1.45 microfiche. Report No. EUR-4866e, 1972. 198 p, 98 fig, 54 tab, 290 ref.

Descriptors: *Radioactive wastes disposal, *Water pollution, *Water pollution sources, *Air pollution, *Radioisotopes, Strontium, Cesium, Cobalt, Ruthenium, Iodine, Water pollution treatment, Assay, Adsorption, Ion exchange, Rivers, Peat.

Identifiers: *Netherlands.

An investigation has been made of the applicability of peat, chemically modified peat, clay and activated carbon for the removal of radionuclides (i.e., of Sr, Cs, Co, Mn, Ce, Ru, I) from aqueous solutions. Natural peat pairs a reasonable ion exchange capacity (approximately 1 meq/g at pH 6) with a pronouced specificity for higher valency cations, especially the cations of the transition elements and of the rare earth elements and for cationic Ru-complexes. Relative to the alkaline elements also the earth alkaline elements are sorbed with a high selectivity. However, within the last group a strong competing action exists between the various species, e.g., peat shows a slight preference for Sr over Ca only at pH <3.5. Strong indications were obtained that humic acids play a predominant part in the behavior of heavy metal ions in aqueous systems. By heating peat in air at a temperature of 120-160 C with either dilute or concentrated H2SO4, a product is obtained with strongly improved properties over the natural peat, i.e., ion exchange capacity (being 2-3.5 meq/g at pH 6), chemical and mechanical stability, swelling properties and selectivity for Sr relative to Ca. Modified peat therefore may be considered as eminently suitable for decontamination of radioactively contaminated aqueous solutions. For the removal of radiocesium some K-fixing Dutch clay deposits, such as Ammerzoden clay, can advantageously be applied. Radioiodine can be removed from aqueous solutions rather effectively by adding activated carbon +C12. (Houser-ORNL)
W73-07960

AGRI-INDUSTRIAL PARK WATER IMPROVEMENTS, SEWAGE TREATMENT FACILITIES, BUSINESS DEVELOPMENT LOAN (FINAL ENVIRONMENTAL IMPACT STATEMENT).
Economic Development Administration, Austin, Tex. Southwestern Region.

Available from the National Technical Information Service as EIS-TX-72-5127-F, $9.00 paper copy, $1.45 microfiche. August 18, 1972. 130 p, 4 fig, 3 map.

Descriptors: *Texas, *Environmental effects, *Treatment facilities, *Groundwater availability, Industrial wastes, Industrial water, Land use, Area redevelopment, Water supply, Sewage treatment, Economic impact, Groundwater resources.
Identifiers: *Environmental Impact Statements, *Cactus (Tex).

This project consists of an agri-industrial park, water improvements, sewage treatment facilities, and a business development loan in Cactus, Texas. Alternatives to this proposed development include a different site for the industrial park and no development. As a result of this project, 701 acres of unimproved range land would be converted to industrial use. Also, 145 acres of cultivated farm land would be converted to the site of a sewage treatment plant. The industries locating in the park would place an additional demand upon the ground water supply of the Ogallala aquifer. Noise, dust, and exhaust emissions would increase in the area both during and after construction. Additional demands would be placed upon the solid waste disposal facilities in the area. (Mockler-Florida)
W73-07978

EFFECT OF INDUSTRIAL WASTES ON OXIDATION POND PERFORMANCE,
For primary bibliographic entry see Field 05C.
W73-08015

INACTIVATION ASSAYS OF ENTEROVIRUSES AND SALMONELLA IN FRESH AND DIGESTED WASTE WATER SLUDGES BY PASTEURIZATION,
For primary bibliographic entry see Field 05C.
W73-08017

WINERY INNOVATES WASTE TREATMENT.

Food Engineering, Vol 44, No 6, p 73-75, June 1972. 3 fig.

Descriptors: *Waste water treatment, Water treatment, *Treatment facilities, *Sludge disposal, Ultimate disposal, Fertilizers, Effluents, Laboratory tests, Aeration, Biological treatment, Activated sludge, Sludge treatment, *New York, Industrial wastes, Water pollution control, Food processing industry.
Identifiers: *Widmer's Wine Cellars, Inc.

At a cost of more than 500,000 dollars, Widmer's Wine Cellars, Inc. located in New York State's Naples Valley, has completed and placed into operation a 3 acre industrial wastewater treatment facility that relies heavily on aeration. Four ten-foot deep aeration ponds are used to encourage bacterial growth and form activated sludge from waste material and natural decomposition. When solids level in the digester must be reduced, digested solids are pumped from the ponds to a tank truck and distributed throughout the vineyard for fertilizer. A new on-site laboratory continually tests effluents moving through the system. (Smith-Texas)
W73-08094

THE ANAEROBIC FILTER FOR THE TREATMENT OF BREWERY PRESS LIQUOR WASTE,
Kentucky Univ., Lexington. Dept. of Civil Engineering.
E. G. Foree, and C. R. Lovan.
Available from the National Technical Information Service as PB-210 924, $3.00 in paper copy, $1.45 in microfiche. Technical Report UKY46-72-CE 12, April, 1972. 52 p, 14 fig, 2 tab, 16 ref.

Descriptors: *Waste water treatment, *Waste treatment, *Industrial wastes, *Sedimentation, Organic wastes, Distillation, Filtration, Anaerobic bacteria, Anaerobic treatment, Filters, Trace elements, Chemical oxygen demand, Organic loading, Dissolved solids, Water pollution control, Water quality.
Identifiers: *Brewery wastes, *Brewery press liquor waste.

Spent grains from the lauter tub and hop strainer are collected in a press where they are concentrated for further drying and processing for cattle feed, in a typical brewing process. The waste liquor from this operation has a very high concentration of dissolved organics and moderately low concentration of suspended organics which can be readily removed by sedimentation and a high temperature, usually 125 to 130F. This investigation was conducted to evaluate the performance of two laboratory scale filters. One was operated at a constant loading of 50 lbs. COD per thousand cu. ft. per day while the other filter load was increased to 100 lb. COD per thousand cu. ft. per day and the concentrations of buffering and trace element solutions were varied. Since COD removal efficiency of 90% or greater was achieved under all conditions and loadings, it was concluded that the anaerobic filter is a feasible means of treating brewery press liquor waste. (Smith-Texas)
W73-08095

USE OF WASTE TREATMENT PLANT SOLIDS FOR MINED LAND RECLAMATION,
Illinois Univ., Urbana.
For primary bibliographic entry see Field 05E.
W73-08096

TERTIARY EFFLUENT DEMINERALISATION,
Permutit Co. Ltd., London (England).
J. Grantham.
Process Biochemistry, Vol 5, No 1, p 31-33 and 38, January 1970. 2 fig, 4 tab, 14 ref.

Descriptors: *Waste water treatment, Water treatment, *Tertiary treatment, *Anion exchange, Treatment facilities, Separation techniques, Detergents, Cost comparison, Cost analysis, Pilot plants, Automation, Capital cost, Sewage treatment, Resins, *Demineralization.
Identifiers: *Anion resins.

This work is based upon current ion exchange demineralizing practice and shows that tertiary treated sewage can be demineralized to a quality suitable for industrial use at a cost comparable in terms of capital investment and running cost to that required for treating water from more conventional sources. A small scale semi-automatic pilot plant was set up and operated for 9 months demineralizing tertiary sewage effluent. The results show that it is practical and economic to abstract filtered sewage effluent direct from the outfall of a sewage works and demineralize it to a quality acceptable for industrial use using conventional water treatment type plants and isoporous anion exchange resins. The only problem of any significance concerns the accumulative effect of detergent on the performance of the anion resins. A cost analysis is included. (Smith-Texas)
W73-08097

'LIQUID FERTILIZER' TO RECLAIM LAND AND PRODUCE CORPS,
Metropolitan Sanitary District of Greater Chicago, Ill.
For primary bibliographic entry see Field 05E.
W73-08098

DESIGN OF TREATMENT FACILITIES FOR THE CONTROL OF PHOSPHORUS,
Robert A. Taft Sanitary Engineering Center, Cincinnati, Ohio.
F. M. Middleton.
Water Research, Vol 6, p 475-476, 1972. 1 fig.

Descriptors: Phosphorus, *Design criteria, *Treatment facilities, *Waste water treatment, *Biological treatment, Flocculation, Chemical reactions, Activated sludge, Sludge treatment, Lime, Alum, Iron compounds, Effluents.
Identifiers: *Chemical treatment, *Phosphorus control.

Biological removal enhanced by lime, alum, or iron compounds can produce effluents that meet new phosphorus standards. Facilities for chemical addition are easily constructed at existing plants. A 30 second flash mix followed by 1-5 minute high energy flocculation and a 5-20 minute low energy flocculation achieves good results. Costs are about 5 cents per 1000 gal for 80% removal. There are both advantages and disadvantages to addition of chemicals to the activated sludge treatment unit. (Anderson-Texas)
W73-08100

FLOW VELOCITIES IN AERATION TANKS WITH MECHANICAL AERATORS,
Emschergenossenschaft, Essen (West Germany).
K-H. Kalbskopf.
Water Research, Vol 6, p 413-416, 1972. 6 fig, 1 ref.

Descriptors: *Settling basins, *Aeration, *Waste water treatment, *Sludge treatment, Domestic wastes, Sewage treatment, Flow rates, Flow characteristics, Flow measurement, Flow profiles, Detergents, Critical flow.
Identifiers: *Mechanical aerators.

Mechanical aerators produce a radial or spiral flow of surface water which then flows down the side of the tank and across the bottom. The flow velocity across the bottom must be rapid enough to prevent sludge settling. The critical velocity is different for municipal and industrial wastes and is affected by the detergent content of the waste. (Anderson-Texas)

W73-08101

OXYGEN DIFFUSION IN WET AIR OXIDATION PROCESSES,
Naval Research Lab., Washington, D.C.
W. W. Willman.
Available from the National Technical Information Service as AD-749 350, $3.00 in paper copy, $1.45 in microfiche. Operations Research Branch Report 72-4, August 1972. 11 p, 6 fig, 14 ref.

Descriptors: *Waste water treatment, *Waste treatment, *Chemical oxygen demand, *Oxygenation, *Dissolved oxygen, Activated sludge, Sludge treatment, Sewage treatment, Aeration, Oxygen demand, Treatment facilities.
Identifiers: *Wet air oxidation processes, *Bubble column reactors.

An essential step in wet air oxidation processes is the diffusion of dissolved oxygen. An investigation was made of the conditions under which this diffusion step becomes a limiting factor for processes which use an air bubble column reactor for sewage treatment. The results show that the importance of oxygen diffusion as a rate limiting step depends mainly on reaction temperature and pressure, chemical oxygen demand, bubble diameter, air supply rate, and reactor height. The implication of these results for shipboard waste treatment processes currently being considered by the Navy are examined. (Smith-Texas)
W73-08102

USAF MOBILITY PROGRAM WASTE WATER TREATMENT SYSTEM,
Illinois Univ., Urbana.
V. L. Snoeyink.
Available from the National Technical Information Service as AD-747 025, $3.00 in paper copy, $1.45 in microfiche. Air Force Weapons Laboratory Technical Report 71-169, April 1972. 210 p, 16 fig, 31 tab, 133 ref.

Descriptors: *Waste water treatment, Water treatment, *Waste treatment, *Waste disposal, Sludge disposal, Reverse osmosis, *Incineration, Industrial wastes, Domestic wastes, Chlorination, Flocculation, Aeration, Activated carbon, Filtration, Flotation, Water quality control, Water pollution control, Brines, Dilution, Settling basins.
Identifiers: *Photographic wastes.

The support systems for the U. S. Air Force Bare Base Mobility program which involves a highly mobile force of 1000 to 6000 men include a waste water treatment system which can treat waste waters to the required degree prior to discharge to the environment. The treatment system involves: (1) separate collection and incineration of human waste and (2) treatment of all waste waters except concentrated photographic wastes in a system which includes chemical clarification, flotation, filtration, activated carbon adsorption and chlorination. The sludge concentrated photographic waste and the skimmings from the aircraft and vehicle washrack waste are incinerated and the ash from the incinerator is disposed of on land. A reclamation system consisting primarily of a reverse osmosis process is recommended for upgrading the quality of the effluent from the waste treatment system such that it is suitable for reuse. The brine from the reverse osmosis treatment is disposed of either by dilution in receiving waters, evaporation from ponds or by transportation from the site. (Smith-Texas)
W73-08103

COST REDUCTION IN SULFATE PULP BLEACH PLANT DESIGN,
Improved Machinery, Inc., Nashua, N.H.
J. K. Perkins.
Tappi, Vol 55, No 10, p 1494-1497, October 1972. 2 fig, 1 tab.

Descriptors: *Design criteria, *Water reuse, *Recycling, *Cost analysis, Cost comparisons, Operating costs, Capital costs, Piping systems, Design standards, Water quality control, Water pollution control, Water pollution sources, *Pulp and paper industry, Treatment facilities.
Identifiers: *Sulfate pulp bleach plant.

A suggested design concept is described for a conventional sulfate bleach plant which will lead to the lowest true cost. First, housing is reduced to a minimum in keeping with climatic conditions. Next, all similar equipment is on the same center line and equally spaced allowing maximum repetition of piping details for reduced prefabrication costs. All mechanical elements except washers and mixers and access to all control valves are on the ground floor. Multiple manifolding is eliminated by tapered inlet boxes that allow straight runs. Finally, stilling type filtrate tanks allow recycling of washwater and better foam control for reduced water usage and treatment costs. (Smith-Texas)
W73-08105

TECHNOLOGIES FOR SHIPBOARD OIL-POLLUTION ABATEMENT: EFFECTS OF OPERATIONAL PARAMETERS ON COALESCENCE.
Naval Ship Research and Development Center, Annapolis, Md.

Available from the National Technical Information Service as AD-749 020, $3.00 in paper copy, $1.45 in microfiche. Report 3598, August 1972. 14 p, 9 fig.

Descriptors: *Separation techniques, *Coalescence, *Filtration, Filters, Oil, Oil wastes, Flow rates, Water quality, Water quality control, Water pollution control, Water pollution, Ships, Research and development, *Waste water treatment.
Identifiers: *Oil/water separator.

The way in which variations of certain operational parameters affected the coalescence subsystem, and the final stage of a three stage oil-water separator system is described. A study was made of cylindrical cartridge type coalescer elements made of resin coated glass fiber. Particulate matter very seriously limited the service life of a coalescer element. Oil viscosity was found to have a very strong effect, with heavier oils decreasing elements life. Increasing oil concentration also decreased the life of the element, but above a certain oil concentration this effect remained the same. No significant effects were noted on coalescer life with variations in the total flow rate in the range of 2 to 8 gallons per minute. (Smith-Texas)
W73-08106

STATE OF THE ART OF WATER FILTRATION,
American Water Works Association, New York. Committee on Filtration Problems.
For primary bibliographic entry see Field 05F.
W73-08107

BUREAU OF RECLAMATION,
Bureau of Reclamation, Denver, Colo. Applied Sciences Branch.
For primary bibliographic entry see Field 05G.
W73-08108

PHILADELPHIA SUBURBAN WATER COMPANY,
Philadelphia Suburban Water Co., Bryn Mawr, Pa.
For primary bibliographic entry see Field 05F.
W73-08109

DETROIT METRO WATER DEPARTMENT,
Detroit Metro Water Dept., Mich.
For primary bibliographic entry see Field 05F.
W73-08110

DALLAS WATER UTILITIES,
Dallas Water Utilities Dept., Tex.
For primary bibliographic entry see Field 05F.
W73-08111

ENGINEERING REPORT ON SHORE DISPOSAL OF SHIP GENERATED SEWAGE AT ACTIVITIES IN THE EASTERN AREA. VOLUME I.
Reynolds, Smith and Hills, Jacksonville, Fla.

Available from the National Technical Information Service as AD-747 998, $3.00 in paper copy, $1.45 in microfiche. Report, June 1969. 179 p, 6 fig, 38 tab. N00025-69-C-0004.

Descriptors: *Waste water treatment, *Treatment facilities, *Cost analysis, *Annual costs, *Construction costs, Cost comparisons, Unit costs, Water quality control, Water pollution control, Sewage treatment, Waste disposal, Economics, Ships, Saline water.
Identifiers: *Naval treatment systems, Engineering estimates, Shore disposal, Ship sewage.

This study was undertaken to provide engineering and cost data for shore disposal of ship generated sewage at 42 Naval activities in the Eastern area of the United States including Puerto Rico. Total capital expenditures required to provide shore disposal of ship sewage in the Eastern area would be $10,078,000. Total cost for handling ship sewage from dockside is estimated at $937,000 per year. The engineering and cost data developed in this study are summarized in a table. This investigation indicates that no serious effects are being experienced with sewage treatment processes exposed to high salt water concentrations introduced by infiltration and sea water flushing systems. The treatment processes acclimate to these high salt water concentrations with very little loss in treatment efficiency. Construction materials and protective coatings should be designed for the salt water environment. (See also W73-08113) (Smith-Texas)
W73-08112

ENGINEERING REPORT ON SHORE DISPOSAL OF SHIP GENERATED SEWAGE AT ACTIVITIES IN THE EASTERN AREA. VOLUME II.
Reynolds, Smith and Hills, Jacksonville, Fla.

Available from the National Technical Information Service as AD-747 999, $3.00 in paper copy, $1.45 in microfiche. Report, June 1969. 213 p, 70 fig. N00025-69-C-0004.

Descriptors: *Sewage treatment, *Sewage disposal, *Sewerage, *Treatment facilities, *Waste water treatment, Waste treatment, Ships, Water pollution sources, Water pollution control, Water quality control, Water pollution, Water quality.
Identifiers: *Sewerage systems.

Maps and drawings show existing conditions, ship berthing and proposed sewerage facilities for 40 activities under this study. Two berthing plans are shown for most of these activities, and are designated: (1) 'Design Maximum Loading of Individual Berthing Facilities', which shows the maximum berthing capability of each pier, wharf, and quay, and (2) 'Typical Berth Employment for a Maximum Day', which shows the maximum ship berthing experienced or expected at the port, except during a national emergency. These berthing plans were used to design the proposed sewerage systems as described in Volume I. Shore facilities to receive ship sewage have been constructed at the Naval shipyard Portsmouth, New Hampshire and Naval Training Center, Great Lakes, Illinois.

All figures contained in this volume are pertinent to and referenced from Volume I, and are arranged by Naval district. (See also W73-08112) (Smith-Texas)
W73-08113

DETECTION, DECONTAMINATION AND REMOVAL OF CHEMICALS IN WATER,
Edgewood Arsenal, Md. Army Development and Engineering Directorate.
For primary bibliographic entry see Field 05A.
W73-08114

ACTIVATED CARBON IN TREATING INDUSTRIAL WASTES,
Westvaco Corp., Covington, Va.
A. W. Loven.
Proc available from the National Technical Information Service as AD-738 544, $3.00 in paper copy, $1.45 in microfiche. In: Proceedings of First Meeting on Environmental Pollution, 15-16 April 1970, Sponsored by American Ordnance Association, Edgewood Arsenal Special Publication EASP 100-78. February 1972, p 83-105. 11 fig, 13 ref.

Descriptors: *Activated carbon, *Filtration, *Waste water treatment, Industrial wastes, Water pollution control, Adsorption, Water quality control, Water quality, Porosity, Design criteria, Pilot plants.
Identifiers: *Industrial waste treatment systems.

Adsorption on activated carbon is perhaps the most promising and widely acceptable process for the removal of dissolved organic compounds from waste water. Activated carbon adsorption principles are reviewed, and activated carbon processes, and activated carbon application to industrial waste water treatment are discussed. (Smith-Texas)
W73-08115

ACTIVATED CARBON IN POLLUTION CONTROL,
Calgon Corp., Pittsburgh, Pa.
G. V. Stone.
Proc available from the National Technical Information Service as AD-738 544, $3.00 in paper copy, $1.45 in microfiche. In: Proceedings of First Meeting on Environmental Pollution, 15-16 April 1970, Sponsored by American Ordnance Association, Edgewood Arsenal Special Publication EASP 100-78. February 1972, p 106-124. 6 fig, 2 tab, 6 ref.

Descriptors: *Activated carbon, *Water pollution control, *Pollution abatement, *Waste water treatment, Industrial wastes, Design criteria, Cost analysis, Biological treatment, Adsorption, Biochemical oxygen demand, Chemical oxygen demand, Water pollution, Water quality control.
Identifiers: *Industrial waste treatment systems.

The economical solution to industrial waste water problem with granular activated carbon is discussed. A polyvinyl chloride and rubber chemical plant was faced with the problem of removing heavy solids loading of polymer particles from the plastic polymer processes effluent and soluble organics from the rubber chemical plant effluent to meet the requirements of the receiving stream. Studies were conducted using 'Filtrasorb' granular activated carbon and biological methods of treatment. The results indicate that this treatment method can be considered as a definite means of treating polyvinyl chloride plastic and rubber chemical plants wastes. A cost analysis and design criteria of the treatment system are included. (Smith-Texas)
W73-08116

HEAVY METALS REMOVAL IN WASTE WATER TREATMENT PROCESSES: PART 1,
Orange County Water District, Santa Ana, Calif.
D. G. Argo, and G. L. Culp.
Water and Sewage Works, Vol 119, No 8, p 62-65, August 1972. 3 tab, 15 ref.

Descriptors: *Waste water treatment, *Heavy metals, *Water reuse, *Recycling, Artificial recharge, Injection wells, Groundwater, Water supply, Water demand, Cost analysis, Water treatment, Water quality, Water pollution, Pilot plants, Trickling filters, *California.
Identifiers: *Orange County Water District (Cal).

The Orange County Water District is investigating the feasibility of waste water reclamation because of increasing water demand and rising cost of imported water delivered to the Southern California area. Studies in waste water reclamation and groundwater recharge through injection wells has been conducted since 1965. Literature is reviewed to determine what is known about the reduction and concentration of heavy metals by waste water treatment and the efficiency of the Orange County Water Districts pilot scale waste water treatment plant for removing heavy metals from trickling effluent is evaluated. (Smith-Texas)
W73-08117

CONVERTING AMD TO POTABLE WATER BY ION EXCHANGE TREATMENT,
Chester Engineers, Inc., Coraopolis, Pa.
W. Zabban, T. Fithian, and D. R. Maneval.
Coal Age, Vol 77, No 7, p 107-111, July 1972. 3 fig, 1 ref.

Descriptors: *Acid mine drainage, *Industrial wastes, *Ion exchange, *Water reuse, *Waste water treatment, Cation exchange, Anion exchange, Flow rates, Demineralization, Treatment facilities, Separation techniques, Costs, Potable water, Water supply, *Pennsylvania.
Identifiers: Counter current ion exchange.

Two small potable water treatment plants have been erected in Pennsylvania, for the purpose of converting acid mine drainage to potable water by ion exchange treatment. These treatment facilities consist of two continuous counter current ion exchange units, one for cation exchange and the other for anion exchange, designed to treat 500,000 gallons per day. Results have proved that this method of demineralization will result in a high quality, low total dissolved solids product costs are included. (Smith-Texas)
W73-08118

A MATHEMATICAL FORMULATION FOR SELECTING HOLDING-TANK-PROCESSOR REQUIREMENTS FOR SHIPBOARD SEWAGE TREATMENT,
M. U. Thomas.
Available from the National Technical Information Service as AD-747 065, $3.00 in paper copy, $1.45 in microfiche. July 1972. 22 p, 2 fig, 4 ref.

Descriptors: *Waste water treatment, *Waste treatment, *Ships, *Sewage treatment, *Mathematical studies, Treatment facilities, Waste disposal, Sewage disposal, Water pollution control, Water pollution sources, Water quality control.
Identifiers: *Shipboard processors, *Shipboard sewage treatment.

A descriptive framework is provided for the decision problem of selecting combined holding-tank-processor systems for shipboard sewage treatment. A mathematical formulation is described for examining tradeoffs between holding-tank capacity and processing rates of any proposed facility subject to the restrictions that all sewage generated must be processed. Following a general formulation of the problem, the first case considered is when the generation of sewage is assumed to be deterministic. In the second and more general case, the arrival streams are considered random, but assumed to follow a Poisson distribution. It is suggested that future studies be addressed to the problem of combining one or more holding-tanks with one or more processors. (Smith-Texas)
W73-08119

WALDHOF TYPE OF FERMENTORS IN DISPOSAL OF FOOD WASTES,
Iowa State Univ., Ames.
G. T. Tsao, and W. D. Cramer.
Chemical Engineering Progress Symposium Series, Vol 67, No 108, (1971), p 158-163. 10 fig, 5 ref. OWRR A-032-IA (10).

Descriptors: *Waste disposal, *Ultimate disposal, Industrial wastes, *Waste water treatment, Treatment facilities, Aeration, *Fermentation, Food processing industry, Foaming, Foam separation.
Identifiers: *Waldhof fermentor, *Aeration rates.

The ability to handle materials that have strong foaming tendencies is one of the important characteristics of a Waldhof type of fermentor. It makes use of the foaming properties by continuous recycling of the foam and by obtaining extra aeration through the air-liquid interface in the foam. A study of the basic mechanisms of the Waldhof type of fermentors is reported. Important factors that affect the operation of a Waldhof type of fermentor and the experimental results, as well as theoretical developments, are included. These operational factors include air pumping by vortex in the system, aeration rate, effects of proteins and dissolved solids on oxygen absorption, foam aeration and foam formation. (Smith-Texas)
W73-08120

SPENT HCL PICKLING LIQUOR REGENERATED IN FLUID BED,
American Lurgi Corp., New York.
P. Marnell.
Chemical Engineering, Vol 79, No 25, p 102-103, November, 1972. 1 fig, 1 tab.

Descriptors: *Industrial wastes, *Waste treatment, *Waste water treatment, Separation techniques, Iron compounds, Water pollution control, Water quality control, Water pollution, Water pollution treatment, Water quality standards.
Identifiers: *Fluid bed combustion, *Acid.

A new regeneration technique known as fluid bed combustion has been adopted in a process developed and licensed by Lurgi Apparate-Technik Gmb H. The fluid bed operation has the advantages over other operations of smoother performance, a relatively dust free iron oxide product, and complete and continuous recovery of clean hydrochloric acid. These features will allow compliance with the stringent anti-pollution regulations currently being adopted by federal, state and local agencies. (Smith-Texas)
W73-08121

CHICAGO DEPARTMENT OF WATER AND SEWERS,
Illinois State Water Survey, Urbana.
H. W. Humphreys.

Descriptors: *Settling basins, *Flow characteristics, *Flow measurement, *Flow control, Flow profiles, Flow system, *Waste water treatment, Water treatment, *Filtration, Water quality, Water quality control, Tracers, Flow rates, Treatment facilities, Separation techniques, *Illinois.
Identifiers: *Chicago.

The Illinois State Water Survey, in cooperation with Chicago, is studying flow characteristics in a hydraulic model of one of the sixteen settling basins in the Central Water Filtration Plant. The objective is to improve flow conditions in the prototype settling basin in order to improve the quality of the settled water applied to the filter. The flow conditions in the model are studied by

Discussions are presented of various water clarification systems used in the aggregate and ready-mixed concrete industries. The overall problem of waste water disposition in each type of plant is studied. An analysis is made of the use of settling ponds, filter ponds and coagulants. Recycling and use of recycled water is discussed with recommendations for further study of the potential use of waste water from ready-mix plants for concrete batch water. Since many aggregate and ready-mix concrete now have effective clarification or recycling systems the overall purpose of the study is to make these systems known throughout the industry so proven systems can be made available to all. The report is based on a review of systems in reported 77 plants and firms plus data obtained from a field trip inspection of 88 plants on the West Coast. The study contains 45 charts and photographs of clarification systems. (EPA)
W73-08126

PHYSICAL-CHEMICAL TREATMENT OF COMBINED AND MUNICIPAL SEWAGE,
Battelle-Northwest, Richland, Wash. Pacific Northwest Lab.
Alan J. Shuckrow, Gaynor W. Dawson, and William F. Bonner.
Copy available from GPO Sup Doc as EP1.23/2:73-149, $2.35; microfiche from NTIS as PB-219 668, $1.45. Environmental Protection Agency, Technology Series, Report EPA-R2-73-149, February 1973. 178 p, 88 fig, 31 tab, 14 ref, append. EPA Project 11020 DSQ. 14-12-519.

Descriptors: *Activated carbon, *Adsorption, *Waste water treatment, *Pilot plants, Alum, Coagulation, Polyelectrolytes, Filtration, Dewatering, Sewage treatment.
Identifiers: *Combined sewage, Alum recovery, Carbon recovery.

The research program included laboratory process development of a unique physical-chemical waste water treatment process followed by design, construction, and field demonstration of a 100,000 gpd mobile pilot plant. In the treatment process, raw waste water is contacted with powdered carbon, coagulated with alum, settled with polyelectrolyte addition and, in some cases, passed through a tri-media filter. The solids from the clarifier, composed of raw sewage solids, powdered carbon, and aluminum hydroxide floc, are readily dewaterable to 20-25 percent solids by direct centrifugation with the powdered carbon acting as a substantial aid to dewatering. The dewatered solids are passed through a fluidized bed furnace developed specifically for powdered carbon regeneration. Alum is recovered by acidifying the regenerated carbon slurry from the furnace to a pH of 2. The recovered carbon and alum are recycled as an acidified slurry and added to the raw sewage with the makeup carbon. The program demonstrated the ability of the treatment process to consistently produce high-quality effluent from raw waste water. Powdered carbon regeneration was highly successful on the pilot scale. Full capacity recovery was achieved with less than two percent carbon loss per regeneration cycle. Alum recovery was also greater than ninety percent. (EPA)
W73-08127

COLOR REMOVAL FROM KRAFT PULP MILL EFFLUENTS BY MASSIVE LIME TREATMENT,
International Paper Co., Springhill, La.
L. Oswalt, and J. G. Land, Jr.
Copy available from GPO Sup Doc as EP1.23/2:73-086, $1.25; microfiche from NTIS as PB-219 594, $1.45. Environmental Protection Agency, Technology Series, Report EPA-R2-73-086, February 1973. 109 p, 21 fig, 14 tab, 10 ref, 4 append. EPA Project 12040 DYD. Grant 135-01 (R-1) (68).

Descriptors: Pulp and paper industry, Effluents, *Waste water treatment, Color reactions, *Lime, *Pilot plants, Operation and maintenance, *Cost analyses, *Pulp wastes, Foaming, Floculation, Sedimentation rates, Chemical precipitation, Water reuse, Biochemical oxygen demand, Calcium carbonate, Feasibility studies, *Capital costs, Bleaching wastes.
Identifiers: *Color removal, Kraft decker effluent, *Recarbonation.

A demonstration plant was installed and operated to determine effectiveness and feasibility of using massive lime treatment (that is, 20,000 ppm lime) to decolor kraft pulp mill effluents. The two most highly colored effluents and mixtures of these treated in the demonstration plant were: (1) the almost black effluent from the caustic extraction stage of pulp bleaching, and (2) the light reddish-brown effluent from the final unbleached pulp washing stage. Objectives were to determine: Effectiveness of color removal, design and performance of massive lime system equipment, effects of normal pulp mill operations, effects on pulp quality, operating costs. Impact of the massive lime system on a hypothetical 1000 tons-per-day bleached kraft pulp and paper mill is described. Using all the lime normally available in such a mill would allow massive lime treatment of four million of the mill's twenty-nine million gallons of effluent. Such treatment would remove 72% of the total mill effluent's color, reducing final effluent color to approximately 740 APHA units at an estimated operating cost of $1.80 per ton of pulp (depreciation, insurance, and taxes included). (EPA)
W73-08128

RENOVATING SEWAGE EFFLUENT BY GROUNDWATER RECHARGE,
Agricultural Research Service, Phoenix, Ariz. Water Conservation Lab.
H. Bouwer, J. C. Lance, and R. C. Rice.
In: Proceedings of the 40th Annual Meeting of the Western Snow Conference, April 18-20, 1972, Phoenix, Ariz: Printed by Colorado State University, Fort Collins, p 9-16, 1972. 3 fig, 2 tab, 14 ref.

Descriptors: *Water reuse, *Water spreading, *Artificial recharge, *Tertiary treatment, Groundwater recharge, Alluvial channels, *Arizona, Hydrogeology.

In the Salt River Valley in central Arizona the performance of a system for renovating waste water by groundwater recharge depends upon the local conditions of climate, soil and groundwater. The hydrogeological conditions in Salt River Valley, central Arizona, are favorable for groundwater recharge with infiltration basins in the river bed. It is estimated that tertiary treatment by this method would cost about $5 per acrefoot or less than one-tenth of equivalent in-plant treatment. This would be an additional source of water to meet future agricultural and other demands. (See also W73-08138) (Knapp-USGS)
W73-08141

WATER QUALITY ASPECTS OF THE STATE OF WASHINGTON, PART B,
Washington State Univ., Pullman.
B. W. Mar, D. A. Nunnallee, W. Mason, and M. Rapp.
Report No. 3B, June 1970. 207 p, 30 fig, 50 tab, 148 ref.

Descriptors: *Surveys, *Data collections, *Water quality, *Waste water treatment, *Washington, Costs, Reviews, Waste water, Industrial wastes, Administration, Municipal wastes, Water pollution control, Waste water disposal, Sewage treatment, Treatment facilities.

As an amplification of An Initial Study of the Water Resources of the State of Washington, this study further examines the quality of State waters. Particular emphasis is placed on three areas: (1) an

examination of water quality and management policics for water quality control, (2) a quantitative compilation of industrial, commercial and municipal wastes in the State, with flow charts of industrial and commercial processes and their relation to waste production, and (3) a comparison of waste treatment efficiencies by type of waste treated, along with costs. In each of the 24 major river basins examined, extensive bacteriologic treatment is necessary in order to meet coliform standards downstream from community discharges. For dissolved oxygen, secondary treatment was not adequate in half the rivers checked. In those cases where secondary treatment is too costly or insufficient, alternative treatment is suggested, along with river monitoring and modelling and a program of research to provide new techniques of technology and management. Eighteen types of industrial and commercial operations, as well as municipal waste treatment, are examined to assist in the development of waste treatment methods. Five methods of water treatment are compared for treatment efficiencies. (Poertner)
W73-08178

WATER, SEWER AND STORM DRAINAGE PLAN FOR THE CUMBERLAND PLATEAU PLANNING DISTRICT.
Thompson and Litton, Inc., Wise, Va.
For primary bibliographic entry see Field 06D.
W73-08179

WATER AND SEWER PLAN.
West Alabama Planning and Development Council, Tuscaloosa.
For primary bibliographic entry see Field 06D.
W73-08181

REGIONAL WATER AND SEWER FACILITIES IMPROVEMENT PROGRAM.
Southeastern Illinois Regional Planning and Development Commission, Harrisburg.
For primary bibliographic entry see Field 06D.
W73-08184

DIGEST OF PROPOSED REGULATION RELATING TO MONITORING AND REPORTING WASTEWATER DISCHARGES AND THEIR EFFECTS UPON RECEIVING WATERS.

North Carolina Dept. of Natural and Economic Resources, Raleigh. (1971), 6 p.

Descriptors: *North Carolina, *Waste water discharges, *Monitoring, *Waste treatment, Waste water treatment, Water pollution sources, Sewage treatment, Treatment facilities, Measurement, Waste water (Pollution), Effluents, Sampling, Testing procedures, Municipal wastes, Industrial wastes, Effluent streams.

A monitoring system must be established for each waste water treatment plant. Such a system must include an effluent flow measuring device and flow recording device. Additionally, samples of the influent and effluent of each waste water treatment plant must be collected and analyzed. Moreover, samples at one or more upstream and downstream sampling points must be collected and analyzed. With reference to the reporting requirements, an annual survey report must be filed for each waste water treatment plant. Also, a monthly monitoring report must be filed for each waste water treatment plant listing the results of samples collected and analyzed in the past month. Several graphs and charts are included. (Mockler-Florida)
W73-08202

CELL REPLICATION AND BIOMASS IN THE ACTIVATED SLUDGE PROCESS,
Illinois Univ., Urbana. Dept. of Sanitary Engineering.
For primary bibliographic entry see Field 05B.
W73-08243

THE MICROBIOLOGY OF AN ACTIVATED SLUDGE WASTE-WATER TREATMENT PLANT CHEMICALLY TREATED FOR PHOSPHORUS REMOVAL,
Pennsylvania State Univ., University Park. Dept. of Microbiology.
J. A. Davis, and R. F. Unz.
Water Research, Vol 7, Nos 1/2, p 325-327, January/February 1973.

Descriptors: *Phosphorus, *Activated sludge, *Microorganisms, *Waste water treatment, Protozoa, Coliforms, Enteric bacteria, Analytical techniques, Phosphates, Cultures, Sewage bacteria, Streptoccocus.
Identifiers: *Alum, Removal, Culture media, Flagellates.

Microbiological research was conducted on a dual, secondary wastewater treatment system which was part of the Pennsylvania State University wastewater treatment plant. Each aeration basin received identical wastewater which was the effluent from a high rate trickling filter. One of the aeration basins was dosed with aluminum sulfate for the purpose of phosphorus removal. The other aeration basin (control) was operated in the conventional manner without alum addition. Plate counts performed on combined chemical-biological sludge and control activated sludge revealed that a higher number of viable microorganisms was contained in the chemical-biological sludge, but the magnitude of difference between the two sludges was significant depending on the culture medium employed. Results suggest the aluminum flocs formed in the chemical-biological treatment enmesh dispersed wastewater microorganisms, some of which are qualitatively unlike those indigenous to natural activated sludge. The combined chemical-biological sludge contained significantly higher numbers of lipolytic, gelatinolytic, and thiosulfate oxidizing microorganisms and, possibly, fewer nitrite oxidizing microorganisms than did control activated sludge. Alum did not appear to affect flagellated protozoa in mixed liquor; however, amoeboid and ciliated protozoa were found less frequently in alum dosed than in control mixed liquor. The settledeffluent from the combined chemical-biological aeration basin generally contained fewer total coliforms, fecal coliforms, and fecal streptoccocci than did counterpart control effluents. (Little-Battelle)
W73-08250

EXPERIENCE WITH ALGAL BLOOMS AND THE REMOVAL OF PHOSPHORUS FROM SEWAGE,
For primary bibliographic entry see Field 05C.
W73-08251

ACTIVATED CARBON TREATMENT OF RAW SEWAGE IN SOLIDS-CONTACT CLARIFIERS,
Westinghouse Electric Corp., Richmond, Va. INFILCO Div.
R. L. Beebe.
Copy available from GPO Sup Doc as EP1.23/2:73-183, $1.25; microfiche from NTIS as PB-219 883, $1.45. Environmental Protection Agency, Technology Series Report EPA-R2-73-183. March 1973. 98 p, 34 fig, 8 tab, 14 ref, append. EPA Project 17050 EGI, 14-12-586.

Descriptors: *Waste water treatment, *Adsorption, *Activated carbon, *Sewage treatment, Flocculation, Filtration, Chemical analysis, Waste treatment, Cost, Settling rates, Pilot plants, Adsorption.
Identifiers: Series clarifier countercurrent adsorption.

Degritted raw municipal sewage was treated with powdered activated carbon in a 28,000-gpd pilot plant. Two high-rate recirculating-slurry solids-contact clarifiers operating in series with countercurrent carbon advance, followed by a gravity polishing filter, produced effluent equal to or better than that produced in a parallel activated sludge plant. TOC and COD removals averaged 88.1 and 88.7 percent, respectively, with higher removals hindered by the concentration of adsorptive-resistant materials present. Filtrable-TOC and -COD removals were 68.0 and 69.9 percent, respectively. Alum and polyelectrolyte flocculated the powdered activated carbon and raw sewage suspended solids into a fast settling floc. Subsidence tests conducted on the solids slurry from the ACCELATOR draft tube indicated ACCLATOR overflow rates equivalent to or greater than 2.5 gpm/ft2. The maximum carbon adsorptive capacity for filtrable COD was 0.50 to 0.55 g COD/g carbon. This capacity was achieved whenever the concentrations of influent COD and carbon matched or exceeded that ratio (adsorptive-resistant COD excluded). Carbon requirements were 55 to 60 percent of theoretical two-stage countercurrent adsorption system requirements. Assuming regeneration cycles 85 percent of the carbon feed, respective treatment cost estimates for 10-mgd and 100-mgd plants were 13.9 cents and 11.2 cents per thousand gallons. (EPA)
W73-08287

THE DISPOSAL OF CATTLE FEEDLOT WASTES BY PYROLYSIS,
Midwest Research Inst., Kansas City, Mo.
W. Garner, and I. C. Smith.
Environmental Protection Agency, Technology Series Report EPA-R2-73-096, January 1973. 99 p, 16 fig, 9 tab, 27 ref, 4 append. EPA Project 13040 EGH; Contract No. 14-12-850.

Descriptors: *Waste disposal, *Farm wastes, Feedlots, *Cattle, Chemical analysis, Organic compounds, Cost analysis, Gases, *Waste treatment.
Identifiers: *Pyrolysis, *Gas condensation.

Beef cattle (steer) manure was obtained from a source that was free of soil contamination, and subsequently dried and pulverized. Replicate batch pyrolyses were carried out in stainless steel, glass, and iron tubes utilizing axial flow, at various levels of elevated temperature, and at atmospheric and lower pressures. Exhausts were carried by inert gas to traps and condensors. Qualitative separations and extractions were performed to determine the presence and quantity of various gases, ash, tar, and organics. Many constituents were extracted, but in such quantities that their value may not pay for the cost of pyrolizing. Larger scale pyrolyzing units should be tested to either confirm or disprove these findings. (EPA)
W73-08290

PILOT PLANT FOR TERTIARY TREATMENT OF WASTE WATER WITH OZONE,
Air Reduction Co., Inc., Murray Hill, N.J. Research and Engineering Dept.
C. S. Wynn, B. S. Kirk, and R. McNabney.
Copy available from GPO Sup Doc as EP1.23/2:73-146, $2.60; microfiche from NTIS as PB-219 877, $1.45. Environmental Protection Agency, Technology Series Report EPA-R2-73-146, January 1973. 229 p, 59 fig, 20 tab, 22 ref, 7 append. EPA Project 17020-DYC, WQO 14-12-597.

Descriptors: *Ozone, *Tertiary treatment, *Waste water treatment, *Pilot plants, *Chemical oxygen demand, Optimization, Water treatment, Oxygen, Evaluation, Cost analysis.
Identifiers: *Washington, D.C.

Tertiary treatment of waste water with ozone in a nominal 50,000 gal./day pilot plant at Blue Plains, Washington, D.C., is described. Plant feeds (10 to 100 ppm COD) were effluents from other pilot

Identifiers: Oxidation ditch, *Anaerobic lagoons, *Food processing wastes, Combined treatment.

Various mixtures of fruit and vegetable cannery wastes, and domestic sewage were treated by anaerobic lagoons followed by an oxidation ditch for a two-year period. The anerobic lagoons consistently achieved BOD reductions of 75 to 85 percent at loadings up to 400 lbs BOD/day/acre provided adequate inorganic nutrients were present. The oxidation ditch reduced the BOD to low levels and was shown to be very stable against overload. Power requirements were less than 0.5 kw.hr/lb of BOD removed and the oxygenation capacity of the rotor was about 30 lbs of BOD per foot of length. (Dostal-EPA)
W73-08293

TREATMENT OF FERROUS ACID MINE DRAINAGE WITH ACTIVATED CARBON,
Bituminous Coal Research, Inc., Pittsburgh, Pa.
C. T. Ford, and J. F. Boyer.
Copy available from GPO Sup Doc as EP1.23/2:73-150, $2.10; microfiche from NTIS as PB-219 826, $1.45. Environmental Protection Agency, Technology Series Report EPA-R2-73-150, January 1973. 123 p, 17 fig. EPA Project 14010 GYH.

Descriptors: *Acid mine water, *Waste water treatment, *Activated carbon, Iron compounds, Oxidation, Adsorption, Ferrobacillus, *Mine drainage, Aeration, Coal mine wastes, Cost analysis.
Identifiers: *Iron removal, Oxidation catalyst.

Laboratory studies were conducted with activated carbon as a catalyst for oxidation of ferrous iron in coal mine water. Batch tests and continuous flow tests were conducted to delineate process variables influencing the catalytic oxidation and to determine the number and types of coal mine water to which this process may be successfully applied. The following variables influence the removal of iron with activated carbon: (a) amount and particle size of the carbon; (b) pH, flow rate concentration of iron, temperature, and total ionic strength of the water; and (c) aeration rate. Adsorption as well as oxidation are the mechanisms involved in iron removal by this process. An evaluation of this process indicated technical feasibility which would permit acid mine drainage neutralization using an inexpensive reagent, such as limestone. The major disadvantage is the cost of the activated carbons since they are rendered inactive after relatively short use by apparently irreversible adsorption of iron. This cost appears to be sufficiently high to prohibit the use of this process for treating coal mine drainage. (EPA)
W73-08296

MICROSTRAINING AND DISINFECTION OF COMBINED SEWER OVERFLOWS--PHASE II,
Crane Co., King of Prussia, Pa. Environmental Systems Div.
G. E. Glover, and G. R. Herbert.
Copy available from GPO Sup Doc as EP1.23/2:73-124, $2.10; microfiche from NTIS as PB-219 879, $1.45. Environmental Protection Agency, Technology Series Report EPA-R2-73-124, January 1973. 116 p, 32 fig, 8 tab, 30 ref, 4 append. EPA Program 11023 FWT.

Descriptors: *Sewers, *Storm runoff, *Filtration, Water pollution control, *Cost comparisons, Water quality, Ozone, Chlorine, Biochemical oxygen demand, Waste water treatment, Screens, Disinfection.
Identifiers: *Microstraining, *Combined sewer overflow, *Suspended solids removal, *Philadelphia.

A microstrainer using a screen with 23 micron apertures reduces the suspended solids of the combined sewer overflow from 50 to 700 mg/l down to 40 to 50 mg/l levels operating at flow rates of 35 to 45 gpm/ft2 of submerged screen. The organic matter as measured by COD and TOC was reduced 25 to 40%. Coliform concentrations were 0.1 to 9 million cells per 100 ml and no reduction was brought about by Microstraining. Stormwater service requires special analytical techniques which are described in detail. The coliform concentrations of both overflow and microstrained overflow were reduced by four or more orders of magnitude by disinfection with 5 mg/l chlorine in specially built, high rate, contact chambers of only 2 minutes contact time. The drainage area served by the combined sewer comprises 11.2 acres of a residential area in the City of Philadelphia having an average dry weather sanitary flow of 1000 gph. The overflow rates recorded were generally 100 times, with a maximum 400 times, the average dry weather flow. The extreme importance of very low - 2 minute - residence volume equipment for suspended solids removal and for disinfection in the very high instantaneous rates encountered with stormwater is shown. The cost of a microstrainer - special chlorine contact chamber installation is cited as $6,750/cfs of peak flow rate capacity less land and engineering. On the basis of 2 cfs instantaneous design overflow per acre this is $13,100/acre. (EPA)
W73-08297

KRAFT EFFLUENT COLOR CHARACTERIZATION BEFORE AND AFTER STOICHIOMETRIC LIME TREATMENT,
Institute of Paper Chemistry, Appleton, Wis. Div. of Natural Materials and Systems.
J. W. Swanson, H. S. Dugal, M. A. Buchanan, and E. E. Dickey.
Copy available from GPO Sup Doc as EP1.23/2:73-141, $1.00; microfiche from NTIS as PB-219 827, $1.45. Environmental Protection Agency, Technology Series Report EPA-R2-73-141, February 1973. Project 12040 DKD.

Descriptors: *Pulp wastes, *Waste water treatment, *Industrial wastes, Pollution abatement, Waste water (Pollution), Effluents, Water reuse, Chemical analysis, Color.
Identifiers: *Lime treatment, *Kraft colors, Kraft effluent, Decker effluent, Kraft decker effluent, *Molecular weights, *Color characterization, Color isolation.

The lime-treatment process was found to remove on an average about 86 percent of the color, 57 percent of the total organic carbon, and 17 percent of total sugars from the waste effluent during the period of approximately 15 months over which the samples were collected. No appreciable change in chloride content was noticed. The 'weight average' molecular weights (Mw) of untreated acid-insoluble fractions varied from < 400 to 30,000 and of the untreated acid-soluble, lime-treated acid-insoluble, and lime-treated acid-soluble fractions from < 400 to 5000. The study shows that color bodies having an apparent Mw of < 400 are not removed by lime treatment and those having Mw of 5000 and above are completely removed. The intermediate range of Mw 400 to 5000 apparently undergoes partial removal. Infrared spectroscopy data indicate that the acid-insoluble color bodies (high Mw) contain a high proportion of conjugated carbonyl groups where conjugation with an aromatic ring is probable. The acid-soluble fractions (low Mw) seem to contain nonconjugated carboxyl groups and may be associated with carbohydrate material. However, color bodies are found to be aromatic in nature (partially degraded lignin), possess a negative charge, and exist primarily as soluble sodium salts in aqueous solutions. The color bodies which are not removed by lime treatment have low Mw, high nonconjugated carboxyl groups, some ligninlike character, and seem to be associated with colorless carbon compounds. (EPA)
W73-08298

PROBLEMS OF COMBINED SEWER FACILITIES AND OVERFLOWS, 1967.
American Public Works Association, Chicago, Ill.
For primary bibliographic entry see Field 05G.
W73-08299

TURBOVENTILATOR FOR THE INPUT OF OXYGEN INTO LIQUIDS,
FMC Corp., San Jose, Calif.
K-H. Kalbskopf.
U. S. Patent No. 3,704,009, 7 p, 18 fig, 1 tab, 8 ref; Official Gazette of the United States Patent Office, Vol 904, No 4, p 531, November 28, 1972.

Descriptors: *Patents, *Waste water treatment, *Aerators, Equipment, *Oxygenation, Ventilation, Pollution abatement, Water quality control.
Identifiers: *Turboventilators.

Water is formed into long lasting aeration-jet streams by means of jet producing devices. The aerator comprises circumferential spaced jet forming arms, each having box-like jet formers mounted to rotate at the surface layer of the liquid. The jet formers are bounded by upstanding, radially spaced vanes. The vanes are parallel and extend between upper and lower closure plates. The jet forming arms are circumferentially spaced providing for unobstructed entry of liquid from between and in front of the arms and for entry of air from above the aerator. The vanes form box-like structures which are inclined from a tangent to their path of motion so that their leading edges have a smaller radius of rotation than their trailing edges. The jet formers accelerate the surface layer of liquid into diverging, horizontal jet streams that embody intrained air bubbles. The jet streams travel onto the walls of the tank whereupon they are diverted downward and hence carry air down into the main body of the liquid. (Sinha-OEIS)
W73-08304

METHOD AND APPARATUS FOR ULTRASONICALLY CLARIFYING LIQUID,
FMC Corp., San Jose, Calif. (assignee).
R. Davidson.
U. S. Patent No. 3,707,230, 4 p, 4 fig, 1 ref; Official Gazette of the United States Patent Office, Vol 905, No 4, p 748, December 26, 1972.

Descriptors: *Patents, *Waste water treatment, *Filtration, *Ultrasonics, Equipment, Water quality control, Pollution abatement.

The effluent is passed through a rotating fabric in a chamber. Solids are deposited on the fabric that serves as a filter. Within the chamber, ultrasonic transducers set up vibrations to remove the solids from the filter. The solids are directed through a discharge pipe for removal. A pressure head is set up between the effluent and the clarified liquid, with the transducers located above the level of the clarified liquid. (Sinha-OEIS)
W73-08306

WASTE WATER CONCENTRATOR WITH PLURAL DISTRIBUTORS,
Sweco, Inc., Los Angeles, Calif. (assignee).
W. J. Talley, Jr.
U. S. Patent No. 3,707,235, 9 p, 5 fig, 3 tab, 3 ref; Official Gazette of the United States Patent Office, Vol 905, No 4, p 749, December 26, 1972.

Descriptors: *Patents, *Sewerage, *Sewage treatment, *Storm runoff, *Storm drains, Water pollution sources, Water quality control, Equipment, Waste water treatment.

The apparatus for use in screening storm water overflows from sewer systems comprises a substantially cylindrical rotary screen disposed for rotation within a housing. A feed device is used to direct influent toward the inner surface of the screen. An outlet receives the effluent which passes through the screen, the concentrate which does not the screen, and backsplash from the screen. (Sinha-OEIS)
W73-08311

EXTENDED AERATION, ACTIVATED SLUDGE PLANT,
BiO2 Systems, Inc., Kansas City, Mo. (assignee).
D. O. Smart, IV, G. B. Pennington, R. M. Plettner, and R. F. Maughan.
U. S. Patent No. 3,709,363, 6 p, 4 fig, 9 ref; Official Gazette of the United States Patent Office, Vol 906, No 3, p 513, January 9, 1973.

Descriptors: *Patents, *Sewage treatment, *Biological treatment, *Activated sludge, *Aerators, *Waste water treatment, Water quality control, Water pollution control, Pollution abatement, *Aerobic bacteria, Microorganisms, Treatment facilities.
Identifiers: Clarifiers.

This plant utilizes many of the basic principles of the activated sludge process. In particular it is designed to efficiently convert organic wastes into oxidation end products which may be safely discharged to the environment. In addition to having an aerator and clarifier, it has a unique airlift arrangement which induces skimmer flow and sludge return flow from the clarifier to the aerator. The air which is employed for inducing recycle flow is ultimately discharged to the aerator for vigorously circulating the mixture and for supplying oxygen for the aerobic microorganisms. (Sinha-OEIS)
W73-08317

METHOD AND APPARATUS FOR DENITRIFICATION OF TREATED SEWAGE,
Dravo Corp., Pittsburgh, Pa. (assignee).
E. S. Savage.
U. S. Patent No. 3,709,364, 4 p, 1 fig, 8 ref; Official Gazette of the United States Patent Office, Vol 906, No 2, p 513, January 9, 1973.

Descriptors: *Patents, *Activated sludge, *Sewage treatment, *Waste water treatment, *Filtration, Bacteria, *Denitrification, Carbon, Nitrates, Nitrites, *Nutrient removal, Water quality control, Water pollution control, Pollution abatement.

Sewage is subjected to an activated sludge treatment and a highly nitrified effluent is discharged from the settling tank of the activated sludge system. The effluent is concurrently clarified and denitrified in a deep bed filter. Backwashing of the filter is controlled so that denitrifying bacteria is either wholly or partially maintained on the filter media or is immediately returned to the media enabling continuous operation of the filter. A supplemental carbon feed is charged to the filter to balance the bacteriological environment. (Sinha-OEIS)
W73-08318

FLOATING SURFACE AERATOR,
Passavant-Werke, Michelbach (West Germany). Michelbacherhutte.
H. Auler, and J. Muskat.
U. S. Patent No. 3,709,470, 4 p, 6 fig, 10 ref; Official Gazette of the United States Patent Office, Vol 906, No 2, p 539, January 9, 1973.

Descriptors: *Patents, *Aerators, Equipment, *Water pollution treatment, Pollution abatement, Water quality control, *Aeration, *Waste water treatment.

This aerating apparatus consists of radially extending blades or shovels which rotate about a vertical axis. The immersed buoyant body may be generally annular in shape. It surrounds the vertical axis. Small stabilizing floats located on the surface of the liquid are placed around the apparatus and are connected with it. A baffle arrangem imparts a screw-shaped torque to the fluid pass through it in the direction opposite to the direct of rotation of the aerator. (Sinha-OEIS)
W73-08319

METHOD AND APPARATUS FOR TREATE SEWAGE,
FMC Corp., San Jose, Calif. (assignee).
Q. L. Hampton.
U. S. Patent No. 3,709,792, 6 p, 3 fig, 4 ref; Of cial Gazette of the United States Patent Offi Vol 906, No 2, p 612-613, January 9, 1973.

Descriptors: *Patents, *Sewage treatment, Trea ment facilities, *Aerobic treatment, Biologic treatment, *Aeration, *Activated sludge, *Was water treatment, Water quality control, Water pc lution control, Biochemical oxygen demand.

Aqueous waste is mixed with aerated, conce trated sludge from a reaeration zone and a supe natant liquor from an aerobic sludge digestic zone to form a mixed liquor. Air is introduced int the aeration zone for mixed liquor, the aerobi digestion zone and the reaeration zone for sludg concentrate. Sludge is settled from the mixe liquor and is discharged into a flow course whic delivers predetermined quantities of the sludg concentrate to the reaeration zone. Aerated sludg and supernatant liquor are discharged into th aeration zone. Control of the character of th sludge being recycled is an important factor ir determining the overall efficiency of treatment o sewage as measured by removal of BOD. The apparatus comprises a tank which is compartmentalized by parallel partitions into a central compartment, a digestion compartment and a reaeration compartment. There is a common longitudinal axis for liquid flow. Conduits are provided in the settling compartment transverse to the longitudinal flow axis. These have airlift devices to induce movement of settled sludge. (Sinha-OEIS)
W73-08320

DOMESTIC APPLICATIONS FOR AEROSPACE WASTE AND WATER MANAGEMENT TECHNOLOGIES,
General Electric Co., Schenectady, N.Y.
For primary bibliographic entry see Field 07B.
W73-08366

REGIONALIZATION AND WATER QUALITY MANAGEMENT,
Camp, Dresser and McKee, Boston, Mass.
K. M. Yao.
Journal Water Pollution Control Federation, Vol 45, No 3, p 407-411, March, 1973. 2 fig, 2 tab, 9 ref.

Descriptors: Streams, *Water quality, Management, *Water pollution control, *Massachusetts, *Connecticut River, *Simulation analysis, Dissolved oxygen, Water utilization, Water resources, Biochemical oxygen demand, Treatment facilities, Water supply, Computer programs, Low flow, Mathematical models, Systems analysis, *Regional analysis.
Identifiers: Regionalization, Treatment plants, Streeter-Phelps equation, Bondi Island, Secondary treatment.

A study is presented which utilizes a realistic example and a simulation technique (a) to demonstrate the effects of regionalization on stream water quality management, and (b) to explore and compare various alternatives from the viewpoint of water quality management as well as that of overall water resources utilization. The study area comprises four municipalities in the Connecticut River basin which have been ordered by the Massachusetts state pollution control agency to upgrade their water pollution abatement facilities for achieving the state stream quality objective on a fixed time schedule. The Streeter-Phelps equation

illustrate the use of the algorithm, a three-stage waste water treatment plant is optimized. To insure meaningful results, it is necessary to include costs for each stage versus the percent removal for that stage. A simple tree diagram is used to illustrate the process. To exemplify the versatility of the algorithm, it is used to show the optimization of a trickling filter using the Eckenfelder equation. The algorithm has many possible applications both within and outside the realm of environmental engineering, and it can be applied to more than three variable systems. (Bell-Cornell)
W73-08386

FLOCCULATING APPARATUS AND METHOD,
Crucible Inc., Pittsburgh, Pa. (assignee)
G. C. Almasi, and W. Slusarczyk.
U. S. Patent No 3,714,037, 3 p, 4 fig, 5 ref; Official Gazette of the United States Patent Office, Vol 906, No 5, p 1722, January 30, 1973.

Descriptors: *Patents, *Flocculation, *Waste water treatment, *Industrial wastes, *Filtration, Water quality control, Water pollution control, Separation techniques, Pollution abatement.
Identifiers: *Magnetic particles, *Magnetization.

Magnetic particles are removed from industrial waste water by magnetizing the particles during flow. The particle containing liquid is introduced into a restricted passage having a large width to height ratio without any change in cross-sectional area. The passage has a magnetic field. Its lines of force are at right angles to the lengthwise liquid flow direction. As the particles are magnetized they agglomerate so that they may be removed by a filter or by settling in a tank. (Sinha-OEIS)
W73-08388

SYSTEMS FOR THE REMOVAL OF POLLUTANTS FROM WATER,
A. J. Shaler, and D. C. McLean.
U. S. Patent No 3,713,542, 9 p, 4 fig, 9 ref; Official Gazette of the United States Patent Office, Vol 906, No 5, p 1606, January 30, 1973.

Descriptors: *Patents, *Activated carbon, *Sewage treatment, *Organic wastes, *Water treatment, Pollution abatement, Water quality control, Water pollution control, Treatment facilities, *Waste water treatment.

This method consists of providing an extensive wall of char or activated carbon, down the height of which the mixture of sewage effluent and unclean water is allowed to flow by gravity. The carbon is continuously regenerated by cycling through a thermal converter. Organically polluted sewage effluent and water from the waterway are continuously mixed and piped to the top of the wall of carbon, passed through it, and the clean water is drained from the bottom. The flow through the system may be reversed to backwash the screens and temporarily partially fluidize the carbon wall to prevent long-term decreases in its permeability. (Sinha-OEIS)
W73-08389

APPARATUS FOR PROCESSING EFFLUENT SLUDGE,
Hazemag G.m.b.H., Muenster (West Germany).
E. von Conrad, K. Rosner, and L. Meyer.
U. S. Patent No 3,712,550, 4 p, 1 fig, 5 ref; Official Gazette of the United States Patent Office, Vol 906, No 4, p 1334, January 23, 1972.

Descriptors: *Patents, *Sludge treatment, Equipment, Separation techniques, *Waste treatment, Water quality control, Water pollution control, Pollution abatement, *Waste water treatment.
Identifiers: *Impact grinders.

The apparatus consists of an impact grinder for converting solid refuse to a particulate stage. The grinder comprises a rotor journalled for rotation about a horizontal axis. There is a supply conduit extending along and above an outlet. A distribution baffle extends along a slot and is inclined downward so that effluent sludge flows in the form of a thin layer over the baffle and onto the ground refuse issuing from the outlet. (Sinha-OEIS)
W73-08390

REVERSE OSMOSIS MODULE,
Westinghouse Electric Corp., Pittsburgh, Pa. (assignee)
G. W. Ellenburg.
U. S. Patent No 3,712,473, 3 p, 3 fig, 4 ref; Official Gazette of the United States Patent Office, Vol 906, No 4, p 1314, January 23, 1973.

Descriptors: *Patents, *Reverse osmosis, *Semipermeable membranes, *Desalination, Water treatment, *Demineralization, Membranes, Water quality control, *Waste water treatment, Water pollution control, Pollution abatement, Separation techniques, *Dissolved solids.

A reverse osmosis module is described for reducing the concentration of dissolved solids in a pressurized liquid by passing the liquid through a semipermeable osmotic membrane. The module is formed from a porous sand bar in which tubular osmotic membranes are placed in longitudinal holes. An enveloping membrane embraces the outer peripheral surface of the bar. The ends of the bar are sealed with epoxy. Only compressive forces are required. (Sinha-OEIS)
W73-08391

PROCESS FOR THE TREATMENT OF PHENOLATED INDUSTRIAL EFFLUENT,
Societe Anonyme pour l'Etude et l'Exploitation des Procedes Georges Claude, Pairs (France). (assignee)
J-P. Zumbrunn, and F. Crommelynck.
U. S. Patent No 3,711,402, 5 p, 3 ref; Official Gazette of the United States Patent Office, Vol 906, No 3, p 1029, January 16, 1973.

Descriptors: *Patents, *Chemical reactions, *Waste water treatment, Chemical wastes, *Liquid wastes, *Phenols, *Industrial wastes, Pollution abatement, Water quality control, Water pollution control.

Industrial effluent polluted by phenolated impurities is treated by conversion of the impurities by the action of an oxidizing reagent containing the HSO5- anion. At least one mole of HSO5- is required per mole of phenol. The oxidizing reagent is a member of the group constituted by monopersulphuric acid and its salts. Eleven examples illustrate possible variations. (Sinha-OEIS)
W73-08394

METHOD OF PURIFYING SEWAGE AND WASTE LIQUIDS AND A DECOCTING SYSTEM FOR CARRYING OUT THE METHOD,
G. E. Lagstrom.
U. S. Patent No 3,711,381, 4 p, 1 fig, 6 ref; Official Gazette of the United States Patent Office, Vol 906, No 3, p 1025, January 16, 1973.

Descriptors: *Patents, *Sewage treatment, *Liquid wastes, *Domestic wastes, *Detergents, *Waste water treatment, Chemical wastes, Water quality control, Water pollution control, Pollution abatement.

Sewage and/or domestic effluent is boiled to produce a foam of the impurities. The steam and foam are removed and separated. The foam is led into a sedimentation tank and the steam either released to the atmosphere or used as a heat source. The sludge obtained in the sedimentation tank as the foam breaks down is removed at intervals and fed back to the boiler. (Sinha-OEIS)

W73-08395

METHOD AND APPARATUS FOR TREATING SEWAGE,
FMC Corp., Chicago, Ill. (assignee)
H. Brociner.
U. S. Patent No 3,710,941, 4 p, 2 fig, 3 ref; Official Gazette of the United States Patent Office, Vol 906, No 3, p 921, January 16, 1973.

Descriptors: *Patents, Pollution abatement, *Waste water treatment, *Sewage treatment, Treatment facilities, *Aeration, Water quality control, Water pollution control, *Liquid wastes.
Identifiers: *Grit.

Grit is removed from liquid sewage in a tank whose floor terminates in an accumulation zone for solids. Air is introduced into the sewage adjacent to the first side wall and above the accumulation zone for solids with uniform distribution along the length of the side wall. A baffle is positioned adjacent to a second side wall and transverse to the direction of liquid flow through the tank. This advances the separation of light weight solids. The sewage suspension flows under the baffle and upward for removal by flow over a weir. (Sinha-OEIS)
W73-08396

SODIUM ALUMINATE FOR REMOVAL OF PHOSPHATE AND BOD,
Babbitt Utilities Commission, Minn.
D. Cole, and P. Tveite.
Public Works, Vol 103, No 10, p 86-87, October, 1972. 1 fig, 1 tab.

Descriptors: *Phosphates, *Biochemical oxygen demand, Domestic wastes, *Waste water treatment, Coliforms, Chlorination, Suspended solids, *Chemical precipitation.
Identifiers: *Sodium aluminate, *Chemical treatment.

In a three month trial at the Babbitt activated sludge treatment plant, 42.5% sodium aluminate was fed to sewage influent to evaluate the chemical for phosphate removal. Sodium aluminate reacts with soluble phosphate to form aluminum phosphate. Sodium aluminate also acts as a coagulant and flocculant, causing both the precipitated aluminum phosphate and other insoluble phosphates in the waste to agglomerate and separate readily in the clarifier or settling tank. Among other benefits of the sodium aluminate treatment were: (1) lower effluent BOD; (2) lower chlorine demand; (3) lower effluent coliforms; (4) reduced air requirements in the aeration chambers; (5) higher effluent dissolved oxygen; and (6) improved primary sludge characteristics. Quantitative results are presented. Babbitt has decided to use the aluminate treatment, and the chemical storage facilities are currently being designed. (Morparia-Texas)
W73-08410

UHDE DETAILS CHLORINE CELL MERCURY RECOVERY PROCESSES.

European Chemical News, Vol 22, No 544, p 22, August 4, 1972.

Descriptors: *Mercury, *Filtration, *Chlorine, Carbon, Chemical reactions, Absorption, *Waste water treatment, Industrial wastes.
Identifiers: Hydrazine sulfide, *Chemical treatment, Caustic soda solution.

Mercury losses from chlorine plants are discussed. Mercury levels which can be obtained through use of new techniques are described. The caustic soda solution can be pre-coat filtered to a mercury content of 0.1 to 0.3 g/m3. Mercury leaving the decomposer can be removed by condensation, Chemical reaction with wet chlorine, or absorption on carbon. Bayer catalyst in place of activated carbon will reduce the mercury content in the hydrogen jar to below 1 microgram/Nm3. Waste water from scrubbing operations can be treated with hydrazine or sulfide. Ventilation air is kept pure by using totally enclosed cells. Handling procedures minimize mercury-air contact time. The resulting mercury loss is 2 to 3 grams per ton of chlorine produced. (Anderson-Texas)
W73-08411

FERMENTATION PATTERNS IN THE INITIAL STAGES OF SEWAGE SLUDGE DIGESTION,
Purdue Univ., Lafayette, Ind.
R. M. Sykes, and E. J. Kirsch.
Developments in Industrial Microbiology, Vol 11, p 357-366, 1969. 5 fig, 1 tab, 16 ref.

Descriptors: *Sludge digestion, *Waste water treatment, Sludge, Activated sludge, *Fermentation, *Sewage treatment, Hydrogen, Carbon dioxide, Amino acids, Humic acids.
Identifiers: Methane, Propionic acids, Butyric acids, *Acetic acid.

The physiological activity of 75 pure culture isolates grown in sterilized raw sewage sludge was compared with that of a mixed culture obtained from a sludge digester. Some of the fermentation products formed were identified and quantitated. In the mixed culture a diphasic pattern of gas and acid production was apparent. During the early stages, acetic, propionic, and butyric acids were produced in nearly equal quantities, while hydrogen and carbon dioxide were evolved in nearly equal quantities. After 3 days, only acetic acid was produced in quantity, hydrogen disappeared, methane was evolved, and carbon dioxide continued to accumulate. Of the pure culture isolates, 25 percent produced acetic, propionic, and butyric acids, hydrogen and carbon dioxide. The remainder produced only acetic acid, propionic acid, and carbon dioxide. None of the pure isolated produced methane. (Murphy-Texas)
W73-08412

BUDDS FARM SEWAGE-TREATMENT WORKS OF HAVANT AND WATERLOO URBAN DISTRICT COUNCIL.

Water Pollution Control, Vol 71, No 4, p 348-350, 1972.

Descriptors: *Waste water treatment, *Treatment facilities, Biological treatment, Settling basins, *Sludge digestion, Sludge, Storage capacity, Capital costs, *Sewage treatment.

The sewage-treatment plant capacity was extended to 145,000 population in 1969. The new units include comminutors, detritor, sedimentation tanks, biological filters with slag medium, and a new ocean outfall. Additional capacity for sludge digestion, storage, and pressing was provided. The standards set by the Hampshire River Authority are 30 mg/l SS and 20 mg/l BOD for flows up to 18 mgd and screening for overflows. A table of construction costs is included. (Anderson-Texas)
W73-08414

THE DISPOSAL OF SOLID WASTES,
For primary bibliographic entry see Field 05E.
W73-08415

FILTER PLANT WASTEWATER TREATMENT,
Pirnie (Malcolm), Inc., Paramus, N.J.
G. P. Westerhoff.
Public Works, Vol 103, No 10, p 79-82, October, 1972. 2 fig.

Descriptors: *Waste water treatment, *Treatment facilities, *Waste disposal, *Tertiary treatment, *Filtration, Sedimentation, *Coagulation, Alum, Chlorination, Flocculation, Settling bas Separation techniques, Data collections, W. water (Pollution).

Water treatment plant wastewater disposal discussed. There are two main sources of wa water: coagulation basin wastewater (referred as alum sludge at plants using alum coagulati and filter backwash wastewater. Water treatm processes used at the Erie County Sturgeon Pc Filtration Plant are aeration, chemical additi mixing, flocculation and sedimentation follow by filtration, pH adjustment, fluoridation a chlorination. Alum recovery from coagulati basin wastewater was proved to hold consideral promise as an effective system for treatme recovery, and disposal of process residue. A stu program has been designed to carry out data co lection and evaluation to meet various goals su as identifying the physical and chemical properti of water treatment plant wastes and the effectiv ness of several methods of liquid-solid separati (both with and without the addition of coagulati aids). The data will reflect overall plant operatic and efficency and the feasibility of each proces Both phases of this study program are describe (Morparia-Texas)
W73-08416

5E. Ultimate Disposal of Wastes

INDUSTRIAL WASTELINE STUDY - A SYSTEM FOR CONTROLLED OCEAN DISPOSAL,
Franklin Inst., Philadelphia, Pa. Labs. for Research and Development.
D. Pindzola, C. T. Davey, and R. A. Erb.
Available from the National Technical Information Service as PB-219 404, $3.00 in paper copy, $1.45 in microfiche. Final Technical Report F-C2577, (1970). 77 p, 12 fig, 23 tab, 33 ref, 2 append. EPA Grant 16070 EOI 08/70.

Descriptors: *Waste disposal, *Pipelines, Baseline studies, Continental shelf, Continental slope, Ultimate disposal, *Outlets, Industrial wastes, Sewage sludge, Water pollution sources, Atlantic Ocean.
Identifiers: *Ocean outfalls, *Ultimate waste disposal, Dredge spoils.

Pipelining waste to the edge of the continental shelf from the Delaware Valley was found to be feasible by defining waste loads, routes, pipe types, conditions of the proposed outfall area and costs. The waste load was examined and projected by type and quantity by percent of total volume as follows: dredge spoils, 24%; sewage sludges, 6%; waste acids and chemicals, 6%; ashes, 0.3%; and dilute industrial waste, 63.7%. Present dilute volume is established at 5,840 million gallons per year. Routes may generally be obtained along power and rail lines currently existing. Of pipe materials considered, polyester-glass reinforced pipe appears best for overland routing and polyolefin for undersea. Estimated installation costs are in the area of $75,000,000. Disposal costs could be reduced from the current $3.50 to $1.85 per 1000 gallons. Ocean studies included otter trawls and direct observation by submersible which showed the biodensity to fall off with increasing depth. The proposed disposal area appears less densely populated than inshore areas and less productive than bays, marshes and estuaries where wastes are currently deposited. Bottom currents were determined to be mainly off shore by current meters and seabed drifters. Of 400 seabed drifters deposited, only two were recovered, these by commercial trawlers. Dissolved oxygen levels were high in the water column. Further examination and baseline data are needed. Actual pilot construction would be desirable. (EPA)
W73-07812

United Kingdom policy in relation to radioactive waste is described and the relevant legislation and methods of control are summarized. Data are given on the amounts of radioactivity discharged as waste from establishments of the United Kingdom Atomic Energy Authority, the nuclear power stations operated by the Electricity Generating Boards and other users of radioactive materials. Studies of the behavior of radioactivity in the environment are reported with particular reference to food chains and other potential sources of irradiation of the public. The results of environmental monitoring are presented and estimates are made of radiation doses received by individual members of the public and larger population groups as a result of waste disposal. It is concluded that the doses received are all within the appropriate limits recommended by the International Commission on Radiological Protection. (Houser-ORNL)
W73-07957

POLLUTION OF SUBSURFACE WATER BY SANITARY LANDFILLS, VOLUME 1,
Drexel Univ., Philadelphia, Pa.
For primary bibliographic entry see Field 05B.
W73-08073

USE OF WASTE TREATMENT PLANT SOLIDS FOR MINED LAND RECLAMATION,
Illinois Univ., Urbana.
T. D. Hinesly, R. L. Jones, and B. Sosewitz.
Mining Congress Journal, Vol 58, No 9, p 66-73, September, 1972. 2 fig, 1 tab, 21 ref.

Descriptors: *Sludge disposal, *Waste disposal, Ultimate disposal, *Land reclamation, Land development, Land resources, Land use, Strip mines, Cost analysis, *Waste treatment, Recycling, Soils, *Illinois.
Identifiers: *Fulton County (Ill), *Stripped mine land reclamation.

It is apparent that the most desirable solution to the problem of sludge disposal is the recycling of the solids to natural biological systems. Fulton County, Illinois, is cited as an example of an area where waste treatment plant solids will be used for land reclamation because of the extensive stripped mine lands. An extensive geologic, biologic, and agricultural overview is presented of the effects expected due to this sludge disposal process. A cost analysis of the entire project is included. (Smith-Texas)
W73-08096

'LIQUID FERTILIZER' TO RECLAIM LAND AND PRODUCE CORPS,
Metropolitan Sanitary District of Greater Chicago, Ill.
B. T. Lynam, B. Sosewitz, and T. D. Hinesly.
Water Research, Vol 6, p 545-549, 1972. 3 tab, 9 ref.

Descriptors: *Sludge disposal, *Land reclamation, *Waste disposal, Nitrogen, Phosphorus, Organic matter, Soils, Soil contamination, Viruses, Fecal coliforms, Agriculture, Fertilizers, *Illinois, Water reuse.
Identifiers: *Chicago, *Nutrient sources.

The Metropolitan Sanitary District of Greater Chicago handles a waste load equivalent of 10 million people. Disposal methods to handle the vast amount of sludge were studied. Land application of liquid sludge was determined to be the cheapest as well as most effective method. Studies by agronomists have shown the value of liquid sludge as a nitrogen source, phosphorus source, organic matter source, and as a soil conditioner. Microbiological studies indicate a rapid die-off rate for virus and fecal coliform. The heavy metal content is small and not expected to cause soil contamination. (Anderson-Texas)
W73-08098

WALDHOF TYPE OF FERMENTORS IN DISPOSAL OF FOOD WASTES,
Iowa State Univ., Ames.
For primary bibliographic entry see Field 05D.
W73-08120

EUROPEAN METHODS OF COMPOSTING AND SLUDGE UTILIZATION,
L. D. Hills.
Compost Science, p 18-19, July-August, 1972.

Descriptors: *Waste treatment, *Land reclamation, *Fertilizers, *Sludge disposal, Ultimate disposal, Waste disposal, Urine, Domestic wastes, Sludge treatment, Organic matter.
Identifiers: *Composting, Clivus toilet.

Methods of composting and sludge utilization in Europe are described. While sludge is primarily used as fertilizer and for land reclamation, composting is done by means of the Clivus Toilet. This is roughly a family Municipal compost plant that takes the excreta and urine and composts them with the kitchen wastes to produce roughly a hundred pounds a year, from a family of three, of a good high potash organic fertilizer. (Smith-Texas)
W73-08123

DISPOSAL OF BRINE EFFLUENTS,
Dow Chemical Co., Walnut Creek, Calif.
R. R. Grinstead, and T. E. Lingafelter.
Available from the National Technical Information Service as PB-215 037, $3.00 in paper copy, $1.45 in microfiche. Office of Saline Water Research and Development Progress Report No. 810, August 1972. 120 p, 18 tab, 33 fig, 23 ref, 5 append. 14-30-2630.

Descriptors: By-products, Solvent extraction, Separation techniques, *Brine disposal, Liquid wastes, *Waste disposal, *Desalination, Salts, Effluents, *Chemical precipitation, *Anion exchange, *Water softening, Recycling, Chlorides, Sulfates.
Identifiers: Metal complexation.

A process is described by which brines from the desalination of brackish water supplies can be converted to either useful products or waste materials which can be readily disposed of without environmental harm. The method involves (1) precipitation of the sulfate with calcium chloride, (2) exchange of chloride for bicarbonate by means of a liquid-liquid anion exchange extraction system, and (3) recycle of the bicarbonate solution to the pretreatment (softening) step. The extraction system is regenerated with lime and carbon dioxide, producing calcium chloride as a product, some of which is used in step (1). In cases where the sodium bicarbonate production is greater than required in pretreatment, a second liquid-liquid extraction system, based on a carboxylic acid cation exchanger, is used to produce sodium carbonate for sale. Preliminary economic estimates indicate that such a process would become competitive with conventional disposal methods (solar ponds or wells) at a production level slightly above 5 mgd of water and would be most attractive in cases where a high sodium carbonate requirement exists for pretreatment. (OSW)
W73-08189

MARINE WASTE DISPOSAL - A COMPREHENSIVE ENVIRONMENTAL APPROACH TO PLANNING,
For primary bibliographic entry see Field 05B.
W73-08247

THE DISPOSAL OF CATTLE FEEDLOT WASTES BY PYROLYSIS,
Midwest Research Inst., Kansas City, Mo.
For primary bibliographic entry see Field 05D.
W73-08290

THE DISPOSAL OF SOLID WASTES,
A. R. Balden.
Industrial Water Engineering, Vol 4, No 8, p 25-27, August, 1967. 6 fig, 2 ref.

Descriptors: *Ultimate disposal, *Industrial wastes, *Waste treatment, Sludge, Sludge disposal, *Incineration, Design, Paints, Chemical wastes, Pilot plants, Tertiary treatment.
Identifiers: *Paint wastes.

Methods of ultimate disposal and their application in handling oily sludges and paint wastes in the automotive industry are reviewed. Five methods from the Environmental Health Series, 'Summary Report, Advanced Waste Treatment Research' are briefly explained. Three variations in incinerator design are described. Each design is capable of burning oily sludges without objectionable emissions. A pit incinerator for treatment of paint sludges, operated by a chemical manufacturer, is described. (Morparia-Texas)
W73-08415

5F. Water Treatment and Quality Alteration

DEVELOPMENT OF MULTIPURPOSE WATER PURIFICATION UNIT FOR ARMY FIELD USE.
Army Mobility Equipment Research and Development Center, Fort Belvoir, Va.
A. Ford, Jr., and D. C. Lindsten.
Available from the National Technical Information Service as AD750 322, $3.00 in paper copy, $1.45 in microfiche. 1972. 15 p, 9 fig, 5 tab, 3 ref.

Descriptors: *Water treatment, *Potable water, *Water supply, *Reverse osmosis, *Membranes, Suspended solids, Chemicals, Botulism, Water quality control, Water pollution treatment, Filtration, Cellulose, Turbidity, Salts, Bacteria, Water purification.
Identifiers: *Ultrafiltration.

The Army Combat Developments Command has established a requirement for research and development to demonstrate the technical feasibility of a single multipurpose unit capable of producing potable drinking water for cullinary, washing, bathing, laundering, and food preparation purposes. Reverse osmosis is capable of removing up to 99% of dissolved salts from water, removing essentially all turbidity from water, removing 99% of chemical agents VX and RZ, 78% of chemical agent GB, 98% of Na2HASO4 and 99.988% of botulinum toxin from water. Furthermore, reverse osmosis is capable of removing essentially all microorganisms from water although chlorination is indicated to protect the product water. The one disadvantage appears to be membrane fouling, which may be solved by feedwater pretreatment, intermittent cleaning or modular replacement. (Smith-Texas)
W73-07833

DESALINIZATION PLANTS, VIRGIN ISLANDS (FINAL ENVIRONMENTAL IMPACT STATEMENT).
Department of Housing and Urban Development, San Juan, Puerto Rico. Region II.
For primary bibliographic entry see Field 03A.
W73-07977

STATE OF THE ART OF WATER FILTRATION,
American Water Works Association, New York. Committee on Filtration Problems.
R. L. Woodward, E. R. Baumann, J. A. Borchardt, N. J. Davoust, and K. A. Dostal.
Journal of the American Water Works Association, Vol 64, No 10, p 662-665, October 1972. 26 ref.

Descriptors: *Filtration, *Filters, *Water treatment, *Water quality control, *Waste water treatment, Water pollution control, Design criteria, Turbidity, *Tertiary treatment, Treatment facilities, Separation techniques.
Identifiers: *Plant operations, Upflow rates, Filtration rates.

Rapid filtration of water is discussed under two main topics: (1) theory of filtration and (2) design and operation of water filters. In the design and operation of filters, consideration is made of filter media, filter aids, filter rates, upflow and byflow filters, turbidity meters, surges and backwashing. Careful collection analysis and reporting of plant operating experience where advanced design or operating features exist is encouraged. (Smith-Texas)
W73-08107

PHILADELPHIA SUBURBAN WATER COMPANY,
Philadelphia Suburban Water Co., Bryn Mawr, Pa.
K. E. Shull.
Journal of the American Water Works Association, Vol 64, No 10, p 647-648, October 1972. 1 fig, 6 ref.

Descriptors: *Water treatment, *Waste water treatment, Research and development, Aluminum, Filtration, Filters, Polyelectrolytes, Anions, Ions, Laboratory tests, Pilot plants, Sludge treatment, Corrosion, *Pennsylvania.
Identifiers: *Philadelphia.

Details are given of various research projects being conducted by the Philadelphia Suburban Water Company. These projects include development of a method for determining small amounts of aluminum in water; use of dual media filters, and polyelectrolyte application; development of a filtrate availability technique; investigations of polyelectrolytes; investigation of the use of cationic, anionic, and non-ionic polyelectrolytes as filter aids; sludge conditioning studies, and laboratory and pilot studies of the use of certain compounds in reducing the corrosive tendencies of water on metal pipes. (Smith-Texas)
W73-08109

DETROIT METRO WATER DEPARTMENT,
Detroit Metro Water Dept., Mich.
G. J. Remus.
Journal of the American Water Works Association, Vol 64, No 10, p 644-645, October 1972.

Descriptors: *Water reuse, *Recycling, Flocculation, Waste water treatment, Water treatment, Corrosion, Corrosion control, Bioassays, Turbidity, Ions, Water pollution control, Water quality control, Michigan.
Identifiers: *Detroit.

Several research programs conducted by the Detroit Metro Water Department are detailed. They include studies of recycling of waste wash water; flocculator tests; dual media tests; studies with polyelectrolytes; fluoride corrosion tests; corrosion studies of copper surfaces; plankton counting and identification reproducibility; bioassay studies; liquid alum feeding; tests of turbidity; study of water samples; and specific ion electrodes. (Smith-Texas)
W73-08110

DALLAS WATER UTILITIES,
Dallas Water Utilities Dept., Tex.
H. J. Graeser.
Journal of the American Water Works Association, Vol 64, No 10, p 638-641, October 1972. 1 fig, 5 tab, 1 ref.

Descriptors: *Waste water treatment, *Waste treatment, *Sewers, *Combined sewers, *Polymers, *Pilot plants, Research and development, Operating costs, Waste disposal, Sewage disposal, Pollution abatement, Water reuse, Treatment facilities, Research facilities, *Texas, Water supply.
Identifiers: Dallas.

The Dallas Water Utilities Department has been able to develop a unified approach to solving problems in water supply and waste water disposal. The City of Dallas has constructed a pilot plant to evaluate different unit processes. Operation of this pilot plant has been costly, but the savings realized in process selection and the operational experience gained prior to start-up to the prototype do much to offset the research cost. The City of Dallas has also been experimenting with developing technological factors of waste water reuse. The two major areas of research are metal and virus removal. Other areas of research now being conducted by the City of Dallas include microbiological studies, studies of the treatment and disposal of infiltration flows in the sanitary sewer system, and experimentation with the use of polymers to increase the hydraulic capacity of sewers. (Smith-Texas)
W73-08111

WATER AND SEWER PLAN.
West Alabama Planning and Development Council, Tuscaloosa.
For primary bibliographic entry see Field 06D.
W73-08181

INVENTORY OF INTERSTATE CARRIER WATER SUPPLY SYSTEMS.
Environmental Protection Agency, Washington, D.C. Water Supply Div.

January 1973. 81 p.

Descriptors: *Water supply, *Environmental sanitation, *Potable water, *Public health, *Water purification, Interstate, Water resources development, Water utilization, Water yield improvement, Resources development, Water supply development, Municipal water, Water quality control, Water quality standards.

In 1970 the Environmental Protection Agency was assigned the responsibility of certifying water supply systems serving interstate carriers, a project formerly accomplished by the Public Health Service. Interstate quarantine regulations were promulgated to control the transmission of communicable disease into the U.S. or between the states; additionally, they contain the standards for acceptable, safe drinking water systems and form the basis for the interstate carrier water supply certification program. Systems in substantial compliance with the standards are classified as approved; those with significant deviations from the quality, surveillance, facilities or operational requirements of the standards are prohibited. Charts and graphs provide all necessary information. The status of all water systems are listed by regional EPA offices as of December 29, 1972. (Mockler-Florida)
W73-08192

A PORTABLE VIRUS CONCENTRATOR FOR TESTING WATER IN THE FIELD,
Baylor Coll. of Medicine, Houston, Tex. Dept. of Virology and Epidemiology.
For primary bibliographic entry see Field 05A.
W73-08262

REGIONALIZATION AND WATER QUALITY MANAGEMENT,
Camp, Dresser and McKee, Boston, Mass.
For primary bibliographic entry see Field 05D.
W73-08383

REAERATION OF WATER WITH TURBINE DRAFT-TUBE ASPIRATORS,

Missouri Univ., Columbia. Dept. of Civil Engineering.

J. J. Cassidy.

Available from the National Technical Information Service as PB-219 263, $3.00 in paper copy, $1.45 in microfiche. Missouri Water Resources Research Center Completion Report, January 5, 1973. 22 p, 8 fig, 32 ref. OWRR A-044-MO (1). 14-31-0001-3525.

Descriptors: *Water quality control, *Reaeration, Dissolved oxygen, Hydraulics, *Draft tubes, Model studies.
Identifiers: *Aspirators.

The rate at which reaeration of water can be accomplished through introduction of air in turbine draft tubes was studied. A laboratory model simulating flow in a draft tube downstream from a turbine was constructed. Independent control of rate of flow of air and rate of flow of water was accomplished. Dissolved oxygen content of flow before and after reaeration was measured. Dimensionless parameters of aeration efficiency, Froude number and air to water content were plotted.
W73-07807

ORGANIC WASTES AS A MEANS OF ACCELERATING RECOVERY OF ACID STRIP-MINE LAKES,

Missouri Univ., Columbia. Dept. of Civil Engineering.

D. L. King, and J. J. Simmler.

Available from the National Technical Information Service as PB-219 264, $3.00 in paper copy, $1.45 in microfiche. Missouri Water Resources Center Completion Report, February 20, 1973. 65 p, 15 fig, 5 tab, 21 ref, append. OWRR A-038-MO (1). 14-31-0001-3225 and 3525.

Descriptors: *Acid mine water, Coal mines, Nutrients, *Aluminum, *Ions, Lakes, *Iron, Mine drainage, Sulfates, Metals, *Organic wastes, Chemical wastes.
Identifiers: *Strip mine lakes.

In the presence of air and water, iron pyrite oxidizes to sulfuric acid and ferric hydroxide. The majority of the hydrogen ions associated with the sulfuric acid never reach the acid strip-mine lake because as they flow over the overburden they are involved in a series of reactions that are responsible for the weathering and dissociation of rocks, clays, and minerals. The majority of the sulfate ions, on the other hand, do reach the strip-mine lake and their concentration in the lake tends to indicate the amount of acid production in the particular watershed. The ferric hydroxide is also washed into the lake with the sulfate ions and settles to the bottom; however, a certain amount redissolves in the lake according to pH and Ksp limitations. Iron, sulfate, and hydrogen ions along with a host of acid dissociated ionic species, including aluminum, manganese, calcium, and magnesium, and allochthonous organic materials are constantly being washed into the acid strip-mine lake. It is these ions and organic matter that characterize the chemistry of these lakes. A small amount of buffer in the acid mine water is from the dissociation of HSO4 (-). Carbon dioxide and hydrogen sulfide gases also contribute considerably to the buffering of the water. However, it is the high concentration of such metals as aluminum and iron that make the greatest contribution to the net buffer capacity of the water. These metal buffers are responsible for the long natural recovery times associated with all acid strip-mine lakes. The amount of such buffers depends upon the amount and type of clays and minerals dissolved on the spoil banks. Depending on the clay type more or less aluminum may be dissolved and allowed to flow into the lake.
W73-07808

ECONOMIC BENEFITS FROM AN IMPROVEMENT IN WATER QUALITY,

Oregon Univ., Corvallis. Dept. of Agricultural Economics.

S. D. Reiling, K. C. Gibbs, and H. H. Stoevener.

Copy available from GPO Sup Doc as EP1.23/3 (73-008), $2.10; microfiche from NTIS as PB-219 474, $1.45. Environmental Protection Agency, Socioeconomic Environmental Studies, Report EPA-R5-73-008, January 1973. EPA Project 16110 FPZ.

Descriptors: *Benefits, Recreation, Water quality, Economics, Lakes, Camping, Sport fishing, *Recreation demand, *Water quality control, *Oregon.
Identifiers: Travel costs, On-site costs, *Klamath Lake (Ore).

A new methodology is introduced and empirically tested for estimating the economic benefits accruing to society from an improved recreational facility. The specific facility under consideration is Upper Klamath Lake, Oregon, which presently has low water quality. The methodology draws upon previous work done in the evaluation of recreational demand; however, it focuses upon the individual recreationist and separates the traditional price variable into on-site costs and travel costs. The model is used to estimate the number of days per visit the recreationist will stay at the site as the water quality improves. Data collected at three other lakes with varied characteristics are used to derive a relationship between the number of visits to a site and the characteristics of the site. This relationship is then used to estimate the increase in visits to Klamath Lake that would be forthcoming with an improvement in water quality. The impact of expanded recreational use of Klamath Lake upon the local economy is also estimated through the use of an input-output model of the Klamath County economy. (EPA)
W73-07813

WATER QUALITY CHANGES IN AN IMPOUNDMENT AS A CONSEQUENCE OF ARTIFICIAL DESTRATIFICATION,

North Carolina Univ., Chapel Hill. School of Public Health.

For primary bibliographic entry see Field 05C.
W73-07818

IMPROVING WATER QUALITY BY REMOVAL OF PESTICIDES WITH AQUATIC PLANTS,

Virginia Polytechnic Inst. and State Univ., Blacksburg. Dept. of Plant Pathology and Physiology.

S. W. Bingham.

Available from the National Technical Information Service as PB-219 389, $4.85 in paper copy, $1.45 in microfiche. Virginia Water Resources Research Institute, Blacksburg, Bulletin 58, March 1973, 94 p, 31 fig, 9 tab, 20 ref. OWRR A-033-VA (3).

Descriptors: *Pesticide removal, Herbicides, Water quality, *Aquatic plants, *Algae, Pesticide residues, Metabolism, *Scenedesmus, Water quality control.
Identifiers: *Herbicide metabolism.

Several species of algae (axenic) and aquatic vascular plants were evaluated for effectiveness in removal of pesticides from water. Pesticides with 14C were utilized to determine plant uptake and molecule degradation. Herbicide concentrations below toxic levels were used; however, these were many times greater than natural residue levels encountered in surface water. It was evident that the plants removed herbicide residues from water, supporting the fact that residues have not yet accumulated in surface water to dangerous levels. Various algae were not equally effective in absorbing pesticides. Scenedesmus was particularly effective, and pH of the medium proved important with 2,4-D. In general, submersed species were

not as effective in herbicide removal as emersed species. This may be related to transpiration and an avenue for translocation through the xylem. Herbicides were metabolized by various aquatic plants. Metabolites as well as the herbicide were released to the culture medium. Metabolism was rapid, particularly with algae. Algae and aquatic vascular plants contribute to the removal of pesticides from water and metabolize the chemicals to less active compounds in the environment.
W73-07821

UNDERSTANDING THE WATER QUALITY CONTROVERSY IN MINNESOTA,
Minnesota Univ., Minneapolis. Water Resources Research Center.
J. J. Waelti.
Available from the National Technical Information Service as PB-219 586, $3.00 in paper copy, $1.45 in microfiche. Minnesota University Agricultural Extension Service, Extension Bulletin 359, 1970. 27 p, 4 fig, 2 tab, 18 ref. OWRR A-019-MINN (1).

Descriptors: *Water quality, *Water quality control, *Water pollution control, Pollutants, Economics, Economic efficiency, Federal government, State governments, Local governments, *Minnesota.

In order to facilitate meaningful, rational debate and better communication between citizen and government, the more important aspects of the water pollution problem are discussed in elementary terms. Pollutants may be classified according to several criteria. For instance, pollutants may be (1) solid or liquid forms, (2) chemical, physical, or biological, (3) degradable or nondegradable, or (4) natural or manmade. Important problem areas discussed include (1) water quality problems in densely populated areas, (2) eutrophication and pollution on lakes, (3) agricultural pollution, and (4) industrial pollution. The losses resulting from water pollution include the direct loss of economic product, the misallocation of resources, and deleterious effects on social and aesthetic values. Next, the role of federal, state, and local governments in water quality control is outlined. The basic problems which a government and its constituents need to resolve include (1) what technical methods should be employed in improving water quality., (2) what level of water quality should be attained., and (3) what institutional means should be used to implement water quality control policy. Numerous technical terms commonly used in discussions of water quality control are carefully defined. (Settle-Wisconsin)
W73-07822

ECONOMIC ASPECTS OF POLLUTION CONTROL,
University of New England, Armidale (Australia).
R. M. Parish.
Australian Economic Papers, Vol 11, No 18, p 32-43, June, 1972. 14 ref.

Descriptors: *Pollutants, *Pollution abatement, *Economics, *Economic efficiency, Equity, Negotiations, Regulation, Control, Cost-benefit analysis, Costs.
Identifiers: Liability.

It is almost exclusively in connection with common-property resources, including the atmosphere and water, that severe pollution and congestion problems arise. With such resources there is often no effective barrier to their being utilized well beyond the optimum level. In these circumstances a social improvement could be effected by either reducing the intensity of use of the resource, or by changing the manner of use. In the absence of transaction costs, externalities will be internalized by negotiation. Furthermore, the outcome of the negotiations will be unaffected by the prevailing liability rules. However, once transactions costs are considered, the particular liability rule affects the final equilibrium. An important cause of high transactions costs is uncertainty regarding the law itself. A government agency charged with regulating polluters will often have a wide range of choices open to it. In general, it must choose: (1) whether to reduce the level of pollution, or mitigate its effects, or both; (2) at what point in the pollution-production or pollution-consumption process to apply controls; and (3) what type of control or mitigation measure to apply. Each of these aspects are briefly examined. (Settle-Wisconsin)
W73-07823

ALTERNATIVE FINANCING FOR STATE AND LOCAL WATER RESOURCES PROJECTS,
Clemson Univ., S.C. Dept. of Economics.
For primary bibliographic entry see Field 06C.
W73-07825

THE ENVIRONMENT AS PROBLEM: II, CLEAN RHETORIC AND DIRTY WATER,
A. M. Freeman, III, and R. H. Haveman.
The Public Interest, No 28, p 51-65, Summer, 1972.

Descriptors: *Water pollution, *Water pollution control, *Regulation, *Standards, Legal aspects, Economic efficiency, Pollution taxes (Charges).
Identifiers: *Laws, Subsidies.

In spite of the several environmental protection laws passed in the last 20 years, indices of environmental quality show that the waste loads imposed on environmental resources have been growing continuously, and rising waste loads mean deteriorating environmental quality. Present federal water pollution control policy has two main elements: (1) a program of federal subsidies to cities for the construction of waste treatment plants, and (2) a procedure for establishing regulations to limit discharges and for enforcing these rules. An examination of these two elements indicates that existing federal water pollution policy has dismally failed to improve the quality of the nation's rivers. The federal subsidies in effect allow polluters to generate and dispose large quantities of wastes without bearing the full cost of their discharges. In fact, the taxpayers' money is used to clean up after them. In addition, the regulatory-enforcement strategy fails because it pits the control authorities against the polluters in an unending sequence of long, drawn out, and often inconclusive battles over licensing and the enforcement of regulations. A desirable alternative may be to impose user or effluent charges on waste dischargers such that the charge is related to the volume discharged. (Settle-Wisconsin)
W73-07827

EPA ASSESSES POLLUTION CONTROL BENEFITS.

Environmental Science and Technology, Vol 6, No 10, p 882-883, October, 1972.

Descriptors: *Water pollution control, *Benefits, *Costs, *Evaluation, Standards, Methodology, Water pollution, Air pollution.
Identifiers: Environmental impact statements, Incentives.

A newly formed division of the Environmental Protection Agency, the Implementation Research Division (IRD), is presently attempting to assess the benefits of air and water pollution control. IRD consists of four branches. The Economic Analysis Branch is on a crash project to assess the benefits associated with pollution control. It will attempt to derive benefit and cost estimates in the human health, materials, animals, vegetation, recreation, and aesthetics areas. The distribution of costs and benefits among the population will also be studied. A second project will estimate the economic damages of water pollution in 1970. A second branch, Ecological Studies and Technology Assessment, will concentrate on the research needed to comment effectively on the environmental impact statements. A third branch, Standards Research, will work on the development of better methodology for evaluating standards, particularly for alternative approaches developed by the Economic Analysis Branch. The fourth branch, Systems Evaluation, has as its main thrust the assignment of identifying incentives and related fiscal inducements to promote pollution control. That is, how can taxes, effluent charges, subsidies, and other fiscal incentives be used to get a better quality environment with minimum costs and inconvenience to society. (Settle-Wisconsin)
W73-07829

MODELS OF INDUSTRIAL POLLUTION CONTROL IN URBAN PLANNING,
RAND Corp., Santa Monica, Calif.
D. P. Tihansky.
Available from the National Technical Information Service as AD742 401, $3.00 in paper copy, $1.45 in microfiche. January, 1972. 47 p, 15 fig, 3 tab, 36 ref.

Descriptors: *Industrial wastes, *Pollution abatement, *Water pollution control, Cities, Economic efficiency, Mathematical models, Profit, Optimization, Standards.

To predict the economic implications of air and water quality management, some simple mathematical models are formulated. The process modeled is one in which urban industries maximize the present value of their aggregate profits over a discretized time span subject to constraints on environmental quality and resource allocation. The model provides quantitative estimates of the responses of the following variables to changes in air and water quality standards: (1) individual profits at each firm and aggregate regional profit; (2) industry output levels; (3) input requirements for manufacturing activities; and (4) resource outlays for gaseous and liquid waste control. Primary effects of enforced waste control on one industry are a diversion of its input resources from manufacturing to pollution control and consequent reduction of output and demand. Secondary effects may involve a decrease in the profits of other industries. The models are based on several restrictive assumptions including (1) a closed regional system with only five industries; (2) linear transport equations; (3) functional relationships based upon recent empirical findings; and (4) exogenously determined prices. These models are not precise enough to indicate the most efficient allocation of resources among firms within a real economy. (Settle-Wisconsin)
W73-07831

SYSTEMS APPROACH TO WATER QUALITY MANAGEMENT,
Water Resources Engineers, Inc., Walnut Creek, Calif.
For primary bibliographic entry see Field 05B.
W73-07922

REPORT OF THE CLINCH VALLEY STUDY, MAY 15-JUNE 2, 1972.
Oak Ridge National Lab., Tenn.

Available from NTIS, Springfield, Va., as ORNL-4835, $3.00 per copy, $1.45 microfiche. Report No ORNL-4835, Jan. 1973. 68 p, 7 ref, 7 append.

Descriptors: *Nuclear powerplants, *Accidents, *Hazards, *Safety, *Population, *Radioactivity, Effluents, Radioisotopes, Strontium, Cesium, Iodine, Safety, Toxicity, Public health, Food chain.
Identifiers: *Emergency procedure.

COPAN LAKE, LITTLE CANEY RIVER, OKLAHOMA (FINAL ENVIRONMENTAL IMPACT STATEMENT).
Army Engineer District, Tulsa, Okla.
For primary bibliographic entry see Field 08A.
W73-07979

SNAGGING AND CLEARING PROJECT ON MILL CREEK AT RIPLEY, WEST VIRGINIA (FINAL ENVIRONMENTAL IMPACT STATEMENT).
Army Engineer District, Huntington, W. Va.
For primary bibliographic entry see Field 04A.
W73-07981

MICHIGAN'S ENVIRONMENTAL PROTECTION ACT OF 1970: A PROGRESS REPORT,
Michigan State Univ., East Lansing.
J. L. Sax, and R. L. Conner.
Michigan Law Review, Vol 20, no 6, p 1003-1106, May 1972. 104 p, 309 ref, 9 append.

Descriptors: *Michigan, *Legislation, *Judicial decisions, *Administrative agencies, Legal aspects, Water pollution control, Remedies, Water quality control, Jurisdiction, Decision making, Eminent domain, Constitutional law.
Identifiers: *Standing, *Citizen suits, Michigan Environmental Protection Act.

Michigan's Environmental Protection Act of 1970 represents a departure from the longstanding tradition under which control of environmental quality has been left almost exclusively in the hands of regulatory agencies, since it gives to ordinary citizens an opportunity to take the initiative in environmental law enforcement. Moreover, the act enlarges the role of the courts because it permits a plaintiff to assert that his right to environmental quality has been violated in much the same way that one has always been able to claim that a property right has been violated. Every significant legal issue, including the Act's constitutionality, remains unresolved by appellate courts. The small number of cases filed under the act shows the statute is not as easily accessible a tool as its supporters had hoped. Two encouraging features in the Act's history are the expedition with which most cases have been handled and the willingness of the courts to face up to the environmental issues that divide the parties. (Mockler-Florida)
W73-07982

ADMINISTRATION OF THE NATIONAL ENVIRONMENTAL POLICY ACT, PART 1.
For primary bibliographic entry see Field 06E.
W73-07983

PROJECTS PROPOSED FOR INCLUSION IN OMNIBUS RIVER AND HARBOR AND FLOOD CONTROL LEGISLATION–1972.

Joint Hearings–Subcomm. on Rivers and Harbors, Subcomm. on Flood Control and Internal Development–Comm. on Public Works, U.S. House of Representatives, 92d Cong, 2d Sess, February 17, 22, 24, 29 and March 2, 1972. 312 p, 3 fig, 20 plate, 1 map, 2 tab, 5 chart.

Descriptors: *Environmental effects, *Rivers and Harbors Act, *Legislation, *Flood control, Water resources development, Federal government, Flood protection, Federal budget, Administrative agencies, Erosion control, Beach erosion, Navigation, Wildlife, Recreation facilities, Flood damages, Channel improvement, Harbors, River regulation.
Identifiers: *Congressional hearings.

The subcommittees heard testimony concerning the environmental effects of twenty-six new flood control and river and harbor projects costing over $210 million. Six of the projects are in the navigation and beach erosion categories, eighteen in flood control and hurricane protection categories, and two involve fish and wildlife conservation at existing projects. All environmental statements required by the National Environmental Policy Act of 1969 have been prepared for each of the projects and the requirement of construction and maintenance of recreational facilities within each project has been fulfilled. The President's 1973 budget proposal providing substantial increases for ongoing water resources projects to avoid the under-financing of past years while limiting new construction to a modest level was emphasized by several speakers. Included are environmental statements for several of the projects. (Beardsley-Florida)
W73-07984

REPORT OF PROCEEDINGS AT PUBLIC HEARINGS RELATING TO APPLICATIONS FILED (TO THE WATER AND AIR QUALITY CONTROL COMMITTEE OF THE NORTH CAROLINA BOARD OF WATER AND AIR RESOURCES).
North Carolina Board of Water And Air Resources, Raleigh. Water and Air Quality Control Committee.

March 30, 1972. 84 p.

Descriptors: *North Carolina, *Water quality standards, *Classification, *Water quality control, Water control, Water zoning, Water resources development, Water utilization, Water pollution control, Adjudication procedure, Recreation, Water supply, Waste water (Pollution).
Identifiers: *Cape Fear River Basin.

Proceedings are reported of the public hearing held on applications requesting the reclassification of certain bodies of water in the Cape Fear River Basin, North Carolina, so that they may be protected and maintained in a suitable condition for the reclassified use. The reclassification was requested in order to protect the waters for fishing and secondary recreation, to protect certain waters as a source of raw water for the University of North Carolina and to protect certain waters for bathing and other water-body contact recreation. A transcript of the hearings containing testimony from interested agencies and citizens is included, as well as prepared reports concerning the specific waters subject to possible reclassification. (Mockler-Florida)
W73-07987

PROPOSED RECLASSIFICATIONS OF CERTAIN WATERS IN THE CHOWAN, NEUSE, PASQUOTANK, ROANOKE, TAR-PAMLICO, AND WHITE OAK RIVER BASINS, ETC.
North Carolina Board of Water and Air Resources, Raleigh. Water and Air Quality Control Committee.

1973. 35 P, 3 TAB.

Descriptors: *North Carolina, *Classification, *Water quality standards, *Water quality control, River basins, River basin development, Regulation, River systems, Water utilization, Water pollution control, Standards, Evaluation.

The proposed reclassification of certain streams in the Chowan, Neuse, Pasquotank, Roanoke, Tar-Pamlico and White Oak River Basins to be considered at public hearings is explained. An evaluation of all surface waters in the above named river basins have been carefully made in accordance with the quality of the waters and the best usage of the waters; the proposal seeks to reclassify them with reference to their best usage and in the best interest of the public. Present classifications and use of the river basins are given, as well as the proposed changes in classification. Comprehensive tables and charts are included, representing an exhaustive and well planned study of the river

basins in their present use as well as pertinent information regarding their use when reclassifed. (Mockler-Florida)
W73-07988

NEW CLASSIFICATION ASSIGNED TO CERTAIN WATERS IN THE CAPE FEAR RIVER BASIN.
North Carolina Board of Water and Air Resources, Raleigh.

Resolutions of March 30, 1972. 3 p.

Descriptors: *North Carolina, *Classification, *Water quality standards, *Water quality control, River basins, Tributaries, Water rights, Water law, Water policy, Water resources development, Water management (Applied), Water pollution control, Adjudication procedure.
Identifiers: *Cape Fear River Basin, *Re-classification (Water quality standards).

The reclassification of certain waters in the Cape Fear River Basin establishes new water quality standards for Buckhorn Creek and its tributaries flowing through Wake County, North Carolina, the waters of Bolin Creek flowing through the town of Chapel Hill, North Carolina, and Cane Creek and its tributaries flowing through Orange County. The reclassification, a result of written applications and public hearings, are discussed along with a description of the concerned subject waters. Any natural, unnamed tributaries in the basin will carry the same classification as that of the segment to which it is tributary. Additionally, those tributaries not specifically covered will continue to carry the same classification. (Mockler-Florida)
W73-07989

(ANNUAL REPORT OF THE DELAWARE RIVER BASIN COMMISSION, 1972).
Delaware River Basin Commission, Trenton, N.J.
For primary bibliographic entry see Field 06E.
W73-07990

ENVIRONMENTAL IMPACT STATEMENTS—A DUTY OF INDEPENDENT INVESTIGATION BY FEDERAL AGENCIES.
For primary bibliographic entry see Field 06E.
W73-07993

PRIVATE REMEDIES FOR BEACH FRONT PROPERTY DAMAGE CAUSED BY OIL POLLUTION OF MISSISSIPPI COASTAL WATERS,
M. Soper.
Mississippi Law Journal, Vol XLIII, No 4, p 516-537, 1972. 161 ref.

Descriptors: *Mississippi, *Oil pollution, *Coasts, *Oil spills, Riparian rights, Legal review, Legislation, Law enforcement, Riparian land, Judicial decisions, Negligence, Penalties (Legal), Water pollution, Beaches, Water pollution sources, Legal aspects, Damages, State government, Federal government, State jurisdiction, Common law, Trespass.
Identifiers: Nuisance (Legal aspects).

Remedies are discussed that a Mississippi citizen owning beach-front property along the Gulf-coast may have for damage caused to his property by the discharge of oil from sea-going vessels. At present there is no Mississippi act which gives a landowner a private right to sue for cleanup or oil damage to his beach property. The state does have legislation, such as the Mississippi Air and Water Pollution Control Act of 1966, which adequately protects state interests; however, existing state law gives no statutory rights to an injured individual landowner. The best avenue presently open to an individual beach property owner for private redress for oil damage is found in the common law actions of trespass, negligence, and nuisance. These remedies are predicated upon the fact that one who owns beach-front property in Mississippi has riparian rights which can be protected through these three common law actions. There are, however, difficulties in maintaining these private civil actions. Federal laws, like Mississippi statutory law, are adequate only to the extent that they protect the national interest. State laws are needed to provide for strict liability for damage to private par ies. (Adams-Florida)
W78-07994

THE ENERGY NEEDS OF THE NATION AND THE COST IN TERMS OF POLLUTION,
Atomic Energy Commission, Washington, D.C.
For primary bibliographic entry see Field 06G.
W73-07995

THE YEAR OF SPOILED PORK: COMMENTS ON THE COURT'S EMERGENCE AS AN ENVIRONMENTAL DEFENDER,
Florida Univ., Gainsville.
W. A. Rosenbaum, and P. E. Roberts.
Law and Society Review, Vol 7, No 1, p 33-60, Fall 1972. 28 ref.

Descriptors: *Judicial decisions, *Legal review, *Decision making, *Administrative agencies, *Administrative decisions, Water resources, Water law, Project planning, Alternate planning, Cost-benefit analysis, Legal aspects, Political aspects, Federal government, Standards, Water pollution, Recreation, Environmental effects, Legislation.
Identifiers: *National Environmental Policy Act, *Sovereign immunity, *Environmental Impact Statements, Standing (Legal).

Federal water resources projects have been consistently challenged by environmentalists because of the fact that Congress frequently votes for projects on the basis of political expediency without considering their ecological impact. The major line of attack upon these projects has been through the courts. There are four crucial legal issues which emerge from environmental lawsuits. First, it must be determined if the government can be sued. This depends upon whether environmentalists have standing, and whether the government enjoys sovereign immunity. Second, can projects which have already been begun be subjected to suit. Third, are benefit-cost calculations susceptible to judicial review. Fourth, what effect will the National Environmental Policy Act (NEPA) and its impact statements have on these lawsuits. Recently a lower court adopted a twin test for the acceptability of NEPA environmental impact statements: they must be submitted in good faith, and they must alert decision-makers to major environmental problems involved in the particular project. If this test is upheld, the effect would be to remove the judiciary from substantive investigations of impact statements, thus diminishing the value of the courts as a means for attacking pork-barrel projects. (Adams-Florida)
W73-07996

F. W. GUEST MEMORIAL LECTURE: RIVER POLLUTION AND THE LAW,
G. H. Newsom.
Otago Law Review, Vol 2, No 4, p 383-392, August 1972. 11 ref.

Descriptors: *Foreign countries, *Penalties (Legal), *Legislation, *Water pollution control, *Riparian rights, Natural flow doctrine, Water law, Legal aspects, Water rights, Banks, Ownership of beds, Riparian waters, Trespass, Common law, Judicial decisions, Fish, Fisheries, Law enforcement.
Identifiers: *New Zealand, *Comparative law, Injunctions (Prohibitory), Nuisance (Legal aspects), England.

This address, delivered in New Zealand in 1971 by a noted English jurist, provides both an account of the English experience in river pollution over the last twenty years, and a critique of New Zealand's legislative efforts to combat river pollution. The most effective weapon utilized in England against river pollution has been, traditionally, common law actions by riparian owners. England has passed a series of acts in order to deal with the problem. These acts, such as the Rivers Act of 1951, and the subsequent Act of 1961, provide for criminal penalties, including imprisonment, for habitual offenders. The laws, both civil and criminal, are now adequate. The problem remains with enforcement; but this can be solved by official resolution and good organization. As for New Zealand's efforts in water pollution control, current legislation contains inadequate penalties. Inclusion of penalties of imprisonment is suggested in order to put teeth into the laws. Another major weapon to be used against river pollution is the organization of public opinion on a massive scale. (Adams-Florida)
W73-07997

CRIMINAL LIABILITY UNDER THE REFUSE ACT OF 1899 AND THE REFUSE ACT PERMIT PROGRAM.
For primary bibliographic entry see Field 06E.
W73-07998

ENVIRONMENTAL LAW: STRICT COMPLIANCE WITH PROCEDURAL REQUIREMENTS OF NEPA–THE AGENCIES MUST PLAY BY THE RULES,
For primary bibliographic entry see Field 06E.
W73-07999

DEPARTMENT OF NATURAL RESOURCES; DIVISION OF WATER POLLUTION CONTROL.
For primary bibliographic entry see Field 06E.
W73-08000

SEEGREN V. ENVIRONMENTAL PROTECTION AGENCY (PETITION FOR HARDSHIP VARIANCE FOR USE OF SANITARY SEWERS).
For primary bibliographic entry see Field 06E.
W73-08008

STALEY MANUFACTURING CO. V. ENVIRONMENTAL PROTECTION AGENCY (REGULATION OF DISCHARGES FROM PRIVATE SEWER INTO MUNICIPAL SEWER).
For primary bibliographic entry see Field 06E.
W73-08009

REEDS CONTROL EUTROPHICATION OF BALATON LAKE,
Research Inst. for Water Resources Development, Budapest (Hungary).
L. Toth.
Water Research, Vol 6, No 12, p 1533-1539, December 1972. 3 fig, 3 tab, 13 ref.

Descriptors: *Sewage treatment, *Effluents, *Water pollution control, Aquatic plants, Water pollution sources, Nitrogen, Phosphorus, *Eutrophication, Water quality control.
Identifiers: *Reeds, *Balaton Lake (Hungary), Macrophytes.

As part of a study on the eutrophication of Lake Balaton, information is presented on the sewage discharged from the sewage purification plants at Tihany, Balatonfuered, and Keszthely into Lake Balaton directly and through stands of reed. Samples of effluent were filtered through Sartorius membrane with a pore size of 0.45 micron and determinations were made of organic nitrogen and phosphorus. The measurements performed at Balatonfuered and Tihany show that in July, i.e. in the

WINERY INNOVATES WASTE TREATMENT.
For primary bibliographic entry see Field 05D.
W73-08094

OXYGEN DIFFUSION IN WET AIR OXIDATION PROCESSES,
Naval Research Lab., Washington, D.C.
For primary bibliographic entry see Field 05D.
W73-08102

TECHNOLOGIES FOR SHIPBOARD OIL-POLLUTION ABATEMENT: EFFECTS OF OPERATIONAL PARAMETERS ON COALESCENCE.
Naval Ship Research and Development Center, Annapolis, Md.
For primary bibliographic entry see Field 05D.
W73-08106

BUREAU OF RECLAMATION,
Bureau of Reclamation, Denver, Colo. Applied Sciences Branch.
W. P. Simmons, L. A. Haugseth, and L. O. Timblin.
Journal of the American Water Works Association, Vol 64, No 10, p 624-627, October 1972. 4 fig.

Descriptors: Water resources, *Water resources development, *Water reuse, Precipitation (Atmospheric), Water management, Water pollution control, Water policy, Water quality, Water quality control, Laboratory tests, Laboratory equipment, Geothermal studies, Surveys.
Identifiers: *Bureau of Reclamation.

The Bureau of Reclamation of the Department of the Interior is concerned with conceiving, evaluating, planning, designing, constructing, operating, and maintaining water resource related projects that maximize social, environmental and economic benefits for the area in the 17 Western States. The Bureau is concentrating present research on new water supplies in the geothermal resources and precipitation management fields. A study is also being made to identify the locations, amounts, and characteristics of various waste waters under the Inventory of Waste Water Reclamation Opportunities Program. Laboratory and field evaluations of techniques, structures, and equipment for use in waste water reclamation projects are also underway. Furthermore, Bureau research involves a continuing effort on methods of water quality determination and evaluation. (Smith-Texas)
W73-08108

HEAVY METALS REMOVAL IN WASTE WATER TREATMENT PROCESSES: PART 1,
Orange County Water District, Santa Ana, Calif.
For primary bibliographic entry see Field 05D.
W73-08117

CONVERTING AMD TO POTABLE WATER BY ION EXCHANGE TREATMENT,
Chester Engineers, Inc., Coraopolis, Pa.
For primary bibliographic entry see Field 05D.
W73-08118

WARWICK MINE NO. 2 WATER TREATMENT,
For primary bibliographic entry see Field 05D.
W73-08124

HUMAN WASTE POLLUTION IN UNITED STATES FORESTS.
Environmental Protection Center, Inc., Inglewood, Calif.
For primary bibliographic entry see Field 05B.
W73-08134

THE APPLICATION OF SNOWMELT FORECASTING TO COMBAT COLUMBIA RIVER NITROGEN SUPERSATURATION PROBLEMS,
Corps of Engineers, Portland, Oreg. North Pacific Div.
For primary bibliographic entry see Field 02C.
W73-08142

CLEAN WATER FOR SAN FRANCISCO BAY.
California State Water Resources Control Board, Sacramento.
For primary bibliographic entry see Field 06E.
W73-08176

REGIONAL WATER AND SEWER GUIDE.
Upper Savannah Planning and Development District, Greenwood, S.C.
For primary bibliographic entry see Field 06D.
W73-08177

THE CLEAN STREAMS LAW OF PENNSYLVANIA.
Pennsylvania Dept. of Environmental Resources, Harrisburg.
For primary bibliographic entry see Field 06E.
W73-08183

INVENTORY OF INTERSTATE CARRIER WATER SUPPLY SYSTEMS.
Environmental Protection Agency, Washington, D.C. Water Supply Div.
For primary bibliographic entry see Field 05F.
W73-08192

THE IMPACT OF THE NATIONAL ENVIRONMENTAL POLICY ACT UPON ADMINISTRATION OF THE FEDERAL POWER ACT.
For primary bibliographic entry see Field 06G.
W73-08195

HOW AN ENFORCER BECOMES THE POLLUTER'S DEFENDER,
W. L. Moonan.
Juris Doctor, p 24, 25, 36, February 1973. 1 photo.

Descriptors: *Legal aspects, *Pollution control, *Pollution abatement, *Water law, Water policy, Industries, Economic impact, Social aspects, Industrial wastes, Industrial production, Administration, Standards, Water pollution sources.

An interview is described with a New Jersey attorney who is one of a growing number of anti-environmental lawyers. Until 1970, he was one of the New Jersey Attorney General's top environmental law enforcers. He expresses concern for the effect of pollution control on the economy, fearing that the costs of cleanup will put many companies out of business and increase unemployment. He is now a partner in a private law firm, and has polluting companies waiting in line for his services. However, he is selective in his cases and will only handle companies who are sincerely interested in pollution control, and are not simply attempting to avoid their responsibility. As he says, he still cares about pollution. (Glickman-Florida)
W73-08196

THE WATER RESOURCES COUNCIL,
National Water Commission, Arlington, Va.
For primary bibliographic entry see Field 06E.
W73-08198

DEPARTMENT OF ECOLOGY.
Washington Natural Resources and Recreation Agencies, Olympia.
For primary bibliographic entry see Field 06G.
W73-08200

DIGEST OF PROPOSED REGULATION RELATING TO MONITORING AND REPORTING WASTEWATER DISCHARGES AND THEIR EFFECTS UPON RECEIVING WATERS.
For primary bibliographic entry see Field 05D.
W73-08202

REPORT OF THE UNITED NATIONS CONFERENCE ON HUMAN ENVIRONMENT, HELD AT STOCKHOLM, 5-16 JUNE, 1972.
For primary bibliographic entry see Field 06G.
W73-08203

LEGAL ASPECTS OF COASTAL ZONE MANAGEMENT IN ESCAMBIA AND SAN ROSA COUNTIES, FLORIDA (ESCAROSA),
For primary bibliographic entry see Field 06E.
W73-08204

IMMINENT IRREPARABLE INJURY: A NEED FOR REFORM,
C. L. Hellerich.
Southern California Law Review, Vol. 45, p. 1025-1061, 1972. 37 p, 145 ref.

Descriptors: *Remedies, *Judicial decisions, *Damages, *Adjudication procedure, Legal aspects, Jurisdiction, Water policy, Local governments, Water pollution control, Environmental effects, Water law, Law enforcement.
Identifiers: *Injunctive relief, Citizens suits, National Environmental Policy Act, Nuisance (Legal aspects).

The most frequently requested remedy in environmental lawsuits is an injunction; however, in order to obtain one the plaintiff usually must show imminent irreparable injury. The requirement that the threatened harm be both immediate and practically certain to occur is no longer acceptable in today's era of over-population, over-industrialization, and new and expanding technology. Also discussed are other needed reforms in the court system, the imminent threat of long-term damage unless procedures are changed, and the authority for reform of the imminent irreparable injury doctrine. The imminent irreparable injury doctrine should be expanded to include future as well as uncertain harm. Various procedural alternatives are set forth with recommendations that the courts reevaluate and restructure their remedies to make them more responsive. (Mockler-Florida)
W73-08205

ENVIRONMENTAL LAW--PRIV.TE CAUSE OF ACTION UNDER THE RIVERS AND HARBORS APPROPRIATION ACT OF 1899 FOR INJURY TO THE ECOLOGY OF NAVIGABLE WATERS,

Texas Law Review, Vol. 50, No 6, p 1255-1264, August 1972. 60 ref.

Descriptors: *Rivers and Harbors Act, *Constitutional law, *Judicial decisions, *Law enforcement, Legislation, Jurisdiction, Legal review, Penalties (Legal), Permits, River regulation, Navigation, Water pollution control, Pollution abatement, Navigable waters, Ecology, Environmental control, Environmental effects, Wildlife conservation, Estuarine fisheries.
Identifiers: *Standing (Legal), *Citizen suits, Injunctions (Prohibitory).

Section 10 of the Rivers and Harbors Act of 1899 makes it unlawful to excavate or fill any navigable water of the United States unless a permit has been obtained from the Army Corps of Engineers. Recently, private citizens have brought suit for damages and injunctive relief under the Act in order to prevent harm to the environment. These suits have raised the question of whether private citizens have standing under the Act. Courts which have allowed citizen suits have imposed two prerequisities: (1) the private plaintiff must be threatened with special injuries, and (2) the special injury must be within the Act's zone of protection. This latter requirement causes attention to focus on the purposes of the 1899 Act. A 1970 case, Zabel v. Tabb, extended the Act to protect the ecology of navigable waters by emphasizing the congressional policy in favor of environmental protection. The first decision to recognize a right to protect the environment through private enforcement of the 1899 Act was based upon the grounds that the plaintiffs would suffer a direct and personal injury from destruction of fisheries and wildlife. (Adams-Florida)
W73-08207

THE NATIONAL ENVIRONMENTAL POLICY ACT OF 1969 SAVED FROM 'CRABBED INTERPRETATION',
For primary bibliographic entry see Field 06E.
W73-08208

SMALL BOAT HARBOR, KING COVE, ALASKA (FINAL ENVIRONMENTAL IMPACT STATEMENT).
Army Engineer District, Anchorage, Alaska.
For primary bibliographic entry see Field 08D.
W73-08209

SMITHVILLE LAKE, LITTLE PLATTE RIVER, MISSOURI (FINAL ENVIRONMENTAL IMPACT STATEMENT).
Army Engineer District. Kansas City, Mo.
For primary bibliographic entry see Field 08A.
W73-08210

WALKER DAM IMPOUNDMENT, AQUATIC PLANT CONTROL PROJECT, NEW KENT COUNTY, VIRGINIA (FINAL ENVIRONMENTAL IMPACT STATEMENT).
Army Engineer District, Norfolk, Va.

Available from the National Technical Information Service as EIS-VA-72-5511-F, $3.75, in paper copy, $0.95 in microfiche. May 15, 1972. 24 p, 2 map, 3 photo, 1 tab.

Descriptors: *Virginia, *Environmental effects, *Aquatic weed control, *Fishkill, Aquatic environment, Water quality, Environmental control, Recreation, Water chemistry, Water conservation, Weed control, Herbicides, Diquat, Aquatic weeds, Impoundments, Reservoirs, Oxygen requirements.
Identifiers: *Egeria, New Kent County (Virginia), *Environmental impact statements.

The proposed project would initiate a program in conjunction with the state, designed to control infestation of egeria, an aquatic weed, through chemical treatment with a 50-50 mixture of herbicide diquat dibromide and potassium endothall. The control of egeria would increase recreational activity, insure adequate water velocities for intake systems and increase fish production on the reservoir. However, adverse environmental effects would include minor fish kills, reduction in available dissolved oxygen associated with bacterial oxidation of dead plants, rendering of reservoir water unsuitable for drinking purposes for approximately two weeks, and potential damage to a tree farm near the upper reaches of the lake. Alternatives to the proposed project include no improvement, alteration of the lake habitat through water level reduction, mechanical control of aquatic vegetation, and removal of nutrient sources. (Mockler-Florida)
W73-08212

VIRGINIA BEACH, VIRGINIA--BEACH EROSION CONTROL AND HURRICANE PROTECTION (FINAL ENVIRONMENTAL IMPACT STATEMENT).
Army Engineer District, Norfolk, Va.
For primary bibliographic entry see Field 08A.
W73-08213

CONSTRUCTION OF ARTIFICIAL REEFS IN THE ATLANTIC OCEAN OFF CAPE HENRY, VIRGINIA (FINAL ENVIRONMENTAL IMPACT STATEMENT).
National Marine Fisheries Service, Beaufort, N.C. Atlantic Coastal Fisheries Center.
For primary bibliographic entry see Field 06G.
W73-08215

WILLOW ISLAND LOCKS AND DAM OHIO RIVER, OHIO AND WEST VIRGINIA (FINAL ENVIRONMENTAL IMPACT STATEMENT).
Army Engineer District, Huntington, W. Va.
For primary bibliographic entry see Field 08A.
W73-08216

PEARL RIVER BASIN, EDINBURG DAM AND LAKE, MISSISSIPPI AND LOUISIANA (FINAL ENVIRONMENTAL IMPACT STATEMENT),
Army Engineer District, Mobile, Ala.
For primary bibliographic entry see Field 08A.
W73-08221

KANAWHA RIVER COMPREHENSIVE BASIN STUDY, NORTH CAROLINA, VIRGINIA, AND WEST VIRGINIA, (FINAL ENVIRONMENTAL IMPACT STATEMENT).
Ohio River Basin Commission, Cincinnati.
For primary bibliographic entry see Field 04A.
W73-08222

LITTLE CALUMET RIVER: LAKE CALUMET HARBOR.
For primary bibliographic entry see Field 06E.
W73-08233

CHANGES IN THE MICROBIAL POPULATIONS OF A RESERVOIR TREATED WITH THE HERBICIDE PARAQUAT,
University of Wales Inst. of Science and Tech., Cardiff.
For primary bibliographic entry see Field 05C.
W73-08239

THE FISH FAUNA OF NERA RIVER AND ITS PROTECTION, (IN RUMANIAN)
Academia R.S.R., Bucharest. Institutul de Biologie.
P. Banarescu, and T. Oprescu.
Ocrotirea Nat. Vol 15, No 2, p 138-148, 1971. Illus. English summary.
Identifiers: *Barbus-Meridionalis, *Chondrostoma-Nasus, Cobitis-Elongata, Fauna, Fish, *Nera River, Protection, River, *Romania, Snails.

The fish-fauna of Nera River (Southern Banat, Romania) include 28 spp. (19 authochthonous ones and 9 ascending the river from the Danube). The most abundant species are Chondrostoma nasus and Barbus meridionalis petenyi. The most interesting species is Cobitis elongata, a preglacial relict, occurring also in some tributaries to the Danube from Yugoslavia and Bulgaria. It is necessary to protect these fish species as well as the whole aquatic fauna of this river, which includes some remarkable prosobranchiate snails, by preventing any pollution of the river or of its tributaries upwards from the gorges. The protection only of the gorges is not sufficient.--Copyright 1972, Biological Abstracts, Inc.
W73-08269

This report was developed from field tests and actual oil spill control experiences of the Marine Division of the New York Fire Department during a twenty-two month period beginning October 8, 1970. The information is intended to assist a community in protecting its area against oil spill damage. Operational procedures described are intended to serve as stop-gap measures, pending the inauguration of clean-up activities by the spiller or responsible Federal Agency. A survey of cities susceptible to oil spills indicates that most responding fire departments are concerned with containing spills as well as dealing with spill-created fire hazards. Research and development which culminated in the production of this manual concentrated on the utilization of existing fire department resources. However, a limited amount of useful ancillary equipment was procured or developed. Such equipment is described and its use is expalined. The manual describes common sources of oil spills and some ecological effects of oil pollution. Pertinent Federal laws and regulations are outlined. Some feasible techniques for dealing with harbor spills are offered. (EPA)
W73-08288

PROBLEMS OF COMBINED SEWER FACILITIES AND OVERFLOWS, 1967.
American Public Works Association, Chicago, Ill.

Federal Water Pollution Control Administration, Water Pollution Control Research Series WP-20-11, December 1, 1967. 183 p, 10 fig, 61 tab, 22 ref. EPA 11020-12/67, FWPCA Contract No. 14-12-65.

Descriptors: *Combined sewers, *Sewerage, *Water pollution, *Urban runoff, Costs, Data collections, Water pollution control, Overflow, Separated sewers, Sanitary engineering, Urban drainage, Water pollution sources, Sewers.
Identifiers: *Combined sewer overflows.

A nationwide survey was made of the effects and means of correcting combined sewer overflows and storm and sanitary sewer discharges. Over 900 communities were surveyed by personal interviews of public officials, and the results were projected for the entire country. Many conclusions were reached and recommendations made on alleviating the problems of flooding and pollution from sewer systems. The communities surveyed did not have adequate information to evaluate the extent and effect of sewer overflows and that sewer overflows are a major part of this country's water pollution problems, contrary to what was previously thought. Separation of existing combined sewers was estimated to cost 48 billion dollars nationally, while alternative methods of control would cost only 15 billion dollars. Few communities were engaged in programs of complete separation of combined sewers. Although separation is the most popular method of control being used, it is usually applied to only a portion of the total sewer system. Additional research is recommended to develop new methods of control and/or treatment of combined sewer overflows as alternatives to sewer separation. Other recommendations were made on improving the knowledge of the frequency and effects of combined sewer overflows and design criteria for sewers to help solve the problems of flooding and overflows. (Poertner)
W73-08299

DESTRUCTION OF OIL SLICKS,
Halliburton Services, Duncan, Okla. (assignee).
R. F. Rensvold.
U. S. Patent No. 3,705,782, 3 p, 4 ref; Official Gazette of the United States Patent Office, Vol 905, No 2, p 353, December 12, 1972.

Descriptors: *Patents, *Oil spills, *Oil pollution, Water quality control, *Pollution abatement.
Identifiers: *Calcium carbide, Acetylene gas, Hydrocarbon gas.

Oil spills are destroyed by increasing the combustibility of the oil film by incorporating an ignitable and combustible gas such as a hydrocarbon gas or hydrogen gas and igniting it in the presence of atmospheric oxygen. The oil film will be converted to carbon dioxide and water and thereby destroyed. Finely divided particles of calcium carbide are deposited on the surface of the oil. The particles settle through the oil film, contact the water and generate acetylene gas. The product of the reaction between calcium carbide and water is calcium hydroxide. The hydroxide is eventually converted to calcium carbonate. The ignition of the oil-hydrocarbon gas mixture may be accomplished by a variety of means, such as floating flares dropped from aircraft, or incendiary projectiles fired from a floating vessel. (Sinha-OEIS)
W73-08305

OIL REMOVAL DEVICE,
R. H. Cross, III.
U. S. Patent No. 3,706,382, 4 p, 4 fig, 3 ref; Official Gazette of the United States Patent Office, Vol 905, No 3, p 526, December 19, 1972.

Descriptors: *Patents, Skimming, *Oil spills, *Oil pollution, Water quality control, *Pollution abatement, Equipment, Separation techniques.

This device includes an H-shaped skimmer consisting of a cross member and two arm portions. Each arm is of rectangular or circular cross section and provides inflow ports. A flexible buoyant suction hose communicates through the structure with the inflow ports for the removal of oil. The skimmer may be suspended from a buoyant frame. The longest dimension of the device is not more than one-fourth the wavelength of the shortest waves whose amplitude might be significant in disturbing the device. (Sinha-OEIS)
W73-08307

SKIMMERS FOR POLLUTION CONTROL DEVICE,
J. W. Harrington, and E. G. Milne.
U. S. Patent No. 3,707,232, 4 p, 2 fig, 3 ref; Official Gazette of the United States Patent Office, Vol 905, No 4, p 748, December 26, 1972.

Descriptors: *Patents, Skimming, *Oil spills, *Oil pollution, *Pollution abatement, Equipment, Water quality control.

This floating skimmer has a top part which can slide on an intake conduit and a bottom which can counteract the internal pressure drop normally incident to the intake operation. An upper cone has a horizontal flange and the lower float has a flat surface which extends slightly past the flange and is positioned close to the flange in normal use. The cone is free to move to achieve adjustment and float on the pollutant surface. Stops may be added on any point of proximity between the top member and the upper plate of the bottom member to maintain a minimum skimming aperture. It is preferred that the upper plate of the bottom member by moderately concave, so as to act somewhat as a small basin. Two variations in the configuration are presented. (Sinha-OEIS)
W73-08309

ANTI-POLLUTION BALLAST CONTAINER,
H. Liles.
U. S. Patent No. 3,707,937, 3 p, 5 fig, 11 ref; Official Gazette of the United States Patent Office, Vol 906, No 1, p 148, January 2, 1973.

Descriptors: *Patents, *Pollution abatement, Equipment, Water pollution control, Water quality control.
Identifiers: *Oil tankers, *Ballast containers.

The container is constructed with a surrounding wall having contraction rings secured at intervals.

The container collapses in folds that are guided by rings secured to the wall of the container intermediate of the folds. This allows the container to expand when ballast fluid is pumped into the container to provide the appropriate quantity of ballast weight within the compartment and yet to be discharged from the tanker without the ballast fluid being contaminated with the oil or causing pollution at the point of discharge. (Sinha-OEIS)
W73-08313

OIL SKIMMER,
Cities Service Oil Co., Tulsa, Okla. (assignee).
E. A. Bell.
U. S. Patent No. 3,708,070, 5 p, 5 fig, 10 ref; Official Gazette of the United States Patent Office, Vol 906, No 1, p 183, January 2, 1973.

Descriptors: *Patents, Skimming, *Oil spills, *Oil pollution, Water quality control, *Pollution abatement, Equipment, Separation techniques, Water pollution control.
Identifiers: *Oil skimmers.

The oil skimming barge has a transition compartment arranged to receive inflow from the surface of the water over the rim of a pivoted floating baffle. Surface liquid inflow to the compartment is determined by the position of the baffle which in turn is determined by the outflow of water from the bottom of the downstream end of the compartment. At the downstream end of the last transition compartment there is an oil recovery compartment having a floating baffle inlet arrangement. Oil is withdrawn at a suitable rate to control the inlet baffle position. The inlet to the recovery compartment faces a downstream direction to render inflow to the recovery compartment substantially independent of perturbations in the liquid resulting from the overall surface flow. (Sinha-OEIS)
W73-08314

SYSTEM AND BARRIER FOR CONTAINING AN OIL SPILL,
Ocean Systems, Inc., Tarrytown, N.Y. (assignee).
R. N. Blockwick.
U. S. Patent No. 3,708,982, 5 p, 18 fig, 3 ref; Official Gazette of the United States Patent Office, Vol 906, No 2, p 418, January 9, 1973.

Descriptors: *Patents, *Barriers, *Oil spills, *Oil pollution, Water pollution control, Water quality control, *Pollution abatement, Equipment, Separation techniques.

Individual barrier modules are coupled end to end for confining a liquid such as oil floating on the surface of a body of water. The barrier module may form a closed loop about a tanker. The barrier may be moored if necessary. The upper section of each module extends above the surface of the water, and the lower section which reaches below the surface serves as ballast and subsurface barrier. They are bound by any water repellent sealing adhesive. Where a porous foam plastic material is used to form the upper section it may be rendered water impervious by sealing the outer periphery. Once immersed into water the mass of the water-absorbing lower section will immediately increase due to entrapped water, providing the necessary ballast. The lower section will exhibit a dynamic response characteristic closely simulating the surface characteristics of the sea itself. The lower section may be fabricated out of a reticulated polyether based polyurethane foam. (Sinha-OEIS)
W73-08315

APPARATUS FOR CONFINING OIL SPILLS,
W. E. Brown, and E. E. Gilbert.
U. S. Patent No. 3,708,983, 4 p, 11 fig, 5 ref; Official Gazette of the United States Patent Office, Vol 906, No 2, p 418, January 9, 1973.

Descriptors: *Patents, Barriers, *Oil spills, *Oil pollution, Water pollution control, Water quality control, Equipment, Separation technique, *Pollution abatement.

The apparatus comprises a series of elongated, hollow, structural units that are connected together to form a closed loop of any desired configuration. Each structural unit is air-tight except for inlet and outlet openings and each is connected to an air line that may be attached to a compressor. Valves on each unit may be actuated simultaneously to allow air to escape from and water to enter to units, causing them to flood and submerge the apparatus. When an oil spill or leakage is to be contained, the outlet valves can be closed and the inlet valves opened so that the compressor can supply air to raise the apparatus to the water surface. Barrier portions extending above and below the hollow air-filled units prevent waves from breaking over the apparatus and oil from passing under it. (Sinha-OEIS)
W73-08316

APPARATUS FOR CONFINING A FLOATABLE LIQUID,
N. Matheson.
U. S. Patent No 3,710,577, 4 p, 9 fig, 10 ref; Official Gazette of the United States Patent Office, Vol 906, No 3, p 829, January 16, 1973.

Descriptors: *Patents, Barriers, *Oil spills, *Oil pollution, *Pollution abatement, Equipment, Water quality control, Water pollution control, Separation techniques.

This apparatus consists of floatable barrier sections, each comprising a flexible water impervious upright curtain. A pair of cables are secured to each curtain and extend longitudinally in spaced relationship one to the other. A pair of buoyant tubes are placed to control the center of buoyancy and develop corrective forces that maintain the barrier in a generally upright stable position. The apparatus has a towing assembly which consists of a pair of outboard buoyancy tanks and a towing bridle that stabilizes the floatable barrier and inhibits yaw, pitch and roll. The barrier can be arranged in a variety of configurations. (Sinha-OEIS)
W73-08321

VARIABLE DISPLACEMENT FENCE FOR OIL SPILL CONTAINMENT AND RECOVERY,
W. M. Davidson, and H. W. Cole, Jr.
U. S. Patent No 3,710,943, 4 p, 8 fig, 5 ref; Official Gazette of the United States Patent Office, Vol 906, No 3, p 922, January 16, 1973.

Descriptors: *Patents, *Oil spills, Barriers, Separation techniques, Water pollution control, Water quality control, *Pollution abatement, *Oil pollution.

A barrier is provided that is responsive to mean wave height rather than to each wave in a heavy sea. This flexible, inflatable barrier is constructed in the form of a tunnel inside of which there is a continuous passage for oil that enters the barrier beneath the water's surface on the upstream side and leaves the barrier on either end of the tunnel. The barrier is weighted at the bottom by suitable ballast means and buoyed at the top by long, continuous air chambers. Lengthwise cables are attached along the top and bottom of the barrier for towing and control purposes. (Sinha-OEIS)
W73-08322

FLOATING BARRAGE,
Pneumatiques, Cacutchouc Manufacture et Plastiques Kleber-Colombes (France).
R. Ducrocq, and C. Moreau.
U. S. Patent No 3,713,410, 3 p, 3 fig, 3 ref; Official Gazette of the United States Patent Office, Vol 906, No 5, p 1573, January 30, 1973.

Descriptors: *Patents, *Oil spills, *Oil poll[illegible], Separation techniques, *Flotsam, Barriers, [illegible] pollution control, Water quality control, *[illegible]tion abatement.

This barrier consists of sections each of [illegible] contains an inflatable bag with panels susp[illegible] from and tangent to it. The panels overlap [illegible] other over a portion of their length. The pane[illegible] connected so as to be able to swing relative to [illegible] other. The barrage formed by an assembly o[illegible]riers has a chain serving as traction means a[illegible] ballast extends underneath the barrage alo[illegible] entire length. The barrage is able to follow [illegible] movements of the surface of seas and ma[illegible] drawn into a variety of configurations. (S[illegible] OEIS)
W73-08323

SYSTEMS FOR THE REMOVAL OF POL[illegible]TANTS FROM WATER,
For primary bibliographic entry see Field 05D.
W73-08389

STANDARDS BASED ON THE QUALITY [illegible] THE RECEIVING WATER,
Trent River Authority (England).
W. F. Lester.
Water Pollution Control, Vol 58, Part 3, p 324-[illegible] 1969. 8 tab, 6 ref.

Descriptors: *River regulation, *Water qu[illegible] standards, Water pollution control, Aerat[illegible] Water law, Regulation, Effluents.
Identifiers: Royal commission river standards.

The Royal Commission established visible st[illegible]dards in 1911 for classifying rivers from 'v[illegible] clean' to 'bad'. In order to establish standa[illegible] based on the quality of the receiving water, m[illegible] subtle factors beyond visual observation must [illegible] considered, such as reareation rate, physi[illegible] parameters, and biological use of the water. Ri[illegible] standards have the disadvantage of requir[illegible] legislative revision. The advantage of river st[illegible]dards is that the river itself determines the qual[illegible] of effluent that should be discharged to it. (And[illegible]son-Texas)
W73-08413

THE EFFECT OF KRAFT PULP MILL E[illegible]FLUENTS ON THE GROWTH OF ZALERI[illegible] MARITIUM,
Simon Fraser Univ., Burnaby (British Columbi[illegible] Dept. of Biological Sciences.
For primary bibliographic entry see Field 05C.
W73-08426

WATER POLLUTION IN SUEZ BAY,
Red Sea Inst. of Oceanography and Fisheries, [illegible] Ghurdaqah (Egypt).
A. H. Meshal.
Bull Inst Oceanogr Fish. 1: p 461-473. 1970. Illu[illegible] Map.
Identifiers: Bays, *Suez Bay, Water pollutio[illegible] treatment, *Oil pollution, *Pollution abatement.

Methods are suggested (and evaluated) for th[illegible] treatment and disposal of floating oil. They are[illegible] covering with fibrous material, sinking with pow[illegible]dered solids, and dispersion with emulsifiers.-[illegible] Copyright 1973, Biological Abstracts, Inc.
W73-08432

THREATENED FRESHWATER FISHES OF THE UNITED STATES,
Michigan Univ., Ann Arbor. Museum of Zoology.
For primary bibliographic entry see Field 02I.
W73-08439

period, a choice of a physical model of the groundwater system that meets the needs of the farm, a ranking of data by the worth to the farm, and a ranking of data by priority for further data collection activities. (Knapp-USGS)
W73-07882

OPTIMIZATION OF DEAD END WATER DISTRIBUTION SYSTEMS,
Roorkee Univ. (India). Dept. of Civil Engineering.
For primary bibliographic entry see Field 04A.
W73-07920

INTEGRATING ALL ASPECTS OF REGIONAL WATER SYSTEMS,
British Columbia Univ., Vancouver. Water Resources Research Centre.
I. K. Fox.
Journal of the Hydraulics Division, American Society of Civil Eningeers, Vol 99, No HY4, Proceedings paper 9662, p 599-603, April, 1973. 5 p, 1 fig.

Descriptors: *Water resources development, *Systems analysis, *Hydraulics, *Economic analysis, *Regional analysis, Social needs, Institutions, Systems, Water quality.
Identifiers: *Wisconsin River, Policies, Behavioral scineces, Cost minimization.

The natural scientist, the economist, and the engineer are concerned with providing better data and information for making decisions about water use, through utilization of water resources system analysis. The behavioral scientist, on the other hand, is concerned with providing information about how people and institutions interact in making water use decisions and how institutions may be altered to produce a different kind of decision. The conceptual and factual foundation utilized by the behavioral scientist has limitations similar to those of the natural scientist and the economist, and systems analysis makes best use of limited data while indicating the areas in which additional knowledge is most urgently needed. The advancement of systems analysis in the natural science, economic, and behavioral science areas, individually and separately, is highly worthwhile. However, a procedural integration of physical-biological subsystems, economic subsystems, and institutional subsystems is necessary for providing an improved foundation for policy and institutional design in water resources development. How such a procedure was followed in a case study of the Upper Wisconsin River is outlined, and the results of using the approach are discussed. (Bell-Cornell)
W73-07921

SYSTEMS APPROACH TO WATER QUALITY MANAGEMENT,
Water Resources Engineers, Inc., Walnut Creek, Calif.
For primary bibliographic entry see Field 05B.
W73-07922

SYSTEMS ANALYSIS IN IRRIGATION AND DRAINAGE,
California Univ., Riverside. Dry-Lands Research Inst.
For primary bibliographic entry see Field 03F.
W73-07923

STATUS OF WATER RESOURCE SYSTEMS ANALYSIS,
Texas Univ., Austin. Center for Research in Water Resources.
L. R. Beard.
Journal of the Hydraulics Division, American Society of Civil Engineers, Vol 99, No HY4, Proceedings paper 9658, p 559-565, April, 1973. 7 p.

Descriptors: *Water resources, *Water management (Applied), *Systems engineering, Hydraulics, *Social needs, *Environmental effects, Ecology, Stochastic processes, Simulation analysis, Optimization, Operations research, Mathematical models.

Although great progress has been made in simulating the physical operation of water resource systems, challenging problems still remain. Primarily, these are multi-objective evaluations of physical output and application of operations research techniques. Conflicting and complementary output functions, stochastic input functions, complex physical, legal and social constraints, and system nonlinearities pose great technical difficulties. Development of an optimum plan of water resources management requires the integration of objectives, such as economic efficiency, environmental protection, ecological management, and social well-being; necessarily, these objectives must be related in terms of a common denominator, or unique objective function. Effective application of operations research techniques, such as linear or dynamic programming, is hindered by the extreme complexity of water resource systems; nonlinearities and interrelationships that change with time and location make optimization particularly difficult. At present, a gradient type of optimization based on detailed system simulation is most useful. Needed is a more realistic and highly sophisticated systems simulation model, capable of accommodating systems of any configuration, inputs, and demand criteria, and containing a framework for operating the system that is sufficiently flexible to respond to all needs. Advances are most promising in the direction of analyzing internal interactions of water resource systems and their impacts on objective functions. (Bell-Cornell)
W73-07924

SYSTEMS APPROACH TO POWER PLANNING,
Bechtel Corp., San Francisco, Calif. Hydro and Community Facilities Div.
For primary bibliographic entry see Field 08C.
W73-07925

EFFECT OF INCLUDING WATER PRICE ON THE CONJUNCTIVE OPERATION OF A SURFACE WATER AND GROUNDWATER SYSTEM,
Plan Organization, Tehran (Iran).
For primary bibliographic entry see Field 04B.
W73-08132

OPTIMAL PATH OF INTERREGIONAL INVESTMENT AND ALLOCATION OF WATER,
Tel-Aviv Univ. (Israel). Dept. of Economics.
U. Regev, and A. Schwartz.
Water Resources Research, Vol 9, No 2, p 251-262, April, 1973. 2 fig, 42 equ, 17 ref.

Descriptors: *Optimization, *Water resources, *Management, *Economics, *Water allocation (Policy), *Investment, *Regions, Seasonal, Water conveyance, Costs, Water storage, Flow, Mathematical models, Systems analysis, Equations.
Identifiers: Shadow prices, Integer programming.

A deterministic model of the optimal allocation of water and investment in a system composed of several water regions is presented. A discrete time control theory is applied to formalize the model in which the main focus is upon the interaction of regional and seasonal considerations. Economic interpretation of the optimal conditions reveals the following price policy implications: Prices at two adjacent regions should differ by (at most) the cost of transportation. The general trend of the water inventory shadow price in present value is increasing over time with a decreasing rate, whereas the seasonal peaks and troughs in water demand produce positive and negative shifts from that

trend and suggest a peak load-pricing system. The marginal productivity of water is related to rental prices of the different equipment types and to capital equipment marginal cost. The latter sets up an upper bound for water prices. Increasing returns to scale are treated by integer programming formulation when setup costs or indivisibility of projects violate the concavity of the objective function. (Bell-Cornell)
W73-08133

PROCEEDINGS OF THE WESTERN SNOW CONFERENCE,
For primary bibliographic entry see Field 02C.
W73-08138

COPING WITH POPULATION GROWTH AND A LIMITED RESOURCE,
Arizona Water Commission, Phoenix.
P. Briggs.
In: Proceedings of the 40th Annual Meeting of the Western Snow Conference, April 18-20, 1972, Phoenix, Ariz: Printed by Colorado State University, Fort Collins, p 6-8, 1972.

Descriptors: *Systems analysis, *Mathematical models, *Water balance, Simulation analysis, Optimization, Water resources development, Water yield, Safe yield, Water management (Applied), *Arizona.

A computerized systems analysis, using a series of optimization and simulation models, was used to prepare recommendations of water allocations in the State of Arizona. The system was divided into three subsystems: economic, hydrologic-engineering, and one which interfaces the first two. These fit together in a closed loop which can be continuously operated until all inputs are internally consistent. This method may be used to develop technical evaluations of specific areas for use by decision makers. (See also W73-08138) (Knapp-USGS)
W73-08140

REGIONALIZATION AND WATER QUALITY MANAGEMENT,
Camp, Dresser and McKee, Boston, Mass.
For primary bibliographic entry see Field 05D.
W73-08383

MODELLING REGIONAL WASTE WATER TREATMENT SYSTEMS,
Michigan Univ., Ann Arbor. School of Public Health.
For primary bibliographic entry see Field 05D.
W73-08385

SIMPLE ALGORITHM FOR COST OPTIMIZATION OF A THREE DECISION VARIABLE SYSTEM,
Wayne State Univ., Detroit, Mich. Dept. of Civil Engineering.
For primary bibliographic entry see Field 05D.
W73-08386

6B. Evaluation Process

A METHODOLOGY FOR PLANNING OPTIMAL REGIONAL WASTEWATER MANAGEMENT SYSTEMS,
Massachusetts Univ., Amherst. Water Resources Research Center.
For primary bibliographic entry see Field 05D.
W73-07805

THE POLITICAL ECONOMY OF A CORPS OF ENGINEERS PROJECT REPORT: THE DELMARVA WATERWAY,
Cornell Univ., Ithaca, N.Y.
For primary bibliographic entry see Field 06E.
W73-07811

ECONOMIC BENEFITS FROM AN IMPROVEMENT IN WATER QUALITY,
Oregon Univ., Corvallis. Dept. of Agricultural Economics.
For primary bibliographic entry see Field 05G.
W73-07813

THE MISSISSIPPI RIVER–A WATER SOURCE FOR TEXAS. (EVALUATION OF A PROPOSED WATER DIVERSION),
Louisiana State Univ., Baton Rouge. Dept. of Civil Engineering.
R. G. Kazmann, and O. Arguello.
Available from the National Technical Information Service as PB-219 362, $3.00 in paper copy, $1.45 in microfiche. Louisiana Water Resources Research Institute, Baton Rouge, Bulletin 9, March 1973. 183 p, 17 fig, 8 tab, 22 ref, 4 append. OWRR A-016-LA (2).

Descriptors: *Diversion, *River forecasting, *Inter-basin transfers, Oultlets, *Water transfer, Water importing, Water costs, Economic feasibility, *Saline water intrusion, Saline water-freshwater interfaces, Mathematical models, Louisiana, *Mississippi River, Interstate, Regional economics.
Identifiers: Texas Water Plan.

Various rates of diversion to Texas were incorporated in computer playbacks of the daily Mississippi River discharges at Vicksburg for the period from 1928 to 1967, assuming that the Old River Control Structure also diverted 25 percent of the flow into the Atchafalaya River at the same time. The frequency of salt-water encroachment in the river at New Orleans (Algiers) during low water would not have changed if the minimum diversion rate to Texas had been 24,000 cfs for 8.5 months per year (annual quantity, 12 million acre-feet). Cost estimates include: construction of new reservoirs in Texas to store 15 million acre-feet for seasonal distribution; 12,000 megawatts of electric power to lift the water 4,000 feet. Total cost of irrigation water ranges from approximately $70 to $100 per acre-foot. Possible future projects in the Mississippi River basin, including proposed navigation improvements and reconstruction of the Old River Control Structure, would necessitate greater diversion rates for shorter periods and, thus, increase water costs. A simple program forecasts the daily discharges at Vicksburg (up to 6 days) and New Orleans (up to 9 days) from daily readings at upstream gaging stations. The estimates are within 11% of recorded flows at least 80% of the time. Salt-water advance up the Mississippi can also be reliably predicted with a subroutine from either the actual or the forecast flows. (Hill-Louisiana)
W73-07816

REGIONAL WATER RESOURCE PLANNING FOR URBAN NEEDS: PART 1,
North Carolina Univ., Chapel Hill. Dept. of City and Regional Planning.
D. H. Moreau, K. Elfers, G. S. Nicolson, and K. Takeuchi.
Available from the National Technical Information Service as PB-219 364, $3.00 in paper copy, $1.45 in microfiche. North Carolina Water Resources Research Institute Report No. 77, March 1973. 159 p, 16 fig, 14 tab, 81 ref. OWRR B-021-NC (2) and B-045-NC (1). 14-31-0001-3313, 14-31-0001-3624.

Descriptors: *Water resources, *Planning, *Alternative planning, *Water supply, *Water pollution control, Urbanization, *Regional analysis, Regional development, North Carolina.
Identifiers: *Urban water resources, Piedmont Triad Region (N.C.).

Objectives of study are three-fold: (1) to demonstrate need for regional water resource planning in emerging urban areas; (2) to examine how regional water resource planning can be realized within the emerging organizational framework of Federal, State, regional, and local planning; and (3) to examine the process, substance, and techniques of field level planning for regional water resource systems. Analysis of present river basin planning and development indicates that problems of urban areas and linkages with river basins have been largely ignored. Bias results from the traditional emphasis on national economic development in formulation of basin plans. Design of municipal water supply and waste disposal systems has been largely single purpose and fractionated with respect to the water resources. Opportunities to achieve economies and resolve externalities through regional planning and management have been ignored. Recent initiatives by Federal and State governments have created an organizational framework within which deficiencies in these processes can be resolved. A water resource planning model and a planning system that can operate within this framework is suggested, but the fragile political basis upon which this framework is founded requires that State and Federal agencies must reinforce the process if it is to resolve the problems of urbanizing regions. The substance of regional water resource planning is examined using a case study of the Piedmont Triad Region of North Carolina.
W73-07819

LOCAL ECONOMIC IMPACT OF RESERVOIR RECREATION,
Tennessee Univ., Knoxville. Water Resources Research Center.
C. B. Garrison.
Available from the National Technical Information Service as PB-219 585, $3.00 in paper copy, $1.45 in microfiche. Research Report No. 27, J[illegible] 1972. 37 p, 15 tab. OWRR A-020-TENN (2). 14-[illegible]-0001-3543.

Descriptors: *Economic impact, Rural areas, *Reservoirs, *Tennessee, Water resources development, *Recreation.
Identifiers: *Norris Reservoir (Tenn), *New Johnsonville (Tenn).

Reservoir recreation affects the economy of the local area and also the larger region in which the reservoir is located. This study estimates the local economic impact of recreation activities at Norris Lake on a three-county rural area of East Tennessee. The impact consists of the primary and secondary effects. The primary effect is determined by estimating the emoloyment and payroll of enterprises which sell goods and services to visitors. Estimation of the secondary effect requires the construction of local income and employment multipliers; these are estimated by the use of economic base theory. The economic effects of reservoir recreation are compared with those due to the establishment of a concentration of water oriented manufacturing industry at New Johnsonville in West Tennessee.
W73-07820

ECONOMIC ASPECTS OF POLLUTION CONTROL,
University of New England, Armidale (Australia).
For primary bibliographic entry see Field 05G.
W73-07823

ALTERNATIVE FINANCING FOR STATE AND LOCAL WATER RESOURCES PROJECTS,
Clemson Univ., S.C. Dept. of Economics.
For primary bibliographic entry see Field 06C.
W73-07825

29, January, 1973; 50 p, 1 fig, 19 tab, 28 ref, append. OWRR C-3277 (3713) (1).

Descriptors: *Population, *Human population, *Mortality, *Urbanization, Fecundity, Migration, Aging, Urban sociology, Water resources development, Income distribution, Water consumption, Water utilization, *Indiana.
Identifiers: *Life expectation, Survival ratios, Lafayette.

Water consumption in a given region is dependent upon variables such as the rate of urbanization, population size, age composition, income, industrial development, recreational opportunities, etc. Estimates of future populations depend upon the calculation of age specific survival ratios for the present population. Mortality is the one stable component in population change, thus the careful calculation of present life tables provides the prerequisite for the development of reliable population projections from a basis of accurate survival ratios. The life tables developed for this research represents a series of probability statements. The data are based upon intervals of observation through the study period, 1950-1970. The reported measures of expectation of life at various ages are presented for the economic regions of Indiana, and like other life tables, are based upon the assumption that age-specific will remain the same in future years as was recorded during the observed periods. The populations in the regions are relatively small, thus the observed mortality rates may not be stable. It is emphasized that change in these rates will result in estimations of expectation of life which are different from those calculated in the study. The study is a part of an inter-disciplinary project entitled 'Systematic Development of Methodologies in Planning Urban Water Resources for Medium Size Communities.'
W73-07964

THE DEVELOPMENT PLAN FOR HARRISTOWN TOWNSHIP, MACON COUNTY, ILLINOIS.
Macon County Regional Plan Commission, Decatur, Ill.

Available from the National Technical Information Service as PB-210 974, $9.35 in paper copy, $1.45 in microfiche. Final Report, February 1972. 143 p, 2 fig, 17 maps, 3 tab. Ill-1-291.

Descriptors: *Land use plan, *Comprehensive planning, *Urban land use, Urbanization, Utilities, Drainage, *Illinois.
Identifiers: Utility extension, Decatur (Ill), Implementation program, Macon County (Ill).

A comprehensive land use plan for Harristown township including several elements such as open space, transportation, commercial, industrial and residential land and a discussion of utilities extension are presented. Discussed are: (1) inventory and analysis including a discussion of land capabili' es for various land uses; (2) growth objectives and the conseqeunces of uncontrolled growth; (3) the proposed land use plan including several maps and policy statements; (4) and the means for implementing the plan. The discussions of utility services, i.e. water supply, sanitary sewers, and the storm sewers, are a minor element of the study, although the extension of these services is a key feature of the implementation of the overall land use plan. In general, the study proposes that the city of Decatur and the Decatur Sanitary District be responsible for extending the services into the township. It is also proposed that casements and drainage ordinances be used to protect the natural drainage ways. (Elfers-North Carolina)
W73-08052

SHORELINE MANAGEMENT PLAN FOR THE ESCAMBIA-SANTA ROSA REGION.
Smith (Milo) and Associates, Tampa, Fla.

Available from the National Technical Information Service as PB-212 438, $12.00 in paper copy, $1.45 in microfiche. Final Report, prepared for Escambia-Santa Rosa Regional Planning Council, Pensacola, Florida. June 1972, 197 p, 25 fig. Fla-P-141.

Descriptors: *Shoreline, *Urbanization, *Planning, *Environmental effects, Design criteria, Management, Coordination, Beaches, Land use, *Florida.
Identifiers: *Shoreline management, Escambia-Santa Rosa Region (Florida), Escambia County (Florida), Santa Rosa County (Florida).

This shoreline management plan is comprehensive study and set of policies and recommendations for the use of the coastal zone around Pensacola, Florida. The study includes sections on the existing urban development patterns, the environmental impacts of these patterns and land uses, and an inventory of physiographic and hydrologic features. The most significant sections include: (1) a discussion of the basic philosophy of shoreland development and the objectives involved; (2) an analysis of potential design concepts and elements for shoreline development, including a morphological approach and an ecological approach; (3) a shoreline utilization plan featuring numerous development policies, standards, and implementation strategies; (4) and a shoreline management program involving protection programs, land use controls, monitoring, coordination of local agencies by the Regional Planning Council, environmental impact statements, and the creation of a special Shoreline Management Committee. Numerous maps and charts are included. (Elfers-North Carolina)
W73-08053

THE INTENSITY OF DEVELOPMENT ALONG SMALL AND MEDIUM SIZED STREAMS IN SURBURBAN PHILADELPHIA,
Regional Science Research Inst., Philadelphia, Pa.
For primary bibliographic entry see Field 03D.
W73-08055

FUNCTIONAL PLANNING AND PROGRAMMING: HOUSING, WATER AND SEWER, RECREATION, LAND USE, ADMINISTRATION.
Mark Twain Regional Planning Commission, Macon, Mo.

Available from the National Technical Information Service as PB-211 603, $11.50 in paper copy, $1.45 in microfiche. Report, June 1972. 180 p. Mo. P. 195 SA/145.

Descriptors: *Planning, *Projects, Water supply, Sewage systems, Priorities, *Missouri.
Identifiers: Citizens Advisory Committee, Mark Twain Region (Missouri), Macon (Missouri).

A water and sewer report is one section of this functionally-oriented planning study for the eight county Mark Twain region in Missouri. The water and sewer section is a summary of two consultant studies of water and sewer needs in the region and of a citizens advisory committee's review of the two studies. The water and sewer needs are outlined according to cities with no public water systems, cities that need additional water supplies, cities that need water treatment improvements and storage improvements, and cities that need sewer systems. These needs are accompanied by some short descriptions and recommendations. Every need or proposed project is then rated as to its priority on the basis of the citizens advisory committee review. (Elfers-North Carolina)
W73-08056

OPTIMAL PATH OF INTERREGIONAL INVESTMENT AND ALLOCATION OF WATER,
Tel-Aviv Univ. (Israel). Dept. of Economics.
For primary bibliographic entry see Field 06A.

W73-08133

WATER SUPPLY PLANNING STUDY OF THE CAPITOL REGION OF CONNECTICUT,
Connecticut Univ., Storrs. Dept. of Civil Engineering.
P. Magyar, P. Renn, A. Shahane, and D. Wall.
Report No. CE 72-57, October 1972. 156 p, 6 fig, 35 tab, 29 ref, 3 append. OWRR A-999-CONN (11).

Descriptors: *Water supply development, *Water demand, *Regional analysis, *Human population, Urbanization, Environmental effects, Social aspects, Economics, Connecticut River, Water reuse, Evaluation, *Connecticut.
Identifiers: *Optimum population, Ethical issues, *Capitol Region (Conn).

This study of water supply schemes for the Capitol Region of Connecticut projected population considering traditional engineering methodologies and ethical points raised by sociologists and demographers. An effort was made to quantify the qualitative concept of ethical issues to arrive at 'optimum population' for the Capitol Region. Total water demand was estimated for the years 1980, 2000, and 2020. To meet the projected demand of 250 MGD, alternative sources, including existing sources, new upland surface sources, the Connecticut River, Quabbin Reservoir and recycling were considered. These combined sources constitute viable alternatives to be evaluated on some rational basis. In the past, benefit-cost analysis has been used. However, public concern called for consideration of social, economic and environmental factors. Since it is relatively difficult to convert these qualitative factors into dollar terms, methods other than benefit-cost analysis were used to put value judgment on these alternatives. Literature review revealed two useful methods for this purpose: the Mason-Moore method, and the weighing factors method adopted by the Connecticut Region Planning Agency. After individual evaluations utilizing these methods, use of the Connecticut River rather than use of new upland sources (reservoirs) was chosen as the recommended alternative. (Edelen-Connecticut)
W73-08302

ORBITAL SURVEYS AND STATE RESOURCE MANAGEMENT,
Battelle Columbus Labs., Ohio. Aerospace Mechanics Div.
For primary bibliographic entry see Field 07B.
W73-08364

SIMPLE ALGORITHM FOR COST OPTIMIZATION OF A THREE DECISION VARIABLE SYSTEM,
Wayne State Univ., Detroit, Mich. Dept. of Civil Engineering.
For primary bibliographic entry see Field 05D.
W73-08386

THE ECONOMICS OF WATER TRANSFER,
Hawaii Univ., Honolulu. Dept. of Agricultural and Resource Economics.
C. Gopalakrishnan.
In: Water Resources Seminar Series No 2, Water Resources Research Center, University of Hawaii, Honolulu, December 1972. 11 p, 2 fig, 1 tab, 12 ref.

Descriptors: *Economics, *Water transfer, Water resources development, *Resource allocation, *Welfare (Economics), Water demand, Prices, Water rights, Optimum development plans.

The economics of water transfer is an aspect of water resource development that has at present generated considerable interest among economists. It deals with the economic implication of transferring water from low-yielding, conventional uses (mainly agricultural) to newly-emerging and more productive uses (industrial and recreational) with a view to enhance the value and productivity of water. The limited supply of water in relation to the fast-expanding demand for it makes this type of transfer almost imperative for the optimum utilization of water resources. The tools of microeconomics and welfare economics are used to study the economic underpinnings of a process of water transfer. To understand the nature of the market for water and the allocation of water among multiple uses, a discussion of the theoretical principles underlying resource allocation is included, examining the concepts of proportionality embodied in the law of variable proportions and the equi-marginal principle. Also examined rather closely are the market for water rights and the problem of pricing water. The three principal characteristics of water development which enable direct application of welfare economics to water resources are described. See also W73-05857) (Bell-Cornell)
W73-08387

6C. Cost Allocation, Cost Sharing, Pricing/Repayment

ECONOMIC EVALUATION OF ALTERNATIVE FARM WATER SOURCES IN THE CLAYPAN AREA OF ILLINOIS,
Illinois Univ., Urbana. Dept. of Agricultural Economics.
For primary bibliographic entry see Field 05G.
W73-07804

DISCHARGES OF INDUSTRIAL WASTE TO MUNICIPAL SEWER SYSTEMS,
Cornell Univ., Ithaca, N.Y. Water Resources and Marine Sciences Center.
For primary bibliographic entry see Field 05D.
W73-07810

SYSTEMS ENGINEERING OF OYSTER PRODUCTION,
Delaware Univ., Newark. Dept. of Mechanical and Aerospace Engineering.
F. A. Costello, and B. L. Marsh.
Available from the National Technical Information Service as COM-72-10698, $3.00 in paper copy, $1.45 in microfiche. College of Marine Studies Publication No 2 EN 066, May, 1972. 55 p, 2 tab, append. Sea Grant No. GH-109.

Descriptors: *Oysters, *Systems analysis, Optimization, Costs, Mathematical models.
Identifiers: *Oyster production, *Uncertainty, *Systems engineering, Marine products.

A stochastic optimization model for a closed-environment oyster production facility is developed. Three types of stochastic uncertainty important to the oyster production system are analyzed: (1) uncertainty in primary system process parameters such as the effect of cell concentration on oyster growth; (2) uncertainty in design variables such as the heat exchanger outlet temperature; and (3) uncertainty in subsystem process functional relations such as algae growth rate versus illumination. The approach used to study the stochastic parameters model is to first develop analytical or empirical distribution functions for the stochastic parameters and then to obtain an expected cost which is a function of deterministic parameters and design variables only. Addition of stochastic design variables to the model requires not only that the distributions for these variables be determined, but also that the distribution for the system cost function be obtained in order to apply the expected value criterion or some other similar objective. The addition of function uncertainty is accomplished through a systems analysis and cost sensitivity approach. The study indicates that algae production rate and cell concentration are of major importance, while illumination intensity and deep tank air requirements are of secondary importance. Other variables affect costs insignificantly. (Settle-Wisconsin)
W73-07824

ALTERNATIVE FINANCING FOR STATE AND LOCAL WATER RESOURCES PROJECTS,
Clemson Univ., S.C. Dept. of Economics.
A. H. Barnett.
South Carolina Water Resources Research Institute, Special Report, November, 1971. 45 p, tab, 13 ref. OWRR B-030-SC (3).

Descriptors: *Water resources development, *Financing, *Economic efficiency, *Equity, Evaluation, Local governments, State governments, Federal government, Pollution taxes (Charges), Bond issues.
Identifiers: User fees.

The criteria of economic efficiency and equity are employed in evaluating various approaches to state financing of local water supply and waste treatment facilities. One approach to such financing is to establish a state fund to be used for grants to local governments. The fund itself could be financed through water use fees or effluent charges. Other alternatives are state general obligation bonds, state bonds financed by local user charges on those receiving the services of the various projects, state loans to municipalities provided so as to meet federal requirements for increased federal aid, and local contributions to a state fund used to provide matching grants to contributing municipalities. The evaluation suggests that any state water resource fund would be difficult to justify. A system requiring benefit taxation or payments would be preferable. In fact, state aid for water projects other than waste treatment seems unjustified except under the unusual conditions where a project's benefits extended state-wide. Given the need to obtain federal aid, it seems that revenue bonds, applicant contributions, and loans to local government provide the most efficient and equitable alternatives. In any case, state sanctions are required to achieve desired water quality. (Settle-Wisconsin)
W73-07825

AN ECONOMIC APPROACH TO LAND AND WATER RESOURCE MANAGEMENT: A REPORT ON THE PUGET SOUND STUDY,
Washington Univ., Seattle. Dept. of Economics.
For primary bibliographic entry see Field 06B.
W73-07826

A PRESENT VALUE-UNIT COST METHODOLOGY FOR EVALUATING WASTEWATER RECLAMATION AND DIRECT REUSE AT A MILITARY BASE OF OPERATION,
Army Mobility Equipment Research and Development Center, Fort Belvoir, Va.
For primary bibliographic entry see Field 05D.
W73-07838

THE FINANCIAL FEASIBILITY OF REGIONALIZATION,
Arkansas Univ., Fayetteville. Dept. of Agricultural Economics and Rural Sociology.
For primary bibliographic entry see Field 03D.
W73-07919

TERTIARY EFFLUENT DEMINERALISATION,
Permutit Co. Ltd., London (England).
For primary bibliographic entry see Field 05D.
W73-08097

COST REDUCTION IN SULFATE PULP BLEACH PLANT DESIGN,
Improved Machinery, Inc., Nashua, N.H.
For primary bibliographic entry see Field 05D.

Greenwood, Laurens, McCormick, and Saluda. Information has been compiled and analyzed in areas of: (1) land use, (2) populations, (3) economics, (4) geography, and (5) existing utilities. With a 1970 population of 159,000 that is expected to grow to 192,000 by 1990, the already inadequate sewer and water services in this region will become even more inadequate as water use rises from 27 mgd to 35 mgd. A major problem in providing these services is the rural nature of the area--more than 80 percent of the land is classified as agricultural or open space. Lack of coordination and political boundaries have also made it difficult to develop adequate facilities. Recommendations include: (1) regionalization of services to provide coordinated, cooperative facilities, (2) planning of future development of systems to control land use, preventing unwanted growth through denial of sewers and water and (3) continuation of the Upper Savannah Regional Planning and Development Council as the areawide planning organization for the six counties. (Poertner)
W73-08177

WATER, SEWER AND STORM DRAINAGE PLAN FOR THE CUMBERLAND PLATEAU PLANNING DISTRICT.
Thompson and Litton, Inc., Wise, Va.

Cumberland Plateau Planning District Commission, Lebanon, Virginia, June 6, 1972. 113 p, 5 tab.

Descriptors: *Planning, *Data collections, *Drainage programs, *Water supply development, *Sewerage, *Virginia, Storm runoff, Storm drains, Water resources, Regional analysis, Public utility districts, Water pollution control, Waste water treatment, Water works, Surveys.
Identifiers: *Cumberland Plateau.

The future development of the Cumberland Plateau Planning District, as well as the present level of development, requires adequate systems of water, sewerage and storm water drainage. Future development can be controlled by the development of these systems to produce growth consistent with the land-use policy of the area, but present development must not be denied adequate facilities even if it doesn't fit into the master plan. As a first step towards providing these services, this report presents a complete inventory of water, sewer and storm drainage in the area, pointing out deficiencies and recommending measures for correcting these short-comings. Each of the four counties comprising the study area has water supply problems. The most severe problems exist in Buchanan and Dickinson Counties where many areas either have no central water systems or inadequate systems. The proposed solution is the development of the John W. Flannagan Reservoir in Dickinson County to provide water for both counties. Similarly, sewerage problems exist. Using a set of criteria, projects have been rated. Those with the highest priority were recommended by the State Water Control Board for Federal and State construction grants. Development of a plan for storm water drainage is still in preliminary stages. (Poertner)
W73-08179

WATER AND SEWER PLAN.
West Alabama Planning and Development Council, Tuscaloosa.

Available from the National Technical Information Service as PB-210 955, $12.50 in paper copy, $1.45 in microfiche. May 1972. 161 p, 30 fig, 10 tab. CPA-AL-04-09-1008.

Descriptors: *Planning, *Regional analysis, *Water resources development, *Water supply, *Sewerage, *Surveys, Data collections, Long-term planning, Administration, Regional development, Water pollution control, Alabama, Forecasting, Public utility districts, Land use, Short-term planning.

The West Alabama Planning and Development Council serves five counties with a total area of 4,154 square miles and a population of 180,000, of which about 120,000 live in one county. Included in the report are: (1) the goals and policies for water and sewer system management, (2) an inventory of existing facilities, (3) an evaluation of these systems, (4) a short range (10 year) plan for water and sewers to meet present and future deficiencies, and (5) a long range (20 year) regional plan. The goals of the region are threefold: (1) to promote a co-ordinated regional approach to sewer and water, (2) to provide adequate water and sewer service to all area residents, and (3) to maximize the use of existing water resources including ground water, surface water and reclaimed waste water. Although population is not expected to change much in the next 20 years, increasingly stringent federal standards will require system up grading, especially for sewer systems. But a higher priority is generally to be assigned for water systems, than for sewer systems, and all improvements should provide maximum service at a minimum cost. County level water systems are recommended with the Council acting as a regional coordinator. An areawide sewer system is seen as unfeasible but sewers are recommended for all incorporated areas, with funding from outside sources. (Poertner)
W73-08181

REGIONAL WATER AND SEWER FACILITIES IMPROVEMENT PROGRAM.
Southeastern Illinois Regional Planning and Development Commission, Harrisburg.

Available from the National Technical Information Service as PB-210 108, $4.65 in paper copy, $1.45 in microfiche. February 1972. 23 p, 4 tab. HUD Ill. P-307 (1-16).

Descriptors: *Water resources, *Regional analysis, *Sewerage, *Water supply, *Illinois, *Planning, Project planning, Project purposes, Evaluation, Water pollution control, Short-term planning.
Identifiers: *Southeastern Illinois.

A previous study (see W72-00643) of this rural, five-county region in southeastern Illinois pointed out extensive municipal water and sewage deficiencies. Strategies are presented of the Southeastern Illinois Regional Planning and Development Commission in overcoming these deficiencies. Criteria for individual projects include: (1) regional economic impact, (2) environmental impact, (3) health factors, (4) public safety factors, (5) size of population served, (6) consistency with regional plans, (7) readiness to proceed and (8) a crisis factor, which rates the project on its need for prevention of a potential catastrophe or to meet state regulations. The Commission has set goals for five one-year phases including the provision of technical assistance to the 7 highest priority projects. Funding is expected to be through utility rates and federal funding, with the Commission serving as the State-appointed clearinghouse for federal funds. The use of bonds is not expected to be required. The Commission also intends to establish a citizen advisory board to deal with all environmental issues in the region. (Poertner)
W73-08184

WATER SUPPLY PLANNING STUDY OF THE CAPITOL REGION OF CONNECTICUT,
Connecticut Univ., Storrs. Dept. of Civil Engineering.
For primary bibliographic entry see Field 06B.
W73-08302

6E. Water Law and Institutions

INTERGOVERNMENTAL RELATIONS AND RESPONSES TO WATER PROBLEMS IN FLORIDA,
Florida Atlantic Univ., Boca Raton. Dept. of Political Science.
R. D. Thomas.
Available from the National Technical Information Service as PB-219 582, $3.00 in paper copy, $1.45 in microfiche. Florida Water Resources Research Center Publication No. 19, 1972. 48 p, 20 tab, 2 append. OWRR A-020-FLA (2). 14-31-0001-3809.

Descriptors: *Governmental interrelations, *Water resources, *Florida, *Attitudes, Water resources development, Water utilization, *State governments.

An exploratory analysis is presented of Florida's legislators' and county commissioners' images (perceptions and attitudes): (1) of eleven selected water problems; (2) of which level or levels of government should have the responsibility for handling and attempting to solve these problems; (3) of the effectiveness of ten selected measures for dealing with water use problems; and (4) of the related factor of growth and development. The data, derived principally from interviews with the legislators and commissioners, showed a basic difference between the legislators and commissioners in their assessment of the severity of water problems; in their assessment of the severity of water problems relative to other public problems such as education, welfare, roads, and health/hospitals; and, in their evaluation of the need to impose controls on growth and development. On the other hand, the data showed considerable agreement between the legislators and commissioners in their evaluation of what solutions would be most effective in dealing with water use problems. (Morgan-Florida)
W73-07803

THE POLITICAL ECONOMY OF A CORPS OF ENGINEERS PROJECT REPORT: THE DELMARVA WATERWAY,
Cornell Univ., Ithaca, N.Y.
L. A. Shahman, P. Willing, D. S. Allee, S. P. Lathrop, and C. Riordan.
Available from the National Technical Information Service as PB-219 403, $3.00 in paper copy, $1.45 in microfiche. New York Water Resources and Marine Sciences Center, Ithaca, Technical Report No. 43, A.E. Res. 72-9, June 1972. 59 p, 1 fig, 4 tab, 35 ref. c-2199; OWRR B-026-NY (3). 14-31-0001-3312; 14-31-0001-3409.

Descriptors: *Decision making, *Planning, Projects, *Delaware, *Maryland, *Virginia, Economies, Political aspects, Federal Government, State governments, Economic analysis, Economic efficiency, Estuaries, Ecosystems, Resources, Cost-benefit ratio.
Identifiers: *Delmarva Waterway, *Information flow, Local interests, Leverage.

A first step in correcting inadequacies in institutional arrangements is to understand such arrangements as they now function. The sequence of events and the system that produced a U.S. Army Corps of Engineers navigation project report are examined. Its orientation is that of seeking rather than solving problems, though some tentative conclusions and suggestions for change are offered. Specific reference is made to the process which led to the Congressional authorization of a waterway on the eastern shore of the Delmarva Peninsula in Delaware, Maryland, and Virginia. The method used to carry out the study involves three related procedures: (1) interviews with principals in the decision-making process and with local people; (2) maintenance of correspondence and memoranda files, studied intensively; and (3) sorting of the resulting data and compiling it according to a framework of theoretical organizing concepts which help illuminate the characteristics of information flow and use in the decision-making process. Information is vital to decision-making, with the result that the content of and power behind information flows largely determine the outcome of the planning process. (Bell-Cornell)
W73-07811

AN APPRAISAL OF FLOODPLAIN REGULATIONS IN THE STATES OF ILLINOIS, INDIANA, IOWA, MISSOURI, AND OHIO,
Illinois Univ., Urbana. Dept. of Landscape Architecture.
For primary bibliographic entry see Field 06F.
W73-07814

REGIONAL WATER RESOURCE PLANNING FOR URBAN NEEDS: PART 1,
North Carolina Univ., Chapel Hill. Dept. of City and Regional Planning.
For primary bibliographic entry see Field 06B.
W73-07819

UNDERSTANDING THE WATER QUALITY CONTROVERSY IN MINNESOTA,
Minnesota Univ., Minneapolis. Water Resources Research Center.
For primary bibliographic entry see Field 05G.
W73-07822

THE ENVIRONMENT AS PROBLEM: II, CLEAN RHETORIC AND DIRTY WATER,
For primary bibliographic entry see Field 05G.
W73-07827

WATER RESOURCE DEVELOPMENT: SOME INSTITUTIONAL ASPECTS,
Hawaii Univ., Honolulu. Coll. of Tropical Agriculture.
C. Gopalakrishnan.
The American Journal of Economics and Sociology, Vol 30, No 4, p 421-428, October, 1971.

Descriptors: *Water resources development, *Institutional constraints, *Water rights, *Prior appropriation, Legal aspects, Economic efficiency, Beneficial use, Conservation, Dams, Water management (Applied), *Montana.

Some of the institutional impediments to Montana's water resource development are examined. In Montana the ownership, control, and use of both surface and groundwater are governed by the doctrine of prior appropriation. Consequently, the system of water rights suffers from some of the following basic weaknesses of the appropriation doctrine: (1) the system is inflexible; (2) the emphasis on priority may result in economic inefficiencies; and (3) actual management of water programs is often vested with local water commissioners who have little expertise in water management. Some drawbacks specific to Montana's system of water rights include: (1) rights determined by private suits in which it is not necessary to join all parties concerned, and (2) rules that define abandonment as the concurrence of relinquishment of possession and intent not to resume it for a beneficial use. Also, state water agencies are handicapped by a dearth of funds and a lack of coordination with local water development groups. Another institutional problem is the attitude of the general public toward federal participation in water projects. Finally, feuds between conservationists and dam-builders have greatly slowed the development of water resources. (Settle-Wisconsin)
W73-07828

SOUTH AFRICA, PROBLEMS OF A QUOTA-CONTROLLED FISHERY.
World Fishing, Vol 21, No 10, p 17, October, 1972.

Descriptors: *Fisheries, Fishing, Regulation.
Identifiers: *Pelagic fish, *Quotas, South Africa, South West Africa.

The west coast of Africa is the last known major pelagic fishing ground in the world; all others have been destroyed. Pelagic fish include pilchards, anchovy, maasbanker, mackerel, redeye, and lanterns. Catches of pelagic shoal fish in South Africa and South West Africa declined during the 1970-71 catching season, and were almost 350,000 tons less than in the previous season. The reduced South West African catch occurred because of the application of relatively severe conservation measures to maintain the pilchard resources. These measures were applied to limit the overall tonnage, and one-third of each quota had to comprise pilchards. When these were landed, fishing operations had to cease. For the first time, a limit was also placed on the South African catch. However, a poor season prevented the industry from even reaching this limit. In addition, the International Commission for the South Atlantic Fisheries will make recommendations to member nations for regulation of fisheries off the west coast of Africa. The Commission predicts its conventions will boost the present two million tons of cultivated pelagic fish off the west coast to twenty million tons by 1980. (Settle-Wisconsin)
W73-07830

MICHIGAN'S ENVIRONMENTAL PROTECTION ACT OF 1970: A PROGRESS REPORT,
Michigan State Univ., East Lansing.
For primary bibliographic entry see Field 05G.
W73-07982

ADMINISTRATION OF THE NATIONAL ENVIRONMENTAL POLICY ACT, PART 1.

For sale by the Superintendent of Documents, U.S. Government Printing Office, Washington, D.C. 20402. Price $5.00. Hearings—Subcomm. on Fisheries and Wildlife Conservation—Comm. on Merchant Marine and Fisheries, United States House of Representatives, 91st Cong, 2d Sess, December 1970. 1270 p, 1 tab, 3 chart, 209 ref, 7 append.

Descriptors: *Administrative agencies, *Federal project policy, *Legislation, *Environmental effects, Governmental interelations, Hydroelectric project licensing, Legal respects, Federal government, Judicial decisions, Law enforcement, Regulation, Federal Power Act, Environmental control, Administrative decisions, Legal review.
Identifiers: *Congressional hearings, *National Environmental Policy Act of 1969.

These hearings were designed to help determine the effectiveness of the National Environmental Policy Act (NEPA), and the adequacy of agency responses to section 102 and 103 of the Act. The hearing also investigated what changes, if any, may be called for to improve the Act. Section 102 requires federal agencies to include detailed environmental statements on every major federal action affecting the environment. Section 103 requires federal agencies to review their operating procedures to collect any inconsistencies which prohibit full compliance with the Act. Inquires as to the status of the environmental impact statement requirement indicates an important start has been made in the implementation of the Act. Included is a mercury report of the Council on Environmental Quality, an analysis of Title 1 of the NEPA, the Draft Environmental Impact Statement for the Trans-Alaskan pipeline, and a detailed analysis of licensing procedure under the Federal Power Act. (Beardlsey-Florida)
W73-07983

graphs, and illustrations are included. (Mockler-Florida)
W73-07990

LIMITING FEDERAL RESERVED WATER RIGHTS THROUGH THE STATE COURTS,
L. B. Craig.
Utah Law Review, Vol 1972, No 1, p 48-59, Spring 1972. 63 ref.

Descriptors: *Reservation doctrine, *Water rights, *Prior appropriation, *Priorities, Legal aspects, Federal jurisdiction, Federal-State water rights conflicts, Jurisdiction, Remedies, Judicial decisions, Legislation, Appropriation, Diversion, Water utilization, Eminent domain, Compensation.
Identifiers: Estoppel.

Much of the surface water currently being appropriated by private individuals in the arid western states is subject to superior federal rights based on the reservation doctrine. The possibility that the government could assert its superior rights to divert private appropriators has discouraged development of water resources in this region. A tracing of the historical development of the reservation doctrine is followed by a discussion of legislation introduced to initiate change. The role of the courts is also discussed; the courts must determine whether justice requires the private appropriator who has acquired water rights pursuant to state law without knowledge of the federal government's prior reserved rights to pay the cost of diversion of water to federal reserved lands or whether the cost would more properly be borne by the public generally. The doctrine of equitable estoppel is examined in relationship to this problem. Condemnation and compensation for private water rights are also discussed. (Mockler-Florida)
W73-07991

WATER LAW—PRIMARY JURISDICTION OF THE BOARD OF CONTROL OVER QUESTIONS OF WATER RIGHTS.

Land and Water Law Review, Vol 7, No 2, p 599-614, 1972. 72 ref.

Descriptors: *Wyoming, *Jurisdiction, *Water rights, *Adjudication procedure, Legal aspects, State jurisdiction, Water law, Abandonment, Legal review, Judicial decisions, Administrative decisions, Watercourses (Legal aspects), Administrative agencies, Water supply.

Plaintiff reservoir company sued to have part of the water rights of defendant reservoir company declared abandoned. The plaintiff contended that the courts have concurrent jurisdiction with the Board of Control and may continue to initially determine abandonment questions; defendant contended that the Board had exclusive jurisdiction in this area. The court held that the Board of Control has primary jurisdiction over questions of abandonment of water rights. Thus while the Board and the courts still have concurrent jurisdiction over abandonment questions, before the lower court will grant relief, the abandonment issue should usually be initially determined by the Board. This comment discussed this case in detail and includes a background discussion, discussion on initiation of proceedings, district court discretion, review by appeals courts, and the impact on other areas of water law. This doctrine of primary jurisdiction will provide much needed flexibility to the adjudication procedure. (Mockler-Florida)
W73-07992

ENVIRONMENTAL IMPACT STATEMENTS—A DUTY OF INDEPENDENT INVESTIGATION BY FEDERAL AGENCIES.
Colorado Law Review, Vol 44, No 1, p 161-172, August 1972. 63 ref.

Descriptors: *Environmental effects, *Legislation, *Administrative decisions, *Federal project policy, Legal aspects, Judicial decisions, Legal review, Administration, Project planning, Water resources development, Watercourses (Legal aspects), Administrative agencies, Adjudication procedure.
Identifiers: *Environmental Impact Statements, *National Environmental Policy Act.

The heart of the National Environmental Policy Act, (NEPA), is the section which requires all agencies of the federal government to include in every recommendation or report on major federal actions significantly affecting the quality of the human environment, a detailed statement on the environmental impact of the proposed action. This comment focuses on the interpretation of this section presented by the Court of Appeals for the Second Circuit in Greene County Planning Board v. Federal Power Commission which held that the environmental impact statement is to be prepared by the agency itself which is proposing the action; the agency cannot satisfy NEPA by preparing the statement on the basis of information provided by others. This case may further the cause of long-range environmental planning. However, NEPA is in serious trouble due to the gap between the goals it sets out and the means and will to implement its procedures. (Mockler-Florida)
W73-07993

PRIVATE REMEDIES FOR BEACH FRONT PROPERTY DAMAGE CAUSED BY OIL POLLUTION OF MISSISSIPPI COASTAL WATERS,
For primary bibliographic entry see Field 05G.
W73-07994

THE YEAR OF SPOILED PORK: COMMENTS ON THE COURT'S EMERGENCE AS AN ENVIRONMENTAL DEFENDER,
Florida Univ., Gainesville.
For primary bibliographic entry see Field 05G.
W73-07996

F. W. GUEST MEMORIAL LECTURE: RIVER POLLUTION AND THE LAW,
For primary bibliographic entry see Field 05G.
W73-07997

CRIMINAL LIABILITY UNDER THE REFUSE ACT OF 1899 AND THE REFUSE ACT PERMIT PROGRAM.

Journal of Criminal Law, Criminal and Police Science, Vol 63, No 3, p 366-376, September 1972. 117 ref.

Descriptors: *Federal Water Pollution Control Act, *Law enforcement, *Penalties (Legal), *Permits, *Legislation, Federal government, State governments, Judicial decisions, Negligence, Legal aspects, Water pollution, Pollution abatement, Water pollution control, Navigable waters, Water law, Rivers and Harbors Act.
Identifiers: *Refuse Act, Absolute liability, Injunctions (Prohibitory), Licenses.

This comment deals with criminal liability for water pollution under the Refuse Act and the Federal Water Pollution Control Act (FWPCA). Section 411 of the Refuse Act provides for fines or imprisonment for discharges of any refuse into navigable water without a permit. However, courts have been reluctant to find corporate officials personally liable for violations caused by their companies. Consequently, the United States Supreme Court has held that injunctive relief is appropriate to insure effectiveness of the statute. The FWPCA does not impose criminal liability, but operates through civil enforcement of state water quality standards. These standards may conflict with liability under the Refuse Act; since the Refuse Act is not superseded by FWPCA legisla-

tion, it is possible that one may be officially in complicance with FWPCA standards and yet be criminally liable under the Refuse Act. In order to resolve this conflict the Refuse Act Permit Program was established which makes the issuance of a permit contingent primarily on the applicant's compliance with state water quality standards. (Adams-Florida)
W73-07998

ENVIRONMENTAL LAW: STRICT COMPLIANCE WITH PROCEDURAL REQUIREMENTS OF NEPA–THE AGENCIES MUST PLAY BY THE RULES,

R. Nielsen.
University of Florida Law Review, Vol 24, No 4, p 814-820, Summer 1972. 74 ref.

Descriptors: *Decision making, *Administrative decisions, *Federal project policy, *Project planning, *Legal review, Environmental effects, Administrative agencies, Permits, Porject benefits, Legal aspects, Federal government, Judicial decisions, Legislation, Water law, Planning, Analytical techniques, Nuclear powerplants, Thermal pollution.
Identifiers: *Environmental Impact Statement, *National Environmental Policy Act, *Licenses.

The National Environmental Policy Act (NEPA) was designed as an agency regulating statute, applicable to all federal agency actions having a potential environmental impact. Substantively, the Act requires the federal government to insure environmental quality in federal activities. It also demands that certain procedural requirements be followed to the fullest extent possible by any federal agency considering environment-affecting projects. Recently, federal agencies have been under attack for lack of compliance with NEPA requirements. In Calvert Cliffs' Coordinating Committee v. Atomic Energy Commission, an environmental protection group brought suit challenging the Commission's approval of a construction permit for two nuclear power plants pending a complete review of NEPA imposed requirements. The U.S. District Court of Appeals held that the Commission's procedures were inadequate because they failed to sufficiently consider environmental values, failed to require environmental impact considerations, and omitted independent project evaluations. The court distinguished substantive duties imposed by NEPA from procedural provisions of the Act on the grounds that the former leave room for agency discretion while the latter must be strictly adhered to. In particular, agencies are required to implement a systematic balancing analysis in each action affecting the environment. (Adams-Florida)
W73-07999

DEPARTMENT OF NATURAL RESOURCES; DIVISION OF WATER POLLUTION CONTROL.

Mass. Ann. Laws ch. 21 secs 1 thru 19, 26 thru 53 (Supp. 1971).

Descriptors: *Legislation, *Water pollution control, *Water management (Applied), *Massachusetts, Eminent domain, Flood control, Public access, Recreation, Aesthetics, Public health (Laws), Permits, Planning, Oil pollution, Administrative agencies, Law enforcement, Impoundments, Water supply, Penalties (Legal), Pollution abatement.

These Massachusetts statutory provisions provide for the creation of a Department of Natural Resources and for the implementation of its purposes. Included within the Department is a Division of Water Resources with various duties of water management and control. The Commission of Water Resources is impowered to purchase, give, lease or acquire by eminent domain any lands and waters to protect water impoundment sites. Reference is made to recreational needs and provisions for source are included. A Division of Water Pollution Control is created and is especially concerned with oil pollution. This division must develop pollution control plans, conduct studies, and adopt water quality standards. The legislation provides for legal sanctions for anyone violating the water quality standards. Municipal corporations may be authorized to construct abatement facilities. Fines are levied upon those who cause pollution in excess of the Commission standards, and orders may be issued to cease. (Smith-Adam-Florida)
W73-08000

WATER RESOURCES RESEARCH PROGRAM.
42 USC secs. 1961 thru 1961a-4 (1972).

Descriptors: *United States, *Grants, *Legislation, *Research and development, *Resources development, Financing, Water resources development, Administration, Federal government, Water law, Water policy, Water utilization, Water resources planning, Institutions, Administrative agencies, Planning, Research priorities, Management.

The Secretary of the Interior is to provide money to each state to assist in establishing and carrying on the work of a competent and qualified water resources research institute at one college or university within the state. It shall be the duty of each institute to plan and conduct or arrange for a component of the college with which it is connected to conduct research in relation to water resources and to provide for training of scientists to conduct such research. The Secretary may also make grants to other institutions or foundations who are qualified to conduct competent water research projects. All grant proposals must be submitted to the President of the Senate and the Speaker of the House. The Secretary will insure that there is no duplication of programs. All results from these programs will be made available to the general public. A cataloging center is to be established for compiling current research in water resources. The President shall clarify agency responsibilities for federal water resources research and provide for interagency coordination of such research. (Glickman-Florida)
W73-08001

REGULATION OF RIVERS, LAKES, AND STREAMS.
111. Ann. Stat. ch. 19 secs. 48 thru 78 (Smith-Hurd 1972).

Descriptors: *Illinois, *Legislation, *Navigation, *Administrative agencies, *Water management (Applied), Legal aspects, Water quality, Regulation, Obstruction to flow, Adjudication procedure, Legal review, Flood control, Penalties (Legal), State jurisdiction, Data storage and retrieval, Docks, Water law, Watercourses (Legal aspects), Lake Michigan, Drainage, Permits, Harbors, Flood plain zoning.
Identifiers: *Obstructions to navigation.

The State Department of Transportation shall have jurisdiction and supervision over all of the rivers and lakes of Illinois, in which the state or the general public shall have any rights or interests. Included within this Chapter are statutes on jurisdiction, data reports, encroachments, navigation, docks and wharves, deep waterways data, duties and powers of the state environmental protection agency, beautifying bodies of water, reclamation and drainage, permitting procedures, taking materials from beds of public waters, building harbors and mooring facilities, flood plain development, water power data, natural resources, fish propagation, flood waters, and rights of the public. Statutes on investigations, hearings, and judicial review are also included. Specific penalties for violations of the statutes are also set out. (Mockler-Florida)
W73-08002

ILLINOIS AND DESPLAINES RIVERS; KASKASKIA RIVER WATERSHED.

111. Ann. Stat. ch. 19 secs. 38 thru 41.1 (Smith-Hurd 1972).

Descriptors: *Illinois, *Legislation, *Watershed management, *Obstructions to flow, *Navigation, Legal aspects, Navigable rivers, Jurisdiction, Flood control, Drainage, Water supply, Watercourses (Legal aspects), Water storage, Water management (Applied), River basin development.

This statute formally recognizes the Desplaines and Illinois Rivers as navigable in law and charged the Governor and Attorney General of Illinois with the responsibility of preventing the erection of any structure in or across the streams without explicit authority. These officials are also directed to take necessary legal action to remove all obstructions now existing that in any way interfere with the intent and purpose of the legislation. Another statute contained herein gives the Illinois Department of Transportation the authority to independently engage in the formulation of plans, acquisition of rights of way, construction, operation and maintenance of any navigation, flood control, drainage, levee, water supply and water storage and other water resource improvements and facilities in connection with the development of the Kaskaskia River Watershed. The Department has jurisdiction and supervision over any and all phases of developments and improvements in the basin. (Mockler-Florida)
W73-08003

REMOVAL OF OBSTRUCTIONS.
111. Ann. Stat. ch. 19 secs 42 thru 47e (Smith-Hurd Supp. 1971).

Descriptors: *Illinois, *Navigation, *Ships, *Legislation, *Flotsam, Legal aspects, Water policy, Obstruction to flow, Penalties (Legal), Transportation, Boats, State jurisdiction, Administrative agencies.

It is unlawful to tie up or anchor vessels or other water craft in public or navigable waters in such manner as to prevent or obstruct in any manner the passage of any vessels or craft. The owner of sunken craft has the duty to immediately mark it with a buoy or beacon during the day and a Lighted lantern at night, and to maintain such marks until the sunken craft is removed or abandoned, and the neglect or failure of the owner to do so is unlawful. If such obstruction is left for more than thirty days, the State Department of Transportation shall remove it in any manner within its discretion. The expense of removal shall be charged to the owner of the vessel. Violations of this act shall be punishable by fines of between five hundred and one thousand dollars and/or by imprisonment of between thirty days and one year. (Mockler-Florida)
W73-08004

HIRSCH V. STEFFEN (LIABILITY FOR OBSTRUCTION OF NATURAL DRAINWAY).
488 S.W.2d 240-245 (Ct. App. Mo. 1972).

Descriptors: *Drainage patterns (Geologic), *Missouri, *Natural drainage, *Obstruction, Legal aspects, Constitutional law, Dams, Water control, Drainage practices, Drainage water, Natural flow, Drainage engineering, Judicial decisions, Drainage effects.

Plaintiffs, upstream riparian landowners, sued to compel defendants, downstream riparian landowners, to remove an obstruction to the natural drainage. Defendants had erected a dam on a dry bed which was usually dry. Hoever, the dam did cause an overflow on upstream lands at times

blocking the flow of surface water. Prior to construction of the dam the old river bed had been the natural drainage flow for the area. The Missouri Court of Appeals held that a natural watercourse or drainway may not be obstructed without liability to those harmed by the obstruction. Moreover, this river bed was a natural drainway and the court could issue a mandatory injunction requiring removal of the obstruction. Furthermore the court could order defendant to restore the old river channel to a near proximation of its previous depth and width prior to defendant's action in damming the old river bed and blocking the channel. (Mockler-Florida)
W73-08005

CONNERY V. PERDIDO KEY, INC. (OWNERSHIP OF LANDS BETWEEN MEANDER LINE AND HIGH WATER MARK).
270 So. 2d 390-394 (1st D.C.A. Fla. 1973).

Descriptors: *High water mark, *Florida, *Meanders, *Boundaries (Property), *Boundary disputes, Legal aspects, Islands, Ownership, Surveys, Mapping, Measurement, Judicial decisions, Water level fluctuations.
Identifiers: *Meander lines.

Plaintiff landowner brought an ejectment action against defendant realty company to determine ownership of a part of an island. Following a government survey and meandering, the island was conveyed by the United States to plaintiff's predecessor in title. The land in dispute lay outside of the government meander line. Defendant contended it was omitted from the survey line either intentionally or as a result of a gross error and thus is not included in the original conveyance. Defendant has filed a homestead application for the unsurveyed land. The District Court of Appeals held that unless lands lying between the meander line and the present high water mark were excluded from the survey intentionally or due to gross error, the meander line does not constitute the boundary; rather the high water mark of the waterbody is the boundary. Meander lines were not designed to be boundaries. The court further held that the evidence was insufficient to support the contention that the lands in dispute were omitted intentionally or because of gross error. Thus, the high water mark is the boundary of plaintiff's lands, not the meander line. The court affirmed the trial court holding for plaintiff. (Mockler-Florida)
W73-08006

IN RE WEST MANAGEMENT DISTRICT (PETITION TO CREATE DRAINAGE DISTRICT).

269 So. 2d 405-407 (2d D.C.A. Fla. 1972).

Descriptors: *Florida, *Public rights, *Local governments, *Drainage districts, Legal aspects, Drainage, governments, Water policy, Drainage systems, Drainage programs, Jurisdiction, Legislation, Judicial decisions, Surface runoff.

Petitioner, a county resident, sought to create a drainage district pursuant to state statute but was opposed by the county. The trial court denied the county's motion to dismiss the petition and also denied the county's right to intervene. The appeals court held that under the rule allowing anyone claiming an interest in pending litigation to assert his right by intervention, the county was entitled to intervene in an action to create a drainage district. The court based its decision on the rationale that the county must be given this right since it had the authority to establish and administer programs concerning drainage and to cooperate with governmental agencies and private enterprises in the development and operation of such programs. The court also indicated that the statute providing that any landowner who did not sign the petition for a drainage district may advocate or resist the organization and incorporation of a drainage district did not preclude the county from intervening in the petition to create the drainage districts. (Mockler-Florida)
W73-08007

SEEGREN V. ENVIRONMENTAL PROTECTION AGENCY (PETITION FOR HARDSHIP VARIANCE FOR USE OF SANITARY SEWERS).
291 N.E.2d 347-349 (Ct. App. Ill. 1972).

Descriptors: *Illinois, *Judicial decisions, *Treatment facilities, *Sewage treatment, Legal aspects, Sanitary systems, Public health, Sewage, Waste disposal, Municipal wastes, Legal review, Sewage effluents, Sewers, Permits, Adjudication procedure, Administrative decisions, Land development, Urbanization.

Petitioner, land developer, sought review of an order of the respondent Illinois Pollution Control Board denying a petition for a hardship variance to permit construction of sewer extensions to a housing development and apartment complex built by petitioner. The Board was under a court order banning further extensions to existing facilities until the treatment plant could adequately treat additional effluents. Petitioner contended that the Board had consistently granted variances and permitted connections to existing facilities where it was shown that substantial investment in and development of the project had occurred prior to the prohibitory order. Petitioner had completed two apartment buildings a month prior to the order. The Illinois Appellate Court held that in view of previous variance policy of the Board, petitioner had proven an unreasonable hardship and the Board's decision was contrary to the weight of evidence. The Court remanded the case to the Board with directions to issue the proper certificates. (Mockler-Florida)
W73-08008

STALEY MANUFACTURING CO. V. ENVIRONMENTAL PROTECTION AGENCY (REGULATION OF DISCHARGES FROM PRIVATE SEWER INTO MUNICIPAL SEWER).
290 N.E.2d 892-897 (Ct. App. Ill. 1972).

Descriptors: *Illinois, *Sewage treatment, *Jurisdiction, *Waste treatment, Legal aspects, Water quality, Water law, Water pollution control, Water pollution sources, Legal review, Administrative decisions, Treatment facilities, Industrial wastes, Effluents.

Petitioner, industrial plant, sought judicial review of a regulation promulgated by the respondent, State Pollution Control Board. Petitioner owned a private sewer which emptied into the municipal treatment system. The disputed regulation would restrict the types, concentrations and quantities of contaminants which could be discharged into municipal sewer systems. Petitioner contended that respondent's jurisdiction extended only to discharges into waters of the state and that this administrative regulation therefore exceeded the Board's statutory authority. The Illinois Appellate Court held that there was a proper nexus between controlling what flows into a sewage treatment plant and what flows out of the plant and into the waters of the state; thus, the Board did have authority to compel the operator of a private sewer system which empties into a municipal sewer to adequately treat its contaminants before they reached the municipal system. The court thereby affirmed the regulation by the Pollution Control Board. (Mockler-Florida)
W73-08009

BOOKER V. WEVER (OWNERSHIP OF RELICTED LAND).
202 N.W.2d 439-443 (Ct. App. Mich. 1972).

Descriptors: *Michigan, *Riparian rights, *Boundaries, *Boundary disputes, Legal aspects, Adjacent landowners, Judicial decisions, Lakes, Riparian land, Watercourses (Legal aspects).
Identifiers: *Reliction.

Plaintiff riparian landowner sought a declaratory judgment against defendant adjacent riparian landowner to determine the ownership of land uncovered by reliction. The issue was raised as a result of the lowering of the water level of the lake. The appeals court held that the relicted land should be divided in proportion to shoreline owned. The shoreline to be used would be that shoreline at the time the lake and surrounding land was patented by the United States to the State of Michigan. The court also added that courses and distances in the deed description would yield to natural monuments such as the shoreline of the lake. (Mockler-Florida)
W73-08010

WILSON CONCRETE COMPANY V. COUNTY OF SARPY (LIABILITY FOR OBSTRUCTION OF NATURAL FLOW).

202 N.W.2d 597-600 (Neb. 1972).

Descriptors: *Nebraska, *Obstruction to flow, *Drainage effects, *Drainage patterns (Geologic), Barriers, Stream flow, Judicial decisions, Highway effects, Drainage engineering, Natural flow, Alteration of flow, Culverts, Drainage water, Drainage practices, Legal aspects.
Identifiers: Injunctive relief.

Plaintiff landowner sued for damages and an injunction requiring the defendant county to provide an adequate drainway under a highway for water of a natural watercourse. In constructing the highway, defendant had placed a single ten-by-ten foot box culvert where there had previously been an open ditch as an outlet for a creek to drain from the plaintiff's land. The Nebraska Supreme Court held that the culvert was inadequate and that it would cause considerable backwater. The court further held that removal of the culvert, thus leaving an open cut, would reduce the flooding, and therefore plaintiff was entitled to a mandatory injunction requiring defendant to remove the obstruction in order to provide for the passage of the surface waters. (Mockler-Florida)
W73-08011

PROCEEDING 1971 TECHNICAL CONFERENCE ON ESTUARIES OF THE PACIFIC NORTHWEST.
Oregon State Univ., Corvallis.
For primary bibliographic entry see Field 05B.
W73-08051

BUREAU OF RECLAMATION,
Bureau of Reclamation, Denver, Colo. Applied Sciences Branch.
For primary bibliographic entry see Field 05G.
W73-08108

FORECAST STUDY FOR PRAIRIE PROVINCES WATER BOARD,
Water Survey of Canada, Calgary (Alberta). Alberta and Northwest Territories District Office.
For primary bibliographic entry see Field 04A.
W73-08143

A WILDERNESS SNOW COURSE,
Forest Service (USDA), Kalispell, Mont. Flathead National Forest.
For primary bibliographic entry see Field 02C.
W73-08148

WILDERNESS IN THE NATIONAL PARKS,
National Park Service, Denver, Colo. Denver Service Center.
For primary bibliographic entry see Field 04C.
W73-08149

CLEAN WATER FOR SAN FRANCISCO BAY.
California State Water Resources Control Board, Sacramento.

April 1971. 16 p, 13 fig.

Descriptors: *Water reuse, *Bays, *Planning, *Administration, *Water pollution control, *Water quality control, California, Water pollution sources, Water pollution effects, Regional analysis, Domestic wastes, Industrial wastes, Waste disposal.
Identifiers: *San Francisco Bay.

In order to preserve and enhance water quality in San Francisco Bay, in view of the inevitable increase in waste loads, the present system of independent decision making by each wastewater discharger must be replaced by an areawide approach. Establishment of a Bay Area service agency, or utility, with powers to plan, finance and construct facilities is needed. Three feasible methods of forming such an agency are: (1) create a permanent agency by special enactment of the California State Legislature (preferred method); (2) create a permanent agency by the voluntary cooperation of all Bay Area wastewater districts; and (3) create a temporary agency by voluntary cooperation until a permanent agency can be set up. While studies show that the Bay is being badly polluted by many sources, including municipal and industrial, and that these pollutants range from nutrients, organics and coliform to toxic materials, these studies have also shown that there is a great potential for wastewater reclamation in the Bay area, especially municipal sewage. Along with maps and pictures showing areas on the Bay which are posted and have yearly fish kills is a map showing the potential reuse of reclaimed water for industrial use, aquifer recharge, landscape irrigation and spray irrigation. The potential market for reclaimed water is estimated at 200 mgd of the 800 mgd currently discharged into the Bay. (Poertner)
W73-08176

BRIDGES, WALLS, FILLS, CHANNEL CHANGES, ETC. (FOR THE INFORMATION OF THOSE INTERESTED IN THE CONSTRUCTION OF).
Pennsylvania Dept. of Environmental Resources, Harrisburg.
For primary bibliographic entry see Field 08A.
W73-08182

THE CLEAN STREAMS LAW OF PENNSYLVANIA.
Pennsylvania Dept. of Environmental Resources, Harrisburg.

1971. 19 P.

Descriptors: *Water quality act, *Water quality control, *Legislation, *Pennsylvania, Water quality standards, Water pollution control, Municipal wastes, Industrial wastes, Legal aspects, Permits, Water pollution sources.

The Clean Streams Law, which became effective January 19, 1971 in Pennsylvania, established the Department of Environmental Resources, the Environmental Quality Board, and a citizen advisory council. Concurrent with this was the abolishment of the Sanitary Water Board and the Department of Mines and Mineral Industries, with the transfer of their duties to the newly created Department of Environmental Resources. Along with this change also came the rewording of laws to correspond to these governmental structural changes and the new law is the subject of this publication. The general rules concerning water pollution are described in ten sections. These ten articles include: (1) general provisions and purpose, (2) sewage pollution, (3) industrial wastes, (4) other pollutions and potential pollutions, (5) domestic water supplies, (6) procedures and enforcement, and (7) existing rights under the old law. Some parts of the old laws were repealed, while additions were made to the new law, including the addition of penalties and other remedies for controlling pollution. (Poertner)
W73-08183

INVENTORY OF INTERSTATE CARRIER WATER SUPPLY SYSTEMS.
Environmental Protection Agency, Washington, D.C. Water Supply Div.
For primary bibliographic entry see Field 05F.
W73-08192

GREAT DISMAL SWAMP AND DISMAL SWAMP CANAL.

Hearing--Subcomm. on Parks and Recreation--Comm. on Interior and Insular Affairs, United States Senate, 92d Cong, 2d Sess, May 9, 1972. 25 p, 1 map.

Descriptors: *Swamps, *Wetlands, *Legislation, *Marsh management, Drainage effects, Legal aspects, Land reclamation, Marshes, Irrigation effects, Recreation, *Virginia, *North Carolina, Boating, Urbanization.
Identifiers: *Congressional hearings.

This hearing took testimony on a bill to authorize a study to determine the most feasible means of protecting and preserving the Great Dismal Swamp and the Dismal Swamp Canal. Encroaching metropolitan areas have threatened and begun to affect the Great Dismal; because it is an area of great national significance the states of Virginia and North Carolina have made efforts to determine the best ways to preserve it. These states decided to apply to all state and federal agencies to obtain aid. This hearing speaks favorably of preserving the Great Dismal, as it is unique in its physical properties, both flora and fauna. Included is a Department of the Interior study and statements by numerous interested parties, including the chairman of the Dismal Swamp Committee, Virginia Division of the Isaac Walton League of America, an Administrative assistant of the Wildlife, National Parks and Conservation Association and senators from the State of Virginia. (Smith-Adam-Florida)
W73-08193

PRESERVATION AND ENHANCEMENT OF THE AMERICAN FALLS AT NIAGARA.
For primary bibliographic entry see Field 06G.
W73-08194

THE IMPACT OF THE NATIONAL ENVIRONMENTAL POLICY ACT UPON ADMINISTRATION OF THE FEDERAL POWER ACT.
For primary bibliographic entry see Field 06G.
W73-08195

HOW AN ENFORCER BECOMES THE POLLUTER'S DEFENDER,
For primary bibliographic entry see Field 05G.
W73-08196

A SURVEY OF STATE REGULATION OF DREDGE AND FILL OPERATIONS IN NON-NAVIGABLE WATERS,
Florida Univ., Gainesville. School of Law.
For primary bibliographic entry see Field 04A.
W73-08197

THE WATER RESOURCES COUNCIL,
National Water Commission, Arlington, Va.
E. Liebman.
Available from the National Technical Information Service as PB-211 443, $6.75 in paper copy, $1.45 in microfiche. May 1972. 224 p, 357 ref.

Descriptors: *Water resources development, *U.S. Water Resources Council, *National Water Commission, *Water Resources Planning Act, *Planning, Water resources, Flood control, Land resources, Interstate compacts, Water pollution, Administration, Legislation, Project planning, Governmental interrelations, River basin commissions, Grants, Government finance, Administrative agencies, Flood plain zoning, Cost sharing.

This report on the Water Resources Council was prepared for the National Water Commission to assist in its deliberations on national water problems. A number of recommendations are made for improving the Council including the appointment of an independent full-time chairman, the addition of an independent board of review, and the provision in future legislation for water resources planning coordination at the state and federal levels. The Water Resources Planning Act and its historical background, the Water Resources Council and its activities, the principles and standards for planning water and related land resources, flood hazards, compacts, water pollution planning, and an evaluation of the assumptions underlying the Water Resources Planning Act are discussed. Portions of vital legislation concerning the Council and its activities are included. (Mockler-Florida)
W73-08198

OCEANOGRAPHIC COMMISSION.
Washington Natural Resources and Recreation Agencies, Olympia.

In: 1972 Annual Report State of Washington, p 47-49, 1972.

Descriptors: *Washington, *Oceanography, *Research and development, *Administrative agencies, Administration, Water law, Water rights, Education, Oil pollution, Water policy, Oceans, Government finance, Oil spills, Data collection, Governmental interrelations.
Identifiers: Hazardous substances (Pollution).

The 1972 annual report of the Oceanographic Commission of the State of Washington is presented. This Commission, created in 1967, is the only state agency specifically mandated to promote, develop, and advise on oceanography. It has neither regulatory powers to enforce state laws nor landlord responsibilities for managing state property. Fiscal year 1972 provided the Commission with a solid economic footing to enable it to meet many of its goals. Programs undertaken by the Commission included: an oil transportation study, a study on hazardous wastes in the area's waters, and oil handling studies. In addition the Commission operated a news bureau, held press conferences, participated in various workshops, and provided speakers for civic groups, schools, and professional societies during the year. In the future the Commission will expand and become involved in such areas as advancement of applied marine research and knowledge, standardization of data, cooperative coordination of state oceanographic and marine programs, and improvement of school and university curricula relating to oceanography. (Mockler-Florida)
W73-08199

DEPARTMENT OF ECOLOGY.
Washington Natural Resources and Recreation Agencies, Olympia.
For primary bibliographic entry see Field 06G.
W73-08200

ties acting alone. Present activities of local governments in coastal zone management in Escambia and Santa Rosa Counties, Florida were examined. Local laws and judicial decisions relevant to county governments, municipal governments, a port authority and the Santa Rosa Island Authority, are examined in relationship to coastal zone management. State laws and federal laws are also discussed in this context. There appear to be a number of opportunities for cooperation between state, county and local bodies which should be explored and developed in order to make Florida coastal zone management fully effective. (Beardsley-Florida)
W73-08204

IMMINENT IRREPARABLE INJURY: A NEED FOR REFORM,
For primary bibliographic entry see Field 05G.
W73-08205

ARTIFICIAL ADDITIONS TO RIPARIAN LAND: EXTENDING THE DOCTRINE OF ACCRETION,
R. E. Lundquist.
Arizona Law Review, Vol 14, No 2, p 315-343, 1972. 134 ref.

Descriptors: *Arizona, *Accretion (Legal aspects), *Riparian rights, *Ownership of beds, *Riparian land, Land tenure, Common law, Easements, Judicial decisions, Riparian waters, Usufructuray right, Submerged Lands Act, Bank erosion, Boundaries (Property), Land reclamation, Navigable rivers, Federal government, Dredging, High water mark, Avulsion, Watercourses (Legal aspects).
Identifiers: Reliction.

The Supreme Court of Arizona in State v. Bonelli Cattle Company, a case of first impression, declined to extend the doctrine of accretion to man-made additions to riparian land. The Bonelli decision is analyzed in terms of its implications for Arizona riparian landowners. In Bonelli the court distinguished between natural and man-made accretions and held that title to lands exposed by dredging projects of the federal government vests in the state. Four arguments are presented in favor of including in the doctrine of accretion man-made additions to riparian land, contrary to the Bonelli holding. This decision fails to consider the certain loss of riparian status due to man-made accretion. Such an omission forbodes the demise of the riparian right of access to navigable waters in Arizona. It is proposed that the riparian owner should be given the right to purchase these accretions with the proceeds going to establish a fund to promote the public use of navigable waters. (Adams-Florida)
W73-08206

ENVIRONMENTAL LAW–PRIV.TE CAUSE OF ACTION UNDER THE RIVERS AND HARBORS APPROPRIATION ACT OF 1899 FOR INJURY TO THE ECOLOGY OF NAVIGABLE WATERS,
For primary bibliographic entry see Field 05G.
W73-08207

THE NATIONAL ENVIRONMENTAL POLICY ACT OF 1969 SAVED FROM 'CRABBED INTERPRETATION',
H. A. Cubell.
Boston University Law Review, Vol 52, No 2, p 425-442, Spring 1972. 104 ref.

Descriptors: *Judicial decisions, *Federal project policy, *Legislation, *Decision making, *Administrative decisions, Legal review, Project planning, Cost-benefit analysis, Federal government, Administrative agencies, Ecology, Environmental effects, Legal aspects, Project feasibility, Administration, Alternate planning, Environment, Economic justification, Comprehensive planning, Water law.
Identifiers: *National Environmental Policy Act.

The judicial interpretation of the National Environmental Policy Act (NEPA) of 1969 was, in Calvert Cliffs' Coordinating Committee v. Atomic Energy Commission, recently expanded. Section 101 of NEPA is a broad policy statement insuring a federal commitment to preserve environmental quality. Section 102 of the Act contains procedural duties which must be met by any agency considering major federal activity which might have an adverse effect on the environment. The court in the Calvert Cliffs' case, after adopting a substantive-procedural dichotomy between Section 101 and 102, held that agencies, in determining the overall desirability of a project, must strike a balance between economic, technical, and environmental costs and benefits. The court acknowledged that the plain language of the Act does not require a balancing analysis; Section 102 only requires that environmental values be given appropriate consideration. However, the court recognized that environmental protection is to be considered as a competing priority to be assessed in view of other essential considerations, which implies a need for a balancing of these values on a case-by-case basis. (Adams-Florida)
W73-08208

ILLINOIS WATERWAY.

Ill. Ann. Stat. Ch. 19 secs. 79 thru 112 (Smith-Hurd 1972).

Descriptors: *Illinois, *Legislation, *Navigable waters, *Navigation, *Canals, Channels, Transportation, Ships, Civil engineering, Inland Waterways, Legal Aspects, Jurisdiction, Eminent domain, Water law.
Identifiers: *Illinois Waterway.

The statutes govern the location, name, general route, dimensions of the channel and locks, plans and specifications for construction, maintenance and operation, lease of surplus waters, rates of toll, rules and regulations, maintenance of power plants and storage facilities, acquisition of property, repair, replacement or reconstruction of public bridges, and sale or lease of lands pertaining to the Waterway. The State Department of Transportation is charged with the overall responsibility of maintaining and improving the Waterway. Other areas covered by statute include permits, contracts, eminent domain procedures, construction procedures, drainage systems, sanitary regulations, and bridges. (Mockler-Florida)
W73-08225

NATURAL RESOURCES DEVELOPMENT BOARD.

Ill. Ann. Stat., ch. 19 secs. 1071 thru 1077.13 (Smith-Hird 1972).

Descriptors: *Illinois, *Legislation, *Water supply, *Water policy, *Water management (Applied), Conservation, Water resources, Water resources development, Water law, Water demand, Water supply development, Flood control, Flood plains, Watershed management, Navigation, Public health, Future planning (Projected), Legal aspects, Governmental interrelations, Administrative agencies, Project planning.

In order to provide sound planning for the proper conservation, development and use of water resources, upon which the health, welfare and economic progress of Illinois depend, the Natural Resources Development Board is created. The Board shall receive staff services from the Department of Business and Economic Development, but when necessary shall receive the assistance of any state educational institution or experiment station.

The Board shall prepare a biennial assessment of the adequacy of the water supplies of the state, and shall recommend to the Governor and State Assembly appropriate policies, legislation and programs necessary to insure the availability of adequate supplies of water. It shall effect maximum coordination between all state agencies in planning and developing water resources, as well as review all proposed legislation, water resources projects and developments, and any state water programs and activities. The Board shall also investigate any project involving navigation improvement, flood control or watersheds of the state. (Glickman-Florida)
W73-08226

UPPER MISSISSIPPI RIVERWAY COMPACT.

Ill. Ann. Stat., ch. 19 secs 1101 thru 1103 (Smith-Hurd 1972).

Descriptors: *Interstate compacts, *River basin development, *Illinois, *Legislation, Resources development, Water management (Applied), Interstate commissions, Mississippi River Basin, Iowa, Minnesota, Wisconsin, Missouri, Natural resources, Conservation, Resource allocation, Research and development, Wildlife conservation, Recreation facilities, Agriculture, Administration, Permits, Planning.
Identifiers: *Upper Mississippi Riverway Compact.

The size of the Upper Mississippi region, the complexity of its economic and social development and the resource needs of its people require a formal instrument for joint and cooperative action in the development and maintenance of a sound and attractive Upper Mississippi region. The purposes of the compact are stated and the geographical area comprising the Upper Mississippi Riverway District is specified within these statutes. A Commission composed of representatives from each party state is established. The structure, powers, duties and responsibilities of the Commission are detailed in the statute. A study of means of preserving and developing the scenic value of both public and private property is to be made by the Commission. The Commission may acquire such easements and make such agreements as may be suitable to preserving or securing patterns or features of land and water use that will be consistent with the terms of the compact. The statute provides that the compact shall enter into force when enacted into law by three of the participating states. (Reed-Florida)
W73-08227

SULLIVAN V. MORENO (WHETHER A SIMPLE RIPARIAN BOUNDARY OPERATES BY LEGAL PRESUMPTION TO CARRY THE RIGHT TO THE SOIL TO THE EDGE OF THE CHANNEL).

19 Fla. 200-231 (1882).

Descriptors: *Ownership of beds, *Florida, *Riparian land, *Riparian rights, Water law, Water rights, Legal aspects, Legislation, Remedies, Water sources, Judicial decisions, Beds under water, Boundary disputes, Adjacent landowners.
Identifiers: Injunctive relief.

This case, decided by the state's supreme court in 1882, involved an action brought by plaintiff landowner against defendant adjacent landowner to determine whether a simple riparian boundary operates by legal presumption to carry the right to the soil at the edge of the channel under the Riparian Act of 1856. Plaintiff contended the possession and enjoyment of his riparian rights had been violated by the defendant's construction of a wharf thereby denying plaintiff access to his wood and coal yard. The court noted that under state law the title to the submerged soil from the channel to the shore is in such riparian proprietor as is contemplated by the act of 1856 and held that this act contemplates riparian proprietorship as requiring a water boundary. An allegation by plaintiff that the party through whom he claims title owned and possessed parcels of land lying on the bay, and that such party had been for thirty years in quiet possession and enjoyment of all riparian rights, does not constitute a sufficient allegation of a riparian proprietorship. As a result plaintiff's injunction was reversed and the case remanded for further proceedings. (Mockler-Florida)
W73-08228

ADAMEK V. CITY OF PORT RICHEY (APPLICATION FOR PERMIT TO CONSTRUCT A DOCK).

214 So. 2d 374-375 (2d D.C.A. Fla. 1968).

Descriptors: *Florida, *Judicial decisions, *Permits, *Docks, Legal aspects, Regulation, State governments, Engineering structures, Riparian rights, Riparian lands, Legislation, Local governments.
Identifiers: Corps of Engineers.

Plaintiff riparian landowner sued defendant city for a judgment compelling the issuance of a permanent permit for construction of a dock. The Circuit Court of Pasco County denied summary judgment to the property owner and an interlocutory appeal was taken. The Florida District Court of Appeals held that denial of summary judgment was proper where the ordinance required the owner to obtain the approval of the Army Corps of Engineers as a condition precedent to the issuance of a permanent permit and that the property owner failed to allege that approval of the Corps of Engineers had been obtained. (Mockler-Florida)
W73-08229

CARTISH V. SOPER (RESERVATION OF RIPARIAN RIGHTS FOR SUBDIVISION LOT OWNERS).

157 So. 2d 150-154 (2d D.C.A. Fla. 1963).

Descriptors: *Florida, *Legal aspects, *Riparian land, *Riparian rights, Water law, Water rights, Remedies, Judicial decisions, Access routes, Easements, Right of way.
Identifiers: Right of Ingress and egress, Water rights (Non-riparians).

Plaintiffs subdivision lot owners sued to obtain a declaration of their riparian rights and the rights of defendants, owners of lots abutting a private parkway, to the parkway and to enjoin the defendants from obstructing the plaintiffs' free use of all parts of the parkway and rebuilding of a dock. The subdivision plat reserved the parkway to allow ingress and egress to an adjacent bay on behalf of all subdivision lot owners. Defendants contend this reservation did not encompass riparian rights. The Florida District Court of Appeals held that riparian rights necessary and incidental to access to the bay were implicit in the plat reservation. Further the court held that the right to build a dock to facilitate the access of the easement holders to the waters was implied. The court thereby affirmed the lower court decision for plaintiffs. (Mockler-Florida)
W73-08230

BLOMQUIST V. COUNTY OF ORANGE (LOCAL RESERVOIR CONSTRUCTION SUBJECT TO STATE APPROVAL).

69 Misc. 2d 1077-1081 (Sup. Ct. N.Y. 1972).

Descriptors: *Municipal water, *Judicial decisions, *Eminent domain, *New York, *Reservoir construction, Reservoirs, Condemnation, Reservoir storage, Reservoir sites, Legislation, Planning, Local governments, Water supply development, State governments, Constitutional law, Legal aspects, Water supply.
Identifiers: *Injunction (Prohibitory).

Plaintiff landowners sued to enjoin the defendant county from acquiring land for a reservoir. Plaintiffs contended inter alia that the county had not submitted its maps and plans to the State Water Resources Commission for approval as required by the Conservation Law. Defendant contended that it was not a public corporation within the meaning of that legislative provision. This law gives the State Commission broad regulatory powers over the potable water supply of the entire state. The New York Supreme Court held that a county is a public corporation within the conservation law and defendant must obtain the state commission's approval prior to condemning land for a reservoir. Since defendant has not complied with the requirements of the Conservation law, it was without authority to acquire land for the reservoir. However, the court denied plaintiff's motion for summary judgment on other grounds. (Smith-Adam-Florida)
W73-08231

STATE V. BLACK RIVER PHOSPHATE CO. (TITLE ASSERTION TO SUBMERGED LANDS ON EDGE OF CHANNEL OF NAVIGABLE STREAM).

13 So. 640-658 (Fla. 1893).

Descriptors: *Ownership of beds, *Florida, *Riparian land, *Riparian rights, Legal aspects, Beds under water, Judicial decisions, Public rights, High water mark, Navigable rivers.
Identifiers: Public trust doctrine.

The plaintiff state sued to enjoin the defendant phosphate company from taking phosphate from the bed of a navigable river. Defendant contended that it owned the bed under the Riparian Act of 1856. The court noted that so long as the land remains submerged the governmental control over it is not lost. The Court held for the state, stating that the Riparian Act of 1856 does not vest in riparian owners an unqualified fee in the lands below the high-water mark, and out to the edge of the channel in navigable streams, bays of the sea or harbors of the state. The court based its decision on the rationale that so long as such submerged lands remained unimproved by the construction of wharves, or unreclaimed by filling in from the shore and converting the water into land the riparian owner, though the legal title is in him has, insofar as the statute is concerned, no greater right to the beneficial use of such submerged land and the waters above them than any other citizen except for the purpose of protecting from invasion the right to improve which the statute gives him. Moreover, the statute was held not to give to the riparian owner the right to take phosphates from the beds of navigable streams, bays of the sea, or harbors, below high-water mark, and out to the edge of the channel, for the purposes of sale. (Mockler-Florida)
W73-08232

LITTLE CALUMET RIVER: LAKE CALUMET HARBOR.

Ill. Ann. Stat. ch. 19 secs 112.1 thru 117 (Smith-Hurd 1972).

Descriptors: *Illinois, *Pollution abatement, *Legislation, *Water pollution control, Administrative agencies, Management, Water law, Water policy, Water pollution, Water quality, Public health, Harbors, Industries, Water pollution treatment, Legal aspects, Excavation.

The interests of the people of Illinois require that the Little Calumet River be cleaned up. The Commission on Operation Little Calumet River i

The objectives were to determine why state statutes and local zoning ordinances are not effectively used in floodplain management, to determine alternative preventive methods available and to analyze the alternatives to determine their suitability for management purposes. The objectives were only partly achieved within the time frame of the study. A repository of enabling legislation, information and inventory reports and ordinances relating to floodplains has been established for Illinois, Indiana, Iowa, Missouri and Ohio. This can be used for monitoring and evaluating changes in regulations and management procedures. An analysis of the floodplain regulations in these five states is presented. It is concluded that local communities are not self-motivated to adopt regulations and require external encouragement. Several areas suitable for regional case studies are identified. A basis for evaluation criteria for a floodplain management program is presented. It is suggested that useful results can be gained through examination of the effectiveness of regional agencies on floodplain management.
W73-07814

NATIONAL PROGRAM FOR MANAGING FLOOD LOSSES, GUIDELINES FOR PREPARATION, TRANSMITTAL, AND DISTRIBUTION OF FLOOD-PRONE AREA MAPS AND PAMPHLETS,
Geological Survey, Washington, D.C.
For primary bibliographic entry see Field 07C.
W73-07849

6G. Ecologic Impact of Water Development

WALKER BRANCH WATERSHED: A STUDY OF TERRESTRIAL AND AQUATIC SYSTEM INTERACTION,
Oak Ridge National Lab., Tenn.
For primary bibliographic entry see Field 04D.
W73-07947

RECTIFICATION OF DEFICIENCIES IN COMPLETED LOCAL PROTECTION PROJECT, WELLSVILLE, NEW YORK (FINAL ENVIRONMENTAL IMPACT STATEMENT).
Army Engineer District, Buffalo, N.Y.
For primary bibliographic entry see Field 08A.
W73-07975

BRANTLEY PROJECT, NEW MEXICO (FINAL ENVIRONMENTAL IMPACT STATEMENT).
Bureau of Reclamation, Denver, Colo.
For primary bibliographic entry see Field 08D.
W73-07976

DESALINIZATION PLANTS, VIRGIN ISLANDS (FINAL ENVIRONMENTAL IMPACT STATEMENT).
Department of Housing and Urban Development, San Juan, Puerto Rico. Region II.
For primary bibliographic entry see Field 03A.
W73-07977

AGRI-INDUSTRIAL PARK WATER IMPROVEMENTS, SEWAGE TREATMENT FACILITIES, BUSINESS DEVELOPMENT LOAN (FINAL ENVIRONMENTAL IMPACT STATEMENT).
Economic Development Administration, Austin, Tex. Southwestern Region.
For primary bibliographic entry see Field 05D.
W73-07978

COPAN LAKE, LITTLE CANEY RIVER, OKLAHOMA (FINAL ENVIRONMENTAL IMPACT STATEMENT).
Army Engineer District, Tulsa, Okla.
For primary bibliographic entry see Field 08A.
W73-07979

LOST CREEK LAKE PROJECT, ROGUE RIVER, OREGON (SUPPLEMENT TO FINAL ENVIRONMENTAL IMPACT STATEMENT).
Army Engineer District, Portland, Oreg.
For primary bibliographic entry see Field 08A.
W73-07980

SNAGGING AND CLEARING PROJECT ON MILL CREEK AT RIPLEY, WEST VIRGINIA (FINAL ENVIRONMENTAL IMPACT STATEMENT).
Army Engineer District, Huntington, W. Va.
For primary bibliographic entry see Field 04A.
W73-07981

ADMINISTRATION OF THE NATIONAL ENVIRONMENTAL POLICY ACT, PART I.
For primary bibliographic entry see Field 06E.
W73-07983

ENVIRONMENTAL IMPACT STATEMENTS—A DUTY OF INDEPENDENT INVESTIGATION BY FEDERAL AGENCIES.
For primary bibliographic entry see Field 06E.
W73-07993

THE ENERGY NEEDS OF THE NATION AND THE COST IN TERMS OF POLLUTION,
Atomic Energy Commission, Washington, D.C.
J. T. Ramey.
Atomic Energy Law Journal, Vol 14, No 1, p 26-58, Spring 1972. 10 fig, 16 ref.

Descriptors: *Water pollution sources, *Energy budget, *Energy dissipation, *Nuclear energy, Oil, Coal, Natural gas, Electricity, Electric power production, Thermal powerplants, Nuclear powerplants, Hydroelectric plants, Alternative planning, Future planning (Projected), Estimated benefits, Estimated cost, Research and development, Environmental effects, Thermal pollution.

In a lecture, the Commissioner of the Atomic Energy Commission described the energy crises now facing the United States and suggested actions to be taken which could resolve the crisis while minimizing further environmental damage caused by increasing energy demands. In 1970, approximately ninety-six per cent of the country's energy needs were being met by a rapidly diminishing supply of coal, oil, and natural gas. Between 1965 and 1970, energy consumption increased at an average rate of five per cent per year. The most important growth has occurred in the use of energy for the production of electricity. Since the production of oil and natural gas is already reaching a critical stage in this country, alternatives must be developed. One alternative to be considered is the expanded use of nuclear reactors for gneration of electricity. In the distant future fusion power, using hydrogen in the oceans as an energy source, may offer an answer to the energy problem. Use of nuclear power will also result in a significant reduction in combustion pollution which at present causes about eight-five per cent of all U.S.air pollution. (Adams-Florida)
W73-07995

THE YEAR OF SPOILED PORK: COMMENTS ON THE COURT'S EMERGENCE AS AN ENVIRONMENTAL DEFENDER,
Florida Univ., Gainsville.
For primary bibliographic entry see Field 05G.
W73-07996

ENVIRONMENTAL LAW: STRICT COMPLIANCE WITH PROCEDURAL REQUIREMENTS OF NEPA—THE AGENCIES MUST PLAY BY THE RULES,
For primary bibliographic entry see Field 06E.

W73-07999

PRESERVATION AND ENHANCEMENT OF THE AMERICAN FALLS AT NIAGARA.

Interim Report to the International Joint Commission, December 1971. 77 p, 21 plate.

Descriptors: *New York, *Environmental effects, *Aesthetics, *Scenery, *International Joint Commission, Rivers and Harbors Act, Legislation, Water quality control, Creativity, Recreation, Erosion control, Flow control, Urbanization, Water levels.
Identifiers: Niagara (New York), American Falls.

Results are summarized of investigation and research authorized under the 1965 Rivers and Harbors Act. The purpose of the study is to consider the nature and extent of measures necessary to preserve and enhance the scenic beauty of American Falls at Niagara, New York. The historical background of the problems is presented the aesthetic factors and physical conditions which must be considered in reaching a solution are discussed. The range of aesthetic and physical options for preserving or enhancing the beauty of the American Falls and for securing the safety of the viewing public are explored. These options are grouped into a few broad alternative courses of action, followed by the Board's conclusions and recommendations. (Mockler-Florida)
W73-08194

THE IMPACT OF THE NATIONAL ENVIRONMENTAL POLICY ACT UPON ADMINISTRATION OF THE FEDERAL POWER ACT.

Land and Water Law Review, Vol 8, No 1, p 93-124, 1973. 32 p, 118 ref.

Descriptors: *Administrative decisions, *Federal Power Act, *Decision making, *Hydroelectric project licensing, Administrative agencies, Federal project policy, Interstate rivers, Legal aspects, Legislation, Hydroelectric plants, Judicial decisions, Project planning, Environmental effects.
Identifiers: *National Environmental Policy Act, Environmental impact statement, Administrative regulations, Licenses.

This article discusses the Federal Power Commission's efforts at compliance with the National Environmental Policy Act of 1969 (NEPA), the problems encountered to date, and some of the conflicts ot be anticipated. The history and the provisions of the Federal Power Act and NEPA are outlined along with the Commission's regulations formulated to comply with NEPA in light of guidelines established by the Council on Environmental Quality. The impact of recent federal court decisions indicates that the Commission's regulations may violate the requirements of NEPA, especially in regard to the granting or renewing of a license for non-federal hydroelectric projects which affect the navigable waterways or lands of the United States or which develop power for transmission to interstate commerce without filing an environmental impact statement prior to a bearing upon application for a license. It is also possible that the court decisions may require environmental statements by the Commission prior to its issuing orders involving wholesale utility rates, interconnections, and the issuance of certain securities by utilities under the regulation of the Federal Power Commission. (Dunham-Florida)
W73-08195

A SURVEY OF STATE REGULATION OF DREDGE AND FILL OPERATIONS IN NON-NAVIGABLE WATERS,
Florida Univ., Gainesville. School of Law.
For primary bibliographic entry see Field 04A.
W73-08197

THE WATER RESOURCES COUNCIL,
National Water Commission, Arlington, Va.
For primary bibliographic entry see Field 06E.
W73-08198

DEPARTMENT OF ECOLOGY.
Washington Natural Resources and Recreation Agencies, Olympia.

In: 1972 Annual Report, State of Washington, p 11-14, 1972.

Descriptors: *Washington, *Environmental effects, *Administrative agencies, *Water management (Applied), Administration, Legal aspects, Recreation, Water quality, Land use, Coasts, Legislation, Water resources development, Government finance, Law enforcement, Oil spills, Planning, Water rights, Flood control, Water quality control, Water pollution control, Water quality standards.
Identifiers: Coastal zone management.

In 1970, three separate state agencies were consolidated into the Washington Department of Ecology to create an environmental quality maintenance department. Information is presented on the Deaprtment's new programs, shoreline management, the state environmental policy act, oil spills, water resources, flood control, water quality, air quality, solid waste disposal management plans, and environmental economics. One of the primary issues is the compatibility of environmental concerns and economic needs; near the end of the fiscal year the Department began work on a major undertaking which could, in the future, resolve this environmental-economic issue. The project is an attempt to develop a line of communication which could lead to the issuance of total environmental authorizations. These authorizations would embody the latest technology to meet all environmental requirements while also meeting industry's economic goals. The primary goal of the Department is to balance environmental concerns with public needs. (Mockler-Florida)
W73-08200

REPORT OF THE UNITED NATIONS CONFERENCE ON HUMAN ENVIRONMENT, HELD AT STOCKHOLM, 5-16 JUNE, 1972.

Available from the National Technical Information Service as PB-211 133, $3.00 in paper copy, $1.45 in microfiche. United Nations General Assembly Report A/CONF.48/14. July 3, 1972. 144 p, 5 append.

Descriptors: *United Nations, *International law, *International commissions, *Environmental control, Public health, Environmental engineering, Water resources, Water conservation, Water resources development, Political aspects, Area redevelopment, Urban renewal, Governments, Foreign countries, Environment, International waters, Legislation.

The Conference adopted this report which contains a declaration on the human environment, an action plan for the human environment and an environmental fund. The fund would be in addition to monies which governments made available to development programs. Certain priorities that require urgent and large-scale action include: water supply sources and ocean and sea pollution. Other areas designated top priority action are understanding and controlling man-produced changes in the ecological systems, acceleration of environmentally sound technology, and encouraging broad international distribution of industrial capacity. Emphasis was also placed on new codes of international law and means for better management of the world's property resources. Included are the following resolutions: World Environment Day, nuclear weapons tests, and the next United Nations Conference on the Human Environment. (Beardsley-Florida)
W73-08203

THE NATIONAL ENVIRONMENTAL POLIC ACT OF 1969 SAVED FROM 'CRABBED I TERPRETATION',
For primary bibliographic entry see Field 06E.
W73-08208

SMALL BOAT HARBOR, KING COV ALASKA (FINAL ENVIRONMENTAL IMPAC STATEMENT).
Army Engineer District, Anchorage, Alaska.
For primary bibliographic entry see Field 08D.
W73-08209

SMITHVILLE LAKE, LITTLE PLATTE RIVE MISSOURI (FINAL ENVIRONMENTAL I PACT STATEMENT).
Army Engineer District. Kansas City, Mo.
For primary bibliographic entry see Field 08A.
W73-08210

BACON CREEK WATERSHED, PLYMOUT AND WOODBURY COUNTIES, IOWA, (FINA ENVIRONMENTAL IMPACT STATEMENT).
Soil Conservation Service, Washington, D.C.
For primary bibliographic entry see Field 08A.
W73-08211

WALKER DAM IMPOUNDMENT, AQUATI PLANT CONTROL PROJECT, NEW KEN COUNTY, VIRGINIA (FINAL ENVIRONME TAL IMPACT STATEMENT).
Army Engineer District, Norfolk, Va.
For primary bibliographic entry see Field 05G.
W73-08212

COW CREEK WATERSHED, STEPHENS AN JEFFERSON COUNTIES, OKLAHOMA (FINA ENVIRONMENTAL IMPACT STATEMENT).
Soil Conservation Service, Washington, D.C.
For primary bibliographic entry see Field 08A.
W73-08214

CONSTRUCTION OF ARTIFICIAL REEFS I THE ATLANTIC OCEAN OFF CAPE HENRY VIRGINIA (FINAL ENVIRONMENTAL IMPAC STATEMENT).
National Marine Fisheries Service, Beaufort, N.C. Atlantic Coastal Fisheries Center.

Available from the National Technical Informa tion Service as EIS-VA-72-5392-F, $3.50, in pape copy, $1.45 in microfiche. October 2, 1972. 23 p, fig.

Descriptors: *Virginia, *Fish attractants, *En vironmental effects, *Reefs, *Sport fishing Recreation, Habitat improvement, Aquati habitats, Fish management, Fish population Fishing.
Identifiers: *Environmental Impact Statement *Artificial reefs, *Cape Henry (Va).

This action consists of the construction of two ar tificial reefs in the Atlantic Ocean fifteen and thi ty miles off Cape Henry, Virginia. The reefs wi be built with surplus navy vessels and will cove approximately 80 and 400 acres. The project wi create habitats attractive to sport fishes in an are where such habitats do not occur naturally. Un derwater observations will be made by the Na tional Marine Fisheries Service to determin changes in the biota on and surrounding the reefs No significant adverse environmental effects ar anticipated as a result of this project. The physica alteration of the ocean that will occur is expecte to create new habitats for many types and substan tial numbers of marine organisms including gam fishes. Other alternative means of increasing gam fish population have been considered but are no considered feasible in this particular area (Mockler-Florida)
W73-08215

WILLOW ISLAND LOCKS AND DAM OHIO RIVER, OHIO AND WEST VIRGINIA (FINAL ENVIRONMENTAL IMPACT STATEMENT).
Army Engineer District, Huntington, W. Va.
For primary bibliographic entry see Field 08A.
W73-08216

DETAILED PROJECT REPORT, INVESTIGATION FOR FLOOD PROTECTION, MUNDAY, TEXAS, BRAZOS RIVER BASIN, TEXAS (FINAL ENVIRONMENTAL IMPACT STATEMENT).
Army Engineer District, Fort Worth, Tex.
For primary bibliographic entry see Field 08A.
W73-08217

KAHULUI HARBOR WEST BREAKWATER REPAIR, MAUI, HAWAII (FINAL ENVIRONMENTAL IMPACT STATEMENT).
Army Corps of Engineers, Honolulu, Hawaii. Pacific Ocean Div.
For primary bibliographic entry see Field 08A.
W73-08218

GILA RIVER BASIN, NEW RIVER AND PHOENIX CITY STREAMS, ARIZONA, DREAMY DRAW DAM, MARICOPA COUNTY, ARIZONA (FINAL ENVIRONMENTAL IMPACT STATEMENT).
Army Engineer District, Los Angeles, Calif.
For primary bibliographic entry see Field 08A.
W73-08219

T OR C WILLIAMSBURG ARROYOS WATERSHED, SIERRA COUNTY, NEW MEXICO (FINAL ENVIRONMENTAL IMPACT STATEMENT).
Soil Conservation Service, Washington, D.C.
For primary bibliographic entry see Field 08A.
W73-08220

PEARL RIVER BASIN, EDINBURG DAM AND LAKE, MISSISSIPPI AND LOUISIANA (FINAL ENVIRONMENTAL IMPACT STATEMENT),
Army Engineer District, Mobile, Ala.
For primary bibliographic entry see Field 08A.
W73-08221

KANAWHA RIVER COMPREHENSIVE BASIN STUDY, NORTH CAROLINA, VIRGINIA, AND WEST VIRGINIA, (FINAL ENVIRONMENTAL IMPACT STATEMENT).
Ohio River Basin Commission, Cincinnati.
For primary bibliographic entry see Field 04A.
W73-08222

SAN LUIS UNIT, CENTRAL VALLEY PROJECT, CALIFORNIA (FINAL ENVIRONMENTAL IMPACT STATEMENT),
Bureau of Reclamation, Sacramento, Calif. Mid-Pacific Regional Office.
For primary bibliographic entry see Field 08A.
W73-08223

CACHE RIVER BASIN FEATURE, MISSISSIPPI RIVER AND TRIBUTARIES PROJECT, ARKANSAS (FINAL ENVIRONMENTAL IMPACT STATEMENT).
Army Engineer District, Memphis, Tenn.

Available from the National Technical Information Service as EIS-AR-72-5350-F, $5.75 in paper copy, $1.45 in microfiche. August 1972. 70 p, 3 plate, 1 map, 1 append.

Descriptors: *Arkansas, *Environmental effects, *Wildlife habitats, *Fish conservation, River basin development, Wildlife, Flood control, Water control, Recreation, Oak trees, Hickory trees, Waterfowl, Hunting, Flood routing, Eminent domain, Land use, Aesthetics, Fish management.
Identifiers: *Environmental Impact Statements, *Cache River Basin (Arkansas), Fish and Wildlife Coordination Act.

This action consists of recommendations to Congress on the modification of the authorized Cache River Basin Project, to acquire approximately 30,000 acres of lands and their development for mitigation of fish and wildlife losses pursuant to general authorities contained in the Fish and Wildlife Coordination Act of 1958. Certain proposed mitigation measures will preserve significant hunting and fishing opportunities; protect biological productivity, aesthetics and other environmental values; and will enhance outdoor recreation opportunities. The Cache River Basin encompasses portions of thirteen counties within the alluvial valley of the Mississippi River in Northeastern Arkansas and Southeastern Missouri. The project itself is for purposes of flood control. (Smith-Adam-Florida)
W73-08224

07. RESOURCES DATA

7A. Network Design

MANAGEMENT MODEL AS A TOOL FOR STUDYING THE WORTH OF DATA,
Geological Survey, Arlington, Va. Water Resources Div.
For primary bibliographic entry see Field 06A.
W73-07882

ERRORS OF THE THERMOMETRIC METHOD OF DEPTH DETERMINATION,
Akusticheskii Institut, Moscow (USSR).
R. D. Sabinin.
Oceanology, Vol 12, No 2, p 285-289, 1972. 1 tab, 2 ref. Translated from Okeanologiya (USSR), Vol 12, No 2, 1972.

Descriptors: *Instrumentation, *Thermometers, *Measurement, *Depth, Water temperature, Thermocline, Time lag, Sampling, Equations.
Identifiers: *USSR, Thermometry.

Errors in the thermometric method of depth determination attributable to differing time lags of protected and unprotected thermometers are discussed. Because of the greater time lag of unprotected thermometers, the thermometric method yields depth values that are too low as ambient water temperature increases and too high as temperature decreases. Formulas are derived for computing these errors, and an error estimate is made for reduced instrument depth during exposure of sampling bottles at various depths and for temperature changes caused by internal waves. In the thermocline of the Arabian Sea these errors may exceed 7.6 m, which precludes use of unprotected thermometers or necessitates equalization of the time lags of protected and unprotected thermometers. (Josefson-USGS)
W73-07911

LAKE SURVEY CENTER 1972 RESEARCH PROGRAM.
National Ocean Survey, Detroit, Mich. Lake Survey Center.
For primary bibliographic entry see Field 02H.
W73-08160

EXPERIMENTAL CURVES AND RATES OF CHANGE FROM PIECEWISE PARABOLIC FITS,
Kentucky Univ., Lexington. Dept. of Mathematics.
P. C. DuChateau, D. L. Nofziger, L. R. Ahuja, and D. Swartzendruber.
Preprint, Journal Paper No. 4442, Purdue University Agricultural Experiment Station, Lafayette, Indiana, 1971. 22 p, 3 fig, 3 tab, 18 equ, 5 ref. OWRR-B-014-IND (2).

Descriptors: Computer programs, Equations, Hydraulics.
Identifiers: *Mathematical analysis, *Data analysis, Sliding parabola, Parabolic splines, Prism method, Form-free curve fitting, Least squares, Nonlinear data, Slope evaluation.

The determination of experimental curves and rates of change for nonlinear data is approached and solved without assuming an artificially restrictive mathematical form for the complete range of the data. This relatively form-free result is achieved by least-squares computer fitting of parabolic segments to short subranges of the experimental data. Two ways of doing this, referred to as the sliding-parabola and parabolic-splines methods, are developed. These are tested on both smooth and scattered data generated basically from the function $y \pm x1/2$, without and with random error, respectively. For smooth data, the sliding-parabola method is slightly better than the parabolic splines, but in general both are subject to only very small errors, and they are also in good agreement with a previously presented graphical prism method. For scattered data wherein the inherent errors of function and slope evaluation are much increased, the parabolic-splines method is distinctly superior to the sliding parabola. Both methods require only relatively short computing times, on the order of 1 sec for 40 data points, and are of utility for determining nonconstant experimental rates of change and for least-squares curve fitting without specification of a complete-range mathematical curve. (Bell-Cornell)
W73-08300

A PROBABILISTIC MODEL FOR STRUCTURING A DRAINAGE NETWORK,
Army Project Mobile Army Sensor Systems Test Evaluation and Review Activity, Fort Hood, Tex.
For primary bibliographic entry see Field 04A.
W73-08380

7B. Data Acquisition

STUDY OF THE HYDROGEOLOGICAL CONDITIONS OF ALLUVIAL FANS BY MULTIPLE LANDSCAPE AND GEOPHYSICAL METHODS,
Moskovskoe Obshchestvo Ispytatelei Prirody (USSR). Geographic Div.
For primary bibliographic entry see Field 02F.
W73-07844

INDICATION OF ROCKS BY DRAINAGE PATTERNS,
Moskovskoe Obshchestvo Ispytatelei Prirody (USSR). Geographic Div.
For primary bibliographic entry see Field 02J.
W73-07846

THE USE OF DRAINAGE PATTERNS FOR INTERPRETING THE MORPHOSTRUCTURES OF THE MOSCOW DISTRICT,
Moskovskoe Obshchestvo Ispytatelei Prirody (USSR). Geographic Div.
For primary bibliographic entry see Field 02J.
W73-07847

EXTRAPOLATION OF INDICATOR SCHEMES WITHIN SALT MARSHES,
Moskovskoe Obshchestvo Ispytatelei Prirody (USSR). Geographic Div.
For primary bibliographic entry see Field 02H.
W73-07848

AN INEXPENSIVE, RECORDING TIDE GAUGE,
Brookhaven National Lab., Upton, N.Y. Dept. of Biology.
N. R. Tempel.
Limnology and Oceanography, Vol 18, No 1, p 178-180, January 1973. 3 fig, 1 ref.

Descriptors: *Tides, *Instrumentation, *Gaging, *Gages, *Water level fluctuations, Data collections, Telemetry, Water levels.
Identifiers: *Tide gages.

An inexpensive, recording tide gage was constructed and tested in a small tidal pond on the north shore of Long Island. The instrument costs less than $50 if a 0-1 mA recorder is already available. (Knapp-USGS)
W73-07852

CHANGES IN CLAY MINERAL ASSEMBLAGES BY SAMPLER TYPE,
George Washington Univ., Washington, D.C. Dept. of Geology.
For primary bibliographic entry see Field 02J.
W73-07868

A TECHNIQUE USING POROUS CUPS FOR WATER SAMPLING AT ANY DEPTH IN THE UNSATURATED ZONE,
Geological Survey, Lubbock, Tex.
For primary bibliographic entry see Field 02G.
W73-07901

ERRORS OF THE THERMOMETRIC METHOD OF DEPTH DETERMINATION,
Akusticheskii Institut, Moscow (USSR).
For primary bibliographic entry see Field 07A.
W73-07911

A NEW SIMPLE WATER FLOW SYSTEM FOR ACCURATE CONTINOUS FLOW TESTS,
Kristinebergs Zoologiska Station, Fiskebackskil (Sweden).
A. Granmo, and S. O. Kollberg.
Water Research, Vol 6, No 12, p 1597-1599, December, 1972. 1 fig, 1 ref.

Descriptors: *Bioassay, *Laboratory equipment, *Design, *Construction, Flow, Research equipment, Flow rates.
Identifiers: *Continuous flow system, Accuracy.

A new simple water flow system is described for use in tests on living organisms in which a continuous controlled flow of water must be utilized. The entire system is made of Perspex plastic. A constant water level is maintained via an overflow and the outlets consist of nozzles screwed onto threaded tubes. By screwing a tube up or down the distance changes and also the velocity of discharge. By this method it is possible to make an adjustment of up to approximately 10 per cent in the rate of flow. Other rates of flow can be obtained simply by replacing the nozzles by other with a cross-section area corresponding to the desired rate. The system can be used for rate of flow ranging from milliters to liters, and has proved to be accurate and easy to handle. A diagram and description of construction of the water flow system are provided. (Holoman-Battelle)
W73-08022

SUPPORT-BONDED POLYAROMATIC COPOLYMER STATIONARY PHASES FOR USE IN GAS CHROMATOGRAPHY,
Applied Automation, Inc., Bartlesville, Okla. Systems Research Dept.
E. N. Fuller.
Analytical Chemistry, Vol 44, No 11, p 1747-1753, September 1972. 8 fig, 3 tab, 17 ref.

Descriptors: *Methodology, *Gas chromatography, Separation techniques, Chemical analysis, Porous media, Efficiencies, Organic compounds.
Identifiers: *Colum preparation, *Polyaromatic copolymers, *Chromatography colums, Aliphatic hydrocarbons, Aromatic hydrocarbons, Styrene, Divinylbenzene, Ethylvinylbenzene, Chromatography peaks, Methane, Ethane, Propane, Isobutane, n-Butane, 2-Methylbutane, n-Pentane, Sample preparation.

The preparation of porous polyaromatic copolymers of divinylbenzene, ethylvinylbenzene, and styrene physically bonded to a solid support is described together with initial results illustrating the utility of these materials as GC column packings. While similar in nature to the widely used porous polymer beads, the support-bonded phases provide more rapid separations and greater column efficiency. Experiments showing the effects of cross-linking and of initial dilution with inert solvent on the resulting copolymer product are also discussed. (Long-Battelle)
W73-08034

ELECTROCHEMICAL CELL AS A GAS CHROMATOGRAPH-MASS SPECTROMETER INTERFACE,
Northgate Lab., Hamden, Conn.
For primary bibliographic entry see Field 02K.
W73-08035

IMPROVED UREA ELECTRODE,
Louisiana State Univ., New Orleans. Dept. of Chemistry.
For primary bibliographic entry see Field 05A.
W73-08042

ACTIVITY MEASUREMENTS AT HIGH IONIC STRENGTHS USING HALIDE-SELECTIVE MEMBRANE ELECTRODES,
State Univ. of New York, Buffalo. Dept. of Chemistry.
For primary bibliographic entry see Field 05A.
W73-08043

ELECTROCHEMICAL CHARACTERISTICS OF THE GOLD MICROMESH ELECTRODE,
Wisconsin Univ., Madison. Dept. of Chemistry.
For primary bibliographic entry see Field 02K.
W73-08044

HYDRAULIC TESTS IN HOLE UAE-3, AMCHITKA ISLAND, ALASKA,
Geological Survey, Lakewood, Colo.
For primary bibliographic entry see Field 04B.
W73-08071

INFLUENCE OF 'BOUND' WATER ON THE CALIBRATION OF A NEUTRON MOISTURE METER,
Ibadan Univ., (Nigeria). Dept. of Agronomy.
For primary bibliographic entry see Field 02G.
W73-08087

APPLICATION OF RHEOLOGICAL MEASUREMENTS TO DETERMINE LIQUID LIMIT OF SOILS,
Central Building Research Inst., Roorkee (India).
For primary bibliographic entry see Field 02G.
W73-08089

AN EXPERIMENTAL STUDY OF THE STRUCTURE, THERMODYNAMICS AND KINETIC BEHAVIOR OF WATER,
Midwest Research Inst., Kansas City, Mo.
For primary bibliographic entry see Field 01A.
W73-08188

DEMOUNTABLE RING-DISK ELECTRODE,
Illinois Univ., Urbana. School of Chemical Sciences.
G. W. Harrington, H. A. Laitinen, and V. Trendafilov.
Analytical Chemistry, Vol 45, No 2, p 433-434, February 1973. 1 fig, 9 ref.

Descriptors: *Construction, *Design, *Methodology, *Efficiencies, Electrochemistry, Laboratory equipment.
Identifiers: *Ring disk electrode, *Platinum, Sensors, Tin oxide-coated glass, Precision.

Although the ring disk electrode has become a valuable tool in studying many electrochemical processes, there are difficulties in fabricating and centering the electrode. The use of different materials requires fabrication of an entirely new electrode for each material. With the present designs, heat treatment of coated electrodes is impossible. An electrode has been specifically designed to permit the use of different materials for the disk and to eliminate problems associated with centering the electrode. Two platinum disk-platinum ring electrodes were constructed and tested with solutions of Cu (II) in 0.5 M KCl; the results agreed exactly with those published previously. Experimental collection efficiency agreed with the theoretical value of 0.345 within less than 2 percent. The electrodes were disassembled and reassembled several times with no changes in the experimental collection efficiency. The procedure, accompanied by a diagram, is given for constructing such an electrode. A tin oxide-coated glass disk-platinum ring electrode and an electrode in which both disk and ring are of tin oxide-coated glass also are described. (Holoman-Battelle)
W73-08260

LINEAR AND NONLINEAR SYSTEM CHARACTERISTICS OF CONTROLLED-POTENTIAL ELECTROLYSIS CELLS,
California Univ., Livermore. Lawrence Livermore Lab.
For primary bibliographic entry see Field 02K.
W73-08261

AN IMPROVED VARIABLE-INTENSITY SPRINKLING INFILTROMETER,
Hebrew Univ., Jerusalem (Israel). Dept. of Soil Science.
E. Rawitz, M. Margolin, and D. Hillel.
Soil Sci Soc Am Proc. Vol 36, No 3, p 533-535. 1972. Illus.
Identifiers: *Instrumentation, Flow, Infiltration, *Infiltrometers, Intensity, Movement, Rain, Simulator, Soils, Sprinkling.

Improvements of the Purdue-Wisconsin infiltrometer are described. A winch was added to the tower to facilitate assembly and the windshield was redesigned. Water distribution was improved by changes in the revolving shutter and the surplus water collection trough. The nozzle mounting was simplified, and water level in the vacuum runoff tank was transmitted to an external recorder. Good uniformity was obtained over a wide range of application rates.—Copyright 1972, Biological Abstracts, Inc.
W73-08340

APPLICATION OF REMOTE SENSING TO SOLUTION OF ECOLOGICAL PROBLEMS,
IBM Federal Systems Div., Bethesda, Md.
A. Adelman.
Proc. available from GPO, Washington, DC 20402 - Price $4.50. In: Space for Mankind's Benefit; Proc of space congress, November 15-19, 1971, Huntsville, Ala: Washington, D C, National Aeronautics and Space Administration Publication NASA SP-313, p 105-108, 1972.

Descriptors: *Remote sensing, *Satellites (Artificial), Water resources, Stream gages, Ecology.

recreation, management, urban and regional planning, and pesticide studies. (Knapp-USGS)
W73-08360

A DATA ACQUISITION SYSTEM (DAS) FOR MARINE AND ECOLOGICAL RESEARCH FROM AEROSPACE TECHNOLOGY,
Mississippi State Univ., Bay Saint Louis. Mississippi Test Facility.
R. A. Johnson.
Proc Available from GPO, Washington, DC 20402, Price $4.50. In: Space for Mankind's Benefit; Proc of space congress, November 15-19, 1971, Huntsville, Ala: Washington, DC, National Aeronautics and Space Administration Publication NASA SP-313, p 149-153, 1972. 3 fig, 2 ref. NASA Grants NGL 25-001-02, 25-001-032, and 25-001-040.

Descriptors: *Telemetry, *Data collections, *Remote sensing, *Instrumentation, *Mississippi, Water resources, Ecology, Aircraft, Water pollution, Air pollution.

A self-contained portable data acquisition system was developed for use in marine and ecological research. The compact, lightweight system is capable of recording 14 variables and is suitable for use in either a boat, pickup truck, or light aircraft. Both self-contained analog recording and a telemetry transmitter are used for real-time digital readout and recording. The prototype system has been utilized in several investigations of air pollution and weather modification. It is currently being used on the Mississippi State University Eco-System Research Project for marine data acquisition. (Knapp-USGS)
W73-08361

SATELLITE OBSERVATIONS OF TEMPORAL TERRESTRIAL FEATURES,
Allied Research Associates, Inc., Concord, Mass.
G. Rabchevsky.
Proc Available from GPO, Washington, DC 20402, Price $4.50. In: Space for Mankind's Benefit; Proc of space congress, November 15-19, 1971, Huntsville, Ala: Washington, DC, National Aeronautics and Space Administration Publication NASA SP-313, p 155-179, 1972. 34 fig, 1 tab, 37 ref.

Descriptors: *Remote sensing, *Satellites (Artificial), *Data collections, Water temperature, Ocean currents, Sea ice, Snow cover, Snowpacks, Lakes, Reservoirs, Floods, Vegetation effects, Mapping, Surveys, Volcanoes, Playas, Deltas, Sedimentation, Forest, Forest fires.

Since the launch of the first orbiting meteorological Television and Infrared Observation Satellite (TIROS I) on April 1, 1960, over 1 million pictures of the earth have been recorded by 25 weather satellites. During the Gemini program, the astronauts took over 2400 color photographs; 2100 pictures were taken during the Apollo Program. Nimbus radiometers sense thermal boundaries of major ocean currents. Some areas of upwelling have been studied in detail. Pack ice boundaries have been established for both polar seas using satellite imagery. Polar (and temperate) ice concentrations can be extracted from satellite data. Changes of the snowline in the upper Missouri-Mississippi River valley were mapped. The seasonal progression of the average surface temperature of Lake Michigan has been obtained using the Nimbus High Resolution Infrared Radiometer. The Nimbus and ITOS television and infrared imaging sensors monitor surface moisture and extent of water bodies. Maps of polar areas have been updated using the Nimbus satellite imagery. Volcanic activity has been recorded by orbital infrared systems. Western U.S. playas have been examined from space. Delta sedimentation plumes have been observed on Nimbus I imagery at the mouth of the Colorado River and at the mouth of the Tigris-Euphrates Rivers. Smoke from large fires in Alaska has been observed on ESSA satellite imagery. (Knapp-USGS)
W73-08362

INTERDISCIPLINARY APPLICATIONS AND INTERPRETATIONS OF REMOTELY SENSED DATA,
Pennsylvania State Univ., University Park.
G. W. Petersen, and G. J. McMurtry.
Proc Available from GPO, Washington, DC 20402, Price $4.50. In: Space for Mankind's Benefit; Proc of space congress, November 15-19, 1971, Huntsville, Ala: Washington, DC, National Aeronautics and Space Administration Publication NASA SP-313, p 181-186, 1972. 2 fig, 2 ref.

Descriptors: *Remote sensing, *Pennsylvania, *Resources, Environment, Soils, Surveys, Forests, Data collections, Recreation, Vegetation, Runoff, Geology, Terrain analysis, Land use, Groundwater, Rivers, Air pollution, Acid mine water.
Identifiers: *Susquehanna River basin.

The use of remote sensing for inventory of natural resources and land use, determination of pollution sources and damage, and analysis of geologic structure and terrain is under investigation. The geographical area of primary interest is the Susquehanna River basin. The specific tasks considered are: identification and characterization of soil; location, inventory, and monitoring of strip-mining spoils; survey and inventory of forest resources; evaluation of potential recreation sites; survey of insect and plant disease epidemics; collection of data for land use management; development of natural resource inventory systems; characterization and analysis of geologic structures and terrain; inventory of mineral resources and mines; detection of groundwater sources from drainage, lineaments, and fracture patterns; determination of runoff; monitoring the environmental effects of power generating plants; detection of sources of acid mine drainage, determination of mixing patterns in surface waters; detection of air pollution damage; and definition and characterization of water quality problems. (Knapp-USGS)
W73-08363

ORBITAL SURVEYS AND STATE RESOURCE MANAGEMENT,
Battelle Columbus Labs., Ohio. Aerospace Mechanics Div.
G. Wukelic, T. L. Wells, and B. R. Brace.
Proc Available from GPO, Washington, DC 20402, Price $4.50. In: Space for Mankind's Benefit; Proc of space congress, November 15-19, 1971, Huntsville, Ala: Washington, DC, National Aeronautics and Space Administration Publication NASA SP-313, p 187-197, 1972. 15 fig, 4 ref.

Descriptors: *Remote sensing, *Ohio, Resources, Surveys, Satellites (Artificial), Data collections, Management.

Ohio, with highly diversified industry, agriculture, and geography, proposes to use orbital survey data and related space capabilities to manage its resources, attack increasing environmental problems, and plan future developments. Short- and long-range benefits are described. The State Government of Ohio foresees opportunities, challenges, and potential benefits in orbital surveys not only for government management responsibility but also for its constituency by providing alternative approaches to resource and environmental problems heretofore unavailable. (Knapp-USGS)
W73-08364

RIVERBED FORMATION,
Colorado State Univ., Fort Collins. Dept. of Civil Engineering.
M. Skinner.
Proc Available from GPO, Washington, DC 20402, Price $4.50. In: Space for Mankind's Benefit; Proc of space congress, November 15-19, 1971, Huntsville, Ala: Washington, DC, National Aeronau-

tics and Space Administration Publication NASA SP-313, p 199-210, 1972. 9 fig, 20 ref.

Descriptors: *Remote sensing, *Aerial photography, *Channel morphology, *Drainage patterns (Geologic), *Alluvial channels, Terrain analysis, Colorado, Beds, Streambeds, River beds, Meanders, Braiding, Erosion, Flood plains, Geomorphology.

The general fluvial processes that work to form a riverbed and produce the characteristic pattern of either meandering, braided, or straight are reviewed. A method for quantification of river pattern and correlation, with the basic hydraulic characteristics of discharge and slope, is presented. Additional characteristics of a river system may be deduced from high-quality photography and imagery obtained from either aircraft or space platforms. (Knapp-USGS)
W73-08365

DOMESTIC APPLICATIONS FOR AEROSPACE WASTE AND WATER MANAGEMENT TECHNOLOGIES,
General Electric Co., Schenectady, N.Y.
F. DiSanto, and R. W. Murray.
Proc Available from GPO, Washington, DC 20402, Price $4.50. In: Space for Mankind's Benefit; Proc of space congress, November 15-19, 1971, Huntsville, Ala: Washington, DC, National Aeronautics and Space Administration Publication NASA SP-313, p 221-230, 1972. 10 fig, 1 tab, 3 ref.

Descriptors: *Water management (Applied), *Research and development, *Waste treatment, *Water utilization, *Water reuse, Technology, Engineering, Systems analysis.
Identifiers: *Aerospace technology.

Tools for solving many pollution problems have been developed by aerospace technologists. These approaches may be used to identify very complex problems, to select the best solution, and to implement vast programs. None of these approaches or technical processes is unique to the aerospace community. Some of the aerospace developments in solid waste disposal and water purification which are applicable to specific domestic problems are described. An overview is presented of the management techniques used in defining the need, in utilizing the available tools, and in synthesizing a solution. Specifically, several water recovery processes are available for domestic applicability, including filtration, distillation, catalytic oxidation, reverse osmosis, and electrodialysis. Also solids disposal methods include chemical treatment, drying, incineration, and wet oxidation. The latest developments in reducing household water requirements and some concepts for reusing water are outlined. (Knapp-USGS)
W73-08366

USING LIGHT TO COLLECT AND SEPARATE ZOOPLANKTON,
Michigan State Univ., East Lansing. Dept. of Fisheries and Wildlife.
J. L. Ervin, and T. A. Haines.
Prog Fish-Cult. Vol 34, No 3, p 171-174, 1972. Illus.
Identifiers: *Light, Plankton, *Zooplankton, Collection device, Design, Construction.

Design and construction details of a zooplantkton trap in expensive materials using either 1 or 2 lamps are given.--Copyright 1973, Biological Abstracts, Inc.
W73-08406

THE BOREAL BIOCLIMATES,
Toronto Univ., (Ontario). Dept. of Geography.
For primary bibliographic entry see Field 02B.
W73-08437

THE EXPENDABLE BATHYOXYMETER,
Oregon State Univ., Corvallis. Dept. of Oceanography.
H. W. Jeter, E. Foyn, M. King, and L. I. Gordon.
Limnol Oceanogr. Vol 17, No 2, p 288-292. 1972. Illus.
Identifiers: *Bathyoxymeter, *Dissolved oxygen, Oxygen, Instrumentation.

A free-fall device is described, analogous to the expendable bathythermograph, which makes dissolved O2 profiles to 500 m in 80 sec.--Copyright 1973, Biological Abstracts, Inc.
W73-08438

SOLUTE POTENTIALS OF SUCROSE SOLUTIONS,
Georgia Univ., Athens. Dept. of Botany.
B. E. Michel.
Plant Physiol. Vol 50, No 1, p 196-198. 1972.
Identifiers: Freezing point depression, Hydrostatic pressure, *Isopiestic method, Osmometer, Solute potential, Sucrose solutions, Vapor pressure.

Comparative solute potentials were calculated using the isopiestic method, hydrostatic pressure, freezing point depression and a vapor pressure osmometer in an attempt to resolve existing discrepancies in the literature.--Copyright 1973, Biological Abstracts, Inc.
W73-08450

7C. Evaluation, Processing and Publication

A MANUAL ON COLLECTION OF HYDROLOGIC DATA FOR URBAN DRAINAGE DESIGN,
Hydrocomp, Inc., Palo Alto, Calif.
R. K. Linsley.
Available from the National Technical Information Service as PB-219 360, $3.00 paper copy, $1.45 microfiche. March 1973. 60 p, 8 fig, 2 tab, 15 ref, append. OWRR X-120 (3750) (1).

Descriptors: *Storm runoff, *Urban drainage, Precipitation, Evaporation, Solar radiation, *Storm drains, Water quality, *Data collections, Management, Publications.

There are very little data on urban storm flow and quality and application of modern methods to urban storm drain management is limited by this lack. This manual is intended to indicate to city and county engineers the types of data which could be gathered by local staff as an aid in the design of future urban runoff management facilities.
W73-07801

THE USE OF LANDSCAPE-INDICATOR METHODS IN HYDROGEOLOGICAL INVESTIGATIONS,
Moskovskoe Obshchestvo Ispytatelei Prirody (USSR). Geographic Div.
For primary bibliographic entry see Field 02F.
W73-07843

STUDY OF THE HYDROGEOLOGICAL CONDITIONS OF ALLUVIAL FANS BY MULTIPLE LANDSCAPE AND GEOPHYSICAL METHODS,
Moskovskoe Obshchestvo Ispytatelei Prirody (USSR). Geographic Div.
For primary bibliographic entry see Field 02F.
W73-07844

LANDSCAPE-INDICATOR INVESTIGATIONS OF KARST,
Moskovskoe Obshchestvo Ispytatelei Prirody (USSR). Geographic Div.
For primary bibliographic entry see Field 02F.
W73-07845

INDICATION OF ROCKS BY DRAINAGE PATTERNS,
Moskovskoe Obshchestvo Ispytatelei Prirod (USSR). Geographic Div.
For primary bibliographic entry see Field 02J.
W73-07846

NATIONAL PROGRAM FOR MANAGING FLOOD LOSSES, GUIDELINES FOR PREPARATION, TRANSMITTAL, AND DISTRIBUTION OF FLOOD-PRONE AREA MAPS AND PAMPHLETS,
Geological Survey, Washington, D.C.
G. W. Edelen, Jr.
Geological Survey open-file report, 1973. 28 p, fig.

Descriptors: *Floods, *Flood plains, *Mapping, *Flood protection, Reviews, Flood profiles, Data processing, Maps, United States, Aerial photographs, Flood frequency, Boundaries (Surfaces), Flood control, Data transmission, Planning, Governments.
Identifiers: *Flood-plain mapping guidelines.

Information is presented to assist Water Resources Division offices in preparing flood-prone area maps and pamphlets. Background and history of the program, legal authority, analytical techniques, printing, distribution, and other operational details are discussed. The instructions and advice should be considered primarily as guidelines. In general, the instructions for map bases, lettering sizes, and reproduction requirements must be followed quite closely whereas instructions pertaining to techniques of delineating flood boundaries may require liberal interpretation in unusual situations. The 89th Congress (1966) in House Document 465 recommended preparation of flood-prone area maps to assist in minimizing flood losses by quickly identifying areas of potential flood hazards. Flood-prone area maps produced to date have been particularly useful during floods in planning the evacuation of areas likely to be flooded. (Woodard-USGS)
W73-07849

MANAGEMENT MODEL AS A TOOL FOR STUDYING THE WORTH OF DATA,
Geological Survey, Arlington, Va. Water Resources Div.
For primary bibliographic entry see Field 06A.
W73-07882

MEASURE OF THREE-DIMENSIONAL DRAINAGE BASIN FORM,
Cambridge Univ. (England). Dept. of Geography.
For primary bibliographic entry see Field 02A.
W73-07892

MULTISITE DAILY FLOW GENERATOR,
Department of the Environment, Ottawa (Ontario). Water Management Service.
For primary bibliographic entry see Field 02A.
W73-07899

THE TRANSIENT FLOW PROBLEM - A ONE-DIMENSIONAL DIGITAL MODEL,
Wyoming Univ., Laramie. Dept. of Civil and Architectural Engineering.
For primary bibliographic entry see Field 02F.
W73-07916

HYDROGEOLOGIC CHARACTERISTICS OF THE VALLEY-FILL AQUIFER IN THE ARKANSAS RIVER VALLEY, BENT COUNTY, COLORADO,
Geological Survey, Washington, D.C.
R. T. Hurr, and J. E. Moore.
Available for sale by USGS, Washington, D.C. 20242, Price $1.00 per set. Geological Survey Hydrologic Investigations Atlas HA-461, 1972. 2 sheet, 2 fig, 4 map, 11 ref.

Descriptors: *Groundwater resources, *Aquifers, *Hydrogeology, *Colorado, Valleys, Rivers, Water wells, Water supply, Water yield, Groundwater recharge, Water levels, Transmissivity, Aquifer characteristics, Irrigation, Hydrologic data, Maps, Groundwater movement.
Identifiers: *Arkansas River valley (Colo), *Bent County (Colo).

This atlas describes the hydrologic characteristics of the valley-fill aquifer in a 36-mile reach of the Arkansas River valley in Bent County, southeastern Colorado. The reach is underlain by saturated valley-fill alluvium consisting of gravel, sand, silt, and clay of Pleistocene to Holocene age. The alluvium occupies a trough eroded in the shale, limestone, and sandstone bedrock of Cretaceous age. In Bent County, the Arkansas River is a gaining stream most of the year owing to groundwater return flow from applied irrigation water. Maximum withdrawals of groundwater in the county occurred in 1964 when about 34,000 acre-feet of water was pumped from 225 irrigation wells and six municipal wells. The yields of the wells range from 100 to 2,500 gpm, and vary considerably from place to place, due mainly to variation in saturated thickness and hydraulic conductivity of the aquifer. The average hydraulic conductivity is about 4,000 gallons per day per square foot. A map of aquifer transmissivity is based on saturated thickness and hydraulic conductivity. (Woodard-USGS)
W73-08068

WATER-RESOURCES RECONNAISSANCE OF THE OZARK PLATEAUS PROVINCE, NORTHERN ARKANSAS,
Geological Survey, Washington, D.C.
A. G. Lamonds.
Available for sale by USGS, Washington, D.C. 20242, Price $1.00 per set. Geological Survey Hydrologic Investigations Atlas HA-383, 1972. 1 sheet, 6 fig, 12 map, 4 tab, 23 ref.

Descriptors: *Water resources, *Surface waters, *Groundwater resources, *Arkansas, Hydrologic data, Water yield, Water quality, Rainfall-runoff-frelationships, Flow characteristics, Flood frequency, Low flow, Aquifers, Hydrogeology, Aquifer characteristics, Water wells, Water utilization, Geology, Hydrology, Basic data collections.
Identifiers: *Ozark Plateaus Province (Ark).

This hydrologic atlas describes the water resources of the Ozark Plateaus province in northern Arkansas. The area is underain by deeply dissected plateaus; the Salem Plateau, the Springfield Plateau, and the Boston Mountains. The Ozark Plateaus in Arkansas encompass all or parts of 22 counties for a total area of about 12,245 square miles. Tributaries to the Arkansas River drain the western and southwestern parts of the plateaus; about 27% of the area. The White River and its tributaries drain about 73% of the area. Geologic units consist mostly of limestone, dolomite, sandstone, and shale. On the basis of geologic and hydrologic similarity, the geologic units are combined into eight hydrologic units. Wells in most of the units generall are less than 300 feet deep and yield less thany 10 gpm. The chemical quality of water from 88 springs and wells is summarized. The quality of groundwater generally is suitable for most uses. A map showing flood-frequency regions and hydrologic areas delineates three regions and six areas in the plateaus. The natural quality of surface water is excellent, and pollution is not a serious problem at present (1969). (Woodard-USGS)
W73-08069

ENTROPY AS A MEASURE OF THE AREAL CONCENTRATION OF WATER-ORIENTED INDUSTRY,
Tennessee Univ., Knoxville. Coll. of Business Administration.
For primary bibliographic entry see Field 03E.
W73-08131

SECTION 22, 'SNOW SURVEY AND WATER SUPPLY FORECASTING,' OF THE SCS NATIONAL ENGINEERING HANDBOOK,
Soil Conservation Service, Portland, Oreg. Water Supply Forecast Unit.
For primary bibliographic entry see Field 02C.
W73-08153

QUALITY OF SURFACE WATERS OF THE UNITED STATES, 1968: PART I. NORTH ATLANTIC SLOPE BASINS.
Geological Survey, Washington, D.C.

Available from Sup Doc GPO, Washington, DC 20402 Price $1.75 (paper cover). Geological Survey Water-Supply Paper 2091, 1972. 373 p, 1 fig, 39 ref.

Descriptors: *Water quality, *Surface waters, *Northeast U.S., *Water temperature, *Sediment transport, Streamflow, Flow rates, Water analysis, Chemical analysis, Water chemistry, Sampling, Basic data collections.
Identifiers: *North Atlantic Slope basins.

During the water year ending September 30, 1968, the U.S. Geological Survey maintained 175 stations on 117 streams in the North Atlantic Slope basins for the study of chemical and physical characteristics of surface water. Samples were collected daily and monthly at 128 of these locations for chemical-quality studies. Samples also were collected less frequently at many other points. Water temperatures were measured continuously at 58 and daily at 39 stations. At chemical-quality stations where data are continuously recorded at the stream site (monitors), the records consist of daily maximum, minimum, and mean values for each constituent measured. Quantities of suspended sediment are reported for 33 stations, and particle-size distributions of sediments for 29 stations. Quality of water stations usually are located at or near points on streams where streamflow is measured. (Woodard-USGS)
W73-08155

RECORDS OF WELLS AND TEST BORINGS IN THE SUSQUEHANNA RIVER BASIN, NEW YORK,
Geological Survey, Albany, N.Y.
A. D. Randall.
New York Department of Environmental Conservation Bulletin 69, 1972. 92 p, 3 fig, 2 plate, 5 tab, 6 ref.

Descriptors: *Groundwater resources, *Water wells, *Well data, *Water quality, *New York, Drillers logs, Aquifer characteristics, Geology, Water levels, Water yield, Water utilization, Chemical analysis, Hydrologic data, Basic data collections.
Identifiers: *Susquehanna River basin (N Y).

Groundwater resources data collected from 1965 through 1968 in the Susquehanna River basin in New York include records of 1,990 wells, chemical analyses of water from 315 wells, detailed logs of 385 wells, and logs of 725 test borings. Well records (except for remarks) and chemical analyses were compiled using automatic data-processing equipment. They are stored by the U.S. Geological Survey on machine cards and magnetic tapes. At least 90% of the records are from the major valleys, which constitute no more than 15% of total basin area. The valleys were emphasized because it is in the valleys that the most productive aquifers occur, most urban development has taken place, and future development is likely to concentrate. A few small upland areas were selected for intensive study. Salt water is known to exist at depth beneath much of the basin. To document its position, salt-water wells were sought and inventoried wherever possible. (Woodard-USGS)
W73-08159

GLOSSARY OF TERMS IN SOIL SCIENCE.
For primary bibliographic entry see Field 02G.
W73-08161

MAP SHOWING GENERAL CHEMICAL QUALITY OF GROUNDWATER IN THE SALINAQUADRANGLE, UTAH,
Geological Survey, Washington, D.C.
D. Price.
Available For Sale by USGS, Washington, D.C. 20242, Price - 75 cents. Geological Survey Miscellaneous Geologic Investigations Maps, Map I-591-K, 1972. 1 sheet, 1 map.

Descriptors: *Water quality, *Groundwater, *Dissolved solids, *Utah, Maps, Locating, Springs, Water wells, Aquifers, Geochemistry, Data collections.
Identifiers: *Salina quadrangle (Utah).

This map of Salina quadrangle, Utah, scale 1:250,000, shows the general chemical quality of water as determined from quality-of-water data collected by the U.S. Geological Survey and cooperating State, local, and Federal agencies. Sources of data include springs, and wells that tap aquifers at depths of less than 1,000 feet. Various colors are used to indicate locations where dissolved solids range from 250 to 1,000 mg/liter, 500 to 1,000 mg/liter, 500 to 3,000 mg/liter, and 1,000 to 3,000 mg/liter. Also shown are locations where dissolved solids are less than 500 mg/liter. (Woodard-USGS)
W73-08172

DRAINAGE AREAS, HARTFORD NORTH QUADRANGLE, CONNECTICUT,
Geological Survey, Washington, D.C.
M. P. Thomas.
Available For Sale by USGS, Washington, D.C. 20242, Price - 75 cents. Geological Survey Miscellaneous Geologic Investigations Maps, Map I-784-J, 1972. 1 sheet, 1 map.

Descriptors: *Drainage area, *Streams, *Reservoirs, *Gaging stations, *Connecticut, Surface waters, Maps, Streamflow, Discharge measurement, Sampling, Sites, Locating, Watersheds (Divides), River basins, Watersheds (Basins), Cities.
Identifiers: *Hartford north quadrangle (Conn).

A map scale 1:24,000) shows drainage areas that contribute streamflow to selected sites on streams in the Hartford north quadrangle, Connecticut. In addition to the drainage area boundary markings, the numbers of square miles drained is shown at stream gaging sites, outlets of surface water impoundments, surface water sampling sites, and mouths of tributary streams. (Woodard-USGS)
W73-08173

MAP OF DEPOSITS ESPECIALLY SUSCEPTIBLE TO COMPACTION OR SUBSIDENCE, PARKER QUADRANGLE, ARAPAHOE AND DOUGLAS COUNTIES, COLORADO,
Geological Survey, Washington, D.C.
J. O. Maberry.
Available For Sale by USGS, Washington, D.C. 20242, Price - 75 cents. Geological Survey Miscellaneous Geologic Investigations Maps, Map I-770-J, 1972. 1 sheet, 1 map, 1 ref.

Descriptors: *Geology, *Geologic units, *Structural engineering, *Sedimentology, *Colorado, Subsidence, Settlement (Structural), Physical properties, Loess, Sands, Alluvium, Sedimentation.
Identifiers: *Arapahoe and Douglas Counties (Colo), Parker quadrangle (Colo).

A map (scale 1:24,000) of Parker quadrangle, Arapahoe and Douglas Counties, Colorado, shows areas of geologic deposits especially susceptible to compaction or subsidence. This information could be important in structural engineering. Colors on the map indicate three different types of geologic

materials: eolian silt (loess), eolian sand, and alluvium. These units are more susceptible to compaction and subsidence or differential settling than other sediments or sedimentary rocks in the quadrangle. Loess occurs in two broad northwest-trending belts west of Cherry Creek and comprises material brought into the area from the Platte River valley by northwesterly winds. It is as much as 18 feet thick. Eolian sand covers the land in a belt along the uplands east of Cherry Creek. It was deposited by northwesterly and westerly winds blowing material out of the Cherry Creek valley and is as much as 40 feet thick. The alluvium considered is the post-Piney Creek alluvium, the youngest in the map area. This alluvium occupies the streambeds of Cherry Creek and its major tributaries. It is made up of loose fine to coarse sand and fine gravel with minor amounts of clay. (Woodard-USGS)
W73-08174

THE REGRESSION OF NET RADIATION UPON SOLAR RADIATION,
Oregon State Univ., Corvallis. Water Resources Research Inst.
For primary bibliographic entry see Field 02D.
W73-08175

AUTOMATED DATA HANDLING USING A DIGITAL LOGGER,
Virginia Polytechnic Inst. and State Univ., Blacksburg. Dept. of Chemistry.
D. G. Larsen.
Analytical Chemistry, Vol 45, No 1, p 217-220, January 1973. 4 fig.

Descriptors: *Digital computers, *Electronic equipment, *Data processing, *Data transmission, Spectrometers, Data storage and retrieval, Design.
Identifiers: *Analog to digital converters, *Data logger.

Data from analytical instruments requiring little or no control may be input to a small computer. Analog data are fed into a Keithley Model 160 Digital Multimeter whose BCD signal is converted to ASCII code by a coupler driver. This code is then fed to an ASR 33 teletype which produces hard copyand paper tape output. The tape is then used as input to the computer. The data logger has proven to be useful for automating digital data collection from a wide variety of spectrometers and other equipment. The logging speed or data rate is limited to a maximum of one data word per second. The logging system has the advantage of low cost and can be built from commercially available equipment. Schematics of the system are included. (Little-Battelle)
W73-08257

EXPERIMENTAL CURVES AND RATES OF CHANGE FROM PIECEWISE PARABOLIC FITS,
Kentucky Univ., Lexington. Dept. of Mathematics.
For primary bibliographic entry see Field 07A.
W73-08300

APPLICATION OF REMOTE SENSING TO SOLUTION OF ECOLOGICAL PROBLEMS,
IBM Federal Systems Div., Bethesda, Md.
For primary bibliographic entry see Field 07B.
W73-08358

USE OF DATA FROM SPACE FOR EARTH RESOURCES EXPLORATION AND MANAGEMENT IN ALABAMA,
Alabama State Oil and Gas Board, Tuscaloosa.
For primary bibliographic entry see Field 07B.
W73-08360

A DATA ACQUISITION SYSTEM (DAS) FOR MARINE AND ECOLOGICAL RESEARCH FROM AEROSPACE TECHNOLOGY,
Mississippi State Univ., Bay Saint Louis. Mississippi Test Facility.
For primary bibliographic entry see Field 07B.
W73-08361

ORBITAL SURVEYS AND STATE RESOURCE MANAGEMENT,
Battelle Columbus Labs., Ohio. Aerospace Mechanics Div.
For primary bibliographic entry see Field 07B.
W73-08364

MAP SHOWING APPROXIMATE GROUNDWATER CONDITIONS IN THE PARKER QUADRANGLE, ARAPAHOE AND DOUGLAS COUNTIES, COLORADO,
Geological Survey, Washington, D.C.
J. O. Maberry, and E. R. Hampton.
For sale by USGS, Washington, D.C. 20242, Price - $0.75. Geological Survey Miscellaneous Geologic Investigations Maps, Map I-770-K, 1972. 1 map, 3 ref.

Descriptors: *Groundwater resources, *Aquifer characteristics, *Water wells, *Colorado, Mapping, Aquifers, Water yield, Water quality, Water utilization, Hydrologic data, Maps.
Identifiers: *Arapahoe and Douglas Counties (Colo), Parker quadrangle.

A map (scale 1:24,000) of the Parker quadrangle, Arapahoe and Douglas Counties, Colorado, shows approximate groundwater conditions. Groundwater is obtained principally from three aquifer systems: stream alluvium and alluvial terraces, relatively shallow bedrock, and relatively deep bedrock units. The greatest amounts of readily available groundwater occur in the sand and gravel alluvial fill of Cherry Creek Valley and upland alluvial and terrace deposits of its major distributaries. The alluvium is as much as 150 feet thick in Cherry Creek Valley. Large-capacity wells producing from alluvium along Cherry Creek yield from 900 to 1,800 gpm and average about 1,200 gpm. Most of these wells are used for municipal water supplies. Dissolved solids in water from the alluvium range from 280 to 380 ppm. (Woodard-USGS)
W73-08369

AVAILABILITY OF GROUNDWATER, HARTFORD NORTH QUADRANGLE, CONNECTICUT,
Geological Survey, Washington, D.C.
R. B. Ryder.
For sale by USGS, Washington, D.C. 20242, Price - $0.75. Geological Survey Miscellaneous Geologic Investigations Maps, Map I-784-K, 1972. 1 sheet, 1 map, 2 ref.

Descriptors: *Groundwater resources, *Groundwater availability, *Aquifer characteristics, *Water wells, *Connecticut, Water yield, Water table, Mapping, Sedimentary rocks, Bedrock, Unconsolidated sediments.
Identifiers: *Hartford (Conn), North Quadrangle.

Groundwater availability in the Hartford north quadrangle, Connecticut, is shown on a map, scale 1:24,000. Areas in which most individual wells can be expected to yield less than 10 gpm are in deposits of till, very fine sand, silt, and clay as well as sand, gravel, and interbedded sand and gravel with a water-saturated thickness of less than 10 feet. Areas in which most individual wells can be expected to yield between 10 and 200 gpm are in deposits of mostly coarse to fine sand. Unconsolidated deposits are underlain by sedimentary bedrock. Properly developed individual bedrock wells can be expected to yield as much as 600 gpm. Locally, it may be necessary to drill through less than 500 feet of low-yielding basalt (traprock) to reach underlying, higher yielding sedimentary bedrock, particularly in the south central map area. (Woodard-USGS)
W73-08370

A PROBABILISTIC MODEL FOR STRUCTURING A DRAINAGE NETWORK,
Army Project Mobile Army Sensor Systems Test Evaluation and Review Activity, Fort Hood, Tex.
For primary bibliographic entry see Field 04A.
W73-08380

SURFACE WATER SUPPLY OF THE UNITED STATES, 1966-1970: PART 6-MISSOURI RIVER BASIN, VOLUME 4-MISSOURI RIVER BASIN BELOW NEBRASKA CITY, NEBRASKA.
Geological Survey, Washington, D.C. Water Resources Div.

Geological Survey Water-Supply Paper 2119, 1972. 901 p, 1 map.

Descriptors: *Streamflow, *Flow measurement, *Missouri River, *Hydrologic data, *Data collections, Stream gages, Surface waters, Gaging stations, Flow rates, Lakes, Reservoirs, Water levels, Average flow, Low flow, Peak discharge, Crest-stage gages, Colorado, Iowa, Kansas, Missouri, Nebraska.
Identifiers: *Missouri River Basin.

This volume of surface water data for the Missouri River basin below Nebraska City, Nebraska, is one of a series of 37 reports presenting records of stage, discharge, and content of streams, lakes, and reservoirs in the United States during the 1966-70 water years. The tables of data include a description of the gaging station, and daily, monthly, and yearly discharges of the stream. The description of the station gives the location, drainage area, records available, type and history of gage, average discharge, extremes of discharge, and general remarks. For most gaging stations on lakes and reservoirs a description of the station and a monthly summary table of stage and contents are given. Data for partial-record stations include measurements at low-flow partial-record stations and annual maximum stage and discharge at crest-stage stations. (Woodard-USGS)
W73-08381

AN INVENTORY AND SENSITIVITY ANALYSIS OF DATA REQUIREMENTS FOR AN OXYGEN MANAGEMENT MODEL OF THE CARSON RIVER,
Nevada Univ., Reno. Dept. of Civil Engineering.
For primary bibliographic entry see Field 05B.
W73-08382

08. ENGINEERING WORKS

8A. Structures

RECTIFICATION OF DEFICIENCIES IN COMPLETED LOCAL PROTECTION PROJECT, WELLSVILLE, NEW YORK (FINAL ENVIRONMENTAL IMPACT STATEMENT).
Army Engineer District, Buffalo, N.Y.

Available from the National Technical Information Service as EIS-NY-72-5586-F, $3.25 in paper copy, $0.95 in microfiche. February 29, 1972. 22 p, 1 map.

Descriptors: *New York, *Environmental effects, *Flood protection, *Channel improvement, Flood control, Bank erosion, Fisheries, Flood plain zoning, Erosion control, River regulation, Dredging, Channel flow, Channel erosion, Flow control.
Identifiers: *Environmental Impact Statements, *Wellsville (N.Y.).

*Oregon, Earthquakes, Dams, Rocks, Geologic investigations, Geological surveys, Earthquake engineering, Reservoirs, Reservoir design, River basins, Structural design, Seismic properties, Structural stability, Slope stability, Landslides, Sedimentation, Turbidity.
Identifiers: *Environmental Impact Statements, *Rogue River (Oregon).

This supplement to the environmental impact statement concerns the Lost Creek Lake Project, a reservoir located in the Rogue River Basin in Oregon. A detailed description of project site geology is provided as well as a summary of reservoir slope stability, area seismicity, and possible earthquake effects on the reservoir. The United States is divided into four seismic-probability zones. Lost Creek Dam site is located in an area classified as Zone 1, a relatively quiet seismic area. Structures within Zone 1 are normally designed for an acceleration of .05g. Lost Creek Dam has been designed for .1g, twice the standard design consideration. Reconnaissance data are included on adjacent slopes regarding possible landslides which might be triggered by an earthquake. The reconnaissance did not locate any potential landslides capable of generating destructive waves or of appreciably reducing reservoir use. In a letter of comment, the U.S. Department of the Interior noted that a temporary adverse effect of the project will be the creation of downstream turbidity and sedimentation during construction. Sedimentation would have an adverse effect on anadromous fish if it occurred during the spawning season. (Adams-Florida)
W73-07980

CASE HISTORIES OF THREE TUNNEL-SUPPORT FAILURES, CALIFORNIA AQUEDUCT,
California State Dept. of Water Resources, Palmdale.
A. B. Arnold.
Bulletin of Association of Engineering Geologists, Vol 9, No 3, p 265-299, Summer 1972. 35 p, 18 fig, 1 tab.

Descriptors: *Collapse, *Construction practices, *Geologic formations, *Tunnels, *California, Rock mechanics, Tunnel failure, Water tunnels (Conveyance), Ribs, Tunnel construction, Tunnel design, Faults (Geology), Fractures (Geology), Swelling pressure, Tunneling, Joints (Geology), Buckling.
Identifiers: *Tunnel supports, *Rock pressure, Liner supports, Swelling pressure, Shields, California Aqueduct, Squeezing pressures, Carley V Porter Tunnel (California), Castaic Dam, California.

Case histories are presented of steel support system failures in 3 tunnels during construction of the California Aqueduct in Southern California. In the Carley V Porter Tunnel, driven by hydraulic shields, the liner plate supports collapsed during the remining of claystone lake bed deposits. This area, constructed several feet below grade, had been excavated and supported a year earlier. The steel supports failed in the Castaic Tunnel during the removal of the bench several months after the top heading was holed through. The collapse was in an area of large crown overbreak and steeply dipping sandstone and clay shale. In the Angeles Tunnel, the arch rib and wallplate support system collapsed in a section of faulted and squeezing siltstones and minor sandstones that had been successfully mined about 2 yrs earlier. Ground support sloughed progressively from beneath the wallplate over a period of time prior to the failure. The tunnel support failures were apparently caused by construction practices that did not fully consider geologic conditions. (USBR)
W73-08059

BRIDGES, WALLS, FILLS, CHANNEL CHANGES, ETC. (FOR THE INFORMATION OF THOSE INTERESTED IN THE CONSTRUCTION OF).
Pennsylvania Dept. of Environmental Resources, Harrisburg.

Form FWWR-23, 1972. 15 p, 5 fig.

Descriptors: *Bridge construction, *Land fills, *Hydraulic structures, *Hydraulic engineering, *Pennsylvania, *Channel improvement, *Regulations, Legal aspects, Permits, Hydraulic design, Bank protection, Stream improvement, Legislation, Administration, Specifications, Engineering structures, Canal construction, State governments.
Identifiers: *Construction requirements.

Information is provided which is necessary for compliance with the rules of the Commonwealth of Pennsylvania concerning the preparation of plans for the construction of certain structures, including bridges, fills, walls along streams, and stream channel changes. Design procedures and sample plans are included, along with general rules for submitting for State approval. Various relevant legislative acts of the Commonwealth are described. (Poertner)
W73-08182

WATER AND LAND RESOURCE ACCOMPLISHMENTS, 1971.
Bureau of Reclamation, Washington, D.C.

Annual Report on Federal Reclamation Projects, 1971. 70 p, 15 fig, 2 maps, 9 tab, 4 chart.

Descriptors: *Land reclamation, *Irrigation, *Reclamation states, *Water management (Applied), *Water supply development, Environmental effects, Water supply, Water pollution control, Federal reclamation law, Irrigated land, Irrigation effects, Municipal water, Industrial water, Recreation, Flood control, Hydroelectric power, Area redevelopment.

For sixty-nine years the Federal Reclamation program has assisted in the management of water resources for the arid West. In 1971 a total of sixteen million people received water service from Reclamation facilities. Accomplishments in irrigation, municipal and industrial water service, hydropower, public recreation, and flood control in 17 western states are described. Moreover, an overview of planning, construction, and research activities and a prospective on the Bureau of Reclamation's future is reported. The industrial impact of construction of Morrow Point Dam and Powerplant in Colorado is described. The purpose of the report is to reflect an overview of the Bureau of Reclamation's role in the Department of the Interior's program of developing and managing natural resources to meet the requirements of present and future societies. Summary data on acreage, yield, production, extension of irrigation to new lands, and the gross value of crops grown on all projects are included. (Mockler-Florida)
W73-08191

SMITHVILLE LAKE, LITTLE PLATTE RIVER, MISSOURI (FINAL ENVIRONMENTAL IMPACT STATEMENT).
Army Engineer District. Kansas City, Mo.

Available from the National Technical Information Service as EIS-MO-72-4723-F, $4.50, in paper copy, $1.45 in microfiche. June 16, 1972. 42 p, 2 plate.

Descriptors: *Water supply, *Flood control, *Missouri, *Environmental effects, *Water quality control, Flood prevention, Dams, Recreation, Wildlife habitats, Flood plain zoning, Levees, Dikes, Dam construction, Impoundments.
Identifiers: *Little Platte River (Mo.), *Environmental impact statements.

This project consists of the construction of a dam and artificial lake on the Little Platte River in Clay and Clinton Counties five miles north of the city limits of Kansas City, Missouri. Desirable environmental effects include flood protection, water supply augmentation, water quality control, and the extensive development of recreational opportunities. Adverse environmental effects include the inundation of 8,040 acres of agricultural land, streams, and fish and wildlife habitat. In addition, the project will cause occasional inundation of up to 4,140 acres of land and the loss of fish and wildlife habitats, in downstream areas. Alternatives to the proposed project include no development at all, a single purpose project for water supply, the development of small flood detention lakes, flood plain zoning, the construction of setback levees, and the obtaining of an adequate water supply through pumping from the Missouri River. (Mockler-Florida)
W73-08210

BACON CREEK WATERSHED, PLYMOUTH AND WOODBURY COUNTIES, IOWA, (FINAL ENVIRONMENTAL IMPACT STATEMENT).
Soil Conservation Service, Washington, D.C.

Available from the National Technical Information Service as EIS-IA-72-5337-F, $3.00 in paper copy, $1.45 in microfiche. May 1972. 37 p, 1 map, 1 tab.

Descriptors: *Flood control, *Sediment control, *Iowa, *Environmental effects, *Watershed management, Water control, Recreation, Fishing, Streams, Reservoirs, Erosion control, Gully erosion, Flood plain zoning, Flood protection, Flood rating, Multiple-purpose projects.
Identifiers: *Environmental Impact Statements, *Bacon Creek Watershed (Iowa).

The Bacon Creek, Iowa, Watershed Project consists of conservation land treatment, 31 grade stabilization structures, 5 floodwater retarding and sediment control structures, one multi-purpose recreation and floodwater retarding structure and basic recreational facilities. Approximately 11,000 acres of the 23.3 square mile project area have been treated by installation of conservation cropping systems, waterways and the like. Gully erosion is by far the most significant problem, and it is proposed that accelerated conservation land treatment be provided for all open land still needing treatment. This project will reduce soil erosion and flood damage and provide an artificial lake for recreational activities. However, approximately ten miles of natural stream channel will be inundated and about seven hundred acres of agricultural land and wildlife habitats will be lost to the project. Since the flood plain is already developed, flood plain zoning is not a feasible alternative. Alternatives include no project or land treatment measures without structural measures. (Smith-Adam-Florida)
W73-08211

VIRGINIA BEACH, VIRGINIA--BEACH EROSION CONTROL AND HURRICANE PROTECTION (FINAL ENVIRONMENTAL IMPACT STATEMENT).
Army Engineer District, Norfolk, Va.

Available from the National Technical Information Service as EIS-VA-72-5322-F, $3.00 in paper copy, $1.45 in microfiche. March 6, 1972. 30 p, 2 plate, 1 map, 2 append.

Descriptors: *Virginia, *Environmental effects, *Beach erosion, Water quantity control, Environmental control, Land management, Erosion rates, Erosion control, Recreation, Beaches, Coasts, Shore protection, Concrete structures, Aesthetics.
Identifiers: *Environmental Impact Statements, *Virginia Beach (Va.).

The Virginia Beach, Virginia, Beach Erosion Control and Hurricane Protection project consists of sheet pile walls capped with concrete, a raising and widening of the beach, and the recommendation of certain non-structural measures. The plan provides for long-term reduction of future damages to structures in the affected area, for reduction in beach erosion, for added recreational opprotunities afforded by the extended beach, and for an impetus to economic stimulation of Virginia Beach. Improvement of the socio-economic climate of the community can be expected with the project; also, enhanced recreational potential by preserving the ocean beaches. Adverse environmental effects include the curtailment of the site as an agricultural area, temporary damage to marine life through the use of dredged material from offshore, and visual restriction of seascape vistas for adjacent residences. (Mockler-Florida)
W73-08213

COW CREEK WATERSHED, STEPHENS AND JEFFERSON COUNTIES, OKLAHOMA (FINAL ENVIRONMENTAL IMPACT STATEMENT).
Soil Conservation Service, Washington, D.C.

Available from the National Technical Information Service as EIS-OK-72-5222-F, $4.50 in paper copy, $1.45 in microfiche. August 1972. 46 p, 1 map, 1 tab.

Descriptors: *Oklahoma, *Environmental effects, *Flood control, *Watershed management, *Erosion control, Flood plain zoning, Recreation, Water management (Applied), Land management, Flood protection, Channel improvement, Stream erosion, Sediment control, Streamflow, Wildlife habitats, Water storage.
Identifiers: *Cow Creek (Oklahoma), *Environmental Impact Statements.

This watershed project on Cow Crrek, Oklahmoa consists of land treatment work, supplemented by 46 floodwater retarding structures, one multipurpose structure with a recreation development, and 2.5 miles of stream channel clearing and snagging. The project will serve to reduce erosion and sediment production on 124,000 acres by 25 per cent, reduce erosion damages to flood plain lands by 56 per cent, reduce floodwater damages on 12,500 acres of flood plain by 70 per cent, and provide a more favorable environment for the residents in and adjacent to the flood plain. Adverse effects include loss of agricultural production on 968 acres covered by sediment pools, loss of 968 acres of game hunting land, and a loss of storage capacity in Lake Texoma. Alternatives include letting the land revert to less intensive uses and developing the flood plain into a publicly owned park or wildlife area. (Mockler-Florida)
W73-08214

WILLOW ISLAND LOCKS AND DAM OHIO RIVER, OHIO AND WEST VIRGINIA (FINAL ENVIRONMENTAL IMPACT STATEMENT).
Army Engineer District, Huntington, W. Va.

Available from the National Technical Information Service as EIS-OH-72-5278-F, $5.25, in paper copy, $1.45 in microfiche. May 16, 1972. 63 p, 3 fig, 1 map, 2 chart.

Descriptors: *Ohio, *West Virginia, *Environmental effects, *Dam construction, *Navigation, Impoundments, Locks, Water quality control, Water utilization , Dredging, Spoil banks, Transportation, Ohio River, Dams.
Identifiers: *Environmental Impact Statements, *Willow Island (Ohio).

The proposed project would involve the construction and operation of a high-lift, non-navigable, gated dam with a main and an auxiliary lock; removal of three existing locks and dams; dredging required to utilize the new structure; and establishment of public access areas at Willow Island on the Ohio River, Ohio. Favorable impacts include conversion of pools of three low-lift dams to a single pool, thereby making the reach of the river compatible with other parts of the Ohio River navigation system. However, adverse environmental effects would include the inundation of land along the riverbank and on islands; deposition of spoil on some islands and riverside land; removal of some trees and shrubs along the shoreline with resultant loss of biotic habitat; and possible modification of dissolved oxygen content in some reaches of the river. Alternatives to the proposed project include modification of existing locks and dams to accommodate larger tows and no action. (Mockler-Florida)
W73-08216

DETAILED PROJECT REPORT, INVESTIGATION FOR FLOOD PROTECTION, MUNDAY, TEXAS, BRAZOS RIVER BASIN, TEXAS (FINAL ENVIRONMENTAL IMPACT STATEMENT).
Army Engineer District, Fort Worth, Tex.

Available from the National Technical Information Service as PB-2106 460-F $3.00 in paper copy, $1.45 in microfiche. April 18, 1972. 28 p, 4 tab.

Descriptors: *Texas, *Flood control, *Environmental effects, *Flood protection, Eminent domain, Channeling, Channel improvement, Flood plain zoning, Scenery, Trees, Wildlife, Flood routing, Chutes, Diversion structures.
Identifiers: *Environmental Impact Statements, *Brazos River Basin (Tex.).

This project would consist of the construction, (on receipt of funds), of a flood control project consisting of a channel, concrete chute, stilling basin, transition sections and bridge replacements in the Brazos River Basin, Knox County, Texas. The impact of the project upon the environment would be minimal. Trees do not exist along the proposed rights of way, and small game, bird-life or fish habitats will not be affected. Although approximately 60 acres of agriculture land would be lost to the project, the advantages of greatly reduced flood damage outweigh those considerations. Comments are included from interested agencies. Alternatives included flood plain zoning and levees, but these were dismissed as only providing minimal benefits, or as being too expensive. (Smith-Adam-Florida)
W73-08217

KAHULUI HARBOR WEST BREAKWATER REPAIR, MAUI, HAWAII (FINAL ENVIRONMENTAL IMPACT STATEMENT).
Army Corps of Engineers, Honolulu, Hawaii. Pacific Ocean Div.

Available from the National Technical Information Service as EIS-HI-72-5007-F, $4.00 paper copy, $1.45 in microfiche. May 17, 1972. 24 p, 4 plate.

Descriptors: *Environmental effects, *Breakwaters, *Hawaii, *Harbors, Water resources development, Maintenance, Navigation, Concrete structures, Air pollution, Coral, Dusts.
Identifiers: *Environmental Impact Statements, *Maui (Hawaii).

This project consists of maintenance and repair work on the breakwater trunk of the Kahului Harbor West in Maui, Hawaii, the only deepwater port on the island of Maui. It is located approximately 94 miles southeast of Honolulu and is the only existing federal navigation project on the island. The repair work would alter the physical condition of the west breakwater by the placement of 19 and 35-ton concrete units on the seaward slope and the construction of concrete ribs on the cap of the breakwater in order to maintain the continued safe

This proposed watershed project for the T or C Williamsburg Arroyos Watershed in Sierra County, New Mexico, consists of conservation land treatment measures, four flood water retarding structures, construction on one mile of channel and pipeline and .6 mile of floodway. Runoff from high intensity summer rains on the steep, unstable and sparsely vegetated watershed causes the flooding, and sediment outwash from the Arroyos is deposited in the valley in alluvial fans. Favorable environmental effects include a 10% reduction of erosion, a 66% reduction of sediment delivered to the Rio Grande, protection of 300 homes and 60 commercial properties from storm consequences, and a 96% reduction in flood damage to 658 acres of urban land and irrigated cropland. Adverse effects include an elimination of 34 acres of rangeland to be used for dams and spillways and periodic interruption of agricultural use and wildlife habitat of the 152 acres in the sediment and detention pools. Alternatives considered include public land acquisition of flood prone areas, conservation land treatment measures alone, or flood proofing of existing fixed improvements and zoning the undeveloped areas for open spaces. (Mockler-Florida)
W73-08220

PEARL RIVER BASIN, EDINBURG DAM AND LAKE, MISSISSIPPI AND LOUISIANA (FINAL ENVIRONMENTAL IMPACT STATEMENT),
Army Engineer District, Mobile, Ala.

Available from the National Technical Information Service as EIS-MS-72-5331-F, $9.00, in paper copy, $1.45 in microfiche. September 1972. 131 p, 1 map, 1 plate.

Descriptors: *Mississippi, *Louisiana, *Environmental effects, *Multiple-purpose reservoirs, *Water management (Applied), Reservoirs, Flood control, Flood protection, Water quality control, Recreation, Fishery habitat, Streamflow, Dam construction, Spillways, Flow augmentation, Wildlife habitats, Land use.
Identifiers: *Environmental Impact Statements, *Pearl River Basin (Miss.).

This project would authorize the construction of a dam and multiple-purpose reservoir on the Pearl River in Neshoba County, Mississippi, for the purposes of flood control, water quality control, general recreation and fish and wildlife enhancement. Alternatives include no development or construction at alternative sites or single-purpose projects which are less efficacious and economically infeasible. The project would consist of the conversion of 16,000 acres of agricultural and forest lands to a lake environment; loss of free-flowing stream habitat; flood protection and low flow augmentation for water quality control; and high quality recreation. Adverse environmental effects include the loss of 16,000 acres of agricultural and forest land and its associated wildlife habitat as well as the loss of free-flowing stream fisheries and the disruption to inhabitants. (Mockler-Florida)
W73-08221

SAN LUIS UNIT, CENTRAL VALLEY PROJECT, CALIFORNIA (FINAL ENVIRONMENTAL IMPACT STATEMENT),
Bureau of Reclamation, Sacramento, Calif. Mid-Pacific Regional Office.

Available from the National Technical Information Service as EIS-CA-72-5404-F, $7.75, in paper copy, $1.45 in microfiche. October 4, 1972. 113 p, 3 map, 1 chart.

Descriptors: *California, *Environmental effects, *Water supply, *Irrigation programs, *Water supply development, Irrigation, Reservoirs, Dams, Drainage systems, Water quali control, Groundwater, Agriculture, Wildlifetyhabitats, Water distribution (Applied).
Identifiers: *Environmental Impact Statements, *San Joaquin Valley (Calif).

This project, the San Luis Unit of the Central Valley Project in California on the west side of the San Joaquin Valley, is designed to deliver supplemental water supplies principally for irrigation of about 600,000 acres of fertile land. The San Luis unit consists of one major dam and reservoir, a forebay dam and reservoir, two detention dams and reservoirs, one pumping plant, two pump generating plants, two major canals, a distribution and drainage collection system, and a major drainage conveyance canal. Alternatives considered included the construction of a desalinization plant and a waste water reclamation project. There is a reasonable basis for concern about the possibility of adverse environmental effects from discharging subsurface agricultural drain flows at the western edge of the delta. Desirable effects include supplemental water supplies for 600,000 acres of fertile land to protect and enhance the existing agricultural economy which previously relied heavily on overdrafting ground water. (Mockler-Florida)
W73-08223

GEOLOGICAL AND GEOHYDROLOGICAL STUDIES FOR ANGOSTURA DAM, CHIAPAS, MEXICO,
Comision Federal de Electricidad, Mexico City.
C. G. Herrera.
Geological Society of America Bulletin, Vol 84, No 5, p 1733-1742, May 1973. 4 fig, 14 ref.

Descriptors: *Hydrogeology, *Karst hydrology, *Damsites, Fractures (Geologic), Grouting, Dam foundations, Grout curtains, Leakage, Reservoir leakage, *Mexico.

A high dam is being constructed in the Angostura Canyon 60 km southeast from the city of Tuxtla Gutierrez, the capital of the State of Chiapas, Mexico. Beds of limestone are present, some of which are soluble and karstic within portions of the proposed reservoir area. Study of the leakage potential along the future reservoir's banks was by means of pressure tests in core holes. Packers were used to isolate sections undergoing investigation. Tests were conducted from the bottom of each hole by progressive intervals upward. Within the limits of the reservoir subject to leakage, eight fractures were located on the reservoir's right bank and five on the left bank. The fractures will be treated by grouting from systems of tunnels to be driven at two levels in the plane of the dam's axis and effect a curtain extending 150 to 200 m into each abutment. These tunnels will serve also as drainage galleries to relieve pressures and intercept water that otherwise would bypass the dam. (Knapp-USGS)
W73-08376

BEACH PROTECTION SYSTEM,
N. P. Rasmussen.
U. S. Patent No 3,712,069, 3 p, 8 fig, 3 ref; Official Gazette of the United States Patent Office, Vol 906, No 4, p 1213, January 23, 1973.

Descriptors: *Patents, *Shore protection, *Beach erosion, Waves (Water), Beaches, Coasts, *Coastal engineering, *Erosion control.

The beach protection system consists of an upright bulkhead extending along the highwater mark. It is fabricated of flat plates in edge to edge coplanar relationship. The lower section is embedded in the beach. Sand fill is placed on the inland side to the edge of the upper bulkhead. A ramp extends downward to the beach level on the seaward side to resist erosion. A brace is placed between the bulkhead and the ramp to reinforce it against abnormally high tides. (Sinha-OEIS)
W73-08393

8B. Hydraulics

IDENTIFICATION OF MULTIPLE REACH CHANNEL PARAMETERS,
California Univ., Los Angeles. Dept. of Engineering Systems.
For primary bibliographic entry see Field 02E.
W73-07887

OPTIMIZATION OF DEAD END WATER DISTRIBUTION SYSTEMS,
Roorkee Univ. (India). Dept. of Civil Engineering.
For primary bibliographic entry see Field 04A.
W73-07920

BOUNDARY EFFECTS ON STABILITY OF STRUCTURES,
Uttar Pradesh Irrigation Research Inst., Roorkee (India).
A. S. Chawla.
Journal of the Hydraulic Division, American Society of Civil Engineers, Vol 98, No HY9, p 1557-1573, Sept 1972. 6 fig, 15 ref, 2 append.

Descriptors: *Seepage, *Structural stability, *Uplift pressure, Water pressure, Bibliographies, Hydraulic structure, Boundary layers, Foundations, Hydraulic gradient, Hydraulics, Permeability, Soil mechanics, Stability, Cutoffs, Mathematical analysis, Confined water, Pervious soils.
Identifiers: India, Transformations.

Design of hydraulic structures founded on permeable soils of finite depth and with finite or infinite pervious reaches on the upstream and downstream sides is controlled by water seepage under the foundations. Water seeping under such structures exerts pressure on the bottom and tends to wash away supporting soil, often causing extensive damage. An exact solution is presented for determining the exit gradient and uplift pressures on the bottom of such structures. Uplift pressures below the structures increase with an increase in the upstream pervious reach and decrease with an increase in the downstream pervious reach. The value of the exit gradient also increases with an increase in the upstream pervious reach; whereas, it decreases with an increase in the downstream pervious reach. Results have been plotted in easy-to-use curves. (USBR)
W73-08058

HYDRAULIC LABORATORY STUDIES OF A 4-FOOT-WIDE WEIR BOX TURNOUT STRUCTURE FOR IRRIGATION USE,
Bureau of Reclamation, Denver, Colo. Engineering and Research Center.
U. J. Palde.
Available from NTIS, Springfield, Va 22151 as PB-213 138, Price $3.00 printed copy; $1.45 microfiche. Report REC-ERC-72-31, September 1972. 12 p, 11 fig, 2 tab, 4 ref.

Descriptors: *Hydraulic models, *Weirs, *Turnouts, *Irrigation design, Model studies, Testing procedures, Baffles, Flow measurement, Hydraulic design, Discharge coefficient, Head loss, Calibrations, Correlation analysis, Design criteria.

A hydraulics laboratory study helped develop the design of a 10-cfs-capacity irrigation weir box structure and determined the head-discharge relationship. A baffle arrangement was developed to adequately distribute the inflow from a circular pipe and provide satisfactory approach flow conditions to the 4-foot-wide measuring weir. Discharge rating calibration was obtained for flows between approximately 2 and 12.5 cfs. The calibration did not agree with accepted equations for suppressed rectangular weirs because of irregular approach flow conditions and the method of measuring head. A discharge equation was derived by combining the Kindsvater-Carter method with a variable effective head concept which considers the disturbances in the approach flow. (Woodard-USGS)
W73-08086

EFFECTIVE HYDRAULIC ROUGHNESS FOR CHANNELS HAVING BED ROUGHNESS DIFFERENT FROM BANK ROUGHNESS,
Army Engineer Waterways Experiment Station, Vicksburg, Miss.
R. G. Cox.
Miscellaneous Paper H-73-2, February 1973. 52 p, 18 plate, 12 tab, 14 ref, 3 append.

Descriptors: *Flow resistance, *Roughness (Hydraulic), *Channel flow, *Channel morphology, *Model studies, Boundaries (Surfaces), Numerical analysis, Laboratory tests, Analytical techniques, Fluid mechanics.
Identifiers: Channel bed roughness, Channel bank roughness.

Two numerical studies are summarized: one laboratory experimental study, and one analytical study of flow resistance in rectangular channels having one type or degree of boundary roughness on the bed and a second type or degree of roughness on the sidewalls. The numerical study showed that usual procedures for estimating resistance coefficients for channels of this type are not adequate. The laboratory study indicated that the relations between the effective Manning's n, the sidewall Manning's n, the bed Manning's n (the n values are developed from flow data in channels of single roughness type), and the channel geometry could be established empirically by extensive laboratory testing. The analytical study proved that laboratory studies using a smooth surface for one of its boundaries had little application to practical design problems. The best solution to the problem probably would result by use of hydraulic roughness values to describe the boundary surfaces. (Woodard-USGS)
W73-08350

8C. Hydraulic Machinery

SYSTEMS APPROACH TO POWER PLANNING,
Bechtel Corp., San Francisco, Calif. Hydro and Community Facilities Div.
J. G. Thon.
Journal of the Hydraulics Division, American Society of Civil Engineers, Vol 99, No HY4, Proceedings paper 9661, p 589-598, April, 1973. 4 fig, 12 ref.

Descriptors: *Electric power industry, *Power system operation, Hydraulics, *Systems analysis, Pumped storage, Thermal powerplants, Planning, Decision making, Mathematical models, Systems engineering.
Identifiers: *Power loads, Site selection, Load flow, National Power Survey (1970), West Region (U.S.).

The development and use of systems engineering techniques in the power industry are reviewed. The power pools of the West Region of the 1970 National Power Survey are used as background for an analysis of system engineering applications in planning for load growth, the sequential position of various types of generating units under the load curve, and interconnection. The place of pumped storage, thermal electric plant siting, and the expected future shifting role of interconnection to reserve status are reviewed. System techniques are of great value to the power industry in estimating reliability and economic aspects of alternative expansion strategies. They are expected to be of increasing value in evaluating environmental effects of alternative expansion strategies and in interacting with the public and regulating agencies; evaluating the effects of alternative social constraints. (Bell-Cornell)
W73-07925

8D. Soil Mechanics

EFFECT OF RESERVOIR FILLING ON STRESSES AND MOVEMENTS IN EARTH AND ROCKFILL DAMS,
California Univ., Berkeley. Col. of Engineering.
For primary bibliographic entry see Field 08D.
W73-07854

EFFECT OF RESERVOIR FILLING ON STRESSES AND MOVEMENTS IN EARTH AND ROCKFILL DAMS,
California Univ., Berkeley. Col. of Engineering.
E. S. Nobari, and J. M. Duncan.
Available from NTIS, Springfield, Va. 22151 as AD-745 216. Price $3.00 printed copy; $1.45 microfiche. Army Engineer Waterways Experiment Station Contract Report S-72-2, January 1972. 186 p, 72 fig, 9 tab, 50 ref, append. USCE Contract DACW39-68-C-0078.

Descriptors: *Impoundments, *Reservoirs, *Stress analysis, *Rockfill dams, *Reviews, Analytical techniques, Evaluation, Engineering structures, Dam foundations, Dam failure, Design criteria, Wettability, Physical properties, Hydrologic data, Hydraulic structures.

The effects of reservoir filling on the stresses and movements in earth and rockfill dams were analyzed. Also included are reviews of the behavior of embankments during reservoir filling, and experimental investigations of the effects of water on the properties of granular materials. Review of the published case histories indicates that reservoir filling has induced significant movements in many well-engineered large dams as well as in small dams. When these movements develop as differential movements in dams and foundations, cracking and consequent erosion and failures have been observed in a number of dams. One of the major factors contributing to the development of differential movements during reservoir filling is the compression due to wetting which occurs in a wide variety of different types of soils. Investigations of the effects of wetting on properties of granular material similar to those used in rock-fill dams showed that moisture had a considerable effect on strength, compressibility, stress-strain, and volume change characteristics of the materials tested. (Knapp-USGS)
W73-07854

BRANTLEY PROJECT, NEW MEXICO (FINAL ENVIRONMENTAL IMPACT STATEMENT).
Bureau of Reclamation, Denver, Colo.

Available from the National Technical Information Service as EIS-NM-72-5263-F, $6.50 paper copy, $1.45 microfiche. 1972. 81 p, 2 map.

Descriptors: *Dam construction, *New Mexico, *Reservoir construction, *Environmental effects, Dams, Flood control, Recreation, Flood protection, Flood routing, Flood plain zoning, Wildlife habitats, Water control, Sport fishing, Water management (Applied).
Identifiers: *Pecos River, *Environmental Impact Statements.

This project consists of the construction of a concrete and earthfill dam and reservoir on the Pecos River in Eddy County, 24 miles upstream from Carlsbad, New Mexico. The Brantley Reservoir will have a permanent pool of 2,000 acre-feet, will eliminate damage to property and threat to life from floods, will result in a net increase in fishing of about 49,000 man-days per year, and will pro-

ide major recreation opportunities. Adverse environmental effects include inundation of an istorical site, a loss of approximately 7,000 acres f wildlife habitat, the partial reduction in value of 7,900 acres, and inundation of a 5-mile stretch of e river and contiguous valley lands, including ome 2,900 acres of agriculturally productive ands. Alternatives to the project include modificaon of McMillan and Avalon Dams, dredging of McMillan Reservoir, enlargement of Avalon Dam, nd flood plain management projects. Addiionally, use of the Carlsbad damsite has been sugested as another alternative. (Mockler-Florida)
W73-07976

ANALYSES OF WACO DAM SLIDE,
Texas Univ., Austin.
S. G. Wright, and J. M. Duncan.
Journal of the Soil Mechanics and Foundations Division, American Society of Civil Engineers, Vol 98, No SM9, p 869-877, Sept 1972. 10 fig, 6 ref, append.

Descriptors: *Anisotropy, *Stability analysis, *Foundation failure, *Dam failure, Dam foundations, Dam stability, Earth dams, Shale, Safety factors, Foundation investigations, Soil mechanics, Soil investigations, Soil strength, Fissures, Clays, Embankments, Shear strength, Sliding.
Identifiers: Progressive failures, Creep rupture strength, Horizontal movement, Waco Dam (Tex), Failure surfaces, Slip surfaces, Embankment subsidence, Slip-circle method.

Significant influence of the anisotropy of Pepper shale on the failure of Waco Dam shows the importance of considering anisotropy in stability analyses of embankments founded on heavily overconsolidated, stiff-fissured clays and clay shales. During construction of Waco Dam, a 1,500-ft-long section of the 85-ft high embankment slid, producing a failure surface that extended horizontally a considerable distance through the Pepper shale and surfaced about 700 ft downstream from the dam axis. Unconsolidated-undrained triaxial tests on the Pepper shale indicated that the strength along an initially horizontal plane was only about 40% of the strength measured in identical tests on vertical specimens. Other tests indicated insignificant strength reduction effects resulting from sustained loading, specimen size, and progressive failure. A stability analysis, which assumed an isotropic, uniform dam foundation with strength values from vertical shale specimens, indicated a safety factor of 1.32 for the most critical circle. When anisotropic soil strengths were used, the safety factor for the new critical circle was reduced to 1.07. Further analyses were made using horizontal failure surfaces through the Pepper shale. (USBR)
W73-08062

STATE-OF-THE-ART REVIEW OF THE DEFINITION OF QUALITY OF A SOIL SAMPLE AND MODERN TECHNIQUES OF SOIL SAMPLING,
Technische Universitaet, Berlin (West Germany).
H. Muhs.
Proceedings, 7th International Conference on Soil Mechanics and Foundations Engineering, Mexico City, p 104, 1969. German Res Assoc Soil Mech, Tech Univ, Berlin, Rep 27, p 41-51 1971. 9 fig, 4 ref.

Descriptors: *Sampling, *Soil investigations, *Subsurface investigations, Samples, Soil mechanics, Boring, Cores, Soil classifications, Core drilling, Reviews, Soils, Soil properties, Test specimens.
Identifiers: *Quality levels, *Soil samplers, Disturbed soils, Swedish foil sampler, Germany, Continuous sample method, Piston samplers, Undisturbed samples, Undisturbed soils, Disturbed samples, Drive samplers.

Classification of soil sample quality should be based on the soil properties that may be accurately determined from the sample rather than on the sampling method. In Germany, the sample quality classification ranges from 1 to 5, depending on the possibility of determining grain size distribution, moisture content, dry density, and shear strength. The quality class obtained is influenced by: (1) boring method, (2) sampling equipment, (3) soil type, and (4) the skill and conscientiousness of the driller. Sampling methods improved after World War II with the Swedish foil sampler which takes continuous cores from 5 to 40 m long and 6.0 to 6.8 cm in dia, but is suitable only for sampling soft soils. The Japanese have developed a modified foil sampler for use in sand and gravel. The Dutch developed 2 continuous core samplers, each having a large inside clearance that fills with fluids to support soft soil and loose sand samples. A third Dutch sampler encases the loose sand sample in a nylon stocking. The German samplers obtain cores 1 m long, and can obtain samples from depths of 40 m in stiff and hard soil. (USBR)
W73-08063

INTERACTION OF SOIL AND POWER PLANTS IN EARTHQUAKES,
Agbabian Associates, Los Angeles, Calif.
J. Isenberg, and S. A. Adham.
Journal of the Power Division, American Society of Civil Engineers, Vol 98, No PO2, p 273-291, Oct 1972. 15 fig, 2 tab, 18 ref, 2 append.

Descriptors: *Structural analysis, *Finite element method, Foundations, Soil dynamics, Computer models, Mathematical models, Structural behavior, Seismic properties, Nuclear powerplants, Dynamics, Bibliographies, Earthquakes, *Seismic tests, Earthquake engineering, Seismology.
Identifiers: Soil-structure interaction, Foundation models, Embedment depth, Elastic models, Computer-aided design, Embedded structures, Earthquake loads, Dynamic response, Dynamic loads.

Bedrock motions are generated from a segment of an earthquake record and applied to a 4-layer elastic finite element model of a soil site. Three different finite element models of an embedded nuclear reactor structure (45 ft below and 85 ft above ground surface) are developed. The least refined model is 1-dimensional, having lumped masses, rigidly connected to the soil and subjected to motions found at 45-ft depth in the free field. A second model is identical to the first, except that springs are interposed between the base of the 1-dimensional model and the soil to represent translational and rotational modes of interaction. The most refined model is a 2-dimensional, dynamic, elastic finite element representation of the soil-embedded foundation, walls, containment, and support structures. Analytical results show the importance of using the refined 2-dimensional model. For the structure analyzed, the lowest dynamic response is obtained by the 1-dimensional model without an interaction mechanism. (USBR)
W73-08065

THE ADAPTATION OF CRITICAL STATE SOIL MECHANICS THEORY FOR USE IN FINITE ELEMENTS,
O. C. Zienkiewicz, and D. J. Naylor.
Proceedings Roscoe Memorial Symposium, Stress-Strain Behavior of Soils, Cambridge University, Great Britain, Contribution 1, p 537-547, Mar 1971. 5 fig.

Descriptors: *Finite element method, *Soil mechanics, Deformation, Theoretical analysis, Mathematical studies, Computer models, Mathematical analysis, Stress analysis, Plasticity, *Soil analysis, Saturated soils, Analytical techniques, Mathematical models, Numerical analysis, Plastic deformation, Clays.
Identifiers: Analytical method, Great Britain, Stress-strain curves.

The critical state theory, developed by the late Kenneth Roscoe and his collaborators for modeling the deformational properties of soil, can be adapted for finite elements. The theory, developed for soft saturated clays, appears to have little application to dense clay or granular materials. The critical state theory, one of several nonlinear soil mechanics theories, has appeal because: (1) application can be made to a wide range of materials from soils to metals; (2) unloading can be more realistically simulated; and (3) relatively few parameters can be used to define the soil stress-strain properties. The theory assumes that creep and recoverable shear strains are negligible. An example presented uses the theory to predict triaxial test deformations. (USBR)
W73-08066

SMALL BOAT HARBOR, KING COVE, ALASKA (FINAL ENVIRONMENTAL IMPACT STATEMENT).
Army Engineer District, Anchorage, Alaska.

Available from the National Technical Information Service as EIS-AK-72-5513-F, $16.00 in paper copy, $1.45 in microfiche. May 1972. 38 p, 1 fig, 1 plate, 5 photo, 12 ref, 1 append.

Descriptors: *Alaska, *Environmental effects, *Harbors, *Navigation, Water quality control, Water pollution, Transportation, Dikes, Shellfish.
Identifiers: *King Cove, Alaska, *Environmental impact statement.

This small boat harbor project at King Cove, Alaska, consists of an earthfill training dike 1,200 feet long, a rock groin 200 feet long, a dredged entrance channel 400 feet long, and an 11-acre anchorage basin. The project would provide a protected mooring area for resident and transient fishing vessels during foul weather and off-season storage. Alternatives to the proposed project include no development, four alternative harbor sites or configurations, and one alternative quarry site. Adverse environmental effects include the loss of or alteration of 23.8 acres of the marine habitat and concomitantly the loss of shellfish, shellfish habitat, and waterfowl habitat. There will also be a temporary increase in water turbidity in the vicinity of the harbor construction. A long-term degradation of water quality within the confines of the harbor will occur due primarily to low-level chronic pollution. (Mockler-Florida)
W73-08209

8E. Rock Mechanics and Geology

CASE HISTORIES OF THREE TUNNEL-SUPPORT FAILURES, CALIFORNIA AQUEDUCT,
California State Dept. of Water Resources, Palmdale.
For primary bibliographic entry see Field 08A.
W73-08059

CONTROLLED FAILURE OF HOLLOW ROCK CYLINDERS IN UNIAXIAL COMPRESSION,
James Cook Univ. of North Queensland, Townsville (Australia).
E. T. Brown.
Rock Mechanics, Vol 4, No 1, p 1-24, June 1972. 11 fig, 5 tab, 41 ref.

Descriptors: *Rock mechanics, *Compressive strength, Rock tests, Bibliographies, Compression tests, Laboratory tests, Cracks, Stress distribution, Servomechanisms, Failure (Mechanics), Fractures, Finite element method, Stress analysis, Test procedures, Scale effect.
Identifiers: *Hollow cylinders, *Uniaxial tests, Uniaxial compression, End restraint, Test results, Platens, Australia, Stress-strain curves, Axial compression, Unconfined compression.

Behavior of hollow rock cylinders loaded in uniaxial compression is discussed. Elastic stress distributions calculated by the finite element method showed radial stresses considerably lower in hollow than in solid cylinders, and that the uniformity of stresses can be improved markedly by using loading platens having the same cross section as the specimen. Servocontrolled (stiff) uniaxial compression tests on solid and thick-walled hollow cylinders of white Tennessee marble showed no essential differences in behavior of the 2 specimen types with similar strengths and fracture phenomena being observed. In both servocontrolled (stiff) and conventional machine tests, the progressive formation of large numbers of short subaxial cracks was followed by the development of macrofractures, such as slabbing and shearing well past the peak of the stress-strain curve. The similar fracture phenomena observed indicated no extraneous fracture features were introduced by the servocontrolled tests. With few exceptions, the 1-in.-od solid and hollow cylinders were stronger than the 2-in-od cylinders, suggesting some degree of strength decrease with size increase. (USBR)
W73-08064

MAP OF DEPOSITS ESPECIALLY SUSCEPTIBLE TO COMPACTION OR SUBSIDENCE, PARKER QUADRANGLE, ARAPAHOE AND DOUGLAS COUNTIES, COLORADO,
Geological Survey, Washington, D.C.
For primary bibliographic entry see Field 07C.
W73-08174

8F. Concrete

SHOTCRETE IN HARD-ROCK TUNNELING,
California Univ., Berkeley.
T. L. Brekke.
Bulletin of the Association of Engineering Geologists, Vol 9, No 3, p 241-264, Summer 1972. 11 fig, 51 ref.

Descriptors: *Tunnel linings, *Tunnel construction, Bibliographies, Tunnels, Pneumatic systems, Concrete mixes, Concrete additives, Aggregates, Tunnel design, Economics, Rock properties, Supports, Quality control, Application equipment, Rock mechanics, Tunneling.
Identifiers: *Sprayed concrete, *Shortcrete, Tunnel supports, Accelerators, Curing compounds, Underground openings.

Shotcrete and composite systems involving shotcrete, both having proved efficient and economical in stabilizing many types of adverse ground conditions, are widely used for underground openings in all parts of the world. Shotcrete as a material, its contribution to tunnel stability, and method of application to a rock surface, are discussed. Satisfactory shotcrete placement requires proper control of aggregate, cement, and water quality, and the cement and accelerator must be compatible. Two basic methods of bringing the mix to the nozzle for application are as a dry or wet mix. Both methods are described. Experience and ability of the nozzleman greatly affects the mix quality and application. Shotcrete stabilizes tunnel openings by locking together adjacent blocks, thus acting as a reinformcement rather than a support. Although dependable mathematical design criteria are not available, well documented case histories from the U S and abroad reveal the circumstances under which shotcrete can contribute to the stability of underground openings. Careful study of these cases by engineering geologists and engineers will aid in using shotcrete correctly. (USBR)
W73-08060

BEHAVIOR OF NONRECTANGULAR BEAMS WITH LIMITED PRESTRESS AFTER FLEXURAL CRACKING,
Leeds Univ. (England).
E. W. Bennett, and N. Veerasubramanian.
Journal of the American Concrete Institute, Proceedings, Vol 69, No 9, p 533-542, Sept 1972. 8 fig, 3 tab, 6 ref, append.

Descriptors: *Cracks, *Deflection, *Prestressing, *Prestressed concrete, *Structural behavior, Beams (Structural), Cracking, Test procedures, Structural design, Laboratory tests, Reinforcing steels, Structural engineering.
Identifiers: *I-beams, *T-beams, *Flexural strength, Test results, Great Britain, India, Prestressed steel, Post-tensioning, Ultimate loads, Pretensioning.

Because of increasing interest in beams with limited prestress, laboratory tests were made to study the flexural behavior after cracking of beams with 4 different cross sections. Deflection, size of cracks, and ultimate strength were studied. The degree to which results obtained from rectangular beams can be applied to beams of nonrectangular sections was also determined. Two methods of calculating the deflection of the beams immediately after cracking were developed and an equation proposed for calculating the width of flexural cracks in beams. Results are compared with current design codes recommended. A numerical example demonstrates both methods of calculating prestressed and nonprestressed reinforcement. (USBR)
W73-08067

WASTEWATER TREATMENT STUDIES IN AGGREGATE AND CONCRETE PRODUCTION,
Smith and Monroe and Gray Engineers, Inc., Lake Oswego, Oreg.
For primary bibliographic entry see Field 05D.
W73-08126

8G. Materials

PHILADELPHIA SUBURBAN WATER COMPANY,
Philadelphia Suburban Water Co., Bryn Mawr, Pa.
For primary bibliographic entry see Field 05F.
W73-08109

DETROIT METRO WATER DEPARTMENT,
Detroit Metro Water Dept., Mich.
For primary bibliographic entry see Field 05F.
W73-08110

GEOLOGICAL AND GEOHYDROLOGICAL STUDIES FOR ANGOSTURA DAM, CHIAPAS, MEXICO,
Comision Federal de Electricidad, Mexico City.
For primary bibliographic entry see Field 08A.
W73-08376

8I. Fisheries Engineering

EXPERIMENTATION OF AN UNIVERSAL TYPE OF ELECTRIC FISH SCREEN ELECTRONICALLY CONTROLLED, USED TO IMPEDE FISH PENETRATION IN HYDROPOWER PLANTS, (IN RUMANIAN),
Institutul de Cercetari si Proiectari Piscicole, Bucharest (Rumania).
M. Niculescu-Duvaz, D. Matei, P. Tabacopol, C. Onu, and N. Reus.
Stud Cercet Piscic Inst Cercet Proiect Aliment. Vol 4, p 363-381, 1971. Illus. English summary.
Identifiers: Electric fish screens, Experimentation, *Fish screens, *Hydropower plants, *Instrumentation.

The most efficient installations electronically-controlled fish screens are those built by the impulse generator. The construction and mechanism system of this fish management tool is described.—Copyright 1972, Biological Abstracts, Inc.
W73-08274

EXPERIMENT IN RAISING COMMERCIAL CARP IN COOLING WATER IN THE ZMIEV AND MIRONOV STATE REGIONAL ELECTRIC POWER PLANTS, (IN RUSSIAN),
V. S. Prosyanyi, Z. A. Makina, and M. N. Mikulina.
Rybn Khoz Resp Mezhved Temat Nauchn Sb. V 13, p 17-19, 1971.
Identifiers: Algae, *Carp, *Cooling water, *Electric power plants, Mironov, Plankton, Retainers, USSR, Zmiev, Zooplankton, Cyanophyta.

Live retainers were put in a warm water discharge canal in the Don fish-raising area. The average 10-day temperature of the water was 25.4-32.0C; during some hours it reached 38 deg C. In the 'Limanskii' nursery of the Kharkov fish raising concern net retainers were placed in the Zmiev Reservoir 400 m from a warm water discharge area. The average 10-day temperature varied from 28-29 deg C. The fish fed on granulated food specially prepared for retainer raising. The natural food base was richer in the Zmiev Reservoir. Studies on the use of food showed that yearling carp in net retainers in the Zmiev Reservoir ate zooplankton; in the canal of the Mironov State Regional Electric Power Plant they ate blue-green algae. The consumption of combined food in the 'Limanskii' fish station was 3.3 g; in the Don fish raising area it was 12.7 g. In the area where the retainers were placed substantial significance in the effectiveness of retainer raising of carp was obtained.—Copyright 1972, Biological Abstracts, Inc.
W73-08275

SOME CHARACTERISTICS OF GENERAL METABOLISM IN CARP YEARLINGS DURING RETAINER RAISING IN WARM WATER, (IN RUSSIAN),
For primary bibliographic entry see Field 05C.
W73-08276

ROTENONE AND ITS USE IN ERADICATION OF UNDESIRABLE FISH FROM PONDS,
Freshwater Fisheries Research Station, Chandpur (Bangladesh).
For primary bibliographic entry see Field 05B.
W73-08279

OCTOMITUS TRUTTAE SCHMIDT, A PARASITE OF YOUNG TROUT IN THE FARM FISHPOND JADRO SOLIN, (IN SERBO-CROATIAN),
Pastrvsko Ribogojilistvo Jadro, Solin (Yugoslavia).
For primary bibliographic entry see Field 05G.
W73-08282

09. MANPOWER, GRANTS AND FACILITIES

9A. Education (Extramural)

WATER RESOURCES RESEARCH PROGRAM.
For primary bibliographic entry see Field 06E.
W73-08001

SCIENTIFIC AND TECHNICAL INFORMATION—Field 10

Preparation of Reviews—Group 10F

SUBJECT INDEX

LEGISLATION

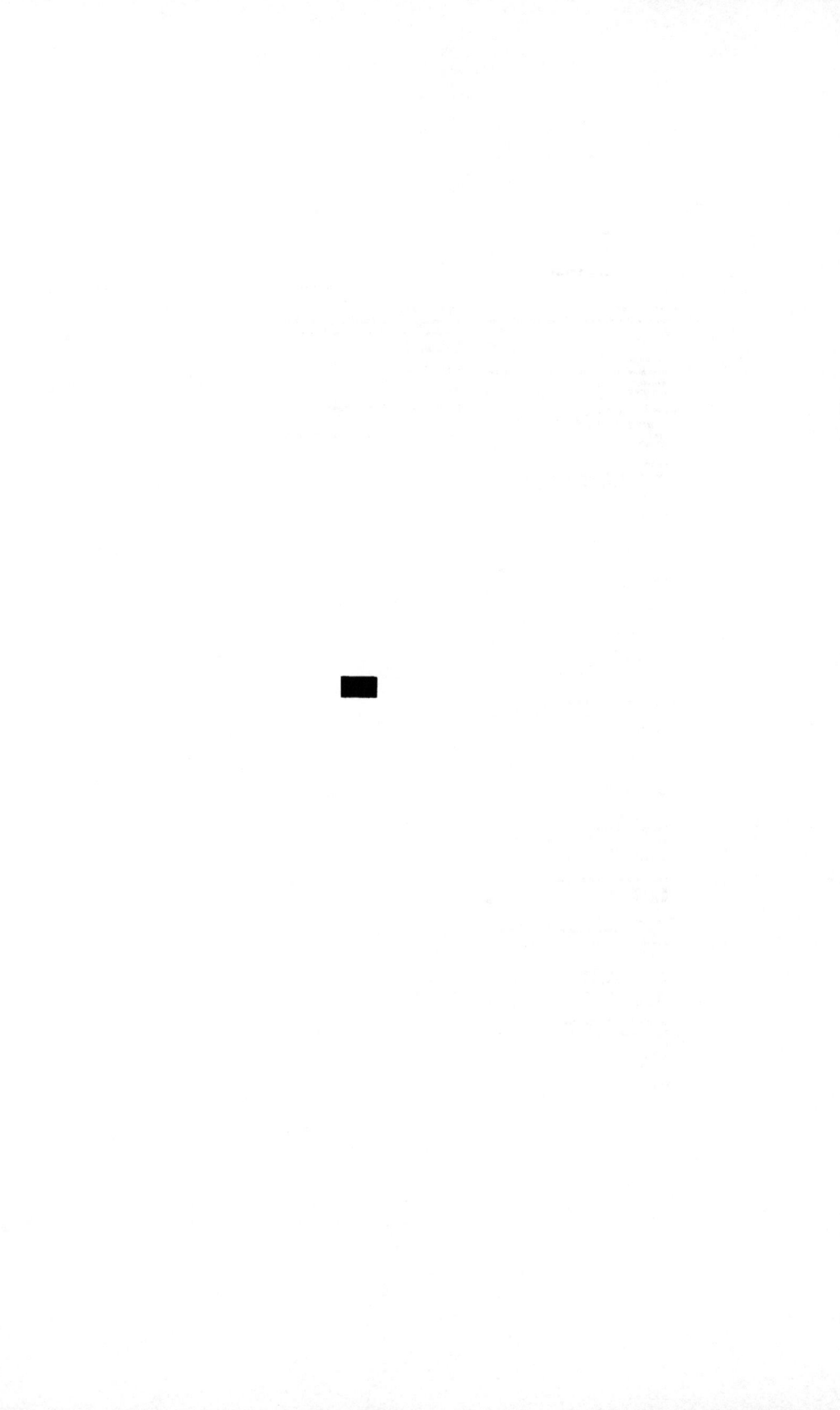

AUTHOR INDEX

ORGANIZATIONAL INDEX

ACCESSION NUMBER INDEX

W73-08113 5D
W73-08114 5A
W73-08115 5D
W73-08116 5D
W73-08117 5D
W73-08118 5D
W73-08119 5D
W73-08120 5D
W73-08121 5D
W73-08122 5D
W73-08123 5E
W73-08124 5D
W73-08125 5D
W73-08126 5D
W73-08127 5D
W73-08128 5D
W73-08129 5A
W73-08130 2H
W73-08131 3E
W73-08132 4B
W73-08133 6A
W73-08134 5B
W73-08135 2J
W73-08136 2H
W73-08137 2L
W73-08138 2C
W73-08139 4C
W73-08140 6A
W73-08141 5D
W73-08142 2C
W73-08143 4A
W73-08144 2C
W73-08145 2B
W73-08146 2C
W73-08147 2C
W73-08148 2C
W73-08149 4C
W73-08150 3B
W73-08151 2A
W73-08152 2C
W73-08153 2C
W73-08154 3B
W73-08155 7C
W73-08156 2L
W73-08157 2K
W73-08158 2I
W73-08159 7C
W73-08160 2H
W73-08161 2G
W73-08162 4A
W73-08163 2F
W73-08164 4C
W73-08165 2C
W73-08166 2C
W73-08167 2C
W73-08168 2C
W73-08169 2C
W73-08170 2C
W73-08171 2C
W73-08172 7C
W73-08173 7C
W73-08174 7C
W73-08175 2D
W73-08176 6E
W73-08177 6D
W73-08178 5D
W73-08179 6D
W73-08180 4A
W73-08181 6D
W73-08182 8A
W73-08183 6E
W73-08184 6D
W73-08185 3A
W73-08186 3A
W73-08187 3A
W73-08188 1A
W73-08189 5E
W73-08190 3A
W73-08191 8A

W73-08192 5F
W73-08193 6E
W73-08194 6G
W73-08195 6G
W73-08196 5G
W73-08197 4A
W73-08198 6E
W73-08199 6E
W73-08200 6G
W73-08201 6E
W73-08202 5D
W73-08203 6G
W73-08204 6E
W73-08205 5G
W73-08206 6E
W73-08207 5G
W73-08208 6E
W73-08209 8D
W73-08210 8A
W73-08211 8A
W73-08212 5G
W73-08213 8A
W73-08214 8A
W73-08215 6G
W73-08216 8A
W73-08217 8A
W73-08218 8A
W73-08219 8A
W73-08220 8A
W73-08221 8A
W73-08222 4A
W73-08223 8A
W73-08224 6G
W73-08225 6E
W73-08226 6E
W73-08227 6E
W73-08228 6E
W73-08229 6E
W73-08230 6E
W73-08231 6E
W73-08232 6E
W73-08233 6E
W73-08234 6E
W73-08235 5C
W73-08236 5A
W73-08237 5A
W73-08238 5B
W73-08239 5C
W73-08240 2D
W73-08241 5C
W73-08242 2I
W73-08243 5B
W73-08244 2H
W73-08245 5A
W73-08246 5C
W73-08247 5B
W73-08248 5C
W73-08249 5C
W73-08250 5D
W73-08251 5C
W73-08252 5C
W73-08253 5C
W73-08254 5C
W73-08255 5C
W73-08256 5A
W73-08257 7C
W73-08258 5C
W73-08259 5C
W73-08260 7B
W73-08261 2K
W73-08262 5A
W73-08263 5A
W73-08264 5A
W73-08265 5A
W73-08266 5A
W73-08267 5A
W73-08268 5C
W73-08269 5G
W73-08270 2H

W73-08271 5B
W73-08272 2I
W73-08273 2I
W73-08274 8I
W73-08275 8I
W73-08276 5C
W73-08277 2H
W73-08278 2H
W73-08279 5B
W73-08280 2I
W73-08281 2I
W73-08282 5G
W73-08283 2H
W73-08284 2I
W73-08285 2H
W73-08286 5G
W73-08287 5D
W73-08288 5G
W73-08289 5A
W73-08290 5D
W73-08291 5D
W73-08292 5D
W73-08293 5D
W73-08294 5B
W73-08295 5A
W73-08296 5D
W73-08297 5D
W73-08298 5D
W73-08299 5G
W73-08300 7A
W73-08301 2C
W73-08302 6B
W73-08303 5B
W73-08304 5D
W73-08305 5G
W73-08306 5D
W73-08307 5G
W73-08308 3A
W73-08309 5G
W73-08310 3A
W73-08311 5D
W73-08312 3A
W73-08313 5G
W73-08314 5G
W73-08315 5G
W73-08316 5G
W73-08317 5D
W73-08318 5D
W73-08319 5D
W73-08320 5D
W73-08321 5G
W73-08322 5G
W73-08323 5G
W73-08324 3F
W73-08325 2I
W73-08326 2H
W73-08327 3F
W73-08328 3F
W73-08329 3F
W73-08330 5C
W73-08331 5C
W73-08332 2I
W73-08333 4A
W73-08334 2I
W73-08335 5C
W73-08336 2H
W73-08337 2D
W73-08338 3F
W73-08339 5C
W73-08340 7B
W73-08341 2G
W73-08342 2K
W73-08343 2I
W73-08344 5C
W73-08345 2I
W73-08346 3F
W73-08347 3C
W73-08348 2D
W73-08349 2G

W73-08350 8B
W73-08351 2K
W73-08352 2G
W73-08353 2G
W73-08354 2G
W73-08355 2J
W73-08356 2G
W73-08357 2A
W73-08358 7B
W73-08359 7B
W73-08360 7B
W73-08361 7B
W73-08362 7B
W73-08363 7B
W73-08364 7B
W73-08365 7B
W73-08366 7B
W73-08367 4B
W73-08368 5B
W73-08369 7C
W73-08370 7C
W73-08371 2J
W73-08372 2J
W73-08373 2J
W73-08374 2J
W73-08375 2J
W73-08376 8A
W73-08377 2K
W73-08378 2J
W73-08379 4B
W73-08380 4A
W73-08381 7C
W73-08382 5B
W73-08383 5D
W73-08384 5B
W73-08385 5D
W73-08386 5D
W73-08387 6B
W73-08388 5D
W73-08389 5D
W73-08390 5D
W73-08391 5D
W73-08392 3A
W73-08393 8A
W73-08394 5D
W73-08395 5D
W73-08396 5D
W73-08397 3A
W73-08398 4A
W73-08399 3F
W73-08400 3F
W73-08401 3B
W73-08402 5C
W73-08403 2L
W73-08404 2H
W73-08405 2I
W73-08406 7B
W73-08407 2I
W73-08408 5B
W73-08409 2G
W73-08410 5D
W73-08411 5D
W73-08412 5D
W73-08413 5G
W73-08414 5D
W73-08415 5E
W73-08416 5D
W73-08417 2I
W73-08418 2I
W73-08419 4A
W73-08420 2I
W73-08421 3F
W73-08422 3B
W73-08423 5B
W73-08424 5A
W73-08425 2J
W73-08426 5C
W73-08427 5C
W73-08428 5C

5C
5B
5C
5G
5C
5C
2I
2G
2B
7B
2I
2G
5C
5A
2L
2I
2L
5C
5B
5C
3F
7B

ABSTRACT SOURCES

Source	Accession Number	Total
A. Centers of Competence		
AEC Oak Ridge National Laboratory, Nuclear Radiation and Safety	W73-07926 — 07929 07931 — 07963 07966 — 07974	46
Battelle Memorial Institute, Methods for Chemical and Biological Identification of Pollutants	W73-08013 — 08019 08021 — 08025 08027 — 08038 08040 — 08051 08235 — 08239 08241, 08243 08245 — 08268	67
Bureau of Reclamation, Engineering Works	W73-08058 — 08060 08062 — 08067	9
Cornell University, Policy Models for Water Resources Systems	W73-07916 — 07925 08131 — 08133 08300 08382 — 08387	20
University of Florida, Eastern U. S. Water Law	W73-07975 — 08011 08191 — 08234	81
University of North Carolina, Metropolitan Water Resources Planning and Management	W73-08052 — 08056	5
University of Texas, Wastewater Treatment and Management	W73-07833 — 07842 08094 — 08098 08100 — 08103 08105 — 08125 08410 — 08416	57
University of Wisconsin, Water Resources Economics	W73-07822 — 07831	10
U. S. Geological Survey, Hydrology	W73-07843 — 07915 08068 — 08093 08134 — 08135 08137 — 08157 08159 — 08174 08349 — 08381	171
B. State Water Resources Research Institutes		
Alabama Water Resources Research Institute	W73-07802	
Arizona Water Resources Research Center	W73-08301	
Connecticut Institute of Water Resources	W73-08302	
Florida Water Resources Research Center	W73-07803	

ABSTRACT SOURCES

Source	Accession Number	To
B. State Water Resources Research Institutes (Cont'd)		
Illinois Water Resources Center	W73-07804, 07814	
Indiana Water Resources Research Center	W73-07815	
Louisiana Water Resources Research Institute	W73-07816	
Massachusetts Water Resources Research Center	W73-07805	
Missouri Water Resources Research Center	W73-07806 — 07808	
Montana Water Resources Research Center	W73-07817	
New York Water Resources and Marine Sciences Center	W73-07809 — 07811	
North Carolina Water Resources Research Institute	W73-07818 — 07819 08303	
Oregon Water Resources Research Institute	W73-08175	
Tennessee Water Resources Research Center	W73-07820	
Virginia Water Resources Research Center	W73-07821	
C. Other		
BioSciences Information Service	W73-07832, 07930 08012, 08020 08026, 08039 08057, 08061 08099, 08104 08130, 08136 08158, 08240 08242, 08244 08269 — 08285 08324 — 08348 08398 — 08409 08417 — 08450	
Engineering Aspects of Urban Water Resources (Poertner)	W73-08176 — 08184 08299	
Environmental Protection Agency	W73-07812 — 07813 07965 08126 — 08129 08286 — 08298	
Ocean Engineering Information Service	W73-08304 — 08323 08388 — 08397	
Office of Saline Water	W73-08185 — 08190	
Office of Water Resources Research	W73-07801, 07964	

CENTERS OF COMPETENCE AND THEIR SUBJECT COVERAGE

- Ground and surface water hydrology at the Water Resources Division of the U.S. Geological Survey, U.S. Department of the Interior.
- Metropolitan water resources planning and management at the Center for Urban and Regional Studies of University of North Carolina.
- Eastern United States water law at the College of Law of the University of Florida.
- Policy models of water resources systems at the Department of Water Resources Engineering of Cornell University.
- Water resources economics at the Water Resources Center of the University of Wisconsin.
- Design and construction of hydraulic structures; weather modification; and evaporation control at the Bureau of Reclamation, Denver, Colorado.
- Eutrophication at the Water Resources Center of the University of Wisconsin, jointly sponsored by the Soap and Detergent Association and the Agricultural Research Service.
- Water resources of arid lands at the Office of Arid Lands Studies of the University of Arizona.
- Water well construction technology at the National Water Well Association.
- Water-related aspects of nuclear radiation and safety at the Oak Ridge National Laboratory.
- Public water supply treatment technology at the American Water Works Association.

Supported by the Environmental Protection Agency in cooperation with WRSIC

- Thermal pollution at the Department of Sanitary and Water Resources Engineering of Vanderbilt University.
- Water quality requirements for freshwater and marine organisms at the College of Fisheries of the University of Washington.
- Wastewater treatment and management at the Center for Research in Water Resources of the University of Texas.
- Methods for chemical and biological identification and measurement of pollutants at the Analytical Quality Control Laboratory of the Environmental Protection Agency.
- Coastal pollution at the Oceanic Research Institute.
- Water treatment plant waste pollution control at American Water Works Association.
- Effect on water quality of irrigation return flows at the Department of Agricultural Engineering of Colorado State University.

Subject Fields

1 NATURE OF WATER

2 WATER CYCLE

3 WATER SUPPLY AUGMENTATION AND CONSERVATION

4 WATER QUANTITY MANAGEMENT AND CONTROL

5 WATER QUALITY MANAGEMENT AND PROTECTION

6 WATER RESOURCES PLANNING

7 RESOURCES DATA

8 ENGINEERING WORKS

9 MANPOWER, GRANTS, AND FACILITIES

10 SCIENTIFIC AND TECHNICAL INFORMATION

INDEXES

SUBJECT INDEX

AUTHOR INDEX

ORGANIZATIONAL INDEX

ACCESSION NUMBER INDEX

ABSTRACT SOURCES

U.S. DEPARTMENT OF COMMERCE
National Technical Information Service
Springfield, Va. 22151

OFFICIAL BUSINESS

PRINTED MATTER

AN EQUAL OPPORTUNITY EMPLOYER

POSTAGE AND FEES PAID
U.S. DEPAR MEN OF COMMERCE

U.S. MAIL

JOHN LITTLEWOOD
.NTS DIVISION
RY

The Library of the

SEP 3 1974

University of Illinois
at Urbana-Champaign

SELECTED WATER RESOURCES ABSTRACTS

VOLUME 6, NUMBER 14
JULY 15, 1973

51 W73-09100

SELECTED WATER RESOURCES ABSTRACTS is published semimonthly for the Water Resources Scientific Information Center (WRSIC) by the National Technical Information Service (NTIS), U.S. Department of Commerce. NTIS was established September 2, 1970, as a new primary operating unit under the Assistant Secretary of Commerce for Science and Technology to improve public access to the many products and services of the Department. Information services for Federal scientific and technical report literature previously provided by the Clearinghouse for Federal Scientific and Technical Information are now provided by NTIS.

SELECTED WATER RESOURCES ABSTRACTS is available to Federal agencies, contractors, or grantees in water resources upon request to: Manager, Water Resources Scientific Information Center, Office of Water Resources Research, U.S. Department of the Interior, Washington, D. C. 20240.

SELECTED WATER RESOURCES ABSTRACTS is also available on subscription from the National Technical Information Service. Annual subscription is $22 (domestic), $27.50 (foreign), single copy price $3. Certain documents abstracted in this journal can be purchased from the NTIS at prices indicated in the entry. Prepayment is required.

SELECTED
WATER RESOURCES ABSTRACTS

'A Semimonthly Publication of the Water Resources Scientific Information Center,
Office of Water Resources Research, U.S. Department of the Interior

MICROFICHE PRICE CORRECTION

The price for all documents listed in this issue as available in microfiche for $1.45 from the National Technical Information Service is in error. The correct price is 95 cents.

VOLUME 6, NUMBER 14
JULY 15, 1973

W73-08451 -- W73-09100

Secretary of the U. S. Department of the Interior has determined that the publication of this periodical ecessary in the transaction of the public business required by law of this Department. Use of funds printing this periodical has been approved by the Director of the Office of Management and Budget ugh August 31, 1973.

As the Nation's principal conservation agency, the Department of the Interior has basic responsibilities for water, fish, wildlife, mineral, land, park, and recreational resources. Indian and Territorial affairs are other major concerns of America's "Department of Natural Resources."

The Department works to assure the wisest choice in managing all our resources so each will make its full contribution to a better United States—now and in the future.

FOREWORD

Selected Water Resources Abstracts, a semimonthly journal, includes abstracts of current and earlier pertinent monographs, journal articles, reports, and other publication formats. The contents of these documents cover the water-related aspects of the life, physical, and social sciences as well as related engineering and legal aspects of the characteristics, conservation, control, use, or management of water. Each abstract includes a full bibliographical citation and a set of descriptors or identifiers which are listed in the **Water Resources Thesaurus.** Each abstract entry is classified into ten fields and sixty groups similar to the water resources research categories established by the Committee on Water Resources Research of the Federal Council for Science and Technology.

WRSIC IS NOT PRESENTLY IN A POSITION TO PROVIDE COPIES OF DOCUMENTS ABSTRACTED IN THIS JOURNAL. Sufficient bibliographic information is given to enable readers to order the desired documents from local libraries or other sources.

Selected Water Resources Abstracts is designed to serve the scientific and technical information needs of scientists, engineers, and managers as one of several planned services of the Water Resources Scientific Information Center (WRSIC). The Center was established by the Secretary of the Interior and has been designated by the Federal Council for Science and Technology to serve the water resources community by improving the communication of water-related research results. The Center is pursuing this objective by coordinating and supplementing the existing scientific and technical information activities associated with active research and investigation program in water resources.

To provide WRSIC with input, selected organizations with active water resources research programs are supported as "centers of competence" responsible for selecting, abstracting, and indexing from the current and earlier pertinent literature in specified subject areas.

Additional "centers of competence" have been established in cooperation with the Environmental Protection Agency. A directory of the Centers appears on inside back cover.

Supplementary documentation is being secured from established discipline-oriented abstracting and indexing services. Currently an arrangement is in effect whereby the BioScience Information Service of Biological Abstracts supplies WRSIC with relevant references from the several subject areas of interest to our users. In addition to Biological Abstracts, references are acquired from Bioresearch Index which are without abstracts and therefore also appear abstractless in SWRA. Similar arrangements with other producers of abstracts are contemplated as planned augmentation of the information base.

The input from these Centers, and from the 51 Water Resources Research Institutes administered under the Water Resources Research Act of 1964, as well as input from the grantees and contractors of the Office of Water Resources Research and other Federal water resources agencies with which the

Center has agreements becomes the information base from which this journal is, and other information services will be, derived; these services include bibliographies, specialized indexes, literature searches, and state-of-the-art reviews.

Comments and suggestions concerning the contents and arrangements of this bulletin are welcome.

Water Resources Scientific Information Center
Office of Water Resources Research
U.S. Department of the Interior
Washington, D. C. 20240

CONTENTS

06 WATER RESOURCES PLANNING
Includes the following Groups: Techniques of Planning; Evaluation Process; Cost Allocation, Cost Sharing, Pricing/Repayment; Water Demand; Water Law and Institutions; Nonstructural Alternatives; Ecologic Impact of Water Development.

07 RESOURCES DATA
Includes the following Groups: Network Design; Data Acquisition; Evaluation, Processing and Publication.

08 ENGINEERING WORKS
Includes the following Groups: Structures; Hydraulics; Hydraulic Machinery; Soil Mechanics; Rock Mechanics and Geology; Concrete; Materials; Rapid Excavation; Fisheries Engineering.

09 MANPOWER, GRANTS, AND FACILITIES
Includes the following Groups: Education—Extramural; Education—In-House; Research Facilities; Grants, Contracts, and Research Act Allotments.

10 SCIENTIFIC AND TECHNICAL INFORMATION
Includes the following Groups: Acquisition and Processing; Reference and Retrieval; Secondary Publication and Distribution; Specialized Information Center Services; Translations; Preparation of Reviews.

SUBJECT INDEX

AUTHOR INDEX

ORGANIZATIONAL INDEX

ACCESSION NUMBER INDEX

ABSTRACT SOURCES

WATER RESOURCES ABSTRACTS

02. WATER CYCLE

2A. General

CHARACTERIZATION OF WATER MOVEMENT INTO AND THROUGH SOILS DURING AND IMMEDIATELY AFTER RAINSTORMS,
Kentucky Water Resources Inst., Lexington.
For primary bibliographic entry see Field 02G.
W73-08464

GROUNDWATER DEVELOPMENT AND ITS INFLUENCE ON THE WATER BALANCE,
Technical Univ., of Denmark, Copenhagen.
For primary bibliographic entry see Field 02F.
W73-08507

HIGH YIELD FROM THE BULL RUN WATERSHED,
British Columbia Dept. of Municipal Affairs, Victoria. Environmental Planning and Management Div.
L. V. Luchin.
Journal of the American Water Works Association, Vol 65, No 3, p 183-186, March 1973.

Descriptors: *Oregon, *Hydrologic cycle, *Evapotranspiration, *Hydrologic budget, Evaporation, Rainfall, Runoff, Soil water, Transpiration, Water balance, Water loss, Hydrologic data, Hydrologic aspects, Mathematical studies, Hydrologic equation, Discharge (Water), Hydrology, Water supply, Soil properties, Watersheds (Basins), Analytical techniques.

The Bull Run Watershed is located within the Mt. Hood National Forest about 25 miles east of Portland, Oregon. In order to determine whether water production is higher or lower than should be expected in the watershed, the magnitude of water loss to components of the hydrologic cycle must be measured. The actual discharge of the watershed is compared with computed yields. The hydrologic cycle is composed of such things as interception, evaporation, transpiration, and soil moisture storage. By utilizing a water balance technique, involving a balance sheet on which water use by the various hydrologic components is budgeted so that the amount of water entering a watershed is compared with the total amount discharged, it is possible to make reasonable estimates of the magnitude of water losses. Data required for computing the water balance are: (1) mean monthly watershed air temperature, (2) mean monthly watershed precipitation, (3) monthly actual and potential evapotranspiration loss, (4) monthly interception loss, (5) monthly evaporation loss, and (6) the amount of water held by the soil at field capacity. Using this technique, it was concluded that in the watershed the annual gross precipitation is greater than the annual water used, and the average annual water runoff is greater than that expected. Water loss constituting the greatest reduction of potential runoff occurred through evapotranspiration. (Adams-Florida)
W73-08667

HYDROGRAPH SIMULATION MODELS OF THE HILLSBOROUGH AND ALAFIA RIVERS, FLORIDA: A PRELIMINARY REPORT,
Geological Survey, Tampa, Fla.
For primary bibliographic entry see Field 04A.
W73-08733

EFFECTS OF COLUMBIA RIVER DISCHARGE ON CHLOROPHYLL A AND LIGHT ATTENUATION IN THE SEA,
Oregon State Univ., Corvallis. Dept. of Oceanography.
For primary bibliographic entry see Field 02L.
W73-08825

WORKING SYMPOSIUM ON SEA-AIR CHEMISTRY: SUMMARY AND RECOMMENDATIONS,
Rhode Island Univ., Kington. Graduate School of Oceanography.
For primary bibliographic entry see Field 02K.
W73-08850

STRUCTURE OF SEA WATER AND ITS ROLE IN CHEMICAL MASS TRANSPORT BETWEEN THE SEA AND THE ATMOSPHERE,
For primary bibliographic entry see Field 02K.
W73-08858

2B. Precipitation

ATMOSPHERIC EFFECTS ON OCEAN SURFACE TEMPERATURE SENSING FROM THE NOAA SATELLITE SCANNING RADIOMETER,
National Oceanic and Atmospheric Administration, Miami, Fla. Physical Oceanography Lab.
For primary bibliographic entry see Field 07B.
W73-08496

FILLING IN GAPS IN RAINFALL RECORDS BY SIMULATED DATA,
Food and Agriculture Organization of the United Nations, Iraklion (Greece). United Nations Development Programme.
B. Samuelsson.
Nordic Hydrology, Vol 4, No 1, p 17-27, 1973. 2 ref, 1 append.

Descriptors: *Simulation analysis, *Rainfall, Data processing, Statistics, Statistical methods, Data collections, Computer programs, Regression analysis, Synthetic hydrology.

The philosophy behind the generation of rainfall data is that historic rainfall records will never be repeated but that their statistical characteristics are preserved. This simulation should be based on historical records over a long enough period of time to include a sufficient number of extreme synoptic situations, in order to provide an acceptable overall statistical representation. The concept of the rainfall coincidence rate is introduced and is used as a tool for correcting the errors in the rainfall records which are due to shifts. The method of simulating missing rainfall data does not smooth the rainfall distribution pattern. Dates are allocated to the generated raindays according to a procedure which takes into account the coincidence rate as well as the occurrence of rainfall over groups of consecutive days. (Knapp-USGS)
W73-08506

THE ATMOSPHERIC EFFECTS OF THERMAL DISCHARGES INTO A LARGE LAKE,
Argonne National Lab., Ill.
For primary bibliographic entry see Field 05C.
W73-08623

THE INFLUENCE OF YEARLY WEATHER VARIATIONS ON SMOKE AND SULPHUR DIOXIDE POLLUTION IN PRETORIA,
National Physical Research Lab., Pretoria (South Africa).
For primary bibliographic entry see Field 05A.
W73-09038

INADVERTENT WEATHER AND PRECIPITATION MODIFICATION BY URBANIZATION,
Illinois State Water Survey, Urbana. Atmospheric Science Section.
For primary bibliographic entry see Field 04C.
W73-09077

2C. Snow, Ice, and Frost

SALINITY VARIATIONS IN SEA ICE,
Dartmouth Coll., Hanover, N.H. Dept. of Earth Sciences.
G. F. N. Cox, and W. F. Weeks.
In: AIDJEX Bulletin No 19, Washington University Division of Marine Resources, Seattle, p 1-17, March 1973. 11 fig, 1 tab, 13 ref.

Descriptors: *Sea ice, *Salinity, Topography, Brines, Water chemistry, Sampling, Cores, Regression analysis, Sea water.

The salinity distribution in multiyear sea ice is dependent on the ice topography and cannot be adequately represented by a single average profile. Cores collected from areas beneath surface hummocks generally show a systematic increase in salinity with depth from 0 at the surface to about 0.4 at the base. Cores collected from areas beneath surface depressions are much more saline with large salinity fluctuations. A strong correlation was found between the average salinity of the ice and the ice thickness. At 0.4 m the dominant brine drainage mechanism changes from brine expulsion to gravity drainage. An annual cyclic variation of the mean salinity probably exists for multiyear sea ice. The mean salinity should reach a maximum at the end of the growth season and a minimum at the end of the melt season. (Knapp-USGS)
W73-08474

1972 AIDJEX INTERIOR FLOW FIELD STUDY, PRELIMINARY REPORT AND COMPARISON WITH PREVIOUS RESULTS,
Washington Univ., Seattle. Dept. of Oceanography.
J. L. Newton, and L. K. Coachman.
In: AIDJEX Bulletin No 19, Washington University Division of Marine Resources, p 19-42, March 1973. 20 fig, 5 tab, 4 ref.

Descriptors: *Ocean currents, *Sea ice, *Arctic Ocean, Winds, Currents (Water), Drifting (Aquatic).
Identifiers: Ice drift, Wind shear.

The interior flow field of the Arctic Ocean was studied by making current and hydrographic measurements simultaneously in an array of fixed dimensions and by observing the ice motion and local weather conditions. The four-week average horizontal currents were in geostrophic balance. Both the four-week average and daily average sea surface slope values were directed approximately 90 deg to the left of the wind stress. As ice motion was within 30 deg of downwind, Coriolis acceleration generally opposed the horizontal pressure gradient force, and their magnitudes were nearly equal. A change in sea surface slope appeared to be generated in one to two days, and the currents at 500 m and 850 m responded directly to these changes. (Knapp-USGS)
W73-08475

MECHANICAL MODELS OF RIDGING IN THE ARCTIC SEA ICE COVER,
Washington Univ., Seattle. Dept. of Aeronautics and Astronautics.
R. R. Parmerter, and M. D. Coon.
In: AIDJEX Bulletin No 19, Washington University Division of Marine Resources, p 59-112, March 1973. 19 fig, 19 ref, 1 append. NSF Grant GV 28807.

Descriptors: *Sea ice, *Arctic Ocean, *Mathematical models, *Stress, *Strain, Computer programs, Deformation, Plastic deformation, Mechanical properties.
Identifiers: *Ice ridges.

In a kinematic model of the formation of pressure ridges in Arctic sea ice, the lateral and vertical motion of ice blocks is combined with a force balance and breaking stress calculation. A computer program encompassing several physical processes simulates ridge formation in ice with thickness from 20 cm to 2 m. Significant dimensionless parameters are identified, and parametric studies are presented of the functional relationships between these parameters. The maximum height, crack location, and required force are functions of the mechanical and geometrical properties of the ice sheet. Typically a force increase of two orders of magnitude is required to make the transition from building a ridge from frozen ice in leads to building it from the parent ice sheet. The large forces are toward the upperend of the range of forces available, in agreement with the observation that most ridges are built from thin iceblocks. When large forces are available and the parent ice is fractured, ridges of limited height but unlimited lateral extent can occur. These correspond to the rubble fields which are occasionally observed. The proposed models agree with observational knowledge and provide predictions which can be tested by field experiments. (Knapp-USGS)
W73-08476

ICEBERGS AS A FRESH WATER SOURCE: AN APPRAISAL,
Cold Regions Research and Engineering Lab., Hanover, N.H.
W. F. Weeks, and W. J. Campbell.
Research Report 200, January 1973. 22 p, 16 fig, 1 tab, 24 ref.

Descriptors: *Icebergs, *Antarctic, *Water supply, South America, Australia, Navigation, Water resources.

The idea of transporting large icebergs to arid regions to provide a fresh water source is reviewed. Only in the Antarctic are supplies of large tabular icebergs available. Data on the size distribution of these icebergs are reviewed, and it is concluded that icebergs of almost any desired size can readily be located. Steady-state towing velocities of different sized icebergs are calculated based on estimates of the drag of the icebergs and pull of tugs. Because drag increases with velocity squared, large icebergs can be towed only at very slow velocities. However, tugs that can be built within the capabilities of current technology are capable of towing extremely large icebergs. Although melting losses are significant and may be excessive for small icebergs, when large icebergs are towed large amounts of ice are left when the iceberg arrives at its destination. Towing trajectories, travel times, and ice delivery rates are calculated for optimum routes between the Amery Ice Shelf and Western Australia and the Ross Ice Shelf and the Atacama Desert. Transit times exceed 107 and 145 days, respectively, with over 50% of the initial ice delivered. After total towing charges are paid, it is possible to deliver ice to Western Australia for 1.3 mills/cu m of water and to the Atacama Desert region for 1.9 mills/cu m. The water delivered by the operation of one super tug alone would irrigate 16,000 sq km. (Knapp-USGS)
W73-08478

AVALANCHES IN SOVIET CENTRAL ASIA (LAVINY SREDNEY AZII).
Sredneaziatskii Nauchno-Issledovatelskii Gidrometeorologicheskii Institut, Tashketn (USSR).

Sredneaziatskiy Nauchno-Issledovatel'skiy Gidrometeorologicheskiy Institut Trudy, No 63 (78), Leningrad, Yu. D. Moskalev, editor, 1972. 137 p.

Descriptors: *Avalanches, *Snow, *Snow cover, *Meteorology, *Topography, Mountains, Slopes, Snowfall, Precipitation (Atmospheric), Temperature, Winds, Synoptic analysis, Crystallography, Ice, Properties, Explosions, Analytical techniques, Forecasting, Seasonal, Equations.
Identifiers: *Soviet Central Asia, *Kazakhstan, *Tien Shan Mountains, Avalanche release, Avalanche hazard, Avalanche control, Snow properties, Snow depth.

Snow and avalanche formations in Soviet Central Asia and Kazakhstan, and avalanche forecasting and control techniques were investigated in the collection of 12 papers prepared by workers at the Central Asian Hydrometeorological Research Institute in Tashkent. The titles of the papers are: (1) mechanism of avalanche formation associated with snowfalls and snowstorms; (2) forms of avalanche release and movement; (3) analysis of present methods of avalanche-hazard evaluation; (4) techniques of avalanche forecasting; (5) quantitative estimates of avalanche formation in Western Tien Shan; (6) snow-avalanche parameters in connection with intentional avalanching and calculation of avalanche motion; (7) some physical properties of newly fallen snow in Western Tien Shan; (8) some observations of avalanching; (9) avalanche formation in Western Tien Shan during large snowfalls in the winter of 1968-69; (10) development of snow transformation processes in the complex mountain zone of Western Tien Shan; (11) methods of anticipating avalanche hazard from rainfall and snowfall intensities in mountains of Zailiski Ala Tau; and (12) redistribution of snow by winds in a mountainous terrain. (See W73-08481 thru W73-08490) (Josefson-USGS)
W73-08480

MECHANISM OF AVALANCHE FORMATION ASSOCIATED WITH SNOWFALLS AND SNOWSTORMS (O MEKHANIZME VOZNIKNOVENIYA LAVIN, SVYAZANNYKH SO SNEGOPADAMI I METELYAMI),
Sredneaziatskii Nauchno-Issledovatelskii Gidrometeorologicheskii Institut, Tashkent (USSR).
K. S. Losev.
In: Laviny Sredney Azii; Sredneaziatskiy Nauchno-Issledovatel'skiy Gidrometeorologicheskiy Institut Trudy, No 63 (78), p 3-11, Leningrad, 1972. 1 fig, 3 tab, 14 ref.

Descriptors: *Avalanches, *Snow, *Snow cover, *Snowfall, *Storms, Winds, Crystallography, Cohesion, Physical properties, Temperature, Forecasting.
Identifiers: *USSR, *Snowstorms, Snow crystals, Snow properties, Slab avalanches, Snow slabs, Wind slabs, Avalanche hazard.

Snow avalanche origin should be studied from the standpoint of snow crystals, since their original form determines the cohesion of newly deposited snow. The factors which influence cohesion of snow cover are type and density of snow crystals, temperature, and wind. Relation of the mechanism of avalanche origin to cohesiveness of newly deposited snow is examined, and recommendations are made for describing cohesiveness of avalanche-forming snow slabs on the basis of genetic, age, and crystallographic characteristics. Additional studies are required of crystallographic properties of deposited snow and of its structural changes in time and after snow deposition; also, a reexamination is needed of methods to determine the end of an avalanche-hazard period in empirical avalanche-forecasting procedures. (See also W73-08480) (Josefson-USGS)
W73-08481

FORMS OF AVALANCHE RELEASE AND MOVEMENT (O FORMAKH OBRUSHENIYA I DVIZHENIYA LAVIN),
Sredneaziatskii Nauchno-Issledovatelskii Gidrometeorologicheskii Institut, Tashkent (USSR).
Yu. D. Moskalev.
In: Laviny Sredney Azii; Sredneaziatskiy Nauchno-Issledovatel'skiy Gidrometeorologicheskiy Institut Trudy, No 63 (78), p 12-17, Leningrad, 1972. 1 tab, 25 ref.

Descriptors: *Avalanches, *Snow, *Movement, *Topography, *Properties, Physical properties, Mechanical properties, Stability, Stress, Deformation, Cohesion, Explosions, Explosives, Seismic waves, Regions, Equations.
Identifiers: *USSR, *Soviet Central Asia, *Tien Shan Mountains, *Avalanche release, Avalanche hazard, Avalanche control, Snow slabs.

The most effective method of avalanche control in avalanche-hazard regions of Western Tien Shan is artificial release by detonation of explosives. To relieve mountain slopes of unstable snow, 20-30 kg of ammonite is usually considered a sufficiently reliable charge. An air wave caused by blasting with high explosives can initiate avalanches of newly fallen snow. Avalanches can be triggered by explosives when the coefficients of snow stability are greater than 1; if the coefficients are greater than 3 or 4, the use of explosives is not always successful. Data on physico-mechanical properties of snow during artificial release in Western Tien Shan are tabulated together with data on maximum distance of flowage of avalanches in the Khibiny Mountains and Soviet Central Asia. (See also W73-08480) (Josefson-USGS)
W73-08484

SOME PHYSICAL PROPERTIES OF NEWLY FALLEN SNOW IN WESTERN TIEN SHAN (NEKOTORYYE FIZICHESKIYE SVOYSTVA SVEZHEVYPAVSHEGO SNEGA V ZAPADNOM TYAN'-SHANE),
Sredneaziatskii Nauchno-Issledovatelskii Gidrometeorologicheskii Institut, Tashkent (USSR).
G. G. Kharitonov.
In: Laviny Sredney Azii; Sredneaziatskiy Nauchno-Issledovatel'skiy Gidrometeorologicheskiy Institut Trudy, No 63 (78), p 62-79, Leningrad, 1972. 6 fig, 3 tab, 6 ref.

Descriptors: *Snow, *Snowfall, *Mountains, *Physical properties, Avalanches, Density, Temperature, Air temperature, Thermocline, Cohesion, Mechanical properties, Measurement, Curves.
Identifiers: *USSR, *Tien Shan Mountains, *Snow properties, Snow stability, Snow crystals.

Density, temperature, and ultimate cohesive strength of newly fallen snow in mountains of Western Tien Shan were investigated in detailed studies carried out at the Naugarzan, Kyzylcha, and Dukant snow-avalanche stations of the Uzbek Administration of the Hydrometeorological Service in 1960-67. To eliminate random errors, measurements of cohesive strength were replicated 20-22 times. At a constant air temperature between two successive periods of observations, cohesion of snow increased by an average of 16.2 kg/sq m. A relation was constructed between the gain in cohesion for each degree of temperature change and negative air temperature. Termination of snowfall and presence of sufficiently high (more than 0.1 deg/cm) temperature gradients in a layer of newly fallen snow are considered basic indicators of the loss of cohesion during periods of reduced air temperatures. Using the methods described, ultimate cohesive strength of newly fallen snow can be computed on the basis of snow density and air temperature data. (See also W73-08480) (Josefson-USGS)
W73-08485

SOME OBSERVATIONS OF AVALANCHING (NEKOTORYYE NABLYUDENIYA NAD SKHODOM LAVIN),
Sredneaziatskii Nauchno-Issledovatelskii Gidrometeorologicheskii Institut, Tashkent (USSR).
A. I. Korolev.
In: Laviny Sredney Azii; Sredneaziatskiy Nauchno-Issledovatel'skiy Gidrometeorologicheskiy Institut Trudy, No 63 (78), p 80-88, Leningrad, 1972. 1 tab, 4 ref.

Descriptors: *Avalanches, *Snow, *Snowfall, *Movement, *Meteorology, Slopes, Cracks, Explosives, Explosions, Roads, Snow removal, Winter.
Identifiers: *USSR, Avalanche release, Avalanche hazard, Avalanche control, Avalanche fracture lines, Snow properties, Snow slabs.

Freezing weather and large snowfalls in the winter of 1968-69 created a serious avalanche hazard in January in the region of the Dukant snow-avalanche station of the Uzbek Administration of the Hydrometeorological Service. Results are presented of a number of avalanche-control measures undertaken to effect artificial release of snow on slopes by the detonation of explosives. (See also W73-08480) (Josefson-USGS)
W73-08486

AVALANCHE FORMATION IN WESTERN TIEN SHAN DURING LARGE SNOWFALLS IN THE WINTER OF 1968-69 (USLOVIYA LAVINOOBRAZOVANIYA V MNOGOSNEZHNUYU ZIMU 1968/69 G. V ZAPADNOM TYAN'-SHANE).
Sredneaziatskii Nauchno-Issledovatelskii Gidrometeorologicheskii Institut, Tashkent (USSR).
M. K. Yefimov.
In: Laviny Sredney Azii; Sredneaziatskiy Nauchno-Issledovatel'skiy Gidrometeorologicheskiy Institut Trudy, No 63 (78), p 89-95, Leningrad, 1972. 1 tab, 8 ref.

Descriptors: *Avalanches, *Snow, *Snowpacks, *Snowfall, *Mountains, Slopes, Meteorology, Anticyclones, Cyclones, Precipitation (Atmospheric), Air temperature, Synoptic analysis, Forecasting, Seasonal, Winter.
Identifiers: *USSR, *Tien Shan Mountains, *Soviet Central Asia, *Kazakhstan, Avalanche release, Avalanche hazard, Snowslides, Snow slabs.

The winter of 1968-69 in Western Tien Shan was characterized by very low air temperatures, exceptionally heavy precipitation, thick snow cover, and intense avalanche activity. Average November-April air temperature in regions of snow-avalanche stations of the Uzbek Administration of the Hydrometeorological Service was minus 3.2 deg C, or 2.1 deg C lower than the long-term average. Precipitation at snow-avalanche stations for November-April varied between 1,300 and 1,500 mm, which was twice the normal precipitation (668 mm) for this period. Snow depth at meteorological plots of the Dukant and Kyzylcha stations reached 242 and 325 cm, respectively. Over 290 avalanches were recorded in the vicinity of the Dukant station, 125 avalanches at the Kyzylcha station, and about 400 avalanches and snowslides in the Naugarzan and Aktash River basins. Avalanching occurred at almost all stations at the same time. The avalanche danger which existed in the winter of 1968-69 arose from intrusions of cold arctic air, wave activity on the polar front, and especially from inrush of southern cyclones into mountainous areas of Soviet Central Asia and Kazakhstan. Provided snow conditions on mountain slopes of different exposure are known, background forecasts of the advent of avalanche-hazard periods can be prepared. (See also W73-08480) (Josefson-USGS)
W73-08487

DEVELOPMENT OF SNOW TRANSFORMATION PROCESSES IN THE COMPLEX MOUNTAIN ZONE OF WESTERN TIEN SHAN (OPYT INDIKATSII PROTSESSOV PREOBRAZOVANIYA SNEGA V SREDNEGORNOY ZONE ZAPADNOGO TYAN'-SHANYA),
Sredneaziatskii Nauchno-Issledovatelskii Gidrometeorologicheskii Institut, Tashkent (USSR).
L. A. Kanayev.

In: Laviny Sredney Azii; Sredneaziatskiy Nauchno-Issledovatel'skiy Gidrometeorologicheskiy Institut Trudy, No 63 (78), p 96-103, Leningrad, 1972. 2 tab, 5 ref.

Descriptors: *Snow, *Snow cover, *Snowfall, *Mountains, *Meteorology, Cyclones, Anticyclones, Synoptic analysis, Weather data, Temperature, Air temperature, Thermocline, Sublimation, Slopes, Valleys, Profiles, Forecasting, Winter.
Identifiers: *USSR, *Tien Shan Mountains, *Complex mountains, Avalanche release, Snow metamorphism.

Observations of snow temperature and stratigraphy were made in 1958-71 in river valleys of the Dukant (elevation--1,500 to 2,100 m) and Kyzylcha (elevation--2,100 to 2,700 m) snow-avalanche stations in the Akhangaran River basin in Western Tien Shan. Based on an analysis of stratigraphic profiles, two types of snow-layer development were distinguished: syngenetic and episyngenetic. A relation was established between the two types of development and the meteorological conditions during a winter period before January 1. A classification of these meteorological conditions can be used to forecast development of a snow layer and possibility of avalanche movement of a particular snow type. (See also W73-08480) (Josefson-USGS)
W73-08488

METHODS OF ANTICIPATING AVALANCHE HAZARD FROM RAINFALL AND SNOWFALL INTENSITIES IN MOUNTAINS OF ZAILISKI ALA TAU (SPOSOBY PROGNOZA LAVINNOY OPASNOSTI PO INTENSIVNOSTI OSADKOV I SNEGOPADOV V GORAKH ZAILIYSKOGO ALATAU),
Sredneaziatskii Nauchno-Issledovatelskii Gidrometeorologicheskii Institut, Tashkent (USSR).
Ye. I. Kolesnikov.
In: Laviny Sredney Azii; Sredneaziatskiy Nauchno-Issledovatel'skiy Gidrometeorologicheskiy Institut Trudy, No 63 (78), p 104-121, Leningrad, 1972. 7 fig, 7 tab, 7 ref.

Descriptors: *Avalanches, *Precipitation intensity, *Rainfall intensity, *Snowfall, *Mountains, Snowpacks, Slopes, River basins, Meteorology, Precipitation (Atmospheric), Humidity, Physical properties, Mechanical properties, Analytical techniques, Curves, Equations, Forecasting, Winter.
Identifiers: *USSR, *Zailiski Ala Tau, *Avalanche hazard, *Snowfall intensity, Avalanche release, Snow stratigraphy.

Investigations of avalanche formation in mountains of Zailiski Ala Tau and of quantitative indicators of developing avalanche hazard during snowfalls were based on observations of snow stratigraphy and physico-mechanical properties and of avalanche movement in the Bol'shaya Almatinka and Malaya Almatinka River basins in the winters of 1966-69. Avalanche activity is greatest in the complex mountain zone on slope angles between 30 and 40 deg at elevations of 1,600 to 3,000 m. Avalanche-forecasting techniques already employed in other parts of the country are examined for possible application in the Tien Shan mountain system. Individual avalanche-forming factors (snow depth, amount and intensity of precipitation) are compared with the time a given slope will avalanche, the release being associated with precipitation in over 70% of the cases. Average difference between computed and actual time of avalanching is plus or minus 0.75-0.85 hrs, and avalanche-hazard is predictable in 94%-95% of the cases. (See also W73-08480) (Josefson-USGS)
W73-08489

REDISTRIBUTION OF SNOW BY WINDS IN A MOUNTAINOUS TERRAIN (PERERASPREDELENIYE SNEZHNOGO POKROVA METELYAMI V USLOVIYAKH GORNOGO REL'YEFA),
Sredneaziatskii Nauchno-Issledovatelskii Gidrometeorologicheskii Institut, Tashkent (USSR).
B. I. Novikov.
In: Laviny Sredney Azii; Sredneaziatskiy Nauchno-Issledovatel'skiy Gidrometeorologicheskiy Institut Trudy, No 63 (78), p 122-131, Leningrad, 1972. 2 fig, 4 tab, 6 ref, append.

Descriptors: *Snow, *Snowpacks, *Snow cover, *Winds, *Mountains, Slopes, Vegetation, Topography, Meteorology, Snowfall, Precipitation (Atmospheric), Avalanches, Curves, Equations, Forecasting.
Identifiers: *USSR, *Carpathian Mountains, *Snowstorms, Snow depth.

Snow accumulation was investigated under various conditions of topography in a mountainous region of the Carpathians in 1966-70. Observations of snow cover were carried out at 16 plots on slopes of different exposure and steepness at elevations of 1,400 to 1,800 m. The amount of snow accumulated in individual areas in winter can best be described by coefficients which relate average or maximum snow depths to similar snow-depth values at a standard plot. The coefficient of snow accumulation for plots on leeward slopes increases with increasing slope angle. Relations are established between increase in snow depth on leeward slopes and amount of precipitation occurring during or immediately before a snowstorm. The empirical relations obtained can be used to determine avalanche activity on slopes of different exposure from the total accumulated depth of snowfall and to calculate incipient avalanching of wind-driven snow from the amount of snow accumulated during a single snowstorm. (See also W73-08480) (Josefson-USGS)
W73-08490

SNOW-WATER EQUIVALENT MEASUREMENT USING NATURAL GAMMA EMISSION,
Geological Survey of Canada, Ottawa (Ontario).
R. L. Grasty.
Nordic Hydrology, Vol 4, No 1, p 1-16, 1973. 5 fig, 4 tab, 13 ref.

Descriptors: *Remote sensing, *Snow surveys, *Snowpacks, *Water equivalent, *Gamma rays, Nuclear moisture meters, Instrumentation, Calibrations, Canada.
Identifiers: *Gamma-ray spectrometers.

The natural gamma radiation emitted by potassium, uranium and thorium is attenuated by snow. This attenuation depends on the water-equivalent of the snow layer. With an airborne gamma-ray spectrometer, snow water-equivalents up to 18 cm could be measured to an accuracy of 2 cm over suitable terrain. The effects of temperature and soil moisture corrections are discussed, together with statistical, instrumental and navigational errors. (Knapp-USGS)
W73-08505

PHOTOSYNTHESIS AND RESPIRATION OF SOME ARCTIC SEAWEEDS,
Scripps Institution of Oceanography, La Jolla, Calif.
For primary bibliographic entry see Field 05C.
W73-08585

THE SNOW REMOVAL CONTROVERSY,
Toronto Univ. (Ontario). Inst. of Environmental Sciences and Engineering.
For primary bibliographic entry see Field 05B.
W73-08601

led to heaving and overstressing in the superjacent turf layers, having been subjected chiefly to disruption and deformation in the southern part of the hummock. The ice streaks having formed in this part raised the earth's surface. However, in the northern part, there occurred a partial settling of the earth's surface; as a result, the individual turf clods became buried under pulverized silty loam. The formation of turf hummocks with an ice core is also explained in terms of hydrostatic pressure, having originated during the freezing of water in the reservoir. The water proved to be in a closed space; as a result, there occurred the upheaval of the turfy reservoir bottom overgrown with sphagnum. (Woodard-USGS)
W73-08735

GREAT LAKES ICE COVER, WINTER 1969-70,
National Ocean Survey, Detroit, Mich. Lake Survey Center.
D. R. Rondy.
Available from NTIS, Springfield, Va 22151 as COM-72-10917; Price $3.00 printed copy; $1.45 microfiche. Technical Memorandum NOS LSC D-3, March 1972. 56 p, 40 plate, 1 tab.

Descriptors: *Ice cover, *Great lakes, *Iced lakes, Winter, Data collections, Aircraft, Ice loads, Lake ice, Freezing, Melting, Snow, Ice-water interfaces, Time, Seasonal.

Thirty-three ice charts were produced from data obtained on 14 visual aerial reconnaissance flights over the Great Lakes during the 1969-70 ice season. Winter temperatures were below normal and were determined from freezing degree-day accumulations. Reported dates of first ice varied from November 3,1969, on southern Lake Huron to December 23, 1969, on eastern Lake Ontario. The period of maximum ice cover varied across the Great Lakes, as did the areal extent of the ice cover. During the periods of maximum ice extent, Lake Superior was estimated to the 85% covered, with a particularly heavy concentration of winter ice in the southeastern portion; Lake Michigan 30% covered, with heavy winter ice confined to the northern end of the lake; Lake Huron 50% covered with differing concentrations around the lake; Lake Erie 95% covered with most of the winter ice containing a drifted snow cover; and Lake Ontario 15% covered with the major concentrations located in the approaches to the St. Lawrence River. The reported dates of last ice ranged from March 31 on eastern Lake Ontario to April 27 on eastern Lake Superior. (Woodard-USGS)
W73-08743

FREEZING EFFECTS ON ARCTIC BEACHES,
Louisiana State Univ., Baton Rouge. Coastal Studies Inst.
A. D. Short, and W. J. Wiseman, Jr.
Available from NTIS, Springfield, Va. 22151, as AD-755 153. Price $3.00 printed copy; $1.45 microfiche. Louisiana State University Coastal Studies Bulletin No 7, p 23-32, Technical Report 128, January 1973. 6 fig, 4 ref.

Descriptors: *Beaches, *Arctic, *Freezing, *Thawing, *Sedimentation, Waves (Water), Erosion, Beach erosion, Stratigraphy, Ice, Snow, Topography, Geomorphology, Sands, Gravels, Sedimentology, Frost action.
Identifiers: *Arctic beaches.

Freezeup on arctic beaches is accompanied by subzero air and water temperatures, which freeze both swash and sediment. Ice in nearshore waters may be deposited on the beach and preserved. Snow may also be preserved in the beach. Beaches with wave activity during the freezeup process contain interbedded snow, frozen swash, ice boulders, and sediment. With spring thaw, the encased ice melts, causing superimposition of the overlying layers and increased deformation of the layers toward the surface. The result is an upper zone (approximately 1.5 meters thick) containing normal beach stratigraphy in addition to irregular, mottled, and complex interbedding, often with lenses, pods, and stringers of sand and gravel. Within the section erosional contacts intersect these beds and occasionally intersect kettle holes and runnels filled with eolian and deformed swash deposits. (Knapp-USGS)
W73-08751

VARIATION IN IONIC RATIOS BETWEEN REFERENCE SEA WATER AND MARINE AEROSOLS,
Centre National de la Recherche Scientifique, Gif-sur-Yvette (France). Centre des Faibles Radioactivites.
For primary bibliographic entry see Field 02K.
W73-08855

CHARACTERISTICS OF HUMMOCKS IN FAST ICE (KHARAKTERISTIKA TOROSOV V PRIPAYE),
Moscow State Univ. (USSR).
V. M. Klimovich.
Meteorologiya i Gidrologiya, No 5, p 80-87, May 1972. 3 fig, 5 tab, 4 ref.

Descriptors: *Sea ice, *Arctic, *Shores, Coasts, Littoral, Winds, Measurement, Height, Spring, Statistical methods, Curves, Equations.
Identifiers: *USSR, *Ice hummocks, *Fast ice, Ice floes, Ice thickness, Icebreakers, Factor analysis.

Height of hummocks and thickness of hummocky floes were measured in an Arctic region of fast ice in the springs of 1962-65. The height distribution of hummocks obeys Maxwell's law of distribution. The amount of hummocked ice and the hummock height are determined by the period of formation of fast ice and by wind conditions. (Josefson-USGS)
W73-08884

VARIATIONS IN SALINITY DURING FORMATION OF ICE IN SHALLOW SEAS (IZMENENIYE SOLENOSTI VODY PRI OBRAZOVANII L'DA VA MELKOVODNYKH MORYAKH),
State Oceanographic Inst., Moscow (USSR).
L. O. Kuzenkova.
Meteorologiya i Gidrologiya, No 5, p 75-79, May 1972. 1 fig, 1 tab, 4 ref.

Descriptors: *Salinity, *Ice, *Shallow water, *Depth, *Variability, Winter, Seasonal, Equations.
Identifiers: *USSR, Ice thickness, Nomograms.

An empirical method for computing salinity in shallow water in winter is based on increase in salinity resulting from changes in water volume during formation of ice. The thickness of ice is considered in relation to the depth of the body of water in which it is formed. Variations in salinity are inversely proportional to the changes in depth of the water body and directly proportional to the changes in ice thickness. (Josefson-USGS)
W73-08885

ICE JAMS IN THE DANUBE DELTA AND SOME MEASURES TO PREVENT FLOODS CAUSED BY THEM (ZATORY L'DA V DEL'TE DUNAYA I NEKOTORYYE MEROPRIYATIYA PO PREDOTVRASHCHENIYU VYZYVAYEMYKH IMI NAVODNENIY),
Vsesoyuznyi Nauchno-Issledovateskii Institut Gidrotekhniki, Leningrad (USSR).
For primary bibliographic entry see Field 04A.
W73-08886

THE ECOLOGY OF SOME ARCTIC TUNDRA PONDS,
Missouri Univ., Columbia.
For primary bibliographic entry see Field 05C.
W73-08909

GLAZE AND RIME PHENOMENA AND ICING OF WIRES IN SOVIET CENTRAL ASIA (GOLOLEDNO-IZMOROZEVYYE YAVLENIYA I OBLEDENENIYE PROVODOV V SREDNEY AZII),
Sredneaziatskii Nauchno-Issledovatelskii Gidrometeorologicheskii Institut, Tashkent (USSR).
G. N. Leukhina.
Trudy, No 7 (88), Leningrad, 1972. 144p.

Descriptors: *Meteorology, *Sleet, *Ice, *Rime, *Transmission lines, Ice loads, Snow, Air temperature, Wind velocity, Synoptic analysis, Topography, Orography, Spatial distribution, Temporal distribution, Variability, Maps.
Identifiers: *USSR, *Soviet Central Asia, *Glaze (Ice), Ice damage.

Investigations of glaze (sleet) and rime phenomena and icing of wires were based on visual and instrument observations carried out at 285 hydrometeorological stations in Soviet Central Asia in 1936-69. The study included the different types and duration of ice formation on wires; space-time distribution of glaze, rime, and wet snow; meteorological and synoptic conditions of glaze formation; formation of especially dangerous ice deposits; and ice loads in transmission lines. Maps were constructed which show space-time variability of the different forms of precipitation. (Josefson-USGS)
W73-08999

DEEP ROTARY CORE DRILLING IN ICE,
Cold Regions Research and Engineering Lab., Hanover, N.H.
G. R. Lange.
Technical Report 94, February 1973. 46 p, 22 fig, 1 tab, 33 ref, 3 append.

Descriptors: *Ice, *Core drilling, *Drilling fluids, *Glaciers, *Polar regions, Glaciology, Equipment, Methodology, Temperature, Depth, Rotary drilling, Drilling samples.
Identifiers: Deep core drilling.

A method of drilling holes and obtaining core to depths of 1500 ft in high polar glaciers was required for International Geophysical Year (IGY) glaciological investigations. Rotary drilling equipment was modified and sent to Greenland and the Antarctic. Principal modifications included the use of chilled compressed air as a circulating fluid and specially designed coring bits. During initial trials in northwest Greenland in 1956 nearly continuous core was obtained to 970 ft but the core was badly cracked at depth. A second season in Greenland produced much more continuous core to 1000 ft, and single coring runs at the 1100-, 1200- and 1300-ft levels produced usable core. Core cracking at depth persisted although it was less severe. During the Antarctic field season of 1957-58 similar equipment was used to take continuous core to 1000 ft at Byrd Station for the IGY glaciological program. The use of a special constant rate of feed device for coring and improved techniques based on experience gained in Greenland resulted in further improvements in core condition and rate of recovery. In 1957-58 the Ross Ice Shelf was penetrated 840 ft at Little America V with equally good results. One core taken with diesel fuel at the bottom of the ice shelf was remarkably free from the cracking which had previously resulted from the rapid release of overburden load in the empty hole. This strongly suggests that increased depths and much better core will be obtained if diesel fuel is used as the circulating fluid. Coring to depths of 10,000 ft should be possible with much heavier ro-

tary drilling equipment and similar techniques, but logistic and other considerations suggest that the feasibility of a wire line method rather than drill pipe be investigated. (Woodard-USGS)
W73-09079

CYCLIC SURGING OF GLACIERS,
Scott Polar Research Inst., Cambridge (England).
G. de Q. Robin, and J. Weertman.
Journal of Glaciology, Vol 12, No 64, p 3-18, 1973. 8 fig, 25 ref.

Descriptors: *Glaciers, *Movement, *Melt water, Regimen, Equations, Rheology, Cryology, Ice, Mathematical studies.
Identifiers: *Surging glaciers, *Glacial surges.

A theory and model are proposed to explain cyclically surging glaciers. During the after-surge portion of a surge cycle the lower portion of a glacier becomes increasingly stagnant. The upper part of the glacier gradually becomes more active as both its thickness and the magnitude of its basal shear stress increase. In the region between these two parts (the trigger zone), the value of the derivative of the basal shear stress in the longitudinal direction of the glacier gradually increases with time. The pressure gradient in the water at the base of a glacier is related to the derivative of the basal shear stress. The pressure gradient decreases as the basal shear-stress gradient increases. The pressure gradient actually cantake on negative values, a condition which produces 'up-hill' water flow at the base of a glacier. A surge is started in the trigger zone when water is dammed there by a zero water-pressure gradient. The zone of fast-sliding velocities propagates up the glacier from the trigger zone with a velocity of the order of a surge velocity. The fast-sliding velocity zone also propagates down the glacier because of increased melt-water production. (Knapp-USGS)
W73-09086

INVERSION OF FLOW MEASUREMENTS FOR STRESS AND RHEOLOGICAL PARAMETERS IN A VALLEY GLACIER,
California Inst., of Tech., Pasadena. Div. of Geological and Planetary Sciences.
C. F. Raymond.
Journal of Glaciology, Vol 12, No 64, p 19-44, 1973. 12 fig, 1 tab, 33 ref.

Descriptors: *Glaciers, *Rheology, *Movement, *Flow, Stress, Strain, Plasticity, Viscosity, Creep, Temperature.

The distribution of stress and effective viscosity in a glacier may be calculated under the assumptions that the ice is quasi-viscous, the flow is time-independent, and acceleration forces are negligible. The differential equations of mechanical equilibrium, expressed in terms of viscosity, strain-rate components, mean stress, and their gradients, are solved for viscosity and mean stress subject to boundary conditions at the free upper surface. For some rectilinear flow patterns, unique distributions of stress and effective viscosity can always be derived. For more complicated flow this is not necessarily so. However, it is possible to choose the best values of rheological parameters, based on the requirement that the residuals of the equations of equilibrium be minimized. The techniques were applied to measurements of internal deformation made in nine boreholes on the Athabasca Glacier. At the centerline the magnitude of the surface-parallel shear stress increases with depth more slowly than would be expected from a standard shape factor correction or the theoretical distribution of Nye. The lateral distribution of lateral shear stress shows the opposite relationships. In the lower one- to two-thirds of the depth, corresponding to a range in effective stress from about 0.5 to 1.2 bars, the gross rheology of the ice is not distinguishably different from the experimentally determined flow law of Glen as generalized by Nye. The upper one- to two-thirds of the glacier constitutes an anomalous zone in which there is either a strong effect from a complex distribution of stress arising from longitudinal stress gradients or more complicated rheology than in a homogeneous power-law material. (Knapp-USGS)
W73-09087

THE CREEP OF ICE SHELVES: THEORY,
Scott Polar Research Inst., Cambridge, (England); and British Antarctic Survey, London (England).
R. H. Thomas.
Journal of Glaciology, Vol 12, No 64, p 45-53, 1973. 2 fig, 12 ref.

Descriptors: *Glaciers, *Rheology, *Movement, *Flow, Stress, Strain, Plasticity, Viscosity, *Creep, Temperature.
Identifiers: *Ice shelves.

A general expression is derived for creep in an ice shelf where the sole restriction is that of zero shear stresses in vertical planes. This is applied to the two special cases: movement of an ice shelf restricted in at least one direction by sea-water pressure only; and movement of an ice shelf flowing between roughly parallel sides. (See also W73-09089) (Knapp-USGS)
W73-09088

THE CREEP OF ICE SHELVES: INTERPRETATION OF OBSERVED BEHAVIOUR,
Scott Polar Research Inst., Cambridge (England); and British Antarctic Survey, London (England).
R. H. Thomas.
Journal of Glaciology, Vol 12, No 64, p 55-70, 1973. 10 fig, 2 tab, 19 ref.

Descriptors: *Glaciers, *Rheology, *Movement, *Flow, Stress, Strain, Plasticity, Viscosity, Creep, Temperature.
Identifiers: *Ice shelves.

A general expression is derived for creep in an ice shelf where the sole restriction is that of zero shear stresses in vertical planes. This is applied to the two special cases: movement of an ice shelf restricted in at least one direction by sea-water pressure only; and movement of an ice shelf flowing between roughly parallel sides. Available measurements of creep rates and dimensions of ice shelves were used to evaluate the flow law parameters at low stresses. The results show good agreement with laboratory work at higher stresses. Adoption of these values can be used to calculate the restraining effects on an ice shelf of obstructions such as areas of grounding. (See also W73-09088) (Knapp-USGS)
W73-09089

A FLOW LAW FOR TEMPERATE GLACIER ICE,
Washington Univ., Seattle. Geophysics Program.
S. C. Colbeck, and R. J. Evans.
Journal of Glaciology, Vol 12, No 64, p 71-86, 1973. 11 fig, 20 ref, append. NSF Grants GA1547 and GA1516 ONR Contract N00014-67-A-0103-0007.

Descriptors: *Glaciers, *Rheology, *Movement, *Flow, Stress, Strain, Plasticity, Viscosity, Creep, Temperature.
Identifiers: *Temperate glaciers.

Uniaxial compressive creep tests were performed on polycrystalline samples of glacier ice at stresses ranging from 0.06 bar to 1.0 bar under conditions similar to those actually occurring in a temperate glacier. Tests were conducted in an ice tunnel on the Blue Glacier, Mt Olympus, Washington, U.S.A., where the temperature remained at the pressure melting point. The flow law proposed as a one-dimensional flow law for temperate glacier ice is a polynomial expression containing linear, cubic and fifth-order terms. This law provides a good fit to the data and is also consistent with the quasi-viscous creep data of Glen. (Knapp-USGS)
W73-09090

RADIO ECHO SOUNDING ON A VALLEY GLACIER IN EAST GREENLAND,
Scott Polar Research Inst., Cambridge (England).
J. L. Davis, J. S. Halliday, and K. J. Miller.
Journal of Glaciology, Vol 12, No 64, p 87-91, 1973. 4 fig, 6 ref.

Descriptors: *Sounding, *Glaciers, *Radar, *Instrumentation, On-site collections, Ice, Radio waves, Depth, Surveys.
Identifiers: *Greenland.

Although radio echo sounding equipment has been used with success for measuring the thickness of ice sheets in the Arctic and Antarctic, a valley glacier poses the additional problems of echoes from the valley walls, which may obscure the bottom echoes, and a high attenuation of radio waves in the ice. During July and August 1970, a study was carried out on Roslin Gletscher in Stauning Alper, East Greenland, to investigate the problems of radio echo sounding on a valley glacier. Reflections from the valley walls may be minimized by using sufficiently directional antennae, but attenuation of the signal in the ice is higher than that in polar ice at the same temperature. Water in and on the ice probably accounts for much of the attenuation, and the use of a lower frequency or measurements before the melt commences should give improved performance. (Knapp-USGS)
W73-09091

ICE VOLCANOES OF THE LAKE ERIE SHORE NEAR DUNKIRK, NEW YORK, U.S.A.,
State Univ. Coll., Fredonia, N.Y. Dept. of Geology.
R. K. Fahnestock, D. J. Crowley, M. Wilson, and H. Schneider.
Journal of Glaciology, Vol 12, No 64, p 93-99, 1973. 7 fig.

Descriptors: *Lake Erie, *Ice, *Lake ice, Beaches, Iced lakes, Great Lakes, Freezing, Ice jams, *New York.
Identifiers: *Ice volcanoes, *Beach ice.

Conical mounds of ice form in a few hours during violent winter storms along the edge of shore-fast ice near Dunkirk, New York. They occur in lines which parallel depth contours, and are evenly spaced in the manner of beach cusps. The height and spacing of mounds and number of rows vary from year to year depending on such factors as storm duration and intensity and the position of the edge of the shore-fast ice at the beginning of the storm. The evenly sloping conical mounds have central channels which increase in width lakeward. The ice between the channels forms headlands above the lake surface. Sprayed-formed levees develop along the headlands and slope gently away from the lake margin. Lake marginal walls of ice are usually vertical. Spray, slush and ice blocks are ejected over the cone as each successive wave is focused by the converging channel walls. Ice blocks, interlayered with frozen slush and dirt, form bedding paralleling the sloping surface of cones, headlands and levees. These features are termed 'ice volcanoes' because their origin is in so many ways analogous to that of true volcanoes. (Knapp-USGS)
W73-09092

SEISMIC REFRACTION AND REFLECTION MEASUREMENTS AT 'BYRD' STATION, ANTARTICA,
Wisconsin Univ., Middleton. Geophysical and Polar Research Center.
H. Kohnen, and C. R. Bentley.

Journal of Glaciology, Vol 12, No 64, p 101-111, 1973. 9 fig, 2 tab, 14 ref. NSF Grants GV-27044 and GAn1706.

Descriptors: *Ice, *Glaciers, *Sea ice, *Antarctic, *Seismic studies, *Sounding, Depth, Surveys, Velocity, Geophysics, Explosives, Travel time.

Seismic refraction and reflection shooting along three profiles about 10 km long, angled 60 deg to one another, near Byrd station, Antarctica, during the 1970-71 field season showed no dependence of velocity upon azimuth, but velocities at 200 or 300 m depth were slightly greater than at a site 30 km away where measurements were made in 1958. The difference can probably be attributed to different ice fabrics arising from a 50% difference in snow accumulation rates at the two sites. The velocity-depth and density-velocity functions at the two sites are also significantly different, but close agreement was found at each site between the depths to significant changes in the velocity gradient and the depths of fundamental change in the densification process. Such agreement may permit density-depth curves, and consequently accumulation rates, to be measured by seismic refraction shooting alone. The reflection shooting on a common reflection-point profile led to a good determination of mean velocity through the ice as a function of angle of incidence. The results agree closely with similar measurements at the 1958 site, and with an anisotropic model based on glaciological and sonic logging observations in the deep drill hole. The mean vertical velocity of 3.90-3.93 km/s through the solid ice is about 2% higher than has commonly been used for determinations of ice thickness from seismic reflection shooting. (Knapp-USGS)
W73-09093

A PHOTOGRAMMETRIC SURVEY OF HOSEASON GLACIER, KEMP COAST, ANTARCTICA,
Ohio State Univ., Columbus. Dept. of Geodetic Science; and Ohio State Univ. Research Foundation, Columbus. Inst. of Polar Studies.
P. J. Morgan.
Journal of Glaciology, Vol 12, No 64, p 113-120, 1973. 4 fig, 1 tab, 9 ref.

Descriptors: *Photogrammetry, *Glaciers, *Antarctic, *Aerial photography, *Movement, Regimen, Surveys, Mapping, Glaciology.
Identifiers: *Hoseason Glacier (Antarctica).

A study of Hoseason Glacier, Antarctica, was made to test whether photogrammetry could yield meaningful glaciological information from repeated series of air photographs without ground control. Large time differences, different camera systems, and the use of sea ice as a level datum do not present insurmountable difficulties in photogrammetric reduction. Surface horizontal movement was determined and level profiles were compiled. Hoseason Glacier is probably in equilibrium. (Knapp-USGS)
W73-09094

MASS TRANSFER ALONG AN ICE SURFACE OBSERVED BY A GROOVE RELAXATION TECHNIQUE,
Cold Regions Research and Engineering Lab., Hanover, N.H.
K. Itagaki, and T. M. Tobin.
Journal of Glaciology, Vol 12, No 64, p 121-127, 1973. 3 fig, 1 tab, 19 ref.

Descriptors: *Mass transfer, *Ice, *Cryology, *Flow, *Evaporation, Freezing, Crystallography, Diffusion.
Identifiers: *Ice structure.

The mass transfer on an ice surface was measured using a groove decay technique at -10 deg C. The evaporation, condensation, and viscous flow terms in Mullins' theory were deduced from the change of decay constant as a function of groove wavelength between 16 and 80 microns. Viscous flow contributes the most to groove decay while evaporation and condensation contribute up to 31.5% of the mass transfer for the shortest wavelength measured. Large discrepancies between the decay constants obtained from the measurements and constants calculated from theory indicate that other mechanisms not considered in Mullins' theory may be responsible for the groove decay. (Knapp-USGS)
W73-09095

GLACIER BORE-HOLE PHOTOGRAPHY,
California Inst. of Tech., Pasadena. Div. of Geological and Planetary Sciences.
W. D. Harrison, and B. Kamb.
Journal of Glaciology, Vol 12, No 64, p 129-137, 1973. 11 fig, 2 ref.

Descriptors: *Photography, *Ice, *Glaciers, *Borehole cameras, Glaciology, Instrumentation, Movement.

A 51 mm diameter bore-hole camera allows observation of subglacial conditions, measurement of basal sliding rates, and study of internal structure and debris in ice at depth. The camera is simple in construction, field operation and maintenance. Water turbidity is a significant problem but it can be overcome by pumping. (Knapp-USGS)
W73-09096

RADIO ECHO SOUNDING: BRINE PERCOLATION LAYER,
Wisconsin Univ., Middleton. Geophysical and Polar Research Center.
J. W. Clough.
Journal of Glaciology, Vol 12, No 64, p 141-143, 1973. 2 fig, 3 ref.

Descriptors: *Sounding, *Radar, *Brines, *Ice, *Antarctic, Percolation, Radio waves, Depth, Surveys.
Identifiers: *Ice shelves.

An abrupt change in radio echo sounding travel-time was observed on the ice shelf near McMurdo station, Antarctica, and was mapped by a zig-zag traverse. This boundary corresponds to the horizontal extent of brine penetration into the edge of the ice shelf. (Knapp-USGS)
W73-09097

ACOUSTIC EMISSION IN SNOW AT CONSTANT RATES OF DEFORMATION,
Montana State Univ., Bozeman. Dept. of Earth Sciences.
W. F. St Lawrence, T. E. Lang, R. L. Brown, and C. C. Bradley.
Journal of Glaciology, Vol 12, No 64, p 144-146, 1973. 1 fig, 3 ref.

Descriptors: *Deformation, *Fractures (Geologic), *Snow, *Compaction, *Sound waves, Strain, Acoustics, Failure (Mechanics), Mechanical properties, Cryology.

Acoustic emissions in the audio spectrum were found in laboratory experiments conducted on snow samples in uniaxial compression. A number of tests show the pattern of acoustic emissions to be a function of the rate of deformation. Over the frequency range 20 to 7,000 Hz acoustic emissions are associated with rates of deformation corresponding to brittle fracture of the snow sample. No acoustic emissions were detected from samples deforming plastically. (Knapp-USGS)
W73-09098

2D. Evaporation and Transpiration

AN APPRAISAL OF POTENTIAL WATER SALVAGE IN THE LAKE MCMILLAN DELTA AREA, EDDY COUNTY, NEW MEXICO,
Geological Survey, Albuquerque, N. Mex.
For primary bibliographic entry see Field 03B.
W73-08493

HIGH YIELD FROM THE BULL RUN WATERSHED,
British Columbia Dept. of Municipal Affairs, Victoria. Environmental Planning and Management Div.
For primary bibliographic entry see Field 02A.
W73-08667

A MODEL FOR SIMULATING TRANSPIRATION OF LEAVES WITH SPECIAL ATTENTION TO STOMATAL FUNCTIONING,
Agricultural Univ., Wageningen (Netherlands). Dept. of Theoretical Production Ecology.
F. W. T. Penning De Vries.
J Appl Ecol. Vol 9, No 1, p 57-77, 1972. Illus.
Identifiers: Carbon, Cells, *Computer models, Diffusion, Leaves, Light, Oxide, Photosynthesis, Simulation, *Stomata, *Transpiration (leaves).

A dynamic model of a water-containing and water-conducting system is described, representing a non-growing, transpiring leaf with an attached root in a nutrient solution. The simulated transpiration rate is determined by environmental conditions and leaf conductivity, the latter being mainly under stomatal control. A hypothesis of stomatal functioning based upon the interaction between guard cells and subsidiary cells is presented. The control mechanism of the guard cells is supposed to be affected both by present and past plant water status, light intensity and CO2 concentration in the leaf, which depends on photosynthesis and diffusion rates. The function of subsidiary cells is taken to be affected only by present and past plant water status. Experiments are simulated to evaluate the model. The model is written in the computer simulation language CSMP (Continuous System Modeling Program).--Copyright 1972, Biological Abstracts, Inc.
W73-09039

2E. Streamflow and Runoff

FLOOD-PRONE AREAS, HARTFORD NORTH QUADRANGLE, CONNECTICUT,
Geological Survey, Washington, D.C.
For primary bibliographic entry see Field 07C.
W73-08472

1972 AIDJEX INTERIOR FLOW FIELD STUDY, PRELIMINARY REPORT AND COMPARISON WITH PREVIOUS RESULTS,
Washington Univ., Seattle. Dept. of Oceanography.
For primary bibliographic entry see Field 02C.
W73-08475

RUNOFF MAP OF SWEDEN: AVERAGE ANNUAL RUNOFF FOR THE PERIOD 1931-60,
Sveriges Meteorologiska och Hydrologiska Institut, Stockholm.
For primary bibliographic entry see Field 07C.
W73-08492

SOME OBSERVATIONS OF TURBULENT FLOW IN A TIDAL ESTUARY,
Naval Research Lab., Washington, D.C. Ocean Sciences Div.
For primary bibliographic entry see Field 02L.

W73-08494

DYNAMICS OF RIP CURRENTS,
Florida State Univ., Tallahassee. Dept. of Mathematics; and Florida State Univ., Tallahassee. Geophysical Fluid Dynamics Inst.
C. K. W. Tam.
Journal of Geophysical Research, Vol 78, No 12, p 1937-1943, April 20, 1973. 7 fig, 16 ref. NSF GP-30136, NOAA NG-3-72.

Descriptors: *Mathematical models, *Rip currents, Surf, Littoral drift, Waves (Water), Equations, Viscosity, Turbulence.

Rip currents were studied using a set of shallow water equations with a horizontal eddy viscosity term. A boundary layer analysis is based on the observed fact that rip currents are rather narrow. Similarity solutions of the model equations give reasonable representations of the velocity profile and other characteristics of rip currents. The mechanisms that are believed to be responsible for the formation of rip heads and turning of rip currents by the longshore momentum carried by the entrained fluid were analyzed. (Knapp-USGS)
W73-08495

A COMPARISON BETWEEN WIND AND CURRENT OBSERVATIONS OVER THE CONTINENTAL SHELF OFF OREGON, SUMMER 1969,
Oregon State Univ., Corvallis. Dept. of Oceanography.
A. Huyer, and J. G. Pattullo.
Journal of Geophysical Research, Vol 77, No 18, p 3215-3220, June 20, 1972. 7 fig, 9 ref. NR 083-102. NSF GA 1435, ONR N00014-67-A-0369-0007.

Descriptors: *Winds, *Currents (Water), *Continental shelf, *Oregon, Waves (Water), Current meters.

Wind and current observations made over the continental shelf off Oregon in the summer of 1969 are compared. The wind and current are related at periods longer than two days. The wind and current data were both filtered to suppress variations of diurnal and higher frequencies. The current behaves similarly to the wind when the wind varies slowly. The amplitude of the current fluctuations depends on the stratification of the water. The current is roughly parallel to the local bottom contours. (Knapp-USGS)
W73-08497

FIELD OBSERVATION OF NEARSHORE CIRCULATION AND MEANDERING CURRENTS,
Louisiana State Univ., Baton Rouge. Coastal Studies Inst.
C. J. Sonu.
Journal of Geophysical Research, Vol 77, No 18, p 3232-3247, June 20, 1972. 13 fig, 30 ref. NR 388 002. ONR N00014-69-A-0211-0003.

Descriptors: *Currents (Water), *Littoral drift, *Rip currents, Surf, Beaches, Meanders, Sand waves, Waves (Water), Topography.

Wave-induced nearshore circulations and meandering longshore currents occur on an undulatory surf-zone bed, under the action of uniform incident waves. Circulations were associated with normal-wave incidence; meandering currents were associated with oblique-wave incidence. Spilling breakers over a shoal underwent greater energy dissipation than plunging breakers in the rip current. Observed streamlines were narrow in the outflow and broad in the inflow, a characteristic that was probably associated with a nonlinear mechanism arising from a steep depression in the rip channel. For a given surf-zone undulation, breaking over the inner bar was essential to the formation of a circulation, and the intensity of breaking, controlled by tide, corresponded with a proportionally stronger circulation. Thus, circulations were generally stronger during low tide than during high tide. Low rip-current velocities at high tide fluctuated with incoming swells, whereas high velocities at low tide tended to fluctuate at surf-beat frequencies. In proportion to increasing rep velocities, the rip pulsation tended toward lower intervals. Mean surface slopes caused by wave setup and setdown agreed with trajectories of neutral-density balls released in the circulation. Meandering currents associated with oblique-wave incidence could be explained as a combined effect of circulation cells and parallel longshore flows. (Knapp-USGS)
W73-08498

ENTRAINMENT OF SHELF WATER BY THE GULF STREAM NORTHEAST OF CAPE HATTERAS,
Naval Oceanographic Office, Washington, D.C. Ocean Science Dept.
A. Fisher, Jr.
Journal of Geophysical Research, Vol 77, No 18, p3248-3255, June 20, 1972. 9 fig, 16 ref.

Descriptors: *Mixing, *Ocean currents, *Remote sensing, Water temperature, Ocean circulation, Salinity, Sea water, Atlantic Ocean.
Identifiers: *Gulf stream, Cape Hatteras.

Entrainment of relatively cold low-salinity water by the Gulf Stream and subsequent formation of a cold filament adjacent to the northern edge of the Gulf Stream were frequently observed during a series of combined ship and aircraft surveys conducted northeastward of Cape Hatteras between October 1968, and May, 1969. Particularly well-defined entrainment was observed during mid-May, 1969, when a band of surface water 10 to 15 km wide and 40 meters thick was observed to reach from the continental shelf to the northern edge of the Gulf Stream. Transport seaward from the outer shelf is of the magnitude 10,000 cu m/sec. Failure to observe entrainment during some of the flights shows the discontinuous nature of the cold filament adjacent to the northern edge. (Knapp-USGS)
W73-08499

TEMPERATURE FLUCTUATIONS AT AN AIR-WATER INTERFACE CAUSED BY SURFACE WAVES,
Argonne National Lab., Ill.
J. Witting.
Journal of Geophysical Research, Vol 77, No 18, p 3265-3269, June 20, 1972. 8 ref.

Descriptors: *Water temperature, *Waves (Water), *Thermal stratification, *Air-water interfaces, Boundary processes, Heat flow.

Surface temperature variation is induced by a plane progressive irrotational linear water wave when a thermal boundary layer exists. When the wave period is sufficiently small (3 sec), the amplitude of the wave-induced temperature fluctuation is proportional to the wave amplitude and to the square root of the wave period. This result is in agreement with experiments. Surface temperature maximums lead wave crests by one-eighth of a period when the average surface temperature is less than the bulk water temperature; surface temperature minimums lead wave crests by one-eighth of a period when the average surface temperature exceeds that of the water below. (Knapp-USGS)
W73-08500

SEA LEVEL VARIATIONS IN THE GULF OF BOTHNIA,
E. Lisitzin.
Nordic Hydrology, Vol 4, No 1, p 41-53, 1973. 5 tab, 6 ref.

Descriptors: *Water level fluctuations, *Sea level, Europe, Tides, Subsidence, Data collections.

Conversely, spring floods destroyed rainbow eggs, thereby enhancing survival of young brook trout. Floods changed the species composition markedly and these changes endured for several years. Adult trout were adversely affected by the worst flood studied, but were unaffected by other floods of lesser magnitude. Effects of floods were not nearly as pronounced or predictable on adult trout as they were on young trout. --Copyright 1973, Biological Abstracts, Inc.
W73-08907

STREAMFLOW DATA USAGE IN WYOMING,
Wyoming Univ., Laramie. Water Resources Research Inst.
For primary bibliographic entry see Field 07C.
W73-09004

FISHES OF THE NORTHEASTERN WOOD BUFFALO NATIONAL PARK REGION, ALBERTA AND NORTHWEST TERRITORIES,
Alberta Univ, Edmonton. Dept. of Zoology.
J. S. Nelson, and M. J. Paetz.
Can Field Nat. Vol 86, No 2, p 133-144. 1972.
Identifiers: Alberta, Canada, *Dace, *Darter, Distribution, Fishes, Habitat, *Minnow (Fathead), National parks, Northwest-Territories, Park, *Stickleback.

Five species of small fishes (northern red-belly dace, fathead minnow, pearl dace, brook stickleback, and Iowa darter) have their known northern distributional limits in landlocked lakes and small creeks in the northern Wood Buffalo National Park region. They are surprisingly abundant along their northern fringe but are poorly known to the immediate south. The majority of individuals of 2 spp. of sickleback (brook and ninespine sticklebacks± fail to develop the pelvic skeleton in several localities. Habitats within the region are unusual, consisting of karst topography (with sinkhole lakes and the possibiltiy of underground drainage connections) and streams of relatively high salinity.--Copyright 1973, Biological Abstracts, Inc.
W73-09008

USE OF DISPOSABLE BEVERAGE CANS BY FISH IN THE SAN JOAQUIN VALLEY,
Fresno State Coll., Calif. Dept. of Biology.
G. M. Kottcamp, and P. B. Moyle.
Trans Am Fish Soc. Vol 101, No 3, p 566. 1972.
Identifiers: Breeding, *California, *Cans (Disposable), *Fish, San-Joaquin Valley.

A surprisingly heavy use of disposable beverage cans, for shelter and breeding, by 6 spp. of fish in streams of the San Joaquin Valley, California, including juveniles of 3 spp. of game fish, is reported.--Copyright 1973, Biological Abstracts, Inc.
W73-09060

FLOODS IN THE IOWA RIVER BASIN UPSTREAM FROM CORALVILLE LAKE, IOWA,
Geological Survey, Iowa City, Iowa.
For primary bibliographic entry see Field 07C.
W73-09084

STABILITY OF ROTATIONAL COUETTE FLOW OF POLYMER SOLUTIONS,
Delaware Univ., Newark. Dept. of Chemical Engineering.
For primary bibliographic entry see Field 08B.
W73-09100

2F. Groundwater

INVESTIGATION OF FACTORS RELATIVE TO GROUNDWATER RECHARGE IN IDAHO,
Idaho Univ., Moscow. Dept. of Hydrology.
For primary bibliographic entry see Field 05D.
W73-08456

MAP SHOWING GENERAL AVAILABILITY OF GROUND WATER IN THE SALINA QUADRANGLE, UTAH,
Geological Survey, Washington, D.C.
For primary bibliographic entry see Field 07C.
W73-08473

MAJOR AQUIFERS AND SAND GRAVEL RESOURCES IN BROWN COUNTY, SOUTH DAKOTA,
Geological Survey, Washington, D.C.
N. C. Koch, W. Bradford, and D. I. Leap.
South Dakota Geological Survey Information Pamphlet No 4, 1973. 9 p, 4 fig, 1 tab.

Descriptors: *Aquifers, *South Dakota, *Glacial drift, Water yield, Groundwater resources, Hydrogeology, Hydrologic data, Sands, Gravels.
Identifiers: *Brown County (S.D.).

The general distribution, quality, and physical characteristics of the major aquifers and the general distribution of sand and gravel resources in Brown County, S. D., are described. Three major glacial aquifers--the Deep James, Middle James, and Elm--are present in the county. The Deep James aquifer underlies an area of about 250 sq mi in Brown County. The aquifer consists of an interconnected system of channels extending from southwestern Brown County in a north and northeast direction across the county. The channels are filled with clay, silt, sand, and gravel, with sand and gravel thickness ranging from 1 to 160 ft. The Deep James aquifer may yield as much as 1,000 gpm to properly constructed wells at depths ranging from 150 to 380 ft. The major bedrock aquifers in Brown County are the Dakota Sandstone and the Fall River and Sundance Formations. (Knapp-USGS)
W73-08477

GROUNDWATER DEVELOPMENT AND ITS INFLUENCE ON THE WATER BALANCE,
Technical Univ., of Denmark, Copenhagen.
E. Hansen, and S. A. Andersson.
Nordic Hydrology, Vol 4, No 1, p 28-40, 1973. 9 fig, 1 tab, 5 ref.

Descriptors: *Surface-groundwater relationships, *Induced infiltration, *Water yield, *Transmissivity, Withdrawal, Recharge, Artesian aquifers, Hydrogeology, Aquifer testing.
Identifiers: *Denmark.

Lowering the head of an artesian aquifer in Zealand, Denmark, by pumping of groundwater affects the regional water balance. Based on known amounts of groundwater being developed, the coefficient of transmissivity was determined and in combination with piezometric maps the net outflow of groundwater was computed for the prepumping period as well as for the pumping period. By lowering the head in the artesian aquifer, groundwater development influences the water balance, especially by causing a decrease in the streamflow as a consequence of an increased infiltration. The decrease in streamflow causes a decrease in the yearly minimum runoff as well as a marked increase in the length of the periods of low flows. The evapotranspiration is approximately 370 mm/per year, uninfluenced by the groundwater development. (Knapp-USGS)
W73-08507

HYDROLOGY OF THE UPPERMOST CRETACEOUS AND THE LOWERMOST PALEOCENE ROCKS IN THE HILIGHT OIL FIELD, CAMPBELL COUNTY, WYOMING,
Geological Survey, Washington, D.C.
For primary bibliographic entry see Field 04B.
W73-08725

GROUNDWATER RESOURCES OF WASHINGTON COUNTY, TEXAS,
Geological Survey, Austin, Tex.
For primary bibliographic entry see Field 04B.
W73-08739

GROUNDWATER RESOURCES OF NAVARRO COUNTY, TEXAS,
Geological Survey, Austin, Tex.
For primary bibliographic entry see Field 04B.
W73-08740

GROUND-WATER RESOURCES OF HARDEMAN COUNTY, TEXAS,
Geological Survey, Austin, Tex.
For primary bibliographic entry see Field 04B.
W73-08741

WATER-LEVEL RECORDS, 1969-73, AND HYDROGEOLOGIC DATA FOR THE NORTHERN HIGH PLAINS OF COLORADO,
Geological Survey, Washington, D.C.
For primary bibliographic entry see Field 07C.
W73-08745

SALINE GROUNDWATER RESOURCES OF MISSISSIPPI,
Geological Survey, Jackson, Miss.
For primary bibliographic entry see Field 07C.
W73-08750

LIMITATIONS OF DARCY'S LAW IN GLASS BEAD POROUS MEDIA,
Oregon State Univ., Corvallis. Dept of Soil Science.
For primary bibliographic entry see Field 02G.
W73-08870

HYDROGEOLOGY AND KARST STUDIES (GIDROGEOLOGIYA I KARSTOVEDENIYE),
Perm State Univ. (USSR). Inst. of Karst Studies and Peleology.
For primary bibliographic entry see Field 02K.
W73-08998

THE PERMEABILITY OF A UNIFORMLY VUGGY POROUS MEDIUM,
Alberta Univ., Edmonton.
G. H. Neale, and W. K. Nader.
Society of Petroleum Engineers Journal, Vol 13, No 2, p 69-74, April 1973. 4 fig, 8 ref.

Descriptors: *Permeability, *Porous media, *Fracture permeability, Equations, Darcys law, Groundwater movement.
Identifiers: *Secondary permeability, *Vugs, *Brinkman equation.

The familiar Darcy equation is inadequate for studies involving liquid flow through porous media containing open spaces. The Brinkman equation was applied to the problem of creeping liquid flow through an isotropic porous medium containing an isolated spherical cavity. An exact solution for the pressure distribution prevailing throughout this system was determined. (Knapp-USGS)
W73-09076

DATA REQUIREMENTS FOR MODELING A GROUND-WATER SYSTEM IN AN ARID REGION,
Geological Survey, Menlo Park, Calif.
F. Kunkel.
Available from NTIS, Springfield, Va. 22151 PB-219 588 Price $3.00 printed copy; $1.45 microfiche. Geological Survey Water-Resources Investigation 4-73, March 1973. 21 p, 12 fig, 11 ref.

Descriptors: *Groundwater resources, *Arid lands, *Hydrology, *Groundwater movement, *Hydrogeology, Model studies, Systems analysis, Input-output analysis, Computer programs, Hydrologic data, Data processing, Water levels, Geology, Transmissivity, Storage coefficient, Groundwater recharge.
Identifiers: Hydrologic models.

The Geological Survey has successfully modeled many groundwater basins in arid regions. In considering the problems of modeling, this paper presents an oversimplified summary. The mathematical formulas are known and the computer capability exists that permits accurate modeling of interrelated surface water and groundwater systems in arid regions if the required data can be assembled. Existing data are often insufficient to provide the needed inputs to construct and verify a model that will be adequate for predicting future water levels. However, even unverified models may be of considerable value in providing the basis for determining the direction and scope of future data-collection programs to insure maximum return. Reports on models for areas in California and for selected areas in Utah and Arizona are given in the list of references. (Woodard-USGS)
W73-09078

2G. Water in Soils

CHARACTERIZATION OF WATER MOVEMENT INTO AND THROUGH SOILS DURING AND IMMEDIATELY AFTER RAINSTORMS,
Kentucky Water Resources Inst., Lexington.
C. T. Haan.
Available from the National Technical Information Service as PB-220 014, $3.00 in paper copy, $1.45 in microfiche. Research Report No 56, December 1972. 34 p, 8 fig, 1 tab, 25 ref. OWRR A-025-KY (3), 14-31-0001-(3217-3517).

Descriptors: *Infiltration, *Mathematical models, *Darcys law, *Rainfall-runoff relationships, Mathematical studies, Diffusivity, Soil water movement, Groundwater movement, Water yield.

Darcy's Law and the continuity equation were studied to find if they could be used to describe watershed infiltration and thus be incorporated into hydrologic models. Even on apparently uniform soils there is a great deal of variability in soil water properties. Handling this variability plus the difficulty of solving the flow equation led to the conclusion that a simpler approach to modelling watershed infiltration is needed. A simple empirical infiltration model was developed and included in a rainfall-runoff model. Tests with the model indicate that it produces satisfactory estimates of monthly runoff from small rural watersheds. (Knapp-USGS)
W73-08464

THE RELATION BETWEEN SOIL CHARACTERISTICS, WATER MOVEMENT AND NITRATE CONTAMINATION OF GROUNDWATER,
Kentucky Water Resources Inst., Lexington.
For primary bibliographic entry see Field 05B.
W73-08465

THE INFLUENCE OF IRON AND ALUMINUM HYDROXIDES ON THE SWELLING OF NA-MONTMORILLONITE AND THE PERMEABILITY OF A NA-SOIL,
Reading Univ. (England). Dept. of Soil Science.
H. M. E. El Rayah, and D. L. Rowell.
Journal of Soil Science, Vol 24, No 1, p 137-144, March 1973. 2 fig, 23 ref.

Descriptors: *Expansive soils, *Expansive clays, *Saline soils, *Montmorillonite, *Permeability, Soil water movement, Aluminum, Iron, Clay minerals, Mineralogy, Ion exchange.

HORIZONTAL INFILTRATION OF WATER IN SOILS: A THEORETICAL INTERPRETATION OF RECENT EXPERIMENTS,
Yale Univ., New Haven, Conn. Dept. of Engineering and Applied Science.
J-Y. Parlange.
Soil Science Society of America Proceedings, Vol 37, No 2, p329-330, March-April 1973. 1 fig, 5 ref.

Descriptors: *Infiltration, *Diffusivity, *Diffusion, *Soil water movement, Soil moisture, Mathematical studies, Equations.
Identifiers: *Horizontal infiltration.

A general theory of water movement is used to describe the horizontal infiltration of water when the diffusivity increases exponentially with moisture content. The results can be expressed in simple analytical form and are in excellent agreement with experimental observations. The results confirm that the theory is accurate because water diffusivity varies with water content. (Knapp-USGS)
W73-08871

WATER MOVEMENT AND CALICHE FORMATION IN LAYERED ARID AND SEMIARID SOILS,
Nevada Univ., Reno.
D. M. Stuart, and R. M. Dixon.
Soil Science Society of America Proceedings, Vol 37, No 2, p 323-324, March-April 1973.

Descriptors: *Caliche, *Leaching, *Soil formation, Weathering, Calcareous soils, Calcium carbonate, Hardpan, Infiltration, Soil water movement.

In soils, downward water movement is restricted when fine-textured materials are underlain by sand or gravel layers. Water accumulates in arid and semiarid soils at the interface of the fine- and coarse-textured materials and may rarely enter the coarse-textured materials. Calcium carbonate, silica, and other salts are deposited at or near the top of the sand or gravel layers as water is removed by evapotranspiration. With time, silica, calcium carbonates, and other salts become cemented or indurated, forming calcareous crusts at these interfaces. The sands and gravels are not wet uniformly when water does enter, because it enters relatively small areas. Calcareous concentrations may form in these small, wetted areas and may be seen as calcareous cemented columns surrounded by noncalcareous sands or gravels. (Knapp-USGS)
W73-08872

ANALYSIS OF MULTIDIMENSIONAL LEACHING,
Commonwealth Scientific and Industrial Research Organization, Wembley (Australia). Div. of Soils.
A. J. Peck.
Soil Science Society of America Proceedings, Vol 37, No 2, p 320, March-April 1973. 4 ref.

Descriptors: *Leaching, *Infiltration, *Steady flow, Soil water movement, Drainage, Saline water, Equations, Saline soils.

A simplified analysis is given for leaching from the soil surface to tile drains while infiltration is steady and uniform. Leachate concentration is predicted to decay exponentially with characteristics determined by physical parameters of the system. Predicted decay characteristics are in good agreement with Mulqueen and Kirkham's experimental data. (Knapp-USGS)
W73-08873

A NOTE ON A THREE-PARAMETER SOIL-WATER DIFFUSIVITY FUNCTION-APPLICATION TO THE HORIZONTAL INFILTRATION OF WATER,
Yale Univ., New Haven, Conn. Dept. of Engineering and Applied Science.
J-Y. Parlange.
Soil Science Society of America Proceedings, Vol 37, No 2, p 318-319, March-April 1973. 2 tab, 5 ref.

Descriptors: *Infiltration, *Diffusivity, *Diffusion, *Soil water movement, Soil moisture, Mathematical studies, Equations.
Identifiers: *Horizontal infiltration.

A realistic water-diffusivity function depending on three parameters described recently by Ahuja and Swartzendruber is solved analitically for the horizontal infiltration of water. The result is expressed in terms of elementary functions and agrees very well with experimental observations. (Knapp-USGS)
W73-08874

THE RELATIONSHIP BETWEEN RAINFALL FREQUENCY AND AMOUNT TO THE FORMATION AND PROFILE DISTRIBUTION OF CLAY PARTICLES,
Illinois Univ., Urbana. Dept. of Agronomy.
T. M. Goddard, E. C. A. Runge, and B. W. Ray.
Soil Science Society of America Proceedings, Vol 37, No 2, p 299-304, March-April 1973. 3 fig, 3 tab, 8 ref.

Descriptors: *Loess, *Clay minerals, *Soil formation, *Soil water, Leaching, Clays, Illinois, Soil chemistry, Infiltration, Weathering.
Identifiers: *Pedogenic clays, *Rainfall frequency.

In Typic Argiudolls in Illinois the amount of pedogenic clay formed is controlled by rainfall frequency. The amount of leaching was not related to the amount of pedogenic clay. This finding may help explain why the clay minerals found in the argillic horizon of some soils are not in the clay mineral stability field predicted by dilute solution chemistry. The distribution of clay within the soil profile is related to the amount and distribution of rainfall (leachable water), depth of leaching, and natural drainage class. (Knapp-USGS)
W73-08875

COLLAPSIBLE LOESS IN IOWA,
Iowa State Univ., Ames.
R. L. Handy.
Soil Science Society of America Proceedings, Vol 37, No 2, p 281-284, March-April 1973. 4 fig, 1 tab, 20 ref.

Descriptors: *Loess, Collapse, *Subsidence, *Compaction, *Soil compaction, Consolidation, Clays, Clay minerals, Soil mechanics, Bulk density, *Iowa.
Identifiers: *Collapsible loess.

The collapsibility, or tendency towards rapid consolidation or settlement upon saturation by water, of many loess soils is pertinent to soil genesis, erosion, and engineering. The potential of a loess soil to collapse may be established from samples saturated while under load in a consolidation testing machine. Loess containing less than 20% 0.005 mm clay is very likely to be collapsible; with 30% clay there is a 50% probability of collapse, and with over 40% clay there is only a small likelihood of collapse. These percentages were compared to published clay content contours to estimate the areal extent of collapsible loess in Iowa. (Knapp-USGS)
W73-08876

EFFECT OF FLOODING ON THE EH, PH, AND CONCENTRATIONS OF FE AND MN IN SEVERAL MANITOBA SOILS,
Manitoba Univ., Winnipeg. Dept. of Soil Science.
M. O. Olomu, G. J. Racz, and C. M. Cho.
Soil Science Society of America Proceedings, Vol 37, No 2, p 220-224, March-April 1973. 9 tab, 19 ref.

Descriptors: *Iron, *Manganese, *Oxidation-reduction potential, *Hydrogen ion concentration, *Saturated soils, Floods, Flooding, Reduction (Chemical), Solubility, Mineralogy, Soil chemistry, Canada, Organic matter.
Identifiers: *Manitoba (Canada).

In flooded soils, the concentrations of Fe and Mn increased with chemical reduction; pH changes were relatively small. The measured concentrations of Fe in the soil solution were much greater than predicted by the observed Eh and pH values. Iron was complexed with organic matter; these complexes were negatively charged and relatively stable. Manganese in soil solution was partly complexed with organic matter in relatively unstable complexes. (Knapp-USGS)
W73-08877

DYNAMIC MEASUREMENT OF SOIL AND LEAF WATER POTENTIAL WITH A DOUBLE LOOP PELTIER TYPE THERMOCOUPLE PSYCHROMETER,
British Columbia Univ., Vancouver. Dept. of Soil Science.
T. L. Chow, and J. De Vries.
Soil Science Society of America Proceedings, Vol 37, No 2, p 181-188, March-April 1973. 11 fig, 14 ref. Canada National Research Council Grant A2556.

Descriptors: *Soil mosture meters, *Water temperature, *Thermometers, Soil temperature, Soil water movement, Calibrations, Temperature, Heat flow, Groundwater potential, Soil-water-plant relationships, Potential flow.
Identifiers: *Psychrometers, *Thermocouples.

Details on the construction, calibration, and performance of a three-terminal double loop thermocouple psychrometer are given. The thermal stability of this psychrometer is about 40 times better than that of the two-terminal psychrometer (Spanner type) for ambient temperature fluctuations with a time rate of change greater than 0.2C/min. The response behaviors of a fritted glass bulb and a ceramic bulb psychrometer were tested for vapor and for liquid phase water movement. For vapor phase flow the fritted glass bulb exhibited a shorter response time then the ceramic bulb psychrometer, whereas the reverse was true when water movement was predominantly in the liquid phase. Water potential measurements carried out on silty clay and silt loam soil samples were within plus or minus 0.4 bar of those from the porous plate extractor. A system that facilitates automatic and continuous in situ measurement of soil water potential using the three-terminal psychrometer is described. (Knapp-USGS)
W73-08878

SOIL WATER FLOWMETERS WITH THERMOCOUPLE OUTPUTS,
Agricultural Research Service, Kimberly, Idaho. Snake River Conservation Research Center.
J. W. Cary.
Soil Science Society of America Proceedings, Vol 37, No 2, p 176-181, March-April 1973. 7 fig, 1 tab, 8 ref.

Descriptors: *Flowmeters, *Soil water movement, *Water temperature, *Instrumentation, Thermometers, Temperature, Soil temperature, Heat flow; Seepage.
Identifiers: *Thermocouples.

Two soil water flowmeters with thermocouple sensors are described. The meter with a sensitivity of 0.1 mm of waterflow per day is recommended for flux measurements in the surface meter of soil when the water matrix potential is greater than 0.8 bar. Calibration factors for three soils with different textures are presented as a family of curves. These curves may be interpolated for using the flowmeter in other soils, possibly without a loss of accuracy greater than the natural water flow variation from place to place in the field. The meter with a sensivity of about 0.5 mm per day will require some additional development and testing before it can be recommended for routine use. It does offer the possibility of making measurements at soil water matric potentials less than 1 bar and at relatively deep soil depths. The thermocouple flow transducer developed for the meters may be used to measure saturated soil waterflow or other liquid flows as small as 1 ml/day. (Knapp-USGS)
W73-08879

A. THE ROLE OF LICHENS IN ROCK WEATHERING AND SOIL FORMATION. B. MERCURY IN SEDIMENTS,
Wisconsin Univ., Madison.
For primary bibliographic entry see Field 05B.
W73-08930

SOIL PHYSICAL CONDITIONS AFFECTING SEEDLING ROOT GROWTH: I. MECHANICAL IMPEDANCE, AERATION AND MOISTURE AVAILABILITY AS INFLUENCED BY BULK DENSITY AND MOISTURE LEVELS IN A SANDY LOAM SOIL,
For primary bibliographic entry see Field 02I.
W73-09029

DIFFUSION COEFFICIENT OF I-IONS IN SATURATED CLAY SOIL,
A. Hamid.
Plant Soil. Vol 36 No 3, p709-711. 1972.
Identifiers: Adsorption, *Clays, Diffusion coefficient, *Iodine ions, Saturated Soils.

The diffusion of I-ions was measured in saturated clay soil using radioactive I131 and I-ions. The measured diffusion coefficient of I-ions was 1.96 x 10-7 cm2/sec in disturbed soil and 0.67 x 10-7 cm2/sec in undisturbed soil. The positive adsorption of I-ions resulted in lower diffusion coefficient of I-ions than that of Cl-ions.–Copyright 1972, Biological Abstracts, Inc.
W73-09030

IRRIGATED AGRICULTURE AS A METHOD OF DESERT SOIL RECLAMATION (IN RUSSIAN),
M. D. Muratov.
Probl Osvoeniya Pustyn'. 1, p 10-15, 1972. English summary.
Identifiers: Agriculture, *Cotton-D, *Desert soils, Irrigation, *Soil reclamation.

Two-year experiments were conducted on desert sandy, grey-brown and takyr soils in the Karshinskaya steppe. In desert sandy soils early cotton germination and ripening was observed. The yield of raw-cotton was 18.5 centners/ha in desert sandy soils, approximately 15 in grey-brown and 34.9 in takyr soils. Desert sandy soils exceed the grey-brown ones in agricultural reclamation properties and productive capacity but, to some extent, are poorer than takyr soils.--Copyright 1972, Biological Abstracts, Inc.
W73-09031

BONDING OF WATER TO ALLOPHANE,
Department of Scientific and Industrial Research, Lower Hutt (New Zealand). Soil Bureau.
For primary bibliographic entry see Field 02K.
W73-09041

INTERLAYER HYDRATION AND THE BROADENING OF THE 10 A X-RAY PEAK IN ILLITE,
Battelle-Northwest, Richland, Wash.
R. C. Routson, J. A. Kittrick, and E. H. Hope.
Soil Sci. Vol 113, No 3, p 167-174, 1972.
Identifiers: Heat, *Hydration, *Illite, *X-ray, Temperature.

X-ray peak broadening was due in part to reversible interlayer hydration. During heating to 300C, 2. K-saturated (based) illites dehydrated gradually, whereas Cs-saturated illites contracted abruptly at lower temperatures. The water reversibly released from K-saturated illites at low temperatures is considered to be physically entrapped due to the closure of dehydrated edges. High temperature water in illites in excess of that predicted from the muscovite formula may also be physically en-trapped.–Copyright 1972, Biological Abstracts, Inc.
W73-09042

THE ECOLOGY OF MORECAMBE BAY: VI. SOILS AND VEGETATION OF THE SALT MARSHES: A MULTIVARIATE APPROACH,
Nature Conservancy, Norwich (England). Coastal Ecology Research Station.
For primary bibliographic entry see Field 02L.
W73-09067

CAUSES OF INFILTRATION INTO YELLOW-BROWN PUMICE SOILS,
Waikato Univ., Hamilton (New Zealand). Dept. of Earth Sciences.
M. J. Selby, and P. J. Hosking.
J Hydrol (Dunedin). Vol 10, No 2, p 113-119. 1971. Illus.
Identifiers: *Infiltration, Soil moisture, Pasture, Prediction equation, *Pumice soils, Vegetation.

Infiltration measurements on 108 plots in the Otutira experimental basin were carried out. Infiltration is correlated with a large number of soil properties, and the best prediction equation gives an explanation of only 57%. Infiltration is least into soils beneath pasture which have been compacted by land use practices, especially when these soils have low pre-existing soil moistures.–Copyright 1972, Biological Abstracts, Inc.
W73-09069

2H. Lakes

ADAPTATION TO EUTROPHIC CONDITIONS BY LAKE MICHIGAN ALGAE,
Wisconsin Univ., Milwaukee. Dept. of Botany.
For primary bibliographic entry see Field 05C.
W73-08463

CARBON DIOXIDE CONTENT OF THE INTERSTITIAL WATER IN THE SEDIMENT OF GRANE LANGSO, A DANISH LOBELIA LAKE,
Copenhagen Univ. (Denmark). Freshwater Biological Lab.
For primary bibliographic entry see Field 05A.
W73-08555

DISTRIBUTION AND FORMS OF NITROGEN IN A LAKE ONTARIO SEDIMENT CORE,
Department of Energy, Mines and Resources, Burlington (Ontario). Canada Center for Inland Waters.
For primary bibliographic entry see Field 05B.
W73-08563

TECHNIQUES FOR SAMPLING SALT MARSH BENTHOS AND BURROWS,
Georgia Univ., Athens. Dept. of Geology.
For primary bibliographic entry see Field 05A.
W73-08582

THE EFFECTS OF ARTIFICIAL AERATION ON THE DEPTH DISTRIBUTION OF THE CRAYFISH ORCONECTES VIRILIS (HAGEN) IN TWO MICHIGAN LAKES,
Michigan State Univ., East Lansing. Dept. of Fisheries and Wildlife.
For primary bibliographic entry see Field 05C.
W73-08583

W73-08917

POLLUTION OF THE 'EL CARPINCHO' POND (PAMPASIC REGION, ARGENTINA) AND ITS EFFECTS ON PLANKTON AND FISH COMMUNITIES,
La Plata Univ. (Argentina). Instituto de Limnologia.
For primary bibliographic entry see Field 05C.
W73-08927

PHOSPHORUS DYNAMICS IN LAKE WATER,
Toronto Univ. (Ontario). Dept. of Zoology.
For primary bibliographic entry see Field 05B.
W73-08931

CESIUM-137 AND STABLE CESIUM IN A HYPEREUTROPHIC LAKE,
Michigan State Univ., East Lansing.
For primary bibliographic entry see Field 05C.
W73-08959

BIOLOGICAL NITROGEN FIXATION IN LAKE MENDOTA,
Wisconsin Univ., Madison.
For primary bibliographic entry see Field 05C.
W73-08962

INITIAL RESPONSES OF PHYTOPLANKTON AND RELATED FACTORS IN LAKE SAMMAMISH FOLLOWING NUTRIENT DIVERSION,
Washington Univ., Seattle.
For primary bibliographic entry see Field 05C.
W73-08973

SOME INFORMATION ON THE CONTAMINATION OF THE VOLGOGRAD WATER RESERVOIR WITH EGGS OF DIPHYLLOBOTHRIUM LATUM, (IN RUSSIAN),
Institute of Medical Parasitology and Tropical Medicine, Moscow (USSR).
For primary bibliographic entry see Field 05C.
W73-09037

HETEROTROPHIC ASSAYS IN THE DETECTION OF WATER MASSES AT LAKE TAHOE, CALIFORNIA,
California Univ., Davis. Dept. of Zoology; and California Univ., Davis. Inst. of Ecology.
For primary bibliographic entry see Field 05C.
W73-09059

STUDIES ON FRESHWATER BACTERIA: ASSOCIATION WITH ALGAE AND ALKALINE PHOSPHATASE ACTIVITY,
Freshwater Biological Association, Windermere (England).
For primary bibliographic entry see Field 05C.
W73-09061

ICE VOLCANOES OF THE LAKE ERIE SHORE NEAR DUNKIRK, NEW YORK, U.S.A.,
State Univ. Coll., Fredonia, N.Y. Dept. of Geology.
For primary bibliographic entry see Field 02C.
W73-09092

2I. Water in Plants

DEGRADATION OF RIPARIAN LEAVES AND THE RECYCLING OF NUTRIENTS IN A STREAM ECOSYSTEM,
Kentucky Water Resources Inst., Lexington.
For primary bibliographic entry see Field 05B.
W73-08471

HYDROPTILA ERAMOSA A NEW CADDIS FLY FROM SOUTHERN ONTARIO (TRICHOPTERA, HYDROPTILIDAE),
Montreal Univ. (Quebec). Dept. of Biological Sciences.
P. P. Harper.
Canadian Journal of Zoology, Vol. 51, No. 3, p 393-394, March 1973. 1 fig, 7 ref.

Descriptors: *Caddisflies, *Aquatic insects, *Systematics, Mature growth stage, Invertebrates, Canada, Speciation.
Identifiers: *Hydroptila eramosa, Trichoptera.

The adult male of Hydroptila eramosa n. sp. is described from Southern Ontario and is briefly compared with other similar species of the 'waubesiana group.' (Holoman-Battelle)
W73-08760

DETERMINATION OF CHLOROPHYLL A AND B IN PLANT EXTRACTS BY COMBINED SUCROSE THIN-LAYER CHROMATOGRAPHY AND ATOMIC ABSORPTION SPECTROPHOTOMETRY,
Texas Tech Univ., Lubbock. Textile Research Center.
For primary bibliographic entry see Field 02K.
W73-08786

INFLUENCE OF REGULATION OF THE LACZANKA RIVER ON THE OCCURRENCE OF HIRUDINEA,
Bialystok Medical Academy (Poland). Dept. of Histology and Embryology.
For primary bibliographic entry see Field 04A.
W73-09019

SOIL PHYSICAL CONDITIONS AFFECTING SEEDLING ROOT GROWTH: I. MECHANICAL IMPEDANCE, AERATION AND MOISTURE AVAILABILITY AS INFLUENCED BY BULK DENSITY AND MOISTURE LEVELS IN A SANDY LOAM SOIL,
B. W. Eavis.
Plant Soil. Vol 36 No 3, p 613-622. 1972 Illus.
Identifiers: Aeration, Bulk, Density, Diffusion, Moisture, Pisum-Sativum-D, *Root growth, *Sandy loam soils, Seedling, Soils.

The role of mechanical impedance, poor aeration and water availability in restricting pea (Pisum sativum L.) seedling root growth in sandy loam soil at 3 bulk densities and 6 matric potentials was studied. Mechanical impedance increased both with bulk density and matric potential. In certain treatments the roots were shorter and thicker as impedance increased but in others shorter, thicker roots were found as impedance decreased and this was attributed to poor aeration. An aeration deficiency index was defined to separate the impedance and aeration effects when the 2 acted in combination. An aeration effect was found in soils having less than 30, 22, and 11% gas-filled pore space at low, medium and high bulk densities, respectively, and the distribution and area of the gas-liquid interface probably influenced diffusion resistance. Effects due to restricted water availability were found only at potentials greater than -3.5 bars, the roots being shorter and thinner as moisture, stress increased.--Copyright 1972, Biological Abstracts, Inc.
W73-09029

METHODS OF DETECTING FUNGI IN ORGANIC DETRITUS IN WATER,
Queen's Univ., Belfast (Northern Ireland). Dept. of Botany.
For primary bibliographic entry see Field 07B.
W73-09057

ON THE ECOLOGY OF HETEROTROPHIC MICROORGANISMS IN FRESHWATER,
Queen's Univ., Belfast (Northern Ireland). Dept. of Botany.
D. Park.
Trans Br Mycol Soc. Vol 58, No 2, p 291-299. 1972.
Identifiers: Classification, Ecology, *Heterotrophic microorganisms, *Fungi.

An outline scheme for the ecological classification of heterotrophic microorganisms (fungi) in freshwater is presented in the hope that it may help to indicate some gaps in our knowledge and stimulate a search for precision and definition in future studies. --Copyright 1972, Biological Abstracts, Inc.
W73-09058

COMPARATIVE STUDIES OF PLANT GROWTH AND DISTRIBUTION IN RELATION TO WATERLOGGING: V. THE UPTAKE OF IRON AND MANGANESE BY DUNE AND DUNE SLACK PLANTS,
Trent Univ., Peterborough (Ontario). Dept. of Biology.
R. Jones.
J Ecol. Vol 60, No 1, p 131-139. 1972. Illus.
Identifiers: Agrostis-Stolonifera-M, Carex-Flacca-M, Carex-Nigra-M, *Dune plants, Festuca-Rubra-M, *Iron, *Manganese, Roots, Sand *Waterlogging, Wetness.

The effect of different waterlogging treatments on the uptake of Fe and Mn by Agrostis stolonifera, Festuca rubra, Carex flacca and C. nigra was studied. Waterlogging resulted in increased Fe and Mn contents of the 4 spp., the greatest increase occurring in the roots. Plants grown in slack sand accumulated more Fe but less Mn than plants grown in dune sand. Field collections of C. flacca from dune slope and slack sites were made. Analysis of this material showed that Fe and Mn content increased with the relative wetness of the collection site, paralleling the results of the pot experiment. A possible protection system of shoots by roots to prevent excessive Fe uptake is discussed.--Copyright 1972, Biological Abstracts, Inc.
W73-09063

COMPARATIVE STUDIES OF PLANT GROWTH AND DISTRIBUTION IN RELATION TO WATERLOGGING: VI. THE EFFECT OF MANGANESE ON THE GROWTH OF DUNE AND DUNE SLACK PLANTS,
Trent Univ., Peterborough (Ontario). Dept. of Biology.
R. Jones.
J Ecol. Vol 60, No 1, p 141-145. 1972. Illus.
Identifiers: Agrostis-Stolonifera-M, Carex-Flacca-M, Carex-Nigra-M, *Dune plants, Festuca-Rubra-M, *Manganese, Toxicity, *Waterlogging.

A solution-culture experiment was undertaken to study the effect of Mn on the growth of Agrostis stolonifera, Festuca rubra, Carex flacca and C. nigra. High Mn concentrations resulted in toxicity symptoms in F. rubra and some plants of C. nigra. Growth of F. rubra was reduced by increasing Mn concentrations while that of A. stolonifera was unaffected. Growth of the Carices was promoted by increasing Mn concentrations. These observations are discussed in relation to the ecology of the respective species. (See also W73-09063)--Copyright 1972, Biological Abstracts, Inc.
W73-09064

PROTEIN QUALITY OF SOME FRESHWATER ALGAE,
North Carolina State Univ., Raleigh.
H. E. Schlichting, Jr.
Econ Bot. Vol 25, No 3, p 317-319. 1971.
Identifiers: *Algae, *Protein quality, Water quality, *Amino acids.

Results are presented of experimental culture of 5 species of freshwater algae. Data on amino acid composition are tabulated. Although the yield and protein quality of the mass cultured freshwater algae were poor, certain amino acids were in greater abundance than in other plants.--Copyright 1972, Biological Abstracts, Inc.
W73-09070

2J. Erosion and Sedimentation

HYDRAULIC AND SEDIMENT TRANSPORT STUDIES IN RELATION TO RIVER SEDIMENT CONTROL AND SOLID WASTE POLLUTION AND ECONOMIC USE OF THE BY-PRODUCTS,
Kentucky Water Resources Inst., Lexington.
B. R. Moore.
Available from the National Technical Information Service as PB-220 016, $3.00 in paper copy, $1.45 in microfiche. Research Report No 59, July 1972. 48 p, 18 fig, 2 tab, 5 ref. OWRR A-034-KY (1), 14-31-0001-3217.

Descriptors: *Sediment transport, *Sands, *Coals, *Bed load, *Scour, Ohio River, Dredging, Water pollution control, Hydraulic models, Streamflow, *Kentucky.
Identifiers: Kentucky River, Big Sandy River.

The distribution of sediments and the conditions of sediment transport were studied in the Kentucky, Big Sandy and Ohio Rivers. Sand and coal were in transport at different flow velocities for the rivers. The principal source of the coal is natural erosion. Deposition of these sediments was a function of the flow conditions at a particular locality. The flow conditions of transport of the sediments were also studied in flumes, as were the hydraulic conditions in model dredge holes, to determine the feasibility of trapping sediment. The conditions of scour and fill were compared with known conditions in a dredge hole in the Ohio River. Flow records from gaging stations were analyzed to determine the periods of sediment transport. Solid waste pollutants can be trapped in dredge holes in some conditions of flow. (Knapp-USGS)
W73-08468

FLUVIAL SEDIMENT IN DOUBLE CREEK SUBWATERSHED NO. 5, WASHINGTON COUNTY, OKLAHOMA,
Geological Survey, Oklahoma City, Okla.
G. A. Bednar, and T. E. Waldrep.
Geological Survey open-file report, 1973. 38 p, 4 fig, 11 tab, 5 ref.

Descriptors: *Sediment yield, *Small watersheds, *Oklahoma, *Reservoir silting, *Trap efficiency, Retention, Water balance.
Identifiers: *Double Creek Watershed (Okla).

Fluvial sediment was measured in Double Creek subwatershed No. 5 in Washington County, Oklahoma. The subwatershed includes 2.39 square miles, and it receives runoff from approximately 5% of the total area of the Double Creek watershed. Approximately three-fourths (47,000 acre-feet) of precipitation is lost by evaporation and transpiration, and a small amount is lost by deep subsurface percolation. Of the total sediment load, 59% was discharged from the reservoir during four major outflow periods representing 34% of the outflow days. Most runoff and sediment yield occur from March through June; 53% of the water discharged and 63% of the sediment yield occurred during this 4-month period. The average annual yield of fluvial sediment from watershed No. 5 was 607 tons per square mile. A total of 21,370 tons of fluvial sediment was transported into reservoir No. 5 and a total of 19,930 tons was deposited. The computed trap efficiency of reservoir No. 5 is 93%. (Knapp-USGS)
W73-08491

THE CONCENTRATION OF COPPER, LEAD, ZINC AND CADMIUM IN SHALLOW MARINE SEDIMENTS, CARDIGAN BAY (WALES),
University Coll. of North Wales, Menai Bridge. Marine Science Labs.
A. S. G. Jones.
Marine Geology, Vol 14, No 2, p M1-M9, February 1973. 2 fig, 3 tab, 15 ref.

Descriptors: *Bottom sediments, *Copper, *Lead, *Zinc, *Cadmium, Bays, Provenance, Distribution patterns, Clays, Sands, Trace elements, Heavy metals.
Identifiers: *Cardigan Bay (Wales).

The concentration of copper, lead, zinc and cadmium in the sediments of the southern Cardigan Bay area were determined. Comparative results are presented for the Pleistocene sediments and the Aberystwyth Grits (Silurian) which outcrop along the shore line. The sediments from the Rheidol and Ystwyth rivers are strongly enriched in lead and zinc. The Pleistocene sediments are enriched in cadmium by a factor of 10 with respect to the seabed sediments in this part of Cardigan Bay. Copper is found in the acid-soluble fraction, 75% of the zinc occurs in the acid-soluble fraction, and 60% of the total cadmium occurs in the acid-soluble fraction. The charted distributions of the four elements did not show any clear dispersal patterns. The minimum concentrations of the four elements are found in the medium to fine sands. (Knapp-USGS)
W73-08501

OBSERVATIONS OF SOME SEDIMENTARY PROCESSES ACTING ON A TIDAL FLAT,
New Hampshire Univ., Durham. Dept. of Earth Sciences; and New Hampshire Univ., Durham. Jackson Estuarine Lab.
F. E. Anderson.
Marine Geology, Vol 14, No 2, p 101-116, February 1973. 11 fig, 12 ref.

Descriptors: *Sediment transport, *Sedimentation, *Suspended load, *Intertidal areas, *New Hampshire, Mud flats, Littoral drift, Waves (Water), Tidal streams, Settling velocity, Sampling.
Identifiers: *Tidal flats.

On tidal flats, sediment resuspension occurs more readily on the floodtide than the ebb. The concentration of suspended sediment follows the water mass distribution and is affected to a lesser degree by tidal currents and small amplitude waves. Deposition occurs during slack water shortly after high tide primarily in the bottom regime (15 cm); it is probably related to coarser particle sedimentation. The water mass distribution was not a simple rise and fall perpendicular to the bottom contours but rather followed a slow clockwise gyre. The net effect on the suspended sediments was to impart a longshore component of drift to the suspended load during the tidal cycle. In order to examine these sedimentary processes, eighteen-foot valves were placed in a small tidal cove in southern New Hampshire. Transport of suspended sediment was determined by comparing concentrations at 15 and 30 cm above the tidal flats throughout a tidal cycle. (Knapp-USGS)
W73-08502

A COMPARISON OF SUSPENDED PARTICULATE MATTER FROM NEPHELOID AND CLEAR WATER,
Lamont-Doherty Geological Observatory, Palisades, N.Y.
M. B. Jacobs, E. M. Thorndike, and M. Ewing.
Marine Geology, Vol 14, No 2, p 117-128, February 1973. 2 fig, 2 plate, 2 tab, 24 ref.

Descriptors: *Suspended load, *Sea water, *Optical properties, *Turbidity, Particle size, Mineralogy.
Identifiers: Nepheloid layers (Sea water).

The total count of suspended particles is a basic difference between water samples from nepheloid layers of the North American and Brazil basins and samples from the clearer water over the Mid-Atlantic Ridge in both the North and South Atlantic. The nepheloid layer of the North American basin has four times and the Brazil basin three times the particle density of clearer water over the mid-ocean ridge. A greater percentage of particles larger than 2 microns prevails in nepheloid water (85-96%) as compared to a range of 76-87% in the clear water. Non-opaque mineral grains form the major constituent. (Knapp-USGS)
W73-08503

A SANDWAVE FIELD IN THE OUTER THAMES ESTUARY, GREAT BRITAIN,
Unit of Coastal Sedimentation, Taunton (England).
For primary bibliographic entry see Field 02L.
W73-08504

THE RESIDUA SYSTEM OF CHEMICAL WEATHERING: A MODEL FOR THE CHEMICAL BREAKDOWN OF SILICATE ROCKS AT THE SURFACE OF THE EARTH,
Guelph Univ. (Ontario). Dept. of Land Resource Science.
For primary bibliographic entry see Field 02K.
W73-08511

UNIT STREAM POWER AND SEDIMENT TRANSPORT,
Illinois Univ., Urgana.
For primary bibliographic entry see Field 08B.
W73-08513

DISTRIBUTION AND FORMS OF NITROGEN IN A LAKE ONTARIO SEDIMENT CORE,
Department of Energy, Mines and Resources, Burlington (Ontario). Canada Center for Inland Waters.
For primary bibliographic entry see Field 05B.
W73-08563

RIVER ECOLOGY AND MAN.
For primary bibliographic entry see Field 06G.
W73-08664

GEOCHEMISTRY AND DIAGENESIS OF TIDAL-MARSH SEDIMENT, NORTHEASTERN GULF OF MEXICO,
Geological Survey, Washington, D.C.
V. E. Swanson, A. H. Love, and I. C. Frost.
Available from GPO, Washington, D.C. 20402, Price $0.55. Geological Survey Bulletin 1360, 1972. 83 p, 11 fig, 7 tab, 81 ref.

Descriptors: *Sediments, *Geomorphology, *Tidal marshes, *Florida, Gulf of Mexico, Sedimentology, Particle size, Sediment distribution, Organic matter, Inorganic compounds, Water chemistry, Chemical analysis, Salinity, Coasts, Estuaries, Tidal effects, Sediment transport, Hydrologic data, Data collections.
Identifiers: *Taylor County (Fla), *Dixie County (Fla).

This investigation was limited to one tidal-marsh area, a belt 1 to 5 miles wide extending 80 miles along the northeastern Gulf of Mexico, from the Aucilla River to the Suwannee River, Fla. Only the near offshore, tidal-stream, and tidal-marsh sediments, with their similarities and differences, are discussed. On the basis of grain size and amount of organic matter, the sediment in the tidal marsh is classified as peaty sandy mud, carbonaceous muddy sand, and relatively clean sand. The range in organic matter content is 15 to 40%, 4 to 15%, and 0.1 to 4%, respectively. Of the abundant organic matter in the tidal marsh sediment 80% is humic material derived from the inland swamp-forest areas and delivered to the Gulf of Mexico by streams. Metals that are enriched in the sediment because of the abundant organic matter are iron, manganese, cobalt, chromium, copper, molybdenum, nickel, and vanadium. Except for iron, each is present as less than 0.1% of the sample. (Woodard-USGS)
W73-08730

EFFECT OF GROUNDWATER ON STABILITY OF SLOPES AND STRUCTURES ERECTED ON THEM ON THAWING OF FROZEN SOILS,
For primary bibliographic entry see Field 02C.
W73-08734

SUSPENDED-SEDIMENT YIELDS OF NEW JERSEY COASTAL PLAIN STREAMS DRAINING INTO THE DELAWARE ESTUARY,
Geological Survey, Trenton, N.J.
L. J. Mansue.
Geological Survey open-file report, 1972. 26 p, 5 fig, 3 tab, 9 ref.

Descriptors: *Sediment transport, *Sediment yield, *Sedimentation, *Delaware River, Data collections, Tributaries, Streams, Estuaries, Bays, Sampling, Particle size, Suspended load, Navigable waters.
Identifiers: *Delaware Bay.

An investigation was begun in 1965 by the U.S. Geological Survey in cooperation with the U.S. Army, Corps of Engineers, for information on sediment transport in the Delaware River basin in relation to the maintenance of navigable channels in the river's estuarine areas. A daily sediment-sampling program at streamflow-gaging stations on Crosswicks Creek, Cooper River, Maurice River, and Raccoon Creek was established. Three partial-record stations were established on the Pennsauken, Big Timber, and Mantua Creeks in 1970 to complete the areal coverage. Records at Delaware River at Trenton and McDonalds Branch in Lebanon State Forest were available also for use in this study. Suspended sediment transported by streams draining into the Delaware estuary and Bay from New Jersey is related to areal variations in geology, physiography, and land use. It is estimated that sediment contribution from the Delaware River mainstem is 700,000 tons per year; that of streams draining the Inner Coastal Plain, 200,000 tons per year; and that of streams draining the Outer Coastal Plain, 9,000 tons per year. Thus, of the approximate 910,000 tons per year transported into the Delaware estuary and Bay, roughly 20% is contributed by the Coastal Plain tributaries. (Woodard-USGS)
W73-08742

FREEZING EFFECTS ON ARCTIC BEACHES,
Louisiana State Univ., Baton Rouge. Coastal Studies Inst.
For primary bibliographic entry see Field 02C.
W73-08751

GEOMORPHIC COASTAL VARIABILITY, NORTHWESTERN AUSTRALIA,
Louisiana State Univ., Baton Rouge. Coastal Studies Inst.
L. D. Wright, J. M. Coleman, and B. G. Thom.
Available from NTIS, Springfield, Va. 22151, as AD-755 154. Price $3.00 printed copy; $1.45 microfiche. Louisiana State University Coastal Studies Bulletin No 7, p 35-64, Technical Report 129, January 1973. 17 fig, 1 tab, 12 ref. ONR-GP Contract N00014-69-A-0211-0003.

Descriptors: *Geomorphology, *Coasts, *Australia, *Tides, *Waves (Water), Erosion, Beaches, Beach erosion, Surf, Littoral drift, Sedimentation, Sediment transport, Structural geology, Tropical regions, Climates.
Identifiers: *Coastal geomorphology.

The coastline from Darwin, Northern Territory, Australia, to the western limit of the Joseph Bonaparte Gulf illustrates multiclass coastal variability within a zonal process environment characterized by an extreme tidal range and a tropical monsoon climate. Ten discrete geomorphic provinces were identified, each possessing a distinct coastal landscape and reflecting varying relative process contributions. The numerous coastal forms and combinations of forms are functions more of local circumstances than of the zonal environment. Because of the feedback between form and process variations, in resistance factors such as geological structure, coastline orientation, and offshore slope, together with variations in local rates of sediment supply create appreciable spatial variability in the nearshore process environment and consequently in the relative contribution of different process factors to the coastal geomorphology. (Knapp-USGS)
W73-08752

PRELIMINARY RIVER-MOUTH FLOW MODEL,
Louisiana State Univ., Baton Rouge. Dept. of Chemical Engineering.
W. R. Waldrop.
Available from NTIS, Springfield, Va. 22151, as AD-755 155, Price $3.00 printed copy; $1.45 microfiche. Louisiana State University Coastal Studies Bulletin No 7, p 67-92, Technical Report 130, January 1973. 11 fig, 27 ref. ONR-GP Contract N00014-69-A-0211-0003.

Descriptors: *Sedimentation, *Deltas, *Numerical analysis, Turbulent flow, Sediment transport, Equations, Mississippi River, Deposition (Sediments), Salinity, Saline water intrusion.

The interaction of the river with its receiving basin near the mouth dominates deltaic growth. The transport phenomena controlling the interaction are extremely complicated. The mixing region involves three-dimensional, viscous, turbulent, three-component, two-phase, incompressible flow. The flow field is steady for a given river stage and tide condition. Turbidity of the river has an insignificant effect upon the fluid mechanics; hence, solution for the transport and deposition of suspended material may be decoupled from the solution of the flow field, and the complications involved in solving two-phase flow can be avoided. Equations describing the fluid mechanics of the region defy analytical solution, but a numerical technique is adequate. Results for two- and three-dimensional test cases are given. A technique is described by which a computed three-dimensional fluid velocity field can be used to calculate the deposition of suspended material from the river. (Knapp-USGS)
W73-08753

CEMENTATION IN HIGH-LATITUDE DUNES,
Louisiana State Univ., Baton Rouge. Coastal Studies Inst.
H. H. Roberts, W. Ritchie, and A. Mather.
Available from NTIS Springfield, Va. 22151, as AD-755 156. Price $3.00 printed copy; $1.45 microfiche. Louisiana State University Coastal Studies Bulletin No 7, p 95-112, Technical Report 131 January 1973.

Descriptors: *Diagenesis, *Sands, *Dunes, *Calcite, Carbonate rocks, Sedimentology, Carbonates, Calcium carbonate.
Identifiers: *Scotland, *Cementation (Diagenesis).

Highly calcareous dune sands are currently being cemented along many coasts of Northern Scotland. These Holocene landforms are stabilized by low Mg-calcite cements with fabrics and geochemical characteristics suggesting precipitation in the vadose meteoric diagenetic environment. Although extensive early carbonate diagenesis is not generally associated with high-latitude environments, this coastal zone cementation is a

widespread phenomenon in Northern Scotland. (Knapp-USGS)
W73-08754

PERIODICITIES IN INTERFACIAL MIXING,
Louisiana State Univ., Baton Rouge. Coastal Studies Inst.
For primary bibliographic entry see Field 02L.
W73-08756

AN AIR-LIFT FOR SAMPLING FRESHWATER BENTHOS,
Reading Univ. (England). Dept. of Zoology.
For primary bibliographic entry see Field 07B.
W73-08795

BOTTOM TOPOGRAPHY AND SEDIMENT TEXTURE NEAR THE COLUMBIA RIVER MOUTH,
Washington Univ., Seattle. Dept. of Oceanography.
For primary bibliographic entry see Field 02L.
W73-08827

DISTRIBUTION OF ORGANIC CARBON IN SURFACE SEDIMENT, NORTHEAST PACIFIC OCEAN,
Washington Univ., Seattle. Dept. of Oceanography.
For primary bibliographic entry see Field 02L.
W73-08828

THE DISTRIBUTION OF MICROBIOGENIC SEDIMENT NEAR THE MOUTH OF THE COLUMBIA RIVER,
Shoreline Community Coll., Seattle, Wash. Div. of Science.
For primary bibliographic entry see Field 02L.
W73-08829

THE RELATIONSHIP BETWEEN RAINFALL FREQUENCY AND AMOUNT TO THE FORMATION AND PROFILE DISTRIBUTION OF CLAY PARTICLES,
Illinois Univ., Urbana. Dept. of Agronomy.
For primary bibliographic entry see Field 02G.
W73-08875

FORMS AND DYNAMICS OF SULFUR IN PEAT AND LAKE SEDIMENTS OF BELORUSSIA (FORMY I DINAMIKA SERY V TORFAKH I OSADKAKH OZER BELORUSSII),
Akademiya Navuk BSSR, Minsk. Institut Geokhimii i Geofiziki.
K. I. Lukashev, V. A. Kovalev, A. L. Zhukhovitskaya, V. A. Generalova, and A. A. Sokolovskaya.
Litologiya i Poleznyye Iskopayemyye, No 3, p 34-47, May-June 1972. 1 fig, 6 tab, 24 ref.

Descriptors: *Sedimentation, *Diagenesis, *Peat, *Lake sediments, *Sulfur, Sulfur compounds, Sulfates, Sulfides, Mineralogy, Inorganic compounds, Organic compounds, Organic matter, Oxidation, Reduction (Chemical), Bogs, Marshes, Coals, Chemical analysis, Geochemistry, Seasonal.
Identifiers: *USSR, *Belorussia, Mineralization.

Diagenesis of sulfur in peat and lake sediments of Belorussia is considered in relation to the seasonal dynamics of the environment and the differences in the supply of mineral and organic matter. Sulfur compounds are present in peat and lake sediments in five forms: monosulfide, free, sulfate, disulfide, and organic. In peat, organic compounds and disulfies predominate, while sulfates and free sulfur are of secondary importance. In lake sediments, the main forms of sulfur are sulfates and monosulfides, followed by disulfides and organic compounds. The quantity of free sulfur in sediments is too small to be determined quantitatively. (Josefson-USGS)
W73-08883

1971 WATER RESOURCES DATA FOR NEW YORK: PART TWO, WATER QUALITY RECORDS.
Geological Survey, Albany, N.Y.
For primary bibliographic entry see Field 07C.
W73-08890

DECOMPOSITION OF OIL POLLUTANTS IN NATURAL BOTTOM SEDIMENTS,
Rutgers - The State Univ., New Brunswick, N.J.
For primary bibliographic entry see Field 05B.
W73-08911

THE DISTRIBUTION, SUBSTRATE SELECTION AND SEDIMENT DISPLACEMENT OF COROPHIUM SALMONIS (STIMPSON) AND COROPHIUM SPINICORNE (STIMPSON) ON THE COAST OF OREGON,
Oregon State Univ., Corvallis.
For primary bibliographic entry see Field 05B.
W73-08912

A. THE ROLE OF LICHENS IN ROCK WEATHERING AND SOIL FORMATION. B. MERCURY IN SEDIMENTS,
Wisconsin Univ., Madison.
For primary bibliographic entry see Field 05B.
W73-08930

SORPTION OF COPPER ON LAKE MONONA SEDIMENTS - EFFECT OF NTA ON COPPER RELEASE FROM SEDIMENTS,
Wisconsin Univ., Madison. Water Chemistry Program.
For primary bibliographic entry see Field 05B.
W73-08946

SUSPENDED-SEDIMENT YIELD AND EROSION FROM MOUNTAINOUS AREAS OF SOVIET CENTRAL ASIA (FORMIROVANYE STOKA VZVESHENNYKH NANOSOV I SNYV S GORNOY CHASTI SREDNEY AZII),
Sredneaziatskii Nauchno-Issledovatelskii Gidrometeorologicheskii Institut, Tashkent (USSR).
O. P. Shcheglova.
Trudy, No 60 (75), Leningrad, 1972. 228 p, 52 fig, 54 tab, 175 ref.

Descriptors: *Sediment yield, *Sediment discharge, *Suspended load, *Erosion, *Mountains, Glaciers, Ice, Snowpacks, Snowmelt, Runoff, Discharge (Water), Rainfall, Raindrops, Air temperature, Erosion rates, Turbidity currents, Provenance, Streams, Watersheds (Basins), Maps.
Identifiers: *USSR, *Soviet Central Asia, Rainwash, Differential erosion.

Methods to separate suspended-sediment yield of mountain streams in Soviet Central Asia into genetic components are based on observations carried out at hydrometeorological stations in the Uzbek, Kirgiz, and Tadzhik Republics and in southern Kazakhstan. Absolute and relative values of the components of suspended-sediment yield from rainfall, snowmelt, and glaciers are evaluated, and total suspended-sediment yield of the rivers of Soviet Central Asia is compiled. Relationships are established between the sources of sediments in mountain streams and absolute basin elevation, and, for the small and receding glaciers of the region, between the rate of glacial erosion and glacier runoff. A map constructed for the Chirchik River Basin in Tashkent Oblast graphically illustrates differential erosion over a basin surface. (Josefson-USGS)
W73-09000

A DEVICE FOR COLLECTING IN-SITU SAMPLES OF SUSPENDED SEDIMENT FOR MICROSCOPIC ANALYSIS,
Johns Hopkins Univ., Baltimore, Md. Chesapeake Bay Inst.
For primary bibliographic entry see Field 07B.
W73-09056

MATHEMATICAL SIMULATION OF THE TURBIDITY STRUCTURE WITHIN AN IMPOUNDMENT,
Army Engineer Waterways Experiment Station, Vicksburg, Miss.
For primary bibliographic entry see Field 05G.
W73-09081

CLAY MINERALOGY OF QUATERNARY SEDIMENTS IN THE WASH EMBAYMENT, EASTERN ENGLAND,
Imperial Coll. of Science and Technology, London (England). Dept. of Geology.
H. F. Shaw.
Marine Geology, Vol 14, No 1, p 29-45, January 1973. 4 fig, 9 tab, 11 ref.

Descriptors: *Clay minerals, *Intertidal areas, *Estuaries, *Tidal marshed, Clays, Illite, Montmorillonite, Kaolinite, Provenance, Mineralogy, Erosion, Sedimentation.

The Wash, England, is an area of extensive intertidal flat sedimentation. Samples were examined from the intertidal zone of deposition and also from areas that are probable sources of the sediments being deposited. The mineralogy of the clay fractions of the sediments is dominated by illite, mixed layer illite-montmorillonite, montmorillonite, kaolinite, and chlorite, with only minor amounts of quartz and calcite. There is little or no significant variation in the relative proportions of the clay minerals in the modern intertidal flat sediments. However, variations are found between the relative proportions of clay minerals present in the probable source materials and the intertidal flat sediments. The main sources of the intertidal sediments are the floors of the North Sea and the Wash, and the exposed coasts of Lincolnshire and Norfolk. Erosion and redeposition of the intertidal flat material is a further important source of clay sediments. (Knapp-USGS)
W73-09083

2K. Chemical Processes

DEVELOPMENT AND TESTING OF A DOUBLE-BEAM ABSORPTION SPECTROGRAPH FOR THE SIMULTANEOUS DETERMINATION OF DIFFERENT CATIONS IN WATER,
Kentucky Water Resources Inst., Lexington.
W. H. Dennen.
Available from the National Technical Information Service as PB-220 017, $3.00 in paper copy, $1.45 in microfiche. Research Report No 53, October 1972. 9 p, 4 fig, 1 tab. OWRR A-036-KY (1), 14-31-0001-3517.

Descriptors: *Water analysis, *Spectroscopy, *Spectrophotometry, *Chemical analysis, Photometry, Analytical techniques, Instrumentation, Spectrometers.
Identifiers: *Double-beam spectrography.

A double-beam absorption spectrographic analysis system using a dc arc multielement source was tested for water analysis. The optical system design brings analytical and reference beams together to illuminate the upper and lower portions of the spectrograph slit. However, a dc arc will not serve as a multielement source for the intended purposes because of excessive thermal broadening of emission lines. A direct excitation dc arc method can be used for the rapid determination of some cations in water. Measurements for Mg, Fe, Ni, Al, Cu, and Si are comparable to known detec-

tion levels. This analytical method is very fast and cheap and could well be applied in reconaissance studies. In operation, spectrographically pure electrodes are simply soaked in water, allowed to dry, and arced at 6 amps for 40 secs. (Knapp-USGS)
W73-08469

DETERMINATION OF TRACE ELEMENTS IN WATER UTILIZING ATOMIC ABSORPTION SPECTROSCOPY MEASUREMENT,
Kentucky Water Resources Inst., Lexington.
For primary bibliographic entry see Field 05A.
W73-08470

SALINITY VARIATIONS IN SEA ICE,
Dartmouth Coll., Hanover, N.H. Dept. of Earth Sciences.
For primary bibliographic entry see Field 02C.
W73-08474

RADIOCHEMICAL MONITORING OF WATER AFTER THE CANNIKIN EVENT, AMCHITKA ISLAND, ALASKA, JULY 1972,
Geological Survey, Lakewood, Colo.
For primary bibliographic entry see Field 05B.
W73-08479

THE CONCENTRATION OF COPPER, LEAD, ZINC AND CADMIUM IN SHALLOW MARINE SEDIMENTS, CARDIGAN BAY (WALES),
University Coll. of North Wales, Menai Bridge. Marine Science Labs.
For primary bibliographic entry see Field 02J.
W73-08501

THE INFLUENCE OF IRON AND ALUMINUM HYDROXIDES ON THE SWELLING OF NA-MONTMORILLONITE AND THE PERMEABILITY OF A NA-SOIL,
Reading Univ. (England). Dept. of Soil Science.
For primary bibliographic entry see Field 02G.
W73-08509

THE RESIDUA SYSTEM OF CHEMICAL WEATHERING: A MODEL FOR THE CHEMICAL BREAKDOWN OF SILICATE ROCKS AT THE SURFACE OF THE EARTH,
Guelph Univ. (Ontario). Dept. of Land Resource Science.
W. Chesworth.
Journal of Soil Science, Vol 24, No 1, p 69-81, March 1973. 6 fig, 2 tab, 35 ref.

Descriptors: *Weathering, *Silicates, Chemical degradation, Hydrolysis, Oxidation, Mineralogy, Iron, Magnesium, Aluminum, Soil formation.

The residua system of chemical weathering is the chemical sink towards which the bulk of earth's surface materials trend during weathering. Chemically the system is defined in terms of four major components: SiO_2, Al_2O_3, Fe_2O_3, H_2O. Evidence is derived from relative solubilities of components under earth-surface conditions, from experiments in which weathering environments have been simulated in the laboratory, and from field studies of rocks and their weathered mantle. The residua system can be used as a framework to construct mineral facies diagrams showing assemblages of the commonest earth-surface minerals that occur within it: quartz, gibbsite, goethite, kaolinite, halloysite. In a qualitative manner the facies diagrams can then be interpreted in terms of the partial pressure of water and the temperature. The usefulness of this approach in systematizing studies of soil genesis is illustrated with reference to the weathering of granite. (Knapp-USGS)
W73-08511

GAS CHROMATOGRAPH INJECTION SYSTEM,
National Aeronautics and Space Administration, Moffett Field, Calif. Ames Research Center.
For primary bibliographic entry see Field 07B.
W73-08528

A REAPPRAISAL: CORRELATION TECHNIQUES APPLIED TO GAS CHROMATOGRAPHY,
Warren Spring Lab., Stevenage (England). Control Engineering Div.
For primary bibliographic entry see Field 05A.
W73-08535

A UNIQUE COMPUTER CENTERED INSTRUMENT FOR SIMULTANEOUS ABSORBANCE AND FLUORESCENCE MEASUREMENTS,
Michigan State Univ., East Lansing. Dept. of Chemistry; and Michigan State Univ., East Lansing. Dept. of Biochemistry.
For primary bibliographic entry see Field 07B.
W73-08541

ON-LINE COMPUTER SYSTEM FOR AUTOMATIC ANALYZERS,
Food and Drug Administration, Chicago, Ill.
For primary bibliographic entry see Field 07C.
W73-08568

APPLICATION OF A LABORATORY ANALOG-DIGITAL COMPUTER SYSTEM TO DATA ACQUISITION AND REDUCTION FOR QUANTITATIVE ANALYSES,
Department of Agriculture, Agassiz (British Columbia). Research Station.
For primary bibliographic entry see Field 07C.
W73-08569

AUTOMATED KJELDAHL NITROGEN DETERMINATION--A COMPREHENSIVE METHOD FOR AQUEOUS DISPERSIBLE SAMPLES,
Miles Lab., Inc., Elkhart, Ind. Marschall Div.
For primary bibliographic entry see Field 05A.
W73-08596

FLOW PROGRAMMING IN COMBINED GAS CHROMATOGRAPHY--MASS SPECTROMETRY,
McDonnell Douglas Research Labs., St. Louis, Mo.
M. A. Grayson, R. L. Levy, and C. J. Wolf.
Analytical Chemistry, Vol 45, No 2, p 373-376, February 1973. 5 fig.

Descriptors: Methodology, Separation techniques, Pollutant identification, Flow rates, Gas chromatography, Mass spectrometry, Chemical analysis, Organic compounds.
Identifiers: *GC-mass spectrometry, *FPGC-mass spectrometry, *Flow programming, Flow programmed gas chromatography, Temperature programmed gas chromatography, Mass spectra, Carrier gas, Molecular effusion separator.

The use of flow programming gas chromatography (FPGC) in combination with GC-mass spectrometry has been studied using a Bendix time-of-flight mass spectrometer coupled to a Beckman gas chromatograph with a molecular effusion type separator. An Analabs flow programmer was used to program the flow rate. The ion source pressure was monitored with a Bayard Alpert ionization gauge. In order to have a molecular separator that operates efficiently over a broad range of carrier gas flow rates the latter of these 3 approaches were used: The outlet of the GC column is directly connected to the molecular separator and the entire column effluent is introduced to the separator; the outlet of the GC column is connected to a 'flow splitter', thus allowing only a fixed flow to enter the separator while the excess flow is vented; the separator is optimized for the highest flow rate encountered in the program. Using the latter, flow programming with GC/MS produced no change in yield during analysis. During an analysis of n-octane, the relatively low background current observed with FPGC (in contrast to TPGC) illustrated the utility of the extension of the useful range of the chromatographic column while minimizing the effect of 'column bleed'. The intensity of the peaks obtained during the elution of n-octane with TPGC indicates the advantage of FPGC-MS in obtaining the net mass spectrum. (Holoman-Battelle)
W73-08597

APPLICATION OF A WAVELENGTH SCANNING TECHNIQUE TO MULTI-ELEMENT DETERMINATIONS BY ATOMIC FLUORESCENCE SPECTROMETRY,
Imperial Coll. of Science and Technology, London (England). Dept. of Chemistry.
For primary bibliographic entry see Field 05A.
W73-08598

GLACIERS AND NUTRIENTS IN ARCTIC SEAS,
Maine Dept. of Sea and Shore Fisheries, West Boothbay Harbor. Fisheries Research Station.
S. Apollonio.
Science, Vol 180, No 4085, p 491-493, May 4, 1973. 1 fig, 11 ref. ONR N00014-70-A-0219-0001.

Descriptors: *Nutrients, *Glaciers, *Fjords, *Arctic, *Canada, Phosphates, Silicates, Nitrates, Scour, Sediments.

Significantly higher concentrations of nitrate and silicate were found in glaciated South Cape Fiord than in unglaciated Grise Fiord, in the Canadian Arctic, or in adjacent Jones Sound. No significant differences in phosphate concentrations were found. Glacial activity apparently enriches the concentrations of those nutrients most critically limiting for arctic phytoplankton requirements. (Knapp-USGS)
W73-08726

INFRARED REFLECTANCE OF SULFATE AND PHOSPHATE IN WATER,
Missouri Univ., Kansas City.
For primary bibliographic entry see Field 05A.
W73-08728

GROUNDWATER RESOURCES OF WASHINGTON COUNTY, TEXAS,
Geological Survey, Austin, Tex.
For primary bibliographic entry see Field 04B.
W73-08739

GROUNDWATER RESOURCES OF NAVARRO COUNTY, TEXAS,
Geological Survey, Austin, Tex.
For primary bibliographic entry see Field 04B.
W73-08740

SALINE GROUNDWATER RESOURCES OF MISSISSIPPI,
Geological Survey, Jackson, Miss.
For primary bibliographic entry see Field 07C.
W73-08750

STANDARD POTENTIALS OF THE SILVER/SILVER-IODIDE ELECTRODE IN AQUEOUS MIXTURES OF ETHYLENE GLYCOL AT DIFFERENT TEMPERATURES AND THE THERMODYNAMICS OF TRANSFER OF HYDROGEN HALIDES FROM WATER TO GLYCOLIC MEDIA,
Jadavpur Univ., Calcutta (India). Dept. of Chemistry.

For primary bibliographic entry see Field 05A.
W73-08757

THE ERROR IN MEASUREMENTS OF ELECTRODE KINETICS CAUSED BY NONUNIFORM OHMIC-POTENTIAL DROP TO A DISK ELECTRODE,
California Univ., Los Angeles. School of Engineering and Applied Science.
W. H. Tiedemann, J. Newman, and D. N. Bennion.
Journal of the Electrochemical Society, Vol. 120, No. 2, p 256-257, February 1973. 1 fig.

Descriptors: *Electrochemistry, Electrodes, Electric currents.
Identifiers: *Disk electrodes, *Errors, *Electrode kinetics, *Current distribution, Potential drop, Accuracy.

Prior studies have shown that nonuniform ohmic, potential drop with rotating disk electrodes will lead to errors in determining kinetic parameters unless special corrections are applied. Linear electrode kinetics have been used to assess the magnitude of the effect. The apparent surface overpotential is taken to be that measured by a reference electrode with the ohmic potential being determined by interruption of the current. Three reference electrode positions--edge, intermediate, and far away--were used. The results show that serious errors in the measurement of electrode kinetic parameters can result from an electrode which is characterized by a nonuniform current distribution. The errors associated with nonuniform current distribution are smallest when the reference electrode is located far from the disk. However, decreasing the error in this manner increases the ohmic drop which must be compensated for by an interrupter technique. Under certain experimental conditions, the compensated ohmic potential can be as large as 90 percent of the total measured overpotential, with a consequent decrease in the accuracy of the surface overpotential. (Little-Battelle)
W73-08758

REVERSED-PHASE FOAM CHROMATOGRAPHY. REDOX REACTIONS ON OPEN-CELL POLYURETHANE FOAM COLUMNS SUPPORTING TETRACHLOROHYDROQUINONE,
Eotvos Lorand Univ., Budapest (Hungary). Inst. of Inorganic and Analytical Chemistry.
For primary bibliographic entry see Field 05A.
W73-08782

ION-EXCHANGE FOAM CHROMATOGRAPHY. PART I. PREPARATION OF RIGID AND FLEXIBLE ION-EXCHANGE FOAMS,
Eotvos Lorand Univ., Budapest (Hungary). Inst. of Inorganic and Analytical Chemistry.
T. Braun, O. Bekeffy, I. Haklits, K. Kadar, and G. Majoros.
Analytica Chimica Acta, Vol 64, No 1, p 45-54, March 1973. 3 fig, 7 tab, 8 ref.

Descriptors: *Construction, Strontium radioisotopes, Pollutant identification, Cation exchange, Separation techniques.
Identifiers: *Ion exchange foam chromatography, *Ion exchange foams, *Foaming, Sr-85, Chloromethylation, Amination, Radiation grafting, Polyurethane foams, Polystyrene foams, Polymerization.

In order to be able to apply the principles of foam chromatography to ion-exchange processes, preparative methods for open-cell ion-exchange foams were investigated. Homogeneous ion-exchange foams were prepared by introducing ion-exchange groups on previously prepared phenol-formaldehyde, polyurethane and polyethylene foams. The maximum capacity of the produced sulfonated phenol-formaldehyde cation-exchange foams was 1.85 meq/g, that of the syrene-polyurethane interpolymer anion-exchange foams was 2.2 meq/g. Weak carboxylic ion-exchange foams were prepared by radiation grafting of polyurethane and polyethylene foams; the maximum capacity of these foams was 4.02 meq/g. Heterogeneous ion-exchange foams were prepared by foaming a fine powder of a commercially available cation exchanger with the precursors of open-cell polyether-type polyurethane foam. The capacity of such a foam containing 26 percent ion-exchange powder was 1.0 meq/g. The kinetics of the cation-exchange process on the heterogeneous foams was measured with Sr-85. (Holoman-Battelle)
W73-08783

ION EXCHANGE IN AQUEOUS AND IN AQUEOUS-ORGANIC SOLVENTS. PART I. ANION-EXCHANGE BEHAVIOUR OF ZR, NB, TA AND PA IN AQUEOUS HCL-HF AND IN HCL-HF-ORGANIC SOLVENT,
Technische Universitaet, Munich (West Germany). Institut fuer Radiochemie.
For primary bibliographic entry see Field 05A.
W73-08784

DETERMINATION OF CHLOROPHYLL A AND B IN PLANT EXTRACTS BY COMBINED SUCROSE THIN-LAYER CHROMATOGRAPHY AND ATOMIC ABSORPTION SPECTROPHOTOMETRY,
Texas Tech Univ., Lubbock. Textile Research Center.
H. Loewenschuss, and P. J. Wakelyn.
Analytica Chimica Acta, Vol 63, No 1, p 230-235, January 1973. 1 fig, 3 tab, 11 ref.

Descriptors: *Methodology, Separation techniques, Reliability, Plant pigments, Solvent extractions, Magnesium, Physical properties.
Identifiers: *Sucrose thin layer chromatography, *Atomic absorption spectrophotometry, *Chlorophyll a, *Chlorophyll b, *Plant extracts, Sample preparation, Sensitivity, Spinach, Pheophytin a, Pheophytin b, Sample size.

A rapid, accurate technique has been described for determining chlorophyll a and b in plant extracts. A plant sample was extracted with MeOH containing $MgCO_3$ at zero degrees for an hour with constant stirring. The $MgCO_3$ was added in order to prevent the chlorphylls from decomposing to pheophytins. The extract was filtered, and a portion evaporated in a stream of N_2 to yield a concentrate for pigment separation by sucrose one-dimensional thin-layer chromatography. The Mg content of those separated compounds was determined by AAS, and was used to determine the ratio of chlorophyll a to b. With this system, chlorophyll a and b are well separated from each other and from other pigments as well. The chlorophyll ratio obtained compared favorably with the published value and spectroscopically determined values. The sample size and sensitivity of the method are limited only by the atomic-absorption spectrophotometer available (0.001 microgram Mg/ml can be measured). The proposed method therefore seems preferable to the normal published techniques since it is fast, reliable, and more convenient. The proposed method can also be used to determine total chlorophyll content of a material. (Holoman-Battelle)
W73-08786

THE PREPARATION AND ANALYTICAL EVALUATION OF A NEW HETEROGENEOUS MEMBRANE ELECTRODE FOR CADMIUM (II),
Rome Univ. (Italy). Istituto di Chimica Analitica.
For primary bibliographic entry see Field 05A.
W73-08787

SEPARATION OF MATRIX ABSORPTION AND ENHANCEMENT EFFECTS IN THE DETERMINATION OF SULFUR IN SEA WATER BY X-RAY FLUORESCENCE,
Hawaii Univ., Honolulu. Dept. of Chemistry; and Hawaii Inst. of Geophysics, Honolulu.
For primary bibliographic entry see Field 05A.
W73-08789

A RADIOREAGENT METHOD FOR THE DETERMINATION OF TRACES OF BISMUTH,
Savannah State Coll., Ga. Dept. of Chemistry.
For primary bibliographic entry see Field 05A.
W73-08790

SPECTROPHOTOMETRIC DETERMINATION OF P.P.B. LEVELS OF LONG-CHAIN AMINES IN WATERS AND RAFFINATES,
Australian Atomic Energy Commission Research Establishment, Lucas Heights.
For primary bibliographic entry see Field 05A.
W73-08793

CHEMICAL BUDGET OF THE COLUMBIA RIVER,
Oregon State Univ., Corvallis. Dept. of Oceanography.
For primary bibliographic entry see Field 02L.
W73-08821

WORKING SYMPOSIUM ON SEA-AIR CHEMISTRY: SUMMARY AND RECOMMENDATIONS,
Rhode Island Univ., Kington. Graduate School of Oceanography.
R. A. Duce, W. Stumm, and J. M. Prospero.
Journal of Geophysical Research, Vol 77, No 27, p 5059-5061. September 20, 1972. 2 ref. NSF GA 30550.

Descriptors: *Air-water interfaces, *Water chemistry, *Sea water, Conferences, Geochemistry, Water cycle.
Identifiers: *Sea-air chemistry.

The Oceanography Section of the National Science Foundation and the American Meteorological Society sponsored the Working Symposium on Sea-Air Chemistry held at Fort Lauderdale, Florida, January 31 through February 3, 1972. Thirty-three papers were presented on a wide variety of subjects associated with sea-air transport processes and the chemistry of the marine atmosphere. The solutions to many problems in air pollution and water pollution as well as a complete assessment of the geochemical cycles of many elements are critically dependent on a better understanding of sea-surface phenomena. Specific areas of investigation include the chemistry of the sea-surface microlayer and its modifications by man, as well as physical and chemical processes involved in bubble formation and bursting and the subsequent production of atmospheric sea-salt nuclei. Also of vital importance is a much more complete understanding of the chemistry of the marine atmosphere in general, including particle-gas interactions and the contribution of continentally derived particulates. (Knapp-USGS)
W73-08850

METAL-ION CONCENTRATIONS IN SEA-SURFACE MICROLAYER AND SIZE-SEPARATED ATMOSPHERIC AEROSOL SAMPLES IN HAWAII,
Hawaii Univ., Honolulu. Dept. of Chemistry.
D. R. Barker, and H. Zeitlin.
Journal of Geophysical Research, Vol 77, No 27, p 5076-5086, September 20, 1972. 1 fig, 3 tab, 36 ref.

Descriptors: *Air-water interfaces, *Aerosols, *Water chemistry, *Sampling, *Water analysis, Geochemistry, Sodium, Potassium, Calcium, Magnesium, Strontium, Copper, Iron, Zinc,

Phosphorus, Nitrates, Organic matter, Boundary processes, *Hawaii, Sea water.

Sea-surface microlayer and marine atmosphere samples from Hawaii were analyzed for sodium, potassin, calcium, magnesium, strontium, copper, iron, zinc, total carbon, total phosphorus, and reactive nitrate. Chemical compositions and ion ratios of most samples differ, often drastically, from normal seawater data. These samples generally have considerably higher total carbon, phosphorus, and nitrate concentrations than samples taken at the same site by at 0.6-meter depth. Microlayer samples are also found to be enriched to a small extent in some alkaline earth metals and to a much larger extent for the transition metals, copper, iron, and zinc. The transition metals were also found to be proportionately more organically associated in the microlayer samples. Aerosol samples studies are enriched in most metals relative to sodium and to the metal sodium ratio found in seawater. Average enrichment values for the transition metals on smaller aerosols approach and exceed 3 and 4 orders of magnitude. Organic coating materials on the aerosols are more concentrated on the smaller aerosols. (Knapp-USGS)
W73-08852

GEOCHEMISTRY OF OCEAN WATER BUBBLE SPRAY,
Negev Inst. for Arid Zone Research, Beersheba (Israel).
M. R. Bloch, and W. Luecke.
Journal of Geophysical Research, Vol 77, No 27, p 5100-5105, September 20, 1972. 3 fig, 1 tab, 22 ref.

Descriptors: *Air-water interfaces, *Water chemistry, *Sea water, *Aerosols, *Bubbles, Sodium, Potassium, Chlorides, Sulfates, Geochemistry, Boundary processes, Adsorption.

The salts in rain, snow, and dew derive from ocean bubble spray. The difference in ion ratios between the spray and the bulk of the solution that is sprayed is in agreement with the requirements of the Gibbs adsorption isotherm if the spray is derived from the surface layer of the bulk solution. Inorganic salts dissolved in ocean water are depleted in the surface of the ocean because the surface tension of their solution rises with their concentration. The gradient of this rise has a characteristic value for each salt and is independent of temperature. For sodium chloride this gradient is 1.85, and for potassium chloride it is 1.50. These and other oceanic salts were tested for changes in ion ratio in the bubble spray of their mixtures. Experiments were performed with seawater and artificial solutions at temperatures ranging from 25 degrees to 130 degrees C. The Gibbs theory predicts quantitatively the ratio changes found in bubble spray at equilibrium conditions only (for ocean environment near 100% humidity). When the sprayed surface is evaporated, the ratio shift is enhanced in favor of potassium ions compared with sodium ions. If condensation occurs at the time of bubble spray generation, however, this change in ratio is diminished. (Knapp-USGS)
W73-08853

FLAME PHOTOMETRIC ANALYSIS OF THE TRANSPORT OF SEA SALT PARTICLES,
National Oceanic and Atmospheric Administration, Hilo, Hawaii. Mauna Loa Observatory.
B. A. Bodhaine, and R. F. Pueschel.
Journal of Geophysical Research, Vol 77, No 27, p 5106-5115, September 20, 1972. 8 fig, 25 ref.

Descriptors: *Aerosols, *Salts, *Water chemistry, *Flame photometry, Analytical techniques, Sea water, Water analysis.
Identifiers: *Flame scintillation photometry.

Flame scintillation spectral analysis was used for the determination of the size distribution of sodium-containing particles in oceanic aerosols. Near the seashore, these aerosols are identical to sea salt particles and amount to approximately 5% of the total Aitken particle count. Measurements made at Mauna Loa Observatory (above the trade wind inversion) indicate a similar size distribution. However, only about 1% of the concentration is transported through the inversion, and removal efficiency tends to favor the removal of larger particles (Knapp-USGS)
W73-08854

VARIATION IN IONIC RATIOS BETWEEN REFERENCE SEA WATER AND MARINE AEROSOLS,
Centre National de la Recherche Scientifique, Gif-sur-Yvette (France). Centre des Faibles Radioactivites.
R. Chesselet, and P. Baut-Menard.
Journal of Geophysical Research, Vol 77, No 27, p 5116-5131, September 20, 1972. 1 fig, 9 tab, 37 ref.

Descriptors: *Aerosols, *Water chemistry, *Geochemistry, *Dusts, *Sea water, Sampling, Water analysis, Chemical analysis, Antarctic, Potassium, Sodium, Air-water interfaces, Salts, Spectrophotometry.

Atmospheric particles were collected over the western Mediterranean, North Atlantic, and Norwegian Sea by air filtration and cascade impactors. The ratios of concentrations of K to Na exhibited marked variation and were greater than the reference ratio in bulk seawater. These concentration ratios are related to chemical fractionation in aerosols produced by the breaking of bubbles at the sea surface. In strong support of this fractionation concept are data based on: laboratory and in situ experiments with artifically produced aerosols; high-volume air sampling over the Antarctic Ocean, where terrigenous dusts are scarce; and coastal region sampling where silicon, aluminum, iron, and potassium provide a corrective index for potassium-rich terrigenous dust concentrations. The Cl/Na ratios measured resemble the seawater values, which could indicate that an enrichment of potassium takes place rather than a depletion of sodium. The Ca/Na ratios often follow the K/Na ratios and suggest identical enrichment of calcium. The enrichments seem to preferentially affect particles of small size (Knapp-USGS)
W73-08855

MOLECULAR ASPECTS OF AQUEOUS INTERFACIAL STRUCTURES,
Miami Univ., Coral Gables, Fla. Dept of Chemistry.
For primary bibliographic entry see Field 01B.
W73-08856

EXPERIMENTAL SEA SALT PROFILES,
National Center for Atmospheric Research, Boulder, Colo.
E. R. Frank, J. P. Lodge, Jr., and A. Goetz.
Journal of Geophysical Research, Vol 77, No 27, p 5147-5151, September 20, 1972. 10 fig, 1 tab, 5 ref.

Descriptors: *Aerosios, *Water chemistry, *Pacific Ocean, *Salts, Chlorides, Sulfates, Winds, Sampling, Particle size, Electron microscopy, Chemicl analysis, Bubbles.

A series of determinations was made of the concentration of airborne chloride particles at heights between 1 and 15 meters above the ocean surface. Samples were taken in the general area between Santa Barbara, California, and Santa Catalina Island. There is no clear gradient in concentration over the height interval measured. Number count, geometric mean, and geometric standard deviation were all extremely similar in al heights. Geometric standard deviation was small, rarely exceeding 2.0, and was often as low as 1.5. Geometric mean diameter, expressed as the diameter of an equivalent sphere of pure sodium chloride, was approximately 2 microns. Concentrations ranged over a wide span from a few thousand to a few million particles per cubic meter. Because of the complex geography of the San Pedro channel and adjacent waters, there was no obvious relationship between local wind velocity and particle concentration. A contributing factor may be the difficulty in making measurements in a small boat at wind velocities that cause significant local surface generation of salt particles. Simultaneous collections of smaller particles for electron microscopy showed phosphate to be a consistent minor ingredient, and sulfuric acid and sulfates to be surprisingly prevalent. (Knapp-USGS)
W73-08857

STRUCTURE OF SEA WATER AND ITS ROLE IN CHEMICAL MASS TRANSPORT BETWEEN THE SEA AND THE ATMOSPHERE,
R. A. Horne.
Journal of Geophysical Research, Vol 77, No 27, p 5170-5176, September 20, 1972. 5 fig, 37 ref.

Descriptors: *Water structure, *Air-water interfaces, *Ion transport, *Mass transfer, *Sea water, Water chemistry, Aerosols, Bubbles, Diffusion, Surface tension, Sodium, Potassium, Vicosity, Hydration, Hydrogen bonding.

Understanding interfacial phenomena associated with liquid water can be aided by a molecular model that envisions the pure bulk liquid as a mixture of flickering clusters and more or less monomeric water. Seawater solutes alter the local water structure into at least two structurally different types of hydration atmospheres, coulombic hydration surrounding ions and polar solutes and hydrophobic hydration enveloping nonpolar solutes and solute segments. The water structure (vicinal water) of the liquid boundary layer at the air-sea interface resembles the hydrophobic hydration, and it may preferentially exclude ionic species and thereby give rise to the Na-K fractionation observed in nature. (Knapp-USGS)
W73-08858

EFFECT OF VARIOUS SALTS ON THE SURFACE POTENTIAL OF THE WATER-AIR INTERFACE,
Naval Research Lab., Washington, D.C.
N. L. Jarvis.
Journal of Geophysical Research, Vol 77, No 27, p 5177-5182, September 20, 1972. 3 fig, 18 ref.

Descriptors: *Chemical potential, *Air-water interfaces, *Sea water, *Aerosols, *Water chemistry, Sulfates, Carbonates, Chlorides, Nitrates, Bromine, Iodine, Potassium, Ammonia, Sodium, Lithium, Barium, Strontium, Magnesium.

Studies of the formation and chemical composition of marine aerosols strongly suggest that some ions may be preferentially excluded from the sea-air interface. Measurements of the surface potential of electrolyte solutions are a useful technique for studying in situ ion exclusion. In a series of surface potential measurements the potential changes due to a variety of Na salts, at a solution concentration of 2 moles, were in the order of decreasing potential: $SO_4 > CO_3 > CH_3COO > Cl > NO_3 > Br > I > SCN$. In general the anion with the smaller hydration energy gave the greater decrease. Regular decreases were observed for the group IA and IIA chlorides at a given concentration: $K \pm NH_4 > Li$ and $Ba > Sr > Mg$. The sign of the surface potential change as well as the magnitude appears to be determined primarily by the anion; this suggests that the anion are preferentially associated with the water molecules at the surface. (Knapp-USGS)
W73-08859

OUR KNOWLEDGE OF THE PHYSICO--CHEMISTRY OF AEROSOLS IN THE UNDISTURBED MARINE ENVIRONMENT,
Max Planck-Institut fur Chemie, Mainz (West Germany).
For primary bibliographic entry see Field 05B.
W73-08860

ADSUBBLE PROCESSES: FOAM FRACTIONATION AND BUBBLE FRACTIONATION,
Cincinnati Univ., Ohio. Dept of Chemical and Nuclear Engineering.
For primary bibliographic entry see Field 05G.
W73-08861

FLOW PATTERNS IN BREAKING BUBBLES,
California Univ., Santa Barbara. Marine Science Inst.
F. MacIntyre.
Journal of Geophysical Research, Vol 77, No 27, p 5211-5228, September 20, 1972. 19 fig, 1 tab, 62 ref.

Descriptors: *Bubbles, *Mass transfer, *Flow, Sea water, Water chemistry, Foaming, Air-water interfaces, Waves (Water), Ocean waves, Surface tension, Viscosity, Aerosols.

Fluid flow during bubble collapse is driven by high pressure gradients associated with high surface curvature. Experiments suggest that the driving force is at a maximum at the surface and preferentially accelerates the surface in boundary-layer shear flow down the cavity wall. An irrotational solitary capillary ripple precedes the main toroidal rim, transporting mass along the surface at about 90% of its phase velocity. The convergence of this flow creates opposed axial jets. About 20% of the boundary-layer thickness flows into the upward jet. The material in the top jet drop was originally spread over the interior bubble surface at an average thickness of some 0.05% of the bubble diameter. (Knapp-USGS)
W73-08862

ELEMENT ENRICHMENT IN ORGANIC FILMS AND FOAM ASSOCIATED WITH AQUATIC FRONTAL SYSTEMS,
Delaware Univ., Newark, College of Marine Studies.
K. H. Szekielda, S. L. Kupferman, V. Klemas, and D. F. Polis.
Journal of Geophysical Research, Vol 77, No 27, p 5278-5282, September 20, 1972. 5 fig, 2 tab, 10 ref.

Descriptors: *Air-water interfaces, *Organic matter, *Foam fractionation, *Ocean circulation, *Trace elements, Chromium, Copper, Iron, Mercury, Lead, Zinc, Ocean currents, Sediments, Suspended load, Water chemistry.
Identifiers: *Fronts (Aquatic).

Surface slicks and foam were collected at frontal convergence zones in Delaware Bay by a method that resulted in the acquisition of gram-sized samples. Concentrations of Cr, Cu, Fe, Hg, Pb, and Zn in these samples were found to be higher by more than 4 orders of magnitude than those in mean ocean water. Inorganic matter comprises more than 80% of the film material. The primary inorganic constituent is quartz. The levels of trace metal enrichment in the slicks and foam are in reasonable agreement with the values predicted from measurements in the surface microlayer in Narragansett Bay. (Knapp-USGS)
W73-08864

A LABORATORY STUDY OF IODINE ENRICHMENT ON ATMOSPHERIC SEA-SALT PARTICLES PRODUCED BY BUBBLES,
Hawaii Univ., Honolulu. Dept of Chemistry.
F. Y. B. Seto, and R. A. Duce.
Journal of Geophysical Research, Vol 77, No 27, p 5339-5349, September 20, 1972, 2 fig, 4 tab, 24 ref.

Descriptors: *Aerosols, *Iodine, *Air-water interfaces, Water chemistry, Sea water, Laboratory tests, Sampling, Salts.

A laboratory model ocean-atmosphere study was undertaken to investigate the mechanisms causing iodine enrichment of up to 500 (compared to the iodine content of seawater) on atmospheric sea-salt particles produced by bubbles in the sea. Experiments were run both by using I-131 tracer in seawater and by using iodine analysis by neutron activation in untreated fresh seawater. Particles produced by bubbling in the model ocean were separated into size fractions with a cascade impactor. Gaseous iodine was collected on activated charcoal. Organically bound iodine probably accounts for an initial iodine enrichment on the particles and may also explain the characteristic U shape of the curve of iodine enrichment versus particle size curve. Gaseous iodine is also a major factor in determining the iodine enrichment on marine atmospheric particulate matter. (Knapp-USGS)
W73-08865

ON THE FLOTATION OF PARTICULATES IN SEA WATER BY RISING BUBBLES,
Naval Research Lab., Washington, D.C.
G. T. Wallace, Jr., G. I. Loeb, and D. F. Wilson.
Journal of Geophysical Research, Vol 77, No 27, p 5293-5301, September 20, 1972. 1 fig, 3 tab, 45 ref.

Descriptors: *Foam fractionation, *Bubbles, *Air-water interfaces, *Sea water, *Water chemistry, Microorganisms, Algae, Aquatic bacteria, Trace elements, Flotation.

A simple batch-type foam separation apparatus was used to determine the degree to which rising bubbles could scavenge and enrich the surface layers of a column of seawater with particulate matter. By comparing the amount of particulated in the original material with that in later recovered material in the enriched surface layers of the column, recoveries were determined for various samples of natural and laboratory origin. Recovery of particulates in natural samples was consistently high, whereas recovery of particulates from unicellular algal cultures varied, the apparent parameter governing the degree of recovery being the age of the culture in question. Bacteria and phytoplankton are known to concentrate trace elements to values larger by orders of magnitude than their ambient sea water concentrations. In turn, bacteria and phytoplankton can be ejected into the atmosphere by the action of breaking bubbles. It seems possible that the trace element enrichment observed in atmospheric sea salt particles may at least in part be due to the biological activity of microorganisms in the sea. (Knapp-USGS)
W73-08866

FRACTIONATION OF THE ELEMENTS F, CL, NA, AND K AT THE SEA-AIR INTERFACE,
Naval Research Lab., Washington, D.C.
P. E. Wilkniss, and D. J. Bressan.
Journal of Geophysical Research, Vol 77, No 27, p 5307-5315, September 20, 1972. 3 fig, 3 tab, 13 ref.

Descriptors: *Aerosols, *Air water interfaces, *Water chemistry, *Sea water, Flourides, Chlorides, Sodium, Potassium, Dusts.

The sea-air interface ion fractionation of F, Cl, Na, and K was studied in the field and the laboratory. Laboratory results show negative enrichement of fluoride and chloride versus Na and a postive enrichment of K. The same results were found in samples that were collected under marine conditions. The fractionation in a marine sea salt aerosol is F ± -0.62, Cl ± +0.12, and K ± +0.40 normalized to Na. The chemistry of marine precipitation is quite different from these values and approaches more the chemistry of the tropospheric background aerosol or of a continental aerosol with the exception of chloride values. (Knapp-USGS)
W73-08867

SMALLER SALT PARTICLES IN OCEANIC AIR AND BUBBLE BEHAVIOR IN THE SEA,
Hawaii Univ., Honolulu. Dept of Oceanography.
A. H. Woodcock.
Journal of Geophysical Research, Vol 77, No 27, p 5316-5321, September 20, 1972. 4 fig, 27 ref.

Descriptors: *Bubbles, *Aerosols, *Water chemistry, Sea water, Sampling, Salts, Air-water interfaces.

Observations of the number and size of the smaller salt particles in the air over Hawaiian and Alaskan seas are presented. Differences in the particle distributions at the two locations are related to differences in the behavior of air bubbles of various sizes near and on a seawater surface. The relationships suggest that a break in the curves among particles of 10 to the -13 to -14 power grams marks a zone of transition from a bubble jet to a bubble film source of droplets and that the Alaskan waters are a relatively poor source of particles. (Knapp-USGS)
W73-08868

RAPID CHANGES IN THE FLUORIDE TO CHLORINITY RATIO SOUTH OF GREENLAND,
Bedford Inst., Dartmouth (Nova Scotia). Atlantic Oceanographic Lab.
J. M. Bewers, G. R. Miller, Jr., D. R. Kester, and T. B. Warner.
Nature Physical Science, Vol 242, No 122, p 142-143, April 30, 1973. 2 fig, 14 ref.

Descriptors: *Chlorides, *Fluorides, *Sea water, *Water chemistry, *Atlantic Ocean, Provenance, Ocean circulation, Ocean currents, Oceanography.
Identifiers: *Greenland.

Fluoride to chlorinity ratios were measured at six stations southeast of Greenland. There were no anomalous fluoride concentrations present at the time these samples were collected. It is unlikely that earlier reported anomalies were the result of sampling or analytical error. Therefore, if the earlier existence of the anomalies is accepted, the present work shows that they are not a steady-state feature of the deep waters of this part of the North Atlantic. The rapid implied change in the deepwater chemistry of this region can only have been caused by an episodic injection of fluoride; but neither volcanism nor mineral dissolution can be responsible. Mineral dissolution rates do not change rapidly enough to explain the temporal changes in water chemistry. Moreover, since the underlying sediments in the area are predominantly calcareous the amounts of fluoride available are insufficient. Therefore, further information regarding the geochemistry of fluoride in the marine environment is required. (Knapp-USGS)
W73-08869

WATER MOVEMENT AND CALICHE FORMATION IN LAYERED ARID AND SEMIARID SOILS,
Nevada Univ., Reno.
For primary bibliographic entry see Field 02G.
W73-08872

ANALYSIS OF MULTIDIMENSIONAL LEACHING,
Commonwealth Scientific and Industrial Research Organization, Wembley (Australia). Div. of Soils.
For primary bibliographic entry see Field 02G.
W73-08873

EFFECT OF FLOODING ON THE EH, PH, AND CONCENTRATIONS OF FE AND MN IN SEVERAL MANITOBA SOILS,
Manitoba Univ., Winnipeg. Dept. of Soil Science.

An ion-selective electrode for calcium was constructed as follows. A membrane 0.5 mm thick was prepared by mixing liquid ion-exchanger and polyvinylchloride and pouring into a polypropylene cap from a 50-ml centrifuge tube. The resulting membrane was cemented to a piece of Tygon tubing. A graphite rod (5 mm dia x 90 mm) was force fitted into the tubing so as to contact the membrane. A copper lead wire was press fitted into the graphite. The performance of the electrode was compared with that of the Orion calcium electrode in calcium solutions prepared by dissolving calcium carbonate in hydrochloric acid, neutralizing excess acid, and diluting with distilled water. The eletrodes were calibrated using these solutions. Selectivity coefficients were determined by additions of Mg, Ba, Ni, Zn, and Pb ions. The results showed that electrode potential varied linearly with the log of calcium activity in the range of 0.1 to 0.00001 M Ca (2 plus). The electrode was also sensitive below 0.00001 M. Both the PVC/graphite and Orion electrodes were useful in the pH range 5.5 to 10.5 at Ca (2 plus) concentrations as low as 0.0001 M. The PVC/graphite electrode was less sensitive to Na than the Orion. Responses were similar for Ni and Zn, less for Mg and Ba with the PVC/graphite electrode, and more for Pb with the PVC/graphite electrode. (Little-Battelle)
W73-08910

INFLUENCE OF PH ON THE RESPONSE OF A CYANIDE ION SELECTIVE MEMBRANE ELECTRODE,
Rome Univ. (Italy). Istituto di Chimica Analitica.
For primary bibliographic entry see Field 05A.
W73-08916

DESIGN AND EVALUATION OF A LOW COST RECORDING SPECTROPOLARIMETER,
Cleveland State Univ., Ohio. Dept. of Chemistry.
For primary bibliographic entry see Field 07B.
W73-08919

(I) OBSERVATIONS OF SMALL PHOTOCURRENTS ON THE DROPPING MERCURY ELECTRODE AND (II) A STUDY OF THE ADSORPTION OF TRIFLUOROMETHYL SULFONATE ON MERCURY,
Georgetown Univ., Washington, D.C.
For primary bibliographic entry see Field 05A.
W73-08926

PULSE POLAROGRAPHY IN PROCESS ANALYSIS. DETERMINATION OF FERRIC, FERROUS, AND CUPRIC IONS,
North American Rockwell Corp., Thousand Oaks, Calif. Science Center.
For primary bibliographic entry see Field 05A.
W73-08952

SOLID STATE ION-SELECTIVE MICROELECTRODES FOR HEAVY METALS AND HALIDES,
State Univ. of New York, Buffalo. Dept. of Chemistry.
For primary bibliographic entry see Field 05A.
W73-08954

GAS-SOLID CHROMATOGRAPHY OF ORGANIC ACIDS AND AMINES USING STEAM CONTAINING FORMIC ACID OR HYDRAZINE HYDRATE AS CARRIER GASES,
Tokyo Univ. of Education (Japan). Inst. for Optical Research.
For primary bibliographic entry see Field 05A.
W73-08956

HYDROGEOLOGY AND KARST STUDIES (GIDROGEOLOGIYA I KARSTOVEDENIYE).
Perm State Univ. (USSR). Inst. of Karst Studies and Peleology.

Vol 4, 1971. 232 p.

Descriptors: *Karst, *Karst hydrology, *Hydrogeology, *Investigations, Analytical techniques, Geology, Structural geology, Fissures (Geologic), Rocks, Carbonate rocks, Limestones, Dolomite, Caves, Groundwater, Subsurface drainage, Erosion, Alluvium, Water types, Reservoirs, Aerial photography.
Identifiers: *USSR, *Karst topography, Karst lakes, Paleokarst, Tectonics.

This collection of 29 papers published by Perm' State University is grouped under four subject headings: karst studies (12 papers); karst hydrogeology and hydrology (9 papers); hydrogeology (5 papers); and geodynamic processes (3 papers). Among the topics discussed are: (1) aerial photographic interpretation of karst areas; (2) karst as an indicator of development of platform structures; (3) fissuration as a condition essential to deep karst development in carbonate rocks; (4) paleokarst development in Carboniferous deposits of the Bashkir and Tatar Republics; (5) buried karst in the Lower Volga region; (6) carbonate breccia in gypsum karst of Dagestan; (7) salt karst on shores of the Toktogul Reservoir (Kirgizia); (8) practical importance of Siberian karst; (9) karst of the Altay Territory; (10) investigation methods and use of karst in Czechoslovakia; (11) characteristics of Cuban karst; (12) geological activity of waters in karst regions of Perm' Oblast; (13) composition of karst-lake deposits in Perm' Oblast; (14) character and intensity of chemical denudation in mountainous regions of the Crimea; (15) hydrogeological significance of clastic karst in the northern Yuryuzano-Sylva depression (Perm' Oblast); (16) use of aqualungs in hydrogeological investigations of karst regions; (17) effects of tectonic structures on subsurface drainage in zone of active water exchange; (18) freshwater and saline-water resources in the Cisural region of Perm' Oblast; (19) exogenic geodynamic processes in the zone of Kama River reservoirs; and (20) dynamic features of alluvium. (Josefson-USGS)
W73-08998

DIFFUSION COEFFICIENT OF I-IONS IN SATURATED CLAY SOIL,
For primary bibliographic entry see Field 02G.
W73-09030

BONDING OF WATER TO ALLOPHANE,
Department of Scientific and Industrial Research, Lower Hutt (New Zealand). Soil Bureau.
N. Wells, and R. J. Furkert.
Soil Sci. Vol 113, No 2, p 110-115, 1972. Illus.
Identifiers: *Allophane, *Bonding, Clusters, *Endothermy, Hydrogen, Physical properties, Infrared.

Modification of the physical properties of allophane by mechanical working has been related to changes induced in the bonding of water. A working action introduced a sharp low temperature endothermic peak. The water in worked samples was able to undergo H--O-H bending at 1620 cm-1 to a greater extend than water in unworked allophane and to a slightly greater extent than in water alone. Worked and unworked allophane taken to wilting point in a pressure membrane apparatus held over 100% by weight (on 110C wt.) in a form that gave a broad high temperature endothermic peak and an IR (H-O-H bending/-O-H stretching) absorbance ratio similar to that given by the unworked, undried allophane. Rewetting to the natural water content did not reproduce the single broad high temperature peak. Rewet allophane could have 3 endothermic peaks: a very

sharp peak at 25C given by surplus free water; a sharp peak at 35C given by single water molecules H bonded to clay surfaces; and a broad peak at 60 to 70C given by H bonded clusters of H_2O molecules cited in microspores. IR evidence suggested that the worked material had the added water mainly present as free water while the unworked allophane had mostly re-incorporated the added water. The unworked allophane taken only part way to wilting point could recover its clusters of water molecules, but the worked material did not recover and had free water present. The large quantity of water present in natural allophane is held in H bonded clusters of H_2O molecules. Mechanically working to produce a slurry breaks the clusters and distributes the water as singly linked molecules on the clay surface.--Copyright 1972, Biological Abstracts, Inc.
W73-09041

INTERLAYER HYDRATION AND THE BROADENING OF THE 10 A X-RAY PEAK IN ILLITE,
Battelle-Northwest, Richland, Wash.
For primary bibliographic entry see Field 02G.
W73-09042

DETAILED TEMPERATURE STRUCTURE OF THE HOT BRINES IN THE ATLANTIS II DEEP AREA (RED SEA),
Bundesanstalt fur Bodenforschung, Hanover, (West Germany).
M. Schoell, and M. Hartmann.
Marine Geology, Vol 14, No 1, p 1-14, January 1973. 5 fig, 4 tab, 13 ref.

Descriptors: *Brines, *Thermal water, *Geothermal studies, *Hot springs, Water temperature, Thermal springs, Thermal stratification, Sampling, Sounding.
Identifiers: *Red Sea.

Semi-continuous temperature profiles in the Atlantis II Deep of the Red Sea reveal a detailed picture of the flow of the recently detected 59 deg C brine. This newly heated brine originates in the southwest basin of the Atlantic II Deep and is spreading to its other basins. In some depressions in the basins, relicts of the previous 56 deg C brine were detected as a bottom layer. In the north basin restricted inflow of the 59 deg C brine led to a maximum temperature of only 58.2 deg C. Detailed bathymetry of the Chain Deep area revealed the existence of three nearly or completely separate brine pools with differing temperature profiles; maximum temperatures recorded are 46.2 and 52.4 deg C, respectively. (Knapp-USGS)
W73-09099

2L. Estuaries

ESTUARINE POLLUTION, A BIBLIOGRAPHY.
Office of Water Resources Research, Washington, D.C.

Available from the National Technical Information Service as PB-220 119, $12.50 in paper copy, $1.45 microfiche. Water Resources Scientific Information Center, Report WRSIC 73-205, April 1973, 477 p, 324 ref.

Descriptors: *Aquatic environment, *Bibliographies, *Estuaries, *Estuarine environment, *Eutrophication, Algae, Benthic fauna, Dissolved oxygen, Environmental effects, Nutrients, Pesticides, Phytoplankton, Radioactivity, Salinity, Thermal pollution, Water pollution effects, Water pollution sources, Water quality control.
Identifiers: *Estuarine pollution.

This bibliography containing 324 abstracts and references, is another in a series of planned bibliographies in water resources to be produced from the information base comprising SELECTED WATER RESOURCES ABSTRACTS (SWRA). At the time of search for this bibliography, the data base had 50,631 abstracts covering SWRA through December 15, 1972 (Volume 5, Number 24). The report contains an author index and extensive subject indexes.
W73-08451

SOME OBSERVATIONS OF TURBULENT FLOW IN A TIDAL ESTUARY,
Naval Research Lab., Washington, D.C. Ocean Sciences Div.
C. M. Gordon, and C. F. Dohne.
Journal of Geophysical Research, Vol 78, No 12, p 1971-1978, April 20, 1973. 6 fig, 26 ref.

Descriptors: *Turbulent flow, *Estuaries, *Tidal effects, *Time series analysis, Tidal streams, Currents (Water), Current meters, Flowmeters, Vortices, Turbulence, Chesapeake Bay.

Time series of tidal current velocities were measured at various depths in an estuary with a pivoted vane current meter. The velocity data were analyzed to obtain mean longitudinal and vertical currents along with their respective turbulent fluctuations. Correlation of the fluctuation components allowed direct measurements of Reynolds stresses, which ranged from near zero at the surface to about 5 dynes/sq cm close to the bottom at 7 meters. Variation of the turbulent kinetic energy with shear stress was approximately linear. The longitudinal and vertical components of the turbulent kinetic energy changed differently with depth and were related to tidal phase. (Knapp-USGS)
W73-08494

THE CONCENTRATION OF COPPER, LEAD, ZINC AND CADMIUM IN SHALLOW MARINE SEDIMENTS, CARDIGAN BAY (WALES),
University Coll. of North Wales, Menai Bridge. Marine Science Labs.
For primary bibliographic entry see Field 02J.
W73-08501

OBSERVATIONS OF SOME SEDIMENTARY PROCESSES ACTING ON A TIDAL FLAT,
New Hampshire Univ., Durham. Dept. of Earth Sciences; and New Hampshire Univ., Durham. Jackson Estuarine Lab.
For primary bibliographic entry see Field 02J.
W73-08502

A SANDWAVE FIELD IN THE OUTER THAMES ESTUARY, GREAT BRITAIN,
Unit of Coastal Sedimentation, Taunton (England).
D. N. Langhorne.
Marine Geology, Vol 14, No 2, p 129-143, February 1973. 10 fig, 1 tab, 19 ref.

Descriptors: *Sand waves, *Estuaries, *Surveys, *Sounding, *Sonar, Navigation, Channel morphology, Sediment transport, Dunes, Currents (Water), Mapping.
Identifiers: *Thames Estuary (England).

A method of presentation in which seabed profiles are scale-corrected and replotted in three-dimensional form was developed to aid the study of complex areas of the seabed. For sandwave research the continuity of crests can be studied, and the positional relationship of individual features compared. By plotting successive surveys of the same area, changes can be detected and significant movement measured. In the sandwave field at Long Sand Head, Thames Estuary, the sandwaves can be grouped in morphological zones, depending upon their amplitude, wavelength and cross-sectional asymmetry. Despite marked asymmetry, which suggests opposing directions of movement in different zones, little progressive movement occurs, but rather a 'flexing' of sinuous crestlines results in displacement of up to 25 m in a year. Side-scan sonar surveys emphasize the complexity of the sandwaves within the field and the presence of less table dune bed forms which often lie at considerable angles to the major crests, suggesting a secondary flow regime. (Knapp-USGS)
W73-08504

PLANS FOR REDUCTION OF SHOALING IN BRUNSWICK HARBOR AND JEKYLL CREEK, GEORGIA,
Army Engineer Waterways Experiment Station, Vicksburg, Miss.
For primary bibliographic entry see Field 08B.
W73-08512

EFFECT OF PULPMILL EFFLUENT ON DISSOLVED OXYGEN IN A STRATIFIED ESTUARY – I. EMPIRICAL OBSERVATIONS,
Fisheries Research Board of Canada, Nanaimo (British Columbia). Biological Station.
For primary bibliographic entry see Field 05C.
W73-08543

EFFECT OF PULPMILL EFFLUENT ON DISSOLVED OXYGEN IN A STRATIFIED ESTUARY - II. NUMERICAL MODEL,
Fisheries Research Board of Canada, Nanaimo (British Columbia). Biological Station.
For primary bibliographic entry see Field 05C.
W73-08544

EFFECTS OF BUNKER C OIL ON INTERTIDAL AND LAGOONAL BIOTA IN CHEDABUCTO BAY, NOVA SCOTIA,
New Brunswick Univ., St. John. Dept. of Biology.
For primary bibliographic entry see Field 05C.
W73-08548

STATEMENT OF FINDINGS--BAL HARBOUR, FLORIDA, PARTIAL BEACH RESTORATION, BEACH EROSION CONTROL AND HURRICANE PROTECTION PROJECT, DADE COUNTY, FLORIDA (FINAL ENVIRONMENTAL IMPACT STATEMENT).
Army Engineer District, Jacksonville, Fla.
For primary bibliographic entry see Field 04A.
W73-08680

GEOCHEMISTRY AND DIAGENESIS OF TIDAL-MARSH SEDIMENT, NORTHEASTERN GULF OF MEXICO,
Geological Survey, Washington, D.C.
For primary bibliographic entry see Field 02J.
W73-08730

SUSPENDED-SEDIMENT YIELDS OF NEW JERSEY COASTAL PLAIN STREAMS DRAINING INTO THE DELAWARE ESTUARY,
Geological Survey, Trenton, N.J.
For primary bibliographic entry see Field 02J.
W73-08742

SUSPENDED SEDIMENT DATA SUMMARY, AUGUST 1969 - JULY 1970 (MOUTH OF BAY TO HEAD OF BAY),
Johns Hopkins Univ., Baltimore, Md. Chesapeake Bay Inst.
For primary bibliographic entry see Field 07C.
W73-08744

SEDIMENTS IN SURFACE WATERS OF THE EASTERN GULF OF GUINEA,
Woods Hole Oceanographic Institution, Mass.
B. D. Bornhold, J. R. Mascle, and K. Harada.

widths seaward of the mouth; (2) periodic seaward repetition of these zones of temperature contrast; (3) abrupt changes in effluent expansion rates associated with the resultant bands of temperature contrast; and (4) well-defined and closely spaced internal wave crests that occur between and parallel to the more widely spaced thermal bands. (Knapp-USGS)
W73-08756

TOOLS FOR COASTAL ZONE MANAGEMENT.
Marine Technology Society, Washington, D.C. Coastal Zone Marine Management Committee.

Marine Technology Society, Conference Proceedings, February 14-15, 1972, Washington D.C., 216 p, $12.00.

Descriptors: *Coasts, *Planning, *Management, *Water resources development, Future planning, Methodology, Research and development, Institutions.
Identifiers: *Conference proceedings.

Various methods and systems for coastal zone management are presented and discussed. Among the topics covered are environmental baseline studies, matrices and inventories, modeling, and information systems. Current information is reviewed and future possibilities explored. (See W73-08803 thru W73-08815) (Ensign-PAI)
W73-08802

ENVIRONMENTAL GEOLOGIC ATLAS OF THE TEXAS COAST: BASIC DATA FOR COASTAL ZONE MANAGEMENT,
Texas Univ., Austin. Bureau of Economic Geology.
C. G. Groat, and L. F. Brown, Jr.
In: Tools for Coastal Zone Management, Feb. 14-15, 1972, Washington, D.C., p 1-15. 5 fig, 3 tab, 2 ref.

Descriptors: *Geologic mapping, *Texas, *Land use, *Cross-sections, Resources, Physical properties, Biological properties, Terrain analysis, Coasts.
Identifiers: *Environmental geologic atlas, *Multifunctional maps, Land use decisions.

The Bureau of Economic Geology of the University of Texas at Austin has produced an environmental geologic atlas of the Texas coastal zone. Through the use of maps, tables and descriptive text, the atlas attempts to provide the necessary physical data for making informed land-use decisions. The atlas contains seven folios, each of which contain, among other maps, an Environmental Geology Map. The units represented on the geologic maps are defined on the basis of physical or biologic characteristics which are meaningful to engineers, archtiects, and planners, as well as to geologists. Maps detailing physical properties, biologic assemblages, current land uses, mineral resources, water systems and topography are also included. (See also W73-08802) (Ensign-PAI)
W73-08803

THE LARGE VARIABILITY OF WATER QUALITY IN COASTAL WATERS AND SUGGESTIONS FOR HOW WE CAN HANDLE THEM,
Massachusetts Inst. of Tech., Cambridge.
E. Mollo-Christensen.
In: Tools for Coastal Zone Management, Feb 14-15, 1972, Washington, D.C., p 17-26. 8 fig, 1 tab, no ref.

Descriptors: *Coasts, *Pollutants, *Physical properties, *Chemical properties, *Variability, Methods, Mixing, Wind, Water pollution control, Path of pollutants.
Identifiers: *Turbulence, *Wind stress, *Observation systems, *Dispersion.

Examples of the variability of chemical and physical properties of coastal waters are presented. The causes of this variability are briefly discussed and methods for coping with coastal pollution are suggested. Mixing processes generated from the turbulence of tides and currents are the chief natural means of dispersing pollutants in rivers, bays, and estuaries. The sources of turbulence available in coastal waters are more limited in number and weaker in force. The main sources of stirring in ocean water are wind stress on the water surface and wave action caused by wind. The study concludes with a recommendation for a program of data acquisition, data handling and interpretation. It is also suggested that the observation system be coupled closely to the generation of computer models for prediction and assessment. (See also W73-08802) (Ensign-PAI)
W73-08804

MODERN APPROACH TO COASTAL ZONE SURVEY,
Michigan Univ., Ann Arbor. Willow Run Labs.
F. C. Polcyn.
In: Tools for Coastal Zone Management, Feb 14-15, 1972, Washington, D.C., p 41-57. 24 fig, no ref.

Descriptors: *Remote sensing, *Spectrometers, *Mapping, *Currents, Water pollution sources.
Identifiers: *Optical scanner, *Multispectral scanner.

A new type of remote sensing instrument is described. The device involves an optical mechanical scanner which operates with mirrors and not lenses so that the scene in any spectral band between the ultraviolet through the visible and into the thermal infrared region can be observed. This kind of sensing instrument is unique in that it uses a spectrometer at the focal point of the energy collecting mirror. Various application areas for the multispectral scanner are discussed -- including its use in pollution studies. The multispectral scanner system could be used to detect the presence of an oil slick and to identify the different kinds of oil involved. Ice monitoring and mapping of shore currents are two other areas of application for the sensing instrument. (See also W73-08802) (Ensign-PAI)
W73-08805

SOME APPLICATIONS OF PHOTOGRAPHY, THERMAL IMAGERY AND X BAND SIDE LOOKING RADAR TO THE COASTAL ZONE,
Grumman Ecosystems Corp., Bethpage, N.Y.
W. C. Coulbourn.
In: Tools for Coastal Zone Management, Feb 14-15, 1972, Washington, D.C., p 59-65.

Descriptors: *Remote sensing, *Aerial photography, *Outfall sewers, *Pollutant identification, Wetlands, Water sampling, Temperature.
Identifiers: *Daedalus Thermal Line Scanner, *Surface photography, Ice monitoring, Water surface patterns.

Results of several water study projects are reported. One study was undertaken to evaluate aerial remote sensing as an effective technique in detecting outfalls into navigable waterways of the New York District. Night and daylight flights were conducted in the study area; a Daedalus Thermal Line Scanner was used to provide thermal imagery in 2 bands. Surface photography was taken from boats, along with surface water sampling and temperature readings. Remote sensing was shown to be useful in detecting outfalls -- 131 outfalls were discovered, 124 of which were clearly visible on the color, and 71 visible on the thermal imagery. Another project was to determine the most efficient scale and photographic film emulsion for use in identifying wetlands vegetation in Maryland. The thermal line scanner proved useful in delineating the waterland or water-vegetation boundary line, but was not used in the identification of wetland vegetation. Remote sensing was also shown

to be helpful in detecting icebergs and ocean surface roughness patterns. (See also W73-08802) (Ensing-PAI)
W73-08806

THE NATIONAL OCEAN SURVEY COASTAL BOUNDARY MAPPING,
National Ocean Survey, Rockville, Md. Coastal Mapping Div.
J. E. Guth.
In: Tools for Coastal Zone Management, Feb 14-15, 1972, Washington, D.C. p 67-69.

Descriptors: *Coasts, *Management, *Boundaries, *Mapping, *Water levels, Remote sensing, State governments, Federal governments.
Identifiers: *National Ocean Survey.

Coastal boundary mapping is an essential prerequisite for coastal zone management. A system for boundary mapping is outlined. Mean low water and high water lines are mapped and used as base lines from which boundary determinations can be made. A twostage system for mapping the wetlands is described whereby remote sensing technique are supplemented with accurate ground truth surveys. It is suggested that the problem of total coastal boundary mapping ought to be handled on the state and federal level, rather than by private sectors. (See also W73-08802) (Ensign-PAI)
W73-08807

METHODS FOR ENVIRONMENTAL PLANNING OF THE CALIFORNIA COASTLINE,
California Univ., Berkeley. Dept. of Landscape Architecture.
R. H. Twiss, and J. C. Sorensen.
In: Tools for Coastal Zone Management, Feb 14-15, 1972, Washington, D.C., P +±*:. + FIG, : REF.

Descriptors: *Coasts, *Planning, *Tourism, *Recreation, *Land use, *Project feasibility, Environmental effects, *California.
Identifiers: Impact identification, Project characteristics.

The Coastal zone planning effort in California is reviewed. Increasing demand for tourist and recreational development of coastal land and all the related problems of this sort of development have made the emphasis in coastal zone planning one of land use rather than coastal water concern. This study outlines a procedure that would allow a local agency to systematically relate the actions of a proposed project to probable changes in environmental conditions. Topics discussed include impact identification, environmental capability, and project characteristics. (See also W73-08802) (Ensign-PAI)
W73-08808

THE CHESAPEAKE BAY INVENTORY SYSTEM,
Johns Hopkins Univ., Baltimore, Md. Dept. of Radiology.
R. F. Beers, Jr.
In: Tools for Coastal Zone Management, Feb 14-15, 1972, Washington, D.C., p 85-93. 1 ref.

Descriptors: *Chesapeake Bay, *Planning, *Management, *Model studies, Data storage and retrieval.
Identifiers: *Interaction model, *Man-environment interface, *Inventory system, Impact studies.

The Chesapeake Bay inventory system is a method for dealing with the planning and management problems of the Bay region. The study's focal point is the interaction between man and his environment. The inventory system is modeled from the Leontief economic input-output model of interacting entities. The interaction model includes biological, physical-geographical, physical-meteorological, physical-hydrological, oceanographical, governmental and societal entities. This system facilitates the assessment of short and long term, direct and indirect consequences of a particular management decision. It can also assist in the identification and description of specific processes which may be controlled or replaced in order to minimize the magnitude of the impact of an interactive event. Lastly, the inventory system delineates areas of ignorance about data essential for the decision maker. (See also W73-08802) (Ensign-PAI)
W73-08809

COASTAL ZONE MANAGEMENT SYSTEM: A COMBINATION OF TOOLS,
Center for the Environment and Man, Inc., Hartford, Conn.
R. H. Ellis.
In: Tools for Coastal Zone Management, Feb 14-15, 1972, Washington, D.C., p 95-111. 6 fig, 1 ref.

Descriptors: *Data storage and retrieval, *Coasts, *Planning, Management, Systems analysis, Synthetis, Analysis, Design.
Identifiers: *MIS, *Long Island.

System techniques for organizing, synthesizing, analyzing and applying information for coastal zone management are discussed. A management information system (MIS) used in Long Island to aid in planning and management of marine coastal resources is described. The components of MIS include: data storage and retrieval, environmental relationships, analytical design, synthesis and analysis and executive control. The importance of information collection, storage, and retrieval is stressed. Ways of combining analytical approaches and models into one management information framework are explored. (See also W73-08802) (Ensign-PAI)
W73-08810

THE MARINE ENVIRONMENT AND RESOURCES RESEARCH AND MANAGEMENT SYSTEM MERRMS,
Virginia Inst. of Marine Science, Glouster Point.
J. B. Pleasants.
In: Tools for Coastal Zone Management, Feb 14-15, 1972, Washington, D.C., p 113-119.

Descriptors: *Data transmission, *Information exchange, *Environment, Resources, *Research and development, *Management, Wetlands, Shores, *Data storage and retrieval, Libraries, Chesapeake Bay.
Identifiers: MERRMS.

Various facets of the Marine Environment and Resources Research and Management System (MERRMS) of the Virginia Institute of Marine Science are described. MERRMS was designed to make relevant data, properly organized and presented, available to advisors, managers and researchers. Sectors covered within the system include: a wetlands, shallows and shorelines inventory, Chesapeake Bay bibliography, data banks, library, and visual displays. MERRMS is presently in the final stages of initial development. (See also W73-08802) (Ensign-PAI)
W73-08811

MARINE TECHNOLOGY TRANSFER DEPENDS UPON INFORMATION TRANSFER,
Environmental Protection Agency, Washington, D.C. Div. of Water Quality Standards.
R. F. Powell.
In: Tools for Coastal Zone Management, Feb 14-15, 1972, Washington, D.C., p 121-143. 3 fig, 1 tab, no ref.

Descriptors: Technology, *Technical writing, *Planning, *Management, *Data transmission, *Information exchange.
Identifiers: TIMP, Coastal zone management.

The Technical Information and Management Planning System (TIMP) of the EPA is discussed as a method of making existing marine technology easily accessible to all. The system has a number of unique characteristics. It recognizes that no technology transfer takes place without a preceding selective, efficiently obtained transfer of information. It can function as a system servicing a system and it offers a means of recognizing specific 'needs statements.' Also the TIMP system functions on standard English language, so no abstractors or interpreters are necessary. It is maintained that with some broadening of data bases and possible system modifications, the TIMP system is highly desirable for the achievement of effective coastal zone management. (See also W73-08802) (Ensign-PAI)
W73-08812

EVALUATION OF COASTAL ZONE MANAGEMENT PLANS THROUGH MODEL TECHNIQUES,
Army Engineer Waterways Experiment Station, Vicksburg, Miss.
R. W. Whalin, and F. A. Herrmann, Jr.
In: Tools for Coastal Zone Management, Feb 14-15, 1972, Washington, D.C., p 149-176. 23 fig.

Descriptors: *Hydraulic models, *Beaches, *Harbors, *Estuaries, *Design, *Management, *Construction, *Planning, Artificial beaches, Shoals, Evaluation.
Identifiers: Hydraulic regimes, Salinity regimes, Shoaling regimes.

Results of several hydraulic model studies are presented. Projects reviewed include studies conducted on estuarine design, estuarine management, beach development, harbor construction and development, and investigations of artificial islands and reefs. Scale models are demonstrated to be useful in evaluating the effects of specific projects on the hydraulic, salinity, and shoaling regimes in estuaries. Modeling is also effective for overall management of coastal zone. (See also W73-08802) (Ensign-PAI)
W73-08813

HYDRAULIC MODELING OF AN ENVIRONMENTAL IMPACT: THE JAMES ESTUARY CASE,
Virginia Inst. of Marine Science, Gloucester Point.
M. M. Nichols.
In: Tools for Coastal Zone Management, Feb 14-15, 1972, Washington, D.C., p 177-184. 4 fig, 8 ref.

Descriptors: *Hydraulic models, *Estuaries, *Channel improvement, *Dredging, *Streambeds, *Oysters, *Productivity, Environmental effects, *Virginia.
Identifiers: *James River Estuary, Flow regime, Salinity regime.

A hydraulic model study was used to predict the effects of channel deepening on oyster production in the James River estuary, Virginia. The model study showed how sensitive the salinity and flow regime of an estuary is to small changes in bottom geometry. The long-term cumulative effects of channel-deepening could be grave. The study also pointed up the need for greater accuracy in verification and testing so that the transient conditions characteristic of natural estuaries can be accomodated. (See also W73-08802) (Ensign-PAI)
W73-08814

PREDICTIVE MODELS OF SALINITY AND WATER QUALITY PARAMETERS IN ESTUARIES,

Massachusetts Inst. of Tech., Cambridge. Dept. of Civil Engineering.
D. R. F. Harleman, M. L. Thatcher, and J. E. Dailey.
In: Tools for Coastal Zone Management, Feb 14-15, 1972, Washington, D.C., p 185-195. 6 fig, 4 ref.

Descriptors: *Estuaries, *Water quality, *Management, *Model studies, *Forecasting, *Mathematical studies, Salinity, Temperature, Dissolved oxygen, Path of pollutants.
Identifiers: Mass transportation, Continuity, Longitudinal momentum equations.

Techniques of mathematical modeling for water quality control and management in mixed estuaries are discussed. A water quality model can be used to study the time-dependent, longitudinal distribution of salinity, temperature, dissolved oxygen or other water quality parameters. The discussion centers on one-dimensional water quality models which utilize and predict information that is related to available or accessible field data. Mass transportation, continuity, and longitudinal momentum equations are presented. Mathematical modeling permits the use of either synoptic or intermittent field data; through its use more attention can be given to the generation, decay and interaction of multiple water quality indicators under estuarine conditions. (See also W73-08802) (Ensign-PAI)
W73-08815

THE COLUMBIA RIVER ESTUARY AND ADJACENT OCEAN WATERS, BIOENVIRONMENTAL STUDIES.

Edited by A.T. Pruter and D. L. Alverson, University of Washington Press, Seattle, Wash., 1972. 876 p. $22.00.

Descriptors: *Columbia River, *Coasts, *Estuaries, *Radioisotopes, Biota, Ecosystems, Environmental effects, Estuarine environment, Physicochemical properties, *Washington, *Oregon, *Pacific Ocean.

The physical, chemical, and biological aspects of the Columbia River estuary and adjacent ocean waters are thoroughly discussed. Special attention is given to the levels and fate of radionuclides in these waters. The material presented represents extensive investigations carried out by Oregon State University, University of Washington, Battelle Memorial Institute, and the National Marine Fisheries Service. It is organized into sections on background and history of the projects, Columbia River-Pacific Ocean relationships, composition and distribution of the biota of the marine environment and radionuclides in the ecosystem. (See W73-08817 thru W73-08849) (Ensign-PAI)
W73-08816

THE HISTORY OF HANFORD AND ITS CONTRIBUTION OF RADIONUCLIDES TO THE COLUMBIA RIVER,

Battelle Memorial Inst., Richland, Wash. Pacific Northwest Labs.
R. F. Foster.
In: The Columbia River Estuary and Adjacent Ocean Waters, University of Washington Press, Seattle, Wash., p 3-18, 1972. 7 fig, 16 ref.

Descriptors: *Columbia River, *Nuclear reactors, *Cooling water, Nuclear wastes, Water pollution sources, Environmental effects, Washington.
Identifiers: Historical survey, Laboratory research, Field research, *Hanford (Wash).

From 1943 to 1945 plutonium-producing reactors were established in Hanford, Washington. Studies of the effects of cooling waters discharged from the plants into the Columbia River were initiated even before the reactors began operations. By the late 1950's scientific and public interest in radioactive wastes and environmental effects had increased significantly enough to support much broader based research in the Columbia River and its estuaries. A short history of these early efforts to monitor the environment is presented. (See also W73-08816) (Ensign-PAI)
W73-08817

PHYSICAL ASPECTS OF THE COLUMBIA RIVER AND ITS ESTUARY,

Oregon State Univ., Corvallis. Dept. of Oceanography.
V. T. Neal.
In: The Columbia River Estuary and Adjacent Ocean Waters, University of Washington Press, Seattle, Wash., p 19-40, 1972. 21 fig, 3 tab, 10 ref. ONR Project NR083-102.

Descriptors: *Columbia River, *Physical properties, *Estuaries, Climatic data, River flow, Tides, Seawater intrusion, Water pollution, Methodology.
Identifiers: Water contaminants, *Flushing, Tidal-prism method, Fraction-of-freshwater method.

Various physical aspects of the Columbia River and its adjacent waters are described. Brief discussions of river flow, navigation, geomorphology, climatic conditions, tides and seawater intrusion are presented. Flushing and pollution distribution in the river estuary (defined as that area of the river subject to salinity intrusion) are examined in greater detail. Some methods used to predict the flushing of contaminants from an estuary are outlined. These include: the classical tidal-prism method, modified tidal-prism method, fraction-of-freshwater method and exponential-decrease method. (See also W73-08816) (Ensign-PAI)
W73-08818

CIRCULATION AND SELECTED PROPERTIES OF THE COLUMBIA RIVER EFFLUENT AT SEA,

Washington Univ., Seattle. Dept. of Oceanography.
C. A. Barnes, A. C. Duxbury, and B. A. Morse.
In: The Columbia River Estuary and Adjacent Ocean Waters, University of Washington Press, Seattle, Wash., p 41-80, 1972. 23 fig, 2 tab, 39 ref. AEC AT (45-1)-1725 and ONR Nonr-477 (37).

Descriptors: *Columbia River, *Climatic data, *Effluents, *Currents, Sediment, Pacific Ocean, Mixing, *Ocean circulation, Seasonal, Path of pollutants.
Identifiers: *Freshwater plume, Ekman transport.

The Columbia River discharges an average of 7300 m3/sec of freshwater into the Pacific Ocean. The areal extent and location of this freshwater plume are the result of seasonal differences in climate. The major seasonal differences in the distribution of river water and the surface and bottom currents of the coastal region are discussed. Two prominent seasonal patterns of effluent distribution prevail; one lies north of the river mouth and inshore during the southerly winds of winter, the other lies south and offshore during the northerly summer winds. Surface currents and inshore bottom currents evidence short-term variability within seasonal trends because they respond rapidly to changes in wind speed and direction. Data gathered on distribution of water properties, movement of seabed drifters, sediment transport, direct current measurements and radionuclide tracer studies are reviewed. The study concludes that there is a consistent pattern of seasonal trend and short-period variability in the distribution and dispersion of the Columbia River water and its dissolved and suspended load. (See also W73-08816) (Ensign-PAI)
W73-08819

REVIEW OF COMMERCIAL FISHERIES IN THE COLUMBIA RIVER AND IN CONTIGUOUS OCEAN WATERS,

National Marine Fisheries Service, Seattle, Wash. Exploratory Fishing and Gear Research Base.
A. T. Pruter.
In: The Columbia River Estuary and Adjacent Ocean Waters, University of Washington Press, Seattle, Wash., p 81-120, 1972. 19 fig, 16 tab, 17 ref.

Descriptors: *Columbia River, *Pacific Ocean, *Fisheries, Fish types, Planning, Fish harvest, Fishing gear, Domestic waters, Foreign waters.
Identifiers: Freshwater (anadromous and marine species).

The important commercial fisheries in the Columbia River and ocean areas are categorized into three units, freshwater, anadromous and marine species, and analyzed in terms of long-range planning and resource development. Representative members of each category of fish are examined with attention to size of harvest, fishing area and types of harvesting gear used. Domestic and foreign ocean fisheries are discussed separately. Some of the fish types studied include the following: freshwater fish -- crawfish and carp; anadromous fish -- salmon, steelhead trout, American shad and sturgeon; and marine species -- Pacific saury, flounders, and northern anchovy. (See also W73-08816) (Ensign-PAI)
W73-08820

CHEMICAL BUDGET OF THE COLUMBIA RIVER,

Oregon State Univ., Corvallis. Dept. of Oceanography.
P. K. Park, C. L. Osterberg, and W. O. Forster.
In: The Columbia River Estuary and Adjacent Ocean Waters, University of Washington, Press, Seattle, Wash., p 123-134, 1972. 6 fig, 4 tab, 13 ref. ONR Nonr 1286 (10) and AEC AT (45-1)-1750; NSF GP-2976 and GP-5317.

Descriptors: *Columbia River, *Estuaries, *Alkalinity, *Salinity, Stream flow, *Phosphates, Water quality, Plankton, Nutrients, *Chemical properties.

Data collected on the chemical features of the Columbia River estuary -- such as alkalinity, pH, dissolved oxygen, phosphate, salinity, nitrate etc. can be used along with stream flow rate to calculate the monthly and annual budgets of the chemicals. Knowing the monthly chemical condition of the estuary is important since it serves as an index of water quality and nutrient condition which is in turn linked to the river's ultimate plankton productivity and distribution. The various economically important estuary organisms depend on a balanced nutrient-plankton relation. Average chemical composition of the river water is 0.5 micro M phosphate, 12 micro M nitrate, 160 micro M silicate, 1.0 meq/liter alkalinity, 1.0 mM carbon dioxide, and 7.4 ml/liter oxygen; average pH is 7.7. (See also W73-08816) (Ensign-PAI)
W73-08821

RIVER-OCEAN NUTRIENT RELATIONS IN SUMMER,

Geological Survey, Menlo Park, Calif. Water Resources Div.
T. J. Conomos, M. G. Gross, C. A. Barnes, and F. A. Richards.
In: The Columbia River Estuary and Adjacent Ocean Waters, University of Washington Press, Seattle, Wash., p 151-175, 1972. 16 fig, 4 tab, 38 ref. AEC AT (45-1)-1725.

Descriptors: *Columbia River, *Pacific Ocean, *Mixing, Upwelling, *Nitrates, *Phosphates, Silicates, Data collections, Sampling, *Seasonal, Nutrients.
Identifiers: Two-stage mixing, Vertical temperature variation.

Mixing processes of Columbia River water and Pacific Ocean water (near the river mouth) were investigated to gain a better understanding of the interaction between rivers and the ocean. Mixing is a two-stage process, controlled primarily by river discharge and tidal action. The first stage occurs in the estuary and involves interaction between freshwater from the river and subsurface ocean water flowing into the estuary above the bottom. River water in the first stage supplies virtually all the silicate and a little phosphate, while the ocean water carries nearly all the nitrate and most of the phosphate. In the second stage of mixing (occurring just seaward of the river mouth) seaward-moving low-salinity water induces upwelling and entrainment of the deeper subsurface ocean water. The study examined mixing processes by assessing data on vertical variations of temperature, salinity, and nutrient concentrations; the data helped distinguish different water parcels and qualitatively determine their movement in the coastal ocean. (See also W73-08816) (Ensign-PAI)
W73-08823

RIVER-OCEAN SUSPENDED PARTICULATE MATTER RELATIONS IN SUMMER,
Geological Survey, Menlo Park, Calif. Water Resources Div.
T. J. Conomos, and M. G. Gross.
In: The Columbia River Estuary and Adjacent Ocean Waters, University of Washington Press, Seattle, Wash., p 176-202, 1972. 17 fig, 4 tab, 36 ref. AEC AT (45-1)-1725.

Descriptors: *Columbia River, *Suspended solids, *Phytoplankton, Particle shape, Particle size, Specific gravity, Currents, Coasts, Seasonal.
Identifiers: Resuspended particles, Particulate pathways, Lithogenous particles, Biogenous particles.

Various processes interact to govern the distribution and dispersal of suspended particulate matter in the Columbia River and coastal ocean. Summer studies conducted on the river and coastal waters indicate that the major sources of suspended particles are the river, phytoplankton growing in the ocean near the river, and particles resuspended from the bottom. The paths followed by suspended sediments depend on such variables as particle morphology, specific gravity, composition and elasped time after discharge from the river mouth. 85 to 95% of particulate matter in the Columbia River consists of lithogenous particles; the bulk of matter found in the ocean is biogenous particles, mainly phytoplankton and detritus. Strong tidal and hydraulic currents near the river mouth serve to resuspend sediment in near-bottom waters. Suspended-particle concentrations are nonconservative properties of seawater; settling of particles is the primary process decreasing particle concentrations. (See also W73-08816) (Ensign-PAI)
W73-08824

EFFECTS OF COLUMBIA RIVER DISCHARGE ON CHLOROPHYLL A AND LIGHT ATTENUATION IN THE SEA,
Oregon State Univ., Corvallis. Dept. of Oceanography.
L. F. Small, and H. Curl, Jr.
In: The Columbia River Estuary and Adjacent Ocean Waters, University of Washington Press, Seattle, Wash., p 203-218, 1972. 5 fig, 3 tab, 18 ref. AEC AT (45-1)-1741 and AT (45-1)-1750.

Descriptors: *Columbia River, *Discharge, *Chlorophyll, *Light penetration, *Nutrient removal, Upwelling, Runoff.
Identifiers: Chlorophyll a, Light extinction, Riley equation.

The effects of Columbia River discharge on chlorophyll a concentration and light attenuation in the sea off Oregon were studied. Although the River acts as a point source for particles and nutrients, its effect on light extinction and chlorophyll a concentration is manifest only out to approximately 45 miles during all seasons. The major factor controlling chlorophyll concentration is thought to be nutrient depletion, which increase with distance from the mouth of the Columbia. Upwelling in summer and coastal runoff in winter generally keep chlorophyll concentrations high inshore in areas away from the river mouth. No seasonal differences within any hydrographic regime could be established for light attentuation. The Riley equation was used to predict mean chlorophyll a and mean light-extinction values. The data was examined in relation to effects of upwellings, coastal runoff other than that from the Columbia River, and oceanic water. (See also W73-08816) (Ensign-PAI)
W73-08825

ASPECTS OF MARINE PHYTOPLANKTON STUDIES NEAR THE COLUMBIA RIVER, WITH SPECIAL REFERENCE TO A SUBSURFACE CHLOROPHYLL MAXIMUM,
Washington Univ., Seattle. Dept. of Oceanography.
G. C. Anderson.
In: The Columbia River Estuary and Adjacent Ocean Waters, University of Washington Press, Seattle, Wash., p 219-240, 1972. 8 fig, 43 ref. AEC AT (45-1)-1725 and ONR Nonr-477 (37).

Descriptors: Washington, Oregon, Coasts, *Phytoplankton, *Primary productivity, Data collections, *Chlorophyll, Nutrients, Theoretical analysis, Columbia River, Seasonal, Pacific Ocean.
Identifiers: Vertical distribution.

Phytoplankton and primary production processes in waters off the Washington and Oregon coasts have been studied since 1961. Findings are reviewed. Areas examined include: size fractionation of phytoplankton populations in different areas and seasons; adaptations of populations to changes in light; excretion of dissolved organic matter by phytoplankton; seasonal and vertical distribution of particulate matter in deep seawater; and the seasonal and areal distributions of phytoplankton concentrations and primary production rates. Data from recent studies on the mechanisms which regulate the supply of nutrients to surface waters and information on the processes responsible for forming chlorophyll concentrations at depth are also reported. The average information synthesized from all prior and on-going studies will be used to define the pathways of radionuclide into the biota. Another equally important use of collected data is in constructing a theoretical quantitative model for phytoplankton production. (See also W73-08816) (Ensign-PAI)
W73-08826

BOTTOM TOPOGRAPHY AND SEDIMENT TEXTURE NEAR THE COLUMBIA RIVER MOUTH,
Washington Univ., Seattle. Dept. of Oceanography.
D. A. McManus.
In: The Columbia River Estuary and Adjacent Ocean Waters, University of Washington Press, Seattle, Wash., p 241-253, 1972. 11 fig, 2 tab, 8 ref. AEC AT (45-1)-1725.

Descriptors: *Columbia River, *Continental shelf, *Sediments, *Topography, Sedimentology, Sampling, Sands.
Identifiers: Sediment textures, Topographical units, Sedimentological units.

The continental shelf area near the Columbia River mouth can be divided into several topographical-sedimentological units, on the basis of descriptions of topography and sediment texture. This study proposes the following units of division: (1) rough topography of relict sediments, (2) outer-shelf silty sand, (3) inner-shelf sand wedge, (4) nearshore sand, (5) Columbia River silt deposit, and (6) shelf-break sand. The first unit occurs mainly in thesouthwest part of the shelf area. The second unit is located on the landward margin of the first unit, both to the north and south of Astoria Canyon. The third unit extends south of the Columbia River and offshore to approximately 73 m water depth, while the fourth unit extends north of the river and out to about 55 m. The fifth unit trends northwesterly along the outer shelf from the mouth of the Columbia and the sixth unit is located at the shelf-break. Sediment textures found in each of these divisions are described in some detail. (See also W73-08816) (Ensign-PAI)
W73-08827

DISTRIBUTION OF ORGANIC CARBON IN SURFACE SEDIMENT, NORTHEAST PACIFIC OCEAN,
Washington Univ., Seattle. Dept. of Oceanography.
M. G. Gross, A. G. Carey, Jr., G. A. Fowler, and L. D. Kulm.
In: The Columbia River Estuary and Adjacent Ocean Waters, University of Washington Press, Seattle, Wash., p 254-264, 1972. 4 fig, 2 tab, 27 ref. ONR NR 083 012. ONR Nonr 1286 (10).

Descriptors: *Pacific Ocean, *Continental shelf, Organic compounds, *Carbon, Sediments, *Sediment transport.
Identifiers: *Organic carbon.

Organic carbon in surface sediment is unevenly distributed in the northeast Pacific Ocean. Observation of carbon-rich and carbon-poor sediments reveals a zoned distribution which roughly parallels the coast. This distribution is believed to be caused by the interaction of several processes, the most important of which seems to be transport of sediment-associated organic matter. Surface and near-bottom currents tend to move particles along the continental shelf rather than across it. The submarine ridge system and seachannels on the ocean floor are both more or less parallel to the coast, which apparently helps restrict the seaward movements of organic carbon in sediments. Relative rates of sediment accumulation and destruction of organic matter are also important. (See also W73-08816) (Ensign-PAI)
W73-08828

THE DISTRIBUTION OF MICROBIOGENIC SEDIMENT NEAR THE MOUTH OF THE COLUMBIA RIVER,
Shoreline Community Coll., Seattle, Wash. Div. of Science.
R. A. Harmon.
In: The Columbia River Estuary and Adjacent Ocean Waters, University of Washington Press, Seattle, Wash., p 265-278, 1972. 4 fig, 14 ref. AEC AT (45-1)-1725.

Descriptors: *Columbia River, *Continental shelf, *Bottom sediments, *Microorganisms, Currents, Hypolimnion.
Identifiers: Biozones, Indigenous organisms, Sediments (Microbiogenic).

The distribution and concentration of microbiogenic hard parts within the bottom sediment on the continental shelf near the Columbia River mouth are described. Based on the microbiogenic components in the sediment, three biozones were established. Boundaries of these zones seemed to be influenced by several factors, including water depth and position relative to shore or the shelf break, nature of coastal substrate or degree of sedimentation, and prevailing currents. In areas of low sediment accumulation, the distribution of biogenic particles reflects either indigenous organisms or possibly, reworked or re-

lict biogenic sediment. It is implied that the distribution of microbiogenic organisms and their living-to-dead ratios in sediments are of potential use for inferring themovement of bottom waters. (See also W73-08816) (Ensign-PAI)
W73-08829

STUDIES OF THE AEROBIC, NONEXACTING, HETEROTROPHIC BACTERIA OF THE BENTHOS,
Georgia Univ., Athens. Dept. of Microbiology.
W. J. Wiebe, and J. Liston.
In: The Columbia River Estuary and Adjacent Ocean Waters, University of Washington Press, Seattle, Wash., p 281-312, 1972. 1 fig, 23 tab, 45 ref. NSF G-19434.

Descriptors: Oregon, Washington, Coasts, *Aerobic bacteria, *Sampling, Continental shelf, *Continental slope, Pseudomonas bacteria, Pacific Ocean.
Identifiers: *Heterotrophic bacteria (Nonexacting), Bacteria counts, Surface sediments, Depth-dependency.

Information on the aerobic, nonexacting heterotrophic bacteria in the surface sediments of the seabed off the Oregon and Washington coasts is presented. Marine sediment samples were obtained with a gravity geological corer. Sampling for aerobic heterotrophic bacteria was performed as soon as possible and never more than a few hours after the sample was first drawn. No statistical difference was found between continental shelf and slope samples; viable counts in the samples averaged 2.0 x 10000 bacteria/ml mud-water slurry. Study results indicate that the bacteria are evenly distributed over the entire region. The results also imply that there are physical and biological pressures which limit bacteria numbers in sediments. Of 366 strains of bacteria examined, gram-negative rods, mainly Pseudomonas types, were predominant. Distribution patterns of most genera and Pseudomonas groups were depth dependent. Biochemical capabilities of isolates at each sampling station were similar. (See also W73-08816) (Ensign-PAI)
W73-08830

DISTRIBUTION OF PELAGIC COPEPODA OFF THE COASTS OF WASHINGTON AND OREGON DURING 1961 AND 1962,
Washington Univ., Seattle. Dept. of Oceanography.
W. K. Peterson.
In: The Columbia River Estuary and Adjacent Ocean Waters, University of Washington Press, Seattle, Wash., p 313-343. 1972. 14 fig, 5 tab, 37 ref. ONR Project NR 083 012. AEC AT (45-1)-1725 and ONR Nonr-477 (37).

Descriptors: Washington, Oregon, Coasts, *Zooplankton, *Sampling, *Copepods, Continental shelf, Continental slope, Seasonal, *Pacific Ocean.
Identifiers: *Plume and ambient water zone.

Zooplankton samples were taken in the upper 200 m off the coasts of Washington and Oregon in order to provide information on the seasonal and geographic distribution of copepods in this area. To gain a better understanding of the distribution of zooplankton, the area sampled was divided into two separate sets of regimes: plume and ambient and shelf, slope, and oceanic zones. Data collected indicate that in summer zooplankton volumes and total number of copepods are greater in ambient than in plume regions and greater in oceanic than in shelf areas. In winter this situation was reversed for total copepods, and shelf water contained more than oceanic water. Possible reasons for these variations are discussed. The study concludes that the Columbia River plume affects the biology of the area studied. Effects on zooplankton are quite large-scale and should probably be considered a general feature of the area, in the same manner that nearshore upwelling is taken as a general feature of the shelf regime in summer. Furthermore, the Columbia River plume can be considered as an extension of an environment similar to inshore water. (See also W73-08816) (Ensign-PAI)
W73-08831

EFFECTS OF THE COLUMBIA RIVER PLUME ON TWO COPEPOD SPECIES,
Oregon State Univ., Corvallis. Dept. of Oceanography.
L. F. Small, and F. A. Cross.
In: The Columbia River Estuary and Adjacent Ocean Waters, University of Washington Press, Seattle, Wash., p 344-350. 1972. 1 fig, 9 ref. AEC AT (45-1)-1751.

Descriptors: *Columbia River, *Discharge, *Copepods, *Sampling, Salinity, Temperature, Water pollution, Environmental effects, Seasonal.
Identifiers: Acartia danae, Centropages mcmurrichi, *Plumes.

Two copepod species, Acartia danae and Centropages mcmurrichi, were examined in order to assess the effect of Columbia River discharge on copepod distribution. The collecting area began at the river mouth, extended south to the Oregon-California border and from the Oregon coast west to 165 nautical miles offshore. Surface net tows were made at each of several sampling stations using modified Clarke-Bumpus quantitative samplers equipped with mesh nets. Temperature and salinity determinations were made concurrently with each copepod collection. The study concludes that Acartia danae is a good indicator of the western and southern limits of Columbia River plume waters. Because A. danae was not found in upwelled coastal waters, no differentiation could be made during the summer between the inshore edge of the plume and upwelled water. Centropages mcmurrichi showed no distributional relation to plume waters. (See also W73-08816) (Ensign-PAI)
W73-08832

DISTRIBUTION AND ECOLOGY OF OCEANIC ANIMALS OFF OREGON,
Oregon State Univ., Corvallis. Dept. of Oceanography.
W. G. Pearcy.
In: The Columbia River Estuary and Adjacent Ocean Waters, University of Washington Press, Seattle, Wash., p 351-377. 1972. 3 fig, 10 tab, 65 ref. AEC AT (45-1)-1750 and NSF GB-5494 and GB-1588.

Descriptors: Oregon, Coasts, *Pacific Ocean, *Aquatic animals, Data collections, Vertical migration, Nekton, Depth, Seasonal.
Identifiers: *Pelagic animals.

Knowledge currently available on nektonic organisms occurring off Oregon is summarized. Pelagic animals identified from collections or observations carried out nearshore and as far as 165 nautical miles off the Oregon coast are listed according to phylum, class and order. The fauna is mainly associated with subarctic and transitional water. The species of animals which dominated the catches (made from 1961 to 1968) remained fairly constant; the abundance of individual species, however, were seasonally variable. Variations in abundance were frequently correlated with seasonal changes in prevailing currents and upwelling. Certain trends discerned from difference in night and day catches strongly suggest diel vertical migrations of small nektonic animals off Oregon. The need for future research on abundance patterns and distribution of pelagic species, depths ranges for animals which migrate vertically, and energy flow, is stressed. (See also W73-08816) (Ensign-PAI)
W73-08833

TECHNIQUES AND EQUIPMENT FOR SAMPLING BENTHIC ORGANISMS,
Oregon State Univ., Corvallis. Dept. of Oceanography.
A. G. Carey, Jr., and H. Heyamoto.
In: The Columbia River Estuary and Adjacent Ocean Waters, University of Washington Press, Seattle, Wash., p 378-408. 1972. 16 fig, 3 tab, 32 ref. AEC AT (45-1)1750 and AT (45-1)1758.

Descriptors: *Sampling, *Reliability, *On-site data collections, *Testing procedures, Quality control.
Identifiers: Biological sampling, Anchor-box dredges, Otter trawls, Corers, Screening devices, Smith-McIntyre grab.

Biological sampling poses certain problems to the investigator, and there is little agreement among scientists on optimum methods for sampling. Most benthic samples are difficult to quantify because each sampling device operates differently in different substrates. Various forms of sampling gear and field methods used by Oregon State University and the U.S. Bureau of Commercial Fisheries are described and discussed. Some of the equipment described include an anchor-box dredge, otter trawls, corers, screening devices and a Smith-McIntyre grab. Each type of sampling device is described and the procedures used to employ the equipment in field situations are discussed. Techniques used to recover, identify, and weigh the catches are outlined; some innovations designed to adapt existing gear to differing conditions are also presented. (See also W73-08816) (Ensign-PAI)
W73-08834

A PRELIMINARY CHECKLIST OF SELECTED GROUPS OF INVERTEBRATES FROM OTTER-TRAWL AND DREDGE COLLECTIONS OFF OREGON,
Oregon State Univ., Corvallis. Dept. of Oceanography.
J. W. McCauley.
In: The Columbia River Estuary and Adjacent Ocean Waters, University of Washington Press, Seattle, Wash., p 409-421. 1972. 1 tab, 20 ref.

Descriptors: *Data collections, Oregon, Coasts, *Aquatic animals, Invertebrates, Distribution pattern, Pacific Ocean.
Identifiers: *Epifaunal species.

A preliminary checklist for epifaunal species of nine phyla occurring off the coast of Oregon is presented. The list was compiled from data collected during some 246 otter-trawl collections and 50 biological-dredge collections. The checklist also provides observed depth ranges from shallowest to deepest occurrence for species within each phylum. (See also W73-08816) (Ensign-PAI)
W73-08835

ECOLOGICAL OBSERVATIONS ON THE BENTHIC INVERTEBRATES FROM THE CENTRAL OREGON CONTINENTAL SHELF,
Oregon State Univ., Corvallis. Dept. of Oceanography.
A. G. Carey, Jr.
In: The Columbia River Estuary and Adjacent Ocean Waters, University of Washington Press, Seattle, Wash., p 422-443. 1972. 8 fig, 5 tab, 24 ref. AEC AT (45-1)-1750 and AT (45-1)-1758.

Descriptors: Oregon, *Pacific Ocean, *Benthic fauna, Sampling, Continental shelf, Biomass, Distribution patterns, Water temperature, Salinity, Oxygen, Sediments, Organic matter.
Identifiers: *Infaunal distribution, *Arthropods, *Polychaetes, Sediment texture.

A short review of some aspects of the ecology of central Oregon shelf benthic invertebrate found is offered. Infaunal samples were obtained through the use of either a Sanders deep-sea anchor dredge

or an anchor-box dredge. Composition of the benthic fauna changes with increasing depth and distance from shore. Macroepibenthos, for instance, changes from a sparse molluscan assemplage to one dominated by numerous echinoderms and arthoropods. Infauna demonstrate a seaward trend in composition: arthropods dominate inshore, polychaetes offshore. Abundance increases seaward, the largest numerical density and biomass occurring at the outer edge of the continental shelf. Water temperature, salinity, and oxygen content all play a part in the abundance and composition of the fauna. The texture and organic content of the shelf sediment is suggested as a major determinant of faunal abundance and composition, but is only one of several interacting factors. (See also W73-08816) (Ensign-PAI)
W73-08836

DISTRIBUTION AND RELATIVE ABUNDANCE OF INVERTEBRATES OFF THE NORTHERN OREGON COAST,
National Marine Fisheries Service, Seattle, Wash. Exploratory Fishing and Gear Research Base.
W. T. Pereyra, and M. S. Alton.
In: The Columbia River Estuary and Adjacent Ocean Waters, University of Washington Press, Seattle, Wash., p 444-474. 1972. 11 fig, 7 tab, 20 ref. AEC AT (49-7)-1971-Mod. No. 1.

Descriptors: Oregon, Coasts, *Benthic fauna, *Sampling, *Distribution patterns, Data collections, Depth, Carbon, Organic compounds, Sediments, *Pacific Ocean, Invertebrates, Mollusks.
Identifiers: *Epibenthic invertebrate fauna, Coelenterates, Arthropods, Echinoderms.

Results of a survey of epibenthic invertebrate fauna conducted in the waters southwest of the mouth of the Columbia River are presented. The distribution and relative abundance of invertebrates was measured along a track line at depths from 50 to 1150 fathoms. The bathymetric availability of major species groups and the relative importance of various segments of invertebrate found are considered. The fauna was sampled by trawl, grab and dredge; trawl sampling was by far the most important source of data. The sampling revealed that the relative abundance of invertebrates was approximately three times greater at depths of 100 to 375 fathoms than at shallower or deeper depths. Coelenterates, arthropods, molluscs and echinoderms made up the bulk of the catches. All species identified are presented in tabular form, along with the depth ranges in which they were found. Study results indicate that the distribution of organic carbon in surface sediments is associated with the changes in invertebrate distribution. (See also W73-08816) (Ensign-PAI)
W73-08837

BATHYMETRIC DISTRIBUTION OF THE ECHINODERMS OFF THE NORTHERN OREGON COAST,
National Marine Fisheries Service, Seattle, Wash. Exploratory Fishing and Gear Research Base.
M. S. Alton.
In: The Columbia River Estuary and Adjacent Ocean Waters, University of Washington Press, Seattle, Wash., p 475-537. 1972. 18 fig, 21 tab, 49 ref.

Descriptors: *Columbia River, Continental shelf, *Continental slope, *Benthic fauna, *Distribution patterns, Density.
Identifiers: Crinoidea, *Echinoidea, Holothuroidea, Benthic zones.

Benthonic fish and invertebrate fauna inhabiting the outer continental shelf and slope southwest of the mouth of the Columbia River were investigated by the Bureau of Commercial Fisheries and the Atomic Energy Commission. A detailed description of the bathymetric distribution of Crinoidea, Echinoidea, and Holothuroidea is presented. Assemblages of echinoderms were related to the following five benthic zones: (1) the outer sublittoral (50-75 fathoms), diversity and catch rates were small here, (2) the upper bathyal (100-250 fathoms), large catches of sea urchins and brittle stars were found, (3) the lower bathyal (275-600 fathoms), large catches of sea stars, (4) the bathyal-abyssal (650-800 fathoms), a transitional zone between the fauna found in lower bathyal depths and abyssal depths, and (5) the abyssal (850-1160 fathoms), many ophiuroids were found. In the upper bathyal zone, several species were found to have circumboreal distributions. The echinoderm fauna from the abyssal zone have multiple species subdominance. Species diversity appeared to be highest at bathyal depths of 187 to 500 fathoms and at abyssal depths of 900 to 1050 fathoms. (See also W73-08816) (Ensign-PAI)
W73-08838

BATHYMETRIC AND SEASONAL ABUNDANCE AND GENERAL ECOLOGY OF THE TANNER CRAB CHIONOECETES TANNERI RATHBUN (BRACHYURA: MAJIDAE), OFF THE NORTHERN OREGON COAST,
National Marine Fisheries Service, Seattle, Wash. Exploratory Fishing and Gear Research Base.
W. T. Pereyra.
In: The Columbia River Estuary and Adjacent Ocean Waters, University of Washington Press, Seattle, Wash., p 538-582. 1972. 10 fig, 9 tab, 58 ref.

Descriptors: *Pacific Ocean, *Continental slope, *Crabs, Life history studies, *Benthic fauna, *Invertebrates, Distribution patterns, Density, Life cycles, Radioisotopes.
Identifiers: Chionoecetes tanneri.

Deepwater investigations conducted near the Columbia River mouth showed Chionoecetes tanneri, the Tanner crab, to be present in substantial numbers on the continental slope. The Tanner crab's relation to the total epibenthic invertebrate biomass, its distribution, seasonal movements, reproduction and life history are all discussed. A life-history model for the crab was developed through a synthesis of all available life-history information. Ecological factors which may serve to determine the life-history patterns observed are explored. One conclusion is that the Tanner crab may serve as a biological transport of radionuclides. Larvae develop at the surface and could transport radiomaterial to the bottom, than developing juveniles could transport it up the slope to the area of adult residency. (See also W73-08816) (Ensign-PAI)
W73-08839

CHARACTERISTICS OF THE DEMERSAL FISH FAUNA INHABITING THE OUTER CONTINENTAL SHELF AND SLOPE OFF THE NORTHERN OREGON COAST,
National Marine Fisheries Service, Seattle, 'Vash. Exploratory Fishing and Gear Research Base.
M. S. Alton.
In: The Columbia River Estuary and Adjacent Ocean Waters, University of Washington Press, Seattle, Wash., p 583-634. 1972. 25 fig, 9 tab, 58 ref. AEC AT (49-7)-1971.

Descriptors: Oregon, Coasts, Fish, Distribution patterns, Density, *Data collections, Bathymetry, *Continental shelf, Continental slope, *Seasonal, Pacific Ocean.
Identifiers: *Demersal fish.

A cooperative study undertaken by the Bureau of Commercial Fisheries and the Atomic Energy Commission provided extensive information on demersal fishes off the Oregon coast. Findings related to the bathymetric occurrences and availability of individual fishes, seasonal changes in their availability and the relative importance of the various species and phylogenetic groups related to depth are discussed. Bathymetric shifts in the availability of certain dominant fishes and size distribution of fish taken, number of species encountered, and species related to depth are also examined. Trawling stations were established at various depths for sampling purposes. Species found in the study area are listed according to phylogenetic groupings. The largest accumulations of fish in the study area were obtained from the bottom depth interval of 50 to 225 fathoms; a secondary zone of fish abundance was found at 250 to 425 fathoms. Changes in fish availability were seasonal and occurred in the outer continental shelf and upper slope regions. These changes were partly attributed to bathymetric and latitudinal shifts in fish populations. (See also W73-08816) (Ensign-PAI)
W73-08840

METHODS FOR THE MEASUREMENT OF HANFORD-INDUCED GAMMA RADIOACTIVITY IN THE OCEAN,
(AEC), Division of Biology and Medicine, Washington, D.C.
C. L. Osterberg, and R. W. Perkins.
In: The Columbia River Estuary and Adjacent Ocean Waters, University of Washington Press, Seattle, Wash., p 637-662. 1972. 12 fig, 1 tab, 32 ref.

Descriptors: *Columbia River, Radioisotopes, *Sampling, *Radioactive wastes, *Analytical techniques, Water pollution sources.
Identifiers: *Radionuclides, *Gamma-ray spectrometry.

Methods for identifying and measuring radionuclides in Columbia River water at sea are discussed. Since not all plants and animals have equal affinities for the same radionuclides, simple sampling and measurement of radionuclide levels in ocean fauna and flora is not accurate. The optimum method appears to be a direct sampling of ocean water for radioactivity. In the past, large samples of water have been processed at sea, with relatively small concentrates being analyzed after return to the laboratory. Recent refinements in sampling methods are discussed. One technique employs sorption beds, which remove radionuclides from many gallons of seawater. The concentrate, when analyzed with multiparameter gamma-ray spectrometry, reveals a number of radionuclides present in the sea. This method has its drawbacks, however, and further research is necessary. (See also W73-08816) (Ensign-PAI)
W73-08841

RADIOACTIVE AND STABLE NUCLIDES IN THE COLUMBIA RIVER AND ADJACENT NORTHEAST PACIFIC OCEAN,
Puerto Rico Nuclear Center, Mayaguez.
W. O. Forster.
In: The Columbia River Estuary and Adjacent Ocean Waters, University of Washington Press, Seattle, Wash., p 663-700. 1972. 17 fig, 21 tab, 28 ref. AEC AT (45-1)-1750.

Descriptors: *Columbia River, *Radioisotopes, Chemical properties, Physical properties, Measurement, Spectrometers, Water pollution sources, *Pacific Ocean, Dilution, *Radioactive wastes.

Results from several individual projects concerning the chemical and physical behavior of several radionuclides in the Columbia River estuary and plum region are presented. All measurements of radioactivity were made by gamma-ray spectrometry. It was discovered that radionuclides associated with particulate matter or found in solution decrease in specific activity as the result of their physical decay rate, biogeochemical removal from the ecosystem, dilution by tributaries, and holdup by dams. (See also W73-08816) (Ensign-PAI)
W73-08842

RADIOECOLOGY OF ZINC-65 IN ALDER SLOUGH, AN ARM OF THE COLUMBIA RIVER ESTUARY,

Oregon State Univ., Corvallis. Dept. of Oceanography.

W. C. Renfro.

In: The Columbia River Estuary and Adjacent Ocean Waters, University of Washington Press, Seattle, Wash., p 755-776. 1972. 8 fig, 3 tab, 18 ref. AEC AT (45-1)-1750.

Descriptors: *Columbia River, *Estuaries, *Radioisotopes, Carriers, *Zinc radioisotopes, *Spectrophotometry, Radiochemical analysis, Environmental effects.
Identifiers: *Zinc-65, Radionuclide transfer.

The cycling of radionuclides was studied in a discrete segment of the Columbia River estuary so that the effects of tides, salinity changes and other environmental variables could be assessed. Zinc-65 levels were measured in water, sediments, algae, emergent vegetation, and in animals in order to understand how 65Zn is transferred through a small estuarine ecosystem. The zinc in the estuary water was concentrated for analysis by a method of coprecipitation; atomic absorption spectrophotometry was used to determine the stable and radioactive zinc in the plants and animals. Study results indicated that zinc activity levels increase during the spring, fall in midsummer, and continue at a low level throughout the remainder of the year. Ecological half-lives were calculated for various organisms. Those in the lowest trophic levels had the shortest ecological half-life, each succeeding trophic level showed a longer half-life. (See also W73-08816) (Ensign-PAI)
W73-08845

SEASONAL AND AREAL DISTRIBUTIONS OF RADIONUCLIDES IN THE BIOTA OF THE COLUMBIA RIVER ESTUARY,

Oregon State Univ., Corvallis. Dept. of Oceanography.

W. C. Renfro, W. O. Forster, and C. L. Osterberg.

In: The Columbia River Estuary and Adjacent Ocean Waters, University of Washington Press, Seattle Wash., p 777-798. 1972. 6 fig, 7 tab, 14 ref. AEC AT (45-1)-1750.

Descriptors: *Columbia River, *Estuaries, *Radioisotopes, *Zinc, Radioisotopes, Aquatic life, Environmental effects, Plankton, *Seasonal, Distribution patterns.
Identifiers: Starry flounder, Sand shrimp, Staghorm sculpin, Radionuclides.

The distributions of radionuclides in organisms from the Columbia River estuary are described quantitatively. The seasonal variances of the radionuclides, as well as their location in the estuary are discussed. The study concentrates on 65 Zn since it can be readily measured; other radionuclides measured in the biota include: 60Co, 54Mn, 51Cr, 46Sc and 32P. Data are presented on the following organisms; plankton, starry flounder, sand shrimp and staghorn sculpin. The plankton exhibited high, but variable, amounts of 65Zn; no seasonal or areal trends could be established. Fish and shrimp had quantities of 65Zn which varied seasonally, reaching a maximum in spring and summer. The starry flounder and staghorn sculpin from upstream showed higher radioactivity levels than those from farther downstream. During the five-year study the zinc-65 specific activity levels in the organisms declined. This appears to be linked to the fact that during the same time span the number of reactors in operation at the Hanford site was reduced from eight to three. (See also W73-08816) (Ensign-PAI)
W73-08846

TOTAL PHOSPHORUS AND PHOSPHORUS-32 IN SEAWATER, INVERTEBRATES, AND ALGAE FROM NORTH HEAD, WASHINGTON,

Washington Univ., Seattle. Coll. of Fisheries; and Washington Univ., Seattle. Lab. of Radiation Ecology.

J. S. Isakson, and A. H. Seymour.

In: The Columbia River Estuary and Adjacent Ocean Waters, University of Washington Press, Seattle, Wash., p 799-818. 1972. 8 fig, 6 tab, 24 ref.

Descriptors: *Columbia River, *Phosphorus radioisotopes, Sampling, *Seawater, *Algae, *Invertebrates, Chemical analysis, Food chains, Environmental effects, Seasonal.
Identifiers: *Phosphorus-32, North Head (Wash).

A biological survey of 32P, a radionuclide released into the waters of the Columbia River by the reactors at Hanford, was made. Monthly samples were taken of seawater, several species of invertebrates, and algae; quantitative determinations were made of 32P and total phosphorus. Seasonal variation in total phosphorus was less pronounced than variation in 32P levels; the peak levels for 32P occurred most frequently in the period from April to June. The values for the abundance of total phosphorus and of 32P in seawater varied independently. The concentration factors for each of the chemicals were different, with 32P being lower. The lesser concentration of 32P can be linked to a non-equilibrium condition for this radionuclide in seawater and biota, as well as the physical decay of 32P in the flood web. Data on phosphorus concentration levels in the tissues of razor clams and muscles are also presented. (See also W73-08816) (Ensign-PAI)
W73-08847

EFFECTS OF STRONTIUM-90 + YTTRIUM-90 ZINC-65, AND CHROMIUM-51 ON THE LARVAE OF THE PACIFIC OYSTER CRASSOSTREA GIGAS,

Washington Univ., Seattle. Coll. of Fisheries; and Washington Univ., Seattle. Lab. of Radiation Ecology.

V. A. Nelson.

In: The Columbia River Estuary and Adjacent Ocean Waters, University of Washington Press, Seattle, Wash., p 819-832. 1972. 1 fig, 3 tab, 15 ref.

Descriptors: *Columbia River, *Bays, *Radioisotopes, *Oysters, Laboratory tests, Lethal limit, Water pollution effects, Environmental effects.
Identifiers: Willapa Bay, Zinc-65, Chromium-51, Crassostrea gigas, Abnormal larvae.

The effects of certain radionuclides, notably, 65Zn, 51Cr, and 90Sr + 90Y, upon Willapa Bay oysters are discussed. The developing eggs and embryos of the oyster Crassostrea gigas, were subjected to the effects of ionizing radiation from radionuclides for 48 hrs. The concentration of radionuclides ranged from 10-8 to 10-2 Ci/liter. Abnormal larvae were defined as those larvae which had incompletely developed shells 48 hr. after fertilization. The results indicated that significant increases in abnormal oyster larvae were present in 65Zn concentrations of 10-4 Ci/liter and greater; the effects of 90Sr + 90Y and 51Cr began at 10-3 Ci/liter. The concentrations of radionuclides necessary to produce abnormal larvae were more than a million times greater than the concentrations which have been measured to date in natural environments. The relationship between laboratory findings and environmental conditions is discussed. (See also W73-08816) (Ensign-PAI)
W73-08848

ZINC-65 IN BENTHIC INVERTEBRATES OFF THE OREGON COAST,

Oregon State Univ., Corvallis. Dept. of Oceanography.

A. G. Carey, Jr.

In: The Columbia River Estuary and Adjacent Ocean Waters, University of Washington Press, Seattle, Wash., p 833-842. 1972. 3 fig, 3 tab, 11 ref. AEC AT (45-1)-1750.

Descriptors: *Pacific Ocean, *Zinc radioisotopes, *Benthic fauna, *Bottom sampling, Trawling, Carriers, Depth, Invertebrates.
Identifiers: *Zinc-65, Sublittoral fauna, Bethyal fauna, Abyssal fauna, Radionuclide uptake.

The spatial distribution of 65Zn in the benthic fauna in the Pacific Ocean adjacent to the Columbia River is discussed. Samples of benthic microfauna and larger epibenthic organisms were gathered through bottom trawling. 65Zn was concentrated by the sublittoral, bathyal and abyssal fauna. The zinc-65 concentrations varied markedly with depth; levels decrease fairly regularly with distance from the Columbia River. 65Zn apparently reaches the fauna through the food web. (See also W73-08816) (Ensign-PAI)
W73-08849

A TEN-YEAR STUDY OF MEROPLANKTON IN NORTH CAROLINA ESTUARIES: AMPHIPODS,
National Marine Fisheries Service, Washington, D.C. Systematics Lab.
A. B. Williams, and K. H. Bynum.
Chesapeake Sci, Vol 13, No 3, p 175-192, 1972. Illus.
Identifiers: Amphipods, *Estuaries, Lunar, *Mero (Plankton), *North Carolina, Plankton.

Occurrence of gammarid and caprellid amphipods in semimonthly nocturnal surface plankton samples taken on flood tide for 10 yr at 1 estuarine station and shorter periods at 11 other stations is summarized on a monthly basis. Fifty-eight spp. are identified from samples taken during the study by no single station produced this variety. The great diversity of species from stations near the sea is contrasted with the sparse and different fauna of meso-oligohaline waters. Seasonal abundance in samples is summarized for all important species on a station basis. Contrasts in density of 8 spp. in surface-bottom tows and 33 spp. in samples taken at alternate lunar phases (new-full) indicate significant behavioral differences. Nocturnal stratification varied with species; about 1/3 showed no apparent abundance in association with lunar phase, another 1/3 showed increase at new, and a final 1/3 at full moon.--Copyright 1973, Biological Abstracts, Inc.
W73-08863

A DESCRIPTION AND ANALYSIS OF COASTAL ZONE AND SHORELAND MANAGEMENT PROGRAMS IN THE UNITED STATES,
Michigan Univ., Ann Arbor. Sea Grant Program.
E. H. Bradley, Jr., and J. M. Armstrong.
Available from NTIS, Springfield, Va 22151 as COM-72 10694 Price $9.00 printed copy; $1.45 microfiche. Technical Report No 20, March 1972. 756 p, 4 tab, 4 ref, 15 append. NOAA MICHU-SG-72-204.

Descriptors: *Coasts, *Management, *Comprehensive planning, *Land management, *Legislation, Water resources development, Legal aspects, Estuaries, Reviews, Beaches, Wetlands, Optimum development plans, Planning.
Identifiers: *Coastal management, *Shoreland management.

All of the existing U.S. coastal management programs are discussed. The programs of 10 states were chosen for close examination as they are representative of the major approaches being undertaken on the state level. An overview is given of the status of shoreland management at the state level in the United States. A historical analysis describes the influence of geographical and political factors on the development of shoreland management programs. Shoreland management programs are described by type. The various issues involved in developing a comprehensive coastal zone management program, and the areas that should be examined in future research are described. (Knapp-USGS)
W73-08887

YAQUINA BAY MARINE DEVELOPMENT PLAN.
Skidmore, Owings and Merrill, Portland, Oreg.

Available from NTIS, Springfield, Va 22151 as COM-72 11305, Price $3.00 printed copy; $1.45 microfiche. Report to Economic Development Administration, Department of Commerce, August 15, 1972. 249 p.

Descriptors: *Water resources development, *Marinas, *Water sports, *Commercial fishing, *Oregon, Bays, Feasibility studies, Economics, Project planning, Land development, Recreation facilities, Tourism, Environmental effects, Ecology, Projections, Evaluation, Costs.
Identifiers: *Yaquina Bay area (Oreg).

This report is Phase III of a continuing action program which is dedicated to promote sound, environmentally compatible economic development of the Yaquina Bay area, Oregon, while preserving and enhancing its natural resources and life style. Its purpose is to establish the feasibility of expanding the marine-oriented economic base to provide a new and more stable employment opportunities. The scope of the study includes the examination of specific interrelated activities: commercial fishing, fish processing, warehousing and storage, moorage and transportation; recreational and sport fishing; supporting commercial-retail facilities, including marine repair and supply; marine-oriented industrial development, including activities related to the Marine Science Center; and transportation bearing directly upon marine-oriented facilities. The culmination of Phase III of this program has resulted: (a) in the recommendation of specific policies to guide the future marine and recreational development of the Yaquina Bay area, and (b) in the citing of specific projects as potential candidates for Federal financial assistance. This material sets a comprehensive framework for healthy environmentally compatible, economic growth, and will allow the Task Force to seek a 'pre-application' interview with Federal authorities for implementative support. (Woodard-USGS)
W73-08888

GRAYS HARBOR ESTUARY, WASHINGTON: REPORT 4, SOUTH JETTY STUDY,
Army Engineer Waterways Experiment Station, Vicksburg, Miss.
For primary bibliographic entry see Field 08B.
W73-08889

MERCURY IN SEDIMENTS FROM THE THAMES ESTUARY,
Institute of Geological Sciences, London (England). Geochemical Div.
For primary bibliographic entry see Field 05A.
W73-08902

RESULT OF FLUVIAL ZINC POLLUTION ON THE ZINC CONTENT OF LITTORAL AND SUB-LITTORAL ORGANISMS IN CARDIGAN BAY, WALES,
University Coll. of Wales, Aberystwyth. Dept. of Zoology.
For primary bibliographic entry see Field 05B.
W73-08925

INTERTIDAL SUBSTRATE RUGOSITY AND SPECIES DIVERSITY,
University of Southern California, Los Angeles.
For primary bibliographic entry see Field 05C.
W73-08928

THE EFFECT OF OIL POLLUTION ON SURVIVAL OF THE TIDAL POOL COPEPOD, TIGRIOPUS CALIFORNICUS,
California State Univ., Northridge. Dept. of Biology.
For primary bibliographic entry see Field 05C.
W73-08929

THE USE OF COLIPHAGE AS AN INDEX OF HUMAN ENTEROVIRUS POLLUTION IN AN ESTUARINE ENVIRONMENT,
New Hampshire Univ., Durham.
For primary bibliographic entry see Field 05A.
W73-08951

ESTUARINE NITRIFICATION,
Rutgers - The State Univ., New Brunswick, N. J.
For primary bibliographic entry see Field 05B.
W73-08971

CONCENTRATION FACTORS OF CHEMICAL ELEMENTS IN EDIBLE AQUATIC ORGANISMS,
California Univ., Livermore. Lawrence Livermore Lab.
For primary bibliographic entry see Field 05B.
W73-08992

THE ECOLOGY OF MORECAMBE BAY: I. INTRODUCTION,
Freshwater Biological Association, Ambleside (England).
J. Corlett.
J Appl Ecol. Vol 9, No 1, p 153-159. 1972. Illus.
Identifiers: Bays, *Ecology, England, *Morecambe Bay (UK).

Between 1968 and 1970 a study of the ecology of Morecambe Bay in north-west England was undertaken as part of a feasibility study for a Morecambe Bay Barrage beign carried out under the authority of the Water Resources Board. The main aspects of the feasibility study are described and an outline of the biological problems is given. This is an introduction to a series of papers describing some of the investigations in more detail. (See also W72-14338 and W72-14339)--Copyright 1972, Biological Abstracts, Inc.
W73-09065

THE ECOLOGY OF MORECAMBE BAY: V. THE SALT MARSHERS OF MORECAMBE BAY,
Nature Conservancy, Norwich (England). Coastal Ecology Research Station.
A. J. Gray.
J Appl Ecol. Vol 9, No 1, p 207-220. 1972. Illus.
Identifiers: Bays, Construction, Ecology, England, Grasses-M, Marshes, *Morecambe Bay (UK), Range, Reservoirs, *Salt marshes, Sheep, Tidal relations, Vegetation.

An account is given of the history, topography, tidal relations and management of the salt marshes in the northern part of Morecambe Bay (England). Their recent history is one of progressive marsh development, a process accelerated by the building of breakwaters and embankments in the upper parts of the estuaries, and by the piecemeal reclamation of land for agriculture or as a result of railway construction. The fluctuations of the low water channels of the rivers draining into the bay have a significant effect on marsh development, producing, at many sites, intermittent phases of erosion and growth. At the present time, 1485 ha of salt marsh in the bay are confined to the upper 2.5 m of a 9.5 m tidal range, the lowest limit of the vegetation being at about 3.9 m A.O.D. (above ordnance datum). Features of the marshes include

directions closely parallel to the basin axis. The effect of axis curvature should be small in the Potomac since the radii of curvature are large compared to the river width (except perhaps at Mathias Pt.). When the model results were applied to the Potomac River, good agreement was obtained with observed mean range and mean high water phase. Manning's roughness coefficient (assumed spacially constant) is near the lower end of tabulated values. (Woodard-USGS)
W73-09080

CLAY MINERALOGY OF QUATERNARY SEDIMENTS IN THE WASH EMBAYMENT, EASTERN ENGLAND,
Imperial Coll. of Science and Technology, London (England). Dept. of Geology.
For primary bibliographic entry see Field 02J.
W73-09083

03. WATER SUPPLY AUGMENTATION AND CONSERVATION

3A. Saline Water Conversion

DESAL ION EXCHANGE FOR DEMINERALIZATION AT SANTEE, CALIFORNIA,
Santee County Water District, Calif.
For primary bibliographic entry see Field 05D.
W73-08975

CARBON ADSORPTION AND ELECTRODIALYSIS FOR DEMINERALIZATION AT SANTEE, CALIFORNIA,
Santee County Water District, Calif.
For primary bibliographic entry see Field 05D.
W73-08976

3B. Water Yield Improvement

AN APPRAISAL OF POTENTIAL WATER SALVAGE IN THE LAKE MCMILLAN DELTA AREA, EDDY COUNTY, NEW MEXICO,
Geological Survey, Albuquerque, N. Mex.
E. R. Cox, and J. S. Havens.
Geological Survey open-file report, March 1973. 71 p, 14 fig, 8 tab, 11 ref.

Descriptors: *Water conservation, *New Mexico, *Evapotranspiration control, *Phreatophytes, *Tomarisk, Reservoirs, Evapotranspiration, Water yield improvement.
Identifiers: *Lake McMillan (N Mex).

The amount of w ter used by saltcedar in the area of the old delta of Lake McMillan, New Mexico, and the amount of water that might be salvaged by eradication of saltcedar by construction of floodways through the delta area are estimated. Saltcedar growth increased from about 13,700 acres in 1952 to about 17,100 acres in 1960, a 25% increase. Most of this increase was in the a real density range near zero to 30%. The estimated average transpiration of phreatophytes in the Artesia to Lake McMillan reach is 29,000 acre-feet of water per year. In the reach from Artesia to the Rio Penasco, where the regional water table is above the Pecos River, saltcedar eradication might salvage from 10,000 to 20,000 acre-feet of water per year for use downstream. From the Rio Penasco to Lake McMillan, the river is perched above the water table; therefore, elimination of the saltcedar would probably not increase flow in the river, nor would drains be effective. Clearing in this reach, however, might increase the flow at Major Johnson Springs below Lake McMillan. Floodways through this reach would eliminate some evapotranspiration but might increase the sediment deposited by floodwaters in Lake McMillan. (Knapp-USGS)
W73-08493

MODIFYING PRECIPITATION BY CLOUD SEEDING,
South Dakota School of Mines and Technology, Rapid City.
A. S. Dennis.
Journal Soil Water Conservation, Vol 25, No 3, p 88-92, May-June 1970. 4 fig, 2 tab, 20 ref.

Descriptors: *Weather modification, *Cloud seeding, Cloud physics, *Precipitation (Atmospheric), *Silver iodide, Snow, Rainfall, Convection, Condensation, Nucleation, Mathematical models, Bibliographies, Orography, On-site tests, Mexico, Great Plains.
Identifiers: Switzerland, Argentina, USSR, Israel, Phillippines, Caribbean, Ice nuclei, Orographic precipitation, Orographic clouds.

Weather modification research has greatly increased understanding of cloud processes and effects of different seeding treatments. The most significant advancement has been the development of numerical models that simulate cloud behavior and predict probable seeding effects. A second advancement has been widespread adoption of stratifying data collected in field experiments, such as according to wind direction, temperature at a selected level, or presence or absence of weather fronts. Every raindrop reaching the ground forms around an embryo. Natural embryos include droplets formed around unusually large condensation nuclei and large droplets resulting from chance collisions among ordinary cloud droplets. Artificial embryos, such as those stimulated by dry ice pellets or silver iodide crystals, are described along with experiments with orographic and convective clouds, the latter on a selective or an area basis. Also discussed are storm abatement projects in switzerland, Argentina, USSR, and the Great Plains of the U S. Orographic seeding can produce precipitation of about 10%, but the effects of seeding convective clouds are still uncertain. (USBR)
W73-08521

3C. Use of Water of Impaired Quality

INVESTIGATION OF FACTORS RELATIVE TO GROUNDWATER RECHARGE IN IDAHO,
Idaho Univ., Moscow. Dept. of Hydrology.
For primary bibliographic entry see Field 05D.
W73-08456

THE FLOURISHING DISTRICT HEATING BUSINESS,
For primary bibliographic entry see Field 06D.
W73-08612

HEATED WATER: A USE FOR A HEADACHE.

Electrical World, Vol 177, No 11, p 26-27, June 1, 1972. 1 fig.

Descriptors: *Heated water, *Experimental farms, *Irrigation programs, *Microclimatology, *Frost protection, Horticulture, Agriculture, Vegetable crops, Demonstration farms, Fruit crops, Multiple-purpose projects, Crop production, Crop response, Growth rates, *Oregon.

An experimental program developed by the Eugene (Oregon) Water and Electric Board, which examines the possible uses of heated water in horticulture, is presented. Twenty-one thousand feet of buried pipeline carry heated water to an irrigation and sprinkler system which services 170 acres of farm land. On this land a large variety of crops

are grown to examine the effects of thermal water upon crop production, growth periods and crops which are usually grown in warmer climates. Thermal water in irrigation is as good as cold water, although flood irrigation can be harmful. Heated water also is an effective form of frost protection and a sun shield when used in a sprinkler system. The difficulties which were encountered in using heated water were minor and could be avoided through the development of techniques in dealing with this new tool. (Jerome-Vanderbilt)
W73-08630

QUALITY STANDARD FOR IRRIGATION WATER IN THE TROPICS,
Hawaii Univ., Honolulu. Dept. of Agronomy and Soil Science.
S. A. El-Swaify.
In: Water Resources Seminar Series No. 1, January 1972. Water Resources Research Center, University of Hawaii, Honolulu, p 53-66.

Descriptors: *Irrigation water, *Tropical regions, *Water quality standards, Soils, Crops, Hawaii, Salts, Leaching, Salinity, Hydraulic conductivity, Anions, Permeability.
Identifiers: Crop tolerance limit, Oxisol, Aridosol.

New water quality standards are needed for irrigating highly weathered soils in the tropics, which should feature higher allowable salinity levels and sodium proportions relative to other cations and stronger emphasis on possible effects of dominant anions, thereby creating an expansion in water resources available for irrigation by putting to use water which is normally considered unsuitable. Just as the U.S. Salinity Laboratory's system proved inadequate for Oxisols and Aridisols, any other attempted generalizations concerning irrigation water quality in the tropics should be avoided due to wide diversity among the soils of these regions. It is important to remember that irrigation water management is a very major factor for modifying the quality of irrigation water. This may be clearly illustrated by the fact that the depth of movement of a given quantity of water--the leaching requirement--into a given soil is strongly dependent on the water application method. Water and accompanying soluble components percolate deeper into soil when unsaturated conditions are maintained during irrigation. It is possible to improve leaching efficiency by proper choice of irrigation method, as it is important to change the allowable level of salinity in irrigation water depending on the irrigation method. (Auen-Wisconsin)
W73-08652

EFFECTS OF MUNICIPAL WASTE WATER DISPOSAL ON THE FOREST ECOSYSTEM,
Pennsylvania State Univ., University Park. School of Forest Resources.
For primary bibliographic entry see Field 05D.
W73-08731

3D. Conservation in Domestic and Municipal Use

MEASURING BENEFITS GENERATED BY URBAN WATER PARKS,
Dornbusch (David M.) and Co., Inc., San Francisco, Calif.
For primary bibliographic entry see Field 06B.
W73-08659

HYDROPNEUMATIC STORAGE FACILITIES.
Illinois State Environmental Protection Agency, Springfield. Div. of Public Water Supplies.
For primary bibliographic entry see Field 05G.
W73-08671

FUNCTIONAL WATER AND SEWERAGE PLAN AND PROGRAM, CATAWBA REGIONAL PLANNING COUNCIL.
LBC and W Associates, Columbia, S.C.
For primary bibliographic entry see Field 06B.
W73-08977

LOWCOUNTRY REGION FUNCTIONAL WATER AND SEWER PLAN AND PROGRAM.
LBC and W Associates, Columbia, S.C.
For primary bibliographic entry see Field 06B.
W73-08978

SOUTH SANTA CLARA COUNTY WATER PLANNING STUDY COMPLETION REPORT.
Santa Clara County Flood Control and Water District, San Jose, Calif.
For primary bibliographic entry see Field 06B.
W73-08979

MALIBU MASTER PLAN OF STORM DRAINS.
Los Angeles County Engineer Dept., Calif.
For primary bibliographic entry see Field 05G.
W73-08980

INITIAL WATER, SEWERAGE AND FLOOD CONTROL PLAN REPORT, JOB 4800.
Duncan and Jones, Berkeley, Calif.
For primary bibliographic entry see Field 06B.
W73-08981

WATER AND SEWER FACILITIES, OSCEOLA COUNTY.
Beck (R. W.) and Associates, Denver, Colo.
For primary bibliographic entry see Field 06B.
W73-08984

WATER SUPPLY PLAN FOR THE SOUTHEASTERN CONNECTICUT REGION: SUMMARY.
Southeastern Connecticut Water Authority, Norwich.
For primary bibliographic entry see Field 06B.
W73-08985

WATER SUPPLY PLAN FOR THE SOUTHEASTERN CONNECTICUT REGION, VOLUME I, INVENTORY.
Southeastern Connecticut Water Authority, Norwich.
For primary bibliographic entry see Field 06B.
W73-08986

STATEWIDE LONG-RANGE PLAN FOR THE MANAGEMENT OF THE WATER RESOURCES OF CONNECTICUT, PHASE I REPORT, 1971.
Connecticut Interagency Water Resources Planning Board, Hartford.
For primary bibliographic entry see Field 06B.
W73-08987

LAND USE AND DEVELOPMENT ALONG THE TENNESSEE RIVER IN THE TARCOG REGION.
Top of Alabama Regional Council of Governments, Huntsville.
For primary bibliographic entry see Field 06B.
W73-08988

INCOG FUNCTIONAL PLAN SUPPLEMENTS (WATER AND SEWER, HOUSING AND OPEN SPACE).
Indian Nations Council of Governments, Tulsa, Okla.
For primary bibliographic entry see Field 06B.
W73-08989

INTERIM RIVER REPORT, NOVEMBER 1970,
Interagency Riverfront Committee, Minneapolis, Minn. Design Team.
For primary bibliographic entry see Field 06B.
W73-08990

ROLE OF WATER IN URBAN PLANNING AND MANAGEMENT,
Geological Survey, Arlington, Va. Water Resources Div.
For primary bibliographic entry see Field 04C.
W73-09082

3E. Conservation in Industry

AN EXAMINATION OF THE ECONOMIC IMPACT OF POLLUTION CONTROL UPON GEORGIA'S WATER-USING INDUSTRIES,
Georgia Inst. of Tech., Atlanta. Engineering Experiment Station.
For primary bibliographic entry see Field 05G.
W73-08453

CRUDE OIL AND NATURAL GAS PRODUCTION IN NAVIGABLE WATERS ALONG THE TEXAS COAST (FINAL ENVIRONMENTAL IMPACT STATEMENT).
Army Engineer District, Galveston, Tex.
For primary bibliographic entry see Field 05G.
W73-08661

3F. Conservation in Agriculture

ESTIMATING THE PRODUCTIVITY OF IRRIGATION WATER FOR SUGARCANE PRODUCTION IN HAWAII,
Hawaii Univ., Honolulu. Water Resources Research Center.
L. B. Rankine, J. R. Davidson, and H. C. Hogg.
Available from the National Technical Information Service as PB-220 007, $3.00 in paper copy, $1.45 in microfiche. Technical Report No. 56, June 1972. 66 p, 9 fig, 7 tab, 14 ref, 3 append. OWRR A-012-HI (2).

Descriptors: *Sugarcane, *Irrigation water, Soil moisture, *Yield equations, *Hawaii, *Crop production, Productivity, Evapotranspiration.
Identifiers: Operational models.

This study concentrated on the development of an operational model. Data were collected from sugar plantations located in two sugar-producing areas chosen to represent the various sugar-production conditions in the state. Area I depended exclusively on water from artesian wells for irrigation. Irrigation in Area II depended mostly on stored surface runoff rainfall and is supplementary in nature. Both areas differ in their physical terrain. The study draws heavily from and builds on the experiences of previous investigations in the general areas of agricultural economics and plant sciences. From an integration of these ideas, the modified composite variable was developed as a measure of water adequacy in cane production. Most of the information required to construct this variable is available from management records. Pan evaporation data were incomplete because standard evaporation pans were only recently installed and then only in a few areas. To overcome this difficulty, techniques were developed to estimate evapotranspiration for the two study areas utilizing available pan data together with other relevant indicators.
W73-08454

AN ANALYTICAL INTERDISCIPLINARY EVALUATION OF THE UTILIZATION OF THE WATER RESOURCES OF THE RIO GRANDE IN NEW MEXICO,
New Mexico State Univ., University Park. Dept. of Agricultural Economics.

For primary bibliographic entry see Field 06B.
W73-08458

IMPROVED METHOD FOR DETERMINATION OF XANTHOPHYLL IN FRESH ALFALFA AND WET FRACTIONS,
Agricultural Research Service, Berkeley, Calif. Western Marketing and Nutrition Research Div.
For primary bibliographic entry see Field 05A.
W73-08602

QUALITY STANDARD FOR IRRIGATION WATER IN THE TROPICS,
Hawaii Univ., Honolulu. Dept. of Agronomy and Soil Science.
For primary bibliographic entry see Field 03C.
W73-08652

DRIP IRRIGATION IN SUGAR CANE,
Hawaii Univ., Honolulu. Dept. of Agronomy and Soil Science.
P. C. Ekern.
In: Water Resources Seminar Series No. 1, January 1972. Water Resources Research Center, University of Hawaii, Honolulu, p 69-72.

Descriptors: *Water conservation, *Irrigation, *Hawaii, Tropical regions, Irrigation practices, Crop production, Evaporation control, Water distribution (Applied).
Identifiers: *Drip irrigation.

Consumptive use of water by sprinkler and drip irrigated sugarcane was measured by four hydraulic load cell lysimeters at the Kunia substation, Hawaii. The original one-eye cane transplants were set in Molokai latosol in October 1968. Water use by the full canopy of sprinkler irrigated cane for both plant crop and ratoon approximated the evaporation from a ground level class A pan and often equalled or exceeded the full equivalent value of the local net radiation, indicative of positive heat advection. The fraction of sunlight converted into net radiation had a strong seasonal change from 0.53 in summer to 0.38 in winter over the cane. Drip replaced sprinkler irrigation in the fall of 1970. With full canopy, both the plant crop and ratoon cane used only 0.75 to 0.8 of the water from that used by sprinkler irrigated cane. Rainfall percolated through the 5 ft depth of soil in the lysimeters and removed approximately 400 lbs of nitrate/acre in the 15 inches of percolate during the winter of 1968-1969, and again in the 30 inches of percolate during the winter of 1970-1971. Yield and growth of cane was satisfactory for both the sprinkler and drip irrigation methods. (Auen-Wisconsin)
W73-08653

CROPPING PATTERN, YIELDS AND INCOMES UNDER DIFFERENT SOURCES OF IRRIGATION (WITH SPECIAL REFERENCE TO IADP DISTRICT ALIGARH, U.P.),
Pant Coll. of Tech., Pantnagar (India).
T. V. Moorti, and J. W. Mellor.
Indian Journal of Agricultural Economics, Vol 27, No 4, p 117-125, October-December 1972. 10 tab, 2 ref.

Descriptors: *Irrigation practices, *Crops, *Income, *Economic impact, Water supply, Agriculture, Farms, Water costs, Water quantity, Irrigation water, Irrigation efficiency, Management, Wheat.
Identifiers: *Cropping pattern, *India, IADP district Aligarh, Economic analysis, Gross returns, Tube-wells, Persian wheels.

Irrigation technology is in a transitional stage in India where both traditional and new methods are being employed simultaneously. Traditional methods utilizing Persian wheels and 'charsa' cannot supply enough water for the new high-yielding varieties of crops. Through use of more recent techniques, these new varieties have greatly increased the returns to irrigation. State and private tube-wells are among the improved sources of irrigation. This transitional stage may have resulted in variations in the cropping pattern, yields, and incomes of farmers. In this study, an attempt is made to assess the changes in the cropping pattern as influenced by increased availability of irrigation. The specific objectives are: (1) to analyze the differences in cropping pattern and yield under different sources of irrigation; and (2) to analyze the differences in gross incomes from various crops under different sources of irrigation. It is concluded that private tube-well farms have better control of water supply in terms of its timely availability in adequate quantity which results in higher cropping intensity, yields, and therefore higher incomes. The two basic factors in irrigation--the quantity and timing of water application--result in the variations in farming pattern ultimately affecting incomes. (Bell-Cornell)
W73-08658

SOIL AND WATER CONSERVATION COMMISSION AND DISTRICTS.
For primary bibliographic entry see Field 06E.
W73-08704

YELLEN V. HICKEL (SUIT TO COMPEL ENFORCEMENT OF THE RECLAMATION LAW).
For primary bibliographic entry see Field 06E.
W73-08707

WATER AND SEWERAGE DEVELOPMENT PLAN, SUMTER COUNTY, SOUTH CAROLINA.
Palmer and Mallard and Associates, Inc., Sumter, S.C.
For primary bibliographic entry see Field 06B.
W73-08982

TRIALS OF N-FERTILIZATION ON DURUM WHEAT WITH REFERENCE TO THE POSSIBILITY OF LEAF DIAGNOSIS, (IN ITALIAN),
Bari Univ. (Italy). Istituto di Agronomia.
A. Patruno.
Riv Agron. Vol 6 No 1, p 13-27. 1972 Illus. (English summary).
Identifiers: Diagnosis, Durum wheat, Fertilization, Leaf, *Nitrogen, *Wheat-M, Yields.

In 2-yr experiments with the cultivar 'Capeiti 8' the maximum grain yield was obtained using 80 kg/ha of N. The fertilization increased straw yield, number of kernels per hectare, and decreased yellow berry, weight of kernels and apparent specific weight. The straw yield was due to increased height, thickness or tillering of culms. The N-content of leaves increased with fertilizer rates. It decreased from jointing until flowering and from the highest leaves to the last 3. However, it varied from year to year and cannot represent a fertility index of the soil. The N-content of the leaves showed several correlations with straw yield, number of kernels per hectare, but not with grain yield. The ratio between the N-content of the leaves harvested at jointing and that of the earing or flowering was correlated with grain yield in the second yr, when the winter rainfall was more than in the first yr and the N-fertilization showed a better effect. Due to the drought the N-content in the kernel was higher in the first year: it increased with N-fertilization in both years, with increasing straw/grain ratio and with decreasing weight of kernels and yellow berry.--Copyright 1972, Biological Abstracts, Inc.
W73-09021

EFFECT OF DIFFERENT NITROGEN FERTILIZERS ON THE PRODUCTIVITY OF FLOODED RICE, (IN PORTUGUESE),
N. Leite, H. Gargantini, A. G. Gomes, and T. Igue.
Bragantia. Vol 29 No 1, p 263-272. 1970. (English summary).
Identifiers: Ammonium, Cal-Nitro, Castor-Bean-D, Fertilizers, Flooded rice, Mead, *Nitrogen, Crop productivity, *Rice-M, Sulfates.

The fertilizers tested were: castorbean mead, ammonium sulfate and Cal-Nitro at 3 different levels. A constant amount of K and P was applied to all treatments. There was a significant increase on rice grain production after application of N fertilizers. Castorbean mead was the most effective in increasing the yield, followed by ammonium sulfate and Cal-Nitro.--Copyright 1972, Biological Abstracts, Inc.
W73-09022

RICE FERTILIZATION: STUDY OF THE EFFECT OF NITROGEN, PHOSPHORUS, MICRONUTRIENTS AND LIME, (IN PORTUGUESE),
N. Leite, H. Gargantini, L. S. Hungria, and T. Igue.
Bragantia. Vol 29 No 1, p 273-285. 1970. (English summary).
Identifiers: Fertilization, Lime, *Nitrogen, Nutrients, *Phosphorus, *Rice-M, Yield.

N and P were applied at levels of 0, 40, and 80 kg/ha. Lime was applied at levels of 5 and 3 t/ha in organic and clayey soils, respectively. Seeds of 'Iguape-agulha' were utilized. Cultivation was under the flooding system. Planting was done directly with seed. A great response to N fertilizer was noted, even at higher doses. In 6 of the 13 experiments, there was some response to P. No response to micronutrients or to liming was noticed.--Copyright 1972, Biological Abstracts, Inc.
W73-09023

EFFECTS OF NITROGENOUS AND PHOSPHOROUS FERTILIZATION ON RICE CULTIVATED IN FLOODED SOILS, (IN PORTUGUESE),
N. Leite, H. Gargantini, and L. S. Hungria.
Bragantia. Vol 29 No 1, p 115-125. 1970. (English summary).
Identifiers: *Brazil, Fertilization, Flooded soils, Nitrogenous, Oryza-Sativa-M, *Phosphorous, *Rice-M, Soils, Yield.

Experiments on Oryza sativa for N were carried out in clayey lowland and for P, in organic lowland soils in Sao Paulo, Brazil. A significant increase in production was obtained with N or P fertilizer applications in most cases.--Copyright 1972, Biological Abstracts, Inc.
W73-09024

IRRIGATION AFTEREFFECT ON THE FIRST AGRICULTURAL CYCLE OF TWO FORAGE GRASSES (DACTYLIS GLOMERATA L. AND FESTUCA PRATENSIS L.), (IN FRENCH),
Institut National de la Recherche Agronomique, Dijon (France).
S. Meriaux.
Ann Agron (Paris). Vol 22 No 1, p 95-111. 1971. (English summary).
Identifiers: Agricultural cycles, Dactylis-Glomerata-M, Festuca-Pratensis-M, *Forage grasses, Fructose, Grasses-M, Irrigation aftereffects, Storage, Stubble, Weight.

The adverse irrigation aftereffect on the yield of grasses during the first agricultural cycle over several years may be due to some depletion of the irrigated plant when subjected to intensive cultivation incompatible with proper build-up of reserves. The trend of regrowth in tilled land, showing the adverse irrigation aftereffects, is reproduced by the biological test, whereas the fructose content in the various parts of the plant changes erratically. Regrowth in tilled land increases with the weight

of the stubble, the tillering sprouts and the soluble-N content in each of these fractions. Regrowth in darkness is unrelated with the fructose weight but correlates with the weight of the stubble, the tillering sprouts and that of the fructoses in the stubble. Irrigation aftereffects on the yield over the first agricultural cycle of the following year may be due to the reduced weight of the stubble and tillering sprouts brought about by such treatment, and also to the lower weight of the stored sugars. The effect can be estimated and forecast by determining the weight of the storage organs.--Copyright 1972, Biological Abstracts, Inc.
W73-09025

IRRIGATED AGRICULTURE AS A METHOD OF DESERT SOIL RECLAMATION (IN RUSSIAN),
For primary bibliographic entry see Field 02G.
W73-09031

THE GROWTH AND STAGNATION OF AN URBAN FRANCE MARKET GARDENING REGION: VIRGINIA, SOUTH AUSTRALIA,
Adelaide Univ. (Australia). Dept. of Geography.
D. L. Smith.
Aust Geogr. Vol 12, No 1, p 35-48. 1972. Illus.
Identifiers: *Australia, Gardening, *Irrigation, Crop production, Stagnation, Urban areas, Vegetables, Water utilization.

The role of irrigation on market gardens in the Virginia Region, the sources of water used, and the irrigation practices adopted were studied. Attention is given to the nature of the water resources of the region, the effects of exploitation upon them, and the efforts which have been made to control their use. The likely effects of these control measures on the future of vegetable production in the region are assessed.--Copyright 1972, Biological Abstracts, Inc.
W73-09068

04. WATER QUANTITY MANAGEMENT AND CONTROL

4A. Control of Water on the Surface

AN APPRAISAL OF POTENTIAL WATER SALVAGE IN THE LAKE MCMILLAN DELTA AREA, EDDY COUNTY, NEW MEXICO,
Geological Survey, Albuquerque, N. Mex.
For primary bibliographic entry see Field 03B.
W73-08493

PLANS FOR REDUCTION OF SHOALING IN BRUNSWICK HARBOR AND JEKYLL CREEK, GEORGIA,
Army Engineer Waterways Experiment Station, Vicksburg, Miss.
For primary bibliographic entry see Field 08B.
W73-08512

SOME DESIGNS AND ANALYSES FOR TEMPORALLY INDEPENDENT EXPERIMENTS INVOLVING CORRELATED BIVARIATE RESPONSES,
California Polytechnic State Univ., San Luis Obispo. Dept. of Computer Science and Statistics.
S-C. Wu.
Biometrics, Vol 28, No 4, p 1043-1061, December 1972. 1 fig, 6 ref, 1 append.

Descriptors: *Statistical methods, *Estimating, *Design, *Runoff, Streamflow forecasting.
Identifiers: Treatment effects, Variance, Continued covariate design, Balanced crossover design, Augmented crossover design, Maximum likelihood.

Designs and analyses are derived for experiments with two experimental units where responses are correlated within and independent among time periods. Fisher's information function for the treatment parameter is maximized to obtain most informative designs for estimating a treatment effect. In the absence of pre-test or historical data on the two units, the most informative design is always a balanced crossover. With historical data, the design is either a continued covariate or an augmented crossover. Efficient analyses of continued covariate and balanced crossover designs are shown to be examples of maximum likelihood (ML) estimation and the analysis of covariance. For the augmented crossover designs, (ML) methods and Wilks' lambda criterion are used to provide efficient large sample procedures. It is shown that use of a most informative design and efficient analysis instead of more familiar designs and analyses can result in a sizable decrease in the variance of the treatment effect estimator. (Little-Battelle)
W73-08546

ORGANIZATION OF LAND AREA IN WASHINGTON FOR WATER AND LAND USE PLANNING,
Washington State Univ., Pullman. Dept. of Agronomy and Soils.
For primary bibliographic entry see Field 06E.
W73-08607

CREATION OF CONDITIONAL DEPENDENCY MATRICES BASED ON A STOCHASTIC STREAMFLOW SYNTHESIS TECHNIQUE,
Arizona Univ., Tucson. Dept. of Systems Engineering.
For primary bibliographic entry see Field 02E.
W73-08608

KNOW WHAT'S HAPPENING IN YOUR WATER SYSTEM,
Parsons, Brinckerhoff, Quade and Douglas, Inc., New York.
H. L. Michel, and J. P. Wolfner.
The American City, Vol 87, No 6, p 80-88, June 1972. 2 fig, 2 tab, 7 ref, 5 p.

Descriptors: *Water distribution (Applied), *Hydraulic models, *Computer programs, *Digital computers, Simulation analysis, Networks, Pipes, Pumps, Head loss, Flow, Water management (Applied), Mathematical models, Systems analysis, Water demand, Analog computers.
Identifiers: *Hardy Cross method, *Newton-Raphson method.

To manage even a moderate-size water utility efficiently, one must be able to estimate the consequence of any operational changes. To do so accurately requires a hydraulic analysis of the distribution network, simulating any number of eventualities. Presented is a digital computer program that overcomes many of the limitations in other hydraulic computer programs, and is easy to use. It assures convergence to a solution, is capable of analyzing a large system, and automatically formulates the initial assumptions required. Computer results are displayed in a format easily used in conjunction with system maps of the utility, thereby eliminating the need to transfer results to special-purpose diagrams. After discussing the disadvantages of the Hardy Cross method and analog computers and the advantages of digital computers in such hydraulic analysis, the program developed herein is described in detail. The program utilizes the Newton-Raphson method which has been proved already in actual practice. Specifically discussed are eleven critical features incorporated into the program, and the problems encountered and overcome in organizing the data. Application of the program to a hypothetical city is included. This program is less costly and time-consuming, and its results are more easily understood by nontechnical personnel. (Bell-Cornell)
W73-08655

VALUATION OF TIMBER, FORAGE AND WATER FROM NATIONAL FOREST LANDS,
Forest Service (USDA), Tucson, Ariz. Rocky Mountain Forest and Range Experiment Station.
For primary bibliographic entry see Field 06B.
W73-08657

NEW ENGLAND RIVER BASINS COMMISSION--1972 ANNUAL REPORT.
New England River Basins Commission, Boston, Mass.
For primary bibliographic entry see Field 06E.
W73-08676

STATEMENT OF FINDINGS--BAL HARBOUR, FLORIDA, PARTIAL BEACH RESTORATION, BEACH EROSION CONTROL AND HURRICANE PROTECTION PROJECT, DADE COUNTY, FLORIDA (FINAL ENVIRONMENTAL IMPACT STATEMENT).
Army Engineer District, Jacksonville, Fla.

Available from National Technical Information Service, Springfield, Virginia, as EIS-FL-72-5591-F. Price: $4.75 in paper copy, $1.45 in microfiche. May 1972. 51 p, 1 plate.

Descriptors: *Florida, *Environmental effects, *Beach erosion, *Recreation, *Artificial beaches, Hurricanes, Borrow pits, Silting, Beaches, Dredging, Aquatic environment, Excavation, Erosion control, Shore protection.
Identifiers: *Environmental Impact Statements, *Coastal zone management, *Bal Harbour (Florida).

This project consists of beach restoration of an 0.85 mile reach of Dade County. The project fill will be obtained from an ocean borrow pit about 1.5 miles offshore in elevations of 36 to 50 feet. About 1.8 million cubic yards of sand will be dredged from an ocean borrow pit and placed along 0.85 mile of beach for restoration of protective and recreational assets. Adverse environmental effects include temporary turbidity and siltation in the borrow and fill areas during construction. Some marine life will be destroyed; however, these populations are expected to become reestablished. Consideration was given to several alternative plans but it was determined the proposed project would be the best and most desirable to preserve the beach on Bal Harbour. Comments on the proposed project were requested and subsequently received from various state, federal, and private agencies. (Mockler-Florida)
W73-08680

ROUGE RIVER FLOOD CONTROL PROJECT. SECTION B FROM I-94 TO MICHIGAN AVENUE ALTERNATIVE NO. IV (SUPPLEMENT TO FINAL ENVIRONMENTAL IMPACT STATEMENT).
For primary bibliographic entry see Field 08A.
W73-08681

FLOOD CONTROL ON SAGINAW RIVER, MICHIGAN AND TRIBUTARIES; FLINT RIVER AT FLINT, SWARTZ, AND THREAD CREEKS (FINAL ENVIRONMENTAL IMPACT STATEMENT).
Army Engineer District, Detroit, Mich.
For primary bibliographic entry see Field 08A.
W73-08682

W73-08690

PERRY COUNTY DRAINAGE AND LEVEE DISTRICTS NOS. 1, 2, AND 3, MISSOURI (FINAL ENVIRONMENTAL IMPACT STATEMENT).
Army Engineer District, St. Louis, Mo.
For primary bibliographic entry see Field 08A.
W73-08691

PROPOSED NEW LOCK--PICKWICK LANDING DAM (FINAL ENVIRONMENTAL IMPACT STATEMENT).
Tennessee Valley Authority, Chattanooga.
For primary bibliographic entry see Field 08C.
W73-08692

REPORT OF THE TEXAS WATER DEVELOPMENT BOARD FOR THE BIENNIUM SEPTEMBER 1, 1970 THROUGH AUGUST 31, 1972.
Texas Water Development Board, Austin.
For primary bibliographic entry see Field 06E.
W73-08732

HYDROGRAPH SIMULATION MODELS OF THE HILLSBOROUGH AND ALAFIA RIVERS, FLORIDA: A PRELIMINARY REPORT,
Geological Survey, Tampa, Fla.
J. F. Turner, Jr.
Geological Survey open-file report 72025, 1972. 107 p, 28 fig, 4 tab, 8 ref.

Descriptors: *Streamflow forecasting, *Mathematical models, *Rainfall-runoff relationships, *Florida, Hydrologic data, Mathematical studies, Equations, Flood control, Watershed management, Hydrographs, Computer programs, Input-output analysis.
Identifiers: *Hillsborough River (Fla), *Alafia River (Fla).

Mathematical (digital) models that simulate flood hydrographs from rainfall records have been developed for the following gaging stations in the Hillsborough and Alafia River basins of west-central Florida: Hillsborough River near Tampa, Alafia River at Lithia, and North Prong Alafia River near Keysville. These models, which were developed from historical streamflow and rainfall records, are based on rainfall-runoff and unit-hydrograph procedures involving an arbitrary separation of the flood hydrograph. Hillsborough and Alafia River flood discharges can be simulated with expected relative errors less than 30% and flood peaks with average relative errors less than 15%. Because of the inadequate rainfall network used in obtaining input data for the North Prong Alafia River model, simulated peaks are frequently in error by more than 40%, particularly for storms having highly variable areal rainfall distribution. The models serve not only as a basis for forecasting floods, but also for simulating hydrologic information needed in flood-plain mapping and delineating and evaluating alternative flood control and abatement plans. (Woodard-USGS)
W73-08733

DELINEATION OF INFORMATION REQUIREMENTS OF THE T.V.A. WITH RESPECT TO REMOTE SENSING DATA,
Tennessee Univ., Knoxville.
For primary bibliographic entry see Field 07B.
W73-08737

STREAM GAUGING INFORMATION, AUSTRALIA, SUPPLEMENT 1970.
Australian Water Resources Council, Canberra.
For primary bibliographic entry see Field 07C.
W73-08738

GEOLOGIC AND HYDROLOGIC MAPS FOR LAND-USE PLANNING IN THE CONNECTICUT VALLEY WITH EXAMPLES FROM THE FOLIO OF THE HARTFORD NORTH QUADRANGLE, CONNECTICUT,
Geological Survey, Washington, D.C.
F. Pessl, Jr., W. H. Langer, and R. B. Ryder.
Geological Survey Circular 674, 1972. 12 p, 2 fig, 21 ref.

Descriptors: *Water resources development, *Land development, *Mapping, *Hydrogeology, *Connecticut, Planning, Land use, Topography, Geology, Land management, Watershed management, Groundwater resources, Streamflow, Water quality, Urbanization, Water quality control.
Identifiers: *Connecticut Valley, *Hartford (Conn).

Decisions that affect land and water use require an understanding of the complex geologic and hydrologic systems that exist within the earth's crust. The U.S. Geological Survey's Connecticut Valley Urban Area Project is developing a flexible resource data base in the form of maps that present a single characteristic or a combination of related characteristics of the land surface, earth materials, or water resources at a common scale and in a simplified format. The maps are prepared by interpretation of existing geologic and hydrologic maps and data available from ongoing U.S. Geological Survey and State programs. Map subjects include unconsolidated materials, depth to bedrock, depth to water table, thickness and distribution of extensive clay deposits, resources of aggregate, slopes, availability of groundwater, flood-prone areas, and quality and quantity of surface water. (Woodard-USGS)
W73-08746

SUMMARIES OF STREAMFLOW RECORDS IN NORTH CAROLINA,
Geological Survey, Raleigh, N.C.
For primary bibliographic entry see Field 07C.
W73-08747

THE CALIFORNIA STATE WATER PROJECT SUMMARY: NINETEEN-SEVENTY-ONE.
California State Dept. of Water Resources, Sacramento.

Bulletin No 132-72, Appendix C, 1972. 31 p.

Descriptors: *Water resources development, *Project purposes, *California, *Reviews, Water utilization, Water supply, Irrigation, Recreation facilities, Flood control, Electric powerplants, Imported water, Rivers, Reservoirs, Costs, Programs, State governments, Hydraulic structures, Project planning.
Identifiers: *California State Water Project, Progress report, Summary report.

The California State Water Project is basically a massive redistribution system designed to help correct an inbalance of water resources. More than 70% of California's available water originates in the State's northern third, while approximately 77% of the demand comes from the southern two-thirds. The State Water Project takes the north's abundant water, much of which now flows unused from the Sacramento-San Joaquin Delta to the Pacific Ocean and delivers it to where it can be productive. For example, in the San Joaquin Valley--the world's top farm region with a $2-billion-a-year agricultural production--the Project provides water to open up a quarter-million barren acres to cultivation and also to give needed relief to another 125,000 acres now irrigated from wells tapping overdrawn groundwater reservoirs. The Project is also designed to supply water to other areas of the State, to control winter and springtime floods, to open dozens of new recreational areas, to save and enhance fisheries and wildlife habitats, and to produce--smog-free--hydroelectric power.

The various aspects of the project are summarized. (Woodard-USGS)
W73-08749

STREAMFLOW FORMATION AND COMPUTATIONS (FORMIROVANIYE I RASCHETY RECHNOGO STOKA).
Ukrainskii Nauchno-Issledovatelskii Gidro-Meteorologicheskii Institut, Kiev (USSR).
For primary bibliographic entry see Field 02E.
W73-08880

ICE JAMS IN THE DANUBE DELTA AND SOME MEASURES TO PREVENT FLOODS CAUSED BY THEM (ZATORY L'DA V DEL'TE DUNAYA I NEKOTORYYE MEROPRIYATIYA PO PREDOTVRASHCHENIYU VYZYVAYEMYKH IMI NAVODNENIY),
Vsesoyuznyi Nauchno-Issledovateskii Institut Gidrotekhniki, Leningrad (USSR).
V. N. Karnovich, and V. I. Sinotin.
Meteorologiya i Gidrologiya, No 4, p 78-85, April 1972. 2 fig, 1 tab, 4 ref.

Descriptors: *Ice jams, *River, *Deltas, *Floods, *Flood control, Channel improvement, Channel morphology, Meteorology, Melting, Ice breakup, Navigation, Explosives.
Identifiers: *USSR, *Danube River, *Ice control, Icebreakers.

Severe ice jams are formed in lower reaches of the Danube once every 7-10 years and are responsible for destruction of dikes, flooding of populated areas, and curtailment of naviatigation. The average time of formation of ice jams is 8-10 hours. The thickness of ice accumulations in severe jams is 7-9 meters. The volume of ice may reach 15 km in length, and the height of ice piles on the banks, 2.5 meters. Ice jams formed in spring last 7-10 days. The cause of formation of ice jams on the Danube is examined in terms of channel morphology, climate, and the water and ice regimes of the river. Methods used to break up ice jams in the Danube delta are described, and recommendations are made for improving ice-jam control by icebreakers, explosives, and channel regulation. (Josefson-USGS)
W73-08886

A DESCRIPTION AND ANALYSIS OF COASTAL ZONE AND SHORELAND MANAGEMENT PROGRAMS IN THE UNITED STATES,
Michigan Univ., Ann Arbor. Sea Grant Program.
For primary bibliographic entry see Field 02L.
W73-08887

DRAINAGE AND FLOOD CONTROL BACKGROUND AND POLICY STUDY.
Nolte (George S.) and Associates, San Francisco, Calif.

San Diego County Comprehensive Planning Organization, San Diego, California, Summary report, May 1970. 20 p, 1 fig. HUD Calif P-294 (G).

Descriptors: *Urban drainage, *Drainage programs, *Flood control, *Computer models, *Regional analysis, Urbanization, Land use, Regional development, Optimum development plans, California, Systems analysis, Data collections, Data processing, Flood plain zoning, Flooding, Costs.
Identifiers: *San Diego County (Calif).

Flooding problems in San Diego County were studied and current practices for flood hazard prevention were inventoried. Although basically a water-short area, floods are becoming a serious problem primarily due to urban development of floodplain lands. The growth of the County is expected to raise the population from the current 1,400,000 to about 2,000,000 by 1990, causing further problems. Over 100 agencies are involved in flood control and storm drainage in the County and over 200 documents were reviewed in this study. Detailed solutions were not offered; rather, a general direction is indicated along with a supply of data on which to base action. Flood control should be placed in the authority of a regional agency on the basis of natural rather than political boundaries. Land use zoning was viewed as a major factor in the prevention of damage from floods. This report, which is a summary of two volumes on flood control and storm drainage, includes a section on definitions of terms for the non-technical reader. This helps bring about community understanding of the problems involved, and citizen cooperation with remedial programs. (Poertner)
W73-08891

STORM DRAINAGE.
Baruth and Yoder, Walnut Creek, Calif.

Available from the National Technical Information Service as PB-211 048, $11.25 in paper copy, $1.45 in microfiche. Mid-Humboldt County Urban Planning Program, Eureka, California, April 1971. 129 p, 22 fig, 21 tab. LPO-P/360-CAS-1.

Descriptors: *Urban drainage, *Drainage systems, *Flood control, *Facilities, *Storm drains, Storm runoff, Flood plain zoning, Costs, Urban runoff, Detention reservoirs, Drainage programs, Drainage districts, Urbanization, Forecasting, Drainage water, Drainage practices, Drainage engineering, Storm water, California.
Identifiers: *Mid-Humboldt County (California).

Problems of storm drainage in mid-Humboldt County, California were studied to determine methods of solution by 2020. The population of the area is expected to grow from the 1970 level of 74,000 to 325,000 by 2020. Concurrently, residential land use is expected to almost triple, commercial acreage will rise to 2 1/2 times its present extent, and industrial land use will more than double. As a consequence, drainage facilities will be required to handle loads considerably larger than existing loads. A detailed master plan for storm drainage is given, covering recommended improvements involving pipes, detention ponds and open channels. Design data are included as is the cost and construction period recommended. A total cost of $32 million was estimated for all construction plus a 30 percent allowance for contingencies, engineering and financing. Financing methods includes general obligation bonds, special assessments, and grants from federal sources. Besides adoption of the master plan, the County should adopt an effective flood plain zoning ordinance to limit development in all flood plains of the study area. (Poertner)
W73-08892

DRAINAGE SYSTEM ALTERNATIVES,
R. K. Brown.
Paper presented at the American Public Works Association Congress and Equipment Show, Minneapolis, Minnesota, September 21-27, 1972. 9 p.

Descriptors: *Urban drainage, *Flooding, *Drainage systems, *Drainage practices, *Drainage programs, Water pollution, Combined sewers, Detention reservoirs, Erosion, Urbanization, Urban runoff, Storm runoff, Canada.
Identifiers: *Ontario (Canada).

Urban development usually has a detrimental effect on natural drainage systems requiring the development of suitable added drainage facilities. Problems of urban drainage include flooding, pollution and erosion and each system studied must include a study of how that system copes with these problems as well as problems of costs, aesthetics, land availability and time needed for implementation. Urban drainage systems are commonly of four types: (1) natural drainage courses, such as rivers which are often over-loaded or polluted from urban drainage; (2) open channels that are natural channels which have been altered by straightening, widening, or deepened to handle the increased urban flow and which can be used in an open space program; (3) trunk storm sewers, which can be very expensive and of limited capacity, needing a means to handle excess flow; and (4) local storm drains, which can be complex, costly and beset by problems of flooding or pollution. Design criteria for drainage facilities are based on the rational formula for runoff rates, but storage of water in ponds and tanks can significantly reduce this rate. Detention ponds for stormwater storage can be incorporated into parks. Proper consideration of all effects of drainage in urban areas is necessary to provide adequate service to the citizens. (Poertner)
W73-08897

LAND DEVELOPMENT AND HEAVY METAL DISTRIBUTION IN THE FLORIDA EVERGLADES,
Florida State Univ., Tallahassee, Dept. of Oceanography.
For primary bibliographic entry see Field 05B.
W73-08906

MALIBU MASTER PLAN OF STORM DRAINS.
Los Angeles County Engineer Dept., Calif.
For primary bibliographic entry see Field 05G.
W73-08980

INITIAL WATER, SEWERAGE AND FLOOD CONTROL PLAN REPORT, JOB 4800.
Duncan and Jones, Berkeley, Calif.
For primary bibliographic entry see Field 06B.
W73-08981

LAND USE AND DEVELOPMENT ALONG THE TENNESSEE RIVER IN THE TARCOG REGION.
Top of Alabama Regional Council of Governments, Huntsville.
For primary bibliographic entry see Field 06B.
W73-08988

INFLUENCE OF REGULATION OF THE LACZANKA RIVER ON THE OCCURRENCE OF HIRUDINEA,
Bialystok Medical Academy (Poland). Dept. of Histology and Embryology.
J. Wilkialis.
Pol Arch Hydrobiol. Vol 18 No 4, p 359-365. 1971 Illus.
Identifiers: Birds, *Hirudinea, Laczanka River, Macrophytes, Mammals, *Poland, *Regulation, Rivers.

As a result of changes of the environment following the regulation of the Laczanka River a decrease in the number of species and of individuals of Hirudinea was observed in the regulated areas. At the regulated stations, the eurytopic species occurred in great numbers, but the rare species which need almost stagnant environments and the presence of prey from the surrounding environments (water birds, mammals) were observed sporadically or not at all. Changes in the number at particular stations of the regulated areas were influenced by both periodical cleaning of the regulated areas and the fall in water level during dry periods. It is believed that the factors favorable for more rapid recolonization of the regulated stations in the river by Hirudinea are accumulation of mud sediments and detritus, the presence of favorable places for attachment, the reappearance of macrophytes and availability of suitable food.–
Copyright 1972, Biological Abstracts, Inc.
W73-09019

MATHEMATICAL SIMULATION OF THE TURBIDITY STRUCTURE WITHIN AN IMPOUNDMENT,
Army Engineer Waterways Experiment Station, Vicksburg, Miss.
For primary bibliographic entry see Field 05G.
W73-09081

FLOODS IN THE IOWA RIVER BASIN UPSTREAM FROM CORALVILLE LAKE, IOWA,
Geological Survey, Iowa City, Iowa.
For primary bibliographic entry see Field 07C.
W73-09084

DEVELOPMENT OF A DISCHARGE MEASUREMENT DEVICE FOR SEWER FLOW,
Illinois Univ., Urbana. Dept. of Civil Engineering.
For primary bibliographic entry see Field 07B.
W73-09085

4B. Groundwater Management

AN ANALYTICAL INTERDISCIPLINARY EVALUATION OF THE UTILIZATION OF THE WATER RESOURCES OF THE RIO GRANDE IN NEW MEXICO,
New Mexico State Univ., University Park. Dept. of Agricultural Economics.
For primary bibliographic entry see Field 06B.
W73-08458

MAP SHOWING GENERAL AVAILABILITY OF GROUND WATER IN THE SALINA QUADRANGLE, UTAH,
Geological Survey, Washington, D.C.
For primary bibliographic entry see Field 07C.
W73-08473

RADIOCHEMICAL MONITORING OF WATER AFTER THE CANNIKIN EVENT, AMCHITKA ISLAND, ALASKA, JULY 1972,
Geological Survey, Lakewood, Colo.
For primary bibliographic entry see Field 05B.
W73-08479

GROUNDWATER DEVELOPMENT AND ITS INFLUENCE ON THE WATER BALANCE,
Technical Univ., of Denmark, Copenhagen.
For primary bibliographic entry see Field 02F.
W73-08507

THE MANAGEMENT OF GROUNDWATER RESOURCE SYSTEMS,
California Univ., Santa Barbara. Dept. of Mechanical Engineering.
For primary bibliographic entry see Field 05G.
W73-08660

LIMESTONE POLICY,
Illinois State Environmental Protection Agency, Springfield. Div. of Public Water Supplies.
For primary bibliographic entry see Field 05G.
W73-08672

HYDROLOGY OF THE UPPERMOST CRETACEOUS AND THE LOWERMOST PALEOCENE ROCKS IN THE HILIGHT OIL FIELD, CAMPBELL COUNTY, WYOMING,
Geological Survey, Washington, D.C.
M. E. Lowry.
Geological Survey open-file report, February 1973. 60 p, 13 fig, 5 plate, 1 tab, 18 ref.

Descriptors: *Groundwater resources, *Geology, *Aquifer characteristics, *Water wells, *Wyoming, Oil fields, Groundwater movement, Hydrogeology, Water yield, Pumping, Drawdown, Hydrologic data, Sandstones, Geologic mapping, Geologic formations.
Identifiers: *Campbell County (Wyo), Hilight oil field (Wyo).

There was a comparatively large increase in the development of groundwater from the Fox Hills Sandstone of Cretaceous age and some of the overlying aquifers in the Powder River basin in 1970-71 for secondary recovery of oil, principally in Campbell County, Wyoming. Recharge to the Lance-Fox Hills aquifer in the Hilight oil field is largely by vertical movement; there is no recharge from the Lance and Fox Hills outcrops on the east side of the basin to the formations in the Hilight area. At the end of the central Hilight water-flood project, the maximum possible drawdown resulting from the pumping of any one well at a distance of 10 miles from the pumped well would be about 15 feet, if the projected pumping were evenly distributed among the project wells. Within a few years after pumping has ceased, water in the project wells will approach the levels present before pumping began. The only irreversible effect of pumping will be the compaction of shale, with attendant subsidence, because the water derived from the shale probably will not be replaced. (Woodard-USGS)
W73-08725

REPORT OF THE TEXAS WATER DEVELOPMENT BOARD FOR THE BIENNIUM SEPTEMBER 1, 1970 THROUGH AUGUST 31, 1972.
Texas Water Development Board, Austin.
For primary bibliographic entry see Field 06E.
W73-08732

GROUNDWATER RESOURCES OF WASHINGTON COUNTY, TEXAS,
Geological Survey, Austin, Tex.
W. M. Sandeen.
Texas Water Development Board, Austin, Report 162, November 1972. 105 p, 21 fig, 9 tab, 52 ref.

Descriptors: *Groundwater resources, *Hydrologic data, *Water wells, *Aquifers, *Texas, Water resources development, Aquifer characteristics, Data collections, Pumping, Water yield, Water utilization, Water quality, Chemical analysis, Groundwater movement, Groundwater recharge, Water level fluctuations, Hydrographs, Hydrogeology.
Identifiers: *Washington County (Tex).

Large quantities of undeveloped fresh water, extending to depths of 1,200 feet below sea level occur in the Catahoula Sandstone, Jaspar aquifer, Evangeline aquifer, and the alluvium of the Brazos River in Texas. In 1968, an estimated 3.2 mgd was pumped from the groundwater reservoir. At lease 8,500 acre-feet per year of fresh groundwater is being transmitted through the Catahoula Sandstone, the Jaspar aquifer, and the Evangeline aquifer, and about 18,700 acre-feet per year of fresh groundwater is being rejected from the outcrops of these units. About 30,700 acre-feet per year of fresh groundwater probably could be withdrawn continuously from the aquifers. About 118,000 acre-feet per year is available for development from the alluvium of the Brazos River. The chemical quality of the groundwater is suitable for most types of uses or can be made suitable with a minimum of treatment. Less than 10 percent of the samples analyzed for dissolved solids contained more than 1,000 mg/liter. In general, the water is very hard (more than 180 mg/liter of hardness) and is slightly alkaline. (Woodard-USGS)
W73-08739

GROUNDWATER RESOURCES OF NAVARRO COUNTY, TEXAS,
Geological Survey, Austin, Tex.
G. L. Thompson.
Available from Texas Water Development Board P.O. Box 13087, Austin, Tex 78711. Texas Water Development Board Report 160, November 1972. 63 p, 12 fig, 8 tab, 47 ref.

Descriptors: *Groundwater resources, *Hydrologic data, *Water wells, *Aquifers, *Texas, Water resources development, Aquifer characteristics, Data collections, Withdrawal, Water yield, Water utilization, Groundwater movement, Groundwater recharge, Precipitation (Atmospheric), Water quality, Chemical analysis, Water levels, Hydrogeology, Hydrographs.
Identifiers: *Navarro County (Tex).

Navarro County, an area of 1,084 square miles, is in the central part of northeastern Texas. About 1.3 mgd of groundwater used in 1968 included public supply, 0.15 mgd; rural domestic, 0.40 mgd; livestock, 0.23 mgd; industry, 0.002 mgd; and irrigation, 0.50 mgd. Most of the water required for public supply and industrial use was supplied by surface water obtained from Navarro Mills and other reservoirs. The principal aquifers and minor water-bearing formations and their approximate quantities of water supplied in 1968 were Woodbine Formation, 0.13 mgd; Nacatoch Sand, 0.05 mgd; Wilcox, Midway, and Navarro Groups (excluding Nacatoch Sand), and Taylor Marl, 0.60 mgd; and alluvial deposits, 0.50 mgd. The Hosston Formation, which is untapped by wells in Navarro County, is potentially a valuable source of groundwater in the western part of the county. Of the samples in which approximate dissolved solids were calculated from specific conductance plus the laboratory determinations of dissolved solids, 61% of all water samples exceeded 500 mg/liter but only 35% exceeded 1,000 mg/liter. (Woodard/USGS)
W73-08740

GROUND-WATER RESOURCES OF HARDEMAN COUNTY, TEXAS,
Geological Survey, Austin, Tex.
M. L. Maderak.
Texas Water Development Board Report 161, November 1972. 44 p, 10 fig, 6 tab, 17 ref.

Descriptors: *Groundwater resources, *Water quality, *Water wells, *Hydrologic data, *Texas, Aquifers, Aquifer characteristics, Water yield, Water level fluctuations, Groundwater movement, Groundwater recharge, Water utilization, Irrigation, Pumping, Hydrogeology, Hydrographs, Maps.
Identifiers: *Hardeman County (Tex).

The Blaine Formation of Permian age and the alluvial terrace deposits of Quaternary age are the most important sources of large quantities of groundwater in Hardeman County, Texas. The quantity of water moving through the alluvium in the vicinity of Chillicothe is estimated at 2,600 acre-feet per year, which is equivalent to about 1 inch or 5% of the average precipitation. Recharge to the Blaine Formation is estimated to be about 17,000 acre-feet per year. Water levels in most wells in the Blaine Formation during the period 1953-69 showed a net decline of 10 to 20 feet; water levels in wells in the alluvium near Chillicothe during the period 1960-69 had an average decline of 3.6 feet, or less than 0.5 foot per year. Withdrawal of groundwater for municipal and domestic supply, livestock use, industrial use, and irrigation is estimated at 11,500 acre-feet in 1968. Water from the alluvium is very hard, but generally has a dissolved-solids content less than 1,000 milligrams per liter. Water from Blaine Formation is more mineralized but has been used successfully for irrigation for many years. (Woodard-USGS)
W73-08741

WATER-LEVEL RECORDS, 1969-73, AND HYDROGEOLOGIC DATA FOR THE NORTHERN HIGH PLAINS OF COLORADO,
Geological Survey, Washington, D.C.
For primary bibliographic entry see Field 07C.
W73-08745

GEOLOGIC AND HYDROLOGIC MAPS FOR LAND-USE PLANNING IN THE CONNECTICUT VALLEY WITH EXAMPLES FROM THE FOLIO OF THE HARTFORD NORTH QUADRANGLE, CONNECTICUT,
Geological Survey, Washington, D.C.
For primary bibliographic entry see Field 04A.
W73-08746

SALINE GROUNDWATER RESOURCES OF MISSISSIPPI,
Geological Survey, Jackson, Miss.
For primary bibliographic entry see Field 07C.
W73-08750

DATA REQUIREMENTS FOR MODELING A GROUND-WATER SYSTEM IN AN ARID REGION,
Geological Survey, Menlo Park, Calif.
For primary bibliographic entry see Field 02F.
W73-09078

4C. Effects on Water of Man's Non-Water Activities

THE SNOW REMOVAL CONTROVERSY,
Toronto Univ. (Ontario). Inst. of Environmental Sciences and Engineering.
For primary bibliographic entry see Field 05B.
W73-08601

DRAINAGE AND FLOOD CONTROL BACKGROUND AND POLICY STUDY.
Nolte (George S.) and Associates, San Francisco, Calif.
For primary bibliographic entry see Field 04A.
W73-08891

STORM DRAINAGE.
Baruth and Yoder, Walnut Creek, Calif.
For primary bibliographic entry see Field 04A.
W73-08892

IMPACT OF HIGHWAYS ON SURFACE WATERWAYS,
Metropolitan Sanitary District of Greater Chicago, Ill.
R. F. Lanyon.
July 14, 1972. 21 p, 3 fig, 1 tab, 9 ref.

Descriptors: *Storm runoff, *Highways, *Detention reservoirs, *Storage, *Highway effects, Water pollution control, Sediment control, Flood control, Flood plains, Water pollution sources, Water pollution, Storm water, Construction, Drainage systems, Drainage effects.

The modern highway has a high efficiency of stormwater drainage which is beneficial to the highway, but detrimental to the waterways into which the runoff is discharged. The associated landscaping near the highway often reduces the natural storage of rainwater and reduces the rate of rainwater infiltration, causing additional problems of increased flow in the waterways. Further problems are encountered when highways are built in floodplains, occupying land which previously was used for stormwater storage. During construction of the roadway, erosion can be a serious problem. Sediment yields have been estimated at 500 to 1,000 tons per year per mile of 100-foot wide construction, or 40 to 80 tons per acre per year. Sediment yields between 0.3 to 0.5 tons per acre per year are cited for undeveloped land. The construction of a roadway across a stream can be a potential flood hazard if debris is allowed to collect at the bridge. Highways can also be detrimental to waterways as a result of pollution from deicing chemicals, oil on the pavement and ordinary surface dirt. Detention storage of stormwater runoff is recommended as a means of controlling many of these problems. Runoff flow rates can be reduced, sediments settled out and additional flood plain storage provided. Detention storage can provide other important benefits such as groundwater replenishment, and recreational and open space facilities. Detention of runoff can also reduce the cost of drainage facilities in comparison to conventional facilities. (Poertner)
W73-08894

TREATMENT OF URBAN RUNOFF.,
Environmental Protection Agency, Washington, D.C. Municipal Pollution Control Section.
For primary bibliographic entry see Field 05G.
W73-08898

PROJECT WAGON WHEEL TECHNICAL STUDIES REPORT NO. 2.
El Paso Natural Gas Co., Tex.
For primary bibliographic entry see Field 05B.
W73-08991

INADVERTENT WEATHER AND PRECIPITATION MODIFICATION BY URBANIZATION,
Illinois State Water Survey, Urbana. Atmospheric Science Section.
S. A. Changnon, Jr.
ASCE Proceedings, Journal of the Irrigation and Drainage Division, Vol 99, No IR1, Paper 9608, p 27-41, March 1973. 14 fig, 2 tab, 10 ref. NSF GA-18781 and GA-28189X.

Descriptors: *Urbanization, *Weather modification, *Precipitation (Atmospheric), Urban hydrology, Rainfall, Cities, Weather, Climates, Air pollution effects, Storms, Runoff.

Urban-industrial complexes produce measurable modification of all weather conditions. Particularly significant are increases in the precipitation conditions of interest to hydrologists and urban planners. In and immediately downwind of major urban areas, the annual precipitation may be increased from 5%-30%, the annual thunderstorm frequency is increased 15%-30%, the heavy daily rainstorm frequencies are increased by 20%-40%. Increases in local runoff may be from 15%-20%. Local crop yields may be increased 2%-10% by this rain modification. (Knapp-USGS)
W73-09077

ROLE OF WATER IN URBAN PLANNING AND MANAGEMENT,
Geological Survey, Arlington, Va. Water Resources Div.
W. J. Schneider, D. A. Rickert, and A. M. Spieker.
Geological Survey Circular 601-H, 1973. 10 p, 2 fig, 1 plate, 1 tab.

Descriptors: *Urbanization, *Urban hydrology, *Hydrologic data, *Data collections, *Planning, Data processing, Rainfall-runoff relationships, Water quality, Water management (Applied), City planning, Land use, Storm drains.
Identifiers: Urban management.

The concentration of people in urban areas has modified the natural landscape, bringing about water problems. Although water resources in urban areas are altered by urbanization, the deleterious effects can be minimized or corrected by comprehensive planning and management. Water-resource information for urban planning is badly needed, but urban planners are not generally able to identify the data that are needed. To help satisfy this need, a water-resources evaluation matrix was developed. The graphic matrix provides a means for organizing the relative importance of water-related problems, and for identifying the data needed to evaluate these problems for the purpose of urban planning. The matrix lists nine subject categories in which water-related urban problems may occur. The matrix also lists 51 possible data inputs for evaluation of the problem areas. The inputs include the standard types of basic hydrologic data as well as information based on interpretation and analysis of these data. The list also includes the factors of climate, land, and culture. The matrix aids in the development of resource evaluation in two ways: first, by promoting interdisciplinary discussion, it leads to a mutual understanding of the water-related problems; second, it serves as a checklist for determining the data needs for evaluation of each problem. (Knapp-USGS)
W73-09082

4D. Watershed Protection

STATEMENT OF FINDINGS--BAL HARBOUR, FLORIDA, PARTIAL BEACH RESTORATION, BEACH EROSION CONTROL AND HURRICANE PROTECTION PROJECT, DADE COUNTY, FLORIDA (FINAL ENVIRONMENTAL IMPACT STATEMENT).
Army Engineer District, Jacksonville, Fla.
For primary bibliographic entry see Field 04A.
W73-08680

OLIVER BOTTOMS RESOURCE CONSERVATION AND DEVELOPMENT PROJECT, SEBASTIAN COUNTY, ARKANSAS (FINAL ENVIRONMENTAL IMPACT STATEMENT).
Soil Conservation Service, Washington, D.C.

Available from National Technical Information Service, U.S. Dep't. of Commerce as EIS-AR-72-5373-F. Price: $3.00 in paper copy, $1.45 in microfiche. September 10, 1972. 16 p.

Descriptors: *Arkansas, *Environmental effects, *Channel improvement, *Erosion control, Wildlife habitats, Water pollution, Water quality control, Flood protection, Flood plain zoning, Watershed management, Land use, Excavation, Vegetation establishment, Bank protection, Bank stabilization, Planting management, Revegetation, Stream erosion, Vegetation effects.
Identifiers: *Environmental Impact Statements, *Oliver Bottoms (Ark).

This watershed project consists of the installation of about 1.4 miles of channel improvement and appurtenant pipe overfall structures for grade stabilization and erosion control in the Oliver Bottoms Watershed, Sebastian County, Arkansas. Excavation of the channel from one side only and plantings of adapted grasses, trees and shrubs to minimize erosion are also planned to provide food and cover for wildlife. Adverse environmental effects include the loss of some productive land resources as a result of widening the channel, removal of some desirable trees to permit construction of the channel, and a minor amount of water pollution during the installation of the project. Floodwater retarding structure sites were considered, several channel locations were also considered, and a third alternative was no action. Other desirable effects include enhanced wildlife food and cover, elimination of stagnant water and mosquito breeding areas, and a reduction in floodwater damage to crops, roads, and land resources. (Mockler-Florida)
W73-08687

PEYTON CREEK, TEXAS--FLOOD CONTROL (FINAL ENVIRONMENTAL IMPACT STATEMENT).
Army Engineer District, Galveston, Tex.
For primary bibliographic entry see Field 08A.
W73-08688

as pathogens. Surface water samples enriched in organic matter were obtained from several recreational lakes in North Georgia. The most probable number test was performed to determine the fluorescent pseudomonads capable of growth at 39 C. Twenty-nine apyocyanogenic strains of fluorescent pseudomonads capable of growth at 39C were isolated for use in comparative studies with a typical pyocyanogenic strain of P. aeruginosa. Among the 29 isolates subjected to selected tests, some strains were clearly distinguishable from P. aeruginosa while others possessed characteristics in common with both P. aeruginosa and certain non-fluorescent aerobic pseudomonads. The pathogenicity to mice of representative strains injected intraperitoneally varied within a range reported in the literature for P. aeruginosa, but differed by at least one order of magnitude from that of a control strain of P. aeruginosa examined. (Holoman-Battelle)
W73-08530

ENVIRONMENTAL FACTORS CORRELATED WITH SIZE OF BACTERIAL POPULATIONS IN A POLLUTED STREAM,
New Mexico Highlands Univ., Las Vegas. Dept. of Biology.
H. Brasfeild.
Applied Microbiology, Vol 24, No 3, p 349-352, September 1972. 2 tab, 14 fig.

Descriptors: Water analysis, Population, *Bioindicators, Water quality, *Streptococcus, *Regression analysis, *Correlation analysis, · Data processing, *Coliforms, Water temperature, Hydrogen ion concentration, Detergents, Nitrates, Nitrites, Sulfates, Chlorides, Phosphates, Bicarbonates, *Pollutant identification, Streams, Water pollution effects, Enteric bacteria, *New Mexico.
Identifiers: *Gallinas River.

Samples of water were taken from a polluted zone of the Gallinas River and analyzed as to numbers of total bacteria, coliforms, and fecal streptococci. Environmental factors measured were temperature, pH and concentrations of detergent, nitrate plus nitrite nitrogen, sulfate, chloride, bicarbonate, and phosphate. Thirty-two observations were made from 12 March through 22 July 1971. Stepwise multiple linear regression analyses of the data were carried out by computer to determine which of the environmental factors were significantly correlated with numbers of bacteria present. A multiple linear regression equation was constructed for each bacteriological parameter as a function of significant variables only. Log total bacteria was correlated positively with bicarbonate, phosphate, and detergent concentrations. Log coliforms was correlated positively with phosphate and sulfate concentrations and negatively with chloride concentration. Log fecal streptococci was correlated positively with bicarbonate and chloride concentration. (Little-Battelle)
W73-08532

THE ISOLATION AND ENUMERATION OF CYTOPHAGAS,
Alberta Univ., Edmonton. Dept. of Soil Science.
P. J. Christensen, and F. D. Cook.
Canadian Journal of Microbiology, Vol 18, No 12, p 1933-1940, December 1972. 1 fig, 9 tab, 22 ref.

Descriptors: *Aquatic bacteria, *Isolation, *Pollutant identification, Detergents, Surfactants, Bactericides, Inhibition.
Identifiers: *Cytophagas, Enumeration, Tween 20, Sodium lauryl sulfate, Culture media, Selective media, Growth.

A comparison of 20 media based on degraded milk, protein or tryptone, yeast extract, and sodium acetate showed that various new formulae were superior to standard media for enumeration of cytophagas from aquatic habitats, for their isolation, and for maximum expression of the spreading characteristic. The use of 0.1 microliter/liter Tween 20 in dilution blanks increased counts of cytophagas up to threefold, but incorporation of Tween 20 into the plating agar had no significant effect on the numbers recovered. Sodium lauryl sulfate (S.L.S.) was examined as a possible screening agent for identification of cytophagas. A concentration of 0.1 percent S.L.S. inhibited the growth of 97 percent of the 66 cytophagas tested, but more than 80 percent of the other 41 organisms tested were also affected. However, 91 percent of the cytophagas were sensitive to S.L.S. and showed proteolysis on skim acetate medium compared to 53 percent of the Flavobacteria and 50 percent of the other organisms tested. None of the third group of organisms could be confused morphologically with the cytophaga group; thus it is suggested that S.L.S. susceptibility performed on skim acetate medium could be a useful screening test for cytophagas. (Holoman-Battelle)
W73-08533

IDENTIFICATION OF VIBRIO CHOLERAE BY PYROLYSIS GAS-LIQUID CHROMATOGRAPHY,
Manchester Univ. (England). Dept. of Bacteriology and Virology.
J. M. Haddadin, R. M. Stirland, N. W. Preston, and P. Collard.
Applied Microbiology, Vol 25, No 1, p 40-43, January 1973. 2 fig, 1 tab, 6 ref.

Descriptors: *Pollutant identification, *Pathogenic bacteria, *Aerobic bacteria, *Chemical analysis, E. coli, Salmonella, Shigella, Varieties.
Identifiers: *Pyrolysis gas liquid chromatography, *Vibrio cholerae, Sample preparation, Aeromonas, Biochemical tests, Proteus, Pseudomonas aeruginosa, Vibrio parahaemolyticus, Vibrio proteus.

Fifty-seven strains of chloera-like vibrios obtained from a variety of sources were grown individually at 37C for 20 hr in a 50-ml static culture of nutrient broth. Formalin was added; the bacteria were harvested by centrifugation, washed three times, and resuspended in distilled water. Samples loaded onto the coil filament of the gas chromatograph were pyrolyzed at 800C for 5 sec. Conventional biochemical tests were also performed to differentiate the biotypes of V. cholerae. Those cholera-like vibrios examined by pyrolysis gas-liquid chromatography could be distinguished from other common aerobic gram-negative bacilli, including oxidase-positive organisms, e.g., Aeromonas. Vibrios in Heiberg group I were subdivided into three types on the basis of differences in one complex in the chromatogram, and these closely corresponded with the identification as classical, El Tor, or 'intermediate' biotypes of Vibrio cholerae by conventional methods. (Holoman-Battelle)
W73-08534

A REAPPRAISAL: CORRELATION TECHNIQUES APPLIED TO GAS CHROMATOGRAPHY,
Warren Spring Lab., Stevenage (England). Control Engineering Div.
G. C. Moss, and K. R. Godfrey.
Instrumentation Technology, Vol 20, No 2, p 33-35, February 1973. 3 fig, 7 ref.

Descriptors: *Gas chromatography, *Pollutants, Pollutant identification, Methodology.
Identifiers: *Correlation techniques, *Trace levels, Detection limits.

The case for applying correlation techniques to gas chromatography has been reexamined following the introduction of on-line cross-correlators. Although there is still little need for these techniques in many chromatographic analyses, a

series of experiments were designed that demonstrate the feasibility of applying cross-correlation to trace gas analysis. The experiments show that correlation offers considerable advantages over the single-sample injection method when minute quantities of a component in a sample gas are being measured. There appears to be much scope for the application of correlation techniques to pollution measurement. (Holoman-Battelle)
W73-08535

EDWARDSIELLA TARDA, A NEW PATHOGEN OF CHANNEL CATFISH (ICTALURUS PUNCTATUS),
Bureau of Sport Fisheries and Wildlife, Stuttgart, Ark. Warmwater Fish Cultural Labs.
F. P. Meyer, and G. L. Bullock.
Applied Microbiology, Vol 25, No 1, p 155-156, January 1973. 3 ref.

Descriptors: *Pathogenic bacteria, *Channel catfish, *Epizootiology, *Enteric bacteria, Fish diseases, Freshwater fish, Fish parasites, Water pollution effects, Toxicity, Coliforms.
Identifiers: *Edwardsiella tarda, *Ictalurus punctatus.

Edwardsiella tarda, an enteric, gram-negative bacterium, causes gas-filled, malodorous lesions in muscle tissue of channel catfish. Incidence and epizootiology of the disease are presented. (Holoman-Battelle)
W73-08536

DEPLETED URANIUM AS CATALYST FOR HYDROCRACKING SHALE OIL,
Bureau of Mines, Laramie, Wyo. Laramie Energy Research Center.
P. L. Cottingham, and L. K. Barker.
Industrial and Engineering Chemistry. Product Research and Development, Vol 12, No 1, p 41-47, March 1973. 11 fig, 1 tab, 11 ref.

Descriptors: Oil, Oil wastes.
Identifiers: *Uranium oxide, *Catalysts, *Gasoline, *Hydrocracking, *Shale oil, Desulfurization, Aromatic hydrocarbons.

Depleted uranium oxide, deposited by the impregnation method on activated alumina and on commercial cobalt molybdate catalyst, was evaluated for catalytic use in hydrocracking crude shale oil at 3000 psig with temperatures from 807 degrees to 1010 degrees F. Adding 10 percent uranium oxide to the catalysts significantly increased the conversion of heavy shale oil to lighter products during hydrocracking. Gasoline produced at the higher conversion levels with the uranium-promoted catalysts were richer in aromatics and had higher octane numbers than those produced with the unpromoted catalysts. Adding uranium to alumina substantially improved the desulfurization activity, but adding uranium to cobalt molybdate did not significantly affect the already high desulfurization activity of this catalyst. (Little-Battelle)
W73-08537

AN APPLICATION OF MULTIVARIATE ANALYSIS TO COMPLEX SAMPLE SURVEY DATA,
North Carolina Univ., Chapel Hill. Dept. of Biostatistics.
For primary bibliographic entry see Field 07B.
W73-08538

PHOTOGRAPHING FUNGI,
Camera M.D. Studios, Inc.
C. M. Weiss.
Industrial Photography, Vol 21, No 11, p 20, 21, 43-45, November 1972.

Descriptors: Methodology, *Pathogenic fungi, *Pollutant identification, *Laboratory equipment, Systematics, Plant morphology, Lighting, Cameras, Color, Films, Cultures.
Identifiers: *Photomicrography, *Photomacrography, Speciation, Color photography, Microsporum canis, Keratinomyces ajelloi, Trichophyton soudanense, Tricophyton gallinae, Microsporum audounii.

Techniques for photographing fungi for species identification include (1) photomacrography used to record gross morphology of fungal cultures, and (2) photomicrography used to record the microscopic morphology of the fungus specimens. Petri dishes (glass or plastic, non-opaque, with shallow sides to minimize shadows) should be used for those cultures to be photographed because they present less difficult problems of lighting, diffraction, and focus than other containers. Badly scratched agar will produce light flare and distortion during photography. Portable equipment can be used but stationary apparatus was used. The equipment consisted of a Polaroid MP-3 camera with 75-, 127-, and 35-mm photomacrography lenses. By using such a fixed setup, those plates requiring serial photography during several weeks of culture growth will not have equipment or photographic variables introduced. Lighting is perhaps the most important photographic requirement and directional lighting is suggested. Since color is a major factor in species identification, several color films were tested for use under controlled photographic conditions and those that work best are listed. (Holoman-Battelle)
W73-08539

NONDESTRUCTIVE INSTRUMENTAL MONITORING OF FUEL OIL FOR VANADIUM, SODIUM, AND SULFUR,
GTE Labs., Inc., Bayside, N.Y.
C. Persiani, and W. D. Shelby.
Environmental Science and Technology, Vol 7, No 2, p 125-127, February 1973. 2 tab, 6 ref.

Descriptors: *Neutron activation analysis, *X-ray fluorescence, *Sulfur, *Sodium, Heavy metals, Oil, Pollutant identification, *Non-destructive tests.
Identifiers: *Fuel oil (No. 6), Detection limits, *Vanadium, Sample preparation, Atomic absorption spectrophotometry, Fuel oil, Oil characterization, Scintillation counting, Precision, Errors.

Samples of No. 6 (bunker) fuel oil were analyzed for sodium and vanadium by neutron activation analysis and for sulfur and vanadium by x-ray fluorescence. Results from the two techniques were compared with those from atomic absorption spectrophotometry. Fuel oil samples for NAA were placed in pressed fit, polyethylene capsules, irradiated, and monitored for gamma rays using a scintillation counter composed of two NaI (Tl) crystals mounted on photomultiplier tubes. Samples for determination of sulfur by x-ray fluorescence were analyzed using the method of successive additions which involves spiking samples with known amounts of amyl ziram and back-calculating original sulfur content. These results were verified using classical methods and served as standards for succeeding samples. For vanadium analysis by x-ray spectrography, samples were ashed, the ash dissolved in HNO_3, dried, triturated, and analyzed. The results show that relative errors for V and Na were 20 and 10 percent, respectively, with NAA. For x-ray spectrographic analysis, the detection limits were 10 and 500 ppm and relative precisions were plus or minus 5.0 and 1.0 percent for V and S, respectively. Interferences with both methods were minor. Since the methods require little sample preparation they should be useful for rapid analysis of fuel oils. (Little-Battelle)
W73-08540

A UNIQUE COMPUTER CENTERED INSTR MENT FOR SIMULTANEOUS ABSORBANC AND FLUORESCENCE MEASUREMENTS,
Michigan State Univ., East Lansing. Dept. Chemistry; and Michigan State Univ., E Lansing. Dept. of Biochemistry.
For primary bibliographic entry see Field 07B.
W73-08541

PERSISTENCE AND REACTIONS OF C-4-CACODYLIC ACID IN SOILS,
Agricultural Research Service, Beltsville, M Agricultural Environmental Quality Inst.
For primary bibliographic entry see Field 05B.
W73-08542

A MALLOMONAS BLOOM IN A BULGARIA MOUNTAIN LAKE,
Copenhagen Univ. (Denmark). Inst. of Pla Anatomy and Cytology.
J. Kristiansen.
Nova Hedwigia, Vol 21, Nos. 2-4, p 877-882, 197 4 fig, 7 ref.

Descriptors: *Electron microscopy, *Polluta identification, Eutrophication, *Phytoplankto Chrysophyta, Aquatic algae, Sampling, Protozoa
Identifiers: *Ultrastructure, *Mallomonas ac roides var. acaroides, *Lake Wapsko (Bulgaria Sample preservation, Sample preparation, *Bu garia, Mallomonas acaroides var. galeata, Flage lates.

Electron microscopic examinations were made samples of Mallomonas taken from Lake Wapsko south of Sofia, Bulgaria. At the time of samplin the water temperature was 10C and the water wa quite turbid due to a mass development of Ma lomonas. The samples were fixed in Bouin's flui and the material for electron microscopy was re peatedly washed in distilled water by means of centrifuge, dried on to carbon-coated grids an shadowcast with chromium. The preparation were examined in a JEM-T 7 electron microscope Other grids with material were coated with gold and examined in a Cambridge Stereoscan Mk II scanning electron microscope. The samples wer shown to contain only one species, Mallomona acaroides var. acaroides (Syn. M. a. var. galeata) and is a confirmation of previously published dat concerning its distribution. (Holoman-Battelle)
W73-08545

CHARACTERIZATION AND IDENTIFICATIO OF SPILLED RESIDUAL FUEL OILS BY GA CHROMATOGRAPHY AND INFRARED SPEC TROPHOTOMETRY,
National Environmental Research Center, Cincinnati, Ohio.
F. K. Kawahara.
Journal of Chromatographic Science, Vol 10, No 10, p 629-636, October 1972. 10 fig, 4 tab, 17 ref.

Descriptors: *Sampling, *Oil spills, *Gas chromatography, Separation techniques, Oil, Density, Pollutant identification.
Identifiers: *Sample preparation, *Oil characterization, *Infrared spectrophotometry, *Aging (Physical), Naphtha, Gasoline, Jet fuel, Kerosene, No. 1 fuel oil, No. 2 fuel oil, No. 5 fuel oil, No. 6 fuel oil, Crude oil, Petroleum jelly, Gas oil, White oil, Motor oil, Asphalt, Grease, Lubricating oil, Preconcentration, Paraffins, Fractionation, Infrared spectra, Chromatograms, Electron capture gas chromatography, Flame ionization gas chromatography.

Analytical methods developed recently at the Analytical Quality Control Laboratory are described for the characterization and identification of heavy residual fuel oil pollutants found in surface waters. Procedures are described for sample collection, extraction, concentration, and analysis. Identity can be made through use of each of

FATTY ACIDS IN SURFACE PARTICULATE MATTER FROM THE NORTH ATLANTIC,

Rhode Island Univ., Kingston. Graduate School of Oceanography.
D. M. Schultz, and J. G. Quinn.

Journal of the Fisheries Research Board of Canada, Vol 29, No 10, p 1482-1486, October 1972. 2 tab, 22 ref.

Descriptors: *Atlantic Ocean, Chemical analysis, *Water analysis, Surface waters, *Pollutant identification, Saline water, Water sampling, Organic acids, Organic matter, Estuaries, Solvent extractions, Separation techniques, Sea water.
Identifiers: *Fatty acids, *Particulate matter, *Narragansett Bay, Sample preparation, Thin layer chromatography, Flame ionization gas chromatography, Gas liquid chromatography.

On R/V Trident Cruise 102 from Reykjavik, Iceland, to Halifax, Nova Scotia (August 1971), samples were taken at seven stations to determine the fatty acid composition of open-ocean surface particulate matter and to compare it with particulate matter in estuarine samples from Narragansett Bay, R. I. Surface samples (20 cm depth) were collected in polyethylene carboys and filtered through Gelman type A glass-fiber filters. The filters were placed in culture tubes and stored at a minus 20C in the ship's freezer for periods up to one month. After transferring the samples to the laboratory for analysis, the following were added (in the order given): an internal standard (17:0), KOH/MeOH, MeOH, and benzene. The samples were saponified and the methyl esters were extracted, and separated and purified using preparative TLC. The esters were then extracted from the stlicic acid and analyzed by GLC. The estuarine samples were collected in glass containers at flood tide and prepared and analyzed as above. Values of total fatty acids (free and esterified) ranged from 4 to 26 micrograms. A high proportion of the fatty acids was long-chain polyunsaturated acids. The low relative abundance of iso and anteiso 15-carbon acids indicated little bacterial contribution to the particulate matter. This fatty acid distribution was in contrast to that found in estuarine particulate matter, which contained only trace amounts of the long-chain polyunsaturated acids and substantial quantities of branched chain acids. (Holoman-Battelle)
W73-08552

CARBON DIOXIDE CONTENT OF THE INTERSTITIAL WATER IN THE SEDIMENT OF GRANE LANGSO, A DANISH LOBELIA LAKE,

Copenhagen Univ. (Denmark). Freshwater Biological Lab.
S. Wium-Andersen, and J. M. Andersen.

Limnology and Oceanography, Vol 17, No 6, p 943-947, November 1972. 4 fig, 1 tab, 9 ref.

Descriptors: *Sediments, *Carbon dioxide, *Connate water, Chemical analysis, *Photosynthesis, Water analysis, Lakes, Sampling.
Identifiers: Sample preparation, *Lobelia dortmanna, *Lake Grane Langso, *Denmark, Litorella, Isoetes.

Sediment samples were collected with Berspex tubes from various depths of Lake Grane Langse for analysis of CO_2 in the interstitial water. Since interstitial water could not be recovered from the sediment, CO_2 was determined directly by a modification of the method of Krogh and Rehberg. In this method, boiling water was added to the sample to drive off free CO_2 and that from HCO_3. The gas was collected in a receiver flask of a still containing a known amount of Ba $(OH)_2$. The amount of $BaCO_3$ formed was calculated by titrating the remaining Ba $(OH)_2$ with HCl. The results showed CO_2 content of the interstitial water to be between 1 and 5 mmole per liter. It is concluded that the sediment must liberate CO_2 to the lake water in amounts sufficient for optimal photosynthesis of Lobelia. Zonation of Isoetes, Littorella, and Lobelia in the lake cannot be correlated with the CO_2 content of the sediment. (Little-Battelle)
W73-08555

IMPROVED TECHNIQUE FOR ANALYSIS OF CARBOHYDRATES IN SEDIMENTS,

Rosenstiel School of Marine and Atmospheric Sciences, Miami, Fla.
S. M. Gerchakov, and P. G. Hatcher.

Limnology and Oceanography, Vol 17, No 6, p 938-943, November 1972. 5 fig, 1 tab, 21 ref.

Descriptors: *Calibrations, *Carbohydrates, *Sediments, *Spectrophotometry, Carbon, Analytical techniques.
Identifiers: Errors, *Total organic carbon, Absorbance, Phenol-sulfuric acid method, Sample preparation, Mangrove Lake, Bermuda.

The phenol-sulfuric acid method for analysis of carbohydrates in sediments has been modified to improve reproductibility and eliminate extraneous color development. Sediment samples were collected from Mangrove Lake, Bermuda for analysis. After filtration, the samples were crushed and dried. Samples to be analyzed were divided into three series. Phenol solution and conc. H_2SO_4 were added to the first; water and conc. H_2SO_4 were added to the second; and phenol and water were added to the third. The absorbance of each solution was determined and the results used to correct for absorbances due to sample/sulfuric acid, sample/phenol, and phenol/sulfuric acid interactions. Carbohydrate content, as glucose, was then determined using a calibration curve constructed using glucose standards. Total organic carbon was also determined by dry combustion of the samples and absorption of the evolved CO_2. Statistical treatment of the absorbances from the three samples indicated that absorbance from sample/phenol or glucose/phenol interactions was insignificant with this sample but may have to be considered with others. Absorbance due to phenol/sulfuric acid interaction can generally be neglected. Correction for absorbance due to sediment/sulfuric acid is necessary with this method. Carbohydrate values obtained with the modified method were generally lower than those obtained with the unmodified method. (Little-Battelle)
W73-08556

A COMPARISON AT SEA OF MANUAL AND AUTOANALYZER ANALYSES OF PHOSPHATE, NITRATE, AND SILICATE,

Oregon State Univ., Corvallis. School of Oceanography.
S. W. Hager, E. L. Atlas, L. I. Gordon, A. W. Mantyla, and P. K. Park.

Limnology and Oceanography, Vol 17, No 6, p 931-937, November 1972. 3 fig, 2 tab, 17 ref.

Descriptors: *Phosphates, *Nitrates, *Silicates, *Colorimetry, Calibrations, *Sea water, Nutrients, Sampling.
Identifiers: Precision, On board analysis, AutoAnalyzer.

Data are presented from a comparison made at sea of automated and manual methods of analysis for phosphate, nitrate plus nitrite, and silicate. An AutoAnalyzer was used for the automated analyses and colorimetric methods, which are described, were used in the manual analyses. Samples were collected in surface buckets and in NIO (Nat. Inst. Oceanogr.) bottles from standard hydrographic depths. Statistical treatment of the results showed that both methods give comparable precision, plus or minus 1 percent, and no significant systematic discrepancies exist between the two technologies for phosphate and silicate analyses. A small discrepancy was found in the nitrate work. (Little-Battelle)
W73-08557

DECOMPOSITION OF DISSOLVED ORGANIC CARBON AND NITROGEN COMPOUNDS FROM LEAVES IN AN EXPERIMENTAL HARD-WATER STREAM,
Michigan State Univ., Hickory Corners. W. K. Kellogg Biological Station.
For primary bibliographic entry see Field 05B.
W73-08558

A METHOD FOR THE IN SITU STUDY OF ZOOPLANKTON GRAZING EFFECTS ON ALGAL SPECIES COMPOSITION AND STANDING CROP,
Yale Univ., New Haven, Conn. Dept. of Biology.
K. G. Porter.
Limnology and Oceanography, Vol 17, No 6, p 913-917, November 1972. 1 fig, 3 tab, 19 ref.

Descriptors: *Grazing, *Zooplankton, *Standing crops, Plant populations, *Phytoplankton, Cyanophyta, Diatoms, Chlorophyta, Chrysophyta, Dinoflagellates, Pyrrophyta, On-site tests, Protozoa, Equipment, Copepods, Chrysophyta.
Identifiers: Diaptomus minutus, Epischura lacustris, Daphnia galeata mendotae, Cyclops scutifer, Macroinvertebrates, Desmids, Flagellates.

A method for in situ study of the effects of grazing on the species composition and standing crop of a natural phytoplankton assemblage is described. Polyethylene enclosures 0.5 cu m in volume are filled with whole lake water or water filtered through 125-micron (No. 10) mesh to remove the major grazers. After 4 days of incubation, contents are sampled and cells counted by species. The bagged whole lake water is also compared with natural lake water. The total number of small algal cells (2-30 microns), small flagellates, large dinoflagellates, large chrysophytes, and ciliates are higher and gelatinous greens are lower in the absence of grazers. There is no detectable experimental effect on the total number of large cells (greater than 30 microns), blue-greens, desmids, small chrysophytes, diatoms, and nonblue-green filaments. (Little-Battelle)
W73-08559

ATP CONTENT OF CALANUS FINMARCHICUS,
Dalhousie Univ., Halifax (Nova Scotia). Inst. of Oceanography.
N. Balch.
Limnology and Oceanography, Vol 17, No 6, pp 906-908, November 1972. 1 tab, 6 ref.

Descriptors: *Biomass, *Food abundance, Carbon, Nitrogen, Copepods, Invertebrates, Statistical methods.
Identifiers: *Calanus finmarchicus, *Adenosine triphosphate, *Organic carbon, Sample preparation.

Measurements of adenosine triphosphate (ATP), carbon, and nitrogen were made on Calanus finmarchicus held in a food-free environment in the laboratory. Over a 23-day starvation period there were significant reductions of absolute amounts of carbon, nitrogen, and ATP, but no significant changes in the relationship between ATP and body carbon. Relative ATP levels in organisms remain sufficiently constant over a range of physiological conditions so as to make them good indicators of biomass. ATP:carbon ratios in lipid-storing copepods are also considered. (Little-Battelle)
W73-08560

TRACE ELEMENTS IN CLAMS, MUSSELS, AND SHRIMP,
Institut Royal des Sciences Naturelles de Belgique, Brussels.
For primary bibliographic entry see Field 05B.
W73-08561

A TECHINQUE FOR THE ESTIMATION OF INDICES OF REFRACTION OF MARINE PHYTOPLANKTERS,
Oregon State Univ., Corvallis. School of Oceanography.
K. L. Carder, R. D. Tomlinson, and G. F. Beardsley, Jr.
Limnology and Oceanography, Vol 17, No 6, p 833-839, November 1972. 4 fig, 2 tab, 24 ref.

Descriptors: *Phytoplankton, *Cultures, *Distribution patterns, *Growth rates, Sea water, Sampling, Chrysophyta, Particle size.
Identifiers: *Isochrysis galbana, *Index of refraction, *Optical systems, Coulter counter.

Measurements of light scattering and particle size distribution were performed on a growing culture of the unicellular phytoplankter Isochrysis galbana. The culture represented a narrow, polydisperse distribution of nearly spherical particles. Assuming the culture to be homogeneous in refractive index, a technique was developed that provided an estimate of the index of refraction of the culture for the wavelengths 5,460 Angstrom and 5,780 Angstrom. The relative index of refraction (relative to that of seawater) of the culture varied with time from 1.026 to 1.036 over a 12-day sampling period and seemed to be related to changes in the surface area:volume ratio of the cells. The indices of refraction determined using green light (5,460 Angstrom) corresponded closely with those found using yellow (5,780 Angstrom), with slight fluctuations perhaps due to changes in cells pigmentation ratios. (Little-Battelle)
W73-08564

USE OF ISOTOPIC ABUNDANCE RATIOS IN IDENTIFICATION OF POLYCHLORINATED BIPHENYLS BY MASS SPECTROMETRY,
Stanford Univ., Pacific Grove, Calif. Hopkins Marine Station.
J. W. Rote, and W. J. Morris.
Journal of the Association of Official Analytical Chemists, Vol 56, No 1, p 188-199, January 1973. 10 fig, 3 tab, 34 ref.

Descriptors: *Polychlorinated biphenyls, *Pollutant identification, *Mass spectrometry, Chemical analysis, *Industrial wastes, Gas chromatography.
Identifiers: *GC-Mass spectrometry, *Isotopic abundance ratios, Aroclor 1232, Aroclor 1254, Aroclor 1268, Isomers, Polychlorinated terphenyls, Chlorinated naphthalenes.

Polychlorinated biphenyls are industrial compounds which are being detected throughout the global ecosystem. Because of the presence of other residues, combined gas chromatography-mass spectrometry must be used for their confirmation. Using isotope abundance ratios, the theoretical probability of the occurrence of ions of different masses in the molecular cluster has been calculated for the polychlorinated biphenyls, polychlorinated terphenyls, and chlorinated naphthalenes. Mass spectra of mono- through decachlorobiphenyl have been taken with a unit resolution, computer-controlled GC-MS system and parent ion clusters were matched with the theoretical isotopic patterns. This method provides the unambiguous identification of polychlorinated biphenyls in the presence of other substances. (Holoman-Battelle)
W73-08565

GAS-LIQUID CHROMATOGRAPHIC SEPARATION OF N-ACETYL-N-BUTYL ESTERS OF AMINO ACIDS,
Agricultural Research Service, Beltsville, Md. Agricultural Marketing Research Inst.
P. G. Vincent, and J. Kirksey.
Journal of the Association of Official Analytical Chemists, Vol 56, No 1, p 158-161, January 1973. 1 fig, 1 tab, 13 ref.

Descriptors: *Separation techniques, *Amino acids.
Identifiers: *Gas liquid chromatography, Proline, Threonine, Homoserine, 4-aminobutyrate, Methionine, Citrulline, Arginine, Aspartate, Phenylalanine, Ornithine, Glutamine, Lysine, Histidine, Tyrosine, Tryptophan, *Esters, N-acetyl-n-butyl esters, Chromatography, Retention time, Elution.

Gas-liquid partition chromatography of a number of N-acetyl-n-butyl esters of amino acids has been investigated. Results obtained with a column packed with methyl silicone grease (DC-200, 12,500 cst viscosity) were better than those reported for Carbowax 1540 or hydrogenated vegetable oil. Amino acid derivatives were distinctly resolved on methyl silicone grease columns. Preliminary determinations were obtained for synthetic mixtures of the acids. The method should prove a useful adjunct to current techniques for the separation and quantitative measurement of amino acids. (Little-Battelle)
W73-08566

COLLABORATIVE STUDY OF THE CATION EXCHANGE CAPACITY OF PEAT MATERIALS,
Michigan Dept. of Agriculture, East Lansing.
For primary bibliographic entry see Field 05B.
W73-08567

ON-LINE COMPUTER SYSTEM FOR AUTOMATIC ANALYZERS,
Food and Drug Administration, Chicago, Ill.
For primary bibliographic entry see Field 07C.
W73-08568

APPLICATION OF A LABORATORY ANALOG-DIGITAL COMPUTER SYSTEM TO DATA ACQUISITION AND REDUCTION FOR QUANTITATIVE ANALYSES,
Department of Agriculture, Agassiz (British Columbia). Research Station.
For primary bibliographic entry see Field 07C.
W73-08569

COLLABORATIVE STUDY OF A SIMPLIFIED HALPHEN PROCEDURE FOR THE QUANTITATIVE DETERMINATION OF LOW LEVELS OF CYCLOPROPENE FATTY ACIDS,
Food and Drug Administration, Washington, D.C. Div. of Chemistry and Physics.
E. C. Coleman.
Journal of the Association of Official Analytical Chemists, Vol 56, No 1, p 82-85, January 1973. 3 tab, 16 ref.

Descriptors: *Pollutant identification, Organic acids, Methodology, Chemical analysis.
Identifiers: *Collaborative studies, Interlaboratory studies, *Corn oil, *Cottonseed oil, Method evaluation, Fatty acids, *Cycloprene fatty acids, Precision, Halphen procedure, Quantitative analysis.

A collaborative study was conducted on a simplified Halphen procedure for the quantitative determination of low levels of cyclopropene fatty acids. Twelve laboratories (some of these participated in a precollaborative study of this procedure) analyzed 9 samples consisting of mixtures of transesterified corn oil and cottonseed oils, whose cyclopropene content ranged from 0.00 to 0.19 percent. The amount of cyclopropene fatty acids in the samples was estimated in the Associate Referee's laboratory from HBr titration data. A statistical evaluation of the collaborative data indicated good precision and accuracy. Evidence for a significant between-laboratory systematic error could not be found. It is recommended that work be continued on development of a suitable calibration procedure and/or acquisition of a reference standard. (Holoman-Battelle)

Identifiers: *Methyl isothiocyanate, *Vorlex, Flame photometric gas chromatography, Conductive gas chromatography, Detection limits, Recovery, Sample preparation, Gas liquid chromatography, Organic solvents, Mineral soils.

Residues of methyl isothiocyanate, the active ingredient (20 percent) in the soil fumigant Vorlex, in soils were extracted with diethyl ether containing 5 percent ethanol, concentrated in a Kuderna-Danish concentrator, and subjected to gas-liquid chromatography without further cleanup. A 6 percent Carbowax 20M column was used in conjunction with either conductivity (nitrogen mode) or flame photometric (sulfur mode) detection systems. Two types of soils, mineral and organic (muck), were used for recovery tests. The soils were air-dried at room temperature, fortified with microgram quantities of CH3NCS, and mixed thoroughly for 60 minutes. Extractions were made immediately and at 3, 7, 14, and 28 days pot-fortification. The observed limits of detection of methyl isothiocyanate were approximately 4.5 ng with the flame photometric detector and 5 ng with the conductivity detector. Recoveries from fortified soils ranged from 80-86 percent using the Kuderna-Danish concentrator. The method is rapid and capable of determining methyl isothiocyanate residues in soils at a level as low as 0.01 ppm. (Holoman-Battelle)
W73-08573

EVALUATION OF CHEMICAL CONFIRMATORY TESTS FOR ANALYSIS OF DIELDRIN RESIDUES,

Department of Agriculture, Ottawa (Ontario). Plant Products Div.
R. B. Maybury, and W. P. Cochrane.
Journal of the Association of Official Analytical Chemists, Vol 56, No 1, p 36-40, January, 1973. 3 fig, 17 ref.

Descriptors: *Pesticide residues, *Dieldrin, *Chemical reactions, *Pollutant identification, Laboratory tests, Evaluation, Chemical analysis, Chlorinated hydrocarbon pesticides, Solvent extractions, Separation techniques, Methodology, Insecticides.
Identifiers: Reagents, Sample preparation, Electron capture gas chromatography, Cleanup, Organic solvents, Gas liquid chromatography, Sensitivity, Derivatization.

Eight of the more commonly used reagents for the confirmation of dieldrin at the residue level were compared for several substrates. After an aliquot of the standard or sample extract was placed in a glass-stoppered centrifuge tube, the solvent was evaporated. The reagent was added, and the tube placed in boiling water bath for 30 min. After cooling, the contents were diluted with distilled H20, mixed, and shaken upon the addition of hexane. The hexane layer was dried and injected into a gas chromatograph. Dry (animal feeds) samples were extracted with hexane, and cleaned up on a column as above. In some instances derivatives were isolated and identified. Two reagents, aqueous HBr solution and BC13/2-chloroethanol, were especially useful because of their application and sensitivity, i.e., 0.0003 ppm for a 10 g dry sample and 0.001 ppm for a 25 g wet sample. The ZnC12/HCl reagent was a practical alternative to the BC13/2-chloroethanol reagent. The HBr/Ac20 reagent gave a more positive confirmation at a lower sensitivity. (Holoman-Battelle)
W73-08574

FLUOROMETRIC DETERMINATION OF OXYTETRACYCLINE IN PREMIXES,

Rutgers - The State Univ., New Brunswick, N. J. Coll. of Agriculture and Environmental Science.
S. E. Katz, and C. A. Fassbender.
Journal of the Association of Official Analytical Chemists, Vol 56, No 1, p 17-19, January, 1973. 3 tab, 10 ref.

Descriptors: *Antibiotics (Pesticides), *Chemical analysis, *Aqueous solutions, *Pollutant identification, Methodology, Fluorometry, Fluorescence, Assay.
Identifiers: *Oxytetracycline, Chemical interference, Recovery, *Premixes, Chlortetracycline, Sample preparation, Sensitivity.

A simple fluorometric procedure is presented for the determination of oxytetracycline in which a highly sensitive fluorometer and ultrapure deionized water are used to eliminate background fluorescence. Standard solutions are mixed with 1 N NH4OH, allowed to stand for 15 min and the fluorescence intensity is measured. Premixes are mixed with an extracting solution for 30 min by shaking; the extract is settled, filtered and diluted with 1 NH4OH for fluorescence intensity measurements. Recoveries from fortified extracts of commercial premixes ranged from 95.0 to 102.5 percent; the ratio of results of microbiological to fluorometric methods ranged from 0.97 to 1.05 percent. The procedure is reasonably specific for oxytetracycline in relationship to interference from chlortetracycline because there is a significant difference between excitation and emission wave lengths for these compounds: 390 and 520 nm for oxytetracycline vs. 350 and 420 nm for chlortetracycline. The main advantage of the fluorometric method over the microbiological method is that the fluorometric method requires less time and can be used effectively for quality assurance and in regulatory laboratories. (Holoman-Battelle)
W73-08575

GAS CHROMATOGRAPHIC DETERMINATION OF TRIARIMOL IN FORMULATIONS AND TECHNICAL MATERIAL,

Lilly (Eli) and Co., Indianapolis, Ind.
R. Frank, O. D. Decker, and E. W. Day, Jr.
Journal of the Association of Official Analytical Chemists, Vol 56, No 1, p 11-13, January, 1973. 2 fig, 1 tab.

Descriptors: *Fungicides, *Assay, Chlorinated hydrocarbon pesticides, Chemical analysis, *Gas chromatography, Diazinon.
Identifiers: *Triarimol, Formulations, Chemical interference, Flame ionization gas chromatography, Sample preparation, Alpha- (2 4-dichlorophenyl)-alpha-phenyl-5-pyrimidine-methanol, Precision, Captan, Disulfoton, Dicofol, Chlorpyrifos, Ethion, Malathion, Methoxychlor, Carbaryl, Maneb.

The triarimol content in formulations and technical materials is determined by flame ionization gas-liquid chromatography after the sample is dissolved in or extracted with chloroform. Pesticides commonly occurring with triarimol do not interfere in this method. Replicate injections of emulsifiable concentrate and wettable powder samples showed good precision. The described procedure has been successfully used for the determination of triarimol in experimental combination formulations containing one of the following pesticides: captan, diazinon, disulfoton, chlorpyrifos, ethion, dicofol, malathion, methoxychlor, carbaryl, and maneb. (Holoman-Battelle)
W73-08576

DEVELOPMENT OF SIMULIUM (PSILOZIA) VITTATUM ZETT (DIPTERA: SIMULIIDAE) FROM LARVAE TO ADULTS AT INCREMENTS FROM 17.0 TO 27.0 C,

Batelle-Pacific Northwest Labs., Richland Wash.
For primary bibliographic entry see Field 05C.
W73-08577

CONFIRMATION OF PESTICIDE RESIDUE IDENTITY. IV. DERIVATIVE FORMATION IN SOLID MATRIX FOR THE CONFIRMATION

OF ALPHA- AND BETA-ENDOSULFAN BY GAS CHROMATOGRAPHY,

Department of the Environment, Ottawa (Ontario). Inland Waters Branch.
A. S. Y. Chau, and K. Terry.
Journal of the Association of Official Analytical Chemists, Vol 55, No 6, p 1228-1231, November, 1972. 4 fig, 8 ref.

Descriptors: *Pesticide residues, *Chlorinated hydrocarbon pesticides, *Pollutant identification, Methodology, Insecticides, Chemical analysis, *Gas chromatography, Solvent extractions.
Identifiers: *Alpha-endosulfan, *Beta-endosulfan, Derivatization, Electron capture gas chromatography, Acetylation, Isomers, Endosulfan diacetate, Gas liquid chromatography, Sample preparation, Organic solvents.

The new solid matrix derivation procedure has now been applied to the confirmation of endosulfan residue identity. This procedure involves the conversion of endosulfans to the same diacetate in alumina impregnated with H_2SO_4 and acetic anhydride. The concentrated sample extract or standard solution is applied to the surface of the solid matrix in the microcolumn. The column is heated for two hours and the eluted with benzene. The eluate is then examined, with and without concentration, by electroncapture gas-liquid chromatography. Comparable yield of the diacetate was obtained from the present approach and from a previously reported acetylation method. Of the various experimental conditions investigated, the best conditions included derivatization for 2 hr at 100C in solid matrix of alumina-acetic anhydride-H_2SO_4 (50 plus 10 plus 5) (W/V/V). The advantage of the present method over the previously reported procedure is discussed, and the applicability of this method to the confirmation of endosulfan in fish and water is illustrated. See also W73-08579 (Holoman-Battelle)
W73-08578

CONFIRMATION OF PESTICIDE RESIDUE IDENTITY. V. ALTERNATIVE PROCEDURE FOR DERIVATIVE FORMATION IN SOLID MATRIX FOR THE CONFIRMATION ALPHA-AND BETA-ENDOSULFAN BY GAS CHROMATOGRAPHY,

Department of the Environment, Burlington (Ontario). Centre for Inland Waters.
A. S. Y. Chau.
Journal of the Association of Official Analytical Chemists, Vol 55, No 6, p 1232-1238, November, 1972. 6 fig, 2 tab, 32 ref.

Descriptors: *Pesticide residues, *Chlorinated hydrocarbon pesticides, *Pollutant identification, Insecticides, Methodology, Chemical analysis, Fish, Sediments, Water analysis, Solvent extractions.
Identifiers: *Alpha-endosulfan, *Beta-endosulfan, Electron capture gas chromatography, Detection limits, Derivatization, Endosulfan ether, Isomers, Sample preparation, Organic solvents.

A quick and sensitive new method has been developed for the confirmation of alpha- and beta-endosulfan by conversion to the same ether in a solid matrix. It seems by far to be the most sensitive method reported to date for positive identification of alpha- and beta-endosulfan. As little as 0.005 ppm of the parent compounds in a 10 g fish or sediment extract and 0.003 ppb in a 2 l water extract can be routinely confirmed. (Holoman-Battelle)
W73-08579

DETERMINATION OF MATACIL AND ZECTRAN BY FLUORIGENIC LABELING, THIN LAYER CHROMATOGRAPHY, AND IN SITU FLUORIMETRY,

Dalhousie Univ., Halifax (Nova Scotia). Dept. of Chemistry.
R. W. Frei, and J. F. Lawrence.
Journal of the Association of Official Analytical Chemists, Vol 55, No 6, p 1259-1264, November, 1972. 6 fig, 21 ref.

Descriptors: *Carbamate pesticides, *Fluorometry, Pollutant identification, Methodology, Chemical analysis, Fluorescence, Nuclear magnetic resonance, Insecticides, Separation techniques, Chemical reactions, Solvent extractions.
Identifiers: *Thin layer chromatography, *Matacil, *Zectran, *Fluorigenic labeling technique, Infrared spectrophotometry, Spectrofluorimetry, Reproducibility, Detection limits, Dansyl chloride, NMR spectra, Infrared spectra, 4-dimethylamino-m-tolyl N-methylcarbamate, 4-dimethylamino-3 5-xylyl N-methylcarbamate, 4-chloro-7-nitrobenzo-2 1 3-oxadiazole, 1-dimethylamino-naphthalene-s-sulfonyl chloride, Organic solvents.

The fluorigenic labeling of Matacil (4-dimethylamino-m-tolyl N-methylcarbamate) and Zectran (4-dimethylamino-3, 5-xylyl N-methylcarbamate) with dansyl chloride (1-dimethylamino-naphthalene-5-sulfonyl chloride) results in 3 fluorescent derivatives, and labeling with NBD-Cl (4-chloro-7-nitrobenzo-2, 1, 3-oxadiazole) produces 2 fluorescent derivatives for each carbamate, all of which can be separated by thin layer chromatography (TLC). These derivatives are identified by nuclear magnetic resonance, infrared, and fluorescence spectroscopy, aided by TLC data. The carbamates are hydrolyzed in dilute base and the resulting amine or phenol hydrolysis products react with the labeling reagents. The derivatives are analyzed by TLC and in situ fluorimetry with a spectrophotometer in the fluorescence mode and a spectrophotofluorometer with the thin layer scanning accessory. Reactions, fluorescence phenomena, and chromatographic properties of the derivatives are investigated for evaluation of the method as a quantitative technique. A reproducibility of 3-5 percent relative standard deviation can be expected in the concentration range from 15 to 300 ng/spot for derivatives of the 2 labeling procedures. The dansyl derivatives are instrumentally detectable as low as 1 ng/spot while the NBD derivatives may be detected at concentrations of less than 0.5 mg/spot. (Holoman-Battelle)
W73-08580

COLORIMETRIC DETERMINATION OF PHENYLAMIDE PESTICIDES IN NATURAL WATERS,

National Research Centre, Cairo (Egypt). Water Pollution Dept.
M. A. El-Dib, and O. A. Aly.
Journal of the Association of Official Analytical Chemists, Vol 55, No 6, p 1276-1279, November, 1972. 4 tab, 23 ref.

Descriptors: *Water analysis, *Pollutant identification, *Colorimetry, Methodology, Carbamate pesticides, Urea pesticides, Chemical analysis, Color reactions, Chlorinated hydrocarbon pesticides, Efficiencies, *Organic pesticides, Solvent extractions, Separation techniques, Spectrophotometry, Monuron, Herbicides, Insecticides, Molluscicides.
Identifiers: Natural waters, *Phenylamide pesticides, Aniline derivatives, Sample preparation, Recovery, Precision, Sensitivity, IPC, CIPC, Diuron, Linuron, Neburon, Propanil, Bayluscide, Vitavax, Karsil, Dicryl, Phenylurea pesticides, Isopropyl N-phenylcarbamate, Isopropyl N- (3-chlorophenyl) carbamate, 3- (4-Chlorophenyl)-1 1-dimethylurea, 3- (3 4-Dichlorophenyl)-1 1-dimethylurea, 3- (3 4-Dichlorophenyl)-1-methoxy-1-methylurea.

Phenylamide pesticides include a great variety of compounds such as N-phenylcarbamates, phenylureas, and anilides. A simple, rapid, and sensitive colorimetric method for their determination in natural waters has been proposed. Phenylamides are rapidly hydrolyzed in acid medium at elevated temperature (150-155 degrees C), yielding the corresponding anilines. The latter compounds are diazotized and coupled with 1-naphthol to yield intensely colored azo dyes. Standard and recovery solutions of specific phenylamides were used in testing the procedure. Aliquots of the solutions were acidified with H_2SO_4, extracted with $CHCl_3$ and evaporated just to dryness. The solution was then refluxed, mixed respectively with $NaNO_2$ solution, sulfanic acid, 1-naphthol, and finally NaOH for maximum color development. Absorbance was read against a reagent blank. Efficiency of the hydrolytic procedure was checked by comparing 100 micrograms of each pesticide with the corresponding theoretical amounts of anilines liberated by complete hydrolysis. Recoveries of the phenylamides from tap and raw river waters ranged from 96 to 100 percent. Using the proposed method, extraction and subsequent determination of the pesticides require about 40 minutes. Both phenylamides and aniline derivatives can be determined in a sample in the presence of each other. The method is sensitive to 0.02 mg/1; coefficients of variation range between 6.0 and 0.2, according to the amount of phenylamide pesticide being measured. (Holoman-Battelle)
W73-08581

TECHNIQUES FOR SAMPLING SALT MARSH BENTHOS AND BURROWS,

Georgia Univ., Athens. Dept. of Geology.
R. W. Frey, P. B. Basan, and R. M. Scott.
The American Midland Naturalist, Vol 89, No 1, p 228-234, January, 1973. 2 fig, 9 ref.

Descriptors: *Sampling, *Benthic fauna, Separation techniques, *Muds, Laboratory equipment, Sediments, Cores, Bottom sampling, Salt marshes, Crabs.
Identifiers: *Burrow casts, *Staining, Sieving, Burrows, Polychaetes, Grass roots, Can cores, Quadrat frames.

Several methods have been developed to reduce the difficulty of sampling salt marsh epibenthos and endobenthos where grass root masses occur. Mobile animals such as crabs were collected using high-walled quadrat frames. These frames are placed while tidewater stands on the marsh. When the water ebbs away, the crabs return to the surface and can be collected. Grass is cropped and removed to eliminate obstructions while collecting the samples. Other studies showed that about 80 percent of the crabs present can be collected this way. Can cores 14x22x33 cm high were used to sample the endobenthos by cutting grass roots around the can and digging the sample out. Time for each sample collected was 2-4 minutes. Organisms were separated from the core samples containing dense grass root by sieving-staining-flotation equipment constructed from a table containing a 1 mm sieve and two No. 3 washtubs, one fitted with a sieve bottom and one fitted with a drain hole. Casts of burrows have been successfully made using polyester plastic and hardener. (Little-Battelle)
W73-08582

A FURTHER CONTRIBUTION TO THE DIATOM FLORA OF SEWAGE ENRICHED WATERS IN SOUTHERN AFRICA,

National Inst. for Water Research, Pretoria (South Africa).
For primary bibliographic entry see Field 05C.
W73-08586

BACTERIOPHAGES RECOVERED FROM SEPTAGE,

Pennsylvania Water and Gas Co., Scranton.
For primary bibliographic entry see Field 05B.
W73-08587

Samples of alfalfa, clovers, and grasses were collected, pooled, chopped, dry ashed or wet digested, analyzed by atomic absorption, and the results subjected to statistical analysis to determine whether the ashing procedures affected results. Twenty-five 2-gram samples were placed in porcelain crucibles and dry ashed at 550 C for 2-3 hours in a muffle furnace. The ashed samples were then boiled in 1 N HCl and diluted with distilled water. The same number of samples were wet ashed by letting them stand overnight in concentrated HNO3. The solution was then boiled, cooled, and HC104 added. After heating, the samples were diluted with distilled water. The prepared samples were analyzed for iron, sodium, potassium, calcium, magnesium, copper, manganese, and zinc by AA. The statistical analysis showed nonsignificant differences in results except for iron concentration in alfalfa and grass. Iron appeared to be higher with dry ashing. A second series of tests using a quartz shield to protect the sample from contamination inside the furnace gave results which were not significantly different. Dry ashing appeared to give better precision, and since it is more rapid than wet digestion, it is suggested that dry ashing may be a more desirable process for sample preparation. (Little-Battelle)
W73-08591

THE GENERATION AND DETERMINATION OF COVALENT HYDRIDES BY ATOMIC ABSORPTION,

Kennecott Copper Corp., Lexington, Mass. Ledgemont Lab.
E. N. Pollock, and S. J. West.
Atomic Absorption Newsletter, Vol 12, No 1, p 6-8, January-February 1973. 3 fig, 2 tab, 10 ref.

Descriptors: *Aqueous solutions, *Heavy metals, Tin, Lead, Laboratory equipment.
Identifiers: *Atomic absorption spectrophotometry, Arsenic, Selenium, Bismuth, Germanium, Antimony, Tellurium, *Hydrides, Sample preparation, Detection limits, Silicon, *Covalent hydrides.

The use of TiC13 with magnesium metal or the use of NaBH4 as reductants and sources of nascent hydrogen for the generation of volatile hydrides from the covalent metals has extended this atomic absorption spectroscopy technique to include Bi, Sb, Te and Ge as well as As and Se. The development of an apparatus compatible with the change in chemical reactions has led to simple precise methods for the determination of these six elements. Further changes in reagents, technique or apparatus may extend the usefulness of this method to the determination of Sn, Si and Pb. The data showed excellent precision to be obtainable in the TiC13-Mg method for determination of As, Se, Sb, Bi, Te at the submicrogram level. Ge determination using NaBH4 was less sensitive and showed poorer precision. Detection limits are shown to be much better with the hydride method than with conventional AA. The method was unsuccessful with Sn, Pb, and Si. (Little-Battelle)
W73-08592

DETERMINATION OF CADMIUM IN BLOOD BY A DELVES CUP TECHNIQUE,

Oklahoma Univ., Oklahoma City. Dept. of Environmental Health.
R. D. Ediger, and R. L. Coleman.
Atomic Absorption Newsletter, Vol 12, No 1, p 3-6, January-February 1973. 2 fig, 3 tab, 6 ref.

Descriptors: *Cadmium, *Heavy metals, Methodology, Chemical analysis, *Pollutant identification.
Identifiers: *Delves cup method, *Atomic absorption spectrophotometry, *Blood, Biological samples, Absorbance, Detection limits, Body fluids.

A simple, rapid method is described for the determination of cadmium in blood utilizing the Delves cup apparatus. Heparinized blood is pipetted into the Delves cups after passage through the air-acetylene flame in order to decontaminate the surface. The cups were placed in their tray, dried at 150 C for 1 min, individually mounted on the Delves cup apparatus, and pushed almost to the center of the flame. Upon cessation of combustion, the cup was immediately pushed to the center of the flame and the Cd absorption peak was recorded. The Deuterium Background Corrector was used to compensate for nonspecific absorption. Standardization was by the method of additions on a normal blood sample spiked with cadmium to yield added concentrations of 0.50, 1.00 and 1.50 micrograms Cd/100 ml. With this procedure, less than 0.02 microgram Cd/100 ml of blood can be detected using a 10-microliter sample. The technique yields values comparable to data obtained with an extraction procedure utilizing the graphite furnace. The effects of cup positioning on absorbance and the use of background correction are discussed. (Holoman-Battelle)
W73-08593

SPECTROPHOTOMETRIC DETERMINATION OF URANIUM (VI) WITH CHROMAZUROL S AND CETYLPYRIDINIUM BROMIDE,

National Inst. for Scientific and Industrial Research, Selangor (Malaysia).
C. L. Leong.
Analytical Chemistry, Vol 45, No 1, p 201-203, January 1973. 3 fig, 5 ref.

Descriptors: *Color reactions, Aqueous solutions, *Spectrophotometry, *Heavy metals, *Pollutant identification, Chemical reactions, Separation techniques, Cations, Hydrogen ion concentration, Solvent extractions, Alkaline earth metals, Calcium, Lead, Manganese, Cobalt, Magnesium, Nickel, Zinc, Titanium, Aluminum, Chromium, Beryllium, Iron, Molybdenum, Methodology.
Identifiers: *Chromazurol, S, *Cetylpyridinium bromide, *Uranium, Absorbance, Ionic interference, Reproducibility, Molar absorptivity, Sensitivity, Chromogenic reagents, Rare earth elements, Organic solvents, Chemical interference, Chloroform, Carbon tetrachloride, Ether, Nitrobenzene, Nitromethane, Isobutyl ketone, Isoamyl alcohol, Benzyl alcohol, Tungsten, Lanthanum, Neodymium, Samarium, Gadolinium, Europium, Bismuth, Ytterbium, Dysprosium, Cerium, Thorium.

Since the uranyl ion has been found to form a soluble blue complex with chromazurol S and cetylpyridinium bromide (CPB), a study has been made of the color reaction with reference to the development of a suitable spectrophotometric method for determining U (VI). A study of the optimum conditions for color development yielded the following: (1) pH 4.8 was chosen for all absorption measurements. (2) At an average temperature of 20 C, maximum absorbance was obtained in 15 minutes and absorbance remained constant for at least 24 hours. (3) The complex was not extractable by chloroform, carbon tetrachloride, ether, nitrobenzene, nitromethane, isobutyl ketone, isoamyl alcohol, and benzyl alcohol. (4) The calibration graph obeyed Beer's law for 0-100 micrograms of U (VI)/50 ml of solution. (5) The color reaction has a molar absorptivity of 99,000 at 625 nm and a relative standard deviation of 3 percent. (6) A 100-fold weight excess of Al (III), Cr (VI), Be (II), Fe (III), Bi (III), Mo (VI), Yb (III), Dy (III), and Ce (III) gave serious interferences, while a 2-fold weight excess of Th (IV) gave interference. Preliminary separation of U (VI) by such methods as ion exchange will be necessary before determination. (Holoman-Battelle)
W73-08594

STABILITY OF METAL DITHIOCARBAMATE COMPLEXES,

Carleton Univ., Ottawa (Ontario). Dept. of Chemistry.
R. R. Scharfe, V. S. Sastri, and C. L. Chakrabarti.
Analytical Chemistry, Vol 45, No 2, p 413-415, February 1973. 1 tab, 13 ref.

Descriptors: *Lead, *Zinc, *Nickel, *Polarographic analysis, *Solvent extractions, Stability, Aqueous solutions, Nitrilotriacetic acid, Chemical reactions, Cations, Heavy metals, Separation techniques, Organic acids, Chelation.
Identifiers: *Dithiocarbamic acids, *Metal complexes, *Stability constants, Ligands, EDTA, Pyrrolidine dithiocarbamate, Diethyl dithiocarbamate, Dimethyl dithiocarbamate, Hexamethylene dithiocarbamate, Piperazine dithiocarbamate.

The objectives were to: (1) to determine the stability constants of complexes of some divalent metal ions with a variety of N-substituted dithiocarbamic acids; and (2) determine the effect of various substituents at the nitrogen atom in disubstituted dithiocarbamic acids on the magnitude of stability constants. Aqueous solutions of the dithiocarbamic acids prepared as the Na salt, of the metal ion (Pb, Ni, Zn), and of the competing ligand (NTA or EDTA) were studied using solvent extraction and polarographic methods. There was fair agreement between the values of the stability constants obtained by those methods; the differences were possibly due to the different ionic strengths, 0.01 and 1.00, respectively, employed in the 2 methods. In the case of Pb, there was a linear relationship between the stability constant values and the pK (basicity) of three dithiocarbamic acids. In the case of Zn and Ni, there was no regular trend in the stability constants with the increasing basicity of dithiocarbamic acids. The order of the stability constants for Zn, Ni, and Pb was hexamethylene DTC greater than pipDTC greater the PyrrDTC. Pyrrolidine DTC forms complexes of the same degree of stability as the widely used diethyl DTC. However, at low pH of 1.0, pyrrolidine DTC is more stable than diethl DTC. Thus, the substitution of pyrrolidine DTC in many existing procedures which call for diethyl DTC will add an additional element of flexibility to many analytical procedures. (Holoman-Battelle)
W73-08595

AUTOMATED KJELDAHL NITROGEN DETERMINATION--A COMPREHENSIVE METHOD FOR AQUEOUS DISPERSIBLE SAMPLES,

Miles Lab., Inc., Elkhart, Ind. Marschall Div.
D. G. Kramme, R. H. Griffen, C. G. Hartford, and J. A. Corrado.
Analytical Chemistry, Vol 45, No 2, p 405-408, February 1973. 2 fig, 1 tab, 10 ref.

Descriptors: *Chemical analysis, *Aqueous solutions, *Carbohydrates, Automation, *Nitrogen, *Pollutant identification, Cations, Heavy metals, Alkaline earth metals, Alkali metals, Methodology, Color reactions, Organic compounds, Amino acids, Chromium, Manganese, Potassium, Iron, Cobalt, Magnesium, Nickel, Zinc, Copper, Cadmium, Mercury, Lead, Ureas.
Identifiers: *Kjeldahl nitrogen, *Kjeldahl procedure (Automated), Ionic interference, Chelating agents, Recovery, Sensitivity, Lithium, Chemical digestion, EDTA, Sodium potassium, tartrate, Autoanalyzer,Chemical concentration, Glucosamine, Casein, Nicotinamide.

A standardized, automated procedure has been developed for the determination of Kjeldahl N without modification regardless of the composition of the sample. A basic Technicon AutoAnalyzer System was used with a special custom made gas aspiration pipet in the continuous digestor. The Technicon procedure was modified to give a tenfold increase in sensitivity and the digester input manifold was changed to increase sample size by 50 percent. The heating stages were reversed to eliminate excess fumes. Reducing the SeO_2 (catalyst) concentration from 0.3 to 0.15 percent eliminated the precipitation problem and maintained good digestion with the reversed heating stages. Various N-containing compounds were analyzed using identical conditions and recoveries were nearly 100 percent. Liquid samples, containing up to 10 percent carbohydrate can be digested without interference or further dilution. Significant interference was observed in tests on color development with 50 ppm Cr (3 plus), Mn (2 plus), Fe (3 plus), and Co (2 plus), and with 400 ppm Li, K, Mg, Ni, Zn, Cu, Cd, Hg, and Pb cations. The use of EDTA or sodium potassium tartrate results in a significant reduction in interference with some metal ions; no change in the interference of others; and with certain other ions, enhances the interference. In this comprehensive method, chelating agents were not included because sample composition is often unknown or widely varied. The automated method was compared with the manual Kjeldahl method by analyzing four unknown samples by both procedures. Good agreement of the two methods was achieved. (Holoman-Battelle)
W73-08596

FLOW PROGRAMMING IN COMBINED GAS CHROMATOGRAPHY–MASS SPECTROMETRY,

McDonnell Douglas Research Labs., St. Louis, Mo.
For primary bibliographic entry see Field 02K.
W73-08597

APPLICATION OF A WAVELENGTH SCANNING TECHNIQUE TO MULTI-ELEMENT DETERMINATIONS BY ATOMIC FLUORESCENCE SPECTROMETRY,

Imperial Coll. of Science and Technology, London (England). Dept. of Chemistry.
J. D. Norris, and T. S. West.
Analytical Chemistry, Vol 45, No 2, p 226-230, February 1973. 4 fig, 2 tab, 18 ref.

Descriptors: *Aqueous solutions, Chemical analysis, *Pollutant identification, *Heavy metals, Nickel, Cobalt, Iron, Manganese, Zinc, Cadmium, Selectivity, Alkaline earth metals, Aluminum, Calcium, Chromium, Copper, Iron, Mercury, Potassium, Magnesium, Sodium, Lead, Strontium, Titanium, Cesium.
Identifiers: *Multielemental analysis, *Atomic fluorescence spectroscopy, *Electrodeless discharge lamps, *Wavelength scanning technique, Selenium, Tellurium, Detection limits, Reproducibility, Chemical interference, Silver, Arsenic, Barium, Bismuth, Cerium, Gallium, Indium, Lithium, Niobium, Silicon, Samarium, Tin, Tantalum, Tellurium, Thorium, Vanadium, Tungsten, Zirconium.

Two dual-element electrodeless discharge lamps, operated from a single microwave generator via a two-port power divider, enable a separated air-acetylene flame to be illuminated with the intense resonance radiation of four elements. The resultant atomic fluorescence may be rapidly measured by scanning over the appropriate wavelength range. Results obtained for the sequential multi-element determination of zinc, cadmium, nickel, and cobalt show that the sensitivity and selectivity are the same by the scanning technique as by conventional atomic fluorescence spectrometry. Wavelength scans are also given for nickel, cobalt, iron, and manganese, and for selenium, tellurium, nickel, and cobalt combinations. (Holoman-Battelle)
W73-08598

IMPROVED METHOD FOR DETERMINATION OF XANTHOPHYLL IN FRESH ALFALFA AND WET FRACTIONS,

Agricultural Research Service, Berkeley, Calif. Western Marketing and Nutrition Research Div.
B. E. Knuckles, E. M. Bickoff, and G. O. Kohler.
Journal of the Association of Official Analytical Chemists, Vol 55, No 6, p 1202-1205, November 1972. 2 tab, 12 ref.

Descriptors: Methodology, *Plant pigments, *Solvent extractions, Separation techniques, Primary productivity, Colorimetry, Color, Analytical techniques.
Identifiers: *Xanthophylls, *Carotene, Organic solvents, Absorbance, Sample preparation, *Alfalfa.

An improved method has been described for the analysis of xanthophyll, as well as carotene, in fresh alfalfa. Samples of the plant are cut and chopped for uniform mixing of necessary. A quantity of the sample is mixed with absolute ethanol and filtered through fluted filter paper. An aliquot of the filtrate is then mixed with 40 percent methanolic KOH and evaporated to dryness under vacuum (60-65 C). The evaporated extract is transferred onto a chromatographic column, vacuum is applied and the pigment eluate is collected in a flask. The eluate is mixed with its eluant, and the volume is diluted with acetone and stored in the dark. In order to determine the pigmenting xanthophylls, ethanolic HCl is added to an aliquot of eluate, diluted with acetone, and mixed. The absorbance can be determined between 5 and 15 min after mixing with any suitably calibrated colorimeter or spectrophotometer. The outlined method is relatively rapid, is suitable for routine use, and measures both carotene and xanthophyll in the same extract. (Holoman-Battelle)
W73-08602

EXPENDABLE BATHYTHERMOGRAPH OBSERVATIONS FROM SHIPS OF OPPORTUNITY,

National Marine Fisheries Service, La Jolla, Calif. Fishery-Oceanography Center.
For primary bibliographic entry see Field 07B.
W73-08629

PHYTOPLANKTON AS A BIOLOGICAL INDICATOR OF WATER QUALITY,

Universidad Central de Venezuela, Caracas. Dept. of Sanitary Engineering.
For primary bibliographic entry see Field 05C.
W73-08634

GLACIERS AND NUTRIENTS IN ARCTIC SEAS,

Maine Dept. of Sea and Shore Fisheries, West Boothbay Harbor. Fisheries Research Station.
For primary bibliographic entry see Field 02K.
W73-08726

INFRARED REFLECTANCE OF SULFATE AND PHOSPHATE IN WATER,

Missouri Univ., Kansas City.
W. P. Nijm.
M Sc Thesis, 1972. 30 p, 4 fig, 19 ref. OWRR A-030-MO (3).

Descriptors: *Water chemistry, *Water properties, *Reflectance, *Optical properties, *Infrared radiation, Remote sensing, Sulfates, Phosphates, Analytical techniques, Pollutant identification.
Identifiers: Infrared reflectance.

The relative specular reflectances of the surfaces of two different aqueous solutions, one containing 0.5 M potassium sulfate and the other 0.5 M ammonium phosphate monobasic were measured throughout the 2-20 micrometer wavelength region of the infrared. Distilled water was the reflectance standard. The infrared radiant flux was incident on the surfaces at an angle of about 70 deg. Relative reflectances were measured for the polarization component having the electric field intensity vector perpendicular to the plane of incidence. Bands attributed to SO_4, PO_4, HPO_4, and H_2PO_4 were observed in the relative reflectance spectra. (Knapp-USGS)

W73-08728

STANDARD POTENTIALS OF THE SILVER/SILVER-IODIDE ELECTRODE IN AQUEOUS MIXTURES OF ETHYLENE GLYCOL AT DIFFERENT TEMPERATURES AND THE THERMODYNAMICS OF TRANSFER OF HYDROGEN HALIDES FROM WATER TO GLYCOLIC MEDIA,
Jadavpur Univ., Calcutta (India). Dept. of Chemistry.
K. K. Kundu, D. Jana, and M. N. Das.
Electrochimica Acta, Vol. 18, No. 1, p 95-103, January 1973. 5 fig, 6 tab, 23 ref.

Descriptors: *Thermodynamic behavior, *Zeta potential, Aqueous solutions, Temperature, Ions, Halides, Entropy, Methodology, Chlorides, Bromides, Iodides, Anions, Hydrogen.
Identifiers: *Mixtures, *Ethylene glycol, *Ion selective electrodes, *Silver/silver iodide electrode, Solvation, Electromotive force, Hydrogen chloride, Hydrogen bromide, Hydrogen iodide, Aqueous solvents.

The standard potentials (E) of the silver/silver-iodide electrode in aqueous mixtures of ethylene glycol (containing 10, 30, 50, 70 and 90 wt-percent glycol) have been determined from emf measurement of the cell. Pt,H2 (g, 1 atm)/HOAc (m1),NaOAc (m2),KI (m3),solvent/AgI/Ag at nine temperatures ranging from 5 to 45C. These E values have been utilized to compute free energy (delta G sub t), entropy (delta S sub t) and enthalpy (delta H sub t) changes accompanying the transfer of HI from water to each of the solvents. These values as well as those for HCl and HBr obtained earlier have also been utilized to evaluate delta S sub t (i) for individual ions by a method of 'simultaneous extrapolation', which in turn furnished the values of delta H sub t (i) for these ions. These quantities and also the 'chemical' contributions for the halide ions as obtained by subtracting the 'electrostatic' contribution computed with the Born equation, have been examined in the light of ion-solvent interactions as well as the structural changes of the solvents. The observed results conform with what is expected from the competitive effects of the preferential solvating capacities of water towards halide ions and that of other solvents towards hydrogen ions, and also of the effects arising from the structural changes of the solvents that are likely to occur in the over-all transfer process. (Holoman-Battelle)
W73-08757

THE ERROR IN MEASUREMENTS OF ELECTRODE KINETICS CAUSED BY NONUNIFORM OHMIC-POTENTIAL DROP TO A DISK ELECTRODE,
California Univ., Los Angeles. School of Engineering and Applied Science.
For primary bibliographic entry see Field 02K.
W73-08758

DEEP-SEA CIRROMORPHS (CEPHALOPODA) PHOTOGRAPHED IN THE ARCTIC OCEAN,
Oregon State Univ., Corvallis. Dept. of Oceanography.
For primary bibliographic entry see Field 07B.
W73-08759

HYDROPTILA ERAMOSA A NEW CADDIS FLY FROM SOUTHERN ONTARIO (TRICHOPTERA, HYDROPTILIDAE),
Montreal Univ. (Quebec). Dept. of Biological Sciences.
For primary bibliographic entry see Field 02I.
W73-08760

THE DISTRIBUTION AND TISSUE RETENTION OF MERCURY-203 IN THE GOLDFISH (CARASSIUS AURATUS),
Wayne State Univ., Detroit, Mich. Dept. of Biology.
M. Weisbart.
Canadian Journal of Zoology, Vol. 51, No. 2, p 121-131, February 1973. 13 fig, 1 tab, 22 ref.

Descriptors: *Mercury, *Water analysis, *Bioassay, Radioactivity techniques.
Identifiers: *Goldfish, *Excretion, *Scintillation counting, *Tissue, Carassius auratus, Biological samples, Gall bladder, Gonads, Spleen, Eyes, Kidney, Intestine, Gills, Heart, Skin, Brain, Liver, Muscle, Swim bladder, Hg-203, Head kidney.

Goldfish weighing between 5.83 and 13.63 g were maintained in aquaria for use in studies of the rate of elimination of mercury. Fish were injected intraperitoneally with 10 microliters of a solution of Hg-203 labelled Hg (NO3)2. At 0 to 672 hours after injection, fish were sacrificed for analysis of mercury residues in whole-body, skin, eyes, brain, gills, heart, gall bladder, liver, gonads, intestines, spleen, kidneys, head kidney, muscle, and swim bladder. Water samples were also analyzed at each period. Hg-203 was determined by gamma counting in a well-type scintillation counter. The results showed that fish lost mercury at an apparent constant rate resulting in a biological half-life of 568 h. Correlated with this loss was a linear increase in the amount of mercury in the water. The mercury-203 content in the tissues displayed four different responses. (1) Gall bladder, gonad, and spleen tissues showed no significant regressions. (2) Eye, kidney, and intestinal tissue manifested significant losses of mercury, but the rate of loss was not significantly different from that of the body as a whole. (3) Gill, heart, skin, and swim bladder tissues lost mercury at rates faster than the body as a whole. (4) Brain, liver, muscle, and head kidney tissues showed no significant losses of mercury. (Little-Battelle)
W73-08761

CONCERNING SOME LITTORAL CLADOCERA FROM AVALON PENINSULA, NEWFOUNDLAND,
Akademiya Nauk SSSR, Moscow. Inst. of Evolutionary Morphology and Animal Ecology.
N. N. Smirnov, and C. C. Davis.
Canadian Journal of Zoology, Vol. 51, No. 1, p 65-68, January 1973. 25 ref.

Descriptors: *Wateflеas, *Littoral, *Systematics, Invertebrates, Marine animals, *Canada, Sampling, Intertidal areas.
Identifiers: Macroinvertebrates, Disparalona (Alonella) acutirostris, Chydorus sphaericus, Pleuroxus laevis, Chydorus sphaericus caelatus, Scapholeberis, Simocephalus, Chydorus piger, Acroperus alonoides, Polyphemus pediculus, Acantholeberis curvirostris, Avalon Peninsula, Streblocerus serricaudatus, Ophryoxus gracilis, Ilyocryptus, Bosmina, Alona guttata, Graptoleberis testudinaria occidentalis, Ophryoxus gracilis spinifera.

During July, 1971, six samples of littoral material were collected by C.C.D. from small bodies of water in the eastern portion of the Avalon Peninsula, Newfoundland. The samples were obtained by scraping through immersed vegetation, or other solid materials, with a small hand net constructed of 100-micron nylon netting. All identifications were made by N.N.S. The Cladocera encountered in the collections are listed. Of the 13 species and varieties that were identified, 10 are new for the island of Newfoundland, 8 are new for the Province of Newfoundland and Labrador, and 5 are new for Canada. Acroperus alonoides Hudendorff and Graptoleberis testudinaria occidentalis Sars have not previously been identified from North America. (Holoman-Battelle)
W73-08762

A CHEMOAUTOTROPHIC AND THERMOPHILIC MICROORGANISM ISOLATED FROM AN ACID HOT SPRING,
New Mexico Bureau of Mines and Mineral Resources, Socorro.
C. L. Brierley, and J. A. Brierley.
Canadian Journal of Microbiology, Vol. 19, No. 2, p 183-188, February 1973. 5 fig, 2 tab, 12 ref.

Descriptors: *Thermophilic bacteria, *Electron microscopy, *Pollutant identification, *Biological properties, *Hot springs, *Acidic water, *Isolation, Heat resistance, Respiration, Cytological studies, Temperature, Oxidation.
Identifiers: *Chemoautotrophic bacteria, Bacterial physiology, Sulfolobus, Culture media, Sample preparation, Culturing techniques, Deoxyribonucleic acid, Manometric techniques, Thiobacillus thiooxidans, Bacillus stearothermophilus, Chemical composition, Biochemical tests, Substrate utilization.

A pleomorphic, acidophilic, and chemoautotrophic microbe is described. The cell is bound by a membrane and a diffuse, amorphous layer. The isolate used either sulfur or iron as a source of energy. Morphological and nutritional similarities as well as corresponding thermophilic and acidophilic requirements suggest a relationship to Sulfolobus. The organism tolerates 80 degrees C for longer than 2 hr, but heat resistance is not attributed to a bacterial spore. The maximum temperature for growth is 70 degrees C; the minimum about 45 degrees C. The DNA base composition is 57 plus or minus 3 mole percent GC. Yeast extract enhances growth of the isolate on iron and sulfur substrates, but does not significantly enhance the isolate's respiration rate on these same substrates. The isolate requires induction by sulfur or iron for maximum respiration on these substrates, respectively. Optimum oxidation of elemental sulfur occurred at pH 2.0 and gave a Q sub 02 of 163; oxidation on iron gave a maximum Q sub 02 of 879. (Holoman-Battelle)
W73-08763

RAPID EXTRACTION AND PHYSICAL DETECTION OF POLYOMA VIRUS,
Sherbrooke Univ. (Quebec). Faculte de Medecine.
H. V. Thorne, and A. F. Wardle.
Canadian Journal of Microbiology, Vol. 19, No. 2, p 291-293, February 1973. 2 fig, 5 ref.

Descriptors: *Pollutant identification, *Separation techniques, *Isolation, *Radioactivity techniques, Viruses, Methodology, Enzymes, Radioactivity, Laboratory tests.
Identifiers: *Polyoma virus, Solubilization, Tissue culture, Purification, Sample preparation, Ribonuclease, Deoxyribonuclease.

In the method for the extraction and purification of polyoma virus from mouse embryo cells in tissue culture, monolayer cultures are infected with the virus and incubated at 37 C in Eagle's medium plus 10 percent calf serum containing H-3-thymidine for 5 days in an atmosphere of 5 percent CO2. The cultures are solubilized with Triton-X-100 (20 percent w/v), then treated with RNase, DNase, and trypsin before equilibrium centrifugation in CsCl and collection of the virus band. Analysis of the virus band showed that Triton-X-100 extraction is a satisfactory alternative to earlier procedures for producing virus suspensions suitable for further purification. The total time, including that for purification, is reduced from about 5 days to less than 24 hr. (Holoman-Battelle)
W73-08764

ULTRASTRUCTURE AND CHARACTERIZATION OF AN ASPOROGENIC MUTANT OF CLOSTRIDIUM BOTULINUM TYPE E,
Manitoba Univ., Winnipeg. Dept. of Microbiology.
R. Z. Hawirko, K. L. Chung, A. C. Emeruwa, and A. J. C. Magnusson.
Canadian Journal of Microbiology, Vol. 19, No. 2, p 281-284, February 1973. 14 fig, 2 tab, 15 ref.

Descriptors: *Pollutant identification, *Electron microscopy, Anaerobic bacteria, Coliforms, Enteric bacteria, Genetics.
Identifiers: *Ultrastructure, *Characterization, *Clostridium botulinum type E, *Pathogenicity, *Biochemical characteristics, *Toxigenic bacteria, Mutants, Sample preparation, Culture media, Cell structures.

The asporogenic mutant, RSpoIIIa, showed septum formation and a nearly completed forespore about 4 hr after onset of sporulation. The cells show defects at a few sites of the forespore membrane, an absence of 'germ cell wall', and within 8 hr lysis of the cytoplasm occurred indicating that the mutant was blocked at stage III. Some aberrant envelopes were seen later. Lysis of the asporogenci mutant was inhibited for up to 36 hr by the addition of 2.4 percent glucose or sucrose to the medium, and 80 percent of the cells showed septum formation. A comparison of the phenotypic characteristics of the asporogenic RSpoIIIa and the sporogenic MSp plus mutants, as well as the wild type, showed the same ultrastructural changes during the development of the forespore with the accumulation of intracellular iodophilic granules. In addition, the mutants showed specific immunofluorescence and precipitin lines of identity with antisera against the wild-type strain, but unlike the toxigenic wild type, the mutants were nontoxigenic by mouse pathogenicity tests. (Holoman-Battelle)
W73-08765

CHEMOTAXONOMIC FATTY ACID FINGERPRINTS OF BACTERIA GROWN WITH, AND WITHOUT, AERATION,
Manchester Univ. (England). Dept. of Bacteriology and Virology.
D. B. Drucker, and I. Owen.
Canadian Journal of Microbiology, Vol. 19, No. 2, p 247-250, February 1973. 2 tab, 16 ref.

Descriptors: *Bacteria, *Aeration, *Oxygen, *Gas chromatography, *Pollutant identification, Separation techniques, Chemical analysis, Cultures, Organic acids, Pathogenic bacteria.
Identifiers: *Fatty acids, *Chemotaxonomy, Chemical composition, Retention time, Sample preparation, Bacillus megaterium, Klebsiella aerogenes, Corynebacterium xerosis, Neisseria catarrhalis, Serratia marcescens, Staphylococcus saprophyticus, Streptococcus faecalis, Palmitate, Myristate, Palmitoleate, Oleate, Isopentadecanoate.

Examinations were made of the differences in qualitative and quantitative fatty acid composition between bacterial species and the extent to which such differences are affected by oxygen availability. Fatty acid fingerprints were obtained for Bacillus megaterium, Corynebacterium xerosus, Klebsiella aerogenes, Neisseria catarrhalis, Serratia marcescens, Staphylococcus saprophyticus, and Streptococcus faecalis. The test organisms were grown with, and without, aeration. Freeze-dried cells were methylated and the fatty acid methyl esters examined gas-chromatographically. Fatty acid fingerprints were similar for all the test organisms when grown without aeration and the major fatty acid esters had the retention characteristics of palmitate, then myristate (17.0), palmitoleate (16.4), and oleate (18.4). The cultures grown with aeration showed fairly dissimilar fatty acid fingerprints; the major fatty acid peak having a similar retention time to palmitate (16.0), oleate, isopentadecanoate (14.5), or three unknown fatty acids of carbon numbers 18.25, 18.7, and 19.25, depending on the species tested. (Holoman-Battelle)
W73-08766

EVALUATION OF THE REDESIGNED ENTEROTUBE--A SYSTEM FOR THE IDENTIFICATION OF ENTEROBACTERIACEAE,
Public Health Service, Savannah, Ga. Technical Development Labs.
K. M. Tomfohrde, D. L. Rhoden, P. B. Smith, and A. Balows.
Applied Microbiology, Vol. 25, No. 2, p 301-304, February 1973. 3 tab, 5 ref.

Descriptors: *Pollutant identification, *Laboratory equipment, *Enteric bacteria, *Separation techniques, E. coli, Salmonella, Shigella.
Identifiers: *Enterotube, *Accuracy, Enterobacter hafniae, Citrobacter, Klebsiella, Serratia, Providencia, Proteus mirabilis, Edwardsiella, Enterobacter liquefaciens, Proteus vulgaris, Enterobacter cloacae, Enterobacter aerogenes, Proteus morganii, Proteus rettgeri.

Because of some discrepancies in analytical results, the Enterotube has been redesigned by the manufacturer. A test for ornithine decarboxylase has been added to one compartment, and lactose has been removed from the lysine decarboxylase compartment. The tests for phenylalanine deaminase and dulcitol fermentation have been combined in a single compartment. An iron salt has been added to this same compartment, thus eliminating the need to add ferric chloride reagent in testing for phenylalanine deaminase activity. To improve the decarboxylase reactions and to allow for the observance of gas production from the fermentation of dextrose, the manufacturer covered the compartments containing tests for dextrose fermentation, lysine decarboxylase, and ornithine decarboxylase with a sterile wax overlay. The redesigned Enterotube has been evaluated with 414 unknown Enterobacteriaceae cultures from the stock culture collection of the Center for Disease Control. When the Enterotube was used as recommended by the manufacturer, an average of 96.4 percent of these cultures were correctly identified. Only two groups (Salmonella and Edwardsiella) were identified with less than 90 percent accuracy (89.2 and 87.5 percent, respectively). The Enterotube now provides a convenient, rapid, and accurate test system for the identification of typically reacting enteric bacteria. (Little-Battelle)
W73-08767

AUXOTAB--A DEVICE FOR IDENTIFYING ENTERIC BACTERIA,
Public Health Service, Savannah, Ga. Technical Development Labs.
D. L. Rhoden, K. M. Tomfohrde, P. B. Smith, and A. Balows.
Applied Microbiology, Vol. 25, No. 2, p 284-286, February 1973. 3 tab, 3 ref.

Descriptors: *Pollutant identification, *Enteric bacteria, *Laboratory equipment, *Separation techniques, E. coli, Salmonella, Shigella.
Identifiers: *Auxotab, Enteric I System, *Accuracy, Citrobacter freundii, Klebsiella pneumoniae, Providencia, Proteus vulgaris, Enterobacter hafniae, Edwardsiella tarda, Serratia marcescens, Proteus morganii, Proteus mirabilis, Proteus rettgeri, Arizona hinshawii, Enterobacter cloacae, Enterobacter liquefacians, Enterobacter aerogenes.

A multitest system called the Auxotab that uses ten dehydrated reagents on a paper card has been evaluated with 417 known stock cultures of Enterobacteriaceae. In double-blind studies with the Auxotab, 87 percent of the strains tested were correctly identified. Results indicate that there is a need for modification of the product in regard to ease of handling, time required for use, and accuracy of identification of enteric bacteria. (Little-Battelle)
W73-08768

SPIRAL PLATE METHOD FOR BACTERIAL DETERMINATION,
Food and Drug Administration, Cincinnati, Ohio. Div of Microbiology.
J. E. Gilchrist, J. E. Campbell, C. B. Donnelly, J. T. Peeler, and J. M. Delaney.
Applied Microbiology, Vol. 25, No. 2, p 244-25[illegible], February 1973. 5 fig, 4 tab, 14 ref.

Descriptors: *Bacteria, *Laboratory equipment, *Pollutant identification, Cultures, Methodology, E. coli, Pathogenic bacteria, Aerobic bacteria, Automation.
Identifiers: *Enumeration, *Spiral plate method, *Culturing techniques, Bacillus subtilis, Pseudomonas aeruginosa, Staphylococcus aureus, Lactobacillus casei, Culture media.

A method is described for determining the number of bacteria in a solution by the use of a machine which deposits a known volume of sample on a rotating agar plate in an ever decreasing amount in the form of an Archimedes spiral. After the sample is incubated, different colony densities are apparent on the surface of the plate. A modified counting grid is described which relates area of the plate to volume of sample. By counting an appropriate area of the plate, the number of bacteria in the sample is estimated. This method was compared to the pour plate procedure with the use of pure and mixed cultures in water and milk. The results did not demonstrate a significant difference in variance between duplicates at the alpha equals 0.01 level when concentrations of 600 to 1,200,000 bacteria per ml were used, but the spiral plate method gave counts that were higher than counts obtained by the pour plate method. The time and materials required for this method are substantially less than those required for the conventional aerobic pour plate procedure. (Holoman-Battelle)
W73-08769

APPLICATION OF THE MOST-PROBABLE-NUMBER PROCEDURE TO SUSPENSIONS OF LEPTOSPIRA AUTUMNALIS AKIYAMI A,
North Carolina Univ., Chapel Hill. Dept. of Environmental Sciences and Engineering.
D. A. Schiemann.
Applied Microbiology, Vol. 25, No. 2, p 235-239, February 1973. 2 fig, 4 tab, 7 ref.

Descriptors: *Population, *Statistical methods, *Estimating, Bacteria, Density, Pathogenic bacteria.
Identifiers: *Most Probable Number test, *Leptospira autumnalis.

Suspensions of washed cells of Leptospira autumnalis Akiyami A were diluted and used to inoculate three series of 120 tubes for use in determining the reliability of the most-probable-number (MPN) procedure for estimating the density of the bacteria. Statistical tests demonstrated that with supplemented Fletchers medium used for cell recovery, the MPN procedure is a reliable technique for estimating the density of suspensions of L. autumnalis. (Little-Battelle)
W73-08770

IN-USE EVALUATION OF A COMMERCIALLY AVAILABLE SET OF QUALITY CONTROL CULTURES,
Public Health Service, Savannah, Ga. Technical Development Labs.
G. W. Douglas, A. Balows, D. Rhoden, K. Tomfohrde, and P. B. Smith.
Applied Microbiology, Vol 25, No 2, p 230-234, February 1973. 6 tab, 14 ref.

Descriptors: *Cultures, *Pathogenic bacteria, E. coli, Enteric bacteria.
Identifiers: *Quality control, *Culture media, *Bact-Chek, *Evaluation, Enterobacter cloacae, Proteus vulgaris, Pseudomonas aeruginosa, Salmonella typhimurium, Staphylococcus aureus, Staphylococcus epidermis, Streptococcus pyogenes.

Increasing awareness of the need for a uniform quality control program prompted an evaluation of a commercially available set (Bact-Chek) of eight organisms. A protocol was designed in which this

ing of maximum absorbance in the UV region to longer wavelengths as the number of fused aromatic rings increased. However, by adjusting the percent of TNF, mobile phase water content, and temperature, it was possible to optimize the column for rapid analysis for the separation of 3-, 4-, 5-, 6- or even 7-fused ring polyaromatic systems. The Corasil I porous layer beads used as support contributed partly to the retention. This effect was markedly reduced by using water saturated heptane as the mobile phase. (Holoman-Battelle)
W73-08773

THE ANALYSIS OF DESIGNED EXPERIMENTS WITH CENSORED OBSERVATIONS,
Unilever Ltd., Sharnbrook (England). Unilever Research Lab.
For primary bibliographic entry see Field 07B.
W73-08777

THE BACTERIA IN AN ANTARTIC PEAT,
University Coll., London (England). Dept. of Botany.
J. H. Baker, and D. G. Smith.
Journal of Applied Bacteriology, Vol 35, No 4, p 589-596, December 1972. 1 tab, 45 ref.

Descriptors: *Peat, *Anarctic, *Pollutant identification, *Separation techniques, *Cultures, *Bacteria, Isolation.
Identifiers: *Sample preparation, Brevibacterium, Arthrobacter, Cellulomonas, Kurthia, Micrococcus, Bacillus, Microbacterium, Sarcina, Flavobacterium, Achromobacter, Alcaligenes, Agrobacterium, Acinetobacter, Corynebacterium, Agars.

Peat samples were collected from 3 depths on the west coast of Signy Island, South Orkney and prepared for analysis by homogenizing, diluting with distilled water, and dropping on to plates of Oxoid Tryptone-Soya Agar. The characters used for identification of the bacteria were: Gram reaction, morphology of old and young cells, pigmentation, motility, position of flagella, endospore formation, action on litmus milk, nitrate reduction, utilization of cellulose, gelatin, starch, chitin, alginate, phenol, alkylamine and alcohol, and fermentation of glucose and lactose. Of a total of 119 strains of bacteria 52 percent belonged to the genus Brevibacterium. Twelve other genera were recorded of which numerically the most important were Arthrobacter, Cellulomonas, Kurthia and Micrococcus. 62 percent of the collection were psychrophilic, but only 4 strains were obligate psychrophiles. No pattern could be established for the various genera from different depths. The fine structure of an obligately psychrophilic pleomorphic rod from the peat is illustrated and discussed. (Little-Battelle)
W73-08778

DETERMINATION OF SOME CARBAMATES BY ENZYME INHIBITION TECHNIQUES USING THIN-LAYER CHROMATOGRAPHY AND COLORIMETRY,
C. E. Mendoza, and J. B. Shields.
Journal of Agricultural and Food Chemistry, Vol 21, No 2, p 178-184, March/April 1973. 5 fig, 7 tab, 22 ref.

Descriptors: *Bioassay, *Pollutant identification, *Carbamate pesticides, *Pesticide residues, *Colorimetry, *Pesticide toxicity, Thiocarbamate pesticides, Spectrophotometry, Bromine, Pollutants, Insecticides, Ultraviolet radiation, Methodology, Chemical analysis.
Identifiers: *Thin layer chromatography, *Tlc-enzyme inhibition techniques, Substrates, Aldicarb, Butacarb, c-8353, Carbaryl, Indophenyl acetate, Formetanate (HCl), Meobal, Mesurol, Methomyl, Promecarb, Enzyme inhibitors, Chemical interference, Detection limits, 5-Bromoindoxyl acetate, Porcine liver esterases.

Detection limits were determined by a thin-layer chromatographic-enzyme inhibition technique for aldicarb, Butacarb, c-8353, carbaryl, formetanate (HCl), Meobal, Mesurol, methomyl, and promecarb. Indophenyl and 5-bromoindoxyl acetates were used as substrates of porcine liver esterases. Effects of ultraviolet irradiation and bromine on the pesticides were also studied. Inhibition of esterase activities by these carbamates was determined by spectrophotometry using indophenyl acetate (IPA) substrate. Percent inhibitions were based on IPA hydrolysis rate in the treated solutions at two different periods vs. that in the control enzyme solutions, and the absorbance of the treated solution at 7.5 or 15 min after the addition of substrate vs. the absorbance of the control solution. The results indicate that tlc and colorimetry using IPA are complimentary techniques, and that the same enzyme sites or system are involved. Both techniques can be used to confirm pesticide residues that are enzyme inhibitors. The data also suggested that Butacarb, promecarb, Mesurol, and carbaryl can be classified as strong inhibitors, and aldicarb and formetanate (HCl) can be classified as weak inhibitors of pig liver esterases catalyzing the hydrolysis of IPA. Since the colorimetric procedure used includes incubation of pesticides before the addition of substrates, it can be used directly to assay enzyme activities from insecticide-treated organisms. IPA, as a substrate, can be used in tlc and colorimetry without involving complex preparations. IPA solution is stable, particularly when stored at 4 degrees. (Holoman-Battelle)
W73-08779

THE DETERMINATION OF ORGANOCARBON IN CLAY MATERIALS,
English Clays Lovering Pochin and Co. Ltd., St. Austell (Enland). Central Labs.
A. P. Ferris, and W. B. Jepson.
Analyst, Vol 97, No 1161, p 940-950, December 1972. 6 fig, 42 ref.

Descriptors: *Clays, *Chemical analysis, *Methodology, *Soil analysis, *Laboratory equipment, Kaolinite, Storage.
Identifiers: *Organic carbon, Kaolin, Dry ashing, Chemical interference, Sulfur dioxide, Silicon tetrafluoride, Hydrogen fluoride, Wet ashing, Precision.

A dry-combustion method is described for the determination of organocarbon in clay materials. The essential features are the use of purified oxygen; slow insertion of the sample into the furnace to avoid rapid dehydroxylation of the kaolinite and attendant loss of pyrolysis fragments in the atmosphere of steam that is generated; and the use of traps to remove gaseous fluorine compounds, steam and sulphur dioxide. Extensive tests on the apparatus with calcium carbonate and with mixtrues of kaolin with glucose and with tannic acid are described. Tests with a commercial kaolin containing organic matter equivalent to about 550 micrograms of carbon per gram of material indicated a standard error of 8 micrograms, or plus or minus 1.5 per cent. A published method in which the kaolin is heated in a stream of oxygen containing 2 per cent V/V of ozone has been re-investigated; the organocarbon was only partially oxidized so that the method has no analytical value. Two wet-combustion methods involving the separate use of potassium dichromate and potassium persulphate, which are used in soil analysis, were evaluated. The organic matter on the clay was again only partially oxidized to carbon dioxide under the experimental conditions. (Holoman-Battelle)
W73-08780

THE ANALYSIS OF ORGANICALLY BOUND ELEMENTS (AS, SE, BR) AND PHOSPHORUS IN RAW, REFINED, BLEACHED AND

HYDROGENATED MARINE OILS PRODUCED FROM FISH OF DIFFERENT QUALITY,
Central Inst. for Industrial Research, Oslo (Norway).
G. Lunde.
Journal of the American Oil Chemists Society, Vol 50, No 1, p 26-28, January 1973. 2 tab, 5 ref.

Descriptors: *Neutron activation analysis, *Biomass, *Phosphorus, Storage.
Identifiers: *Fish oil, *Arsenic, *Selenium, Bleaching, Oil refining, Hydrogenation, Oil characterization, Spoilage, *Marine oils.

The purpose was to study how the selenium, arsenic, phosphorus, and bromine contents varied in oils produced from raw fish materials of differing quality and to follow these elements in the oils during the alkaline-refining, the bleaching, and the hydrogenation processes. Herring, mackerel, and capelin oils, were produced from raw materials stored for varying lengths of time and consequently were of varying quality. Neutron activation analysis of samples showed that when the raw material deteriorates during storage, an increase in the selenium and phosphorus content in the oils produced from these materials is observed, whereas the bromine and the arsenic content is nearly constant. During the refining the arsenic and phosphorus disappear almost completely from the oils, whereas the selenium content is reduced to about two-thirds and the bromine content is nearly unaffected. In the hydrogenation step the selenium disappears relatively fast and the bromine more slowly. (Little-Battelle)
W73-08781

REVERSED-PHASE FOAM CHROMATOGRAPHY. REDOX REACTIONS ON OPEN-CELL POLYURETHANE FOAM COLUMNS SUPPORTING TETRACHLOROHYDROQUINONE,
Eotvos Lorand Univ., Budapest (Hungary). Inst. of Inorganic and Analytical Chemistry.
T. Braun, A. B. Farag, and A. Klimes-Szmik.
Analytica Chimica Acta, Vol 64, No 1, p 71-76, March 1973. 2 fig, 1 tab, 5 ref.

Descriptors: *Chemical reactions, *Heavy metals, *Pollutant identification, Iron, Cations, Temperature, Flow rates, Reduction (Chemical), Oxidation.
Identifiers: *Reversed phase foam chromatography, *Open cell polyurethane foam, *Support columns, *Chloranil, Polyurethane foams, Cerium, Vanadium, Tetrachlorohydroquinone.

The use of elastic polyurethane foam as a support for chloranil in reversed-phase foam chromatography was proved successful. Reductions of cerium (IV), vanadium (V) and iron (II) on foam-filled columns were carried out quantitatively and rapidly. The effect of flow-rate and temperature on the reduction of each metal ion was examined in detail. Cerium (IV) was reduced quantitatively on passing through the foam-redox column at flow-rates of 2-11 ml/min at room temperature. The reduction of vanadium (V) and iron (III) was slower; complete reduction occurred only at flow-rates up to 4 and 2 ml/min for V (V) and Fe (III), respectively. At 35 degrees, however, it was possible to use flow-rates of 7 and 6 ml/min for the quantitative reduction of V (V) and Fe (III) respectively. (Holoman-Battelle)
W73-08782

ION-EXCHANGE FOAM CHROMATOGRAPHY. PART I. PREPARATION OF RIGID AND FLEXIBLE ION-EXCHANGE FOAMS,
Eotvos Lorand Univ., Budapest (Hungary). Inst. of Inorganic and Analytical Chemistry.
For primary bibliographic entry see Field 02K.
W73-08783

ION EXCHANGE IN AQUEOUS AND IN AQUEOUS-ORGANIC SOLVENTS. PART I. ANION-EXCHANGE BEHAVIOUR OF ZR, NB, TA AND PA IN AQUEOUS HCL-HF AND IN HCL-HF-ORGANIC SOLVENT,
Technische Universitaet, Munich (West Germany). Institut fuer Radiochemie.
J. I. Kim, H. Lagally, and H-J. Born.
Analytica Chimica Acta, Vol 64, No 1, p 29-43, March 1973. 9 fig, 20 ref.

Descriptors: *Anion exchange, *Radioactivity techniques, Adsorption, Separation techniques, Aqueous solutions, Pollutant identification, Heavy metals.
Identifiers: *Aqueous solvents, *Organic solvents, *Zirconium, *Niobium, *Protactinium, *Tantalum, Ion exchange resins, Gamma spectrometry, Pa-233, zr-95, Nb-95, Ta-182, Pa-231, Hydrochloric acid, Hydrofluoric acid, Acetic acid, Isopropanol, Methanol, Protactinium radioisotopes, Tantalum radioisotopes, Niobium radioisotopes, Zirconium radioisotopes.

Ion-exchange behaviors of Zr (IV, Nb (V), Ta (V), and Pa (V) on the anion exchanger Dowex 1-x8 were investigated first in a wide variety of aqueous mixed hydrochloric acid-hydrofluoric acid and secondly in the same aqueous system mixed with various organic solvents. Equilibrium adsorptions of these four elements as a function of hydrochloric acid concentration as well as hydrofluoric acid concentration, or both acid concentrations, were strongly differentiated. This fact can be utilized for convenient separations of these elements from each other. Based on the equilibrium results, the possible complex formation of the metal ions and the separation possibilities for the elements are discussed. (Holoman-Battelle)
W73-08784

THE SIMULTANEOUS DETERMINATION OF TRACES OF COBALT, CHROMIUM, COPPER, IRON, MANGANESE AND ZINC BY ATOMIC FLUORESCENCE SPECTROMETRY WITH PRECONCENTRATION BY AN AUTOMATED SOLVENT EXTRACTION PROCEDURE,
Imperial Coll. of Science and Technology, London (England). Dept. of Chemistry.
M. Jones, G. F. Kirkbright, L. Ranson, and T. S. West.
Analytica Chimica Acta, Vol 63, No 1, p 210-215, January 1973. 2 fig, 2 tab, 11 ref.

Descriptors: *Heavy metals, *Aqueous solutions, *Chemical analysis, *Pollutant identification, Water analysis, Solvent extractions, Trace elements, Cobalt, Chromium, Copper, Iron, Manganese, Zinc, Sea water, Automation, Methodology, Efficiencies, Separation techniques.
Identifiers: *Sample preparation, Preconcentration, *Automated solvent extraction, *Atomic fluorescence spectroscopy, *Multielemental analysis, Detection limits, Recovery, Accuracy, Organic solvents.

A description is given of the assembly and operation of an automated solvent extraction system for preconcentration; the subsequent simultaneous determination of Co, Cr, Cu, Fe, Mn and Zn by atomic fluorescence spectrometry in a manner similar to that described elsewhere is also reported. The application of this system to the rapid determination of trace amounts of Cu, Fe, Mn and Zn in sea-water is also described. Detection limits and extraction efficiencies were obtained for all 6 elements in aqueous and organic solutions. Except for chromium, an improvement of 6-10-fold in the detection limit was achieved by the extraction procedure. The analysis rate of the spectrometer is limited by the rate of the automated solvent extraction process; under the conditions used the maximum analysis rate was 25 samples per h. Recovery of the analyte elements from the sea water samples was checked by making standard additions of each of the elements to an analyzed sample; recoveries exceeded 90 percent, indicating that the accuracy sufficed for the routine determination of these metals (less than ca. 0 ppm) in sea water. The automated solvent extraction method developed can be successfully employed for simultaneous determinations of six elements with a multichannel atomic fluorescence spectrometer. Improved detection limits are obtained, but only a 4:1 sample:solvent mixing ratio is tolerable if the system is to be successfully interfaced with the nebulizer of the flame spectrometer. (Holoman-Battelle)
W73-08785

DETERMINATION OF CHLOROPHYLL A AND B IN PLANT EXTRACTS BY COMBINED SUCROSE THIN-LAYER CHROMATOGRAPHY AND ATOMIC ABSORPTION SPECTROPHOTOMETRY,
Texas Tech Univ., Lubbock. Textile Research Center.
For primary bibliographic entry see Field 02K.
W73-08786

THE PREPARATION AND ANALYTICAL EVALUATION OF A NEW HETEROGENEOUS MEMBRANE ELECTRODE FOR CADMIUM (II),
Rome Univ. (Italy). Istituto di Chimica Analitica.
M. Mascini, and A. Liberti.
Analytica Chimica Acta, Vol 64, No 1, p 63-70 March 1973. 7 fig, 1 tab, 7 ref.

Descriptors: *Calibrations, *Aqueous solutions *Hydrogen ion concentration, *Cadmium, Design Silver, Mercury, Copper, Lead, Zinc, Cobalt Nickel, Iron.
Identifiers: *Ion selective electrodes, *Detection limits, Sensitivity, *Chemical interfere.

A new heterogeneous membrane electrode for cadmium ion was prepared by thermomoulding mixtures of cadmium and silver sulfides and polyethylene. Electrodes were prepared with eight different salt preparations. Calibrations were obtained with pure cadmium nitrate solutions and with sodium nitrate additions to the solutions. Only electrodes prepared using Ag_2S-CdS mixtures precipitated from acidic solutions with hydrogen sulfide and heat treated exhibited Nerstian behavior. This method of preparation, therefore, was chosen. The electrodes prepared in this manner responded linearly down to 0.00001 M Ca (2 plus) and indicated changes in concentrations down to 0.0000001 M. The effects of mixed solvents, pH variations, and interference from Ag, Hg, Cu, H, Pb, Zn, Co, Ni, and Fe were also investigated. The electrode was shown to be useful in both aqueous and nonaqueous solutions. (Little-Battelle)
W73-08787

GRAPHITE ROD ATOMIZATION AND ATOMIC FLUORESCENCE FOR THE SIMULTANEOUS DETERMINATION OF SILVER AND COPPER IN JET ENGINE OILS,
Florida Univ., Gainesville. Dept. of Chemistry.
B. M. Patel, and J. D. Winefordner.
Analytica Chimica Acta, Vol 64, No 1, p 135-138, March 1973. 2 fig, 1 tab, 9 ref.

Descriptors: *Copper, *Chemical analysis, *Methodology, Heavy metals, Oil, Pollutant identification.
Identifiers: *Atomic fluorescence spectroscopy, *Silver, *Jet engine oil, Spectrochemical Oil Analysis Program, Fluorescent spectra, Accuracy.

Silver and copper can be determined simultaneously by atomic fluorescence without altering any conditions for sample atomization. Both of the elements are atomized simultaneously from a heated graphite rod at an intermediate temperature (1800 K), and the two separate fluorescence peaks

vert Bi ion to radioactive tetraiodobismuthic acid by treatment with excess of I-131. The active product is extracted into n-butyl acetate and the gamma-ray activity of a portion of the extract is determined. The relationship between the net activity of the reaction product and the concentration of bismuth was linear over the range 1-25 micrograms of bismuth. The sensitivity of the proposed method appears to be better than that of any other method except the dithizone method, which has been applied only to biomaterials. Elements such as Cd, Pb, Cu, Ag, Hg, Sv and V interfere with the direct determination of bismuth, but bismuth can still be determined directly in solutions containing iron, nickel, or chromium, or in most biological samples. The results of the analyses of the N.B.S. reference standards with the preliminary extraction were in fairly good agreement with the values certified by N.B.S. (Holoman-Battelle)
W73-08790

A TECHNIQUE FOR FAST AND REPRODUCIBLE FINGERPRINTING OF BACTERIA BY PYROLYSIS MASS SPECTROMETRY,

Stichting voor Fundamenteel Onderzoek der Materie, Amsterdam (Netherlands). Instituut voor Atoom en Molecuulfysica.
H. L. C. Meuzelaar, and P. G. Kistemaker.
Analytical Chemistry, Vol 45, No 3, p 587-590, March 1973. 4 fig, 1 tab, 22 ref.

Descriptors: *Bacteria, Mass spectrometry, Pathogenic bacteria, Laboratory equipment, Pollutant identification.
Identifiers: *Pyrolysis mass spectrometry, *Sample preparation, *Freeze drying, *Characterization, Neisseria sicca, Leptospira, Neisseria meningitides, Biological samples, Mass spectra.

Neisseria sicca, N. meningitides and Leptospira strains were cultured, washed in isotonic salt solutions, centrifuged, resuspended in distilled water, and freeze-dried for analysis by pyrolysis mass spectrometry. The freeze-dried samples were suspended in carbon disulfide and small drops applied to the ferromagnetic Curie-point pyrolysis wires which were rotated to evenly distribute the sample. The mass spectrometer used was a fast scanning quadrupole mass filter with a multi-channel signal averager for recording the spectra. The pyrolysis wires operated at 510 C, and mass spectra were obtained by the accumulation of 70 scans of 0.5 sec duration. Although the results are preliminary, the method proved to be rapid, reproducible, and able to differentiate the three bacteria. Work is being continued using low-voltage, electron-impact ionization. (Little-Battelle)
W73-08792

SPECTROPHOTOMETRIC DETERMINATION OF P.P.B. LEVELS OF LONG-CHAIN AMINES IN WATERS AND RAFFINATES,

Australian Atomic Energy Commission Research Establishment, Lucas Heights.
T. M. Florence, and Y. J. Farrar.
Analytica Chimica Acta, Vol 63, No 2, p 255-261, February 1973. 3 fig, 2 tab, 3 ref.

Descriptors: *Chemical analysis, *Water analysis, *Spectrophotometry, *Methodology, Pollutant identification, Rivers, Sea water, Solvent extractions, Industrial wastes, Organic compounds, Separation techniques, Chlorides, Anions, Cations, Copper, Iron, Magnesium, Heavy metals, Chromium, Zinc, Manganese, Nickel, Nitrates, Phosphates, Freshwater, Nitrogen compounds, Water sampling, Equipment, Storage.
Identifiers: *Uranium processing raffinates, *Aliphatic amines, *Alamine 336, Ionic interference, Detection limits, Accuracy, Precision, Aliquat 336, Primene JM-T, Amberlite LA-1, Uranium, Absorbance, Trace levels, Organic solvents.

A method has been developed for the determination of traces of long-chain amines in river and sea water and uranium processing raffinates. The amine is extracted as an ion-association complex with chromate from sulfuric acid solution into chloroform and the extracted Cr (VI) is determined spectrophotometrically with diphenylcarbazide. Alamine 336, a mixture of tri-n-octyl and tri-n-decylamines, was studied using this method. Tests for adsorption losses showed that polyethylene containers are unsuitable for collection or storage but glass containers are safe for up to a week. Several metal ions (Cu (II), Fe (III), Mg (II), Mn (II), Ni (II), U (VI) and Zn (II)) were found to have no effect on the determination of Alamine 336 by the chromate method, even when present at 500 times the amine concentration. Chloride, nitrate, and phosphate at concentrations of 25, 5, and 250 ppm, respectively, could be tolerated in determining 1 ppm Alamine 336 by a single chromate extraction. The limits of detection for Alamine 336 with a 100-ml and a 250-ml sample were 15 and 7 ppb, respectively. The relative standard deviation of the method at 1.5 ppm Alamine 336 level was 1.9 percent. The application of the chromate method to the determination of some other commercial amines (Aliquat 336, Primene JM-T, Amberlite LA-1) indicates that primary, secondary, and quatenary long-chain amines can also be determined. (Holoman-Battelle)
W73-08793

TRACE DETERMINATION OF MERCURY IN BIOLOGICAL MATERIALS BY FLAMELESS ATOMIC ABSORPTION SPECTROMETRY,

Central Inst. for Industrial Research, Oslo (Norway).
S. H. Omang.
Analytica Chimica Acta, Vol 63, No 2, p 247-253, February 1973. 3 tab, 17 ref.

Descriptors: *Mercury, *Chemical analysis, *Methodology, Heavy metals, Iodides, Bromides, Anions, Pollutant identification.
Identifiers: *Sample preparation, *Wet ashing, *Biological samples, Flameless atomic absorption spectrophotometry, Chemical interference, Trace levels.

A new, rapid, wet digestion procedure which utilizes a mixture of hydrobromic acid and nitric acid has been used for preparing biological materials for analysis for the presence of mercury. The samples are boiled for 5-10 min in a mixture of nitric and hydrobromic acid under reflux. Mercury is then determined by the flameless cold-vapor, atomic absorption technique and complete release is shown to occur even when the fat present is not fully decomposed. The results are discussed with regard to the normal bromine and iodine content of marine samples. The enhancement obtained with bromine can easily be compensated by digesting the standards. Iodine, however, reduces the peak heights and the content of this element must be judged separately for samples other than fish, which can always be safety analyzed without interference. (Holoman-Battelle)
W73-08794

AN AIR-LIFT FOR SAMPLING FRESHWATER BENTHOS,

Reading Univ. (England). Dept. of Zoology.
For primary bibliographic entry see Field 07B.
W73-08795

COMPARATIVE HABITAT DIVERSITY AND FAUNAL RELATIONSHIPS BETWEEN THE PACIFIC AND CARIBBEAN DECAPOD CRUSTACEA OF PANAMA,

Miami Univ., Fla.
L. G. Abele.
Availale from Univ. Microfilms, Inc., Ann Arbor, Mich. 48106 Order No. 73-5817. Ph D Dissertation, 1972, 124 p.

Descriptors: *Distribution patterns, *Habitats, Beaches, *Mangrove swamps, *Dominant organisms, Biological communities, Pacific Ocean, Sampling, Population, Benthic fauna, Sands, Muds, Rocks, Crustaceans.
Identifiers: *Species diversity, *Decapods, Caribbean Sea, Substrates, *Panama.

The community structure of the decapod crustacean fauna of seven tropical, shallow water, marine habitats was examined. Sandy beaches, mangrove swamps and rocky intertidal habitats on both the Pacific and Caribbean coasts of Panama were sampled. The Pocillopora coral habitat of the Bay of Panama was also sampled. The species composition and their relative abundances were noted for each habitat. Four thousand three hundred and seventy-two specimens representing 247 species were collected from the seven habitats. An index of faunal similarity was calculated for each pair (Pacific-Caribbean) of habitats. This is the number of ecologically and taxonomically similar species which occurred in both habitats expressed as a percentage of the total number of species present in pair of habitats. A few specialized species dominated each of the communities. It is demonstrated that the number of species present in a habitat increases as the structural complexity of the habitat increases. Structural complexity of the habitats was approximated by the number of different types of substrates present. (Little-Battelle)
W73-08796

SIMPLE METHOD FOR CULTURING ANAEROBES,
California Univ., San Diego, La Jolla. Clinical Microbiology Lab.
C. E. Davis, W. J. Hunter, J. L. Ryan, and A. I. Braude.
Applied Microbiology, Vol 25, No 2, p 216-221, February 1973. 1 fig, 2 tab, 14 ref.

Descriptors: *Anaerobic bacteria, *Methodology, *Pathogenic bacteria, *Pollutant identification, Laboratory tests, Water pollution sources, Enteric bacteria, Isolation.
Identifiers: *Culturing techniques, *Selective media, Culture media, Pathogenicity, Actinomyces spp, Vibrio sputorum, Propionobacterium, Fusobacterium spp, Peptococcus, Sarcina, Catenabacterium spp, Streptococcus spp, Ramibacterium spp, Bacteroides spp, Gentamicin sulfate, Lactobacillus, Veillonella, Corynebacteria, Sphaerophorus necrophorus, Clostridium spp, Gaffkya anaerobia, Eubacterium spp.

A simple, effective method is needed for growing obligate anaerobes in the clinical laboratory. A pre-reduced anaerobic bottle is described that can be used for direct inoculation, provides a flat agar surface for evaluation of number and morphology of colonies, and can be incubated in conventional bacteriological incubators. Its efficiency was evaluated by testing it on a routine basis in the clinical laboratory by three criteria: (i) recovery of bacteria that did not grow aerobically but were observed in Gram stains; (ii) isolation of a wide range of pathogenic anaerobic bacteria; and (iii) comparison to a GasPak jar closed immediately after the inoculation of a single specimen and not opened for 48 hr. Each anaerobic culture set consisted of two bottles containing brain heart infusion agar and CO2. Gentamicin sulfate (50 micrograms/ml) was added to one of these to inhibit facultative enteric bacilli. Comparison of the anaerobic bottles with an identical aerobic bottle which was also routinely inoculated permitted early identification of anaerobic colonies. Representative species of most anaerobic genera of proven pathogenicity for man have been isolated from this system during 10 months of routine use. (Holoman-Battelle)
W73-08798

FLUORESCENT ANTIBODY AS A METHOD FOR THE DETECTION OF FECAL POLLUTION: ESCHERICHIA COLI AS INDICATOR ORGANISMS,
North Texas State Univ., Denton. Dept. of Biology.
R. L. Abshire, and R. K. Guthrie.
Canadian Journal of Microbiology, Vol 19, No 2, p 201-206, February 1973. 6 tab, 23 ref.

Descriptors: *E. coli, *Bioindicators, *Pollutant identification, Methodology, Water pollution, Enteric bacteria, Reliability, Waste water (Pollution), Water pollution sources, Water quality, Aerobic bacteria, Pathogenic bacteria.
Identifiers: *Fecal pollution, *Fluorescent antibody techniques, Fecal coliforms, Culture media, Sample preparation, Heterologous bacteria, Biochemical tests, Fluorescent microscopy, Aerobacter aerogenes, Klebsiella pneumoniae, Arizona arizonae, Salmonella spp, Shigella spp, Proteus vulgaris, Pseudomonas aeruginosa, Alicalgenes faecalis, Providencia, Staphylococcus aureus, Streptococcus faecalis, Serratia marcescens, Streptococcus pyogenes.

A study was made to determine whether fluorescent antibody (FA) techniques, with E. coli as an indicator organism, offer a potential as a tool for the very rapid detection of fecal pollution in water. Detection of Escherichia coli of fecal origin by this method was compared to IMVIC typing and to detection by the E C Broth method. Reliability of this method was tested by reaction of specific E. coli antisera with heterologous bacterial species and with isolates from polluted waste water and umpolluted water and soil sources. There was no attempt to detect individual strains or to differentiate between pathogenic and nonpathogenic strains. Detection of E. coli was possible by this method in 2 to 3 hr after filtration of a sample of wastewater effluent and placing the filter in lactose broth at 37 degrees C. (Holoman-Battelle)
W73-08799

SELECTRODE (TM)--THE UNIVERSAL ION-SELECTIVE ELECTRODE. PART IV. THE SOLID-STATE CADMIUM (II) SELECTRODE IN EDTA TITRATIONS AND CADMIUM BUFFERS,
Technical Univ. of Denmark, Lyngby. Kemisk Laboratorium A.
J. Ruzicka, and E. H. Hansen.
Analytica Chimica Acta, Vol 63, No 1, p 115-128, January 1973. 4 fig, 3 tab, 13 ref.

Descriptors: *Aqueous solutions, *Cadmium, *Volumetric analysis, *Evaluation, Storage, Calibrations.
Identifiers: *Selectrode, *Ion selective electrodes, *Sensitivity, Conditioning, Cadmium nitrate.

A series of cadmium buffers was prepared and used for calibration of a newly developed cadmium Selectrode activated with CdS/Ag2S. The electrode exhibits Nernstian response with a sensitivity close to the theoretical limit imposed by the conditional solubility product of cadmium sulphide, i.e., the response is linear up to pCd 9 at pH 6.7 and up to pCd 11 at pH 9. It is suitable for direct measurements of cadmium ion activities as well as for direct compleximetric titrations of cadmium ions. Prolonged storage or use in acid solutions below pH 3 increases the standard potential value and causes a gradual loss of sensitivity, because of dissolution of cadmium sulphide. When damaged, the ionsensitive surface can, however, be readily renewed as with any other Selectrode. (See also W73-01079) (Little-Battelle)
W73-08800

MODERN APPROACH TO COASTAL ZONE SURVEY,
Michigan Univ., Ann Arbor. Willow Run Labs.
For primary bibliographic entry see Field 02L.
W73-08805

SOME APPLICATIONS OF PHOTOGRAPHY, THERMAL IMAGERY AND X BAND SIDE LOOKING RADAR TO THE COASTAL ZONE,
Grumman Ecosystems Corp., Bethpage, N.Y.
For primary bibliographic entry see Field 02L.
W73-08806

THE COLUMBIA RIVER ESTUARY AND ADJACENT OCEAN WATERS, BIOENVIRONMENTAL STUDIES.
For primary bibliographic entry see Field 02L.
W73-08816

STUDIES OF THE AEROBIC, NONEXACTING, HETEROTROPHIC BACTERIA OF THE BENTHOS,
Georgia Univ., Athens. Dept. of Microbiology.
For primary bibliographic entry see Field 02L.
W73-08830

METHODS FOR THE MEASUREMENT OF HANFORD-INDUCED GAMMA RADIOACTIVITY IN THE OCEAN,
(AEC), Division of Biology and Medicine, Washington, D.C.
For primary bibliographic entry see Field 02L.
W73-08841

METAL-ION CONCENTRATIONS IN SEA-SURFACE MICROLAYER AND SIZE-SEPARATED ATMOSPHERIC AEROSOL SAMPLES IN HAWAII,
Hawaii Univ., Honolulu. Dept. of Chemistry.
For primary bibliographic entry see Field 02K.
W73-08852

RAPID CHANGES IN THE FLUORIDE TO CHLORINITY RATIO SOUTH OF GREENLAND,
Bedford Inst., Dartmouth (Nova Scotia). Atlantic Oceanographic Lab.
For primary bibliographic entry see Field 02K.
W73-08869

FORMS AND DYNAMICS OF SULFUR IN PEAT AND LAKE SEDIMENTS OF BELORUSSIA (FORMY I DINAMIKA SERY V TORFAKH I OSADKAKH OZER BELORUSSII),
Akademiya Navuk BSSR, Minsk. Institut Geokhimii i Geofiziki.
For primary bibliographic entry see Field 02J.
W73-08883

CESIUM KINETICS IN A MONTANE LAKE ECOSYSTEM,
Colorado State Univ., Fort Collins.
For primary bibliographic entry see Field 05B.
W73-08901

MERCURY IN SEDIMENTS FROM THE THAMES ESTUARY,
Institute of Geological Sciences, London (England). Geochemical Div.
J. D. Smith, R. A. Nicholson, and P. J. Moore.
Environmental Pollution, Vol 4, No 2, p 153-157, February 1973. 2 fig, 1 tab, 3 ref.

Descriptors: *Mercury, *Sediments, *Soil analysis, *Chemical analysis, *Estuaries, Aquatic soils, Heavy metals, Saline soils, Sampling, Pollutant identification.
Identifiers: *Thames estuary, *Flameless atomic absorption spectrophotometry, Sample preparation.

Fresh unconsolidated surface sediment samples from the outer Thames estuary and samples from a

MICRODETERMINATION OF VOLATILE ORGANICS BY GALVANIC COULOMETRY,
Minnesota Univ., Minneapolis. Lab. for Biophysical Chemistry.
A. Anusiem, and P. A. Hersch.
Analytical Chemistry, Vol 45, No 3, p 592-594, March 1973. 3 fig, 2 tab, 7 ref.

Descriptors: *Organic compounds, *Volatility, *Aqueous solutions, Electrolytes, Oxygen, Pollutant identification, Chemical analysis, Methodology.
Identifiers: *Galvanic coulometry, *Trace levels, *Hydrocarbons, Nitrobenzene, Sensitivity.

The principle of determing organic gases and vapors carried by an inert gas stream upon adding a constant proportion of O2 to the stream, passing the stream through a hot tube for complete combustion, and determing the O2 left over in the effluent, can be extended to hydrocarbons and other slightly soluble species in an aqueous sample. Experiments conducted using stock solutions prepared from recrystallized nitrosobenzene confirm that combustion of trace organics followed by galvanic coulometry of the residual oxygen yields accurate analyses in the microgram range with only modest demands on time, training, and equipment. The range can probably be extended to much smaller quantities, given electronic amplification of the signal, before limitations by various types of noise are encountered. Ultimate sensitivity is expected to exceed that of flame ionization which contributes about 1 in 100,000 carbon atoms to the signal. Coulometry utilizes every carbon and hydrogen atom combusted. This principle should be applicable to effluent analysis in both gas and liquid chromatography and to many micro-analytical requirements of organic and biological chemistry. (Holoman-Battelle)
W73-08908

CALCIUM ION-SELECTIVE ELECTRODE IN WHICH A MEMBRANE CONTACTS GRAPHITE DIRECTLY,
Lever Bros. Co., Edgewater, N.J. Research and Development Div.
For primary bibliographic entry see Field 02K.
W73-08910

DECOMPOSITION OF OIL POLLUTANTS IN NATURAL BOTTOM SEDIMENTS,
Rutgers - The State Univ., New Brunswick, N.J.
For primary bibliographic entry see Field 05B.
W73-08911

THE DISTRIBUTION, SUBSTRATE SELECTION AND SEDIMENT DISPLACEMENT OF COROPHIUM SALMONIS (STIMPSON) AND COROPHIUM SPINICORNE (STIMPSON) ON THE COAST OF OREGON,
Oregon State Univ., Corvallis.
For primary bibliographic entry see Field 05B.
W73-08912

ANALYSIS OF HALOGENATED BIPHENYLS BY PULSED SOURCE-TIME RESOLVED PHOSPHORIMETRY,
Florida Univ., Gainesville. Dept. of Chemistry.
C. M. O'Donnell, K. F. Harbaugh, R. P. Fisher, and J. D. Winefordner.
Analytical Chemistry, Vol 45, No 3, p 609-611, March 1973. 2 fig, 1 tab, 6 ref.

Descriptors: *Chemical analysis, *Methodology, Polychlorinated biphenyls, Organic compounds, Pollutant identification Separation techniques.
Identifiers: *Halogenated biphenyls, *Quantitative analysis, *Pulsed source time resolved phosphorimetry, *Mixtures, Errors, Precision, Phosphorescence spectra, 2-chlorobiphenyl, 4-chlorobiphenyl, 4-bromobiphenyl, 4-iodobiphenyl.

Pulsed source, time resolved phosphorimetry has been used for the quantitative analysis of several mixtures of halogenated biphenyls. The procedure followed was identical to that of Fisher and Winefordner (1972) except for changes involving delay and sweep times with reference to recording the phosphorescence decay curve. The sweep time was chosen to temporally compress the shorter-lived species with respect to the longest-lived phosphor; the resulting phosphorescence signal of the longest-lived species extrapolated to zero delay time was subtracted from the sum of the two longest-lived phosphors to determine the phosphorescence signal of the intermediate lived species. Analysis of a 4-halobiphenyl mixture in ethanolic solvent resulted in a maximum absolute error in concentration of 5 percent or less for 4-chlorobiphenyl and 4-bromobiphenyl and 14 percent for the shortest-lived species, 4-iodobiphenyl. A mixture of 2-chlorobiphenyl and 4-chlorobiphenyl similarly measured resulted in errors of 5.4 and 10 percent, respectively. With this method it is possible to analyze structurally and spectrally similar molecules. Quantitative information and qualitative identification of species aided by phosphorescence lifetimes can be obtained. For very complex mixtures, a simple thin layer or gas chromatographic separation prior to analysis may also be needed, but overall, the time for analysis of a multicomponent mixture should be considerably reduced as compared to methods previously used. (Holoman-Battelle)
W73-08913

EFFECTS OF SIZE-SELECTIVE PREDATION ON COMMUNITY STRUCTURE IN LABORATORY AQUATIC MICROCOSMS,
Texas Univ., Austin.
For primary bibliographic entry see Field 05C.
W73-08914

INFLUENCE OF PH ON THE RESPONSE OF A CYANIDE ION SELECTIVE MEMBRANE ELECTRODE,
Rome Univ. (Italy). Istituto di Chimica Analitica.
M. Mascini.
Analytical Chemistry, Vol 45, No 3, p 614-615, March 1973. 1 fig, 7 ref.

Descriptors: *Hydrogen ion concentration, *Iodides, *Chemical reactions, Aqueous solutions, Electrochemistry.
Identifiers: *Ion selective electrodes, *Cyanides, Accuracy, Errors, Reaction products.

Chemical reactions are given in contrast with those given previously to better describe the effect of pH on the response of a cyanide ion-selective electrode membrane. Values of iodide and cyanide concentrations as a function of pH obtained with the new equation show that the results from previous determinations agree only above pH 9. It is concluded that the response of the cyanide electrode is affected by undissociated HCN which can generate I ions. The same considerations can be used to explain the behavior of cyanide selective membrane electrodes in the presence of a metal ion that binds CN (-) in strong or weak complexes. (Little-Battelle)
W73-08916

SIGNIFICANCE OF CELLULOSE PRODUCTION BY PLANKTONIC ALGAE IN LACUSTRINE ENVIRONMENTS,
Massachusetts Univ., Amherst.
For primary bibliographic entry see Field 05C.
W73-08917

RAPID METHYLATION OF MICRO AMOUNTS OF NONVOLATILE ACIDS,
Pittsburgh Univ., Pa. School of Medicine.
M. J. Levitt.
Analytical Chemistry, Vol 45, No 3, p 618-620, March 1973. 1 fig, 1 tab, 11 ref.

Descriptors: *Organic acids, *Laboratory equipment, Chemical analysis, Research equipment, Design, Pollutant identification, Methodology.
Identifiers: *Methylation, *Nonvolatile acids, *Esterification, *Sample preparation, *Diazomethane generating system, Electron capture gas chromatography, Methyl esters, Lithocholic acid.

A Diazomethane Generating System is described which permits microgram or smaller amounts of complex organic acids to be quantitatively esterified without detectable side-product formation. Multiple samples can be processed in sequence at two-minute intervals. The methyl esters produced are suitable for further analysis by electron-capture gas chromatography. The apparatus is constructed from components present in most laboratories, and may be dismantled easily for storage. Details are given on the construction and operation of the esterification apparatus. (Holoman-Battelle)
W73-08918

DESIGN AND EVALUATION OF A LOW COST RECORDING SPECTROPOLARIMETER,
Cleveland State Univ., Ohio. Dept. of Chemistry.
For primary bibliographic entry see Field 07B.
W73-08919

IMPROVED EXPERIMENTAL TECHNIQUE FOR REVERSE ISOTOPE DILUTION METHOD,
Rohm and Haas Co., Bristol, Pa. Bristol Research Labs.
W. H. Graham, and W. E. Bornak.
Analytical Chemistry, Vol 45, No 3, p 623-624, March 1973. 2 tab, 1 ref.

Descriptors: *Radioactivity techniques, *Methodology, *Pollutant identification, *Radiochemical analysis, *Pollutants, Chemical analysis, Pesticides, Pesticide residues, Radioactivity, Laboratory equipment, Radioisotopes.

The carrier-addition or reverse isotope dilution (RID) method is a useful technique for both qualitative and quantitative identification of radioactive compounds in complex mixtures. Described are the principles involved, the apparatus and procedures used, and the results obtained in trials of a modified RID technique. In the modified RID method, the test of constant specific radioactivity is determined differently. Instead of isolating the crystalline material after each recrystallization, it is necessary to radioassay only the mother liquor. Two consecutive recrystallizations of a chemically pure but radiolabeled material will contain the same amount of radioactivity in the same volume of mother liquor at constant temperature. At that point, the crystalline material may be isolated and radioassayed to determine its specific activity and complete the analysis. The advantages of the method are speed and simplicity; the small amount of cold carrier and radioactive material needed; ease of handling sensitive carrier compounds; and increased boiling range of solvent systems, including mixed solvents, available through use of pressure vials. The RID method is particularly useful in pesticide residue analysis because trace levels of multiple degradation products can be determined quantitatively from the radioactive parent pesticide. (Holoman-Battelle)
W73-08920

DATA PROCESSING ESTIMATIONS FOR SAMPLE SURVEYS,
Georgia Univ., Athens.
For primary bibliographic entry see Field 07C.
W73-08921

DIRECT QUALITATIVE AND QUANTITATIVE INFRARED SPECTROSCOPIC ANALYSIS OF ZINC-DIALKYLDITHIOCARBAMATE WITHOUT PRIOR SEPARATION, (DIREKTE QUALITATIVE UND QUANLITATIVE INFRAROTSPEKTROSKOPISCHE ANALYSE VON ZINK-DIALKYLDITHIOCARBAMIDATEN OHNE VORHERGEHENDE AUFTRENNUNG),
Technische Hochschule, Vienna (Austria). Institut fuer Analytische Chemie und Mikrochemie.
W. Sztark, H. Malissa, and R. Kellner.
Analytica Chimica Acta, Vol 63, No 2, p 285-293, February 1973. 13 fig, 2 tab, 19 ref.

Descriptors: *Pollutant identification, Chemical analysis, Heavy metals, Methodology.
Identifiers: *Mixtures, *Zinc-dialkyldithiocarbamates, *Infrared spectrophotometry, Infrared spectra, Detection limits, Quantitative analysis, Zinc-dimethyldithiocarbamate, Zinc-diaethyldithiocarbamate, Zinc-dibutyldithiocarbamate, Zinc-dibenzyldithiocarbamate, Organometallic compounds.

A new method for the i.r. analysis of zinc-dialkyldithiocarbamate mixtures has been developed. The identification and detection limits were improved, compared to the standard method, by the use of microtechniques (identification limit 2 micrograms, detection limit 30-80 micrograms). Binary and ternary mixtures can be analyzed without preliminary separation, because the other components of the mixtures have very little influence on the measured absorption band. Beer's law was obeyed for the binary and two ternary mixtures. The relative standard deviations and the coefficients of correlation were calculated for the macro- and micromethods. A comparison of statistical data shows that i.r. spectroscopy is suitable for quantitative analysis even in the region of 80-10 micrograms of substance. (See also W73-08923) (Holoman-Battelle)
W73-08922

A METHOD FOR THE QUALITATIVE AND QUANTITATIVE INFRARED SPECTROSCOPIC ANALYSIS OF METAL-TETRAMETHYLENEDITHIOCARBAMATE MIXTURES WITHOUT PRIOR SEPARATION (EINE METHODE ZUR QUALITATIVEN UND QUANTITATIVEN INFRAROTSPEKTROSKOPISCHEN ANALYSE VON METALL-TETRA-METHYLENDITHIOCARBAMIDAT-GEMISCHEN OHNE VORHERGEHENDE AUFTRENNUNG),
Technische Hochschule, Vienna (Austria). Institut fuer Analytische Chemie und Mikrochemie.
H. Malissa, and R. Kellner.
Analytica Chimica Acta, Vol 63, No 2, p 263-275, February 1973. 10 fig, 3 tab, 11 ref.

Descriptors: *Chemical analysis, *Methodology, *Pollutant identification, *Trace elements, *Aqueous solutions, Chelation, Heavy metals, Mercury, Iron, Cobalt, Nickel, Sodium, Alkali metals.
Identifiers: *Metal complexes, *Infrared spectrophotometry, *Quantitative analysis, *Mixtures, Infrared spectra, Metal chelates, Detection limits, Lead-tetramethylenedithiocarbamate, Mercury-tetramethylenedithiocarbamate, Iron-tetramethylenedithiocarbamate, Cobalt-tetramethylenedithiocarbamate, Nickel-tetramethylenedithiocarbamate.

A method for the direct i.r. analysis of metal-tetramethylenedithiocarbamate mixtures without preceding separation is proposed. Because of the remarkable shifts of the M-S-stretching vibrations and the small losses of energy by stray light, the region beyond 400/cm is the most useful for a determination in potassium bromide. Model analyses with standard mixtures of Pb (II)- and Hg (II)-TMDTC, and Fe (III)-, Co (III)- and Ni (II)-TMDTC are described. The method can be used for the determination of trace amounts of heavy metals in aqueous solutions after complexation with Na-TMDTC and extraction with chloroform. The detection limit for application of the 13-mm standard KBr method and a routine spectrometer (without ordinate expansion) is about 10 micrograms of metal. (See also W73-08922) (Holoman-Battelle)
W73-08923

STATISTICAL METHODS FOR THE DETECTION OF POLLUTION UTILIZING FISH MOVEMENTS,
Virginia Polytechnic Inst. and State Univ., Blacksburg.
For primary bibliographic entry see Field 05B.
W73-08924

RESULT OF FLUVIAL ZINC POLLUTION ON THE ZINC CONTENT OF LITTORAL AND SUB-LITTORAL ORGANISMS IN CARDIGAN BAY, WALES,
University Coll. of Wales, Aberystwyth. Dept. of Zoology.
For primary bibliographic entry see Field 05B.
W73-08925

(I) OBSERVATIONS OF SMALL PHOTOCURRENTS ON THE DROPPING MERCURY ELECTRODE AND (II) A STUDY OF THE ADSORPTION OF TRIFLUOROMETHYL SULFONATE ON MERCURY,
Georgetown Univ., Washington, D.C.
J. C. Kreuser.
Available from Univ. Microfilms, Inc., Ann Arbor, Mich. 48106. Order No. 72-34,182. Ph D Dissertation, 1972. 332 p.

Descriptors: *Adsorption, *Aqueous solutions, *Electrochemistry, Light.
Identifiers: *Photocurrents, *Dropping mercury electrode, *Trifluoromethyl sulfonate, Ion selective electrodes.

Part I deals with the photoemission of electrons from mercury in aqueous solutions. An instrument was built to measure photocurrents, and experimental results obtained with it are described. Mathematical methods used to analyze the data are shown in detail. Threshold potentials E sub T obtained by extrapolation of current-voltage curves do not show the expected dependence on light energy E sub L, i.e., d E sub T/d E sub L is not equal to 1. Several possible sources of error have been investigated, but no satisfactory explanation has been found. Part II deals with the specific adsorption of the trifluoromethyl sulfonate (CF3SO3 (-)) anion on the dropping mercury electrode, using a constant ionic strength approach. Data were taken with the system described in Part I, and analyzed directly by computer. Results indicated adsorption of CF3SO3 (-) is similar to that of Cl04 (-) and NO3 (-). (Little-Battelle)
W73-08926

SOUND VELOCIMETERS MONITOR PROCESS STREAMS,
NUSonics, Inc. Paramus, N.J.
E. M. Zacharias, Jr., and D. W. Franz.
Chemical Engineering, Vol 80, No 2, p 101-108, January 22, 1973. 6 fig, 1 tab, 9 ref.

Descriptors: *Monitoring, *Aqueous solutions, *Flowmeters, *Flow rates, *Specific gravity, *Density, *Suspended solids, *Compressibility, Measurement, Flow measurement, Electronic equipment.
Identifiers: *Sound velocimeters, Sound waves.

Sound velocimeters have been developed which can be used to accurately monitor changes in density, specific gravity, bulk modulus, concentration of solutions, solids content, and percent conversion of monomers to polymers in process streams. The instrumemts employ an electro-acoustic amplifier, which acts as a transmitter and receiver,

Nitella stuartii, Nitella myriotricha, Nitella leptostachys, Nitella tasmanica, Nitella furcata, Nitella pseudoflabellata, Nitella subtilissima, Nitella hyalina.

A floristic study of the Characeae of Australia was made based upon extensive collecting in 1960-61 throughout the continent together with a re-examination of available herbarium specimens. Twenty-nine species are treated in accordance with the author's revised classification; and a key, brief descriptions, figures, and lists of specimens cited are provided. Five genera are represented--Chara, Lamprothamnium, Lychnothamnus, Nitella, and Tolypella. The most common species in decreasing order of frequency of occurrence are C. corallina, N. pseudofalbellata, Lamp. papulosum, N. cristat, and N. lhotzkyi. Examples of species widespread in the world which are least common in Australia are C. zeylanica and N. acuminata. Eight species are endemic--C. leptopitys, N. congesta, N. cristat, N. lhotzkyi, N. penicillata, N. subtilissima, N. tumida, and N. verticillata. Of these, all but the last are known to be dioecious. The results generally support the author's earlier revision, but newly-rediscovered type material (MEL) requires several changes and reinterpretations of the literature. In addition, four new varieties are described, and seven new varietal combinations are made. (Holoman-Battelle)
W73-08937

THE DYNAMICS OF BROWN TROUT (SALMO TRUTTA) AND SCULPIN (COTTUS SPP.) POPULATIONS AS INDICATORS OF EUTROPHICATION,

Michigan State Univ., East Lansing.
W. L. Smith.
Available from Univ. Microfilms, Inc., Ann Arbor, Mich. 48106 Order No. 73-5491. Ph D Dissertation, 1972. 53 p.

Descriptors: *Brown trout, *Sculpins, *Bioindicators, *Eutrophication, Fish populations, Coldwater fish, Freshwater fish, Fish reproduction, Fish physiology, Survival, Michigan, Pollutant identification.
Identifiers: Salmo trutta, Cottus spp, Jordan River, Au Sable River, Population density.

Brown trout and sculpin populations were studied in three variously perturbed stream sites in northern Michigan. Intraspecific comparisons were made of several aspects of population dynamics including the intrinsic rate of natural increase (r). The upper Jordan River, nearly pristine and with high population densities, exhibited r values judged adequate for maintenance of the populations. The other sites were compared with this baseline. The moderately perturbed lower Jordan River had less population densities and survival but greater mean fecundities for both species. This resulted in a positive r for the trout but the birth rate of the sculpins could not compensate for the death rate and the population was declining. The Au Sable River, the most eutrophic, also had lesser population densities and greater mean fecundities than the upper Jordan. Survivorship of the sculpins was sufficient to yield a positive r. The low survival of the trout resulted in a strongly negative r, suggesting inability of the population to sustain itself. The intrinsic rate of natural increase of short-lived, coldwater fish species could be a useful tool in monitoring water quality, especially if studies were continued through several generations. (Holoman-Battelle)
W73-08940

ISOLATION OF BACTERIA CAPABLE OF UTILIZING METHANE AS A HYDROGEN DONOR IN THE PROCESS OF DENITRIFICATION,

National Inst. for Water Research, Pretoria (South Africa).
T. R. Davies.
Water Research, Vol 7, No 4, p 575-579, April 1973. 2 tab, 14 ref.

Descriptors: *Methane bacteria, *Denitrification, Laboratory tests, *Isolation, *Pollutant identification, Methane, Nitrogen, Hydrogen, Water pollution sources, Anaerobic conditions.
Identifiers: *Substrate utilization, Energy sources, *Enrichment, *Culture media, Fate of pollutants, Growth, Methanol, Ethanol, Malate, Lactate, Methanomonas, Culturing techniques, Alcaligenes spp, Achromobacter spp, Pseudomonas spp, Bacillus, Micrococcus denitrificans, Substrates.

Bacteria were isolated which are able to denitrify when methane is supplied as the sole carbon source. These isolates were not found to be specific to methane, but could use other carbon compounds as hydrogen donors. They were capable of using nitrate both as a source of cell nitrogen and as an alternative terminal electron acceptor to oxygen. Results obtained from a laboratory scale dentrifying unit indicated that denitrification with methane as the energy source could become an attractive commercial proposition. (Holoman-Battelle)
W73-08941

MOLECULAR ARCHITECTURE OF THE CHLOROPLAST MEMBRANES OF CHLAMYDOMONAS REINHARDI AS REVEALED BY HIGH RESOLUTION ELECTRON MICROSCOPY,

California Univ., Los Angeles.
For primary bibliographic entry see Field 05C.
W73-08942

SORPTION OF COPPER ON LAKE MONONA SEDIMENTS - EFFECT OF NTA ON COPPER RELEASE FROM SEDIMENTS,

Wisconsin Univ., Madison. Water Chemistry Program.
For primary bibliographic entry see Field 05B.
W73-08946

MASS AND MONOXENIC CULTURE OF VORTICELLA MICROSTOMA ISOLATED FROM ACTIVATED SLUDGE,

Tokyo Univ. (Japan). Inst. of Applied Microbiology.
R. Sudo, and S. Aiba.
Water Research, Vol 7, No 4, p 615-621, April 1973. 3 fig, 1 tab, 13 ref.

Descriptors: *Protozoa, Growth rates, *Activated sludge, *Isolation, Animal growth, Cultures, Sewage bacteria, Biomass, Animal populations, Invertebrates, Pollutant identification.
Identifiers: *Mass cultures, *Monoxenic cultures, *Vorticella microstoma, Alcaligenes faecalis, Culture media.

Following previous work on the monoxenic culture of Vorticella microstoma isolated from activated sludge, Sudo and Aiba (1971), a mass culture of the protozoa was attempted. An appropriate range of phosphate buffer concentration in cultivating the protozoa, using sludge bacteria as food, was from 1/150 to 1/75 M, and the optimum pH value of the culture medium ranged from 6.5 to 7.5. For a measurement of dry mass of single cells of Vorticella microstoma, and also for an assessment of some relationship between protozoan growth and bacterial consumption, a pure culture of Alcaligenes faecalis was given as food. The specific growth rate, mu (equals 1.5-1.8/day) observed with the protozoa was independent of the inoculum size, provided the bacterial concentration was less than 200 mg/l (360,000,000/ml). The yield on conversion from Alcaligenes faecalis to Vorticella microstoma was 0.47. (Holoman-Battelle)
W73-08950

THE USE OF COLIPHAGE AS AN INDEX OF HUMAN ENTEROVIRUS POLLUTION IN AN ESTUARINE ENVIRONMENT,
New Hampshire Univ., Durham.
J. M. Vaughn.
Available from Univ. Microfilms, Inc., Ann Arbor, Mich. 48106 Order No. 73-4361. Ph D Dissertation 1972. 68 p.

Descriptors: *Bioindicators, *Estuaries, Viruses, Sewage effluents, Oysters, E. coli, Enteric bacteria.
Identifiers: *Shellfish waters, *Coliphages, *Enterovirus, Survival, Characterization, *Biological samples, Fecal pollution, Isolation, Macroinvertebrates.

Parallel examinations of sewage effluents, shellfish and shellfish growing waters for coliphage and enteric virus indicated a wide dissemination of coliphage throughout the estuary, generally occurring in the absence of detectable enteric virus activity. A majority of the enteric virus isolations were observed in samples yielding no coliphage activity. Under controlled conditions, oysters were observed to accumulate more coliphage than enteric virus. Replication of coliphage in the estuary during the summer months was shown to occur when proper host cell was present. Two major coliphage types were observed in field samples based on their reactivity with different Escherichia coli strains. Survival times of coliphage and enteric virus in estuarine waters along with retention values in oysters were shown to be similar with a slight advantage shown by coliphage. Inability to correlate accurately coliphage and enteric virus occurrence in field samples along with the potential for the presence of more than one dominant coliphage type indicated the serious shortcomings of the coliphage indicator system as a method of enteric virus detection. A secondary characterization study was performed on one of the two dominant bacteriophage types occurring in field samples. Nutritional studies revealed an absolute requirement for copper ions. (Little-Battelle)
W73-08951

PULSE POLAROGRAPHY IN PROCESS ANALYSIS. DETERMINATION OF FERRIC, FERROUS, AND CUPRIC IONS,
North American Rockwell Corp., Thousand Oaks, Calif. Science Center.
E. P. Parry, and D. P. Anderson.
Analytical Chemistry, Vol 45, No 3, p 458-463, March 1973. 6 fig, 1 tab, 7 ref.

Descriptors: *Polarographic analysis, *Chemical analysis, *Pollutant identification, Sampling, Cations, Iron, Copper, Electrolytes, Methodology, Automation, Kinetics.
Identifiers: *Pulse polarography, *Process streams, Chemical interference, Pyrophosphate, Sensitivity, Reproducibility.

A form of pulse polarography was described which can be used for automated analysis of process streams. The technique is rapid, has good sensitivity, and is not severely affected by small amounts of oxygen. For the automated analysis, a sample (5-10 microliters) was drawn from the sample stream and added to the polarographic cell. Sulfuric acid was added, the solution bubbled for 3 minutes, and the potential pulse was applied to determine Cu. Pyrophosphate solution adjusted to 0.1 M and 0.002 percent in Triton X-100 was added and the ferric and ferrous ions were determined. Only one 50-millisecond pulse to the diffusion plateau for each species to be analyzed was required for the analysis after suitable calibration. The choice of supporting electrolyte (s) is of great importance in the successful application of the technique. For determination of two different oxidation states of the same species (or where both oxidation states can be present), it is necessary that the pulse polarographic wave be irreversible. The kinetic parameters are discussed. Pyrophosphate solution was shown to be a suitable medium for the simultaneous determination of ferric and ferrous ions and the pulse polarographic behavior of these ions, as well as cupric ion in this medium, is described in detail. (Holoman-Battelle)
W73-08952

SOLID STATE ION-SELECTIVE MICROELECTRODES FOR HEAVY METALS AND HALIDES,
State Univ. of New York, Buffalo. Dept. of Chemistry.
J. D. Czaban, and G. A. Rechnitz.
Analytical Chemistry, Vol 45, No 3, p 471-474, March 1973. 4 fig, 11 ref.

Descriptors: *Heavy metals, *Halides, *Anions, *Cations, Copper, Cadmium, Lead, Chlorides, Bromides, Iodides, Design, Selectivity, Physical properties, Electrical properties.
Identifiers: *Microelectrodes, *Ion selective electrodes, Silver, Sensors, Solid state electrodes.

Ion-selective microelectrodes with tip diameters of 100-150 microns of the type (1) metal sulfide/silver sulfide, selective towards Ag, Cu, Pb, and Cd cations, and (2) silver halide/silver sulfide, selective towards sulfide, bromide, chloride, and iodide anions were constructed and evaluated in solution volumes of 0.5-1.0 microliters. Details are given for the preparation of materials, forming of membranes, machining of microtips, and construction of probe electrodes. Successful microelectrodes for Ag, Cu, Cd, Pb, chloride, bromide and iodide are reported. (Holoman-Batelle)
W73-08954

ON THE MICROBIOLOGY OF SLIME LAYERS FORMED ON IMMERSED MATERIALS IN A MARINE ENVIRONMENT,
Hawaii Univ., Honolulu.
For primary bibliographic entry see Field 05C.
W73-08955

GAS-SOLID CHROMATOGRAPHY OF ORGANIC ACIDS AND AMINES USING STEAM CONTAINING FORMIC ACID OR HYDRAZINE HYDRATE AS CARRIER GASES,
Tokyo Univ. of Education (Japan). Inst. for Optical Research.
A. Nonaka.
Analytical Chemistry, Vol 45, No 3, p 483-487, March 1973. 4 fig, 13 ref.

Descriptors: *Organic acids, Methodology, *Organic compounds, Acidity, Alkalinity, Laboratory equipment.
Identifiers: *Amines, *Steam carrier gas solid chromatography, Gas solid chromatography, Carrier gas, Formic acid, Hydrazine hydrate, Adsorbents, Flame ionization detector, Sample preparation, Chromatography columns, Fatty acids, Stearic acid, Ethylenediamine, Ethanolamine, n-propylamine, n-butylamine, Amylamine, n-hexylamine, n-octylamine, Laurylamine, Stearylamine, Acetic acid, Propionic acid, n-caproic acid, Enanthic acid, Lauric acid, Myristic acid, Palmitic acid.

Gas-solid chromatography can be carried out for samples of free organic acids and amines using carrier steam containing 10 percent formic acid or 10 to 20 percent hydrazine hydrate. The mixed carrier vapors are introduced into the column by pumping aqueous solutions of formic acid or hydrazine hydrate into a vaporizing port set in the GC system. The adsorbents, such as diatomaceous firebrick and porous glass beads, can be used as stationary solids, without any coating. The FID can be employed as a detector with the mixed carriers. The effect of stream and the added polar vapors is so significant that acid and amine samples are eluted very rapidly without any marked tailing. By changing the carrier vapor from the acidic to a basic one, acid and amine samples can be analyzed on the same column. Chromatograms of lower and higher fatty acids and their alkali salts, lower and higher fatty amines, etc. is reported. (Holoman-Battelle)
W73-08956

PERIPHYTON PRODUCTION AND GRAZING RATES IN A STREAM MEASURED WITH A P-32 MATERIAL BALANCE METHOD,
Oak Ridge National Lab., Tenn.
J. W. Elwood, and D. J. Nelson.
Oikos, Vol 23, No 3, p 295-303, 1972. 6 fig, 2 tab, 29 ref.

Descriptors: *Standing crops, *Grazing, *Periphyton, *Biomass, *Radioactivity techniques, *Primary productivity, Phosphates, Streams, Cycling nutrients, Sampling, Water analysis, Snails, Mollusks, Secondary productivity.
Identifiers: Goniobasis clavaeformis, Macroinvertebrates.

Net production rates and standing crops of periphyton and grazing rates on periphyton were measured in a small, woodland stream in southeastern United States using a material balance method. The material balance of radioactive phosphorus was followed at three times during the year in periphyton, consumer organisms and stream water for up to six weeks following a one-hr release of P-32 labelled PO4 to the stream. Rates of decrease of P-32 in periphyton per unit weight,per unit area of substrate, and in the entire study reach of stream were used to compute biomass turnover rates. Periphytonstanding crops in July, September, and November were estimated at 200, 198, and 658 mg ash-free dry weight per sq m respectively, while estimates of net production rates were 22, 24, and 16 mg ash-free dry weight per sq m per day, respectively. Estimated grazing rates on periphyton during these periods were 23, 15, and 14 mg ash-free dry wt per sq m per day, respectively. Biomass turnover rates of periphyton suggest that grazing limited periphyton production rates in this stream by controlling the standing crop of periphyton. The method is of wide applicability since less than maximum permissible concentrations of radioactive phosphorus were used in the spike releases. (Little-Battelle)
W73-08958

DEOXYRIBONUCLEIC ACID IN ECOSYSTEMS,
North Carolina Univ., Chapel Hill.
M. J. Canoy.
Available from Univ. Microfilms, Inc., Ann Arbor, Mich. 48106 Order No. 73-4806. Ph D Dissertation, 1972. 355 p.

Descriptors: *Trophic level, *Biomass, *Primary productivity, *Respiration, *Ecosystems, Stress, Aquatic plants, Algae, *Chlorophyll, Model studies, Secondary productivity.
Identifiers: *DNA, Chlorophyll a, Thalassia.

Thirteen temperate and tropical ecosystems were sampled for DNA content per gram dry weight of major species making up 90 percent or more of the biomass. Mean DNA/g/ species data were used to derive mean DNA/sq m concentrations. Tropical and temperate producers were found to have a range of DNA contents from 0.02 mg/g to 0.72 mg/g with a mean of 0.36 mg/g for aquatic plants and a range of 0.22 mg/g to 1.20 mg/g for terrestrial producers with a mean of 0.46 mg/g. Mean DNA content was very similar for all systems studied. That of producer systems was 0.40-0.50 mg/g and that of consumers was 4.50-6.00 mg/g. Mean DNA concentration per square meter was also similar for normal systems of similar type. Attached aquatic systems had 83.00-420.00 mg/sq m. Calculations of DNA by trophic level indicate that for mature systems, where most major elements are considered, the trophic level biomass pyramid can

Descriptors: *Electron microscopy, *Slime, Cultures.
Identifiers: *Freeze etching, *Klebsiella pneumoniae, *Enterobacter aerogenes, Characterization, Cell structures.

The capsule of Klebsiella pneumoniae and slime of Enterobacter aerogenes A3 (SL) were examined by electron microscopy using the freeze-etch technique, which permits observation of hydrated specimens. The capsule of K. pneumoniae was composed of several layers containing many fibers 10 nm thick; while the polysaccharide slime of E. aerogenes A3 (SL) was composed of a diffuse network of fibrils. This work represents the first time the image of a hydrated bacterial capsule or slime has been observed in the electron microscope. The slime of E. aerogenes A3 (SL) resembled the layered structure of the capsule of K. pneumoniae. Freeze-etching of the bacterial plasma membrane revealed a mosaic of glycoprotein granules (10 nm in diameter) on the fracture surface. A capsule of phagocytized K. pneumoniae was observed to be a layered structure resembling the freeze-etched preparations of pure cultures of K. pneumoniae. (Little-Battelle)
W73-08966

THE ISOLATION AND CHARACTERIZATION OF A HITHERTO UNDESCRIBED GRAM-NEGATIVE BACTERIUM,
North Texas State Univ., Denton.
C. B. Lassiter.
Available from Univ. Microfilms, Inc., Ann Arbor, Mich. 48106 Order No. 73-2916. Ph D Dissertation, 1972. 144 p.

Descriptors: *Cultures, Toxicity, Heavy metals, Salinity, *Pollutant identification, Separation techniques, *Pseudomonas.
Identifiers: Characterization, Pseudomonas multiflagella, Culture media, *Isolation.

A unique undescribed gram-negative rod is extensively characterized. The most distinguishing characteristic of the water isolate is a polar tuft of 35-40 flagella that aggregate to function as a single organelle which is visible under phase contrast. Aging cells deposit poly-beta-hydroxybutyric acid granules. It also possesses an unusual exterior membrane outside the cell wall which contains large fibrils of protein. The organism grows optimally at 37 C, under well aerated conditions. The optimum beginning pH is 7.0. This organism does not ferment carbohydrates nor can it utilize them as a source of carbon and energy. It is catalase negative and reduces nitrates to nitrites. It does not liquefy gelatin or attack starch. It is oxidase-positive but does not produce indole or hydrogen sulfide. It has no vitamin or amino acid requirement and grows well when many of the amino acids in the aspartic acid family are offered as a substrate. It was only able to utilize sodium acetate and malic acid as sources of carbon and energy. It is best cultivated in the laboratory in a medium containing 0.2 percent sodium acetate, 0.01 percent calcium chloride, and 0.05 percent sodium acetate. It cannot survive sodium chloride concentrations above 0.5 percent. Metal ions were found to be toxic to the organism, with the divalent metal ions being approximately twice as toxic as the monovalent ions. After careful evaluation of the findings, the new bacterium was subsequently named Pseudomonas multiflagella. (Little-Battelle)
W73-08968

LEUCOTHRIX MUCOR AS AN ALGAL EPIPHYTE IN THE MARINE ENVIRONMENT,
Indiana Univ., Bloomington.
J. A. Bland.
Available from Univ. Microfilms, Inc., Ann Arbor, Mich. 48106 Order No. 73-2688. Ph D Dissertation, 1972. 179 p.

Descriptors: *Cultures, *Rhodophyta, *Artificial substrates, *Hosts, Marine Bacteria, Sea water, On-site tests.
Identifiers: *Leucothrix mucor, *Survival, Epiphytes, Bangia, Porphyra.

This study was concerned with the marine bacterium Leucothrix mucor as an algal epiphyte and showed (1) that the algal-bacterial relationship is essential for the survival of leucothrix; and (2) that Leucothrix is an algal epiphyte throughout the intertidal zone. These conclusions were verified by first selecting and field testing a suitable artificial substrate so that the role of the natural living host might be evaluated more easily. Polypropylene strips were chosen, and found to accumulate the benthic flora of the algae which served as a substrate for attachment by L. mucor. The second technique developed was a method to quantitate populations of Leucothrix on any given substrate. The technique is based on the optical design of the microscope being used. Experimental results showed that Leucothrix would attach and grow on the artificial substrate, but was crowded out rapidly by the other attaching benthic forms. When L. mucor was attached to the strips in the laboratory and then incubated in natural environments, growth occurred, but was poor. When given a choice between natural and artificial substrates, Leucothrix attached preferentially to the natural host. Thus the conclusion that the natural host is essential for the survival of Leucothrix was supported. Particular species of red algae, Bangia and Porphyra, were found to be the preferred hosts. (Little-Battelle)
W73-08970

AERIAL REMOTE SENSING, A BIBLIOGRAPHY.
Office of Water Resources Research, Washington, D.C.
For primary bibliographic entry see Field 07B.
W73-08974

STUDIES ON SOME EGYPTIAN FRESHWATER ISOLATES OF UNICELLULAR GREEN ALGAE,
Ain Shams Univ., Cairo (Egypt). Dept. of Botany.
S. M. Taha, S. A. Z. Mahmoid, M. A. Abdel-Halim, and M. M. El-Haddad.
J Microbiol U A R. Vol 3 No 2, p 113-121. 1970 Illus.
Identifiers: *Algae, Carbon dioxide, Cells, *Chlorella, *Chlorococcum, *Egypt (UAR), Light, Oxides, Temperature, Chlorophyta.

Unicellular green algae were isolated from different localities in the United Arab Republic. The isolates were purified and identified to genus. Six isolates of the genus Chlorella and 3 isolates of Chlorococcum were finally obtained. Isolates were screened to select the most active organism. The cultivation of the isolates was attained under controlled conditions of light (2500 lux), temperature (30 C) and a stream of 5% CO_2 in air mixture by using a specially designed apparatus.--Copyright 1972, Biological Abstracts, Inc.
W73-09020

CHROMATOGRAPHIC SEPARATION AND DETERMINATION OF BENZENE AND ISOPROPYLBENZENE IN WATER, (IN RUSSIAN),
Nauchno-Issledovatelskii Institut Gigieny, Moscow (USSR).
Z. N. Boldina.
Gig Sanit. Vol 36 No 9, p 69-70. 1971.
Identifiers: *Benzene, *Chromatographic separation, *Isopropyl benzene, Pollutant identification.

The methods proposed for investigating benzene and isopropylbenzene in water are based on the ability of these compounds to be nitrated and yield a colored compound following the addition of al-

kali and acetone. The minimum determinable concentration of isopropylbenzene on the chromatogram was 10 microgram and of benzene 5 microgram.--Copyright 1972, Biological Abstracts, Inc.
W73-09028

USE OF C14 FOR ACCELERATED DETERMINATION OF THE NUMBER OF ESCHERICHIA COLI IN WATER, (IN RUSSIAN),
Institute of General and Municipal Hygiene, Moscow (USSR).
L. E. Korsh, O. I. Yurasova, A. G. Nikonova, and M. A. Motova.
Gig Sanit. Vol 36, No 9, p 78-81, 1971. Illus.
Identifiers: Carbon dioxide, *Carbon-14, Metabolic rate, Oxides, Water pollution, *E. coli, *Pollutant identification.

The proposed method for rapid determination of the number of E. coli in waste waters is based on the quantity of metabolic carbon dioxide evolved as a result of their activity from media labeled with 14C. Calibration curves of metabolic carbon dioxide plotted against the number of bacteria in the sample were constructed for rapid determination of the number of bacteria.--Copyright 1972, Biological Abstracts, Inc.
W73-09032

USE OF GAS CHROMATOGRAPHY IN HYGIENIC INVESTIGATIONS, (IN RUSSIAN),
Nauchno-Issledovatelskii Institut Gigieny, Moscow.
V. A. Voronenko, and Y. V. Novikov.
Gig Sanit. Vol 36, No 9, p 81-87, 1971.
Identifiers: Air pollution, Chromatography, Food, *Gas chromatography, Hygienic studies, Industry, Water pollution, *Pollutant identification.

Gas-chromatographic analysis can be used for investigating atmospheric pollution, pollution of the air of industrial rooms, water pollution, for establishing the quality and nutritive value of food products and the presence of extraneous chemical impurities in them and for detecting traces of medicinal preparations, hormones, pesticides, food colors, etc.--Copyright 1972, Biological Abstracts, Inc.
W73-09033

GAS CHROMATOGRAPHY IN ENVIRONMENTAL SAFETY: A CONTRIBUTION ON GAS CHROMATOGRAPHY INVESTIGATION OF WATER, (IN GERMAN),
G. Goeke.
Fette Seifen Anstrichm. Vol 74, No 3, p 168-172, 1971, Illus. English summary.
Identifiers: Chromatography, Effluents, Environmental effects, *Gas chromatography, Industrial wastes, Lakes, Rivers, *Pollutant identification.

The detection of mineral oil, gasoline, petroleum, various organic solvents and pesticides is reported along with specific mention of instrumental parameters. The analytical chemist is often faced with a difficult task, because of the complexity of possible impurities in water. Besides the constant pollution of rivers and lakes with more or less well-clarified water from the civic communities, pollution is mainly caused periodically by industrial effluents due to disposal of toxic substances from factories, either intentionally or negligently. In addition, materials stored in underground tanks, basements and garbage depots pose a standing threat to groundwater.--Copyright 1972, Biological Abstracts, Inc.
W73-09034

THE INFLUENCE OF YEARLY WEATHER VARIATIONS ON SMOKE AND SULPHUR DIOXIDE POLLUTION IN PRETORIA,
National Physical Research Lab., Pretoria (South Africa).
E. Kemeny, and E. C. Halliday.
Arch Meteorol Geophys Bioklimatol Ser B Klimatol Umweltmeteorol Strahlungsforsch. Vol 20, No 1 p 49-78, 1972. Illus.
Identifiers: Oxide, *Air pollution, *Pretoria, Smoke, South Africa, *Sulfur dioxide, Weather, Yearly.

In Pretoria, South Africa, the atmospheric pollution by smoke and SO2 shows 2 distinct seasons: one (Oct.-March) during which the concentrations are low; and another during which the concentrations go through 3 phases: general increase (April-May), maximum concentration (June-July) and general decrease (Aug.-Sept.). The 2 different seasons can be explained by the fact that 7 main weather types occur over Pretoria and the South African highveld. Between Oct. and March rain and atmospheric instability predominate whereas between May and Sept. weather characterized by atmospheric stability prevails. April is a transition period between these extremes. The annual and diurnal variations in the concentrations of smoke and SO2 in the atmosphere which are mainly due to climatic factors are discussed. Long-term trends indicate that smoke levels remained fairly constant at the 2 sites where they were measured over periods of 13 and 7 yr, respectively, whereas the levels of SO2 decreased markedly. The long-term trends are due to human factors.--Copyright 1972, Biological Abstracts, Inc.
W73-09038

WATER-AFFECTED FRACTION OF NATURAL 1.5-9 MICRO DIAMETER AEROSOL PARTICLES,
National Center for Atmospheric Research, Boulder, Colo.
J. Rosinski, and C. T. Nagamoto.
J Colloid Interface Sci. Vol 40, No 1, p 116-120. 1972. Illus.
Identifiers: *Aerosol particles, *Colorado, Microns, Particles, Pollution, *Hydrosol particles.

The water-affected fraction of a given size class of an aerosol population reflects the change in particle concentration when that population becomes hydrosolized. The water-affected fraction was determined for natural aerosol particles in the 1.5-9 micro diameter size range. The largest change was observed in the 1.5-3 micro size group in a polluted atmosphere, especially in the presence of scattered rain and overcast sky. The aggregates formed through evaporation of cloud droplets seem to be the principal source of smaller hydrosol particles. Parallel determination of changes in air concentration of chloride particles and water-affected fraction of aerosol particles, coupled with analysis of air parcel trajectories, showed excellent agreement in identification of the movement of maritime air parcels over the central part of the continent (Colorado).--Copyright 1973, Biological Abstracts, Inc.
W73-09055

ON THE ECOLOGY OF HETEROTROPHIC MICROORGANISMS IN FRESHWATER,
Queen's Univ., Belfast (Northern Ireland). Dept. of Botany.
For primary bibliographic entry see Field 02I.
W73-09058

HETEROTROPHIC ASSAYS IN THE DETECTION OF WATER MASSES AT LAKE TAHOE, CALIFORNIA,
California Univ., Davis. Dept. of Zoology; and California Univ., Davis. Inst. of Ecology.
For primary bibliographic entry see Field 05C.
W73-09059

PROTEIN QUALITY OF SOME FRESHWATER ALGAE,
North Carolina State Univ., Raleigh.
For primary bibliographic entry see Field 02I.
W73-09070

5B. Sources of Pollution

ESTUARINE POLLUTION, A BIBLIOGRAPHY,
Office of Water Resources Research, Washington D.C.
For primary bibliographic entry see Field 02L.
W73-08451

STREAM POLLUTION FROM CATTLE BARN LOT (FEEDLOT) RUNOFF,
Ohio State Univ., Columbus. Dept. of Agricultural Engineering.
R. K. White.
Available from the National Technical Information Service as PB-220 010, $3.00 in paper copy, $1.45 in microfiche. Ohio Water Resources Center, Columbus, Project Completion Report No. 393X, December, 1972, 33 p, 14 fig, 5 tab, [illegible] ref. OWRR A-023-OHIO (1).

Descriptors: *Agricultural runoff, *Feed lots, *Cattle, Water pollution sources, Water quality, Livestock, *Biochemical oxygen demand, *Chemical oxygen demand, *Ohio, *Farm wastes, *Solid wastes.

This project has established that solids, BOD and COD transport in barnlot runoff are significant. (A barnlot, as distinguished from a feedlot, has less than 100 head of cattle wintered in a lot with access to a barn for feeding, watering and/or sleeping. The barnlot is typical for about two-thirds, 700,000, of the beef cattle raised in Ohio.) Runoff from a 0.17-hecatre, unpaved barnlot at the North Appalachian Experimental Watershed, Coshocton, Ohio, with 60 beef steers, was monitored. Thirty-nine events were sampled and analyzed in the period, March, 1970 to April, 1972. Total solids in the runoff varied from 0.10 to 1.20%. The BOD concentration varied from 9 mg/l (Jan, 1970) to 640 mg/l (Feb, 1972). Runoff usually occurs with rainfall of 1.3 cm (0.5 in.) or more. Antecedent soil moisture conditions significantly affect the amount of solids, BOD and COD in the runoff, with increased amounts following extended periods without rainfall. Graphs show that BOD concentrations and transport were higher in the winter and significantly less in the summer. A significant reduction of solids and BOD in the runoff was effected by using a grassed waterway or runoff collection pond and irrigation.
W73-08459

THE RELATION BETWEEN SOIL CHARACTERISTICS, WATER MOVEMENT AND NITRATE CONTAMINATION OF GROUNDWATER,
Kentucky Water Resources Inst., Lexington.
G. W. Thomas.
Available from the National Technical Information Service as PB-220 015, $3.00 in paper copy, $1.45 in microfiche. Research Report No 52, September 1972. 35 p, 4 fig, 9 tab, 14 ref. OWRR A-024-KY (1), 14-31-0001-3217.

Descriptors: *Leaching, *Nitrates, *Fertilizers, *Water pollution sources, Infiltration, Soil water movement, Return flow, Ion transport, *Kentucky.

Soils from several areas in Kentucky were placed in columns and leached with Ca (NO3)2. Subsoils high in iron oxide were found to retard the leaching of nitrate very significantly. In other soils, the nitrate moved through as fast as or slightly faster than the water. Field application of nitrogen to corn was most efficient in the spring or summer near the time that corn takes it up. The one exception to this was a red soil, where fall application of nitrogen resulted in little loss due to the retarding effect. Soils on which a sod or cover crop is killed and in which corn is planted lose little water by evaporation. Because of this, much more water movement occurs and nitrate moves out of the soil during the summer. In contrast, the soil

The U.S. Geological Survey established a sampling program on Amchitka Island in 1967 in cooperation with the U.S. Atomic Energy Commission. Water samples are analyzed routinely for tritium, gross alpha, and gross beta/gamma content. Frequency of sampling was semiannual prior to the Cannikin underground nuclear explosion and has been bimonthly since the Cannikin event (November 6, 1971). The sampling frequency will be quarterly for the remainder of the first year (through November 1972). Sampling frequency will be changed to an annual basis approximately 1 1/2 years after the event. Radiochemical data are presented from samples collected during July 1972. The gross alpha activity, as U equivalent, ranged from less than 0.3 to 3.0 pCi/liter (picocuries per liter). The gross beta activity in water as Cs-137 equivalent ranged from 2.0 to 39 pCi/liter. The only water samples having detectable tritium activity that is greater than 640 pCi/liter (200 tritium units) were those collected at the Long Shot site. The gross alpha and gross beta determinations showed no trends when compared with previously determined ranges for fresh waters on Amchitka Island. (Woodard-USGS)
W73-08479

BIODEGRADATION OF PETROLEUM IN SEA-WATER AT LOW TEMPERATURES,
Rutgers - The State Univ., New Brunswick, N.J. Dept. of Biochemistry and Microbiology.
R. M. Atlas, and R. Bartha.
Canadian Journal of Microbiology, Vol 18, No 12, p 1851-1855, December 1972. 5 fig, 14 ref.

Descriptors: *Microbial degradation, *Biodegradation, *Sea water, *Water temperature, Gas chromatography, Water sampling, Inhibition, Weathering, Monitoring, Chemical analysis.
Identifiers: *Crude oil, *Fate of pollutants, Carbon dioxide evolution technique, Mineralization, Degradation rates, Sweden crude oil, Coastal waters.

To evaluate the significance of biodegradation in the removal of polluting oil from cold oceans, freshly collected seawater samples were treated with petroleum and were incubated at controlled temperatures between 5 and 20C. Biodegradation was monitored by the measurement of CO_2 evolution and by quantitative gas chromatographic analysis. Low water temperatures not only resulted in slower degradation rates, but caused increasing lag periods that preceded the onset of measurable biodegradation. A substantial portion of these lag periods was eliminated when, instead of fresh petroleum, a 'weathered' sample was used. The results suggest that some volatile components of petroleum that are inhibitory to oil-degrading microorganisms evaporate only very slowly at low temperatures, and thus retard biodegradation. (Holoman-Battelle)
W73-08531

PERSISTENCE AND REACTIONS OF C-14-CACODYLIC ACID IN SOILS,
Agricultural Research Service, Beltsville, Md. Agricultural Environmental Quality Inst.
E. A. Woolson, and P. C. Kearney.
Environmental Science and Technology, Vol 7, No 1, p 47-50, January 1973. 4 fig, 4 tab, 18 ref.

Descriptors: *Arsenicals (Pesticides), *Pesticide kinetics, *Soil types, Persistence, Chemical reactions, *Soil contamination, Herbicides, Soil analysis, Chemical analysis, Degradation (Decomposition), Soil chemical properties, Soil physical properties, Pesticide residues, Clay loam, Aerobic conditions, Biodegradation, Loam, Radioactivity techniques.
Identifiers: *Cacodylic acid, Chemical distribution, Hagerstown silty clay loam, Lakeland loamy sand, Fate of pollutants, Christiana clay loam, Hydroxydimethylarsine oxide, Arsenic, Recovery, Carbon dioxide evolution.

Carbon-14-labeled cacodylic acid (hydroxydimethyl-arsine oxide) was prepared by reacting C-14 methyl iodide with methyl dichloroarsine. Concentrations of 1, 10, and 100 ppm of cacodylic acid were established in three soils of varying iron and aluminum content. At 2, 4, 8, 16, 24, and 32 weeks, soils were analyzed for C-14 and total arsenic in the water-soluble (ws), calcium (Ca), iron (Fe), and aluminum (Al) fractions. Initially, cacodylic acid was distributed in the following fractions: ws much greater than Al greater than Fe greater than Ca. After 32 weeks, the distribution was ws greater than Al greater than Fe greater than Ca. In contrast, inorganic arsenate (5 plus) was largely present in the Fe and Al fractions. Cacodylic acid persistence was a function of soil type and after 32 weeks the following amounts of C-14 were recovered in each soil type by combustion: Christiana (23 percent), Hagerstown (53 percent), Lakeland (62 percent). A decrease in both total C-14 and total arsenic occurred in all soils with time. A pungent garlic odor was detected in soils receiving 100 ppm, suggesting the production of a volatile alkyl arsine. The loss of arsenic suggests that one route of cacodylic acid loss from aerobic and anaerobic soils is by alkyl arsine volatility. Degradation under aerobic conditions also occurred by cleavage of the C-As bond, presumably yielding CO_2 and AsO_4 (3 minus). This degradation is presumably due to microbiological action. (Holoman-Battelle)
W73-08542

EFFECT OF PULPMILL EFFLUENT ON DISSOLVED OXYGEN IN A STRATIFIED ESTUARY -- I. EMPIRICAL OBSERVATIONS,
Fisheries Research Board of Canada, Nanaimo (British Columbia). Biological Station.
For primary bibliographic entry see Field 05C.
W73-08543

EFFECT OF PULPMILL EFFLUENT ON DISSOLVED OXYGEN IN A STRATIFIED ESTUARY - II. NUMERICAL MODEL,
Fisheries Research Board of Canada, Nanaimo (British Columbia). Biological Station.
For primary bibliographic entry see Field 05C.
W73-08544

FATTY ACIDS IN SURFACE PARTICULATE MATTER FROM THE NORTH ATLANTIC,
Rhode Island Univ., Kingston. Graduate School of Oceanography.
For primary bibliographic entry see Field 05A.
W73-08552

CARBON DIOXIDE CONTENT OF THE INTERSTITIAL WATER IN THE SEDIMENT OF GRANE LANGSO, A DANISH LOBELIA LAKE,
Copenhagen Univ. (Denmark). Freshwater Biological Lab.
For primary bibliographic entry see Field 05A.
W73-08555

DECOMPOSITION OF DISSOLVED ORGANIC CARBON AND NITROGEN COMPOUNDS FROM LEAVES IN AN EXPERIMENTAL HARD-WATER STREAM,
Michigan State Univ., Hickory Corners. W. K. Kellogg Biological Station.
R. G. Wetzel, and B. A. Manny.
Limnology and Oceanography, Vol 17, No 6, p 927-931, November 1972. 3 fig, 20 ref.

Descriptors: Water pollution sources, *Decomposing organic matter, *Microbial degradation, *Leaves, Water analysis, Model studies, Path of pollutants, Degradation (Decomposition), Biodegradation, Bacteria, Cultures, Population, Leaching, Chemical analysis, Spectrophotometry.
Identifiers: *Dissolved organic nitrogen, *Dissolved organic carbon, Natural organics, Dissolved organic matter.

To study decomposition of natural organic substances in water, dry hickory and maple leaves, enclosed in nylon mesh, were immersed in one of two 12-m recirculating streams simulating natural hard-water woodland streams. Particulate organic carbon was determined by sulfuric-dichromate oxidation and spectrophotometry; total dissolved organic carbon was determined by persulfate oxidation of filtrates and infrared CO_2 analysis; and total dissolved organic nitrogen was fractionated into labile and refractory components on the basis of UV combustion and spectrophotometry. Bacteria were also counted. Addition of leaves for 30 hr increased dissolved organic carbon (DOC) levels nearly 10-fold. Bacterial populations which developed rapidly decomposed labile organic carbon and nitrogen compounds in the leaf leachate within 72 hr. Bacteriologically labile and refractory dissolved organic carbon fractions, with T1/2 decomposition rates of 2 and 80 days, respectively, were present in the leachate. Most refractory dissolved organic nitrogen compounds persisted unmodified for at least 24 days. The processing capacity of woodland streams for natural dissolved organic compounds is much greater than previously believed. (Little-Battelle)
W73-08558

TRACE ELEMENTS IN CLAMS, MUSSELS, AND SHRIMP,
Institut Royal des Sciences Naturelles de Belgique, Brussels.
K. K. Bertine, and E. D. Goldberg.
Limnology and Oceanography, Vol 17, No 6, p 877-884, November 1972. 4 tab, 11 ref.

Descriptors: *Trace elements, *Clams, *Mussels, *Shrimp, *Neutron activation analysis, *Path of pollutants, Shellfish, Chemical analysis, Invertebrates, Mollusks, Crustaceans, Iron, Cobalt, Chromium, Zinc, Mercury.
Identifiers: *Shells, Tissue, Antimony, Scandium, Selenium, Silver, Ensis ensis, Ensis arcuatus, Ensis siligua, Ensis vagsina, Mytilus edulis, Mytilus afer, Mytilus galloprovincialis, Crangon crangon, Arthropods.

Compositional changes in the trace element content of shells of mussels and clams that might be related to man's influence on the composition of inshore marine waters over the past hundred years were sought but not found. The elements Rb, Fe, Co, Sb Sc, Ag, Cr, Zn, Se, and Hg were analysed in freshly caught and museum specimens by instrumental neutron activation analysis. Contamination from the preservatives was evident in the shells and tissues of museum specimens. Elemental concentrations in the calcareous shells, both aragonitic and mixed aragonitic-calcitic, are similar and may reflect the composition of the waters in which they lived, rejection by the organisms, or surface associated features. The proteinaceous molts of shrimp contained high levels of these elements, in agreement with previous investigations. With 20-25 molts per shrimp per year, the molting of shrimp can cause a redistribution of these elements within the marine environment. (Holoman-Battelle)
W73-08561

MERCURY IN A MARINE PELAGIC FOOD CHAIN,
Stanford Univ., Pacific Grove, Calif. Hopkins Marine Station.
G. A. Knauer, and J. H. Martin.
Limnology and Oceanography, Vol 17, No 6, p 868-876, November 1972. 2 fig, 6 tab, 25 ref.

Descriptors: *Mercury, *Zooplankton, *Phytoplankton, Marine fish, *Food chains, Chemical analysis, Path of pollutants, Pollutant identification, Heavy metals, Methodology, Sampling, Copepods, Diatoms, Protozoa, *California.
Identifiers: *Organic mercury, *Monterey Bay (Calif), Atomic absorption spectrophotometry, Flameless atomic absorption spectrophotometry, Sample preparation, Anchovy, Monterey Bay, Myctophidae, Engraulis mordax, Euphausiids, Gills, Muscle, Skin, Liver, Gonads, Diaphus theta, Tarletonbeania crenularis, Stenobrachius leucopsaris, Lampanyctus ritteri.

Phytoplankton, zooplankton, and anchovies collected in Monterey Bay, California, over a 10-month period were analyzed for total mercury. Known wet weight or dry weight quantities of homogenized samples were weighed into disposable ampoules, a 2:1 solution of concentrated H_2SO_4 and HNO_3 added (wet samples were chilled over ice), and the samples heated overnight at 80C. After acid digestion, the ampoules were chilled in ice and a 6 percent $KMnO_4$ solution added for the development of color. Excess permanganate was reduced with 30 percent H_2O_2 and the solution was back titrated with the permanganate solution to yield a pink color (this avoided suppression of peaks). The sample digest was drawn, a reductant added, and the Hg partitioned on a vortex mixer. The resulting vapor was injected into an AA spectrophotometer. A few samples of anchovies and phytoplankton were analyzed for organic Hg by flameless AAS after extracting the homogenate with benzene and cysteine acetate. The cysteine/Hg complex was treated as for total Hg except no HNO_3 or back titration was necessary. In general, mercury levels were low and no evidence of food chain amplification was observed. Temporal variations of Hg concentrations in phytoplankton were greater than those for zooplankton; however, no seasonal trends were observed for either group. Mercury levels were approximately equal in inshore and offshore zooplankton. The highest average mercury concentration was found in phytoplankton-net samples from the open ocean that contained radiolarians and other small zooplankton forms as well as diatoms. (Holoman-Battelle)
W73-08562

DISTRIBUTION AND FORMS OF NITROGEN IN A LAKE ONTARIO SEDIMENT CORE,
Department of Energy, Mines and Resources, Burlington (Ontario). Canada Center for Inland Waters.
A. L. W. Kemp, and A. Mudrochova.
Limnology and Oceanography, Vol 17, No 6, p 855-867, November 1972. 3 fig, 4 tab, 34 ref.

Descriptors: *Nitrogen, *Lake sediments, *Cores, *Lake Ontario, *Sediment-water interface, *Spatial distribution, Amino acids, Soil analysis, Chemical analysis, Nitrates, Nitrites, Path of pollutants, Bottom sediments, Sampling, Aquatic soils, Hydrogen ion concentration, Mud, Hypolimnion, Denitrification, Ammonification.
Identifiers: Ammonium, Hexosamine, Fate of pollutants, Surface sediments, Organic nitrogen, Nitrification.

Fixed ammonium nitrogen is the dominant inorganic form of nitrogen in the sediments of Lake Ontario. The fixed ammonium nitrogen concentration is around 300 micrograms/g of sediment at the top of a 10-m core and increases gradually to 525 micrograms at 150 cm, below which it remains constant. As nitrification is precluded in the reduced sediments below 3 cm, the ammonium ion is either fixed within the sediment clay lattices or migrates upward in the sediment interstitial waters. The uniform fixed and exchangeable ammonium concentrations below 150 cm in the core indicate that the sediment is saturated with respect to ammonium fixation, and the decrease in these concentrations above 150 cm in the core suggest that equilibrium is not attained with the ammonium ion. The deeper sediments probably do not regenerate nitrogen to Lake Ontario, but most of the nitrogen released to the hypolimnion by the sediments is from nitrification, denitrification, and ammonification reactions at the sediment-water interface. A minimum of 20 percent of the organic nitrogen input to the sediments is regenerated to the lake from the top 6 cm of sediments. About [illegible] percent of the nitrogen in the surface muds is organic: 28-46 percent as amino acid-N, 4-7 percent as hexosamine-N, and 21-31 percent as hydrolyzable unidentified-N. From 29-57 percent of the total nitrogen could not be accounted for as amino acids, hexosamines, fixed and exchangeable ammonia, nitrate, and nitrite in the surface sediments. (Holoman-Battelle)
W73-08563

COLLABORATIVE STUDY OF THE CATION EXCHANGE CAPACITY OF PEAT MATERIALS,
Michigan Dept. of Agriculture, East Lansing.
V. A. Thorpe.
Journal of the Association of Official Analytical Chemists, Vol 56, No 1, p 154-157, January 1973. [illegible] tab, 5 ref.

Descriptors: *Peat, *Humus.
Identifiers: *Collaborative studies, *Cation exchange capacity, *Reed-sedge, Sample preparation, Sphagnum moss, Statistical methods.

To provide a measure of the total amount of exchangeable cations that can be held by peat expressed as mequiv./100 g air-dried peat, the modified method of Puustjarvi for cation exchange capacity has been proposed and studied collaboratively. The statistical treatment of the collaborators' results indicates a satisfactory degree of precision and accuracy for the 3 products considered, moss, humus, and reed-sedge. The method for cation exchange capacity of peat materials, with the description of the transfer technique included, has been adopted as official first action. The 7 ASTM methods have been adopted as procedures. (Little-Battelle)
W73-08567

BACTERIOPHAGES RECOVERED FROM SEPTAGE,
Pennsylvania Water and Gas Co., Scranton.
J. F. Calabro, B. J. Cosenza, and J. J. Kolega.
Journal Water Pollution Control Federation, Vol 44, No 12, p 2355-2358, December, 1972. 3 fig, 1 tab, 7 ref.

Descriptors: *Bacteriophage, *Electron microscopy, Pollutant identification, Septic tanks, Waste water (Pollution), Assay, E. coli, Coliforms, Cultures, Pollutants, Hosts, Bioindicators, Filtration, Pseudomonas, Detergents, Linear alkylate sulfonates.
Identifiers: *Coliphages, *Septage, *Citrobacter freundii, *Salmonella typhimurium, *Shigella flexneri, Host specificity, Sample preparation, Culturing techniques, Plaque counts, Alcaligenes, Streptococcus fecalis, Mima spp, Herella spp, Archomobacter spp.

A study was made to (a) determine the presence of bacteriophage in septage where numbers of coliforms are markedly reduced, (b) examine their morphology and host specificity, and (c) detect any biological effects of linear alkyl sulfonate (LAS) on phage attachment in the system. Three active filtrates were recovered, and two distinct morphological types were observed with an electron microscope. The short-tail phage infected Citrobacter freundii, Escherichia coli, and cell-wall mutants of Salmonella typhimurium, and a long-tail phage was specific for Shigella flexneri. Host specificity suggested three different types. Only rough strains of hosts were found to be susceptible to phage. Concentrations of LAS normally encountered in septage showed no deleterious effect on the phage and host model system used. (Holoman-Battelle)
W73-08587

possible should be left at roadside. Heavy metal content of snow does not appear to be above that of the receiving waters. The fate of oils, greases, and toxic organics is only partially understood at the present time. (Little-Battelle)
W73-08601

A SURVEY OF LENTIC WATER WITH RESPECT TO DISSOLVED AND PARTICULATE LEAD,

Alaska Univ., College. Inst. of Water Resources.
D. Nyquist, L. A. Casper, and J. D. LaPerriere.
Available from the National Technical Information Service as PB-219 994, $3.00 in paper copy, $1.45 in microfiche. Completion Report No IWR-30, November 1972. 33p, 3 fig, 9 tab, 5 ref, append. OWRR A-035-ALAS (1).

Descriptors: *Lead, *Heavy metals, *Lentic environment, Bottom sediments, Plankton, Aquatic plants, *Alaska, *Ice fog, Lakes, Water pollution sources.
Identifiers: Fairbanks (Alas), Atomic adsorption spectrophotometry, Super cooled fog, Automobile combustion wastes.

Lead associated with automobile combustion is deposited on snow after adsorption onto ice fog particles in the Fairbanks, Alaska area. Fifty selected bodies of lentic water in the area were sampled for dissolved and particulate lead and for lead concentrations in the muds, plankton, and aquatic plants. Lead analysis was accomplished using a Perkin-Elmer Model 303 atomic adsorption spectrophotometer. Statistical analyses were carried out to determine correlations existing between the lead concentrations in the different fractions of each lake. Most correlations were highly significant (at the 99% probability level) except those between aquatic plant lead levels and the dissolved fraction, and between the dissolved fraction and the lead concentration in the nearby snow; these were both significant (at the 95% probability level) correlations. Additionally, no significant correlation was found between the lead levels in the net plankton and in the aquatic plants. An attempt was made to examine the lead concentrations for each fraction statistically, grouping the lakes according to areas of ice fog severity, to see if the fallout of the lead and its partitioning into the various fractions related to ice fog formation. Some evidence was found that this does occur. The limitations of the project are discussed in relation to the implications to human health.
W73-08604

ANALYSIS OF SELECTED EXISTING WATER QUALITY DATA ON THE CONNECTICUT RIVER,

Massachusetts Univ., Amherst. Dept. of Public Health.
A. J. Gross, and J. R. Folts.
Available from the National Technical Information Service as PB-219 981, $3.00 in paper copy, $1.45 in microfiche. Massachusetts Water Resources Research Center, Amherst, Publication no. 27, Completion Report (1973). 30p, 6 fig, 14 tab, 11 ref. OWRR A-055-MASS (1).

Descriptors: *Water data, Statistical methods, Water quality, *Regression analysis, Stream pollution, *Connecticut River, *Dissolved oxygen, Model studies, *Time series analysis, Computer programs, *Coliforms, Sampling.
Identifiers: Wilder (Vermont), Enfield (Conn), and Northfield (Mass), *Variance analysis.

A variety of statistical procedures was used to analyze data which measure water quality in the Connecticut River. The procedures included analysis of variance, regression and stepwise regression analyses and analyses of time series. The following observations may be made on the basis of the conclusions of the study: (1) Dissolved oxygen levels vary significantly among days and among sampling stations along the Connecticut River, (2) there was no significant variation in dissolved oxygen levels at either of the two sampling depths within a sampling station although there was a significant variation between the two depths, and (3) several different time series models and their requisite computer programs were developed in order to determine which model best fitted the six years of weekly coliform data available. From the time series analysis the coliform content is shown to be increasing from year to year. Finally, a model was developed using regression analysis to predict fecal coliform content from total coliform content.
W73-08605

HEATED DISCHARGE FROM FLUME INTO TANK - DISCUSSION,

Vanderbilt Univ., Nashville, Tenn. Dept. of Environmental and Water Resources Engineering.
For primary bibliographic entry see Field 08B.
W73-08610

A NEW TYPE OF WATER CHANNEL WITH DENSITY STRATIFICATION,

Johns Hopkins Univ., Baltimore, Md. Dept. of Mechanics and Materials Science.
G. M. Odell, and L. S. Kovasznay.
Journal of Fluid Mechanics, Vol 50, part 3, p 535-543, 1971. 5 fig, 7 ref.

Descriptors: *Stratification, *Density stratification, *Open channel flow, Velocity, Density, Conductivity, Mixing, Laboratory tests, Fluid mechanics, Flow characteristics, Distribution patterns.
Identifiers: *Flow visualization, Velocity profiles.

Existing laboratory facilities for experiments with stratified fluids are mainly of the density-stratified towing tank or wind tunnel type. The novel idea presented is based on using a disk pump to drive each fluid layer independently round a closed-return density-stratified water channel. A small-scale model was built and tested to demonstrate the feasibility of the idea. A hydrogen-bubble technique was used for both flow visualization and velocity measurement. Density measurements were made with an electrical conductivity probe developed by the authors. The tests have demonstrated the usefulness of such a facility, especially for longsteady-state experiments in density-stratified flows. (Oleszkiewicz-Vanderbilt)
W73-08615

TOPLYR-II A TWO-DIMENSIONAL THERMAL-ENERGY TRANSPORT CODE,

Hanford Engineering Development Lab., Richland, Wash.
D. C. Kolesar, and J. C. Sonnichsen, Jr.
Report HEDL-TME 72-46, March 1972. 49 p, 16 fig, 13 ref, 2 append. AEC-AT (45-1)-2170.

Descriptors: *Mathematical models, *Thermal pollution, *Heat transfer, *Heat, *Temperature, Thermal pollution, Water pollution, Energy, Algorithms, Meteorology, Diffusivity, Water temperature, Numerical analysis, Dispersion, Mixing.
Identifiers: *TOPLYR-II±MODEL, Eddy diffusivity, Thermal discharges, North Platte River, Thermal energy conservation.

TOPLYR-II describes, in certain types of rivers and lakes, the time varying transport of energy caused by the discharge and subsequent dissipation of condenser cooling water from electric power plants. The model flow field possesses a negligible vertical temperature gradient, and is located where the momentum source due to the plant discharge no longer affects the flow. TOPLYR-II is a modification of the TOPLYR code. As in TOPLYR, the mathematical formulation of the law of conservation of thermal-energy is employed to equate accumulation to convective

(via mean current), to diffusive and to source and sink terms. TOPLYR-II differs from TOPLYR primarily in that equations accurate to second-order rather than first-order have been employed to reduce errors due to numerical dispersion. The fluid depth is allowed to vary with position. The heat source is introduced as a temperature boundary condition which is permitted to change with time. An alternating-direction implicit algorithm is employed to solve the finite difference equations once velocity distribution, thermal eddy diffusivities, meteorological conditions, bottom contour, initial fluid temperature distribution and inlet fluid temperature (a function of time) are supplied. TOPLYR-II results compare favorably with field data. (Oleszkiewicz-Vanderbilt)
W73-08618

ENVIRONMENTAL ENGINEERING PROGRAMS QUARTERLY TECHNICAL PROGRESS REPORT JANUARY, FEBRUARY, MARCH 1972.
Hanford Engineering Development Lab., Richland, Wash.

Progress Report HEDL-TME 72-58, 1972. 25 p, 15 fig, 1 tab, 6 ref. AEC-AT (45-1)-2170.

Descriptors: *Temperature, *Mathematical models, *Heat, *Thermal powerplants, Rivers, Water pollution, Heat transfer, Mixing, Diffusion, Cooling, Heat transfer.
Identifiers: *Thermal discharges, *Cooling ponds, *Cooling capacity, Roanoke River, Pearl River, Tombigbee River.

Technical progress made by Westinghouse Hanford Company and its subcontractors during January, February and March 1972 is summarized. Programs covered include: (1) Simulation Modeling of Environmental Effects of Thermal Generation, (2) Regional Modeling of Surface Water Temperature from Projected Power Growth, and (3) Nuclear Plant Site Selection - Engineering Evaluation of Environmental Aspects. Previously published results of a sensitivity study associated with the undefined parameters of the TOPLYR-II program are updated. The program was modified to include variable thermal eddy diffusivity. Extension of TOPLYR-II to Potomac River data has necessitated consideration of local disturbances to the flow field such as islands. Low flow cooling capacities for the major river basins of the Eastern Gulf Coast and Southeastern Atlantic states were computed using the MAXPWR code. Techniques to aid in fog abatement and cooling pond technology are described briefly. (Oleszkiewicz-Vanderbilt)
W73-08619

ENVIRONMENTAL ENGINEERING PROGRAMS QUARTERLY TECHNICAL PROGRESS REPORT JULY, AUGUST, SEPTEMBER 1971.
Hanford Engineering Development Lab., Richland, Wash.

Report, HEDL-TME 71-165, November 1971. 45 p, 32 fig, 1 tab, 15 ref. AEC-AT (45-1)-2170.

Descriptors: *Temperature, *Mathematical models, *Heat, *Thermal powerplants, Rivers, Water pollution, Heat transfer, Mixing, Diffusion, Columbia River, Cooling.
Identifiers: *Thermal discharges, *Assimilative capacity, *Cooling capacity, Direct cooling capacity, Roanoke River, Tombigbee River, Pearl River.

Technical progress made by WADCO and WADCO subcontractors during July, August and September 1971 is summarized. The program is divided into three sections: (1) Simulation Modeling of Environmental Effects of Thermal Generation, (2) Regional Modeling of Surface Water Temperature from Projected Power Growths, and (3) Hydro-Environmental Evaluation of Nuclear Power Stations. A comparison of TOPLYR-II predictions with field data has demonstrated the capability of the code to simulate actual river data on the North Platte River. A transient flow routine was tested using data collected in the Columbia River Estuary and was added to the COLHEAT river simulation system. The MAXPWR code, used in calculating the cooling capacities of national waterways, has been modified to reduce the effort required to assemble input data and interpret program output and permit problems of greater complexity to be handled. Total and assimilative cooling capacities were calculated for 15 major rivers in the Eastern Gulf Coast and Southeastern Atlantic states. Thermal power plant intake structures and biological samplers are discussed briefly. (Oleszkiewicz-Vanderbilt)
W73-08620

THE COLHEAT RIVER SIMULATION MODEL.
Hanford Engineering Development Lab., Richland, Wash.

Report HEDL-TME 72-103, August 1972. 80 p, 20 fig, 4 tab, 23 ref. 3 append. AEC-AT (45-1)-2170.

Descriptors: Rivers, *Heated water, *Mathematical models, *Computer models, Electric power production, Thermal powerplants, *Thermal pollution, Water pollution, Discharge (Water), Theoretical analysis, Model studies, Computers, *Simulation analysis, Temperature, Flow, Meteorological data, Measurements, Evaluation, Heat budget, Heat transfer, Evaporation, Density.

The underlying theory, the development of the computer model, the result of various model tests and a brief user's guide for the Colheat River Simulation Model are presented. The model is formulated in terms of the far field resulting from the coarse grid normally employed. However, it has been proven quite flexible and capable of providing much of the fine structure. Routine operation of the system requires specification of: river or reservoir dimension reduced to equivalent non-parallel trapezoidal cross section; water temperature at the upstream end of the reach under study; stream flow information; and meteorological data. Special data are required for use of the transient flow routines and certain mixing options. Although the model was originally designed to simulate the thermal properties of a flowing water body, it has been used to model other pollutants. (Jerome-Vanderbilt)
W73-08624

RIVER JET DIFFUSER PLANNED.
For primary bibliographic entry see Field 05G.
W73-08631

URBAN STORMWATER QUALITY AND ITS IMPACT ON THE RECEIVING SYSTEM,
Rex Chainbelt, Inc., Milwaukee, Wis. Ecology Div.
For primary bibliographic entry see Field 05C.
W73-08635

THE MANAGEMENT OF GROUNDWATER RESOURCE SYSTEMS,
California Univ., Santa Barbara. Dept. of Mechanical Engineering.
For primary bibliographic entry see Field 05G.
W73-08660

GEOCHEMISTRY AND DIAGENESIS OF TIDAL-MARSH SEDIMENT, NORTHEASTERN GULF OF MEXICO,
Geological Survey, Washington, D.C.
For primary bibliographic entry see Field 02J.
W73-08730

THE DISTRIBUTION AND TISSUE RETENTION OF MERCURY-203 IN THE GOLDFISH (CARASSIUS AURATUS),
Wayne State Univ., Detroit, Mich. Dept. of Biology.
For primary bibliographic entry see Field 05A.
W73-08761

DURATION OF VIABILITY AND THE GROWTH AND EXPIRATION RATES OF GROUP E STREPTOCOCCI IN SOIL,
Missouri Univ. Columbia. School of Veterinary Medicine.
J. A. Schmitz, and L. D. Olson.
Applied Microbiology, Vol 25, No 2, p 180-183, February 1973. 1 fig, 2 tab, 9 ref.

Descriptors: *Feedlots, *Streptococcus, *Soils, Separation techniques, Pollutant identification, Isolation, Water pollution sources.
Identifiers: *Survival, *Irradiation.

In irradiated and nonirradiated feedlot and pasutre soils inoculated with group E streptococci, the organism was not recovered 17 days postinoculation from either the irradiated or nonirradiated feedlot soils incubated at 37C, but survived in the irradiated pasture soils for 24 and 31 days postinoculation. The streptococci survived in irradiated and nonirradiated soils incubated at 4C for 116 days and in one irradiated feedlot soil for 165 days. The population of streptococci did not increase in either irradiated or nonirradiated soil, and the expiration rate was greater in the soils incubated at 37 and 25 C than at 4C. With the relatively prolonged druation of viability of group E streptococci in soil at 4C, it is suggested that soil contaminate with exudate from draining abscesses of infected swine could act as a source of infection during the colder season. (Little-Battelle)
W73-08774

MICROBIAL RESPONSE IN LOW TEMPERATURE WASTE TREATMENT,
Toronto Univ. (Ontario).
J. G. F. Henry.
Available in microfilm from National Library of Canada at Ottawa, Ph D Dissertation, 1971.

Descriptors: *Temperature, *Population, *Bacteria, *Separation techniques, *Sewage sludge, Growth rates, Model studies, Pathogenic bacteria, Sewage bacteria, Sewage treatment, Food abundance, Pseudomonas, Waste treatment, Waste water treatment.
Identifiers: Flavobacterium, Achromobacter, Vibrio, Mesophilic bacteria, Psychrophilic bacteria, Substrates, Gram stain.

The relationships between low temperature, psychrophilic population and microbial activity are clarified. Full scale treatment plants, batch studies and continuous laboratory models were used. Batch studies on pure cultures of 26 psychrophiles and 4 mesophiles were conducted at temperatures of 1,4,9.5, and 18 C with food (nutrient broth) in excess. A comparison of the temperature coefficients for the psychrophiles and the mesophiles revealed a significant difference between them. Continuously fed laboratory models with detention times of 2.4, 9.6, 24 and 96 hours were also investigated at the same temperatures as used in the batch tests. The proportion of psychrophilic bacteria in the continuous models proved to be a function mainly of termperature but also of detention. Microscopic examination, colonial characteristics, Gram-staining and other diagnostic tests were used on the mixed population from the models to show the variation in microorganisms with temperature and detention and also to identify to genus level the bacterial cultures isolated for the pure culture studies. Psychrophilic Pseudomonas, Flavobacterium, Achromobacter and Vibrio all occurred. In the batch tests, with unlimited food, yield decreased at colder temperatures. However, in the continuous experiments, where food was

ASPECTS OF MARINE PHYTOPLANKTON STUDIES NEAR THE COLUMBIA RIVER, WITH SPECIAL REFERENCE TO A SUBSURFACE CHLOROPHYLL MAXIMUM,
Washington Univ., Seattle. Dept. of Oceanography.
For primary bibliographic entry see Field 02L.
W73-08826

DISTRIBUTION OF ORGANIC CARBON IN SURFACE SEDIMENT, NORTHEAST PACIFIC OCEAN,
Washington Univ., Seattle. Dept. of Oceanography.
For primary bibliographic entry see Field 02L.
W73-08828

STUDIES OF THE AEROBIC, NONEXACTING, HETEROTROPHIC BACTERIA OF THE BENTHOS,
Georgia Univ., Athens. Dept. of Microbiology.
For primary bibliographic entry see Field 02L.
W73-08830

RADIOACTIVE AND STABLE NUCLIDES IN THE COLUMBIA RIVER AND ADJACENT NORTHEAST PACIFIC OCEAN,
Puerto Rico Nuclear Center, Mayaguez.
For primary bibliographic entry see Field 02L.
W73-08842

RADIONUCLIDE DISTRIBUTION IN COLUMBIA RIVER AND ADJACENT PACIFIC SHELF SEDIMENTS,
Puerto Rico Nuclear Center, Mayguez.
For primary bibliographic entry see Field 02L.
W73-08843

SEDIMENT-ASSOCIATED RADIONUCLIDES FROM THE COLUMBIA RIVER,
State Univ. of New York, Stony Brook. Marine Sciences Research Center.
For primary bibliographic entry see Field 02L.
W73-08844

SEASONAL AND AREAL DISTRIBUTIONS OF RADIONUCLIDES IN THE BIOTA OF THE COLUMBIA RIVER ESTUARY,
Oregon State Univ., Corvallis. Dept. of Oceanography.
For primary bibliographic entry see Field 02L.
W73-08846

TOTAL PHOSPHORUS AND PHOSPHORUS-32 IN SEAWATER, INVERTEBRATES, AND ALGAE FROM NORTH HEAD, WASHINGTON,
Washington Univ., Seattle. Coll. of Fisheries; and Washington Univ., Seattle. Lab. of Radiation Ecology.
For primary bibliographic entry see Field 02L.
W73-08847

ORGANIC FILMS ON NATURAL WATERS: THEIR RETRIEVAL, IDENTIFICATION, AND MODES OF ELIMINATION,
Cornell Aeronautical Lab., Inc., Buffalo, N.Y.
R. E. Baier.
Journal of Geophysical Research, Vol 77, No 27, p 5062-5075, September 20, 1972. 5 fig, 1 tab, 16 ref.

Descriptors: *Air-water interfaces, *Path of pollutants, *Oily water, *Bubbles, Biodegradation, Water analysis, Gas chromatography, Spectrophometry, Water quality, Infrared radiation, Ultraviolet radiation, New York.
Identifiers: *Lake Chautauqua (NY).

Air-water interfacial films were sampled and analyzed by infrared spectroscopy. A field program on Lake Chautauqua in western New York correlated the buildup of oily films over the lake surface with peak boating activity and demonstrated the rapid elimination of such pollutant layers by natural mechanisms. Bubble breaking and its concomitant ejection of film fragments into the air was by far the most efficient process, especially when simultaneous irradiation accelerated the formation of polar moieties in the films. Chemical fractionation also occurs by this mechanism, with the more polar organic components and their associated inorganic elements being stripped from the interface preferentially. Thus, air-sea interfacial films are implicated in sea-to-air material transport and in the chemistry of marine aerosols. (Knapp-USGS)
W73-08851

EFFECT OF VARIOUS SALTS ON THE SURFACE POTENTIAL OF THE WATER-AIR INTERFACE,
Naval Research Lab., Washington, D.C.
For primary bibliographic entry see Field 02K.
W73-08859

OUR KNOWLEDGE OF THE PHYSICO-CHEMISTRY OF AEROSOLS IN THE UNDISTURBED MARINE ENVIRONMENT,
Max Planck-Institut fur Chemie, Mainz (West Germany).
C. E. Junge.
Journal of Geophysical Research, Vol 77, No 27, p 5183-5200, September 20, 1972. 11 fig, 3 tab, 48 ref.

Descriptors: *Aerosols, *Sea water, *Salts, Geochemistry, Distribution patterns, Dusts, Path of pollutants, Particle size, Sampling.

A survey is given of present knowledge of marine aerosols. On the basis of the aerosol size distribution in undisturbed marine environments, five different components of the aerosol are discussed: particles with radii larger than 20 microns sea spray particles, tropospheric background particles, mineral dust particles, and particles with radii smaller than 0.03 microns. The troposphere over the oceans is filled with a fairly uniform background aerosol on which the sea spray aerosol is superimposed only within the lowest kilometers above the sea surface. The sea spray component is discussed in detail with respect to size distribution, production, and chemical composition. The various components of the background aerosol seem to be either of continental origin or produced within the atmosphere. New data are presented on the mineral dust component over the Atlantic Ocean that originates from the Sahara. (Knapp-USGS)
W73-08860

1971 WATER RESOURCES DATA FOR NEW YORK: PART TWO, WATER QUALITY RECORDS.
Geological Survey, Albany, N.Y.
For primary bibliographic entry see Field 07C.
W73-08890

TREATMENT OF URBAN RUNOFF.,
Environmental Protection Agency, Washington, D.C. Municipal Pollution Control Section.
For primary bibliographic entry see Field 05G.
W73-08898

WATER QUALITY TRENDS IN INSHORE WATERS OF SOUTHWESTERN LAKE MICHIGAN,
Metropolitan Sanitary District of Greater Chicago, Ill.
W. G. Schmeelk, E. W. Knight, and C. Lue-Hing.
Presented at the Illinois Institute of Technology Symposium on Water Pollution in Metropolitan

Areas, Chicago, Illinois, November 30, 1972. 26 p, 9 fig, 10 ref.

Descriptors: *Water quality, *Lake Michigan, *Data collections, Water pollution, Chemical analysis, Water quality control, Water pollution control, Domestic wastes, Water analysis, Water pollution sources.
Identifiers: *Water quality trends.

The quality of Lake Michigan water is of great concern to public health officials and scientists, especially to the Metropolitan Sanitary District of Greater Chicago. With the establishment of the Sanitary District by State of Illinois Statue, the District was given authority to prevent discharges of any type into the Lake in its geographical area of jurisdiction. Waste discharges flow to the Chicago River, a former tributary of Lake Michigan, but the flow of the River has been reversed to prevent waste discharges to the Lake. An extensive testing program of Lake water quality has been set up by the District to determine water quality deterioration from discharges outside the Districts authority. Over 20 such monitoring stations are in existence and thirty water quality parameters are measured at each station. The data of four such stations are examined in depth with emphasis on plankton and diatom counts. Water quality of Lake Michigan is changing with time. Sulfates and chlorides are increasing, calcium levels are remaining almost constant, and sodium and potassium levels are declining somewhat. Domestic wastes are occasionally a problem and nitrogen and phosphorus are sufficient to support large algae growths. (Poertner)
W73-08899

CESIUM KINETICS IN A MONTANE LAKE ECOSYSTEM,
Colorado State Univ., Fort Collins.
T. E. Hakonson.
Available form Univ. Microfilms, Inc., Ann Arbor, Mich. 48106 Order No. 73-2791. Ph D Dissertation, 1972. 166 p.

Descriptors: *Path of pollutants, *Kinetics, Ecosystems, *Neutron activation analysis, *Radiochemical analysis, Water analysis, Lake sediments, Seston, Amphipoda, Zooplankton, Trout, *Colorado, Aquatic plants, Aquatic life, Absorption, Bottom sediments, Tracers.
Identifiers: *Cs-133, *East Twin Lake (Colo), Montane Lake, Bioaccumulation, Cesium radioisotopes, Cs-137, Scuds, Fate of pollutants.

The kinetics of cesium were determined in East Twin Lake, a 5 hectare, natural semi-drainage lake which lies at an elevation of 2880 meters in the north-central Colorado Rockies. One kilogram of Cs-133 was introduced into the water of East Twin Lake on September 15, 1970. Samples of water, seston, sediment, amphipods, zooplankton, trout and three species of vegetation were obtained over a 393 day period following the dosing event. Stable cesium was measured by neutron activation analysis. Loss of Cs-133 from the water occurred in a rapid phase, which resulted in the loss of 60 percent of the cesium from water and had a loss half-time of 0.5 days and a slow phase which had a loss half-time of 130 days. The seston fraction of each water sample contained from 25 percent to 80 percent of the Cs-133 present in each liter of unfiltered water. Bottom sediments were identified as the major site of deposition of the Cs-133 dose. Amphipods and zooplankton reached equilibrium with the water within about three weeks after the Cs-133 administration and achieved concentration factors of about 700 and 150, respectively. Trout accumulated the Cs-133 more slowly than the invertebrates and reached a maximum concentration factor of about 5500 some 260 days following the dosing event. The simulation of the kinetics of Cs-133 in the East Twin Lake ecosystem was approached by utilizing the observed in-growth data to solve for the intercompartmental transfer rate constants. The constant coefficient model described the general behavior of the observed data very well. The use of Cs-133 as a tracer for Cs-137 was both reliable and feasible. Absolute quantities as small as 0.01 micrograms of Cs-133 were detectable. Concentrations of Cs-133 and fallout Cs-137 in trout muscle were significantly correlated which indicated that the kinetic behavior of the isotopes was similar. (Holoman-Battelle)
W73-08901

LAND DEVELOPMENT AND HEAVY METAL DISTRIBUTION IN THE FLORIDA EVERGLADES,
Florida State Univ., Tallahassee. Dept. of Oceanography.
G. J. Horvath, R. C. Harriss, and H. C. Mattraw.
Marine Pollution Bulletin, Vol 3, No 12, p 182-184, December 1972. 1 fig, 3 tab, 13 ref.

Descriptors: *Manganese, *Cobalt, *Copper, *Zinc, *Cadmium, *Lead, *Hydrologic systems, *Estuaries, Agricultural runoff, Water pollution sources, Water analysis, Land development, Heavy metals, Water sampling, Estuarine environment, Pollutant identification, Florida, Chemical analysis.
Identifiers: *Florida Everglades, Atomic absorption spectrophotometry, Barron River, Chokoloskee Bay, Lastman Bay, Barron River Canal, Sample preparation, Precision, Accuracy.

In order to make a quantitative appraisal of Big Cypress land development on the distribution and abundance of heavy metals in the Everglades estuaries, a study was conducted to elucidate the geochemistry of manganese, cobalt, copper, zinc, cadmium and lead in the canal-estuary system of the region. The hydrological system chosen for the study consists of farmed lands to the north connected by the Barron River canal system of Chokoloskee Bay to the south. Water samples were collected during moderate to high flow conditions, iced, filtered, and subjected to nitric acid-persulfate oxidation. Total dissolved metals were concentrated using Chelex-100 ion-exchange resin columns with an efficiency of 80-100 percent. Metal concentrations were determined by atomic absorption spectrophotometry, using the standard addition method for all samples. Analytical precision was between plus or minus 2 and 8 percent. The highest concentration of total dissolved metals occurred in the portion of the canal adjacent to cultivated areas. Data demonstrate that Chokoloskee Bay is enriched in the metals investigated by factors of 2.6-4.8. A comparison of dissolved metal concentrations in the developed canal area with water from estuarine areas indicates that copper, cadmium, and lead exhibit maximum concentrations in Chokoloskee Bay. Both of the Everglades estuaries investigated have higher dissolved Mn, Co, Cd, and Pb contents than have been previously reported for coastal marine waters. (Holoman-Battelle)
W73-08906

DECOMPOSITION OF OIL POLLUTANTS IN NATURAL BOTTOM SEDIMENTS,
Rutgers - The State Univ., New Brunswick, N.J.
T. B. Shelton.
Available from Univ. Microfilms, Inc., Ann Arbor, Mich. 48106. Order No. 73-4778. Ph D Dissertation, 1972. 165 p.

Descriptors: *Degradation (Decomposition), *Oil, *Bottom sediments, Pollutants, Rivers, *Aerobic conditions, *Anaerobic conditions, Soil analysis, Aquatic soils, Persistence, Chromatography, Pollutant identification, Consolidation, Oil pollution, Organic compounds, Separation techniques, Dissolved oxygen, Biochemical oxygen demand, Chemical oxygen demand, Hydrogen ion concentration, Color, Hydrogen sulfide, Nitrogen, Phosphorus, Solvent extractions.
Identifiers: *Fate of pollutants, Infrared spectrophotometry, Chemical composition, Degradation products, Hydrocarbons, Organic carbon Total organic carbon, Organic solvents, Hexane Benzene, Volatile solids.

An aerobic and anaerobic experimental system was designed to study the decomposition of oil pollutants in natural bottom sediments and to answer 3 questions: (1) What is the relative persistence of oils compared with other organics from a natural bottom sediment containing oil. The relative persistence of aliphatic, aromatic and oxy-compound components of the oil were also determined; (2) How is this relative persistence influenced by anaerobic conditions in the overlying water, and (3) Is organic carbon released from a natural bottom sediment during the decomposition process. The aerobic system was a constant flow apparatus and the anaerobic system operated on a fill and draw basis with nitrogen gas used to achieve anaerobic conditions. A suitable bottom sediment containing approximately 8 percent oil on a dry weight basis was used. Appropriate samples of the sediment were placed on the aerobic and anaerobic experimental systems and the decomposition followed for a 33 week period. At weekly intervals dissolved oxygen, pH, TOC, BOD were measured on the aerobic system. H2S, pH, COD, TOC, and color were determined on the anaerobic system. Dry solids, volatile solids, nitrogen, phosphorus, hexane extractables, benzene extractables were measured on the bottom sediment. The hexane extractable oils were subjected to chromatographic separation using a silica gel technique. These fractions were examined by infrared spectrophotometry for changes in functional groups. It was found that: (1) the oxygen utilization rate of the sediment varied from 1.5 to 1.73 g/sq m/day; (2) oils are lost and there is a steady release of organic C from natural bottom sediments with time with both aerobic and anaerobic conditions present in the overlying water; (3) under aerobic conditions oils persist in the bottom sediment longer than do other organics. Under anaerobic conditions, the reverse seems to be true; (4) natural bottom sediments containing oil consolidate under their own weight with time; (5) oils which reach the bottom sediments of rivers and other water bodies have lost all their lighter fractions corresponding to a carbon chain length of 20; and (6) there is an increased percentage of partially oxidized hydrocarbons in oils extracted from natural bottom sediments. (Holoman-Battelle)
W73-08911

THE DISTRIBUTION, SUBSTRATE SELECTION AND SEDIMENT DISPLACEMENT OF COROPHIUM SALMONIS (STIMPSON) AND COROPHIUM SPINICORNE (STIMPSON) ON THE COAST OF OREGON,
Oregon State Univ., Corvallis.
J. E. McCarthy.
Available from Univ. Microfilms, Inc., Ann Arbor, Mich. 48106. Order No. 73-7840. Ph D Dissertation, 1973. 68 p.

Descriptors: *Distribution patterns, Habitats, *Sediments, Laboratory tests, Amphipoda, Salinity, Population, Sands, *Oregon.
Identifiers: *Corophium spinicorne, *Corophium salmonis, *Substrate selection, Substrates, Macroinvertebrates, Scuds, Arthropods.

The amphipoda, Corophium salmonis (Stimpson) and C. spinicorne (Stimpson) were examined from three aspects. The first study involved the distribution and natural habitat of these species on the Oregon coast. Both species were found in fresh-water (less than 0.4 percent salinity) and in estuarine water up to 1.0 percent salinity (C. spinicorne) and 2.9 percent salinity (C. salmonis). Both species disappeared in winter months, reoccurring in increasing numbers during spring months and reaching highest densities in the

summer. Under natural conditions C. salmonis was found most frequently in mud sediments while C. spinicorne occurred more often in sand. Both species formed burrows in sediments and tubes upon submerged surfaces. The second study was to determine how specific these animals were in selecting substrates. The animals were exposed simultaneously to their native sediments, acid-cleaned sand and a foreign mud or sand. These multiple choice experiments demonstrated that these species of Corophium can readily determine their native substrate. Both species, when occurring in sand, could distinguish the native sand from similar sand of other areas. The third study determined the amount of material removed when the animal established. The average for both species was 0.07977 gm (0.032 ml)/burrow. (Little-Battelle)
W73-08912

NUTRIENT PATHWAYS IN SMALL MOUNTAIN STREAMS,

Georgia Univ., Athens.
W. R. Woodall, Jr.

Available from Univ. Microfilms, Inc., Ann Arbor, Mich. 48106. Order No. 73-5809. Ph D Dissertation, 1972. 139 p.

Descriptors: *Nutrients, *Biodegradation, *Decomposing organic matter, *Leaves, *Cycling nutrients, *Benthic fauna, Detritus, Potassium, Calcium, Magnesium, Crayfish, Salamanders, Path of pollutants, Streams, Leaching, Invertebrates, Surface runoff.
Identifiers: Biotransformation, Macroinvertebrates.

Population sizes of benthic organisms in streams draining four small watersheds were estimated. The watersheds, which were located in the southern Appalachian mountains, were each covered in a different vegetation type. Microflora were not included in the model which was devised. Grazers performed a minor function in Coweeta's small, shaded streams. Crayfish and salamanders were responsible for most of the standing crops biomass in the detritivore and predator compartments respectively and also account for most of the fluxes in their compartments. They are important in small stream nutrient flow by forming a sink in the remineralization process. An increase in potassium concentrations and a decrease in calcium and magnesium concentrations was associated with an increase in trophic levels. Since the food material was richer in calcium and magnesium than potassium, detritivores concentrated proportionately more potassium than calcium or magnesium. Some insects showed a positive correlation between calcium concentrations and degree of sclerotization. An inverse relationship between potassium concentration and degree of sclerotization was even more pronounced. Most taxa, however, showed neither the positive correlation with calcium nor the inverse relation with potassium. The principal mechanism of potassium release from detritus was through leaching. For calcium and magnesium, which are chemically bound in leaf tissue, the principal mechanism for release was the feeding activity of detritivores. Streams are important in the watersheds because they are the primary mechanism for nutrient removal from the system. (Little-Battelle)
W73-08915

STATISTICAL METHODS FOR THE DETECTION OF POLLUTION UTILIZING FISH MOVEMENTS,

Virginia Polytechnic Inst. and State Univ., Blacksburg.
J. W. Hall.

Available from Univ. Microfilms, Inc., Ann Arbor, Mich. 48106. Order No. 73-5872. Ph D Dissertation, 1973. 114 p.

Descriptors: *Zinc, *Statistical methods, *Monitoring, *Pollutant identification, *Fish behavior, Movement, Heavy metals, Stochastic processes, Mathematical studies, Water pollution effects, Fish physiology.
Identifiers: Camp-Paulson approximation.

A statistical technique was developed for using fish movement patterns measured by light beam interruptions to continuously monitor water for the presence of zinc. The sequence of light beam interruptions per hour, which is the realization of a stochastic process, was studied by analyzing some of Waller's (1971) data. It was determined that the mean of the process varied with time, having a twenty-four hour cycle and that the observations were serially correlated. Although a shortage of data prevented a firm conclusion, it was decided provisionally that the observations for a single hour were distributed negative binomially. Using the Camp-Paulson approximation as a transformation and making some additional assumptions allowed the process to be transformed to a strictly stationary one for which critical values could easily be calculated. The procedure was compared with Waller's method both on the data which had been analyzed and on some additional data. The procedure worked as well as Waller's method for low concentration of zinc and detected the presence of zinc faster when the concentration was high. The procedure allowed some statement about the probability of the errors involved whereas Waller's method did not. (Holoman-Battelle)
W73-08924

RESULT OF FLUVIAL ZINC POLLUTION ON THE ZINC CONTENT OF LITTORAL AND SUB-LITTORAL ORGANISMS IN CARDIGAN BAY, WALES,

University Coll. of Wales, Aberystwyth. Dept. of Zoology.
M. P. Ireland.

Environmental Pollution, Vol 4, No 1, p 27-35, January 1973. 2 fig, 4 tab, 19 ref.

Descriptors: Chemical analysis, *Zinc, *Path of pollutants, *Pollutant identification, *Ecological distribution, Water analysis, Kelp, Littoral, Tidal effects, Mollusks, Crustaceans, Food habits, Sampling, Methodology, Sea water, Marine animals, Marine algae, Water sampling, Heavy metals, Scuba diving, Invertebrates, Clams, Snails, Gastropods, Mussels, Flow, Freshwater.
Identifiers: Atomic absorption spectrophotometry, Macroinvertebrates, Sponges, Sublittoral, Sample preparation, Fucus vesiculosus, Barnacles, Littorina littorea, Mytilus edulis, Actinia equina, Thais lapillus, Balanus balanoides, Halichondria panicea, Tunicates, Botryllus schlosseri, *Wales, *Cardigan Bay.

A study has been made of the distribution of fluvial zinc in sea water, five littoral animals and one species of seaweed, together with two sub-littoral animals at varying distances from the source of pollution. Water samples were taken 15 cm below the surface, and filtered. Fresh water samples were concentrated 50 times by boiling and acidifying with HCl (Aristar, BDH) to a concentration of 0.06 N. Sea water samples were concentrated with organic solvents, as described by Willis (1961). Littoral animals were collected in the area where kelp was located and sublittoral species were collected at a low spring tide by SCUBA divers. Allowed to purge themselves overnight, the animal samples were rinsed in distilled water, dried at 105C to a constant weight, pulverized, and digested in 6 N HCl. Whole-bodied specimens were used except for a species of barnacle and the kelp. Zinc was estimated by AAS at 214 nm. The Zn content of sea water was indicative of polluted water becoming progressively more dilute. The distribution of Zn in the various species was found to be related to tidal flow, diet, and species specificity for the metal. (Holoman-Battelle)
W73-08925

A. THE ROLE OF LICHENS IN ROCK WEATHERING AND SOIL FORMATION. B. MERCURY IN SEDIMENTS,

Wisconsin Univ., Madison.
I. K. Iskandar.

Available from Univ. Microfilms, Inc., Ann Arbor, Mich. 48106. Order No. 72-24,886. Ph D Dissertation, 1972. 177 p.

Descriptors: *Path of pollutants, *Mercury, *Sediments, *Soils, *Lichens, Lakes, Sewage, Iron, Heavy metals, Water pollution sources, Aluminum, Calcium, Magnesium, Solubility, *Wisconsin.
Identifiers: Mobilization, Sample preparation, *Atomic absorption spectrophotometry, Chemical recovery.

Results showed that low but significant amounts of lichen compounds dissolve in water. Lichen compounds and water solutions of lichen compounds formed soluble metal complexes, frequently colored, when shaken with silicate materials for 96 hours, as shown by spectrophotometric and chemical analyses of the extracts. Release of cations from the silicate materials resulted largely from metal-complex formation rather than from reactions directly involving hydrogen ions. Similar amounts of Fe, Al, Ca, and Mg were released from the silicates by water solutions of lichen compounds and by solid lichen compounds. Contrary to popular belief, lichen compounds are sufficiently soluble in water to form soluble metal complexes and to cause chemical weathering of minerals and rocks. A highly sensitive and precise procedure was developed for the quantitative determination of total Hg in sediments and soils. Undried samples were treated with concentrated HNO_3/H_2SO_4, $KMnO_4$, and $K_2S_2O_8$ for digestion and oxidation of all forms of Hg to Hg^{2+} plus, which is subsequently determined by flameless atomic absorption spectrophotometry. Recovery of Hg ranged from 98 to 105 percent. The developed procedure extracted more Hg from sediments and soils than did extraction with concentrated HNO_3. The vertical distribution of Hg in sediment cores from a range of hard- and soft-water lakes in Wisconsin was evaluated in terms of potential sources of Hg during the nineteenth and twentieth centuries. For the Madison lakes, the trends in Hg distribution were related to variations in sewage inputs during the last 80 years. It is unlikely that either inputs of sewage or erosional products are responsible for the observed accumulation of Hg. Background levels varied from 0.06 to 0.24 ppm of Hg in precultural sediments from the Wisconsin lakes investigated. (Little-Battelle)
W73-08930

PHOSPHORUS DYNAMICS IN LAKE WATER,

Toronto Univ. (Ontario). Dept. of Zoology.
D. R. S. Lean.

Science, Vol 179, No 4074, p 678-680, February 16, 1973. 2 fig, 14 ref.

Descriptors: *Cycling nutrients, *Phosphorus, *Metabolism, *Plankton, *Organophosphorus compounds, Nutrients, Phosphates, Absorption, Lakes, Separation techniques, Path of pollutants.
Identifiers: *Excretion, *Gel filtration, Heart Lake, Canada, Colloids, Biotransformation.

Radioactive phosphate was used in conjunction with gel filtration to identify biologically important forms of P in lake water and the rate constants for their formation. Water samples from Heart Lake (Ontario), a small eutrophic lake, were spiked with tagged phosphate, filtered, and analyzed. The experiments revealed that a steady state is rapidly attained between tagged P and lake water in the summer with the following composition: a particulate fraction containing the bulk of the P plus small amounts of a soluble high-molecular-weight organic P compound; a low-molecular-weight organic P compound; and soluble inorganic phosphate, the turnover time of which is very

short. It is concluded that an exchange mechanism exists in lake water between phosphate and plankton, but the excretion of an organic phosphorus compound by the plankton is also a significant process. It results in the extracellular formation of a colloidal substance, and most of the nonparticulate phosphorus in lake water is in this form. (Little-Battelle)
W73-08931

MICROBIAL DEGRADATION OF PARATHION,
Mississippi State Univ., State College.
W. L. Gibson.
Available from Univ. Microfilms, Inc., Ann Arbor, Mich. 48106 Order No. 73-168. Ph D Dissertation, 1972. 125 p.

Descriptors: *Microbial degradation, *Phosphothioate pesticides, *Metabolism, Biochemistry, Biodegradation, Organophosphorus pesticides, Phenols, Alcohols, Hydrogen ion concentration, Temperature, Pollutant identification, Aerobic bacteria.
Identifiers: *Metabolites, *Pseudomonas aeruginosa, Substrate utilization, *Parathion, Fate of pollutants, Ethanol, Growth, Bacterial physiology, Thin layer chromatography, Paraoxon, p-Nitrophenol, p-Aminophenol, Organic solvents.

An organism capable of utilizing parathion as the sole carbon and energy source was isolated by enrichment culture techniques. The bacterium was characterized and tentatively classified as Pseudomonas aeruginosa. A pH of 7.0 - 7.5 and temperature of 30 C were optimum for the consumption of parathion. Virtually no oxygen utilization was observed with resting cell suspensions when nonsolubilized parathion was employed. The use of ethanol as solvent for parathion in resting cell studies or preincubation of cells in ethanol obviated this problem and rapid parathion oxidation was demonstrable. Approximately 80 percent of the parathion consumed by resting cells was present terminally as carbon dioxide. Permeability of the cell to parathion or its metabolites was contingent upon the use of ethanol as either solvent or denaturant. The nature of the enzyme system in parathion utilization was constitutive. Enhanced parathion utilization was noted when grown in a carbon-dioxide-enriched atmosphere. The isolate was capable of growth on parathion, paraoxon, p-nitrophenol, p-aminophenol, and phenol. Metabolites were tentatively identified by thin layer chromatography. Thin layer chromatographic data generally supported the growing evidence for the proposed pathway of parathion metabolism. (Holoman-Battelle)
W73-08932

GAGING AND SAMPLING INDUSTRIAL WASTEWATERS,
Calgon Corp., Pittsburgh, Pa.
For primary bibliographic entry see Field 05A.
W73-08935

REGROWTH OF COLIFORMS AND FECAL COLIFORMS IN CHLORINATED WASTEWATER EFFLUENT,
Hadassah Medical School, Jerusalem (Israel). Environmental Health Lab.
For primary bibliographic entry see Field 05C.
W73-08939

ISOLATION OF BACTERIA CAPABLE OF UTILIZING METHANE AS A HYDROGEN DONOR IN THE PROCESS OF DENITRIFICATION,
National Inst. for Water Research, Pretoria (South Africa).
For primary bibliographic entry see Field 05A.
W73-08941

A NOTE ON THE IRON-ORGANIC RELATIONSHIP IN NATURAL WATER,
Wisconsin Univ., Madison. Water Chemistry Program.
R. H. Plumb, Jr., and G. F. Lee.
Water Research, Vol 7, No 4, p 581-585, April 1973. 2 fig, 3 tab, 7 ref. EPA-5T1-WP-22.

Descriptors: *Iron, Separation techniques, *Organic compounds, *Colorimetry, Heavy metals, Water analysis, Organic matter, *Wisconsin.
Identifiers: *Fractionation, *Gel filtration, Preconcentration, *Lake Mary (Wis), Sample preparation.

Water samples were collected at the 15 m depth of Lake Mary, a small, highly colored meromictic lake in Wisconsin, for use in studies of iron-organic relationships. Most of the samples were fractionated on a Sephadax G-25 fine column. Iron concentrations were determined by a colorimetric method using 2,4,6-tripyridal-s-triazine (TPTZ). Organic matter was monitored with a fluorometer. The iron elution pattern for the Sephadex column was investigated using two approaches. The first consisted of passing ionic iron through the column and determining recovery of iron in the effluent. The second consisted of tying up the iron with a complexing agent and determining the amount that would pass through the column. The results showed that complexed iron, that associated with organic matter, could not pass through the column. A second approach in which natural water was treated with TPTZ strengthened the observation. It is concluded that gel filtration is potentially useful for evaluating organic-inorganic interactions. However, the use of concentrated samples distorts and overemphasizes the importance of the iron-organic systems in the environment. The use of unconcentrated samples with Sephadex columns should be a means for evaluating the amount of iron associated with organic matter. (Little-Battelle)
W73-08945

SORPTION OF COPPER ON LAKE MONONA SEDIMENTS - EFFECT OF NTA ON COPPER RELEASE FROM SEDIMENTS,
Wisconsin Univ., Madison. Water Chemistry Program.
I. Sanchez, and G. F. Lee.
Water Research, Vol 7, No 4, p 587-593, April 1973. 6 fig, 2 tab, 4 ref. EPA-5T1-WP-22.

Descriptors: *Sediments, Iron, *Copper, Manganese, *Nitrilotriacetic acid, Carbonates, *Leaching, *Sorption, Heavy metals, Sulfides, Alkalinity, Calcium, Magnesium, *Wisconsin.
Identifiers: Binding capacity, Mobilization, *Lake Monona (Wis).

Sediment samples from Lake Monona, Wisconsin were dried and solutions prepared in flasks by adding distilled water and a standard solution of Cu. The solutions were shaken for 24 hr, left to stand for 12 hr, and/or centrifuged through membrane filters and the filtrate analyzed for copper. Wet sediment was also aerated by shaking for one month to oxizide Fe, Mn, sulfide, and organics to determine their effect on the binding capacity of the sediment for copper. Additional studies were conducted with large quantities of NTA added to sediment solutions to determine its effect on the release of heavy metals. Copper, iron, and manganese were determined by AA after 24 hr of contact. The results show that the binding capacity of Cu by Lake Monona sediments is primarily related to the alkalinity (Ca, Mg carbonates present in the sediments). Apparently organics and sulfides play a minor role in binding copper to sediments; in the case of sulfides most probably for reasons of low content rather than for reasons of solubility of cupric sulfides. Over very dilute copper solutions, sulfide in the sediment should control the binding mechanism. The sediments investigated show a relatively large binding capacity amounting to approximately 26 mg of copper being fixed per gram of dry sediments. The addition of large amounts of NTA to Lake Monona sediments resulted in an increase in the amounts of iron and manganese leached from the sediments. However, the copper released from the sediments decreased with increasing NTA. (Little-Battelle)
W73-08946

TRACE METAL RELATIONSHIPS IN A MARINE PELAGIC FOOD CHAIN,
Stanford Univ., Calif.
For primary bibliographic entry see Field 05C.
W73-08947

TOXIC EFFECTS OF CUPRIC, CHROMATE, AND CHROMIC IONS ON BIOLOGICAL OXIDATION,
Calgary Univ. (Alberta). Dept. of Chemical Engineering.
For primary bibliographic entry see Field 05C.
W73-08948

MASS AND MONOXENIC CULTURE OF VORTICELLA MICROSTOMA ISOLATED FROM ACTIVATED SLUDGE,
Tokyo Univ. (Japan). Inst. of Applied Microbiology.
For primary bibliographic entry see Field 05A.
W73-08950

THE USE OF COLIPHAGE AS AN INDEX OF HUMAN ENTEROVIRUS POLLUTION IN AN ESTUARINE ENVIRONMENT,
New Hampshire Univ., Durham.
For primary bibliographic entry see Field 05A.
W73-08951

STUDIES ON THE OCCURRENCE, PHYSIOLOGY, AND ECOLOGY OF BIOLUMINESCENCE IN DINOFLAGELLATES,
Oregon State Univ., Corvallis.
For primary bibliographic entry see Field 05C.
W73-08953

A COMPARATIVE STUDY OF THE DECOMPOSITION OF CELLULOSIC SUBSTRATES BY SELECTED BACTERIAL STRAINS,
Massachusetts Univ., Amherst.
M. C. Segal.
Available from Univ. Microfilms, Inc., Ann Arbor, Mich. 48106 Order No. 73-6723. Ph D Dissertation, 1972. 141 p.

Descriptors: *Microbial degradation, *Cellulose, *Degradation (Decomposition), Cultures.
Identifiers: *Viscometric techniques, Cellulomonas, Cellvibrio, Cytophaga, Bacillus, *Bacterium, Pseudomonas, Arthrobacter, *Substrates, Vibrios, Cotton cellulose, Cotton linters, Wood cellulose, Hydroxethyl cellulose, Carboxymethyl cellulose.

The decomposition of cellulosic materials by cultures of aerobic mesophilic bacteria was investigated using viscometric techniques. Substrates for decomposition studies included insoluble cotton celluloses of various Degrees of Polymerization, raw cotton linters, wood cellulose, hydroxethyl cellulose and soluble carboxymethyl celluloses of various Degrees of Polymerization and Degrees of Substitution. Viscometric methodologies were readily adaptable for quantitative determinations using whole cells. The most active cultures included species of Cellulomonas, Cellvibrio and Cytophaga. Cultures showing less activity than the above species included strains of Bacillus, Bacterium, Cytophaga. Pseudomonas. Cultures showing little or no activity included strains of Arthrobacter, Bacillus, and Cellvibrio. (Little-Battelle)
W73-08957

CESIUM-137 AND STABLE CESIUM IN A HYPEREUTROPHIC LAKE,
Michigan State Univ., East Lansing.
For primary bibliographic entry see Field 05C.
W73-08959

ACETYLENE REDUCTION IN SURFACE PEAT,
Dundee Univ. (Scotland). Dept. of Biological Sciences.
G. J. Waughman, and D. J. Bellamy.
Oikos, Vol 23, No 3, p 353-358, 1972. 3 fig, 2 tab, 19 ref.

Descriptors: *Peat, *Nitrogen fixation, *Assay, Ecosystems, *Path of pollutants, Gas chromatography, Mass spectrometry, Chemical reactions, Nitrogen, Chemical analysis, Enzymes, Mosses, Radioactivity techniques, Organic soils, Aquatic plants, Chlorophyta.
Identifiers: *Acetylene reduction, *Mire, Ethylene, N-15, Nitrogenase, Enzyme kinetics, Fate of pollutants, Nitrogen radioisotopes, Myricagale, Macrophytes, Nuphar lutea, Bryum pseudotriquetrum, Drepanocladus revolvens, Chara, Potamogeton polygonifolius, Carex spp, Scorpidium scorphiodes, Utricularia minor, Campylium stallatum, Riccardia pingius, Potentilla palustris, Molinia caerulea, Phragmites communis, Sphagnum spp, Aulaconnium palustre, Narthecium ossifragum, Calluna vulgaris, Rhynchospora alba, Drosera rotundifolia, Erica tetralix, Hylocomium splendens, Eriophorum spp.

An investigation was carried out between April and August 1971, using the acetylene reduction assay for nitrogen fixation. Ten samples were collected from three mire types and from blanket peat and flushed with a gas phase of O2, CO2 and Ar. Acetylene was injected into the sample chambers and the ethylene content was estimated at intervals by gas chromatography. Other time course experiments were carried out to determine the effect of glucose, carbon monoxide and a high ambient oxygen tension on acetylene reduction by the peat. Anaerobic nitrogenase activity and N-15 uptake were also measured using samples of one of the mire types. Nitrogen fixation in the mire system was confirmed by uptake of N-15. The rates were very low and near to the limits of detection by mass spectrometry. Nitrogenase activity measurements indicate that a relationship does exist between mire type and such activity, the highest rate being in rheophilus mire and the lowest in ombrophilous mire. (Holoman-Battelle)
W73-08960

DEOXYRIBONUCLEIC ACID IN ECOSYSTEMS,
North Carolina Univ., Chapel Hill.
For primary bibliographic entry see Field 05A.
W73-08961

EXCRETION MEASUREMENTS OF NEKTON AND THE REGENERATION OF NUTRIENTS NEAR PUNTA SAN JUAN IN THE PERU UPWELLING SYSTEM DERIVED FROM NEKTON AND ZOOPLANKTON EXCRETION,
Washington Univ., Seattle.
T. E. Whitledge.
Available from Univ. Microfilms, Inc., Ann Arbor, Mich. 48106 Order No. 73-3803. Ph D Dissertatiou, 1972. 126 p.

Descriptors: Water analysis, Fish, *Ammonia, *Nitrates, *Silicates, *Ureas, *Primary productivity, *Cycling nutrients, *Plankton, Bacteria, Sea water, Fluorescence, Phosphates, Zooplankton, Phytoplankton, Limiting factors, Diatoms, Food abundance, Upwelling, Oxygen demand, Automation.
Identifiers: *Creatine, *Excretion, Feces, Peruvian anchoveta, Northern anchovy, Tubesnout, Three-spined stickleback, Piceral, Rainbow wrasse, Stargazer, Detection limits, Biological samples, Orthophosphates, Sample preparation.

The excretion of nutrients and organic substances was examined for eight species of fish from several regions of the world's oceans. The excreted substances measured were ammonia, nitrate, orthophosphate, silicate, creatine, and urea. In addition soluble organic nitrogen excretion, oxygen utilization, and physical parameters were determined. Automated and manual methods for the determination of creatine in seawater were developed. The analysis consists of the addition of (1.2-cyclohexylenedinitrilo)tetraacetic acid, ninhydrin, and potassium hydroxide after which the fluorescence of the solution is measured. Concentrations of 0.05 to 30 microgram-at/creatine-N/liter were measured with the analysis. Regenerated production derived from the uptake of ammonia nitrogen was 28 percent of the nitrogen utilized by phytoplankton near Punta San Juan in Peru while new production derived from nitrate was used for the remaining 72 percent of the nitrogen. Of the ammonia nitrogen required for primary production 71 percent was estimated to be supplied by the excretion of anchoveta and 8 percent by zooplankton. The bacterial uptake of creatine was two times the quantity excreted by anchoveta. Phytoplankton uptake of urea was estimated to be three times the combined excretion of nekton and zooplankton. Excretion produces 37 and 4 percent respectively of the phosphorus and silica requirements of primary production in the Peru upwelling area. Silicate may limit the growth of diatoms. Anchoveta consume 72 to 80 percent of primary production. (Little-Battelle)
W73-08963

UTILIZATION OF CRUDE OIL HYDROCARBONS BY MIXED CULTURES OF MARINE BACTERIA,
Florida State Univ., Tallahassee.
H. I. Kator.
Available from Univ. Microfilms, Inc., Ann Arbor, Mich. 48106 Order No. 73-194. Ph D Dissertation, 1972. 257 p.

Descriptors: *Marine bacteria, *Microbial degradation, Sea water, Oily water, Degradation (Decomposition), Aromatic compounds, Organic compounds, Chemical analysis.
Identifiers: *Crude oil, *Substrate utilization, *Paraffins, *Fate of pollutants, Bacterial physiology, Column chromatography, Activated silica gel, Aliphatic hydrocarbons, Biosynthesis, Fatty acids, Aromatic hydrocarbons, Naphthalene, Anthracene, Alkyl benzenes, Pristane, Phytane, Culture media, Louisiana crude oil, Kuwait crude oil, Venezuelan crude oil, Enrichment.

Mixed populations of marine bacteria were isolated from a variety of marine and estuarine locations and maintained on a medium consisting of a sterile crude oil in a nutrient-salts enriched seawater (ESW). They degraded a variety of pure n-paraffins of chain lengths ranging from C-10 to C-30. While visible growth was not evident on selected examples of branched paraffins, aromatics, and heterocyclic compounds, a substrate transfer experiment revealed that some examples of alkyl benzenes, napthalene, and anthracene supported growth. A variety of crude oils of different composition (Louisiana, Kuwait, and Venezuelan crude oils) were degraded (in ESW) with consistent utilization of n-paraffins C-10 to C-30. Generally, the rates of utilization were inversely related to the chain lengths and directly related to the amount of each n-paraffin added to the growth flask. Sequential utilization of n-paraffins was followed by utilization of the branched paraffins pristane and phytane. Rates of utilization (20 C) for mixed cultures were 2 to 22 times larger in ESW than in unenriched seawater. Activated silica gel column chromatography of degraded crude oils revealed a preferential utilization of saturated paraffins in Kuwait, Venezuelan and Louisiana crude oils. Relative to the absolute weights of saturated paraffins utilized, aromatic-naphthenic utilization in whole crude oils was limited. Aromatic-naphthenic utilization in a crude oil from which normal paraffins were removed (using a molecular sieve) was small but significant. Intracellular fatty acids produced by bacterial cells grown on crude oils were similar to fatty acid profiles found in nonhydrocarbon grown bacterial cells. De novo fatty acid synthesis, rather than direct incorporation of the oxidized hydrocarbons as fatty acids, was therefore implied. (Holoman-Battelle)
W73-08967

ESTUARINE NITRIFICATION,
Rutgers - The State Univ., New Brunswick, N. J.
B. J. Berdahl.
Available from Univ. Microfilms, Inc., Ann Arbor, Mich. 48106, Order No. 73-4719. Ph D Dissertation, 1972. 334 p.

Descriptors: *Estuarine environment, *Ammonia, *Model studies, *Kinetics, *Estuaries, Salinity Nitrates, Nitrites, Equations, Chemical reactions, Oxidation, Mathematical studies, Dissolved oxygen, Water temperature, Hydrogen ion concentration, Conductivity, Saline water, Nitrogen.
Identifiers: *Nitrification, Fate of pollutants.

Nitrification, the oxidation of ammonia to nitrate by specific autotrophic bacteria, was studied in a model estuary system having a total flow (flushing) time of 30 days. During this time interval, the salinity was being increased exponentially until it reached that of coastal (100 percent) seawater. Ammonia-nitrogen was added to this salinity gradient at three points (day 0, day 10 and day 20 flow) corresponding to 2, 6, and 34 percent seawater. The concentrations ranged from 1 to 9 mg/l NH3-N. Daily measurements of ammonia, nitrite and nitrate-nitrogen along with conductivity (salinity), dissolved oxygen, temperature and pH were taken. Only two representative ammonia-nitrogen concentrations, 3 and 5 mg/l, were studied. Upon discharge at 2 and 6 percent seawater, NH3 was oxidized in two definite kinetic stages that were best described by an autocatalytic first-order equation (Stage 1) and a simple first-order equation (Stage 2). At 34 percent seawater, there was only sufficient time (10 days) for the first (autocatalytic) stage to develop. The 6 percent seawater starting salinity reduced (inhibited) the quantity of oxidized ammonia compared to 2 and 6 percent seawater. Low dissolved oxygen concentrations of 0.5-1/0 mg/l existed for at least 24 hours without any apparent effect upon nitrification. (Holoman-Battelle)
W73-08971

INVERTEBRATE DRIFT IN AN OHIO STREAM AND ITS UTILIZATION BY WARM-WATER FISHES,
Ohio State Univ., Columbus.
M. E. Sisk, Jr.
Available from Univ. Microfilms, Inc., Ann Arbor, Mich. 48106, Order No. 73-2128. Ph D Dissertation, 1972. 80 p.

Descriptors: Streams, *Food habits, *Fish, *Mayflies, *Diptera, Food abundance, Dominant organisms, Population, Invertebrates, Shiners, Sunfishes, Darters, Biomass, Sampling, Secondary productivity, Ohio.
Identifiers: *Drift organisms, Alum Creek, Baetis, Caenis.

The drift pattern of aquatic and terrestrial invertebrates in a Central Ohio stream (Alum Creek) supporting a warm-water fish fauna of shiners, darters and sunfishes was investigated. The study was intended to ascertain what fishes fed on drift and to what extent, whether or not the drift-food resource was shared by the fishes, whether or not the feeding activity of fishes was correlated with greatest drift activity, how abundant drift organisms were on a diel and seasonal basis, what portion of the drift was allochthonous in origin and what taxa were represented in the drift. Eleven 24-

hour collections were taken at a riffle site during the period of August, 1969 and July, 1970. Each monthly collection consisted of 10-minute drift net samples taken every two hours to determine quality and quantity of drift, four collections of fishes for stomach analysis and two bottom samples for determination of benthic biomass. Maximum drift of aquatic invertebrates in numbers and biomass occurred at 2300 hours. Baetidae was represented in the drift by five genera. The Chironomidae exhibited the most regular drift pattern and occurred in greatest numbers. Maximum drift in numbers and biomass of aquatic species occurred in May. Terrestrial adults contributed 74 percent in biomass and 24 percent by numbers to the annual drift and were collected mostly during daylight hours. The fishes fed heavily on both aquatic and terrestrial organisms but stomach contents of the fishes did not correlate with aquatic drift peaks. (Little/Battelle)
W73-08972

PROJECT WAGON WHEEL TECHNICAL STUDIES REPORT NO. 2.
El Paso Natural Gas Co., Tex.

Available from NTIS, Springfield, Va., as PNE-WW-13; $3.00 in paper copy, $1.45 microfiche. Report PNE-WW-13, Oct. 1972. L. A. Rogers, editor. 215 p, 75 fig, 72 tab, 118 ref.

Descriptors: *Radioactivity effects, *Radioecology, *Nuclear explosions, *Monitoring, Water wells, Well data, Texas, Baseline studies, Feasibility studies, Public health, Natural gas, Seismic properties, Seismic waves, Water pollution effects, Air pollution, Desert plants, Sagebrush, Semiarid climates, On-site investigations, Path of pollutants, Nuclear wastes, Biota, Radioisotopes, Absorption, Hazards.

Aspects of the proposed project (stimulation of natural gas by five 100-kiloton nuclear explosions) are evaluated in separate reports on effects of predicted ground motion, environmental radiation, off-site water-well and natural-spring documentation, temperature of the chimney, initial composition of the gas, release of radioactivity, stresses in the gas reservoir, and ecology of the environs. Environmental-radiation monitoring programs and equipment were tested and base-line data were obtained by measurements of samples of airborne particles, surface water, municipal and well water, precipitation, milk, soil, urine, vegetation, bottom sediment, fish, and animals. Sixty-two water wells and natural springs were identified within a 10-mile radius, 363 within a 20-mile radius and many were surveyed for later comparison with post-event data. Negligible environmental impact due to radioactivity from burning of the gas during well testing is shown. Base-line data on native vegetation were provided by inspection at 43 locations. Several varieties of sagebrush and grass are the principal vegetation in the immediate area. Environmental effects from farming, ranching, construction, and other activities were evident. (Bopp-ORNL)
W73-08991

CONCENTRATION FACTORS OF CHEMICAL ELEMENTS IN EDIBLE AQUATIC ORGANISMS,
California Univ., Livermore. Lawrence Livermore Lab.
S. E. Thompson, C. A. Burton, D. J. Quinn, and Y. C. Ng.
Available from NTIS, Springfield, Va., as UCRL-50564 Rev. 1; $3.00 paper copy, $1.45 microfiche. Report UCRL-50564 Rev. 1, Oct 1972. 73 p, 6 tab, 300 ref.

Descriptors: *Radioisotopes, *Absorption, *Food chains, *Reviews, *Bibliographies, Public health, Path of pollutants, Fallout, Nuclear wastes, Freshwater, Sea water, Estuarine environment, Marine animals, Marine plants, Marine fish, Marine algae, Freshwater fish, Aquatic plants, Aquatic animals, Aquatic environment, Radioecology, Trace elements, On-site data collections, Data collections.

In the compilation many of the estimated values of UCRL-50564 are replaced by measured values. As a result, certain values increased more than an order of magnitude (for uptake of Ta and Po by marine invertebrates; Th by marine fish; Ra and W by freshwater plants; Li, B, F, V, and Po by freshwater invertebrates; I, W, and Re by freshwater fish) and other values decreased more than an order of magnitude (Sb and U by marine plants; Zr, Sb, and U by marine invertebrates; Sc and Br by marine fish; Cl, Cr, and Br by freshwater plants; Cr, Zr, and Sb by freshwater invertebrates; Be, Co, Zr, Ag, Cd, and Sb by freshwater fish). For a discussion of the uses and limitation of concentration factors see W72-04484, W72-04491, W72-4490, and W73-00821. (Bopp-ORNL)
W73-08992

FINAL ENVIRONMENTAL STATEMENT RELATED TO OPERATION OF CALVERT CLIFFS NUCLEAR POWER PLANT UNITS 1 AND 2.
Directorate of Licensing (AEC), Washington, D.C.

Available from NTIS, Springfield, Va., as Docket 50317-87. $6.00 per copy, $1.45 microfiche. Dockets 50317-87 and 50318-84, April 1973. 309 p, 16 fig, 30 tab, 95 ref, 4 append.

Descriptors: *Nuclear powerplants, Effluents, Environment, Administrative agencies, *Comprehensive planning, *Sites, Geology, Investigations, Hydrology, Seismology, Climatology, Meteorology, Ecology, Radioactive wastes, Water pollution, Water pollution sources, Radioactive effects, Monitoring, Public health, Transportation, Beneficial use, Cost-benefit analysis, *Maryland, Chesapeake Bay, *Thermal pollution.
Identifiers: Atomic Energy Commission, *Pressurized water reactors, Lusby (Maryland), *Environmental impact statements.

This final environmental statement was prepared in compliance with the National Environmental Policy Act and relates to the continuation of construction and operation of the Calvert Cliffs nuclear power plants. Each of these units will employ a pressurized-water reactor cooled by salt water from and discharged to the Chesapeake Bay. The site is near Lusby, Maryland. Environmental impacts are assessed and after consideration of alternatives an environmental benefit-cost summary was compiled. Some environmental factors considered include climate, hydrology (surface water and ground water), ecology including aquatic life, cooling-water supply and discharge, cooling towers, cooling lakes, spray ponds, radioactive chemical and sanitary wastes, amount of dissolved oxygen and toxic chemicals in effluent water. The conclusion is to continue the construction permits and issue operating licenses for the facility subject to the following conditions: (1) establish a comprehensive biological monitoring program; (2) establish a hydrological monitoring program; (3) establish a radiological monitoring program to include sampling and analysis of milk and forage; (4) if harmful effects appear, a program of remedial action is to be taken immediately. (Houser-ORNL)
W73-08993

FINAL ENVIRONMENTAL STATEMENT FOR THE EDWIN I. HATCH NUCLEAR PLANT UNITS I AND 2.
Directorate of Licensing (AEC), Washington, D.C.

Available from NTIS, Springfield, Va., as Docket 50321-54; $3.00 as paper copy, $1.45. Docket 50321-54, Oct 1972. 194 p, 13 fig, 22 tab, 65 ref, 2 append.

Descriptors: *Nuclear powerplants, Effluents, Environment, Administrative agencies, *Comprehensive planning, *Site, Geology, Investigations, Hydrology, Seismology, Climatology, Meteorology, Ecology, Radioactive wastes, Water pollution, Water pollution sources, Radioactive effects, Monitoring, Public health, Transportation, Beneficial use, Cost-benefit analysis, Rivers, *Georgia, *Thermal pollution.
Identifiers: Atomic Energy Commission, *Boiling water reactors, *Altamaha River (Geo), *Environmental impact statements.

This final environmental statement was prepared in compliance with the National Environmental Policy Act and relates to the proposed issuance of operating license for Unit No. 1 and a construction permit for Unit No. 2 of the Edwin I. Hatch Nuclear Plant located at the Altamaha River, near Baxley, Georgia. The two boiling water reactors will be cooled by a closed-cycle cooling system using mechanical-draft cooling towers discharging waste to the atmosphere. Makeup water will be taken from the Altamaha River and blowdown returned to it. Environmental impacts are assessed and after consideration of alternatives an environmental benefit-cost summary was compiled. Some environmental factors considered include climate, hydrology (surface water and ground water), ecology including aquatic life, cooling-water supply and discharge, cooling towers, cooling lakes, spray ponds, radioactive chemical and sanitary wastes, amount of dissolved oxygen and toxic chemicals in effluent water. USAEC will issue an operating license for Unit 1 and a construction permit for Unit 2, subject to the following conditions: Applicant shall (1) establish preoperational base line data distribution of aquatic species, (2) minimize damage due to construction of transmission lines, (3) establish prior to operation a monitoring program for liquid effluents, (4) establish a program to monitor the ecosystems, (5) establish a radiological monitoring program, (6) provide a course of action to alleviate detrimental effects found, (7) provide analysis of alternates of venting gaseous effluents from the turbine building. (Houser-ORNL)
W73-08994

FINAL ENVIRONMENTAL STATEMENT RELATED TO OPERATION OF SHOREHAM NUCLEAR POWER STATION.
Directorate of Licensing (AEC), Washington, D.C.

Available from NTIS, Springfield, Va., as Docket 50322-30; $6.00 per copy, $1.45 microfiche. Docket 50322-30, September 1972. 585 p, 37 fig, 52 tab, 503 ref, 8 append.

Descriptors: *Nuclear powerplants, Effluents, Environment, Administrative agencies, *Comprehensive planning, *Sites, Geology, Investigations, Hydrology, Seismology, Climatology, Meteorology, Ecology, Radioactive wastes, Water pollution, Water pollution sources, Radioactive effects, Monitoring, Public health, Transportation, Beneficial use, Cost-benefit analysis, *New York, *Thermal pollution.
Identifiers: Atomic Energy Commission, *Boiling water reactors, *Brookhaven (N.Y.), *Environmental impact statements.

This final environmental statement was prepared in compliance with the National Environmental Policy Act and relates to the proposed construction of the Shoreham Nuclear Power Station, located at Brookhaven, New York. The Shoreham Station will consist of a boiling-water reactor with waste heat dissipated by once-through cooling seawater drawn from and discharged to Long Island Sound. Environmental impacts are assessed and after consideration of alternatives an environmental benefit-cost summary was compiled. Environmental factors considered include hydrology (surface water and ground water), ecology including aquatic life, cooling-water supply and discharge,

cooling towers, cooling lakes, spray ponds, radioactive chemical and sanitary wastes, amount of dissolved oxygen and toxic chemicals in effluent water. The conclusion is to issue a construction permit subject to the following conditions for protection of the environment: (1) protect Wading River marsh from blowing sand; (2) conduct additional biological studies to evaluate potential impact; (3) consider alternative intake systems and modifications to present design; (4) monitor accumulation and/or erosion of beach sand and gravel in vicinity of the intake jetties. (Houser-ORNL)
W73-08995

FINAL ENVIRONMENTAL STATEMENT RELATED TO OPERATION OF THREE MILE ISLAND NUCLEAR STATION UNITS 1 AND 2.
Directorate of Licensing (AEC), Washington, D.C.

Available from NTIS, Springfield, Va., as Docket 50289-80. $3.00 per copy, $1.45 microfiche. Dockets 50289-80 and 50320-38, Dec 1972. 251 p, 19 fig, 24 tab, 60 ref, 4 append.

Descriptors: *Nuclear powerplants, Effluents, Environment, Administrative agencies, *Comprehensive planning, *Sites, Geology, Investigations, Hydrology, Seismology, Climatology, Meteorology, Ecology, Radioactive wastes, Water pollution, Water pollution sources, Radioactive effects, Monitoring, Public health, Transportation, Beneficial use, Cost-benefit analysis, Rivers, *Pennsylvania, *Thermal pollution.
Identifiers: Atomic Energy Commission, *Pressurized water reactors, Susquehanna River, *Environmental impact statements, *Harrisburg (Penn).

This final environmental statement was prepared in compliance with the National Environmental Policy Act and relates to the proposed construction and operation of the Three Mile Island Nuclear Station Units 1 and 2. The station is comprised of two pressurized water reactor units having four natural draft cooling towers for dissipating waste heat from the closed cycle cooling water system and is located near Harrisburg, Pennsylvania. Cooling water will be discharged to the Susquehanna River. Environmental impacts are assessed and after consideration of alternatives an environmental benefit-cost summary was compiled. Some environmental factors considered include hydrology (surface water and ground water), ecology including aquatic life, cooling-water supply and discharge, cooling towers, cooling lakes, spray ponds, radioactive chemical and sanitary wastes, amount of dissolved oxygen and toxic chemicals in effluent water. The conclusion is to continue the construction permits and issue operating license for units 1 and 2 subject to certain specified monitoring programs for the protection of the environment. If harmful effects appear, the applicant is required to provide an analysis of expected impacts and a course of action to minimize these impacts. (Houser-ORNL)
W73-08996

FINAL ENVIRONMENTAL STATEMENT RELATED TO THE LA SALLE COUNTY NUCLEAR STATION.
Directorate of Licensing (AEC), Washington, D.C.

Available from NTIS, Springfield, Va., as Docket 50373-32. $3.00 per copy, $1.45 microfiche. Dockets 50373-32 and 50374-31, Feb 1973. 267 p, 14 fig, 22 tab, 60 ref, 2 append.

Descriptors: *Nuclear powerplants, Effluents, Environment, Administrative agencies, *Comprehensive planning, *Sites, Geology, Investigations, Hydrology, Seismology, Climatology, Meteorology, Ecology, Radioactive wastes, Water pollution, Water pollution sources, Radioactive effects, Monitoring, Public health, Transportation, Beneficial use, Cost-benefit analysis, Lakes, *Illinois, *Thermal pollution.
Identifiers: Atomic Energy Commission, *Boiling water reactors, *Brookfield (Ill), *Environmental impact statements.

This final environmental statement was prepared in compliance with the National Environmental Policy Act and relates to a decision on the applications for construction permits for two reactors near Brookfield, Illinois. The two boiling water reactors are to be cooled by water obtained from and discharged to an artificial lake to be constructed. These facilities are part of the La Salle County Nuclear Station. Environmental impacts are assessed and after consideration of alternatives an environmental benefit-cost summary was compiled. Some environmental factors considered include climate, hydrology (surface water and ground water), ecology including aquatic life, cooling-water supply and discharge, cooling towers, cooling lakes, spray ponds, radioactive chemical and sanitary wastes, amount of dissolved oxygen and toxic chemicals in effluent water. The conclusion is to issue construction permits for the facilities subject to the following conditions: (1) applicant to submit a lake management program; (2) cleanup and resoration of transmission line construction; (3) locate intake to the blowdown system in the cooler part of the lake; and (4) implement a monitoring program to determine environmental effects of construction and operation. (Houser-ORNL)
W73-08997

DISTRIBUTION OF ICHTHYOPLANKTON IN THE WARM WATER DISCHARGE ZONE OF THE NOVOROSSIISK THERMAL ELECTRIC POWER PLANT, (IN RUSSIAN),
Rostov-on-Don State Univ. (USSR).
For primary bibliographic entry see Field 05C.
W73-09018

RARE CASE OF POISONING BY WATER POLLUTED WITH TOXIC AGRICULTURAL CHEMICALS, (IN RUSSIAN),
Khorezmskii Pedagogicheskii Institut, Urgench (USSR).
For primary bibliographic entry see Field 05C.
W73-09026

5C. Effects of Pollution

ESTUARINE POLLUTION, A BIBLIOGRAPHY.
Office of Water Resources Research, Washington, D.C.
For primary bibliographic entry see Field 02L.
W73-08451

BIOLOGICAL AND PHOTOBIOLOGICAL ACTION OF POLLUTANTS ON AQUATIC MICROORGANISMS,
Auburn Univ., Ala. Dept. of Chemistry.
W. C. Neely, R. C. Smith, R. M. Cody, J. R. McDuffie, and J. A. Lansden.
Available from the National Technical Information Service as PB-220 167, $5.45 paper copy, $1.45 microfiche. Alabama Water Resources Research Institute, Auburn Bulletin 9, February 1973. 122 p, 23 fig, 5 tab, 148 ref. OWRR A-017-ALA (2).

Descriptors: Pollutants, *Heavy metals, *Pesticide residues, Water pollution effects, Aquatic microorganisms, E. coli, Insecticides, Ions, *Pesticide toxicity, Photoactivation.
Identifiers: *Photosensitization, *Mutagens, Polycyclic amines, Metal ions, *Phototoxicity.

The various aspects of the biological and photobiological action of pollutants on aquatic microorganisms have been explored in a series of studies covering a wide range of chemical species and two classes of aquatic microorganisms, paramecia and bacteria. Certain metal ions, insecticides, polycyclic amines and mycotoxins are capable of drastic alteration of the life processes in test strains of either bacteria or paramecia or both. In some cases concurrent exposure to light and the pollutant was necessary and in others the toxic and/or mutagenic reactions were independent of light exposure. In particular the uranyl ion, beta naphthyl amine, and aflatoxin B1 were phototoxic while the insecticide, phygon, was strongly darktoxic. All agents affected Paramecium caudatum and all but aflatoxin B1 affected Escherichia coli. In addition, the uranyl ion caused morphological mutations in E. coli. The public health hazard connected with such alterations in life-cycles resulting from water-horne pollutants is difficult to assess from laboratory data but the possibility should be further investigated.
W73-08452

AQUATIC FUNGI OF THE LOTIC ENVIRONMENT AND THEIR ROLE IN STREAM PURIFICATION,
Virginia Polytechnic Inst. and State Univ., Blacksburg. Dept. of Biology.
R. A. Paterson, and D. F. Farr.
Available from the National Technical Information Service as PB-220 012, $3.00 in paper copy, $1.45 in microfiche. Virginia Water Resources Research Center, Blacksbury, Completion Report, March 1973, 40 p, 18 fig, 10 ref. OWRR A-043-VA (1).

Descriptors: *Aquatic fungi, Running waters, Water pollution effects, *Water purification, *Toxicity, Zinc, Surfactants, *Virginia.
Identifiers: Pure culture studies, Mannitol, Cyanide, *Phycomycetes.

Samples were collected from three rivers and streams of varying physical and chemical characteristics. The samples were examined for aquatic Phycomycetes. These fungi were found in all samples collected. The filamentous forms such as Achlya, Saprolegnia, and Pythium were abundant in all three rivers. However, the unicellular forms such as Rhizophydium and Phlyctochytrium were infrequently encountered or absent in the eutrophic river but very common in the oligotrophic situations. Samples collected immediately below the introduction of the effluent from an Army ammunition plant showed a greatly reduced flora of aquatic Phycomycetes. Achlya flagellata, a Rhizidium sp. and a Phlyctochytrium sp. were isolated from natural collections and their responses in terms of growth and reproduction to zinc, mannitol, cyanide, and surfactants were observed. These studies showed the filamentous form, Achlya flagellata, to be more tolerant to the toxic effects of these chemicals than were the unicellular Phlyctochytrium and Rhizidium.
W73-08462

ADAPTATION TO EUTROPHIC CONDITIONS BY LAKE MICHIGAN ALGAE,
Wisconsin Univ., Milwaukee. Dept. of Botany.
K. C. Lin, and J. L. Blum.
Available from the National Technical Information Service as PB-220 013, $3.00 in paper copy, $1.45 in microfiche. Wisconsin Water Resources Center, Madison, Technical Report WIS-WRC 73-03, 1973. 24 p, 4 tab, 5 fig, 18 ref. OWRR A-041-WIS (1), 14-31-0001-3550.

Descriptors: *Aquatic algae, *Sessile algae, Nutrients, Tertiary treatment, *Phosphorus compounds, *Phosphates, Phosphorus, Plankton, Municipal wastes, *Lake Michigan, Eutrophication, *Wisconsin, Cladophora.
Identifiers: Milwaukee River, Milwaukee Harbor, *Ulothrix sp, *Cladophora glomerata.

Algae, when supplied with sufficient orthophosphate, can absorb phosphorus in quantities far in excess of the amount needed for optimal growth. This surplus phosphorus is then available for the algae to continue to grow when the external supply of orthophosphate is depleted. Studies were conducted on Cladophora glomerata and Ulothrix sp. to determine whether algae could utilize polyphosphate. Results showed that the extractable orthophosphate was inversely proportional to the ability of the algae species to hydrolyze either pyrophosphate or tripolyphosphate. Field investigations indicated that algae growing near the Milwaukee harbor area received a sufficient supply of orthophosphate, whereas the algae located some distance from the harbor did not. It is likely that the algae accumulated the soluble orthophosphate supplied by the waste water effluent and non-point sources discharged by the Milwaukee River. At the same time, the insoluble polyphosphate and organic phosphates dispersed into the open lake. The algae growing furthest from the harbor provided phosphatases to hydrolyze the polyphosphates, because of the lack of orthophosphate. This method of phosphorus assimilation by the algal metabolism produced less substantial growth. Field samples were collected in 1971 from four sample sites in the Milwaukee Harbor and five sites in Lake Michigan with a five kilometer radius of the harbor. (Kerrigan-Wisconsin)
W73-08463

BENTHAL OXYGEN DEMANDS AND LEACHING RATES OF TREATED SLUDGES,
Manhattan Coll., Bronx, N.Y.
J. A. Mueller, and W. J-l. Su.
Journal Water Pollution Control Federation, Vol 44, No 12, p 2303-2315, December 1972. 15 fig, 8 tab, 14 ref.

Descriptors: *Biochemical oxygen demand, *Cultures, *Leaching, *Sewage sludge, *Sludge disposal, Water pollution effects, *Chemical oxygen demand, Laboratory equipment, Heavy metals, Laboratory tests, Nutrients, Sludge treatment, Sea water, Phosphates, Nitrogen, Ammonia, Aluminum, Cadmium, Chromium, Copper, Iron, Lead, Magnesium, Manganese, Nickel, Zinc, Titanium, Benthos.
Identifiers: Arsenic.

The effects of sludge dumping in coastal waters, such as the New York Bight, were studied in the laboratory in continuous flow reactors supplied with raw, digested, heat treated, and wet oxidized sludges. The sludges were concentrated, analyzed, and seeded using sea water overlying sludge deposits. The reactors consisted of plexiglas cylinders 2 3/4 inch i.d. by 5 inches high mounted on a square base. A plastic screen covered the sludge to maintain stability during continuous air diffusion. The reactors were placed in a water bath maintained at 15 C. For each sludge, one reactor was used for determination of oxygen uptake and effluent analyses during a 45-day incubation period. A second reactor was used for sludge analysis after 21 days. After 45 days, sludge in a third reactor was mixed with its supernatant and oxygen uptake monitored for several days. Test results indicated that treatment of sludges before disposal reduced benthal demands and nutrient leaching rates. Benthal demands could be reduced 50 percent by heat treatment, low-pressure wet oxidation, or digestion, and over 90 percent by intermediate and high-pressure wet oxidation. Wet oxidation also provided the best reduction of volatile solids in the sludges. (Little-Battelle)
W73-08527

EFFECT OF PULPMILL EFFLUENT ON DISSOLVED OXYGEN IN A STRATIFIED ESTUARY - I. EMPIRICAL OBSERVATIONS,
Fisheries Research Board of Canada, Nanaimo (British Columbia). Biological Station.
R. R. Parker, and J. Sibert.
Water Research, Vol 7, No 4, p 503-514, April 1973. 8 fig, 22 ref.

Descriptors: *Canada, *Dissolved oxygen, *Pulp wastes, Water pollution effects, *Stratification, *Estuaries, On-site investigations, On-site tests, Laboratory tests, Chemical analysis, Water analysis, Industrial wastes, Water sampling, Salinity, Color, Light penetration, Saline water, Oxygen sag, Solar radiation, Primary productivity, Biomass, Nutrients.
Identifiers: *Alberni Inlet (Canada).

That pulpmill effluent discharged at the water's surface does suppress photosynthesis, thus oxygen production, in the stratum beneath the halocline by restricting light penetration has been examined empirically in the Alberni Inlet, Canada. Measurements were made of salinity, DO, primary productivity, color, light transmission, and solar radiation. The data indicate that the pulpmill effluent adversely affects DO levels in the estuary, and several possible causes are discussed. Net primary productivity was diminished by the light-absorbing stain. The situation cannot be remedied by removal of BOD from the effluent, rather the staining properties must be diminished or removed. (See also W73-08544) (Holoman-Battelle)
W73-08543

EFFECT OF PULPMILL EFFLUENT ON DISSOLVED OXYGEN IN A STRATIFIED ESTUARY - II. NUMERICAL MODEL,
Fisheries Research Board of Canada, Nanaimo (British Columbia). Biological Station.
J. Sibert, and R. R. Parker.
Water Research, Vol 7, No 4, p 515-523, April 1973. 3 fig, 2 tab, 16 ref.

Descriptors: *Canada, *Mathematical models, *Dissolved oxygen, *Estuaries, *Stratification, *Pulp wastes, Color, Light intensity, Nutrients, Inlets (Waterways), Inhibition, Photosynthesis, Respiration, Model studies, Biochemical oxygen demand, Industrial wastes, Primary productivity, Biomass, Bacteria, Algae, Growth rates, Saline water, Turbidity.
Identifiers: Pollutant removal, *Alberni Inlet (Canada), Substrate utilization.

A numerical model is presented which simulates the biological and physical processes of oxygen supply in a stratified inlet. The results of the model show that the introduction of pulpmill effluent into the upper layer of the inlet causes a decrease in the dissolved oxygen concentration of the water due to blockage of photosynthesis in the stratum of water immediately below the halocline. Removal of the biochemical oxygen demand from the effluent has very little effect on the oxygen concentration in the upper layer since the supply of oxygen from the lower layer is blocked by the stain. Removal of 90 percent of the stain from the effluent restores the oxygen concentrations to near normal. (See also W73-08543) (Holoman-Battelle)
W73-08544

A MALLOMONAS BLOOM IN A BULGARIAN MOUNTAIN LAKE,
Copenhagen Univ. (Denmark). Inst. of Plant Anatomy and Cytology.
For primary bibliographic entry see Field 05A.
W73-08545

EFFECTS OF BUNKER C OIL ON INTERTIDAL AND LAGOONAL BIOTA IN CHEDABUCTO BAY, NOVA SCOTIA,
New Brunswick Univ., St. John. Dept. of Biology.
M. L. H. Thomas.
Journal of the Fisheries Research Board of Canada, Vol 30, No 1, p 83-90, January 1973. 30 ref.

Descriptors: Water pollution effects, *Oil pollution, *Oil spills, Intertidal areas, *Lagoons, *Biota, Marine plants, Marine animals, Mortality, Water pollution sources, Marine algae, Kelp, Phaeophyta, Clams, Invertebrates, Mussels, Crustaceans, Mollusks, Fishkill, Marine fish, Sculpins, Gastropods, Snails, *Canada.
Identifiers: *Bunker C oil, Isle Madame, *Chedabucto Bay (Nova Scotia), Mya arenaria, Spartina alterniflora, Pucus spiralis, Barnacles, Arthropods, Macroinvertebrates, Ulvaria subbifurcata, Myoxocephalus scorpius, Liparis atlanticus, Pholis gunnellus, Rock eel, Ulva fish, Balanus balanoides, Littorina saxatilis, Littorina littorea, Littorina obtusata, Zostera marina, Eel grass, Cord grass, Macrophytes.

In February 1970, a large spill of Bunker C oil occurred in Chedabucto Bay, Nova Scotia. Observations were made on the effects of the oil on the intertidal and lagoonal biota from seven sites in the Isle Madame area where oiling was heavy and repeated. Transects delineated the observational area at each station. Initial effects of oil involved minor smothering of fauna and tearing loose of algae. Longer term effects involved extensive mortalities of Pucus spiralis on rocky shores and Mya arenaria and Spartina alterniflora in lagoons. Other biota were not visually affected. In all three affected species, mortalities took place either continuously or only in the second year of pollution. Causes of death are unknown. It is recommended that in all intertidal areas very heavy oil deposits should be mechanically removed and the remainder of the oil left to natural degradation. (Holoman-Battelle)
W73-08548

ENDRIN TOXICOSIS IN RAINBOW TROUT (SALMO GAIRDNERI),
Bureau of Sport Fisheries and Wildlife, Columbia, Mo. Fish-Pesticide Research Lab.
B. F. Grant, and P. M. Mehrle.
Journal of the Fisheries Research Board of Canada, Vol 30, No 1, p 31-40, January 1973. 2 fig, 4 tab, 26 ref.

Descriptors: *Rainbow trout, *Endrin, *Pesticide toxicity, *Fish physiology, *Bioassay, Water pollution effects, Insecticides, Chlorinated hydrocarbon pesticides, Lethal limit, Survival, Laboratory tests, Electrolytes, Animal growth, Proteins.
Identifiers: *Salmo gairdneri, Biochemical tests, Serum, Cholesterol, Cortisol, Lactate, Glucose, Glycogen, Liver, Fat tissue.

In mature rainbow trout (Salmo gairdneri) receiving sublethal doses of endrin (4.3-145 micrograms/kg body wt/day in 0.215-7.25 mg/kg of food) for 163 days and then forced to swim for 1 hr, the insecticide affected serum electrolytes, osmolality, total protein, cholesterol, cortisol, lactate, glucose, liver glycogen, and growth. Forced swimming alone altered 9 of 16 serum parameters examined. Apparent increases in serum Na and Cl and significant increases in osmolality and liver glycogen were directly related to dosage. A biphasic distribution of phosphate, total protein, and cholesterol with dosage was apparent. Glucose was increased about 50 percent by 145 micrograms/kg but was unaffected by lower doses. Variance analysis of zone electrophoretic patterns disclosed an interaction between serum protein distribution and dose. Mobilization of liver glycogen was apparently inhibited by low doses and almost totally blocked by high doses. Correspondingly, trout given 14.5 micrograms/kg or more had lowered serum cortisol levels whereas the lowest dose elevated cortisol. Growth was inhibited appreciably by 145 micrograms/kg but not by lower doses. Visceral fat accumulated 4.8-8.7 micrograms endrin/g tissue in the 43 and 145 micrograms/kg exposures. It is concluded that endrin caused dysfunction of physiologic processes critical to survival. (Holoman-Battelle)
W73-08549

collections, Animal physiology, Crustaceans, Invertebrates, Dissolved oxygen, Lake morphometry, Eutrophication, Sampling.
Identifiers: Orconectes virilis, *Hemlock Lake (Michigan), *Section Four Lake (Michigan), Macroinvertebrates, Arthropods, Decapods.

Crayfish (Orconectes virilis) usually exhibit seasonal depth distributions in certainnorthern Michigan lakes based on sex, age and water temperature. After releasing attached young in shallow water, the adult females typically migrate to deeper cold water while the adult males remain in the warm shallow water. This pattern was thought to be related to the sexual maturation cycle. However, when two lakes were artificially aerated and destratified with compressed air, both sexes distributed throughout the lakes. It is, therefore, postulated that under normal conditions of thermal stratification, the social aggression of the larger males forces the females into deeper, colder water and that this aggression is temperature-related. If oxygen or some other factor is not limiting, 10 C seems to be the lowest temperature selected by O. virilis during the summer. (Holoman-Battelle)
W73-08583

CHANGES IN CHLOROPLAST STRUCTURE DURING CULTURE GROWTH OF PERIDINIUM CINCTUM FA. WESTII (DINOPHYCEAE),
Tel-Aviv Univ., (Israel). Lab. for Electron Microscopy.
G. Messer, and Y. Ben-Shaul.
Phycologia, Vol 11 Nos 3/4, p 291-299, December 1972. 2 fig, 1 tab, 18 ref.

Descriptors: *Aging (Biological), *Dinoflagellates, *Cytological studies, Electron microscopy, Pyrrophyta, Cultures, Protoza, Plant growth, Animal growth, Plant pigments, Spectrophotometry, Statistical methods.
Identifiers: *Peridium cinctum westii, *Unialgal cultures, Chlorophyll a, Chlorophyll c, Chloroplasts, Ultrastructure, Thylakoids, Flagellates, Culture media.

Observations are presented which indicate that chlorophyll content and chloroplast ultrastructure may serve as parameters of aging during the growth of Peridinium cinctum westii in laboratory culture. Unialgal cultures were grown in liquid medium Carefoot medium: Lake Tiberias water, 1:4 at pH 7.5 under constant illumination of 1800 lux at 19C. Samples were taken for determinations of cell number, mean cell volume, and chlorophyll, and for examination by electron microscopy. These analyses were performed at intervals during 75 days of culture maintenance. Cell size and chlorophyll content increased during early growth and decreased with culture aging. Chloroplasts contained fewer thylakoids, were narrower, and showed greater association with endoplasmic reticulum and cytoplasmic ribosomes in cells at stationary phases than in those during exponential phase. Chloroplast divisions at early growth phases showed that bands comprising three thylakoids are continuous between daughter chloroplasts and may be synthesized prior to division. (Holoman-Battelle)
W73-08584

PHOTOSYNTHESIS AND RESPIRATION OF SOME ARCTIC SEAWEEDS,
Scripps Institution of Oceanography, La Jolla, Calif.
F. P. Healey.
Phycologia, Vol 11, Nos 3/4, p 267-271, December, 1972. 1 fig, 2 tab, 15 ref.

Descriptors: *Photosynthesis, *Respiration, *Plant physiology, *Marine algae, On-site tests, Laboratory tests, Phaeophyta, Kelp, Chlorophyta, Light intensity, Plant growth, Water temperature, *Alaska, Arctic, Rhodophyta.
Identifiers: Chaetomorpha, Halosaccion glandiforme, Fucus, Laminaria complanata, Corallina officinalis var. chilensis, Heterochordaria abietina, Rhodomela larix, Ahnfeltia spp, Gigartina, *Arctic seaweeds.

Both laboratory and in situ measurements of photosynthesis and respiration were made of some Artic seaweeds in order to determine: (1) the ability of the algae to show positive net photosynthesis under natural or simulated conditions, and (2) growth of the specimens suspended beneath ice as well as adaptations to low temperatures. Conventional manometric techniques were used for laboratory measurements and clear and darkened BOD bottles were used in situ measurements. The temperature optima of these reactions were similar to those of temperate seaweeds and much higher than the local temperatures. However, reaction rates at low temperatures (as a percentage of the maximum rate) were higher than those reported for plants grown or collected at higher temperaturesLaboratory and in situ measurements resulted in similar estimates of compensating intensities. They also showed that little or no growth could be expected beneath snow-covered ice, but growth could occur beneath about 70 cm of ice cleared of snow. (Holoman-Battelle)
W73-08585

A FURTHER CONTRIBUTION TO THE DIATOM FLORA OF SEWAGE ENRICHED WATERS IN SOUTHERN AFRICA,
National Inst. for Water Research, Pretoria (South Africa).
F. R. Schoeman.
Phycologia, Vol 11, Nos 3/4, p 239-245, December, 1972. 10 fig, 2 tab, 36 ref.

Descriptors: *Diatoms, *Eutrophication, Water pollution, *Systematics, *Sewage, Ecology, Aquatic algae, Chrysophyta, Water analysis, Chemical analysis, Physical properties, Chemical properties, Biochemical oxygen demand, Chemical oxygen demand, Ammonia, Nitrogen, Nitrates, Hydrogen ion concentration, Dissolved solids, Alkalinity, Sampling, Hardness (Water), Sodium, Chlorides, Sulfates, Bioindicators.
Identifiers: *Walvis Bay, *South Africa, Organic nitrogen, Amphora subacutiuscula, Amphora tenerrima, Anomoeoneis sphaerophora, Cyclotella meneghiniana, Cymbella pusilla, Fragilaria pinnata, Navicula pseudohalophila, Nitzschia apiculata, Nitzchia fonticola, Nitzschia thermalis, Synedra tabulata.

The systematics and autecology of the diatoms observed in an algal sample from a maturation pond of the Walvis Bay (South West African coast) sewage works are discussed. Fourteen diatom species are recorded. A new species, Amphora subacutiuscula, is described. The diatom association was subjected to a statistical analysis to determine the relative abundance of the different species in the association. Since the structure of the diatom association is the result of environmental conditions, it is possible to employ the association to determine the ecological conditions prevailing in the water. The chemical results of the maturation pond water suggest eutrophic, alkaline water with a high concentration of dissolved solids. Therefore, species favoring these conditions are expected to occur in this water. Except for the new species, Amphora subacutiuscula, whoe autecology is still unknown, all the other species recorded in this sample are alkaline water inhabitants. Many of these are brackish species or are able to tolerate certain fluctuations in osmotic pressure. The dominant species, Amphora tenerrima, suggests brackish conditions. Cyclotella meneghiniana, a sub-dominant species, is known to be nitrogen heterotrophic and, therefore, grew well in the pond. (Holoman-Battelle)
W73-08586

DIATOM COMMUNITY RESPONSE TO PRIMARY WASTEWATER EFFLUENT,
Montana State Univ., Bozeman. Dept. of Botany and Microbiology.
L. L. Bahls.
Journal Water Pollution Control Federation, Vol 45, No 1, p 134-144, January, 1973. 2 fig, 7 tab, 27 ref.

Descriptors: Water pollution effects, *Biological communities, *Diatoms, Sewage treatment, *Waste water (Pollution), Water analysis, Chemical analysis, Nutrients, Bioindicators, Nitrogen, *Montana, Plant populations, Water sampling, Alkalinity, Sulfates, Fluorides, Chlorides, Nitrites, Nitrates, Ammonia, Silica, Turbidity, Specific conductivity, Hydrogen ion concentration, Chrysophyta, Aquatic algae, Water pollution, Sewage effects, Outlets, Physical properties, Chemical properties.
Identifiers: *East Gallatin River (Wyo), Species diversity, Nitzschia dissipata, Atomic absorption spectrophotometry, Nitzschia epiphytica, Navicula cryptocephala var. veneta, Achnanthes minutissima, Navicula pelliculosa, Navicula tripunctata, Navicula minuscula, Gomphonema olivaceum, Nitzschia palea, Navicula minima, Cymbella ventricosa, Navicula viridula, Surirella ovata, Amphora ovalis var. pediculus, Nitzschia communis, Diatoma vulgare, Navicula cryptocephala, Cocconeis placentula var. euglypta, Nitzschia romana.

A study was made of the diatom community in the East Gallatin River above and below the Bozeman, Montana, wastewater treatment plant. Water samples for chemical and physical analyses and diatom collections from natural substrates were made monthly. The water quality parameters measured were water temperature, pH, Ca, Mg, alkalinity, sulfate, orthophosphate, fluoride, chloride, nitrite, nitrate, ammonia, silica, turbidity, specific conductivity, Na, and K. The diatom taxa identified were correlated with the nutrients that were present in the forms of phosphate, NH3-N, and nitrate-nitrogen. Of the 12 major taxa identified, Nitzschia dissipata was the most frequently occurring diatom. Just below the wastewater outfall a nitrogen heterotroph, Nitzschia epiphytica, was most abundant. Together these two taxa accounted for 50 percent of the mean abundance of all the taxa in the river. Calculations were made to determine simple and multiple correlation coefficients between diatom diversity and individual taxa and the various nutrients. Overall, diversity correlated negatively with ammonia and positively with phosphate. (Holoman-Battelle)
W73-08588

WHAT'S KNOWN ABOUT PENTACHLOROPHENOLS,
Environmental Protection Service, Halifax (Nova Scotia).
For primary bibliographic entry see Field 05B.
W73-08599

THE EFFECTS OF WASTE DISCHARGES FROM RADFORD ARMY AMMUNITION PLANT ON THE BIOTA OF THE NEW RIVER, VIRGINIA,
Virginia Polytechnic Inst. and State Univ., Balcksburg. Dept. of Biology.
J. Cairns, and K. L. Dickson.
Available from the National Technical Information Service as PB-219 982, $3.00 in paper copy, $1.45 in microfiche. Virginia Water Resources Research Center, Blacksburg, Bulletin 57, April 1973. 58 p, 13 tab, 5 fig, 9 ref. OWRR B-017-VA (10).

Descriptors: Water pollution effects, *Fish, Minnows, Perches, Sucker, Sunfishes, Catfishes, *Biota, Baseline studies, *Waste assimilative capacity, *Virginia, Chemical wastes, Liquid wastes, Industrial wastes, Effluents, Organic wastes, Thermal pollution, Acids, *Benthic fauna, Aquatic plants, Aquatic animals, *Invertebrates, Nitrogen compounds, *Algae, Diatoms.
Identifiers: Stream survey, Biological survey, *Munition wastes, TNT wastes, Nutrient enrichment, *New River (Va), Ash wastes, Waste discharge, *Pollution recovery (River), Bacillariophyta.

The effects of waste discharges from Radford Army Ammunition Plant (RAAP), Radford, Virginia, on fish, bottom fauna, algae, and higher aquatic plants in the New River were investigated in June 1971. Sampling stations were established upstream and downstream from the RAAP discharges. These stations were located to evaluate the individual effects of the various discharges (acid wastes, organic solvent wastes, thermal effluents, nitrogenous TNT wastes, and ash) on the fauna and flora. This was possible because there were a number of waste streams discharged at various points along the New River rather than a single combined waste discharge. An independent analysis of the data for each taxonomic group is presented along with general conclusions which integrate the findings of all the investigators. The waste discharges from the plant caused localized damage to the fauna and flora of the New River; however, the river had recovered in the five miles contained within the boundaries of the plant property.
W73-08606

THE EFFECTS OF TEMPERATURE AND IONS ON THE CURRENT-VOLTAGE RELATION AND ELECTRICAL CHARACTERISTICS OF A MOLLUSCAN NEURONE,
Massachusetts Eye and Ear Infirmary, Boston. Dept. of Ophthalmology.
M. F. Marmor.
Journal of Physiology, Vol 218, No 3, p 573-598, 1971. 12 fig, 52 ref.

Descriptors: *Semipermeable membranes, *Temperature, *Ion transfer, *Mollusks, *Electrical properties, Biochemistry, Physiology, Cytological studies, Metabolism, Structure, Chemical analysis, Chemical reactions, Water pollution effects.

This study continues experiments performed on the gastro-oesophageal giant neurone (G cell) of the marine mollusc Anisodoris nobilis MacFarland. It demonstrates, first, that inward-going rectification in the G cell depends upon the temperature and the external K ion concentrations. Secondly, temperature and Ca ions influence a high conductance state which develops when the cell is strongly hyperpolarized. Thirdly, temperature, membrane potential, and external ions affect the time and resistance constants of the membrane. Finally observations that the conductance of the G cell soma is dominated by the axon are confirmed under conditions where inward-going rectification does not complicate the analysis. (Jerome-Vanderbilt)
W73-08621

THE ATMOSPHERIC EFFECTS OF THERMAL DISCHARGES INTO A LARGE LAKE,
Argonne National Lab., Ill.
J. E. Carson.
Journal of the Air Pollution Control Association, Vol 22, No 7, p 523-528, July 1972. 32 ref, 1 append.

Descriptors: *Nuclear powerplants, *Heated water, *Environmental effects, *Meteorology, *Fog, Electric power production, Thermal powerplants, Discharge (Water), Atmosphere, Lakes, *Lake Michigan, Precipitation, Evaluation.

Potential meteorological effects of waste heat from nuclear power plants with once-through cooling systems are discussed with the primary emphasis upon the atmospheric changes over a large lake like Lake Michigan. The amount of heat energy in the cooling water is quite small compared to the natural heat processes and since it will be dissipated over a large area the changes in weather will be small and probably impossible to isolate in the natural variability of weather elements. An exception is the increase in steam fog at the point of discharge during fall and winter. Alternate cooling systems, such as cooling towers, cooling ponds and spray canals release their energy rapidly over a small area, therefore, the potential for local weather changes is increased. From a meteorologist's view, once-through cooling is preferred over other methods. Assumptions and calculations for setting up a meteorological model are presented. (Jerome-Vanderbilt)
W73-08623

THE EFFECT OF TEMPERATURE ACCLIMATION UPON SUCCINIC DEHYDROGENASE ACTIVITY FROM THE EPAXIAL MUSCLE OF THE COMMON GOLDFISH (CARASSIUS AURATUS L.)-I. PROPERTIES OF THE ENZYME AND THE EFFECT OF LIPID EXTRACTION,
Nebraska Univ., Lincoln. Dept. of Zoology.
J. R. Hazel.
Comparative Biochemistry and Physiology, Vol 43B, No 4B, p 837-861, 1972. 10 fig, 3 tab, 66 ref.

Descriptors: *Adaptation, Fish, *Enzymes, *Catalysts, *Lipids, Biochemistry, *Water temperature, Chemical reactions, Physiology, Environmental effects, Metabolism, Kinetics, Inhibitors, Measurement, Laboratory tests, Water pollution effects.
Identifiers: *Succinic dehydrogenease, *Goldfish.

The objectives were twofold: to examine the properties of a single enzyme system which contributes to the maintenance of a temperature independent rate of catalysis over three distinct time course periods; and to determine whether or not the temperature induced changes in lipid composition which occur during temperature acclimation may be responsible for changes in respiratory enzyme activity. Evidence strongly suggests that goldfish SHD is non-isozymal and is present in the same molecular form at all temperatures. It is also suggested that temperature-induced change in membrane composition may influence the catalytic activity of membrane-bound enzymes. The SDH system exhibits compensation of catalytic function by virtue of an increased specific activity following a period of cold acclimation, by an E (a) function which correlates with environmental temperature during evolutionary adaptation, and by several inherent properties of the enzymatic protein which result in temperature independent catalysis over the short term (immediate compensation). (See also W73-08626) (Jerome-Vanderbilt)
W73-08625

THE EFFECT OF TEMPERATURE ACCLIMATION UPON SUCCINIC DEHYDROGENASE ACTIVITY FROM THE EPAXIAL MUSCLE OF THE COMMON GOLDFISH (CARASSIUS AURATUS L.)-II. LIPID REACTIVATION OF THE SOLUBLE ENZYME,
Nebraska Univ., Lincoln. Dept. of Zoology.
J. R. Hazel.
Comparative Biochemistry and Physiology, Vol 43B, No 4B, p 863-882, 1972. 8 fig, 3 tab, 46 ref.

Descriptors: Fish, *Metabolism, *Water temperature, *Biological membranes, *Enzymes, *Lipids, Biochemistry, Animal physiology, Adaptation, Respiration, Environmental effects, Morphology, Kinetics, Laboratory tests, Chemical analysis, Water pollution effects.
Identifiers: *Goldfish.

The properties of lipids extracted from the mitochondria of goldfish acclimated to 5C and 25C are compared in terms of their ability to reactivate the soluble (lipid-free) succinic dehydrogenase

from both acclimated groups. The materials and methods of investigation in these experiments are described and results are presented and discussed. The specific activity of soluble succinic dehydrogenase was increased by incubation with soluble lipid preparations. A total mitochondrial lipid extract from 5C - acclimated goldfish was a more effective reactivator of the enzyme than a comparable extract from 25C - acclimated goldfish; this difference was referable to the greater degree of unsaturation of the 5C lipid. It was indicated that the affinity of the enzyme for lipid was largely dependent upon the phosphatide species, whereas the magnitude of the reactivation was dependent primarily upon the degree of unsaturation of the fatty acid residues. The results may be interpreted as indicating that the increased specific activity of succinic dehydrogenase following a period of cold acclimation can largely be explained by underlying changes in the membrane composition rather than the quantity of the enzyme. (See also W73-08625) (Jerome-Vanderbilt)
W73-08626

TEMPERATURE DEPENDENCE OF THE ATP-ASE ACTIVITIES IN BRAIN HOMOGENATES FROM A COLD-WATER FISH AND A WARM-WATER FISH,

Turku Univ. (Finland). Zoophysiological Lab.
J. Kohonen, R. Tirri, and K. Y. H. Lagerspetz.
Comparative Biochemistry and Physiology, Vol 44B, No 3B, p 819-821, 1973. 1 fig, 5 ref.

Descriptors: Biochemistry, Fish, *Chemical reactions, *Water temperature, Animal physiology, Metabolism, Enzymes, Chemical properties, Bioassay, Laboratory studies, Water pollution effects.
Identifiers: *ATPase, *Roach, *Gouramis.

A species of cold water fish (Leuciscus rutilus) and a species of warm water fish (Helostoma temmincki) were chosen as test animals for this study which investigated the brain adenosinetriphosphatases as they relate to the different temperature requirements of the two species. The gouramis were kept at 28C while the roach were maintained at 5C. The fish were decapitated and the brains were excised and homogenized. Assays of the Mg (++) and the Na (+)-K (+) ATPase activities in the homogenates were performed. The Mg (++) ATPase activity in the brain homogenates from the cold water fish was less sensitive to temperature and showed maximum activity and a discontinuity point at temperatures 5C lower than the Mg (++) ATPase from warm water fish. The brain Na (+)-K (+) ATPase activity in the roach showed a curvilinear temperature dependence with activity maxima at 41C and 20C, while the gourami had one maximum at 45C and was relatively insensitive to temperature between 35C and 20C. It is suggested that there are present at least two Na (+)-K (+) isozymes with different temperature characteristics. (Jerome-Vanderbilt)
W73-08627

EFFECT OF TEMPERATURE ON MEMBRANE POTENTIAL AND IONIC FLUXES IN INTACT AND DIALYSED BARNACLE MUSCLE FIBRES,

Instituto Venezolano de Investigaciones Cientificas, Caracas.
R. Dipolo, and R. Latorre.
Journal of Physiology, Vol 225, No 2, p 255-273, 1972. 2 tab, 8 fig, 23 ref.

Descriptors: *Aquatic animals, *Water temperature, *Ion transport, *Membranes, Physiology, Biochemistry, Environmental effects, Metabolism, Inhibitors, Radioisotopes, Chemical reactions, Dialysis, Water pollution effects, *Crustaceans.
Identifiers: *Barnacles.

The temperature-dependent component of resting potential is insensitive to inhibitors of the sodium pump. The large size of barnacle muscle fiber permits measurements of ionic fluxes with radioactive isotopes by means of a dialysis technique at different temperatures in the same muscle fiber. The experiments indicate that the phenomenon under study is probably related to a variation in the passive permeability rather than to active transport. The maintenace of specimens and dissection of single fibers, dialysis procedure, membrane potential measurements, flux determinations, solutions, and calculation methods are discussed. Addition of strophanthidin, removal of the external potassium, and replacement of the external sodium by lithium all strongly reduced the sodium efflux in barnacle muscle fibers, but none of these conditions affects the variation of resting potential with temperature. In contrast, the removal of the external chloride in barnacle muscle cells produces a substantial decrease in the voltage shift induced by temperature. The effect of changes in external chloride concentration on the temperature-dependent component of the resting potential remains unclear and needs further study. (Jerome-Vanderbilt)
W73-08628

EFFECTS OF THERMAL ACCLIMATION ON PHYSIOLOGICAL RESPONSES TO HANDLING STRESS, CORTISOL AND ALDOSTERONE INJECTIONS IN THE GOLDFISH, CARASSIUS AURATUS,

Cincinnati Univ., Ohio. Dept. of Biological Sciences.
B. L. Umminger, and D. H. Gist.
Comparative Biochemistry and Physiology, Vol 44A, p 967-977, 1973. 3 fig, 1 tab, 52 ref.

Descriptors: Fish, Water temperature, *Metabolism, *Stress, Biochemistry, Physiology, Adaptation, Environmental effects, Chemical reactions, Laboratory tests, Measurements, Evaluation, Water pollution effects.
Identifiers: *Cortisol, *Aldosterone, *Goldfish.

As part of a study to evaluate the actions of steroid hormones, cortisol and aldosterone, on carbohydrates and electrolyte regulation in goldfish acclimated to several different temperatures, this investigations was performed to assess the stressful effects of the injection procedure. Eighty-four goldfish were divided into groups of 28 which were held at 20 C, 32C and 10C. Fish at each temperature were injected with cortisol or subjected to sham injection procedures and analysed along with fish from control groups maintained at the same temperature. The fish subjected to stress and sham injection responded with a hyperglycemia and a decline in serum chloride and sodium concentrations. The stressful effects were most pronounced in fish acclimated to 10 C and least apparent in fish acclimated to 32 C. Injections of cortisol further lowered serum electrolyte levels in warm acclimated fish and increased the serum glucose level of fish acclimated to 20 C. (Jerome-Vanderbilt)
W73-08632

PHYTOPLANKTON AS A BIOLOGICAL INDICATOR OF WATER QUALITY,

Universidad Central de Venezuela, Caracas. Dept. of Sanitary Engineering.
I. Villegas, and G. de Giner.
Water Research, Vol 7, No 3, p 479-487, 1973. 4 fig, 3 tab, 9 ref.

Descriptors: *Phytoplankton, *Bioindicators, *Water quality, Biological communities, Diatoms, Chlorophyta, Cyanophyta, Scenedesmus, Rivers, Ecological distribution, Water pollution effects, Systematics.
Identifiers: *San Pedro River (Venezuela), Species diversity, Ulothrix.

The San Pedro River, Venezuela, was chosen for study of the magnitude of ecological variations in a river in relation to water quality. Five samples stations were selected on the basis of the agricultural, industrial, and domestic wastes entering the river or on the basis of topographical changes of the land. Routine sampling and analyses were carried on the 12 consecutive months at three times per week. The parameters more frequently used in evaluating the sanitary water quality were determined by: coliforms, fecal coliforms, bacterial density, dissolved oxygen demand, organic nitrogen, ammonia nitrogen, nitrites, nitrates, total residue, fixed residue, volatile residue, filtrable residue, non-filtrable residue and pH. A qualitative and quantitative study was made of the phytoplankton oriented towards calculation of abundance of a group as percentage of total phytoplankton population of each sample. The quantitative and qualitative study of phytoplankton related to non-biological values indicating pollution shows that the ecological condition of this biological group appears to bear a direct relation to the sanitary state of water. Use of phytoplankton as biological community indicators of pollution has value when comprehensive ecological study is made. (Jones-Wisconsin)
W73-08634

URBAN STORMWATER QUALITY AND ITS IMPACT ON THE RECEIVING SYSTEM,

Rex Chainbelt, Inc., Milwaukee, Wis. Ecology Div.
E. H. Bryan.
In: Proceedings of the 20th Southern Water Resources and Pollution Control Conference, April 1-2, 1972, University of North Carolina, Durham, p 38-51. 2 fig, 4 tab, 10 ref.

Descriptors: *Water pollution sources, *Urban drainage, *Storm water, *Water quality, *Municipal wastes, North Carolina, Basins, Biochemical oxygen demand, Chemical oxygen demand, Chlorides, Coliforms, Phosphates, Lead, Rainfall-runoff relationships, Combined sewers, Sediment load, Urban runoff, Storm runoff.
Identifiers: Durham (N.C.), Total solids, Volatile total solids, Tulsa (Okla.).

Results of a study undertaken to characterize urban stormwater from a typical drainage basin in a North Carolina community are presented and compared with those from other locations. The drainage area, a section of the City of Durham, included commercial business districts, an industrial plant, housing districts, shopping center, public recreational park, schools, churches, and a cemetery; the downtown section of an expressway was under final construction. Seventeen storm series provided data representative of rainfall-runoff characteristics for an entire year's precipitation-runoff events. Urban stormwater was concluded to be a significant source of pollutional constituents. The annual contribution of BOD by stormwater was estimated to equal that of the sanitary wastewater effluent from secondary treatment. The contribution of total organic matter, as measured by chemical oxygen demand, was greater than that which would be obtained by discharge of raw, sanitary sewage from a strictly residential, average urban area of the same size. The total solids contribution was substantially larger than would be expected from such discharge of average raw domestic sewage to the stormwater collection and drainage system. Some indications of treatment possibilities are emerging, including alternatives to separation of combined sewer systems. (Jones-Wisconsin)
W73-08635

NONTHERMAL EFFECTS OF MICROWAVE ON ALGAE,

Manitoba Univ., Winnipeg. Dept. of Electrical Engineering; and Manitoba Univ., Winnipeg. Dept. of Botany.
M. A. K. Hamid, and S. Badour.

In: Digest of Papers of the 4th Canadian Medical and Biological Engineering Conference, September 7-9, 1972, Winnipeg, Manitoba, p 48.

Descriptors: *Plant growth, *Microwaves, *Algae, Chlamydonomas, Photosynthesis, Metabolism, Heat, Plant growth regulators, Inhibition.

The thermal and nonthermal effects of microwaves on algae are reported. Algal samples were exposed to microwaves at 2450 MHz for varying periods using appropriately designed waveguide applicators, resulting in either inhibition of photosynthesis or stimulation of the processes depending on culture age of developmental stage. Since microwave treatment is associated with increase in temperature and this injurious effect on photosynthesis is solely due to the thermal effect on microwaves, heated sample control were used which consistently exhibited decrease in photosynthetic capacity of the algae as compared to the control (unheated sample); providing the temperature increases were the same, decreases in photosynthesis were significantly higher than caused by microwaves. Microwaves stimulated photosynthesis, though such stimulation was marked by the injurious effects of heat. Study of the effect of microwaves on developmental stages of the algae to specify the stage which preferentially responds to the ameliorating nonthermal microwave effects and at the same time resists harmful effect of resulting heat is of basic importance. Results provide evidence that microwaves artificially applied or already available in the natural environment may exert a stimulatory effect on plants (primary producers) as concluded from primary results of experiments conducted on Chlamydomonas segnis. (Jones-Wisconsin)
W73-08636

NITROGEN-LIMITED GROWTH OF MARINE PHYTOPLANKTON--II. UPTAKE KINETICS AND THEIR ROLE IN NUTRIENT LIMITED GROWTH OF PHYTOPLANKTON,

Hawaii Inst. of Marine Biology, Honolulu.
J. Caperon, and J. Meyer.
Deep-Sea Research, Vol 19, No 9, p 619-632, 1972. 6 fig, 4 tab, 16 ref.

Descriptors: *Absorption, *Plant growth, *Nutrients, *Marine phytoplankton, *Kinetics, Ammonia, Mathematical models, Growth rates, Sea water, Nitrogen, Culture, Limiting factors.
Identifiers: Luxury uptake, Chemostate cultures, Monochrysis lutheri, Dunaliella tertiolecta, Coccochloris stagnina, Cyclotella nana.

To clarify the relationship between nutrient uptake kinetics and the nutritional experience of the population, these experiments provide reliable estimates of uptake parameters under different nutrient preconditioning regimes. Nitrate uptake experiments were conducted with nitrate-preconditioned Monochrysis lutheri, Dunaliella tertiolecta, Coccochloris stagnina, and Cyclotella nana. Ammonium uptake experiments were conducted with Monochrysis and Dunaliella preconditioned on ammonium. In six of these experiments the growth chamber was inoculated with nitrate and ammonium simultaneously, and the uptake of both was monitored. All cases are well described by the Michaelis-Menten hyperbolas. The two sets of half-saturation constants do not show significant interspecific variation, nor do they show intraspecific variation with different steady-state preconditioning growth rates. The maximum uptake rate parameter shows marked intraspecific variability that is positively correlated with the preconditioning population growth rate. This linear function, combined with the limiting nutrient uptake hyperbola and the relation between growth rate and the nutritional state of the population, as previously developed, provides a model that relates steady-state growth rate to environmental substrate concentration. The evidence suggests that this model would be applicable to field situations for a mixed phytoplankton population in either a nitrate- or ammonium-limited environment. (Jones-Wisconsin)
W73-08638

ON NITROGEN DEFICIENCY IN TROPICAL PACIFIC OCEANIC PHYTOPLANKTON. II. PHOTOSYNTHETIC AND CELLULAR CHARACTERISTICS OF A CHEMOSTAT-GROWN DIATOM,

California Univ., San Diego, La Jolla. Inst. of Marine Resources.
W. H. Thomas, and A. N. Dodson.
Limnology and Oceanography, Vol 17, No 4, p 515-523, 1972. 2 fig, 2 tab, 28 ref.

Descriptors: *Nitrogen, *Productivity, *Deficient elements, *Tropic, Pacific Ocean, Photosynthesis, Diatoms, Growth rates, Chlorophyll, Pigments, Carbon, Cultures, Ammonia, Nutrient requirements, Nutrients, Equations.
Identifiers: *Chemostats, Chaetoceros gracilis, Assimilation ratios.

Varying degrees of nitrogen deficiency for a tropical Pacific diatom, Chaetoceros gracilis, were established and assessed. Cells were grown in a chemostat at varying percentages of the maximum growth rate; at each limited submaximal growth rate, assimilation ratio and various intracellular parameters, such as C:N, carbon:chlorophyll, and carotenoid:chlorophyll ratios were expressed as functions of growth rate; cells may be grown in a chemostat at a constant degree of deficiency. Assimilation ratio (photosynthesis at light saturation per unit chlorophyll) increased with increasing growth rate. Cellular C:chlorophyll ratios decreased with increasing growth rate, but carotenoid: chlorophyll ratios showed no obvious trend. The C:N ratio decreased and chlorophyll: cell increased with increasing growth rate. Parameters measurable at sea are assimilation ratio, C:Chl ratio, and carotenoid:Chl ratio. Most cellular parameters, such as C:N ratio, are difficult to measure in natural communities because of detrital contamination. Steady-state cell numbers were not constant at different growth rates, but decreased as growth rates increased. Growth rates seemed to be controlled by internal nitrogen supplies and the apparent half saturation constant decreased with increasing growth rate. (Jones-Wisconsin)
W73-08639

NITROGEN, PHOSPHORUS AND OTHER INORGANIC MATERIALS IN WATERS IN A GRAVITY-IRRIGATED AREA,

Idaho Univ., Moscow. Dept. of Agricultural Engineering; and Idaho Univ., Moscow. Dept. of Soils.
D. W. Fitzsimmons, G. C. Lewis, D. V. Naylor, and J. R. Busch.
Transactions of the American Society of Agricultural Engineering, Vol 15, No 2, p 292-295, 1972. 5 fig, 5 tab, 12 ref.

Descriptors: *Water pollution, *Nitrogen, *Inorganic compounds, *Irrigation, Phosphorus, Surface waters, Groundwater, Nitrates, Ammonia, Idaho, Fertilizers, Leaching.
Identifiers: *Gravity irrigation, Boise Valley (Idaho), Total solids.

The kinds and amounts of inorganic materials were determined in surface and groundwaters in an intensively-farmed gravity-irrigated area (Boise Valley, Idaho). Water samples, taken throughout the crop growing season were analyzed for nitrate-nitrogen, ammonia, and organic nitrogen, two forms of phosphorus, and total solids. Water samples were collected at 2-week intervals throughout the 1970 season from 79 sites which included 29 farms in four irrigation districts. The mean concentrations of all three forms of nitrogen were found to be relatively low. The mean concentration of each nitrogen form was greater in the surface runoff than in the headwater; the groundwater contained more nitrate-nitrogen than the other water sources, possibly an indication that nitrate is being leached from the soil by percolating irrigation water; this material could also come from feedlots, dairies and septic tank drain fields in the area. The surface runoff also contained the largest concentrations of total phosphorus and total solids. The groundwater was found to contain a relatively large concentration of both ortho and total phosphorus. (Jones-Wisconsin)
W73-08640

LEACHING OF P AND N FROM OHIO SOILS,

Ohio Agricultural Research and Development Center, Wooster.
T. J. Logan, E. O. McLean, B. L. Schmidt, and M. E. Kroetz.
Ohio Report on Research and Development, Vol 51, No 5, p 74-76, 1972. 3 fig, 2 tab.

Descriptors: *Leaching, *Phosphorus, *Nitrogen, *Ohio, Soils, Nutrients, Fertilization, Groundwater.
Identifiers: Wauseon sandy loam, Toledo clay loam, Rossmoyne silt loam.

Studies on extent to which nitrogen and phosphorus are lost from soil and factors affecting this loss reported. Fertilizer phosphate was tagged with P-32, applied at various rates, and subjected to varying rates and intensities of leaching in laboratory studies. Phosphate leaching was investigated in three Ohio soils, sandy loam, clay loam, and silt loam. Since phosphate is strongly adsorbed by soil, leaching is quite restricted. Results suggest that the rate at which phosphorus moves through the soil is dependent on ability of the soil to increase or decrease phosphorus concentration in the soil solution, and on rate at which soil water moves through the soil. In general, only on well-fertilized sandy or organic soils low in phosphorus-fixing capacity would phosphorus leaching be of significance. Field experiments were conducted to determine soil management practices for minimizing nitrogen leaching. Nitrate, unlike phosphate, is not adsorbed by soil, and completely soluble in water, is subject to extensive leaching. The possibility of substantial nitrate leaching from some sandy soils indicates need of careful prediction of nitrogen crop requirements to avoid nitrate excesses in the soil which may subsequently contribute to groundwater contamination. (Jones-Wisconsin)
W73-08641

ALGAL GROWTH IN RELATION TO NUTRIENTS.

Water Pollution Research Lab., Stevenage (England).

In: Pollution of Fresh Waters - Report of the Director 1971, p 25-31. 6 fig, 3 tab, 3 ref.

Descriptors: *Eutrophication, *Water pollution, *Freshwater, *Nutrients, Nitrogen compounds, Silica, Phosphorus compounds, Limiting factors, Oxidation lagoons, Cyanophyta, Reservoirs, Diatoms, Chlorophyll, Ammonia, Chlorophyta, Carbon dioxide, Hydrogen ion concentration, Alkalinity, Temperature, Bicarbonates, Suspended solids, Biochemical oxygen demand, Carbon, Scenedesmus.
Identifiers: Grafham Water (England), Aphanizomenon, Microcystis.

Marked seasonal changes in average concentrations of oxidized nitrogen, silicon, and soluble orthophosphate in the water columns have been observed, particularly during algal blooms and when reservoir was thermally stratified. During the bloom of blue-green algae in July and August 1970, concentrations of oxidized nitrogen and phosphorus fell. During the subsequent diatom

Future seasonal and permanent populations will be controlled by zoning regulations to some predetermined number as a measure of eutrophication control of Lake Tahoe. The Lake Tahoe Area Council was formed in the 1950's with the basic function of conducting research and promoting orderly and environmentally sound development around the lake. A consultants' study concluded that municipal sewage posed the greatest threat to the clarity of the lake, prompting a program of effluent export from the basin. After lengthy negotiations, Alpine County agreed to accept the effluent in a proposed 160-acre reservoir located outside of the Tahoe Basin if the effluent met drinking water quality standards; the reservoir would be used for recreations and as a storage facility for irrigation water. One of the most advanced waste water treatment plants, operated by the South Lake Tahoe Public Utilities District (the several processes are listed) purifies the effluent before it is pumped over a 1500-ft. elevation to the Indian Creek Reservoir. A 1966 FWPCA demonstration grant and a 1969 EPA grant resulted in intensive studies (which are described) of the basin. Lake Tahoe should not be the basic criteria for limiting development as when detectable degradation is observed environmental degradation will have already occurred. (Auen-Wisconsin)
W73-08647

ORGANOCHLORINE POLLUTANTS OF THE MARINE ENVIRONMENTS OF HAWAII,
Hawaii Univ., Honolulu. Dept. of Agricultural Biochemistry.
A. Bevenue.
In: Water Resources Seminar Series No. 1, January 1972. Water Resources Research Center, University of Hawaii, Honolulu, p 45-50.

Descriptors: *Pesticides, *Hawaii, Organic pesticides, Water pollution sources, DDT, Chlorinated hydrocarbon pesticides, Dieldrin.
Identifiers: Lindane, Chlordane.

Various water, sediment, algae and fish specimen samples were obtained on Oahu, Hawaii, Kauai and Maui from which pesticide residues were isolated. Preliminary cleanup studies of the biota samples indicated the absence of any polychlorobiphenyl contaminants. Hawaiian rainwater contained p,p'-DDT, dieldrin, and lindane at parts per trillion (ppt) levels. Drinking water contained chlordane, DDT, dieldrin, and lindane in the very low ppt range and about 1/5th, or less, the amounts noted in rainwater. Pesticide residues in non-potable waters were all at the low ppt level, ranging from 0.1 to 64 ppt, with p,p'-DDT representing the highest level. An attempt to classify the sampled areas into two categories of rural and urban sections of the state indicated that the only major difference in residues in the waters was dieldrin, which averaged about seven times higher in the urban areas. Amounts of chlordane and dieldrin found in the waters and sediments and the pentachlorophenol residue observed in the sewage suggests that these pesticides originated primarily in the urban areas where many households and home builders use these materials for termite control. Based on the present pesticide usage pattern, pollution of Hawaii's waters by organochlorine pesticides does not occur to a significant degree. (Auen-Wisconsin)
W73-08651

EFFECTS OF MUNICIPAL WASTE WATER DISPOSAL ON THE FOREST ECOSYSTEM,
Pennsylvania State Univ., University Park. School of Forest Resources.
For primary bibliographic entry see Field 05D.
W73-08731

CONCERNING SOME LITTORAL CLADOCERA FROM AVALON PENINSULA, NEWFOUNDLAND,
Akademiya Nauk SSSR, Moscow. Inst. of Evolutionary Morphology and Animal Ecology.
For primary bibliographic entry see Field 05A.
W73-08762

CULTURED RED ALGA TO MEASURE POLLUTION,
Durham Univ. (England). Dept. of Botany.
For primary bibliographic entry see Field 05A.
W73-08772

DURATION OF VIABILITY AND THE GROWTH AND EXPIRATION RATES OF GROUP E STREPTOCOCCI IN SOIL,
Missouri Univ. Columbia. School of Veterinary Medicine.
For primary bibliographic entry see Field 05B.
W73-08774

VERTEBRATE INSECTICIDE RESISTANCE: IN VIRON AND IN VITRO ENDRIN BINDING TO CELLULAR FRACTIONS FROM BRAIN AND LINER TISSUES OF GAMBUSIA,
Mississippi State Univ., State College. Dept. of Zoology.
M. R. Wells, and O. Yarbrough.
Journal of Agricultural and Food Chemistry, Vol 20, No 1, p 14-16, January-February, 1972. 2 fig, 1 tab, 11 ref.

Descriptors: Fish, *Insecticides, *Endrin, *Resistivity, Animal physiology, Cytological studies.
Identifiers: *Mosquitofish, *Gambusia affinis, *Binding patterns, *Cellular fractions, Brain tissue, Liner tissue.

A comparison was made of the in viro and in vitro binding patterns of endrin -14C in susceptible and resistant Gambusia affinis brain and liner cellular fractions. The resistant fish cell membrane fractions were found to bind more endrin than the cell membrane fractions of susceptible fish. The resistant mitochondria binds less endrin than susceptible fish mitochondria. Possible reasons for endrin resistance in mosquitofish are: differences between endrin uptake in susceptible and resistant fish; retention of endrin by brain cell membranes; a blood-brain barrier; and a structural difference in myelin. (Ensign-PAI)
W73-08801

THE COLUMBIA RIVER ESTUARY AND ADJACENT OCEAN WATERS, BIOENVIRONMENTAL STUDIES.
For primary bibliographic entry see Field 02L.
W73-08816

ASPECTS OF MARINE PHYTOPLANKTON STUDIES NEAR THE COLUMBIA RIVER, WITH SPECIAL REFERENCE TO A SUBSURFACE CHLOROPHYLL MAXIMUM,
Washington Univ., Seattle. Dept. of Oceanography.
For primary bibliographic entry see Field 02L.
W73-08826

DISTRIBUTION OF PELAGIC COPEPODA OFF THE COASTS OF WASHINGTON AND OREGON DURING 1961 AND 1962,
Washington Univ., Seattle. Dept. of Oceanography.
For primary bibliographic entry see Field 02L.
W73-08831

EFFECTS OF THE COLUMBIA RIVER PLUME ON TWO COPEPOD SPECIES,
Oregon State Univ., Corvallis. Dept. of Oceanography.
For primary bibliographic entry see Field 02L.
W73-08832

DISTRIBUTION AND ECOLOGY OF OCEANIC ANIMALS OFF OREGON,
Oregon State Univ., Corvallis. Dept. of Oceanography.
For primary bibliographic entry see Field 02L.
W73-08833

A PRELIMINARY CHECKLIST OF SELECTED GROUPS OF INVERTEBRATES FROM OTTER-TRAWL AND DREDGE COLLECTIONS OFF OREGON,
Oregon State Univ., Corvallis. Dept. of Oceanography.
For primary bibliographic entry see Field 02L.
W73-08835

ECOLOGICAL OBSERVATIONS ON THE BENTHIC INVERTEBRATES FROM THE CENTRAL OREGON CONTINENTAL SHELF,
Oregon State Univ., Corvallis. Dept. of Oceanography.
For primary bibliographic entry see Field 02L.
W73-08836

DISTRIBUTION AND RELATIVE ABUNDANCE OF INVERTEBRATES OFF THE NORTHERN OREGON COAST,
National Marine Fisheries Service, Seattle, Wash. Exploratory Fishing and Gear Research Base.
For primary bibliographic entry see Field 02L.
W73-08837

BATHYMETRIC DISTRIBUTION OF THE ECHINODERMS OFF THE NORTHERN OREGON COAST,
National Marine Fisheries Service, Seattle, Wash. Exploratory Fishing and Gear Research Base.
For primary bibliographic entry see Field 02L.
W73-08838

BATHYMETRIC AND SEASONAL ABUNDANCE AND GENERAL ECOLOGY OF THE TANNER CRAB CHIONOECETES TANNERI RATHBUN (BRACHYURA: MAJIDAE), OFF THE NORTHERN OREGON COAST,
National Marine Fisheries Service, Seattle, Wash. Exploratory Fishing and Gear Research Base.
For primary bibliographic entry see Field 02L.
W73-08839

CHARACTERISTICS OF THE DEMERSAL FISH FAUNA INHABITING THE OUTER CONTINENTAL SHELF AND SLOPE OFF THE NORTHERN OREGON COAST,
National Marine Fisheries Service, Seattle, Wash. Exploratory Fishing and Gear Research Base.
For primary bibliographic entry see Field 02L.
W73-08840

RADIOECOLOGY OF ZINC-65 IN ALDER SLOUGH, AN ARM OF THE COLUMBIA RIVER ESTUARY,
Oregon State Univ., Corvallis. Dept. of Oceanography.
For primary bibliographic entry see Field 02L.
W73-08845

TOTAL PHOSPHORUS AND PHOSPHORUS-32 IN SEAWATER, INVERTEBRATES, AND ALGAE FROM NORTH HEAD, WASHINGTON,
Washington Univ., Seattle. Coll. of Fisheries; and Washington Univ., Seattle. Lab. of Radiation Ecology.
For primary bibliographic entry see Field 02L.
W73-08847

EFFECTS OF STRONTIUM-90 + YTTRIUM-90 ZINC-65, AND CHROMIUM-51 ON THE LARVAE OF THE PACIFIC OYSTER CRASSOSTREA GIGAS,
Washington Univ., Seattle. Coll. of Fisheries; and Washington Univ., Seattle. Lab. of Radiation Ecology.
For primary bibliographic entry see Field 02L.
W73-08848

ZINC-65 IN BENTHIC INVERTEBRATES OFF THE OREGON COAST,
Oregon State Univ., Corvallis. Dept. of Oceanography.
For primary bibliographic entry see Field 02L.
W73-08849

A TEN-YEAR STUDY OF MEROPLANKTON IN NORTH CAROLINA ESTUARIES: AMPHIPODS,
National Marine Fisheries Service, Washington, D.C. Systematics Lab.
For primary bibliographic entry see Field 02L.
W73-08863

SECONDARY PRODUCTION OF SELECTED INVERTEBRATES IN AN EPHEMERAL POND,
Texas A and M Univ., College Station.
P. R. Becker.
Available from Univ. Microfilms, Inc., Ann Arbor, Mich. 48106 Order No. 73-3511. PhD Dissertation, 1972. 154 p.

Descriptors: *Trophic level, *Succession, Temporary pond stage, *Amphipoda, *Rotifers, *Copepods, *Isopods, *Insects, Phosphates, Nitrates, Secondary productivity, Stability, Ponds, Rain, Surface runoff, Mosquitoes.
Identifiers: Species diversity, Orthophosphates, Streptocephalus seali, Diaptomus clavipes, Asellus militaris, Anostraca, Fairy shrimp, Sow bugs.

Between September, 1970 and August, 1971, an ecological study was made of a small ephemeral pond located in Brazos County, south-central Texas. The pond exhibited swift succession of physico-chemical characteristics begween filling and drying. As drying proceeded, there was a general decrease in the total amount of dissolved materials. Concentrations of most materials increased; however, the concentrations of orthophosphate and nitrate decreased. Rainfall and surface runoff modified the above patterns by diluting the concentrations of most materials, but increasing nitrate and orthophosphate. Biotic succession was rapid. Typical ephemeral pond species such as rotifers, copepods, anostracans, and mosquito larvae appeared first. These were followed by amphipods, isopods, insects, and other forms more characteristic of permanent ponds. Secondary production of Streptocephalus seali, Diaptomus clavipes, and Asellus militaris appeared high when compared to published values for similar organisms. The ratio of daily production to standing biomass was initially high, then decreased as the pond dried. It is proposed that the ephemeral pond is maintained in early low maturity stages of succession by the external physical disturbance of drying. The physico-chemical patterns of the pond coupled with the low species diversity and the high ratio of production to standing biomass indicates a system of low maturity and instability. (Little-Battelle)
W73-08903

EFFECTS OF PHENOLIC INHIBITORS ON GROWTH, METABOLISM, MINERAL DEPLETION, AND ION UPTAKE IN PAUL'S SCARLET ROSE CELL SUSPENSION CULTURES,
Oklahoma Univ., Norman.
M. L. Croak.
Available from Univ. Microfilms, Inc., Ann Arbor, Mich. 48106 Order No. 73-4940. PhD Dissertation, 1972. 67 p.

Descriptors: *Inhibitors, *Phenols, *Bioassay, *Cytological studies, Metabolism, Absorption, Ions, Radioactivity techniques, Organic acids, Magnesium, Calcium, Phosphorus, Potassium, Cations, Iron, Manganese, Trace elements, Molybdenum, Inhibition, Carbon radioisotopes.
Identifiers: *Rose cell suspension cultures, Pollutant effects, Growth, Cinnamic acid, p-Coumaric acid, Ferulic acid, Chlorogenic acid, Scopoletin.

Cinnamic, p-coumaric, and ferulic acids were very inhibitory to the growth of rose-cell suspension cultures when present in low concentrations, while chlorogenic acid and scopoletin were effective as inhibitors only at high concentrations. Treatment of 5-day cells with 0.0001 M ferulic acid and 0.00001 M cinnamic acid resulted in altered patterns of incorporation of C-14 from glucose-UL-C-14 into the following cell constituents: amino acids, organic acids, protein, and lipids. The effect of incubation of cells with 0.0001 M ferulic acid on the rate of depletion of Mg, Ca, K, P, Fe, Mn, and Mo ions from the medium during the 14-day growth cycle varied with age of the cells and the ion under consideration. In general, rates of uptake were higher than control rates in older cells and in very young cells and less than control rates in cells 3-5 days old. The degree of inhibition of uptake of Rb-86 also varied with age in cells treated with 0.0001 M ferulic acid. Young (4-5 day) cells showed approximately 50 percent inhibition at higher concentrations of RbCl (system 2) and approximately 25 percent inhibition at lower concentrations of RbCl (system 1). In contrast, the rate of Rb-86 uptake in 10-day cells was not significantly altered by incubation in ferulic acid at either system 1 or system 2. (Holoman-Battelle)
W73-08905

THE ECOLOGY OF SOME ARCTIC TUNDRA PONDS,
Missouri Univ., Columbia.
D. A. Kangas.
Available from Univ. Microfilms, Inc., Ann Arbor, Mich. 48106 Order No. 73-7045. PhD Dissertation, 1972. 343 p.

Descriptors: *Tundra, *Ponds, *Arctic, *Secondary productivity, *Ecology, Alaska, Standing crops, Crustaceans, Aquatic insects, Invertebrates, Biomass, Diptera, Water chemistry, Aquatic plants, Regression analysis, Trophic level, Efficiencies, Correlation analysis, Lake Morphometry, Physical properties, Chemical properties, Waterfleas, Alkalinity, Magnesium, Chlorides, Sodium, Calcium, Conductivity, Mathematical studies, Air temperature, Daphnia, Copepods, Zooplankton, Water temperature, Wind velocity, Hydrogen ion concentration, Carbonates, Phosphates.
Identifiers: Macroinvertebrates, Species density, Macrophytes, Anostraca, Ostracods, Arthropods, Artemia salina, Daphnia middendorffiana, Branchinecta paludosa, Polyartemiella hazeni, Artemiopsis bungei, Arctophila fulva, Ranunculus pallasii, Insolation, Tannin, Orthophosphates, Chlorophyll a.

Seventeen tundra ponds in the northernmost tip of Alaskan tundra were surveyed. Morphometric data were collected for seven of these ponds, and physical and chemical data were gathered for waters in all of them. Daily thermal oscillations were recorded in two ponds and predictive equations were derived by multiple linear correlation between U.S. Weather Bureau data on air temperature, wind direction and speed and insolation. Six water chemistry factors including alkalinity, magnesium, chloride, sodium, calcium and conductivity were significantly correlated by linear regression. Arctophila fulva and Ranunculus pallasii were the more common vascular plants which invaded these ponds. Some 28 genera of phytoplankton were identified in these ponds. Chlorophyll a standing crops ranged between 0.3 and 37 micrograms/l. Numerical standing crop

are drawn between the changes in system behavior found in this study and those resulting from predator manipulations in both computer models and natural systems. (Little-Battelle)
W73-08914

NUTRIENT PATHWAYS IN SMALL MOUNTAIN STREAMS,
Georgia Univ., Athens.
For primary bibliographic entry see Field 05B.
W73-08915

SIGNIFICANCE OF CELLULOSE PRODUCTION BY PLANKTONIC ALGAE IN LACUSTRINE ENVIRONMENTS,
Massachusetts Univ., Amherst.
J. Rho.
Available from Univ. Microfilms, Inc., Ann Arbor, Mich. 48106. Order No. 73-6704. Ph D Dissertation, 1972. 112 p.

Descriptors: *Phytoplankton, *Cellulose, *Aquatic algae, *Aquatic environment, Lakes, Chlorophyta, Chrysophyta, Pyrrophyta, Eutrophication, Oligotrophy, Dystrophy, Ponds, Carbohydrates, Dinoflagellates, Plant tissues, Vascular tissues, Bodies of water.
Identifiers: *Fate of pollutants, Particulate matter, Vertical distribution, Absorption spectra.

Numbers and types of cellulose-containing algae were observed in an eutrophic pond, a dystrophic pond and an oligotrophic lake. The cellulose-containing algae, members of Chlorophyceae, Dinophyceae and Chrysophyceae, were estimated to be approximately 70 percent of the total phytoplankton population observed. Fourteen algal species representing the dominant cellulose-containing phytoplankton were investigated for cellulose content which accounted for 2-39 percent on a dry weight basis of the cells studied. The amounts of cellulose present in the water column ranged from 110-1,185 micrograms/l, which accounted for 4-50 percent of the total dry weight of particulate matter. A comparison of absorption spectra of the extracts from algae, higher plants, particulate matter from bodies of water studied and purified cellulose, indicated that the cellulosic material present in particulate matter was of algal origin. It was estimated that less than 30 percent of cellulose in the water column was actually contributed by the viable algal cells present at the time of sampling. There was more cellulose found in the water column than could be accounted for by the number of algal cells observed. (Holoman-Battelle)
W73-08917

POLLUTION OF THE 'EL CARPINCHO' POND (PAMPASIC REGION, ARGENTINA) AND ITS EFFECTS ON PLANKTON AND FISH COMMUNITIES,
La Plata Univ. (Argentina). Instituto de Limnologia.
L. Freyre.
Environmental Pollution, Vol 4, No 1, p 37-40, January 1973. 3 tab, 2 ref.

Descriptors: Water pollution effects, *Waste water (Pollution), *Industrial wastes, *Zooplankton, Freshwater fish, Chemical analysis, Water pollution sources, Dairy industry, Catfishes, Silversides, Biological communities, Lead, Iron, Nitrates, Phosphates, Heavy metals, Pollutants, Chlorides, Biochemical oxygen demand, Odor, Sodium, Organic wastes, South America, Lakes.
Identifiers: *El Carpincho Pond, *Basilichthys bonariensis, *Parapimelodus valenciennesi, Plastic bag factory, *Argentina, Sulfites, Species diversity index.

A pond in the Southern Pampasic region of Argentina, 'El Carpincho', with highly developed fishing and other recreational activities, has been polluted since 1969 by the effluents from three industrial plants (milk processing plant, plastic bag factory, and an establishment for washing cattle trucks). The effects of this on the plankton and fish communities were investigated. Chemical analyses showed an increase in Na and sulfite, nitrates and phosphates. Lead and iron were detected in the water for the first time, and there was a high BOD and chloride demand, with an objectionable odor in wastes from the milk industry. Changes in the Mg/Ca ratio caused a reduction in the density of zooplankton, and in the numbers of the freshwater argentine silverside fish (Basilichthys bonariensis). On the other hand, an explosive increase in catfish (Parapimelodus valenciennesi) was observed. The diversity index for the relative composition of the fish fauna rose from 0.3093 in 1966 to 1.9384 in 1970. (Holoman-Battelle)
W73-08927

INTERTIDAL SUBSTRATE RUGOSITY AND SPECIES DIVERSITY,
University of Southern California, Los Angeles.
M. J. Risk.
Available from Univ. Microfilms, Inc., Ann Arbor, Mich. 48106. Order No. 73-763. Ph D Dissertation, 1972. 92 p.

Descriptors: *Intertidal areas, Standing crops, Aquatic habitats, Water levels, Tropical regions, California, Virgin Islands, Indicators, Aquatic populations, Tidal waters.
Identifiers: *Species diversity, *Rugosity, *Substrates, Panama, Marine environment.

A study of the relationship in the middle and upper intertidal between small-scale rugosity and species diversity was undertaken, restricting sample substrates to basalt flows. A total of fifty-five samples was taken in Panama and the Virgin Islands, and on the coasts of California and Baja, California. Species diversity was calculated both by numbers of individuals and by ash-free dry weights. Duplicates of the rock surfaces were produced by a technique involving field molding with an alginate impression material, casting a positive with plaster of paris, and finally producing a negative with latex compound. Rugosity was determined by calculating the variance of the differences between profile heights on the original surface and smoothed profile heights produced by taking a moving average of the original profile heights. Values for tidal height, total surface area exposed, standing crop (both in numbers and in weights) and latitude were also obtained for each sample area. Results indicate that the measured environmental parameters are much better predictors of weight species diversity than of number species diversity. Both measures of diversity correlate strongly with rugosity, and not as strongly (or not at all) with total surface area exposed. Rugosity is seen to increase with a decrease in tidal height, perhaps as a result of wave action and biological boring and rasping activities. Tropical intertidal diversity is low, and tropical areas seem to have lower species-carrying abilities. The tropical intertidal seems then to be an exception to the generally noted diversity increase in the tropics. (Holoman-Battelle)
W73-08928

THE EFFECT OF OIL POLLUTION ON SURVIVAL OF THE TIDAL POOL COPEPOD, TIGRIOPUS CALIFORNICUS,
California State Univ., Northridge. Dept. of Biology.
J. E. Kontogiannis, and C. J. Barnett.
Environmental Pollution, Vol 4, No 1, p 69-79, January 1973. 6 fig, 19 ref.

Descriptors: *Water pollution effects, *Oil pollution, Copepods, Bioassay, Toxicity, Crustaceans, Invertebrates, Sea water, Mortality, Oxygen, Air-water interface, Inhibition, Diffusion, Laboratory tests.

Identifiers: *Crude oil, *Mineral oil, *Tigriopus californicus, *Survival, Harpacticoids.

An investigation was undertaken to determine if the survival of Tigriopus californicus, a harpacticoid copepod, is affected by crude oil contamination and, if so, to determine whether death is caused mainly by oil acting as a physical barrier to O2 transfer between air and water or because of the presence of toxic chemicals in the oil itself. The experiments conducted involved subjecting adult copepods to (1) crude or mineral oil in sea water, with or without bubbling, at 17.5C in the dark for 1-7 days; and (2) crude or mineral oil as a surface film on partially deoxygenated sea water. A 1.5 mm thick layer of crude oil on the water surface caused the death of all animals within three days, while a similar layer of mineral oil resulted in complete mortality in five days. When oxygen was added to the water containing crude oil, total mortality was delayed by two days. Approximately 100 percent mortality was extended to seven days when crude oil was enclosed in a dialysis membrane bag immersed in the water and oxygen was supplied. When mineral oil was used and air provided, the animals survived as well as the controls. It was concluded that death resulted because the oil acts as a barrier to oxygen transfer between air and water, and because it contains substances toxic to Tigriopus. (Holoman-Battelle)
W73-08929

PHOSPHORUS DYNAMICS IN LAKE WATER,
Toronto Univ. (Ontario). Dept. of Zoology.
For primary bibliographic entry see Field 05B.
W73-08931

GROWTH AND SURVIVAL OF ANURAN TADPOLES (BUFO BOREAS AND RANA AURORA) IN RELATION TO ACUTE GAMMA RADIATION, WATER TEMPERATURE, AND POPULATION DENSITY,
Washington State Univ., Pullman.
J. L. Fish.
Available from Univ. Microfilms, Inc., Ann Arbor, Mich. 48106 Order No. 73-44. PhD Dissertation, 1972. 91 p.

Descriptors: *Bioassay, *Water temperature, *Gamma rays, *Growth rates, Water pollution effects, Frogs, Toads, Laboratory tests, Radioactivity effects, Thermal stress, Amphibians, Radiation, Electromagnetic waves, Animal populations, Larval growth stage.
Identifiers: *Tadpoles, Rana aurora, Bufo boreas, Survival, *Population density, Continuous flow techniques.

Effects on growth and survival of anuran tadpoles (Bufo boreas and Rana aurora) in relation to acute gamma radiation, water temperature, and population density were studied in continuous-flow laboratory test chambers utilizing a factorial randomized block design. Approximately one week after hatching, tadpoles were exposed to radiation doses ranging from 0-2000 Roentgen (R), then reared for 28 days at three temperatures (15, 20, or 25 C) and at four densities (5, 15, 35, or 100 animals). Survivors were measured (total length) and counted weekly. Comparisons between species indicated that R. aurora was more radio-sensitive than B. boreas. Densities greater than 35 animals in a half gallon of water significantly slowed growth and adversely affected survival. There appeared to be increased radiation damage at greater densities. Growth and survival were enhanced by higher temperatures, and radiation damage developed faster at higher temperatures. A temperature of 25 C with densities greater than 35 animals slowed growth in B. boreas. The three-way interaction of these variables was not conclusively demonstrated. (Holoman-Battelle)
W73-08934

GROWTH AS RELATED TO INGESTION IN A STREAM ISOPOD, LIRCEUS BRACHYURUS (HARGER) FED A MACROPHYTE OR PERIPHYTON,
Pennsylvania State Univ., University Park.
C. L. Hawkes.
Available from Univ. Microfilms, Inc., Ann Arbor, Mich. 48106 Order No. 72-33,170. Ph D Dissertation, 1972. 57 p.

Descriptors: *Food habits, *Growth rates, *Isopods, *Periphyton, Invertebrates, Crustaceans, Digestion, Radioactivity techniques, Cobalt radioisotopes, Aquatic plants, Aquatic weeds, Animal growth.
Identifiers: Lirceus brachyurus, Elodea canadensis, *Ingestion rates, *Macrophytes, Macroinvertebrates, Waterweeds.

The roles of a macrophyte Elodea canadensis and periphyton as food for the isopod Lirceus brachyurus were compared by investigating growth as related to ingestion. Ingestion rates, growth rates, percentage growth rates, and the ratios of growth rates to ingestion rates were obtained for isopods fed these two plant types. Ingestion was measured using C-60 labeled plants with a method involving a one hour feeding period and a direct counting of the whole live animal with the radioactive meal. Growth was based on the area of the frontal outline of each isopod calculated by multiplying the length by the width of a calibrated photographic image of the animal. Ingestion rates for small and medium sized isopods feeding on the macrophyte were significantly higher than those on periphyton with the medium sized animals having the largest difference. The highest growth rates and percentage growth rates were obtained when the isopods were fed periphyton. The ratios of growth rates to ingestion rates were higher for all size classes of iospods fed periphyton. (Holoman-Battelle)
W73-08936

SEASONAL SUCCESSION OF PHYTOPLANKTON AND A MODEL OF THE DYNAMICS OF PHYTOPLANKTON GROWTH AND NUTRIENT UPTAKE,
Wisconsin Univ., Madison.
J. F. Koonce.
Available from Univ. Microfilms, Inc., Ann Arbor, Mich. 48106 Order No. 72-24,888 Ph D Dissertation, 1972. 204 p.

Descriptors: *Phytoplankton, *Mathematical models, *Growth rates, *Absorption, *Nutrients, Plant populations, Aquatic environment, Lentic environment, Carbon, Nitrogen, Phosphorus, Marine algae, Aquatic algae, Model studies, Cytological studies, *Wisconsin, Ecosystems, Water temperature, Light intensity, Mathematical studies, Plant physiology, Plant morphology, Chlorophyta.
Identifiers: *Seasonal succession, Marine environment, Silicon, *Lake Wingra (Wisconsin), Selenastrum capricornutum, Unialgal culture, Desmids.

Seasonal succession in phytoplankton associations has been observed in freshwater and marine lentic environments. The appearance and disappearance of algal populations has been correlated with variations in physical, chemical, and biological factors in aquatic ecosystems. To study the succession of phytoplankton, a series of weekly observations for one year were made at Lake Wingra, Madison, Wisconsin. In addition to enumeration and identification of phytoplankton species, analyses were conducted for major algal nutrients (Si, C, N, and P), and light and temperature were measured. The results revealed species' succession and nutrient depletion patterns similar to other lakes. Proceeding from an assumption that growth optimization was the dominant process in the observed succession pattern, a mathematical model was derived from theoretical considerations of algal physiology. The prediction of the model was rate of algal growth, and model parameters were related to efficiencies of biological processes (such as nutrient uptake). The model was tested against unialgal cultures of Selenastrum capricornutum and was found to predict adequately the time course of algal growth and uptake on N and P in batch cultures of algae. The verified model was then related to morphological variation and physiological adaptation of phytoplankton species. Through appropriate use of mathematical optimization routines, model parameters could be varied to simulate maximum growth rates under an imposed set of nutrient concentrations (such as low P and high N or low N and high P concentrations). By relating values of model parameters to cell morphology and physiological adaptation such procedure suggests the possibility of predicting morphological variation--this succession of phytoplankton populations. The derived model is useful only for variation in concentration of N and P. (Holoman-Battelle)
W73-08938

REGROWTH OF COLIFORMS AND FECAL COLIFORMS IN CHLORINATED WASTEWATER EFFLUENT,
Hadassah Medical School, Jerusalem (Israel). Environmental Health Lab.
H. I. Shuval, J. Cohen, and R. Kolodney.
Water Research, Vol 7, No 4, p 537-546, April 1973. 7 fig, 14 ref.

Descriptors: *Coliforms, *Waste water (Pollution), *Chlorination, *Sewage effluents, Environmental effects, On-site investigations, Laboratory tests, Water quality control, Enteric bacteria, Chlorine, Water sampling, Water analysis, Disinfection, Bactericides, Reservoir storage, Aerobic bacteria, Pathogenic bacteria.
Identifiers: *Fecal coliforms, *Regrowth, Survival, Biochemical tests, Growth.

Observations made both in the field in chlorinated effluent, and in laboratory experiments show that coliforms and fecal coliforms are capable of regrowth in chlorinated wastewater. Under field conditions regrowth of coliforms in chlorinated effluent held in a storage reservoir for about 3 days appeared inversely correlated to: (1) the residual chlorine in the storage reservoir and (2) the number of coliforms surviving chlorination. In the laboratory experiments regrowth occurred after initial doses as high as 11 ppm total chlorine even when there was no chemical inactivation of the chlorine. Fecal coliforms did not generally show regrowth to the same extent as coliforms. Regrowth occurred even when coliforms were not detectable in 10-ml of samples after chlorination. Since coliforms and fecal coliforms are capable of regrowth in chlorinated sewage effluent and adminstures of it, the sanitary significance of the number of coliforms after storage or in receiving bodies of water is difficult to interpret. Thus standards might be based on the number of coliforms, or fecal coliforms detected in effluents immediately after chlorination. However, this would not be justified if in addition to coliforms, pathogenic bacteria can regrow in chlorinated effluents. (Holoman-Battelle)
W73-08939

THE DYNAMICS OF BROWN TROUT (SALMO TRUTTA) AND SCULPIN (COTTUS SPP.) POPULATIONS AS INDICATORS OF EUTROPHICATION,
Michigan State Univ., East Lansing.
For primary bibliographic entry see Field 05A.
W73-08940

MOLECULAR ARCHITECTURE OF THE CHLOROPLAST MEMBRANES OF CHLAMYDOMONAS REINHARDI AS REVEALED BY

IIGH RESOLUTION ELECTRON MICROSCO-'Y,
'alifornia Univ., Los Angeles.
'. L. Kretzer.
vailable from Univ. Microfilms, Inc., Ann Ar-or, Mich. 48106 Order No. 73-1711. Ph D Disser-ation, 1972. 78 p.

)escriptors: *Molecular structure, *Cytological tudies, *Biological membranes, Electron nicroscopy, Chlorophyta, Aquatic algae, 'rotozoa, Methodology.
dentifiers: *Chlamydomonas reinhardi, *High esolution electron microscopy, *Chloroplasts, ample preparation, Ultrastructure, Flagellates.

'he chloroplast membranes of Chlamydomonas einhardi were prepared for thin section, high esolution electron microscopy with the specific oal of minimally denaturing the protein and with he least lipid solubilization possible. This recluded all conventional procedures (OsO4, (MnO4, high concentrations of glutaraldehyde, nd acetone or ethanol dehydration). The scheme irst used by Sjostrand and Barajas was adapted to his system: membrane proteins were first cross-nked with a very short exposure to 1 percent glu-araldehyde, dehydrated a short time in cold thylene glycol, and infiltrated with Vestopal. After such a preparatory procedure, the partition nembranes were 270A thick, nearly three times he thickness of the membranes preserved by con-entional proceudres. The 270A-thick partition nembranes showed a heterogeneous (10-100A) lobular substructure. It is proposed that the sub-tructure is the enzymatic and structural protein of he thylakoids. The parameters of the original jostrand-Barajas preparatory scheme were aried, and the chloroplast membranes responded ke a system sensitive to the phenomena of rotein denaturation. The membrane varied con-istently from 100-270A depending on the denatur-ıg potential of the 'fixative' and the hydrophobic haracter of the dehydrating agent. (Holoman-Bat-elle)
V73-08942

CUTE TOXICITY OF HEAVY METALS TO OME MARINE LARVAE,
finistry of Agriculture, Fisheries and Food, urnham-on-Crouch (England). Fisheries Lab.
'. M. Connor.
larine Pollution Bulletin, Vol 3, No 12, p 190-192,)ecember 1972. 3 fig, 3 tab, 6 ref.

)escriptors: *Toxicity, *Heavy metals, *Larval rowth stage, *Shellfish, *Crustaceans, *Mol-ısks, Larvae, Copper, Mercury, Zinc, Oysters, hrimp, Crabs, Invertebrates, Bioassay, Lethal mit, Lobsters, Mortality, Mature growth stage, larine animals, Laboratory tests.
dentifiers: *Median tolerance limit, Sensitivity,)strea edulis, Crangon crangon, Carcinus maenas, lomarus gammarus, Macroinvertebrates.

he toxicity of copper, mercury and zinc to the ırvae of oysters, shrimp, crab and lobsters has een examined over periods of up to 64 hours. lercury was found to be more toxic than copper nd zinc, which had similar levels of toxicity. Over ıe experimental period, the relationship between ›xicity and concentration was linear. Larvae were 'om 14 to 1,000 times more susceptible than dults of the same species. The median lethal con-entrations (LC50) of each metal to the most sensi-ve species of larvae, tested over a 48 hour period, xceeded the concentrations found in natural sea ater by a factor of 100. For longer test periods, ıe LC50 would be considerably less and this fac-ır would then be considerably reduced. Hence ıe continued addition of these metals to confined aters should give cause for concern. (Holoman-attelle)
'73-08943

EFFECTS OF DIELDRIN ON WALLEYE EGG DEVELOPMENT, HATCHING AND FRY SURVIVAL,
Ohio State Univ., Columbus.
E. M. Hair.
Available from Univ. Microfilms, Inc., Ann Arbor, Mich. 48106 Order No. 73-2012. Ph D Dissertation, 1972, 52 p.

Descriptors: *Bioassay, Absorption, *Fish eggs, *Dieldrin, *Mortality, Animal physiology, Fry, *Walleye, Pesticide toxicity, Water pollution effects, Laboratory equipment, Growth rates, Hatching, Fish physiology.
Identifiers: *Bioaccumulation, Biological magnification.

The objectives were to determine whether walleye eggs absorb dieldrin from water during incubation, and if so, whether uptake is concentration-dependent, and whether absorbed dieldrin affects embryo hatchability, or fry development and survival through the yolk sac stage. Eggs were reared in a miniature jar hatchery and were continuously exposed to 0.0, 1.0, 10.0, or 100.0 ppb dieldrin from spawning until hatching. Water and egg samples were taken daily to monitor HEOD concentration and embryo development. After hatching fry were held in glass battery jars until yolk sac absorption was complete. Samples were taken daily to monitor their development and HEOD content. Eggs absorbed dieldrin from the water at all levels of exposure. The amount absorbed depended on the concentration in water and varied with the duration of exposure. Corresponding with the appearance of eyes in the embryos there was a sharp increase in dieldrin uptake. After 5 days of exposure eggs had concentrated levels in the water by factors of 800 (exposure to 1 ppb), 1210 (10 ppb), and 914 (100 ppb). Dieldrin absorbed by developing walleye eggs did not affect embryo development, rate of development or hatchability. At the higher levels of exposure (10 and 100 ppb) fry survival was affected. After exposure to 10 or 100 ppb, fry became pigmented sooner and were more active than the controls. (Little-Battelle)
W73-08944

TRACE METAL RELATIONSHIPS IN A MARINE PELAGIC FOOD CHAIN,
Stanford Univ., Calif.
G. A. Knauer.
Available from Univ. Microfilms, Inc., Ann Arbor, Mich. 48106 Order No. 73-4534. Ph D Dissertation, 1972, 156 p.

Descriptors: *Heavy metals, *Food chains, *Phytoplankton, *Zooplankton, *Fish, Absorption, Water analysis, Sea water, Cadmium, Copper, Manganese, Lead, Zinc, Sodium, Magnesium, Calcium, Potassium, Strontium, Aluminum, Iron, Nickel, Titanium, Organic compounds.
Identifiers: *Biological magnification, *Bioaccumulation, Atomic absorption spectrophotometry, Biological samples, Barium, Silver, Anchovy, Silicon.

Surface water samples collected over a one year period from Monterey Bay, California were analyzed for Cd, Cu, Mn, Pb and Zn. Mixed phytoplankton and zooplankton samples were analyzed for Na, Mg, Ca, K, Sr, Si (phytoplankton), Ba, Al, Zn, Fe, Cu, Mn, Ni, Ti, Ag, Cd, and Pb. Samples of the northern anchovy Engraulis mordax were analyzed for Pb, Cd, Ag, Ni, Mn, Cu, Fe, Zn, Al, and Ba. Analysis was by conventional atomic absorption spectrometry. Phytoplankton samples were separated into acid soluble and silica fractions before analysis. The acid soluble fraction contained large amounts of Na, K, Mg, Ca and Si and low concentrations of the remaining elements. Titanium usually was not detected in this fraction. In the silica fraction, Na, K and Mg were found to have the highest concentrations followed by Al, Ca, Fe and Sr. The lowest levels found in this fraction were for the elements Cu and Zn with Ba, Mn and Cd detected only occasionally. The metals Pb, Cd, Ag, Ni, Mn, Cu, Fe, Zn, Al and Ba were also compared in phytoplankton, zooplankton and anchovy samples in order to determine existing relationships in this simple food chain. In general, concentrations of biologically active metals were relatively constant in all three trophic levels. For non-biologically active metals (e.g., Cd, Ni, Pb) little evidence for food chain amplification was found. Phytoplankton concentration factors (relative to seawater) were found to be highest for Pb, Fe, Si, Cd, Al, and Ti respectively while relatively low values were obtained for Ba, Zn, Cu, Mn, Ni and Ag. Except for K, Sr, and Ba, the alkali and alkaline earth metals were not concentrated relative to seawater. Mercury levels in this food chain were also found to be low. (Little-Battelle)
W73-08947

TOXIC EFFECTS OF CUPRIC, CHROMATE AND CHROMIC IONS ON BIOLOGICAL OXIDATION,
Calgary Univ. (Alberta). Dept. of Chemical Engineering.
A. Lamb, and E. L. Tollefson.
Water Research, Vol 7, No 4, p 599-613, April 1973. 13 fig, 3 tab, 10 ref.

Descriptors: *Toxicity, *Cations, *Chromium, *Copper, *Activated sludge, Laboratory tests, Suspended solids, Sewage bacteria, Microbial degradation, Chemical analysis, Heavy metals, Aqueous solutions, Biodegradation, Sludge digestion, Biochemical oxygen demand, Dissolved oxygen, Nutrients.
Identifiers: Pollutant effects, *Biological oxidation, Substrate utilization, Substrate concentration, Fate of pollutants, Atomic absorption spectrophotometry, Glucose, Total carbon analyzer.

Using a completely mixed, continuously operated, lightly loaded, laboratory activated sludge system, the toxic effects of cupric, chromic and chromate ions under conditions of shock loading were observed. These were determined with the aid of a total carbon analyzer and simple mass balance techniques in terms of conversion of the organic nutrient fed. The distribution of the metal ion between aqueous solution and suspended solids was measured using atomic absorption spectrophotometry. Toxic effects were in the order: cupric greater than chromate greater than chromic while the reductions in conversion were 90, 50 and 20 percent, respectively, for concentrations of 5 ppm metal ion. Cupric ion toxicity was directly proportional to the weight of copper absorbed per unit mass of suspended matter within the total copper concentration range (0-5.5 ppm) studied. This toxicity decreased markedly with increased suspended solids concentration: an 80 percent decrease in conversion at 210 ppm suspended solids was reduced to a negligible quantity (3 percent) by increasing the suspended solids to 4000 ppm. At 210 ppm suspended solids, 34 percent of the added copper was removed by the sludge in 7 h. The results suggest that the toxic effect of metal ions on a sewage plant activated sludge system could be reduced by rapidly increasing the suspended solids concentration, possibly by the addition of dried sludge. This work supports a mechanism involving rapid adsorption of the cupric ion by both viable and dead sludge followed by a slower rate determining step resulting in the toxic effect. The first order rate constant for substrate utilization was found to be 1.07 plus or minus 0.6/hr. (Holoman-Battelle)
W73-08948

THE EFFECT OF PASSAIC RIVER BENTHAL DEPOSITS ON DEOXYGENATION IN OVERLYING WATERS,
Rutgers - The State Univ., New Brunswick, N.J.
For primary bibliographic entry see Field 05G.
W73-08949

STUDIES ON THE OCCURRENCE, PHYSIOLOGY, AND ECOLOGY OF BIOLUMINESCENCE IN DINOFLAGELLATES,
Oregon State Univ., Corvallis.
W. E. Esaias.
Available from Univ. Microfilms, Inc, Ann Arbor, Mich. 48106 Order No. 73-7826. Ph D Dissertation, 1973. 87 p.

Descriptors: *Dinoflagellates, *Bioluminescence, *Environmental effects, *Physiological ecology, Pyrrophyta, Marine algae, Protozoa, *Oregon, Plant physiology, Animal physiology, Ecology.
Identifiers: *Yaquina Bay, Flagellates, Ceratium fusus, Gonyaulax digitale, Gonyaulix catenella, Gonyaulax acatenlla, Gonyaulax tamarensis, Peridinium depressum, Gonyaulax triacantha, Photoinhibition.

To provide further information on the occurrence and geographical variations of bioluminescent capabilities of marine dinoflagellates, forty species, representing twelve genera, of dinoflagellates from Yaquina Bay, Oregon, were examined for bioluminescence as single cell isolates. Seventeen species from the genera Ceratium (1 sp.), Gonyaulax (3 sp.), and Peridinium (13 sp.) were found to be bioluminescent. Ceratium fusus was the only member of the genus found to emit light; G. triacantha was found to be non-bioluminescent. The total photon emission of each luminescent species is reported. Values ranged from 10,500,000,000 photons per P. depressum to 21,000,000 photons per G. digitale. As a taxon, the genus Peridinium emitted more light by an order of magnitude than did Ceratium or Gonyaulax. The mechanically stimulable bioluminescence of members of the Gonyaulax catenella group can be photoinhibited completely by exposure to as little as 10 trillion quanta/sq cm delivered as a pulse of width between 0.1 and 10 seconds. There is an initial time lag of one minute, followed by a first order decay to approximately one percent of the bioluminescence of unexpected controls. The half time of this decay is only 50 seconds. Action spectra for photoinhibition in Gonyaulax catenella, G. acatenalla, and G. tamarensis revealed a single absorption band with a maximum of 562 nm. Grazing experiments were conducted with three calanoid copepods and three species of bioluminescent dinoflagellates, using procedures which yielded samples of cultures with high and low capacities for mechanically stimulable bioluminescence. In all cases the ingestion rates were lower for the high bioluminescent capacity samples than for the samples having a reduced bioluminescent capacity. These results indicate that dinoflagellate bioluminescence has survival value as a defense against copepod grazing. (Holoman-Battelle)
W73-08953

ON THE MICROBIOLOGY OF SLIME LAYERS FORMED ON IMMERSED MATERIALS IN A MARINE ENVIRONMENT,
Hawaii Univ., Honolulu.
G. E. Sechler.
Available from Univ. Microfilms, Inc., Ann Arbor, Mich. 48106 Order No. 73-5276. Ph D Dissertation, 1972. 114 p.

Descriptors: Sampling, *Artificial substrates, *Sea water, *Diatoms, *Bacteria, *Microscopy, *Pollutant identification, Adsorption, Zinc, Aluminum.
Identifiers: Glass, Plexiglass, Wood, Stainless steel, Steel, Monel, Phosphor bronze, Teflon overlay technique, Parlodion filming technique.

Three separate techniques were used to investigate the development of microorganisms in primary films found on various materials immersed in marine waters. These included microbial viable counts by the traditional swabbing method, and total visible counts by two new direct microscopic observation techniques developed during this study: the Teflon overlay technique and the Parlodion filming technique. A variety of test materials were investigated, including glass, plexiglass, wood, zinc, stainless steel, steel, Monel, aluminum, and phosphor-bronze. Test panels were immersed for intervals ranging from 1 hour to 40 days. Although similar varieties of aerobic heterotrophic bacteria were found on all surfaces regardless of their chemical nature, viable population levels were characteristic of each test material during the first few days following immersion. Wood accumulated the greatest number of bacteria in the shortest period of time (within 3 days). Bacteria were found to attach to Teflon membranes over-lying test surfaces by 1 day and to proliferate in situ by at least 4 days. Using the Parlodion filming technique it was found that the number of bacteria per sq cm of test panel could be accurately determined within the first 10 days after immersion. Diatom counts were accurate for 24 days, while extraneous particle counts were valid at least 75 days. No stable diatom population was recognized on any test surface until 6 days after immersion, indicating that bacterial growth may prepare the test surface for the development of unicellular algae. The more chemically passive test panels (glass, plexiglass, and stainless steel) consistently exhibited the highest diatom population level up to 16 days after immersion. (Little-Battelle)
W73-08955

A COMPARATIVE STUDY OF THE DECOMPOSITION OF CELLULOSIC SUBSTRATES BY SELECTED BACTERIAL STRAINS,
Massachusetts Univ., Amherst.
For primary bibliographic entry see Field 05B.
W73-08957

PERIPHYTON PRODUCTION AND GRAZING RATES IN A STREAM MEASURED WITH A P-32 MATERIAL BALANCE METHOD,
Oak Ridge National Lab., Tenn.
For primary bibliographic entry see Field 05A.
W73-08958

CESIUM-137 AND STABLE CESIUM IN A HYPEREUTROPHIC LAKE,
Michigan State Univ., East Lansing.
L. D. Eyman.
Available from Univ. Microfilms, Inc., Ann Arbor, Mich. 48106 Order No. 73-5367. Ph D Dissertation, 1972. 113 p.

Descriptors: Ecosystems, Lakes, *Eutrophication, *Path of pollutants, *Stable isotopes, *Radiochemical analysis, *Cesium, Water analysis, Aquatic plants, Aquatic algae, Zooplankton, Freshwater fish, Lake sediments, Aquatic environment, Soil analysis, Aquatic soils, Trophic level.
Identifiers: *Cs-137, Cesium radioisotopes, Macrophytes, Fate of pollutants.

The inputs of Cs-137 and stable Cs and their distribution among the various components of an aquatic ecosystem were studied in a lake exhibiting an advanced stage of eutrophy. Components sampled and analyzed for these two isotopes of cesium included water, sediments, macrophytes, filamentous algae, zooplankton, and several species of fish. Most of the cesium pool (87 percent Cs-137; 98 percent Cs) was associated with the sediments. A trend of increased Cs-137 concentration at higher trophic levels was demonstrated for those fish that are free-ranging limnetic feeders. No such trend was evident for stable Cs. Forms closely associated with sediments had higher Cs-137 concentrations than expected based on their feeding habits. On the dates samples were collected, specific activity in limnetic fishes was constant but was variable in other forms. The degree of association of biotic forms with sediments was reflected in their specific activity. (Holoman-Battelle)
W73-08959

BIOLOGICAL NITROGEN FIXATION IN LAKE MENDOTA,
Wisconsin Univ., Madison.
M. L. S. Torrey.
Available from Univ. Microfilms, Inc., Ann Arbor, Mich. 48106 Order No. 72-27,352. Ph D Dissertation, 1972. 452 p.

Descriptors: *Nitrogen fixation, *Bacteria, *Phytoplankton, *Cyanophyta, Water temperatures, Dissolved oxygen, Hydrogen ion concentration, Sediments, Phosphates, Nitrogen, Ammonia, Eutrophication, Depth, Light, Statistical methods, Algae, Lakes, *Wisconsin.
Identifiers: *Acetylene reduction, *Lake Mendota (Wis), Orthophosphates, Wisconsin.

The study investigated the validity of prior estimates of N fixation in Lake Mendota, the factors influencing algal fixation, and the possibility of heterotrophic fixation in the sediments and the water column. All nitrogen fixation studies were performed with acetylene. In addition the ratio of acetylene reduced to nitrogen was estimated. Physical, chemical, and biological analyses were performed at each depth sampled. Short, medium, and long-term studies all showed that heterocyst content was significantly related to the activity and efficiency of acetylene reduction. In long-term studies, temperature was positively correlated with acetylene reduction activity and efficiency. The depth at which the sample was collected was negatively correlated with acetylene reduction activity and efficiency. Dissolved oxygen had a positive regression coefficient in the multiple linear regression analysis of acetylene reduction. pH also was related to acetylene reduction. Available data do not distinguish whether dissolved oxygen and pH were affecting acetylene reduction or whether these were a result of changes in water quality caused by active photosynthesis. Bacterial acetylene reduction activity in the waters from which colonial phytoplankton had been removed was very low. Sediment acetylene reduction activities were also low, but can be significant over a long period of time. Nitrogen fixation contributed about ten percent or about 39,800 kg of the annual input of nitrogen to Lake Mendota during 1970 and 1971. While this is much less significant than groundwater, rural runoff, or dry fallout, most of it is added at a period of time when combined nitrogen is depleted and thus may be significant during the summer months in maintaining blue-green algal nuisances in the surface waters. (Little-Battelle)
W73-08962

EXCRETION MEASUREMENTS OF NEKTON AND THE REGENERATION OF NUTRIENTS NEAR PUNTA SAN JUAN IN THE PERU UPWELLING SYSTEM DERIVED FROM NEKTON AND ZOOPLANKTON EXCRETION,
Washington Univ., Seattle.
For primary bibliographic entry see Field 05B.
W73-08963

CHARACTERIZATION OF SEVERAL THYMINE-REQUIRING MUTANTS OF ESCHERICHIA COLI Y MEL,
New Hampshire Univ., Durham.
J. H. Maryanski.
Available from Univ. Microfilms, Inc., Ann Arbor, Mich. 48106 Order No. 73-4362. Ph D Dissertation, 1972. 236 p.

Descriptors: *Cultures, *E. coli, Separation techniques, Pollutant identification, Growth rates.
Identifiers: *Culture media, Thymidine, Thymidilic acid, Agars, Tryptophan, Thymine, Mutants.

Escherichia coli Y melT (-), E. coli B27T (-), and E. coli B9T (-) were isolated from E. coli Y mel, E. coli B27, and E. coli B9, respectively, by trimethoprim selection. E. coli B27 and E. coli B9 were tryptophan delection mutants derived from E. coli Y mel. Mutants, E. coli Y melT (-), E. coli

marinus under Ni stress. Scanning electron microscopy by the freeze-etching technique indicated multivacuolation characterized by vacuole membranes, polybetahydroxybutyric acid granules, and a loss of cell envelope protein granules under Ni stress. An ecological survey of three seawater environments, open ocean, coastal, and estuarine, indicated the open ocean heterotrophic microbial population as the most Ni tolerant. (Little-Battelle)
W73-08969

LEUCOTHRIX MUCOR AS AN ALGAL EPIPHYTE IN THE MARINE ENVIRONMENT,
Indiana Univ., Bloomington.
For primary bibliographic entry see Field 05A.
W73-08970

INITIAL RESPONSES OF PHYTOPLANKTON AND RELATED FACTORS IN LAKE SAMMAMISH FOLLOWING NUTRIENT DIVERSION,
Washington Univ., Seattle.
R. M. Emery.
Available from Univ. Microfilms, Inc., Ann Arbor, Mich. 48106, Order No. 72-28594. Ph D Dissertation, 1972. 244 p.

Descriptors: *Trophic level, *Nutrients, *Phytoplankton, *Primary productivity, *Biomass, Water pollution effects, Phosphorus, Nitrogen, Lakes, Sewage, Urban runoff, Carbon, *Washington.
Identifiers: *Lake Sammamish (Wash), *Sewage diversion, Restoration, Lake Washington, Silicon.

A two-year study on Lake Sammamish was carried out to evaluate the responses of phytoplankton and related factors to a sewage diversion project completed in September, 1968. Trophic indices of pre-diversion years were compared to those in nearby Lake Washington to determine the relative extent of eutrophication. Post-diversion trophic indices were compared to those of pre-diversion years to determine the extent of recovery. In addition, post-diversion changes of indices were compared to those in Lake Washington, a lake with an established pattern of response to diversion. This study shows that, so far, only minor changes which suggest recovery have occurred in Lake Sammamish since diversion. This lack of early response is not consistent with the predicted recovery time of 2.8 year, calculated using a lake restoration model. Concentrations of P and N in Lake Sammamish have not changed significantly since 1968, with one exception. Seasonal levels of phytoplankton production and biomass have not decreased significantly since nutrient diversion, although there was a noticeable change in the composition of phytoplankton populations. Statistical multivariate analyses indicate that Si and C, rather than P and N, have the most influence on algal activity. Results indicate that urban runoff was not a significant source of nutrient enrichment. (Little-Battelle)
W73-08973

DISTRIBUTION OF ICHTHYOPLANKTON IN THE WARM WATER DISCHARGE ZONE OF THE NOVOROSSIISK THERMAL ELECTRIC POWER PLANT, (IN RUSSIAN),
Rostov-on-Don State Univ. (USSR).
L. P. Kostyuchenko.
Vopr Ikhtiol. Vol 11 No 6, p 987-992. 1971 Illus.
Identifiers: Discharge zone, Distribution, Electric powerplants, *Ichthyo plankton, *Novorossiisk Bay (USSR), Plankton, *Thermal pollution, USSR, Warm water.

Information is given on the distribution of eggs and larvae of fishes in the warm water zone of Novorossiisk bay formed by the local power plant. The species composition and quantitative development of ichthyoplankton are analyzed in individual areas of the investigated region as the distance increases from the origin of discharge of warm waters. Data on the change of the spawning time of certain species of fishes as a function of temperature are given. A comparative characterization of ichthyoplankton is presented.--Copyright 1972, Biological Abstracts, Inc.
W73-09018

STUDIES ON SOME EGYPTIAN FRESHWATER ISOLATES OF UNICELLULAR GREEN ALGAE,
Ain Shams Univ., Cairo (Egypt). Dept. of Botany.
For primary bibliographic entry see Field 05A.
W73-09020

RARE CASE OF POISONING BY WATER POLLUTED WITH TOXIC AGRICULTURAL CHEMICALS, (IN RUSSIAN),
Khorezmskii Pedagogicheskii Institut, Urgench (USSR).
B. Kh. Khasanov.
Gig Sanit. Vol 36 No 9, p 96. 1971.
Identifiers: *Agricultural chemicals, *Arsenic, Poisoning, Polluted water, Toxicity, Water pollution effects.

The case history of As poisoning of several farm workers who drank water from a well dug at the site where a storehouse for agricultural chemicals had stood 15 yr before is presented. Excavation around the well revealed 2 corroded drums at a distance of 3.5 m from the well at a depth of 90 cm. The drums contained sodium arsenate which had percolated into the well water.--Copyright 1972, Biological Abstracts, Inc.
W73-09026

HYGIENIC EVALUATION OF NEW POLYMERS USED IN THE DRINKING WATER SUPPLY SYSTEMS, (IN RUSSIAN),
Vsesoyuznyi Nauchno-Issledovatelskii Institut Gigieni i Toksikologii Pestitsidov, Kiev (USSR).
Z. M. Tsam.
Gig Sanit. Vol 36 No 9, p 100-101. 1971.
Identifiers: Hygienic studies, *Polymers, Toxicity, *Potable water, Water supply, Organoleptic properties.

Data are given on the effect of certain polymer materials used in drinking water supply systems on the toxic, physicochemical, organoleptic and quality of the water.--Copyright 1972, Biological Abstracts, Inc.
W73-09027

POLLUTION OF DRINKING WATER AND INFECTIONS BY PSEUDOMONAS AERUGINOSA IN A PEDIATRIC SECTION,
Siena Univ. (Italy). Istituto di Clinica Pediatria.
C. Panero, A. Grasso, C. LaCauza, F. Ragazzini, and G. Bosco.
G Batteriol Virol Immunol Ann Osp Maria Vittoria Torino. Vol 63, No 9/10, p 517-527, 1970.
Identifiers: Infections, *Italy, Nate, Pediatrics, Pollution, *Pseudomonas-aeruginosa, Siena, Water pollution effects, *Potable water.

Because of possible serious infections or its resistance to numerous antibiotics, P. aeruginosa plays an important role during the first year of human life and especially in immature neonates. To prevent infection, efficient prophylaxis including scrupulous disinfection, systematic bacterial control and location of contact areas are important. A series of pediatric cases involving P. aeurginosa in Siena may have been caused by a contaminated drinking water reservoir.--Copyright 1972, Biological Abstracts, Inc.
W73-09035

HYGIENIC PROBLEMS ARISING DUE TO DISCHARGE OF WASTE WATERS OF WOODWORKING PLANTS INTO OPEN WATERS, (IN RUSSIAN),

Institutul de Medicina si Farmacia, Iasi (Rumania).
G. Zamfir, S. Apostol, M. Filipyuk, L. Akeksa, and K. Melinte.
Gig Sanit. Vol 36, No 9, p 91-92, 1971. Illus.
Identifiers: Discharge, *Hygienic studies, Rivers, Toxicity, Treatment, Waste waters, *Woodworking industry, Water treatment.

Analysis of the content of rivers serving as receivers of waste waters of woodworking plants and experimental investigations indicate the complex composition, difficulty of decomposition and high toxicity of cellulose waste. The adverse and long effect on water quality in rivers necessitates an increase of the capacity of treatment plants and of their effectiveness for treating the waste waters.--Copyright 1972, Biological Abstracts, Inc.
W73-09036

SOME INFORMATION ON THE CONTAMINATION OF THE VOLGOGRAD WATER RESERVOIR WITH EGGS OF DIPHYLLOBOTHRIUM LATUM, (IN RUSSIAN),

Institute of Medical Parasitology and Tropical Medicine, Moscow (USSR).
A. S. Artamoshin.
Med Parazitol Parazit Bolezn. Vol 41, No 1, p 87-90, 1972. English summary.
Identifiers: Contamination, Diphyllobothrium-latum, eggs, Fish, Infestation, Reservoir, USSR, *Volgograd reservoir, Water pollution effects, *Helminth.

The significance of the contamination of a water body with eggs of D. latum is discussed. Diphyllobothrial invasion was determined on the basis of catching sites and the extent of infestation of fry of some fish with plerocercoids of this helminth.--Copyright 1972, Biological Abstracts, Inc.
W73-09037

HETEROTROPHIC ASSAYS IN THE DETECTION OF WATER MASSES AT LAKE TAHOE, CALIFORNIA,

California Univ., Davis. Dept. of Zoology; and California Univ., Davis. Inst. of Ecology.
H. W. Paerl, and C. R. Goldman.
Limnol Oceanogr. Vol 17, No 1, p 145-148. 1972. Illus.
Identifiers: Assays, Bacterial studies, California, Cells, *Heterotrophis, Lakes, *Lake Tahoe (Calif), Autoradiographs.

A sensitive biological method for water mass detection was successfully tested in the extremely oligotrophic waters of Lake Tahoe, California. A trace amount of 14C dissolved acetate is added to water samples and microbial incorporation monitored; assimilation rates demonstrate the presence of water masses originating from stream outflow. The same technique also shows microbial stimulation of Tahoe water when stream water is added in culture experiments. Autoradiographs indicate that bacterial cells, sometimes associated with detrital matter, are responsible for the acetate uptake.--Copyright 1972, Biological Abstracts, Inc.
W73-09059

STUDIES ON FRESHWATER BACTERIA: ASSOCIATION WITH ALGAE AND ALKALINE PHOSPHATASE ACTIVITY,

Freshwater Biological Association, Windermere (England).
J. G. Jones.
J Ecol. Vol 60, No 1, p 59-75. 1972. Illus.
Identifiers: *Algae, Alkaline, *Bacteria, Biomass, Diatoms, Dinoflagellates, *Phosphatase activity.

A survey of Windermere North and South Basins and Esthwaite Water at 2 sites at the south end of the lake was made in an attempt to establish the degree of association between the phytoplankton and the bacteria. Positive correlations were established, when all samples were considered, between chlorophyll alpha estimates and bacterial counts. This association varied with the species involved and the degree of attachment by bacteria was seen to diminsih along the series colonial green and blue-green algae > filamentous green and blue-green algae > diatoms > dinoflagellates. This trend was observed when using 'viable' as well as direct counting methods. An increase in the degree of attachment to moribund algal cells when compared with healthy cells was also noted. The positive correlation of bacterial estimates ('viable' and direct counts) with water temperature was observed when all samples were included in the analysis. Alkaline phosphatase activity of paper-filtered samples correlated positively with both algal and bacterial numbers; activity of membrane-filtered samples was significantly correlated only with numbers of phosphatase-producing bacteria. Further examination is required before the origin and control of the enzymes can be established. Estimates of algal and bacterial biomass were made in an attempt to determine their respective roles in the CO2 balance of the lakes. It was noted that indirect estimates of bacterial CO2 evolution were considerably lower than similar estimates of the net uptake of the algal biomass.--Copyright 1972, Biological Abstracts, Inc.
W73-09061

TOLERANCE OF JUVENILE SOCKEY SALMON AND ZOOPLANKTON TO THE SELECTIVE SQUAWFISH TOXICANT 1, 1' -METHYLENEDI-2-NAPHTHOL,

International Pacific Salmon Fisheries Commission, New Westminister (British Columbia).
J. M. Johnston.
Prog Fish-Cult. Vol 34, No 3, p 122-125. 1972.
Identifiers: Costia, Diet, Fishery, Juvenile, Methylene-D1-2, Naphthol, Oncorhynchus-Nerka, Plankton, Ptychocheilul-Oregonensis, *Salmon (Sockeye), Squawfish toxicant, Tolerance, Toxicants, Zooplankton.

The selective squawfish toxicant, 1, 1' - methylenedi-2-naphthol, (Squoxin) was tested for its toxicity to 2 races of juvenile sockeye salmon (Oncorhynchus nerka) at various stages of development, and to 10 spp. of zooplankton occurring in their diet. Static 96-hr bioassays were used to determine the highest concentration of Squoxin that was lethal to 0% of the sockeye (LC0) and compared with the lowest concentration that was lethal to 100% (LC100) of northern squawfish (Ptychocheilus oregonensis). The maximum LC0 values for sockeye depended mainly on their stage of development, ranging from 0.09 to 0.10 ppm for alevins with yolk, to as high as 0.90 ppm for fry and fingerlings. Temperatures between 4.4 degrees and 12.8 degrees C. had little influence on sockeye tolerance to Squoxin, but infection with the protozoan Costia reduced the LC0 of fingerlings to 0.30 ppm. No differences werenoted between the 2 races of sockeye studied. The minimim LC100 for northern squawfish varied with temperature, decreasing from 0.10 ppm at 4.4 degrees C to 0.01 ppm at 12.8 degrees C. Comparison of the 2 spp. indicated that newly hatched sockeye alevins would be subjected to lethal concentrations of Squoxin if squawfish eradication were attempted at temperatures of 4.4 degrees C or less. After yolk absorption, sockeye could tolerate 3 to 90 times the Squoxin concentration lethal to squawfish. No mortalities occurred among the zooplankton tested in static bioassays for 7 days at 12.8 degrees C, even at concentrations of Squoxin 10 times greater than needed to kill northern squawfish at the same temperature.--Copyright 1973, Biological Abstracts, Inc.
W73-09062

COMPARATIVE STUDIES OF PLANT GROWTH AND DISTRIBUTION IN RELATION TO WATERLOGGING: VI. THE EFFECT OF MANGANESE ON THE GROWTH OF DUNE AND DUNE SLACK PLANTS,

Trent Univ., Peterborough (Ontario). Dept. of Biology.
For primary bibliographic entry see Field 02I.
W73-09064

THE ECOLOGY OF MORECAMBE BAY: I. INTRODUCTION,

Freshwater Biological Association, Ambleside (England).
For primary bibliographic entry see Field 02L.
W73-09065

DDT INTOXICATION IN RAINBOW TROUT AS AFFECTED BY DIELDRIN,

Bureau of Sport Fisheries and Wildlife, Columbia, Mo. Fish-Pesticide Research Lab.
F. L. Mayer, Jr., J. C. Street, and J. M. Neuhold.
Toxicol Appl Pharmacol. Vol 22, No 3, p 347-354. 1972. Illus.
Identifiers: Adipose, Brain, *DDT, *Dieldrin, Intoxication, Mortality, *Rainbow trout, Tissue, Trout, Water pollution effects.

Rainbow trout were dosed orally with 5 mg DDT alone and in combination with 0.04 mg or 0.20 mg dieldrin on alternate days, 4 treatments in all. Fish were also pretreated on 4 alternate days with 0.20 mg dieldrin/dose followed by 4 alternate day treatments of 5 mg DDT/dose. The presence of dieldrin caused reduced mortality from DDT intoxication. The time until death after demonstrating signs of intoxication was also much longer in fish receiving dieldrin in addition to DDT than in those receiving DDT only. DDT increased with time in adipose tissue, but the lethal brain levels of DDT were nearly constant. Rainbow trout die at a much lower brain level of DDT than reported for rats or birds.--Copyright 1972, Biological Abstracts, Inc.
W73-09072

RETENTION OF 14C-DDT IN CELLULAR FRACTIONS OF VERTEBRATE INSECTICIDE-RESISTANT AND SUSCEPTIBLE FISH,

Middle Tennessee State Univ., Murfreesboro. Dept. of Biology.
M. R. Wells, and J. D. Yarbrough.
Toxicol Appl Pharmacol. Vol 22, No 3, p 409-414. 1972.
Identifiers: *Carbon-14, Cellular, *DDT, Fish, Gambusia-Affinis, Insecticides, *Mosquito-fish, Vertebrates, Water pollution effects.

DDT resistant and susceptible populations of mosquitofish (Gambusia affinis) were assayed for DDT retention in particulate fractions of livers and brains following in vivo and in vitro treatment with 14C-DDT. Livers and brains were homogenized, and particulate fractions were prepared by differential centrifugation and counted for radioactivity. The in vivo studies indicate a significant difference in retention between brains from susceptible and resistant fish and point to a membrane barrier to DDT in tissues from resistant fish. There is a more effective bloodbrain barrier in resistant fish than in susceptible fish. There was a difference in total distribution of DDT based on percent retention in brain tissue of resistant and susceptible fish. The cell membranes of resistant fish retain DDT, thus reducing the amount entering the cell. The opposite effect is seen in the mitochondrial membrane as DDT is apparently prevented from entering rather than being membrane bound. In the in vivo liver treatments the heavy microsome fraction was the only significantly different cell component in which the heavy microsomes from resistant fish retained more 14C-DDT than those from susceptible fish. For the in vitro treatments all liver fractions from susceptible fish showed a significantly greater retention than did those from resistant fish.--Copyright 1972, Biological Abstracts, Inc.

W73-09073

5D. Waste Treatment Processes

OZONE TREATMENT OF WASTE EFFLUENT,
Idaho Univ., Moscow. Water Resources Research Inst.
R. R. Furgason.
Available from the National Technical Information Service as PB-220 008, $3.00 in paper copy, $1.45 in microfiche. Completion Report, April 1973, 14 p, 5 fig, 1 tab, 5 ref. OWRR A-037-IDA (1) 14-01-001-3212.

Descriptors: *Waste water treatment, *Ozones, Oxidation, *Pulp wastes, *Idaho, Odor, Color, Biodegradation, Waste treatment, Water treatment, Chemical oxygen demand, Sulfates, Iron, Manganese.
Identifiers: Ozonation, Sulfate liquor treatment, Moscow (Ida).

A portable ozone test unit was designed and built for use in treating liquid materials with ozone. Due to its powerful oxidation properties, ozone has a great deal of potential as a treatment system for liquid waste materials and water supplies. The portable test unit was constructed so that it could easily be transported to a field site for testing. Several applications used the ozone test unit. The dark brown, odorous waste effluent from a kraft pulp mill could be satisfactorily decolorized and deodorized by the ozone. In addition, the material after ozonation was more biodegradable than before ozonation. The ozone test unit was also used to treat potable water supplies to remove iron and manganese and to reduce off-odors and off-tastes. Tests run on the Moscow, Idaho, water supply demonstrated that ozone could effectively remove high concentrations of both iron and manganese (1 - 5 mg/l Fe and 0.1 - 1 mg/l Mg).
W73-08455

INVESTIGATION OF FACTORS RELATIVE TO GROUNDWATER RECHARGE IN IDAHO,
Idaho Univ., Moscow. Dept. of Hydrology.
R. E. Williams, A. T. Wallace, D. Eier, D. Wallace, and O. Shadid.
Available from the National Technical Information Service as PB-220 009, $3.00 in paper copy, $1.45 in microfiche. Idaho Water Resources Research Institute, Moscow, Completion Report, June 1972. 140 p, 14 fig, 26 ref, 6 append. OWRR A-028-IDA (1), 14-01-001-3212.

Descriptors: *Groundwater movement, *Idaho, Canneries, Sweet corn, *Recharge, Mathematical models, Soil chemical properties, *Water reuse, *Waste water disposal.
Identifiers: Camas Prairie (Idaho), Buhl (Idaho).

This study examined hydrogeologic environments as related to groundwater flow systems, recharge and the reuse of waste waters. Parameters utilized to delineate areas most suitable for waste water reuse include thickness of unconsolidated material, thickness of unsaturated material, regional slope and whether or not the land is being irrigated and farmed. This project has consisted of four parts: (1) An evaluation and application of methods for delineating large areas which can be safely utilized for the terrestrial disposal of waste water; (2) An examination of chemical changes in soil properties resulting from the application of waste water, particularly sweet corn canning process effluent; (3) An investigation of flow system in a mountain valley, including the application of a mathematical model to the groundwater flow system in order to evaluate its contribution to the hydrologic budget of the valley; and (4) Investigation of the effect of a terrestrial disposal operation on soil water and groundwater in the vicinity of a sweet corn canning operation near Buhl, Idaho. (Trihey-Idaho)
W73-08456

USE OF WATER SOFTENING SLUDGE IN SEWAGE TREATMENT,
Kansas Univ., Lawrence. Dept. of Civil Engineering.
W. J. O'Brien, and J. A. Moore.
Available from the National Technical Information Service as PB-220 120, $6.00 paper copy, $1.45 in microfiche. Kansas Water Resources Institute, Manhattan, Contribution No. 108, December 1972, 183 p, 28 fig, 43 tab, 12 append. OWRR A-031-KAN (1), 14-31-0001-3016.

Descriptors: *Waste water treatment, Lime, Sludge, Suspended solids, Costs, *Kansas, Water quality, *Sewage treatment, Economics, *Water softening, Biochemical oxygen demand, Chemical oxygen demand.
Identifiers: *Lime softening sludge, *Lawrence (Kan).

This investigation was conducted to determine whether sludge produced by the excess lime softening process could be used in sewage purification. The study was performed at the water and waste water treatment plants located in Lawrence, Kansas. The chemical composition of the water softening sludge was approximately 85% $CaCO_3$ and 4% $Mg(OH)_2$. About 600 mg (dry weight) of sludge was available for treating each liter of sewage. The average BOD5 of the raw sewage was 290 mg/l. The total COD was 470 mg/l and the total phosphorous was 11.6 mg/l. These values are greater than those found in typical municipal sewage. The total suspended solids were 188 mg/l. This value is slightly lower than expected. Only 19% of the BOD5 and 46% of the suspended solids were removed by conventional primary sedimentation. The addition of softening sludge to a pilot plant consisting of a mixing basin followed by sedimentation with sludge recycle increased the BOD5 removal to 83%, the suspended solids removal to 81%, and the total phosphorous removal to 45%. The optimum concentration of total suspended solids in the mixing chamber was about 2500 mg/l. The annual cost of capital, labor and supplies, for a 10 MGD activated sludge plant using a solids contact basin for primary treatment and vacuum filtration for sludge dewatering was slightly less than the cost of providing separate sewage treatment and softening sludge disposal facilities for Lawrence. Use of the softening sludge for sewage treatment was very economical if additional phosphorous removal was obtained by use of alum, ferric iron, or ferrous iron.
W73-08457

SLUDGE PRODUCTION AND DISPOSAL FOR SMALL, COLD CLIMATE BIO-TREATMENT PLANTS,
Alaska Univ., College. Inst. of Water Resources.
T. Tilsworth.
Available from the National Technical Information Service as PB-219 980, $3.00 in paper copy, $1.45 in microfiche. Completion Report No IWR-32, December 1972. 43 p, 11 fig, 5 tab, 54 ref, 2 append. OWRR A-033-ALAS (2).

Descriptors: *Freezing, *Refrigeration, *Sludge disposal, Ultimate disposal, Activated sludge, Cost analysis, *Dewatering, *Sludge treatment, Waste water treatment.
Identifiers: Natural refrigeration, Package sewage treatment plants, Extended aeration, Freeze-thaw technique, Sand drying beds.

A sludge disposal process consisting of modified sand drying beds in combination with the freeze-thaw technique utilizing natural refrigeration was evaluated. Special consideration was directed at designing a process that was simple to operate and easy to maintain. The purpose of freeze-thawing of sewage sludge is principally to condition the sludge such that it is readily dewaterable and, subsequently, results in a reduced volume of solids to be further processed. The study consisted of three separate evaluations using three model drying beds for each respective run. Two of the beds were studies at sludge depths of six inches, and the third bed was evaluated at an eighteen-inch sludge depth. Settled activated sludge for placement on the model beds was obtained from the sedimentation tank of a local activated sludge plant. Field evaluations were conducted from February to June, 1971. The use of natural refrigeration for conditioning waste activated sludge by the freeze-thaw method was very effective. Freezing of the sludge substantially improved its settleability and dewatering characteristics. The sludge handling characteristics were greatly improved, the sludge being granular and earthy with little to no odor. Design of a prototype operation should consider two operational phases including summertime conventional sand drying and wintertime lagoon freezing. The depth of the sludge to be frozen is dependent on the local climatology. Estimated costs for the process, based on a depth of twelve inches for freezing, compare favorably with the alternatives available for sludge dewatering. Capital and operating costs are estimated at approximately $50 per ton of sludge processed as compared to $70 per ton using vacuum filtration, centrifugation or incineration and $100+ per ton for sand drying beds.
W73-08603

ENVIRONMENTAL ENGINEERING PROGRAMS QUARTERLY TECHNICAL PROGRESS REPORT JANUARY, FEBRUARY, MARCH 1972.
Hanford Engineering Development Lab., Richland, Wash.
For primary bibliographic entry see Field 05B.
W73-08619

ENVIRONMENTAL ENGINEERING PROGRAMS QUARTERLY TECHNICAL PROGRESS REPORT JULY, AUGUST, SEPTEMBER 1971.
Hanford Engineering Development Lab., Richland, Wash.
For primary bibliographic entry see Field 05B.
W73-08620

FROM PHOSPHATE AND ALGAE TO FISH AND FUN WITH WASTE WATER--LANCASTER, CALIFORNIA,
F. D. Dryden, and G. Stern.
In: Report No 24, June 1969, University of Arizona, Tuscon, 'Water Reclamation by Tertiary Treatment Methods,' p 1-40. 12 fig, 2 tab.

Descriptors: *Water reuse, *Water quality standards, *Recreation demand, *Reclaimed water, *California, Oxidation lagoons, Turbidity, Coagulation, Costs, Simulation analysis, Pilot plants, Laboratory tests, Wastewater treatment, Settling basins, Testing procedures, Filtration, Chlorination.
Identifiers: *Lancaster (Calif.).

The most economical method of renovating waste water for use in recreational lakes was determined and pilot studies conducted. Criteria on turbidity, phosphate concentration, pH value, low concentration of COD and BOD, algal counts, and coliform and virus determinations were established. Basic data on the character of the water to be treated were collected. The most effective coagulants and coagulant aids, their required dosages, and the best flocculating conditions were determined in laboratory studies. Various treatment processes were compared and reasonable costs estimated. A practical tertiary process was developed to supply boating and fishing lakes near Lancaster, California. The process selected and piloted features treatment with alum, sedimentation, filtration, and chlorination. Phosphate and algal counts are consistently reduced. Product water has been evaluated in the laboratory and in large simulated lakes stocked with fish. Algae slowly grow to high concentrations in the water but the rate of biological activity does not deplete

the oxygen supply, interfere with fish life, or cause nuisances. Algal growth rates are inhibited at phosphate concentrations less than 0.5 mg/l and almost stop at less than 0.05 mg/l. The ecology of the simulated lake is typical of an oligotrophic fresh water lake. (Jones-Wisconsin)
W73-08645

EFFECTS OF MUNICIPAL WASTE WATER DISPOSAL ON THE FOREST ECOSYSTEM,
Pennsylvania State Univ., University Park. School of Forest Resources.
W. E. Sopper, and L. T. Kardos.
Pennsylvania Institute for Research on Land and Water Resources Reprint Series No 31, 1972. 6 p, 4 photo, 1 tab, 16 ref. (Reprint from Journal of Forestry, Vol 70, No 9, September 1972.) OWRR B-047-PA (1).

Descriptors: *Waste water disposal, *Sprinkler irrigation, *Irrigation effects, *Forests, *Water reuse, Forest management, Water pollution control, Sewage disposal, Recharge, Municipal wastes, Environmental effects.

Diversion of wastewater to the forest ecosystem should help to eliminate or alleviate many disposal and pollution problems. In some instances it might even provide such secondary benefits as increased recharge of groundwater reservoirs, increased growth of vegetation, and amelioration of barren unproductive sites. Satisfactory renovation of municipal waste water was achieved when the sewage effluent was applied at rates of 1, 2, and 4 inches per week in forested areas during the period April to November. Soil samples analyzed indicated no detrimental effects on the chemical status of the upper 5 feet of soil. Irrigation of red pine with 1 inch sewage effluent per week did not significantly increase diameter growth. Two inches per week was detrimental. Height growth of red pine was increased with 1 inch of effluent and decreased with 2 inches per week. Diameter growth of the mixed hardwood species was not affected by the 1-inch per week application but was significantly increased by 2 and 4 inches per week. Height growth of white spruce saplings was significantly increased by irrigation with sewage effluent at the rate of 2 inches per week. Irrigation caused significant changes in the herbaceous ground cover in the old-field area. Height growth, density, and dry matter production were all significantly increased. About 90% of the waste water applied at 2 inches per week was recharged to the groundwater reservoir. Annual recharge averaged 1.6 million gallons per acre during the April to November irrigation period. (Knapp-USGS)
W73-08731

MICROBIAL RESPONSE IN LOW TEMPERATURE WASTE TREATMENT,
Toronto Univ. (Ontario).
For primary bibliographic entry see Field 05B.
W73-08797

WATER POLLUTION CONTROL, A STATE-OF-THE-ART REVIEW,
Port of New York Authority, N.Y.
D. P. Costa.
In:Actual Specifying Engineer, Vol 24, No 1, 1970. p 1-10. Reprint, BIF, Providence, Rhode Island, January 1971.

Descriptors: *Water pollution control, *Reviews, *Sewage treatment, *Treatment facilities, *Watewater treatment, Water pollution, Trickling filters, Activated sludge, Water quality control, Domestic wastes, Industrial wastes, Technology.

Waste water treatment facilities, manpower, and funds have not kept pace with water pollution. Water consumption averages about 155 gallons per capita per day and, because of this high consumption, disposal methods that rely on the capacity of the environment to absorb and dilute pollutants are ineffective in urban areas. Pollution from waste water can be reduced by three methods: dilution, treatment, or the creation of less potentially harmful material. Dilution is rarely a feasible solution because waste sources are concentrated and large volumes are present. Treatment plants usually provide primary and/or secondary treatment. Primary treatment consists of mechanical separation to remove the larger, floating, suspended and undissolved particles. Such processes remove only visible pollutants and particulate matter. Reliance on primary treatment alone can result in pollution of the receiving waters. Secondary treatment employs biochemical processes to remove pollution. Bacteria react with the organic matter through absorption, digestion, oxidation, assimilation or decomposition. The processes most commonly used are (1) trickling filter and (2) activated sludge. Trickling filters are stone beds onto which primary treated waters are sprinkled. The seeping water comes into contact with biological growth on the stones and, in the presence of oxygen, receives treatment. Activated sludge accomplishes the same task in a completely wet system, with the primary effluent entering a tank with sludge from a settling basin. The sludge is rich with biological growths and provides treatment to the incoming waters. Schematic sketches are included to illustrate processes. (Poertner)
W73-08900

DESAL ION EXCHANGE FOR DEMINERALIZATION AT SANTEE, CALIFORNIA,
Santee County Water District, Calif.
H. Filar, Jr.
Available from the National Technical Information Service as PB-220 123, $3.00 in paper copy, $1.45 in microfiche. Environmental Protection Agency Report EPA-R2-73-239, May 1973. 117 p, 19 fig, 14 tab, 7 ref, 5 append. EPA Project 17080 EDV. WPRD 5-01-67.

Descriptors: *Ion exchange, *Demineralization, *Desalination, Water purification, *Resins, Anion exchange, Cation exchange, Potable water, Water quality, Tertiary treatment, *California, Cost analysis, Waste water treatment.
Identifiers: *Desal ion exchange process, Santee (Calif).

A 50,000 gpd Desal ion exchange pilot plant for the demineralization of lime treated, dual media filtered, tertiary effluent has been built and tested at Santee, California. The plant removed up to 86 percent of the total dissolved solids (TDS) as CaCo3. Influent TDS ranged from 700 to 1000 mg/l during the two month operating period (All TDS values were the sum of the chloride, sulfate, and alkalinity concentrations as CaCO3). Anion resin (45 cu ft of IRA-68) capacities of 11.4 and 9.7 Kgr (as CaCO3)/cu ft were obtained with a tertiary effluent flow of 1 gal./cu ft/min and an operating pressure of 40 psig. The system was designed to demineralize a portion of the influent stream and blend it with influent water to make a product TDS equal to or better than the Colorado River drinking water supplied to Santee. Because of the short operating time, emphasis was placed on the demineralization process, especially the anion IRA-68 resin performance. The estimated total cost, based on Santee, for 64,370 gpd of 500 mg/l (as CaCO3) blended effluent is $1.00 per 1000 gallons. This estimate calls for activated carbon adsorption influent, two complete operating cycles per day, and a properly functioning blending system. (See also W73-08976) (EPA)
W73-08975

CARBON ADSORPTION AND ELECTRODIALYSIS FOR DEMINERALIZATION AT SANTEE, CALIFORNIA,
Santee County Water District, Calif.
H. Filar, Jr.
Available from the National Technical Information Service as PB-220 360, $6.00 in paper copy, $1.45 in microfiche. Environmental Protection Agency, Report EPA-R2-73-240, May 1973. 437 p[illegible], 37 fig, 34 tab, 20 ref, 5 append. EPA Project 170[illegible] FKG. 14-12-444.

Descriptors: *Demineralization, *Desalination, *Electrodialysis, *Activated carbon, Carbon, Membranes, Membrane process, Dissolved solids, Pilot plants, Filtration, Fouling, Reclaimed water, Potable water, Tertiary treatment, *California, Cost analysis, Waste water treatment.
Identifiers: Santee (Calif), Salt removal, *Carbon adsorption.

A 100,000 gpd activated carbon adsorption pilot plant followed by a 50,000 gpd electrodialysis pilot plant for the demineralization of lime treated tertiary effluent, has been built and tested at Santee, California. The average total chemical oxygen demand (COD) removal, by the entire carbon column, was 57 percent with an average 30.9 mg/l influent and an average 13.4 mg/l effluent concentration. The average dissolved COD removal was 58 percent with an average 26.9 mg/l influent and an average 11.4 mg/l effluent concentration. The cost of the 100,000 gpd activated carbon plant, including the replacement of expended carbon with virgin activated carbon, was 35.9 cents per 1000 gallons of product. The electrodialysis pilot plant produced water near the Federal Public Health Service Drinking Water Standards. Typical operation on clean membranes was 60 percent removal of the total dissolved solids with an 1135 mg/l influent and a 455 mg/l effluent concentration. Sulfuric acid flushes, enzymatic detergent flushes, and plant shutdowns were effective in retarding the organic (fouling) and inorganic (scaling) buildups on the membrane surfaces. An estimate for a 1.5 MGD electrodialysis plant, excluding brine disposal costs, is 28.6 cents per 1000 gallons of product. (See also W73-08975) (EPA)
W73-08976

FINAL ENVIRONMENTAL STATEMENT RELATED TO OPERATION OF CALVERT CLIFFS NUCLEAR POWER PLANT UNITS 1 AND 2.
Directorate of Licensing (AEC), Washington, D.C.
For primary bibliographic entry see Field 05B.
W73-08993

FINAL ENVIRONMENTAL STATEMENT FOR THE EDWIN I. HATCH NUCLEAR PLANT UNITS 1 AND 2.
Directorate of Licensing (AEC), Washington, D.C.
For primary bibliographic entry see Field 05B.
W73-08994

FINAL ENVIRONMENTAL STATEMENT RELATED TO OPERATION OF SHOREHAM NUCLEAR POWER STATION.
Directorate of Licensing (AEC), Washington, D.C.
For primary bibliographic entry see Field 05B.
W73-08995

FINAL ENVIRONMENTAL STATEMENT RELATED TO OPERATION OF THREE MILE ISLAND NUCLEAR STATION UNITS 1 AND 2.
Directorate of Licensing (AEC), Washington, D.C.
For primary bibliographic entry see Field 05B.
W73-08996

FINAL ENVIRONMENTAL STATEMENT RELATED TO THE LA SALLE COUNTY NUCLEAR STATION.
Directorate of Licensing (AEC), Washington, D.C.
For primary bibliographic entry see Field 05B.

A mathematical formulation employing the kinetic model of Willimon and Andrews is used to simulate conventional and contact anaerobic digestion processes consisting of two stages. Presented is a system of two completely mixed digesters in which the process is carried out without recycle (conventional) and with recycle (contact). Performance equations of the system are developed by means of mass balance. A unified analysis of the steady states of the system is made because of the possible existence of multiple steady states. Two specific steady-state operations--normal steady-state and washout steady-state--are considered in detail. The critical flow rates that cause washout of some or all species are investigated. To further clarify the results of the steady-state and washout analyses, numerical simulations are carried out. Finally, the optimal design policy for a two-stage continuous anaerobic digester system is determined by joining an economic model to the process model. The optimization problem is nonlinear and is performed using the simplex search technique. Optimization is carried out by constructing an economic objective function and the optimal first- and second-stage mean holding times are obtained. Increases in recycle ratio decrease treatment cost and improve efficiency in an optimum system. (Bell-Cornell)
W73-09007

NEW TECHNIQUES IN WATER POLLUTION CONTROL,
House, Washington, D.C.
For primary bibliographic entry see Field 06E.
W73-09014

AEROSPACE VEHICLE WATER-WASTE MANAGEMENT,
National Aeronautics and Space Administration, Washington, D.C.
For primary bibliographic entry see Field 05G.
W73-09017

5E. Ultimate Disposal of Wastes

BENTHAL OXYGEN DEMANDS AND LEACHING RATES OF TREATED SLUDGES,
Manhattan Coll., Bronx, N.Y.
For primary bibliographic entry see Field 05C.
W73-08527

SLUDGE PRODUCTION AND DISPOSAL FOR SMALL, COLD CLIMATE BIO-TREATMENT PLANTS,
Alaska Univ., College. Inst. of Water Resources.
For primary bibliographic entry see Field 05D.
W73-08603

5F. Water Treatment and Quality Alteration

OZONE TREATMENT OF WASTE EFFLUENT,
Idaho Univ., Moscow. Water Resources Research Inst.
For primary bibliographic entry see Field 05D.
W73-08455

USE OF WATER SOFTENING SLUDGE IN SEWAGE TREATMENT,
Kansas Univ., Lawrence. Dept. of Civil Engineering.
For primary bibliographic entry see Field 05D.
W73-08457

PUBLIC WATER SUPPLY SYSTEMS RULES AND REGULATIONS.
Illinois State Dept. of Health, Springfield. Div. of Sanitary Engineering.
For primary bibliographic entry see Field 06E.
W73-08893

PRESENT DAY ENGINEERING CONSIDERATIONS AS APPLIED TO WATER LINE DESIGN,
Woolpert (Ralph L.) Co., Dayton, Ohio.
For primary bibliographic entry see Field 08A.
W73-08896

HYGIENIC EVALUATION OF NEW POLYMERS USED IN THE DRINKING WATER SUPPLY SYSTEMS, (IN RUSSIAN),
Vsesoyuznyi Nauchno-Issledovatelskii Institut Gigieni i Toksikologii Pestitsidov, Kiev (USSR).
For primary bibliographic entry see Field 05C.
W73-09027

POLLUTION OF DRINKING WATER AND INFECTIONS BY PSEUDOMONAS AERUGINOSA IN A PEDIATRIC SECTION,
Siena Univ. (Italy). Istituto di Clinica Pediatria.
For primary bibliographic entry see Field 05C.
W73-09035

HYGIENIC PROBLEMS ARISING DUE TO DISCHARGE OF WASTE WATERS OF WOOD-WORKING PLANTS INTO OPEN WATERS, (IN RUSSIAN),
Institutul de Medicina si Farmacia, Iasi (Rumania).
For primary bibliographic entry see Field 05C.
W73-09036

5G. Water Quality Control

AN EXAMINATION OF THE ECONOMIC IMPACT OF POLLUTION CONTROL UPON GEORGIA'S WATER-USING INDUSTRIES,
Georgia Inst. of Tech., Atlanta. Engineering Experiment Station.
W. G. Dodson, and R. B. Cassell.
Available from the National Technical Information Service as PB-220 006, $3.00 in paper copy, $1.45 in microfiche. Georgia Environmental Resources Center Report No. ERC-0173, February 1973, 43 p, 8 tab, 114 ref, 5 append. OWRR B-064-GA (1).

Descriptors: *Industrial wastes, Food processing industry, Pulp and paper industry, Textile industry, Kaolin industry, Electric power industry, Institutions, *Water quality standards, *Georgia, *Attitudes, *Economic impact, Administrative agencies.
Identifiers: *Water quality management.

The study used interviews of management personnel in 28 companies and of state and Federal water pollution control officials to assess the economic impact of existing water pollution control programs upon Georgia's large water-using industries including pulp and paper manufacturing, textiles, food processing, kaolin mining and processing, and electric power generation. The goals were to evaluate industry acceptance of and support for water pollution control programs to propose methods for improving such acceptance. The interviews revealed extensive voluntary cooperation between government and industry in addressing pollution problems. Companies believed the basic legislation to be fair even though they reported profit losses, but they perceived a failure of officials to consider differing waste characteristics and were concerned by pressures by environmentalists to legislate unrealistic standards. Officials were concerned over the lack of available information for them to use in assessing the economic impact of alternative control policies. Community leaders were unaware of the effects of pollution control programs on the ability of a community to attract industry. Many companies have difficulty in passing pollution control costs on to the consumer, and small businesses particularly need technical and financial help. More effective com-

munications channels between enforcement officials and industry are also important and should be used to develop improved predictions of the effects of water quality management programs on local economies. (James-Georgia)
W73-08453

STREAM POLLUTION FROM CATTLE BARNLOT (FEEDLOT) RUNOFF,
Ohio State Univ., Columbus. Dept. of Agricultural Engineering.
For primary bibliographic entry see Field 05B.
W73-08459

EFFECTS OF LONG CHAIN POLYMERS ON THE SIZE DISTRIBUTION OF OIL-IN-WATER EMULSIONS,
Brown Univ., Providence, R.I.
R. I. Tanner, and R. W. Fisk.
Available from the National Technical Information Service as PB-220 011, $3.00 in paper copy, $1.45 in microfiche. R. I. Water Resources Center Completion Report, (1973). 13 p, 7 fig, 1 ref. OWRR A-043-RI (2).

Descriptors: *Polymers, *Emulsions, *Oil-water interfaces, *Particle size, Water pollution control.
Identifiers: Long chain polymers, *Coulter counter, Polyethylene oxide, Separan, Cetane.

The effects of long chain, turbulent drag reducing polymers on oil-in-water emulsions were studied using the Coulter Counter. Both 50 ppm polyethylene oxide and 500 ppm Separan MGL proved to reduce particle counts while 500 ppm polyethylene oxide had the adverse effect of increasing the density. Solutions of 4.4% polyisobutylene in cetane and 50 ppm Separan MGL had no noticeable effect on the size distribution. The inflection point diameter could not be found in each emulsion. Therefore it was recommended that future testing be continued investigating smaller particles through the use of a smaller (30 micro) aperture tube.
W73-08460

USE OF ADVANCED WATER RESOURCES PLANNING TECHNIQUES IN THE DEVELOPMENT OF REGIONAL WATER QUALITY MANAGEMENT PROGRAMS,
Clemson Univ., S.C. Dept. of Environmental Systems Engineering.
B. C. Dysart, III.
Available from the National Technical Information Service as PB-220 121, $9.00 in paper copy, $1.45 in microfiche. Report No. 34, South Carolina Water Resources Research Institute, Clemson, S.C. December 1972. 319 p, 62 fig, 19 tab, 57 ref, 5 append. OWRR B-017-SC (8), 14-31-0001-3127.

Descriptors: *Planning, *Water management (Applied), *Water quality control, *Decision making, *Mathematical models, Dynamic programming, Institutinal constraints, Economic efficiency, Hydrologic models, Public opinion, Algal nutrients, Eutrophication, Social aspects, Interbasin water transfers, *South Carolina.
Identifiers: *Greenville (SC), *Reedy River (SC), *Public participation, Alternate management plans, Lake Greenwood (SC).

Results are presented of investigations of water resources planning and management techniques of four types. The first section considered basinwide water quality modeling from a practical viewpoint. Several alternate management policies, e.g., an optimal policy based upon dynamic programming and a uniform treatment policy, were derived for a basin. The economic and water quality ramifications were presented for use by dischargers and regulatory agencies in selecting the best policy. Individual treatment as well as regional plants and multiple outfalls were considered. The second section dealt with nutrient and eutrophication problems in the Reedy River basin and a reservoir. Sources, effects, and possible remedial measures were proposed. The third section presented the development of an improved water yield model and application to the upper Reedy River basin and the Greenville, S.C. area. The utility of the model in describing or predicting effects of interbasin diversions on the hydrology of an area was presented. Such information is necessary for water quality management planning. The model has as inputs 5-day precipitation and produces runoff for 5-day periods. The final section reported on an investigation of public preferences for water quality in the Greenville, S.C. area. The public desired improved water quality and were willing to pay a rather substantial amount to achieve this.
W73-08461

AQUATIC FUNGI OF THE LOTIC ENVIRONMENT AND THEIR ROLE IN STREAM PURIFICATION,
Virginia Polytechnic Inst. and State Univ., Blacksburg. Dept. of Biology.
For primary bibliographic entry see Field 05C.
W73-08462

HYDRAULIC AND SEDIMENT TRANSPORT STUDIES IN RELATION TO RIVER SEDIMENT CONTROL AND SOLID WASTE POLLUTION AND ECONOMIC USE OF THE BY-PRODUCTS,
Kentucky Water Resources Inst., Lexington.
For primary bibliographic entry see Field 02J.
W73-08468

SOME DESIGNS AND ANALYSES FOR TEMPORALLY INDEPENDENT EXPERIMENTS INVOLVING CORRELATED BIVARIATE RESPONSES,
California Polytechnic State Univ., San Luis Obispo. Dept. of Computer Science and Statistics.
For primary bibliographic entry see Field 04A.
W73-08546

AIR COOLED HEAT EXCHANGERS-NEW GROWTH,
T. C. Elliott.
Power, Vol 115, No 8, p 88-90, August 1971. 3 fig.

Descriptors: *Heat exchangers, *Cooling, *Economics, Environmental effects, Thermal powerplants, Thermal pollution, Electric power demand, Temperature, Corrosion, Common control, Economic justification, Refrigeration.
Identifiers: *Air cooling, Natural draft, Mechanical draft.

Air cooling is discussed from the standpoint of economics and its environmental attractiveness. The biggest disadvantage is cost caused by the outsize system needed to offset the poor heat transfer capabilities of air. Air-cooled heat exchangers cost from 15-37 $/KW (mechanical draft to 25-65 $/KW (Natural draft) as compared to evaporative cooling towers 4-14 $/KW (mechanical draft) or 6-20 $/KW (Natural draft). Because of sensible heat transfer, dry bulb temperature may limit cooling capability. Air-side fouling is negligible, however, when compared to problems of scale and corrosion in water cooling systems. Air is the best choice when the overall heat transfer coefficient is low, process fluid temperatures are high and water usage is costly, especially in the western part of the United States. (Oleszkiewicz-Vanderbilt)
W73-08611

THE FLOURISHING DISTRICT HEATING BUSINESS,
For primary bibliographic entry see Field 06D.
W73-08612

UNDERGROUND NUCLEAR POWER PLA[illegible] SITING,
Aerospace Corp., San Bernardino, Calif.
For primary bibliographic entry see Field 06G.
W73-08617

UTILITY BATTLES: YOU CAN'T WIN '[illegible] ALL.

Electrical World, Vol 176, No 7, p 27-28, Oct 19[illegible] 1 fig.

Descriptors: *Thermal powerplants, *Thern[illegible] pollution, *Regulation, *Nuclear powerplan[illegible] *Permits, Electric power production, Envir[illegible] mental effects, Legal aspects, Administrat[illegible] agencies, Control systems, Management, Wa[illegible] policy, Water treatment, Heated water, Judic[illegible] decisions.
Identifiers: Atomic Energy Commission.

Strict licensing criteria set up by the Atomic En[illegible] gy Commission threaten to slow down or suspe[illegible] construction schedules for some 51 nuclear plan[illegible] A test suit between the Interior Department a[illegible] the Florida Power and Light Company failed [illegible] clearly resolve whether heated water could be co[illegible] sidered under the 1899 Refuse Act but, FPLC h[illegible] been required to build 165-185 miles of paral[illegible] canals in order to operate under conditions whi[illegible] are not harmful to the environment. A D.C. A[illegible] peals Court directive has instructed the AEC [illegible] consider the environmental impact when licensi[illegible] new facilities. This may result in further delays [illegible] power plant operation which could have serio[illegible] consequences for the national power supply. T[illegible] AEC's only concession was to allow plants who[illegible] hearings for operating licenses were already pen[illegible] ing to operate at 20% of full power. Concern h[illegible] been expressed over 11 plants with approximate[illegible] 7500 Mw potential which are nearing completic[illegible] (Jerome-Vanderbilt)
W73-08622

RIVER JET DIFFUSER PLANNED.

Electrical World, Vol 176, No 5, p 25, Septemb[illegible] 1, 1971. 1 photo.

Descriptors: *Thermal pollution, *Dischar[illegible] (Water), *Heated water, *Jets, *Regulation, Ele[illegible] tric power production, Hydraulic engineering, E[illegible] fluents, Environmental effects, *Mississip[illegible] River, Design specifications, Structural desig[illegible] Standards, Fish barriers, Temperature, *Illinois.

The plans of Commonwealth Edison Company [illegible] build a $6 million multi-jet diffuser in the river b[illegible] of the Mississippi River at Cordova, Illinois, a[illegible] discussed. The proposed system would invol[illegible] two pipes extending 1,180 ft. across with orific[illegible] at 20 ft. intervals. It would discharge 2,270 c[illegible] ft/sec. of effluent and 12 billion B.T.U.'s/hr. [illegible] heat into average river flow of 45,000 cu. ft./se[illegible] The high velocity of discharge would me[illegible] complete mixing within 40 ft. with a temperatu[illegible] increase of less than 4.3 F. Extensive testing a[illegible] design studies have been completed, but approv[illegible] must be obtained from the Environmental Prote[illegible] tion Agency and state agencies. Questions ha[illegible] been raised about the true mixing zone distance, [illegible] request for temporary exclusion from month[illegible] maximum temperature regulations, and the pos[illegible] bility that the diffuser will present a barrier to fi[illegible] passage. (Jerome-Vanderbilt)
W73-08631

A SLOW AND PAINFUL PROGRESS TOWARD MEETING OUR WATER POLLUTION CO[illegible] TROL GOALS,
For primary bibliographic entry see Field 06E.
W73-08650

ing concentration of pollutants within the groundwater domain, so that one can formulate control procedures to prevent degradation of both the possible source of potable water and the soil matrix which affects this water. After a brief introduction to groundwater basins, types of groundwater pollutants and how they penetrate the groundwater environment are discussed. Next, five controls for groundwater pollution are described: (1) controls supplied by the natural environment of the basin; (2) controls implemented through logic in knowing the imminent dangers present in the environment; (3) controls eliminating or reducing the pollutant at its source; (4) in situ controls for pollution above the tolerable toxic limit; and (5) computer simulation. Finally, two prototype systems are presented and analyzed in detail, illustrating cases where knowledge of concentration of a pollutant is vital for prevention of groundwater degradation. The Indus Valley and the Oxnard Plain in California are used as case examples. (Bell-Cornell)
W73-08660

CRUDE OIL AND NATURAL GAS PRODUCTION IN NAVIGABLE WATERS ALONG THE TEXAS COAST (FINAL ENVIRONMENTAL IMPACT STATEMENT).
Army Engineer District, Galveston, Tex.

Available from National Technical Information Service, U.S. Dept. of Commerce as EIS-TX-72-5069-F, $12.50 paper copy, $1.45 microfiche. August 9, 1972. 198 p, 1 plate.

Descriptors: *Permits, *Drilling, *Oil industry, *Texas, *Environmental effects, Pipelines, Turbidity, Oil spills, Estuaries, Dredging, Drilling equipment, Recreation, Navigation, Navigable waters, Aesthetics, Lagoons, Land subsidence.
Identifiers: *Environmental Impact Statements, Coastal waters.

This action concerns the continued issuance of permits for erection of structures and construction of ancillary facilities associated with exploration for and production of crude petroleum and natural gas within the coastal waters, lagoons and estuaries of Texas, subject to imposition of certain additional limitations and restrictions based on environmental considerations. The prospective drilling will have only a minimal impact on the environment if associated discharges are properly controlled. The placing of pipelines, dredging and spoiling may harm the marine habitat, create water turbidity and siltation and damage archaeological sites in the affected areas. It is unlikely that large scale production potential will be encountered in these shallow formations. Additionally large concentrations of well structures may pose a navigation hazard and detract from the aesthetic value of the seascape. (Smith-Adam-Florida)
W73-08661

WATER QUALITY IN A STRESSED ENVIRONMENT, READINGS IN ENVIRONMENTAL HYDROLOGY.

Burgess Publishing Company, Minneapolis, Minnesota, 1972. W.A. Pettyjohn, editor. 309 p, 19 fig, 16 map, 19 tab, 11 chart, 63 ref.

Descriptors: *Water pollution, *Water pollution control, *Surface-groundwater relationships, *Water law, Surface waters, Ground waters, Hydrology, Water resources, Water conservation, Water consumption, Water control, Water demand, Water rights, Governments, Judicial decisions, Legal aspects, Water pollution effects, Trace elements, Water pollution sources, Water sources, Water supply, Water utilization, Water management (Applied).
Identifiers: Hazardous substances (Pollution), U.S. Army Corps of Engineers.

The selected papers deal primarily with groundwater and surface water contamination in an attempt to point out a few significant occurrences in water pollution that have been reported over the last twenty-five years. The book is divided into six major headings: (1) The Water We Drink, (2) Sources of Surface-Water Pollution, (3) Geologic Controls and Ground-Water Pollution, (4) Examples of Ground-Water Pollution, (5) Trace Elements--A New Factor in Water Pollution, and (6) Water Pollution and Legal Controls. The readings have been drawn mainly from scientific journals and governmental publications. The three articles on water law have wide application in environmental problems including legal approaches to water rights, enforcement of the 1899 Refuse Act through citizen action, and the role of the Army Corps of Engineers in preventing the destruction and pollution of our waters and wetlands. An epilogue looks into the future of environmental monitoring. (Dunham-Florida)
W73-08662

RIVER ECOLOGY AND MAN.
For primary bibliographic entry see Field 06G.
W73-08664

HEARING FOR THE REVIEW AND REVISION OF MISSOURI'S WATER QUALITY STANDARDS.
Missouri Clean Water Commission, Jefferson City.

March 21, 1973. 15 p.

Descriptors: *Water quality standards, *Federal Water Pollution Control Act, *Water pollution control, *Missouri, Environmental effects, Water pollution, Administrative agencies, Governmental interrelations, Water sources, Legislation, Administrative decisions, Regulations, Water management (Applied), Water quality, Public health, Water utilization, Flow, Streams, Lakes, Rivers, Ecology, Institutions, State governments.
Identifiers: Missouri Clean Water Law.

In an attempt to comply with the Federal Water Pollution Control Act of 1972, the following proposed revisions and changes to Missouri's existing Water Quality Standards are made. The water quality characteristics of Missouri's streams will be published and referenced in the introduction. Existing statutory authority and approval standards will be revised to reflect the authority granted in the Missouri Clean Water Law. Additions or deletions resulting from a hearing will be made to water uses and stream classification. Streams and lakes will be classified according to their use as provided in an attached use classification. Water quality criteria will be revised in relation to general criteria, water quality, flow, stream sampling and analytical testing. Programs to control water pollution will be summarized including a list of regulations to be promulgated by the Clean Water Commission. The Water Quality Surveillance Program will replace present compliance and time schedules and will be published as a separate document. (Dunham-Florida)
W73-08668

WATER SUPPLIES.

Rules--State of Florida Department of Health and Rehabilitative Services, Division of Health, ch. 10D-4 secs 4.01 thru 4.15 (1971).

Descriptors: *Florida, *Water quality control, *Water supply, *Potable water, Water purification, Legal aspects, Regulation, Water utilization, Public health, Municipal water, Water demand, Wells, Well permits, Well regulations.

Rules concerning water supplies as promulgated by the State of Florida Department of Health and

Rehabilitative Services are presented. The rules cover such areas as water supply in general, potable water standards, approval of public water supply systems, application for well drilling permits, abandonment of water supply wells, potable water in places serving the public, location and construction of water supply wells, and fluoridation of public water supply systems. Under the rules, a public water supply system is defined as a system serving more than twenty-five persons or otherwise making water available to public groupings or the public in general; including works and auxiliaries for collection, treatment, storage, and distribution of water from a source or sources of supply to the free-flowing outlet of the ultimate consumer. Graphic, charts, chemical analysis techniques, and other pertinent information are included. (Mockler-Florida)
W73-08670

HYDROPNEUMATIC STORAGE FACILITIES.
Illinois State Environmental Protection Agency, Springfield. Div. of Public Water Supplies.

Technical Release No. 10-8, (1972). 2 p.

Descriptors: *Illinois, *Water storage, *Pumped storage, *Storage capacity, Storage, Storage tanks, Surplus water, Water distribution (Applied), Water supply, Water tanks, Water shortage, Water supply development, Flow rates, Flow measurement, Pumps, Pumping.
Identifiers: Illinois Environmental Protection Act, Illinois Public Water Supply Control Law.

Following the enactment of the Illinois Public Water Supply Control Law of 1951 and the Environmental Protection Act, many existing subdivision water supplies are being classified as public water supplies. Many subdivisions must develop their own water supply facilities, and in many of these supplies, hydropneumatic storage facilities exist or are proposed. A study was done on these tanks, and a total gross volume of 35 gallons per capita was recommended as the optimum size. The maximum flow rate was calculated to be 62.5 g.p.m. in a water system consuming 15,000 gallons a day. An air compressor of proper size should be provided to maintain the air cushion in the pressure tank. Finally, the actual capacity of the well pump should be greater than the maximum hourly rate of consumption. (Glickman-Florida)
W73-08671

LIMESTONE POLICY.
Illinois State Environmental Protection Agency, Springfield. Div. of Public Water Supplies.

Technical Release No. 10-6, (1972). 2 p.

Descriptors: *Groundwater resources, *Illinois, *Chlorination, *Water supply treatment, Limestones, Supply treatment, Water purification, Water quality control, Water treatment, Sewage treatment, Water pollution sources, Water quality, Dependable supply, Groundwater, Water table aquifers, Water wells, Water resources, Groundwater basins.

Certain standards are recommended in protecting the safety of water supplies in the Illinois limestone areas. In heavily populated areas, a fifty foot thickness of glacial drift must exist over the bedrock limestone before the water should be used without chlorination. In rural areas, only thirty feet is necessary between the bedrock and the lowest points of pollution. A well may be in danger of contamination if it draws water from a limestone bed which crops out less than one fourth of a mile away. Where it is evident that the limestone supply is contaminated with sewage, use of the water will not be considered satisfactory unless it is treated. If water is obtained from a supply completely below the limestone, it should be chlorinated before use. (Glickman-Florida)
W73-08672

AN ANALYSIS OF ENVIRONMENTAL STATEMENTS FOR CORPS OF ENGINEERS WATER PROJECTS,
Stanford Univ., Calif. Dept. of Civil Engineering.
For primary bibliographic entry see Field 06G.
W73-08673

LEGISLATIVE NEEDS IN WATER RESOURCES MANAGEMENT IN ALABAMA,
Alabama Univ., University. Natural Resources Center.
For primary bibliographic entry see Field 06E.
W73-08674

CALIFORNIA LAWYERS LEAD ENVIRONMENTAL FIGHT,
For primary bibliographic entry see Field 06E.
W73-08675

NEW ENGLAND RIVER BASINS COMMISSION--1972 ANNUAL REPORT.
New England River Basins Commission, Boston, Mass.
For primary bibliographic entry see Field 06E.
W73-08676

CITIZEN ALERT: PUBLIC PARTICIPATION IN THE FEDERAL WATER POLLUTION CONTROL ACT AMENDMENTS OF 1972.
Natural Resources Defense Council, Washington, D.C.

Project on Clean Water, 1972. 6 p.

Descriptors: *Water pollution control, *Legislation, *Water quality control, *Federal Water Pollution Control Act, Administrative agencies, Decision making, Legal aspects, Governments, Planning, Standards, Water policy, Administrative decisions, Law enforcement, Permits, Adjudication procedure.
Identifiers: *Federal Water Pollution Control Act Amendments of 1972.

The recently enacted Federal Water Pollution Control Act Amendments of 1972 establish a broad policy of public participation in all aspects of its implementation. This newsletter describes the measures proposed by the federal government regarding public participation and presents a section-by-section analysis of the proposed regulations implementing the Amendments. Since these regulations and the laws they are designed to implement have potentially far-reaching implications, it is vital that the public take advantage of this opportunity to effect the degree and form of public participation in the national effort to clean up public waterways. It is indicated that the proposed guidelines contain inadequate provisions on public participation in enforcement and legal proceedings. This newsletter also recommends changes to the proposed regulations and presents a detailed analysis of such recommendations. Finally, all interested citizens are urged to write the Environmental Protection Agency to encourage adoption of these recommendations. (Mockler-Florida)
W73-08677

PUBLIC PARTICIPATION IN WATER POLLUTION CONTROL PROGRAMS.
For primary bibliographic entry see Field 06E.
W73-08678

A VETERAN VIEWS THE POLLUTION FIGHT,
V. J. Yannacone, Jr.
Juris Doctor, p 28-29, February 1973. 2 p, 1 photo.

Descriptors: *Pollution abatement, *Administrative agencies, *Judicial decisions, *Public heal[illegible] Legal aspects, Legislation, Administrative de[illegible] sions, Regulation, Institutions, Public rights, C[illegible] stitutional law, Jurisdiction, Penalties (Legal), [illegible] vironmental effects.
Identifiers: National Environmental Policy Act.

The fundamental environmental issues are bec[illegible] ing lost in a maze of procedural legal formali[illegible] such as standing, jurisdiction and existence o[illegible] cause of action. If these trends continue [illegible] general public will be the eventual loser. The N[illegible] tional Environmental Policy Act cannot assure [illegible] vironmental quality. In fact the law can't solve [illegible] environmental problems because it in part crea[illegible] them. As medical evidence establishes the dam[illegible] ing effects of environmental degradation on l[illegible] and health, environmental personal injury actio[illegible] will increase and class actions seeking mon[illegible] damages will become more common. These cl[illegible] actions will also recognize that environmen[illegible] rights are constitutionally protected. Lawy[illegible] must master a complex scientific background [illegible] the issues to be tried. The vast bulk of enviro[illegible] mental litigation will concern administrative [illegible] tion, but current administrative agencies are [illegible] capable of considering environmental matters w[illegible] the requisite degree of ecological sophisticatio[illegible] Thus the goal of all environmental lawyers shou[illegible] be the drafting of responsible and feasible legis[illegible] tion. (Glickman-Florida)
W73-08679

AQUATIC PLANT CONTROL AND ERADIC[illegible] TION PROGRAM, STATE OF TEXAS (FIN[illegible] ENVIRONMENTAL IMPACT STATEMENT).
Army Engineer District, Galveston, Tex.
For primary bibliographic entry see Field 04A.
W73-08683

UNITED STATES V. GRANITE STA[illegible] PACKING COMPANY (VIOLATION [illegible] REFUSE ACT OF 1899).
For primary bibliographic entry see Field 06E.
W73-08695

BANKERS LIFE AND CASUALTY COMPAN[illegible] V. VILLAGE OF NORTH PALM BEAC[illegible] FLORIDA (SUIT BY RIPARIAN LANDOWNE[illegible] TO COMPEL CORPS OF ENGINEERS T[illegible] RENEW DREDGE AND FILL PERMITS).
For primary bibliographic entry see Field 06E.
W73-08696

COWELL V. COMMONWEALTH [illegible] PENNSYLVANIA DEPARTMENT OF TRAN[illegible] PORTATION (CHALLENGE TO CONDEMN[illegible] TION PROCEEDINGS INSTITUTED TO PR[illegible] VIDE CHANNEL IMPROVEMENT).
For primary bibliographic entry see Field 06E.
W73-08697

CHARLES COUNTY SANITARY DISTRIC[illegible] INC. V. CHARLES UTILITIES, INC. (ACTIO[illegible] BY WATER AND SEWER CORPORATIO[illegible] SEEKING JUDICIAL VALIDATION OF [illegible] FRANCHISE).
For primary bibliographic entry see Field 06E.
W73-08699

SOIL AND WATER CONSERVATION COMMI[illegible] SION AND DISTRICTS.
For primary bibliographic entry see Field 06E.
W73-08704

BOOTH V. CORN PRODUCTS CO. (TA[illegible] PAYER'S SUIT FOR RECOVERY OF COSTS O[illegible] TREATING EXCESS WASTES).
For primary bibliographic entry see Field 06E.
W73-08705

segregated by the foam. Even in the absence of foam, a bubble fractionation process may occur in the liquid, the process yielding a concentration gradient. Besides chemical factors, both processes are influenced by hydrodynamic and diffusional factors. (Knapp-USGS)
W73-08861

IMPACT OF HIGHWAYS ON SURFACE WATERWAYS,
Metropolitan Sanitary District of Greater Chicago, Ill.
For primary bibliographic entry see Field 04C.
W73-08894

INDUSTRIAL WASTE MANUAL, A GUIDE FOR THE PREPARATION OF APPLICATIONS, REPORTS AND PLANS.
Pennsylvania Dept. of Environmental Resources, Harrisburg. Bureau of Water Quality Management.
For primary bibliographic entry see Field 06E.
W73-08895

TREATMENT OF URBAN RUNOFF.,
Environmental Protection Agency, Washington, D.C. Municipal Pollution Control Section.
F. J. Condon.
Paper presented at the American Public Works Association Congress and Equipment Show, Minneapolis, Minnesota, September 21-27, 1972. 18 p, 27 ref.

Descriptors: *Urban drainage, *Water pollution, *Water pollution treatment, *Storm runoff, Water pollution sources, Urban runoff, Urban hydrology, Storm water, Erosion, Treatment.

Urban runoff has been found to be an important factor in producing water pollution. Control of this pollution is necessary to protect our water resources. A great deal of technology exists to deal with the problem of combined sewer overflow, the major point source of pollution from urban runoff. There are three major problems in dealing with the control of pollution from urban runoff; these are: (1) a lack of total system management, (2) a lack of accurate information about the quality of urban runoff, and (3) a lack of accurate knowledge about the hydrology of urban runoff. The total system of urban drainage must be studied to incorporate the interaction of causes, sources and effects of different pollutants, along with the effects on receiving waters of various corrective actions. A model of urban drainage must be developed to study the impact of various alternatives. The quality of urban runoff is a big unknown, and studies made across the country give contradictory results. Standard methods of measurement, as well as definition of pollutants, seems to be needed. Detailed information on urban hydrology on a small scale is also badly needed to be able to provide adequate facilities for control of urban runoff pollution. Control of pollutants at the source is a primary goal. It involves controlling erosion from land under development, better methods of street cleaning, reducing peak flows of storms and the use of stormwater rather than merely disposing of it. (Poertner)
W73-08898

WATER POLLUTION CONTROL, A STATE-OF-THE-ART REVIEW,
Port of New York Authority, N.Y.
For primary bibliographic entry see Field 05D.
W73-08900

THE EFFECT OF PASSAIC RIVER BENTHAL DEPOSITS ON DEOXYGENATION IN OVERLYING WATERS,
Rutgers - The State Univ., New Brunswick, N.J.
O. M. Donovan.
Available from Univ. Microfilms, Inc., Ann Arbor, Mich. 48106 Order No. 73-4738. Ph D Dissertation, 1972, 161 p.

Descriptors: *Reaeration, *Oxygen sag, Water pollution effects, Water pollution, Ammonia, Nitrates, Phosphates, Streamflow, Discharge (Water), Water analysis, cChemical analysis, Dissolved oxygen, Oxygen demand, Rates, *New Jersey.
Identifiers: *Passaic River, *Benthal deposits, Reaeration coefficients.

Field work was carried out on natural benthal deposits on the upper Passaic River using a benthal respirometer. The deposits showed an average areal oxygen demand of 2.22 grams/sq m/day, with a range of from 1.13 grams/sq m/day to 3.79 grams/sq m/day. Studies on the areal extent of the deposits allowed calculation of the average benthal demand on the overlying water as a function of the stream discharge rate. This demand ranged from 0.26 mg/l/day at 200 cfs to 0.12 mg/l/day at a discharge rate of 1400 cfs. Benthal deposits from the Passaic River, along with samples of the overlying water, were brought back to the laboratory for examination under more carefully controlled conditions. Resulting oxygen sag curves in the laboratory apparatus were examined and provided graphic examples of the capacity of pollutional sediments in the causation of adverse effects on stream oxygen balance. Quantitative studies with oxygen sag curves generated in the laboratory apparatus showed that a very close fit could be obtained between predicted and observed data. This indicated that the methods used to experimentally determine constants for use in the oxygen sag expression were reasonably accurate and could be put to use in engineering examinations of polluted rivers and streams. Corollary studies on NH_3, nitrate, and phosphate concentrations in the water overlying the deposits showed that the benthal deposits released ammonia to the overlying waters and sorbed nitrate from the overlying waters. Comparative studies on the reaeration rates of highly polluted Passaic River waters as opposed to clean waters revealed no significant differences in the reaeration coefficients observed. (Holoman-Battelle)
W73-08949

FUNCTIONAL WATER AND SEWERAGE PLAN AND PROGRAM, CATAWBA REGIONAL PLANNING COUNCIL.
LBC and W Associates, Columbia, S.C.
For primary bibliographic entry see Field 06B.
W73-08977

LOWCOUNTRY REGION FUNCTIONAL WATER AND SEWER PLAN AND PROGRAM.
LBC and W Associates, Columbia, S.C.
For primary bibliographic entry see Field 06B.
W73-08978

MALIBU MASTER PLAN OF STORM DRAINS.
Los Angeles County Engineer Dept., Calif.

March, 1969. 54 p, 12 tab, 12 maps, photos. Urban Planning Project No. California P-267.

Descriptors: *Planning, *Storm drains, *Drainage, Storm runoff, Drainage basins, Floodplains, Urbanization, Topography, Maps, Storm water, *California.
Identifiers: *Stormwater drainage, Malibu, *Los Angeles County (Calif).

This report is intended to serve as a guide to both public agencies and private developers for stormwater drainage improvements, including design data and relations to county ordinances, e.g. subdivision and floodplain regulations. Stormwater drainage is important in this area of the southwestern corner of Los Angeles County

because of recent urbanization and the mountainous topography causing rapid runoff. The report and proposed drainage improvements are closely related to the General Plan for the Malibu area. Some engineering-oriented explanation of the planning procedure and some administrative and financial information is included. However, the key elements are the numerous maps, charts, and photographs. Each sub-basin in the area is shown on large fold-out maps which include existing and proposed drains, major watercourses, and natural channels. Photographs of some of the areas to be served by major storm drainage lines are also included. (Elfers-North Carolina)
W73-08980

INITIAL WATER, SEWERAGE AND FLOOD CONTROL PLAN REPORT, JOB 4800.
Duncan and Jones, Berkeley, Calif.
For primary bibliographic entry see Field 06B.
W73-08981

WATER AND SEWER FACILITIES, OSCEOLA COUNTY.
Beck (R. W.) and Associates, Denver, Colo.
For primary bibliographic entry see Field 06B.
W73-08984

WATER SUPPLY PLAN FOR THE SOUTHEASTERN CONNECTICUT REGION: SUMMARY.
Southeastern Connecticut Water Authority, Norwich.
For primary bibliographic entry see Field 06B.
W73-08985

STATEWIDE LONG-RANGE PLAN FOR THE MANAGEMENT OF THE WATER RESOURCES OF CONNECTICUT, PHASE I REPORT, 1971.
Connecticut Interagency Water Resources Planning Board, Hartford.
For primary bibliographic entry see Field 06B.
W73-08987

INCOG FUNCTIONAL PLAN SUPPLEMENTS (WATER AND SEWER, HOUSING AND OPEN SPACE).
Indian Nations Council of Governments, Tulsa, Okla.
For primary bibliographic entry see Field 06B.
W73-08989

POLLUTION CONTROL FINANCING IN THE UNITED KINGDOM AND EUROPE,
Thames Conservancy, Reading (England).
H. Fish.
Journal Water Pollution Control Federation, Vol 45, No 4, p 734-741, April, 1973.

Descriptors: *Financing, Water pollution control, *Planning, Regional analysis, *Management, *Waste water disposal, Costs, *Pollution taxes (Charges), *Europe.
Identifiers: United Kingdom, Netherlands, France, Germany.

Four basic methods for financing pollution control facilities apply in part to procedures for waste water disposal financing in the United Kingdom, the Netherlands, France, and Germany. They are: (1) payment of all costs for waste water disposal by the discharger; (2) same as number 1 except that a credit is given recognizing the reuse value of the waste water discharged; (3) charges related to the volume and quantity of the waste water levied on the discharger in order to meet the costs of a public authority; and (4) a single bill of charge relating to the volume and quality of the waste water levied to the discharger by a public authority responsible for providing these services. Present and future positions are considered, and the organization of regional water authorities for the control of all regulatory and management aspects of pollution control is presented as the most practicable method of financing. (Bell-Cornell)
W73-09005

A MODEL OF A PRIVATE ENVIRONMENTAL AGENCY,
For primary bibliographic entry see Field 06E.
W73-09009

THE ALLOCATION OF TROUBLESOME MATERIAL,
T. A. Ferrar, and A. Whinston.
(1972), 19 p, 7 fig, 5 ref. OWRR-B-020-IND (16).

Descriptors: *Resource allocation, *Urban sociology, *Waste disposal, *Numerical analysis, *Optimization, Water resources, Land, Solid wastes, Costs, Recycling, Supply, Prices, Decision making, Constraints, Mathematical models, Systems analysis.
Identifiers: *Nondisposable materials, Material property rights, Urban centers, Land areas, Iterative processes, Incentive-feedback algorithm, Government commissions, Environmental standards, Allocation programs, Synthetic materials, Raw materials.

Orientation to convenient containers and the production and sale of virtually nondisposable packaging materials for containerization threaten the land areas surrounding urban centers. The supply price of these materials incorrectly reflects the cost burden on society—the true cost of disposal of these materials is presently shielded from the packaging industry; thus, we experiencing an oversupply of containers made of these materials. It is recommended that an optimal quantity of these materials be permitted to enter the urban region per period; this optimal quantity should reflect the particular waste disposal capacity of the region for these materials and the preference structure of the population for these materials relative to substitute packaging means. A model is developed for determining the allocation of material-usage property rights to the distributors of packaged products that is consistent with government-specified standards. To enforce these environmental standards, a management agency would be created, with the task of allocating the legal material usage rights in accordance with the provided goal function among the distributers of the community. The method utilized to accomplish this task involves an incentive-feedback algorithm that iteratively allocates the supplies among the distributors. (Bell-Cornell)
W73-09010

JIM KEE SPEAKS ON WATER RESOURCES,
House, Washington, D.C.
For primary bibliographic entry see Field 06E.
W73-09011

NEW CONCEPTS IN SITING OF NUCLEAR POWER PLANTS,
House, Washington, D.C.
For primary bibliographic entry see Field 06E.
W73-09012

HERCULES CONTINUES TO FIGHT POLLUTION,
House, Washington, D.C.
For primary bibliographic entry see Field 06E.
W73-09015

BIOASTRONAUTICS DATA BOOK, SECOND EDITION.

Washington, D.C.: U.S. Government Printing Office, $7.50. 1973, 930 p.

Descriptors: *Water quality standards, *Wa
quality control, Potable water, Water supp
Recycling, Water reuse, Pollutant identification
Identifiers: *Space craft water quality standard

Chapter 10, Toxicology, section on Water Qua
Standards for Space Missions, (p478-487) sta
that requirements for space craft water qua
standards are based on aesthetics or phys
criteria, trace chemical content, and microbiol
cal impurities. It was decided that the water qu
ty standards need not be as stringent as those u
for municipal water supplies since the latter w
established to protect a broad spectrum of
population. Two tables are included in the rep
giving space craft water quality standards
physical properties and space craft water qua
standards for chemical content respectively. (
also W73-09017) (Smith-Texas)
W73-09016

AEROSPACE VEHICLE WATER-WAS
MANAGEMENT,
National Aeronautics and Space Administrati
Washington, D.C.
J. N. Pecoraro.
In: Bioastronautics Data Book, Second Editi
Chapter 20, National Aeronautics and Space A
ministration SP-3006 p 915-922, 1973. 3 fig, 1 tab
ref.

Descriptors: *Waste storage, *Waste dispo
*Wastes, Air pollution, Water quality contr
Water reuse, Recycling, Ultimate disposal, Sc
waste, Liquid wastes.
Identifiers: *Waste management, *Aerospace p
gram.

Various aerospace vehicle waste managem
systems are evaluated. In a space craft, was
must be collected and transported to stor
and/or recovery process equipment in such
manner that they do not contaminate the crew a
the internal environment of the vehicle. If
wastes are to be stored on board for an extend
period of time, they should be sterilized to av
contamination if the storage vessel fails. The
jective of the urine and liquid waste collection a
transport subsystem is to provide a means for c
lecting and transporting these wastes to the wa
management subsystem, where treating a
processing are performed. Biodegradatic
vacuum thermal drying, and incineration are so
of the currently considered methods for disposi
of solid waste. The results of the processing m
be either stored in an aesthetic and sanit
manner or dumped overboard to protect the cr
from microorganisms and noxious gases. The si
plest and most versatile method for storing
types of wastes is vacuum drying to remove t
water necessary for biological mobility a
growth. This technique does not destroy t
microorganisms, but it does inactivate them suf
ciently to permit storage in a plastic bag f
periods of 120 days or more. (See also W73-0901
(Smith-Texas)
W73-09017

PLASTIC PACKAGING AND COASTAL PO
LUTION,
Aston Univ., Birmingham (England). Dept.
Chemistry.
G. Scott.
Int J Environ Stud. Vol 3, No 1, p 35-36. 1972.
Identifiers: *Coastal pollution, *Plastic packagin
Pollution wastes, Water pollution control.

Pollution of shore line due to plastic packaging i
result of the deposition of seaborne waste. T
waste is a by-product of international commer
and not due to irresponsibility of casual visito
Common packaging plastics vary markedly in th
resistance to environmental breakdown. Low de
sity polyethylene which is used for agricultu
sacks, wrapping film and detergent bottles su
vives unchanged for many years, whereas hi

ganizational structure and the talents which have made it a great institution without tearing down any ivy or cracking any marble. It can be done while giving the individual student the freedom to pursue his particular area of interest throughout his college years and yet insure that he will emerge with a broad understanding of the interrelationship of man and his environment plus an understanding of how to bring some special areas of knowledge to bear upon environmental problems in concert with other areas of knowledge. A university can emphasize environmental problems without neglecting other areas of emphasis and a way can and should exist for every student to develop his area of interest continuously throughout his educational program. The field of environmental management is too broad to be encompassed by any single program of studies, as all disciplines are relevant to environmental management and in the real world the 'manager' may come from any disciplinary background. Every individual must have some cohesive body of knowledge that he can apply to environmental problems thus dispensing with an environmental generalist. (Auen-Wisconsin)
W73-08648

CONCEPTS FOR ANALYSIS OF MASSIVE SPILL ACCIDENT RISK IN MARITIME BULK LIQUID TRANSPORT,
Coast Guard, Washington, D.C. Office of Research and Development.
For primary bibliographic entry see Field 05G.
W73-08654

KNOW WHAT'S HAPPENING IN YOUR WATER SYSTEM,
Parsons, Brinckerhoff, Quade and Douglas, Inc., New York.
For primary bibliographic entry see Field 04A.
W73-08655

VALUATION OF TIMBER, FORAGE AND WATER FROM NATIONAL FOREST LANDS,
Forest Service (USDA), Tucson, Ariz. Rocky Mountain Forest and Range Experiment Station.
For primary bibliographic entry see Field 06B.
W73-08657

THE CHESAPEAKE BAY INVENTORY SYSTEM,
Johns Hopkins Univ., Baltimore, Md. Dept. of Radiology.
For primary bibliographic entry see Field 02L.
W73-08809

COASTAL ZONE MANAGEMENT SYSTEM: A COMBINATION OF TOOLS,
Center for the Environment and Man, Inc., Hartford, Conn.
For primary bibliographic entry see Field 02L.
W73-08810

PREDICTIVE MODELS OF SALINITY AND WATER QUALITY PARAMETERS IN ESTUARIES,
Massachusetts Inst. of Tech., Cambridge. Dept. of Civil Engineering.
For primary bibliographic entry see Field 02L.
W73-08815

SOUTH SANTA CLARA COUNTY WATER PLANNING STUDY COMPLETION REPORT.
Santa Clara County Flood Control and Water District, San Jose, Calif.
For primary bibliographic entry see Field 06B.
W73-08979

MALIBU MASTER PLAN OF STORM DRAINS.
Los Angeles County Engineer Dept., Calif.
For primary bibliographic entry see Field 05G.
W73-08980

ANALYSIS AND OPTIMIZATION OF TWO-STATE DIGESTION,
Kansas State Univ., Manhattan. Dept. of Chemical Engineering.
For primary bibliographic entry see Field 05D.
W73-09007

COLLECTIVE UTILITY IN THE MANAGEMENT OF NATURAL RESOURCES: A SYSTEMS APPROACH,
Arizona Univ., Tucson. Dept. of Systems Engineering.
E. G. Dupnick.
Technical Reports on Hydrology and Water Resources. Technical Report No 5, June 1971, Arizona University, Tucson. 172 p, 8 fig, 112 ref. OWRR A-024-ARIZ (5). 14-31-0001-3503.

Descriptors: *Planning, Operations, *Natural resources, *Management, Economics, Systems analysis.
Identifiers: Multicriteria, Cost effectiveness, *Collective utility, Externalities.

The main purpose is to develop an economic theory, along the lines of the Bergson-Samuelson social welfare theory, to regulate the utilization of natural resources in the long-term interest of a political-economic group of individuals and firms. The theory, called Collective Utility, qualifies as a 'systems approach' because of its inherent flexibility, generality, and comprehensiveness. Collective Utility is a function of individual satisfactions and firm revenues, which are, in general, contingent upon the actions of other individuals and/or firms. Such interactions are called externalities. Efficient management of natural resources will follow from efficient control of externalities. A taxation-subsidy structure is suggested as an efficient control and the complete mathematics of determining and implementing such a structure are provided. Finally, the idea of externalities is integrated within the framework of Collective Utility to form an optimal policy for the utilization of natural resources using the techniques of calculus of variations.
W73-09074

6B. Evaluation Process

AN ANALYTICAL INTERDISCIPLINARY EVALUATION OF THE UTILIZATION OF THE WATER RESOURCES OF THE RIO GRANDE IN NEW MEXICO,
New Mexico State Univ., University Park. Dept. of Agricultural Economics.
R. R. Lansford, S. Ben-David, T. G. Gebhard, Jr., W. Brutsaert, and B. J. Creel.
Available from the National Technical Information Service as PB-220 070, $3.00 in paper copy, $1.45 in microfiche. New Mexico Water Resources Research Institute, Las Cruces, Completion Report 020, March 1973, 152 p, 18 fig, 39 tab, 91 ref, 2 append. OWRR B-026-NMEX (1), B-019-NMEX (1), and B-016-NMEX (1).

Descriptors: *New Mexico, *Water demand, Economics, *Water resources development, Management, *Economic prediction, *Surface-groundwater relationships, Natural resources, Water requirements, Resource allocation, River basin, Groundwater management, Water law, Interstate compacts, International compacts, Treaties, Litigation, Water quality, *Water utilization, Human population, Employment, Industrial water, Recreation, Water management (Applied), Linear programming, Model studies.
Identifiers: *Rio Grande Basin, *Socio economic models, *Interdisciplinary, Groundwater appropriation, Input-output coefficients, Conjunctive use models, Economic land classification, Groundwater models, Irrigation water diversions.

An interdisciplinary approach to the solution of the water resource problems of the Rio Grande region in New Mexico was conducted. The primary objective was the evaluation of the social and economic impacts of alternative water-use policies. A socio-economic model was developed to represent the New Mexico economy, with special emphasis placed upon the Rio Grande region. Inputs into the socio-economic model were obtained from separate studies covering the hydrological, agricultural, municipal, and industrial areas. Three sets of alternatives were considered: (1) growth without a water constraint; (2) growth, holding surface water constraint; (3) growth, holding both surface and groundwater constraint. Without a water constraint, both production and depletions are expected to exhibit the largest increase (59.2 percent and 49.6 percent, respectively). When a surface water constraint is imposed, the value of production is reduced by only $5.6 million in the year 2000, and by $14.2 million in 2020; water depletions are expected to decrease about 27 percent by 2020. When a total water constraint is imposed, the value of production is decreased $2.7 million below that expected when using only a surface water constraint, and water depletions are reduced only slightly. (Creel-New Mexico State)
W73-08458

ORGANIZATION OF LAND AREA IN WASHINGTON FOR WATER AND LAND USE PLANNING,
Washington State Univ., Pullman. Dept. of Agronomy and Soils.
For primary bibliographic entry see Field 06E.
W73-08607

WHERE WILL NUCLEAR POWER BE IN 1980,
Bechtel Corporation, Michigan.
For primary bibliographic entry see Field 06D.
W73-08613

VALUATION OF TIMBER, FORAGE AND WATER FROM NATIONAL FOREST LANDS,
Forest Service (USDA), Tucson, Ariz. Rocky Mountain Forest and Range Experiment Station.
P. F. O'Connell.
The Annals of Regional Science, Vol vi, No 2, p 1-14, December 1972. 3 fig, 18 ref.

Descriptors: Forest watersheds, *Management, *National forests, Water resources, *Forages, *Evaluation, Methodology, Decision making, Linear programming, Constraints, Prices, Economics, Water yield, *Arizona, Mathematical models, Systems analysis, *Lumber.
Identifiers: *Salt-Verde Basin (Arizona), *Tradeoff values.

Conservation is no longer adequate by itself for justifying the kind of expenditures required on public lands. With more people and increasing affluence, greater demand is being placed on limited forest resources. Extensive management for timber and forage will be replaced gradually by intensive management for outdoor recreation, landscape esthetics, water, and wildlife, in addition to intensive timber and forage management. This trend is especially noticeable in the Salt-Verde Basin in central Arizona, where water users have been urging authorities to increase the water yield in the basin. To make a decision regarding the extent of watershed treatments, tradeoff criteria are required. A methodology is discussed for determining the relative tradeoff values for timber, forage, and water; the data used are specifically from the Salt-Verde Basin, but the procedure is appliable to any National Forest lands. The use of economic criteria is the most logical way to judge the relative worth to society of timber, forage and consumptive water. Since each of these three resources is used as an input into a production process, a range of resource values is derived from the price of lumber, livestock, and feed grains, using a valuation model. A multiple-use linear programming model for analyzing the range of values is then discussed. Results are presented in detail. (Bell-Cornell)
W73-08657

CROPPING PATTERN, YIELDS AND INCOMES UNDER DIFFERENT SOURCES OF IRRIGATION (WITH SPECIAL REFERENCE TO IADP DISTRICT ALIGARH, U.P.),
Pant Coll. of Tech., Pantnagar (India).
For primary bibliographic entry see Field 03F.
W73-08658

MEASURING BENEFITS GENERATED BY URBAN WATER PARKS,
Dornbusch (David M.) and Co., Inc., San Francisco, Calif.
A. H. Darling.
Land Economics, Vol XLIX, No 1, p 22-34, February 1973. 1 fig, 9 tab, 3 equ, 6 ref.

Descriptors: Water resources, *Parks, *Benefits, *Property values, Evaluation, Lakes, *Linear programming, *California, Mathematical models, Estimating, Recreation, Real property, Methodology, Ponds, Demand, *Urbanization.
Identifiers: *Water parks, Interviews, Residents, Lake Merritt (Cal.), Lake Murray (Cal.), Santee Lakes (Cal.).

A number of projects have outputs that are not sold in the open market, and these outputs require some measure other than price to value their benefits. Public park project outputs are of this type, consisting primarily of recreation and aesthetic quality. Presented is a method for evaluating urban water parks benefits, which regards increases in the value of property near urban parks as a measure of the benefits derived. Results are reported of an attempt to measure the benefits of three urban water parks in California using a property value model similar to that of Kitchen and Hendon, and using an interview technique similar to that developed by Davis for the Maine woods. The three water parks chosen are: (1) located within urban areas; (2) significantly developed for recreation; and (3) analytically tractable. Characteristic differences between the sites provide a basis for testing the efficiency of the method for a wide range of water resources and urban characteristics. It is concluded that: (1) the value of an urban water park is measurable; (2) the value of an urban water resource is apparently large; and (3) regardless of the limited sample size available in this study, the techniques presented have merit for evaluating such intangible benefits. (Bell-Cornell)
W73-08659

FEDERAL DECISIONMAKING FOR WATER RESOURCE DEVELOPMENT, IMPACT OF ALTERNATIVE FEDERAL DECISION-MAKING STRUCTURES FOR WATER RESOURCES DEVELOPMENT,
Michigan State Univ., East Lansing. Dept. of Agricultural Economics.
A. A. Schmid.
Available from the National Technical Information Service as PB-211 441, $5.45 in paper copy, $1.45 in microfiche. Report prepared for National Water Commission, December 1971. 99 p, 1 fig, 43 ref.

Descriptors: *Water resources development, *Decision making, *Administrative decisions, *Adjudication procedure, Administration, Governments, Coordination, Comprehensive planning, Legislation, Legal aspects, Water management (Applied), Governmental interrelations, Institutional constraints.

An extensive and comprehensive discussion is presented of the impact of alternative federal decision-making structures and the effects of federal decision making on water resources development. The important criteria for choice among decision making structures are described and the impacts of a number of alternative structures are assessed. The general discussion is divided into three parts. First, the pros and cons of movements toward consolidation of agencies are discussed. Second, external bargaining rules are specified with respect to agency-clientele bargaining, interagency bargaining, state-federal and state-state bargaining and finally market bargaining. The third part is concerned with bargaining and management internal to the agency. Detailed consideration must be given to organizational changes that affect negotiation rules and the rules that shape the kind of information available to various interested groups. (Mockler-Florida)
W73-08669

AN ANALYSIS OF ENVIRONMENTAL STATEMENTS FOR CORPS OF ENGINEERS WATER PROJECTS,
Stanford Univ., Calif. Dept. of Civil Engineering.
For primary bibliographic entry see Field 06G.
W73-08673

THE CALIFORNIA STATE WATER PROJECT SUMMARY: NINETEEN-SEVENTY-ONE.
California State Dept. of Water Resources, Sacramento.
For primary bibliographic entry see Field 04A.
W73-08749

TOOLS FOR COASTAL ZONE MANAGEMENT.
Marine Technology Society, Washington, D.C. Coastal Zone Marine Management Committee.
For primary bibliographic entry see Field 02L.
W73-08802

METHODS FOR ENVIRONMENTAL PLANNING OF THE CALIFORNIA COASTLINE,
California Univ., Berkeley. Dept. of Landscape Architecture.
For primary bibliographic entry see Field 02L.
W73-08808

MARINE TECHNOLOGY TRANSFER DEPENDS UPON INFORMATION TRANSFER,
Environmental Protection Agency, Washington, D.C. Div. of Water Quality Standards.
For primary bibliographic entry see Field 02L.
W73-08812

EVALUATION OF COASTAL ZONE MANAGEMENT PLANS THROUGH MODEL TECHNIQUES,
Army Engineer Waterways Experiment Station, Vicksburg, Miss.
For primary bibliographic entry see Field 02L.
W73-08813

REVIEW OF COMMERCIAL FISHERIES IN THE COLUMBIA RIVER AND IN CONTIGUOUS OCEAN WATERS,
National Marine Fisheries Service, Seattle, Wash. Exploratory Fishing and Gear Research Base.
For primary bibliographic entry see Field 02L.
W73-08820

A DESCRIPTION AND ANALYSIS OF COASTAL ZONE AND SHORELAND MANAGEMENT PROGRAMS IN THE UNITED STATES,
Michigan Univ., Ann Arbor. Sea Grant Program.
For primary bibliographic entry see Field 02L.
W73-08887

siderations, and environmental problems, a section on water resource data, sections on goals and design criteria, and county-by-county analysis of existing water supply and wastewater disposal facilities. The plan and implementation program is also presented county-by-county and is divided into short range improvements and long range policies. (Elfers-North Carolina)
W73-08978

SOUTH SANTA CLARA COUNTY WATER PLANNING STUDY COMPLETION REPORT.
Santa Clara County Flood Control and Water District, San Jose, Calif.

Available from the National Technical Information Service as PB-213 231, \$3.50 in paper copy, \$1.45 in microfiche. Final report, August 1972. 24 p, 1 append. Calif. P-277. Calif. P2-77 (G).

Descriptors: *Planning, *Water supply, *Optimization, *Water allocation, Linear programming, Dynamic programming, Scheduling, Water demand, Cost analysis, Systems analysis, *California, Urbanization.
Identifiers: *Santa Clara County (Calif).

Water supply and allocation for the fast growing southern part of Santa Clara County is discussed in this planning report which is a summary of a computer oriented water resource management study. The basic objective was to meet water supply demands to 2020 via the optimal allocation of water and staging of the system elements. Projections of population and land use together with information on geology, hydrology, existing reservoirs, and economic relationships were used to formulate optimal water management systems. Linear and dynamic programming techniques were used to arrive at the optimal combinations of system elements (e.g. reservoirs, wastewater reclamation, importing of water) and their timing. Such computer tools are not only useful for the present study, but the programs can be used for future allocations and plan modifications. (Elfers-North Carolina)
W73-08979

INITIAL WATER, SEWERAGE AND FLOOD CONTROL PLAN REPORT, JOB 4800.
Duncan and Jones, Berkeley, Calif.

Summary report prepared for the San Diego County Comprehensive Planning Organization, San Diego, California, June 1972. 82 p, 5 maps, 4 tab. HUD P-384 (g).

Descriptors: *Planning, *Water supply, *Sewerage, *Flood control, Environmental effects, Coordination, Administration, Water reuse, *California, Urbanization.
Identifiers: *Utility extension planning, Extension policies, Urban development patterns, *San Diego County (Calif).

Plans and policies for the extension of water supply, sanitary sewerage and flood control facilities are presented. One of the main purposes is to establish a set of policies to guide water resources planning and management in San Diego County. These policies particularly relate to carrying out water resources planning as an integral part of the overall planning for the county and to use the water systems extensions to help guide urban development patterns. The report is divided into five sections: (1) an introduction which explains the relation of water resources planning to other planning activities; (2) goals and policies, particularly extension policies, emphasizing environmental quality and relations to land use patterns; (3) the adequacy of the existing water resource systems (much greater detail on this is presented in a technical report of the same title); (4) the proposed improvements program presented via long lists of each project, its costs and priority; and (5) a proposed program and organization for implementation wherein governmental responsibilities and organization are discussed. Five large fold-out maps are included. (Elfers-North Carolina)
W73-08981

WATER AND SEWERAGE DEVELOPMENT PLAN, SUMTER COUNTY, SOUTH CAROLINA.
Palmer and Mallard and Associates, Inc., Sumter, S.C.

Report prepared for Sumter County Planning Board, Sumter, South Carolina, July 1968. 100 p, 26 tab, 21 map, 11 ref, 1 append.

Descriptors: *Planning, Water demand, *Water supply, *Sewerage, Agriculture, *South Carolina, *Rural areas.
Identifiers: *Utility extension planning, Inventory, Land use plan, Sumter County (South Carolina).

A water supply and sewerage plan for Sumter County is outlined in very general form on maps of the ten planning districts in the county. The planning period extends to 1985. Both the water and sewerage plans consist of several small systems since the county is not yet very urbanized. The report focuses on background data for the plan and the relation of water and sewerage services to the land use plan. Thus, there are sections on population, the economic base, agricultural land patterns, natural resources, public facilities, water resources, and a county land use plan. The study was sponsored by the Farmers Home Administration, which gives assistance in developing small, rural-oriented water resource systems. More of a comprehensive collection of planning data than a detailed water and sewerage plan, one section is devoted to proposed water and sewerage systems with detailed maps provided for individual areas. A proposed Sumter County Master Plan which will complement this water and sewerage plan is now in process. (Elfers-North Carolina)
W73-08982

WATER AND SEWER FACILITIES, OSCEOLA COUNTY.
Beck (R. W.) and Associates, Denver, Colo.

Report prepared for Orange-Seminole-Osceola Planning Commission, Orlando, Florida, November 1969. 104 p, 4 map, 21 tab, 6 append. HUD Fla. P-92.

Descriptors: *Planning, *Water supply, *Sewerage, Comprehensive planning, Environmental effects, Urbanization, Financing, Administration, *Florida, Urbanization.
Identifiers: *Utility extension planning, *Osceola County (Florida), East Central Florida Region.

Long-range water supply and sewerage plans for Osceola County are part of the overall Comprehensive Development Plan for the county and are based on the previously completed plan elements of land use, transportation, recreation, and open space. These plan elements were used to determine service areas, demands, and line and treatment plant locations. The plans use a 'wholesaling concept' in which the County provides the major facilities throughout the area. The main objectives are to provide good service at minimum cost, to project the environmental quality of the area, especially the lakes, and to promote orderly development. The study is divided into four main parts: (1) an analysis of existing facilities; (2) long-range (to 1990) water and sewerage plans; (3) financial considerations including water and sewer charges; and (4) a discussion of organization and implementation measures. There are also appendices on hydrology, geology, design

criteria, treatment, and a glossary. (Elfers-North Carolina)
W73-08984

WATER SUPPLY PLAN FOR THE SOUTHEASTERN CONNECTICUT REGION: SUMMARY.
Southeastern Connecticut Water Authority, Norwich.

February 1970. 11 p, 3 fig, 3 map, 4 tab. U.P.A. Conn. P-104.

Descriptors: *Planning, *Water supply, Administration, Financing, Water demand, Water sources, *Connecticut, *Regional development.
Identifiers: Inventory, Southeastern Connecticut, Service areas, Load centers.

This summary covers the major sections of a comprehensive two volume study with a few descriptive paragraphs, a number of charts, and several maps. Volume I is basically an inventory of the 74 water supply systems in the region in 1969. Included in the inventory are the service areas, the rates of use, sources of supply, treatment, transmission and distribution facilities, and storage. Volume II presents the recommended water supply plan for the region through the year 2020, and the analysis leading to the formulation. This includes projections of population growth and water demands, analysis of surface water and groundwater supplies, determination of construction costs, and considerations about administration and coordination. The recommended plan is framed in terms of a central system and several outlying 'load centers' and is planned in four construction phases. The creation of the Southeastern Connecticut Water Authority is a key element in the implementation of the Plan. (See also W73-08986) (Elfers-North Carolina)
W73-08985

WATER SUPPLY PLAN FOR THE SOUTHEASTERN CONNECTICUT REGION, VOLUME I, INVENTORY.
Southeastern Connecticut Water Authority, Norwich.

September 1969. 167 p, 22 fig, 3 tab, 1 append. U.P.A.Conn. P-104.

Descriptors: *Planning, Water supply, Water quality, Water rates, Operation and maintenance, Water demand, *Connecticut, *Regional development.
Identifiers: *Inventory, Southeastern Connecticut, Service areas.

A detailed inventory of all existing water supply systems in Southeastern Connecticut is presented. The inventory is the first phase of a water supply study and plan formulation being carried out jointly by the newly formed Southeastern Connecticut Water Authority (1967) and the regional planning agency. The Water Authority has broad powers to construct, operate, and maintain regional water supply systems. The inventory covers each of the 74 existing systems in the region in great detail including numerous maps. Data are presented relating to various physical, institutional, and economic base aspects of the specific service area, the population served, source of water supply, plant operating characteristics, water quality, and water rates. Schematic diagrams of many of the systems are also presented. (See also W73-08985) (Elfers-North Carolina)
W73-08986

STATEWIDE LONG-RANGE PLAN FOR THE MANAGEMENT OF THE WATER RESOURCES OF CONNECTICUT, PHASE I REPORT, 1971.
Connecticut Interagency Water Resources Planning Board, Hartford.

May 1971. 152 p, 54 fig, 27 tab, 4 append. HUD P-128.

Descriptors: *Planning, *Long-term planning, *Regional development, *Water resources development, Water supply, Water quality control, Flood control, Recreation, Water demand coordination, Environmental effects, Industrial water, *Connecticut, Urbanization.

The state of Connecticut is very concerned about the orderly development of its land and water resources and, in particular, the resolution of conflicts among various uses of its water resources such as for water supply, wastewater disposal, and recreation. This report, the initial findings of a comprehensive, statewide water resources planning program based on an act passed in 1967 (Public Act 477), is divided into four sections: (1) an introduction covering the background and organization of the study; (2) a description of various physical aspects of the state such as population, land use, climate, geology, and water resources; (3) an inventory of existing water resource uses and facilities including wastewater treatment facilities, water-based recreation, water supplies, industrial water use, and flood control structures; and (4) projections of future requirements for water uses, based on the Connecticut Interregional Planning Program's year 2000 projections. A glossary, a great number of maps and tables plus appendices on water quality standards and flood control protective devices are included. (Elfers-North Carolina)
W73-08987

LAND USE AND DEVELOPMENT ALONG THE TENNESSEE RIVER IN THE TARCOG REGION.
Top of Alabama Regional Council of Governments, Huntsville.

Technical Paper No. 4, April 1972. 21 p, map, 7 ref. CPA-AL-04-09-1006.

Descriptors: *Comprehensive planning, *Land use, Urbanization, Regional development, Coordination, Recreation, *Alabama, *Tennessee River.
Identifiers: *Shorelands, Land use trends, Land use conflicts, *TARCOG region (Alabama).

Urban development trends and planning recommendations for an area along the Tennessee River in northern Alabama are presented. The focus is on the land adjacent to the Tennessee River because of its overriding economic and recreational influence in the region. The purpose of the report is to create a public awareness of the potential conflicts among various uses of the land along the river and the need for comprehensive planning. An inventory of current uses along the river, a discussion of future trends in these uses and the potential conflicts, and some planning recommendations are included. At present the four main uses of riverfront land (industrial areas, residential subdivisions, recreational areas, and wildlife areas) are each concentrated in a single county. However, as industrial and residential areas continue to expand there will arise numerous conflicts without proper planning. The formulation of a regional development plan, the creation of strong county planning agencies, local coordination programs, and the increased use of the regional clearinghouse for review and coordination of development are recommended for optimal use of the land. (Elfers-North Carolina)
W73-08988

INCOG FUNCTIONAL PLAN SUPPLEMENTS (WATER AND SEWER, HOUSING AND OPEN SPACE).
Indian Nations Council of Governments, Tulsa, Okla.

Available from the National Technical Information Service as PB-213 156, $7.50 in paper copy $1.45 in microfiche. Final report. June 1972. 99 p 13 tab, append. CPA OK 06 56 1012.

Descriptors: *Planning, *Evaluation, *Water supply, *Sewerage, Legislation, Waste disposal, *Oklahoma.
Identifiers: *Plan review, Implementation, Policies, Capital improvements, *INCOG.

A review and update of three elements of the Indian Nations Council of Governments comprehensive plan is presented. The review is based upon the objective of maintaining the planning program as a continuing process. The water and sewer plan element was completed in 1970 and the first review and evaluation is described in Chapter II, p. 9-30. The review is conducted by the INCOG staff and the Water and Sewer Technical Advisory Committee. The review is organized into four areas: (1) plan effectiveness, which is measured largely in terms of whether recommended projects were either implemented or are in the design stage; (2) water and wastewater disposal policy analysis in which goals and implementation policies are reviewed and possibly changed; (3) legislative analysis of new federal acts and programs relevant to the plan; (4) and capital improvement program updating in which specific project proposals and priorities are reassessed. The review also includes a revised table of capital improvements for the next five years. (Elfers-North Carolina)
W73-08989

INTERIM RIVER REPORT, NOVEMBER 1970.
Interagency Riverfront Committee, Minneapolis, Minn. Design Team.

November 1970. 114 p, illus, 31 maps.

Descriptors: *City planning, *Community development, Aesthetics, Urban renewal, Railroads, Architecture, *Design, *Minnesota.
Identifiers: *Urban riverfront, *Minneapolis, *Mississippi River.

The interim report on the development opportunities for the Mississippi riverfront in Minneapolis is primarily intended to serve as a basis for community discussion and participation. It points out the importance of immediate action while there is great national interest in environmental quality and favorable local circumstances such as the planned merger of several railroads, the expansion of the University of Minnesota, and several existing redevelopment projects. The importance of community values and goals, the aesthetic and amenity potential of the riverfront, the special character of riverfront activities, and the design of riverfront buildings and land uses are stressed. Much of the report is devoted to suggesting design concepts for the various districts or distinctive areas along the river. The river is seen as a key element in the overall development of the city, particularly as a means to help integrate various areas and activities within the city. (Elfers-North Carolina)
W73-08990

ROLE OF WATER IN URBAN PLANNING AND MANAGEMENT,
Geological Survey, Arlington, Va. Water Resources Div.
For primary bibliographic entry see Field 04C.
W73-09082

6C. Cost Allocation, Cost Sharing, Pricing/Repayment

ROUGE RIVER FLOOD CONTROL PROJECT. SECTION B FROM I-94 TO MICHIGAN AVENUE ALTERNATIVE NO. IV (SUPPLE-

Nuclear power plants must be located with equal emphasis on environmental design and the conventional design aspects of plant siting. Advanced technology environmental abatement systems must continue to be developed to assure man's requirement for electrical energy can be met while preserving or enhancing the environment. Advanced generation systems must continue to be developed and demonstrated during the '70s to assure economic energy generation in the decades ahead. (Oleszkiewicz-Vanderbilt)
W73-08613

METHODS FOR ENVIRONMENTAL PLANNING OF THE CALIFORNIA COASTLINE,
California Univ, Berkeley. Dept. of Landscape Architecture.
For primary bibliographic entry see Field 02L.
W73-08808

FUNCTIONAL WATER AND SEWERAGE PLAN AND PROGRAM, CATAWBA REGIONAL PLANNING COUNCIL.
LBC and W Associates, Columbia, S.C.
For primary bibliographic entry see Field 06B.
W73-08977

WATER SUPPLY PLAN FOR THE SOUTHEASTERN CONNECTICUT REGION, VOLUME I, INVENTORY.
Southeastern Connecticut Water Authority, Norwich.
For primary bibliographic entry see Field 06B.
W73-08986

STATEWIDE LONG-RANGE PLAN FOR THE MANAGEMENT OF THE WATER RESOURCES OF CONNECTICUT, PHASE I REPORT, 1971.
Connecticut Interagency Water Resources Planning Board, Hartford.
For primary bibliographic entry see Field 06B.
W73-08987

6E. Water Law and Institutions

AN EXAMINATION OF THE ECONOMIC IMPACT OF POLLUTION CONTROL UPON GEORGIA'S WATER-USING INDUSTRIES,
Georgia Inst. of Tech., Atlanta. Engineering Experiment Station.
For primary bibliographic entry see Field 05G.
W73-08453

ORGANIZATION OF LAND AREA IN WASHINGTON FOR WATER AND LAND USE PLANNING,
Washington State Univ., Pullman. Dept. of Agronomy and Soils.
W. A. Starr.
Available from the National Technical Information Service as PB-219 983, $3.00 in paper copy, $1.45 in microfiche. Washington Water Research Center, Pullman, Report no. 12, November 1973. 97 p, 7 fig, 13 tab, 14 ref, 3 append. OWRR A-051-WASH (1).

Descriptors: *Planning, *Land use, *Management, *Washington, *Land tenure, *Local governments, Institutions, Organizations, Water policy.

Three levels of planning and management units for the state of Washington have been examined. They are the Land and Water Management Region, the Watershed, and the Local Planning (hydrologic) Unit. An evaluation of the distribution of land and water resources was made at each level of planning and related to the distribution of ownership of the land. The king and number of local improvement and public purpose districts were also associated with the planning units. Eleven Land and Water Management Regions and 87 Watersheds were considered. They were compared with regions and watersheds identified in other planning concepts and public action programs. Their utility for planning, administration, and regulatory functions were evaluated. The use of Local Planning (hydrologic) Units was evaluated and a method for construction of local units within a watershed developed. The utility of the local units for local analysis of land and water problems and developments, for land and water plan integration, and as a means of involvement of local people in the planning process have been considered.
W73-08607

UTILITY BATTLES: YOU CAN'T WIN 'EM ALL.
For primary bibliographic entry see Field 05G.
W73-08622

MOBILIZING FOR A CLEAN ENVIRONMENT: WHO SHOULD BE DOING WHAT, HOW AND WHY,
L. R. Freeman.
In: Water Resources Seminar Series No 1, January 1972. Water Resources Research Center, University of Hawaii, Honolulu, p 27-36.

Descriptors: *Environment, *Comprehensive planning, *Governments, *Pollution abatement, Decision making, Institutions, Social aspects, Industry, Air environment, Aquatic environment, Hawaii.

The goal is to better understand the nature of the system within which we are striving for a better environment. The federal government recommends broad, general policy and general means for implementing that policy; the state augments and interprets general policy and translates this into a broadly defined plan of action, and finally, local government and industry are made responsible for taking action to abate pollution. Accomplishing the goal of 'protection and enhancement' of the environment requires two principal actions--one is the abatement of existing problems, and the other is prevention of future problems; a third category is program support activities, such as surveillance, systems analysis, etc. The most dramatic example of preventing future problems is the 'environmental impact statement'. Two widely divergent approaches are possible in dealing with the mechanics of administering a pollution control program. The problem approach focuses on identifying problems and dealing with them. Pollution prevention characteristically focuses on planning or analyses designed to forecast future problem areas and to head these problems off. An alternative approach is to focus on action requirements which is based on the observation that many problems can be prevented or controlled by controlling the cause. (Auen-Wisconsin)
W73-08649

A SLOW AND PAINFUL PROGRESS TOWARDS MEETING OUR WATER POLLUTION CONTROL GOALS,
S. Soneda.
In: Water Resources Seminar Series No 1, January 1972. Water Resources Research Center, University of Hawaii, Honolulu, p 39-41.

Descriptors: *Water pollution control, *Hawaii, *Regulation, Pollutant identification, Legal aspects, Water quality standards, Water policy, State governments, Management, Tariff.
Identifiers: Mixing zone.

Chapter 37-A of Hawaii's Public Health Regulations classifies the state waters in accordance with the uses to be protected and delineates basic water quality standards applicable to all water areas and

specific standards applicable, depending upon the water area classification. Due to the stringency of the standards a provision was made for a limited zone of mixing for the assimilation of waste discharges which were first required to be subjected to 'the best practicable treatment or control or such lesser degree of treatment or control as will provide for a water quality commensurate with the classified use of the waters outside the zone of mixing'. Recently the regulation has been changed to read 'the best practicable treatment or control'. Those involved considered the term 'best practicable' as including an economic consideration; others are now insisting that this term sould read as 'the best state of the art', exclusive of economic considerations. As Hawaii is committing its financial resources to back the construction of water pollution control facilities, it should consider the next, logical step--the utility business: that part of the business dealing with the treatment-resource recover-disposal of waste water as the primary mechanism to control pollution. (Auen-Wisconsin)
W73-08650

WATER QUALITY IN A STRESSED ENVIRONMENT, READINGS IN ENVIRONMENTAL HYDROLOGY.
For primary bibliographic entry see Field 05G.
W73-08662

COMMENTARY: FILLING AND DREDGING IN MICHIGAN--SOME FURTHER THOUGHTS,
Wayne State Univ., Detroit, Mich.
R. W. Bartke.
Wayne Law Review, Vol 18, p 1515-1526, 1972. 51 ref.

Descriptors: *Judical decisions, *Public rights, *Riparian land, *Riparian rights, *Dredging, Eminent domain, Riparian waters, Navigable waters, Michigan, Highway effects, Construction, Environmental control, Water policy, Law enforcement, Non-navigable waters, Ownership of beds, Environmental effects.
Identifiers: Standing, Injunctions (Prohibitory), Estoppel, Public trust doctrine, Navigability tests, Fill permits.

This comment is an extension and continuation of an earlier note in this Review. The addition has become necessary due to several significant judicial decisions since the first article. Of particular concern is the effect that dredge and fill operations may have upon privately owned lakes and waters adjoining riparian land. The avenues of relief open to such landowners are discussed in detail in connection with recent decisions by the Supreme Courts of Michigan and the United States. The decisions are very dissatisfying because too much emphasis was placed upon navigability of the waters and the public's right to use than upon the crucial concern of the effect of filling and dredging upon the lakes' natural conditions. Numerous legal remedies are available, including injunctions and claims for estoppel. If riparian landowners themselves wish to fill land they should not be automatically permitted to do so since the title they obtain is burdened by a public trust. Bold action on all fronts, legislative, administrative and judicial, is needed to meet the ecological crisis. (Smith-Adam-Florida)
W73-08665

ENVIRONMENTAL IMPACT STATEMENTS--A DUTY OF INDEPENDENT INVESTIGATION BY FEDERAL AGENCIES.
For primary bibliographic entry see Field 06G.
W73-08666

HEARING FOR THE REVIEW AND REVISION OF MISSOURI'S WATER QUALITY STANDARDS.
Missouri Clean Water Commission, Jefferson City.
For primary bibliographic entry see Field 05G.
W73-08668

FEDERAL DECISIONMAKING FOR WATER RESOURCE DEVELOPMENT, IMPACT OF ALTERNATIVE FEDERAL DECISION-MAKING STRUCTURES FOR WATER RESOURCES DEVELOPMENT,
Michigan State Univ., East Lansing. Dept. of Agricultural Economics.
For primary bibliographic entry see Field 06B.
W73-08669

WATER SUPPLIES.
For primary bibliographic entry see Field 05G.
W73-08670

HYDROPNEUMATIC STORAGE FACILITIES.
Illinois State Environmental Protection Agency, Springfield. Div. of Public Water Supplies.
For primary bibliographic entry see Field 05G.
W73-08671

LIMESTONE POLICY.
Illinois State Environmental Protection Agency, Springfield. Div. of Public Water Supplies.
For primary bibliographic entry see Field 05G.
W73-08672

LEGISLATIVE NEEDS IN WATER RESOURCES MANAGEMENT IN ALABAMA,
Alabama Univ., University. Natural Resources Center.
D. M. Grubbs, and H. Cohen.
Available from National Technical Information Service, U.S. Department of Commerce as PB-211 121. Price: $3.00 in paper copy, $1.45 in microfiche. Report prepared for Alabama Development Office, November 1971. 86 p, 82 ref.

Descriptors: *Alabama, *Water management (Applied), *Water resources development, *Water allocation (Policy), *State jurisdiction, *Legislation, Water supply development, Water quality standards, Alternative water use, Water conservation, Water rights, Water law, Legal aspects, Administrative agencies, Governmental interrelations, Riparian rights, Potential water supply, Marshland management, Groundwater resources, Flood plain zoning, Eminent domain, Constitutional law, Planning.
Identifiers: Coastal waters, New York Conservation Law.

This study identifies legislative action needed in the planning methodology to place responsibility and authority in the hands of Alabama state agencies charged with development and conservation of water resources. The dilemma in water supply and water use is discussed with particular emphasis on problems of riparian rights, groundwater use, diffused surface water, control of watercourses, and conflicting claims and rights to water in Alabama. Alternative legislative measures of various states are outlined. Finally, there is a proposed plan of legislative action for long-range development of Alabama's water resources divided into interim legislative needs, primarily concerning conservation and pollution control, and legislation requiring a constitutional amendment aimed at establishing priorities in water use. Legislative action has now become critical to the formulation of long-range plans for the development and management of Alabama's water resources. (Dunham-Florida)
W73-08674

CALIFORNIA LAWYERS LEAD ENVIRO MENTAL FIGHT,
J. Adler.
Juris Doctor, p 23, February 1973. 1 p, 1 fig.

Descriptors: *California, *Legal aspects, *Poll tion control, *Legislation, Administrative de sions, Regulation, Institutions, Public rights, E vironmental effects, Conservation, Resource Natural resources, Land use, Land developme Air pollution, Coasts.
Identifiers: Coastal zone management.

California is the scene of a great deal of enviro mental law activity, which may well preview th to come in other states. At the polls in Novembe environmentalists stunned the large oil, land a utility interests with the passage of the Coastal I itiative, which creates regional commissioners draw up long range plans for the conservation the coastline. Even the most conservative sta legislators are concerned about air emmission and drastic steps may be politically feasible. La use planning is another matter, however, and e vironmentalists and developers are gearing up fo a fight over this element of conservation. Despi intense lobbying, the Environmental Quality A was passed virtually intact. This provided th decisions by local officials on building and cond tional use permits and zoning variances for priva construction require environmental impact stat ments. (Glickman-Florida)
W73-08675

NEW ENGLAND RIVER BASINS COMMI SION--1972 ANNUAL REPORT.
New England River Basins Commission, Bosto Mass.

Report 1972. 36 p, 6 fig, 1 map, 13 photo, 1 chart, append.

Descriptors: *New England, *River basins, *Rive basin development, *River basin commission *Comprehensive planning, Offshore platform Water resources, Water resources developmen Water management (Applied), Water resource Planning Act, Intergovernmental relations, O pollution, Connecticut River, Regional analysis.

Under the Water Resources Planning Act, th New England River Basins Commission serves a the principal agency in its region for the coordina tion of plans for the use and development of wate and related land resources. Results are presente of special studies carried out with reference t Boston Harbor, hydroelectric power, offshore oil and the Nashua River as well as a comprehensiv analysis of the northern states' guide plan pro gram. In addition, an analysis of three separat comprehensive planning programs for water an related land resources is discussed. During 1972 four Connecticut River Basin states and federa agencies approved a comprehensive plan fo management of the water and related lan resources of the basin through 1980. Significan progress was also made toward developing com prehensive programs for protection and manage ment of water and related land resources i southeastern New England and Long Islan Sound. Detailed analyses of specific resource con ditions and needs were undertaken in both areas a a basis for formulating preservation, developmen and management programs. Illustrations, graphs charts, and the Commission's annual financia statement are also included. (Mockler-Florida)
W73-08676

CITIZEN ALERT: PUBLIC PARTICIPATION I THE FEDERAL WATER POLLUTION CON TROL ACT AMENDMENTS OF 1972.
Natural Resources Defense Council, Washington D.C.
For primary bibliographic entry see Field 05G.
W73-08677

pellee denied both contentions. The U.S. Supreme Court held that the grant of a right-of-way of 200 feet on either side of the pipeline violates MLA which only authorizes 25 feet. The court refused to reach claims under NEPA concluding they were not yet ripe for adjudication. The Court instructed the District Court to enter a decree enjoining appellee from issuing permits inconsistent with MLA restrictions. (Dunham-Florida)
W73-08693

DALY V. VOLPE (ENVIRONMENTAL STUDIES REQUIRED FOR HIGHWAY CONSTRUCTION).

250 F. Supp. 252-261 (W. D. Wash. 1972).

Descriptors: *Washington, *Administrative decisions, *Legal review, *Highway effects, *Environmental effects, Judicial decisions, Highways, Legal aspects, Road construction, Environmental control, Decision making, Federal project policy, Wildlife habitats, Marsh management.
Identifiers: *National Environmental Policy Act, *Environmental Impact Statements, Federal Aid Highway Act.

Plaintiffs, residents and landowners, sued to enjoin the construction of a federally funded highway by defendants, government officials. The highway was to go through a wildlife and waterfowl refuge which was privately owned. Plaintiffs maintained that the construction would violate the Federal Aid Highway Act, since dfendants failed to make required environmental findings prior to routing a federal aid highway through a wildlife and waterfowl refuge, and the National Environmental Policy Act (NEPA) because the eivironmental impact statement was inadequate. The U.S. District Court held: (1) the Federal Aid Highway Act is not violated since the wildlife, waterfowl refuge is privately and not publicly owned; (2) that NEPA is applicable to federal aid highway projects; and (3) the NEPA has been violated because the planning report inadequately discussed the adverse environmental effects construction would have upon the wildlife, waterfowl refuge. The court therefore enjoined further construction until an environmental impact statement was filed which met the NEPA required procedural and substantive requisites. (Mockler-Florida)
W73-08694

UNITED STATES V. GRANITE STATE PACKING COMPANY (VIOLATION OF REFUSE ACT OF 1899).

470 F.2d 303-304 (1st Cir. 1972).

Descriptors: *Rivers and Harbors Act, *Water pollution sources, *United States, *Sewage effluents, New Hampshire, Navigable rivers, Legislation, Culverts, Open channels, Water pollution, Industrial wastes, Organic wastes, Environmental sanitation, Local governments, Municipal wastes, Sewage, Environmental effects, Judicial decisions, Legal aspects, Watercourses (Legal aspects), Permits.
Identifiers: *Refuse Act.

Defendant meat packing firm appeals from prosecution for violating the Refuse Act of 1899 brought by the United States government. Appellant's predecessor in occupation constructed a stone culvert to carry away waste discharges. Later, the city connected a sewer to the culvert. Since then appellant and the city have discharged effluents which mingle in the culvert and flow into the Merrimack River, a navigable river covered by the Refuse Act. Appellant contended that the city's extended use of the culvert made it a part of the city's sewer system and that appellant is therefore freed of its liability. The United States Court of Appeals, First Circuit, held that if a party deposits an impermissible substance in a municipal sewer, knowing that the sewer leads directly into navigable water, it causes, suffers, or procures the substance to be discharged into the stream. Moreover, the statute is not restricted to direct deposits. The fact that the Refuse Act empowers the Secretary of the Army to grant exceptions does not aid appellant's argument because the statute makes receipt of a permit a condition precedent to relieve liability from the statutory prohibition. (Dunham-Florida)
W73-08695

BANKERS LIFE AND CASUALTY COMPANY V. VILLAGE OF NORTH PALM BEACH, FLORIDA (SUIT BY RIPARIAN LANDOWNERS TO COMPEL CORPS OF ENGINEERS TO RENEW DREDGE AND FILL PERMITS).

469 F.2d 994 (5th Cir. 1972).

Descriptors: *United States, *Administrative decisions, *Riparian rights, *Permits, Administrative agencies, Water law, Legal aspects, Legislation, Navigable waters, Water conservation, Lakes, Lake shores, Dredging, Environmental effects, Land development, Governmental interrelations, Judicial decisions, State governments, Legal review.
Identifiers: *Army Corps of Engineers, National Environmental Policy Act, Fish and Wildlife Coordination Act, Fill permits.

Plaintiff riparian landowner sued the defendant, Army Corps of Engineers, to compel renewal or issuance of a new dredge and fil. permit. The defendant refused to renew the permit so long as an objection was filed by a state agency involved. It also refused to issue the permit on ecological grounds. Plaintiff maintained that the objection by the state agency was not binding on appellants and since permit criteria did not include ecological considerations when the original permit was issued, such criteria should not be considered now. The United States Court of Appeals held that, where plaintiff waited over five years to object to appellant's denial of an extension, he was in no position to complain if, during such period, requirements for issuance of a permit became more stringent concerning the effect on the environment. The court further held that the Corps could not be kept from complying with subsequent requirements merely because of their stating to plaintiff that an extension of the permit would be granted upon withdrawal of a deferral request by the state agency involved. Finally, the court held that defendant properly denied the permit when so requested by a concerned state agency. (Dunham-Florida)
W73-08696

COWELL V. COMMONWEALTH OF PENNSYLVANIA DEPARTMENT OF TRANSPORTATION (CHALLENGE TO CONDEMNATION PROCEEDINGS INSTITUTED TO PROVIDE CHANNEL IMPROVEMENT).

297 A.2d 529 (Commonwealth Ct. Pa. 1972).

Descriptors: *Pennsylvania, *Eminent domain, *Drainage programs, *Environmental effects, Legal aspects, Judicial decisions, Condemnation, Administrative agencies, Channel improvement, Water pollution, Flood control , Erosion control, Highway effects, Legal review.
Identifiers: Clean Streams Act (Pennsylvania), Administrative Code Act (Pennsylvania).

Plaintiff-condemnee sought to prevent a taking of a portion of her land by eminent domain to improve a highway drainage channel. The channel, arising on one side of the highway, carried water through a pipe beneath the highway and then across plaintiff's land and into a creek. The defendant, highway department, contended that additional condemnation of plaintiff's land was necessary to provide an improved channel resulting in

better protection of the highway from flooding and erosion. Plaintiff contended that defendant failed to comply with Pennsylvania's Administrative Code Act (ACA) requiring an environmental effect report and hearing; that there was no proper public purpose for taking; and that the project contravenes policies set forth in Pennsylvania's Clean Streams Law. The Commonwealth Court of Pennsylvania found: (1) the project did not constitute a transportation route or program within the meaning of the ACA, (2) evidence of continued flooding and erosion unless the channel was altered was a sufficient public purpose, and (3) evidence that the commonwealth had not caused existing pollution indicated the project did not contravene the Clean Streams Law. (Dunham-Florida)
W73-08697

UNITED STATES V. 100 ACRES OF LAND, MORE OR LESS, IN MARIN COUNTY, STATE OF CALIFORNIA (TITLE DISPUTE INVOLVED IN FEDERAL CONDEMNATION PROCEEDINGS OF SMALL PENINSULA).

468 F.2d 1261 (9th Cir. 1972).

Descriptors: *United States, *Boundaries (Property), *Condemnation, *High water mark, Sand spit, Tidal effects, Water levels, Pacific Coast Region, Water law, boundary disputes, Seashores, Sea level, Sand bars, Real property, Legal aspects, Condemnation value, Judicial decisions, Islands, Meanders, Patents, Surveys.

Condemnee landowner claimed title to a small peninsula on the coast of California and sought compensation from the United States government in this federal eminent domain proceeding. The government conceded that appellee was successor in interest to the peninsula, but contended that the parcel of land in question was a small offshore island, and that at the time of the original government patent it was not within the high tide line of the condemnee's land. The government argued the patent was granted only to a portion of the peninsula and that the land in question belonged to the United States, and therefore the United States did not own compensation to appellee. The United States Court of Appeals, Ninth Circuit, held that on consideration of the entire record there was ample evidence to support the findings and conclusions of the trial court including the fact that appellee did possess title to the land in question. (Dunham-Florida)
W73-08698

CHARLES COUNTY SANITARY DISTRICT, INC. V. CHARLES UTILITIES, INC. (ACTION BY WATER AND SEWER CORPORATION SEEKING JUDICIAL VALIDATION OF IT FRANCHISE).
298 A.2d 419 (Ct. App. Md.1973).

Descriptors: *Maryland, *Sewage districts, *Judicial decisions, Sewage treatment, Legal aspects, Governments, Sewage disposal, Liquid wastes, Waste water treatment, Water pollution, Treatment facilities, Administrative agencies.

This case involved an action brought by the complainant water and sewer corporation seeking declaratory and injunctive relief against a sanitary district which refused to recognize or give its approval to a water and sanitary district franchise to the corporation beyond the boundaries of a certain area. The main issue of the case was whether a franchise had already been indirectly granted. The court held that the state tax commission's acceptance of the incorporation papers and the amendments thereto stating the corporation's purpose of operating a water and sewerage system did not constitute the grant of a franchise to the corporation. The court emphasized the granting of a franchise is a legislative function and that the public service commission did not have such authority. Moreover, the court also indicated an order of the state department of health ordering several commercial establishments to hook up to the sewer and water corporation's facilities did not constitute the grant of a franchise to the corporation. (Mockler-Florida)
W73-08699

EXCHANGE NATIONAL BANK OF CHICAGO V. BEHREL (CITY NOT REQUIRED TO SUPPLY WATER TO LANDOWNER).
292 N.E.2d 164 (Ct. App. Ill. 1972).

Descriptors: *Illinois, *Judicial decisions, *Water supply, *Water rights, Legal aspects, Permits, Water utilization, Water policy, Competing uses, Local governments, Water delivery, Water distribution (Applied), Municipal water.

This case involved a declaratory judgment proceeding instituted by plaintiff landowner against the defendant city concerning whether the city was required to supply water to the plaintiff. Previously, the municipality had entered into a contract with an improvement association to supply water. This contract provided that no other parties were allowed to connect to the water main without the joint consent of the parties to the contract. Subsequently, plaintiff's predecessor in title paid the required amount to tap into the water main and was allowed to do so. When plaintiff acquired the property he paid the necessary amount to continue the water rights but the city refused to grant the requested permission. The appeals court affirmed the lower court and held the municipality was within its discretion in refusing to allow plaintiff to tap into the water main. (Mockler-Florida)
W73-08700

TAKING OYSTERS FROM BROADKILN, MISPILLION OR MURDERKILL RIVERS OR THEIR TRIBUTARIES.
Del. Code Ann. tit. 7, secs 2131 thru 2134 (Supp. 1970).

Descriptors: *Delaware, *Commercial fishing, *Oysters, *Legislation, Shellfish farming, Aquatic animals, Aquatic life, Commercial shellfish, Invertebrates, Marine animals, Shellfish, Mollusks, Aquiculture, Industries, Ecology, Environmental effects, Governments, Water resources development, Water law, Administration, Regulation.
Identifiers: Broadkiln River, Mispillion River, Murderkill River, Del.

No oysters less than three inches in length shall be taken from the Broadkiln River or its tributaries, and all oysters taken shall be culled in the stream of water of at least two feet deep at mean low water. Oysters used solely for planting purposes are excepted from this provision. A limit of 25 bushels from the Broadkiln or its tributaries, and a limit of 15 bushels of oysters from the Mispillion and Murderkill Rivers per boat per day is imposed with an allowance of 20% of the total catch made for waste. Taking of oysters from the three rivers and their tributaries is limited to a period between September first and the following March thirty-first. The State Board of Health may permit oysters to be taken from the Broadkiln River for seeding purposes only in home waters. Seed oysters may not be taken from the Mispillion or Murderkill Rivers or their tributaries. (Dunham-Florida)
W73-08701

CRABS AND CLAMS.

Del. Code Ann. tit. 7, secs. 2301 thru 2306 (Supp. 1970).

Descriptors: *Delaware, *Commercial fishi *Legislation, *Crabs, *Clams, Delaware Riv Conservation, Protection, Wildlife conservati Aquatic animals, Aquatic life, Commercial shel ish, Invertebrates, Shellfish, Mollusks, Aquic ture, Marine animals, Ecology, Environmental fects, Governments, Water resources devel ment, Water law, Administration, Regulation.
Identifiers: Hard-shell crabs, Soft-shell cra Peeler crabs, Big Assawoman Bay.

Sections 2301-2306, Title 7, Delaware Code A notated, provide that no person shall take a female crabs bearing eggs or any female crab fr which the egg pouch or bunion has been remove Except as otherwise provided, any U.S. citiz may take hard shell crabs of not less than fo inches in any of the state's tidal waters withou license. A non-resident is limited to 50 crabs a may not use more than four hand lines. Taking hard shell crabs for commercial purposes is limit to the Delaware River and Bay and Big A sawoman Bay, and no person shall have in his p session more than 1 bushel of hard shell cra other than peeler crabs, less than 5 inches acro The taking of soft shell or peeler crabs is limited the same areas with a 3 1/2 inch minimum for s shell crabs and a 3 inch minimum for peelers. Pe sons operating boats using a dredge or rake mu be licensed. It is unlawful to take crabs by dred from March 16 to December 15 of any year. N person shall take any clams in Rehoboth Bay or I dian River Bay by motor powered devices hauled by a boat propelled by motor power. (Du ham-Florida)
W73-08702

LOBSTERS.

Del. Code Ann. tit. 7, secs 2501 thru 2504 (Sup 1970).

Descriptors: *Delaware, *Commercial fishin *Legislation, *Lobsters, Crustaceans, Conserva tion, Protection, Wildlife conservation, Aquati animals, Aquatic life, Commercial shellfish, Inve tebrates, Shellfish, Shellfish farming, Aquicultur Marine animals, Ecology, Environmental effect Governments, Water resources developmen Water law, Administration, Regulation.
Identifiers: Lobster pot.

No person shall take or attempt to take lobster from the waters of Delaware between Septembe first of each year and the last day of the followin April, or between the hours of 2 p.m. every Satu day and 12 midnight of the following Sunday. N person may take lobsters other than by a lobste pot, trap or set; and each person is limited to n more than 50 such devices at one time. Lobster measuring less than 3 1/8 inches or any spawnin lobsters may not be taken from the public water of the state; and lobsters less than the minimu length, whether caught within state jurisdiction limits or not, may not be possessed or offered fo sale. Lobsters of any kind may not be taken fro state waters by non-residents or aliens. Penaltie for violation of these provisions are provide (Dunham-Florida)
W73-08703

SOIL AND WATER CONSERVATION COMMIS SION AND DISTRICTS.

Del. Code Ann. tit. 7 secs. 3901 thru 3912 (Sup 1970).

Descriptors: *Delaware, *Legislation, *Soil co servation, *Water resources development, Lan management, Land resources, Land use, Natur resources, Water resources, Water law, Wate rights, Water quality, Water pollution, Wate management (Applied), Water pollution contro Ecology, Environmental effects, Government Administration, Administrative agencies, Regula tion, Water districts, Conservation.

second largest earth and rock-filled dam in the United States. The appellants contended that the appellees, representatives of the U.S. government, have failed to comply with the provisions of the National Environmental Policy Act (NEPA) requiring an environmental impact statement (EIS), and that they have failed to obtain permits from the proper California agency as required by the California Water Code. The government denied both contentions. The United States District Court held that although the EIS required further supplementation to deal with alternatives to uses which the conservation yield might be put, balancing of the equities required that commencement of construction of the dam not be preliminarily enjoined pending preparation of the supplemented EIS. The court also held that federal agencies were required to obtain proper permits under the California Water Code, and that appellees should apply for the permits prior to the preparation of the NEPA statement. (Dunham-Florida)
W73-08706

YELLEN V. HICKEL (SUIT TO COMPEL ENFORCEMENT OF THE RECLAMATION LAW).

352 F.Supp. 1300-1319 (S.D. Cal. 1972).

Descriptors: *United States, *Judicial decisions, *Federal Reclamation Law, *Land development, *Water control, Legal aspects, Water utilization, Competing uses, Water loss, Water supply, Legislation, Water rights, Irrigation districts.

Plaintiff landowners sought a writ of mandamus to compel the Secretary of the Interior to enforce the Reclamation Law of 1902, which provides that no right to use water for land in private ownership shall be sold for a tract exceeding 160 acres to any one landowner, and no such sale shall be made to any one landowner unless he be an actual bona fide resident thereon. The court held that the Reclamation Law Section was applicable to a contract between an irrigation district and the federal government whereby the district agreed to repay the construction costs of a project, despite the contention of nonapplicability because there was no sale and because a district rather than individual landowners was involved. Moreover the court held this section does not limit residency to the threshold requirement and that a water right can validly be conditioned upon continued residence. In addition, this section of the 1902 Reclamation Law was applicable despite application of the Boulder Canyon Project Act. (Mockler-Florida)
W73-08707

BELLE FOURCHE IRRIGATION DISTRICT V. SMILEY (ACTION TO RESTRAIN RIPARIAN OWNER FROM DIVERTING WATER).
204 N.W.2d 105 (So. Dak. 1973).

Descriptors: *Water rights, *South Dakota, *Water law, *Riparian rights, Prior appropriation, Legal aspects, Judicial decisions, Legislation, Alteration of flow, Diversion, Domestic water, Civil law, Irrigation water, Water policy, Riparian land, Prescriptive rights, Eminent domain, Beneficial use, Water utilization.

Plaintiff irrigation district sought to restrain the defendant riparian landowner from diverting water from a river for irrigation purposes. The defendant contended that he had a vested riparian right to use and divert water for domestic and irrigation purposes by virtue of his ownership of land contiguous to the river, and that right he asserted, became an inseparable incident of the land; to deny him such right deprived him of property without just compensation and due process of law. The South Dakota Supreme Court held for the plaintiff because of a comprehensive water law enacted in 1955. The act limited the riparian doctrine to accrued rights thereunder. Thus, mere ownership of land contiguous to a stream did not carry with it a vested right to divert unlimited amounts of water. There had to be appropriation and application to a beneficial use, and that measure of use was the amount of water divertable by a riparian owner for irrigation. Defendant was allowed to divert water for domestic purposes. (Glickman-Florida)
W73-08708

CARPENTIER V. ELLIS (LANDOWNER'S RIGHT TO LATERAL SUPPORT).
489 S.W.2d 388 (Ct. Civ. App. Texas 1973).

Descriptors: *Texas, *Water law, *Judicial decisions, *Negligence, *Adjacent landowners, Structures, Legal aspects, Water policy, Accidents, Risks, Wave action, Shore protection, Soils, Retaining walls, Abutments, Bulkheads.

Plaintiffs and defendants were landowners of adjacent lots on a lake that was not filled in until after each had purchased his lot. Plaintiffs constructed a bulkhead in front of their lot on the lake side, and later defendants likewise built a retaining wall that abutted plaintiffs. As the lake filled, wave action caused defendants' retaining wall to collapse, and plaintiffs alleged that because of the loss of lateral support, damage was caused to their wall. Plaintiffs contended that defendants owed them a duty of lateral support, and sought injunctive relief to compel defendants to contruct a proper retaining wall as well as money damages. The appellate court held for the defendants, stating that although an owner of land had an absolute right to the lateral support of adjoining land, this applied only to soil and not to buildings or other structures placed upon the land. As to the latter, it was essential to show negligence to recover. Plaintiffs made no allegations or proof of negligence, and thus could not obtain relief. (Glickman-Florida)
W73-08709

AYERS V. TOMRICH CORPORATION (LIABILITY FOR ALTERATION OF SURFACE RUNOFF FLOW).
193 S.E.2d 764 (Ct. App. North Carolina 1973).

Descriptors: *Alteration of flow, *North Carolina, *Drainage effects, *Drainage water, *Judicial decisions, Water law, Water policy, Legal aspects, Diversion structures, Flood control, Natural flow doctrine, Reasonable use, Relative rights, Water spreading, Earthworks, Diversion.

Plaintiff landowner sued defendant, a land developer, for damages to plaintiff's land and destruction of a bridge thereon. Defendant had piled dirt 30 feet high immediately adjacent to plaintiff's property, and following a heavy rainwater, rocks and mud flowed onto plaintiff's land causing the damage. The court held that a lower landowner is not required to receive such material which in the natural condition of the lands would not be carried by the normal flow of surface waters from the upper to the lower tracts. The court did not uphold the claim for destruction of the bridge, because it was merely destroyed by water, and a lower tract of land is burdened with an easement to receive waters from the upper tract, which naturally flows therefrom. Further, the owner of the higher tract may even increase the natural flow of water, so long as he does not divert it. However, the court remanded the case for a new trial on the issue of damages to the land itself resulting from the runoff of dirt and rocks from defendant's embankment. (Glickman-Florida)
W73-08710

INTERSTATE MOTELS, INC. V. BIERS (SUIT TO ENJOIN INTERFERENCE WITH ACCESS AND DRAINAGE RIGHTS).
193 S.E.2d 658 (Virginia 1973).

Descriptors: *Virginia, *Riparian rights, *Riparian land, *Water rights, *Judicial decisions, Tidal streams, Drainage, Water law, Legal aspects, Severance, Riparian waters, Water sources, Low water mark, High water mark, Outlets, Drainage practices, Watercourses (Legal aspects).
Identifiers: Subdivision plats.

Complainant and respondent owned adjoining lots. Complainant sought to enjoin respondent from damming a cove of a river which crossed respondent's lot and extended into complainant's, and claimed riparian and drainage rights in the cove. Since the purchase of his lot complainant had used the cove as an access to the river. Respondent contended that the original owner of both lots had manifested on the subdivision plat an intention to sever from complainant's lot the riparian rights and drainage rights in the cove in question. The court granted complainant an injunction, and held that if the original owner had intended to sever the riparian rights appurtenant to complainant's lot, it was necessary that he clearly manifest his intent to do so. The court felt that, to the contrary, he had displayed the intent that the riparian rights appurtenant to the lot were to be enjoyed by its owner, as indicated by the plat, which depicted the cove as extending into complainant's lot. (Glickman-Florida)
W73-08711

DOLPHIN LANE ASSOCIATES, LTD. V. TOWN OF SOUTHAMPTON (RIPARIAN OWNER'S RIGHTS IN INTERTIDAL AREAS).

339 N.Y.S.2d 966 (Sup. Ct. N.Y. 1971).

Descriptors: *New York, *Intertidal areas, *High water mark, *Public rights, Legal aspects, Zoning, Fluctuations, Water levels, Easements, Navigable waters, Water law, Local governments, Judicial decisions, Riparian rights, Riparian land, Boundaries (Property), Public access.
Identifiers: *Public trust doctrine.

Plaintiff landowner commenced this action to declare unconstitutional certain zoning ordinances enacted by the defendant municipality. The ordinance claimed on behalf of the people in the municipality superior title and/or interest to certain portions of the real property claimed to be owned by the plaintiff. The court held that a conveyance by a governmental agency of land fronting on a navigable body of water conveys title only to the high-water line and the applicable act of 1818 reserved to the inhabitants of the municipality a public easement over the beach land above the high water mark. This mark was defined by the court as a line marked by periodic flow of the tide, excluding any advance of water caused by winds, storms, and unusual conditions. The court went on to point out that the existence of certain flora was evidence of the high water mark, since it is a type of grass which thrives naturally in saltwater areas only if the soil from which it grows is regularly inundated twice a day by tidal flow. The court therefore held the ordinance was valid. (Mockler-Florida)
W73-08712

PIESCO V. FRANCESCA (VALIDITY OF WATERFRONT ZONING ORDINANCE).

338 N.Y.S.2d 286 (Sup. Ct. N.Y. 1972).

Descriptors: *New York, *Water zoning, *Land use, *Riparian rights, Zoning, Legal aspects, Local governments, Public rights, Water law, Marinas, Constitutional law, Judicial decisions, Legislation, Legal review, Non-structural alternatives, Riparian land.

This was an action brought by plaintiff landowner against defendant municipality seeking a judgment declaring a zoning ordinance passed by the municipality invalid insofar as it affected the plaintiff's waterfront property. The ordinance required plaintiffs to obtain special permits in order to build a marina. The court held that the upland lot owner was not entitled to a declaratory judgment that the zoning ordinance was confiscatory or invalid unless there was a factual showing to permit a determination of whether the plaintiff had suffered any significant economic injury. The court also indicated it was not discriminatory to allow other owners of waterfront property having nonconforming marina uses at the time of enactment of the zoning ordinance to continue such use and at the same time to disallow other owners a nonconforming use when application was made after enactment of the zoning regulation. Thus, the effect on plaintiff's riparian rights was held not invalid or discriminatory. (Mockler-Florida)
W73-08713

MCQUINN V. TANTALO (EXISTENCE OF EASEMENT OF NECESSITY WHERE PLAINTIFF'S LAND WAS ACCESSIBLE BY NAVIGABLE WATER).
339 N.Y.S.2d 541 (Sup. Ct. N.Y. 1973).

Descriptors: *New York, *Riparian land, *Easements, *Adjacent landowners, Equity, Judicial decisions, Right-of-way, Trespass, Water rights, Water law, Legal aspects.

Plaintiff riparian landowner brought suit to establish a right-of-way across the defendant adjacent landowner's lot to afford access to his lot. Plaintiff had access to his lot only via a navigable waterway. The court held that where the plaintiff's land was accessible by navigable water which plaintiff had the right to use, no way of necessity existed that would support an implied easement across the defendant's lot to afford plaintiff's access to his lot. (Mockler-Florida)
W73-08714

TEMPLETON V. HUSS (LIABILITY FOR ALTERATION OF SURFACE DRAINAGE).
292 N.E.2d 530 (Ct. App. Ill. 1973).

Descriptors: *Illinois, *Surface drainage, *Natural flow, *Alteration of flow, Judicial decisions, Legal aspects, Water rights, Diversion, Reasonable use, Relative rights, Surface runoff, Drainage effects, Drainage practices, Land development.

Plaintiff landowner sued for an injunction and damages against defendant subdividers alleging that in constructing a subdivision the defendants changed the natural course of the surface drainage and brought water from a different watershed into the natural watershed in which the plaintiff's land was located. Even though expert testimony was introduced, it failed to prove water from a different watershed was diverted onto plaintiff's land. The court held that owners of servient land have no cause of action against the owner of a dominant tract who alters its condition in such a manner that water, in the natural course of surface drainage, flows onto the servient tract at an increased rate and in much greater quantities. However, the court emphasized in dicta that the owner of a dominant tract cannot divert the natural course of drainage to bring in water from another watershed or remove natural barriers, thus allowing water which would not otherwise flow in such a direction to be deposited on the servient owner's land. Since plaintiff failed to prove this the relief sought was denied. (Mockler-Florida)
W73-08715

ZABEL V. PINELLAS COUNTY WATER AND NAVIGATION CONTROL AUTHORITY (GROUNDS FOR DENIAL OF DREDGE AND FILL PERMIT).
171 So. 2d 376-388 (Fla. 1965).

remove or arrange for the removal of the pollutant. The person responsible for the discharge shall be liable to such governmental body for the costs incurred in the cleanup operation. That person may still remain liable for damage to the property resulting from the discharge. (Glickman-Florida)
W73-08720

WATER SUPPLY AND SEWAGE SYSTEM.

Ill. Ann. Stat. ch. 24 secs 11-125-1 thru 11-139-11 (Smith-Hurd Supp. 1971).

Descriptors: *Illinois, *Water supply, *Sewage systems, *Legislation, Water works, Eminent domain, Condemnation, Water conveyance, Water control, Municipal wastes, Legal aspects, Bond issues, Costs, Local governments, Government finance, Municipal water, Water rates, Water management (Applied).

In order to construct a waterworks system or sewage system, each municipality has the power to levy special assessments on all property benefited therefrom within one mile from its corporate limits. The proceeds of the tax may be used solely for waterworks planning or construction. A municipality may also pay for such waterworks by the issuance and sale of revenue bonds. The issuance of these bonds shall be provided for by ordinance passed by the corporate authorities of the municipality. The city has the power of eminent domain. Several municipalities may jointly acquire and operate a waterworks which shall be run by a joint water commission. This commission is also authorized to issue revenue bonds to finance the project. Water may be sold to non-party municipalities or other political subdivisions, corporations or persons by the Commission. Any municipality may acquire and operate a combined waterworks and sewerage system and may finance it by issuing revenue bonds or applying federal monies. The municipalities may enact any rules and regulations necessary to maintain the system and preserve the public health. (Glickman-Florida)
W73-08721

STATE V. HARDEE (OWNERSHIP OF TIDELANDS AND SUBMERGED LANDS).
193 S.E.2d 497-509 (So. Car. 1972).

Descriptors: *South Carolina, *Intertidal areas, *Boundaries (Property), *Highwater mark, Tidal streams, Low water mark, Water law, Judicial decisions, Estuaries, Legal aspects, Water policy, Streambeds, Beds underwater, Beds, Navigable waters, Boundary disputes.
Identifiers: *Public trust doctrine.

The respondent, state, alleged in its complaint ownership of all tidelands, that being the area between the usual high and low water marks, and all submerged lands, that being the area below the usual low water mark. The state further alleged that it held title in trust for the people of the state. The appellant riparian landowner alleged ownership of the land in question down to the usual low water mark by virtue of a deed. The Supreme Court of South Carolina upheld the decision of the trial court in favor of the state. The court first stated that in determining the extent of the boundary of a body of land a different rule applied to tidal navigable streams such as the one in question than to nonnavigable streams. As to the latter, the boundary was generally conceded to be the middle of the stream, while as to the former it was deemed to be the high water mark in the absence of language to the contrary in a deed or plat. The appellant here failed to show such specific language in her deed, and thus the portion of land between the high and low water marks remained in ownership of the state for the benefit of the people. (Glickman-Florida)
W73-08722

SOLID WASTE DISPOSAL.

42 USC secs 3251 thru 3259 (1972).

Descriptors: *United States, *Waste treatment, *Waste disposal, *Solid wastes, Waste storage, Governments, State governments, Governmental interrelations, Water pollution sources, Pollution abatement, Water treatment, Legislation, Water law, Water policy, Radioactive waste disposal, Wastes, Local governments.

While the collection and disposal of solid wastes should continue to be primarily the function of state, regional and local agencies, the problems of waste disposal have become a matter national in scope which necessitates federal action through financial and technical assistance and leadership in the development of new and improved methods to reduce the amount of waste and unsalvagable materials and to provide for proper solid waste disposal practices. The Secretary of HEW shall conduct and render assistance to other agencies in the conduct of research and investigations to implement the purpose of this section, and shall also make public any information obtained through these experiments. The Secretary shall encourage cooperative activities by the states and local governments in connection with solid waste disposal programs. The Secretary may authorize grants to various administrative agencies to carry out the purposes of this act. The Secretary shall recommend to appropriate agencies guidelines for solid waste recovery, collection and separation systems which shall be consistent with the public health and welfare, and shall submit to Congress a comprehensive report for the creation of a system of national disposal sites for hazardous wastes. (Glickman-Florida)
W73-08723

CONTROL OF FLOODS AND CONSERVATION OF WATER.
Ill. Ann. Stat. ch. 19 secs. 121 thru 128.3 (Smith-Hurd 1972).

Descriptors: *Illinois, *Flood protection, *Flood control, *Watershed management, *Legislation, Administrative agencies, Eminent domain, Condemnation, Public health, Flood damage, Flood forecasting, Water policy, Watersheds (Basins), Floods, Water law, Rivers, Regulation, Flow control, Dams, Hydroelectric power.

The unregulated flow of the rivers and waters of Illinois constitutes a menace to the general welfare of the people of Illinois. Regulation is thus a proper activity of the state as are investigations and improvements of the rivers and waters for the purpose of controlling floods and low waters. The Illinois Department of Transportation is authorized to make the necessary examinations and constructions of improvements, either by doing the work itself or letting contracts to do the work. The Department may enter into cooperative agreements with federal and municipal agencies to implement the purpose of this section. Where surplus waters become available in the construction of works, the Department may operate a power plant and dispose of the power so generated. The Department may exercise the power of eminent domain if necessary. In the case of emergencies due to floods, the Department may make expenditures for the purpose of flood relief. Watershed protection is found to offer a sound approach to flood protection. The Illinois Department of Agriculture is authorized to plan for works of improvement in any approved watershed in the state. (Glickman-Florida)
W73-08724

REPORT OF THE TEXAS WATER DEVELOPMENT BOARD FOR THE BIENNIUM SEPTEMBER 1, 1970 THROUGH AUGUST 31, 1972.
Texas Water Development Board, Austin.

Report, 1973. 75 p, 31 fig, append.

Descriptors: *Water resources development, *Groundwater resources, *Surface waters, *Texas, Planning, Projections, Evaluation, Water quality, Water yield, Water users, Water demand, Hydrologic data, Reviews, Sedimentation, Aquifers, Streamflow, Water utilization, Water conservation, Well data, Estuaries, Land use, Legal aspects, Watershed management.

The primary goal of the Texas Water Development Board is to assure that the present and future water requirements of the people of Texas are met. In the biennium, September 1, 1970 through August 31, 1972, the Board oriented its program and activities toward solution of present or immediately foreseeable water supply problems. Some internal agency reorganization was initiated to improve functional operations. Techniques of program monitoring and control were employed to improve program performance and budget efficiency. The Board's programs and organization and some issues involving water resource development are described. Approximately 50 reports published during the biennium are listed in the appendix. (Woodard-USGS)
W73-08732

THE CALIFORNIA STATE WATER PROJECT SUMMARY: NINETEEN-SEVENTY-ONE.
California State Dept. of Water Resources, Sacramento.
For primary bibliographic entry see Field 04A.
W73-08749

A DESCRIPTION AND ANALYSIS OF COASTAL ZONE AND SHORELAND MANAGEMENT PROGRAMS IN THE UNITED STATES,
Michigan Univ., Ann Arbor. Sea Grant Program.
For primary bibliographic entry see Field 02L.
W73-08887

PUBLIC WATER SUPPLY SYSTEMS RULES AND REGULATIONS.
Illinois State Dept. of Health, Springfield. Div. of Sanitary Engineering.

Illinois Department of Public Health, Division of Sanitary Engineering, Springfield, (1970). 24 p.

Descriptors: *Administration, *Water works, *Water treatment, *Illinois, *Regulation, *Municipal water, Permits, Water quality standards, Legal aspects, Construction, Public health, Water supply, Water distribution (Applied), Water supply development, Engineering, Safety, State governments.
Identifiers: Water supply rules.

The safeguarding of public health through water supply system control was set forth by the State of Illinois Department of Public Health in July 1959. In 1970, the establishment of the Illinois Environmental Protection Agency was accompanied by the transfer of the Division of Public Water Supplies from the Department of Public Health to the Illinois Environmental Protection Agency. The rules and regulations were transferred unchanged. There are six main articles in the rules, namely: (1) definitions, (2) submission of plan documents, (3) design of waterworks, (4) protection of water supplies during repair work or construction, (5) operation and maintenance, and (6) rules of practice in administrative hearings. Article three, pertaining to waterworks design, covers eight topics; these are (1) surface water supplies, (2) groundwater supplies, (3) pumping stations, (4) storage reservoirs, (5) distribution systems, (6) chlorination, (7) cross-connections and (8) disinfection. (Poertner)
W73-08893

INDUSTRIAL WASTE MANUAL, A GUIDE FOR THE PREPARATION OF APPLICATIONS, REPORTS AND PLANS.
Pennsylvania Dept. of Environmental Resources, Harrisburg. Bureau of Water Quality Management.

Publication Number 14, 1971. 35 p.

Descriptors: *Industrial wastes, *Administration, *Regulation, *Pennsylvania, Water quality control, Water pollution control, Permits, Legal aspects, State governments.
Identifiers: *Industrial waste manual.

Because streams serve as both sources of water supply and as receivers and carriers of waste water from homes, farms, factories and mines, conflicts of interest between water users and waste dischargers are inevitable. The Pennsylvania Department of Environmental Resources has been empowered by law to abate surface and groundwater pollution and has written this manual as a guide for the preparation of waste discharge applications, reports, and plans from industrial sources. Minimum standards are set, with the understanding that further revision and more stringent standards may be forthcoming. This guide is divided into four parts, namely: (1) general information on the Clean Streams Law of Pennsylvania, the Bureau of Water Quality Management and the role of other agencies in waste discharges; (2) the procedure for obtaining a permit; (3) waste treatment requirements; and (4) pollution incident prevention. It is emphasized that this publication is intended only as a guide and that specific instances may require special regulations. Acid mine drainage from coal or clay mines are not covered in this manual, but this is covered in a separate report. All other industrial processes are covered by this guide. (Poertner)
W73-08895

INCOG FUNCTIONAL PLAN SUPPLEMENTS (WATER AND SEWER, HOUSING AND OPEN SPACE).
Indian Nations Council of Governments, Tulsa, Okla.
For primary bibliographic entry see Field 06B.
W73-08989

A MODEL OF A PRIVATE ENVIRONMENTAL AGENCY,
T. A. Ferrar, and A. Whinston.
(1972), 28 p, 2 fig, 9 ref. OWRR-B-020-IND (17). 14-31-0001-3080.

Descriptors: Environment, *Management, Quality control, *Governments, Economics, *Pollution abatement, *Water quality, *Social needs, Resource allocation, Economies of scale, Marginal costs, Optimization, Constraints, Waste disposal, Air pollution, Numerical analysis, Equations, Reach (Streams), Environmental sanitation.
Identifiers: *Private agencies, *Environmental use permits, Environmental property rights, Environmental assimilative capacity, Free enterprise, Market equilibrium price.

The efficacy of creating government-regulated private agencies to control a region's air and water quality is considered. Assumed is the formation of a government commission responsible for specifying environmental quality standards as a measure of the legal supply of naturally available environmental waste disposal capacity. This environmental commission would establish a private environmental agency giving it the responsibility of maintaining the environmental quality standards in the given region, and giving it the right to sell user permits or environmental property rights to potential municipal or industrial polluters; the revenue obtained from the sale of these rights would be invested in environmental improvement facilities, reducing the burden on the established municipal and industrial structure while protecting the en- vironment to the specified level. Specifically, the paper portrays the existence of two key priva public sector linking parameter sets, and outli the welfare trade-off existing between envir mental use permit production and capital resou efficient utilization. The framework presented p mits the criterion efficiency to modify the in politically related structure to one reflecti economies of scale, thus restoring the competiti efficiency characteristics of free enterprise ope tion to the area of environmental manageme (Bell-Cornell)
W73-09009

JIM KEE SPEAKS ON WATER RESOURCES,
House, Washington, D.C.
J. Wright.
Congressional Record, Vol. 118, p E8861-E88 (daily ed.) Oct. 18, 1972. 2 p.

Descriptors: *Bank protection, *Legislatio *Ohio River, *Flood control, *Water polluti control, Federal government, West Virginia, Ju cial decisions, Water pollution, Navigable water Cost-benefit analysis, Legal aspects, Water la Administrative agencies, Natural resource United States, Supervisory control (Powe Floods, Flood protection.
Identifiers: Standing (Legal), National Enviro mental Policy Act, Federal Water Pollution Co trol Act Amendments of 1972, Injunctio (Prohibitory).

This statement concerns legislation coming out o the 92nd Congress. First, the omnibus rivers an harbors and flood control bill is discussed. Th legislation contains an authorization for bank pr tection works along the Ohio River from upstrea New Matamoras to Cincinnati and local flood pr tection works at Chillicothe prior to the beginnin of construction of the Mill Creek Reservoir. The there is the Federal Water Pollution Control A Amendments of 1972. Coupled with legislation a ready passed this legislation will go far in reducin and eliminating pollution to rivers and stream The question of standing and the use of injuncti relief is reviewed in relationship to Sierra Club v Morton. The court held: a person has standin only if he can show that he himself has suffered o will suffer injury, whether economic or otherwis Proposed legislation for the 93rd Congress o limiting increases on interest rates used in figurin cost-benefit ratios is discussed. The National En vironmental Policy Act of 1969 is also discusse (Tolle-Florida)
W73-09011

NEW CONCEPTS IN SITING OF NUCLEA POWER PLANTS,
House, Washington, D.C.
J. L. Evins.
Congressional Record, Vol. 118, p. E8820-E882 (daily ed.) October 18, 1972. 3 p, 15 ref.

Descriptors: *Nuclear energy, *Nuclear power plants, *Environmental effects, *Power system operation, Breeder reactors, Desalination Desalination plants, Offshore platforms, Electrici ty, Water resources development, Rivers Streams, Oceans, Thermal pollution, Nuclea wastes, Water pollution sources, Energy.
Identifiers: *Atomic Energy Act of 1954.

If the needs and goals of our energy-dependen society are to be met, this country must efficiently tap all its available resources. However, publi concern over possible environmental effects, and a growing competition for available land and wate resources have made it more difficult to obtai nuclear power plant sites. The Commissioner of the Atomic Energy Commission addressed himself to this problem in a speech inserted in the Con gressional Record. The Commissioner discussed the status of nuclear power in the United States

ıd the uses to which it could be put in desalina- ɔn processes. The dual purpose vector might give ıe to the energy center. An energy center means ı integrated complex in which the large scale ·oduction of energy is effectively combined with ɔnsuming industries at a single site or contiguous ·ea. These centers might also form the nucleus of new city. Power plants might be located on large arges or man-made islands. The power plant park ıd its attendant difficulties are also discussed. urrent research on other potential sources of lectric power is also mentioned. (Reed-Florida)
/73-09012

'OREIGN POLICY ISSUES AND THE LAW OF 'HE SEA CONFERENCE,
Iouse, Washington, D.C.
. M. Pelly.
Congressional Record, Vol. 118, p. E8931-E8934 daily ed.) October 25, 1972. 4 p.

Descriptors: *International law, *International waters, *Law of the sea, *United Nations, United States, Federal Government, Natural resources, Investment, Oceans, Commercial fishing, Pacific Northwest U.S., Legal aspects, Jurisdiction, Political aspects, Washington, Marine fisheries.
Identifiers: International Law of the Sea Conference.

This address was delivered on October 11, 1972, at a meeting at the University of Washington in Seattle on the local aspects and interest in United Nations' Meetings on the International Law of the Sea. The speaker was the Hon. John R. Stevenson, the legal advisor to the Department of State. The general outline of issues and the difficulties in establishing agreement on new and up-to-date international conventions were presented. Because of the divergence of view regarding what is permitted under current international law, there is too great a tendency to overlook the gradual convergence of ideas on what effect of the Law of the Sea Conference should be. The overwhelming majority of states from all regions are supporting agreement on a 12-mile territorial sea. Also there is widespread support for the establishment of an international regime for the seabeds in the area beyond coastal state economic jurisdiction. The major problems to be resolved include: (1) free transit of straits used for international navigation; (2) the nature of coastal state rights over resources beyond the territorial sea; (3) treaty standards protecting the integrity of investment in developing coastal countries; and (4) scientific research, particularly in areas of coastal state resource jurisdiction. (Tolle-Florida)
W73-09013

NEW TECHNIQUES IN WATER POLLUTION CONTROL,
House, Washington, D.C.
G. Gude.
Congressional Record, Vol. 118, p E8947 (daily ed.) October 25, 1972. 1 p.

Descriptors: *Nutrient removal, *Water pollution control, *Phosphorous, *Sewage treatment, Sanitary engineering, Great Lakes, Sewage effluents, Canada, Sewage, Illinois, Waste water (Pollution), Sewerage, Pumping plants, Environmental engineering, Public health, Biological treatment, Treatment facilities, Water quality control.

A new biological process was developed to successfully remove phosphorus from sewage. Biospherics Incorporated reports that its PhoStrip method reduced sewage effluent phosphorus levels to meet regulations imposed by the State of Illinois this year on the Great Lakes and the Fox River. Similar regulations have been adopted by Canada and the United States to curb pollution in the Great Lakes. Tests conducted in Chicago lasted 22 days, and represented the third sewage treatment plant where the Biospherics PhoStrip process was pilot tested. The others were the Washington, D.C. Water Pollution Control Plant, and the Piscataway Plant in southwestern Prince George's County, Maryland. The phosphorus removed from the sewage effluent was concentrated into a waste flow of approximately five percent of the total plant flow resulting in proportionately lower treatment costs. The phosphorus could be chemically removed from this relatively small volume by currently available standard methods. Complete details of the process will be published in the October 1973 issue of the Journal of the Water Pollution Control Federation. (Tolle-Florida)
W73-09014

HERCULES CONTINUES TO FIGHT POLLUTION,
House, Washington, D.C.
E. J. Patten.
Congressional Record, Vol. 118, p. E8988 (daily ed.) October 25, 1972. 1 p.

Descriptors: *Pollution abatement, *Industrial wastes, *Pollutant identification, *Discharge (Water), Regulation, Communication, Annual costs, Waste water (Pollution), Water pollution control, Water pollution effects, Air pollution, Air pollution effects, New Jersey, Industrial plants, Water quality control, Waste treatment, Treatment facilities.

Those companies doing outstanding jobs of trying to clean up the environment are commended. Such a company is Hercules, Inc., operating one of its plants in Parlin, New Jersey. A formal corporate policy was developed by Hercules in 1967 to: 'do more than just comply with (pollution) regulations. We want to fully carry out our responsibilities for air and water quality in the communities of which we are a part'. In order to carry out this policy, Hercules set up an Environmental Health Committee. The Committee, formed in late 1967, monitors air, water and solid waste pollution abatement practices in 48 principal domestic plants located in 25 states. Although the Committee sees to it that each Hercules operating department pays attention to abatement problems and regulations, the plant manager is directly responsible for pollution control in his plant. At Hercules, each plant has a pollution abatement coordinator responsible for all pollution control activities. In a study made last year an estimated 35 percent of the operating expenses of pollution control facilities was returned to the plant by more efficient plant operations, the recovery of waste materials, and the reduction in water usage and other utilities. (Tolle-Florida)
W73-09015

6F. Nonstructural Alternatives

DOLPHIN LANE ASSOCIATES, LTD. V. TOWN OF SOUTHAMPTON (RIPARIAN OWNER'S RIGHTS IN INTERTIDAL AREAS).
For primary bibliographic entry see Field 06E.
W73-08712

PIESCO V. FRANCESCA (VALIDITY OF WATERFRONT ZONING ORDINANCE).
For primary bibliographic entry see Field 06E.
W73-08713

6G. Ecologic Impact of Water Development

UNDERGROUND NUCLEAR POWER PLANT SITING,
Aerospace Corp., San Bernardino, Calif.
M. B. Watson, W. A. Kammer, L. A. Selzer, R. L. Beck, and N. P. Langley.
California Institute of Technology, Environmental Quality Laboratory, Pasadena, EQL Report No 6, September 1972. 42 fig, 4 tab, 10 ref, 7 append.

Descriptors: *Nuclear power plants, *Structural engineering, *Estimated costs, *Safety, *Underground powerplants, Electric power production, Engineering, Geology, Earthquakes, Estimating, Evaluation, Estimated benefits, Model studies, Systems analysis, Structural design, Underground structures, Rock mechanics.

This study is part of a larger evaluation which examines the problems associated with the siting of nuclear power plants. It is based on four partially buried European facilities and the discussion is restricted to light-water nuclear power plants of about 1000 Mwe, using present designs for components of the nuclear steam system. The postulated advantages to underground siting would be greater safety, reduced area requirements and improved aesthetics. The basic structural configurations and equipment packaging are presented. Three different underground plants are examined; two which represent the straight-forward adaptation of surface PWR and BWR plants and the third represents a reconfigured BWR plant in which the pressure supression emergency system has been eliminated. Structural liners and the depth of burial of the plants are discussed. Typical rock media should provide adequate insulation against leakage of radioactive material. A rough estimate of the cost penalty associated with underground siting is presented, based on 1970-1972 construction costs. (Jerome-Vanderbilt)
W73-08617

CRUDE OIL AND NATURAL GAS PRODUCTION IN NAVIGABLE WATERS ALONG THE TEXAS COAST (FINAL ENVIRONMENTAL IMPACT STATEMENT).
Army Engineer District, Galveston, Tex.
For primary bibliographic entry see Field 05G.
W73-08661

WATER QUALITY IN A STRESSED ENVIRONMENT, READINGS IN ENVIRONMENTAL HYDROLOGY.
For primary bibliographic entry see Field 05G.
W73-08662

ENVIRONMENTAL DATA BANK.
For primary bibliographic entry see Field 07C.
W73-08663

RIVER ECOLOGY AND MAN.

Proceedings of an International Symposium on River Ecology and the Impact of Man, held at the Univ. of Mass., June 20-23, 1971. Academic Press, New York, 1972, R.T. Oglesby, C.A. Cadson and J.A. McCann, editors. 465 p, 17 fig, 2 map, 12 photo, 9 tab, 5 chart, 164 ref.

Descriptors: *Geomorphology, *Sedimentology, *Aquatic life, *Pesticide residues, *Radioactive wastes, Ecosystems, Water pollution sources, Poisons, Balance of nature, Water pollution effects, Thermal pollution, Disharge (Water), Chemical analysis, Marine geology, Environmental effects, Marine plants, Marine animals, Rivers, Columbia River, Delaware River, Waste discharges, Nutrients.
Identifiers: Illinois River.

In a series of symposium papers, the term river is described and defined in terms of the flora, fauna, chemical composition and geomorphometry by specialists in each field. Case histories of specific rivers, past and present, throughout the world are given with special emphasis placed on man's interaction with them. There is a consideration of the effects of various river uses on morphometry,

discharge and sedimentation; and analyses of the impact of heat, radionuclides, pesticides and industrial wastes on river ecosystems are also presented. Finally, the economic, environmental and political implications of the multiple use of rivers in an attempt to avoid disastrous side effects in the future are explored. (Dunham-Florida)
W73-08664

ENVIRONMENTAL IMPACT STATEMENTS--A DUTY OF INDEPENDENT INVESTIGATION BY FEDERAL AGENCIES.

University of Colorado Law Review, Vol 44, p 161-172, 1972. 63 ref.

Descriptors: *Legislation, *Administrative decisions, *Decision making, *Federal project policy, *Administrative agencies, Judicial decisions, Federal jurisdiction, Legal review, Environmental control, Permits, Law enforcement, Project planning, Adjudication procedure, Environmental effects, Planning, Water resources development, Legal aspects.
Identifiers: *National Environmental Policy Act, Environmental Impact Statements.

The Second Circuit Court of Appeals has recently held that the detailed environmental impact statements required by the National Environmental Policy Act (NEPA) must be prepared and developed by the agency itself which is proposing the action. The agency cannot satisfy NEPA by preparing the statement on the basis of information provided by others. This case, Green County Planning Board v. Federal Power Commission, makes it clear that federal agencies have an affirmative duty not only to consider environmental factors and include them in the record, but to do so at every stage of the process. This duty cannot be delegated. The practical impact of the holding is to increase substantially the agency's work load. The practical ability of the federal agencies to supply the type of analysis required is doubted, unless further funds are made available for NEPA implementation. The thrust of the decision is to neutralize technological bias by prohibiting industry participation in the planning stages. NEPA is in trouble due to the distance between the goals it sets and the means and will available to implement its procedures. However, this decision may do a great deal to further long-range environmental planning under NEPA. (Smith-Adam-Florida)
W73-08666

AN ANALYSIS OF ENVIRONMENTAL STATEMENTS FOR CORPS OF ENGINEERS WATER PROJECTS,
Stanford Univ., Calif. Dept. of Civil Engineering.
L. Ortolano, and W. W. Hill.
Available from National Technical Information Service as AD-747 374, Price: $3.00 in paper copy, $1.45 in microfiche. Report to Army Engineer Institute for Water Resources, June 1972. 135 p, 15 tab.

Descriptors: *Coastal structures, *Inland waterways, *Engineering structures, *Environmental effects, Jetties, Breakwaters, Dredging, Coastal engineering, Spoil banks, Groins, Dikes, Barriers, Ecology, Channels, Reservoirs, Levees, Construction, Conservation, Water resources development, Water pollution, Federal government, Administrative agencies, Dam construction, Federal project policy, Channeling.
Identifiers: *Army Corps of Engineers, *National Environmental Policy Act, *Environmental Impact Statements, Revetments.

The first 234 environmental impact statements prepared by the Army Corps of Engineers through August of 1971, in response to requirements of the National Environmental Policy Act (NEPA), are analyzed. Included is a catalog of impact statements along with a summary and discussion of the statements in two main areas: (1) projects on coastal waters, and (2) projects on inland waters. The coastal waters projects are subdivided into: (a) dredging, (b) spoil disposal, (c) breakwaters, (d) jetties and groins, and (e) revetments, dikes and barriers. The inland waters projects are subdivided into: (a) channelization, (b) dams and reservoirs, (c) levees, (d) dredging, (e) spoil disposal, (f) construction activities, and (g) miscellaneous structures and activities. There is a discussion of alternatives, short term uses and long term productivity, irreversible and irretrievable resource commitments, and controversial issues. Finally there are suggestions for improving the content of future impact statements to better comply with NEPA and Council on Environmental Quality guidelines. (Dunham-Florida)
W73-08673

LEGISLATIVE NEEDS IN WATER RESOURCES MANAGEMENT IN ALABAMA,
Alabama Univ., University. Natural Resources Center.
For primary bibliographic entry see Field 06E.
W73-08674

CALIFORNIA LAWYERS LEAD ENVIRONMENTAL FIGHT,
For primary bibliographic entry see Field 06E.
W73-08675

STATEMENT OF FINDINGS--BAL HARBOUR, FLORIDA, PARTIAL BEACH RESTORATION, BEACH EROSION CONTROL AND HURRICANE PROTECTION PROJECT, DADE COUNTY, FLORIDA (FINAL ENVIRONMENTAL IMPACT STATEMENT).
Army Engineer District, Jacksonville, Fla.
For primary bibliographic entry see Field 04A.
W73-08680

ROUGE RIVER FLOOD CONTROL PROJECT. SECTION B FROM I-94 TO MICHIGAN AVENUE ALTERNATIVE NO. IV (SUPPLEMENT TO FINAL ENVIRONMENTAL IMPACT STATEMENT).
For primary bibliographic entry see Field 08A.
W73-08681

FLOOD CONTROL ON SAGINAW RIVER, MICHIGAN AND TRIBUTARIES; FLINT RIVER AT FLINT, SWARTZ, AND THREAD CREEKS (FINAL ENVIRONMENTAL IMPACT STATEMENT).
Army Engineer District, Detroit, Mich.
For primary bibliographic entry see Field 08A.
W73-08682

AQUATIC PLANT CONTROL AND ERADICATION PROGRAM, STATE OF TEXAS (FINAL ENVIRONMENTAL IMPACT STATEMENT).
Army Engineer District, Galveston, Tex.
For primary bibliographic entry see Field 04A.
W73-08683

EDWARDS UNDERGROUND RESERVOIR, GUADALUPE, SAN ANTONIO, AND NUECES RIVERS AND TRIBUTARIES, TEXAS (FINAL ENVIRONMENTAL IMPACT STATEMENT).
Army Engineer District, Fort Worth, Tex.
For primary bibliographic entry see Field 08A.
W73-08684

SOUTH BRANCH, RAHWAY RIVER, NEW JERSEY, FLOOD CONTROL PROJECT, RAHWAY, NEW JERSEY (FINAL ENVIRONMENTAL IMPACT STATEMENT).
Army Engineer District, New York.
For primary bibliographic entry see Field 08A.
W73-08685

SANTA PAULA CREEK CHANNEL, VENTU[RA] COUNTY, CALIFORNIA (FINAL ENVIRO[N]MENTAL IMPACT STATEMENT).
Army Engineer District, Los Angeles, Calif.
For primary bibliographic entry see Field 08A.
W73-08686

OLIVER BOTTOMS RESOURCE CONSER[VA]TION AND DEVELOPMENT PROJEC[T], SEBASTIAN COUNTY, ARKANSAS (FINAL [EN]VIRONMENTAL IMPACT STATEMENT).
Soil Conservation Service, Washington, D.C.
For primary bibliographic entry see Field 04D.
W73-08687

PEYTON CREEK, TEXAS--FLOOD CONTR[OL] (FINAL ENVIRONMENTAL IMPACT STAT[E]MENT).
Army Engineer District, Galveston, Tex.
For primary bibliographic entry see Field 08A.
W73-08688

FLOOD CONTROL PROJECT, THOMPSON CREEK, HOLLAND PATENT, NEW YOR[K] (FINAL ENVIRONMENTAL IMPACT STAT[E]MENT).
Army Engineer District, New York.
For primary bibliographic entry see Field 08A.
W73-08689

CAMP GROUND LAKE, SALT RIVER BASI[N] KENTUCKY (FINAL ENVIRONMENTAL IM[PACT STATEMENT]).
Army Engineer District, Louisville, Ky.
For primary bibliographic entry see Field 08A.
W73-08690

PERRY COUNTY DRAINAGE AND LEVEE DIS[TRICTS] NOS. 1, 2, AND 3, MISSOURI (FINA[L] ENVIRONMENTAL IMPACT STATEMENT).
Army Engineer District, St. Louis, Mo.
For primary bibliographic entry see Field 08A.
W73-08691

PROPOSED NEW LOCK--PICKWICK LAND[]ING DAM (FINAL ENVIRONMENTAL IMPAC[T] STATEMENT).
Tennessee Valley Authority, Chattanooga.
For primary bibliographic entry see Field 08C.
W73-08692

WILDERNESS SOCIETY V. MORTON (IN[]JUNCTIVE RELIEF SOUGHT AGAINST TH[E] ISSUANCE OF LAND USE PERMIT FOR CON[]STRUCTION PURPOSES ON THE TRANS -ALASKAN PIPELINE PROJECT).
For primary bibliographic entry see Field 06E.
W73-08693

DALY V. VOLPE (ENVIRONMENTAL STUDIES REQUIRED FOR HIGHWAY CONSTRUCTION).
For primary bibliographic entry see Field 06E.
W73-08694

ENVIRONMENTAL DEFENSE FUND, INC. V ARMSTRONG (ACTION FOR INJUNCTIV[E] AND OTHER RELIEF WITH RESPECT TO TH[E] NEW MELONES DAM PROJECT).
For primary bibliographic entry see Field 06E.
W73-08706

LAND USE AND DEVELOPMENT ALONG TH[E] TENNESSEE RIVER IN THE TARCOG RE[]GION.
Top of Alabama Regional Council of Govern[]ments, Huntsville.
For primary bibliographic entry see Field 06B.
W73-08988

Descriptors: *Seismic studies, *Borehole geophysics, *Rock mechanics, *Poisson ratio, Foundation rocks, Mechanical properties, Rock properties, Elasticity (Mechanical), Shear strength, Compressibility, Subsurface investigations.

The methods of seismic refraction surveying may be adapted to the accurate determination of depth and undulation of a rock surface where the depth to rock is less than 50 feet. Four seismic field methods and a laboratory method were used to determine shear wave propagation velocities and shear moduli for two sites. The four seismic methods are standard seismic refraction survey, downhole shooting refraction survey, transient Rayleigh wave survey, and crosshole shooting survey. A torsional resonant column apparatus was used for the laboratory tests. The crosshole shooting method gave the best results because direct measurements were made. Methods which measure compression wave velocity give inconsistent results because the conversion to shear wave velocity is very sensitive to Poisson's ratio. Laboratory tests data gave consistently low values. Strength reduction due to sampling was one cause. Laboratory tests also showed increase in values with time. None of the seismic methods discussed should be the sole subsurface investigative tool when engineering properties are desired. They must be used in conjunction with conventional boring, sampling, and laboratory testing techniques to gain a more complete picture of existing subsurface conditions. (See also W73-08466) (Knapp-USGS)
W73-08467

DEVELOPMENT AND TESTING OF A DOUBLE-BEAM ABSORPTION SPECTROGRAPH FOR THE SIMULTANEOUS DETERMINATION OF DIFFERENT CATIONS IN WATER,
Kentucky Water Resources Inst, Lexington.
For primary bibliographic entry see Field 02K.
W73-08469

DETERMINATION OF TRACE ELEMENTS IN WATER UTILIZING ATOMIC ABSORPTION SPECTROSCOPY MEASUREMENT,
Kentucky Water Resources Inst., Lexington.
For primary bibliographic entry see Field 05A.
W73-08470

METHODS OF ANTICIPATING AVALANCHE HAZARD FROM RAINFALL AND SNOWFALL INTENSITIES IN MOUNTAINS OF ZAILISKI ALA TAU (SPOSOBY PROGNOZA LAVINNOY OPASNOSTI PO INTENSIVNOSTI OSADKOV I SNEGOPADOV V GORAKH ZAILIYSKOGO ALATAU),
Sredneaziatskii Nauchno-Issledovatelskii Gidrometeorologicheskii Institut, Tashkent (USSR).
For primary bibliographic entry see Field 02C.
W73-08489

ATMOSPHERIC EFFECTS ON OCEAN SURFACE TEMPERATURE SENSING FROM THE NOAA SATELLITE SCANNING RADIOMETER,
National Oceanic and Atmospheric Administration, Miami, Fla. Physical Oceanography Lab.
G. A. Maul, and M. Sidran.
Journal of Geophysical Research, Vol 78, No 12, p 1909-1916, April 20, 1973. 4 fig, 1 tab, 17 ref.

Descriptors: *Remote sensing, *Satellites (Artificial), *Water temperature, *Sea water, Calibrations, Cloud cover, Data processing, Humidity.
Identifiers: *Radiometers.

The effects of atmospheric state, nadir angle, cloud amount, cloud height, and random noise on temperature data from the NOAA series satellites were investigated. These satellites have a dual-channel (visible and infrared) scanning radiometer. Temperature departures for nadir viewing range from 22 deg K for a dry winter atmosphere to 10.5 deg K under moist subtropical summer conditions. An 8 deg K temperature difference at the sea surface when viewed at zero nadir angle through the 30 deg N July standard atmosphere registers less than 3 deg K at the satellite and is further compressed to 1 deg K when viewed at a 60 deg nadir angle. A 10% cloud cover can introduce errors that range from 0.5 deg K to 4 deg K depending on cumuloform cloud height; hence clouds must be completely eliminated in the analysis. Random noise in radiosonde data can introduce errors greater than 1 deg K for any given sounding; however, compositing and smoothing will eliminate most of this error source. A correction scheme for cloud-free conditions is developed that is essentially free of bias. A reasonable error estimate for the total system is of the order of 2 deg K. (Knapp-USGS)
W73-08496

SNOW-WATER EQUIVALENT MEASUREMENT USING NATURAL GAMMA EMISSION,
Geological Survey of Canada, Ottawa (Ontario).
For primary bibliographic entry see Field 02C.
W73-08505

A METHOD FOR STUDYING SOIL-WATER PROPERTIES OF SOIL PROFILES,
Agricultural Research Council, Cambridge, (England). Unit of Soil Physics.
For primary bibliographic entry see Field 02G.
W73-08510

GAS CHROMATOGRAPH INJECTION SYSTEM,
National Aeronautics and Space Administration, Moffett Field, Calif. Ames Research Center.
G. E. Pollock, M. E. Henderson, and R. W. Donaldson, Jr.
Available from the National Technical Information Service as N92-21433, $3.00 in paper copy, $1.45 in microfiche. Report Nos NASA-CASE-ARC-10344-1, US-PATENT-APPL-SN-180962, September 16, 1971. 13 p.

Descriptors: *Gas chromatography, *Laboratory equipment, Research equipment, Instrumentation.
Identifiers: *Injection systems.

This patent covers a gas chromatograph injection system which consists of a sample chamber instead of a septum. The sample is placed in the chamber at room temperature and pressure and the solvent removed by evaporation. After the chamber is closed, the position of the carrier gas control value is changed, the chamber is heated, and the sample is volatilized and swept into the analysis apparatus. Disadvantages of syringe injection methods are obviated. (Little-Battelle)
W73-08528

A REAPPRAISAL: CORRELATION TECHNIQUES APPLIED TO GAS CHROMATOGRAPHY,
Warren Spring Lab., Stevenage (England). Control Engineering Div.
For primary bibliographic entry see Field 05A.
W73-08535

AN APPLICATION OF MULTIVARIATE ANALYSIS TO COMPLEX SAMPLE SURVEY DATA,
North Carolina Univ., Chapel Hill. Dept. of Biostatistics.
G. G. Koch, and S. Lemeshow.
Journal of the American Statistical Association, Vol 67, No 340, p 780-782, December 1972. 3 tab, 12 ref.

Descriptors: *Sampling, *Estimating, Statistical methods.

Identifiers: *Multivariate analysis, *Variance, Replication, Balanced repeated replication, Univariate analysis.

A standard method of multivariate analysis is adapted to a highly complex sampling design utilizing the method of balanced repeated replication for calculating valid and consistent estimates of variance. An example illustrates that by doing univariate tests no significant differences are found between two groups. However, the multivariate approach yields a significant result because the directions of the differences between two groups with respect to two positively correlated variables are reversed. (Little-Battelle)
W73-08538

A UNIQUE COMPUTER CENTERED INSTRUMENT FOR SIMULTANEOUS ABSORBANCE AND FLUORESCENCE MEASUREMENTS,
Michigan State Univ., East Lansing. Dept. of Chemistry; and Michigan State Univ., East Lansing. Dept. of Biochemistry.
J. F. Holland, R. E. Teets, and A. Timnick.
Analytical Chemistry, Vol 45, No 1, p 145-153, January 1973. 6 fig, 5 tab, 29 ref.

Descriptors: *Instrumentation, *Data processing, *Fluorescence, *Digital computers, Rhodamine, Automation, Computer programs, Automatic control, Absorption, Research equipment, Laboratory equipment, Design, Sulfates.
Identifiers: *Spectrophotometers, *Spectrofluorimeters, *Absorbance, Quinine bisulfate, Anthracene, Absorption corrected fluorescence, Quantum efficiency.

A computer centered spectrophotometer-spectrofluorimeter combination instrument has been fabricated that will allow simultaneous absorption and fluorescence measurements. The dedicated computer applies corrections to the data collected for many of the instrumental and photophysical variables of fluorescence measurements. From data collected during an excitation scan, the computer can output absorbance, fluorescence, quanta corrected fluorescence, and a unique quantity, partial quantum efficiency. From an emission scan, absorbance, fluorescence, quantum corrected fluorescence, and total quantum efficiency can be obtained. Absorption corrected fluorescence is the name proposed for measurements which correct for the attenuation of excitation beam by absorption processes in the sample cell. The computer program is detailed and results are presented to evaluate the system for the various absorption and fluorescence measurements. Samples of quinine bisulfate, anthracene, and Rhodamine B were analyzed with the equipment. (Little-Battelle)
W73-08541

IMPROVED TECHNIQUE FOR ANALYSIS OF CARBOHYDRATES IN SEDIMENTS,
Rosenstiel School of Marine and Atmospheric Sciences, Miami, Fla.
For primary bibliographic entry see Field 05A.
W73-08556

EXPENDABLE BATHYTHERMOGRAPH OBSERVATIONS FROM SHIPS OF OPPORTUNITY,
National Marine Fisheries Service, La Jolla, Calif. Fishery-Oceanography Center.
J. F. T. Saur, and P. D. Stevens.
Mariners Weather Log, Vol 16, No 1, p 1-8, January 1972. 4 tab, 6 fig.

Descriptors: *Oceans, *Water temperature, *Bathythermographs, Oceanography, On-site testing, Sampling, Instrumentation, Aquatic environment, Physical properties, Currents, Thermal stratification, Thermocline, Geophysics, Marine biology.

A relatively new expendable bathythermograph called the XBT has made it possible to take ocean temperature readings to a depth of 760 m from cooperating nonoceanographic vessels. This is significant because temperature is one of the most basic oceanographic factors in such marine biological and physical problems as the relation of marine organisms to their environment; migration patterns and group behavior; the monitoring of ocean currents; the relationship of large bodies of water to weather; and the propagation of sound in the sea. The bathythermograph system which consists of a launcher, a recorder and expendable probes and signal wires, is described and discussed. The simple observational procedures, which generally take no longer than 20 min. and do not interfere with the ship's operation, are reviewed. Programs for XBT observations on ships of opportunity are supported by separate or cooperative projects among several government agencies and some of these are discussed briefly. Further studies should result in a more complete picture of ocean currents and temperatures. (Jerome-Vanderbilt)
W73-08629

DELINEATION OF INFORMATION REQUIREMENTS OF THE T.V.A. WITH RESPECT TO REMOTE SENSING DATA,
Tennessee Univ., Knoxville.
J. B. Rehder.
Available from NTIS, Springfield, Va 22151 as PB-210 114 Price $3.00 printed copy; $1.45 microfiche. Geological Survey Interagency Report USGS-235, December 1971. 20 p, 12 ref.

Descriptors: *Tennessee Valley Authority, *Data collections, *Aerial photography, *Satellites (Artificial), *Project planning, Water resources development, Land use, Agriculture, Recreation facilities, Land development, Land management, Investigations, Evaluation, Methodology.
Identifiers: Data needs, Interviews.

The information needs and interests of T.V.A. (Tennessee Valley Authority) that can be supplied by hyperaltitude imagery and future ERTS-A satellite imagery were determined by interviews. Two series of interviews were used to obtain specific data requirements keyed to the topics of information needed, the variation of scales and coverage involved between aircraft and satellite platforms, and temporal considerations. (Woodard-USGS)
W73-08737

MATHEMATICAL ANALYSIS: STILL-WATER GAGE,
Louisiana State Univ., Baton Rouge. Dept. of Chemical Engineering.
T. F. Dominick, B. Wilkins, Jr., and H. W. Roberts.
Available from NTIS, Springfield, Va. 22151, as AD-755 157. Price $3.00 printed copy; $1.45 microfiche. Louisiana State University Coastal Studies Bulletin No 7, p 115-124, Technical Report 132, January 1973. 6 fig, 7 ref. ONR-GP Contract N00014-69-A-0211-0003.

Descriptors: *Waves (Water), *Water level fluctuations, *Gages, Instrumentation, Calibrations.
Identifiers: *Wave gages.

The response of a stillwater gage to wave action was analyzed. The mean water level in the gage coincides with the mean level of the impinging waves for symmetric wave shapes. For asymmetric wave shapes the mean level in the gage may not coincide with the mean level of the impinging waves. Deviations between the mean levels are a function of height, shape, period of the wave, and the dimensions of the gage. Prediction of the order of magnitude of the error for asymmetric waves is possible with given figures, which may also be used to choose gage dimensions to minimize the inherent error. (Knapp-USGS)
W73-08755

DEEP-SEA CIRROMORPHS (CEPHALOPOI PHOTOGRAPHED IN THE ARCTIC OCEAN,
Oregon State Univ., Corvallis. Dept. of Ocea raphy.
W. G. Pearcy, and A. Beal.
Deep-Sea Research and Oceanographic Abstrac Vol. 20, No. 1, p 107-108, January 1973. 2 fig, tab, 4 ref.

Descriptors: *Mollusks, *Arctic Ocean, Inv tebrates, Marine animals, Systematics, Dee water habitats, Deep water, Animal populatio On-site data collections, Photography.
Identifiers: *Cephalopods, *Cirromorp Macroinvertebrates, Underwater cameras.

Cirromorphs (cephalopods) were photograph just above the sea floor in the Arctic Ocean duri the summer of 1965. Photographs were obtain with a 35 mm EG and G Deep-Sea Camera fro the USS Staten Island (AGB-5) at 6 locatio north of Point Barrow, Alaska. The camera w positioned above the bottom using a Precisi Depth Recorder and pinger attached to t camera. The usual distance above the bottom w six meters. Over 2900 photos were examined. C romorphs appeared in 21 photos representing maximum of 12 separate individuals. The c romorphs photographed had a prominent ov head, large fins inserted near its well-develop eyes, a web or umbrella that extended to the ti of the arms and long cirri on the arms. A second ry web (a web extending from the arms to the ma web) was not evident. (Holoman-Battelle)
W73-08759

THE USE OF MATCHED SAMPLING AN REGRESSION ADJSTMENT TO REMOVE BIA IN OBSERVATIONAL STUDIES,
Harvard University, Cambridge, Mass. Dept c Statistics.
D. B. Rubin.
Biometrics, Vol 29, No 1, p 185-203, March 1973. fig, 6 tab, 5 ref.

Descriptors: *Sampling, Statistical method Regression analysis.
Identifiers: *Bias, *Matched sampling, *Regre sion adjustment, Bias reduction.

The ability of matched sampling and linear regre sion adjustment to reduce the bias of an estimat of the treatment effect in two sample observ tional studies is investigated for a simple matchin method and five simple estimates. Monte Carl results are given for moderately linear exponenti response surfaces and analytic results ar presented for quadratic response surfaces. Th conclusions are (1) in general both matched sam pling and regression adjustment can be expecte to reduce bias, (2) in some cases when the varianc of the matching variable differs in the two popul tions both matching and regression adjustment ca increase bias, (3) when the variance of th matching variable is the same in the two popul tions and the distributions of the matching variabl are symmetric the usual convariance adjusted est mate based on random samples is almost unbiase and (4) the combination of regression adjustme in matched samples generally produces the lea biased estimate. (Little-Battelle)
W73-08775

MATCHING TO REMOVE BIAS IN OBSERVA TIONAL STUDIES,
Harvard Univ., Cambridge, Mass. Dept. c Statistics.
D. B. Rubin.
Biometrics, Vol 29, No 1, p 159-183, March 1973. fig, 7 tab, 18 ref, 3 append.

Descriptors: Statistical methods, Sampling.
Identifiers: *Bias, *Matched sampling, Mea matching, Pair matching, Bias reduction.

needs to be regulated, a circular metal disc with a large central hole through which the tube of the air-lift passes, may be fitted. Information is given on the procedure for sampling the various types of sediments. The main advantages of this apparatus are that it is light, cheap to construct, and can be used to sample most freshwater sediments. A preliminary sorting of material can be carried out while the sample is being taken. That the air-lift can sample unionid mssels and flints of approximately 40 mm diameter is an indication of the size of particle that can be sampled. Test results are presented from a comparison between the air-lift and a Maitland corer. (Holoman-Battelle)
W73-08795

MODERN APPROACH TO COASTAL ZONE SURVEY,
Michigan Univ., Ann Arbor. Willow Run Labs.
For primary bibliographic entry see Field 02L.
W73-08805

SOME APPLICATIONS OF PHOTOGRAPHY, THERMAL IMAGERY AND X BAND SIDE LOOKING RADAR TO THE COASTAL ZONE,
Grumman Ecosystems Corp., Bethpage, N.Y.
For primary bibliographic entry see Field 02L.
W73-08806

TECHNIQUES AND EQUIPMENT FOR SAMPLING BENTHIC ORGANISMS,
Oregon State Univ., Corvallis. Dept. of Oceanography.
For primary bibliographic entry see Field 02L.
W73-08834

METHODS FOR THE MEASUREMENT OF HANFORD-INDUCED GAMMA RADIOACTIVITY IN THE OCEAN,
(AEC), Division of Biology and Medicine, Washington, D.C.
For primary bibliographic entry see Field 02L.
W73-08841

FLAME PHOTOMETRIC ANALYSIS OF THE TRANSPORT OF SEA SALT PARTICLES,
National Oceanic and Atmospheric Administration, Hilo, Hawaii. Mauna Loa Observatory.
For primary bibliographic entry see Field 02K.
W73-08854

DYNAMIC MEASUREMENT OF SOIL AND LEAF WATER POTENTIAL WITH A DOUBLE LOOP PELTIER TYPE THERMOCOUPLE PSYCHROMETER,
British Columbia Univ., Vancouver. Dept. of Soil Science.
For primary bibliographic entry see Field 02G.
W73-08878

SOIL WATER FLOWMETERS WITH THERMOCOUPLE OUTPUTS,
Agricultural Research Service, Kimberly, Idaho. Snake River Conservation Research Center.
For primary bibliographic entry see Field 02G.
W73-08879

DESIGN AND EVALUATION OF A LOW COST RECORDING SPECTROPOLARIMETER,
Cleveland State Univ., Ohio. Dept. of Chemistry.
S. J. Simon, and K. H. Pearson.
Analytical Chemistry, Vol 45, No 3, p 620-623, March 1973. 6 fig, 3 tab.

Descriptors: *Aqueous solutions, Laboratory equipment.
Identifiers: *Recording spectropolarimeter, *Errors, *Optical rotatory dispersion, Sucrose, Neodymium-D (-)PTDA, Tris (D- (-)-1 2-propylenediamine)cobalt (III) iodide, Spectropolarimetry.

A Perkin-Elmer Model 141 polarimeter was modified to obtain continuous optical rotatory dispersion data over the entire spectral region of 650-240 nanometers. Standard solution of sucrose in deionized water, neodymium-D (-)PDTA, and Tris (D- (-)-1,2-propylenediamine)cobalt (III) iodide were prepared and the errors between observed and calculated ORD values determined. The results show that the equipment performed satisfactorily. (Little-Battelle)
W73-08919

SOUND VELOCIMETERS MONITOR PROCESS STREAMS,
NUSonics, Inc. Paramus, N.J.
For primary bibliographic entry see Field 05A.
W73-08933

AERIAL REMOTE SENSING, A BIBLIOGRAPHY.
Office of Water Resources Research, Washington, D.C.

Available from the National Technical Information Service as PB-220 163, $10.60 in paper copy, $1.45 in microfiche. Water Resources Scientific Information Center Report WRSIC 73-211, March 1973. 482 p. (edited by Donald B. Stafford, Clemson University)

Descriptors: *Remote sensing, *Bibliographies, *Aerial photography, *Data collections, *Earth resources, Hydrologic data, Infrared radiation, Instrumentation, Mapping, Microwaves, Monitoring, Oil spills, Photogrammetry, Pollutant identification, Sattelites (Artificial), Telemetry, Water temperature.

This report, containing 272 abstracts, is another in a series of planned bibliographies in water resources to be produced from the information base comprising SELECTED WATER RESOURCES ABSTRACTS (SWRA). At the time of search for this bibliography, the data base had 50,631 abstracts covering SWRA through December 15, 1972 (Volume 5, Number 24). Author and subject indexes are included.
W73-08974

COMPARATIVE CHARACTERISTICS OF CONTINUOUS COUNTING AND SAMPLING METHODS FOR REGISTERING YOUNG PACIFIC SALMON, (IN RUSSIAN),
E. T. Nikolaeva.
Izv Tikhookean Nauchno-Issled Inst Rybn Khoz Okeanogr. 78, p 73-80, 1970. English summary.
Identifiers: Counting methods, *Pacific salmon, Registering, *Salmon, *Sampling.

A series of experiments were done to verify the effectiveness of widely used sampling methods for registering migrating salmon young in the Far East. The effectiveness of the traps depends on the dimensions of the down-stream migrants, so that established coefficients of trapping capacity vary for different species of juveniles. The effectiveness of sampling methods can increase by replacement of short sturdy traps with shorter finer ones.--Copyright 1972, Biological Abstracts, Inc.
W73-09043

A DEVICE FOR COLLECTING IN-SITU SAMPLES OF SUSPENDED SEDIMENT FOR MICROSCOPIC ANALYSIS,
Johns Hopkins Univ., Baltimore, Md. Chesapeake Bay Inst.
J. R. Schubel, and E. W. Schiemer.
J Mar Res. Vol 30, No 2, p 269-273. 1972. Illus.

Identifiers: Aquatic sampler, Collecting device, Sediments, *Suspended sediment, *Sampling, Instrumentation.

An in-situ sampler for collecting small samples of suspended sediment for microscopic analysis has been built and tested. The device rapidly freezes a thin layer of water entrapping all of the suspended particles in it; when the sampler is recovered, the disc of ice is placed on a suitable substrate and freeze-dried. The particles can then be examined in an undisturbed state with a light microscope or with an electron microscope.--Copyright 1972, Biological Abstracts, Inc.
W73-09056

METHODS OF DETECTING FUNGI IN ORGANIC DETRITUS IN WATER,
Queen's Univ., Belfast (Northern Ireland). Dept. of Botany.
D. Park.
Trans Br Mycol Soc. Vol 58, No 2, p 281-290. 1972.
Identifiers: Ecology, *Fungi, *Ireland, Isolation, *Organic detritus, Streams, Unpolluted water, Microoganisms.

Observations were made of the fungi in 8 unpolluted streams in Northern Ireland, using different techniques of isolation, and the results are recorded. It is suggested that the systematic use of several methods of isolation can yield valuable information on the ecology of micro-organisms.--Copyright 1972, Biological Abstracts, Inc.
W73-09057

DEVELOPMENT OF A DISCHARGE MEASUREMENT DEVICE FOR SEWER FLOW,
Illinois Univ., Urbana. Dept. of Civil Engineering.
H. G. Wenzel, B. E. Burris, and C. D. Morris.
Paper presented at 8th American Water Resources Conference, October 30 - November 2, 1972, St. Louis, Missouri, 1972. 20 p, 9 fig, 1 tab, 6 ref. OWRR B-063-Ill (1).

Descriptors: *Flow measurement, *Flowmeters, *Discharge measurements, Venturi meters, *Urban runoff, Storm drains, Venturi flumes, Sewers, Instrumentation.

A suitable meter for continuous in-system flow measurement should be reasonably accurate throughout a range of flow measurement including free surface and pressurized flow conditions, cause minimum reduction in flow capacity, require a minimum of field maintenance, be capable of continuous, real time measurement with remote data transmission, and could be built at reasonable cost. A Venturi device was designed to act as a Venturi type flume under free surface or part full conditions and as a conventional pipe Venturi meter under pressurized of full flow conditions. The constriction should be designed for a specific installation with the acceptable reduction in capacity and error in discharge prescribed. (Knapp-USGS)
W73-09085

RADIO ECHO SOUNDING ON A VALLEY GLACIER IN EAST GREENLAND,
Scott Polar Research Inst., Cambridge (England).
For primary bibliographic entry see Field 02C.
W73-09091

SEISMIC REFRACTION AND REFLECTION MEASUREMENTS AT 'BYRD' STATION, ANTARTICA,
Wisconsin Univ., Middleton. Geophysical and Polar Research Center.
For primary bibliographic entry see Field 02C.
W73-09093

A PHOTOGRAMMETRIC SURVEY OF HOSEASON GLACIER, KEMP COAST, ANTARCTICA,
Ohio State Univ., Columbus. Dept. of Geodetic Science; and Ohio State Univ. Research Foundation, Columbus. Inst. of Polar Studies.
For primary bibliographic entry see Field 02C.
W73-09094

GLACIER BORE-HOLE PHOTOGRAPHY,
California Inst. of Tech., Pasadena. Div. of Geological and Planetary Sciences.
For primary bibliographic entry see Field 02C.
W73-09096

RADIO ECHO SOUNDING: BRINE PERCOLATION LAYER,
Wisconsin Univ., Middleton. Geophysical and Polar Research Center.
For primary bibliographic entry see Field 02C.
W73-09097

ACOUSTIC EMISSION IN SNOW AT CONSTANT RATES OF DEFORMATION,
Montana State Univ., Bozeman. Dept. of Earth Sciences.
For primary bibliographic entry see Field 02C.
W73-09098

7C. Evaluation, Processing and Publication

FLOOD-PRONE AREAS, HARTFORD NORTH QUADRANGLE, CONNECTICUT,
Geological Survey, Washington, D.C.
M. P. Thomas.
For sale by USGS, Washington, D C 20242 Price - 75 cents. Geological Survey Miscellaneous Geologic Investigations Maps, Map I-784-M, 1972. 1 sheet, 1 map.

Descriptors: *Flood frequency, *Flood recurrence interval, *Connecticut, Mapping, Hydrologic data, Cities, Flood plains, Reservoirs, Streams, Streamflow.
Identifiers: *Hartford (Conn), North quadrangle.

A map (scale 1:24,000) shows flood-prone areas in the Hartford North quadrangle, Connecticut. Areas shown on the map have a 1-in-100 chance on the average of being flooded during any year. The areas were delineated for natural conditions but, where necessary, were adjusted for the effects of storage in flood-control reservoirs and channel improvements that existed in 1972. Smaller areas (not shown separately on map) within those shown have a greater than 1-in-100 chance on the average of being flooded during any year; some areas immediately adjacent to streams may be flooded almost every year. Larger areas outside those shown have a less than 1-in-100 chance on the average of being flooding during any year. Depth and duration of flooding are not indicated. (Woodard-USGS)
W73-08472

MAP SHOWING GENERAL AVAILABILITY OF GROUND WATER IN THE SALINA QUADRANGLE, UTAH,
Geological Survey, Washington, D.C.
D. Price.
For sale by USGS, Washington, D C 20242 Price - 75 cents. Geological Survey Miscellaneous Geologic Investigations Maps, Map I-591-M, 1972. 1 sheet, 1 map.

Descriptors: *Groundwater resources, *Water wells, *Water yield, *Utah, Maps, Locating, Aquifers, Hydrologic data.
Identifiers: *Salina quadrangle (Utah).

This map of Salina quadrangle, Utah, scale 1:250,000, shows the general availability of groundwater as determined from well records collected by the U.S. Geological Survey and cooperating State, local, and other Federal agencies. Data are for aquifers that in most places are less than 1,000 feet below the land surface. In areas of few or no wells, potential well yields are inferred from geological data. Various colors are used to indicate locations where well yields range from less than 1 to 10 gpm, 5 to 50 gpm, and 50 to 1,000 gpm. (Woodard-USGS)
W73-08473

RUNOFF MAP OF SWEDEN: AVERAGE ANNUAL RUNOFF FOR THE PERIOD 1931-60,
Sveriges Meteorologiska och Hydrologiska Institut, Stockholm.
O. Tryselius.
Meddelanden (Communications, Series C, No 7), Stockholm, 1971. 2 p, 1 map, 9 ref.

Descriptors: *Runoff, *Average runoff, *Hydrologic data, Maps, Foreign countries, Precipitation (atmospheric), Data collections, Rain gages, Snowfall, Melt water, Meteorological data, Water balance, Hydrology, Streamflow.
Identifiers: *Sweden.

A map (scale 1:3,000,000) showing the normal average annual runoff in Sweden from 1931-60 is intended for hydrological purposes and therefore the runoff is given in liters per second per square kilometer. The runoff map is mainly based on point values of precipitation minus evaporation, but adjustments and corrections have also been made where actual runoff values from small drainage basins are known. A pure statistical computation of the results from the precipitation stations gives a normal average annual precipitation of 588 mm for Sweden during the period 1931-60. (Woodard-USGS)
W73-08492

FILLING IN GAPS IN RAINFALL RECORDS BY SIMULATED DATA,
Food and Agriculture Organization of the United Nations, Iraklion (Greece). United Nations Development Programme.
For primary bibliographic entry see Field 02B.
W73-08506

SOME DESIGNS AND ANALYSES FOR TEMPORALLY INDEPENDENT EXPERIMENTS INVOLVING CORRELATED BIVARIATE RESPONSES,
California Polytechnic State Univ., San Luis Obispo. Dept. of Computer Science and Statistics.
For primary bibliographic entry see Field 04A.
W73-08546

ON-LINE COMPUTER SYSTEM FOR AUTOMATIC ANALYZERS,
Food and Drug Administration, Chicago, Ill.
M. W. Overton, L. L. Alber, and D. E. Smith.
Journal of the Association of Official Analytical Chemists, Vol 56, No 1, p 140-146, January 1973. 3 fig, 1 tab.

Descriptors: *Digital computers, *Data processing, Automation, Automatic control, Programming languages.
Identifiers: *Autoanalyzer, Spectrophotometers.

A computer system has been developed for acquisition and reduction of data obtained from an Autoanalyzer. An 8K core memory computer was interfaced with the Autoanalyzer through an operational amplifier manifold and plug-in components. The interfacing circuitry is based on the use of a retransmitting slidewire mounted on the pen cable pulley of the spectrophotometer. The program, which is included was written in

ENVIRONMENTAL DATA BANK.

Hearings--Subcomm. on Fisheries and Wildlife Conservation--Comm. on Merchant Marine and Fisheries, U.S. House of Representatives, 91st Cong, 2d Sess, June 2, 3, 25, 26, 1970. 395 p, 17 fig, 8 photo, 5 tab, 2 chart, 7 ref.

Descriptors: *Legislation, *Information retrieval, *Data collections, *Data storage and retrieval, Legal aspects, United States, Environmental control, Documentation, Data processing, Computers, Computer models, Data transmission, Statistics, Environment.
Identifiers: *Congressional hearing, *National Environmental Policy Act.

These hearings took testimony on an amendment to the National Environmental Policy Act of 1969 which would provide a powerful tool for implementing the letter and spirit of the Act, a national environmental data bank. The data bank would serve as the central national depository of all information relating to the environment. Technical experts reported that such a system could be operative within a two or three year period. This system would be useful for many planning decisions. Included within this report are typical data which have been stored in a model of the proposed data bank. The model system under development is geared to meet the informational needs of not only scientists and planners but also laymen. The system will contain data programs both in quantity and quality. The base of the system will come in part from existing information sources of various governmental agencies but the total data base is still not defined. Included are several data bank proposals and existing information storage systems of various governmental agencies. (Beardsley-Florida)
W73-08663

TERRESTRIAL SURVEYING AND GEOLOGICAL MAPPING OF THE ROCK FACE OF THE DEWATERED AMERICAN NIAGARA FALLS,
Department of the Environment, Ottawa (Ontario). Inland Waters Branch.
I. A. Reid.
1972. 5 SHEETS, 6 MAPS, 2 REF.

Descriptors: *Waterfalls, *United States, *Canada, *Geologic mapping, *Geological surveys, Methodology, Dewatering, Channels, Engineering structures, Cofferdams, Aerial photography, Photogrammetry, Terrain analysis, Rock properties, Bedrock, Planning.
Identifiers: *Niagara Falls, *American Falls, *Canadian Horseshoe Falls, *Waterfall deterioration control, Rock face.

The crest lines of Niagara Falls (the American Falls and the Canadian Horseshoe Falls) have been receding steadily throughout history, due to the eroding action of flowing water on the rock structure of the falls. The channel bed at the falls consists primarily of hard dolomite underlain by shale. Shale, more susceptible to weathering than dolomite, erodes more readily. Thus, the harder dolomitic rocks are undermined, and rockfalls result at the crest. Descriptions of the bedrock formation exposed on the rock face are shown on maps. In addition, there is a selection of 28 profiles of the rock face, which are identified by profile identification numbers. To carry out mapping and drilling operations, the channel was dewatered June 12, 1969 by constructing a cofferdam at the head of the American Falls Channel between Goat Island and the U.S. mainland, and diverting the flow through existing power canals. The surveying and mapping of the channel and rock face was a cooperative venture involving the Canadian Department of the Environment and the Buffalo District Office of the United States Army Corps of Engineers. (Woodard-USGS)
W73-08727

HYDROGRAPH SIMULATION MODELS OF THE HILLSBOROUGH AND ALAFIA RIVERS, FLORIDA: A PRELIMINARY REPORT,
Geological Survey, Tampa, Fla.
For primary bibliographic entry see Field 04A.
W73-08733

STREAM GAUGING INFORMATION, AUSTRALIA, SUPPLEMENT 1970.
Australian Water Resources Council, Canberra.

Australian Government Publishing Service, Canberra, 1972. 39 p.

Descriptors: *Stream gages, *Gaging stations, *Australia, Flow measurement, Streamflow, Sites, Indexing, Surface waters, River basins, Data collections, Hydrologic data.

This publication is a supplement to Stream Gauging Information, Australia, December 1969 published by the Australian Government Publishing Service and is intended for use in conjunction with that publication. Details of 150 new gaging stations and other stations established throughout Australia during the year ended December 31, 1970 are given. In addition 30 stations were upgraded through improvements in control or through installation of more advanced instrumentation and equipment. The supplement also lists 12 stations which had been omitted from the previous publication. The 22 authorities which undertake stream gaging are listed with the number of stream gaging stations operated by each of them in December 1970. The total number of stations was 2,405. In comparison the number of stream gaging stations operated by these authorities in December 1969 was 2,275. The gaging information is presented by drainage divisions. The numbering system is in accordance with that used in Stream Gauging Information, Australia, December 1969. Comments are provided in the columns headed 'Variation' and 'Remarks'. Where information on the purpose of the station has been provided this is shown in the 'Remarks' column. (Woodard-USGS)
W73-08738

GROUNDWATER RESOURCES OF WASHINGTON COUNTY, TEXAS,
Geological Survey, Austin, Tex.
For primary bibliographic entry see Field 04B.
W73-08739

GROUNDWATER RESOURCES OF NAVARRO COUNTY, TEXAS,
Geological Survey, Austin, Tex.
For primary bibliographic entry see Field 04B.
W73-08740

GROUND-WATER RESOURCES OF HARDEMAN COUNTY, TEXAS,
Geological Survey, Austin, Tex.
For primary bibliographic entry see Field 04B.
W73-08741

GREAT LAKES ICE COVER, WINTER 1969-70,
National Ocean Survey, Detroit, Mich. Lake Survey Center.
For primary bibliographic entry see Field 02C.
W73-08743

SUSPENDED SEDIMENT DATA SUMMARY, AUGUST 1969 - JULY 1970 (MOUTH OF BAY TO HEAD OF BAY),
Johns Hopkins Univ., Baltimore, Md. Chesapeake Bay Inst.
J. R. Schubel, C. H. Morrow, W. B. Cronin, and A. Mason.
Available from NTIS, Springfield, Va 22151 as COM-72-11184 Price $3.00 printed copy; $1.45

microfiche. Special Report 18, (Reference 70-10), November 1970. 39 p, 1 fig, 1 tab, 1 ref.

Descriptors: *Sediment transport, *Suspended load, *Organic matter, *Salinity, *Chesapeake Bay, Water temperature, Weather data, Data collections, Methodology, Estuaries, Bays.

Monthly surveys (August 1969 - July 1970) were conducted to determine the spatial and temporal variations of a number of the characteristic propeties of suspended sediment in the Chesapeake Bay. These properties include the concentration of total suspended solids (suspended sediment) and the percent of the concentration of total suspended solids made up of combustible organic matter. The concentrations of total suspended solids (suspended sediment) were determined by filtration of measured volumes of water through preweighed filters. The concentrations of total particulate organic matter were estimated by determining the loss of mass of the total solids on ignition. Salinities were computed from measurements of conductivity and temperature made with the Chesapeake Bay Institute Induction Conductivity Temperature Indicator. Weather conditions at the time of sampling were noted. (Woodard-USGS)
W73-08744

WATER-LEVEL RECORDS, 1969-73, AND HYDROGEOLOGIC DATA FOR THE NORTHERN HIGH PLAINS OF COLORADO,
Geological Survey, Washington, D.C.
W. E. Hofstra, and R. R. Luckey.
Colorado Water Resources Basic-Data Release No 28, 1973. 52 p, 1 fig, 2 tab.

Descriptors: *Groundwater resources, *Water levels, *Water wells, *Aquifer characteristics, *Colorado, Water utilization, Irrigation, Water level fluctuations, Well data, Hydrogeology, Hydrologic data, Basic data collections.
Identifiers: *Northern High Plains (Colo).

The northern High Plains of Colorado is bounded by the State line on the east, the South Platte River on the norht, the Arkansas River on the south, and extends about 8 miles west beyond Lincoln County. The 9,500 square-mile area includes all or part of 11 counties. Irrigation supplies are derived almost entirely from wells tapping the Ogallala Formation. The Ogallala, of Pliocene and locally of Miocene age, consists of sand, gravel, silt, and clay, and ranges in thickness from a few feet on the west to as much as 400 feet in places near the state line. Wells tap the following geologic units: Quaternary deposits undivided, Ogallala Formation, White River Group, and the Niobrara Formation. Water level measurements were made during the winter prior to the 1973 irrigation season in more than 1,000 wells in the northern High Plains of Colorado. Measurements for the 4 preceding winters are also included to serve as references illustrating declining or rising water levels. Records of 386 large-capacity wells (yields greater than 100 gpm) inventoried from 1971 through 1972 are tabulated. This table supplements 'Basic-Data Release No. 23' which includes records of all large-capacity wells inventoried before 1971. (Woodard-USGS)
W73-08745

GEOLOGIC AND HYDROLOGIC MAPS FOR LAND-USE PLANNING IN THE CONNECTICUT VALLEY WITH EXAMPLES FROM THE FOLIO OF THE HARTFORD NORTH QUADRANGLE, CONNECTICUT,
Geological Survey, Washington, D.C.
For primary bibliographic entry see Field 04A.
W73-08746

SUMMARIES OF STREAMFLOW RECORDS IN NORTH CAROLINA,
Geological Survey, Raleigh, N.C.
N. O. Thomas.
North Carolina Department of Natural and Economic Resources, Raleigh, Office of Water and Air Resources, 1973. 303 p, 1 plate, 30 ref.

Descriptors: *Streamflow, *Flow rates, *Gaging stations, *Hydrologic data, *North Carolina, Stream gages, River basins, Low flow, Peak discharge, Average runoff, Water yield, Basic data collections, Reviews, Computer programs, Water resources.
Identifiers: *Streamflow records (NC).

Summaries of daily streamflow records collected at continuous-record gaging stations in North Carolina include: tables of highest mean discharges for 1 day and for 3, 7, 15, 30, 60, 90, 120, and 183 consecutive-day periods for each water year ending September 30; lowest mean discharges for 1 day and for 3, 7, 14, 30, 60, 90, 120, and 183 consecutive-day periods for each climatic year ending March 31; and a duration-of-flow table for each water year and for the period of record. A computer program was utilized in preparing the summaries. The summaries were prepared for all active and discontinued stream-gaging stations for which 5 or more water years of record were available as of September 30, 1968. Each set of summary tables is preceded by a station description giving the location of the gage and information on the station's operation. The summaries are fundamental in analyzing streamflow records. Further analyses and studies that may be undertaken to develop key long-term streamflow and runoff values needed in water-availability studies and in the design and management of water-related facilities are briefly described by reference to a few selected reports. The techniques employed in the reference reports are summarized. (Woodard-USGS)
W73-08747

SALINE GROUNDWATER RESOURCES OF MISSISSIPPI,
Geological Survey, Jackson, Miss.
G. J. Dalsin.

Descriptors: *Groundwater resources, *Saline water, *Water quality, *Chemical analysis, *Mississippi, Water wells, Aquifer characteristics, Dissolved solids, Water yield, Desalination, Projections, Water utilization, Hydrologic data, Maps.

This one-sheet atlas describes the saline groundwater resources of Mississippi. Saline groundwater occurs under all or most of the State of Mississippi and is a potentially important natural resource because it may someday be used for water supplies, either directly or after desalination. Also, as an aid in pollution control the less mineralized saline-water sections can serve as a buffer between the fresh-water section and the highly mineralized water section into which wastes may be injected. For these reasons the State has a policy of protecting from manmade pollution the aquifers containing water with less than 10,000 mg/liter of dissolved solids. A well of the size commonly constructed for public and industrial water supplies in the artesian aquifers of Mississippi can be expected to yield 25-30 gpm for each foot of water-level drawdown over a long period of continuous pumping. Chemical analyses of saline groundwater from selected wells are given in tables, and the wells are shown on maps. The water is a sodium bicarbonate type. (Woodard-USGS)
W73-08750

APPLICATION OF THE MOST-PROBABLE-NUMBER PROCEDURE TO SUSPENSIONS OF LEPTOSPIRA AUTUMNALIS AKIYAMI A,
North Carolina Univ., Chapel Hill. Dept. of Environmental Sciences and Engineering.
For primary bibliographic entry see Field 05A.
W73-08770

PRACTICAL CONSIDERATIONS FO DIGITIZING ANALOG SIGNALS,
Alberta Univ., Edmonton. Dept. of Chemistry.
P. C. Kelly, and G. Horlick.
Analytical Chemistry, Vol. 45, No 3, p 518-527, March 1973. 10 fig, 3 tab, 13 ref.

Descriptors: *Sampling, *Data processing, Mathematical studies.
Identifiers: *Errors, *Analog to digital conversion *Signal reconstruction, *Digitization, Precision Regeneration.

Quantitative measures for the quality of digital data can be obtained by attempting to regenerate an analog signal from digital data. The resulting signal can be compared to the original analog signal. Regeneration may be carried out by use of simple operations on the Fourier transform of a continuous representation of digital data. The effects of sampling interval, sampling duration, quantization, digitization time, aperture time, and random variations in sampling interval are examined. The maximum sampling interval and minimum number of samples needed to digitize triangular, exponential, Lorentzian, and Gaussian peaks for given values of maximum absolute error are tabulated. (Little-Battelle)
W73-08791

TOOLS FOR COASTAL ZONE MANAGEMENT.
Marine Technology Society, Washington, D.C. Coastal Zone Marine Management Committee.
For primary bibliographic entry see Field 02L.
W73-08802

ENVIRONMENTAL GEOLOGIC ATLAS OF THE TEXAS COAST: BASIC DATA FOR COASTAL ZONE MANAGEMENT,
Texas Univ., Austin. Bureau of Economic Geology.
For primary bibliographic entry see Field 02L.
W73-08803

THE NATIONAL OCEAN SURVEY COASTAL BOUNDARY MAPPING,
National Ocean Survey, Rockville, Md. Coastal Mapping Div.
For primary bibliographic entry see Field 02L.
W73-08807

THE CHESAPEAKE BAY INVENTORY SYSTEM,
Johns Hopkins Univ., Baltimore, Md. Dept. of Radiology.
For primary bibliographic entry see Field 02L.
W73-08809

COASTAL ZONE MANAGEMENT SYSTEM: A COMBINATION OF TOOLS,
Center for the Environment and Man, Inc., Hartford, Conn.
For primary bibliographic entry see Field 02L.
W73-08810

THE MARINE ENVIRONMENT AND RESOURCES RESEARCH AND MANAGEMENT SYSTEM MERRMS,
Virginia Inst. of Marine Science, Glouster Point.
For primary bibliographic entry see Field 02L.
W73-08811

MARINE TECHNOLOGY TRANSFER DEPENDS UPON INFORMATION TRANSFER,
Environmental Protection Agency, Washington, D.C. Div. of Water Quality Standards.
For primary bibliographic entry see Field 02L.
W73-08812

1971 WATER RESOURCES DATA FOR NEW YORK: PART TWO, WATER QUALITY RECORDS.
Geological Survey, Albany, N.Y.

Geological Survey Basic Data Report, 1973. 254 p, 10 fig, 4 tab.

Descriptors: *Water quality, *Chemical analysis, *Surface water, *Groundwater, *New York, Water analysis, Water chemistry, Water resources, Sampling, Basic data collections, Sediment transport, Water temperature, Streamflow, Flow rates.

Water resources data for the 1971 water year for New York include records for the chemical and physical characteristics of surface water, groundwater and precipitation. These data represent that portion of the National Water Data System collected by the U.S. Geological Survey and cooperating state, local, and federal agencies in New York. Water quality information is presented for chemical quality, microbiology, water temperature, and fluvial sediment. Chemical quality includes concentrations of individually dissolved constituents and certain properties or characteristics such as hardness of water, specific conductance, and pH. Microbiological information includes quantitative identification of certain bacteriological indicator organisms. Data include a description of the sampling station and tabulations of the samples analyzed. The description of the sampling station gives location, drainage area, periods of record for the various water quality data, extremes of the pertinent data, and general remarks. (Woodward-USGS)
W73-08890

DRAINAGE AND FLOOD CONTROL BACKGROUND AND POLICY STUDY.
Nolte (George S.) and Associates, San Francisco, Calif.
For primary bibliographic entry see Field 04A.
W73-08891

DATA PROCESSING ESTIMATIONS FOR SAMPLE SURVEYS,
Georgia Univ., Athens.
A. Y-S. Huong.
Available from Univ. Microfilms, Inc., Ann Arbor, Mich. 48106. Order No. 73-5717. Ph D Dissertation, 1972. 150 p.

Descriptors: *Data processing, *Computer programs, *Regression analysis, Sampling, Statistical methods, Estimating.
Identifiers: *Data processing estimation, Variance, Covariance, Errors.

A new method for summarizing sample surveys called Data Processing Estimation is based upon a regression type estimation technique. This method, which generalizes many of the commonly used standard estimation procedures, utilizes regression relationships of the response variable on the vector of descriptive variables to estimate all unsampled units of the finite population and their corresponding sampling variance-covariances factors. As such, this approach allows these individual estimates to be processed so as to be able to evolve an estimate of the mean of any specified subpopulation by only averaging the sum of the sampled and unsampled units of this particular subpopulation. Since this technique also determines for each estimate the associated sampling error factors, an estimated variance of the estimated means can also be determined by simply processing the variances and covariances factors. The advantages of the technique are: the generality of the approach, the dynamic nature of the procedure and the simplicity of the estimation method. (Little-Battelle)
W73-08921

AERIAL REMOTE SENSING, A BIBLIOGRAPHY.
Office of Water Resources Research, Washington, D.C.
For primary bibliographic entry see Field 07B.
W73-08974

STREAMFLOW DATA USAGE IN WYOMING,
Wyoming Univ., Laramie. Water Resources Research Inst.
V. E. Smith.
Water Resources Series No 33, February, 1973. 9 p, 4 tab, 17 ref.

Descriptors: *Streamflow, *Hydrologic data, *Wyoming, *Surveys, *High-flow, Estimating, Planning, Stream gages, Networks, Low flow, Hydrologic systems, Forecasting, Mathematical models.
Identifiers: Data usage, Hydrologic networks, Streamflow observations, Peak flow data, Daily flows, Water availability.

To aid in future planning of hydrologic networks, a survey of federal, state, local government and private organizations was conducted to ascertain streamflow data, a principal hydrologic component, usage in Wyoming. Survey findings are discussed in detail. While several kinds of hydrologic studies are commonly made, high-flow studies are the most numerous. Of the 19 public agencies and two private firms that provide the input for these findings, all but five make high-flow studies; several make low-flow, water availability, or forecast studies as well or instead. Several models of runoff are in existance, but they are still in the development stage. Pointing to the need for more streamflow observations, numerous methods of estimating streamflow are used; the frequency of sue of these estimation methods is summarized. Since economic considerations will continue to be a constraint on any hydrologic network, estimations will continue therefore to be a necessary part of hydrologic studies. (Bell-Cornell)
W73-09004

A MODEL FOR SIMULATING TRANSPIRATION OF LEAVES WITH SPECIAL ATTENTION TO STOMATAL FUNCTIONING,
Agricultural Univ., Wageningen (Netherlands). Dept. of Theoretical Production Ecology.
For primary bibliographic entry see Field 02D.
W73-09039

FLOODS IN THE IOWA RIVER BASIN UPSTREAM FROM CORALVILLE LAKE, IOWA,
Geological Survey, Iowa City, Iowa.
A. J. Heinitz.
Geological Survey open-file report, February 1973. 75 p, 16 fig, 2 tab, 7 ref, 1 append.

Descriptors: *Floods, *Flood profiles, *Flood data, *Iowa, Flood damage, Flood control, Historic floods, Peak discharge, Streamflow, Gaging station, Hydrologic data, Basic data collections, Flow rates, Low flow engineering structures, Bridge design, Design criteria, Flow characteristics, Flood frequency.
Identifiers: *Iowa River basin.

Flood information is reported for 207 miles of the main stem, 23 miles on the West Branch, and 23 miles on the East Branch of the Iowa River. The information will be of use to those concerned with the design of bridges and other structures and the conduct of various operations on the flood plains of the streams. Included are flood-peak records, gaging-station records, and low-water profiles. Outstanding floods treated in this report are the 1954 flood which is the greatest known in the upper reaches of the Iowa River, the 1969 flood which is the greatest flood in recent years in the central reach of the river, and the 1947 flood which is the greatest flood recorded, excepting the 1918 historic flood, in the lower reaches of the river. Selected data are also given for the 1918 flood, the greatest flood recorded on the Iowa River. Flood profiles for the main stem include those for the 1947, 1954, 1969, the computed 25- and 50-year floods, and a partial profile for the June 1972 flood. (Woodard-USGS)
W73-09084

08. ENGINEERING WORKS

8A. Structures

CULVERT VELOCITY REDUCTION BY INTERNAL ENERGY DISSIPATORS,
Virginia Polytechnic Inst. and State Univ., Blacksburg.
For primary bibliographic entry see Field 08B.
W73-08514

UNDERGROUND NUCLEAR POWER PLANT SITING,
Aerospace Corp., San Bernardino, Calif.
For primary bibliographic entry see Field 06G.
W73-08617

RIVER JET DIFFUSER PLANNED.
For primary bibliographic entry see Field 05G.
W73-08631

AN ANALYSIS OF ENVIRONMENTAL STATEMENTS FOR CORPS OF ENGINEERS WATER PROJECTS,
Stanford Univ., Calif. Dept. of Civil Engineering.
For primary bibliographic entry see Field 06G.
W73-08673

ROUGE RIVER FLOOD CONTROL PROJECT. SECTION B FROM I-94 TO MICHIGAN AVENUE ALTERNATIVE NO. IV (SUPPLEMENT TO FINAL ENVIRONMENTAL IMPACT STATEMENT).

Available from National Technical Information Service, Springfield, Virginia, as EIS-MI-72-5462-S. Price: $4.25 in paper copy, $1.45 in microfiche. 1972. 42 p, 6 fig, 4 tab.

Descriptors: *Michigan, *Environmental effects, *Levees, *Flood control, Recreation, Engineering structures, Flood protection, Water quality control, Channel improvement, Wildlife habitat.
Identifiers: *Environmental Impact Statements, *Rouge River (Mich).

A study was undertaken to consider an optimized levee system and flood control plan for the Rouge River in Michigan. The levee protection system with associated interior drainage facilities is considered feasible from a structural engineering standpoint, and will be designed to provide protection against floods. The major benefit that would accrue from a levee project is the prevention of flood damages. Various recreational benefits would also be derived from the project. The project will have an adverse effect on the sewer systems because of the surcharging that would result. However, the project would preserve 19 acres of wildlife habitat, 5 acres of river, and 3 acres of woods. Alternative structural measures were considered to provide the same degree of protection as the proposed channel improvement plan. The adverse effects of such alternatives would be similar to those of the proposed project. In addition to the above information, numerous charts, graphs, illustrations, channel profiles, and economic evaluation summaries are included. (Mockler-Florida)
W73-08681

FLOOD CONTROL ON SAGINAW RIVER, MICHIGAN AND TRIBUTARIES; FLINT RIVER AT FLINT, SWARTZ, AND THREAD CREEKS (FINAL ENVIRONMENTAL IMPACT STATEMENT).
Army Engineer District, Detroit, Mich.

Available from National Technical Information Service, as EIS-MI-72-5186-F. Price: $10.25 in paper copy, $1.45 in microfiche. May 12, 1972. 83 p, 9 map.

Descriptors: *Michigan, *Environmental effects, *Flood control, *Channel improvements, Flood protection, Biological communities, Channel flow, Stream improvement, Regulated flow, Water management (Applied), Water control, Excavation, Levees.
Identifiers: *Environmental Impact Statements, *Flint River (Mich).

This flood control project for the Flint River and Swartz and Thread Creeks, in Genesee County, Michigan, consists of channel realignments and modifications of approximately 11,000 feet of the main stem of the Flint River and of approximately 8,900 feet of Swartz and Thread Creeks. The completed project will provide flood protection for approximately 455 acres of land adjacent to the Flint River. This area, subject to periodic flooding, includes a concentrated industrial complex for automotive manufacturing and other commercial enterprises. The proposed actions would have an adverse impact upon the existing stream channels and the immediate, contiguous areas by displacing biological communities through the excavations and the construction of a concrete channel. Alternatives include not to continue with the proposed structural techniques. This would reduce the effectiveness of those segments already completed and would provide no further protection. (Mockler-Florida)
W73-08682

EDWARDS UNDERGROUND RESERVOIR, GUADALUPE, SAN ANTONIO, AND NUECES RIVERS AND TRIBUTARIES, TEXAS (FINAL ENVIRONMENTAL IMPACT STATEMENT).
Army Engineer District, Fort Worth, Tex.

Available from National Technical Information Service, as EIS-TX-72-5323-F. Price: $6.25 in paper copy, $1.45 microfiche. November 19, 1971. 79 p, 1 tab.

Descriptors: *Texas, *Multiple purpose reservoirs, *Environmental effects, *Flood control, *Water supply, Flood protection, Levees, Irrigation, Reservoir storage, Recreation, Fish, Wildlife, Lakes, Wildlife habitats, Agriculture, Flood plains, Vegetation, Dams, Estuarine environment, Water importing, Archaeology, Dam construction, Flow control, River regulation, Groundwater resources, Groundwater recharge.
Identifiers: *Environmental Impact Statements, *Blanco River (Tex).

This project consists of a multiple purpose reservoir at Cloptin Crossing on the Blanco River in south central Texas. The reservoir would provide storage for flood control, municipal and industrial water supply, general recreation, and fish and wildlife enhancement. The impoundment created would inundate 6,060 acres of wildlife habitat and affect 1,670 additional acres of land by periodic inundation during flood control stages. Environmental impacts on the area, in addition to a reduction of flood damages and promotion of increased agricultural development in the flood plain, would include losses of natural stream, loss of natural vegetation and loss of wildlife habitats. This impact would be due to the construction of a dam, creation of a lake, and regulation of downstream flows. Curtailment of flows would deteriorate stream fish habitat. Adverse environmental effects might also include deterioration of coastal estuarine environment. Alternatives considered included other dam locations, water recycling, water importation, and no project. Since there are significant archeological values in the project area, the Department of Interior requested a study to determine possible adverse impact. (Adams-Florida)
W73-08684

SOUTH BRANCH, RAHWAY RIVER, NEW JERSEY, FLOOD CONTROL PROJECT, RAHWAY, NEW JERSEY (FINAL ENVIRONMENTAL IMPACT STATEMENT).
Army Engineer District, New York.

Available from National Technical Information Service, as EIS-NJ-72-5587-F. Price: $3.75 in paper copy, $1.45 in microfiche. May 1972. 30 p, 9 plate.

Descriptors: *New Jersey, *Environmental effects, *Flood control, Channel improvements, Land fills, Levees, Floodwalls, Flood protection, Check structures, Multiple-purpose projects.
Identifiers: *Environmental Impact Statements, *Rahway (NJ).

The project consists of construction of local flood protection facilities in Rahway. These flood protection facilities will include channel improvement, levees, floodwalls, land filling, a roadway bridge, and interior drainage facilities. A pump station will also be constructed. Beautification features are provided for in project plans. The project will result in the reduction of flood damage threat for a residential area that is presently planned as the site for an urban renewal project. Adverse environmental effects include construction disturbances and the loss of tidal mud flats. Alternatives to the proposed project include reservoir flood detention, greater emphasis on channel enlargement, providing no flood protection, and flood plain regulation measures. Implementation of the flood control improvements is essential to the program of urban redevelopment and many economic and social gains would accrue. (Mockler-Florida)
W73-08685

SANTA PAULA CREEK CHANNEL, VENTURA COUNTY, CALIFORNIA (FINAL ENVIRONMENTAL IMPACT STATEMENT).
Army Engineer District, Los Angeles, Calif.

Available from National Technical Information Service, U.S. Dep't. of Commerce, EIS-CA-72-5334-F. Price: $6.25 in paper copy, $1.45 in microfiche. March 1972. 37 p, 2 plate, 3 illus, 8 photo.

Descriptors: *California, *Environmental effects, *Channel improvements, *Flood protection, *Flood control, Water quality control, Recreation, Water pollution, Flood plain zoning, Water storage, Area redevelopment, Water control.
Identifiers: *Environmental Impact Statements, *Santa Paula Creek (Calif).

This flood control project on the Santa Paula Creek channel in Ventura County, California, would consist of the construction of two debris basins, one each on Santa Paula and Mud Creeks and construction of channel improvements along both creeks. The proposed project would provide protection for the city of Santa Paula and adjacent agricultural and urban areas from floods originating in the Santa Paula Creek drainage area. The recommended improvements would also protect the existing local water supply and provide for additional recreational facilities. The natural landscape would be altered. Other adverse environmental effects include loss of existing natural streambeds; increased requirements for water and sewer facilities; and increased air, water, and noise pollution caused by urbanization and other development. Alternatives to the project include the use of upstream locations, no action, flood plain management, and use of other concrete channel sections on Santa Paula Creek. (Mockler-Florida)
W73-08686

OLIVER BOTTOMS RESOURCE CONSERVATION AND DEVELOPMENT PROJECT SEBASTIAN COUNTY, ARKANSAS (FINAL ENVIRONMENTAL IMPACT STATEMENT).
Soil Conservation Service, Washington, D.C.
For primary bibliographic entry see Field 04D.
W73-08687

PEYTON CREEK, TEXAS—FLOOD CONTROL (FINAL ENVIRONMENTAL IMPACT STATEMENT).
Army Engineer District, Galveston, Tex.

Available from National Technical Information Service, as EIS-TX-72-5006-F. Price: $6.00 in paper copy, $1.45 in microfiche. June 1972. 29 p, 7 tab.

Descriptors: *Texas, *Channel improvement, *Flood control, *River basin development, *Environmental effects, Controlled drainage, Soil conservation, Watersheds (Basins), Marsh management, Drainage effects, Agricultural runoff, Depositions (Sediments), Water pollution control, Aquatic environment, Watershed management, Sediment control, Dredging, Spoil banks, Water quality control.
Identifiers: *Environmental Impact Statements.

In the interest of flood control and drainage, this project proposes channel enlargement and the use of marshes as spoil areas in the Peyton Creek basin. The project would remove the threat of serious flooding in the urban area of Bay City, Texas, reduce flood damage to agricultural lands adjacent to the project, increase agricultural lands adjacent to the project, increase agricultural productivity by improved drainage, and help control future development with the designated floodway areas. The project would result in the conversion of about 82 acres of estuarine marsh area either for channel rights-of-way or placement of excavated material. There will be no significant effect on the water quality of Peyton Creek; there will be less sediment deposition in Lake Austin than is occurring presently. Channel enlargement and diversion channels in areas other than the proposed plan, along with a combination of non-structural improvements, were considered as alternatives; but they were felt to have no distinct developmental or environmental advantage over the recommended proposal. Comments from interested agencies expressed concern over the possibility of a deteriorating effect on the ecology in the lower reaches of Peyton Creek and Lake Austin. (Dunham-Florida)
W73-08688

FLOOD CONTROL PROJECT, THOMPSON'S CREEK, HOLLAND PATENT, NEW YORK (FINAL ENVIRONMENTAL IMPACT STATEMENT).
Army Engineer District, New York.

Available from National Technical Information Service, as EIS-NY-72-5285-F. Price: $3.25 in paper copy, $1.45 in microfiche. June 28, 1972. 15 p, 1 fig, 1 map.

Descriptors: *Flood control, *New York, *Channel improvement, *Drainage systems, *Environmental effects, Overflow, Planning, Flood damage, Flood protection, Flood proofing, Nonstructural alternatives, Cost-benefit ratio, Fish, Channels, Flood discharge, Diversion structures, Drainage programs, Culverts.
Identifiers: *Environmental Impact Statements, *Ninemile Creek Basin (NY).

This project consists of flood control improvements proposed for the Ninemile Creek Basin at Holland Patent, New York, necessitated by periodic overflow of Thompson's Creek, a tributary of Ninemile Creek which flows through Holland Patent. The overflowing affects approximately twelve acres of residential and commercial properties in the town. Since the principal cause of the overflow is the inadequate capacity of Thompson's Creek's channel, the proposed project will consist of the construction of a diversion channel from Thompson's Creek to Willow Creek and another waterway flowing through the town and having greater channel capacity. Other features of the project include culverts, drainage ditches and beautification measures. Implementation of the improvement will minimize future flood damage to the village, enhance land values, and will not significantly affect fish and wildlife resources in the area. Several alternatives were considered. Structural alternatives included impounding the flood discharges in a detention basin and diversion of flood discharges to Willow Creek. A nonstructural alternative considered was flood proofing. However, none of these alternatives is as economically efficient as the proposed project. (Adams-Florida)
W73-08689

CAMP GROUND LAKE, SALT RIVER BASIN, KENTUCKY (FINAL ENVIRONMENTAL IMPACT STATEMENT).
Army Engineer District, Louisville, Ky.

Available from National Technical Information Service, as EIS-KY-72-5447-F. Price: $3.00 in paper copy, $1.45 in microfiche. September 1972. 24 p, 2 map, 1 append.

Descriptors: *Multiple purpose reservoirs, *Dam construction, *Kentucky, *Flood protection, *Environmental effects, Dams, Reservoirs, Ecology, Flooding, Recreation, Fish, Wildlife habitats, Flood damage, Flood plains, Flood plain zoning, Flood plain insurance, Flood control, Flood forecasting, Water supply, Stream flow, Alteration of flow, Channels, Non-structural alternatives, Project planning, Water quality control.
Identifiers: *Environmental Impact Statements, *Camp Ground Lake (Ky).

The Camp Ground Lake project, located in the Salt River Basin, Kentucky, would consist of construction of a multi-purpose reservoir comprised of a dam and appurtenances. Project purposes include flood control, water supply, water quality control, and general fish and wildlife recreation. By regulating the amount of water released from the dam during periods of flooding, the project would provide flood damage reduction to about 20,000 acres of downstream lands. At seasonal pool elevation, approximately 5,070 acres of land and 50 miles of free-flowing stream will be inundated by the project. Aquatic habitat will be altered from free-flowing to slack water, causing an adverse impact. Non-structural alternatives considered included evacuation, acquisition of flood prone lands, and flood plain zoning. Flood plain zoning would have minimal impact on the environment. However, responsibility for this method generally lies with state and local governments, and furthermore, it would have no effect on reducing damage to development in flood prone areas. Structural alternatives, such as levees, were also considered. All alternatives were considered economically infeasible, or confining to area development. (Adams-Florida)
W73-08690

PERRY COUNTY DRAINAGE AND LEVEE DISTRICTS NOS. 1, 2, AND 3, MISSOURI (FINAL ENVIRONMENTAL IMPACT STATEMENT).
Army Engineer District, St. Louis, Mo.

Available from National Technical Information Service as EIS-MO-72-5279-F. Price $4.25 in paper copy; $1.45 in microfiche. May 1972. 14 p.

Descriptors: *Missouri, *Environmental effects, *Flood control, *Mississippi River, *Drainage districts, Levee districts, Pumps, Drainage engineering, Drainage systems, Drainage programs, Flood protection, Flood damage, Levees, Flooding, Agriculture, Ditches, Annual flood, Cultivated lands, Wildlife habitats, Environmental effects, Non-structural alternatives.
Identifiers: *Environmental Impact Statements, *Perry County (Mo.), *Randolph County (Ill.).

A high degree of flood protection from the Mississippi River was provided by the Perry County Drainage and Levee Districts Numbers 1, 2, and 3, located in Perry County, Missouri, and Randolph County, Illinois, upon completion of the existing levee in 1968. However, repetitive and substantial flood damages have occurred in the area due to blocked drainage through the levee. This project concerns a proposed improvement, recommended as a modification of the existing levee project, which is designed to provide relief from interior flooding on the agricultural lands within the districts. The recommended plan of improvement consists of the construction of four pumping stations ranging in size from 60 c.f.s. to 130 c.f.s., and two new drainage ditches totaling 13,900 feet in length. Improvements will reduce the average area flooded annually from 5,850 acres to 2,420 acres. With better drainage, some of the districts' 280 acres of scattered tracts of brush and forest land may eventually be converted to cropland. Since the area consists of highly developed agricultural land, there is a minimal wildlife population and the improvement will have little adverse effect on ecological systems in the districts. (Adams-Florida)
W73-08691

PRESENT DAY ENGINEERING CONSIDERATIONS AS APPLIED TO WATER LINE DESIGN,
Woolpert (Ralph L.) Co., Dayton, Ohio.
J. A. Herman.
Paper presented at American Water Works Association Annual Conference, Denver, Colorado, 1971. 26 p, 10 fig, 1 tab, 4 ref.

Descriptors: *Pipelines, *Water conveyance, *Municipal water, *Design criteria, *Water distribution (Applied), Water supply, Water delivery, Water requirements, Water works, Underground structures, Tunnels.

The design of a water line, or a system of water lines, includes many factors such as the following: peak flow rates generated by maximum instantaneous usages of water, types of materials to be utilized in the construction of the pipelines, desired ranges of system operating pressures, allowable losses of head as determined by flow velocities and pipe characteristics, types of joints to be utilized, possible effects of corrosion to the pipeline, the restraint of thrust in pipelines, etc. The application of design criteria has not changed greatly during the past few decades but conditions and environments for water lines have changed considerably from the past. Two such problems are peak system demand and thrust restraint. Peak demand is rising ever faster, especially in predominately residential areas, where lawn watering, air conditioners, etc., require more water than in the past. Large industries, although they demand large quantities of water, reduce the ratio of peak flow to average flow due to their continuous operation. Thrust-restraint of pipelines is an important consideration with the changing environmental conditions being imposed on all utilities, especially with utility tunnels and corridors. More positive restraints will be needed than are in use today. (Poertner)
W73-08896

TREATMENT EFFICIENCY OF WASTE STABILIZATION PONDS,
South Dakota School of Mines and Technology, Rapid City.
For primary bibliographic entry see Field 05D.

W73-09001

8B. Hydraulics

PLANS FOR REDUCTION OF SHOALING IN BRUNSWICK HARBOR AND JEKYLL CREEK, GEORGIA,
Army Engineer Waterways Experiment Station, Vicksburg, Miss.
F. A. Herrmann, Jr., and I. C. Tallant.
Available from NTIS, Springfield, Va., 22151, as AD-751-749 Price $3.00 printed copy; $1.45 microfiche. Technical Report H-72-5, September 1972. 202 p, 16 fig, 117 plate, 15 photo, 13 tab, 2 append.

Descriptors: *Harbors, *Navigation, *Channel improvement, *Georgia, *Coastal engineering, Model studies, Design criteria, Estuaries, Data collections, Atlantic Ocean, Engineering structures, Environmental effects, Tidal effects, Sand bars, Channel morphology, Shoals.
Identifiers: *Brunswick Harbor (Ga), *Jekyll Creek (Ga), *Shoaling reduction.

A fixed-bed model of Brunswick Harbor area, Georgia constructed to scales of 1:500 horizontally and 1:100 vertically, reproduced approximately 67 square miles. The model was equipped with the necessary appurtenances for the accurate reproduction of tides, tidal currents, shoaling patterns, and other significant prototype phenomena. The model study tested the effectiveness of proposed plans for the reduction of shoaling in Brunswick Harbor and Jekyll Creek. Closure of the East River channel was the only scheme tested which could effect significant reduction of shoaling. Schemes to increase current velocities or turbulence were unsuccessful in reducing channel shoaling in the critical reach of East River. Deepening the navigation channel from 30 ft to 36 ft will reduce shoaling in East River and increase shoaling in Turtle and Brunswick Rivers, even if no additional improvement works are constructed. A plan consisting of 10,150 ft of impermeable training walls was developed for reducing shoaling in Jekyll Creek. This reduced overall shoaling in the channel and shifted substantial amounts of shoal material from areas of high-cost maintenance to areas of low-cost maintenance. Dye dispersion tests were conducted in the model to determine the dispersion of wastes from the Brunswick Pulp and Paper Company into Turtle River and from the Brunswick sewage treatment plant into Academy Creek. (Woodard-USGS)
W73-08512

UNIT STREAM POWER AND SEDIMENT TRANSPORT,
Illinois Univ., Urgana.
C. T. Yang.
Journal of the Hydraulics Division, American Society of Civil Engineers, Vol 98, No HY10, p 1805-1826, Oct 1972. 9 fig, 3 tab, 28 ref, 3 append.

Descriptors: *Sediment transport, *Concentration, *Sediment discharge, *Statistical analysis, *Alluvial channels, Hydraulics, Sediments, Rivers, Open channel flow, Open channels, Sediment load, Turbulence, Particle size, Velocity, On-site data collections, Laboratory tests, Bibliographies.
Identifiers: Bed profiles, Unit stream power, Water surface profiles, Test results.

Sediment transport equations previously used often provided misleading results because of unrealistic assumptions made in their derivation. Unit stream power, defined as the time rate of potential energy expenditure per unit weight of water in an alluvial channel, is shown to dominate the total sediment concentration. Statistical analyses of 1225 sets of laboratory flume data and 50 sets of field data indicate the existence and generality of the linear relationship between the

logarithm of total sediment concentration and the logarithm of the effective unit stream power. The coefficients in the proposed equation are shown as being related to particle size and water depth, or particle size and width-depth ratio. An equation generalized from Gilbert's data can be applied to natural streams for accurately predicting total sediment discharge. (USBR)
W73-08513

CULVERT VELOCITY REDUCTION BY INTERNAL ENERGY DISSIPATORS,
Virginia Polytechnic Inst. and State Univ., Blacksburg.
J. M. Wiggert, and P. D. Erfle.
Concrete Pipe News, Vol 24, No 5, p 87-94, Oct 1972. 10 fig, 4 tab.

Descriptors: *Culverts, Energy dissipation, On-site tests, Obstruction to flow, Hydraulic jump, Laboratory tests, Concrete pipes, Plastic pipes, *Design criteria, Velocity, Graphical analysis, Turbulent flow.
Identifiers: *Energy dissipators, Tumbling flow, Test results, Roughness projections, Spacing, Rings, *Critical slopes, Free flow, Critical depth.

Laboratory and field studies were made on culverts on steep slopes to develop parameters for design of internal energy dissipators. The purpose was to eliminate the need for costly outlet structures. The report describes free flow and full flow conditions for culverts on 3 steep slopes with inlet control. Laboratory tests confirmed in the field showed that 4 annular rings properly spaced in the culvert would produce tumbling flow and the required dissipation of energy. Air entrained at the inlet because of vortex action had an insignificant effect on capacity. Larger diameter pipe is needed for free flow design to compensate for the rings, but for full flow design, no increase in pipe diameter is required. The suggested design is based on 1 large ring followed by 3 smaller ones, with double spacing between the large ring and the first smaller one. Laboratory tests were made in a 6-in.-dia plastic pipe, and field tests in an 18-in.-dia concrete pipe. Tables, graphs, and a nomograph summarize the test data, and provide aids to design. Two sample problems are solved in step-by-step detail. (USBR)
W73-08514

CONTROL OF SURGING IN LOW-PRESSURE PIPELINES,
Bureau of Reclamation, Denver, Colo.
J. J. Cassidy.
Report REC-ERC-72-28, Sept 1972. 35 p, 27 fig, 3 tab, 8 ref, append.

Descriptors: *Closed conduit flow, *Surges, On-site tests, Hydraulic gradient, Graphical analysis, Pipelines, Pipe flow, Pipe tests, Flow, Hydraulic transients, Head losses, Mathematical studies, Darcy-Weisbach equation, Unsteady flow, Computer programs, Fluid mechanics, Hydraulic design.
Identifiers: *Pipeline surges, Canadian River Project (Tex), Computer-aided design.

Pipelines designed for a hydraulic gradient nearly parallel to the ground profile are inherently underdamped when operating at less than design discharge. Consequently, any unsteadiness in the flow may be amplified, depending upon the characteristics of the system. The study was performed to establish parameters for use in design of low-pressure pipe systems by predicting amplitude and frequency of unsteady flow occurring at flow rates below the design value. A system must be designed so that small disturbances in one reach are not amplified in a succeeding reach; when the natural period of a downstream section exceeds the period of the incoming disturbance by a factor of 1.43, then incoming surges are not likely to occur. An analytical study of surge dynamics was conducted. Tabulations display the results, and numerous curves are given to show dimensionless discharge variations versus a function of dimensionless time and dimensionless discharge cutback ratio. Others show surge amplitudes as a function of a resistance coefficient. An appendix gives computer programs for dimensionless aqueduct parameters and surge characteristics. (USBR)
W73-08522

HEATED DISCHARGE FROM FLUME INTO TANK - DISCUSSION,
Vanderbilt Univ., Nashville, Tenn. Dept. of Environmental and Water Resources Engineering.
B. A. Benedict.
Journal of the Sanitary Engineering Division, American Society of Civil Engineers Vol 97, No. SA4, p 537-540, August 1971. 2 fig, 3 ref.

Descriptors: *Jets, *Heated water, *Temperature, *Velocity, Laboratory tests, Water temperature, Mathematical models, Mixing, Entrainment, Distribution patterns.
Identifiers: Flow establishment zone, Boussinesq assumption.

The original article by H. Stefan (See W71-05132) is timely experimental evidence, since the motion of heated discharges into lakes and rivers is of real concern to environmentalists. Of particular value is the measurement of the velocity field as well as the temperature field. The discussion concentrates on the reported data for the decline of center line surface water temperature. A form of the Motz-Benedict model is applied to these data to yield added insight into the problem. It follows from the graphical presentation that the theoretical results fit the data reasonably well. It appears that an entrainment mechanism is generally adequate to describe centerline decay of a discharge into the tank. It may not then be so necessary to consider different zones of discharge. The author also points out the importance of differentiating between the basic entrainment mechanisms for stagnant and flowing ambient streams. (Oleszkiewicz-Vanderbilt)
W73-08610

A NEW TYPE OF WATER CHANNEL WITH DENSITY STRATIFICATION,
Johns Hopkins Univ., Baltimore, Md. Dept. of Mechanics and Materials Science.
For primary bibliographic entry see Field 05B.
W73-08615

A TEMPERATURE ADJUSTMENT PROCESS IN A BOUSSINGESQ FLUID VIA A BUOYANCY-INDUCED MERIDIONAL CIRCULATION,
Kyoto Univ. (Japan). Dept. of Aeronautical Engineering.
T. Sakurai, and T. Matsuda.
Journal of Fluid Mechanics, Vol 54, part 3, p 1-6, August 1972. 4 ref.

Descriptors: *Buoyancy, *Temperature, *Hydraulics, Fluid mechanics, Laboratory tests, Mathematical models, Mathematical studies.
Identifiers: *Boussinesq fluid, Prandtl Number.

The existence of a new time scale expressed by means of Prandtl number, the thermometric conductivity, a typical length and the Brunt-Vaisala frequency, is clarified for a temperature adjustment process of a Boussinesq fluid in a circular cylinder. Relevant physical processes are discussed together with the in-depth mathematical treatment of the subject. (Oleskiewicz-Vanderbilt)
W73-08616

EVALUATION OF COASTAL ZONE MANAGEMENT PLANS THROUGH MODEL TECHNIQUES,
Army Engineer Waterways Experiment Station, Vicksburg, Miss.
For primary bibliographic entry see Field 02L.
W73-08813

HYDRAULIC MODELING OF AN ENVIRONMENTAL IMPACT: THE JAMES ESTUARY CASE,
Virginia Inst. of Marine Science, Gloucester Point.
For primary bibliographic entry see Field 02L.
W73-08814

GRAYS HARBOR ESTUARY, WASHINGTON: REPORT 4, SOUTH JETTY STUDY,
Army Engineer Waterways Experiment Station, Vicksburg, Miss.
N. J. Brogdon, Jr.
Available from NTIS, Springfield, Va 22151 as AD-749 257, Price $6.00 printed copy; $1.45 microfiche. Technical Report H-72-2, September 1972. 300 p, 260 plate, 16 photo, 4 tab.

Descriptors: *Engineering structures, *Coastal engineering, *Harbors, *Jetties, *Washington, Navigation, Shore protection, Breakwaters, Model studies, Design criteria, Planning, Estuaries, Tidal effects.
Identifiers: *Grays Harbor Estuary (Wash).

This is the fourth report in a series to be published on the results of model tests on the Grays Harbor, Washington, Estuary model conducted for the Army Engineer District, Seattle. This study was conducted during the period November 1970 to June 1971. The entrance to the harbor is protected by two converging stone jetties which are about 1.23 miles apart at their outer ends. The condition of the south jetty influences the stability of Point Chehalis and the surrounding areas. The conclusions and plans developed during the previous model study were put into effect and did result in stabilizing Point Chehalis. The navigation channel, however, continued to migrate south until the south jetty is now in danger of being undermined. For the tests in this report the entire model was in a fixed-bed condition. The model was constructed to linear scales of 1:500 horizontally and 1:100 vertically, which resulted in the following model-to-prototype scales based on the Proudian relations: velocity 1:10, time 1:50, discharge 1:500,000, volume 1:25,000,000, and slope 5:1. The salinity scale was 1:1. As a basis for selecting the best south jetty plan, hydraulic, salinity, dye dispersion, and shoaling characteristics were investigated. (Woodard-USGS)
W73-08889

STABILITY OF ROTATIONAL COUETTE FLOW OF POLYMER SOLUTIONS,
Delaware Univ., Newark. Dept. of Chemical Engineering.
M. M. Denn.
American Institute of Chemical Engineers Journal, Vol 18, No 5, p 1010-1016, September 1972. 2 fig, 1 tab, 18 ref.

Descriptors: *Rheology, *Viscosity, *Critical flow, *Aqueous solutions, *Rotational flow, Turbulence, Vortices, Shear stress.
Identifiers: Taylor number, Secondary flow, *Polymer solutions.

The onset of secondary flow between rotating cylinders (Taylor vortices) was observed for a dilute polymer solution whose viscometric flow properties were characterized rheogoniometrically. The critical Taylor number was predicted accurately by linear stability theory with a stress constitutive equation describing viscous behavior. The cell spacing differed significantly from that predicted by linear theory. A nonlinear analysis shows that linear theory will predict the ultimate cell size only for an inelastic liquid. For an elastic liquid, closer spacing is a lower energy configuration than the linear theory spacing. This is consistent with experiment. (Knapp-USGS)
W73-09100

:. Hydraulic Machinery

MONSTON PUMPING PLANT: NATION'S GHTIEST,
lifornia State Dept. of Water Resources, Sacramento.
B. Jansen.
vil Engineering, Vol 42, No 10, p 67-71, Oct 72. 4 fig, 5 photo.

escriptors: *Pumping plants, *Pumps, Pumping, ydraulic machinery, Seismic design, Surge nks, Structural design, Mechanical equipment, lectric motors, Electrical equipment, Discharge es, Control, California, Structural engineering, utomatic control, Manual control, Design iteria, Hydraulic design, Remote control.
Identifiers: *High head, A. D. Edmonston Pump lant (California), California Water Plan, arthquake zones.

he Edmonston pumping plant, a key element in e California Water Project, lifts Northern alifornia water over the Tehachapi Mountains orth of Los Angeles. With a discharge capacity of 410 cfs and a lift of 1926 ft, this plant is the Nation's largest in combined lift and capacity. The pecial 4-stage European-type pumps selected for his plant are described. Each pump has a capacity of 315 cfs, requiring an 80,000-hp synchronous notor, for a total power requirement for all 14 pumps of more than 1 million hp. The unique motor-starting features are discussed. The pumping operation is computer- or operator-controlled from the plant room and, ultimately, can be controlled from a remote Area Control Center. As a precautionary measure against earthquake hazards prevelant in this area, structures were designed for a seismic acceleration of 0.5 g. The surge tank, a key feature of the discharge system, is described. (USBR)
W73-08523

ELECTRICAL EFFECTS OF AC TRANSMISSION LINES ON PIPELINES,
Corrosion Control Technologists, Houston, Tex.
M. A. Riordan.
Materials Protection Performance, Vol 11, No 10, p 26-31, Oct 1972. 4 fig, 3 tab, 13 ref.

Descriptors: *Pipelines, *Induced voltage, *Alternating current, Safety, Transmission lines, Hazards, Electric fields, Electrical faults, Ground currents, Corrosion, Electrical grounding.
Identifiers: Potential gradients, Electric shock, *Induced currents, Ground return, Earth resistivity (Electrical).

A metallic pipeline represents a single-line conductor and as such will respond to the electrical field when a pipeline is located near an overhead power transmission line. As the availability of easements decreases, coexistence of powerlines and pipelines increases. With parallel coexistence, electrical exposures may develop dangerous voltages on the metallic portions of the pipeline system. Since pipelines are usually undergound, they are not as responsive as aerial cable to inductive effects from transmission lines, but may be more responsive to ground voltage effects during fault conditions on the transmission lines. The pipeline engineer should be aware of these electrical influences that could develop into undesirable conditions; both human exposure and mechanical hazards must be considered. Human tolerances, fault conditions, modes of energy transfer, normal and abnormal powerline operation, and the magnitude of induced voltages are discussed with respect to pipeline safety for installation, operation, and maintenance. (USBR)
W73-08524

PROPOSED NEW LOCK--PICKWICK LANDING DAM (FINAL ENVIRONMENTAL IMPACT STATEMENT).
Tennessee Valley Authority, Chattanooga.

Available from the National Technical Information Service $3.25 in paper copy, $1.45 in microfiche. Tennessee Valley Authority, Office of Health and Environmental Science, TVA-OHES-EIS-72-7, September 13, 1972. 17 p.

Descriptors: *Tennessee, *Tennessee River, *Tennessee Valley Authority, *Locks, *Environmental effects, Navigation, Navigable rivers, Dams, Dam construction, Concrete structures, Damsites, Excavation, Federal project policy, Federal government, Administrative agencies, Water pollution, Natural resources, Transportation, Dredging, Turbidity.
Identifiers: *Environmental Impact Statements, Water traffic control.

The proposed project is the construction and operation of a new main lock at Pickwick Landing Dam in Hardin County, Tennessee. The new lock will improve conditions for water traffic and reduce congestion and the possibility of accidents. The site is part of the present dam reservation and was provided for the proposed lock in the original design of the dam. The environmental impact involves excavation of some 2.7 million cubic yards of earth and rock, including some 450,000 cubic yards of dredging, the disposal of excavated material on low-lying land near the lock, and the placement of about 425,000 cubic yards of concrete. Adverse impacts include a temporary increase in turbidity in the Tennessee River and the temporary loss of about 30 acres of wildlife habitat. Extension of the existing lock and development of bypass methods were considered as alternatives. (Dunham-Florida)
W73-08692

DEEP ROTARY CORE DRILLING IN ICE,
Cold Regions Research and Engineering Lab., Hanover, N.H.
For primary bibliographic entry see Field 02C.
W73-09079

8D. Soil Mechanics

FOUNDATION AND ABUTMENT TREATMENT FOR HIGH EMBANKMENT DAMS ON ROCK.
American Society of Civil Engineers, New York. Committee on Embankment Dams and Slopes.
For primary bibliographic entry see Field 08E.
W73-08515

ENVIRONMENTAL DEFENSE FUND, INC. V. ARMSTRONG (ACTION FOR INJUNCTIVE AND OTHER RELIEF WITH RESPECT TO THE NEW MELONES DAM PROJECT).
For primary bibliographic entry see Field 06E.
W73-08706

8E. Rock Mechanics and Geology

LOCATION OF SOLUTION CHANNELS AND SINKHOLES AT DAM SITES AND BACKWATER AREAS BY SEISMIC METHODS: PART I--ROCK SURFACE PROFILING,
Kentucky Water Resources Inst., Lexington.
V. P. Drnevich.
Available from the National Technical Information Service as PB-220 063, $3.00 in paper copy, $1.45 in microfiche. Research Report No 54, 1972. 45 p, 14 fig, 11 ref, 2 append. OWRR A-026-KY (3), 14-31-0001-3217.

Descriptors: *Seismic studies, *Subsurface investigations, *Bedrock, Profiles, Subsurface mapping, Borehole geophysics, Computer programs, Model studies.

Sledge hammer seismic refraction surveys and a modified version called downhole shooting are useful for rock surface profiling. Advantages of downhole shooting are calibration at the end points of the survey, measurement of vertical wave propagation velocities directly, and having a refracted wave ray path for almost the entire survey length. The downhole method was simulated with the digital computer. The method can handle any rock surface profile and generates corresponding travel time curves for the forward and reverse profile surveys. This program was used to study the effects of anomalies on the travel time curves. A method of data reduction was developed that enables an estimate of the rock surface profile to be made from the travel time data. Field tests were performed at four sites having soil and rock characteristics different from each other. Rock surface profiles were estimated from the travel time curves using the procedure developed and these were compared with the depth to rock by proof drilling. The sources of error and some limitations of use are discussed. For the sledge hammer method to be used for rock surface profiling, the rock surface should be within 25 to 30 ft of the soil surface and the minimum width of solution channel that can be sensed with this method is on the order of two feet. (See also W73-08467) (Knapp-USGS)
W73-08466

LOCATION OF SOLUTION CHANNELS AND SINKHOLES AT DAM SITES AND BACKWATER AREAS BY SEISMIC METHODS: PART II--CORRELATION OF SEISMIC DATA WITH ENGINEERING PROPERTIES,
Kentucky Water Resources Inst., Lexington.
For primary bibliographic entry see Field 07B.
W73-08467

FOUNDATION AND ABUTMENT TREATMENT FOR HIGH EMBANKMENT DAMS ON ROCK.
American Society of Civil Engineers, New York. Committee on Embankment Dams and Slopes.

Journal of the Soil Mechanics and Foundations Division, American Society of Civil Engineers, Vol 98, No SM10, p 1115-1128, October 1972. 4 tab.

Descriptors: *Dam foundations, *Abutments, *Rock foundations, Grout curtains, Earth dams, Rockfill dams, Treatment, Excavation, Grouting, Instrumentation, Topography, Geologic investigations, Slopes, Cutoffs, Dam construction, Foundations, Foundation investigations, Dam design, Design practices, Surface, Drainage.
Identifiers: Surface treatment, Surface preparation, Site selection.

Current and past practices of abutment and foundation treatment for high embankment dams founded on rock are reviewed. Organizations contributing information on 23 dams include: Bechtel Corp, British Columbia Hydro and Power Authority, Harza Engineering Company, Snowy Mountains Authority, State of California Department of Water Resources, Bureau of Reclamation, and US Army Corps of Engineers. Topics include: (1) geology and topography of the abutments and foundations--rock treatment and dam location and design, (2) excavation of the abutment and foundation areas--permissible slopes and quality of acceptable rock surfaces for both core and shell portions of the dam, (3) foundation preparation, including methods for normal and poor rock treatment, drainage, and grouting, and (4) adequacy of abutment and foundation treatment. (USBR)
W73-08515

TERRESTRIAL SURVEYING AND GEOLOGICAL MAPPING OF THE ROCK FACE OF THE DEWATERED AMERICAN NIAGARA FALLS,
Department of the Environment, Ottawa (Ontario). Inland Waters Branch.
For primary bibliographic entry see Field 07C.
W73-08727

8F. Concrete

THE WINDSOR PROBE,
Gulick-Henderson Labs., Inc., Whitestone, N.Y.
L. D. Long, and C. Gordon.
Paper presented to American Concrete Institute Seminar, Atlanta, Georgia, Sept 1972. 5 p.

Descriptors: *Probes (Instruments), *Concrete technology, Compressive strength, *Nondestructive tests, *Concrete tests, Test equipment, Test procedures, Hardness tests, Quality control, Inspection, On-site tests, Aggregates, Coarse aggregates, Concretes, Concrete mixes, Materials testing.
Identifiers: *Windsor probe test system, *Probes (Instruments).

The recently developed Windsor probe is considered by some to be one of the most reliable nondestructive methods of measuring concrete strength in a structure. The probe measures compressive strength by the resistance of the mortar and by the force required to shatter a piece of coarse aggregate. This instrument, if used correctly within its limits, is a convenient and valuable tool. A favorable feature is that the probe can be left in place, if required. Factors affecting concrete strength, such as size, type, amount, mix, and condition of the aggregates, which may not be reflected by probe tests, should be considered when using the probe. Operator qualifications include: (1) experience in the use of this instrument, (2) expertise in concrete technology, (3) capable of sight recognition of commonly used aggregates, and (4) ability to test and identify such aggregates. (USBR)
W73-08517

CONCRETE NONDESTRUCTIVE TESTING,
Law Engineering Testing Co., Birmingham, Ala.
R. Muenow.
Paper presented at American Concrete Institute Seminar, Atlanta, Georgia, Sept 1972. 21 p.

Descriptors: *Nondestructive tests, *Concrete tests, *Test equipment, *Test procedures, *Concrete technology, Electrical properties, Magnetic properties, Electrical resistance, Sound waves, Materials tests, Quality control, Inspection, On-site tests.
Identifiers: Resonance frequency method, Sonic devices, Ultrasonic tests, Nuclear density meters, Radiographic inspection, Infrared spectroscopy.

Because of tradition, outdated quality control techniques are still used even though useful information is no longer derived. Today, many new methods of testing are available for obtaining information about construction materials or the condition of structures without destroying any part of them. Now, properties of materials in a structure can be measured, flaws and early signs of deterioration detected, and tests performed more quickly and accurately. The operating principles and uses of 8 unusual instruments in facilitating construction repairs are reviewed. Methods described are: (1) nuclear activation and absorption, (2) low-frequency ultrasonics, (3) resonant frequency, (4) radiographic or isotope, (5) infrared, (6) inspection of steel by magnetic or electrical methods, (7) piling inspection, and (8) sonic emission. (USBR)
W73-08519

EVALUATION OF LOW STRENGTH CONCRETE,
Law Engineering Testing, Co., Birmingham, Ala.
F. A. Kozeliski.
Paper presented at American Concrete Institute Seminar, Atlanta, Georgia, September 1972. 5 p, 6 ref.

Descriptors: *Concrete tests, *Non-destructive tests, *Compressive strength, Concrete technology, Concretes, Test equipment, Test procedures, Materials tests, Inspection, Quality control, On-site tests, Electronic equipment, Concrete mixes, Evaluation.
Identifiers: *Pulse velocity tests, *Soniscopes, Pulse generators, Pulse method, Sonic velocity, Transducers.

The soniscope with transducers is one of the most acceptable methods of nondestructive testing of concrete. With this method the pulse velocity of the concrete is measured to evaluate the strength and workmanship of concrete throughout the structure. The pulse velocity method is an accepted ASTM procedure under ASTM designated C597-71. Description of the apparatus and the method of measuring the pulse velocity are presented. Some factors affecting the pulse velocity are: (1) cement content, (2) type of aggregate, (3) age of concrete, and (4) curing temperature. The soniscope can also be used in the field to monitor concrete strength development, determine acceptable times to remove formwork, and to transfer prestress forces to concrete. Other uses include evaluating quality control of concrete placement, construction workmanship, and the strength of old buildings. (USBR)
W73-08525

DUCTILITY OF REINFORCED CONCRETE COLUMN SECTIONS IN SEISMIC DESIGN,
Canterbury Univ., Christchurch (New Zealand).
R. Park, and R. A. Sampson.
Journal of the American Concrete Institute, Proceedings, Vol 69, No 9, p 543-551, Sept 1972. 5 fig, 1 tab, 17 ref, append.

Descriptors: *Columns, *Earthquake-resistant structures, *Reinforced concrete, *Structural design, Bibliographies, Deformation, Reinforcing steels, Seismic design, Concrete structures, Theoretical analysis.
Identifiers: Curvature, New Zealand, Axial loads, Eccentric loading, Limited design, Spirals, Stress-strain curves.

Ductility required of eccentrically loaded reinforced concrete column sections in seismic design is discussed. A method is suggested for determining the amount of special transverse steel required for ductility. The method is based on a theoretical study using stress-strain curves for concrete confined by rectangular hoops and for steel including the effect of strain hardening. Factors considered were: (1) required ultimate curvature, (2) level of axial load, (3) longitudinal steel content, and (4) material strengths. Present code recommendations for transverse steel may be less than appropriate in some instances of high axial load and low longitudinal steel content, and could be relaxed in others. (USBR)
W73-08526

8G. Materials

LOCATION OF SOLUTION CHANNELS AND SINKHOLES AT DAM SITES AND BACKWATER AREAS BY SEISMIC METHODS: PART I--ROCK SURFACE PROFILING,
Kentucky Water Resources Inst., Lexington.
For primary bibliographic entry see Field 08E.
W73-08466

THE CORROSION ENGINEER'S LOOK . STRESS CORROSION,
Texas Univ., El Paso.
D. P. Redzie, and F. E. Rizzo.
Mater Prot Performance, Vol 11, No 10, p 53- Oct 1972. 4 fig, 2 tab, 20 ref.

Descriptors: *Corrosion, *Brittle fractur Cathodic protection, Heat treatment, Bibliog phies, Failure, Corrosion control, Cracking, Te sile stress, Fabrication, Time, Materials engine ing, Materials testing.
Identifiers: *Stress corrosion, *Brittle fracture Ion concentration, Notch effect, Microstructu Stress concentration, Materials failure.

Stress corrosion cracking can occur in almost engineering materials. Recent research indicat that failure is continuous on a seconds time sca Failure takes place in a brittle rather than a duct fashion, and is usually perpendicular to the loa The characteristic appearance includes a lack deformation with relatively small amounts general corrosion; at the microscopic level, it c be transangular or intergranular. Stress corrosi cracking is more probable as the strength of material increases, is related to heat treatment a fabrication, and is dependent upon a specific i whose concentration may be low. Time to failu is decreased by: (1) using a susceptible alloy, he treatment, or fabrication method; (2) presence cuts, sharp corners, or notches; (3) high conce trations of the specific ion; and (4) high stre levels. Stress corrosion cracking can be avoid by eliminating a susceptible material, a specific e vironment, time, or tensile stress. Cathodic pr tection can also be used to control stress corrosic cracking. No unified theory of stress corrosic cracking has been accepted by workers in th field. (USBR)
W73-08516

IMPROVED CORROSION RESISTANCE, RE LIABILITY, AND INTEGRITY OF CAS STAINLESS ALLOYS,
Duriron Co., Inc., Dayton, Ohio.
W. A. Luce, M. G. Fontana, and J. W. Cangi.
Corrosion, Vol 28, No 4, p 115-128, Apr 1972. 1 fig, 4 tab, 4 ref.

Descriptors: *Alloys, Castings, Materials Moldings, Stainless steel, Interfaces, Graphic analysis, Applied research, Laboratory tests.
Identifiers: *Cast metals, *Corrosion resistance Carburizing, Cast steel, Carbon steel, Corrosio tests, Accelerated tests.

Lower carbon content improves general corrosio resistance of stainless alloys and minimize susceptibility to intergranular attack. Producers o alloy castings have the problem of surface carbu rization occurring at the metal-mold interfac when the mold contains organic or carbonaceou material. The study was undertaken to: (1) defin the nature and extent of surface carburization oc curring on stainless steel alloy castings poured i carbon-containing and carbon-free mold environ ments, and (2) determine effects of surface carbu rization on corrosion rate. Carbon enters liqui metal by gaseous transfer or direct contact wit carbon or carbide residues. A carbon concentra tion gradient is established at the metal-mold inter face. Castings were made in resin shell, ceramic and green sand molds. Chemical composition wa determined from samples cut from the flange Results are given for CF-3, CF-8, and CN-7M castings. Numerous electron microscope photo graphs of casting surfaces and graphs of carbo profiles are included. The degree of carburizatio and reduction in corrosion resistance were mor pronounced in the CF-3 than the other 2 alloys Carburization is inversely proportional to allo carbon content. Green sand and ceramic molds im part little carbon to the casting. Pronounced carbu rization may occur with resin shell molds. (USBR)
W73-08518

USE OF MICRO-ALGAE SUSPENSION IN FEEDING BALTIC SALMON JUVENILES, (IN RUSSIAN),
T. M. Aronovich.
Tr Vses Nauchno-Issled Inst Prudovogo Rybn Khoz. 18: p 11-16. 1971. (English summary).
Identifiers: *Fish diets, Algae, Baltic, Chlorella, Juveniles, Salmo salar, Salmon, Scenedesmus, Suspension.

Supplements of micro-algae (Chlorella and Scenedesmus cultures) in the food of juvenile Baltic salmon (Salmo salar) during raising in fish hatcheries were highly effective.--Copyright 1972, Biological Abstracts, Inc.
W73-08554

DOWNSTREAM MIGRATION OF VOBLA, BREAM AND WILD CARP FROM ARAKUMSK BODIES OF WATER, (IN RUSSIAN),
K. A. Adzhimuralov.
Sb Nauchn Soobshch Kafedry Fiziol Cheloveka Zhivotn Zool Biol Khim Dagest Univ. 4. p 49-52. 1970.
Identifiers: *Fish management, Arakumsk, Bream, Carp, Downstream, *Fish migration, USSR, Vobla.

In Arakumsk bodies of water in the Dagestan ASSR, 2 forms of downstream migration by juveniles from the spawning grounds were observed: passive and active. The passive form was observed in April as result of unfavorable factors: hunger, disturbances in the temperature and the changes in the water level. Active migration was observed in May-July, when the juveniles reached the downstream-migrant stage. For normal downstream migration regulation of the water supply and increases in its flow at the spawning ground in the direction of the fishway are necessary.--Copyright 1972, Biological Abstracts, Inc.
W73-08600

STUDY OF IMPROVING THE QUALITATIVE COMPOSITION OF CARP FOOD. METHODS OF FEEDING, (IN RUSSIAN),
F. M. Sukhoverkhov.
Tr Vses Nauchno-Issled Inst Prudovogo Rybn Khoz. 17 p 36-46. 1971. English summary.
Identifiers: *Fish diets, *Carp, Composition, Feeding, Methods, Yeast.

Methods for experimental feeding of carp are discussed, including the use of duplicate experimental ponds, production checks, correct selection of ponds, stocking material and dietary requirements. The optimal norms of food supplements of dietary yeasts, tissue preparation, terramycin, carotene, vitamin B12, phosphatides and microelements (C0) are established.--Copyright 1972, Biological Abstracts, Inc.
W73-08609

THE INDUCED SPAWNING OF THE PHYTO- AND PLANKTOPHAGOUS FISHES FROM THE EXPERIMENTAL STATION NUCET DURING 1966-1970, (IN RUMANIAN),
Institutul de Cercetari si Proiectari Aliment., Bucharest (Rumania).
A. Nicolau, S. Luscan, and E. Nichiteanu.
Stud Cercet Piscic Inst Cercet Proiect Aliment. 4: 273-298, Illus, 1971. English summary.
Identifiers: *Fish reproduction, Aristichthys nobilis, Ctenopharyngodon idella, Fishes, Hypophthalmichthys molitrix, Nucet, Phytophagous fish, Pituitary treatment, Planktophagous fish, Romania, *Spawning.

Results are presented on induced spawning of the grass carp (Ctenopharyngodon idella), the silver carp (Hypophythalmichthys molitrix) and Aristichthys nobilis during a period of 5 yr. The factors affecting and determining the variations of the percentage of ripened gonads from females subject to pituitary treatment, the fecundity, the survival rate from ova until 5 days old, and the number of 5-day old fry were analyzed. The ponds and the local climate offer to phytophagous and planktonophagous fishes favorable conditions for growth and spawning. H. molitrix and A. nobilis had great quantities of natural food. The induced spawning was influenced by complex internal and external factors. The physiological conditions of the spawners are discussed.--Copyright 1972, Biological Abstracts, Inc.
W73-08614

THE NATURAL AND SUPPLEMENTARY FEEDING ECOLOGY OF POND CARP, (IN RUMANIAN),
Inst. Cercet. Project. Aliment., Bucuresti, Rom. Institutul de Cercetari si Proiectari Aliment., Bucharest (Rumania).
M. Niculescu-Duvaz.
Stud Cercet Piscic Inst Cercet Project Aliment. 4 p 59-251. Illus. 1971. English summary.
Identifiers: *Carp, *Fish diets, Ecology, Feeding, Fish growth, Production.

Detailed analysis is presented on the importance of supplementary feeding and the necessary pond conditions for increased growth and production of this valuable fish.--Copyright 1972, Biological Abstracts, Inc.
W73-08633

COREGONOID FISH OF SIBERIA. BIOLOGICAL BASIS ON COMMERCIAL EXPLOITATION AND REPRODUCTION OF RESOURCES, (IN RUSSIAN),
B. K. Moskalenko.
Pishch. Prom-st': Moscow. 1971. 184p Illus.
Identifiers: *Cisco, *Commercial fish, Biology, Dynamics, Exploitation, Fish population, Fish reproduction, Siberia, USSR.

The descent and distribution of coregonoid fish of Eurasia, migration, reproduction, feeding, nutritional interrelationships, population dynamics and productivity are covered. A biological characterization of commercial species including 'muksun,' cisco, omul, 'pizhyan,' broad whitefish, 'peled' and tugun, evaluation of reserves, and regulation of fishing and artificial propagation are included.--Copyright 1972, Biological Abstracts, Inc.
W73-08637

EFFICIENCY OF PROTEINS IN RELATION TO THEIR CONCENTRATION IN THE RAINBOW TROUT DIET, (IN FRENCH),
P. Luquet.
Ann Hydrobiol. Vol 2 No 2, p 175-186. 1971. Illus. English summary.
Identifiers: Concentration, *Fish diets, Herring, Maize M, *Proteins, *Rainbow trout.

Four semi-synthetic diets based on herring meal and raw maize starch containing 30-60% protein were given ad lib to 4 groups of rainbow trout for 36 wk. The body weight composition, food consumption and protein utilization were studied. A weight gain of 200 g in 36 wk was obtained. It is estimated that trout require less than 30% protein in their diet under similar conditions.--Copyright 1972, Biological Abstracts, Inc.
W73-08642

INTRODUCTION OF SALMONIDS IN THE KERGUELEN ISLANDS: I. FIRST RESULTS AND PRELIMINARY OBSERVATIONS, (IN FRENCH),
Station d'Hydrobiologie Continentale, Biarritz (France).
R. Lesel, Y. Therezien, and R. Vibert.
Ann Hydrobiol. Vol 2, No 2, p 275-304. 1971. Illus. English summary.

Identifiers: *Fish establishment, *Fish diets, Condition, Insects, Kerguelen, Islands, Plankton, Salmonids.

Preliminary observations indicate that the condition factor for lake trout living in rivers is moderate, whereas that of fish living in lakes is good. The study of food shows 2 diets: one essentially based on plankton in lakes, the other on insect larvae of exogenous origin in streams and rivers.--Copyright 1972, Biological Abstracts, Inc.
W73-08643

CONDITION AND FATNESS OF THE GRASS CARP CTENOPHARYNGODON IDELLA (VAL.) IN THE AMUR BASIN (IN RUSSIAN),
Tikhookeanskii Nauchno-Issledovatelskii Institut Rybnogo Khozyaistva i Okeanografii, Khabarovsk (USSR).
E. I. Gorbach.
Vopr Ikhtiol, Vol 11, No 6, p 1002-1013, 1971, Illus.
Identifiers: *Fish physiology, Amur Basin, *Carp, Condition, Ctenopharyngodon idella, Fatness, USSR, Fish diets.

The condition and fatness of C. idella as a function of size, age, sex, stage of gonadal maturity, season, and food supply were investigated. It was established that the changes of the condition and fatness of this species are governed by the physiological state of the organism and by the effect of the food supply.--Copyright 1972, Biological Abstracts, Inc.
W73-08822

COLLECTION OF JUVENILE SALMON AND STEELHEAD TROUT PASSING THROUGH ORIFICES IN GATEWELLS OF TURBINE INTAKES AT ICE HARBOR DAM,
National Marine Fisheries Service, Seattle, Wash. Biological Lab.
D. L. Park, and W. E. Farr.
Trans Am Fish Soc. Vol 101, No 2, p 381-184. 1972. Illus.
Identifiers: Collection, Dam, Gatewells, Ice Harbor Dam, Juvenile, Orifices, Passing, *Salmon, *Steelhead trout, Turbine intakes, *Washington.

A total of 172,785 juvenile salmon and steelhead trout were trapped over a 2-mo period by the collection system at Ice Harbor Dam, Washington, and the ability of the system to capture and hold a large number of fish is demonstrated. Its applicability to other dams where orifices are constructed is indicated. This provides sufficient numbers of the fish in good condition for transportation studies.--Copyright 1973, Biological Abstracts, Inc.
W73-08983

STUDIES ON GILL PARASITOSIS OF THE GRASSCARP (CTENOPHARYNGODON IDELLA) CAUSED BY DACTYLOGYRUS LAMELLATUS ACHMEROW, 1952: II. EPIZOOTIOLOGY,
K. Molnar.
Acta Vet Acad Sci Hung. Vol 21, No 4, p 361-375. 1971. Illus.
Identifiers: *Carp, Ctenopharyngodon-Idella, Dactylogyrus-Lamellatus, Epizootiology, Gill, *Parasitosis, Temperature.

The gill parasite D. lamellatus affected grasscarp of all ages and sizes throughout the year. A high stocking density favors its spread, and thus the parasite threatens the populations primarily of the fry-rearing ponds, nursery ponds and wintering ponds. The fatal outcome of dactylogyrosis depends on the parasite count of the host, but a warm temperature and a low O2 content of the water accelerate its onset. Three epizootiologically different forms of D. lamellatus infection were observed in ponds farms: acute dactylogyrosis of the fry, and subacute and chronic dactylogyrosis of 1 and 2 summer grasscarps. A continuous low-degree infestation with D. lamellatus protected the fishes against a not too massive reinfection by larval stages of the parasite. This kind of host-parasite equilibrium seems to depend on the individual resistance of the host. (See also W73-09003)--Copyright 1972, Biological Abstracts, Inc.
W73-09002

STUDIES ON GILL PARASITOSIS OF THE GRASSCARP (CTENOPHARYNGODON IDELLA) CAUSED BY DACTYLOGYRUS LAMELLATUS ACHEMEROW, 1952: III. THERAPY AND CONTROL,
K. Molnar.
Acta Vet Acad Sci Hung. Vol 21, No 4, p 377-382. 1971.
Identifiers: *Carp, Ctenopharyngodon-Idella, Dactylogyrus-lamellatus, Ditrifon, E, Flibol, Garoona, Gill, Host, Nuvanol, Parasit-drugs, *Parasitosis, USSR.

Experimental and field observations showed that the establishment of D. lamellatus infection among grasscarp fry can be prevented by supplying the water from reservoir ponds not populated with older infected fishes and by removal of the growing fry to larger ponds before the population becomes too high. Bathing of the hosts in a 2.5% NaCl solution kills D. lamellatus and the organic phosphoric acid ester solutions (Ditrifon 50, Flibol E, Nuvanol, Gardona) tested for long-term action were 100% effective when applied to a concentration of 1 ppm for 48 hr. Ditrifon 50, also tested for short-term action at a concentration of 100 ppm, killed all parasites in 0.5-4 hr. The combination of breeding technological and chemotherapeutic methods ensures effective control of grasscarp dactylogyrosis. (See also W73-09002)--Copyright 1972, Biological Abstracts, Inc.
W73-09003

PISCIVOROUS ACTIVITIES OF BROWN BULLHEADS IN LOCKHART POND, ONTARIO, CANADA,
Guelph Univ., (Ontario). Dept. of Zoology.
J. W. Moore.
Prog Fish-Cult. Vol 34, No 3, p 141-142, 1972. Illus.
Identifiers: *Bullheads (Brown), Canada, Ictalurus-nebulosus, Lepomis-gibbosus, Lockhart pond, Notemigonus-crysoleucas, Ontario, *Piscivorous, Ponds.

Brown bullhead (Ictalurus nebulosus) greater than 21.5 cm, collected from Lockhard Pond, Ontario, Canada, fed mainly on fish (88% by volume). Pumpkinseeds (Lepomis gibbosus) comprised 55% of the total stomach volume, golden shiners (Notemigonus crysoleucas) 15%, and unidentified fish remains, 18%. Invertebrates, primarily larval Chironomidae, comprised the remaining 12%. High population density of bullheads, high incidence of encounter betwen bullheads and other fish species, and semidystrophic limnological conditions may be responsible for the unusual feeding habits.--Copyright 1973, Biological Abstracts, Inc.
W73-09040

ENERGY METABOLISM AND FOOD RATIONS OF STEELHEAD TROUT UNDER CONDITIONS OF THE CHERNAYA RACHKA TROUT INDUSTRY, (IN RUSSIAN),
E. P. Skazkina.
Tr Vses Nauchno-Issled Inst Morsk Rybn Khoz Okeanogr. 76, p 130-134, 1970. English summary.
Identifiers: Chernaya Rechka, Food, *Metabolism, Rations, *Steelhead trout, USSR, *Trout.

Properties of metabolism are discussed for early steelhead juveniles which were received from the USA and raised on Chernorechenskaya trout. The larval rations were computed by respiration method. The obtained values for growth and energy metabolism were very small. This was evident due to the constant low water temperature in t streams of Chernaya Rechka.--Copyright 197 Biological Abstracts, Inc.
W73-09044

DIETARY CHARACTERISTICS OF SILVE CARP AND GRASS CARP LARVAE AT VAR OUS STAGES OF DEVELOPMENT, (IN RU SIAN),
A. I. Strelova.
Tr Vses Nauchno-Issled Inst Prudovogo Ryt Khoz. 18, p 188-194, 1971. English summary.
Identifiers: *Carp, Cyclops, *Dietary studi (Carp), Grass carp, Larvae, *Rotifers, Silver car

Experiments in aquariums and pond observatior showed that rotifers were the basic food com ponent of silver carp and grass carp at 24 stage development. When there was an insufficiency rotifers in the food of both species, the fish bega to consume small Cyclops. Larvae of the 2 spp. ar competitors for rotifers in the first stage an copepodid cyclops in the second and third.--Copy right 1972, Biological Abstracts, Inc.
W73-09045

CONDITION OF RESOURCES, SOME BIOLOG ICAL TRAITS AND PERSPECTIVES O REPRODUCTION OF THE AMUR STURGEO AND STERLET, (IN RUSSIAN),
V. G. Svirskii.
Uch Zap Dal'nevost Univ. Vol 15, No 31, 1971.
Identifiers: Amur sturgeon, Biological studies *Reproduction, *Sterlet, *Sturgeon, USSR.

Kaluga, Amur sturgeon, green sturgeon and sterle inhabit the Amur River basin. Introduction of ster let was begun in 1956. The dynamics of catches o sturgeon were studied for the period 1891-1964 The large-scale operation of artificial breeding was registered. In all 30,000 juveniles of the same age were put into the Amur.--Copyright 1972, Biological Abstracts, Inc.
W73-09046

SEASONAL RACES OF PINK SALMON ON THE KURIL ISLANDS, (IN RUSSIAN),
V. N. Ivankov.
Uch Zap Dal'nevost Univ. Vol 15, No 3, p 34-43, 1971.
Identifiers: Fecundity, Islands, *Kuril Islands, Races, *Salmon (Pink),Seasonal, *Spawning, USSR.

In the Slavnaya River on Iturup Island 499 juveniles and 10,250 adult fish were analyzed. The existence of summer and autumn races differing by the traveling time to spawning grounds, spawning ground location, dimensions, growth rate, fecunity, condition of the gonads and period of spawning migration was confirmed. Juvenile pink salmon from each race migrated downstream at different times: the peak for the summer race occurred during the third 5-day period of May, in the autumn race it was the first-second 5-day period in June. Differences in the duration of the migration were reflected in the scale structure. The scales in the first yr zone had a greater number of sclerites in the summer variety. In an area where both races occurred, the spawning area was used more extensively, which increased the reproductive efficiency of this salmon. The distribution of silver salmon in the Kuril Islands was mosaic-like. The center of intraspecies biological differentiation in this region was on Iturup Island. Mostly summer silver salmon reproduced in its northern area and autumn salmon in the south.--Copyright 1972, Biological Abstracts, Inc.
W73-09047

CONDITION OF THE BROOD STOCK OF AUTUMN CHUM IN THE AMUR RIVER LIMAN

W73-09051

INCREASING THE USE OF FOOD RESERVES IN BODIES OF WATER BY MEANS OF AGE-RELATED STRUCTURES (IN RUSSIAN),
M. V. Zheltenkova.
Tr Vses Nauchno-Issled Inst Morsk Rybn Khoz Okeanogr. 79 p 72-77. 1971. (English summary).
Identifiers: *Age (Fish), *Fish size, Food reserves, Structures.

The food spectrum increased along with increases in the size of fish. The younger specimens consumed small organisms with little mobility while the older ones fed on large, quickly moving organisms and those with powerful defenses. Feeding and the extent of migration increase with size. As a result, the amount of food accessible to the whole population increased.--Copyright 1972, Biological Abstracts, Inc.
W73-09052

WINTERING OF THE BELUGA STERLET HYBRIDS IN WESTERN GEORGIA (IN RUSSIAN),
E. N. Fedoseeva.
Sb Nauchno-Issled Rad Vses Nauchno-Issled Inst Prudovogo Rybn Khoz. 4 p 127-132. 1970. (English summary).
Identifiers: *Fish diets, Georgia (USSR), Hybrids, *Sterlet (Beluga), USSR, Wintering.

In the mild climate of Western Georgia wintering of the hybrids beluga X sterlet took place in normal carp ponds of the summer type. Two-yr old hybrids weighting 200-300 g did not stop feeding and in the majority of cases gained weight during the winter. Fish weighing over 400 g gained only in 30 cases. Management of food during the winter is recommended for these hybrids.--Copyright 1972, Biological Abstracts, Inc.
W73-09053

EFFECT OF A TISSUE PREPARATION ON CARP GROWTH,
F. M. Sukhoverkhov, G. B. Gribanova, N. V. Pechnikova, and M. R. Sidel'nikov.
Tr Vses Nauchno-Issled Inst Prudovogo Rybn Khoz. 20 p 152-160. 1971. (English summary).
Identifiers: *Carp growth, *Tissue preparation, Fish food.

Experiments on the introduction of a tissue preparation into the food to increase growth and productivity of carp were studied. The tissue preparation stimulated growth in yearlings and underyearlings. Different doses were tested. The best results for the growth of yearlings and for a decrease in food consumption per unit weight increase were obtained with 7 g/kg daily. This dose increased growth by 17.4 and decreased food consumption per unit of weight increase by 9.4%. For underyearlings a dose of 3.5 g/kg had the best growth stimulating effect, increasing growth by 17% compared to the control. --Copyright 1972, Biological Abstracts, Inc.
W73-09054

QUINNAT SALMON INVESTIGATIONS: REVIEW OF INVESTIGATIONS OF TECHNICAL FIELD SERVICE OF THE COUNCIL OF SOUTH ISLAND ACCLIMATISATION SOCIETIES: FEBRUARY 1969,
Marine Dept., Christchurch (New Zealand). Fisheries Lab.
C. J. Hardy.
N Z Mar Dep Fish Tech Rep. 84. p 1-38. 1972. Illus.
Identifiers: Acclimatization, Field, Islands, New Zealand, *Quinnat salmon, *Reviews, *Salmon, Service.

Quinnat salmon were introduced into New Zealand from California at the turn of the century and by the early 1920's good catches of quinnat were being reported from all the major South Island east coast rivers from the Waiau to the Clutha. The utilization and present status of quinnat salmon stocks are described. Progress is reviewed of investigations being undertaken by the Technical Field Service of the Council of South Island Acclimatization Societies principally on the Glenariffe Stream which is a tributary of the Rakaia River.--Copyright 1973, Biological Abstracts, Inc.
W73-09075

09. MANPOWER, GRANTS AND FACILITIES

9A. Education (Extramural)

ACADEMIC PROGRAMS IN ENVIRONMENTAL STUDIES,
California Univ, Berkeley. Sanitary Engineering Research Lab.
For primary bibliographic entry see Field 06A.
W73-08648

10. SCIENTIFIC AND TECHNICAL INFORMATION

10A. Acquisition And Processing

ENVIRONMENTAL DATA BANK.
For primary bibliographic entry see Field 07C.
W73-08663

10C. Secondary Publication And Distribution

ESTUARINE POLLUTION, A BIBLIOGRAPHY.
Office of Water Resources Research, Washington, D.C.
For primary bibliographic entry see Field 02L.
W73-08451

PROBLEMS AND TECHNIQUES OF KARST RESEARCH (PROBLEMY IZUCHENIYA KARSTA I PRAKTIKA),
For primary bibliographic entry see Field 02K.
W73-08881

AERIAL REMOTE SENSING, A BIBLIOGRAPHY.
Office of Water Resources Research, Washington, D.C.
For primary bibliographic entry see Field 07B.
W73-08974

CONCENTRATION FACTORS OF CHEMICAL ELEMENTS IN EDIBLE AQUATIC ORGANISMS,
California Univ., Livermore. Lawrence Livermore Lab.
For primary bibliographic entry see Field 05B.
W73-08992

10F. Preparation of Reviews

WATER POLLUTION CONTROL, A STATE-OF-THE-ART REVIEW,
Port of New York Authority, N.Y.
For primary bibliographic entry see Field 05D.
W73-08900

SUBJECT INDEX

POLLUTANT IDENTIFICATION

AUTHOR INDEX

ORGANIZATIONAL INDEX

ACCESSION NUMBER INDEX

2C
2L
5G
4C
2J
7C
7B
2C
2C
2C
2C
2C
2C
2C
2C
2C
2C
2C
2C
2C
2K
8B

ABSTRACT SOURCES

ırce	Accession Number	Total
Centers of Competence		
AEC Oak Ridge National Laboratory, Nuclear Radiation and Safety	W73-08991 -- 08997	7
Battelle Memorial Institute, Methods for Chemical and Biological Identification of Pollutants	W73-08527 -- 08528 08530 -- 08552 08555 -- 08599 08601 -- 08602 08757 -- 08800 08901 -- 08906 08908 -- 08973	188
Bureau of Reclamation, Engineering Works	W73-08513 -- 08519 08521 -- 08526	13
Cornell University, Policy Models for Water Resources Systems	W73-08654 -- 08655 08657 -- 08660 09004 -- 09007 09009 -- 09010	12
Oceanic Research Institute, Coastal Pollution	W73-08801 -- 08821 08823 -- 08849	48
University of Florida, Eastern U.S. Water Law	W73-08661 -- 08724 09011 -- 09015	69
University of North Carolina, Metropolitan Water Resources Planning and Management	W73-08977 -- 08982 08984 -- 08990	13
University of Texas, Wastewater Treatment and Management	W73-09016 -- 09017	2
University of Wisconsin, Eutrophication	W73-08634 -- 08636 08638 -- 08641 08644 -- 08653	17
U.S. Geological Survey, Hydrology	W73-08464 -- 08512 08725 -- 08756 08850 -- 08862 08864 -- 08890 08998 -- 09000 09076 -- 09100	149
Vanderbilt University, Thermal Pollution	W73-08610 -- 08613 08615 -- 08632	22
State Water Resources Research Institutes		
Alabama Water Resources Research Institute	W73-08452	
Alaska Institute for Water Research	W73-08603 -- 08604	2
Arizona Water Resources Research Center	W73-08608, 09074	2
Georgia Environmental Resources Center	W73-08453	
Hawaii Water Resources Research Center	W73-08454	
Idaho Water Resources Research Institute	W73-08455 -- 08456	2

ABSTRACT SOURCES

Source	Accession Number	Total
B. State Water Resources Research Institutes (Cont'd)		
Kansas Water Resources Research Institute	W73-08457	
Massachusetts Water Resources Research Center	W73-08605	
New Mexico Water Resources Research Institute	W73-08458	
Ohio Water Resources Center	W73-08459	
Rhode Island Water Resources Center	W73-08460	
South Carolina Water Resources Research Institute	W73-08461	
Utah Center for Water Resources Research	W73-09001	
Virginia Water Resources Research Center	W73-08462, 08606	2
Washington Water Research Center	W73-08607	
Wisconsin Water Resources Center	W73-08463	
C. Other		
BioSciences Information Service	W73-08520, 08529 08553 -- 08554 08600, 08609 08614, 08633 08637 08642 -- 08643 08656, 08822 08863, 08907 08983 09002 -- 09003 09008 09018 -- 09073 09075	76
Engineering Aspects of Urban Water Resources (Poertner)	W73-08891 -- 08900	10
Environmental Protection Agency	W73-08975 -- 08976	2
Office of Water Resources Research	W73-08451, 08974	2

☆ U. S. GOVERNMENT PRINTING OFFICE; 1973 OL - 542-949(1)

CENTERS OF COMPETENCE AND THEIR SUBJECT COVERAGE

- ▶ Ground and surface water hydrology at the Water Resources Division of the U.S. Geological Survey, U.S. Department of the Interior.
- ▶ Metropolitan water resources planning and management at the Center for Urban and Regional Studies of University of North Carolina.
- ▶ Eastern United States water law at the College of Law of the University of Florida.
- ▶ Policy models of water resources systems at the Department of Water Resources Engineering of Cornell University.
- ▶ Water resources economics at the Water Resources Center of the University of Wisconsin.
- ▶ Design and construction of hydraulic structures; weather modification; and evaporation control at the Bureau of Reclamation, Denver, Colorado.
- ▶ Eutrophication at the Water Resources Center of the University of Wisconsin, jointly sponsored by the Soap and Detergent Association and the Agricultural Research Service.
- ▶ Water resources of arid lands at the Office of Arid Lands Studies of the University of Arizona.
- ▶ Water well construction technology at the National Water Well Association.
- ▶ Water-related aspects of nuclear radiation and safety at the Oak Ridge National Laboratory.
- ▶ Public water supply treatment technology at the American Water Works Association.

Supported by the Environmental Protection Agency in cooperation with WRSIC

- ▶ Thermal pollution at the Department of Sanitary and Water Resources Engineering of Vanderbilt University.
- ▶ Water quality requirements for freshwater and marine organisms at the College of Fisheries of the University of Washington.
- ▶ Wastewater treatment and management at the Center for Research in Water Resources of the University of Texas.
- ▶ Methods for chemical and biological identification and measurement of pollutants at the Analytical Quality Control Laboratory of the Environmental Protection Agency.
- ▶ Coastal pollution at the Oceanic Research Institute.
- ▶ Water treatment plant waste pollution control at American Water Works Association.
- ▶ Effect on water quality of irrigation return flows at the Department of Agricultural Engineering of Colorado State University.

Subject Fields

1 NATURE OF WATER

2 WATER CYCLE

3 WATER SUPPLY AUGMENTATION AND CONSERVATION

4 WATER QUANTITY MANAGEMENT AND CONTROL

5 WATER QUALITY MANAGEMENT AND PROTECTION

6 WATER RESOURCES PLANNING

7 RESOURCES DATA

8 ENGINEERING WORKS

9 MANPOWER, GRANTS, AND FACILITIES

10 SCIENTIFIC AND TECHNICAL INFORMATION

INDEXES

SUBJECT INDEX

AUTHOR INDEX

ORGANIZATIONAL INDEX

ACCESSION NUMBER INDEX

ABSTRACT SOURCES

U.S. MAIL

POSTAGE AND FEES PAID
U.S. DEPARTMENT OF COMMERCE
COM 211

AN EQUAL OPPORTUNITY EMPLOYER

U.S. DEPARTMENT OF COMMERCE
National Technical Information Service
Springfield, Va. 22151

OFFICIAL BUSINESS

PRINTED MATTER

SELECTED WATER RESOURCES ABSTRACTS

VOLUME 6, NUMBER 15
AUGUST 1, 1973

9101 – W73-09750

SELECTED WATER RESOURCES ABSTRACTS is published semimonthly for the Water Resources Scientific Information Center (WRSIC) by the National Technical Information Service (NTIS), U.S. Department of Commerce. NTIS was established September 2, 1970, as a new primary operating unit under the Assistant Secretary of Commerce for Science and Technology to improve public access to the many products and services of the Department. Information services for Federal scientific and technical report literature previously provided by the Clearinghouse for Federal Scientific and Technical Information are now provided by NTIS.

SELECTED WATER RESOURCES ABSTRACTS is available to Federal agencies, contractors, or grantees in water resources upon request to: Manager, Water Resources Scientific Information Center, Office of Water Resources Research, U.S. Department of the Interior, Washington, D. C. 20240.

SELECTED WATER RESOURCES ABSTRACTS is also available on subscription from the National Technical Information Service. Annual subscription is $22 (domestic), $27.50 (foreign), single copy price $3. Certain documents abstracted in this journal can be purchased from the NTIS at prices indicated in the entry. Prepayment is required.

SELECTED
WATER RESOURCES ABSTRACTS

'A Semimonthly Publication of the Water Resources Scientific Information Center,
Office of Water Resources Research, U.S. Deoartment of the Interior

MICROFICHE PRICE CORRECTION

The price for all documents listed in this issue as available in microfiche for $1.45 from the National Technical Information Service is in error. The correct price is 95 cents.

VOLUME 6, NUMBER 15
AUGUST 1, 1973

W73-09101 -- W73-09750

The Secretary of the U. S. Department of the Interior has determined that the publication of this periodical is necessary in the transaction of the public business required by law of this Department. Use of funds for printing this periodical has been approved by the Director of the Office of Management and Budget through August 31, 1978.

As the Nation's principal conservation agency, the Department of the Interior has basic responsibilities for water, fish, wildlife, mineral, land, park, and recreational resources. Indian and Territorial affairs are other major concerns of America's "Department of Natural Resources."

The Department works to assure the wisest choice in managing all our resources so each will make its full contribution to a better United States—now and in the future.

FOREWORD

Selected Water Resources Abstracts, a semimonthly journal, includes abstracts of current and earlier pertinent monographs, journal articles, reports, and other publication formats. The contents of these documents cover the water-related aspects of the life, physical, and social sciences as well as related engineering and legal aspects of the characteristics, conservation, control, use, or management of water. Each abstract includes a full bibliographical citation and a set of descriptors or identifiers which are listed in the **Water Resources Thesaurus.** Each abstract entry is classified into ten fields and sixty groups similar to the water resources research categories established by the Committee on Water Resources Research of the Federal Council for Science and Technology.

WRSIC IS NOT PRESENTLY IN A POSITION TO PROVIDE COPIES OF DOCUMENTS ABSTRACTED IN THIS JOURNAL. Sufficient bibliographic information is given to enable readers to order the desired documents from local libraries or other sources.

Selected Water Resources Abstracts is designed to serve the scientific and technical information needs of scientists, engineers, and managers as one of several planned services of the Water Resources Scientific Information Center (WRSIC). The Center was established by the Secretary of the Interior and has been designated by the Federal Council for Science and Technology to serve the water resources community by improving the communication of water-related research results. The Center is pursuing this objective by coordinating and supplementing the existing scientific and technical information activities associated with active research and investigation program in water resources.

To provide WRSIC with input, selected organizations with active water resources research programs are supported as "centers of competence" responsible for selecting, abstracting, and indexing from the current and earlier pertinent literature in specified subject areas.

Additional "centers of competence" have been established in cooperation with the Environmental Protection Agency. A directory of the Centers appears on inside back cover.

Supplementary documentation is being secured from established discipline-oriented abstracting and indexing services. Currently an arrangement is in effect whereby the BioScience Information Service of Biological Abstracts supplies WRSIC with relevant references from the several subject areas of interest to our users. In addition to Biological Abstracts, references are acquired from Bioresearch Index which are without abstracts and therefore also appear abstractless in SWRA. Similar arrangements with other producers of abstracts are contemplated as planned augmentation of the information base.

The input from these Centers, and from the 51 Water Resources Research Institutes administered under the Water Resources Research Act of 1964, as well as input from the grantees and contractors of the Office of Water Resources Research and other Federal water resources agencies with which the

Center has agreements becomes the information base from which this journal is, and other information services will be, derived; these services include bibliographies, specialized indexes, literature searches, and state-of-the-art reviews.

Comments and suggestions concerning the contents and arrangements of this bulletin are welcome.

Water Resources Scientific Information Center
Office of Water Resources Research
U.S. Department of the Interior
Washington, D. C. 20240

CONTENTS

For primary bibliographic entry see Field 02F.

A mathematical model was developed to simulate watersheds in flat areas with many surface depressions and indicate the response due to watershed changes. The model, which simulates the watershed discharge and soil moisture status continuously throughout the crop season is based on a single-storm hydrologic model and uses daily pan evaporation and time-depth precipitation records as input. It simulates the processes of interception, surface storage, infiltration, surface runoff, soil profile storage, percolation to the water table, subsurface tile drainage, soil moisture redistribution, evapotranspiration, depression routing, tile-main routing, and drainage ditch routing. The resulting outputs are daily evapotranspiration, soil moisture status in the crop root zone, and watershed discharge. The hydrology of the East Fork Hardin Creek watershed near Jefferson, Iowa (with a drainage area of 24 sq mi) was simulated. The model satisfactorily simulated the watershed discharge and soil moisture status. (Knapp-USGS)
W73-09101

THE TRANSIENT BEHAVIOR OF RECHARGE-DISCHARGE AREAS IN REGIONAL GROUND-WATER SYSTEMS,
Maryland Univ., College Park. Dept. of Civil Engineering.
For primary bibliographic entry see Field 02F.
W73-09105

A BAYESIAN APPROACH TO AUTOCORRELATION ESTIMATION IN HYDROLOGIC AUTOREGRESSIVE MODELS,
Massachusetts Inst. of Tech., Cambridge. Ralph M. Parsons Lab. for Water Resources and Hydrodynamics.
R. L. Lenton, I. Rodriguez-Iturbe, and J. C. Schaake, Jr.
Available from the National Technical Information Service as PB-220 354, $3.00 in paper copy, $0.95 in microfiche. Partial Completion Report 163, January 1973. 121 p, 24 fig, 12 tab, 32 ref, 2 append. OWRR C-4118 (9021) (1).

Descriptors: Statistical methods, *Correlation analysis, *Time series analysis, Stochastic processes, *Synthetic hydrology, Monte Carlo method, Risks, Uncertainty, *Model studies.
Identifiers: *Bayes estimators, Estimation theory, Loss function.

Three general approaches leading to the marginal posterior probability distribution function for the autocorrelation coefficient of the first order annual autoregressive model are presented, based on varying assumptions about the incidental parameters of the model. The performance of the Bayes estimators for the quadratic, symmetric, linear and asymmetric linear loss functions is evaluated by Monte Carlo methods, and compared to the performance of some classical estimators, under the expected risk criterion and for conditions of limited data. The robustness of the Bayes estimator under changes of the loss function is also determined. The general framework for the derivation of a loss function for a hydrologic design problem is presented.
W73-09120

WATER BALANCE IN NORTH AMERICA (VODNYY BALANS SEVERNOY AMERIKI),
Akademiya Nauk SSSR, Moscow. Institut Geografii.
N. N. Dreyer.
Akademiya Nauk SSSR Izvestiya, Seriya Geograficheskaya, No 1, p 108-118, January-February 1972. 7 fig, 1 tab, 16 ref.

Descriptors: *Water balance, *Water resources, *North America, United States, Canada, Mexico, Hydrologic cycle, Inflow, Discharge (Water), Precipitation (Atmospheric), Runoff, Evaporation, River basins, Maps, Estimating.
Identifiers: *USSR.

Water resources of North America have been studied in considerable detail except for northern Alaska, the Arctic Archipelago, and northwestern regions of Canada. On the basis of American and Canadian water-supply papers and Soviet maps prepared from American, Canadian, and Mexican atlases, water-balance equations were developed for 80 river basins in Canada and for 135 river basins in the United States, including 10 in Alaska. The components of the water balance examined are precipitation; total runoff, including outflow due to subsurface seepage; total evaporation; and total moisture. Rough estimates of the water resources of North and Central America are tabulated. (Josefson-USGS)
W73-09158

WATER RESOURCES INVESTIGATIONS IN TEXAS, FISCAL YEAR 1973.
Geological Survey, Austin, Tex.
For primary bibliographic entry see Field 07C.
W73-09278

CONCEPTS OF KARST DEVELOPMENT IN RELATION TO INTERPRETATION OF SURFACE RUNOFF,
Geological Survey, Raleigh, N.C.

W73-09294

DESCRIPTION AND HYDROLOGIC ANALYSIS OF TWO SMALL WATERSHEDS IN UTAH'S WASATCH MOUNTAINS,
Forest Service, (USDA), Logan, Utah. Intermountain Forest and Range Experiment Station.
R. S. Johnston, and R. D. Doty.
USDA Forest Serv. Res. Pap. INT-127, July 1972. 53 p, Illus.

Descriptors: *Hydrologic data, Transpiration, *Streamflow, *Climatic data, *Water yield improvement, Rainfall-runoff relationships, Snowmelt, Snow packs, Air temperature, Water temperature, Water chemistry, Sediment yields, Evapotranspiration, Vegetation, Geology, Soils, Soil water, *Utah, *Small watersheds.
Identifiers: Experimental watersheds, Aspen.

The climate, geology, soils, and vegetation are included in a description of two small watersheds characteristic o the high-elevation aspen type of northern Utah. Precipitation, soil-water use, evapotranspiration, and quantity and quality of streamflow on these relatively undisturbed catchments are graphically illustrated and discussed. These data permit pretreatment calibration of these watersheds. This thorough inventory will allow a sensitive, multi-resource analysis of planned treatments designed to increase water yields.
W73-09378

MORPHOMETRY OF THREE SMALL WATERSHEDS, BLACK HILLS, SOUTH DAKOTA AND SOME HYDROLOGIC IMPLICATIONS,
Forest Service, (USDA), Fort Collins, Colo. Rocky Mountain Forest and Range Experiment Station.
T. Yamamoto, and H. K. Orr.

Descriptors: *Geomorphology, *Hydrogeology, *Geologic control, *Water yield, *Dimensional analysis, Hydrology, Peak discharge, Topography, *Small watersheds, Drainage basins.
Identifiers: *Forested watersheds, *Ponderosa pine, Black Hills.

Morphometry of three small contiguous watersheds together with their hydrologic character is analyzed in terms of dimensional anal-

ysis and similitude concepts. Within this framework it has been found that maximum length of master watershed and average relief of first order basins scale, most nearly the same as volume yield and stormflow peaks. Therefore these variables are better indicators of relative water yield, for example, than is 'area' on the hard rock of the laccolith on which the study watersheds are located. These relationships need to be further tested under conditions of different parent rock and climate.
W73-09382

A LOSING DRAINAGE BASIN IN THE MISSOURI OZARKS IDENTIFIED ON SIDE-LOOKING RADAR IMAGERY,
Geological Survey, Rolla, Mo.
For primary bibliographic entry see Field 07B.
W73-09446

WORLD WATER BALANCE (MICROVOY VODNYY BALANS),
Akademiya Nauk SSSR, Moscow. Institut Geografii.
M. I. L'vovich.
Izvestiya Vsesoyuznogo Geograficheskogo Obshchestva, Vol 102, No 4, p 314-324, July-August 1970. 6 tab, 83 ref.

Descriptors: *Water balance, *Hydrologic cycle, *Water resources, Precipitation (Atmospheric), Runoff, Evaporation, Oceans, Land, Water utilization, Water conservation, Conferences.
Identifiers: *USSR, *Global water balance, Hydrosphere.

Discussion of general problems of global water balance and tabulation of the total amount of water in the Earth system and of its movement through the world hydrological cycle are based on a paper presented at the International Symposium on Water Balance of the World, held in Reading, England, 1970. A rough approximation of world water resources for the year 2000 shows that the greatest threat to water conservation lies in the utilization of rivers, lakes, and reservoirs for waste-water removal and treatment. Adoption of more rational principles of water conservation and use can provide a basis for the protection of water from depletion and lessen the prospects of a water crisis. (Josefson-USGS)
W73-09459

PAPERS BY YOUNG SCIENTISTS (DOKLADY MOLODYKH SPETSIALISTOV),
Gosudarstvennyi Gidrologicheskii Institut, Leningrad (USSR).

Gosudarstvennyy Gidrologicheskiy Institut Trudy, No 204, Grushevskiy, M. S., and Ivanova, A. A., editors, Leningrad, 1972. 152 p.

Descriptors: *Hydrology, *Investigations, Channels, Rivers, Estuaries, Surges, Dunes, Sediments, Sediment transport, Meteorology, Atmosphere, Ice cover, Heat balance, Soil moisture, Evaporation, Evaporation pans, Measurement, Aerial photography, Correlation analysis, Equations.
Identifiers: *USSR, Heat exchange, Isobaths, Nomograms.

This collection contains papers presented at a conference of young scientists, held at the State Hydrologic Institute in Leningrad, April 12-14, 1971. The papers present the results of theoretical, laboratory, and field investigations in different branches of hydrology. The subjects examined include: (1) optimal spatial interpolation at various distances between precipitation stations; (2) quantitative criteria of channel macroforms; (3) similitude criteria of channel deformations in a shoal type of channel process; (4) dune movement of coarse sediments; (5) evaporation from soil surfaces; (6) evaporation measurement by pans; (7) heat exchange in rivers during cooling and freezing; and (8) influence of man on changes in river regime and natural conditions. (Josefson-USGS)
W73-09463

GENERAL REPORT ON WATER RESOURCES SYSTEMS: OPTIMAL OPERATION OF WATER RESOURCES SYSTEMS,
Illinois Univ., Urbana. Dept. of Hydraulic Engineering.
V. T. Chow.
Paper presented at International Symposium on Mathematical Models in Hydrology, July 26-31, 1971, Warsaw, Poland: International Association of Scientific Hydrology, 1971. 9 p, 4 ref. OWRR B-060-ILL (1).

Descriptors: *Hydraulic models, *Hydrologic data, *Data collections, Analytical techniques, Synthetic hydrology, Model studies, Reservoir operation, Streamflow, Channel morphology, Flood control, Mathematical studies, Systems analysis, Input-output analysis, Statistical methods, Methodology, Reviews, Conferences.

For the optimization of a tangible water resources system, two categories of modelling techniques may be distinguished; namely, the analytical technique and the simulation technique. The analytical technique is to convert the optimization problem to a mathematical model which can be solved by an analytical optimization method. Structuring of the model, or modelling, begins with the selection of the physical variables for which an optimum quantity is desired, such as reservoir size, flood channel capacity, and hydropower plant scale. Mathematical expressions consisting of objective functions and constraints are then formulated to express the physical relationships among the variables and limit the range of values that given variables may assume. The simulation technique is a direct simulation of the system and its functioning, therefore offering great freedom and flexibility in modelling. Systems sensitivity to a particular input, transfer function, or policy feature is readily ascertained. The simulation technique thus avoids the difficulty of the analytical technique in condensing mathematical formulations describing the functioning of a prototype system into a form solvable by available analytical optimization methods. Five papers are reviewed in this general report. (Woodard-USGS)
W73-09588

MAKARA, IHD EXPERIMENTAL BASIN NO 6, 1968-1970,
National Water and Soil Conservation Organization, Wellington (New Zealand).
M. E. Yates.
New Zealand Ministry of Works Hydrological Research Annual Report No 23, 1972. 54 p, 14 fig, 18 tab, 3 ref, 5 append.

Descriptors: *Land management, *Soil-water-plant relationships, *Hydrologic data, *Vegetation, Foreign countries, International Hydrological Decade, Projects, Reviews, Data collections, Grasses, Growth rates, Trees, Soil types, Soil moisture, Precipitation (Atmospheric), Temperature, Land development.
Identifiers: *Makara experimental basin (N.Z.), *New Zealand.

Under the auspices of the International Hydrological Decade the Ministry of Works is establishing a network of experimental basins to study the hydrological characteristics of important soil and vegetation complexes of New Zealand. Studies of the effect of cultural change on these characteristics are included. This is the second annual research report for Makara, the first begin published in 1970. The Makara experimental basin is situated on the west coast of the southern tip of the North Island and is 8 kilometers in a direct line west-northwest of Willington City. In broad terms the objective at Makara is the investigation of the effect of land use and land management practices on the hydrological regimen of stable central yellow-brown earths. In 1970, projects included work with unimproved exotic pasture; improved pasture--by oversowing and topdressing; improved pasture--by cultivation, sowing and topdressing; cropping; exotic forest; and retirement for regeneration of badly depleted indigenous forest. Data are included concerning types of vegetation, vegetation growth rates, soil types, topography, rainfall, runoff, winds, temperature, and other related information. (Woodard-USGS)
W73-09596

MOUTERE, IHD EXPERIMENTAL BASIN NO 8, 1971,
National Water and Soil Conservation Organization, Wellington (New Zealand).
F. Scarf.
New Zealand Ministry of Works Hydrological Research Annual Report No 22, 1972. 16 p, 11 fig, 5 tab, 1 ref, 1 append.

Descriptors: *Land management, *Soil-water-plant relationships, *Vegetation, Plant growth, *Soil environment, Foreign countries, International Hydrological Decade, Rainfall, Runoff, Soil moisture, Climatic data, Data collections, Correlation analysis.
Identifiers: *Moutere experimental basin (N.Z.), New Zealand.

Under the auspices of the International Hydrological Decade the Ministry of Works is establishing a network of experimental basins to study the hydrological characteristics of important soil and vegetation complexes of New Zealand. Studies of the effect of cultural change on these characteristics are included. This is the sixth annual research report for Moutere. The present catchment cultural change program has been in operation since 1970. Pictures, tables of data, and illustrations describe vegetation types, vegetation growth rates, rainfall, runoff, soil moisture, and other related data concerned with land management projects in the area. (Woodard-USGS)
W73-09597

REMOTE SENSING AND DATA HANDLING--THEIR APPLICATION TO WATER RESOURCES,
International Business Machines Corp., Gaithersburg, Md.
For primary bibliographic entry see Field 07B.
W73-09673

2B. Precipitation

WATER RESOURCE OBSERVATORY CLIMATOLOGICAL DATA WATER YEAR 1972.
Wyoming Univ., Laramie. Water Resources Research Inst.
For primary bibliographic entry see Field 07C.
W73-09107

WEATHER MODIFICATION: PRECIPITATION INDUCEMENT, A BIBLIOGRAPHY.
Office of Water Resources Research, Washington, D.C.
For primary bibliographic entry see Field 03B.
W73-09114

A FIELD INVESTIGATION AND NUMERICAL SIMULATION OF COASTAL FOG,
Cornell Aeronautical Lab., Inc., Buffalo, N.Y.
E. J. Mack, W. J. Eadie, C. W. Rogers, and R. J. Pilie.

Journal of Geophysical Research, Vol 77, No 12, p 2159-2165, April 20, 1972. AEC (45-1)-1830.

Descriptors: *Radioisotopes, *Air circulation, *Fallout, Air-water interfaces, Cloud physics, Precipitation, Model studies, Climatic data, *Washington.
Identifiers: *Cosmogenic radionuclides, Bomb debris, Cloud scavenging.

Many natural trace elements are picked up by cloud water and deposited in precipitation. Bomb debris in precipitation, through in-cloud scavenging, can be a threat to man by similar process. This paper describes a study of the problem. Results are presented from a model developed to evaluate deposition of cosmogenic radionuclides picked up by cloud water. The results are an in-cloud scavenging coefficient of less than 0.0001 per sec and less than 0.001 per sec for Cl-38 and Cl-39. (Lang-USGS)
W73-09146

A SELECTED ANNOTATED BIBLIOGRAPHY OF THE CLIMATE OF THE GREAT LAKES,
National Oceanic and Atmospheric Administration, Silver Spring, Md. Environmental Data Service.
For primary bibliographic entry see Field 02H.
W73-09149

ROLE OF SEA SALTS IN THE SALT COMPOSITION OF RAIN AND RIVER WATERS,
Akademiya Nauk SSSR, Moscow. Institut Okeanologii.
For primary bibliographic entry see Field 02K.
W73-09152

CHARACTER OF SUMMER PRECIPITATION IN EAST SIBERIA (O PRIRODE LETNIKH OSADKOV VOSTOCHNOY SIBIRI),
Akademiya Nauk SSSR, Moscow. Institut Geografii.
G. N. Vitvitskiy.
Akademiya Nauk SSSR Izvestiya, Seriya Geograficheskaya, No 1, p 93-99, January-February 1972. 3 tab.

Descriptors: *Precipitation (Atmospheric), *Summer, *Meteorology, Atmosphere, Water vapor, Winds, Air circulation, Atmospheric pressure, Isobars, Mapping.
Identifiers: *USSR (East Siberia), Isohypses.

Most precipitation in East Siberia is associated with the warm season of the year, with about 40% of the annual total occurring in July and August. In the lower half of the troposphere, the transport of water vapor eastward and northward predominates, with the eastward transport generally greater than the northward. The predominance of eastward transport is further proof that water vapor from the adjoining seas of Okhotsk and Japan contributes little to moisture circulation over East Siberia. (Josefson-USGS)
W73-09157

RAINFALL EFFECT ON SHEET FLOW OVER SMOOTH SURFACE,
Colorado State Univ., Fort Collins. Dept. of Civil Engineering.
For primary bibliographic entry see Field 02E.
W73-09273

THE CLIMATE OF AN ORANGE ORCHARD: PHYSICAL CHARACTERISTICS AND MICROCLIMATE RELATIONSHIPS,
Commonwealth Scientific and Industrial Research Organization, Canberra (Australia). Div. of Land Research.
For primary bibliographic entry see Field 03F.
W73-09320

A RAINFALL CLIMATOLOGY OF HILO, HAWAII,
Hawaii Univ., Hilo. Cloud Physics Observatory.
C. M. Fullerton.
Available from the National Technical Information Service as PB-220 573, $3.00 in paper copy, $0.95 in microfiche. Hawaii University Water Resources Research Center, Honolulu, Technical Report No 61 (UHMET 72-03), December 1972. 34 p, 20 fig, 3 tab, 8 ref. OWRR B-024-HI (1).

Descriptors: *Rainfall, *Meteorology, *Rainfall intensity, *Hawaii, Rainfall disposition, Data collections, Rain gages, Frequency analysis, Time series analysis.
Identifiers: *Hilo (Hawaii).

Annual, monthly, and hourly rainfall data for Hilo, Hawaii, are plotted and analyzed as a preliminary phase of the study of space-time variations in high intensity rainfall. A summary of the Hilo climate is provided. Annual rainfall amounts are highly variable, while the annual rainfall frequency remains relatively constant at about 33%. Monthly rainfall and maximum 24-hour rainfall amounts are displayed in a monthly rainfall expectancy graph. Hourly rainfall amounts are divided into four intensity categories: 0.01-0.24, 0.25-0.49, 0.50-0.99, and> 1.00 inches per clock hour. The percentage distribution of annual and monthly rainfall amounts and frequencies by rainfall intensity category are plotted and discussed. Hourly data are displayed in a series of monthly diurnal distributions by rainfall intensity category. (Woodard-USGS)
W73-09585

CLIMATE OF THE POTOMAC RIVER BASIN AND CLIMATOLOGICAL SUMMARIES FOR SPRUCE KNOB, WARDENSVILLE, ROMNEY AND MARTINSBURG, WEST VIRGINIA,
West Virginia Univ., Morgantown.
R. O. Weedfall, W. H. Dickerson, M. S. Baloch, and E. N. Henry.
Available from NTIS, Springfield, Va 22151 COM-72-11334 Price $3.00 printed copy; 95 cents microfiche. West Virginia University Water Research Institute, Morgantown, Information Report 6, 1972. 15 p, 3 fig, 7 tab.

Descriptors: *Climatology, *Potomac River, *Climatic data, *West Virginia, Precipitation (Atmospheric), Temperature, Evaporation, Topography, Elevation, Data collections, Weather data, Meteorological data.

The general climate of the Potomac River basin is classified as semi-humid continental because of its relatively high and generally evenly distributed precipitation and its marked temperature contrasts between summer and winter. The West Virginia part of the basin includes a small part of the broad, fertile Shenandoah Valley. Eight of the States 55 counties cover a 3490.26 square mile area in the basin. Climatological summaries are presented for Spruce Knob, Wardensville, Romney and Martinsburg, West Virginia. Temperature and precipitation means and extremes are shown for the period 1931-1970 or portions thereof. Data tables also show total monthly precipitation at Wardensville (1931-1970) and average temperature and average pan evaporation for the same area. In addition, departures in inches from normal growing season rainfall and probabilities of low temperatures in spring and fall at selected locations are tabulated. (Woodard-USGS)
W73-09602

SNOW CRYSTAL FORMS AND THEIR RELATIONSHIP TO SNOWSTORMS,
National Oceanic and Atmospheric Administration, Boulder, Colo. Atmospheric Physics and Chemistry Lab.
For primary bibliographic entry see Field 02C.
W73-09610

Group 2B—Precipitation

A RELATIONSHIP BETWEEN SNOW ACCUMULATION AND SNOW INTENSITY AS DETERMINED FROM VISIBILITY,
National Weather Service, Garden City, N.Y. Eastern Region.
For primary bibliographic entry see Field 02C.
W73-09612

APPROACHES TO MEASURING 'TRUE' SNOWFALL,
National Weather Service, Silver Spring, Md. Office of Hydrology.
For primary bibliographic entry see Field 02C.
W73-09613

GLOBOSCOPE - A NEW LOOK AROUND,
West Virginia Univ., Morgantown. Div of Forestry.
D. A. Morgan.
West Virginia Agriculture and Forestry, Vol 4, No 4, p 2-5, September 1972. 7 fig, 5 ref. OWRR A-020-WVA (1).

Descriptors: *Local precipitation, *Gaging stations, *Cameras, *Terrain analysis, Rainfall, Winds, Climatology, Rain gages, *West Virginia, Instrumentation, Measurement, Ecology, Forests.
Identifiers: Site description, *Gage exposure, Photographic devices, Stereographic projections, Paraboloidal mirrors.

A unique photographic device was developed to provide a routine objective method of recording and evaluating the exposure of precipitation gages. The photographs are precise stereographic projections of the celestial hemisphere taken from a paraboloidal mirror. Vertical and horizontal angles to obstructions in the vicinity of a gage (to 12 degrees below the horizontal) can be measured directly from the projections. In conjunction with a zenith angle grid system, or a sun path chart, useful application possibilities for globescope technology exist in ecology, forestry, architecture, climatology, and related fields. (Chang-West Virginia)
W73-09620

ELEVATION IN WEST VIRGINIA,
West Virginia Univ., Morgantown.
R. Lee, M. Chang, and R. Calhoun.
West Virginia Agriculture and Forestry, Vol 5, No 1, p 5-9, January 1973. 5 fig, 3 tab, 5 ref. OWRR A-020-WVA (2).

Descriptors: *Topography, *Elevation, *Terrain analysis, Geomorphology, Climatology, Mountains, *Climatic zones, Orography, *West Virginia, Topographic mapping, *Appalachian Mountain Region.
Identifiers: Terrain climatology.

The vertical distribution of land in West Virginia was determined from a systematic sample of elevations at 2.5-minute intervals of latitude and longitude. The mean, median, and extreme elevations, and the total relief were computed for the State, and for each of its 55 counties. The data were shown to be useful in the delineation of climatological zones, in the estimation of mean annual temperature and precipitation, and in the assessment of terrain influences in agriculture, forestry, hydrology, and related endeavors.
W73-09621

2C. Snow, Ice, and Frost

IRON IN SURFACE AND SUBSURFACE WATERS, GRIZZLEY BAR, SOUTHEASTERN ALASKA,
Alaska Univ., College. Dept. of Geology.
For primary bibliographic entry see Field 05A.
W73-09111

SEDIMENTATION AND PHYSICAL LIMNOLOGY IN PROGLACIAL MALASPINA LAKE, ALASKA,
Massachusetts Univ., Amherst. Coastal Research Center.
For primary bibliographic entry see Field 02J.
W73-09141

HOW TO GET THE SNOW WHERE YOU WANT IT TO GO,
Forest Service (USDA), Fort Collins, Colo. Rocky Mountain Forest and Range Experiment Station.
R. A. Schmidt, Jr.
Skiing Area News 6 (3): 20-21, 43-44, Summer 1971.

Descriptors: *Snow management, *Snow removal, *Skiing, *Slopes, Snowpacks, Snow cover, Structures, *Sublimation, Environmental engineering, Blizzards.
Identifiers: *Snowfences, *Blowing snow.

Through examples, the article shows how general principles of snow fence design and specific data may be applied to certain wind-caused snow control problems on ski areas.
W73-09380

PROCESSING SIZE, FREQUENCY, AND SPEED DATA FROM SNOW PARTICLE COUNTERS,
Forest Service (USDA), Fort Collins, Colo. Rocky Mountain Forest and Range Experiment Station.
R. A. Schmidt, Jr.

Descriptors: Snow, *Particle Size, *Data processing, Snowpacks, Snow management, *Sublimation, Suspended load, Electronic equipment, Instrumentation, *Blizzards, Frequency.
Identifiers: Particle counters, particle speed, Blowing snow.

This note describes techniques for electronically processing magnetic tape records from a photoelectric snow particle counter. Examples of the resulting particle size distributions, particle frequency plots, and measurements of particle speed are included.
W73-09381

GLACIATION NEAR LASSEN PEAK, NORTHERN CALIFORNIA,
Geological Survey, Denver, Colo.
D. R. Crandell.
Available from Sup Doc, GPO, Washington, D C 20402 - Price $3.00. In: Geological Survey Research 1972, Chapter C; U S Geological Survey Professional Paper 800-C, p C179-C188, 1972. 4 fig, 2 tab, 19 ref.

Descriptors: *Glaciation, *Pleistocene epoch, *California, *Weathering, *Stratigraphy, Glacial drift, Dating, Volcanoes, Glaciers, Sediments.
Identifiers: *Lassen Volcanic National Park.

Lassen Volcanic National Park in northern California was mostly covered by icecap glaciers during the Tahoe and Tioga Glaciations, and at least once during pre-Tahoe time. Glacial deposits of various ages are subdivided chiefly by weathering profiles in till, as shown by the relative amounts of clay formation, the presence and thickness of weathered rinds on volcanic stones at or near the ground surface, the depth of oxidation, and the color hues of the oxidized zone. During each glaciation the major ice-accumulation area in the western part of Lassen Volcanic National Park was on the slopes of the old Brokeoff Volcano, a Pleistocene volcano which covered an area of more than 100 square miles. Lassen Peak erupted during late Tioga time, about 11,000 years ago. In latest Tioga time, small glaciers formed on the flanks of Lassen Peak and on adjacent mountains at altitudes above 8,000 feet, whereas during the Tioga icecap glaciation the regional snowline near Lassen Peak was probably at about 7,000 feet. (Knapp-USGS)

;e. The weight of the snow on the pillow con- uration causes an increase in the internal pres- re of the pillow fluid. This pressure is measured means of a pressure transducer. Data are trans- tted by radios via repeater links, as required, to central base station. There the data are con- rted to digital form. The data can then be dis- ıyed or retransmitted by use of microwave or ısed-line teletype circuits. Relative humidity and ıer hydrometeorologic parameters can be emetered by using appropriate sensors such as ecipitation, temperature, wind speed, and rection. (Knapp-USGS)
73-09611

RELATIONSHIP BETWEEN SNOW ACCU- ULATION AND SNOW INTENSITY AS ETERMINED FROM VISIBILITY,
ational Weather Service, Garden City, N.Y. ıstern Region.
E. Wasserman, and D. J. Monte.
: Proceedings of the 29th Annual Eastern Snow onference, February 3, 5, 1972, Oswego, New ork: Eastern Snow Conference Publication, Vol , p 47-52, 1972. 4 tab, 2 ref.

escriptors: *Precipitation intensity, *Snowfall, stimating, *Precipitation (Atmospheric), Data llections, Meteorological data, Weather data, ew York.
entifiers: Visibility, Snowfall rate, Snowfall in- nsity, *New York City (NY).

ourly snow accumulation may be estimated from servations of snow intensity or visibility. The ocedure was tested at La Guardia Field, New ork City. The relationship between reported ıow intensity and estimated snowfall was shown ith independent data to be reliable. Results ıould aid in forecasting additional snow accumu- tion and in estimating previous snowfall when ıis information is not available from direct mea- ırements. (Knapp-USGS)
/73-09612

PPROACHES TO MEASURING 'TRUE' NOWFALL,
ational Weather Service, Silver Spring, Md. Of- ce of Hydrology.
.. W. Larson.
ı: Proceedings of the 29th Annual Eastern Con- erence, February 3,4, 1972, Oswego, New York: astern Snow Conference Publication, Vol 17, p 5-76, 1972. 4 fig, 2 tab, 27 ref.

escriptors: *Precipitation gages, *Snowfall, *In- trumentation, Data collections, Calibrations, Wyoming, *Vermont, Gaging stations, Winds, now.

he catch of precipitation gages is affected by ıany variables. Wind has the strongest influence nd an increase in wind speed will generally result ı a decrease in catch. Proper cite protection can educe the turbulence near the gage and will result ı a more consistent and reliable measurement. nowfall measured at a well-protected site is robably quite close to the actual point snowfall. he Wyoming project has as its major goal the in- estigation of shielding by the use of artificial ind barriers. Alter shields are used in conjunc- on with various fence combinations. The objec- ve is to duplicate, as nearly as possible, the recipitation catch at a well-protected site in a orest opening with the catch of a gage located in n open windy area. The Vermont project has four ites. Two of these sites are suitable for use in valuating the dual-gage approach. Well designed rtificial wind barriers can be used to provide site rotection for precipitation gages approximately qual to a well-protected natural site. The dual- age approach and the profiling method for calcu- ıting ground true precipitation also have con- iderable merit. (Knapp-USGS)
V73-09613

ALBEDO OF SIMULATED SNOW SURFACES RELATED TO ROUGHNESS,
New Brunswick Univ., Fredericton. Faculty of Forestry.
R. B. B. Dickison.
In: Proceedings of the 29th Annual Eastern Snow Conference, February 3, 5, 1972, Oswego, New York: Eastern Snow Conference Publication, Vol 17, p 77-82, 1972. 4 fig, 2 tab, 3 ref.

Descriptors: *Albedo, *Snow cover, *Forests, Model studies, Snow, Snowpacks, Interception, *Canada, Snow surveys, Runoff forecasting, Snowmelt, Solar radiation.
Identifiers: *New Brunswick (Canada).

To investigate the effect of forest density on albedo in winter, a 12-foot square model forest was constructed of inverted conical paper cups and pipe cleaners. Artificial snow was sprayed onto the model to simulate intercepted snow on a forest canopy. Reflected solar radiation measurements were taken over the model as crown closure was reduced in a random block pattern from 100% to 10%. Albedo increased from about 0.66 to 0.88 as the density of the model forest decreased. (Knapp-USGS)
W73-09614

WIND EFFECTS ON SKI TRAILS,
New Hamshire Univ., Durham. Dept. of Mechanical Engineering.
R. J. A. Lof, R. W. Alperi, and C. K. Taft.
In: Proceedings of the 29th Annual Eastern Snow Conference, February 3, 5, 1972, Oswego, New York: Eastern Snow Conference Publication, Vol 17, p 83-97, 1972. 6 fig, 9 ref.

Descriptors: *Wind erosion, *Snowpacks, *Skiing, *Snow management, Model studies, Winds, Forests.
Identifiers: Ski trails, Snow fences.

Ski trails facing into the wind direction and those that are above tree line are particularly susceptible to being scoured of their snow cover. To find novel wind barrier techniques to reduce the scouring effect of the wind and induce snow deposition, an open channel model through which air is forced was designed. Tests on models simulating ski trails indicate that narrow, winding trails with short runs into the prevailing winds can greatly reduce wind erosion. A wind barrier in the form of a lifting body shape similar to a delta winged aircraft, properly oriented, was tested. This barrier has the ability to induce snow deposition into drifts of larger volume than conventional snow fences. Depending on the orientation and the number of barriers, one, two or three drifts are possible. Tests of larger scale models under actual wind conditions are being conducted. (Knapp-USGS)
W73-09615

THE MAPPING OF SNOWFALL AND SNOW COVER,
Atmospheric Environment Service, Ottawa (Ontario).
G. A. McKay.
In: Proceedings of the 29th Annual Eastern Snow Conference, February 3, 5, 1972, Oswego, New York: Eastern Snow Conference Publication, Vol 17, p 98-110, 1972. 2 fig, 43 ref.

Descriptors: *Mapping, *Maps, *Snowfall, *Snow cover, *Reviews, Data collections, Streamflow forecasting, Runoff forecasting, Climatology, Weather, Surveys.

Maps provide an excellent means of presenting and storing information on snowfall and snow cover. The history and status of snow mapping are reviewed. Most maps suffer from the deficiencies which have long been recognized to exist in snow data; this is, lack of precision and unrepresentativeness of measurements. Within recent years networks for the measurement of snow have been substantially enhanced. The accumulated data and new sources of information now make possible the preparation of maps of increased variety and detail. Until recently snowfall and snow-cover maps have had a more academic than practical value. Generally they were mean-value maps prepared on the basis of a skeletal network so as to obtain rough comparisons of snowiness and the duration of winter. Over the past 25 years more precise and detailed information has made possible a greater variety of maps. Snow maps are used or can be used in water resource planning, development, and operations; in the determination of snow loads; in the planning and development stages of tourism and recreation; in the planning of fertilizer sales; in land-use evaluation; in the allocation of funds for highway maintenance; and in evaluating the probabilities of survival and shaping conservation procedures for plants and wildlife. (Knapp-USGS)
W73-09616

LYSIMETER SNOWMELT AND STREAMFLOW ON FORESTED AND CLEARED SITES,
Forest Service (USDA), Durham, N.H. Northeastern Forest Experiment Station.
C. W. Martin, and J. W. Hornbeck.
In: Proceedings of the 29th Annual Eastern Snow Conference, February 3, 5, 1972, Oswego, New York: Eastern Snow Conference Publication, Vol 17, p 111-118, 1972. 6 fig, 1 tab, 8 ref.

Descriptors: *Snowmelt, *Lysimeters, *Runoff forecasting, *New Hampshire, Streamflow forecasting, Peak discharge, Melt water, Snow surveys, Snowpacks.
Identifiers: *Snowmelt lysimeters.

Snowmelt lysimeters were operated for two years on cleared and forested watersheds in central New Hampshire in an experiment to improve measurements for determining how specific variables affect snowmelt-streamflow relationships. The lysimeters readily detected advance in melt water contribution caused by forest clearing. On a daily basis, maximum melt water contribution at the soil surface occurred about one hour ahead of daily streamflow peaks. Peak melt water flow is moderated during passage through the soil. The continuous record of melt water at the base of the snowpack appears to be more useful and convenient for snowmelt studies than the interval measurement of snowpack changes obtained by the widely used snow-tube samplers. (Knapp-USGS)
W73-09617

2D. Evaporation and Transpiration

EVAPOTRANSPIRATION BY SUGAR CANE AS INFLUENCED BY METEOROLOGICAL FACTORS,
Hawaii Univ., Honolulu. Dept of Geography.
P. C. Prasad.
Ph D Thesis, December, 1971. 293 p, 38 fig, 46 tab, 173 ref.

Descriptors: *Evapotranspiration, *Lysimeters, *Hawaii, *Sugarcane, Solar radiation, Advection, Consumptive use, Irrigation effects.

Daily consumptive water use by springkler irrigated sugarcane was measured in four 100-ft square x 5-ft deep weighing lysimeters. About 4 months of the daily evapotranspiration was included in the analysis for the plant cane (age 127-254 days), a period when the consumptive water use for the crop increases to its maximum value. For the same incident radiation, a considerably higher pan evaporation occurred from August to December than did from January to July, a situation which reflected significant transport of advected energy. For the fullycanopied sprinkler-irrigated plant cane, the correlation coefficient

between evapotranspiration and the pan evaporation was 0.66, whereas for the entire ratoon cane it was 0.62. For the entire study period, which included many varied situations, 42.1% of the cane water use was explained by the pan evaporation. Incident radiation accounted for only 48.1% of the variations of water use for the sprinkler-irrigated plant cane with complete canopy, and 19.9% for the entire ratoon cane. The poor relationship was explained by the advection of energy from the surrounding fields. (Knapp-USGS)
W73-09137

AN ELECTRICAL ANALOGUE OF EVAPORATION FROM, AND FLOW OF WATER IN PLANTS,
Australian National Univ., Canberra. Dept. of Environmental Biology.
I. R. Cowan.
Planta (Berl). Vol 106, No 3, p 221-226. 1972.
Identifiers: *Plant physiology, Electrical analogue, *Evaporation, *Flow, Heat, Loss, Plants, Transport, Vapor.

The currents generated in the analogue circuit represent vapor loss from leaves, heat loss from leaves, and liquid flow in plant and soil. The plant and soil resistances are defined in such a way that they are consistent with the resistances to transport of vapor in the atmosphere and there is continuity of potential at the analogue liquid: air interface in the leaves. The action of the environment on plant water movement is treated as an application of Thevenin's theorem of electric circuits.--Copyright 1973, Biological Abstracts, Inc.
W73-09420

EVALUATION OF TURBULENT TRANSFER LAWS USED IN COMPUTING EVAPORATION RATES,
Geological Survey, Bay Saint Louis, Miss. Water Resources Div.
H. E. Jobson.
Open-file report, April 1973. 169 p, 33 fig, 26 ref.

Descriptors: *Evaporation, *Ponds, *Mathematical models, *Meterorological data, *Oklahoma, Analytical techniques, Equations, Correlation analysis, Mass transfer, Measurement, Estimating, Forecasting.
Identifiers: *Evaporation rates, Aerodynamic method.

The aerodynamic method of computing evaporation rates has a number of significant advantages over other methods. In order to use the aerodynamic method some functional form describing the variation of wind velocity with elevation must be assumed. Although the logarithmic law appears to be an adequate description when atmospheric conditions are neutrally stable, no wind law has been found which is satisfactory under all conditions of atmospheric stability. A combined logarithmic and linar law was designed to extend the applicability of the log law to conditions which are nearly neutrally stable. Data collected at Hefner, Oklahoma, were used to evaluate the theoretical correctness and practicality of the log-linear law for computing evaporation rates by the aerodynamic method. The theoretical correctness of the log-linear law was evaluated by comparing its results with those obtained by use of the logarithmic law. The mass-transfer method was used as a reference from which the practicality of the log-linear law can be judged. Provided that the measurement errors in the velocities are averaged out in a prescribed manner, the log-linear law can be expected to provide monthly evaporation rates which are accurate to within 17%. This accuracy approaches that which can be expected from the mass-transfer method. (Woodard-USGS)
W73-09434

ECOLOGY OF CORTICOLOUS LICHENS: III. A SIMULATION MODEL OF PRODUCTIVITY AS A FUNCTION OF LIGHT INTENSITY AND WATER AVAILABILITY,
McMaster Univ., Hamilton (Ontario). Dept. of Biology.
For primary bibliographic entry see Field 02I.
W73-09512

A WATER BALANCE MODEL FOR RAIN--GROWN, LOWLAND RICE IN NORTHERN AUSTRALIA,
Commonwealth Scientific and Industrial Research Organization, Darwin (Australia). Coastal Plains Research Station.
For primary bibliographic entry see Field 03F.
W73-09553

EFFECT OF AERIAL ENVIRONMENT AND SOIL WATER POTENTIAL ON THE TRANSPIRATION AND ENERGY STATUS OF WATER IN WHEAT PLANTS,
Washington State Univ., Pullman. Dept. of Agronomy and Soils.
For primary bibliographic entry see Field 03F.
W73-09570

CLIMATE OF THE POTOMAC RIVER BASIN AND CLIMATOLOGICAL SUMMARIES FOR SPRUCE KNOB, WARDENSVILLE, ROMNEY AND MARTINSBURG, WEST VIRGINIA,
West Virginia Univ., Morgantown.
For primary bibliographic entry see Field 02B.
W73-09602

EVAPOTRANSPIRATION OF EPIPHYTIC VEGETATION ON THE LOWER PART OF OAK (QUERCUS ROBUR) AND MAPLE-TREE (ACER CAMPESTRE) TRUNKS IN THE MIXED OAK FOREST AT VIRELLES-BLAIMONT (IN FRENCH),
Brussels Univ. (Belgium). Laboratoire de Botanique Systematique et d'Ecologie.
G. Schnock.
Bull Soc R Bot Belg, Vol 105, No 1, p 143-150, 1972. Illus. English summary.
Identifiers: Acer-campestre-D, Belgium, Biomass, *Epiphytic vegetation, *Evapotranspiration, Forests, *Maple-D, *Oak-D, Quercus-robur-D, Rain, Transpiration, Tree trunks, Vegetation, Virelles-Blaimont.

In the mixed oak forest at Virelles-Blaimont (Belgium), the biomass of epiphytic layer overgrowing the bark at the lower part of the trunks (between 0 and 1 m) of O. robur and A. campestre amounts 355, 1 kg/ha in early year 1970. Its water saturation capacity, reached with an 8.0 mm rainfall during the leafy phenophase, is estimated at 0.133 mm. During this period, the water content of epiphytes (We) is linked with the atmospheric evaporating power (Ep ± Piche) by the following relation: We ± 0.1542 Ep-0.64. With the number, the size and the periodicity of showers, the water content of the epiphytic layer can be estimated using this relation, and the evapotranspiration may be calculated by subtraction between 2 successive showers. The calculations show that the mean evapotranspiration by shower is equal to 0.059 mm. Extrapolation to the whole leafy phenophase (mean of 3 periods: 1966, 1967 and 1968) suggests that the epiphytic vegetation of the lower part of the trunks may be assumed to evapotranspirate about 5.0 mm (0.92% of the total rainfall or 1.18% of the phytocenose evapotranspiration). This work shows that there is a slight overestimation of the infiltration water by omitting the evapotranspiration of the epiphytic vegetation overgrowing the trunks under run-off channels.--Copyright 1972, Biological Abstracts, Inc.
W73-09677

PERIODICAL FLUCTUATIONS OF THE MOISTURE CONTENT OF PRINCIPAL SPECIES COMPOSING THE HERBACEOUS LAYER AND MASS STRATA DURING 1969 (IN FRENCH),
Brussels Univ. (Belgium). Laboratoire de Botanique Systematique et d'Ecologie.
For primary bibliographic entry see Field 02I.
W73-09679

2E. Streamflow and Runoff

EFFECTS OF RIVER CURVATURE ON A RESISTANCE TO FLOW AND SEDIMENT DISCHARGES OF ALLUVIAL STREAMS,
Iowa Univ., Iowa City. Inst. of Hydraulic Research.
Y. Onishi, S. C. Jain, and J. F. Kennedy.
Available from the National Technical Information Service as PB-220 249, $3.00 in paper copy, $0.95 in microfiche. Iowa State Water Resources Research Institute Completion Report ISWRRI-46, December 1972. 150 p, 37 fig, 3 tab, 19 ref, append. OWRR A-029-IA (3) 14-01-0001-1634, 14-31-0001-3015 and 3215.

Descriptors: *Sediment transport, *Bed load, *Sediment discharge, *Channel morphology, *Meanders, Discharge (Water), Streamflow, Alluvial channels, Hydraulic models, Profiles, Sedimentology.
Identifiers: *Bend loss (Sediment transport).

The effects of channel meandering and of width of sinuous streams on the sediment loads and the friction factors of alluvial streams were investigated in laboratory flumes. Experiments were conducted in the subcritical flow domain in straight and meandering laboratory channels with sand beds and rigid walls. For given flow conditions, the sediment discharge per unit width in the full-width meandering channel was greater than sediment discharge in the straight flume, which in turn was greater than sediment discharge in the half-width meandering channel. The bend-loss coefficient, defined as the head loss due to channel curvature, normalized by the mean-velocity head, increases with Froude number, the ratio of bed hydraulic radius to median sand diameter, and the ratio of width to centerline radius of curvature. The bed forms, especially the point bars along the inner banks, play a dominant role in regulating the quasi-steady characteristics of meandering alluvial streams. (Knapp-USGS)
W73-09102

WATER RESOURCES OBSERVATORY STREAMFLOW DATA WATER YEAR 1972,
Wyoming Univ., Laramie. Water Resources Research Inst.
For primary bibliographic entry see Field 07C.
W73-09109

A BAYESIAN APPROACH TO AUTOCORRELATION ESTIMATION IN HYDROLOGIC AUTOREGRESSIVE MODELS,
Massachusetts Inst. of Tech., Cambridge. Ralph M. Parsons Lab. for Water Resources and Hydrodynamics.
For primary bibliographic entry see Field 02A.
W73-09120

HYDROLOGIC INVESTIGATIONS OF PRAIRIE POTHOLES IN NORTH DAKOTA, 1959-68,
Geological Survey, Washington, D.C.
For primary bibliographic entry see Field 02H.
W73-09130

arrangement of the ridges corresponds to the direction of the tidal currents in a given area, and the ratio of the distance between ridges to their height varies between 200 and 650. Data are presented on the structure and composition of ridges and of the bottom waters above them, and various hypotheses are advanced to explain the mechanism of ridge formation and ridge dynamics. (Josefson-USGS)
W73-09153

EXPERIMENTAL STUDY OF HORIZONTAL TURBULENCE IN THE LITTORAL ZONE USING OCEAN CURRENT DATA FROM AN AERIAL SURVEY,
State Oceanographic Inst., Moscow (USSR).
I. Z. Konovalova.
Oceanology, Vol 12, No 3, p 443-450, 1972. 5 fig, 3 tab, 15 ref. Translated from Okeanologiya (USSR), Vol 12, No 3, p 527-534, 1972.

Descriptors: *Oceanography, *Ocean currents, *Littoral, *Turbulence, *Aerial photography, Surveys, Floats, Velocity, Distance, Fluctuations, Variability, Curves.
Identifiers: *USSR, *Black Sea, Turbulent diffusion.

Computation of horizontal turbulent diffusion coefficients and determination of some statistical turbulence parameters from current velocity fluctuations were based on aerial photographic surveys of floats off the Caucasian coast of the Black Sea in the summers of 1961-62. Variability of the horizontal diffusion coefficients obtained was analyzed as a function of the scale of the phenomenon and the distance from the coast. Turbulence characteristics obtained from aerial photo survey data are reliable and can be used in general estimates of horizontal exchange in a coastal zone and in various practical computations. (Josefson-USGS)
W73-09154

SURFACE WATER SUPPLY OF THE UNITED STATES, 1966-70: PART 3, OHIO RIVER BASIN--VOLUME 2, OHIO RIVER BASIN FROM KANAWHA RIVER TO LOUISVILLE, KENTUCKY.
Geological Survey, Washington, D.C.
For primary bibliographic entry see Field 07C.
W73-09160

SUSPENDED SEDIMENT DISCHARGE IN WESTERN WATERSHEDS OF ISRAEL,
Hebrew Univ., Jerusalem (Israel). Dept of Geography.
For primary bibliographic entry see Field 02J.
W73-09163

SOME MECHANISMS OF OCEANIC MIXING REVEALED IN AERIAL PHOTOGRAPHS,
Lamont-Doherty Geological Observatory, Palisades, N.Y.
For primary bibliographic entry see Field 07B.
W73-09168

PRECISION REQUIREMENTS FOR A SPACECRAFT TIDE PROGRAM,
National Oceanic and Atmospheric Administration, Miami, Fla. Atlantic Oceanographic and Meteorological Labs.
For primary bibliographic entry see Field 07B.
W73-09170

ABUNDANCE AND HARVEST OF TROUT IN SAGEHEN CREEK, CALIFORNIA,
California Univ., Berkeley. School of Forestry and Conservation.
R. Gard, and D. W. Seegrist.
Trans Am Fish Soc. Vol 101, No 3, p 463-477. 1972. Illus.

Identifiers: *California, Fish management, Sagehen, Creek (Calif), Salmo-Gairdneri, Salmo-Trutta, Salvelinus-Fontinalis, *Trout, *Fish harvest.

Between 1952 and 1961 standing crops of brook (Salvelinus fontinalis), rainbow (Salmo gairdneri), and brown (Salmo trutta) trout were determined in Aug. for a 5.7 mile section of Sagehen Creek in east-central California. In 1953 a creel census was initiated which continued through 1961. The purpose of the study was to determine if a moderately productive wild trout stream could support a substantial angler harvest over an extended period without augmentation with hatchery-reared trout. The 10-yr average standing crop of all trout was 1578 (37 lb)/acre. Standing crops declined somewhat during the study period, largely due to habitat deterioration rather than to fishing. Floods and abandonment of beaver impoundments were the primary adverse influences on trout habitat. Catch did not decline significantly over the period. Fishermen annually removed 23-47% (average, 33%) of all trout over 99 mm in length, but recruitment replaced the loss. Although the natural fishery has proven to be viable, the trout population is not characterized by many large fish. Recommended is an experimental management program aimed at restoring some of the brown trout habitat by construction of low dams similar to the beaver dams that formerly were so productive, and increasing the number of larger trout in the stream by regulation of the take. If each large trout is subject to multiple capture and release, additional high-quality sport would be provided. Future studies could evaluate the effect on the fishery of these changes in management.--Copyright 1973, Biological Abstracts, Inc.
W73-09179

FLEXIBLE ROUGHNESS IN OPEN CHANNELS,
Waterloo Univ. (Ontario). Dept. of Civil Engineering.
For primary bibliographic entry see Field 08B.
W73-09271

RAINFALL EFFECT ON SHEET FLOW OVER SMOOTH SURFACE,
Colorado State Univ., Fort Collins. Dept. of Civil Engineering.
H. W. Shen, and R.-M. Li.
ASCE Proceedings, Journal of the Hydraulics Division, Vol 99, No HY5, Paper 9733, p 771-792, May 1973. 7 fig, 2 tab, 9 ref, append. OWRR Grant 14-01-0001-1435.

Descriptors: *Sheet flow, *Fluid friction, *Impact (Rainfall), *Statistical methods, *Shear stress, Mathematical models, Boundary layers, Numerical analysis, Overland flow, Hydraulics, Turbulence, Rainfall intensity.

The variation of friction factor for sheet flow influenced by various flow and rainfall conditions was established by statistical analysis. Boundary shear stress was also directly measured by hot-film anemometry. The friction factor is a function of both the flow Reynolds number and the rainfall intensity for a flow Reynolds number of less than 900. For a flow Reynolds number greater than 2,000, the friction factor is a function only of Reynolds number. A numerical model and a simplified procedure to predict the water surface profiles and boundary shear stresses for sheet flow with rainfall are presented. The computations of flow depths and boundary shear stresses are not too sensitive to the effect of uncertainties in selecting friction factor. (Knapp-USGS)
W73-09273

FLOODS IN MAPLE PARK QUADRANGLE, NORTHEASTERN ILLINOIS,
Geological Survey, Washington, D.C.
For primary bibliographic entry see Field 07C.
W73-09298

FLOODS IN RILEY QUADRANGLE, NORTHEASTERN ILLINOIS,
Geological Survey, Washington, D.C.
For primary bibliographic entry see Field 07C.
W73-09299

LINEAR STABILITY OF PLAIN POISEUILLE FLOW OF VISCOELASTIC LIQUIDS,
Delaware Univ., Newark. Dept. of Chemical Engineering.
K. C. Porteus, and M. M. Denn.
Transactions of the Society of Rheology, Vol 16, No 2, p 295-308, 91972. 9 fig, 17 ref.

Descriptors: *Rheology, *Critical flow, *Laminar flow, *Turbulent flow, Reynolds number, Viscosity.
Identifiers: Viscoelastic fluids.

The stability of plane Poiseuille flow to infinitesmal perturbations was studied for the second order and Maxwell fluid rheological models. When the disturbance propagation (based on the Deborah number) is small, the second order fluid is described by a consistent constitutive equation, and the two models give identical results. This occurs for elasticity number less than 0.0005. At higher values, the second order fluid model cannot be used. The critical Reynolds number is a slowly decreasing function of elasticity up to 0.0001 and a rapidly decreasing function subsequently, in the region where fluid relaxation effects become important. For a sufficiently elastic fluid the flow transition is described by a new mode of the Orr-Sommerfeld equation and differs qualitatively from that for a Newtonian liquid. Instability is likely at low Reynolds numbers in highly elastic liquids. (Knapp-USGS)
W73-09431

WATER-RESOURCE EFFECTS,
Geological Survey, Washington, D.C.
For primary bibliographic entry see Field 02F.
W73-09432

INVESTIGATIONS OF MORPHOLOGY AND HYDRAULICS OF RIVER CHANNELS, FLOOD PLAINS, AND BODIES OF WATER FOR STRUCTURAL DESIGN (ISSLEDOVANIYA MORFOLOGII I GIDRAVLIKI RECHNYKH RUSEL, POYM I VODOYEMOV DLYA NUZHD STROITEL' NOGO PROYEKTIROVANIYA).
Gosudarstvennyi Gidrologicheskii Institut, Leningrad (USSR).
For primary bibliographic entry see Field 08B.
W73-09460

HYDROMORPHOLOGICAL INVESTIGATIONS OF FLOOD-PLAIN AND CHANNEL PROCESSES (GIDROMORFOLOGICHESKIYE ISSLEDOVANIYA POYMENNOGO I RUSLOVOGO PROTSESSOV).
Gosudarstvennyi Gidrologicheskii Institut, Leningrad (USSR).
For primary bibliographic entry see Field 08B.
W73-09462

THE CHEMICAL TYPE OF WATER IN FLORIDA STREAMS,
Geological Survey, Tallahasse, Fla.
For primary bibliographic entry see Field 07C.
W73-09594

REPORTS FOR CALIFORNIA BY THE GEOLOGICAL SURVEY WATER RESOURCES DIVISION,
Geological Survey, Menlo Park, Calif. Water Resources Div.
J. S. Bader, and F. Kunkel.
Open-file report, 1969. 95 p.

Descriptors: *Bibliographies, *Publications, *Water resources, *California, Documentation, Reviews, Indexing, Information retrieval, Computer programs.

This bibliography lists alphabetically by author all water-resources reports prepared by the U.S. Geological Survey for public agencies in California through September 1969. It also lists selected nationwide reports that contain information pertinent to California. In addition, general hydrologic and research reports by personnel of the California district office of the Pacific Coast regional office at Menlo Park, Calif., are included, although those reports make no specific reference to California. The bibliographic list is indexed according to hydrologic area, county, and subject. A report may be indexed under one or more hydrologic areas, counties, and (or) subjects. The indexes list only the first line of the bibliographic citation; for the complete citation, refer to the bibliographic list. The computer program for which this compilation was prepared can be expanded to include additional subject categories and states. (Woodard-USGS)
W73-09603

FLOODFLOW CHARACTERISTICS OF EAST FORK HORSEHEAD CREEK AT INTERSTATE HIGHWAY 40, NEAR HARTMAN, ARKANSAS,
Geological Survey, Little Rock, Ark.
For primary bibliographic entry see Field 04C.
W73-09608

LYSIMETER SNOWMELT AND STREAMFLOW ON FORESTED AND CLEARED SITES,
Forest Service (USDA), Durham, N.H. Northeastern Forest Experiment Station.
For primary bibliographic entry see Field 02C.
W73-09617

FLOODS IN THE RIO GUANAJIBO VALLEY, SOUTHWESTERN PUERTO RICO,
Geological Survey, Washington, D.C.
For primary bibliographic entry see Field 07C.
W73-09618

2F. Groundwater

THE TRANSIENT BEHAVIOR OF RECHARGE-DISCHARGE AREAS IN REGIONAL GROUNDWATER SYSTEMS,
Maryland Univ., College Park. Dept. of Civil Engineering.
M-T. Tseng, R. M. Ragan, and Y. M. Sternberg.
Maryland Water Resources Research Center, Technical Report No 12, December 1972. 102 p, 15 fig, 3 tab, 47 ref, 6 append. OWRR A-017-MD (1) 14-31-0001-3220.

Descriptors: *Mathematical models, *Groundwater movement, *Recharge, *Unsteady flow, Numerical analysis, Hydraulic models, Laplaces equation, Artificial recharge, Drawdown, Water table, Water yield, Computer programs.

Mathematical models were developed for theoretical analysis of unsteady fluid motion in an unconfined homogeneous, isotropic aquifer subject to local recharge. The models apply to one- and two-dimensional analysis of the flow problems within the saturated zone. In the one-dimensional analysis, the system of partial differential equations is quasi-linear and of the hyperbolic type. In the two-dimensional analysis, the problem reduces to the solution of the Laplace equation satisfying the boundary conditions, including that of the complete nonlinear free surface. Numerical solutions were obtained for the outflow hydrograph and the temporal variation of the free surface for the one-dimensional case. For the two-dimensional analysis, the temporal and spatial variations

C. W. Poth.
Pennsylvania Geological Survey, 4th Series, Bulletin W 30, 1972. 52 p, 5 fig, 1 plate, 7 tab, 15 ref.

Descriptors: *Groundwater resources, *Aquifers, *Well data, *Water quality, Pennsylvania, Pumping, Water yield, Drawdown, Specific capacity, Groundwater recharge, Groundwater movement, Chemical analysis, Water analysis, Water utilization, Aquifer characteristics, Hydrogeology, Hydrologic data.
Identifiers: *Lehigh County (Penn), *Northampton County (Penn).

The Martinsburg Formation underlies the northern half of Lehigh and Northampton Counties, and is of Middle and Late Ordovician age. It is bounded on the south by older Ordovician limestone formations and on the north by a ridge-forming conglomerate of Silurian age. Recent mapping has supported a three-part division of the Martinsburg into a lower thin-bedded slate (Bushkill Member), a middle graywacke-bearing unit (Ramseyburg Member), and an upper thick-bedded slate (Pen Argyl Member). Data were collected on 332 wells in Lehigh County and 402 wells in Northampton County. Fifty-four wells were test pumped for 1 hour. The depths of the nondomestic wells average about twice those of the domestic wells (240 feet versus 125 feet in Lehigh County and 225 feet versus 112 feet in Northampton County). Also, the nondomestic wells yield about three to five times as much water as the domestic wells: 36 gpm versus 13 gpm in Lehigh County and 75 gpm versus 15 gpm in Northampton County. The specific capacities were greatest in the Pen Argyl Member and least in the Bushkill Member. The formation as a whole had a specific capacity of about 0.5 gpm per foot drawdown. Chemical analyses show that the groundwater has a median dissolved-solids content of 166 mg/liter. The principal dissolved constituents are calcium, magnesium, bicarbonate, and sulfate ions. (Woodard-USGS)
W73-09286

GEOHYDROLOGY OF THE NEEDLES AREA, ARIZONA, CALIFORNIA, AND NEVADA,
Geological Survey, Washington, D.C.
D. G. Metzger, and O. J. Loeltz.
For sale by Sup Doc, GPO, Washington, D.C. 20402 Price $1.75. Geological Survey Professional Paper 486-J, 1973. 54 p. 25 fig, 3 plate, 11 tab, 31 ref.

Descriptors: *Groundwater resources, *Water quality, *Hydrogeology, *Colorado River, Interstate, Arizona, California, Nevada, Water wells, Aquifer chracteristics, Pumping, Water yield, Drawdown, Specific capacity, Groundwater movement, Groundwater recharge, Irrigation, Water level fluctuations, Precipitation (Atmospheric), Hydrologic data.
Identifiers: *Needles Area (Colorado River).

The Needles area, which includes Mohave and Chemehuevi Valleys, extends from Davis Dam (57 miles south of Hoover Dam) southward to Parker Dam. It is in Mohave County, Ariz., San Bernardino County, Calif., and Clark County, Nev. Wells in the area that tap a sufficient thickness of Colorado River gravels have specific capacities as high as 400 gallons per minute per foot of drawdown. Groundwater in the Colorado River alluviums occurs under water-table conditions. Groundwater may occur under artesian conditions in or below the Bouse Formation. Sources of recharge to the groundwater reservoir are the Colorado River, unused irrigation water, runoff from precipitation, and underflow from bordering areas. Of these, the Colorado River is by far the principal source. Groundwater in the Needles area is of better quality than that in other parts of the lower Colorado River. Of the 95 samples analyzed, 46 had dissolved solids of less than 1,000 mg/liter and six had less than 500; the smallest concentration was 314 mg/liter. On the other extreme, six analyses had dissolved-solids content of more than 2,000 mg/liter; the largest concentration was 3,290 mg/liter. (Woodard-USGS)
W73-09287

CHANGE IN POTENTIOMETRIC HEAD IN THE LLOYD AQUIFER, LONG ISLAND, NEW YORK,
Geological Survey, Mineola, N.Y.
For primary bibliographic entry see Field 04B.
W73-09293

CONCEPTS OF KARST DEVELOPMENT IN RELATION TO INTERPRETATION OF SURFACE RUNOFF,
Geological Survey, Raleigh, N.C.
H. E. LeGrand, and V. T. Stringfield.
Journal of Research of the U S Geological Survey, Vol 1, No 3, p 351-360, May-June 1973. 8 fig, 32 ref.

Descriptors: *Karst hydrology, *Surface-groundwater relationships, *Streamflow, Caves, Subsurface drainage, Springs, Sinks, Topography, Geomorphology, Hydrogeology, Carbonate rocks.

In regions underlain by carbonate rocks the streamflow characteristics are related to processes of karstification, which is dependent on circulation of subsurface water and solution of the rock. Cavernous and permeable unsaturated zones tend to keep the water table depressed below land surface in many karst regions, a condition that leads to a low density of perennial streams. The uneven distribution of permeability beneath surface karst streams causes them to lose or gain water, depending on the position of the water table with reference to stream level. The conventional techniques of interpolation and extrapolation that have been reasonably successful in approximating streamflow of ungaged sites in nonkarstic regions have only limited use in karst regions. An understanding of principles of karstification and an understanding of the hydrogeologic framework of a carbonate terrane provide a useful basis for evaluating the streamflow characteristics. (Knapp-USGS)
W73-09294

EVALUATING THE RELIABILITY OF SPECIFIC-YIELD DETERMINATIONS,
Geological Survey, Tucson, Ariz.
For primary bibliographic entry see Field 04B.
W73-09297

WATER-RESOURCE EFFECTS,
Geological Survey, Washington, D.C.
A. O. Waananen, and W. R. Moyle, Jr.
Available from Sup Doc, GPO, Washington, D.C. 20402, Price $6.25. In: The Borrego Mountain Earthquake of April 9, 1968: Geological Survey Professional Paper 787, p 183-189, 1972. 4 fig, 2 tab, 6 ref.

Descriptors: *Earthquakes, *California, *Water level fluctuations, Groundwater, Surface waters, Seiches, Turbidity, Water quality.
Identifiers: *Borrego Mountain earthquake (Calif).

The principal effect of the Borrego Mountain, (Calif.), earthquake of 1968 on water resources was seismic in nature and almost entirely transient, as shown by the charts from recorders on groundwater observations wells in California, Arizona, and Nevada, and at gaging stations in California. Hydroseisms were recorded at several sites nearly 500 km from the epicenter. At most sites, the effects were limited to water-level or water-surface fluctuations, with some minor and insignificant net changes in water level at many wells; changes in streamflow were observed at only two sites. Hydroseisms were observed also in many swimming pools near the epicenter, in San Diego County, and near Los Angeles. The effects

on water quality were minor, and no significant problems were reported. (Knapp-USGS)
W73-09432

REGIONAL RATES OF GROUND-WATER MOVEMENT ON LONG ISLAND, NEW YORK,
Geological Survey, Mineola, N.Y.
O. L. Franke, and P. Cohen.
Available from Sup Doc, GPO, Washington, D C 20402 - Price $3.00. In: Geological Survey Research 1972, Chapter C; U S Geological Survey Professional Paper 800-C, p C271-C277, 1972. 5 fig, 1 tab, 14 ref.

Descriptors: *Groundwater movement, *New York, *Analog models, Saline water intrusion, Water balance, Path of pollutants, Septic tanks, Sewers, Water pollution control, Urban hydrology.
Identifiers: *Long Island (N Y).

Regional rates of groundwater movement on Long Island, N.Y., computed with the aid of a steady-state electrical analog model, indicate that near the boundary between Nassau and Suffolk Counties the length of time required for groundwater recharge to move seaward of the barrier beaches is about 800 years for water entering the Magothy aquifer and 3,000 years for water entering the Lloyd aquifer. These computations are based upon an assumed rate of natural recharge of 21 inches per year and upon the configuration of the natural groundwater flow net associated with that rate of recharge. About 25-30 years is the maximum time required for water to drain from one of the shallow groundwater subsystems into East Meadow Brook. If the dissolved substances are assumed to move at the same rate as the water, then these lengths of time indicate the orders of magnitude of the times required for groundwater containing substances of sewage origin (largely derived from cesspools and septic tanks) to be flushed from the groundwater system after completion of planned wide-scale sanitary sewering systems in Nassau and Suffolk Counties. (Knapp-USGS)
W73-09441

DISTORTION OF THE GEOTHERMAL FIELD IN AQUIFERS BY PUMPING,
Geological Survey, Washington, D.C.
R. Schneider.
Available from Sup Doc, GPO, Washington, D C 20402 - Price $3.00. In: Geological Survey Research 1972, Chapter C; U S Geological Survey Professional Paper 800-C, p C267-C270, 1972. 3 fig, 10 ref.

Descriptors: *Geothermal studies, *Water temperature, *Hydrogeology, *Withdrawal, Drawdown, Aquifers, Aquifer characteristics, Heat flow, Sand aquifers, Karst, Artesian aquifers.

The extent and nature of distortion of geothermal fields in aquifers by pumping were assessed by comparing observed geothermal profiles with reconstructed natural profiles for three areas of contrasting geohydrologic conditions in three Tertiary aquifers. In the Sparta Sand at Fordyce, Ark., a negative departure of about 0.8 deg C, results from induced downward flow of cool water. A positive departure of about 1.9 deg C in a shallow zone of the heavily pumped artesian carbonate-rock aquifer in Brunswick, Ga., near the center of an extensive potentiometric depression, results from horizontal flow of warm water derived in part from a distant region where water moves upward. A more complex condition was found in the intensively developed '500-foot' sand (Claiborne Group) in Memphis, Tenn. A positive departure of about 0.8 deg C near the lower part of the screened interval of a municipal well and a negative departure of about 0.4 deg C near the upper part suggest simultaneous movement of warm water from below and cool water from above. In the confining beds above the artesian aquifer downward infiltration of warm water from a shallow water-table aquifer is suggested by the positive 1.2 deg C departure. (Knapp-USGS)
W73-09442

GEOLOGY AND GROUND-WATER RESOURCES OF PRATT COUNTY, SOUTH-CENTRAL KANSAS,
Geological Survey, Lawrence, Kans.
For primary bibliographic entry see Field 04B.
W73-09453

SUBSURFACE WASTEWATER INJECTION, FLORIDA,
Geological Survey, Tallahassee, Fla.
For primary bibliographic entry see Field 05E.
W73-09456

FLUORIDE CONCENTRATIONS IN WATERS OF THE MOLDAVIAN ARTESIAN BASIN (K VOPROSU OB USLOVIYAKH FORMIROVANIYA FTORSODERZHASHCHIKH VOD MOLDAVSKOGO ARTEZIANSKOGO BASSEYNA),
Leningrad State Univ. (USSR).
For primary bibliographic entry see Field 02K.
W73-09457

GROUNDWATER OF THE KURA-ARAKS LOWLAND (GRUNTOVYYE VODY KURA-ARAKSINSKOY NIZMENNOSTI),
G. Yu. Israfilov.
Izdatel'stvo 'Maarif', Baku, 1972. 208 p.

Descriptors: *Groundwater, *Hydrogeology, Geomorphology, Stratigraphy, Petrology, Aquifers, Surface-groundwater relationships, Groundwater recharge, Infiltration, Water sources, Water table, Water levels, Water level fluctuations, Water chemistry, Water types, Confined water, Water temperature, Geologic time, Profiles, Maps.
Identifiers: *USSR, *Azerbaydzhan, *Kura-Araks lowland.

The physiography and geological structure of the Kura-Araks lowland in eastern Azerbaydzhan are described with particular emphasis given to changes in the hydrogeological conditions of groundwater aquifers in the last 35-40 years. Hydrogeological maps show the depth of occurrence and chemistry of groundwater and the delineation of the lowland according to groundwater types. (Josefson-USGS)
W73-09458

DETERMINATION OF THE AGE OF GROUNDWATER IN THE FERGANA ARTESIAN BASIN (OB OPREDELENII VOZRASTA PODZEMNYKH VOD FERGANSKOGO ARTEZIANSKOGO BASSEYNA),
Institute of Hydrogeology and Engineering Geology, Tashkent (USSR).
A. N. Sultankhodzhayev, G. Yu. Azizov, and S. U. Latipov.
Uzbekskiy Geologicheskiy Zhurnal, No 2, p 72-75, 1971. 1 tab, 4 ref.

Descriptors: *Groundwater, *Groundwater basins, *Artesian wells, *Radioactive dating, *Age, Geologic time, Stratigraphy, Rocks, Gases, Potassium, Deuterium, Confined water, Radioactive well logging, Estimating, Equations.
Identifiers: *USSR, *Fergana, Uranium, Thorium, Radium.

To estimate the age of groundwater in the Fergana artesian basin, the helium-argon method was used. Rock samples of different composition and age were analyzed to determine the average content of uranium, thorium, and radium in the rocks. The age of groundwater in Upper Neogene deposits of the basin varies between 0.56 and 6.0 million years, and the age of groundwater in Lower Neogene deposits ranges from 3.5-5 million years in the central part of the basin to 7.37-9.8 million years in the low foothills bordering the depression. (Josefson-USGS)
W73-09466

GROUND-WATER RESOURCES OF BRAZORIA COUNTY, TEXAS,
Geological Survey, Austin, Tex.
For primary bibliographic entry see Field 04B.
W73-09590

GROUND-WATER RESOURCES OF DONLEY COUNTY, TEXAS,
Geological Survey, Austin, Tex.
For primary bibliographic entry see Field 04B.
W73-09591

COMPUTER MODEL FOR DETERMINING BANK STORAGE AT HUNGRY HORSE RESERVOIR, NORTHWESTERN MONTANA,
Geological Survey, Menlo Park, Calif.
For primary bibliographic entry see Field 04A.
W73-09600

HYDROGEOLOGIC CHARACTERISTICS OF THE VALLEY-FILL AQUIFER IN THE STERLING REACH OF THE SOUTH PLATTE RIVER VALLEY, COLORADO,
Geological Survey, Lakewood, Colo.
For primary bibliographic entry see Field 04B.
W73-09605

HYDROGEOLOGIC CHARACTERISTICS OF THE VALLEY-FILL AQUIFER IN THE BRUSH REACH OF THE SOUTH PLATTE RIVER VALLEY, COLORADO,
Geological Survey, Lakewood, Colo.
For primary bibliographic entry see Field 04B.
W73-09606

2G. Water in Soils

HYDROLOGIC SIMULATION OF DEPRESSIONAL WATERSHEDS,
Iowa State Water Resources Research Inst., Ames.
For primary bibliographic entry see Field 02A.
W73-09101

USE OF NATURALLY IMPAIRED WATER, A BIBLIOGRAPHY.
Office of Water Resources Research, Washington, D.C.
For primary bibliographic entry see Field 03C.
W73-09116

SUBSURFACE DRAINAGE SOLUTIONS BY GALKERIN'S METHODS,
Colorado State Univ., Fort Collins. Dept of Civil Engineering.
For primary bibliographic entry see Field 04A.
W73-09132

THE EFFECT OF SURFACE TREATMENTS ON SOIL WATER STORAGE AND YIELD OF WHEAT,
South Australian Dept. of Agriculture, Adelaide.
For primary bibliographic entry see Field 03F.
W73-09241

WEATHERING OF MICACEOUS CLAYS IN SOME NORWEGIAN PODZOLS,
Norges Landbrukshoegskole, Vollebekk. Inst. of Soil Science.

sively drained, well drained, moderately well drained, imperfectly or somewhat poorly drained, and poorly drained. These are based on topography, soil color, water table and the presence, depth and abundance of mottles. There is usually a relationship between the abundance of mottling and soil color. Topography and ground water are also related to the determination of soil drainage classes.--Copyright 1973, Biological Abstracts, Inc.
W73-09421

INFLUENCE OF LAND-USE ON CERTAIN SOIL CHARACTERISTICS IN THE AFFORESTED AREA OF KHUNTI IN CHOTANAGPUR,
Ranchi Agricultural Coll., Kanke (India).
P. K. Roy, and B. P. Sahi.
Indian J Agric Sci. Vol 41, No 11, p 971-980. 1971.
Identifiers: *Soil properties, *Land use, Aggregates, Chotanagpur, Conductivity, Forests, Hydraulics, India, Khunti, Root penetration, Soils, Hydrogen ion concentration.

A comparative study of certain soil characteristics like root penetration, its abundance, hydraulic conductivity, volume of expansion and total water-stable aggregates was made under forest land, deforested barren land and cultivated land. The values of these characters declined on deforestation and cultivation. There was appreciable degradation in larger aggregates (2mm) to deforestation, which further degraded on cultivation. No differentiations in texture, pH and other physical constants were observed, probably because of sufficient weathering of soils. Organic matter and N were found in the descending order deforested land > forest land > cultivated land; this was probably due to thick grasses and shrubs in the deforested land. Most of the differences were observed in the surface layers and at the same site.--Copyright 1973, Biological Abstracts, Inc.
W73-09424

MOISTURE CONTENT OF SOILS FOR MAKING SATURATION EXTRACTS AND THE EFFECT OF GRINDING,
Commonwealth Scientific and Industrial Research Organization, Canberra (Australia). Div. of Soils.
J. Loveday.
Aust Commonw Sci Ind Res Organ Div Soils Tech Pap. 12, p 3-9. 1972. Illus.
Identifiers: *Soil moisture, *Analytical techniques, Extracts, Grinding, Moisture, Saturation, Soils, Wetting.

Details are given of the capillary wetting method for determining the moisture content at which to make saturated pastes for extracting the soluble ions. For other routine analysis it is recommended that samples be ground to 2 mm. The effect of different grinding machines was considerable, and it is suggested that laboratories using this wetting technique should standardize on a particular type of grinding machine.--Copyright 1973, Biological Abstracts, Inc.
W73-09425

EFFECTS OF SOIL AND MANAGEMENT FACTORS ON FERTILIZER RESPONSES AND YIELD OF IRRIGATED WHEAT IN IRAN,
For primary bibliographic entry see Field 03F.
W73-09436

DETERMINATION OF SILVER IN SOILS, SEDIMENTS, AND ROCKS BY ORGANIC-CHELATE EXTRACTION AND ATOMIC ABSORPTION SPECTROPHOTOMETRY,
Geological Survey, Denver, Colo.
For primary bibliographic entry see Field 02K.
W73-09451

EFFECT OF SEEPAGE STREAM ON ARTIFICIAL SOIL FREEZING,
University Engineers, Inc., Norman, Okla.
H. T. Hashemi, and C. M. Sliepcevich.
ASCE Proceedings, Journal of the Soil Mechanics and Foundations Division, Vol 99, No SM3, Paper 9630, p 267-289, March 1973. 9 fig, 2 tab, 16 ref, 1 append. Army CRREL Grant DA-AMC-27-021-65-G19

Descriptors: *Freezing, *Frozen soils, *Groundwater movement, *Soil water movement, Numerical analysis, Soil mechanics, Frost, Thawing, Ice, Seepage, Heat transfer.

A numerical technique is presented for predicting the two-dimensional transient temperature distribution in freezing or thawing wet soil. The technique may be used for estimating the influence of seepage stream velocity on an artificial free-draining wet soil by a row of freeze pipes. Small groundwater flows perpendicular to the vertical axis of a freeze pipe have no significant effect on the rate of growth of the ice boundary in the horizontal direction. In fact, a flow of 0.01 cu ft per hr per sq ft to possibly 0.02 cu ft per hr per sq ft may even enhance the rate of propagation of the ice boundary. Increases in flow above those limits diminish the rate of growth of the ice boundary markedly. At a flow rate of 0.50 cu ft per hr per sq ft, closure of the ice boundary between a row of 6-inch freeze pipes with a center-to-center spacing of 8.6 ft will never occur. (Knapp-USGS)
W73-09454

CHEMICO-OSMOTIC EFFECTS IN FINE-GRAINED SOILS,
California Univ., Berkeley. Dept. of Civil Engineering.
J. K. Mitchell, J. A. Greenberg, and P. A. Witherspoon.
ASCE Proceedings, Journal of the Soil Mechanics and Foundations Division, Vol 99, No SM4, Paper 9678, p 307-322, April 1973. 5 fig, 2 tab, 12 ref, 1 append.

Descriptors: *Soil water movement, *Soil stabilization, *Osmosis, *Chemical potential, *Clays, Soils, Chemical reactions, Groundwater movement, Pore water, Seepage, Soil chemical properties, Soil physical properties, Soil physics, Soil treatment.

A theory describes the simultaneous coupled diffusional flow of salt and water in soils. The effects of coupling should increase as soil void ratio decreases and soil compressibility and salt concentration differences increase. Test results show that chemico-osmotic effects exist, and the rates of solution flow are in reasonable accord with theoretical predictions. Chemico-osmotic consolidation is likely to be small for most soils, except for very fine-grained active clays like bentonite. However, because chemico-osmotic coupling is capable of moving soil pore water and dissolved salts, chemico-osmotic soil stabilization might be feasible in some cases. (Knapp-USGS)
W73-09455

TRACE ELEMENTS IN SOILS AND FACTORS THAT AFFECT THEIR AVAILABILITY,
Macaulay Inst. for Soil Research, Aberdeen (Scotland).
For primary bibliographic entry see Field 05B.
W73-09468

DISTRIBUTION OF TRACE ELEMENTS IN THE ENVIRONMENT AND THE OCCURRENCE OF HEART DISEASE IN GEORGIA,
Geological Survey, Denver, Colo.
For primary bibliographic entry see Field 05B.
W73-09469

DETERMINATION OF MERCURY IN SOILS BY FLAMELESS ATOMIC ABSORPTION SPECTROMETRY,
Department of Scientific and Industrial Research, Petone (New Zealand).
For primary bibliographic entry see Field 05A.
W73-09471

SELENIUM DEFICIENCY IN SOILS AND ITS EFFECT ON ANIMAL HEALTH,
Oregon State Univ., Corvallis. Dept. of Animal Science.
For primary bibliographic entry see Field 02K.
W73-09476

LEAD IN SOILS AND PLANTS: A LITERATURE REVIEW,
Colorado State Univ., Fort Collins. Dept. of Botany and Plant Pathology.
For primary bibliographic entry see Field 05B.
W73-09479

PYROPHOSPHATE HYDROLYSIS IN SOIL AS INFLUENCED BY FLOODING AND FIXATION,
Manitoba Univ., Winnipeg. Dept. of Soil Science.
G. J. Racz, and N. K. Savant.
Soil Sci Soc Am Proc. Vol 36, No 4, p 678-682, 1972. Illus.
Identifiers: Fertilizers, Fixation, Flooding, *Hydrolysis, Phosphatase, *Pyrophosphate, *Soils.

The rate of hydrolysis of pyrophosphate (K4P2O7) in a noncalcareous soil maintained at field capacity moisture content or under flooded conditions was determined. Attempts were also made to study the influence of fixation of added pyrophosphate by soil on the rate of pyrophosphate hydrolysis. The initial rate of pyrophosphate hydrolysis was rapid and approximately the same for both flooded soil and soil maintained at field capacity moisture content. The first-order reaction rate constants, obtained by plotting the log of the concentration of unhydrolyzed pyrophosphate vs. time were 0.000814/min and 0.000837/min for soil at field capacity and under flooded conditions, respectively. However, after about 3 days of incubation, flooded soil hydrolyzed pyrophosphate at a relatively greater rate (k ± 0.000020/min) than did soil maintained at field capacity (k-0.000013/min). This difference appeared to be due to an increase in pyrophosphatase activity in the flooded soil. The half-life of the added pyrophosphate was approximately 2 days. Pyrophosphate adsorbed by an anion exchange resin or fixed by soil hydrolyzed at a slower rate than did pyrophosphate in solution. The ratio of orthophosphate to total phosphate (orthophosphate ± pyrophosphate) extracted from phosphated soils by various extractants decreased in the order; water, O1N H2SO4, and 0.5N2H4SO. Water-soluble or loosely bound pyrophosphate hydrolyzed at a much greater rate than did water-insoluble or strongly bonded pyrophosphate.–Copyright 1973, Biological Abstracts, Inc.
W73-09516

POTENTIOMETRIC TITRATION OF SULFATE IN WATER AND SOIL EXTRACTS USING A LEAD ELECTRODE,
Agricultural Research Service, Riverside, Calif. Salinity Lab.
For primary bibliographic entry see Field 02K.
W73-09518

GASEOUS LOSSES OF NITROGEN FROM FRESHLY WETTED DESERT SOILS,
Wisconsin Univ., Madison. Dept. of Soil Science.
A. N. Macgregor.
Soil Sci Soc Am Proc. Vol 36, No 4, p 594-596, 1972.
Identifiers: Denitrification, Desert soils, Gaseous losses, *Nitrogen, Nitrogen-15, *Soils, Spectrometry, Volatilization, Wetted soils.

On the basis of manometric data and mass spectrometer analyses, N2, and N2O were formed when samples from the dry surface of 2 semiarid soils were moistened and incubated under for 22-42 hr at 38 degC. Of the endogeneous soil nitrate-N present in each soil, approximately 60% was volatilized by Sonoita sandy loam (virgin desert soil). When either soil was amended with organic C, more gaseous N (N2 and N2O) was evolved than could be accounted for in terms of the initial level of endogenous nitrate. In 15N-experiments, N2O was derived primarily from nitrate. However, although significant amounts of N2 were evolved by soils, much of the N2 was unlabelled further suggesting nonnitrate sources play a role in volatilization of N from the soils under study.–Copyright 1973, Biological Abstracts, Inc.
W73-09519

HYDROLYSIS OF PROPAZINE BY THE SURFACE ACIDITY OF ORGANIC MATTER,
Agricultural Research Service, Beltsville, Md. Soils Lab.
For primary bibliographic entry see Field 05B.
W73-09520

CHARACTERIZATION OF THE OXIDIZED AND REDUCED ZONES IN FLOODED SOIL,
Louisiana State Univ., Baton Rouge. Dept. of Agronomy.
W. H. Patrick, Jr., and R. D. Delaune.
Soil Sci Soc Am Proc. Vol 36, No 4, p 573-576. 1972. Illus.
Identifiers: Ammonium, Flooded soil, Iron, Manganese, Nitrates, Oxygen, Soils, Sulfides, Zones, *Redox potential.

The oxidized and reduced layers in flooded soil were characterized by vertical distribution of the oxidation-reduction (redox) potential and concentrations of manganous Mn, ferrous Fe, sulfide, nitrate, and ammonium. Redox potential was measured with a special motor-driven assembly which advanced a Pt electrode at a rate of 2 mn/hr through the flooded soil profile. Vertical distribution of reduced forms of Mn, Fe and S and of nitrate and ammonium was determined by freezing and slicing the flooded soil into segments 1 or 2 mm thick. The apparent thickness of the oxidized layer was different when evaluated by the distribution of the various components in the profile, with the sulfide profile indicating the thickest oxidized zone, and Mn profile indicating the thinnest oxidized zone, and the Fe profile showing an intermediate thickness. The thickness of the oxidized layer increased with duration of flooding.–Copyright 1973, Biological Abstracts, Inc.
W73-09522

INFILTRATION AND SOIL PHYSICAL PROPERTIES,
Department of Agriculture, Ashburton (New Zealand). Winchmore Irrigation Research Station.
P. D. Fitzgerald, G. G. Cossens, and D. S. Rickard.
J Hydrol (Dunedin). Vol 10, No 2, p 120-126. 1971.
Identifiers: *Infiltration, Physical properties, *Regression analysis, *Soil properties.

An investigation was carried out into the relationship between soil physical properties and various infiltration parameters. The results of both simple and multiple regression analyses showed that there was little association. The technique of measuring infiltration by the double ring method is responsible for this lack of association.–Copyright 1972, Biological Abstracts, Inc.
W73-09530

RICE RESPONSE TO ZN IN FLOODED AND NONFLOODED SOIL,
Tennessee Valley Authority, Muscle Shoals, Ala. Soils and Fertilizer Research Branch.
For primary bibliographic entry see Field 03F.
W73-09551

EFFECT OF AERIAL ENVIRONMENT AND SOIL WATER POTENTIAL ON THE TRANSPIRATION AND ENERGY STATUS OF WATER IN WHEAT PLANTS,
Washington State Univ., Pullman. Dept. of Agronomy and Soils.
For primary bibliographic entry see Field 03F.
W73-09570

LONG-TERM GRAZING EFFECTS ON STIPA-BOUTELOUA PRAIRIE SOILS.
Department of Agriculture, Lethbridge (Alberta) Research Station.
S. Smoliak, J. F. Dormarr, and A. Johnston.
J Range Manage. Vol 25, No 4, p 246-250, 1972. Illus.
Identifiers: *Canada (Alberta), Carbon, *Grazing. Moisture, *Prairie soils, Saccharides, Sheep. Soils, *Stipa-Bouteloua, Hydrogen ion concentration.

The effects of grazing on prairie soils in Alberta were evaluated after 19 yr of continuous summer use by sheep at 3 stocking intensities. Analysis of the soils under the heavy grazing treatment showed lower values for pH and percent spring moisture but higher values for total C, alcohol/benzene-extractable C, alkaline-soluble C, polysaccharides, and belowground plant material than the soil under light or no grazing. The results were attributed to changes in amounts and kinds of roots due to species changes caused by grazing and to increased amounts of manure depositied by sheep on fields grazed at a higher intensity. Shallow-rooted species replaced the deeper-rooted ones on the drier environment induced by heavy grazing.–Copyright 1973, Biological Abstracts, Inc.
W73-09579

DEVELOPMENT OF GRASS ROOT SYSTEMS AS INFLUENCED BY SOIL COMPACTION,
Agricultural Research Service, Big Springs, Tex.
For primary bibliographic entry see Field 03F.
W73-09580

EFFECT OF SOIL MANAGEMENT ON CORN YIELD AND SOIL NUTRIENTS IN THE RAIN FOREST ZONE OF WESTERN NIGERIA,
Iban Univ. (Nigeria). Dept. of Agronomy.
For primary bibliographic entry see Field 03F.
W73-09581

SOIL MANAGEMENT FACTORS AND GROWTH OF ZEA MAYS L. ON TOPSOIL AND EXPOSED SUBSOIL,
For primary bibliographic entry see Field 03F.
W73-09582

STUDIES ON THE PERMEABILITY OF SOME TYPICAL SOIL PROFILES OF RAJASTHAN IN RELATION TO THEIR PORE SIZE DISTRIBUTION AND MECHANICAL COMPOSITION,
Udaipur Univ. (India). Coll. of Agriculture.
K. S. Singh, and G. Singh.
Ann Arid Zone. Vol 10, No 2/3, p 105-110. 1971.
Identifiers: Distribution, *India (Rajasthan), Mechanical composition, *Permeability, *Pore size, Soil profiles, Size.

The permeability of the soil profiles in general decreased downwards. The coefficient of permeability of Dungargarh soil was highest and of Kota was lowest at all depths. Capillary porosity and silt plus clay content of the soil samples were negatively correlated, whereas non-capillary porosity, total pore space and sand content were positively correlated with the coefficient of permeability. Capillary porosity, silt plus clay percent, non-capillary porosity and sand percent were significantly correlated with permeability. The relationship between the coefficient of permeability and percent total pore space was not significant.–Copyright 1973, Biological Abstracts, Inc.

sizable summer thunderstorms in the Southwest. (Popkin-Arizona)
W73-09683

FACTORS INFLUENCING INFILTRATION AND EROSION ON CHAINED PINYON-JUNIPER SITES IN UTAH,
Utah State Univ., Logan. Dept. of Range Science.
G. Williams, G. F. Gifford, and G. B. Coltharp.
Journal of Range Management, Vol 25, No 3, May 1972, p 201-205. 1 fig, 1 tab, 13 ref. Bureau of Land Management 14-11-0008-2837.

Descriptors: *Infiltration, *Erosion, *Pinyon-juniper trees, *Juniper trees, *Hydrologic properties, Limiting factors, Vegetation, Soil properties, Infiltrometers, Simulated rainfall, Controls, Sediments, Runoff, Soil texture, Bulk density, Organic matter, Porosity, Moisture content, Regression analysis, Forecasting, Variability, Time, Semiarid climates, *Utah, Great Basin.
Identifiers: Woodland.

Many factors influence infiltration in semiarid lands. Defining them is requisite to understanding hydrologic behavior of the 61.4 million acres of pinyon-juniper woodland in the western United States. This study reports the influence of vegetal and edaphic factors on infiltration and sediment production rates of pinyon-juniper sites in Utah. A Rocky Mountain infiltrometer was used to simulate high intensity rainfall on 2.5 sq ft plots for 550 trials on 28 chain-treated and 28 untreated field sites. Plots were pre-wet, runoff and sediment were measured and soil surface characteristics were determined. Parameters of interest include vegetal crown cover, soil texture, bulk density, organic matter, soil porosity and moisture content. Multiple regression analysis was performed. Ability to predict infiltration rates varied with time and location. Time of event, location, infiltration rates and erosion were important considerations. The predictive regression equations are not given in the paper, though the percent variance and other statistical properties are discussed. Factors influencing sediment discharge were so variable that no consistent relationship was found. The complexity that exists with regard to hydrologic phenomenon is recognized. Geographic areas, parameters of interest, and timing of an event should be considered. (Popkin-Arizona)
W73-09684

2H. Lakes

HYPOLIMNION REAERATION OF SMALL RESERVOIRS AND LAKES,
Missouri Univ., Rolla.
For primary bibliographic entry see Field 05G.
W73-09103

CIRCULATION PATTERNS IN LAKE SUPERIOR,
Wisconsin Univ., Madison. Dept. of Civil and Environmental Engineering; and Wisconsin Univ., Madison. Dept. of Meteorology.
J. A. Hoopes, R. A. Ragotzkie, S. L. Lien, and N. P. Smith.
Available from the National Technical Information Service as PB-220 244, $3.00 in paper copy, $0.95 in microfiche. Wisconsin Water Resources Center, Madison, Technical Report, April 1973. 80 p, 1 tab, 79 fig, 64 ref. OWRR B-009-WIS (5) 14-01-0001-1057.

Descriptors: *Lake Superior, Air circulation, Lake breezes, Lakes, Mixing, *Water circulation, Stratification, Winds, *Bathythermography, *Water temperature, *Currents (Water), Coriolis force, Model studies, *Mathematical models, *Energy transfer, Movement.
Identifiers: Keweenaw Peninsula, Isle Royale.

Lake circulation studies were directed toward Lake Superior for the purpose of broadening the quantitative understanding of large-scale circulation patterns induced by wind and atmospheric pressure variations over lakes under homogeneous and stratified conditions. One phase considers the numerical and physical modelling of lake circulation under uniform density conditions. The effects of lake geometry and wind stress distributions on the movement of water are considered. Coriolis forces were considered in the models. The second phase involves the analysis and interpretation of field observations of currents and temperatures for the central region of Lake Superior, between the Keweenaw Peninsula and Isle Royale during seasons when lake stratification conditions predominate. Bathythermography data and airborne infrared surface temperature surveys were collected at the same time to permit the investigators to make inferences about the temperature structure of the interior of the lake, given only the surface temperature pattern. By using a diverse approach of combining many sampling techniques and methods of analysis, it was possible to detect and define the nature of the temperature and current patterns for the study area. (Kerrigan-Wisconsin)
W73-09106

A PROPOSAL FOR IMPROVING THE MANAGEMENT OF THE GREAT LAKES OF THE UNITED STATES AND CANADA.
Cornell Univ., Ithaca, N.Y. Water Resources and Marine Sciences Center.
For primary bibliographic entry see Field 06E.
W73-09110

CHEMICAL QUALITY OF SURFACE WATER IN THE FLAMING GORGE RESERVOIR AREA, WYOMING AND UTAH,
Geological Survey, Washington, D.C.
For primary bibliographic entry see Field 02K.
W73-09127

HYDROLOGIC INVESTIGATIONS OF PRAIRIE POTHOLES IN NORTH DAKOTA, 1959-68,
Geological Survey, Washington, D.C.
W. S. Eisenlohr, Jr.
Geological Survey Professional Paper 585-A, 1972. 102 p, 92 fig, 3 plate, 34 tab, 48 ref.

Descriptors: *Potholes, *Ponds, *Water balance, *Salinity, *Wetlands, Evapotranspiration, Glacial drift, Hydrogeology, Evaporation, Seepage, Habitats, Topography, Grasslands, *North Dakota.
Identifiers: *Prairie potholes.

Prairie potholes, a result of glacial processes, are depressions capable of storing water. Large numbers of them have been drained for agricultural use. This report discusses prairie potholes in the Coteau du Missouri, the eastern part of the glaciated northern Great Plains region in North Dakota. It is a rolling upland covered with glacial drift. Seasonal evaporation from potholes clear of vegetation was found to very nearly equal the generalized evaporation values published by the U.S. Weather Bureau. The effect of hydrophytes in potholes was twofold; their presence reduced evaporation from the water surfaces; and at the height of the growing season their transpiration rate, added to the reduced evaporation rate, was frequently greater than the evaporation rate from potholes clear of vegetation. Net seepage outflow from potholes was generally very small--less than 0.01 foot per day per unit of water surface. This rate of seepage was not insignificant, however, because it often amounted to more than one-fourth of the total seasonal loss of water from a pothole. The source of water was primarily precipitation on the water surface of a pothole pond. Augmenting this supply was basin inflow, including overland flow, flow in channels, and seepage inflow. Where

there is no overflow or seepage outflow, there is no mechanism for the removal of dissolved solids brought to the pond by basin inflow. Such potholes are saline. The species of vegetation are excellent indicators of water quality and permanence of ponds. (Knapp-USGS)
W73-09130

SEDIMENTATION AND PHYSICAL LIMNOLOGY IN PROGLACIAL MALASPINA LAKE, ALASKA,
Massachusetts Univ., Amherst. Coastal Research Center.
For primary bibliographic entry see Field 02J.
W73-09141

A SELECTED ANNOTATED BIBLIOGRAPHY OF THE CLIMATE OF THE GREAT LAKES,
National Oceanic and Atmospheric Administration, Silver Spring, Md. Environmental Data Service.
H. Hacia.
Available from NTIS, Springfield, Va 22151 as COM-72 10830, Price $3.00 printed copy; $1.45 microfiche. Technical Memorandum EDS BS-7, March 1972. 70 p, 201 ref.

Descriptors: *Climatology, *Great Lakes, *Bibliographies, *Abstracts, *Publications, Meteorology, Precipitation (Atmospheric), Storms, Weather modification, Temperature, Water balance, Water levels, Winds, Waves, Solar radiation, Vapor pressure, Humidity.

This Selected Annotated Bibliography (201 entries) of the Climate of the Great Lakes has been compiled from readily available materials, particularly the American Meteorological Society's Meteorological and Geoastrophysical Abstracts (MGA), January 1960 through April 1971, and National Oceanic and Atmospheric Administration (NOAA) Atmospheric Sciences Library (DAS) files. Atmospheric Sciences Library call numbers are recorded with entry if available. (Woodard-USGS)
W73-09149

CONCENTRATIONS OF BIOGENIC ELEMENTS IN THE MOZHAYSK RESERVOIR (ZAPASY BIOGENNYKH ELEMENTOV V MOZHAYSKOM VODOKHRANILISHCHE),
Moscow State Univ. (USSR). Chair of Hydrology.
M. A. Khrustaleva.
Vestnik Moskovskogo Universiteta, Seriya V, Geografiya, No 4, p 77-80, 1972. 4 tab, 2 ref.

Descriptors: *Trace elements, *Reservoirs, *Water chemistry, Water sampling, Water analysis, Nitrogen, Phosphorus, Iron, Inorganic compounds, Nitrates, Nitrites, Seasonal.
Identifiers: *USSR, *Mozhaysk Reservoir, Silicon.

Investigations of the content of biogenic elements in waters of the Mozhaysk Reservoir in Moscow Oblast were based on analysis of 562 water samples collected during different seasons in 1968-70. The elements examined were nitrogen, phosphorus, iron, and silicon. Large concentrations of the elements observed in the dry year of 1969 were 1.7 times those in the moderately wet year of 1968 and 1.9 times those in the wet year of 1970. (Josefson-USGS)
W73-09151

THE EIGHTEENTH INTERNATIONAL LIMNOLOGY CONGRESS AT LENINGRAD (XVIII MEZHDUNARODNYY LIMNOLOGICHESKIY KONGRESS V LENINGRADE),
Akademiya Nauk SSSR, Moscow. Institut Geografii.
L. L. Rossolimo.
Akademiya Nauk SSSR Izvestiya, Seriya Geograficheskaya, No 1, p 151-156, January-February 1972.

Descriptors: *Conferences, *Limnology, *Bodies of water, *Lakes, Reservoirs, Aquatic environment, Aquatic life, Benthos, Aquatic productivity, Eutrophication, Ecology, Ecosystems, Microbiology, Fish management, Parasitism, Water chemistry, Water quality, Water pollution, Self-purification, Meteorology.
Identifiers: *USSR.

The Eighteenth International Congress on Limnology, held in Leningrad, August 19-26, 1971, was attended by 820 scientists from 34 countries, including 425 from the Soviet Union. Of the 300 papers presented at the Congress, 75 were delivered by Soviet delegates. The papers focused on general limnology, applied limnology, hydrometeorology, primary production, primary production and phytoplankton, hydrochemistry, microbiology, fish biology and fish management, ecology of aquatic organisms, zooplankton, benthos, and reservoirs. Plenary sessions were devoted to problems of general and regional limnology; bodies of water with regulated flow; heightened fish productivity in inland bodies of water by acclimatization of fishes and food invertebrates and by other means; and water quality and self purification of bodies of water. Symposia dealt with such specific topics as interactions between land and water; paleolimnology; parasites and productivity of fishes in fresh bodies of water; and regional productivity of the Vimba vimba species. (Josefson-USGS)
W73-09159

OCCURRENCE OF FREE AMINO ACIDS IN POND WATER,
Polish Academy of Science, Krakow. Zaklad Biologii Wod.
For primary bibliographic entry see Field 05C.
W73-09174

WISCONSIN LAKES RECEIVING SEWAGE EFFLUENTS,
Wisconsin Univ., Madison. Water Resources Center.
For primary bibliographic entry see Field 05B.
W73-09176

SOLAR RADIATION, WATER TEMPERATURE AND PRIMARY PRODUCTION IN CARP PONDS,
Polish Academy of Sciences, Golysz. Zaklad Biologii Wod.
For primary bibliographic entry see Field 05C.
W73-09181

PHYTOPLANKTON OF FINGERLING PONDS,
Polish Academy of Sciences, Krakow. Zaklad Biologii Wod.
For primary bibliographic entry see Field 05C.
W73-09183

BIOMASS OF ZOOPLANKTON AND PRODUCTION OF SOME SPECIES OF ROTATORIA AND DAPHNIA LONGISPINA IN CARP PONDS,
Polish Academy of Sciences, Golysz. Zaklad Biologii Wod.
For primary bibliographic entry see Field 05C.
W73-09184

CILIATA IN BOTTOM SEDIMENTS OF FINGERLING PONDS,
Polish Academy of Sciences, Krakow. Zaklad Biologii Wod.
For primary bibliographic entry see Field 05C.
W73-09185

TERUS SALMOIDES) AT MERLE COLLINS RESERVOIR,
California State Dept. of Fish and Game, Sacramento.
R. A. Rawstron, and K. A. Hashagen, Jr.
Calif Fish Game. Vol 58, No 3, p 221-230, 1972.
Identifiers: *Bass (Large mouth), Exploitation, Merle-Collins, Micropterus-Salmoides, *Mortality, Reservoirs, *Survival, Tagged fish, *California.

A tagging study from 1965 through 1969 revealed that exploitation rates of largemouth bass increased after the first season following impoundment, reaching a high of 0.65 in 1968 and 1970. Annual survival rates generally increased and stabilized near 0.2. Natural mortality declined. A combination of the highest reported exploitation rates, reduced annual catches, lowered catch/hour, increased bluegill populations,and competition with smallmouth bass and threadfin shad indicate possible continued depletion and overexploitation of largemouth bass.--Copyright 1973, Biological Abstracts, Inc.
W73-09277

DIFFERENTIAL SEASONAL MIGRATION OF THE CRAYFISH, ORCONECTES VIRILIS (HAGEN), IN MARL LAKES,
Ohio State Univ., Columbus. Dept. of Zoology.
H. Gowing.
Ecology. Vol 53, No 3, p 479-483. 1972. Illus.
Identifiers: *Crayfish, Gonads, Lakes, *Marl lakes, Maturation, *Michigan, Migration, Orconectes-virilis, *Seasonal migration.

In an earlier study during 1962 and 1963, movement of female O. virilis from shallow to deeper water in summer was noted in West Lost Lake, Otsego County, Michigan. The present additional observations on O. virilis, during 3 summers and in 3 lakes, confirm this migration. Evidence from other areas suggest that this movement is a general phenomenon of the species throughout most of its range. Migration of females appears to be associated essentially with maturation of the gonads.--Copyright 1973, Biological Abstracts, Inc.
W73-09303

VITAL STATISTICS, BIOMASS, AND SEASONAL PRODUCTION OF AN UNEXPLOITED WALLEYE (STIZOSTEDION VITREUM VITREUM) POPULATION IN WEST BLUE LAKE, MANITOBA,
Ministry of Natural Resources, Port Dover (Ontario).
J. R. M. Kelso, and F. J. Ward.
J Fish Res Board Can. Vol 29, No 7, p 1043-1052. 1972. Illus.
Identifiers: *Biomass, *Canada (Manitoba), Lakes, Seasonal production, Statistics, Stizostedion-vitreum-vitreum, *Walleye, *West Blue Lakes.

The population of S. vitreum vitreum, (25 cm and greater) in West Blue Lake, Man., in May 1969 decreased from 1090-819 in May 1970. During summer and fall, the intial population was supplemented by new recruits, age II+; reaching a maximum of 3451 in Sept. 1969. The total population in May 1970 was 2037. Mean daily mortality for all ages was small for the year, 0.0045, but loss rate was greater in fall and winter. Growth in length was greatest for age II+ followed by older fish. Growth in weight was similar for ages III-V and lowest for age II+. Mean biomass, approximately 800 kg, was similar in May 1969 and May 1970. Production, 340 kg, coincided with the growing season (June to Oct.) and was greatest from June to Sept. Greatest contribution to walleye production and biomass was made by age II+ fish. During winter and spring, zero or negative production occurred.--Copyright 1973, Biological Abstracts, Inc.
W73-09309

COPEPODS OF MARINE AFFINITIES FROM MOUNTAIN LAKES OF WESTERN NORTH AMERICA,
Alaska Univ., College. Inst. of Marine Science.
For primary bibliographic entry see Field 05C.
W73-09325

ECOLOGICAL OBSERVATIONS ON HETEROTROPHIC, METHANE OXIDIZING AND SULFATE REDUCING BACTERIA IN A POND,
Hydrobiologisch Instituut, Nieuwersluis (Netherlands).
For primary bibliographic entry see Field 05A.
W73-09337

LIMNOLOGICAL INVESTIGATIONS IN THE AREA OF ANVERS ISLAND, ANTARCTICA,
Virginia Polytechnic Inst. Blacksburg. Dept. of Biology.
For primary bibliographic entry see Field 05C.
W73-09339

ON THE QUANTITATIVE CHARACTERISTICS OF THE PELAGIC ECOSYSTEM OF DALNEE LAKE (KAMCHATKA),
For primary bibliographic entry see Field 05C.
W73-09340

TROUT IN LLYN ALAW, ANGLESEY, NORTH WALES: II. GROWTH,
Liverpool Univ. (England). Dept. of Zoology.
P. C. Hunt, and J. W. Jones.
J Fish Biol. Vol 4, No 3, p 409-424. 1972. Illus.
Identifiers: Anglesey, Growth, *Llyn-Alaw (No. Wales), *Trout growth, *Wales.

The annual and seasonal growth, length/weight relationship, changes in seasonal condition and specific growth rates of 17 types of trout in the newly flooded reservoir, Llyn Alaw, in Anglesey, North Wales, were investigated. Scales and length-weight data from 2076 trout caught by angling and netting during the first 4 yr after impoundment were analyzed to show the age, origins and growth of all fish both before and after flooding.--Copyright 1973, Biological Abstracts, Inc.
W73-09342

FINE STRUCTURE OF SOME BRACKISH-POND DIATOMS,
Rhode Island Univ., Kingston. Narragansett Marine Lab.
For primary bibliographic entry see Field 05A.
W73-09353

FIRST FOOD OF LARVAL YELLOW PERCH, WHITE SUCKER, BLUEGILL, EMERALD SHINER, AND RAINBOW SMELT,
National Water Quality Lab., Duluth, Minn.
R. E. Siefert.
Trans Am Fish Soc, Vol 101, No 2, p 219-225, 1972, Illus.
Identifiers: *Eutrophic lake, Bluegill, Bosmina-coregoni, Catostomus-commersoni, Cyclops-bicuspidatus, Food, *Larval stages (Fish), Lepomis-macrochirus, Notropis-atherinoides, Osmerus-mordax, Perca-flavescens, Perch, Polyarthra, Shiner, Smelt, Sucker, Trichocera, *Copepod.

Entire digestive tract contents from larval yellow perch (Perca flavescens (Mitchill)), white sucker (Catostomus commersoni (Lacepede)), bluegill (Lepomis macrochirus Rafinesque), emerald shiner (Notropis atherinoides Rafinesque), and rainbow smelt (Osmerus mordax (Mitchill)), along with plankton samples taken concurrently with fish samples, were identified and counted. Food selection was calculated by use of Ivlev's electivity index. First food of yellow perch from an eutrophic lake were copepod nauplii and cyclopoid

copepods; after fish reached 11 mm in length Bosmina coregoni was the dominant food. Yellow perch from an oligotrophic lake fed first on the rotifer Polyarthra, and to a lesser extent on copepod nauplii; after fish reached 11 mm cyclopoid copepods became the most important food items. Rotifers contributed an important portion of the early diet of white sucker; copepod nauplii and Cyclops bicuspidatus provided the remainder of the diet. Cladocerans increased in numbers after the fish reached 14 mm. Food of bluegills at the initiation of feeding consisted of Polyarthra and copepod nauplii. When fish reached 7 mm other rotifers and cyclopid copepods appeared in the diet, and at 8mm cladocerans become the dominant food and remained so for larger fish. Rotifers, mainly Trichocera, were the dominant food of 1st-feeding emerald shiners, and they remained an important food of larger fish. Copepod nauplii became a dominant food when fish reached 10 mm, and Cyclops bicuspidatus and cladocerans were important foods in fish 7 mm and larger. Algae were found in fish of all sizes, and calanoid copepods were found in fish 13 mm and larger. First-feeding rainbow smelt contained C. bicuspidatus, copepod nauplii, diatom, and green algae; C. bicuspidatus was the principal food until fish reached 21 mm. Calanoid copepods succeeded as the dominant food for larger fish. Electivity indices showed the following to be highly selected prey of fish at the initiation of feeding: yellow perch-Polyarthra and cyclopoid copepods; bluegill-Polyarthra; emerald shiner-rotifers and copepod nauplii; and rainbow smelt-C. biscuspidatus and copepod nauplii. Selection of food among the fish species studied was similar. Necessary criteria of a desirable prey include small size, the inability to escape capture, close proximity to the predator by swimming movements.--Copyright 1973, Biological Abstracts, Inc.
W73-09372

WINTER FOOD OF TROUT IN THREE HIGH ELEVATION SIERRA NEVADA LAKES,
Bureau of Sport Fisheries and Wildlife, Bishop, Calif.
G. V. Elliot, and T. M. Jenkins, Jr.
Calif Fish Game. Vol 58, No 3, p 231-237. 1972.
Identifiers: Fish food, Lakes, Salmo-Gairdneri, Salvelinus-Fontinalis, *Sierra-Nevada lakes, *Trout (Brook), Winter, *Trout (Rainbow).

A year-round analysis of trout stomach contents from 3 high elevation Sierra Nevada lakes showed that brook trout (Salvelinus fontinalis) and rainbow trout (Salmo gairdneri) fed on a variety of aquatic prey throughout the winter. However, under-ice feeding was poor relative to summer feeding even when the loss of surface forms is accounted for. Since fish were active and readily caught with bait at temperatures of 1C, poor feeding was due to reduced availability of important aquatic prey species rather than to a change in trout activity or readiness to feed.--Copyright 1973, Biological Abstracts, Inc.
W73-09417

DEVELOPMENT OF A SPORT FISHERY FOR LANDLOCKED SALMON IN A RECLAIMED POND,
Maine Dept. of Inland Fisheries and Game, Bangor.
For primary bibliographic entry see Field 08I.
W73-09419

REMOTE SENSING OF NEW YORK LAKES,
Geological Survey, Albany, N.Y.
J. M. Whipple.
Available from Sup Doc, GPO, Washington, D C 20402 - Price $3.00. In: Geological Survey Research 1972, Chapter C; U S Geological Survey Professional Paper 800-C, p C243-C247, 1972. 3 fig, 6 ref.
Descriptors: *Remote sensing, *Lakes, *New York, Infrared radiation, Water temperature, Thermal stratification, Limnology, Lake Ontario, Air-water interfaces, Mixing, Turnovers, Lake ice.
Identifiers: *Cayuga Lake (N Y), *Onondaga Lake (N Y).

Reverse flow through the outlet of Onondaga Lake and thermal activity associated with the spring mixing of Cayuga Lake were discerned on thermal-infrared imagery. Thermal radiances of natural and artificial discharges into Lake Ontario were measured, delineating the thermally pulsating nature of discharges into the open lake and thermal relationships between water masses. Quantitative imagery (radiometry) is useful in defining energy exchanges at the air-water interface. (Knapp-USGS)
W73-09447

SOME ALGAE COLLECTED FROM MT. TAYUAN, ILAN,
T-P Chang.
Quart J Taiwan Mus, Vol 23, No 1/2, p 127-134, 1970, Illus, Map.
Identifiers: *Algae (Alpine), Daphnia, Ilan, Mountains, Mt. Tayuan, Plankton, Species, *Taiwan, Zooplankton, *Green Peak Lake (Taiwan).

Alpine algae were collected at Green Peak Lake, which is 10 ha in area. Because the lake is surrounded by peaks (about 2300 m above sea-level), the lake accumulates mountain water from the creeks draining them. As the collection was procured in the dry season, a bare shore remained between the water margin and the grassy bank. The water color was due to the richness of zooplankton, especially Daphnia. Mt. Ali is the only Taiwanese locality to have algal samples studied. Although almost all the alga species listed are new to Taiwan, they are taxa commonly found in alpine habitats in other countries.--Copyright 1973, Biological Abstracts, Inc.
W73-09470

PHYTOPLANKTON SUCCESSION IN A EUTROPHIC LAKE WITH SPECIAL REFERENCE TO BLUE-GREEN ALGAL BLOOMS,
Academy of Natural Sciences of Philadelphia, Pa. Dept. of Limnology.
For primary bibliographic entry see Field 05C.
W73-09534

AN EXPERIMENTAL STUDY OF FEEDING BEHAVIOR AND INTERACTION OF COASTAL CUTTHROAT TROUT (SALMO CLARKI CLARKI) AND DOLLY VARDEN (SALVELINUS MALMA),
Department of Fisheries and Forestry, Vancouver (British Columbia). Fisheries Service.
D. C. Schutz, and T. G. Northcote.
J Fish Res Board Can. Vol 29, No 5, p 555-565, 1972. Illus.
Identifiers: *Fish feeding behavior, Coastal lakes, *Dolly Varden, Salmo-Clarki-Clarki, Salvelinus-Malma, *Trout (Cutthroat).

Feeding behavior of S. clarki clarki and Dolly Varden S. malma cohabiting a small coastallake was studies experimentally to examine the importance of food exploitation as a mechanism for spatial and food segregation observed there. The hypothesis that Dolly Varden could feed more successfully on benthic prey, and cutthroat on surface prey was supported by laboratory experiments involving isolated individuals and interspecies pairs exposed to food in benthic, surface, and both locations. Differences between the species were found in resting and orientation positions and in behavior associated with food searching, location, and capture. Dolly Varden also were more successful than cutthroat in capturing benthic prey at low light intensities. Observed differences in feeding between the species were fully expressed in solitary individuals and did not appear to be magnified by interaction. It is suggested that these differences in sympatric stocks of the 2 species may be inherent and, therefore, that segregation is largely selective rather than interactive, even though the populations still retain considerable plasticity, enabling them to change diets or habitats when necessary or advantageous.--Copyright 1972, Biological Abstracts, Inc.
W73-09544

PHYTOPLANKTON, PHOTOSYNTHESIS, AND PHOSPHORUS IN LAKE MINNETONKA, MINNESOTA,
Minnesota Univ., Minneapolis. Limnological Research Center.
For primary bibliographic entry see Field 05C.
W73-09546

MICROBIOLOGICAL STUDIES OF THERMAL HABITATS OF THE CENTRAL VOLCANIC REGION, NORTH ISLAND, NEW ZEALAND,
Wisconsin Univ., Madison. Dept. of Bacteriology.
For primary bibliographic entry see Field 05C.
W73-09554

SOME FACTORS ASSOCIATED WITH FLUCTUATION IN YEAR-CLASS STRENGTH OF SAUGER, LEWIS, AND CLARK LAKE, SOUTH DAKOTA,
Bureau of Sport Fisheries and Wildlife, Yankton, S. Dak. North Central Reservoir Investigations.
C. H. Walburg.
Trans Am Fish Soc. Vol 101, No 2, p 311-316, 1972. Illus.
Identifiers: Fluctuation, Lakes, *Sauger, *South Dakota, Spawning, Stizostedion-Canadense, Year, *Lewis and Clark Lake (S. Dak), Year-class strength.

The strength of sauger, Stizostedion canadense (Smith), year classes in Lewis and Clark Lake has fluctuated widely since formation of the reservoir in 1956. Studies suggest that more than 80% of the variability in year-class strength measured from the catch of fish older than age 0 and can be predicted from knowledge of water level change over the spawning ground, June reservoir water temperature, and reservoir water exchange rate. These factors were measured and probably function during the first 2 mo. of fish life. Sauger year classes of better than average strength can be expected in most years if both power-peaking operations at Fort Randall Dam during the fish spawning and incubation period, and the reservoir water exchange rate during June are minimized.--Copyright 1973, Biological Abstracts, Inc.
W73-09558

DIATOMS FOUND IN A BOTTOM SEDIMENT SAMPLE FROM A SMALL DEEP LAKE ON THE NORTHERN SLOPE, ALASKA,
For primary bibliographic entry see Field 05C.
W73-09622

THE VARIATIONS IN THE DISSOLVED OXYGEN AND TOTAL PHOSPHORUS CONCENTRATIONS IN LAKE GENEVA DURING THE YEARS 1969 AND 1970 (IN FRENCH),
Societe Vaudoise des Sciences Naturelles, Lausanne (Switzerland).
For primary bibliographic entry see Field 05B.
W73-09653

THE MER BLEUE SPHAGNUM BOG NEAR OTTAWA: II. SOME ECOLOGICAL FACTORS, (IN FRENCH),
Montreal Univ. (Quebec). Dept. of Biological Sciences.
For primary bibliographic entry see Field 02I.

Identifiers: Animals, Marshes, *Mice, Mus-musculus, *Salt marshes, Water sources.

An attempt was made to determine the natural water sources of feral house mice, Mus musculus Linnaeus 1758, from salt marshes of southern California by comparing the urine concentrations of field-collected samples with those obtained in a previous laboratory study. Two mice in the field had high urine chloride concentrations (620 and 895 mEq Cl/l) similar to those produced by feral mice drinking seawater in the laboratory. All 11 field samples had low urine osmotic concentrations (1580 to 2820 milliosmolals/kg) similar to those from feral mice drinking water ad lib in the laboratory. Plant and animal matter, dew, and seawater serve as possible sources of water. It appears that feral mice are not stressed routinely for water in salt marshes, even during seasonal periods of aridity.--Copyright 1973, Biological Abstracts, Inc.
W73-09194

SPECIES DIVERSITY OF STREAM INSECTS ON FONTINALIS SPP. COMPARED TO DIVERSITY ON ARTIFICIAL SUBSTRATES,
Plymouth State Coll, N. H. Dept. of Natural Sciences.
J. M. Glime, and R. M. Clemons.
Ecology. Vol 53, No 3, p 458-464. 1972. Illus.
Identifiers: Fontinalis-Spp, *Insects (Stream), *Species diversity, Streams, Substrates, *Mosses (Artificial).

Two types of artificial mosses (string and plastic) were compared with Fontinalis spp. for their insect inhabitants. These communities were sampled on 6 dates at 8 stream sites and compared by community coefficients, information theory analysis, and rank correlations. These results showed that those insects which were abundant on moss were also abundant on the artificial substrates. Fewer species and fewer individuals were present on the latter.--Copyright 1973, Biological Abstracts, Inc.
W73-09414

DOWNSTREAM MOVEMENTS OF WHITE SUCKER, CATASTOMUS COMMERSONI, FRY IN A BROWN-WATER STREAM OF ALBERTA,
Alberta Univ., Edmonton. Dept. of Zoology.
H. F. Clifford.
J Fish Res Board Can. Vol 29, No 7, p 1091-1093. 1972. Illus.
Identifiers: Biomass, *Brown water streams, *Canada (Alberta), Catostomus-Commersoni, Fry, Streams, *Sucker (White).

In a brown-water stream of Alberta, C. commersonii fry moved downstream almost entirely at night. During June and July when the fry were moving downstream, their nocturnal 'drift' pattern was more pronounced than that of any of the drifting invertebrates; and at this time their total biomass crossing a width transect per 24 hr exceeded the combined total biomass of the drifting invertebrate taxa. As the fry became larger, many more moved downstream near the surface of the water than near the stream's substrate.--Copyright 1973, Biological Abstracts, Inc.
W73-09426

THE RESPONSE OF CYANOGENIC AND ACYANOGENIC PHENOTYPES OF TRIFOLIUM REPENS TO SOIL MOISTURE SUPPLY,
Sheffield Univ. (England). Dept. of Botany.
W. Foulds, and J. P. Grime.
Heredity. Vol 28, No 2, p 181-187. 1972. Illus.
Identifiers: Acyanogenic, Cyanogenic, *Soil moisture, Phenotypes, *Trifolium repens.

Under experimental conditions of severe drought, fatalities in Ac phenotypes of T. repens exceeded those in ac phenotypes by a factor of 3. Under sublethal moisture stresses the yields of Ac and ac phenotypes were not significantly different. In moist conditions ac phenotypes attained higher yields than the Ac plants suggesting a possible linkage between the Ac gene and genes affecting vegetative vigor. Cyanogenic plants displayed a low sexual reporductive vigor in all treatments. Moisture stress inhibited flowering completely under conditions in which flowers were produced in all 3 acyanogenic phenotypes.--Copyright 1973, Biological Abstracts, Inc.
W73-09428

OSCILLATIONS IN STOMATAL CONDUCTANCE AND PLANT FUNCTIONING ASSOCIATED WITH STOMATAL CONDUCTANCE: OBSERVATIONS AND A MODEL,
Australian National Univ., Canberra. Dept. of Environmental Biology.
I. R. Cowan.
Planta (Berl). Vol 106, No 3, p 185-219. 1972.
Identifiers: *Plant physiology, Capacitors, Circadian rhythm, *Stumatal conductance, *Cotton, Illumination, Models, Oscillations, Permeability, Plants, Transpiration.

Measurements of transpiration, leaf water content, and flux of water in a cotton plant exhibiting sustained oscillations in stomatal conductance were presented, and a model of the mechanism causing this behavior was developed. The dynamic element of the model was capacitors--representing the change of water content with water potential in mesophyll, subsidiary and guard cells--interconnected by resistances representing flow paths in the plant. Increase of water potential in guard cells caused an increase in stomatal conductance. Increase of water potential in the subsidiary cells had the opposite effect and provided the positive feed-back which caused stomatal conductance to oscillate. The oscillations were shown to have many of the characteristics of free-running oscillations in real plants. The behavior of the model was examined, using an analogue computer, with constraints and perturbations representing some of those which could be applied to real plants in physiological experiments. Aspects of behavior which were simulated are opening and closing of stomata under the influence of changes in illumination, transient responses due to step changes in potential transpiration, root permeability and potential of water surrounding the roots, the influence of these factors on the occurrence and shape of spontaneous oscillations, and modulation of sustained oscillations due to a circadian rhythm in the permeability of roots.--Copyright 1973, Biological Abstracts, Inc.
W73-09429

CLASSIFICATION AND EVALUATION OF FRESHWATER WETLANDS AS WILDLIFE HABITAT IN THE GLACIATED NORTHEAST,
Massachusetts Univ., Amherst. Dept. of Biology.
F. C. Golet.
Ph D Thesis, October 1972. 179 p, 15 fig, 27 plate, 3 tab, 112 ref. OWRR B-012-Mass (3) and B-023-Mass (3).

Descriptors: *Wetlands, *Wildlife, *Freshwater marshes, *Classification, *Northeast U.S., Data collections, Massachusetts, Publications, Reviews, Marshes, Swamps, Vegetation, Bodies of water, Aquatic environment, Shallow water, Bogs, Water chemistry, Aquatic plants, Aquatic animals.

During the summer of 1970, qualitative data on habitat were collected at over 100 wetlands located throughout Massachusetts. Thirty-eight wetlands of high value to wildlife were selected for study during the fall of 1971. An extensive literature review of the habitat requirements of wetland wildlife species supplemented data gathered in the field. This dissertation presents the findings in two parts. Part I describes a detailed classification system for freshwater wetlands in the glaciated Northeast. Wetland vegetation is classified into

five life forms and 18 subforms. Other components of the classification system include: 5 size categories, 6 site types, 8 cover types, 3 vegetative interspersion types, and 6 surrounding habitat types. Methods and general considerations for application of the system are outlined also. Part II presents a list of 10 criteria for evaluating wetlands as wildlife habitat. These criteria are based largely upon the classification system in order of most to least important: (1) wetland class richness, (2) dominant wetland class, (3) size category, (4) subclass richness, (5) site type, (6) surrounding habitat type (s), (7) cover type, (8) vegetative interspersion type, (9) wetland juxtaposition, and (10) water chemistry. (Woodard-USGS)
W73-09430

AVAILABILITY OF MANGANESE AND IRON TO PLANTS AND ANIMALS,
West Virginia Univ., Morgantown. Dept. of Animal Industries and Veterinary Science.
For primary bibliographic entry see Field 02K.
W73-09474

ECOLOGY OF CORTICOLOUS LICHENS: III. A SIMULATION MODEL OF PRODUCTIVITY AS A FUNCTION OF LIGHT INTENSITY AND WATER AVAILABILITY,
McMaster Univ., Hamilton (Ontario). Dept. of Biology.
G. P. Harris.
J Ecol. Vol 60, No 1, p 19-40, 1972. Illus.
Identifiers: Canopy, Carbon, *Corticolouslichens, Ecology, Evaporation, *Lichens productivity, Light intensity, Oak-D, Parmelia-Caperata, Productivity, Rainfall, Simulation models, Through-Fall, Trees.

A mathematical simulation model for predicting epiphytic corticolous lichen productivity was built relating lichen productivity to environmental conditions at Shaugh, S. Devon, (England). Net C assimilation rates were calculated for Parmelia caperata (L) Ach. at 6 heights in a model oak tree, for a simulated time period of 2 yr. Weekly totals of net C assimilation were calculated for lichens at each height. The model has been constructed as a series of blocks each of which relates to, and was tested against, observed field data. It included a main program-responsible for setting up arrays, reading data, and calling sub-routines, and a radiation intensity sub-routine-using input parameters of time of year, time of day, cloud cover, and radiation drop inside the tree canopy; radiation levels were calculated for each of the 6 height zones in the tree. A wetting-up sub-routine used rates of wetting of different parts of the tree during rain, observed in the field using paper-grid conductivity sensors and a portable 12-track chart recorder. A new technique for modeling rainfall/throughfall was used to simulate the wetting-up system and was tested against observed data, with good agreement. Measurements of evaporation were carried out in the field using the paper-grid conductivity sensors and small capillary evaporimeters. Calculations of evaporation by Penman's method were used to estimate evaporation rates from the entire tree canopy. Various methods of apportioning the evaporation rates within the canopy are discussed. Calculations of thallus physiology are based on data published in a previous paper in this series. The model appears to give a calculated result which compares favorably with the vertical distribution of P. caperata measured in the field. Variation in weekly net C assimilation totals are shown to be highly variable and are thought to be correlated with water availability in the environment. Changes in the vertical evaporation gradient are shown to have a strong effect on the vertical distribution of this species.--Copyright 1972, Biological Abstracts, Inc.
W73-09512

THE BOTTOM FAUNA OF THE SUCHA WODA STREAM (HIGH TATRA MTS) AND ITS ANNUAL CYCLE,
Polish Academy of Sciences, Krakow. Zaklad Biologii Wod.
M. Kownacka.
Acta Hydrobiol. Vol 13, No 4, p 415-438, 1971. (Polish summary).
Identifiers: *Annual cycles, Baetis-Alpinus, *Bottom fauna, *Insects, Caddisflies, Diamesa-Latitarsis, Eukiefferiella, Fauna, Mountains, Orthocladius-Rivicola, Orthocladius-Thienemanni, Parorthocladius-Nudipennis, *Poland, Rhitrogena-Loyolaea, Stone-Flies, Streams, Sucha Woda (Pol).

The bottom fauna in the stream Sucha Woda was chiefly represented by larvae of insects, especially of flies (dominant Chironomidae), mayflies, stoneflies, and caddisflies. The distribution of these groups showed a distinct 'etagement,' as well as quantitative and qualitative differences according to the height and the velocity of the current. Above 1550 m the fauna was poor both qualitatively and quantitatively. The dominant forms were Diamesa gr. latitarsis. One quantitative peak was observed in Sept., brought about chiefly by D. gr. latitarsis and by some stoneflies, and Simuliidae. At altitude 1550-1000 m the number of species and individuals greatly increased in the stream. Eukiefferiella minor and Parorthocladius nudipennis, as well as Baetis alpinus and Rhitrogena loyolaea predominated. A thrice repeated increase in fauna was noted in winter (the greatest), in spring, and summer. Below 1000 m a number of species characteristic of rivers with a small gradient and a higher content of Ca appeared in the stream. A large number of individuals were noted. The dominant forms were Orthocladius rivicola and O. thienemanni. The fauna increased 3 or 4 times in number. The smallest number of individuals was noted in winter and the greatest in summer.--Copyright 1972, Biological Abstracts, Inc.
W73-09517

PENETRATION OF STOMATA BY LIQUIDS: DEPENDENCE ON SURFACE TENSION, WETTABILITY, AND STOMATAL MORPHOLOGY,
Michigan State Univ., East Lansing. Dept. of Horticulture.
J. Schoenherr, and M. J. Bukovac.
Plant Physiol. Vol 49, No 5, p 813-819. 1972. Illus.
Identifiers: Cuticular ledges, *Leaf surfaces, Liquids, Morphology, *Stomata, Stomatal, *Surface tension, Vestibule, *Wettability, Zebrina-purpusii-m.

Wettability of the leaf surface, surface tension of the liquid, and stomatal morphology control penetration of stomata by liquids. The critical surface tension of the lower leaf surface of Zebrina purpusii Bruckn. was estimated to be 25-30 dyne cm -1. Liquids having a surface tension less than 30 dyne cm -1 gave zero contact angle on the leaf surface and infiltrated stomata spontaneously while liquids having a surface tension greater than 30 dyne cm -1 did not wet the leaf surface and failed to infiltrate stomata. Considering stomata as conical capillaries, it was possible to show that with liquids giving a finite contact angle, infiltration depended solely on the relationship between the magnitude of the contact angle and the wall angle of the aperture. Generally, spontaneous infiltration of stomata will take place when the contact angle is smaller than the wall angle of the aperture wall. The degree of stomatal opening (4, 6, 8, or 10 micro m) was of little importance. Cuticular ledges present at the entrance to the outer vestibule and between the inner vestibule and substomatal chamber resulted in very small if not zero wall angles, and thus played a major role in excluding water from the intercellular space of leaves. It is shown why the degree of stomatal opening cannot be assessed by observing spontaneous infiltration of stomata by organic liquids of low surface tension.--Copyright, 1972, Biological Abstracts, Inc.
W73-09531

PATHOGENICITY OF PYRICULARIA AND COCHLIO-BOLUS CONIDIA COMPRESSED BY HYDROSTATIC PRESSURE (EFFECT OF PHYSICAL FACTORS UPON PLANT DISEASES: V.). (IN JAPANESE),
Gifu Univ. (Japan). Faculty of Agriculture.
H. Ikegami, and I. Takagi.
Ann Phytopathol Soc Jap. Vol 38, No 2, p 95-99. 1972. Illus. English summary.
Identifiers: Cochliobolus-miyabeanus, *Conidia, Germination, *Hydrostatic pressure, *Pathogenicity, Piricularia-Oryzae, Plant diseases, *Rice-M.

When conidia of Pyricularia oryzae were compressed at 0, 500, 1000, 1500, and 1800 kg/cm2 for 10 min., germination was reduced from 97% to 16%. The infectivity of these conidia to rice leaves was correspondingly reduced. At 0-1000 kg/cm2 the number of conidia with appressoria exceeded those with ordinary germ tubes, whereas at 1500 and 1800 kg/cm2 conidia with appressoria were less than those with germ tubes. With increasing pressures (0-1800 kg/cm2) the number of susceptible type lesions decreased 7.5-0.5/10 cm of rice leaf, whereas the number of resistant type lesions remained approximately constant at 1/10 cm. When conidia of Cochliobolus miyabeanus were exposed to increasing pressures, the germination and infection were reduced, but not as much as with p. oryzae. Conidia exposed to pressures above 1500 kg/cm2 formed smaller lesions on rice leaves than those exposed to lower pressures.--Copyright 1972, Biological Abstracts, Inc.
W73-09536

THE INTERNAL WATER STATUS OF THE TEA PLANT (CAMELLIA SINENSIS): SOME RESULTS ILLUSTRATING THE USE OF THE PRESSURE CHAMBER TECHNIQUE,
Tea Research Inst. of East Africa, Kericho (Kenya).
For primary bibliographic entry see Field 03F.
W73-09541

EFFECTS OF WATER DEFICITS ON FIRST PERIDERM AND XYLEM DEVELOPMENT IN FRAXINUS PENNSYLVANICA,
Wisconsin Univ., Madison. Dept of Forestry.
G. A. Borger, and T. T. Kozlowski.
Can J For Res. Vol 2 No 2, p 144-151, 1972. Illus.
Identifiers: Desiccation, Ethylene, *Fraxinus-Pennsylvanica-D, Glycol, *Periderm, Physiological studies, Seed, Water deficit, *Xylem.

Three-day-old seedlings of F. pennsylvanica were transferred to soil undergoing periodic drying or continuous subirrigation. Seedlings were harvested after 24 days. Germinating seeds of F. pennsylvanica were exposed to 1, 10, 20, or 30% polyethylene glycol (PEG) for 35 days and 1-day-old seedlings were exposed to 1, 10, or 20% PEG for 35 days. Seedlings grown on drying cycles or in 20% PEG had smaller periderm and xylem increments than seedlings grown in continuously subirrigated soil or 1 or 10% PEG, respectively. Xylem increment was affected more than periderm increment by water deficits. Germinating seeds exposed to 10 or 20% PEG did not develop periderm or secondary xylem and did not expand foilage. Thirty percent PEG caused desiccation and death or germinating seeds. Seeds exposed to 1% PEG germinated and developed normally. When water deficits occurred during germination the foilage failed to expand and periderm formation was thereby prevented. Water deficits subsequent to seed germination reduced periderm increment less than they reduced xylem increment, indicating difference in physiological controls.--Copyright 1972, Biological Abstracts, Inc.

centage of the moss layer (internal plus external water) depends more on the rainfall distribution than on the phenology of the species: 60% in dry period and 470% after rainy weather. The mean water content of the ground flora is lower than 1 mm, and shows considerable variations which reach 50% (0.4 mm) of the mean value. These variations are due to extreme values observed i January 1972 to discuss public rights and private use of beaches along the Texas coast. This report summarizes Texas law of the beaches and reports by participants which suggested necessary amendments to the Open Beaches Act (Act) to make it more protective of the public rights. Although the conference was not aimed at reaching any specific conclusions, the discussions did point out that the Texas Open Beaches Act may not guarantee the public a right to use all beaches; but, where such rights exist, state and local regulations are desirable to assure orderly development of public use. The Open Beaches Act was enacted to reduce to statutory language some common law concepts that recognize certain public rights to use the beaches. The act directs the Attorney General to protect the people's right, defined as a right of ingress and egress to that portion of the beach owned by the state and also to that part of the beach impressed with a presumption of a right of use by the people. Moreover, the Act provides for a presumption of prescriptive right to the area between low tide and, generally, the vegetation mark. There is also a discussion of the proposed National Open Beaches Act and actions taken by other coastal states. (Mockler-Florida)
W73-09679

ON NEONEMURA ILLIESI NOV. SPEC. AND SOME OTHER STONEFLIES FROM CHILE (INS., PLECOPTERA),

Max-Planck-Institut fuer Limnologie, Schlitz (West Germany). Limnologische Flussstation.
P. Zwick.

Stud Neotrop Fauna, Vol 7, No 1, p 95-100. 1972. Illus.

Identifiers: Araucanioperla, C, *Chile, Insects, Larval, *Neonemura-illiesi, Plecoptera, *Stoneflies.

Neoneumura illiesi is described from Chile, together with larval type C of Araucanioperla (.), found under stones in a small stream. Notes are also given on 6 other species of stoneflies from Chile.--Copyright 1972, Biological Abstracts, Inc.
W73-09680

THE MER BLEUE SPHAGNUM BOG NEAR OTTAWA: II. SOME ECOLOGICAL FACTORS, (IN FRENCH),

Montreal Univ. (Quebec). Dept. of Biological Sciences.
R. Joyal.

Can J Bot. Vol 50, No 6, p 1209-1218, 1972. Illus. English summary.

Identifiers: *Bogs, Climate, *Ecological studies, Humidity, Mer-bleue, Moss, Ombro, Peat, Picea-mariana-G, Polytrichum, Sphagnum bogs, Spruce-G, Temperature, Tropic level, *Canada (Ottawa).

Mer Bleue is an ombrotrophic Sphagnum and black spruce (Picea mariana) bog near Ottawa, Ontario. Its 5000 acres offer a vegetation composed of 10 associations and 3 sub-associations. Day temperature in the bog is 2.7F higher and night temperature 3.2F lower than at Uplands Airport 7 mi away. Relative humidity is also higher. The black spruce stand is the most humid and the coolest of all plant associations and its water table is the highest, while the Polytrichum association is the dryest and the warmest. Black spruce stands also have the most stable climate. Temperature at 1 ft depth in the peat moss reaches a minimum of 55F in summer. At 2 ft depth a maximum of 47C occurs in Sept. and a minimum of 24F in Feb.--Copyright 1972, Biological Abstracts, Inc.
W73-09702

THE OCCURRENCE, LIFE CYCLE, AND PATHOGENICITY OF ECHINURIA UNCINATA (RUDOLPHI, 1819) SOLOVIEV, 1912 (SPIRURIDA, NEMAT ODA) IN WATER FOWL AT DELTA, MANITOBA,

Naval Medical Research Inst., Bethesda, Md.
F. G. Austin, and H. E. Welch.

Can J Zool. Vol 50, No 4, p 385-393. 1972. Illus.

Identifiers: *Parasites (Waterfowl), Canada (Manitoba), Ceriodaphnia species, Chirocephalopsis-bundyi, *Daphnia-magna, *Deltas, *Duck, *Echinuria-uncinata, Eurycercus-lamellatus, Gammarus-lacustris, Geese, Hyallela-azteca, Lynceus-brachyurus, Mallard, Moina-macrocopa, *Nematoda, Pathogenicity, Simocephalus-vetulus.

In the Delta Marsh, third-stage E. uncinata juveniles were found in Daphnia magna, D. pulex, Simocephalus vetulus, and Gammarus lacustris. D. magna, the major host, were infected from late May to early Nov. with a peak of 108 parasites/100 Daphnia in early August. Experimentally, D. magna, D. pulex, Ceriodaphnia reticulata, C. acanthina, S. vetulus, Moina macrocopa, Eurycercus lamellatus, G. lacustris, Hyallela azteca, Chirocephalopsis bundyi, and Lynceus brachyurus became infected when exposed to E. uncinata eggs. Parasites developed to the infective stage in D. magna and D. pulex in 30 days at 15 C and in 10 days at 20-24 C. In mallard ducks, E. uncinata completed the fourth molt 20 days after infection; male worms were sexually mature after 30 days and females oviposited 40 days after infection. Parasites grew faster in 1 wk-old Delta mallards than in 2 and 3 mo.-old birds. Adult nematodes were located beneath the mucosal layer at the junction of the proventriculus and gizzard where granulomas formed after 30 days. The number of granulomas was correlated with the number of parasites. Mallards, pintails, gadwalls, lesser scaup, common eiders, and domestic geese were more susceptible to Echinuria infection than were shovellers, blue-winged teal, redheads, ruddy ducks, and American coots. Parasite eggs died when frozen but 50% survived 85 days when dried on filter paper. Echinuria uncinata can survive winter in resident mallards.--Copyright 1972, Biological Abstracts, Inc.
W73-09749

2J. Erosion and Sedimentation

EFFECTS OF RIVER CURVATURE ON A RESISTANCE TO FLOW AND SEDIMENT DISCHARGES OF ALLUVIAL STREAMS,

Iowa Univ., Iowa City. Inst. of Hydraulic Research.

For primary bibliographic entry see Field 02E.
W73-09102

RATES OF HILLSLOPE LOWERING IN THE BADLANDS OF NORTH DAKOTA,

North Dakota Univ., Grand Forks. Dept. of Geology.
L. Clayton, and J. R. Tinker, Jr.

North Dakota University Water Resources Research Institute Brookings Report WI-221-012-71, December, 1971. 36 p, 20 fig, 9 tab, 21 ref.

Descriptors: *Erosion, *Slopes, *Sediment transport, *Sediment yield, *North Dakota, Small watersheds, Valleys, Topography, Geomorphology, Hydrology, Runoff, Sedimentation.
Identifiers: *Badlands (N DAK), Buttes, Slopewash.

Measurements in a small drainage basin in the Little Missouri Badlands of western North Dakota indicate an average rate of hillslope lowering by slopewash of 0.41 inch per year on the west-facing hillslopes underlain by the Sentinel Butte Formation, 0.14 inch per year on the southwest-facing hillslopes underlain by the Tongue River Formation, and 0.11 inch per year on the northeast-facing hillslopes underlain by the Tongue River Forma-

tion. Soil creep occurs mainly on the Tongue River Formation and is mostly restricted to the northeast-facing hillslopes where the average rate of soil creep parallel to the hillslope surface is 0.23 inch per year in the upper 2.5 inches of surficial sediment. Erosion perpendicular to the face of seepage steps is 0.29 inch per year. The Sentinel Butte Formation has a lower rate of infiltration and percolation, which results in a higher rate of surface runoff than on the Tongue River Formation. The lowering of the hillslopes by slopewash contributes 99.9% of the 43,000 cubic feet of sediment per year from the hillslopes in the study areas. Comparison of the hillslope sediment yield with the rates of valley-bottom deposition from June to July, 1969, indicated that approximately 62% of the hillslope sediment left the drainage basin. (Woodard-USGS)
W73-09121

OBSIDIAN HYDRATION DATES GLACIAL LOADING,
Geological Survey, Denver, Colo.
I. Friedman, K. L. Pierce, J. D. Obradovich, and W. D. Long.
Science, Vol 180, No 4087, p 733-734, May 18, 1973. 1 fig, 1 tab, 6 ref.

Descriptors: *Hydration, *Dating, *Glaciation, *Weathering, Pleistocene epoch, Lava, Igneous rocks, Rhyolites.
Identifiers: *Obsidian.

The thickness of hydration rinds on the surface of obsidian (volcanic glass of rhyolitic composition) increases with age and is used for dating obsidian artifacts that were manufactured 200 to 250,00 years ago. Hydration rinds are also used to measure the age of rhyolite volcanic flows. In an attempt to relate hydration thicknesses on a rhyolite flow containing obsidian to the age of the flow as determined by the K-Ar method obsidian was collected in a small cavelike structure in the face of Obsidian Cliff, Yellowstone National Park, Wyoming, about 30 m below the flow surface. The average thickness of the thickest (oldest) group of hydration rinds is 16.3 micrometers and can be related to the original emplacement of the flow 176,000 years ago. In addition to these original surfaces, most thin sections show cracks and surfaces which have average hydration rind thicknesses of 14.5 and 7.9 micrometers. These later two hydration rinds compare closely in thickness with those on obsidian pebbles in the Bull Lake and Pinedale terminal moraines in the west Yellowstone basin, which are 14 to 15 and 7 to 8 micrometers thick, respectively. The later cracks are thought to have been formed by glacial loading during the Bull Lake and Pinedale glaciations, when as estimated 800 meters of ice covered the Obsidian Cliff flow. (Knapp-USGS)
W73-09125

SEDIMENTATION AND PHYSICAL LIMNOLOGY IN PROGLACIAL MALASPINA LAKE, ALASKA,
Massachusetts Univ., Amherst. Coastal Research Center.
T. C. Gustavson.
Available from NTIS, Springfield, Va. 22151 as AD-750 782, Price $3.00 printed copy; $0.95 microfiche. Technical Report No 5-CRC, August 1972. 48 p, 16 fig, 3 tab, 16 ref. Contract No N00014-67-A-0230-0001.

Descriptors: *Sedimentation, *Lakes, *Limnology, *Glacial sediments, *Alaska, Data collections, Glaciology, Lake morphology, Lake sediments, Varves, Streamflow, Melt water, Inflow, Discharge (Water), Density currents, Turbidity currents, Glaciohydrology.
Identifiers: *Proglacial lake, *Malaspina Lake (Alaska), Glaciolacustrine sediements.

Glaciolacustrine varved sediments are presently being deposited in Malaspina Lake, which lies along the southeastern margin of the Malaspina Glacier in Alaska. Malaspina Lake water exhibit a reserve thermal stratification. Water near the temperature of maximum density (3.94 deg C) occurs close to the lake surface, while water as cold as 0.3 deg C occurs at the lake bottom. The lake is normally stratified, however, with respect to suspended sediment content, which ranges from 0.1 g/liter at the surface to 0.7 g/liter at a depth of 45 m. Two large surface streams, Russell Stream and Tarr Stream, flow into the lake and their combined discharge is 140 cm. Discharge from the lake is as much as 600 cm. thus subglacial and englacial streams apparently discharge as much as 460 cm into the lake. Bottom topography of the lake is quite irregular except where the selective infilling of basins by turbidity currents has produced flat to gently sloping topography. Cores taken of these flat areas contain varved sediments. Varves, deposited at depths of 50 m or more, contain normal and reverse graded beds and horizontal beds. The current-bedded portion of the varve was deposited from an underflow or turbidity current. (Woodard-USGS)
W73-09141

RESEARCH IN THE COASTAL AND OCEANIC ENVIRONMENT.
Delaware Univ., Newark.
For primary bibliographic entry see Field 02L.
W73-09144

QUANTITATIVE ESTIMATES OF MUDFLOW HAZARD (O KOLICHESTVENNOY OTSENKE SELEOPASNOSTI),
Moscow State Univ. (USSR). Problemnaya Laboratoriya Nezhnykh Lavin i Selei.
S. M. Fleyshman.
Vestnik Moskovskogo Universiteta, Seriya V, Geografiya, No 4, p 26-31, 1972. 1 tab.

Descriptors: *Erosion, *Mass wasting, *Mudflows, Estimating, Evaluation, Mountains, Valleys, Slopes, Basins, Priorities.
Identifiers: *USSR, *Mudflow hazard.

A quantitative criterion for mudflow-hazard evaluation in mountainous areas is based on the relation between frequency of mudflow occurrence (mudflow activity) and volume of material transported by a mudflow during mass movement. Problems are considered which relate mudflow danger to the order of urgency of remedial measures for its prevention and control in the national economy. (Josefson-USGS)
W73-09150

SUSPENDED SEDIMENT DISCHARGE IN WESTERN WATERSHEDS OF ISRAEL,
Hebrew Univ., Jerusalem (Israel). Dept of Geography.
M. Negev.
Israel Hydrological Service Hydrological Paper No 14, 1972. 73 p, 15 fig, 9 tab, 6 ref, 3 append.

Descriptors: *Sediment transport, *Streamflow, *Suspended load, *Sediment yield, Forecasting, Methodology, Regression analysis, Floods, Equations, Hydrologic data, Foreign countries, Runoff, Correlation analysis, Evaluation.
Identifiers: *Israel.

Suspended sediment data were collected during a 3-year period on several streams in Israel. Two methods of analysis were used; one was a regression of sediment discharge on water discharge and other variables, and the other was a regression of sediment loads transported by single floods on flood volumes and other variables. In the first method, prediction equations were formulated based on the first year of data and then examined for competence with data for the last 2 years. This analysis reveals stable relationships in four of the streams, and suggests curtailing costly sediment sampling activities in these streams. The second method provides a simple and convenient tool for predicting sediment loads; however, there is less precision than with the first method. (Woodard-USGS)
W73-09163

CLAY MOBILITY IN RIDGE ROUTE LANDSLIDES, CASTAIC, CALIFORNIA,
Stanford Univ., Palo Alto, Calif.
P. F. Kerr, and I. M. Drew.
The American Association of Petroleum Geologists Bulletin, Vol 56, No 11 (Part I and II), p 2168-2184, November 1972. 17 fig, 6 tab, 14 ref.

Descriptors: *Landslides, *Clay, *California, *Embankments, Mass wasting, Clay minerals, Slopes, Stratigraphy.
Identifiers: *Castaic (Calif).

Landslides along clay-bearing bedding planes in road cuts along Interstate 5, where it crosses the western extension of the San Gabriel Mountains in southern California between Castaic and Gorman, occur where the strike of the clay-bearing strata in road cuts is parallel with the highway, and the dip is toward the roadway at an angle less than the slope of the cut. Slides also occur on dip slopes. Montmorillonite and illite are abundant minerals in the clay-bearing strata. When the water content of the clay that contains these minerals exceeds 505, the cone shear resistance drops to 0.001 to 0.1 tsf, whereas at 20%-35% water content, the shear remains in excess of 1.0 tsf. A large highway fill across West Liebre Gulch slid badly. South of the gulch, strata strike approximately normal to the line of the freeway and dip north toward the bottom of the gulch. Clay-bearing zones lie just beneath the surface. An old landslide, formed long before construction was contemplated, extended across the line chosen for the freeway. Fill, emplaced when the freeway was consructed, slid when the old slide, partly buried beneath the fill, was reactivated following heavy rains both in 1967 and 1969. (Knapp-USGS)
W73-09164

DEEP SEA SEDIMENTS AND THEIR SEDIMENTATION, GULF OF MEXICO,
Missouri Univ., Columbia. Dept. of Geology.
D. K. Davies.
The American Association of Petroleum Geologists Bulletin, Vol 56, No 11, p 2212-2239, November 1972. 34 fig, 2 tab, 31 ref. USGS 14-08-0001-11990; ONR N0014-038 (0002).

Descriptors: *Bottom sediments, *Gulf of Mexico, *Turbidity currents, *Sedimentation, *Provenance, Mississippi River, Stratigraphy, Sedimentology, Distribution patterns, Sediment distribution, Sediment transport.
Identifiers: *Deep-sea sediments.

Five distinct sources have contributed terrigenous sediment to the Gulf of Mexico during the Quaternary: (1) the Apalachicola, (2) the Mississippi, (3) the central Texas Rivers, (4) the Rio Grande, and (5) the rivers of northeast Mexico. Sands and silts originating from the Mississippi, Rio Grande, and northeast Mexico are mapped from the continental shelf to the abyssal plain. Sediments from the other sources are so adulterated by adjacent inputs as to be unrecognizable,.or remained on the shelf and slope. The provinces of the Campeche shelf and the Florida shelf contributed carbonates. Differential pelagic settling is responsible for the deposition of much of the clay-size sediments. Settling has produced parallel lamination and parallel bedding. Turbidity currents are the most important mechanism for the transport of sands and silts. The most significant sources of abyssal turbidites were the Mississippi River and Campeche shelf. Bottom currents have had little effect on the sands

and silts of the deep gulf. The bottom currents are sporadic, lowsflow phenomena, local, and of short duration. (Knapp-USGS)
W73-09165

HYDROLOGY AND EFFECTS OF CONSERVATION STRUCTURES, WILLOW CREEK BASIN, VALLEY COUNTY, MONTANA, 1954-68,
Geological Survey Washington, D.C.
For primary bibliographic entry see Field 04A.
W73-09167

THE ROLE OF SEDIMENTATION IN THE PURIFICATION OF RESIDUAL WATER, (NOTE SUR LA ROLE DE LA SEDIMENTATION DANS L'EPURATION DES EAUX RESIDUAIRES),
Centre d'Etudes et de Recherches de Biologie et d'Oceanographie Medicale, Nice (France).
For primary bibliographic entry see Field 05G.
W73-09205

SEDIMENT, FISH, AND FISH HABITAT,
Agricultural Research Service, Oxford, Miss. Sedimentation Lab.
For primary bibliographic entry see Field 05C.
W73-09213

RUNOFF AND SOIL LOSSES ON HAGERSTOWN SILTY CLAY LOAM: EFFECT OF HERBICIDE TREATMENT,
Pennsylvania State Univ., University Park.
J. K. Hall, and M. Pawlis.
Journal of Soil and Water Conservation, Vol 28, No 2, p 73-76, March-April 1973. 3 fig, 1 tab, 2 ref.

Descriptors: *Water yield, *Sediment yield, *Runoff, *Soil erosion, *Herbicides, Erosion, Path of pollutants, Soil conservation.
Identifiers: *Atrazine.

Runoff, sediment losses, and entrained herbicide residues were measured in field plots during the first two years of a corn-corn-oats rotation. The plots, situated on Hagerstown silty clay loam (14% slope), were treated in 1967 with atrazine at 0.5, 1, 2, 4, 6, and 8 pounds per acre. Although chemical losses in runoff and soil sediments were minor, the magnitude of soil and water losses, which increased generally with the rate of herbicide application, suggests the need for continued and expanded use of adequate agronomic and conservation systems and management practices. (Knapp-USGS)
W73-09279

HYDRAULIC EQUIVALENT SEDIMENT ANALYZER (HESA),
Massachusetts Univ., Amherst.
For primary bibliographic entry see Field 07B.
W73-09282

VOLUME WEIGHT OF RESERVOIR SEDIMENT IN FORESTED AREAS,
Forest Service (USDA), Boise, Idaho. Intermountain Forest and Range Experiment Station.
For primary bibliographic entry see Field 04D.
W73-09373

EFFECTS OF LOGGING AND LOGGING ROADS ON EROSION AND SEDIMENT DEPOSITION FROM STEEP TERRAIN,
Forest Service (USDA), Boise, Idaho. Intermountain Forest and Range Experiment Station.
For primary bibliographic entry see Field 04C.
W73-09374

AQUATIC SEDIMENT AND POLLUTION MONITOR,
For primary bibliographic entry see Field 05A.
W73-09390

QUALITY OF STREAM WATERS OF THE WHITE CLOUD PEAKS AREA, IDAHO,
Geological Survey, Washington, D.C.
For primary bibliographic entry see Field 07C.
W73-09435

DEGRADATION OF THE EARTHQUAKE LAKE OUTFLOW CHANNEL, SOUTHWESTERN MONTANA,
Geological Survey, Helena, Mont.
M. V. Johnson, and R. J. Omang.
Available from Sup Doc, GPO, Washington, D C 20402 - Price $3.00. In: Geological Survey Research 1972, Chapter C; U S Geological Survey Professional Paper 800-C, p C253-C256, 1972. 4 fig.

Descriptors: *Degradation (Stream), *Stream erosion, *Montana, *Landslides, Lakes, Erosion, Degradation (Slope).
Identifiers: *Earthquake Lake (Montana), *Madison River (Montana).

The Madison River is cutting though the Madison Slide, which was caused by the Hebgen Lake earthquake of August 17, 1959. Since July 15, 1960, the crest of the outlet channel of Earthquake Lake, on the slide, has been lowered about 4 feet by degradation. The outlet channel has been degraded as much as 19 feet in one short reach. Degradation of the channel was as much as 8.5 feet during the 1971 spring and summer runoff. Vertical and horizontal changes in the channel have been accompanied by sloughing of the banks. Degradation is expected to continue until the slope of the channel is adjusted to provide only the velocities necessary to transport the available sediment. Future adjustments in the reach of the channel through the landslide may endanger some works of man and cause environmental changes to the river downstream from the landslide. (Knapp-USGS)
W73-09445

ROLE OF ATMOSPHERIC PRECIPITATION AND GROUNDWATER IN THE FORMATION OF LANDSLIDES ON THE LEFT BANK OF THE CHIRCHIK RIVER (O ROLI ATMOSFERNYKH OSADKOV I PODZEMNYKH VOD V FORMIROVANII OPOLZNEY LEVOBEREZH-'YA R. CHIRCHIK),
Institute of Hydrogeology and Engineering Geology, Tashkent (USSR).
G. A. Mavlyanov, and K. A. Artykov.
Uzbekskiy Geologicheskiy Zhurnal, No 2, p 68-71, 1971. 1 fig, 3 ref.

Descriptors: *Landslides, *Precipitation (Atmospheric), *Groundwater, Loess, Moisture content, Wetting, Pressure, Drainage.
Identifiers: *USSR, *Chirchik River.

New landslides occur each year in upper reaches of the Chirchik River, particularly in the Bel'dersay, Gel'vasay, and Chimgansay basins. The main factor responsible for landslides in the region is the high natural moisture content of loess, which is continuously replenished by groundwater. Despite a relation between rainfall and landslide frequency, precipitation plays a secondary role. Landslide control to reduce the natural moisture content of loess includes drainage of the slide area provided the interstitial water under pressure at the contact between bedrock and loess is drained. (Josefson-USGS)
W73-09465

STRATIGRAPHY OF THE JACKSON GROUP IN EASTERN GEORGIA,
Georgia Univ., Athens. Dept. of Geology.
R. E. Carver.
Southeastern Geology, Vol 14, No 3, p 153-181, September 1972. 6 fig, 1 tab, 31 ref, 1 append. OWRR A-006-GA (10).

Descriptors: *Geology, *Sediments, *Geological surveys, *Georgia, Data collections, Geologic formations, Geomorphology, Geochemistry, Geologic history, Sediment transport, Sediment distribution, *Stratigraphy, Sedimentary structures.
Identifiers: *Jackson Group (Ga).

The depositional strike of the Jackson Group in eastern Georgia, based on locations of known outcrops of the Albian member of the Barnwell formation, is approximately N80 deg E. Because the trend of the fall line between Macon and Augusta is about N65 deg E, exposures of the Jackson Group near Macon represent more typically marine facies than do outcrops near Augusta. The Jackson Group is predominantly Upper Eocene in age. It is a transgressive-regressive sequence with a thin, extensively developed transgressive sand and a much thicker, more complex fine to course clastic regressive phase. In down-dip areas the group is represented by the Ocala Limestone; in up-dip areas, by fluviatile sediments indistinguishable from the late Cretaceous to possibly Middle Eocene Middendorf Formation. Between the down-dip marine limestone facies and the up-dip fluviatile facies occurs a lithologically complex near-shore facies, the Barnwell Formation. While general patterns of lighologic distribution can be recognized and the formation roughly divided into members, individual lithologic units are lenticular or deeply channeled and can not be traced over distances of more than a few miles. (Woodard-USGS)
W73-09586

ANALYTICAL METHODS FOR RUTHENIUM-106 IN MARINE SAMPLES,
Institute of Public Health, Tokyo (Japan). Dept. of Radiological Health.
For primary bibliographic entry see Field 05A.
W73-09648

ENVIRONMENTAL ASPECTS OF DREDGING IN ESTUARIES,
Skidaway Inst. of Oceanography, Savannah, Ga.
For primary bibliographic entry see Field 05C.
W73-09651

THE REGULARITIES OF VERTICAL DISTRIBUTION OF BENTHOS IN BOTTOM SEDIMENTS OF THREE MASURIAN LAKES,
Polish Academy of Sciences, Warsaw. Inst. of Ecology.
For primary bibliographic entry see Field 02H.
W73-09750

2K. Chemical Processes

IRON IN SURFACE AND SUBSURFACE WATERS, GRIZZLEY BAR, SOUTHEASTERN ALASKA,
Alaska Univ., College. Dept. of Geology.
For primary bibliographic entry see Field 05A.
W73-09111

ECOLOGICAL STUDIES OF THE SURFACE WATERS OF THE WHITEWATER CREEK WATERSHED, WALWORTH, ROCK AND JEFFERSON COUNTIES, WISCONSIN,
For primary bibliographic entry see Field 04C.
W73-09117

CHEMICAL QUALITY OF SURFACE WATER IN THE FLAMING GORGE RESERVOIR AREA, WYOMING AND UTAH,
Geological Survey, Washington, D.C.
R. J. Madison, and K. M. Waddell.
Geological Survey Water-Supply Paper 2009-C, 1973. 18 p, 8 fig, 1 plate, 3 tab, 4 ref.

Descriptors: *Water quality, *Dissolved solids, *Reservoirs, *Wyoming, *Utah, Leaching, Evaporation, Water chemistry, Sulfates, Bicarbonates.
Identifiers: *Flaming Gorge Reservoir, Green River.

The major inflow to the Flaming Gorge Reservoir, Wyoming and Utah, is from the Green River, which contributes an average of 81% of the water and 59% of the inflow load of dissolved solids. Together, Blacks Fork and Henrys Fork contribute about 16% of the water and about 23% of the dissolved-solids load, whereas minor tributaries contribute approximately 3% of the total inflow water to the reservoir, but about 18% of the total incoming load of dissolved solids. The concentration of dissolved solids in the reservoir in October 1966 was about 150 mg/liter greater than the concentration of the 1962-66 inflow. The increased concentration is due mostly to leaching of minerals from the reservoir bottom. The major difference between the chemical composition of the inflow during 1963-66 and that of the reservoir in 1966 is an increase in sulfate and a decrease in bicarbonate. Impoundment caused the concentration of dissolved solids in the river system to increase by about 32%. Evaporation accounted for an increase of 15 mg/liter, and leaching accounted for an increase of 115 mg/liter. (Knapp-USGS)
W73-09127

GROUNDWATER RESOURCES OF THE NORCO AREA, LOUISIANA,
Geological Survey, Baton Rouge, La.
For primary bibliographic entry see Field 04B.
W73-09128

DIFFUSION IN DILUTE POLYMETRIC SOLUTIONS,
Delaware Univ., Newark. Dept of Chemical Engineering.
For primary bibliographic entry see Field 01B.
W73-09134

OXYGEN DISTRIBUTION IN THE PACIFIC OCEAN,
Bedford Inst., Dartmouth (Nova Scotia).
For primary bibliographic entry see Field 02E.
W73-09147

CONCENTRATIONS OF BIOGENIC ELEMENTS IN THE MOZHAYSK RESERVOIR (ZAPASY BIOGENNYKH ELEMENTOV V MOZHAYSKOM VODOKHRANILISHCHE),
Moscow State Univ. (USSR). Chair of Hydrology.
For primary bibliographic entry see Field 02H.
W73-09151

ROLE OF SEA SALTS IN THE SALT COMPOSITION OF RAIN AND RIVER WATERS,
Akademiya Nauk SSSR, Moscow. Institut Okeanologii.
V. D. Korzh.
Oceanology, Vol 12, No 3, p 355-362, 1972. 1 fig, 3 tab, 39 ref. Translated from Okeanologiya (USSR), Vol 12, No 3, p 423-430, 1972.

Descriptors: *Oceanography, *Water chemistry, *Salts, *Rain water, *Rivers, Runoff, Hydrologic cycle, Meteoric water, Precipitation (Atmospheric), Atmosphere, Oceans, Sea water, Chlorides, Trace elements, Aerosols, Equations.
Identifiers: *USSR.

A quantitative determination of the role of the ocean in supplying salts to atmospheric precipitation and river water is based on the assumption that salts are carried to continents in the same proportion as they occur in atmospheric moisture over the ocean. The closeness of chloride ratios computed for Li, Na, Rb, Cs, B, and I at the ocean surface and determined from their concentrations in global river runoff indicate that B, I, Cl, Li, Na, Rb, and Cs are mainly cyclic, i.e., they reach land from the ocean through the atmosphere and return to the ocean with river runoff. The most likely proportion of cyclic salts in rainwater over the USSR is 14%-28% and in global river runoff, 6%-15%. According to computations, the actual values should approximate the upper limits. (Josefson-USGS)
W73-09152

RECOMMENDED METHODS FOR WATER DATA ACQUISITION.
For primary bibliographic entry see Field 07A.
W73-09161

WORLDWIDE SULFUR POLLUTION OF RIVERS,
Yale Univ., New Haven, Conn. Dept. of Geology and Geophysics.
For primary bibliographic entry see Field 05B.
W73-09169

INSTRUMENTATION FOR ENVIRONMENTAL MONITORING: WATER.
California Univ., Berkeley. Lawrence Berkeley Lab.
For primary bibliographic entry see Field 07B.
W73-09268

SURFACE WATER TEMPERATURES AT SHORE STATIONS, UNITED STATES WEST COAST, 1971.
Scripps Institution of Oceanography, La Jolla, Calif.

Availabile from NTIS, Springfield, Va 22151 AD-749 031 Price $3.00 printed copy; $0.95 cents microfiche. Data Report (SIO Reference 72-62), June 1972. 21 p. ONR Contract N00014-69-A-0200-6006.

Descriptors: *Water temperature, *Salinity, *Pacific Ocean, *Coasts, *United States, Sampling, Measurement, Data collections, Washington, Oregon, California, Surface waters, Methodology.

Temperature and salinity data are presented from observations made during 1971 at shoreline stations along the west coast of the United States from the Strait of Juan de Fuca, Washington, to La Jolla, California. The data consist of monthly means, ranges and standard deviations based on daily observations. Daily temperature and salinity values from which the means were derived are reproduced by computer for limited distribution and are available, upon request, for the cost of reproduction. The agencies that participated are: Hopkins Marine Station, National Oceanic and Atmospheric Administration/National Ocean Survey, U. S. Coast Guard, the California State Park System, Oregon State University, Pacific Gas and Electric Company, and Scripps Institution of Oceanography of the University of California, San Diego. Temperature readings and water samples were obtained from surf and sandy beaches, off rocky cliffs and ledges, over the sides of lightships and off piers, depending upon the station location. All stations, excluding those of NOAA/NOS and those reporting to Oregon State University, are maintained in cooperation with Scripps Institution of Oceanography. (Woodard-USGS)
W73-09269

WEATHERING OF MICACEOUS CLAYS IN SOME NORWEGIAN PODZOLS,
Norges Landbrukshoegskole, Vollebekk. Inst. of Soil Science.
For primary bibliographic entry see Field 02G.
W73-09275

dssolved solids. Conditions at the time of the survey generally represent average June-September conditions. At the time of the survey, most of the water in the drainage system upstream from Woodruff Warm Springs was diverted for irrigation. (Knapp-USGS)
W73-09448

NATURAL BACKGROUND CONCENTRATION OF MERCURY IN SURFACE WATER OF THE ADIRONDACK REGION, NEW YORK,
Geological Survey, Albany, N.Y.
W. Buller.
Available from Sup Doc, GPO, Washington, D C 20402 - Price $3.00. In: Geological Survey Research 1972, Chapter C; U S Geological Survey Professional Paper 800-C, p C233-C238, 1972. 2 fig, 2 tab, 17 ref.

Descriptors: *Water chemistry, *Mercury, Public health, Data collections, Water quality, Heavy metals, Surveys, *New York.
Identifiers: *Adirondack region (N Y).

Natural background concentrations of mercury in water supplies are not well known. Establishment of these natural background concentrations requires the study of areas relatively free of man's influence on the environment. The Adirondack region of New York State is such an area. Determinations of mercury concentration of samples collected from streams and lakes of the Adirondack region in fall, winter, and spring 1970-71 indicate that the natural background concentration of mercury is less than 0.5 micrograms per liter, which is less than one-tenth the 5 micrograms per liter limit recommended for drinking water in New York. (Knapp-USGS)
W73-09449

DETERMINATION OF SILVER IN SOILS, SEDIMENTS, AND ROCKS BY ORGANIC-CHELATE EXTRACTION AND ATOMIC ABSORPTION SPECTROPHOTOMETRY,
Geological Survey, Denver, Colo.
T. T. Chao, J. W. Ball, and H. M. Nakagawa.
Analytica Chimica Acta, Vol 54, p 77-81, 1971. 2 fig, 2 tab, 6 ref.

Descriptors: *Chemical analysis, *Metals, Sediments, *Soil chemistry, *Trace elements, Rocks, Analytical techniques, Organic matter, Acids, Spectrophotometry, Spectroscopy.
Identifiers: *Silver determinations, Atomic absorption spectrophotometry, Triisooctyl thiophosphate, Methyl isobutyl ketone, Nitric acid.

A method for the determination of silver in soil, sediment, and rock samples in geochemical exploration is presented. The sample is digested with concentrated nitric acid, and the silver extracted with triisooctyl thiophosphate (TOTP) in methyl isobutyl ketone (MIBK) after dilution of the acid digest to approximately 6 moles. The extraction of silver into the organic extractant is quantitative and not affected by the nitric acid concentration from 4 to 8 moles, or by different volumes of TOTP-MIBK. The extracted silver is stable and remains in the organic phase for several days. The silver concentration is determined by atomic absorption spectrophotometry. (Woodard-USGS)
W73-09451

GEOLOGY AND GROUND-WATER RESOURCES OF PRATT COUNTY, SOUTH-CENTRAL KANSAS,
Geological Survey, Lawrence, Kans.
For primary bibliographic entry see Field 04B.
W73-09453

CHEMICO-OSMOTIC EFFECTS IN FINE-GRAINED SOILS,
California Univ., Berkeley. Dept. of Civil Engineering.
For primary bibliographic entry see Field 02G.
W73-09455

FLUORIDE CONCENTRATIONS IN WATERS OF THE MOLDAVIAN ARTESIAN BASIN (K VOPROSU OB USLOVIYAKH FORMIROVANIYA FTORSODERZHASHCHIKH VOD MOLDAVSKOGO ARTEZIANSKOGO BASSEYNA),
Leningrad State Univ. (USSR).
Ye. V. Petrakov, E. V. Kozlova, and N. A. Sargsyants.
Vestnik Leningradskogo Universiteta, No 12, Geologiya-Geografiya, No 2, p 69-72, June 1972. 1 fig, 3 tab, 5 ref.

Descriptors: *Water properties, *Water quality, *Fluorides, *Groundwater basins, *Artesian aquifers, Rocks, Sedimentary rocks, Igneous rocks, Mineralogy, Salts, Leaching, Geologic time, Chemical analysis.
Identifiers: *USSR, *Moldavia.

Fluoride concentrations in groundwater of the Moldavian artesian basin are extremely high, reaching 20 mg/liter. An analysis of the conditions of formation of these waters shows a clear genetic relation of fluorides to both marine sedimentary rocks and igneous rocks. (Josefson-USGS)
W73-09457

DETERMINATION OF THE AGE OF GROUNDWATER IN THE FERGANA ARTESIAN BASIN (OB OPREDELENII VOZRASTA PODZEMNYKH VOD FERGANSKOGO ARTEZIANSKOGO BASSEYNA),
Institute of Hydrogeology and Engineering Geology, Tashkent (USSR).
For primary bibliographic entry see Field 02F.
W73-09466

HAIR AS A BIOPSY MATERIAL,
Agricultural Research Service, Grand Forks, N. Dak. Human Nutrition Lab.
For primary bibliographic entry see Field 05A.
W73-09467

DETERMINATION OF MERCURY IN SOILS BY FLAMELESS ATOMIC ABSORPTION SPECTROMETRY,
Department of Scientific and Industrial Research, Petone (New Zealand).
For primary bibliographic entry see Field 05A.
W73-09471

AVAILABILITY OF MANGANESE AND IRON TO PLANTS AND ANIMALS,
West Virginia Univ., Morgantown. Dept. of Animal Industries and Veterinary Science.
D. J. Horvath.
Geological Society of America Bulletin, Vol 83, p 451-462, February 1972. 10 tab, 72 ref.

Descriptors: *Iron, *Manganese, *Trace elements, *Soil, *Plants, *Animals, Agriculture, Animal pathology, Hydrogen ion concentration, Geochemistry, Grazing, Nutrients, Fertilizers, Domestic animals, Zinc, Copper, Nitrogen, Heavy metals, Potassium, Soil investigations, Soil chemistry.

Mn and Fe are essential to both plants and animals and are transferred successfully from the soil in all situations in which life persists. At the soil:plant interface pH and Eh are dominant factors and are frequently regulated by agronomic technology. Of the plant macronutrients used in fertilizer, P is most likely to reduce directly the availability of Fe

and Mn. Availability of Fe and Mn of plants to animals is characteristically described as low, in the order of 1/10. However, availability increases during deficiency and, in the case of Fe, following hemorrhage. Fe deficiency is considered one of the most common trace-element deficiencies of man. Intensive agronomic practices will have little effect, however, on the Fe or Mn status of grazing animals; and, consequently are likely to be damped. (Oleszkiewicz-Vanderbilt)
W73-09474

SELENIUM DEFICIENCY IN SOILS AND ITS EFFECT ON ANIMAL HEALTH,
Oregon State Univ., Corvallis. Dept. of Animal Science.
J. E. Oldfield.
Geological Society of America Bulletin, Vol 83, No 1, p 173-180, January 1972. 52 ref.

Descriptors: *Heavy metals, *Soils, *Animal pathology, *Pathology, *Trace elements, Irrigation, Agriculture, Nutrient requirements, Nutrients, Animal physiology, Forages, Chemical reactions, Hazards, Soil chemistry, Soil investigations.
Identifiers: *Selenium.

Selenium in minute quantities has been shown to be a dietary essential for animal life, and soil-plant-animal relations have been identified in the distribution of the element. In some cases, soils are frankly deficient in selenium - most particularly those derived from igneous rocks, and the deficiency in surface layers may be aggravated by intensive irrigation. Analytical surveys have revealed also that considerable variation exists among plant species in their abilities to take up and retain selenium from the soil. Some experiments have investigated the effectiveness of additions of selenium to the soil in overcoming selenium deficiency among farm animals. Protection for 2 years has been achieved by this technique; however, the various factors influencing the soil-plant-animal relations of selenium direct caution in its application. (Oleszkiewicz-Vanderbilt)
W73-09476

NICKEL EXPLORATION BY NEUTRON CAPTURE GAMMA RAYS,
Geological Survey, Washington, D.C.
For primary bibliographic entry see Field 05A.
W73-09480

DETERMINATION OF MERCURY IN NATURAL WATERS AND EFFLUENTS BY FLAMELESS ATOMIC ABSORPTION SPECTROPHOTOMETRY,
Central Inst. for Industrial Research, Oslo (Norway).
For primary bibliographic entry see Field 05A.
W73-09481

SPECTROPHOTOMETRIC DETERMINATION OF TRACE AMOUNTS OF MERCURY (II) BY EXTRACTION WITH BINDSCHEDLER'S GREEN,
Tottori Univ. (Japan). Faculty of Education.
For primary bibliographic entry see Field 05A.
W73-09482

NATURAL WATERS IN AMAZONIA: V. SOLUBLE MAGNESIUM PROPERTIES,
Instituto Nacional de Pesquisas da Amazonia, Manaus (Brazil).
W. L. F. Brinkmann, and A. Dos Santos.
Turrialba. Vol 21, No 4, p 459-465. 1971. Illus.
Identifiers: *Amazonia, *Brazil, *Magnesium, Microorganisms, Natural waters, Solubility, Trees.

In the Tertiary region of Amazonia, near Manaus, the principal sources of Mg in the natural waters are the water that flows over the trunks of trees and all the forest water in general. The Mg content comes from 'washings' of foliage, tops of trees, trunks and buds by rain water and, up to a certain point, the solution of metabolic products of macro-and micro-organisms. Soil water contained some Mg, but in rain water and brooks only traces were found. Mg can be regarded as an element circulating within the closed system of the humid tropical forest. In comparison with the Mg content of natural waters in the world as a whole, the concentrations found in the circulating waters of tertiary Amazonia are extremely low.--Coyright 1972, Biological Abstract, Inc.
W73-09497

NOVEL WET-DIGESTION PROCEDURE FOR TRACE-METAL ANALYSIS OF COAL BY ATOMIC ABSORPTION,
Bureau of Mines, Pittsburgh, Pa. Pittsburgh Mining and Safety Research Center.
For primary bibliographic entry see Field 05A.
W73-09506

NOTES ON THE DETERMINATION OF MERCURY IN GEOLOGICAL SAMPLES,
Imperial Coll. of Science and Technology, London (England). Dept. of Geology; and Imperial Coll. of Science and Technology, London (England). Applied Geochemistry Research Group.
For primary bibliographic entry see Field 05A.
W73-09509

POTENTIOMETRIC TITRATION OF SULFATE IN WATER AND SOIL EXTRACTS USING A LEAD ELECTRODE,
Agricultural Research Service, Riverside, Calif. Salinity Lab.
J. O. Goertzen, and J. D. Oster.
Soil Sci Soc Am Proc. Vol 36, No 4, p 691-693, 1972. Illus.
Identifiers: Electrodes, Lead, *Potentiometric titration, Soils, *Sulfates, Titration, Water pollution sources.

Sulfate concentrations in natural waters and soil-water extracts were determined semiautomatically using a pb-ion electrode to indicate the solution potential change at the endpoint, a constant flow device to deliver the titrant, a pH-mV meter to measure the potential, and a strip chart recorder to plot the solution potential thus indicating the endpoint. Sulfate concentrations as low as 0.5 meq/liter in the sample solution were determined by the potentiometric titration system. This semiautomated direct titration of sulfate has the advantages of increased sensitivity and speed as compared with the precipitation method.--Copyright 1973, Biological Abstracts, Inc.
W73-09518

GASEOUS LOSSES OF NITROGEN FROM FRESHLY WETTED DESERT SOILS,
Wisconsin Univ., Madison. Dept. of Soil Science.
For primary bibliographic entry see Field 02G.
W73-09519

HYDROLYSIS OF PROPAZINE BY THE SURFACE ACIDITY OF ORGANIC MATTER,
Agricultural Research Service, Beltsville, Md. Soils Lab.
For primary bibliographic entry see Field 05B.
W73-09520

CHARACTERIZATION OF THE OXIDIZED AND REDUCED ZONES IN FLOODED SOIL,
Louisiana State Univ., Baton Rouge. Dept. of Agronomy.
For primary bibliographic entry see Field 02G.
W73-09522

THE GROWTH, COMPOSITION AND NUTRIENT UPTAKE OF SPRING WHEAT: EFFECTS OF FERTILIZER-N, IRRIGATION AND CCC ON DRY MATTER AND N, P, K, CA, MG, AND NA,
Rothamsted Experimental Station, Harpenden (England).
For primary bibliographic entry see Field 03F.
W73-09552

A NEW METHOD FOR THE GAS CHROMATOGRAPHIC SEPARATION AND DETECTION OF DIALKYLMERCURY COMPOUNDS-APPLICATION TO RIVER WATER ANALYSIS,
National Environmental Research Center, Cincinnati, Ohio. Analytical Quality Control Lab.
For primary bibliographic entry see Field 05A.
W73-09584

CATALYTIC OXIDATION OF PHENOL OVER COPPER OXIDE IN AQUEOUS SOLUTION,
Delaware Univ., Newark. Dept. of Chemical Engineering.
For primary bibliographic entry see Field 05B.
W73-09587

GROUND-WATER RESOURCES OF BRAZORIA COUNTY, TEXAS,
Geological Survey, Austin, Tex.
For primary bibliographic entry see Field 04B.
W73-09590

GROUND-WATER RESOURCES OF DONLEY COUNTY, TEXAS,
Geological Survey, Austin, Tex.
For primary bibliographic entry see Field 04B.
W73-09591

AN EVALUATION OF APHA METHOD FOR DETERMINING ARSENIC IN WATER,
Environmental Health Lab., McClellan AFB, Calif.
For primary bibliographic entry see Field 05A.
W73-09593

THE CHEMICAL TYPE OF WATER IN FLORIDA STREAMS,
Geological Survey, Tallahasse, Fla.
For primary bibliographic entry see Field 07C.
W73-09594

GEOHYDROLOGY OF SUMTER, DOOLY, PULASKI, LEE, CRISP, AND WILCOX COUNTIES, GEORGIA,
Geological Survey, Washington, D.C.
For primary bibliographic entry see Field 07C.
W73-09619

ANALYTICAL METHODS FOR RUTHENIUM--106 IN MARINE SAMPLES,
Institute of Public Health, Tokyo (Japan). Dept. of Radiological Health.
For primary bibliographic entry see Field 05A.
W73-09648

CHEMICAL COMPOSITION OF ZARAFSHAN RIVER WATER (IN RUSSIAN),
Selskokhozyaistvennyi Institut, Samarkand (USSR).
Ch. N. Takhtamyshev.
Probl Osvoeniya Pustyn'. 3, p 73-77. 1972. Illus. English summary.
Identifiers: *Chemicals, Rivers, USSR, *Zarafshan River.

In the Zarafshan upper reaches carboniferous waters predominate. Middle-stream they change into carbonate-sulfuric and sulfuric-hydrocarbonic waters and down-stream-into sulfurous ones.

established in the area. More than 80 years of hydrologic data could be provided which would supply a dependable basis for predicting future conditions. (Woodard-USGS)
W73-09139

RESEARCH IN THE COASTAL AND OCEANIC ENVIRONMENT.
Delaware Univ., Newark.

Available from NTIS, Springfield, Va. 22151 as AD-752 503, Price $3.00 printed copy; $0.95 microfiche. Annual Status Report, November, 1972. 43 p, 9 fig, 12 ref. ONR N00014-69-A0407.

Descriptors: *Sediment transport, *Coasts, *Littoral drift, *Delaware, Projects, Data collections, Beaches, Ocean currents, Ocean waves, Environment, Winds, Soils, Vegetation, Shores, Estuaries, Bays, New Jersey, Maryland, Atlantic Ocean, Shore protection, Tidal effects.
Identifiers: *Delaware Bay.

Accomplishments are presented for the third year of an interdisciplinary investigation of the coastal processes and salient physical characteristics of a section of the Atlantic seacoast comprising Delaware Bay and adjoining areas in New Jersey and Maryland. This study, sponsored by the Geography Programs of the Office of Naval Research, is intended to provide thorough basic knowledge of this representative coastal zone that can be applied to similar littoral areas throughout the world in order to identify offshore, tidal zone, beach, and backshore surface and subsurface features from limited and preferably remote observations. Studies that were completed in 1972 include: the testings, correlation and alteration of mechanical properties of soils; vivianite as a cementitious material in soil stabilization; generation of water waves by wind; microclimatological investigations in coastal areas; rooted aquatics and their interaction with sublittoral sedimentary processes; and soil and vegetation characteristics of a low-lying marsh fringe-sand barrier coastline. (Woodard-USGS)
W73-09144

HYDROGRAPHIC OBSERVATIONS IN ELKHORN SLOUGH AND MOSS LANDING HARBOR, CALIFORNIA, OCTOBER 1970 TO NOVEMBER 1971,
Moss Landing Marine Labs., Calif.
For primary bibliographic entry see Field 05B.
W73-09162

DEEP SEA SEDIMENTS AND THEIR SEDIMENTATION, GULF OF MEXICO,
Missouri Univ., Columbia. Dept. of Geology.
For primary bibliographic entry see Field 02J.
W73-09165

PRECISION REQUIREMENTS FOR A SPACECRAFT TIDE PROGRAM,
National Oceanic and Atmospheric Administration, Miami, Fla. Atlantic Oceanographic and Meteorological Labs.
For primary bibliographic entry see Field 07B.
W73-09170

MICROBIOLOGICAL SEA WATER CONTAMINATION ALONG THE BELGIAN COAST, I - GEOGRAPHICAL CONSIDERATIONS,
Belgian Army Military Hospital, Ostend. Lab. for Sea Microbiology.
For primary bibliographic entry see Field 05B.
W73-09203

UTILIZATION OF CARD-FLOATS FOR THE STUDY OF SURFACE DRIFT AND APPLICATION IN THE PREVENTION OF COASTAL POLLUTION (UTILISATION DE CARTES-FLOTTEURS POUR L'ETUDE DES DERIVES DE SURFACE ET APPLICATION A LA PREVISION DES P OLLUTIONS CATIERES),
Service Hydrographique de la Marine, Paris (France).
For primary bibliographic entry see Field 05B.
W73-09210

THE POLLUTION OF THE COASTS AND EFFLUENTS IN THE SEA (LA POLLUTION DES COTES ET LES REJETS EN MER),
For primary bibliographic entry see Field 05G.
W73-09215

EFFECTS OF MAN'S ACTIVITIES ON ESTUARINE FISHERIES,
Florida State Univ., Tallahassee. Dept. of Oceanography.
For primary bibliographic entry see Field 05C.
W73-09216

A TRACER SIMULATION OF WASTE TRANSPORT IN THE MUDDY CREEK-RHODE RIVER ESTUARY, MARYLAND,
Geological Survey, Washington, D.C. Water Resources Div.
For primary bibliographic entry see Field 05B.
W73-09217

ANTIBIOTIC-RESISTANT COLIFORMS IN FRESH AND SALT WATER,
Alabama Univ., Birmingham. Medical Center.
For primary bibliographic entry see Field 05B.
W73-09218

POLLUTION CONTROL ASPECTS OF THE BAY MARCHAND FIRE,
For primary bibliographic entry see Field 05G.
W73-09231

MERCURY IN THE NETHERLANDS ENVIRONMENT, (KWIK IN HET NEDERLANDSE MILIEU (SLOT)),
For primary bibliographic entry see Field 05B.
W73-09237

SURFACE WATER TEMPERATURES AT SHORE STATIONS, UNITED STATES WEST COAST, 1971.
Scripps Institution of Oceanography, La Jolla, Calif.
For primary bibliographic entry see Field 02K.
W73-09269

EXCHANGE PROCESSES IN SHALLOW ESTUARIES,
Miami Univ., Fla. Sea Grant Institutional Program.
T. N. Lee, and C. Rooth.
Available from NTIS, Springfield, Va 22151 COM-72 10671 Price $3.00 printed copy; $1.45 microfiche. Special Bulletin No 4, January 1972. 33 p, 13 fig, 3 ref, 1 append. NOAA Sea Grant No 2-35147.

Descriptors: *Flow characteristics, *Estuaries, *Shallow water, *Model studies, Forecasting, Mixing, Tidal effects, Salinity, Water circulation, Winds, Bays, *Florida, Currents (Water), Analytical techniques, Dye releases, Dispersion, Diffusion, Tracking techniques, Inflow, Freshwater, Saline water intrusion.
Identifiers: *Biscayne Bay (Fla).

A modular approach to the analysis of mixing and flow cjaracteristics in shallow tidal estuaries is presented using South Florida's Biscayne Bay as an example. The method depends on isolating rela-

tively simple characteristic flow regimes in different parts of an estuary. These can be considered as building blocks, which when recombined in different configurations are capable of yielding a qualitative model for a specific estuary. Such models are of immediate value in preliminary assessments of estuarine water quality and interaction problems. This method can provide an effective base for further studies where more precise information is needed. Wind effects were found to have a great influence on exchange processes in the study estuary. Wind effects mix the estuary horizontally and vertically and can set up a mean circulation that advects interior water into the direct exchange region of the tidal inlets, thereby substantially decreasing the basin's flushing time. (Woodard-USGS)
W73-09283

POLYCHAETOUS ANNELIDS COLLECTED BY 'UMITAKA-MARU' FROM THE ARABIAN GULF,
Kuwait Univ. Dept. of Zoology.
For primary bibliographic entry see Field 05A.
W73-09341

USE OF A COMMERCIAL DREDGE TO ESTIMATE A HARDSHELL CLAM POPULATION BY STRATIFIED RANDOM SAMPLING,
Rhode Island Dept. of Natural Resources, Providence. Div. of Fish and Wildlife.
For primary bibliographic entry see Field 05B.
W73-09348

AN INSTRUMENT FOR MEASURING CONDUCTIVITY PROFILES IN INLETS,
British Columbia Univ., Vancouver. Inst. of Oceanography.
For primary bibliographic entry see Field 05A.
W73-09349

TRANSITION OF POLLUTION WITH MERCURY OF THE SEA FOOD AND SEDIMENTS IN MINAMATA BAY (IN JAPANESE),
Kumamoto Univ. (Japan). Dept. of Hygiene.
For primary bibliographic entry see Field 05B.
W73-09416

GRAYS HARBOR ESTUARY, WASHINGTON: REPORT 3, WESTPORT SMALL-BOAT BASIN STUDY,
Army Engineer Waterways Experiment Station, Vicksburg, Miss.
For primary bibliographic entry see Field 08B.
W73-09438

REGIONAL RATES OF GROUND-WATER MOVEMENT ON LONG ISLAND, NEW YORK,
Geological Survey, Mineola, N.Y.
For primary bibliographic entry see Field 02F.
W73-09441

PRELIMINARY STUDIES OF COLLOIDAL SUBSTANCES IN THE WATER AND SEDIMENTS OF THE CHESAPEAKE BAY,
Geological Survey, Washington, D.C.
For primary bibliographic entry see Field 02K.
W73-09443

COMMON MARSH, UNDERWATER AND FLOATING-LEAVED PLANTS OF THE UNITED STATES AND CANADA,
For primary bibliographic entry see Field 02I.
W73-09548

ON THE POPULATION ECOLOGY OF THE COMMON GOBY IN THE YTHAN ESTUARY,
Fisheries Research Board of Canada, Winnipeg (Manitoba).
M. C. Healey.
J Nat Hist. Vol 6, No 2, p 133-145, 1972, Illus.
Identifiers: Ecology, Estuaries, Gobius-Microps, *Goby, Population ecology, *Scotland, *Ythan estuary.

The abundance of common gobies (Gobius microps) was measured in 2 areas of the Ythan estuary, Scotland, at different time of the year. Information on their growth, reproduction and food habits is presented.--Copyright 1972, Biological Abstracts, Inc.
W73-09557

THE OCCURRENCE AND FOOD HABITS OF TWO SPECIES OF HAKE, UROPHYCIS REGIUS AND U. FLORIDANUS IN GEORGIA ESTUARIES,
Georgia Univ., Sapelo Island. Marine Inst.
W. B. Sikora, R. W. Heard, and M. D. Dahlberg.
Trans Am Fish Soc. Vol 101, No 3, p 513-525. 1972. Illus.
Identifiers: *Fish diets, Amphipoda, Estuaries, *Georgia, *Hake, Macrura, Mysidacea, Natantia, Urophycis floridanus, Urophycis regius.

From 1967-1970 a total of 2683 spotted hake, U. regius and 470 southern hake, U. floridanus were collected and found to exhibit migratory patterns in Georgia similar to northern and Gulf populations of these fish. The food habits of inshore juvenile populations of these 2 species of hake collected from coastal salt march-estuarine areas near Sapelo Island, Georgia were examined. The 341 spotted hake and 192 southern hake examined contained identifiable food items. These were analyzed for the numbers of individual food organisms, percent frequency of occurrence, and percent biomass. The most important group in occurrence and biomass was the Crustacea with Natantia most important gravimetrically, Amphipoda and Mysidacea most frequently occurring. These data when combined with the habits of the food organisms established these 2 hakes as species which use the estuary as 'nursey grounds'.--Copyright 1973, Biological Abstracts, Inc.
W73-09565

DISTRIBUTION AND ABUNDANCE OF YOUNG-OF-THE-YEAR STRIPED BASS, MORONE SAXATILIS, IN RELATION TO RIVER FLOW IN THE SACRAMENTO-SAN JOAQUIN ESTUARY,
California State Dept. of Fish and Game, Stockton. Anddromous Fisheries Branch.
J. L. Turner, and H. K. Chadwick.
Trans Am Fish Soc. Vol 101, No 3, p 442-452, 1972. Illus.
Identifiers: Abundance, *Bass (Striped), *California, Distribution, Estuaries, Morone-saxatilis, Rivers, *Salinity, *San-Joaquin estuary.

Annual distribution and abundance of young-of-the-year striped bass were measured from 1959 to 1970 in the Sacramento-San Joaquin estuary. Annual abundance of young bass in late summer was closely related to the amount of river flow in June-July into the estuary ($r \pm 0.89$). Highly significant correlations existed between striped bass abundance and salinity and water diverted from the estuary, both of which were mutually related to the amount of river flow. Six mechanisms which may control these relationships are discussed. Annual striped bass distribution in the estuary was also related to river flow ($r \pm -0.93$) and salinity ($r \pm 0.88$) with bass being farther upstream in years of low runoff and high salinity.--Copyright 1973, Biological Abstracts, Inc.
W73-09568

CIRCULATION AND BENTHIC CHARACTERIZATION STUDIES--ESCAMBIA BAY, FLORIDA.
Environmental Protection Agency, Athens, Ga. Southeast Water Lab.
For primary bibliographic entry see Field 05B.
W73-09592

A REPORT ON THE PROTOTYPE CURRENT VELOCITY AND SALINITY DATA COLLECTED IN THE UPPER CHESAPEAKE BAY FOR THE CHESAPEAKE BAY MODEL STUDY,
Johns Hopkins Univ., Baltimore, Md. Chesapeake Bay Inst.
J. C. Klepper.
Chesapeake Bay Institute Special Report 27 (Reference 72-12), Johns Hopkins University, December 1972. 44 p, 10 fig, 6 tab, 2 ref, 6 append. Contracts DACW31-70-C-0078 and DACW31-70-C-0077.

Descriptors: *Estuaries, *Chesapeake Bay, *Saline water intrusion, *Currents (Water), Flow measurement, Current meters, Velocity, Salinity, Sampling, Boats, Analytical techniques, Data collections, Hydrologic data, Tidal effects.

This report describes current velocity and salinity data collected in the Upper Chesapeake Bay during the Chesapeake Bay Model Study by the Chesapeake Biological Laboratory, Natural Resources Institute, of the University of Maryland and by the Chesapeake Bay Institute of The Johns Hopkins University. The joint field program consisted of same slack salinity runs from January 1971 through December 1972 and four five-day deployments. The same slack salinity runs were made each month along the main channel of the Bay to the upper limit of intrusion of sea-derived salinity. The 1381 Braincon histogram current meter was used to obtain current velocity measurements. Salinity measurements were made with CBI Induction Conductivity Temperature Indicators and Beckman RS-5-3 Industrial Salinometers. A 7094/1401 IBM computer was used to tabulate these data into a useable format. (Woodard-USGS)
W73-09595

SEA GRANT NEWSLETTER INDEX, 1968-71,
Rhode Island Univ., Narragansett. Pell Marine Science Library.
P. K. Weedman, S. Scott, and R. E. Bunker.
Available from NTIS Springfield, Va 22151 Com-72 11479, Price $3.00 printed copy; $0.95 microfiche. National Oceanic and Atmospheric Administration Technical Memorandum EDS ESIC-6, September 1972. 83 p, 1 fig, 4 tab. NOAA 2-35473.

Descriptors: *Oceanography, *Estuaries, *Publications, *Documentation, *Information retrieval, Indexing, Marine fish, Marine biology, Marine geology, Marine microorganisms, Marine plants, Marine animals, Ecology.
Identifiers: *Marine resources.

The Sea Grant Newsletter Index contains all issues of newsletters that have been produced with Sea Grant support, received by the National Sea Grant Depository (NSGD), and dated 1971 or earlier. Most of the articles in the newsletter are indexed. Some exceptions are those that merely list new publications or describe future meetings. In addition, continuing features which contain the same type of information in each issue are only indexed the first time they appear. An indication of this will be found in the notes line of the entry in the document listing. The listing is in alphanumeric order by document number. All available bibliographic information for each newsletter is given. The entries for the individual newsletter articles are brief, normally consisting of document number, date, title, and citation. If other information, such as personal author and author affiliation, is available or comments are needed, these are included. (Woodard-USGS)
W73-09598

TIDAL DATUM PLANES AND TIDAL BOUNDARIES AND THEIR USE AS LEGAL BOUNDA-

Results of hydrological investigations (temperature, salinity, O2) from 1961 to 1965 in the Bay of Marseilles are discussed. The most frequently observed salinity is about 37.9% and it alternates with a salinity higher than 38%. The latter can be explained by the vertical mixing in winter and by the upwelling of subsurface water. Slight dilutions of surface water are clearly more frequent than rare strong dilutions due to the influence of the Rhone riverwater. Hydrological features show a strong vertical mixing in winter. Average values of the summer temperatures can be expected to be the lowest of Mediterranean coasts. This must be explained by the upwelling of subsurface water induced by strong N-W winds (Mistral). At the end of summer a thermocline at 50 m depth separates an upper layer of warm water with salinity (S%>38) from a deeper layer, which shows a salinity minimum and an O2 maximum in its upper part. Hydrological conditions are very similar to those observed in Villefranche-sur-Mer. The principal difference between the 2 regions is a more important cooling of coastal water near Marseilles by strong winds from the N-W; this cooling is also favored by the continental shelf.--Copyright 1972, Biological Abstracts, Inc.
W73-09690

MESOBENTHIC COMPLEXES IN THE BAYS OF THE KILIYAN DELTA OF THE DANUBE, (IN RUSSIAN),
Akademiya Nauk URSR, Kiev. Institut Hidrobiologii.
M. N. Dekhtyar.
Gidrobiol Zh. Vol 8, No 1, p 44-49, 1972. Illus. English summary.
Identifiers: Bays, *Benthos, *Danube River, *Deltas, Fauna, Kiliyan delta, Salinity, USSR.

Five mesobenthic found complexes are described on different grounds of the Danube delta. The dependence of their formation on the salinity value frequency and range of variations is shown. These complexes preserve the typical structure of the delta under unstable conditions because of their ecological heterogeneity.--Copyright 1972, Biological Abstracts, Inc.
W73-09703

PROTECTING AMERICA'S ESTUARIES: PUGET SOUND AND THE STRAITS OF GEORGIA AND JUAN DE FUCA.
For primary bibliographic entry see Field 06E.
W73-09712

PROTECTING AMERICA'S ESTUARIES: PUGET SOUND AND THE STRAITS OF GEORGIA AND JUAN DE FUCA (20TH REPORT).
For primary bibliographic entry see Field 06E.
W73-09713

03. WATER SUPPLY AUGMENTATION AND CONSERVATION

3A. Saline Water Conversion

A TRANSPORT MECHANISM IN HOLLOW NYLON FIBER REVERSE OSMOSIS MEMBRANES FOR THE REMOVAL OF DDT AND ALDRIN FROM WATER,
Tennessee State Univ., Nashville. Dept. of Civil Engineering.
For primary bibliographic entry see Field 05D.
W73-09302

THERMAL REGENERATION ION EXCHANGE PROCESS WITH TRIALLYLAMINE POLYMERS,
ICA Australia Ltd., Melbourne. (Assigne).
H. A. J. Battaerd.
U. S. Patent No. 3,716,481, 10 p, 7 tab, 2 ref; Official Gazette of the United States Patent Office, Vol 907, No 2, p 480, February 13, 1973.

Descriptors: *Patents, *Demineralization, *Polymers, Water quality control, Chemical reactions, *Radiation, *Ion exchange, Desalination, Waste water treatment.
Identifiers: Triallylamine polymers.

Experimental results and tabulated data are given to illustrate the process set forth in this disclosure. It sets forth the process for the demineralization of water by a mixed bed of weak acid and weak base type ion exchange resins in which the mixed bed is regenerated by eluting it with water or a saline solution at a temperature exceeding that used during the adsorption stage. The weak base resin is at least one solid polymer comprising 3 and 100 mole percent of triallylamine and a balance of at least one allylamine comer. The polymer used is prepared by exposing the allylic monomer or monomers to a total dose of from 3 to 20 megarads of high enerby radiation at a temperature between -80 degC and+12 degC. (Sinha-OEIS)
W73-09386

ION EXCHANGE DEMINERALIZING SYSTEM,
CCI Aerospace Corp., Van Nuys, Calif. (Assignee).
A. M. Johnson.
U. S. Patent No 3,715,287, 7 p, 5 fig, 4 ref; Official Gazette of the United States Patent Office, Vol 907, No 1, p 172, February 6, 1973.

Descriptors: *Patents, *Demineralization, *Desalination, *Ion exchange, *Resins, Carbon, Salt, Salinity, Anion exchange, Cation exchange, Separation techniques, Water purification, Water quality control, *Waste water treatment.

This water treatment system consists of a preconditioner unit and a salinity buffer-storage tank unit which can be used separately or in a combination, and also in conjuntion with a demineralizing unit. The preconditioner unit consists of two containers which operate in push-pull arrangement. These tanks contain strong acid resins arranged with one in service to soften the raw incoming water and one in regeneration, short duty cycles. A combination storage tank-salinity buffer is provided for use with a demineralization system. This tank contains a mixture of anion responsive carbon particles and cation responsive carbon particles. These act as a buffer. The extent of salt take-up by the carbons is a function of the saltiness of the surrounding water in the storage tank, which receives water from the demineralizer unit. (Sinha-OEIS)
W73-09395

TUBULAR OSMOTIC MEMBRANE,
E. A. G. Hamer.
U. S. Patent No 3,715,036, 5 p, 6 fig, 2 ref; Official Gazette of the United States Patent Office, Vol 907, No 1, p 110, February 6, 1973.

Descriptors: *Patents, *Desalination apparatus, *Reverse osmosis, *Membranes, Equipment, Separation techniques.

This invention relates to osmotic membranes in a tubular form rendering them particularly useful as components in desalination apparatus and to methods for preparing such tubular membranes. The membrane is formed from a flat membrane which is curled and lap seamed. An adhesive is employed to seal the lap seam. The adhesive does not permanently bond together the lapped surfaces of the membrane until after the membrane is positioned in contact with the support tube and fluid pressure is applied to force the membrane into firm and conforming contact with the support member. (Sinha-OEIS)
W73-09396

VAPOR FREEZING TYPE DESALINATION METHOD AND APPARATUS,
Pacific Lighting Service Co., Los Angeles, Calif. (assignee)
R. E. Peck.
U. S. Patent No. 3,714,791, 6 p, 1 fig, 6 ref; Official Gazette of the United States Patent Office, Vol 907, No 1, p 44, February 6, 1973.

Descriptors: *Patents, *Desalination, *Evaporation, Heat exchangers, *Flash distillation, Sea water, Potable water, Equipment, Saling water, *Condensation, Brine.
Identifiers: *Vacuum freezing.

Cooled sea water is introduced into a low pressure chamber to form water vapor, ice, and cold brine. In the first step of this batch-type process there is a continous freezing of water vapor to form ice in a second chamber. In the second step relatively warmer sea water is introduced into a low pressure region to form water vapor and cold brine. The resulting water vapor is contacted with ice from the previous operation for condensing the water vapor and melting the ice. The apparatus comprises a vacuum freezing chamber, a spray nozzle near its upper portion and at its bottom a conduit controlled by a valve to lead brine from this chamber through a heat exchanger. A screen near the bottom of the chamber retains the ice formed in it. There is also provided a condensation chamber that is connected to the vacuum freezing chamber by means of a vapor conduit. This chamber contains refrigeration tubes that cause condensation of the water vapor. The refrigerant may be liquified natural gas or cold natural gas. After passing through the refrigeration tubes, the natural gas passes to an expansion turbine where the expansion produces a substantial amount of shaft horsepower used in the pumps needed in the operation. In the evaporation chamber, incoming seawater is flash evaporated and the resultant brine somewhat cooled and discharged. The water vapor formed here flows into the condensation and vacuum freezing chambers. (Sinha-OEIS)
W73-09402

FLASH EVAPORATOR,
P. R. Bom.
U. S. Patent No 3,713,989, 3 p, 4 fig, 8 ref; Official Gazette of the United States Patent Office, Vol 906, No 5, p 1711, January 30, 1973.

Descriptors: *Patents, *Flash distillation, Separation techniques, *Evaporators, *Distillation, *Water treatment, *Water purification, Water quality control, *Desalination.

The flash evaporator consists of an elongated vessel divided by partitions into a series of connected chambers. Cold raw water to be preheated is conveyed in cooling pipes successively through the vapor condensing portions of the chamber and then flashed in reversed flow through. The vapors which result are condensed and fall into a receptacle in each chamber. An exterior chamber is longitudinally divided into compartments by transverse partitions. A hood in each chamber has an opening over each receptacle to form a collecting space for non-condensable gases which may be present in the vapor. The non-condensable gases are removed from closed channels by a special conduit. It is possible to put one chamber out of operation for maintenance and repairs without interrupting the entire operation. (Sinha-OEIS)
W73-09404

GEOTHERMAL RESOURSE INVESTIGATIONS, IMPERIAL VALLEY, CALIFORNIA: DEVELOPMENTAL CONCEPTS.
Bureau of Reclamation, Washington, D.C.

January 1972. 58 p, 2 fig, 12 plate, 3 photo, 4 tab.

Descriptors: *Desalination, *Geothermal studies, *Water purification, *Thermal water, *California, Methodology, Steam, Groundwater, Surface water, Imported water, Injection wells, Subsidence, Water supply, Water quality, Costs, Electric power, Planning, Water utilization.
Identifiers: *Imperial Valley (Calif).

This report describes the need for augmenting the Colorado River with high quality water, and shows how the hot geothermal brines of the Imperial Valley, California, could be used to supply this need through multipurpose development of the resource. The hot brine will flash into a mixture of steam and brine when pressures are relieved by a well. This mixture will flow to the surface and can be used to produce not only electric power, but also desalted water and possibly mineral by-products. The program would be developed in three stages. The Research and Development Stage would, through extensive geological, geophysical, and water chemistry investigations, determine the potential and extent of the geothermal resource. This program would require expenditures of about $16 million over a 7-year period. The Demonstration Stage would demonstrate the feasibility of large-scale development. The concept would use a local salt or brackish water supply for replacement fluids such as Salton Sea, Wellton-Mohawk Drain, or groundwater. The magnitude of the development would be about 100,000 acre-feet of fresh water per year and about 400-500 megawatts of electric power. Cost of desalted water produced and delivered would range from $85 to $130 per acre-foot and electric energy would be produced at 3 to 5 mills per kwh. The Large-Scale Development Stage would augment the Colorado River by delivering as much as 2.5 million acre-feet of desalted water annually with electric power production of about 10,000 mw. This would require importation of Pacific Ocean or Gulf of California water for replacement fluids. Cost of desalted water produced and delivered would range from $100 to $150 per acre-foot, and electric energy would be produced at 3 to 5 mills per kwh. (Woodard-USGS)
W73-09439

ECONOMIC STUDY OF THE VACUUM FREEZING EJECTOR ABSORPTION DESALTING PROCESS,
Colt Industries, Inc., Beloit, Wis.
J. Koretchko.
Available from the National Technical Information Service as PB-217 616, $6.00 in paper copy, $0.95 in microfiche. Office of Saline Water Research and Development Progress Report No 833, July 1972. 162 p, 22 fig, 5 tab, 5 ref, 3 append. 14-30-2988.

Descriptors: *Desalination, Desalination apparatus, Freezing, Economics, *Cost Analysis, Crystallization, Absorption.
Identifiers: *Vacuum freezing.

The work effort consisted of the following: (1) A 1.0 MGD reference desalting plant was designed in sufficient detail to demonstrate plant operability and allow accurate estimating of the capital cost and the water cost. Assuming a power rate of 8.5 mills/Kw-hr, a steam rate of 50 cents/1,000,000 Btu, a fixed charge rate of seven pervent, and a 90 percent plant load factor, the following table shows the cost of desalted water for the 1.0 MGD reference plant as the feed salinity and temperature are varied: Feed Salinity, ppm 5,000, 20,000, 35,000 Feed Temperature, F 60 to 80, 60 to 80, 60 to 80, Water Cost Range, cent/Kgal 63 to 72, 68 to 79, 72 to 84. (2) Nomographs were developed to obtain the minimum water cost for a 1.0 MGD plant with the following parametric ranges: 60 to 80 F feed temperature, 0.5 to 1.2 cents/Kw-hr power rate, 30 to 80 cents/1,000,000 Btu steam rate, 7 to 14% fixed charge rate, 60 to 90% plant load factor, 0.5 to 3.5% feed salinity. For these parametric ranges, the water cost range is 53 to 125 cents/Kgal. (OSW)
W73-09623

POROUS MATERIALS FOR REVERSE OSMOSIS MEMBRANES: THEORY AND EXPERIMENT,
Massachusetts Inst. of Tech., Cambridge.
G. Jacazio, R. F. Probstein, A. A. Sonin, and D. Yung.
Available from the National Technical Information Service as PB-214 972, $3.00 in paper copy, $0.95 in microfiche. Office of Saline Water Research and Development Progress Report No 809, December 1972. 43 p, 7 fig, 18 ref. 14-30-2575.

Descriptors: *Desalination, *Membranes, *Porous media, *Pore Radius, *Reverse osmosis.
Identifiers: *Porous membranes, Compacted clay, Debye length, Peclet No., Salt rejection, Zeta potential, Filtration velocity.

A theory is presented for the salt rejecting characteristics of porous membranes, whose pore size is large compared with molecular dimensions, under reverse osmosis conditions where the saline solution is moved through the membrane pores by an applied pressure gradient. The salt rejection mechanism is assumed to be an electrokinetic one resulting from charge built up on the interior surfaces of the material when in contact with the salt solution. The performance of the porous material is shown to depend on three parameters: the ratio of Debye length to effective pore radius; a dimensionless wall potential related to the zeta potential; and a Peclet number based on the filtration velocity through the pore, the membrane thickness and the diffusion coefficient of the salt in the water. A universal correlation is given for the fractional salt rejection in terms of the three membrane parameters. Experiments were carried out on the salt rejecting properties of compacted clay, through which saline solutions were forced under high pressures. Absolute comparisons between the experimentally determined and theoretically predicted rejection characteristics are shown to be excellent. A comparison of the theory with experimental data of Michelsen and Harriott on cellophane also shows very good agreement. (OSW)
W73-09624

EVAPORATION OF WATER FROM AQUEOUS INTERFACES,
Rochester Inst. of Tech., N.Y.
K. Hickman, I. White, and W. Kayser.
Available from the National Technical Information Service as PB-215 184, $3.00 in paper copy, $0.95 in microfiche. Office of Saline Water Research and Development Progress Report, No 808, April, 1972. 110 p, 39 fig, 9 tab, 48 ref. 14-30-2572.

Descriptors: *Air-water interface, *Condensation, *Mass transfer, *Evaporation, *Boule flotation, *Desalination, Distillation, Films Surfaces.
Identifiers: Surface purification, Transfer coefficients.

The effects of trace contaminants on the thermal properties of superheated water were measured. When water covered by steam is maintained at the boiling point, it super-heats to a consistent value for a given input of additional heat energy. This value was lowest for the purest water and increased when trace amounts of long chain organic substances were added. Typifying the additives, cis-12-docosenoic acid (erucic) and n-tetracosenoic acid (lignoceric) gave a significant (11%) increase in superheat when one molecule of fatty acid was supplied for each 4600 A2 of exposed water surface. With 25 times the quantity, i.e., 1 molecule for each 180A2 of water surface, a limiting increase of 40% - 47% was attained. Suppression of evaporation appears to be due to stabilization of the sub-surface, which prevents evaporatively cooled layers descending, rather than to a reduction of molecular permeability to steam. Experiments were also carried out on the schizoid

surface, i.e., the separation of an evaporating liquid surface into two distinct boiling modes, one surface seething rapidly, the other scarcely at all. In addition measurements are reported on the electrical conductivity of water near the normal boiling point. (OSW)
W73-09625

DEVELOPMENT OF IMPROVED POROUS SUPPORT FOR REVERSE OSMOSIS MEMBRANES,
Uniroyal, Inc., Wayne, N.J.
H. P. Smith, and K. K. Sirkar.
Available from the National Technical Information Service as PB-215 077, $3.00 in paper copy, $0.95 in microfiche. Office of Saline Water Research and Development Progress Report, No 807, December, 1972. 55 p, 17 tab, 4 fig, 3 ref. 14-30-2564.

Descriptors: *Reverse Osmosis, *Membranes, Permeability, *Desalination, Tensile strength, Compressive strength, Heat treatment, Permselective membranes.
Identifiers: *Membrane support, *Porous polyethylene, *Flux, *Salt rejection, Braided wire reinforcement, Extrusion of tublar support.

The composite support that evolved from the contract work consists of a porous polyethylene tube with an external braid of stainless steel wire. The porous tube is made by mixing powdered polyethylene and plasticized polyethylene oxide, extruding the compound on a mandrel in the form of a tube, wrapping the tube with a braid of stainless steel wire, sintering the tube, removing the mandrel, and extracting the polyethylene oxide to leave open pores in the polyethylene. After application of end fittings, a membrane is inserted to form the complete desalination unit. A composite support of 1/2 inch inside diameter containing a CA membrane heat treated in situ was subjected to desalination testing at pressures of 275, 600, 700, 900, 1200 and 1500 psi. Salt rejection and water flow were measured. At 1500 psi the flow was 7.7 GFD with 53.6% salt rejection. When the pressure was dropped back to 300 psi, flow and salt rejection returned to approximately the same values observed at that pressure previously, i.e., before using the elevated pressures. It is concluded that there was no gross damage to the membrane in the test of the reinforced support. It is difficult to conclude how much of the non-ideal behavior of the composite is due to deficiencies in the membrane itself. (OSW)
W73-09626

ANNUAL REPORT (FY 1971) WEBSTER TEST FACILITY AND ELECTRODIALYSIS TEST-BED PLANT, WEBSTER, SOUTH DAKOTA,
Mason-Rust, Lexington, Ky.
J. S. Nordin, N. Call, R. A. Ackerman, D. R. Bogue, and J. E. Gugeler.
Available from the National Technical Information Service as PB-215 183, $3.00 in paper copy, $0.95 in microfiche. Office of Saline Water Research and Development Progress Report, No 805, February, 1972. 130 p, 20 fig, 23 tab, 11 ref. 14-01-0001-2263.

Descriptors: *Electrodialysis, *Desalination, Economics, *Brackish water, *Water purification, South Dakota, Potable water.

The Asahi Chemical Industry (ACI)-designed Test Bed Plant completed its ninth year of operation as the only Office of Saline Water Test Bed Plant for converting brackish water to potable water by the electrodialysis process. The plant has a production capacity of 325,000 gallons per day. Since May 1969, the feed water has been pretreated using lime softening to reduce the level of iron, maganese, and hardness. During FY 70, the performance of the Test Bed Plant was evaluated on the basis of anion membranes subject to organic fouling, using carbon filtration pretreatment for feed water organic removal to compensate for the anion fouling. The major goal during this report period was to evaluate performance of the Test Bed Plant using an organic foulant-resistant membrane without carbon pretreatment. The membrane combination tested was the Tokuyama Soda Company, Ltd. (Japan) AF-4T anion and CH-2T cation permeable membranes. These membranes were chosen on the basis of successful pilot plant results completed in FY 1970. (OSW)
W73-09627

MODEL STUDIES OF OUTFALL SYSTEMS FOR DESALTING PLANTS (PART III - NUMERICAL SIMULATION AND DESIGN CONSIDERATIONS),
Dow Chemical Co., Freeport, Tex.
For primary bibliographic entry see Field 05E.
W73-09628

A STUDY OF THE EFFECT OF DESALINATION PLANT EFFLUENTS ON MARINE BENTHIC ORGANISMS,
Dow Chemical Co., Freeport, Tex.
E. F. Mandelli, and W. F. McIlhenny.
Available from the National Technical Information Service as PB-215 182, $3.00 in paper copy, $0.95 in microfiche. Office of Saline Water Research and Development Progress Report, No 803, October, 1971. 154 p, 56 tab, 30 fig, 64 ref. 14-01-0001-2253.

Descriptors: *Brine disposal, Discharge (Water), *Desalination plants, *Copper, *Benthos, Copper compounds, Thermal stress, Salinity, Marine animals, Marine fish, Shrimp, Oysters, Effluents, Toxicity, Larvae.
Identifiers: *Marine benthic organism.

The objective was to determine the effects of increased salinities, temperatures and copper concentrations on marine benthic organisms as a result of desalination brine disposal into coastal and estuarine environments. In order to evaluate the effects of the seawater-brine mixtures on typical benthic fauna, the American oyster, Crassostrea virginica (Gmelin) and the penaeid shrimps, Penaeus aztecus (Ives) and Penaeus duorarum (Burkenroad), were selected and long-term multivariate, seasonal experiments on juvenile and adult speciments were conducted. The effects of the mixtures of brine and seawater were also tested on the eggs and larval stages of these marine benthic organisms, using short-term bioassay tests. Seawater-brine dilutions containing 0.02 ppm of total copper (with about 0.01 ppm present as ionic copper) had an acute toxic effect on C. virginica larvae. The mixture was also harmful to juvenile and adult specimens due to the copper accumulation in their tissues that impaired the metabolic functions of the organism and lowered the tolerance level to other environmental stresses. With normal copper concentrations for seawater, an increase of temperature of salinity to 10 percent above the environmental values enhanced the incidence level of the pathogenic fungus, Labyrinthomyxa marina, letha to adult oysters. Penaeid shrimp exhibited a greater tolerance to the presence of the desalination brine in the coastal seawater. Larval stages of shrimp were able to develop normally in the presence of 0.025 ppm of dissolved copper, at salinities up to 35,000 ppm and temperatures of 81.5F. A copper concentration of 0.05 ppm in the seawater-brine mixtures was lethal to the nauplial, protozoal and mysis stages of P. aztecus and P. duorarum. (OSW)
W73-09629

A PROGRAMMED INDIRECT FREEZING PROCESS WITH IN-SITU WASHING AND IN-SITU MELTING OPERATIONS,
Denver Research Inst., Colo.
Chen-yen Cheng, Gary Van Riper, and V. Grant Fox.
Available from the National Technical Information Service as PB-215 181, $3.00 in paper copy, $0.95 in microfiche. Office of Saline Water Research and Development Progress Report, No 802, December 1972. 59 p, 34 fig, 4 ref, 2 tab. 14-30-2526.

Descriptors: *Desalination, *Freezing, *Ice, Crystallization.
Identifiers: Indirect freeze desalination.

A programmed indirect freezing process with the following features has been studied: (1) It is a cyclic process to be conducted in a unified freezer-melter and each cycle consists of a feeding step, a freezing step, an in-situ washing step and an in-situ melting step. Because of the in-situ washing and in-situ melting steps, problems associated with scraping of ice and transfer of ice have been eliminated. (2) By an appropriate coordination of the design of a unified freezer-melter and the control of the freezing step, a consolidated ice bed of a structure similar to that of a hydraulic ice washing column used in a direct contact freezing process is obtained at the end of each freezing step. Therefore, an efficient washing of ice bed can be accomplished with a small loss of wash water. (3) New ways of accomplishing heat reuse, such as the use of a working medium which undergoes solid equal liquid transformations and the method based on low pressure freezing and high pressure melting operations, have been incorporated in the process to significantly reduce the energy requirement. (OSW)
W73-09630

DEVELOPMENT OF LOW COST, CORROSION RESISTANT ALLOYS AND OF ELECTROCHEMICAL TECHNIQUES FOR THEIR PROTECTION IN DESALINATION SYSTEMS,
Tyco Labs., Inc., Waltham, Mass.
For primary bibliographic entry see Field 08G.
W73-09631

MODIFIED POLYOLEFIN MEMBRANES FOR USE IN REVERSE OSMOSIS OR PRESSURE DIALYSIS,
Southwest Research Inst., San Antonio, Tex.
H. F. Hamil.
Available from the National Technical Information Service as PB-215 180, $3.00 in paper copy, $0.95 in microfiche. Office of Saline Water Research and Development Progress Report, No 800, July 1972. 45 p, 19 tab. 14-30-2719.

Descriptors: *Membranes, *Reverse osmosis, *Ion transport, Cation Exchange, Anion exchange, Permeability, *Desalination, Polyelectrolytes.
Identifiers: *Polyolefin membranes, Pressure dialysis, Mosaic membranes, Grafting, Polysalt membranes, Hydrophilic membranes.

Preparation of crude mosaic membranes by selective areal grafting of acidic and basic monomers onto polyethylene film was accomplished. Preparative procedures evaluated included direct grafting, indirect grafting, and peroxidation grafting using gamma irradiation, beta irradiation, and both high and low energy electrons. The peroxidation technique using electron irradiation was most successful. A better procedure for mosaic membrane preparation involved chemical addition of cationic and anionic sites to a styrene grafted polyethylene film using a masking technique to obtain the desired control of charge domain size and spacing. Efforts to prepare neutral hydrophilic membranes by irradiation grafting of polyethylene film with hydrophilic monomers failed to give membranes with acceptable reverse osmosis characteristics. Preparation of random polysalt membranes by sequential grafting of polyethylene with acid and basic monemers afforded membranes which displayed selective transport porperties for different ions when evaluated with mixed ionic species in the feed. (OSW)
W73-09632

FREEZING PROCESS STUDIES: ANISOTROPIC ICE BEDS, BUTANE FLASHING, AND BUTANE BOILING,
Syracuse Univ., N.Y. Dept. of Chemical Engineering and Materials Science.
J. S. Huang, A. V. Naimpally, Y. F. Rmissirlis, P. A. Rice, and A. J. Barduhn.
Available from the National Technical Information Service as PB-215 178, $3.00 in paper copy, $0.95 in microfiche. Office of Saline Water Research and Development Progress Report, No 797, December 1972. 126 p, 30 fig, 15 tab, 64 ref. 14-30-2578.

Descriptors: *Desalination, *Freezing, *Ice, *Crystallization, Permeability.
Identifiers: *Anisotropy, *Ice beds, Butane.

This report covers research on subjects: Anisotropy in Beds of Ice Crystals, Removal of Dissolved Butane from Water by Vacuum Flashing, and the Rate of Evaporation of Butane Droplets Suspended in Water. Ice beds tested showed 'apparent' anisotropy in which the effective ratio of horizontal to vertical permeability was about 1.2. It is also shown that the effect on wash column design of increased ratio (horizontal/vertical) permeability is to decrease the slurry pumping cost slightly and to give more operating flexibility by lowering the brine crown height. It does not increase the capacity of the column to wash ice, however. Experimental and mathematical analysis of vacuum flashing of butane from water solutions is reported. One model assumed the drop completely mixed with the entire resistance to the butane evaporation residing in the gas phase. Experimental evaporation rates were much slower than predicted by this mixed drop model. The second model assumed no gas phase resistance and the drop to be stagnant with liquid diffusion controlling. The experimental rates were much higher than predicted by this stagnant drop model. A model with internal drop circulation was not attempted but has a better chance of agreeing with the experimental data. The first stages of evaporation of a butane drop formed beneath the tip of a vertical capillary tube submerged in warm water were observed by cinematography and the rates of evaporation, areas, delta T driving forces, and heat transfer coefficients were measured as a function of time. The process was observed until 1% of the drop was evaporated. Bubble growth rates varied as the square root of time and heat transfer coefficients were as large as 200 to 1400 kcal/ (cm2hr. deg C) depending on delta T. Lower heat ransfer coefficients corresponded to higher delta T's. (OSW)
W73-09633

NEW BORON-NITROGEN POLYIONS AND RELATED COMPOUNDS FOR ION EXCHANGE,
Texas Christian Univ., Fort Worth.
H. C. Kelly.
Available from the National Technical Information Service as PB-215 177, $3.00 in paper copy, $0.95 in microfiche. Office of Saline Water Research and Development Progress Report, No 794, September, 1972. 10 p, 2 tab, 7 ref. 14-01-0001-2175.

Descriptors: *Ion exchange, *Anion exchange, *Desalination, Ions, Salts, Boron, *Electrolysis.
Identifiers: *Boron-nitrogen polyions.

Both heterogeneous nucleophilic displacement and vinyl polymerization have proven successful as routes for the synthesis of salts of boron containing cations and polymeric boron (n+) ion salts. Although the study was limited in the number of electrolytic species synthesized, the potential for synthesis is great and a wide variety of cationic species are possible through both preparative routes. In general, the products obtained are high melting and relatively chemically inert. The question of water solubility is of primary consideration for any potential application of such boron cation salts as membrane components. It has been found that iodide salts of the cations have a high solubility in aqueous media, however, relatively insoluble derivatives can be obtained in the form of hexafluorophosphate, tetraphenylborate or other salts. Even if halide salts should generally be found to exhibit significant solubility it is feasible that they could be incorporated into an appropriate matrix in a membrane permitting utilization of the cationic system for anion exchange. A study of the evaluation of such species as potential components of ion exchange membranes was not within the scope of the study. (OSW)
W73-09634

EFFECT OF TURBULENCE PROMOTERS ON LOCAL MASS TRANSFER - SECOND REPORT,
Ionics, Inc., Watertown, Mass.
L. Marincic, and F. B. Leitz.
Availablefrom the National Technical Information Service as PB-215 176, $3.00 in paper copy, $0.95 in microfiche. Office of Saline Water Research and Development Progress Report, No 793, July 1972. 150 p, 112 fig, 6 tab, 4 ref. 14-01-0001-2174.

Descriptors: Desalination, *Turbulence, Reynolds number, *Electrodialysis, Laminar flow, *Mass transfer.
Identifiers: *Turbulence promoters, Segmented electrode, Polarization.

An instrument has been developed for detailed measurement of mass transport in channels similar to those in electrodialysis equipment. Mass transport promoters are commonly used to increase the usable current density. The final objective of this investigation is the development of techniques for optimization of such promoters. The apparatus consists of an electrochemical cell containing a segmented cathode, which is so connected that current through each segment can be measured separately. The aggregate of these measurement provides a mass transfer profile over the surface of this cathode. To test the apparatus, runs were made under conditions which matched as closely as possible the conditions for which a mathematical solution to the mass transfer equations could be obtained. The conditions selected were well developed laminar flow in a rectangular channel of constant cross-section with the concentration gradient beginning at the leading edge of the electrode. The channel width was 25 times the channel height and data only from the middle portion of the channel were used so this could reasonably be compared to flow between parallel plates of infinite width. The data yielded a diffusion constant which correlated closely to values reported in the literature. A full investigation was made of single rectangular, circular, and triangular turbulence promoters placed against both polarized and nonpolarized surfaces. Reynolds number ranged across most of the laminar flow region. The data thus obtained made it possible to construct a unified picture of flow and mass transfer with each promoter. Each promoter shape has its own characteristics. These data show an unexpected phenomenon. Ahead of the rectangular promoter, when it is placed against the polarized surface, the current density reaches a minimum about 10% below the unpromoted value. This minimum was observed in all runs with a promoter having a bluff leading edge and is apparently caused by pulling away of the streamlines from the polarized surface in a region where current passes unimpeded through the solution. This behavior is not observed with the triangular promoter or with promoters in other locations. (OSW)
W73-09635

SYNTHESIS OF BLOCK POLYMERS,
Iowa Univ., Iowa City.
J. K. Stille, M. Kamachi, and M. Kurihara.
Available from the National Technical Information Service as PB-215 172, $3.00 in paper copy, $0.95 in microfiche. Office of Saline Water Research and Development Progress Report, No 792, February 1972. 81 p, 43 fig, 9 tab, 26 ref. 14-01-0001-2157.

Descriptors: *Desalination, *Membranes, *Reverse osmosis, Permeability, Cation exchange, Anion exchange.
Identifiers: *Block copolymers, *Polymerization, Pressure dialysis, 'Living' polymers, Salt rejection, Polymer properties, Film forming polymers.

Block copolymers of 2-vinylpyridine and various methacrylic and acid esters were synthesized by 'living' anionic polymerization at low temperature. Block copolymers having uniform block lengths of controlled molecular weight and with a narrow molecular weight distribution were prepared by this method. Base hydrolysis of the ester sequence produced the block copolymer of 2-vinylpyridine and methacrylic acid. Quaternization of the 2-vinylpridine block followed by dehydrohalogenation yields a polymer containing anionic and cationic blocks. The physical properties and salt rejection of these polymers has also been studied. (OSW)
W73-09636

THE USE OF POLYMERIC GELS TO REDUCE COMPACTION IN CELLULOSE ACETATE REVERSE OSMOSIS MEMBRANES,
Carnegie-Mellon Univ., Pittsburg, Pa.
S. L. Rosen, C. Irani, and L. Baayens.
Available from the National Technical Information Service as PB-215 175, $3.00 in paper copy, $0.95 in microfiche. Office of Saline Water Research and Development Progress Report, No 791, December 1972. 43 p, 15 fig, 17 tab. 14-01-0001-2125.

Descriptors: *Reverse osmosis, *Membranes, *Compaction, Permeability, Saline water, Pressure, *Desalination.
Identifiers: *Cellulose acetate membranes, Crosslinking, Polymeric gels, Flux decline, Salt rejection, Flux, Emulsion gels, Solution gels, Membrane hydrodynamics.

Polymeric gels were synthesized by crosslinking cellulose acetate in emulsion and solution reactions. These gels were used to replace a portion of the linear cellulose acetate in a standard Manjikian-type asymmetric membrane, with the aim of reducing long-term flux decline. Membranes were tested on a 3.5% NaCl solution at 1500 psi for up to 400 hours. The solution polymerization technique was capable of producing a wide range of gel characteristics. Appropriate solution polymerized gels reduced the compaction slope by about a factor of two, without seriously comprising salt rejections, and by casting somewhat thinner membranes, the fluxes could be maintained at the control level. Experiments on a series of dense cellulose acetate membranes of different thickness revealed that essentially all the hydrodynamic resistance in an asymmetric membrane is in the dense, salt-rejecting layer, and further indicate that this is the locus of flux decline processes. (OSW)
W73-09637

DEVELOPMENT OF HOLLOW FINE FIBER REVERSE OSMOSIS DESALINATION SYSTEMS,
Monsanto Research Corp., Durham, N.C.
T. A. Orofino.
Available from the National Technical Information Service as PB-215 179, $3.00 in paper copy, $0.95 in microfiche. Office of Saline Water Research and Development Progress Report, No 789, April 1972. 82 p, 18 fig, 40 tab, 7 ref. 14-30-2585.

Descriptors: *Desalination, *Reverse osmosis, *Membranes, Membrane process, Hollow fibers, *Desalination apparatus.
Identifiers: *Fiber Bundles, Laminar-Flow hydrodynamics.

gel. The final phase of the work involved the extension of the grafting technology to other base polymer systems. Acrylic acid, ethylacrylate, and 2-vinyl pyridine were grafted to poly (vinylchloride) films. (OSW)
W73-09639

HYPERFILTRATION IN POLYELECTROLYTE MEMBRANES,
Weizmann Inst., of Science, Rehoveth (Israel).
D. Vofsi, and O. Kedem.
Available from the National Technical Information Service as PB-215 173, $3.00 in paper copy, $0.95 in microfiche. Office of Saline Water Research and Development Progress Report, No 787, December 1972. 36 p, 9 tab, 6 fig, 10 ref. 14-01-0001-961.

Descriptors: *Reverse osmosis, *Membranes, *Permeability, *Ions, Ion transport, Desalination.
Identifiers: *Hyperfiltration, *Polyelectrolyte membranes, Salt rejection, Charged membranes, Streaming potential, Asymmetric membranes, Maleinated acetyl-cellulose, TMS model.

Hyperfiltration (Reverse osmosis) research in the following areas is reported: asymmetric membranes prepared from maleinated acetyl cellulose; hyperfiltration in charged membranes: prediction of salt rejection from equilbrium measurements; and streaming potential during hyperfiltration. (OSW)
W73-09640

SPRAY FREEZING, DECANTING, AND HYDROLYSIS AS RELATED TO SECONDARY REFRIGERANT FREEZING,
Avco Corp, Wilmington, Mass. Avco Space Systems Div.
Wallace E. Johnson, James H. Fraser, Walter E. Gibson, Anthony P. Modica, and Gershon Grossman.
Available from the National Technical Information Service as PB-215 036, $3.00 in paper copy, $0.95 in microfiche. Office of Saline Water Research and Development Progress Report No. 786, June 1972. 137 p, 20 tab, 48 fig, 57 ref. 14-30-2770.

Descriptors: *Desalination, *Crystallization, *Freezing, *Ice, Solubility, Hydrolysis.
Identifiers: Secondary refrigerant freezing, Spray freezer, Decanter, Freon 113, Freon 114.

This report describes research performed on a secondary refrigerant desalination process, specifically freezing, decanting and Freon hydrolysis. Research on a spray freezer is reported. High rates of heat-transfer between feed and refrigerant are effected in a finely dispersed well-mixed spray of the two fluids. Reasonable crystal quality is obtained. The performance of a triplet nozzle configuration is studied from the viewpoint of production rate and permeability. Theory is developed to correlate and explain the experimental data. A theoretical model describes decanting of a stationary, initially uniform, mixture of Freon and water in a vertical column. The analysis provides a simple experimental method to determine droplet size distribution in a liquid-liquid dispersion band and measurement of its decanting rate. Mixtures of water and Freon 113 were studied experimentally. Droplet size distribution was defined in terms of three characteristic parameters. The solubility and hydrolysis rate of Freon 114 have been measured over a range of temperatures and pressures. The apparent first order hydrolysis rate constant of Freon 114 is shown to be sufficiently low that hydrolysis losses can be neglected. The rate of hydrolysis in saline water was somewhat accelerated by the presence of aluminum. A general theory was developed to predict the first order rate constant for hydrolysis of Freon refrigerants. (OSW)
W73-09641

3B. Water Yield Improvement

WEATHER MODIFICATION: PRECIPITATION INDUCEMENT, A BIBLIOGRAPHY.
Office of Water Resources Research, Washington, D.C.

Available from the National Technical Information Service as PB-220 348, $6.75 in paper copy, $0.95 in microfiche. Water Resources Scientific Information Center Report WRSIC 73-212, March 1973. 246 p.

Descriptors: *Weather modification, *Bibliographies, *Artificial precipitation, *Cloud seeding, *Precipitation (Atmospheric), Cloud physics, Meteorology, Climatology, Rainfall, Water yield improvement.

This report, containing 169 abstracts, is another in a series of planned bibliographies in water resources to be produced from the information base comprising SELECTED WATER RESOURCES ABSTRACTS (SWRA). At the time of search for this bibliography, the data base had 53,230 abstracts covering SWRA through February 15, 1973 (Volume 6, Number 4). Author and subject indexes are included.
W73-09114

INTERCEPTION LOSS IN LOBLOLLY PINE STANDS OF THE SOUTH CAROLINA PIEDMONT,
Forest Service (USDA), Asheville, N.C. Southeastern Forest Experiment Station.
W. T. Swank, B. G. Norbert, and J. D. Helvey.
Journal of Soil and Water Conservation, Vol 27, p 160-164, July-August 1972. 5 tab, 17 ref.

Descriptors: *Interception, *Streamflow, *Throughfall, *Loblolly pine trees, *Mixed forests, *South Carolina, Water resources, Watershed management, Water yield improvement, Regression analysis.
Identifiers: South Carolina Piedmont.

Annual interception loss was measured in 5-, 10-, 20-, and 30-year-old loblolly pine stands and in a mature hardwood-pine forest in the Piedmont of South Carolina. Interception loss for the loblolly pine stands was estimated to be 14, 22, 18, and 18 percent of annual precipitation (54 inches). Annual interception loss from the hardwood-pine stand was similar to that of the pine stands. However, on the average, the loss of water intercepted annually by loblolly pine appeared to be about 4 inches greater than the loss estimated from a number of hardwood studies. Where extensive conversions of hardwood to loblolly pine occur, significant reductions in the amount of water available for streamflow or groundwater should be expected.
W73-09383

GEOTHERMAL RESOURSE INVESTIGATIONS, IMPERIAL VALLEY, CALIFORNIA: DEVELOPMENTAL CONCEPTS.
Bureau of Reclamation, Washington, D.C.
For primary bibliographic entry see Field 03A.
W73-09439

CRITICAL SOIL MOISTURE LEVELS FOR FIELD PLANTING FOURWING SALTBUSH,
Forest Service (USDA), Albuquerque, N. Mex. Rocky Mountain Forest and Range Experiment Station.
For primary bibliographic entry see Field 02G.
W73-09683

3C. Use of Water of Impaired Quality

USE OF NATURALLY IMPAIRED WATER, A BIBLIOGRAPHY.
Office of Water Resources Research, Washington, D.C.

Available from the National Technical Information Service as PB-220 350, $6.00 in paper copy, $0.95 in microfiche. Water Resources Scientific Information Center Report WRSIC 73-217, May 1973. 364 p.

Descriptors: *Saline water, *Saline soils, *Salinity, *Irrigation water, *Bibliographies, *Impaired water quality, Salt tolerance, Alkaline soils, Arid lands, Crop production, Desalination, Economic evaluations, Drainage, Groundwater, Land reclamation, Leaching, Osmotic pressure, Plant growth, Sodium.

This report, containing 246 abstracts, is another in a series of planned bibliographies in water resources to be produced from the information base comprising SELECTED WATER RESOURCES ABSTRACTS (SWRA). At the time of search for this bibliography, the data base had 53,230 abstracts covering SWRA through February 15, 1973 (Volume 6, Number 4). Author and subject indexes are included.
W73-09116

PRODUCTION OF EIGHT CROPPING SYSTEMS ON SALINE AND GLEYSOLIC ALLUVIUM SOIL, THE PAS, MANITOBA,
E. D. Spratt, B. J. Gorby, and W. S. Ferguson.
Can J Soil Sci. Vol 52, No 2, p 187-193. 1972. Illus.
Identifiers: Alfalfa, Brome, *Canada, Clover, *Gleysolic alluvium soils, Grass, Manitoba, Pas, *Saline soil, Wheat, Crop production.

In 1961, 8 crop rotations were established on a recently drained alluvium soil (5 ha in size). Within the rotation strips the average initial electrical conductivity (ECe) of the 0-30-cm depth ranged from 0.3-28.6 mmhos/cm. Three rotations contained only wheat (with various amounts of summer fallow), 3 contained only forage crops (alfalfa or bromegrass or both), and 2 contained both wheat and forage crops (clover or brome-alfalfa). From 1961-1969 the wheat-clover and continuous alfalfa rotations gave the best yields, thus giving the highest net returns in dollars. In 1969, better yields of wheat were obtained after the forage crop rotations than after the grain rotations. Ground water levels and subsurface soil salinity remained relatively constant throughout the study.--Copyright 1973, Biological Abstracts, Inc.
W73-09238

FUNDAMENTAL PRINCIPLES IN THE STUDY OF WATER AND SALT BALANCES OF GROUNDWATER IN IRRIGATED REGIONS (OSNOVNYYE PRINTSIPY IZUCHENIYA VODNO-SOLEVOGO BALANSA GRUNTOVYKH VOD OROSHAYEMYKH TERRITORIY),
Institute of Hydrogeology and Engineering Geology, Tashkent (USSR).
For primary bibliographic entry see Field 07B.
W73-09464

3D. Conservation in Domestic and Municipal Use

GEOLOGY AND LAND RECLAMATION,
Kent State Univ., Ohio. Dept. of Geology.
For primary bibliographic entry see Field 04A.
W73-09533

ENVIRONMENTAL AND ECONOMIC IMPACT OF RAPID GROWTH ON A RURAL AREA: PALM COAST,
Southeastern Environmental Services, Jacksonville, Fla.
For primary bibliographic entry see Field 05G.
W73-09656

SHORTCUT PIPELINE MODIFICATION CONTRA COSTA CANAL UNIT, CENTRAL VALLEY PROJECT, CALIFORNIA (FINAL ENVIRONMENTAL STATEMENT).
Bureau of Reclamation, Washington, D.C.
For primary bibliographic entry see Field 08A.
W73-09696

3E. Conservation in Industry

THE GEOPHYSICAL DIRECTORY, (TWENTY-EIGHTH EDITION 1973).

Globe Universal Sciences, Inc., Houston, Texas, 1973, 405 p.

Descriptors: *Exploration, Oil industry, *Geophysics, Seismology, Explosives, Drilling, Core drilling, Drilling equipment, Drill holes, Mineral industry, Soil analysis, Gravimetric analysis, Radioactivity, Radioactive well logging, *Logging (Recording).
Identifiers: Consulting services, Contractors.

A listing of all companies and individuals directly connected with, or engaged in, geophysical exploration is presented. Included are listings in the fields of: Aircraft Services, Boats and Marine Equipment, Consultants (Geophysical), Core Drilling, Data Exchange Service, Drilling Bits, Electronic Supplies, Explosives, Geophysical Instrumentation, Logging services, Mining, Non-Dynamite Energy Source Equipment, Oceanographic Instruments, Radioactive Surveying, Seismology, Soil Analysis, and Supplies (Geophysical). Names, addresses, and company affiliations are given (in the case of individuals.) (Campbell-NWWA)
W73-09256

INTERINDUSTRY ECONOMIC MODELS AND REGIONAL WATER RESOURCES ANALYSES,
For primary bibliographic entry see Field 06B.
W73-09409

THE ECONOMICS OF INDUSTRIAL WASTE TREATMENT,
Environmental Control Consultancy Services Ltd., London (England).
For primary bibliographic entry see Field 05D.
W73-09410

THE EFFECT OF MIXING AND GRAVITATIONAL SEGREGATION BETWEEN NATURAL GAS AND INERT CUSHION GAS ON THE RECOVERY OF GAS FROM HORIZONTAL STORAGE AQUIFERS,
New Mexico Inst. of Mining and Technology, Socorro.
For primary bibliographic entry see Field 04B.
W73-09589

MINERALS AND DEPOSITS ON STATE LANDS--SALE OR LEASE.
For primary bibliographic entry see Field 06E.
W73-09735

Identifiers: Copper, Environment, Manganese, Moisture, Oil, Phosphorus, Potassium, Soils, *Soybeans, Yield, *Soil temperature.

Under comparable field conditions of air temperature and solar radiation, yield of soybeans increased 43.4% when the day-degrees units of the soil were raised from a seasonal value of 859 to 1822 (>5 degree C) and yield decreased 82.4% when the day-degrees for the same period were lowered to 408 for the 20-cm depth. These heat values from July 10 to September 21 correspond to mean daily soil temperatures of 11.2, 17.7, and 31.2 degrees C. The change in yield represents a reduction on the cold soils of 208 kg/ha (3.1 bu/acre)/1 degree C below the seasonal temperature, and an increase on the warm plots of 54 kg/ha/1 degree C (unfertilized) above the seasonal temperature. In general, yield was related linearly to the reciprocal of the temperature and of the day-degrees. Oil concentration varied little among the 3 soil temperatures, although the I number decreased and the percentage protein increased with higher soil temperature. Coinciding with the high yields on the warm soils was a high concentration of P in the foliage, e.g., 0.15% P at 11.2 degrees C and 0.42% at 31.9 degrees C. During early growth, concentration of K in the plant material increased and that of Mn and Cu was reduced with high soil temperature. This resulted in a greater removal of P and K from the warm soil than from the cool soil, and little difference in removal of Mn and Cu between the low and high temperatures.--Copyright 1973, Biological Abstracts, Inc.
W73-09307

THE CLIMATE OF AN ORANGE ORCHARD: PHYSICAL CHARACTERISTICS AND MICROCLIMATE RELATIONSHIPS,
Commonwealth Scientific and Industrial Research Organization, Canberra (Australia). Div. of Land Research.
J. D. Kalma, and G. Stanhill.
Agric Meteorol. Vol 10, No 3 p 185-201p. 1972. Illus.
Identifiers: Canopy, *Climate, Crop. Humidity, Models, *Orange orchard, Physical properties, Pressure, Radiation, Rainfall, Temperature, Vapor, Wind, *Israel.

Measurements are given of radiative and aerodynamic characteristics of a typical mature orange orchard in Israel's coastal plain. Net radiation to the canopy can be calculated adequately for weekly periods and longer from global radiation and 3 radiative crop characteristics. The role of aerodynamic crop characteristics, e.g., in crop water use studies, is discussed. Rainfall, wind run, air temperature and humidity of the air were measured at equivalent heights both in a climate station outside the orchard and above the tree canopy at the experimental orchard site, and the microclimatic modifications associated with the above characteristics are discussed. Rainfall differences were negligible, whereas wind run at the climate station is shown to be significantly under estimated. Absolute differences in temperature never exceeded 2 degrees C, while relative humidity of the air above the canopy was almost constantly 8% higher. Relationships are given for calculating the relevant crop and climate parameters from standard climatic data. The distribution of net radiation, wind flow, air temperature and vapor pressure throughout the day is illustrated. The canopy structure of the orchard is not homogeneous horizontally. The horizontal variations especially affect the transfer of mass, energy and momentum within the canopy. Attempts to simulate these processes within the stand are mostly based on 1-dimensional models, assuming horizontal homogeneity. One-dimensional models of Cowan (1968) for the absorption of net radiation and the alternation of windflow within the canopy were tested. Reasons are given for model breakdown and the need for general 3-dimensional models is stressed.--Copyright 1973, Biological Abstracts, Inc.
W73-09320

A PREDICTION TECHNIQUE FOR SNAP BEAN MATURITY INCORPORATING SOIL MOISTURE WITH THE HEAT UNIT SYSTEM,
National Weather Service, Clemson, S.C.
A. J. Kish, W. L. Ogle, and C. B. Loadholt.
Agric Meteorol. Vol 10 No 3, p 203-209. 1972. Illus.
Identifiers: *Beans (Snap), *Maturity, Prediction, Soils, Technique, *Soil moisture.

The growing-degree hour method was found to be unreliable in predicting the maturity for 3 plantings of 'Harvester' and 'Tendercrop' snap beans. Other environmental factors, in addition to temperature, affected maturity. The available soil moisture for each of the 3 plantings varied greatly. Because of the unreliability of the heat unit method, the available soil moisture parameter was integrated into the degree hour method. The formula that gave the smallest coefficient of variation was one using the daily heat unit multiplied by ratio of the daily available soil moisture to a constant soil moisture. Optimum available soil moisture level to be used as the constant soil moisture value in the formula was in the 57-59% range. Predicting the maturity of snap beans was improved by integrating available soil moisture into the heat unit system.--Copyright 1973, Biological Abstracts, Inc.
W73-09336

THE GROWTH AND PERFORMANCE OF RAIN-GROWN COTTON IN A TROPICAL UPLAND ENVIRONMENT: I. YIELDS, WATER RELATIONS AND CROP GROWTH,
Commonwealth Scientific and Industrial Research Organization, Kununurra (Australia). Div. of Land Research.
A. B. Hearn.
J Agric Sci, Vol 79, No 1, p 121-135, 1972, Illus.
Identifiers: *Cotton, *Crop production, Density, Environment, Genotype, Growth, Irrigation, Leaf, Light, Matter, Rain, Tropical, Upland, Yields.

The effects of environment and genotype on growth and yield of cotton were studied in 3 experiments from 1966-1969. Treatments were date of sowing, variety, fertilizer, plant population and water. Variety BPA66 sown in June at 4-10 plants m-2 out-yielded other varieties, sowing dates and population densities. Compound fertilizer at 1.25 t ha-1 increased yield by 15% and irrigation increased yield by 38%. The soil water deficit (CSWD) was calculated from meteorological data, and the relative water content (RWC) of the plants was measured. CSWD did not affect growth until a critical value (CD) was reached, which increased from 20 to 50 mm as the crop aged. When CD was reached RWC was 0.94 at dawn and 0.83 at 1400 hr. Growth stopped when CSWD > or equal to CD, except while any rain, insufficient to make CSWD < CD, was being consumed. Days while such rain was being consumed and days when CSWD < CD were added to give the effective numbers of growing days which accounted for differences in numbers of mainstem nodes caused by sowing date and spacing, and for differences in plant dry weight and leaf area caused by sowing date. Variation in light transmitted by the crop canopy depended on leaf area index (L) alone; spacing, fertilizer and CSWD had no independent effects. The measured extinction coefficient was 1.1 compared with 0.9 predicted by de Wit's (1965) model. Measured value of crop growth rate (C) agreed with values predicted by the de Wit model for the vegetative phase. Spacing and fertilizer only affected C through L. During the reproductive phase C became much less dependent on L, and the form of the relationship changed. Some varieties including BPA66 had a greater net assimilation rate. Maximum C was expected when L is equivalent or equal to 3, but L was seldom >2. Crops sown at current and previously recommended spacing had sparse canopies and did not fully use light available for dry matter production.--Copyright 1973, Biological Abstracts, Inc.
W73-09379

EFFECT OF TIMING OF IRRIGATION ON TOTAL NONSTRUCTURAL CARBOHYDRATE LEVEL IN ROOTS AND ON SEED YIELD OF ALFALFA (MEDICAGO SATIVA L.),
Volcani Inst. of Agricultural Research, Bet-Dagan (Israel). Dept. of Soil and Water.
Y. Cohen, H. Bielorai, and A. Dovrat.
Crop Sci. Vol 12, No 5, p 634-636. 1972.
Identifiers: *Irrigation efficiency, *Crop production, Alfalfa, Carbohydrate, Irrigation, Medicago sativa, Roots, Seeds, Timing.

The effect of certain soil moisture regimes, manipulated by timing of irrigation, on the seed yield of alfalfa was assessed. Sprinkler-irrigated alfalfa (cv 'Hairy Peruvian') was cut once and the regrowth was left for seed production or cut periodically for forage during the spring to mid-summer period. Seed plots were irrigated either at day 10 or 23, or both (following cutting) to 180-cm soil depth. Forage plots were irrigated once immediately after each cutting. Irrigation at day 23 resulted in a considerably lower rate of foliar regrowth during the first 3 wk following cutting than when irrigation was applied at day 10. Slow regrowth was conducive to greater accumulation of total nonstructural carbohydrate (TNC) in roots than quick regrowth. High TNC level in the roots was associated with high percentage of pods set and with high seed yield. Irrigation both at day 10 and also at day 23 increased the flowering intensity that was associated with a lower percentage of pods set than the other treatments. The seasonal decline of TNC in the roots of plants harvested for seed was smaller than that in the roots of plants cut for forage. It was concluded that timing of irrigation that leads to an increase in the root reserves during the period of initial regrowth is likely to increase the seed yield potential of alfalfa.--Copyright 1973, Biological Abstracts, Inc.
W73-09418

FLOW-METER FOR UNDERGROUND IRRIGATION PIPELINES,
Indian Agricultural Research Inst., New Delhi. Div. of Agricultural Engineering.
For primary bibliographic entry see Field 07B.
W73-09423

FACTORS RELATED TO THE IRRIGATION OF MARYLAND TOBACCO: I. AGRONOMIC EFFECTS,
Virginia Polytechnic Inst. and State Univ., Blacksburg. Coll. of Agriculture.
G. W. Brown, and O. E. Street.
Tob Sci. 16, p 55-60. 1972. Illus.
Identifiers: Agronomic effects, Fertilization, Irrigation, *Maryland, Nitrogen, Plant population, *Tobacca, Yield.

The effect of irrigation, N level, plant population, and 2 varieties on the yield and value of Maryland tobacco was studied. Irrigation increased the yield in 2 of 4 yr tested. The value/acre was increased only in 1966, a very dry year. Increasing the N rate from 90-120 lb/acre significantly increased the yield. The plant populations evaluated resulted in no significant alterations of yield or value. The 'Wilson' cultivar outyielded 'Catterton.' Yield increased as N rate increased for irrigated and non-irrigated tobacco. Value/acre was highest at the 120 lb N rate on irrigated plots. On non-irrigated plots, tobacco which received the 150 lb N rate had a higher value/acre than that from the 90 lb N rate but was the same as for tobacco fertilizer with 120 lb of N. When irrigation was applied, 'Wilson' produced the greater yield. When no irrigation was applied, no difference in yield and value occurred between the 2 cultivars suggesting that this result was a function of the water alone. Reduced growth because of water caused reduced yields with each of the 3 N rates.--Copyright 1973, Biological Abstracts, Inc.
W73-09427

OSCILLATIONS IN STOMATAL CONDUCTANCE AND PLANT FUNCTIONING ASSOCIATED WITH STOMATAL CONDUCTANCE: OBSERVATIONS AND A MODEL,
Australian National Univ., Canberra. Dept. of Environmental Biology.
For primary bibliographic entry see Field 02I.
W73-09429

EFFECTS OF SOIL AND MANAGEMENT FACTORS ON FERTILIZER RESPONSES AND YIELD OF IRRIGATED WHEAT IN IRAN,
T. P. Abraham, and A. Hoobakht.
Exp Agric. Vol 8, No 3, p 195-202. 1972.
Identifiers: Fertilizers, *Iran, Irrigation, Management, Soils, *Wheat, Yield.

The yield data obtained from fertilizer experiments conducted at 195 locations for 3 yr were examined to study the relations between yield, response to N and P,soil test values, and number of irrigations, using multiple regression technique. Irrigation accounted for 20-36% of the yield variation, whereas soil factors did not show appreciable effects. N response was influenced by irrigation, while some soil factors affected response to P.--Copyright 1973, Biological Abstracts, Inc.
W73-09436

INFLUENCE OF THE MID-SUMMER DRAINAGE ON THE ACTIVITY OF ROOTS OF PADDY RICE (IN JAPANESE),
Tokyo Univ. of Agriculture and Technology (Japan). Faculty of Agriculture.
T. Tanabe.
J Agric Sci Tokyo Nogyo Daigaku, Vol 16, No 1, p 35-43, 1971, Illus, English summary.
Identifiers: *Drainage (Paddy rice), Guttation, Naphthylamine, Oxidation, *Rice, *Roots (Rice), Seasonal.

Experimental plots were drained at 55-45 days, at 45-35 days and at 35-25 days before heading time. In addition to its influence on root activity, the amount of alpha-naphthylamine oxidation by roots, amount of guttation from the leaf and the number of water drops/number of green leaves were measured. By mid-summer drainage, the root activity was vigorous and remained so until the later part of growth. The measurement of guttation and of the number of water drops/number of green leaves were correlated with alpha-naphthylamine oxidation, and root activity could be assessed from the shoot growth conditions. The most effective time of the mid-summer drainage was 45-35 days before heading.--Copyright 1973, Biological Abstracts, Inc.
W73-09472

WATER USE BY IRRIGATED COTTON IN SUDAN: III. BOWEN RATIOS AND ADVECTIVE ENERGY,
Projet Rech. Agron, Bassin Fleuve Senegal, Dakar.
D. A. Rijks.
J Appl Ecol, Vol 8, No 3, p 643-663, 1971, Illus.
Identifiers: *Advective energy, *Cotton, Energy, Humidity, Irrigation, *Sudan, Temperature, Wind, *Bowen ratios.

Wet-and dry-bulb temperatures and wind speed were measured over irrigated cotton at various heights and distances from the leading edge of a field lying downwind from a bare, fallow area. Together with radiation balance data, they were used to estimate vertical (Ev) and horizontal (delta E sub h) fluxes of water vapor during day and night. Vertical fluxes were estimated both by the Bowen ratio method and from the Thornthwaite-Holzman formula. Under near-neutral conditions the 2 estimates did not agree well. On theoretical grounds the Bowen ratio estimate was preferred. Horizontal fluxes below the effective level of measurement of Ev were obtained from wind speed and the humidity gradient in the x-direction at various heights. Cross-wind variation in profiles of temperature and humidity was small at heights greater than 40 cm above the crop. In a check on the estimated income and expenditure of energy over a complete irrigation cycle, a discrepancy from 5 to 15% (depending on fetch and crop development) was found. Mean Bowen ratios were calculated to illustrate the variation in temperature and humidity above the crop with fetch, stage in irrigation cycle and crop development. The estimates of total evaporation suggested that irrigation water was unevenly distributed over the field; that evaporation rates early in an irrigation cycle were smaller than in the middle of a cycle, perhaps because the roots were less active, an effect more apparent early in the season than late; that evaporation was less than the potential rate towards the end of the irrigation cycles in the first 1/2 of the season, probably because soil water was limiting; that the rate of evaporation from an actively growing crop, not short of water, was often as much as 1.8 times as large as the rate at which net radiant energy became available to the crop, if both local and general advection were important, and as much as 1.5 times the rate of net radiant energy supply if general advection alone was important; that local advection was not very noticeable beyond about 200 m from the leading edge, but that general advection was noticeable over the whole field. (See also W69-06756)--Copyright 1973, Biological Abstracts, Inc.
W73-09473

THE STUDY ON SOME METEOROLOGICAL CONDITIONS AND PRODUCTIVITY OF RICE IN KOREA (IN KOREAN),
Seoul National Univ., Suwon (Republic of Korea). Coll. of Agriculture.
E. W. Lee.
Res Rep Off Rural Dev (Crop) (Suwon). 14, p 7-31, 1971, English summary.
Identifiers: *Korea, *Meteorological conditions, *Rice, Crop production.

The highest rice yield and ripening ratio were obtained when the rice cultivars headed about Aug. 15 at Suwon, about Aug. 20 at Milyang and about Aug. 25 at Iri, respectively. Highly significant correlations between the climatic conditions during the period of 20 days before and after heading and rice yield or its components, were noted.--Copyright 1973, Biological Abstracts, Inc.
W73-09475

RECOVERY PROCESSES AFTER LEAF CUTTING IN CEREALS AND FORAGE GRASSES: IV. SOME PHYSIOLOGICAL CHANGES IN RICE PLANTS FOLLOWING DEFOLIATION, (IN JAPANESE),
Tohoku Univ., Sendai (Japan). Faculty of Agriculture.
K. Sato.
Proc Crop Sci Soc Jap. Vol 39, No 1, p 15-20. 1970. Illus. English summary.
Identifiers: Carbohydrates, *Cereals, Chlorophyll, Defoliation, *Forage grasses, Grasses-M, *Leaf cutting, Nitrogen, Physiological changes, *Rice-M, Roots, Transpiration.

In defoliated plants, water and N content increased with a concomitant increase in transpiration per unit weight of shoot. Recovery was most rapid in leaf formation and N absorption with the reduction of carbohydrate content in the stem and leaf-sheath. Chlorophyll content was higher in the treated plant with a higher photosynthesis. During the 10 days following defoliation, root activity was lower due to the cessation of new root formation. After that as the new roots emerged and elongated, the activity of the treated plants surpassed that of the control. The higher growth rate in the later half of the life of the defoliated plant was considered to be caused by the physiological juvenilization induced from the later formation of

new roots and their persistence.--Copyright 1972, Biological Abstracts, Inc.
W73-09483

CHANGES IN CYTOKININ ACTIVITY ASSOCIATED WITH THE DEVELOPMENT OF VERTICILLIUM WILT AND WATER STRESS IN COTTON PLANTS,
California Univ., Davis. Dept. of Plant Pathology.
I. Misaghi, J. E. Devay, and T. Kosuge.
Physiol Plant Pathol. Vol 2, No 3, p 187-196. 1972. Illus.
Identifiers: Chromatography, *Cotton-D, Gossypium-hirsutum-D, Tracheal, *Verticillium wilt, Water stress, *Cytokinin activity.

Tracheal fluid and extract from leaves and stems of both healthy and Verticillium-infected cotton plants (Gossypium hirsutum L. cv. 'Acala SJ-1') contain 3 cytokinin-active substances which can be separated from each other by thin-layer chromatography. There are significant reductions in the levels of cytokinins in tracheal fluid and extract from leaves and stems of Verticillium-infected cotton plants compared to that of healthy plants. A time course study showed that the changes in cytokinin activity in Verticillium-infected cotton occurred after development of leaf symptoms. The leaves and stems of cotton plants exposed to water stress contained less cytokinins than those from normal plants. However, there was a greater loss of cytokinin activity in extract from leaves and stems in Verticillium-infected plants than in plants subjected to water stress. The TLC Rf values of 2 of the factors from cotton were similar to those of zeatin mononucleotide and zeatin; however, their structural relatedness to zeatin or zeatin mononucleotide is unknown. The chromatographic behavior of the third factor from cotton was different from the known cytokinins.--Copyright 1972, Biological Abstracts, Inc.
W73-09532

PATHOGENICITY OF PYRICULARIA AND COCHLIO-BOLUS CONIDIA COMPRESSED BY HYDROSTATIC PRESSURE (EFFECT OF PHYSICAL FACTORS UPON PLANT DISEASES: V.). (IN JAPANESE),
Gifu Univ. (Japan). Faculty of Agriculture.
For primary bibliographic entry see Field 02I.
W73-09536

THE INTERNAL WATER STATUS OF THE TEA PLANT (CAMELLIA SINENSIS): SOME RESULTS ILLUSTRATING THE USE OF THE PRESSURE CHAMBER TECHNIQUE,
Tea Research Inst. of East Africa, Kericho (Kenya).
M. K. V. Carr.
Agric Meteorol. Vol 9, No 5, p 447-460, 1972. Illus.
Identifiers: Camellia-Sinensis-D, Depletion, Illustrating, Internal, Moisture, Plant, *Pressure chamber technique, Soils, *Sap tension, Stomata, *Tea-D, Tension, Wind.

There is a diurnal change in the sap tension of shoots taken from both irrigated and unirrigated tea plants. Early in the morning the sap tension is low; it increases until about 10:00 hr and then remains almost constant, decreasing from about 16:00 hr. Stomatal closure appear to limit the increase in sap tension during the middle of the day, even when the tea is growing in wet soil. The stomata of 2 genetically distinct populations of tea differ in their sensitivity to increasing sap tension. Sap tension measurements confirm that sheltered tea can deplete soil water faster than tea more exposed to the wind.--Copyright 1972, Biological Abstracts, Inc.
W73-09541

WINTER AND SUMMER GROWTH OF PASTURE SPECIES IN A HIGH RAINFALL AREA OF SOUTHEASTERN QUEENSLAND,
Commonwealth Scientific and Industrial Research Organization, St. Lucia (Australia). Div. of Tropical Pastures.
For primary bibliographic entry see Field 04A.
W73-09545

THE RESPONSE OF A PANGOLA GRASS PASTURE NEAR DARWIN TO THE WET SEASON APPLICATION OF NITROGEN,
Northern Territory Administration, Darwin (Australia). Animal Industry Agricultural Branch.
K. Hendy.
Trop Grassl. Vol 6, No 1, p 25-32. 1972. Illus.
Identifiers: *Australia (Darwin), Digitaria-Decumbens-m, *Grass-m, *Nitrogen, *Pangola pasture, Protein, Wet season.

The response of a pangola grass (Digitaria decumbens) pasture to N fertilizer at 0, 100, 200, 400, 600 lb N/acre in split applications during the wet season was measured. There was a response in dry matter production up to the highest N level, but the differences in response occurred mainly during the early showers and not during the monsoonal rains of the wet season. Crude protein content and crude protein production/acre increased and the N recovery decreased with increasing levels of N fertilizer over the whole growing season. The P and K contents of the pasture were low.--Copyright 1972, Biological Abstracts, Inc.
W73-09549

ECOLOGY OF PASTORAL AREAS IN THE ARID ZONE OF RAJASTHAN,
Central Arid Zone Research Inst., Jodhpur (India).
For primary bibliographic entry see Field 04A.
W73-09550

RICE RESPONSE TO ZN IN FLOODED AND NONFLOODED SOIL,
Tennessee Valley Authority, Muscle Shoals, Ala. Soils and Fertilizer Research Branch.
P. M. Giordano, and J. J. Mortvedt.
Agron J. Vol 64, No 4, p 521-524, 1972. Illus.
Identifiers: Cultivars, Fertilization, Flooded soils, Minerals, Nonflooded soils, Nutrients, Oryza-Sativa-M, Phosphorus, *Rice-M, *Soils, *Zinc.

Rice (Oryza sativa L. cv. 'Bluebelle') was grown in greenhouse pots on flooded and moist, limed Crowley oil (pH 7.5). P, as concentrated superphosphate, was applied at rates of 20 and 200 ppm and Zn as ZnSO4, was applied at rates of 1, 4, and 16 ppm. Dry matter production and P uptake of immature plants were doubled and uptake of Zn was up to 5 times as great when the plants were grown on flooded rather than moist soil. The Zn concentration in the tissue decreased at the high P rate on moist but not flooded soil. The percentage recovery of fertilizer Zn by the crop was greater from flooded than moist soil, especially at the high P rate. Mixed and surface applications of Zn were comparable for flooded rice, whereas middepth placement (8 cm below soil surface) was less effective. Both surface and middepth placement were less effective than mixing under nonflooded soil conditions. The mobility of Zn applied as ZnSO4 was comparable in flooded and moist soil. The amount of diethylenetriamine pentaacetate (DTPA)-extractable Zn tended to be higher in moist than in flooded soil. This suggests that DTPA may not provide a reliable measure of Zn availability in flooded soils and that differences in Zn uptake may be related to root distribution or some other physiological or morphological characteristic of rice roots growing in moist or flooded soil.--Copyright 1972, Biological Abstracts, Inc.
W73-09551

THE GROWTH, COMPOSITION AND NUTRIENT UPTAKE OF SPRING WHEAT: EFFECTS OF FERTILIZER-N, IRRIGATION AND CCC ON DRY MATTER AND N, P, K, CA, MG, AND NA,
Rothamsted Experimental Station, Harpenden (England).
J. K. R. Gasser, and M. A. P. Thorburn.
J Agric Sci. Vol 78, No 3, p 393-404, 1972. Illus.
Identifiers: Ammonium, Calcium, Chlorides, *Fertilizers, Growth, *Irrigation, Magnesium, *Nitrogen, Nutrients, Phosphorus, Potassium, Sodium, *Wheat-M, Dry matter.

Kloka spring wheat grown on a sandy-loam soil was given 50, 100, 151 or 201 kg N/ha as fertilizer, was irrigated during drought or not irrigated, and was sprayed with CCC (2-chloroethyltrimethylammonium chloride) or not sprayed. Samples, taken approximately weekly from brairding to heading and less frequently to maturity measured the production of dry matter. The dried samples were analyzed for percentage N, P, K, Ca, Mg and Na and the uptakes of these were calculated to measure the effects of treatments on the composition of the crop and the weights of nutrients it contained. Increasing amounts of fertilizer-N up to 151 kg N/ha increased the maximum weight of straw without irrigation and up to 201 kg N/ha with irrigation; similarly, maximum weights of ears were with 100 kg N/ha without irrigation and 151 kg N/ha with irrigation. The maximum weight of straw was at flowering and of ears at maturity. Irrigation increased yields of straw and ears, more with the 2 larger than with the 2 smaller amounts of fertilizer-N. Spraying with CCC decreased the yield of straw, and did not affect the yield of ears. Increasing amounts of fertilizer-N increased percentage nutrients in the green crop and in the straw, and of N, P and Mg in the ears. Effects of irrigation and spraying with CCC on composition differed between nutrients and between ears and straw, sometimes increasing, sometimes decreasing, and sometimes having no effect, on percentage in dry matter. Increasing fertilizer-N increased the maximum weight of N, K, Ca and Na in the crop. P and Mg increased with up to 151 kg N/ha, but was no more with 201 kg N/ha. Maximum weight of N. P., Mg and Na was found at or near maturity, of K at heading and of Ca at flowering.--Copyright 1972, Biological Abstracts, Inc.
W73-09552

A WATER BALANCE MODEL FOR RAIN-GROWN, LOWLAND RICE IN NORTHERN AUSTRALIA,
Commonwealth Scientific and Industrial Research Organization, Darwin (Australia). Coastal Plains Research Station.
A. L. Chapman, and W. R. Kininmonth.
Agric Meteorol. Vol 10, No 1/2, p 65-82, 1972. Illus.
Identifiers: *Australia, *Evapotranspiration, Model studies, Northern, Rain, Rainfall, *Rice-M, Transpiration, Water balance model.

The rice farming system is based on short duration (100 days) photoperiod-insensitive varieties. The seed bed is prepared in the flooded state, and the crop is sown during mid-Jan.-early Feb. when rainfall and tillage criteria are satisfied. The model, based mainly on field measurements of evapotranspiration, was used to estimate the success frequency of rain-grown, lowland rice at Darwin, Koolpinyah and Humpty Doo with 93, 48 and 24 yr of rainfall records, respectively. Daily estimates of soil water storage and depth of ponded water for each wet season, Sept. 1-May 31 were obtained. Rainfall was regarded as adequate providing there were at least 14 days pondage plus or minus 3 in. between Dec 31 and Feb. 18 for wet tillage before sowing, at least 80 days between sowing date and the last date at which ponded water was present on the field, not more than 10 consecutive zero pondage days between 50 and 80 days after sowing, corresponding to the stage of

panicle initiation through flowering. A 23-yr comparison showed that on the average, the water availability was similar at the 3 stations, although differences sometimes occurred within individual seasons. The expectation of crop failure due to inadequate rainfall at Darwin is about 1 ye in 30. In addition there is an expectation of major yield reduction of about 1 yr in 10. The seasons of crop failure and low yield tended to occur in groups after relatively long runs of seasons of adequate rainfall. Designing for a maximum field pondage greater than 10 in. did not reduce the estimated number of crop failures at any station. The results are dicussed in relation to practical farming operations.--Copyright 1972, Biological Abstracts, Inc.
W73-09553

WHEAT RESPONSE TO DIFFERENT SOIL WATER-AERATION CONDITIONS,
California Univ., Riverside. Dept. of Soil Science and Agricultural Engineering.
M. G. Anaya, and L. H. Stolzy.
Soil Sci Soc Am Proc. Vol 36, No 3, p 485-489, 1972. Illus.
Identifiers: *Aeration, Grain Oxygen, Protein, Soils, Triticum-Aestivum-M, *Wheat-M, *Soil water.

A graphical yield surface was constructed for wheat (Triticum aestivum L.) grown on Yolo silt loam with 13 different soil water-aeration combinations. The experiment was conducted in a growth chamber under controlled environmental conditions. O_2 over the soil surface was maintained at different percent O_2 concentrations 0.9-21%. Various soil water regimes were obtained by irrigation at different soil suctions 8-99 centibars (cbars). The highest grain yields were obtained in 2 treatments, one of 9.6% O_2 watered at a soil suction of 15 cbars, and another treatment of 4.3% watered at cbars. The lowest production was in the treatment of 0.9% O_2 watered at 99 cbars, and the difference between the highest and lowest yields was 347%. From the data, the regression equation for grain yield was $Y \pm 15.94 - 0.1324\,X1 + 3.1813\,X2 \pm 0.1297X22$, where $Y \pm$ grain yield in g/pot, $X1 \pm$ soil suction in cbars, and $X2 \pm$ percent O_2. The maximum predicted yield calculated from the equation is at a level of 12.0% of O_2 irrigated at a soil suction of 8 cbars. However, the maximum production recorded was obtained at 9.6% O_2. There was a high correlation coefficient (0.94) between water consumption and grain production. An inverse relationship existed between grain yield and protein content. The highest level of grain protein content (22.6%) was obtained in the lowest producing treatment, while the lowest level of grain protein content (16.5%) occurred in the highest producing treatment. The difference in grain protein content between treatments was 37%.--Copyright 1972, Biological Abstracts, Inc.
W73-09555

METHODS OF PLANTING RICE (ORYZA SATIVA L.) IN PUDDLED SOIL,
Central Rice Research Station, Shoranur (India).
R. R. Nair, G. R. Pillai, P. N. Pisharody, and R. Gopalakrishnan.
Indian J Agric Sci. Vol 41, No 11, p 948-951, 1971.
Identifiers: Oryza-Sativa, Planting, *Rice, Seeds, *Soils puddled, Sprouts, Transplanting.

Transplanting 25-day-old seedlings from wet nursery, raising seedlings according to 'dapog' system and transplanting at the age of 14 days, dibbling sprouted seeds, and broadcasting sprouted seeds on mud and in standing water were compared. Under conditions of adequate water control, dibbling sprouted seeds gave a yield equal to that given by transplanting. These 2 methods were superior to the rest. Dibbling was also the most economical method.--Copyright 1973, Biological Abstracts, Inc.
W73-09559

INFLUENCE OF WATER REGIMES AND STARCH ON THE GROWTH, YIELD AND NUTRIENT ABSORPTION BY RICE (ORYZA SATIVA L.),
Fertilizer Corp. of India, Sindri. Planning and Development Div.
J. D. Naphade, and B. P. Ghildyal.
Indian J Agric Sci. Vol 41, No 11, p 963-966. 1971.
Identifiers: Absorption, Growth, Nutrients, Oryza-sativa, *Rice, Roots, *Starch, Toxic substances, *Water regime, Yield.

In lateritic sandy clay-loam soil, plant growth and yield were significantly higher under shallow flooding than at a soil-moisture tension of 60 cm and deep flooding, possibly due to higher root proliferation and greater uptake of N, P and Fe, and a lower uptake of Mn. Starch application to the soil increased the absorption of N, P, Fe and Mn in the plant because of highly reducing conditions and their consequent increased in the soil. Starch had an adverse effect on the growth and yield of rice; this was probably due to intense reducing conditions and the formation of toxic substances in the soil.--Copyright 1973, Biological Abstracts, Inc.
W73-09562

WATER REQUIREMENTS OF DWARF WHEAT (TRITICUM AESTIVUM L.) UNDER WEST BENGAL CONDITIONS,
Jute Agricultural Research Inst., Barrackpore (India).
S. R. Gupta, and K. S. Dargan.
Indian J Agric Sci. Vol 41, No 11, p 958-962. 1971.
Identifiers: Climate, *India (Bengal), Irrigation, Triticum-aestivum, *Water requirements, *Wheat (Dwarf).

The optimum water requirement was 382.04 mm in 1967-68 and 221.70 mm in 1968-69. The optimum regime to schedule irrigation was 50% of the available moisture; but it decreased to 10% of available moisture when the climate was wet and cool, as in 1968-69. 'PV 18' gave 1.68 q/ha more yield (mean for 2 yr) than 'Sonora 64'. The application of 80 and 120 kg N over 40 kg N/ha, 3/4 at sowing and 1/4 at first irrigation, increased the grain yield by 28.1 and 51.2% in 1967-68 and 27.8 and 45.7% in 1968-69, the mean increase being 27.9 and 48.2% respectively. The peak period of the rate of water-use occurred at the earing and grain-filling stages.--Copyright 1973, Biological Abstracts, Inc.
W73-09563

EFFECTS OF MOISTURE REGIMES AND NITROGEN LEVELS ON GROWTH AND YIELD OF DWARF WHEAT VARIETIES,
Rockefeller Foundation, New Delhi (India).
N. P. Singh, and N. G. Dastane.
Indian J Agric Sci. Vol 41, No 11, p 952-958. 1971.
Identifiers: Growth, Moisture, *Nitrogen, Triticum-aestivum, *Wheat (Dwarf), Yield, Cultivars, Fertilization.

Three dwarf wheat (Triticum aestivum L.) cultivars were studied for their response to 3 moisture regimes and 3 N levels. 'Kalyan Sona' gave the maximum yield (49.9 q/ha), followed by 'Safed Lerma' (45.9 q/ha) and 'Sonalika' (42.5 q/ha) when sown on equal-grain-weight basis. 'Sonalika' had the boldest grain and the yield was best on per-plant basis. Irrigation at 0.25 atm tension measured at a depth of 22 cm was the optimum. Use of water was the lowest at 0.75 atm tension. N at 150 kg/ha produced the highest yield in the first yr when the initial N status of the soil was 0.038%. In the second year, no differential response was noted when the initial N content of the soil was 0.065%. For economic utilization of fertilizer resources, therefore, N application should be commensurate with initial N status of the soil.--Copyright 1973, Biological Abstracts, Inc.
W73-09566

RICE RESEARCH IN FIJI, 1960-1970: III. WEED CONTROL,
Department of Agriculture, Suva (Fiji). Research Div.
N. P. Patel.
Fiji Agric J. Vol 34, No 1, p 27-34. 1972.
Identifiers: Dicots, *Fiji, Grass, *Herbicides, Phenoxy, Propanil, *Rice, *Weed control, Grasses.

In 1967-1971 herbicides were evaluated for use in short-stature, close-spaced dryland rice. Propanil applied post emergence at 3 lb active ingredient/acre (3.4 kg/ha), when weeds are at the 2-3 leaf stage and growing actively, gave good control of grasses and common dicots. Trials in 1969 and 1970 showed that phenoxy herbicides can be safely used to control broad-leafed weeds at a later stage. In transplanted wetland rice, grass weeds are no problem and other weeds generally are not serious, particularly under controlled irrigation (as opposed to rain-fed) conditions. Phenoxy herbicides applied 3-4 wk planting give adequate control. With direct seeding (broadcasting) on to puddled surface, followed by flooding, grass weeds are a serious problem. Propanil is useful but does not give complete control.--Copyright 1973, Biological Abstracts, Inc.
W73-09569

EFFECT OF AERIAL ENVIRONMENT AND SOIL WATER POTENTIAL ON THE TRANSPIRATION AND ENERGY STATUS OF WATER IN WHEAT PLANTS,
Washington State Univ., Pullman. Dept. of Agronomy and Soils.
S. J. Yang, and E. DeJong.
Agron J. Vol 64, No 5, p 574-578, 1972. Illus.
Identifiers: Aerial environment, Capillary conductivity, *Energy status, Evaporation, Humidity, Soils, Temperature, Texture, *Transpiration, Triticum-aestivum, *Wheat, Wilting, *Soil-water potential.

Thatcher wheat (Triticum aestivum L.) was grown under 4 combinations of air temperature and relative humidity in a loam and a clay soil. The relationship between transpiration rate and soil water potential depended on evaporative demand and soil texture. The decline in the transpiration rate from its maximum commenced at higher soil water potentials under conditions of higher evaporative demand and was more gradual on the clay than on the loam soil, presumably due to the higher capillary conductivity of the former. Permanent wilting occurred at soil water potentials of -20 to -25 bars on the loam soil and at -45 to -50 bars on the clay soil. At these potentials the capillary conductivities of both soils were about equal. Resistance to water flow in the plant decreased with increased temperature, while changes in relative humidity had no consistent effect. The total resistance to water movement in the plant and the soil increased with decreasing soil water potential and decreasing air temperature. The relationship between leaf water potential and relative water content was affected by aerial environment and soil texture.--Copyright 1973, Biological Abstracts, Inc.
W73-09570

INFLUENCE OF WINDBREAKS ON LEAF WATER STATUS IN SPRING WHEAT,
Agricultural Research Service, Mandan, N. Dak. Northern Great Plains Research Center.
A. B. Frank, and W. O. Willis.
Crop Sci. Vol 12, No 5, p 668-672. 1972.
Identifiers: *Leaf water status, Shelterbelts, Stomatal diffusion, Triticum-aestivum, *Wheat, Wind, *Windbreaks, Water potential.

Spring wheat (Triticum aestivum L.) was grown for 2 years with 3 shelter conditions: (a) exposed, (b) surrounded by a slat barrier, and (c) adjacent to tree shelterbelts. Leaf water potential, stomatal diffusion resistance, and meteorological factors were monitored on selected days. The shelterbelt

as much as 75% at 12C. Shoot initiation occurred at 7 to 10% higher water content than radicle initiation. The rate of radicle elongation was independent of the shoot but removal of the radicle reduced shoot growth by 15% at 20C. The apical meristem of an emerged seedling at 20C remained below the soil surface and was a function of the depth of planting. The equations derived for the processes and events have been incorporated into a mathematical model to predict germination and emergence of corn.--Copyright 1973, Biological Abstracts, Inc.
W73-09573

AERIAL THERMAL SCANNER TO DETERMINE TEMPERATURES OF SOILS AND OF CROP CANOPIES DIFFERING IN WATER STRESS,
Florida Univ., Gainesville. Dept. of Climatology.
For primary bibliographic entry see Field 07B.
W73-09575

EFFECTS OF NORTH-AND SOUTH-FACING SLOPES ON YIELD OF KENTUCKY BLUEGRASS (POA PARTENSIS L.) WITH VARIABLE RATE AND TIME OF NITROGEN APPLICATION,
Agricultural Research Service, Morgantown, W. Va. Soil and Water Conservation Research Div.
O. L. Bennet, E. L. Mathias, and P. R. Henderlong.
Agron J. Vol 64, No 5, p 630-635, 1972.
Identifiers: *Kentucky bluegrass, *Nitrogen, Nutrition, Poa-pratensis, Radiation, *Slopes, Temperature, Yield.

Kentucky bluegrass was treated with 0, 112, 224, 448, and 672 kg/ha rates of N applied singly and in split applications. Slopes were approximately 35%. Soil on both slopes was Gilpin silt loam. The rate of growth, total yield, seasonal distribution of yield, and N recovery in plants were influenced greatly by slope orientation. Production on the north-facing slope was more than twice that on the south-facing slope. Splitting the N applications tended to produce a higher yield and more uniform seasonal distribution. A combination of high soil temperatures and low soil moisture levels appeared to limit growth on the south-facing slope. Maximum soil temperature 2.5 cm (1 in.) below the soil surface on the north-facing slopes were 8-10C less than on the south-facing slope. Yield differences due to slope orientation completely overshadowed any yield responses from specific fertility treatments.--Copyright 1973, Biological Abstracts, Inc.
W73-09576

GROWTH ANALYSIS OF TWO SAINFOIN (ONOBRYCHIS VICIAEFOLIA SCOP.) GROWTH TYPES,
Agricultural Research Service, Bozeman, Mont. Plant Science Research Div.
C. S. Cooper.
Agron J. Vol 64, No 5, p 611-613, 1972.
Identifiers: Assimilation, *Growth, Irrigation, Leaf, Moisture, Onobrychis-viciaefolia, Radiation, *Sainfoin, Soil types, Yields.

Growth of 'Remont,' a '2-cut' type cultivar, and 'Eski,' a '1-cut' type cultivar was quantitatively analyzed in relation to adaptation to dryland or to areas with limited irrigation water. Yields of 'Remont' were significantly higher than those of 'Eski' when sampled on May 12, but because of a greater relative growth rate (RGR, g/g/wk), 'Eski' surpassed 'Remont' in yield when sampled on June 21. The greater RGR of Eski was associated with a significantly greater leaf area ratio (LAR, dm sq/g). The LAR of both species decreased with advancing maturity as a result of increasing stem-leaf ratio. Net assimilation rate (g/dm sq/wk) of these 2 cultivars was not significantly different. Crop growth rate (CGR, g/m sq/da) increased as the season advanced and appeared to be greatest for 'Eski.' Both species produced the majority of their leaves early with maximum leaf area index occurring on or before June 8. Specific leaf weight (cm sq/g) of cultivars was similar at the first 3 sampling dates, but was significantly greater for 'Remont' at the last sampling date. A late-maturing '1-cut' type sainfoin would most efficiently utilize soil moisture and incident radiation in areas where soil moisture is limiting after July 1.--Copyright 1973, Biological Abstracts, Inc.
W73-09577

EFFECT OF WET GROWING SEASONS ON HERBAGE GROWTH ON SOILS WITH IMPEDED DRAINAGE,
Foras Taluntais, Ballinamore (Ireland). Soil Physics Field Station.
J. MulQueen.
Ir J Agric Res. Vol 11, No 1, P 120-124, 1972. Illus.
Identifiers: *Drainage, Farming, Grasslanu, *Wet growing, *Herbage growth, *Ireland, Seasons, Soils.

Dry matter production from grassland on a soil with impeded drainage at Castle Archdale, Co. Fermanagh, was substantially reduced in the wet growing seasons of 1965-66. Reductions of the order of 16% were common on fertilized treatments and were as high as 37% on unfertilized treatments. The significance of these results in grassland farming systems is discussed.--Copyright 1973, Biological Abstracts, Inc.
W73-09578

LONG-TERM GRAZING EFFECTS ON STIPA-BOUTELOUA PRAIRIE SOILS.
Department of Agriculture, Lethbridge (Alberta). Research Station.
For primary bibliographic entry see Field 02G.
W73-09579

DEVELOPMENT OF GRASS ROOT SYSTEMS AS INFLUENCED BY SOIL COMPACTION,
Agricultural Research Service, Big Springs, Tex.
D. W. Fryrear, and W. G. McCully.
J Range Manage. Vol 25, No 4, p 254-257, 1972. Illus.
Identifiers: Bouteloua-curtipendula, *Grass roots, *Soil compaction, *Seedlings (Sideout grama).

The roots of 'Premier' sideoats grama (Bouteloua curtipendula (Michx.) Torr.) seedlings do not penetrate a shallow compacted layer the first year. This restrictive layer, commonly found in cultivated fields being converted to grass, can be modified by tillage to permit grass roots to expoit the soil beneath these compacted layers to obtain nutrients and water.--Copyright 1973, Biological Abstracts, Inc.
W73-09580

EFFECT OF SOIL MANAGEMENT ON CORN YIELD AND SOIL NUTRIENTS IN THE RAIN FOREST ZONE OF WESTERN NIGERIA,
Iban Univ. (Nigeria). Dept. of Agronomy.
A. A. Agboola, and A. A. Fayemi.
Agron J. Vol 64, No 5, p 641-644. 1972.
Identifiers: Calopgonium-Muccunoides, Conservation, *Corn yield, Forest zone, *Nigeria, Nutrients, Phaseolus-Aureus, Phosphorus, Potassium, Rain, Crop rotation, *Soil nutrients, Vigna-Sinensis, Zea-Mays, *Leume crops.

An experiment involving both interplanted legume crops in rotation plus fertilizer was conducted in Ibadan, which has an annual rainfall of 140 cm and average temperature of 25 C. The experiment was located on Iwo soil series with unusually high available P (100 kg/ha) and exchangeable K (600 kg/ha), soil pH of 6.5, and cation exchange capacity (CEC) of 15 meg/100g. Fertilizer (55-10-55 kg/ha of N-P-K) increased yields of corn (Zea mays L.)

from 1510 to 2460 kg/ha. Legumes (greengram (Phaseolus aureus), cowpea (Vigna sinensis), and calopo (Calopogonium mucunoides)) either interplanted or rotated also increased the yield of corn. The magnitude of the increase for calopo was equivalent to that for the fertilizer. Legumes tended to conserve available P and exchangeable K in the surface soil.--Copyright 1973, Biological Abstracts, Inc.
W73-09581

SOIL MANAGEMENT FACTORS AND GROWTH OF ZEA MAYS L. ON TOPSOIL AND EXPOSED SUBSOIL,
A. R. Batchelder, and J. N. Jones, Jr.
Agron J. Vol 64, No 5, p 648-652. 1972.
Identifiers: Exposed soils, Fertilizers, *Corn, Growth, Irrigation, Lime, *Soil management, Mulch, Runoff, Soils, Subsoil, Surface tension, Topsoil, Zea-Mays, *Water storage.

Corn (Zea mays L.) was grown for 4 yr on a Groseclose clay loam surface soil and exposed clay subsoil to study the effects of selected management practices on subsoil productivity and the available water storage in exposed subsoils. Half of the exposed subsoil plots were covered with straw mulch, and half of all plots were irrigated. Lime was added the first 2 yr and fertilizer was added each year to alleviate the growth-limiting effects of low pH and low fertility. The treatments were reflected in surface water runoff from the 10-12% slopes, in soil water tension relationships, in plant growth, and in grain yields. Mulching reduced the surface water runoff, resulting in greater net water input. The soil water tensions to a depth of 45 cm were generally lower under the straw mulch without irrigation than with either of the topsoil treatments. Top growth and grain yields generally increased on the subsoil treatments each year, and in the fourth yr yields from the irrigated subsoil without mulch and both the irrigated and nonirrigated subsoil with mulch were equivalent to those of the irrigated topsoil, and significantly greater than those of topsoil without irrigation. In the last 2 yr, when soil tests indicated that the low pH and low fertility had been mitigated, relative yields were markedly affected by the number of days during which the soil water tension at the 45-cm depth exceeded 700 millibar. --Copyright 1973, Biological Abstracts, Inc.
W73-09582

AGRICULTURAL RESPONSE TO HYDROLOGIC DROUGHT,
Colorado State Univ., Fort Collins. Dept. of Civil Engineering.
V. J. Bidwell.
Colorado State University Press, Hydrology Papers, No 53, July 1972. 25 p, 2 fig, 3 tab, 24 ref.

Descriptors: *Droughts, *Mathematical models, *Least squares method, *Planning, *Crop response, Agriculture, Hydrology, Dry farming, Crop production, Evaluation, Physiological ecology, Plant physiology, Mathematical studies.
Identifiers: Drought prediction.

Of all the detrimental effects ascribed to droughts, the total or partial loss of agricultural production is probably the most serious. Within this context is included both crops for direct harvesting and those sued as pasture for animal grazing. This paper deals with the effects of hydrologic drought on the yield of dryland farming and the evaluation of these effects. Some of the physiological principles involved in crop growth are reviewed in the context of a system which relates reduction in crop production to deficits in the water input. A possible set of non-linear functions is developed to describe this system by using the physical properties of crop production as constraints on the mathematics. The method of non-linear least squares regression is described as a suitable technique for solving the system function. One of the main purposes of the model is to reduce the uncertainty in estimation of the statistical properties of agricultural drought over large areas. Application of the analytical methods to a set of corn yield data is demonstrated with several types of system functions, one of which is suggested as a suitable practical model. The predictability of the models is not high because of restriction imposed by the availability of data. Some improvements are suggested for further development when more suitable data are available. (Black-Arizona)
W73-09678

GREAT BASIN EXPERIMENT STATION COMPLETES 60 YEARS,
Technical Writers' Services, Ogden, Utah.
For primary bibliographic entry see Field 04A.
W73-09681

RICE CALENDAR OF THAILAND,
Meteorological Dept. Bangkok (Thailand).
V. Rasmidatta.
Thai J Agric Sci. Vol 5 No 1, p 63-76 1972.
Identifiers: Armyworm, Bacteria, Diseases, Fungi, Jassids, Mealy bugs, Rainfall, *Rice-M, *Thailand, Thrips.

Rainfall is the most important meteorological factor in rice cultivation in Thailand. Drought from mid-June to July obstructs growth of plants. Flood in Sept. not only damages rice crops directly but also causes the outbreak of climbing cutworms. The infestation of armyworms, thrips, jassids and mealybugs are heaviest in the period of poor rains. Heavy rainfall causes blast, stem rot, collar rot, and oriental sheath and leaf spot diseases. The spread of bacterial leaf stripe diseases is greatest during the storm season, and brown spot during cold spells.--Copyright 1972, Biological Abstracts, Inc.
W73-09682

ARIZONA AGRICULTURE - CURRENT STATUS AND OUTLOOK.
Arizona Univ., Tucson. Coll. of Agriculture.

Arizona Agricultural Experiment Station and Cooperative Extension Service, Bulletin A-74, April 1973. 24 p, 41 fig, 3 tab.

Descriptors: *Arizona, *Agriculture, *Statistics, *Crop production, *Groundwater, Water levels, Pumping, Land use, Livestock, Irrigation.

General topics affecting Arizona Agriculture during 1972 such as the general economy, international trade, farm income, agricultural policies and programs, rural development, water, land use, agricultural credit and purchase inputs are detailed. Trends in specific crops such as sugar beets, safflower and cottonseed, cotton, grains, alfalfa, vegetables and melons, deciduous tree crops, and citrus fruit are analyzed, followed by similar treatment for livestock, including sheep, lambs and wool, hogs, and beef cattle, as well as poultry. Tables display total surface water diversions and groundwater pumping over a 7-year period, 1965-1971, land ownership in Arizona, 1972, and farm and ranch loans in Arizona by lender group over the period 1960-1973. Total water diversions and pumping for the period 1965-1972 were 60.1 million acre-feet, with 90 percent being used in irrigated agriculture. Figures reveal comparative statistics over varying periods of time during the last decade. A companion report, Arizona Agricultural Statistics, is abstracted separately. (Paylore-Arizona)
W73-09686

1972 ARIZONA AGRICULTURAL STATISTICS,.
Arizona Crop and Livestock Reporting Service, Phoenix.

Bulletin S-8, April 1973. 82 p.

Results of hydrologic observations in Willow Creek basin in northeastern Montana are presented for the period 1954-68. Frequency curves for runoff and precipitation are provided. The average annual precipitation for the 14-year period 1954-67 was about 0.5-inch less than the 30-year average; also, the average for the first half of the period was 3.4 inches less than that for the second half. The average seasonal runoff volume measured at Northwest Burnett, Cactus Flat, and Sheepshed reservoirs was 93.9, 96.5, and 37.5 acre-feet per square mile, respectively, for the 10-year period 1958-67. Sediment yields from Northwest Burnett, Cactus Flat, Sheepshed, and Triple Crossing basins varied from 0.1 to about 2.0 acre-feet per square mile. During the study period numerous reservoirs and waterspreaders were constructed in the basin. This upstream regulation of flow has caused a decrease in both the channel dimensions and the suspended-sediment load at the gaging station near the mouth of the basin. The peak stream discharge recorded at the gaging station is estimated to be about 45% less than it would have been had no reservoirs or waterspreaders been in the basin. The annual runoff volume recorded at the gaging station is estimated to be 18% less than it would be if the basin were totally uncontrolled. The capacity of the reservoir system is being reduced at an estimated rate of 1.6%, owing to sedimentation. (Woodard-USGS)
W73-09167

PHOSPHATE NUTRITION OF THE AQUATIC ANGIOSPERM, MYRIOPHYLLUM EXALBESCENS FERN.,
Cornell Univ., Ithaca, N.Y. Dept. of Agronomy.
For primary bibliographic entry see Field 05C.
W73-09201

REMOTE SENSING AS AN AID TO THE MANAGEMENT OF EARTH RESOURCES,
California Univ., Berkeley. Space Sciences Lab.
For primary bibliographic entry see Field 07B.
W73-09262

HYDROLOGY OF SMOOTH PLAINLANDS OF ARID AUSTRALIA.
Department of National Development, Canberra (Australia).
For primary bibliographic entry see Field 07C.
W73-09276

INTERCEPTION LOSS IN LOBLOLLY PINE STANDS OF THE SOUTH CAROLINA PIEDMONT,
Forest Service (USDA), Asheville, N.C. Southeastern Forest Experiment Station.
For primary bibliographic entry see Field 03B.
W73-09383

STUDY FOR DETERMINATION OF SOIL DRAINAGE CLASSES IN THE SOIL SURVEY (IN KOREAN),
Institute of Plant Environment, Suwon (Korea).
For primary bibliographic entry see Field 02G.
W73-09421

INFLUENCE OF LAND-USE ON CERTAIN SOIL CHARACTERISTICS IN THE AFFORESTED AREA OF KHUNTI IN CHOTANAGPUR,
Ranchi Agricultural Coll., Kanke (India).
For primary bibliographic entry see Field 02G.
W73-09424

CLASSIFICATION AND EVALUATION OF FRESHWATER WETLANDS AS WILDLIFE HABITAT IN THE GLACIATED NORTHEAST,
Massachusetts Univ., Amherst. Dept. of Biology.
For primary bibliographic entry see Field 02I.
W73-09430

A LOSING DRAINAGE BASIN IN THE MISSOURI OZARKS IDENTIFIED ON SIDE-LOOKING RADAR IMAGERY,
Geological Survey, Rolla, Mo.
For primary bibliographic entry see Field 07B.
W73-09446

GEOLOGY AND LAND RECLAMATION,
Kent State Univ., Ohio. Dept. of Geology.
M. R. McComas.
Ohio J Sci., Vol 72, No 2, p 65-74. 1972. Illus.
Identifiers: Drainage, Flooding, Geology, *Land reclamation, Landslides, Water pollution control.

Expansion of the population in the periphery of urban centers creates increased need for land for food, housing, recreation and waste disposal. As much of land is occupied, derelict lands such as floodplains, swamps and bogs, abandoned mineral working and steep slopes are subjected to reclamation. Reclamation of wetlands for agriculture in north-central USA is accomplished simply by land drainage. Reclamation of derelict land (wetlands, floodplains, stripped land and quarries, and landslide-prone slopes) for municipal, industrial, or residential needs requires study by environmentally oriented scientists as well as engineers, to prevent environmental disruption. In Illinois and Ohio, reclamation or intensive use of some derelict lands has precipitated problems of flooding, surface-water pollution, ground-water pollution, and landslides. Study of the geology and hydrology of the area considered for reclamation prior to construction is highly desirable for making more effective reclamation and for deciding on the best land use after reclamation. Low-intensity land uses such as parks on floodplains, agriculture in old strip-mined areas, and ski runs on unstable slopes may be, at least for the present, the highest value land-uses available for these sites.--Copyright 1972, Biological Abstracts, Inc.
W73-09533

THE RELATION OF FORESTS AND FORESTRY TO WATER RESOURCES,
Georgia Univ., Athens. School of Forest Resources.
J. D. Hewlett.
S Afr For J. 75. p 4-8. 1970. Illus.
Identifiers: Ecology, Erosion, Evaporation, *Forestry management, Forests, *Land management, Water resources, *Water utilization.

The various users and planners of a basin's water resource must recognize and submit to a reasonable distribution of the total water available, taking in account the inevitable surplus and deficit periods, and recognizing both on-site and downstream uses of water as necessary to optimize benefits to the community as a whole. The effects of land use on erosion, evaporation, water yield, and production coefficients for water must be considered if an effective plan for watershed management is ever to be enacted. Therefore, an intensive and well-supported program of research into the hydrologic and ecologic processes in the headwaters should be maintained until all the salient facts are known and agreed on.--Copyright 1972, Biological Abstracts, Inc.
W73-09537

WINTER AND SUMMER GROWTH OF PASTURE SPECIES IN A HIGH RAINFALL AREA OF SOUTHEASTERN QUEENSLAND,
Commonwealth Scientific and Industrial Research Organization, St. Lucia (Australia). Div. of Tropical Pastures.
M. C. Rees.
Trop Grassl. Vol 6, No 1, p 45-54, 1972.
Identifiers: *Australia (Queensland), Desmodium-Intorum-D, Desmodium-Uncinatum-D, Glycine-Wightii-D, Panicum-Coloratum-M, *Pasture species, Phalaris-M, Rainfall, Setaria-M-Spp, Setaria-Sphacelata-M, Trifolium-Repens-D, *Seasonal growth.

The persistence and seasonal distribution of growth of 134 tropical and temperate grasses and legumes were studied without irrigation over 4 yr in a high rainfall area. Grasses were heavily fertilized with N in addition to the molybdenized superphosphate which was applied to all species tested. Most tropical grasses were capable of high summer dry matter production, but Setaria spp. and Panicum coloratum (Kabulabula) appeared most suitable, combining persistence with production. Setaria sphacelata CPI 33452 from which Narok setaria was selected, persisted better and gave higher winter dry matter yields than Nandi and Kazungula setaria and most temperate grasses. Nandi setaria survived only when no N was applied in winter. Following winter rains, temperate grasses outyielded setaria and are probably superior for winter production where irrigation is available. Phalaris hybrid was the only temperate grass to persist for 4 yr. Silverleaf desmodium (Desmodium uncinatum) and Cooper glycine (Glycine wightii) were the most persistent of the 31 tropical legumes tested and silverleaf desmodium and greenleaf desmodium (D. intortum) were the highest yielding. Of 53 temperate legumes tested ladino and Louisiana white clovers (Trifolium repens) were the best. Annual legumes failed to reestablish in the second season due to competition from existing naturalized white clover.--Copyright 1972, Biological Abstracts, Inc.
W73-09545

ECOLOGY OF PASTORAL AREAS IN THE ARID ZONE OF RAJASTHAN,
Central Arid Zone Research Inst., Jodhpur (India).
R. K. Gupta.
Ann Arid Zone. Vol 10, No 3, p 136-157, 1971. Illus.
Identifiers: Animals, Arid zones, Climates, Drought, *Ecology, Grassland, *India (Rajasthan), Land, Management, *Pastoral population, Range, Soils, Stocking.

The area, land use and climate of pastoral areas of arid Rajasthan are reviewed. Grasslands are divided into ephemeral drought evading, perennial drought evading, perennial drought resistant, and perennial drought resisting semi-succulent types. The distribution of the main grassland types with regard to climate and soil types is discussed. The interrelation of grassland, animal population and stocking rate is given. Suggestions for collecting data on consumers of the native vegetation and decomposers are highlighted for a better understanding of the range ecosystem in the arid regions of Rajasthan.--Copyright 1972, Biological Abstracts, Inc.
W73-09550

RICE RESEARCH IN FIJI, 1960-1970: III. WEED CONTROL,
Department of Agriculture, Suva (Fiji). Research Div.
For primary bibliographic entry see Field 03F.
W73-09569

GENERAL REPORT ON WATER RESOURCES SYSTEMS: OPTIMAL OPERATION OF WATER RESOURCES SYSTEMS,
Illinois Univ., Urbana. Dept. of Hydraulic Engineering.
For primary bibliographic entry see Field 02A.
W73-09588

MAKARA, IHD EXPERIMENTAL BASIN NO 6, 1968-1970,
National Water and Soil Conservation Organization, Wellington (New Zealand).
For primary bibliographic entry see Field 02A.
W73-09596

MOUTERE, IHD EXPERIMENTAL BASIN NO 8, 1971,
National Water and Soil Conservation Organization, Wellington (New Zealand).
For primary bibliographic entry see Field 02A.
W73-09597

APPRAISAL OF THE WATER AND RELATED LAND RESOURCES OF OKLAHOMA.
Oklahoma Water Resources Board, Oklahoma City, Region Ten.

Available from NTIS, Springfield, Va 22151 COM-72 11470 Price $3.00 printed copy; $0.95 microfiche. Publication No 40, 1972. 137 p.

Descriptors: *Water resources, *Land development, *Oklahoma, *Water quality, *Water utilization, Streamflow, Groundwater, Hydrology, Climatology, Flood control, Watershed management, Geology, Economics, Oil, Recreation, Wildlife conservation, Water demand.

This is the ninth in a series of 12 reports which are part of the first phase of gathering information basic to the eventual Oklahoma statewide water plan. Region X in north central Oklahoma definitely has water problems; even in areas where water is available, quality is often too poor to allow use for beneficial purposes. Various aspects of the needs and assets of this area are described and discussed. Information is given on the geology, soils, hydrology, climatology, surface water, watershed protection and flood prevention, groundwater water quality, agriculture, manufacturing and industry, oil and gas, power and fuel, and recreation and wildlife. Maps, charts and tables provide data summaries for the region. (Woodard-USGS)
W73-09599

COMPUTER MODEL FOR DETERMINING BANK STORAGE AT HUNGRY HORSE RESERVOIR, NORTHWESTERN MONTANA,
Geological Survey, Menlo Park, Calif.
T. H. Thompson.
Open-file report, 1972. 33 p, 7 fig, 4 tab, 8 ref.

Descriptors: *Bank storage, *Reservoirs, *Montana, *Simulation analysis, Computer programs, Mathematical models, Surface-groundwater relationships, Water yield.
Identifiers: *Hungry Horse Reservoir (Mont).

A mathematical model computes bank storage at Hungry Horse Reservoir in northwestern Montana. The model uses daily reservoir elevations as input. Monthly accumulated bank-storage volumes for the period October 1951 through September 1972 were calculated. About 5.8% of the usable reservoir storage volume would be available from bank storage for at-site power generation and downstream benefits if the reservoir were subjected to a long term, cyclic drawdown. The model is most sensitive to changes in aquifer width on an annual basis and to changes in the storage coefficient on a seasonal basis. The computer model can be used to compute bank-storage volumes whenever historical or assumed elevations are available for Hungry Horse Reservoir. The parameters can be changed to estimate bank-storage volumes at other reservoirs having similar geologic, physiographic, and hydrologic conditions. (Knapp-USGS)
W73-09600

REPORTS FOR CALIFORNIA BY THE GEOLOGICAL SURVEY WATER RESOURCES DIVISION,
Geological Survey, Menlo Park, Calif. Water Resources Div.
For primary bibliographic entry see Field 02E.
W73-09603

AN ECONOMIC APPROACH TO LAND AND WATER RESOURCE MANAGEMENT: A REPORT ON THE PUGET SOUND STUDY,
Washington Univ., Seattle. Dept. of Economics.
For primary bibliographic entry see Field 06B.
W73-09646

GREAT BASIN EXPERIMENT STATION COMPLETES 60 YEARS,
Technical Writers' Services, Ogden, Utah.
W. M. Keck.
Journal of Range Management, Vol 25, No 3, May 1972, p 163-166. 4 fig.

Descriptors: *Watershed management, *Range management, *Climatology, *Planting management, *History, Summer, Floods, Air temperature, Precipitation (Atmospheric), Evaporation, Pressure, Wind velocity, Rainfall, Runoff, Vegetation, Erosion, Sediments, Storms, Revegetation, Ecosystems, Environmental effects, Semiarid climates, *Utah, Great Basin.

The Great Basin Experimental Range in Utah completed 60 years of continuous ecological research in 1972. Research pioneered include watershed management, range management, climatology and plant ecology. A history of the Experiment Station is reviewed in brief. The Station was established in 1912 in response to damages caused by summertime floods in the West. Climate studies include collection and interpretation of air temperature, sunshine, precipitation, evaporation, barometric pressure and wind velocity data. Watershed studies include monitoring rainfall, runoff and vegetatal cover relations with regard to erosion and sediment production due to high intensity summer storms. Plant studies include revegetation efforts and planting management. Aspen studies and game range restoration are also performed by the Station. The role of range science in ecological and environmental research is summarized. (Popkin-Arizona)
W73-09681

FACTORS INFLUENCING INFILTRATION AND EROSION ON CHAINED PINYON-JUNIPER SITES IN UTAH,
Utah State Univ., Logan. Dept. of Range Science.
For primary bibliographic entry see Field 02G.
W73-09684

EXPERIENCE IN USING HERBICIDE SIMAZINE FOR CONTROLLING PLANTS IN WATER BODIES, (IN RUSSIAN),
T. A. Oktyabr'skaya, and I. G. Shishkov.
Med Parazitol Parazit Bolezn. Vol 41 No 1, p 64-66, 1972. English summary.
Identifiers: Duckweed-M, *Herbicides, Horsetail-P, Mace-M, Reeds, *Simazine, Water pollution control, *Weed control.

A greater part of the vegetation was controlled by simazine in 3-6 wk. The following year partial overgrowth occurred predominantly with duckweed, as well as with reedmace and horsetail. Summer treatment was more effective during the period of active growth of hydrophytes. For destruction of the latter in summer a dose of 10 kg/ha was calculated for the active substance (20 kg/ha of 50% moistened powder). A repeated treatment in the season is required for elimination of perennial herbs. When the compound is used in early spring, its dosage should be increased to 12-15 kg/ha. Granulated form is particularly effective. Treatment of water bodies may be done both with hand-operated sprayer and mechanical sprayers. Particularly convenient is MRZh-12 sprayer mounter on a lorry.--Copyright 1972, Biological Abstracts, Inc.
W73-09685

WAHKIAKUM COUNTY CONSOLIDATED DIKING DISTRICT NO. 1, WAHKIAKUM

CONSERVANCY DISTRICTS.
For primary bibliographic entry see Field 06E.
W73-09721

INDIANA DRAINAGE CODE--DRAINAGE BOARDS--POWERS AND DUTIES.
For primary bibliographic entry see Field 06E.
W73-09725

DRAINAGE, SYSTEM OF COUNTY DRAINAGE.
For primary bibliographic entry see Field 06E.
W73-09739

4B. Groundwater Management

WATER QUALITY AS RELATED TO POSSIBLE HEAVY METAL ADDITIONS IN SURFACE AND GROUNDWATER IN THE SPRINGFIELD AND JOPLIN AREAS, MISSOURI,
Missouri Water Resources Research Center, Rolla.
For primary bibliographic entry see Field 05A.
W73-09104

THE TRANSIENT BEHAVIOR OF RECHARGE--DISCHARGE AREAS IN REGIONAL GROUNDWATER SYSTEMS,
Maryland Univ., College Park. Dept. of Civil Engineering.
For primary bibliographic entry see Field 02F.
W73-09105

PROCEEDINGS OF CONFERENCE ON TOWARD A STATEWIDE GROUNDWATER QUALITY SUBCOMMITTEE CITIZENS ADVISORY COMMITTEE, GOVERNORS ENVIRONMENTAL QUALITY COUNCIL.
For primary bibliographic entry see Field 05B.
W73-09113

GROUND-WATER LEVELS IN THE LOWER ARKANSAS RIVER VALLEY OF COLORADO, 1969-73,
Geological Survey, Denver, Colo.
For primary bibliographic entry see Field 07C.
W73-09123

APPRAISAL OF GROUNDWATER FOR IRRIGATION IN THE LITTLE FALLS AREA, MORRISON COUNTY, MINNESOTA,
Geological Survey, Washington, D.C.
J. O. Helgesen.
Geological Survey Water-Supply Paper 2009-D, 1973. 40 p, 19 fig, 1 plate, 4 tab, 17 ref.

Descriptors: *Groundwater resources, *Minnesota, *Glacial drift, Groundwater, Hydrogeology, Aquifers, Water yield, Safe yield, Model studies, Analog models.
Identifiers: *Morrison County (Minn).

Geologic conditions cause groundwater availability to vary widely in the area near Little Falls, Morrison County, Minnesota. The largest and most readily available groundwater source is the glacial outwash sand and gravel from which the soils were derived. Saturated outwash is as much as 50-100 feet thick where the outwash fills a probable former melt-water channel. Transmissivity of the thicker parts of the aquifer approaches or exceeds 100,000 gallons per day per foot, and probable well yields should exceed 1,000 gallons per minute. In about two-thirds of the study area, a saturated thickness of less than 40 feet generally limits well yields to less than 300 gallons per minute. Response to pumping was studied through electric analog analyses by stressing the model aquifer system in accordance with areal variations in expected well yields. Most of the sustained pumpage would be obtained from intercepted base flow and evapotranspiration. Simulated withdrawals totaling 18,000 acre-feet of water per year for 10 years resulted in little adverse effect on the aquifer system. Simulated larger withdrawals, assumed to represent denser well spacing, caused greater depletion (excessive in some areas) of aquifer storage, streamflow, and lake volumes. (Knapp-USGS)
W73-09126

GROUNDWATER RESOURCES OF THE NORCO AREA, LOUISIANA,
Geological Survey, Baton Rouge, La.
R. L. Hosman.
Louisiana Geological Survey and Department of Public Works Water Resources Bulletin, No 18, 1972. 61 p, 7 fig, 3 plate, 1 tab, 19 ref.

Descriptors: *Groundwater resources, *Water wells, *Hydrogeology, *Water quality, *Louisiana, Groundwater movement, Pumping, Water yield, Water utilization, Chemical analysis, Groundwater recharge, Gulf of Mexico, Mississippi River, Aquifer characteristics, Data collections, Hydrologic data, Saline water intrusion, Maps, Saline water-freshwater interfaces.
Identifiers: *Norco area (La).

The Norco area, Louisiana, a site of industrial development on the Mississippi River east of New Orleans, is hydrologically complex. The hydraulic interplay between the major aquifers and the Mississippi River offers possibilities for areal or regional water-management planning, and much of the potential of the aquifer system has already been demonstrated. Relations between freshwater and saltwater in the area are ever changing because of water movement generated by heavy pumping. Salty water is moving into the Norco aquifer from the Gramercy aquifer through the convergence. This salty water is, in turn, being replaced by fresh, very hard water moving into the Gramercy aquifer through contact with the Mississippi River via point-bar deposits. A sizable area in the Gramercy aquifer had already been flushed in this manner, and this water will eventually reach the Norco aquifer is present pumping is continued. Total groundwater pumpage in the area is about 16 million gallons per day. Most of the water is brackish or salty and is used for cooling purposes. The Norco aquifer yields about 9.5 million gallons per day and is capable of yielding much more and fresher water. (Woodard-USGS)
W73-09128

RESPONSE OF KARST AQUIFERS TO RECHARGE,
Agricultural Research Service, Athens, Ga. Southeast Watershed Research Center.
For primary bibliographic entry see Field 02F.
W73-09129

EXPERIMENTAL STUDY OF UNDERGROUND SEEPAGE TOWARD WELLS,
Rhode Island Univ., Kingston. Dept of Engineering.
For primary bibliographic entry see Field 05B.
W73-09131

SAFE YIELD IN WEST MAUI BASAL AQUIFERS,
Hawaii Univ., Honolulu.
F. L. Peterson.
Termination Report, 1972. 2 p, 2 ref. OWRR A-033-HI- (1).

Descriptors: *Groundwater, *Water supply, *Safe yield, *Hawaii, *Aquifers, Mathematical models, Equations, Water yield, Groundwater recharge, Correlation analysis, Evaluation, Groundwater movement.
Identifiers: West Maui Basal Aquifers (Hawaii).

To estimate recharge and safe yield of the Waikapu basal aquifer of West Maui, Hawaii, a correlation was made between rainfall, draft, and head changes for use on basal aquifers on Oahu, Hawaii, in the vicinity of Honolulu. Draft is correlated with a rainfall index, head, and three coefficients. Loss is proportional to the first power of the head, and there is a gain or loss proportional to the first power of the function of rainfall, in values which fluctuate on either side of zero depending on whether the rainfall accumulation is excessive or deficient. Thus, the equation is a linear regression of draft on the values for head and for rainfall. (Woodard-USGS)
W73-09135

ARTIFICIAL RECHARGE IN THE WATERMAN CANYON-EAST TWIN CREEK AREA, SAN BERNARDINO COUNTY, CALIFORNIA,
Geological Survey, Menlo Park, Calif.
J. W. Warner, and J. A. Moreland.
Geological Survey open-file report, November 16, 1972. 26p, 13 fig, 2 plate, 7 ref.

Descriptors: *Artificial recharge, *Groundwater recharge, *Imported water, *California, Feasibility studies, Test wells, Aquifer characteristics, Faults (Geologic), Hydrogeology, Groundwater movement, Water level fluctuations, Infiltration rates, Hydrologic data.
Identifiers: *San Bernardino (Calif), Waterman Canyon (Calif), East Twin Creek area (Calif).

The feasibility of recharging 30,000 acre-feet a year of imported water from northern California in the Waterman Canyon-East Twin Creek area in San Bernardino County, Calif., will depend on the effectiveness of fault K as a barrier to groundwater movement near the land surface. Fault K, about three-fourths of a mile southwest of the south branch of the San Andreas fault, is nearly parallel to the San Andreas fault. This fault is not evident from surface expression but is postulated from geophysical and water-level data. Results of test drilling and infiltration tests reveal that the subsurface material at the spreading grounds is permeable enough to allow recharged water to percolate to the water table. Data indicate that fault K extends into the Waterman Canyon-East Twin Creek area and may impede the lateral movement of recharged water. Fault K has no known surface expression and therefore probably does not affect the highly permeable younger alluvium. If that is so, fault K will be less effective as a barrier to groundwater movement as the recharge mound rises. (Woodard-USGS)
W73-09140

AQUIFER TEST, SOLDOTNA, ALASKA,
Geological Survey, Anchorage, Alaska. Water Resources Div.
G. S. Anderson.
Geological Survey open-file report, April 1972. 17 p, 3 fig, 8 tab.

Descriptors: *Groundwater resources, *Test wells, *Pumping, *Hydrologic data, *Alaska, Water wells, Water yield, Groundwater recharge, Aquifer characteristics, Groundwater movement, Driller s logs, Drawdown, Well data.
Identifiers: *Soldotna (Alaska).

The results of an aquifer test on a production well drilled by Soldotna Drilling Company for the city of Soldotna, Alaska, are presented. The well was tested on July 23, 1971, by Adams, Corthell, Lee, Wince and Associates. The U.S. Geological Survey, in cooperation with the city of Soldotna and the Kenai Penisula Borough, assisted in the test as a part of a boroughwide water-resources study. Drawdown and recovery data obtained by measuring water levels in the Parker well and the Soldotna test well during and following the pumping period are plotted against time on logarithmic paper. The geologic and hydrologic setting of the Soldotna test well and the results of the aquifer test indicate leaky artesian conditions. The production well is deriving water from storage in the main aquifer, storage in the semiconfining units, and induced leakage through the semiconfining units. The values of the leakage and storage parameter, the transmissivity, and storage coefficient of the main aquifer are summarized. (Woodard-USGS)
W73-09143

PLANNING AND DEVELOPING GROUND WATER SUPPLIES,
Dames and Moore, Inc., New York.
R. B. Ellwood, and J. M. Heckard.
Engineering Bulletin No 35, p 1-4, 1969. 3 fig.

Descriptors: Wells, Water wells, Groundwater, *Water utilization, *Planning, *Groundwater availability, Water quality, Exploration, Surveys, Data collections, Aquifer testing, Pump testing, Hydrologic properties.
Identifiers: Well development, Well design, Exploration boring.

Wells and ground water supply systems are shown to present a number of problems which have to be analyzed carefully to enable an optimum use of a ground water resource and an optimum design of a well system. It is demonstrated that the knowledgeable consultant, familiar with the basic principles of ground water occurrence, can be of considerable help to his clients in developing the most suitable, most economical source of water supply. Four phases of ground water supply investigations are outlined, showing successive stages: reconnaissance, survey, exploration, testing, development, design and installation, rehabilitation, and maintenance. Emphasis is placed on good design, since the cost of proper design, high-quality materials, correct installation, and development is repaid many times over because of the resulting lower costs of long-term pumping. It is also emphasized that close attention to drilling techniques, methods of setting well screens, and developing the wells can make the difference between the mediocre well and an excellent well. (Campbell-NWWA)
W73-09243

WATER SUPPLY AND GROUND WATER UTILIZATION IN FOREIGN COUNTRIES,
Dames and Moore, Salt Lake City, Utah.
W. E. Mead.
Engineering Bulletin No 35, p 5-7, 1969. 4 fig.

Descriptors: Wells, Water wells, Africa, Asis, South America, Urban hydrology, Water supply, *Water resources development, Water treatment, United Nations, Surveys, *Population.
Identifiers: *Developing countries, World Health Organization.

Study by the World Health Organization revealed serious need for expanded water supplies. Population projections indicate that, although not an immediate threat, a point will ultimately be reached when all of the easily available, usable fresh water in the world will have been developed. Present urban water supply conditions in developing countries as rated by WHO regional offices show that most countries studied have unsatisfactory to grossly unsatisfactory water supplies. Although sometimes these conditions have resulted from inadequate distribution systems, it is shown that most often the quantity of water available is either unknown or the water resource has not been developed to a sufficient degree. It is further alleged that the exploitation of ground water resources by industry, municipalities, and agriculture has barely begun throughout the world. A basic premise is that large underground reserves of water can be used directly, with little or no treatment, and that large production wells, though expensive, are usually cheaper than the treatment of contaminated surface water. (Smith-NWWA)
W73-09244

WATER WELL CONSTRUCTION AN DESIGN,
Dames and Moore, San Francisco, Calif.
For primary bibliographic entry see Field 08A.
W73-09245

EVALUATING AND MANAGING GROUN WATER SUPPLIES WITH MODELS AND COM PUTERS,
Dames and Moore, Los Angeles, Calif.
J. R. Mount.
Engineering Bulletin 35, p 16-19, 1969. 3 fig.

Descriptors: Model studies, Groundwat *Groundwater availability, Aquifer chara teristics, *Aquifer management, Compute *Computer models, Pump testing, Water suppl Theis equation, Drawdown, Well spacing.
Identifiers: *Ground water management, *Electr analog models.

In planning a water supply system, great pai may be taken with almost every aspect of t development; but all too often one critical eleme is overlooked. This is the crucial factor of wat supply dependability - the capacity to sustain t long-term demand. In the investigation plannin and maintenance of reliable water supplies, preliminary estimate of the available ground wat supply - subsequently refined by application collected data - is extremely valuable. The im portance of periodically reviewing and appraisi collected data in order to learn more about aquifer is emphasized. Predicting the changes water levels in response to pumping through t use of models is explained. Models are designed establish a reasonable correspondence betwee their properties and those of the prototyp aquifers; they can simulate ground water flov storage, recharge, pumping, natural discharge, an impervious boundaries. (Smith-NWWA)
W73-09246

DEEP WELL INJECTION OF LIQUID WASTS,
Dames and Moore, New York.
For primary bibliographic entry see Field 05G.
W73-09247

ARSENIC INTOXICATION FROM WEL WATER IN THE UNITED STATES,
Minnesota Dept. of Health, Minneapolis.
For primary bibliographic entry see Field 05B.
W73-09248

CEMENTING GEO-THERMAL STEAM WELLS
Dow Chemical Co., Tulsa, Okla. Dowell Div.
For primary bibliographic entry see Field 08F.
W73-09252

GROUNDWATER HYDRAULICS OF EXTEN SIVE AQUIFERS,
J. H. Edelman.
International Institute for Land Reclamation an Improvement, Wageningen, The Netherlands 1972. 216 p.

Descriptors: Groundwater, *Groundwater move ment, Porous media, *Hydraulics, Aquifers Mathematical studies, *Mathematical models Aquifer characteristics, Engineering geology *Numerical analysis.
Identifiers: *Groundwater hydraulics, Sche matized models, Laws (Physical).

A series of strongly schematized problems i analyzed with a view to applying the results i groundwater engineering. Mathematical deriva tions are given in full, starting from the fundamen tal physical laws. As a rule, a problem is discusse in four stages: posing the problem, formulating th solution, deriving the formulae and analyzing th result. The term extensive aquifers is used t denote aquifers whose horizontal dimensions ar

CHANGE IN POTENTIOMETRIC HEAD IN THE LLOYD AQUIFER, LONG ISLAND, NEW YORK,

Geological Survey, Mineola, N.Y.
G. E. Kimmel.
Journal of Research of the U S Geological Survey, Vol 1, No 3, p 345-350, May-June 1973. 6 fig, 18 ref.

Descriptors: *Water levels, *Drawdown, *New York, *Aquifers, *Water yield, *Withdrawal, Artesian aquifers, Leakage, Saline water intrusion, Hydrogeology.
Identifiers: *Long Island (NY).

The potentiometric surface of the Lloyd aquifer in 1970 locally was as much as 40 feet lower than in 1900. During this period, withdrawal of water from wells was estimated to exceed 300 billion gallons, and the amount of water released from aquifer storage by compressive forces was estimated to be 1.6 billion gallons (about 0.5% of the withdrawal). The remainder of the withdrawal was derived from downward leakage through overlying aquifers and confining layers and by the displacement of freshwater in the aquifer by landward movement of salty groundwater. (Knapp-USGS)
W73-09293

EVALUATING THE RELIABILITY OF SPECIFIC-YIELD DETERMINATIONS,

Geological Survey, Tucson, Ariz.
R. L. Hanson.
Journal of Research of the U S Geological Survey, Vol 1, No 3, p 371-376, May-June 1973. 5 fig, 2 tab, 12 ref.

Descriptors: *Specific yield, *Drawdown, *Soil water movement, *Aquifer testing, *Moisture content, Storage coefficient, Infiltration, Percolation, Unsaturated flow, Saturated flow, Water level fluctuations.

The specific yield of the alluvial aquifer in the Gila River flood plain in southeastern Arizona was determined using both the time-drawdown method and the soil-moisture-content method. Time-drawdown data measured at 17 observation wells during a 3.5-day aquifer test define an average apparent specific yield of 0.13. Soil-moisture-content data measured at nine access holes during the aquifer test indicated that complete gravity drainage had not been attained in the cone of depression by the end of the drawdown period. The moisture-content data were therefore extrapolated with time to define an average specific yield in the range 0.13 to 0.15. The results obtained by the two methods are in close agreement. However, the significantly lower standard deviation of the results from the moisture-content analysis indicates that extrapolation of the apparent values derived by this method may provide a more reliable estimate of the true specific yield than the apparent values derived by the time-drawdown method. Reliable estimates of specific yield in the zone of seasonal water-level fluctuations are also possible from an evaluation of the soil-moisture change in the spring and summer recession period. (Knapp-USGS)
W73-09297

MANAGEMENT OF GROUND WATER,

Nebraska Univ., Lincoln. Dept. of Agricultural Economics.
L. K. Fischer.
Nebraska Journal of Economics and Business, Vol 11, No 3, p 23-32, Summer, 1972.

Descriptors: *Optimum development plans, *Water supply development, Irrigation water, Aquifiers, Water management (Applied), Benefits, Economic feasibility, *Nebraska.
Identifiers: Stock, Flow, Institutional framework, Incentives.

Using empirical data from Nebraska, groundwater management is considered primarily in terms of irrigation, since 80 percent of the groundwater withdrawn from aquifiers in Nebraska is used for this purpose. However, the principles discussed are peculiar neither to Nebraska nor to irrigation. On the contrary, they generally apply wherever the rate of recharge, natural or induced, is substantial but less than the expected rates of withdrawal. It was assumed that development of water is justified only to the extent that utilization will yield net benefits, and the objective of the utilization of water is to maximize over time the total net benefits derived. Analysis of the history of the present system of incentive and restraints for the development of groundwater and of projections for development over the next decade, however, indicate that this objective can only be achieved by rapidly developing an effective institutional framework for public intervention. This public policy would place restraints on the current rates of development which, if unhindered, will lead to inevitable and substantial overdevelopment. Such overdevelopment, though beguiling in the short run, would be inimical to the public interest in the long run. (Weaver-Wisconsin)
W73-09413

DISTORTION OF THE GEOTHERMAL FIELD IN AQUIFERS BY PUMPING,

Geological Survey, Washington, D.C.
For primary bibliographic entry see Field 02F.
W73-09442

GEOLOGY AND GROUND-WATER RESOURCES OF PRATT COUNTY, SOUTH-CENTRAL KANSAS,

Geological Survey, Lawrence, Kans.
D. W. Layton, and D. W. Berry.
Kansas Geological Survey Bulletin 205, February 1973. 33 p, 15 fig, 8 tab, 47 ref.

Descriptors: *Groundwater resources, *Water quality, *Basic data collections, *Geology, *Kansas, Water wells, Aquifer characteristics, Withdrawal, Water utilization, Groundwater movement, Groundwater recharge, Streams, Surface-groundwater relationships, Topography, Water analysis, Chemical analysis, Mineralogy, Hydrology, Irrigation.
Identifiers: *Pratt County (Kans).

A continuing program of groundwater investigations was begun in 1937 to define and describe the occurrence, availability, and chemical quality of groundwater in Kansas. Pratt County comprises an area of 729 square miles in south-central Kansas. The county is underlain by unconsolidated deposits of Pleistocene age that, in some areas, could yield as much as 3,000 gallons of water per minute to wells. About 26,000 acre-feet of groundwater is withdrawn annually for municipal, industrial, and irrigation uses. This withdrawal probably could be doubled without causing serious depletion of storage in the groundwater reservoir or of flow in the South Fork Ninnescah River. The water generally is suitable chemically for all the common uses. However, water in deposits of Pleistocene age in a small area near Cairo is contaminated by saline water from underlying rocks of Permian age. The dissolved-solids discharge in a 7.7-mile reach of the South Fork Ninnescah River near Cairo increases from 30 to 185 tons per day. (Woodard-USGS)
W73-09453

FUNDAMENTAL PRINCIPLES IN THE STUDY OF WATER AND SALT BALANCES OF GROUNDWATER IN IRRIGATED REGIONS (OSNOVNYYE PRINTSIPY IZUCHENIYA VODNO-SOLEVOGO BALANSA GRUNTOVYKH VOD OROSHAYEMYKH TERRITORIY),

Institute of Hydrogeology and Engineering Geology, Tashkent (USSR).
For primary bibliographic entry see Field 07B.

W73-09464

THE EFFECT OF MIXING AND GRAVITATIONAL SEGREGATION BETWEEN NATURAL GAS AND INERT CUSHION GAS ON THE RECOVERY OF GAS FROM HORIZONTAL STORAGE AQUIFERS,
New Mexico Inst. of Mining and Technology, Socorro.
A. Kumar, and O. K. Kimbler.
Preprint of paper (SPE 3866) to be published in Journal of Petroleum Technology, American Institute of Mining, Metallurgical, and Petroleum Engineers, 1972. 12 p, 9 fig, 1 tab, 11 ref. OWRR A-002-LA (7) A-011-LA (3).

Descriptors: *Injection wells, *Natural gas, *Storage capacity, *Hydrofracturing, *Aquifers, Model studies, Mathematical studies, Secondary recovery (Oil), Oil reservoirs, Aquifer characteristics, Pores.
Identifiers: *Inert cushion gas, Water-gas interfaces.

The underground storage of natural gas during the summer months is widely used by the natural gas industry to supplement the increased winter demand in urban areas. Furthermore, it enables pipe lines to be operated at nearly a 100% load factor throughout the year. Gas injected from the surface under sufficient pressure will desaturate the pore space and provide storage for the gas. Ultimately, the water saturation in the gas-swept region may approach an irreducible minimum value. A calculation procedure, which considers the miscible boundary only, has been developed to calculate the gas recovery efficiency. The calculation procedure was validated by displacements in a horizontal three-dimensional radial consolidated sand model utilizing miscible liquids of differing densities. Only the natural gas-inert gas boundary is considered, which is tantamount to assuming that sufficient inert gas has been injected so that the aquifer water neither contacts nor closely approaches the mixed zone between the gases. Thus, the water-inert gas interface is considered to have no effect on the natural gas-cushion gas boundary. The percent of injected gas recovered during a particular storage cycle should increase with the number of years of operation of the storage project. (Woodard-USGS)
W73-09589

GROUND-WATER RESOURCES OF BRAZORIA COUNTY, TEXAS,
Geological Survey, Austin, Tex.
J. B. Wesselman.
Texas Water Development Board Report 163, February 1973. 199 p, 29 fig, 10 tab, 45 ref.

Descriptors: *Groundwater resources, *Water quality, *Water wells, *Aquifer characteristics, *Texas, Well data, Withdrawal, Pumping, Water yield, Drawdown, Groundwater recharge, Land subsidence, Hydrogeology, Hydrologic data, Basic data collections, Chemical analysis, Water levels, Salinity, Water utilization.
Identifiers: *Brazoria County (Tex).

The Chicot and Evangeline aquifers are the only hydrologic units bearing fresh (less than 1,000 milligrams per liter dissolved solids) or slightly saline water (1,000-3,000 mg per liter dissolved solids) in Brazoria County, Texas. These aquifers are composed of gravel, sand, silt, and clay of Pliocene, Pleistocene, and Holocene age. The average permeability of the Evangeline aquifer is about 250 gpd per square foot. The Chicot aquifer has a range of permeability from 130 to 1,655 gpd per square foot. A large cone of depression occurs in the water-level surface as a result of pumping from the upper part of the Chicot in the Brazosport area of southern Brazoria County. Land-surface subsidence of more than 1.5 feet, attributed mostly to groundwater removal, has taken place in northeast Brazoria County. Groundwater pumpage for all uses in 1967 was about 43 mgd. Of this, 22.6 mgd was used for irrigation, 12.7 for industrial use, and 7.7 mgd for public and domestic supplies. The fresh groundwater potential of the Brazosport area is fully developed or overdeveloped while in some areas in northern Brazoria County, it is relatively undeveloped. (Woodard-USGS)
W73-09590

GROUND-WATER RESOURCES OF DONLEY COUNTY, TEXAS,
Geological Survey, Austin, Tex.
B. P. Popkin.
Texas Water Development Board Report 164, February 1973. 75 p, 9 fig, 9 tab, 18 ref.

Descriptors: *Groundwater resources, *Water wells, *Water quality, *Aquifer characteristics, *Texas, Well data, Withdrawal, Pumping, Water yield, Drawdown, Groundwater recharge, Hydrogeology, Hydrologic data, Basic data collections, Water levels, Chemical analysis, Water utilization, Irrigation.
Identifiers: *Donley County (Tex).

The principal source of groundwater in Donley County, Texas, is the Ogallala Formation. The Whitehorse Group supplies small quantities of water for livestock. The allivium supplies small quantities for domestic use and irrigation. In 1967, about 38,000 acre-feet of groundwater was used in the county. Of this amount, 36,000 acre-feet was pumped for irrigation. Nearly all the water was from the Ogallala Formation. Recharge from precipitation on the outcrop of the Ogallala is estimated at 10,000 acre-feet per year. Water levels in the vicinity of Clarendon have declined as much as 20 feet since 1942 as a result of concentrated pumping for irrigation. Water in the Ogallala Formation is a very hard, calcium bicarbonate water that generally contains less than 1,000 milligrams per liter dissolved solids. About 3 million acre-feet of freshwater is in storage in the Ogallala Formation, but only half of this amount is considered to be available to wells. Water in the Whitehorse Group is a highly mineralized, calcium plus sulfate water that is suitable for livestock, supplemental irrigation, and some industrial uses. Water in the alluvium varies in quality according to the source of recharge. (Woodard-USGS)
W73-09591

APPRAISAL OF THE WATER AND RELATED LAND RESOURCES OF OKLAHOMA.
Oklahoma Water Resources Board, Oklahoma City, Region Ten.
For primary bibliographic entry see Field 04A.
W73-09599

REPORTS FOR CALIFORNIA BY THE GEOLOGICAL SURVEY WATER RESOURCES DIVISION,
Geological Survey, Menlo Park, Calif. Water Resources Div.
For primary bibliographic entry see Field 02E.
W73-09603

HYDROGEOLOGIC CHARACTERISTICS OF THE VALLEY-FILL AQUIFER IN THE STERLING REACH OF THE SOUTH PLATTE RIVER VALLEY, COLORADO,
Geological Survey, Lakewood, Colo.
R. T. Hurr, and P. A. Schneider, Jr.
Open-file report, 1972. 2 p, 1 fig, 6 plate, 4 ref.

Descriptors: *Groundwater resources, *Water wells, *Aquifer characteristics, *Hydrogeology, *Colorado, Valleys, Rivers, Hydrologic data, Maps, Contours, Bedrock, Water table, Saturation, Transmissivity.
Identifiers: *South Platte River Valley (Colo), Valley-fill aquifer, Stream-depletion factor.

This illustrated report of the Sterling reach of the South Platte River Valley in Colorado presents data in 6 maps, scale one mile to the inch, showing location of wells in the reach, bedrock configuration beneath the valley-fill aquifer, water-table contours of the valley-fill aquifer, saturated thickness of the aquifer, transmissivity of the aquifer, and stream-depletion factor of the aquifer. (Woodard-USGS)
W73-09605

HYDROGEOLOGIC CHARACTERISTICS OF THE VALLEY-FILL AQUIFER IN THE BRUSH REACH OF THE SOUTH PLATTE RIVER VALLEY, COLORADO,
Geological Survey, Lakewood, Colo.
R. T. Hurr, and P. A. Schneider, Jr.
Open-File report, 1972. 2 p, 1 fig, 6 plate, 4 ref.

Descriptors: *Groundwater resources, *Water wells, *Aquifer characteristics, *Hydrogeology, *Colorado, Valleys, Rivers, Hydrologic data, Maps, Contours, Bedrock, Water table, Saturation, Transmissivity.
Identifiers: *South Platte River Valley (Colo), Valley-fill aquifer, Stream-depletion factor.

This illustrated report of the Brush reach of South Platte River valley in Colorado presents data in 6 maps, scale one mile to the inch, showing location of wells in the reach, bedrock configuration beneath the valle-fill aquifer, water-table contours of the valley-fill aquifer, saturated thickness of the aquifer, transmissivity of the aquifer, and stream-depletion factor of the aquifer. (Woodard-USGS)
W73-09606

GEOHYDROLOGY OF SUMTER, DOOLY, PULASKI, LEE, CRISP, AND WILCOX COUNTIES, GEORGIA,
Geological Survey, Washington, D.C.
For primary bibliographic entry see Field 07C.
W73-09619

ARIZONA AGRICULTURE - CURRENT STATUS AND OUTLOOK.
Arizona Univ., Tucson. Coll. of Agriculture.
For primary bibliographic entry see Field 03F.
W73-09686

REGULATIONS FOR PREVENTION OF POLLUTION FROM WELLS TO SUBSURFACE WATERS OF THE STATE PRESENTED AT HEARING ON MARCH 21, 1973.
Missouri Clean Water Commission, Jefferson City.
For primary bibliographic entry see Field 06E.
W73-09694

4C. Effects on Water of Man's Non-Water Activities

ECOLOGICAL STUDIES OF THE SURFACE WATERS OF THE WHITEWATER CREEK WATERSHED, WALWORTH, ROCK AND JEFFERSON COUNTIES, WISCONSIN,
S. G. Smith.
Available from the National Technical Information Service as PB-220 351, $3.00 in paper copy, $0.95 in microfiche. Wisconsin Water Resources Center, Madison, Technical Report, 1973. 76 p, 9 tab, 8 fig, 12 ref, append. OWRR A-021-WIS (1) 14-01-0001-1870.

Descriptors: *Nutrients, *Water temperature, *Electrical conductance, Freshwater fish, Fish types, *Ecotypes, *Vegetation, *Aquatic animals, *Aquatic life, *Aquatic plants, *Agricultural watersheds, Sodium chloride, *Waste water

Roadbanks, Soil stabilization, Sedimentation rates, Watershed management.
Identifiers: Logging (Forest).

Erosion plots and sediment dams were used to evaluate the effects of jammer and skyline logging systems on erosion and sedimentation in steep, ephemeral drainages in the Idaho Batholith of central Idaho. Plot data collected periodically over a 5-year period indicated that no difference in erosion resulted from the two skidding systems as applied in the study. Concurrent sediment dam data were used to show that the logging operations alone (excluding roads) increased sediment production only about 0.6 times over the natural sedimentation rate. Roads associated with the jammer logging system increased sediment production an average of about 750 times over the natural rate for the 6-year period following construction.
W73-09374

CHANGES IN THE FOREST FLOOR UNDER UPLAND OAK STANDS AND MANAGED LOBLOLLY PINE PLANTATIONS,
Forest Service (USDA), Oxford, Miss. Southern Forest Experiment Station.
B. P. Dickerson.
J. For. 70 (9): 560-562, Sept 1972, Illus.

Descriptors: *Loblolly pine trees, *Clearcutting, *Erosion control, *Soil stabilization, Forest management, Soil erosion, Surface runoff, Forest soils, Mississippi.
Identifiers: *Litter, *Litterfall, *Forest floor, Thinning.

Three years after all standing timber in a loblolly pine (Pinus taeda L.) plantation in northern Mississippi was removed, the forest floor there averaged 3 tons per acre less than under uncut stands. Reducing pine basal area from 130 to 75 square feet per acre caused a 1-1/2-ton-per-acre decline in forest floor weight and an 18-percent decline in annual litterfall. The forest floor under deadened low-grade oak stands deteriorated rapidly and would have been completely gone in six years had not other vegetation invaded the area.
W73-09376

GROWTH AND HYDROLOGIC INFLUENCE OF EUROPEAN LARCH AND RED PINE 10 YEARS AFTER PLANTING,
Forest Service, (USDA), La Crosse, Wis. Watershed Lab.
R. S. Sartz, and A. R. Harris.
USDA Forest Serv. Res. Note NC-144, 1972. 4 p.

Descriptors: *Flood protection, *Water conservation, *Land management, *Vegetation establishment, Soil properties, Forest management, Land use, Infiltration.
Identifiers: *Litter, *Soil water storage, Soil bulk density.

Ten years after planting, European larch and red pine diameters averaged 11.2 and 9.6 cm., and heights averaged 9.7 and 5.1 m. Litter on the larch plots was twice as heavy as on the pine and unplanted control plots. Organic carbon content of the top 5 cm. of soil appeared to reflect vegetation differences, but soil bulk density did not. The amount of water depleted by the two species was about the same, and it was about twice the amount depleted by a grass and weed cover.
W73-09377

WATER-RESOURCE EFFECTS,
Geological Survey, Washington, D.C.
For primary bibliographic entry see Field 02F.
W73-09432

QUALITY OF STREAM WATERS OF THE WHITE CLOUD PEAKS AREA, IDAHO,
Geological Survey, Washington, D.C.
For primary bibliographic entry see Field 07C.
W73-09435

CLEAR-CUTTING AND ITS EFFECT ON THE WATER TEMPERATURE OF A SMALL STREAM IN NORTHERN VIRGINIA,
Geological Survey, Arlington, Va.
E. J. Pluhowski.
Available from Sup Doc, GPO, Washington D C 20402 - Price $3.00. In: Geological Survey Research 1972, Chapter C; U S Geological Survey Professional Paper 800-C, p C257-C262, 1972. 5 fig, 1 tab, 8 ref.

Descriptors: *Clear-cutting, *Water temperature, *Virginia, *Urbanization, *Urban hydrology, Storm runoff, Vegetation effects, Solar radiation.
Identifiers: Reston (Va).

Tree and shrub removal from a reach of 1,100 feet at the lower end of Colvin Run near Reston, Va., has altered stream-temperature patterns. Because increased solar radiation, especially in summer, maximum water temperature at the lower end of the reach is frequently 1 deg to 3.5 deg C higher than that observed at the upper end. An energy budget, prepared for the period 1415-1500 hours, July 15, 1969, quantifies the principal energy sources controlling stream temperature in the reach. (Knapp-USGS)
W73-09444

FLOODFLOW CHARACTERISTICS OF EAST FORK HORSEHEAD CREEK AT INTERSTATE HIGHWAY 40, NEAR HARTMAN, ARKANSAS,
Geological Survey, Little Rock, Ark.
J. L. Patterson, and J. N. Sullavan.
Open-file report, November 1972. 8 p, 3 fig.

Descriptors: *Floods, *Backwater, *Obstruction to flow, *Arkansas, Routing, Highway effects, Bridges, Abutments, Streamflow, Flow profiles.

The Arkansas State Highway Department requested a study of the East Fork of Horsehead Creek bridge site on Interstate Highway 40, approximately 5 miles west of the junction of U.S. Highway 64 and I-40 at Clarksville, Ark. This report includes a study of the magnitude and duration of the December 10, 1971, flood and the amount of backwater caused by the highway bridge and embankment. The flood of Dec. 10, 1971 which was slightly smaller than the 50-year flood, and has a recurrence interval of about 40 years, was used in the hydraulic analysis. By using routing procedures, it was determined that the elevation in the vicinity of the buildings, 600 feet upstream from the bridge, would have been about 424.6 feet before the highway was built as compared with 426.2 feet with the highway in place. The water elevation would be above 423 feet for about 2 hours without backwater and about 3 hours under present conditions. (Knapp-USGS)
W73-09608

4D. Watershed Protection

HYDROLOGY AND EFFECTS OF CONSERVATION STRUCTURES, WILLOW CREEK BASIN, VALLEY COUNTY, MONTANA, 1954-68,
Geological Survey Washington, D.C.
For primary bibliographic entry see Field 04A.
W73-09167

EFFECT OF TALL VEGETATIONS ON FLOW AND SEDIMENT,
Colorado State Univ., Fort Collins. Dept. of Civil Engineering.
For primary bibliographic entry see Field 04C.
W73-09274

VOLUME WEIGHT OF RESERVOIR SEDIMENT IN FORESTED AREAS,
Forest Service (USDA), Boise, Idaho. Intermountain Forest and Range Experiment Station.
W. F. Megahan.
Journal of the Hydraulics Division, ASCE, Vol 98, No HY8, Proc Paper 9129, August 1972, p 1335-1342.

Descriptors: *Sedimentation, *Reservoir silting, *Organic matter, *Mountain forests, Reservoir storage, Particle size, Watersheds (Basins), Bulk density, Idaho.
Identifiers: *Experiment watersheds, *Idaho batholith, *Volume weight, Undisturbed forest.

Volume weight samples of deposited sediments were collected from small sediment retention dams in up to twelve forested watersheds in Idaho during 1969 and 1970. Volume weights varied considerably among the various watersheds and within a given impoundment. The percentage of organic matter by weight in the samples provided a reliable estimate of sediment volume weight--better than the more conventional approach of relating volume weight to sediment particle size. The organic matter content of sediments may represent a substantial volume of storage loss for reservoirs on forested or range watersheds. In addition, organic sediments may have important ecological implications.
W73-09373

SOIL WATER DISTRIBUTION ON A CONTOUR-TRENCHED AREA,
Forest Service (USDA), Logan, Utah. Intermountain Forest and Range Experiment Station.
For primary bibliographic entry see Field 02G.
W73-09375

CHANGES IN THE FOREST FLOOR UNDER UPLAND OAK STANDS AND MANAGED LOBLOLLY PINE PLANTATIONS,
Forest Service (USDA), Oxford, Miss. Southern Forest Experiment Station.
For primary bibliographic entry see Field 04C.
W73-09376

DESCRIPTION AND HYDROLOGIC ANALYSIS OF TWO SMALL WATERSHEDS IN UTAH'S WASATCH MOUNTAINS,
Forest Service, (USDA), Logan, Utah. Intermountain Forest and Range Experiment Station.
For primary bibliographic entry see Field 02A.
W73-09378

THE RELATION OF FORESTS AND FORESTRY TO WATER RESOURCES,
Georgia Univ., Athens. School of Forest Resources.
For primary bibliographic entry see Field 04A.
W73-09537

FACTORS INFLUENCING INFILTRATION AND EROSION ON CHAINED PINYON-JUNIPER SITES IN UTAH,
Utah State Univ., Logan. Dept. of Range Science.
For primary bibliographic entry see Field 02G.
W73-09684

CORNUDAS, NORTH AND CULP DRAWS WATERSHED, HUDSPETH COUNTY, TEXAS AND OTERO COUNTY, NEW MEXICO (FINAL ENVIRONMENTAL IMPACT STATEMENT).
Soil Conservation Service, Washington, D.C.
For primary bibliographic entry see Field 08A.
W73-09699

WATERSHED FIELD INSPECTIONS–1971.
For primary bibliographic entry see Field 06E.
W73-09704

ALABAMA WATERSHED MANAGEMENT ACT.
For primary bibliographic entry see Field 06E.
W73-09748

05. WATER QUALITY MANAGEMENT AND PROTECTION

5A. Identification of Pollutants

WATER QUALITY AS RELATED TO POSSIBLE HEAVY METAL ADDITIONS IN SURFACE AND GROUNDWATER IN THE SPRINGFIELD AND JOPLIN AREAS, MISSOURI,
Missouri Water Resources Research Center, Rolla.
P. Proctor, G. Kisvarsanyi, E. Garrison, and A. Williams.
Available from the National Technical Information Service as PB-220 245, $3.00 in paper copy, $0.95 in microfiche. Completion Report, April 1973. 56 p, 18 fig, 2 tab, 22 ref, append. OWRR B-054-MO (1) 14-31-0001-3606.

Descriptors: Water quality, *Heavy metals, Cadmium, Mercury, Copper, *Lead, *Zinc, Iron, Surface waters, Groundwater, Springs, Mining, Trace elements, Distribution patterns, *Missouri, Potable water, Water quality standards, Water wells, *Water analysis.
Identifiers: Source rocks, Joplin (Mo), Springfield (Mo).

Some 165 seasonal water samples were collected and analyzed for heavy metals from surface and subsurface sources in a one hundred mile area around Springfield and Joplin, Missouri, respectively. Joplin is in a former large zinc mining district. Springfield is 72 miles east. Locally, cadmium, lead, zinc and iron exceed acceptable PHS standards for drinking water, but the majority of water samples are well within the established limits. Yet, ten percent of the water wells sampled in the Springfield area and twenty-five percent of those sampled in the Joplin area approached or exceeded the PHS limits of the one or more heavy metals for drinking water. High zinc values are related to known zinc-lead mineralization in both areas. Average cadmium values are slightly higher in Joplin, copper content is similar for both areas, and lead content is slightly higher near Joplin. Surface waters in Joplin are 17 times higher in average zinc content than in Springfield, though shallow wells for both areas are similar in zinc content. Iron is higher and more variable in Joplin. Mercury, in very low quantity in both areas, is somewhat higher in the Springfield area. Some seasonal variation occurs in the heavy metal content in both areas. Alternate sources of water are suggested for those areas having heavy metal content in excess of PHS standards. Effects on living systems within areas containing anomalous heavy metal content are unknown.
W73-09104

IRON IN SURFACE AND SUBSURFACE WATERS, GRIZZLEY BAR, SOUTHEASTERN ALASKA,
Alaska Univ., College. Dept. of Geology.
C. M. Hoskins, and R. M. Slatt.
Available from the National Technical Information Service as PB-220 248, $3.00 in paper copy, $0.95 in microfiche. Alaska Institute of Water Resources, Fairbanks, Publication No IWR-29, August 1972, 15 p, 3 fig, 1 tab, 39 ref. OWRR A-019-ALAS (3), and 026-ALAS (4).

Descriptors: *Iron, *Alaska, *Spectrophotometry, Melt water, Pollutant identification, *Glacial drift, Iron compounds.
Identifiers: *Atomic absorption spectrophotometry, Norris Glacier (Alas), Iron precipitates.

tional Training Center, FWQA, Cincinnati, Ohio, are incorporated and applied. An introduction to the importance of sampling, a basic equipment list for reference use, description of preparations and uses for specific pieces of equipment, plus detailed accounts of sampling techniques are given in the section on Sampling. The section on Lab Preparation lists photo illustrated descriptions of equipment used, plus various laboratory media and reagents needed to perform the three basic indicator organism tests. Other sections deal with Laboratory testing (including fecal coliform, fecal strep, and FC/FS ratio), Field Testing, Analysis of Swimming Pool Waters, General Discussions (indicator organisms) and Appendices (Detection of Enteric pathogens, monitoring of non-sanitary microorganisms). (Campbell-NWWA)
W73-09254

INSTRUMENTATION FOR ENVIRONMENTAL MONITORING: WATER.
California Univ., Berkeley. Lawrence Berkeley Lab.
For primary bibliographic entry see Field 07B.
W73-09268

WATER RESOURCES INVESTIGATIONS IN TEXAS, FISCAL YEAR 1973.
Geological Survey, Austin, Tex.
For primary bibliographic entry see Field 07C.
W73-09278

HYDRAULIC EQUIVALENT SEDIMENT ANALYZER (HESA),
Massachusetts Univ., Amherst.
For primary bibliographic entry see Field 07B.
W73-09282

PROGRESS REPORT ON WATER-RESOURCES INVESTIGATIONS IN PALM BEACH COUNTY, FLORIDA,
Geological Survey, Tallahassee, Fla.
For primary bibliographic entry see Field 07C.
W73-09284

THE FRACTIONATION OF HUMIC ACIDS FROM NATURAL WATER SYSTEMS,
Geological Survey, Lakewood, Colo.
For primary bibliographic entry see Field 02K.
W73-09295

FLUORITE EQUILIBRIA IN THERMAL SPRINGS OF THE SNAKE RIVER BASIN, IDAHO,
Geological Survey, Menlo Park, Calif.
For primary bibliographic entry see Field 02K.
W73-09296

DETERMINATION OF RESIDUES FROM HERBICIDE N- (1,1-DIMETHYLPROPYNYL)-3, 5-DICHLOROBENZAMIDE BY ELECTRON CAPTURE GAS-LIQUID CHROMATOGRAPHY.
Rohm and Haas Co., Bristol, Pa. Bristol Research Labs.
I. L. Adler, C. F. Gordon, L. D. Haines, and J. P. Wargo, Jr.
Journal of the Association of Official Analytical Chemists, Vol 55, No 4, p 802-805, July 1972. 2 fig, 4 tab, 4 ref.

Descriptors: *Pesticide residues, Urine, Separation techniques, Herbicides, Gas chromatography, Pollutant identification, Water pollution sources, Alfalfa, Lettuce, Tomatoes, Legumes, Soils, Milk, Solvent extractions, Radiochemical analysis, Chlorinated hydrocarbon pesticides, Organic pesticides, Halogenated pesticides, Digestion, Soil analysis, Sugar beets.
Identifiers: N- (1 1-dimethylpropynyl)-3 5-dichlorobenzene, *Gas liquid chromatography, Sample preparation, *Biological samples, *Electron capture gas chromatography, Liver, Kidneys, Muscle, Fat, Metabolites, Egg white, Egg yolk, Feces, Petroleum ether, Benzene, Hexane, Methanol, Florisil, C-14, Chemical recovery, Detection limits, Co-distillation, Method validation.

Terminal residues of n- (1,1-dimethylpropynyl)-3,5-dichlorobenzamide and its metabolites were determined by an electron capture gas chromatographic analysis of various biological samples which had been predigested with sulfuric acid and methanol to convert the residues to methyl 3,5-dichlorobenzoate (MDCB). In preparing the digested biological samples (including alfalfa, eggs, fat, feces, liver, kidney, muscle, lettuce, tomatoes, milk, soil, beets, urine) for chromatographic analysis, the MDCB was first codistilled from the reaction mixture and purified by chromatography on Florisil. This GLC analytical procedure proved sensitive to residues of 0.01 ppm. The procedure was validated by recovering labelled C-14 MDCB which had been added to samples prior to the digestion step; recoveries were determined by measuring the peak due to MDCB on the gas chromatogram as well as the radioassay of the final eluant. (Long-Battelle)
W73-09300

DETECTION OF HEXACHLOROBENZENE RESIDUES IN DAIRY PRODUCTS, MEAT FAT, AND EGGS,
Australian Customs Lab., Seaton.
R. J. Smyth.
Journal of the Association of Official Analytical Chemists, Vol 55, No 4, p 806-808, July 1972. 3 tab, 6 ref.

Descriptors: *Pesticide residues, *Chlorinated hydrocarbon pesticides, *Pollutant identification, *Gas chromatography, Heptachlor, DDE, Dieldrin, DDD, DDT, Aldrin, Endrin, Separation techniques, Organic pesticides, Dairy products, Fungicides, Wheat, Chemical analysis.
Identifiers: *Biological samples, Thin layer chromatography, Detection limits, Sample preparation, Hexachlorobenzene, Florisil, Hexane, Alpha-benzene hexachloride, Fat, Eggs, Lindane, Heptachlor epoxide, Butter, Cheese, Chemical recovery, Cleanup.

A gas chromatographic procedure, using deactivated Florisil for cleanup of fatty acids, is employed for the detection of hexachlorobenzene residues extracted with hexane from dairy products, meat fat, and eggs. The extracted residues, including alpha-benzene hexachloride, hexachlorobenzene, lindane, heptachlor, aldrin, heptachlor epoxide, DDE, dieldrin, endrin, DDD, and DDT, were eluted from the column containing a mixture of DC-200 and QF-1. The retention times for these pesticides are tabulated in reference to the retention time of aldrin. The pesticides detected by this method were confirmed by thin-layer chromatography; the lower detection limit of the pesticides examined was about 0.002 ppm based on a 1 g sample of fat. (Long-Battelle)
W73-09301

ORTHOPHOSPHATE DETERMINATIONS USING PREMEASURED REAGENTS,
Missouri Univ., Columbia. Dept. of Microbiology.
R. C. Baskett.
Water and Sewage Works, Vol 120, No 1, p 47, January 1973. 1 tab, 2 ref.

Descriptors: *Spectrophotometry, *Water analysis, Colorimetry, Phosphates.
Identifiers: Accuracy, *Orthophosphates, Reagents.

A simple and convenient method for determining orthophosphate in water involves using a spectrophotometer, premeasured powder pillows of reagent (Murphy and Riley formula), and 5 ml water

samples. The spectrophotometer is operated at a wavelength setting of 710 millimicrons. The percent transmittance can be converted to mg/l phosphate by means of a standard curve. The test has proved to be rapid and well within the established limits of accuracy for conventional orthophosphate measurements. (Little-Battelle)
W73-09305

HYDROGEN PEROXIDE AIDS IN MEASURING SLUDGE OXYGEN UPTAKE RATES,
Iowa State Univ., Ames. Dept. of Civil Engineering.
For primary bibliographic entry see Field 05B.
W73-09308

FLUORESCAMINE: A REAGENT FOR ASSAY OF AMINO ACIDS, PEPTIDES, PROTEINS, AND PRIMARY AMINES IN THE PICOMOLE RANGE,
Roche Inst. of Molecular Biology, Nutley, N.J.
S. Udenfriend, S. Stein, P. Bohlen, W. Dairman, and W. Leimgruber.
Science, Vol 178, No 4063, p 871-872. November 24, 1972. 3 fig, 1 tab, 15 ref.

Descriptors: *Amino acids, *Proteins, *Peptides, *Assay, *Aqueous solutions, Fluorescence, Pollutant identification, Hydrogen ion concentration, Organic acids.
Identifiers: *Primary amines, *Fluorescamine, Reagents, Organic solvents, Thin layer chromatography, Gamma-aminobutyric acid, Beta-alanine, Histamine, Catecholamines, Amphetamine, Spermine, Amino sugars, Spermidine.

Fluorescamine is a new reagent for the detection of primary amines in the picomole range. Its reaction with amines is almost instantaneous at room temperature in aqueous media. The products are highly fluorescent, whereas the reagent and its degradation products are nonfluorescent. An important application has been the assay of proteins during protein purification procedures. Assays carried out manually can be used to determine 0.5 microgram of protein. With a semi-automated procedure (similar to the one for amino acid assay) and a Bio-Gel column to separate nonprotein material, as little as 0.05 microgram of protein can be assayed. Fluorescamine can be used as an amine reagent not only in aqueous solution but also in organic solvents and on solids. It has been used as a spray to detect amino acids and peptides on thin layer chromatograms. As little as 20 pmole of each can be detected. Fluorescamine yields intense fluorescence with other primary amines of biological importance. Thus, gamma-aminobutyric acid, beta-alanine, histamine, catecholamines, amphetamine, amino sugars, spermine, and spermidine also yield the characteristic fluorophore. Procedures for extracting and assaying the two polyamines in milligram quantities of tissue have been found feasible. (Holoman-Battelle)
W73-09310

ANALYTICAL METHOD FOR DETERMINATION OF INORGANIC NITRATES IN NATURAL STREAM WATER,
Naval Ordnance Lab., White Oak, Md.
D. Sam.
Report No NOLTR 72-295, December 26, 1972. 10 p, 2 fig, 2 tab, 1 ref. Contract No ORD332005-201-23.

Descriptors: *Nitrites, Methodology, *Water analysis, *Natural streams, *Pollutant identification, Chemical analysis, Spectrophotometry, Anion exchange, Reliability.
Identifiers: Diazotization, Azo dyes, Ion exchange resins, Detection limits, Sample preparation, Absorbance, Sensitivity, Preconcentration.

A sensitive and reliable procedure has been developed which is applicable for determination of inorganic nitrites in natural stream water in concentrations as low as 0.1 micromoles/liter. Samples are concentrated by an ion exchange procedure prior to the spectrophotometric measurement of the nitrite in the form of an azo dye. The spectrophotometric measurement of the azo dye is found to conform to Beer's law. The nitrite sample is treated with sulfanilic acid and N- (1-Naphthyl)-ethylenediamine dihydrochloride to form a red azo dye. The dye is passed through a column of Dowex 1-x8, 50-100 mesh, anion exchange resin and then it is eluted with 60 percent acetic acid. The resulting effluent is measured in a Cary 16 spectrophotometer at 550 nm in a cell of 10 cm optical path. The application of this method to the determination of natural stream water is discussed. (Holoman-Battelle)
W73-09311

THALASSIOSIRA ECCENTRICA (EHRENB.) CLEVE, T. SYMMETRICA SP. NOV., AND SOME RELATED CENTRIC DIATOMS,
Texas A and M Univ., College Station. Dept. of Oceanography.
For primary bibliographic entry see Field 05B.
W73-09312

DISSOLVED FREE AMINO ACIDS IN SOUTHERN CALIFORNIA COASTAL WATERS,
California Inst. of Tech., Pasadena. W. M. Keck Lab. of Engineering Materials.
For primary bibliographic entry see Field 05B.
W73-09324

PREPARATION OF MEMBRANE FILTER SAMPLES FOR DIRECT EXAMINATION WITH AN ELECTRON MICROSCOPE,
Texas A and M Univ., College Station. School of Geosciences.
J. E. Harris, T. R. McKee, and R. C. Wilson, Jr.
Limnology and Oceanography, Vol 17, No 5, p 784-787, September 1972. 3 fig, 7 ref.

Descriptors: *Electron microscopy, *Laboratory equipment, Distribution patterns, Aquatic microorganisms, Clay minerals.
Identifiers: *Membrane filters, *Sample preparation, Cellulose membranes, Particulate matter.

A simple apparatus is described for transferring samples of suspended matter collected on Millipore filters to specimen grids for examination with an electron microscope. The apparatus is made from glassware available in most laboratories and is relatively inexpensive. Samples of suspended matter prepared with this method have been used to obtain the particle size distribution and shapes for particles between 5 and 0.05 microns. The distribution of coccoliths and other nanoorganism remains in the water column of the Gulf of Mexico are being studied. Clay minerals have been studied and successfully identified using selected area electron diffraction on samples prepared with this apparatus. This method should also be useful in preparing samples for light microscopy. (Holoman-Battelle)
W73-09326

FRESHWATER HARPACTICOID COPEPODS OF NEW ZEALAND, I. ATTHEYELLA ELAPHOIDELLA (CANTHOCAMPTIDAE),
Auckland Univ. (New Zealand). Dept. of Zoology.
M. H. Lewis.
New Zealand Journal of Marine and Freshwater Research, Vol 6, Nos 1 and 2, June 1972. 14 fig, 14 ref.

Descriptors: *Copepods, *Crustaceans, *Aquatic animals, *Systematics, Invertebrates, Ecological distribution, Habitats, Mosses, Peat.
Identifiers: *Harpacticoids, *Speciation, Attheyella (Delachauxiella) spp., Canthocamptus

A gamma spectroscopic method, using a sodium iodide (thallium-activated) crystal, was evaluated by 25 collaborators for the determination of cesium-137 in milk. Triplicate analyses were performed on 2 milk samples with high and low activities. The values obtained were statistically analyzed by the use of laboratory mean values. The statistical analysis showed that the collaborators were able to determine levels of activity in the two samples satisfactorily. There was a greater dispersion of results for the low level sample which was not surprising and was to be expected. The percent error found for the low level of Cs-137 activity was 10.2 and that for the high level was 5.9. Overall average recoveries of 295 plus or minus 18.1 and 53 plus or minus 5.3 pCi/L were obtained for samples containing 305 and 52 pCi/L, respectively. The presence of iodine-131 and barium-140 did not interfere with the analyses. This method has been recommended to be adopted as official first action since: (1) the slight bias on the low side and the greater dispersion of results for the low level sample were not considered serious; and (2) the method is acceptable for the analysis of various levels of Cs-137 activity in fluid milk. (Holoman-Battelle)
W73-09333

DETERMINATION OF STRONTIUM-90 IN WATER: COLLABORATIVE STUDY,

Environmental Protection Agency, Winchester, Mass. Quality Control Service.
E. J. Baratta, and F. E. Knowles, Jr.
Journal of the Association of Official Analytical Chemists, Vol 56, No 1, p208-212, January 1973. 3 tab, 5 ref.

Descriptors: *Water analysis, *Pollutant identification, Methodology, Strontium radioisotopes, Chemical analysis, Aqueous solutions, Chemical precipitation.
Identifiers: *Collaborative studies, *Interlaboratory studies, *Sr-90, Method evaluation, Precision, Y-90, Yttrium radioisotopes, Accuracy.

A procedure involving carbonate precipitation and purification with nitric acid was evaluated by 11 collaborators for the determination of strontium-90 in water. Four 1050-ml samples were analyzed by each collaborator. Unknown to the collaborators, these samples consisted of 2 sets of similar, but not identical, samples. The request was to perform single analyses, using the procedure without modification. The ingrowth level of yttrium-90 was counted and used to calculate the strontium-90 activity present. The accuracy of the method ranged from 9 to 16 percent and within-laboratory precision ranged from 13 to 19 percent for strontium-90 levels of 28.2 to 99.6 pCi/L. A slight bias on the low side, caused by individual laboratory techniques, was not considered serious. The method has been adopted as official first action. (Holoman-Battelle)
W73-09334

ION EXCHANGE DETERMINATION OF STRONTIUM-89 AND STRONTIUM-90 IN MILK: COLLABORATIVE STUDY,

Environmental Protection Agency, Winchester, Mass. Quality Control Service.
E. J. Baratta, and F. E. Knowles, Jr.
Journal of the Association of Official Analytical Chemists, Vol 56, No 1, p 213-218, January 1973. 5 tab, 2 ref.

Descriptors: *Ion exchange, *Milk, *Pollutant identification, Methodology, Strontium radioisotopes, Laboratory tests, Chemical analysis, Aqueous solutions, Separation techniques, Radioactivity.
Identifiers: *Sr-89, *Sr-90, *Collaborative studies, Precision, Method evaluation, Interlaboratory studies, Accuracy, Yttrium radioisotopes, Y-90, Recovery, Errors.

A method for the determination of strontium-89 and strontium-90 in milk, involving ion exchange and purification with nitric acid, was evaluated by 11 collaborators. The accuracy of the method for strotium-89 ranged from 2 to 56 percent and the within-laboratory precision ranged from 11 to 76 percent for a concentration range of 68-480 pCi/L; values for strontium-90 in a concentration range of 50.2-116.2 pCi/L were 6.4 to 10 and 4 to 11 percent, respectively. The slight bias on the high side for strontium-90 was not considered serious. The collaborators had difficulty with strontium-89 analyses, as shown by the high precision and accuracy estimation values. (Holoman-Battelle)
W73-09335

ECOLOGICAL OBSERVATIONS ON HETEROTROPHIC, METHANE OXIDIZING AND SULFATE REDUCING BACTERIA IN A POND,

Hydrobiologisch Institutt, Nieuwersluis (Netherlands).
Th. E. Cappenberg.
Hydrobiologia, Vol 40, No 4, p 471-485, December 15, 1972. 7 fig, 2 tab, 22 ref.

Descriptors: *Aquatic bacteria, *Sulfur bacteria, Ecological distribution, *Temporal distribution, *Methane bacteria, *Seasonal, Dissolved oxygen, Sulfates, Hypolimnion, Methane, Epilimnion, Aerobic bacteria, Hydrogen ion concentration, Eutrophication, Anaerobic bacteria, Water sampling, Water temperature, Gas chromatography, Volumetric analysis, Spring, Summer, Autumn, Winter, Oxidation, Reduction (Chemical).
Identifiers: *Heterotrophic bacteria, Vertical distribution, Metalimnion, Substrate utilization, Microaerophiles, Lake Vechten, Valas Test Water Collector, Culture media, Membrane filters, Counting, Oxygen electrode, Flame ionization gas chromatography.

Numbers of heterotrophic, methane oxidizing and sulfate reducing bacteria were counted in Lake Vechten and a dynamic distribution pattern was found. A maximum of heterotrophs (numbers of greater than or equal to one billion bact./l) occured in the deepest part of the lake in spring and in the metalimnion during summer-stratification. These bacteria use nearly all available oxygen in the hypolimnion. It was found that the concentration of available organic material and the oxygen tension caused the numbers of heterotrophs in the metalimnion to be high. The maximal numbers of methane oxidizers (number of greater than or equal to 500,000 bact./l) were found at a depth of maximal methane concentration: the de-oxygenated hypolimnion. Preliminary evidence indicated that these organisms were facultative methane oxidizers and must be regarded as microaerophyllics. By oxidizing methane they removed the residual oxygen under the metalimnion. The sulfate reducing bacteria could be observed in the hypolimnion. The sulfate reducing bacteria could be observed in the hypolimnion only. Decreased SO4- (2-) concentration and increased numbers of bacteria were found in the bottom water. An association between the methane oxidizers and the sulfate reducers could be deduced. It was assumed that favourable redox requirements for obligate anaerobic sulfate reducers were the results of the activities of the methane oxidizing bacteria. The dynamic distribution equilibrium of the investigated groups of bacteria was disturbed by the autumn turn-over. The heterotrophic and methane oxidizing bacteria decreased in number and were equally distributed; no sulfate reducers could be detected in the free water. (Holoman-Battelle)
W73-09337

LIMNOLOGICAL INVESTIGATIONS IN THE AREA OF ANVERS ISLAND, ANTARCTICA,

Virginia Polytechnic Inst. Blacksburg. Dept. of Biology.
For primary bibliographic entry see Field 05C.
W73-09339

POLYCHAETOUS ANNELIDS COLLECTED BY 'UMITAKA-MARU' FROM THE ARABIAN GULF,

Kuwait Univ. Dept. of Zoology.
M-B. M. Mohammad.

Hydrobiologia, Vol 40, No 4, p 553-560, December 15, 1972. 10 ref.

Descriptors: *Annelids, *Invertebrates, *Systematics, Marine animals, Water temperature, Salinity, Dissolved oxygen.
Identifiers: *Polychaetes, *Arabian Gulf, *Pelagic animals, Speciation, Lepidonotus carinulatus, Euphrosine foliosa, Plotohelmis capitata, Tomopteris, Leocrates claparedii, Syllis gracilis, Autolytus cf. A. lonistaffi, Aglaophamus, Glycera prashadi, Eunice antennata, Eunice indica, Palola siciliensis, Lysidice ninetta, Chaetopterus variopedatus, Ammotrypane aulogaster, Pista cristata, Polycirrus coccineus.

Seventeen species of polychaetous annelids, belonging to 12 families, have been identified from a collection made by the Japanese research 'Umitaka-Maru' in December 1968 from the Arabian Gulf. The material was collected from 12 sites along the coasts of Kuwait, Bahrain, Qatar, and the Trucial States. During the collection period, the mean sea water temperature, salinity, and dissolved oxygen near the bottom (depth, 15 to 20 metrics) at the stations at which the polychaetes were collected were in the neighborhood of 24.1 degrees C, 41.0 ppt, and 4.6 cc/1, respectively (Tokyo University of Fisheries and Kuwait Institute for Scientific Research, 1969). Two pelagic families (Alciopidae, Tomopteridae) and five species (Plotohelmis capitata, Tomopteris sp., Autolytus cf. A. longistaffi, Lysidice ninetta, Pista cristata) have been recorded new to the fauna of the Arabian Gulf. (Holoman-Battelle)
W73-09341

DIATOMS FROM THE DEVIL'S HOLE CAVE FIFE, SCOTLAND,

Elm Bank, Hawick (Scotland).
J. R. Carter.

Nova Hedwigia, Vol 21, Nos. 2-4, p657-681, 1971. 113 fig, 37 ref.

Descriptors: *Diatoms, *Chrysophyta, Aquatic algae, *Systematics, Ecological distribution, *Marine algae, Pollutant identification.
Identifiers: *Speciation, *Scotland, *Devil's Hole Cave (U.K), Achnanthes spp., Amphora granulata, Caloneis spp., Cymbella spp., Denticula tenuis, Diploneis ovalis, Eunotia spp., Fragilaria spp., Gomphonema spp., Licmophora oedipus, Melosira dickiei, Navicula spp., Nitzschia spp., Pinnularia spp., Pleurosigma rigidum, Stauroneis leguman, Synedra spp.

Material collected by the late Mr. George West of the University College, Dundee, in the Devil's Hole Cave in Kincraig Hill, Fife, Scotland has been examined for the presence of diatoms. The preparations yielded an abundance of small and delicately structured forms, several of which could not be identified with certainty. A total of 94 species and other subcategories are reported, five of which are new to science. Caloneis borealis, Cymbella diavola, Navicula variolinea, Navicula vula, and Nitzschia disputata nov. spp. are described. The types of the new species are lodged in the British Museum (No. 77777 and 77778). (Holoman-Battelle)
W73-09343

ADDITIONS TO THE MARINE ALGAL FLORA OF GHANA I.

Ghana Univ., Legon. Dept. of Botany.
D. M. John, and G. W. Lawson.

Nova Hedwigia, Vol 21, Nos. 2-4, p817-842, 1971. 14 fig, 94 ref.

Descriptors: *Marine algae, Plant morphology, Ecological distribution, Spatial distribution, *Systematics, *Chlorophyta, *Phaeophyta, *Rhodophyta, Atlantic Ocean, Africa, Pollutant identification.
Identifiers: *Ghana.

A list is given of 36 new records of marine algae from the Ghanaian coast. It includes information in the annotations on their morphology and other features of taxonomic interest together with notes on their localities and world distributions. A large proportion of these new records are of plants confined to the sublittoral (21) and this reflects the flora of new areas where collecting has now been made possible by aqua lung diving. Thirteen of these new records are of species not previously known from West Africa and 5 of these plants are new to the Atlantic Ocean. (Holoman-Battelle)
W73-09344

THE LIFE-HISTORY OF SPHACELARIA FURICIGERA KUTZ. (PHAEOPHYCEAE) II. THE INFLUENCE OF DAYLENGTH AND TEMPERATURE ON SEXUAL AND VEGETATIVE REPRODUCTION,

Groningen Rijksuniversiteit, Haren (Netherlands). Afdeling Plantenoecologie.
For primary bibliographic entry see Field 05C.
W73-09345

USE OF A COMMERCIAL DREDGE TO ESTIMATE A HARDSHELL CLAM POPULATION BY STRATIFIED RANDOM SAMPLING,

Rhode Island Dept. of Natural Resources, Providence. Div. of Fish and Wildlife.
For primary bibliographic entry see Field 05B.
W73-09348

AN INSTRUMENT FOR MEASURING CONDUCTIVITY PROFILES IN INLETS,

British Columbia Univ., Vancouver. Inst. of Oceanography.
D. M. Farmer, and T. R. Osborn.

Journal of the Fisheries Research Board of Canada, Vol 29, No 12, p 1767-1769, December 1972. 6 fig, 1 ref.

Descriptors: *Conductivity, *Estuaries, *Monitoring, *Instrumentation, *Profiles, *Measurement, *Density stratification, *Inlets (Waterways), Automation, On-site data collections, Physical properties, Operation, Design, Calibration.
Identifiers: Sensors, Digital chart recorder.

A description is given of an instrument for monitoring conductivity profiles in the upper few meters of stratified estuarine waters. Watertight containers of P.V.C. tubing separately house the batteries and the electronics. Four low specific gravity 6-volt, lead-acid batteries provide the power. Each probe was calibrated by immersing it in a large container of salt water of known temperature, for which the conductivity could be determined accurately by other means. After noting the output, the cell was removed from the salt solution and a conducting loop in series with a potentiometer was passed through the probe. In this way, it was possible to find a cell constant, Kc, in terms of the solution conductivity, Sc, and the 'equivalent resistance', Req, of the potentiometer that yielded the same output as that obtained with the cell in solution. The instrument measures the electrical conductivity of seawater at 14 depths by successively interrogating, for 1 min at a time, each of the 14 conductivity probes on the chain. There is a 15th probe inside the instrument housing which serves as a check on any drift in the response of the electronics. Output from the conductivity measuring circuit is fed to a Rustrak chart recorder. A converted fiberglass marker buoy with a lid at one end and an inner sheath of plastic foam affords shock protection and flotation for the battery and instrument cases. The instrument runs for one week between chart replacements. Contours of constant conductivity derived from the chart records have provided a grap description of wind effects in Alberni Inl (Holoman-Battelle)
W73-09349

MERCURY REMOVAL FROM FISH PROTE CONCENTRATE,

Fisheries Research Board of Canada, Halif (Nova Scotia). Halifax Lab.
L. W. Regier.

Journal of the Fisheries Research Board Canada, Vol 29, No 12, p 1777-1779, Decemb 1972. 3 tab, 4 ref.

Descriptors: *Mercury, *Solvent extractior *Separation techniques, Methodology, Hea metals, Alcohols, Fish, Marine fish.
Identifiers: *Fish protein concentrate, *Swor fish, Halifax propanol extraction method, Orgar solvents, Ethyl alcohol, Isopropyl alcohol, Biolo ical samples, Xiphias gladius, Hydrochloric acid.

The potential reclamation of mercury co taminated proteinaceous food by an extracti procedure such as employed in marking fi protein concentrate was investigated. Pieces frozen swordfish (Xiphias gladius) known to co tain high levels of mercury were thawed in air room temperature, chopped, and deboned. A po tion of the deboned meat was refrozen polyethylene bags and kept at -18 C. Th remainder of the meat was used to make FPC i the pilot plant by the Halifax Isopropanol (IPA extraction method (Power 1962). Samples we taken of the filter cake after each of the three e tractions and of the product after drying. Thes samples, together with a sample of the debone meat, were analyzed for mercury by the atomi absorption method (Uthe et al., 1970). Since th regular process did not remove the Hg, th complete FPC was re-extracted with alcoho acidified with HC1. The addition of hydrocholori acid to the isopropanol extractant was found t give removals as high as 93 percent from dr swordfish protein concentrate. Preliminary studie of the variables indicated acid concentration, kin of alcohol (ethanol and isopropanol), alcohol con centration, extractant volume, and number of ex tractions were important in the extraction (Holoman-Battelle)
W73-09351

FINE STRUCTURE OF SOME BRACKISH -POND DIATOMS,

Rhode Island Univ., Kingston. Narraganset Marine Lab.
R. E. Hargraves, and M. Levandowsky.

Nova Hedwigia, Vol 21, Nos 2-4, p 321-336, 1971 34 fig, 35 ref.

Descriptors: *Electron microscopy, *Diatoms *Brackish-water, *Chrysophyta, *Systematics *Ecological distribution, Marine algae, Nannoplankton, Eutrophication, Water temperature Salinity, Hydrogen ion concentration, Phytoplankton, Water properties, Summer, Pollutant identification.
Identifiers: *Ultranstructure, Sample preparation, Speciation, Achnanthes hauckiana, Amphiprora hyalina, Amphora tenuissima, Bellerochea Cyclotella cryptica, Cyclotella meneghiniana Cylindrotheca closterium, Fragilaria pinnata Fragilaria virescens var. subsalina, Navicula cryptocephala, Navicula cryptolyra, Nitzcschia frustulum, Nitzschia ovalis, Skeletonema costatum Thalassionema nitzschioides, Thalassiosira alleni Thalassiosira pseudonana.

Samples of diatoms ere taken during summer from a brackish pond on the northern shore of Long Island, New York, 0.25 mile west of Mt. Sinai Harbor mouth. The diatoms were examined in the Philips EM-200 electron microscope at 80 kv as carbon replicas and as direct preparations of acid cleansed unreplicated frustules. Some of the taxa

likelihood (ML) estimates of the unknown parameters is then described and the proposed methods are illustrated by an example of the analysis of a quantal assay of trypanosomes involving a mixture of two distinct populations. The calculation of the parameter estimates involved an iterative procedure which is only practicable using a digital computer, and it is pointed out that the quantity and quality of the experimental data must conform to very stringent criteria if an analysis is to be worthwhile. (Little-Battelle)
W73-09363

BAYESIAN ESTIMATION AND DESIGN OF EXPERIMENTS FOR GROWTH RATES WHEN SAMPLING FROM THE POISSON DISTRIBUTION,
American Cyanamid Co., Stamford, Conn.
D. W. Behnken, and D. G. Watts.
Biometrics, Vol 28, No 4, p 999-1009, December 1972. 2 fig, 3 tab, 6 ref.

Descriptors: *Estimating, *Growth rates, *Algae, *Statistical methods, Bioassay, Sampling.
Identifiers: *Data interpretation, *Experimental design, Poisson distribution, Selenastrum capricornutum.

Data on algal growth rates were used to investigate the problem of estimating the growth rate parameter, beta, for a process with an expected value lambda sub i equal to alpha e to the beta t sub i power. This equation produces data from a Poisson distribution p of lambda sub i at time t sub i (i equals 1,2, ..., n). The posterior distribution for beta is derived, and an approach to designing experiments in this situation is suggested. (Little-Battelle)
W73-09364

MICROBIAL POPULATION OF FEEDLOT WASTE AND ASSOCIATED SITES,
Agricultural Research Service, Peoria, Ill. Northern Regional Research Lab.
For primary bibliographic entry see Field 05B.
W73-09366

ENTEROBACTERIA IN FEEDLOT WASTE AND RUNOFF,
Agricultural Research Service, Peoria, Ill. Northern Regional Research Lab.
For primary bibliographic entry see Field 05B.
W73-09367

WATER POLLUTION MONITORING,
California Univ., Berkeley. Dept. of Geoscience Engineering.
U. Conti.
Industrial Photography, Vol 21, No 7, p 30-31, 49, July 1972.

Descriptors: *Water pollution, *Monitoring, *Automation, *Photography, *Water properties, *Mechanical equipment, Water quality, Physical properties, Chemical properties, Cameras, Reliability, Depth, Dissolved oxygen, Water temperature, Salinity, Hydrogen ion concentration, Chlorides, Ions, Sulfides, Ambient light.
Identifiers: *Sensors.

Members of the Geoscience Engineering Department at the University of California, Berkeley, are testing an automatic, continuous, in situ water pollution monitoring system which promises to be both more economical and easier to operate than a comparable system already in use. This system consists of a towed vehicle about four feet long, capable of following a given vertical path (constant depth of constant distance from the bottom) and housing a number of sensors. The parameters measured are depth, dissolved oxygen, temperature, salinity, pH, chloride ion activity, sulfide ion activity and ambient light. The system is powered by its own battery pack and does not require an electrical cable incorporated in the tow line. The problem of recording the eight parameters (plus two internal checks and time for a total of 11 parameters) has been solved very economically with a movie camera by translating the various parameters to voltages, which are displayed sequentially in alphanumeric form on a Weston Model 1292 digital voltmeter. A Bell and Howell 16mm magazine camera takes single-frame pictures of the digital voltmeter, of an electric watch and of the scanner readout which shows which parameter is being measured. The camera has been modified to be motor driven at a speed of 20 frames per minute. As no changes have been made on the shutter mechanism, each frame is exposed for about 1.5 seconds. A standard 50-foot 16mm magazine contains 2000 frames or data points, and supplies more than 1.5 hours of continuous record. The film is read on a small standard editor, and the data from one 50-foot reel can be transcribed onto tables or computer cards in about two hours. The main advantages of this system are great reliability, ease of operation and low cost. The entire system has been tested extensively at sea in Mexico and in the San Francisco Bay. The recording system worked flawlessly, recording very small changes in the measured parameters. (Holoman-Battelle)
W73-09369

CALIBRATING PLATINUM RESISTANCE THERMOMETERS,
Westinghouse Electric Corp., Pittsburgh, Pa.
R. P. Benedict, and R. J. Russo.
Instruments and Control Systems, Vol 45, No 10, p 55-56, October 1972. 1 fig, 3 tab, 5 ref.

Descriptors: *Calibrations, *Temperature, Measurement, Thermometers.
Identifiers: *Platinum thermometers, Resistance thermometers.

Replacement of the International Practical Temperature Scale of 1948 (IPTS-48) with IPTS-68 requires that the calibration of platinum resistance thermometers be changed. Consequently, the Callender interpolation equation has been revised for determining calibration constants from experimental data in the temperature range between the ice and antimony points. A correction factor is applied to the equation to satisfy the new scale. (Little-Battelle)
W73-09370

TEMPERATURE COEFFICIENTS AND THEIR COMPENSATION IN ION-SELECTIVE SYSTEMS,
Foxboro Co., Mass.
L. E. Negus, and T. S. Light.
Instrumentation Technology, Vol 19, No 12, p 23-29, December 1972. 2 fig, 4 tab, 16 ref.

Descriptors: *Thermal properties, *Instrumentation, *Mathematical studies, Temperature, Laboratory equipment, Research equipment, Temperature control, Electrochemistry.
Identifiers: *Ion selective electrodes, *Temperature coefficients, *Temperature compensation, Sensors, Silver electrode, Chloride electrode, Bromide electrode, Iodide electrode, Sulfide electrode, Cyanide electrode, Copper electrode, Hydrogen electrode, Sodium electrode, Fluoride electrode.

Temperature effects in ion-selective electrode systems are discussed as well as compensation techniques for these effects. The electrodes discussed fit in either of the following categories: 1. The cell potential and thermal characteristics are independent of any internal filling solution (solid-state membrane electrodes for Ag, chloride, bromide, iodide, sulfide, cyanide, and Cu). 2. The cell potential and thermal characteristics are dependent upon the internal filling solution (elec-

trodes for pH, Na, fluoride, Ca, Mg, and water hardness). Application of such electrodes is aided if there is exact information available on their thermal characteristics. It has been shown that those characteristics can be derived from the Nernst equation and from experimental data, and how various thermal effects can be minimized. (Holoman-Battelle)
W73-09371

AQUATIC SEDIMENT AND POLLUTION MONITOR,
R. Y. Anderson.
U. S. Patent No 3,715,913. 6 p, 8 fig, 2 ref; Official Gazette of the United States Patent Office, Vol 907, No 2, p 339, February 13, 1973.

Descriptors: *Patents, *Sampling, Pollutants, Sediments, *Monitoring, *Instrumentation, Water pollution, Water quality, *Pollutant identification.

This invention relates to the collection and measurement of the natural materials and polluting substances that accumulate in bodies of water. The device is an elongated, vertically alignable collecting tube having an open upper end and a closed lower end for collecting over a period of time. A funnel-shaped magnifying cone is positioned with the small diameter end extending into the open end to magnify the collected matter. A baffle prevents large organisms from entering the tube and helps to minimize turbulence within the tube. Marking material settles as a distinct layer on the settled material. The marking material is held in a rotatable magazine which has many dispensing chambers. There is a thermostat which activates a different dispensing device every time the temperature passes a different preset level. After a given period of time, which may be several months or several years, the monitor may be recovered for study. The collecting tube will contain layers of material that correspond to the preset interval of a timing mechanism. (Sinha-OEIS)
W73-09390

WATER SAMPLING DEVICE,
Department of the Navy, Washington, D.C. (assignee)
For primary bibliographic entry see Field 07B.
W73-09401

FLIGHT TEST OF AN OCEAN COLOR MEASURING SYSTEM,
TRW Systems Group, Redondo Beach, Calif. Display and Imaging Dept.
For primary bibliographic entry see Field 07B.
W73-09440

NATURAL BACKGROUND CONCENTRATION OF MERCURY IN SURFACE WATER OF THE ADIRONDACK REGION, NEW YORK,
Geological Survey, Albany, N.Y.
For primary bibliographic entry see Field 02K.
W73-09449

DETERMINATION OF SILVER IN SOILS, SEDIMENTS, AND ROCKS BY ORGANIC-CHELATE EXTRACTION AND ATOMIC ABSORPTION SPECTROPHOTOMETRY,
Geological Survey, Denver, Colo.
For primary bibliographic entry see Field 02K.
W73-09451

HYDROGEOLOGICAL CRITERIA FOR EVALUATING SOLID-WASTE DISPOSAL SITES IN JAMAICA AND A NEW WATER SAMPLING TECHNIQUE,
Geological Survey of Jamica, Kingston.
For primary bibliographic entry see Field 05B.
W73-09452

HAIR AS A BIOPSY MATERIAL,
Agricultural Research Service, Grand Forks, N. Dak. Human Nutrition Lab.
L. M. Klevay.
Archives of Environmental Health, Vol 26, p 169-172, April 1973. 1 fig, 3 tab, 28 ref.

Descriptors: *Biopsy, *Lead, *Spectrometry, *Pollutant identification, Biochemistry, Analytical techniques, Laboratory tests, Measurements, Evaluation, Pathology, Path of pollutants.

The possibility is presented that hair could be used as an easily acquired, efficient biopsy material for the determination of exposure to quantities of lead. This study was appended to the 1967 Nutritional Survey of the Republic of Panama. Hair was collected and analyzed by atomic absorption spectrometry. For ten replicate analyses on a single hair sample with a mean lead concentration of 4.0 micrograms/gm, the standard deviation was 1.0 microgram/gm and the standard error was 0.3 microgram/gm. Data was separated and analyses for variance according to sex, age and location of residence were performed. Location appeared to be the most important factor in lead concentration differences. Hair and bone lead concentration correlations from other studies were discussed. It was concluded that if hair is used as a biopsy material for assessment of human exposure to lead in the environment, only individuals or groups which have been matched for age, sex, and place of residence may be compared. (Jerome-Vanderbilt)
W73-09467

DETERMINATION OF MERCURY IN SOILS BY FLAMELESS ATOMIC ABSORPTION SPECTROMETRY,
Department of Scientific and Industrial Research, Petone (New Zealand).
B. G. Weissberg.
Economic Geology, Vol 66, No 7, p 1042-1047, 1971. 2 fig, 4 tab, 16 ref.

Descriptors: *Mercury, *Analytical techniques, *Soil, *Soil investigations, Soil chemistry, Laboratory equipment, Spectrophotometry, Sampling, Testing, Pollutant identification, Metals, Absorption, Heavy metals.
Identifiers: *Atomic absorption spectroscopy, Organomercurials, *New Zealand.

Mercury analyses of acceptable accuracy for geochemical prospecting were obtained on ten New Zealand soil samples, containing 2 to 10% of total organic matter, using a simple single beam atomic absorption spectrometer combined with a cold gold filter that removed mercury from the vapors released on heating the sample. A modification of the analytical procedure, using two gold filters heated to 170 degrees C during mercury collection, gave results lower than the values obtained by an acid oxidizing digestion of the samples with subsequent evolution of mercury from the solutions by reduction with stannous chloride. These low results suggest that as much as 30% of the mercury in the soils tested is present as volatile mercury-organic compounds. (Oleszkiewicz-Vanderbilt)
W73-09471

AVAILABILITY OF MANGANESE AND IRON TO PLANTS AND ANIMALS,
West Virginia Univ., Morgantown. Dept. of Animal Industries and Veterinary Science.
For primary bibliographic entry see Field 02K.
W73-09474

DETERMINATION OF SUBMICROGRAM QUANTITIES OF MERCURY IN PULP AND PAPERBOARD BY FLAMELESS ATOMIC ABSORPTION SPECTROMETRY,
Weyerhaeuser Co., Longview, Wash. Analytical and Test Dept.
D. C. Lee, and C. W. Laufmann.
Analytical Chemistry, Vol 43, No 8, p 1127-1129, July 1971. 2 fig, 2 tab, 5 ref.

Descriptors: *Evaluation, *Mercury, *Pulp and paper industry, *Spectrophotometry, Chemistry, Chemical analysis, Analytical techniques, Absorption, Instrumentation, Solvents, Acids, Laboratory tests.

A procedure for analyzing pu‑ ‑ and paperboard for concentrations of mercury is presented in which the loss of mercury is virtually eliminated. The wet oxidation mixture of nitric acid (70%) and hydrochloric acid (37%) is prepared in a 1:1 ratio forming aqua regia. Aqua regia will react with pulp or paperboard to break down inorganic forms of mercury as well as the organic materials, to form soluble mercuric compounds, leaving behind cellulose fibers. The sensitivity of the proposed method will depend primarily on the dimensions of the absorption cell used. Using a 10 cm cell with the Perkin-Elmer Model 303 Atomic Absorption Spectrophotometer, as little as 5 nanograms of mercury can be detected with very little difficulty. The apparatus, reagents, instrument settings, calibration and experimental procedure are presented. Inasmuch as the determination of mercury in this method is based upon peak height rather than peak area, it is very important that all conditions of analysis be carefully controlled. (Jerome-Vanderbilt)
W73-09477

NICKEL EXPLORATION BY NEUTRON CAPTURE GAMMA RAYS,
Geological Survey, Washington, D.C.
F. E. Senftle, P. F. Wiggins, D. Duffey, and P. Philbin.
Economic Geology, Vol 66, No 4, p 583-590, 1971. 8 fig, 2 tab, 14 ref.

Descriptors: *Analytical techniques, *Neutron activation analysis, *Nickel, *Exploration, Chemical analysis, Metals, Geology, Measurements, Evaluation, Boreholes, Geophysics, Subsurface investigations, Gamma rays.

An in situ analytical technique which appears feasible for nickel detection is presented in this report. This method depends on the very high energy gamma rays emitted when nickel absorbs a thermal or low energy neutron and is made possible by the advent of large lithium drifted germanium, Ge (Li), detectors with their excellent resolution. Analysis down to 0.1 percent nickel generally compared well with chemical analysis. Measurements using the technique as a detection method were made on an artificially prepared surface deposit of low grade (about 1%) nickel ore and also in a test bore hole in the same ore. Nickel was easily detected in both cases. (Jerome-Vanderbilt)
W73-09480

DETERMINATION OF MERCURY IN NATURAL WATERS AND EFFLUENTS BY FLAMELESS ATOMIC ABSORPTION SPECTROPHOTOMETRY,
Central Inst. for Industrial Research, Oslo (Norway).
S. H. Omang.
Analytica Chimica Acta, Vol 53, p 415-419, 1971. 1 tab, 10 ref.

Descriptors: *Spectrophotometry, *Water analysis, *Mercury, Analytical techniques, Chemical analysis, Absorption, Water pollution, Effluents, Discharge (Water), Sampling, Laboratory tests, Evaluation, Wastes.
Identifiers: *Atomic absorption spectroscopy.

Analyses of water and effluents for mercury content have been reported, but the sensitivity of the methods used has been rather low. Modifications and refinements developed in order to reach a detection limit of 0.02 micrograms of mercury per

phenylmercury acetate and phenylmercury chloride examined, but methoxyethylmercury chloride and ethoxyethylmercury chloride samples contained as many as five mercury-containing impurities and up to 12 percent of inorganic mercury compounds. (Oleszkiewicz-Vanderbilt)
W73-09487

DETERMINATION OF METHYLMERCURY IN FISH AND IN CEREAL GRAIN PRODUCTS,
Food and Drug Directorate, Ottawa (Ontario). Reasearch Labs.
W. H. Newsome.
Journal of Agriculture and Food Chemistry, Vol 19, No 3, p 567-569, March 1971. 2 tab, 5 ref.

Descriptors: *Mercury, *Gas chromatography, *Analytical techniques, *Pollutant identification, Separation techniques, Fish, Foods, Fish toxins, Water pollution, Laboratory equipment.
Identifiers: *Methylmercury, *Organomercurials, Cereal grain products.

Methylmercury was determined in fish by gas-liquid chromatography following a modification of existing extraction procedures. The method obviated centrifugation to assist phase separation by incorporating a filtration step and by the use of hydrobromic rather than hydrochloric acid. Overall recoveries were 94% plus or minus 6% for whitefish and 98% plus or minus 6% for cod of methylmercury. Methylmercury was also determined in wheat flour and ground oats by extraction with a benzene-formic acid mixture followed by purification and gas-liquid chromatography. Interfering substances were removed from the extracts by column chromatography on silicic acid and partitioning with cysteine acetate solution. The method is sensitive in the 0.01 - 0.90 ppm range, with a mean recovery generally greater than 95%. (Oleszkiewicz-Vanderbilt)
W73-09488

MERCURY IN THE ATMOSPHERE,
For primary bibliographic entry see Field 05B.
W73-09489

ZINC UPTAKE IN NEOCOSMOSPORA VASINFECTA,
Queen's Univ., Kingston (Ontario). Dept. of Biology.
W. H. N. Paton, and K. Budd.
Journal of General Microbiology, Vol 72, Part I, p 173-184, 1972. 7 fig, 20 ref.

Descriptors: *Zinc, *Heavy metals, *Fungi, Adsorption, Metabolism, Growth rates, Temperature, Cytological studies, Nutrients, Model studies, Laboratory tests, Path of pollutants.
Identifiers: Zinc uptake, Nucleus, *Neocosmospora vasinfecta.

Mycelia of Neocosmospora vasinfecta harvested in mid-logarithmic phase absorb Zn (+2) from dilute solutions in the absence of growth. Zn (+2) uptake involves two phases: a rapidly established phase I believed to represent adsorption to negatively charged groups in the hyphal surface-membrane and a slowly established phase 2 which represents transport into the cytoplasm. Phase I is not influenced by low temperature, NaN3 or anaerobiosis applied for short periods, conforms to the Langmuir adsorption equation and is reduced in the presence of various other divalent cations. Phase 2 is strongly inhibited by low temperature, NaN3 and anaerobiosis and exhibits carrier-type kinetics. Electron microscopy of unstained material indicates that part at least of phase 2 Zn is deposited in the cytoplasm and nucleus. A model for the metabolic uptake of Zn (2+) involving phase I binding as a requisite preliminary process is suggested. (Oleszkiewicz-Vanderbilt)
W73-09491

PARTICLES CONTAINING LEAD, CHLORINE, AND BROMINE DETECTED ON TREES WITH AN ELECTRON MICROPROBE,
Connecticut Agricultural Experiment Station, New Haven. Dept. of Ecology.
G. H. Heichel, and L. Hankin.
Environmental Science and Technology, Vol 6, No 13, p 1121-1122, December 1972. 2 tab, 7 ref.

Descriptors: *Waste identification, *Lead, *Plant morphology, *Trees, Air pollution, Chemistry, Toxins, Gasoline, White pine trees, Public health, Spectrophotometry.
Identifiers: Automobile exhaust, Airborne lead particles.

Samples of branches 55 mm in diameter from a white pine tree situated 2 to 3 meters from densely traveled roadway, and samples of bark 4 mm thick from an elm tree .75 meter from a city street, were collected to analyze the amount and composition of the lead particulate matter imbedded on them. Analyses by atomic absorption spectrophotometry determined the lead content on the pine twig to be 120 micrograms of lead per gram of dry weight, while on the elm bark it was 108 micrograms of lead per gram of dry weight. Further analysis with an electron probe microanalyzer showed that although the lead content in the two samples differed greatly, the lead/chlorine/bromine ratio was similar. The lead composition of the particles imbedded on the tree bark was much the same as particles in automobile exhaust. Knowledge of the form and composition of lead residues is important to determine how lead affects plants, as some forms of lead are more toxic than others. (Jerome-Vanderbilt)
W73-09494

CONCENTRATIONS OF ARSENIC FROM WATER SAMPLES BY DISTILLATION,
McGill Univ., Montreal (Quebec). Dept. of Chemical Engineering.
E. J. Farkas, R. C. Griesbach, D. Schachter, and M. Hutton.
Environmental Science and Technology, Vol 6, No 13, p 1116-1117, December 1972. 1 tab, 11 ref.

Descriptors: *Analytical techniques, *Separation techniques, *Water quality, *Arsenic, Chemistry, Chemical engineering, *Distillation, Water pollution, Sampling.

Determination of the amount of arsenic in samples of drinking water is facilitated if the arsenic in the sample can first be concentrated into a smaller volume. The utility of the distillation method described in standard reference works was investigated. Optimum conditions for application of this method were determined based on experiments in which percentage recovery of the arsenic present in the original sample was measured as a function of the amounts of the reagents added to the original sample and of the volume of distillate collected. The acid concentration in particular influences the results quite markedly. Improvements in experimental technique are also described. The acid concentration necessary could be achieved by bubbling HCl gas into the sample. (Jerome-Vanderbilt)
W73-09495

SOME GENERAL AND ANALYTICAL ASPECTS OF ENVIRONMENTAL MERCURY CONTAMINATION,
Hope Coll., Holland, Mich.
D. H. Klein.
Journal of Chemical Education, Vol 49, No 1, January 1972 p 7-10, 1 tab, 10 ref.

Descriptors: *Analytical techniques, *Mercury, *Laboratory tests, Reviews, Chemistry, Heavy metals, Environment, Diseases, Evaluation, Spectrophotometry, Atmosphere, Air pollution, Water pollution, Food chains, Biota.

The object is to encourage chemists in undergraduate institutions to develop programs in which students can participate in the collection of much needed data on the presence of mercury and other heavy metals in the environment. In order to further this goal the author has conducted a broad discussion of relatively simple analytical methods which can be performed with inexpensive equipment by small groups on readily available samples. The reduction-aeration-flameless atomic absorption technique of analysis is discussed briefly along with modifications in digestion procedures which make it applicable to the analysis of fish tissue or brines, sludges, air, water and other substances. Problems of initially erratic results seem to be overcome by operator experience and breaking in of the apparatus. Preparation of standards, reagent contamination and absorption and volatility of the sample are discussed. (Jerome - Vanderbilt)
W73-09502

NOVEL WET-DIGESTION PROCEDURE FOR TRACE-METAL ANALYSIS OF COAL BY ATOMIC ABSORPTION,
Bureau of Mines, Pittsburgh, Pa. Pittsburgh Mining and Safety Research Center.
A. M. Hartstein, R. W. Freedman, and D. W. Platter.
Analytical Chemistry, Vol 45, No 3, p 611-614, March 1973. 1 fig, 3 tab, 8 ref.

Descriptors: *Chemical analysis, *Trace elements, *Separation techniques, *Spectrophotometry, *Laboratory tests, Chemistry, Analytical techniques, Testing procedures, Sampling, Evaluation, Research facilities, Instrumentation.

A procedure is presented which permits the rapid and reliable quantitative determination of trace elements in coal at sample weights as low as 10 mg. The organic matter in coal is destroyed by fuming nitric acid, and the dissolution of siliceous material is accomplished by hydrofluoric acid. Thus a relatively concentrated solution is provided having little matrix interference for atomic absorption and several other types of analysis. Instrumental conditions are presented for determination of Be, Cd, Ca, Co, Cu, Li, Mg, Mn, Ni and K. The accuracy in the experiments performed ranges from an average of 97% for beryllium to an average of 106% for nickel and is obtained by averaging the percent recovery of a spike of 0.25 ppm, 0.50 ppm and 0.75 ppm for each element. (Jerome - Vanderbilt)
W73-09506

NOTES ON THE DETERMINATION OF MERCURY IN GEOLOGICAL SAMPLES,
Imperial Coll. of Science and Technology, London (England). Dept. of Geology; and Imperial Coll. of Science and Technology, London (England). Applied Geochemistry Research Group.
M. Koksoy, P. M. D. Bradshaw, and J. S. Tooms.
Institution of Mining and Metallurgy Transactions, London, Vol 76, p B-121-B-124, 1967. 3 fig, 3 tab, 3 ref.

Descriptors: *Mineralogy, *Analytical techniques, *Mercury, *Evaluation, Chemistry, Laboratory tests, Instrumentation, Separation techniques, Sampling, Gases, Water pollution.
Identifiers: *Turkey.

Various aspects of preparation and analysis of naturally occurring materials was conducted during an investigation of the primary and secondary dispersion of mercury from cinnabar and stibnite deposits in Turkey. The instrument used for all analyses was the mercury vapor meter described by James and Webb, with only minor modifications. The data presented demonstrate that to obtain comparable and consistent results it is essential that the time and temperature to which samples are heated be standardized; the samples be ground to very fine powder; the drying of samples be carried out at room temperature; and anomalous and background samples be stored separately or the samples be analyzed after a very limited storage life. (Jerome - Vanderbilt)
W73-09509

SELECTIVE ATOMIC-ABSORPTION DETERMINATION OF INORGANIC MERCURY AND METHYLMERCURY IN UNDIGESTED BIOLOGICAL SAMPLES,
Medical Research Council, Carshalton (England). Toxicology Unit.
L. Magos.
Analyst, Vol 96, p 847-853, December 1971. 3 fig, 2 tab, 10 ref.

Descriptors: *Mercury, *Analytical techniques, *Foods, *Pollutant identification, Spectrophotometry, Laboratory equipment, Path of pollutants, Fish, Fish toxins, Toxicity.
Identifiers: *Atomic absorption spectroscopy, *Methylmercury, Organomercurials, Biological samples.

A simple method for the determination of total mercury in biological samples contaminated with inorganic mercury and methylmercury is described. The method is based on the rapid conversion of organomercurials first into inorganic mercury and then into atomic mercury suitable for aspiration through the gas cell of a mercury vapour concentration meter, by a combined tin (II) chloride-cadmium chloride reagent. It was found that if 100 mg of tin (II) chloride alone were added instead of the tin (II) chloride-cadmium chloride reagent, only the release of inorganic mercury influenced the peak deflection of the potentiometer, thus permitting the selective determination of inorganic mercury in the presence of methylmercury. It was possible first to release inorganic mercury then, after re-acidification of the reaction mixture, methylmercury, by adding the tin (II) chloride-cadmium chloride reagent and sodium hydroxide. When total mercury and inorganic mercury were determined separately, the difference between results gave the methylmercury content of the sample. (Oleszkiewicz-Vanderbilt)
W73-09513

DETERMINATION OF THE COPPER CONCENTRATIONS TOXIC TO MICROORGANISMS,
Moscow State Univ. (USSR). Dept. of Soil Biology.
Z. A. Avakyan, and I. L. Rabotnova.
Trans from Mikrobiologiya, Vol 35, No 5 p 682-687, September 1966. 7 tab, 6 ref.

Descriptors: *Copper, *Toxicity, *Microorganisms, *Metabolism, Biochemistry, Heavy metals, Water pollution, Environmental effects, Nutrients, Phosphorus compounds, Acidity, Chemical reactions.

An investigation to determine the true concentration of copper in solution after the introduction of copper sulfate into various nutrient media and to determine the copper concentration toxic to Torula utilis, was undertaken. It was shown that on media with glycerin, sucrose, or mannitol, copper is contained in solution simultaneously with phosphate only at pH<5.5. Therefore studies of the effects of copper ions can be conducted only with microorganisms that grow well at an acid pH. Heavy metals produce a phosphorus deficiency in the medium. It was found that on T. utilis, the copper ion begins to exert an inhibiting effect at a concentration of 20-30 mg/liter, while at 40 mg/liter growth stops completely. After 50 passages on glycerin medium with copper, the resistance could not be increased. (Jerome - Vanderbilt)
W73-09515

THE LEACHING OF TOXIC STABILIZERS FROM UNPLASTICIZED PVC WATER PIPE: II. A SURVEY OF LEAD LEVELS IN UPVC DISTRIBUTION SYSTEMS,
R. F. Packham.
Water Treat Exam. Vol 20, No 3, p 144-151, 1971.
Identifiers: Chlorides, Distribution, Leaching, *Lead, Pipes, *Polyvinylchloride pipes, *Toxic stabilizers, United-Kingdom, Water analysis.

Water samples were collected from many uPVC (unplasticized polyvinylchloride) distribution systems in the United Kingdom and analyzed for Pb. Pb was also determined where no uPVC contact had taken place. Pb concentrations sampled were generally very low, usually identical to content prior to contact with uPVC. The use of stabilizers in uPVC pipe does not seem to constitute a health hazard to domestic water consumers.--Copyright 1973, Biological Abstracts, Inc.
W73-09521

A ROTARY SYSTEM FOR CALIBRATING FLOWMETERS USED FOR PLANKTON SAMPLING,
Lamont-Doherty Geological Observatory, Palisades, N.Y.
For primary bibliographic entry see Field 07B.
W73-09523

VITAL STAINING TO SORT DEAD AND LIVE COPEPODS,
Virginia Inst. of Marine Science, Gloucester Point.
For primary bibliographic entry see Field 07B.
W73-09539

A NEW METHOD FOR THE GAS CHROMATOGRAPHIC SEPARATION AND DETECTION OF DIALKYLMERCURY COMPOUNDS-APPLICATION TO RIVER WATER ANALYSIS,
National Environmental Research Center, Cincinnati, Ohio. Analytical Quality Control Lab.
R. C. Dressman.
J Chromatogr Sci. Vol 10, No 7, p 472-475. 1972. Illus.
Identifiers: Pollution detection, *Chromatography, Gas, *Mercury compounds, Rivers, Separation techniques, Water analysis.

A method of selectively analyzing for dialkylmercury compounds at the ng level without converting these compounds to their chloride salts for electron capture detection was developed. An homologous series of 4 dialkylmercury compounds beginning with dimethylmercury was successfully separated using a 6 ft x 2 mm, i.d. (internal diameter), glass column packed with 5% DC-200 + 3% QFI (copacked) on Gas Chrom Q (80/100 mesh). The separation was accomplished by programming the column oven temperature from 70 deg C, after a 2 min hold, to 180 deg C at a rate of 20 deg C/min. The separated compounds were combusted in an FID (flame ionization detector) and the resultant free Hg was passed into a cold vapor Hg detector. Practical absolute sensitivity for Hg was 0.1 ng using a Colman Mercury Analyzer MAS-50. This combined separation and detection technique provides a unique, specific and highly sensitive method of direct identification of dialkylmercury compounds.--Copyright 1973, Biological Abstracts, Inc.
W73-09584

AN EVALUATION OF APHA METHOD FOR DETERMINING ARSENIC IN WATER,
Environmental Health Lab., McClellan AFB, Calif.
E. G. Robles, Jr.
Available from NTIS, Springfield, Va 22151 AD-752 526, Price $3.00 printed copy; $0.95 microfiche. Air Force Environmental Health Laboratory Professional Report No 70M-8, March 1970. 19 p, 2 fig, 1 tab, 7 ref, 6 append.

animals, *Analytical techniques, *Chemical analysis, Radioactivity, Bottom sediments, Environmental effects, Pollutant identification.
Identifiers: *Ruthenium-106.

Various methods for analyzing ruthenium-106 in marine samples are presented. Ways of treating suspended materials in sea water preserving water samples and procedures for extracting ruthenium completely from organisms and sediments are discussed in detail. In this study marine organisms and sediments were ashed at 400-500 deg C and then fused with a mixture of potassium hydroxide and potassium nitrate. Ruthenium was extracted with carbon tetrachloride as ruthenium tetroxide, and then back-extracted with sodium hydroxide solution containing a reducing agent. The loss of ruthenium throughout ashing and chemical procedures was found to be negligible. Data concerning chemical and radiochemical yields, sensitivity, and contamination factors in the analyses are included. Optimum methods for radiochemical analysis of shellfish and seaweed, bottom sediments, and sea water are described step-by-step. (Ensign-PAI)
W73-09648

ENVIRONMENTAL ASPECTS OF DREDGING IN ESTUARIES,
Skidaway Inst. of Oceanography, Savannah, Ga.
For primary bibliographic entry see Field 05C.
W73-09651

5B. Sources of Pollution

WATER QUALITY AS RELATED TO POSSIBLE HEAVY METAL ADDITIONS IN SURFACE AND GROUNDWATER IN THE SPRINGFIELD AND JOPLIN AREAS, MISSOURI,
Missouri Water Resources Research Center, Rolla.
For primary bibliographic entry see Field 05A.
W73-09104

CIRCULATION PATTERNS IN LAKE SUPERIOR,
Wisconsin Univ., Madison. Dept. of Civil and Environmental Engineering; and Wisconsin Univ., Madison. Dept. of Meteorology.
For primary bibliographic entry see Field 02H.
W73-09106

PROCEEDINGS OF CONFERENCE ON TOWARD A STATEWIDE GROUNDWATER QUALITY SUBCOMMITTEE CITIZENS ADVISORY COMMITTEE, GOVERNORS ENVIRONMENTAL QUALITY COUNCIL.

Available from the National Technical Information Service as PB-220 390, $3.00 in paper copy, $0.95 in microfiche. University of Minnesota, Water Resources Research Center, St. Paul, Mimeographed Rept., February 1973. 238 p. OWRR A-999-MINN (27) 14-31-0001-3823.

Descriptors: *Groundwater, *Water quality, *Information exchange, *Hydrogeology, Water pollution, Solid wastes, Waste disposal, Agricultural runoff, *Minnesota.
Identifiers: *Hydrogeologic framework, *Reactive system, *Transport system.

The objectives of the Conference on Toward A Statewide Groundwater Quality Information System were: to document, publicize, and promote the need for a statewide groundwater quality information system; to consider possible institutional arrangements for designing the system; and to review the factors to be considered in designing the system. The following topics were discussed: the natural quality of groundwater in Minnesota, the use of groundwater in Minnesota, hydrogeologic framework for deterioration in groundwater quality, groundwater pollution problems in Minnesota, establishing the impact of agricultural practices on groundwater quality, spray disposal of sewage effluent, solid waste disposal, needs and uses for a groundwater quality data system, water well records and information system needs, subsurface geologic information system in Minnesota, groundwater quality information system experiences in other states, Federal water information systems, and relation of groundwater quality information system and other systems in Minnesota. (Walton-Minnesota)
W73-09113

MATHEMATICAL MODEL FOR SCREENING STORM WATER CONTROL ALTERNATIVES,
Massachusetts Inst. of Tech., Cambridge. Ralph M. Parsons Lab. for Water Resources and Hydrodynamics.
P. H. Kirshen, D. H. Marks, and J. C. Schaake, Jr.
Available from the National Technical Information Service as PB-220 353, $3.00 in paper copy, $0.95 in microfiche. Partial Completion Report 157, October 1972. 125 p, 10 fig, 9 tab, 47 ref. OWRR C-2137 (3403) (4).

Descriptors: *Storm water, Urban runoff, Urban hydrology, *Waste storage, Detention, *Combined sewers, Drainage systems, Linear programming, *Mathematical models, *Waste water disposal, *Ohio, Simulation analysis.
Identifiers: Screening models, *Cincinnati (Ohio).

The general problem of pollution from combined sewer systems is discussed and control alternatives are described. A linear programming model for screening the sizes and operating policies of storage tanks, pipes, and treatment plants is formulated. The thesis also discusses a storm water simulation model and shows how it can be used interactively with the screening model to plan for the control of combined sewer overflow and local flooding in the Bloody Run Drainage Basin, Cincinnati, Ohio. The results of this case study indicate that the screening model and the planning method are reliable.
W73-09119

EXPERIMENTAL STUDY OF UNDERGROUND SEEPAGE TOWARD WELLS,
Rhode Island Univ., Kingston. Dept of Engineering.
C-F. Tsai.
M Sc Thesis, 1972. 38 p. 15 fig, 2 tab, 20 ref.

Descriptors: *Groundwater movement, *Mathematical models, *Path of pollutants, *Hydraulic models, Soil disposal fields, Water wells, Hydrogeology, Septic tanks, Seepage, Drawdown.

A mathematical solution for groundwater flow problems was studied experimentally using a test model of 8 feet in diameter and 6 feet in height. The mathematical solution agreed satisfactorily with the measured data. The mathematical solution was then used to establish relationships among such parameters as the rate of seepage from an infiltration area to a well and distance between the infiltration area and the well for various hydraulic boundary conditions. The relationships are helpful for determination of a safe distance between a water supply and a waste disposal field provided that the properties of pathogenic substances in the waste are known. (Knapp-USGS)
W73-09131

DREDGE SPOIL DISPOSAL IN RHODE ISLAND SOUND,
Rhode Island Univ., Kingston. Dept. of Oceanography; and Rhode Island Univ., Kingston. Dept. of Zoology.
For primary bibliographic entry see Field 05E.
W73-09142

HYDROGRAPHIC OBSERVATIONS IN ELKHORN SLOUGH AND MOSS LANDING HARBOR, CALIFORNIA, OCTOBER 1970 TO NOVEMBER 1971,
Moss Landing Marine Labs., Calif.
W. W. Broenkow, and R. E. Smith.
Available from NTIS, Springfield, Va 22151 COM-72 11273 Price $3.00 printed copy; $0.95 cents microfiche. Technical Publication 72-3, July 1972. 74 p, 1 fig, 66 tab, 10 ref. NOAA Grant No 2-35137.

Descriptors: *Water quality, *Chemical analysis, *Harbors, *Estuaries, *California, Basic data collections, Water analysis, Water chemistry, Sampling, Boats, Water pollution sources, Path of pollutants, Water resources development, Planning, Environmental effects, Dredging.
Identifiers: *Elkhorn Slough (Calif), *Moss Landing Harbor (Calif), Deepwater port planning.

In Ocotber 1970, Moss Landing Marine Laboratories began an observational program to determine the seasonal changes in the water chemistry of Elkhorn Slough and Moss Landing Harbor, California. This report contains the first year of data (October 1970 to November 1971). These data are of immediate interest in determining the flushing and mixing mechanisms of the slough and in establishing the effect that local domestic and industrial effluents have on the distribution of chemical parameters. In recent years, various plans have been suggested for the further development of Elkhorn Slough ranging from the construction of a deepwater port to the development of commercial shellfish production. Data in this report should aid those agencies ultimately responsible for the development of the area. Samples were taken simultaneously from two outboard motor boats, one for Elkhorn Slough at five stations and one for four Harbor stations. The parameters measured were water temperature, salinity, dissolved oxygen, and nutrient ions. (Woodard-USGS)
W73-09162

SOME MECHANISMS OF OCEANIC MIXING REVEALED IN AERIAL PHOTOGRAPHS,
Lamont-Doherty Geological Observatory, Palisades, N.Y.
For primary bibliographic entry see Field 07B.
W73-09168

WORLDWIDE SULFUR POLLUTION OF RIVERS,
Yale Univ., New Haven, Conn. Dept. of Geology and Geophysics.
R. A. Berner.
Journal of Geophysical Research, Vol 76, No 27, p 6597-6600, September 20, 1971. 3 tab, 13 ref. NSF GA-1441.

Descriptors: *Sulfur, *Water pollution sources, *Rivers, Path of pollutants, Sulfates, Weathering, Industrial wastes, Acid mine water.

The amount of sulfur that has been contributed to the world average of river water from different sources is estimated. From two independent calculations the contribution due to pollution is estimated to be 27% and 28% although both are probable minimum values. Other sources are weathering of sedimentary rocks, 35%; volcanic emanations and hot springs, 7%; and cyclic sulfur carried inland from the oceans, 30%. (Knapp-USGS)
W73-09169

WISCONSIN LAKES RECEIVING SEWAGE EFFLUENTS,
Wisconsin Univ., Madison. Water Resources Center.
J. P. Wall, P. D. Uttormark, and M. J. Ketelle.
Technical Report WIS-WRC 73-1. January 1973. 26 p. EPA R-801363.

Descriptors: *Water pollution sources, *Wisconsin, *Lakes, *Sewage effluents, Eutrophication, Nutrients, Phosphorus, Nutrient removal, Industrial wastes, Urban runoff, Agricultural runoff, River basins, Basins.

Presumably some Wisconsin lakes could be improved if nutrient contributions from wastewater sources were eliminated or substantially reduced, thus as the first step, the lakes which receive sewage effluent discharges are listed. The drainage basin surveys and pollution investigation reports, conducted comprehensively for over twenty years and published by the Wisconsin Department of Natural Resources, furnished the data and material on which this presentation is based. Data exist concerning sewage treatment plants, whereas information about industrial effluent or urban runoff or agricultural drainage is scanty. The sources of discharge are identified in each watershed with accompanying descriptions of the type of treatment, quantity of output, quality of effluent, and recommendations for improvement where appropriate. The receiving stream is named and associated chemical and biological data are presented as available. The lakes fed by these streams have been identified and pertinent data presented. Somewhat greater emphasis is given to those lakes with surface areas of 100 acres or greater, but smaller lakes and ponds receiving effluent are identified as encountered. Nine of the lakes named are among the fifteen largest bodies of water in the state. (Jones-Wisconsin)
W73-09176

ATMOSPHERIC CONTRIBUTIONS OF NITROGEN AND PHOSPHORUS,
Wisconsin Univ., Madison. Water Resources Center.
J. D. Chapin, and P. D. Uttormark.
Technical Report WIS-WRC 73-2, February 1973. 35 p. EPA R-801343.

Descriptors: *Atmosphere, *Nitrogen, *Phosphorus, Reviews, Water quality, Lakes, Eutrophication, Nutrients, Precipitation (Atmospheric), Rainfall, Snowfall, Fallout, United States, Geographical regions, Land use, Seasonal, Birds, Dusts, Soil erosion, Industries, Smoke, Ammonia, Nitrates, Seeds, Pollen, Bibliographies, Water pollution sources.
Identifiers: Dry fallout.

Estimates are presented, compiled from a literature review, of atmospheric contributions of nitrogen and phosphorus via bulk precipitation, which includes rainfall, snowfall, and dry fallout. Data are converted to kg/ha/yr, and, to a large extent, are site specific. The contribution of phosphorus via bulk precipitation has been studied less intensively than nitrogen because other sources of phosphorus are likely to be more important. Nitrogen contribution by precipitation are mapped for the United States. Although considerable differences are noted between findings of various investigators, it is generally agreed that storms and prevailing winds off the ocean are low in phosphorus content and that the major source of phosphorus in rainfall is from dust generated over the land. Urbanization and industrialization, soil erosion, industrial ash, and smoke are commonly listed. Data interpretation is hampered by the failure of authors to specify the form of nitrogen and phosphorus measured or the analytical procedure used. Atmospheric contributions of nitrogen and phosphorus appear more significant than is generally recognized with respect to control of lake eutrophication and they are likely to be larger in the future. (Jones--Wisconsin)
W73-09177

MICROBIOLOGICAL SEA WATER CONTAMINATION ALONG THE BELGIAN COAST, I - GEOGRAPHICAL CONSIDERATIONS,
Belgian Army Military Hospital, Ostend. Lab. for Sea Microbiology.
J. Pinon, J. Pijck, and C. Van Cauwenberghe.
Revue Internationale d'Oceanographie Medicale Nice, Vol 27, p 5-15, 1972. 2 fig, 5 ref.

Descriptors: *Coasts, *Water pollution sources *Microorganisms, Sedimentation, Sands, Sand bars, *Tides, *Currents (Water), Wind, Temperature, Density, Irradiation, Turbidity, Human population, Seasonal.
Identifiers: *Belgian Coast, *North Sea.

The area under study on the Belgian Coast is homogeneous and 65 km long. The southern part of the North Sea is usually less than 50 m deep with a bottom of fine sand and has sand banks parallel to the coast. Size and direction of the tides and tidal currents are described along with the dominating wind direction. Water temperature density, sun irradation, coastal populations at the varying seasons, water turbidity and evacuation points are presented as background to the microbial contamination study. (See also W73-09204) (Ensign-PAI)
W73-09203

MICROBIOLOGICAL SEA WATER CONTAMINATION ALONG THE BELGIAN COAST, II - TECHNIQUES - NORMS - PRELIMINARY RESULTS,
Belgian Army Military Hospital, Ostend. Lab. for Sea Microbiology.
J. Pinon, and J. Pijck.
Revue Internationale d'Oceanographie Medicale, Nice, Vol 27, p 17-40, 1972. 7 fig, 4 tab, 16 ref.

Descriptors: *Coasts, Water pollution sources, Data collections, *Bioindicators, *Bacteria, Microorganisms, *Coliforms, Sampling, Cultures, Human population, Temperature, Currents, Sites, Time, Data processing.
Identifiers: *Belgian Coast, *North Sea.

A statistical investigation on the pollution indicator-organisms was carried out for a better understanding of, and the prediction of fluctuations of microbial pollution along the Belgian coasts. A mean fecal coliform concentration greater than 10/ml is considered dangerous. For bacterial counting, under experimental conditions, the standard error for the 'total microbes, 37 degrees C' f.i. is approximately 20%. Technical details on sampling, culture and counting are given. Indicator-organisms, chosen were the group of organisms growing on a rich medium at 37 degrees C, the total coliforms and the fecal coliforms. Nine points of observation were chosen along the coast, one identical to one of the nine pollution source points. Data on the 20 parameters gathered are processed by a computer. (See also W73-09203) (Ensign-PAI)
W73-09204

T-90, CRITERION OR CHIMAERA,
Aarhus Univ. (Denmark). Inst. of Hygiene.
G. J. Bonde.
Revue Internationale d'Oceanographie Medicale, Nice, Vol 25, p 5-15, 1972. 2 fig, 1 tab, 8 ref, 1 append.

Descriptors: *Bacteria, *Degradation (Decomposition), Planning, Sewage disposal, *Outlets, Tracers, Radioisotopes, *Bioindicators, *E. coli, Mathematical studies, Salinity, Dilution, Sedimentation, Mortality.
Identifiers: *T-90, *Bacterial decay.

Estimating bacterial decay is a serious problem in both field and laboratory experiments. The determination of T-90 (the time necessary for a 90% decrease in numbers) by field experiments was used as an aid in planning sewage outlet sites. The T-90 estimates are sometimes computationally difficult being situated in the asymptotic region of the reduction curve. A computation based on the dose-response relationship and a transformation to probits is advocated along with an estimation of T-

(UNTERSUCHUNG DER VERMISCHUNGS- UND VERDUNNUNGSPROZESSE VON VERUNREINIGUNGEN IN ABWASSEREINLEITUNGSGEBIETEN IM MEER MIT HILFE FLUORESZIERENDER INDIKATOREN),
Institute of Biology of the Southern Seas, Sevastopol (USSR).
V. I. Zac, and M. S. Nemirovskij.
Beitrage zur Meereskunde, No 30-31, p 163-172, 1972. 3 fig, 1 tab, 7 ref.

Descriptors: Coasts, *Sewage, *Outlets, Diffusion, *Waste dilution, *Mixing, *Fluorescent dye, Fluorometry, Path of pollutants.
Identifiers: *USSR, *Crimean Coast, Uranin, Vertical dilution, Horizontal dilution.

Measurements of diffusion intensity, mixing, and dilution of sewage in the Crimean coastal waters were taken with fluorescent indicators. Special fluorometer, light filter, and rheostat were used for calculations near sewage outfalls. Uranin was introduced into the sewers along the coast and dye concentrations were measured at various depths and distances along the jet. At deep-lying sewage outfall sites, vertical dilution was 10-100 times higher than horizontal dilution. (Ensign-PAI)
W73-09212

THE POLLUTION OF THE COASTS AND EFFLUENTS IN THE SEA (LA POLLUTION DES COTES ET LES REJETS EN MER),
For primary bibliographic entry see Field 05G.
W73-09215

A TRACER SIMULATION OF WASTE TRANSPORT IN THE MUDDY CREEK-RHODE RIVER ESTUARY, MARYLAND,
Geological Survey, Washington, D.C. Water Resources Div.
N. Yotsukura, R. L. Cory, and K. Murakami.
Chesapeake Science, Vol 13, No 2, p 101-109, June 1972. 6 fig, 1 tab, 7 ref.

Descriptors: *Estuaries, *Discharge, *Sewage effluents, *Dye releases, Rhodamine, Dye concentrations, *Path of pollutants, Water pollution, Treatment facilities, *Maryland.
Identifiers: *Muddy Creek (Md), *Rhodamine WT dye.

A dye injection study was conducted in the Muddy Creek estuary, Maryland to determine the effects of discharge from a proposed sewage treatment plant. Rhodamine WT dye was injected near the head of the estuary and concentration of the dispersed dye was periodically measured at a number of fixed sampling stations downstream. The principle of superposition in a linear system was used in the extimation of dye concentration buildup in the study area. The results of the study suggest that the water of the north fork of the estuary would be contaminated as much as the south fork, even if the effluent site were located on the south fork. (Ensign-PAI)
W73-09217

ANTIBIOTIC-RESISTANT COLIFORMS IN FRESH AND SALT WATER,
Alabama Univ., Birmingham. Medical Center.
T. W. Feary, A. B. Sturtevant, Jr., and J. Lankford.
Archives of Environmental Health, Vol 25, No 3, p 215-220, September 1972. 1 fig, 4 tab, 12 ref.

Descriptors: *Bacteria, *Coliforms, *Antibiotics, *Public health, Water sampling, Water pollution effects, *Enteric bacteria, *Alabama.
Identifiers: R factors, *Infectious drug resistance, *Mobile Bay, *Tombigbee River system (Ala).

Infectious drug resistance, mediated by episomal elements (R factors), is an important factor in the spread of multiple resistance among enteric bacteria. Results of a study of the incidence of R factors among antibiotic-resistant coliforms isolated from the Black-Warrior-Tombigbee River system and Mobile Bay, are presented. Surface water samples were collected from 20 sites on the river system and 16 sites within the Bay. Antibiotic-resistant coliforms were detected in nearly all of River samples and in about 50% of the bay samples. Of the 194 strains tested, 20% contained demonstrable R factors; the majority of the R factors conferred resistance to two or more antibiotics and in addition, 53% conferred resistance to three or more antibiotics. Potential health hazardous to man and domestic animals are discussed. (Ensign-PAI)
W73-09218

INCREASING ORGANIC POLLUTION IN KINGSTON HARBOUR, JAMAICA,
University of the West Indies, Kingston (Jamaica). Dept. of Zoology.
B. A. Wade, L. Antonio, and R. Mahon.
Marine Pollution Bulletin, Vol 3, No 7, p 106-110, July 1972. 4 fig, 5 tab, 5 ref.

Descriptors: Harbors, Organic compounds, *Nutrients, *Bioindicators, Water pollution sources, Sewage, Industrial wastes, Runoff, Environmental effects.
Identifiers: *Kingston Harbor, *Jamaica, *Spiochaetopterus oculatus, *Capitella capitata.

The current status of Kingston Harbor, Jamaica, with respect to pollution, is reported. The harbor is a major deep water port and serves as a center for transportation, trade and commerce. It also serves as a receptacle for large volumes of primary treated domestic sewage, raw industrial wastes and agricultural wastes. The natural freshwater runoff of river and gully courses is a major source of high organic and nutrient input into the harbor. Two known indicators of high organic pollution, Spiochaetopterus oculatus and Capitella capitata, are present in increasing numbers in Kingston Harbor. The study concludes that if the present rate of environmental deterioration is allowed to continue unchecked complete destruction of the benthic fauna over most of the harbor may be imminent. (Ensign-PAI)
W73-09219

OIL POLLUTION OF THE SEA,
N. Pilpel.
Ecologist, Vol 2, No 3, 3 p, March 1972. 1 fig.

Descriptors: *Oil pollution, *Oceans, *Bacteria, Water pollution sources, Laboratory tests, On-site investigations, Dispersion.
Identifiers: *Oil dispersion processes, *Decomposition.

The nature and scope of oil pollution at sea is reviewed. Certain natural processes which help disperse spilled oil are discussed, along with activities which prolong the existence of oil. Research on oxidation and bacterial decomposition is summarized. Preliminary results from lab experiments indicate that many species of bacteria are capable of utilizing crude oil as a source of food and energy. Field work done at sea on the processes involved in the decomposition of oil is also discussed. The question of whether or not man can continue to rely on natural sources to clean up the oceans he pollutes remains unanswered; indications are, though, that he cannot. (Ensign-PAI)
W73-09222

AN EXPERIMENTAL EVALUATION OF OIL SLICK MOVEMENT CAUSED BY WAVES,
Missouri Univ., Rolla. Dept. of Mechanical and Aerospace Engineering.
D. J. Alofs, and R. L. Reisbig.

Journal of Physical Oceanography, Vol 2, No 4, p 439-443, October 1972. 6 fig, 1 tab, 11 ref. USCG DOT-CG12196-A.

Descriptors: *Oil spills, *Waves, *Laboratory tests, Surface waters, Aquatic drift, Floats, Forecasting, Water pollution sources, *Path of pollutants, Stokes law.
Identifiers: *Oil spill movement, Surface drift velocities.

The influence of waves on the direction of a oil spill movement was studied experimentally using mechanically generated gravity waves in a water tank. Results of the experiment showed that the measured surface drift velocities were in all cases higher than those predicted by the Stokes theory of deep water waves. It was also concluded that drift velocity is insensitive to float size when the float length is larger than one wavelength. Thin plastic floats were found to have the same drift speed as similarly sized oil lenses. Possible adaptation of this experimental data to predict oil spill movement in the open ocean is discussed. (Ensign-PAI)
W73-09233

MERCURY IN THE NETHERLANDS ENVIRONMENT, (KWIK IN HET NEDERLANDSE MILIEU (SLOT)),
J. E. Carriere.
Water, Vol 56, No 1, p 18-21, January 1972.

Descriptors: *Mercury, Water pollution sources, Coasts, *Fish, Rivers, Runoff, Environment.
Identifiers: *Netherlands Coast, *Rhine River, *Seals.

Investigations of the mercury content in fresh and salt water fish and in the coastal waters of the Netherlands were conducted. The mercury content in fish was between 0.01 and 0.1 ppm. Seals were found to have the highest mercury content. The northern and southern areas of the Netherlands coast showed higher concentrations of mercury than other areas. Thus contamination was thought to be related to this element being carried by the Rhine into the littoral. (Ensign-PAI)
W73-09237

ARSENIC INTOXICATION FROM WELL WATER IN THE UNITED STATES,
Minnesota Dept. of Health, Minneapolis.
E. J. Feinglass.
The New England Journal of Medicine, Vol 288, No 16, April 19, 1973, p 828-830. 1 tab, 11 ref.

Descriptors: Wells, *Water wells, *Arsenicals (Pesticides), Arsenic compounds, Toxins, Toxicity, Water pollution, Shallow wells, Poisons, Pesticides, *Pesticide toxicity, Cores, Minnesota.
Identifiers: Gastrointestinal illness, *Arsenic intoxication.

In 11 cases reported from a small town in western Minnesota, acute and subacute arsenic intoxication was related to ingestion of well water. In May, 1972, a local building contractor expanded his facilities by building a warehouse and office structure on the outskirts of the town. A well was driven, water from which was used by 13 people working on the premises. During the next 2 1/2 months, eleven of the thirteen persons reported positive symptoms of arsenic poisoning. Specimens from the well water in question contained 21,000 and 11,800 ppb arsenic (maximum allowable in the U.S. is 50 ppb). In the course of interviews with area residents, it was learned that grasshopper bait composed of arsenic, bran, and sawdust had been kept on the ground on property now owned by the contractor. Although there were no records of these events, it was believed that the bait was buried in that area. Burial of arsenical material results in high concentrations of the substance in a circumscribed area. Contamination of wells dug directly through such areas is not surprising; although the likelihood of repetitions of these events is remote, the importance of throughtful planning in the disposition of potentially toxic material must be stressed. (Campbell-NWWA)
W73-09248

INFANTILE METHEMOGLOBINEMIA CAUSED BY CARROT JUICE,
Saint Louis Children's Hospital, Mo.
For primary bibliographic entry see Field 05C.
W73-09249

BIOLOGICAL ANALYSIS OF WATER AND WASTEWATER.
Millipore Corp., Bedford, Mass.
For primary bibliographic entry see Field 05A.
W73-09254

LIMITATIONS TO LENGTH OF CONTAINED OIL SLICKS,
Technical Univ. of Denmark, Copenhagen. Inst. of Hydrodynamics and Hydraulic Engineering.
D. L. Wilkinson.
ASCE Proceedings, Journal of the Hydraulics Division, Vol 99, No HY5, Paper 9711, p 701-712, May 1973. 9 fig, 2 ref, append.

Descriptors: *Oily water, *Oil spills, *Oil-water interfaces, *Water pollution treatment, *Hydraulics, Path of pollutants, Stratified flow, Froude number, Waves (Water).

The drag forces exerted by a current passing beneath a contained oil slick cause it to thicken in the direction of flow. In rivers and channels of finite depth this process cannot continue indefinitely; ultimately, if the slick is of sufficient length an instability develops at the oil-water interface, making the retention of more oil impossible. The ultimate or critical length, the volume, and the profile of a contained oil slick may be expressed as a nondimensional function of the Froude number of the river flow and the interfacial and boundary friction coefficients. Laboratory experiments verify the dynamics of the theoretical model and enable magnitudes of the interfacial friction coefficient to be calculated. The interfacial stress is markedly increased if waves develop at the interface. (Knapp-USGS)
W73-09270

PREDICTION OF WATER QUALITY IN STRATIFIED RESERVOIRS,
Stone and Webster Engineering Corp., Boston, Mass. Environmental Engineering Div.
M. Markofsky, and D. R. F. Harleman.
ASCE Proceedings, Journal of the Hydraulics Division, Vol 99, No HY5, Paper 9730, p 729-745, May 1973. 7 fig, 1 tab, 10 ref, append.

Descriptors: *Water quality, *Reservoirs, *Stratified flow, *Thermal stratification, *Mathematical models, Numerical analysis, Forecasting, Dissolved oxygen, Withdrawal, Water temperature.
Identifiers: *Water quality forecasting.

A transient mathematical model which couples the conservation of heat and mass equations was developed and solved numerically. Vertical temperature and concentration profiles, outlet temperature, and concentrations for conservative and nonconservative substances may be obtained. The results of the predictive model were compared to laboratory measurements of a conservative dye tracer and to field measurements of dissolved oxygen in the Fontana Reservoir during 1 yr of operation. The trends shown for dissolved oxygen as a function of time were satisfactorily reproduced. (Knapp-USGS)
W73-09272

aquae, Asterionella formosa, Substrate utilization, Glycolate, C-14, Release rates, Chemical composition, Axenic cultures, Gel filtration, Inorganic carbon, Organic carbon, Heart Lake, Grenadier Pond, Filtrates, Culture media, Fate of pollutants.

In short-term experiments using cultures of Chlorella pyrenoidosa, Anabaena flos-aquae, Asterionella formosa, and Navicula pelliculosa, both the proportion of photosynthetic products released from cells and the composition of these products altered with age. In the first 3 species, percentage extracellular release values increased with increasing growth rates, but the reverse trend was shown by Navicula. Fractionation of filtrates using Sephadex indicated that, in general, larger molecular weight compounds became predominant as cultures aged. Also a time-dependent shift in a similar direction occurred in cultures of all ages. In several lakes a predominance of large molecular weight compounds was apparent in filtrates even from short-term experiments. Filtrates of mixed cultures of planktonic bacteria growing on C-14 glycolate were found to contain large molecular weight organic compounds. It was demonstrated that in nonaxenic cultures of algae and in lake water, bacteria utilize low molecular weight extracellular metabolites of algae origin and larger molecular weight compounds are formed. (Holoman-Battelle)
W73-09315

THE UPTAKE OF GLUCOSE BY CHLAMYDOMONAS SP.,
North Carolina State Univ., Raleigh. Dept. of Zoology.
M. E. Bennett, and J. E. Hobbie.
Journal of Phycology, Vol 8, No 4, p 392-398, December 1972. 4 fig, 4 tab, 31 ref.

Descriptors: *Absorption, *Chlamydomonas, *Aquatic algae, *Velocity, Organic compounds, Carbohydrates, Phytoplankton, Protozoa, Invertebrates, Aquatic animals, Light intensity, Photosythesis, Primary productivity, Path of pollutants, Radioactivity techniques, Carbon radioisotopes.
Identifiers: *Heterotrophy, *Glucose, *Substrate utilization, Cellobiose, Maltose, Fructose, Substrate concentration, C-14.

The glucose uptake of a species of Chlamydomonas was studied at various concentrations of D-glucose plus glucose-1-C-14 (0.003-10.0 mg/liter) and at various light levels (0-220 ft-c). The alga grows at 4 C either in the light or in the dark with added glucose, cellobiose, maltose, or fructose. Uptake of glucose could be described by the Michaelis-Menten equation, and both the maximum velocity of uptake and the half-saturation constant increased when the cells were exposed to glucose in the dark. However, the high value of the half-saturation constant (5 mg glucose/liter) compared with the low levels of glucose in nature (5-10 micrograms/liter) makes it unlikely that a transport system is effective under natural conditions. Even if a total of 10.0 mg/liter of glucose plus other organic compounds were available as substrate, the rate of photosynthesis would still be more than 10 times higher (at 220 ft-c) than the rate of organic substrate uptake. Light had no effect on the total uptake of glucose but did reduce the percentage of C-14O2 evolved from 61 percent of the total C-14 taken up in the dark to 0 percent at 220 ft-c. This decrease could be due to either preferential use of the C-14O2 in photosynthesis or of the photosynthate in respiration. (Holoman-Battelle)
W73-09317

TURBULENT DIFFUSION OF OIL IN THE OCEAN,
Louisiana State Univ., Baton Rouge. Coastal Studies Inst.
S. P. Murray.
Limnology and Oceanography, Vol 17, No 5, p 651-660, September 1972. 4 fig, 2 tab, 18 ref.

Descriptors: *Oceans, *Oil wells, *Mathematical studies, *Oil spills, *Theoretical analysis, Oceans, Marine environment, Oil wastes, Eddies, Diffusivity, Equations, Water pollution sources, Dispersion, Currents (Water).
Identifiers: *Turbulent diffusion, *Petroleum oil, Diffusion coefficients, Surface tension theory, Taylor turbulent diffusion theory, Fickian diffusion theory.

On-site observations of oil slick geometries and current speeds during the Chevron spill of March, 1970, in the Gulf of Mexico have allowed a comparative evaluation of the role of large-scale turbulence (in the form of a horizontal eddy diffusivity) and surface tension effects in the spreading of oil from a continuously emitting well into a steady current. The initial outline of the slick (roughly the first 50 percent of slick length) follows the laws of expansion as predicted by Taylor's turbulent diffusion theory. The gross size and overall shape (neglecting details of outline) of this type of slick are well represented by a solution of the Fickian diffusion equations which predict approximate slick geometry as a function of current speed, horizontal eddy diffusivity, the oil discharge rate, and an empirically determined constant (the boundary concentration). Under the conditions observed the effect of surface tension confined to within the first few hundred meters downslick and can probably be neglected for practical purposes under moderate oil discharge rates and current speeds as low as even 5 cm/sec. (Holoman-Battelle)
W73-09319

NITROGEN FIXATION IN CLEAR LAKE, CALIFORNIA. I. SEASONAL VARIATION AND THE ROLE OF HETEROCYSTS,
California Univ., Davis. Div of Environmental Studies.
For primary bibliographic entry see Field 05C.
W73-09321

NITROGEN FIXATION IN CLEAR LAKE, CALIFORNIA. II. SYNOPTIC STUDIES ON THE AUTUMN ANABAENA BLOOM,
California Univ., Davis. Inst. of Ecology.
For primary bibliographic entry see Field 05C.
W73-09322

THE UPTAKE OF UREA BY NATURAL POPULATIONS OF MARINE PHYTOPLANKTON,
Scripps Institution of Oceanography, La Jolla, Calif.
For primary bibliographic entry see Field 05C.
W73-09323

DISSOLVED FREE AMINO ACIDS IN SOUTHERN CALIFORNIA COASTAL WATERS,
California Inst. of Tech., Pasadena. W. M. Keck Lab. of Engineering Materials.
M. E. Clark, G. A. Jackson, and W. J. North.
Limnology and Oceanography, Vol 17, No 5, p749-758, September 1972. 3 fig, 4 tab, 33 ref.

Descriptors: *Amino acids, *Outlets, *Sewage effluents, *Water analysis, *Sea water, *Pollutant identification, Water sampling, Surface waters, Hypolimnion, Scuba diving, Separation techniques, Spectrophotometry, Chemical analysis, Connate water, Saline water, Organic acids, Solvent extractions, *California, Methodology.
Identifiers: *Coastal water, Sample preparation, Thin layer chromatography, Scintillation counting, Optical density, Organic solvents, Ornithine, Glycine, Alanine, Serine, Glutanic acid, Arginine, Lysine, Histidine, Aspartic acid, Ninhydrin, Valine, Threonine, Isoleucine, Leucine, Proline, Tryosine, Phenylalanine, Cystine, Methionine.

The dissolved free amino acids (DFAA) in seawater samples from an inshore area near a large

submarine sewage outfall and from an inshore area relatively remote from any major outfall were compared. Twenty samples were analyzed; all were collected from a small boat and immediately filtered on board using Millipore HA (0.45-micron) membrane filters to remove bacteria, plankton, and sediments. Precautions were taken to avoid contamination by ninhydrin-positive materials. Samples were stored on dry ice for transit to the laboratory, then frozen at minus 60 C to await analysis. Surface samples were taken directly over the side of the boat. Bottom and interstitial waters were collected by SCUBA divers using cylindrical Plexiglas containers of about 6 liters capacity. The amino acids were initially separated from salts by ligand exchange on Chelex-100 resin (BioRad) converted to the Cu-NH3 form and analyzed by thin-layer chromatography after complete desalting and eluting with NH4OH. Individual amino acids were estimated by the method of Clark (1968). After two-dimensional separation and color development with ninhydrin, individual spots were identified, cut out, and eluted in 2 ml of 50 percent aqueous n-propanol. Optical densities were determined at 570 nm on a microsample spectrophotometer. Quantities of amino acids were estimated from standard curves of chromatograms of a mixture of amino acids (Cal Biochem). Subsamples (1 ml) were taken for scintillation counting at each step in the procedure, so that overall loss of amino acids at each step was known. At both sites, surface waters contained DFAA levels of about 115 micrograms/liter (ca. 1 micromole/liter), about twice the value found in surface waters at both sites, with significantly more (290 micrograms or ca. 2.4 micromoles/liter) near an outfall than away from one (170 micrograms or ca. 1.6 micromoles/liter). Interstitial waters near the outfall contained about 350 micrograms or 3.0 micromoles DFAA/liter. The composition of the DFAA in all samples was similar to that obtained by other workers elsewhere. (Holoman-Battelle)
W73-09324

DIRECTION OF DRIFT OF SURFACE OIL WITH WIND AND TIDE,
Department of Scientific and Industrial Research, Wellington (New Zealand). Oceanographic Inst.
N. W.Ridgway.
New Zealand Journal of Marine and Freshwater Research, Vol 6, Nos 1 and 2, p 178-184, June 1972. 4 fig, 11 ref.

Descriptors: *Oil spills, Movement, *Wind velocity, *Surface waters, Forecasting, Methodology, *Currents (Water), Aquatic drift, *Path of pollutants, Tidal streams, Mathematical studies, Oil wastes, Oily water.
Identifiers: *Vectorial addition, Wellington Harbor, New Zealand, Whangarei Harbor, Errors, Error sources.

The directions of surface oil slicks spreading from the wreck of t.e.v. Wahine during April 1968 are compared with predicted movements of surface water obtained by the vectorial addition of wind-induced surface currents and tidal currents. Wind factors of 1 percent, 2 percent, 3 percent, 4 percent, and 5 percent of recorded surface wind speeds were used to calculate wind-induced currents which were assumed to be in the same direction as the wind. The observed oil slick movements correspond best with the resultants obtained using a 3-5 percent wind factor, 60-64 percent of these falling within plus or minus 20 degrees of the observed directions of the oil slicks. Possible sources of error in the method are discussed and an example is given to illustrate how the method might be used to predict movements of surface oil resulting from an oil spillage. (Holoman-Battelle)
W73-09328

STANDARDS FOR FAECAL COLIFORM BACTERIAL POLLUTION,
Otago Univ., Dunedin (New Zealand). Dept of Microbiology.
For primary bibliographic entry see Field 05A.
W73-09331

ON THE QUANTITATIVE CHARACTERISTICS OF THE PELAGIC ECOSYSTEM OF DALNEE LAKE (KAMCHATKA),
For primary bibliographic entry see Field 05C.
W73-09340

MERCURY CONCENTRATION IN RELATION TO SIZE IN SEVERAL SPECIES OF FRESHWATER FISHES FROM MANITOBA AND NORTHWESTERN ONTARIO,
Fisheries Research Board of Canada, Winnipeg (Manitoba). Freshwater Inst.
D. P. Scott, and F. A. J. Armstrong.
Journal of the Fisheries Research Board of Canada, Vol 29, No 12, p 1685-1690, December 1972. 1 fig, 2 tab, 5 ref.

Descriptors: *Absorption, *Mercury, *Freshwater fish, *Size, *Mathematical studies, *Regression analysis, Heavy metals, Canada, Yellow perch, Sauger, Walleye, Lake trout, Statistical methods, Path of pollutants.
Identifiers: White sucker, Burbot, Northern pike, Sheepshead, Lake whitefish, Tullibee, Goldeye, Catostomus commersoni, Lota lota, Perca flavescens, Stizostedion canadense, Stizostedion vitreum vitreum, Esox lucius, Freshwater drum, Aplodinotus grunniens, Coregonus clupeaformis, Coregonus artedii, Salvelinus namaycush, Hiodon alosoides.

Statistical analysis of 53 samples of 11 species of fishes from a number of areas of Manitoba and northwestern Ontario indicated that in general there was a positive correlation between mercury concentration and length. There also appeared to be a more variable positive relation between mercury concentration and fish condition (fatness). It was possible to predict, for 31 of the 53 samples, the range of lengths within which there was a 95 percent probability of all fish containing less than 0.5 ppm mercury. However, within species the relation between mercury concentration and length was not consistent. The results show that, for certain of the populations closed to commercial fishing because of mercury contamination, selection (by fishing technique or otherwise) of certain sizes should provide fish of acceptably low mercury concentration. (Holoman-Battelle)
W73-09346

USE OF A COMMERCIAL DREDGE TO ESTIMATE A HARDSHELL CLAM POPULATION BY STRATIFIED RANDOM SAMPLING,
Rhode Island Dept. of Natural Resources, Providence. Div. of Fish and Wildlife.
H. J. Russell, Jr.
Journal of the Fisheries Research Board of Canada, Vol 29, No 12, p 1731-1735, December 1972. 2 fig, 1 tab, 9 ref.

Descriptors: *Aminal populations, *Estimating, Statistical methods, Clams, Mollusks, Invertebrates, Marine animals, Efficiencies, Dredging.
Identifiers: *Stratified random sampling, *Mercenaria mercenaria, *Rocking-chair dredge, Performance evaluation, Random sampling, Sampling design, Quahog, Macroinvertebrates, Data interpretation.

Results are presented of the application of a Fall River rocking-chair dredge in a stratified random sampling design in which a limited number of samples sufficed to provide a reasonably accurate population estimate of M. mercenaria in Narragansett Bay, Rhode Island, Stratified random sampling was designed such that the preliminary dredge survey served to identify centers o abundance and provide the information necessar to construct contours enclosing areas of equ abundance. Subsequent sampling was the directed towards approaching equal sampling frac tions based on areas between contours, or strat Because of the nonrandom distribution pattern o M. mercenaria, treatment of the data with norma statistics would indicate a mean of 1.64 bu pe dredge tow with a variance of 1.80. Stratifie sampling, due to its inherent weighting feature provided a similar estimate of 1.36 bu per tow wit a variance of 0.033. (Holoman-Battell)
W73-09348

SALMON-CANNING WASTE WATER AS A MICROBIAL GROWTH MEDIUM,
Fisheries Research Board of Canada, Vancouve (British Columbia). Vancouver Lab.
For primary bibliographic entry see Field 05C.
W73-09350

HYPHOMYCETES UTILIZING NATURAL GAS
University of Western Ontario, London. Dept. o Plant Sciences; and University of Western Ontario, London. Faculty of Engineering Science.
J. S. Davies, A. M. Wellman, and J. E. Zajic.
Canadian Journal of Microbiology, Vol 19, No 1, p 81-85, January 1973. 3 fig, 1 tab, 20 ref.

Descriptors: *Aquatic fungi, *Sewage sludge *Isolation, *Natural gas, *Pollutant identification Systematics, Activated sludge, Methodology Aquatic microorganisms.
Identifiers: *Hyphomycetes, Substrate utilization Sewage fungi, Ethane, Propane, n-butane, Graphium, Phialophora jeanselmei, Acremonium, Continuous cultures, Culturing techniques, Culture media, Aliphatic hydrocarbons, Coty's agar.

Twleve cultures of Hyphomyetes capable of growth upon the ethane component of natural gas were isolated from raw sewage by a continuous enrichment technique. Five continuous enrichment systems were operated simulataneously Each consisted of a closed 4-liter Erlenmeyer flask containing Coty's mineral salts medium; the pH was adjusted with 2 N HC1. A mixture of 50 percent natural gas and 50 percent air was bubbled through each system. Agitation of the medium was effected by means of magnetic stirrers which also aided in dispersion of the gas bubbles. During the enrichment procedure fresh medium was pumped in daily by means of a multi-channel peristaltic pump and the effluent was harvested by positive gas pressure at the same rate, thus effecting dilution. Each system received an inoculum of 5 percent by volume of raw sewage, taken from several points in an activated sludge-sewage treatment plant. All five systems were run continuously for 6 weeks; after the first 2 weeks, attempts were made at weekly intervals to isolate individual fungi from the effluent of each system by the following method. Samples of the enrichment effluent were treated for 30 s in a waring Blendor to break up hyphal clumps into smaller fragments. Each sample was washed by suspending the hyphae in sterile Coty's medium (pH 5.5) followed by centrifugation at 100 g for 2 min. Tenfoled dilutions in sterile Coty's medium were made from the resulting suspension and 0.1-ml aliquots were spread over the surface of Petri plates of Coty's agar. Replicate plates of each dilution of the sample were incubated at room temperature (22-25 C) in desiccators, gassed manometrically as follows: (1) Methand 40 percent:air 60 percent, (2) Ethane 40 percent: air 60 percent, (3) Air 100 percent (control). The plates were examined regularly for growth; when significant growth of fungal colonies occurred, serial transfers were made to fresh medium until pure cultures were obtained. All isolates also grew on propane and n-butane but not methane. These cultures are described and their toxanomic position discussed. Four were tentatively placed in Graphium; three were identified as

Canadian Journal of Microbiology, Vol 19, No 1, p 43-45, January 1973. 16 ref.

Descriptors: *Marine bacteria, *Microbial degradation, *Electron microscopy, *Cytological studies, Aerobic bacteria.
Identifiers: *Ultrastructure, *Crude oil, Flavobacterium, Brevibacterium, Heterotrophic bacteria bacteria, Light microscopy, Sample preservation, Sample preparation.

Two species of marine bacteria with the ability to degrade crude oil were compared ultrastructurally after growing in the presence and absence of oil. The procedure of negative staining was used to evaluate the specimens by electron microscopy. For observations by light microscopy, the organism cells were stained with Sudan black B according to Burdon's method of fat staining. Large electron-dense inclusions, which were located predominantly at the cell terminus, characterized species of Flavobacterium and Brevibacterium when growing on oil. Cells of Flavobacterium sp. had smaller inclusions when grown on marine agar, while inclusion bodies were not found in Brevibacterium sp. grown on marine agar. Sudan black B staining indicated the inclusions are stored lipids. (Holoman-Battelle)
W73-09357

A BAYESIAN APPROACH TO BIOASSAY,
Oregon State Univ., Corvallis.
For primary bibliographic entry see Field 05C.
W73-09358

ESTIMATION OF MICROBIAL VIABILITY,
Department of Scientific and Industrial Research, Palmerston North (New Zealand). Applied Mathematics Div.
V. J. Thomas, N. A. Doughty, R. H. Fletcher, and J. G. Robertson.
Biometrics, Vol 28, No 4, p 947-958, December 1972. 3 tab, 8 ref.

Descriptors: *Estimating, *Viability, *Microorganisms, *Mathematical studies, *Model studies, Probability, Soil fungi, Water pollution effects, Environmental effects.
Identifiers: *Stochastic models, *Data interpretation, Slide culture techniques, Nocardia corallina, Cluster analysis, Random distribution, Pollutant effects.

Determinations of microbial viability, using the slide culture technique on suspensions in which organisms are aggregated to form clusters, give higher values than would be obtained if all organisms were suspended singly. Formulae have been derived for estimating the viability of individual organisms from the viability of clusters and the distribution of organisms among clusters of different sizes. With the limited stochastic model used it was assumed that clusters of a given size are randomly distributed over a slide and that the distributions of numbers of different sized clusters are statistically independent. An additional assumption was made to the effect that the probability of any organism being alive is constant with value pi, and that life or death for any one organism is an independent binomial variate with parameter pi. Depending on the method of sampling used there are two reasonable approaches for estimating the viability of individual organisms from the cluster viability and the cluster size distribution. (1) The first approach depends on a sampling scheme in which a count is made of the number of live and dead clusters for each cluster size over a number of separate samples. In practice this could be carried out by counting the same field of view on a microscope slide before and after growth of the clusters. (2) If preparation of a sample for counts of live clusters differs from, and is incompatible with, preparation for counts of cluster sizes, two separate sets of samples are used. The latter method was used for estimating viability on suspensions of Nocardia corallina. A difference of 8.8 percent was obtained between the viability of clusters and that of individual organisms. This value was considerably greater than the degree of incertainty in both estimates. Itwas concluded that clustering does have a considerable effect on viability determination and it should be considered especially where interpertations of intracellular materials on maintenance of viability are to be made. (Holoman-Battelle)
W73-09360

INVERSE BINOMIAL SAMPLING AS A BASIS FOR ESTIMATING NEGATIVE BINOMIAL POPULATION DENSITIES,
For primary bibliographic entry see Field 07B.
W73-09362

MICROBIAL DEGRADATION OF CRUDE OIL: FACTORS AFFECTING THE DISPERSION IN SEA WATER BY MIXED AND PURE CULTURES,
Tel Aviv Univ. (Israel). Dept. of Microbiology.
A. Reisfeld, E. Rosenberg, and D. Gutnick.
Applied Microbiology, Vol 24, No 3, p 363-368, September 1972. 4 fig, 4 tab, 14 ref.

Descriptors: *Microbial degradation, *Sea water, *Dispersion, *Kinetics, *Environmental effects, Oil spills, Oil wastes, Marine bacteria, Aerobic bacteria, Hydrogen ion concentration, Inhibition, Nitrogen, Phosphorus, Nutrient requirements, Isolation, Oil pollution.
Identifiers: *Crude oil, *Fate of pollutants, Enrichment, Arthrobacter, Mixed cultures, Pure cultures, Culturing techniques, Hexadecane, Heterotrophic bacteria, Substrate utilization, Emulsification, Azide, Culture media.

By means of the enrichment culture technique, a mixed population of microorganisms was obtained which catalyzed the dispersion of crude oil in supplemented sea water. From this enrichment culture, eight pure cultures were isolated and studied. Only one of the isolates (RAG-1) brought about a significant dispersion of crude oil. RAG-1 has been tentatively characterized as a member of the genus Arthrobacter. The other seven isolates gave rise to colonies on supplemented oil agar, but were neither able to disperse oil nor to stimulate the dispersion catalyzed by RAG-1. The dispersion of crude oil by either RAG-1 or the enrichment culture was absolutely dependent on exogenous sources of nitrogen and phosphorus and completely inhibited by 0.01 M azide. The increase in cell number of RAG-1 was directly proportional to the concentration of crude oil added to the medium over the range 0.05 to 1.0 mg/ml. Within this linear region, 1.0 mg of crude oil yielded 90 billion cells and approximately 65 percent of the oil was converted into a nonbenzene extractable form. Accompanying the emulsification was a decrease in the pH from 7.6 to 5.0. Acidic conditions, however, were neither necessary nor sufficient for oil dispersion. When sea water was supplemented with 0.029 mM K_2HPO_4 and 3.8 mM $(NH_4)_2SO_4$ and inoculated with RAG-1, oil dispersion occurred within 1 day. This dispersion could also be brought about by the supernatant following separation of the cells from the medium. Similarly, the supernatant obtained following growth of RAG-1 on hexadecane was capable of emulsifying crude oil in 60 min. (Holoman-Battelle)
W73-09365

MICROBIAL POPULATION OF FEEDLOT WASTE AND ASSOCIATED SITES,
Agricultural Research Service, Peoria, Ill. Northern Regional Research Lab.
R. A. Rhodes, and G. R. Hrubant.
Applied Microbiology, Vol 24, No 3, p 369-377, September 1972. 4 fig, 1 tab, 14 ref.

Descriptors: *Farm wastes, *Feed lots, *Cattle, *Runoff, *Pollutant identification, *Microorganisms, Confinement pens, Coliforms, Anaerobic bacteria, Yeasts, Fungi, Water pollution sources, Sampling, Methodology, Bacteria, Isolation, Soil disposal fields, Domestic animals, Ruminants, Analytical techniques.
Identifiers: Sample preparation, Culture media, Streptomycetes, Enumeration.

A quantitative determination was made every 2 months for a year of the microflora of beef cattle waste and runoff at a medium-sized midwestern feedlot. Counts were obtained for selected groups of organisms in waste taken from paved areas of pens cleaned daily and, therefore, reflect the flora of raw waste. Overall, in terms of viable count per gram dry weight, the feedlot waste contained 10 billion total organisms, one billion anaerobes, 100,000,000 gram-negative bacteria, 10,000,000 coliforms, 1,000,000 sporeformers, and 100,000 yeasts, fungi, and streptomycetes. The specific numbers and pattern of these groups of organisms varied only slightly during the study in spite of a wide variation in weather. Data indicate that little microbial growth occurs in the waste as it exists in the feedlot. Runoff from the pens contained the same general population pattern but with greater variation attributable to volume of liquid. Comparable determinations of an associated field disposal area (before and after cropping), stockpiled waste, and elevated dirt areas in the pens indicate that fungi, and especially streptomycetes, are the aerobic organisms most associated with final stabilization of the waste. Yeasts, which are the dominant type of organism in the ensiled corn fed the cattle, do not occur in large numbers in the animal waste. Large ditches receiving runoff and subsurface water from the fields have a population similar to the runoff but with fewer coliforms. (Holoman-Battelle)
W73-09366

ENTEROBACTERIA IN FEEDLOT WASTE AND RUNOFF,
Agricultural Research Service, Peoria, Ill. Northern Regional Research Lab.
G. R. Hrubant, R. V. Daugherty, and R. A. Rhodes.
Applied Microbiology, Vol 24, No 3, p 378-383, September 1972. 6 tab, 12 ref.

Descriptors: *Cattle, *Feed lots, *Runoff, *Enteric bacteria, *Farm wastes, *Pollutant identification, Isolation, Aerobic bacteria, E. coli, Domestic animals, Ruminants, Pseudomonas, Sampling, Salmonella, Coliforms.
Identifiers: *Drainage ditches, Enterobacteriaceae, Enrichment, Culture media, Plate counts, Selective media, Citrobacter, Klebsiella, Enterobacter cloacae, Proteus vulgaris, Proteus morganii, Proteus mirabilis, Proteus rettgeri, Providencia stuartii, Bacillus spp, Arizona, Enterobacter aerogenes, Enterobacter liquefaciens, Providencia alcalifaciens.

Samples of beef cattle feedlot waste (FLW), runoff from the pens, and water from a large drainage ditch at the feedlot were examined for Enterobacteriaceae. The drainage ditch receives the runoff but contains primarily subsurface drainage from fields on which FLW is spread for disposal. Plating and enrichment techniques with seven different media were used to isolate 553 cultures of enterobacteria. FLW contains about 50 million enterobacteria/g dry weight. More than 90 percent of these were Escherichia coli, none of which were enteropathogenic types as determined with multivalent sera. Citrobacter and Enterobacter cloacae were other organisms present in moderate numbers. Application of enrichment techniques broadened the spectrum of enterobacteria isolates to include the four Proteus spp., both Providencia spp., Klebsiella, Enterbacter aerogenes, Arizona, and a single isolate of Salmonella (serological group C2). Shigella was not isolated. The wide spectrum of enterobacteria in FLW may be a hazard if unsterilized waste is refed. Fewer enterobacteria occurred in the runoff and in the drainage ditch; the most numerous species in FLW also were most numerous at these sites. However, neither Salmonella nor Arizona was isolated from runoff or drainage-ditch waters. (Holoman-Battelle)
W73-09367

WATER POLLUTION MONITORING,
California Univ., Berkeley. Dept. of Geoscience Engineering.
For primary bibliographic entry see Field 05A.
W73-09369

ACCUMULATION AND PERSISTENCE OF DDT IN A LOTIC ECOSYSTEM,
Maine Univ., Orono. Dept. of Entomology.
J. B. Dimond, A. S. Getchell, and J. A. Blease.
J Fish Res Board Can. Vol 28, No 12, p 1877-1882. 1972.
Identifiers: *Path of pollutants, Accumulation, Birds, *DDT, Ecosystem, Fish, Invertebrates, Plants, *Maine.

DDT persisted in streams of several small watersheds in Maine for at least 10 yr following light applications to the forest. Residues declined sharply within 2 or 3 yr after application, but after 10 yr were still well above concentrations detected in untreated streams. This was true of all components studied: muds, plants, invertebrates, fish, and fish-eating birds. The prolonged persistence led to cumulative residue levels in streams sprayed more than once. Concentration of residues through the food chain was evident.--Copyright 1972, Biological Abstracts, Inc.
W73-09415

TRANSITION OF POLLUTION WITH MERCURY OF THE SEA FOOD AND SEDIMENTS IN MINAMATA BAY (IN JAPANESE),
Kumamoto Univ. (Japan). Dept. of Hygiene.
K. Irukayama, M. Fujiki, S. Tajima, and S. Omori.
Jap J Public Health. Vol 19, No 1, p 25-32. 1972. Illus. English summary.
Identifiers: *Path of pollutants, Industrial wastes, Waste disposal, Effluents, Fish, Hair, Japan, *Mercury, Minamata Bay, Water pollution, Seafood, Sediments, Shellfish, Water pollution effects.

Since the end of 1959, a chemical factory has disposed of effluents from its acetaldehyde and vinyl chloride plant, in which Hg was used as catalyst, into Minamata Bay. Methods of waste disposal were revised several times, but in May, 1968, the production of acetaldehyde was stopped, and in June, 1971, the method of production of vinylchloride was changed to use acetylene instead of ethylene. At present no Hg is used in the factory. Since the beginning of 1960, the Hg contents in the sea food from Minamata Bay have gradually decreased. In 1970, Hg contents in the fish from Minamata Bay showed below 0.5 micro g/g of wet weight and Hg contents in the hair of inhabitants in Minamata City was below 5 micro g/g in average. Abundant sediments containing Hg were deposited in the bottom of Minamata Bay, almost all in the form of inorganic Hg compounds, although a trace of methyl mercury was detected.--Copyright 1972, Biological Abstracts, Inc.
W73-09416

TRITIUM AND ITS EFFECTS IN THE ENVIRONMENT--A SELECTED LITERATURE SURVEY,
Battelle Memorial Inst., Columbus, Ohio.
For primary bibliographic entry see Field 05C.
W73-09433

GRAYS HARBOR ESTUARY, WASHINGTON: REPORT 3, WESTPORT SMALL-BOAT BASIN STUDY,
Army Engineer Waterways Experiment Station, Vicksburg, Miss.
For primary bibliographic entry see Field 08B.
W73-09438

REGIONAL RATES OF GROUND-WATER MOVEMENT ON LONG ISLAND, NEW YORK,
Geological Survey, Mineola, N.Y.
For primary bibliographic entry see Field 02F.
W73-09441

SPECIFIC-CONDUCTANCE SURVEY OF THE MALAD RIVER, UTAH AND IDAHO,
Geological Survey, Logan, Utah.
For primary bibliographic entry see Field 02K.
W73-09448

HYDROGEOLOGICAL CRITERIA FOR EVALUATING SOLID-WASTE DISPOSAL SITES IN JAMAICA AND A NEW WATER SAMPLING TECHNIQUE,
Geological Survey of Jamica, Kingston.
R. M. Wright.
Journal of the Scientific Research Council of Jamaica, Vol 3, No 2, p 59-92, 1972. 6 fig, 4 tab, 53 ref.

Descriptors: *Waste disposal, *Solid wastes, *Sanitary engineering, *Landfills, *Water pollution sources, Water quality, Monitoring, Instrumentation, Specific conductivity, Leaching, Path of pollutants, Groundwater, Groundwater movement.
Identifiers: *Jamaica.

The hydrogeological factors for evaluating sanitary landfill sites in Jamaica are described. Favorable topography includes flat upland areas, heads of gullies and ravines, disused quarries, and dry strip mines. Unfavorable topography includes lower reaches of gullies, depressions where water accumulates, stream flood plains, and sites where leachate might discharge into water. Favorable bedrocks are shale and granodiorite. Unfavorable bedrocks are sandstone, cavernous limestone, and cavernous dolomite. Favorable unconsolidated deposits are clays and silts. Unfavorable materials are sand and gravel less than 50 feet thick. Thickness requirements vary depending on the nature and drainage characteristics of unconsolidated materials. Local sources (and potential sources) of water are favorable in deep bedrock wells; in sand, gravel and alluvium wells with thick impermeable cover over aquifers; and in dug wells if more than 500 feet from the landfill site; and unfavorable in shallow bedrock wells and in sand, gravel and alluvium wells with thin cover. (Woodard-USGS)
W73-09452

SUBSURFACE WASTEWATER INJECTION, FLORIDA,
Geological Survey, Tallahassee, Fla.
For primary bibliographic entry see Field 05E.
W73-09456

TRACE ELEMENTS IN SOILS AND FACTORS THAT AFFECT THEIR AVAILABILITY,
Macaulay Inst. for Soil Research, Aberdeen (Scotland).
R. L. Mitchell.
Geological Society of America Bulletin, Vol 83, No 4, p 1069-1076, April 1972. 6 fig, 18 ref.

Descriptors: *Trace elements, *Soils, *Agriculture, Animal pathology, Human pathology, Heavy metals, Metals, Drainage, Irrigation, Precipitation, Drainage effects, Plants, Nutrients, Geochemistry, Air pollution, Water pollution, Plant uptake, Chelation, Soil investigations.

content of plants grown in natural soils is generally less than 10 ppm d.w. When taken up by a plant, lead may be found principally in the roots, with little translocation to the shoot. Lead uptake is enhanced by low soil pH and can be reduced by liming the soil. (Jerome-Vanderbilt)
W73-09479

MERCURY IN THE ATMOSPHERE,
S. H. Williston.
Journal of Geophysical Research, Vol 73, No 22, p 7051-7055, November 18, 1968. 2 fig, 3 ref.

Descriptors: *Mercury, *Air pollution, *Analytical techniques, Sampling, Testing, Laboratory equipment, Wind velocity, Temperature, Smog, Atmosphere.
Identifiers: *Mercury pollution.

High-sensitivity mercury-vapor analysis has been automated to provide a continuous monitoring of mercury in the earth's atmosphere over the last 2-year period. In the San Francisco Bay area (Los Altos) winter concentrations range from slightly >1/2 to 25 nanograms of mercury per cubic meter of air. Summer concentrations range from slightly over 1 to 50 nanograms per cubic meter. Concentration of mercury seems to depend primarily on wind direction, wind speed, and seasonal temperature variations. Numerous other variables also affect it. It is shown that high mercury levels always coincide with high smog levels. (Oleszkiewicz-Vanderbilt)
W73-09489

PARTICLES CONTAINING LEAD, CHLORINE, AND BROMINE DETECTED ON TREES WITH AN ELECTRON MICROPROBE,
Connecticut Agricultural Experiment Station, New Haven. Dept. of Ecology.
For primary bibliographic entry see Field 05A.
W73-09494

RADIOCHEMICAL STUDIES ON EFFECT OF BATHING: II. EFFECT OF SERIAL BATHING ON DISTRIBUTION OF MERCURY COMPOUNDS IN MICE, (IN JAPANESE),
Kyushu Univ., Beppu (Japan). Inst. of Balneotherapeutical Research I.
For primary bibliographic entry see Field 05C.
W73-09496

METHYLMERCURY IN NORTHERN PIKE (ESOX LUCIUS): DISTRIBUTION, ELIMINATION, AND SOME BIOCHEMICAL CHARACTERISTICS OF CONTAMINATED FISH,
Fisheries Research Board of Canada, Winnipeg (Manitoba). Freshwater Inst.
W. L. Lockhart, J. F. Uthe, A. R. Kenney, and P. M. Mehrle.
Journal of the Fisheries Research Board of Canada, Vol 29, No 11, p 1519-1523, 1972. 4 tab, 18 ref.

Descriptors: *Mercury, *Pikes, *Biopsy, Distribution, Water pollution sources, Biochemistry, Organic compounds, Inorganic compounds, Fish, Lakes, Sampling, Evaluation, Environmental effects, On-site data collection.
Identifiers: *Methylmercury.

Northern pike heavily contaminated with methylmercury were captured from Clay Lake, Ontario, and released in Heming Lake, Manitoba, an area relatively free of mercury. Mercury levels in muscle biopsy samples at the time of transfer and at subsequent recaptures indicated that only 30% was eliminated in one year. Distribution among various body tissues was essentially unchanged, those organs most heavily contaminated being lens, kidney, and liver in decreasing order. Biochemical profiles of blood serum constituents showed several differences between samples from the two lakes, especially in levels of inorganic phosphate, total protein, alkaline phosphatase, and cortisol. Serum values for transplanted fish tended toward those in the clean lake and we have concluded that biochemical profiles were sensitive to the environmental change. (Jerome-Vanderbilt)
W73-09498

ECOLOGICAL STUDIES IN THE MAMMOTH CAVE SYSTEM OF KENTUCKY: II. THE ECOSYSTEM,
Kentucky Univ., Lexington. Inst. of Speleology.
For primary bibliographic entry see Field 05C.
W73-09503

DISTRIBUTION AND EXCRETION RATE OF PHENYL- AND METHYLMERCURY NITRATE IN FISH, MUSSELS, MOLLUSCS AND CRAYFISH,
Helsinki Univ. (Finland). Dept. of Radiochemistry.
For primary bibliographic entry see Field 05C.
W73-09508

MONITORING ZINC CONCENTRATIONS IN WATER USING THE RESPIRATORY RESPONSE OF BLUEGILLS (LEPOMIS MACROCHIRUS RAFINESQUE),
Virginia Polytechnic Inst. and State Univ., Blacksburg. Dept. of Biology; and Virginia Polytechnic Inst. and State Univ., Blacksburg. Center for Environmental Studies.
R. E. Sparks, J. Cairns, Jr., R. A. McNabb, and G. Suter II.
Hydrobiologia, Vol 40, No 3, p 361-369, 1972. 7 fig, 5 ref. EPA Project 18050 EDQ.

Descriptors: *Sunfish, *Respiration, *Zinc, *Measurement, *Analytical techniques, Biochemistry, Physiology, Fish, Metabolism, Mortality, Toxicity, Water pollution, Evaluation, Heavy metals.

Pressure changes in both the buccal and opercular cavities of bluegill sunfish exposed to lethal and sublethal concentrations of zinc, were measured according to a method developed by Schaumburg, Howard, and Walden. The object was to determine whether this measuring technique might be useful in monitoring heavy metal pollution. Fish were exposed to varying concentrations of $ZnSO_4 \cdot 7H_2O$ and respiration and coughing rates were measured along with cavity pressure levels. Zinc induced coagulation and sloughing of mucus in bluegills and the function of coughing might be to reverse water flow through the gills in order to free them of mucus. The coughing rates increase proportionally to the concentration of zinc in the water. Although useful physiological information was obtained by recording pressure in both the buccal and opercular cavities, it appears that cannulation of the mouth alone would provide sufficient information on minor toxicants. Measuring cough frequency of bluegills is a good monitoring technique for zinc. (Jerome - Vanderbilt)
W73-09510

SIGNIFICANCE OF MERCURY IN THE ENVIRONMENT,
Department of Agriculture, Saskatoon (Saskatchewan). Research Station.
J. G. Saha.
Residue Reviews, Vol 42, p 103-163, 1972. 15 tab, 229 ref.

Descriptors: *Mercury, *Path of pollutants, *Public health, *Toxins, Biochemistry, Chemistry, Pollution, Biological communities, Ecological distribution, Industrial production, Environmental effects, Industrial wastes, Food chains, Aquatic populations, Fish, Diseases, Heavy metals, *Reviews, *Bibliographies.

A general review of mercury in the environment and its effects on the biological community in general and humans in particular is presented. There is a natural circulation of mercury in the environment called the background mercury level, which ranges from less than 1 ppb in the atmosphere to approximately 1 or 2 ppm in soil; and which seems relatively harmless to man. In addition to the background mercury, man's activities in manufacturing and agriculture contribute considerable amounts of mercury to the environment. Levels of mercury in a large variety of materials are listed. Occupational exposure to mercury and high concentrations of mercury in the food chain are capable of affecting the central nervous system and causing genetic defects with toxicity depending on the type of compound. Two estimates of allowable daily intakes are presented and discussed. The allowable level of mercury in food in the U.S. and Canada is 0.5 ppm, which makes it unlikely that fish eaters in these countries will ever have harmful consumption levels of this element. (Jerome - Vanderbilt)
W73-09511

MICROBIAL UPTAKE OF LEAD,
Colorado State Univ., Fort Collins. Dept. of Microbiology.
For primary bibliographic entry see Field 05C.
W73-09514

PYROPHOSPHATE HYDROLYSIS IN SOIL AS INFLUENCED BY FLOODING AND FIXATION,
Manitoba Univ., Winnipeg. Dept. of Soil Science.
For primary bibliographic entry see Field 02G.
W73-09516

POTENTIOMETRIC TITRATION OF SULFATE IN WATER AND SOIL EXTRACTS USING A LEAD ELECTRODE,
Agricultural Research Service, Riverside, Calif. Salinity Lab.
For primary bibliographic entry see Field 02K.
W73-09518

GASEOUS LOSSES OF NITROGEN FROM FRESHLY WETTED DESERT SOILS,
Wisconsin Univ., Madison. Dept. of Soil Science.
For primary bibliographic entry see Field 02G.
W73-09519

HYDROLYSIS OF PROPAZINE BY THE SURFACE ACIDITY OF ORGANIC MATTER,
Agricultural Research Service, Beltsville, Md. Soils Lab.
D. C. Nearpass.
Soil Sci Soc Am Proc. Vol 36, No 4, p 606-610, 1972. Illus.
Identifiers: Acidity, Adsorption, Herbicides, *Hydrolysis, *Organic matter, *Propazine, Water pollution sources.

Propazine (2-chloro-4, 6 bis (isopropylamino)-s-triazine) hydrolysis in acidic aqueous soil-free systems was pH dependent, increasing with lower pH values. At a given pH, degradation followed first-order kinetics. At 23.5 degC, the relationship was log t1/2 (days) ± 0.59 pH-0.21. Adsorption of propazine by organic matter prepared from Michigan peat was also pH dependent, but both molecular and cationic adsorption occurred. An increased rate of degradation in the presence of organic matter was postulated as due to hydrolysis by ionized surface H. Increasing the Ca saturation resulted in decreased hydrolysis. Increasing the CaCl2 concentration of the propazine-organic matter-aqueous system had no effect on acid hydrolysis. In the salt-amended systems, an increased hydrolysis, to be expected from the lower pH in the ambient solution, was apparently offset by a decreased hydrolysis due to the lowered amounts of ionized surface H.--Copyright 1973, Biological Abstracts, Inc.
W73-09520

CHARACTERIZATION OF THE OXIDIZED AND REDUCED ZONES IN FLOODED SOIL,
Louisiana State Univ., Baton Rouge. Dept. of Agronomy.
For primary bibliographic entry see Field 02G.
W73-09522

GEOGRAPHICAL INFLUENCE ON COOLING PONDS,
Vanderbilt Univ., Nashville, Tenn. Dept. of Environmental and Water Resources Engineering.
E. L. Thackston, and F. L. Parker.
J Water Pollut Control Fed. Vol 44, No 7, p 1334-1351. 1972. Illus.
Identifiers: *Cooling ponds, Geographical effects, *Thermal pollution, *Energy budget.

The energy budget approach was outlined and applied to cooling ponds. Monthly average weather data from 88 stations throughout the USA were used to calculate equilibrium temperatures, heat exchange coefficients, and amount of cooling in various-sized ponds receiving the effluent from a standard power plant of 1000-Mw capacity, both for average and extreme weather conditions. The data for each station were plotted on a chart, and the variation of these results across the USA is depicted by a series of maps of the USA with contours connecting equal values of the parameters. The maps disclose variations across the USA on a given date, of up to 55 degree F (30.5 degree C) in equilibrium temperature, up to 100% difference in heat exchange coefficients, up to 50% difference in heat lost from a given pond size, and up to 200% difference in the size of a pond necessary to produce an equal cooling effect. The results may also be used to estimate cooling pond performance for power plants of other sizes.--Copyright 1972, Biological Abstracts, Inc.
W73-09525

BIODEGRADATION OF ANIMAL WASTE BY LUMBRICUS TERRESTRIS,
Georgia Univ., Athens. Dept. of Dairy Science.
For primary bibliographic entry see Field 05G.
W73-09528

EFFECT OF THE WASTE-WATER OF SUGAR-WORKS ON NATURAL HISTORY OF THE ZAGYVA RIVER,
Authority Water Conservancy Middle Course Tisza, Szolnok (Hungary).
J. Hamar.
Tiscia. Vol 6, p 109-128. Illus. 1970/1971.
Identifiers: History, *Hungary, Rivers, Sugar industry, *Waste water pollution, *Zagyva River.

The chemical composition and natural history of the Zagyva river, as well as changes as a result of various waste-water are analyzed. The effect of standing depends upon the water output, while the influence of periodical pollutions (waste-water waves) depends primarily on the storage time of waste-water.--Copyright 1972, Biological Abstracts, Inc.
W73-09538

PHYTOPLANKTON, PHOTOSYNTHESIS, AND PHOSPHORUS IN LAKE MINNETONKA, MINNESOTA,
Minnesota Univ., Minneapolis. Limnological Research Center.
For primary bibliographic entry see Field 05C.
W73-09546

CATALYTIC OXIDATION OF PHENOL OVER COPPER OXIDE IN AQUEOUS SOLUTION,
Delaware Univ., Newark. Dept. of Chemical Engineering.
A. Sadana.

approach to further research is stressed. (Ensign-PAI)
W73-09647

UNSTRUCTURED MARINE FOOD WEBS AND 'POLLUTANT ANALOGUES',
Scripps Institution of Oceanography, La Jolla, Calif.
J. D. Isaacs.
Fishery Bulletin, Vol 70, No 3, p 1053-1059, July 1972. 1 fig, 1 tab, 1 ref.

Descriptors: *Cesium, Fish, *Oceans, *Food chains, *Food webs, Trace elements, Trophic level, Water pollution sources, Marine animals.
Identifiers: *Gulf of California, *Salton Sea.

Cesium levels in several species of fish living in the Gulf of California differ substantially from levels of the same species living in the Salton Sea. Simplified trophic models of the two environments are developed by applying known information about the food chain in the Salton Sea to the more uncertain food web in the Gulf of California. Data on an unstructured food web matrix and equations are presented. It is believed that the study of concentrations of cesium and other natural trace elements in marine organisms may lead to an understanding of the trophic position of the organisms. Such an understanding could yield fuller knowledge of the existing or potential distribution of pollutants in marine organisms. (Ensign-PAI)
W73-09649

PLUTONIUM AND POLONIUM INSIDE GIANT BROWN ALGAE,
Scripps Institution of Oceanography, La Jolla, Calif.
For primary bibliographic entry see Field 05C.
W73-09652

THE VARIATIONS IN THE DISSOLVED OXYGEN AND TOTAL PHOSPHORUS CONCENTRATIONS IN LAKE GENEVA DURING THE YEARS 1969 AND 1970 (IN FRENCH),
Societe Vaudoise des Sciences Naturelles, Lausanne (Switzerland).
Pierre Revelly.
Bull Soc Vaudoise Sci Nat. Vol 71, No 4, p 211-215, 1972.. English summary.
Identifiers: *Dissolved oxygen, *Lake Geneva, Oxygen, *Phosphorus, Switzerland, Seasonal.

Resultsol analytical tests on the water of lake Leman Switzerland for the years 1968-1970 show a sudden increase in total P and dissolved O2 in the lower layers. However, it cannot yet be concluded that this variation is characteristic of the evalou-tion of the lake as a whole.--Copyright 1972, Biological Abstracts, Inc.
W73-09653

THE OCCURRENCE, LIFE CYCLE, AND PATHOGENICITY OF ECHINURIA UNCINATA (RUDOLPHI, 1819) SOLOVIEV, 1912 (SPIRURIDA, NEMAT ODA) IN WATER FOWL AT DELTA, MANITOBA,
Naval Medical Research Inst., Bethesda, Md.
For primary bibliographic entry see Field 02I.
W73-09749

5C. Effects of Pollution

THE EIGHTEENTH INTERNATIONAL LIMNOLOGY CONGRESS AT LENINGRAD (XVIII MEZHDUNARODNYY LIMNOLOGICHESKIY KONGRESS V LENINGRADE),
Akademiya Nauk SSSR, Moscow. Institut Geografii.
For primary bibliographic entry see Field 02H.
W73-09159

PHOTOSYNTHETIC PROPERTIES OF PERMEAPLASTS OF ANACYSTIS,
Texas Univ., Austin. Dept. of Botany; and Texas Univ., Austin. Dept. of Zoology.
B. Ward, and J. Myers.
Plant Physiology, Vol 50, No 5, p 574-550, 1972. 3 fig, 1 tab, 17 ref. NIH GM 11300.

Descriptors: *Cytological studies, *Photosynthesis, Plant physiology, Plant morphology, Algae.
Identifiers: *Permeaplstts, *Anacystis nidulans, Electron transport.

For preparation of photosynthetic procaryotic cells, functionally analogous to chloroplasts of eucaryotes, a procedure using lysozyme and EDTA reproducibly gives intact but permeable cells (permeaplasts) of Anacystis nidulans with isolated photosystems 1 and 2 of high ability (although with lowered rates) of complete photosynthesis. Rates of electron transport from water to carbon dioxide, ferricyanide, 2, 6-dichlorophenol indophenol, benzoquinone, and methyl viologen, and from reduced indophenol to methyl viologen were measured as a function of treatment time. Oxygen evolution rates in complete photosynthesis and electron flow from water to methyl viologen showed rapid and parallel decline with time. Electron flow from water to ferricyanide and from reduced indophenol to methyl viologen increased during the first half hour (phase 1) to 60 to 80% of original photosynthetic rate. Longer treatment (phase 2) resulted in decreased ferricyanide reduetion rate but not in rate of methyl viologen reduction from indophenol. Electron flow from water to quinone was two to three times higher than for complete photosynthesis in intact cells remaining high during phase 1 and declining during phase 2. Phase 1 permeaplasts apparently retain high activity for photosystems 1 and 2 photoreactions. (Jones-Wisconsin)
W73-09171

BLOOMS OF SURF-ZONE DIATOMS ALONG THE COAST OF THE OLYMPIC PENINSULA, WASHINGTON. I. PHYSIOLOGICAL INVESTIGATIONS OF CHAETOCEROS ARMATUM AND ASTERIONELLA SOCIALIS IN LABORATORY CULTURES,
Washington Univ., Seattle. Dept. of Oceanography.
J. Lewin, and D. Mackas.
Marine Biology, Vol 16, No 2, p 171-181, 1972. 14 fig, 7 ref. NSF GB7373

Descriptors: *Adaptation, *Marine algae, *Plant physiology, *Coasts, *Diatoms, Cultures, Light intensity, Nutrients, Temperature, Salinity, Growth rates, Iron.
Identifiers: *Surf zone, *Chaetoceros armatum, *Asterionella socialis.

Virtually all plant biomass in the surf zone along the Olympic Peninsula, Washington, consists of two diatoms, Chaetoceros armatum and Asterionella socialis, which were isolated, grown in culture, and their physiological behavior studied. A. socialis displayed the greater maximum growth rate. Both species tolerated a wide range of salinities and temperatures and were favored by aeration. In response to increasing light intensity, C. armatum reached its maximum growth rate at 750 to 1000 lux (at 13C) whereas A. socialis only reached its maximum at 4000 lux. In its natural habitat, C. armatum is far the most important of the two, since its blooms often consist of almost pure stands. A mucilage envelope, with its accompanying clay particles, is invariably present surrounding C. armatum chains in their natural habitat, but the diatom did not produce this envelope under any culture conditions. The absence of the mucilage envelope (and clay particles) may explain the iron nutrition peculiarities observed in culture experiments, that is, a dependence on iron as NaFe-EDTA or ferric citrate, and failure to utlize ferric chloride as a source of Fe for growth. (Jones-Wisconsin)
W73-09172

PRODUCTION OF CARBOHYDRATES BY THE MARINE DIATOM CHAETOCEROS AFFINIS VAR. WILLEI (GRAN) HUSTEDT. I. EFFECT OF THE CONCENTRATION OF NUTRIENTS IN THE CULTURE MEDIUM,
Norwegian Inst. of Seaweed Research, Trondheim.
S. Myklestad, and A. Haug.
Journal of Experimental Marine Biology and Ecology, Vol 9, No 2, p 125-136, 1972. 7 fig, 6 tab, 16 ref.

Descriptors: *Cytological studies, *Plant physiology, *Carbohydrates, *Marine algae, Diatoms, Nutrients, Cultures, Chemical properties, Nitrates, Proteins, Phosphates, Fjords.
Identifiers: *Chaetoceros affinis, Glucan, Polysaccharide, Trondheimsjord (Norway).

Influence of the concentration of nitrate and phosphate on the chemical composition of the marine diatom, Chaetoceros affinis var. willei was studied, particularly production of carbohydrates, not only total amounts, but also amounts of glucan and of extracellular polysaccharide. The diatom, collected from the Trondheimsjord, was isolated from an enrichment culture and made bacteria-free by treatment with antibiotics. Chemical changes with age of the culture were followed. The cells' chemical composition was markedly influenced by nutrient concentration in the medium. In the logarithmic growth phase the cellular glucan content was relatively low, but in the stationary phase the glucan content showed a rapid increase which seemed to coincide with depletion of nitrate from the medium, which in turn led to very pronounced variations in the ratio of protein to carbohydrate. This ratio can easily be determined and seems to be a sensitive and convenient parameter for characterizing physiological state of the diatom cells. It is evident that the extracellular polysaccharide which was produced in the stationary phase, especially in a medium with a high proportion of nitrate to phosphate, accounts for a considerable part of the total cell production. (Jones-Wisconsin)
W73-09173

OCCURRENCE OF FREE AMINO ACIDS IN POND WATER,
Polish Academy of Science, Krakow. Zaklad Biologii Wod.
J. Zygmuntowa.
Acta Hydrobiologica, Vol 14, No 1, p 317-325, 1972. 3 fig, 1 tab, 36 ref.

Descriptors: *Amino acids, *Ponds, *Organic matter, Fish farming, Hypolimnion, Temporal distribution, Epilimnion, Indicators, Eutrophication, Chromatography.
Identifiers: Fish ponds, Cystine, Aspartic acid, Serine, Glutamic acid, Valine, Leucine, Alanine, Tyrosine, Tryptophan, Methionine, Proline, Lysine, Arginine, Threonine, Phenylalanine.

Free amino acid content in pond water was studied to check the qualitative and quantitative variations at various depths and times of day. Samples were taken from similar carp breeding ponds, differing only in amount of foodstuffs supplied to the fish. Temperature, oxygen content, and hydrogen ion concentration were determined. Besides the chemical analysis, the qualitative composition of phytoplankton was determined as the variations in the content of free amino acids on the surface and at the bottom seemed to correspond with the vertical migration of the phytoplankton. The thin-layer-chromatographic method proved successful because it permits the identification of substances in a relatively small amount of the basic sample. Appearing most frequently were: cystine, aspartic acid, serine with glycine, glutamic acid, alanine, valine, and leucine. There seemed to be accumulation of free amino acids at the bottom in the morning indicating sinking of organic matter and a decomposition process taking place. The concentration of free amino acids dissolved in pond water depends, to a certin extent, on the fertility of the pond, the time of day, and the layer in the vertical section. The content of free amino acids in waters of a similar type may reflect their fertility. (Jones-Wisconsin)
W73-09174

WATER QUALITY SIMULATION MODEL,
Birmingham Univ. (England). Dept. of Civil Engineering.
N. Green, M.D.
Journal of the Sanitary Engineering Division, Proceedings of the American Society of Civil Engineers, Vol 98, Nd SA4, p 669-670,1972. 2 ref.

Descriptors: *Equations, *Mathematical models, *Simulation analysis, Reaeration, Biochemical oxygen demand, Hydraulic properties, Streamflow, Dissolved oxygen.
Identifiers: Sensitivity analysis, Great Britain.

The mathematical model presented by Grantham, Schaake and Pyatt (See W71-13978) appears to have a number of inadequacies and a more accurate representation, describing a natural process subject to variation, is suggested. A daily pollution model was developed for a river system in Great Britain where there is a single effluent discharge point at the top end of a reach and the critical DO level within the reach is printed out for each time increment. This model was used to carry out a sensitivity test for each of the input parameters. First, the sensitivity of the daily critical DO level within the reach, averaged over four years, using a synthetic input, was determined with respect to changes in the mean value of each input parameter. Second, the sensitivity of the system in terms of the change in the standard deviation of the output was measured by changing the standard deviation of the inputs. Two important points regarding the application of the results of a sensitivity study are significant: First, the results of a sensitivity analysis are only applicable to a particular system or to systems that are close to the original; secondly, a sensitivity study can be used to guide further fieldwork. (Jones-Wisconsin)
W73-09175

WISCONSIN LAKES RECEIVING SEWAGE EFFLUENTS,
Wisconsin Univ., Madison. Water Resources Center.
For primary bibliographic entry see Field 05B.
W73-09176

ATMOSPHERIC CONTRIBUTIONS OF NITROGEN AND PHOSPHORUS,
Wisconsin Univ., Madison. Water Resources Center.
For primary bibliographic entry see Field 05B.
W73-09177

DDT AND MALATHION: EFFECT ON SALINITY SELECTION BY MOSQUITOFISH,
Environmental Protection Agency, Gulf Breeze, Fla. Gulf Breeze Lab.
D. J. Hansen.
Trans Am Fish Soc. Vol 101, No 2, p 346-350. 1972. Illus.
Identifiers: *DDT, *Malathion, *Mosquitofish, Salinity.

It is believed that salinity selection by test mosquitofish (Gambusia affinis) is probably the combined result of their preferences as wild fish, effects of acclimation salinity and effects of the pesticide.--Copyright 1973, Biological Abstracts, Inc.
W73-09178

EMERSON ENHANCEMENT OF CARBON FIXATION BUT NOT OF ACETYLENE REDUC-

ponds. A relationship between the species composition and amount of algae and type of fertilization was confirmed. It seems that biological analyses by no means always strictly correspond with chemical analyses, which sometimes leads even to different classifications of reservoirs. It is probable that the age of the ponds was the factor underlying the plankton composition as the type of succession observed was characteristic during the first years of the filling of the ponds. (Jones-Wisconsin)
W73-09183

BIOMASS OF ZOOPLANKTON AND PRODUCTION OF SOME SPECIES OF ROTATORIA AND DAPHNIA LONGISPINA IN CARP PONDS,
Polish Academy of Sciences, Golysz. Zaklad Biologii Wod.
M. Lewkowicz.
Polskie Archiwum Hydrobiologii, Vol 18, No 2, p 215-223, 1971. 4 fig, 3 tab, 17 ref.

Descriptors: *Biomass, *Zooplankton, *Productivity, Rotifers, Daphnia, Fish hatcheries, Fertilization, Nitrogen, Phosphorus, Temperature, Crustaceans, Copepods.
Identifiers: *Rotatoria, *Daphnia longispina, Golysz (Poland), Polyarthra trigla vulgaris, Brachionus calyciflorus, Filinia longiseta, Cyclops.

Relationship of fertilization to zooplankton biomass and production of Rotatoria species and of Daphnia longispina was assessed at the Experimental Farm at Golysz, Poland, June 28-October 9, 1967 in three fingerling carp ponds which were unfertilized, fertilized with nitrogen + phosphorus, and fertilized with phosphorus alone. Fertilization increased zooplankton biomass. In the pond fertilized with nitrogen + phosphorus, zooplankton developed more rapidly than in the pond fertilized with phosphorus only. Rotatoria production in the fertilized ponds was several times greater than in the unfertilized. Production of Brachionus calyciflorus was found to be the highest in the pond fertilized with nitrogen + phosphorus. A particularly large production increase under the influence of fertilization was found in Polyarthra trigla vulgaris, Brachionus calyciflorus, and Filinia longiseta. Production of Daphia longispina in the fertilized ponds was much higher than in the unfertilized, the highest production always observed after a rapid decline of organic matter. No large differences between seasonal production in the pond fertilized with nitrogen + phosphorus and in that fed with phosphorus alone were found. Immediately after filling the unfertilized pond, Daphnia longispina appeared on a mass scale, perhaps associated with intensive destruction processes occurring on the bottom. (Jones-Wisconsin)
W73-09184

CILIATA IN BOTTOM SEDIMENTS OF FINGERLING PONDS,
Polish Academy of Sciences, Krakow. Zaklad Biologii Wod.
E. Grabacka.
Polskie Archiwum Hydrobiologii, Vol 18, No 2, p 225-233, 1971. 2 fig, 1 tab, 8 ref.

Descriptors: *Protozoa, *Bottom sediments, Fish hatcheries, Eutrophication, Fertilization, Nitrogen, Phosphorus, Productivity, Daphnia, Systematics.
Identifiers: *Ciliates, Golysz (Poland), Coleps hirtus, Loxocephalus.

Ciliata are detritus feeders contributing much to organic matter reduction on bottom mud and are an element in the food chain. The development of Ciliata (mainly bacteriophagous) populations in bottom mud of fingerling ponds was investigated as part of a survey at the Experimental Farm, Golysz, Poland. The ponds were unfertilized, fertilized with phosphorus supplied in superphosphate, and fertilized with phosphorus and nitrogen, supplied in Norway saltpetre and superphosphate. The number of Ciliata species was rather small and the percentage of occurrence in all samples was computed for each species; quantitative dynamics in the ponds varied. The richest microfauna developed in the pond fertilized with nitrogen and phosphorus, in contrast to the pond fertilized with phosphorus only, and the control, where the amounts of Ciliata were markedly smaller. A marked increase in Ciliata numbers, particularly Coleps hirtus and Loxocephalus was observed soon after the maximum production in plankton, that is, during increased sinking of these crustacean organic remnants to the bottom, which was most remarkable in the pond fertilized with nitrogen and phosphorus. Ciliata feeding basically on bacteria prevailed while those feeding on algae were in the minority. (Jones-Wisconsin)
W73-09185

PRODUCTION OF MACROBENTHOS IN FINGERLING PONDS,
Polish Academy of Sciences, Krakow. Zaklad Biologii Wod.
J. Zieba.
Polskie Archiwum Hydrobiologii, Vol 18, No 2, p 235-246, 1971. 1 fig, 2 tab, 17 ref.

Descriptors: *Diptera, *Productivity, Benthos, *Eutrophication, Fish hatcheries, Phosphates, Ammonia, Nitrates, Biomass Larvae.
Identifiers: Golysz (Poland).

As part of a survey on productivity of new ponds at the Experimental Farm, Golysz, Poland, production of benthos biomass in three ponds--control, fertilized with phosphorus, and fertilized with phosphorus + nitrogen--was estimated. For comparability of data, each pond was stocked equally with fish. Among the most abundant larvae in the macrobenthos were Chironomus f.l. thummi and Ch. f.l. plumosus while Glyptotendipes ex gr. gripekoveni were less numerous; Chironomus larvae formed nearly 100% of the total community biomass. Soon after the ponds were filled, an abundance of Chironomus f.l. thummi appeared. After a few weeks, the large form, Chironomus f.l. plumosus, prevailed in fertilized ponds for a longer period than in the control. Highest production of Chironomus f.l. thummi and of total benthos production was recorded in the nitrogen + phosphorus fertilized pond. Larvae developed mainly on food from decomposed water plants on the bottom. The influence of fertilization was expressed in the occurrence of the second predominant species, Ch. f.l. plumosus. In the phosphorus fertilized and in the control ponds, biomass production values were similar. The production/biomass coefficient was lowest in the phosphorus + nitrogen fertilized pond and highest in the control. (Jones-Wisconsin)
W73-09186

BIOMASS AND PRODUCTION OF CARP FRY IN DIFFERENTLY FERTILIZED FINGERLING PONDS,
Polish Academy of Sciences, Krakow. Zaklad Biologii Wod.
J. M. Wlodek.
Polskie Archiwum Hydrobiologii, Vol 18, No 2, p 247-264, 1971. 7 fig, 4 tab, 10 ref.

Descriptors: *Biomass, *Fry, *Productivity, *Fish hatcheries, Carp, Fertilization, Animal growth, Mathematical models, Nitrogen, Phosphorus, Phosphates, Mortality, Fish management, Fish reproduction.
Identifiers: Golysz (Poland).

The influence of mineral fertilization of ponds at the Experimental Farms, Golysz, Poland, was estimated during the vegetation season of 1967, and a mathematical model for fish survival

developed. Variability of fish weight and its correlation with production were also studied. Twelve ponds were investigated using four unfertilized controls, four fertilized with phosphorus, and four fertilized with nitrogen + phosphorus in the form of superphosphate and Norway saltpetre. Biomass was calculated on the basis of fish measurements from random samples and on the basis of the hypothetical densities of the sampled populations. Density was calculated at the time of sampling taking alternative seven elementary mathematical functions as a regularity in mortality; therefore the changes in the biomass with time were represented by clusters of curves. The greatest biomass and greatest production were observed in the nitrogen + phosphorus fertilized ponds. Time variability of five biometrical features was also examined by means of the coefficients of variation. In the unfertilized ponds there was found a high and statistically significant negative correlation between production and relative variability. There was a statistically significant fertilization influence on the individual fish growth in weight. (Jones-Wisconsin)
W73-09187

STUDIES ON CULTURAL PROPERTIES OF PLANKTONIC, BENTHIC AND EPIPHYTIC BACTERIA,

Nicolas Copernicus Univ. of Torun (Poland). Lab. of Microbiology.
E. Strzelczyk, W. Donderski, and A. Mielczarek.
Zeszyty Naukowe Uniwersytetu Mikolaja Kopernika, Nauki Matematyczno-Przyrodnicze, Zeszyt 28 (Limnological Paper No 7), p 3-12, 1972. 6 tab, 16 ref.

Descriptors: *Cultures, *Bacteria, *Plant growth, Plankton, Benthos, Aquatic bacteria, Biochemistry, Biological communities, Plant morphology, Carbon, Nitrogen, Habitats, Distribution.
Identifiers: *Epiphytic bacteria, Heterotrophic bacteria, Lake Jeziorak (Poland).

Heterotrophic bacterial morphology was studied in cultures derived from samples of surface waters, mud, and Canadian pondweed, which were grown in liquid iron pepton medium. Ability of bacteria to utilize various carbon compounds, growth at various temperatures, the effect of pH on growth, and comparative bacterial growth were analyzed. Bacteria inhabiting bottom deposits differ from those occurring in water and on the surface of submersed plants in regard to morphology and nutrition. Gram negative rods predominate amoud epiphytic bacteria while among mud bacteria, the predominant types were pleomorphic forms (Coryne-bacterium, Arthrobacter, and Nocardia). The epiphytic bacteria show best growth in the presence of organic nitrogen whereas the majority of water and mud isolates respond almost equally to mineral nitrogen (ammonium nitrogen). A wide range of carbon compounds was utilized by the bacterial isolates studied. The largest number of slow growing bacteria was found among the mud isolates. Optimal temperatures were 26C for the epiphytic isolates and 20C for the mud and water isolates. Bacteria developed well over pH 5.5-9, however pH 7 was optimal for most isolates derived from all sources mentioned. (Jones-Wisconsin)
W73-09188

THE HORIZONTAL DIFFERENTIATION OF THE BOTTOM FAUNA IN THE GOPLO LAKE,

Nicolas Copernicus Univ. of Torun (Poland). Lab. of Hydrobiology.
A. Gizinski, and S. Kadulski.
Zeszyty Naukowe Uniwersytetu Mikolaja Kopernika, Nauki Matematyczno-Przrodnicze, Zeszyt 28 (Limnological Paper No 7), p 57-76, 1972. 1 fig, 5 tab, 21 ref.

Descriptors: *Systematics, *Benthic fauna, Habitats, Lakes, Oligochaetes, Spatial distribution, Distribution, Diptera, Mollusks, Water pollution effects, Profundal zone.
Identifiers: *Goplo Lake (Poland), Dreissena polymorpha, Asellus aquatcus, Chaoborus.

General characterization of bottom fauna of Goplo Lake, Poland, especially considering horizontal qualitative and quantitative benthic composition, is given. An attempt is made to establish to what degree samples, taken at a particular location, may be representative of bottom fauna of the whole lake to support findings regarding faunistic typology. Lake Goplo's trophic level is intermediate, between the oligo--and mezo-plumosus-Chaoborus type. The environmental and faunistic character of 14 stations are given. At stations along the longest axis a steady increase of the relative quantity of Chironomus f.l. plumosus larvae were found. The percent of Oligochaeta and Chaoborus larvae decreased as they did also in proportion to the proximity to the town of Kruszwica while Chironomus plumosus larvae tended to increase. The station at the town showed the poorest fauna, both qualitatively and quantitatively due to industrial and communal pollution. The most representative profundal fauna, qualitatively and quantitatively, were found at a depth approaching the maximum but in another part of the lake. The station with the abundant Einfeldia ex gr. carbonaria larvae showed the highest coefficient of difference from the other midlake stations. (Jones-Wisconsin)
W73-09189

ZOOPLANKTON OF THE BAY PART OF GOPLO LAKE,

Nicolas Copernicus Univ. of Torun (Poland). Lab. of Hydrobiology.
A. Adamska, and D. Bronisz.
Zeszyty Naukowe Uniwersytetu Mikolaja Kopernika, Nauki Matematyczno-Przyrodnicze, Zeszyt 28 (Limnological Paper No 7), p 39-55, 1972. 4 fig, 4 tab, 24 ref.

Descriptors: *Zooplankton, *Shallow water, *Lakes, *Systematics, Dominant organisms, Ecological distribution, Spatial distribution, Crustaceans, Copepods, Rotifers, Seasonal.
Identifiers: *Goplo Lake (Poland).

Qualitative and quantitative zooplankton compostion in the eutrophic bay of Goplo Lake, Poland, were examined and are classified. The most important species, horizontal and vertical distribution of Cladocera, Copepods, and Rotatoria, periodic fluctuations in composition, numbers and distribution, and numbers and proportionate share of Rotatoria, Cladocera, and Copepoda during a year's cycle are delineated. A description of the lake and the various stations with temperature, transparency, and oxygen content is given. Copepoda predominate (average number of individuals 3763/10 1). The second place belongs to Rotatoria (21 species, 3602 individuals/10 1). Cladocera occur less numerously (1182 individuals/10 1), but they are the richest in species and are represented by 34 species. Populations differed between the pelagic stations near the East Notec River, which flows through the lake, and the stations containing aquatic vegetation. Noticeable differences of zooplankton were found in various layers, with the most abundant in the surface and near the bottom. The maximum growth of Rotatoria and Cladocera in summer was recorded. Seasonal fluctuations of Copepoda differ at various stations. (Jones-Wisconsin)
W73-09190

BOTTOM FAUNA OF THE BAY PART OF GOPLO LAKE,

Nicolas Copernicus Univ. of Torun (Poland). Lab. of Hydrobiology.
A. Gizinski, and E. Toczek-Boruchowa.
Zeszyty Naukowe Uniwersytetu Mikolaja Kopernika, Nauki Matematyczno-Przyrodnicze, Zeszyt 28 (Limnological Paper No 7), p 77-93, 1972. 1 fig, 4 tab, 11 ref.

Descriptors: *Benthic fauna, *Shallow water, *Lakes, Habitats, Seasonal, Eutrophication, Systematics, Baseline studies, Oligochaetes, Environmental gradient, Diptera, Mollusks.
Identifiers: *Lake classification, Ephemeroptera, Hirundenea.

Continuing studies of shallow, extremely eutrophic lakes, the bay of Goplo Lake, Poland, is a pond type and classified according to the fauna of its deeper parts as oligo--or mesoplumosus-Chaoborus type, was investigated. The general characterization of the bottom fauna habitats, with the list of fauna, are given together with the dynamics of seasonal fluctuations in composition (qualitative and quantitative). Prevailing forms at littoral stations and at deeper stations, similar to profundal, are listed. January proved to be the best period for taking samples which are representative of the fauna. The profundal samples taken that month showed the greatest similarity to the average estimated on a base of samples taken during the whole year. The greatest divergence from the average was shown in samples taken just before freezing of the lake and just after retreat of the ice cover. Increase in fauna numbers occurred at the shallower stations at the beginning of the freeze suggesting the possibility of development from eggs or from younger larval states, smaller 0.5 mm, in winter. (Jones-Wisconsin)
W73-09191

THE BOTTOM FAUNA OF THE WATER RESERVOIRS WHICH NEWLY CAME INTO THE NEIGHBOURHOOD OF THE KORONOWO DAM RESERVOIR,

Nicolas Copernicus Univ. of Torun (Poland). Lab. of Hydrobiology.
A. Gizinski, and A. Paliwoda.
Zeszyty Naukowe Uniwersytetu Mikolaja Kopernika, Nauki Matematyczno-Przyrodnicze, Zeszyt 28 (Limnological Paper No 7), p 97-108, 1972. 2 fig, 5 tab, 15 ref.

Descriptors: *Benthic fauna, *Ponds, *Systematies, Diptera, Habitats, Environmental gradient.
Identifiers: *Koronowo Reservoir (Poland), Brda River (Poland), Chaoborus, Tendipes, Tanytarsus.

In damning the Brda River, Poland, for the new Koronwo Reservoir in 1960-1962, small, permanent water basins were formed in ground depressions. The fauna of some of these basins were examined, listed, and the dominant forms are described together with the physical features of the basins. Despite the fact that the basins were only recently formed (1961-1962) the bottom fauna was relatively rich in quality and quantity. During the fourfold sampling in the four seasons, 5392 specimens from 67 taxons were collected in 70 samples. The larvae of Chaoboridae and Chironomidae were the most abundant and the most frequently caught representatives of the benthos. Abundance of bottom fauna was quite varied in the different habitats and at various times. The great variability at the same stations during the year proves the unstabilized, yet, environmental and biocenotic conditions. Only the eurytopic forms, such as the larvae of the Chaoborus and Chironomus genera, occurred in relatively constant and large numbers. The ecologic conditions, typical of 'old' basins, may be stabilized in several years when a stratum of mud will be deposited at the bottom. (Jones-Wisconsin)
W73-09192

EFFECT OF THE CHEMICAL COMPOSITION OF THE MEDIUM ON THE GROWTH OF CHLORELLA PYRENOIDOSA,

Instytut Zootechniki, Oswiecim (Poland). Samodzielna Pracownia Biologii Ryb i Strodowiska Wodnego.

J. Zarnowski.
Acta Hydrobiologica, Vol 14, No 3, p 215-223, 1972. 1 fig, 4 tab, 18 ref.

Descriptors: *Algae, *Cultures, *Chemical properties, *Plant growth, Chlorella, Biomass, Ureas.
Identifiers: *Chlorella pyrenoidosa, *EDTA, Tamiya medium.

The growth intensity of Chlorella pyrenoidosa on several media, selected from the available literature, was observed and the effect of different EDTA concentrations in the medium on the rate of growth of algal cultures, measured by the number of cells in 1 ml and by the weight of dry matter, was investigated. Each medium was used in four combinations: ethylenediaminetraacetic acid (EDTA) without and with 7, 37, and 100 mg of EDTA in 1 liter of the culture. The temperature ranged from 19 to 22C. Each combination was carried out in five replications. Results of growth on 16 mineral media are tabulated. Results of utilization of basic mineral nutrients, nitrogen, phosphorus, and potassium are also presented suggesting that degree of utilization of different compounds is variable. The greatest increases in biomass were obtained on the media containing complex compounds: EDTA and sodium citric, besides the mineral macro- and micro-elements. The greatest amounts of nutrient were utilized by the algal culture on the Tamiya medium. The additions of EDTA at a rate of up to 100 mg/l of the medium stimulated the growth of the alga. (Jones-Wisconsin)
W73-09193

FIXATION OF ATMOSPHERIC NITROGEN BY AZOTOBACTER SP., AND OTHER HETEROTROPHIC OLIGONITROPHILOUS BACTERIA IN THE ILAWA LAKES,
Wyzsza Szkola Rolnicza, Olsztyn-Kortowo (Poland). Instytut Inzynierii i Biotechnolgii Zywnosci.
S. Niewolak.
Acta Hydrobiologica, Vol 14, No 3, p 287-305, 1972. 8 fig, 3 tab, 33 ref.

Descriptors: *Nitrogen fixation, *Azotobacter, *Nitrogen fixing bacteria, *Lakes, Bottom sediments, Distribution, Water pollution, Seasonal, Plankton, Pseudomonas, Clostridium, Eutrophication, Limiting factors.
Identifiers: *Ilawa Lakes (Poland), Azotobacter beijerinckii, Aeromonas, Vibrio, Achromobacter, Flavobacterium, Corynebacterium, Clostridium pasteurianum, Aerobacter aerogenes.

The occurrence of bacteria fixing atmospheric nitrogen in water, bottom sediments, and netplankton of the Ilawa Lakes, Poland, especially strains recently isolated, the dynamics of their development in the annual cycle, and the intensity of molecular nitrogen fixed by them were studied. The choice of sampling points was determined by their distance from pollution sources and the character of the bottom. In the collected samples determinations were made of the number of bacteria of the genus Azotobacter fixing atmospheric nitrogen, the number of bacteria of the Pseudomonas-Achromobacter group, and Clostridium pasteurianum. Considerable differences were noted in the bacteria development, according to degree of pollution or mineralization of the water, season of the year, and type of the bottom. The greatest quantities of bacteria fixing atmospheric nitrogen occurred in the netplankton and in the bottom sediments. One maximum occurred in water and bottom sediments in May or June; and another in September and November. Azotobacter beijerinckii fixed the greatest quantity of atmospheric nitrogen in pure cultures. Yield of strains of Aeromonas, Vibrio, Achromobacter, Flavobacterium, and Corynebacterium was within the range typical of nitrogen fixing bacteria. (Jones-Wisconsin)
W73-09196

PSEUDOMONAS CYTOCHROME C PEROXIDASE. IV. SOME KINETIC PROPERTIES OF THE PEROXIDATION REACTION, AND ENZYMATIC DETERMINATION OF THE EXTINCTION COEFFICIENTS OF PSEUDOMONAS CYTOCHROME C-551 AND AZURIN,
Helsinki Univ. (Finland). Dept. of Biochemistry.
R. Soininen, and N. Ellfolk.
Acta Chemica Scandinavica, Vol 26, No 3, p 861-872, 1972. 8 fig, 3 tab, 29 ref.

Descriptors: *Pseudomonas, *Enzymes, *Pigments, *Biochemistry, Kinetics, Cytological studies, Aquatic bacteria.
Identifiers: *Cytochrome, Peroxydation reaction, Peroxidase, Hydrogen peroxide, Azurin.

Location in the cell and some physicochemical properties of the peroxidation reaction catalyzed by Pseudomonas cytochrome c peroxidase were studied previously. The kinetics of the reaction are now reported. The optimum pH of the peroxidation of Pseudomonas ferrocytochrome c-551 was found to be 6.0 and the optimum ionic strength 0.01 in sodium acetate and sodium phosphate buffers. Pseudomonas cytochrome c peroxidase reacts specifically with two Pseudomonas respiratory chain pigments, ferrocytochrome c-551 and reduced azurin. Horse heart ferrocytochrome c, yeast ferrocytochrome c, and conventional low molecular weight peroxidase substrates such as pyrogallol, guajacol, sodium ascorbate, ferrocyanide, and reduced 2,6 dichlorobenzenone-indo-3-chlorophenol were found to be poor donors in the peroxidation. Ethyl hydroperoxide could not replace hydrogen peroxide as an electron acceptor in the peroxidation reaction. Maximal turnover rates for Pseudomonas ferrocytochrome c-551 and reduced azurin were found. The extinction coefficients of the peroxidase substrates Pseudomonas cytochrome c-551 and azurin were determined enzymatically by the reaction catalyzed by Pseudomonas cytochrome c peroxidase. Values were obtained for cytochrome c-551 and for azurin. (Jones-Wisconsin)
W73-09198

COMPARATIVE STUDY OF THE LIMNOLOGY OF TWO SMALL LAKES ON ROSS ISLAND, ANTARCTICA,
California Univ., Davis. Div. of Environmental Studies; and California Univ., Davis. Inst. of Ecology.
Charles R. Goldman, David T. Mason, and Brian J. B. Wood.
In: Antarctic Terrestrial Biology, p 1-50. American Geophysical Union, Washington, D.C., 1972. 57 fig, 16 tab, 63 ref, 2 append.

Descriptors: *Limnology, *Lakes, *Antarctic, Lake Morphometry, Periphyton, Meteorology, Birds, Cyanophyta, Light penetration, Heat budget, Light intensity, Productivity, Conductivity, Seston, Temperature, Lake morphology, Winds, Solar radiation, Hydrogen ion concentration, Phosphates, Alkalinity, Benthos, Diatoms, Standing crop, Chemical analysis, Organic matter, Plant physiology, Plankton, Primary productivity, Water chemistry, Trace elements, Biology.
Identifiers: *Ross Island (Antarctica), McMurdo Sound (Antarctica).

Investigations of two small antarctic lakes on Ross Island, Cape Evans, Antarctica during the austral summers of 1961-1962 and 1962-1963 were directed toward energy flows through their unique physical and biological conditions with Algal and Skua lakes illustrating extreme conditions. Skua Lake is linked to productivity of the sea by avian (skua) fertilization and responds with seasonal phytoplanktonic blooms comparable to eutrophic temperate waters. The chemical characteristics of these lakes reflect proximity to the ocean and the kenyte lava surrounding them. Rates and efficiencies of productivity are remarkably similar, although benthic production is more important than planktonic. Solar heating of the benthos early in the season opens large productive areas before a truly planktonic community can exist. The overall plankton production and chlorophyll estimates indicate that Skua Lake is about eight times more productive than Algal Lake, whereas other sestonic crop estimates indicate that Skua Lake is only five times more productive, suggesting a greater portion of the total seston of turbid Skua Lake is actively producing, a phenomenon perhaps largely attributable to great light inhibition in the clear water of Algal Lake. (Jones-Wisconsin)
W73-09199

PROFILES OF CHLOROPHYLL CONCENTRATIONS BY IN VIVO FLUORESCENCE: SOME LIMNOLOGICAL APPLICATIONS,
Kinneret Limnological Lab., Tiberias (Israel).
T. Berman.
Limnology and Oceanography, Vol 17, No 4, p 616-618, 1972. 1 fig, 9 ref.

Descriptors: *Analytical techniques, *Chlorophyll, *Fluorometry, Phytoplankton, Distribution, Algae, On-site tests, Sampling, Vertical migration.
Identifiers: Lake Kinneret (Israel), Water column.

To ensure representative sampling, an in vivo fluorometric technique of chlorophyll-a sensing enables accurate and rapid delineation of vertical phytoplankton profiles. Work in Lake Kinneret, Israel, illustrates some potential limnological applications of this technique, especially in situations where microstratification or vertical patchiness of algae occur. This procedure leads to a rapid determination of important hetero-geneities in vertical algal distribtuion that are often missed by sampling at fixed depths and permits rational sampling for chemical and biological parameters. In addition to this primary exploratory function, continuous chlorophyll profiles can obviously provide valuable information on the amounts of this pigment in the water column. In Lake Kinneret, eight representative stations were monitored to follow the seasonal changes in total lake chlorophyll concentrations. During the peak of the winter-spring bloom, when more than 95% of the algal biomass consisted of Peridinium westii, fluorescence could be calibrated directly in cells/ml. In such cases, or where reasonable estimations of the ratio of algal carbon to chlorophyll are available, phytoplankton biomass carbon can be rapidly measured by fluorometry which can thus complement more exact, but exceedingly tedious, microscopic observations. (Jones-Wisconsin)
W73-09200

PHOSPHATE NUTRITION OF THE AQUATIC ANGIOSPERM, MYRIOPHYLLUM EXALBESCENS FERN.,
Cornell Univ., Ithaca, N.Y. Dept. of Agronomy.
D. O. Wilson.
Limnology and Oceanography, Vol 17, No 4, p 612-616, 1972. 1 fig, 1 tab, 12 ref.

Descriptors: *Phosphates, *Aquatic plants, *Plant growth, Nutrients, Aquatic weeds, Rooted aquatic plants.
Identifiers: *Myriophyllum exalbescens, Phosphorus storage, Watermilfoil.

In order to control and decrease the spread of watermilfoil in aquatic environments, the effect of external concentrations of inorganic phosphate on Myriophyllum exalbescens growth was evaluated. A vegetative sterilization method was developed by treatment with a sodium hypochlorite solution followed by rinses with sterile distilled water. A sufficiently high concentration of sodium hypochlorite for bleaching the leaves completely was necessary. Surface algae were killed and an algal-free axillary bud formed. The concentration and time of exposure to sodium hypochlorite de-

pended on size and age of the stem segment, but concentrations of 0.2-1.0% for 10 to 20 minutes were usually satisfactory. Growth and tissue phosphorus concentrations were measured in these algal-free plants grown in nutrient solutions with phosphate maintained at concentrations from 0.1 to 100 micromoles. No major growth differences were observed over 5 weeks. Tissue phosphorus concentrations (dry-wt basis) ranged from 0.062-1.38% with no apparent signs of phosphorus deficiency or toxicity. This species can store large quantities of phosphorus which it uses when external sources are low. It is apparent that lowering the phosphate concentration of the water would not be effective in controlling the spread of this aquatic angiosperm. (Jones-Wisconsin)
W73-09201

OCCURRENCE OF STEROLS IN THE BLUE-GREEN ALGA, ANABAENA-CYLINDRICA,
Kagoshima Univ. (Japan). Faculty of Fisheries.
S. Teshima, and A. Kanazawa.
Bulletin of the Japanese Society of Scientific Fisheries, Vol 38, No 10, p 1197-1202, 1972. 1 fig, 2 tab, 15 ref.

Descriptors: *Biochemistry, *Anabaena, *Cytological studies, Enzymes, Cyanophyta, Algae.
Identifiers: *Anabaena cylindrica, Sterols.

Occurrence and origin of sterols in blue-green algae was studied, especially their composition and biosynthesis. Anabaena cylindrica was grown aseptically on a modified Detmer's medium containing only the salts and without sterols, from which sterols were isolated by the digitonin-precipitation method. Identification of sterols was performed by gas-liquid chromatography and mass spectrometry. They were composed of brassicasterol (90%), cholesterol (8%), 22-dihydrobrassicasterol (2%), and 22-dehydrocholesterol (less than 1%). The occurrence of sterols in A. cylindrica was further confirmed by the fact that the incubation of its cells with acetate-1-C14 gave radioactive squalene and sterols. The sterol content of A. cylindrica (0.253 mg/g of fresh cells) obtained in the present study was very high as compared with that of Phormidium luridum (0.03 mg/g of fresh cells) which was reported by earlier workers. This may be attributed to the discrepancy in extraction methods for sterols from the cells. There is almost no doubt that cells of blue-green algae as well as other algal cells contain sterols. (Jones-Wisconsin)
W73-09202

MECHANISM OF RESPIRATORY EXCHANGES IN AQUATIC ENVIRONMENT. A GENERAL REVIEW INCLUDING PERSONAL RESULTS,
Amiens Univ. (France). Laboratoire de Physiologie.
P. Harichaux, A. Poizot, and M. Freville.
Revue Internationale d'Oceanographic Medicale, Nice, Vol 27, p 71-84, 1972. 4 fig, 1 tab, 9 ref.

Descriptors: *Water pollution effects, *Chemical wastes, *Toxicity, Fish, *Respiration, Weight, Temperature, *Metabolism, *Oxygen, Phosphorus.
Identifiers: *Gardonus rutilus.

Respiratory exchanges in the aquatic environment are reviewed. A method for continuous measuring of PO2 and PCO2 variations in the aquatic environment of the fish is presented. This device computes oxygen consumption in relation to species weight, ambient temperature and the chemical factors of pollution. The method permits, among other things, the study of metabolic variations after artificial poisoning of the medium and also serves as a test of pollution (experiments conducted with 2-4 dinitrophenol provoked spontaneous death for a dose of 0.2 mg/ml of the bath.) It is assumed that the different metabolic conditions or the manner of supply of O2 in the blood explains the existing differences in the limiting values of PO2 of the medium which assures survival. (Ensign-PAI)
W73-09206

DISTRIBUTION OF POLYCHAETA IN POLLUTED WATERS,
Modena Univ. (Italy). Inst. of Zoology.
For primary bibliographic entry see Field 05B.
W73-09208

RESEARCH OF THE ORGANOCHLORINE PESTICIDES IN THE LITTORAL ENVIRONMENT (RECHERCHE DES PESTICIDES ORGANOCHLORES DANS LES MILIEUX LITTORAUX),
J. Fougeras-Lavergnolle.
Revue des Travaux de l'Institut des Peches Maritimes. Paris. Vol 35, No 3, p 367-371, October 1971. 1 fig, 1 tab, 8 ref.

Descriptors: *Littoral, *Oysters, *Chromatography, *Pesticide toxicity, *Pesticides, *Insceticides, Aldrin, Dieldrin, Heptachlor, DDE, DDT.
Identifiers: *Crassostrea angulata, Hexachlorocyclohexanes, TDE.

The effect of pesticides on oysters in the coastal zone was studied. The oyster tissue accumulated much higher concentrations of DDT than the surrounding waters. This concentration varied according to the concentration in the water, the temperature, and the exposure time, but was possible to reach a level of 70,000 times the environmental waters. If the concentration in the water was less than 0.002 ppm the oysters growth and behavior were normal. Crassostrea angulata, were used to test the insecticides hexachlorocyclohexanes, heptachlor, heptachlor epoxide, aldrin, dieldrin, DDE, DDT, and TDE. Analysis was done by ebromatography. None of the insecticides heptachlor, heptachlor epoxide, aldrin and dieldrin were identified in the samples analysed. The hexachlorocyclohexanes and were found in all samples. (Ensign-PAI)
W73-09209

SEDIMENT, FISH, AND FISH HABITAT,
Agricultural Research Service, Oxford, Miss. Sedimentation Lab.
J. C. Ritchie.
Journal of Soil and Water Conservation, Vol 27, No 3, p 124-125, May-June 1972. 24 ref. AEC 49-7 (3029).

Descriptors: *Sediments, *Turbidity, *Aquatic life, *Fish populations, *Suspended solids, Water pollution, Environmental effects, Water pollution effects.
Identifiers: *Spawning grounds, Survival rates.

The effects of sediment and/or turbidity on aquatic ecosystems are discussed. Some of the areas briefly examined include changes in light penetration, reduction of oxygen in water, effects on survival rates of eggs and alevins, reduction in the number and kind of bottom organisms and alteration in the depth and duration of streamflow. Fish population changes are also described. Studies of turbidity have led to the conclusion that a definite relationship exists between the period of survival of fish and their exposure to high concentrations of suspended material. Sedimentation can destroy fish and oyster spawning grounds, as evidenced by the ruin of spawning grounds in the upper estuary of the Chesapeake Bay. This study maintains that quantitatively, sediment is the greatest single pollutant in U.S. waters. (Ensign-PAI)
W73-09213

TRAGEDY AT NORTHPORT,
National Marine Fisheries Service, Highlands N.J. Sandy Hook Lab.
M. J. Silverman.
Bulletin of the American Littoral Society, Vol 7 No 2, p 15-18. 4 fig. Fall - 1971.

Descriptors: *Thermal pollution, Fish, *Water temperature, *Fish migration, Habitats, Water pollution effects, *Fishkill, *New York.
Identifiers: *Long Island Sound, *Bluefish (Pomatomus saltatrix).

The effect of the Northport, Long Island power plant discharge on the bluefish, Pomatomus saltatrix, is reported. Young bluefish migrating from Long Island Sound to warmer Atlantic waters encountered water warmed by the power plant discharge. The fish wintered in this artifically warmed habitat until gusting winds mixed the cold Sound water into the warmer water in the plant's discharge basin. It has been estimated that as many as 10,000 bluefish died from cold shock. The kill was the largest one known to have occurred at the power plant. (Ensign-PAI)
W73-09214

EFFECTS OF MAN'S ACTIVITIES ON ESTUARINE FISHERIES,
Florida State Univ., Tallahassee. Dept. of Oceanography.
R. W. Menzel.
Bulletin of the American Littoral Society, Vol 7, No 2, p 19-31. 6 fig. Fall - 1971.

Descriptors: *Estuaries, *Aquatic animals, Aquatic populations, *Water pollution effects, *Environmental effects, Dredging, Dams, Industrial wastes, Thermal pollution, *Sea basses, *Shrimps, *Oysters.
Identifiers: *Sea trout, Environmental changes.

Estuaries are described and discussed with respect to a variety of forms of environmental degradation. The slow but cumulative sub-lethal environmental changes that reduce organismic populations are explored. Dredging and filling, the damming of rivers, industrial effluents and thermal pollution are among topics examined. Some species of estuarine fauna are studied in light of the effects of pollution. Oysters, for instance, are susceptible to all types of pollution since they are sessile all their adult life. Results of various studies conducted on the effects of different kinds of pollution on oysters are summarized. Shrimp, spotted bass, and sea trout are also studied; current knowledge concerning population levels and pollution effects are presented for these organisms as well. (Ensign-PAI)
W73-09216

ANTIBIOTIC-RESISTANT COLIFORMS IN FRESH AND SALT WATER,
Alabama Univ., Birmingham. Medical Center.
For primary bibliographic entry see Field 05B.
W73-09218

INCREASING ORGANIC POLLUTION IN KINGSTON HARBOUR, JAMAICA,
University of the West Indies, Kingston (Jamaica). Dept. of Zoology.
For primary bibliographic entry see Field 05B.
W73-09219

INFLUENCE OF VENEZUELAN CRUDE OIL ON LOBSTER LARVAE,
Guelph Univ. (Ontario). Dept. of Zoology.
P. G. Wells.
Marine Pollution Bulletin, Vol 3, No 7, p 105-106, July 1972. 7 ref.

Descriptors: *Oil wastes, *Toxicity, *Lobsters, *Bioassays, Laboratory tests, Larvae, Water pollution effects, *Lethal limit.

J. P. Keating, M. E. Lell, A. W. Strauss, H. Zarkowsky, and G. E. Smith.
The New England Journal of Medicine, Vol 228, No 16, April 19, 1973, p 824-826. 1 tab, 19 ref.

Descriptors: Nitrogen, Nitrogen compounds, *Nitrites, *Nitrates, *Carrots, Reduction (Chemical), Toxicity, Fertilizers, Pollutants, Water pollution effects.
Identifiers: *Methemoglobinemia, Cyanosis, Health hazards.

A two-week old black male infant was found to have symptoms of methemoglobinemia after ingestion of fresh carrot juice. Analysis of similar carrots, of the same brand and bought at the same store as the suspect carrots, showed large amount of nitrate and nitrite present. It was calculated that the patient had ingested approximately 104 mg. of nitrite and 70 mg. of nitrate per kilogram (toxic effects may be expected if 2 mg. of nitrite per kilogram is ingested). Nitrate is a health hazard if conditions permit its reduction to nitrite, which when ingested, will cause methemoglobinemia. This potentially toxic reduction, in this case, occurred in the carrots after harvest without definite bacterial contamination. Studies in spinach suggest that plant reductases are responsible and that initial high nitrate content may be important to the reaction. The index carrots contained approximately 1300 ppm of nitrate at harvest--an amount far in excess of that usually found. Because nitrogen is essential plant nutrient, elimination of nitrates from soil or food is not feasible. However, it is recommended that the ingestion of vegetables of known high nitrate content be restricted in infants under three months of age, who are particularly susceptible to nitrite poisoning. (Smith-NWWA)
W73-09249

CORRELATION TO EVALUATE THE EFFECTS OF WASTE WATER PHOSPHORUS ON RECEIVING WATERS,

Wisconsin Univ., Madison. Water Resources Center.
G. P. Fitzgerald, S. L. Faust, and C. R. Nadler.
Water and Sewage Works, Vol 120, No 1, p 48-55, January 1973. 7 tab, 12 ref.

Descriptors: *Bioassay, Growth rates, *Phosphates, *Chemical analysis, Water pollution sources, *Limiting factors, *Cladophora, Lakes, Rivers, On-site tests, Water analysis, Water pollution effects, Wisconsin, Absorption.
Identifiers: *Orthophosphate, Lake Mendota, Lake Monona, Lake Wingra, Lake Wanbesa, Lake Kegonsa, Lake Koshkonong, Badfish Creek, Yahara River, Rock River, Crawfish River, Bark River, Rhizoclonium, Selanastrum capricornutum.

Algal growth tests were conducted with water from several rivers and lakes in Wisconsin to determine the amounts of available phosphorus and the effect and sources of the phosphorus. The methods used were short- and long-term bioassays for measuring available phosphorus, in situ bioassay to determine limiting or surplus phosphorus, and chemical analyses to determine soluble ortho and total phosphorus. Cladophora sp., Rhizoclonium sp., and Selenastrum capricornutum were used in the bioassays. The water system investigated was Lake Mendota, Lake Monona, Lake Wingra, Lake Wanbesa, Lake Kegonsa, Lake Koshkonong and their connecting and feeding rivers. The results of the bioassays and the chemical analyses were compared to ascertain their validity. The results show that soluble ortho PO4-P analyses correlate with bioassays for available phosphorus, and total phosphorus analyses do not correlate. This indicates that either method (bioassay or chemical ortho PO4-P analysis) could be used to measure the phosphorus available for the growth of algae. There was close agreement of results between the two types of bioassay: the one-day sorption-extraction test versus the long-term growth assay, even though the two methods differ greatly. However, since the growth of algae on the phosphorus in water samples is ultimate proof of the availability of phosphorus, the longer growth tests must still be used occasionally to check on the results obtained with the short-term bioassays or chemical analyses of soluble ortho PO4-P. The significance of the tests are discussed with regard to the sources and effects of phosphorus in the lakes and rivers analyzed. (Little-Battelle)
W73-09306

TOXICITY OF COPPER TO THALASSIOSIRA PSEUDONANA IN UNENRICHED INSHORE SEAWATER,

National Marine Water Quality Lab., West Kingston, R.I.
S. J. Erickson.
Journal of Phycology, Vol 8, No 4, p 318-323, December 1972. 6 fig, 3 tab, 16 ref.

Descriptors: *Toxicity, *Copper, *Water pollution effects, *Diatoms, *Bioassay, Heavy metals, Marine algae, Chrysophyta, Sea water, Growth rates, Absorption, Carbon radioisotopes, Water sampling, Water storage, Bacteria, Salinity, Hydrogen ion concentration, Tides.
Identifiers: *Thalassiosira pseudonana, Cyclotella nana, C-14, Population density, Enrichment.

Toxicity of copper to T. pseudonana (formly Cyclotella nana, clone 13-1) was examined in inshore seawater using a 96-hr bioassay method. Raw unenriched seawater was filtered through a 0.22-micron membrane filter and then pasteurized for 30 min at 60 C. Following this treatment, samples contained 0.68-1.14 micrograms Cu/liter. Copper was added as the chloride in 5-microgram increments over the range of 5 to 30 micrograms/liter (about 0.1-0.5 microM). Population densities, mean cell volume, and C-14 bicarbonate uptake were measured. Population growth and C-14 uptake by T. pseudonana displayed inhibition over the entire range of added copper. Growth rate constant (k) to T. pseudonana decreased with increasing copper concentration and during the course of growth of each concentration. Correspondingly, mean cell volumes increased with copper concentration and time. Copper toxicity varied in different water samples. The presence of decomposed natural plankton and detritus decreased toxicity. In the absence of enrichment, bacteria had little effect on copper toxicity. Results were influenced by glassware treatment, collection and storage of seawater, and absence of enrichment. (Holoman-Battelle)
W73-09313

THE RELATIONSHIP OF LIGHT, TEMPERATURE, AND CURRENT TO THE SEASONAL DISTRIBUTION OF BATRACHOSPERMUM (RHODOPHYTA),

Pennsylvania State Univ., University Park. Dept. of Biology.
D. E. Rider, and R. H. Wagner.
Journal of Phycology, Vol 8, No 4, p 323-331, December 1972. 7 fig, 1 tab, 27 ref.

Descriptors: *Limiting factors, Ecological distribution, *Rhodophyta, *Laboratory tests, *On-site investigations, Environmental effects, Light intensity, Water temperature, Aquatic algae, Seasonal, Currents (Water), Separation techniques, Water analysis, Growth rates, Degradation (Decomposition), Carbon dioxide, Solvent extractions, Chlorophyll, Chemical analysis, Hydrogen ion concentration, Iron, Hardness, Sulfates, Spectrophotometry.
Identifiers: *Batrachospermum vagum, *Batrachospermum moniliforme, Chlorophyll a, Mathanol, Organic solvents, Absorbance.

Batrachospermum vagum and B. moniliforme, collected in a dark-water and a clear-water stream,

respectively, were studied in order to elucidate species variation, growth phenomena, and the rate and causes of disintegration. To permit manipulation of environmental factors and simulate the natural habitat of Batrachospermum, a series of chambers was designed and constructed which allowed water in a trough to flow from one compartment to the other at a rate that could be adjusted by changing the slope of the trough. The water used in the sytem, transported from Black Moshannon Spring, was chosen because of its transparency. Carbon dioxide levels were maintained by filling an inverted 2-liter Erlenmeyer flask with pure CO_2 once or twice a day and allowing it to diffuse slowly through the water. Chemical characteristics of the water (CO_2, pH, iron, hardness, and sulfate) determined, using a Hach kit, after 3 weeks of running the water system with algal samples in it, remained unchanged. Light intensity was varied by raising or lowering lights above the chambers; readings were taken with a YSI-Kettering Model 65 Radiometer and a Weston Model 756 Sunlight Illumination Meter. Samples of B. vagum and B. moniliforme were maintained simultaneously in the trough under constant illumination for periods of 2 or 3 weeks. Chlorophyll was extracted from each sample after complete dehydration. All washes but one were made with methanol and the resulting lump of material was ground with mortar and pestle using MeOH until the wash was colorless and the cell remains were brown. Chlorophyll absorption and the concentration of Chl a were determined using data obtained with the Beckman DB recording spectrophotometer. Observations of the 2 growth sites and the laboratory experiments performed on B. vagum and B. moniliforme indicate that B. moniliforme is better suited to higher intensity illumination than is B. vagum. Without the protection of dark water, the B. moniliforme disintegrated under the high light intensities of summer. In the laboratory, B. moniliforme retained a healthy macroscopic condition over a full range of light-intensity treatments; above 250 ft-c B. vagum was increasingly disintegrated. It was found that within the limitations caused by water depth, current velocity can become limiting to the growth of B. vagum. Although the evidence for environmental factors affecting the growth of B. moniliforme in Smay's Run is less extensive, light intensity and current velocity in this site also appear to interact to control the growth pattern of the algae. (Holoman-Battelle)
W73-09314

GROWTH AND EXCRETION IN PLANKTONIC ALGAE AND BACTERIA,

Scarborough Coll., Toronto (Ontario).

For primary bibliographic entry see Field 05B.
W73-09315

THE EFFECT OF GROWTH ILLUMINATION ON THE PIGMENTATION OF A MARINE DINOFLAGELLATE,

Organization of American States, Washington, D.C.
E. F. Mandelli.

Journal of Phycology, Vol 8, No 4, p 367-369, December 1972. 2 fig, 2 tab, 20 ref.

Descriptors: *Dinoflagellates, *Light intensity, *Plant pigments, *Environmental effects, Pyrrophyta, Marine algae, Protozoa, Solvent extractions, Cultures, Invertebrates, Marine animals, Separation techniques, Phytoplankton.
Identifiers: *Amphidinium klebsii, Chlorophyll a, Chlorophyll c, Carotenoids, Peridinin, Diadinoxanthin, Biosynthesis, Absorption spectra, Infrared spectra, Organic solvents, n-hexane, Ethanol, Carbon disulfide, Acetone, Infrared spectrophotometry, Absorption spectrophotometry, Culture media.

Variations in pigment concentration, especially within the carotenoid pigment system, were investigated in the marine dinoflagellate Amphidinium klebsii in relation to light intensity. Cultures of the dinoflagellate were grown in Guillard's medium 'f' under constant temperature and light intensities of 0.003, 0.032, 0.051, and 0.129 ly/min. Day-night conditions were simulated by alternating 12-hr periods of light and darkness. Cultures were harvested during the light photoperiod at mid-log growth phase; cell densities were determined using a hemacytometer. The pigments were extracted using 90 percent acetone, separated using diethyl ether, hydrated, and spotted on silica gel G. The developing solvent contained petroleum ether, ethyl acetate, and diethylamine (58:30:12v/v). Estimation of pigment concentration and identification of carotenoid pigments by determining absorption spectra in various solvents were carried out using a Beckman spectrophotometer. Amphidinium klebsii cultures grown under different light intensities exhibited similar chlorophyll a content per cell. Among the accessory pigments, chlorophyll c concentration decreased slightly in cells exposed to increasingly light intensities up to 0.129 ly/min. The concentration of the 2 major xanthophylls present in A. Klebsii cells--peridinin and diadinoxanthin--however, varied according to the light background of the cells. Some biochemical pathways in the formation of peridinin in dinoflagellates are discussed. (Holoman-Battelle)
W73-09316

INHIBITION OF PHOTOSYNTHESIS IN ALGAE BY DIMETHYL SULFOXIDE,

York Univ., Toronto (Ontario). Dept. of Biology.
K. H. Cheng, B. Grodzinski, and B. Colman.

Journal of Phycology, Vol 8, No 4, p 399-400, December 1972. 1 fig, 11 ref.

Descriptors: *Photosynthesis, Water pollution effects, *Aquatic algae, *Photosynthetic oxygen, *inhibition, Plant physiology, *Laboratory tests, Bioassay, Chlorophyta, Cyanophyta, Radioactivity techniques, Carbon radioisotopes, Cultures.
Identifiers: Dimethyl sulfoxide, Pollutant effects, Oscillatoria, Anabaena flos-aquae, Anacystis nidulans, Chlorella pyrenoidosa, Organic solvents, C-14.

Bacteria-free cultures of the blue-green algae Oscillatoria sp., Anabaena flos-aquae, and Anacystis nidulans were grown and harvested according to a previously described method, and Chlorella pyrenoidosa was grown on the Watt and Fogg medium (1966) and harvested as were the blue-greens. The rate of photosynthesis was determined at 30 C by measuring O_2 evolution with a Gilson differential respirometer fitted with photoflood lamps which gave a light intensity of 10 klux at the flask level, and by measuring the incorporation of C-14 from C-14-bicarbonate under the same conditions. Cells were suspended to 2.7 ml of 0.01 M phosphate buffer, pH 8, containing 35 micromoles $NaHCO_3$ and O_2 evolution was measured for 30 min. Dimethyl sulfoxide (DMSO), a polar solvent which increases the permeability of cell membranes, of the appropriate concentration was added and O_2 evolution measured for an additional 30 min, at which time 0.1 ml of C-14-sodium bicarbonate (0.6 microCi) was added and O_2 evolution measured for an additional 30 min. A sample (100 microliters) of cell suspension was removed and C-14-incorporation determined. DMSO at concentrations above one percent inhibited photosynthetic oxygen evolution and C-$14O_2$ fixation by all 4 species of algae. (Holoman-Battelle)
W73-09318

NITROGEN FIXATION IN CLEAR LAKE, CALIFORNIA. I. SEASONAL VARIATION AND THE ROLE OF HETEROCYSTS,

California Univ., Davis. Div of Environmental Studies.
A. J. Horne, and C. R. Goldman.

Limnology and Oceanography, Vol 17, No 5, p 678-692, September 1972. 4 fig, 5 tab, 43 ref.

Descriptors: *Eutrophication, *Nitrogen fixation, Nitrogen cycle, Energy budget, *Path of pollutants, Nitrates, Ammonia, Phosphates, Phosphorus, Cyanophyta, Aquatic algae, Phytoplankton, Primary producuvity, Cycling nutrients, Methodology, Chemical analysis, Nuisance algae, *California, Colorimetry, Spectrophotometry, Radioactivity techniques, Solvent extractions, Water sampling, Denitrification, Nitrogen, Photosynthesis, Chlorophyta, Fresh-water.
Identifiers: Acetylene reduction, *Heterocysts, *Seasonal variation, *Clear Lake (Calif), Ethylene, Aphanizomenon flos-aquae, Anabaena circinalis, C-14, Chlorophyll a, Gas liquid chromatography, Multiple regression analysis, Van Dorn bottles, Kjeldahl procedure, Oocystis.

Details are given of more than 2200 acetylene reduction measurements of aquatic N_2 fixation for a 7-month period and their relationship to a correspondingly large number of measurements of other relevant physical and chemical variables. Samples were taken at 0.5- or 1.0-m intervals throughout the water column with a Van Dorn bottle. Samples from each depth were pooled to give a representative sample of the whole basin at that depth. Acetylene reduction was measured essentially by the original technique of Stewart et al. (1967). The main modifications were the exclusive use of unconcentrated phytoplankton samples, larger bottles (60 ml here) and in situ incubation (Horne 1969). Gas samples were analyzed for ethylene by GLC. Daily and hourly rates of reduction were calculated; algae, and heterocysts and cells were counted using a microscopic technique. Heterocysts were expressed either as a percentage of the total number of vegetative cells per species or as the total heterocyst numbers in the entire water column. Spectrophotometric determinations of nitrate, ammonia, and chlorophyll a were made; total-P and PO_4-P were estimated colorimetrically. Carbon fixation was measured using the C-14 technique, and total-N using a standard micro-Kjeldahl method. The nitrogen budget was calculated from regular measurements of the nitrogen content of the inflowing and outflowing streams and rainfall, including storm periods. The amount of nitrogen deposited in the sediments was found by direct measurement. Denitrification was estimated from changes in lake nitrogen when conditions were suitable for denitrification. The annual contribution of N_2 fixation to Clear Lake in 1970 was about 550 tonnes, 500 Mg (megagrams) or 18 kg/ha, 43 percent of the lake's yearly nitrogen inflow. Biological N_2 fixation is implicated as the main cause of large algal nuisance blooms on this lake. A sustained spring peak of fixation was associated with a simultaneous bloom of Aphanizomenon flow-aquae in all three basins, an autumn peak with an ephemeral bloom of Anabaena circinalis occurring at a different time in each basin. A stepwise multiple regression analysis showed that fluctuations in N_2 fixation were best described by variations in heterocysts, quantities of blue-green algae, PO_4-P, NO_3-N, and temperature. Annual rates of N_2 fixation were correlated with the proportion of heterocysts in Anabaena. Previous nitrogen budgets for Clear Lake have shown a large excess of nitrogen in outflow over inflow, which is accounted for by the levels of N_2 fixation measured. (See also W73-09322) (Holoman-Battelle)
W73-09321

NITROGEN FIXATION IN CLEAR LAKE, CALIFORNIA. II. SYNOPTIC STUDIES ON THE AUTUMN ANABAENA BLOOM,

California Univ., Davis. Inst. of Ecology.
A. J. Horne, J. E. Dillard, D. K. Fujita, and C. R. Goldman.

Limnology and Oceanology, Vol 17, No 5, p 693-703, September 1972. 4 fig, 4 tab, 18 ref.

Descriptors: *Nitrogen fixation, *Eutrophication, *Seasonal, Aquatic alage, Cyanophyta, Primary productivity, Nitrogen cycle, Cycling nutrients,

Water pollution effects, Nitrogen, Iron, Phytoplankton, Nitrates, *California, Water sampling, Protozoa, Chrysophyta, Chlorophyta, Photosynthesis, Ammonia, Biomass.
Identifiers: *Anabaena circinalis, *Heterocysts, *Clear Lake (Calif), Chlorophyll a, Ammonium, Dissolved organic nitrogen, Particulate carbon, Van Dorn bottles, Aphanizomenon, Microcistis, Oscillatoria, Melosira, Cryptomonads, Ciliates, Acetylene reduction, Coscinodiscus, Schroederia.

Nitrogen fixation at three stages of an autumal bloom of Anabaena circinalis was measured after almost simultaneous collection at up to 32 stations in Clear Lake and algal heterocysts, phytoplankton cell numbers, NO3-N, NH4-N, dissolved organic-N, PO4-P, Fe, primary production, particulate carbon, and chlorophyll a were also measured. Nitrogen fixation was significantly and positively correlated to Anabaena heterocyst numbers (P greater 0.0001), negatively correlated to NO3-N (P greater 0.01), and positively correlated to dissolved organic-N (P greater 0.01) and PO4-P (P greater 0.05). A negative correlation with NH4 is probable; no significant relationship was found with the other variables measured. An explanation of the apparent restriction of substantial cyanophycean N2 fixation to nonoligotrophic waters is proposed. The results are consistent with the theory that heterocysts are responsible for N2 fixation in situ under oxic conditions. (See also W73-09321) (Holoman-Battelle)
W73-09322

THE UPTAKE OF UREA BY NATURAL POPULATIONS OF MARINE PHYTOPLANKTON,
Scripps Institution of Oceanography, La Jolla, Calif.
J. J. McCarthy.
Limnology and Oceanography, Vol 17, No 5, p 738-748, September 1972. 2 fig, 5 tab, 25 ref.

Descriptors: *Absorption, *Marine algae, *Phytoplankton, *Ureas, Radioactivity techniques, Primary productivity, Nitrates, Carbon, Chemical analysis, Sea water, Secondary productivity, Water sampling, California, Light intensity, Mass spectrometry, Carbon radioisotopes, Nutrients.
Identifiers: *On board analysis, Ammonium, Van Dorn bottles, PVC sampler, Enrichment, Particulate nitrogen, C-14, N-14, N-15, Nitrogen radioisotopes.

Nitrogen-15 isotopes were used to study the uptake of nitrate, ammonium, and urea by natural phytoplankton populations in 36 samples collected at nine stations off the coast of southern California. Samples for measurement of nitrate, ammonium, and urea productivity and for chemical analyses were collected with a 50-liter PVC sampler or an 8-liter Van Dorn bottle from depths corresponding to the 87, 43, 20, 7, 4, and 1 percent surface light intensity as determined by either submarine photometer or a Secchi disk. Subsamples were filtered through 183-micron nylon mesh to remove the larger zooplankton and were then used to fill 4-liter Pyrex bottles covered with neutral density filters calibrated to simulate the light level at the depth from which the samples were collected. Additions of the N-15-labeled compounds were made to different bottles from each sample; the bottles were plugged with silicone rubber stoppers and placed under natural light in Plexiglas incubators on the deck of the research vessel. The bottles were shaken every few hours and after 24 hr the contents were filtered, the particulate samples dessicated under partial vacuum over silica gel, combusted in a Coleman nitrogen analyzer, and the gaseous product swept into a glass vacuum system. A single beam Nier sector-type mass spectrometer was used to determine the N-15: N-14 ratio of each sample. Portions of sea water samples were analyzed for nitrate, ammonium, urea, particulate nitrogen, and carbon productivity. The percentage of the total phytoplankton nitrogen productivity accounted for by urea varied from less than 1 percent to greater than 60 percent and for the entire study averaged 28 percent. The percentage of total available nitrogenous nutrient (ambient nitrate, ammonium, and urea plus the N-15 additions) utilized per day varied among the stations from a minimum of 5 percent at station 4 (12 km off San Diego) to a maximum of 46 percent at station 19 (off White Point). The average C:N uptake ratio was 12.4. (Holoman-Battelle)
W73-09323

COPEPODS OF MARINE AFFINITIES FROM MOUNTAIN LAKES OF WESTERN NORTH AMERICA,
Alaska Univ., College. Inst. of Marine Science.
M. S.Wilson.
Limnology and Oceanography, Vol 17, No 5, p 762-763, September 1972. 11 ref.

Descriptors: *Copepods, *Salt tolerance, *Lakes, Crustaceans, Invertebrates, Aquatic animals, Zooplankton, Deep water, Systematics.
Identifiers: *Euryhaline, Limnocalanus macrurus, Senecella calanoides, Mesochra rapiens, Huntemannia lacustris, Macroinvertebrates, Harpacticoids, Bear Lake, Arthropods, Waterton Lake, Cedar Lake.

Occurrences of Limnocalanus macrurus, Senecella calanoides, a variation of Mesochra rapiens and Huntemannia lacustris, all copepods either euryhaline or related to marine forms, are recorded from lakes in the Cascade and Rocky Mountains, western North America. (Holoman-Battelle)
W73-09325

FRESHWATER HARPACTICOID COPEPODS OF NEW ZEALAND, I. ATTHEYELLA ELAPHOIDELLA (CANTHOCAMPTIDAE),
Auckland Univ. (New Zealand). Dept. of Zoology.
For primary bibliographic entry see Field 05A.
W73-09327

RECOVERY OF SHIGELLA UNDER ACIDIC CONDITIONS,
Food and Drug Administration, Washington, D.C. Div. of Microbiology.
For primary bibliographic entry see Field 05A.
W73-09332

ECOLOGICAL OBSERVATIONS ON HETEROTROPHIC, METHANE OXIDIZING AND SULFATE REDUCING BACTERIA IN A POND,
Hydrobiologisch Institutt, Nieuwersluis (Netherlands).
For primary bibliographic entry see Field 05A.
W73-09337

ON THE CHANGES IN THE STRUCTURE OF TWO ALGAL POPULATIONS: SPECIES DIVERSITY AND STABILITY,
F. Symons.
Hydrobiologia, Vol 40, No 4, p 499-502, December 15, 1972. 4 ref.

Descriptors: *Aquatic algae, *Plant populations, *Stability, Aquatic plants, Diatoms, Chrysophyta, Chlorophyta, Mathematical studies.
Identifiers: *Species diversity, Synedra acus, Synedra affinis, Synedra ulna, Synedra ulna var. oxyrhynchus, Navicula cryptocephala var. veneta, Navicula cuspidata, Navicula gracilis, Amphora ovalis var. pedunculus, Cymbella affinis, Nitzschia acicularis, Nitzschia recta, Nitzschia sigmoidea, Surirella robusta var. splendida, Pediastrum boryanum, Pediastrum duplex var. clathratum, Pediastrum tetras, Scenedesmus abundans var. brevicauda, Scenedesmus dimorphus, Scenedesmus obliquus var. intermedius, Scenedesmus quadricauda var. maximus, Scenedesmus quadricauda var. westii.

Two algal populations sampled in the same pond were compared and two assumptions were made: (1) When comparing two sample places in the same pond, the one taken nearer to the water edge will show a more diversified population structure. (2) The population taken nearer to the water edge having a more diversified population structure will be more self-regulatory than the one taken from the open water. For each of those two assumptions a quantitative measure was proposed. The calculated coefficients did seem to confirm the assumptions. (Holoman-Battelle)
W73-09338

LIMNOLOGICAL INVESTIGATIONS IN THE AREA OF ANVERS ISLAND, ANTARCTICA,
Virginia Polytechnic Inst. Blacksburg. Dept. of Biology.
G. L. Samsel, Jr., and B. C. Parker.
Hydrobiologia, Vol 40, No 4, p 505-511, December 15, 1972. 2 tab, 13 ref.

Descriptors: Limnology, *Aquatic algae, *Primary productivity, Water properties, *Iced lakes, Photosynthesis, Chlorophyll, Phytoplankton, Aquatic plants, Antarctic, Physical properties, Chemical properties, Chlorophyta, Cyanophyta, Chrysophyta, Biomass, Aquatic bacteria, Protozoa, Crustaceans, Lake sediment, Lake morphology, Water analysis, Chlorides, Nitrates, Water temperature, Chlorella, Chlamydomonas, Hydrogen ion concentration, Turbidity, Dissolved oxygen, Alkalinity, Hardness (Water), Nitrites, Phosphates, Silicates, Calcium, Iron, Sulfates, Diatoms, Radioactivity techniques, Carbon radioisotopes, Radiation.
Identifiers: *Antarctica, Bryophytes, Anvers Island, Lichens, Snow algae, Ammonium, Chroococcus, Oscillatoria, Orthophosphates, Aphanocapsa, Merismopedia, Phormidium, Scotiella, Chlorosarcina, Palmellopsis, Dactylococcopsis, Navicula, Ellipsoidion, Trachychloron, Trochisicia, Monostroma, Prasiola, Chrysococcocus, Chrysapsis, Xanthoria spp, Calliergidium austrostramineum, Polytrichum juniperinum, Caloplaca, C-14.

Comparisons were made of primary productivity, physical features, and chemical and biological composition of two small lakes possessing different 'trophic states' during January, 1970 at Anvers Island, Antarctica. Measurements were made of radiation, water temperature, pH, primary productivity via a C-14 technique, extractable chlorophyll, and carbon fixation. Both lakes, less than 500 meters apart, had partial ice cover the entire season and were underlain with a similar silica-rich granite. Striking dissolved chemical differences were chloride (7.5 and 35.0), ammonium-N (0.1 and 2.5), and total phosphate-P (0.03 and 1.7 mg/l), respectively, for lake nos. 1 and 2. Extractable total chlorophyll in subsurface water ranged from 15-41 mg/sq m in lake no. 1 and 35-112 mg/cu m in lake no. 2 during the three week study period. Ranges in net photosynthesis were 0.78-3.5 (Lake no. 1) and 9.0-72.0 mgC/sq m/hr (Lake no. 2). Diel ranges for chlorophyll and carbon fixation also fell within these values. It is hypothesized that enrichment of lake no. 2 with phosphate-P and ammonium-N may account for its higher 'trophic state'. (Holoman-Battelle)
W73-09339

ON THE QUANTITATIVE CHARACTERISTICS OF THE PELAGIC ECOSYSTEM OF DALNEE LAKE (KAMCHATKA),
Yu. I. Sorokin, and E. B. Paveljeva.
Hydrobiologia, Vol 40, No 4, p 519-552, December 15, 1972. 22 fig, 5 tab, 26 ref.

Descriptors: *Ecosystems, *Trophic level, *Biomass, *Primary productivity, *Zooplankton, *Phytoplankton, *Metabolism, Food habits, Ecological distribution, Vertical migration, Organic matter, Aquatic bacteria, Aquatic microorganisms, Eutrophication, Dominant organisms,

Stratification, Predation, Aquatic algae, Copepods, Rotifiers, Diatoms, Carbon radioisotopes.
Identifiers: *Pelagic animals, Energy flow, *Lake Dalnee, Stephanodiscus astraea, Vertical distribution, C-14, Biosynthesis, Nauplii, Cyclops scutifer, Flagellates, Ciliates, Tintinnopsis, Strombidium, Asplanchna priodonata, Desmids, Neutrodiaptomus angustilobis, Keretalla, Asterionella, Ankistrodesmus, Staurastrum.

Data are presented on the biomass, production, metabolism, and trophic relations of the components of the ecosystem of Lake Dalnee (Kamchatka). Water samples were taken with a plastic water bottle of the Suslijaev type from depths selected according to the stratification of the water column. Primary productivity of phytoplankton, trophic characteristics of zooplankters, and bacterial production were measured by a C-14 technique. Counts of phytoplankton, colorless flagellates, and zooplankton were made. The carbon content of the predominant diatom, Stephanodiscus astraea, was measured by wet combustion of its suspension obtained in culture. The data on the spectrum of feeding, on rations and on optimal food concentrations were ascertained. All these data, together with the observations in the lake, were used for the construction of a scheme of energy flow. The scheme shows that the ecosystem of the lake receives the third part of its energy from the land as allochtonous organic matter via microbial biosynthesis. The main part of energy accessible to the animals of the second trophic level is used by protozoa, and of a third part by the predatory rotifer Asplanchna. (Holoman-Battelle)
W73-09340

THE LIFE-HISTORY OF SPHACELARIA FURICIGERA KUTZ. (PHAEOPHYCEAE) II. THE INFLUENCE OF DAYLENGTH AND TEMPERATURE ON SEXUAL AND VEGETATIVE REPRODUCTION,
Groningen Rijksuniversiteit, Haren (Netherlands). Afdeling Plantenoecologie.
F. Colijn, and C. van den Hoek.
Nova Hedwigia, Vol 21, Nos. 1-4, p 899-922, 1971. 7 tab, 11 ref.

Descriptors: *Marine algae, *Phaeophyta, *Life history studies, Reproduction, Cultures, Temperature.
Identifiers: *Sphacelaria furcigera, Culture media.

Cultures of gametophytes and sporophytes of Sphacelaria furcigera from Hoek van Holland formed abundant propagules under 12 C long day (equals 16 hours daylight) and 17 C long day conditions. This agrees with field data from temperature regions which indicate multiplication by propagules in the summer half year. No propagules were formed under 17 C short day, 12 C short day, 4 C long day, and 4 C short day conditions. Plurilocular macrogametangia were abundantly formed under 4 C long day and 12 C long day conditions. Under 17 C long day conditions plurilocular macrogametangia were initially formed, but were soon succeeded by propagules. Unilocular sporangia of the diploid generation were formed at 12 C and 4 C. There are some indications that short day conditions might have an an influence on the formation of these structures, but the evidence is inconclusive. In nature (in temperate regions) unilocular zoidangia and plurilocular gametangia are known mainly from the winter half year, but the scarce data do not indicate whether the daylength might also play a role in their induction. (Holoman-Battelle)
W73-09345

MERCURY CONCENTRATION IN RELATION TO SIZE IN SEVERAL SPECIES OF FRESHWATER FISHES FROM MANITOBA AND NORTHWESTERN ONTARIO,
Fisheries Research Board of Canada, Winnipeg (Manitoba). Freshwater Inst.
For primary bibliographic entry see Field 05B.
W73-09346

EFFECTS OF VARIOUS METALS ON SURVIVAL, GROWTH, REPRODUCTION, AND METABOLISM OF DAPHNIA MAGNA,
National Water Quality Lab., Duluth, Minn.
K. E. Biesinger, and G. M. Christensen.
Journal of the Fisheries Research Board of Canada, Vol 29, No 12, p 1691-1700, December 1972. 1 fig, 7 tab, 21 ref.

Descriptors: *Toxicity, *Heavy metals, *Alkali metals, *Alkaline earth metals, *Bioassay, Water pollution effects, Animal physiology, Crustaceans, Zooplankton, Aquatic animals, Proteins, Enzymes, Statistical methods, Laboratory tests, Water analysis, Cations, Growth rates, Reproduction, Metabolism, Sodium, Calcium, Magnesium, Potassium, Strontium, Chromium, Iron, Manganese, Aluminum, Zinc, Gold, Nickel, Lead, Copper, Cobalt, Mercury, Cadmium, Sulfides, Chemical analysis.
Identifiers: *Daphnia magna, Data interpretation, Biochemical tests, Macroinvertebrates, Arsenic, Barium, Tin, Platinum, Glutamic oxalacetic transaminase.

The toxicities of various metals to Daphnia magna was evaluated on the basis of a 48-hr 50 percent lethal concentration (LC50), a 3-week LC50, and a 16 percent decrease in the number of young born (reproductive impairment). The 3-week 16 percent reproductive impairment concentrations (in micrograms per liter) for the metal ions tested were: Na (I), 680,000; Ca (II), 116,000; Mg (II), 82,000; K (I), 53,000; Sr (II), 42,000; Ba (II), 5800; Fe (III), 4380; Mn (II), 4100; As (V), 520; Sn (II), 350; Cr (III), 330; Al (III), 320; Zn (II), 70; Au (III), 60; Ni (II), 30; Pb (II), 30; Cu (II), 22; Pt (IV), 14; Co (II), 10; Hg (II), 3.4; and Cd (II), 0.17. At metal concentrations permitting survival but impairing reproduction, daphnids weighed less than control animals. Amounts of total protein and glutamic oxalacetic transaminase activity varied with the different metals. The negative logarithm of the solubility product constant (pK sub sp) of the metal sulfides, electronegativity, and the logarithm of the equilibrium constant (log K sub eg) of the metal-ATP complex were positively correlated with toxicity to D. magna. Other physicochemical properties were considered, but no additional correlations were found. (Holoman-Battelle)
W73-09347

SALMON-CANNING WASTE WATER AS A MICROBIAL GROWTH MEDIUM,
Fisheries Research Board of Canada, Vancouver (British Columbia). Vancouver Lab.
G. A. Strasdine, and J. M. Melville.
Journal of the Fisheries Research Board of Canada, Vol 29, No 12, p 1769-1771, December 1972. 2 tab, 9 ref.

Descriptors: *Industrial wastes, Fish handling facilities, *Microbial degradation, Waste water (Pollution), *Bacteria, Canneries, Nutrient requirements, Carbohydrates, Biodegradation, Nitrogen.
Identifiers: *Culture media, *Substrate utilization, *Salmon cannery effluents, Fate of pollutants, Nutrient sources, Sorangium, Bacillus, Streptococcus faecalis, Glucose, Trypticase, Bacto-peptone, Polypeptone, Phytone, Acidicase, Beef extract, Pseudomonas putrefaciens, Aerobacter aerogenes, Lactobacillus plantarum.

Salmon-canning waste water was shown to support the growth of six species of bacteria both as a complete growth medium and as a supplementary source of available nitrogen. Salmon-canning effluent (SCE) was collected from the outfall of a local cannery located on the Fraser River, B.C. The crude waste water was filtered through glass wool and 50-ml samples dispensed into 250-ml Erlenmeyer flasks. Six species of bacteria which may be of commercial value were cultured on the SCE-salts media, and on media using the following available nitrogen sources: trypticase, Bacto-peptone, polypeptone, phytone, acidicase, and beef extract. Total carbohydrate, free glucose, reducing sugar, and total nitrogen were determined by previously described methods. All six species of bacteria showed relatively good growth in the SCE-salts media; with the exception of polypeptone and possibly phytone, this was not true for the remaining nitrogen sources. Based on the results obtained, two general approaches may be considered for the microbial utilization of salmon-canning waste water. The first and probably the simplest of these would include those possibilities in which the SCE serves as an inexpensive source of available nitrogen for the microbial degradation and/or utilization of nitrogen-deficient wastes. One such waste is the spent liquor from the sulfite pulping process. The second approach for SCE would include those methods designed for the microbial utilization of the waste per se as a pollution abatement system and for the production of a marketable product. (Holoman-Battelle)
W73-09350

SOME PHYSIOLOGICAL CONSEQUENCES OF HANDLING STRESS IN THE JUVENILE COHO SALMON (ONCORHYNCHUS KISUTCH) AND STEELHEAD TROUT (SALMO GAIRDNERI),
Bureau of Sport Fisheries and Wildlife, Seattle, Wash. Western Fish Disease Lab.
G. Wedemeyer.
Journal of the Fisheries Research Board of Canada, Vol 29, No 12, p 1780-1783, December 1972. 2 fig, 1 tab, 8 ref.

Descriptors: *Fish physiology, *Coho salmon, *Rainbow trout, Juvenile fish, Environmental effects, Water balance, Chemical analysis, Bioassay, hardness (Water), Metabolism, Chlorides, Calcium.
Identifiers: Handling stress, *Biochemical tests, *Blood, Oncorhynchus kisutch, Steelhead trout, Salmo gairdneri, Plasma glucose, Cholesterol, Ascorbate.

The stress of handling juvenile coho salmon (Oncorhynchus kisutch) and steelhead trout (Salmo gairdneri) in soft water and in water with added salts was evaluated using blood and tissue chemistry fluctuations as indices of metabolic and endocrine function. Changes in plasma glucose, chloride, calcium, and cholesterol levels indicated that significant osmoregulatory and metabolic dysfunctions can occur and persist for about 24 hr after handling in soft water. Pituitary activation, as judged by lack of interrenal ascorbate depletion, did not occur. Increasing the ambient NaCl and Ca (2 plus) levels to about 100 milliosmols and 75-120 ppm, respectively, partially or completely alleviated the hyperglycemia and hypochloremia indicating that the stress of handling had been reduced. (Holoman-Battelle)
W73-09352

FINE STRUCTURE OF SOME BRACKISH-POND DIATOMS,
Rhode Island Univ., Kingston. Narragansett Marine Lab.
For primary bibliographic entry see Field 05A.
W73-09353

A BAYESIAN APPROACH TO BIOASSAY,
Oregon State Univ., Corvallis.
F. L. Ramsey.
Biometrics, Vol 28, No 3, p 841-858, September 1972. 7 fig, 11 ref.

Descriptors: *Bioassay, *Toxicity, *Statistical methods, *Estimating, Lethal limits, Water pollution effects, Mathematical studies.
Identifiers: *Bayesian method, *Median tolerance limit, *Data interpretation, Potency curves, Experimental design.

A prior distribution for the class of continuous, non-decreasing potency curves is intorduced. The Bayes posterior distribution resulting from an assay experiment with quantal responses is discussed. Several examples are presented where a posterior modal function is used to summarize the posterior distribution. The examples illustrate the value of obtaining smooth estimates of potency and the value of experimental designs using many doses with few observations per dose. The methods suggested may be used to analyze quantal response from experiments with one observation per dose, situations in which the standard probit and logit methods cannot be used. It is even suggested that one observation per dose may be in fact the best design for estimating ED50. (Holoman-Battelle)
W73-09358

SOME DISTRIBUTION-FREE PROPERTIES OF THE ASYMPTOTIC VARIANCE OF THE SPEARMAN ESTIMATOR IN BIOASSAYS,
California Univ., Los Angeles. School of Public Health.
P. C. Chang, and E. A. Johnson.
Biometrics, Vol 28, No 3, p 882-889, September 1972. 3 tab, 11 ref.

Descriptors: *Bioassay, *Lethal limits, Statistical methods, Laboratory tests, Toxicity, Resistance, Mathematical studies.
Identifiers: *Experimental design, *Spearman estimator, *Asymptotic variance, *Precision, *Data interpretation, Quantal assays, Median tolerance limit.

Some distribution-free properties of the asymptotic variance of the Spearman estimator in bioassay have been investigated. The integral of the function, F (1-F) dx was investigated for distribution functions with varying skewness and kurtosis. It was shown that for various shapes of the tolerance distribution, the asymptotic variance of the Spearman estimator can be expressed approximately in terms of either the standard deviation or the distance between the 5th and the 95th percentile of the tolerance distribution. It was observed that K prime sub 60, K prime sub 80, and K sub sigma all depend upon skewness and kurtosis of distribution functions more than does K prime sub 90. K prime sub 90 equals approximately 0.17 for all those distribution functions except for the t distribution with 1 degree of freedom. These distribution-free properties can be used in planning quantal assays of required precision. (Holoman-Battelle)
W73-09359

ESTIMATION OF MICROBIAL VIABILITY,
Department of Scientific and Industrial Research, Palmerston North (New Zealand). Applied Mathematics Div.
For primary bibliographic entry see Field 05B.
W73-09360

QUANTAL RESPONSE ANALYSIS FOR A MIXTURE OF POPULATIONS,
Exter Univ. (England).
For primary bibliographic entry see Field 05A.
W73-09363

FIRST FOOD OF LARVAL YELLOW PERCH, WHITE SUCKER, BLUEGILL, EMERALD SHINER, AND RAINBOW SMELT,
National Water Quality Lab., Duluth, Minn.
For primary bibliographic entry see Field 02H.
W73-09372

ACCUMULATION AND PERSISTENCE OF DDT IN A LOTIC ECOSYSTEM,
Maine Univ., Orono. Dept. of Entomology.
For primary bibliographic entry see Field 05B.
W73-09415

TRANSITION OF POLLUTION WITH MERCURY OF THE SEA FOOD AND SEDIMENTS IN MINAMATA BAY (IN JAPANESE),
Kumamoto Univ. (Japan). Dept. of Hygiene.
For primary bibliographic entry see Field 05B.
W73-09416

TRITIUM AND ITS EFFECTS IN THE ENVIRONMENT—A SELECTED LITERATURE SURVEY,
Battelle Memorial Inst., Columbus, Ohio.
A. W. Rudolph, T. E. Carroll, and R. S. Davidson.
Available from NTIS, Springfield, Va. 22151, BMI-171-203, Price $6.00 printed copy; $0.95 in microfiche. Battelle Memorial Institute Columbus Laboratories Report BMI-171-203, June 30, 1971. 378 p. USAEC Contract AT (26-1)-171 and W-7405-eng-92.

Descriptors: *Bibliographies, *Abstracts, *Tritium, Path of pollutants, Water pollution sources, Water pollution effects.

This bibliography surveys the literature on tritium and its effects in the environment, with emphasis on man. Abstracts are included to describe the content of the references without relating specific details. All index terms associated with each article are listed in the compilation immediately following the abstract. A subject, author, and report number index are provided. Each entry consists of an alphabetically ordered index term, full title of the reference, and all significant terms by which the article is indexed. (Knapp-USGS)
W73-09433

DISTRIBUTION OF TRACE ELEMENTS IN THE ENVIRONMENT AND THE OCCURRENCE OF HEART DISEASE IN GEORGIA,
Geological Survey, Denver, Colo.
For primary bibliographic entry see Field 05B.
W73-09469

LEAD IN SOILS AND PLANTS: A LITERATURE REVIEW,
Colorado State Univ., Fort Collins. Dept. of Botany and Plant Pathology.
For primary bibliographic entry see Field 05B.
W73-09479

EMBROYO-FETOTOXIC EFFECT OF SOME ORGANIC MERCURY COMPOUNDS,
Nagoya Univ. (Japan). Research Inst. of Environmental Medicine.
U. Murakami.
Annual Report of the Research Insitute of Environmental Medicine, Nagoya University, Vol 19, p 61-68, 1972. 5 tab, 14 ref.

Descriptors: *Toxicity, *Human diseases, *Animal pathology, *Mercury, Biology, Biochemistry, Pathology, Water pollution effects, Public health, Laboratory tests, Cytological studies, Heavy metals.

Information is updated on cases of Minamata disease given in the 1971 annual report and presents briefly several new experiments using the organic mercury compounds, methylmercuric chloride, methylmercuric sulfide and methylmercuric dicyandiamide. The autopsy findings in human cases and results of animal experiments showed incidence of malocclusion and congenital malformation of the teeth such as irregular size, which when compared with incidence in the general population suggest that alkyl mercuries may show embryotoxicity. Several animal experiments using organic mercury seem to show evidence of embryo-toxicity and teratogenicity. In studies of the effect of organic mercury compounds upon mitotis it was observed that chromosome breakage and c-mitosis occurred, resulting in polyploid and aneuploid cells, which were morphologically manifest as tumors. It can be concluded that there is the possibility that organic mercury compounds might cause not only fetopathy, but also embryopathy and gametopathy. This should be an urgent subject for study. (See also W73-09485) (Jerome-Vanderbilt)
W73-09484

EFFECT OF METHYLMERCURIC CHLORIDE ON EMBRYONIC AND FETAL DEVELOPMENT IN RATS AND MICE,
Nagoya Univ. (Japan). Research Inst. of Environmental Medicine.
M. Inouye, K. Hoshino, and U. Murakami.
Annual Report of the Research Institute of Environmental Medicine Nagoya University, Vol 19, p 69-74, 1972. 4 fig, 3 tab, 6 ref.

Descriptors: *Animal pathology, *Mercury, *Environmental effects, Biology, Biochemistry, Pathology, Animal physiology, Laboratory tests, Analytical techniques, Water pollution effects, Mammals, Diseases, Toxicity, Growth rates, *Rodents.
Identifiers: *Methylmercury.

Rats and mice were used in experiments to investigate the effects of methylmercuric chloride upon embryonic and fetal development. A daily dosage of 5 mg/kg/day of methylmercuric chloride was administered to rats on days 0-12, days 7-14, and days 12-20 of pregnancy while a single 30 mg/kg dose was issued to mice on each day from 6 to 13 of pregnancy. Apparent teratogenicity was found in the mouse fetus while there was no apparent teratogenic effect shown in the rat. The effect of methylmercury on brain development in the rat and the mouse was similar. Tissue defects were found in the brain mantle, corpus callosum, fornix, nucleus caudatus-putamen, hypothalamus, pedunculus cerebri, and cerebellar primordium. The ventricles were enlarged and ependymal lining was exfoliated. In mice there were no differences in the brain lesion among the groups treated in different developmental stages. These findings support the opinion that alkylmercury might not influence the immature brain but might cause greater damage when it has developed. (See also W73-09484) (Jerome-Vanderbilt)
W73-09485

EXCRETION RATE OF METHYL MERCURY IN THE SEAL (PUSA HISPIDA),
Helsinki Univ. (Finland). Dept. of Radiochemistry.
M. Tillander, J. K. Miettinen, and I. Koivisto.
Food and Agriculture Organization of the United Nations, Bulletin Number FIR:MP/70/E-67, October 21, 1970. 6 p, 3 fig, 1 tab, 7 ref.

Descriptors: *Aquatic animals, *Mercury, *Distribution patterns, Biology, Biochemistry, Bioassay, Water pollution effects, Food chains, Diseases, Furbearers, Toxins, Analytical techniques, Radiochemical analysis.
Identifiers: *Methylmercury.

The retention and excretion rates of a methylmercury compound in the body of a seal were investigated using 203 Hg-labeled methylmercury and the whole body counting technique. CH3 203 Hg-proteinate containing 6.5 microcuries 203 Hg and 12 micrograms stable Hg was packed into two gelatin capsules which were concealed in herrings and fed to a 9 month old female seal weighing approximately 40 kg. A third capsule was sealed in a glass tube and used as a standard. The seal suffered no ill effects from this treatment. A whole body count was taken on a daily basis and a Baird

Atomics single channel analyzer was used to measure the 203 Hg activity. The 203 Hg initially showed an apparent increase due to redistribution of the methylmercury into fatty tissue near the surface. After 27 days the activity decreased. Retention was estimated to be between 80 and 100%. About 45% of the CH3 203 Hg administered as a single dose of methylmercury was excreted with the very long biological half-time of 500 days, but most was excreted with a half-time of 3 weeks. (Jerome-Vanderbilt)
W73-09486

TOXICITY AND DISTRIBUTION OF MERCURY IN PIGS WITH ACUTE METHYLMERCURIALISM,

Washington State Univ., Pullman. Dept. of Veterinary Pathology.
R. C. Piper, V. L. Miller, and E. O. Dickinson.

American Journal of Veterinary Research, Vol 32, No 2, p 264-273, February 1971. 2 tab, 35 ref.

Descriptors: *Mercury, *Toxicity, *Laboratory animals, Heavy metals, Toxins, Analytical techniques, Distribution patterns, Foods, Animal metabolism, Animal pathology.
Identifiers: *Methylmercury, Organomercurials, *Pigs.

Acute toxicity, clinical course, and concentration and distribution of mercury in tissues were determined in 20 pigs given 1 inoculation (oral route) of methylmercury dicyandiamide (MMD) at dose levels of 2.5 to 160 mg./kg. of body weight. Chemical analyses of tissues for mercury were performed by a modified dithizone procedure. Three of 4 pigs given 20 mg of MMD/kg. and all pigs given larger doses died or were euthanatized in extremis. Signs of methylmercurialism, but not deaths, were observed in 6 of 8 pigs, given 5, 10, or 15 mg. of MMD/kg. Toxicosis was not observed in 2 pigs given 2.5 mg./kg. Mercury was diffusely distributed trough the body; concentrations were greatest in the gastrointestinal tract at postinoculation (PI) hours 12 to 28. At PI days 7 to 35, concentrations of mercury were greatest in the renal cortex and then in the liver. Above-average concentrations were also seen in renal medulla, skeletal muscles, and cerebrum; lesser amounts were in all other tissues examined. A latency period of up to 3 weeks was observed in some pigs between administration of MMD and onset of clinical signs. (Oleszkiewicz-Vanderbilt)
W73-09490

LEAD POISONING: RAPID FORMATION OF INTRANUCLEAR INCLUSIONS,

Rochester Univ. Medical Center, N.Y.
D. D. Choie, and G. W. Richter.

Science, Vol 177, No 4055, p 1194-5, September 29, 1972. 1 fig, 1 tab, 9 ref.

Descriptors: *Lead, *Heavy metals, *Toxicity, *Animal pathology, Pathology, Cytological studies, Laboratory animals, Laboratory tests, Animal metabolism.
Identifiers: *Lead poisoning, Intranuclear inclusions.

A single dose of lead (0.05 milligram per gram of body weight) induced characteristic intranuclear inclusions in the epithelium of proximal tubules in rat kidney within 1 to 6 days. The development of the intranuclear inclusions is thus an acute manifestation of lead poisoning, not a delayed one, as has been thought hitherto. Cytoplasmic structures resembling the intranuclear inclusions and situated in the vicinity of endoplasmic reticulum were regularly found in cells bearing the pathognomonic intranuclear inclusions. The latter and the cytoplasmic structures may be derived from a common precursor, perhaps a soluble protein-lead complex. (Oleszkiewicz-Vanderbilt)
W73-09492

RADIOCHEMICAL STUDIES ON EFFECT OF BATHING: II. EFFECT OF SERIAL BATHING ON DISTRIBUTION OF MERCURY COMPOUNDS IN MICE, (IN JAPANESE),

Kyushu Univ., Beppu (Japan). Inst. of Balneotherapeutical Research I.
H. Kawamura.

J Hyg Chem. Vol 17, No 4, p 287-297. 1971. Illus. English summary.
Identifiers: Alkyl mercury, Distribution, *Japan, *Mercury compounds, Mice, Radiochemical studies, Shellfish, *Toxicosis, Methylmercury.

To elucidate the action mechanism of S sludge bath, which showed a specific effect in accelerating urinary excretion of Hg during spa treatment for alkylmercury toxicosis, methyl methylmercuric sulfide, isolated from a poisonous shellfish caught in Minamata Bay, Japan, was labeled with 203 Hg and the Me-Me203HgS and 203 Hg acetate were administered to mice. The effect of hot spring bath on the biological distribution of these labeled Hg compounds was examined. Both labeled Hg compounds showed a marked accumulation in the kidneys after i.p. administration. Systemic distribution of the sulfide was more gradual and homogeneous than that of the acetate. Disappearance of the acetate and sulfide from the organs after i.p. administration was accelerated by continued bathing in S sludge for 3 wk but this action appeared more slowly in the sulfide. This accelerative effect was not significant when the bathing was in simple spa water. After bathing in water containing 203Hg, the amount of Hg in various organs increased up to the 6th day but the increase became insignificant thereafter. The amount of Hg in the animals given simple spa baths was twice that in the animals given S sludge baths. When the mice given Me-Me203HgS were bathed in S sludge labeled with 197Hg acetate, decrease of 203Hg from the liver was much faster than that from the kidneys and the accumulation of 197Hg was extremely small. Concurrent measurement of 197Hg and 203Hg in the liver and kidneys was performed with a 400-channel pulseheight analyzer. The large amount of Hg excreted into the urine of patients with alkylmercury toxicosis by continued bathing in S sludge is the Hg already accumulated in the body and not due to turnover of Hg in the S sludge.--Copyright 1972, Biological Abstracts, Inc.
W73-09496

METHYLMERCURY IN NORTHERN PIKE (ESOX LUCIUS): DISTRIBUTION, ELIMINATION, AND SOME BIOCHEMICAL CHARACTERISTICS OF CONTAMINATED FISH,

Fisheries Research Board of Canada, Winnipeg (Manitoba). Freshwater Inst.

For primary bibliographic entry see Field 05B.
W73-09498

ULTRASTRUCTURAL EVIDENCE FOR NEPHROPATHY INDUCED BY LONG-TERM EXPOSURE TO SMALL AMOUNTS OF METHYL MERCURY,

Oregon Univ., Portland. Dept. of Pathology.
B. A. Fowler.

Science, Vol 175, No 4023, p 780-781, February 18, 1972. 3 fig, 14 ref.

Descriptors: *Mercury, *Metabolism, *Rodents, *Animal pathology, Biology, Chemistry, Physiology, Structure, Mammals, Laboratory tests, Evaluation, Environmental effects, Pollutants, Disease.
Identifiers: *Methylmercury.

Proteinuria has been observed in persons who are occupationally exposed to organo-mercury compounds. To obtain further information on this phenomenon, the author investigated the effects of long term exposure of low doses of methyl mercuric chloride (CH3 Hg Cl) on rat nephron. Wistar rats were fed a high protein diet, and starting at the age of 28 days mercury at 2 ppm as CH3HgCl was

J. K. Miettinen, M. Tillander, K. Rissanen, V. Miettinen, and Y. Ohmomo.
Proceedings of the 9th Japan Conference on Radioisotopes, Tokyo, p 474-478, May 13-15, 1969, Japan Atomic Industrial Forum, Inc. 4 fig, 4 tab, 6 ref.

Descriptors: *Metabolism, *Mercury, Biochemistry, Toxicity, Analytical techniques, Testing, Laboratory studies, Radioisotopes, Physiology, Distribtuion, Seawater, Freshwater, Water pollution effects, *Mollusks, *Mussels, *Crayfish.

This study was performed to ascertain whether there was any difference in the toxicity, excretion rate, and distribution pattern of radioactive methylmercury when administered by mouth, either as the ionic form CH3 Hg (+) or bound to protein. Organ activity was determined aftervarious time lapses by whole body counting of the live animals. No significant differences were noted between the fish in brackish water and in fresh water. Ten to fifty percent of the administered activity was excreted within one or two days. The remainder was excreted more slowly with phenylmercury nitrate having a shorter biological half time than methylmercury nitrate. After a span of about three weeks the highest concentrations of mercury were found in the flesh, stomach, liver, ovary and kidneys. Although the experiments are continuing, the results show that there is no essential difference in the toxicity, distribution or excretion rate between the ionic and protein-bound forms of methylmercury. From experiments with molluscs, mussels and crayfish it is evident that when mercury compounds are bound to the muscle protein (mercury compound injected) they have a slower excretion rate than when bound to the digestive organs (administered by mouth). (Jerome - Vanderbilt)
W73-09508

MONITORING ZINC CONCENTRATIONS IN WATER USING THE RESPIRATORY RESPONSE OF BLUEGILLS (LEPOMIS MACROCHIRUS RAFINESQUE),
Virginia Polytechnic Inst. and State Univ., Blacksburg. Dept. of Biology; and Virginia Polytechnic Inst. and State Univ., Blacksburg. Center for Environmental Studies.
For primary bibliographic entry see Field 05B.
W73-09510

SIGNIFICANCE OF MERCURY IN THE ENVIRONMENT,
Department of Agriculture, Saskatoon (Saskatchewan). Research Station.
For primary bibliographic entry see Field 05B.
W73-09511

ECOLOGY OF CORTICOLOUS LICHENS: III. A SIMULATION MODEL OF PRODUCTIVITY AS A FUNCTION OF LIGHT INTENSITY AND WATER AVAILABILITY,
McMaster Univ., Hamilton (Ontario). Dept. of Biology.
For primary bibliographic entry see Field 02I.
W73-09512

MICROBIAL UPTAKE OF LEAD,
Colorado State Univ., Fort Collins. Dept. of Microbiology.
T. G. Tornadene, and H. W. Edwards.
Science, Vol 176, No 4041, p 1334-1335, June 23, 1972. 15 ref.

Descriptors: *Lead, *Heavy metals, *Cytological studies, Microbial degradation, Microbiology, Laboratory tests, Bioassay, Bacteria.
Identifiers: Cell count, Lead uptake, *Micrococcus luteus, *Azotobacter sp.

Micrococcus luteus and Azotobacter sp. cells grown in broth in contact with a dialysis membrane containing lead bromide were found to immobilize 4,900 and 3,100 milligrams of lead per gram of whole cells, on a dry weight basis, respectively. Culture turbidity and cell count measurements on these and other cell cultures show that lead bromide, lead iodide, and lead bromochloride in concentrations approaching solubility limits have no detectable effect on overall growth rate and cell viability. Analyses of cellular subfractions reveal that fractions of cell wall plus membrane contain 99.3 and 99.1 percent of the lead found associated with Micrococcus luteus and Azotobacter sp., respectively. The remainder is found associated with the cytoplasmic fractions. (Oleszkiewicz-Vanderbilt)
W73-09514

DETERMINATION OF THE COPPER CONCENTRATIONS TOXIC TO MICROORGANISMS,
Moscow State Univ. (USSR). Dept. of Soil Biology.
For primary bibliographic entry see Field 05A.
W73-09515

HEALTH ASPECTS OF TOXIC MATERIALS IN DRINKING WATER,
Environmental Health Service, Rockville, Md.
J. H. McDermott, P. W. Kabler, and H. W. Wolf.
Am J Public Health. Vol 61, No 11, p 2269-2276. 1971.
Identifiers: *Public health, Potable water, Pollution, *Toxic materials, Water pollution effects.

Toxic materials in drinking water are a major problem of water pollution. The development of this situation and the actions being taken to deal with it are reviewed.--Copyright 1972, Biological Abstracts, Inc.
W73-09524

HEALTH ASPECTS OF WATER QUALITY,
Environmental Protection Service, Ottawa (Ontario).
S. K. Krishnaswami.
Am J Public Health. Vol 61, No 11, p 2259-2268. 1971.
Identifiers: *Epidemiology, *Public health, *Human diseases, Water pollution effects, Potable water, Water quality.

The epidemiologic significance of continual human exposure to low-level concentrations of a plethora of organic chemicals and some inorganics in water is not understood. The nutritional significance of water constituents has received little consideration. The safety to various organic and inorganic chemicals purposefully added to drinking water during treatment and distribution should be evaluated and regulation by laws is recommended. The health aspects of other water uses (public recreation, agriculture and food and soft-drink manufacture) should be studied epidemiologically.--Copyright 1972, Biological Abstracts, Inc.
W73-09526

PHYTOPLANKTON SUCCESSION IN A EUTROPHIC LAKE WITH SPECIAL REFERENCE TO BLUE-GREEN ALGAL BLOOMS,
Academy of Natural Sciences of Philadelphia, Pa. Dept. of Limnology.
C. K. Lin.
Hydrobiologia, Vol 39, No 3, p 321-334. 1972. Illus.
Identifiers: Anabaena, Aphanizomenon-flos-aquae, Asterionella-formosa, Canada, Cyclotella-meneghiniana, *Eutrophication, Lakes, Microcystis-aeruginosa, Nutrients, Oxygen, *Phytoplankton, Prairies, Silica, Temperature, Water pollution effects, *Cyanoplyta.

An investigation of phytoplankton in Astotin Lake was made between mid-May of 1966 and Sept. of 1967, with particular attention to the ice-free seasons. Astotin Lake is a typical, small eutrophic, kettle lake with shallow, landlocked, hard water in the Canadian prairies. High concentrations of nutrients supported heavy blooms of blue-green algae throughout the summer. The spring communities were dominated by Asterionella formosa in 1966 and by Cyclotella meneghiniana in 1967. Oxygen depletion under ice cover probably explains the failure of an A. formosa population to appear in 1967. Deficiency of silica and a rise in water temperature apparently caused the decline of the spring pulses of diatoms. Relatively high summer water temperature favored the blue-green algal blooms and resulted in a high concentration of organic matter. The decomposition of dead Anabaena cells played an important part in the development of subsequent waterblooms, i.e., Microcystis aeruginosa and Aphanizomenon flos-aquae. The sequence of waterblooms of those species was closely related to the change in water temperature. A. flos-aquae became incompatible with M. aeruginosa when the temperature fluctuated in a wide range. Most of the non-blue-green algae apparently were inhibited by these cyanophyte blooms. Great species diversity appeared intermittently between blooms and a few species of the Scenedesmaceae and the Oocystaceae were relatively compatible to these blooms.--Copyright, 1972, Biological Abstracts, Inc.
W73-09534

PHYTOPLANKTON, PHOTOSYNTHESIS, AND PHOSPHORUS IN LAKE MINNETONKA, MINNESOTA,
Minnesota Univ., Minneapolis. Limnological Research Center.
R. O. Megard.
Limnol Oceanogr. Vol 17, No 1, p 68-87, 1972. Illus.
Identifiers: Anabaena, Lakes, *Minnesota, *Lakes, Minnetonka, *Phosphorus, *Photosynthesis, *Phytoplankton.

The temporal and spatial variations of chlorophyll concentrations (n) and integral photosynthesis (p) are described for a large, morphometrically complex lake. The daily rate of photosynthesis at the average depth where illumination is optimal (p-max) increases, whereas the depth where illumination begins to be saturating (zi) decreases, as chlorophyll concentrations increases. These opposing effects influence the geometric proportions (expressed by the ratio p:pmax) of photosynthesis depth profiles systematically. The regression of the ratio p:pmax) (±zi) on chlorophyll is fitted to an exponential equation with which zi can be estimated from the chlorophyll concentration. Values of zi so estimated are similar to those expected from the reciprocal relationship between zi and chlorophyll required by Lambert's law. Therefore daily integral photosynthesis at any locality can be calculated from measurements of the chlorophyll concentration and the rate of photosynthesis in a single water sample incubated for a half day at the average depth where illumination is optimal. Chlorophyll concentrations increase as concentrations of total P increase, but generally no relationship exists between P concentrations and maximum specific rates of photosynthesis (Pmax). The linear relationship between P and chlorophyll occurs only when Anabaena, a nitrogen-fixing alga, is an important component of the phytoplankton. The quantity of chlorophyll above the depth zi (± ni) represents ca. 60-80% of the chlorophyll in the photic zone. The value of ni where chlorophyll concentrations are 60 mg/m3 (the maximum concentration at most localities in Lake Minnetonka) is about 100 mg Chl/m2. The usual limit for integral photosynthesis at 20C is therefore 5 g C m-2 day-1.--Copyright 1972, Biological Abstracts, Inc.
W73-09546

MICROBIOLOGICAL STUDIES OF THERMAL HABITATS OF THE CENTRAL VOLCANIC REGION, NORTH ISLAND, NEW ZEALAND,
Wisconsin Univ., Madison. Dept. of Bacteriology.
T. D. Brock.
N Z J Mar Freshwater Res. Vol 5, No 2, p 233-258, 1971. Illus.
Identifiers: *Bacteria, *Algae, Cyanidium-Caldarium, Ephydrella-Thermarum, Islands, Mastigocladus- Laminosus, Microbiological studies, *New Zealand, Phormidium-Sp, Synechococcus-Sp, Thermal water regions, Volcanic ash, Zygogonium-Sp.

Studies on the algae and bacteria of North Island thermal areas recorded temperature, pH and species found in these microbial habitats with special attention to organisms living at the highest temperatures. Thermal features were studied at Rotorna (Whakarewarewa and Ohinemutu), Waiotapu (Tourist Reserve and Lady Knox Geyser), Orakei Korako, Taupo Spa, Waikite Springs, Wairakei thermal valley, Wairakei geothermal field, Tikitere, Ketetahi, Lake Rotokawa (Taupo region), Waimangu, De Brett Thermal Hotel (Taupo). The upper temperature limit for blue-green algae in New Zealand is 60-65 deg C, and the species living at the thermal limit is generally Mastigocladus laminosus, although in some cases Phormidium sp. or Synechococcus sp. was found. The Synechococcus sp. characteristic of high temperatures (73-74 deg C) present in North America was not found in New Zealand. In virtually all boiling pools (99-101 deg C) with pH values in the neutral and alkaline range bacteria were found, but in acidic boiling pools, bacteria were absent. The presence in New Zealand of the eucaryotic algae Cyanidium caldarium and Zygogonium sp. is reported for the first time. Further records for the hot spring brine fly Ephydrella thermarum and other ephydrids are given. The observations are compared with previous data on thermal habitats in Yellowstone Park, in Iceland, and in other parts of the world.--Copyright 1972, Biological Abstracts, Inc.
W73-09554

EVALUATION OF TWO METHOD OF AERATION TO PREVENT WINTERKILL,
Virginia Polytechnic Inst. and State Univ., Blacksburg. Div. of Forestry and Wildlife Science.
For primary bibliographic entry see Field 05G.
W73-09560

THE EFFECT OF ANNUAL RECHANNELING ON A STREAM FISH POPULATION,
Tennessee Univ., Knoxville. Dept. of Zoology.
D. A. Etnier.
Trans Am Fish Soc. Vol 101, No 2, p 372-375. 1972.
Identifiers: Annual, *Channeling, *Fish population, Streams, *Rechanneling (Annual).

Substrate instability and decreased variability of the physical habitat are believed to be the most significant factors responsible for the changes in the fish fauna in the altered stream.--Copyright 1973, Biological Abstracts, Inc.
W73-09564

THE FEEDING BEHAVIOUR OF SOME COMMON LOTIC INSECT SPECIES IN TWO STREAMS OF DIFFERING DETRITAL CONTENT,
National Inst. for Physical Planning and Construction Research, Dublin (Ireland).
E. Fahy.
J Zool Proc Zool Soc Lond. Vol 167, No 3, p 337-350. 1972. Illus.
Identifiers: Algae, Cannibalism, Detrital, Content (Streams), *Feeding habits (Insects), *Insects, Invertebrates, *Lotic insect species, Plecoptera, Prey, Streams.

The trophic relationships of the benthic invertebrates at 2 places of differing detrital content in an oligotrophic stream system were examined during a 12 mo. period. The sites are described and their detrital contents and other abiotic features compared. The faunal list at each site is fairly similar but certain species are more common at or restricted to one or other site. The reasons for this are briefly discussed. The predators and omnivores are in 2 groups: the first feed only on certain prey and are thus limited by the distribution of their prey. The others consume a wide range of prey and have wider distributions. The predator species are non-selective in their choice of prey which are devoured in an intensity which cannot however be related to their density in the benthos. In poorer trophic conditions predators adapt by consuming a greater quantity of algae and detritus; cannibalism and interpredator-predation occur. The prey species are more intensely consumed under poor trophic conditions. The ecology of 2 species of predatory Plecoptera is examined and the results indicate that competition probably occurs between them.--Copyright 1973, Biological Abstracts, Inc.
W73-09567

FLUOROSIS OF SHEEP CAUSED BY THE HEKLA ERUPTION IN 1970,
Iceland Univ., Reykjavik. Inst. for Experimental Pathology.
G. Georgsson, and G. Petursson.
Fluoride. Vol 5, No 2, p 58-66, 1972. Illus.
Identifiers: Ash, Bones, *Fluorosis, Iceland (Hekla eruption), *Sheep, Teeth, *Volcanic ash, Water.

Samples of ash from the erupted volcano contained up to 2,000 ppm of F. Values decreased to 10% within 2 wk and to 2% within 3 wk. During the first week stagnant surface water contained up to 70 ppm F, running water up to 10 ppm. Grass analyzed under layers of ash 10 mm thick, showed 4,300 ppm F on the second day after eruption, and less than 30 ppm after 35-40 days, partly due to rainfall. Acute fluorosis killed 3% of the sheep and 8-9% of lambs in the affected area during the first few weeks through convulsive seizures, pulmonary edema and kidney and liver changes. No skeletal fluorosis was noted except for slight periosteal thickening in less than 0.25% of 400 animals which were x-rayed. In the bones of lambs, F increased 4 fold the normal amount but in adult sheep only 50%. Dental fluorosis occurred in 25.3% of the third incisors which erupted 4 to 9 mo. later and in 8.6% of the second incisors which erupted 9-13 mo. after the volcanic eruption.--Copyright 1973, Biological Abstracts, Inc.
W73-09574

DIATOMS FOUND IN A BOTTOM SEDIMENT SAMPLE FROM A SMALL DEEP LAKE ON THE NORTHERN SLOPE, ALASKA,
N. Foged.
Nova Hedwigia, Vol 21, Nos 2-4, p 923-1035, 1971. 331 fig, 58 ref.

Descriptors: *Diatoms, *Chrysophyta, *Lake sediments, *Aquatic algae, Alaska, Bottom sediments, Systematics.
Identifiers: Achnanthes species, Amphipleura pellucida, Amphora species, Asterionella formosa, Caloneis species, Campylodiscus echineis, Campylodiscus hibernicus, Ceratoneis arcus, Cocconeis species, Cyclotella species, Cymatopleura species, Cymbella species, Denticula species, Diatoma species, Diploneis species, Epithemia species, Epithemia zebra var. saxonica, Eunotia species, Fragilaria species, Gomphonema species, Gyrosigma species, Hantzschia species, Melosira granulata, Melosira teres, Navicula species, Neidium species, Nitzschia species, Oestrupia zachariasis, Opephora martyi, Pinnularia species, Rhoicosphenia curvata, Rhopalodia gibba, Stauroneis species, Stephanodiscus species, Surirella species, Synedra species, Tabellaria species.

n a sediment sample from 15.5 m depth in an unamed freshwater lake of 1.25 sq. km on the Northern Slope, Alaska (70 degrees 01 min N. lat., 153 degrees 36 min W. Gr.) about 400 forms of liatoms belonging to 36 genera were identified (19 nov. spec., 5 nov. var., and 6 nov. fo. are lescribed). The flora of diatoms is characterized, n addition to the richness of species, by the coniderable number of forms, which previously often vere found in interglacial or postglacial deposits in Europe or Asia. Especially remarkable are the inds of a number of species of Surirella which is urprisingly great for arctic regions. The lake must e very old, possibly from the Tertiary, since the rea has neither been covered by inland ice since hat time nor exposed to transgressions by the sea. Holoman-Battelle)
V73-09622

IOLOGICAL ASPECTS OF OIL POLLUTION N THE MARINE ENVIRONMENT, A REVIEW,
McGill Univ., Montreal (Quebec). Marine Sciences Centre.
M. J. A. Butler, and F. Berkes.
M.S.C. Manuscript Report No. 22, December 972. 122 p, 174 ref.

Descriptors: *Oil spills, *Oil pollution, Organic ompounds, Solvents, Marine plants, Marine nimals, On-site investigations, *Environmental ffects, *Bibliographies, *Reviews, Water pollu-ion effects.
dentifiers: *Literature review, Torrey Canyon, Arrow, *Biological effects.

A synthesis of some of the most significant esearch done on oil pollution and its biological efects is presented. The report includes fairly exensive sections on petroleum hydrocarbons, solent-emulsifiers, effects of oil pollution on birds, nammals, fish, sediments, and marine communiies and case studies of oil pollution including the Torrey Canyon' disaster, the Santa Barbara oil pill and the 'Arrow' accident. The geography of il pollution is explored in such areas as the Baltic ea, North Sea and Arctic Ocean. The study also ims to analyze the environmental implications of series of hypothetical incidents that would be asociated with activities involving oil exploration, xploitation, export and import, coastal movement nd marine transportation activities and facilities. Ensign-PAI)
W73-09643

NSTRUCTURED MARINE FOOD WEBS AND POLLUTANT ANALOGUES',
cripps Institution of Oceanography, La Jolla, alif.
or primary bibliographic entry see Field 05B.
W73-09649

NVIRONMENTAL ASPECTS OF DREDGING N ESTUARIES,
kidaway Inst. of Oceanography, Savannah, Ga.
I. L. Windom.
ournal of the Waterways, Harbors and Coastal ngineering Division, Vol 98, No WW4, p 475-87, November 1972. 9 fig, 4 tab, 2 ref.

escriptors: *Salt marshes, *Estuaries, *Atlantic oastal plain, *Dredging, *Water quality, Water ollution, Environmental effects, Balance of nare, Chemical degradation, Sampling.

esults of the initial phase of a project designed to etermine the nature and magnitude of environental changes resulting from dredging activities estuarine salt marshes along the southeastern tlantic coast are reported. The chemical response salt marsh sediments to the deposition of edged materials and the water quality response dredging in the estuarine environment are scussed. Two salt marsh areas were studied bere and after dredging activities; sediment core mples were taken. Research findings indicate that the depth of spoil material deposited on a marsh will determine what parts of the spoil bank may ultimately return to production as well as dictating the time necessary for chemical rebalancing so that artificial seeding or sprigging could be successfully carried out to accelerate marsh regeneration. Water quality was tested by collecting water samples hourly during 24-hr periods at several stations surrounding each area before, during, and after dredging. It was found that no significant changes in water quality occurred during or after dredging activities. It was concluded that in order to evaluate water quality changes due to dredging specific information must be obtained about the sediments in the area to be studied. No general criteria can be set up for dredging in marine waters. (Ensign-PAI)
W73-09651

PLUTONIUM AND POLONIUM INSIDE GIANT BROWN ALGAE,
Scripps Institution of Oceanography, La Jolla, Calif.
K. M. Wong, V. F. Hodge, and T. R. Folsom.
Nature, Vol 237, No 5356, p 460-462, June 23, 1972. 2 fig, 3 tab, 15 ref.

Descriptors: *Radioisotopes, *Sea water, Pollutants, *Algae, Absorption, *Trace elements, Water pollution efffects, Laboratory tests, *Phaeophyta.
Identifiers: *Plutonium, *Polonium, Pelagophycus porra, Biological indicators.

The giant brown alga, Pelagophycus porra, was sampled to determine plutonium and polonium levels in sea water. The samples were analysed by alpha spectrometry for 239Pu and 210Po content. Polonium concentrations in the outermost thin layer of the algae were found to be 1,000 times greater than those found in the inner layers; the highest concentration of plutonium was also found in outer layers. These differences in sensitivity within a single species suggest that, when comparing two different environments, identical sample tissue from a certain species should be compared. Sampling thin parts of brown algae, or thin outer layers should provide great sensitivity for detecting early changes in the extremely small traces of plutonium which are expected to accumulate in the oceans due to reactor effluents, nuclear fuel processing and fallout. (Ensign-PAI)
W73-09652

5D. Waste Treatment Processes

THE ROLE OF SEDIMENTATION IN THE PURIFICATION OF RESIDUAL WATER, (NOTE SUR LA ROLE DE LA SEDIMENTATION DANS L'EPURATION DES EAUX RESIDUAIRES),
Centre d'Etudes et de Recherches de Biologie et d'Oceanographie Medicale, Nice (France).
For primary bibliographic entry see Field 05G.
W73-09205

THE NAVY AND THE EFFLUENT SOCIETY,
For primary bibliographic entry see Field 05G.
W73-09223

USE OF WASTES FOR IRRIGATION IN THE DONBAS, (OROSHEVIYE STOCHNYMI VODAMI V DONBASSE),
Ukrainskii Nauchno-Issledovatelskii Institut Gidrotekhniki i Melioratsii, Kiev (USSR).
S. I. Repetun.
NWWA Translation, Typescript, No 8, p 10, August, 1972. Trans. from (Gidrotekhnika i Melioratsya).

Descriptors: Irrigation, Irrigated land, Irrigation efficiency, *Irrigation practices, Gypsum, Alkalinity, Crops, Crop production, Sewage, *Sewage treatment, *Industrial wastes, Phosphates, Calcium sulfate, Sodium, Domestic wastes.
Identifiers: Soviet Union (Ukraine), *Yield (Crops), Sulfuric acid, Field studies.

The natural purification of waste waters in the fields in combination with artificial and economic purification was found to be the most effective, reliable and economic measure for protection of water resources. Conclusions reached after two years of testing at an experimental station were: (1) Industrial waste water from sulfuric acid and superphosphate production can be used for irrigation provided it is diluted with biologically purified domestic sewage in the ratio of 1:4; (2) The best method of waste water irrigation is to water along the ridges at the rate of 2000-2400 m3/hectare (4 growing season waterings of 500-600 m3/hectare each); (3) Potato yield was increased by 24% and corn yield by 54%. Gypsum, when added to the waste-water (in concentrations below 2.t/hectare) had no effect on yield and prevented absorption of sodium by the soil; and (4) Quality, as well as quantity, of the harvest was improved by irrigation with wastewater. (Campbell-NWWA)
W73-09258

IRRIGATION BY SEWAGE, (OROSHENIYE STOCHNYMI VODAMI),
Y. I. Tikhonov.
NWWA Translation, Typescript, No 9, p 5, September, 1972. Trans. from (Gidrotekhnika i Melioratsya).

Descriptors: Irrigation, *Irrigation practices, Crop production, *Sewage treatment, *Fertilization, Fertilizers, Economics, *Economic impact, Waste water treatment.
Identifiers: *Yield (Crops), Soviet Union, Field studies.

In the area under study, considerable yield increase of field crops was realized by irrigation with the sewage from a local town, supplemented in the summer with water from a nearby river. Harvest of the main crops, i.e., cabbage, carrots, beets, corn, and hay were in some cases more than doubled in irrigated fields over the yield in dryland. Profits were increased to the extent that capital investments in the irrigation system were paid for in less than five years. It was also found that mixing mineral fertilizers with the wastewater used for irrigation made it possible to spread them more evenly than by scattering them from a tractor and increased their fertilization effectiveness in terms of crop yield. Shortcomings in the method include frequent interruptions of the water supply, making it difficult to irrigate the fields on schedule; and much hand labor in irrigation was needed, since no machines were available for irrigation with sewage. (Campbell-NWWA)
W73-09259

USE OF SEWAGE FOR IRRIGATION, (ISPOLZOVAIYE STOCHNYKH VOD DLYA OROSHENIA),
V. E. Fedotov, and F. G. Gorkopenko.
NWWA Translation, Typescript, No 9, p 4, September, 1972. Trans. from (Gidrotekhnika i Melioratsya).

Descriptors: Irrigation, *Irrigation practices, *Crop production, Sewage, *Sewage treatment, Fertilizers, Waste water treatment, Sewage bacteria, Sewage lagoons, Irrigation water, Industrial wastes, *Aerobic treatment.
Identifiers: Soviet Union, Yield (Crops).

The experiences of a collective farm with irrigation by means of sewage from a nearby community are given. The farm is located in a dry area, and production of field crops would be almost impossible without irrigation. Yields of crops irrigated with sewage were found to be higher than the

yields of those irrigated with usual irrigation water (river water). Since the river water required the introduction of mineral fertilizers, the cost of these was also saved with irrigation by sewage. Irrigation of treated waste water (mineral and organics, helminth larvae and pathogenic microflora removed) was found to be better than irrigation with the untreated sewage since the materials in the untreated sewage clogged the soil causing rotting of the crops in the fields. Treatment of sewage was carried out directly at the farm in a series of settling ponds. A recent sharp increase in the percentage of industrial waste waters in the sewage was found to adversely affect crop yield. (Smith-NWWA)
W73-09260

WATER AND WASTEWATER PROBLEMS IN RURAL AMERICA.
Commission on Rural Water, Washington, D.C.
For primary bibliographic entry see Field 05G.
W73-09264

WASTE WATER TREATMENT SYSTEMS FOR RURAL COMMUNITIES,
System Sciences, Inc., Bethesda, Md.
S. N. Goldstein, and W. J. Moberg.
Commission on Rural Water, Washington, D.C., 1973. 340 p, 29 fig, 23 tab, 39 ref, 3 append.

Descriptors: *Rural areas, *Waste treatment, Waste water (Pollution), Waste water treatment, *Sewage disposal, *Septic tanks, Biochemical oxygen demand, Microorganisms, Coliforms, Sewage lagoons, Soil disposal fields, Soil filters, Soil microbiology, Disinfection, Irrigation, *Maintenance.
Identifiers: Septage, Package plants, Central systems.

A guide to systems and components available for treating waste waters in rural areas is presented, assisting designers and planners of waste water treatment facilities for rural communities in laying out preliminary system plans, and providing the potential client with sufficient information on available alternatives, to enable him to exercise an informed choice among those presented for his consideration. Basic concepts of domestic sewage and treatment processes appropriate to rural settings are given in Chapter One. In Chapter Two, the role and use of soils in waste water treatment and disposal are described. Traditional systems and design approaches for waste water treatment systems are reviewed in Chapter Three. Chapter Four describes new or unusual methods for waste water collection, conveyance, treatment, and disposal where traditional approaches are not appropriate. The traditionally poor record of maintenance and service of small plants is described in Chapter Five; types of management organization that can provide proper plant maintenance and service are suggested. A basis for anticipating the costs of components and systems for treating waste water in rural communities is provided in Chapter Six. Three Appendices are included, presenting: A-an approach for determining building lot sizes on the basis of soil suitability for waste water disposal; B-the role of the National Sanitation Foundation in Component testing; and C-a representative selection of equipment which can be used in small waste water treatment systems. (Campbell-NWWA)
W73-09266

A TRANSPORT MECHANISM IN HOLLOW NYLON FIBER REVERSE OSMOSIS MEMBRANES FOR THE REMOVAL OF DDT AND ALDRIN FROM WATER,
Tennessee State Univ., Nashville. Dept. of Civil Engineering.
L. A. Abron, and J. O. Oshurn.
Water Research, Vol 7, No 3, p 461-477, March 1973. 18 fig, 6 tab, 8 ref.

Descriptors: *DDT, *Aldrin, *Pesticide removal, *Aqueous solutions, *Semipermeable membranes, *Reverse osmosis, Chlorinated hydrocarbon pesticides, Membrane processes, Separation techniques, Water quality control, Infiltration, Solubility, Diffusivity, Adsorption, Pollutant identification, Physical properties, *Waste water treatment.

The DuPont hollow nylon fiber reverse osmosis membranes were investigated for use in the removal of DDT and Aldrin from aqueous solution. Because these membranes were developed for the demineralization of brackish waters the removal characteristics for the pesticides were compared with the removal characteristics for the ions commonly found in brackish and hard waters. These membranes rejected 85-95 percent of the brackish-producing ions. Such inorganic ions with respect to the membrane were classified as membrane-non-interacting solutes. DDT and Aldrin were classified as membrane-interacting solutes, and thus the concept of solute rejection by the membrane was different from that which had been developed for the inorganic ions. (Holoman-Battelle)
W73-09302

DENITRIFICATION. STUDIES WITH LABORATORY-SCALE CONTINUOUS-FLOW UNITS,
National Inst. for Water Research, Pretoria (South Africa).
P. J. Du Toit, and T. R. Davies.
Water Research, Vol 7, No 3, p 489-500, March 1973. 1 fig, 9 tab, 11 ref.

Descriptors: *Denitrification, Methodology, *Sewage, Domestic wastes, Laboratory tests, Nitrates, Nitrites, Suspended solids, Chemical oxygen demand, Ammonia, Nitrogen compounds, Gas chromatography, Alkylbenzene sulfonates, Sampling, Chemical analysis, Hydrogen ion concentration, *Waste water treatment.
Identifiers: *Lactate, *Methanol, Continuous flow system, Substrate utilization, *Denitrifying bacteria, Orthophosphates, Fatty acids, Enzymatic techniques, Culture media, Enrichment culture.

Settled domestic sewage, lactate, and methanol were all investigated as carbon sources for denitrification in different types of denitrifying units. The denitrifying units consisted of a suspended growth (SG) unit and a PC unit. The SG units were operated at a hydraulic residence time of 40 h. The stone PC unit was operated at a hydraulic residence time of 10 h. The coke PC unit was operated at different residence times to determine the minimum time required for efficient denitrification of the nitrified effluent. Samples were withdrawn, at the different sampling ports (PC units) at each stage of the experiment after at least three hydraulic residence times had elapsed. Gas samples were periodically taken for analysis by gas chromatography. The units were operated in a controlled temperature room at 20 C. General chemical analysis (COD, total-. and orthophosphate, total- and NH3-nitrogen, volatile fatty acids and ABS determinations) was perfomed by means of a Technicon autoanalyzer. Nitrates and nitrites were analyzed by previously described methods and lactate was determined enzymatically. Settled domestic sewage is unsuitable as a carbon source for inducing denitrification, although efficient COD removals can be obtained in the system. The fact that ammonia nitrogen flows freely through this system results in poor total nitrogen removal values. Lactate is an efficient additive for achieving denitrification. Efficient removal of nitrate by lactate addition can be achieved effectively by a C:N ratio of less than 1-5:1, where N represents -NO3 or NO2 nitrogen. Methanol is also suitable for inducing an acceptable rate of denitrification. The packed column type unit appears to be a very effective system for use as a denitrifying unit, particularly when packed with coke to increase the surface are. A denitrifying unit packed with coke and receiving methanol as hydrogen donor used in series with a bacterial disc unit appears to provide an efficient unit for use in water reclamation schemes. (Holoman-Battelle)
W73-09304

PROCESS AND APPARATUS FOR RECOVERING CLEAN WATER FROM DILUTE SOLUTIONS OF WASTE SOLIDS,
Carver-Greenfield Corp., East Hanover, N.J. (Assignee)
C. Greenfield, R. E. Casparian, and A. J. Bonanno.
U. S. Patent No 3,716,458, 10 p, 1 fig, 6 ref; Official Gazette of the United States Patent Office, Vol 907, No 2, p 474, February 13, 1973.

Descriptors: *Patents, *Waste water treatment, *Waste dilution, *Solid wastes, *Evaporation, *Condensation, Water pollution control, Water quality control, Pollution abatement, Equipment.

Dilute solutions of waste solids are concentrated by heat evaporation to yield water vapor and concentrated solutions holding waste solids. A tank is adapted to receive a stream of the dilute solution. It has a stirring or mixing device to homogenize the dilute solution. A conduit extends from this tank to the first evaporator and another conduit extends to the first condenser. After the dilute solution is heated and water vapor forms, the water vapor flows into the condenser for condensation and recovery. There is a combustion apparatus which supplies heat to the evaporator or evaporators. The waste solids are mixed with oil and subjected to drying by evaporation. If desired, the waste solids may be separated from the oil which may then be recycled. (Sinha-OEIS)
W73-09385

THERMAL REGENERATION ION EXCHANGE PROCESS WITH TRIALLYLAMINE POLYMERS,
ICA Australia Ltd., Melbourne. (Assigne).
For primary bibliographic entry see Field 03A.
W73-09386

PROCESS FOR REMOVING ORGANIC SUBSTANCES FROM WATER, '10A. RENNER.
Ciba-Geigy, Basel (Switzerland). (Assignee).
For primary bibliographic entry see Field 05G.
W73-09387

PROCESS FOR SUBSTANTIAL REMOVAL OF PHOSPHATES FROM WASTEWATERS,
Michigan Chemical Corp., St. Louis. (Assignee).
P. A. Lincoln, and P. G. Delamater.
U. S. Patent No 3,716,484, 5 p, 3 tab, 6 ref; Official Gazette of the United States Patent Office, Vol 907, No 2, p 480, February 13, 1973.

Descriptors: *Patents, *Waste water treatment, Chemical reactions, *Sewage treatment, *Phosphates, *Nutrient removal, *Pollution abatement, Water pollution control, *Water quality control.
Identifiers: Primary treatment, Chemical treatment.

Tabulated results of two experiments are offered as support for the process advancing the substantial removal of phosphates from waste waters and sewage effluents. Soluble phosphates may be removed in the primary treatment stages by means of chemical precipitation utilizing calcium chloride at a pH of 8 - 9.5, or a combination of calcium chloride and calcium hydroxide. (Sinha-OEIS)
W73-09388

WASTE TREATMENT SYSTEM,
Westinghouse Electric Corp., Pittsburgh, Pa. (Assignee).
A. B. Turner.

Identifiers: *Cyanide, *Metal finishing wastes, Chemical treatment.

The apparatus and process provide a system for continuously treating and destroying toxic hexavalent chromium and cyanide constituents in dilute waste streams. A first stage pH electrode and an oxidation-reduction potential cell are mounted in contact with the solution flowing from the first compartment to the second compartment. The ORP meter operates to control the injection of sulfur dioxide or equivalent reagents, such as sodium metabisulfite for admixture with the solution. The pH electrode controls the addition of a mineral acid such as sulfuric acid. A two step reaction occurs under appropriate pH conditions and the solution containing all of the hexavalent chromium in the form of trivalent chromic sulfate overflows into a rectangular duct of a second tank. In the second tank an alkali hydroxide such as caustic raises the pH to 7 or 8. Trivalent chromic sulfate is converted to chromic hydroxide which can be removed by settling. For the removal of cyanide alkaline chlorination is used at a pH of 10.5 or more. The resultant cyanogenchloride (tear gas) is rapidly converted to sodium cyanate by hydrolysis. Sodium cyanate is further heated by oxidation form nitrogen and carbon dioxide, sodium chloride and water. (Sinha-OEIS)
W73-09392

PROCESS FOR SEPARATING WATER FROM POLLUTANTS IN WATER,
M. W. Mar.
U. S. Patent No 3,715,306, 11 p, 4 fig, 2 ref; Official Gazette of the United States Patent Office, Vol 907, No 1, p 177, February 6, 1973.

Descriptors: *Patents, Organic compounds, *Wood preservatives (Pesticides), Creosote, *Pesticides, *Wood wastes, Construction materials, Industrial wastes, *Waste water treatment, *Filtration, Solvents, Pollution abatement.
Identifiers: Aromatic halogen compounds, Aliphatic halogen compounds, Pentachlorophenol, Bunker oil.

This process involves the removal of pollutants such as wood preservatives or insecticides from the effluent of a wood treating facility. Common preservatives are creosote alone, or creosote in conjunction with pentachlorophenol, or a mixture of creosote and bunker oil. In the treatment of wood with preservatives some of the organic compounds in the wood may be dissolved by the preservatives and these may be discharged into the water and act as pollutants. Twenty-five examples are presented to illustrate variations in the removal of these pollutants. The polluted water is treated with a heterocyclic compound containing nitrogen and a solvent selected from the group consisting of an aromatic halogen compound and an aliphatic halogen compound. The treated water is passed through a filter which is so constructed as to allow continuous filtration. (Sinha-OEIS)
W73-09393

METHOD OF CONTROLLING THE TEMPERATURE IN A SEWAGE TREATMENT SYSTEM, AND APPARATUS THEREFOR,
H-P. Hefermehl.
U. S. Patent No 3,715,304, 3 p, 1 fig, 6 ref; Official Gazette of the United States Patent Office, Vol 907, No 1, p 176, February 6, 1973.

Descriptors: *Patents, *Flocculation, *Filtration, *Aerobic treatment, Temperature, *Farm wastes, Precipitation (Chemical), *Waste water treatment, Polymers, *Sewage treatment, Pollution abatement, Biological treatment, Polyelectrolytes.
Identifiers: Multivalent metals, Chemical treatment, Polyacrylic acis, Ionogenic polyelectrolytes.

This method and apparatus are used for the biological conversion of animal waste products. After separating a phase containing the large solid particles from the other which comprises liquid and small particles, the liquid phase is treated so as to cause flocculation and precipitation of small or suspended particles. The clear liquid is then treated for aerobic biological purification. Hot vapor obtained downstream from the drying of the large particle phase passes into the liquid pahse where the temperature rises to about 20 deg - 50 degC to increase the rate of aerobic purification. Precipitation of the suspended particles present after preliminary treatment is achieved in a range of pH of 4.5 to 7 in conjunction with suitable flocculants and precipitants. Suitable flocculants are particularly high polymerized compounds such as homopolymerized and/or copolymerized polyacrylic acids as ionogenic polyelectrolytes and non-ionogenic anionic and cationic polymers. Recommended precipitants are salts of multivalent metals. (Sinha-OEIS)
W73-09394

ION EXCHANGE DEMINERALIZING SYSTEM,
CCI Aerospace Corp., Van Nuys, Calif. (Assignee).
For primary bibliographic entry see Field 03A.
W73-09395

HIGH CAPACITY PORTABLE WATER PURIFIER,
For primary bibliographic entry see Field 05F.
W73-09397

FLUID TREATMENT DEVICES,
S. Nicko.
U. S. Patent No 3,715,032, 6 p 7 fig, 4 ref; Official Gazette of the United States Patent Office, Vol 907, No 1, p 109, February 6, 1973.

Descriptors: *Patents, *Filters, *Filtration, *Water treatment, Equipment, *Water softening, Water quality control, *Valves, Waste water treatment.

The object is to provide a fluid treatment device with a unit that is removable and disposable. The unit may comprise a filter, strainer, water softener, or the like. It is mounted on a valve body and has inlet and outlet ports to be connected to the fluid supply system. An automatic valve mechanism is provided in the valve body to disconnect the discharge port from the inlet port when the treatment unit is removed from the body. (Sinha-OEIS)
W73-09399

SEWAGE DISPOSAL DEVICE,
W. E. Dorn.
U. S. Patent No 3,714,914, 4 p, 4 fig, 7 ref; Official Gazette of the United States Patent Office, Vol 907, No 1, p 77, February 6, 1973.

Descriptors: *Patents, *Sewage treatment, *Filtration, Solids, *Sewage disposal, Incineration, Separation techniques, *Pollution abatement, Water pollution control, Water quality control, *Waste water treatment, Equipment, Waste disposal wells.

An outer concrete shell houses a primary settling chamber, a liquid treating chamber, and an access well. A separation and filtration bucket moves on rails between the primary settling chamber and the equipment holding chamber. A drying and separation plate is located directly below the separation and filtration bucket. There is a cast iron grate at the bottom of the combustion chamber and below the grate is an ash pit. The liquid that has been filtered into the primary settling chamber passes over a baffle wall and into a liquid holding chamber where it undergoes final treatment before being discharged. The solids are incinerated in the combustion chamber. (Sinha-OEIS)
W73-09400

METHOD AND APPARATUS FOR STABILIZING ACCUMULATIONS IN AERATED LAGOONS,
Atara Corp., Montreal (Canada). (assignee)
R. W. Slater.
U. S. Patent No 3,714,036, 3p, 5 fig, 2 ref; Official Gazette of the United States Patent Office, Vol 906, No 5, p 1722, January 30, 1973.

Descriptors: *Patents, *Aeration, *Waste water treatment, Temperature, *Aerated lagoons, *Sewage treatment, *Aerobic treatment, Water pollution control, Water quality control, Pollution abatement, Treatment facilities.

Raw sewage is fed successively through two pools in series when the temperature is below 10 degrees C. Both pools are lightly aerated allowing decomposable material to settle in the first pool. When the temperature rises to above about 15 degrees C, the sewage is fed directly into the second pool while this pool is aerated more heavily than previously. After three to six weeks bypassing and aerating the feeding of raw sewage through the two pools in series is resumed. (Sinha-OEIS)
W73-09403

ACTIVATED SEWAGE PLANT,
Dravo Corp., Pittsburgh, Pa. (assignee)
D. F. Heaney.
U. S. Patent No 3,713,543, 5 p, 5 fig, 9 ref; Official Gazette of the United States Patent Office, Vol 906, No 5, p 1607, January 30, 1973.

Descriptors: *Patents, *Treatment facilities, *Sewage treatment, *Waste water treatment, *Filtration, *Aerobic treatment, Chlorination, Water pollution control, Pollution abatement, Water quality control.

Compact disposal units can be installed in locations or neighborhoods where plants requiring lagoons or usual filters would not be acceptable. This is achieved by the use of an in-depth filter through which the effluent liquid from the settling tank flows to the clearwell in which it is collected and chlorinated. Using two filters, the sewage plant can operate continously, one being used while the other is being backwashed and prepared for reuse. The compact package consists of two concentric tanks, the outer one divided into several compartments. The inner tank serves as a settling zone. The outer comprises in rotation, a contact zone, a stabilization zone, an aerobic digester, the backwash well, and a chlorinating and effluent discharge zone. (Sinha-OEIS)
W73-09405

WATER AERATOR,
For primary bibliographic entry see Field 05G.
W73-09406

APPARATUS FOR TREATING WASTE MATERIALS,
FMC Corp., San Jose, Calif. (assignee)
R. Davidson, and F. F. Sako.
U. S. Patent No 3,713,540, 4 p, 4 fig, 5 ref; Official Gazette of the United States Patent Office, Vol 906, No 5, p 1606, January 30, 1973.

Descriptors: *Patents, *Waste water treatment, *Sewage treatment, *Filtration, Separation techniques, Equipment, Pollution abatement, *Aeration, Water pollution control, Water quality control, Biochemical oxygen demand.

The apparatus may consist of a tank for the sewage, a perforate drum mounted for partial submergence in the sewage and a continuous screen which is guided through the tank in contact with the drum and acts as a filter. Sewage is introduced into the tank and brought into contact with the drum and the clarified liquid is piped out. An air conduit supplies air for reducing the BOD. Control over the volume of liquid in the tank is coordinated with the flow control device which is associated with the outlet pipe drawing off the clarified effluent. When operating with a relatively high solids concentration in the liquid under aeration, a screen with a nominal hole size of 20 microns will produce a filtrate of 40 ppm to 80 ppm of solids. This system can be used in a tank designed for concentrating solids, in an aeration tank of an aerobic digestion system, or in the aeration tank of an activated sludge system. (Sinha-OEIS)
W73-09407

FILTER CARTRIDGE,
Ecodyne Corp., Chicago, Ill. (Assignee)
A. J. Soriente.
U. S. Patent No 3,715,033, 4 p, 2 fig, 2 ref; Official Gazette of the United States Patent Office, Vol 907, No 1, p 109, February 6, 1973.

Descriptors: *Patents, *Liquid wastes, *Waste water treatment, *Filters, Filtration, Pollution abatement, Water pollution control, Water quality control, *Ion exchange, Resins.

This invention provides a precoat filter cartridge comprising a rigid tubular core with holes forming an open area in the range 3-9 percent of the total surface area of the core element. A layer of coarse screen in the size 6-10 mesh is wrapped around it. A layer of fine mesh screen approximately 50 x 250 mesh is wrapped around the coarse screen. By uniformly reducing the percent of open area of the surface of the core element, the resulting increase in pressure drop causes the precoat slurry to flow up towards the top of the filter cartridge thereby resulting in a uniform distribution of the precoat medium. The precoat is a layer of ion exchange resin particles in the size range of 60 to 400 mesh. (Sinha-OEIS)
W73-09408

THE ECONOMICS OF INDUSTRIAL WASTE TREATMENT,
Environmental Control Consultancy Services Ltd., London (England).
D. Anderson.
The Chemical Engineer, No 267, p 422-425, November, 1972. 3 tab, 6 ref.

Descriptors: *Waste disposal, *Effluents, *Industrial wastes, *Waste water treatment, Optimization.
Identifiers: Plant selection and design, Waste parameters, In-plant treatment, Sewer disposal.

Problems of industrial waste management differ from those of sewage treatment because they are more complex and their solution may be more flexibly approached. Milk, yeast and phenolic wastes were used as examples of management problems. It was found that the nature, composition, volume, manner, location and timing of industrial waste production are all controllable by the industry, which can then optimize the problem in terms of production economics. Examples illustrate the options open to industry and the need for carefully evaluating the total problem through a multi-discipline approach. Specifically, two such options were considered: (1) discharge of industrial wastes to sewers, and (2) in-plant treatment. Based on United Kingdom data for 1970 and 1972, typical costs of each option were estimated using standard methods. For sewer disposal, oxygen demand characteristics and volume flow were the typical waste parameters having the greatest influence on disposal costs. Representative problems connected with in-plant treatment were discussed since there are no typical effluents and therefore no typical costs. (Weaver-Wisconsin)
W73-09410

ANIMAL WASTES, THEIR HEALTH HAZARDS AND TREATMENT,
Applied Scientific Research Corp. of Thailand, Bangkok.
S. M. A. Durrani.
Thai J Agric Sci. Vol 4, No 4, p 265-270. 1971.
Identifiers: *Farm wastes, *Waste treatment, Animals, Fungi, Hazards, Public health, Methods, Parasites Salmonella.

Animal wastes are a source of many diseases communicable to man including salmonellosis and other bacterial infections, fungus and parasitic infections. Potentially dangerous chemicals are excreted through animal feces and urine. A treatment process must check water pollution caused by waste disposal and must be economical to install. Anaerobic digestion, anaerobic and aerobic lagoons and incinerators are discussed.—Copyright 1973, Biological Abstracts, Inc.
W73-09422

RECYCLABLE COAGULANTS LOOK PROMISING FOR DRINKING WATER TREATMENT,
For primary bibliographic entry see Field 05F.
W73-09478

CHEMICAL TREATMENT OF PLATING WASTE FOR ELIMINATION OF CHROMIUM, NICKEL AND METAL IONS,
Lancy Labs., Hamden, Conn.
J. J. Martin, Jr.
Journal New England Water Pollution Control Association, Vol 6, No 1, p 53-72, June 1972. 8 fig, 4 tab.

Descriptors: *Heavy metals, *Chromium, *Nickel, *Waste water treatment, *Industrial wastes, Metals, Recycling, Economics, Water pollution, Settling, Hydrogen ion concentration, Cehmical precipitation, Neutralization.
Identifiers: *Electroplating industry.

It has been found that chemical rinsing of dragout contaminates and chemical batch treatment of spent processing solutions is a practical treatment solution for electroplating installations. Equipment as well as control of treatment processes is of a type familiar to electroplating and plant maintenance departments and does not require special maintenance skills or procedures. Cyanide is treated with sodium hydroxide and sodium hypochlorite; nickel is precipitated as the hydroxide or carbonate; chromium is reduced by sodium bisulfite and precipitated as chromium hydroxide; copper and zinc solutions are neutralized and hydrazine causes the metallic hydroxide to be precipitated. Treatment costs have not materially affected the competitive position of the company in marketing. Chemical rinsing plus water reuse reduces water requirements by three-quarters or from 90 gpm to 20 gpm. A bleed-off of about one-quarter of the total water keeps the dissolved salts low enough to provide adequate rinsing in selected areas. Chemical rinsing and water treatment of spent solutions provides an effluent that can be safely discharged to the stream. (Oleszkiewicz-Vanderbilt)
W73-09493

POTATO STARCH FACTORY WASTE EFFLUENTS: II. DEVELOPMENT OF A PROCESS FOR RECOVERY OF AMINO ACIDS, PROTEIN AND POTASSIUM,
Agricultural Research Service, Philadelphia, Pa. Eastern Marketing and Nutrition Research Div.
E. G. Heisler, J. Siciliano, and S. Krulick.
J Sci Food Agric. Vol 23, No 6, p 745-762, 1972. Illus.
Identifiers: *Amino-Acids, Effluents, Fertilizers, Oxygen, *Potassium, Potato wastes, Protein, Starch, *Waste water treatment, Industrial wastes.

Problems associated with the secondary waste effluents from potato starch factories and the basic studies leading to a proposed ion-exchange process for the treatment of such wastes to decrease the biochemical oxygen demand (b.o.d.) by removal of protein and amino acids are

Paper presented at 63rd Annual Meeting, The American Institute of Chemical Engineers, November 29 - December 3, 1970. 21 p, 18 ref. EPA Grant 17030 DGQ.

Descriptors: *Biodegradation, *Waste water treatment, *Adsorption, Kinetic, Organic compounds. Identifiers: Biological kinetics, Biological uptake, Slime layer.

A fixed biologically active slime layer was used to determine the kinetics of removal ofremoval of organic substrates. Uptake by slime surfaces acclimated to single substrates appeared to be mass transfer limited at concentrations below 100 mg/l and reaction limited at higher concentrations. Comparison of the rates of carbon uptake versus glucose uptake shows that glucose is almost consumed. By contrast, only a fraction of the degraded starch is consumed, less than 40%. These observations are consistent with the concept of a surface which has catalytic activity for both hydrolysis and carbon uptake and when the active sites are independent and physically separated.
W73-09688

USE OF FILM-FLOW REACTORS FOR STUDYING MICROBIOLOGICAL KINETICS,
Minnesota Univ., Minneapolis. Dept. of Civil and Mineral Engineering.
W. J. Maier.
Applied Microbiology, Vol 16, No 7, p 1095-1097, July 1968. 2 fig. EPA Grant 17030 DGQ.

Descriptors: Biodegradation, *Waste water treatment, Adsorption, Kinetics, Trickling filters. Identifiers: *Biological kinetics, Biological uptake, *Film flow reactors, Slime layers.

The use of a film-flow reactor for studying the kinetics of substrate removal is described. The reactor provides a means of maintaining a uniform layer of biologically-active slime which is stationary while the feed solution is made to flow over the slime surface in a thin film. The study was designed to evaluate process limitations in the trickling filter process which is designed for the biological removal of waste matter from dilute solutions.
W73-09689

SIMULATION OF THE TRICKLING FILTER PROCESS,
Minnesota Univ., Minneapolis. Dept. of Civil and Mineral Engineering.
W. J. Maier, V. C. Behn, and C. D. Gates.
Journal of the Sanitary Engineering Division, Proceedings of the American Society of Civil Engineers, Vol 93, No SA4, p 91-112, August 1967. 7 fig, 10 tab, 32 ref. EPA Grant 17030 DGQ.

Descriptors: *Trickling filters, *Biological treatment, *Waste water treatment, Kinetics. Identifiers: Biological kinetics, Slime layers.

A simple physical model consisting of a flat surface covered with biologically-active slime is used to simulate trickling filter operations. Process variable studies were conducted using glucose as the sole source of carbon in a mixed salt solution. Liquid feed rate has a marked effect on the rate of glucose use at low liquid rates. The results show that mass transfer in the liquid film limits nutrient removal; the top surface of the slime is the major site of metabolic activity.
W73-09691

REGULATIONS FOR SINGLE RESIDENCE WASTE WATER TREATMENT DEVICES PRESENTED AT HEARING ON MARCH 21, 1973.
Missouri Clean Water Commission, Jefferson City.
For primary bibliographic entry see Field 06E.

W73-09695

5E. Ultimate Disposal of Wastes

DREDGE SPOIL DISPOSAL IN RHODE ISLAND SOUND,
Rhode Island Univ., Kingston. Dept. of Oceanography; and Rhode Island Univ., Kingston. Dept. of Zoology.
S. B. Saila, S. D. Pratt, and T. T. Polgar.
Available from NTIS, Springfield, Va. 22151 as COM-72-11537, Price $3.00 printed copy; $0.95 microfiche. Marine Technical Report No 2, 1972. 48 p, 23 fig, 4 tab, 96 ref. Sea Grant URI 98-20-6012.

Descriptors: *Dredging, *Sludge disposal, *Sounds, *Rhode Island, Environmental effects, Sediment transport, Data collections, Bathymetry, Benthic fauna, Channel improvement, Waste disposal, Sediments, Physical properties. Identifiers: *Rhode Island Sound, Dredging disposal, Sea dumping.

Between December 1967 and September 1970 a total of 8.2 million cubic yards of dredge spoil from the Providence River was deposited on an offshore site in Rhode Island Sound. The spoil was carried to sea in scows of 2,000- and 3,000-cubic yard capacity and discharged within a 1-square mile dumping area approximately 4 miles south of Newport, Rhode Island. The depths were between 96 and 106 feet before dumping began. An investigation carried out between March and September 1970 was designed to update physical observations. Although the emphasis of this study was on aspects of spoil disposal which relate to the management of a specific offshore area, information of general ecological importance was obtained on the composition of the benthic assemblages of Rhode Island Sound and on the colonization of new sea bottom by benthic animals. The disposition of spoil was determined by two bathymetric surveys and two series of sediment samples. Patches of spoil were found as much as a mile outside the dump site toward the northwest and the southwest. Over 30 benthic species were found in the lower Providence River indicating a relatively low level of pollution in that part of the spoil source area. (Woodard-USGS)
W73-09142

OIL POLLUTION OF THE SEA,
For primary bibliographic entry see Field 05B.
W73-09222

SOUND WASTE MANAGEMENT CUTS OFFSHORE COSTS,
Weston (Roy F.), Inc., West Chester, Pa. Marine Waste Management Div.
For primary bibliographic entry see Field 05G.
W73-09224

DEEP WELL INJECTION OF LIQUID WASTS,
Dames and Moore, New York.
For primary bibliographic entry see Field 05G.
W73-09247

DEEP WELL DISPOSAL GAINING FAVOR.
For primary bibliographic entry see Field 05G.
W73-09255

HYDRAULIC FRACTURING AND THE EXTRACTION OF MINERALS THROUGH WELLS,
Wisconsin Univ., Madison. Dept. of Minerals and Metals Engineering.
For primary bibliographic entry see Field 08E.
W73-09288

INITIATION AND EXTENSION OF HYDRAULIC FRACTURES IN ROCKS,
Minnesota Univ., Minneapolis.
For primary bibliographic entry see Field 08E.
W73-09289

REAL STRESSES AROUND BOREHOLES,
Wisconsin Univ., Madison.
For primary bibliographic entry see Field 08E.
W73-09290

STATIC FRACTURING OF ROCK AS A USEFUL TOOL IN MINING AND ROCK MECHANICS,
Wisconsin Univ., Madison.
For primary bibliographic entry see Field 08E.
W73-09291

MICROBIOLOGICAL ASPECTS OF GROUND-WATER RECHARGE--INJECTION OF PURIFIED UNCHLORINATED SEWAGE EFFLUENT AT BAY PARK, LONG ISLAND, NEW YORK,
Geological Survey, Menlo Park, Calif.
For primary bibliographic entry see Field 05B.
W73-09292

SEWAGE DISPOSAL DEVICE,
For primary bibliographic entry see Field 05D.
W73-09400

SUBSURFACE WASTEWATER INJECTION, FLORIDA,
Geological Survey, Tallahassee, Fla.
M. I. Kaufman.
ASCE Proceedings, Journal of the Irrigation and Drainage Division, Vol 99, No IR1, Paper 9598, p 53-70, March 1973. 10 fig, 4 tab, 20 ref.

Descriptors: *Waste disposal wells, *Florida, *Injection wells, *Underground waste disposal, *Hydrogeology, Path of pollutants, Permeability, Aquifer characteristics, Water quality, Aquifers, Brines, Environmental effects, Transmissivity, Limestones.

Deep-well injection of liquid waste is being evaluated in Florida as a management option to help alleviate environmental deterioration of fresh and estuarine waters and increasing problems of waste disposal. Extensive areas of Florida are underlain by permeable saline-aquifer systems that are separated from overlying freshwater aquifers by low-permeability confining materials consisting of clay, evaporites, or dense carbonate rocks. These deep saline zones are the subject of current research designed to assess the environmental impact of subsurface waste storage. Three active deep-well disposal systems currently exist, and two additional systems are under construction. Industrial and municipal wastes are injected into subsurface environments of different hydrogeologic characteristics at depths ranging from 1,400 ft to 3,000 ft. Transmissivities of the receiving carbonate aquifers range from 6,500 gpd per ft in northwest Florida to more than 17,000,000 gpd per ft for the cavernous Boulder Zone in southern peninsular Florida. (Knapp-USGS)
W73-09456

MODEL STUDIES OF OUTFALL SYSTEMS FOR DESALTING PLANTS (PART III - NUMERICAL SIMULATION AND DESIGN CONSIDERATIONS),
Dow Chemical Co., Freeport, Tex.
M. A. Zeitoun, W. F. McIlhenny, R. O. Reid, and T. M. Mitchell.
Available from the National Technical Information Service as PB-215 171, $3.00 in paper copy, $0.95 in microfiche. Office of Saline Water Research and Development Progress Report, No 804, December 1972. 79 p, 24 fig, 10 tab, 25 ref. 14-01-0001-2169, Amdt. 2.

Descriptors: *Brine disposal, *Model studies, Desalination plants, Effluents, Outlets, Waste disposal.
Identifiers: *Marine ecology, *Effluent dispersion, Diffusers, Model flume, Outfall, Designs.

Although at coastal locations, the saline dense effluent from a desalting plant is potentially harmful to the marine organisms, the environmental effects can be minimized or eliminated if the effluent is properly mixed with and diluted by the receiving water. Conceptual designs of outfall systems were previously developed on the basis of laboratory tests and are presented in OSW R and D Progress Report No. 550. Experimental verification of the conceptual designs by the U.S. Army Corps of Engineers was published as Parts I and II of this report (R and D Nos. 714 and 736). The data obtained in the flume study were used to calibrate several existing models for the jet regime. The model chosen gives a reasonable simulation of the geometry of the jet axis and the dilution at the peak of the jet, but was found to oversimulate the lateral spreading factor. Application of equations to the design of diffusers is presented. Numerical simulation of the dispersion of a dense effluent discharge into a homogeneous stream is included. (OSW)
W73-09628

SHIPBOARD WASTE TREATMENT SYSTEMS.
For primary bibliographic entry see Field 05D.
W73-09654

5F. Water Treatment and Quality Alteration

HIGH CAPACITY PORTABLE WATER PURIFIER,
G. H. Teeple, Jr., and J. W. Welsh.
U. S. Patent No 3,715,035, 2 p, 4 fig, 4 ref; Official Gazette of the United States Patent Office, Vol 907, No 1, p 110, February 6, 1973.

Descriptors: *Patents, *Water purification, *Filtration, *Resins, *Activated carbon, Potable water, Sediments, Suspended solids, Bacteria, Metals, Salts, Pollution abatement, Separation techniques, Water treatment.

This portable water purifier comprises a two container filter assembly. The containers are arranged one above the other and are separated by a supporting structure. The upper container includes a flexible bag that is adapted to receive and hold liquid to be purified. The bag extends below the level of the lower container inlet. The bag interior communicates with that inlet so that the liquid in the bag may flow to the inlet in response to the elevation of the bag. The bag may be suspended to extend closely about a substantially rigid duct forming portion of the lower container thus forming a sediment trap. Activated charcoal and deionizing resin may be held within compartments of the duct. A particle entrapping porous filter media may separate the charcoal and resin compartments and may be located at the entrance and exit of the rigid duct whereby additional filtering as well as flow regulation may be achieved. The second flexible bag may be suspended by the apparatus to receive purified water discharged from the duct. (Sinha-OEIS)
W73-09397

FLASH EVAPORATOR,
For primary bibliographic entry see Field 03A.
W73-09404

TION IN THE PREVENTION OF COASTAL POLLUTION (UTILISATION DE CARTES-FLOTTEURS POUR L'ETUDE DES DERIVES DE SURFACE ET APPLICATION A LA PREVISION DES P OLLUTIONS CATIERES),
Service Hydrographique de la Marine, Paris (France).
For primary bibliographic entry see Field 05B.
W73-09210

THE POLLUTION OF THE COASTS AND EFFLUENTS IN THE SEA (LA POLLUTION DES COTES ET LES REJETS EN MER),
J. Garancher.
Houille Blanche, Vol 26, No 8, p 779-783, 1971.

Descriptors: *Coasts, *Sewerage, *Sewage disposal, *Outlets, *Water pollution control, Disposal, Dispersion, Control systems, Regulation, Human population, Community development, Water quality control.

Sewerage control measures in coastal towns are insufficient to prevent sea pollution. Appropriate regulations for better control are considered and it is concluded that control of effluent discharge into the sea alone is not sufficient for good coastal water quality. The theoretical aspects of discharge into the sea and the prediction of effluent distribution after discharge are discussed. Measures recently put into effect are evaluated. Consideration is given to the possibility of varying the method of disposal to accomodate the needs of the different coastal populations. (Ensign-PAI)
W73-09215

A SURFACE-ACTIVE CHEMICAL SYSTEM FOR CONTROLLING AND RECOVERING SPILLED OIL FROM THE OCEAN,
Shell Pipe Line Corp., Houston, Tex.
E. A. Milz, and J. P. Fraser.
Journal of Petroleum Technology, p 255-262, March 1972. 9 fig, 6 tab, 5 ref.

Descriptors: *Oil spills, *Chemical wastes, *Surface tension, Water pollution treatment, Water pollution control.
Identifiers: *Oil recovery, *Containment, *Polyurethane foam.

Results of research conducted to devise a system for oil-slick control and oil spill recovery that is independent of the limitations that weather imposes on mechanical containment and recovery devices are presented. The system devised involves the use of a surface-tension modifier. The introduction of a chemical to the water which has a greater spreading force than that of the oil inhibits the spread of the slick. Weather conditions do not significantly alter the effectiveness of this kind of treatment. For oil recovery, the use of polyurethane foam is suggested. The foam has excellent oil sorption properties and it can be generated at the site of the spill. The foam can then be removed through the use of large mesh fish nets; simple squeezing devices will wring the oil from the foam. Tests conducted on this kind of use of chemicals indicate that their ecological effect is less harmful than the effect of the oils they eliminate. (Ensign-PAI)
W73-09220

THE NAVY AND THE EFFLUENT SOCIETY,
T. R. Colemon, and B. M. Kopec.
Sea Power, p 18-21, September 1972. 4 fig.

Descriptors: *Oceans, Ships, *Waste disposal, *Waste storage, *Waste treatment, *Oil spills, Water pollution treatment, Water pollution control, Costs, Design criteria.
Identifiers: *Pollution control programs, Dumping.

The role of the U.S. Navy in the fight against ocean pollution is depicted. Their anti-pollution program is a two-headed one. One facet involves holding refuse on board while ships are in territorial waters (to be dumped later at sea or disposed of on land); the other approach involves developing methods of waste treatment and disposal aboard ship. Cost and design problems associated with the development of shipboard waste disposal systems are outlined. The problem of accidental oil spills and clean-up projects is also discussed. Results of experiments which the Navy has conducted to test the effectiveness of oil barriers, skimmers, and a new chemical treatment called Sea Beads are reported. (Ensign-PAI)
W73-09223

SOUND WASTE MANAGEMENT CUTS OFFSHORE COSTS,
Weston (Roy F.), Inc., West Chester, Pa. Marine Waste Management Div.
M. W. Hooper.
World Oil, p 41-44, September, 1972. 2 fig, 8 ref.

Descriptors: *Governments, *Regulations, *Legislation, Ships, *Waste disposal, *Oceans, Channels, Oil industry, Water quality standards, Water pollution control.
Identifiers: Offshore operators, Platform regulations.

Various state, federal and foreign regulations regulating the disposal of solid wastes into waterways and oceans are reviewed. Jurisdictional conflicts and differences in standards among the regulatory agencies involved are pointed out. The discrepencies between standards for platform operations and seagoing vessels are discussed. For instance, a crewboat located 20 miles offshore may discharge raw sewage into the Gulf of Mexico; a production platform in the same field must process its sewage. Existing sewage treatment systems for vessels and platforms are described. The report concludes that further research ought to be conducted to determine actual and potential effect of solid and liquid waste discharges on marine environment. (Ensign-PAI)
W73-09224

RESCUE OPERATIONS FOR OILED SEABIRDS,
Newcastle-upon-Tyne Univ. (England). Dept. of Zoology.
R. B. Clark, and J. P. Croxall.
Marine Pollution Bulletin, Vol 3, No 8, p 123-127, August 1972. 1 fig, 2 tab, 9 ref.

Descriptors: Oil wastes, *Oil spills, *Birds, *Cleaning, *Detergents, Water pollution control, Environmental effects, Treatment facilities.
Identifiers: *Sea birds, *Water-repellency.

New methods for rescuing sea birds oiled by accidental tanker spills is described. These cleaning methods protect the water repellency of the birds' feathers. However, in some cases, the methods devised are shown to be either impermanent or prohibitively costly. The best cleaning agents developed are believed to be the simple liquid detergents found in most homes. Used at a concentration of one percent and a temperature from 40 deg to 45 deg the detergents were reliable and efficient at removing contaminating oil, and also easily washed off the birds without leaving behind harmful residues. Birds cleaned with detergent can be released within 2-3 weeks of capture. Restoring sea birds damaged by oil entails some essential facilities of a permanent establishment, so it is suggested that such centers be established in areas where they can serve large sectors of coast. (Ensign-PAI)
W73-09225

SEA FISHERIES COMMITTEES,
Manchester Univ. (England). Faculty of Law.

J. McLoughlin.
Marine Pollution Bulletin, Vol 3, No 8, p 118-122, August 1972.

Descriptors: *Foreign countries, *Regulations, *Water policy, *Water quality control, Coasts, Estuaries, Water pollution control, Legal aspects, History, Jurisdiction.
Identifiers: *Great Britain, *Pollution control authorities.

The role of the Sea Fisheries Committees of England and Wales as pollution control authorities is examined. A brief history of the origin and purposes of these committees is presented. Since they were primarily established to control sea fishing within fishery limits, the Sea Fisheries Committees have only a limited jurisdiction over pollution. However, the critical need for ecological protection of coastal areas has led some of the committees to interpret their bylaws freely and act as regulatory agencies for controlling pollution. The major sources of pollution in Fisheries Areas are direct discharges to the sea by local authorities and/or industry, estuarial discharges, and pollution flowing into the sea from rivers. The need for greater dissemination of information to the public on polluting discharges is stressed. Under existing British law there can be no actions taken against polluters unless they are generally known as public nuisances. (Ensign-PAI)
W73-09226

ECOLOGICAL EFFECTS OF DREDGED BORROW PITS,
Research and Ecology Projects, Inc., San Francisco, Calif.
For primary bibliographic entry see Field 05C.
W73-09227

WHAT HAPPENS WHEN LNG SPILLS,
Phillips Petroleum Co., Bartlesville, Okla. Research Div.
W. W. Crouch, and J. C. Hillyer.
Chemtech, p 210-215, April 1972. 5 fig, 14 ref.

Descriptors: *Natural gas, Chemical properties, Ships, Design standards, *Safety factors, *Burning, Water pollution control.
Identifiers: *Liquified natural gas, *Tankers, *Accidental spills, Containment, Recovery.

Plans for transportaing liquified natural gas (LNG) from producing countries to gas-short nations are in the making. Since the transportation of LNG could lead to accidents and spillage it is important to know how the gas behaves during and after a spill. Modern tankers carrying LNG will be double-hull membrane vessles which can withstand a severe collision without rupturing the inner LNG-containing wall. Furthermore, the ships will employ extensive fire fighting equipment for the engine, pump rooms, upper deck and pipe lines. If LNG should be accidentally spilled on water it will evaporate at a steady rate; with large spills this may take several minutes. The principle hazard in a large LNG spill lies in the formation of a combustible vapor that will envelop the spill, drift downwind, and remain near the surface until it is diluted and warmed. If this vapor should come in contact with an ignition source a fire would occur, propagating backwards through the combustible region. (Ensign-PAI)
W73-09228

CALIFORNIA TIGHTENS CLAMP ON EFFLUENTS,
H. M. Wilson.
Oil and Gas Journal, July 24, 1972, p 22-23. 1 tab.

Descriptors: *California, *Water quality control, *Legislation, *Industrial wastes, Water quality standards, Governments, Costs, *Regulation.
Identifiers: *Compliance time tables.

New water-effluent regulations released by the California State Water Resources Control Board are reviewed. The controls apply mainly to water discharged into the ocean by industries and municipalities. Surveys conducted by the industries affected indicate that none of them can presently meet the required standards in all categories. Compliance time tables will be established by regional boards. Some companies fear that the total cost of meeting the new regulations will force them into dire economic straits. The new limits for chromium discharge, for example, will be particularly difficult for most companies to meet. (Ensign-PAI)
W73-09230

POLLUTION CONTROL ASPECTS OF THE BAY MARCHAND FIRE,
W. L. Berry.
Journal of Petroleum Technology, p 241-249, March 1972. 8 fig, 1 tab, 8 ref.

Descriptors: *Oil spills, *Burning, Oil fields, Water pollution treatment, Beaches, Water pollution control, Pollution abatement, *Louisiana.
Identifiers: *Bay Marchand fire, *Oil recovery, *Skimmers, *Oil scoops.

Various methods used to contain and recover the unburned fraction of oil from the Bay Marchand fire are described. An open-seas recovery fleet employed a skimmer system to recover unburned oil as it was emitted from Platform 'B'. Skimming operations were conducted 24 hrs. a day, sea and weather conditions permitting. Oil Scoop systems employed are discussed in some detail. Measures taken to protect the shoreline and to remove oil from beaches and shallow water are reviewed, along with methods for waterfowl protection. The use of chemical dispersants like Shell Oil's Oil Herder is mentioned. (Ensign-PAI)
W73-09231

SCRAMBLE,
J. R. Burkhardt.
Petroleum Today, No 1, p 20-23, 1972. 3 fig.

Descriptors: *Oil spills, *Massachusetts, Equipment.
Identifiers: *Citizen groups, *Oil clean-up, *Oil recovery, Ocean 80, Boston Harbor, Chelsea Creek.

Efforts by the Coast Guard, oil companies and concerned citizens in the Boston area have helped improve Boston Harbor and surrounding coastal waters. Oil clean-up and recovery operations conducted when the barge Ocean 80 ran aground in Chelsea Creek are described. The Boston cooperative team provides strong evidence that oil spill disasters can be avoided through the use of good equipment, organization, manpower and experience. (Ensign-PAI)
W73-09232

EFFECT OF STEEPNESS OF REGULAR AND IRREGULAR WAVES ON AN OIL CONTAINMENT BOOM,
Shell Oil Co., Houston, Tex.
G. E. Walker, Jr.
Journal of Petroleum Technology, p 1007-1013, August 1972. 11 fig, 7 ref.

Descriptors: *Oil spills, Model studies, *On-site investigations, Equipment, *Waves, *Mathematical models.
Identifiers: *Oil containment booms, Splashover.

The effectiveness of oil containment booms in the presence of waves was evaluated. Commercially available booms were experimentally tested in wave tanks, on site in the Gulf of Mexico, and with scale models. A mathematical modeling of boom response in waves was attempted. The tests conducted with production type and scale model

LIMITATIONS TO LENGTH OF CONTAINED OIL SLICKS,
Technical Univ. of Denmark, Copenhagen. Inst. of Hydrodynamics and Hydraulic Engineering.
For primary bibliographic entry see Field 05B.
W73-09270

A TRANSPORT MECHANISM IN HOLLOW NYLON FIBER REVERSE OSMOSIS MEMBRANES FOR THE REMOVAL OF DDT AND ALDRIN FROM WATER,
Tennessee State Univ., Nashville. Dept. of Civil Engineering.
For primary bibliographic entry see Field 05D.
W73-09302

LIQUID SURFACE SWEEPING APPARATUS,
JBF Scientific Corp., Burlington, Mass. (Assignee).
R. A. Bianchi.
U. S. Patent No 3,716,142, 4 p, 11 fig, 8 ref; Official Gazette of the United States Patent Office, Vol 907, No 2, p 400, February 13, 1973.

Descriptors: *Patents, *Flotsam, *Oil spills, *Oil pollution, Skimming, *Pollution abatement, Water pollution control, Water quality control, Separation techniques, Equipment.

The sweeping apparatus consists of a pair of triangular shaped floats or pontoons secured to the bow of the skimmer. The floats when viewed from above have the shape of a right triangle with the hypotenuse being the sweeping face. Efficient sweeping is achieved because the curved underbody of the pontoons causes water flowing under them to be substantially free of turbulence. Since the floats are buoyant, they will follow wave action, lifting with the waves and thus will contain the material floating on the wave surface. The two arms of the float are located with their planar surfaces opposing each other and converging in the direction of water flow. (Sinha-OEIS)
W73-09384

PROCESS FOR REMOVING ORGANIC SUBSTANCES FROM WATER, '10A. RENNER.
Ciba-Geigy, Basel (Switzerland). (Assignee).

U. S. Patent No 3,716,483, 3 p, 4 ref; Official Gazette of the United States Patent Office, vol 907, No 2, p 480, February 13, 1973.

Descriptors: *Patents, *Industrial wastes, Chemical wastes, *Organic compounds, Waste water treatment, Paints, *Oil wastes, Oily water, Dyes, Resins, *Polymers, *Pollution abatement, Water quality control, Water pollution control, Oil spills, Oil pollution.
Identifiers: Melamine-formaldehyde, Urea-formaldehyde.

Contaminated water is brought into contact with a highly disperse, solid, water-insoluble organic polymer of average molecular weight greater than 1000 and a specific surface area greater than 5 m2/g and the polymer charged with the impurity is separated from the water. The polymer can be brought into contact with the contaminated water by stirring or vibration. The polymer is referred to as an absorbent and may be an insoluble melamine-formaldehyde resin or a urea-formaldehyde resin. Five examples are given. The polymer, absorbent may be used for removing dissolved, emulsified or suspended organic substances such as fat, oil, or dyestuff. To remove oil film from a water surface, the polymer may be sprinkled or sprayed as uniformly as possible onto the liquid surface. (Sinha-OEIS)
W73-09387

DEVICE FOR REMOVING OIL SLICKS,
A. Ivanoff.
U. S. Patent No 3,715,034, 6 p, 23 fig, 9 ref; Official Gazette of the United States Patent Office, Vol 907, No 1, p 110, February 6, 1973.

Descriptors: *Patents, *Oil spills, *Oil pollution, Skimming, *Pollution abatement, Water pollution control, Water quality control, Separation techniques.
Identifiers: *Water craft, Skimmers.

A flat bottomed water craft is constructed to have a broad squared-off bow section that slants sternward from the deck toward the water line. Below the water line in or near the bow section, it has a transverse elongate opening for the ingress of the oil flowing as a continuous layer above the water. The craft has openings preferably near the stern for discharging water which may have entered with the scooped-up oil. Conduits may be used to transfer the collected oil to a mother ship. (Sinha-OEIS)
W73-09398

WATER AERATOR,
E. P. Aghnides.
U. S. Patent No 3,712,548, 5 p, 14 fig, 6 ref; Official Gazette of the United States Patent Office, Vol 906, No 4, p 1333, January 23, 1973.

Descriptors: *Patents, *Aeration, Equipment, *Waste water treatment, *Aerobic treatment, *Sewage treatment, *Pollution abatement, Water pollution control, Water quality control.
Identifiers: *Aerators.

This aerator has a relatively thin diaphragm with square cross-section holes. There is a bridge arrangement over the holes on the upstream side to form either one or two water entrances to each hole. Where there are two entrances they may be of different size to permit the issuing turbulent jet of water to diverge or converge. The casing has an annular enlargement at the downstream end. (Sinha-OEIS)
W73-09406

POLLUTION CONTROL--USES OF CORRECTIVE TAXES RECONSIDERED,
Arizona Univ., Tucson. Dept. of Economics.
J. T. Wenders.
Natural Resources Journal, Vol 12, No 1, p 76-82, January, 1972.

Descriptors: *Water pollution control, *Taxes, *Pollution taxes (Charges), Pollution abatement, Economic efficiency.
Identifiers: *Corrective tax, Equiproportionate abatement rule.

Professor Colin Wright (Natural Resources Journal, 1969) argued that in certain situations the equiproportionate abatement rule is superior to a single corrective tax as a pollution control device. The present author claims that Wright's analysis is incorrect and that the optimal level of pollution can be achieved by a single, constant per unit tax on pollutants emitted. Such a tax insures pollution abatement according to the marginal rule for efficiency, which requires that marginal cost of abatement be the same for each polluter at each successive level of total abatement. Furthermore, a corrective tax would always be superior to the equiproportionate abatement rule, which was found to be more expensive for the same amount of abatement. Administrative costs in the two cases differed insignificantly and thus were irrelevant to the selection of a pollution control device. Two arguments for corrective taxes were presented: (a) they provide an efficient source of revenue for the pollution control board, allowing pollution control programs a chance for quick implementation; and (b) the expected rate of improvement in pollution abatement technology is higher where they are used because the innovating firm experiences a reduction in both tax liability and direct cost of abatement. (Weaver-Wisconsin)

W73-09411

FLUORIDE AVAILABILITY AND THE PREVALENCE OF CARIES-FREE NAVAL RECRUITS: A PRELIMINARY TEN YEAR REPORT,
Naval Dental Research Inst., Great Lakes, Ill.
H. J. Keene, G. H. Rovelstad, S. Hoffman, and W. R. Shiller.
Arch Oral Biol. Vol 16, No 3, p 343-346. 1971.
Identifiers: Epidemiology, *Fluoridation, Fluorides, *Dental health, *Caries, Public health, Potable water.

Epidemiologic and geographic data on caries-free recruits at Great Lakes, Illinois, were collected from 1960-1969. The prevalence of caries-free recruits, as related to the availability of naturally occurring fluoride in public water supplies, is evaluated by correlation analysis. A caries-free recruit was one without clinical or radiographic evidence of dental caries upon entrance into the Navy. A home-state determination was made, based on residence during the major years of tooth development (from birth to 12 yr). The number of caries-free men/1000 recruits was calculated, based on the geographic distribution of a 10% sample of 50,000 men.--Copyright 1972, Biological Abstracts, Inc.
W73-09527

BIODEGRADATION OF ANIMAL WASTE BY LUMBRICUS TERRESTRIS,
Georgia Univ., Athens. Dept. of Dairy Science.
O. T. Fosgate, and M. R. Barr.
J Dairy Sci. Vol 55, No 6, p 870-872. 1972.
Identifiers: *Animal wastes, *Degradation (Biological), Fecal matter, Lumbricus-terrestris, Meal, Soils, Waste recycling, *Earthworms.

The possibility of recycling animal waste through the common earthworm, L. terrestris was studied. Earthworms were raised in beds and fed only raw feces and water with lime added as a buffer. The conversion of kilograms of fecal dry matter to kilograms of live earthworms was 2:1. The excretion (castings) of the earthworm was a loose, friable humus type of soil containing 3.0% N. Earthworm meal dry matter analyzed 58% protein and 2.8% fat and proved to be very palatable when fed to domestic cats. Worm dirt was equal to greenhouse potting soil for the production of flowering plants. An added advantage is that the worm dirt weighs only about 50% as much as normal potting soil.--Copyright 1972, Biological Abstracts, Inc.
W73-09528

GEOLOGY AND LAND RECLAMATION,
Kent State Univ., Ohio. Dept. of Geology.
For primary bibliographic entry see Field 04A.
W73-09533

STUDIES ON PRODUCTION AND FOOD CONSUMPTION BY THE LARVAL SIMULIIDAE (DIPTERA) OF A CHALK STREAM,
Freshwater Biological Association, Wareham (England). River Lab.
M. Ladle, J. A. B. Bass, and W. R. Jenkins.
Hydrobiologia. Vol 29, No 3, p 429-448. 1972. Illus.
Identifiers: Chalk, *Diptera, Food, *Larval stages, Production, Ranunculus-D, Simuliidae, Simulium-equinum, Simulium-ornatum, Stream, *United Kingdom.

Large populations of Simulium ornatum and S. equinum inhabit Ranunclus plants in the Bere Stream at Bere Heath. A weekly sampling program obtained both larvae and pupae. The results of weekly sampling indicated 4 annual generations of S. ornatum and 3 of S. equinum. Quantitative samples of larvae, obtained at least 3 times in each generation gave estimates of production ranging from 0.26 g to 2.75 g dry weight m-2 per generation of larvae of S. equinum ornatum. Retention times of food by larvae, under field conditions were in the order of 20 to 30 min. Following determination of the weight of the gut contents for each size class of larvae, the activity of feeding larvae was calculated to achieve complete theoretical clearance of suspended material in a distance of 0.6 km, at peak population densities in the summer months.--Copyright 1972, Biological Abstracts, Inc.
W73-09535

EFFECT OF THE WASTE-WATER OF SUGAR--WORKS ON NATURAL HISTORY OF THE ZAGYVA RIVER,
Authority Water Conservancy Middle Course Tisza, Szolnok (Hungary).
For primary bibliographic entry see Field 05B.
W73-09538

SEASONAL VARIATIONS IN THE SIZE OF CHIRONOMUS PLUMOSUS L, (IN GERMAN),
Bern Univ. (Switzerland). Zoologisches Institut.
Anna Maria Kloetzli, F. Roemer, and S. Rosin.
Rev Suisse Zool. Vol 78, No 3, p 587-603. 1971. English and French summary.
Identifiers: *Chironomus-Plumosus, Populations, *Seasonal, *Size, *Mosquitos.

During 2 yr measurements of the winglength were carried out on adults of C. plumosus. The material originated from emergence traps and from net-trappings of swarming mosquitoes. A gradual decline of the size leads in both sexes to 3 relatively uniform levels of size during the summer. In autumn a 4th slightly higher level is observed in some places. The seasonal fluctuations of adults and the apparition of the grades of size can be explained by the shallow-water population of the Wohlensee near Berne which develops in 2 generations. Hibernated larvae emerge at the end of April until end of July. This generation extends to the 21st levels of size. The 10 to 15% smaller specimens of the 2nd generation emerge particularly in Aug. and Sept.--Copyright 1972, Biological Abstracts, Inc.
W73-09540

ECOLOGY. THE EXPERIMENTAL ANALYSIS OF DISTRIBUTION AND ABUNDANCE,
For primary bibliographic entry see Field 06G.
W73-09547

EVALUATION OF TWO METHOD OF AERATION TO PREVENT WINTERKILL,
Virginia Polytechnic Inst. and State Univ., Blacksburg. Div. of Forestry and Wildlife Science.
R. T. Lackey, and M. Levandowsky.
Prog Fish-Cult. Vol 34, No 3, p 175-178, 1972. Illus.
Identifiers: *Aeration, Fish, Fishery, *Winterkill, *Dissolved oxygen.

Continuous winter aeration and commencing aeration when dissolved O2 reached low levels were compared in their effect on temperature, dissolved O2, and Fe. Continuous winter aeration lowered water temperatures, eliminated thermal gradients, maintained dissolved O2 at high levels, and eliminated Fe buildups. Commencing aeration when dissolved O2 was low had little effect on water temperature, but caused an immediate lowering of dissolved O2 and an increase in Fe. Dissolved O2 slowly increased in concentration after this initial decline. Commencing aeration when dissolved O2 is low, to prevent fish winterkill, should be done cautiously.--Copyright 1973, Biological Abstracts, Inc.
W73-09560

THE 1971 VIRGINIA PESTICIDE STUDY, (PURSUANT TO HOUSE JOINT RESOLUTION 51).
Virginia Dept. of Agriculture and Commerce, Richmond.
(1972), 126 p, 36 tab, 6 append.

Descriptors: *Virginia, *Pesticides, *Pesticide residues, *Monitoring, *DDT, *Water pollution control, *Systems analysis, Public health, Environmental effects, Birds, Legislation.

The Virginia General Assembly initiated an in-depth study of pesticides used in Virginia and their impact on the resources of the State; the full report is presented. Pesticide residue monitoring programs indicate widespread presence of DDT and/or its metabolites. The levels of pesticide residues in water and air did not exceed health standards, however. Occupational hazards for applicators, field workers and people engaged in the manufacture and formulation of pesticides were found to be high. Reproductive failure in several avian species were correlated with DDE and Dieldrin. The report recommends a review system for use of pesticides. Availability of substitute chemical and non-chemical data from residue monitoring and long-term toxicology studies would be readily accessible for decision-making. A systems approach would also be valuable for controlling the use of potentially hazardous pesticides. (Ensign-PAI)
W73-09642

AN ANALOGUE OF THE ARTIFICIAL AERATION OF BASINS (IN RUSSIAN),
Akademiya Nauk URSR, Kiev. Institut Kibernetiki.
A. G. Ivakhnenko, and N. V. Gulyan.
Gidrobiol Zh. Vol 8, No 1, p 75-81. 1972. English summary.
Identifiers: *Aeration, *Artificial aeration, Equations, Oxygen, Regression analysis, *River basins.

The obtained regression equation (equation of control) permits calculating the quantity of air which is necessary for each cubic meter of the basin in order to obtain necessary content of O2 in water. For this purpose some characteristics should be given of the basin state for the previous 8 hrs. The equations are true for all basins, which by hydrobiological indices and specific characteristics of aeration systems (specific density of distribution and design of bubblers) are similar to the given one.--Copyright 1972, Biological Abstracts, Inc.
W73-09644

NEW SANITATION RULES: WHAT THEY MEAN,
Marland Environmental Systems, Inc., Wayne, Pa.
W. F. Roberts.
Marine Engineering/Log, p 62-63, 103, November 1972. 2 fig, 2 ref.

Descriptors: *Water quality standards, *Water quality control, Sewage disposal, Ships, *Legislation, Design, Waste treatment, Administrative agencies.
Identifiers: *Compliance time tables.

The Environmental Protection Agency's sanitation standards for the nation's waterways are reviewed. The new regulations are designed to reduce and eventually eliminate the dumping of sewage into rivers, lakes and channels. New vessels must meet the 'no discharge' final standard after two years; existing vessels have a full year compliance time table. Some design methods for sewage storage and/or treatment shipboard are discussed. The quantity of waste water to be handled is a prime consideration in each of the disposal procedures described. The report concludes that the help of a marine engineer/architect or pollution control expert would be advisable since many of the EPA requirements are complex. (Ensign-PAI)
W73-09650

LLUTION REMEDIES A CONTINENT ART.

orld Ports, Vol 34, No 8, p 32, September 1972. 1
.

scriptors: *Oil spills, Oil pollution, *Skimming, hips, *Pollution abatement, Technology, search and development.
ntifiers: *Oil skimmers, Clean-up boats, Oil covery.

new oil-skimming device employing a concept led Dynamic Inclined Plane (DIP) is described. e machine works in wind, waves and current, lects sorbents as well as oil and recovers the ac- l spilled oil rather than oily water. Also cussed is the recent development of a self-con- ned, automated harbor clean-up craft. The ship a catamaran with a steel hull and a hydraulic sket at the bow for scooping floating debris out the water. The boat also has facilities for con- ning small oil spills. (Ensign-PAI)
73-09655

VIRONMENTAL AND ECONOMIC IMPACT RAPID GROWTH ON A RURAL AREA: LM COAST,
utheastern Environmental Services, Jackson- le, Fla.
P. Bird.
vironmental Affairs, Vol 2, No 1, p 154-171, ring 1972. 2 fig, 12 ref.

scriptors: *Florida, *Coasts, *City planning, nvironmental effects, *Saline water intrusion, man population, Economics, Ecology, Legisla- n, Urban sociology.
ntifiers: *Palm Coast (Fla).

e Palm Coast city development is assessed as a ve environmental threat. Problems of salt water rusion, open-ended canals, tree preservation, hway construction and mosquito control are all iewed. The ITT Community Development Cor- ation (which is building Palm Coast) is charged h poor planning, misinformation, and possible gal activities all connected with their proposed . The social impact of Palm Coast on Flagler unty is also examined. The question of tax cture is discussed, along with school facilities accreditation. The study concludes that much re information is necessary concerning the en- nmental impact of Palm Coast and that a mora- um on further development should be con- red. (Ensign-PAI)
3-09656

E ROLE OF THE UN IN ENVIRONMENTAL BLEMS,
mbia Univ., New York.
primary bibliographic entry see Field 06G.
-09659

ING THE TIME DIMENSION TO EN- ONMENTAL POLICY,
ed P. Sloan School of Management, Cam- ge, Mass.
primary bibliographic entry see Field 06G.
-09660

ROLE OF SPECIAL PURPOSE AND NON- ERNMENTAL ORGANIZATIONS IN THE IRONMENTAL CRISIS,
national Council of Scientific Unions, Rome,). Special Committee on Problems of the En- ment.
rimary bibliographic entry see Field 06G.
-09662

THE ROLE OF THE WMO IN ENVIRONMENTAL ISSUES,
World Meteorological Organization, Geneva (Switzerland).
For primary bibliographic entry see Field 06G.
W73-09663

INTERNATIONAL INSTITUTIONS FOR ENVIRONMENTAL MANAGEMENT,
Council on Environmental Quality, Washington, D.C.
For primary bibliographic entry see Field 06G.
W73-09664

EXPERIENCE IN USING HERBICIDE SIMAZINE FOR CONTROLLING PLANTS IN WATER BODIES, (IN RUSSIAN),
For primary bibliographic entry see Field 04A.
W73-09685

WATER USE CLASSIFICATIONS AND WATER QUALITY STANDARDS.
Georgia Dept. of Natural Resources, Atlanta.

Rules of the Environmental Protection Division, Ga. Code Ann. ch 730-3, 1972. 11 p.

Descriptors: *Water supply development, *Water quality standards, *Georgia, *Water pollution control, *Water quality control, Legal aspects, Administration, Governments, Water treatment, Public health, Water conservation, Reasonable use, Jurisdiction, Environmental effects, Ecology, Administrative agencies, Administrative decisions, Water utilization, Water management (Applied).
Identifiers: *Administrative regulations.

Promulgated for the express purpose of establishing water quality standards, these rules regulate water quality enhancement, water use classifications, general criteria for all waters, specific criteria for classified water usage, natural water quality, waste treatment requirements, stream flows, and mixing zones for effluents. The specific criteria for classified water areas deal with drinking water supplies, recreation, aquatic life, agriculture, industry, navigation, wild rivers, scenic rivers, and urban streams. The purposes and intent of the state of Georgia in establishing water quality standards are to provide enhancement of water quality and prevention of pollution; to protect the public health or welfare in accordance with the public interest for drinking water supplies, conservation of fish, game and other beneficial aquatic life, and agricultural, industrial, recreational, and other beneficial use. The rules indicate that in applying these policies and requirements, the state of Georgia will recognize and protect the interest of the federal government in interstate waters. Toward this end the state will consult and cooperate with the Environmental Protection Agency on all matters affecting the federal interest. (Mockler-Florida)
W73-09692

DEFINITIONS FOR CLEAN WATER COMMISSION REGULATIONS PRESENTED AT HEARING ON MARCH 21, 1973.
Missouri Clean Water Commission, Jefferson Cit .
Foprimary bibliographic entry see Field 06E.
W73-09693

REGULATIONS FOR PREVENTION OF POLLUTION FROM WELLS TO SUBSURFACE WATERS OF THE STATE PRESENTED AT HEARING ON MARCH 21, 1973.
Missouri Clean Water Commission, Jefferson City.
For primary bibliographic entry see Field 06E.
W73-09694

REGULATIONS FOR SINGLE RESIDENCE WASTE WATER TREATMENT DEVICES PRESENTED AT HEARING ON MARCH 21, 1973.
Missouri Clean Water Commission, Jefferson City.
For primary bibliographic entry see Field 06E.
W73-09695

SHORTCUT PIPELINE MODIFICATION CONTRA COSTA CANAL UNIT, CENTRAL VALLEY PROJECT, CALIFORNIA (FINAL ENVIRONMENTAL STATEMENT).
Bureau of Reclamation, Washington, D.C.
For primary bibliographic entry see Field 08A.
W73-09696

LOCAL PROTECTION PROJECT, LOG JAM REMOVAL ON SALAMONIE RIVER NEAR BLUFFTON IN WELLS COUNTY, INDIANA (FINAL ENVIRONMENTAL STATEMENT).
Army Engineer District, Louisville, Ky.
For primary bibliographic entry see Field 04A.
W73-09700

THE BEACHES: PUBLIC RIGHTS AND PRIVATE USE (PROCEEDINGS OF A CONFERENCE).
Texas Law Inst. of Coastal and Marine Resources, Houston.
For primary bibliographic entry see Field 06E.
W73-09701

A REPORT TO CONGRESS ON WATER POLLUTION CONTROL MANPOWER DEVELOPMENT AND TRAINING ACTIVITIES.
For primary bibliographic entry see Field 06E.
W73-09708

OMNIBUS WATER RESOURCES AUTHORIZATIONS--1972, PARTS I AND II.
For primary bibliographic entry see Field 06E.
W73-09709

ENVIRONMENTAL PROTECTION ACT OF 1971.
For primary bibliographic entry see Field 06E.
W73-09711

PROTECTING AMERICA'S ESTUARIES: PUGET SOUND AND THE STRAITS OF GEORGIA AND JUAN DE FUCA.
For primary bibliographic entry see Field 06E.
W73-09712

PROTECTING AMERICA'S ESTUARIES: PUGET SOUND AND THE STRAITS OF GEORGIA AND JUAN DE FUCA (20TH REPORT).
For primary bibliographic entry see Field 06E.
W73-09713

WATER IMPROVEMENT COMMISSION.
For primary bibliographic entry see Field 06E.
W73-09714

CONSERVANCY DISTRICTS.
For primary bibliographic entry see Field 06E.
W73-09721

WATER SUPPLY.
For primary bibliographic entry see Field 06E.
W73-09724

UNITED STATES V. SKIL CORPORATION (CRIMINAL ACTION AGAINST CORPORATION FOR WATER POLLUTION).
For primary bibliographic entry see Field 06E.

W73-09732

ALKIRE V. CASHMAN (CONSTITUTIONAL CHALLENGE TO STATE STATUTE REQUIRING FLUORIDATION OF MUNICIPAL WATER SUPPLIES).
For primary bibliographic entry see Field 06E.
W73-09733

GEORGIA WATER QUALITY CONTROL ACT.
For primary bibliographic entry see Field 06E.
W73-09738

DELAWARE RIVER BASIN COMPACT.
For primary bibliographic entry see Field 06E.
W73-09746

06. WATER RESOURCES PLANNING

6A. Techniques of Planning

SIMULATION THEORY APPLIED TO WATER RESOURCES MANAGEMENT - PHASE III, DEVELOPMENT OF OPTIMAL OPERATING RULES,
Nevada Univ., Reno. Desert Research Inst.
J. W.Fordham.
Available from the National Technical Information Service as PB-220 352, $3.00 in paper copy, $0.95 in microfiche. Technical Report Series H-W, Publication No 13, 1972. 45 p, 7 fig, 4 tab, 3 append. OWRR C-2153 (3372) (2).

Descriptors: *Planning, *Mathematical models, *Decision making, *Simulation analysis, Reservoir operation, Regression analysis, Synthetic hydrology, Inter-basin transfers, Water allocation, Linear programming, *Nevada, California, Model studies, Water demand, Optimal development plants.
Identifiers: *Truckee River, *Carson River, Out-of-kilter Algorithm.

This project evaluates simulation as a planning and management tool for water resources. This tool was evaluated in the context of Truckee and Carson River System in Nevada and California. A simulation model of the two-river system was constructed which was verified using historical flows and management practices. This simulation model was then embodied in an optimization algorithm to develop 'optimum' operating rules for the system as a whole. Since the demands on the system were incomencerate in economic terms and are greater than the available resource, the problem was resolved into one of allocation of the resource among the various demands. To accomplish this, the problem was formulated as a capacitated flow network and solved using the 'out-of-kilter' algorithm. The reservoir releases and diversions from several flow traces were then subject to multiple regression analysis to determine 'optimal' operating rules for the five reservoirs and for diversions within the system. From the results of the particular runs made it can be said that this two-river multiple time period problem can be approached in this manner and operating rules can be derived which significantly improve overall system operation and satisfy demands to as great as an extent possible. (See also W71-05042)
W73-09118

INTERINDUSTRY ECONOMIC MODELS AND REGIONAL WATER RESOURCES ANALYSES,
For primary bibliographic entry see Field 06B.
W73-09409

ADDING THE TIME DIMENSION TO ENVIRONMENTAL POLICY,
Alfred P. Sloan School of Management, Cambridge, Mass.
For primary bibliographic entry see Field 06G.
W73-09660

6B. Evaluation Process

POSSIBLE WAYS OF DEVELOPING IRRIGATED AGRICULTURE IN SOVIET CENTRAL ASIA AND THE FUTURE OF THE ARAL SEA (VOZMOZHNYYE PUTI RAZVITIYA OROSHAYEMOGO ZEMLEDELIYA SREDNEY AZII I BUDUSHCHEYE ARAL'SKOGO MORYA),
All-Union Designing, Surveying and Scientific Research Inst. Hydroproject, Moscow (USSR).
For primary bibliographic entry see Field 03F.
W73-09155

STRUCTURE OF WATER CONSUMPTION IN THE UNITED STATES AND ITS SPATIAL CHARACTERISTICS (STRUKTURA VODOPOTREBLENIYA V SSHA I YEYE TERRITORIAL'NYYE OSOBENNOSTI),
Akademiya Nauk SSSR, Moscow. Institut Geografii.
For primary bibliographic entry see Field 06D.
W73-09156

REMOTE SENSING AS AN AID TO THE MANAGEMENT OF EARTH RESOURCES,
California Univ., Berkeley. Space Sciences Lab.
For primary bibliographic entry see Field 07B.
W73-09262

INTERINDUSTRY ECONOMIC MODELS AND REGIONAL WATER RESOURCES ANALYSES,
E. M. Lofting.
In: Economics of Engineering and Social Systems, John Wiley and Sons--Interscience, New York, New York, p 235-257, 1972. 1 tab, 51 ref.

Descriptors: *Leontief models, *Economic justification, *Regional development, *Regional analysis, Investment, *Decision making, Input-output analysis.
Identifiers: *Interindustry models, Regional investment, Decisions, Investment criteria, Interindustry demand.

Regional investment decisions require a knowledge of the structure of the regional economy, the magnitude and value of commodity imports and exports, and the real value of regional resources that may be necessary to develop. In a survey of methods for regional investment decision-making, the author reviewed the static, open interindustry model as a framework for studying water resource investment. This input-output model requires a regional accounting scheme which can then be used to isolate the dynamic processes of change within a regional economy. This knowledge forms the foundation for investment decisions by indicating the extent to which public projects should be undertaken to meet the input demands of the regional economy. If the satisfaction of effective demand for inputs is considered a benefit, then the input-output framework may be considered a tool for benefit-cost analysis. Methods and feasibility of specific applications were discussed. (Weaver-Wisconsin)
W73-09409

A VALUE-ORIENTED APPROACH TO WATER POLICY OBJECTIVES,
Colorado State Univ., Fort Collins. Dept. of Economics.
A. R. Dickerman.
Land Economics, Vol 48, No 4, p 398-403, November, 1972.

water resources regions in the United States in 1965, five (North Atlantic Region, Great Lakes Region, Ohio Region, Columbia-North Pacific Region, and California Region) accounted for 58.5% of the nation's water withdrawals; three regions (South Atlantic-Gulf, Texas-Gulf, and Missouri) accounted for 21%; and the remaining 12 regions for 20.5% of the total withdrawal. Withdrawals for the first five regions are expected to grow to 67% by 2020. In 1965, streams supplied 67%, groundwater aquifers 22%, and saline sources, most of which were from the ocean, 11% of the nation's water withdrawals. Groundwater supplies are used primarily for irrigated agriculture (66%), urban water supply (14%), and industrial water supply (14%). The three most important users of water are irrigated agriculture (41%), the thermal electricity generating industry (31%), and the manufacturing industries (17%). A division of the conterminous United States into 6 regions according to water resources adequacy and development needs is based on comparison of current water supplies with projections of water withdrawal requirements. (Josefson-USGS)
W73-09156

THE RELATION OF FORESTS AND FORESTRY TO WATER RESOURCES,
Georgia Univ., Athens. School of Forest Resources.
For primary bibliographic entry see Field 04A.
W73-09537

PUBLIC ACCESS TO RESERVOIRS TO MEET GROWING RECREATION DEMANDS.
For primary bibliographic entry see Field 06E.
W73-09710

6E. Water Law and Institutions

A LOOK BACK AT THE FIRST SEVEN YEARS OF WYO. WRRI (1965-1971),
Wyoming Univ., Laramie. Water Resources Research Inst.
For primary bibliographic entry see Field 09A.
W73-09108

A PROPOSAL FOR IMPROVING THE MANAGEMENT OF THE GREAT LAKES OF THE UNITED STATES AND CANADA.
Cornell Univ., Ithaca, N.Y. Water Resources and Marine Sciences Center.

Available from the National Technical Information Service as PB-220 359, $3.00 in paper copy, $0.95 in microfiche. Canada--United States University Seminar, 1971-1972, Technical Report No 62, January 1973. OWRR C-2145 (3370) (1).

Descriptors: *Great Lakes Region, *Institutions, *International waters, *Canada, *United States, *International Joint Commission, Comprehensive planning, International law, Land management, Multiple purpose, Water management (Applied), Organizations.
Identifiers: Large lakes.

A study of the management of the Great Lakes was undertaken by the Canada-United States University Seminar in 1971-1972. The Seminar, a binational group composed of faculty members representing some twenty Canadian and U.S. universities in the Great Lakes region, explored ways in which the management of the water and related land resources of the Great Lakes Basin might be strengthened to the mutual advantage of both countries. The investigation was concerned with determining what kinds of government organizations are needed and how these organizations should be related to one another in order to achieve effective management of the natural resources of the Great Lakes Basin at the lowest possible economic, political and social costs. The report proposes that the United States and Canada must act individually and jointly, to remove such obstacles. The report considers: The current resource use situation in the Great Lakes; the International Joint Commission (IJC), its structure and functions, and an assessment of the Commission; general specifications for an enhanced institutional structure for the Great Lakes; and alternative institutional arrangements. Resulting from the Seminar was general agreement on the need for substantial changes in the present institutional framework. Two alternatives were identified: (1) to introduce organizational improvements within a considerably strengthened International Joint Commission; and (2) to establish by treaty new organizational arrangements distinct from the IJC. (Bell-Cornell)
W73-09110

DIRECTORY OF FACULTY ENGAGED IN WATER RESEARCH AT CONNECTICUT UNIVERSITIES AND COLLEGES,
Connecticut Univ., Storrs. Inst. of Water Resources.
For primary bibliographic entry see Field 09A.
W73-09112

SOUND WASTE MANAGEMENT CUTS OFFSHORE COSTS,
Weston (Roy F.), Inc., West Chester, Pa. Marine Waste Management Div.
For primary bibliographic entry see Field 05G.
W73-09224

SEA FISHERIES COMMITTEES,
Manchester Univ. (England). Faculty of Law.
For primary bibliographic entry see Field 05G.
W73-09226

CALIFORNIA TIGHTENS CLAMP ON EFFLUENTS,
For primary bibliographic entry see Field 05G.
W73-09230

WATER AND WASTEWATER PROBLEMS IN RURAL AMERICA.
Commission on Rural Water, Washington, D.C.
For primary bibliographic entry see Field 05G.
W73-09264

WATER RESOURCES RESEARCH AT THE UNIVERSITY OF MARYLAND, 1965-1972,
Maryland Univ., College Park. Water Resources Research Center.
For primary bibliographic entry see Field 09A.
W73-09267

MANAGEMENT OF GROUND WATER,
Nebraska Univ., Lincoln. Dept. of Agricultural Economics.
For primary bibliographic entry see Field 04B.
W73-09413

TIDAL DATUM PLANES AND TIDAL BOUNDARIES AND THEIR USE AS LEGAL BOUNDARIES -- A STUDY WITH RECOMMENDATIONS FOR VIRGINIA,
Virginia Inst. of Marine Science, Gloucester Point.
For primary bibliographic entry see Field 02L.
W73-09609

THE 1971 VIRGINIA PESTICIDE STUDY, (PURSUANT TO HOUSE JOINT RESOLUTION 51).
Virginia Dept. of Agriculture and Commerce, Richmond.
For primary bibliographic entry see Field 05G.
W73-09642

NEW SANITATION RULES: WHAT THEY MEAN,
Marland Environmental Systems, Inc., Wayne, Pa.
For primary bibliographic entry see Field 05G.
W73-09650

THE UNITED NATIONS AND THE ENVIRONMENT,
Conference on the Human Environment, New York.
For primary bibliographic entry see Field 06G.
W73-09658

THE UNITED NATION'S INSTITUTIONAL RESPONSE TO STOCKHOLM: A CASE STUDY IN THE INTERNATIONAL POLITICS OF INSTITUTIONAL CHANGE,
Sussex Univ., Brighton (England). Inst. for the Study of International Organization.
For primary bibliographic entry see Field 06G.
W73-09661

THE ROLE OF SPECIAL PURPOSE AND NONGOVERNMENTAL ORGANIZATIONS IN THE ENVIRONMENTAL CRISIS,
International Council of Scientific Unions, Rome, (Italy). Special Committee on Problems of the Environment.
For primary bibliographic entry see Field 06G.
W73-09662

INTERNATIONAL INSTITUTIONS FOR ENVIRONMENTAL MANAGEMENT,
Council on Environmental Quality, Washington, D.C.
For primary bibliographic entry see Field 06G.
W73-09664

WATER USE CLASSIFICATIONS AND WATER QUALITY STANDARDS,
Georgia Dept. of Natural Resources, Atlanta.
For primary bibliographic entry see Field 05G.
W73-09692

DEFINITIONS FOR CLEAN WATER COMMISSION REGULATIONS PRESENTED AT HEARING ON MARCH 21, 1973,
Missouri Clean Water Commission, Jefferson City.

Hearing of Clean Water Commission--Definition for Regulations presented March 21, 1973. 2 p.

Descriptors: *Water quality, *Administration, *Water quality standards, *Missouri, Water pollution, Water law, Waste water, Water pollution treatment, State governments, Administrative agencies, Administrative decisions, Regulation, Water pollution control, Wells, Aquifer, Cesspools, Effluents, Sewerage, Sinks, Waste water (Pollution), Waste water treatment.
Identifiers: *Administrative regulations, *Definitions, Engineers, Well plugging, State waters.

This document contains definitions for Clean Water Commission regulations as promulgated by the Missouri Clean Water Commission. One of the more important terms defined is pollution, described as such contamination or other alteration of the physical, chemical or biological properties, of any waters of the state, including change in temperature, taste, color, turbidity, or odor of the waters, or such discharge of any liquid, gaseous, solid, radioactive, or other substance into any waters of the state as will or is reasonably certain to create a nuisance or render such waters harmful, detrimental or injurious to public health, safety or welfare, or to domestic, industrial, agricultural, recreational, or other legitimate beneficial uses, or to wild animals, birds, fish or other aquatic life. Other defined terms include abandoned well, agency, aquifer, cesspool, effluent, engineer, plugging, sewer system, single family residence, single residence waste water treatment device, sinkhole, waste waste, waste water treatment facility, waters of the state, well, and well owner. (Mockler-Florida)
W73-09693

REGULATIONS FOR PREVENTION OF POLLUTION FROM WELLS TO SUBSURFACE WATERS OF THE STATE PRESENTED AT HEARING ON MARCH 21, 1973,
Missouri Clean Water Commission, Jefferson City.

Regulations of March 21, 1973. 1 p.

Descriptors: *Missouri, *Regulation, *Oil wastes, *Waste water (Pollution), *Waste water disposal, Subsurface waters, Wells, Water pollution, Legal aspects, Drainage, Underground streams, Aquifers, Water quality control, mining, Water pollution control, Administrative agencies, Administrative decisions, Water resources development, Water management (Applied), Well regulation.
Identifiers: *Administrative regulations, *Well plugging.

No person shall release any waste water into the subsurface waters of the state, or store or dispose of such waste water in any way which causes or permits it to enter the subsurface waters of the state directly or indirectly. Moreover, no waste water shall be introduced into sinkholes, caves, fissures, or other openings in the ground which do or are reasonably certain to drain into aquifers. Whenever any well is abandoned, it shall be plugged or sealed as necessary to prevent pollution of subsurface waters of the state. An affidavit detailing the procedure used in plugging is required to be filed with the Missouri Clean Water Commission. It shall not be a violation to return oil field wastes to oil producing formations from which they were obtained or backfill cavities as an integral part of a mining operation with aggregate or other material obtained from that operation to either reduce accumulation of wastes on the surface or to provide additional ground support in the mined out area or to inundate such cavities with water devoid of toxic materials. (Mockler-Florida)
W73-09694

REGULATIONS FOR SINGLE RESIDENCE WASTE WATER TREATMENT DEVICES PRESENTED AT HEARING ON MARCH 21, 1973,
Missouri Clean Water Commission, Jefferson City.

Regulations of March 21, 1973. 2 p.

Descriptors: *Regulation, *Waste water treatment, *Waste water disposal, *Missouri, *Water pollution control, Legal aspects, Sewage disposal, Water treatment, Water pollution sources, Waste disposal, Water management (Applied), Administrative agencies, Effluents, Waste water (Pollution), Sanitary engineering, Sewerage, Environmental effects.
Identifiers: *Administrative regulations.

When a single family residence is located in an area where a public or privately owned sewer system is not available, the owner must construct a waste water treatment device in accordance with the Missouri Clean Water Commission (Commission) regulations. If such a sewer system becomes available within 150 feet of the owner's property line, his waste water treatment system must be abandoned and a connection made to the sewer system within one year thereafter. The waste water treatment device must be a minimum of 50 feet from the nearest property line. If the owner wishes to install a waste water treatment device on a lot less than three acres in size, permission must be obtained from the Commission, and in no event may such a device be installed on a lot less than one acre in size. Effluents from these waste water treatment devices shall be contained within the owner's property line and may not be released into waters of the state. Special requirements relating to residences in subdivisions, construction requirements, and location of waste water treatment devices are also mentioned. (Mockler-Florida)
W73-09695

THE BEACHES: PUBLIC RIGHTS AND PRIVATE USE (PROCEEDINGS OF A CONFERENCE).
Texas Law Inst. of Coastal and Marine Resources, Houston.

Available from National Technical Information Service, U.S. Department of Commerce, PB-212 901, for $3.00 paper copy and $0.95 microfiche. Proceedings of a conference held January 15, 1972. 74 p, 12 ref.

Descriptors: *Texas, *Legislation, *Beaches, *Coastal land use, *Public rights, Shoreline boundary, Legal aspects, Water resources development, Competing uses, Project planning, Water law, Relative rights, Recreation, Recreation facilities, Public access, Public lands, Public benefits, Easements, Water utilization, Water policy, Prescriptive rights.
Identifiers: *Coastal zone management, Ingress, Egress.

A conference was held in January 1972 to discuss public rights and private use of beaches along the Texas coast. This report summarizes Texas law of the beaches and reports by participants which suggested necessary amendments to the Open Beaches Act (Act) to make it more protective of the public rights. Although the conference was not aimed at reaching any specific conclusions, the discussions did point out that the Texas Open Beaches Act may not guarantee the public a right to use all beaches; but, where such rights exist, state and local regulations are desirable to assure orderly development of public use. The Open Beaches Act was enacted to reduce to statutory language some common law concepts that recognize certain public rights to use the beaches. The act directs the Attorney General to protect the people's right, defined as a right of ingress and egress to that portion of the beach owned by the state and also to that part of the beach impressed with a presumption of a right of use by the people. Moreover, the Act provides for a presumption of prescriptive right to the area between low tide and, generally, the vegetation mark. There is also a discussion of the proposed National Open Beaches Act and actions taken by other coastal states. (Mockler-Florida)
W73-09701

WATERSHED FIELD INSPECTIONS--1971.

Hearings--Subcomm. on Conservation and Watershed Development--Comm. on Public Works, U.S. House of Representatives, 92d Cong, 1st Sess, 1971. 545 p, 2 fig, 4 tab, 4 chart.

Descriptors: *Flood control, *Watershed management, *Water resources development, *Water utilization, *Water management (Applied), Federal government, Erosion control, Land management, River basin development, Soil management, Water conservation, Water storage, Water supply, Water yield improvement, Watersheds (Basins), Water sources, Channel improvement, Legislation, Project planning, Environmental effects, Recreation basins.

Hearings were held in various communities between July and November of 1971 located in

fourth bill represents an alternative and substantial departure from the Golden Ealge program. Texts of the bills; reports from federal agencies; statements, testimony, and letters from interested groups within and outside the government are included along with a series of tables related to federal recreation fees and their use in land and water conservation. (Dunham-Florida)
W73-09706

A REPORT TO CONGRESS ON WATER POLLUTION CONTROL MANPOWER DEVELOPMENT AND TRAINING ACTIVITIES.

Report--Environmental Protection Agency, Office of Water Programs, March 1972. 209 p, 4 fig, 16 tab, 9 chart.

Descriptors: *Federal Water Pollution Control Act, *Manpower, *Training, *Forecasting, Federal government, Administrative agencies, Estimating, Evaluation, Future planning (Projected), Employment opportunities, Personnel, Occupations, Personnel management, Scientific personnel, Professional personnel, Schools (Education), Water pollution control, Comprehensive planning, Control, Regulation.

This report focuses on the manpower development activities of all organizations, both public and private, that are concerned with water pollution control. The report is made in response to a requirement of the Federal Water Pollution Control Act. It is divided into three main areas: (1) manpower planning; (2) manpower recruitment, retention and utilization; and (3) manpower training. Actions taken to develop a forecasting system for manpower and training needs and interim estimates of future needs are included. There is a discussion of the extent and effectiveness of both the Environmental Protection Agency, (EPA), and non-EPA training programs in the field of water pollution control. A projection of manpower and training needs until 1976 in the three principal water pollution control manpower categories of operators, technicians, and professionals is also included. Various charts, graphs and tables accompany the report. (Dunham-Florida)
W73-09708

OMNIBUS WATER RESOURCES AUTHORIZATIONS--1972, PARTS I AND II.
Hearings--Subcomm. on Flood Control--River and Harbors--Comm. on Public Works, U.S. Senate, 92d Cong, 2d Sess, June 1972. 821 p, 6 fig, 3 map, 41 photo, 19 tab, 11 chart, 2 append.

Descriptors: *Federal government, *Water resources development, *Water policy, *Water utilization, Administration, Comprehensive planning, Conservation, Legislation, Multiple-purpose projects, Natural resources, Optimum development plans, Potential water supply, Project purposes, Research and development, Water allocation (Policy), Water conservation, Water requirements, Water resources, Water supply, Flood control, Recreation facilities, Environmental effects.

These hearings were initiated for the purpose of determining the necessity of legislation authorizing the construction of federal water resources projects for navigation, water supply, flood control, recreation facilities, conservation, beach erosion control and water utilization. The Army Corps of Engineers presented reports of on-going projects and proposals for the further development and conservation of the Nation's water resources. Also included are statements and testimony of witnesses at the hearing, texts of proposed legislation, letters and commentaries by various groups and individuals interested either in particular projects or future water resources development at the federal level. Charts, maps, and tables are interspersed through the report. (Dunham-Florida)
W73-09709

PUBLIC ACCESS TO RESERVOIRS TO MEET GROWING RECREATION DEMANDS.
House of Representatives Report No. 279, 92d Cong, 1st Sess, October 21, 1971. 38 p.

Descriptors: *Federal project policy, *Artificial lakes, *Reservoir design, *Public access, *Recreational facilities, Federal government, Administrative agencies, Reservoir construction, Reservoir sites, Reservoirs, Reservoir operation, Competing uses, Recreation, Water management (Applied), Wildlife management, Multiple-purpose reservoirs, Conservation, Lakes, Recreation demands, Access routes, Public benefits, Public lands, Legislation, Administrative decisions.
Identifiers: Environmental impact statements.

Based on hearings held before the U.S. House Subcommittee on Conservation and Natural Resources, this report examines whether federal agencies which constrict or finance the construction of reservoir lakes are complying with congressioual outdoor recreation policy of providing public access to such reservoirs. After brief analyses of efforts by several federal agencies to comply with the above policy the report concludes that the changes made by the U.S. Army Corps of Engineers (Corps) concerning the 1962 Joint Reservoir Land Acquisition Policy should be promptly rescinded. No changes in the joint policy or its implementation should be made either by the Corps or the Bureau of Reclamation without (1) providing public notice of the proposed change; (2) holding public hearings; and (3) following the requirements of the National Environmental Policy act by preparing an environmental impact statement and consulting with agents having special interest or expertise with respect to the effects of such changes. (Dunham-Florida)
W73-09710

ENVIRONMENTAL PROTECTION ACT OF 1971.

Hearing--Subcomm. on the Environment--Comm. on Commerce, United States Senate, 92d Cong, 1st Sess, September 27, 1971. 294 p.

Descriptors: *Legislation, *Environmental protection, *Federal government, *Legal review, Governments, Water law, Water resources development, Water pollution, Air pollution, Regulation, Legal aspects, Federal jurisdiction, Pollution abatement, Pollutants, Ecology, Environmental control, Public health, Public benefits.
Identifiers: *Environmental Protection Act of 1971.

Hearings were held in September of 1971 to take testimony on the Environmental Protection Act of 1971. The bill, which seeks to facilitate environmental lawsuits, overcomes the defense of sovereign immunity and provides a remedy for all citizens when the administrative process does not give the environment the protection it requires. The bill would serve to promote and protect the free flow of interstate commerce without unreasonable damage to the environment, assure that activities which affect interstate commerce will not unreasonably injure environmental rights, provide a right of action for relief for protection of the environment from unreasonable infringement by activities which affect interstate commerce and to establish the right of all citizens to the protection, preservation, and enhancement of the environment. Witnesses appearing before the subcommittee included Ralph Nader and representatives of Common Cause and General Electric Co. Additional articles, letters, and statements provided by individuals, private groups, and federal agencies interested in the proposed legislation are included. (Mockler-Florida)
W73-09711

PROTECTING AMERICA'S ESTUARIES: PUGET SOUND AND THE STRAITS OF GEORGIA AND JUAN DE FUCA.

Hearings--Subcomm. of the Comm. on Government Operations, United States House of Representatives, 92d Cong, 1st Sess, December 10, 11, 1971. 727 p, 14 fig, 1 map, 17 tab.

Descriptors: *Environmental effects, *Resources development, *Administrative agencies, *Estuaries, *Estuarine environment, Sounds, Water law, Competing uses, Governmental interrelations, Laws, Dredging, Flood control, Ecology, Wildlife conservation, Wildlife management, Federal government, Comprehensive planning, Administrative decisions, Regulation, Oil fields.
Identifiers: *Puget Sound, Coastal zone management, Administrative regulations.

Hearings were held before the House Subcommittee on Government Operations with reference to protecting the Puget Sound estuary. The report indicates the Puget Sound region is probably the most important resource of the Pacific Northwest. It abounds with wildlife and is the principal wintering area for waterfowl from British Columbia, the Northwest Territories, Alaska, and eastern Russia. At the hearings, the subcommittee testimony on whether the various federal agencies are responsibly, imaginatively, economically, and efficiently carrying out their responsibilities under existing laws, executive orders and directives to protect the environmental quality of the Puget Sound area. One matter of concern to the Subcommittee was the increasing use of the Puget Sound area for the transportation and production of oil. Moreover, the Subcommittee wanted to learn the extent to which dredging and filling, including Army Corps of Engineers maintenance dredging, has caused problems. The Subcommittee was also interested in learning the extent to which dredging in harbors is being used as a substitute for effective treatment of wastes. (See also W73-09713) (Mockler-Florida)
W73-09712

PROTECTING AMERICA'S ESTUARIES: PUGET SOUND AND THE STRAITS OF GEORGIA AND JUAN DE FUCA (20TH REPORT).

Comm. on Government Operations, U.S. House of Representatives, Report No. 731, 92d Cong, 2d Sess, September 18, 1972. 83 p.

Descriptors: *Estuarine environment, *Federal government, *Water pollution control, *Water pollution sources, Legislation, Water pollution, Waste water disposal, Industrial wastes, Environmental effects, Aquatic habitats, Sounds, Washington, Estuaries, Oil pollution, Oil spills, Pulp wastes, Sewage treatment, Ecology, Administrative agencies, Water resources development, Water management (Applied).
Identifiers: *Puget Sound.

This current report is directed to the question of whether federal agencies, in cooperation with state and local agencies, are giving adequate consideration to the effects of their activities in the Puget Sound area and whether they are developing means to avoid or mitigate harmful effects wherever possible. Specifically the report finds a heightened increase in the risk of oil spills due to an increase in the petroleum industry and failure of federal agencies to carry out existing law and policy, regulations regarding sewage discharge from vessels have not been formulated as required, a state agency unilaterally extended pollution control schedules for pulpmills in violation of federal water quality standards, the Refuse Act has not been enforced adequately or uniformly, and the state of Washington is developing legislation which may help minimize damage to Puget Sound and other state shorelines and estuaries. A list of recommendations for the various agencies involved is included. (See also W73-09712) (Mockler-Florida)
W73-09713

WATER IMPROVEMENT COMMISSION.

Ala. Code tit. 22, secs. 140 (1) thru 140 (12g) (Supp 1971).

Descriptors: *Alabama, *Legislation, *Water pollution control, *Administrative agencies, Water resources development, Water utilization, Water permits, Water pollution treatment, Water quality, Water purification, Water quality standards, Federal Water Pollution Control Act, Water management (Applied), Permits, Governments, Regulation, Administrative decisions, Environmental effects, Adjudication procedure, Legal review.
Identifiers: Notice, Administrative regulations.

To carry out the declared public policy to improve and conserve the waters of the state and to prevent water pollution, the Water Improvement Commission (Commission) is created. The Commission is empowered to investigate, research and prepare programs pertaining to the purity and conservation of state waters. The Commission is authorized to promulgate rules, regulations and orders in furtherance of state policy and to issue licenses or permits to insure water quality standards. Provisions for the composition and administration of the Commission are listed. The Commission is required to hold hearings prior to taking specific action with an appeal procedure provided for affected parties. Penalties are provided of orders or permit requirements. The Commission is designated the state water pollution control agency as required by the Federal Water Pollution Control Act, and is authorized to work in cooperation with agencies within the federal government on matter of mutual interest. (Dunham-Florida)
W73-09714

COMPREHENSIVE LAND MANAGEMENT AND USE PROGRAM IN FLOOD PRONE AREAS.

Ala. Code tit. 12, secs. 341 thru 364 (Supp 1971).

Descriptors: *Alabama, *Legislation, *Flood control, *Land management, *Flood plain zoning, Zoning, Flood damage, Flood protection, Water management (Applied), Building codes, Local governments, Land management, Land resources, Land use, Ecology, Administration, Administrative decisions, Regulation, Legal aspects, Adjudication procedure, Legal review, Penalties (Legal), Comprehensive planning.
Identifiers: Subdivision regulation.

To reduce the damaging effects of flooding, each county in the state may be provided with a comprehensive land-use management plan by: (1) regulating the development of land in flood-prone areas; (2) guiding the development of proposed construction away from locations threatened by flood hazards; (3) assisting in reducing damage caused by floods; and (4) improving long-range management and use of flood prone areas. Each county governing body is authorized to adopt zoning ordinances and building codes for such flood prone areas in accordance with the above policies. The county government may also create a planning commission for enforcing these provisions. The organizational structure and powers of the commission are listed. Anyone desiring to develop flood prone lands may be required to submit plans to the county. Provisions for hearings and appeal of decisions to a board of admustment, which is required to be set up by the county government to adjudicate problems arising in implementing these provisions, are provided. Special provisions for regulation of subdivisions in flood prone areas are also included. (Dunham-Florida)
W73-09715

WATER CONSERVATION AND IRRIGATION AGENCIES.

Ala. Code tit. 12, secs. 297 (1) thru297 (14) (Supp 1971).

Descriptors: *Alabama, *Legislation, *Water conservation, *Irrigation districts, Irrigation, Irrigated land, Irrigation efficiency, Water delivery, Water requirements, Irrigation systems, Irrigation programs, Irrigation water, Water allocation (Policy), Water utilization, Water supply development, Water resources development, Administration, Governments, Permits, Bonds, Governmental interrelations.

In the interest of water conservation and land irrigation the legislature authorizes the establishment of an incorporated state development agency constituting an irrigation district or districts for the state. The membership of the corporation shall consist of title holders to the land irrigated or proposed to be irrigated within the boundaries of the district or districts to be established; such members shall elect a board of directors. Provisions for incorporation and the management structure are set forth. The powers of the corporation are enumerated. The corporation is required to obtain permits from the Department of Conservation prior to any project construction. Contracts awarded by the corporation must be on the basis of competitive bids. The corporation is authorized to issue bonds to finance proposed projects. No officer of the corporation may have an interest in any contract awarded. The corporation is required to cooperate with other state agencies in the development of irrigation and water conservation. (Dunham-Florida)
W73-09716

WATER RESOURCES RESEARCH INSTITUTE.
Ala. Code tit. 8, secs. 266 thru 269 (Supp 1971).

Descriptors: *Research and development, *Research facilities, *Alabama, *Legislation, *Water resources development, *Investigations, Water resources, Water sources, Water supply, Natural resources, Optimum development plans, Project purposes, Water allocation (Policy), Water conservation, Water policy, Administration, Governments, Ecology, Grants.

In order to help assure the state policy of the legislature to provide an abundance of water, both as to quality and quantity, the Board of Trustees of Auburn University may establish a water resources research institute or center which is designated as the state agency to accept federal funds for the use of such centers. The institute will be under the control of the trustees and president of Auburn University. It shall be the duty of the institute to research, investigate and experiment in the field of water and resources as they affect water, and to encourage the training of scientists in this field. The result of the institute's research shall be published as determined by the director of the institute. The provisions are not intended to restrict or duplicate programs conducted by the Geological Survey of Alabama or the University of Alabama in the area of water resources. (Dunham-Florida)
W73-09717

SEAFOODS.
Ala. Code tit. 8, secs. 130 (1) thru 164 (5) (Supp 1971).

Descriptors: *Alabama, *Legislation, *Shellfish farming, *Regulation, Marine animals, Mussels, Oysters, Mollusks, Shrimp, Shellfish, Commercial shellfish, Aquatic animals, Crustaceans, Commercial fishing, Beds, Reefs, Baits, Permits, Administration, Administrative agencies, Administrative decisions, Legal aspects, Governments, Penalties (Legal).

dissolution of the Commission upon completion of its work is included. (Dunham-Florida)
W73-09720

CONSERVANCY DISTRICTS.
Ind. Ann. Stat. secs. 27-1501 thru 27-1599 (1970).

Descriptors: *Water conservation, *Water districts, *Legislation, *Indiana, *Flood control, Water resources development, Legal aspects, Drainage, Irrigation, Water quality standards, Sewage disposal, Economics, Jurisdiction, Administrative agencies, State governments, Comprehensive planning, Regulation, Administration, Financing, Governmental interrelations, Legal review, Water pollution control.

Conservancy districts may be established for flood prevention and control, improving drainage, providing for irrigation, providing water supply (including treatment and distribution for domestic, industrial, and public use), and providing for the collection, treatment and disposal of sewage and other liquid wastes. The express policy of this act is to enable the owners of land to organize special districts with power to tax and with other necessary powers to effect maximum beneficial utilization that can be made of the water resources of the state. The purposes of the act may be accomplished by cooperating with federal and state agencies whose programs are designed to accomplish any of the purposes of the district. Provisions relating to the establishment, administration, membership of the district's board and project financing are included with guidelines and requirements for hearings and appeal procedures by landowners affected either by the creation of a conservancy district or projects approved by the district's board of directors. (Mockler-Florida)
W73-09721

OHIO RIVER AREA COORDINATING COMMISSION.
Ind. Ann. Stat. secs. 68-1301 thru 68-1304 (Supp. 1972).

Descriptors: *Indiana, *Ohio River, *Legislation, *Comprehensive planning, *Governmental interrelations, Navigation, Project purposes, Interstate rivers, Evaluation, Economics, Social aspects, Water pollution control, Environmental effects, Economic impact, Social impact, Legal aspects, Regulation, Planning, Rivers, Administrative agencies, State governments, Public benefits.
Identifiers: *Economic opportunities.

The Ohio River Area Coordinating Commission (Commission) is established, and the means by which its members are to be selected is specified. The Commission is to make annual reports to the state legislature of the problems and economic opportunities arising from the modernization of navigation facilities on the Ohio River along the southern border of Indiana. The statute also provides that the Commission shall encourage and attempt to coordinate the activities of the state of Indiana and agencies thereof with the activities of federal agencies, agencies of other states bordering on the Ohio River, private agencies and private individuals interested in promoting the economy and welfare of the entire Ohio River area. (Reed-Florida)
W73-09722

INDIANA PORT COMMISSION.

Ind. Ann. Stat. secs. 68-1201 thru 68-1238 (Supp. 1972).

Descriptors: *Indiana, *Legislation, *Port authorities, *Channels, Administrative agencies, Administrative decisions, Construction, Erection, Administration, Legal Aspects, Water utilization, Transportation, State governments, Federal government, Planning, Permits, Regulation, Economics, project planning, Leases, Canals, Public benefits, Harbors.
Identifiers: *Foreign trade zones, *Public ports, Administrative regulations.

The Indiana Port Commission (Commission) is authorized to construct, maintain and operate, in cooperation with the federal government, or otherwise, at such location on Lake Michigan and/or the Ohio River, and/or the Wabash River, as shall be approved by the governor, public ports with terminal facilities and traffic exchange points for all forms of transportation. The powers and duties of the Commission and the general powers to prepare and operate a port are specified. The Commission is authorized to acquire by purchase any interest in land as it may deem necessary or convenient for the performance of its responsibilities. The Commission is authorized to apply for the privilege of establishing, operating and maintaining foreign-trade zones. The Commission is empowered to construct new canals or improve any existing canal, river or other waterway. The statute provides several means by which the Commission may finance its activities. (Reed-Florida)
W73-09723

WATER SUPPLY.

Ind. Ann. Stat. secs. 35-2921 thru 35-2931 (Supp 1972).

Descriptors: *Treatment facilities, *Indiana, *Legislation, *Water supply, Water pollution control, Water yield, Environmental sanitation, Water conservation, Water distribution (Applied), Water management (Applied), Water policy, Water supply development, Water utilization, Water works, Water treatment, Waste water treatment, Permits, Regulation, Administrative agencies, State governments.
Identifiers: *Licenses.

It shall be unlawful for any person, firm or corporation, both municipal and private, to operate a water or wastewater treatment plant, or a water distribution system unless the operator in charge is duly certified, and it is also unlawful for any person to perform the duties of such an operator without being so certified. The State Board of Health is authorized to adopt rules and regulations with regard to the basis for classification of water and water treatment systems, provisions establishing qualifications for operator applicants and procedures for examination of candidates, and such other provisions necessary to enforce this chapter. The State Health Commissioner is to issue certificates attesting to the competency of operators; provisions relating to application, fees, notice of revocation and expiration are also set forth. (Dunham-Florida)
W73-09724

INDIANA DRAINAGE CODE--DRAINAGE BOARDS--POWERS AND DUTIES.

Ind Ann Stat secs, 27-2004 thru 27-2016 (Supp. 1972).

Descriptors: *Indiana, *Legislation, *Drainage programs, *Land management, *Drainage districts, Water management (Applied), Drainage, Drainage area, Drainage systems, Multiple purpose projects, Soil conservation, State governments, Local governments, Administrative agencies, Water control, Flood control, Drainage engineering, Project planning, Controlled drainage, Governmental interrelations.

The county commissioners or a board expressly appointed by them shall constitute the County Drainage Board, (Board). The Board is authorized to employ an attorney to represent and advise it. The Board is required to employ either the county

surveyor or a qualified deputy to investigate and formulate necessary construction or reconstruction of drainage facilities. When it appears that projected programs may affect lands in more than one county, a joint meeting of affected drainage boards is required. The membership and administration of such a joint board meeting is provided with a possibility of waiver by the affected county's Board under certain circumstances. Board members having an interest in any construction, reconstruction or maintenance of drains by reason of ownership of lands affected by such drains must disqualify themselves in that proceeding. (Dunham-Florida)
W73-09725

WATER RIGHTS.
Ind Ann Stat secs, 27-1401 thru 27-1409, (1970).

Descriptors: *Indiana, *Legislation, *Riparian rights, *Water rights, *Water allocation (Policy), Legal aspects, Water law, Administrative agencies, Competing uses, Irrigation, Regulation, Water contracts, Water utilization, Water resources development, Water policy, Comprehensive planning, State governments, Water management (Applied), Water conservation, Consumptive use, Non-consumptive use, Public health, Surface waters, Diversion.

In promoting and expanding the beneficial uses of surface water resources in the interest of the general welfare of the people of the state, the surface waters of the state are declared to be public waters and subject to regulation by the Indiana general assembly. Users of such waters who institute withdrawal of water for artificial uses or who increase such uses shall be subject to any regulations enacted by the legislature. Riparian owners' rights, insofar as use of public water for domestic purposes and/or impounding such waters for irrigation, are set forth. The Indiana Flood Control and Water Resources Commission, (Commission), is authorized to mediate disputes arising between the users of surface waters. Such mediation is not binding and does not preclude redress to the courts. The Commission is also empowered to contract to dispose and sell water stored in state financed projects with the governor's approval. Proceeds from such contracts will go into the Water Resources Development Fund, and will be used by the state to further water resources development. (Dunham-Florida)
W73-09726

DRAINAGE MAINTENANCE AND REPAIR DISTRICTS; MISCELLANEOUS PROVISIONS.
Ind Ann Stat secs, 27-247, 27-248, 27-411 thru 27-420 (Supp. 1972).

Descriptors: *Indiana, *Legislation, *Drainage programs, *Drainage districts, *Assessments, Drainage engineering, Controlled drainage, Drainage systems, Water policy, Land use, Soil conservation, Flood control, Multiple purpose projects, Water management (Applied), Water control, Taxes, Artificial watercourses, Ditches, Channels, Penalties (Legal).

The Commissioners of Drainage Maintenance and Repair Districts appointed by the court shall keep and maintain dredge ditches and drains in proper condition and shall have full power and authority to do all acts necessary and incident thereto. The procedure for levying assessments to raise drainage maintenance funds is set forth, and no obligations by the commissioners may be made until the assessments are levied. The commissioners are required to keep a complete record of expenses incurred and must file itemized statements with the county auditor. Regulations regarding compensation for and appointment of commissioners are provided. Provisions are also included for the construction of channels by private persons. No such channel may be constructed without the written approval of the Commission. (Dunham-Florida)
W73-09727

CANAL AUTHORITY V. MILLER (CHALLENGE TO COURT'S GRANTING A PERPETUAL EASEMENT RATHER THAN A FEE SIMPLE IN EMINENT DOMAIN PROCEEDINGS).
230 So 2d 193-196 (1st D.C.A. Fla. 1970).

Descriptors: *Florida, *Canals, *Eminent domain, *Adjacent landowners, Judicial decisions, Public rights, Legal aspects, Boundaries, Adjudication procedure, Administrative agencies, Public benefits, Navigable waters, Legislation, Condemnation, Easements.
Identifiers: *Cross-Florida Barge Canal, *Legislative interpretation.

This eminent domain proceeding resulted in the granting of a perpetual easement over appellee landowner's property through which the appellant canal authority, as a delegate of the state, planned construction of the Cross-Florida Barge Canal. Chapters 73, 74 and 374 of the Florida Statutes vested appellant with the power of eminent domain. Appellant was denied the order of taking for a fee simple title to such properties. The appeals court affirmed holding that an acquiring authority will not be permitted to take a greater quantity of property, or a greater interest or estate therein, than is necessary to serve the particular public use for which the property is being acquired. Evidence presented at the hearing was deemed adequate that the Canal Authority be vested with a perpetual easement to control water and land usage; it was not necessary that fee simple title be granted to the authority by the sovereign power of eminent domain. (Smith-Adam-Florida)
W73-09728

THIESEN V. GULF, F. AND A. RY. CO. (EXTENT OF LAND OWNERSHIP OF RIPARIAN INTERESTS ALONG NAVIGABLE WATERS).

78 So 491-507 (Fla. 1917).

Descriptors: *Florida, *Navigable waters, *Reparian rights, *Judicial decisions, Compensation, Condemnation, Landfills, Legal aspects, Banks, Legislation, Adjacent landowners, Eminent domain, Legal review, Common law, Navigable waters, Public rights, Docks, High water mark, Low water mark, Watercourses (Legal aspects), Penalties (Legal), Boundaries (Property).
Identifiers: Fill permits, Public trust.

Plaintiff landowner attempted to recover damages when defendant railroad company, in accordance with powers granted it by the state, filled in areas from the shoreline towards the channel opposite plaintiff's land. Plaintiff relied upon Florida Statutes purporting to vest title to such submerged lands in riparian owners. The trial court directed a verdict for defendant, holding that plaintiff had not adequately proven that his land was bounded by the low water mark of the bay. After originally affirming the directed verdict, upon rehearing the Florida Supreme Court reversed holding that since there was room for difference of opinion as to whether the boundaries extended to the high water mark of the bay, and since plaintiff had exercised rights of ingress and egress over the bay for a sufficient time, that plaintiff landowner had common law rights as riparian owners. These rights at common law could not be taken away without just compensation. Though some courts had deemed riparian rights subordinate to public rights, the railroad company had utilized the water front only incidentally for public purposes; thus the plaintiff's claim for damages was a proper one for adjudication. (Smith-Adam-Florida)
W73-09729

SUBAQUEOUS LANDS.

Del. Code Ann tit 7, secs, 6451 thru 6462 (Supp. 1970).

Descriptors: *Beds, *Beds under water, *Delaware, *Legislation, *Public lands, Ownership of beds, Leases, Permits, Legal aspects, Administrative agencies, Regulation, Easements, State governments, Administrative decisions, State jurisdiction, Water law, Water policy, Legal review, Governments, Water utilization, Water management (Applied).
Identifiers: Licenses, Public hearings.

Except as otherwise provided, the Commission and the Governor shall have exclusive jurisdiction to lease or grant permits or easements in or over subaqueous lands of the state and no use of such lands shall be undertaken except pursuant to a permit or grant. Each applicant for a lease, permit or grant under these provisions must file a detailed request with the Commission. A public hearing on such application shall be held: (1) if use of subaqueous lands is sought for a period of over 10 years; (2) if the Commission determines a hearing to be in the public interest; or (3) if written objection to the application is filed. The Commission may appoint a board to investigate any application. Provisions for the membership and administration of such a board are included. The Commission is to report its recommendation to the Governor who, at his discretion, may execute or refuse any lease, permit, deed or grant. (Dunham-Florida)
W73-09734

MINERALS AND DEPOSITS ON STATE LANDS--SALE OR LEASE.

Ark Stat Ann secs, 10-1001 thru 10-1013 (Supp. 1971).

Descriptors: *Arkansas, *Legislation, *Beds under water, *Mining, Ownership of beds, Leases, Permits, Administrative agencies, Lakes, Navigable waters, Mineralogy, State governments, Regulation, State jurisdiction, Water law, Public lands, Legal aspects, Lake beds, River beds, Oil, Gases, Sands, Gravels, Artificial watercourses.
Identifiers: Recording statutes.

This legislation provides for the regulation of the removal of sand, gravel, oil, coal and other minerals from the navigable rivers and lakes or lands of the state. The Commissioner of Revenues, (Commissioner), is authorized to regulate such activities by the granting of permits and/or leases. Specific details as to the types of permits issuable are provided. The Commissioner is authorized to establish and maintain a permanent record of all leases and agreements for the extraction of such minerals from navigable waters or lands of the state, and anyone receiving a lease is required to record it in the permanent ledger. The state shall not acquire title to oil, gas or minerals in and under artificially created navigable waters created by agencies of the United States or the state where such minerals are not compensated for. The private ownership of minerals in lands covered by artificially created navigable waters are subservient to the public use of such waters. Procedures for record proof of ownership in mineral rights by the owner are set forth. (Dunham-Florida)
W73-09735

GEORGIA WATERWAYS COMMISSION.

17 Ga. Code Ann secs, 301 thru 306, secs 401 thru 403 (Supp 1972).

Descriptors: *Georgia, *Legislation, *Negotiations, *Water resources development, *River basin development, Flood control, Legal aspects, Comprehensive planning, Governments, Planning, Water utilization, Governmental interrelations, State governments, Federal government, Multiple-purpose projects, Watershed management, Land management, Drainage systems, Administrative agencies.

The Georgia Waterways Commission (Commission) is hereby created, consisting of a chairman appointed by the Governor and six appointed members from various river basins of the state. The statutes provide that the commission shall negotiate with the proper authorities of the United States from time to time concerning the development of the rivers of Georgia, and particularly with reference to flood control. Moreover, the Commission shall be privileged to appear before committees of Congress concerning appropriations for river development, to gather and disseminate information, and to cooperate with the federal and state authorities toward the development of Georgia's rivers. However, the Commission shall have no power to make any contract binding upon the State of Georgia or to incur any obligation against the State. Information pertinent to the development of Georgia's rivers as may be obtained by the Commission may be filed from time to time with the Department of Industry and Trade and with the Department of Archives and History. (Mockler-Florida)
W73-09736

NAVIGATION COMPANIES.
17 Ga Code Ann secs, 201 thru 213 (Supp 1972).

Descriptors: *Legislation, *Navigation, *Water policy, *Georgia, *Transportation, Reasonable use, Administration, Legal aspects, Navigable waters, Water law, Harbors, Non-consumptive use, Ships, Public benefits, Water allocation (Policy), State governments, Social needs, Social mobility, Boats.
Identifiers: *Corporations, *Navigation companies.

This chapter provides for the establishment and creation of navigation companies, how they may become incorporated, regulations concerning petition for incorporation, certificates of incorporation, organization of the companies, and the powers of incorporation. The statutes provide that any number of persons, not less than five, who desire to be incorporated for that purpose, may form a company by filing a written petition, addressed to the Secretary of State and stating the names and residences of each of the persons desiring to form said corporation. Other required information includes the name of the navigation company they desire to have incorporated, the amount of the proposed capital stock, the number of years it is to continue and the place where its principal office is to be located. The statutes further provide that said company shall be empowered to acquire, purchase, hold, and operate all such real and personal property as may be necessary or convenient for the maintenance and operation of its said business, to provide services incident to a general navigation business, to erect structures to accommodate freight and passengers, to regulate fees and shipping schedules, and to borrow funds to finance its operations. (Mockler-Florida)
W73-09737

GEORGIA WATER QUALITY CONTROL ACT.

17 Ga Code Ann secs, 501 thru 530 (Supp. 1972).

Descriptors: *Georgia, *Legislation, *Water pollution, *Water quality control, *Water purification, *Water supply development, Legal aspects, Governments, Water resources development, Administrative agencies, Water law, Water policy, Water utilization, Comprehensive planning, Public health, State governments, Water pollution control, ecology, Environmental effects, Administrative decisions.
Identifiers: Administrative regulations.

It is declared to be the policy of the state that its water resources shall be utilized prudently to the maximum benefit of the people in order to restore and maintain a reasonable degree of water purity and to require, where necessary, reasonable treatment of sewage, industrial wastes, and other wastes prior to their discharge into the waters of the state. The act requires that the Water Quality Control Board (Board) be created within the State Department of Public Health and that this agency be charged with the duty and authority to require the use of reasonable methods, after having considered the technical means available, for the reduction of pollution and the economic factors involved to prevent and control pollution. Moreover, the Act provides administrative facilities and procedures within the executive branch and to confer discretionary administrative authority upon the Board to establish procedures designed to best protect the public interest. The powers and duties of the board are listed along with provisions for its administration and budgeting. (Mockler-Florida)
W73-09738

DRAINAGE, SYSTEM OF COUNTY DRAINAGE.

23 Ga Code Ann secs, 2501 thru 2573 (Supp. 1972).

Descriptors: *Georgia, *Drainage systems, *Legislation, *Water management (Applied), Legal aspects, Water quality control, Administration, Easements, Canals, Dams, Economics, Remedies, Water control, Drainage programs, Water utilization, Flood control, Water conservation, Land use, Project planning, Administrative agencies, Administrative decisions, Legal review, Financing, Bond issues, State governments, Local governments.
Identifiers: Administrative regulations.

In an attempt to fulfill the state's policy of draining and reclaiming wet, swamp, or overflowed lands for the public benefit, these provisions authorize each county to establish and maintain a system of lowland drainage and acquire the right-of-way and other easements necessary for the construction of structures incident to such a system. The right-of-way may be acquired from the landowners, with their consent, for the consideration of benefits conferred upon them by the drainage established. Procedures are established to initiate drainage districts, maintain established districts, and provide for their administration and budgeting. There is provided a framework for hearings and appeals for those landowners affected by the creation, administration or project planning of a drainage district. Provisions relating to financing through tax assessments and bonds with remedies for bond holders are also included. (Mockler-Florida)
W73-09739

CONDEMNATION BY COUNTIES FOR WATERSHEDS.

36 Ga Code Ann secs, 401 thru 1405 (Supp 1972).

Descriptors: *Georgia, *Eminent domain, *Condemnation, *Watershed management, *Legislation, Flood prevention, Easements, Legal aspects, Public access, Right-of-way, Access routes, Flood protection, Flood control, Local governments, Recreation facilities.

Every county of the State of Georgia is granted the power of eminent domain for the purpose of taking by condemnation necessary easements and property rights to enable the county to complete small watershed projects and improve watershed protection and flood control. The right of eminent domain may also be used to acquire lands for certain public recreation facilities to be developed in connection with watershed projects. The power of eminent domain includes the right to condemn land to allow for rights of ingress and egress from any such project. The procedures to be followed for such takings are those set forth by the pertinent eminent domain statutes of the state. The condemnor shall be required to condemn the fee simple title to any land which will be covered by permanent flooding. (Glickman-Florida)
W73-09740

WATER RIGHTS, NAVIGABLE STREAMS, ETC.

85 Ga. Code Ann. secs. 1301 thru 1312 (Supp. 1972).

Descriptors: *Georgia, *Beds, *Legislation, *Streams, Beds under water, Ownership of beds, Streambeds, Low water mark, Navigable waters, Tidal waters, Tidal streams, Water law, Dams, Embankments, Bridges, Legal aspects, State governments, Boundaries (Property), Riparian rights, Adjacent landowners, Shores, Accretion (Legal aspects), Boundary disputes, Water rights.
Identifiers: *Non-navigable streams.

Running water belongs to the owner of the land it is on, but he has no right to divert it or interfere with the enjoyment of it by the next owner. The beds of nonnavigable streams belong to the owner of the adjacent land, with the center of the main current the dividing line between two landowners. Accretions of land on either side belong to the owner. The rights of the owner of lands adjacent to navigable streams extend to the low water mark of the streambed. The owner of a nonnavigable stream is entitled to exclusive possession thereof. Owners of land on both sides of nonnavigable streams may construct dams across them. The title to beds of all tide waters shall belong to the owner of the adjacent land. The rights of the owners shall extend to the low water mark in the bed. Owners of lands on any watercourses are authorized to ditch and embank their land to protect it from overflows of the watercourse. The right to construct a private bridge is appurtenant to the land, but the right to establish a public bridge is a franchise of the state. (Glickman-Florida)
W73-09741

DRAINAGE OF LANDS ADJACENT PUBLIC ROADS,

Del. Code Ann. tit. 17, sec. 901 (Supp. 1970).

Descriptors: *Delaware, *Surface drainage, *Drainage, *Controlled drainage, Drainage programs, Administration, Legal aspects, Legislation, Water law, Water policy, Drainage engineering, Contours, Ditches, Drainage area, Highway effects, Alteration of flow, Surface runoff.

The Department of Highways and Transportation shall cause to be drained by sufficient ditches or conduits, waters which have backed upon lands along the right of way of any road or highway caused by changing the natural contour lines and levels, along and over the area confined within such right of way, whereby the natural flow and drainage from such lands is hindered or stopped. (Glickman-Florida)
W73-09742

FLOOD LOSS PREVENTION.

Ark. Stat. Ann. secs. 21-1901 thru 21-1904 (Supp. 1971).

Descriptors: *Flood plain zoning, *Land management, *Arkansas, *Legislation, *Flood control, Regulation, Floods, Flood damage, Building codes, Flood protection, Flood plains, Public health, Drainage systems, Floodproofing, Zoning, Land development, Land use, Penalties (Legal), Local governments.
Identifiers: Nuisance (Legal aspects), Injunctive relief.

lands, flood control, and water conservation, the court of probate of any county within the state is given jurisdiction and authority to establish water management districts upon entertaining a petition by a specified number of landowners within the proposed district. Upon the organization of a district the court is authorized to appoint a board of water management commissioners with replacement also by appointment. The commissioners are authorized to appoint a district engineer to aid them in their work. Specific provisions relating to the powers and duties of the commissioners acting with the county probate court are set forth including such areas as: eminent domain, levying taxes and issuing bonds to finance district programs, contractual authority, and annexing land to the district. Also included are provisions for notice-giving, public hearings, and appeal for persons affected by decisions or programs of the commissioners. (Dunham-Florida)
W73-09748

6G. Ecologic Impact of Water Development

FORESTS, WETLANDS AND WATERFOWL POPULATIONS IN THE TURTLE MOUNTAINS OF NORTH DAKOTA AND MANITOBA,
North Dakota State Univ., Fargo.
For primary bibliographic entry see Field 02I.
W73-09136

ECOLOGY. THE EXPERIMENTAL ANALYSIS OF DISTRIBUTION AND ABUNDANCE,
C. J. Krebs.
Harper and Row, Publishers: New York, N.Y., London, Engl, 1972. 694 p. Illus. Maps, Pr. $14.95.
Identifiers: Abundance, *Animals, Books, Distribution, *Ecology, *Mathematical models, *Plants, Water pollution control.

Special interest is planned on problems in ecology and substantiating these problems with examples from diverse representatives from the plant and animal kingdoms. The first section, what is ccology, contains a brief introduction to the science of ecology and its basic problems and approach to these problems. Section 2, the problem of distribution in populations, discusses the following limitations on distribution: dispersal, behavior, interrelations with other organisms, temperature, moisture, other chemical and physical factors; and adaptation complications. Section 3, the problem of abundance in populations, describes the following aspects: population parameters; demographic techniques; population growth; species interactions, including competition and predation; natural regulation of population size; examples of population studies; applied problems, including optimum yield and biological control. Part 4, distribution and abundance at the community level, covers the following topics: community parameters; the nature of the community; community change; community metabolism, both primary and secondary production; species diversity; community organization; and evolutionary ecology. The final section, human ecology, discussed a variety of aspects of human population and food production. Two appendixes (estimation of the size of the marked population in capture-recapture studies, and instantaneous and finit rates), a glossary, a list of mathematical symbols, an extensive bibliography and a species index are included.--Copyright 1972, Biological Abstracts, Inc.
W73-09547

ENVIRONMENTAL AND ECONOMIC IMPACT OF RAPID GROWTH ON A RURAL AREA: PALM COAST,
Southeastern Environmental Services, Jacksonville, Fla.
For primary bibliographic entry see Field 05G.
W73-09656

THE UNITED NATIONS AND THE ENVIRONMENT,
Conference on the Human Environment, New York.
M. F. Strong.
World Eco-Crisis is available, as W-105, from University of Wisconsin Press, Madison, for $2.50. In: World Eco-Crisis: International Organization in Response. University of Wisconsin Press, Madison, Wisconsin. p 3-6. June 1972.

Descriptors: *United Nations, *Environmental control, *Planning, *Documentation.

The complex nature of environmental problems is explained. The concept of 'human environment' is developed to encompass natural and man-made pollution, use and misuse of natural resources, population, food production, etc. The unity of the natural environment creates a chain of cause and effect relationships which link local and global concerns. The role of the U.N. in gathering information and implementing action is discussed. (See W73-09659 thru W73-09664) (Ensign-PAI)
W73-09658

THE ROLE OF THE UN IN ENVIRONMENTAL PROBLEMS,
Columbia Univ., New York.
R. N. Gardner.
In: World Eco-Crisis: International Organizations in Response. University of Wisconsin Press, Madison, Wisconsin. p 69-86, June 1972.

Descriptors: *United Nations, *Environmental control, *Pollution abatement, *Planning.
Identifiers: *Stockholm conference.

The state of the world environment and particnlarly the problems of air and water pollution are topics of concern for the United Nations. The UN is the only framework available for unilateral cooperation among nations. Although pollution problems have been created largely by the technologically 'advanced' countries of the world, the current and future actions of less developed states can have serious effects on the global environment. The UN's efforts to plan a conference on the Human Environment in Stockholm are reviewed. The basic organizational principle underlying the Conference, as well as the range of organizational choices, are discussed in some detail. The success of the Stockholm conference is contingent upon the ability of the participating nations to devise a workable pattern of international cooperation for dealing with environmental problems. (See also W73-09658) (Ensign-PAI)
W73-09659

ADDING THE TIME DIMENSION TO ENVIRONMENTAL POLICY,
Alfred P. Sloan School of Management, Cambridge, Mass.
D. L. Meadows, and J. Randers.
In: World Eco-Crisis: International Organization in Response. University of Wisconsin Press, Madison, Wisconsin. p 47-66, June 5, 1972. 5 fig, 1 tab.

Descriptors: *DDT, *Solid wastes, *Systems analysis, *Environmental effects, *Planning, Economics, Projections, Decision making.
Identifiers: Policy considerations, Time factors.

Problems of environmental deterioration are by their nature longterm and not always clearly apparent. As such, they do not force immediate corrective action. Data gathered on two studies, one concerning DDT in the environment, the other on solid waste are presented in some detail. Systems-analysis was used to bring together existing information about DDT and solid waste flows and then to assess the dynamic implications of that information in order to instigate and/or influence long-term policy considerations. In each of the two pollutant studies discussed, the negative factor of

time delays is repeatedly mentioned. The environmental costs of pollution in each case extend for a much longer time period than do the benefits derived from the polluting activity. However, once a pollution limiting policy has been established it may take as many as 10 years before its first benefits become obvious. Any new policies must be based on an active commitment to the future. (See also W73-09658) (Ensign-PAI)
W73-09660

THE UNITED NATION'S INSTITUTIONAL RESPONSE TO STOCKHOLM: A CASE STUDY IN THE INTERNATIONAL POLITICS OF INSTITUTIONAL CHANGE,
Sussex Univ., Brighton (England). Inst. for the Study of International Organization.
B. Johnson.
In: World Eco-Crisis: International Organization in Response. University of Wisconsin Press, Madison, Wisconsin. p 87-134, June 1972. 2 append.

Descriptors: *United Nations, *Environmental control, *Monitoring, *Governments, *Decision making.
Identifiers: National enforcement, Public opinion.

Attention is focussed on the implications of international response to environmental problems. An attempt is made to clarify the present distribution of environmental responsibilities among the major international organizations. Certain principles and possible approaches to the problem of organizing the world's environmental monitoring and protection activities are discussed. Attitudes of various governments, scientific bodies, and the general public are explored. George Kennan's proposal for a world ecological agency is reviewed in light of power distributions, the role of emerging countries, etc. Practical institutional alternatives are proposed; the optimum design developed remits execution of international policy primarily to national governments. At the international level, institutions and policies should be flexible, pragmatic, and cautious. (See also W73-09658) (Ensign-PAI)
W73-09661

THE ROLE OF SPECIAL PURPOSE AND NONGOVERNMENTAL ORGANIZATIONS IN THE ENVIRONMENTAL CRISIS,
International Council of Scientific Unions, Rome, (Italy). Special Committee on Problems of the Environment.
J. E. Smith.
In: World Eco-Crisis: International Organizations in Response. University of Wisconsin Press, Madison, Wisconsin. p 135-159, June 1972. 1 tab.

Descriptors: *Environmental control, *International commissions, *Inter-agency cooperation, *Water quality, *Planning, Natural resources.

Special purpose and nongovernmental organizations can serve important functions in the area of environmental improvement. This study provides information on such groups as the International Council of Scientific Unions (ICSU), the World Health Organization (WHO), the Food and Agriculture Organization (FAO), and the World Meteorological Organization (WMO). These bodies serve to promote international collaborative inquiry into the state of the environment; investigate the quality and quantity of man-made environmental changes; and note the effects of such changes on man and living ecosystems. Some of the specific projects undertaken in areas such as water quality and biological investigations are discussed. Attention is given to ways in which the present institutional arrangements might be better utilized and areas in which new processes are necessary for better coordination, communication, and implementation of research. (See also W73-09658) (Ensign-PAI)
W73-09662

THE ROLE OF THE WMO IN ENVIRONMENTAL ISSUES,
World Meteorological Organization, Geneva (Switzerland).
D. A. Davies.
In: World Eco-Crisis: International Organizations in Response. University of Wisconsin Press, Madison, Wisconsin. p 161-170, June 1972.

Descriptors: *Environmental control, *Monitoring, *Air pollution, *Water pollution, Path of pollutants, Sampling, *Inter-agency cooperation, Pollution abatement.
Identifiers: *World Meteorological Organization, Pollution control.

The role of the World Meteorological Organization (WMO) in identifying and alleviating world pollution problems is reviewed. Areas such as monitoring the atmosphere, research on climatic changes, studies of pollutant transport and water resource development are examined. Atmospheric monitoring, for example, is carried on by the WMO's World Weather Watch which surveys the atmosphere on a daily basis and supplies all the countries of the world with necessary meteorological data. The organization and routine processes of the World Weather Watch are described. The importance of cooperation between the WMO and other agencies is acknowledged. (See also W73-09658) (Ensign-PAI)
W73-09663

INTERNATIONAL INSTITUTIONS FOR ENVIRONMENTAL MANAGEMENT,
Council on Environmental Quality, Washington, D.C.
G. J. MacDonald.
In: World Eco-Crisis: International Organizations in Response. University of Wisconsin Press, Madison, Wisconsin. p 207-235, June 1972. 2 tab.

Descriptors: *Environmental control, *Governments, Pollutants, *Pollution abatement, *Economics, *Legislation, *Management.
Identifiers: Worldwide institutions, National control.

Various environmental problems facing the nations of the world are outlined. Pollutants are ordered by references to their lifetime or resident time in air, water, or land, where they are introduced, and what their potential effects are. In this way it can be better established which problems are best treated by bilateral, regional, or worldwide institutions. The economic consequences of environmental measures are also studied. Since the additional cost of production due to pollution control measures is or will be reflected in higher prices which will in turn effect world trade, international mechanisms must be developed for reaching fair agreements. The use and misuse of natural resources, including the land itself, is also considered. The report concludes that environmental problems are many-faceted and are usually best dealt with at a national, rather than international level. Furthermore, the lack of tradition and capabilities of enforcement by international organizations support the argument for national measures. (See also W73-09658) (Ensign-PAI)
W73-09664

SHORTCUT PIPELINE MODIFICATION CONTRA COSTA CANAL UNIT, CENTRAL VALLEY PROJECT, CALIFORNIA (FINAL ENVIRONMENTAL STATEMENT).
Bureau of Reclamation, Washington, D.C.
For primary bibliographic entry see Field 08A.
W73-09696

WAHKIAKUM COUNTY CONSOLIDATED DIKING DISTRICT NO. 1, WAHKIAKUM COUNTY, WASHINGTON (FINAL ENVIRONMENTAL IMPACT STATEMENT).
Army Engineer District, Portland, Oreg.
For primary bibliographic entry see Field 08D.
W73-09697

PROSPERITY LAKE, SPRING RIVER AN[D] TRIBUTARIES, MISSOURI-KANSA[S]-OKLAHOMA (FINAL ENVIRONMENTA[L] STATEMENT).
Army Engineer District, Tulsa, Okla.
For primary bibliographic entry see Field 08A.
W73-09698

CORNUDAS, NORTH AND CULP DRAW WATERSHED, HUDSPETH COUNTY, TEXA[S] AND OTERO COUNTY, NEW MEXICO (FINA[L] ENVIRONMENTAL IMPACT STATEMENT).
Soil Conservation Service, Washington, D.C.
For primary bibliographic entry see Field 08A.
W73-09699

LOCAL PROTECTION PROJECT, LOG JA[M] REMOVAL ON SALAMONIE RIVER NEA[R] BLUFFTON IN WELLS COUNTY, INDIAN[A] (FINAL ENVIRONMENTAL STATEMENT).
Army Engineer District, Louisville, Ky.
For primary bibliographic entry see Field 04A.
W73-09700

ENVIRONMENTAL PROTECTION ACT O[F] 1971.
For primary bibliographic entry see Field 06E.
W73-09711

07. RESOURCES DATA

7A. Network Design

RECOMMENDED METHODS FOR WATER DATA ACQUISITION.

Preliminary Report of the Federal Interagency Work Group on Designation of Standards for Water Data Acquisition: Geological Survey, Office of Water Data Coordination, December 1972. 415 p.

Descriptors: *Water resources, *Data collections, *Methodology, *Testing procedures, *Water sampling, Surface waters, Groundwater, Fluvial sediments, Water quality, Biological properties, Inorganic compounds, Physicochemical properties, Monitoring, Automation.
Identifiers: *Water data acquisition methods.

This report documents results of interagency efforts during 1970-72 to designate recommended methods for acquisition of water data. Methods specified were identified as the most acceptable of those currently in use. The methods represent the findings of six task groups established by the Federal Interagency Work Group on Designation of Standards for Water Data Acquisition. This Work Group was established under the auspices of the Geological Survey's Office of Water Data Coordination. The task groups consisted of more than 60 technical experts in the field of water data from six departments and two independent agencies of the federal government. Task groups covered the following areas of concern: (1) surface-water stage and quantity; (2) chemical (inorganic) and physical quality; (3) biologic, bacteriologic, and chemical (organic) quality; (4) sediment; (5) groundwater; and (6) automatic water-quality monitors. The findings of the task groups are presented as separate chapters. (Woodard-USGS)
W73-09161

HYDROLOGY OF SMOOTH PLAINLANDS OF ARID AUSTRALIA.
Department of National Development, Canberra (Australia).
For primary bibliographic entry see Field 07C.

Descriptors: *Mixing, *Currents (Water), *Oceans, *Aerial photography, *Tracking techniques, Ocean currents, Diffusion, Dispersion, Winds, Waves (Water).
Identifiers: Langmuir cells, Ekman spirals.

Circulation rates within the mixed layer of the ocean were made based on aerial photography of dye injections and floating cards in the surface and near-surface levels under different wind, sea state, and thermal profile conditions. Except under calm conditions, the first few meters of the sea are subjected to helical flow of small-size Langmuir cells with 3 to 6 meter spacing between the convergence lines. Under moderate to strong wind conditions (10-30 knots), a hierarchy of larger-size Langmuir cells is developed. The maximum horizontal spacing between the larger cells is approximately the same as the depth of the mixing layer. The spacing of the convergence zones between the largest cells was 280 meters, and they were accompanied by medium and small-size cells of approximately 35 and 5 meters, respectively. Estimates of the apparent vertical diffusion associated with the large and medium Langmuir cells are thousands and hundreds of sq cm/sec, respectively. A stratified flow resembling an Ekman spiral was observed under moderate conditions. The transition from the Ekman to the Langmuir regime occurs with turbulence Reynolds numbers of approximately 100. Under calm conditions following a period of moderate winds and thermal instability, convergence zones have anticyclonic rotation. (Knapp-USGS)
W73-09168

PRECISION REQUIREMENTS FOR A SPACECRAFT TIDE PROGRAM,
National Oceanic and Atmospheric Administration, Miami, Fla. Atlantic Oceanographic and Meteorological Labs.
B. D. Zetler, and G. A. Maul.
Journal of Geophysical Research, Vol 76, No 27, p 6601-6605, September 20, 1971. 3 fig, 2 tab, 10 ref.

Descriptors: *Satellites (Artificial), *Water level fluctuations, *Tides, Data collections, Radar, Statistics.

A tidal analysis of sea-surface elevations from an orbiting altimeter will include errors due to the instrument and the orbital determinations. Furthermore, the results may be somewhat degraded by the small amount of data and their random distribution in time. These effects were evaluated in a numerical feasibility investigation with the use of observed hourly heights at two tide stations and a hypothetical satellite orbit with a 30 deg inclination and a period of about 95 min. For a station at 13 deg N with a water elevation assumed to be obtained each time the altimeter transits a 1 deg square centered at the station, 86 observations are obtained in a year. For a station at 22 deg N, areas of 1 deg to 5 deg square were used, and the number of observations in a year varied from 112 to 551. Degradation of the signal was studied by introducing sequentially increased levels of white noise to the observations. The harmonic constants for the smaller constituents deteriorate rather quickly as white noise is added, but the largest constituents appear to be reasonably determined even when random numbers with extreme absolute values as large as the amplitude are added to the data. The results indicate that previously stated minimum precision requirements for a spacecraft tide program can be relaxed by roughly an order of magnitude. (Knapp-USGS)
W73-09170

MECHANISM OF RESPIRATORY EXCHANGES IN AQUATIC ENVIRONMENT. A GENERAL REVIEW INCLUDING PERSONAL RESULTS,
Amiens Univ. (France). Laboratoire de Physiologie.
For primary bibliographic entry see Field 05C.
W73-09206

REMOTE SENSING AS AN AID TO THE MANAGEMENT OF EARTH RESOURCES,
California Univ., Berkeley. Space Sciences Lab.
R. N. Colwell.
American Scientist, Vol 61, No 2, March-April, 1973, p 175-183, 9 fig, 1 tab, 6 ref.

Descriptors: *Remote sensing, *Resources, *Radar, Infrared radiation, Electromagnetic waves, Satellites (Artificial), California, Vegetation, *Photography, Agriculture, Management.
Identifiers: *Renewable resources, Nonrenewable resources, *Imagery analysis, Spacecraft, Iuventory (Resources).

In the field of remote sensing--acquiring information through the use of cameras and related devices, such as radar and thermal infrared sensors, operated from aircraft and spacecraft--several useful developments have taken place which aid in the acquisition of better and more timely resource information and hence in better resource management. Remote sensing is expected to be of greatest use to mankind in the monitoring of renewable resources, allowing, for example, agriculturists to predict each year the best time to mobilize the labor force, equipment, and food-processing facilities that will be needed for the harvest and to determine how much of each component will be needed. Multiband, multidate, and multistage photographs and their uses are explained, along with a series of color plates illustrating the various types of photographs and their interpretation. Thermal infrared and radar imagery techniques are also outlined. It is shown that a considerable savings in sampling and surveying costs may be realized by the use of remote sensing methods, by greatly reducing the number of on-the-ground observations. (Campbell-NWWA)
W73-09262

INSTRUMENTATION FOR ENVIRONMENTAL MONITORING: WATER.
California Univ., Berkeley. Lawrence Berkeley Lab.

Lawrence Berkeley Laboratory LBL-1, Vol 2, February 1, 1973. 236 p. NSF AG-271.

Descriptors: *Water quality, *Chemical analysis, *Monitoring, *Instrumentation, Analytical techniques, Water pollution sources, Water quality control, Remote sensing, Research equipment, Research and development, Path of pollutants.

A survey of instrumentation for environmental monitoring is being carried out by the Lawrence Berkeley Laboratory, University of California, under a grant from the National Science Foundation. Instruments being investigated are those useful for measurements in air quality, water quality, radiation, and biomedicine related to environmental research and monitoring. Consideration is given to instruments and techniques presently in use and to those developed for other purposes but having possible applications to this work. The results are given as descriptions of the physical and operating characteristics of available instruments, critical comparisons among instrumentation methods, and recommendations of promising methodology and development of new instrumentation. The survey material will be compiled in four looseleaf volumes which can be periodically updated. This is volume 2. Emphasis in this volume is on monitoring instruments suitable for freshwater and estuarine water. Chemical and biological contaminants have been included as well as physical parameters. (Woodard-USGS)
W73-09268

HYDRAULIC EQUIVALENT SEDIMENT ANALYZER (HESA),
Massachusetts Univ., Amherst.
F. S. Anan.
Available from NTIS, Springfield, Va 22151 AD-748 652 Price $3.00 printed copy; $1.45 microfiche.

Coastal Research Center Technical Report No 3-CRC, July 1972. 38 p, 4 fig, 6 tab, 23 ref.

Descriptors: *Sediments, *Particle size, *Sediment yield, *Instrumentation, Design, Analytical techniques, Data collections, Methodology, Automation, Data transmission, Charts, Sedimentation, Settling velocity, Measurement.
Identifiers: *Sediment analyzer (HESA).

In 1966 construction began on the sediment settling tube (HESA) now being used by the Coastal Research Center, University of Massachusetts. The tube was built to circumvent the disadvantages and inadequacies of sieves, as well as some of the existing settling tubes. The HESA is a sensitive relay-recording settling tube that provides the following features: (1) a continuous cumulative curve based on the hydraulic equivalence of the sediment particles; (2) a long fall distance, 200 cm, which permits adequate separation of the grains; (3) a large internal diameter of the tube, 20.32 cm, which, in conjunction with a venetian blind gate, allows the use of a representative sample of about 20 grams; (4) a chart ten inches wide which enables detection of minor variations in the cumulative curve; (5) an initial weighing station which provides the use of the full width of the chart for each sample, regardless of its weight, and determination of the exact amount of the fine material remaining in suspension in the tube at the end of each analysis; and (6) a sensitive measurement system based on weight rather than pressure. (Woodard-USGS)
W73-09282

UNDERWATER CAMERA SYSTEM FOR DEEP-SEA BOTTOM PHOTOGRAPHY,
Department of Scientific and Industrial Research, Wellington (New Zealand). Oceanographic Inst.
For primary bibliographic entry see Field 05A.
W73-09329

DEVICE FOR DETECTING AND MEASURING ACTIVITY OF LARGE MARINE CRUSTACEANS,
Marine Dept., Wellington, (New Zealand). Fisheries Research Div.
R. F. Coombs.
New Zealand Journal of Marine and Freshwater Research, Vol 6, Nos 1 and 2, p 194-205, June 1972. 8 fig, 12 ref.

Descriptors: *Crustaceans, *Marine animals, *Laboratory equipment, *Measurement, Invertebrates, Electrodes, Lobster, Bioassay, Research equipment, Animal physiology, Animal behavior, Voltage regulations, Water temperature, Dissolved oxygen, Electrolytes, Sea water.
Identifiers: Locomotor activity, *Sensors, *Macroinvertebrates, Detectors, Jasus edwardsii, Decapods, Arthropods, Sensitivity, Digital recording system, Recorders.

A device is described for detecting and measuring the locomotor activity of large marine crustaceans in captivity. The device consists of two electrodes: one of stainless steel, which acts as a sensing surface, and the other of aluminum immersed in a tank containing sea water. The sensor is essentially a simple galvanic cell, of which the electrolyte is the sea water in which the animal is maintained. When the cell circuit is initially closed a current flows which rapidly decreases as the cell polarizes mainly because of a build-up of hydrogen gas on the cathode. Any disturbance in the water displaces gas and the current produced by the cell increases. The activity sensor described was devised to record the activity of the spiny lobster Jasus edwardsii, and operates by detecting movements in the water and by contact of the animal with a sensing surface. It is effective in a large tank, and may be built at modest cost. It can, however, be used only for marine animals. A punched-tape recording system used with this device is also described. (Holoman-Battelle)
W73-09330

AN INSTRUMENT FOR MEASURING CONDUCTIVITY PROFILES IN INLETS,
British Columbia Univ., Vancouver. Inst. of Oceanography.
For primary bibliographic entry see Field 05A.
W73-09349

SOME DISTRIBUTION-FREE PROPERTIES OF THE ASYMPTOTIC VARIANCE OF THE SPEARMAN ESTIMATOR IN BIOASSAYS,
California Univ., Los Angeles. School of Public Health.
For primary bibliographic entry see Field 05C.
W73-09359

INVERSE SAMPLING OF A POISSON DISTRIBUTION,
Commonwealth Scientific and Industrial Research Organization, Adelaide (Australia). Div. of Mathematical Statistics.
H. Weiler.
Biometrics, Vol 28, No 4, p 959-970, December 1972. 1 fig, 1 tab, 6 ref.

Descriptors: *Sampling, *Statistical methods, Quality control, Probability.
Identifiers: *Confidence limits, *Poisson distribution, *Sample size, Precision.

Given a situation where an observation or 'trial' may result in a number X equals 0, 1, 2, ... of events or 'successes', with X a poisson variate of mean mu, a series of independent trials is conducted, resulting in X1, X2, ... successes respectively. The series is terminated as soon as the total number X1 plus X2 plus ... plus Xn of successes has reached or exceeded a preassigned number k. Let Nk be the number of trials required to reach k successes. The moments, of Nk over k to m are derived for both positive and negative integer values of m. Exact confidence limits for mu, based on Nk, are derived, and the width of the confidence interval in relation to the size of the trial unit is discussed. In applications to counts of particles in suspension, the most suitable dilution and size of the trial unit are investigated to ensure an approximate Poisson distribution. (Little-Battelle)
W73-09361

INVERSE BINOMIAL SAMPLING AS A BASIS FOR ESTIMATING NEGATIVE BINOMIAL POPULATION DENSITIES,
D. J. Gerrard, and R. D. Cook.
Biometrics, Vol 28, No 4, p 971-980, December 1972. 2 fig, 2 tab, 11 ref.

Descriptors: *Sampling, *Estimating, *Statistical methods, *Population, *Distribution pattern, Probability.
Identifiers: *Data interpretation, Poisson distribution, Variance.

A sequential sampling procedure has been developed which yields minimum-variance, unbiased estimates of a mean density population and relies solely upon the presence or absence of individuals in sampling units. Using inverse binomial samples of presence or absence, the population densities of negative binomially distributed organisms are characterized by a known parameter k. The estimator is shown to be the only unbiased, uniformly minimum variance estimator obtainable from inverse binomial samples. A comparison of its cost efficiency with that of conventional sampling indicates that the proposed technique offers a promising, economical alternative whenever direct enumeration of populations is either difficult or impractical. (Little-Battelle)
W73-09362

QUANTAL RESPONSE ANALYSIS FOR A MIXTURE OF POPULATIONS,
Exter Univ. (England).
For primary bibliographic entry see Field 05A.
W73-09363

BAYESIAN ESTIMATION AND DESIGN OF EXPERIMENTS FOR GROWTH RATES WHEN SAMPLING FROM THE POISSON DISTRIBUTION,
American Cyanamid Co., Stamford, Conn.
For primary bibliographic entry see Field 05A.
W73-09364

WATER POLLUTION MONITORING,
California Univ., Berkeley. Dept. of Geosciences Engineering.
For primary bibliographic entry see Field 05A.
W73-09369

TEMPERATURE COEFFICIENTS AND THEIR COMPENSATION IN ION-SELECTIVE SYSTEMS,
Foxboro Co., Mass.
For primary bibliographic entry see Field 05A.
W73-09371

AQUATIC SEDIMENT AND POLLUTION MONITOR,
For primary bibliographic entry see Field 05A.
W73-09390

WATER SAMPLING DEVICE,
Department of the Navy, Washington, D.C. (assignee)
R. S. Keir.
U. S. Patent No. 3,714,830, 3 p, 1 fig, 4 ref; Official Gazette of the United States Patent Office, Vol 907, No 1, p 55, February 6, 1973.

Descriptors: *Patents, *Sampling, Water analysis, Laboratory tests, Water quality, Pollutant identification, Water pollution.

This water sampling device utilizes an Erlenmeyer flask, which has openings at the top and bottom to permit the passage of water as it is lowered to the desired depth. The openings are regulated so as to close at the selected water depth to capture a water sample at that level. When the sampler is raised to the surface the laboratory analysis can be made in the same container. The avoidance of the need to transfer the sample to another container minimizes the likelihood of coatmination. (Sinha-OEIS)
W73-09401

FLOW-METER FOR UNDERGROUND IRRIGATION PIPELINES,
Indian Agricultural Research Inst., New Delhi. Div. of Agricultural Engineering.
D. R. Arora, and M. S. Rao.
In J Agric Sci. Vol 41, No 11, p 914-919. 1971. Illdian

Identifiers: *Flow measurement, Flow, *Irrigation, Meters, Pipelines, Underground, Instrumentation.

A flow-meter for measuring the discharge through riser pipes of an underground-irrigation water-distribution system has been designed and developed. It is a simple instrument consisting of a metallic float with guide-sleeve, a central shaft made of brass, an indicator and a calibrated discharge, indicating plate mounted over the central shaft. Flow is directly measured in terms of ha cm/hr both under free- and submerged-flow conditions. The meter is simple to install and does not restrict the outflow from the riser.--Copyright 1973, Biological Abstracts, Inc.
W73-09423

For primary bibliographic entry see Field 02H.
W73-09447

HYDROGEOLOGICAL CRITERIA FOR EVALUATING SOLID-WASTE DISPOSAL SITES IN JAMAICA AND A NEW WATER SAMPLING TECHNIQUE,
Geological Survey of Jamica, Kingston.
For primary bibliographic entry see Field 05B.
W73-09452

FUNDAMENTAL PRINCIPLES IN THE STUDY OF WATER AND SALT BALANCES OF GROUNDWATER IN IRRIGATED REGIONS (OSNOVNYYE PRINTSIPY IZUCHENIYA VODNO-SOLEVOGO BALANSA GRUNTOVYKH VOD OROSHAYEMYKH TERRITORIY),
Institute of Hydrogeology and Engineering Geology, Tashkent (USSR).
M. S. Alimov.
Uzbekskiy Geologicheskiy Zhurnal, No 2, p 23-28, 1971. 1 tab.

Descriptors: *Groundwater, *Water balance, *Salt balance, *Irrigated land, Water table, Groundwater recharge, Infiltration, Evaporation, Analytical techniques, Evaluation, Forecasting, Lysimeters, Maps.
Identifiers: *USSR, Uzbekistan, Golodnaya Steppe.

Several methods were employed in the 'Malek' test area of the Golodnaya Steppe hydrogeological station in Uzbekistan to study evaporation and groundwater recharge by infiltration. These included the lysimeter, heat-balance, and chloride-cobalt methods and the method of cut samples. Techniques were developed for a regional evaluation of the sources of groundwater supply and discharge for irrigated lands of the steppe and for other similar regions in the Republic. (Josefson-USGS)
W73-09464

NICKEL EXPLORATION BY NEUTRON CAPTURE GAMMA RAYS,
Geological Survey, Washington, D.C.
For primary bibliographic entry see Field 05A.
W73-09480

NOVEL WET-DIGESTION PROCEDURE FOR TRACE-METAL ANALYSIS OF COAL BY ATOMIC ABSORPTION,
Bureau of Mines, Pittsburgh, Pa. Pittsburgh Mining and Safety Research Center.
For primary bibliographic entry see Field 05A.
W73-09506

A ROTARY SYSTEM FOR CALIBRATING FLOWMETERS USED FOR PLANKTON SAMPLING,
Lamont-Doherty Geological Observatory, Palisades, N.Y.
S. M. Harrison, and A. W. H. Be.
Limnol Oceanogr. Vol 17, No 1, p 152-156. 1972. Illus.
Identifiers: Calibration, *Flow meters, *Plankton sampling.

A calibration system is described consisting of a circular tank with a rotating arm from which a flowmeter is suspended and towed at speeds that can be varied up to 250 cm/sec (5 knots). The velocity of the water movement that is generated by the friction of the correct calibration factor.--Copyright 1972, Biological Abstracts, Inc.
W73-09523

VITAL STAINING TO SORT DEAD AND LIVE COPEPODS,
Virginia Inst. of Marine Science, Gloucester Point.
D. M. Dressel, D. R. Heinle, and M. C. Grote.
Chesapeake Sci. Vol 13, No 2, p 156-159. 1972.
Identifiers: *Cladocera, *Copepods, *Sorting, Staining, Vital staining.

Intra vitam staining with neutral red dye vividly stains live copepods, providing a rapid method of sorting dead copepods from live. Organisms stained and preserved in acidic media retain coloration for several days, allowing large numbers of organisms to be leisurely sorted following bioassays. Cooling the preserved samples increases color retention. Copepods and cladocerans filter and ingest carmine particles providing an indication of feeding activity. Staining with neutral red at concentrations of 3:200,000 to 1:200,000 is relatively independent of temperature over a range of 5 to 30 C.--Copyright 1972, Biological Abstracts, Inc.
W73-09539

AERIAL THERMAL SCANNER TO DETERMINE TEMPERATURES OF SOILS AND OF CROP CANOPIES DIFFERING IN WATER STRESS,
Florida Univ., Gainesville. Dept. of Climatology.
J. F. Bartholic, L. N. Namken, and C. L. Wiegand.
Agron J. Vol 64, No 5, p 603-608, 1972. Illus.
Identifiers: *Aerial thermal scanner, *Crop canopies, Gossypium-hirsutum, *Soils, *Temperatures, Water stress.

An airplane-mounted thermal scanner was used to measure irradiance in the 8- to 14-micrometer wavelength interval over an extensively instrumented agricultural area. The area included soils differing in water and tillage condition, and replicated cotton (Gossypium Hirsutum L.) plots with a wide range of plant water stress. The scanner data were recorded on analog magnetic tape and on 70-mm film. The film densities of the various soil and cotton treatments and film calibration information were determined with a microdensitometer. The observed irradiances corresponded to cotton plant canopy temperature differences up to 6C between the most and the least water-stressed plots. The irradiance data from soils showed large differences as a function of time after tillage and irrigation. Thermal imagery offers potential as a useful aid for delineating water-stressed and nonstressed fields, evaluating uniformity of irrigation, and evaluating surface soil water conditions.--Copyright 1973, Biological Abstracts, Inc.
W73-09575

A NEW METHOD FOR THE GAS CHROMATOGRAPHIC SEPARATION AND DETECTION OF DIALKYLMERCURY COMPOUNDS-APPLICATION TO RIVER WATER ANALYSIS,
National Environmental Research Center, Cincinnati, Ohio. Analytical Quality Control Lab.
For primary bibliographic entry see Field 05A.
W73-09584

AUTOMATIC DATA ACQUISITION IN THE WESTERN UNITED STATES,
Soil Conservation Service, Portland, Oreg. Water Supply Forecast Unit.
For primary bibliographic entry see Field 02C.
W73-09611

APPROACHES TO MEASURING 'TRUE' SNOWFALL,
National Weather Service, Silver Spring, Md. Office of Hydrology.
For primary bibliographic entry see Field 02C.
W73-09613

GLOBOSCOPE - A NEW LOOK AROUND,
West Virginia Univ., Morgantown. Div of Forestry.
For primary bibliographic entry see Field 02B.
W73-09620

REMOTE SENSING AND DATA HANDLING--THEIR APPLICATION TO WATER RESOURCES,
International Business Machines Corp., Gaithersburg, Md.
P. A. Castruccio.
Paper, 13th Meet Panel Sci Tech, p 209-220, 1972. 12 p, 18 fig.

Descriptors: *Remote sensing, *Data collections, Data processing, Watershed management, Research and development, Economics, Water resources, Application methods, Conservation, Hydrographs, Model studies, Rainfall, Runoff forecasting, Infiltration rate, Evapotranspiration, Water sources, Water demand.
Identifiers: Environmental quality.

Remote sensing has been a useful and growing technology since the beginning of the century. Black and white, color, and infrared photography and, recently, radar imaging, have added to the technology. A potential application for remote sensing is in conservation and better utilization of water resources. The search for new water sources has been intensified but a better, more rapid means of transferring collected data into useful watershed information is needed. The technique of transferring items and quantities measured directly from imagery onto the parameters of a model is applicable to the entire spectrum of hydrological models, from the simple ones used by local governments, to the complex formulations for large watersheds. The point of applying remote sensing to determining the hydrologic regime of watersheds is 2-fold: (1) improving the predictive accuracy of instrumented and modeled watersheds; and (2) determining the hydrologic regimes of unknown watersheds, with reductions in time, labor, and costs over present methods. Watershed management is only one example of the results of applying remote sensing to environmental conservation. Finding ways to accelerate the technology evolution process, rather than the technological development of remote sensing itself, is the challenge. (USBR)
W73-09673

7C. Evaluation, Processing and Publication

WATER RESOURCE OBSERVATORY CLIMATOLOGICAL DATA WATER YEAR 1972.
Wyoming Univ., Laramie. Water Resources Research Inst.

Available from the National Technical Information Service as PB-220 252, $3.00 in paper copy, $0.95 in microfiche. Water Resources Series No. 34, February 1973. 291 p. OWRR A-001-WYO (63).

Descriptors: *Data processing, *Data collections, *Humidity, *Air temperature, *Wyoming, *Climatic data.

Temperature and relative humidity data that have been reduced from hygrothermograph charts and precipitation data from recording and nonrecording precipitation gages are presented in tabular form. Four readings per day at 0600, 1200, 1800, and 2400 hours and the maximum and minimum are presented. The mean, maximum, and minimum temperatures are also shown graphically. The reduced data are transferred to punch cards for computation and tabulating by the University of Wyoming's digital computer. The card format and computer program are presented in WyoWRRI Water Resources Series No. 8.
W73-09107

WATER RESOURCES OBSERVATORY STREAMFLOW DATA WATER YEAR 1972,
Wyoming Univ., Laramie. Water Resources Research Inst.
L. E. Allen, and V. E. Smith.

Available from the National Technical Information Service as PB-220 250, $3.00 in paper copy, $0.95 in microfiche. Water Resources Series No 32, February 1973. 29 p. OWRR A-001-WYO (62).

Descriptors: *Streamflow, Data processing, *Wyoming, *Data collections, *Flow rates, *Water level recorders.

Data from water level recorders and rated stream sections are presented for stations operated by the Wyoming Water Resources Research Institute. The reduced data are processed through the University of Wyoming's digital computer. Program output presented herein includes station identification, mean daily flows in second-feet, total monthly flows in second-foot-days, mean monthly flows in second-feet, monthly discharges in acre-feet, and annual flow in acre-feet for each water year.
W73-09109

PROCEEDINGS OF CONFERENCE ON TOWARD A STATEWIDE GROUNDWATER QUALITY SUBCOMMITTEE CITIZENS ADVISORY COMMITTEE, GOVERNORS ENVIRONMENTAL QUALITY COUNCIL.
For primary bibliographic entry see Field 05B.
W73-09113

GROUND-WATER LEVELS IN THE LOWER ARKANSAS RIVER VALLEY OF COLORADO, 1969-73,
Geological Survey, Denver, Colo.
O. J. Taylor, and R. R. Luckey.
Colorado Water Resources Basic-data Release No 29, 1973. 28 p, 1 fig.

Descriptors: *Groundwater, *Water wells, *Water levels, *Colorado, Basic data collections, Water level fluctuations, Hydrologic data, Aquifers, Well data.
Identifiers: *Lower Arkansas River Valley (Colo).

Water levels are tabulated for about 600 wells from data collected during the spring prior to the 1973 irrigation season in the lower Arkansas River Valley of Colorado. Measurements made each spring during the 4 preceding years are included to serve as references illustrating declining or rising water levels. Previous water-level measurements in about 450 other wells are also included. The lower Arkansas River valley of Colorado extends from Pueblo, Colo., to the Kansas State line, a distance of 150 miles, and occupies 450 square miles in parts of five counties. The valley-fill aquifer which is adjacent to, underlies, and is in hydraulic connection with the Arkansas River yields groundwater which is used as a supplemental supply for irrigation in the Arkansas River valley. The valley fill is as much as 300 feet thick and consists of gravel, sand, silt, and clay of Pleistocene to Holocene ages. The fill ranges from 1 to 14 miles in width and rests in a broad trough cut into the bedrock. The bedrock consists of the Pierre Shale, the Niobrara Formation, the Carlile Shale, the Greenhorn Limestone, the Graneros Shale, and the Dakota Sandstone, all of Cretaceous age. (Woodard-USGS)
W73-09123

PROCESSING FOR SPACEBORNE SYNTHETIC APERTURE RADAR IMAGERY,
Computer Sciences Corp., Huntsville, Ala.
For primary bibliographic entry see Field 07B.
W73-09124

SURFACE WATER SUPPLY OF THE UNITED STATES, 1966-70: PART 3, OHIO RIVER BASIN-VOLUME 2, OHIO RIVER BASIN FROM KANAWHA RIVER TO LOUISVILLE, KENTUCKY.
Geological Survey, Washington, D.C.

the 4 preceding years are included to serve as reference. The aquifer supplies nearly all of the groundwater used in the South Platte River valley as a supplemental supply for irrigation. The valley fill ranges in thickness from 0 to about 300 feet; in much of the area, however, the thickness is about 50 to 200 feet. The valley fill consists of gravel, sand, silt, and clay of Pleistocene to Holocene ages. The valley fill ranges from 2 to 10 miles in width and rests in a broad trough cut into the bedrock. In northeastern Colorado bedrock consists of pre-Pleistocene sedimentary formations. A list of 29 Colorado water resources basic-data releases is included. (Woodard-USGS)
W73-09280

PROGRESS REPORT ON WATER-RESOURCES INVESTIGATIONS IN PALM BEACH COUNTY, FLORIDA,
Geological Survey, Tallahassee, Fla.
L. F. Land.
Open-file report (72014), 1972. 40 p, 19 fig, 1 tab, 3 ref.

Descriptors: *Water resources development, *Water supply, *Water quality, *Hydrologic data, *Florida, Hydrology, Groundwater, Surface waters, Canals, Lakes, Water utilization, Water levels, Water yield, Tidal effects, Saline water intrusion, Water table, Water wells, Hydrographs, Chemical analysis, Water pollution sources, Rainfall-runoff relationships, Urbanization.
Identifiers: *Palm Beach County (Fla).

The entire southeastern coast of Florida, of which Palm Beach County is a part is experiencing a rapid growth in population. To assure that the fresh-water supply will remain adequate, much hydrologic information is needed for water-resource management and development decisions. A hydrologic network was established in 1970 to collect data on canal discharge; water-level fluctuations in wells and canals; quality of water in wells, canals, and lakes; geologic and hydrologic characteristics of the shallow aquifer system; sea-water intrusion; and tidal fluctuations. The network provides the basic information from which hydrologic interpretations can be made. In May 1971 the water-table decline in the vicinity of some well fields in the extreme northeast and southeast coastal parts of the county was severe enough to permit salty water to contaminate some wells. The water levels of canals and lakes in Palm Beach County are extensively controlled. The water quality in the shallow aquifer in the county is generally acceptable for domestic, agricultural, and industrial uses in the coastal ridge area. Dissolved solids in goundwater ranges from 250 to 300 mg/liter. (Woodard-USGS)
W73-09284

FLOODS IN MAPLE PARK QUADRANGLE, NORTHEASTERN ILLINOIS,
Geological Survey, Washington, D.C.
R. T. Mycyk, and G. L. Walter.
For sale by USGS, Washington, D C 20242 Price 75 cents. Hydrologic Investigations Atlas HA-458, 1972. 1 sheet, 9 fig, 1 map, 2 ref.

Descriptors: *Floods, *Flood data, *Streamflow, *Illinois, Hydrologic data, Maps, Flood plains, Flood frequency, Flood profiles, Flood peak, Flood stages, Curves, Topographic mapping, Flood control.
Identifiers: Maple Park (Ill).

This one-sheet atlas presents hydrologic data that can be used to evaluate the extent and depth of flooding that affect the economic development of flood plains in the Maple Park quadrangle, northeastern Illinois. The approximate areas inundated by floods are delineated on a topographic map. Inundated areas for the flood of October 1954 are shown along Union ditch No. 2, Union ditch No. 3, Virgil ditch No. 1, Virgil ditch No. 2, Virgil ditch No. 3, and several unnamed streams. This flood was reported by local residents to be the highest observed in the past 65 years. The flood boundaries on the map provide a record of historic fact that reflect channel conditions existing when the floods occurred. (Woodard-USGS)
W73-09298

FLOODS IN RILEY QUADRANGLE, NORTHEASTERN ILLINOIS,
Geological Survey, Washington, D.C.
R. T. Mycyk, and R. S. Grant.
For sale by USGS, Washington, D C 20242 Price $1.00. Hydrologic Investigations Atlas HA-464, 1972. 1 sheet, 10 fig, 1 map, 1 tab.

Descriptors: *Floods, *Flood data, *Streamflow, *Illinois, Hydrologic data, Maps, Flood plains, Flood frequency, Flood profiles, Flood peak, Flood stages, Curves, Topographic mapping, Flood control.
Identifiers: *Riley (Ill).

This one-sheet atlas presents hydrologic data that can be used to evaluate the extent, depth, and frequency of flooding that affect the economic development of flood plains in the Riley quadrangle, northeastern Illinois. The areas inundated by floods are delineated on a topographic map. Local residents reported that the flood of June 1970 was one of the highest observed in the last 70 years on Coon Creek. The flood of June 1967 was about the highest observed in the last 50 years on Spring Creek. Greater floods than those whose boundaries are shown on the map are possible. The flood boundaries shown provide a record of historic fact that reflect channel conditions existing when the floods occurred. Other illustrated data summaries include flood height, flood discharge, flood frequency, and flood profiles. (Woodard-USGS)
W73-09299

ESTIMATION OF MICROBIAL VIABILITY,
Department of Scientific and Industrial Research, Palmerston North (New Zealand). Applied Mathematics Div..
For primary bibliographic entry see Field 05B.
W73-09360

CYCLIC ANALOG-TO-DIGITAL CONVERSION,
Motorola, Inc., Franklin Park, Ill.
J. Barnes.
Instruments and Control Systems, Vol 46, No 3, p 35-36, March 1973. 3 fig, 1 tab.

Descriptors: *Data processing, *Data transmission, *Electronic equipment, Sampling.
Identifiers: *Analog to digital converters, Converters.

Cyclic converters are continuous devices in which lists are calculated serially and an output word is always available. The conversion may be implemented in a series of successive approximation stages. To determine whether the bit at each stage is ONE or ZERO, the reference voltage, which equals half the full-scale of the converter, is subtracted from the unknown input. The difference is multiplied by -2, and is applied to a sign comparator. For the most significant bit (MSB) a negative value causes the sign comparator to generate a ONE and a positive value results in a ZERO. For all subsequent bits, the polarities are reversed. The output of the reference comparator is also applied to an absolute value circuit, to create a residue for use as the input to the next stage. The analog portion of each converter stage can be implemented with operational amplifiers. The reference comparator, sign comparator, and absolute value circuit are described. (Little-Battelle)
W73-09368

QUALITY OF STREAM WATERS OF THE WHITE CLOUD PEAKS AREA, IDAHO,
Geological Survey, Washington, D.C.
W. W. Emmett.
Water Resources Investigations 6-72, 1972. 5 sheets, 8 fig, 2 tab.

Descriptors: *Water quality, *Streams, *Sediment transport, *Idaho, *Chemical analysis, Sampling, Mountains, Discharge (Water), Low flow, Average flow, High flow, Correlation analysis, Maps, Graphical analysis, Drainage area, Topography, Geology, Water analysis, Inorganic compounds, Hydrogen ion concentration, Turbidity.
Identifiers: *White Cloud Peaks area (Idaho).

Water quality in the White Cloud Peaks area of Idaho is depicted on a series of maps. The area lies in the upper drainage basin of the Salmon River in south-central Idaho. In the center of this area are located the White Cloud Peaks, an aggregation of rugged mountains considered as part of the Sawtooth Mountain Range. Data for chemical and sediment quality of stream waters were collected in 1971. Each of the 39 water data stations was visited at least three times from late May to early October to determine water-quality conditions at times of low-, medium-, and high-water runoff. This determination included measurements of discharge and temperature and collection of water samples for analysis of dissolved chemical constituents and suspended sediment. Water quality in the study area is generally excellent. Deviations in the quality can be related to the geology, topography, and to a certain extent man-induced environmental impacts. Man's impact is graphically illustrated by the high suspended sediment concentrations at a stream site near poor road maintenance practices. (Woodard-USGS)
W73-09435

A RAINFALL CLIMATOLOGY OF HILO, HAWAII,
Hawaii Univ., Hilo. Cloud Physics Observatory.
For primary bibliographic entry see Field 02B.
W73-09585

GROUND-WATER RESOURCES OF BRAZORIA COUNTY, TEXAS,
Geological Survey, Austin, Tex.
For primary bibliographic entry see Field 04B.
W73-09590

GROUND-WATER RESOURCES OF DONLEY COUNTY, TEXAS,
Geological Survey, Austin, Tex.
For primary bibliographic entry see Field 04B.
W73-09591

THE CHEMICAL TYPE OF WATER IN FLORIDA STREAMS,
Geological Survey, Tallahasse, Fla.
M. I. Kaufman.
Florida Bureau of Geology, Map Series No 51, 1972. 1 sheet, 4 fig, 1 map, 8 ref.

Descriptors: *Water quality, *Chemical analysis, *Water types, *Florida, *Streams, Data collections, Sampling, Maps, Curves, Correlation analysis, Streamflow, Flow rates, Low flow, High flow, Inorganic compounds, Water chemistry.

This map report portrays inorganic chemical quality characteristics of Florida's streams and delineates broad regions of the State where stream waters are of similar chemical type (ionic composition) during low flow, and correlates chemical quality of surface waters with changes in streamflow and with certain natural and manmade environmental influences. The information should aid in determining areal coverage, frequency, and required types of analyses for a Statewide chemical-quality-sampling network representative of the differing hydrogeochemical environments in Florida. Water samples were collected from 1940 through 1967. Examination of long-term chemical data for selected streams indicates no significant changes in water-quality characteristics. Low flow refers to the lower 25% of flow based on the flow-duration data of Heath and Wimberly (1971). Five chemical types are described in terms of the dominant cations and anions, as: (A) calcium and magnesium bicarbonate, (B) sodium bicarbonate and chloride, (C) mixed type,--no dominant cation or anion, (D) sodium chloride and (E) calcium and magnesium sulfate. (Woodard-USGS)
W73-09594

A REPORT ON THE PROTOTYPE CURRENT VELOCITY AND SALINITY DATA COLLECTED IN THE UPPER CHESAPEAKE BAY FOR THE CHESAPEAKE BAY MODEL STUDY,
Johns Hopkins Univ., Baltimore, Md. Chesapeake Bay Inst.
For primary bibliographic entry see Field 02L.
W73-09595

COMPUTER MODEL FOR DETERMINING BANK STORAGE AT HUNGRY HORSE RESERVOIR, NORTHWESTERN MONTANA,
Geological Survey, Menlo Park, Calif.
For primary bibliographic entry see Field 04A.
W73-09600

CLIMATE OF THE POTOMAC RIVER BASIN AND CLIMATOLOGICAL SUMMARIES FOR SPRUCE KNOB, WARDENSVILLE, ROMNEY AND MARTINSBURG, WEST VIRGINIA,
West Virginia Univ., Morgantown.
For primary bibliographic entry see Field 02B.
W73-09602

HYDROGEOLOGIC CHARACTERISTICS OF THE VALLEY-FILL AQUIFER IN THE STERLING REACH OF THE SOUTH PLATTE RIVER VALLEY, COLORADO,
Geological Survey, Lakewood, Colo.
For primary bibliographic entry see Field 04B.
W73-09605

HYDROGEOLOGIC CHARACTERISTICS OF THE VALLEY-FILL AQUIFER IN THE BRUSH REACH OF THE SOUTH PLATTE RIVER VALLEY, COLORADO,
Geological Survey, Lakewood, Colo.
For primary bibliographic entry see Field 04B.
W73-09606

TIDAL DATUM PLANES AND TIDAL BOUNDARIES AND THEIR USE AS LEGAL BOUNDARIES -- A STUDY WITH RECOMMENDATIONS FOR VIRGINIA,
Virginia Inst. of Marine Science, Gloucester Point.
For primary bibliographic entry see Field 02L.
W73-09609

THE MAPPING OF SNOWFALL AND SNOW COVER,
Atmospheric Environment Service, Ottawa (Ontario).
For primary bibliographic entry see Field 02C.
W73-09616

FLOODS IN THE RIO GUANAJIBO VALLEY, SOUTHWESTERN PUERTO RICO,
Geological Survey, Washington, D.C.
W. J. Haire.
For sale by Geological Survey Price 75 cents, Washington, D.C. 20242. Hydrologic Investigations Atlas HA-456, 1972. 1 sheet, 7 fig, 2 ref.

Descriptors: *Floods, *Flood plains, *Strean flow, *Hydrologic data, *Puerto Rico, Floo damage, Maps, Flood frequency, Flood profile Flood peak, Flood stages, Flood data, Topc graphic mapping.
Identifiers: *Rio Guanajibo Valley (Puerto Rico).

This one-sheet atlas provides factural and in terpretive information that will aid administrators planners, engineers, and other interested person concerned with future development in areas sub ject to flooding in the Rio Guanajibo basin, Puert Rico. The study area lies in the southwestern par of the Puerto Rico and encompasses about 9 square miles of land area. Floods that have inun dated sizable areas since 1928 and for which som water-surface elevation data are available oc curred in 1943, 1945, 1952, 1954, 1958, 1960, 1963 1967, and 1968; but records are too fragmentary t determine accurate water-surface profiles for any flood except the flood of July 30, 1963. The 196 flood is the 7th highest of record. The area inun dated by the flood is delineated on a topographic map. The flood boundary on the map was developed from historical data and reflects the channel and flood-plain conditions that existed a that time. Fourteen highway bridges are in the val ley, 11 of which cross Rio Guanajibo. The water surface during this flood was above low steel on five of these bridges, and two of the five bridges were completely inundated. (Woodard-USGS)
W73-09618

GEOHYDROLOGY OF SUMTER, DOOLY, PULASKI, LEE, CRISP, AND WILCOX COUNTIES, GEORGIA,
Geological Survey, Washington, D.C.
R. C. Vorhis.
For sale by Geological Survey Price $1.00 per set, Washington, D.C. 20242. Hydrologic Investigations Atlas HA-435, 1972. 2 sheets, 3 fig, 2 tab, 13 ref.

Descriptors: *Groundwater resources, *Aquifers, *Water yield, *Water quality, *Georgia, Water wells, Test wells, Pumping, Drawdown, Groundwater recharge, Aquifer characteristics, Mapping, Geology, Hydrogeology, Hydrologic data, Water resources development.
Identifiers: *Sumter Co (Ga), *Dooly Co (Ga), Pulaski Co (Ga), Lee Co (Ga), Crisp Co (Ga), Wilcox Co (Ga).

The increasing use of Interstate Highway 75 through Georgia has generated the drilling of many wells to supply water for light industry, motels, and restaurants. Furthermore, a recent increase in groundwater use for supplementary irrigation has created a need for information on the required depths of wells, the amount of water to be expected, and the quality of the obtainable water. This 2-sheet atlas was prepared to make such information publicly available. The illustrations provide information on the geologic formations from which water is obtainable, their depth, and the probable quantity and quality of the water. The total water recharge per year for the 2,266 square-mile area is about 80 billion gallons. In order to use this annual recharge to the maximum, ideally spaced wells would have to be pumped continuously at a rate of 150,000 gallons per minute. An estimate of water use suggests that less than 10% seeable future, water quantity will not be a limiting factor in the economic growth of the six-county area in southwest Georgia. (Woodard-USGS)
W73-09619

ELEVATION IN WEST VIRGINIA,
West Virginia Univ., Morgantown.
For primary bibliographic entry see Field 02B.
W73-09621

mineral incrustation are: (1) Use screens with the most possible open area; (2) Reduce drawdown through maximization of well efficiency; (3) Reduce drawdown by limitation of pumping rate; (4) Provide periodic chemical treatment; and (5) Keep oxygen away from the screen section of the well by the use of packers or vacuum seals. Since incrustation from manganese, magnesium carbonate, and calcium sulfate are rare in water wells, types emphasized are the calcium carbonate and ferrous or ferric hydroxide incrustations. The problem of incrustation by growth of iron bacteria Gallionella, Crenothrix, and Leptothrix is outlined. (Campbell-NWWA)
W73-09250

WAVE DAMPING EFFECTS OF FIBROUS SCREENS,
Army Engineer Waterways Experiment Station, Vicksburg, Miss.
For primary bibliographic entry see Field 08B.
W73-09437

A TALK ON THE FINITE ELEMENT METHOD AND GEOTECHNICAL ENGINEERING IN THE BUREAU OF RECLAMATION,
Bureau of Reclamation, Denver, Colo.
For primary bibliographic entry see Field 08G.
W73-09668

SHORTCUT PIPELINE MODIFICATION CONTRA COSTA CANAL UNIT, CENTRAL VALLEY PROJECT, CALIFORNIA (FINAL ENVIRONMENTAL STATEMENT).
Bureau of Reclamation, Washington, D.C.

Available from National Technical Information Service, U.S. Dept. of Commerce as EIS-CA-72-5493-F, for $3.75 paper copy, $0.95 microfiche. October 16, 1972. 15 p, 1 map, 2 append.

Descriptors: *California, *Pipelines, *Water supply, *Project planning, *Environmental effects, Closed conduits, Piping systems (Mechanical), Water utilization, Pipes, Pollution abatement, Economics, Programs, Water allocation (Policy), Water delivery, Project purposes, Zoning, Recreation, Canals, Water supply development, Water shortage, Water management (Applied), Water distribution (Applied).
Identifiers: *Socioeconomic development, *Environmental impact statement, *Central Valley Project (Calif.).

The existing Contra Costa Canal has not the capacity to supply the present demand for water in the Martinez, California, portion of its service area. To meet this need, the construction of a buried pipeline, through the urban Concord area, has been begun. The pipeline will supply the increasing needs of the Martinez urban and industrial areas. The right-of-way will be available for development as a green belt or part of a recreational trail system. The pollution potential of increased urbanization and industralization of the area has been recognized, and discharges are under the regulation set by the California Water Pollution Control Board. Alternatives to the present plan which were considered were either: (1) the enlargement of the existing canal; or (2) no charge in the existing situation. Without the pipeline, existing and planned economic development in the area would be restricted and the socioeconomic environment of the area impaired. The area adjoining almost all of the pipeline right-of-way has been committed to industrial development. The subsurface of the right-of-way is committed to pipeline use. Water rights associated with the service area are already in existence. The comments of interested agencies are included in the statement. (Reed-Florida)
W73-09696

PROSPERITY LAKE, SPRING RIVER AND TRIBUTARIES, MISSOURI-KANSAS-OKLAHOMA (FINAL ENVIRONMENTAL STATEMENT).
Army Engineer District, Tulsa, Okla.

Available from National Technical Information Service, U.S. Dept. of Commerce as EIS-MO-72-5277-F, for $3.75 paper copy, $0.95 microfiche. April 14, 1971. 33 p, 2 tab.

Descriptors: *Flood control, *Artificial lakes, *Federal project policy, *Water supply, *Environmental effects, Missouri, Kansas, Oklahoma, Water policy, Dams, Flood damage, Reservoirs, Water resources development, Water quality, Recreation, Water sports, Recreation facilities, Project planning, Project purposes, Ecology.
Identifiers: *Spring River (Mo.-Kan.-Okla.), *Environmental impact statement, National Environmental Policy Act.

The proposed project would result in the construction of a dam and the creation of an artificial lake along a valley bottom located on the northwestern flank of the Ozark uplift in the Ozark Plateau. Flood control, water supply and recreation are the project's purposes. The project will provide millions of gallons of good quality water daily to the residents of the Joplin-Carthage area and will increase the agricultural productivity of the area, thereby raising the living standard of the rural residents. The project will also provide an opportunity for water-related recreational development. The project would necessitate the innundation of several thousand acres of land and the relocation of one cemetery and various utilities. Considered as alternatives to the proposed project were levees, an exclusive upstream flood control program, flowage easements, and alternative lake sites. The only known irreversible commitment of resources would be the covering of 1,880 acres of land and streams and their associated ecosystems with the project structures and water. The comments of interested federal and state agencies are included in the statement. (Reed-Florida)
W73-09698

CORNUDAS, NORTH AND CULP DRAWS WATERSHED, HUDSPETH COUNTY, TEXAS AND OTERO COUNTY, NEW MEXICO (FINAL ENVIRONMENTAL IMPACT STATEMENT).
Soil Conservation Service, Washington, D.C.

Available from National Technical Information Service, U.S. Dept. of Commerce as EIS-NM-72-5708-F, for $5.00 paper copy, $0.95 microfiche. November 29, 1972. 29 p, 1 plate, 1 tab, 1 append.

Descriptors: *Texas, *New Mexico, *Environmental effects, *Watershed management, Flood prevention, Erosion, Runoff, Irrigation, Wildlife habitat, Agricultural land, Flood control, Flood plain management, Land use, Soil conservation, Land resources, Land development, Watersheds (Basins), Water supply, Habitat improvement, Governmental interrelations.
Identifiers: *Environmental impact statement, Hudspeth County (Texas), Otero County (N.M.), Interstate projects.

The project proposes land treatment measures within the existing watershed supplemented by three floodwater retarding structures for the purpose of watershed protection and flood prevention in Hudspeth County, Texas, and Otero County, New Mexico. The project will reduce erosion and runoff; reduce the waste of irrigation water; improve big game habitat; reduce floodwater damages on about 14,173 acres by 84%; reduce sediment deposition damage by 88%; reduce scour damage by 65%; provide protection to Dell City from all floods up to and including the 100-year event; provide an average annual 2,700 acre-feet of groundwater recharge; and reduce future energy needs associated with water desalinization in

Dell City. Adverse effects include losting 483 acres of agricultural land and interrupting agricultural use on an additional 811 acres. Alternatives considered to the proposed project include less intensive use of the flood plain, an alternate combination of a larger number of smaller structures, and the construction of channels to carry released floodwater across the highly developed flood plain. Comments of interested state and federal agencies are included. (Mockler-Florida)
W73-09699

OMNIBUS WATER RESOURCES AUTHORIZATIONS--1972, PARTS I AND II.
For primary bibliographic entry see Field 06E.
W73-09709

8B. Hydraulics

EFFECTS OF RIVER CURVATURE ON A RESISTANCE TO FLOW AND SEDIMENT DISCHARGES OF ALLUVIAL STREAMS,
Iowa Univ., Iowa City. Inst. of Hydraulic Research.
For primary bibliographic entry see Field 02E.
W73-09102

ELACTIC EFFECTS IN FLOW OF VISCOELASTIC LIQUIDS,
Delaware Univ., Newark. Dept of Chemical Engineering.
M. M. Denn, and K. C. Porteous.
Chemical Engineering Journal, Vol 2, p 280-286, 1971. 5 fig, 28 ref.

Descriptors: *Fluid mechanics, *Fluid friction, *Viscosity, *Pipe flow, *Elasticity (Mechanical), Flow, Turbulence, Shear drag, Mathematical models, Equations, Reynolds Number, Design criteria.
Identifiers: *Fluid elasticity.

In design for systems involving viscoelastic liquids it is necessary to have available simple estimates of when deviations from inelastic behavior might be expected. The traditional engineering approach is through dimensional analysis. A basic approach to dimensional analysis estimation in viscoelastic liquid flows is based upon considerations of information transport in such systems. The shear wave velocity is the relevant elastic parameter. As a result of this approach it seems possible to bring within a common framework the appearance and correlation of a diversity of phenomena broadly attributable to fluid elasticity, including low Reynolds Number instability in melts and turbulent drag reduction in dilute polymer solutions. The influence of fluid elasticity on information transport is best illustrated by the development of velocity profile between infinite parallel plates, both initially at rest, when one is suddenly set in motion at a constant velocity. (Woodard-USGS)
W73-09133

HORIZONTAL FORCES DUE TO WAVES ACTING ON LARGE VERTICAL CYLINDERS IN DEEP WATER,
Naval Undersea Center, San Diego, Calif.
E. R. Johnson.
Available from NTIS, Springfield, Va 22151 AD-751 574 Price $3.00 printed copy; $0.95 microfiche. Report NUC TP 322, October 1972. 15 p, 4 fig, 2 tab, 31 ref.

Descriptors: *Waves (Water), *Deep water, *Ocean waves, *Floats, Model studies, Mathematical models, Equations, Correlation analysis, Hydrologic data, Methodology, Design criteria, Engineering structures.
Identifiers: *Ocean platforms, *Floating platforms.

Wave forces on vertical cylinders are due to both viscous and inertial effects. The problem appears to be considerably simplified when the forces are predominantly inertial. In general, the inertial forces predominate as cylinder diameter and water depth increase. A column-stabilized floating ocean platform presents such a case. An expression for the maximum horizontal wave force on large-diameter circular cylinders mounted vertically in deep water was analytically derived. Experimental model studies were also conducted and the resulting measured forces were within 20% of predicted forces. The maximum horizontal force due to waves is proportional to the square of the cylinder diameter when the maximum force is inertial. The force distribution is concentrated near the surface. About one-half the possible maximum force on the cylinder occurs in the first one-tenth wavelength of depth from the water surface. For the first one-tenth wavelength (200 ft for a 20-sec wave) of depth, the force distribution is almost linear with depth. For platform buoyancy columns, the horizontal forces due to waves are about proportional to the column length and the square of the diameter. Therefore, wave force considerztions should not be a factor in column proportioning since for a given column buoyancy, the force would be about the same regardless of the length-to-diameter ratio. However, the magnitude and distribution of the wave forces on the vertical cylinders do need to be known for the structural design. (Woodard-USGS)
W73-09166

PLANNING AND DEVELOPING GROUND WATER SUPPLIES,
Dames and Moore, Inc., New York.
For primary bibliographic entry see Field 04B.
W73-09243

WATER SUPPLY AND GROUND WATER UTILIZATION IN FOREIGN COUNTRIES,
Dames and Moore, Salt Lake City, Utah.
For primary bibliographic entry see Field 04B.
W73-09244

WATER WELL CONSTRUCTION AND DESIGN,
Dames and Moore, San Francisco, Calif.
For primary bibliographic entry see Field 08A.
W73-09245

EVALUATING AND MANAGING GROUND WATER SUPPLIES WITH MODELS AND COMPUTERS,
Dames and Moore, Los Angeles, Calif.
For primary bibliographic entry see Field 04B.
W73-09246

GROUNDWATER HYDRAULICS OF EXTENSIVE AQUIFERS,
For primary bibliographic entry see Field 04B.
W73-09263

FLEXIBLE ROUGHNESS IN OPEN CHANNELS,
Waterloo Univ. (Ontario). Dept. of Civil Engineering.
N. Kouwen, and T. E. Unny.
ASCE Proceedings, Journal of the Hydraulics Division, Vol 99, No HY5, Paper 9723, p 713-728, May 1973. 8 fig, 1 tab, 12 ref, append.

Descriptors: *Open channel flow, *Roughness (Hydraulic), *Vegetation, *Hydraulic models, Flumes, Aquatic plants, Flow resistance, Fluid friction, Roughness coefficient.
Identifiers: *Flexible roughness (Hydraulic).

The logarithmic flow formula for open channels is valid for flow over a roughness cover made up of flexible plastic strips representing vegetation. The deflection of such roughness directly influences the relative roughness. Dimensionless parameters relate the amount of bending to the stiffness of the roughness and the boundary shear. Three regimes of boundary behavior were observed for flexible plastic roughness, leading to two resistance functions. The flexible plastic roughness was produced by thin plastic strips of various thickness on the bed of a flume. (Knapp-USGS)
W73-09271

WAVE DAMPING EFFECTS OF FIBROUS SCREENS,
Army Engineer Waterways Experiment Station, Vicksburg, Miss.
G. H. Keulegan.
Available from NTIS, Springfield, Va. 22151, AD-751 133 Price $3.00 printed copy; $0.95 in microfiche. Army Engineer Waterways Experiment Station Research Report H-72-2, September 1972. 105 p, 35 fig, 18 tab, 9 ref.

Descriptors: *Waves (Water), *Harbors, *Control structures, *Hydraulic models, Design criteria, Engineering structures, Analytical techniques, Materials, Testing procedures, Filters, Resistivity, Mathematical studies, Equations.
Identifiers: *Wave absorbers, *Wave filters, Fibrous materials.

This is the second of three reports planned on the design of filters and wave absorbers of the type used in harbor wave-action models as a part of the Civil Works Investigation, Engineering Study 833, sponsored by the Office, Chief of Engineers, U.S. Army. Expressions are derived for the damping of essentially translatory waves by screens prepared from fibrous materials such as aluminum wool, rubberized hair, and polyurethane foam. The derivations involve either the resistance coefficients or the permeabilities of these materials. With the aluminum wool the sampling is exponential and yields expressions admitting logarithmic decrements. A comparison is made with the Biesel theory of decay in porous media. With the foams the damping is also exponential; on the other hand, with the rubberized hair the damping is best described in terms of a power formula. (Woodard-USGS)
W73-09437

GRAYS HARBOR ESTUARY, WASHINGTON: REPORT 3, WESTPORT SMALL-BOAT BASIN STUDY,
Army Engineer Waterways Experiment Station, Vicksburg, Miss.
N. J. Brogdon, Jr.
Available from NTIS, Springfield, Va. 22151, AD-748 814, Price $3.00 printed copy; $0.95 in microfiche. Army Engineer Waterways Experiment Station Technical Report H-72-2, September 1972. 152 p, 116 plate, 22 photo, 1 tab.

Descriptors: *Coastal engineering, *Harbors, *Recreation facilities, *Washington, *Estuaries, Model studies, Channel improvement, Environmental effects, Ecology, Data collections, Hydraulic models, Currents (Water), Shoals, Path of pollutants, Sewage, Outfall sewers.
Identifiers: *Grays Harbor (Wash).

The existing, comprehensive fixed-bed model of the Grays Harbor estuary was used to determine current velocities, surface current patterns, flushing and dispersion characteristics, and qualitative shoaling rates and patterns for four small-boat basin plans located near Westport, Wash. With the increased interest in sport fishing in this area and expansion of the charter boat and commercial fishing industry, this small-boat harbor no longer has the capacity to serve the ever-increasing number of boats desiring anchorage; therefore, there is a need for expanding this facility. Model tests indicate that only insignificant local changes in current velocities and surface current patterns outside

of ice; (7) construction of ice-engineering installations on rivers in northern Siberia; (8) formation of ice on foreign surfaces; (9) stability of frozen peat; (10) extension of navigation periods on inland waterways; (11) aerodynamics of cooling towers; (12) remote-control measurement of water discharge in large-diameter pipes; and (13) measurement of small flow velocities by pressure floats. (Josefson-USGS)
W73-09461

HYDROMORPHOLOGICAL INVESTIGATIONS OF FLOOD-PLAIN AND CHANNEL PROCESSES (GIDROMORFOLOGICHESKIYE ISSLEDOVANIYA POYMENNOGO I RUSLOVOGO PROTSESSOV).
Gosudarstvennyi Gidrologicheskii Institut, Leningrad (USSR).

Gosudarstvennyy Gidrologicheskiy Institut Trudy, No 195, Snishchenko, B. F., editor, Leningrad, 1972. 99 p.

Descriptors: Investigations, *Flood plains, *Channels, *Channel morphology, *Hydraulics, Floods, Bodies of water, Meanders, Braiding, Water levels, Flow, Currents (Water), Reservoirs, Statistical methods, Probability, Aerial photography, Analytical techniques, Instrumentation, Digital computers, Model studies, Laboratories.
Identifiers: *USSR, Profilographs.

Results are presented of research conducted by scientific workers and graduate students at the State Hydrologic Institute's Department of Channel Processes, Investigations of the morphology and hydraulics of channels, flood plains, and bodies of waters are based on the use of aerial photography and digital computers. A description is given of an ultrasonic profilograph for use in erodible-model studies in channel laboratories. (Josefson-USGS)
W73-09462

GENERAL REPORT ON WATER RESOURCES SYSTEMS: OPTIMAL OPERATION OF WATER RESOURCES SYSTEMS,
Illinois Univ., Urbana. Dept. of Hydraulic Engineering.
For primary bibliographic entry see Field 02A.
W73-09588

THE EFFECT OF MIXING AND GRAVITATIONAL SEGREGATION BETWEEN NATURAL GAS AND INERT CUSHION GAS ON THE RECOVERY OF GAS FROM HORIZONTAL STORAGE AQUIFERS,
New Mexico Inst. of Mining and Technology, Socorro.
For primary bibliographic entry see Field 04B.
W73-09589

CONSIDERATIONS OF PIPELINE TENSION TIE PROBLEMS,
Santa Clara County Flood Control and Water District, San Jose, Calif.
For primary bibliographic entry see Field 08G.
W73-09671

ORIFICE LOSSES FOR LAMINAR APPROACH FLOW,
Indian Inst. of Science, Bangalore (India).
N. S. L. Rao, and K. Sridharan.
Proc Am Soc Civ Eng, J Hydraul Div, Vol 98, No HY11, p 2015-2034, Nov 1972. 20 p, 13 fig, 19 ref, 2 append.

Descriptors: *Orifices, *Orifice flow, *Laminar flow, Turbulent flow, Reynolds number, Research and development, Basic research, Pipe flow, Fluid mechanics, Curves, Bibliographies.
Identifiers: India, Foreign studies.

Very little information is available on variability of loss coefficient with changing Reynolds number for valves, contractions, expansions, and bends. The reported studies deal with laminar approach flow to several types of orifices, including sharp-edged concentric, eccentric and segmental, quadrant-edged concentric, and square-edged long. Flow medium was oil in a 36.6-mm-dia pipe. Reynolds number was varied from 10 to 2,000. The ratio of orifice diameter to pipe diameter varied from 0.2 to 0.8. The loss coefficient curves for several orifices give an indication of the critical Reynolds number at which flow immediately downstream of the orifices becomes turbulent even though the approach flow is laminar. The effects of Reynolds number, orifice-to-pipe diameter ratio, edge radius, eccentricity, segmental opening, and orifice length are studied. Numerous curves of loss coefficient versus Reynolds number are given for each orifice. (USBR)
W73-09672

8C. Hydraulic Machinery

FLUID TREATMENT DEVICES,
For primary bibliographic entry see Field 05D.
W73-09399

USE OF CARRIER LINE TRAPS WITH NO TUNING PACKS,
Tennessee Valley Authority, Chattanooga.
A. B. Bagwell, and H. I. Dodson.
Inst Electra Electron Eng Trans Power Appar Syst, Vol PAS-91, No 6, p 2305-2312, Nov-Dec 1972. 8 p, 15 fig, 4 ref, disc.

Descriptors: *Powerline carriers, *Transmission lines, Carrier-current, Carriers, Power operation and maintenance, Communication, Protective relaying, Tests, Extra high voltage, Instrumentation.
Identifiers: *Line traps, Tuning.

On construction projects involving changes to existing facilities, removing power transmission lines from service often becomes necessary so that carrier coupling can be modified or line traps retuned. Even more frequently necessary is taking powerline outages for line trap maintenance because of defective tuning packs. Ever since line traps were first used with powerline carrier, tuning packs have been the weakest link. A tuning scheme that uses a self-resonant line trap coil as 1 leg of a pi-section bandpass filter to eliminate line trap tuning pack problems is described and compared with other approaches. Results illustrate that a carrier circuit can be coupled successfully to a powerline without susceptibility to line trap tuning pack failure and without degrading the characteristics of the carrier channel. (USBR)
W73-09665

8D. Soil Mechanics

EFFECT OF SEEPAGE STREAM ON ARTIFICIAL SOIL FREEZING,
University Engineers, Inc., Norman, Okla.
For primary bibliographic entry see Field 02G.
W73-09454

EARTHQUAKE RESISTANCE OF EARTH AND ROCK-FILL DAMS: REPORT 2. ANALYSIS OF RESPONSE OF RIFLE GAP DAM TO PROJECT RULISON UNDERGROUND NUCLEAR DETONATION,
Army Engineer Waterways Experiment Station, Vicksburg, Miss. Soils and Pavements Lab.
J. E. Ahlberg, J. Fowler, and L. W. Heller.
Available from NTIS, Springfield, Va 22151 AD-744 420 Price $3.00 printed copy; 95 cents microfiche. Miscellaneous Paper S-71-17, June 1972. 132 p, 22 fig, 66 plate, 4 tab, 10 ref.

Descriptors: *Dams, *Movement, *Nuclear explosions, *Underground, *Colorado, Correlation analysis, Model studies, Earthquakes, Mathematical models, Data collections, Forecasting, Earth dams, Rockfille dams, Design criteria, Engineering structures, Hydraulic structures.
Identifiers: *Rifle Gap Dam (Colo).

The motion of Rifle Gap Dam in Colorado was measured in September 1969 during the Project Rulison underground nuclear explosion. The observed response was then compared with the response computed in a mathematical model. Observed and computed responses were similar. From this study it appears that the mathematical models used are applicable to the design and analysis of soil structures, at least for ground motion intensities comparable to those observed at Rifle Gap Dam. The type of material used, sand or clay, determined the relationship used to modify the material properties for shear strain level. For the foundation material at Rifle Gap Dam, better agreement with the observed data was obtained by considering the material as sand. The granular material was present in both Bu Rec borings, but cohesive material was also present and results comparable to those observed could be obtained only by multiplying the cohesive modulus by a factor of 1.875, which actually gave about the same modulus as that for the sand curves at that shear strain level. (Woodard-USGS)
W73-09604

FINITE ELEMENT ANALYSES OF STRESSES AND MOVEMENTS IN DAMS, EXCAVATIONS AND SLOPES (STATE-OF-THE-ART),
California Univ., Berkeley.
J. M. Duncan.
Paper, Appl Finite Elem Method Geotech Eng, Proc Symp, Vicksburg, Miss, May 1972, p 267-326, Sept 1972. 60 p, 16 fig, 67 ref, 2 append.

Descriptors: *Finite element analysis, *Stress analysis, *Slopes, *Slope stability, Earth dams, Rockfill dams, Deformation, Settlement (Structural), Excavation, Earth pressure, Pore pressure, Costs, Reviews, Bibliographies, Stability, California.
Identifiers: Buena Vista Pump Plt (Calif), Scammonden Dam, Great Britain, Hydraulic fracturing, Nonlinear properties, Oroville Dam (Calif).

Until the finite element method (FEM) was devised, geotechnical problems involving nonhomogeneous and nonlinear materials, complex boundary conditions, and sequential loading were considered impossible to analyze realistically. In analyzing the behavior of embankment and excavation slopes, FEM can be used to predict: (1) movements and earth pressures, (2) locations likely to crack, (3) locations susceptible to hydraulic fracturing, (4) pore pressure changes for undrained conditions, and (5) local failure and stability. The most significant limitation on the accuracy of geotechnical FEM analyses is caused by a limited ability to represent the stress-strain behavior of soils and rocks. Methods for modeling the stress-strain behavior include: linear elastic, successive approximations, successive increments, bilinear, hyperbolic, and spline functions. FEM examples given are: (1) a slope stability analysis of the Buena Vista Pumping Plant excavation in alluvium, (2) slope stability analysis of the failure of a canal constructed in highly plastic clay, (3) movement prediction and observation at Scammonden Dam, (4) movement prediction for a soft clay test fill loaded to failure, and (5) movement prediction and observation of Oroville Dam during construction and reservoir filling. (USBR)
W73-09666

APPLICATIONS OF FINITE ELEMENT ANALYSES IN SOIL DYNAMICS (STATE-OF-THE-ART),
California Univ., Berkeley.
H. B. Seed, and J. Lysmer.
Paper, Appl Finite Elem Method Geotech Eng, Proc Symp, Vicksburg, Miss, May 1972, p 885-912, Sept 1972. 28 p, 11 fig, 35 ref, append.

Descriptors: *Soil dynamics, *Finite element analysis, *Seismic design, *Damping, Earth dams, Slope stability, Stress analysis, Seismic waves, Vibrations, Seismology, Engineering geology, Bibliographies, Dynamics, Faults (Geologic), Earthquake engineering.
Identifiers: Computer-aided design, Caracas (Venezuela), Blast loads, Soil-structure interaction, *Damping.

This review of finite element analyses emphasizes efforts in the fields of earthquake engineering, seismology, blast effects on structures, and foundation vibrations. To obtain meaningful results from the analyses, correct representations are needed of the soil stress-strain and damping properties. Seismic stability analyses of earth dams and slopes can be used to compute stress time histories for both rigid and traveling wave base motions. Other earthquake studies analyze surface accelerations of soil deposits underlain by sloping rock boundaries. Recent seismological analyses include the ground vibrations produced by a fault rupture and the influence of structural irregularities on the propagation of surface waves. Classical analyses of foundation vibration problems have been limited to rigid circular footings resting on a homogeneous half-space, but with the finite element method, footings can be analyzed that are resting on layered soil deposits or embedded below the ground surface. After considering several methods of representing the stress-strain and damping properties of soil, a method using equivalent linear moduli with variable Rayleigh damping is considered best for a wide range of situations. (USBR)
W73-09667

STABILITY OF ROCKFILL IN END-DUMP RIVER CLOSURES,
Alberta Univ., Edmonton.
B. P. Das.
Proc Am Soc Civ Eng, J Hydraul Div, Vol 98, No HY11, p 1947-1967, Nov 1972. 21 p, 6 fig, 3 tab, 22 ref, 2 append.

Descriptors: *Construction, *Hydraulics, *Rock fills, Stability, Rockfill dams, Cofferdams, Alluvial channels, Flow, Laboratory tests, River flow, Velocity, Constrictions, Bibliographies, Canada.
Identifiers: *River closures, Foreign studies.

Closure of a river is necessary to divert the flow prior to any construction on the bed. One method is to end-dump loose material, such as rockfill, from one or both banks to constrict the waterway progressively and ultimately block the channel. However, the usual methods of determining stable closure material size by correlation with either mean velocity through the gap or velocity near the boundary are not realistic. Since most major hydraulic structures are sited at the submountain stage, where the river has an alluvial bed, stability of materials on end-dump closure dams built into alluvial channels was investigated. Closure material stability was defined numerically by a parameter efficiency of the material, denoting the ratio of the volume of material remaining in the body of the dam to the volume dumped at any stage. Analysis of all relevant parameters indicated that essential interrelated nondimensional variables describing stability are: (1) Froude number of approach flow, (2) contraction ratio, (3) ratio of closure material size to the approach flow depth, and (4) efficiency of closure material. Functional relationships of these variables were established by laboratory studies. Practicable design curves to determine stable size of closure material are presented. (USBR)
W73-09670

LIQUID FAILURE.
American Wood Preservers Inst., McLean, Va.

Wood Preserving, Vol 50, No 8-9, p 7-15, Aug-Sept 1972. 9 p, 1 map, 6 photo, 2 tab, 9 chart, 4 ref.

Descriptors: *Soil compaction, Soil stability, Sands, Soil density, Soil investigations, Soil engineering, Soil mechanics, Cohesionless soils, Seismic studies, *California.
Identifiers: Niigata Earthquake (Japan), Vibroflotation, *Compaction piles, Soil liquefaction, Compaction piles, Treasure Island (Calif), Earthquake damage.

At Treasure Island, a hydraulically placed sandfill was determined to be in danger of liquefaction failure in event of a major earthquake. The sandfill extends from the ground surface to a depth of about 45 ft and the ground-water level is 6 ft below the ground surface. Between depths of 15 to 25 ft, some sand was at low relative densities of 25 to 30%, considerably below the 70% desired for safety from liquefaction. Methods considered to increase sand density included: (1) increasing the confining stress by adding more fill, (2) increasing the confining stress by lowering the ground-water level, (3) grouting, (4) excavating and compacting, (5) using sand compaction piles, and (6) vibroflotation. Sand compaction piles used at one building site were found unsatisfactory because sand density was both nonuniform and difficult to measure. At other sites, 20-ft-long nonstructural displacement piles were driven completely below the water table. Sand was compacted by conventional methods above the water table. The nonstructural piles method was simpler, easier to specify, open to all bidders, easier to determine density increases, and less expensive than other effective methods. (USBR)
W73-09675

WAHKIAKUM COUNTY CONSOLIDATED DIKING DISTRICT NO. 1, WAHKIAKUM COUNTY, WASHINGTON (FINAL ENVIRONMENTAL IMPACT STATEMENT).
Army Engineer District, Portland, Oreg.

Available from National Technical Information Service, U.S. Dept. of Commerce as EIS-WA-74-5280-F, for $5.50 paper copy, $0.95 microfiche. June 1972. 37 p, 1 map, 1 photo.

Descriptors: *Washington, *Flood control, *Levees, *Project planning, *Environmental effects, Levee districts, Diversion structures, Ditches, Dikes, Flood damage, Flood protection, Project purposes, Stagnant water, Wildlife, Wildlife habitats, Ecology, Dredging, Landfills, Water management (Applied), Wildlife conservation.
Identifiers: *Columbia River (Wash.), *Environmental impact statement, Borrow ditches.

The proposed action is construction of various improvements to correct deficiencies in the existing levee system on Puget and Little Islands in the Columbia River, Washington. The project's purpose is to secure a higher degree of safety from flooding for the diking district. The project will prevent flooding in the project area, reduce flood damages resulting from levee failure and seepage, and eliminate stagnant water areas. The project would also eliminate the habitat of ducks, pheasants, furbearers and small birds with a consequent reduction in population of these animals. Alternatives considered were a levee of a different height and a levee in a different location, but neither were felt to provide sufficient flood control. Implementation of the proposed action is not expected to alter existing agricultural practices; however, residential development will probably accelerate. Reduction of wildlife population will be long-term. The only commitment of resources is the conversion of wet borrow ditches created at the time of original levee construction into dry

sand fill. However, the ditches could be redredged to create a habitat similar to that which is destroyed. The comments of various governmental and private agencies are included in the statement. (Reed-Florida)
W73-09697

8E. Rock Mechanics and Geology

USE OF ACOUSTIC LOGGING TECHNIQUE FOR DETERMINATION OF INTERGRAIN CEMENTATION PROPERTIES,
Continental Oil Co., Houston, Tex.
For primary bibliographic entry see Field 08G.
W73-09242

HYDRAULIC FRACTURING AND THE EXTRACTION OF MINERALS THROUGH WELLS,
Wisconsin Univ., Madison. Dept. of Minerals and Metals Engineering.
B. Haimson, and E. J. Stahl.
In: Third Symposium on Salt, Vol 2; Northern Ohio Geological Society Publication, p 421-432, 1970. 19 fig, 2 tab, 8 ref.

Descriptors: *Rock mechanics, *Injection wells, *Stress, *Strain, *Cracks, Fractures (Geologic), Rock properties, Elasticity (Mechanical), Deformation, Rupturing, Mechanical properties, Failure (Mechanics).
Identifiers: *Hydraulic fracturing.

Hydraulic fracturing is a method used to create artificial fractures for the purpose of increasing flow capacity around a well or enhancing communication between two adjacent wells. The fractures are theoretically assumed to be tensile ruptures extending in a plane perpendicular to the direction of the smaller horizontal principal compressive stress. The pressures required to initiate and extend vertical fractures depend on the principal tectonic stresses, the porous-elastic parameters of the rock and on tensile strength. Experimental work on simulated wells in laboratory rock samples under triaxial loading is described. Results confirm theoretical predictions of fracture type, fracture initiation pressure and fracture orientation. In the field, oriented impression packers were used to determine the fracture azimuth at the well bore. Wells belonging to a common field yielded hydraulic fractures of approximately same orientation. This substantiates the theoretical and laboratory conclusion that the smallest tectonic horizontal compressive stress direction determines the fracture orientation. (Knapp-USGS)
W73-09288

INITIATION AND EXTENSION OF HYDRAULIC FRACTURES IN ROCKS,
Minnesota Univ., Minneapolis.
B. Haimson, and C. Fairhurst.
Society of Petroleum Engineers Journal, Vol 7, No 3, p 310-318, September 1967. 7 fig, 14 ref.

Descriptors: *Rock mechanics, *Injection wells, *Stress, *Strain, *Cracks, Fractures (Geologic), Rock properties, Elasticity (Mechanical), Deformation, Rupturing, Mechanical properties, Failure (Mechanics).
Identifiers: *Hydraulic fracturing.

A criterion is proposed for the initiation of vertical hydraulic fracturing taking into consideration the three stress fields around the wellbore. These fields arise from (1) nonhydrostatic regional stresses in earth, (2) the difference between the fluid pressure in the wellbore and the formation fluid pressure, and (3) the fluid flow from the wellbore into the formation. The wellbore fluid pressure required to initiate a fracture is a function of the elastic constants of the rock, the horizontal principal regional stresses, the tensile strength of the rock, and the formation fluid pressure. A constant injection rate will extend the fracture to a point where equilibrium is reached and then, to keep the fracture open, the pressure required is a function of the elastic constants of the rock, the component of the regional stress normal to the plane of the fracture, the formation fluid pressure, and the dimensions of the crack. The same expression may also be used to estimate the vertical fracture width, provided all other variables are known. The equations for the initiation and extension pressures in vertical fracturing may be employed to solve for the two horizontal, regional, principal stresses in the rock. (Knapp-USGS)
W73-09289

REAL STRESSES AROUND BOREHOLES,
Wisconsin Univ., Madison.
B. C. Haimson, and T. M. Tharp.
Preprint of Paper SPE 4241; prepared for 6th Conference on Drilling and Rock Mechanics January 22-23, 1973, Austin, Tex: Society of Petroleum Engineers Preprints, p 107-116, 1973. 5 fig, 3 tab, 12 ref.

Descriptors: *Rock mechanics, *Injection wells, *Stress, *Strain, *Cracks, Fractures (Geologic), Rock properties, Elasticity (Mechanical), Deformation, Rupturing, Mechanical properties, Failure (Mechanics).
Identifiers: *Hydraulic fracturing.

Generally the internal pressure required to rupture a borehole is higher than the expected tensile strength of the rock. The tangential stresses around holes as developed according to linear elasticity appear to be deceptively high. Rock is nonlinear in its elastic behavior. When linearized for purpose of mathematical analysis the slope of the stress-strain curve in compression is most often higher than in tension. Recalculation of the stresses around a circular hole based on the assumption of bilinearity reveals severe discrepancies between the real stresses developed and those apparent from linear elastic solutions. The results obtained can be directly applied to calculations of allowable pressures in wellbores (during drilling or hydraulic fracturing), pressure tunnels and boreholes. Stresses around circular holes in bilinear rock are a function of the material elastic properties, unlike the linear case. (Knapp-USGS)
W73-09290

STATIC FRACTURING OF ROCK AS A USEFUL TOOL IN MINING AND ROCK MECHANICS,
Wisconsin Univ., Madison.
B. C. Haimson.
In: Rock Fracture; Proceedings of International Symposium on Rock Mechanics, October 4-6, 1971, Nancy, France: International Society for Rock Mechanics, Part II, Paper II-30, 1971. 10 fig, 2 tab, 10 ref.

Descriptors: *Rock mechanics, *Injection wells, *Stress, *Strain, *Cracks, Fractures (Geologic), Rock properties, Elasticity (Mechanical), Deformation, Rupturing, Mechanical properties, Failure (Mechanics).
Identifiers: *Hydraulic fracturing.

Static or hydraulic fracturing of rock by means of borehole pressurization is a versatile technique of stimulating oil production, improving salt and potash extraction, disposing of waste materials, storing gas, and determining in-situ stresses at any depth. To understand this fracturing phenomenon, a theory relating fracturing fluid pressures and fracture orientation to the principal in-situ stresses and directions was developed. Laboratory testing results are given which verify most of the theoretical assumptions and criteria. Results from oil-field and underground mine tests show that the method is suitable for in-situ stress determination. Results of routine hydraulic fracturing jobs in an oil field could be used to estimate in-situ stresses. Experimental work indicates that mechanically induced notches play an important role in directing the fracture at least in the neighborhood of the borehole. This result could be used in enhancing communication between wells in solution mining. (Knapp-USGS)
W73-09291

8F. Concrete

CEMENTING GEO-THERMAL STEAM WELLS,
Dow Chemical Co., Tulsa, Okla. Dowell Div.

1972. 8 P, 5 TAB, APPEND.

Descriptors: Wells, *Grouting, Silica, *Cement grouting, Cements, Steam, Slurries, Technology, Well data, Compressive strength, Density.
Identifiers: *Geothermal wells, *Portland cement, Pressure grouting, Bottom-hole temperature.

Cement, once in place in a geo-thermal well, is expected to support the pipe, preventing channelling by forming a hydraulic seal, and remain competent under producing conditions. It is recommended that a minimum of 30% fine silica sand be used in all cement systems placed in geo-thermal wells, in order to minimize the decrease in compressive strength and increase in permeability of portland cement suffered while in high temperature use, a phenomenon commonly referred to as strength retrogression. Guidelines for selecting a cement system and appropriate technique are given; good cementing practices are listed, including the use of centralizers, hole conditioning, cooldown, bottom plugging, etc. Tables are given which contain typical data on primary cement systems recommended for use in cementing geo-thermal wells. It is recommended, however, that laboratory tests under actual well conditions on the materials to be used should be the basis for the final job design. (Campbell-NWWA)
W73-09252

CEMENTING FUNDAMENTALS,
Dow Chemical Co., Tulsa, Okla. Dowell Div.

1972. 11 P.

Descriptors: *Cements, *Cement grouting, Cellulose, *Additives, Wells, Materials, Grouting, Sealants, Slurries, Equipment, Drilling, Rotary drilling, Technology.
Identifiers: Barite, Accelerators (Cement), Retarders (Cement), Antifoam agents.

The overall mechanism of cementing pipe in a well appears to be rather simple; however, the tremendous amount of technology which has been developed and the research which is continually in progress proves that cementing wells is more a science than an art. Additives, which are used to alter the properties of the basic cement-water mixture known as neat cement are discussed, including weighting materials, extenders, accelerators, retarders, lost circulation materials, and special additives. Special additives include: fine sand to control strength loss in high-temperature applications, antifoam agents, dispersants, radioactive materials (for logging purposes) and fluid loss materials. Pieces of auxiliary equipment are listed and the purpose of each is explained. These include: float collars and shoes, centralizers, scratchers, casing wiper plugs, and cement heads. (Smith-NWWA)
W73-09253

SYSTEMATIC STRESS ANALYSIS OF CIRCULAR PIPE,
Naigai Engineering Co., Tokyo (Japan).
J. Katoh.

Proc Am Soc Civ Eng, Transp Eng J, Vol 98, No TE4, p 1039-1063, Nov 1972. 25 p, 10 fig, 6 tab, 3 append.

Descriptors: *Loads (Forces), *Stress analysis, *Concrete pipes, Deformation, Pipelines, Pipes, Structural analysis, Stress, Numerical analysis, Shear, Cracks.
Identifiers: *Pipe design, Foreign studies, Japan.

Despite increasing demands for concrete pipes in sewer and irrigation systems, stress analysis of pipes usually depends on approximate formulas because of complex rational analysis. For small pipes, sufficient factors of safety may cover deficiencies and cause little difference in results. However, for larger pipes, probability of crack growth inside the crown and bottom of large reinforced concrete pipes requires closer analysis of lateral deformation of pipe and behavior of lateral soil reaction. The principles described facilitate a more rational design by permitting simple, systematic analysis and calculation of the section force or deformation. Using tables presented of arbitrary load and reaction systems, simple calculations can be made to obtain values for the bending moment, shear, and axial force at any point in the pipe. Pipe deformation is useful for analyzing lateral resistance of the fill, evaluating concrete strength in the pipes, and predicting crack growth. Methods of computing deformation are given which closely follow general derivation. (USBR)
W73-09669

8G. Materials

ANALYSIS OF CATHODIC PROTECTION CRITERIA,
Hawaii Univ., Honolulu.
D. A. Jones.
Corrosion, Vol 28, No 11, p 421-423, November, 1972. 4 fig, 8 ref.

Descriptors: Corrosion, *Corrosion control, *Cathodic protection, Electrochemistry, Anodes, Reduction (Chemical), Cathodes, Steel structures, *Graphical analysis, Electrolytes.
Identifiers: Corrosion cell, *Aqueous corrosion, Polarization potential (Electrical), Tafel constant, Nernst equation.

The criterion of cathodic protection is that level of cathodic potential change (polarization) which is sufficient to give adequate protection for a given application. Ideally, it is preferred to polarize to the reversible potential of the anodic reaction, but this potential is often impossible to determine by experiment or to calculate from theory for complex corrosive electrolytes. Polarization to a specific potential (e.g.-0.85 volts) or to various potential changes (e.g., 100 to 300 mv) have also been proposed but none have been universally accepted. Electrochemical kinetic theory has been very successful in predicting and correlating the corrosion behavior of metals. However, the theory has not been applied to cathodic protection until recently. It has been shown that the anodic Tafel constant determines the cathodic polarization necessary to effect a given reduction in corrosion rate. Thus, polarization required to obtain a given degree of protection can be predicted if the anodic Tafel constant is known. Determining the anodic Tafel constant may be difficult for complex structures in service. However, even an estimate from a small sample in the required corrosive would be very helpful in making the criterion of cathodic protection more quantitative. (Campbell-NWWA)
W73-09239

CHEMISTRY OF GRAIN BOUNDARIES AND ITS RELATION TO INTERGRANULAR CORROSION OF AUSTENTITIC STAINLESS STEEL,
Brookhaven National Lab., Upton, N.Y.
A. Joshi, and D. F. Stein.
Corrosion, Vol 28, No 9, p 321-330, September 1972. 10 fig, 2 tab, 37 ref.

Descriptors: *Corrosion, Steel, *Stainless steel, Electronic equipment, *Chromium, Nickel, Nitrogen, Phosphorous, Alloys, Sulfur, Silicon, Heat treatment.
Identifiers: *Intergranular corrosion, *Electron spectroscopy, Annealing, Nitric acid, Fractographs.

The basis of intergranular corrosion in austenitic stainless steels in examined by relating the grain boundary composition to the corrosion properties. The technique of Auger electron spectroscopy has been used to obtain the chemistry of intergranular fracture surfaces. It was found that the chromium depleted zones exist and that the depletion theory was valid for tests in weakly oxidizing solutions. In highly oxidizing solutions, however, the impurity segregation and not chromium depletion best explains the deterioration of corrosion properties. Impurity elements such as sulfur, silicon, nitrogen, and phosphorous were observed in the various steels examined. An attempt is made to explain the observed corrosion properties on the basis of chromium depletion and solute segregation theories combined with an electrochemical mechanism. (Smith-NWWA)
W73-09240

USE OF ACOUSTIC LOGGING TECHNIQUE FOR DETERMINATION OF INTERGRAIN CEMENTATION PROPERTIES,
Continental Oil Co., Houston, Tex.
A. M. Zanier, and H. L. Overton.
Paper presented at Eleventh Annual Logging Symposium, p K1-K12, 1970. Society of Professional Well Log Analysts, Houston, Texas, 7 fig, 2 tab, 12 ref.

Descriptors: Logging (Recording), Wells, *Acoustics, Shear, Shear stress, Porosity, Elasticity (Mechanical), Compressibility, *Rock properties, Shear strength, Well data.
Identifiers: *Dilatational waves, *Wave velocity, Cementation, Formation evaluation.

The acoustic velocity of the dilatational and the shear waves in rocks increases as the hydrostatic pressure increases. However, the rate of increase of velocity is different for dilatational than for shear waves. For dilatational waves, the rate of velocity is a function of packing of the grains, while for shear, the rate of increase is a function of shear coupling. The shear waves are sensitive to the cementing material in the rocks, while the dilatational waves are mainly dependent upon the grains' properties. A theoretical exponent has been derived and tested experimentally in the laboratory on various rocks. Dilatational and shear wave test have been made and bulk compressibility and shear strength recorded under confining conditions. The tests indicate that a relation exists between this exponent and the 'cementationfactor'. From the scope picture of the acoustic pulse, it is possible to determine the exponent, along with porosity and elastic properties of the rock in place. (Campbell-NWWA)
W73-09242

ROLE OF THE ALLOYING ELEMENTS AT THE STEEL/RUST INTERFACE,
Centro Sperimentale Metallurgico S.p.A., Rome (Italy).
R. Bruno, A. Tamba, and G. Bombara.
Corrosion, Vol 29, No 3, March, 1973, p 95-99, 8 fig, 1 tab, 19 ref.

Descriptors: *Alloys, Corrosion, Analysis, *Steel, *Rusting, Copper, Chromium, Phosphorus, Nickel, *Oxidation, Oxides, Steel structures, Materials.
Identifiers: Atmospheric moisture, Polarization, Carbon steel, *Self-protection.

Three low alloy steels and an ordinary carbon steel were investigated as to the distribution of the alloy elements (Cu, Cr, P, Ni) in rust coverings produced either by weathering or by slight anodic oxidation. Electron microprobe examinations of rust showed, besides an overall enrichment in the alloy elements, the presence of sharp concentration peaks in layers some ten um wide. The more resistant the steel to atmospheric corrosion, the higher the rust to steel concentration gradient and the closer the concentration peaks to the steel/rust interface. Corrosion potential and polarization measurements in 0.1M Na2SO4 on rusted specimens evidenced a marked effect of the interfacial enrichment in increasing the anodic polarizability and the tendency to a coverage passivation. This leads to looking upon the corrosion generated steel/rust interface as directly contributing, in addition to the rust itself, to the self protection of low alloy steels during atmospheric corrosion. (Smith-NWWA)
W73-09261

DEVELOPMENT OF LOW COST, CORROSION RESISTANT ALLOYS AND OF ELECTROCHEMICAL TECHNIQUES FOR THEIR PROTECTION IN DESALINATION SYSTEMS,
Tyco Labs., Inc., Waltham, Mass.
F. H. Cocks, A. H. Taylor, and S. B. Brummer.
Available from the National Technical Information Service as PB-215 265, $6.75 in paper copy, $0.95 in microfiche. Office of Saline Water Research and Development Progress Report, No 801, December 1972. 227 p, 123 fig, 25 tab, 104 ref. 14-01-0001-1730.

Descriptors: *Desalination, *Corrosion, Metallurgy, Electrochemistry, *Copper Alloys, *Corrosion control, *Alloys.
Identifiers: Heat-treatment, Yield strengths, Anodic protection.

A corrosion - resistant alloy of the Cu-Zn-Ni-Al family has been developed for desalination systems. Modified by the addition of Fe and Mn, the alloys can be fabricated soft and thermally treated to impart a yield strength in excess of 140 000 psi. In a 268 deg F saline solution an improvement in corrosion resistance over arsenical aluminum brass by a factor of three was noted. Also studied were means to protect the alloy, and others commercially available, by anodic protection techniques. (OSW)
W73-09631

FINITE ELEMENT ANALYSES OF STRESSES AND MOVEMENTS IN DAMS, EXCAVATIONS AND SLOPES (STATE-OF-THE-ART),
California Univ., Berkeley.
For primary bibliographic entry see Field 08D.
W73-09666

APPLICATIONS OF FINITE ELEMENT ANALYSES IN SOIL DYNAMICS (STATE-OF-THE-ART),
California Univ., Berkeley.
For primary bibliographic entry see Field 08D.
W73-09667

A TALK ON THE FINITE ELEMENT METHOD AND GEOTECHNICAL ENGINEERING IN THE BUREAU OF RECLAMATION,
Bureau of Reclamation, Denver, Colo.
H. W. Anderson.
Paper, Appl Finite Elem Method Geotech Eng, Proc Symp, Vicksburg, Miss, May 1972, p 131-154, Sept 1972. 24 p, 10 fig, 14 ref.

Descriptors: *Finite element analysis, Engineering geology, Foundations, Soil mechanics, Rock mechanics, Stress analysis, Concrete dams, Earth dams, Rockfill dams, Structural analysis, Flow nets, Computer models, Dynamics, Rock foundations, Rock excavation, Deformation.

Identifiers: Grand Coulee Dam, Auburn Dam (Calif), Nonlinear properties, Foundation models, Computer graphics, Computer-aided design.

The Bureau of Reclamation began using the finite element method in 1962, when a stability analysis was performed for the underground chamber of Morrow Point Dam Powerplant. As the years passed, the Bureau applied the method to many analytical problems including in situ rock, concrete dams, earth and rockfill dam foundations, structures and flow nets. Results of Grand Coulee Forebay Dam foundation analyses were useful in determining the amount of concrete treatment necessary in a fault zone under the dam, for checking the adequacy of the proposed foundation drainage system, and for determining the stability of the dam foundation along a major joint system. Static analyses of stresses and deformations in earth and rockfill dams are made using a program that models the sequential construction of embankments and the nonlinear, stress-dependent behavior of the materials. As applications to larger problems grew, methods to simplify input of data improved with automatic generation of the finite element mesh and much of the input data. Improving the materials analysis to allow truly representative values and comparing actual structural behavior with predicted behavior are areas in which more work should be done. (USBR)
W73-09668

SYSTEMATIC STRESS ANALYSIS OF CIRCULAR PIPE,
Naigai Engineering Co., Tokyo (Japan).
For primary bibliographic entry see Field 08F.
W73-09669

CONSIDERATIONS OF PIPELINE TENSION TIE PROBLEMS,
Santa Clara County Flood Control and Water District, San Jose, Calif.
L. C. Fowler, R. P. Lundahl, and R. W. Purdie.
Proc Am Soc Civ Eng, Transp Eng J, Vol 98, No TE4, p 969-984, Nov 1972. 16 p, 7 fig, 8 tab, append.

Descriptors: *Pipelines, *Hydrostatic pressure, Joints (Connections), Shear, Strain, Tension, Structural design, Tensile stress, Strain gages, *California.
Identifiers: *Bolted joints, *Pipe joints, Santa Clara County (Calif), Lugs, Couplings, Flexible couplings.

The Santa Clara County Flood Control and Water District recently completed construction of more than 44 mi of pressure pipeline, ranging from 20 to 84 in. in diameter. The joint harness type tension tie was used extensively on these pipelines at flexible coupling locations where unbalanced internal forces occurred. During preliminary hydrostatic tests, 2 types of joint failures occurred at 5 locations. The first resulted when the tension tie nut and washer pulled through the back plate of the harness lug. The joint was repaired by replacing all plate lugs with new 1/2-in.-thick plate lugs twice as thick as the originals. In the second, the joint harness lug tore a section of pipe wall at the heel of the lug along the periphery of the lug to pipe weld. Satisfactory repairs were made by installing a 3/8-in. steel wrapper plate over the failed section of the pipe and welding new tension tie lugs on the wrapper plate. These failures, their possible causes, and recommendations for minimizing similar occurrences are discussed. Strain gage data obtained from field tests of 3 joint harness connections and calculated strains and stresses are presented in an appendix. (USBR)
W73-09671

LABORATORY LOAD TESTS ON BURIED FLEXIBLE PIPE,
Bureau of Reclamation, Denver, Colo.
A. K. Howard.
Journal of the American Water Works Association, Vol 64, No 10, Part 1, p 655-662, Oct 1972. 8 p, 12 fig, 3 tab, 8 ref.

Descriptors: *Laboratory tests, *Steel pipes, Deflection, Elastic deformation, Deformation, Structural behavior, Soil density, Clays, Research and development, Soil pressure, Strain measurement, Buckling.
Identifiers: *Flexible pipes, *Buried pipes, *Pipe tests.

Investigations were undertaken to determine soil-structure interaction of buried pipe. Steel pipe 18, 24, and 30 in. in diameter with wall thicknesses of 7, 10, and 14 gage for each diameter was buried in a large soil container for loading with a large universal testing machine. Nine pipes were tested with backfill soil placed at 90% Proctor maximum dry density, and 5 were tested at 100% Proctor. Four pipes buckled elastically at vertical deflections of less than 8%. Tests on 5 others, after the first 10-psi surcharge, produced good correlations with the Iowa formula for flexible pipe design. Under deflection, pipe took either an elliptical or rectangular shape, depending upon the relationship between pipe stiffness and soil stiffness. Elliptical pipe deformation created high compressive strains on the inner pipe surface at the horizontal diameter. Rectangular pipe deformation showed such strains at 45, 135, 225, and 315 deg points from the vertical axis. Pressures at the sides of the pipe were about 100-150% of the surcharge pressure for elliptically deformed pipe and 50-100% for rectangularly deformed pipe. (USBR)
W73-09674

TOTAL THERMAL ANALYSIS: POLYMERS AS A CASE IN POINT,
Owens-Illinois, Inc., Toledo, Ohio.
G. W. Miller.
Materials Research Standards, Vol 12, No 10, p 10-21, Oct 1972. 13 p, 25 fig, 27 ref.

Descriptors: *Differential thermal analysis, *Polymers, *Materials testing, Heat, Heat treatment, Temperature, Microscopy, Quality control, Measurement, Crystallization, Thermal expansion, Tensil properties, Bolumetric analysis, Deformation, Laboratory tests, Bibliographies, Calorimeters.
Identifiers: Transition, Thermogravimetric method, Dilatometers.

Thermal analysis has been identified with the techniques of differential thermal analysis (DTA), differential scanning calorimetry (DSC), and thermogravimetric analysis (TGA). Application of linear heating rates, increased amplification, and reliable quantification to these techniques have made them extremely useful to the materials scientist. Availability of highly sensitive transducers has promoted commercialization of reliable thermal mechanical analyzers. New thermal analysis techniques using polymers are discussed and correlated with old methods. Each new technique produces a specific response for further defining the structural behavior of polymers. Calorimetric or thermometric measurements complement these techniques in monitoring and controlling material properties for routine applications. To reliably interpret polymer behavior, specific modes of thermal analysis should be used. (USBR)
W73-09676

8I. Fisheries Engineering

BIOMASS AND PRODUCTION OF CARP FRY IN DIFFERENTLY FERTILIZED FINGERLING PONDS,
Polish Academy of Sciences, Krakow. Zaklad Biologii Wod.
For primary bibliographic entry see Field 05C.
W73-09187

BIASED HARVEST ESTIMATES FROM A POSTAL SURVEY OF A SPORT FISHERY,
Wisconsin Dept. of Natural Resources, Waupaca.
R. F. Carline.
Trans Am Fish Soc. Vol 101, No 2, p 262-266. 1972. Illus.
Identifiers: Fishery, *Postal surveys, Salvelinus-fontinalis, *Sport fishery, Surveys, *Brook trout.

A partial creel census and a postal card survey were simultaneously employed to determine the harvest of wild brook trout (Salvelinus fontinalis) from several ponds. A comparison of the catch per effort ratios (c/e) from interviews and postcards revealed a positive bias in the estimates obtained from returned postcards. The c/e ratios from postcards were double the c/e ratios from interviews. This discrepancy was partly due to the reluctance of unsuccessful anglers to return postcards. There was a significantly higher proportion of unsuccessful anglers from interviews than from postcard returns. Positive errors in the c/e ratios also resulted from a nonresponse of successful anglers and from more than one fisherman reporting on a single card. Estimates of fishing success from postcards seriously overestimated the season's harvest of trout and the exploitation rates.--Copyright 1973, Biological Abstracts, Inc.
W73-09229

EVALUATION OF THE CONTRIBUTION OF SUPPLEMENTAL PLANTINGS OF BROWN TROUT SALMO TRUTTA (L.) TO A SELF-SUSTAINING FISHERY IN THE SYDENHAM RIVER, ONTARIO, CANADA,
Guelph Univ. (Ontario). Dept. of Zoology.
T. J. Millard, and H. R. MacCrimmon.
J Fish Biol. Vol 4, No 3, p 369-384. 1972. Illus.
Identifiers: *Canada (Ontario), Fishery, Rivers, Salmo-Trutta, *Sydenham River (Ont.), *Trout (Brown).

An autumn planting of 4000 tagged yearling S. trutta (L.) in 1969 resulted in an over-winter survival of 26%, an angler recovery the following year of 8.1% and made up 22% of the March, 1970 standing population of the species. Aug. standing populations of brown trout increased from 142 trout/ha (17.6 kg/ha) in 1969 to 360 trout/ha (39.3 kg/ha) in 1970 while angler harvest of the species increased from 61 trout/ha (12.7 kg/ha) at a rate of 0.26 fish/h to 89 trout/ha (18.5 kg/ha) at a rate of 0.34/h. Using angler recovery and standing population as criteria the planting contributed substantially to the fishery. Actual contribution of stocked trout however, is questioned after detailed analysis of resident population structure and the potential of natural recruitment. It is suggested that the true benefit of stocked trout may be measured by the presence of those stocked fish in excess of the number of resident trout of that size predictable from a normal length distribution curve in waters with self-sustaining populations. Complexities in evaluating the merits of supplemental plantings of hatchery-reared brown trout to existing stream fisheries are examined.--Copyright 1973, Biological Abstracts, Inc.
W73-09285

DEVELOPMENT OF A SPORT FISHERY FOR LANDLOCKED SALMON IN A RECLAIMED POND,
Maine Dept. of Inland Fisheries and Game, Bangor.
K. Warner.
Prog Fish-Cult. Vol 34, No 3, p 133-140. 1972.
Identifiers: *Fish management, Development, Fishery, Osmerus mordax, Ponds, Salmo salar, *Salmon, *Maine.

A 108-acre mesotrophic lake in northern Maine was reclaimed with toxaphene in 1961 to evaluate the possibility of establishing a landlocked salmon (Salmo salar) fishery, with rainbow smelts (Osmerus mordax) as forage in ponds previously considered too small for salmon management, and

to evaluate factors affecting success. Initial salmon plantings failed to survive, but salmon were established by 1964, the third year after reclamation. After 5 unsuccessful eyed-egg plantings were made, smelts became established by 1968. Of 13 spp. present prior to reclamation, 7 spp. were eradicated and 2 others are probably no longer present. Because heavy salmon stocking and late establishment of smelts threatened salmon growth rate, length and bag limits were liberalized in 1966, resulting in a heavy harvest of salmon. Salmon growth improved to above average by 1969-71, attributable to establishment of an abundant smelt population, reduced salmon stocking, and a moderate salmon harvest. A high quality salmon fishery is presently being provided under existing conditions. The advantages and constraints of salmon management in similar habitats are discussed.--Copyright 1973, Biological Abstracts, Inc.
W73-09419

RELATION BEWTEEN FOOD CONSUMPTION AND FOOD SELECTIVITY IN FISHES: II. EXPERIMENTAL STUDY ON THE FOOD SELECTION BY JUVENILE GUPPY (POECILIA RETICULATA) (IN JAPANESE),
Kawaga Univ., Takamatsu (Japan). Inst. of Biology.
T. Sunaga.
Jap J Ecol. Vol 21, No 1/2, p 67-70, 1972. Illus. (English summary).
Identifiers: Cyclops-Vicinus, Daphnia-Pulex, *Fishes (Food), *Juvenile, Guppy, *Poecilia-Reticulata.

A series of experiments using the juvenile P. reticulata was made under laboratory conditions to examine how the feeding rate of a fish can be affected by the intensity of food selection. Throughout the experiments, the guppy was placed in a 2-l nylon screen basket suspended into a glass aquarium. Alive specimens of Daphnia pulex cultured in the laboratory and Cyclops vicinus collected from a reservoir were filtrated with a nylon screen net (1.2 mm in mesh). The filtrated plankters were mixed in various ratios and supplied to the fish. Juvenile guppy seemed to prefer D. pulex more than C. vicinus. The feeding rate (number of food specimen in the fish intestine/fish body weight/hr) for D. pulex increased significantly with an increase in number of the plankters supplied. In the case of C. vicinus such an increase of feeding rate as in the former species was not observed, and the feeding rate increased in accordance with the decrease of number of D. pulex given together. The total feeding rate (sum of D. pulex and C. vicinus) of guppy increased accompanied by the increase of the total number of plankters supplied. The greater the proportion of D. pulex to C. vicinus in plankters supplied, the higher the total feeding rate occurred.--Copyright 1972, Biological Abstracts, Inc.
W73-09504

SPAWNING OF THE AMERICAN SHAD, ALOSA SAPIDISSIMA IN THE LOWER CONNECTICUT RIVER,
Essex Mar. Lab., Conn.
C. Barton, Jr.
Chesapeake Sci. Vol 13, No 2, p 116-119, 1972. Illus.
Identifiers: Alosa-Sapidissima, Connecticut River, Rivers, *Shad (American), *Spawning.

Eggs were collected using plankton nets in the lower 54 mi of the Connecticut River during 1967 and 1968 to locate spawning areas of American shad and to determine environmental factors influencing spawning success. Spawning occurred between 13 and 27 June, the peak being June in 1967 and May in 1968. Shad spawned throughout the study area, but primarily above mile 28. Based on egg age and river flows, most eggs could have traveled 1 to 4 mi from the spawning sites. Shad arrived earlier in 1968 and spawned earlier and further downstream than in 1967, possibly due to warmer temperatures in May 1968. Lower water temperature in June 1968 may have contributed to the higher egg mortality observed in that year (18% in 1967 and 30% in 1968). Most eggs in 1968 (65%) were collected in temperatures lower than 16.0C, as opposed to only 3% in 1967. Other investigators have shown that shad eggs developing at these low temperatures have a prolonged development and reduced survival; the paucity of juveniles collected in 1968 gave supporting evidence for a weakened year class.--Copyright 1972, Biological Abstracts, Inc.
W73-09543

EVIDENCE ON THE INFLUENCE OF THE COPEPOD (SALMINCOLA CALIFORNIENSIS) ON THE REPRODUCTIVE PERFORMANCE OF A DOMESTICATED STRAIN OF RAINBOW TROUT (SALMO GAIRDNERI),
California Univ., Davis. Dept. of Animal Science.
G. A. E. Gall, E. L. McClendon, and W. E. Schafer.
Trans Am Fish Soc. Vol 101, No 2, p 345-346, 1972.
Identifiers: *Copepods, *Rainbow trout, Reproduction (Trout), Salmincola-Californiensis, Salmo-Gairdneri, *Fish eggs (Trout).

Egg number and egg size of hatchery-raise rainbow trout before and after eradication of this copepod parasite are analyzed.--Copyright 1973, Biological Abstracts, Inc.
W73-09556

SOME FACTORS ASSOCIATED WITH FLUCTUATION IN YEAR-CLASS STRENGTH OF SAUGER, LEWIS, AND CLARK LAKE, SOUTH DAKOTA,
Bureau of Sport Fisheries and Wildlife, Yankton, S. Dak. North Central Reservoir Investigations.
For primary bibliographic entry see Field 02H.
W73-09558

MINERAL COMPOSITION OF OREGON PELLET PRODUCTION FORMULATIONS,
Oregon State Univ., Astoria. Seafoods Lab.
D. L. Crawford, and D. K. Law.
Prog Fish-Cult. Vol 34, No 3, p 126-130. 1972.
Identifiers: *Fish diet, Fish hatchery, *Minerals, Oregon, *Pellet production.

The mineral composition of 2 Oregon pellet production formulations was determined. Four variations in the wet fish fraction composition for each formulation were included. Formulations were charcterized as to proximate composition. Levels of total ash, P, K, Ca, Mg, Na, Al, Sr, Fe, B, Cu, Zn, and Mn were determined. The success of these formulations under hatchery conditions seems to support the supposition that essential minerals in required amounts are available from these formulations for good growth.--Copyright 1973, Biological Abstracts, Inc.
W73-09561

09. MANPOWER, GRANTS AND FACILITIES

9A. Education (Extramural)

A LOOK BACK AT THE FIRST SEVEN YEARS OF WYO. WRRI (1965-1971),
Wyoming Univ., Laramie. Water Resources Research Inst.
P. A. Rechard.
Available from the National Technical Information Service as PB-220 251, $3.00 in paper copy, $0.95 in microfiche. Water Resources Series No 35, March 1973. 51 p, 2 append. OWRR A-001-WYO (64).

Descriptors: *Research and development, *Water resources Institute, Publications, *Education, Administration, *Wyoming, Training, Colleges, *Universities.

The organization, role, program, and accomplishments of the University of Wyoming Water Resources Research Institute are reviewed. The individuals involved with the WRRI are recognized and the publications resulting from the first seven years of operations are listed.
W73-09108

DIRECTORY OF FACULTY ENGAGED IN WATER RESEARCH AT CONNECTICUT UNIVERSITIES AND COLLEGES,
Connecticut Univ., Storrs. Inst. of Water Resources.
W. C. Kennard, and M. S. Williams.
Available from the National Technical Information Service as PB-220 246, $3.00 in paper copy, $0.95 in microfiche. Report No 17, October 1972, 36 p, 2 tab, append. OWRR A-999-CONN (12)

Descriptors: *Education, *Universities, *Connecticut, *Scientific personnel, Research and development, Water resources.
Identifiers: *Directories.

In order to determine the extent of interest at academic institutions in water-related research, a survey of the 21 colleges and universities in the state of Connecticut was conducted. Results of the survey were used to compile this directory. Eighteen out of the 21 reporting institutions had faculty in water research, representing a total of 186 individuals. Faculty was classified according to department, institution and subject matter. Of the 38 academic departments represented, the social sciences of law, economics and political science comprise less than 10%, the biological sciences more than 25%, the engineering sciences, along with the earth sciences and chemistry, approximately 14% each. Lesser percentages from a wide range of subjects such as Art, Pathobiology, Journalism, Statistics and Elementary Education formed the remaining 23%. The directory will be of value to people throughout the state who are concerned with the proper management and development of water resources in Connecticut. In addition, greater exchange of information and ideas among staff members at academic institutions in the state should occur. (Edelen-Connecticut)
W73-09112

WATER RESOURCES RESEARCH AT THE UNIVERSITY OF MARYLAND, 1965-1972,
Maryland Univ., College Park. Water Resources Research Center.
R. L. Green.
Water Resources Resèarch Center Research Report, 1972. 38 p, 3 fig, 5 tab, 25 ref. OWRR A-999-MD (7).

Descriptors: *Water Resources Research Act, *Water resources institute, *Universities, *Maryland, Research and development, Chesapeake Bay, Bibliographies, Reviews.

The Water Resources Research Center (WRRC) was established at the University of Maryland in the spring of 1965. Its Interdisciplinary Advisory Committee includes a chemical engineer, a soil physicist, a recreationist, a sociologist, a biologist and an economist. The water oriented activities of the state are primarily in the eastern shore, southern Maryland areas on the waterfront of the Chesapeake Bay, other smaller bays and the Potomac River and their tributaries. The demands on Maryland water resources are from the competitive uses for transportation, recreation, commercial fishing, agriculture and industrial uses including cooling water for electric power generators. Research by the Water Resources Research Center is summarized, and a bibliography of completed reports is presented. (Knapp-USGS)

SEA GRANT NEWSLETTER INDEX, 1968-71,
Rhode Island Univ., Narragansett. Pell Marine Science Library.
For primary bibliographic entry see Field 02L.
W73-09598

REPORTS FOR CALIFORNIA BY THE GEOLOGICAL SURVEY WATER RESOURCES DIVISION,
Geological Survey, Menlo Park, Calif. Water Resources Div.
For primary bibliographic entry see Field 02E.
W73-09603

10F. Preparation of Reviews

LEAD IN SOILS AND PLANTS: A LITERATURE REVIEW,
Colorado State Univ., Fort Collins. Dept. of Botany and Plant Pathology.
For primary bibliographic entry see Field 05B.
W73-09479

SIGNIFICANCE OF MERCURY IN THE ENVIRONMENT,
Department of Agriculture, Saskatoon (Saskatchewan). Research Station.
For primary bibliographic entry see Field 05B.
W73-09511

EVALUATION OF MEASUREMENT METHODS AND INSTRUMENTATION FOR ODOROUS COMPOUNDS IN STATIONARY SOURCES. VOLUME 1—STATE OF THE ART,
Esso Research and Engineering Co., Linden, N.J. Government Research Div.
For primary bibliographic entry see Field 05A.
W73-09601

BIOLOGICAL ASPECTS OF OIL POLLUTION IN THE MARINE ENVIRONMENT, A REVIEW,
McGill Univ., Montreal (Quebec). Marine Sciences Centre.
For primary bibliographic entry see Field 05C.
W73-09643

FINITE ELEMENT ANALYSES OF STRESSES AND MOVEMENTS IN DAMS, EXCAVATIONS AND SLOPES (STATE-OF-THE-ART),
California Univ., Berkeley.
For primary bibliographic entry see Field 08D.
W73-09666

APPLICATIONS OF FINITE ELEMENT ANALYSES IN SOIL DYNAMICS (STATE-OF-THE-ART),
California Univ., Berkeley.
For primary bibliographic entry see Field 08D.
W73-09667

SUBJECT INDEX

AUTHOR INDEX

ORGANIZATIONAL INDEX

ACCESSION NUMBER INDEX

ACCESSION NUMBER INDEX

ABSTRACT SOURCES

Source	Accession Number	Total
A. Centers of Competence		
Battelle Memorial Institute, Methods for Chemical and Biological Identification of Pollutants	W73-09300 -- 09302 09304 -- 09306 09308 09310 -- 09319 09321 -- 09335 09337 -- 09341 09343 -- 09371 09622	67
Bureau of Reclamation, Engineering Works	W73-09665 -- 09676	12
National Water Well Association, Water Well Construction Technology	W73-09239 -- 09240 W73 09242 -- 09250 09252 -- 09266	26
Oceanic Research Institute, Coastal Pollution	W73-09203 -- 09228 09230 -- 09237 09642 -- 09643 09645 -- 09652 09654 -- 09656 09658 -- 09664	54
University of Arizona, Arid Land Water Resources	W73-09678, 09681 09683 -- 09684 09686 -- 09687	6
University of Florida, Eastern U. S. Water Law	W73-09692 -- 09701 09704 -- 09748	55
University of Wisconsin, Eutrophication	W73-09171 -- 09177 09180 -- 09194 09196 09198 -- 09202	28
University of Wisconsin, Water Resources Economics	W73-09409 -- 09413	5
U. S. Geological Survey, Hydrology	W73-09101 -- 09103 09105 09121 -- 09170 09267 -- 09276 09278 -- 09280 09282 -- 09284 09286 -- 09299 09430 -- 09435 09437 -- 09466 09585 -- 09619	155
Vanderbilt University, Metals Pollution	W73-09467 -- 09469 09471, 09474 09476 -- 09482 09484 -- 09495 09498 -- 09502 09505 -- 09506 09508 -- 09511 09513 -- 09515	38

ABSTRACT SOURCES

Source	Accession Number	Tot
B. State Water Resources Research Institutes		
Alaska Institute for Water Research	W73-09111	
Connecticut Institute of Water Resources	W73-09112	
Minnesota Water Resources Research Center	W73-09113	
Missouri Water Resources Research Center	W73-09104	
West Virginia Water Research Institute	W73-09620 -- 09621	
Wisconsin Water Resources Center	W73-09106, 09117	
Wyoming Water Resources Research Institute	W73-09107 -- 09109	
C. Other		
BioSciences Information Service	W73-09178 -- 09179 09195, 09197 09229, 09238 09241, 09251 09277, 09281 09285, 09303 09307, 09309 09320, 09336 09342, 09372 09379, 09414 09415 -- 09429 09436, 09470 09472 -- 09473 09475, 09483 09496 -- 09497 09503 -- 09504 09507, 09512 09516 -- 09584 09644, 09653 09657, 09677 09679 -- 09680 09682, 09685 09690 09702 -- 09703 09749 -- 09750	
Environmental Protection Agency	W73-09688 -- 09689 09691	
Forest Service	W73-09373 -- 09378 09380 -- 09383	
Ocean Engineering Information Service	W73-09384 -- 09408	
Office of Saline Water	W73-09623 -- 09641	
Office of Water Resources Research	W73-09110 09114 -- 09116 09118 -- 09120	

☆ U. S. GOVERNMENT PRINTING OFFICE: 1973 O! - 542-949/2)

CENTERS OF COMPETENCE AND THEIR SUBJECT COVERAGE

- ▶ Ground and surface water hydrology at the Water Resources Division of the U.S. Geological Survey, U.S. Department of the Interior.
- ▶ Metropolitan water resources planning and management at the Center for Urban and Regional Studies of University of North Carolina.
- ▶ Eastern United States water law at the College of Law of the University of Florida.
- ▶ Policy models of water resources systems at the Department of Water Resources Engineering of Cornell University.
- ▶ Water resources economics at the Water Resources Center of the University of Wisconsin.
- ▶ Design and construction of hydraulic structures; weather modification; and evaporation control at the Bureau of Reclamation, Denver, Colorado.
- ▶ Eutrophication at the Water Resources Center of the University of Wisconsin, jointly sponsored by the Soap and Detergent Association and the Agricultural Research Service.
- ▶ Water resources of arid lands at the Office of Arid Lands Studies of the University of Arizona.
- ▶ Water well construction technology at the National Water Well Association.
- ▶ Water-related aspects of nuclear radiation and safety at the Oak Ridge National Laboratory.
- ▶ Public water supply treatment technology at the American Water Works Association.

Supported by the Environmental Protection Agency in cooperation with WRSIC

- ▶ Thermal pollution at the Department of Sanitary and Water Resources Engineering of Vanderbilt University.
- ▶ Water quality requirements for freshwater and marine organisms at the College of Fisheries of the University of Washington.
- ▶ Wastewater treatment and management at the Center for Research in Water Resources of the University of Texas.
- ▶ Methods for chemical and biological identification and measurement of pollutants at the Analytical Quality Control Laboratory of the Environmental Protection Agency.
- ▶ Coastal pollution at the Oceanic Research Institute.
- ▶ Water treatment plant waste pollution control at American Water Works Association.
- ▶ Effect on water quality of irrigation return flows at the Department of Agricultural Engineering of Colorado State University.

Subject Fields

1 NATURE OF WATER

2 WATER CYCLE

3 WATER SUPPLY AUGMENTATION AND CONSERVATION

4 WATER QUANTITY MANAGEMENT AND CONTROL

5 WATER QUALITY MANAGEMENT AND PROTECTION

6 WATER RESOURCES PLANNING

7 RESOURCES DATA

8 ENGINEERING WORKS

9 MANPOWER, GRANTS, AND FACILITIES

10 SCIENTIFIC AND TECHNICAL INFORMATION

INDEXES

SUBJECT INDEX

AUTHOR INDEX

ORGANIZATIONAL INDEX

ACCESSION NUMBER INDEX

ABSTRACT SOURCES

U.S. DEPARTMENT OF COMMERCE
National Technical Information Service
Springfield, Va. 22151

OFFICIAL BUSINESS

PRINTED MATTER

AN EQUAL OPPORTUNITY EMPLOYER

POSTAGE AND FEES PAID
U.S. DEPARTMENT OF COMMERCE
COM 211

U.S. MAIL

R A JENSEN
WATER RESOURCE SCI INFO CTR
DEPT OF THE INTERIOR
WASHINGTON DC 20240

SELECTED WATER RESOURCES ABSTRACTS

THE LIBRARY OF THE
SEP 10 1973
UNIVERSITY OF ILLINOIS
AT URBANA-CHAMPAIGN

VOLUME 6, NUMBER 16
AUGUST 15, 1973

W73-09751 -- 10400

SELECTED WATER RESOURCES ABSTRACTS is published semimonthly for the Water Resources Scientific Information Center (WRSIC) by the National Technical Information Service (NTIS), U.S. Department of Commerce. NTIS was established September 2, 1970, as a new primary operating unit under the Assistant Secretary of Commerce for Science and Technology to improve public access to the many products and services of the Department. Information services for Federal scientific and technical report literature previously provided by the Clearinghouse for Federal Scientific and Technical Information are now provided by NTIS.

SELECTED WATER RESOURCES ABSTRACTS is available to Federal agencies, contractors, or grantees in water resources upon request to: Manager, Water Resources Scientific Information Center, Office of Water Resources Research, U.S. Department of the Interior, Washington, D. C. 20240.

SELECTED WATER RESOURCES ABSTRACTS is also available on subscription from the National Technical Information Service. Annual subscription is $22 (domestic), $27.50 (foreign), single copy price $3. Certain documents abstracted in this journal can be purchased from the NTIS at prices indicated in the entry. Prepayment is required.

SELECTED
WATER RESOURCES ABSTRACTS

'A Semimonthly Publication of the Water Resources Scientific Information Center,
Office of Water Resources Research, U.S. Deoartment of the Interior

MICROFICHE PRICE CORRECTION

The price for all documents listed in this issue as available in microfiche for $1.45 from the National Technical Information Service is in error. The correct price is 95 cents.

VOLUME 6, NUMBER 16
AUGUST 15, 1973

W73-09751 -- 10400

retary of the U. S. Department of the Interior has determined that the publication of this periodical is neces- the transaction of the public business required by law of this Department. Use of funds for printing this pe- has been approved by the Director of the Office of Management and Budget through August 31, 1978.

As the Nation's principal conservation agency, the Department of the Interior has basic responsibilities for water, fish, wildlife, mineral, land, park, and recreational resources. Indian and Territorial affairs are other major concerns of America's "Department of Natural Resources."

The Department works to assure the wisest choice in managing all our resources so each will make its full contribution to a better United States—now and in the future.

FOREWORD

Selected Water Resources Abstracts, a semimonthly journal, includes abstracts of current and earlier pertinent monographs, journal articles, reports, and other publication formats. The contents of these documents cover the water-related aspects of the life, physical, and social sciences as well as related engineering and legal aspects of the characteristics, conservation, control, use, or management of water. Each abstract includes a full bibliographical citation and a set of descriptors or identifiers which are listed in the **Water Resources Thesaurus**. Each abstract entry is classified into ten fields and sixty groups similar to the water resources research categories established by the Committee on Water Resources Research of the Federal Council for Science and Technology.

WRSIC IS NOT PRESENTLY IN A POSITION TO PROVIDE COPIES OF DOCUMENTS ABSTRACTED IN THIS JOURNAL. Sufficient bibliographic information is given to enable readers to order the desired documents from local libraries or other sources.

Selected Water Resources Abstracts is designed to serve the scientific and technical information needs of scientists, engineers, and managers as one of several planned services of the Water Resources Scientific Information Center (WRSIC). The Center was established by the Secretary of the Interior and has been designated by the Federal Council for Science and Technology to serve the water resources community by improving the communication of water-related research results. The Center is pursuing this objective by coordinating and supplementing the existing scientific and technical information activities associated with active research and investigation program in water resources.

To provide WRSIC with input, selected organizations with active water resources research programs are supported as "centers of competence" responsible for selecting, abstracting, and indexing from the current and earlier pertinent literature in specified subject areas.

Additional "centers of competence" have been established in cooperation with the Environmental Protection Agency. A directory of the Centers appears on inside back cover.

Supplementary documentation is being secured from established discipline-oriented abstracting and indexing services. Currently an arrangement is in effect whereby the BioScience Information Service of Biological Abstracts supplies WRSIC with relevant references from the several subject areas of interest to our users. In addition to Biological Abstracts, references are acquired from Bioresearch Index which are without abstracts and therefore also appear abstractless in SWRA. Similar arrangements with other producers of abstracts are contemplated as planned augmentation of the information base.

The input from these Centers, and from the 51 Water Resources Research Institutes administered under the Water Resources Research Act of 1964, as well as input from the grantees and contractors of the Office of Water Resources Research and other Federal water resources agencies with which the

Center has agreements becomes the information base from which this journal is, and other information services will be, derived; these services include bibliographies, specialized indexes, literature searches, and state-of-the-art reviews.

Comments and suggestions concerning the contents and arrangements of this bulletin are welcome.

Water Resources Scientific Information Center
Office of Water Resources Research
U.S. Department of the Interior
Washington, D. C. 20240

CONTENTS

runoff as a primary goal must consider basic hydrologic elements such as soil type, basin configuration and slopes, channel characteristics, antecedent moisture, season, vegetation cover, and the significance of varying rainfall intensities. (See also W73-09773) (Woodard-USGS)
W73-09777

MODELLING PROCESSES FOR STUDY AND EXPLORATION OF LAKE ECOLOGICAL SYSTEM (IN RUSSIAN),
Akademiya Nauk SSSR, Leningrad. Institut Evolyutsionnoi Fiziologii i Biokhimii.
For primary bibliographic entry see Field 02H.
W73-09821

WATER RESOURCES DEVELOPMENT IN THE MULLICA RIVER BASIN PART II: CONJUNCTIVE USE OF SURFACE AND GROUNDWATERS OF THE MULLICA RIVER BASIN,
Rutgers - The State Univ., New Brunswick, N.J. Water Resources Research Inst.
For primary bibliographic entry see Field 04B.
W73-09870

THE RELATIONSHIP BETWEEN RAINFALL, GEOMORPHIC DEVELOPMENT AND THE OCCURRENCE OF GROUNDWATER IN PRECAMBRIAN ROCKS OF WESTERN AUSTRALIA,
For primary bibliographic entry see Field 02F.
W73-09958

MATERIALS OF THE ANNIVERSARY SESSION OF THE ACADEMIC COUNCIL (MATERIALY YUBILEYNOY SESSII UCHENOGO SOVETA).
Kazakhskii Nauchno-Issledovatelskii Gidrometeorologicheskii Institut, Alma-Ata (USSR).

Kazakhskiy Nauchno-Issledovatel'skiy Gidrometeorologicheskiy Institut Trudy, No 51, Kh. A. Akhmedzhanov, and A. P. Agarkova, editors, Leningrad, 1971. 200 p.

Descriptors: *Investigations, *Climatology, *Meteorology, Weather, Atmosphere, Air temperature, Snowpacks, Precipitation (Atmospheric), Evaporation, Runoff, Water balance, Water resources, Orography, Mountains, Mudflows, Model studies, Forecasting, Curves, Equations, Maps.
Identifiers: *USSR, *Kazakhstan.

Development of scientific investigations at the Kazakh Hydrometeorological Scientific Research Institute (KazNIGMI) is discussed in this commemorative issue marking the 20th anniversary of the Institute's founding. Topics in this collection of 18 papers include: (1) results of 20 years of research by KazNIGMI; (2) 20 years of research on the problem of long-range weather forecasting; (3) models of mudflow processes; (4) glacial mudflows and forecasting techniques; (5) mudflow classification and distribution on the northern slope of Zailiski Alatau; (6) observations and computation of mudflows in the Kokcheka River basin; (7) spatial distribution of atmospheric precipitation, streamflow, and maximum water equivalent of snowpacks in Kazakhstan; (8) calculation of water-balance components for mountain basins on the northern slope of Zailiski Alatau; (9) complex use of water resources in the Aral Sea basin; (10) synoptic and meteorologic conditions in development of wheat rust in northern Kazakhstan; (11) advance forecasting of amount and duration of storm precipitation for the northern slope of Zailiski Alatau; (12) comparison of methods for the determination of average long-term annual evaporation in southeastern Kazakhstan. (Josefson-USGS)
W73-10048

WATER BALANCE IN THE GEORGIAN SSR (VODNYY BALANS GRUZII),
Akademiya Nauk Gruzinskoi SSR, Tiflis. Institut Geografii.
L. A. Vladimirov.
Meteorologiya i Gidrologiya, No 6, p 86-88, June 1972. 1 tab, 6 ref.

Descriptors: *Water balance, *Orography, Mountains, Elevation, Precipitation (Atmospheric), Runoff, Surface runoff, Subsurface runoff, Subsurface flow, Runoff coefficient, Evaporation.
Identifiers: *USSR, *Georgian SSR.

Annual precipitation for the Georgian SSR is 96.9 cu km, of which 65.3 cu km occurs in western Georgia and 31.6 cu km in eastern Georgia. Annual runoff of streams is 56.9 cu km, including 42.5 cu km from the western part and 14.4 cu km from the eastern part. Total streamflow, including flow derived from outside the Republic, is 65.3 cu km. The runoff coefficient is highest (0.85) in the high-mountain zone of the Western Caucasus and lowest (0.20) in the lowlands of the Khrami River basin. Methods are given for studying items in the water balance in the high-mountain zone and for determining subsurface flow into streams. (Josefson-USGS)
W73-10050

WATER-TABLE FLUCTUATION IN SEMIPERVIOUS STREAM-UNCONFINED AQUIFER SYSTEMS,
California Univ., Davis.
For primary bibliographic entry see Field 02F.
W73-10058

A REVIEW OF SOME MATHEMATICAL MODELS USED IN HYDROLOGY, WITH OBSERVATIONS ON THEIR CALIBRATION AND USE,
Institute of Hydrology, Wallingford (England).
R. T. Clarke.
Journal of Hydrology, Vol 19, No 1, p 1-20, May 1973. 91 ref.

Descriptors: *Statistical models, *Mathematical models, *Model studies, Variability, Probability, Least squares method, Regression analysis, Correlation analysis, Stochastic processes, Systems analysis.

Terms that are frequently used (and misused) where mathematical models are applied to solve hydrological problems are reviewed, and some of the many available models are classified into four main groups. It is suggested that models with parameters estimated by a least-squares objective function represent an application of nonlinear regression theory to situations in which the assumptions in this theory are seldom valid. The correction required is not the use of objective functions, but the use of more realistic assumptions concerning the stochastic structure of the model residuals. Interdependence between model parameters necessitates extensive exploration of the sum-of-squares surface in the neighborhood of its minimum even when regression assumptions are valid; this is particularly true where the model is to be used to examine the likely effects of a proposed physical change to the catchment, since the complexity of such a change will not generally be represented by a change in one (or even some) of the model's parameters, regardless of the remainder. (Knapp-USGS)
W73-10061

METHOD OF EVALUATING WATER BALANCE IN SITU ON THE BASIS OF WATER CONTENT MEASUREMENTS AND SUCTION EFFECTS (METHODE D'EVALUATION DU BILAN HYDRIQUE IN SITU A PARTIR DE LA MESURE DES TENEURS EN EAU ET DES SUCCIONS),
Grenoble Univ. (France).

J-F. Daian, and G. Vachaud.
In: Proceedings of Symposium on the Use of Isotopes and Radiation in Research on Soil-Plant Relationships Including Applications in Forestry, Dec 13-17, 1971, Vienna (IAEA-SM-151), International Atomic Energy Agency, Vienna, 1972, p 649-660. 5 fig, 1 tab, 8 ref.

Descriptors: *Unsaturated flow, *Hydraulic conductivity, *Water balance, *Groundwater recharge, Moisture meters, Radioactivity techniques, Tensiometers, Soil water movement, Darcys law, Capillary conductivity, Hydraulic gradient, Precipitation (Atmospheric), Recharge, Evaporation.

Knowledge of the entire hydrologic region as well as strictly local factors are required to predict the water balance (including aquifer recharge). In 1970, measurements (including daily determination with a neutron moisture meter of soil moisture at 10-120 cm depth) indicated that 66% to 91% of rainfall reached 120 cm depth at a site near the Isere River. Uncertainties in interpretation included estimation of seepage from the river, spatial inhomogeneity in local rainfall, neglect of lateral flow, and assumption that air in soil voids was at atmospheric pressure. (See also W72-12358 and W72-10679) (Bopp-ORNL)
W73-10197

2B. Precipitation

PROCEEDINGS--SOUTHERN PLAINS REGION CLOUDTAP CONFERENCE.
For primary bibliographic entry see Field 03B.
W73-09773

RESULTS FROM AN APPLICATIONS PROGRAM OF HAIL SUPPRESSION IN TEXAS,
Atmospherics, Inc., Fresno, Calif.
For primary bibliographic entry see Field 03B.
W73-09779

FIFTH ANNUAL SURVEY REPORT ON THE AIR WEATHER SERVICE, WEATHER MODIFICATION PROGRAM (FY 1972),
Air Weather Service, Scott AFB, Ill.
For primary bibliographic entry see Field 03B.
W73-09785

THE MOKOBULAAN RESEARCH CATCHMENTS,
For primary bibliographic entry see Field 03B.
W73-09910

SAMPLING, HANDLING AND COMPUTER PROCESSING OF MICROMETEOROLOGICAL DATA FOR THE STUDY OF CROP ECOSYSTEMS,
For primary bibliographic entry see Field 07C.
W73-10001

ICE NUCLEATION: ELEMENTAL IDENTIFICATION OF PARTICLES IN SNOW CRYSTALS,
National Oceanic and Atmospheric Administration, Boulder, Colo. Atmospheric Physics and Chemistry Lab.
For primary bibliographic entry see Field 03B.
W73-10037

THEORETICAL APPROXIMATION TOWARD TWO FUNDAMENTAL MODELS OF CLIMATE ON EARTH,
S. De J. Soria.
Acta Agron (Palmira). Vol 21, No 4, p 165-177. 1971. Illus.
Identifiers: Biomes, *Climates, Earth, *Model studies, Theoretical studies.

Climatic conditions prevailing under an equinoxial and a solstitial model of the earth are compared, using an extraterrestrial energy curve in which the ordinate represents Langleys per day and the abscissa the months of the year. The equinoxial curve is taken at latitude 0 deg (the equator) and solstital curves are shown at latitudes ranging from latitudes 23 deg, 30, 45 and 90. The astronomic fundamentals are defined and many authors are quoted as to the effect of the various conditions on atmospheric circulation, rain, evaporation and evapotranspiration, as well as the resulting biomes, which are the major biological systems morphologically characterized by the equinoxial belt.--Copyright 1973, Biological Abstracts, Inc.
W73-10154

AIR FLOW LAND-SEA-AIR-INTERFACE: MONTEREY BAY, CALIFORNIA - 1971,
Moss Landing Marine Labs., Calif.
For primary bibliographic entry see Field 05B.
W73-10223

A REVIEW OF METHODS FOR PREDICTING AIR POLLUTION DISPERSION,
National Aeronautics and Space Administration, Hampton, Va. Langley Research Center. National Aeronautics and Space Administration, Langley Station, Va. Langley Research Center.
For primary bibliographic entry see Field 05A.
W73-10324

2C. Snow, Ice, and Frost

FIFTH ANNUAL SURVEY REPORT ON THE AIR WEATHER SERVICE, WEATHER MODIFICATION PROGRAM (FY 1972),
Air Weather Service, Scott AFB, Ill.
For primary bibliographic entry see Field 03B.
W73-09785

EVALUATION OF NORTH WATER SPRING ICE COVER FROM SATELLITE PHOTOGRAPHS,
McGill Univ., Montreal (Quebec). Dept. of Meteorology.
P. G. Aber, and E. Vowinckel.
Available from NTIS, Springfield, Va 22151 N72-29300 Price $3.25 printed copy; $0.95 microfiche. Publication in Meteorology No 101, June 1971. 19 p, 4 fig, 1 tab, 4 ref. Defense Res Bd. of Canada Grant 9511-107.

Descriptors: *Ice cover, *Mapping, *Aerial photography, *Arctic, Satellites (Artificial), Methodology, Reviews, Evaluation, Data collections, Climatology, Ice, Melting, Freezing, Distribution patterns, Remote sensing.
Identifiers: *Baffin Bay, North water.

Satellite photographs for two years (March-September) have been used to study ice cover in the polynya called 'North Water' and to determine whether reliable ice maps can be made from satellite data without computer analysis. After early July the clouds become opaque and distinction between cloud and ice is impossible. Ice distribution for short periods can best be obtained by careful human photo interpretation. The most persistant open water is at the northern edge, at about 78 deg N. The southern ice edge is diffuse. The changes in ice cover in the North are mainly caused by freezing and melting, while ice transport is important in the southern area. (Woodard-USGS)
W73-09800

MANIFESTATION OF EROSION IN PERMAFROST ON THE YAMAL PENINSULA (PROYAVLENIYE EROZII V MNOGOLETNEMERZLYKH PORODAKH NA YAMALE),
For primary bibliographic entry see Field 02J.

Exploratory work was designed to evaluate the storage-duration index as related to inventory-prediction variables of forest density, potential insolation (slope steepness and aspect), and elevation. The study utilized 75 sample plots in the ponderosa pine type on the Beaver Creek Watershed in Arizona for 1967-68 and 1968-69. The storage-duration index provides an integrated single estimate of initial snow storage and subsequent melt rates, and was developed for each time period by adding together snowpack water equivalent measurements made in successive surveys for the period. A multiple regression analysis showed similar relationships between storage-duration index values and inventory-prediction variables for both years of study; such a relationship is useful for identifying sites with desired snowpack storage conditions. On the Beaver Creek Study Area large initial storage followed by slow melting was associated with low forest densities, low potential insolation values, and high elevations, as sites exhibiting all three characteristics possessed maximum index values. Low initial storage followed by rapid melting was associated with the opposite site characteristics. (White-Arizona)
W73-10152

WATER UPTAKE OF PINES FROM FROZEN SOIL,
Finnish Forest Research Inst., Helsinki (Finland).
For primary bibliographic entry see Field 02D.
W73-10189

EFFECTS OF OIL UNDER SEA ICE,
Massachusetts Inst. of Tech., Cambridge. Fluid Mechanics Lab.
For primary bibliographic entry see Field 05B.
W73-10246

2D. Evaporation and Transpiration

EVAPOTRANSPIRATION OF DIFFERENT CROPS IN A MEDITERRANEAN SEMI-ARID REGION, (IN FRENCH),
Office de la Recherche Scientifique et Technique Outre-Mer, Abidjan (Ivory Coast). Centre d'Adiopodoume.
B. A. Monteny.
Agric Meteorol. Vol 10, No 1/2, p 19-38, 1972. Illus. English summary.
Identifiers: *Advection, *Arid lands, Crops, Density, Energy balance, *Evapotranspiration, Humidity, Index, Irrigation, Leaf, Mediterranean areas, Stomata, Temperature, Transpiration, Winds.

The value of the maximum evapotranspiration in an arid region depends on the kind of crop chosen. The energy balance method has shown the importance of a source of energy in this semi-arid region (Tunisia)-advection. Under conditions of large scale advection (oasis effect), the ratio between maximum evapotranspiration and net radiation exceeds generally unity. The water loss from the different crops receiving optimum irrigation treatment varies widely. The factors which could influence the maximum crop evapotranspiration include: the air characteristics (temperature, humidity, wind-speed) flowing from a dry area into the experimental irrigated area; the solar energy interception and the crop architecture; the extent of the transpirating surface (leaf development, leaf area index) and its characteristics (geometry, stomatal density, tropism). For a large area cultivated with a defined crop the measured evapotranspiration from this crop would be a more accurate method of irrigation control. It is difficult in those arid conditions to evaluate the water requirements for a cropped surface by climatological formulae because of the importance of the physiological factors.--Copyright 1972, Biological Abstracts, Inc.
W73-09836

GROWTH AND WATER USE OF LOLIUM PERENNE: I. WATER TRANSPORT,
Rothamsted Experimental Station, Harpenden (England).
For primary bibliographic entry see Field 02I.
W73-09973

THE EFFECT OF TOTAL WATER SUPPLY AND OF FREQUENCY OF APPLICATION UPON LUCERNE: I. DRY MATTER PRODUCTION,
Reading Univ. (England). Dept. of Agricultural Botany.
For primary bibliographic entry see Field 03F.
W73-09975

THE EFFECT OF DIFFERENT LEVELS OF WATER SUPPLY ON TRANSPIRATION COEFFICIENT AND PHOTOSYNTHETIC PIGMENTS OF COTTON PLANT,
National Research Centre, Cairo (Egypt). Plant Physiology Dept.
For primary bibliographic entry see Field 03F.
W73-09979

PRELIMINARY RESULTS ON THE EFFECT OF POLYETHYLENE FOIL APPLIED AS A BARRIER LAYER, (IN GERMAN),
Rostock Univ. (East Germany). Institut fuer Meliorationswesen.
H. Koenker.
Arch Acker-Pflanzenbau Bodenkd. Vol 16, No 4/5, p 407-413. 1972. Illus. English summary.
Identifiers: *Barrier layers, Crops, *Evapotranspiration, Polyethylene foils, Retention, Transpiration, Water retention.

The increase of the precipitation retention of a sandy soil by means of a polyethylene foil barrier applied at a depth of 1 m was studied in a field experiment. The water that was retained above the barrier led to a rise of soil moisture up to closely below the topsoil. In dry periods, water evapotranspiration on the barrier variants was about 4 times as high as that on the control plot. Deep-rooted crop plants made good use of the water that was retained above the barrier, while it was less available for plants with a shallow root system.--Copyright 1973, Biological Abstracts, Inc.
W73-09983

ON THE HYDROPHOBIZATION OF A SOIL LAYER WITH THE VIEW OF INTERRUPTING CAPILLARY WATER MOVEMENT, (IN GERMAN),
Dsutsche Akademie der Landwirtschaftswissenschaften zu Berlin, Muencheberg (East Germany). Forschungszentrum Bodenfruchtbarkeit.
H. Hergehan.
Arch Acker-Pflanzenbau Bodenkd. Vol 16, No 4/5, p 399-405. 1972. Illus. English summary.
Identifiers: *Evaporation, *Hydrophobization, Soil layers, *Water movement (Capillary).

The wettability of soil samples may be reduced by means of substances that lower the interfacial tension (structural hydrophobization) and by partial sealing of the aggregate pores by means of emulsions with a naturally hydrophobic disperse phase (indirect hydrophobization). The extent of hydrophobization depends on the applied amount of the substance and largely on the length of the hydrophobic hydrocarbon and on the charge of the active group. Hydrophobization of a soil layer leads to an interruption of the capillary water movement and thus, in the laboratory experiments, to a reduction of unproductive evaporation from the soil samples.--Copyright 1973, Biological Abstracts, Inc.
W73-09984

MODES OF ACTION AND METHODS OF APPLICATION OF SOIL CONDITIONERS FOR IMPROVING PHYSICAL PROPERTIES OF THE SOIL, (IN GERMAN),
Deutsche Akademie der Landwirtschaftswissenschaften zu Berlin, Muenchberg (East Germany). Forschungszentrum Bodenfruchtbarkeit.
For primary bibliographic entry see Field 02G.
W73-09985

PATTERNS OF GROUNDWATER DISCHARGE BY EVAPORATION AND QUANTITATIVE ESTIMATES OF THE FACTORS AFFECTING EVAPORATION AND CONVECTIVE HEAT EXCHANGE (ZAKONOMERNOSTI RASKHODA GRUNTOVYKH VOD NA ISPARENIYE I KOLICHESTVENNYY UCHET F AKTOROV, VLIYAYUSHCHIKH NA ISPARENIYE I KONVEKTIVNYY TEPLOOBMEN),
Zakavkazskii Nauchno-Issledovatelskii Gidrometeorologicheskii Institut, Tiflis (USSR).
A. M. Mkhitaryan, A. S. Akopyan, and M. G. Dagestanyan.
Zakavkazskiy Nauchno-Issledovatel'skiy Gidrometeorologicheskiy Institut Trudy, No 51 (57), Leningrad, 1972. 180 p.

Descriptors: *Estimating, *Groundwater, *Evaporation, *Convection, *Heat transfer, Heat balance, Temperature, Radiation, Air circulation, Atmosphere, Turbulence, Mixing, Water balance, Water table, Zone of aeration, Drainage, Irrigation, Lysimeters, Geomorphology, Measurement.
Identifiers: *USSR, *Armenia, *Ararat Plain, *Heat exchange, Moisture exchange.

Moisture exchange in the zone of aeration and evaporation from the groundwater surface at different water-table depths were investigated on the Ararat Plain in Armenia. Physiographic conditions and heat and water regimes of the plain are examined together with techniques of evaporation measurement. Quantitative estimates are made of the influence of different factors on the processes of evaporation and convective exchange of heat with the atmosphere. (Josefson-USGS)
W73-10056

ON THE ESTIMATION OF EVAPORATION AND CLIMATIC WATER BALANCE FOR PRACTICAL FARMING, (IN GERMAN),
For primary bibliographic entry see Field 03F.
W73-10072

TRANSPIRATION OF REED (PHRAGMITES COMMUNIS TRIN.),
Polish Academy of Sciences, Warsaw. Dept. of Applied Limnology.
J. Krolikowska.
Pol Arch Hydrobiol. Vol 18 No 4, p 347-358. 1971. Illus.
Identifiers: Aquatic plants, Humidity, Infection, Leaves, Panicle, Phragmites-Communis, Radiation, *Reed, Seasonal, Temperature, *Transpiration, Ustilago-Grandis, *Poland, *Mikolajskie Lake.

Transpiration of P. communis in a reed-belt of Mikolajskie Lake was measured at 2 sites: one terrestrial and one aquatic during the vegetation season of 1969. The intensity of reed transpiration increases with the increase of its fresh weight; it also increases with the rise of air temperature and of solar radiation values, but decreases with the rise of relative humidity; transpiration rate rises rapidly in the morning and gradually decreases in the afternoon--maximal values were found between 11 a.m. and 1 p.m. The transpiration intensity increases from spring till mid-summer, then decreases; the maximum was found in late end of July and early August. The reed growing on land transpires less than the one in the lake, and is also shorter. The leaves of the upper part of shoot (younger ones) transpire more than the lower (older) ones. The formation of panicle slows down the transpiration of reed leaves. The infection by Ustilago grandis Fr. lowers the intensity of reed transpiration. Transpiration of reed is of an intensity which can influence the water balance of the lake.--Copyright 1973, Biological Abstracts, Inc.
W73-10160

WATER UPTAKE OF PINES FROM FROZEN SOIL,
Finnish Forest Research Inst., Helsinki (Finland).
M. Leikola, and E. Paavilainen.

Descriptors: *Tracers, *Coniferous forests, *Transpiration, *Cold regions, Scotch pine trees, Soil-water-plant relationships, Available water, Moisture stress, Moisture uptake, Moisture availability, Plant physiology, Freezing, Neutron activation analysis, Radioactivity techniques, Radioecology.
Identifiers: Scandium radioisotopes.

Although a control peatland plot thawed in May, a second plot was caused to remain partially frozen until July by shoveling off the snow in winter and by insulating with straw in spring. Transpiration of young trees was measured by analysis of the needles for a tracer that had been injected into the trunk. The presence of tracer in the needles of trees on the colder plot showed that transpiration occurred at 0 dec C; however, about 5.5 times more tracer was present in needles of trees on the warmer plot for a May 8 injection (measured 2-3 days after injection), and about 3.4 times more for a May 28 injection. (Bopp-ORNL)
W73-10189

USE OF TRITIATED WATER FOR DETERMINATION OF PLANT TRANSPIRATION AND BIOMASS UNDER FIELD CONDITIONS,
Argonne National Lab., Ill.
J. R. Kline, M. L. Stewart, C. F. Jordan, and P. Kovac.
In: Proceedings of Symposium on the Use of Isotopes and Radiation in Research on Soil-Plant Relationships Including Applications in Forestry, Dec. 13-17, 1971, Vienna (IAEA-SM-151), International Atomic Energy Agency, Vienna, 1972, p 419-437, 3 fig, 8 tab, 13 ref.

Descriptors: *Water requirements, *Trees, *Grasslands, *Tritium, Tracers, Water stress, Transpiration, Soil-water-plant relationships, Jack Pine Trees, Corn (Field), Oats, Coniferous trees, Coniferous forests, Analytical techniques, Radioecology.

The tracer method was corroborated by other methods. Transpiration ranged from 6 ml/hr (small saplings) to 900 ml/hr (large conifers). Biomass ranged from 167 g to 400 kg. The feasibility of using tracer to measure the mean water turnover in grasslands was tested by labeling by the vapor exchange method. This method appears to be inherently more precise than tritium injection that must be used for trees and large plants. The method has been applied to field studies of grasslands as will be described in a future report. The high water turnover rate in grasslands makes measurements feasible in a few hours as compared with days and weeks for trees. (See also W72-10679) (Bopp-ORNL)
W73-10190

ESTIMATE OF WATER BALANCE IN AN EVERGREEN FOREST IN THE LOWER IVORY COAST (ESTIMATION DU BILAN HYDRIQUE DANS UNE FORET SEMPERVIRENTE DE BASSE COTE-D'IVOIRE),
Office de la Recherche Scientifique et Technique Outre-Mer, Abidjan (Ivory Coast). Centre d'Adiopodoume.
C. Huttel.

Identifiers: *Belgium, Biological studies, Chemistry, Conservation, Creeks, Ecological studies, *Fauna, Flanders, Invertebrates, Malacological fauna, Planktonic, Rivers, *Scheldt River.

In the north of the province of east Flanders, extending from some 30 km north-west of Ghent to north-west of Antwerp, and also along the river Scheldt, are many inland waters called 'kreken' (creeks) which are the remains of consecutive dike ruptures. The water of the creeks is slightly brackish: the chloride content generally varies between 0.6 and 1.8 g Cl/l. One southern creek however, more distant from the river, may be considered as a freshwater biotope (0.14 g Cl/l), while the northwest group is more brackish (2.2-3.65 g Cl/l). The creeks have well-buffered waters and are productive. As the waters mostly are oligohaline, many freshwater euryhaline species are present. Most characteristic are the invertebrates which are more or less confined to the brackish biotopes. Physico-chemical data obtained by water analysis and its importance in relation to animal life are added; finally a plea is made for conservation of these unique natural sites.—Copyright 1972, Biological Abstracts, Inc.
W73-10086

WATER RESOURCES OF THE RED RIVER OF THE NORTH DRAINAGE BASIN IN MINNESOTA,
Geological Survey, St. Paul, Minn.
R. W. Maclay, T. C. Winter, and L. E. Bidwell.
Available from NTIS, Springfield, Va 22151, PB-218 965, Price $3.00 printed copy; $0.95 microfiche. Geological Survey Water-Resources Investigations 1-72, November 1972. 129 p, 49 fig, 5 tab, 77 ref.

Descriptors: Water resources, *Groundwater resources, *Water quality, *Minnesota, Surface waters, Groundwater, Water chemistry, Water yield, Water supply, Hydrogeology, Aquifers, Streamflow, Water pollution, Floods, Water balance, Glacial drift.
Identifiers: *Red River of the North (Minn).

Water problems in the Red River of the North basin in Minnesota include flooding, pollution, and water shortages. In the morainal area, problems generally are absent; but in the flat plain of former Glacial Lake Agassiz, they can be severe. About 5.1 million acre-feet of water is perennially available. Average annual flow in streams tributary to the Red River equals 1.7 inches over the basin. Runoff ranges from less than 1 to more than 4 inches. Glacial sand and gravel in the morainal and in the Halma-Lake Bronson areas are potentially large aquifers. Presently, other sand and gravel aquifers, generally local, provide most supplies. Much saline water occurs at depth in sedimentary rocks in the northwest part of the basin. Regionally, groundwater moves westward from morainal area to lake plain or Red River. Locally, groundwater in the morainal area moves from high areas to adjacent lowlands. Much deep moving water discharges at the east edge of the lake plain. Groundwater in recharge areas is generally calcium magnesiun bicarbonate type. Sodium bicarbonate and sodium chloride type waters occur in association with Cretaceous and sedimentary rocks (in the northwest part of the basin), respectively. (Knapp-USGS)
W73-10321

REGIONAL ANALYSES OF STREAMFLOW CHARACTERISTICS,
Geological Survey, Washington, D.C.
H. C. Riggs.
Available from GPO, Washington, D.C. 20402, Price $0.45. Geological Survey Techniques of Water-Resources Investigations, Book 4, Chapter B3, 1973. 15 p, 4 fig, 2 tab, 34 ref.

Descriptors: *Streamflow, *Runoff, *Flow rates, *Regional analysis, *Data collections, Analytical techniques, Stochastic processes, Statistical methods, Low flow, Peak discharge, Average flow, Regression analysis, Correlation analysis, Hydrologic data, Stream gages, Gaging stations.
Identifiers: *Streamflow characteristics.

Regional analysis is concerned with extending records in space as differentiated from extending them in time. Because streamflow records are collected at only a few of the many sites where information is needed, gaging-station information must be transferred to ungaged sites. A regional analysis provides a tool for doing this. This manual describes various ways of generalizing streamflow characteristics and evaluates the applicability and reliability of each under various hydrologic conditions. Several alternatives to regionalization are briefly described. The manual is one of a series of manuals on techniques describing procedures for planning and executing specialized work in water-resources investigations. Material is grouped under major subject headings called books and further subdivided into sections and chapters; section B of book 4 is on surface water. (Woodard-USGS)
W73-10323

FIELD OBSERVATIONS OF EDGE WAVES,
Institute of Coastal Oceanography and Tides, Birkenhead (England).
D. A. Huntley, and A. J. Bowen.
Nature, Vol 243, No 5403, p 160-161, May 18, 1973. 2 fig.

Descriptors: *Waves (Water), *Surf, *Beaches, Littoral drift, Frequency analysis, Time series analysis, Fourier analysis.
Identifiers: *Edge waves.

Edge wave solutions are used for wave motion trapped against a shoaling beach. Their amplitude varies sinusoidally along the shore and diminishes rapidly seawards from the shoreline. Each solution is characterized by a mode number which gives the number of zero crossings in the seaward decay of amplitude; each mode has a dispersion relation linking the longshore wave number with the edge wave frequency. The existence of edge waves on natural shorelines may be inferred from the periodic spacing of rip currents and the existence of beach cusps and crescentic bars. Measurements of waves were made during August 1972 at Slapton Beach, South Devon, England. Spectral analyses of four pairs of 15-minute time series were made using the fast Fourier transform technique and smoothing filter. An unusual feature of the spectrum is the rather sharp peak at about 0.1 Hz, about one half of the predominant wind wave frequency. This energy peak occurs in each of the spectra and for both onshore and longshore components of the flow. The variation with offshore distance of this energy at 0.1 Hz suggests that it is due to the presence of an edge wave. The ratio of onshore to longshore current amplitudes is approximately constant, independent of offshore position, at 4.6. This is consistent only with the $n \pm 0$ edge wave mode and would be inconsistent with any higher order mode. Because the value of the ratio is not unity, as would be expected for a progressive edge wave, the currents must be associated with a trapped standing edge wave. (Knapp-USGS)
W73-10327

REGIMES OF FORCED HYDRAULIC JUMP,
Alberta Dept. of Environment, Edmonton. Design and Construction Div.
For primary bibliographic entry see Field 08B.
W73-10331

EXPECTED NUMBER AND MAGNITUDES OF STREAM NETWORKS IN RANDOM DRAINAGE PATTERNS,
California Univ., Irvine.
For primary bibliographic entry see Field 04A.
W73-10335

2F. Groundwater

GROUNDWATER CONSITIONS IN THE CENTRAL VIRGIN RIVER BASIN, UTAH,
Geological Survey, Salt Lake City, Utah.
For primary bibliographic entry see Field 04B.
W73-09791

GROUNDWATER RESOURCES OF MOTLEY AND NORTHEASTERN FLOYD COUNTIES, TEXAS.
Geological Survey, Austin, Tex.
For primary bibliographic entry see Field 04B.
W73-09792

GROUNDWATER RESOURCES OF HALL AND EASTERN BRISCO COUNTIES, TEXAS.
Geological Survey, Austin, Tex.
For primary bibliographic entry see Field 04B.
W73-09793

GROUND-WATER FAVORABILITY AND SURFICIAL GEOLOGY OF THE LOWER AROOSTOOK RIVER BASIN, MAINE,
Geological Survey, Washington, D.C.
For primary bibliographic entry see Field 07C.
W73-09801

INVESTIGATIONS AND FORECAST COMPUTATIONS FOR THE CONSERVATION OF GROUNDWATER (ISSLEDOVANIYA I PROGNOZNYYE RASHETY DLYA OKHRANY PODZENMYKH VOD),
For primary bibliographic entry see Field 05B.
W73-09802

ROLE OF SUBSURFACE FLOW IN THE FORMATION OF THE SALT COMPOSITION OF THE CASPIAN SEA (O ROLI PODZEMNOGO STOKA V FORMIROVANII SOLEVOGO SOSTAVA KASPIYSKOGO MORYA),
Akademiya Nauk SSSR, Moscow. Institut Vodnykh Problem.
R. G. Dzhamalov, I. S. Zektseri, N. V. Rodionov, and A. V. Meskheteli.
Meteorologiya i Gidrologiya, No 10, p 70-76, October 1972. 1 tab, 15 ref.

Descriptors: *Subsurface flow, *Subsurface runoff, *Groundwater movement, *Salt balance, *Sea water, Salinity, Ions, Aquifers, Surface runoff, Estimating.
Identifiers: *USSR, *Caspian Sea.

Quantitative estimates of subsurface discharge of dissolved ions are based on hydrodynamic calculations of subsurface flow into the Caspian Sea from shallow aquifers. Subsurface flow on the west coast of the Caspian is 1,159,389,500 cu m/yr, and the discharge of dissolved ions is 4,486,000 metric tons/yr. On the east coast, the volume of subsurface flow is 59,330,000 cu m/yr, and the discharge of dissolved ions is 588,300 metric tons/yr. On the Iranian coast of the Caspian, subsurface flow is 62,830,000 cu m/yr, and the discharge of dissolved ions is 314,000 metric tons/yr. Total subsurface flow into the sea reaches 1.3 cu km/yr, and total discharge of dissolved ions is 5,388,300 metric tons/yr. The amount of salts carried into the sea by surface runoff and subsurface flow is compared, and the importance of subsurface flow in the formation of the salt balance of seawater is discussed. (Josefson-USGS)
W73-09808

COMPUTER SIMULATION OF GROUND WATER AQUIFERS OF THE COASTAL PLAIN OF NORTH CAROLINA,
North Carolina Univ., Chapel Hill. Dept. of Environmental Sciences and Engineering.
J. K. Sherwani.
Available from the National Technical Information Service as PB-220 756, $3.00 in paper copy, $0.95 in microfiche. North Carolina Water Resources Research Institute, Raleigh, UNC-WRRI Report No. 75, April 1973. 126 p, 19 fig, 7 tab, 25 ref, 2 append. OWRR B-005-NC (3). 14-01-0001-1931.

Descriptors: *Simulation analysis, Hydrologic models, Analog models, *Computer models, *Groundwater, Aquifers, *North Carolina, Model studies, Coastal plains, Pumping, Water quality, Saline water intrusion.
Identifiers: *Castle Hayne aquifer (NC), *Leaky aquifers, Sensitivity analysis.

The determination of the hydrologic characteristics of a leaky confined aquifer is approached from several viewpoints: (a) analysis of pumping tests, (b) fitting of steady-state mathematical models, (c) analog simulation, and (d) digital computer simulation. The Castle Hayne aquifer in eastern North Carolina is used as a case study. The aquifer is analyzed as a heterogeneous isotropic system. The inputs and outputs of the system are considered; the hydraulic connection between the aquifer and the underlying and overlying aquifers is examined. The verified digital model is used for prediction purposes under hypothetical strategies of development. The groundwater quality is considered to be of major importance in the optimal management of the aquifer on a regional bases. The vertical movement of the estuarine water and the lateral flow from lenses of poor quality water within the aquifer are examined using simple mathematical relationships. Single-valued estimates based on specific values of controlling parameters can be subject to large errors because of the accuracy of the data. The possible range of variation is investigated by carrying out sensitivity analysis. The conceptual problems in the measurement of social benefits and costs are outlined, and possible regulation techniques are explored. A possible framework for the regional management of the aquifer taking quality of water and economic consequences into account is presented.
W73-09871

DRAWDOWN AND RECOVERY OF AN ARTESIAN SAND AND GRAVEL AQUIFER, LONDON, ONTARIO, CANADA,
Ontario Water Resources Commission, Toronto. Div. of Water Resources.
D. N. Jeffs, and A. A. Mellary.
Proc available from 24th Int Geological Congress, 601 Booth St, Ottawa, Canada KIA OE8 Price $7.00. In: Proceedings of the 24th Session of the International Geological Congress, Section 11, Hydrogeology; Montreal, Canada, 1972: International Geological Congress Publication, p 5-15, 1972. 6 fig, 1 tab, 8 ref.

Descriptors: *Drawdown, *Recharge, *Withdrawal, *Artesian aquifers, Groundwater movement, Canada, Hydrogeology, Aquifer characteristics, Sands, Gravels, Glacial drift, Water levels, Water yield.
Identifiers: *London (Ontario).

Drawdown and recovery were studied in the White Oak aquifer, London, Ontario, using six high-capacity wells operated from 1959 to 1967. Production from the wells totaled 7,500 million imperial gallons. The aquifer consists of sand and gravel, interbedded with lenses of silt, clay and till, and is overlain generally by till and local surface deposits of clay, silt and sand. It has a maximum thickness of approximately 150 feet. In an observation well about 6,000 feet east of the production area, the maximum drawdown was 74 feet, and the residual drawdown 4 years after pumping ceased was 53 feet. The production dewatered a significant portion of the aquifer. Most of the water released from storage was from the dewatered portion. The volume of water which was released from the dewatered portion and from the confined portions of the aquifer was compared with the total production and the difference attributed to recharge. Storage coefficients range from 0.03 to 0.04 in the dewatered portion of the aquifer. Recharge rate was 40,000 to 50,000 imperial gallons per day per square mile during the production period. (Knapp-USGS)
W73-09948

OBSERVATIONS ON MESOZOIC SANDSTONE AQUIFERS IN SAUDI ARABIA,
Ministry of Agriculture and Water, Riyadh (Saudi Arabia).
G. Otkun.
Proc available from 24th Int Geological Congress, 601 Booth St, Ottawa, Canada KIA OE8 Price $7.00. In: Proceedings of the 24th Session of the International Geological Congress, Section 11, Hydrogeology; Montreal, Canada, 1972: International Geological Congress Publication, p 28-35, 1972. 1 fig, 4 tab, 8 ref.

Descriptors: *Aquifers, *Water yield, *Hydrogeology, Sandstones, Water balance, Groundwater movement, Water quality.
Identifiers: *Saudi Arabia.

In Saudi Arabia all water requirements are supplied from groundwater. Because the rainfall is very low and irregular, and temperatures and evaporation very high, the floods which occur once in a while either infiltrate underground and/or evaporate before reaching the sea. The extent, water movement, and chemical and hydraulic characteristics of two main sandstone aquifers of Mesozoic age are discussed along with the results obtained from isotopic analyses of the samples collected from deep boreholes. These aquifers are the Minjur of the Upper Triassic and the Wasia of the Middle Cretaceous. The Minjur formation is sandstone with some shale of littoral-continental facies. It is productive almost everywhere, but varies in quality and aquifer characteristics. The age of its water ranges between 10,000 and 35,000 years. The Wasia formation is also sandstone and shale of marine and nonmarine origin. It can be followed from north to south over a length of 1,450 km and, despite variation in quality, it is one of the country's most prolific aquifers. (Knapp-USGS)
W73-09949

ESTIMATING THE YIELD OF THE UPPER CARBONATE AQUIFER IN THE METROPOLITAN WINNIPEG AREA BY MEANS OF A DIGITAL MODEL,
F. W. Render.
Proc available from 24th Int Geological Congress, 601 Booth St, Ottawa, Canada KIA OE8 Price $7.00. In: Proceedings of the 24th Session of the International Geological Congress, Section 11, Hydrogeology; Montreal, Canada, 1972: International Geological Congress Publication, p 36-45, 1972. 6 fig, 8 ref.

Descriptors: *Safe yield, *Mathematical models, *Simulation analysis, *Hydrogeology, Withdrawal, Water levels, Drawdown, Limestones, Permeability, Fracture permeability, Storage coefficient, Canada.
Identifiers: *Winnipeg (Manitoba), Secondary permeability.

An extensive confined aquifer, pumped at a rate of 3 thousand million gallons per year, is in the fractured and jointed upper 100 feet of the thick Paleozoic carbonate rock sequence underlying metropolitan Winnipeg. Groundwater withdrawals have created a major drawdown come in the central industrial area. Pumpage varies from 5 mgd in the winter months to 10 mgd during the summer.

guished within the deeper lying groundwater resources: (1) isolated groundwater reservoirs largely sealed by aquifuges; and (2) groundwater reservoirs in the old massive crystalline rocks, taking part in the regional groundwater migration. Groundwater of drinking-water quality is only to be found within orographic uplift areas, where favorable slope conditions led to an increased velocity of flow of the groundwater in the aquifer; great distances to the ground surface are a protection against evaporation from the aquifer, and a thin soil cover facilitates a direct and rapid infiltration. Within regions of quartzite ridges and inselbergs there are isolated groundwater resources of freshwater quality. (Knapp-USGS)
W73-09952

THE METHODS OF EXPLORATION OF KARST UNDERGROUND WATER FLOWS IN POPOVO POLJE, YUGOSLAVIA,
Trebisnica Hydrosystem, Trebinje (Yugoslavia).
P. Milanovic.
Proc available from 24th Int. Geological Congress, 601 Booth St., Ottawa, Canada, KIA OE8. Price $7.00. In: Proceedings of the 24th Session of the International Geological Congress, Section 11, Hydrogeology; Montreal, Canada, 1972: International Geological Congress Publication, p 80-88, 1972. 8 fig, 2 ref.

Descriptors: *Karst hydrology, *Exploration, *Hydrogeology, Borehole geophysics, Investigations, Seismic studies, Groundwater movement, Aquifer characteristics.
Identifiers: *Yugoslavia.

Development of water power at Trebisnjica, Yugoslavia, required detailed hydrogeological exploration of the lowest part of the karst terrain of Popovo Polje, in which construction of a storage compensation reservoir was planned. After geological and hydrogeological maps were prepared and the results of exploratory drilling were obtained, other conventional methods such as geoelectrical mapping and sounding, dye tracers, and speleology were applied. The complexity of the problems required application of new methods, including gamma-gamma logging, radioactive tracers, TV-cameras, diving speleology, time-bomb seismic studies, and stereo photography. The hypothesis that the anticlinal dolomite core functions as a hydrogeological barrier was proved. Also the spatial position of underground water flows was determined. Important data on high velocities of the groundwater flow in the zone of swallow holes were obtained. Radioactive isotopes, time-bomb and stereo photo-camera studies proved to be especially useful. (Knapp-USGS)
W73-09953

DETERMINATION OF HYDROGEOLOGICAL PROPERTIES OF FISSURED ROCKS,
Department of the Environment, Ottawa (Ontario). Inland Waters Branch.
For primary bibliographic entry see Field 02J.
W73-09954

TIDAL FLUCTUATIONS IN A CONFINED AQUIFER EXTENDING UNDER THE SEA,
Vrije Universiteit, Amsterdam (Netherlands). Inst. of Earth Sciences.
G. Van der Kamp.
Proc available from 24th Int. Geological Congress, 601 Booth St., Ottawa, Canada KIA OE8. Price $7.00. In: Proceedings of the 24th Session of the International Geological Congress, Section 11, Hydrogeology; Montreal, Canada, 1972: International Geological Congress Publication, p 101-106, 1972. 3 fig, 3 ref.

Descriptors: *Water level fluctuations, *Tides, *Artesian aquifers, *Confined water, Barometric efficiency, Compressibility, Hydraulic gradient, Surface-groundwater relationships.

Tidal fluctuations in sea level can induce corresponding fluctuations of groundwater pressure in underlying confined aquifers. Two different phenomena are involved: the initiation of the fluctuations by compression of the subsea portion of the aquifer, and their propagation landwards. Measurement of such fluctuations can yield information on the aquifer parameters. The continuity of flow between the subsea and inland regions is taken into account for the case of a completely confined aquifer. The existing theory was modified. The new theory allows an acceptable treatment of cases involving adjacent land-sea boundaries such as that of a narrow channel. Some empirical data are presented which support the theoretical results. (Knapp-USGS)
W73-09955

CONCEPTUAL MODELS FOR EVALUATION OF THE SUBTERRANEAN WATER CYCLE IN PALEOZOIC HIGHLANDS,
Bonn Univ. (West Germany). Institut fuer Geologie.
K. U. Weyer.
Proc available from 24th Int. Geological Congress, 601 Booth St., Ottawa, Canada KIA OE8. Price $7.00. In: Proceedings of the 24th Session of the International Geological Congress, Section 11, Hydrogeology; Montreal, Canada, 1972: International Geological Congress Publication, p 107-117, 1972. 12 fig, 2 ref.

Descriptors: *Water balance, *Groundwater movement, *Hydrologic cycle, *Hydrogeology, Model studies, Systems analysis, Rainfall-runoff relationships, Surface-groundwater relationships, Hydrograph analysis, Recharge, Infiltration.
Identifiers: Rheinisches Schiefergebirge.

Several conceptual models are described which are useful for the evaluation of soil moisture and groundwater flow systems in mountainous terranes such as the Rheinisches Schiefergebirge. In these mountains groundwater flows in the fractures, joints and cleavage planes of the bedrock. In both bedrock and residual soil the water flows within two distinct hydraulic networks. In residual soil this involves movement of water through pores smaller than about 50 micrometers, and movement through pores larger than about 50 micrometers (root holes, cracks, decayed-root channels and earthworm holes). In bedrock water moves through joints, bedding joints, cleavage planes and hydraulically ineffective faults, and through hydraulically very effective open faults. Under normal conditions the main groundwater recharge is not decisively influenced by vegetation, because this recharge does not occur during the vegetation period. Because the relationship between groundwater and base runoff is usually linear and time invariant, the hydrologic pattern of the bedrock and groundwater recharge can be satisfactorily determined from the discharge of the channels. (Knapp-USGS)
W73-09956

ASSESSMENT OF GROUNDWATER RECHARGE IN THE UPPER TARNAK BASIN OF AFGHANISTAN,
United Nations Development Program, New York.
V. R. Dixon, and D. Duba.
Proc available from 24th Int. Geological Congress, 601 Booth St., Ottawa, Canada KIA OE8. Price $7.00. In: Proceedings of the 24th Session of the International Geological Congress, Section 11, Hydrogeology; Montreal, Canada, 1972: International Geological Congress Publication, p 130-138, 1972. 1 fig, 2 tab.

Descriptors: *Recharge, *Withdrawal, *Water balance, Groundwater basins, Hydrogeology, Alluvium, Water levels, Groundwater movement, Groundwater recharge.
Identifiers: *Afghanistan.

The physiography, geology, and hydrometeorology of the semiarid Upper Tarnak basin of Afghanistan are briefly described. The main aquifers are in the alluvial and colluvial sediments of Recent age. Phreatic conditions exist in places, particularly in the fanheads and along the valley flanks, whereas leaky artesian conditions are evident in the middle and lower fans and valley center. Groundwater levels are more than 10 m below the ground surface throughout most of the valley and in places exceed 50 m. Measured permeabilities vary from 6 m/day to more than 170 m/day. Infiltration of runoff from the mountains is by far the most important source of groundwater recharge. The direct infiltration of precipitation, and infiltration from streams in the valley center, appears to be negligible. Large groundwater withdrawals by means of karez amount to 190 million cu m annually. The amount is equivalent to about 14% of the total precipitation on the basin for an average year, 44% of the precipitation falling on the mountains during the December-April season, and 95% of the latter precipitation minus basin evaporation. (Knapp-USGS)
W73-09957

THE RELATIONSHIP BETWEEN RAINFALL, GEOMORPHIC DEVELOPMENT AND THE OCCURRENCE OF GROUNDWATER IN PRECAMBRIAN ROCKS OF WESTERN AUSTRALIA,
K. H. Morgan.
Proc available from 24th Int. Geological Congress, 601 Booth St., Ottawa, Canada KIA OE8. Price $7.00. In: Proceedings of the 24th Session of the International Geological Congress, Section 11, Hydrogeology; Montreal, Canada, 1972: International Geological Congress Publication, p 143-152, 1972. 3 fig, 2 tab, 7 ref.

Descriptors: *Hydrogeology, *Water quality, *Geomorphology, *Australia, Water balance, Leaching, Laterites, Salinity, Groundwater basins, Erosion, Climates.
Identifiers: *Western Australia.

In Western Australia, groundwater quality and quantity in regions underlain by crystalline rock are related primarily to the climatic and geomorphic regimes. Where precipitation is confined to the summer months, the groundwater is less saline than comparable regions with winter rain. Areas with misty winter rains tend to have poorer quality groundwater. The two geomorphic divisions are areas of regenerated erosion and those of ancient landscape with recognizable elements of staged development. Permeability, storage, and rainfall intake can be related to the stages of weathering, leaching, lateritization, and erosion. Laterite at the headwaters of the ancient river systems has undergone several stages of leaching and is more permeable than younger laterites in higher rainfall areas. Post-Miocene erosion in the Eastern Goldfields stripped the lower reaches of the ancient drainages to expose less-leached relatively impermeable parts of the laterite profile. The high groundwater salinities are derived from ancient accumulations of saline water as well as from salinity buildup caused by low rainfall on the low-permeability soils. (Knapp-USGS)
W73-09958

PROPERTIES AND MANIFESTATIONS OF REGIONAL GROUNDWATER MOVEMENT,
Research Council of Alberta, Edmonton.
J. Toth.
Proc available from 24th Int. Geological Congress, 601 Booth St., Ottawa, Canada KIA OE8. Price $7.00. In: Proceedings of the 24th Session of the International Geological Congress, Section 11, Hydrogeology; Montreal, Canada, 1972: International Geological Congress Publication, p 153-163, 1972. 11 fig, 29 ref.

Descriptors: *Groundwater movement, *Groundwater basins, *Hydrogeology, Aquifers, Confined water, Artesian aquifers, Subsurface waters, Surface-groundwater relationships, Water balance, Water yield, Water chemistry.
Identifiers: *Regional groundwater movement.

It is possible to evaluate theoretically the general properties of the regional movement of groundwater and to make prognoses of its natural manifestations. Analysis of flow patterns suggests that manifestations of regional movement are: areas of inflow, lateral flow, and outflow showing water levels declining, constant, and rising, respectively; flowing well conditions in areas of outflow; thinning of the phreatic belt from the divide toward the valley bottom; change in the rate of water transfer between the atmosphere and saturated zone from a maximum demand at the divide to a maximum yield at the valley bottom; relatively low mineralization of groundwater in recharge areas, and relatively high concentrations in discharge areas; abrupt differences in water types across flow-system boundaries; accumulation of dissolved, suspended, emulsified and colloidal matter at stagnant zones; humid conditions in areas of outflow relative to areas of inflow, reflected by differences in vegetation and by distribution and intensity of springs, seeps, soil erosion and salinization; near-surface reversals of flow direction; anomalous groundwater levels reflecting permeability differences at depth; and negligible direct groundwater contribution to streams. (Knapp-USGS)
W73-09959

AQUIFER PROPERTIES OF THE BUNTER SANDSTONE IN NOTTINGHAMSHIRE, ENGLAND,
Bristol Univ. (England). Dept. of Geology.
B. P. J. Williams, R. A. Downing, and P. E. R. Lovelock.
Proc available from 24th Int. Geological Congress, 601 Booth St., Ottawa, Canada KIA OE8. Price $7.00. In: Proceedings of the 24th Session of the International Geological Congress, Section 11, Hydrogeology; Montreal, Canada, 1972: International Geological Congress Publication, p 169-176, 1972. 3 fig, 9 ref.

Descriptors: *Hydrogeology, *Groundwater movement, *Sandstones, *Aquifer characteristics, Cores, Laboratory tests, Aquifer testing, Permeability, Hydraulic conductivity, Artificial recharge, Correlation analysis.
Identifiers: *England, Bunter sandstone.

The Bunter Sandstone (Lower Triassic) is a fluvial red-bed sequence of sandstones and conglomerates with occasional thin mudstones. It is an important aquifer in England. A detailed laboratory investigation of the physical and hydrological properties of core samples was made and the results compared with field data derived from pumping tests. The purpose was to assess the hydraulic conductivity of the Sandstone, using small-diameter boreholes to assist in the selection of sites for artificial recharge. The relationships between physical and hydrological properties were influenced by the laminated nature of the Sandstone and by cementation; correlation coefficients were low. The best relationships were between vertical hydraulic conductivity and median grain size, and between vertical hydraulic conductivity and the smaller grain sizes. The laboratory results were compared with field pumping tests. Values for vertical hydraulic conductivity were similar, but horizontal hydraulic conductivities determined in the laboratory were significantly lower than those derived from pumping tests because groundwater movement is controlled by fissures accentuated by subsidence due to the mining of coal in the underlying rocks. (Knapp-USGS)
W73-09960

REGIONALIZATION OF HYDROGEOLOGIC PARAMETERS FOR USE IN MATHEMATICAL MODELS OF GROUNDWATER FLOW,
IBM Watson Research Center, Yorktown Heights, N.Y.
R. A. Freeze.
Proc available from 24th Int. Geological Congress, 601 Booth St., Ottawa, Canada KIA OE8. Price $7.00. In: Proceedings of the 24th Session of the International Geological Congress, Section 11, Hydrogeology; Montreal, Canada, 1972: International Geological Congress Publication, p 177-190, 1972. 13 fig, 26 ref.

Descriptors: *Mathematical models, *Simulation analysis, *Groundwater movement, Hydraulic conductivity, Moisture content, Hydrogeology, Saturated flow, Unsaturated flow, Statistical methods, Numerical analysis, Anisotropy.

The weakest links in the use of deterministic mathematical models of groundwater flow lie in the scarcity of available data, and in the absence of reliable techniques to extract representative parameters from the data which do exist. For steady-state, saturated-unsaturated flow systems, the necessary hydrogeologic parameters are the pressure-head-dependent functions of hydraulic conductivity and moisture content. It is possible to effectively reduce these complex functions to a two-parameter representation in saturated hydraulic conductivity and a parameter that can be determined from the moisture content curve. Input values for mathematical simulations are best determined from statistical analyses of the available measured values. The hydraulic conductivity data commonly form a lognormal frequency distribution. From these data, it is possible to calculate a probability distribution for the geometric mean value of hydraulic conductivity for a given formation, or for the ratio of geometric means of two adjacent formations. This allows presentation of quantitative results in the form of statistical probability functions, rather than as single-valued solutions of unknown reliability. (Knapp-USGS)
W73-09961

ON THE GEOMETRICAL BASIS OF INTRINSIC HYDROGEOLOGIC PARAMETERS IN POROUS MEDIA FLOW,
Minnesota Univ., Minneapolis. Dept. of Geology and Geophysics.
H-O. Pfannkuch.
Proc available from Int. Geological Congress, 601 Booth St., Ottawa, Canada KIA OE8. Price $7.00. In: Proceedings of the 24th Session of the International Geological Congress, Section 11, Hydrogeology; Montreal, Canada, 1972: International Geological Congress Publication, p 191-200, 1972. 4 fig, 1 tab, 26 ref.

Descriptors: *Porous media, *Groundwater movement, *Pores, Porosity, Hydrogeology, Mass transfer Electrolytes, Interstices, Diffusion, Dispersion, Hydraulic conductivity, Permeability.
Identifiers: Formation factor.

Intrinsic geometric properties of porpous media were defined on a microscopic scale. These properties are then compared to characteristic lengths of interest in transport processes. Characteristic length dimensions of pores are analyzed in thin sections and through transport processes in consolidated materials. The parameters discussed are porosity, permeability, relative roughness of pore channels, dispersivity, tortuosity and the formation factor concept. A very precise assessment even of secondary geometric effects such as internal surface roughness and overgrowths is necessary. The discrepancy between the microscopic geometric parameters and measured expected results of transport experiments based on these lengths may lead to a better insight into the transport process itself. For electrolyte conductance the influence of the internal specific surface area is demonstrated, while the discrepancy between

precipitated from the brines due to decreases in temperature and pressure. (Knapp-USGS)
W73-09964

GROUNDWATER FLOW SYSTEMS, PAST AND PRESENT,
Department of the Environment (Alberta). Inland Waters Branch.
L. D. Delorme.
Proc available from 24th Int. Geological Congress, 601 Booth St., Ottawa, Canada KIA OE8. Price $7.00. In: Proceedings of the 24th Session of the International Geological Congress, Section 11, Hydrogeology; Montreal, Canada, 1972: International Geological Congress Publication, p 222-226, 1972. 2 tab, 3 ref.

Descriptors: *Paleohydrology, *Groundwater movement, *Lakes, *Surface-groundwater relationships, Springs, Recharge, Discharge (Water), Land reclamation, Climates, Water quality.

The hydrological regime of a lake or pond is a function of time, climate and geology. Geology can be considered to be constant for the history of such a surface-water body whereas climate can be shown, through the use of fossils, not to be constant in time. As climate influences the nature of groundwater flow systems to and from the lake or pond, the present groundwater flow system is not necessarily the one that has always been there. A correlation can be shown between permanency, chemistry and the depth and direction of the groundwater flow system. Shelled organisms, such as ostracodes and snails provide a means to determine the state of permanency and the chemistry of the lake or pond during specific times in the past. They can, therefore, be used with other evidence to decipher the changing nature of the groundwater flow systems through time. The knowledge gained from such an analysis could help in making land reclamation decisions. (Knapp-USGS)
W73-09965

BASIC STUDIES IN HYDROLOGY AND C-14 AND H-3 MEASUREMENTS,
Niedersaechsisches Landesamt fuer Bodenforschung, Hanover (West Germany).
M. A. Geyh.
Proc available from 24th Int. Geological Congress, 601 Booth St., Ottawa, Canada KIA OE8. Price $7.00. In: Proceedings of the 24th Session of the International Geological Congress, Section 11, Hydrogeology; Montreal, Canada, 1972: International Geological Congress Publication, p 227-234, 1972. 6 fig, 1 tab, 24 ref.

Descriptors: *Paleohydrology, *Carbon radioisotopes, *Tritium, *Glaciation, *Recharge, Saline water-freshwater interfaces, Europe, South America, Karst, Groundwater movement, Paleoclimatology, Lakes, Sedimentation, Geochemistry, Radioactive dating.

Interrelations among the initial C-14 concentration of groundwater, geologic features of catchment areas and aquifers, as well as palaeoclimatic fluctuations, were studied using 624 groundwater samples from central Europe and Brazil. Groundwaters recently recharged in the karst of southern Germany, in regions covered by loess, and in the crystalline rocks of northeast Brazil have diverse initial C-14 concentrations as a consequence of specific geochemical features of the catchment areas and different mean residence times. Variations of the groundwater recharge rate over the last 30,000 years are reflected in the C-14 age distribution of the groundwater in a confined aquifer in South Germany, in the C-14 age sequence of the saltwater-freshwater interface on the North Sea coast, as well as in the C-14 histogram based on 103 dates for central European groundwaters. The groundwater recharge rate of the Holocene may have been three times greater than that of the time before 20,000 B.P. No groundwater was recharged during the period from 20,000 to 13,000 B.P. in central Europe, simultaneous with the recession in the formation of speleothems in the production of organic matter. (Knapp-USGS)
W73-09966

A PRELIMINARY STUDY OF THE GEOCHEMICAL CHARACTERISTICS OF THERMAL WATERS IN NEW MEXICO,
New Mexico Inst. of Mining and Technology, Socorro.
W. K. Summers, and R. A. Deju.
Proc available from 24th Int. Geological Congress, 601 Booth St., Ottawa, Canada KIA OE8. Price $7.00. In: Proceedings of the 24th Session of the International Geological Congress, Section 11, Hydrogeology; Montreal, Canada, 1972: International Geological Congress Publication, p 241-250, 1972. 3 fig, 3 tab, 9 ref.

Descriptors: *Water chemistry, *Thermal water, *New Mexico, Leaching, Equilibrium, Hot springs, Calcium, Sodium, Potassium Magnesium, Geothermal studies, Thermal springs.

The geologic and hydrologic settings of nine thermal springs in New Mexico that discharge groundwater from igneous rocks are described. These spring areas constitute a representative cross section of igneous terrains. In laboratory experiments, rock particles were leached in a dilute HCl solution. The following factors have an effect on the water chemistry: the initial pH of the solution; the atmosphere under which the experiment is conducted; the presence of extraneous or free ions such as sulfur; and the surface area of the particles. The total amount of monovalent cations plus calcium and magnesium entering the solution increased in proportion to increases of these constituents in the rock matrix. The ratios of Ca to Na and K to Mg in the groundwater discharging at Ponce de Leon Warm Spring, Hot Spring Rio San Antonio, and Warm Spring Rio San Antonio are essentially the same as those of the rock at each spring. No similar relation exists at the other sites investigated. It seems that the water and the rock in these three areas are nearly in ionic equilibrium, whereas the others are not. (Knapp-USGS)
W73-09967

THE APPLICATION OF THE TRANSPORT EQUATIONS TO A GROUNDWATER SYSTEM,
Geological Survey, Denver, Colo.
For primary bibliographic entry see Field 05B.
W73-09968

SOURCE DETERMINATION OF MINERALIZATION IN QUATERNARY AQUIFERS OF WEST-CENTRAL KANSAS,
Kansas State Dept. of Health, Dodge City.
For primary bibliographic entry see Field 05B.
W73-09969

VARIATIONS OF GROUNDWATER CHEMISTRY BY ANTHROPOGENIC FACTORS IN NORTHWEST GERMANY,
Niedersaechsisches Landesamt fuer Bodenforschung, Hanover (West Germany).
For primary bibliographic entry see Field 05B.
W73-09970

HYDROGEOLOGIC CRITERIA FOR THE SELF-PURIFICATION OF POLLUTED GROUNDWATER,
Hessiches Landesamt fuer Bodenforschung, Wiesbaden (West Germany).
For primary bibliographic entry see Field 05B.
W73-09971

GROUNDWATER IN PERSPECTIVE,
Geological Survey, Washington, D.C.
For primary bibliographic entry see Field 04B.

W73-09994

THE REGIONAL SETTING,
Institute of Geological Sciences, London (England).
For primary bibliographic entry see Field 08A.
W73-10002

SUBSURFACE WASTE STORAGE--THE EARTH SCIENTIST'S DILEMMA,
Geological Survey, Denver, Colo.
For primary bibliographic entry see Field 05E.
W73-10022

INJECTION WELLS AND OPERATIONS TODAY,
Bureau of Mines, Bartlesville, Okla. Bartlesville Energy Research Center.
For primary bibliographic entry see Field 05B.
W73-10023

SUBSIDENCE AND ITS CONTROL,
Geological Survey, Sacramento, Calif.
For primary bibliographic entry see Field 04B.
W73-10024

RESPONSE OF HYDROLOGIC SYSTEMS TO WASTE STORAGE,
Geological Survey, Washington, D.C. Water Resources Div.
For primary bibliographic entry see Field 05B.
W73-10025

GEOMETRY OF SANDSTONE RESERVOIR BODIES,
Shell Oil Co., Houston, Tex.
R. J. LeBlanc.
In: Underground Waste Management and Environmental Implications, Proc of Symposium held at Houston, Tex, Dec 6-9, 1971: Tulsa, Okla, American Association of Petroleum Geologists Memoir 18, p 133-189, 1972. 35 fig, 4 tab, 568 ref.

Descriptors: *Aquifer characteristics, *Sedimentation, *Sandstones, Hydrogeology, Stratigraphy, Alluvium, Beaches, Turbidity currents, Deltas, Dunes, Aquifers, *Bibliographies.
Identifiers: *Sandstone aquifers.

Sandstones are important natural underground reservoirs capable of containing water, petroleum, and gases. The principal sandstone-generating environments are (1) fluvial environments such as alluvial fans, braided streams, and meandering streams; (2) distributary-channel and delta-front environments; (3) coastal barrier islands, tidal channels, and chenier plains; (4) desert and coastal eolian plains; and (5) deeper marine environments. The most common sandstone reservoirs are of deltaic origin. They consist of two types: delta-front or fringe sands and abandoned distributary-channel sands. Fringe sands are sheetlike, and their landward margins are abrupt. Seaward, these sands grade into the finer prodelta and marine sediments. Distributary-channel sandstone bodies are narrow, they have abrupt basal contacts, and they decrease in grain size upward. They cut into, or completely through, the fringe sands and also connect with upstream fluvial sands. Some of the more porous and permeable sandstone reservoirs consist of well-sorted beach and shoreface sands associated with barrier islands and tidal channels. Barrier sand bodies are long and narrow, are aligned parallel with the coastline, and are characterized by an upward increase in grain size. The most porous and permeable sandstone reservoirs are products of wind activity in coastal and desert regions. Eolian sands are very well sorted, highly crossbedded, and they occur as extensive sheets. (Knapp-USGS)
W73-10026

APPLICATION OF TRANSPORT EQUATIONS TO GROUNDWATER SYSTEMS,
Geological Survey, Denver, Colo.
For primary bibliographic entry see Field 05B.
W73-10027

CIRCULATION PATTERNS OF SALINE GROUNDWATER AFFECTED BY GEOTHERMAL HEATING--AS RELATED TO WASTE DISPOSAL,
Alabama Univ., University. Dept. of Civil and Mineral Engineering.
For primary bibliographic entry see Field 05B.
W73-10028

CHEMICAL EFFECTS OF PORE FLUIDS ON ROCK PROPERTIES,
Texas A and M Univ., College Station. Coll. of Geosciences.
For primary bibliographic entry see Field 02K.
W73-10029

NATURAL AND INDUCED FRACTURE ORIENTATION,
Geological Survey, Washington, D.C.
M. K. Hubbert.
In: Underground Waste Management and Environmental Implications, Proc of Symposium held at Houston, Tex, Dec 6-9, 1971: Tulsa, Okla American Association of Petroleum Geologists Memoir 18, p 235-238, 1972. 6 ref.

Descriptors: *Rock mechanics, *Fractures (Geologic), *Injection wells, *Underground waste disposal, *Geologic control, Water pressure, Strength, Stress, Strain, Fracutre permeability, Rock properties, Hydrogeology, Aquifer characteristics, Waste disposal wells.
Identifiers: *Hydraulic fracturing (Rocks).

Fractures may be induced in rocks by fluid pressures applied through wells. Hydraulically induced fractures tend to follow surfaces parallel with the greatest and intermediate principal compressive stresses and perpendicular to the least stress. In tectonically relaxed regions characterized by normal faulting, the greatest principal stress in nearly vertical and the intermediate and least principal stresses are nearly horizontal. In such a region, the preferred orientation of hydraulic fractures is vertical, perpendicular to the least principal stress, and parallel with the strike of the local normal faults. Hydraulic fracture orientation may also be influenced by anisotropy or planar inhomogeneities in the rock such as bedding, schistose cleavage, or a system of parallel joints. If such a planar system does not depart too far from perpendicularity to the axis of least stress, hydraulic fractures may follow such a zone of weakness. In this case, provided the rocks are also stressed tectonically, slippage along the fracture and possible resultant earthquakes are possible consequences of increasing the fluid pressure in the rock. (Knapp-USGS)
W73-10030

MECHANICS OF HYDRAULIC FRACTURING,
Geological Survey, Washington, D.C.
M. K. Hubbert, and D. G. Willis.
In Underground Waste Management and Environmental Implications, Proc of Symposium held at Houston, Tex, Dec 6-9, 1971: Tulsa, Okla, American Association of Petroleum Geologists Memoir 18, p 239-257, 1972. 26 fig, 22 ref.

Descriptors: *Rock mechanics, *Fractures (Geologic), *Injection wells, *Underground waste disposal, *Geologic control, Water pressure, Strength, Stress, Strain, Fracture permeability, Rock properties, Hydrogeology, Aquifer characteristics, Waste disposal wells.
Identifiers: *Hydraulic fracturing (Rocks).

GEOCHEMICAL EFFECTS AND MOVEMENT OF INJECTED INDUSTRIAL WASTE IN A LIMESTONE AQUIFER,
Geological Survey, Tallahassee, Fla.
For primary bibliographic entry see Field 05B.
W73-10036

RECENT SUBSIDENCE RATES ALONG THE TEXAS AND LOUISIANA COASTS AS DETERMINED FROM TIDE MEASUREMENTS,
National Ocean Survey, Rockville, Md.
For primary bibliographic entry see Field 02L.
W73-10045

USE OF INFRARED AERIAL PHOTOGRAPHY IN IDENTIFYING AREAS WITH EXCESS MOISTURE AND GROUNDWATER OUTFLOW (OB ISPOL'ZOVANII INFRAKRASNOY AEROS'YEMKI PRI VYYAVLENII UCHASTKOV IZBYTOCHNOGO UVLAZHNENIYA I VYKHODOV PODZEMNYKH VOD),
Nauchno-Issledovatelskii Institut Geologii Arktiki, Leningrad (USSR).
For primary bibliographic entry see Field 07B.
W73-10054

PROBLEMS OF SUBSURFACE FLOW INTO SEAS (PROBLEMA PODZEMNOGO STOKA V MORYA),
Akademiya Nauk SSSR, Moscow. Institut Vodnykh Problem.
B. I. Kudelin, I. S. Zektser, A. V. Meskheteli, and S. A. Brusilovskiy.
Sovetskaya Geologiya, No 1, p 72-80. January 1971. 48 ref.

Descriptors: *Subsurface flow, *Groundwater movement, *Oceans, Lakes, Analytical techniques, Estimating, Hydrogeology, Hydrology, Geophysics, Geochemistry.
Identifiers: *USSR.

Investigations of subsurface flow into oceans, seas, and lakes are important for solution of a number of hydrogeological and hydrological problems. The problems include (1) the role of subsurface flow in the hydrologic cycle; (2) the influence of subsurface flow on water, salt, temperature, and hydrobiological regimes of seas; and (3) the role of subsurface flow in the formation of mineral resources on the sea floor. Hydrogeological, hydrological, geophysical, and geochemical methods of investigation are examined for quantitative estimates of subsurface flow into seas and inland bodies of water. (Josefson-USGS)
W73-10055

PATTERNS OF GROUNDWATER DISCHARGE BY EVAPORATION AND QUANTITATIVE ESTIMATES OF THE FACTORS AFFECTING EVAPORATION AND CONVECTIVE HEAT EXCHANGE (ZAKONOMERNOSTI RASKHODA GRUNTOVYKH VOD NA ISPARENIYE I KOLICHESTVENNYY UCHET F AKTOROV, VLIYAYUSHCHIKH NA ISPARENIYE I KONVEKTIVNYY TEPLOOBMEN),
Zakavkazskii Nauchno-Issledovatelskii Gidrometeorologicheskii Institut, Tiflis (USSR).
For primary bibliographic entry see Field 02D.
W73-10056

HORIZONTAL SOIL-WATER INTAKE THROUGH A THIN ZONE OF REDUCED PERMEABILITY,
Purdue Univ., Lafayette, Ind.
For primary bibliographic entry see Field 02G.
W73-10057

WATER-TABLE FLUCTUATION IN SEMIPERVIOUS STREAM-UNCONFINED AQUIFER SYSTEMS,
California Univ., Davis.
M. A. Marino.
Journal of Hydrology, Vol 19, No 1, p 43-52, May 1973. 4 fig, 7 ref.

Descriptors: *Surface-groundwater relationships, *Water level fluctuations, Recharge, Infiltration, Alluvial channels, Permeability, Drawdown, Equations.

Available expressions which describe the water-table fluctuation in a stream-aquifer system are based primarily on the assumption that the bed of the stream is as permeable as the aquifer it completely cuts through. New analytical expressions were developed, in terms of the head averaged over the depth of saturation, to take into account the semiperviousness of the streambed. The situations considered the finite and semi-infinite aquifer systems in which the water level in the semipervious stream is suddenly lowered below its initial elevation, and suddenly raised above its initial elevation, and maintained constant thereafter. As a by-product, solutions are also obtained for finite and semi-finite aquifer systems in which the bed of the stream is as permeable as the aquifer. (Knapp-USGS)
W73-10058

INTERSTITIAL WATER CHEMISTRY AND AQUIFER PROPERTIES IN THE UPPER AND MIDDLE CHALK OF BERKSHIRE, ENGLAND,
Institute of Geological Sciences, London (England). Dept. of Hydrogeology.
For primary bibliographic entry see Field 02K.
W73-10060

METHOD OF EVALUATING WATER BALANCE IN SITU ON THE BASIS OF WATER CONTENT MEASUREMENTS AND SUCTION EFFECTS (METHODE D'EVALUATION DU BILAN HYDRIQUE IN SITU A PARTIR DE LA MESURE DES TENEURS EN EAU ET DES SUCCIONS),
Grenoble Univ. (France).
For primary bibliographic entry see Field 02A.
W73-10197

TEST-OBSERVATION WELL NEAR PATERSON, WASHINGTON: DESCRIPTION AND PRELIMINARY RESULTS,
Geological Survey, Tacoma, Wash.
For primary bibliographic entry see Field 04B.
W73-10322

GROUND WATER IN COASTAL GEORGIA,
Geological Survey, Atlanta, Ga.
For primary bibliographic entry see Field 04B.
W73-10326

AQUIFER SIMULATION MODEL FOR USE ON DISK SUPPORTED SMALL COMPUTER SYSTEMS,
Illinois State Water Survey, Urbana.
T. A. Prickett, and C. G. Lonnquist.
Illinois State Water Survey Circular 114, 1973. 21 p, 15 fig, 6 ref.

Descriptors: *Groundwater movement, *Aquifer characteristics, *Simulation analysis, *Computer models, *Hydrogeology, Computer programs, Systems analysis, Input-output analysis, Mathematical studies, Equations, Water wells, Pumping, Groundwater recharge, Evapotranspiration, Digital computers, Forecasting.

A generalized digital computer program listing is given that can simulate two-dimensional flow of groundwater in heterogeneous aquifers under nonleaky and/or leaky artesian conditions. A special feature of the program is that a large simulation (up to an estimated 10,000 node problem) can be accomplished on a disk supported small computer with only 4,000 words (32 bits per word) of core storage. The computer program can handle time varying pumpage from wells, natural or artificial recharge rates, the ralationships of water exchange between surface waters and the groundwater reservoir, the process of groundwater evapotranspiration, and the mechanism of flow from springs. A finite difference approach is used to formulate the equations of groundwater flow. A modified alternating direction implicit method is used to solve the set of resulting difference equations. The discussion of the digital technique includes the necessary mathematical background, the documented program listing, sample computer input data, and explanations of job setup procedures. The program is written in FORTRAN and will operate with any consistent set of units. (Woodard-USGS)
W73-10328

EFFICIENCY OF PARTIALLY PENETRATING WELLS,
Maryland Univ., College Park. Dept. of Civil Engineering.
For primary bibliographic entry see Field 04B.
W73-10336

2G. Water in Soils

THE INFLUENCE OF NITROGEN LEACHING DURING WINTER ON THE NITROGEN SUPPLY FOR CEREALS, (IN GERMAN),
Kiel Univ. (West Germany). Institut fuer Pflanzenbau und Pflanzenzuechtung.
For primary bibliographic entry see Field 03F.
W73-09874

CRITERIA OF DETERMINATION OF INDUSTRIAL LANDS, (IN POLISH),
For primary bibliographic entry see Field 04C.
W73-09947

CONCEPTUAL MODELS FOR EVALUATION OF THE SUBTERRANEAN WATER CYCLE IN PALEOZOIC HIGHLANDS,
Bonn Univ. (West Germany). Institut fuer Geologie.
For primary bibliographic entry see Field 02F.
W73-09956

REACTIONS OF THE MONOCALCIUM PHOSPHATE MONOHYDRATE AND ANHYDROUS DIACALCIUM PHOSPHATE OF VOLCANIC SOILS, (IN SPANISH),
Chile Univ., Santiago.
J. Urrutia, and K. Igue.
Turrialba. Vol 22, No 2, p 144-149. 1972. Illus. English summary.
Identifiers: Allophane, Anhydrous compounds, Calcium, *Fertilizers, *Hydrates, Liming, *Phosphates, Soils, *Volcanic soils.

Reactions of monocalcium phosphate monohydrate (MCPM), ordinary superphosphate (MCP-S), and dicalcium phosphate anhydrous (DCPA) were studied in 3 highly P fixing volcanic soils. Soil columns were used to reproduce the micorenvironment around the fertilizer zone. The movement and reaction products of fertilizers in different soil layers in the column were determined. MCPM and MCP-S dissolved better than DCPA. Most of the P was retained in the upper layer. P movement through the column was inversely related to allophane content. The ratio of Al-P/F3-P decreased with increasing distance from the site of application. Liming allophanic soil up to 16 tons/ha did not show a marked increase in Ca-P. There was no marked effect of fertilizer upon occluded and reductant soluble forms of P under the experimental conditions.--Copyright 1973, Biological Abstracts, Inc.
W73-09982

ON THE HYDROPHOBIZATION OF A SOIL LAYER WITH THE VIEW OF INTERRUPTING CAPILLARY WATER MOVEMENT, (IN GERMAN),

Dsutsche Akademie der Landwirtschaftswissenschaften zu Berlin, Muencheberg (East Germany). Forschungszentrum Bodenfruchtbarkeit.

For primary bibliographic entry see Field 02D.

W73-09984

MODES OF ACTION AND METHODS OF APPLICATION OF SOIL CONDITIONERS FOR IMPROVING PHYSICAL PROPERTIES OF THE SOIL, (IN GERMAN),

Deutsche Akademie der Landwirtschaftswissenchaften zu Berlin, Muenchberg (East Germany). Forschungszentrum Bodenfruchtbarkeit.
J. Lehfeldt, A. Kullman, K. Steinbrenner, D. Kleinhempel, and H. Lindner.

Arch Acker-Pflanzenbau Bodenkd. Vol 16, No 4/5, p 281-300. 1972. English summary.
Identifiers: Evapotranspiration, Hydrophobing, *Soil physical properties, *Soil conditioners, Temperature, Transpiration.

The mode of action and the application of different soil conditioners primarily for improving the physical properties of the soil are described on the basis of recent literature. For structure stabilization, linear polymers are particularly suitable, but dispersions of organic polymers and hydrophobing products have also been used. Evapotranspiration may best be reduced if the surface pores are sealed by a mulch cover. Organic dispersions, such as bituminous emulsion, are suitable. Total or partial sealing of the pores at the depth of 60 to 100 cm or the establishment of barriers may improve the water storage capacity of light soils. The water storage capacity of light soils is also improved by the application of soil conditioners, such as open-cell foam materials and gels. Soils rich in colloids are lossened by hydrophobic, decomposition-resistant and mechanically stable materials, such as polystyrene foam material. Surface-applied dyed soil conditioners (e.g. bituminous emulsion) reduce the short-wave reflex radiation and thus raise soil temperature at the depth of sowing. Finally, the low sorptive capacity of many soils may be raised by adding highly sorptive products.--Copyright 1973, Biological Abstracts, Inc.

W73-09985

ON THE EFFECT OF A BITUMINOUS MULCH COVER ON WATER INFILTRATION, AIR COMPOSITION AND CONDENSATION PROCESSES IN THE SOIL, (IN GERMAN),

Humboldt-Universitaet sur Berlin (East Germany). Institut fuer Bodenkunde und Pflanzenernaehrung.
W. Krueger, and H. Benkenstein.

Arch Acker-Pflanzenbau Bodenkd. Vol 16, No 4/5, p 298-391. 1972. Illus. English summary.
Identifiers: Air composition, Bituminous mulch, Carbon, Condensation, *Infiltration, *Mulch cover, Oxides, Oxygen, Processes, Soil surfaces.

Infiltration was reduced by bituminous mulch covers if the bituminous emulsion was applied to a smooth soil surface. The rate of infiltration declined as the applied quantity went up. If a finely aggregated soil surface was stabilized with a bituminous mulch cover, the rate of infiltration was higher even at an input of 6 t/ha than on untreated soil. Results from sandy and loamy sand soils had the same trend. On a light loam sandy soil with an unvegetated, smooth surface application of 6 t bituminous emulsion/ha changed the CO_2 and O_2 contents of the soil air insignificantly as compared with the unmulched variant. Significant differences appeared only at an input of 12 t/ha. Results from a silty loam in Mitscherlich pots showed the same trend. On sunny days, the different temperature of mulched and untreated plots resulted in higher vapor pressure values under the mulch cover. During a period of 4 days diffusion from soil 50 cm in depth was less by about 1 l of water /m3 less in the mulched soil than in the untreated one.--Copyright 1973, Biological Abstracts, Inc.

W73-09987

RELATION OF PHYSICAL AND MINERALOGICAL PROPERTIES TO STREAMBANK STABILITY,

Agricultural Research Service, Chickasha, Okla.

For primary bibliographic entry see Field 02J.

W73-09993

STUDIES ON THE RECLAMATION OF ACID SWAMP SOILS,

Ministry of Agriculture and Lands, Kuala Lumpur (Malaysia). Div. of Agriculture.
K. Kanapathy.

Malays Agric J. Vol 48, No 2, p 33-46. 1971. Illus.
Identifiers: Crops, Drainage (Soils), Field tests, Laboratory tests, *Liming, Palm-M, *Soil reclamation, *Soils (Acid swamp).

Tests were carried out in the laboratory, green house and in 2 areas in the field. The results obtained show that drainage and liming could be effectively used to reclaim acid swamp soils as well as improve crops growing on them. The oil palm was a very tolerant crop to acidity.--Copyright 1972, Biological Abstracts, Inc.

W73-10017

HORIZONTAL SOIL-WATER INTAKE THROUGH A THIN ZONE OF REDUCED PERMEABILITY,

Purdue Univ., Lafayette, Ind.
L. R. Ahuja, and D. Swartzendruber.

Journal of Hydrology, Vol 19, No 1, p 71-89, May 1973. 8 fig, 29 ref.

Descriptors: *Infiltration, *Percolation, *Soil water movement, *Unsaturated flow, Saturated flow, Numerical analysis, Hydraulic conductivity, Wetting.
Identifiers: *Horizontal infiltration.

One-dimensional, horizontal soil-water absorption through a thin zone of constant nonzero hydraulic resistance was examined theoretically as well as a similarity reduction of the problem for early to intermediate times. The flow equations were transformed by introducing a dimensionless parameter which enables the solution for any value of thin-zone resistance to be obtained from the solution for a given known thin-zone resistance. At the inlet boundary between the thin zone and the soil column, the soil-water content increases with time to approach the saturated value. The cumulative absorption of water by the soil column increases more than proportionally with the square root of time for early and intermediate times, and approaches a square-root-or-time proportionality at large times. For both the soil-water content at the inlet boundary and the cumulative water absorption by the soil column, simple expressions arise from the similarity-reduction analysis, which is based on specific functional forms of the soil-water diffusivity and suction head. For early to intermediate times of flow, the similarity-reduction analysis describes adequately the calculated numerical-solution flow data for Yolo soil, as well as the measurements obtained experimentally on Salkum silty clay loam. (Knapp-USGS)

W73-10057

MAJOR SOILS IN THE UPPER VALLEY OF THE MANSO RIVER, RIO NEGRO, (IN GERMAN),

Universidad Nacional del Sur, Bahia Blanca (Argentina).
H. A. Laya.

Pev Invest Agropecu Ser 3 Clima Suelo. Vol 8, No 4, p 171-183, 1971. Illus. English summary.
Identifiers: Alluvial, *Argentina (Manso River), Ash deposits, Glacial deposits, Rio-Negro, Rivers, *Soil types, Valley, Volcanic.

The principal soil groups are described and their geographical distribution given. The soils were classified using the NORTH-COTE key, and also the Natural Classification and 7th Approximation. Brief meteorological, vegetational and geological data are included. Some geomorphological elements of this narrow glacial valley, which display exceptional characteristics, are explained in more detail. Without considering the lithosols and skeletal soils of the higher slopes and the summits of mountains, the main soils are as follows. Glacial deposits form non calcic lithosols with small distribution. Volcanic ash deposits of 'O' layer common on the lower valley slopes, are superimposed over rocky formations or glacial and fluvioglacial deposits. Those soils that are Ando-like. A thin recent volcanic ash layer is commonly present in the upper part of the soils. Alluvial deposits include higher terrace and levee: Alluvial regosolic hydromorphic soils; very frequent stoniness at different depths and lower terrace: Peat Bog soils, with more than 1 m of decomposed plant remains. These soils are covered by water most or all of the year. These are the dominant soils in the valley bottoms.--Copyright 1973, Biological Abstracts, Inc.

W73-10067

NITROGEN IN NATURE AND SOIL FERTILITY, (IN RUSSIAN),

Akademiya Nauk SSSR, Moscow. Institut Mikrobiologii.
E. N. Mishustin.

Izv Akad Nauk SSSR Ser Biol. 1. p 5-22, 1972. Illus. English summary.
Identifiers: *Soil fertility, Fertilizers, Fixation, *Nitrogen, Soils, Symbiosis.

Abiotic processes take place in the atmosphere causing the formation of oxidized and reduced N compounds. These substances due to sorption and via rain water penetrate into the soil. The general enrichment of the soil in N by means of these processes is rather low. The soil becomes slightly enriched in N through the activity of free-living N fixators. Symbiotic N fixation whose exploitation in agricultural practice yields substantial results appears to be far more effective. Mineral N fertilizers are best applied to soils whose fertility has reached its maximum due to utilization of biological processes. While considering the N balance of USSR soils an insufficient utilization of biological N is observed. Future possibilities of a more intense N assimilation leading to an increase of soil fertility are discussed.--Copyright 1973, Biological Abstracts, Inc.

W73-10068

TEXTURAL ASPECT OF SOIL POROSITY, (IN FRENCH),

Institut National de la Recherche Agronomique, Versailles (France).
C. J. Fies.

Ann Agron (Paris). Vol 22, No 6, p 655-685, 1971. English summary.
Identifiers: Illite, Kaolinite, Model studies, Montmorillonite, *Soil porosity, Soils, *Texture (Soils).

The relationship between the overall porosity of soils and their texture is undertaken on the soil fabric level from models consisting of clay and soil skeleton mixture. The latter are selected from loams, silts and sands in combination of different proportions of clay, a kaolinite, a montmorillonite and a mixture (extracted from soil) of those two, with illite. These mixtures are made according to a record whereby the samples obtained were massive and structureless. After measuring the overall porosities of the various samples, the data are compared with those computed by means of a theoretical structural model. This comparison

highlights the disturbance brought about in the arrangement of the skeletal particles on initiation of the clay phase. This physical effect suggested taking the apparent density of the argillaceous phase and the specific surface of the grain skeleton as a basis for an equivalent soil fabric model. In the latter, the available clay contributes both to coating the skeletal particles (chlamydate clay) and filling-in of the intergranular voids (intertextic clay). This arrangement was defined with dry samples but is applicable to water-saturated samples. The parametric values, defining the skeletal arrangement (matrix porosity) and the percentage of clay in the intergranular voids was then derived from determination of the swelling of pure clay.--Copyright 1973, Biological Abstracts, Inc.
W73-10069

ATMOSPHERIC MOISTURE CONDENSATION IN SOILS, (IN RUSSIAN),
V. I. Tochilov.
Probl Osvoeniya Pustyn'. 3, p 10-16. 1972. Illus. English summary.
Identifiers: Atmospheric moisture, Condensation, Equations, Laplace equation, Moisture, Soils, Thomson equation.

From the Laplace and Thomson equations relative critical molecular mositure was obtained. Critical molecular moisture is lower than that of capillary air. When the values are equal, capillary condensation occurs on which thermal condensation is superimposed.--Copyright 1973, Biological Abstracts, Inc.
W73-10173

DIFFUSION IN COMPLEX SYSTEMS,
UT-AEC Agricultural Research Lab., Oak Ridge, Tenn.
For primary bibliographic entry see Field 02K.
W73-10183

THE USE OF NUCLEAR TECHNIQUES IN SOIL, AGROCHEMICAL AND RECLAMATION STUDIES. (PRIMENENIE METODOV ATOMNOI TEKHNIKI V POCHVENNO-AGROKHIMICHESKIKH I MELIORATIVNIKH ISSLEDOVANIYAKH),
Moskovskaya Selskokhozyaistvennaya Adademiya (USSR).
V. V. Rachinskii.
In: Proceedings of Symposium on the Use of Isotopes and Radiation in Research on Soil-Plant Relationships Including Applications in Forestry, Dec. 13-17, 1971, Vienna (IAEA-SM-151), International Atomic Energy Agency, Vienna, 1972, p 123-130. 71 ref.

Descriptors: *Reviews, *Soil water movement, *Soil-water-plant relationships, *Tracers, Carbon radioisotopes, Tritium, Soil bacteria, Soil investigations, Soil science, Podzols, Soil moisture, Cycling nutrients, Moisture uptake, Decomposing organic matter, Radioactivity techniques, Radioisotopes, Humic acids, Analytical techniques, Forest soils, Absorption, Ion exchange, Diffusion, Groundwater movement, Radioecology.

Work at the Timiryazevsk Academy in the use of radioactivity techniques is reviewed. Chromatographic equations predicted sorption of phosphates by soil columns, by non-ideal behavior was shown by Ca and Sr. Fe mobility was affected by the presence of organic compounds; phosphate mobility, only by organic compounds containing Fe or Al. The absorption capacity of podzolic soil was 0.2 mg sulfate/g. Humic acid formation from C14 labeled pine needles occurred with a 50-60 year half-time in field studies. Decomposition of C14 labeled amino acid incorporated 1-2% of the C14 into humic acid. The rate constant for decomposition of urea by urobacilli was 10/mega-second; the temperature dependence was described by the Arhenius equation. Quasi-diffusion coefficients of 0.001-0.01 sq cm/sec were measured at velocities where hydrodynamic factors play a role. Soil moisture was measured by neutron techniques. (See also W72-10679) (Bopp-ORNL)
W73-10187

MOVEMENT OF ZN65 INTO SOILS FROM SURFACE APPLICATION OF ZN SULPHATE IN COLUMNS,
Haile Sellassie I Univ., Alem Maya (Ethiopia). Coll. of Agriculture.
M. Abebe.
In: Proceedings of Symposium on the Use of Isotopes and Radiation in Research on Soil-Plant Relationships Including Applications in Forestry, Dec 13-17, 1971, Vienna (IAEA-SM-151), International Atomic Energy Agency, Vienna, 1972, p 507-515.

Descriptors: Agriculture, *Soil-water-plant relationships, *Radioecology, *Tracers, *Soil chemistry, Soil science, Cycling nutrients, Cation exchange, Tracers, Path of pollutants.

Precipitation of Zn sulfate in base-saturated soils required the application of a larger amount in order to reach a given depth than predicted from only the cation exchange capacity of the soil. Competition with displaced cations for exchangeable sites decreased Zn retained to a depth of 6 cm (with an application of 55 millequivalents to a column of orchard soil, 7.5 cm in diameter and 30 cm long). At greater depth Zn was occluded by precipitated Ca sulfate. Zn concentration was nearly uniform at 4-10 cm depth; than it dropped to very low values. (See also W72-10679) (Bopp-ORNL)
W73-10191

STUDY ON THE SENSITIVITY OF SOIL MOISTURE MEASUREMENTS USING RADIATION TECHNIQUES, NYLON RESISTANCE UNITS AND A PRECISE LYSIMETER SYSTEM,
Ghent Rijksuniversiteit (Belgium). Dept. of Agricultural Science.
M. de Boodt, H. Verplancke, and A. V. Laksmipathy.
In: Proceedings of Symposium on the Use of Isotopes and Radiation in Research on Soil-Plant Relationships Including Applications in Forestry, Dec 13-17, 1971, Vienna (IAEA-SM-151), International Atomic Energy Agency, Vienna, 1972, p 605-619. 7 fig, 4 tab, 8 ref.

Descriptors: *Soil moisture meters, *Radioactivity techniques, *Electrical resistance, *Lysimeters, Pervious soils, Sands, Loam, Analytical techniques.

Moisture was measured in-situ in loamy sand soils (1) by the change in weight of the soil in a lysimeter indicated by a pressure transducer, (2) by the electrical resistance of nylon blocks buried in the soil, and (3) by a neutron moisture meter. The neutron meter, which responded rapidly, was about 10-16% higher than the lysimeter (within the variation expected from soil inhomogeneity). The electrical resistance blocks agreed better with the lysimeter, but a 5-day period was required for equilibration. (See also W72-10679) (Bopp-ORNL)
W73-10194

MONITORING THE WATER UTILIZATION EFFICIENCY OF VARIOUS IRRIGATED AND NON-IRRIGATED CROPS, USING NEUTRON MOISTURE PROBES (CONTROLE DE L'UTILISATION ET DE L'EFFICIENCE DE L'EAU CHEZ DIVERSES CULTURES IRRIGUEES OU NON, AU MOYEN D'HUMIDIMETRES A NEUTRONS),
For primary bibliographic entry see Field 03F.
W73-10195

OPTIMISATION OF SAMPLING FOR DETERMINING WATER BALANCES IN LAND UNDER CROPS BY MEANS OF NEUTRON MOISTURE METERS (OPTIMISATION DE L'ECHANTILLONNAGE POUR DETERMINER LES BILANS HYDRIQUES SOUS CULTURE AU MOYEN DE L'HUMIDIMETRE A NEUTRONS),
Institut Technique des Cereales et Fourrages, Paris (France).
For primary bibliographic entry see Field 03F.
W73-10196

ESTIMATE OF WATER BALANCE IN AN EVERGREEN FOREST IN THE LOWER IVORY COAST (ESTIMATION DU BILAN HYDRIQUE DANS UNE FORET SEMPERVIRENTE DE BASSE COTE-D'IVOIRE),
Office de la Recherche Scientifique et Technique Outre-Mer, Abidjan (Ivory Coast). Centre d'Adiopodoume.
For primary bibliographic entry see Field 02D.
W73-10198

GROUND MOISTURE MEASUREMENT IN AGRICULTURE (BODENFEUCHTEMESSUNG IN DER LANDWIRTSCHAFT),
European Atomic Energy Community, Brussels (Belgium). Bureau Eurisotop.
For primary bibliographic entry see Field 03F.
W73-10209

STUDIES ON THE PERMEABILITY OF SOME TYPICAL SOIL PROFILES OF RAJASTHAN IN RELATION TO THEIR PORE SIZE DISTRIBUTION AND MECHANICAL COMPOSITION,
Udaipur Univ. (India). Coll. of Agriculture.
K. S. Singh, and G. Singh.
Annals of Arid Zone, Vol 10, No 2-3, p 105-110, 1972. 1 tab, 7 ref.

Descriptors: *Permeability, *Soil investigations, *Porosity, *Soil classification, *Soil texture, Infiltration, Percolation, Soil profiles, Distribution, Soil physical properties, Agriculture, Planning, Depth, Capillary fringe, Sampling, Soil structure.

Information about permeability is essential to the functional classification of the soil. It is needed for adequate planning of drainage and irrigation systems, and in planning flood control measures. Permeability, pore size distribution and mechanical composition of some profiles of Rajasthan were determined. The permeability of the soil profiles in general decreased downwards. The coefficient of permeability of Dungargarh soil was highest and of Kota was lowest at all depths. The capillary porosity and silt plus clay content of the soil samples were negatively correlated, whereas, non-capillary porosity, total pore space and sand content were positively correlated with the coefficient of permeability. Capillary porosity, silt plus clay percent, non-capillary porosity and sand percent had significant correlation with the permeability. The relationship between coefficient of permeability and percent total pore space was not significant. (Black-Arizona)
W73-10210

ROCK MULCH IS REDISCOVERED,
Arizona Univ., Tucson. Dept. of Soils, Water and Engineering.
For primary bibliographic entry see Field 03B.
W73-10212

MATHEMATICAL MODELING OF SOIL TEMPERATURE,
Arizona Univ., Tucson. Office of Arid-Lands Research.
K. E. Foster, and M. M. Fogel.
Progressive Agriculture in Arizona, Vol 25, No 1, p 10-12, Jan-Feb 1973. 2 fig, 2 ref.

Descriptors: *Soil temperature meters, *Mathematical models, *Soil temperature, *Microenvironment, *Energy budget, Canopy, Temperature, Correlation analysis, Heat budget, Arizona.

A data collection program was initiated in the summer of 1970 at the Santa Rita Experimental Range, in southern Arizona, to test a mathematical model developed to simulate soil surface temperature and subsurface temperature profiles to any desired depth on an hourly basis. While a surface energy budget for a bare soil can be written, the need for soil temperature simulation beneath a variable vegetation cover is desirable, so an equation was written to account for the energy partitioning between the plant canopy and the soil surface, resulting in an energy balance equation for both plant canopy and soil surface. Programming on a CDC 6400 computer produced calculations of hourly soil surface and subsurface temperature variation at 5 cm increments down to 70 cm, and model results were verified in the field by 2 Honeywell Tektronik 24-channel recorders, with a natural state control plot and a cleared plot. Observed and calculated temperatures agreed in most cases within 2 degrees F. During late mornings and early afternoons, vegetation intercepted at least 30 percent of the incoming solar radiation. This study successfully simulates a complex natural phenomenon occurring at the earth's surface. Soil temperature variation was the variable measured in the field which was a measure of the model's accuracy. The model also simulated hourly the latent heat flux, soil heat flux, and sensible heat flux. An additional use of the model could simulate energy budget response under varying plant canopies and initial soil conditions. (Paylore-Arizona)
W73-10213

RELATIVE LEACHING POTENTIALS ESTIMATED FROM HYDROLOGIC SOIL GROUPS,
Agricultural Research Service, Beltsville, Md. Hydrograph Lab.
C. B. England.
Water Resources Bulletin, Vol 9, No 3, p 590-597, June 1973. 2 fig, 2 tab, 7 ref.

Descriptors: *Leaching, *Soil water movement, *Soil groups, Recharge, Groundwater movement, Infiltration, Soil properties, Path of pollutants, Percolation.

Leaching of soils with water can be both beneficial and hazardous at the same time, by removing salts harmful to plants and contributing dissolved substances to groundwater. The leaching potential of a given soil is difficult to assess, even with complex instrumentation. The final infiltration rates associated with the Hydrologic Soil Groups used by the USDA Soil Conservation Service in watershed planning may provide a useful guide in estimating quantities of leaching water moving through soil profiles. (Knapp-USGS)
W73-10333

THE EFFECT OF TEMPERATURE AND MOISTURE ON MICROFLORA OF PEAT-BOGGY SOIL, (IN RUSSIAN),
For primary bibliographic entry see Field 02I.
W73-10381

FIELD TRIALS ON IMPROVEMENT OF SALINE SOIL IN THE SIBARI PLAIN, (IN ITALIAN),
Bari Univ. (Italy).
For primary bibliographic entry see Field 03C.
W73-10382

POLLUTION AND DEGRADATION OF SOILS: A SURVEY OF PROBLEMS,
State Coll. of Agronomical Science, Gembloux (Belgium).
For primary bibliographic entry see Field 05B.
W73-10384

2H. Lakes

STUDIES ON BENTHIC NEMATODE ECOLOGY IN A SMALL FRESHWATER POND,
Auburn Univ., Ala. Dept. of Botany and Microbiology.
For primary bibliographic entry see Field 05A.
W73-09751

NUTRIENT TRANSPORT BY SEDIMENT-WATER INTERACTION,
Illinois Univ., Chicago. Dept. of Materials Engineering.
For primary bibliographic entry see Field 05C.
W73-09754

CLADOPHORA AS RELATED TO POLLUTION AND EUTROPHICATION IN WESTERN LAKE ERIE,
Ohio State Univ., Columbus. Dept. of Botany.
For primary bibliographic entry see Field 05C.
W73-09757

AIR INJECTION AT LAKE CACHUMA, CALIFORNIA,
Geological Survey, Menlo Park, Calif. Water Resources Div.
For primary bibliographic entry see Field 05G.
W73-09799

LIMNOLOGY OF DELTA EXPANSES OF LAKE BAYKAL (LIMNOLOGIYA PRIDEL'TOVYKH PROSTRANSTV BAYKALA).

Akademiya Nauk SSSR Sibirskoye Otdeleniye, Limnologicheskiy Institut Trudy, No 12 (32), Galaziy, G. I., editor, Leningrad, 1971. 295 p.

Descriptors: *Limnology, *Lakes, *Deltas, *Aquatic life, Fish, Aquatic algae, Aquatic bacteria, Benthic fauna, Seston, Plankton, Biomass, Primary productivity, Benthos, Bottom sediments, Organic matter, Inorganic compounds, Metals, Mineralogy, Meteorology, Topography.
Identifiers: *USSR, *Lake Baykal, Oozes, Saprophytes, Infusoria.

This collection contains 22 papers devoted to recent investigations by the Baykal Limnological Institute in shallow waters of Lake Baykal. The papers are grouped under four subject headings: hydrometeorological elements (3 papers); bottom topography and chemical composition of bottom sediments (5 papers); benthos and food sources for benthos-feeding fish (5 papers); and plankton and food sources for plankton-feeding fish (9 papers). (Josefson-USGS)
W73-09806

PRESENT AND FUTURE WATER AND SALT BALANCES OF SOUTHERN SEAS IN THE USSR (SOVREMENNYY I PERSPEKTIVNYY VODNYY I SOLEVOY BALANS YUZHNYKH MOREY SSSR).
State Oceanographic Inst., Moscow (USSR).

Gosudarstvennyy Okenograficheskiy Institut Trudy, No 108, Simonov, A. I., and Goptarev, N. P., editors, Moscow, 1972. 236 p.

Descriptors: *Lakes, *Water balance, *Salt balance, *Water properties, Salinity, Water temperature, Ice, Water levels, Water chemistry, Hydrology, Hydrologic cycle, Inflow, Discharge (Water), Runoff, Precipitation (Atmospheric), Evaporation, Currents (Water), Bottom sediments, Diversion, Regulation.
Identifiers: *USSR, *Sea of Azov, *Caspian Sea, *Aral Sea, Nomograms.

Computations of present and future hydrologic and hydrochemical regimes of the Azov, Caspian, and Aral Seas were based on investigations carried out by the State Oceanographic Institute in 1966-70. Scientifically sound proposals are suggested for maintaining and developing the natural resources of these southern seas. For the Sea of Azov, these include regulation of water and salt balances in the Kerch Strait for optimal salinity and water levels without costly diversions of freshwater from other river basins; for the Caspian Sea, diversion of part (20-25 cu km/yr) of the surplus flow of northern rivers into the Volga-Kama basin, regulation of flow of Caspian waters into Kara-Bogaz-Gol at 5 cu km annually, and reduction of evaporation by separation of the shallow northeastern part of the sea by an earthen dike; and for the Aral Sea, irrigation and flooding of lands of Soviet Central Asia and Kazakhstan, and maintenance of the sea level by diversion of part of the discharge of the Ob' and Yenisey. (Josefson-USGS)
W73-09807

ROLE OF SUBSURFACE FLOW IN THE FORMATION OF THE SALT COMPOSITION OF THE CASPIAN SEA (O ROLI PODZEMNOGO STOKA V FORMIROVANII SOLEVOGO SOSTAVA KASPIYSKOGO MORYA),
Akademiya Nauk SSSR, Moscow. Institut Vodnykh Problem.
For primary bibliographic entry see Field 02F.
W73-09808

RATIO OF ORGANIC AND MINERAL COMPONENTS OF THE SOUTH BAIKAL SUSPENDED WATERS (IN RUSSIAN),
For primary bibliographic entry see Field 05B.
W73-09812

VERTICAL DISTRIBUTION AND ABUNDANCE OF THE MACRO- AND MEIOFAUNA IN THE PROFUNDAL SEDIMENTS OF LAKE PAIJANNE, FINLAND,
Jyvaskyla Univ. (Finland). Dept. of Biology.
J. Sarkka, and L. Paasivirta.
Ann Zool Fenn. Vol 9, No 1, p 1-9. 1972. Illus.
Identifiers: Chaoborus-flavicans, *Fauna, *Finland, Lakes, *Lake Paijanne, Peloscolex-ferox, Pisidium-conventus, Primary productivity, *Profundal, Sediments, Tubifex-tubifex, Vertical distribution, Lake sediments.

In the bottom sediments of the profundal of Lake Paeijaenne the maximum of all species or groups of the macro- and meiofauna at depths below 30 m seems to be near the surface of the sediment. On an average 58.4% of the macrofauna was in the 2 uppermost and 86.7% in the 6 uppermost cm. For the meiofauna the corresponding figures were 78.3 and 99.1%. In the more polluted areas the macrofauna lives deeper in the sediment than in the cleaner parts of the lake. It is suggested that here the redox potential is less favorable near the surface of the sediment than in the deeper layers. Differences in numbers of individuals of the bottom macrofauna and meiofauna in different parts of the lake are discussed, as well as the relations between the numbers of bottom fauna and the primary productivity. In the most polluted northern part of the lake dominant forms of the macrofauna are Tubifex tubifex and Chaoborus flavicans, in the clean southern part Pisidium conventus, Peloscolex ferox, Tanytarsini, Tanypodinae, Lumbriculidae, and T. tubifex. In the northern part the dominant forms of the meiofauna are Cyclopoida, in the southern part Nematoda, Cyclopoida and Harpacticoida.--Copyright 1972, Biological Abstracts, Inc.
W73-09818

MODELLING PROCESSES FOR STUDY AND EXPLORATION OF LAKE ECOLOGICAL SYSTEM (IN RUSSIAN),
Akademiya Nauk SSSR, Leningrad. Institut Evolyutsionnoi Fiziologii i Biokhimii.

ITS EFFECT ON PISCICULTURE, (IN RUMAINIA),
Institutul Agronomic Dr. Petru Groza, Cluj (Rumania).
L. Egri, and C. Man.
Inst Agron Dr. Petru Groza Cluj Lucr Stiint Ser Zooteh. 26. p 253-260. 1970. English summary.
Identifiers: Fish, Minerals, Oxygen, Physicochemical studies, *Pisciculture, Ponds, *Romania, Water chemistry, *Dissolved oxygen.

Physicochemical analyses of some Romanian pond waters were periodically made with special regard to the elements of fishery importance e.g. color, turbidity, temperature, alkalinity, Ca, Mn, chlorides, sulfates, H2S, Fe, ammonia, nitrates, nitrities, dissolved O2 and water oxidability. It was concluded that the pond waters furnish favorable ecological conditions for profitable fish production.--Copyright 1972, Biological Abstracts, Inc.
W73-09882

GROUNDWATER FLOW SYSTEMS, PAST AND PRESENT,
Department of the Environment (Alberta). Inland Waters Branch.
For primary bibliographic entry see Field 02F.
W73-09965

FORMATION AND DIAGENESIS OF INORGANIC CA MG CARBONATES IN THE LACUSTRINE ENVIRONMENT,
Heidelberg Univ. (West Germany). Sediment Research Lab.
G. Mueller, G. Irion, and U. Foerstner.
Naturwissneschaften, Vol 59, No 4, p 158-164, 1972, Illus.
Identifiers: Calcium, *Carbonates, *Diagenesis, Environment, *Lacustrine formation, *Magnesium, Lakes, Minerals (Inorganic).

Lacustrine formation of primary carbonate minerals (calcite, high-Mg calcite, aragonite, hydrous Mg carbonates) and of secondary carbonates (dolomite, huntite and magnesite), as determined from 25 lakes with differing hydrochemistry and salinity, is dependent on the Mg/Ca ratio of the lake or pore water, respectively. Secondary carbonates are found only in lake sediments with high-Mg calcite and a Mg/Ca ratio of more than 7. The only difference between nonmarine (including speleothems) and marine carbonates is the absence of calcite in the marine environment.--Copyright 1972, Biological Abstracts, Inc.
W73-10003

THE PATHWAYS OF PHOSPHORUS IN LAKE WATER,
Wyzsza Szkola Rolnicza, Wroclaw (Poland).
For primary bibliographic entry see Field 05B.
W73-10004

INVESTIGATIONS ON CRUSTACEAN AND MOLLUSCAN REMAINS IN THE UPPER SEDIMENTARY LAYER OF LAKE BALATON,
Magyar Tudomanyos Akademia, Tihany. Biological Research Inst.
J. E. Ponyi.
Ann Inst Biol (Tihany) Hung Acad Sci. 38, p 183-197, 1971, Illus, Maps.
Identifiers: *Lake Balaton (Hungary), Bosmina-longirostris, *Candona, *Crustacea, Darwinula, Dreissena, Lakes, *Molluscan remains, Sediments, Siltation.

The frequency and relative amounts of Mollusca and Ostracoda shells rapidly increased northward especially between the depth axis of the lake and the northern shore. The relative amount of Cladocera fragments is lower in the cores of the Keszthely-Bay than in the other investigated sections. The ratio of Candona and Darwinula shells compared to each other seems to be connected with the amount of detritus. The decrease of the latter causes a change in the ratio in favor of Candona while an increase brings an opposite effect. The population of Bosmina longirostris might have been significant in the southwestern basin of Lake Balaton during the last 70-90 yr. On the basis of occurrence of Dreissena shells in depth, the average siltation of Lake Balaton can be estimated as 1.7 mm/yr. The siltation seems to be most intense in the central regions of the lake.--Copyright 1972, Biological Abstracts, Inc.
W73-10005

BLUE-GREEN ALGAE FROM THE KREMENCHUG RESERVOIR AS A BASE FOR GROWING MICROORGANISMS, (IN RUSSIAN),
Akademiya Nauk URSR, Kiev. Inst. of Microbiology and Virology.
For primary bibliographic entry see Field 05C.
W73-10013

GEOGRAPHY OF LAKES IN THE BOL'SHEZEMEL'SKAYA TUNDRA (GEOGRAFIYA OZER BOL'SHEZEMEL'SKOY TUNDRY),
L. P. Goldina.
Izdatel'stvo 'Nauka', Leningrad, 1972. 103 p.

Descriptors: Lakes, *Lake morphology, *Lake morphometry, *Lake basins, *Tundra, Lake sediments, Lake ice, Lake fisheries, Meteorology, Geology, Geomorphology, Hydrology, Water chemistry, Physicochemical properties, Water balance, Water levels, Temperature, Bathymetry, Mapping, Curves.
Identifiers: *USSR, *Geography, Lake classification.

Physiographic, hydrologic, and hydrochemical characteristics of lakes were investigated in 1962-63 in the eastern part of the Bol'shezemel'skaya tundra in northeastern European Russia. The number of lakes in the region, which exceeds 1,500 sq km, is more than 6,000. The importance of the lakes to the national economy of the Komi Republic is discussed, and a system of lake classification is proposed, based on morphometric parameters, thermal properties, origin of lake basins, physicochemical characteristics, and certain biological indices. (Josefson-USGS)
W73-10047

TRACER RELATIONS IN MIXED LAKES IN NON-STEADY STATE,
Weizmann Inst. of Science, Rehovot (Israel). Isotope Dept.
A. Nir.
Journal of Hydrology, Vol 19, No 1, p 33-41, May 1973. 6 ref.

Descriptors: *Tracers, *Tracking techniques, *Mixing, *Lakes, *Limnology, Water chemistry, Evaporation, Radioisotopes, Stable isotopes, Thermal stratification.

In limnological applications of tracer data, the concentration of a tracer in a completely mixed reservoir is studied when the concentration of the tracer in the inflow is a varying function of time. Tracer concentration in mixed lakes is usually calculated under assumption of steady hydrologic state with constant input, output, and volume. Departures from steady state are treated by the use of average flow or weighting the concentration by inflow values. An exact analytical solution indicates the limits of validity of the approximations. The exact solution can be adapted for multiple inputs and outputs, exchange with atmospheric moisture, evaporation with isotope fractionation and formation of epilimnion. The solution is simplified for certain types of connection between outflow and volume. (Knapp-USGS)
W73-10059

RETENTION LAKES OF LARGE DAMS IN HOT TROPICAL REGIONS, THEIR INFLUENCE ON ENDEMIC PARASITIC DISEASES,
For primary bibliographic entry see Field 05C.
W73-10070

DEPENDENCE OF PRODUCTION OF BACTERIA ON THEIR ABUNDANCE IN LAKE WATER,
Polish Academy of Sciences, Warsaw. Inst. of Ecology.
For primary bibliographic entry see Field 05C.
W73-10082

THE UTILIZATION OF BOGS IN CZECHOSLOVAKIA,
For primary bibliographic entry see Field 04A.
W73-10131

DETERMINATION OF THE YIELD BY LIQUID SCINTILLATION OF THE 14C IN THE MEASUREMENT OF PRIMARY PRODUCTION IN A LAGOON ENVIRONMENT, (DETERMINATION DES RENDEMENTS DU COMPTAGE PAR SCINTILLATION LIQUIDE DU 14C DANS LES MESURES DE PRODUCTION PRIMAIRE EN MILIEU LAGUNAIRE).
For primary bibliographic entry see Field 05C.
W73-10137

ENRICHMENT EXPERIMENTS IN THE RIVERIS RESERVOIR,
Bonn Univ. (Germany). Zoological Inst.
For primary bibliographic entry see Field 05C.
W73-10157

GLOEOTRICHIA PISUM THURET IN LAKE TRASIMENO, (IN ITALIAN),
Perugia Univ. (Italy). Istituto di Botanica Generale.
B. Granetti.
Riv Idrobiol. Vol 8 No 1/2, p 3-15. 1969. Illus. (English summary).
Identifiers: *Gloeotrichia-Pisum, *Italy, Lakes, *Lake Trasimeno.

A vast G. pisum station which seems to have found excellent growth conditions in Lake Trasimeno was recorded. The physical and chemical characteristics of the water from a summer analysis are mentioned. The alga and its annual growth cycle are described.--Copyright 1973, Biological Abstracts, Inc.
W73-10166

RESERVOIR RELEASE RESEARCH PROJECT.
For primary bibliographic entry see Field 05G.
W73-10261

CONTRIBUTION CONCERNING THE AQUATIC AND PALUSTRAL VEGETATION OF DUDU SWAMP AND LAKE MOGOSOAIA, (IN RUMANIAN),
Bucharest Univ. (Rumania).
G. A. Nedelcu.
An Univ Bucur Biol Veg. 18, p 235-255, 1969.
Identifiers: Aquatic vegetation, Carex, Dudu swamp, Elodea, Hydrocharis, Lakes, Lemna, *Lake Mogosoaia, Myriophyllum, Palustral vegetation, Phragmites, Potamogeton, *Romania, Salvinia, Scirpus, *Swamps, Typha, *Vegetation.

In a study on the aquatic and palustral vegetation of 2 aquatic basins of the Romanian plain, Dudu Swamp and Lake Mogosoaia, 17 associations were discussed. Some such as Lemno-Salvinietum natantis, Potametum lucentis, Typhetum angustifoliae, Caricetum riparioacutiformis are common for the 2 basins. Hydrocharitetum morsus-ranae, Myriophyllo-Potametum, Typhetum laxmanii grow at Dudu, while Riccietum fluitantis, Elodeetum canadensis, Phragmitetum natantis, grow at Mogosoaia. The subassociation Typhosum laxmannii Ubrizsy, 1961, was elevated to the rank of an association. Its ecology differs from that of the Scirpo-Phragmitetum association and it has its own characteristic species. The species characteristic of the Phragmitetum natantis, Borza, 1960 association were described for the first time.--Copyright 1973, Biological Abstracts, Inc.
W73-10353

A METHOD OF CONCENTRATING TRACES OF HEAVY METAL IONS FROM LAKE WATERS,
Wyzsza Szkola Rolnicza, Olsztyn-Kortowa (Poland).
F. Wieclawski.
Acta Hydrobiol. Vol 14, No 1, p 57-66, 1972. Illus.
Identifiers: *Heavy metals, *Ions, Lakes, *Trace elements, Water analysis.

The usefulness of the dithizone column for separating and initially concentrating traces of heavy metal ions (Pb, Cd, Zn, Mn, Cu, and Co) from lake waters was tested. The fully described methods applied include the collection of water samples, the preparation of the dithizone extraction column, and the determination of trace heavy metals by the polarographic method. The periodically great biological production of lake waters and the variable O2 conditions prevailing in them make it difficult to make full use of all the advantages of the tested method.--Copyright 1972, Biological Abstracts, Inc
W73-10373

WATER MICROFLORA OF KAPCHAGAJ RESERVOIR DURING THE FRIST YEAR OF ITS FILLING, (IN RUSSIAN),
Akademiya Nauk Kazakhskoi SSR, Alma-Ata. Inst. of Microbiology and Virology.
N. K. Gulaya, and N. L. Tyuten'kova.
Mikrobiologiya. Vol 41, No 2, p 349-355, 1972. English summary.
Identifiers: *Bacteria, Flora, Fungi, *Kapchagaj Reservoir, *Microflora, Photosynthesis, Reservoirs, USSR, Year, Yeast.

The first microflora determination of the Kapchagaj reservoir was performed in July, 1970. The total bacterial number was the highest in waters of the shoal at the leftbank flood-lands of the river, and the lowest in the open part of the reservoir. The bacterial number correlated with the amount of photosynthesis. The determination of the rate of water self-purification showed that 48-62% of easily oxidized organic matter was decomposed each day in shallow waters and only 5-12% in the open part of the reservoir. Numerous sulfate-reducing, thionic, cellulose-decomposing bacteria, yeasts and fungi were also detected.--Copyright 1972, Biological Abstracts, Inc.
W73-10375

POPULATIONS OF CLADOCERA AND COPEPODA IN DAM RESERVOIRS OF SOUTHERN POLAND,
Jagellonian Univ., Krakow (Poland). Dept. of Hydrobiology.
K. Starzykowa.
Acta Hydrobiol. Vol 14, No 1, p 37-55, 1972. Illus.
Identifiers: *Cladocera, *Copepoda, Dams, Plankton, *Poland, Reservoirs, *Rotatoria, Southern, Zooplankton.

Populations of planktonic crustaceans were investigated in 10 dam reservoirs on rivers lying in the Vistula basin and 90 spp. were found (37 Rotatoria, 32 Cladocera, and 21 Copepoda). The most varied crustacean plankton occurred in water steps, where the diversity index was 23-25. The most uniform was in the lowland reservoir at Kozlowa Gora and in Boznow. The magnitude of zooplankton production depended on the rate of water exchange and on the age of the reservoir. Three types of curves of the development of populations in the course of the season were distinguished. In the majority of reservoirs seasonal maxima were simultaneously attained by 2 populations. In Cladocera populations mature individuals prevailed.--Copyright 1972, Biological Abstracts, Inc.
W73-10376

PATTERN AND PROCESS IN DUNE SLACK VEGETATION ALONG AN EXCAVATED LAKE IN THE KENNEMER DUNES, (IN DUTCH),
G. Londo.
Verb Rijksinst Natuurbeheer. 2. p 9-279. 1971. Illus. English summary.
Identifiers: Biomass, Calamagrostis-epigejos, Carex-flacca, *Dunes, Eleocharis-palustris-spp. palustris, Elodea-spp, Excavated, Hippophae, Infiltration, Juncus-spp, Kennemer, Lakes, Lime, Lycopus-europaeus, Mentha-aquatica, Morphology, Mosses, *Netherlands, Nitrogen, Phosphorus, Phragmites-communis, Salix-repens, Seepage, *Vegetation (Dune slack).

General features of the specific dune environment are discussed, followed by discussion of several external and internal habitat factors regarding the lake and its shores. The lake is in the most calcareous Dutch dune area where Hippophae scrubs and Calamagrostis epigejos grasslands dominate. The lake is on a slope of the phreatic surface; consequently, on the western and northern shore, seepage, and on the eastern and southern shores, infiltration, occur. Besides a high lime content (4-9%), a relatively high amount of N and P is found, probably from bird excrement. They hygrosere is discussed as an ecosystem where many factors interact. The investigation involved successive vegetation mapping (in 1956, 1963 and 1968), detailed mapping of species populations, and analysis of permanent quadrats. The methods are discussed. Succession is discussed in each of the hygrosere zones. In the aquatic community, Chara communities were first replaced by Elodea canadensis and E. nuttallii, but then regained dominance. A succession from Juncus articulatus and Juncus bufonius to Eleocharis palustris ssp. palustris, then to Phragmites communis, is described for the lowest shore zone. In the higher part of the hygrosere zonation, a Salix repens vegetation prevails with the mosses Drepanocladus aduncus and Calliergonella cuspidata developing in the latter part of the study. The mesosere Galio-Koelerion-Hippophae vegetation and the transitional mesosere-hygrosere community are discussed. In the latter, Carex flacca was the original dominant, but due to flooding with nutrient-rich water, tall forb thickets comprising C. epigejos, Lycopus europaeus, and Mentha aquatica developed. With time, sharp borders among community types deteriorated, leaving more gradual transitions. A negative correlation between biomass and species diversity was shown. Floristic inventories are summarized in a differentiated list of all vascular plant species observed in the period 1956-1968 along and in the Grote Vogelmeer. Details are given for each species. The vegetation development along the Grote Vogelmeer runs roughly parallel with that along and in other excavated lakes and slacks in the calcareous dunes; it deviates in various aspects from the succession as described for the Wadden district (dunes poor in lime). The reasons no Junco-Schoenetum development along the Grote Vogelmeer are discussed. Guidelines for creation and management of habitats for dune slack vegetation are outlined.--Copyright 1973, Biological Abstracts, Inc.
W73-10392

CONTRIBUTIONS TO THE KNOWLEDGE OF THE ALGAE OF THE ALKALINE ('SZIK') WATERS OF HUNGARY. III. THE

vesiculosa, Utricularia minor, U. intermedia, and Nymphaea candida. Only 1 sp. each of Litorella uniflora (observed some years ago), Isolepis setacea, Isoetes lacustris, and Petasites spurius was also encountered.--Copyright 1973, Biological Abstracts, Inc.
W73-10398

2I. Water in Plants

DISTRIBUTION OF MICROARTHROPODS IN SPRUCE-FORESTS OF THE CHONKEMIN RIVER (TYAN-SHAN), (IN RUSSIAN),
Leningrad State Univ. (USSR). Dept. of Entomology.
P. P. Vtorov, and E. F. Martynova.
Zool Zh. Vol 51, No 3, p 370-375. 1972. English summary.
Identifiers: *Arthropods (Distribution), Biomass, China, *Chon-Kemin River, Forests, Mites, Rivers, Soils, Springtails, *Spruce-G, Tyan-Shan.

The upper horizons of the litter and soil were studied for assessing microarthropod numbers and biomass; the animals were distributed into size-weight classes. In all the habitats (under the spruce-tree crown, under the edge of the crown, moss-shade-herb spruce forest, herboso-graminetum meadow in spruce-forest) mites predominated as to numbers. As a biomass, springtails predominated. The biomass of mites ranged from 520 to 1905 mg/m2, that of collembolans from 300 to 2473/m2. Twenty spp. of springtails were obtained in the samples.--Copyright 1972, Biological Abstracts, Inc.
W73-09794

THE INFLUENCE OF ANNUAL FLOODING ON POPULATION CHANGES OF SOIL ARTHROPODS IN THE CENTRAL AMAZONIAN RAIN FOREST REGION,
Ruhr-Universitaet Bochum (West Germany).
L. Beck.
Pedobiologia. Vol 12, No 2, p 133-148, 1972. Illus. English summary.
Identifiers: *Arthropods (Soil), *Flooding (Annual), Forests, Population, *Rain forests, Rostrozetes-foveolatus, Soils.

The soil arthropods of the flood forests are subjected to extreme changes which are caused by annual flooding. The macro-arthropods seek refuge from these in the more elevated woods. The meso-arthropods remain in flooded soil. The oribatid mites of the flood forests investigated in more detail do not differ in their ecological requrements from those of the surrounding terra firma. They can only live in the flood area because their developmental cycle is adapted to the annual rhythm of the flooding. In general, this is achieved by a developmental cycle as short as 3-5 mo. Rostrozetes foveolatus is the only oribatid species of the flood forests with a post-embryonic development of more than 5 mo. The other species survive flooding in the egg, whereas R. foveolatus tolerates these conditions as an adult. Its eggs grow during the submerged phase and are laid just after the water has receded. This developmental cycle is characterized by parthenogenesis, which is a necessary prerequisite for a terra firma species such as R. foveolatus to invade the flood areas.--Copyright 1972, Biological Abstracts, Inc.
W73-09860

AFRICAN JOURNAL OF TROPICAL HYDROBIOLOGY AND FISHERIES.
East African Freshwater Fisheries Research Organization, Jinja (Uganda).

Vol I, No 1, 1971. Semi-annually. East African Literature Bureau: P.O. Box 30022, Nairobi, Kenya. Pr. $10.00.
Identifiers: *Africa, Fisheries, *Hydrobiology, Tropical, Publications, Documentation.

This journal will accept original manuscripts which are relevant to the African aquatic environment. Topics will include: fish and fisheries, freshwater and marine biology, limnology, parasitology, fisheries economics, fish processing, aquatic weeds, rural fishing development, fishing gear technology and development and other fields in the aquatic sciences. Short notes, reviews of hydrobiological and fisheries and book reviews are also solicited. Its quality will conform to International standards and will be published in English and French, with English being the preferred language. A short resume in English must accompany articles in French.--Copyright 1972, Biological Abstracts, Inc.
W73-09903

STUDY OF THE BIOCENOSIS OF THE BOTTOM OF THE RIVER PO AT TRINO VERCELLESE IN THE YEARS 1967-68 (IN ITALIAN),
Milan Univ. (Italy). Laboratorio di Zoologia.
V. Parisi, P. Magnetti, M. Dotti, M. Michelangeli, and A. P. DiChiara.
Ist Lombardo Accad Sci Lett Rend Sci Biol Med B., Vol 105, No 1, p 3-56, 1971. Illus. English summary.
Identifiers: Amphipoda, *Biocenosis, Bottom, *Italy, Mollusca, Oligochaeta, *Po River, Trichoptera, Trino, Vercellese.

A sketch of the bottom communities of the Po river shows 88 spp. of animals. The most common were species of Oligochaeta Amphipods, Trichoptera and Mollusca.--Copyright 1972, Biological Abstracts, Inc.
W73-09912

STUDIES ON THE RELATIVE GROWTH RATE OF SOME FRUTICOSE LICHENS,
Turku Univ. (Finland). Dept. of Botany.
L. Karenlampi.
Ann Univ Turku Ser A II Biol Geogr. 46, p 33-39. 1971. Illus.
Identifiers: Cetraria-Nivalis, Cladonia-Alpestris, Cladonia-Mitis, Cladonia-rangiferina, *Fruticose lichens, *Growth rate (Plants), Lapland, *Lichens, Rainfall, Stereocaulon-Paschale.

Observations were made of the relative growth rate (RGR) of some lichens that were grown in plastic boxes in a natural environment. Weight growth was determined in the laboratory on air-dry material, a correction to oven-dry weight being made. Length growth was measured from photographs. The material included Cladonia alpestris (different sizes and different parts), C. rangiferina, C. mitis, Stereocaulon paschale and Cetraria nivalis. Growing in boxes seemed to give a good starting point for productivity calculations. Daily rainfall was found to be the dominant controlling climatic factor of the RGR in this area of Lapland. The distribution of RGR of weight in the thallus was found to correspond to the results of earlier studies. No significant differences could be shown between the RGRs calculated from the weights of the reindeer lichen species, but these values seemed to be higher than those of Stereocaulon and Cetraria.--Copyright 1972, Biological Abstracts, Inc.
W73-09972

GROWTH AND WATER USE OF LOLIUM PERENNE: I. WATER TRANSPORT,
Rothamsted Experimental Station, Harpenden (England).
D. W. Lawlor.
J Appl Ecol. Vol 9, No 1, p 79-98. 1972. Illus.
Identifiers: Absorption, Conductivity, Distribution, *Lolium-perenne-M, *Plant growth, Soils, *Transpiration, Transport, Vertical, Water potential (Plants).

Water absorption and transpiration are considered in relation to the gradients of water potential between the soil and the leaves of L. perenne

grown in a small volume of sandy soil. The vertical distribution of water potential within the soil during a drying cycle was estimated and a calculated value of soil-water potential related to water loss. Small changes in soil-water potential diminished water loss, because the stomata closed when plant-water potential decreased by 2-3 bar. Plant-water potential decreased from a maximum of -3 bar when the integrated soil-water potential decreased to -1 bar. With more severe stress, plant-water potential decreased faster than soil-water potentials, because of increased plant resistance. The gradients of water content and potential between the bulk soil and the root surface were calculated. The gradients were very small when all the roots were considered and the soil-water potential was above wilting. Larger gradients were predicted at more severe plant water stress if water flowed into the larger roots only. The total resistance of the plant and soil (Rp + Rs) increased from 20 bar day cm-1 at -3 bar plant-water potential to 300 at -16 bar. Within the range from full to zero turgor, Rp greatly exceeds Rs, although Rs increased much faster than Rp. Rp increased as the plant-water potential decreased. Some difficulties of estimating the gradients of soil-water potential and the relative magnitudes of Rs and Rp are considered and also the relation of Rs to soil-water conductivity. (See also W73-09974)--Copyright 1972, Biological Abstracts, Inc.
W73-09973

GROWTH AND WATER USE OF LOLIUM PERENNE: II. PLANT GROWTH,
Imperial Coll. of Science and Technology, London (England). Dept. of Botany.
D. W. Lawlor.
J Appl Ecol. Vol 9, No 1, p 99-105. 1972.
Identifiers: Dry matter, Leaves, *Lolium-perenne-M, Matter, *Plant growth, *Turgor, Water potential (Plants), Leaves.

The effects of decreasing water potential of the soil and plant on growth of L. perenne were studies. Leaf and root extension growth decreased by 50% when plant water potential decreased from -3 bar to -6 bar and growth stopped at -16 bar. Dry matter accumulation and leaf area also decreased becaus of slower expansion growth and senescence of the older leaves. Growth cannot be related to water potential with accuracy because of the rapid changes of soil- and plant-water potential but a small decrease in soil-water potential has a large effect on growth because of the more rapid decrease in plant water potential due to a large plant resistance. The importance of turgor in growth is considered. The causes of increased plant resistance and decreased growth are briefly discussed. (See also W73-09973)--Copyright 1972, Biological Abstracts, Inc.
W73-09974

EFFECT OF SOIL MOISTURE ON THE GROWTH OF ANNUAL SEEDLINGS OF THE GREEN ALDER, ALNUS VIRIDIS, AND THE GREEN ASH, FRAXINUS VIRIDIS, (IN BYELORUSSIAN),
M. D. Nesterovich, T. F. Deryugina, and B. S. Aliker.
Vyestsi Akad Navuk B SSR Syer Biyal Navuk. 2. p 5-11. 1972. Illus.
Identifiers: Alder, Alnus-Viridis, Ash, Fraxinus-Viridis, *Green alder, Seedlings, *Soil moisture, Transpiration, *Plant growth, *Green ash.

The development, anatomical and physiological characters of the seedlings cultivated in soil of 20, 40, 60, 80% humidity were studied. The humidity of 20% was not enough for growth seedlings of green alder. The decrease in soil humidity from 80% to 20% decreased seedling height by 23.8-61.3%, main root length by 22.2-51.9%, stem diameter by 28.2-48.7% and assimilation intensity by 8.1-27.9%. Decreases in total and free water in the leaves and the transpiration coefficient were also observed in the same conditions.--Copyright 1973, Biological Abstracts, Inc.
W73-09977

DISTRIBUTION OF THE PHYTOPHYLLIC INVERTEBRATES AND METHODS OF THEIR QUANTITATIVE EVALUATION. (1), (IN RUSSIAN),
Akademiya Nauk URSR, Kiev. Instytut Hidrobiologii.
L. I. Zimbalevskaya.
Gidrobiol Zh, Vol 8, No 2, p 49-55, 1972, Illus, English summary.
Identifiers: Density, *Invertebrates (Distribution), *Phytophyllic invertebrates, Quantitative factors, Weight, Sampling.

Quantitative treatment of samples of phytophyllic invertebrates, obtained with a hydrobiological net, showed the relationships on the number of samples taken, their weight, and population density. The optimal weight of samples for different ecological vegetation groups-submerged and semi-submerged was different, and was determined by the morphological structure peculiarities. The variability coefficient is proposed for use to establish aggregative and mosaic distribution of invertebrates.--Copyright 1972, Biological Abstracts, Inc.
W73-10007

BIOHYDRODYNAMIC DISTINCTION OF PLANKTON AND NEKTON (IN RUSSIAN),
Institute of Biology of the Southern Seas, Sevastopol (USSR).
Yu. G. Aleev.
Zool Zh, Vol 51, No 1, p 5-12, 1972, Illus, English summary.
Identifiers: Hydrodynamics, *Nekton, *Plankton, Reynolds number.

Biohydrodynamic distinctions of plankton and nekton are considered. Indices for distinguishing between plankton and nekton obtained in experimental studies are given. Plankton and nekton are regarded as eco-morphological types. The ecological importance of the factor characterized by the Reynolds number (Re) is noted.--Copyright 1972, Biological Abstracts, Inc.
W73-10008

ECOPHYSIOLOGICAL STUDIES ON HEAVY METAL PLANTS IN SOUTHERN CENTRAL AFRICA,
Munester Univ. (West Germany).
W. Ernst.
Kirkia. Vol 8, No 2, p 125-145. 1972. Illus.
Identifiers: Africa, Bioassay, Cells, Chromium, Copper, Eco, Grasses-M, *Heavy metals, Indigofera-D, Lead, Leaves, Nickel, Nutrients, *Physiological studies (Plants), Rhodesia, Sedges-M, Shrubs, Trees, Zambia, Zinc.

Investigations into the ecology and ecophysiology of heavy metal plants were carried out in south central Africa, one site on the Zambian Copperbelt, the remainder in Rhodesia. Correlated with the gradient in the concentrations of plant-available heavy metals there is a well-established zonation; on the highest metal contaminated areas there is a treeless zone of sedges and grasses, followed by small shrubs and stunted trees. All the normal macronutrients are as available as in normal soils. The uptake of heavy metals depends on their solubility status in the soil. It was shown by a bioassay that, for instance, Cu is present as an organic plant-available compound. The uptake of heavy metals is specific for each species, and within 1 species, for different tissues. The greatest accumulation of heavy metals is in the leaves, as well as in the cortex of the shoots and roots; the xylem or wood of both shoots and roots has only small amounts. In relation to the very high levels of plant-available Cu in some soils, the Cu uptake by plants is comparatively lower than that of Zn Cr, Ni and other heavy metals. Extractions of the tissues of leaves, shoots, and roots has revealed that the binding of heavy metals within the cells is specific to the individual metal. Cr and Pb are very tightly bound in contrast to Zn and Cu. However there are great differences among various tissues The plasmatic resistance of various populations of Indigofera against heavy metals is highly specific for individual heavy metals as well as being related to the amount of plant-available heavy metals in the soil.--Copyright 1972, Biological Abstracts Inc.
W73-10012

ASPECTS ON PRODUCTIVITY IN REED-FEEDING INSECTS, (IN GERMAN),
Vienna Univ. (Austria). Zoologisches Institut.
W. Waitzbauer.
Verb D Tsch Zool Ges. 65: p 116-119. 1971. English summary.
Identifiers: Diptera, *Insects (Herbivorous), Lepidoptera, Phragmites-Communis, *Productivity (Plants), *Reeds, Rhynchota.

The present work investigates the influence of herbivorous insects on the productivity of Phragmites communis. The investigations were done in the reed-belt on the westshore of the Lake 'Neusiedler See' in eastern Austria. Several groups of Phragmites-consumers of the orders of Lepidoptera, Diptera and Rhynchota were distinguished: gall-formers on the top or in the inside of the plant; stem-miners; shoot-suckers and leaf-suckers. Not even the stem-miners with an extensive distribution in the reed-belt and great consumption-rate influenced the growth of Phragmites. The gall-formers damage the plants heavily by destroying their vegetation tips. They have a local distribution with suboptimal stock of reeds. Stem-suckers and leaf-suckers are abundant and are parasites of Phragmites. It seems therefore that herbivorous insects have an insignificant influence on the productivity of living Phragmites.--Copyright 1973, Biological Abstracts, Inc.
W73-10064

METHODS OF MEASURING DRIFT FOR PRODUCTION BIOLOGY STUDIES IN RUNNING WATERS, (IN GERMAN),
Max-Planck-Institut fuer Limnologie, Schlitz (West Germany). Limnologische Flussstation.
M. P. D. Meijering.
Verb D Tsch Zool Ges. 65: p 69-73, 1971. Illus. English summary.
Identifiers: *Drift measuring, *Gammarus, Measuring methods, Production, Running waters.

All gammarids leaving a 10 m long section of a woodland tributary by drifting downstream, together with those moving upstream, were counted over a period of 28 days. Losses from both ends of the controlled section halved its Gammarus-population within 20 days. The daily losses from drifting remained fairly constant at between 1 and 2% of the standing crop. During the first and second week those moving upstream were the more numerous, but later these movements were almost entirely stopped.--Copyright 1973, Biological Abstracts, Inc.
W73-10065

PRODUCTION ECOLOGY OF GLACIER BROOKS IN ALASKA AND LAPLAND, (IN GERMAN),
Bundesanstalt fuer Land- und Forstwirtschaft, Berlin (West Germany). Institut fuer Zool. Biol.
A. W. Steffan.
Verb Dtsch Zool Ges. 65, p 73-78. 1971. Illus. English summary.
Identifiers: *Alaska, Animals, Brooks, Diamesa, Ecology, Gracier brooks, *Lapland, Plants, *Productivity studies, Prosimulium, Scandinavia.

and predator-prey relationships between A. brightwelli and B. rubens which had been detected in former studies. In each case the association between 2 of the species is modified by the presence of the third. The causes (alternative prey, destruction of the competitors etc.) are discussed.--Copyright 1973, Biological Abstracts, Inc.
W73-10153

RESEARCH ON THE FOREST ECOSYSTEM. SERIES C: THE OAK PLANTATION CONTAINING GALEOBDOLON AND OXALIS OF MESNIL-EGLISE (FERAGE). CONTRIBUTION NO. 17: ECOLOGICAL GROUPS, TYPES OF HUMUS, AND MOISTURE CONDITIONS, (IN FRENCH),
A. Froment, G. Schnock, and M. Tanghe.
Bull Soc R Bot Belg. Vol 103 No 2, p 293-310. 1970. Illus. (English summary).
Identifiers: *Belgium, Coryletum, Drainage, Ecological groups, Ecosystems, Ferage, Forests, Galeobdolon, Humus, *Mesnil-Eglise, Moisture, Number, *Oak, Oxalis, Querceto, Soil types, *Flora (Ground).

The ecological groups of the Querceto-Coryletum ground flora from Ferage (Mesnil-Eglise near Dinant, Belgium) are distributed in distinct concentric areas. The spreading of the groups is connected with the different humus forms of mull type. The study of the soil water content and the observation of the water-level shows that these areas have different water regime; the infiltrating water flows laterally in the pedological layers. The lateral drainage is due to the slope (about 3%) and to the higher clay content of the B and B/C horizons of the soil. In the Ferage forest, water regime of the soil is an important ecological factor for the generation and the distribution of the humus types and their corresponding ecological groups of the ground flora.--Copyright 1973, Biological Abstracts, Inc.
W73-10161

THE ROLE OF ALGAE AND MOSSES IN CREATING THE LIME CONTENT OF SPRINGS IN THE BUKK MOUNTAIN RANGE, (IN HUNGARIAN),
H. Attila.
Bot Kozlem. Vol 57 No 3, p 233-244. 1970. Illus.
Identifiers: Achnanthes, *Algae, Brachythecium-Rutabulum, *Bukk Mountain range, Carbon dioxide, Concephalum, Cratoneuron-Commutatum, Cratoneuron-Felicinum, Diurnal, Eupatorium = Cannabinum, Gomphonema, *Hungary, Leptothrix-Ochracea, *Lime, *Mosses, Navicula, Nitzschia, Pelonema-Tenue, Petasites-Albus, Petasites-Hybrides, Springs.

The CO2 absorption of mosses and algae is a major factor in the deposition of lime formations in mountain springs. In the Karst streams of the Bukk mountains, the algae promote the formation of CaCO3. The preponderance and large number of species of some pebble moss facilies (Navicula, Nitzschia, Achnanthes, Gomphonema) is rather remarkable. Calcified algal colonies of the Monosbeli Vizfo form pieces that go into building the stony masses of cauliflower-like shape in that area. There are also iron bacteria (Leptothrix ochracea) among the algal colonies of the Haromkuti-volgi Valley. One of these is Pelonema tenue in the Szalajka-brook. Among the moss species of the lime building brooks of the mountain, the most common are Brachythecium rutabulum (found in 14 locations), Cratoneuron filicinum (in 11), C. commutatum (in 8), Conocephalum conicum (in 5), and Mnium undulatum (in 5). The frequent occurrence of Brachythecium rutabulum in freshwater lime-forming streams was not previously reported. In the mountain area one also finds both species of Cratoneurum, and in some places these two are the only mosses found. The mossland built by C. commutatum created the Bryolith Steps of the Pacsirta Spring in Tal Szentleleki-volgy. Vegetative rather than generative (spore) multiplication is probable here, with detached stalk parts or other propagating amalgamations. As algae and mosses diminish the CO2 content of the flowing water, the travertine formation in such areas shows a diurnal rhythm. Among the species of mosses of high order of the Bukk mountains Eupatorium cannabinum, Petasites hybrides and P. albus are the most common. The prevalence of P. albus at the Feheracsas Spring of Tal Nagyszallas-volgy, and the colonies of Bolhasi Wasserschlinger indicates a further spreading of P. albus in the Bukk mountains. Aside from the production of limestone, the form and structure of the calcareous tufa is also determined by the vegetation.--Copyright 1973, Biological Abstracts, Inc.
W73-10162

CONTRIBUTIONS TO THE STUDY OF THE BIOLOGY OF THE ASSOCIATION FESTUCETUM VALESIACAE BURD. ET AL., 1956 IN THE MOLDAVIAN FOREST STEPPE, (IN RUMANIAN),
Institutul Agronomic, Iasi (Rumania).
D. Rosca.
Inst Agron Ion Ionescu De La Brad Iasi Luer Stiint I. Agron-Hort. 1970, p 233-254. 1970. Illus. Map. (English summary).
Identifiers: Festucetum-Valesiacae, Forests, Humus-Rich, *Moldavian Forest steppe, Slopes, Soils, Steppe, *Xerophilous variants, *Romania.

Two ecological variants were determined: a xerophilous variant of dry stations and a mesox mesoxerophilous variant of stations with more favorable water conditions. The conclusion is reached that this association is encountered especially on NE or NW facing slopes, on deep soils, rich in humus, with slightly acid or slightly alkaline reaction, not salinized, and with a high exchange capacity.--Copyright 1973, Biological Abstracts, Inc.
W73-10163

PRELIMINARY DATA ON THE VEGETATIVE REPRODUCTION AND ORGANOGENESIS OF SOME AQUATIC PLANTS, (IN RUMANIAN),
Bucharest Univ. (Rumania).
M. Andrei.
An Univ Bucur Biol Veg. 19. p 163-182. 1970. Illus.
Identifiers: *Aquatic plants, Batracium, Ceratophyllum-Demersum, Hydrocharis-Morsus-Ranae, Myriophyllum-Spicatum, Nymphaea-Alba, Nymphoides-Peltata, *Organogenesis, Phragmites-Communis, Potamogeton-Crispus, Potamogeton-Perfoliatus, *Romania, Rorippa-Amphibia-F-Auriculata, Rorippa-Sylvestris, Stratiotes-Aloides, Typha-Angustifolia, Utricularia-Vulgaris, *Vegetative reproduction.

Vegetative reproduction was discussed for Batrachium trichophyllum, Nymphaea alba, Ceratophyllum demersum, Rorippa amphibia f. auriculata, R. sylvestris, Myriophyllum spicatum, Utricularia vulgaris, Nymphoides peltata, Stratiotes aloides, Hydrocharis morsus-ranae, Potamogeton perfoliatus, P. crispus, Typha angustifolia and Phragmites communis. Reproduction by layering is predominant and almost general in aquatic plants. The formation of turions is determined by a decrease in the temperature of the habitat. The common aquatic plants of the Crapina-Jijila (Romania) lake complex reproduce by natural layering.--Copyright 1973, Biological Abstracts, Inc.
W73-10164

VEGETATION AND STRATIGRAPHY OF PEATLANDS IN NORTH AMERICA,
H. Osvald.
Acta Univ Ups Nova Acta Regiae Soc Sci Ups Ser V C. 1, p 7-94 1970. Illus.
Identifiers: Chamaedaphne, Climates, Larix, Ledum-Groenlandicum, Ledum-Palustre, *North

America, *Peatlands, Picea, Pinus, *Stratigraphy, *Vegetation.

In North America there are types of mires corresponding to most of the European ones. The most striking difference is that types similar to the west European types of bog, with stagnation and erosion complexes, have not yet been described. The bogs close to the east coast are even more similar to the continental bogs in Europe than to bogs belonging to the western maritime type. In the flat bogs of Nova Scotia Chamaedaphne has a prominent role, while this dwarf-shrub is lacking on all bogs in western Europe. In western North America Ledum groenlandicum is the predominant dwarf-shrub on the bogs close to the coast (even on Queen Charlotte Islands), while in Europe L. palustre only forms communities on bogs far away from the ocean coast. There are trees (Picea, Larix, or Pinus) on all raised bogs in North America, but not in the westernmost parts of Europe. Bogs strongly influenced by the maritime climate in North America show several features which in Europe are found only on bogs of a fairly continental type. It seems probable that these differences are due to differences in the summer climate. The bog areas in North America have a considerably warmer summer climate than the corresponding areas in Europe. The exceptions from this rule, Newfoundland and the extreme maritime zone of the Pacific coast, are not discussed. Fires have played and are still playing a prominent role in the development of the mires in North America and affect the composition of the plant communities and the boundaries between them. This phenomenon is characteristic not only of peat-land vegetation but also of the vegetation on mineral soil, such as heaths, meadows, grasslands and woodlands. Since areas with vegetation that may be ravaged by fires are much more common in North America than in Europe, it is often difficult to separate the different plant communities from each other. Instead of fairly clear boundaries between the plant communities, there is quite frequently a gradual change in the composition from one plant community to an adjacent one. The same phenomenon may also be due to influences other than fires, for instance, grazing.--Copyright 1973, Biological Abstracts, Inc.
W73-10165

AUTORADIOGRAPHY AS A TOOL IN PRIMARY PRODUCTION RESEARCH,
Brussels Univ. (Belgium).
J. P. Mommaerts.
Neth J Sea Res. Vol 5 No 4, p 437-439. 1972.
Identifiers: *Autoradiography, *Phyto-plankton, *Primary production, Photosynthesis.

Autoradiography of C-14 labeled phytoplankton collected onto a millipore filter allows the determination of the number of photosynthetically active cells in a water sample and thus gives useful information on the standing crop in primary production studies.--Copyright 1973, Biological Abstracts, Inc.
W73-10168

SYSTEMS ANALYSIS OF MINERAL CYCLING IN FOREST ECOSYSTEMS,
Goettingen Univ. (West Germany).
For primary bibliographic entry see Field 02K.
W73-10188

BEHAVIOUR OF RU IN AN ESTABLISHED PASTURE SOIL AND ITS UPTAKE BY GRASSES,
Centre d'Etude de l'Energie Nucleaire, Mol (Belgium). Laboratoires.
For primary bibliographic entry see Field 05B.
W73-10192

DISTRIBUTION OF RADIOSTRONTIUM AND RADIOCAESIUM IN THE ORGANIC AND MINERAL FRACTIONS OF PASTURE SOILS AND THEIR SUBSEQUENT TRANSFER TO GRASSES,
Centre d'Etude de l'Energie Nucleaire, Mol (Belgium). Laboratoires.
For primary bibliographic entry see Field 05B.
W73-10193

A PRELIMINARY LIST OF INSECTS AND MITES THAT INFEST SOME IMPORTANT BROWSE PLANTS OF WESTERN BIG GAME,
Forest Service (USDA), Moscow, Idaho. Forestry Sciences Lab.
M. M. Furniss.
USDA Forest Service Research Note INT-155. February 1972. 16 p, 26 ref.

Descriptors: *Insects, *Mites, *Shrubs, *Wildlife, *Forages, Idaho, Pacific Northwest U.S., Systematics, Willow trees, Big game, Entomology.
Identifiers: Serviceberry, Bitterbrush, Mountain mahogany, Ceanothus.

This list is organized alphabetically by scientific names of the host plants and species, the latter by order, family, genus, and species. Parts of plants damaged are indicated when known. Omitted are all parasitic and predacious species. Most records are from Idaho and adjacent states. Collections have been most intensive on bitterbrush, a shrub attacked by large numbers and many kinds of insects. Other genera important as browse for western big game, such as deer, elk, moose, antelope, bighorn sheep, and mountain goat, that are cited in this list include ceanothus, mountain mahogany, willow, and serviceberry. (Paylore-Arizona)
W73-10218

PRELIMINARY ANNOTATED LIST OF DISEASES OF SHRUBS ON WESTERN GAME RANGES,
Forest Service (USDA), Ogden, Utah. Intermountain Forest and Range Experiment Station.
R. G. Krebill.
USDA Forest Service Research Note INT-156. February 1972. 8 p, 29 ref.

Descriptors: *Plant diseases, *Shrubs, *Forages, *Wildlife, Air pollution, Bacteria, Fungi, Viruses, Sagebrush, Willow trees, Pacific Northwest U.S., Southwest U.S., Rocky Mountain region, Big game.
Identifiers: Chokecherry, Ceanothus, Mountain mahogany, Bitterbrush.

Lists 7 genera of shrubs browsed by big game in the western U.S., excluding Alaska, Hawaii, and Texas, with indications of diseases that attack the various species of each, and the locations of observed attacks. Diseases caused by fungi, bacteria, viruses, and parasitic plants, as well as common physiogenic problems such as winter injury, are included. Only those organisms known to be pathogenic to browse hosts are cited. Air pollution as it affects certain genera is cited. (Paylore-Arizona)
W73-10219

ALGAE FLORA IN THE UPPER PART OF GULCHA RIVER (IN RUSSIAN),
B. K. Karimova.
Uzb Biol Zh. Vol 15, No 5, p 30-32. 1971.
Identifiers: *Algae, *Diatoms, *Gulcha River, Oligosaprobic water, Rivers, Cyanophyta, USSR, *Agnatic plants.

The study was carried out during the summer-autumn period of 1969. The water temperature on July 22 and Aug. 12 at 5 PM was recorded to be 10.2 deg C. Altogether 106 spp., varieties or forms of algae were detected. Of these one belonged to

Identifiers: Algae, Fish, Nutrition, Plankton, *Rhodeus-sericeus-Amarus, Zooplankton, *Hungary.

On the basis of analysis of the stomach contents of 113 specimens of the fish from several rivers in Hungary, it was determined that the species is genuinely phytophagous and stenophagous. The characteristic shows itself at the age of 2 mo. after a short-time use of zooplankton, when its phytophagy is decidedly restricted to algae. The use of higher plants by this species was not observed. The index of gut-fullness is unusally high, in yearly average attaining 38.6%. This is connected with the phytophagous nature of the species, and rises continually during the summer until it reaches 64.1% in Sept., becoming lower in fall and winter. The fish feed also in winter. Feeding intensity varies with age and size: up to 30 mm body length (less caudal fin) it increases, but thereafter it decreases in steps, reflecting the life-periods of the animal.--Copyright 1973, Biological Abstracts, Inc.
W73-10361

ECOPHYSIOLOGICAL STUDIES ON DESERT PLANTS: VII. WATER RELATIONS OF LEPTADENIA PYROTECHNICA (FORSK.) DECNE. GROWING IN THE EGYPTIAN DESERT,
Cairo Univ., Giza (Egypt). Dept. of Botany.
A. M. Migahid, A. M. Abdel Wahab, and K. H. Batanouny.
Oecologia (Berl). Vol 10, No 1, p 79-91, 1972. Illus.
Identifiers: Climates, Deserts, Ecology, *Egypt, *Leptadenia-pyrotechnica, Morphology, *Osmotic pressure, Physiological studies, Seasons, *Transpiration.

L. pyrotechnica (Forsk.) Decne. grows in valleys of the Eastern Desert in Egypt where the climatic and edaphic drought is very severe. The transpiration rate is low in winter months, and rises by the onset of summer. The rise in transpiration rate is not comparable with that of the evaporation. The maximum transpiration rate is generally attained earlier than the maximum of the evaporating factors, specially in summer. The transpiration curves show a rapid decrease after this maximum, despite the continuous rise of the evaporating power of the atmosphere. This points to an effective stomatal regulation. The osmotic pressure of Leptadenia is relatively low. The osmotic pressure values show narrow daily and seasonal fluctuations. L. pyrotechnica is one of the most drought resistant plants. Its xeromorphic leafless habit implies reduction in the transpiring surface, while its deep extensive roots and low transpiration rate ensure a favorable water balance. The mechanisms of drought resistance in this plant are discussed.--Copyright 1973, Biological Abstracts, Inc.
W73-10364

HUMIDITY REACTIONS OF THE YOUNG OF THE TERRESTRIAL ISOPODS PORCELLIO SCABER LATR. AND TRACHEONISCUS RATHKEI (BRANDT),
Turku Univ. (Finland). Zoophysiological Lab.
O. V. Lindqvist.
Ann Zool Fenn. Vol 9, No 1, p 10-14, 1972. Illus.
Identifiers: *Humidity, *Isopods (Terrestrial), Porcellio-Scaber, Tracheoniscus-Rathkei.

The postlarval young of P. scaber, when taken out of the marsupium of the female, are able to react to environmental humidities. Their preference is for high humidities and the reaction intensity is related to the steepness of the gradient. A low temperature (12degC) does not interfere with this ability. If the female is desiccated in dry air, the young which have been lying against the oostegites and were thus desiccated showed stronger preference for the humid side than those which had been located next to the body wall of the female and were thus protected against water loss. The young of both P. scaber and T. rathkei after their first molt show an about equal preference for high humidities. It is assumed that all developmental stages of these 2 spp. are able to react to external humidities, which may be a factor in their success in colonization of dry land. The role and origin of the marsupial fluid in terrestrial isopods is discussed.--Copyright 1972, Biological Abstracts, Inc.
W73-10371

FRENCH SPEAKING AFRICA SOUTH OF SAHARA AND MADAGASCAR, INLAND WATER PRODUCTION, (IN FRENCH),
Centre Technique Forestier Tropical, Nogent-sur-Marne (France).
J. Bard.
Bois For Tropi. 140, p 3-12, 1971. Illus. English summary.
Identifiers: *Africa, Cultures, *Fisheries, Inland waters, Madagascar, Sahara desert.

A brief description of the waters and of their fauna is followed by a review of fisheries in natural waters of intensive fish culture and of extensive (or restocking) fish culture. The possibilities in fisheries as well as fish culture are still to be fully exploited.--Copyright 1972, Biological Abstracts, Inc.
W73-10372

THE SELECTION OF ALGAE FOR MASS CULTURE PURPOSES,
Instytut Zootechniki, Oswiecim (Poland). Zaklad Doswiadczalnej Zator.
T. Bednarz, and M. Nowak.
Acta Hydrobiol. Vol 14, No 1, p 1-18, 1972. Illus.
Identifiers: *Algae, Ankistrodesmus, Chlamydomonas, Chlorella, Coelastrum, Culture, *Protein, Protococcus, Scenedesmus, Stichococcus, Tetraedron.

A total of 950 algae strains belonging to the genera Chlorella, Scenedesmus, Ankistrodesmus, Tetraedron, Coelastrum, Protococcus, Stichococcus and Chlamydomonas were isolated and subjected to a 3-stage selection for a high content of crude protein. Fifty-six strains were obtained, belonging to the genera Chlorella, Scenedesmus and Ankistrodesmus; the crude protein content varied from 43 to 65%.--Copyright 1972, Biological Abstracts, Inc.
W73-10374

CHIRONOMIDAE IN STREAMS OF THE POLISH HIGH TATRA MTS.,
Polish Academy of Sciences, Krakow. Zaklad Biologii Wod.
A. Kownacki.
Acta Hydrobiol. Vol 13, No 4, p 439-464, 1971. Illus.
Identifiers: Algae, *Chironomidae, Insect larvae, Moss, Streams, Tatra Mountain streams, *Bottom fauna, *Poland.

The main components of the bottom fauna in streams of the High Tatra Mts are insect larvae, especially Chironomidae (40-100% of the total number of animals). On the basis of the dominant species it is possible to distinguish 5 types of Chironomidae, distributed in accordance with changes in altitude. It was possible to show differentiation of Chironomidae associations in individual habitats (stones, moss, algae) also within particular localities, as well as the dependence of the course of seasonal changes on altitude.--Copyright 1972, Biological Abstracts, Inc.
W73-10377

GENERAL CHARACTERISTICS OF THE BIOCOENOSIS IN THE POLISH HIGH TATRAS,
Polish Academy of Sciences, Krakow. Zaklad Biologii Wod.
B. Kawecka, M. Kownacka, and A. Kownacki.

Acta Hydrobiol. Vol 13, No 4, p 465-476, 1971. Illus.
Identifiers: *Biocoenosis, Chamesiphon-Polonicus, Diamesa-Latitarsis, Diamesa-Steinboecki, Diatoma-Hiemale, Homoeothrix-Janthina, Hydrurus-Foetidus, Orthocladius-Rivicola, Parorthocladius-Nudipennis, Streams, *Poland.

On the basis of algae and bottom-fauna communities in the streams of the High Tatra Mts the following zones were distinguished: the zone of high-mountain streams (at an altitude of 1550-2100 m), with community composed of Cyanophyceae (Chamaesiphon polonicus) and larvae of Chironomidae (Diamesa steinboecki, D. gr. latitarsis); the zone of the montane streams (at an altitude of 1000-1550 m) of which Hydrurus foetidus, Homoeothrix janthina, Diatoma hiemale, Baetis alpinus, Rhitrogena loyolaea, and Parorthocladius nudipennis on substratum or Orthocladius rivicola on granite substratum are typical; the zone of the submontane streams and rivers (at an altitude of 500-1000 m) in which Diatoma vulgare var. ehrenbergii, D. vulgare var. capitulatum, Cymbella affinis, Synedra ulna, Ulothrix zonata dominate with Orthocladius rivicola, O. thienemanni, and Simuliidae.--Copyright 1972, Biological Abstracts, Inc.
W73-10378

THE EFFECT OF TEMPERATURE AND MOISTURE ON MICROFLORA OF PEAT-BOGGY SOIL, (IN RUSSIAN),
E. Z. Tepper, and T. V. Pushkareva.
Dokl Mosk S-Kh Akad Im K A Timiryazeva. 169, p 160-165, 1971.
Identifiers: *Microflora, Moisture, Peat, *Soils (Peat-boggy), Temperature.

The effect of incubation temperatures (10, 20-30 deg) and various moisture levels (40, 60-85% of total water capacity) on microflora of lowland peat-bog soil and the accumulation of mineral forms of N was studied. The soil was enriched with compost for 100 days and analyzed every 10 days. Nitrate forms of N predominated at 20-30 deg C and ammonia N at 10 deg, especially with 85% moisture. In the latter case, the activity of nitrifying agents was suppressed and N immobilization decreased since the development of aerobic cellulose-decomposing bacteria was inhibited. Excess ammonia N (due to the low absorption capacity of peat) could be lost through water drainage. Variations in the development of cellular mass of various groups of microorganisms were characterized by systematic changes in the increase and decrease of cellular mass, which were especially abrupt at higher temperatures and moisture levels.--Copyright 1972, Biological Abstracts, Inc.
W73-10381

CONTRIBUTION TO THE STUDY OF THE ALGAL FLORA OF ALGERIA,
Algiers Univ. (Algeria). Faculty of Science.
R. Baudrimont.
Bull Soc Hist Nat Afr Nord. Vol 61, No 3-4, p 155-168. 1970. Illus.
Identifiers: *Algal flora, *Algeria, *Artesian well water, Cyanophyceae, Diatoms, Sahara, Salinity, Springs, Temperature, Wells.

Artesian well water in the west Chott ech Chergui are strongly mineralized, hard, mesohaline (over 400 mg/l Cl), rich in $CaSO_4$ and weakly alkaline. The temperature varied from 22 to 44 C, depending on the well studied. Forty-one genera and 64 spp., principally Cyanophyceae (17 spp.) and some diatoms (24 spp.), generally mesohalobes, thermophilic and alkaliphilic. These algae are common in the upper littoral in continental salt waters of the central and northern Sahara and in hot springs of Algeria. Three different diatom associations were defined in springs whose physico-chemical composition is stable during the year.--Copyright 1973, Biological Abstracts, Inc.
W73-10399

2J. Erosion and Sedimentation

KINETICS OF CARBONATE-SEAWATER INTERACTIONS,
Hawaii Inst. of Geophysics, Honolulu.
For primary bibliographic entry see Field 02K.
W73-09790

GEOLOGIC PROCESS INVESTIGATIONS OF A BEACH-INLET-CHANNEL COMPLEX, NORTH INLET, SOUTH CAROLINA,
South Carolina Univ., Columbia. Dept. of Geology.
For primary bibliographic entry see Field 02L.
W73-09795

COBEQUID BAY SEDIMENTOLOGY PROJECT: A PROGRESS REPORT,
McMaster Univ., Hamilton (Ontario). Dept. of Geology.
For primary bibliographic entry see Field 02L.
W73-09796

THE COASTAL GEOMORPHOLOGY OF THE SOUTHERN GULF OF SAINT LAWRENCE: A RECONNAISSANCE,
South Carolina Univ., Columbia. Dept. of Geology.
For primary bibliographic entry see Field 02L.
W73-09797

MANIFESTATION OF EROSION IN PERMAFROST ON THE YAMAL PENINSULA (PROYAVLENIYE EROZII V MNOGOLETNEMERZLYKH PORODAKH NA YAMALE),
I. I. Shamanova.
Akademiya Nauk SSSR Izvestiya, Seriya Geograficheskaya, No 2, p 92-98, March-April 1971. 4 fig, 4 ref.

Descriptors: *Erosion, *Permafrost, *Geomorphology, Terraces (Geologic), Gullies, Ice, Temperature, Solifluction.

Observations were conducted in the summer of 1969 to study erosion processes in the central part of the Yamal Peninsula in Northwestern Siberia. The region is a distinctly terraced, gently undulating aggradational plain with erosion-thermokarst rills and numerous thermokarst lakes. Maximum absolute elevations (50-60 m) are confined to the fourth marine terrace and minimum elevations (5-7 m), to flood-plain terraces and low seacoast plains. The region has a severe climate with a long-term average air temperature of minus 9.5 deg C. Annual precipitation is 480 mm, most of which occurs during the cold period of the year. Snow depth on the surface of terraces generally does not exceed 20-30 cm, and snow density varies between 0.2 and 0.3 g/cu cm. A polygenetic type of continuous permafrost prevails in the region, and the depth to the bottom of permafrost is generally greater than 300-400 m. In many cases, the presence of low-temperature permafrost near the surface intensifies erosion at high geomorphological levels (third and fourth terraces). This intensification is associated with specific characteristics of mechanical erosion and thermal degradation in regions of perennially frozen ground. (Josefson-USGS)
W73-09805

VERTICAL DISTRIBUTION AND ABUNDANCE OF THE MACRO- AND MEIOFAUNA IN THE PROFUNDAL SEDIMENTS OF LAKE PAIJANNE, FINLAND,
Jyvaskyla Univ. (Finland). Dept. of Biology.
For primary bibliographic entry see Field 02H.
W73-09818

was applied to analyze on a regional scale the influence of fissures on the bedrock permeability on Prince Edward Island. The range of directional permeabilities agrees in magnitude with average values determined by pumping tests and tidal analyses, and the orientations of the principal axes are similar, nearly vertical in the case of major axes and nearly horizontal in the case of other axes. (Knapp-USGS)
W73-09954

RELATION OF PHYSICAL AND MINERALOGICAL PROPERTIES TO STREAMBANK STABILITY,
Agricultural Research Service, Chickasha, Okla.
D. W. Goss.
Water Resources Bulletin, Vol 9, No 1, p 140-144, February 1973. 1 tab, 10 ref.

Descriptors: *Bank stability, *Stream erosion, *Bank erosion, *Clays, *Particle shape, Soil mechanics, Soil properties, Mineralogy, Clay minerals, Soil physical properties, Alluvium, Alluvial channels, *Oklahoma.
Identifiers: *Washita River (Okla).

Clay mineralogy and bulk density do not appear to be factors contributing to the relative stability of streambanks on the Washita River, Oklahoma. The Washita River basin does not have a large enough variety of clay minerals to determine the effect of clay mineralogy on streambank stability. The total clay content may contribute to stability. Sand grain shape and clay distribution seem to have the greatest potential in explaining some of the relative differences in streambank stability. The sand-sized grains of the stable areas were less rounded than those of the unstable areas. This somewhat angular shape of the grains may have produced an interlocking between grains that added stability to the bank material. Also, clay coatings on the sand-sized grains from the stable areas may have caused cementation between grains. (Knapp-USGS)
W73-09993

SILTATION PROBLEMS IN RELATION TO THE THAMES BARRIER,
Hydraulics Research Station, Wallingford (England).
For primary bibliographic entry see Field 08A.
W73-09997

AVALANCHE MODE OF MOTION: IMPLICATIONS FROM LUNAR EXAMPLES,
Geological Survey, Menlo Park, Calif.
K. A. Howard.
Science, Vol 180, No 4090, p 1052-1055, June 8, 1973. 3 fig, 19 ref.

Descriptors: *Landslides, *Avalanches, *Mass wasting, Rockslides, Degradation (Slope), Movement.
Identifiers: *Moon.

A large avalanche (21 square kilometers) at the Apollo 17 landing site moved out several kilometers over flat ground beyond its source slope. If not triggered by impacts, then it was as 'efficient' as terrestrial avalanches attributed to air-cushion sliding. Evidently lunar avalanches are able to flow despite the lack of lubricating or cushioning fluid. (Knapp-USGS)
W73-10038

BEACH RIDGE SLOPE ANGLES VS AGE,
Florida State Univ., Tallahassee. Dept. of Geology.
W. F. Tanner, and J. C. Hockett.
Southeastern Geology, Vol 15, No 1, p 45-51, April 1973. 2 fig, 2 ref.

Descriptors: *Beaches, *Geomorphology, *Dating, *Topography, *Degradation (Slope), Erosion, Slopes, Florida, Pleistocene epoch, Recent epoch.

Sixteen beach ridge sets, including 76 Holocene beach ridges, were studied in northwest Florida. All ridge sets at any one age have the same side slopes; older ridges have gentler side slopes. No ridges, on sand of the same size and under the same climate, should be visible after about 170,000 years. Correlation of slopes and ages can be used for estimating ridge age where materials are not too course. (Knapp-USGS)
W73-10039

DISTRIBUTION OF THE COREY SHAPE FACTOR IN THE NEARSHORE ZONE,
Virginia Univ., Charlottesville. Dept. of Environmental Sciences.
D. Poche, A. Jones, and B. Taylor.
Southeastern Geology, Vol 15, No 1, p 29-36, April 1973. 4 fig, 2 tab, 10 ref.

Descriptors: *Particle shape, *Sands, *Settling velocity, Surf, Beaches, Alluvium, Sedimentology, Distribution patterns.
Identifiers: Shape factor (Sands).

The Corey shape factor for several thousand sand grains sampled from the nearshore zone had a mean value of 0.66. This is lower than the previously assumed value of 0.70 which is commonly used for rapid sediment analysis. This difference could cause about 5% error in the estimation of the nominal size of any sand sample. Comparison with the shape factors of river sands indicate no significant differences between surf abrasion and fluvial abrasion for particles of sand size. Histograms of the data are presented. (Knapp-USGS)
W73-10040

A MARKOV MODEL FOR BEACH PROFILE CHANGES,
Louisiana State Univ., Baton Rouge. Coastal Studies Inst.
C. J. Sonu, and W. R. James.
Journal of Geophysical Research, Vol 78, No 9, p 1462-1471, March 20, 1973. 6 fig, 3 tab, 9 ref. NR 388002. ONR Contract N00014-69-A-0211-0003.

Descriptors: *Statistical models, *Mathematical models, *Beach erosion, *Markov processes, Numerical analysis, Simulation analysis, Profiles, Beaches, Sedimentation, Geomorphology, Waves (Water), Surf, Probability.
Identifiers: *Beach profiles.

The history of beach geometry can be modeled as a specific case of first-order Markov process. Under the assumptions that the profile transition is controlled only by random excitations from waves and that the transition probability is identical for all the possible states of beach profile, it is demonstrated that a beach profile time series contains cycles having negative binomial distribution. A simplified case in which the transition probability is taken as one-half (i.e., equal probability for either erosional or accretional transition for any profile state) is derived through both numerical simulation and theoretical derivation. The result shows reasonable agreement with field observations. (Knapp-USGS)
W73-10043

A MODEL FOR FORMATION OF TRANSVERSE BARS,
Florida State Univ., Tallahassee. Geophysical Fluid Dynamics Inst.
A. I. Barcilon, and J. P. Lau.
Journal of Geophysical Research, Vol 78, No 15, p 2656-2664, May 20, 1973. 4 fig, 3 tab, 22 ref. ONR N-00014-68-0159, NOAA NG-3-72.

Descriptors: *Sand bars, *Sand waves, *Beaches, *Sediment transport, Mathematical models, Sedimentary structure, Hydraulic models, Surf, Beach erosion, Dunes, Ripple marks, Littoral drift, Currents (Water).

Subtle rhythmic transverse bars more or less normal to the beach are distinct features of low-energy beaches. A simple theoretical model accounts for the formation of these bars. Linearized potential flow equations, an empirical formula for the bed load discharge, a continuity equation for the sediment, and the artifice of the lag distance between fluid and bed waves are used to formulate the model. The resulting loose boundary instability mechanism is interpreted as being related to transverse sand bar formation. Although in nature the transverse bars are not perpendicular to the coast and the currents do not run parallel to the shore as they do in the model, an estimate can be made of the spacing between transverse bars. Field observations for five locations are compared with model results. Sustained longshore currents are responsible for transverse sand bar formation. The spacing between the bars depends on the inverse of the beach slope and on the squares of the drift velocity across the bars. (Knapp-USGS)
W73-10046

MODIFICATIONS IN THE SOLUBILITY AND PRECIPITATION BEHAVIOR OF VARIOUS METALS AS A RESULT OF THEIR INTERACTION WITH SEDIMENTARY HUMIC ACID,
Bedford Inst. of Oceanography, Dartmouth (Nova Scotia). Atlantic Geoscience Center.
For primary bibliographic entry see Field 02K.
W73-10062

PROVENANCE AND ACCUMULATION RATES OF OPALINE SILICA, AL, TI, FE, CU, MN, NI, AND CO IN PACIFIC PELAGIC SEDIMENTS,
Rosenstiel School of Marine and Atmospheric Sciences, Miami, Fla.
K. Bostrom, T. Kraemer, and S. Gartner.
Chemical Geology, Vol 11, No 2, p 123-148, April 1973. 19 fig, 3 tab, 68 ref.

Descriptors: *Bottom sediments, *Pacific Ocean, *Provenance, *Silica, *Trace elements, Aluminum, Titanium, Iron, Manganese, Copper, Nickel, Cobalt, Water chemistry, Weathering, Volcanoes, Sediment transport, Sedimentation.

Accumulation rates and chemical compositions of pelagic sediments were studied at 73 locations in the Pacific Ocean and 11 in the Indian Ocean. In the Pacific, many elements accumulate rapidly close to the continents and slowly in the central part of the ocean. This pattern is interrupted by two major zones of relatively high accumulation rates, one along the Equator and one along the East Pacific Rise. Deposition of opaline silica is almost completely restricted to areas of high biological productivities at the Equator and at very high latitudes. Cu and Ni show stronger tendencies than Fe and Mn to precipitate with opaline silica. The highest accumulation rates of Fe and Mn in the open Pacific occur along the East Pacific Rise. To some extent Cu and Ni are enriched there due to volcanic processes. Al and Ti show high accumulation rates only close to the continents; these elements appear to be almost completely terrigenous. Significant quantities of basaltic matter (oceanic crust) are incorporated into the sediments only in areas of very low total sedimentation rates in the vicinity of oceanic island groups such as Polynesia and Hawaii, whereas hydrothermal processes act as a major sediment source only on the East Pacific Rise. Sediments in the north as well as in the southernmost part of the Pacific are nearly entirely terrigenous. Between 75% and 95% of all sediments in the Pacific are terrigenous, and submarine weathering and submarine exhalations each account for only a small fraction of the sediments. (Knapp-USGS)
W73-10063

ALPHA-RECOIL THORIUM-234: DISSOLUTION INTO WATER AND THE URANIUM-234/-URANIUM-238 DISEQUILIBRIUM IN NATURE,
Gakushuin Univ., Tokyo (Japan). Dept. of Chemistry.
For primary bibliographic entry see Field 05B.
W73-10325

THE DISCHARGE/WAVE-POWER CLIMATE AND THE MORPHOLOGY OF DELTA COASTS,
Louisiana State Univ., New Orleans.
For primary bibliographic entry see Field 02L.
W73-10334

2K. Chemical Processes

KINETICS OF CARBONATE-SEAWATER INTERACTIONS,
Hawaii Inst. of Geophysics, Honolulu.
F. Pesret.
Available from NTIS, Springfield, Va., 22151, as AD-747 979; Price $3.00 printed copy; $0.95 microfiche. Hawaii Institute of Geophysics Report HIG-72-10, August 1972. 64 p, 14 fig, 30 ref, 3 append. ONR Contract N00014-70-A-0016-0001.

Descriptors: *Water chemistry, *Sea water, *Calcite, Solubility, Aqueous solutions, Chemical properties, Analytical techniques, Correlation analysis, Organic matter, Dissolved solids, Magnesium, Strontium, Calcium, Sulfates, Phosphates.

Most of the calcium carbonate removed from the oceans is precipitated out by pelagic organisms living in the upper layers of the world's oceans. However, only a small fraction of that amount accumulates on the ocean floor as sediments. Thus, there is the question of where the dissolution takes place. Using the spinning disk method, an experimental setup was devised to study the rate of dissolution of calcite in aqueous solutions. Different models were developed to describe the reaction and to estimate what chemical processes may take place. The object of this study was to compare the relative influence of individual seawater constituents such as Mg, Sr, Ba, Ca, SO4, PO4 and dissolved organic matter, on the rate of calcite solution. The effect of temperature was studied by the same method. Calcium ions, followed by magnesium ions and dissolved organic matter, were found to have the greater influence on the solution-rate constant. (Woodard-USGS)
W73-09790

GROUNDWATER CONSITIONS IN THE CENTRAL VIRGIN RIVER BASIN, UTAH,
Geological Survey, Salt Lake City, Utah.
For primary bibliographic entry see Field 04B.
W73-09791

GROUNDWATER RESOURCES OF MOTLEY AND NORTHEASTERN FLOYD COUNTIES, TEXAS.
Geological Survey, Austin, Tex.
For primary bibliographic entry see Field 04B.
W73-09792

GROUNDWATER RESOURCES OF HALL AND EASTERN BRISCO COUNTIES, TEXAS.
Geological Survey, Austin, Tex.
For primary bibliographic entry see Field 04B.
W73-09793

HYDROCHEMICAL TYPING OF RIVER BASINS OF THE UKRAINE,
Akademiya Nauk URSR, Kiev. Instytut Hidrobiologii.
For primary bibliographic entry see Field 05B.
W73-09811

DETERMINATION OF THE TOTAL AMOUNT OF CARBOHYDRATES IN OCEAN SEDIMENTS,
For primary bibliographic entry see Field 05A.
W73-09884

DETERMINATION OF TRACES OF IRON, COBALT AND TITANIUM IN SEA WATER AND SUSPENDED MATTER (IN SAMPLES FROM THE BALTIC SEA AND ATLANTIC OCEAN),
For primary bibliographic entry see Field 05A.
W73-09895

A GUIDE TO MARINE POLLUTION.
For primary bibliographic entry see Field 05A.
W73-09915

OBSERVATIONS ON MESOZOIC SANDSTONE AQUIFERS IN SAUDI ARABIA,
Ministry of Agriculture and Water, Riyadh (Saudi Arabia).
For primary bibliographic entry see Field 02F.
W73-09949

PROPERTIES AND MANIFESTATIONS OF REGIONAL GROUNDWATER MOVEMENT,
Research Council of Alberta, Edmonton.
For primary bibliographic entry see Field 02F.
W73-09959

GEOCHEMICAL PROCESSES IN SHALLOW GROUNDWATER FLOW SYSTEMS IN FIVE AREAS IN SOUTHERN MANITOBA, CANADA,
Waterloo Univ. (Ontario). Dept. of Earth Sciences.
For primary bibliographic entry see Field 02F.
W73-09964

A PRELIMINARY STUDY OF THE GEOCHEMICAL CHARACTERISTICS OF THERMAL WATERS IN NEW MEXICO,
New Mexico Inst. of Mining and Technology, Socorro.
For primary bibliographic entry see Field 02F.
W73-09967

REACTIONS OF THE MONOCALCIUM PHOSPHATE MONOHYDRATE AND ANHYDROUS DIACALCIUM PHOSPHATE OF VOLCANIC SOILS, (IN SPANISH),
Chile Univ., Santiago.
For primary bibliographic entry see Field 02G.
W73-09982

CHARACTERISTICS OF DISTRIBUTION OF FLUORINE IN WATERS OF THE STEPPE REGIONS OF KAZAKHSTAN IN RELATION TO ENDEMIC FLUOROSIS, (IN RUSSIAN),
For primary bibliographic entry see Field 05B.
W73-09990

INTERNAL WATERS: MONOGRAPHS FROM LIMNOLOGY AND NEIGHBORING AREAS, VOL. 25. BIOLOGY OF BRACKISH WATER,
A. Remane, and C. Schlieper.
E. Schweizerbart'sche Verlagsbuchhandlung (Naegele u. Obermiller): Stuttgart, West Germany; John Wiley and Sons, Inc: New York, N.Y., 2nd Revised Ed. 372 p, Illus, Maps. 1971. Pr. $21.75.
Identifiers: Animals, Biotopes, Books, *Brackish water, Distribution, *Limnology, Monographs, Organisms, Salinity.

This book represents the translation of the 1958 German publication on brackish water; it has been completely revised and covers a much broader field of knowledge. The book confines its presentation to the important ecological features peculiar to the region of brackish water and the problems

INTERSTITIAL WATER CHEMISTRY AND AQUIFER PROPERTIES IN THE UPPER AND MIDDLE CHALK OF BERKSHIRE, ENGLAND,
Institute of Geological Sciences, London (England). Dept. of Hydrogeology.
W. M. Edmunds, P. E. R. Lovelock, and D. A. Gray.
Journal of Hydrology, Vol 19, No 1, p 21-31, May 1973. 4 fig, 1 tab, 15 ref.

Descriptors: *Pore water, *Water chemistry, *Limestones, *Hydrogeology, Aquifer characteristics, Weathering, Diagenesis, Fracture permeability, Groundwater movement.
Identifiers: *Chalk formation (England).

Interstitial water from core material from the Chalk of southern England has a variable chemical composition and is more mineralized by as much as a factor of 10 compared with the water from fissures samples in the borehole column. Measurements of porosity and permeability of the Chalk in the same core material show that there is a correlation between the physical and chemical parameters. The mineralized interstitial water in the lower zone may well represent pore water modified considerably since the Cretaceous by dilution with meteoric water, and the composition appears to have been influenced by diagenetic changes in the Chalk itself. The increase in the total mineralization appears to be associated with a decrease in the intrinsic permeability and porosity of the formation, suggesting that the rate of mass transfer is directly related to physical properties of the Chalk. The interstitial water composition in the upper zone of the borehole is more difficult to explain, particularly because there appears to be no correlation with changes in physical properties. The base exchange reactions which appear to have occurred in this zone are probably related to processes operating prior to any development pumping. The chemical composition may somehow be related to the former presence of Tertiary cover. Reading Beds of Eocene age occur at the surface within a few hundred meters of the site, and the abnormal compositions found in the upper zone may be related to migration of sodium rich waters derived from the overlying formations before erosion led to the exposure of the Chalk. (Knapp-USGS)
W73-10060

MODIFICATIONS IN THE SOLUBILITY AND PRECIPITATION BEHAVIOR OF VARIOUS METALS AS A RESULT OF THEIR INTERACTION WITH SEDIMENTARY HUMIC ACID,
Bedford Inst. of Oceanography, Dartmouth (Nova Scotia). Atlantic Geoscience Center.
M. A. Rashid, and J. D. Leonard.
Chemical Geology, Vol 11, No 2, p 89-97, April 1973. 2 fig, 4 tab, 15 ref.

Descriptors: *Humic acids, *Water chemistry, *Chelation, Leaching, Sedimentation, Chemical precipitation, Weathering, Diagenesis, Solubility.

Sedimentary humic acid or its acid-hydrolysate, consisting of various amino acids, is effective in dissolving unusually large quantities of metals (up to 682 mg/g of organic matter) from their insoluble salts. The presence of humic acid in a reaction media which had favorable conditions for the precipitation of metals as carbonates, hydroxides or sulfides prevented the formation of insoluble metal salts. Infrared analysis suggests that the metals added to various anionic systems and humic acid do not react with the anion, but apparently a complex formation between metals and organic matter keeps the metal in solution. The enhanced solubility and consequent decrease in precipitation of metals under the influence of humic compounds plays a leading role in the accumulation of metals in sedimentary deposits. (Knapp-USGS)
W73-10062

PROVENANCE AND ACCUMULATION RATES OF OPALINE SILICA, AL, TI, FE, CU, MN, NI, AND CO IN PACIFIC PELAGIC SEDIMENTS,
Rosenstiel School of Marine and Atmospheric Sciences, Miami, Fla.
For primary bibliographic entry see Field 02J.
W73-10063

A FUNDAMENTAL STUDY OF THE SORPTION FROM TRICHLOROETHYLENE OF THREE DISPERSE DYES ON POLYESTER,
Auburn Univ., Ala. Dept. of Textile Engineering.
For primary bibliographic entry see Field 03E.
W73-10151

DIFFUSION IN COMPLEX SYSTEMS,
UT-AEC Agricultural Research Lab., Oak Ridge, Tenn.
G. E. Spalding.
Available from NTIS, Springfield, Va., as ORO-695; $5.45 in paper copy, $0.95 in microfiche. In: Report ORO-695, p 71-73. July 1972.

Descriptors: *Soil chemistry, *Diffusion, *Ion transport, *Computer programs, Numerical analysis, Electrodialysis, Electrolytes, Equations, Cation exchange, Soil-water-plant relationships, Path of pollutant.

A computer program was developed which will numerically solve a set of time-dependent Nernst-Planck equations and thus predict the rates and paths of diffusion when only self-diffusion coefficients are known. The program was written for 3 mobile ionic species and simultaneously solves 2 nonlinear second-order partial differential equations. Initial and boundary conditions can be changed. For (1) the interdiffusion of NaCl and Sr (Cl)2 in aqueous solution and (2) the exchange between a H (+) saturated phenosulfonic membrane and NaCl solution predictions agreed with experiment generally; however systematic deviations suggested that the electric field is reduced by ion-pair formation. (Bopp-ORNL)
W73-10183

POTENTIALITIES AND LIMITS OF MULTIELEMENT NEUTRON ACTIVATION ANALYSIS OF BIOLOGICAL SPECIMENS WITH OR WITHOUT SEPARATION FROM AN ACTIVE MATRICE, (POSSIBILITIES ET LIMITES DE L'ANALYSE PAR ACTIVATION NEUTRONIQUE MULTIELE MENTAIRE D'ECHANTILLONS BIOLOGIQUES AVEC OU SANS SEPARATION DE LA MATRICE ACTIVABLE.),
Commissariat a l'Energie Atomique, Saclay (France). Centre d'Etudes Nucleaires.
For primary bibliographic entry see Field 05A.
W73-10184

SYSTEMS ANALYSIS OF MINERAL CYCLING IN FOREST ECOSYSTEMS,
Goettingen Univ. (West Germany).
B. Ulrich, and R. Mayer.
In: Proceedings of Symposium on the Use of Isotopes and Radiation in Research on Soil-Plant Relationships Including Applications in Forestry, Dec. 13-17, 1971, Vienna (IAEA-SM-151), International Atomic Energy Agency, Vienna, 1972, p 329-339. 1 tab, 2 fig, 16 ref.

Descriptors: *Deciduous forests, *Radioecology, *Soil-water-plant relationships, *Tracers, Balance of nature, Cycling nutrients, Ecology, Soil moisture meters, Moisture uptake, Root zone, Transpiration, Radioactivity techniques, Nitrogen cycle, Radioecology, Soil science.

Use of radioisotopes for answering questions in systems analysis of mineral cycling is described briefly. Bioelement fluxes (H, Na, K, Ca, Mg, Fe, Mn, Al, N, P, S, Cl) for a 125-year-old beech stand were measured: (1) input into the ecosystem by

precipitation; (2)- (4) soil input by through-fall (by passage of rains through the tree crown), stem flow and litter, respectively; (5)- (7) output from the humus layer, from the intensive root layer in a 50-cm soil depth and from the extensive root layer in a 100-cm soil depth. From an annual balance of bioelement fluxes, information is drawn about the steady-state behavior of the ecosystem, plant uptake of bioelements, separation of plant uptake into mass flow and diffusion transport mechanisms, and ion-uptake selectivity. Transfer functions expressing the fluxes as functions of compartment inputs are derived from monthly flux balances. (See also W72-10679) (Bopp-ORNL)
W73-10188

USE OF TRITIATED WATER FOR DETERMINATION OF PLANT TRANSPIRATION AND BIOMASS UNDER FIELD CONDITIONS,
Argonne National Lab., Ill.
For primary bibliographic entry see Field 02D.
W73-10190

MOVEMENT OF ZN65 INTO SOILS FROM SURFACE APPLICATION OF ZN SULPHATE IN COLUMNS,
Haile Sellassie I Univ., Alem Maya (Ethiopia). Coll. of Agriculture.
For primary bibliographic entry see Field 02G.
W73-10191

BEHAVIOUR OF RU IN AN ESTABLISHED PASTURE SOIL AND ITS UPTAKE BY GRASSES,
Centre d'Etude de l'Energie Nucleaire, Mol (Belgium). Laboratoires.
For primary bibliographic entry see Field 05B.
W73-10192

GEOTHERMAL WATER AND POWER RESOURCE EXPLORATION AND DEVELOPMENT FOR IDAHO,
Boise State Coll., Idaho. Dept. of Geology.
For primary bibliographic entry see Field 06B.
W73-10217

THE FLUXES OF MARINE CHEMISTRY,
Scripps Institution of Oceanography, La Jolla, Calif.
E. D. Goldberg.
Proceedings of the Royal Society of Edinburg. Section B: Biology, Vol 72, p 357-364, 1971/72. 21 ref.

Descriptors: *Oceans, Chemistry, *Wastes, Air-earth interfaces, *Air-water interfaces, *Earth-water interfaces, Transfer, Water pollution control.

The development of chemical oceanography as a scientific discipline is reviewed, and transfer of materials from oceans to continents to the atmosphere and back to continents is described. (Ensign-PAI)
W73-10225

PUBLIC WATER SUPPLIES OF SELECTED MUNICIPALITIES IN FLORIDA, 1970,
Geological Survey, Tallahassee, Fla.
For primary bibliographic entry see Field 04B.
W73-10319

WATER RESOURCES OF THE RED RIVER OF THE NORTH DRAINAGE BASIN IN MINNESOTA,
Geological Survey, St. Paul, Minn.
For primary bibliographic entry see Field 02E.
W73-10321

NON-EQUILIBRIUM PHENOMENA IN CHEMICAL TRANSPORT BY GROUND-WATER FLOW SYSTEMS,
Nevada Univ., Reno. Desert Research Inst.
For primary bibliographic entry see Field 05B.
W73-10329

VEGETATION OF PRAIRIE POTHOLES, NORTH DAKOTA, IN RELATION TO QUALITY OF WATER AND OTHER ENVIRONMENTAL FACTORS,
Geological Survey, Washington, D.C.
For primary bibliographic entry see Field 02I.
W73-10330

GROUNDWATER QUALITY IN THE CORTARO AREA NORTHWEST OF TUCSON, ARIZONA,
For primary bibliographic entry see Field 05B.
W73-10332

RELATIVE LEACHING POTENTIALS ESTIMATED FROM HYDROLOGIC SOIL GROUPS,
Agricultural Research Service, Beltsville, Md. Hydrograph Lab.
For primary bibliographic entry see Field 02G.
W73-10333

TURBULENCE, MIXING AND GAS TRANSFER,
Georgia Inst. of Tech., Atlanta. School of Civil Engineering.
For primary bibliographic entry see Field 05B.
W73-10338

RELATIVE GAS TRANSFER CHARACTERISTICS OF KRYPTON AND OXYGEN,
Georgia Inst. of Tech., Atlanta. School of Civil Engineering.
For primary bibliographic entry see Field 05B.
W73-10339

FIELD TRACER PROCEDURES AND MATHEMATICAL BASIS,
For primary bibliographic entry see Field 05B.
W73-10340

LABORATORY PROCEDURES,
For primary bibliographic entry see Field 05B.
W73-10342

OXYGEN BALANCE OF THE SOUTH RIVER,
For primary bibliographic entry see Field 05B.
W73-10344

2L. Estuaries

PHYSICAL AND CHEMICAL PROPERTIES OF THE COASTAL WATERS OF GEORGIA,
Georgia Univ., Sapelo Island. Marine Inst.
R. Kuroda, and F. C. Marland.
Available from the National Technical Information Service as PB-220 680, $3.00 in paper copy, $0.95 in microfiche. Georgia Environmental Resources Center, Atlanta, Report ERC-0373, April 1973, 82 p 25 fig, 10 tab, 11 ref, append. OWRR B-035-GA (1). 14-0001-1892.

Descriptors: *Ocean circulation, Ocean waves, Winds, *Salinity, *Conductivity, *Turbidity, *Dissolved oxygen, *Georgia, Sea water, Seasonal, Atlantic Ocean, Physicochemical properties, Estuaries.
Identifiers: *Sapelo Island (Georgia), Offshore waters, Ocean water masses.

Previously unpublished data are presented on the circulation and seasonal movements of the ocean waters, waves, and wind off the Georgia coast, especially in the vicinity of Sapelo Island. The total surface circulation was found to be strongly influenced by or at least coincident with prevailing winds. Analyses of the sea water for temperature, salinity, conductivity, acidity, turbidity, and dissolved oxygen content and for seasonal changes in these properties were used to classify the offshore waters. The consistent water masses were classified, proceeding from the open sea toward the coast, as North Atlantic Central Water, Mixing Water, Shelf Water, and Coastal Water. Each mass was mapped so that its seasonal movements and distributions could be traced. These original data on sea water conditions form a general benchmark for future investigations into the hydrographic conditions of the Georgia coast and a specific benchmark for a companion report on circulation within a Georgia Coastal Estuary. Good hydrographic information on an estuary is essential to effective water pollution control (James--Georgia Tech).
W73-09752

REPORT ON POLLUTION OF THE ST. MARYS AND AMELIA ESTUARIES, GEORGIA-FLORIDA,
Environmental Protection Agency, Athens, Ga. Southeast Water Lab.
For primary bibliographic entry see Field 05B.
W73-09788

GEOLOGIC PROCESS INVESTIGATIONS OF A BEACH-INLET-CHANNEL COMPLEX, NORTH INLET, SOUTH CAROLINA,
South Carolina Univ., Columbia. Dept. of Geology.
R. J. Finley.
Maritime Sediments, Vol 8, No 2, p 65-67, September 1972. 3 fig, 1 ref.

Descriptors: *Beaches, *Tidal waters, *Geomorphology, *South Carolina, *Marshes, Inlets (Waterways), Tides, Sediment transport, Beach erosion, Littoral drift, Sedimentary structures, Sand bars, Sand waves, Sand spits, Sedimentation.
Identifiers: *North Inlet (SC).

The coastal area adjacent to North Inlet, Georgetown County, South Carolina, was used in a study of short-term geological and geochemical variables because it is relatively unaffected by man's activities. Climatic factors, physical and chemical characteristics of the tidal prism, substrate composition, bedform morphology, and beach and inlet processes were investigated. The dominant drift is to the south. A northerly drift just south of the inlet may be the result of wave refraction around the ebbtidal delta. Mean tidal range at North Inlet is 1.4 m. No temperature or salinity stratification was found. There is no major freshwater influx to the tidal channels. Summer water temperatures ranged from 26.5 deg C to 29 deg C, and variations with depth up to 8 m rarely exceeded a few tenths of a degree Centigrade. The waters carry substantial suspended matter, both organic and inorganic. (Knapp-USGS)
W73-09795

COBEQUID BAY SEDIMENTOLOGY PROJECT: A PROGRESS REPORT,
McMaster Univ., Hamilton (Ontario). Dept. of Geology.
R. J. Knight.
Maritime Sediments, Vol 8, No 2, p 45-60, September 1972. 15 fig, 6 tab, 24 ref.

Descriptors: *Sedimentology, *Channel morphology, *Sand waves, *Tidal waters, *Ripple marks, Sediment transport, Tides, *Canada, Tidal effects, Tidal bores, Tidal streams.
Identifiers: *Bay of Fundy (Cobequid Bay).

COMPUTER SIMULATION OF GROUND WATER AQUIFERS OF THE COASTAL PLAIN OF NORTH CAROLINA,
North Carolina Univ., Chapel Hill. Dept. of Environmental Sciences and Engineering.
For primary bibliographic entry see Field 02F.
W73-09871

BARTOW MAINTENANCE DREDGING: AN ENVIRONMENTAL APPROACH,
For primary bibliographic entry see Field 05G.
W73-09881

HYDRAULIC MODELS PREDICT ENVIRONMENTAL EFFECTS.
For primary bibliographic entry see Field 08B.
W73-09883

DEVELOPMENT PROPOSAL FOR LAGOONS,
Rhode Island Univ., Kingston. Dept. of Economics.
For primary bibliographic entry see Field 03C.
W73-09886

ARTIFICIAL RADIOACTIVITY IN FRESH WATER AND ESTUARINE SYSTEMS,
Ministry of Agriculture, Fisheries and Food, Lowestoft (England). Fisheries Radiobiological Lab.
For primary bibliographic entry see Field 05B.
W73-09893

SEDIMENTARY PHOSPHORUS IN THE PAMLICO ESTUARY OF NORTH CAROLINA,
North Carolina Univ., Chapel Hill. Dept. of Environmental Sciences and Engineering.
For primary bibliographic entry see Field 05B.
W73-09897

STUDIES ON THE COASTAL OCEANOGRAPHY IN THE VICINITY OF FUKUYAMA, HIROSHIMA PREF., I. DISTRIBUTION PATTERNS OF TEMPERATURE, CHLORINITY, PH AND INORGANIC NUTRIENT (PHOSPHATE-P, AMMONIA-N, NITRITE-N, NITRATE-N) CONTENTS OF SEA WATER IN EARLY FEBRUARY, 1968, (IN JAPANESE),
Hiroshima Univ. (Japan). Dept. of Fisheries.
For primary bibliographic entry see Field 05B.
W73-09899

NATURAL AND MANMADE HYDROCARBONS IN THE OCEANS,
Texas Univ., Port Aransas. Marine Science Inst.
For primary bibliographic entry see Field 05B.
W73-09907

THE KENT COAST IN 1971,
British Museum of Natural History, London. Marine Algal Section.
For primary bibliographic entry see Field 05C.
W73-09909

WASTE DISPOSAL AND WATER QUALITY IN THE ESTUARIES OF NORTH CAROLINA,
North Carolina State Univ., Raleigh. Dept. Civil Engineering.
For primary bibliographic entry see Field 05B.
W73-09911

COAGULATION IN ESTUARIES,
North Carolina Univ., Chapel Hill. Dept. of Environmental Sciences and Engineering.
J. K. Edzwald.
Sea Grant Publication UNC-SG-72-06, July 1972. 204 p. 37 fig, 25 tab, 131 fig, 4 append.

Descriptors: *Coagulation, *Flocculation, *Estuaries, Clay minerals, Sediments, Kaolinite, Illite, Montmorillonite, *North Carolina.
Identifiers: *Pamlico Estuary (NC).

A model in which the stability of clay minerals is dependent on the chemistry of the system is proposed and tested. It is assumed that clay suspensions are destabilized by compression of the electrical double layer when these suspensions are transported from fresh waters into estuaries. The stability factor of three clay minerals was determined in the laboratory in solutions at various ionic strengths. Illite was found more stable than Kaolinite and this more stable than Montmorillonite. Particle destabilization increased with salinity; it similarly improved when solutions with divalent cations were used instead of solutions with monovalent solutions. Selected sediments from the Pamlico Estuary of North Carolina were studied, their composition being determined by x-ray diffraction. Kaolinite was abundant in the upper estuary and decreased towards the mouth, Illite increased towards the mouth and Montmorillonite was found in minor amounts, the rest comprised chlorite and chlorite-like intergrade clay. The stability factor decreased downstream. The coagulation and deposition of clays in estuaries can affect water quality via adsorption and release of soluble pollutants. It is concluded from adsorbed phosphorus data that sediments entering the estuary from upstream would lose phosphorus as salinity increases. Depending upon the strength of the solid-solute interaction, coagulated materials could serve as a source or sink for a soluble substance in estuarine waters. (Ensign-PAI)
W73-09925

TIDAL FLUCTUATIONS IN A CONFINED AQUIFER EXTENDING UNDER THE SEA,
Vrije Universiteit, Amsterdam (Netherlands). Inst. of Earth Sciences.
For primary bibliographic entry see Field 02F.
W73-09955

RIVER THAMES--REMOVABLE FLOOD BARRIERS,
Bruce White, Wolfe Barry and Partners, London (England).
For primary bibliographic entry see Field 08A.
W73-09996

SILTATION PROBLEMS IN RELATION TO THE THAMES BARRIER,
Hydraulics Research Station, Wallingford (England).
For primary bibliographic entry see Field 08A.
W73-09997

POLLUTION PROBLEMS IN RELATION TO THE THAMES BARRIER,
Water Pollution Research Lab., Stevenage (England).
For primary bibliographic entry see Field 08A.
W73-09998

INTERNAL WATERS: MONOGRAPHS FROM LIMNOLOGY AND NEIGHBORING AREAS, VOL. 25. BIOLOGY OF BRACKISH WATER,
For primary bibliographic entry see Field 02K.
W73-10006

COMPUTATION OF SALINITY AND TEMPERATURE FIELDS IN RIVER MOUTHS NEAR THE SEA (IN RUSSIAN),
Akademiya Nauk SSSR, Kaliningrad. Institut Okeanologii.
For primary bibliographic entry see Field 02K.
W73-10009

GEOMETRY OF SANDSTONE RESERVOIR BODIES,
Shell Oil Co., Houston, Tex.
For primary bibliographic entry see Field 02F.
W73-10026

DISTRIBUTION OF THE COREY SHAPE FACTOR IN THE NEARSHORE ZONE,
Virginia Univ., Charlottesville. Dept. of Environmental Sciences.
For primary bibliographic entry see Field 02J.
W73-10040

THE EFFECT OF HYDROLOGIC FACTORS ON THE PORE WATER CHEMISTRY OF INTERTIDAL MARSH SEDIMENTS,
South Carolina Univ., Columbia. Dept. of Geology.
For primary bibliographic entry see Field 02K.
W73-10041

RECENT SUBSIDENCE RATES ALONG THE TEXAS AND LOUISIANA COASTS AS DETERMINED FROM TIDE MEASUREMENTS,
National Ocean Survey, Rockville, Md.
R. L. Swanson, and C. I. Thurlow.
Journal of Geophysical Research, Vol 78, No 15, p 2665-2671, May 20, 1973. 3 fig, 3 tab, 11 ref.

Descriptors: *Subsidence, *Land subsidence, *Gulf Coastal Plain, *Texas, *Louisiana, Sedimentation, Withdrawal, Structural geology, Sea level, Surveys.

Sea level and tide level rose at several locations along the Louisiana and Texas coasts during the years 1959-1970, from as little as 0.5 cm/yr at Port Isabel, Texas, to a maximum of 4.3 cm/yr at South Pass, Louisiana. These rates were determined by comparing tide records along the Louisiana and Texas coasts with the long term tide record at Pensacola, Florida, a location assumed to be stable for the purpose of this study. At locations for which observations were available before 1959 the rise in sea (or tide) level was greater from 1959 to 1970 than during the decade prior to 1959. This rise in sea level is attributed to land subsidence. Subsidence in deltaic regions of the Mississippi River is related to combinations of geosynclinal downwarping, compaction of pore spaces by increased sediment burden, and lateral displacement by plastic flow of fine-grained sediments. West Bay has undergone subsidence caused by faulting and withdrawal of oil and water. Subsidence at Freeport is believed to reflect the deltaic environment of the Brazos River. The data were collected in the old river outlet that was abandoned in favor of a dredged channel in the 1930's. Since then the delta at the old outlet has receded. Consequences of withdrawal of oil and water from coastal substrata must also be considered. (Knapp-USGS)
W73-10045

PROBLEMS IN HYDROLOGY OF ESTUARINE REACHES OF SIBERIAN RIVERS (PROBLEMY GIDROLOGII UST'YEVYKH OBLASTEY SIBIRSKIKH REK).
Arkticheskii i Antarkticheskii Nauchno-Issledovatelskiy Institut, Leningrad (USSR).

Arkticheskiy i Antarkticheskiy Nauchno-Issledovatel 'skiy Institut Trudy, No 297, V. S. Antonov, editor, Leningrad, 1972. 147 p.

Descriptors: *Hydrology, *Estuaries, *Rivers, *Arctic, *Ice, Ice jams, Ice breakup, Melting, Snow, Discharge (Water), Runoff, Flow, Water levels, Water level fluctuations, Navigation, Hydrography, Mapping, Measurement, Instrumentation, Forecasting.
Identifiers: *USSR, *Siberia, Bars (Coastal).

Hydrologic and ice regimes of lower reaches and estuaries of major Siberian rivers are investigated in this collection of 15 papers published by the Arctic and Antarctic Scientific Research Institute. Specific topics include: (1) hydrologic regime of lower reaches of the Yenisey after regulation of flow; (2) computation of unsteady flow in estuaries of large Siberian rivers; (3) long-range forecasts of water levels on the Kolyma River bar; (4) melting of ice cover in arctic river mouths; (5) flow of Ob' River water into the sea and long-term variability of runoff; (6) water-level fluctuations of the Pur River; (7) runoff redistribution in the Yenisey delta; (8) distribution of spring ice jams on rivers in the Siberian Arctic; (9) effect of ice jams on breakup of ice cover at the mouth of the Yenisey; (10) a new map of drainage density for the Lena River basin; and (11) water-level measurements in river mouths by the compensation method. (Josefson-USGS)
W73-10053

PROBLEMS OF SUBSURFACE FLOW INTO SEAS (PROBLEMA PODZEMNOGO STOKA V MORYA),
Akademiya Nauk SSSR, Moscow. Institut Vodnykh Problem.
For primary bibliographic entry see Field 02F.
W73-10055

COASTAL ZONE POLLUTION MANAGEMENT, PROCEEDINGS OF THE SYMPOSIUM.
For primary bibliographic entry see Field 05G.
W73-10098

APPLICATIONS OF ESTUARINE WATER QUALITY MODELS,
Manhattan Coll., Bronx, N.Y.
For primary bibliographic entry see Field 05B.
W73-10108

EFFECT OF POLLUTION ON THE MARINE ENVIRONMENT, A CASE STUDY,
Florida State Univ., Tallahassee. Marine Lab.
For primary bibliographic entry see Field 05C.
W73-10110

NUMERICAL MODEL FOR THE PREDICTION OF TRANSIENT WATER QUALITY IN ESTUARY NETWORKS,
Massachusetts Inst. of Tech., Cambridge.
For primary bibliographic entry see Field 05B.
W73-10112

ESTUARIES AND INDUSTRY,
University of Strathclyde, Garelochhead (Scotland). Marine Lab.
For primary bibliographic entry see Field 05B.
W73-10133

STUDIES ON THE WATER POLLUTION AND PRODUCTIVITY IN KYONGGI BAY IN SUMMER SEASON (3): CLASSIFICATION OF THE PHYTOPLANKTON (III), (IN KOREAN),
Seoul National Univ. (Republic of Korea). Dept. of Botany.
For primary bibliographic entry see Field 05C.
W73-10159

EFFECTS OF SR90 (Y90), ZN65, AND CR51 ON THE LARVAE OF THE PACIFIC OYSTER CRASSOSTREA GIGAS,
Washington Univ., Seattle. Coll. of Fisheries.
For primary bibliographic entry see Field 05C.
W73-10172

BIOENVIRONMENTAL FEATURES: AN OVERVIEW.
National Marine Fisheries Service, Seattle, Wash.

STUDY OF THE OCEANOGRAPHIC CONDITIONS IN THE RIA DE AROSA, IN SUMMER, (IN SPANISH),
Spanish Inst. of Oceanography, Madrid.
J. Gomez Gallego.
Bol Inst Esp Oceanogr. 147, p 1-39, 1971. Illus.
Identifiers: Oceanographic studies, *Ria-De-Arosa, Salinity, *Spain, Summer, Water temperature, *Dissolved oxygen.

The Ria de Arosa is an estuary in northwestern Spain between the provinces of La Coruna and Pontevedra. The sources of the water which it deposits into Atlantic Ocean are rain water and the outflow of the rivers Ulla and Umia. The study comprises statistical findings of temperatures, salinity, dissolved O2, phosphates and the direction and intensity of currents. Numerous maps and charts provide the results of the investigation, conducted at various stations and at various times, and depths. The thermal gradient is very high and reaches 1C/m on occasions. The currents are relatively weak. No reference is made to biological subjects.--Copyright 1973, Biological Abstracts, Inc.
W73-10363

03. WATER SUPPLY AUGMENTATION AND CONSERVATION

3A. Saline Water Conversion

WATER TREATMENT BY MEMBRANE ULTRAFILTRATION,
North Carolina State Univ., Raleigh. School of Engineering.
For primary bibliographic entry see Field 05D.
W73-09758

HEMIHYDRATE SCALING THRESHOLD ENHANCEMENT BY MAGNESIUM ION AUGMENTATION,
California Univ., Los Angeles. School of Engineering and Applied Science.
K. S. Murdia, J. Glater, and J. W. McCutchan.
California Water Resources Center Desalination Report No. 49, April 1972. 109 p, 18 fig, 29 tab, 42 ref, 2 append. OSW Contract 14-30-2785.

Descriptors: *Desalination, *Sea water, *Desalination processes, *Descaling, Water Chemistry, Chemical reactions, Methodology, Calcium compounds, Sulfates, Mixing, Additives, Magnesium, Chlorides, Ion exchange, Equipment, California.

Development of low cost water by distillation technology is severely limited by the deposition of mineral scale. Calcium sulfate scaling thresholds in natural seawater can be markedly enhanced by enrichment with magnesium chloride. The effect of magnesium ion on calcium sulfate solubility was studied in natural seawater concentrates. Experiments were carried out at magnesium concentrations up to 3 times the ambient level, concentration factors between 1 and 4, and temperatures up to 340 F. Equilibrium solubilities of hemihydrate were measured in teflon-lined pressure vessels under nonboiling, nonevaporating conditions. The effect of salinity and magnesium ion concentration on hemihydrate-anhydrite phase transition was also studied. This work carried out at temperatures and concentrations normally encountered in high temperature evaporators is most relevant to the scaling problem. A process for recovery of magnesium chloride from distilling plant blowdown is proposed. Sample calculation of material balances for this system are presented at varying levels of magnesium ion, blowdown concentrations and recycle ratios. (Woodard-USGS)
W73-09789

METHOD OF IMPROVING THE SALT REJECTION OF SEMIPERMEABLE REVERSE OSMOSIS MEMBRANES,
Westinghouse Electric Corp., Pittsburgh, Pa. (assignee).
R. R. Stana.
U.S. Patent No 3,721,623, 4 p, 1 fig, 1 tab, 1 ref; Official Gazette of the United States Patent Office, Vol 908, Nos 3, p 692, March 20, 1973.

Descriptors: *Patents, *Desalination, Membranes, *Semipermeable membranes, *Reverse osmosis, Separation techniques, Sea water, Potable water, Detergents, Additives.
Identifiers: Sodium phosphate, Alkyl sulfonates, Alkylaryl sulfonates.

To improve the salt rejection capability of a semipermeable reverse osmosis membrane, the membrane is treated with sodium phosphate and alkyl sulfonates and/or alkylaryl sulfonates. The detergent additive may be added to the feed liquid. One example is cited and results are tabulated. (Sinha-OEIS)
W73-09927

FORWARD-OSMOSIS SOLVENT EXTRACTION,
W. T. Hough.
U.S. Patent No 3, 721, 621, 4 p, 3 ref; Official Gazette of the United States Patent Office, Vol 908, No. 3, p 692, March 20, 1973.

Descriptors: *Patents, *Desalination, *Osmosis, *Membranes, *Filtration, Sea water, Potable water, Separatin techniques, Hydrogen ion concentration.

Pressure is applied to a quantity of seawater to facilitate the forward osmosis of the sea water in passing through a membrane having pores of maximum diameter. The water moves into a solution of more concentrated removable solute having a pH which may render the removable solute insoluble. The insoluble solute is then filtered and removed allowing the palatable water to be removed. Either an osmotic membrane or a non-osmotic membrane may be used, as long as the membrane is characterized by its microscopic pores. (Sinha-OEIS)
W73-09930

WATER HEATING AND PURIFICATION SYSTEM,
Pollution Research and Control Corp., Glendale, Calif. (Assignee).
A. V. Sims.
US Patent No 3,718,544, 6 p, 6 fig, 10 ref; Official Gazette of the United States Patent Office, Vol 907, No 4, p 1033, February 27, 1973.

Descriptors: *Patents, *Water purification, *Distillation, *Condensation, *Heat exchangers, Water treatment.

An improved system is provided for supplying hot undistilled water together with distilled water employing a non-scale heat exchanger for condensation of water vapor. The apparatus comprises a vessel having an upper and lower portion. Water is heated in the lower portion. To minimize heat loss in the lower portion a coating or sheath of insulating material is sandwiched between the inner wall and the outer wall. The metal wall of the upper portion is not insulated to facilitate heat transfer. A glass or Teflon lining may be added to the wall in the upper portion to facilitate dropwise condensation while at the same time preventing corrosion. (Sinha-OEIS)
W73-09936

PARTICLE SEPARATOR AND METHOD,
For primary bibliographic entry see Field 05F.
W73-09937

DEVICE FOR RECLAIMING SWEET WATER FROM SEA WATER OR BRACKISH WATER,
Siemens A.G., Munich (West Germany). (Assignee).
K. Ruthrof.
US Patent No 3,717,554, 4 p, 3 fig, 9 ref; Official Gazette of the United States Patent Office, Vol 907, No 3, p 780, February 20, 1973.

Descriptors: *Patents, *Desalination, *Vaporization, *Centrifugation, Sea water, Brackish water, Potable water, Separation techniques, Water treatment, *Water purification.

Sea water or brackish water is heated and fed into a centrifugal force field so that vaporization, due to low pressure occurring in a condenser located downstream of the centrifugal force field, takes place at the inner surface of a water ring formed by the centrifugal force field. The device comprises an outer stationary chamber and an inner chamber which is rotationally symmetrical. The gaseous medium is produced by vaporization in the inner chamber by means of an axial discharge device. The unwanted residue accumulates in the space between the two chambers and is led out by a special outlet tube. (Sinha-OEIS)
W73-09941

INVESTMENT SEQUENCING RECOGNIZING EXTERNALITIES IN WATER DESALTING,
Massachusetts Univ., Amherst. Dept. of Agricultural and Food Economics.
C. E. Willis, and G. C. Rausser.
Water Resources Bulletin, Vol 9, No 1, p 54-72, February, 1973. 1 fig, 13 tab, 13 ref.

Descriptors: Water resources, *Investment, *Decision making, *Project planning, *Desalination plants, Construction, Timing, Size, Project feasibility, Project benefits, Estimating, Economics, *Water costs, Water conveyance, Computer programs, Equations, Mathematical models, Operations research, *California, Risks.
Identifiers: Learning functions, *Externalities, Bayesian procedures, Investment sequencing, Learning benefits.

The failure to recognize the learning process in new technologies such as desalting may lead to incorrect water resource investment decisions for two reasons. First, to neglect cost reductions stemming from 'learning by doing' implies an overestimation of desalting costs. Second, since learning in a particular plant may result in external (learning) benefits to other plants, these may serve as the basis for a subsidy intended to internalize such benefits. An investment decision model is presented in which these aspects are explicitly acknowledged. The criterion function for the model is net benefits (i.e., internal plus external benefits less costs); hence the components of the decision framework include an external benefits (subsidy) model, internal benefits measured in terms of an opportunity cost (importation of fresh water) which values water quality differences, and desalting costs. The research reported herein includes an estimation of learning functions for desalting and the results of a formulation designed to measure external benefits on the basis of these functions. These results are then incorporated into the decision framework for water resource investments which recognizes uncertainty in determining optimal timing of desalting construction; considered also are capacity and discount and growth rates. The region investigated comprises the San Luis Obispo and Santa Barbara Counties in California. (Bell-Cornell)
W73-10018

3B. Water Yield Improvement

PROCEEDINGS--SOUTHERN PLAINS REGION CLOUDTAP CONFERENCE.

Proceedings of Southern Plains Region Cloudtap Conference, October 16, 1972, Dallas, Texas: Texas Water Development Board, 1972. 165 p.

Descriptors: *Weather modification, *Cloud seeding, *Artificial precipitation, *Conferences, *Texas, Publications, Information retrieval, Great plains, Programs, Project purposes, Cloud physics, Precipitation (Atmospheric), Methodology, Aircraft, Silver iodide, Fertilizers, Radar, Data collections, Crop response, Correlation analysis, Economics, Costs, Governments.
Identifiers: *Symposium proceedings, Dallas (Tex).

The Texas Water Development Board, the Texas Water Conservation Association, and the Oklahoma State Department of Agriculture jointly arranged a symposium on weather modification projects and programs currently being carried out in Texas, Oklahoma, New Mexico, Kansas, and elsewhere. The Southern Plains Region Cloudtap Conference, which convened in Dallas, Texas, on October 16, 1972, had as its objective the bringing together of scientists involved in weather modification activities, and farmers, ranchers, businessmen, and public officials. Because of the wide interest in weather modification in the Southern Plains Region and the potential that cloud seeding may hold as a source of supplemental water supplies, the 11 papers presented at the symposium have been compiled and reproduced in this volume to make them available for distribution. The introduction made by the Honorable David Hall, Governor of Oklahoma, and comments made by participants of the symposium are also included. (See W73-09774 thru W73-09784) (Woodard-USGS)
W73-09773

RECLAMATION'S GREAT PLAINS WEATHER MODIFICATION PROGRAM--A STUTUS REPORT,
Bureau of Reclamation, Denver, Colo. Div. of Atmospheric Water Resources Management.
A. M. Kahan.
In: Proceedings of Southern Plains Region Cloudtap Conference, October 16, 1972, Dallas, Texas: Texas Water Development Board, p 5-24, 1972. 3 fig, 1 tab.

Descriptors: *Weather modification, *Cloud seeding, *Artificial precipitation, *Precipitation (Atmospheric), *Great plains, Aircraft, Federal government, Projects, Methodology, Snow, Cloud physics, Silver iodide, Salts, Ammonium salts, Nitrates, Ureas, Costs, Reviews, Evaluation, Conferences.
Identifiers: *Bureau of Reclamation, Skywater projects, Hygroscopic materials.

Weather modification is rapidly acquiring a transitional status from strictly research to full application. Under the Bureau of Reclamation's leadership, a program called Project Skywater has grown from a budget of $100,000 covering 3 contracts to an annual budget of $6,600,000 covering 49 contracts. From the beginning, attention was focused on the Great Plains Region. A table lists the contracts involved in the Great Plains Region and the completion dates. About 90% of Project Skywater funds are spent in the field by the participating contractors, many of whom are in private sector. A bureau staff of 25, including people experienced in weather modification, meteorology, other sciences, engineering, and administration, is centrally located in Denver to plan and direct the program. Hygroscopic seeding is a relatively new type of seeding being developed which promises to be very effective in modifying clouds unaffected by silver iodide. In hygroscopic seeding, such materials as common salt and a mixture of ammonium nitrate/urea may be used. They have the unique ability to absorb water vapor from the surrounding air. It is the Bureau of Reclamation's contention that scientifically sound local, state, and federal cooperative programs can be implemented to provide both the necessary learning and the immediate benefits of additional water. (See also W73-09773) (Woodard-USGS)
W73-09774

THE SOUTHERN PLAINS SKYWATER PROJECT,
Weather Science, Inc., Norman, Okla.

In: Proceedings of Southern Plains Region Cloudtap Conference, October 16, 1972, Dallas Texas: Texas Water Development Board, p 25-30, 1972. 1 fig.

Descriptors: *Weather modification, *Cloud seeding, *Artificial precipitation, *Precipitation (Atmospheric), *Southwest U.S., Oklahoma, Texas, Projects, Methodology, Silver iodide, Ammonium salts, Nitrates, Ureas, Fertilizers, Cloud physics, Aircraft, Correlation analysis, Reviews, Evaluation.
Identifiers: Southern Plains, Skywater project, Hygroscopic materials.

The Southern Plains Skywater Project is sponsored by the Division of Atmospheric Water Resources Management, Bureau of Reclamation, Denver, Colorado. The project was formed during the spring of 1972 and is planned to continue for three consecutive years. The project will develop better methods for increasing usable precipitation from cloud seeding. In addition, the project is designed to provide increased water supplies for the people in the target area as a co-product of the experimental program. The operational area includes nine counties of southwestern Oklahoma and three counties of the Texas Panhandle. Experimental cloud seeding is to be conducted primarily during the spring, summer, and fall when convective or cumulus cloud systems deliver most of the precipitation which falls annually in this area. Since the Skywater Project is developing operationally useful techniques, the constraints are similar to those which are commonly experienced by operational cloud seeders. This includes the use of applicable aircraft, radar and other equipment and a 24-hour, seven-day-per-week alert. (See also W73-09773) (Woodard-USGS)
W73-09775

AREAS OF NEEDED EMPHASIS TO BRING WEATHER MODIFICATION TO AN ACCOUNTABLE OPERATIONAL STATE,
Meteorology Research, Inc., Altadena, Calif.
J. R. Stinson.
In: Proceedings of Southern Plains Region Cloudtap Conference, October 16, 1972, Dallas, Texas: Texas Water Development Board, p 31-38, 1972.

Descriptors: *Weather modification, *Cloud seeding, *Precipitation (Atmospheric), *Programs, Decision making, Governments, Regulation, Methodology, Planning, Reviews, Evaluation.
Identifiers: *Weather modification accountability.

This paper outlines some of the areas of needed emphasis to make weather modification operations more accountable to: (1) public at large, (2) program sponsor, (3) scientific and engineering fields on which these operations are based, and (4) the environment in general. The basic point of view is that greater accountability is required of the regulators, sponsors, and operators. Those programs with inadequate conceptual design, concern for public safety, decision systems, engineering execution, evaluation, and environmental concern should be expected to be denied. The respon-

silver iodide at cloudbase in the inflow areas considered to be appropriate to those particular volumes of clouds which give birth to and allow growth of hailstones. The crop-hail insurance loss data indicate that the loss-cost values in the target county area of 1970 and 1971 were below average, and that most control counties had below normal annual loss-cost in these two years. Pertinent data are illustrated. (See also W73-09773) (Woodard-USGS)
W73-09779

SEEDING CLOUDS FROM THE GROUND IN OKLAHOMA,
Water Resources Development Corp., Palm Springs, Calif.
I. P. Krick.
In: Proceedings of Southern Plains Region Cloudtap Conference, October 16, 1972, Dallas, Texas: Texas Water Development Board, p 67-87, 1972. 10 fig, 1 plate, 1 tab, 17 ref.

Descriptors: *Weather modification, *Cloud seeding, *Artificial precipitation, *Oklahoma, Crop response, Cloud physics, Silver iodide, Rainfall, Meteorological data, Correlation analysis, Hydrologic data, Maps, Contours.
Identifiers: Silver iodide dispensing ground generator.

Oklahoma's Governor Hall established an executive committee for the study of weather modification in all of its aspects in July 1971. The work of this broad-based group, under the chairmanship of Mr. Ferdie Deering, editor and manager of the Farmer-Stockman, has been effective in producing far-reaching benefits to Oklahoma and awakening widespread interest in the subject throughout the nation. The programs described were first initiated in Oklahoma during April of 1972. A short and long range weather forecasting laboratory is maintained for the development of operational design and for calculating control data, so that dispersion patterns can be accurately computed continuously from ongoing surface and upper air information. This technique permits one to key the operational impact, either from single cloud formations or cloud bands associated with storm fronts, to specified target areas. Under generally similar airflow conditions and meteorological characteristics, it is possible to calculate the required ground-based generator sites to use in seeding a given target with silver iodide for optimum amounts of precipitation. With an output of one-half gram per hour, the optimum effects are achieved around 30 to 50 miles downwind from the generator in typical Oklahoma summer rainfall situations. Pertinent data are illustrated. (See also W73-09773) (Woodard-USGS)
W73-09780

A RANDOMIZED CLOUD SEEDING PROJECT IN NORTHERN NEW MEXICO,
New Mexico State Univ., University Park. Jemez Atmospheric Water Resources Research Project.
C. G. Keyes, Jr.
In: Proceedings of Southern Plains Region Cloudtap Conference, October 16, 1972, Dallas, Texas: Texas Water Development Board, p 89-100, 1972. 3 fig, 10 ref. AWRM 14-06-D-6903 and 14-06-D-6144.

Descriptors: *Weather modification, *Cloud seeding, *Artificial precipitation, *New Mexico, Silver iodide, Rainfall, Clouds, Water yield improvement, Meteorological data, Projects, Planning, Methodology, Precipitation (Atmospheric), Snowfall.
Identifiers: *Santa Fe area (N Mex), *Cuba area (N Mex), Silver iodide dispensing ground generator.

The Jemez Atmospheric Water Resources Research Project of New Mexico State University is developing a precipitation management system in northern New Mexico. Since the fall of 1968, randomized ground-based AgI seeding operations have been conducted during each winter season, November through March, from in or near the town of Cuba, New Mexico. The primary target area is in the Sierra Nacimiento Mountains just east of Cuba, and the secondary target area is located in the Sangre de Cristo Mountains just east of Santa Fe, New Mexico. The Mesoscale observation network and other supporting facilities operated during each winter season included: ground-based AgI generators; a network of 25 or more continuous recording Belfort precipitation gages in the immediate target area and 4 or more gages in the downwind area; upper-air measurements during seedable units; weather radar and ice nuclei counter equipment; and other support facilities, including radio communications, over-snow vehicles, a vehicle maintenance section, and a time-share computer system for real-time and post evaluation of the project. A few significant increases (and/or decreases) of precipitation have occurred during some of the seeding periods of the past four years. Pertinent data are illustrated. (See also W73-09773) (Woodard-USGS)
W73-09781

FEASIBILITY STUDIES OF PRECIPITATION ENHANCEMENT IN ILLINOIS,
Illinois State Water Survey, Urbana. Atmospheric Science Section.
S. A. Changnon, Jr.
In: Proceedings of Southern Plains Region Cloudtap Conference, October 16, 1972, Dallas, Texas: Texas Water Development Board, p 101-109, 1972. 3 fig.

Descriptors: *Weather modification, *Cloud seeding, *Artificial precipitation, *Illinois, Feasibility studies, Project purposes, Crop response, Methodology, Aircraft, Silver iodide, Clouds, Agriculture, Economics, Costs, Computer models, Meteorological data, Correlation analysis.

Illinois is considered by many to have adequate rainfall. However, studies show that in 9 out of 10 years crop yields would benefit from increased July-August rainfall. Weather modification benefits to agriculture and water supplies have potential in all parts of Illinois. However, the greatest economic benefits from weather modification would probably be realized in southern Illinois because of its thinner soils and greater dependence on surface water supplies. To know whether an economically beneficial rain enhancement program is feasible in Illinois, it must be determined if the clouds will respond to an artificial stimulus and how often suitable clouds may be expected. Answers about the clouds will be obtained by flying an instrumented airplane beneath and through clouds also observed by radar during the various seasons of the year. A computer will be used to simulate the clouds and cloud changes on the mesoscale. It will utilize measures obtained from radiosondes of temperature, relative humidity, and pressure. It would be premature to say that a cloud seeding program could be carried out within the state. This question will be resolved by the various scientific studies described. (See also W73-09773) (Woodard-USGS)
W73-09782

PROGRESS IN THE SOUTH DAKOTA WEATHER MODIFICATION PROGRAM,
South Dakota Weather Control Commission, Pierre.
M. C. Williams.
In: Proceedings of Southern Plains Region Cloudtap Conference, October 16, 1972, Dallas, Texas: Texas Water Development Board, p 111-126, 1972. 5 fig.

Descriptors: *Weather modification, *Cloud seeding, *Artificial precipitation, *South Dakota, Project purposes, State governments, Contracts, Data

collections, Precipitation (Atmospheric), Clouds, Aircraft, Radar, Silver iodide, Rain gages, Crop response, Conferences.
Identifiers: *Hail suppression.

Design, organization and initiation of a statewide program of weather modification was undertaken by the South Dakota Weather Control Commission in September 1971. The program was designed to provide for increased rainfall and decreased hail by cloud seeding. These efforts led to the development of a limited scale field program during the summer of 1972. State funding for the summer field activities was provided by the 1972 legislature in a $250,000 appropriation. Contracts with private operators for equipment and personnel were awarded in March in preparation for field activities during the period May 1 through August 31. Aircraft acquired under these contracts were of the turbocharged, light, twin-engine type, fully equipped for high altitude and instrument operations. Each aircraft was equipped with both acetone-silver iodide type generators with a capability of dispensing silver iodide at a rate of 150 to 300 grams per hour and pyrotechnic devices capable of dispensing 200 grams per minute. Thus, a wide range of capabilities was provided for two types of seeding materials to permit either rainfall increase or hail suppression activities. Seeding results for 1972 operations were not available in time for this report. (See also W73-09773) (Woodard-USGS)
W73-09783

GROMET II, RAINFALL AUGMENTATION IN THE PHILIPPINE ISLANDS,
Naval Weapons Center, China Lake, Calif.
P. St-Amand, D. W. Reed, T. L. Wright, and S. D. Elliott.
In: Proceedings of Southern Plains Region Cloudtap Conference, October 16, 1972, Dallas, Texas: Texas Water Development Board, p 127-160, 1972. 54 fig, 11 tab.

Descriptors: *Weather modification, *Cloud seeding, *Artificial precipitation, *Islands, Pacific Ocean, Crop response, Methodology, Aircraft, Silver iodide, Rainfall, Agriculture, Economics, Data collections, Costs, Evaluation, Governments.
Identifiers: *Philippine Islands.

A severe drought in the Philippine Islands during 1968 and 1969 led the Philippine Government to try cloud seeding as a means of rainfall augmentation. With the help of the United States, a silver iodide seeding project, GROMET II, was conducted over the entire archipelago from the end of April through mid-June 1969. Pyrotechnically generated silver iodide was released in updrafts in growing clouds, and through judicial placement and timing of seeding events individual clouds were organized into larger cloud systems. Rainfall estimated as at least 3 x 10 to the tenth power cubic meters of water fell from seeded clouds. Benefits derived from the project included marked improvement in the agriculture, increased sugar production amounting to 43 million U.S. dollars, and augmented crops of rice and corn sufficient to make anticipated importation unnecessary. In addition, local personnel were trained in seeding techniques. Because of the success of GROMET II the Government of the Philippines conducted a similar operation during 1970 and planned another for 1971. (See also W73-09773) (Woodard-USGS)
W73-09784

FIFTH ANNUAL SURVEY REPORT ON THE AIR WEATHER SERVICE, WEATHER MODIFICATION PROGRAM (FY 1972),
Air Weather Service, Scott AFB, Ill.
H. S.Appleman, T. S. Cress, R. I. Sax, and K. M. Weickmann.
Technical Report 249, December 1972. 16 p, 10 fig, 7 tab, 3 ref.

Descriptors: *Weather modification, *Cloud seeding, *Fog, *Ice, Methodology, Aircraft, Clouds, Projects, Costs, Propane.
Identifiers: *Ice-fog dispersal, Downdraft.

This annual report of the weather modification activities of the Air Weather Service covers the projects, their operation and results, undertaken during FY 1972. Its primary purpose is to inform AWS field personnel of the progress in weather modification made during the year. Details of the projects are only briefly discussed as such details are published elsewhere. COLD FLAKE, COLD WAND, COLD CLEAR, COOL VIEW, and the Elmendorf ground-based propane system are the activities covered. COLD WAND and COLD FLAKE were ground-based techniques at Fairchild AFB and Hahn AB, respectively, and COLD CLEAR was airborne technique in Alaska and Europe. In addition, a ground system was installed at Elmendorf AFB, Alaska, to replace airborne seeding there. Project COOL VIEW was an AWS field test of ice-fog dissipation techniques at Eielson AFB, Alaska. A brief description of procedures, equipment, and results of the test is given. (Woodard-USGS)
W73-09785

THE MOKOBULAAN RESEARCH CATCHMENTS,
U. W. Nanni.
S Afr For J. 78, p 5-13, 1971. Illus.
Identifiers: *Catchments, Flow, Forests, Hydrology, *Mokobulaan, Hydrologic project, *Rainfall, Stream.

The Mokobulaan Hydrologic Project, consisting of 3 small catchments, is described. Two catchments will be afforested with 2 different exotic tree species within the next few years while the third catchment will remain under its present management, conservative grazing by sheep. Monthly rainfall and streamflow data for 12 years are tabulated.--Copyright 1972, Biological Abstracts, Inc.
W73-09910

WEATHER MODIFICATION: AN ECONOMIC ALTERNATIVE FOR AUGMENTING WATER SUPPLIES,
South Dakota State Dakota State Univ., Brookings. Dept. of Economics.
R. K. Rudel, H. J. Stockwell, and R. G. Walsh.
Water Resources Bulletin, Vol 9, No 1, p 116-128, February 1973. 2 fig, 5 tab, 18 ref.

Descriptors: *Colorado River Basin, *Weather modification, *Cost-benefit analysis, Economic impact, Economic feasibility, Water supply, Water resources development, Cloud seeding, *Alternative planning, *Colorado.
Identifiers: Extra-market, *Water supply augmentation, San Juan Mountains (Colorado).

Evaluation of weather modification as a competitive or supplemental means of augmenting water supplies depends on quantitative estimates of benefits and costs, and the full consideration of extra-market values. Experimental results of cloud seeding in Southwestern Colorado suggest that runoff can be increased by 25 percent over a 3,300 mile area. The economic consequences in the Colorado River Basin are estimated using benefit-cost analysis. Results suggest that weather modification is an economically feasible means to provide additional water for the basin, and it appears to be one of the least cost alternatives compared with other proposed means of augmenting water supplies. A very low proportion of weather modification costs are fixed; thus the program is easily reversible. Also, a relatively small increase in daily precipitation covers the direct costs of operation. Benefits of water produced by weather modification include power production and irrigation of forage crops. In the long run, if water is used for higher valued fruit and vegetable production, or for domestic and industrial purposes, its value would rise sharply. Preliminary investigation of extra-market costs and benefits suggests that while they have little effect on the benefit-cost ratio, they may be very important to individuals and groups affected. (Bell-Cornell)
W73-10021

ICE NUCLEATION: ELEMENTAL IDENTIFICATION OF PARTICLES IN SNOW CRYSTALS,
National Oceanic and Atmospheric Administration, Boulder, Colo. Atmospheric Physics and Chemistry Lab.
F. P. Parungo, and R. F. Pueschel.
Science, Vol 180, No 4090, p 1057-1058, June 8, 1973. 2 fig, 4 ref.

Descriptors: *Ice, *Cloud seeding, *Electron microscopy, *X-ray analysis, *Nucleation, Silver iodide, Crystal growth, Chemistry of precipitation, Instrumentation.

Information about the elemental composition of atmospheric nuclei is necessary for a quantitative evaluation of deliberate and inadvertent weather modification. In seeding experiments on supercooled clouds, AgI is most often used as the seeding agent. Because the presence of silver in seeded snow can be due to nucleation, scavenging, or sedimentation, analyzing the silver concentration in the snow provides information about the spatial and temporal distribution of AgI in a particular seeding area. Only by examining individual AgI particles in relation to individual snow crystals and identifying AgI as the nucleus is it possible to make an objective analysis of seeding experiments. A scanning field-emission electron microscope combined with an x-ray analyzer may be used to locate the ice nucleus within a three-dimensional image of a snow crystal and determine the chemical composition of the nucleus. This method can also be used for elemental analysis of natural ice nuclei and condensation nuclei. (Knapp-USGS)
W73-10037

ROCK MULCH IS REDISCOVERED,
Arizona Univ., Tucson. Dept. of Soils, Water and Engineering.
W. H. Fuller.
Progressive Agriculture in Arizona, Vol 25, No 1, p 8-9, January-February 1973, 5 fig.

Descriptors: *Mulching, *Rocks, *Rock fill, *Soil stabilization, Slope stability, *Erosion control, Surfaces, Rainfall-runoff relationships, Arid lands, Landscaping.

An example is presented of modern application of the ancient practice of rock mulching to conserve soil moisture for plant growth and soil erosion control. Illustrations show how 10 years experimentation have stabilized the steep slope originally characterized by lack of vegetation and erodible soil. Advantages cited are: soil erosion control, water conservation, aesthetic value, inexpensive land surface, requires little care, weed control relatively easy, establishment of native plants easy, and economy of maintenance. Since a high proportion of world deserts become paved with stones and rocks naturally, and are subject to high intensity rainfall, this practice provides effective protection. (Paylore-Arizona)
W73-10212

3C. Use of Water of Impaired Quality

PRESENT AND FUTURE WATER AND SALT BALANCES OF SOUTHERN SEAS IN THE USSR (SOVREMENNYY I PERSPEKTIVNYY

VODNYY I SOLEVOY BALANS YUZHNYKH MOREY SSSR).
State Oceanographic Inst., Moscow (USSR).
For primary bibliographic entry see Field 02H.
W73-09807

DEVELOPMENT PROPOSAL FOR LAGOONS,
Rhode Island Univ., Kingston. Dept. of Economics.
H. C. Lampe.
Maritimes, p 13-14, May 1972. 1 fig.

Descriptors: *Lagoons, *Brackish water, *Management, Research and development, *Social impact.
Identifiers: *Food sources, *Community-environment interaction.

Management schemes for brackish water lagoons are discussed. Proper development of lagoons may provide additional sources of high protein foods for increasing human populations. Lagoon environments and ecosystems are not fully understood at this point. Any plans to develop them effectively will entail more basic research. Part of this research must focus on the needs, aspirations, abilities and activities of the people affected by lagoon development. An Institute of Lagoon Studies has been proposed to coordinate research efforts by such varied groups as economists, sociologists, biologists and engineers. (Ensign-PAI)
W73-09886

PROBLEMS OF GROUNDWATER EXPLORATION IN METAMORPHIC ROCKS FOR DOMESTIC WATER SUPPLIES IN NORTHEASTERN BRAZIL,
Bundesforschungsanstalt fuer Bodenforschung, Hanover (West Germany).
For primary bibliographic entry see Field 02F.
W73-09952

STUDIES ON THE RECLAMATION OF ACID SWAMP SOILS,
Ministry of Agriculture and Lands, Kuala Lumpur (Malaysia). Div. of Agriculture.
For primary bibliographic entry see Field 02G.
W73-10017

CULTIVATION OF BARLEY FOR EVALUATING DESALINATION RATE OBTAINED BY LEACHINGS BY CASPIAN WATER, (IN RUSSIAN),
I. S. Rabochev, and V. N. Kravchuk.
Probl Osvoeniya Pustyn'. 2, p 51-53. 1972. English summary.
Identifiers: *Barley, Caspian waters, Cultivation, Evaluation, Leaching, *Saline soils.

Field tests were undertaken to prove Caspian water effective for desalinization of solonchaks. Barley may be cultivated with salt concentrations of 27 g/l in the top meter of soil and a chloride concentration of 13.5 g/l. Barley appears to be of biological value in desalting.--Copyright 1973, Biological Abstracts, Inc.
W73-10075

A POSSIBLE NEW METHOD FOR IMPROVING PLANT GROWTH IN SALINE AND CALCAREOUS ENVIRONMENTS,
University Coll. of North Wales, Bangor. Dept. of Biochemistry and Soil Science.
R. G. Wyn Jones.
In: Proceedings of Symposium on the Use of Isotopes and Radiation in Research on Soil-Plant Relationships Including Applications in Forestry, Dec. 13-17, 1971, Vienna (IAEA-SM-151), International Atomic Energy Agency, Vienna, 1972, p 109-122. 4 fig, 6 tab, 30 ref.

Descriptors: *Saline soils, *Salt tolerance, *Corn (Field), *Ornamentals, Shrubs, Soil properties, Alkalinity, Plant growth regulators, Root systems, Plant growth, Radioecology, Sodium compounds.
Identifiers: Rubidium, Sodium radioisotopes.

Root application of choline chloride to crop plants and ornamental shrubs may prove useful in certain saline and calcareous environments. Possible causes of growth stimulation are: (1) increased root system, (2) root acid-production increasing nutrient availability as outlined, and (3) alteration in organic acid metabolism. Although growth was less than for low-salt controls, the shoot-growth inhibition during the first 24 hours was decreased with maize plants. It appeared that choline promoted tolerance only to the sodium ion and that the maximum effect was short-lived. In a small field trial on commercial rhododendron shrubs growing in a pH-7.5 alluvial soil, leaf chlorophyll contents increased three-to-four fold although the response of different varieties varied widely. (See also W72-10679) (Bopp-ORNL)
W73-10186

FIELD TRIALS ON IMPROVEMENT OF SALINE SOIL IN THE SIBARI PLAIN, (IN ITALIAN),
Bari Univ. (Italy).
V. V. Bianco, A. Patruno, and L. Cavazza.
Riv Agron. Vol 6, No 1, p 3-12, 1972. Illus. English summary.
Identifiers: Crops, Cultivation, Fertilizers, Field trials, *Italy (Sibari plain), Plain, *Saline soils, Soils.

Field trials on reclamation of a saline soil rich in Mg was attempted by deep ploughing, different crops, cultural practice and manure, S, and gypsum application. A negative relationship between salinity and pH was found; both were influenced by the amount of rainfall during the month before drawing samples. Soil reaction was weakly influenced by deep ploughing and rotation, but not in the same way every year. None of the techniques or materials used influenced the soil salinity. Probably the best way to improve this type of soil will be to lower and stabilize the water table that now is 150-200 cm deep. Among the plants used, sugar beet and alfalfa failed to grow and only sorghum and horse-bean gave good results. A weak positive correlation between barley yield and soil salinity was found. On wheat yields a negative action was shown by burying the barley straw and by gypsum application; good results were obtained by deep ploughing (50 cm) or green manure of horse-bean combined with shallow ploughing (25 cm). Yield of corn silage was enhanced by deep ploughing. Horse-bean yield was better with deep ploughing, manure and possible gypsum application.--Copyright 1972, Biological Abstracts, Inc.
W73-10382

3D. Conservation in Domestic and Municipal Use

ESTIMATING THE YIELD OF THE UPPER CARBONATE AQUIFER IN THE METROPOLITAN WINNIPEG AREA BY MEANS OF A DIGITAL MODEL,
For primary bibliographic entry see Field 02F.
W73-09950

3E. Conservation in Industry

WATER AND WASTEWATER MANAGEMENT IN DAIRY PROCESSING,
North Carolina State Univ., Raleigh. Dept. of Food Science.
For primary bibliographic entry see Field 05B.
W73-09755

THERMAL EFFECTS OF PROJECTED POWER GROWTH: SOUTH ATLANTIC AND EAST GULF COAST RIVER BASINS,
Hanford Engineering Development Lab., Richland, Wash.
For primary bibliographic entry see Field 05C.
W73-09848

SYSTEMS APPROACH TO POWER PLANNING,
Bechtel Corp., San Francisco, Calif. Hydro and Community Facilities Div.
For primary bibliographic entry see Field 06A.
W73-09859

A U.S. ENERGY POLICY NOW.
For primary bibliographic entry see Field 06D.
W73-09862

LOW GRADE HEAT FROM THERMAL ELECTRICITY PRODUCTION, QUANTITY, WORTH AND POSSIBLE UTILIZATION IN SWEDEN,
Swedish Board for Technical Development, Studvik (Sweden).
For primary bibliographic entry see Field 05G.
W73-09865

A FUNDAMENTAL STUDY OF THE SORPTION FROM TRICHLOROETHYLENE OF THREE DISPERSE DYES ON POLYESTER,
Auburn Univ., Ala. Dept. of Textile Engineering.
S. Perkins, and D. M. Hall.
Textile Research Journal, Vol 43, No 2, p 115-120, February 1973. 9 fig, 4 tab, 12 ref. OWRR A-020-ALA (1).

Descriptors: *Dyes, Analytical techniques, Enthalpy, Polarity, Solvents, *Textiles, *Sorption, Dye dispersion, Equilibrium, Films, Plastics, Polymers, Industrial production, Chemical industry.
Identifiers: *Polyester film, *Trichloroethylene, Dyeing rates, Diffusion coefficient, *Solvent processing.

Dyeing of polyester film from trichloroethylene with three highly purified disperse dyes is reported. Rate of sorption of each dye by polyester increases greatly with increasing temperature. Equilibrium sorption measurements indicate that two of the dyes, C.I. Disperse Red 15 and C.I. Disperse Orange 3, give constant partition coefficients. C.I. Disperse Orange 3 gives Freundlich type isotherms, possibly due to aggregation of dye in solution which increases with increasing concentration of dye in solution and with decreasing temperature. The tendency of dye to transfer from solution to fiber as expressed by the free energy of sorption increases with increasing polarity of the dye molecules. The enthalpy of dyeing is used to quantitatively express the effect of temperature on equilibrium sorption of dye by the substrate.
W73-10151

GEOTHERMAL WATER AND POWER RESOURCE EXPLORATION AND DEVELOPMENT FOR IDAHO,
Boise State Coll., Idaho. Dept. of Geology.
For primary bibliographic entry see Field 06B.
W73-10217

A STUDY OF THE ECONOMIC IMPACT ON THE STEEL INDUSTRY OF THE COSTS OF MEETING FEDERAL AIR AND WATER POLLUTION ABATEMENT REQUIREMENTS, PART 2.
Council on Environmental Quality, Washington, D.C.
For primary bibliographic entry see Field 05G.
W73-10312

FISH PROTEIN CONCENTRATE: A REVIEW OF PERTINENT LITERATURE WITH AN EMPHASIS ON PRODUCTION FROM FRESH-WATER FISH,
State Univ., Coll., Buffalo, N.Y. Great Lakes Lab.
E. B. Robson.
Special Reports No. 3 and No. 5, January and March, 1970. 188 p, 10 fig, 17 tab, 206 ref.

Descriptors: Freshwater fish, *Proteins, Technology, *Reviews, Bibliographies, *Fisheries.
Identifiers: *Fish protein concentrate, Fish flour, Manufacturing.

The state of the art of fish protein concentrate (FPC) technology is discussed. Such information is essential background for establishing an FPC factory. Topics discussed in the survey include (1) chemical and biological aspects of FPC, (2) practical and theoretical problems of FPC manufacture, (3) methods of FPC manufacture, (4) the possibility of producing FPC from non-hake-like fish, and (5) by-products of FPC manufacture. In addition, studies on the use of FPC in human nutrition are reviewed and a list of selected references is provided. Recent developments involved with the manufacture of FPC are summarized in a supplement. These include issues in regulation of the manufacture of FPC, socio-economic factors behind the use of FPC, and discussion of new technologies not previously mentioned. (Weaver-Wisconsin)
W73-10313

3F. Conservation in Agriculture

INVESTMENT PLANNING FOR IRRIGATION DEVELOPMENT PROJECTS,
Boston Univ., Mass.
T. S. Glickman, and S. V. Allison.
Socio-Economic Planning Sciences, Vol 7, No 2, p 113-122, April 1973. 19 tab, 3 ref, 1 appen.

Descriptors: *Irrigation, *Project planning, *Investment, *Planning, *Optimization, Agriculture, *Benefits, Capital costs, Cost-benefit analysis, Project planning, Dynamic programming, Mathematical models, Operations research, Groundwater, Replacement costs, Operation and maintenance, Drilling, Wells, Decision making, Dug wells, Water distribution (Applied), Pipes, Electric power.
Identifiers: *Tubewells, Developing countries, Bangladesh, Investment schedules, Diesal power.

This paper considers the most economical selection of available technologies which should be employed and the schedule which should be followed in implementing them for irrigation by tubewells to obtain an investment plan which maximizes the net agricultural benefits from a proposed project in a developing country. Cost and benefit relationships are derived and incorporated into a mathematical model which is solved using a modification of the dynamic programming procedure for solving the knapsack problem. The optimization model is intended to assist in planning groundwater development for irrigation by means of drilled wells (Tubewells), as opposed to dug wells. To generate optimal investment schedules within the range of available technologies, the model employs data from engineering/economic relationships representing conditions encountered in Bangladesh. Each technology is characterized by a particular combination of drilling method, component material, and discharge capacity. Costs attributed to the project are capital costs of well construction and costs of operation, maintenance and replacement during the projected 20-year economic lefe of the project. Project benefits are the increased agricultural revenues during this time horizon resulting from improvement in crop yields. The optimal schedule is seen to favor small capacity wells, drilled by indigenous methods, with supplementary water distribution systems. (Bell-Cornell)
W73-09831

EVAPOTRANSPIRATION OF DIFFERENT CROPS IN A MEDITERRANEAN SEMI-ARID REGION, (IN FRENCH),
Office de la Recherche Scientifique et Technique Outre-Mer, Abidjan (Ivory Coast). Centre d'Adiopodoume.
For primary bibliographic entry see Field 02D.
W73-09836

THE INFLUENCE OF NITROGEN LEACHING DURING WINTER ON THE NITROGEN SUPPLY FOR CEREALS, (IN GERMAN),
Kiel Univ. (West Germany). Institut fuer Pflanzenbau und Pflanzenzuechtung.
J. Koehnlein.
Landwirtsch Forsch. Vol 25, No 1, p 1-15. 1972. English summary.
Identifiers: *Cereals, *Leaching, *Nitrogen, Potato-D, Summer, Wheat-M, Winter, Soils.

The leaching of N dressings in autumn was studied for 4 yr in crop rotations on loamy Parabrownearth. The experimental crop was summer wheat after potatoes dressed with organic manure. The accessibility of rainfall during winter was controlled by using PE-Folies placed on the plots (30 m2) at different periods. The following plots were compared high N supply in autumn and normal winter rainfall; high N supply in autumn and reduced winter rainfall; high N supply in autumn without any rainfall during winter; high N supply in spring; reduced N supply in spring; without N supply and with normal rainfall; in 2 yr: no N supply and no rainfall. When the accessibility of rainfall during winter was stopped, the effect of a N-fertilizer dressing in autumn showed no difference from that given in spring. When the accessibility of the rainfall during winter was reduced, the effect of N-fertilizer dressed in autumn appeared later than that given in spring and also later than the N-fertilizer dressed in autumn without rainfall. The effect in the yield lies between the N-fertilizer given in spring and the N-fertilizer given in autumn with the whole rainfall during winter. The effect of N-fertilizer in autumn with normal rainfall fluctuates between 'no effect' and the effect obtained with reduced amount of rainfall during winter.--Copyright 1972, Biological Abstracts, Inc.
W73-09874

COMPLEX OF SOIL-DWELLERS IN SUGAR--BEET FIELDS ON FLOODLANDS, (IN RUSSIAN),
Rostov-on-Don State Univ. (USSR).
V. A. Minoranskii.
Zool Zh. Vol 51, No 2, p 300-303. 1972. English summary.
Identifiers: *Beet-D, *Floodlands, Pests (Soil), Phytophages, Predators, Saprophages, Soils, Sugar beets.

The complex of soil-dwellers of sugar beet fields located on floodlands is characterized by a higher density and variety as compared with that in terrace fields. In floodland fields the numbers of most of predators and saprophages are higher. Of phytophages, hygro- and mesophilous species predominate. Some of them may become dangerous pests. At the same time, some insects in floodland fields are of no agricultural importance in river valleys.--Copyright 1972, Biological Abstracts, Inc.
W73-09888

THE EFFECT OF TOTAL WATER SUPPLY AND OF FREQUENCY OF APPLICATION UPON LUCERNE: I. DRY MATTER PRODUCTION,
Reading Univ. (England). Dept. of Agricultural Botany.
R. W. Snaydon.
Aust J Agric Res. Vol 23, No 2, p 239-251. 1972. Illus.
Identifiers: *Dry matter, *Evaporation (Pan), Irrigation, *Lucerne-D, Medicago-Sativa-D, Water supply (Plants).

The total quantity of water supplied to an established lucerne (Medicago sativa) stand, and the frequency of water application, were varied independently. The total quantity of water supplied during summer, autumn, and early winter was expressed as a proportion of class A pan evaporation (E sub pan), ranged from 0.4 to 3.5 E sub pan. Dry matter production, at total water supply equal to E sub pan, varied between 8.6 kg per ha per day in early winter (May-June) and 97 kg per ha per day in summer (Jan.-Feb.), and was correlated (r ± 0.99) with mean minimum temperature. The form of the response curve of log sub e dry matter production to water supply, relative to E sub pan, was similar during all seasons. Dry matter production increased approximately logarithmically with increasing water supply up to the equivalent of 0.5 E sub pan, reached a maximum at about 1.5 E sub pan, and declined above this. A response surface was constructed, based upon total water supply and mean minimum temperature. Frequency of application, at a given total water supply, had no significant effect on dry matter production when total supply exceeded 0.5 E sub pan in any season. At a total water supply equal to 0.2 E sub pan, frequent small applications of water (5mm) produced 50% more dry matter than less frequent large applications (20 and 80 mm) during summer, but only 30% as much as the less frequent applications in the subsequent autumn. Water use efficiency (ratio of dry matter production to total water supplied) was greatest at 0.5 E sub pan at all frequencies of application in summer. During the 6 mo. after irrigation treatments ceased, dry matter production was generally related to the amount of available water in the soil, but production was less, even 18 mo. later, on plots that had previously received the heaviest irrigation (130 mm per month). Frequency of application had no significant effect upon subsequent yield, but frequent small applications in 1 summer reduced the capacity of lucerne to extract soil water at low water potentials in the following summer. (See also W73-09976)--Copyright 1972, Biological Abstracts, Inc.
W73-09975

THE EFFECT OF TOTAL WATER SUPPLY, AND OF FREQUENCY OF APPLICATION, UPON LUCERNE: II. CHEMICAL COMPOSITION,
Reading Univ. (England). Dept. of Agricultural Botany.
R. W. Snaydon.
Aust J Agric Res. Vol 23, No 2, p 253-256. 1972. Illus.
Identifiers: Chemical composition, Digestibility, Evaporation (Pan), *Lucerne-D, Medicago-Sativa-D, *Nitrogen, *Phosphorus, Water supply (Plants).

Total water supply, expressed as a proportion of class A pan evaporation (E sub pan), and frequency of water application were varied independently during summer. The P concentration in the shoots of lucerne (Medicago sativa) increased by 35% when the total water supply was increased from 0.1 to 1.0 E sub pan; the N concentration was not significantly affected. The in vitro digestibility of leaf and stem fractions decreased with increasing total water supply, and the proportion of highly digestible fractions (leaf and flower) also decreased, so that total shoot digestibility decreased from 65% at 0.24 E sub pan to 55% at 0.58 E sub pan. Frequency of water application had no significant effect upon P or N concentra-

tion or in vitro digestibility. (See also W73-09975)--Copyright 1972, Biological Abstracts, Inc.
W73-09976

INFLUENCE OF THE NUTRIENT SOLUTION CONCENTRATION UPON THE WATER ABSORPTION OF PEA PLANTS, (IN BULGARIAN),
Institut po Furazhite, Pleven (Bulgaria).
Iv Kalaidjiev, V. Petkov, and V. Radeva.
Rastenievd Nauki. Vol 9, No 3, p 3-10. Illus. 1972. English summary.
Identifiers: Absorption (Plants), *Plant nutrients, *Pea plants.

Pea plants absorb water at a different intensity during their development. The first water absorption maximum occurs at the mass bloom stage and the second and greatest one occurs at the filling and ripening of seeds. Plants raised in a nutrient solution of a higher concentration have a larger content of bound and a lesser content of free water than plants raised in a normal nutrient solution. The water potential of the cells increases with the advancement of their development. It is greater in plants raised in a nutrient solution of a higher concentration.--Copyright 1973, Biological Abstracts, Inc.
W73-09978

THE EFFECT OF DIFFERENT LEVELS OF WATER SUPPLY ON TRANSPIRATION COEFFICIENT AND PHOTOSYNTHETIC PIGMENTS OF COTTON PLANT,
National Research Centre, Cairo (Egypt). Plant Physiology Dept.
M. T. El-Saidi, and A. A. El-Mosallami.
Z Acker Pflanzenbau. Vol 135, No 2, p 135-142. 1972.
Identifiers: Carotenoids, Chlorophyll, *Cotton plants, *Photosynthesis, Pigments, *Transpiration coefficient, Water supply (Plants).

Different levels of soil moisture had large effect on the cotton yield, transpiration coefficient and productivity of transpiration and the photosynthetic pigments. This effect is related mainly to which phase of cotton development was exposed to drought or excess of water supply. Excess of water supply in different period of plant growth increased the amount of chlorophyll a, especially in the treatment with excess of water from seedlings to budding, in which chlorophyll a remained at a higher level than the control. Excess of water in the flowering stage decreased the amount of carotenoids. Water stress increased the total chlorophyll contents a + b in green leaves but not in yellow and brown ones. Productivity of transpiration decreased at water stress and excess of water supply, but the transpiration coefficient was high. This means that a larger amount of water was lost than in control.--Copyright 1973, Biological Abstracts, Inc.
W73-09979

NEW METHODS OF SOIL RECLAMATION IN ARID LANDS, (IN RUSSIAN),
I. S. Zonn.
Probl Osvoeniya Pustyn'. 1, p 46-55. 1972. Illus. English summary.
Identifiers: *Arid lands, Irrigation, *Literature surveys, Reviews, *Soil reclamation.

A survey of the world literature is given on soil reclamation in arid lands, new irrigation techniques and constructing controlled environment systems to grow agricultural plants.--Copyright 1973, Biological Abstracts, Inc.
W73-09981

PRELIMINARY RESULTS ON THE EFFECT OF POLYETHYLENE FOIL APPLIED AS A BARRIER LAYER, (IN GERMAN),
Rostock Univ. (East Germany). Institut fuer Meliorationswesen.
For primary bibliographic entry see Field 02D.
W73-09983

THE EFFECTIVITY OF NITROGEN FERTILIZATION UNDER DIFFERENT CONDITIONS OF PRODUCTION, (IN GERMAN),
Deutsche Akademie der Landwirtschaftswissenschaften zu Berlin, Leipzig (East Germany). Institut fuer Mineralduengung.
R. Jauert, H. Goerlitz, O. Hagemann, R. Breternitz, and H. Ansorge.
Arch Acker-Pflanzenbau Bodenkd. Vol 16, No 4/5, p 349-360. 1972. Illus. English summary.
Identifiers: Fertilization, Grass, Irrigation, Maize, *Nitrogen fertilizer, Oats, Potatoes, Rye, *Sprinkler irrigation, Sugarbeets.

On Tieflehm-Fahlerde (Sl 4D 33/30), longterm combination experiments with varied N fertilization and additional clean-water sprinkler irrigation of sugarbeet (O to 400 kg N/ha), oats (O to 200 kg N/ha), field grass (O to 700 kg N/ha), potatoes (O to 300 kg N/ha), winter rye (O to 250 kg N/ha) and ensilage maize (O to 300 kg N/ha) showed very good effects of combinations of fertilization and sprinkler irrigation at a high productivity per kg N. The economic effect was highest with sugarbeet and potatoes. With sugarbeet, oats, field grass, potatoes, winter wheat and ensilage maize that were grown on black earth (L 2 Lo 94/96) and dressed with the same amounts of fertilizer, the effectiveness of N fertilization was only slightly improved by additional sprinkler irrigation.--Copyright 1973, Biological Abstracts, Inc.
W73-09986

SAMPLING, HANDLING AND COMPUTER PROCESSING OF MICROMETEOROLOGICAL DATA FOR THE STUDY OF CROP ECOSYSTEMS,
For primary bibliographic entry see Field 07C.
W73-10001

THE DAILY COURSE OF STOMATA APERTURES IN OATS (AVENA SATIVA L.) OF DIFFERENT AGE OF THE PLANTS RESPECTIVELY LEAVES,
Leipzig Univ. (East Germany).
I. Frommhold.
Biochem Physiol Pflanz (BPP). Vol 163, No 2, p 216-224, 1972. Illus.
Identifiers: Age, Avena-Sativa, Carbon dioxide, Daily, Growth, *Leaves, *Oats, Oxide, Plants, *Stomata apertures.

During the growing of oat, the daily course of stomatal opening was measured at different stages of plant growth, and the opening capacity of the stomata in CO2-diminished air was checked. No direct correlation was found between the age of the plant leaf and the degrees of aperture of the stomata. Oats generally showed small apertures during day-time, but after application of CO2-diminished air the stomata opened wide. Relatively large openings seem to occur only during intensive growing of the leaves or plants.--Copyright 1973, Biological Abstracts, Inc.
W73-10066

WATER UPTAKE FROM FOLIAR-APPLIED DROPS AND ITS FURTHER DISTRIBUTION IN THE OAT LEAF,
Instituut voor Toepassing van Atoomenergie in de Landbouw, Wageningen (Netherlands).
D. H. Ketel, W. G. Dirkse, and A. Ringoet.
Acta Bot Neerl. Vol 21, No 2 p 155-166. 1972. Illus.
Identifiers: Calcium chloride, Cells, Cuticle, Ectodesmata, Foliar, Humidity, Mesophyll, *Oat-M, Radiography, Stomata, Xylem, *Leaves (Oat).

A fraction (12%) of the water from tritium-labeled foliar-applied drops rapidly penetrates through the cuticle and, after transport through the mesophyll, is distributed in an acropetal direction and probably through the xylem, over the oat leaf. The initial rapid penetration is favorably affected by a water 'continuum' that is always formed between tissue and drop, immediately after drop application. It is suggested that the pathway for the creation of this water continuum are the ectodesmata in the guard cells of the stomata. Increasing the CaCl2 concentration of the applied drops has little influence on the water uptake from these drops. Higher salt content of particular parts of the oat leaf increases the water retention by these parts. In the present experimental conditions (60-70% relative humidity) aqueous vapor uptake from the ambient air is small (1.7% of the amounts evaporated), as compared to the uptake from applied drops. Tritium from foliar-absorbed water does not combine with stable compounds (mainly organic) in the leaf.--Copyright 1972, Biological Abstracts, Inc.
W73-10071

ON THE ESTIMATION OF EVAPORATION AND CLIMATIC WATER BALANCE FOR PRACTICAL FARMING, (IN GERMAN),
U. Wendling.
Arch Acker-Pflanzenbau Bodenkd. Vol 16, No 3, p 199-205. 1972. Illus. English.
Identifiers: *Climatic data, Cover, Equation, *Evaporation, *Farming, Germany, Grass, Precipitation, Water balance.

A 7-yr series of measurements of the potential evapotranspiration of grass-covered areas with maximal water supply is compared with methods of evaporation estimation that are based on meteorological data. The method according to Klatt yields too high values for evaporation. The method according to Turc is most suitable. A simplified form of this equation of estimation is given, which is adapted to the climatic conditions in the German Democratic Republic (GDR). A map with weekly precipitation values and the potential evaportranspiration from the observations made at 35 meteorological stations of the GDR is presented.--Copyright 1973, Biological Abstracts, Inc.
W73-10072

AUTOMATIC WICK IRRIGATION OF POT EXPERIMENTS, (IN GERMAN),
Deutsche Akademie der Landwirschaftswissenschaften zu Berlin (East Germany). Institut fuer Pflanzenzuechtung.
H. Griess.
Arch Acker-Pflanzenbau Bodenkd. Vol 16, No 3, p 185-198. 1972. Illus. English summary.
Identifiers: *Irrigation methods, *Pot experiments, Wick, Automation.

Different materials, sizes and forms of wicks in pots of different sizes with sand or standard substratum, with or without plants and with different equipment for maintaining a constant water level were tested. Fiberglass wicks that were inserted in the pots from the bottom and whose lower end dipped into a channel with a constant water level, proved to be suitable for automatic irrigation in the optimal range of 60-80% moisture capacity. The desired value may be pricisely adjusted with a variation $< 5\%$ moisture capacity via the height between the lower bottom of the pot and the water level. This height depends on the kind of substratum and the height of the pot (height of the substratum) and must be determined once. The stability of the control system is so high that under normal experimental conditions the interfering item 'evapotranspiration' is almost completely balanced over several months. On an average of several pots deviations of less than or equal to plus or minus 5% moisture capacity and in single cases total deviations of less than or equal to plus or minus 10% moisture capacity are to be expected.

Thus wick irrigation is clearly superior to watering. However, wick irrigation is not recommended for the range < 50% moisture capacity. The horizontal uniformity of moisture in the pot was good, while the vertical uniformity was not. Shallow pots are recommended and technical details are described.--Copyright 1973, Biological Abstracts, Inc.
W73-10073

CONSIDERING THE SITE AND THE REACHED LEVEL OF FERTILIZATION WHEN CALCULATING THE OPTIMAL NITROGEN FERTILIZATION OF GRAIN CROPS IN THE MODEL FOR FERTILIZATION SYSTEMS, (IN GERMAN),
Deutsche Akademie der Landwirtschaftswissenschaften zu Berlin, Leipzig (East Germany). Institut fuer Mineraldueng.
O. Hagemann, H. Ansorge, R. Jauert, and H. Goerlitz.
Arch Ackerpflanzenbau Bodenkd. Vol 16, No 2, p 123-132. 1972. English summary.
Identifiers: Crops, *Fertilization, Grain, Irrigation, Models, *Nitrogen fertization, Nutrients, Optimization, Sites.

The optimal N fertilization rates of grain crops that were determined on the basis of optimization calculations, in dependence on the site (with and without sprinkler irrigation) are given. Information is given on N fertilizers of both 'unlimited' and 'limited' availability.--Copyright 1973, Biological Abstracts, Inc.
W73-10074

SUNFLOWER DROUGHT RESISTANCE AT VARIOUS GROWTH STAGES, (IN RUMANIAN),
Institutul de Cercetari Pentru Cereale si Plante Tehnice, Fundulea (Rumania).
L. Pirjol, C. I. Milica, and V. Vranceanu.
An Inst Cercet Pentru Cereale Plante Teh Fundulea Ser C Amelior Genet Fiziol Tehnol Agric. 37, p 191-208. 1969. Illus. English summary.
Identifiers: *Drought, Growth, Seed oil, *Sunflowers, Yield.

The experiments were carried out in a greenhouse, drought being induced for 10 days at different periods (vegetative growth, inflorescence growth, anthesis, seed formation). Drought induced at anthesis proved most detrimental and had irreversible effects, reducing plant metabolic activity and causing a significant yield decrease by forming a small number of flowers with normal development but only partial fertilization. When drought was induced at seed formation, a high number of flowers and seeds developed, but the filling process stagnated, giving high percentage of empty seeds and low seed oil content. This was due to the fact that after the drought a great part of the leaf area dried out. Drought induced during the vegetative growth period was the least detrimental to quantitative and qualitative yield, affecting only growth processes. The plants remained small in size and root, but with a big inflorescence 'SH 53' was more resistant to drought than 'Record'.--Copyright 1973, Biological Abstracts, Inc.
W73-10076

IRRIGATION EXPERIMENTS WITH OIL FLAX 1965-1970,
A. Nordestgaard.
Tidsskr Planteavl. Vol 75, No 6, p 788-792. 1971. English summary.
Identifiers: *Irrigation, *Oil flax, Seed, Straw, Yield, Drought.

Nine irrigation experiments with oil flax on loamy soil were conducted. The effect of drought during growth elongation showed a positive influence on seed yield. Irrigation during this period often influenced seed yield negatively, while the straw yield on the contrary is influenced positively. Drought during and after flowering damaged both seed and straw yield. If maximum seed yield is desired during this period oil flax should be irrigated.--Copyright 1973, Biological Abstracts, Inc.
W73-10077

THE DOWNY MILDEW OF ONION (PERONOSPORA DESTRUCTOR (BERK.) CASP.) IN ISRAEL,
Ministry of Agriculture, Tel-Aviv (Israel).
J. Palti, S. Brosh, M. Stettiner, and M. Zilkha.
Phytopathol Mediterr. Vol 11, No 1, p 30-36. 1972. Illus.
Identifiers: Age, Density, Humidity, Irrigation, *Israel, *Mildew, *Onions, Peronospora-destructor, Plants.

Factors conducive to epidemics of the onion downy mildew include those affecting humidity: stand density, direction of the rows and plant age. Sprinkler irrigation favored the disease more than furrow or trickle irrigation. The number and timing of sprinklings applied in spring to crops sown in autumn or winter greatly influenced the extent of mildew development and yield. On crops sown in Sept.-Oct. the mildew can be expected to appear in late Dec. or early Jan. in years with (a) mean minimum Nov. temperatures not exceeding 12 degrees C and at least 7 rainy days in Oct.-Nov. In fields sown in Nov. the mildew does not usually appear before Feb. Low temperatures in Dec.-Jan. (minima of 3-4 degrees C in the screen) delay development of the disease in winter.--Copyright 1973, Biological Abstracts, Inc.
W73-10083

STUDY OF THE WATER STATUS IN PEA STEMS BY THE METHOD OF SUPERHIGH--FREQUENCY DIELECTRIC MEASUREMENTS, (IN RUSSIAN),
N. N. Ishmukhametova, and A. V. Startseva.
Dokl Akad Nauk SSR Ser Biol. Vol 200, No 4, p 996-998. 1971.
Identifiers: Frequency, Measurements, *Pea stems, *Dielectric measurement.

The water status in pea stems was studied by measuring their dielectric characteristics at superhigh frequencies. Reproduction of the results was very good for specimens with a high water content. The value of the dielectric constant depended on the total water content in the tissue. When measuring the dielectric constant of stems experiencing dehydration, scattering of the data occurred.--Copyright 1973, Biological Abstracts, Inc.
W73-10085

TECHNICAL AND ECONOMIC FACTORS OF VARIOUS TECHNIQUES OF RANGELAND IRRIGATION, (IN RUSSIAN),
A. Ataev, and S. Kerimberdiev.
Probl Osvoeniya Pustyn'. 3, p 66-68. 1972. English summary.
Identifiers: Boreholes, Catchments, Distillation, Economic aspects, *Irrigation, *Rangelands, Techniques, Wells.

Dug wells, boreholes, artificial water catchment sites, water transportation by tankers and distillation installations are discussed and compared.--Copyright 1973, Biological Abstracts, Inc.
W73-10090

IRRIGATION TRIALS WITH ONIONS,
Horticultural Research Inst., Pretoria (South Africa).
F. J. Van Eeden, and J. Myburgh.
Agroplantae. Vol 3, No 4, p 57-62. 1971.
Identifiers: *Irrigation practices, *Onions-M, Plants, Spacing, Trials, Soils, Evaporation.

A good yield was obtained if water was applied whenever 60% of the available moisture in the 0-30 cm soil-zone had been depleted. Yield was depressed by allowing the soil moisture depletion to reach 90% of the available moisture during the second half of the growing season. The yield loss as a result of the dry regime was, to a certain extent, compensated for by using a narrow spacing of 22.9 cm (9 in.) between rows instead 38.1 cm (15 in.). A good yield/unit water applied was obtained with a close spacing combined with a dry regime. Irrigation scheduling according to evaporation from the Class A pan is possible under local irrigation practices. It is recommended that the following Esubt/Esubo-ratios (Et ± soil moisture depletion from top soil, Eo ± pan evaporation) be used for each consecutive 2-weekly period after transplanting: 0.3; 0.4; 0.5; 0.6 and 0.7 for the rest of the growing season.--Copyright 1972, Biological Abstracts, Inc.
W73-10139

METHODS OF ALFALFA IRRIGATION UNDER CONDITIONS OF HIGH GROUNDWATER TABLE, (IN BULGARIAN),
Academy of Agricultural Sciences, Pazardzhik (Bulgaria). Experiment Station of Irrigated Agriculture.
L. Delchev.
Rasteniev'd Nauki. Vol 9 No 1, p 51-61. 1972. Illus.
Identifiers: *Alfalfa, Groundwater, *Irrigation methods, Water table.

Watering by gravity in strips in a soil with an 80-100 cm deep active soil layer, sprinkling in the same type of soil, and sprinkling in soil with a 50 cm deep soil level were compared with a non-irrigated culture of alfalfa. Irrigation considerably increased growth. The first variant of the experiment with gravity flow seemed best.--Copyright 1973, Biological Abstracts, Inc.
W73-10169

TECHNICAL AND ECONOMICAL EFFICIENCY OF FORAGE PLANT CULTIVATION IN KAZAKHSTAN IRRIGATED DESERT RANGES (IN RUSSIAN),
N. V. Danil'chenko.
Probl Osvoeniya Pustyn'. 1, p 56-61. 1972. English summary.
Identifiers: Alfalfa, Corn, Cultivation, Deserts, Economics, *Fertilizers, *Forage plants, Grass, Irrigation, *Kazakhstan, Plant, Ranges, Silage, Sudan.

Under desert conditions 100-120 c/ha of alfalfa dry hay, 400-500 c/ha of corn silo and 90-100 c/ha of sudan can be obtained from an irrigated hectare treated with manure (20-30 ton/ha) and mineral nutrients (5-6 c/ha). The presowing moisture content of the root layer must not be less than 50-60% of saturation.--Copyright 1973, Biological Abstracts, Inc.
W73-10170

INFLUENCE OF THE GROWING NITROGEN DOSES ON THE YIELDS OF MIXTURES OF LUCERNE WITH GRASSES AS WELL AS ON THOSE OF LUCERNE AND GRASSES IN PURE CULTURES, (IN POLISH),
Instytut Uprawny Nowozenia i Gleboznawstwa, Baborow (Poland).
H. Burczyk, W. Cwojdzinski, and E. Naglik.
Pamiet Pulawski. 52, p 159-181. 1972. Illus. English summary.
Identifiers: Brome grass, Cocksfoot, Cultures, *Grasses, Growing, *Lucerne, Mixtures, *Nitrogen fertilizers, Nutrients, Oat grass, Rainfall, Yield, Yields.

Three series of experiments on the fertilizing of mixtures of lucerne with grasses as well as of grasses (cocksfoot, awnless brome and oat grass) and lucerne 'Miechowska' in pure culture were

carried out. The following N doses were applied: 0, 120, 240, 360 and 480 kg/ha on the constant background of P (64 P2O5 kg/ha and K (160 K2O kg/ha). The nutrient uptake was greater in wet years than in the dry ones. When rainfall during the vegetation period was sufficient, the yield of grasses intensively fertilized with N was equal to that of the mixture. Under conditions of seasonal drought the yields of the mixture were higher.--Copyright 1973, Biological Abstracts, Inc.
W73-10171

THE USE OF NUCLEAR TECHNIQUES IN SOIL, AGROCHEMICAL AND RECLAMATION STUDIES. (PRIMENENIE METODOV ATOMNOI TEKHNIKI V POCHVENNO-AGROKHIMICHESKIKH I MELIORATIVNIKH ISSLEDOVANIYAKH),
Moskovskaya Selskokhozyaistvennaya Adademiya (USSR).
For primary bibliographic entry see Field 02G.
W73-10187

MONITORING THE WATER UTILIZATION EFFICIENCY OF VARIOUS IRRIGATED AND NON-IRRIGATED CROPS, USING NEUTRON MOISTURE PROBES (CONTROLE DE L'UTILISATION ET DE L'EFFICIENCE DE L'EAU CHEZ DIVERSES CULTURES IRRIGUEES OU NON, AU MOYEN D'HUMIDIMETRES A NEUTRONS),
C. Maertens, and J.-R. Marty.
In: Proceedings of Symposium on the Use of Isotopes and Radiation in Research on Soil-Plant Relationships Including Application in Forestry, Dec 13-17, 1971, Vienna (IAEA-SM-151), International Atomic Energy Agency, Vienna, 1972. p 621-630. 6 fig, 3 ref.

Descriptors: *Soil moisture meters, *Radioactivity techniques, *Water requirements, *Agronomic crops, Irrigation, Crop response, Corn (Field), Sorghum, Fescues, Alfalfa, Soybeans, Clovers.

Efficiency of water utilization (quantity of dry matter produced/volume of water consumed) was measured for various crops (meadow fescue, Italian rye grass, alfalfa, corn, sorghum, soya, and sunflower) in southwestern France, using a neutron moisture meter to determine soil moisture in-situ. The efficiency decreased in summer as compared with spring in connection with increased transpiration in the instance of rye-red clover but was unchanged for alfalfa. For fescue meadows, the efficiency was increased by irrigations; but when transpiration exceeded 3 mm/day, the efficiency dropped with increasing deficiency of climatic moisture. The summertime efficiency was better for corn and sorghum than for fescue meadows. The soil moisture available to certain plants was decreased by the relatively shallow root structure resulting when these plants were irrigated. For corn the availability of soil moisture was decreased 25-30% throughout the summer months; for sorghum, sunflower, and soya, the reduction was 50% for early September. (See also W72-10679) (Bopp-ORNL)
W73-10195

OPTIMISATION OF SAMPLING FOR DETERMINING WATER BALANCES IN LAND UNDER CROPS BY MEANS OF NEUTRON MOISTURE METERS (OPTIMISATION DE L'ECHANTILLONNAGE POUR DETERMINER LES BILANS HYDRIQUES SOUS CULTURE AU MOYEN DE L'HUMIDIMETRE A NEUTRONS),
Institut Technique des Cereales et Fourrages, Paris (France).
P. Peyremorte, G. Philippeau, and J. Marcesse.
In: Proceedings of Symposium on the Use of Isotopes and Radiation in Research on Soil-Plant Relationships Including Applications in Forestry, Dec 13-17, 1971, Vienna (IAEA-SM-151), International Atomic Energy Agency, Vienna, 1972, p 631-647. 6 tab.

Descriptors: *Radioactivity techniques, *Agronomic crops, *Soil moisture meters, *Water requirements, Underground, Water balance, Statistical methods, Variability, Reliability, On-site tests, Instrumentation, Soil analysis, Optimization, Non-destructive tests, Testing procedures, Cultivation, Alfalfa, Sorghum, Corn (Field), Soil water.

Work at the Greoux agricultural research station showed the importance of a representative sampling to account for the heterogeneous nature of the soil-water-plant medium. Optimization of equipment use was achieved, reducing the counting time from about 1 minute to 20 seconds and increasing the number of measuring points. Horizontal implantation of the tubes to provide access of the neutron meter to underground locations was advantageous as compared to vertical emplacement since less interference with cultivation resulted and since the installation labor was reduced 50%. Tube lengths of tens of meters may be used to traverse several experimental plots or for large plots such as vineyards or orchards. A precision within 10 mm moisture with 95% probability resulted using 3 horizontal tubes at 20, 60, and 100 cm depths. The use of vertically emplaced tubes required 12 to achieve the same precision. Analyses of variance are given for water balances for alfalfa, sorghum, corn, and sunflower crops. (See also W72-10679) (Bopp-ORNL)
W73-10196

GROUND MOISTURE MEASUREMENT IN AGRICULTURE (BODENFEUCHTEMESSUNG IN DER LANDWIRTSCHAFT),
European Atomic Energy Community, Brussels (Belgium). Bureau Eurisotop.
H. Dreiheller.
Available from NTIS, Springfield, Va., as EURISOTOP-67; $4.00 in paper copy, $0.95 in microfiche. Report EURISOTOP-67, May 1972, 32 p, 15 fig, 10 ref.

Descriptors: *Soil moisture meters, *Neutron absorption, *Radioactivity techniques, *Irrigation, Water table, Soil-water-plant relationships, Instrumentation, Gamma rays, Root zone, Water requirements, Analytical techniques, Reviews, Research and development.

Advantages are indicated of radiometric techniques over classical methods in irrigation studies for measurement of (1) average soil moisture over 0-60 cm depth, (2) groundwater at greater depth, and (3) soil moisture at the depth of the root zone of various plants. Neutron meters are generally preferable to density measurements using gamma radiation and are suitable for automatic recording and control. Special designs give resolution in the depth ranges of plant root zones. An equilibration time that is required by other methods is not involved in the neutron-meter measurement of water table, and a separate accounting of the soil porosity is not required. (Bopp-ORNL)
W73-10209

REMOTE CONTROL OF IRRIGATION SYSTEMS,
Akademiya Nauk Kirgizskoi SSR, Frunze. Institut Automatiki i Telemekhaniki.
V. I. Kurotchenko, and N. I. Babanin.
Available from the National Technical Information Service as TT N-53085 , $3.00 in paper copy, $0.95 in microfiche. Translation from Russian prepared for the National Science Foundation and Dept. of Interior, Bureau of Reclamation, by Dr. Ismail Saad. 1971. 164 p, 58 fig, 13 tab, 73 ref.

Descriptors: *Irrigation systems, *Irrigation engineering, Sprinkler irrigation, Irrigation operation and maintenance, *Mathematical models, Analytical techniques, *Telemetry, *Remote control, Automatic control, Installation, Equipment, Instrumentation.
Identifiers: U.S.S.R.

A pulse-code device, labelled BKT-62, for telemetering technological parameters in irrigation systems was constructed in 1962. In addition to metering the levels, it also positions the sluices at irrigation headworks. Fifteen time channels are required for the transmission of telemechanical information, achieved by means of a selected 2-line cable and using step-by-step synchronization. Experimental tests with sprinklers over a 20-month period to 1965 were conducted under different conditions, with the system functioning satisfactorily in the execution of over 50 million measurements. The work was not affected by fluctuations either in the supply lines or in temperature variations from -20 degrees C to more than 50 degrees C. This detailed description of the complicated components of the system is illustrated with numerous figures, equations, and specifications. (Paylore-Arizona)
W73-10214

INFLUENCE OF TEMPERATURE AND PRECIPITATION ON THE DEVELOPMENT OF GROWTH PHASES OF MILLET (PANICUM MILIACEUM L.), (IN CZECH),
Vyzkumny Ustav Rastlinnej Vyroby, Piestanoch (Czechoslovakia).
J. Sinska.
Ved Pr Vysk Ustavu Rastlinnej Vyroby Piestanoch. 9. p 35-42. 1971. English summary.
Identifiers: Growth, *Millet (Growth), Panicum-Miliaceum, *Precipitation (Atmospheric), *Temperature.

Cool dry weather lengthened the period from emergence to flowering. Cultivars with a longer vegetation period were more affected. Higher than average temperature and high precipitation lengthened the period from emergence to flowering, but the length of the vegetation period as a whole was considerably shorter than for cool, dry years.--Copyright 1973, Biological Abstracts, Inc.
W73-10224

EFFECT OF WATER SUPPLY ON THE DAILY WATER CONSUMPTION OF COTTON OF THE SPECIES GOSSYPIUM HIRSUTUM L. AND GOSSYPIUM BARBADENSE L. (IN RUSSIAN),
Akademiya Nauk Uzbekskoi SSR, Tashkent, Inst. of Experimental Plant Biology.
Kh. S. Samiev, and O. T. Kurbaev.
Uzb Biol Zh. Vol 15, No 6, p 14-16, 1971.
Identifiers: *Cotton, Daily, Gossypium-Barbadense, Gossypium-Hirsutum, Species, *Water uptake (Plants), Biomass.

Irrespective of the water supply Upland cotton (G. hirsutum) consumes more water daily than Sea Island cotton (G. barbadense) up to the end of flowering. The biomass accumulates more intensely in the former species than in the latter up to the end of flowering. Thereafter daily water consumption is higher in G. barbadense than in G. hirsutum.--Copyright 1973, Biological Abstracts, Inc.
W73-10229

PHOTOSYNTHESIS PRODUCTIVITY OF THE COTTON PLANT, VARIETY 152-F, DEPENDING ON THE LEVEL OF WATER SUPPLY, (IN RUSSIAN),
Akademiya Nauk Uzbekskoi SSR, Tashkent. Inst. of Experimental Plant Biology.
Kh. S. Samiev, E. E. Uzenbaev, and U. S. Sidikov.
Uzb Biol Zh. Vol 15, No 5, p 17-19. 1971.
Identifiers: *Cotton, Cultivars, Moisture, *Photosynthesis, Plant productivity, Soils, Water supply, Irrigation moisture.

In serozem soils with a low groundwater level, best productivity was obtained when irrigation moisture was maintained at a level not lower than 60% of the moisture capacity of the field.--Copyright Biological Abstracts, Inc.
W73-10230

THE EFFECT OF HERBICIDE ON THE DEVELOPMENT OF WHEAT VARIETIES AND THEIR CLIMATIC ECOLOGICAL BEHAVIOR, (IN HUNGARIAN),
University of Agriculture, Debrecen (Hungary). Novenytani Novenyelettani Tansz.
G. Mandy.
Bot Kozl. Vol 58, No 4, p 221-227. 1971. Illus. English summary.
Identifiers: Climatic conditions, Ecological studies, Flowering, *Herbicides, Metoxuron, Ripening, Straw, *Wheat cultivers, *Doseanex.

The effect of the herbicide Dosanex (metoxuron) on 3 winter wheat cultivars, (Bezostaya 1., Fertodi 293., and Multibraun) under extreme weather conditions was studied. Dosanex eradicated the wheat population under unfavorable weather conditions and had a strong effect on the ecological capacity for adaptation. The Dosanex treatment increased the phenological phenomena causing an uneven ripening. The herbicide also decreased the length of straw by 5-19%.--Copyright 1973, Biological Abstracts, Inc.
W73-10244

WESTLANDS WATER DISTRICT.
For primary bibliographic entry see Field 06E.
W73-10274

CULTIVATION OF CARP IN RICE PADDIES OF WESTERN FERGANA AND THE EFFECT OF FISH ON RICE YIELD, (IN RUSSIAN),
E. A. Tashpulatov.
Uzb Biol Zh. Vol 15, No 6, p 62-63, 1971.
Identifiers: *Carp cultivation, Cyprinus-carpio, *Fergana, Fish, Paddies, *Rice paddies, Stocking, USSR, Yield.

Experiments demonstrated that one of the ways to increase the yield of rice is to grow yearling carp (Cyprinus carpio) in the paddies at a stocking rate of 500-550 fishes/ha.--Copyright 1973, Biological Abstracts, Inc.
W73-10359

ON THE POSSIBILITES OF GROWING TOMATO PLANTS FOR INDUSTRIAL PURPOSES IN ANGOLA: A GRAPHIC ANALYSIS, (IN PORTUGUESE),
P. Martins, and A. Carlos.
Rev Agron (Lisb). Vol 53, No 1, p 11-28, 1970. Illus. English summary.
Identifiers: *Angola, Climates, *Graphic analysis, Plants, Rainfall, Temperature, *Tomato.

Several meteorological stations of Angola were studied in order to determine if the weather would be favorable for growing processing tomatoes. The following climatic conditions are suitable for successful growth: 2 or 3 dry months average monthly maximum temperature under 30C and average monthly minimum temperature over 12C. Vila Arriaga appears to have favorable conditions, in Vila Pereira d'Eca the dry season is too cold, in Luanda, Carmona and Cela low sunlight and blight diseases are problems, in Cela, Nova Lisboa, Luso e Sa' da Bandeira the dry season is too short and cold and in Forte Rocadas it may be too cold at times.--Copyright Biological Abstracts, Inc.
W73-10362

CORRELATIONS BETWEEN NITROGEN FERTILIZATION AND SPRINKLER IRRIGATION WITH FORAGE CROPS OF SEVERAL CUTS GROWN ON HEAVY SOILS, (IN GERMAN),
D. Roth.
Arch Acker-Pflanzenbau Bodenkd. Vol 16, No 2, p 149-159, 1972. Illus. English summary.
Identifiers: Clay, Crops, *Fertilization, Forage, Grasses, Irrigation, Legumes, *Nitrogen fertilizer, Soils, *Sprinkler irrigation.

Combined N fertilization and sprinkler irrigation experiments with intensively used grassland, legume-grass-mixtures and field grass were carried out on 3 sites with heavy soils (>30% clay). Increasing rates of N application resulted in substantial yield increase with all the tested forage crops. In case of pure field grass, sprinkler irrigation led to considerably higher surplus yields than with legume-grass-mixtures and grassland. A positive correlation between additional water supply and N fertilization was found with pure grass stands only.--Copyright 1973, Biological Abstracts, Inc.
W73-10369

PATCHES IN CULTIVATED SPECIES IN THE CENTRAL REGION OF SANTA FE PROVINCE. I. SOIL-CLIMATIC FACTORS, (IN SPANISH),
J. L. Panigatti, A. Pineiro, and F. P. Mosconi.
Rev Invest Agropecu Ser 3 Clima Suelo. Vol 8, No 4, p 141-154, 1971. Illus. English summary.
Identifiers: *Argentina (Santa Fe), Crops, Patches, Species, Yield, *Water balance, *Soil-climatic factors (Crops).

Water balance in the area, and the influence that external (climate) and internal (soil) factors have in water economy and crop development were important. Those factors may produce patches in the crops with significant differences in quality and quantity in the crop yield.--Copyright 1973, Biological Abstracts, Inc.
W73-10383

INFLUENCE OF ATMOSPHERIC PRECIPITATION, MEAN TEMPERATURE AND RELATIVE HUMIDITY OF AIR ON CROP OF GRAIN, STRAW AND ON TECHNOLOGICAL VALUE OF SPRING BARLEY, (IN CZECH),
Vyzkumny Ustav Raslinnej Vyrbby, Piestanoch (Czechoslovakia).
J. Kandera.
Ved Pr Vysk Ustavu Rastlinnej Vyroby Piestanoch. No 9, p 147-154, 1971. English summary.
Identifiers: Crops, Grain, Humidity, Mean temperature, *Precipitation (Atmospheric), *Relative humidity, Spring barley, Straw, Temperature, *Fertilization.

The influence of increased rates of N together with normal fertilization with P2O5 and K2O was investigated over a 6 yr period using 'Slovensky 802' introduced into the crop rotation after sugarbeet fertilized by manure on soils of degraded czernozem. The results of 12 combinations of fertilizing were evaluated as a whole in the relation to atmospheric precipitations, mean temperatures and to the relative humidity of the air during the vegetation period by computing correlation coefficients. Positive, significant correlations were seen between the atmospheric precipitation and the waste (size of the grains under 2.2 mm), between mean temperature in deg.C and grain to straw ratio, the hectolitre weight of the grain, the absolute weight of the grain, the size of grain with a diameter of 2.5 mm, the starch content and the malt extract. A significant positive correlation was also seen between relative air humidity and absolute weight of the grain, the protein and starch content.--Copyright 1973, Biological Abstracts, Inc.
W73-10388

04. WATER QUANTITY MANAGEMENT AND CONTROL

4A. Control of Water on the Surface

THE HYDROLOGY OF FOUR SMALL SNOW ZONE WATERSHEDS,
Forest Service (USDA), Berkeley, Calif. Pacific Southwest Forest and Range Experiment Station.
For primary bibliographic entry see Field 02A.
W73-09765

A NEW METHOD OF DETECTING LEAKS IN RESERVOIRS OR CANALS USING LABELLED BITUMEN EMULSIONS,
Centre d'Etude de l'Energie Nucleaire, Grenoble (France).
For primary bibliographic entry see Field 08G.
W73-09770

HYDROLOGIC IMPLICATIONS OF THE SAN ANGELO CUMULUS PROJECT,
Texas A and M Univ., College Station. Dept. of Meteorology.
For primary bibliographic entry see Field 02A.
W73-09777

SOME PROBLEMS IN THE CONSTRUCTION OF RESERVOIRS IN THE USSR (NEKOTORYYE PROBLEMY SOZDANIYA VODOKHRANILISHCH V SSSR),
Akademiya Nauk SSSR, Moscow. Institut Vodnykh Problem.
For primary bibliographic entry see Field 06G.
W73-09803

PRESENT AND FUTURE WATER AND SALT BALANCES OF SOUTHERN SEAS IN THE USSR (SOVREMENNYY I PERSPEKTIVNYY VODNYY I SOLEVOY BALANS YUZHNYKH MOREY SSSR).
State Oceanographic Inst., Moscow (USSR).
For primary bibliographic entry see Field 02H.
W73-09807

SHORT-RANGE FORECAST OF MAXIMUM FLOOD DISCHARGES ON THE KYZYLSU RIVER AND ITS TRIBUTARIES (KRATKOSROCHNYY PROGNOZ MAKSIMAL'NYKH RASKHODOV PAVODKOV NA R. KYZYLSU I YEYE PRITOKAKH),
Yu. I. Lisitsyn.
Meteorologiya i Gidrologiya, No 10, p 100-101, October 1972. 3 ref.

Descriptors: *Flood forecasting, *Flood discharge, *Peak discharge, *Rivers, *Tributaries, Air masses, Meteorology, Synoptic analysis, Orography, Probability, Equations.
Identifiers: *USSR, *Tadzhikistan, *Kyzylsu River.

A method is described for preparing a short-range forecast of maximum flood discharges on the Kyzylsu River in southern Tadzhikistan. The basis for the forecast is the aerosynoptic method, which considers the characteristics of air masses and the wetting indices of the river basin. The probability of achieving the allowable forecast error is 78%. (Josefson-USGS)
W73-09809

INTERNATIONAL HYDROLOGICAL DECADE (MEZHDUNARODNOYE GIDROLOGICHESKOYE DESYATILETIYE),
For primary bibliographic entry see Field 02E.
W73-09810

systems. The case study approach, applied to a prospective system to serve the needs of several urbanizing areas in the Piedmont of North Carolina is employed to demonstrate the practicality of methods developed. In the mathematical modeling stage, problem reduction techniques are considered and a preliminary set of rational simplification theories are constructed. Then, the problem of finding appropriate decisions that minimize current and expected future losses to economic efficiency within each month is formulated as a convex piecewise-linear programming problem. That mathematical problem contains an unknown function of end-of-month reservoir levels that yields the expected future losses. To estimate that unknown function, the linear programming is coupled with the backward algorithm of dynamic programming; the resultant problem is referred to as a DCL problem. Two computational procedures for obtaining approximate solutions based on the DCL technique are applied to the resource system under investigation, and their theoretical and computational soundness is tested by simulation. The DCL technique is concluded to be practical for realistic systems and to be satisfactory in achieving the objectives of the dissertation. (Bell-Cornell)
W73-09835

CONTROL OF THE OPERATING REGIMES OF WATER RESOURCES SYSTEMS,
A. Sh. Reznikovskii.
Hydrotechnical Construction, No 1, p 56-61, January, 1972. 7 p, 11 ref. Translated from Gidrotekhnicheskoe Stroitel'stvo, No 1, p 34-37, January, 1972.

Descriptors: *Water management (Applied), *Multiple-purpose reservoirs, *Operation and maintenance, *Methodology, *Hydroelectric plants, River basins, Optimum development plans, Comprehensive planning, Hydraulics, Economics, Reservoir releases, Streamflow, Water consumption, Equations.
Identifiers: *Optimal control rules, Probabilistic method, Distribution curves, *USSR.

Formation and development of large water resources systems characterize a relatively new phase of water management in the Soviet Union, including the operation, construction, and design of reservoirs which adapt the nonuniform and asynchronous flow of different river systems to the time-variable demands of the different water users of the system. At present, comprehensive plans have not yet been developed for operating large reservoir systems covering entire river basins. The combined operation of several reservoirs provides increases in the guaranteed yield of the reservoirs and in the effective utilization of streamflow. However, difficulties in creating a methodology for formulating optimal control rules arise from a present lack of probabilistic descriptions of streamflow variations and water consumption, and in the evaluation of economic characteristics which constitute a special-purpose function to optimization, wherein it is necessary to accomplish a complex analysis of not only the multiple-purpose water resources system, but also of several allied systems (power, transport, etc.). This article examines some aspects of the methodology for formulation of control rules for the regimes of systems of reservoirs of hydroelectric plants operating in an energy system and satisfying the given requirements of the participants in the water management complex. (Bell-Cornell)
W73-09868

AN INVESTIGATION INTO THE LONGTERM PERSISTENCE OF U. S. HYDROLOGIC RECORDS,
IBM Watson Research Center, Yorktown Heights, N.Y.
J. R. Wallis.

Available from the National Technical Information Service as PB-220 738, \$4.50 paper copy, \$0.95 microfiche. Completion Report, January 1973. 54 p, 7 fig, 42 ref. OWRR C-3135 (3691) (1).

Descriptors: *Reservoir yield, *Hydrologic data, Seasonal, *Synthetic hydrology, Long-term planning, Regional analysis, *Monte Carlo method, *Stochastic processes, Optimum development plants.
Identifiers: ARIMA process, Fractional noises, Rescaled range.

Longterm persistence in hydrologic records influences the concept of reserovir firm yield. This study attempted to assess regional and seasonal persistence in U. S. hydrologic records. The attempt was not successful because of the nature of the hydrologic data base and the power of the relevant statistics to separate longterm from shorterm persistence given records of normal hydrologic length.
W73-09869

WATER RESOURCES DEVELOPMENT IN THE MULLICA RIVER BASIN PART II: CONJUNCTIVE USE OF SURFACE AND GROUNDWATERS OF THE MULLICA RIVER BASIN,
Rutgers - The State Univ., New Brunswick, N.J. Water Resources Research Inst.
For primary bibliographic entry see Field 04B.
W73-09870

MOVABLE DAM AND METHOD OF OPERATION,
For primary bibliographic entry see Field 08A.
W73-09938

WATER REGIME OF THE BANK SANDS OF KURTLI STORAGE LAKE, (IN RUSSIAN),
A. Nurmukhamedov.
Probl Osvoeniya Pustyn'. 1, p 71-74. Illus. 1972. English summary.
Identifiers: Banks, Cultivation, *Desert shouls, *Kurtli storage, Lake, Sands, *Mesophyte trees, USSR.

The banks consist of fine-grained, non- and weakly saline sands. The underground water depth varies from 0.7-5.0 m according to season. The water regime of the root soil horizons depends upon the ground water depth and the rainfall. These areas may be used for cultivation of desert shrubs and mesophyte tree species.—Copyright 1973, Biological Abstracts, Inc.
W73-09980

HOW FREQUENTLY WILL FLOODS OCCUR,
Pennsylvania State Univ., University Park. Dept. of Civil Engineering.
B. M. Reich.
Water Resources Bulletin, Vol 9, No 1, p 187-188, February 1973. 1 fig, 2 ref.

Descriptors: *Flood forecasting, *Probability, *Planning, *Flood frequency, *Flood recurrence interval, Flood protection, Water management (Applied), Warning systems, Floods, Flood discharge.

It has become customary for engineers to refer to floods in terms of their return period or recurrence interval. While this common practice provides a means for comparing the relative severity of floods on different sized rivers, it often leads to misunderstandings. The only simple interpretation of return period is that its reciprocal is the probability of obtaining a flood that size or greater in any one year. Thus the 100-year flood peak has a probability of being exceeded in any one year of 0.01. It is necessary to consider the probability that the selected flood discharge will be exceeded in 25, 50 years, or another design lifetime. For in-

stance, there is a probability of 0.22 that the 100-year flood will be exceeded at least once in 25 years. The percentage change of getting floods of that or higher levels during a 100-year design lifetime is 64%. A table gives up to five equivalent interpretations for a specific return period. Lifetimes from 1 to 100 years are considered. Return periods are spaced according to the commonly used extreme value paper of Powell or Gumbel. Besides their use on graph paper for plotting observed floods and an extrapolated frequency curve, the values could provide the basis for a table in which estimates are given in terms of the discharge of progressively rarer events. Alternatively flood elevations or costs of alternative protection schemes could be listed alongside these six columns. Decision makers could see at a glance the change of failure over various planning horizons for all options proposed. (Knapp-USGS)
W73-09992

RIVER THAMES--REMOVABLE FLOOD BARRIERS,
Bruce White, Wolfe Barry and Partners, London (England).
For primary bibliographic entry see Field 08A.
W73-09996

NAVIGATION ON THE THAMES,
Port of London Authority (England).
For primary bibliographic entry see Field 08A.
W73-09999

RAIN-FOREST CONSERVATION IN AFRICA,
T. T. Struhsaker.
Primates. Vol 13, No 1, p 103-109. 1972. Illus.
Identifiers: *Africa, *Conservation, Forests, *Rain forests, Ecosystem.

Outlined are some of the causes of rain-forest ecosystem destruction in Africa, reasons why it should be conserved, brief descriptions of a few specific forests and suggestions on how forest conservation might be implemented.--Copyright 1972, Biological Abstracts, Inc.
W73-10016

HYDRO-POWER SYSTEM SIMULATION TECHNIQUE,
Corps of Engineers, Portland, Oreg. North Pacific Div.
C. E. Abraham.
Water Resources Bulletin, Vol 9, No 1, p 34-40, February, 1973. 3 fig, 1 plate, 5 ref.

Descriptors: *Simulation analysis, *Computer programs, *Reservoirs, *Regulation, *Multiple-purpose projects, *Hydrologic systems, Scheduling, Hydroelectric power, Dams, Columbia River, Optimum development plans, Project benefits, Grand Coulee Dam, Systems analysis, Mathematical models.
Identifiers: Operating constraints.

The non-power requirements of a large hydro-power system of multiple-purpose projects often conflict with the best power peaking operation. In order to schedule an optimum multiple-purpose operation, advanced procedures that necessitate computers are required. One of the family of programs used by the North Pacific Division of the Corps of Engineers for its reservoir regulation and power scheduling activities is described. This program, Hydro-Power System Regulation Analysis, is used to regulate the Columbia System of hydro projects, and is designed for short incremental time periods from one hour to one day. Special techniques are used in a computer program that provide the ability to define operating constraints in order of priority. Emphasis is given to rapid turn-around time yielding the maximum useful information while requiring as little manual input as possible. These techniques are easily adapted to practical reservoir regulation problems so that the program is useful in daily reservoir regulation scheduling. (Bell-Cornell)
W73-10019

MATERIALS OF THE ANNIVERSARY SESSION OF THE ACADEMIC COUNCIL (MATERIALY YUBILEYNOY SESSII UCHENOGO SOVETA).
Kazakhskii Nauchno-Issledovatelskii Gidrometeorologicheskii Institut, Alma-Ata (USSR).
For primary bibliographic entry see Field 02A.
W73-10048

THE UTILIZATION OF BOGS IN CZECHOSLOVAKIA,
H. Broekke.
Tidsskr Skogbruk. Vol 79, No 2, p 214-228. 1971. Illus.
Identifiers: *Bogs, *Czechoslovakia, Water utilization, Forestry, Timber industry.

Czechoslovakia has about 2000 bogs with a total of 33,000 ha, which is not much compared with northern countries. However, due to industry's pressure on resources, much scientific and practical interest is devoted to bog lands. The Czechs consider production of quality timber and the forest's social significance as the basis of their forestry.--Copyright 1972, Biological Abstracts, Inc.
W73-10131

DESCRIBING ARIZONA SNOWPACKS IN FORESTED CONDITION WITH STORAGE-DURATION INDEX,
Arizona Univ., Tucson. Dept. of Watershed Management.
For primary bibliographic entry see Field 02C.
W73-10152

PROPER USE AND MANAGEMENT OF GRAZING LAND.
Arizona Interagency Range Committee.

1972. P 48, 7 FIG, 3 TAB.

Descriptors: *Range management, *Grazing, *Wildlife management, *Arizona, *Natural resources, Land use, Grasses, Forage grasses, Livestock, Cost-benefit analysis, Carrying capacity, Droughts.

The first of a series of publications to be issued by an interagency committee composed of federal, state and University personnel, this one begins by recognizing the proper use of Arizona's grazing lands that now make up a large percentage of the State's land area. Since at present most of this land is producing vegetation far below its carrying capacity, the Committee has addressed itself to a consideration of the following topics: plant factors and proper grazing use, animal factors bearing upon proper use, procedures for determining grazing use, and practices to facilitate proper use. The figures show the growth of various species, effect of clipping, seasonal livestock gains, and a cost-return evaluation of proper range use. Practices recommended to effect proper use include development features such as wells, ponds, and pipelines; plant control, seeding, fire, and fertilization; and control of human use to minimize disruptions detrimental to proper use. (Paylore-Arizona)
W73-10211

A PRELIMINARY LIST OF INSECTS AND MITES THAT INFEST SOME IMPORTANT BROWSE PLANTS OF WESTERN BIG GAME,
Forest Service (USDA), Moscow, Idaho. Forestry Sciences Lab.
For primary bibliographic entry see Field 02I.
W73-10218

PRELIMINARY ANNOTATED LIST OF DISEASES OF SHRUBS ON WESTERN GAME RANGES,
Forest Service (USDA), Ogden, Utah. Intermountain Forest and Range Experiment Station.
For primary bibliographic entry see Field 02I.
W73-10219

SUBTERRANEAN PART CHARACTERISTICS OF SOME NATURAL GRASS COMMUNITIES ON EROSIVE SLOPES, (IN RUSSIAN),
Avademiya Nauk Litovskoi SSR, Vilnius. Inst. of Botany.
N. A. Lapinskene.
Liet Tsr Moksiu Akad Darb Ser B. 3 p 13-24. 1971. Illus. English summary.
Identifiers: Agrostis-tenuis, Anthoxanthum-odoratum, Briza-media, *Erosive slopes, Festuca-ovina, Festuca-rubra, *Grass, Hieracium-pilosella, Medicago-lupulina, Moisture, Morphology, Nutrition, Poa-angustifolia, Poa-compressa, Roots, Soils, Trifolium-medium, Trifolium-montanum, Weather, *Lithuania.

In 1965-1966 the communities typical of erosive slopes in Ignalina district (Northeast Lithuania) and in Silale district (West Lithuania), Anthoxanthum odoratum + Agrostis tenuis, Anthoxanthum odoratum + Briza media, Festuca ovina-Trifolium montanum, F. ovina + Hieracium pilosella, Festuca rubra-Medicago lupulina, F. rubra + Agrostis tenuis, Poa angustifolia + F. rubra, Poa compressa-Trifolium medium were investigated. The interrelations between the subterranean part of these communities and the soils which are reflected by the morphological properties of the roots and by the distribution of their mass in separate horizons of sod-podzolic loam and sandy loam soil were established. The subsurface part of the communities was influenced by the physical and chemical properties of the soil. Dryness and a bad water-weather-nutritional status in the deeper soil horizons have a negative influence. The main mass of the subsurface part of the investigated communities is in the surface humic horizon, in the depth of 0-20 cm, where it forms turf of about 3 cm thick. It strengthens the soil against the erosion. Only separate individual plant roots penetrate to the depth of 50-100 cm.--Copyright 1973, Biological Abstracts, Inc.
W73-10245

SIMON RUN WATERSHED PROJECT, POTTAWATTAMIE COUNTY, IOWA (FINAL ENVIRONMENTAL STATEMENT).
Soil Conservation Service, Des Moines, Iowa.
For primary bibliographic entry see Field 04D.
W73-10263

RESTORATION OF HISTORIC LOW FLOW DIVERSION TO ARROYO COLORADO; LOWER RIO GRANDE FLOOD CONTROL PROJECT, TEXAS (FINAL ENVIRONMENTAL IMPACT STATEMENT).
Corps of Engineers, Dallas, Tex. Southwestern Div.

Available from National Technical Information Service, U.S. Dept. of Commerce as EIS-TX-72-5578-F. $4.50 in paper copy; $0.95 in microfiche. Report prepared for U.S. Section International Boundary and Water Commission, October 1972. 42 p, 1 fig.

Descriptors: *Texas, *Environmental effects, *Diversion structures, *Alteration of flow, Arroyos, Channels, Navigation, Water management (Applied), Obstruction to flow, Flow augmentation, Earthworks, Drainage engineering.
Identifiers: *Environmental Impact Statements, *Rio Grande Flood Control Project (Tex).

The proposed action is a modification of the existing Lower Rio Grande Flood Control System to divert flood flows to the Arroyo Colorado before

flows begin in the north floodway. This will be accomplished by construction of a low earth sill in the entrance to the North Floodway. The low flow diversion will restore drainage conditions to lands outside of and draining into the North Floodway. The sill will require less than one-acre and will not affect fishery in the lake. The increase in diversion to the Arroyo Colorado will not cause either significant sedimentation or velocities that would adversely affect navigation in the barge channel. The infrequent increased flow should not cause any major changes in the salinity of brackish waters of the Arroyo; it will have only a slight temporary effect on nursery and feeding areas for marine species. An alternative structural measure to accomplish the same result would require additional concrete openings through the divisor dike and a new channel requiring 34 acres of pasture and farmland. (Mockler-Florida)
W73-10264

DREDGING AND FILLING, COWIKEE STATE PARK, LAKEPOINT RESORT, WALTER F. GEORGE LAKE, CHATTAHOOCHEE RIVER, ALABAMA (FINAL ENVIRONMENTAL IMPACT STATEMENT).
Army Engineer District, Mobile, Ala.

Available from National Technical Information Service, U.S. Department of Commerce, EIS-AL-72-5555-F, $4.00 in paper copy; $0.95 in microfiche. June 1972. 33 p, 1 map.

Descriptors: *Alabama, *Project planning, *Environmental effects, *Artificial beaches, *Recreation facilities, Dredging, Recreation, Projects, Public benefits, Comprehensive planning, Land use, Multiple-purpose projects, Recreation demand, Water sports, Campsites, Fishing, Lakes, Land use, Marinas, Ecosystems, Social aspects, Swamps.
Identifiers: *Environmental impact statement, *Eufaula, Ala.

The site of the proposed project is in southeastern Alabama in Barbour County near the city of Eufaula on Walter F. George Lake on the Chattahoochee River. The project consists of the removal of 745,000 cubic yards of material by hydraulic dredge and placing the material in a low swampy area adjacent to the lake and within an area occupied by the summer pool. The filled area will be used for the construction of a marina, bathing beach, roads and parking facilities. Some land and water bottom will be covered by the dredged material. The ecological character of the terrestrial system will be permanently altered. A no-action alternative would result in continuing overuse of limited recreation area. Although the project requires the removal of some low relief area, the project will make available a much larger acreage for man's long-term recreational use. Implementation of the project would result in the permanent alteration of 42 acres of upland and 21 acres of water bottom. These resources would be committed to a park, recreation and nature area. The comments of various governmental and private agencies are included. (Reed-Florida)
W73-10265

CONTROL OF EURASIAN WATERMILFOIL (MYRIOPHYLLUM SPICATUM L.) IN TVA RESERVOIRS (FINAL ENVIRONMENTAL IMPACT STATEMENT).
Tennessee Valley Authority, Chattanooga.

Available from the National Technical Information Service as EIS-TN-72-3397-F, $6.50 paper copy, $0.95 in microfiche. Office of Health and Environmental Science, Report TVA-OHES-EIS-72-8, September 29, 1972. 82 p, 1 fig, 2 photo, 9 tab, 1 chart, 12 ref.

Descriptors: *Aquatic weed control, *Environmental effects, *Tennessee Valley Authority, *Water quality, Weeds, Weed control, Water level fluctuations, Reservoir operation, Water pollution sources, 2-4-D, Tennessee River, Federal project policy, Rivers, Waterfowl, Herbicides, Water pollution control, Recreation, Fishing, Plant growth, Aquatic plants.
Identifiers: *Eurasian watermilfoil, *Environmental Impact Statements.

This action consists of water level management and application of 2,4-D herbicide in order to achieve and maintain control of watermilfoil to the degree necessary to protect public health and to assure that economic and recreational values of the Tennessee Valley Authority reservoir system are not materially impaired. Beneficial impacts of the project include socioeconomic benefits resulting from returning reservoirs and contiguous lands to maximum potential for recreational and other land uses, increased production of sport fish as a result of water level drawdown and the removal of shelter for forage fish. Potential adverse effects include low concentration of herbicides in water supplies, minor damage to nontarget aquatic and terrestrial plants, minor loss of food and shelter for some fish species and waterfowl and decreased fish spawning. Alternatives to this program included mechanical control, biological control, water level management, use of 2,4-D alone and use of other herbicides. (Reed-Florida)
W73-10266

BRANTLEY PROJECT.
For primary bibliographic entry see Field 06E.
W73-10270

PUBLIC ACCESS TO RESERVOIRS TO MEET GROWING RECREATION DEMANDS.
For primary bibliographic entry see Field 06E.
W73-10273

SAN ANGELO PROJECT, TEXAS.
For primary bibliographic entry see Field 06E.
W73-10275

DRAINAGE DISTRICTS, COMMISSION COMPOSITION AND POWERS.
For primary bibliographic entry see Field 06E.
W73-10301

REGIONAL WATER DISTRIBUTION DISTRICTS.
For primary bibliographic entry see Field 06E.
W73-10302

ARKANSAS RIVER BASIN COMPACT.
For primary bibliographic entry see Field 06E.
W73-10305

TIME PERIODS AND PARAMETERS IN MATHEMATICAL PROGRAMMING FOR RIVER BASINS,
Massachusetts Inst. of Tech., Cambridge. Ralph M. Parsons Lab. for Water Resources and Hydrodynamics.
J. A. Poblete, and R. T. McLaughlin.
Report No 128, September 1970. 325 p, 7 fig, 34 tab, 57 ref, 7 append.

Descriptors: *Linear programming, *Mathematical models, *River basin development, *Synthetic hydrology, *Reservoir operation, Model studies.
Identifiers: Time periods, Integer programming, Deterministic hydrology.

The preliminary analysis of a river-basin development is repeated with several different mathematical models of increasing complexity and with several different values for important parameters included in the models. The results of these various models are compared for their effects upon practical insights and decision-making. Five models employed linear programming with deterministic hydrology in which various time periods were used to describe the hydrology and reservoir operation. In addition, an integer program was used to allow discrete values of reservoir size and to permit reservoir costs to reflect advantages of scale. This model was cyclical with 4 periods of 3 months. The parameters varied within these models were both economic and physical. Simple mathematical programming models were found to be both valuable and fairly sensitive tools. Various methods were discussed for imporving the outcomes of these models. Mixed integer programming was found to introduce significantly more 'engineering reality' into the approach. (Weaver-Wisconsin)
W73-10317

BENEFIT-COST ANALYSIS FOR WATER RESOURCE PROJECTS,
Tennessee Valley Authority, Muscle Shoals, Ala.
For primary bibliographic entry see Field 06B.
W73-10318

NAVIGATION CHANNEL IMPROVEMENT--GASTINEAU CHANNEL, ALASKA,
Army Engineer Waterways Experiment Station, Vicksburg, Miss.
For primary bibliographic entry see Field 08B.
W73-10320

EXPECTED NUMBER AND MAGNITUDES OF STREAM NETWORKS IN RANDOM DRAINAGE PATTERNS,
California Univ., Irvine.
C. Werner.
Proceedings of the Association of American Geographers, Vol 3, p 181-185, 1971. 1 fig, 2 tab, 7 ref. NSF GS-2989.

Descriptors: *River basins, *Drainage area, Drainage patterns (Geologic), *Topography, *Networks, Mathematical models, Equations, Hydrologic data, Correlation analysis, Kentucky, Geology, Geomorphology, Streams.
Identifiers: *Stream networks, Random drainage patterns, Stream merger distribution.

How does a set of first order streams behave, i.e., what kind of resulting drainage pattern can be expected if all topologically different patterns of stream mergers (including no merger) are equally likely. For this assumption deductions are made for mathematical expressions of the number of topologically different patterns possible, the frequencies of networks by magnitude, the expected number of networks to develop, and their distribution with regard to magnitude. The expected number of networks emerging from any sequence of first order streams is always smaller than 3, and one of them is always identical with one of the original first order streams. The theoretical findings are tested by using samples of drainage network data from an area in eastern Kentucky that is essentially free of geologic controls. Chisquare tests show that empirical data follow closely to the predicted figures. (Woodard-USGS)
W73-10335

PATTERN AND PROCESS IN DUNE SLACK VEGETATION ALONG AN EXCAVATED LAKE IN THE KENNEMER DUNES, (IN DUTCH),
For primary bibliographic entry see Field 02H.
W73-10392

AN ANALYSIS OF ECOLOGICAL FACTORS LIMITING THE DISTRIBUTION OF A GROUP OF STIPA PULCHRA ASSOCIATIONS,
Kon-Kuk Univ., Seoul (South Korea). Dept. of Biology.

R. H. Robinson.
Korean J Bot. Vol 14, No 3, p 1-20. 1971. Illus.
Identifiers: Avena-fatua, *California (Monterey), *Distribution patterns (Plants), Ecological studies, Foothill, Forbs, Grasslands, Moisture, Phosphorus, Soils, *Stipa-pulchra, Valley, Woodlands.

Ecological factors limiting S. pulchra have been determined in experimental gardens and at several sites in the hills south and east of Monterey, California. The S. pulchra facies of Valley Grassland communities were found to be dominated by that species, though a total of 36 grasses and forbs were collected and identified. Basal area was not large, but aerial cover by Stipa alone averaged over 50%. Across an ecotone between a Stipa association and the California Annual Type a sudden and dramatic change was recorded. Soil measurements there, and in other nearby areas, showed a much higher clay content with more available water and elemental P at the Stipa sites. Germination of Stipa seeds was high under all laboratory and field conditions, though growth of seedlings was highly variable. Seedlings grown in Stipa soil with an abundance of water were vigorous and reached anthesis the first year. In other soils they grew less, and when grown in competition with Avena fatua they scarcely grew at all. These findings indicate that when established on desirable soils, Stipa competes well and apparently precludes the dominance of A. fatua and other large annual grasses. On the other hand, because of a lack of vigor in its seedlings, Stipa cannot reinvade the rich more friable soils on which it was once found, andon which it was shown to grow satisfactorily. This supports the contention that S. pulchra was the dominant grass through much of the Walley Greenland and Foothill Woodland, but also indicates that well-drained soils and those poor in mineral nutrients probably never supported such associations.--Copy 1973, Biological Abstracts, Inc.
W73-10400

4B. Groundwater Management

GROUNDWATER CONSITIONS IN THE CENTRAL VIRGIN RIVER BASIN, UTAH,
Geological Survey, Salt Lake City, Utah.
R. M. Cordova, G. W. Sandberg, and W. McConkie.
Utah Department of Natural Resources, Salt Lake City, Technical Publication No 40, 1972. 64 p, 10 fig, 3 plate, 18 tab, 12 ref, 1 append.

Descriptors: *Groundwater resources, *Well data, *Aquifer characteristics, *Water quality, *Utah, Hydrogeology, Water wells, Water levels, Withdrawal, Pumping, Water yield, Water utilization, Groundwater recharge, Chemical analysis, Hydrologic data, Basic data collections.
Identifiers: *Virgin River basin (Utah).

The central Virgin River basin, in Washington and Iron Counties, Utah, includes about 1,000 square miles in the drainage basin of the Virgin River downstream from Hurricane Cliffs. Aquifers in both consolidated and unconsolidated rocks supply water for public supply, irrigation, stock, industry, and domestic uses. The chief unconsolidated-rock aquifers are alluvial fans and channel-fill deposits, which supply about 80% of the water withdrawn by wells in the basin. The chief consolidated-rock aquifers include the Moenkopi, Chinle, Moenave, and Kayenta Formations, the Navajo Sandstone, basalt, and Tertiary igneous rocks of the Pine Valley Mountains. Average annual recharge to the aquifers of the central Virgin River basin is estimated to be 100,000 acre-feet. Discharge from wells averaged 6,600 acre-feet annually for the years 1968-70. Water-level hydrographs give no indication that withdrawals of groundwater to date have had any significant effect on the amount of groundwater in storage. The dissolved-solids concentration in the water differs considerably from aquifer to aquifer and from place to place. The aquifers that are most likely to yield water containing less than 1,000 milligrams per liter are the Navajo Sandstone and basalt. The Chinle and Moenkopi Formations are most likely to yield water containing more than 3,000 milligrams per liter. (Woodard-USGS)
W73-09791

GROUNDWATER RESOURCES OF MOTLEY AND NORTHEASTERN FLOYD COUNTIES, TEXAS.
Geological Survey, Austin, Tex.
J. T. Smith.
Texas Water Development Board, Austin, Report 165, March 1973. 66 p, 8 fig, 10 tab, 45 ref.

Descriptors: *Groundwater resources, *Water wells, *Water quality, *Aquifer characteristics, *Texas, Well data, Withdrawal, Pumping, Water yield, Drawdown, Groundwater recharge, Hydrogeology, Hydrologic data, Basic data collections, Water levels, Chemical analysis, Water utilization.
Identifiers: *Motley County (Tex), Floyd County (Tex).

The principal sources of groundwater in Motley County and northeastern Floyd County, Texas, are the alluvial deposits of Quaternary age, the Ogallala Formation of Tertiary age, and the upper part of the Dockum Group (Trujillo Formation) of Triassic age. Rocks of Permian age supply small amounts of slightly saline to very saline water. The alluvial deposits and the upper part of the Dockum Group are the most prolific aquifers. The average yield of irrigation wells tapping these units is about 400 gpm. The Permian rocks usually yield less than 100 gpm to wells that range in depth from about 50 to over 300 feet. The alluvial deposits cover about 25% of the area but supply a large part of the water needs. The Permian rocks are used almost entirely as a source of water for domestic and stock supplies. In 1968, about 11,200 acre-feet of groundwater was pumped for municipal supply, industrial use, and irrigation; of the 11,200 acre-feet pumped, about 9,400 acre-feet was used to irrigate 6,823 acres. Generally, the water from the Ogallala Formation and the Dockum Group is the least mineralized. The water from the Permian rocks is more highly mineralized and in some parts of the area it is unfit for domestic use. The quality of the water in the alluvium depends upon the source of recharge. (Woodard-USGS)
W73-09792

GROUNDWATER RESOURCES OF HALL AND EASTERN BRISCO COUNTIES, TEXAS.
Geological Survey, Austin, Tex.
B. P. Popkin.
Texas Water Development Board, Austin, Report 167, April 1973. 84 p, 8 fig, 7 tab, 16 ref.

Descriptors: *Groundwater resources, *Water wells, *Water quality, *Aquifer characteristics, *Texas, Well data, Withdrawal, Pumping, Water yield, Drawdown, Groundwater recharge, Hydrogeology, Hydrologic data, Basic data collections, Water levels, Chemical analysis, Water utilization, Irrigation, Water supply.
Identifiers: *Hall County (Tex), *Eastern Brisco County (Tex).

Hall and eastern Brisco Counties are in the southeastern part of Texas Panhandle. Nearly all the water used in the area is from wells and springs. Most of the water for irrigation and all water for public supply in Hall County is from alluvial deposits. In eastern Brisco County, most of the water for irrigation and public supply is pumped from aquifers in the Ochoa Series of Permian age. The total groundwater withdrawal in 1968 was approximately 28,700 acre-feet. About 65% of this amount was from the alluvial deposits and nearly 20% from the Ochoa Series. Water levels in most of the irrigation areas declined during the period 1960-69 from less than 1.0 to 29.8 feet. Water from the Permian units generally is of the calcium sulfate type and has a dissolved-solids content that ranges from less than 1,000 to more than 10,000 milligrams per liter. The quality of water from the Quaternary alluvium and terrace deposits varies widely, depending upon the source of recharge. Where the alluvium is recharged from Permian rocks, the water usually contains calcium and sulfate as the major constituents. Where the recharge is mostly from direct infiltration of rainfall, the water contains calcium and bicarbonate as the major constituents. (Woodard-USGS)
W73-09793

GROUND-WATER FAVORABILITY AND SURFICIAL GEOLOGY OF THE LOWER AROOSTOOK RIVER BASIN, MAINE,
Geological Survey, Washington, D.C.
For primary bibliographic entry see Field 07C.
W73-09801

INVESTIGATIONS AND FORECAST COMPUTATIONS FOR THE CONSERVATION OF GROUNDWATER (ISSLEDOVANIYA I PROGNOZNYYE RASHETY DLYA OKHRANY PODZENMYKH VOD),
For primary bibliographic entry see Field 05B.
W73-09802

WATER RESOURCES DEVELOPMENT IN THE MULLICA RIVER BASIN PART II: CONJUNCTIVE USE OF SURFACE AND GROUNDWATERS OF THE MULLICA RIVER BASIN,
Rutgers - The State Univ., New Brunswick, N.J. Water Resources Research Inst.
M. L. Granstrom, G. H. Nieswand, and R. Ahmed.
Available from the National Technical Information Service as PB-220 739, $3.00 in paper copy, $0.95 microfiche. Partial Completion Report, March 1973. 59 p, 8 fig, 20 tab, 32 ref. OWRR B-014-NJ (5) and B-018-NJ (3). 14-01-0001-1529, 14-31-0001-3105.

Descriptors: *Surface-groundwater relationships, Hydrology, *Optimization, Water supply, Water yield, Groundwater, Subsurface waters, *Conjunctive use, *New Jersey, *Balance of nature, *Water transfer, Linear programming, Model studies.
Identifiers: *Mullica River basin (N.J.).

The potential transfer of water from the Mullica River Basin prompted a joint study by biologists and engineers to determine (1) the minimum average monthly streamflow necessary to maintain an appropriate ecological balance in the river-bay system and (2) the optimum combination of groundwater-surface water pumping to maximize yields and to meet the ecological criteria. This report describes the geological and hydrological characteristics of the Mullica River Basin and explores engineering means of withdrawing water supply under somewhat unusual circumstances. The report is based upon the procedural methods developed in two Ph.D. theses in Engineering at Rutgers University. A change constrained linear programming model for the conjunctive use of ground- and surface water was used, along with groundwater aquifer-stream system equations to obtain rates of optimum withdrawal of surface waters and groundwater consistent with specified objective functions and sets of physical and operational constraints. The physical constraints are those given in terms of minimum streamflow residuals. Optimization was obtained using linear programming techniques. (See also W72-12706) (Whipple-Rutgers)
W73-09870

WELL PUMPING APPARATUS FOR POLLUTED WATER,
Mowid Anstalt, Vaduz (Liechtenstein). (Assignee).
For primary bibliographic entry see Field 05G.
W73-09935

DRAWDOWN AND RECOVERY OF AN ARTESIAN SAND AND GRAVEL AQUIFER, LONDON, ONTARIO, CANADA,
Ontario Water Resources Commission, Toronto. Div. of Water Resources.
For primary bibliographic entry see Field 02F.
W73-09948

GROUNDWATER RESEARCH IN THE OPEN-CAST MINING AREA OF THE SOUTHERN NIEDERRHEIN-BUCHT (WEST GERMANY),
Geologisches Landesamt Nordrhein-Westfalen, Krefeld (West Germany).
For primary bibliographic entry see Field 02F.
W73-09951

ASSESSMENT OF GROUNDWATER RECHARGE IN THE UPPER TARNAK BASIN OF AFGHANISTAN,
United Nations Development Program, New York.
For primary bibliographic entry see Field 02F.
W73-09957

GROUNDWATER IN PERSPECTIVE,
Geological Survey, Washington, D.C.
R. L. Nace.
Water Resources Bulletin, Vol 9, No 1, p 18-24, February 1973. 1 fig, 1 tab, 8 ref.

Descriptors: *Aquifers, *Groundwater resources, Water storage, Water management (Applied), Hydrogeology, Aquifer characteristics, Conjunctive use, Water resources development.

Because of their enormous capacity, groundwater reservoirs are at least equal in importance to the groundwater itself. As regulators of water movement in the hydrological cycle, these reservoirs surpass all lakes and reservoirs combined. While many aquifers are not well understood, data on many others are adequate for long-range broad-scale planning. An example is the basalt aquifer of the Snake River Plain in Idaho. However, the area has managerial problems which concern the time, the place, and the feasibility of manipulations of water. All continents of the world contain great aquifers. For every huge aquifer, however, hundreds of smaller ones occur, and even these contain large amounts of water. Aquifers in the Ohio River basin of the United States are good examples. Management of total water resources is a difficult problem at many places. But many problems could be met and many water shortages alleviated or eliminated by use of aquifers, not merely as sources of water but as reservoirs for management of water. (Knapp-USGS)
W73-09994

SUBSIDENCE AND ITS CONTROL,
Geological Survey, Sacramento, Calif.
J. F. Poland.
In: Underground Waste Management and Environmental Implications, Proc of Symposium held at Houston, Tex, Dec 6-9, 1971: Tulsa, Okla, American Association of Petroleum Geologists Memoir 18, p 50-71, 1972. 20 fig, 2 tab, 50 ref.

Descriptors: *Land subsidence, *Withdrawal, *Compaction, *Overdraft, Subsidence, Groundwater, Hydrogeology, Oil fields, Injection wells, Aquifer characteristics, Pore pressure, Drawdown.

Land subsidence caused by fluid withdrawal most commonly affects overdrawn groundwater basins, but subsidence of serious proportions also has occurred in several oil and gas fields. The San Joaquin Valley in California is the area of the most intensive land subsidence in the United States. Surface-water imports to subsiding areas have reduced groundwater extractions and raised the artesian head, causing subsidence rates to decrease. Wilmington oil field, in the harbor area of Los Angeles and Long Beach, California, is not only the oil field of maximum subsidence (29 ft) in the United States but also the outstanding example of the subsidence control by injection and repressuring. Large-scale repressuring was begun in 1958 by use of injection water obtained from shallow wells. Subsidence at some bench marks was stopped by 1960. In evaluation of potential land subsidence due to the fluid withdrawal, an essential parameter is the compressibility of aquifers. When effective (grain-to-grain) stress exceeds maximum prior stress, the compaction is primarily inelastic and nonrecoverable. If fluid pressures in an elastic, compacting, confined system are increased sufficiently, subsidence will stop. If fluid pressures continue to increase, the system will expand elastically. (Knapp-USGS)
W73-10024

NATURAL AND INDUCED FRACTURE ORIENTATION,
Geological Survey, Washington, D.C.
For primary bibliographic entry see Field 02F.
W73-10030

MECHANICS OF HYDRAULIC FRACTURING,
Geological Survey, Washington, D.C.
For primary bibliographic entry see Field 02F.
W73-10031

EARTHQUAKES AND FLUID INJECTION,
National Center for Earthquake Research, Menlo Park, Calif.
For primary bibliographic entry see Field 02F.
W73-10033

TECHNICAL AND ECONOMIC FACTORS OF VARIOUS TECHNIQUES OF RANGELAND IRRIGATION, (IN RUSSIAN),
For primary bibliographic entry see Field 03F.
W73-10090

CLOSED BASIN DIVISION, SAN LUIS VALLEY PROJECT, COLORADO.
For primary bibliographic entry see Field 06E.
W73-10268

CONSERVATION--LICENSING POWERS.
For primary bibliographic entry see Field 06E.
W73-10279

ALABAMA WATER WELL STANDARDS.
For primary bibliographic entry see Field 06E.
W73-10291

WATER WELL CONSTRUCTION ACT.
For primary bibliographic entry see Field 06E.
W73-10304

PUBLIC WATER SUPPLIES OF SELECTED MUNICIPALITIES IN FLORIDA, 1970,
Geological Survey, Tallahassee, Fla.
H. G. Healy.
Florida Department of Natural Resources, Bureau of Geology, Information Circular No 81, 1972. 213 p, 28 fig, 4 tab, 79 ref.

Descriptors: *Potable water, Water resources, *Water supply, *Groundwater, *Surface waters, *Florida, Cities, Water utilization, Water wells, Well data, Aquifers, Streams, Lakes, Chemical analysis, Water users, Human population, Urbanization, Municipal water, Water quality, Water analysis, Basic data collections, Hydrologic data, Desalination.
Identifiers: *Florida cities.

Of the 138 Florida municipalities in this report, 119 use groundwater and 19 use surface water either wholly or partly. The groundwater supply constitutes 87%, 635 million gallons per day (mgd), of the total demand of the 138 municipalities, and surface water supplies 13%, or 91 mgd. The principal sources of groundwater used by the 138 cities and the amounts are: the Biscayne aquifer, 305 mgd; the Floridan aquifer, 277 mgd; the sand-and-gravel aquifer, 21 mgd; and the shallow sand aquifer, 20% mgd. Other minor aquifers yielded 10 mgd. Most of the surface-water supply, 55 mgd, was diverted from rivers. Lakes supplied 31 mgd and reservoir 4 mgd. An infiltration gallery and desalinization plant together supplied 3 mgd. Of the rivers, Hillsborough River supplied the most, 45 mgd to Tampa; the Myakka-Hatchee River supplied the least, 0.25 mgd to North Port Charlotte. Per capita use of water in the 138 municipalities ranged from 40 to 390 gallons per day and averaged 167. Of the 162 chemical analyses listed in the report, 148 represent water quality during 1967-72 and 14 before 1967. The analyses represent both raw and treated water, virtually all from groundwater sources. Groundwater supplies are typically hard to very hard except in extreme northwestern Florida where water from the sand-and-gravel aquifer is very soft. (Woodard-USGS)
W73-10319

WATER RESOURCES OF THE RED RIVER OF THE NORTH DRAINAGE BASIN IN MINNESOTA,
Geological Survey, St. Paul, Minn.
For primary bibliographic entry see Field 02E.
W73-10321

TEST-OBSERVATION WELL NEAR PATERSON, WASHINGTON: DESCRIPTION AND PRELIMINARY RESULTS,
Geological Survey, Tacoma, Wash.
H. E. Pearson.
Geological Survey Water-Resources Investigations 9-73, 1973. 23 p, 5 fig, 4 tab, 3 ref.

Descriptors: *Test wells, *Well data, *Washington, Aquifer characteristics, Pumping, Water yield, Water quality, Hydrogeology, Specific capacity, Drawdown, Observation wells, Irrigation wells, Chemical analysis.
Identifiers: *Paterson (Wash).

Initial irrigation development in the Horse Heaven Hills area started near the towns of Paterson and Plymouth, Washington, along the Columbia River and has relied on surface water from the river. The first stages of irrigation of land farther from the river created a need for more complete data on groundwater conditions than that provided from the few existing irrigation wells in the area. A 860-foot test-observation well, 10 to 6 inches in diameter, was drilled in the Horse Heaven Hills about 8 miles north of Paterson in the valley of the East Branch of Glade Creek. The well, drilled by the air-rotary method, penetrates productive aquifer zones at the 735- and 860-foot depths. Test pumping showed that the well had specific capacities of 58 gpm per foot of drawdown at the 735-foot depth and 148 gpm per foot of drawdown at the 860-foot and final depth. The tests indicate that large amounts of water occur only at these, and possibly greater, depths. The water is soft and is suitable for all common uses. (Woodard-USGS)
W73-10322

ALPHA-RECOIL THORIUM-234: DISSOLUTION INTO WATER AND THE URANIUM-234/-URANIUM-238 DISEQUILIBRIUM IN NATURE,
Gakushuin Univ., Tokyo (Japan). Dept. of Chemistry.
For primary bibliographic entry see Field 05B.
W73-10325

GROUND WATER IN COASTAL GEORGIA,
Geological Survey, Atlanta, Ga.
R. E. Krause.
The Georgia Operator, p 12-13, Winter 1973. 2 fig.

Descriptors: *Groundwater resources, *Aquifer characteristics, Hydrogeology, *Georgia, *Water wells, Coasts, Pumping, Withdrawal, Water yield, Water utilization, Industrial water, Drawdown, Water level fluctuations, Grundwater recharge, Groundwater movement, Water quality, Saline water intrusion, Reviews, Water supply, Water demand.

The coastal area of Georgia draws most of its water from the 'principal artesian aquifer,' which is a limestone deposited more than 25 million years ago. The aquifer, or water-bearing formation, is more than 500 feet thick and lies from 150 to 600 feet below land surface. Groundwater withdrawal has increased markedly since the first wells were laborously drilled into the limestone in the late 1800's. Single industrial wells with diameters of over 20 inches now pump millions of gallons of water daily, around the clock, throughout the year. Pumpage from the aquifer in coastal Georgia now totals well over a quarter of a billion gallons per day. Pumping is centered mainly in Brunswick, Savannah, and Jesup, and over 80 percent of that used is for industry. The water level in this aquifer is declining as a result of groundwater withdrawal. The decline in water level and resultant decrease in area where flowing wells exist is noted on two maps. The groundwater is of good quality and is generally not treated except for aeration and chlorination when used for municipal supply. (Woodard-USGS)
W73-10326

NON-EQUILIBRIUM PHENOMENA IN CHEMICAL TRANSPORT BY GROUND-WATER FLOW SYSTEMS,
Nevada Univ., Reno. Desert Research Inst.
For primary bibliographic entry see Field 05B.
W73-10329

EFFICIENCY OF PARTIALLY PENETRATING WELLS,
Maryland Univ., College Park. Dept. of Civil Engineering.
Y. M. Sternberg.
Ground Water, Vol 11, No 3, p 5-8, May-June 1973. 3 fig, 2 tab, 5 ref.

Descriptors: *Drawdown, *Water yield, *Water wells, Transmissivity, Aquifer characteristics, Storage coefficient, Discharge (Water).
Identifiers: *Well efficiency, *Partially penetrating wells.

A graphical solution is presented for the determination of total drawdown of partially penetrating wells. This simplified solution agrees with the more complicated exact solution and can be used to evaluate the efficiency of such wells. (Knapp-USGS)
W73-10336

CONTRIBUTION TO THE STUDY OF THE ALGAL FLORA OF ALGERIA,
Algiers Univ. (Algeria). Faculty of Science.
For primary bibliographic entry see Field 02I.
W73-10399

4C. Effects on Water of Man's Non-Water Activities

STREAM CHANNEL ENLARGEMENT DUE TO URBANIZATION,
Regional Science Research Inst., Philadelphia, Pa.
T. R. Hammer.
Water Resources Research, Vol 8, No 6, p 1530-1540, December 1972. 3 tab, 7 ref.

Descriptors: *Urbanization, *Channel erosion, Channel morphology, Land use, Watersheds, *Regression analysis, Sewers, *Pennsylvania, Erosion, Urban hydrology, Storm runoff, Storm drains, Rainfall-runoff relationships, Topography, Slopes, Storm water.
Identifiers: *Impervious land cover, *Philadelphia, Stream channel enlargement.

Results are summarized of an empirical study of the effect of urbanization on stream flows and stream channel enlargements. Land use data and stream channel cross-sectional areas were obtained for 78 small watersheds near Philadelphia. Regression analyses were used to relate the variables. Some of the findings are: land areas containing sewered streets and major impervious areas such as parking lots had the greatest channel enlargement effects, residential areas without sewered streets have little effect, and urban areas over 30 years old have little effect. The influence of urbanization on channel enlargement is significantly related to the topography of the watershed, the location of the urban area within the watershed, and man imposed drainage alterations. In particular, a critical factor is basin slope. Channel enlargement was considered to be an important phenomenon to study because of its effects on the aesthetic and recreational value of the stream. (Elfers - North Carolina)
W73-09878

CRITERIA OF DETERMINATION OF INDUSTRIAL LANDS, (IN POLISH),
L. Langhamer.
Pamiet Pulawski. 48, p 31-39, 1971. Illus. English summary.
Identifiers: Chemistry, Criteria, Deformation (Land), Dust, Flooding, *Industrial land use, Mechanical industry, Mining, *Soils.

Industrial lands arise in consequence of industrial activity, especially of mining, building and infrastructural factors upon the soils. The change of the natural properties and in consequence deformation and devastation of soils is the result of these activities. Factors forming industrial lands are: mechanical deformations, covering with dust, overdrying and flooding. The direct influence of mining depends on the deformation of the surface causing changes in the water relationships of soils. Mechanical deformation or devastation of soils results from the direct influence of industry and building. The indirect effect of industry, mining, building and infrastructure is reduced to hydrological changes in soils and chemical changes in soils. Industrial lands occur on the territories of different areas, around industrial centers, agglomeration of towns, in the valleys of industrial sewage and along the tracks of communication. In most cases industry, especially mining and building ground development is unfavorable for agriculture changes in soils and reduces the agricultural productive area.--Copyright 1972, Biological Abstracts, Inc.
W73-09947

4D. Watershed Protection

ROCK MULCH IS REDISCOVERED,
Arizona Univ., Tucson. Dept. of Soils, Water and Engineering.
For primary bibliographic entry see Field 03B.
W73-10212

SIMON RUN WATERSHED PROJECT, POTTAWATTAMIE COUNTY, IOWA (FINAL ENVIRONMENTAL STATEMENT).
Soil Conservation Service, Des Moines, Iowa.

Available from National Technical Information Service, Department of Commerce, EIS-USDA-SCS-ES-WS (ADM)-73-12 (F). $3.00 in paper copy, $0.95 in microfiche. September 1972. 17 p, 1 plate, 1 tab.

Descriptors: *Iowa, *Watershed management, *Environmental effects, *Erosion control, *Flood control, Corn Belt, Land management, River basin development, Soil management, Water conservation, Water management (Applied), Water storage, Dams, Spillways, Ponds, Gully erosion, Watersheds (Basins), Water utilization, Voids.
Identifiers: *Environmental Impact Statements, *Simon Run Watershed, Iowa.

The project proposes conservation land treatment and a series of eight grade stabilization structures for flood prevention and gully erosion control. The project will eliminate gully erosion on 904 acres, reduce flood damage 52% and inundate 55 acres of various land types to create permanent pools. Two and one-half miles of intermittent stream channels will be inundated by the pools. Twenty-five acres of cropland, gully and pasture are expected to change to grasses and aquatic vegetation. A system of structural measures only was rejected because it does not provide upstream treatment. A land treatment program only was also considered but rejected because it would not stop land voiding and depreciation or flood control. Use of the watershed for agricultural production is not expected to change and the project should improve productivity. Approximately 70 acres of land will be used for dams, spillways, and permanent pools. (Dunham-Florida)
W73-10263

NAVIGATION CHANNEL IMPROVEMENT--GASTINEAU CHANNEL, ALASKA,
Army Engineer Waterways Experiment Station, Vicksburg, Miss.
For primary bibliographic entry see Field 08B.
W73-10320

05. WATER QUALITY MANAGEMENT AND PROTECTION

5A. Identification of Pollutants

STUDIES ON BENTHIC NEMATODE ECOLOGY IN A SMALL FRESHWATER POND,
Auburn Univ., Ala. Dept. of Botany and Microbiology.
T. W. Merritt, Jr.
Available from the National Technical Information Service as PB-220 679, $3.00 in paper copy, $0.95 in microfiche. Alabama Water Resources Research Institute Auburn Bulletin 8, February 1973. 73 p, 28 fig, 19 tab, 27 ref. (partial completion report). OWRR A-004-ALA (1).

Descriptors: *Nematodes, *Bacteria, *Fungi, Population, Distribution patterns, Habitats, Dominant organisms, Ponds, *Alabama, *Benthic fauna, Dissolved oxygen, Water temperature, Hydrogen ion concentration, *Bottom sediments.

Benthic nematode population fluctuations and biological, chemical and physical factors of the nematode environment were studied over a nine months period at ten selected stations in a small farm pond situation. Nematode community structure was investigated over a three months period.

Changes in temperature and pH of the bottom muds and in dissolved oxygen of the overlying water were not found to correlate with nematode concentration changes. Peaks in nematode numbers were observed in December 1966 and in April 1967. A definite correlation in numbers of nematodes, numbers of bacteria, and numbers of fungi with the thickness of the detritus layer overlying the mud surface was observed. Selective deposit feeding nematodes was the predominant feeding type found at the ten stations during the three months period.
W73-09751

PHYSICAL AND CHEMICAL PROPERTIES OF THE COASTAL WATERS OF GEORGIA,
Georgia Univ., Sapelo Island. Marine Inst.
For primary bibliographic entry see Field 02L.
W73-09752

A CRITICAL TEST OF METHODS FOR ISOLATION OF VIRUSES FOR USE IN CONTROL OF NUISANCE ALGAE,
Illinois Univ., Chicago. Dept. of Biological Sciences.
H. N. Guttman.
Available from the National Technical Information Service as PB-220 681, $3.00 in paper copy $0.95 in microfiche. Illinois, Water Resources Center, Urbana, Research Report No 63, April 1973. 17 p, 4 fig, 4 ref. OWRR A-043-ILL (1). 14-31-0001-3513.

Descriptors: *Algae, *Viruses, *Cyanophyta, *Plant viruses, *Isolation, Algal control, Biocontrol, Pollutant identification, Illinois, *Nuisance algae.
Identifiers: Temperate viruses, *Plectonema boryanum, Chicago, Virus removal, Lysogenesis.

The objects of this study were (a) to compare the two major methods for isolation of blue-green algae viruses in order to determine whether the finding of viruses in any one local depended upon the isolation method used; (b) to isolate an array of virulent blue-green algae viruses which could be used for field control of these algal polluters. It was found that the large sample plus concentration method of Padan et al, is useful for detecting a high incidence of temperate viruses to blue-green algae (at the initiation time of our study temperate viruses to blue-green algae were unknown). Two temperate viruses to Plectonema boryanum was studies in detail. It was found that algae infected with temperate viruses, like our type T-5, grow well in the Spring and Fall (moderate water temperature) but not in the summer when ambient water temperature rises. Since infection of algae with temperature viruses immunizes the algae to infection with closely related virulent viruses, we concluded that use of virulent viruses to remove blue-green algae from waterways in the greater Chicago metropolitan area is not practical. Virus removal of blue-green algae is only practical in (a) areas in which the natural incidence of temperature viruses is low or (b) cases in which noncross reacting virulent viruses are available. Virus removal of algae remains an attractive concept because of its specificity but, at the present, may require too much economic investment to locate appropriate viruses.
W73-09753

AUTOMATION OF THE BACTERIOLOGICAL ANALYSIS OF WATER, I, STUDY OF A NEW SPECIFIC TEST FOR FECAL CONTAMINATION AND THE OPTIMAL CONDITIONS FOR ITS DETERMINATION,
Institut Pasteur, Lille (France). Laboratoire d'-Hydrobiologie.
P. A. Trinel, and H. Leclerc.
Water Research, Vol 6, No 12, p 1445-1458, December 1972. 13 fig, 1 tab, 14 ref.

Descriptors: *Synthesis, *Measurement, *Automation, Methodology, Bacteria, E. coli, Enzymes, *Water analysis, Amino acids, Separation techniques, Analytical techniques.
Identifiers: *Glutamic acid decarboxylase, Autoanalyzer.

With the aid of the Technicon auto-analyzer, the optimal conditions for synthesizing and measuring glutamic acid decarboxylase in suspensions of Escherichia coli have been defined. The proposed system allows for a simple, rapid and precise search for the glutamic acid decarboxylase in bacterial suspensions. It can also be used for the bacteriological analysis of water. (Holoman-Battelle)
W73-09772

POLLUTION RELATED STRUCTURAL AND FUNCTIONAL CHANGES IN AQUATIC COMMUNITIES WITH EMPHASIS ON FRESHWATER ALGAE AND PROTOZOA,
Virginia Polytechnic Inst. and State Univ., Blacksburg. Dept. of Biology.
For primary bibliographic entry see Field 05C.
W73-09875

DETERMINATION OF THE TOTAL AMOUNT OF CARBOHYDRATES IN OCEAN SEDIMENTS,
V. Ye. Artem'yev, L. N. Krayushkin, and Ye. A. Romankevich.
Oceanology, Vol 11, No 6, p 934-936, August 1972. 3 tab, 8 ref.

Descriptors: *Carbohydrates, *Sediments, *Sea water, *Inorganic compounds, *Organic compounds, Ions, Proteins, Amino acids, Phenols, Pollutant identification.
Identifiers: *Phenol method.

The applicability of the phenol method for the determination of carbohydrates in oceanic sediments was studied. Using the phenol method, various organic and inorganic compounds and combinations at concentrations greatly exceeding their content in bottom sediments were found to have no effect on the determination of carbohydrates. The maximum permissible concentrations of some inorganic compounds were determined in sediments. (Ensign-PAI)
W73-09884

MERCURY IN SEA-FOOD FROM THE COAST OFF BOMBAY,
Tata Inst. of Fundamental Research, Bombay (India).
For primary bibliographic entry see Field 05C.
W73-09891

DETERMINATION OF TRACES OF IRON, COBALT AND TITANIUM IN SEA WATER AND SUSPENDED MATTER (IN SAMPLES FROM THE BALTIC SEA AND ATLANTIC OCEAN),
Ye. M. Yemel'yanov, I. K. Blazhis, R. Yu. Yuryavichyus, R. I. Payeda, and Ch. A. Valyukyavichyus.
Oceanology, Vol 11, No 6, p 924-933, August 1972. 1 fig, 11 tab, 16 ref.

Descriptors: *Iron, *Cobalt, *Titanium, Sea water, Suspended solids, Membrane filtration, Colorimetry, *Atlantic Ocean, Pollutant identification.
Identifiers: *Baltic Sea, *North Sea, Wet combustion, Extraction-kinetic method, Sensitivity.

Iron, cobalt and titanium were determined in 250 small samples of marine suspended matter (1 to 3 mg), collected by membrane filtration on filters with pores 0.5 to 0.7Micro in diameter. The samples were dissolved by 'wet combustion' and iron was then determined colorimetrically with violet pyrocatechol, titanium colorimetrically with tiron and cobalt by the extraction-kinetic method. The sensitivity of the methods used was 2 Microgram/liter for iron, 2.0.01 Microgram/liter for cobalt and 10 Microgram/liter for titanium. In the Baltic Sea, North Sea and Northeast Atlantic 5.7 to 2595.4 Microgram/liter of Fe, 2.6.0.0001 to 3.07.0.01 Microgram/liter of Co were determined. The amounts of 0.29 to 24.81 and 1.10-5 to 1.38.10-3%, make up the per cents of dry weight respectively. In the Baltic Sea, 0.00 to 17.88 Microgram/liter or 0.00 to 0.53% of Ti was determined. (Ensign-PAI)
W73-09895

SEDIMENTARY PHOSPHORUS IN THE PAMLICO ESTUARY OF NORTH CAROLINA,
North Carolina Univ., Chapel Hill. Dept. of Environmental Sciences and Engineering.
For primary bibliographic entry see Field 05B.
W73-09897

INDIGENOUS AND PETROLEUM-DERIVED HYDROCARBONS IN A POLLUTED SEDIMENT,
Woods Hole Oceanographic Institution, Mass.
M. Blumer, and J. Sass.
Marine Pollution Bulletin, Vol 3, No 6, p 92-94, June, 1972. 1 fig, 1 tab, 16 ref. ONR CO-241, NSF GA-19472.

Descriptors: *Bays, *Oil wastes, Chemical analysis, Marine animals, Marine plants, *Sediments, Pollutant identification, Sampling, Laboratory tests, *Massachusetts.
Identifiers: *Buzzard's Bay (Mass), *Paraffins.

Chemical analysis was used to determine the presence of petroleum and petroleum products in Buzzard's Bay (Mass.) organisms and sediments. Core samples made of sediments at various locations in the bay show evidence of bacterial depletion of fuel oils in the upper 2.5 cm of the smaple. Sediments in lower sections (7.5 cm and below) of the core exhibit a strong predominance of odd carbon number paamffins. Spilled oil exhibits no odd carbon predominance, so these core analyses suggest that fuel oils have not penetrated the sediment beyond the 7.5 cm level. The area between 2.5 cm and 7.5 cm is an intermediate one, showing some signs of polluted oil as well as some natural hydrocarbon content. Spilled oil in the uppermost section of the core samples has been partially dissolved and degraded by bacteria; still, after 2 years, the fuel oil hydrocarbon in this top segment exceeds the indigenous hydrocarbons in the lower section by 1-1/2 orders of magnitude. The study maintains that chemical analysis by gas chromatography is a valid method for distinguishing between indigenous and pollutant hydrocarbons in recent sediments. (Ensign-PAI)
W73-09898

STUDIES ON THE COASTAL OCEANOGRAPHY IN THE VICINITY OF FUKUYAMA, HIROSHIMA PREF., I. DISTRIBUTION PATTERNS OF TEMPERATURE, CHLORINITY, PH AND INORGANIC NUTRIENT (PHOSPHATE-P, AMMONIA-N, NITRITE-N, NITRATE-N) CONTENTS OF SEA WATER IN EARLY FEBRUARY, 1968, (IN JAPANESE),
Hiroshima Univ. (Japan). Dept. of Fisheries.
For primary bibliographic entry see Field 05B.
W73-09899

MERCURY IN FUR SEALS,
National Marine Fisheries Service, Seattle, Wash. Marine Mammal Biological Lab.
R. E. Anas.
Paper presented at Mercury in the Western Environment Workshop, Feb. 24-26, 1971. 4 fig, 8 ref.

Descriptors: *Mercury, *Mammals, *Laboratory tests, Animal physiology, Water pollution effects,

Analytical techniques, *Neutron activation analysis, Pollutant identification.
Identifiers: *Fur seals, Atomic absorption testing.

Fur seals (Callorhinus ursinus) off the Pribilof Islands, Alaska, were tested by flameless atomic absorption for mercury contamination. Liver, muscle and brain tissues were sampled from pup, young male and adult female seals. All of the samples contained mercury, though levels were higher in the liver tissue than muscle or brain. Mercury content in livers of the pups was 0.20 ppm, 10.8 ppm for young males and 67.2 ppm for the adult females. While increased mercury levels in liver tissue were significantly correlated with age, muscle tissue showed no such correlation. Brain tissues were analyzed by neutron activation; mercury values were higher using this method than those determined by atomic absorption. The sources of mercury contamination off the Pribilof Islands have not yet been identified. (Ensign-PAI)
W73-09901

NATURAL AND MANMADE HYDROCARBONS IN THE OCEANS,
Texas Univ., Port Aransas. Marine Science Inst.
For primary bibliographic entry see Field 05B.
W73-09907

A GUIDE TO MARINE POLLUTION.

Papers presented at Food and Agriculture Organization Seminar, December 4-10, 1970. Rome, Italy 174 p. Published by Gordon and Beach, New York, 1972.

Descriptors: *Water pollution, *Pollutants, *Pollutant identification, Organic compounds, Chemicals, Nutrients, *Radioisotopes, *Monitoring, Analytical techniques.

The papers included in this seminar deal with problems of marine pollution. They present current knowledge of what substances are significant polluters and how to measure these substances. Broad categories of pullutants like hydrocarbons, organic and inorganic chemicals, nutrients, and radionuclides are discussed. Analytical techniques for identifying and quantifying these materials are reviewed. The usefulness and feasibility of small and large-scale monitoring systems are debated. The need for expanded and continued research in all areas of marine pollution is repeatedly stressed. (See W73-09916 thru W73-09924) (Ensign-PAI)
W73-09915

HALOGENATED HYDROCARBONS,
Canadian Wildlife Service, Ottawa (Ontario).
R. W. Risebrough, E. Huschenbeth, S. Jensen, and J. E. Portmann.
In: A Guide to Marine Pollution, papers presented at Food and Agriculture Organization Seminar, December 4-10, 1970, p 1-17, 1972. 26 ref.

Descriptors: *Organic compounds, *Pollutant identification, Marine plants, Marine animals, *Analytical techniques, *Monitoring, Water pollution, Research and development, *Reviews, *Bioindicators.

The current state of knowledge concerning the distribution and accumulation of halogenated hydrocarbons in the sea is reviewed. It is now believed that some form of equilibrium exists between the level of organo-chlorine pollutants in water and their level in marine animals. However, present methodology is not capable of detecting the quantities of these compounds on a routine basis in the open sea. Techniques now in use to analyze the hydrocarbons are outlined and briefly described. Areas in which greater research is needed are enumerated. A discussion of existing monitoring systems is presented; criteria for expanded systems are developed within the context of the current state of methodology. (See also W73-09915) (Ensign-PAI)
W73-09916

PETROLEUM,
Woods Hole Oceanographic Institution, Mass.
M. Blumer, P. C. Blokker, E. B. Cowell, and D. F. Duckworth.
In: A Guide to Marine Pollution, papers presented at Food and Agriculture Organization Seminar, December 4-10, 1970, p 19-40, 1972. 4 fig, 1 tab, 71 ref.

Descriptors: *Oceans, *Oil, *Oil pollution, *Distribution, *Organic compounds, Analytical techniques, Pollutant identification, Monitoring, Environmental effects.
Identifiers: *Marine ecosystems.

The complexity of the problem of petroleum pollution in sea water is discussed. Little information is available on the extent of pollution in different marine ecosystems; concentrations of pollutants vary from bulk oil to oil at very low concentration. Existing methods of hydrocarbon analysis of water, sediments and organisms are reviewed. Some of the techniques discussed include passive tagging, active tagging, mass spectrometry, and isolation of hydrocarbons. The possibility of determining a limited range of hydrocarbon standards and using automated analytical techniques are explored for the future. The concept of environmental monitoring is also covered. It is noted that a rapid method for examining large numbers of sites produces useful data more rapidly than very detailed counts in a limited number of areas. (See also W73-09915) (Ensign-PAI)
W73-09917

INORGANIC CHEMICALS,
Goeteborg Univ. (Sweden). Dept. of Analytical Chemistry.
D. Dyrssen, C. Patterson, J. Ui, and G. F. Weichert.
In: A Guide to Marine Pollution, papers presented at Food and Agriculture Organization Seminar, December 4-10, 1970, p 41-58, 1972. 2 tab, 89 ref.

Descriptors: *Oceans, *Metals, *Mercury, *Lead, *Toxicity, Aquatic life, Distribution, Pollutant identification, Environmental effects, Beryllium, Iron, Cadmium, Copper, Zinc, Water pollutants effects.

The presence of heavy metals in marine waters is discussed. Since mercury and lead are considered to be the most threatening pollutants they are given special attention. The concentration of lead in the surface waters of the oceans in the Northern Hemisphere has increased by a factor of 5 above natural levels. Mercury concentrations are not as high; however, the tendency of mercury compounds to decompose into methyl mercury which then accumulates in marine organisms through the food chain makes it an especially hazardous metal. Other toxic metals are investigated, such as beryllium, iron, cadmium, copper and zinc. Methods for detecting the presence of metals in ocean water are described briefly. The need for improved analytical methods in area like multielement and automatic analysis is noted. (See also W73-09915) (Ensign-PAI)
W73-09918

ORGANIC CHEMICALS,
Stockholm Univ. (Sweden). Institutionen for Analytisk Kemi.
J. G. Widmark, W. D. Garrett, and K. H. Palmork.
In: A Guide to Marine Pollution, papers presented at Food and Agriculture Organization Seminar, December 4-10, 1970, p 59-80, 1972. 5 fig, 4 tab, 26 ref.

Descriptors: *Oceans, *Organic compounds, *Pollutant identification, *Analytical techniques, *Sampling, Data collections, Chemicals, Water pollution effects.
Identifiers: Soluble pollutants, Insoluble pollutants, Biological response, Pollutant distribution.

The general problems associated with organic chemical pollutants are discussed. The difficulty of establishing the presence of pollutants against a background of naturally occurring carbon compounds is noted. The fact an organic compound may differ in response to various biological situations makes an exact definition of a pollutant difficult. Three primary categories of distribution of chemical pollutants in the sea are identified: water-soluble compounds, partly soluble but highly surface active compounds, hydrophobic and largely water insoluble compounds. Analytical techniques used to identify pollutants are reviewed. One method for the determination of soluble organic carbon is presented in some detail. A method for the collection of filmforming materials from the sea surface is separately discussed. (See also W73-09915) (Ensign-PAI)
W73-09919

NUTRIENT CHEMICALS (INCLUDING THOSE DERIVED FROM DETERGENTS AND AGRICULTURAL CHEMICALS),
Liverpool Univ. (England). Dept. of Oceanography.
J. P. Riley, K. Grasshoff, and A. Voipio.
In: A Guide to Marine Pollution, papers presented at Food and Agriculture Organization Seminar, December 4-10, 1970, p 81-110, 1972. 5 fig, 13 ref.

Descriptors: *Oceans, *Nutrients, Water Pollution, *Pollutant identification, Methodology, Design, Mechanical equipment, *Detergents, *Agricultural chemicals.
Identifiers: *AutoAnalyzer, Micronutrients, Pollutant levels.

Several automated procedures using an AutoAnalyzer to determine the levels of nutrient elements in sea water are described. The methods used to identify micronutrients are outlined along with the means for determining levels of dissolved phosphorus and polyphosphates. Methodology involved in identifying nitrites, nitrates, nitrogen and silicon is discussed. The AutoAnalyzer, manufactured by Technicon Instruments Corp. offers a speedy method of nutrient analysis which can be adapted to large-scale surveys and detailed monitoring. The instrument consists of a number of modules in which samples are circulated and pumped with reagents. The mixed stream is circulated through glass helices to assist mixing of the sample and the reagents; no new reagents are added until the first reaction has proceeded to completion. Typical use of the AutoAnalyzer with silicates, phosphates, nitrites and nitrates is described. (See also W73-09915) (Ensign-PAI)
W73-09920

SUSPENDED SOLIDS AND TURBIDITY,
Copenhagen Univ. (Denmark). Inst. of Physical Oceanography.
N. G. Jerlov, H. Postma, and B. Zeitschel.
In: A Guide to Marine Pollution, papers presented at Food and Agriculture Organization Seminar, December 4-10, 1970, p 111-127, 1972. 20 ref.

Descriptors: *Oceans, *Suspended solids, *Turbidity, *Pollutant identification, Methodology, *Analytical techniques, *Reviews.

River transport, effluent discharges and dumping are the major pathways by which pollutants reach the ocean. Reliable identification of these wastes in turbid waters is possible only if the pollutants have specific characteristics, which they often do not. No method has been developed yet to identify and quantify only those particles is suspended

sediments which have been introduced by pollution. Techniques for determining suspended solids such as filtering, size distribution and various types of optical methods are described. Detection of pollutants in suspended matter must involve an identification of the substances by their physical and chemical nature. Current techniques are not sufficiently sophisticated to accomplish this. (See also W73-09915) (Ensign-PAI)
W73-09921

RADIOACTIVITY,
Ministry of Agriculture, Fisheries and Food, Lowestoft (England). Fisheries Radiobiological Lab.
A. Preston, R. Fukai, H. L. Volchok, N. Yamagata, and J. W. R. Dutton.
In: A Guide to Marine Pollution, papers presented at Food and Agriculture Organization Seminar, December 4-10, 1970, p 129-146, 1972. 2 fig, 4 tab, 25 ref.

Descriptors: *Oceans, *Radioactivity, *Radioactive wastes, *Sampling, Aquatic life, Analytical techniques, Water pollution effects, Water pollution control, Monitoring, *Power plants.
Identifiers: Nuclear testing, *Gamma spectrometry, Environmental impact.

Environmental background levels of radiation have increased steadily as a result of nuclear testing and radioactive wastes from power plants. Nuclear fallout has resulted in low levels of radioactive contamination over wide ocean areas. Disposal of wastes from nuclear fission has resulted in high concentrations of radioisotopes in localized areas. The detection and measurement of radionuclides in the marine environment is discussed. For samples of seawater, biota, and sediments the final determination of the majority of radionuclides concerned can be performed by gamma-spectrometry. Chemical methods of separation for radionuclides which do not emit gamma radiation are described. Currently available methods for the detection and measurement of artificial radionuclides are adequate for the control of radioactive waste disposal. Fallout from nuclear explosions will not be of concern as long as testing continues at its present low level. (See also W73-09915) (Ensign-PAI)
W73-09922

TEST, MONITORING AND INDICATOR ORGANISMS,
Environmental Protection Agency, Gulf Breeze, Fla. Gulf Breeze Lab.
P. Butler, L. E. Andren, G. J. Bonde, A. B. Jerneloev, and D. J. Reish.
In: A Guide to Marine Pollution, papers presented at Food and Agriculture Organization Seminar, December 4-10, 1970, p 147-159, 1972. 67 ref.

Descriptors: *Oceans, Marine animals, *Bioindicators, *Laboratory tests, *Monitoring, Marine plants, Pollutant identification, Water pollution effects.

Selected species from a marine environment useful for the detection and evaluation of pollution are discussed. The distinctions between monitoring, indicator, and test organisms are noted. Monitoring organisms are used primarily to quantify pollution levels; indicator organisms are used to identify pollutants; and test organisms are monitoring and/or indicator organisms which are small and easy to handle and culture. The usefulness of a particular species depends upon the type of information desired. Various species such as bacteria, phytoplankton, algae, molluscs, fish, birds and mammals are reviewed in terms of the roles they can play in environmental monitoring. The need for further research on test, monitoring, and indicator organisms is stressed. Information on life histories, sensitive life stages, trophic webs, population fluctuations, etc. is inadequate at present. (See also W73-09915) (Ensign-PAI)
W73-09923

DESIGN OF A WORLD MONITORING SYSTEM,
Ministry of Agriculture, Fisheries and Food, Lowestoft (England). Fisheries Lab.
A. J. Lee, E. Eriksson, I. J. Haahtela, B. G. Lundholm, and S. Mizuno.
In: A Guide to Marine Pollution, papers presented at Food and Agriculture Organization Seminar, December 4-10, 1970, p 161-168, 1972. 16 ref.

Descriptors: *Oceans, *Pollutant identification, *Monitoring, *Planning, *Management, *Water pollution control, Data collections.

There are three possible types of monitoring, local, regional and global. The structure of a global system is discussed. Such a system would include a network of base-line stations, of impact stations, remote sensing systems and special systems. Base-line stations would chronicle large-scale changes; impact stations would measure the most marked changes and be sited in areas where such changes would be harmful to the marine environment; remote sensing systems, such as satellites would allow some feature of the ocean to be viewed on a global scale; and special systems would include the use of species to indicate the level of certain pollutants in various geographical areas. Implementing a global monitoring system would entail a preliminary exploratory survey, a determination of specific substances to be monitored, a decision on the frequency of sampling, and a free exchange of data among the nations of the world. (See also W73-09915) (Ensign-PAI)
W73-09924

SAMPLER-CULTURE APPARATUS FOR THE DETECTION OF COLIFORM BACTERIA IN POTABLE WATERS,
A. A. Hirsch.
US Patent No 3,708,400, 4 p, 6 fig, 1 ref; Official Gazette of the United States Patent Office, Vol 906, No 1, p 260, January 2, 1973.

Descriptors: *Patents, *Pollutant identification, *Coliforms, Bacteria, *Potable water, *Public health, Laboratory equipment.

Improvements are described for a specialized apparatus for determining the presence of coliform bacteria in potable waters, raw reservoir water, shellfish beds, and in other aqueous environments in which low density of these bacteria is imperative. A go-on-go gage determines directly whether bacterial quality of a drinking water sample meets the U.S. Public Health Service Standards; all laboratory manipulations are eliminated from sampling to gas observations. The apparatus is creep-proof; it avoids exposures and operations are precise. Gas from fermentation shows in a Durham vial held diagonally in a screw cap bottle either by being clamped in a cage or by a tailrod, both methods eliminating all axial and lateral motion. A confirmatory Brilliant Green Bile broth tube is seeded by pressing into the screw cap and inverting, thus obviating all extraneous utensils throughout the procedure. (Sinha-OEIS)
W73-09945

A SYSTEM FOR REMOTE TELEMETRY OF WATER QUALITY DATA,
Maine Univ., Orono. Dept. of Electrical Engineering.
For primary bibliographic entry see Field 07B.
W73-09946

EFFECTS OF FLUORINE IN CATTLE: A SUMMARY OF RESULTS OF STUDIES OVER A NUMBER OF YEARS IN THE NEIGHBORHOOD OF A HYDROFLUORIC ACID FACTORY AND AN ALUMINUM FACTORY (IN GERMAN),
Tieraerztlichen Hochschule, Hanover (West Germany).
For primary bibliographic entry see Field 05C.
W73-09991

IN SITU DISSOLVED OXYGEN MEASUREMENTS IN THE NORTH AND WEST ATLANTIC OCEAN,
Rhode Island Univ., Kingston. Graduate School of Oceanography.
R. B. Lambert, Jr., D. R. Kester, M. E. Q. Pilson, and K. E. Kenyon.
Journal of Geophysical Research, Vol 78, No 9, p 1479-1483, March 20, 1973. 2 fig, 13 ref. NSF Grant GA27272 ONR Grant N00014-68-A-0215-0003.

Descriptors: *Dissolved oxygen, *Atlantic Ocean, Mixing, Ocean currents, Ocean circulation, Profiles, Mass transfer, Density stratification, Water temperature.

Continuous vertical profiles of dissogved oxygen were obtained between Iceland and Nove Scotia and between Bermuda and Cape Hatteras. Small-scale variability was observed between the main oxygen minimum and the surface layer. Continuous profiling of dissolved oxygen is useful in studying small-scale mixing as well as large-scale water mass transport. (Knapp-USGS)
W73-10042

WATER QUALITY OF COLORADO RIVER (PROVINCE BUENOS AIRES) USED IN IRRIGATION, (IN SPANISH),
Universidad Nacional del Sur, Bahia Blanca (Argentina).
For primary bibliographic entry see Field 05B.
W73-10089

ESTUARINE SURVEYS IN THE SOUTHEAST,
Environmental Protection Agency, Athens, Ga. Southeast Environmental Research Lab.
For primary bibliographic entry see Field 05B.
W73-10106

PETROLEUM HYDROCARBONS IN OYSTERS FROM GALVESTON BAY,
Kiel Univ. (West Germany). Institut fuer Meereskunde.
M. Ehrhardt.
Environmental Pollution, Vol 3, p 257-271, 1972. 11 fig, 32 ref.

Descriptors: *Oil wastes, Water pollution effects, Environmental effects, *Oysters, Chromatography, *Gas chromatography, *Texas, Pollutant identification.
Identifiers: *Hydrocarbons, Aliphatic hydrocarbons, Aromatic hydrocarbons, *Galveston Bay.

Oysters from Galveston Bay, Texas were analyzed for petroleum-derived hydrocarbons. The lipids and hydrocarbons were Soxhlet-extracted with benzene/methanol and then partitioned into n-pentane. Separation of hydrocarbons from lipids was done by column chromatography on a bed of silicagel covered by alumina. These absorbents were deactivated to prevent the formation of artifacts from sensitive components of the lipid fraction. The column effluent was then resolved into aliphatic, mono-, di-, and tri- aromatic hydrocarbons by preparative TLC on activated silicagel. Identification of compounds and compound types was made from their gas chromatographic retention indices, mass spectra, and UV spectra. Severe oil contamination of the oysters was apparent. Aliphatic hydrocarbon concentration distribution was similar to the distribution found in a crude oil. Contaminated oysters had higher concentrations of alicyclic and aro-

matic hydrocarbons than the uncontaminated oysters. (EnsignnPAI)
W73-10117

DETERMINATION OF THE YIELD BY LIQUID SCINTILLATION OF THE 14C IN THE MEASUREMENT OF PRIMARY PRODUCTION IN A LAGOON ENVIRONMENT, (DETERMINATION DES RENDEMENTS DU COMPTAGE PAR SCINTILLATION LIQUIDE DU 14C DANS LES MESURES DE PRODUCTION PRIMAIRE EN MILIEU LAGUNAIRE).
For primary bibliographic entry see Field 05C.
W73-10137

TRACE METALS IN CORES FROM THE GREAT MARSH, LEWES, DELAWARE,
Delaware Univ., Newark. Dept. of Geology.
R. N. Strom, and R. B. Biggs.
College of Marine Studies Report No. 2 GL-105, December, 1972. 29 p, 12 fig, 1 tab, 7 ref. NOAA 2-35223.

Descriptors: *Bays, *Sediments, *Trace elements, Cores, Sampling, Zinc, Copper, Chromium, Iron, Cadmium, *Delaware, *Metals.
Identifiers: *Great Marsh (Del).

A study was undertaken to determine the areal and vertical changes in trace metal concentrations in sediments deposited before the industrialization of the Delaware Bay watershed. Sediment core samples were taken by hand driving plastic pipe sections into the sediment. The samples were analyzed for zinc, copper, chromium, iron, lead and cadmium. The levels of lead and cadmium were less than 1 ppm in the sediment sample. No significant changes with depth could be found for copper or chromium, while zinc concentrations increased slightly with depth. The concentration of trace metals in the <63 micrometer fraction and the >63 micrometer fraction did not differ significantly, contrary to expectations. (Ensign-PAI)
W73-10141

AERIAL EXPERIMENTS OF RADIOMETRIC DETECTION OF HYDROCARBON DEPOSITS IN THE SEA, (EXPERIENCES AERIENNES DE DETECTION RADIOMETRIQUE DE NAPPES D'HYDROCARBURES EN MER).

Bulletin d'Information, Centre National Pour l'Exploitation des Oceans, Republique Francaise, No 40, p 15-16, April 1972.

Descriptors: *Organic compounds, *Oil spills, Research and development, Programs, Monitoring, Ships, Accidents, Discharge measurement, *Remote sensing, Pollutant identification.
Identifiers: *Oil spill detection, *Radiometric detection, Radiometers, Barns, Super-Cyclope.

Experiments with aerial radiometric detection of hydrocarbon spills in the sea were conducted by the Centre National pour l/exploitation des Oceans and a research group from the ministry of National Defense. Two types of radiometers were used: Barns and Super-Cyclope. The experiments constituted the first phase of a research and development program to assure the surveillance of oil spills originating from shipwrecks of petroleum ships as well as the detection of fraudulent discharges made by ships in operation. (Ensign-PAI)
W73-10148

A NEW APPROACH TO THE MEASUREMENT OF RATE OF OXIDATION OF CRUDE OIL IN SEAWATER SYSTEMS,
University Coll. of North Wales, Menai Bridge. Marine Science Labs.
For primary bibliographic entry see Field 05B.
W73-10150

RADIATION DATA, SECTION II, WATER.
Office of Radiation Programs, Washington, D.C.
For primary bibliographic entry see Field 05B.
W73-10181

POTENTIALITIES AND LIMITS OF MULTIELEMENT NEUTRON ACTIVATION ANALYSIS OF BIOLOGICAL SPECIMENS WITH OR WITHOUT SEPARATION FROM AN ACTIVE MATRICE, (POSSIBILITIES ET LIMITES DE L'ANALYSE PAR ACTIVATION NEUTRONIQUE MULTIELE MENTAIRE D'ECHANTILLONS BIOLOGIQUES AVEC OU SANS SEPARATION DE LA MATRICE ACTIVABLE.),
Commissariat a l'Energie Atomique, Saclay (France). Centre d'Etudes Nucleaires.
B. Maziere, A. Gaudry, W. Stanilewicz, and D. Comar.
Available from NTIS, Springfield, Va., as 721010-26; $3 in paper copy, $0.95 in microfiche. Report CONF-721010-26, 1972. 22 p, 4 fig, 4 tab, 8 ref.

Descriptors: *Radioactivity techniques, *Analytical techniques, *Neutron activation analysis, *Bioindicators, *Pollutant identification, *Trace elements, Gamma rays, Spectrometry, Separation techniques, Ion exchange, Radiochemical analysis.

Eight essential-nutrient trace elements were determined in blood serum and powdered liver. Chemical separation was not required for Fe, Co, Zn and Se; chemical-separation methods requiring a few hours were used with Mn and Cu; more extensive chemical methods were used with Mn and Cr. A total of 43 trace elements were determined of which precise values were obtained for 12, and upper limits for the remainder. The determination of Yb, Th, Ir, Cd, Au, Hf, Ru, Sb, Sc, Tb, Ta, and Eu was not as sensitive as by other methods. (Bopp-ORNL)
W73-10184

ENVIRONMENTAL SURVEILLANCE FOR RADIOACTIVITY IN THE VICINITY OF THE CRYSTAL RIVER NUCLEAR POWER PLANT, AN ECOLOGICAL APPROACH,
Florida Univ., Gainesville. Dept. of Environmental Engineering Sciences.
For primary bibliographic entry see Field 05B.
W73-10185

ANNUAL ENVIRONMENTAL MONITORING REPORT: CALENDAR YEAR 1972,
Mound Lab., Miamisburg, Ohio.
For primary bibliographic entry see Field 05B.
W73-10201

ENVIRONMENTAL LEVELS OF RADIOACTIVITY IN THE VICINITY OF THE LAWRENCE LIVERMORE LABORATORY: 1972 ANNUAL REPORT,
California Univ., Livermore. Lawrence Livermore Lab.
For primary bibliographic entry see Field 05B.
W73-10202

ENVIRONMENTAL MONITORING AT ARGONNE NATIONAL LABORATORY - ANNUAL REPORT FOR 1972,
Argonne National Lab., Ill.
J. Sedlet, N. W. Golcbert, and T. L. Duffy.
Available from NTIS, SPringfield, Va., as Report No ANL-8007; $3.00 per copy, $0.95 microfiche. Report No ANL-8007, March 1973. 68 p, 5 fig, 26 tab, 22 ref, 1 append.

Descriptors: *Monitoring, *Environment, *Safety, *Public health, *Radioactivity, Air pollution, Water pollution, Soil contamination, Water pollution sources, Fallout, Effluents, Nuclear wastes, Aquatic algae, Aquatic animals, Milk, Food chains.

The environmental monitoring program at Argonne National Laboratory for 1972 is described and the results are presented. To evaluate the effect of Argonne operations on the environment, measurements were made for a variety of radionuclides in air, surface water, soil, grass, benthos, and milk; for a variety of chemical constituents in surface and Argonne effluent water; and for the environmental penetrating radiation dose. Sample collections and measurements were made both on and off the Argonne site for comparison purposes. The results of the program are interpreted in terms of the sources and origin of the radioactive and chemical substances (natural, fallout, Argonne, and other) and are compared with accepted environmental quality standards. (Houser-ORNL)
W73-10206

RADIOACTIVITY OF THE WATER OF THE NORTHEASTERN PART OF THE ATLANTIC OCEAN,
Akademiya Nauk URSR, Sevastopol. Marine Hydrophysics Inst.
For primary bibliographic entry see Field 05B.
W73-10220

A STUDY OF THE COST EFFECTIVENESS OF REMOTE SENSING SYSTEMS FOR OCEAN SLICK DETECTION AND CLASSIFICATION,
New Hampshire Univ., Durham. Dept. of Electrical Engineering.
G. C. Gerhard.
Sea Grant Programs Report No UNHSG-101, EDAL Report No 112, April 1972. 26 p, 3 fig, 5 ref. UNHSG-2-35244.

Descriptors: *Water pollution, *Pollutant identification, *Oil spills, *Remote sensing, Surveys, Monitoring, Instrumentation, *Costs.

A 5-month investigation concerning the uses of, and the need for, remote sensing devices and techniques to be applied to the oil slick detection and classification problem is reported. The current U.S. Coast Guard surveillance system, consisting of daytime visual spottings by patrol helicopter pilots, is evaluated. A sensor system which allows only a 1 to 5% improvement in detection capability can in a 1-yr period justify reasonable costs for technical feasibility studies and prototype development work. Other needs such as improved oil slick thickness monitoring capabilitiies and nighttime identification techniques warrant further study and development to realize significant improvements in the cost effectiveness of sensor packages. The type of instruments and sensor packages based on Coast Guard requirements and availability, as well as the best obtainable cost estimates, are reviewed. (Ensign-PAI)
W73-10228

THE SOURCE IDENTIFICATION OF MARINE HYDROCARBONS BY GAS CHROMATOGRAPHY,
Woods Hole Oceanographic Institution, Mass.
M. Ehrhardt, and M. Blumer.
Environmental Pollution, London, Vol 3, p 179-194, July 1972. 8 fig, 2 tab, 21 ref. NSF-GA-19472, EPA-18050 EBN.

Descriptors: *Pollutant identification, *Organic compounds, *Chromatography, *Seawater, *Oil spills, Oil, Degradation (Decomposition), Evaporation, Bacteria, Weathering.
Identifiers: Dissolution, Oil compositional parameters.

A method for the source identification of oil spills by gas chromatographic analysis is described. The compositional parameters for eight different crude oils are given. The environmental degradation of oil by evaporation, dissolution, and bacteria decomposition is discussed. Distinguishing compositional features are still recognizable after more

Available from NTIS, Springfield, Va. 22151. NASA SP-322 Price $3.00.

Descriptors: *Air pollution, *Air pollution effects, *Pollution abatement, Control, Model studies, *Dispersion, Testing procedures, *Reviews, Evaluation, Diffusion, Aircraft, Meteoric water, Bibliographies.
Identifiers: *Atmospheric pollution dispersion, Aircraft emissions.

A survey was performed to identify problem areas in modeling air pollution dispersion and to summarize the current state of the art for these areas. Included are a summary of previous urban dispersion models, a review of the underlying concepts and assumptions, the techniques employed, and recommended research efforts which potentially can improve the ability to predict the dispersion of pollutants in the atmospheric boundary layer. Specific recommendations are made for utilizing these models to study the effects of pollutant contributions resulting from aircraft operations. Emphasis is on areas which are of direct concern to NASA or to which NASA's expertise could readily be applied. (Woodard-USGS)
W73-10324

BIOXIDATION OF WASTE IN ACTIVATED SLUDGE: I. MEASURING OXYGEN RESPIRATION OF ACTIVATED SLUDGE BY MEANS OF POLAROGRAPHIC AND OXYGEN PROBE METHODS,
G. Buraczewski, and A. Rotowska.
Acta Microbiol Pol Ser B Microbiol Appl. Vol 3, No 2, p 121-124, 1971. Illus.
Identifiers: *Activated sludge, Electrochemical methods, Measuring methods, *Polarographic method, Respiration, Sludge, Waste water treatment, *Oxygen probe, *Bioxidation.

Two electrochemical methods, using polarography and an oxygen probe, were tested. Experiments and apparatus are described. Both methods can be used for respiration experiments of activated sludge. The oxygen probe by Todt or a modified measuring technique using the polargraphic method should be used because smaller discrepancies in measuring results. Micro devices allow for a constant flow of activated sludge synchronized with the movement of the register tape inside the polarograph.--Copyright 1972, Biological Abstracts, Inc.
W73-10387

BACTERIOLOGICAL FINDINGS IN LAKE, RIVER AND WELL WATER SUPPLIES OF RANGOON,
Institute of Medicine (I), Rangoon (Bruma). Dept. of Microbiology.
S. Hla-Gyaw, K. Sann-Myint, H. Chen, and M. Tu.
Union Burma J Life Sci. Vol 4, No 1, p 95-103. 1971. Illus.
Identifiers: Alcaligenes-faecalis, Baxillus-sp, *Bacteriological studies, *Burma (Rangoon), Citrobacter-freundii, Clostridium-perfringens, Enterobacter-aerogenes, Escherichia-coli, Klebsiella-aerogenes, Proteus-mirabilis, Proteus-morganii, Proteus-vulgaris, Providencia, Pseudomonas-aeruginosa, Pseudomonas-fluorescens, Streptococcus-faecalis, *Water supply.

Water samples, both unchlorinated and chlorinated, from 20 natural water sources in Rangoon towns comprising lakes, river and wells were examined during the period July 1966 to July 1968 for the Presumptive Coliform Count, the Presumptive Enterococcus Count and the presence of presumptive Clostridium perfringens. Bacteria isolated from MacConkey bile salt lactose peptone water and sodium azide medium primary cultures were identified. Using as criteria the Presumptive Coliform Count, the Presumptive Enterococcus Count and the isolation of Escherichia coli and/or Klebsiella aerogenes and/or Streptococcus faecalis, 18 out of 20 samples were found unsatisfactory for drinking purposes. The bacteria isolated were E. coli, K. aerogenes, Citrobacter freundii, Enterobacter aerogenes, Alcaligenes faecalis, a Bacillus sp., Proteus mirabilis, P. morganii, P. vulgaris, a Providencia strain, Pseudomonas aeruginosa, P. fluorescens and S. faecalis.--Copyright 1972, Biological Abstracts, Inc.
W73-10391

5B. Sources of Pollution

PHYSICAL AND CHEMICAL PROPERTIES OF THE COASTAL WATERS OF GEORGIA,
Georgia Univ., Sapelo Island. Marine Inst.
For primary bibliographic entry see Field 02L.
W73-09752

NUTRIENT TRANSPORT BY SEDIMENT-WATER INTERACTION,
Illinois Univ., Chicago. Dept. of Materials Engineering.
For primary bibliographic entry see Field 05C.
W73-09754

WATER AND WASTEWATER MANAGEMENT IN DAIRY PROCESSING,
North Carolina State Univ., Raleigh. Dept. of Food Science.
R. E. Carawan, V. A. Jones, and A. P. Hansen.
Available from the National Technical Information Service as PB-220 704 $3.00 in paper copy, $0.95 in microfiche. North Carolina Water Resources Research Institute, Raleigh, UNC-WRRI Report No 79, December 1972. 132 p, 19 fig, 20 tab, 21 ref, 4 append. OWRR A-058-NC (1). 14-31-0001-3533.

Descriptors: *Dairy Industry, *Food Processing Industry, Water conservation, *Waste disposal, *Municipal wastes, Industrial water, *North Carolina, *Water utilization, Biochemiacal oxygen demand, Effluents, Industrial wastes.
Identifiers: *Milk processing.

A typical North Carolina multi-product plant was evaluated for water use and wastewater effluent quantities and characteristics. The maximum rate of milk usage was 40,000 gal/day. Water use totaled 325.5 gal/1000 lb total product with the processing areas requiring 311.4 gal/1000 lb total product. The largest water use operation was utilities which accounted for 38.8% of the total process water. The effluent was sampled at two locations in the process drain system with the frozen products area contributing 54% of the BOD and 69% of the fat for only 11% of the plant production. Waste parameters identified included BOD, 7.34 lb/1000 lb product (2257 mg/l); TSS, 3.59 lb/1000 lb product (1104 mg/l) and Fat, 2.34 lb/1000 lb product (720 mg/l). Effluent temperatures were monitored and averaged 95-100 F for the fluid products and by-products drain and 85-90 F for the frozen products drain. The average BOD/COD ratio was 0.64. BOD/TOC values ranged from 0.74-2.22 for limited data. Wide variation was observed in the day-to-day water use, waste discharge and production of products. A program is presented to help dairies reduce their water use and waste generation. This program stresses the need for minimum water use and waste discharge to achieve economic and environmental goals.
W73-09755

REPORT ON POLLUTION OF THE ST. MARYS AND AMELIA ESTUARIES, GEORGIA-FLORIDA,
Environmental Protection Agency, Athens, Ga. Southeast Water Lab.
T. P. Gallager.

Available from NTIS, Springfield, Va. 22151, PB-213 349; Price $3.00 printed copy; $0.95 microfiche. EPA report, August 1971. 51 p, 7 fig, 4 tab, 2 append.

Descriptors: *Water pollution sources, *Estuaries, *Georgia, *Florida, Chemical analysis, Pollutant identification, Industrial wastes, Municipal wastes, Sampling, Water analysis, Dissolved oxygen, Biochemical oxygen demand, Coliforms, Water quality.
Identifiers: *St. Marys Estuary, *Amelia Estuary.

Survey results indicate that pollution of the St. Marys and Amelia Estuaries in Georgia and Florida have resulted in deleterious water quality. Data are based on the results of water quality surveys conducted by the Federal Water Pollution Control Administration in 1967. More than 982,000 population equivalents (PE) of 5-day biochemical oxygen demand were discharged into these waters--710,000 PE (72%) into Florida waters and 272,000 PE (28%) into Georgia waters. Untreated wastes discharged from industries and municipalities caused dissolved oxygen in the Amelia River to be depleted to a mean value of 2.0 mg/liter with a minimum observed value of 0.6 mg/liter. Total coliform bacteria densities in the Amelia River exceeded 1,000/100 ml which was considered minimal for direct contact recreation. Pollution from waste discharges limits the use of the river for recreation and fishing activities. (Woodard-USGS)
W73-09788

KINETICS OF CARBONATE-SEAWATER INTERACTIONS,
Hawaii Inst. of Geophysics, Honolulu.
For primary bibliographic entry see Field 02K.
W73-09790

INVESTIGATIONS AND FORECAST COMPUTATIONS FOR THE CONSERVATION OF GROUNDWATER (ISSLEDOVANIYA I PROGNOZNYYE RASHETY DLYA OKHRANY PODZENMYKH VOD),
Ye. L. Minkin.
Izdatel'stvo 'Nedra', Moscow, 1972. 110 p.

Descriptors: *Hydrogeology, *Groundwater, *Water conservation, *Water pollution, *Path of pollutants, Water pollution sources, Pollutants, Pollutant identification, Water pollution control, Water quality, Water chemistry, Aquifers, Water wells, Groundwater mining, Withdrawal, Industrial plants, Mathematics, Equations, Maps, Investigations, Forecasting.
Identifiers: *USSR, Mineralization.

Sources of pollution and pathways by which pollutants move through groundwater are examined for a classification of groundwater contamination conditions. Hydrogeological investigations to justify measures for the protection of groundwater from pollution are described, and hydrogeological computations are given for forecasting possible changes in groundwater quality in regions of withdrawals. (Josefson-USGS)
W73-09802

HYDROCHEMICAL TYPING OF RIVER BASINS OF THE UKRAINE,
Akademiya Nauk URSR, Kiev. Instytut Hidrobiologii.
A. D. Konenko, and N. M. Kuz'menko.
Hydrobiological Journal, Vol 8, No 1, p 1-10, 1972. 1 fig, 1 tab, 24 ref.

Descriptors: *Water chemistry, *Water quality, *River basins, Rivers, Streams, Ponds, Organic compounds, Inorganic compounds, Salts, Salinity, Ions, Water pollution, Investigations.
Identifiers: *USSR, *Ukraine.

Tabulation of hydrochemical characteristics of river basins is based on long-term, large-scale investigations of the hydrochemistry of Ukrainian rivers and ponds. The salt composition of the waters of small rivers is reflected in the different types of basins described. The influence of man is reflected in the sporadic increase in chloride and sulfate concentrations and in the sharp increase in concentrations of biogenic and organic substances. Small rivers of the Ukraine are polluted by household wastes, which results in eutrophication of reservoirs and ponds. Municipal sewage discharge and runoff from fertilized fields result in higher ammonium, nitrate, and phosphorus concentrations and in a high oxidation potential of the water. Manmade salinization is most evident in rivers of the Dnieper steppe, Dnieper-Donetsk steppe, Donets steppe, and Azov region. Man-induced salinization is less apparent in regions with abundant moisture. (Josefson-USGS)
W73-09811

RATIO OF ORGANIC AND MINERAL COMPONENTS OF THE SOUTH BAIKAL SUSPENDED WATERS (IN RUSSIAN),
E. N. Tarasova.
Gidrobiol Zh. Vol 8, No 1, p 17-25. 1972. Illus. English summary.
Identifiers: *Lake Baikal, Melosira, *Minerals, *Organic matter, Suspended solids, Synedra, USSR, Water pollution sources.

Data are presented for the first time on the range of total amount of suspensions in the surface layer of Lake Baikal. Seasonal changes in ratios of mineral and organic components of the suspension are considered by the vertical line and the role of the Baikal endemic diatoms (Melosira and Synedra) is estimated in suspended organic matter of waters in the open part of the lake. Increased concentrations of suspended organic matter were found near the littoral zone and in deep water they were found in May and Aug. of 1968. The ratio of mineral and organic components of the suspension undergoes distinct seasonal changes. The phytoplankton in the period of its mass development (spring) is not constant in suspended organic matter.--Copyright 1972, Biological Abstracts, Inc.
W73-09812

METROPOLITAN COMPREHENSIVE WATER, SEWAGE, AND SOLID WASTE PLANNING STUDY.
Bi-State Metropolitan Planning Commission, Rock Island, Ill.
For primary bibliographic entry see Field 06A.
W73-09829

THE DISSIPATION OF EXCESS HEAT FROM WATER SYSTEMS,
Geological Survey, Bay St. Louis, Miss.
H. E. Jobson.
Journal of the Power Division, American Society of Civil Engineers, Vol 99, No P01, Proceedings paper 9702, p 89-103, May 1973. 1 fig, 1 tab, 35 equ, 16 ref.

Descriptors: *Thermal pollution, *Water temperature, *Mathematical models, *Heat transfer, Predicting, Equations, *Diffusion, Humidity, Air temperature, Water resources.
Identifiers: Heat transmission, Wind speed.

A model for the prediction of excess temperature is shown to have the form of the classical diffusion equation. The excess temperature in this model is defined as the difference between the actual water temperature and the water temperature which would have occurred provided that a particular heat source, such as a power plant, had never existed. This definition of excess temperature is convenient because it directly represents the incremental effect that a heat source has on a water system and, provided that historical water temperature records are available, the excess temperature can be predicted without the use of solar or atmospheric radiation data. A prediction equation for the surface transfer coefficient for excess heat illustrates that this coefficient is primarily dependent upon the water temperature and wind speed, and that it is almost independent of the humidity and temperature of the air. (Bell-Cornell)
W73-09839

THERMAL CAPACITY OF OUR NATION'S WATERWAYS,
Westinghouse Hanford Co., Richland, Wash.
For primary bibliographic entry see Field 05G.
W73-09840

OPTIMIZATION OF CIRCULATING WATER SYSTEM,
Worcester Polytechnic Inst., Mass.
For primary bibliographic entry see Field 05G.
W73-09842

QUALITATIVE COMPOSITION OF ORGANIC SUBSTANCE IN WATERS OF THE DNIEPER RESERVOIRS, (IN RUSSIAN),
Akademiya Nauk URSR, Kiev. Institut Hidrobiologii.
G. A. Enaki.
Gidrobiol Zh. Vol 8, No 1, p 26-31, 1972. English summary.
Identifiers: *Amino-Acids, *Carbohydrates, *Dnieper reservoirs, Organic matter, Qualitative analysis, Reservoirs, USSR, Water pollution sources.

The amino acid and carbohydrate contents in the Dnieper cascade reservoirs which have arid and dry climate was higher than in the Kiev reservoir. The tendency to increase in the content of biochemically available compounds in the reservoir waters down the cascade was observed.--Copyright 1972, Biological Abstracts, Inc.
W73-09843

THERMAL EFFECTS OF PROJECTED POWER GROWTH: SOUTH ATLANTIC AND EAST GULF COAST RIVER BASINS,
Hanford Engineering Development Lab., Richland, Wash.
For primary bibliographic entry see Field 05C.
W73-09848

THE CASE FOR THERMAL POLLUTION,
Wisconsin Univ., Madison.
For primary bibliographic entry see Field 05G.
W73-09864

SIMPLIFIED MATHEMATICAL MODEL OF TEMPERATURE CHANGES IN RIVERS,
Associated Water and Air Resources Engineers, Inc., Nashville, Tenn.
V. Novotny, and P. A. Krenkel.
Journal Water Pollution Control Federation, Vol 45, No 2, p 239-248, February 1973. 3 fig, 7 ref, 2 append. EPA Grant 1642-11.

Descriptors: *Mathematical models, Equations, Flow, *Heat transfer, Streams, Hydraulic models, Hydraulic engineering, Model studies, Fluid mechanics, Analytical techniques, Theoretical analysis, Movement, Energy, Radiation, Temperature, *Thermal pollution, Heated water, Discharges (Water), Rivers, *Water temperature.

Mathematical models for describing stream temperature changes which are based on studies of evaporation phenomena in lakes are seen as inadequate because they fail to take the dynamic behavior of the air-water interface, water body size differences or topographical differences into account. Theoretical equations are presented for

as turbulence, entrainment, buoyancy, and heat transfer. The analytical results are supported by experimental data and demonstrate the usefulness of the theory for estimating the location and size of the effluent with respect to the discharge point. The capability of predicting jet flow properties, as well as two- and three-dimensional jet paths, was enhanced by obtaining the jet cross-sectional area during the solution of the conservation equations (a number of previous studies assume a specific growth for the area). Realistic estimates of temperature in the effluent were acquired by accounting for heat losses in the jet flow due to forced convection and to entrainment of free-stream fluid into the jet.
W73-09876

ARTIFICIAL RADIOACTIVITY IN FRESH WATER AND ESTUARINE SYSTEMS,
Ministry of Agriculture, Fisheries and Food, Lowestoft (England). Fisheries Radiobiological Lab.
A. Preston.
Proceedings of the Royal Society of London, Series B, Vol 180, p 421-436, 1972. 6 fig, 6 tab, 54 ref.

Descriptors: *Estuaries, Freshwater, *Radioactivity, Behavior, *Nuclear wastes, *Waste disposal, Water pollution control, Data collections, Control systems, Safety, Standards, *Estuarine environment.
Identifiers: *United Kingdom.

The nature and origin of artificial radioactivity in freshwater and estuarine environments are described. Man-made sources, such as atmospheric nuclear tests, residues associated with nuclear weapon production and electric power generation from nuclear fission are considered. The situation in the United Kingdom, particularly as regards the expansion of nuclear power to the end of the century is reviewed. The policy in the U.K. concerning the disposal of radioactive waste is stated and the system of control used to achieve the objectives of this policy (based on the international standards for safe exposure to members of the public) is described. Artificial radionuclide behavioral data, on which systems of control are initially based, are reviewed and the current radiological status of freshwater and estuarine disposal areas of the United Kingdom is presented. (Ensign-PAI)
W73-09893

SEDIMENTARY PHOSPHORUS IN THE PAMLICO ESTUARY OF NORTH CAROLINA,
North Carolina Univ., Chapel Hill. Dept. of Environmental Sciences and Engineering.
J. B. Upchurch.
Sea Grant Publication UNC-SG-72-03, May 1972. 39 p, 12 fig, 1 tab, 29ref, 2 append.

Descriptors: *Estuaries, *Sediments, *Phosphorus, Freshwater, *Salinity, Separation techniques, Iron, Manganese, Silt, Clay, *North Carolina.
Identifiers: Ferric oxides, *Pamlico estuary (N.C.).

Observations along the 30 mile length of the Pamlico Estuary on the amount of 'available' phosphorus extracted from sediment samples indicated a decrease from 1.6 mg P/gm sediment in fresh water to 0.3 mg P/gm sediment in water with a salinity of 18 ppt. A modification of the HCl-H2SO4 (pH 1.1) acid extraction procedure measured the available phosphorus. A high degree of correlation (r±0.99) between oxalate-extractable iron and available phosphorus was found in the upper reach of the estuary (salinity less than 1 ppt). The Fe-P correlation decreased (r±0.86) in the lower part of the estuary. The decrease in the available P and in the Fe-P correlation along the length of the estuary are consistent with the theory that P is held to suspended sediments by some type of Fe-inorganic P complex of limited stability. The amount of phosphorus would decrease when sediments entering the estuary in the freshwater inflow were transported through the more saline waters to the estuary mouth. (Ensign-PAI)
W73-09897

STUDIES ON THE COASTAL OCEANOGRAPHY IN THE VICINITY OF FUKUYAMA, HIROSHIMA PREF., I. DISTRIBUTION PATTERNS OF TEMPERATURE, CHLORINITY, PH AND INORGANIC NUTRIENT (PHOSPHATE-P, AMMONIA-N, NITRITE-N, NITRATE-N) CONTENTS OF SEA WATER IN EARLY FEBRUARY, 1968, (IN JAPANESE),
Hiroshima Univ. (Japan). Dept. of Fisheries.
H. Koyama, and H. Ochiai.
Journal of the Faculty of Fisheries and Animal Husbandry, Hiroshima University, Vol 11, No 1 p 65-77, July 1972. 8 fig, 3 tab. English summary.

Descriptors: Estuaries, *Coasts, Oceanography, Pollutants, Water temperature, *Chlorinity, Hydrogen ion concentration, *Phosphates, *Ammonia, Nitrites, *Nitrates, Chemical industry, Waste water (Pollution), Outlets, Circulation, Cooling, Nutrients, Water pollution sources.
Identifiers: *Japan, *Ashida River estuary.

Measurements were taken of water temperature, chlorinity, pH and phosphate-P, ammonia-N, nitrite-N and nitrate-N contents of seawater off the estuary of the Ashida River, at 14 stations at both high and low water in February, 1968. The water temperature was within a range of 6-8 deg, its vertical variation was very slight reflecting the vertical circulation of water due to convective cooling. Chlorinity also demonstrated a slight vertical variation due to the same cause everywhere except near the river mouth and an outfall from a chemical plant. Normal pH values were encountered everywhere except in the vicinity of the chemical plant where values as low as 2.4-8.0 pH were measured. Nutrient content in seawater was low but was somewhat higher near the Ashida River estuary, and abnormally high in water samples near waste water outlets from the chemical plant. (Ensign-PAI)
W73-09899

SEWAGE DISPOSAL IN THE TURKISH STRAITS,
Woodward-Envicon, Inc., San Diego, Calif.
C. G. Gunnerson, E. Sungur, E. Bilal, and E. Ozturgut.
Water Research, Vol 6, p 763-774, 1972. 4 fig, 1 tab, 14 ref.

Descriptors: *Waste disposal, *Sewage, *Oceanography, *Circulation, *Currents (Water), *Mixing, Anaerobic conditions, Sanitary engineering, Public health, Coliforms, Bacteria, Outlets, Design criteria, Treatment, Construction.
Identifiers: *Turkish Straits, Mediterranean Sea, Aegean Sea, Dardanelles, Sea of Marmara, Bosporus Sea, Black Sea.

The Turkish Straits extend approximately 300 km from the Aegean Sea through the Dardanelles, Sea of Marmara, and Bosporus to the Black Sea. There is a stable two layer current system where the Mediterranean's heavier highly saline water seeks its own level and flows north to the anaerobic lower portion of the Black Sea. The brackish surface waters carrying river runoff from the Black Sea flows southward. Due to the circulation patterns of these straits it is possible for Istanbul and nearby communities to safely discharge, after flotation, sewage into the lower layer where it will mix with Mediterranean waters flowing into the anaerobic lower portion of the Black Sea. The oceanography of the area is described in detail along with the sanitary engineering factors such as waste characteristics, public health, and coliform bacteria persistence. Functional design criteria

such as depth of discharge, outfall and treatment requirements, vertical flux of sewage, suburban locations and construction priorities are considered. (Ensign-PAI)
W73-09904

MARINE POLLUTION,
For primary bibliographic entry see Field 05G.
W73-09905

NATURAL AND MANMADE HYDROCARBONS IN THE OCEANS,
Texas Univ., Port Aransas. Marine Science Inst.
K. Winters, and P. L. Parker.
1972, 25 P. 5 FIG, 11 TAB, NO REF.

Descriptors: *Organic compounds, Oceans, Aquatic life, Oil, *Oil pollution, *Gulf of Mexico, *Laboratory tests, Bioindicators, Plankton, Sampling.
Identifiers: *Petroleum paraffins.

The abundance of various paraffins in marine materials was measured in order to establish base line data concerning the presence of petroleum derived organic matter in the Gulf of Mexico and other selected locations. The samples were solvent extracted for 12 hours using freshly distilled hexane in a continuous extractor. Laboratory testing yielded the following results: (1) normal paraffin concentrations are higher in waters over the shelf than in open ocean waters, (2) plankton are useful petroleum pollution indicators, and (3) higher organisms from the open sea do not appear to be highly polluted with petroleum. It is also noted that the experimental method used has a direct influence on the data obtained. The problem of transporting samples is a serious one and must be overcome before extensive petroleum pollution research can be undertaken. (Ensign-PAI)
W73-09907

FURTHER INVESTIGATIONS OF THE TRANSFER OF BOMB 14C TO MAN,
Scottish Research Reactor Centre, Glasgow (Scotland). Radiocarbon Unit.
D. D. Harkness, and A. Walton.
Nature, Vol 240, p 302-303, December 1, 1972. 2 fig, 2 tab, 20 ref.

Descriptors: *Carbon radioisotopes, *Atmosphere, *Diet, *Food chains, *Human pathology, Transfer, Absorption, Tracers, Kinetics, Air pollution.
Identifiers: *Bomb 14C, Residence time.

Concentration of 'bomb 14C' reported for human tissues have differed slightly from atmospheric 14C variations. Two factors have been considered as the major contributors to this damping effect: the selective composition of the human diet and the finite residence time for carbon in human tissues. The mean dietary intake of 14C by the population of the United Kingdom is evaluated and the possible application of temporal changes in 'bomb 14C' concentrations as a kinetic tracer for human metabolic processes is assessed. (Ensign-PAI)
W73-09908

WASTE DISPOSAL AND WATER QUALITY IN THE ESTUARIES OF NORTH CAROLINA,
North Carolina State Univ., Raleigh. Dept. Civil Engineering.
C. D. Riley.
School of Engineering, Summer Project CE 598, Summer 1972. 68p, 1 append.

Descriptors: *Estuaries, *Water quality, *Waste disposal, *Water pollution sources, *Water quality control, Municipal wastes, Urban runoff, Agricultural runoff, Industrial wastes, Water utilization, Hydrology, *North Carolina.

An investigation into waste disposal and pollution in the marine waters of North Carolina is reported. The purpose is to obtain a feeling for the scope and magnitude of the pollution problem of these areas. Inventories of municipal and industrial waste discharges are developed. Pollution associated with diffuse sources such as marine traffic, agricultural runoff, and urban runoff is investigated. The magnitude of the solid waste and air pollution problem is discussed. A series of maps indicate the 'best use' of these waters as established by the North Carolina Department of Water and Air Resources. Restricted shellfish areas as determined by the North Carolina Department of Public Health are also indicated. The general hydrological regime of the geographic area of concern is discussed. (Ensign-PAI)
W73-09911

REVIEW ON THE STATE OF MARINE POLLUTION IN THE MEDITERRANEAN SEA.
Rome, February 1972. 47 p, 78 ref.

Descriptors: *Reviews, Water pollution sources, *Transportation, *Sewage, *Industrial wastes, Oil, *Domestic wastes, Pesticides, Radioactive wastes, Thermal pollution, Plastics, Polychlorinated biphenyls, Suspended load, Water pollution effects, Water quality control, Regulation, International law.
Identifiers: *Mediterranean Sea.

The study on the state of pollution in the Mediterranean Sea covers the following subjects: domestic sewage, industrial pollution, oil pollution, pesticides and other types of pollution. The data were collected sending a questionnaire to all countries bordering the Mediterranean. The following conclusions are presented: all the main urban centers still discharge waste water directly into the sea. The level of organic load by industrial waste and the impact of toxic substances on the water in the eastern part of the Mediterranean are not yet dangerous. The eastern Mediterranean is mainly a region of oil production, and there are only a few big and well defined industrial centers contributing to the contamination by oil. In the western region the main sources of oil pollution are the industrial oil processing enterprises and the trans-shipment of their product. Pollution by pesticides was a local problem, the adverse effects have been found primarily in coastal zones. Radioactive substances, thermal pollution, plastics, suspended matter and polychlorinated biphenyls have not yet seriously affected the water quality in the Mediterranean. The second part of the report is a review of the regulations of Mediterranean countries governing pollution of fresh water resources and the seas. (Ensign-PAI)
W73-09913

A GUIDE TO MARINE POLLUTION.
For primary bibliographic entry see Field 05A.
W73-09915

HALOGENATED HYDROCARBONS,
Canadian Wildlife Service, Ottawa (Ontario).
For primary bibliographic entry see Field 05A.
W73-09916

SUSPENDED SOLIDS AND TURBIDITY,
Copenhagen Univ. (Denmark). Inst. of Physical Oceanography.
For primary bibliographic entry see Field 05A.
W73-09921

RADIOACTIVITY,
Ministry of Agriculture, Fisheries and Food, Lowestoft (England). Fisheries Radiobiological Lab.
For primary bibliographic entry see Field 05A.
W73-09922

A SYSTEM FOR REMOTE TELEMETRY OF WATER QUALITY DATA,
Maine Univ., Orono. Dept. of Electrical Engineering.
For primary bibliographic entry see Field 07B.
W73-09946

THE APPLICATION OF THE TRANSPORT EQUATIONS TO A GROUNDWATER SYSTEM,
Geological Survey, Denver, Colo.
J. D. Bredehoeft, and G. F. Pinder.
Proc available from 24th Int. Geological Congress, 601 Booth St., Ottawa, Canada KIA OE8. Price $7.00. In: Proceedings of the 24th Session of the International Geological Congress, Section 11, Hydrology; Montreal, Canada, 1972: International Geological Congress Publication, p 255-263, 1972. 5 fig, 7 ref.

Descriptors: *Groundwater movement, *Path of pollutants, *Numerical analysis, Equations, Mathematical models, Model studies, Hydrogeology, Mixing, Water chemistry, Mass transfer, Ion transport, *Georgia.
Identifiers: *Brunswick (Ga).

Predicting changes in groundwater quality in a complex hydrologic system generally requires simulation of the field problem through the use of a deterministic model. In the most general case, a complete physical-chemical description of moving groundwater requires the simultaneous solution of the differential equations that describe the transport of mass, momentum, and energy. The difficulties in solving this set of equations requires simplified subsets of equations. The equation of motion for single-component groundwater flow was solved for many different initial and boundary conditions. To describe the transport of miscible fluids of differing density, the mass-transport equation and the equation of motion were coupled and solved numerically. Numerical solutions were also obtained for the heat transport equation and the equation of motion, particularly for convection problems. A case history of groundwater contamination at Brunswick, Georgia, illustrates the use of the mass- and momentum-transport equations in predicting movement of contaminants. (Knapp-USGS)
W73-09968

SOURCE DETERMINATION OF MINERALIZATION IN QUATERNARY AQUIFERS OF WEST-CENTRAL KANSAS,
Kansas State Dept. of Health, Dodge City.
W. R. Bryson.
Proc available from 24th Int. Geological Congress, 601 Booth St., Ottawa, Canada KIA OE8. Price $7.00. In: Proceedings of the 24th Session of the International Geological Congress, Section 11, Hydrogeology; Montreal, Canada, 1972: International Geological Congress Publication, p 264-269, 1972. 4 ref.

Descriptors: *Path of pollutants, *Groundwater movement, *Hydrogeology, *Kansas, Aquifer characteristics, Malenclaves, Saline water intrusion, Water pollution sources, Brines.

The unconsolidated Quaternary aquifers of west-central Kansas have problems in areas where mineralized waters adversely affect the natural chemical quality. Basic geologic data such as (1) slope and paleotopography of the underlying bedrock surface, (2) permeability changes within the aquifer and (3) the lithology and hydrostatic pressures imposed by the underlying rock are more useful in pinpointing sources of mineralized water than are ratio comparisons of various inorganic constituents. Chemical analyses of water samples collected at bedrock depth from test holes, and at prescribed intervals above the basal collection depth, reveal trends by which areas of mineralized solutions can be outlined. These same test-hole data can be useful in determining the point at which the mineralization is entering the

of organic matter by microorganisms, by chemical precipitation and coprecipitation, by sorption on soil particles, on bacterial slimes, and on hydroxides of iron and manganese, by exchange of gases, and by dilution. Sorptive bacterial slimes develop at the surface of the solid particles of aquifers in the polluted zones rich in nutrients. Hydroxides of iron and manganese are typically precipitated in the oxidation zone downstream from polluted areas. (Knapp-USGS)
W73-09971

CHARACTERISTICS OF DISTRIBUTION OF FLUORINE IN WATERS OF THE STEPPE REGIONS OF KAZAKHSTAN IN RELATION TO ENDEMIC FLUOROSIS, (IN RUSSIAN),
T. M. Belyakova, and G. S. Dzyadevich.
Vestn Mosk Univ Geogr. 6, p 79-85. 1971.
Identifiers: Dental caries, *Fluorine distribution, Fluorosis (Endemic), *Kazakhstan (Steppes), USSR, Minerals.

Fluorine enables fixation of Ca and P in the bone tissue; surplus Fl causes fluorosis; insufficient Fl causes dental caries. Results of geochemical studies in Pavlodar and Karaganda Regions, USSR clarify migration and accumulation of Fl. Mean Fl level was 1.5-2.5 mg/l in drinking water. There was no correlation between the degree of mineralization and Fl content. Fl concentration in ocean waters reached 8 mg/l. There was a close relation between the content of chemical elements in underground and surface water and the geological structure of the region.--Copyright 1973, Biological Abstracts, Inc.
W73-09990

POLLUTION PROBLEMS IN RELATION TO THE THAMES BARRIER,
Water Pollution Research Lab., Stevenage (England).
For primary bibliographic entry see Field 08A.
W73-09998

THE PATHWAYS OF PHOSPHORUS IN LAKE WATER,
Wyzsza Szkola Rolnicza, Wroclaw (Poland).
J. B. Golachowska.
Pol Arch Hydrobiol, Vol 18, No 4, p 325-345, 1971, Illus.
Identifiers: Algae, Bacteria, Lakes, *Phosphorus, Water pollution sources, *Path of pollutants.

The circulation of P in lakes presented shows that the production ability of a water body is determined not only by the absolute level of P content but also the form in which it occurs. In lakes of different types total P in the summer period consists of 40-60% dissolved orthophosphates, 16-33% dissolved organic P, 3-20% particulate organic P and 4.2-11.7% particulate mineral P. These 4 forms have a basic significance for the interpretation of biological processes. Attention is given to the fact that in some lakes in the summer a significant position in total P balance is played by dissolved organic P. It is supposed that the sources of this form are algae and bacteria. It is shown that humus compounds, Fe and C strongly influence the phytoplankton production. These compounds may act either as inhibitors or stimulators. Significance loss of P from lake water in the autumn was caused by sorption by bottom deposits of the 'mineral' type. Deposits of 'organic' type show very weak sorption and, therefore, in some conditions they release phosphates to the water.--Copyright 1972, Biological Abstracts, Inc.
W73-10004

THE DISTRIBUTION OF CHITIN- AND N-ACETYLGLUCOSA-MINE-DECOMPOSING BACTERIA IN AQUATIC ENVIRONMENTS,
Kagawa Univ., Takamatsu (Japan). Dept. of Food Science.
K. Okutani, and H. Kitada.
Kagawa Daigaku Nogakubu Gakuzyutu Hokoku, Vol 23, No 1, p 127-136, 1971, Illus.
Identifiers: *Bacteria, Decomposition (Bacteria), Distribution, Aquatic environments.

Investigations of chitin- and N-acetylglucosamine (NGA)-decomposing systems by bacteria in aquatic environments which may be the first step of ammonia splitting systems from chitin were carried out. The chitin-decomposing bacteria were distributed in sea water and the bottom muds at 10 cells/ml and 1000-10,000 cells/gm of wet muds at coastal regions. In freshwater environments, the chitin-decomposing bacteria were distributed at 1000-10,000 cells/g of wet muds and 10 cells/ml of water. The population of chitin-decomposing bacteria was much smaller than that of the heterotrophic bacteria. The NGA-decomposing bacteria were detected in sea water at 100-1000 cells/ml and 10,000-100,000 cells in the bottom muds at coastal regions. At the open-sea, 0-1000 cells/100 ml of the NGA-decomposing bacteria were detected in sea water and 1000 cells/g of the bottom muds. NGA-decomposing bacteria were distributed in freshwater environments at 10 cells/ml of water.--Copyright 1972, Biological Abstracts, Inc.
W73-10010

THE ENVIRONMENTAL MATRIX: INPUT-OUTPUT TECHNIQUES APPLIED TO POLLUTION PROBLEMS IN ONTARIO,
Department of Energy, Mines and Resources, Burlington (Ontario). Canada Center for Inland Waters.
For primary bibliographic entry see Field 05G.
W73-10020

INJECTION WELLS AND OPERATIONS TODAY,
Bureau of Mines, Bartlesville, Okla. Bartlesville Energy Research Center.
E. C. Donaldson.
In: Underground Waste Management and Environmental Implications, Proc of Symposium held at Houston, Tex, Dec 6-9, 1971: Tulsa, Okla, American Association of Petroleum Geologists Memoir 18, p 24-46, 1972. 11 fig, 8 tab, 21 ref, 1 append.

Descriptors: *Waste disposal wells, *Injection wells, *Underground waste disposal, Water chemistry, Kinetics, Dispersion, Water pollution effects, Water pollution sources.

The feasibility and limitations of the underground injection of industrial wastes were studied by observing installations at industrial plants, cities, and oil fields. The chemical industry is using about 175 deep wells to inject approximately 30 million gal per day of waste solutions. The wastes are inorganic salts, mineral and organic acids, basic solutions, chlorinated and oxygenated hydrocarbons, and municipal sewage. The wells, ranging from 300-2, 440 m deep, are completed in three general types of formations: unconsolidated sand, consolidated sandstone, and vugular carbonate rock. The chemical and physical characteristics of the formation and waste dictate the design of the injection system and govern its operation. Commonly, underground injection is the most economical method for disposal of liquid wastes not amenable to surface treatment. Operating costs are lower for pretreatment and subsurface disposal than for surface treatment systems, and plant area requirements are less. Chemical treatment is minimal, and generally the only physical treatment required for underground injection is filtration. (Knapp-USGS)
W73-10023

RESPONSE OF HYDROLOGIC SYSTEMS TO WASTE STORAGE,
Geological Survey, Washington, D.C. Water Resources Div.
J. G. Ferris.

In: Underground Waste Management and Environmental Implications, Proc of Symposium held at Houston, Tex, Dec 6-9, 1971: Tulsa, Okla, American Association of Petroleum Geologists Memoir 18, p 126-132, 1972. 1 ref.

Descriptors: *Underground waste disposal, *Waste disposal wells, *Path of pollutants, *Water pollution effects, Hydrogeology, Aquifer characteristics, Groundwater movement, Groundwater basins.

The demand for cleanup of lakes and streams is literally driving pollution underground. The use of sanitary landfills for disposal of solid wastes, the use of spray irrigation for disposal of partly treated sewage effluent, and the use of deep-well injection for disposal of certain industrial wastes are increasing rapidly. Aquifers are dynamic flow systems that undergo change whenever a new stress is imposed. Attendant upon the injection of fluid into an aquifer is increase in hydraulic head which ultimately influences the hydrologic regime throughout the entire flow system. Disposal to shallow aquifers, which are generally sources of water supply, poses a threat not only to present and future well developments but also to lakes and streams that are sustained by groundwater seepage. In deep-lying confined aquifers, the hydraulic transmissivity is generally small; consequently, the pressures required for significant rates of injection are large. In marked contrast to the very slow migration of the waste, pressure is propagated outward very rapidly. Thus, evaluation of the consequences of waste injection requires not only consideration of the effects of the advancing waste but also the far-reaching effects of the pressure increase. (Knapp-USGS)
W73-10025

APPLICATION OF TRANSPORT EQUATIONS TO GROUNDWATER SYSTEMS,
Geological Survey, Denver, Colo.
J. D. Bredehoeft, and G. F. Pinder.
In: Underground Waste Management and Environmental Implications, Proc of Symposium held at Houston, Tex, Dec 6-9, 1971: Tulsa, Okla, American Association of Petroleum Geologists Memoir 18, p 191-201, 1972. 6 fig, 17 ref.

Descriptors: *Groundwater movement, *Path of pollutants, *Numerical analysis, Equations, Mathematical models, Model studies, Hydrogeology, Mixing, Water chemistry, Mass transfer, Ion transport.
Identifiers: Brunswick (Ga).

Predicting changes in groundwater quality in a complex hydrologic system generally requires simulation of the field problem through the use of a deterministic model. In the most general cases a complete physical-chemical description of moving groundwater requires the simultaneous solution of the differential equations that describe the transport of mass, momentum, and energy. The difficulties in solving this set of equations requires simplified subsets of equations. The equation of motion for single-component groundwater flow has been solved for many different initial and boundary conditions. To describe the transport of miscible fluids of differing density, the mass-transport equation, and the equation of motion have been coupled and solved numerically. Numerical solutions have also been obtained for the heat transport equation and the equation of motion, particularly for convection problems. A case history of groundwater contamination at Brunswick, Georgia, illustrates the use of the mass- and momentum-transport equations in predicting movement of contaminants. (Knapp-USGS)
W73-10027

CIRCULATION PATTERNS OF SALINE GROUNDWATER AFFECTED BY GEOTHERMAL HEATING-AS RELATED TO WASTE DISPOSAL,
Alabama Univ., University. Dept. of Civil and Mineral Engineering.
H. R. Henry, and F. A. Kohout.
In: Underground Waste Management and Environmental Implications, Proc of Symposium held at Houston, Tex, Dec 6-9, 1971: Tulsa, Okla, American Association of Petroleum Geologists Memoir 18, p 202-221, 1972. 26 fig, 43 ref.

Descriptors: *Groundwater movement, *Path of pollutants, *Waste disposal wells, *Hydrogeology, *Convection, Florida, Water circulation, Hydraulic models, Numerical analysis, Equations, Underground waste disposal, Geothermal studies.

The fate of the waste liquids entrained in saline aquifer systems depends upon the state of motion in the aquifer before injection and the modification of this state by the injection process. The Floridan aquifer underlying peninsular Florida is more than 2,000 ft thick and provides a field situation for comparison with mathematical and hydraulic models. The aquifer is exposed to cold seawater where truncated by the deep trenches of the Gulf of Mexico and the Florida Strait. Natural upwellings of warm saline water and observations of temperature and salinity in wells suggest that the seawater flows inland at depth then upward into shallow parts of the aquifer, and, after mixing with freshwater it flows seaward again to form a large, geothermally heated, convective flow cycle. A hydraulic sand model was built to simulate a saline aquifer, a geothermal source, freshwater recharge, and waste-injection wells. The hydraulic-flow equation, the diffusion equation for salt and injected contaminants, and the heat-diffusion equation were solved simultaneously on a high-speed digital computer. Theoretical solutions for the hydraulic model qualitatively correspond to observed distributions of temperature and salinity in the Floridan aquifer. (Knapp-USGS)
W73-10028

HYDRODYNAMICS OF FLUID INJECTION,
California Univ., Berkeley.
P. A. Witherspoon, and S. P. Neuman.
In: Underground Waste Management and Environmental Implications, Proc of Symposium held at Houston, Tex, Dec 6-9, 1971: Tulsa, Okla, American Association of Petroleum Geologists Memoir 18, p 258-272, 1972. 11 fig, 19 ref, 1 append.

Descriptors: *Injection wells, *Artesian aquifers, *Hydrogeology, *Waste disposal wells, Equations, *Path of pollutants, Hydrodynamics, Groundwater movement, Fractures (Geologic), Porosity, Permeability, Stratigraphy, Anisotropy, *Leakage, Injection, Confined water.

A theory for the flow of slightly compressible fluids in multilayered porous systems predicts the effects of injecting into a permeable layer and also the consequent effects in adjacent confining beds. Pressure buildup in a multilayered system depends on the degree of communication that develops within the system between aquifer and aquitard. This communication can be characterized in terms of the permeabilities, storage coefficients, and bed thicknesses. By considering multilayered systems under quasi-steady state conditions, a conservative method of evaluating leakage through an aquitard can be made. (Knapp-USGS)
W73-10032

WATER-MINERAL REACTIONS RELATED TO POTENTIAL FLUID-INJECTION PROBLEMS,
Geological Survey, Menlo Park, Calif. Water Resources Div.
I. Barnes.
In: Underground Waste Management and Environmental Implications, Proc of Symposium held at Houston, Tex, Dec 6-9, 1971: Tulsa, Okla, American Association of Petroleum Geologists Memoir 18, p 294-297, 1972. 3 tab, 5 ref.

Descriptors: *Waste disposal wells, *Water chemistry, *Mineralogy, *Water pollution effects, *Chemical reactions, Injection wells, Thermodynamics, Water quality, Carbonates, Silicates, Clay minerals, Geochemistry, Chemical precipitation, Corrosion.
Identifiers: *Water compatibility (Injection).

Reinjection of formation waters creates few chemical problems if no large change in temperature has occurred, no gas or vapor has separated, and access of air has been prevented. The fluids already have had an opportunity to react with the minerals to the point of compatibility. Injection of incompatible fluids, however, may cause chemical problems; thus, prediction of fluid-mineral reactions should be attempted. Reactions should be studied in both reaction directions. All the species in a solution generated by complete solution of the solid must be considered. Using incongruent reactions (reactions producing a new solid directly from the reactant solid) introduces the unwarranted assumption of equilibrium. In general, hydroxide species such as amorphous silica, limonite, and brucite, as well as simple carbonates, have been found to dissolve or precipitate in natural systems. More complex silicates including serpentine, kenyaite, and magadiite may not precipitate from solutions. Sulfides such as covellite, chalcocite, and pyrite may not precipitate even where supersaturation exceeds 30 kcal, although some sulfide minerals will dissolve readily where oxidation of sulfur can take place. (Knapp-USGS)
W73-10034

NATURAL-MEMBRANE PHENOMENA AND SUBSURFACE WASTE EMPLACEMENT,
Geological Survey, Washington, D.C.
B. B. Hanshaw.
In: Underground Waste Management and Environmental Implications, Proc of Symposium held at Houston, Tex, Dec 6-9, 1971: Tulsa, Okla, American Association of Petroleum Geologists Memoir 18, p 308-317, 1972. 6 fig, 23 ref.

Descriptors: *Waste disposal wells, *Injection, *Aquitards, *Membrane processes, *Clay minerals, Shales, Osmosis, Clays, Reverse osmosis, Adsorption, Semipermeable membranes, Water pollution effects, Path of pollutants, Injection wells, Ion exchange.

In any plan to emplace liquid waste in an aquifer system, the membrane behavior of shale must be taken into account. Clay minerals are natural semipermeable membranes capable of retarding the passage of charged species through its micropores. If a chemical, electrical, or thermal gradient is imposed across a semipermeable membrane, the result is a movement of water tending to equalize the chemical potential of water on the two sides of the membrane. If liquid wastes are injected into an aquifer system, the emplacement will likely upset the dynamic equilibrium; it may cause (1) chemical reactions with the existing fluid and rocks, (2) thermal changes, and (3) increased pressure on the aqueous phase. If a shale capable of behaving as a membrane is an aquitard, its membrane characteristics must be taken into account. An osmotic cell may be set up between a nearby aquifer and the emplacement aquifer with the intervening shale acting as a membrane. A pressure increase beyond that anticipated could occur across the shale membrane. If it exceeds the pressure required for osmotic balance, ultrafiltration can result. The effect would be to cause flow across the shale and increase the chemical concentration of the filtrate in the aquifer beyond the planned account. (Knapp-USGS)
W73-10035

Inst Nac Tecnol Agropecu Suelos Publ. 122, p 57-69. 1971. Illus. English summary.
Identifiers: Adsorption, *Buenos-Aires, Calcium, Chemical analysis, Conductivity (Electrical), *Irrigation water, Rivers, Salinity, Sodium, Water quality, Hydrogen ion concentration.

A study was made of the quality of irrigation water of 'rio Colorado' (Lower valley in Buenos Aires). Five places were selected for sampling. The samples were taken at the beginning and middle of each month, between Nov. 1967 and April 1968. Determined were: electrical conductivity, pH, Ca, Mg, Na, K, bicarbonate, sulfate, chloride, dissolved solids, B and Na-adsorption-ratio (S.A.R.). The results show high and medium salinity hazard with low Na (alkali) hazard. The B concentrations permits normal growth of sensitive and semitolerant crops. Bicarbonate ion concentration is safe for irrigation water. Variations were observed in different dates in the same place and there was a ratio between the values in the same place and other observation points.—Copyright 1973, Biological Abstracts, Inc.
W73-10089

CHEMICAL POLLUTION AND MARINE TROPHODYNAMIC CHAINS, (POLLUTIONS CHIMIQUES ET CHAINES TROPHODYNAMIQUES MARINES),
Centre d'Etude et de Recherches de Biologie et d'Oceanographie Medicale, Nice (France).
For primary bibliographic entry see Field 05C.
W73-10091

UTILIZATION OF A TROPHODYNAMIC CHAIN OF A PELAGIC TYPE FOR THE STUDY OF TRANSFER OF METALLIC POLLUTION, (UTILISATION D'UNE CHAINE TROPHODYNAMIQUE DE TYPE PELAGIQUE POUR L'ETUDE DES POLLUTIONS METALLIQUES),
Centre d'Etudes et de Recherches de Biologie et d'Oceanographie Medicale, Nice (France).
For primary bibliographic entry see Field 05C.
W73-10092

STUDY OF THE TOXICITY OF EFFLUENTS FROM PAPER MILLS IN THE MARINE ENVIRONMENT, (ETUDE DE LA TOXICITE D'EFFLUENTS DE PAPETERIE EN MILIEU MARIN),
Centre d'Etude et de Recherches de Biologie et d'Oceanographie Medicale, Nice (France).
For primary bibliographic entry see Field 05C.
W73-10093

MERCURY CONTAMINATION OF FISH OF THE SEA, (CONTAMINATION DES POISSONS DE MER PAR LE MERCURE),
G. Cumont, G. Viallex, H. Lelievre, and P. Bobenrieth.
Revue Internationale d'Oceanographie Medicale, Vol 28, p 95-127, 1972. 1 tab, 7 histograms, 6 graphs, 8 ref.

Descriptors: *Mercury, *Fish, Water pollution sources, *Toxicity, *Water quality standards, Food chain.
Identifiers: *Concentration levels.

Mercury determination in the muscle of various species of fish is researched. Results showed concentrations of this metal generally to be below the norms admissable for today's standards. The accumulation of mercury is always proportional to the weight of the fish. Consequently, the larger individuals at the end of the food chain are apt to have concentrations higher than these norms. Fish from certain areas showed abnormally high levels indicating a higher rate of mercury pollution in those specific areas. (Ensign-PAI)
W73-10094

THE MARINE ENVIRONMENT AND ORGANIC MATTER, (LE MILIEU MARIN ET LES MATIERES ORGANIQUES),
Centre d'Etude et de Recherches de Biologie et d'Oceanographie Medicale, Nice (France).
For primary bibliographic entry see Field 05C.
W73-10097

COASTAL ZONE POLLUTION MANAGEMENT, PROCEEDINGS OF THE SYMPOSIUM.
For primary bibliographic entry see Field 05G.
W73-10098

WE ARE KILLING THE OCEANS,
Senate, Washington, D.C.
E. F. Hollings.
In: Coastal Zone Pollution Management, Proceedings of the Symposium, Charleston, South Carolina. February 21-22, 1972. Clemson University, Clemson, S.C. 1972. p 1-9.

Descriptors: *Coasts, Water pollution control, Sewage, *Domestic wastes, *Industrial wastes, *Agricultural runoff, *Radioactive wastes, *Oil, Mixing, Natural resources, Fish, *Legislation, Government supports, Programs.

The global aspects of coastal and ocean pollution are reviewed, and instances of pollution from domestic sewage, agricultural runoff, industrial and radioactive wastes, oil, and undersea mining and exploration are reported. Depletion of fish populations and other ocean resources is discussed and legislation and government programs to protect the coastal zone are considered. (See also W73-10098) (Ensign-PAI)
W73-10099

REGIONAL COASTAL POLLUTION CONTROL,
Environmental Protection Agency, Atlanta, Ga. Region IV.
For primary bibliographic entry see Field 05G.
W73-10101

INVOLVEMENT OF NOAA IN COASTAL POLLUTION MANAGEMENT,
National Oceanic and Atmospheric Administration, Rockville, Md. Office of Ecology and Environmental Conservation.
For primary bibliographic entry see Field 05G.
W73-10104

ESTUARINE SURVEYS IN THE SOUTHEAST,
Environmental Protection Agency, Athens, Ga. Southeast Environmental Research Lab.
J. E. Hagan, III.
In: Coastal Zone Pollution Management, Proceedings of the Symposium, Charleston, South Carolina, February 21-22, 1972, Clemson University, Clemson, S.C. 1972. p 155-174.

Descriptors: *Estuaries, *Coasts, Water pollution, *Ecosystems, *Analytical techniques, Pollutants, *Water quality, Odor, Electric powerplants, Fish-kill, Bacteria, Pulp wastes, Waste disposal, Fisheries, Shellfish, Coliforms, Chemicals, Mining, Florida, Georgia, South Carolina, Alabama, *Southeast US.
Identifiers: Spectral analysis, Dye tracer studies, Brown foam.

This is a review of selected estuarine studies which have been conducted in recent years along the southeastern Atlantic and Gulf Coast, and an attempt to draw some generalization concerning the types of pollution situations encountered, study techniques which have been particularly useful (or not useful), and some of the problems which seem to recur with the greatest frequency. Studies are selected which attempt to deal with fairly large segments of the ecosystem, demonstrate particular techniques, highlight recurring

problems, and were of particular personal interest to the author. The effects of rediversion of the Cooper River on the water quality of Charleston Harbor, South Carolina, are considered. Other topics include: obnoxious odors in Hillsborough Bay, Florida; the influence of a steam-electric generating station on the shores of lower Biscayne Bay, Florida; fish kills in Escambia Bay, Florida; the effect of a paper mill on Perdido Bay, Alabama; bacterial pollution of Mobile Bay Alabama; waste disposal in the Savannah River Estuary, Georgia; water quality in Waccasassa and Fernholloway Estuaries, Florida; pollutant problems in Lake Worth, Florida; the effects of chemical disposal on Port Royal Sound, South Carolina and strip mining in the area of the Georgia coastal marsh. (See also W73-10098) (Ensign-PAI)
W73-10106

OCEANOGRAPHIC SURVEYS FOR DESIGN OF POLLUTION CONTROL FACILITIES,
Engineering-Science, Inc., Berkeley, Calif.
For primary bibliographic entry see Field 05E.
W73-10107

APPLICATIONS OF ESTUARINE WATER QUALITY MODELS,
Manhattan Coll., Bronx, N.Y.
D. J. O'Connor, and R. V. Thomann.
In: Coastal Zone Pollution Management, Proceedings of the Symposium, Charleston, South Carolina, February 21-22, 1972, Clemson University, Clemson, S.C. 1972. p 215-226. 5 fig.

Descriptors: *Estuaries, *Water quality, *Mathematical models, Dissolved oxygen, Biochemical oxygen demand, Nitrogen, Delaware River, Continuity equation, Waste water disposal, Waste water treatment.
Identifiers: Time stepping, Finite differences, *Delaware River estuary.

A mathematical model for the concentration of a material in estuarine waters based on the continuity equation is discussed. Various approximative techniques are given for both the stationary and non-stationary cases, these include subdivision of the area of interest in subregions with constant coefficient equations as well as a frequency response analysis on time stepping. The possibility of sources or sinks of material due to effluents and chemical or biochemical oxygen demand for the Delaware Estuary are computed for various conditions of water discharge and waste treatment. The steady state solution is compared with a time response technique and with a segmentation technique. The results compare well. (See also W73-10098) (Ensign-PAI)
W73-10108

SUBMERGED DIFFUSERS IN SHALLOW COASTAL WATERS,
Massachusetts Inst. of Tech., Cambridge. Dept. of Civil Engineering.
For primary bibliographic entry see Field 05G.
W73-10109

COASTAL ZONE POLLUTION MANAGEMENT. A SUMMARY,
Clemson Univ., S.C. Dept. of Civil Engineering.
For primary bibliographic entry see Field 05G.
W73-10111

NUMERICAL MODEL FOR THE PREDICTION OF TRANSIENT WATER QUALITY IN ESTUARY NETWORKS,
Massachusetts Inst. of Tech., Cambridge.
J. E. Dailey, and D. R. F. Harleman.
Sea Grant Project Office, Report No. MITSG 72-15, October 1972. 226 p, 56 fig, 8 tab, 42 ref, 2 append. DSR 72602, 73479 and 73986. NOAA-GH-88 and 2-35150. EPA-R-800429.

Descriptors: *Estuaries, *Water quality, *Forecasting, *Mathematical models, Model studies, Numerical analysis, *Tides, Salinity, Temperature, Biochemical oxygen demand, Dissolved oxygen, Turbulence, *Dispersion, Waste disposal, Eddies, Diffusion, Mixing, *Path of pollutants.
Identifiers: James River Model, Cork Harbour (Ireland).

A one-dimensional network model that can be applied to geometrically complex estuaries is developed for the prediction of transient estuarine water quality. The time scale of the model is on the order of minutes and is called tidal time and is used to model longitudinal dispersion. Longitudinal dispersion is associated with tidal motion where a constituent is spread out by eddy diffusion and velocity distribution interaction. Salinity, temperature, BOD and DO are the constituents included in the model. The concentration distribution of these constituents are obtained by solving the conservation of mass equations including transport, mixing and reaction processes. The tidal time permits formulation of an ocean boundary condition related to the direction of tidal flow. The numerical solution procedures were based on the method of weighted residuals, an application technique of the finite element method. Computed results were compared with available data from dye test experiments in the James River Model and field data from dye tests in Cork Harbour, Ireland to demonstrate the model's predictive capabilities. (Ensign-PAI)
W73-10112

A PRELIMINARY STUDY OF BACTERIAL POLLUTION AND PHYSIOGRAPHY OF BEIRUT COASTAL WATERS,
American Univ., Beirut (Lebanon). Dept. of Biology.
H. Kouyoumjian.
Revue Internationale d'Oceanographie Medicale, Vol 26, p 5-26, 1972. 5 fig, 2 tab, 28 ref.

Descriptors: *Coasts, Water pollution sources, Sewage effluents, *Sewage bacteria, *Bacteria, Salinity, Hydrogen ion concentration, Temperature, Sewers, *Outlets, Waste dilution, Climatology, Oceanography, Sewage treatment, Reclamation, Waste water treatment.
Identifiers: *Beirut, Republic of Lebanon, *Mediterranean Coast.

Twenty-four sampling stations were fixed with reference to the major sewer outlets and swimming beaches in the highly polluted coastal waters. During a twelve month period in 1969-70, regular seawater samples were taken for general bacteriological studies. Salinity, pH, and temperature were determined for each station. The climatology, oceanography, and sewerage in Lebanon are surveyed and the bacterial quality of Beirut's coastal waters is discussed. The placing of a major sewer for effective bacterial dilution is discussed and the advantages and disadvantages of two possible locations are considered. Sewage reclamation is recommended as a permanent solution to both water pollution and water resources scarcity. (Ensign-PAI)
W73-10119

RESTRUCTURING OF SHORES AND SECONDARY POLLUTION. STUDY OF THE EUTROPHICATION OF HARBOR ZONES. (RESTRUCTURATION DES RIVAGES ET POLLUTIONS SECONDAIRES. ETUDE DE L'EUTROPHISATION DE ZONES PORTUAIRES),
Centre d'Etudes et de Recherches de Biologie et d'Oceanographie Medicale, Nice (France).
For primary bibliographic entry see Field 05C.
W73-10120

A SHORT SURVEY OF THE ENVIRONMENT AT THE DUMPING SITE FOR SANTA CRUZ HARBOR DREDGING,
Moss Landing Marine Labs., Calif.
R. E. Arnal.
Technical Publication No. 72-6, August 1972. 20 p, 5 fig, 2 tab, 8 ref. NOAA 2-35137.

Descriptors: *Harbors, *Dredging, *Sediments, Marine plants, Marine animals, On-site investigations, Environmental effects, Water pollution control, *California.
Identifiers: *Santa Cruz (Calif), *Environmental impact survey, *Dumping sites.

A survey of the environment in the Santa Cruz Harbor District was conducted prior to a proposed expansion of the small craft harbor to provide additional, 400 boat berths. Local surface currents, bottom topography, sediments in the dredge site, sediments in the disposal site, and local benthic communities were all investigated. A very high organic content was discovered in the sediments at the dredge site, regardless of particle size; and there was a high biomass value at the disposal site and surrounding area. The best dumping site was determined to be on the 122 deg 00 min west longitude meridian at the intersection with the 5 fathom contour. Dredge spoil was delivered to the dumping site by means of a 3,000 foot floating pipeline. (Ensign-PAI)
W73-10124

THE CHANGING CHEMISTRY OF THE OCEANS,
Goteborg Univ. (Sweden). Dept. of Analytical Chemistry.
D. Dyrssen.
Ambio, Vol 1, No 1, p 21-25, February 1972. 6 fig.

Descriptors: Conferences, *Oceans, *Chemicals, *Aerosols, *Monitoring, *Water pollution, Water pollution control, Foreign countries, Foreign research, Reviews.
Identifiers: *Research reviews.

Some key ideas presented at Nobel Symposium 20 on the 'Changing Chemistry of the Oceans' are recapped. Scientists stress that the natural fluctuations in atmospheric and oceanic variables must be understood before man-induced effects can be accurately ascertained. Some areas of concern such as the increase of carbon dioxide in the atmosphere, marine aerosols, and coastal oil pollution are briefly discussed; the effects of synthetic organics are also reviewed. The restricted use of pollutants such as chlorinated hydrocarbons is advised. Monitoring systems are suggested as a means of guarding threatened ecosystems from distruction by lethal levels of pollutants. It is also recommended that information on the production and use of potential pollutants be made easily available worldwide. (Ensign-PAI)
W73-10125

SOME ADDITIONAL MEASUREMENTS OF PESTICIDES IN THE LOWER ATMOSPHERE OF THE NORTHERN EQUATORIAL ATLANTIC OCEAN,
Rosenstiel School of Marine and Atmospheric Sciences, Miami, Fla.
J. M. Prospero, and D. B. Seba.
Atmospheric Environment, Vol 6, No 5, p 363-364, May 1972. 1 fig, 7 ref. NSF GA-25916, GA-22033.

Descriptors: *Aerosols, *Marine air masses, *Pesticides, *Dusts, Water pollution sources, *Air pollution, *Atlantic Ocean, Air-water interfaces.
Identifiers: Trade winds, West Indies, Pollutant transport.

Marine aerosols collected from the trade winds at Barbados, West Indies during July, 1970, were measured for dust and pesticide levels. The

MARINE POLLUTION, CONCENTRATING ON THE EFFECTS OF HYDROCARBONS IN SEAWATER,

Massachusetts Inst. of Tech., Cambridge. Dept. of Mechanical Engineering.
D. P. Hoult.
In: Law of the Sea Workshop, Canadian-U.S. Maritime Problems, June 15-17, 1971, Toronto, Canada, University of Rhode Island, Kingston, Law of the Sea Institute, p 29-31, 1972. LM. Alexander and G. R. S. Hawkins, editors.

Descriptors: *Oceans, Oil pollution, Oil spills, *Surface tension, *Viscosity, *Gravity, *Mathematical models, Analytical techniques, Path of pollutants.
Identifiers: *Oil spill control, Skimmers, Barriers.

The growing concern over oil pollution is discussed. The spread of oil once spilled is the result of four forces, gravity, surface tension, viscous drag and inertia. A mathematical statement representing the spread of oil into a thin film (due to surface tension) is presented. The rate at which oil spreads into a film appears to be independent of the volume of oil originally released. There is not, then, a direct correlation between the amount of oil spilled and the size of the resultant slick. Mathematical analyses are also applied to techniques for controlling oil spills. All mechanical barriers are subject to a restricted usefulness dependent upon the size of the water current. Skimmers are also limited--if the velocity at the orifice used to suck up the oil is too great, either water or oil will be entrained. (Ensign-PAI)
W73-10144

DISPOSAL OF CHINA CLAY WASTE IN THE ENGLISH CHANNEL,

Marine Biological Association of the United Kingdom, Plymouth (England). Lab.
N. A. Holme.
In: Colloque sur la Geologic de la Manche, Symposium, January 14-15, 1971, Paris, France. Memoires, Bureau de Recherches Geologiques et Minieres No 79, 1972, 2 p. 1 ref.

Descriptors: *Industrial wastes, *Waste disposal, Pipelines, *Outfalls, *Dispersion, Water pollution effects.
Identifiers: *Tracer studies, *China clay wastes.

A method for disposing of fine micaceous waste from the china clay industry is outlined. The scheme involves the use of a pipeline which would run overland to a point near the coast, then tunnel to an outfall position about a mile offshore. A number of tests were done to determine the best location of the outfall, based on the waste dispersion pattern. One such study involved the use of radioactive isotopes as tracers. An experimental slurry was prepared, including the tracers, and then discharged off the coast from a temporary pipeline. Most of the slurry settled initially on the sea bed close to the outfall pipe and subsequent monthly surveys indicated that it only slowly dispersed from this area. It is believed that greater quantities of the actual clay residues would behave in a similar fashion. (Ensign-PAI)
W73-10145

A NEW APPROACH TO THE MEASUREMENT OF RATE OF OXIDATION OF CRUDE OIL IN SEAWATER SYSTEMS,

University Coll. of North Wales, Menai Bridge. Marine Science Labs.
C. F. Gibbs.
Chemosphere, Vol 1, No 3, p 119-124, May, 1972. 3 fig, 8 ref.

Descriptors: *Oil, Oil spills, *Biodegradation, *Measurement, Marine microorganisms, *Laboratory tests, Nutrients, Pollutant identification, Water pollution sources.
Identifiers: *Decomposition rates, *Winkler analysis.

A more accurate method for measuring the rate of decomposition of oil by microorganisms in a natural marine environment is proposed. This approach involves a semi-closed system with oxygen concentrations measured by Winkler titration on abstracted samples. Kuwait crude oil was used in the experiments, mixed into aged seawater. The results of a series of lab tests indicate that the rate of decomposition of oil by marine organisms is very low in unenriched seawater; levels of nitrogen and/or phosphorous nutrients appear to be important factors in degradation rates. The problem of approximately natural nutrient levels in a laboratory situation remains to be solved. (Ensign-PAI)
W73-10150

BIOENVIRONMENTAL FEATURES: AN OVERVIEW.

National Marine Fisheries Service, Seattle, Wash.
D. L. Alverson.
In: The Columbia River Estuary and Adjacent Ocean Waters, Bioenvironmental Studies (A.T. Pruter and D.L. Alverson, eds.), University of Washington Press, Seattle, Washington 1972. p 845-857, 1 fig, 7 ref.

Descriptors: *Columbia River, *Nuclear wastes, *Radioisotopes, *Radioecology, Environmental effects, Path of pollutants, Model studies, Theoretical analysis, Public health, Waste assimilative capacity, Disperson, Estuarine environment, Effluents, Water pollution, Water pollution effects.

The Columbia flows through the Hanford site, some 375 miles upstream from its mouth. Features of the Columbia and its estuary are described. In all, over 60 radionuclides have been measured in the radioactive effluent at Hanford. The more abundant, longer-half-life radionuclides are reconcentrated by biological systems and can be detected in the river and adjacent ocean system. Some speculative models of diel, seasonal, and annual transport patterns are discussed. Various authors noted that the discharged radionuclides have proved a powerful tool, but they did not attempt to evaluate the decision made 3 decades ago to discharge the cooling effluent. The studies provide a wealth of new information which is of value in enhancing capacity to interpret the likely downstream features of radioactive and other pollutants accidentally or purposely discharged into the river, and the impact on the associated biota. (Bopp-ORNL)
W73-10174

FALLOUT PROGRAM. QUARTERLY SUMMARY REPORT (DEC. 1, 1972 THROUGH MARCH 1, 1973),

New York Operations Office (AEC), N.Y. Health and Safety Lab.
E. P. Hardy, Jr.
Available from NTIS, Springfield, Va., as Rpt. No HASL-273; $3.00 per copy, $0.95 microfiche. Report No. HASL-273, April 1, 1973. 227 p, 24 fig, 39 tab, 87 ref, 1 bib.

Descriptors: *Radioactivity, *Radioisotopes, *Fallout, *Water pollution, Marine algae, Marine animals, Aquatic life, Water pollution sources, Air pollution, Soil contamination, Assay, Measurement, Food chains, Public health, Milk, Water, Strontium, Lead.

Current data are reported from the HASL Fallout Program; The Laboratory of Radiation Ecology, University of Washington; and the EURATOM Joint Nuclear Research Centre at Ispra, Italy. The initial section consists of interpretive reports on radium daughter products and lead in marine organisms, inventories of radionuclides in the stratosphere, strontium-90 in diet, and the tropospheric baseline concentration of lead. Subsequent sections include tabulations of radionuclide levels in fallout, surface air, stratospheric air, foods, milk,

and tap water. A bibliography of recent publications related to radionuclide studies is also presented. (See also W73-10176) (Houser-ORNL)
W73-10175

APPENDIX TO HEALTH AND SAFETY LABORATORY FALLOUT PROGRAMS - QUARTERLY SUMMARY REPORT. (DEC. 1, 1972 THROUGH MARCH 1, 1973),
New York Operations Office (AEC), N.Y. Health and Safety Lab.
E. P. Hardy, Jr.
Available from NTIS, Springfield, Va., as Rpt. HASL-273 Append., $6 per copy, $0.95 microfiche. Report No. HASL-273 Appendix, April 1, 1973. 451 p.

Descriptors: *Data collections, *Fallout, *Radioactivity, *Radioisotopes, *Strontium, Sampling, Atlantic Ocean, Air pollution, Lead, Water pollution, Water pollution sources, Assay, Milk, Water table.

Current data from HASL Fallout Program are reported in six appendices, as follows: Appendix A, Sr-90 and Sr-89 in monthly deposition at world land sites. Appendix B, Radiostrontium deposition at Atlantic Ocean weather stations. Appendix C, Radionuclides and lead in surface air. Appendix D, Radiostrontium in milk and tap water. Appendix E, Table of conversion factors. Appendix F, Table of radionuclides. (See also W73-10175) (Houser-ORNL)
W73-10176

FINAL ENVIRONMENTAL STATEMENT RELATED TO OPERATION OF SALEM NUCLEAR GENERATING STATION UNITS 1 AND 2.
Directorate of Licensing (AEC), Washington, D.C.

Available from NTIS, Springfield, Va., as Docket 50272-57, $3.00 per copy, $0.95 microfiche. Dockets 50272-57 and 50311-57, April 1973. 264 p, 21 fig, 33 tab, 95 ref, 6 append.

Descriptors: *Nuclear powerplants, Effluents, Environment, Administrative agencies, *Comprehensive planning, *Sites, Climatology, Meteorology, Ecology, Radioactive wastes, Water pollution, Water pollution sources, Radioactive effects, Monitoring, Public health, Transportation, Beneficial use, Cost-benefit analysis, Delaware River, *New Jersey, *Thermal pollution.
Identifiers: Pressurized water reactors, *Salem (N.J.), *Environmental impact statements.

This final environmental statement was prepared in compliance with the National Environmental Policy Act and relates to the proposed continuation of the provisional construction permits and operation licenses of the Salem Nuclear Generating Station Units 1 and 2. The station is located on Artificial Island in the Delaware River, state of New Jersey, near Salem. The two units will be pressurized water reactors cooled by once-through flow of brackish water from and discharged to the Delaware River. Environmental impacts are assessed and after consideration of alternatives an environmental benefit-cost summary was complied. Some environmental factors considered include climate, hydrology (surface water and ground water), ecology including aquatic life, cooling-water supply and discharge, cooling towers, cooling lakes, spray ponds, radioactive chemical and sanitary wastes, amount of dissolved oxygen and toxic chemicals in effluent water. The conclusion is to continue the construction permits and issue operating licenses for the facility subject to the following conditions: (1) Establish baseline studies of plankton and zooplankton losses and a continuing program of monitoring for protection of aquatic life; (2) Measurement of chlorine concentration in discharge; (3) Establish necessary radiological and environmental monitoring practices; and (4) Propose a course of action to alleviate other detectable harmful effects. (Houser-ORNL)
W73-10177

FINAL ENVIRONMENTAL STATEMENT RELATED TO OPERATION OF PEACH BOTTOM ATOMIC POWER STATION UNITS 2 AND 3.
Directorate of Licensing (AEC), Washington, D.C.

Available from NTIS, Springfield, Va., as Docket 50277-83; $6.00 per copy, $0.95 microfiche. Dockets 50277-83 and 50278-77, April 1973. 530 p, 44 fig, 69 tab, 333 ref, 14 append.

Descriptors: *Nuclear powerplants, Effluents, Environment, Administrative agencies, *Comprehensive planning, *Site, Climatology, Ecology, Radioactive wastes, Water pollution, Water pollution sources, Radioactive effects, Monitoring, Public health, Transportation, Beneficial use, Cost-benefit analysis, *Pennsylvania, *Thermal pollution.
Identifiers: Boiling water reactors, *Susquehanna River, *Environmental impact statements.

This final environmental statement was prepared in compliance with the National Environmental Policy Act and relates to the proposed construction and operation of the Peach Bottom Station, Units 2 and 3, located on the Susquehanna River, Peach Bottom, Pennsylvania. The exhaust steam will be cooled by once-through flow of water obtained from and discharged to the Susquehanna River and also by forced-draft towers when needed. Environmental impacts are assessed and after consideration of alternatives an environmental benefit-cost summary was compiled. Environmental factors considered include hydrology (surface water and ground water), ecology including aquatic life, cooling-water supply and discharge, cooling towers, cooling lakes, spray ponds, radioactive chemical and sanitary wastes, amount of dissolved oxygen and toxic chemicals in effluent water. The conclusion is to continue the construction permits and issue an operating license subject to the following conditions for the protection of the environment: (1) Carry out both radiological and nonradiological monitoring programs for assessment of impacts; (2) require a closed-cycle cooling system after Nov. 1, 1975, and furnish an impact evaluation of the system; (3) restriction of chlorine concentration; (4) reduce iodine releases to as low as practicable; (5) furnish plan of action to reduce significantly any harmful effects. (Houser-ORNL)
W73-10178

FINAL ENVIRONMENTAL STATEMENT RELATED TO THE PROPOSED SAN ONOFRE NUCLEAR GENERATING STATION - UNITS 2 AND 3.
Directorate of Licensing (AEC), Washington, D.C.

Available from NTIS, Springfield, Va., as Docket No. 50361-37. $6.00 per copy, $0.95 microfiche. Docket Number 50361-37. March 1973. 459 p, 50 fig, 53 tab, 366 ref, 6 append.

Descriptors: *Nuclear powerplants, Effluents, Environment, Administrative agencies, *Comprehensive planning, *Sites, Climatology, Ecology, Radioactive wastes, Water pollution, Water pollution sources, Radioactive effects, Monitoring, Public health, Transportation, Beneficial use, Cost-benefit analysis, Pacific Ocean, Plankton, Fishkill, *California, *Thermal pollution.
Identifiers: *Pressurized water reactors, *San Onofre (Calif), *Environmental impact statements.

The final environmental statement was prepared in compliance with the National Environmental Policy Act and relates to the proposed construction of the San Onofre Nuclear Generating Station, Units 2 and 3, pressurized water reactors using once-through water from the Pacific Ocean, Camp Pendleton, California. Environmental impacts are assessed and after consideration of alternatives an environmental benefit-cost summary was compiled. Environmental factors considered include climate, hydrology (surface water and ground water), ecology including aquatic life, cooling-water supply and discharge, cooling towers, cooling lakes, spray ponds, radioactive chemical and sanitary wastes, amount of dissolved oxygen and toxic chemicals in effluent water. It is concluded that construction permits should be issued for the facilities subject to certain requirements affecting cooling water discharges, chlorine concentration discharges, construction practices, and expansion of the current environmental monitoring program. (Houser-ORNL)
W73-10179

CALCULATIONS OF ENVIRONMENTAL RADIATION EXPOSURES AND POPULATION DOSES DUE TO EFFLUENTS FROM A NUCLEAR FUEL REPROCESSING PLANT,
Office of Radiation Programs, Washington, D.C.
J. A. Martin, Jr.
Available from U.S. Govt. Printing Office, Washington, D.C.; $1.00 per copy. Radiation Data and Reports, Vol 14, No 2, p 59-76, Feb 1973. 6 fig, 12 tab, 33 ref, 1 bib.

Descriptors: *Nuclear energy, *Industrial plants, Effluents, *Nuclear wastes, *Air pollution, Water pollution, Fallout, Water pollution sources, Assay, Measurement, Survey, Sampling, Public health, Monitoring, Radioisotopes, Iodine, Strontium, Cesium, Tritium, Krypton, Radioisotopes, Regulation, Path of pollutant, Food chains, *New York.
Identifiers: Ruthenium, Concentration, West Valley, Fuel reprocessing, *Nuclear fuel services, Buttermilk Creek, Chattaraugas Creek.

Effluent and environmental data pertinent to the Nuclear Fuel Services nuclear fuel reprocessing plant in West Valley, N.Y., were analyzed to determine doses to sample populations. Individual doses to the maximally exposed population group were considered. Population doses (man-rems) from air, fish, deer, and water pathways were calculated. The most significant radionuclides contributing to doses were tritium, krypton-85, strontium-90, cesium-134, and cesium-137. It was concluded that the impact of this facility upon humans was well below applicable guides in 1971. (Houser-ORNL)
W73-10180

RADIATION DATA, SECTION II, WATER.
Office of Radiation Programs, Washington, D.C.

Available from U.S. Supt. of Documents Washington, D.C.; $1.00 per copy. Radiation Data and Reports, Vol 14, No 2, p 102-114, Feb 1973. 1 fig, 18 tab, 26 ref.

Descriptors: *Monitoring, *Sampling, *Assay, *Radioactivity, Environment, *Data collections, Air pollution, Water pollution, Water pollution sources, Radioisotopes, Nuclear explosions, Fallout, Nuclear powerplants, Effluents, Nuclear wastes, Regulation, Federal Government, Radioisotopes, Food chains, Public health, *Kansas, *Tritium.

Data are provided by federal, state, and foreign governmental agencies and other cooperating organizations. Data reported are accumulated from surveillance programs concerning: radioactivity in Kansas surface waters, January-December 1971; tritium surveillance system, July-September 1972; and water surveillance programs, Nevada test site area, April-June 1972. Also given are environmen-

organic components of the soil, but migration to the 10-25 cm depth was continuing after two months. About 9.5% of the activity was in the aerial parts two days following deposition; the half-time in the field for retention by plant matter was about 8 days. Less than 0.0006% of the activity ingested by cows in a day was secreted in one liter of milk, much less than for Sr85 (0.06%) and Cs134 (0.4%). (See also W72-10679) (Bopp-ORNL)
W73-10192

DISTRIBUTION OF RADIOSTRONTIUM AND RADIOCAESIUM IN THE ORGANIC AND MINERAL FRACTIONS OF PASTURE SOILS AND THEIR SUBSEQUENT TRANSFER TO GRASSES,
Centre d'Etude de l'Energie Nucleaire, Mol (Belgium). Laboratoires.
T. J. D'Souza, R. Kirchmann, and J. J. Lehr.
In: Proceedings of Symposium on the Use of Isotopes and Radiation in Research on Soil-Plant Relationships Including Applications in Forestry, Dec 13-17, 1971, Vienna (IAEA-SM-151), International Atomic Energy Agency, Vienna, 1972, p 595-604. 7 fig, 4 tab, 8 ref.

Descriptors: *Soil-water-plant relationships, *Grasslands, *Pastures, *Fallout, Absorption, *Strontium, Radioisotopes, Food chains, Path of pollutants, Moisture availability, Monthly, Clay loam, Organic soils, Soil contamination, Crop response, Radioecology, Soil science.
Identifiers: *Cesium radioisotopes, *Belgium.

Pasture soils from four different locations in Belgium, artifically contaminated, were studied. The Sr/Ca ratio in the organic component of the four soils decreased from about 2.7 to 1.3 times that in the mineral component as the clay component (particle size less than 2 microns) increased from 2.2 to 15.8%. Both Sr85 and Cs134 were more readily available to plants when present in the organic matter as compared with the clay. Greenhouse studies of the effect of varying the content of organic matter in sand-clay mixtures will be reported in the future. (See also W72-10679) (Bopp-ORNL)
W73-10193

ANNUAL ENVIRONMENTAL MONITORING REPORT: CALENDAR YEAR 1972,
Mound Lab., Miamisburg, Ohio.
D. G. Carfagno, and W. H. Wastendorf.
Available from NTIS, Springfield, Va., as Report No. MLM-2028; $3.00 per copy, $0.95 microfiche. Report No. MLM-2028, March 15, 1973. 32 p, 3 fig, 18 tab, 5 ref, 1 append.

Descriptors: Environment, Monitoring, Radioactivity, Radioisotopes, Assay, Measurement, Tritium, Regulation, Public health, Sampling, Analytical techniques, Air pollution, Water pollution, Soil contamination, Toxicity, Administrative agencies, Water pollution sources.
Identifiers: *Polonium, *Plutonium.

The environment surrounding Mound Laboratory was monitored and reported on for calendar year 1972. Samples analyzed and reported on included air, water, foodstuffs, soil and silt. For radioactive species, the average concentrations of polonium-210, plutonium-238 and tritium detected were well within the stringent standards adopted by the Atomic Energy Commission and the Environmental Protection Agency. Data concerning non-radioactive species in air and water are also presented. Data for these indicate that Mound Laboratory operations have neglibible effect on the environment. (Houser-ONRL)
W73-10201

ENVIRONMENTAL LEVELS OF RADIOACTIVITY IN THE VICINITY OF THE LAWRENCE LIVERMORE LABORATORY: 1972 ANNUAL REPORT,
California Univ., Livermore. Lawrence Livermore Lab.
P. H. Gudiksen, C. L. Lindeken, J. W. Meadows, and K. O. Hamby.
Available from NTIS, Springfield, Va., as Rept. UCRL-51333; $3.00 per copy, $0.95 microfiche. Report No. UCRL-51333, March 7, 1973. 42 p, 11 fig, 26 tab, 8 ref, 1 append.

Descriptors: *Monitoring, *Radioactivity, *Radioisotopes, Tritium, Uranium, Plutonium, Sampling, Analytical techniques, Air pollution, Water pollution sources, Soil contamination, Milk, Food chains, Public health, Assay, Measurement.
Identifiers: *Lawrence Livermore Laboratory.

The Lawrence Livermore Laboratory continuously monitors the levels of radioactivity within the Livermore Valley and Site 300. Results of analyses performed during 1972 for gross radioactivity and for specific radioisotopes of interest in a variety of environmental samples are presented. In all cases, the levels of radioactivity were below the concentration-guide values in AEC Manual Chapter 0524. Sampling included airborne particles, soil, vegetation, and water. Assessment of the radiation doses to an individual from the observed environmental activities indicates the contribution from artificially produced radionuclides is small in comparison with the approximately 100-mrem/yr dose received from natural sources. (Houser-ORNL)
W73-10202

AQUATIC RADIOLOGICAL ASSESSMENTS AT NUCLEAR POWER PLANTS,
Portland General Electric Co., Oreg.
J. W. Lentsch.
In: Water Quality Considerations for Siting and Operating Nuclear Power Plants, Oct 1-4, 1972, Washington, D.C., Atomic Industrial Forum, Inc., New York, New York, 36 p, 2 fig, 46 ref.

Descriptors: *Nuclear wastes, *Nuclear powerplants, *Effluents, *Public health, Path of pollutants, Monitoring, Radioisotopes, Absorption, Fish diets, Estuarine environment, Environmental effects, Forecasting, Systems analysis.

Assurance that dose guidelines are not exceeded requires evaluation and continual environmental monitoring of possible pathways (utilization of water supplies including irrigation and recreational facilities, and consumption of edible biota). Which pathways require detailed consideration are generally indicated by approximations with regard to dilution, biological concentration, and sedimentation. For predicting radionuclide uptake by aquatic organisms, the 'specific activity' method (see W72-04484) is known to be adequate for Sr90, Cs137, Zn65, and Mn54; but in the use of 'concentration factors' more conservatism is required with estuarine ecosystems as compared with freshwater or marine ecosystems. Methods are outlined for predicting radionuclide uptake by sediments, soil, irrigated crops, and cattle flesh; and for assessing internal and external doses. (Bopp-ORNL)
W73-10204

FINAL ENVIRONMENTAL STATEMENT RELATED TO CONTINUATION OF CONSTRUCTION AND OPERATION OF UNITS 1 AND 2 AND CONSTRUCTION OF UNITS 3 AND 4 OF NORTH ANNA POWER STATION.
Directorate of Licensing (AEC), Washington, D.C.

Available from NTIS, SPringfield, Va., as Docket 50338-60; $6.00 per copy, $0.95 microfiche. Docket No. 50338-60. April 1973. 474 p, 43 fig, 40 tab, 208 ref, 4 append.

Descriptors: *Nuclear powerplants, Effluents, Environment, Administrative agencies, *Comprehensive planning, *Sites, Climatology, Ecology, Radioactive wastes, Water pollution, Water pollution sources, Radioactive effects, Monitoring, Public health, Transportation, Beneficial use, Cost-benefit analysis, Lakes, *Virginia, *Thermal pollution.
Identifiers: *Pressurized water reactors, Lake Anna, *Anna River, Environmental Impact Statements.

This final environmental statement was prepared in compliance with the National Environmental Policy Act and relates to the proposed construction and operation of the North Anna Power Station. This station is located on the North Anna River, Virginia, near the town of Mineral. The four units will employ pressurized water reactors and be cooled by once-through flow of water obtained from and discharged to Lake Anna. Environmental impacts are assessed and after consideration of alternatives an environmental benefit-cost summary was compiled. Environmental factors considered include climate, hydrology (surface water and groundwater), ecology including aquatic life, cooling-water supply and discharge, cooling towers, cooling lakes, spray ponds, radioactive chemical and sanitary wastes, amount of dissolved oxygen and toxic chemicals in effluent water. The conclusions reached are: (1) continue construction permits for units 1 and 2; (2) issue operating license for units 1 and 2; (3) issue construction permits for units 3 and 4; (4) necessary monitoring programs are to be continued and modified to assess the impact of construction and operation; (5) if harmful effects appear, the applicant will propose action to alleviate the problem. (Houser-ORNL)
W73-10205

ENVIRONMENTAL MONITORING AT ARGONNE NATIONAL LABORATORY - ANNUAL REPORT FOR 1972,
Argonne National Lab., Ill.
For primary bibliographic entry see Field 05A.
W73-10206

RADIOACTIVITY OF THE WATER OF THE NORTHEASTERN PART OF THE ATLANTIC OCEAN,
Akademiya Nauk URSR, Sevastopol. Marine Hydrophysics Inst.
I. P. Baksheyeva, A. D. Zemlyanoy, V. N. Markelov, and B. A. Nelepo.
Oceanology, Vol 11, No 6, p 861-867, 4 fig, 1 tab, 22 ref. Translated from Okeanologiia, Vol 11, No 6, 1971.

Descriptors: *Water pollution sources, *Radioactive wastes, Industries, *Nuclear powerplants, *Nuclear explosions, Testing, *Strontium, Distribution patterns, *Atlantic Ocean.
Identifiers: Irish Sea, Hebrides, Orkneyes.

Radioactive wastes from atomic industrial facilities along with radioactive fission products from nuclear weapon tests discharged into the ocean are an important source of radioactive contamination of some regions of the world's oceans. Spatial distribution studies in the northeastern Atlantic indicated that the content of 90Sr at the end of 1965 and the beginning of 1966 was higher than the average level for the ocean and amounted to about 53 counts/min-100 1 in the surface layer. The most intense transport of artificial radioactive products out of the Irish Sea was detected in the northern and northeastern directions along the Hebrides and the Orkneyes. (Ensign-PAI)
W73-10220

POLYCHLORINATED BIPHENYLS-ENVIRONMENTAL IMPACT. A REVIEW BY THE PANEL ON HAZARDOUS TRACE SUBSTANCES. MARCH 1972,
Minnesota Univ., St. Paul. Coll. of Veterinary Medicine.
P. B. Hammond, I. C. I. Nisbet, A. F. Sarofim, W. H. Drury, and N. Nelson.
Environmental Research, Vol 5, No 3, p 249-362, September 1972. 15 fig, 32 tab, 234 ref.

Descriptors: *Polychlorinated biphenyls, *Properties, Industrial production, Use rates, Transfer, *Toxicity, *Environmental effects, Analytical techniques, Model studies, Chemicals, Testing, Laboratory tests, On site tests, *Trace elements, *Reviews, Bibliographies.
Identifiers: Transport, Transformation, *Hazardous substances.

A review of the literature concerning polychlorinated biphenyls (PCBs) is presented. The properties, production, and uses of PCBs; their transport and transformation in the environment; environmental occurrence and human exposure to PCBs; toxic impurities; biological effects; and analytical methods are discussed. Systems models need to be developed to describe environmental transport of PCBs and to predict the effects of varying inputs of the chemicals in sensitive areas. Testing procedures should be established to effectively screen potentially hazardous chemicals before they become large scale environmental pollutants. Further study needs to be done on the effects of pollutants on field populations, so that laboratory results can be correlated with observations in the natural environment. These needs apply not only to PCBs but to a wide range of chemicals. (Ensign-PAI)
W73-10221

AIR FLOW LAND-SEA-AIR-INTERFACE: MONTEREY BAY, CALIFORNIA - 1971,
Moss Landing Marine Labs., Calif.
R. G. Read.
Technical Publication No 72-4, 1972. 28 p, 15 fig, 4 ref. NOAA 2-35137.

Descriptors: *Coasts, *Bays, *Air pollution effects, Transport, *Waste dilution, Trapping, *Air-Earth interfaces, *Air-water interfaces, Winds, Flow, Velocity, Industries, Urbanization, Planning, Atmosphere, *California, Dilution.
Identifiers: *Monterey Bay (Calif).

The atmospheric conditions that determine the transport, dilution, and trapping of air pollutants in the coastal areas of Monterey Bay and the Salinas Valley are greatly influenced by the air flow at the land-sea-air interface. Analysis of the hourly air flow on a daily and monthly basis indicates patterns of stagnation from midnight to noon of the following day with moderate to strong air flow during the period 1:00 to 10:00 p.m. Whenever flow was >5mph, the prevailing wind direction was onshore and from a westerly direction throughout 1971. Suggestions for urbanization and industrialization are made on the basis of an understanding of the atmospheric conditions which lead to trapping and dispersal of atmospheric waste. (Ensign-PAI)
W73-10223

THE FLUXES OF MARINE CHEMISTRY,
Scripps Institution of Oceanography, La Jolla, Calif.
For primary bibliographic entry see Field 02K.
W73-10225

THE IMPACT OF ARTIFICIAL RADIOACTIVITY ON THE OCEANS AND ON OCEANOGRAPHY,
Ministry of Agriculture, Fisheries and Food, Lowestoft (England). Fisheries Radiobiological Lab.
A. Preston, D. S. Woodhead, N. T. Mitchell, and R. J. Pentreath.
Proceedings of the Royal Society of Edinburgh. Section B: Biology, Vol 72, p 411-423, 1971/72. 1 fig, 5 tab, 85 ref.

Descriptors: *Oceans, *Radioisotopes, *Nuclear explosions, *Radioactive wastes, *Fallout, *Distribution patterns, Toxicity, Environmental effects, Natural resources, Water pollution control.

A total inventory of artificial radionuclides introduced into the world oceans is presented. Weapon fallout is the largest artificial component, but it is spread throughout surface waters and occurs at very low concentrations. Radioactive waste is extremely restricted in its geographic distribution and therefore occurs at relatively high concentrations. Artificial radionuclides have widely varying radiotoxicities and their significance in a particular environment depends not only on their radiotoxicities but also upon the components of that environment and the uses of the radionuclides leading to radiation exposure. The impact of radioactive waste disposal on marine resources is assessed, and control measures are discussed. Physical, chemical, and biological parameters determining the distribution of various radionuclides introduced into the oceans are examined. (Ensign-PAI)
W73-10226

CURRENT STATUS OF KNOWLEDGE OF THE BIOLOGICAL EFFECTS OF HEAVY METALS IN THE CHESAPEAKE BAY,
Johns Hopkins Univ., Baltimore, Md. Dept. of Environmental Medicine.
For primary bibliographic entry see Field 05C.
W73-10232

CURRENT STATUS OF KNOWLEDGE CONCERNING THE CAUSE AND BIOLOGICAL EFFECTS OF EUTROPHICATION IN CHESAPEAKE BAY,
Maryland Univ., Solomons. Natural Resources Inst.
For primary bibliographic entry see Field 05C.
W73-10233

BIOLOGICAL CRITERIA OF ENVIRONMENTAL CHANGE IN THE CHESAPEAKE BAY,
Virginia Inst. of Marine Science, Gloucester Point. Dept. of Ecology-Pollution.
R. C. Swartz.
Chesapeake Science, Vol 13, (Supplement), p S17-S41, December 1972. 1 fig, 5 tab, 121 ref.

Descriptors: *Chesapeake Bay, *Biota, Biology, *Estuarine environment, *Environmental effects, Analytical techniques, Bioassay, Application methods, Aquatic populations, Biological communities, Structure, Metabolism, Bioindicators, Habitats, Seasonal, Bays.
Identifiers: Change of conditions.

Biological phenomena in the Chesapeake Bay are expectionally complex. Simple, unequivocal standards for characterizing biotic conditions are not available. A review of techniques for the determination of direction and rate of change of conditions in the Bay is provided. Specific bioassays and their applicability to situations in the Bay are discussed. Condition indices, population dynamics, community structure, and community metabolism are also sensitive indicators of environmental disturbance. For the immediate future it is desirable to rely upon several different procedures to document biological changes. The method selected will depend upon the habitat, season, and nature of the environmental alteration. (Ensign-PAI)
W73-10235

(BIOTA OF THE CHESAPEAKE BAY), CONCLUSIONS AND RECOMMENDATIONS,
Maryland Univ., Prince Frederick, Hallowing Point Field Station.

the Delaware River, and Mg, Cr, Cu, and Sr have predominately seaborne sources. Water currents seem to be the primary factor influencing distribution of these materials, regardless of source. A 'hot spot' of high concentrations of Cr, Cu, Pb, and Sr appears associated with the mouth of the Cohansey River, and there seems to be a sink of trace metals being formed offshore from the mouths of the Murderkill and St. Jones rivers. (Ensign-PAI)
W73-10250

RISK ANALYSIS OF THE OIL TRANSPORTAION SYSTEM.

Pacific Northwest SEA, Vol. 5, No 4, 22 p, 1972.

Descriptors: *Pacific Northwest U.S., *Washington, *Oil spills, *Transportation, *Hazards, Equipment, Operations research, Training, Manpower, Accidents, Data collections, Data transmission, Regulations.
Identifiers: *Risk analysis.

A program for assessing the risks of oil spills in Washington State waters is discussed, and hazards inherent in transportation operations are considered. Fault tree analysis is described, and ways of reducing risk levels are suggested, including improving equipment, operations and maintenance procedures, and crew selection and training; diverting petroleum crudes and products to transportation modes and systems other than ships; improving accident data acquisition and distribution; and improving system controls, regulations, and enforcement actions. (Ensign-PAI)
W73-10251

INDICATOR SPECIES--A CASE FOR CAUTION,
University Coll. of North Wales, Bangor. Marine Science Labs.
R. A. Eagle, and E. I. S. Rees.
Marine Pollution Bulletin, London, Vol 4, No 2, p 25, February 1973.

Descriptors: *Bioindicators, *Outlets, Construction, *Sewage, Sands, Silts, Clay, Habitats, Water pollution sources, Path of pollutants.
Identifiers: *Polychaetes, Capitella capitata.

The polychaete worm Capitella capitata is one of the best indicators of organic enrichment in coastal seas. It is often found in abundance areound sewage outfalls. A case is presented in which this polychaete colonized an outfall area before sewage discharge. The cause of colonization is attributed to construction of the outfall. Sand, silt, and clay displaced during construction fell close to the pipe providing an adequate habitat for colonization. Populations of C. capitata may be due to temporary disturbance of the environment rather than to sewage effluent; thus, their use as bioindicators should be viewed with caution. (Ensign-PAI)
W73-10252

POLLUTION AND THE OFFSHORE OIL INDUSTRY,
British Petroleum Co. Ltd., Sunbury-on-Thames (England).
T. E. Lester, and L. R. Beynon.
Marine Pollution Bulletin, London, Vol 4, No 2, p 23-25, February 1973.

Descriptors: *Exploration, *Exploitation, *Oil pollution, *Water pollution, Oil wells, Accidents, Environmental effects, Planning.
Identifiers: *North Sea, Gulf of Mexico, Santa Barbara.

Exploration and exploitation of the North Sea oil field has caused concern about the possibility of major oil pollution. Three major accidents to offshore wells, Santa Barbara in Jan. 1969, Gulf of Mexico in spring 1970, and Gulf of Mexico in Dec. 1970, are examined to discover the risk and likely consequences of a blow-out in the North Sea. Biological effects, spillage estimates, and contingeney plans are discussed. (Ensign-PAI)
W73-10253

ENVIRONMENTAL RADIOACTIVITY IN MICHIGAN, 1970,
Office of Radiation Programs, Washington, D.C.
For primary bibliographic entry see Field 05A.
W73-10262

ALPHA-RECOIL THORIUM-234: DISSOLUTION INTO WATER AND THE URANIUM-234/-URANIUM-238 DISEQUILIBRIUM IN NATURE,
Gakushuin Univ., Tokyo (Japan). Dept. of Chemistry.
K. Kigoshi.
Science, Vol 173, No 3991, p 47-48, July 2, 1971. 1 fig, 1 tab, 5 ref.

Descriptors: *Water pollution sources, *Groundwater, *Radioactivity, *Uranium radioisotopes, Sediments, Path of pollutants, Analytical techniques, Water analysis, Soil types, Soil properties, Infiltration, Leaching, Groundwater movement.
Identifiers: Residence time.

In laboratory leaching experiments on fresh igneous rocks or minerals, solutions containing uranium-234 have been found to contain less than 1.3 times the equilibrium amount of uranium-238. The rate of ejection of alpha-recoil thorium-234 into solution from the surface of zircon sand gives an alpha-recoil range of 550 angstroms. The alpha-recoil thorium-234 atoms ejected into the groundwater may supply excess uranium-234. In pelagic sediments, ejected alpha-recoil thorium-234 may contribute to the supply of mobile uranium-234 in the sedimentary column. (Woodard-USGS)
W73-10325

NON-EQUILIBRIUM PHENOMENA IN CHEMICAL TRANSPORT BY GROUND-WATER FLOW SYSTEMS,
Nevada Univ., Reno. Desert Research Inst.
C. L. Carnahan, and J. V. A. Sharp.
In: Proceedings of Fall Annual Meeting of the American Geophysical Union, December 5, 1972, San Francisco, California: Preprint No 72, Paper H-16, 1972. 13 p, 7 fig, 10 ref. OWRR A-046-NEV (1).

Descriptors: *Path of pollutants, *Groundwater movement, *Water chemistry, *Thermodynamics, *Ion transport, Equilibrium, Entropy, Free energy, Heat flow, Diffusion, Translocation, Chemical reactions.

The methods of irreversible thermodynamics provide a self-consistent and unifying approach to study of simultaneous, coupled, non-equilibrium processes affecting chemical mobilization and transport in groundwater systems, where the systems are acted upon by a variety of gradients and forces. Although the methods are applicable to any groundwater flow system, certain cases of special interest can be identified. These include chemical processes in regions of large thermal gradients and transient interactions of contaminants introduced into aquifers. (Knapp-USGS)
W73-10329

GROUNDWATER QUALITY IN THE CORTARO AREA NORTHWEST OF TUCSON, ARIZONA,
K. D. Schmidt.
Water Resources Bulletin, Vol 9, No 3, p 598-606, June 1973. 2 fig, 3 tab, 12 ref.

Descriptors: *Water quality, *Groundwater, *Path of pollutants, *Arizona, *Waste disposal, Sewage disposal, Infiltration, Nitrates, Coliforms, Alluvium, Malenclaves, Hydrogeology, Groundwater movement.
Identifiers: *Tucson (Ariz).

The Cortaro area is used for disposal of much of the liquid waste from the city of Tucson, Arizona. In the past, more than one-half of the sewage effluent was used for crop irrigation. However, since 1970 virtually all of the sewage effluent has been percolated in the normally dry Santa Cruz River channel. Nitrate and chloride contents are monitored in water samples from about 20 large-capacity irrigation wells. Contents and seasonal trends for these constituents are closely related to the disposal of sewage effluent. Water quality problems other than nitrate include total dissolved solids, boron, coliform, and lead. High lead contents in the area appear to be a natural phenomenon and the coliform contents are related to poor well construction. The other quality problems are primarily caused by sewage effluent. (Knapp-USGS)
W73-10332

RELATIVE LEACHING POTENTIALS ESTIMATED FROM HYDROLOGIC SOIL GROUPS,
Agricultural Research Service, Beltsville, Md. Hydrograph Lab.
For primary bibliographic entry see Field 02G.
W73-10333

SYMPOSIUM ON DIRECT TRACER MEASUREMENT OF THE REAERATION CAPACITY OF STREAMS AND ESTUARIES.

Copy available from GPO Sup Doc as EP1.16:16050 for 01/72, $1.00; microfiche from NTIS as PB-218 478, $0.95. Environmental Protection Agency Water Pollution Control Research Series 16050 for 01/72, January 1972. 193 p, 72 fig, 12 tab, 58 ref. Project No. 16050 for 01/72.

Descriptors: *Mixing, *Turbulence, *Reaeration, *Tracers, Pollutants, Estuaries, Detergents, Domestic wastes, Fluorescent dyes, Oxygen, Streams, Analytical techniques, Rivers, Hydraulic properties, Water pollution effects, Tritium, On-site data collections, Measurement, Model studies, Radioisotopes, Dissolved oxygen, Biochemical oxygen demand, Radioactivity techniques, Sampling, Linear alkylate sulfonates.
Identifiers: *Gas transfer, *Krypton, Kr-85, Reaeration capacity, Oxygen balance, Yaquina River, South River, Chattahoochee River, James River, Flint River, Patuxent River.

Papers presented at a symposium at the Georgia Institute of Technology, Atlanta, during July 7-8, 1970, reflect the state-of-the-art of measuring and predicting the reaeration capacity of streams and estuaries. The papers presented were as follows: 'Turbulence, Mixing and Gas Transfer' by E. C. Tsivoglou; 'Relative Gas Transfer Characteristics of Krypton and Oxygen' by E. C. Tsivoglou; 'Field Tracer Procedures and Mathematical Basis' by J. R. Wallace; 'Field Hydraulic Studies' by J. R. Wallace and D. E. Hicks; 'Laboratory Procedures' by R. J. Velten; 'Reaeration Capacity of the Flint, South and Patuxent Rivers' by E. C. Tsivoglou; 'Oxygen Balance of the South River' by A. G. Herndon; 'Reaeration Studies of the Chattahoochee River' by J. R. Wallace; 'Model Study of Reaeration Capacity of the James River Estuary (Virginia)' by M. W. Lammering; 'Field Studies in Yaquina River Estuary of Surface Gas Transfer Rates' by D. J. Baumgartner, M. H. Feldman, L. C. Bentsen, and T. L. Cooper; 'Radiological Safety' by Jon R. Longstin; 'Effect of Hydraulic Properties on Reaeration' by Edward L. Thackston; 'Pollutant Effects on Reaeration' by L. A. Neal; 'Observed vs. Calculated Reaeration Capacities of Several Streams' by J. R. Wallace; 'Relationships Between Hydraulic Properties and Reaeration' by E. C. Tsivoglou. (See W73-10338 thru W73-10352) (Snyder-Battelle)
W73-10337

TURBULENCE, MIXING AND GAS TRANSFER,
Georgia Inst. of Tech., Atlanta. School of Civil Engineering.
E. C. Tsivoglou.
In: Proceedings of a Symposium on Direct Tracer Measurement of the Reaeration Capacity of Streams and Estuaries; Georgia Institute of Technology, Atlanta, July 7-8, 1970: Environmental Protection Agency Water Pollution Control Research Series 16050 FOR 01/72, p 5-18, January 1972. 4 fig. 8 ref.

Descriptors: *Reaeration, *Dissolved oxygen, *Tracers, *Water chemistry, Biodegradation, Biochemical oxygen demand, Path of pollutants, Turbulence, Mixing, Oxygen sag, Self-purification, Diffusion.

A brief outline is given of the principles of gas transfer in turbulent streams. Stream self-purification involves the depletion of dissolved oxygen by bacterial degradation of organic wastes and replenishment of the dissolved oxygen by absorption of oxygen from the atmosphere. Reaeration is a direct function of turbulence. The purpose of recent research has been to define the hydraulic properties that determine the reaeration coefficient in real streams; this is tantamount to attempting to define the rate of surface replacement in terms of measurable hydraulic properties such as velocity, depth, time of flow, hydraulic gradient, hydraulic radius, flow, roughness, and wetted perimeter. (See also W73-10337) Knapp-USGS)
W73-10338

RELATIVE GAS TRANSFER CHARACTERISTICS OF KRYPTON AND OXYGEN,
Georgia Inst. of Tech., Atlanta. School of Civil Engineering.
E. C. Tsivoglou.
In: Proceedings of Symposium on Direct Tracer Measurement of the Reaeration Capacity of Streams and Estuaries; Georgia Institute of Technology, Atlanta, July 7-8, 1970: Environmental Protection Agency Water Pollution Control Research Series 16050 FOR 01/72, p 19-30, January 1972. 6 fig, 1 ref.

Descriptors: *Tracers, *Reaeration, *Krypton radioisotopes, *Radioactivity techniques, *Dissolved oxygen, Water chemistry, Mixing, Oxygen sag, Aeration, Dispersion, Self-purification, Diffusion, Path of pollutants, Turbulence.

Accurate evaluation of stream reaeration capacity is a problem of major importance in water pollution control. A gaseous tracer for dissolved oxygen can be used to circumvent problems of measurement that have proved to be otherwise insurmountable. The relative gas transfer capability of krypton-85 and oxygen, measured as the ratio of reaction rate coefficients, is 0.83, the transfer of krypton-85 being the slower process. The transfer ratio is not significantly affected by temperature over the range 10 deg C to 32 deg C, by the degree of turbulent mixing over the range studied, by the direction in which the gases transfer (into or out of the water), or by the presence of a broken water surface. (See also W73-10337) (Knapp-USGS)
W73-10339

FIELD TRACER PROCEDURES AND MATHEMATICAL BASIS,
J. R. Wallace.
In: Proceedings of Symposium on Direct Tracer Measurement of the Reaeration Capacity of Streams and Estuaries; Georgia Institute of Technology, Atlanta, July 7-8, 1970: Environmental Protection Agency Water Pollution Control

haps 50 yards in length and containing several small hydraulic jumps. Below, there are two old mill ponds, each followed in turn by a waterfall 12 feet or so high. (See also W73-10337) (Knapps-USGS)
W73-10343

OXYGEN BALANCE OF THE SOUTH RIVER,
A. G. Herndon.
In: Proceedings of Symposium on Direct Tracer Measurement of the Reaeration Capacity of Streams and Estuaries; Georgia Institute of Technology, Atlanta, July 7-8, 1970: Environmental Protection Agency Water Pollution Control Research Series 16050 FOR 01/72, p 83-87, January 1972. 1 fig, 1 tab.

Descriptors: *Reaeration, *Self-purification, Oxygen sag, Biochemical oxygen demand, Channel morphology, *Georgia, Path of pollutants.
Identifiers: South River (Ga).

The Federal Water Quality Administration, in cooperation with the Georgia Water Quality Control Board, has conducted studies on the South River, Atlanta, Ga., to determine pollution loads and assimilation characteristics. These studies included measurements of dissolved oxygen, BOD, ammonia, and nitrates. Because the flow in the upper end of the South River is only 13 cfs and it receives 13,700 pounds of BOD over a distance of about 15 miles, this river has a heavy pollution load. A very broad range of deoxygenation or BOD reduction rates were found in the South River. Some sections have wide, sandy bottoms and other sections have rocky, turbulent areas. It appears that in some sections of this river the BOD is being removed without requiring an equivalent oxygen demand; or else, additional oxygen is being supplied, other than that of normal reaeration. The general appearance of the stream would indicate that the rocky areas which have sandy, shallow bottoms. (See also W73-10337) (Knapp-USGS)
W73-10344

REAERATION STUDIES OF THE CHATTAHOOCHEE RIVER,
J. R. Wallace.
In: Proceedings of Symposium on Direct Tracer Measurement of the Reaeration Capacity of Streams and Estuaries; Georgia Institute of Technology, Atlanta, July 7-8, 1970: Environmental Protection Agency Water Pollution Control Research Series 16050 FOR 01/72, p 89-91, January 1972.

Descriptors: *Tracers, *Reaeration, *Krypton radioisotopes, *Radioactivity techniques, *Dissolved oxygen, Water chemistry, Mixing, Oxygen sag, Aeration, Dispersion, Self-purification, Diffusion, Path of pollutants, Turbulence, *Georgia.
Identifiers: *Chattahoochee River (Ga).

The tracer techniques for measuring reaeration originally developed for small streams were applied to a much larger stream. Flows in the Flint River, Ga., are typically 10 to 20 cubic feet per second; on the South River, Ga., the flows are between 50 to 100 cfs; and on the Chattahoochee River, Ga., flows are around 1000 to 2000 dfs. The hydraulic studies on the Chattahoochee were similar to the hydraulic studies on the Flint and the South. Reaeration coefficients are in the range of 0.01 to 0.04, values which are considerably lower than those reported on the Flint and South. (See also W73-10337) (Knapp-USGS)
W73-10345

MODEL STUDY OF REAERATION CAPACITY OF THE JAMES RIVER ESTUARY (VIRGINIA),
M. W. Lammering.
In: Proceedings of Symposium on Direct Tracer Measurement of the Reaeration Capacity of Streams and Estuaries; Georgia Institute of Technology, Atlanta, July 7-8, 1970: Environmental Protection Agency Water Pollution Control Research Series 16050 FOR 01/72, p 93-114, January 1972. 12 fig, 2 tab, 5 ref.

Descriptors: *Hydraulic models, *Reaeration, *Tracers, *Radioactivity techniques, Dissolved oxygen, Krypton radioisotopes, Tritium, Mathematical models, Dye releases, Diffusion, Mixing, Dissolved oxygen, *Virginia.
Identifiers: *James River (Va).

Radio-tracer reaeration studies were conducted in the hydraulic model of the James River estuary. This study was conducted to obtain the reaeration data required for a mathematical model of the estuary. Reaeration rate constants for the prototype were calculated by applying a reduction factor of 1/100th to the model values. The reaeration rate constants were used successfully in a mathematical model of the prototype for oxygen-demanding wastes. The average value of the reaeration constant was 0.069. The use of gaseous krypton-85 for the direct measurement of reaeration capacity was successful. (See also W73-10337) (K-napp-USGS)
W73-10346

FIELD STUDIES IN YAQUINA RIVER ESTUARY OF SURFACE GAS TRANSFER RATES,
D. J. Baumgartner, M. H. Feldman, L. C. Bentsen, and T. L. Cooper.
In: Proceedings of Symposium on Direct Tracer Measurement of the Reaeration Capacity of Streams and Estuaries; Georgia Institute of Technology, Atlanta, July 7-8, 1970: Environmental Protection Agency Water Pollution Control Research Series 16050 FOR 01/72, p 115-137, January 1972. 13 fig, 1 tab, 13 ref.

Descriptors: *Tracers, *Reaeration, *Krypton radioisotopes, *Radioactivity techniques, *Dissolved oxygen, Water chemistry, Mixing, Oxygen sag, Aeration, Dispersion, Self-purification, Diffusion, Path of pollutants, Turbulence, Fluorescent dye, Tritium, Estuaries, *Oregon.
Identifiers: Yaquina River (Oreg).

The Yaquina River estuary near Toledo, Oregon, was chosen for a reaeration study because it offers a range of natural conditions necessary for collection of data for completely general application, yet it is small enough for ease of study. Tests under a variety of density distribution conditions may provide data useful for relating gas transfer rates to stream hydrodynamics. Current meters are set in place up-current from the injection point, a vertical profile of salinity and temperature is made, surface current drogues are released, and approximately 400 millicuries each of krypton-85 and tritium water mixed with rhodamine WT and local water are discharged into the center of the stream. Samples are taken until the dye peak has passed. Samples are returned to the lab, transferred to replicate counting vials containing scintillation fluid and each counted twice. (See also W73-10337) (Knapp-USGS)
W73-10347

RADIOLOGICAL SAFETY,
J. P. Longtin.
In: Proceedings of Symposium on Direct Tracer Measurement of the Reaeration Capacity of Streams and Estuaries; Georgia Institute of Technology, Atlanta, July 7-8, 1970: Environmental Protection Agency Water Pollution Control Research Series 16050 FOR 01/72, p 139-146, January 1972. 1 fig, 2 tab, 3 ref.

Descriptors: *Radioactivity techniques, *Safety, *Public health, *Tracers, *Reaeration, Krypton radioisotopes, Tritium.

The radiation safety aspects of the double tracer reaeration technique are radiation exposure to the general population in the area of the reaeration study and radiation exposure to the personnel performing the study. Procedures are recommended to minimize radiation exposure to personnel and to the general population. While the total quantity of radioactivity involved in the reaeration measurement technique can be fairly large, the rapid dispersion in the stream coupled with the low relative hazards associated with tritiated water and krypton-85 insure that, with proper planning and execution, the exposures to the general population and to the personnel conducting the study can be kept well below those permitted by the AEC (See also W73-10337) (Knapp-USGS)
W73-10348

EFFECT OF HYDRAULIC PROPERTIES ON REAERATION,
E. L. Thackston.
In: Proceedings of Symposium on Direct Tracer Measurement of the Reaeration Capacity of Streams and Estuaries; Georgia Institute of Technology, Atlanta, July 7-8, 1970: Environmental Protection Agency Water Pollution Control Research Series 16050 FOR 01/72, p 147-164, January 1972. 4 fig, 2 tab, 24 ref.

Descriptors: *Reaeration, *Equations, *Hydraulics, Mathematical models, Discharge (Water), Turbulence, Water temperature, Oxygen, Dissolved oxygen, Aeration.

Measurement of the reaeration coefficient by the tracer method is accurate and reliable, but it is also expensive and time-consuming, and it requires highly skilled personnel and a major investment in instrumentation. Most organizations interested in obtaining reaeration coefficients cannot justify this investment. However, many times basic hydraulic data (discharge, depth, slope, velocity, temperature) are available for the reach of stream in question, or they can be obtained for less expenditure than a tracer test would require. It is possible that acceptably accurate average values of the hydraulic data could be obtained with significantly fewer measurements. With these data a mathematical model may be used for prediction of the reaeration coefficient. Methods used to predict reaeration are reviewed. The Thackston-Krenkel equation is slightly more accurate than any of the others. All the equations apply to fully developed turbulent shear flow with a regular vertical velocity profile and do not apply to laminar flow. (See also W73-10337) (Knapp-USGS)
W73-10349

POLLUTANT EFFECTS ON REAERATION,
L. A. Neal.
In: Proceedings of Symposium on Direct Tracer Measurement of the Reaeration Capacity of Streams and Estuaries; Georgia Institute of Technology, Atlnta, July 7-8, 1970: Environmental Protection Agency Water Pollution Control Research Series 16050 FOR 01/72, p 165-177, January 1972. 4 fig, 3 tab, 5 ref.

Descriptors: *Reaeration, Water pollution effects, *Alkylbenzene sulfonates, Laboratory tests, Self-purification, Rivers, *Georgia, Water chemistry, Detergents.
Identifiers: *Chattahoochee River (Ga).

Some pollutants can alter the reaeration capacity of a stream. In order to measure the effect of pollutants on the gas transfer rate coefficient, an open top reactor was constructed. Each pollutant effect test consisted of two reactor runs. The first run was conducted on clean water and a krypton-85 transfer rate determined. The second run was conducted on the polluted sample under conditions identical to those of the clean water run. The reaeration rate is significantly reduced by the surface active agent, LAS. The reaeration rates in the highly polluted reaches of the South and Chattahoochee Rivers, Georgia, are lower than they would be in the absence of pollution. (See also W73-10337) (Knapp-USGS)
W73-10350

OBSERVED VS. CALCULATED REAERATION CAPACITIES OF SEVERAL STREAMS,
J. R. Wallace.
In: Proceedings of Symposium on Direct Tracer Measurement of the Reaeration Capacity of Streams and Estuaries; Georgia Institute of Technology, Atlanta, July 7-8, 1970: Environmental Protection Agency Water Pollution Control Research Series 16050 FOR 01/72, p 179-183, January 1972. 1 fig, 1 tab.

Descriptors: *Reaeration, *Hydraulics, *Water measurement, *Channel morphology, Discharge measurement, Flow rates, Gaging stations, Hydrographs, Streamflow, On-site data collections, Hydrologic data .

Measured reaeration values and values computed from several of the formulas that are available are compared. In the equations, three different parameters used are depth, velocity, and slope. The velocity used in these equations is simply the length of the reach divided by the time of passage as measured from dye studies. The depth flow is the average of the average depth at each cross section. Slope is the difference in elevation at the upstream and downstream end of the reach divided by the length. Mathematical models provide predictions that are closer to the observed reaeration values than values from prediction equations in uniform reaches. The equations, to some extent, are based upon assumptions about the uniformity of the channel and on the assumption of unbroken surfaces. (See also W73-10337) (Knapp-USGS)
W73-10351

RELATIONSHIPS BETWEEN HYDRAULIC PROPERTIES AND REAERATION,
Georgia Inst. of Tech., Atlanta. School of Civil Engineering.
For primary bibliographic entry see Field 05G.
W73-10352

STUDY OF THE OCEANOGRAPHIC CONDITIONS IN THE RIA DE AROSA, IN SUMMER, (IN SPANISH),
Spanish Inst. of Oceanography, Madrid.
For primary bibliographic entry see Field 02L.
W73-10363

POLLUTION AND DEGRADATION OF SOILS: A SURVEY OF PROBLEMS,
State Coll. of Agronomical Science, Gembloux (Belgium).
G. Manil.
Ann Gembloux, Vol 77, No 3, p 175-182, 1971.
Identifiers: *Degradation, *Soil properties, Surveys, Water pollution sources.

This is the first paper of a conference program on pollution of soils. Certain laws which govern the properties of soils are outlined, such as the specific rhythm, the biodynamic character indicating the state of fertility, and the specific behavior. In the latter case physical functions (heat, water and air capacity), chemical functions (e.g., nutritive elements), physicochemical functions (acidity, oxidation, reduction potential), and regulatory functions of the biochemistry are to be considered. Pollution is the endogenous or exogenous incorporation of deleterious substances into the soil. Changes in the acidity may lead to increase in the edaphic milieu of Al+++, toxic to non-acidophilic plants. Pesticides, antiparasitic material, growth control substances and radioactive fallout are examples of exogenous pollution.--Copyright 1972, Biological Abstracts, Inc.
W73-10384

A CONTRIBUTION TO THE CHEMICAL AND MICROBIOLOGICAL STUDY OF WATER-CATCHMENT SYSTEMS IN CONNECTION WITH DECREASED YIELD DURING OPERATION, (IN SERBO CROATIAN),
For primary bibliographic entry see Field 05C.
W73-10386

THE UTILIZATION AND CONSUMPTION OF MERCURY IN FRANCE: OUTLINE OF THE RESULTING HEALTH PROBLEMS, (IN FRENCH),
Laboratoire Central de Recherches Veterinaires Alfort (France).
G. Cumont.
Recueil Med Vet Ec Alfort. Vol 148, No 4, p 427-442. 1972. English summary.
Identifiers: Air pollution, Food contamination, *France, *Health aspects, Mercury, Pollution, Soils, Water pollution sources, Industrial wastes.

The pollution of the air, water, soil and food by increasing quantities of Hg is alarming. This pollution stems from industrial wastes and agricultural use. Highly industrialized countries should take measures to reduce these pollution sources.--Copyright 1972, Biological Abstracts, Inc.
W73-10389

PHENOL-AND RHODANE DESTRUCTING MICROFLORA AS A FACTOR FOR NATURAL PURIFICATION OF THE DNIEPER RESERVOIRS, (IN RUSSIAN),
Dnepropetrovskii Gosudarstvennyi Universitet (USSR).
For primary bibliographic entry see Field 05C.
W73-10394

5C. Effects of Pollution

STUDIES ON BENTHIC NEMATODE ECOLOGY IN A SMALL FRESHWATER POND,
Auburn Univ., Ala. Dept. of Botany and Microbiology.
For primary bibliographic entry see Field 05A.
W73-09751

A CRITICAL TEST OF METHODS FOR ISOLATION OF VIRUSES FOR USE IN CONTROL OF NUISANCE ALGAE,
Illinois Univ., Chicago. Dept. of Biological Sciences.
For primary bibliographic entry see Field 05A.
W73-09753

NUTRIENT TRANSPORT BY SEDIMENT-WATER INTERACTION,
Illinois Univ., Chicago. Dept. of Materials Engineering.
C. A. Moore, and M. L. Silver.
Available from the National Technical Information Service as PB-220 682 $3.00 in paper copy, $0.95 in microfiche. Illinois Water Resources Center, Urbana, Research Report No 65, April 1973. 51 p, 12 fig, 1 tab. OWRR A-053-ILL (2). 14-31-0001-3213.

Descriptors: *Eutrophication, *Lake beds, *Phosphates, Great lakes, *Nutrients, *Sediment-water interfaces, Water pollution effects, Mathematical models, Laboratory tests, Analytical techniques.
Identifiers: P-32 techniques, Phosphate-sediment exchange.

This report presents the results of a series of laboratory tests to investigate phosphate transport in sediments subjected to one dimensional consolidation type loading. P-32 techniques are employed. The results indicate that measurable transport occurs for phosphate concentrations on the order of 2 mg p/gm dry soil or more and for loads

RATIO OF ORGANIC AND MINERAL COMPONENTS OF THE SOUTH BAIKAL SUSPENDED WATERS (IN RUSSIAN),
For primary bibliographic entry see Field 05B.
W73-09812

SOME EFFECTS OF A POWER PLANT ON MARINE MICROBIOTA,

Florida Univ., Gainesville. Dept. of Environmental Engineering Sciences.
J. L. Fox, and M. S. Moyer.

Chesapeake Science, Vol 14, No 1, p 1-10, March 1973. 15 fig, 1 tab, 12 ref.

Descriptors: *Electric power production, *Environmental effects, *Heated water, Marine microorganisms, Analytical techniques, Evaluation, Discharge (Water), Effluents, Water cooling, Temperature, Toxicity, *Florida, Sampling, On-site-tests, Thermal powerplants, Thermal pollution, Water pollution effects, *Biota.

Various analytical methods, including primary productivity, chlorophyll a, adenosine triphosphate (ATP), bacterial counts, total and dissolved solids, temperature and dissolved oxygen were used to evaluate the direct and indirect effects of power plant cooling water on marine microbiota. Both 'shock' effects and changes occurring as organisms remained exposed to the warmer effluent were determined by following and sampling the heated water as it flowed out the discharge canal. The results showed that increased water temperature is having an effect on the marine organisms present. Effects are most profound immediately following heat exposure and their severity seems to be proportional to the temperature of the intake water. Primary productivity dropped an average of 25.9 percent. ATP and bacterial populations generally increased. Chlorophyll a showed wide fluctuations. The results showed that some organisms, such as phytoplankton, may be killed (or at least hindered in their ability to assimilate carbon) whereas other organisms, such as bacteria, survive condenser tube passage and may even increase in numbers as a result of prolonged exposure to increased heat. (Jerome-Vanderbilt)
W73-09845

SWIMMING PERFORMANCE OF THREE WARMWATER FISHES EXPOSED TO A RAPID TEMPERATURE CHANGE,

Virginia Polytechnic Inst. and State Univ., Blacksburg. Dept. of Biology.
C. H. Hocutt.

Chesapeake Science, Vol 14, No 1, p 11-16, March 1973. 4 fig, 2 tab, 21 ref.

Descriptors: *Heated water, Discharges (Water), *Fish migration, Fish, Biology, Thermal powerplants, *Thermal pollution, Effluents, Flow, Environmental effects, Metabolism, Bass, Channel catfish, Shiners, Laboratory tests, *Water temperature, *Pennsylvania, *Susquehanna River.
Identifiers: *Fish (Swimming habits), *Conowingo Reservoir (Penn).

The influence of rapid temperature change on the swimming performance of some warmwater fishes common to Conowingo Reservoir, Susquehanna River, Pennsylvania was tested. Largemouth bass (Micropterus Salmoides), spotfin shiner (Notropis spilopterus), and channel catfish (Ictalurus punctatus) were acclimated to temperatures ranging from 15 to 35 C at 5 C intervals for 14 to 20 hours. For a given test, a single specimen was dipped out of the holding tank and placed in the respirometer. After a conditioning period to accustom the fish to the current flow, the velocity of the current was gradually increased. Results of 55 tests are reported. The largemouth bass remained tranquil upon introduction to the respirometer and proved to be a consistant swimmer. The spotfin shiners were excitable in the test chamber and maximum swimming speed was decidedly higher than for the bass or the catfish. As with the bass and the shiner the catfish swimming performance increased with temperature, from 15 to 30 C, but decreased thereafter. It is felt that the methods presented simulate conditions near industrial discharges where temperature gradients may occur. (Jerome - Vanderbilt)
W73-09846

THERMAL EFFECTS OF PROJECTED POWER GROWTH: SOUTH ATLANTIC AND EAST GULF COAST RIVER BASINS,

Hanford Engineering Development Lab., Richland, Wash.
S. L. Engstrom, G. F. Bailey, P. M. Schrotke, and D. E. Peterson.

Report HEDL TME 72-131, November 1972. 157 p, 76 fig, 26 tab, 30 ref, 3 append. AEC-AT (45-1)-2170.

Descriptors: *Electric power production, *Heated water, *Thermal pollution, *River basins, Thermal powerplants, Discharge (Water), Heat flow, Heat transfer, Cooling water, Waste water, Forecasting, *Southeast U.S., River flow, Standards, Regulation, Temperature.

The direct cooling capacities of twelve major southeastern river systems were calculated for average and low flow conditions. Total regional cooling capacities considering not only the direct heating of the rivers, but also their ability to dissipate heat to the atmosphere, ranged from a high of 328 GW (T) to a low of 65 GW (T). The Alabama, Apalochicola, Altamaha and Savannah rivers contribute essentially all of the capacity based on low flow conditions. Considering that roughly 1.7 MW (T) are rejected regionally for each MWe generated, the available cooling capacity will support the equivalent of 38 GWe at one time. A projection estimates that approximately 65 GWe could be accommodated in this area. Although improvements in plant efficiency could increase this to 70 or 80 GWe, analysis of projected power generation patterns indicates that this is insufficient to cool, on a once through basis, the anticipated growth in the electrical generation industry. Recourse to supplemental cooling techniques will be required throughout much of the Southeastern United States. Jerome - Vanderbilt)
W73-09848

THE EFFECT OF TEMPERATURE ON THE RESPIRATORY FUNCTION OF COELACANTH BLOOD,

Bristol Univ. (England). Research Unit for Comparative Animal Respiration.
G. M. Hughes, and Y. Itazawa.

Experientia, Vol 28, No 10, p 1247, October 15, 1972. 1 fig, 11 ref.

Descriptors: Fish, *Respiration, Oxygen, *Laboratory tests, Biochemistry, Chemical analysis, *Fish physiology, Metabolism, *Temperature, Environmental effects.
Identifiers: *Coelacanth blood, *Oxygen content analyzer, *Latimeria chalumnae.

This study used samples of blood from Latimeria chalumnae caught at Anjouan in January and Iconi, Grande Comore in March 1972, in order to investigate the oxygen content of coelacanth blood equilibrated with various gas mixtures at temperatures of 28C and 15C. The analysis was performed with a Lex Oxygen Content analyzer. The blood showed a lower affinity for oxygen at 28C than at 15C. The pH is also lower at higher temperatures. These results show agreement with results obtained from other species of fish. Differing oxygen affinity of the blood may explain some survival differences of specimens. (Jerome-Vanderbilt)
W73-09856

TEMPERATURE ADAPTATION OF EMBRYONIC DEVELOPMENT RATE AMONG FROGS,
Dalhousie Univ., Halifax (Nova Scotia). Dept. of Biology.
I. A. McLaren, and J. M. Cooley.
Physiological Zoology, Vol 45, No 3, p 223-228, 1972. 3 fig, 1 tab, 20 ref.

Descriptors: *Adaptation, *Growth rates, *Frogs, *Embryonic growth stage, *Mathematical studies, Biology, Physiology, Amphibians, Environmental effects, *Temperature, Analytical techniques, Evaluation, Measurements, Estimating.

Analysis of temperature adaptation of Rana pipiens is extended in this study to available data on the genus Rana in North America, Europe and Japan. Where the relationship between development rate and temperature is monotonic and slightly curvilinear within natural temperature ranges, it may well be described by Belehradek's temperature function. This function was applied to data from the literature on the effects of temperature on the development rate of embryos of Rana spp. between first cleavage and beginning of gill circulation. Development rates at temperatures which produced abnormalities or death of more than 50% of the embryos have been excluded. Rates for 14 species of Rana are presented. It is suggested that more precise measurement of temperatures at spawning might lead to an even closer correlation of the relationships between various species. It is concluded that deeper inquiry into the biological meaning of the parameters of Belehradek's temperature function holds promise of useful analytical and predictive tools for ecologists. (Jerome-Vanderbilt)
W73-09858

METABOLIC RATE INDEPENDENT OF TEMPERATURE IN MOLLUSC TISSUE,
Memorial Univ. of Newfoundland, St. John's. Marine Sciences Research Lab.
J. A. Percy, and F. A. Aldrich.
Nature, Vol 231, No 5303, p 393-394, June 11, 1971. 2 fig, 1 tab, 8 ref.

Descriptors: *Temperature, *Oysters, *Animal metabolism, Respiration, Animal behavior, Aniaml physiology, *Seasonal, Invertebrates.
Identifiers: Bivalves.

There is considerable controversy about the possible occurrence of temperature--insensitive regions in poikilotherm rate-temperature (R-T) curves. Investigations of seasonal shifts of metabolism in excised tissues of the American oyster yielded results that support the concept of temperature-independent metabolism on a tissue level homeostatic mechanism. The adaptive significance of the temperature independent respiration of the muscle tissue may be related to the fact that the adductor muscle is one of the principal defensive mechanisms of the bivalves and as such, should ideally be relatively independent of environmental fluctuations. (Oleszkiewicz-Vanderbilt)
W73-09863

EFFECTS OF TEMPERATURE AND SALINITY ON THERMAL DEATH IN POSTLARVAL BROWN SHRIMP, PENAEUS AZTECUS,
Galveston Marine Lab., Tex.
L. M. Wiesepape, D. V. Aldrich, and K. Strawn.
Physiological Zoology, Vol 45, No 1, p 22-23, 1972. 4 fig, 1 tab, 27 ref. NSF Grant GH-59.

Descriptors: *Aquiculture, *Shrimp, *Heat resistance, *Salinity, *Thermal pollution, Biology, Animal physiology, Aquatic animals, Temperature, Mortality, Growth stages, Environmental effects, Testing, Electric power production, Thermal powerplants, *Texas, *Gulf of Mexico.

Interest has been expressed in the possibilities of commercial mariculture of brown shrimp which naturally occur in bays in the Northwest Gulf of Mexico. The lethal temperature for shrimp acclimated at the same salinity and different temperatures was determined. The thermal resistance of postlarvae transferred to a higher temperature with two salinities was measured at timed intervals to determine differences in acclimation. To determine the effects of changes in temperature and salinity at the same time, a large number of postlarval shrimp acclimated at various temperature-salinity combinations were subjected to various other temperature-salinity conditions. The level of thermal resistance does not seem to be affected by reduced salinity. It now seems possible to raise brown shrimp from the egg to the postlarval stage under hatchery conditions. The effects of heated effluents on postlarval brown shrimp depend primarily on the temperature of the effluent and the temperature at which the shrimp are acclimated. Secondary considerations are salinity and acclimation salinity. (Jerome-Vanderbilt)
W73-09866

EFFECTS OF INDUSTRIAL EFFLUENTS ON PRIMARY PHYTOPLANKTON INDICATORS,
Christian Bros. Coll., Memphis, Tenn.
R. J. Staub, J. W. Appling, and J. Haas.
Available from the National Technical Information Service as PB-220 741, $3.00 in paper copy, $0.95 in microfiche. Tennessee Water Resources Research Center, Knoxville, Research Report No 26, April 1973. 19 p, 5 tab, 6 ref. OWRR B-011-TENN (1) 14-31-001-3334.

Descriptors: *Phytoplankton, *Effluents, Pollution, *Eutrophication, Chromium, Copper, Fluoride, Iron, Manganese, Mercury, Salt, Diatoms, *Tennessee, *Mississippi River, *Chlamydomonas, Scenedesmus, Industrial wastes, Hydrogen ion concentration, Waste dilution, Metals, Water pollution effects.
Identifiers: *Memphis (Tenn.), Cyclotella, Fragilaria, Melosira, Navicula, Synedra, EDTA.

A method is presented whereby an industry can evaluate its effluent and decide whether it (1) is suitable for emptying into the public waterways, (2) can be made satisfactory merely by adjusting the pH, (3) can be made satisfactory merely by adjusting the pH and reasonable dilution with water, or (4) is unsatisfactory and can be made suitable only by some special treatment. Cultured phytoplankton are inoculated into a series of flasks containing varying amounts of effluent mixed with nutrient medium; the amount of growth after two weeks is recorded and interpreted. A suitable or properly treated effluent is one which will support the growth of representative genera of tolerant phytoplankton, previously isolated from local waterways, without leading to rapid eutrophication. Examples of all four cases were found in the effluent samples tested. Data are presented on tolerance levels of a few local algae to some metallic and other ions.
W73-09873

POLLUTION RELATED STRUCTURAL AND FUNCTIONAL CHANGES IN AQUATIC COMMUNITIES WITH EMPHASIS ON FRESHWATER ALGAE AND PROTOZOA,
Virginia Polytechnic Inst. and State Univ., Blacksburg. Dept. of Biology.
J. Cairns, Jr., G. R. Lanza, and B. C. Parker.
Proceedings of the Academy of Natural Sciences of Philadelphia, Vol 124, No 5, p 79-127, December 29, 1972. 9 fig, 4 tab, 237 ref.

Descriptors: Water pollution effects, Assessments, *Monitoring, *Aquatic algae, Aquatic environment, *Bioassay, *Bibliographies, *Protozoa.
Identifiers: *Microbial bioassays, Environmental impact, Diversity indices.

Past and present approaches to microbial assessment of water pollution are discussed particularly the uses of algal and protozoan community structure and function to assay pollution. Types of pollution and their potential effect on algal and protozoan community structure and function are discussed under 11 different headings: Organic Pollution, Nitrogen, Phosphorus, Roles of Nitrogen and Phosphorus in Eutrophication, Importance of Available Carbon, Other Nutrients, Interactions of Ions and Other Factors, Culture Studies on Nutrient Requirements and Metabolism of Protozoa, Vitamins and Other Organic Requirements, TemperatureEffects on Community Structure and Function, and Toxic Substances.
W73-09875

SUPPOSED TOXICITIES OF MARINE SEDIMENTS. BENEFICIAL EFFECTS OF DREDGING TURBIDITY,
Resources and Ecology Projects, Inc., San Francisco, Calif.
For primary bibliographic entry see Field 05G.
W73-09880

FACTORS CAUSING ENVIRONMENTAL CHANGES AFTER AN OIL SPILL,
University of Southern California, Los Angeles.
D. Straughan.
Journal of Petroleum Technology, p 250-254, March 1972. 1 tab, 29 ref.

Descriptors: Oil, *Oil spills, *Environmental effects, Marine plants, Marine animals, *On-site investigations, Forecasting, Water pollutant effects.
Identifiers: Biological impact, *Environmental factors, Oil pollution control.

Factors which govern the biological impact of an oil spill are discussed. Nine categories of factors are proposed: (1) the type of oil spilled, (2) the dose of oil, (3) the physiography of the area of the spill, (4) weather conditions at the time of spill, (5) the biota of the area, (6) the season of the spill, (7) previous exposure of the area to oil, (8) exposure to other pollutants, and (9) the treatment of the spill. Since any of the areas may interact with another, the environmental impact of an oil spill can only be predicted in a general way. The type of oil spilled is probably the single most important factor; thin, brown oil is the most toxic. Various oil spill disasters, such as the one occurring in Santa Barbara, Torrey Canyon and Tampico Maru are reviewed within the framework of the nine categories. The psychological perspective of a spill observed is also important. The optimist searches for signs of recovery, the pessimist concentrates on the damage done. (Ensign-PAI)
W73-09885

SLUDGE DUMPING AND BENTHIC COMMUNITIES,
Bordeaux Univ., Arcachon (France). Inst. of Marine Biology.
I. R. Jenkinson.
Marine Pollution Bulletin, Vol 3, No 7, p 102-105, July 1972. 2 fig, 1 tab, 16 ref.

Descriptors: *Coasts, *Waste dumps, *Waste disposal, *Sewage sludge, Water pollution, Aquatic life, Environmental effects.
Identifiers: *Needles Spoil Ground (U.K.), *Dumping.

A survey of existing conditions in dumping areas off the south coast of England was conducted. The properties and immediate fate of sewage sludge dumped near the Southampton-Portsmouth region were examined; the ways in which dumping has affected the fauna of the Needles Spoil Ground were also investigated. Laboratory and field tests were undertaken. Despite settling out by the sludge, neither the presence nor the absence of any species could be attributed to the dumping. However, further study is necessary before it can be stated

with certainty whether the current rate of dumping is without deleterious effect. (Ensing-PAI)
W73-09887

BACTERIAL POLLUTION OF THE BRISTOL CHANNEL,
Bristol Univ. (England). Dept. of Bacteriology.
G. C. Ware, A. E. Anson, and Y. F. Arianayagam.
Marine Pollution Bulletin, Vol 3, No 6, p 88-90, June 1972. 2 fig, 7 ref.

Descriptors: *Industrial wastes, *Channels, *Sampling, *Aerobic bacteria, Computer models, Water circulation, Water pollution effects, *Coliforms.
Identifiers: *Bristol Channel (UK), Population density.

The effects of industrial and urban effluents on the marine and estuarine environment of the Bristol Channel, England, were investigated. Over 150 samples of water were collected from various points in the Channel. Total aerobic organisms found in the samples ranged from 10/ml to 5.8 x 10,000/ml, depending somewhat upon the area of the Channel from which the samples were drawn. The overall movement of the water in Bristol Channel was also investigated; the midstream surface water body was found to have a general easterly movement. A computer program was developed to help estimate the water movement as well as to predict sampling locations at standard tide height. Bacterial organisms are either surviving longer than normal (coliform bacteria normally die in a matter of hours when exposed to salt water) or actually multiplying in the Channel. (Ensign-PAI)
W73-09889

STRESSED CORAL REEF CRABS IN HAWAII,
Naval Biomedical Research Lab., Oakland, Calif.
L. H. DiSalvo.
Marine Pollution Bulletin, Vol 3, No 6, p 90-91, June 1972. 1 fig, 8 ref.

Descriptors: Marine animals, *Stress, *Crabs, *Sedimentation, *Sewage, Water pollution effects, Animal physiology, Environmental effects, *Hawaii, Coral.
Identifiers: *Kaneohe Bay (Hawaii), Shell distress, Encrustation, *Chitinoclastic bacteria.

A brief account of shell distress in several species of xanthid crabs inhabiting coral reefs in Kaneohe Bay, Hawaii, is presented. The crabs exhibited several kinds of defects, including encrustation by sessile organisms, shell disease (pitting) and overt physical damages (cracks, holes). Heavy sedimentation and the presence of numerous chitinoclastic bacteria in the Bay sediments probably promoted shell disease among the crabs living inshore. Pitting is believed to arise from microzonation of the distribution of chitinoclastic bacteria in the bacterial flora of the shell. The overt physical damage of the crabs observed is probably the result of encounters with stomatopod crustaceans. It is concluded that the recruitment of juveniles produced in clean reef habitats could provide for replacement of stressed inshore stocks of xanthid crabs. (Ensign-PAI)
W73-09890

MERCURY IN SEA-FOOD FROM THE COAST OFF BOMBAY,
Tata Inst. of Fundamental Research, Bombay (India).
B. L. K. Somayajulu.
Current Science, Vol 41, No 6, p 207-208, March 20, 1972. 2 tab, 13 ref.

Descriptors: *Coasts, *Mercury, Marine animals, *Toxicity, Laboratory tests, Water pollution, Water quality standards, Water pollution effects, Neutron activation analysis.
Identifiers: *Bombay (India), *Seafood, Beta-gamma counting.

Specimens of common seafood, including lobster, pomfret, Bombay duck and prawns were collected off the coast of Bombay, India, and analyzed for mercury content. Neutron activation and beta-gamma coincidence counting were used to measure the mercury levels in the samples. The results of the testing indicate a mercury concentration of about 100 ppb for all analyzed organisms, except salmon, which evidenced higher levels. The upper limit of mercury allowed by U.S. Food and Drug Administration is 500 ppb; the samples all fell considerably below that limit. Preliminary evaluations must be augmented by further research before conclusive statements on mercury pollution can be made. (Ensign-PAI)
W73-09891

PROBLEMS OF THE CASPIAN,
V. Rich.
Marine Pollution Bulletin, Vol 3, No 6, p 84-85, June, 1972. 7 ref.

Descriptors: Sea water, *Water level, *Salinity, Water pollutant ion effects, *Industrial effluents, Fisheries, Water pollution control, Environmental control, *Exploitation.
Identifiers: *Caspian Sea, Environmental protection.

The problems caused by pollution in the waters of the Caspian Sea are briefly discussed. The hazards of a falling water level and increasing salinity have been further complicated by off-shore oil drilling and pollution from transport ships and fishing fleets. The Soviet Union has been attempting to alleviate some of these problems by diverting water from the Siberian rivers (via canal) towards the Caspian. Anti-pollution controls have been instigated to cut down on the amount of petroleum products, phenols, copper, zinc and untreated sewage which are carried to the Sea by feeder rivers. The construction of a new water purification plant at the West Caspian oil center of Neftyanye Kamni is also proposed. The system will be designed to deal with oil leakage and wastes and also discharges from bilges and storm damaged tankers. These plans, and other less ambitious ones, suggest that the ecological importance of the Caspian is at last taking precedence over its geological significance. (Ensign-PAI)
W73-09896

INDIGENOUS AND PETROLEUM-DERIVED HYDROCARBONS IN A POLLUTED SEDIMENT,
Woods Hole Oceanographic Institution, Mass.
For primary bibliographic entry see Field 05A.
W73-09898

MERCURY IN FUR SEALS,
National Marine Fisheries Service, Seattle, Wash. Marine Mammal Biological Lab.
For primary bibliographic entry see Field 05A.
W73-09901

ACUTE TOXICITY OF POLYOXYETHYLENE ESTERS AND POLYOXYETHYLENE ETHERS TO S. SALAR AND G. OCEANICUS,
Fisheries Research Board of Canada, St. Andrews (New Brunswick). Biological Station.
D. J. Wildish.
Water Research, Vol 6, p 759-762, 1972. 2 fig, 1 tab, 5 ref.

Descriptors: *Surfactants, *Toxicity, Aquatic animals, Water pollution effects.
Identifiers: *Salmo salar, *Gammarus oceanicus, *Polyoxyethylene esters, *Polyoxyethylene ethers, Ethylene oxide groups.

Results show a close linear relationship between the degree of polymerization of the polyoxyethylenic chain and toxicity. The fewer ethylene oxide groups per molecule, the greater the toxicity. The data also showed that polyoxyethylene ethers are generally more toxic. S. salar polyoxyethylene (4) lauryl ether is 9 times more toxic at the incipient lethal levels (ILL), than polyoxyethylene (14) monolaurate, and in G. oceanicus polyoxyethylene (4) lauryl ether is 4000-14,000 times more toxic at the 96-h LC50 than polyoxyethylene esters. This difference may be due to the relative ease of cleavage of the ester linkage by lipid metabolizing enzymes in comparison with the more resistant ether linkage. No significant mortality occurred in controls of either species and any mortality that did occur was probably due to the effects of the nonionic surfactants. (Ensign-PAI)
W73-09906

FURTHER INVESTIGATIONS OF THE TRANSFER OF BOMB 14C TO MAN,
Scottish Research Reactor Centre, Glasgow (Scotland). Radiocarbon Unit.
For primary bibliographic entry see Field 05B.
W73-09908

THE KENT COAST IN 1971,
British Museum of Natural History, London. Marine Algal Section.
I. Tittley.
Marine Pollution Bulletin, London, Vol 3, No 9, p 135-138, September 1972. 2 tab, 15 ref.

Descriptors: *Coasts, *Estuaries, Water pollution, *Oil wastes, Oil, Accidents, Ships, Cleaning, *Detergents, Water pollution effects, *Aquatic populations, Marine animals, Vegetation.
Identifiers: *United Kingdom (Kent coast), Thames river estuary.

Oil pollution due to accidental and/or unknown causes during 1971 on the Kent coast of the United Kingdom is reviewed. Three major oil tanker accidents that produced heavy bouts of pollution by oil and detergent along 100 km of Kent coast are described in regard to their effect on the flora and fauna. The methods used in clearing the oil slicks from sea and shore are described as well as the detrimental effects of the detergents used. Other minor oil pollution incidents, some of them of unknown causes are analyzed. The fact that a recovery of biological activity (algae and fish) has been observed in some portions of the Thames river estuary that were previously completely deserted is interpreted as a sign of the overall improvement in the state of the river water. (Ensign-PAI)
W73-09909

PETROLEUM,
Woods Hole Oceanographic Institution, Mass.
For primary bibliographic entry see Field 05A.
W73-09917

INORGANIC CHEMICALS,
Goeteborg Univ. (Sweden). Dept. of Analytical Chemistry.
For primary bibliographic entry see Field 05A.
W73-09918

ORGANIC CHEMICALS,
Stockholm Univ. (Sweden). Institutionen for Analytisk Kemi.
For primary bibliographic entry see Field 05A.
W73-09919

TEST, MONITORING AND INDICATOR ORGANISMS,
Environmental Protection Agency, Gulf Breeze, Fla. Gulf Breeze Lab.

For primary bibliographic entry see Field 05A.
W73-09923

DANGERS OF MERCURY AND ITS COMPOUNDS,.

Vet Ital. Vol 22, No 11/12, p 715-721. 1971.
Identifiers: Contamination, Ecological effects, *Mercury, Mercurials, Reviews.

Ecological contamination and the biological aspects of inorganic and organic mercurials were reviewed and discussed.--Copyright 1973, Biological Abstracts, Inc.
W73-09988

SOME METAL TRACE ELEMENTS IN THE BLOOD OF PATIENTS WITH ENDEMIC FLUOROSIS, (IN RUSSIAN),
V. I. Nikolaev, V. I. Sidorkin, and K. G. Kas'yanova.
Vestn Akad Med Nauk SSSR. Vol 26, No 10, p 77-80. 1971. Illus. English summary.
Identifiers: Barium, Blood, Cobalt, Cobaltemia, *Flurosis (Endemic), Hyper cobaltemia, *Metals, Nickel, *Trace elements, USSR, Zinc.

The Co, Ni, Zn and Ba blood content was determined by spectral analysis in patients with endemic fluorosis and in healthy residents of 3 settlements in the Apsheron peninsula, USSR, where the water of native sources contains fluorine. Fluorosis-afflicted patients had a markedly elevated absolute and relative Ba and Ni coneentration in their blood, along with a drop of the Zn level. Some patients suffered from hypercobaltemia.--Copyright 1973, Biological Abstracts, Inc.
W73-09989

EFFECTS OF FLUORINE IN CATTLE: A SUMMARY OF RESULTS OF STUDIES OVER A NUMBER OF YEARS IN THE NEIGHBORHOOD OF A HYDROFLUORIC ACID FACTORY AND AN ALUMINUM FACTORY (IN GERMAN),
Tieraerztlichen Hochschule, Hanover (West Germany).
H-D. Gruender.
Zentralbl Veterinaermed Reihe A, Vol 19, No 4, p 265-309, 1972, Illus, English summary.
Identifiers: Aluminum, *Cattle, Fluoric-acid, *Fluorine, Hydrofluoric acid, *Air pollution effects, Industrial wastes (Air).

Pollution of the air brought about by civilization in centuries of industrial expansion produces a serious danger for plants, animals and man. The natural F2 content of soil, surface water and air in the inhabited areas of the earth is continually increasing as a result of F produced by an almost all combustion process. In addition to certain organic F compounds with a high toxicity for mammals, damage can also result from the erosive effect of free hydrofluoric acid or from acute or chronic increased absorption of F ions. In cattle, a species particularly susceptible to F2, various forms of chronic absorptive fluoride poisoning occur (geofluorosis, alimentary fluorosis, hydrofluorosis, industrial fluorosis) which result from increased uptake of F2 accumulating over periods of mo. or yr in the hard tissues of the body, a distinction can be made between fluorosis of the teeth without loss of production (tolerance dose) and teeth and/or bone fluorosis with reduced production (toxic dose). Studies were carried out between 1962 and 1967 in the area near a factory producing hydrofluoric acid and another producing Al. The studies involved a total of 36 cattle kept under control conditions on 2 farms. In relation to the early recognition, reduction and prevention of the damage caused by fluorosis, various important factors are considered which influence the occurrence and extent of F effects in cattle kept under ordinary conditions of management and feeding. Attention was also paid to the regression of damage caused by fluorosis outside this area. Damage in cattle can be largely prevented by paying sufficient attention to the various factors related to the way in which animals are housed, managed and fed.--Copyright 1973, Biological Abstracts, Inc.
W73-09991

THE DISTRIBUTION OF CHITIN- AND N-ACETYLGLUCOSA-MINE-DECOMPOSING BACTERIA IN AQUATIC ENVIRONMENTS,
Kagawa Univ., Takamatsu (Japan). Dept. of Food Science.
For primary bibliographic entry see Field 05B.
W73-10010

ECOLOGY OF A EUTROPHIC ENVIRONMENT: MATHEMATICAL ANALYSIS OF DATA, (IN FRENCH),
Centre Universitaire de Luminy, Marseille (France). Laboratoire d'Hydrobiologie Marine.
F. Blanc, M. Leveau, M. C. Bonin, and A. Lauree.
Mar Biol (Berl). Vol 14, No 2, p 120-129. 1972. Illus. English Summary.
Identifiers: Ecology, *Eutrophic environment, France, Mathematical analysis, Phyto plankton, *Rhone River (Mouth), Salinity, Zoo plankton.

Numerous data on physical, chemical and biological parameters in the dilution layer of the Rhone mouth have been studied by multivariate techniques: principal-component analysis, part correlation. A new technique of cluster analysis is also proposed. By these means, a very euryhaline group of zooplankton species was isolated and the extremely low sensibility by phytoplankton towards salinity was shown. However, temperature seems to be the most important ecological factor. Instability and eutrophy of this area do not appear to disturb the phytoplankton cycle, which occurs with its usual successions. Chlorophyll a and organic matter do not seem of value for estimation of the biomass in the area studied.--Copyright 1972, Biological Abstracts, Inc.
W73-10011

BLUE-GREEN ALGAE FROM THE KREMENCHUG RESERVOIR AS A BASE FOR GROWING MICROORGANISMS, (IN RUSSIAN),
Akademiya Nauk URSR, Kiev. Inst. of Microbiology and Virology.
E. I. Kvasnikov, I. P. Stognii, T. P. Travchuk, I. F. Shchelokova, and T. M. Klyushnikova.
Gidrobiol Zh. Vol 7, No 6, p 80-83. 1971.
Identifiers: Algae, Aphanizomenon-Flos-Aquae, Fermentation, Growing, *Kremenchug Reservoir, Microcystis-Aeruginosa, Microcystis-Wesenbergii, *Microorganisms, Reservoirs, USSR, *Cyanophyta.

Desiccated Microsystis aeruginosa, M. wesenbergii and Aphanizomenom flos-aquae which were collected in 1968-69 in the Kremenchug Reservoir were studied to determine their suitability for use as a cheap base for the cultivation of heterotrophic microorganisms in the food and fermentation industries. Blue-green algae contained all the necessary components, including substantial amounts of polysaccharides which were difficult and easy to hydrolyze, and rich amounts of protein. There were sufficient amounts of ash elements, though they were poor in P. There was a wide assortment of group B vitamins.--Copyright 1972, Biological Abstracts, Inc.
W73-10013

EFFECT OF ARTIFICIAL AERATION ON SOME HYDROCHEMICAL INDICES OF WATER QUALITY, (IN RUSSIAN),
Akademiya Nauk URSR, Kiev. Institut Hidrobiologii.
For primary bibliographic entry see Field 05G.
W73-10014

BEHAVIOR OF IODINE 131 IN FRESHWATER BODIES, (IN RUSSIAN),
V. N. Gus'kova, A. N. Bragina, B. N. Il'in, A. A. Zasedatelev, and V. M. Kupriyanova.
Gig Sanit. Vol 36, No 9, p 46-49. 1971. Illus.
Identifiers: Fish, *Iodine-131, Pollution, Protozoa, Water pollution effects, Organism.

Experimental investigations were made of the effect of I131 on freshwater organisms, from protozoa to fishes, under conditions of 1-time pollution of the water with I131 in concentrations of 10 to the minus 7th power, 10 to the minus 6th power, 10 to the minus 5th power, 10 to the minus 4th power, and 10 to the minus 3rd power Ci/l. The results indicated that concentrations of I131 in a body of water of 10 to the minus 6th power Ci/l and less do not disturb its normal biological activity.--Copyright 1972, Biological Abstracts, Inc.
W73-10015

RETENTION LAKES OF LARGE DAMS IN HOT TROPICAL REGIONS, THEIR INFLUENCE ON ENDEMIC PARASITIC DISEASES,
R. Deschiens.
Bull Soc Pathol Exot. Vol 63, No 1, p 35-51, 1970. English summary.
Identifiers: Africa, Arthropods, Bilharziasis, Dams, *Diseases endemic, Lakes, Malaria, Mollusks, Onchocercosis, *Parasitic diseases, *Retention lakes, Tropical endemics.

A freshwater environment, complicated by industry and agriculture provokes, particularly in warm countries, geographic, climate, floristic and faunistic alterations, which may influence public health. Tropical endemics in Africa include malaria (3,000,000 fatal cases/yr in the world), bilharziasis (200-300 million patients in the world, and onchocercosis (severe African and American cases of blindness), which are dangerous parasitic infestations, whose vectors (arthropods, mollusks) are aquatic. The increase of freshwater areas provokes an extension of biocenoses comprising arthropods and their larvae and freshwater mollusks. A multiplication of malaria, bilharziasis and onchocercosis vectors, and an increase in diffusion possibilities of these parasitic infestations result. Epidemiologists and health services possess efficient defenses against such pathogenic complexes and sanitary hazards: controlled epidemiologic and demographic investigations and surveys of the pathogenic agents, clinical aspects of the diseases, and invertebrate vector (location, topographical distribution) and ethnic studies. Such documentation determines the specific prophylaxis to be used: drug chemoprophylaxis and control of vector invertebrates by either insecticide and molluskicide agents and pertinent sanitary education.--Copyright 1972, Biological Abstracts, Inc.
W73-10070

INFLUENCE OF FOOD QUALITY AND QUANTITY ON THE POPULATION DYNAMICS OF THE PLANKTONIC ROTIFER BRACHIONUS CALYCIFLORUS IN LABORATORY EXPERIMENTS AND IN THE FIELD, (IN GERMAN),
Munich Univ. (West Germany). Zoologisch-Parasitologisches Institut.
U. Halbach.
Verh D Tsch Zool Ges. 65, p 83-88. 1971. English summary.
Identifiers: Biomass, Brachionus-calyciflorus, Chlorella-pyrenoidosa, Food quality, Oocystis-parva, *Planktonic population, *Rotifers, Scenedesmus-obliquus, *Algae.

The influence of food quantity (Chlorella pyrenoidosa) upon duration of life, fertility, intrinsic rate of natural increase, and mean population density (environmental capacity) of B. calyciflorus was investigated by laboratory experiments. All these parameters show optima in relation to food density. This is the consequence of 2 different influences of the algae: at low food densities the

parameters are reduced by starvation, whereas at high densities they are reduced by detrimental effects of algal metabolites. In 15 field tanks (0.3 to 50 m3) strong fluctuations of the animals' population parameters as well as algal biomass were recorded. The algal densities covered all experimental ranges. Analysis showed a correlation of phytoplankton biomass with egg-ratio and population growth of Brachionus, but no correlation with population density. The relationships are mainly caused by the nutritional effect of Oocystis parva and Scenedesmus obliquus.--Copyright 1973, Biological Abstracts, Inc.
W73-10079

ECOLOGICAL FACTORS THAT INDUCE SEXUAL REPRODUCTION IN ROTIFERS, (IN GERMAN),
Munich Univ. (West Germany). Zoologisch-Parasitologisches Institut.
U. Halbach, and G. Halbach-Keup.
Oecologia (Berl). Vol 9, No 3, p 203-214. 1972. English summary.
Identifiers: Brachionus-Angularis, Brachionus-Calyciflorus, Brachionus-Rubens, Competition, Ecological factors, Phytoplankton, *Reproduction, *Rotifers.

Monogonont rotifers reproduce parthenogenetically or sexually. The proportion of sexual females in a population (rate of mixis) can be modified by external factors. Published data about these factors are inconsistent and in part even contradictory. In summer 1967 quantitative plankton studies were done in 15 tanks (0.3-50m3). The following parameters were recorded every third day: population density, egg rate (eggs/female), and rate of mixis of the 3 rotifer species Brachionus calyciflorus, B. rubens and B. angularis; pH, temperature, rainfall, and phytoplankton biomass (dry weight). The latter was subdivided into 3 categories: ultra, nanno and micro-plankton. A correlation analysis of the environmental factors revealed many intercorrelations. The coefficients of correlation between each rate of mixis and all other parameters are given. A most striking result is the absence of significant correlations among the rates of mixis of the 3 species. This means that the periods of sexuality of the 3 related species are independent of one another in the same biotope. No 1 factor shows a consistent positive or negative correlation with the rates of mixis of all 3 species. But there are no contradictions, i.e., none of the parameters is correlated positively in 1 species and negatively in another species. Positive correlations (or none) are demonstrated with temperature, changing of temperature, and micro-phytoplankton; negative correlations (or none) with total phytoplankton, ultra-phytoplankton, nanno-phytoplankton, eggs/female, and population density of the competing Brachionus species, in no case are significant correlations found with pH and rainfall. That factors with significant correlations do not show these correlations with all species could be due to different theshold values of the mixis-inducing factors in the 3 rotifer species. This study in no case shows a significant positive correlation between the rate of mixis and the population growth rate, or the rate of eggs/female. Some cases show a significant negative correlation. It is difficult to decide whether the significant factors influence the rate of mixis directly or indirectly. The intercorrelations of the factors suggest that in many, if not most, cases these influences are indirect.--Copyright 1973, Biological Abstracts, Inc.
W73-10080

DEPENDENCE OF PRODUCTION OF BACTERIA ON THEIR ABUNDANCE IN LAKE WATER,
Polish Academy of Sciences, Warsaw. Inst. of Ecology.
W. A. Godlewska-Lipowa.
Bull Acad Pol Sci Biol. Vol 20, No 5, p 305-308. 1972. Illus.
Identifiers: Abundance, *Bacteria, Lakes, *Productivity.

The effect of artificial decrease or increase of a natural bacterial community in lake water was studied. The lower the initial numbers, the higher was the percentage increase and the shorter was the generation time. All variants with lower densities, and the variant with the maximum density, displayed similar changes in number, beginning with 12-hr exposure. Generation time was much longer in all experimental variants (with the exception of the variant with the greatest density) than in the control sample.--Copyright 1973, Biological Abstracts, Inc.
W73-10082

CHEMICAL POLLUTION AND MARINE TROPHODYNAMIC CHAINS, (POLLUTIONS CHIMIQUES ET CHAINES TROPHODYNAMIQUES MARINES),
Centre d'Etude et de Recherches de Biologie et d'Oceanographie Medicale, Nice (France).
M. Aubert.
Revue Internationale d'Oceanographie Medicale, Vol 28, p 9-25, 1972. 33 ref.

Descriptors: Water pollution, Industrial wastes, *Chemical wastes, *Detergents, *Pesticides, Metals, *Food chains, *Trophic level, Toxicity, Metabolism, Retention, Transfer, Biodegradation.
Identifiers: *Pelagic chain, Neritic chain, Benthic chain.

Chemical pollutants from industrial wastes, detergents and pesticides in the oceans are now not only hazardous to the flora and fauna in the marine environment but retransferred along the food chain to man. These chemical substances are metabolized by organisms living in the sea and by some phenomenon of concentration are retained in their tissues and then transferred along the various food chains. Some of these pollutants, particularly heavy metals, are very toxic to the human population. These phenomena were studied using several marine trophodynamic chains, the pelagic chain (phytoplankton, zooplankton, fish, and mammifers), neritic chain with mollusks (phytoplankton, mollusks, and mammifers), neritic chain with crustaceans, (Marine microorganisms, invertebrates, crustaceans; and mammifers), and the benthic chain, (Marine microorganisms, invertebrates, benthic fish and mammifers). Results of the research on toxicity thresholds, pollutant biodegradability and eventural toxicity transmission are presented. (Ensign-PAI)
W73-10091

UTILIZATION OF A TROPHODYNAMIC CHAIN OF A PELAGIC TYPE FOR THE STUDY OF TRANSFER OF METALLIC POLLUTION, (UTILISATION D'UNE CHAINE TROPHODYNAMIQUE DE TYPE PELAGIQUE POUR L'ETUDE DES POLLUTIONS METALLIQUES),
Centre d'Etudes et de Recherches de Biologie et d'Oceanographie Medicale, Nice (France).
M. Aubert, R. Bittel, F. Laumond, M. Romeo, and B. Donnier.
Revue Internationale d'Oceanographie Medicale, Vol 28, p 27-52, 1972. 22 fig, 8 tab, 19 ref.

Descriptors: Water pollution effects, Metals, *Heavy metals, *Copper, *Chromium, *Zinc, *Lead, *Mercury, Toxicity, Food chains, *Trophic level, Transfer.
Identifiers: *Pelagic chain.

Discharge of mineral products into the sea is becoming more frequent and in larger amounts making it a necessity to assess the hazards of contamination to man through the consumption of marine products. This concentration phenomenon is studied in vitro. Through the use of marine food chains we can measure the factors of transfer of the different metals at each level of the pelagic food chain. Five heavy metals: copper, zinc, chromium, lead and mercury in an ionic or complex form, are studied and the results are presented in tables and the conclusions discussed. (Ensign-PAI)
W73-10092

STUDY OF THE TOXICITY OF EFFLUENTS FROM PAPER MILLS IN THE MARINE ENVIRONMENT, (ETUDE DE LA TOXICITE D'EFFLUENTS DE PAPETERIE EN MILIEU MARIN),
Centre d'Etude et de Recherches de Biologie et d'Oceanographie Medicale, Nice (France).
B. Donnier.
Revue Internationale d'Oceanographie Medicale, Vol 28, p 53-93, 1972. 1 fig, 4 tab, 31 graphs, 14 ref.

Descriptors: Water pollution effects, Pulp and paper industry, *Pulp wastes, *Toxicity, *Food chains, Trophic level, Transfer, Sewage, Degradation (Decomposition), Fish, Mollusks, Crustaceans.
Identifiers: *Pelagic chain, *Neritic chain, *Benthic chain.

The hazards incurred by pollution of the marine environment from paper mill wastes are determined by studying the marine food chains to the final consumer. The study of direct toxicity before and after degradation of sewage in the sea and the transmission of toxic products was conducted on four chains: pelagic chain, neritic chain with mollusks, neritic chain with crustaceans and the benthic chain with fish. Results are discussed and illustrated in tables and graphs. (Ensign-PAI)
W73-10093

MERCURY CONTAMINATION OF FISH OF THE SEA, (CONTAMINATION DES POISSONS DE MER PAR LE MERCURE),
For primary bibliographic entry see Field 05B.
W73-10094

EFFECTS OF CHEMICAL POLLUTION WITH RESPECT TO TELEMEDIATORS OCCURRING IN MICROBIOLOGICAL AND PLANKTONIC ECOLOGY IN THE MARINE ENVIRONMENT, (EFFETS DES POLLUTIONS CHIMIQUES VIS-A-VIS DE TELEMEDIATEURS INTERVENANT DAN S L'ECOLOGIE MICROBIOLOGIQUE ET PLANCTONIQUE EN MILIEU MARIN),
Centre de'Etudes et de Recherches de Biologie et d'Oceanographie Medicale, Nice (France).
M. Aubert, M. Gauthier, B. Donnier, D. Pesando, and J-M. Pincemin.
Revue Internationale d'Oceanographie Medicale, Vol 28, p 129-165, 1972. 43 tab, 22 ref.

Descriptors: Industrial wastes, *Chemical wastes, Aquatic environment, *Phytoplankton, *Bacteria, *Inhibitors, Diatoms, Pseudomonas, Pesticides, Detergents, Oil, Pulp wastes, Electricity, Fabrics.
Identifiers: *Telemediators, *Antibiotic action, Asterionella japonica.

The chemical pollutant's action, from industrial wastes, on primary telemediators acting on phytoplanktonic and bacterian relations was studied. For some pollutants, an inhibition or antibiotical activity was observed in the diatom Asterionella japonica and in two strains of marine bacteria belonging to the genus Pseudomonas. (Ensign-PAI)
W73-10095

LARVAL DEVELOPMENT OF MYTILUS EDULIS (L.) IN THE PRESENCE OF ANTIBIOTICS, (DEVELOPPEMENT LARVAIRE DE MYTILUS EDULIS (L.) EN PRESENCE D'ANTIBIOTIQUES),
Universite de Bretagne-Occidentale, Brest (France). Laboratoire de Zoologie.

M. Le Pennec, and O. Prieur.
Revue Internationale d'Oceanographie Medicale, Vol 28, p 167-180, 1972. 8 plates, 10 ref. English summary.

Descriptors: *Bacteria, *Antibiotics, *Mollusks, *Larvae, Toxicity, Water pollution effects.
Identifiers: *Lamellibranch, *Mytilus edulis, Chloramphenicol, Aureomycine, Erythromycine, Sulfamerazine.

The influence of four antibiotics at several concentrations: chloramphenicol, aureomycine, erythromycine, and sulfamerazine on the larvae of Mytilus edulis, laboratory reared from the straight hinge stage to metamorphosis was investigated. The number of larvae alive in function of time for each of twenty-eight cultures is presented in diagrams and discussed. Effective and non-toxic concentrations of these antibiotics are shown. (Ensign-PAI)
W73-10096

THE MARINE ENVIRONMENT AND ORGANIC MATTER, (LE MILIEU MARIN ET LES MATIERES ORGANIQUES),
Centre d'Etude et de Recherches de Biologie et d'Oceanographie Medicale, Nice (France).
M. Aubert, J. Aubert, and M. Gauthier.
Revue Internationale d'Oceanographie Medicale, Vol 28, p 181-193, 1972. 10 ref. English Summary.

Descriptors: *Oceans, *Organic matter, Natural resources, *Biomass, Evolution, *Degradation (Decomposition), *Recycling, Resource development, Aquatic populations, Waste disposal, *Eutrophication, Productivity, Balance of nature, Ecology.

The processes of organic matter cycles in the sea: sources, evolution, degradation and recycling are reviewed. Man's influence on the natural cycle of the sea by depleting the biomass and the demographical expansion leading to an increase in organic matter production from wastes effect the balance of nature in the sea. Data on these processes are assessed and the consequences considered with emphasis on the phenomenon of eutrophication. The question of what man takes from the ocean in utilizing and developing the natural resources, balancing that which man puts into the ocean by discharge and the resulting production is discussed. (Ensig-PAI)
W73-10097

WE ARE KILLING THE OCEANS,
Senate, Washington, D.C.
For primary bibliographic entry see Field 05B.
W73-10099

TRENDS IN MARINE WASTE DISPOSAL, GOOD AND BAD,
Engineering-Science, Inc., Pasadena, Calif.
For primary bibliographic entry see Field 05E.
W73-10105

EFFECT OF POLLUTION ON THE MARINE ENVIRONMENT, A CASE STUDY,
Florida State Univ., Tallahassee. Marine Lab.
R. C. Harriss, H. Mattraw, J. Alberts, and A. R. Honke.
In: Coastal Zone Pollution Management, Proceedings of the Symposium, Charleston, South Carolina, February 21-22, 1972, Clemson University, Clemson, S.C. 1972. p 249-264. 3 fig, 1 tab, 17 ref.

Descriptors: *Estuaries, Water pollution effects, *Industrial wastes, Waste disposal, *Mining, *Phosphates, *Domestic wastes, Envrionmental effects, Plankton, Aquatic animals, Nitrogen, Eutrophication, *Florida.
Identifiers: *Charlotte Harbor Estuary.

The impact of industrial activities on the Charlotte Harbor Estuary in Florida is considered. The estuary was subjected to waste discharges from phosphate mining and processing operations for 84 years, and extensive short term damage occurred from periodic failures in waste facilities. Long-term exposure to low-level, partially-treated, phosphate waste discharges did not measurably affect most plankton and animal communities, but the combined effects of phosphates and domestic wastes containing nitrogen are producing eutrophic conditions. A system that could determine the capacity of an aquatic environment to assimilate waste discharges is proposed. (See also W73-10098) (Ensign-PAI)
W73-10110

CHOICE OF PRODUCTS FOR CONTROL AGAINST POLLUTION OF THE MARINE ENVIRONMENT BY HYDROCARBONS. II. EFFICACY OF ANTIPETROLEUM PRODUCTS, (CHOIX DE PRODUITS POUR LUTTER CENTRE LA POLLUTION DU MILIEU MARIN LES HYDROCARBURES. II. EFFICACITE DES PRODUITS ANTIPETROLE),
Institut Scientifique et Technique des Peches Maritimes, Nantes (France).
For primary bibliographic entry see Field 05G.
W73-10114

CHOICE OF PRODUCTS FOR CONTROL AGAINST POLLUTION OF THE MARINE ENVIRONMENT BY HYDROCARBONS. III. THE RELATIVE TOXICITY OF ANTIPETROLEUM PRODUCTS ON TWO MARINE ORGANISMS, (CHOIX DE PRODUITS POUR LUTTER CONTRE LA POLLUTION DU MILIEU MARIN PAR LES HYDROCARBURES. III. TOXICITE RELATIVE DE PRODUITS ANTIPETROLE SUR DEUX ORGANISMES MARINS),
Institut Scientifique et Technique des Peches Maritimes, Nantes (France).
C. Alzieu.
Revue des Travaux de l'Institut des Peches Maritimes, Vol 36, No 1, p 103-119, March 1972. 19 fig, 2 tab, 27 ref.

Descriptors: Water pollution control, *Organic compounds, *Oil pollution, *Cleaning, *Emulsifiers, Detergents, Environmental effects, Wildlife, *Oysters, *Phytoplankton, *Toxicity, Seawater, Dispersion, Pollution abatement, Water pollution effects.
Identifiers: *Agglomerants, Precipitants, Polyglycol ethers, Crassostrea angulata, Phaeodactylum tricornutum.

The influence of toxic antipetroleum agents on marine fauna and flora is discussed and various methods to determine the toxicity and LD on different organisms are surveyed. Immediate toxicity was tested on oysters (Crassostrea angulata) and a phytoplanktonic alga (Phaeodactylum tricornutum) on which it feeds, to study effects on oyster closing and phytoplankton growth. Various concentrations of emulsifiers were tested and agglomerating and precipitating products were tested for dispersion in seawater. All products containing anionic detergents or polyglycol ethers reduced the coefficient of valvate activity. Polyglycol ether appeared less toxic to phytoplankton than to oysters. (See also W73-10113) (Ensign-PAI)
W73-10115

CHOICE OF PRODUCTS FOR CONTROL AGAINST POLLUTION OF THE MARINE ENVIRONMENT BY HYDROCARBONS. IV. THE RELATIVE TOXICITY OF SEVEN ANTIPETROLEUM EMULSION PRODUCTS, (CHOIX DE PRODUITS POUR LUTTER CONTRE LA POLLUTION DU MILIEU MARIN PAR LES HYDROCARBURES. IV. TOXICITE RELATIVE DE SEPT PRODUITS EMULSIONNANTS ANTIPETROLE),
Institut Scientifique et Technique des Feches Maritimes, Nantes (France).
P. Maggi.
Revue des Travaux de l'Institut des Peches Maritimes, Vol 36, No 1, p 121-124, March 1972. 2 tab, 7 ref.

Descriptors: *Emulsifiers, *Toxicity, Water pollution effects, Aquatic animals, *Gastropods, Crustaceans, Fish, *Pollution abatement.
Identifiers: *Lamellibranchs, *Anguilla anguilla.

The relative toxicity of seven antipetroleum emulsifiers (E22, E30, E41, E46, E47, E52, and E53) was tested on three gastropods (Gibbula umbilicalis, Purpura lapillus and Littorina littorea), one lamellibranch (Mytilus edulis), three crustaceans (Artemia salina, Penaeus kerathurus and Clinabarius misanthropus), and one fish (Anguilla anguilla). The experiments were conducted on lots of 20-25 animals. Emulsifer E47 appeared the least toxic. (See also W73-10113) (Ensign-PAI)
W73-10116

PETROLEUM HYDROCARBONS IN OYSTERS FROM GALVESTON BAY,
Kiel Univ. (West Germany). Institut fuer Meereskunde.
For primary bibliographic entry see Field 05A.
W73-10117

EFFECTS OF CHLORINATED ALIPHATIC HYDROCARBONS ON LARVAL AND JUVENILE BALANUS BALANOIDES (L.),
Swedish Water and Air Pollution Research Lab., Goteborg (Sweden).
R. Rosenberg.
Environmental Pollution, Vol 3, p 313-318, 1972. 2 fig, 1 tab.

Descriptors: *Chlorinated hydrocarbons, Environmental effects, *Lethal limit, Larval growth stage, Juvenile growth stage, Life cycle, Analytical techniques, Water pollution effects, *Growth stages.
Identifiers: *Balanus balanoides (L.), Vinyl chloride, *North Sea, Norway coast.

Dumping of by-products from vinyl chloride-production (EDC-tar) was observed in the North Sea and off the west coast of Norway in concentrations of one-tenth the amount known to cause acute biological effects. The sublethal concentrations of these chlorinated aliphatic hydrocarbons on larval and juvenile Balanus balanoides (L) were determined. Reduction in swimming activity of the stage II nauplii occurred in an EDC-concentration of 2 ppm. Stage V and VI nauplii were affected in concentrations around 4 ppm. Newly-settled, 2-3 week old specimens, showed about ten times greater tolerance than did the stage H nauplii to EDC-tar. Tolerance seems to gradually increase with age in B. balanoides even when the species is exposed to abnormal conditions, such as heat, surfactants, detergents, or EDC-tar. (Ensign-PAI)
W73-10118

RESTRUCTURING OF SHORES AND SECONDARY POLLUTION. STUDY OF THE EUTROPHICATION OF HARBOR ZONES. (RESTRUCTURATION DES RIVAGES ET POLLUTIONS SECONDAIRES. ETUDE DE L'EUTROPHISATION DE ZONES PORTUAIRES),
Centre d'Etudes et de Recherches de Biologie et d'Oceanographie Medicale, Nice (France).
M. Aubert, J. Aubert, J-M. Pincemin, N. Desirotte, and J-Ph. Brettmeyer..
Revue Internationale d'Oceanographie Medicale, Vol 26, p 53-64, 1972. 3 tab, 6 ref. English summary.

Identifiers: *Laminairie hyperborea, *Species diversity, Long-term effects, *North Sea.

A spatial and temporal study of the neotenous, pollution-tolerant community found in the holdfast habitats of Laminaria hyperborea off the coast of Britain was conducted. Random samples of 1 to 7 year old holdfasts of the kelp plant were collected at each of fifteen stations, in the polluted and non-polluted waters along the shores of northeast England and southwest Scotland. Species diversity was consistently greater at non-polluted stations. Mytilus edulis, Asterias rubens, Nereis pelagica and Sabellaria spinulosa werbellaria spinulosa were the most abundant species in polluted areas. It is noted that this kind of reduction in species diversity leaves only one possible detritus food chain in the habitat. Possible long-term effects of pollution on kelp forests are discussed. (Ensign-PAI)
W73-10123

A SHORT SURVEY OF THE ENVIRONMENT AT THE DUMPING SITE FOR SANTA CRUZ HARBOR DREDGING,
Moss Landing Marine Labs., Calif.
For primary bibliographic entry see Field 05B.
W73-10124

THE CHANGING CHEMISTRY OF THE OCEANS,
Goteborg Univ. (Sweden). Dept. of Analytical Chemistry.
For primary bibliographic entry see Field 05B.
W73-10125

THE OCEAN: OUR COMMON SINK,
Norske Videnskaps-Akademi i Oslo.
T. Heyerdahl.
Environment this Month, Vol 1, No 1, p 6-10, July 1972. 3 fig.

Descriptors: *Oceans, Pollutants, *Environmental effects, Water pollution effects, Waste disposal.
Identifiers: *Kon-Tiki, *Ra Expeditions.

Detrimental changes in ocean conditions are discussed; observations from an expedition in 1947 are compared to those made in 1970. The condition of coastal rocks and cliffs (where much marine life centers) near islands like Malta and Cyprus in the Mediterranean is deplorable. Fins, plastic containers, and oil slicks float visibly between boulders and in crevices. The high content of insecticides and detergents flowing into the oceans from industry and agriculture are beginning to sterilize the coastline. The oceans are being greatly over-taxed by growing amounts of 'unnatural' pollutants; the ocean cannot filter all of these wastes, consequently we are in danger of depleting one of the earth's greatest resources. (Ensign-PAI)
W73-10126

THE EFFECTS OF SUSPENDED 'RED MUD' ON MORTALITY, BODY WEIGHT, AND GROWTH OF THE MARINE PLANKTONIC COPEPOD, CALANUS HELGOLANDICUS,
California Univ., San Diego. Inst. of Marine Resources.
G-A. Paffenhoefer.
Water, Air, and Soil Pollution, Vol 1, No 3, p 314-321, July 1972. 2 fig, 2 tab, 7 ref. AEC AT (11-1) GEN 10, P.A. 20.

Descriptors: *Industrial wastes, *Aluminum, *Copepods, Growth, *Dumping, Water pollution effects, Laboratory tests, Toxicity, Sediments, Mud.
Identifiers: *Red mud, Calanus helgolandicus.

'Red mud' is the residual substance after the extraction of Al from bauxite; it is a waste product associated with aluminum production. The effects of red mud on growth, body weight and mortality of the copepod Calanus helgolandicus were investigated. Laboratory tests were conducted using copepods which ingested large quantities of red mud and control groups. Growth of the experimental copepods was delayed in comparison to the control group; the red mud group also had a higher mortality rate than the controls. The small amount of phytoplankton present in red mud weakens the animals; although they ingest large quantities of mud, they were unable to attain a sufficient amount of nutritive material. Adult copepods in the red mud group exhibited sluggish swimming movements and slow reactions which would make them easy prey for their predators. The study concludes that the proposed dumping of red muds into the North Sea will create a serious problem in the nutrition, survival, and reproduction of planktonic copepods. (Ensign-PAI)
W73-10127

GULF STATES OYSTER SEMINAR REVIEWS RECENT AREA SCIENTIFIC DEVELOPMENTS.

Fish Boat, Vol 17, No 11, p 16, 34-36, November 1972.

Descriptors: *Gulf of Mexico, *Oysters, *Fisheries, *Dredging, *Sewage effluents, Disease, Pollutants, Water quality standards, Water pollution effects.

Some problems facing the oyster industry along the Gulf Coast are discussed. Scientists from Texas, Louisiana, Florida, Alabama and Mississippi briefly outline the sorts of problems which represent the most significant threats to oysters in their state waters. Some common complaints include pollution caused by dredging, by dumping raw sewage, and by increased mercury levels. Troubles attributed to hurricanes, siltage and oxygen depletion were also mentioned. The spread of diseases carried by oysters, such as hepatitis, is potentially dangerous, unless strict standards of sanitation are enforced within the oyster industry. The need for greater research into areas of pollution, disease etc., is stressed. (Ensign-PAI)
W73-10130

ESTUARIES AND INDUSTRY,
University of Strathclyde, Garelochhead (Scotland). Marine Lab.
For primary bibliographic entry see Field 05B.
W73-10133

SPECIAL PROBLEMS OF THE ARCTIC ENVIRONMENT,
Arctic Inst. of North America, Washington, D.C.
M. E. Britton.
In: Canadian-U.S. Maritime Problems, Law of the Sea Workshop, June 15-17, 1971, Toronto, Canada, University of Rhode Island, Kingston, Law of the Sea Institute, p 9-28. L.M. Alexander and G.R.S. Hawkins, editors, 6 fig, 18 ref. 1972.

Descriptors: *Arctic, *Arctic Ocean, *Ice, *Water pollution, *Air pollution, Oil industry, Oil fields, Tundra, Ecosystems, Balance of nature, Water pollution effects.

A discussion of Arctic environments is presented, with emphasis on the special problems which are unique to the area. Atmospheric conditions, basic features of the Arctic Ocean basin, the role of the continental shelves, the marine ecosystem and the tundra ecosystem are some of the areas discussed. Pollution of the Arctic atmosphere and/or ice could upset present climatic and oceanographic conditions and begin a potentially disasterous warming trend. About two-thirds of the Arctic Ocean bottom is continental shelf, in which lies oil. However, ice and shallow waters make it difficult to extract and transport. Oil resources on the tundra are more feasible; the possible effects of oil

exploitation on the tundra ecosystem and the complex nature of permafrost are investigated. The marine ecosystem is also discussed in some detail; although the ocean has few numbers of species, the individuals of each species are abundant. All primary production occurs in a very shallow ocean layer where light and heat are appropriate. Any alteration of this layer by pollutants could cause far-reaching effects on the ecosystem. (Ensing-PAI)
W73-10134

A CONTINUOUS CULTURE OF DESULFOVIBRIO ON A MEDIUM CONTAINING MERCURY AND COPPER IONS,
Netherlands Inst. for Sea Research, Den Helder.
J. H. Vosjan, and G. J. Van der Hoek.
Netherlands Journal of Sea Research, Vol 5, No 4, p 440-444, May, 1972. 2 tab, 7 ref.

Descriptors: *Bacteria, *Mercury, *Copper, *Cultures, *Biodegradation, Toxicity, Water pollution effects, Growth.
Identifiers: *Desulfovibrio, Continuous cultures.

A continuous culture of Desulfovibrio was used to study the effects of mercury and copper. It was determined that once a continuous culture of Desulfovibrio is in progress the toxic effects of Hg and Cu are nullified and do not affect the growth of the organisms. Atomic absorption analyses showed that the elements had been removed from the liquid. Similar experiments carried out on batch cultures of Desulfovibrio caused the cessation of all growth. It appears from this study that microbiological processes of the sort that occur in continuous cultures, could be used to render toxic metals innocuous, to remove them from polluted waters, and to fix them as sulphides. (Ensign-PAI)
W73-10135

INFLUENCE OF AEROSOLS IN THE MARINE ENVIRONMENT (INFLUENCE DES AEROSOLS MARINS DANS L'ENVIRONMENT),
Centre d'Etudes et de Recherches de Biologie et d'Oceanographie Medicale, Nice (France).
For primary bibliographic entry see Field 05B.
W73-10136

DETERMINATION OF THE YIELD BY LIQUID SCINTILLATION OF THE 14C IN THE MEASUREMENT OF PRIMARY PRODUCTION IN A LAGOON ENVIRONMENT, (DETERMINATION DES RENDEMENTS DU COMPTAGE PAR SCINTILLATION LIQUIDE DU 14C DANS LES MESURES DE PRODUCTION PRIMAIRE EN MILIEU LAGUNAIRE).

Revue Internationale d'Oceanographie Medicale, Vol 26, p 27-41, 1972. 3 fig, 12 ref. English summary.

Descriptors: *Lagoons, *Carbon radioisotopes, Environmental effects, *Primary productivity, Cytological studies, Solubility, Solvents, Pollutant identification, Water pollution effects.
Identifiers: *Liquid scintillation, *Counting efficiency.

Liquid scintillation was used to calculate 14C in primary productivity measurements. Lagoons presented non-homogeneous samples for determination of counting efficiency. A partial solubilization of cells by organic solvent was indicated from the results of this experiment. The channel ratio should be used for a correct counting efficiency determination. (Ensign-PAI)
W73-10137

EFFECT OF PESTICIDES ON CHOLINESTERASE FROM AQUATIC SPECIES: CRAYFISH, TROUT AND FIDDLER CRAB,
Louisiana State Univ., New Orleans. Dept. of Chemistry.
G. G. Guilbault, R. L. Lozes, W. Moore, and S. S. Kaun.
Environmental Letters, Vol 3, No 4, p 235-245, 1972. 2 fig, 3 tab, 12 ref. NIH ES-00426-01.

Descriptors: Pesticides, *Pesticide residues, Oceans, *Crabs, *Crayfish, *Trout, Laboratory tests, *Enzymes, Water pollution effects.
Identifiers: *Tolerance levels, *Cholinesterase enzymes.

Purified cholinesterose enzymes were isolated in samples of trout, crayfish and fiddler crabs; nineteen pesticides were tested for inhibition of cholinesterase activity. Laboratory analyses show that trout cholinesterase is insensitive to all chlorinated and most organophosphorous pesticides. Crayfish have a high sensitivity to paraoxon and Mesurol and fiddler crabs are selectively inhibited by six pesticides. Overall, chlorinated pesticides seem to have little inhibitory effect on trout or crab cholinesterase; carbamates strongly inhibit all three enzymes. The experimental method of isolating the purified enzyme from each sampled species permitted a very sensitive and selective assay. (Ensign-PAI)
W73-10138

PHENYLALANINE METABOLISM ALTERED BY DIETARY DIELDRIN,
Bureau of Sport Fisheries and Wildlife, Columbia, Mo. Fish-Pesticide Research Lab.
P. M. Mehrle, M. E. DeClue, and R. A. Bloomfield.
Nature, Vol 238, No 5365, p 462-463, August 25, 1972. 3 fig, 16 ref.

Descriptors: *Trout, *Dieldrin, *Metabolism, Insecticides, Enzymes, Amino acids, Water pollution effects.
Identifiers: *Salmo gairdneri, *Liver, Blood, Urine samples.

The chronic effects of dieldrin on metabolic pathways in the rainbow trout, Salmo gairdneri, are described. Specific areas studied in the trout were liver phenylaline hydroxylase activity, serum phenylalanine, and phenylpyruvic acid concentrations in urine. The fish were fed daily portions of food containing differing doses of dieldrin equivalent to 4% of their body weight for 300 days; then liver, blood, and urine samples were taken and analyzed. Whole body residues were also investigated in some of the fish samples. The results showed that liver phenylalanine hydroxylase activity was significantly decreased by all doses of dieldrin; there was a negative correlation between dieldrin dose and enzyme activity. Serum phenylalanine concentration was increased by all doses of dieldrin; urinary phenylpyruvic acid increased in dieldrin. It is concluded that dieldrin has a marked effect on phynylalanine metabolism and can induce the biochemical manifestations of phenylketonuria. (Ensign-PAI)
W73-10140

NATURE AND POISONOUS CHEMICALS, (IN RUSSIAN),
D. F. Rudnev, and N. E. Kononova.
Lesn. prom'st Moscow. 1971. 141 p, Illus.
Identifiers: Ant, Bee, Bird, Chemicals, Fish, Humans, *Insecticides, Mammals, *Poisonous chemicals, *Entomofauna, *Fauna.

Information is given on the effect of insecticides on the useful fauna. There is a discussion of the different types of insecticides, the effect of chemical treatment on the soil of the forest and the soil fauna, on the entomofauna, the ants, bees, fishes, birds, mammals, and man.—Copyright 1973, Biological Abstracts, Inc.
W73-10155

ON THE PHYTOPLANKTON OF WATERS POLLUTED BY A SULPHITE CELLULOSE FACTORY,
Jyvaskyla Univ. (Finland). Dept. of Biology.
P. Eloranta.
Ann Bot Fenn. Vol 9 No 1, p 20-28. 1972. Illus.
Identifiers: Carbon dioxide, Cellulose, *Finland *Phytoplankton, Polluted water, *Sulfites, Water pollution effects, *Pulp wastes, Industrial wastes.

In summer 1970 a study was made of the species composition and biomass of the phyto-plankton in the northern part of Kuorevesi, in the northeast of the Kokemaenjoki basin, Finnish Lake District whose waters are polluted by the effluents of a sulfite cellulose factory. Observations were also made on the concentrations of O2, CO2 and on the pH.—Copyright 1973, Biological Abstracts, Inc.
W73-10156

ENRICHMENT EXPERIMENTS IN THE RIVERIS RESERVOIR,
Bonn Univ. (Germany). Zoological Inst.
H. Schnitzler.
Hydrobiologia. Vol 39 No 3, p 383-404. 1972. Illus. (English summary).
Identifiers: *Germany, *Nitrates, *Phosphates, Phytoplankton, Reservoirs, *Riveris reservoir, Nutrients.

From March 1967 to Oct. 1968 enrichment experiments with PO4 (3-) and NO3 (2-) nitrate were carried out in the Riveris reservoir near Trier (Germany). Plastic bags filled with 70 1 of the reservoir water were placed into the lake after NO3 (2-) and PO4 (3-) was added. When only NO3 (2-) was added, no enhancement of phytoplankton grown could be established. The phytoplankton in those bags to which PO4 (3-) was added in most cases increased according to the amount of added phosphate. Only in a few cases was an increase of plankton seen when NO3 (2-) and PO4 (3-) were added. These results show that in the Riveris reservoir phosphate is a limiting factor for the phytoplankton.—Copyright 1973, Biological Abstracts, Inc.
W73-10157

DEPENDENCE OF PRIMARY PRODUCTION OF PHYTOPLANKTON ON BIOGENIC ELEMENTS AND LIGHT, (IN RUSSIAN),
Akademiya Nauk SSSR, Moscow. Institut Okeanologii.
Yu. G. Kabanova, and Yu. E. Ochakovskii.
Dokl Akad Nauk SSSR Ser Biol. Vol 201 No 5, p 1227-1230. 1971. Illus.
Identifiers: Biogenic elements, Light, *Phytoplankton, *Primary production, *Photosynthesis.

A direct relation between the increment of primary production and concentration of elements at all levels of light energy was observed for natural populations of phytoplankton. The optimal illumination for photosynthesis of natural populations of phytoplankton was on the average at an energy of 46 W/m2 at a depth of 25 m. On the surface, in the case of a subsurface energy exceeding 300 W/m2, photosynthesis was inhibited by an excess of light, and at a depth of 50 m at an energy of about 13 W/m2, by a shortage of energy.—Copyright 1973, Biological Abstracts, Inc.
W73-10158

STUDIES ON THE WATER POLLUTION AND PRODUCTIVITY IN KYONGGI BAY IN SUMMER SEASON (3): CLASSIFICATION OF THE PHYTOPLANKTON (III), (IN KOREAN),
Seoul National Univ. (Republic of Korea). Dept. of Botany.
Y. H. Chung, J. H. Shim, and M. J. Lee.
Identifiers: Bays, Chrysophyta, Classification, *Korea, *Kyonggi Bay, *Phytoplankton, Pollution, *Productivity, Pyrrophyta, Seasonal, Species, Water pollution effects.

FINAL ENVIRONMENTAL STATEMENT RELATED TO THE PROPOSED SAN ONOFRE NUCLEAR GENERATING STATION - UNITS 2 AND 3.
Directorate of Licensing (AEC), Washington, D.C.
For primary bibliographic entry see Field 05B.
W73-10179

CALCULATIONS OF ENVIRONMENTAL RADIATION EXPOSURES AND POPULATION DOSES DUE TO EFFLUENTS FROM A NUCLEAR FUEL REPROCESSING PLANT,
Office of Radiation Programs, Washington, D.C.
For primary bibliographic entry see Field 05B.
W73-10180

RESEARCH ON THE MARINE FOOD CHAIN, PROGRESS REPORT JULY 1971-JUNE 1972.
California Univ., San Diego, La Jolla. Inst. of Marine Resources.

Available from NTIS, Springfield, Va., as UCSD 10P20-121; $9.00 in paper copy, $0.95 in microfiche. Report UCSD 10P20-121, 1972. 728 p.

Descriptors: *Oceanography, *Phytoplankton, *Estuarine environment, *California, Path of pollutants, Carbon radioisotopes, Nitrogen fixation, *Nutrients, Upwelling, Organic matter, Carbon cycle, Nitrogen, Phosphorus, Europhication, Polar regions, Research and development, Vitamins, Vitamin B.

Progress reports on current research programs, manuscripts of completed work that will be submitted to the open literature, and data on nutrients and phytoplankton in waters off southern California from a July 1970 cruise are presented. Topics include factors affecting eutrophication in coastal waters; the content in polar seas of N, P, dissolved organic C, and vitamins; fixation of dissolved nitrogen gas; and the radioactivity of dissolved and particulate organic C. (Bopp-ORNL)
W73-10200

A STATISTICAL STUDY OF RADIATION DOSE FOR FISHERMEN,
Battelle-Pacific Northwest Labs., Richland, Wash.
T. H. Beetle.
Available from NTIS, Springfield, Va., as BNWL-SA-4449; $4.00 in paper copy, $0.95 in microfiche. Report BNWL-SA-4449, Dec 1972, 19 p, 10 fig.

Descriptors: *Columbia River, *Radioactivity effects, *Public health, *Food chains, Radioisotopes, Absorption, Fish diets, Statistical models, Surveys, Statistical methods, Water pollution effects, Nuclear wastes, Path of pollutants, Environmental effects.

Statistical concepts are applied in an investigation of an environmental disturbance (release of radionuclides to the Columbia River) on a human population. The design of multistage sampling decreased the standard-deviation error of the estimate obtained with a given time allocation for the total number of interviews. It is estimated that the average internal dose for a Tri-City eater of Columbia River fish in 1966-67 was 8.25 mrem/yr and that the average external dose of a fisherman was 0.6 mrem/yr. (Bopp-ORNL)
W73-10203

AQUATIC RADIOLOGICAL ASSESSMENTS AT NUCLEAR POWER PLANTS,
Portland General Electric Co., Oreg.
For primary bibliographic entry see Field 05B.
W73-10204

FINAL ENVIRONMENTAL STATEMENT RELATED TO CONTINUATION OF CONSTRUCTION AND OPERATION OF UNITS 1 AND 2 AND CONSTRUCTION OF UNITS 3 AND 4 OF NORTH ANNA POWER STATION.
Directorate of Licensing (AEC), Washington, D.C.
For primary bibliographic entry see Field 05B.
W73-10205

ENVIRONMENTAL MONITORING AT ARGONNE NATIONAL LABORATORY - ANNUAL REPORT FOR 1972,
Argonne National Lab., Ill.
For primary bibliographic entry see Field 05A.
W73-10206

A COMPARATIVE STUDY OF TWO METHODS OF ESTIMATING HUMAN INTERNAL EXPOSURE HAZARDS RESULTING FROM THE INGESTION OF SEA FOODSTUFFS,
Commissariat a l'Energie Atomique, Fontenay-aux-Roses (France). Centre d'Etudes Nucleaires.
R. Bittel.
Available from NTIS, Springfield, Va., as CONF-720503-5; $3.00 in paper copy, $0.95 in microfiche. CONF-720503-5, 1972. 5 p, 1 tab, 5 ref.

Descriptors: *Public health, *Food chain, *Radioisotopes, *Nuclear wastes, *Fallout, Path of pollutants, Radioecology, Mathematical models, Systems analysis, Marine biology, Marine animals, Marine fisheries, Radiosensitivity, Absorption, Metabolism, Fish diets, Water pollution effects, Radioactivity effects, Radioactivity techniques.

The maximum permissible concentration of radionuclides in sea water was calculated on the basis of the dose to the critical organ from an annual diet consisting of 12 kg of fish, 13 of mollusks, and 3 of crustaceans. Ratios of the values obtained by the specific-activity method (w72-04484) to those obtained by the concentration-factor method were: Ca45, 50; Cs137, 15; Mn54, 12; I131, 10. Close agreement by the two methods was obtained for Na24, P32, Fe55, Co60, Zn65, and Mo99. It is considered that either approach is justified when the elements have metabolic regulation, and that the choice between the two depends upon the availability of reliable values of the required parameters. Different methods are required when the critical organ is the gastrointestinal tract. (Bopp-ORNL)
W73-10208

POLYCHLORINATED BIPHENYLS-ENVIRONMENTAL IMPACT. A REVIEW BY THE PANEL ON HAZARDOUS TRACE SUBSTANCES. MARCH 1972,
Minnesota Univ., St. Paul. Coll. of Veterinary Medicine.
For primary bibliographic entry see Field 05B.
W73-10221

THE IMPACT OF ARTIFICIAL RADIOACTIVITY ON THE OCEANS AND ON OCEANOGRAPHY,
Ministry of Agriculture, Fisheries and Food, Lowestoft (England). Fisheries Radiobiological Lab.
For primary bibliographic entry see Field 05B.
W73-10226

CURRENT STATUS OF RESEARCH ON THE BIOLOGICAL EFFECTS OF PESTICIDES IN CHESAPEAKE BAY,
Westinghouse Ocean Research Lab., Annapolis, Md.
T. O. Munson, and R. J. Huggett.
Chesapeake Science, Vol 13, (Supplement), p S 154-S 156, December 1972.

Descriptors: *Chesapeake Bay, *Pesticides, *Water pollution effects, *Ecosystem, Fish, Gulls, Clams, Research, Programs, *Monitoring, Planning, Management, Virginia, Maryland, Federal government, Bays.

Monitoring and research programs for Chesapeake Bay carried out by Virginia, Maryland, and federal agencies are discussed. Apart from the EPA monitoring program on fish, no Bay-wide pesticide monitoring is presently being done. Sea gulls and soft-shelled clams would be important and representative species to monitor in a Bay-wide program. Several programs could be initiated to give management a better estimate of both existing conditions and probable pesticide effects on the ecosystem. (Ensign-PAI)
W73-10231

CURRENT STATUS OF KNOWLEDGE OF THE BIOLOGICAL EFFECTS OF HEAVY METALS IN THE CHESAPEAKE BAY,
Johns Hopkins Univ., Baltimore, Md. Dept. of Environmental Medicine.
J. M. Frazier.
Chesapeake Science, Vol 13, (Supplement), p S 149-S 153, December 1972. 35 ref.

Descriptors: *Chesapeake Bay, *Estuarine environment, Water pollution sources, *Heavy metals, *Distribution patterns, *Water pollution, Environmental effects, *Biota, Productivity, Toxicity, Food chains, Trophic level, Temperature, Salinity, Dissolved oxygen, Sediments, Bays.
Identifiers: Bioavailability.

The knowledge concerning the bioavailability, effects on productivity, and contamination problems of heavy metals in the Chesapeake Bay is discussed. The investigation of the bioavailability of heavy metals in estuarine systems has barely scratched the surface. The effects of heavy metals on the productivity of higher trophic levels cannot be predicted at this time. Adequate information concerning biological contamination by these metals is still lacking in many areas. Specific areas where research is required are listed. (Ensign-PAI)
W73-10232

CURRENT STATUS OF KNOWLEDGE CONCERNING THE CAUSE AND BIOLOGICAL EFFECTS OF EUTROPHICATION IN CHESAPEAKE BAY,
Maryland Univ., Solomons. Natural Resources Inst.
D. A. Flemer.
Chesapeake Science, Vol 13, (Supplement), p S144-S149, December 1972. 28 ref.

Descriptors: *Chesapeake Bay, *Estuarine environment, *Eutrophication, *Phytoplankton, *Nutrients, Water pollution control, Primary productivity, Standing crops, Photosynthesis, Plant physiology, Nutrient requirements, Protein, Phosphorus, Carbon, Nitrogen, Hydrogen, Oxygen, Research, Planning, Bays.

The basic terms and concepts of eutrophication are briefly presented. Conditions in the Chesapeake Bay are discussed with reference to phytoplankton blooms and guidelines on nutrient requirements and supply. Solutions for the control of the effects of eutrophication are discussed and general research guidelines for the Chesapeake Bay and other waters are listed. (Ensign-PAI)
W73-10233

CURRENT STATUS OF THE KNOWLEDGE OF THE BIOLOGICAL EFFECTS OF SUSPENDED AND DEPOSITED SEDIMENTS IN CHESAPEAKE BAY,
Maryland Univ., Prince Frederick, Hallowing Point Field Station.
J. A. Sherk, Jr.
Chesapeake Science, Vol 13, (Supplement), p S137-S144, December 1972. 47 ref.

Descriptors: *Chesapeake Bay, *Sedimentation, *Deposition (Sediments), *Suspended loads, Inorganic compounds, Organic matter, Estuarine environment, Environmental effects, Aquatic life, Protection, Projects, Bays.

Suspended particulate organic and inorganic material and substratum changes associated with its deposition may be expected to affect estuarine organisms in a number of ways. The nature of the sediment input into the Chesapeake Bay is reviewed. The effects of suspended particles and deposited sediments on organisms in the Bay are discussed. Criteria for protection of the Bay are presented. Areas for consideration when assessing a projects effects on sedimentation are listed. (Ensign-PAI)
W73-10234

BIOLOGICAL CRITERIA OF ENVIRONMENTAL CHANGE IN THE CHESAPEAKE BAY,
Virginia Inst. of Marine Science, Gloucester Point. Dept. of Ecology-Pollution.
For primary bibliographic entry see Field 05B.
W73-10235

(BIOTA OF THE CHESAPEAKE BAY), CONCLUSIONS AND RECOMMENDATIONS,
Maryland Univ., Prince Frederick, Hallowing Point Field Station.
For primary bibliographic entry see Field 05B.
W73-10236

POSSIBLE EFFECTS OF THERMAL EFFLUENTS ON FISH: A REVIEW,
Washington, Univ., Seattle. Coll. of Fisheries.
J. R. Sylvester.
Environmental Pollution, London, Vol 3, p 205-215, July 1972. 2 fig, 1 tab, 75 ref.

Descriptors: *Thermal pollution, Temperature, Environmental effects, *Fish physiology, *Metabolism, *Reproduction, Growth rates, Lethal limit, Spawning, Fry, Heat resistance, Water pollution effects, *Reviews.
Identifiers: *Thermal shock, Acclimation, Sensing, Temperature preferences, Synergism.

Literature on specific considerations of thermal effluents (metabolism; thermal shock; reproduction and development; acclimation and lethal limits; fluctuations and sensing; temperature preferences; and synergism) is reviewed. Fish physiology is affected generally and specifically by thermal stress. Over a species-specific temperature range, increasing temperatures generally increase metabolic rates and other activity. A cessation of spawning behavior or an increase of abnormal fry can be caused by deviations from optimal temperature. The final level of temperature acclimation is a function of species, size, condition, and past environment. Most fish can acclimate faster to higher temperatures than to lower temperatures. Fluctuating temperatures increase thermal resistance in some species. Temperature preferences are a function of acclimation temperature and of the species-specific optimum temperature. Slight increases in temperature can be lethal because of synergism. (Ensign-PAI)
W73-10238

KILLING OF MARINE LARVAE BY DIESEL OIL,
Alberta Univ., Edmonton. Dept. of Zoology.
F-S. Chia.
Marine Pollution Bulletin, London, Vol 4, No 2, p 29-30, February 1973.

Descriptors: *Oil, Water pollution effects, *Larvae, *Animal physiology, *Mortality.
Identifiers: *Diesel oil (No 2), Echinodermata, Mollusca, Annelida, Arthropoda, Urochordata.

Laboratory observations are given on the effect of Texaco No 2 diesel oil on 14 species of 5 phyla of pelagic larvae. All larvae died at different durations (from 3-72 hours) after being placed in 0.5% oiled seawater, except Crossaster which survived eight days. Symptoms of contact with oil are acute contractions of the gut, asynchronization and sluggish beat of cilia or setae, and occasional violent movement of the body. Species specificity in terms of survival may be related to larval size. (Ensign-PAI)
W73-10241

EFFECTS OF OIL UNDER SEA ICE,
Massachusetts Inst. of Tech., Cambridge. Fluid Mechanics Lab.
For primary bibliographic entry see Field 05B.
W73-10246

UPPER POTOMAC ESTUARY EUTROPHICATION CONTROL REQUIREMENTS,
Environmental Protection Agency, Washington, D.C. Office of Research and Monitoring.
N. A. Jaworski, L. J. Clark, and K. D. Feigner.
Environmental Protection Agency, Annapolis Field Office, Technical Report No 53, April 1972. 45 p, 6 fig, 5 tab, 11 ref.

Descriptors: *Chesapeake Bay, *Potomac River, *Estuaries, *Water pollution control, *Eutrophication, *Nutrient removal, Programs, Planning, Management, Costs, Treatment, Information exchange, Treatment facilities.

Identification of the needs, costs, and mechanisms for controlling eutrophication in the Potomac Estuary was made and an attempt at implementing the program has begun. With capital cost for nutrient removal of over $250,000,000, a need exists for continuous efforts to improve eutrophication control, treatment methods, cost estimates, and institutional arrangements. Maintenance of free-flowing continuous exchange of information among the various agencies conducting the removal requirement studies, designing and facilities, and planning the overall management needs is also necessary. These interactions are the basis for successful management planning. (Ensign-PAI)
W73-10247

INTERTIDAL ALGAL SPECIES DIVERSITY AND THE EFFECT OF POLLUTION,
Sydney Univ. (Australia). School of Biological Sciences.
M. A. Borowitzka.
Australian Journal of Marine and Freshwater Research, Vol 23, No 2, p 73-84, December 1972. 2 fig, 4 tab, 35 ref.

Descriptors: *Sewage, *Outfalls, *Water pollution effects, *Intertidal areas, *Algae, Sites, Growth rates, Water analysis.
Identifiers: *Species composition, Species diversity, Species reduction.

Determination of the species diversity of the larger intertidal algae at 3 sites along the coastline of Sydney, N.S.W. was made. The height from mean of low water (MLW), distance from the edge of the rock platform at MLW, and distance from a sewer outfall (i. e., the degree of pollution) were correlated with changes in species composition and species diversity. Algal species number reduced in the vicinity of the outfall. This reduction was most evident in the Phaeophyceae and the Rhodophyceae. The maximum value of algal species diversity was also reducted at higher levels above MLW, near the outfalls and away from the platform edge. (Ensign-PAI)
W73-10249

INDICATOR SPECIES--A CASE FOR CAUTION,
University Coll. of North Wales, Bangor. Marine Science Labs.
For primary bibliographic entry see Field 05B.
W73-10252

POLLUTION AND THE OFFSHORE OIL INDUSTRY,
British Petroleum Co. Ltd., Sunbury-on-Thames (England).
For primary bibliographic entry see Field 05B.
W73-10253

SECOND ANNUAL REPORT OF THE ADVISORY COMMITTEE ON OIL POLLUTION OF THE SEA. RESEARCH UNIT ON THE REHABILITATION OF OILED SEABIRDS.
Newcastle-upon-Tyne Univ. (England). Dept. of Zoology.
For primary bibliographic entry see Field 05G.
W73-10255

RESPIRATION AND RESISTANCE TO ANAEROBIOSIS OF FRESH-WATER ALGAE, (IN GERMAN),
Vienna Univ. (Austria).
M. Dokulil.
Int Rev Gesamten Hydrobiol. Vol 56, No 5, p 751-768, 1971. English summary.
Identifiers: *Algae, *Anaerobiosis, Necrosis, *Respiration, Temperature, Water pollution effects.

Several species of unicellular and filamentous freshwater algae were studied. Both cultured forms and species gathered in the field were utilized. Respiration was measured by means of a Warburg apparatus. The effect of anaerobiosis on subsequent respiration was examined. Anaerobiosis was obtained by bubbling N2 through the medium. Most of the algae adapted their intensity of respiration to sudden changes in the concentration of environmental O2. After a considerable period without O2, respiratory rates up to 500% above that of the controls were measured. When anaerobic conditions prevailed too long, however, the algae became necrotic, and did not recover. The blue-green algae tested could exist for 10 days without O2, but none of the other types could exist more than 48 hr, and most of them by no means this long. When temperatures were elevated, the resistance to anaerobiosis was lessened.--Copyright 1973, Biological Abstracts, Inc.
W73-10365

STATE OF THE CARDIOVASCULAR SYSTEM OF MEN AFTER LONG-TERM USE OF CHLORIDE DRINKING WATER, (IN RUSSIAN),
Institute of General and Municipal Hygiene, Moscow (USSR).
A. I. Bokina, V. K. Fadeeva, and E. M. Vikhrova.
Gig Sanit. Vol 37, No 3, p 10-14, 1972. Illus. English summary.
Identifiers: *Cardio vascular system, *Chlorides (Drinking water), Diseases, Hypertension, Salination, Water pollution effects, *Potable water.

The state of the cardiovascular system was studied in persons drinking highly mineralized water for a long time. The constant use for drinking purposes of water was high content of NaCl is hazardous for the onset of hypertensive disease. If the chloride content of water is above the level of 1 g/l (chlorine-ion), the mineral composition of water should be corrected by desalination.--Copyright 1973, Biological Abstracts, Inc.
W73-10370

ON THE QUESTION OF ECOLOGICAL MAIN FACTORS IN TESTACEA ASSOCIATIONS,
W. Martin.
Limnologica. Vol 8, No 2, p 357-363, 1971.
Identifiers: Ecological studies, Moisture, *Testacea, *Biotopes, Reviews.

Areview is given of the literature on whether moisture or food is the major determining factor in the spread of Testacea, particular in secondary biotopes. Nutrition may be the major factor in determining their presence in such associations with moisture as a limiting factor. The Testacea of the secondary biotope are euryplastic with respect to moisture. Detritus is the main food of the moss and swamp Testacea.--Copyright 1972, Biological Abstracts, Inc.
W73-10379

EXPERIMENTAL STUDIES OF SUBSTANTIATING HYGIENIC STANDARD VALUES OF M-DIISOPROPYLBENZOL AND P-DIISOPROPYLBENZOL IN WATER BODIES, (IN RUSSIAN),
Institute of General and Municipal Hygiene, Moscow (USSR).
A. M. Sologub, and T. P. Bogdanova.
Gig Sanit. Vol 36, No 9, p 18-21, 1972. Illus. English summary.
Identifiers: *Benzols (Diisopropyl), *Hygienic standards, Water quality standards, *Isomers, Mice.

Isomers impart a specific smell to water but have no significant effect on the sanitation of water bodies. The effect of the substances on mice with long-term intake of water manifested itself mainly in pathologies of the central nervous system and the protein forming functions of the liver. The hyeienic standard level for both substances in water should be set at a level of 0.05 mg/l.--Copyright 1972, Biological Abstracts, Inc.
W73-10380

A CONTRIBUTION TO THE CHEMICAL AND MICROBIOLOGICAL STUDY OF WATER-CATCHMENT SYSTEMS IN CONNECTION WITH DECREASED YIELD DURING OPERATION, (IN SERBO CROATIAN),
V. Vajgand, J. Vucetic, M. Stojanovic, D. Babac, and M. Zdraveski.
Mikrobiologija (Belgr). Vol 7, No 2, p 139-148, 1970.
Identifiers: Catchments, Chemical studies, *Gallionella-Sp, *Leptothrix-Sp, Microbiological studies, *Siderocapsa-Sp, *Iron bacteria.

The incrustation of drains with iron from the water as a result of growth of iron-sustained bacteria (Leptothrix sp., Gallionella sp. and Siderocapsa sp.) decreased their specific yield. The presence of these bacteria is due to the combination of increased ferrous ions and CO2, with decreased oxygen in wells.--Copyright 1972, Biological Abstracts, Inc.
W73-10386

THE WATER DIVIDER AND REPRODUCTIVE CONDITIONS OF VOLGA STURGEON, (IN RUSSIAN),
Kaspiiskii Nauchno-Issledovatelskii Institut Rybnogo Khozyaistva, Astrakhan (USSR).
For primary bibliographic entry see Field 08I.
W73-10393

PHENOL-AND RHODANE DESTRUCTING MICROFLORA AS A FACTOR FOR NATURAL PURIFICATION OF THE DNIEPER RESERVOIRS, (IN RUSSIAN),
Dnepropetrovskii Gosudarstvennyi Universitet (USSR).
A. K. Stolbunov.
Gidrobiol Zh. Vol 7, No 2, p 11-19. 1971. Illus. English summary.
Identifiers: *Dnieper reservoirs, *Micro flora, Phenols, Reservoirs, Rhodane, USSR, *Self-purification, *Bacterial.

Investigations carried out in summer 1968-1969 showed that phenol- and rhodane-destructing bacteria are constantly present in the basin. They possess the highest biochemical activity in places of discharge into the basin of sewage from the sources of phenol and rhodanide contamination. Moving off the source of contamination during the processes of natural self-purification, the phenol-destructing ability of water decreased up to the backround values in 50 km at an average and rhodane-destructing ability of water in 10-20 km. Mathematical-statistical analysis showed the positive moderate correlation of phenol-destructing ability of water and content of phenol-destricting bacteria in it, especially in the upper and most contaminated plots of reservoirs as well as the positive strong correlation of an amount of saprophytic and phenol-destructing bacteria in water. Phenol- and rhodane-destructing microflora and the physicochemical processes are important factors of self-purification of basins from contamination by phenols and rhodanides.--Copyright 1972, Biological Abstracts, Inc.
W73-10394

LETHAL DROPS IN HYDROSTATIC PRESSURE FOR JUVENILES OF SOME FRESH-WATER FISH, (IN RUSSIAN),
Akademiya Nauk SSSR, Moscow. Inst. of Evolutionary Morphology and Animal Ecology.
V. I. Tsvetkov, D. S. Pavlov, and V. K. Nezdolii.
Vopr Ikhtiol. Vol 12, No 2, p 344-356. 1972. Illus.
Identifiers: *Hydrostatic pressure, *Lethal limit, Acipenser-haeri, Carassius-carassius, Coregonus-albula, *Fish (Juveniles), Leucaspius-delineatus, Perca-fluviatilis, Phoxinus-phoxinus, Rutilus-rutilus, Water pollution effects.

The effect of the amount and rate of drops in hydrostatic pressure were investigated for Leucaspius delineatus, Phoxinus phoxinus, Carassius carassius, Perca fluviatilis, Acipenser guldenstadti, Acipenser baeri, Rutilus rutilus, Coregonus albula, Dalmo salar, Alburnus alburnus and Thymalus thymallus. Sharp drops in pressure caused death not only in physoclistics, but also in physostomous fishes which had adapted to conditions of increased pressure. The rate of decrease was more critical for survival than the absolute decrease in pressure. An increase in the pressure drop above the rate of decompression caused severe trauma and death. The maximum rate of decompression was close to 0.9 atm/sec for the physostomous fishes and was much lower for physoclistics. Young juveniles with a swim bladder were more sensitive to pressure drops than older juveniles. It was concluded that the death of a large number of juveniles (11-67%) during downstream migration across the turbines of a high pressure hydroelectric power station resulted from decreased pressure in the turbines, especially in fishes which had adapted to increase pressure conditions in the head bay.--Copyright 1973, Biological Abstracts, Inc.
W73-10395

CONTRIBUTIONS TO THE KNOWLEDGE OF THE ALGAE OF THE ALKALINE ('SZIK') WATERS OF HUNGARY. III. THE PHYTOSESTON OF THE ALKALINE PONDS NEAR KUNFEHERTO,
Damjanich Muzeum, Szolnok (Hungary). Laboratorium Tisza-Forsch.
For primary bibliographic entry see Field 02H.
W73-10397

5D. Waste Treatment Processes

WATER AND WASTEWATER MANAGEMENT IN DAIRY PROCESSING,
North Carolina State Univ., Raleigh. Dept. of Food Science.
For primary bibliographic entry see Field 05B.
W73-09755

WATER TREATMENT BY MEMBRANE ULTRAFILTRATION,
North Carolina State Univ., Raleigh. School of Engineering.
J. A. Palmer, H. B. Hopfenberg, and R. M. Felder.
Available from the National Technical Information Service as PB-220 685, $3.00 in paper copy, $0.95 in microfiche. Environmental Protection Agency, Report EPA-R2-73-109, May 1973. 66 p, 18 fig, 2 tab, 39 ref. EPA Project 17010 EDR.

Descriptors: *Waste water treatment, *Surfactants, *Reverse osmosis, *Surface tension, *Semipermeable membrane, Membrane processes, Water treatment, Separation techniques, Sulfonates, Soaps, Detergents.
Identifiers: *Concentration polarization, *Flux-limiting effects, *Solute asymmetry, Ultrafiltration.

The effect of solute asymmetry and in turn solute surface activity on flux-limiting membrane interactions was studied in a 6 cell, duplex, continuous flow-through ultrafiltration test loop. Triton X-100, sodium dodecylbenzenesulfonate, and Carbowax 600 were chosen as the model nonionic surfactant, anionic surfactant, and nonionic symmetrical solute respectively. The presence of low concentrations of surface active agents in aqueous ultrafiltration feeds significantly reduced the steady-state trans-membrane flux. The nonionic surfactant and the anionic surfactant interacted with a polysalt membrane, causing maximum flux reductions of 24% and 20% respectively. The nonionic surfactant interacted with a cellulose acetate membrane, causing a maximum flux decline of 25%. The smallest flux reduction, of the order of 5%, was observed for the anionic surfactant interacting with the cellulose acetate membrane. In all but one case the distilled water flux returned to its original value consequent to a distilled water rinse following ultrafiltration in the presence of surface-active solute. There was essentially no observable flux decline associated with the ultrafiltration of Carbowax 600 solutions with the cellulose acetate membrane. A 5% flux reduction was observed during ultrafiltration experiments on these same solutions with the polysalt complex membrane. These composite results suggest that at least for the specific systems studied, solute asymmetry and in turn surface activity contributes specifically to flux-limiting membrane interactions. (EPA)
W73-09758

COST ESTIMATING GUIDELINES FOR WASTEWATER TREATMENT SYSTEMS,
Bechtel Corp., San Francisco, Calif.
J. W. Porter, G. W. Brothers, and W. B. Whitton.
Environmental Protection Agency Report EPA-R2-73-237, May 1973. 92 p, 1 fig, 7 tab, 48 ref. EPA Project 17090 DRU.

Descriptors: *Estimated costs, *Annual costs, *Capital costs, *Maintenance costs, *Operating costs, *Waste water treatment, Cost analysis, Cost trends, Electric power costs, Fuels, Financial analysis, Government finance, Interest rate, Wages, Water costs.
Identifiers: *Guidelines, Manuals.

This manual provides guidelines for the cost estimator concerned with various waste water treatment systems now under investigation by the Federal Water Quality Administration. Direction is given for the uniform presentation of capital and annual cost data now being developed. Individual sections deal with suggested forms for submitting reports, including an outline, and method and format for preparing capital and annual cost estimates for FWQA studies. The following specific data are appended: Summary of cost indices useful in scaling capital costs of previous years to present Description of typical operating cost data; sources of additional data Discussion of municipal bonds and financing Discussion of levels of estimates and expected accuracies. (EPA) .
W73-09759

PRELIMINARY WATER AND WASTE MANAGEMENT PLAN.
Metcalf and Eddy, Inc., Boston, Mass.
For primary bibliographic entry see Field 06A.
W73-09813

WATER AND WASTEWATER FOR BROWARD COUNTY, FLORIDA.
Benham-Blair and Affiliates of Florida, Inc. Fort Lauderdale.
For primary bibliographic entry see Field 06B.
W73-09816

HUD SEWER POLICY IN URBAN RENEWAL AREAS,
Paterson Redevelopment Agency, N.J.
For primary bibliographic entry see Field 06E.
W73-09820

COOLING PONDS - A SURVEY OF THE STATE OF THE ART,
Hanford Engineering Development Lab., Richland, Wash.
J. C. Sonnichsen, Jr., S. L. Engstrom, D. C. Kolesar, and G. C. Bailey.
Report HEDL-TME 72-101 September 1972. 109 p, 14 fig, 16 tab, 129 ref, append. AEC-AT (45-1)-2170.

Descriptors: *Cooling water, *Heat transfer, *Ponds, *Design criteria, *Model studies, *Cost analysis, Electric power production, Thermal powerplants, Heated water, Discharge (Water), Thermal pollution, Cooling, Structural engineering, Stratification, Design, Theoretical analysis, Analytical techniques, Mathematical models, Evaluation, Spraying, Economics, Environment, *Reviews, Wast water treatment.

A review of cooling pond technology is presented in which major emphasis was placed on examination of engineering and environmental aspects of design, mathematical and physical modeling, use of sprays, and economics. The design of a cooling pond is affected by the local climatic, topographic, and hydrologic characteristics of the specific site. Construction methods rely on embankments to form an artificial water body or floating partitions to establish a pond in a larger body of water. Surface area, volume and depth are the main considerations of design as heat loss to the atmosphere is the main cooling factor. Methods for both theoretical and practical modeling are presented along with the equations necessary. Advantages associated with cooling ponds are service as a settling basin for suspended solids and operation for extended periods without make-up water. The economic aspects of cooling ponds include the commercial and recreational possibilities of such facilities. (Jerome - Vanderbilt)
W73-09849

LARGE SCALE SPRAY COOLING,
Richards of Rockford, Inc., Ill.
R. B. Kelley.
Industrial Water Engineering, Vol 8, No 7, p 18-20, Aug-Sept 1971. 3 fig.

Descriptors: *Spraying, *Heated water, *Heat transfer, Hydraulic engineering, *Cooling, Cooling water, Sprays, *Water cooling, Ponds, Electric power production, Thermal power plants, Air-water interfaces, Structural design, Temperature, Model studies, Theoretical analysis, Equations, Waste water treatment.

The use of spray cooling of heated water is discussed and compared to cooling towers for economy and efficiency. It has been shown that spray devices which produce fine droplets (less than 100 microns in diameter) achieve greater thermal dissipation than cooling towers. Previously, the energy which was required to operate spray systems made them uneconomical. But, now, proper nozzle and pump selection can create more economical systems. The performance of a spray system depends upon droplet size. Drift loss must be considered in design. Some pump and head combinations are discussed. To calculate comparative efficiency, wet bulb temperature must be converted into enthalpy units and certain saturation conditions must be assumed. Equations for calculating the number of transfer units for a mechanical spray device and the total heat dissipation for a given spray nozzle are given. (Jerome - Vanderbilt)
W73-09850

SPINNING DISCS: A BETTER WAY TO COOL POND WATER,
J. Papamarcos.
Power Engineering, Vol 75, No 9, p 54-57, September 1971. 4 fig.

Descriptors: *Cooling, *Heated water, *Thermal pollution, Thermal powerplants, Powerplants, Heat transfer, Waste water treatment, Temperature, Prototypes, Ponds, *Sprays.
Identifiers: *Cooling ponds, *Spinning discs cooling system.

A new cooling system shows promise of being effective and versatile enough to meet many different operating requirements. The cooling device was developed by Cherne Industrial, Inc., and 23 utilities have joined with the company to help finance a demonstration project. Spray is formed by a spinning disc which is 24 in. in diameter by 1/4 in. thick. In an operating system it is expected that up to 50 discs would be assembled on a single shaft to make up what is called a Thermal Rotor. Several of these rotors might be joined together to be driven by a single electric motor. Variable-speed hydraulic motors, rather than electric motors, might also be used to drive the rotors. This would make it possible to vary the cooling rate and to reduce drift during high-wind conditions. The demonstration project is set up in a holdup basin near Stillwater, Minn. with nine floating units which will operate year-round with the rotors in various modes. (Oleszkiewicz-Vanderbilt)
W73-09853

REMOVAL OF CHROMATE FROM COOLING TOWER BLOWDOWN BY REACTION WITH ELECTRO-CHEMICALLY GENERATED FERROUS HYDOXIDE,
Los Alamos Scientific Lab., N. Mex.
E. I. Onstott, W. S. Gregory, and E. F. Thode.
Environmental Science and Technology, Vol 7, No 4, p 333-337, April 1973. 5 fig, 4 tab, 16 ref.

Descriptors: *Waste water treatment, *Cooling towers, *Chemical precipitation, *Chromium, *Electrolysis, Chemical engineering, Thermal power plants, Chemical reaction, Electrochemistry, Effluents, Phosphates, *Thermal pollution.

A new method of treating cooling tower blowdown water utilizing electrochemically generated ferrous hydroxide as a reducing agent is reported. This method quantitatively reduced Cr (VI) to Cr (III) and concurrently precipitated it without pH adjustment. Concentrations of Cr (VI) of <0.005 mg/l. were achieved. Ferrous hydroxide treatment also precipitated a large fraction of the phosphate and other anions to improve product effluent quality significantly with respect to total dissolved solids. Steady-state generation of ferrous hydroxide in blowdown was accomplished with high Faraday efficiency in a flow cell which utilized sacrificial low carbon steel anodes and stainless steel cathodes. Colloids were formed under most electrolysis conditions, and special procedures were required to obtain filterable precipitates. (Jerome-Vanderbilt)
W73-09854

LOW GRADE HEAT FROM THERMAL ELECTRICITY PRODUCTION, QUANTITY, WORTH AND POSSIBLE UTILIZATION IN SWEDEN,
Swedish Board for Technical Development, Studvik (Sweden).
For primary bibliographic entry see Field 05G.
W73-09865

PROCESS FOR TREATING WASTE EFFLUENT,
Minnesota Mining and Manufacturing Co., St. Paul. (assignee).
R. Fisch, and N. Newman.
U.S. Patent No 3,721,624, 2 p, 2 ref; Official Gazeytt of the United States Patent Office, Vol 908, No 3, p 692, March 20, 1973.

Descriptors: *Patents, *Toxicity, Pollution abatement, *Waste water treatment, Water quality control, Water pollution control, Biochemical oxygen demand.
Identifiers: *Chemical treatment, Oxidizing agents, Peroxy compounds, Hydrogen peroxide, Peroxydisulfate.

An oxidizing agent is added to the photographic fix solution to oxidize thiosulfate to sulfate. The oxidized fix solution is added to the spent color developer. The phenylene diamine is thereby isolubilized, forming a sludge which is easily removed. Three oxidizing agents were tested: perozy compounds, hydrogen peroxide, and potassium peroxydisulfate. (Sinha-OEIS)
W73-09926

PROCESS FOR THE BIO-OXIDATION OF NITROGEN DEFICIENT WASTE MATERIALS,
R. K. Finn, and A. L. Tannahill.
U.S. Patent No 3,721,622, 6 p, 4 ref; Official Gazette of the United States Patent Office, Vol 908, No 3, p 692, March 20, 1973.

Descriptors: *Patents, Liquid wastes, Organic wastes, *Waste water treatment, Pollution abatement, Water quality control, *Nitrogen fixation, Molybdenum, *Aeration, Microorganisms, Nutrients.
Identifiers: *Bio-oxidation, Vanadium.

Basically the process for the bio-oxidation of nitrogen-deficient aqueous organic waste consists of feeding the waste into an aeration zone containing: an active culture comprising at least 75 percent by weight of cells of aerobic, free-living nitrogen fixing microorganisms; a sufficient amount of nutrients to maintain growth of the cell (0.01 ppm molybdenum or vanadium); and at least 0.01 atmospheres of dissolved oxygen. The aeration zone is maintained at a temperature of about 20C to 40C and a pH of about 6 to 9. Retention time might range from 3 to 12 hours. Ten examples are presented to illustrate variations in this invention. (Sinha-OEIS)
W73-09928

MICROBIAL BIOLYSIS PROCESS,
E. G. Smith, and J. W. Hood.
US Patent No 3,718,582, 5 p, 4 fig, 5 ref; Official Gazette of the United States Patent Office, Vol 4, p 1042, February 27, 1973.

Descriptors: *Patents, Liquid wastes, Solids, *Heat treatment, *Biological treatment, *Microbial degradation, Water pollution control, Water quality control, Pollution abatement, *Waste water treatment.
Identifiers: *Biolysis.

This system includes an aerobic biological oxidation step wherein a mixed liquor is formed that contains purified suspending liquid and biologically active solid matter suspended within it. Purified suspending liquid is the effluent from the process. Portions of the sludge are heated to 140 to 212F and held at that temperature to effect nontoxic microbial biolysis. The material is made biologically inactive and is usable as food for the remaining active organisms. The portion subjected to biolysis is so regulated that substantially all of the biodegradable matter is consumed and only a relatively small quantity is removed to final disposal. (Sinha-OEIS)
W73-09929

PROCESS FOR TREATING AQUEOUS CHEMICAL WASTE SLUDGES AND COMPOSITION PRODUCED THEREBY,
Corson (G. and W. H.), Inc., Plymouth Meeting, Pa. (Assignee).
C. L. Smith, and W. C. Webster.
US Patent No 3,720,609, 7 p, 1 fig, 10 ref; Official Gazette of the United States Patent Office, Vol 908, No 2, p 431, March 13, 1973.

Descriptors: *Patents, *Chemical wastes, *Sludge treatment, *Sludge disposal, Pollution abatement, *Ions, *Waste water treatment, Landfills, Sulfates, Aluminum, Iron, Calcium, Magnesium.

Residual reactive materials, including sulfate ions, aluminum ions, iron ions, calcium ions and magnesium ions or sources of these ions present in various forms in the sludge, are reacted with ion yielding aluminum compounds, lime and soluble sulfates, added when necessary to raise the concentrations of these materials in the sludge above certain minimum levels, to produce a composition which will harden over a period of days or weeks. This material may be used as a land fill or base material or be otherwise easily disposed. (Sinha-OEIS)
W73-09932

METHOD AND APPARATUS FOR CONDITIONING AND DISPOSING OF ALUM SLUDGE FROM WATER TREATMENT,
Bonham, Grant and Brundage Ltd., Columbus, Ohio. (Assignee).
J. D. Stauffer.
US Patent No 3,720,608, 3 p, 2 fig, 4 ref; Official Gazette of the United States Patent Office, Vol 908, No 2, p 431, March 13, 1973.

Descriptors: *Patents, *Sludge treatment, *Dewatering, Equipment, Water quality control, Water pollution control, Pollution abatement, *Waste water treatment.
Identifiers: *Alum sludges.

This apparatus consists of a holding tank which receives solids from clarification settling tanks, a system of heat exchangers for heat recovery and auxiliary heat application to raise the temperature of the incoming sludge, a detention vessel to provide for a required reaction time, a decant tank, and a dewatering unit for separation of the solids from the conditioned sludge. The alum sludge is subjected to a temperature of about 212F for a period of about 30 minutes. The solids are separated from the liquid to produce a low moisture content sludge cake. (Sinha-OEIS)
W73-09933

DEODORIZING AND SEWAGE TREATMENT FORMULATION,
Biogenics Co., Inc., Greensboro, N.C. (Assignee).
R. E. Horney, and H. T. Jackson.
US Patnet No 3,720,606, 4 p, 5 ref; Official Gazette of the United States Patent Office, Vol 908, No 2, p 431, March 13, 1973.

Descriptors: *Patents, *Sewage treatment, *Waste water treatment, *Bacteria, Odor, *Aerobic treatment.
Identifiers: *Odor suppressants, Bacillus.

The formulation combines the features of an improved odor suppressing or masking agent with a

novel bacterial agent, the latter induces aerobic decomposition of the odor producing matter to a form free of odor. The masking agent includes a perfumant dissolved in a film-forming carrier to form a barrier or shield between the waste and the atmosphere. The formulation comprises a solution containing one or more aerobic, mesophilic, spore-forming bacterial agents chosen from Group 1 of the genus Bacillus to induce aerobic decomposition. (Sinha-OEIS)
W73-09934

PERIPHERAL FEED AND EFFLUENT SYSTEM FOR SEDIMENTATION TANKS,
Rex Chainbelt Inc., Milwaukee, Wis. (Assignee).
W. H. Boyle.
US Patent No 3,717,257, 5 p, 8 fig, 2 ref; Official Gazette of the United States Patent Office, Vol 907, No 3, p 710, February 20, 1973.

Descriptors: *Patents, *Waste water treatment, Sedimentation, Hydraulic gradients, Pollution abatement, Water pollution control, Water quality control, *Settling basins.

Adjoining feed and effluent channels are of complementary dimensions. Their combined width is uniform but their diminishing widths provide the required hydraulic gradients. The sedimentation tank has a single horizontal shelf projecting inward. The shelf is of uniform width and an upright wall at its inner periphery forms the overflow weir of the tank. An inner scum baffle is uniformly spaced in the weir and is concentric with the vertical axis of the rotating sludge removal mechanism at the center of the tank. (Sinha-OEIS)
W73-09942

LIQUID CLARIFICATION UNIT,
Bauer Bros. Co., Springfield, Ohio. (Assignee).
R. P. Rowland, and D. P. Michel.
US Patent No 3,717,255, 6 p, 3 fig, 2 ref; Official Gazette of the United States Patent Office, Vol 907, No 3, p 710, February 20, 1973.

Descriptors: *Patents, *Water purification, Separation techniques, *Settling basins, *Waste water treatment, *Pollution abatement, Water quality control, Water pollution control.

This invention relates to a process of clarifying contaminated liquids and to the simplified separation of liquids from solids. It consists of a single compartment open tank, having a feed trough along one side to provide for the infeed. Turbulence is avoided for quick settling of heavy solids. Bridging the lateral extent of the tank are riffly plates mounted on a continuous chain. The plates are continuously moved along the bottom to pick up the solid contaminants and lead them away for separate discharge. Clarified liquid is directed from the tank under pressure. (Sinha-OEIS)
W73-09943

METHOD AND APPARATUS FOR FILTERING SOLIDS,
Q. L. Hampton.
US Patent No 3,717,251, 9 p, 15 fig, 5 ref; Official Gazette of the United States Patent Office, Vol 907, No 3, p 709, February 20, 1973.

Descriptors: *Patents, *Sewage treatment, *Waste water treatment, *Filtration, Pollution abatement, Water quality control, Water pollution control, Equipment, Filters.

A washing system utilizes the hydraulic characteristics created by the use of two greatly different volumes of water flow when filtering and washing within a discharge trough of fixed dimensions. Different velocities created in the discharge of the water carrying trough above the filter media produce two different flow trajectories and proper location of a baffle in a discharge receiving area permits complete separation of filtered water flow from washwater flow. The filter is cleaned by increasing the rate at which the water flows through the filter media. No devices are required to control fluidizing of the bed except volumetric flow to the unit during washing. Dispersion of air bubbles in the washwater supply to the filter during washing operations does not agitate the filter bed to the extent that filter media is lost in the washwater. Upon leaving the filter media, the filtered liquid flows over a weir and into an effluent trough from where it is directed outward for suitable disposition. (Sinha-OEIS)
W73-09944

NEUTRALIZATION OF WASTE WATERS BY BETTER EXPLOITATION OF RAW MATERIAL DURING PRODUCTION OF POTATO STARCH,
J. Haeusler, and J. Malcher.
Staerke. Vol 24, No 7, p 229-235. 1972. Illus. English summary.
Identifiers: *Fermentation, *Neutralization, *Potato starch, Industrial wastes, Waste water treatment, Coagulation.

Neutralization of fruit juice released during potato starch production can be efficiently accomplished in 2 steps. The first is based on protein coagulation at 85 degrees C in an acidic medium. The acid fermentation step in methane fermentation is used. Heterogeneous lactic acid fermentation is the main process occuring here. Coagulated protein may be efficiently processed into feed, making waste water purification more profitable. The second step is methane fermentation. At loads higher than the fermentation tank's maximum load of 1.5 kg/m3, the period of fermentation is shortened enough to make anaerobic processes impossible. During fermentation, 15.25 m3 of fermentation gas/m3 raw water is produced, with a methane content of 54%. This quantity should be sufficient for heating the waste water to the temperature needed for coagulation. Fruit juice fermentation must be followed by a 1-hr period of mud sedimentation to make purification more effective. With this method, the fruit juice concentration is decreased enough that it can be further purified, along with the processing water, in a biological activation filter plant.--Copyright 1973, Biological Abstracts, Inc.
W73-10087

CHEMICAL AND PHYSICAL ALTERATIONS OF HOUSEHOLD WASTE DURING BIOTHERMAL DISINFECTION IN EXPERIMENTAL DIGESTION CELLS AT THE INSTITUTE FOR MUNICIPAL ADMINISTRATION, DRESDEN, (IN GERMAN),
P. Czerney.
Z Gesamte Hyg Grenzgeb. Vol 18, No 2, p 90-93. 1972. Illus.
Identifiers: Digestion cells, *Disinfection (Biothermal), Germany, Microorganisms, Thermal, *Waste treatment, *Domestic wastes (Household).

Continuous temperature measurements and gas analyses, pH, dry substance, water content and capacity, loss on burning, carbon and nitrogen content, and density were determined. Household waste was ground and sieved and put in stirred digestion cells where the action of microorganisms developed heat and produced CO2. The fresh material was weakly alkaline, 35% water and 26% organic with 0.8% N. The output was more homogeneous, had less water and C but the same N content. The dry weight decreased 10%. The density was higher together with the water capacity. Disinfection succeeds best with low pH and lower water content.--Copyright 1973, Biological Abstracts, Inc.
W73-10088

ACTIVITIES OF THE CORPS OF ENGINEERS IN MARINE POLLUTION CONTROL,
Corps of Engineers, Atlanta, Ga. South Atlantic Div.
For primary bibliographic entry see Field 05G.
W73-10103

A PRELIMINARY STUDY OF BACTERIAL POLLUTION AND PHYSIOGRAPHY OF BEIRUT COASTAL WATERS,
American Univ., Beirut (Lebanon). Dept. of Biology.
For primary bibliographic entry see Field 05B.
W73-10119

STUDY OF VARIOUS POSSIBLE DESIGNS FOR WATER INTAKE AND OUTLET SYSTEMS OF LARGE POWER STATIONS, CONSIDERING HYDRUALIC PLANT OPERATING PROBLEMS AND SITE SUITABILITY (QUELQUES CONSIDERATIONS SUR LE CHOIX DU SYSTEME P RISE-REJET DES GRANDS AMENAGEMENTS THERMIQUES SELON LEUR LOCALISATION),
Laboratoire National d'Hydraulique, Chatou (France).
For primary bibliographic entry see Field 05G.
W73-10147

WHY CORROSION PRODUCTS MUST BE CONTROLLED IN POWER PLANT WATERS,
Battelle-Pacific Northwest Labs. Richland, Wash.
For primary bibliographic entry see Field 08G.
W73-10182

REMOVAL OF CHROMATES FROM PROCESS COOLING WATER BY THE REDUCTION AND PRECIPITATION PROCESS,
Union Carbide Corp., Paducah, Ky. Paducah Plant.
E. W. Richardson.
Available from NTIS, Springfield, Va., as Ky-641; $4.00 in paper copy, $0.95 in microfiche. Report KY-641, Sept 1972. 22 p, 3 fig, 4 tab, 2 ref.

Descriptors: *Chromium, *Cooling towers, *Waste water treatment, Pollution abatement, *Reduction (Chemical), Filtration, *Chemical precipitation, Flocculation, Zinc, Phosphates, Feasibility studies, Cost analysis.

Pilot plant tests favored either ferrous sulfate or sodium sulfite rather than an iron-filing bed for reduction of Cr (6+) to Cr (3+). The advantages were a smaller system size (to provide adequate retention time and surface area) and elimination of problems with regard to channeling and passivation. Passivation resulted in part from collection of hydrogen gas on the surface of the filings. After reduction at pH 2.2-3.0, hydrolytic precipitation by lime to adjust pH to 8.5-9.0 reduced Cr and phosphates to less than 0.1 ppm and Zn to less than 1 ppm. Average water content of the filtered sludge was 93%. (Bopp-ORNL)
W73-10199

CONDITIONS FOR THE REMOVAL OF CHROMATE FROM COOLING WATER,
Paducah Gaseous Diffusion Plant, Ky.
A. J. Lemonds.
Available from NTIS, Springfield, Va., as KY-642; $4.00 in paper copy, $0.95 in microfiche. Report KY-642, July 1972. 10 p, 2 fig, 4 tab, 2 ref.

Descriptors: *Chromium, *Cooling towers, *Waste water treatment, *Pollution abatement, Reduction (Chemical), Filtration, Chemical precipitation, Zinc, Phosphates, Feasibility studies, Cost analysis, Chemical engineering, Laboratory tests, *Thermal pollution.

W73-10387

5E. Ultimate Disposal of Wastes

DEEP WELL DISPOSAL STUDY FOR BALDWIN, ESCAMBIA, AND MOBILE COUNTIES, ALABAMA.
Geological Survey, University, Ala.

Prepared for the South Alabama Regional Planning Commission. Geological Survey of Alabama Circular 58, June 1970. 49 p, 8 fig, 10 maps, 1 tab, 1 append. HUD Ala. P-63 (G).

Descriptors: *Deep wells, *Waste water disposal, *Injection wells, Geology, Hydrology, Environmental effects, Alabama.
Identifiers: Baldwin County (Alabama), Escambia County (Alabama), Mobile County (Alabama).

A study of the potential use of deep wells for the disposal of liquid wastes is presented. The area is a 3-county region including the highly industrialized Mobile metropolitan area. This area is underlain by sedimentary rocks of the type used for deep well disposal in other areas of the country. The study describes the theory of deep well injection of liquid wastes, looks at the geology of the area in detail, and discusses other important factors such as hydrology, chemical aspects, the nature of the wastes, and the economics involved. There are several potential zones for deep-well injection for certain types of industrial wastes. These zones meet the criteria for safe waste disposal and generally contain water that is not suitable for human use. There is still a need for more data, especially for more analyses of water from deep wells in these zones. Deep well injection proposals must be studied carefully to prevent environmental damage. (Elfers-North Carolina)
W73-09817

ST. LOUIS REGION WATER AND SEWERAGE FACILITIES SUPPLEMENT NUMBER 1 - PROGRAMMING,
East-West Gateway Coordinating Council, St. Louis, Mo.
L. D. Leitner.
June 1972. 59 p, 3 fig, 4 append. HUD EWG-LL-0199.08.0.

Descriptors: *Planning, *Water supply, *Sewerage, Financing, Coordination, Priorities, Missouri.
Identifiers: St. Louis Region, Policy review.

This report is one in a series of water and sewerage reports from a continuing planning program being carried out by the East-West Gateway Coordinating Council for the St. Louis region. It is basically a revision of a previous report, St. Louis Region Water and Sewerage Facilities. The emphasis is on establishing a priority rating system for existing and proposed water supply and sewage treatment facilities. The ratings are based on projecting the year that the facilities would become overloaded. Cost estimates for proposed facilities and some general observations and recommendations are also discussed. The major problems noted include a great shortage of funds for construction, the probable inability to formulate the federally required Water Quality Management Plan by the deadline, and the great number of local agencies which complicate review and coordination procedures. There is no indication that the situation revealed by the evaluation of water and sewer facilities within the region will improve in the future. Appendices provide data on water supply and sewage treatment inventories, evaluations of existing facilities, and impacts of proposed projects. (See also W71-00421) (Elfers-North Carolina)
W73-09826

METROPOLITAN COMPREHENSIVE WATER, SEWAGE, AND SOLID WASTE PLANNING STUDY.
Bi-State Metropolitan Planning Commission, Rock Island, Ill.
For primary bibliographic entry see Field 06A.
W73-09829

STUDY ON: ATOMIC RESIDUES IN THE DEEP SEA, (PREOCUPACAO: RESIDUOS ATOMICOS NO FUNDO DO MAR),
C. R. Schroth.
Revista Nacional da Pesca. Sao Paulo. Vol 12, No 102, p 14-16, November 1970.

Descriptors: *Bottom sediments, *Radioisotopes, *Underground storage, *Safety, Water pollution sources, Nuclear energy, Byproducts, Brine disposal, Water pollution control.

Dangerous materials are enclosed in lead boxes and placed in the salt deposits on the ocean floor. The subterranean salt deposits will eventually be transformed into radioactive residue because of the by-products given off in the production of atomic energy. It was concluded in 1965-67 by North American scientists that radioactive materials can be stored safely in ocean floor salt deposits. (Ensign-PAI)
W73-09900

LOCAL SCOUR UNDER OCEAN OUTFALL PIPELINES,
Montgomery (James M.), Inc., Pasadena, Calif.
For primary bibliographic entry see Field 08A.
W73-09902

SEWAGE DISPOSAL IN THE TURKISH STRAITS,
Woodward-Envicon, Inc., San Diego, Calif.
For primary bibliographic entry see Field 05B.
W73-09904

SUBSURFACE WASTE STORAGE--THE EARTH SCIENTIST'S DILEMMA,
Geological Survey, Denver, Colo.
R. W. Stallman.
In: Underground Waste Management and Environmental Implication, Proc of Symposium held at Houston, Tex, Dec 6-9, 1971: Tulsa, Okla, American Association of Petroleum Geologists Memoir 18, p 6-10, 1972. 20 ref.

Descriptors: *Waste disposal wells, *Injection wells, *Underground waste disposal, Water chemistry, Kinetics, Dispersion, Water pollution effects, Water pollution sources.

Some of the possible consequences of underground waste injection are (1) groundwater pollution, (2) surface-water pollution, (3) changes in rock permeability, (4) subsidence, (5) earthquakes, and (6) mineral-resource pollution. Although much work has been done to predict some of the effects, the current state of knowledge is not adequate to estimate them accurately. Knowledge of dispersion, nonlinear relations between rock stress and strain, and interrelations of hydraulics, heat, chemistry, and rock mechanics at macroscale is especially deficient for application to the waste problem. Advances in chemical thermodynamics and kinetics of geochemistry may provide improved technology. (Knapp-USGS)
W73-10022

INJECTION WELLS AND OPERATIONS TODAY,
Bureau of Mines, Bartlesville, Okla. Bartlesville Energy Research Center.
For primary bibliographic entry see Field 05B.
W73-10023

RESPONSE OF HYDROLOGIC SYSTEMS TO WASTE STORAGE,
Geological Survey, Washington, D.C. Water Resources Div.
For primary bibliographic entry see Field 05B.
W73-10025

CIRCULATION PATTERNS OF SALINE GROUNDWATER AFFECTED BY GEOTHERMAL HEATING--AS RELATED TO WASTE DISPOSAL,
Alabama Univ., University. Dept. of Civil and Mineral Engineering.
For primary bibliographic entry see Field 05B.
W73-10028

NATURAL-MEMBRANE PHENOMENA AND SUBSURFACE WASTE EMPLACEMENT,
Geological Survey, Washington, D.C.
For primary bibliographic entry see Field 05B.
W73-10035

GEOCHEMICAL EFFECTS AND MOVEMENT OF INJECTED INDUSTRIAL WASTE IN A LIMESTONE AQUIFER,
Geological Survey, Tallahassee, Fla.
For primary bibliographic entry see Field 05B.
W73-10036

TRENDS IN MARINE WASTE DISPOSAL, GOOD AND BAD,
Engineering-Science, Inc., Pasadena, Calif.
H. F. Ludwig, J. R. Thoman, and P. N. Storrs.
In: Coastal Zone Pollution Management, Proceedings of the Symposium, Charleston, South Carolina, February 21-22, 1972, Clemson University, Clemson, S.C., 1972. p 105-130, 9 ref, 1 tab.

Descriptors: *Waste disposal, *Environmental effects, *Planning, *Economics, Temperature, Dissolved oxygen, Suspended solids, Oil, Nutrients, Bacteria, Pesticides, Sedimentation, Benthos, Light, Water quality standards, Regulations, Legislation.

The overall objective in planning for marine waste disposal is to provide an economical means of disposal with minimum adverse effects on the environment. The first problem is essentially technical in nature: the problem of defining the effects of waste waters and their constituents on the marine environment, and the paucity of scientific parameters available for making such assessments. A second problem is defining 'adverse effects' in any given situation. Temperature, DO, pH, suspended solids, floatables, oil, grease, nutrients, metals, bacteria and pesticides are considered along with changes in light transmission, benthic organisms and sediment characteristics. The current trend in marine water quality standards appears to shortcut essential investigation and research and to establish the answers by legislative or regulatory fiat. In most cases this will result either in uneconomical utilization of resources or in less than maximal protection of environmental resources. (See also W73-10098) (Ensign-PAI)
W73-10105

OCEANOGRAPHIC SURVEYS FOR DESIGN OF POLLUTION CONTROL FACILITIES,
Engineering-Science, Inc., Berkeley, Calif.
D. L. Feuerstein.
In: Coastal Zone Pollution Management, Proceedings of the Symposium, Charleston, South Carolina, February 21-22, 1972, Clemson University, Clemson, S.C. 1972. p 175-183.

Descriptors: *Surveys, *Oceanography, Water pollution control, *Waste water disposal, Facilities, *Outlets, Application methods, Dispersion, Dilution, Ecosystems, Environmental effects, *Design.
Identifiers: Predesign, Predischarge, Postdischarge.

Oceanographic surveys for waste water disposal systems can be placed in one of three general categories: predesign, predischarge and postdischarge. Three objectives must be satisfied in the conduct of a predesign oceanographic survey: must determine dispersion of diluting characteristics of the receiving waters; must provide sufficient information on the ecosystem in the proposed discharge area to assure that biologically significant or sensitive areas will not be adversely affected by the disposal system; and must determine foundation conditions for outfall and diffuser placement prior to preparation of engineering plans and specifications of the waste disposal system. Predischarge surveys are conducted to establish the baseline or natural conditions of the receiving waters, receiving sediments and adjacent shoreline prior to discharge of waste waters. Postdischarge surveys are monitoring programs conducted for the purpose of determining the effects of a discharge on the receiving environment. (See also W73-10098) (Ensign-PAI)
W73-10107

THE OCEAN: OUR COMMON SINK,
Norske Videnskaps-Akademi i Oslo.
For primary bibliographic entry see Field 05C.
W73-10126

GROUNDWATER QUALITY IN THE CORTARO AREA NORTHWEST OF TUCSON, ARIZONA,
For primary bibliographic entry see Field 05B.
W73-10332

5F. Water Treatment and Quality Alteration

PRELIMINARY WATER AND WASTE MANAGEMENT PLAN.
Metcalf and Eddy, Inc., Boston, Mass.
For primary bibliographic entry see Field 06A.
W73-09813

COMPREHENSIVE AREA-WIDE WATER AND SEWER PLAN.
Robert and Co. Associates, Atlanta, Ga.
For primary bibliographic entry see Field 06A.
W73-09814

WATER AND WASTEWATER FOR BROWARD COUNTY, FLORIDA.
Benham-Blair and Affiliates of Florida, Inc. Fort Lauderdale.
For primary bibliographic entry see Field 06B.
W73-09816

ST. LOUIS REGION WATER AND SEWERAGE FACILITIES SUPPLEMENT NUMBER 1 - PROGRAMMING,
East-West Gateway Coordinating Council, St. Louis, Mo.
For primary bibliographic entry see Field 05E.
W73-09826

MUNICIPAL WATER SYSTEMS - A SOLUTION FOR THERMAL POWER PLANT COOLING,
Washington Univ., Seattle. Dept. of Civil Engineering.
R. G. Hansen, C. R. Knoll, and B. W. Mar.
Journal of the American Water Works Association, Vol 65, No 3, p 174-181, March 1973. 5 fig, 1 tab, 14 ref.

Descriptors: *Thermal power plants, *Heated water, *Water distribution, *Potable water, *Heat transfer, Discharges (Water), Theoretical analysis, Computer models, Model studies, Temperature, Water temperature, Water quality control, Impoundments, Piping systems, Thermal pollution.

An analysis of the possibility of discharging power plant waste heat into a drinking water system was conducted for the transmission system and the distribution grid. The hypothesis was that heated water could be discharged into drinking water transmission mains. General heat loss calculations were made for both flowing and stagnant water in buried pipes. Except with very low flows, large diameter and medium diameter pipes experienced very little temperature loss. Results indicate that major changes are required before an existing water-supply and distribution system can be employed as a sink for waste heat from a thermal power generating facility. There would be marginal costs or benefits from integrating the water supply system with power plant cooling if the heat was rejected at less than 100F (40C). High rejecting temperatures would not create benefits for the water user. (Jerome-Vanderbilt)
W73-09851

WATER HEATING AND PURIFICATION SYSTEM,
Pollution Research and Control Corp., Glendale, Calif. (Assignee).
For primary bibliographic entry see Field 03A.
W73-09936

PARTICLE SEPARATOR AND METHOD,
A. S. King.
US Patent No 3,718,256, 3 p, 2 fig, 6 ref; Official Gazette of the United States Patent Office, Vol 907, No 4, p 964, February 27, 1973.

Descriptors: *Patents, Ions, *Nucleation, *Crystallization, *Electric fields, *Water purification, *Demineralization, Polarity, Water treatment.

Particle separation is provided by using electric fields to create polarizing action within the water solution to free unwanted ions from water molecule clusters to permit formation of ionic crystals by nucleation or coagulation. An electrically conductive surface is maintained at ground potential at a temperature which differs from the fluid to be treated. The fluid is directed against the surface and the particles collect on it. The particles are thereafter washed off and then separated by settling as a sediment which may then be removed by scraping. (Sinha-OEIS)
W73-09937

DEVICE FOR RECLAIMING SWEET WATER FROM SEA WATER OR BRACKISH WATER,
Siemens A.G., Munich (West Germany). (Assignee).
For primary bibliographic entry see Field 03A.
W73-09941

SAMPLER-CULTURE APPARATUS FOR THE DETECTION OF COLIFORM BACTERIA IN POTABLE WATERS,
For primary bibliographic entry see Field 05A.
W73-09945

CHARACTERISTICS OF DISTRIBUTION OF FLUORINE IN WATERS OF THE STEPPE REGIONS OF KAZAKHSTAN IN RELATION TO ENDEMIC FLUOROSIS, (IN RUSSIAN),
For primary bibliographic entry see Field 05B.
W73-09990

PUBLIC WATER SUPPLIES OF SELECTED MUNICIPALITIES IN FLORIDA, 1970,
Geological Survey, Tallahassee, Fla.
For primary bibliographic entry see Field 04B.
W73-10319

STATE OF THE CARDIOVASCULAR SYSTEM OF MEN AFTER LONG-TERM USE OF

Descriptors: *Thermodynamic behavior, *Aeration, *Lakes, *California, *Thermocline, Destratification, Limnology, Reservoirs, Thermal stratification, Data collections, Evaporation, Water quality, Chemical analysis.
Identifiers: *Lake Cachuma (Calif), Air injection in lakes, Air injection effects.

A 4-year study was undertaken to determine the effects of air injection on Lake Cachuma, Santa Barbara County, Calif. Lake Cachuma has a capacity of 205,000 acre-feet and was nearly full through most of the study period. The preliminary investigations at Lake Cachuma and studies at other lakes indicated that the use of air injection would be economically feasible in reducing undesirable tastes and odors, increasing dissolved oxygen in the hypolimnion, and reducing evaporation rates. The primary purpose of air injection--elimination or reduction of thermal stratification--was achieved only partially. The greatest effect of the air injection was on temperature distribution, but the test did not produce a totally destratified reservoir. The density barriers to complete mixing were not eliminated. The dissolved-oxygen stratification was only moderately affected, and the nutrient distribution was unaffected by the air injection. No undesirable phytoplankton blooms occurred. No change could be detected in evaporation rates. The air injection did not have an adverse effect on the physical, chemical, or biological conditions of the lake. Possibly a different and larger scale injection program might prove beneficial. (Woodard-USGS)
W73-09799

PROJECTIONS OF WATER USE AND CONSERVATION IN THE USSR, 2000 (ORIYENTIROVOCHNYY PROGNOZ ISPOL'ZOVANIYA I OKHRANY VODNYKH RESURSOV SSSR NA UROVNE 2000 G.),
Akademiya Nauk SSSR, Moscow. Institut Geografii.
For primary bibliographic entry see Field 06B.
W73-09804

COMPREHENSIVE AREA-WIDE WATER AND SEWER PLAN.
Robert and Co. Associates, Atlanta, Ga.
For primary bibliographic entry see Field 06A.
W73-09814

A QUALITY CONTROL STUDY FOR A NEW WATERSHED,
Denver Water Board, Colo.
C. G. Farnsworth.
Public Works, Vol 102, No 11, p 61-62, November 1971. 2 photos, 1 tab.

Descriptors: *Planning, *Water supply, *Watersheds, *Water quality, Diversion, Diversion tunnels, Colorado.
Identifiers: Denver (Colorado).

A water quality study for potential watersheds to be used for water supply for Denver, Colorado, is described. Denver and its metropolitan area have developed most of the nearby surface water and are now planning to use watersheds in the Rocky Mountains. The watersheds in question, the Eagle, Piney, and Ten Mile, are on the western slope of the Rockies, and the water would have to be diverted through the mountains in tunnels. The study is particularly concerned about the water quality in these watersheds as compared to the water in the receiving reservoirs of the Denver system. Field studies were made of the water quality covering such parameters as pH, turbidity, chlorides, hardness, and coliforms. The new streams were of better quality than the potential receiving reservoir. Continual testing is necessary to monitor possible changing conditions. (Elfers-North Carolina)
W73-09815

PROGRAM DESIGN FOR PLANNING WATER QUALITY MANAGEMENT IN KENTUCKY RIVER BASINS.
Schimpeler-Corradino Associates, Louisville, Ky.; and Schimpeler-Corradina Associates, Lexington, Ky.
For primary bibliographic entry see Field 06B.
W73-09825

INITIAL WATER, SEWERAGE AND FLOOD CONTROL PLAN,
Duncan and Jones, Berkeley, Calif.; and Yoder-Trotter-Orlob and Associates, Berkeley, Calif.
For primary bibliographic entry see Field 06D.
W73-09828

COMPUTER LANGUAGES FOR ENGINEERS,
New Mexico State Univ., University Park. Dept. of Chemical Engineering.
H. G. Folster, and D. B. Wilson.
Journal of the American Water Works Association, Vol 65, No 4, p 248-254, April 1973. 4 fig, 5 tab, 35 ref.

Descriptors: Engineering, *Computer programs, *Waste water (Pollution), *Simulation analysis, Mathematical models, Systems analysis, Digital computers.

Now through computer simulations, many variations, process configurations and cases can be examined in the engineering analysis of waste water systems. Many engineers desire to use the capabilities of the computer but do not because they are unfamiliar with machine language. Moreover, with so many computer languages available today, the engineer needs guidance in choosing the best one for a particular situation. This article examines the types of computer languages and their use in waste water system analysis. The information flow through a digital computer--which occurs in three phases: programming, translation, and execution--is described to aid in understanding the types of machine interaction. The programming phase probably has been the principal deterent to widespread use of the computer by non-computer oriented engineers since it involves learning a new language. The various levels of programming languages and computer languages at each level are discussed, including procedure- and problem-oriented languages operations using numerical quantities or alphabetical character strings, comparision of problem-oriented languages classed as either 'computer aids' to design or 'computer-aided' design languages, and finally, nomenclature and glossary terms. For engineers interested in learning to use computers, references are made to instructional manuals for the various levels of computer languages in use today. (Bell-Cornell)
W73-09834

ESTIMATION THEORY APPLICATIONS TO DESIGN OF WATER QUALITY MONITORING SYSTEMS,
Massachusetts Inst. of Tech., Cambridge. Dept. of Civil Engineering.
S. F. Moore.
Journal of the Hydraulics Division, American Society of Civil Engineers, Vol 99, No HY 5, Proceedings paper No 9755, p 815-831, May 1973. 8 fig, 6 tab, 9 equ, 12 ref.

Descriptors: *Water quality control, *Monitoring, *Stochastic processes, *Data collections, *Estimating, Hydraulics, Water quality standards, Economics, Optimization, Costs, Management, Mathematical models, *Design, Systems analysis, Equations, Aquatic environment.
Identifiers: *Data systems, Filtering techniques.

The objectives of water quality control are quantified in water quality standards, and enforcement of standards is the feedback mechanism for quality control. To achieve effective control, engineers need a knowledge of the state of the aquatic

ecosystem. The necessary information is obtained from monitoring or data collection programs and an understanding of the phenomena involved (a model). A quantitative methodology, utilizing Kalman filtering techniques, is developed for designing water quality monitoring systems. A basis is established for: (1) Improvement of current practices of specification and enforcement of water quality standards; and (2) evaluating the economic trade-off between temporal and spatial frequency of sampling. Monitoring systems are characterized by spatial and temporal frequency of sampling and the variables to be measured. Utilizing a dynamic model of the aquatic environment and estimates of the uncertainty in model error and measurement error, a best sampling program is selected from a set of feasible sampling programs by sequentially minimizing a specified measurement system cost function. An optimal solution is not guaranteed. The power of the technique is based on the unique combination of model and data obtained from filtering techniques. The major shortcomings are: (1) The need for a model of the systems and (2) high computer costs. (Bell-Cornell)
W73-09837

THERMAL CAPACITY OF OUR NATION'S WATERWAYS,
Westinghouse Hanford Co., Richland, Wash.
D. E. Peterson, J. C. Sonnichsen, Jr, S. L. Engstrom, and P. M. Schrotke.
Journal of the Power Division, American Society of Civil Engineers, Vol 99, No PO1, Proceedings paper No 9736, p 193-204, May 1973. 6 fig, 3 tab, 16 ref.

Descriptors: *Simulation analysis, River basins, *Thermal pollution, Watersheds (Basins), *Computer programs, Environmental engineering, Water pollution, Numerical analysis, Cooling, Mississippi River, Missouri River, Colorado River, Systems analysis, Mathematical models, Water temperature.
Identifiers: *Thermoelectric power generation, *Once-through cooling systems.

A study made to assess the thermal capacity of rivers within the contiguous United States is described. Formats used by both the Geological Survey and Federal Power Commission delineate study basins; within each basin, projected power growth is compared to computed capacities. The siting of powerplants and the use of once-through cooling systems are subjects of major controversy. In some regions there still exists a considerable capacity for once-through cooling of steam electric power stations. A methodology for assessing this capacity is discussed in hopes of placing the potential for once-through cooling in the proper perspective. Capacities are computed using simulation models. The capacity calculations are based upon conditions of low and average flow. The calculations are contrained by imposing the appropriate water temperature standards. The effect of maximum temperatures is considered by analyzing long-term temperature records wherever available. A number of basins are found to possess suitable capacity for siting additional thermoelectric power plants. More detailed examination may prove this capacity is nonexistent. Projected power through the year 1990 is considered. (Bell-Cornell)
W73-09840

OPTIMIZATION OF CIRCULATING WATER SYSTEM,
Worcester Polytechnic Inst., Mass.
R. S. Gupta, and B. R. Webber.
Journal of the Power Division, American Society of Civil Engineers, Vol 99, No PO1, Proceedings paper No 9756, p 235-247, May 1973. 5 fig, 5 tab, 9 equ, 6 ref, 5 appen.

Descriptors: Computers, *Nuclear powerplants, *Cooling towers, *Condensers, *Optimization, *Thermal pollution, Water supply, Water temperature, Flow, Pumps, Cooling water, Rivers, Atmosphere, Vermont, Connecticut River, Steam.
Identifiers: *Circulating water systems, Iterative procedures.

Due to strict thermal standards, utilities are forced to look for efficient and economic methods of removing waste heat from steam electric generating plants. A code is presented for optimizing the circulating water system such that requirements imposed by state and Federal Authorities are not violated. Two large cooling towers, a 135-acre cooling pond and the circulating water system are coupled together. The major savings realized as a result of this code are: (1) a minimization of the number of operating circulating water pumps, circulating water booster pumps, and cooling water fans; (2) a maximization (within limits) of the amount of heat rejected to the river; and (3) a minimization of unit heat rate consistent with existing conditions and limitations. (Bell-Cornell)
W73-09842

MUNICIPAL WATER SYSTEMS - A SOLUTION FOR THERMAL POWER PLANT COOLING,
Washington Univ., Seattle. Dept. of Civil Engineering.
For primary bibliographic entry see Field 05F.
W73-09851

THE CASE FOR THERMAL POLLUTION,
Wisconsin Univ., Madison.
G. F. Lee, and C. Stratton.
Industrial Water Engineering, Vol 9, No 6, p 12-16, Oct-Nov 1972. 3 tab.

Descriptors: *Thermal pollution, *Effluents, *Water quality standards, *Cooling towers, Water pollution, Discharges (Water), Electric power production, Thermal powerplants, Cooling, Cooling water, *Lake Michigan, Potable water, Aquatic life, Chemical analysis, Water treatment, *Beneficial use, Reasonable use.

A discussion is presented of the expected chemical composition of cooling tower blowdown water from a thermoelectric plant, and the potential water quality problems that may exist should cooling towers be required to meet the thermal effluent standards. Data gathered from Commonwealth-Edison from the Lake Michigan waters near the Zion installation are used as being representative of cooling tower make-up water, and consideration is given to the chemicals which are added for treatment of cooling tower water. Ammonia, boron, cadmium, chloride, copper, iron, mercury, nitrate, hexane soluble oil, phenols, phosphorus, sulfate, suspended solids and zinc are all discussed briefly, and seen as approaching the limits allowed for lake and effluent water as a result of cooling tower operation. Although cooling towers will relieve the problem of thermal pollution it is clear that others will result from periodic blowdowns. In many cases the thermal difficulties may be much smaller than those related to their solution. (Jerome-Vanderbilt)
W73-09864

LOW GRADE HEAT FROM THERMAL ELECTRICITY PRODUCTION, QUANTITY, WORTH AND POSSIBLE UTILIZATION IN SWEDEN,
Swedish Board for Technical Development, Studvik (Sweden).
J. Christensen.
Aktiebolaget Atomenergi Report AE 448, April 1972. 101 p, 8 fig, 9 tab, 115 ref.

Descriptors: *Heated water, *Heat flow, *Water utilization, *Beneficial use, *Electric power production, Wastes, Cooling water, *Thermal pollution, Thermal powerplants, Cost analysis, Economics, Equations, Sewage treatment, Distillation, Evaporation, Circulation, Aquiculture.

WATER QUALITY STUDIED ON TEXAS GULF COAST,
Texas A and M Univ., College Station. Dept. of Civil Engineering.
R. W. Hann, Jr., and J. F. Slowey.
World Dredging and Marine Construction, Vol 8, No 13, p 30-34, December 1972. 7 fig, 2 tab.

Descriptors: *Texas, *Estuaries, *Water quality control, *Dredging, Oil, Sedimentation rate, Heavy metals, Copper, Zinc, Lead, Chromium, Cadmium, Mercury, Gulf of Mexico, *Metals.
Identifiers: Environmental modification, Environmental pollution, Grease, Oxygen demanding organic material, Organic carbon.

The Texas Gulf Coast is an estuarine system which has undergone substantial environmental modifications and which in places is subjected to excessive levels of environmental pollution. The study has as its first objectives the quantitative and qualitative evaluation of sediment materials, determining the overall mass balance of oxygen demanding organic material, the rate of sedimentation and the oil and grease content. Additional quality parameters considered are: heavy metal content of the sediments, including copper, chromium, zinc, lead, cadmium and mercury and total organic carbon. Data were obtained for the Houston Ship Channel, and five additional estuaries including the Neches, Sabine, Brazos, the San Bernard Estuaries, and the Corpus Christi Ship Channel. Recommendation is made that the dredging industry more carefully document its position, with regard to environmental modification and environmental pollution with special consideration given to coastal bays, marshes and estuaries. (Ensign-PAI)
W73-09894

PROBLEMS OF THE CASPIAN,
For primary bibliographic entry see Field 05C.
W73-09896

STUDY ON: ATOMIC RESIDUES IN THE DEEP SEA, (PREOCUPACAO: RESIDUOS ATOMICOS NO FUNDO DO MAR),
For primary bibliographic entry see Field 05E.
W73-09900

MARINE POLLUTION,
A. J. O'Sullivan.
Water Pollution Control, Vol 71, No 3, p 312-323, 1972. 39 ref.

Descriptors: *Coast, Pollutant identification, *Path of pollutants, *Water pollution control, Governments, Legislation, *Reviews.
Identifiers: *Legal jurisdiction, *United Kingdom (England).

The problem of coastal pollution in England is reviewed. The origin and growth of river and subsequently, coastal water, pollution is outlined. Major pollutants and their sources are described, including domestic and agricultural waste, industrial waste, oil, radioactive material, and heat. Current knowledge concerning the environmental effects of each of these same pollutants is summarized. Measures and methods used to control coastal pollution are discussed. The changing roles of the river authorities and the sea fisheries committees are traced. It is suggested that the control and regulation of all activities affecting the marine environment might be best handled by a single environmental agency. (Ensign-PAI)
W73-09905

WASTE DISPOSAL AND WATER QUALITY IN THE ESTUARIES OF NORTH CAROLINA,
North Carolina State Univ., Raleigh. Dept. Civil Engineering.
For primary bibliographic entry see Field 05B.
W73-09911

REVIEW ON THE STATE OF MARINE POLLUTION IN THE MEDITERRANEAN SEA.
For primary bibliographic entry see Field 05B.
W73-09913

ENGINEERING COMMITTEE ON OCEANIC RESOURCES PROCEEDINGS OF THE FIRST GENERAL ASSEMBLY.
For primary bibliographic entry see Field 06E.
W73-09914

DESIGN OF A WORLD MONITORING SYSTEM,
Ministry of Agriculture, Fisheries and Food, Lowestoft (England). Fisheries Lab.
For primary bibliographic entry see Field 05A.
W73-09924

COAGULATION IN ESTUARIES,
North Carolina Univ., Chapel Hill. Dept. of Environmental Sciences and Engineering.
For primary bibliographic entry see Field 02L.
W73-09925

PROCESS FOR THE BIO-OXIDATION OF NITROGEN DEFICIENT WASTE MATERIALS,
For primary bibliographic entry see Field 05D.
W73-09928

COLLECTION AND DISPOSAL OF SHIP'S SEWAGE,
Department of the Navy, Washington, D.C.
C. W. Walker.
US Patent No 3,721,207, 2 p, 3 fig, 3 ref; Official Gazette of the United States Patent Office, Vol 908, No 3, p 589, March 20, 1973.

Descriptors: *Patents, *Sewage disposal, Ships, Water pollution control, Water quality control, Pollution abatement, *Waste water disposal.

A system is provided for collecting ships' discharges by running the discharges by gravity through a floating pipe into a submerged sewage pumping plant or a holding station from which it is pumped into a sewer system on shore. (Sinha-OEIS)
W73-09931

WELL PUMPING APPARATUS FOR POLLUTED WATER,
Mowid Anstalt, Vaduz (Liechtenstein). (Assignee).
L. Bood.
US Patent No 3,720,488, 3 p, 1 fig, 2 ref; Official Gazette of the United States Patent Office, Vol 908, No 2, p 405, March 13, 1973.

Descriptors: *Patents, Water pollution treatment, *Pumps, *Wells, *Pollution abatement, Water pollution control, Water quality control.

This invention comprises a well with one or more supply openings for polluted water and a discharge line for the water to be pumped and a pump. The pump housing is provided with a suction opening at the lower side which is connected with the polluted water collected in the well and it also has a discharge line. There is a removable portion (for cleaning) which consists of the electric motor, motor housing and an impeller. (Sinha-OEIS)
W73-09935

WAVE RIDING WATER BARRIER,
J. D. Harper.
US Patent No 3,718,001, 2 p, 4 fig, 4 ref; Official Gazette of the United States Patent Office, Vol 907, No 4, p 897, February 27, 1973.

Descriptors: *Patents, *Oil spills, *Oil pollution, *Pollution abatement, Barriers, Water quality control, Water pollution control.

A floating water barrier is provided that is flexible not only in both vertical and horizontal planes. It is flexible in the sense of being able to ride wave slopes without tilting sharply from a vertical disposition. It consists of a strip of flexible material acting as a curtain, a ballast mass, and a pair of flexible buoyancy pockets formed in opposite sides of the strip at the edge opposite the ballast. The tops and bottoms of the pockets are secured to the strip. A float is placed in each of the pockets. The floats are formed of closed cell foamed plastics and are rectangular in cross section with the long side of the rectangle lying adjacent to the strip. (Sinha-OEIS)
W73-09939

REMOVAL OF PETROLEUM HYDROCARBONS FROM WATER SURFACES,
Pfizer Inc., New York. (Assignee).
J. Warren.
US Patent No 3,717,573, 3 p, 2 ref; Official Gazette of the United States Patent Office, Vol 907, No 3, p 784, February 20, 1973.

Descriptors: *Patents, *Iron oxides, Oil spills, *Oil pollution, *Pollution abatement, Water quality control, Water pollution control, Separation techniques, Wetting.
Identifiers: *Oleophilic substances, Magnetism.

A process of scavenging an oil layer on a body of water is disclosed. A comminuted ferromagnetic oleophilic substance is spread on the oil, and the agglomerates formed are removed by magnetic attraction. Seven examples are given to suggest variations in experimental application of the process. (Sinha-OEIS)
W73-09940

GROUNDWATER RESEARCH IN THE OPEN-CAST MINING AREA OF THE SOUTHERN NIEDERRHEIN-BUCHT (WEST GERMANY),
Geologisches Landesamt Nordrhein-Westfalen, Krefeld (West Germany).
For primary bibliographic entry see Field 02F.
W73-09951

PROBLEMS OF GROUNDWATER EXPLORATION IN METAMORPHIC ROCKS FOR DOMESTIC WATER SUPPLIES IN NORTHEASTERN BRAZIL,
Bundesforschungsanstalt fuer Bodenforschung, Hanover (West Germany).
For primary bibliographic entry see Field 02F.
W73-09952

EFFECT OF ARTIFICIAL AERATION ON SOME HYDROCHEMICAL INDICES OF WATER QUALITY, (IN RUSSIAN),
Akademiya Nauk URSR, Kiev. Institut Hidrobiologii.
A. K. Ryabov, B. I. Nabivanets, Zh. M. Aryamova, E. M. Palamarchuk, and I. S. Kozlova.
Gidrobiol Zh. Vol 8, No 1, p 63-67. 1972. Illus. English summary.
Identifiers: *Aeration (Artificial), Algae, *Hydrochemical indices, USSR, Water quality, *Cyanophyta.

Data are presented on the effect of artificial aeration on O2 content, carbonic acid, pH, organic substances, organic C and N, mineral forms of N, and oxidizability of water. Aeration of the basin in the first turn results in destratification of O, creating favorable conditions for mineralization of organic substances in bottom layers of water. As a result of aeration mineralization of organic substances containing N and the development of blue-green algae are suppressed.--Copyright 1972, Biological Abstracts, Inc.
W73-10014

THE ENVIRONMENTAL MATRIX: INPUT-OUTPUT TECHNIQUES APPLIED TO POLLUTION PROBLEMS IN ONTARIO,
Department of Energy, Mines and Resources, Burlington (Ontario). Canada Center for Inland Waters.
T. R. Lee, and P. D. Fenwick.
Water Resources Bulletin, Vol 9, No 1, p 25-33, February, 1973. 5 tab, 13 ref.

Descriptors: *Input-output analysis, *Canada, *Heavy metals, Economic impact, Environmental effects, Industries, Water pollution control, Ecology, Mathematical models, Systems analysis, Water pollution sources.
Identifiers: *Ontario (Canada), Economic-ecologic linkages, *Residuals management, Compounds, Trace wastes, Iterative processes, Environmental quality.

The increased intensity of concern with problems of environmental quality and, particularly in Canada, the occurrence of a number of problems with specific industrial waste discharges of mercury, fluorides, oil, and others, has led to the focussing of attention on what might be termed trace wastes. Monitoring of the environment and wildlife has produced a partial account of the rapid increase in the amounts of many substances. But no real accounting of the rate of residual production has been made. Recently, emphasis is being placed upon the concept of waste production or residuals management, and many studies are focussing on constructing economic (particularly input-output) models. These studies attempt to link interrelationships between alternative pollution control measures and economic development with a key variable being the levels of residuals entering the environment. The environmental impact of various economic activities in Ontario has been investigated by means of input-output tables with emphasis on the use of certain materials in the production processes, rather than on the discharge of wastes from these processes. A brief discussion of the general utility of the input-output technique is made. Considerable emphasis is given to the background and methodology of this study, and specific deficiencies and strengths are examined. (Bell-Cornell)
W73-10020

EMERGENCY CONTROL, A NEW METHOD FOR PRODUCTIVITY STUDIES IN RUNNING WATERS, (IN GERMAN),
Max-Planck-Institut fuer Limnologie, Schlitz (West Germany). Limnologische Flusstation.
J. Illies.
Verh D Tsch Zool Ges. 65, p 65-69. 1971. English summary.
Identifiers: *Biomass, Emergency control, *Insects, Method, *Productivity studies, Running waters.

Quantitative records of the number and biomass of insect adults from a given area of running water courses were obtained by setting greenhouses over small streams. Comparing the results of emergence in 2 brooks one observes an outstanding equality of biomass, in spite of striking differences in numbers and species composition. The faunal differences are significant of the ecological character of the brooks. Emergence records of 3 subsequent years in the same place show a remote fluctuation of individuals in many species and sudden alterations in the abundance of others. The new method seems of considerable value for productivity studies and faunistic analysis in running waters.--Copyright 1973, Biological Abstracts, Inc.
W73-10078

COASTAL ZONE POLLUTION MANAG MENT, PROCEEDINGS OF THE SYMPOSIUM
Available from Office of Industrial and Municip Relations, College of Engineering, Clems University, Clemson, South Carolina 29631. Pric $8.00. Symposium held in Charleston, Sou Carolina, February 21-22, 1972. Clemson Unive sity, Clemson, South Carolina, B.Y. Edge, editc 1972. 281 p.

Descriptors: *Coasts, *Southeast U.S., *Estu ries, *Water pollution control, *Managemen Waste disposal, Water pollution sources, Pr grams, Environmental effects.

The purpose of the symposium was to bri together experts in coastal zone pollution manag ment from throughout the United States so th there could be a mutual sharing of knowledge wi those participants from the Southeast U.S. Tl symposium was intended to identify the importa problems that the Southeast U.S. faces in coast zone pollution management, explain the role of n tional, regional, state and local government agencies, and indicate available techniques fc solution of these problems. (See W73-10099 thr W73-10111)
W73-10098

NATIONAL COASTAL POLLUTION PRC GRAM,
Environmental Protection Agency, Washingtor D.C. Office of Water Programs.
E. T. Jensen.
In: Coastal Zone Pollution Managemen Proceedings of the Symposium, Charleston, Sout Carolina, February 21-22, 1972. Clemson Univer sity, Clemson, S.C. 1972. p 11-19.

Descriptors: *Coasts, *Estuaries, *Water pollu tion control, *Pollution abatement, *Biologica communities, *Balance of nature, Land, Fresh water, Saline water, Energy, Water utilization Classification, Water quality standards, Stan dards.
Identifiers: Citizen awareness.

An important environmental concern is the pollu tion of estuarine areas and the open oceans. Many unique biosystems are found in estuarine areas Estuarine systems have four distinct natural ele ments that must be kept in balance: land, fresh anc salt waters, biosystems, and energy. The many and varied uses of our estuarine systems must alsc be kept in balance. The proper balance may be maintained by classifying each estuary as to the way it is used and an appropriate level of water quality established. The sources of estuarine pollution are man-made; therefore, it is man's decision whether or not abatement of estuarine pollution is possible. The environmental Protection Agency's Water Quality Standards Program is the basis of the nation's water pollution control effort; this program classifies bodies of water according to use, establishes pollutant limits, and sets implementation plans to abate further water pollution. Polluters are required to file pollutants discharged with the EPA. Public awareness and involvement both locally and nationally is a key factor to abatement progress. (See also W73-10098) (Ensign-PAI)
W73-10100

REGIONAL COASTAL POLLUTION CONTROL,
Environmental Protection Agency, Atlanta, Ga. Region IV.
J. E. Ravan.
In: Coastal Zone Pollution Management. Proceedings of the Symposium, Charleston, South Carolina, February 21-22, 1972, Clemson University, Clemson, S.C. 1972. p 21-35.

Descriptors: *Coasts, *Estuaries, *Water pollution control, Pollutants, Water pollution sources.

ıdustrial wastes, Environmental effects, search, Training, Manpower, *Southeast U.S. entifiers: *Charleston (So.Car.), Gulf Coast ver Basin.

general view of the estuarine and coastal pollu- n problems confronted by the Region IV outheast) of the Environmental Protection ency (EPA) is presented. Even if this region es not have as high a population concentration others, it has an important industrial pollution oblem compounded by its special physical aracteristics: a very wide and shallow continen- platform that precludes the fast dispersion of llutants to the marine environment. Among the oblem pollutants in waste discharge are the fol- ving: decomposable organic materials; thogenic organisms; inorganic plant nutrients; lustrial waste toxic materials; and heat, floating d sedimentary materials. Each of these has dif- ent deleterious effects on the estuarine en- onment. Studies of the Charleston area as well of other smaller cities in the Gulf Coast River sin were conducted by the EPA, identifying the ajor pollution sources in local industries. The commendations of a 1969 study of the IV Region EPA on estuarine pollution are enumerated, all them are concerned with the need for research the basic physical, chemical and biological ocesses in estuaries. The need of continued sup- rt of academic research to guarantee the availa- ity of trained manpower is emphasized. (See so W73-10098) (Ensign-PAI)
73-10101

LUTION TO CONFLICT IN THE COASTAL NE: PART I--A NATIONAL COASTAL ZONE ANAGEMENT PROGRAM,
rginia Inst. of Marine Science, Gloucester Point.
. J. Hargis, Jr.
: Coastal Zone Pollution Management, oceedings of the Symposium, Charleston, South rolina, February 21-22, 1972, Clemson Univer- y, Clemson, S.C. 1972. p 37-65, 22 ref.

scriptors: *United States, *Coasts, *Manage- nt, *Programs, *Planning, Legislation, Stan- rds, Monitoring, Research and Development, gineering, Manpower, Water quality standards.

e history of coastal zone management is iewed, and the need for the development of a ional coastal zone program is stressed. A na- al program should comprise planning, legisla- , regulations, standards, monitoring, enforce- nt, basic and applied research, engineering elopment, technical assistance and advisory vices. The governmental structure, manage- nt strategy, organization, and manpower of h a program are considered. (See also W73- 98) (Ensign-PAI)
-10102

IVITIES OF THE CORPS OF ENGINEERS ARINE POLLUTION CONTROL,
s of Engineers, Atlanta, Ga. South Atlantic

Tabb.
Coastal Zone Pollution Management, eedings of the Symposium, Charleston, South lina, February 21-22, 1972, Clemson Univer- Clemson, S.C. 1972. p 67-84.

riptors: *Harbors, *Coasts, *Estuaries, r pollution control, Engineering, Water rces, Dredging, Waste disposal, *Treatment ties, *Oil spills, *Cleaning, Thermal pollu- Navigable waters, Regulation, Civil engineer- Water works, Construction, Design, Legisla- Rivers and Harbors Act.

J.S. Army Corps of Engineers is involved in tal concept of water resources management concerned with certain aspects of pollution coastal zone. The Corps interest has a twofold aspect: that which is involved in studies, design and construction of Federal Civil Works projects and the regulation responsibilities in navigable waters of the United States. Model studies and investigations on sanitary waste disposal, floating waste treatment facilities, oil spills, harbor cleanup, thermal pollution and ocean dumping are reported. Laws regulating pollution of navigable waters are outlined, including the 1899 River and Harbor Act. One engineering activity of particular concern to the coastal environment is dredging because it modifies the physical hydrology, changes the environment available to organisms and modifies the quality of water to some degree. Refuse discharge permits are also discussed. (See also W73-10098) (Ensign-PAI)
W73-10103

INVOLVEMENT OF NOAA IN COASTAL POLLUTION MANAGEMENT,
National Oceanic and Atmospheric Administration, Rockville, Md. Office of Ecology and Environmental Conservation.
W. Aron.
In: Coastal Zone Pollution Management, Proceedings of the Symposium, Charleston, South Carolina, February 21-22, 1972, Clemson University, Clemson, S.C., 1972. p 85-103. 8 fig.

Descriptors: *Coasts, Environment, *Ecosystems, *Forecasting, *Environmental control, Fisheries, *Mapping, Charts, Tides, Satellites, Remote sensing, *Management, Water pollution control.

NOAA's role of describing and predicting the environment is reviewed as an essential element in understanding pathological problems in the coastal zone and as a basis for both improving such problems, and providing understanding for environmental improvement in this area. The responsibilities and capabilities within NOAA are described including knowledge of coastal fisheries, mapping, charting, tidal and current predictions and ability to employ the techniques of satellite sensing to develop the knowledge required for management of the coastal regions. The belief that the problems of the coastal zone cannot be treated separately but must be examined as a total ecosystem is emphasized. (See also W73-10098) (Ensign-PAI)
W73-10104

TRENDS IN MARINE WASTE DISPOSAL, GOOD AND BAD,
Engineering-Science, Inc., Pasadena, Calif.
For primary bibliographic entry see Field 05E.
W73-10105

ESTUARINE SURVEYS IN THE SOUTHEAST,
Environmental Protection Agency, Athens, Ga. Southeast Environmental Research Lab.
For primary bibliographic entry see Field 05B.
W73-10106

OCEANOGRAPHIC SURVEYS FOR DESIGN OF POLLUTION CONTROL FACILITIES,
Engineering-Science, Inc., Berkeley, Calif.
For primary bibliographic entry see Field 05E.
W73-10107

APPLICATIONS OF ESTUARINE WATER QUALITY MODELS,
Manhattan Coll., Bronx, N.Y.
For primary bibliographic entry see Field 05B.
W73-10108

SUBMERGED DIFFUSERS IN SHALLOW COASTAL WATERS,
Massachusetts Inst. of Tech., Cambridge. Dept. of Civil Engineering.
D. R. F. Harleman.
In: Coastal Zone Pollution Management, Proceedings of the Symposium, Charleston, South Carolina, February 21-22, 1972, Clemson University, Clemson, S.C. 1972. p 227-247. 10 fig, 14 ref.

Descriptors: *Coasts, *Shallow water, *Diffusion, *Waste disposal, *Thermal pollution, Nuclear powerplants, Fossil fuels, Mixing, Momentum transfer, Flow, Nozzles, Theoretical analysis, Equipment, Water pollution effects.
Identifiers: Diffusers (Submerged).

Experimental and analytical results have been presented for submerged multi-port diffusers in shallow waters, typical of coastal waters on the East Coast and in the Great Lakes. The motivation for the research came from the need to find effective ways of disposing of degradable wastes and waste heat from fossil and nuclear power generation. A one-dimensional theory, similar to the momentum theory for propellers, was developed and compared with laboratory experiments. The separate effects of mixing induced by the momentum input of the diffuser and mixing induced by an ambient cross flow in the receiving waters are shown. The nozzles may be pointed in the direction of the cross current, in opposition, or in alternating directions. (See also W73-10098) (Ensign-PAI)
W73-10109

EFFECT OF POLLUTION ON THE MARINE ENVIRONMENT, A CASE STUDY,
Florida State Univ., Tallahassee. Marine Lab.
For primary bibliographic entry see Field 05C.
W73-10110

COASTAL ZONE POLLUTION MANAGEMENT. A SUMMARY,
Clemson Univ., S.C. Dept. of Civil Engineering.
B. L. Edge.
In: Coastal Zone Pollution Management, Proceedings of the Symposium, Charleston, South Carolina, February 21-22, 1972, Clemson University, Clemson, S.C., 1972, p 265-278. 2 fig, 1 tab.

Descriptors: *Coasts, *Water pollution control, *Management, *Surveys, *Environmental effects, *Outlets, Sites, Waste disposal, Industrial wastes, Chemical wastes, Sewage, Solid wastes, Incineration, Dredging, Recycling.

Some of the more basic research efforts that must be undertaken for effective marine pollution control are: conduct baseline surveys (monitor physical, chemical and biological parameters) of waste disposal sites, to find out what damage has been done; study the impact of both treated and untreated effluents on estuarine and ocean waters; select the offshore sites most suitable for disposal of compatible wastes and study the best ways to place them; study the rate at which sewage sludge can be safely assimilated on the continental shelf; and establish quantitative receiving-water values, values that reflect the relative biological value of inland, estuarine, coastal and deep ocean waters. These will allow engineers to select the best locations for outfalls and dumping sites. Waste disposal should be considered as a total system, including ocean dumping, incineration, land disposal, and, most important, recycling. (See also W73-10098) (Ensign-PAI)
W73-10111

CHOICE OF PRODUCTS FOR CONTROL AGAINST POLLUTION OF THE MARINE ENVIRONMENT BY HYDROCARBONS. I. DETERMINING SOME CHOICES, (CHOIX DE PRODUITS POUR LUTTER CONTRE LA POLLUTION DU MILIEU MARIN PAR LES HYDROCARBURES. I. MOTIF DU CHOIX),
Institut Scientifique et Technique des Peches Maritimes, Nantes (France).
F. Soudan.
Revue des Travaux de l'Institut des Peches Maritimes, Vol 36, No 1, p 81-83, March 1972.

Descriptors: Water pollution control, *Organic compounds, *Cleaning, *Emulsifiers, *Solvents, *Detergents, Chemicals, Toxicity, Efficiencies, *Pollution abatement.

A variety of products to combat marine pollution by hydrocarbons is discussed. A test of these various products determined their efficacy and innocuousness. A product that might be noxious, but that could be employed in weak doses due to its efficiency, was preferred to a relatively mild product employed in large quantities. Emulsifiers with a weak density were preferred for reducing surface tension, and solvents with low boiling points can emulsify viscous petroleum. Relative solubility in seawater was not related to efficacy. Anionic detergents and ethers of polyglycol or perchlorethylene should be avoided because of their toxicity. (See W73-10114 thru W73-10116) (Ensign-PAI)
W73-10113

CHOICE OF PRODUCTS FOR CONTROL AGAINST POLLUTION OF THE MARINE ENVIRONMENT BY HYDROCARBONS. II. EFFICACY OF ANTIPETROLEUM PRODUCTS, (CHOIX DE PRODUITS POUR LUTTER CENTRE LA POLLUTION DU MILIEU MARIN LES HYDROCARBURES. II. EFFICACITE DES PRODUITS ANTIPETROLE),
Institut Scientifique et Technique des Peches Maritimes, Nantes (France).
P. Michel.
Revue des Travaux de l'Institut des Peches Maritimes, Vol 36, No 1, p 85-102, March 1972. 12 fig, 3 tab, 11 ref.

Descriptors: *Oil pollution, *Cleaning, Efficiencies, *Stability, *Emulsifiers, Toxicity, Environmental effects, *Pollution abatement, Water pollution control.
Identifiers: *Agglomerants, Precipitants.

One hundred and one antipetroleum products were studied to select the most effective. According to their method of action against hydrocarbons, three categories were established: Emulsifiers (liquids), agglomerants (powder), and precipitants (powder). The products were tested on petroleum of three different origins. The emulsifying power, stability, cleaning power, and efficacy coefficient were determined. There is no universal antipetroleum product; emulsifiers are the easiest to apply in all conditions at sea and on beaches, but their toxicity makes them unsuitable where flora and fauna must be protected. Few products are effcient in all cases, especially on viscous petroleum. (See also W73-10113) (Ensign-PAI)
W73-10114

CHOICE OF PRODUCTS FOR CONTROL AGAINST POLLUTION OF THE MARINE ENVIRONMENT BY HYDROCARBONS. III. THE RELATIVE TOXICITY OF ANTIPETROLEUM PRODUCTS ON TWO MARINE ORGANISMS, (CHOIX DE PRODUITS POUR LUTTER CONTRE LA POLLUTION DU MILIEU MARIN PAR LES HYDROCARBURES. III. TOXICITE RELATIVE DE PRODUITS ANTIPETROLE SUR DEUX ORGANISMES MARINS),
Institut Scientifique et Technique des Peches Maritimes, Nantes (France).
For primary bibliographic entry see Field 05C.
W73-10115

CHOICE OF PRODUCTS FOR CONTROL AGAINST POLLUTION OF THE MARINE ENVIRONMENT BY HYDROCARBONS. IV. THE RELATIVE TOXICITY OF SEVEN ANTIPETROLEUM EMULSION PRODUCTS, (CHOIX DE PRODUITS POUR LUTTER CONTRE LA POLLUTION DU MILIEU MARIN PAR LES HYDROCARBURES. IV. TOXICITE RELATIVE DE SEPT PRODUITS EMULSIONNANTS ANTIPETROLE),
Institut Scientifique et Technique des Peches Maritimes, Nantes (France).
For primary bibliographic entry see Field 05C.
W73-10116

COAST GUARD ANTI-POLLUTION ENGINEERING,
Coast Guard, Washington, D.C.
R. D. Parkhurst.
The Military Engineer, Vol 64, No 421, p 309-311, September-October, 1972. 3 fig.

Descriptors: *Ships, *Design, *Waste treatment, Water pollution control, *Pollution abatement.

Pollution prevention and control operations conducted by the U. S. Coast Guard on its own vessels are reviewed. Pending the completion of an effective shipboard waste treatment system, a method for retaining waste on board for eventual disposal on shore has been adopted. The complex problems surrounding the disposal of bulge and ballast wastes are discussed. At present, no suitable device has been found to effectively and economically treat these wastes. The Air Delivered Anti-Pollution Transfer System (ADAPTS), a method for reducing the quantity of oil spilled in tanker accidents, is another project which the Coast Guard has undertaken to reduce the problems of pollution in the marine environment. The design history of ADAPTS is briefly stated. Other related activities by the Coast Guard include investigations into air fallout caused by sandblasting and painting activities at shore bases, testing oil harvesting equipment, and the development of an airborne all-weather sensor system to detect, map, quantify, and classify oil spills. (Ensign-PAI)
W73-10122

THE ECONOMICS OF OIL TRANSPORTATION IN THE ARCTIC,
Toronto Univ. (Ontario). School of Business.
G. D. Quirin, and R. N. Wolff.
In: Canadian - U. S. Maritime Problems, Law of the Sea Workshop, June 15-17, 1971, Toronto, Canada, University of Rhode Island, Kingston, Law of the Sea Institute, p 32-46, 1972, L. M. Alexander and G. R. S. Hawkins, editors. 10 tab, 3 ref.

Descriptors: *Oil, Oil fields, *Oil industry, *Transportation, *Arctic, *Environmental effects, *Costs, Legal aspects.
Identifiers: *Oil transport, *Pollution potentials.

The possible ecological impact of future transportation of oil down to Canada and the U.S. from Arctic areas is discussed. Alternative energy sources to oil are examined, including hydroelectric energy, natural gas and coal. It is concluded that the need for oil will not be significantly reduced for the next 2 to 6 decades. Transporting oil from the Arctic is viewed as less potentially dangerous to the environment than shipping oil across large ocean areas, the other alternative. Factors such as frequency of pollution, severity of pollution, initial cost of crude oil, and transport costs are compared and contrasted between using northern sources or overseas reserves. The various modes of transporting oil from Arctic areas—pipeline, tanker, and airlift system, etc., are each discussed in some detail. Legal issues, and the issue of national security are also investigated. Shipping oil from northern areas will incur overall pollution costs which will be significantly lower than those arising from using alternative sources of supply. (Ensign-PAI)
W73-10129

OIL SPILLS RESEARCH: A NEW SKIMMER.

Ecolibrium, Vol 1, No 2, p 8, Fall, 1972. 2 fig.

INE POLLUTION, CONCENTRATING ON EFFECTS OF HYDROCARBONS IN SEA-ER,
achusetts Inst. of Tech., Cambridge. Dept. of anical Engineering.
rimary bibliographic entry see Field 05B.
10144

OSAL OF CHINA CLAY WASTE IN THE LISH CHANNEL,
ie Biological Association of the United King- Plymouth (England). Lab.
rimary bibliographic entry see Field 05B.
10145

EFFECT OF MARINE FOULING ON COR-ON OF METALS ANTI-CORROSION INIQUES (INFLUENCE DE LA SALIS-; MARINE SUR LA CORROSION DES AUX ET MOYENS DE PREVENTION),
ricite de France, (Chatou). Service Etudes et ts Thermiques et Nucleaires.
rimary bibliographic entry see Field 08G.
10146

IY OF VARIOUS POSSIBLE DESIGNS FOR ER INTAKE AND OUTLET SYSTEMS OF GE POWER STATIONS, CONSIDERING RUALIC PLANT OPERATING PROBLEMS SITE SUITABILITY (QUELQUES CON-RATIONS SUR LE CHOIX DU SYSTEME P -REJET DES GRANDS AMENAGEMENTS RMIQUES SELON LEUR LOCALISA-I),
ratoire National d'Hydraulique, Chatou ce).
Lebreton, and R. Bonnefille.
lle Blanche, Vol 205, No 2-3, p 199-202, 1972.

riptors: Coasts, Estuaries, *Lagoons, *En-mental effects, *Powerplants, Hydraulic ms, *Intakes, *Outlets, *Cooling, *Design, Sedimentation, Silting, Suspended solids, nal pollution, Mixing, Currents.
ifiers: Artificial islands.

us designs are possible for water intake and systems of large nuclear power stations. For on an estuary site, cooling capacity must be d as well as the disturbances to estuarine rium likely to result from heated effluents. g characteristics between different density in the estuary must be known. Further ms relate to sediment behavior and cross-ts that can be created by the intake and out-tem. For coastal plants the recycle problem Various solutions were studied for tidal d for sites in lagoons. Problems with silting spended particles can be avoided by putting nt on an artificial island. Intake and outlet t-sited plants are simpler because they may ed in separate pools or docks not connected other. (Ensign-PAI)
147

OIL DEVELOPS NEW OIL BARGE FOR SPILL RECOVERY.

e, Vol 32, No 11, p 106, October, 1972.

tors: *Ships, *Equipment, *Oil spills, pollution control, Research and develop-ollution abatement.
rs: *Oil skimmers.

il skimming device developed by Shell Oil scribed. The skimming barge is equipped system that allows various barge com-to flex with waves and currents. The ship collect oil from the water surface with Oil Herder, a biodegradable chemical n reduce the size of a floating slick and owards a mechanical diverter. This new skimming barge is capable of recovering oil from water at speeds up to 2.75 feet per second, with 95% efficiency. (Ensign-PAI)
W73-10149

CONDITIONS FOR THE REMOVAL OF CHROMATE FROM COOLING WATER,
Paducah Gaseous Diffusion Plant, Ky.
For primary bibliographic entry see Field 05D.
W73-10207

MARINE RESOURCE DEVELOPMENT, A BLUEPRINT BY CITIZENS FOR THE NORTHEAST PACIFIC.
Washington Univ., Seattle. Div. of Marine Resources.

Pacific Sea Grant Advisory Program, Seattle, Report No PASGAP 4, August 1972. 24 p.

Descriptors: *Pacific Ocean, *Coasts, *Management, *Resources, Fisheries, Recreation, Tourism, *Water pollution control, Ships, Water pollution control, Conservation, Education, Training, Publications.
Identifiers: *Northeast Pacific coast.

Marine resource problems along the coast of the Northeast Pacific Ocean and suggestions for solutions are summarized. Fisheries, shipping, and coastal zone management are considered, along with pollution, recreation, and tourism. Education, training, and public dissemination of information are suggested, and various methods of conserving marine resources are proposed. (Ensign-PAI)
W73-10222

RECOMMENDED TREATMENT OF OILED SEABIRDS.
Newcastle-upon-Tyne Univ. (England). Dept. of Zoology.

Research Unit on the Rehabilitation of Oiled Seabirds, Advisory Committee on Oil Pollution of the Sea, England Report. June 1972. 11 p.

Descriptors: *Oil pollution, Water pollution effects, Birds, *Cleaning, *Detergents, *Treatment facilities, Transportation, Marking techniques, Treatment, Wettability.
Identifiers: *Seabirds.

A method of treating oiled seabirds which allows most birds to be returned to sea a few weeks after capture is described. Essential facilities of a cleaning center are enumerated along with methods of capture, preliminary treatment, and transportation to cleaning centers. Oiled seabirds should not be washed immediately after capture. The first priority is to keep birds warm, quiet, and well fed. Cleaning involves washing birds with a warm solution (detergent). Birds should be thoroughly rinsed with hot water and then dried. Aftercare procedures are also explained. Birds can be released as soon as they are strong and healthy and have completely water-repellent plumage. Records should be kept on each bird. (Ensign-PAI)
W73-10227

CURRENT STATUS OF RESEARCH ON THE BIOLOGICAL EFFECTS OF PESTICIDES IN CHESAPEAKE BAY,
Westinghouse Ocean Research Lab., Annapolis, Md.
For primary bibliographic entry see Field 05C.
W73-10231

THE PROTECTED OCEAN, HOW TO KEEP THE SEAS ALIVE,
W. Marx.
Coward, McCann, and Geoghegan, Inc., New York. 1972. 95 p, 19 ref.

Descriptors: *Oceans, Oceanography, Chemistry, Biology, *Water pollution sources, *Waste disposal, Fisheries, Construction, Oil pollution, Harbors, Coasts, Protection, Technology, Legislation, *Legal aspects, Water pollution control.

Physical, chemical, and biological characteristics of the oceans are reviewed. Activities of man that are destroying the oceans, such as waste disposal, coastal development, barbor building, fish harvesting, and oil pollution, are described. Legal and technical methods for protecting the oceans from further degradation and for restoring them to a healthy state are summarized. (Ensign-PAI)
W73-10243

UPPER POTOMAC ESTUARY EUTROPHICATION CONTROL REQUIREMENTS,
Environmental Protection Agency, Washington, D.C. Office of Research and Monitoring.
For primary bibliographic entry see Field 05C.
W73-10247

DEEPWATER OFFSHORE PETROLEUM TERMINALS,
Esso Research and Engineering Co., Florham Park, N.J.
J. Mascenik.
Paper presented at American Society of Civil Engineers National Transportation Meeting, July 17-21, 1972, Milwaukee, Wisconsin. Meeting preprint 1748, 27 p, 1972. 8 fig, 5 ref.

Descriptors: *Oil, *Offshore platforms, *Design, Operations, *Buoys, Island, *Water pollution control, Sites, Accidents.
Identifiers: *Mooring.

Requirements for a reliable offshore oil facility are discussed, and design considerations, berth operations, and environmental limitations of sea islands, multi-buoy moorings, and single point moorings are described. Offshore petroleum terminals located away from congested ship channels and ports will minimize pollution from collisions and groundings. Single point mooring is most suitable for installation in unprotected waters and is reliable in severe operating environments. Multi-use of offshore terminals should be avoided. Industry's role in preventing pollution is reviewed. (Ensign-PAI)
W73-10248

RISK ANALYSIS OF THE OIL TRANSPORTAION SYSTEM.
For primary bibliographic entry see Field 05B.
W73-10251

SECOND ANNUAL REPORT OF THE ADVISORY COMMITTEE ON OIL POLLUTION OF THE SEA. RESEARCH UNIT ON THE REHABILITATION OF OILED SEABIRDS.
Newcastle-upon-Tyne Univ. (England). Dept. of Zoology.

1971. 38 P, 9 TAB, 7 REF.

Descriptors: *Oil pollution, *Birds, *Cleaning, *Wettability, *Detergents, Water pollution effects, Mortality.
Identifiers: *Plumage-cleaning materials, *Water-repellency, *Seabirds.

Characteristics of the naturally waterproof plumage of aquatic birds were investigated to devise cleaning methods that would avoid destroying this waterproof property. Laboratory tests showed that minute traces of oil or the cleaning agent on feathers left them wettable. Residual traces of detergent are particularly hard to remove and seemed likely cleaned birds. A considerable range of cleaning agents and detergents were tested. Most were unsuitable for cleaning the oiled plumage but a number of useful preparations were

identified. Desirable properties of any plumage-cleaning material can be specified. Results of autopsies performed on various birds that died after contamination by oil or cleaners are given. (Ensign-PAI)
W73-10255

REGULATION OF ACTIVITIES AFFECTING BAYS AND ESTUARIES: A PRELIMINARY LEGAL STUDY.
Texas Law Inst. of Coastal and Marine Resources, Houston.
For primary bibliographic entry see Field 06E.
W73-10256

SUMMARY OF SELECTED LEGISLATION RELATING TO THE COASTAL ZONE.
Texas Law Inst. of Coastal and Marine Resources, Houston.
For primary bibliographic entry see Field 06E.
W73-10257

STORAGE FACILITIES (FOR HAZARDOUS SUBSTANCES) REGULATION PRESENTED AT HEARING ON MARCH 21, 1973.
Missouri Clean Water Commission, Jefferson City.

March 21, 1972. 1 p.

Descriptors: *Storage requirements, *Safety, *Water pollution control, *Missouri, Legal aspects, Water management (Applied), Water quality control, Water pollution sources, Agriculture, Oil, Fertilizers, Pesticides, Herbicides, Public health, Ecology, Administrative agencies, Administrative decisions, Accidents, Hazards, Damages, Regulation.
Identifiers: *Administrative regulations, *Hazardous substances (Pollution), Catchment areas, Entrapment dikes, Spillage.

Any owner or operator of storage facilities for any hazardous materials such as, but not limited to, petroleum products, fertilizers, pesticides, herbicides, cyanide or cyanogen compounds, which are hazardous to health and welfare, or are capable of causing pollution if accidentally released, shall locate and construct such facilities so as to prevent any spillage which might result in pollution of the waters of the state of Missouri, or result in substantial damage to a sewer system or wastewater treatment facility. Such persons shall install any facilities necessary to entrap spillage, such as catchment areas, relief vessels, or entrapment dikes, and take all other measures necessary to prevent accidental pollution of the waters of the state. Site locations, facilities and measures to entrap spillage may be inspected by the agency for compliance with the requirements of this regulation and any other requirements of the Missouri Clean Water Law and regulations, orders, and permits thereunder. (Mockler-Florida)
W73-10258

(EMERGENCY) ENFORCEMENT REGULATION (TO PROTECT STATE WATERS) PRESENTED AT HEARING ON MARCH 21, 1973.
Missouri Clean Water Commission, Jefferson City.

March 21, 1973. 1 p.

Descriptors: *Missouri, *Legal aspects, *Water pollution control, *Water quality control, *Remedies, Law enforcement, Public rights, Local governments, Administration, Water law, Water quality standards, Water pollution control, Water management (Applied), Administrative agencies, Administrative decisions, Regulation, Environmental effects.
Identifiers: *Administrative regulations, *Emergency action (Pollution).

Upon telephonic approval of any four members of the Clean Water Commission, the Executive Secretary may immediately requiest the Attorney General, any prosecuting attorney, or any other counsel, to bring any legal action authorized in the Missouri Clean Water Law. Notwithstanding the foregoing, where in the opinion of the Executive Secretary, immediate action must be taken to protect the waters of the state and he is unable to contact the necessary members of the Commission, or if time prevents such contacts, he may immediately request the Attorney General, any prosecuting attorney or any other counsel to bring an action for injunctive relief to protect the waters of the state or prohibit any violation of the Missouri Clean Water Law. The definitions as set forth in the Missouri Clean Water Law and Commission Definitions Regulation shall apply to those terms when used in this regulation, unless the context clearly requires otherwise. (Mockler-Florida)
W73-10259

MISSOURI WATER POLLUTION CONTROL BOARD REGULATIONS FOR TURBID WATER DISCHARGES, (TO PREVENT INTERMITTENT OR ACCIDENTAL POLLUTION OF WATERS OF THE STATE).
Missouri Water Pollution Board, Jefferson City.

January 19, 1967. 2 p.

Descriptors: *Missouri, *Administrative agencies, *Water pollution sources, *Water pollution effects, *Turbidity, State governments, Administrative decisions, Legal aspects, Legislation, Water management (Applied), Water policy, Water resources, Density, Water quality, Public health, Safety, Waste disposal, Regulation.
Identifiers: *Administrative regulations, *Spillage, Jackson Candle Turbidimeter, Missouri Water Pollution Law.

The turbidity of any waste discharge from any washing operation into clean waters of the state shall not exceed 50 turbidity units at the point of discharge, as measured by approved methods outlined in standard methods for the examination of water and waste water, twelfth edition. The resultant turbidity of waste discharge mixed with waters of the receiving stream shall not exceed turbidity stipulated by standards in accordance with the Missouri Water Pollution Law. Storage facilities for materials which are hazardous to health and welfare or capable of casuing pollution shall be located so as to prevent spillage into state waters, sewage collection systems or waste treatment works that might result in pollution or substantial damage. Measures to entrap spillage shall be installed to prevent accidental pollution of state waters. Site locations and measures to entrap spillage may be inspected by the Missouri Water Pollution Board. (Dunham-Florida)
W73-10260

RESERVOIR RELEASE RESEARCH PROJECT.

Sport Fishing Institute Bulletin, No 243, p 1-4, April 1973. 1 fig.

Descriptors: *Oxygen demand, *Reservoir releases, *Reservoir fisheries, *Fish populations, *Aquatic environment, Reservoir construction, Discharge (Water), Fish management, Surface waters, Water management (Applied), Fish management, Impoundments, Water storage, Aquatic life, Fish conservation, Research and development, Sport fish, Ecology.
Identifiers: Hypolimnial discharge, Epilimnial discharge.

A five-year research project dealing with the effect of hypolimnial and epilimnial discharge regimes of reservoirs as they affect dissolved oxygen distribution concludes that reservoir-outlet locations do not appear to affect significantly either sport fish catch or biological productivity. However, hypolimnial discharge made possible substantial additional coldwater fisheries in tailwaters without detriment to existing warmwater fisheries. Researchers recommended, to insure maximum vertical distribution within the reservoir system of biologically significant quantities of dissolved oxygen, that reservoirs be discharged from the deepest possible water strata. Field data were collected from three experimental reservoirs to measure any significant differences in reservoir productivity indices. An outline of the experiment and methods of gathering data are included. (Napolitano-Florida)
W73-10261

DREDGING AND FILLING, COWIKEE STATE PARK, LAKEPOINT RESORT, WALTER F. GEORGE LAKE, CHATTAHOOCHEE RIVER, ALABAMA (FINAL ENVIRONMENTAL IMPACT STATEMENT).
Army Engineer District, Mobile, Ala.
For primary bibliographic entry see Field 04A.
W73-10265

CONTROL OF EURASIAN WATERMILFOIL (MYRIOPHYLLUM SPICATUM L.) IN TVA RESERVOIRS (FINAL ENVIRONMENTAL IMPACT STATEMENT).
Tennessee Valley Authority, Chattanooga.
For primary bibliographic entry see Field 04A.
W73-10266

ELIMINATION OF HAZARDOUS OPEN CANALS.
For primary bibliographic entry see Field 06E.
W73-10267

THIRD ANNUAL REPORT OF THE COUNCIL ON ENVIRONMENTAL QUALITY.
For primary bibliographic entry see Field 06E.
W73-10272

WATER POLLUTION.
For primary bibliographic entry see Field 06E.
W73-10277

REGULATION OF WATER RESOURCES.
For primary bibliographic entry see Field 06E.
W73-10278

CONSERVATION—GENERALLY.
For primary bibliographic entry see Field 06E.
W73-10280

NATURAL AND SCENIC RIVERS SYSTEM.
For primary bibliographic entry see Field 06E.
W73-10282

PROHIBITIONS ON CERTAIN DETERGENTS.
For primary bibliographic entry see Field 06E.
W73-10283

PROSPECTING AND DEVELOPMENT OF PETROLEUM PRODUCTS IN RIVER BEDS, LAKE BEDS, AND SHORELANDS—PREFERENTIAL RIGHTS.
For primary bibliographic entry see Field 06E.
W73-10284

PLUGGING WELLS—DECLARATION OF POLICY, AND ADMINISTRATIVE REGULATION.
For primary bibliographic entry see Field 06E.
W73-10285

Available from the National Technical Information Service as PB-211 918, $5.45 in paper copy, $0.95 in microfiche. July 27, 1972. 103 p, 5 tab, 30 exhibits.

Descriptors: *Water pollution control, Acidic water, Effluents, Federal Water Pollution Control Act, Pollutant identification, Economic feasibility, *Economic impact, *Pollution abatement, Water pollution sources.
Identifiers: Industry study, *Steel industry.

This volume is Part II of a final report on the potential economic impact on the steel industry of meeting Federal air and water pollution abatement requirements. The volume includes (1) an evaluation of the steel industry's growth patterns, operating practices, and competitive environment; (2) an assessment of the industry's recent financial performance; and (3) a description of the process of manufacturing steel in order to identify significant pollutants generated at each production stage along with abatement techniques to control them. It was found that air pollution is more difficult to control and requires more advanced technology than that required to treat waste water. Relative quantities of water effluent from various operations should be evaluated in light of the percentage of product that is processed by each operation. (Weaver-Wisconsin)
W73-10312

ON TAXATION AND THE CONTROL OF EXTERNALITIES,
Princeton Univ., N.J. Dept. of Economics.
W. J. Baumol.
The American Economic Review, Vol 62, No 3, p 307-322, June 1972. 3 fig, 22 ref.

Descriptors: *Taxes, *Pollution taxes (Charges), Economic efficiency, Water pollution control, Optimization, *Water allveation.
Identifiers: *Pigouvian tax-subsidy programs, *Optimal resource allocation, Externalities, Public goods, Economic theory.

In a critical analysis of the theory of Pigouvian tax-subsidy programs, it is argued that the prescriptions of the Pigouvian tradition are generally those required for an optimal allocation of resources. In addition, where an externality is of the public goods variety, neither compensation to nor taxation of those who are affected by it is compatible with optimal resource allocation. This is an important consideration with respect to the typical pollution problem. Pigouvian taxes (subsidies) upon the generator of the externality are all that is required. The Pigouvian proposals suffer from a number of serious shortcomings as operational criteria when they are implemented precisely as they emerge from theory. Thus, a modified approach is discussed. This approach consists of two steps: (1) the setting of standards for levels of pollution that are considered tolerable, and (2) the design of taxes and effluent charges whose rates are shown by experience to be sufficient to achieve the selected standards of acceptibility. (Weaver-Wisconsin)
W73-10314

SYMPOSIUM ON DIRECT TRACER MEASUREMENT OF THE REAERATION CAPACITY OF STREAMS AND ESTUARIES.
For primary bibliographic entry see Field 05B.
W73-10337

REAERATION CAPACITY OF THE FLINT, SOUTH AND PATUXENT RIVERS,
Georgia Inst. of Tech., Atlanta. School of Civil Engineering.
For primary bibliographic entry see Field 05B.
W73-10343

REAERATION STUDIES OF THE CHATTAHOOCHEE RIVER,
For primary bibliographic entry see Field 05B.
W73-10345

MODEL STUDY OF REAERATION CAPACITY OF THE JAMES RIVER ESTUARY (VIRGINIA),
For primary bibliographic entry see Field 05B.
W73-10346

FIELD STUDIES IN YAQUINA RIVER ESTUARY OF SURFACE GAS TRANSFER RATES,
For primary bibliographic entry see Field 05B.
W73-10347

RADIOLOGICAL SAFETY,
For primary bibliographic entry see Field 05B.
W73-10348

EFFECT OF HYDRAULIC PROPERTIES ON REAERATION,
For primary bibliographic entry see Field 05B.
W73-10349

POLLUTANT EFFECTS ON REAERATION,
For primary bibliographic entry see Field 05B.
W73-10350

OBSERVED VS. CALCULATED REAERATION CAPACITIES OF SEVERAL STREAMS,
For primary bibliographic entry see Field 05B.
W73-10351

RELATIONSHIPS BETWEEN HYDRAULIC PROPERTIES AND REAERATION,
Georgia Inst. of Tech., Atlanta. School of Civil Engineering.
E. C. Tsivoglou.
In: Proceedings of Symposium on Direct Tracer Measurement of the Reaeration Capacity of Streams and Estuaries; Georgia Institute of Technology, Atlanta, July 7-8, 1970: Environmental Protection Agency Water Pollution Control Research Series 16050 FOR 01/72, p 185-194, January 1972. 2 fig.

Descriptors: *Reaeration, *Hydraulics, *Water measurement, *Channel morphology, Discharge measurement, Flow rates, Gaging stations, Hydrographs, Streamflow, On-site data collections, Hydrologic data.

The basic relationships between the reaeration capacity of a stream and its hydraulic properties were studied with the practical aim of developing the capability to predict reaeration on the basis of field measurement of appropriate hydraulic properties. The rate of reaeration is directly proportional to the rate of water surface replacement. The rate of water surface replacement is proportional to the rate of energy dissipation in open channel flow. Therefore, gas transfer in a turbulent natural stream is dependent only upon the change in water surface elevation. At a given water temperature the amount of tracer gas that will be lost to the atmosphere in a specific length of stream channel, or the amount of DO deficit that will be satisfied, can be predicted on the basis solely of the change in water surface elevation. All of the individual tracer gas loss data obtained from five separate tests in the Patuxent River have been plotted, with the observed percent loss of tracer gas shown as a function of the change in water surface elevation between sampling points. The relationship predicted is clearly demonstrated by the data. (See also W73-10337) (Knapp-USGS)
W73-10352

THE UTILIZATION AND CONSUMPTION OF MERCURY IN FRANCE: OUTLINE OF THE RESULTING HEALTH PROBLEMS, (IN FRENCH),
Laboratoire Central de Recherches Veterinaires, Alfort (France).
For primary bibliographic entry see Field 05B.
W73-10389

HYGIENIC SUBSTANTIATION OF PERMISSIBLE HUMIDITY LEVEL IN CASE OF HIGH TEMPERATURE, (IN RUSSIAN),
Institut Gigieny Truda i Professionalnykh Zabolevanii, Moscow (USSR).
E. F. Medvedeva, and N. A. Fedotova.
Gig Sanit. Vol 36, No 12, p 23-25, 1971. Illus. English summary.
Identifiers: Humans, *Humidity, *Hygienic standards, Regulation, Temperature, *Thermo regulation strain.

The investigations detected the most pronounced strain in human body thermoregulation under conditions of chamber air temperature amounting to 28 deg., that is the maximum permissible level for warm season in accordance with sanitary standards. Under conditions of 40% relative air humidity the changes in physiological reactions correspond to moderate thermoregulation strain; in more humid air (55-65%) there was intense strain. Alterations in sanitary standards are recommended.--Copyright 1972, Biological Abstracts, Inc.
W73-10390

06. WATER RESOURCES PLANNING

6A. Techniques of Planning

PRELIMINARY WATER AND WASTE MANAGEMENT PLAN.
Metcalf and Eddy, Inc., Boston, Mass.

Report to Sacramento Regional Area Planning Commission, Sacramento, California, November 1969. 118 p, 6 fig, 20 tab.

Descriptors: *Planning, *Water supply, *Sewerage, *Drainage, Project planning, Groundwater, Water reuse, Waste water treatment, Financing, Administration.
Identifiers: *Utility extension planning, Consolidation, Wastewater reclamation, Sacramento, California.

Preliminary plans for water supply, sewerage, and drainage facilities for the Sacramento region are presented in this report. The plans are meant to be a guide to what is feasible and logical in terms of water resource projects and are framed in both long-range plans and short-range projects. Since financing is an important consideration the recommended short-range projects are those of highest priority and are intended to alleviate existing problems or deficiencies. The report emphasizes a planning approach that includes planning objectives, identification of problems, evaluation of system adequacy, and programs to be implemented by local agencies under regional review. The main sections of the report are land use and population, water systems, sewerage systems, and drainage. Estimated costs for remedial and short range projects are given. Some of the recommendations include the need for more data on groundwater, an emphasis on using reclaimed wastewater especially for agriculture and groundwater recharge, consolidation of facilities to eliminate a number of small treatment plants, and consideration of advanced wastewater treatment facilities or transport of treated effluent. (Elfers-North Carolina)
W73-09813

COMPREHENSIVE AREA-WIDE WATER AND SEWER PLAN.
Robert and Co. Associates, Atlanta, Ga.

Prepared for the Middle Georgia Planning Commission, Macon. Phase I, July 1969. 79 p, 1 photo, 4 maps, 3 tab, 1 append. HUD UPA Ga. P-61.

Descriptors: *Planning, Water supply, Sewers, Project planning, Financing, Coordination.
Identifiers: *Utility extension planning, *Inventory, Macon, Georgia, Zoning controls, Rural water systems, Middle Georgia Area.

The existing systems around Macon, Georgia were analyzed. The seven-county area is described in terms of land use, population, transportation, geology, drainage basins, and water quantity and quality. Existing water and sewer systems are described and projected improvements discussed. Means of financing the improvements and related zoning and subdivision regulations are presented. The basic recommendations include the need for a 5-10 year improvement program, a program for area-wide coordination and continued planning, a storm drainage and flood control plan, and a number of specific feasibility studies for expansion projects. (Elfers-North Carolina)
W73-09814

LAKEFRONT NEW ORLEANS, PLANNING AND DEVELOPMENT, 1926-1971,
Louisiana State Univ., New Orleans. Urban Studies Inst.
J. A. Filipich, and L. Taylor.
53 P, 36 FIGS, 10 REF. 1971.

Descriptors: *Planning, Land use, Levee, Long-range planning, Financing, Design, Land reclamation.
Identifiers: *Lakefront development, New Orleans (Louisiana), Lake Pontchartrain, Master plan.

The land reclamation and lakefront development that has occurred on Lake Pontchartrain since 1926 and some insights for future development are presented as described. The New Orleans Lakefront experience can provide useful information and guidelines for other urban areas. Much of the report traces the history of the lakefront development including levee construction, financing, parks, several subdivisions, the lakefront airport, recent expansions, and new design concepts. The importance of long-range planning is stressed. Several key factors for successful implementation are: (1) the formulation of a master plan with compatible land uses; (2) the creation of an operating agency to supervise the development according to the master plan; (3) the provision of a strong financial and jurisdictional base for the operating agency; (4) use of feedback information to provide flexibility in the overall program; (5) strong control over the development to prevent speculative intrusions from private developers. (Elfers-North Carolina)
W73-09822

PLANNING FOR ENVIRONMENTAL QUALITY, PHASE I,
North Carolina State Dept. of Administration, Raleigh. State Planning Div.
R. A. Paul.
Available from the National Technical Information Service as PB-213 147, $7.25 in paper copy, $0.95 in microfiche. Prepared in cooperation with the Interagency Task Force on Environment and Land Use Planning. Report No. NCP-146.09. June 1972. 81 p, 8 fig, 2 append. HUD NCP-146. CPA-NC-04-00-0146.

Descriptors: *Environmental effects, Water quality, Land use, Natural resources, Waste water disposal, North Carolina.
Identifiers: *Environmental planning, *Environmental indicators.

This report is the first of a planned series of documents with the purpose of articulating the concern of the State of North Carolina for preserving its environmental quality. The series is intended to cover such areas as environmental quality goals, policies, legislation and research activities and to communicate this information to state agencies and the public. This first report reviews the environmental planning process in North Carolina and discusses the concept and use of environmental indicators. The two key planning bodies are the N.C. Council on State Goals and Policy and the Interagency Task Force on Environmental and Land Use Planning. Some of the environmental indicators discussed relate to air quality, water quality, wildlife abundance, housing quality, recreational areas, and natural resources. The environmental indicators are also related to broader areas of environmental quality such as energy and land use policies. Water supply, wastewater disposal, and solid waste disposal systems are also used as environmental indicators. (Elfers-North Carolina)
W73-09824

METROPOLITAN COMPREHENSIVE WATER, SEWAGE, AND SOLID WASTE PLANNING STUDY.
Bi-State Metropolitan Planning Commission, Rock Island, Ill.

Summary. March 31, 1970. 30 p, Illus, 5 map, 4 fig.

Descriptors: *Planning, Water supply, Sewerage, Drainage systems, Solid wastes, Pollution abatement, Project planning, Water management, Iowa, Illinois.
Identifiers: *Utility extension planning, Davenport (Iowa), Rock Island (Ill), Moline (Ill), Scott County (Iowa), Rock Island County (Ill).

This summary of a three volume report on the planning and development of water supply, sanitary sewer, storm water drainage, and solid waste disposal systems in the two county area which includes the cities of Rock Island, Moline, and Davenport, is oriented to a quick and interesting presentation of the principal concepts and plan elements of the main report. It is divided into the following basic section: (1) an introduction including a discussion of pollution problems; (2) a listing of goals and objectives plus an outline of the planning process; (3) the basic plan concepts and elements of the water supply, sewerage, solid waste disposal, and storm drainage system for the central urbanized area; (4) and similar concepts and plans for the outlying communities in the region. The planning summary is to be used as a public information document, but not as a source of detailed information on the actual plans. (Elfers-North Carolina)
W73-09829

A MARGINAL ANALYSIS - SYSTEM SIMULATION TECHNIQUE TO FORMULATE IMPROVED WATER RESOURCES CONFIGURATIONS TO MEET MULTIPLE OBJECTIVES,
Massachusetts Inst. of Tech., Cambridge. Dept. of Civil Engineering.
E. A. McBean, and J. C. Schaake, Jr.
Doctoral thesis, Report No. 166, Ralph M. Parsons Laboratory for Water Resources and Hydrodynamics, Department of Civil Engineering, MIT, Cambridge, Mass., February 1973. 366 p, 114 fig, 29 tab, 178 sym, 138 ref, 5 appen.

Descriptors: *Simulation analysis, Water resources development, *Planning, *Reservoirs, *Multiple-purpose projects, *Markov processes, Analytical techniques, *Optimization, Estimating, River basins, Streamflow, Reservoir releases, Reservoir storage, Optimum development plans, Benefits, Equations, Water supply, Design, Systems analysis, *Mathematical models, Computer models.

development of the hydroelectric potential of a river basin or the supply of water to a growing city. A sensitivity testing procedure is given for computing the amount by which the cost of any project would have to change to enable it to occupy any position in the optimal sequence. This procedure forms the base for an efficient planning strategy for the case in which the engineer is particularly interested in correctly selecting the next few projects, but being sure that they are compatible with the overall scheme of development. It is simple to use and it provides results in a form that can be readily assimilated, an important consideration in complex planning problems dealing with many intangibles. (Bell-Cornell)
W73-09838

THE DISSIPATION OF EXCESS HEAT FROM WATER SYSTEMS,
Geological Survey, Bay St. Louis, Miss.
For primary bibliographic entry see Field 05B.
W73-09839

THERMAL CAPACITY OF OUR NATION'S WATERWAYS,
Westinghouse Hanford Co., Richland, Wash.
For primary bibliographic entry see Field 05G.
W73-09840

AN APPROACH TO ANALYSIS OF ARCH DAMS IN WIDE VALLEYS,
Southampton Univ. (England). Dept. of Civil Engineering.
For primary bibliographic entry see Field 08A.
W73-09841

OPTIMIZATION OF CIRCULATING WATER SYSTEM,
Worcester Polytechnic Inst., Mass.
For primary bibliographic entry see Field 05G.
W73-09842

SYSTEMS APPROACH TO POWER PLANNING,
Bechtel Corp., San Francisco, Calif. Hydro and Community Facilities Div.
J. G. Thon.
Journal of the Hydraulics Division, American Society of Civil Engineers, Vol 99, No HY4, p 589-598, April 1973. 4 fig, 12 ref.

Descriptors: Computers, *Transmission (Electrical), *Electrical networks, *Planning, *Coordination, Electric power production, Thermal power plants, Hydroelectric power, Pumped storage, *Systems analysis, Dynamic programming, Design, Management, Estimating, Power system operation, Comprehensive planning.

A broad discussion of systems planning for the electric power industry is presented. The main geographic area which is considered is the Western Region of the United States. This region had the first power pooling agreements in the 1940's and is likely to see a large amount of growth in the future. Today interconnection has been accomplished for nearly all systems in the U.S. and Canada. The feasibility of bulk power transmission over long distances and the projected increases and accompanying energy requirement increases have made coordination of planning unavoidable. Weekly load curves and daily peak requirements on large power pools must be met economically and efficiently. Combinations of base load thermal plants, less efficient oil-fired plants and fast response hydroelectric facilities are necessary. Stronger connections with adjacent areas represent another alternative. A great many factors, such as new energy sources and technology, pollution controls, engineering obstacles, generation interconnection, fuel location, plant siting and others must be considered in any short or long range planning. Planning models are discussed briefly. (Jerome-Vanderbilt)
W73-09859

CONTROL OF THE OPERATING REGIMES OF WATER RESOURCES SYSTEMS,
For primary bibliographic entry see Field 04A.
W73-09868

AN INVESTIGATION INTO THE LONGTERM PERSISTENCE OF U. S. HYDROLOGIC RECORDS,
IBM Watson Research Center, Yorktown Heights, N.Y.
For primary bibliographic entry see Field 04A.
W73-09869

HOW FREQUENTLY WILL FLOODS OCCUR,
Pennsylvania State Univ., University Park. Dept. of Civil Engineering.
For primary bibliographic entry see Field 04A.
W73-09992

HYDRO-POWER SYSTEM SIMULATION TECHNIQUE,
Corps of Engineers, Portland, Oreg. North Pacific Div.
For primary bibliographic entry see Field 04A.
W73-10019

THE ENVIRONMENTAL MATRIX: INPUT-OUTPUT TECHNIQUES APPLIED TO POLLUTION PROBLEMS IN ONTARIO,
Department of Energy, Mines and Resources, Burlington (Ontario). Canada Center for Inland Waters.
For primary bibliographic entry see Field 05G.
W73-10020

TIME PERIODS AND PARAMETERS IN MATHEMATICAL PROGRAMMING FOR RIVER BASINS,
Massachusetts Inst. of Tech., Cambridge. Ralph M. Parsons Lab. for Water Resources and Hydrodynamics.
For primary bibliographic entry see Field 04A.
W73-10317

6B. Evaluation Process

THE TEXAS WATER PLAN AND ITS INSTITUTIONAL PROBLEMS,
Texas A and M Univ., College Station. Water Resources Inst.
C. W. Jensen, and W. L. Trock.
Available from the National Technical Information Service as PB-220 683, $3.00 in paper copy $0.95 in microfiche. Technical Report 37, January 1973, 73 p, 64 ref. OWRR B-025-TEX (10). 14-01-0001-1558.

Descriptors: *Institutions, *Water rights, Costs benefit analysis, *Cost sharing, Financing, Operating costs, Jurisdiction, Indirect costs, Texas, Water districts, Water supply, Water allocation, Planning, Decision making, *Imported water.
Identifiers: *Texas Water Plan, Master water districts.

An expansion of the supply of water greater efficiency in its use are necessary for the future economic development of the state of Texas. Imported water, supplemental to that available in the state, is an important part of development plans as outlined in the Texas Water Development Board. Implementing the Board's plan to reallocate water supplies and improve the efficiency of land and water use will raise many serious problems. Solutions will be required in a wide array of institu-

tional problems that will extend to such areas as the interstate diversion and interbasin transfers of water, doctrines or water rights and legislated water use-priorities, acreage restrictions established in federal reclamation law, comingling public and private water, construction financing, revenue production through a system of taxation and water sales, and the organizing of new institutions for governing the entire system. As the need for master or other special districts is faced, decisions will be required as to whether to organize for centralized control from the state level or with emphasis on control by the local area. Reorganizing institutions, or their formalized cooperation, will be necessary to permit local control, yet be able to induce the desired efficiency in resource use that will make the Texas Water System succeed. Failure to achieve efficiency in the functioning of institutions may result in an institutional overhead so high as to prohibit realization of the anticipated System benefits. (Runkles-Texas A andM)
W73-09756

PROJECTIONS OF WATER USE AND CONSERVATION IN THE USSR, 2000 (ORIYENTIROVOCHNYY PROGNOZ ISPOL'ZOVANIYA I OKHRANY VODNYKH RESURSOV SSSR NA UROVNE 2000 G.),
Akademiya Nauk SSSR, Moscow. Institut Geografii.
M. I. L'vovich, and N. I. Koronkevich.
Akademiya Nauk SSSR Izvestiya, Seriya Geograficheskaya, No 2, p 35-47, March-April 1971. 5 tab, 40 ref.

Descriptors: *Projections, *Water utilization, *Alternative water use, *Water conservation, *Water resources development, Water supply, Water demand, Water users, Consumptive use, Waste water (Pollution), Waste water disposal, Waste water treatment, Water pollution control, Hydrologic budget.
Identifiers: *USSR.

Several alternatives are considered in developing a plan for utilization and conservation of water resources in the USSR for the year 2000. Two alternatives call for disposal of waste water in rivers, based either on present pollution control techniques or on improvements in water utilization and waste-water treatment. A third alternative, based on water utilization and conservation principles developed by the Institute of Geography of the USSR Academy of Sciences, calls for nationwide stoppage of waste-water disposal in rivers, and a fourth alternative, for limited stoppage affecting only the most heavily populated regions of the country. The fourth alternative will solve the water problem as successfully as the third at a one-third reduction in capital costs. (Josefson-USGS)
W73-09804

WATER AND WASTEWATER FOR BROWARD COUNTY, FLORIDA.
Benham-Blair and Affiliates of Florida, Inc. Fort Lauderdale.

Prepared for Broward County Planning Board, Fort Lauderdale, Florida, February 1970. 256 p, 37 fig, 25 tab, 7 append.

Descriptors: *Wastewater disposal, *Planning, *Administration, *Water supply, Sewerage, Groundwater, Pollution abatement, Project planning, Water management, Financing, Florida.
Identifiers: *Utility extension planning, Broward County (Florida), Fort Lauderdale (Florida).

Water supply and wastewater disposal are discussed for the eastern part of Broward County, which includes Fort Lauderdale. Because of the great growth in this area, there is an urgent need to plan for the optimum use of the groundwater resources, to begin a concerted effort to control water pollution, and to centralize the administration of the county's water resources. Detailed information is presented on land use and water demands, geology and groundwater, water pollution and water quality criteria, and financial and organizational data. The basic recommendations are that water be supplied to the whole planning area via 3 districts using optimally located wells to prevent salt water intrusion and that a single county-wide sewerage agency collect wastewater and dispose it out in the Atlantic Ocean after primary treatment and chlorination at 2 central treatment plants. These recommendations are largely based upon least-cost financial considerations and an objective of administrative efficiency. (Elfers-North Carolina)
W73-09816

HUD SEWER POLICY IN URBAN RENEWAL AREAS,
Paterson Redevelopment Agency, N.J.
For primary bibliographic entry see Field 06E.
W73-09820

PROGRAM DESIGN FOR PLANNING WATER QUALITY MANAGEMENT IN KENTUCKY RIVER BASINS.
Schimpeler-Corradino Associates, Louisville, Ky.; and Schimpeler-Corradina Associates, Lexington, Ky.

Available from the National Technical Information Service as PB-212 957, $12.25 in paper copy, $0.95 in microfiche. Prepared for Kentucky Development Office and Kentucky Department of Health, Division of Water Pollution Control, June 1972. 138 p, 11 fig, photos, 20 tab, 9 append. HUD - CPA-KY-04-00-0088.

Descriptors: *Water quality control, Water quality standards, River basins, Environmental effects, Coordination, Financing, Kentucky, Water pollution sources, Management.
Identifiers: *Water quality management.

This study was to interpret the Environmental Protection Agency's quidelines for water quality management and to design a program for the formulation of water quality management plans for the ten major river basins in Kentucky. The EPA guidelines, issued in 1971, set certain policies and procedures for water quality planning and management and the formulation of the plans is necessary to qualify for federal wastewater treatment facility grants. Part I is a step-by-step procedure for preparing basin water quality management plans and covers basin characteristics, pollution sources, water resources, planning alternatives, and a discussion of management strategies. Part II covers the implementation of the water quality plans and strategies in Kentucky river basins and includes descriptions of the ten river basins, priorities for the planning efforts, costs of formulating the plans, the institutional aspects involved, and the means of financing the management programs. The report also covers water quality standards, information systems, and environmental assessment. (Elfers-North Carolina)
W73-09825

REGIONAL BENEFITS IN FEDERAL PROJECT EVALUATIONS,
Michigan Univ., Ann Arbor. School of Natural Resources.
G. Schramm.
The Annals of Regional Science, Vol VI, No 1, p 84-95, June 1972. 1 fig, 15 ref.

Descriptors: Water resources, *Federal government, Projects, *Income, Regions, *U.S. Water Resources Council, *Benefits, Costs, Feasibility, *Regional analysis, Regional development, Evaluation, Project benefits.
Identifiers: Project selection, Regional benefits.

personal freedom due to environmental control and willingness to pay for higher environmental quality indicated a high degree of environmental consciousness among both year-round and seasonal residents. However, respondents opposed restriction of time periods during which recreational facilities could be used as well as regulations on architectural design. Community attitude surveys such as the present one were said to be of great value to planners in constructing locally supported programs. (Weaver-Wisconsin)
W73-10310

ON TAXATION AND THE CONTROL OF EXTERNALITIES,
Princeton Univ., N.J. Dept. of Economics.
For primary bibliographic entry see Field 05G.
W73-10314

CAUSAL FACTORS IN THE DEMAND FOR OUTDOOR RECREATION,
University of New England, Armidale (Australia).
For primary bibliographic entry see Field 06D.
W73-10315

EVALUATION OF RECREATIONAL BENEFITS ACCRUING TO RECREATORS ON FEDERAL WATER PROJECTS--A REVIEW ARTICLE,
Clemson Univ., S.C.
R. M. Pope, Jr.
The American Economist, Vol 16, No 2, p 24-29, Fall 1972. 6 fig, 1 tab, 12 ref.

Descriptors: *Cost-benefit analysis, *Estimated benefits, *Indirect benefits, *Intangible benefits, Recreation, Reviews, *Federal project policy.
Identifiers: Flood Control Act of 1936, Simulated demand analysis.

The Flood Control Act of 1936 requires justification of water resource projects on the basis of cost-benefit criterion. However, cost-benefit analysis is complicated by the existence of intangible and incommensurable benefits. The author demonstrates and discusses several methods for attaching dollar value to such benefits. Simulated demand curves are shown to involve problems of determining whether the entire expenditures of recreators are in fact measures of benefits received from the use of the project. These demand curves, however, are much more adequate than other techniques, such as home interviews, trade-offs, and costs of providing recreational services. (Weaver-Wisconsin)
W73-10316

BENEFIT-COST ANALYSIS FOR WATER RESOURCE PROJECTS,
Tennessee Valley Authority, Muscle Shoals, Ala.
H. Hinote.
Center for Business and Economic Research, Tennessee University, Knoxville, June 1969.

Descriptors: *Cost-benefit analysis, *Water resources development, *Land development, *Bibliographies, Flood control, Navigation, Water quality, Recreation, Water supply, *Projects.
Identifiers: Forecasting demand.

This bibliography provides a convenient guide and source document in the field of economic analysis of water resource projects. Works cited are concerned with the current state of the art of benefit-cost analysis, and emphasis is placed on the following project purposes: (1) Flood control, (2) Navigation, (3) Water quality (Pollution), (4) Recreation, and (5) Water supply. Value enhancement of land surrounding a water resource project is also emphasized. The literature reviewed is classified into the following basic steps followed in performing benefit-cost analysis: (1) definition of the problem or the costs and benefits, (2) forecasting demand, (3) benefit measurement or cost determination, (4) evaluation techniques, and (5) decision criteria. Some works incorporate more than one of the above steps, and some may not be classified under any of the steps. In the latter case, the works are classified under the title of Basic Works. (Weaver-Wisconsin)
W73-10318

ATTITUDES TOWARDS WATER FLOURIDATION IN TURKU, (IN FINNISH),
Turku Univ. (Finland).
For primary bibliographic entry see Field 05F.
W73-10385

6C. Cost Allocation, Cost Sharing, Pricing/Repayment

THE TEXAS WATER PLAN AND ITS INSTITUTIONAL PROBLEMS,
Texas A and M Univ., College Station. Water Resources Inst.
For primary bibliographic entry see Field 06B.
W73-09756

COST APPORTIONMENT PLANNING REPORT.
Metcalf and Eddy, Inc., Boston, Mass.

Prepared fot the Metropolitan Council of the Twin Cities Area, Minnesota, November 1968. 99 p, 7 fig, 18 tab.

Descriptors: *Cost allocation, *Financing, Planning, Sewerage, Coordination, Administration, Minnesota.
Identifiers: *Metropolitan Council of the Twin Cities (Minnesota), Service areas, Policy plans.

A study of cost apportionment for the major sewerage works outlined in Metropolitan Sewer Plan, A Preliminary Concept Plan (1968) is presented. The purpose is to assist the Metropolitan Council in formulating cost apportionment and financing procedures for the construction and operation of a coordinated metropolitan-wide sewerage system. This is being undertaken within a governmental framework of the Metropolitan Council which is an overall policy-making body and the Sewer Board, appointed by the Council, which carries out detailed planning. The report includes chapters on variations of cost apportionment methods, their use in other metropolitan areas, and their application to specific service areas in the Twin Cities area. In some areas the Metropolitan Council may oversee operation and management of treatment plants which would then be treated as metropolitan works. Recommendations include the grouping of communities into service areas, the charging of costs to communities rather than individual households or firms, that charges be based on sewage flows, the creation of metropolitan loan funds to help communities construct facilities, and the creation of a metropolitan grant fund for communities with special costs because of metropolitan-wide decisions. (Elfers-North Carolina)
W73-09827

INITIAL WATER, SEWERAGE AND FLOOD CONTROL PLAN,
Duncan and Jones, Berkeley, Calif.; and Yoder-Trotter-Orlob and Associates, Berkeley, Calif.
For primary bibliographic entry see Field 06D.
W73-09828

INVESTMENT SEQUENCING RECOGNIZING EXTERNALITIES IN WATER DESALTING,
Massachusetts Univ., Amherst. Dept. of Agricultural and Food Economics.
For primary bibliographic entry see Field 03A.
W73-10018

THE ECONOMICS OF OIL TRANSPORTATION IN THE ARCTIC,
Toronto Univ. (Ontario). School of Business.
For primary bibliographic entry see Field 05G.
W73-10129

A STUDY OF THE COST EFFECTIVENESS OF REMOTE SENSING SYSTEMS FOR OCEAN SLICK DETECTION AND CLASSIFICATION,
New Hampshire Univ., Durham. Dept. of Electrical Engineering.
For primary bibliographic entry see Field 05A.
W73-10228

A STUDY OF THE ECONOMIC IMPACT ON THE STEEL INDUSTRY OF THE COSTS OF MEETING FEDERAL AIR AND WATER POLLUTION ABATEMENT REQUIREMENTS, PART 2.
Council on Environmental Quality, Washington, D.C.
For primary bibliographic entry see Field 05G.
W73-10312

ON TAXATION AND THE CONTROL OF EXTERNALITIES,
Princeton Univ., N.J. Dept. of Economics.
For primary bibliographic entry see Field 05G.
W73-10314

6D. Water Demand

PROJECTIONS OF WATER USE AND CONSERVATION IN THE USSR, 2000 (ORIYENTIROVOCHNYY PROGNOZ ISPOL'ZOVANIYA I OKHRANY VODNYKH RESURSOV SSSR NA UROVNE 2000 G.),
Akademiya Nauk SSSR, Moscow. Institut Geografii.
For primary bibliographic entry see Field 06B.
W73-09804

INITIAL WATER, SEWERAGE AND FLOOD CONTROL PLAN,
Duncan and Jones, Berkeley, Calif.; and Yoder-Trotter-Orlob and Associates, Berkeley, Calif.

Prepared for the San Diego County Comprehensive Planning Organization, California, June 1972. Technical report. 68 p, 5 fig, 6 tab, 5 append. HUD P-384 (g).

Descriptors: *Planning, *Water supply, *Sewerage, *Flood control, Storm runoff, Water demand, Financing, Water quality control, California.
Identifiers: San Diego County (Calif), Utility extension planning, Land suitability.

Background and technical information used to prepare the Initial Water, Sewerage, and Flood Control Plan, Summary Report, is presented in the Technical Report. The technical information is divided into sections: (1) an evaluation of the adequacy of the water resources in the region including the availability of water supplies to meet future demands, the potential problems from wastewater discharges, and the relation between future water quality objectives and water supply requirements; (2) a forecast of 1980 water supply demands, wastewater loads, and storm water runoff; (3) an evaluation of the adequacy of existing water resources facilities in the region to meet future demands and loads; (4) an inventory and description of various financing methods including bonds, taxes, user charges and grants; and (5) a land suitability analysis to determine where urban development should occur. The report contains appendixes on beneficial use of water, water quality objectives, wastewater sources, and drainage structures. (Elfers-North Carolina)
W73-09828

SEARCH TECHNIQUE FOR PROJECT SEQUENCING,
British Columbia Hydro and Power Authority, Vancouver.
For primary bibliographic entry see Field 06A.
W73-09838

THERMAL EFFECTS OF PROJECTED POWER GROWTH: SOUTH ATLANTIC AND EAST GULF COAST RIVER BASINS,
Hanford Engineering Development Lab., Richland, Wash.
For primary bibliographic entry see Field 05C.
W73-09848

A U.S. ENERGY POLICY NOW.

Chemical Engineering, Vol 78, No 14, p 50-54, June 28, 1971.

Descriptors: *Oil, *Natural gas, *Electric power demand, Natural resources, Fossil fuels, Economics, Legislation.
Identifiers: *Energy crisis.

Oil company executives and a financial expert stress the necessity for a clear national policy that would assure increased domestic production of oil and gas. Sharp increases in taxes on production, proposals to scrap the import quota system, on-again off-again leasing policies, and delays in approval of the Alaskan pipeline undermine the basis for national planning by the oil companies, and rational investment by outsiders. The nation will not be able to resolve its energy problems by continuing to manage individual energy resources each in isolation from the other. The emergence of an energy-supply gap that can endanger the progress and security of this nation is noted along with the progressing dependence on foreign supplies. (Oleszkiewicz-Vanderbilt)
W73-09862

PUBLIC ACCESS TO RESERVOIRS TO MEET GROWING RECREATION DEMANDS.
For primary bibliographic entry see Field 06E.
W73-10273

ATTITUDES ON ENVIRONMENTAL QUALITY IN SIX VERMONT LAKESHORE COMMUNITIES,
Vermont Univ., Burlington.
For primary bibliographic entry see Field 06B.
W73-10310

CAUSAL FACTORS IN THE DEMAND FOR OUTDOOR RECREATION,
University of New England, Armidale (Australia).
R. L. Ranken, and J. A. Sinden.
The Economic Record, Vol 47, No 119, p 418-426, September 1971. 4 tab.

Descriptors: *Regression analysis, *Recreation demand, *Motivation, Water resources development, Land development, *Australia, Water demand.
Identifiers: Demand analysis.

The demand for outdoor recreation should be a major consideration in the development of land and water resources. In order that planning meet this demand, an understanding of the factors that promote recreational activities and the development of a model for predictive purposes is required. Using Australia as an example, relationships between recreational activities and family characteristics were studied. Specifically, the influence of factors other than price on the consumption of various recreation services was estimated using a sample drawn from Armidale, New South Wales, and the statistical technique of regression analysis. Various classes of recreation were used as alternative dependent variables while independent variables included measures of household income, educational level, age and sex composition, and number of holidays per year. Several results were pointed out as different from those obtained from similar analyses in the United States. (Weaver-Wisconsin)
W73-10315

6E. Water Law and Institutions

THE TEXAS WATER PLAN AND ITS INSTITUTIONAL PROBLEMS,
Texas A and M Univ., College Station. Water Resources Inst.
For primary bibliographic entry see Field 06B.
W73-09756

HUD SEWER POLICY IN URBAN RENEWAL AREAS,
Paterson Redevelopment Agency, N.J.
F. J. Blesso.
Journal of the Urban Planning and Development Division, Proceedings of the American Society of Civil Engineers, Vol 98, No UP1, July 1972. p 33-43.

Descriptors: *Urban renewal, *Sewerage, *Financing, Comprehensive planning, Water quality control, Legislation.

HUD sewer policy in urban renewal areas has been an important influence in the construction of sanitary and storm sewers since 1949. The Housing Act of 1949 which established the basic urban renewal program provided grants for many facilities including sewers. Since then there have been many revisions to the program and several related water and sewer programs. One key policy statement was Local Public Agency Letter 415 (1967) which covered such issues as secondary treatment, separation of combined sewers, construction costs, and extensions of the sewer system beyond the renewal area. The main benefit of the policy and program has been the construction of badly needed sewer systems, while a key drawback has been the complexity of the grant procedures and the adherence to federal requirements that may produce second-best local solutions. Under revenue sharing procedures many of the drawbacks can be eliminated. Urban renewal programs should be used to explore new and innovative technology for sewer design and construction. (Elfers-North Carolina)
W73-09820

REGIONAL BENEFITS IN FEDERAL PROJECT EVALUATIONS,
Michigan Univ., Ann Arbor. School of Natural Resources.
For primary bibliographic entry see Field 06B.
W73-09830

MARINE POLLUTION,
For primary bibliographic entry see Field 05G.
W73-09905

REVIEW ON THE STATE OF MARINE POLLUTION IN THE MEDITERRANEAN SEA.
For primary bibliographic entry see Field 05B.
W73-09913

ENGINEERING COMMITTEE ON OCEANIC RESOURCES PROCEEDINGS OF THE FIRST GENERAL ASSEMBLY.

(1972). 100 p.

Descriptors: Engineering, Resources, *International commissions, *Oceans, *Environmental control, United States.
Identifiers: *Marine affairs.

Selected problems of governmental management of the coasll zone are discussed. The major legal problem encountered by plans for management of the coastal zone is the Fourteenth Amendment's prohibition against the taking of life, liberty, or property without due process of law. This problem is exemplified in legislative efforts in several states to preserve the ecology of coastal wetlands. Comprehensive coastal planning, called for in recent wetlands acts, should prevent most private owners from later complaining about restrictions on the use of their property. For those cases which are challenged, the regulatory authorities should be prepared to muster evidence that in its natural state, as regulated, the property has some significant utility to the owner or evidence that the proposed modification of the area will cause direct and immediate harm to a legal interest which is conventionally protected. (Ensign-PAI)
W73-10254

REGULATION OF ACTIVITIES AFFECTING BAYS AND ESTUARIES: A PRELIMINARY LEGAL STUDY.
Texas Law Inst. of Coastal and Marine Resources, Houston.

Available from National Technical Information Service, U.S. Dep't. of Commerce as PB 212 902. $3.00 in paper copy; $0.95 in microfiche. March 1972. 25 p. NSF-GT-26.

Descriptors: *Texas, *Coastal structures, *Coastal engineering, *Governmental interrelationships, *Coasts, Beach erosion, Shores, Shore protection, Coastal plains, Bays, Estuaries, Environmental effects, Legal aspects, Dredging, Excavation, Waste disposal, Construction, Canals, Channels, Dredging, Spoil banks, Drainage, Flood control, Wells, Water pollution sources.
Identifiers: *Coastal zone management.

Federal, state, and local agencies which regulate or oversee coastal activities in Texas identified by the Bay and Estuarine Management Study are indicated by charts and graphs in this preliminary study; the study reveals gaps and overlaps in institutional authority supervising coastal zone activities. The particular areas subject to regulation are: liquid waste disposal, gaseous waste disposal, solid waste disposal, offshore and coastal construction, land canals, offshore channels, dredging and soil disposal, excavation, drainage, filling, flood control, wells, fertilizers and biocides, and traversing with vehicles. A summary of selected legislation relating to the coastal zone is also included. (Dunham-Florida)
W73-10256

SUMMARY OF SELECTED LEGISLATION RELATING TO THE COASTAL ZONE.
Texas Law Inst. of Coastal and Marine Resources, Houston.

Available from National Technical Information Service, U.S. Dept. of Commerce as PB-212 885. $3.00 in paper copy, $0.95 in microfiche. February 1972. 111 p, 36 ref. NSF-GT-26.

Descriptors: *Legislation, *Governments, *Federal-state water rights conflicts, *Water resources development, Federal jurisdiction, State jurisdiction, Law enforcement, Public rights, Regulation, Competing uses, Governmental interrelations, Coasts, Coastal engineering, Coastal structures, Administration, Comprehensive planning, Federal government, State governments, Water management (Applied), Water pollution control, Environmental effects.
Identifiers: *Coastal zone management, Coastal waters.

A preliminary summary is presented of selected legislation relating to federal and state regulation of the coastal zone, in terms of authorizing legislation, planning, financing, and enforcement which will provide the basis for detailed studies of the areas covered in further reports. The federal government, acting under the commerce clause of the Federal Constitution, has preempted substantially regulations of interstate air transportation; is mainly responsible for irrigation projects under the Federal Reclamation Laws; has the biggest role in flood control, reservoir construction, and channel improvements; and has stricter and more comprehensive legislation regarding air pollution. However, states and their political subdivisions have the primary responsibility in controlling urban runoff; have primary responsibility and authority to manage fish and wildlife; and have retained extensive control over surface transportation. Other areas of legislation studied included those affecting pollution, minerals, air pollution, commercial fishing, recreation, and ports. (Mockler-Florida)
W73-10257

STORAGE FACILITIES (FOR HAZARDOUS SUBSTANCES) REGULATION PRESENTED AT HEARING ON MARCH 21, 1973.
Missouri Clean Water Commission, Jefferson City.
For primary bibliographic entry see Field 05G.
W73-10258

(EMERGENCY) ENFORCEMENT REGULATION (TO PROTECT STATE WATERS) PRESENTED AT HEARING ON MARCH 21, 1973.
Missouri Clean Water Commission, Jefferson City.
For primary bibliographic entry see Field 05G.
W73-10259

MISSOURI WATER POLLUTION CONTROL BOARD REGULATIONS FOR TURBID WATER DISCHARGES, (TO PREVENT INTERMITTENT OR ACCIDENTAL POLLUTION OF WATERS OF THE STATE).
Missouri Water Pollution Board, Jefferson City.
For primary bibliographic entry see Field 05G.
W73-10260

ELIMINATION OF HAZARDOUS OPEN CANALS.

Hearing—Subcomm. on Water and Power Resources—Comm. on Interior and Insular Affairs, U.S. Senate, 92d Cong, 2d Sess, June 5, 1972. 72 p.

Descriptors: *Legislation, *Safety factors, *Canal design, *Open channels, *Cost sharing, Federal government, Canals, Irrigation canals, Concrete-lined canals, Canal construction, Channels, Conduits, Conveyance structures, Ditches, Drains, Engineering structures, Water conveyance, Public health, Safety, Public benefits, Drowning, Hazards, Design, Cost allocation.
Identifiers: *Congressional hearings.

Hearings were held to take testimony on a proposed bill to authorize federal cost sharing in promoting public safety through the elimination of hazardous open canals by converting them to closed conduits and by fencing. The purpose of the legislation would be to establish cost sharing assistance for water using entities in financing protective measures for canals adjacent to urban areas. The Secretary of the Interior would be authorized to contribute up to 50% of the costs of constructing such protective devices. A text of the bill, statements and testimony of witnesses appearing at the hearing and communications relating to the proposed legislation are included for the record. Much of the criticism by opponents to the proposed bill was that the cost to the federal government was not warranted by the claimed

benefits of public health and safety. (Dunham-Florida)
W73-10267

CLOSED BASIN DIVISION, SAN LUIS VALLEY PROJECT, COLORADO.

Hearing--Subcomm. on Water and Power Resources--Comm. on Interior and Insular Affairs, U. S. Senate, 92d Cong. 2d Sess, June 26, 1972. 115 p.

Descriptors: *Legislation, *Groundwater resources, *Multiple-purpose projects, *Groundwater mining, Federal government, Groundwater, Dug wells, Groundwater availability, Groundwater basins, Groundwater potential, Surface-groundwater relationships, Water sources, Water supply, Water wells, Water law, Mexican Water Treaty, Rio Grande River, Project planning, Water management (Applied), Water distribution (Applied), Water resources development.
Identifiers: *Congressional hearings.

Hearings were held in June of 1972 to take testimony on a proposed bill authorizing the construction, operation and maintenance of the Closed Basin division, San Luis Valley project, Colorado. The project is a multipurpose water resource development which, by means of shallow wells and conveyance facilities, would collect groundwater presently being wasted and deliver it to the Rio Grande River for beneficial use. The project would assist in fulfilling Colorado's obligation under the Rio Grande Compact to deliver water at the Colorado-New Mexico state line and in fulfilling U.S. treaty obligations to deliver water to Mexico. A new wildlife refuge would also be established. A text of the bill, statements and testimony of witnesses at the hearing, and communications relating to the proposed project are included. (Dunham-Florida)
W73-10268

THE LAW OF THE SEA CRISIS--AN INTENSIFYING POLARIZATION.

Report--Comm. on Interior and Insular Affairs Part 2, US Senate, 92d Cong, 2d Sess, May 1972. 54 p.

Descriptors: *United Nations, *International law, *Law of the sea, *Governmental interrelations, *Mining, Water law, Governments, Organizations, International waters, Legal aspects, Competing uses, Public rights, Water policy, Mineralogy, Mineral industry, Oceans, Continental shelf, Beds under water.
Identifiers: *Seabed mining, Coastal waters.

This report outlines developments at the March 1972 session of the United Nations Seabed Committee and supplements a prior report on the 1971 session of the same committee. The present law of the sea, major law of the sea issues under present consideration and a history of the UN Seabed Committee are described before a more detailed account of the activities of the 1972 Committee session is presented. Present U.S. rights to mine the ocean floor are in jeopardy due to the increasing militance of a caucus of developing countries of Africa, Asia, and Latin America. Pending legislation in the U.S. Senate may soon need to be acted upon in order to prevent U.S. forfeiture of freedom of the seas, including the right to mine minerals of the seabed lying beyond the continental margin of the U.S. (Dunham-Florida)
W73-10269

BRANTLEY PROJECT.

Hearing--Subcomm. on Water and Power Resources, Comm. on Interior and Insular Affairs, U.S. Senate, 92d Cong, 2d Sess, January 25, 1972. 84 p.

Descriptors: *Flood control, *Legislation, *Multiple-purpose projects, *River basin development, New Mexico, Dams, Diversion, Diversion structures, Erosion control, Flood plains, Flood protection, Reservoirs, Water distribution (Applied), Water management (Applied), Water policy, Project feasibility, Project planning, Project purposes, Administration, Water resources development, Federal government.
Identifiers: *Congressional hearings, *Brantley Project, Pecos River basin.

Hearings were held in January of 1972 to take testimony on a proposed bill authorizing the Secretary of the Interior to construct, operate and maintain the Brantley Project in the Pecos River Basin of New Mexico. The project was recommended by the Bureau of Reclamation to improve existing reclamation projects in that area which, because of reservoir silting and inadequate spillways, created a potential flood hazard. Replacement storage for the existing projects, irrigation water supplies, recreation, and conservation of fish and wildlife resources would also be provided by construction of the Brantley Project. The statements and testimony of interested persons both within and outside the federal government are reported. Communications relating to the project sent to the Senate committee are also included. (Dunham-Florida)
W73-10270

DEVELOPMENT OF HARD MINERAL RESOURCES OF THE DEEP SEABED.

Hearing--Subcomm. on Minerals, Materials, and Fuels--Comm. on Interior and Insular Affairs, U.S. Senate, 92d Cong, 2d Sess, June 2, 1972. 77 p.

Descriptors: *Legislation, *Mining, *Beds, Oceans, Mineral industry, Legal aspects, Water resources development, International law, United Nations, Beds under water, Law of the sea, International waters, Governmental interrelations, Penalties (Legal), Permits.
Identifiers: *Seabed mining, *Congressional hearings, Licenses, Water rights (Non-riparians).

S.2801 would provide the Secretary of the Interior with the authority to promote the conservation and orderly development of the hard mineral resources of the deep seabed, pending adoption of an international regime therefor. A great deal of criticism was directed toward this proposed bill. Some witnesses indicated that the bill would fly in the face of United Nations Resolutions, would negate basic principles of the U.S. Draft Seabed Treaty, would damage the U.S. position in negotiations at the United Nations Seabed Committee, would threaten the world's best present opportunity to move toward the rule of just law, and would injure the interests of the U.S. in peace, world order, the environment and an equitable sharing of the world's resources. Alternatives to the proposed bill would include no exploitation at all, the unregulated and virtually unlimited exploitation by corporations and nations of these resources, or international supervision and issuance of licenses to exploit a certain amount of minerals over a period of years. Overall, the testimony presented recommended defeat of the bill. (Mockler-Florida)
W73-10271

THIRD ANNUAL REPORT OF THE COUNCIL ON ENVIRONMENTAL QUALITY.

Hearing--Subcomm. on Public Health and Environment--Comm. on Interstate and Foreign Commerce, U.S. House of Representatives, 92d Cong, 2d Sess, August 16, 1972. 30 p.

Descriptors: *Water pollution control, *Legislation, *Environmental control, *Pollution abatement, Federal government, Ecology, Environment, Social aspects, Environmental effects, Environmental sanitation, Air pollution, Pollutants, Public health, Radioactive wastes, Toxins, Solid wastes, Waste disposal, Water pollution, Water quality, Natural resources, Comprehensive planning.
Identifiers: *Congressional hearings, *Council on Environmental Quality, Noise pollution, Hazardous substances (Pollution).

This hearing was held to receive testimony of members of the Council on Environmental Quality, (CEQ), with respect to the third annual report on the state of the environment by the CEQ. The subcommittee chairman asked to hear comments relating both to an overview of the total environmental situation in the United States and on the effects of recently passed environmental laws relating to air pollution, solid waste pollution, radiation and noise. Statements and testimony of various CEQ members are included along with additional material submitted for the record by the CEQ. These additional materials relate to proposed toxic waste disposal control act, CEQ cost estimates from 1971 to 1980, pollution control at federal facilities, status of the Army's hybrid combustion engine program, and additional funding to carry out the responsibilities of the National Oil and Hazardous Substances Pollution Contingency Plan. (Dunham-Florida)
W73-10272

PUBLIC ACCESS TO RESERVOIRS TO MEET GROWING RECREATION DEMANDS.

Hearing--Subcomm. of the Comm. on Government Operations, U.S. House of Representatives, 92d Cong, 1st Sess, June 15, 1971. 192 p.

Descriptors: *Federal government, *Reservoir operation, *Recreation facilities, *Public access, Administration, Reservoirs, Recreation, Water management, Administrative agencies, Coordination, Water resources development, Social impact, Recreation demand, Fringe benefits, Water allocation (Policy), Water utilization, Water policy, Public health, Reasonable use, Multiple-purpose projects.
Identifiers: *Congressional hearings.

Hearings held before a House Subcommittee of the Committee on Government Operations indicate that the Nations's population growth, coupled with increased leisure time, has led to an ever-expanding demand for public recreational opportunities, such as fishing, hunting, swimming, boating, and camping. Since reservoirs at federal or federally-aided water resource projects offer an excellent opportunity to meet these mounting recreational demands, the Subcommittee held hearings to determine whether the federal agencies involved in reservoir projects are making these valuable recreational resources available for the benefit of the American public. The subcommittee heard from representatives of the Interior Department, Corps of Engineers; Tennessee Valley Authority; the Soil Conservation Service; the Kansas Forestry, Fish, and Game Commission; and several conservation organizations. The report also indicates the extent of the subcommittee's long and continuing interest in these federal agencies' policies as they relate to public access to reservoirs financially aided by federal funds. (Mockler-Florida)
W73-10273

WESTLANDS WATER DISTRICT.

Hearing--Subcomm. on Water and Power Resources--Comm. on Interior and Insular Affairs, U.S. Senate, 92d Cong, 1st Sess, December 6, 1971. 25 p.

Descriptors: *California, *Legislation, *Land management, *Agriculture, *Government finance, Irrigation, Land use, River basin development, Farm management, Reclamation, Financing, Cost sharing, Water management (Applied).

Identifiers: *Congressional hearings.

H. R. 1682 would provide for deferment of construction charges payable by the Westlands Water District attributable to lands of the Naval Air Station in Lemoore, California. The bill would provide authorization for the Secretary of the Navy to lease lands owned by the federal government for grazing or agricultural purposes in accordance with sound land management practices. The proceeds from the leases would go to reclamation funds. Acreage limitations would be imposed and direct charges for water would be paid by the various lessees to the Westlands Water District on a cost plus expenses basis. In order to assure that the contemplated national defence purposes are served, the bill would give the Secretary of the Navy the authority to include in the leases various provisions on cancellation, use of the land, and other terms of the lease. The lands covered by the bill consist of 19,752 acres, 12,000 of which acres have already been leased for agricultural purposes since 1958. (Mockler-Florida)
W73-10274

SAN ANGELO PROJECT, TEXAS.

Hearing--Subcomm. on Water and Power Resources--Comm. on Interior and Insular Affairs, United States Senate, 92d Cong, 1st Sess, September 9, 1971. 19 p.

Descriptors: *Legislation, *Reservoirs, *Government finance, *Water supply development, Flood control, Planning, Water resources development, Regulation, Water supply, Impoundments, Water yield, *Texas, Water management (Applied), Municipal water, Water shortage, Droughts, Cost sharing.
Identifiers: *Congressional hearings, *San Angelo (Tex).

S.1151 would authorize the Secretary of the Interior to revise a repayment contract with the San Angelo Water Supply Corporation in San Angelo, Texas. Specifically, the bill would provide financial relief to the municipality and extend the repayment contract from 40 to 50 years. A portion of the operating costs would be credited against the corporations debt to flood control, fish and wildlife conservation and make them nonreimbursible, and otherwise use any and all available funds to carry out the project. The Twin Buttes Reservoir is the principal feature of the San Angelo project. Its active capacity is 632,000 acre-feet, of which 177,800 acre-feet are active conservation storage and 454,000 acre-feet are flood control space. The significance of the proposal is due to the fact that the San Angelo area has experienced severe drought in recent years and has not been able to build up a reserve of water for the future. (Mockler-Florida)
W73-10275

UNDERWATER LANDS.

Del. Code Ann. Tit. 7, secs. 6401 thru 6440 (Supp. 1970).

Descriptors: *Delaware, *Legislation, *Mining, *Leases, Mineral industry, Dredging, Drilling, Exploration, Ownership of beds, Exploitation, Oil fields, Investigations, Oil wells, Research and development, Seismic studies, Sites, Subsurface investigations, Water pollution, Water pollution sources, Water pollution control, Oil, Natural gas, Permits, Administrative agencies, Environmental effects.

The Governor and the Water and Air Resources Commission, (Commission), have exclusive jurisdiction to lease for mineral exploration and exploitation all ungranted submerged tidelands owned by the state of Delaware. The Commission is empowered to administer and control all such lands and to lease such lands as provided. Specifically, the Commission is to exercise control over surveys, forms of leases and permits, royalties, rental of leased lands, pollution, joint exploration, and interests in leases. The Commission is authorized to promulgate reasonable rules, regulations, and orders necessary to administer operations to remove minerals from submerged state lands. All leases and permits must be subject to prior federal approval from the Department of Defense. The Commission may delegate the State Geologist to act in its behalf under these provisions. (Dunham-Florida)
W73-10276

WATER POLLUTION.

Del. Code Ann. tit. 7, secs. 6301 thru 6307 (Supp. 1970).

Descriptors: *Delaware, *Legislation, *Water pollution control, *Sewage treatment, Standards, Water pollution, Impaired water quality, Industrial wastes, Municipal wastes, Pollutants, Public health, Sewage, Treatment facilities, Waste treatment, Water pollution sources, Water pollution treatment, Water quality standards, Water resources, Water utilization, Administrative agencies, Governments, Legal aspects, Environmental effects, Adjudication procedure.
Identifiers: Injunctive relief, Administrative regulations.

In carrying out state policy of maintaining a reasonable quality of water consistent with public health and enjoyment, protection of wildlife, and industrial development of the state of Delaware, it is the purpose of these provisions to control any existing pollution and prevent new pollution. The Water and Air Resources Commission, (Commission), is empowered to regulate water pollution through rules, regulations and orders after notice and hearing. The Department of Natural Resources and Environmental Control, (Department), is empowered to exercise supervision over administration and enforcement of state water pollution laws, and may issue special orders after notice and hearing to secure control of pollution of state waters. On request, political entities and industrial concerns are required to furnish the Commission or the Department with information pertinent to the discharge of their duties. The Commission may take summary action prior to a hearing to protect the public health and may seek injunctive relief against violations of its final orders. All plans for construction or alteration of sewage systems must be submitted to the Department for approval prior to commencement. (Dunham-Florida)
W73-10277

REGULATION OF WATER RESOURCES.

Del. Code Ann. tit. 7 secs 6101 thru 6106 (Supp. 1970).

Descriptors: *Delaware, *Legislation, *Water resources development, *Water management (Applied), Water resources, Water rights, Water control, Water distribution (Applied), Water law, Governments, Administration, Administrative agencies, Water pollution, Water pollution control, Water sources, Water quality, Public health, Safety, Natural resources, Environmental effects, Water quality control, Delaware River Basin Commission, Adjudication procedure, Comprehensive planning.
Identifiers: Injunctive relief, Administrative regulations.

In fulfilling a policy to plan for and regulate the water resources of the state to assure optimal beneficial use, the Delaware Water and Air Resources Commission (Commission) shall formulate and adopt a statewide comprehensive master plan for development and use of state water resources after holding public hearings. The Commission is empowered to issue rules, regulations and orders to carry out its duties. The Department of Natural Resources and Environmental Control shall advise the Delaware River Basin Commission, study and investigate matters concerning water use in the state, approve the allocation and use of state waters, and approve plans for water facilities. No increase in the amount of water used within the state shall be made without prior approval of the Department, excepting municipal users supplying nonindustrial subscribers or for emergency purposes. The Department shall designate by regulation those increases which will be approved by simple notice and those requiring a public hearing. The Commission is provided authority to seek injunctive relief for violations of its final orders. (Dunham-Florida)
W73-10278

CONSERVATION--LICENSING POWERS.

Del. Code Ann. tit. 7, secs. 6021 thru 6024 (Supp. 1970).

Descriptors: *Delaware, *Legislation, *Well regulations, *Well permits, *Water wells, Water resources development, Water resources, Water law, Water rights, Water quality, Drilling, Legal aspects, Administrative agencies, Water conservation, Wells, Regulation, Permits, Drill holes, Pumping, Pumps, Water sources, Test wells, Water supply, Governments, Groundwater resources, Groundwater mining.
Identifiers: *Licenses.

No well water contractor shall contract for or engage in the construction, repair or installation of a water well, test wells or pumping equipment until he has received a license from the Delaware Water and Air Resources Commission, (Commission); any activity related to the above shall be under the supervision of a licensed water well contractor. The grant or denial of a license shall be based on an examination to determine whether applicant has sufficient knowledge of the proper method of constructing water wells and the rules and regulations of the Commission pertaining to water wells. The Commission is empowered to enact rules and regulations to effectuate the above provisions. A water well contractor is defined as an individual firm or corporation with whom a contract is made for constructing or repairing a well or test well or installing pumping equipment in a well or test well. (Dunham-Florida)
W73-10279

CONSERVATION--GENERALLY.

Del. Code Ann. tit. 7, secs 6001 thru 6017 (Supp. 1970).

Descriptors: *Delaware, *Legislation, *Water resources development, *Water pollution control, Air environment, Air pollution, Air pollution effects, Water pollution, Water resources, Water law, Water quality, Water management (Applied), Water pollution sources, Environmental effects, Public health, Safety, Conservation, Administrative agencies, Adjudication procedure, Legal review, Permits, Penalties (Legal), Law enforcement, Jurisdiction.
Identifiers: Administrative regulations.

In an attempt to fulfill the policy of preservation and conservation of water and air resources within the state of Delaware, the development, utilization and control of these resources are placed in the hands of the Water and Air Resources Commission, (Commission), and the Department of Natural Resources and Environmental Control, (Department). The organization and procedure for the appointment of Commission members is set forth. The Secretary of the Department is charged with attempting to correct violations of rules or regulations of the Department, and provisions for a public hearing are included in the event the person

charged does not comply. Individuals whose interests are affected by any action or order of the Secretary may appeal to the Commission and then the Superior Court. The Secretary is also authorized to hold hearings with respect to permits, leases or grants. The Commission is authorized to adopt rules and regulations in accord with state policy and issue general and specific orders after public hearings. Again, an appropriate appeal procedure is set forth. The Commission is authorized to issue temporary, emergency orders when no procedure to control hazardous operations exists prior to enacting more permanent procedure. (Dunham-Florida)
W73-10280

DRAINAGE OF LANDS; TAX DITCHES--PRELIMINARY PROVISIONS.

Del. Code Ann. tit. 7, secs. 4101 thru 4104 (Supp. 1970).

Descriptors: *Drainage districts, *Delaware, *Legislation, *Controlled drainage, *Flood control, Drainage area, Drainage effects, Drainage systems, Arable land, Ditches, Land management, Soil water movements, Surface waters, Watersheds (Basins), Irrigation effects, Soil-water-plant relationships, Administrative agencies, Public health, Environmental effects, Flood prevention, Drainage programs.

It is declared that the drainage and prevention of flooding of susceptible lands shall be considered a public benefit and conducive to public health, safety and welfare. This chapter provides a basis for a uniform system for establishing, financing, administering, and maintaining drainage organizations in the state of Delaware under the supervision of the Department of Natural Resources and Environmental Control in order to conserve the state's resources. Drainage means water management, by drainage areas or watersheds, to safely remove or control excess surface flood waters and damaging, excess sub-surface waters. Landowners who desire their lands to be drained or protected from flooding and landowners served by a drainage organization prior to June 1, 1951 established under state law may petition for establishment of a Tax Ditch. If a drainage organization already exists the present assets for liabilities may be transferred to the Tax Ditch as provided. (Dunham-Florida)
W73-10281

NATURAL AND SCENIC RIVERS SYSTEM.
La. Rev. Stat. secs. 56:1841 thru 56:1849 (Supp. 1973).

Descriptors: *Louisiana, *Legislation, *Aesthetics, *Environmental control, *River basin development, Water conservation, Wildlife conservation, Land conservation, Preservation, Protection, Geologic controls, Streams, Administrative agencies, Recreation, Fish management, Scenery, Project planning, Ecology, Natural streams, Stream improvement, Wildlife management.
Identifiers: *Scenic rivers.

This chapter provides for the establishment of the Louisiana Natural and Scenic River System to be administered by the Louisiana Wildlife and Fisheries Commission. The system is to be administered for the purposes of preserving, protecting, developing reclaiming and enhancing the wilderness qualities, scenic beauties and ecological regimen of certain free-flowing streams or segments thereof. The provisions call for full and equal consideration to be given to potential natural and scenic river areas. All river basin reports and project plans should discuss any such potential and all economic evaluations should consider aesthetic as well as monetary values. No state agency may authorize or concur in plans of local or federal agencies that would detrimentally affect a natural or scenic river or upon which full and equal consideration as above stated has not been discussed or evaluated. Channelization, clearing and snagging, channel realignment and reservoir construction of those rivers and streams included within the system is prohibited. Included is a list of natural and scenic rivers. Recognizing that some streams recommended for inclusion may be privately owned, the legislature encourages such owners to grant to the system administrator scenic easements and surface easements. (Napolitano-Florida)
W73-10282

PROHIBITIONS ON CERTAIN DETERGENTS.

Ind. Ann. Stat. secs 35-5101 thru 35-5107 (Supp. 1972).

Descriptors: *Indiana, *Legislation, *Detergents, *Alkylbenzene sulfonates, *Phosphorus, Water pollution control, Water pollution, Water pollution sources, Chemical wastes, Chemicals, Soaps, Sulfonates, Phosphates, Water pollution treatment, Water quality, Water quality control, Permits, Legal aspects, Penalties (Legal), State governments, Administrative agencies, Administrative decisions, Water policy, Nutrients.
Identifiers: Injunctive relief.

It is unlawful to use, sell or otherwise dispose of any detergent containing alkylbenzene sulfonate (ABS) within the state of Indiana or into the boundary waters of this state. Prior to January 1, 1973, it is unlawful to use, sell or dispose of detergent containing more than 8.7% phosphorus by weight as determined by current applicable standards; after that date it shall be unlawful to use, sell or dispose of detergent containing any amount phosphorus by weight. An exception as to the effective date is provided for institutional or commercial users. These provisions do not apply to a detergent or cleaning compound contained in fuel or lubricating oil. The Stream Pollution Control Board shall enforce these provisions and is authorized to adopt reasonable rules and regulations consistent with this chapter. Penalties, including the availability of injunctive relief, are provided for violations of these provisions. (Dunham-Florida)
W73-10283

PROSPECTING AND DEVELOPMENT OF PETROLEUM PRODUCTS IN RIVER BEDS, LAKE BEDS, AND SHORELANDS--PREFERENTIAL RIGHTS.

Ind. Ann. Stat. secs. 46-1620 (Supp. 1972).

Descriptors: *Indiana, *Exploration, *Legislation, *Oil, Natural resources, Natural gas, Banks, River beds, Lake beds, Drilling, Oil industry, Subsurface investigation, Shores, Land use, Oil wells, Administration, Legal aspects, Permits, Regulation, Leases, Adjacent landowners, Contracts.
Identifiers: *Petroleum products.

In the case of any application for prospecting permits or leases to river beds, lake beds, and shorelands, the owners of the right to prospect for and develop and produce petroleum from the abutting lands shall have a preferential right for a period of 30 days, after he has received notice of such application, to a prospecting permit, or if petroleum has been discovered in commercial quantities, in any structure underlying such abutting lands, a lease to the portion of such river bed, lake bed and shorelines as adjoin such abutting lands upon the terms and conditions provided. The authorized state agency dealing with these applications has the power to withold any tracts of public land from prospecting or leasing for petroleum purposes if the best interest of the state will be served by doing so. All leases, permits or other contracts of public lands for the development of oil and gas entered into by the state or any of its political subdivisions subsequent to April 11, 1944, are declared to be in full force and effect. (Reed-Florida)
W73-10284

PLUGGING WELLS--DECLARATION OF POLICY, AND ADMINISTRATIVE REGULATION.

Ind. Ann. Stat. secs.33 thru 46-1740 (Supp. 1972).

Descriptors: *Indiana, *Water pollution control, *Well regulations, *Oil wells, *Legislation, Oil spills, Drilling, Injection wells, Test wells, Saline water intrusion, Exploration, Oil industry, Legal aspects, Penalties (Legal), Water pollution, Subsurface investigation, Legal review, Administrative decisions, Water conservation.
Identifiers: *Well plugging, Administrative regulations.

If, after notice and hearing, the Indiana Department of National Resources finds that a well drilled for the exploration, development, or production of oil or gas, or as an injection or salt water disposal well, geological structure test well, gas storage well or gas storage observation well, has been abandoned and is leaking or may result in the leaking of deleterious substances into any fresh water formation or onto the surface of the land in the vicinity of the well, and if the situation has not been remedied within a specified time, then the department or any person or agency duly authorized by the department may repair or plug the well. In the case of an emergency no notice or hearing is required. Any person who acts pursuant to the authority granted in the statute shall not be held liable for any damages resulting from reasonably necessary operations, nor shall he be held liable for future remedial work. Any person who acted under this statute shall have a cause of action, against the person who was obligated by law to repair the well, for the reasonable costs and expenses incurred for such repairs. (Reed-Florida)
W73-10285

FERRIES ON STREAMS BORDERING STATE.

Ind. Ann. Stat. secs. 36-2506 thru 36-2512 (Supp. 1972).

Descriptors: *Indiana, *Legislation, *Ferry boats, *Navigation, *Permits, Transportation, Adjudication procedure, Penalties (Legal), Regulation, State governments, Legal aspects, Local governments.
Identifiers: *Licenses.

No person or corporation is permitted to maintain a public ferry across any lake, river or stream running through or bounding any city or town in the state of Indiana without first having obtained a license from the common council or board of town trustees of the proper city or town for that purpose. The requirements that must be met prior to the granting of the license are specified herein. The rates of ferriage, reasonable rules for operation and the schedule upon which such ferry shall be operated is to be fixed by the common council or board of town trustees. If the ferrykeeper, or any user of such ferry, be aggrieved by the establishing of such rates, rules or schedule, he has the right to appeal to the circuit court of the proper county. The statute provides a penalty for the violation of the conditions of the license. (Reed-Florida)
W73-10286

AN ACT TO PROVIDE FURTHER FOR WATER POLLUTION CONTROL, ESTABLISHING A NEW WATER IMPROVEMENT COMMISSION AND PRESCRIBING ITS JURISDICTION,

POWERS, AND DUTIES.

Act No. 1260, Acts of Alabama, p. 2175-2190, 1971.

Descriptors: *Alabama, *Water quality standards, *Permits, *Water pollution control, *Legislation, Water quality control, Public health, Administrative agencies, Pollution abatement, Administrative decisions, Waste disposal, Law enforcement, Legal review, Penalties (Legal), Damages, Federal Water Pollution Control Act.
Identifiers: Notice.

The Water Improvement Commission is charged with controlling water pollution within Alabama. Water quality standards shall be set following public notice and a hearing by the Commission. Such standards must take into account present and future uses, the public interest and other policies relating to existing or proposed future pollution. The Commission shall issue permits for all waste discharges into waters of the state. Any person discharging, or applying to discharge pollution into waters of the state may be required to maintain records, file reports and use monitoring equipment. Only the Commission, the Attorney General or a District Attorney may bring an action to enforce this statute. Those found in violation of the act may be liable for criminal fines and for civil damages covering clean up costs as well as expenses for restocking any fish killed or replenishing any wildlife killed. Provisions are included on membership of the Commission and judicial review of orders or decisions of the Commission. (Mockler-Florida)
W73-10287

TOMBIGBEE VALLEY DEVELOPMENT AUTHORITY.

Ala. Code tit. 38, secs. 126 (1) thru 126 (15) (Supp. 1971).

Descriptors: *Alabama, *Legislation, *Inland waterways, *Bond issues, *Administrative agencies, Channels, Surface waterways, Regulation, Administrative decisions, Water management (Applied), Comprehensive planning, Water resources, Water supply, Water supply development, Water utilization, Governments, Surface water availability, Conservation, Agriculture, Industry, Recreation, Tennessee River, Transportation.
Identifiers: *Tombigbee River.

To implement the construction of a navigable waterway between Demopolis and the Tennessee River utilizing the channel of the Tombigbee River to promote water conservation and supply, flood control, irrigation, industrial development and public recreation in the Tombigbee River basin, the legislature authorizes the formation of a public corporation. The corporation, in conjuction with other state and federal agencies, is authorized to undertake all obligations and perform all actions necessary to complete the project. Provisions for the establishment and administration of the corporation are provided, and the corporation is authorized to issue bonds and collect the proceeds therefrom. A provision for dissolution of the public corporation when no duties or obligations remain is included. (Dunham-Florida)
W73-10288

DISCHARGE OF SEWAGE AND LITTER FROM WATERCRAFT.

Ala. Code tit. 38, secs. 97 (36) thru 97 (47) (Supp. 1971).

Descriptors: *Alabama, *Legislation, *Recreation wastes, *Water pollution sources, *Boats, Administrative agencies, Boating, Litter, Wastes, Disposal, Environmental sanitation, Waste disposal, Water pollution, Water quality, Public health, Sewage disposal, Recreation, Water utilization, Water policy, Safety, Recreation facilities, Law enforcement, Legal aspects, Penalties (Legal), Ecology.
Identifiers: Marine toilets, Hazardous substances (Pollution), Nuisance (Legal aspects).

No person shall deposit or discharge into the waters of this state materials from watercraft detrimental to the public health or to the enjoyment of water for recreational purposes. The Board of Health is authorized to formulate rules and regulations to effectively carry out these provisions. Provisions specifically related to marine toilets, on-shore trash receptacles, and a public education program are included. To enforce rules, regulations and orders of the Board of Health, provisions for prosecution and penalties are provided. Definitions of watercraft, sewage, litter, and marine toilet are listed. (Dunham-Florida)
W73-10289

SEWAGE COLLECTION, ETC., FACILITIES.

Ala. Code tit. 22, secs. 140 (13) thru 140 (18) (Supp. 1971).

Descriptors: *Alabama, *Legislation, *Sewage disposal, *Sewage treatment, Administrative agencies, Regulation, Permits, Administrative decisions, Project planning, Legal aspects, Penalties (Legal), Sewage, Sanitary engineering, Sewage districts, Water pollution, Water pollution sources, Wastes, Sewerage, Cities, Plumbing, Public health, Ecology, Environmental effects, Water management (Applied).

It is unlawful to build, maintain, or use insanitary sewage facilities within the State excluding plumbing within structures located within and regulated by municipal corporations. The State Board of Health and/or county boards are empowered to require proper sewage facilities conforming with specifications, rules and regulations adopted by these boards. All plans and specifications applying to sewage collection, treatment, and disposal must be approved by the state and/or county boards in conformance with their standards. Municipal corporations are empowered to issue permits relating to the installation within structures located in their jurisdiction; but permits relating to the installation of plumbing serving structures outside their jurisdictions must be issued in accordance with state and/or county board requirements. Penalties are provided for violation of these provisions. (Dunham-Florida)
W73-10290

ALABAMA WATER WELL STANDARDS.
Ala. Code Tit. 22, secs 140 (19) thru 140 (29) (Supp. 1971).

Descriptors: *Alabama, *Legislation, *Water wells, *Well permits, *Well regulations, Legal aspects, Permits, Regulation, Drilling, Wells, Casings, Water conservation, Drill holes, Excavation, Groundwater, Pumping, Well spacing, Penalties (Legal), Law enforcement, Administrative agencies, Administrative decisions, Administration, Governments, Water management (Applied).
Identifiers: *Licenses, Administrative regulations.

Every person who intends to drill water wells in Alabama shall annually apply for and obtain a license. The Alabama Water Wells Standard Board is empowered to promulgate and publish rules and regulations in this regard; other provisions included herein set forth the Board's membership and authority. No person may attempt to drill or repair a water well unless under the supervision of a licensed driller. The criteria for refusal, suspension, revocation and renewal of licenses are provided with a requirement of Board notice and a hearing on the matter if requested. Any person aggrieved by an act or decision of the Board may appeal to the circuit court. Each application for a well driller's license is to be accompanied by a surety bond for $1,000. Penalties for violation of these provisions or for violation of rules and regulations established by the Board are included. (Dunham-Florida)
W73-10291

ALLISON V. FROEHLKE (CHALLENGE TO ENVIRONMENTAL IMPACT STATEMENT OF RESERVOIR PROJECT).

470 F.2d 1123-1128 (5th Cir. 1972).

Descriptors: *Texas, *Judicial decisions, *Environmental effects, *Dam construction, Dams, Reservoirs, Water law, Water policy, Legal aspects, Legislation, Remedies, Flood control, Eminent domain, Wildlife conservation, Wildlife habitats, Wildlife, Legal review.
Identifiers: *Environmental Impact Statements, *National Environmental Policy Act.

Plaintiff landowners sought a permanent injunction against the construction of three reservoirs and dams in the Brazos River Basin by the defendant Army Corps of Engineers. The plaintiffs contended that the project as presently planned had not been so authorized by Congress and violated specific environmental statutes, including the National Environmental Policy Act. The Fifth Circuit Court of Appeals affirmed the lower court's denial of the injunction and held that the project was authorized, even though the plans did not conform exactly to those contained in the original Congressional approval. Further, neither the fact that the environmental impact statement, (EIS), relating to the project failed to refer to certain specific animals, nor the absence of a botanist, zoologist, demographer or economist on the staff writing the EIS renders it statutorily defective. (Glickman-Florida)
W73-10292

HARELSON V. PARISH OF EAST BATON (SUIT FOR DAMAGES FOR PARISH'S FAILURE TO COMPLETE DRANAGE PROJECT).

272 So. 2d 382 (Ct. App. La. 1973).

Descriptors: *Louisiana, *Drainage, *Spoil banks, *Damages, Drainage engineering, Drainage programs, Local governments, Water law, Water policy, Legal aspects, Judicial decisions, Easements, Remedies, Erosion control, Farm management, Erosion, Slope protection.

Plaintiff landowner granted a drainage servitude to the defendant parish for the purposes of widening a major drainage artery. Defendant allegedly breached the contract by not distributing spoil from the excavation in accordance with the agreement; the spoil was to be distributed so as to prevent erosion of the banks. The trial court held for the plaintiff, and the appellate court affirmed but modified the decision. It awarded damages first in the amount that it would cost the plaintiffs to regrade the channel bank to the condition in which it should have been delivered to them, and additional damages for the defendant's failure to burn brush piles created during the performance of the widening task. Finally plaintiffs were awarded damages for the loss of use of their entire tract for pasture purposes because of the defendant's failure to properly fence and grade the project in accordance with the contract. (Glickman-Florida)
W73-10293

BOARD OF TRUSTEES OF THE INTERNAL IMPROVEMENT TRUST FUND V. MADEIRA BEACH NOMINEE, INC. (OWNERSHIP OF LAND ACCRETED DUE TO PUBLIC IMPROVEMENT PROJECT).
272 So. 2d 209 (Fla. App. Ct. 1973).

Descriptors: *Florida, *Riparian rights, *Accretion (Legal aspects), *Water rights, Sea walls, Water law, Water policy, Legal aspects, Riparian waters, Bank erosion, Boundaries (Property), Retaining walls, Shore protection, Erosion control, Erosion, Judicial decisions, Riparian land, Boundary disputes.
Identifiers: Public trust doctrine.

Plaintiff state agency sued to enjoin the construction of a seawall by defendant riparian landowner on land accreted due to a municipal erosion control project. Defendant cross-claimed to quiet title to the accreted land. Plaintiff contended that defendant did not become the owner of land which accreted due to a public improvement project. The Florida District Court Appeals held that title to accreted lands vests in the riparian owners of abutting lands unless the riparian himself causes the accretion. The court also refused to give retroactive application to a Florida statute which purported to vest title to accretions caused by public works in the state. Thus, the judgment of the trial court was affirmed and the title to the accreted land was vested in defendant. However, the court noted that if the state could show that erection of the seawall would endanger the erosion control project, an injunction would be proper to prevent its construction. (Glickman-Florida)
W73-10294

ADAM V. GREAT LAKES DREDGE AND DOCK COMPANY (SUIT TO RECOVER DAMAGES FOR DESTRUCTION OF OYSTER BED).

273 So. 2d 60 (Ct. App. La. 1973).

Descriptors: *Louisiana, *Oysters, *Dredging, *Beds under water, Shellfish farming, Streambeds, Beds, Judicial decisions, Water law, Water policy, Legal aspects, Civil law, Water rights.

Plaintiff owner of a lease of certain water bottoms for oyster harvesting, sued defendants, a dredging company, the Parish owning the waterway dredged, and the state of Louisiana, from whom he had obtained the oyster lease, for damages resulting from trespassing upon the leasehold and destroying the oysters. The complaint was dismissed as to the state by the trial court, but on appeal it was held that a cause of action against the state was stated by the allegation that the plaintiff had obtained from the state the lease of water bottoms for oyster purposes, and that the dredging was done by a third party under the supervision of the state. The court therefore remanded the case with the state as a party defendant. (Glickman-Florida)
W73-10295

CALE V. WANAMAKER (ACTION FOR DECLARATORY JUDGMENT GRANTING RIGHT OF WAY OVER DEFENDANTS' LANDS WHERE THE ONLY ACCESS TO PLAINTIFFS' PROPERTY WAS BY NAVIGABLE WATER).
296 A.2d 329 (N.J. 1972).

Descriptors: *New Jersey, *Easements, *Right-of-way, *Judicial decisions, Legal aspects, Bays, Adjacent landowners, Boundaries, Boundary disputes, Riparian lands.

Plaintiffs owned a lot surrounded by lots belonging to defendants and a bay. They sought a declaratory judgment to determine whether they had a right-of-way over defendants' land, either by way of necessity or by an implied grant of quasi-easement. Plaintiffs had used as access to their lot the land of a former railroad right-of-way, but that use was restrained, and the plaintiffs only lawful means of access was by way of water. The court held that an implied grant of quasi-easement did not exist, because there was no pre-existing use upon which to support it. However, they further decided that a way of necessity does not rest on a pre-existing use, but on a need for a way across property. As long as there had at one time been a unity of title, a way of necessity may pass through several transfers and be exercised at any time. The court found an implied way of necessity across defendants' land in favor of plaintiffs. (Glickman-Florida)
W73-10296

MUNICIPAL WATER AND SEWER AUTHORITY OF CENTER TOWNSHIP, BUTLER COUNTY V. NORTHVUE WATER COMPANY, INC. (ACTION TO DIRECT SEWER AND WATER AUTHORITY TO SUPPLY WATER TO UTILITY COMPANY).
298 A.2d 677 (Commonwealth Ct. Pa. 1972).

Descriptors: *Pennsylvania, *Water districts, *Water supply, *Adjudication procedure, Legal aspects, Rates, Administrative decisions, Local governments, Water law, Water allocation (Policy), Equity, Water rates, Municipal water, Water demand, Legislation, Judicial decisions.

Complainant public utility water company filed a complaint requesting a mandatory injunction directing defendant municipal sewer and water authority to supply necessary water at reasonable rates. The state department of health had ordered the complainant to find another water supply; complainant applied to and was refused service by the defendant. The appeals court held that the complainant's request that defendant supply water to complainant for service to complainant's customers within the municipality came within the meaning of the word service in the applicable governing statute. This statute provided procedures for challenging the reasonableness of the defendant's services and the court held complainant had to follow this exclusive procedure provided for in the statute and could not in the alternative bring a suit in equity in order to rectify the situation. (Mockler-Florida)
W73-10297

PROPERTY OWNERS ASSOCIATION OF BALTIMORE CITY, INC. V. MAYOR AND CITY COUNCIL OF BALTIMORE (LANDLORD'S SUIT TO HAVE WATER SERVICE IN TENANT'S NAME AND NOT HIS OWN).
299 A.2d 824 (Ct. App. Md. 1973).

Descriptors: *Water supply, *Maryland, *Water rights, *Local governments, Legal aspects, Water utilization, Rates, Water law, Judicial decisions, Municipal water, Water costs, Water rates.

This case involved an action on a bill of complaint brought by plaintiff landlord and others against the defendant city seeking a declaratory judgment with respect to an agreement of a tenant to pay directly to the city for all water charges. Plaintiff also sought a writ of mandamus against the mayor or the city council requiring the placement of the tenant's name on the water account and the mailing of water bills in the name of the tenant. The appeals court held that where a landlord stood in the shoes of an owner who had signed and filed with the city bureau of water supply an application for water supply service before execution of the agreement with the tenant, the city's decision not to place the name of such tenant on the water bills but rather to deal with the landlord was not unreasonable and did not operate to discriminate against the tenant. The landlord's action was supported by other landlords in the city who sought the same relief. (Mockler-Florida)
W73-10298

CITY OF TEXARKANA V. TAYLOR (ACTION FOR DAMAGES CAUSED BY SEWAGE BACKUP).
490 S.W.2d 191 (Ct. Civ. App. Texas 1973).

Descriptors: *Texas, *Judicial decisions, *Local governments, *Sewage districts, Negligence, Legal aspects, Water law, Water policy, Accidents, Sewers, Governments, Municipal wastes, Sewage.
Identifiers: *Sovereign immunity, *Nuisance (Legal aspects).

Plaintiffs were owners of a residence and of an apartment house that were damaged when the defendant city's sanitary sewer system backed up. The plaintiffs contended that the defendant was negligent in the operation of a segment of the sewer system and the jury found for the plaintiffs. The appellate court reversed however, and held that a municipality was immune from liability for negligence in the performance of governmental functions, and that maintenance of a sanitary sewer system was clearly a governmental function. While it was recognized that governmental immunity did not apply to private nuisances, that doctrine was held merely to be a classification of a type of damage inflicted, and not a cause of damage as was negligence. Private nuisance therefore did not provide a grounds for plaintiffs recovery. (Glickman-Florida)
W73-10299

CAPT. SOMA BOAT LINE, INC. V. CITY OF WISCONSIN DELLS (PUBLIC RIGHT TO UNOBSTRUCTED NAVIGATION IN NAVIGABLE RIVER).

203 N.W.2d 369 (Wis. 1973).

Descriptors: *Wisconsin, *Legislation, *Public rights, *Navigable waters, *Navigation, Bridges, Judicial decisions, Legal aspects, Water law, Water policy, Recreation, Common law, Constitutional law, Local governments, Water rights.
Identifiers: *Obstruction to navigation, *Water rights (Non-riparians), Injunctive relief.

Plaintiff, a boat tourline owner sought to abate the maintenance of a bridge by the defendant city as a public nuisance. Plaintiff contended that the bridge was never authorized to be constructed or maintained by law and constituted an obstruction and hazard to the use of a navigable stream. The trial court granted defendant a motion for summary judgment and the Wisconsin Supreme Court affirmed the decision. The court held that the statute plaintiff sought to enforce applied to bridges constructed by individuals and not by cities. Further, the proper remedy under that statute against the maintenance of a bridge was a forfeiture of liquidated damages, and not abatement. The court went on to note that the state held in trust the public right to the unobstructed use of navigable waters for recreation and commercial use, but stated that the legislature could authorize municipalities to construct bridges, provided that the bridges maintained did not obstruct navigable waters. The judgment entered in this case did not preclude the plaintiff from pursuing any other available remedy. (Glickman-Florida)
W73-10300

DRAINAGE DISTRICTS, COMMISSION COMPOSITION AND POWERS.

Ark. Stat. Ann. secs. 21-505.1 thru 21-553 (Supp. 1971).

Descriptors: *Arkansas, *Legislation, *Drainage districts, *Drainage area, Levee districts, Governments, Federal government, State governments, Bond issues, Financing, Contracts, Legal aspects, Procurement, Bids, Performance, Administrative agencies, Governmental interrelations, Administrative decisions, Legal review.

The membership of the Drainage District Board of Commissioners for any drainage district may be increased from three to five by a majority vote of

the real property owners of the district. This can be accomplished by presenting a petition to the Court of the district signed by more than 50% of the real property owners. The number of commissioners may also be increased by order of the court having jurisdiction in the drainage district. The terms of office shall be 9 years for Commissioners in three-men districts and 10 years for those in five-men districts. The Board of Commissioners shall have control of the construction of improvements in their districts. They may advertise for contract bids. If the work is to be done in cooperation with the federal government, the federal agency involved shall let and carry out the contracts. All contractors shall be required to give bond to insure the performance of their contracts. The Board may borrow money and issue bonds to pay for the improvements. (Glickman-Florida)
W73-10301

REGIONAL WATER DISTRIBUTION DISTRICTS.

Ark. Stat. Ann. secs. 21-1407 thru 21-1408 (Supp. 1971).

Descriptors: *Arkansas, *Water district, *Districts (Water), *Administrative agencies, *Legislation, Governments, Local governments, Water law, Water policy, Financing, Bond issues, Budgeting, Capital, Loans, Interest, Financial feasibility, Economic efficiency, Water utilization, Water management (Applied), Project planning.

All powers granted a water district under this act shall be executed by a board of directors composed of three registered voters of that district. The directors shall be elected by the voters of the district as part of the General Election. Each district shall have power to borrow money and otherwise contract indebtedness, to issue its obligations therefor, and to secure the payment thereof by mortgage. The bonds must be issued by resolution of the board of directors. The directors may also fix, regulate and collect rates, fees, rents or other charges for water or any other facilities or services furnished by the water district. If any district distributes water to consumers outside of the district, their fees shall be calculated to pay the cost of such distribution. (Glickman-Florida)
W73-10302

WATER CONSERVATION COMMISSION.

Ark. Stat. Ann. secs. 21-1302 thru 21-1331 (Supp. 1971).

Descriptors: *Arkansas, *Legislation, *Dams, *Water conservation, Impoundments, Dam construction, Administrative agencies, Regulation, Legal aspects, Water management (Applied), Water policy, Water resources, Water resources development, Land resources, Land development, Natural resources, Bond issues, Legal review, Permits, Water utilization, Water allocation (Policy).
Identifiers: Administrative regulations.

The Water Conservation Commission (Commission) shall have the power to issue permits for the construction of dams to impound water on streams. Before any person shall have the right to construct or own a dam or impound water, he must obtain a permit from the Commission. Before the permit is granted, the Commission will hold a hearing to see if there are any objections to the issuance of the permit. However, any person owning land or having the right to possess land shall have the right to impound water on it without obtaining a permit. Any person affected by any rule or order made by the Commission may go to court to contest it. Any person diverting water from any stream, lake or pond must register with the Commission. The Soil and Water Conservation Commission has the duty of formulating and engaging in a comprehensive program for the development and management of the state's water and related land resources. The Commission may establish and maintain a fund to pay for any water development project. The Commission may also issue revenue bonds to finance projects. (Glickman-Florida)
W73-10303

WATER WELL CONSTRUCTION ACT.

Ark. Stat. Ann. sec. 21-2001 thru 21-2020 (Supp. 1971).

Descriptors: *Arkansas, *Public health, *Water wells, *Legislation, *Water supply development, Regulation, Water sources, Wells, Water supply, Water resources, Water management (Applied), Water law, Legal aspects, Water policy, Natural resources, Water yield improvement, Water utilization, Permits, Public benefits, Administrative agencies, Water resources development, Legal review.
Identifiers: *Licenses, *Administrative regulations, Emergency powers.

In that there is an increasing demand for water necessitating the construction of water wells, it is imperative that underground water supplies be developed in such a way as to protect the general health, safety and welfare while still providing sufficient potable water supplies. Because it is essential that those engaged in waterwell drilling cooperate with the state in such development, a Committee on Water Well Construction, (Committee), is established to oversee such cooperation. The Committee shall adopt rules and regulations governing the construction, repair and abandonment of water wells. The Committee is authorized to inspect and either approve or disapprove any abandoned or operative water well. All water well contractors shall be required to have a rig permit issued by the Committee before any rig shall be operated. Further, every person who wishes to engage in the business of water well contracting must obtain a license from the Committee to conduct such business. Procedural requirements for obtaining permits, licenses and for registration are set forth along with provisions dealing with the membership and administration of the Committee. Penalties for violation of this chapter are included. (Glickman-Florida)
W73-10304

ARKANSAS RIVER BASIN COMPACT.

Ark. Stat. Ann. sec. 21-2101 (Supp. 1971).

Descriptors: *Legislation, *Arkansas, *Oklahoma, *Water resources development, *Interstate compacts, Water yield, Administrative agencies, Water utilization, Watershed management, Watersheds (Basins), Water law, Legal aspect, Water policy, Water pollution abatement, Water resources, River basin development, Water allocation (Policy), Water management (Applied), Interstate commissions, Equitable apportionment, Regulation.
Identifiers: *Arkansas River.

Arkansas and Oklahoma have agreed on a compact respecting the waters of the Arkansas River and its tributaries. The compact's major purposes are to fairly apportion the waters of the Arkansas River between the two states and to promote its development; to provide an agency to apportion the water to be known as the Arkansas-Oklahoma Arkansas River Compact Commission (Commission); to encourage the maintenance of a pollution abatement program in the two states; and to facilitate cooperation of water administrative agencies in both states to develop and manage the water resources of the Arkansas River Basin. Each state may construct and maintain for its own needs water storage reservoirs in the state. The Commission shall have the power to adopt rules and regulations governing its operations and implementing this compact, as well as authority to enforce those rules. It shall also collect and analyze data on stream flows and water quality and continue research for developing means of effective pollution abatement and determining total basin yields. (Glickman-Florida)
W73-10305

NAVIGABLE WATERWAYS, RIVER CROSSING PROPRIETORSHIP.

Ark. Stat. Ann. secs. 73-2201 thru 73-2216 (Supp. 1971).

Descriptors: *Arkansas, *Legislation, *Navigable waters, *Bridge, Water law, Legal aspects, Water policy, Ownership of beds, Navigable rivers, Structures, Land tenure, Legal review, Jurisdiction, Administrative agencies, Judicial decisions.
Identifiers: *Navigable water crossing, Administrative regulations, Administrative hearings.

The Public Service Commission, (Commission), shall have jurisdiction over all navigable water crossings. It shall have the power to require that a crossing be constructed or operated in a manner consistent with public safety and that causes no interference with some paramount use of the Navigable Waterway. Any person who wants to operate a crossing must file a petition for approval with the Commission. The Commission shall then hold a hearing to determine if there are valid objections to the petition, and if there are none, shall grant it. Rights granted to the River Crossing Proprietor are perpetual. The applicant shall be required to pay all costs and expenses of a proceeding under this Act. Any party to this proceeding may file a petition in the Circuit Court to have the decision reviewed, and then appeal the final action from there to the Supreme Court of Arkansas. (Glickman-Florida)
W73-10306

CESSPOOLS, PRIVY WELLS AND DRAINAGE SYSTEMS.

Del. Code Ann. tit. 16, sec. 1506 (Supp. 1970).

Descriptors: *Delaware, *Cesspools, *Legislation, *Drainage systems, *Underground waste disposal, Sumps, Water control, Water management (Applied), Water sources, Legal aspects, Water law, Water policy, Administrative agencies, Public health, Drains, Water pollution control, Waste water disposal, Drainage wells, Waste water (Pollution), Sewage disposal.
Identifiers: Privy wells.

Before construction or alteration of cesspools, privy wells or drainage systems, the plans for these water supply systems shall be submitted to the Delaware Department of Natural Resources and Environmental Control for its approval. That Department shall not grant approval if the Department of Health and Social Services disapproves the plans within thirty days of their submission. (Glickman-Florida)
W73-10307

WATERCOURSES; DIVERSION, OBSTRUCTION, OR POLLUTION AS CONSTITUTING TRESPASS.

105 Ga. Code Ann. secs. 1407 thru 1410 (Supp. 1972).

Descriptors: *Georgia, *Watercourses (Legal aspects), *Legislation, *Trespass, Natural streams, Perennial streams, Non-navigable waters, Legal aspects, Water law, Water policy, Riparian rights, Right-of-Way, Water pollution, Diversion, Obstruction to flow, Reasonable use, Relative rights, Diversion structures, Groundwater, Base flow.

Identifiers: Natural flow doctrine.

The owner of land through which nonnavigable watercourses flow is entitled to have that water come to his land in its natural and usual flow subject only to such lessening as may be caused by a reasonable use of it by other riparian owners. Diverting the stream obstructing it so as to impede its course or cause it to overflow, or polluting it shall be a trespass on that landowner's property. Trespass may not be brought for any supposed interference with the rights of a proprietor of an underground stream because of the difficulties of ascertaining its course. However, the owner of realty owns title upward and downward indefinitely, and any unlawful interference with those rights gives him a cause of action. Unlawful interference with a right of way is a trespass against the party entitled to it. (Glickman-Florida)
W73-10308

WATER SUPPLY QUALITY CONTROL.

88Ga. Code Ann. secs. 2601 thru 2618 (Supp. 1972).

Descriptors: *Georgia, *Legislation, *Public health, *Water quality control, *Water supply, Legal aspects, Regulation, Water purification, Water pollution, Administrative agencies, Dependable supply, Water tanks, Water storage, Water supply development, Water sources, Water policy, Water law, Storage requirements, Water utilization, Water management (Applied), Water conservation, Water yield.
Identifiers: *Administrative regulations, Certification, Licenses.

The water supplies of Georgia shall be used to the maximum benefit of the people. The Board of Health, (Board), is designated to establish and maintain a water quality control program adequate for present and future needs. The Board shall have the duties to establish water quality standards, as well as policies and standards governing the purification, treatment and storage of water for public water supply systems. The Board has the power to enforce its policies, as well as authorize research to update and better its programs. No person shall operate any public water system without first securing approval from the Board of the source of the water supply, the treatment and storage methods, the specifications for construction of the facility and its operation. Members of the Board shall be permitted access at all reasonable times to any property for the purpose of investigating conditions related to the furnishing of water to the public. In the case of an emergency affecting public health, the Board may order action taken without notice or a hearing. It shall be a misdemeanor to violate the Board's rulings. (Glickman-Florida)
W73-10309

6G. Ecologic Impact of Water Development

SOME PROBLEMS IN THE CONSTRUCTION OF RESERVOIRS IN THE USSR (NEKOTORYYE PROBLEMY SOZDANIYA VODOKHRANILISHCH V SSSR),
Akademiya Nauk SSSR, Moscow. Institut Vodnykh Problem.
A. B. Avakyan.
Akademiya Nauk SSSR, Izvestiya, Seriya Geograficheskaya, No 4, p 56-63, July-August 1972. 10 ref.

Descriptors: *Reservoirs, *Reservoir construction, Reservoir operation, Environmental effects, Water users, Economics, Agriculture, Land use, Irrigation, Water supply, Flood control, Electric power, Fish management, Navigation, Regulation.
Identifiers: *USSR.

Arguments against the view that reservoirs are generally undesirable are based on the contention that the positive impact of manmade lakes has received less attention than the negative effects. The development of a set of principles is urged to guide future reservoir development. These principles would evaluate reservoirs as complex structures with both positive and negative impacts on the environment and on various sectors of the economy. Techniques are needed to predict these impacts and to calculate the overall economic effect. Expansion of reservoir-focused research will require the organization of permanent research stations and field expeditions at selected reservoirs in a variety of physical settings and the use of these reservoirs for experiments designed to test theoretically sound proposals and recommendations. (Josefson-USGS)
W73-09803

URBAN PLANNING AND DEVELOPMENT OF RIVER CITIES,
Vollmer Associates, New York.
A. H. Vollmer.
Journal of the Urban Planning and Development Division, American Society of Civil Engineers, Vol 96, No UP1, p 59-64, March 1970.

Descriptors: *Planning, *Urbanization, Recreation, Rivers, Land use, Water quality Aesthetics.
Identifiers: *Riverfront planning, *Riverfronts, Waterfronts.

A general discussion of the reasons cities have located on rivers and the evolution of their riverfront areas is presented. Cities have located on rivers largely for transportation and trade purposes and the waterfronts have been predominantly used for industrial and commercial activities. However, as our life style and economy have changed, many of these areas have become obsolete and deteriorating. With our present concern for the environment and demands for water-based recreation there arises significant conflicts over the use of the river and its riverfront land. Considerations of aesthetics and compatible functions must be key factors in future planning for urban waterfront areas. The main objective should be to meet the new demands for the urban river yet allow for order and meaning. (Elfers-North Carolina)
W73-09819

BARTOW MAINTENANCE DREDGING: AN ENVIRONMENTAL APPROACH,
For primary bibliographic entry see Field 05G.
W73-09881

THE SAVANNAH RIVER PLANT SITE,
Savannah River Ecology Lab., Aiken, S.C.
T. M. Langley, and W. L. Marter.
Available from NTIS, Springfield, Va., as Report No. DP-1323; $3 per copy, $0.95 microfiche. Report No. DP-1323, Jan. 1973. 175 p, 44 fig, 21 tab, 66 ref, 9 append.

Descriptors: *Sites, *South Carolina, *Savannah River, Laboratories, *Research facilities, Administrative agencies, Water supply, Geology, Hydrology, Meteorology, Biology, Ecology, Streams, Surface waters, Groundwater, Animal groupings, Plant groupings, Earthquakes, Atlantic Ocean.
Identifiers: *Savannah River Plant (AEC).

On June 20, 1972, the Atomic Energy Commission designated nearly 200,000 acres of land near Aiken, South Carolina, as the nation's first National Environmental Research Park. The designated land surrounds the AEC Savannah River Plant production complex. The site, which borders the Savannah River for 22 miles, includes swampland, pine forests, abandoned town sites, a large man-made lake for cooling water impoundment, fields, streams, and watersheds. The geological, hydrological, meteorological, and biological characteristics of the Savannah River Plant site are described. Discussed in detail are the streams, ground and surface water, climatology (including hurricanes), earthquakes, soil, forests, and wildlife. (Houser-ORNL)
W73-10167

FINAL ENVIRONMENTAL STATEMENT RELATED TO OPERATION OF SALEM NUCLEAR GENERATING STATION UNITS 1 AND 2.
Directorate of Licensing (AEC), Washington, D.C.
For primary bibliographic entry see Field 05B.
W73-10177

FINAL ENVIRONMENTAL STATEMENT RELATED TO OPERATION OF PEACH BOTTOM ATOMIC POWER STATION UNITS 2 AND 3.
Directorate of Licensing (AEC), Washington, D.C.
For primary bibliographic entry see Field 05B.
W73-10178

FINAL ENVIRONMENTAL STATEMENT RELATED TO THE PROPOSED SAN ONOFRE NUCLEAR GENERATING STATION - UNITS 2 AND 3.
Directorate of Licensing (AEC), Washington, D.C.
For primary bibliographic entry see Field 05B.
W73-10179

FINAL ENVIRONMENTAL STATEMENT RELATED TO CONTINUATION OF CONSTRUCTION AND OPERATION OF UNITS 1 AND 2 AND CONSTRUCTION OF UNITS 3 AND 4 OF NORTH ANNA POWER STATION.
Directorate of Licensing (AEC), Washington, D.C.
For primary bibliographic entry see Field 05B.
W73-10205

SIMON RUN WATERSHED PROJECT, POTTAWATTAMIE COUNTY, IOWA (FINAL ENVIRONMENTAL STATEMENT).
Soil Conservation Service, Des Moines, Iowa.
For primary bibliographic entry see Field 04D.
W73-10263

RESTORATION OF HISTORIC LOW FLOW DIVERSION TO ARROYO COLORADO; LOWER RIO GRANDE FLOOD CONTROL PROJECT, TEXAS (FINAL ENVIRONMENTAL IMPACT STATEMENT).
Corps of Engineers, Dallas, Tex. Southwestern Div.
For primary bibliographic entry see Field 04A.
W73-10264

DREDGING AND FILLING, COWIKEE STATE PARK, LAKEPOINT RESORT, WALTER F. GEORGE LAKE, CHATTAHOOCHEE RIVER, ALABAMA (FINAL ENVIRONMENTAL IMPACT STATEMENT).
Army Engineer District, Mobile, Ala.
For primary bibliographic entry see Field 04A.
W73-10265

CONTROL OF EURASIAN WATERMILFOIL (MYRIOPHYLLUM SPICATUM L.) IN TVA RESERVOIRS (FINAL ENVIRONMENTAL IMPACT STATEMENT).
Tennessee Valley Authority, Chattanooga.
For primary bibliographic entry see Field 04A.
W73-10266

SHORT-RANGE FORECAST OF MAXIMUM FLOOD DISCHARGES ON THE KYZYLSU RIVER AND ITS TRIBUTARIES (KARATKOSROCHNYY PROGNOZ MAKSIMAL'NYKH RASKHODOV PAVODKOV NA R. KYZYLSU I YEYE PRITOKAKH),
For primary bibliographic entry see Field 04A.
W73-09809

A SYSTEM FOR REMOTE TELEMETRY OF WATER QUALITY DATA,
Maine Univ., Orono. Dept. of Electrical Engineering.
W. E. Daniels, and W. W. Turner.
In: Water Quality Instrumentation, Vol 1; Selected Papers from International Symposia presented by Instrument Society of America: Pittsburgh, Penn, Instrument Society of America, p 166-175, 1972. 9 fig, 2 ref. OWRR B-003-ME (5), Me R 1088-12 and Me R 1094-12.

Descriptors: *Water quality, *Instrumentation, *Telemetry, *Monitoring, Sampling, Data collections, Water analysis, Water pollution, Dissolved oxygen, Water temperature, Salinity.

An instrumentation system capable of measuring dissolved oxygen, temperature, and salinity at a remote location and transmitting this information to a central computer center can be modified for use in either fresh or ocean water or in measuring other parameters. All the measuring equipment is in a 10-foot decked boat anchored at the desired site. Water is pumped from the various depths in sequence, and all measurements for a particular depth are made simultaneously at the surface by appropriate sensors. The sensor outputs are used to modulate an audio frequency tone associated with each parameter. In addition the frequency of another audio tone is changed each time the pump changes to a new depth. These tones are summed and the composite signal frequency modulates a radio transmitter. The radio signal is received by a shore station and demodulated, and the regained composite audio signal is sent by leased telephone line to the computer center. The audio signal is then filtered to produce separate audio frequencies. These frequencies are demodulated and the resulting d-c signals are sent to a computer through appropriate converters. (Knapp-USGS)
W73-09946

THE METHODS OF EXPLORATION OF KARST UNDERGROUND WATER FLOWS IN POPOVO POLJE, YUGOSLAVIA,
Trebisnica Hydrosystem, Trebinje (Yugoslavia).
For primary bibliographic entry see Field 02F.
W73-09953

BASIC STUDIES IN HYDROLOGY AND C-14 AND H-3 MEASUREMENTS,
Niedersaechsisches Landesamt fuer Bodenforschung, Hanover (West Germany).
For primary bibliographic entry see Field 02F.
W73-09966

MICROWAVE RADIOMETRIC INVESTIGATIONS OF SNOWPACKS--FINAL REPORT,
Aerojet-General Corp., El Monte, Calif. Microwave Div.
For primary bibliographic entry see Field 02C.
W73-09995

DISTRIBUTION OF THE PHYTOPHYLLIC INVERTEBRATES AND METHODS OF THEIR QUANTITATIVE EVALUATION. (I), (IN RUSSIAN),
Akademiya Nauk URSR, Kiev. Instytut Hidrobiologii.
For primary bibliographic entry see Field 02I.
W73-10007

SEA ICE OBSERVATION BY MEANS OF SATELLITE,
Alaska Univ., College. Geophysical Inst.
G. Wendler.
Journal of Geophysical Research, Vol 78, No 9, p 1427-1448, March 20, 1973. 8 fig, 1 tab, 32 ref.

Descriptors: *Sea ice, *Remote sensing, *Mapping, Albedo, Surveys, Satellites (Artificial), Arctic, Alaska, Aerial photography.

Ice conditions were examined in detail by satellite imagery in a small area (about 120 km X 150 km) of the Arctic Ocean off the shore of north Alaska. Brightness was distinguished by computer for each data point. To suppress the transient cloudiness, minimum brightness composites were used. The ice conditions could be mapped in great detail for 5-day periods for the summer of 1969. The ice conditions depend strongly on the wind direction. Offshore wind moved the ice out, and wind toward the shore brought it back. A good correlation was found with data taken from the weather maps, although the data from a ground-based climatological station were not always in good agreement with the ice movement. Comparison with ice charts mapped by more conventional methods showed good agreement in most cases. The satellite picture, however, gave much greater detail. Monthly mean albedo maps were constructed from the brightness composites for the 4 months from mid-May to mid-September. (Knapp-USGS)
W73-10044

USE OF INFRARED AERIAL PHOTOGRAPHY IN IDENTIFYING AREAS WITH EXCESS MOISTURE AND GROUNDWATER OUTFLOW (OB ISPOL'ZOVANII INFRAKRASNOY AEROS'YEMKI PRI VYYAVLENII UCHASTKOV IZBYTOCHNOGO UVLAZHNENIYA I VYKHODOV PODZEMNYKH VOD),
Nauchno-Issledovatelskii Institut Geologii Arktiki, Leningrad (USSR).
B. V. Shilin, N. A. Gusev, and Ye Ya Karizhenskiy.
Sovetskaya Geologiya, No 1, p 155-160, January 1971. 7 fig, 8 ref.

Descriptors: *Aerial photography, *Infrared radiation, *Excess water (Soils), *Discharge (Water), *Groundwater, Hydrogeology, Soil moisture, Moisture content, Analytical techniques.
Identifiers: *USSR.

Examples are given of diverse natural and technical conditions for the use of infrared aerial photography in hydrogeological investigations. Refinements in techniques stem from the need to determine optimal conditions for the application of infrared photography to hydrogeology in different geological regions and climatic zones. Application of the techniques is especially important in locating groundwater in desert and semidesert zones. (Josefson-USGS)
W73-10054

TRACER RELATIONS IN MIXED LAKES IN NON-STEADY STATE,
Weizmann Inst. of Science, Rehovot (Israel). Isotope Dept.
For primary bibliographic entry see Field 02H.
W73-10059

AERIAL EXPERIMENTS OF RADIOMETRIC DETECTION OF HYDROCARBON DEPOSITS IN THE SEA, (EXPERIENCES AERIENNES DE DETECTION RADIOMETRIQUE DE NAPPES D'HYDROCARBURES EN MER),
For primary bibliographic entry see Field 05A.
W73-10148

USE OF TRITIATED WATER FOR DETERMINATION OF PLANT TRANSPIRATION AND BIOMASS UNDER FIELD CONDITIONS,
Argonne National Lab., Ill.

For primary bibliographic entry see Field 02D.
W73-10190

STUDY ON THE SENSITIVITY OF SOIL MOISTURE MEASUREMENTS USING RADIATION TECHNIQUES, NYLON RESISTANCE UNITS AND A PRECISE LYSIMETER SYSTEM,
Ghent Rijksuniversiteit (Belgium). Dept. of Agricultural Science.
For primary bibliographic entry see Field 02G.
W73-10194

MONITORING THE WATER UTILIZATION EFFICIENCY OF VARIOUS IRRIGATED AND NON-IRRIGATED CROPS, USING NEUTRON MOISTURE PROBES (CONTROLE DE L'UTILISATION ET DE L'EFFICIENCE DE L'EAU CHEZ DIVERSES CULTURES IRRIGUEES OU NON, AU MOYEN D'HUMIDIMETRES A NEUTRONS),
For primary bibliographic entry see Field 03F.
W73-10195

OPTIMISATION OF SAMPLING FOR DETERMINING WATER BALANCES IN LAND UNDER CROPS BY MEANS OF NEUTRON MOISTURE METERS (OPTIMISATION DE L'ECHANTILLONNAGE POUR DETERMINER LES BILANS HYDRIQUES SOUS CULTURE AU MOYEN DE L'HUMIDIMETRE A NEUTRONS),
Institut Technique des Cereales et Fourrages, Paris (France).
For primary bibliographic entry see Field 03F.
W73-10196

GROUND MOISTURE MEASUREMENT IN AGRICULTURE (BODENFEUCHTEMESSUNG IN DER LANDWIRTSCHAFT),
European Atomic Energy Community, Brussels (Belgium). Bureau Eurisotop.
For primary bibliographic entry see Field 03F.
W73-10209

MATHEMATICAL MODELING OF SOIL TEMPERATURE,
Arizona Univ., Tucson. Office of Arid-Lands Research.
For primary bibliographic entry see Field 02G.
W73-10213

A STUDY OF THE COST EFFECTIVENESS OF REMOTE SENSING SYSTEMS FOR OCEAN SLICK DETECTION AND CLASSIFICATION,
New Hampshire Univ, Durham. Dept. of Electrical Engineering.
For primary bibliographic entry see Field 05A.
W73-10228

7C. Evaluation, Processing and Publication

TRANSMISSION INTERCHANGE CAPABILITY--ANALYSIS BY COMPUTER,
Commonwealth Edison Co., Chicago, Ill.
For primary bibliographic entry see Field 08C.
W73-09764

THE DIGITAL PHOTO MAP,
Soft Ware Oriental Co. Ltd., Tokyo (Japan).
R. Kamiya.
Photogramm Eng., Vol 38, No 10, p 985-988, Oct 1972. 2 fig.

Descriptors: *Topographic mapping, *Photogrammetry, *Mapping, Design data, Contours, Maps, Topography, Terrain analysis, Photography, Aerial photography, Data processing.
Identifiers: *Photomaps, Data preparation, Automatic plotters, Computer-aided design, Stereoscopic map plotters, Analog to digital converters, Photographic analysis.

Use of a new type of photographic map--a digital photo map--is proposed instead of a topographic contour map for preparing layout designs. This proposal allows designers to use original photographs with essential digital information, eliminating the need for contour maps. In stereoscopic observation of a digital photo map, a stereoscopic model of the terrain and a grid mesh in 3 dimensions closely following the terrain surface can be seen. The ground elevations are indicated digitally at each grid point. The designer can draw a freehand design on the photographs under stereoscopic observation, and the digital figures required can be easily obtained from terrain data stored in the computer. The major operations required in making a digital photo map are very simple and almost automatic, requiring very little operator skill. A flow diagram showing principal operations in making a digital photo map is included. (USBR)
W73-09767

LOW-FLOW CHARACTERISTICS OF STREAMS IN THE PACIFIC SLOPE BASINS AND LOWER COLUMBIA RIVER BASIN, WASHINGTON,
Geological Survey, Tacoma, Wash.
For primary bibliographic entry see Field 02E.
W73-09786

STREAM CATALOG--PUGET SOUND AND CHEHALIS REGION,
Washington State Dept. of Fisheries, Olympia.
R. W. Williams, R. M. Laramie, and J. J. Ames.
Available from NTIS, Springfield, Va. 22151 COM-72-11326; Price $3.00 printed copy; $0.95 microfiche. Annual Progress Report March 1972. 71 p. NOAA 14-17-0001-2156.

Descriptors: *Fisheries, *Salmon, *Streams, *Washington, *Habitats, Water quality, Tributaries, Indexing, Data collections, Mapping, Hydrologic data, River basins, Physical properties, Land use, Water utilization, Spawning, Environmental effects, Ecology.
Identifiers: *Stream Catalog, *Puget Sound (Wash), *Chehalis Region (Wash).

A Stream Catalog of the Puget Sound and Chehalis Basins, Washington, has been compiled in a rough draft form by the Aquatic Environmental Management Section of the Washington Department of Fisheries. This catalog brings together all available physical descriptions and data concerning the individual rivers, tributary streams, and adjacent lands, as well as present status data through 1970, associated with the various salmon species and stream habitat utilization by each species. This interrelationship of stream physical conditions and present status of fish populations for each watershed is information vitally needed for sound management of these natural resources to insure perpetuation under the heavy demands projected beyond 1980. A numbering system for all Puget Sound rivers and streams was developed as an outgrowth of the needs of this catalog. This system also provides a highly useful code index adaptable for IBM identification. All federal, state, county and local water management agencies will have specific needs for this compilation of data in future planning. (Woodard-USGS)
W73-09787

GROUND-WATER FAVORABILITY AND SURFICIAL GEOLOGY OF THE LOWER AROOSTOOK RIVER BASIN, MAINE,
Geological Survey, Washington, D.C.
G. C. Prescott, Jr.
For sale by Geological Survey, price $0.75, Washington, D.C. 20243. Hydrologic Investigation Atlas HA-443, 1972. 1 sheet, 9 fig, 9 ref.

TIME PERIODS AND PARAMETERS IN MATHEMATICAL PROGRAMMING FOR RIVER BASINS,
Massachusetts Inst. of Tech., Cambridge. Ralph M. Parsons Lab. for Water Resources and Hydrodynamics.
For primary bibliographic entry see Field 04A.
W73-10317

REGIONAL ANALYSES OF STREAMFLOW CHARACTERISTICS,
Geological Survey, Washington, D.C.
For primary bibliographic entry see Field 02E.
W73-10323

AQUIFER SIMULATION MODEL FOR USE ON DISK SUPPORTED SMALL COMPUTER SYSTEMS,
Illinois State Water Survey, Urbana.
For primary bibliographic entry see Field 02F.
W73-10328

08. ENGINEERING WORKS

8A. Structures

SEISMIC RISK AND DESIGN RESPONSE SPECTRA,
Massachusetts Inst. of Tech., Cambridge.
For primary bibliographic entry see Field 08G.
W73-09761

SOME PROBLEMS IN THE CONSTRUCTION OF RESERVOIRS IN THE USSR (NEKOTORYYE PROBLEMY SOZDANIYA VODOKHRANILISHCH V SSSR),
Akademiya Nauk SSSR, Moscow. Institut Vodnykh Problem.
For primary bibliographic entry see Field 06G.
W73-09803

AN APPROACH TO ANALYSIS OF ARCH DAMS IN WIDE VALLEYS,
Southampton Univ. (England). Dept. of Civil Engineering.
J. R. Rydzewski, and R. W. Humphries.
Journal of the Power Division, American Society of Civil Engineers, Vol 99, No PO1, Proceedings paper No 9710, p 165-174, May 1973. 8 fig, 4 tab, 9 ref.

Descriptors: *Arch dams, *Embankments, *Stress analysis, *Design, *Water resources, *Computer programs, *Finite element analysis, River flow, Spillways, Valleys, Foundations, Rock fill, Costs, Mathematical models, Systems analysis, Flood plains.
Identifiers: *South Africa, Artificial abutments.

The use of arch dams in combination with earth or rock fill embankments for wide valley sites has arisen from the need to provide adequate spillway capacity on certain rivers in southern Africa. The stress analysis of the resulting structure, consisting of an arch dam with artificial abutments can be achieved by means of finite element techniques. However, cost factors would tend to discourage their use during preliminary stages of the design process. This paper investigates the behavior of several arch dam shapes in wide valleys of different elastic properties and then proceeds to suggest a way in which the Southampton University Arch Dam (SUAD) computer program, based on thinshell approximation of the arch dam, can be extended to cover this type of problem. A comparison with a finite element solution indicates that the SUAD program can be used with confidence for preliminary design studies of arch dams with artificial abutments. Once the final shape of the dam and abutments has been obtained, the total structure should be checked using finite element methods. (Bell-Cornell)
W73-09841

COOLING PONDS - A SURVEY OF THE STATE OF THE ART,
Hanford Engineering Development Lab., Richland, Wash.
For primary bibliographic entry see Field 05D.
W73-09849

HELLER DISCUSSES HYBRID WET/DRY COOLING,
Eotvos Lorand Univ., Budapest (Hungary).
For primary bibliographic entry see Field 05D.
W73-09855

CONTROL OF THE OPERATING REGIMES OF WATER RESOURCES SYSTEMS,
For primary bibliographic entry see Field 04A.
W73-09868

LOCAL SCOUR UNDER OCEAN OUTFALL PIPELINES,
Montgomery (James M.), Inc., Pasadena, Calif.
J. L. Chao, and P. V. Hennessy.
Journal Water Pollution Control Federation, Vol 44, No 7, p 1443-1447, July 1972. 4 fig, 18 ref.

Descriptors: *Waste water disposal, *Outlets, *Pipelines, *Scour, *Currents (Water), Flow, Velocity, Shear stress, Sediments, Particle size, Analytical techniques.

An analytical method is derived for estimating the maximum depth of scour caused by a subsurface current under an ocean outfall pipeline. From the potential and stream functions governing the flow, the velocity in the scour hole and the discharge through the hole per unit length of pipe can be computed. The maximum boundary shear stress can be calculated from the velocity; and the maximum scour depth, by matching boundary shear stress with critical tractive stress of the sand grain size composing the ocean floor. The method provides an order-of-magnitude evaluation of hole depth for analysis of the pipeline's structural stability. (Ensign-PAI)
W73-09902

ENGINEERING COMMITTEE ON OCEANIC RESOURCES PROCEEDINGS OF THE FIRST GENERAL ASSEMBLY.
For primary bibliographic entry see Field 06E.
W73-09914

MOVABLE DAM AND METHOD OF OPERATION,
J. Aubert.
US Patent No 3,718,002, 5 p, 21 fig, 3 ref; Official Gazette of the United States Patent Office, Vol 907, No 4, p 897, February 27, 1973.

Descriptors: *Patents, *Dams, *Engineering structures, Movalbe dams, *Floodgates, Flood control, River regulation, Water control, Surface waters.

A movable dam is discussed which is intended to be built across waterways of different types for the purpose of regulating the water level at predetermined locations upstream of the structure. The dam has a fixed portion which is anchored in the bed of the watercourse and a movable gate system for regulating the cross-sectional area provided for the flow of water through the dam. It differs from others by having at least one closure element which can be regulated under a heavy load and another which can be operated only 'in the dry' or under a so-called light load. (Sinha-OEIS)
W73-09938

RIVER THAMES–REMOVABLE FLOOD BARRIERS,
Bruce White, Wolfe Barry and Partners, London (England).
A. H. Beckett.
Royal Society of London Philosophical Transactions, Mathematical and Physical Sciences, Vol 272, No 1221, p 259-274, May 4, 1972. 16 fig.

Descriptors: *Barriers, *Surges, *Wind tides, *Estuaries, *Gates, *Navigation, Cost-benefit analysis, Flood control, Floodgates.
Identifiers: *Thames estuary (England).

A removable flood barrier in Long Reach of the Thames is the basis of a flood defense system compatible with the navigation interests, yet avoiding the high cost of bank raising in the metropolis. Three barrier designs were developed, each embodying two 150-m-wide navigation openings. The lesser use by shipping of reaches above the Royal Docks permitted narrower openings. Schemes for some six different sites and over 40 variations in span arrangement were investigated and led to a proposal for four 60-m navigation openings in Woolwich Reach which might be closed by a form of rising section gate. This has proved to be the cheapest, most reliable, and quickest to install of all the schemes investigated and is now the basis of design for contract. (Knapp-USGS)
W73-09996

SILTATION PROBLEMS IN RELATION TO THE THAMES BARRIER,
Hydraulics Research Station, Wallingford (England).
M. P. Kendrick.
Royal Society of London Philosophical Transactions, Mathematical and Physical Sciences, Vol 272, No 1221, p 223-243, May 4, 1972. 21 fig, 1 tab, 5 ref.

Descriptors: *Silting, *Estuaries, *Barriers, *Flood control, Hydraulic models, Floodgates, Mathematical models, Tides, Wind tides, Surges, Sedimentation, Sediment transport, Bed load, Suspended load, Silts, Sands.
Identifiers: *Thames estuary (England).

A well-designed storm-surge barrier in the Thames used only to exclude surges should produce no insuperable siltation problems. Both mathematical and physical model results show that continuous tide control leads to increased siltation, the zone of greatest deposition depending on the barrier site. Methods adopted in tackling the problem include four large-scale estuarine surveys; the establishment of stations for continuously monitoring the suspended solids content of the river; field and laboratory tests to determine the properties of Thames silt; the development of a mathematical model to study the effect of the tide control on the distribution of silt which is in the navigation channel likely to follow barrier construction and continuous half-tide control. Current velocity, suspended silt concentration, salinity, and temperature at different depths along the estuary change throughout a spring and a neap tide during both high and low river flows. (Knapp-USGS)
W73-09997

POLLUTION PROBLEMS IN RELATION TO THE THAMES BARRIER,
Water Pollution Research Lab., Stevenage (England).
M. J. Barrett, and B. M. Mollowney.
Royal Society of London Philosophical Transactions, Mathematical and Physical Sciences, Vol 272, No 1221, p 213-221, May 4, 1972. 10 fig, 5 ref.

Descriptors: *Path of pollutants, *Estuaries, *Tidal streams, *Barriers, Flow control, Water level fluctuations, Streamflow, Tides, Water pollution control, Mathematical models, Hydraulic models, Surges, Wind tides, Saline water intrusion, Floodgates.
Identifiers: *Thames estuary (England).

Although the storm-surge barrier on the Thames, which is to be constructed about 15 km seaward of London Bridge, need only be closed to exclude exceptionally high tides to fulfill its primary function as a flood-prevention device, it could also be operated as a tide-control structure. A theoretical study to assess the effect of operating the barrier on the Thames estuary on a regular basis to prevent water levels from falling below a fixed datum is outlined. Studies on a large-scale physical model indicate that tide control would bring about fundamental changes in the estuary. The tidal range would be reduced, and current velocities would fall to almost zero. There would be a consequent reduction in tidal range and hence, in the tidal excursion. There would also be changes in salinity. The upstream limit of the saltwater would be displaced some 4 km farther seaward. When tide control was discontinued the original salinity distribution would quickly be reestablished. A one-dimensional, time-dependent numerical model, which encompasses the ebb and flow of the tide, was also developed. (Knapp-USGS)
W73-09998

NAVIGATION ON THE THAMES,
Port of London Authority (England).
G. R. Rees, and P. F. C. Satow.
Royal Society of London Philosophical Transactions, Mathematical and Physical Sciences, Vol 272, No 1221, p 201-212, May 4, 1972. 4 fig.

Descriptors: *Surges, *Wind tides, *Estuaries, *Barriers, *Flood control, Floods, Flood protection, Gates, Storms, Seiches, Tides, Navigation, Floodgates.
Identifiers: *Thames estuary (England).

Navigation on the Thames will be affected by the size of the opening of the storm-tide barrier. Shipping needs the widest possible channel, and any constriction should be as far upriver as practicable. During construction, special arrangements for navigation will be required to ensure the minimum of disruption to traffic. The closure of the barrier will halt all vessels bound through it and is likely to delay them for up to 12 hours. It will also have some effect on vessels bound for docks below the barrier. Special arrangements will be required to enable all such vessels to wait. Acceptance of the Woolwich site was governed by the forecast of the likely rate of closures to be initially no more than 2 per year. With opening widths of 61 m, a long straight approach is needed for navigation, and the piers of the barrier must be aligned to the direction of the tidal stream. To minimize disruption of navigation, the level of the sill must be low enough to ensure that vessels are not hindered. Sill level also needs to be considered in relation to the regime of the river. Use of the barrier other than for flood control could create problems of siltation and pollution, and it is essential before using the barrier for other than its prime purpose to establish that there is no such possibility. (Knapp-USGS)
W73-09999

CURRENT PROPOSALS FOR THE THAMES BARRIER AND THE ORGANIZATION OF THE INVESTIGATIONS,
Department of Public Health Engineering, London (England).
R. W. Horner.
Royal Society of London Philosophical Transactions, Mathematical and Physical Sciences, Vol 272, No 1221, p 179-185, May 4, 1972. 5 fig, 5 ref.

Descriptors: *Surges, *Wind tides, *Estuaries, *Barriers, *Flood control, Floods, Flood protection, Gates, Storms, Seiches, Tides, Navigation, Floodgates.
Identifiers: *Thames estuary (England).

A considerable area of London close to the River Thames floods when the water in the river reaches an exceptionally high level. The very high waters due to exceptional meteorological conditions, are steadily increasing in level. The possible ways of meeting this threat are by raising the existing flood defenses, by constructing a barrage across the mouth of the estuary, or by constructing a removable flood barrier to be closed only when a dangerous surge tide is liable to occur. A site in Woolwich Reach was chosen for a flood barrier with four main openings of 61 m. The rising sector gate will be used for the main openings. (Knapp-USGS)
W73-10000

THE REGIONAL SETTING,
Institute of Geological Sciences, London (England).
K. C. Dunham.
Royal Society of London Philosophical Transactions, Mathematical and Physical Sciences, Vol 272, No 1221, p 81-86, May 4, 1972. 2 fig, 20 ref.

Descriptors: *Subsidence, *Sea level, Pleistocene epoch, Floods, Silting, Submergence.
Identifiers: *England.

Post-Pleistocene subsidence is still apparently continuing in the Thames Estuary and London. Holocene times have been a period of subsidence of the land relative to sea level, the total extent being over 20 m. The current rate of subsidence is about 100 mm per century. In areas like the Thames Estuary, the activities of man have undoubtedly affected the situation. (Knapp-USGS)
W73-10002

ACTIVITIES OF THE CORPS OF ENGINEERS IN MARINE POLLUTION CONTROL,
Corps of Engineers, Atlanta, Ga. South Atlantic Div.
For primary bibliographic entry see Field 05G.
W73-10103

SOLAR ENERGY AT OSC, (OPTICAL SCIENCES CENTER).
Arizona Univ., Tucson. Optical Sciences Center.

Optical Sciences Center Newsletter, Vol 6, No 3, p 65-67, December 1972. 4 photos.

Descriptors: *Solar radiation, *Heat transfer, *Thermal conductivity, *Design data, Arizona, Prototype tests, Research and development, Power system operation, *Energy conversion, Energy loss, Energy transfer, Energy budget, Model studies.
Identifiers: *Solar energy collector.

The first thermal test bed model of a solar energy collector on display at the University of Arizona's Optical Sciences Center demonstrates at full scale all essential features of the collector system as designed by A. B. and Marjorie P. Meinel. The model was built by Helio Associates, Inc., and funded through four western utility companies. Its purpose is to establish through experimentation that the concept will work as predicted, and to use it to obtain baseline engineering data on the collecting and heat transfer subsystems. Two basic types of data will be derived from the measurements, a dynamic analysis or determination of heating and cooling rates, and heat transfer characteristics whereby gas transfer and absorber temperatures are determined as a function of the flow rate, pressure, etc. It is hoped that a 10 megawatt experimental solar power farm can be operational near Yuma, Arizona, by 1976. For long-range plans, see W73-10216. (Paylore-Arizona)
W73-10215

A HARVEST OF SOLAR ENERGY,
Arizona Univ., Tucson. Optical Sciences Center.
A. B. Meinel, and M. P. Meinel.
Optical Sciences Center Newsletter, Vol 6, No 3, p 68-75, December 1972. 13 fig.

Descriptors: *Solar radiation, *Heat transfer, *Thermal conductivity, *Design data, Arizona, Prototype tests, Research and development, Power system operation, Energy conversion, Model studies, Environmental engineering, Environmental effects, Water yield improvement, Economic feasibility, Projections, Rainfall-runoff relationships.
Identifiers: Solar energy collectors.

Of several energy options open for research and development, the major contender is solar power farms, and they can be operational by 1985. A thermal conversion system was designed as a prototype that allows surface collectors to reach the high temperatures required by optical thin film coatings that absorb sunlight but prevent heat loss in the infrared. Material used, such as silicon, is opaque (absorbent) to sunlight but transparent to infrared radiation. Costs of such film is now low enough to make their use feasible. To reach desired operating temperature, sunlight is concentrated onto the coatings through lens that focus on a glass pipe and a following series of transfers. In describing a proposed solar energy farm, the designers point out that additional rainfall runoff from extensive collectors can materially improve the ability of an arid land to support growth of grasses of commercial importance to ranchers. Prime land for such a prototype farm appears to be 5000 sq. miles of collectors over the approximately 130,000 sq. miles of desert in the southwestern U.S. and northern Mexico. Environmentally, the energy balance will remain unchanged. If the world is to be prepared for alternative sources of energy by the time conventional sources are depleted, the solar energy farms here proposed may be the answer. (See also W73-10215) (Paylore-Arizona)
W73-10216

8B. Hydraulics

FLOW MEASUREMENT OF PUMP-TURBINES DURING TRANSIENT OPERATION,
Central Research Inst. of Electric Power Industry, Tokyo (Japan).
H. Suzuki, H. Nakabori, and A. Kitajima.
Water Power, Vol 24, No 10, p 366-370, Oct 1972.

Descriptors: *Flow measurement, *Pump turbines, *Discharge (Water), *Unsteady flow, Electronic equipment, Flowmeters, Ultrasonics, Oscillographs, Penstocks, Reynolds number, Hydraulic machinery, Foreign research, Velocity, Fluid mechanics, Instrumentation.
Identifiers: *Ultrasonic flow measurement, Japan, Probes (Instruments).

During transient operations of pump turbines, such as startup, input-rejection when pumping, and changing mode from power generation to pumping, the guidevane opening, rotational speed of the machines, discharge and pressure in the penstock all vary. Recording the changing conditions of the pump-turbine components on site is sometimes desirable to study flow characteristics under transient operating conditions. Since recording is difficult, discharge changes in a penstock have never before been measured in situ. Ultrasonic flowmeters used for steady-state flow measurements were adapted for measuring transient discharge with good results. An electromagnetic oscillograph records output from ultrasonic devices while measuring transient discharges. The theory and measuring apparatus used to determine flow in pump turbines during transient operating conditions are described. (USBR)
W73-09771

VERTICAL TURBULENT MIXING IN STRATIFIED FLOW - A COMPARISON OF PREVIOUS EXPERIMENTS,
California Univ., Berkeley. Coll. of Engineering.
J. E. Nelson.
Waste Heat Management Report No WHM-3, December 1972. 33 p, 3 fig, 8 tab, 21 ref. NSF-EI-34 932.

Descriptors: *Mixing, *Stratified flow, *Mathematical models, Velocity, Density, Hydraulics, Temperature, Momentum transfer, Momentum equation, Turbulence.
Identifiers: *Turbulent mixing, Richardson numbers, Transfer coefficients.

The decrease in both mass and momentum transfer coefficients with increasing stability has been clearly shown. Mathematical formulations to describe changes in the transfer coefficients usually presume only one independent non-dimensioual parameter, having the form of a Richardson number. Large quantities of data have been gathered in both atmospheric and hydrologic environments with a variety of measuring techniques, and mass and momentum transfer coefficients have been calculated over a wide range of Richardson numbers. Unfortunately, the scatter of data is so great that almost no direct correlation can be deduced between the stability as defined by a Richardson number and any transfer coefficient. It is clear that none of the presently available mathematical formulations represent an acceptable match to all the experimental data. Evidently a Richardson number per se does not provide a complete description of the effect of stratification on vertical mixing. (Oleszkiewicz-Vanderbilt)
W73-09852

ANALYSIS OF THE INJECTION OF A HEATED, TURBULENT JET INTO A MOVING MAINSTREAM, WITH EMPHASIS ON A THERMAL DISCHARGE IN A WATERWAY,
Virginia Polytechnic Inst. and State Univ., Blacksburg. Dept. of Aerospace and Ocean Engineering.
For primary bibliographic entry see Field 05B.
W73-09876

HYDRAULIC MODELS PREDICT ENVIRONMENTAL EFFECTS.

World Dredging and Marine Construction, Vol 8, No 13, p 38-42, December 1972. 6 fig.

Descriptors: *Channels, *Estuaries, *Hydraulic models, *Model studies, *Environmental effects, *Hydraulic engineering, Engineering, Facilities, Project purposes.
Identifiers: *Waterways Experiment Station, Vicksburg (Miss), Mobile (Ala), James River, Hudson River.

A general description of the facilities and work of the U.S. Army Engineer Waterways Experiment Station (WES) in Vicksburg, Miss., is given. The WES Hydraulic Laboratory has about 60 hydraulic models in operation at any one time. Among them, the following are mentioned: A model of Mobile Bay, Alabama, to test the effects of the proposed Theodore Ship Channel on the general environment; an estuary model of the James River in the Chesapeake Bay to determine the effect of proposed dredging of the navigation channel; a model for the Federal navigation channels in the Hudson River between the Battery and the George Washington Bridge to study the improvement of present necessary dredging operations. The studies undertaken by WES are of a multidisciplinary character ranging from ecological and biological to engineering aspects of waterways problems. The staff of WES covers a wide range of specialities. A system analysis approach is emphasized in integrating different aspects of a problem, using results from models, field measurements, and other data sources. Instrumentation for monitoring of environmental variables as well as remote sensing and profiling are developed. Different techniques are developed to study particular waterways problems such as tidal modeling, dispersion of effluents, shoaling of channels and protection from hurricane surges. (Ensign-PAI)
W73-09883

LOCAL SCOUR UNDER OCEAN OUTFALL PIPELINES,
Montgomery (James M.), Inc., Pasadena, Calif.
For primary bibliographic entry see Field 08A.
W73-09902

CURRENT PROPOSALS FOR THE THAMES BARRIER AND THE ORGANIZATION OF THE INVESTIGATIONS,
Department of Public Health Engineering, London (England).
For primary bibliographic entry see Field 08A.
W73-10000

NAVIGATION CHANNEL IMPROVEMENT--GASTINEAU CHANNEL, ALASKA,
Army Engineer Waterways Experiment Station, Vicksburg, Miss.
F. A. Herrmann, Jr.
Available from the National Technical Information Service as AD-753 337, $3.00 in paper copy, $0.95 in microfiche. Technical Report H-72-9, November 1972. 170 p, 18 fig, 48 plate, 68 photo, 8 tab.

Descriptors: *Navigation, *Channel improvement, *Shoals, *Erosion control, *Alaska, Model studies, Design criteria, Engineering structures, Dikes, Sediment transport, Sediment distribution, Sediment control, Channels, Hydraulic models, Testing procedures.
Identifiers: *Gastineau Channel (Alaska).

The existing Federal project through the Gastineau Channel, Alaska, provides for a navigation channel 4 ft deep at mllw (including overdepth dredging) with a bottom width of 75 ft. The channel was constructed in 1959-60 through an area with a prevailing bottom elevation of plus 10 to plus 15 mllw and soon experienced rapid shoaling at several locations. No maintenance dredging has been performed, primarily because of the large cost of moving a dredge to this remote area. A model study was conducted to determine the best means of resolving the shoaling problem. The model, constructed to linear scale ratios of 1:500 horizontally and 1:100 vertically, reproduced about 7 miles of Gastineau Channel from Fritz Cove on the west to 1 mile north of Juneau, Alaska, on the east. It was equipped to reproduce and study prototype tides, tidal currents, freshwater inflow, and shoaling. The model tests showed that any one of several impermeable dikes with a top elevation above high water and located along the north side of the navigation channel would reduce shoaling by 80 to 85%. Diversion of Fish Creek away from the navigation channel would result in an additional 5% reduction. The shortest dike tested was 17,250 ft long, and the shoaling reduction for this plan was essentially the same as that for longer dikes. (Woodard-USGS)
W73-10320

FIELD OBSERVATIONS OF EDGE WAVES,
Institute of Coastal Oceanography and Tides, Birkenhead (England).
For primary bibliographic entry see Field 02E.
W73-10327

REGIMES OF FORCED HYDRAULIC JUMP,
Alberta Dept. of Environment, Edmonton. Design and Construction Div.
V. Mura Hari.

Water Resources Bulletin, Vol 9, No 3, p 613-617, June 1973. 1 fig, 8 ref.

Descriptors: *Hydraulic jump, *Supercritical flow, *Critical flow, Turbulent flow, Froude number, Open channel flow, Regime.

The various types of forced hydraulic jumps formed with the assistance of baffle walls and baffle blocks and methods of analysis and research required to understand these flow configurations are briefly discussed. In a laboratory flume, to form a hydraulic jump at a particular location, presence of adequate tailwater is necessary in the basin. In the field, the tailwater depths are due to the incoming supercritical flow rather than being present beforehand. Once a certain discharge is let into a canal or a stream in the form of a supercritical jet, the development of the tailwater depends on the hydraulic properties of the downstream channel. (Knapp-USGS)
W73-10331

FIELD HYDRAULIC STUDIES,
Georgia Inst. of Tech., Atlanta. School of Civil Engineering.
For primary bibliographic entry see Field 05B.
W73-10341

OBSERVED VS. CALCULATED REAERATION CAPACITIES OF SEVERAL STREAMS,
For primary bibliographic entry see Field 05B.
W73-10351

RELATIONSHIPS BETWEEN HYDRAULIC PROPERTIES AND REAERATION,
Georgia Inst. of Tech., Atlanta. School of Civil Engineering.
For primary bibliographic entry see Field 05G.
W73-10352

8C. Hydraulic Machinery

TRANSMISSION INTERCHANGE CAPABILITY--ANALYSIS BY COMPUTER,
Commonwealth Edison Co., Chicago, Ill.
G. L. Landgren, H. L. Terhune, and R. K. Angel.
Inst Electr on Eng Trans Power Appar Syst, Vol PAS-91, No 6, p 2405-2414, Nov-Dec 1972. 10 p, 6 fig, 6 ref, disc.

Descriptors: *Computer programs, Electric power, Energy transfer, Extra high voltage, Electric power production, Transmission (Electrical), Power system operations, Reliability, Transmission lines.
Identifiers: Interchanges, Outages, *Load flow, *Power interchange, Interconnected systems, Overloads.

Until recently, interchange capability has not been widely used, since few high-capacity transmission ties existed between power systems. With the growth in the number of EHV interconnections, as well as generating unit size, the choice of suitable key and limiting facilities for examination of an interchange capability study has become more complex. Many more conditions must be investigated now because of system size and the number of interconnections. These investigations have resulted in large expenditures of computer, engineering, and clerical time. A program developed for determining the transmission system interchange capability and those outages and elements which could limit the transfer of energy between interconnected power systems is discussed. The method has been extensively tested during interchange capability studies of the Mid-America interpool Network (MAIN). The program is an effective planning and operating tool, reducing both engineering and computer time required to analyze interchange limitations, yet maintaining accuracy within a 5% tolerance. (USBR)
W73-09764

OVERVOLTAGES AND HARMONICS ON EHV SYSTEMS,
Westinghouse Electric Corp., Pittsburgh, Pa.
E. R. Taylor.
Inst Electra Electron Eng Trans Power Appar Syst, Vol PAS-91, No 6, p 2537-2544, Nov-Dec 1972. 10 fig, 4 tab, 5 ref, disc.

Descriptors: *Extra high voltage, *Transmission lines, Frequency, Capacitors, Electric currents, Networks, Digital computers, Electrical power, Alternating currents, Transmission (Electrical).
Identifiers: *Overvoltage, *Harmonics, Series compensation, Line charging, Series capacitors.

One system condition that produces steady state 60-hz overvoltages is a relatively long, lightly loaded or unloaded EHV line connected to the system at only one end, the sending end of the line. Under no-load or a lightly loaded system the transmission line acts as a capacitance at the fundamental frequency of 60 hz. The capacitive charging current of the line causes a voltage rise from the system voltage source to the sending end of the line and from the sending end to the receiving end. To reduce the high overvoltages and absorb excess vars produced under these conditions, shunt reactors are usually added to the line. In practice, shunt reactors directly connected to the line usually do not compensate entirely all line capacitance. Therefore, a long shunt compensated line acts as a capacitance to some degree at the 60-hz frequency and may experience high steady state overvoltages in the line when fed by a weak source. Discussed are results of digital and transient network analyzer (Anacom) studies made to analyze the effect of system parameters on harmonics in steady state overvoltages. Factors influencing harmonic distortion of steady state voltages were emphasized. (USBR)
W73-09768

EFFECT OF LOAD CHARACTERISTICS ON THE DYNAMIC STABILITY OF POWER SYSTEMS,
Servicos de Electricidade S.A., Sao Paulo (Brazil).
A. Semlyen.
Inst Electra Electron Eng Trans Power Appar Syst, Vol PAS-91, No 6, p 2295-2304, Nov-Dec 1972. 8 fig, 12 ref, 3 append, disc.

Descriptors: *Power system operation, *Digital computers, *Stability, Dynamics, Electric generators, Electric motors, Equations, Resistance, Foreign research, Electrical impedance, Canada.
Identifiers: *Power loads, Brazil, Reactive loads.

Small signal stability (also called steady state or dynamic stability) is receiving increasing attention from power engineers. Improvement in equipment design, resulting in the introduction of sophisticated control systems and generators with higher reactances, makes knowledge of the limits of stable operation equally important in steady state and under transient conditions. New concepts in control theory, powerful computer systems, and mathematical algorithms have provided accurate models for representing synchronous generators, both in steady state and transient stability studies. Nonsynchronous loads have been considered either as constant impedances, or as static elements whose active and reactive powers are related to voltage through algebraic nonlinear equations. A proposed method based on the state space theory, can be applied to any type of terminal equipment to more accurately represent steady state loads in stability investigations. The influence of dynamic load behavior on the stability limit of a power system is shown. State space methods of application are exemplified. (USBR)
W73-09769

FLOW MEASUREMENT OF PUMP-TURBINES DURING TRANSIENT OPERATION,
Central Research Inst. of Electric Power Industry, Tokyo (Japan).
For primary bibliographic entry see Field 08B.
W73-09771

LARGE SCALE SPRAY COOLING,
Richards of Rockford, Inc., Ill.
For primary bibliographic entry see Field 05D.
W73-09850

MUNICIPAL WATER SYSTEMS - A SOLUTION FOR THERMAL POWER PLANT COOLING,
Washington Univ., Seattle. Dept. of Civil Engineering.
For primary bibliographic entry see Field 05F.
W73-09851

REMOTE CONTROL OF IRRIGATION SYSTEMS,
Akademiya Nauk Kirgizskoi SSR, Frunze. Institut Automatiki i Telemekhaniki.
For primary bibliographic entry see Field 03F.
W73-10214

8D. Soil Mechanics

TARBELA DAM CONSTRUCTION REACHES HALF-WAY MARK,
Tippetts-Abbett-McCarthy-Stratton, New York.
L. A. Lovell, J. Lowe, III, and W. V. Binger.
Water Power, Vol 24, No 9-10, Part I-II, p 317-325, Sept-Oct 1972. 20 p, 18 fig, 8 photo, 2 tab.

Descriptors: *Earth dams, *Dam construction, Rockfill dams, Switchyards (Electrical), Reservoir silting, Hydroelectric plants, Outlet works, Intake structures, Hydraulic gates, Hydraulic structures, Bulkhead gates, Tunnel construction, Dam foundations, Cofferdams, Tunnel linings, Foreign projects, Spillways, Diversion structures.
Identifiers: *Tarbela Dam (Pakistan), *Pakistan, Austrian tunneling method.

Tarbela Dam, the world's largest fill dam, has risen 225 ft of the planned height of 470 ft. The dam design has a cobble gravel and sandy silt sloping core supported on the upstream and downstream sides by well-graded granular shells. A foundation of mainly deep (up to 700 ft) cobble gravel and fine sand necessitated extending the dam core 5700 ft upstream. The usable reservoir capacity of 9.3 million acre-ft is expected to be lost to sediment after 50 or 60 yrs, but the power-generating capacity will remain. Construction began in 1968 with excavation of a 3-mi-long diversion channel, 4 tunnels, and placement of a small section of the embankment located between the diversion channel and the natural river channel. In October 1970, the river was diverted into the diversion channel, and construction began on the entire length of the main embankment, except the section where the diversion channel flows. Future closing of a buttress dam built across the diversion channel will divert the riverflow through the 4 tunnels. Details of the project are discussed, including: the main embankment, auxiliary dams, materials processing and handling, spillways, tunnels, intake and outlet structures, gate operation structures, and mechanical and electrical equipment. (USBR)
W73-09762

DYNAMIC DESIGN OF DAMS FOR EARTHQUAKES,
F. P. Jaecklin, and W. A. Wahler.
Int Civ Eng Vol 2, No 9, p 420-426, Mar 1972. 7 p, 9 fig, 9 ref.

Descriptors: *Dam design, *Seismic design, Earth dams, Earthquakes, Dynamics, Dams, Soil dynamics, Deformation, Vibrations, Structural design.
Identifiers: *Dam ability, Computer-aided design, Earthquake loads, Earthquake-resistant structures, Stability analysis, Seismic stability.

In the past, the pseudostatical method, which assumes a particular horizontal acceleration, was used in earthquake design of dams in Europe and other parts of the world. Dam deformation may be more realistically predicted by considering specific earthquake hazards of the damsite, dynamic strength properties of the foundation and fill, and the dynamic deformation characteristics of the dam. This study shows that earthquakes may induce considerably higher stresses in a dam than any other load, and that a dam may be less safe than predicted by earlier methods of analysis. Particularly unfavorable conditions exist when the natural frequency of the dam is close to the expected earthquake vibration frequencies. Experience gained in the application of this method leads to the conclusion that not only is a higher effective safety ensured, but weak points can be detected in the design stage. This dynamic method influences determination of the face slopes, degree of compaction of the fill, height of the crest above the headrace level, and dimensions of the filter and drainage system. (USBR)
W73-09766

AN APPROACH TO ANALYSIS OF ARCH DAMS IN WIDE VALLEYS,
Southampton Univ. (England). Dept. of Civil Engineering.
For primary bibliographic entry see Field 08A.
W73-09841

RESTORATION OF HISTORIC LOW FLOW DIVERSION TO ARROYO COLORADO; LOWER RIO GRANDE FLOOD CONTROL PROJECT, TEXAS (FINAL ENVIRONMENTAL IMPACT STATEMENT).
Corps of Engineers, Dallas, Tex. Southwestern Div.
For primary bibliographic entry see Field 04A.
W73-10264

8E. Rock Mechanics and Geology

STUDY OF EXPANSION-TYPE ROCK BOLTS IN MINE-GALLERY AND TUNNEL CONSTRUCTION,
For primary bibliographic entry see Field 08H.
W73-09760

8F. Concrete

UNDERWATER CONCRETE CONSTRUCTION,
B. C. Gerwick, Jr.
Mech Eng, Vol 94, No 11, p 29-34, Nov 1972. 6 p, 4 illus, 3 photo, 3 dwg, 6 ref.

Descriptors: *Underwater construction, *Concrete construction, *Concrete placing, Grouting, Concrete mixes, Concrete technology, Construction, Concretes, Reinforcement, Bentonite, Concrete control, Concrete testing, Bonding.
Identifiers: Grout mixes, Underwater structures, Heat of hydration, Laitance, *Tremie concrete.

Underwater concrete construction is reliable and proven. Techniques exist for placement at all depths at which other construction can be performed. Four basic methods described are: tremie, bucket, grout-intruded aggregate, and grouting. The essential element of tremie concrete placement is that the concrete must be discharged beneath the surface of previously placed concrete in a manner that least disturbs that surface. Concrete dropped directly through sea water will segregate into sand and gravel, with the cement going into collodial suspension. High-quality, durable, water-resistant structures can be achieved if procedures are rigidly followed to avoid errors which are far more serious underwater than with conventional dry concrete. Selection of the proper method to use must consider: (1) depth; (2) working conditions, such as rough seas; (3) logistics (supply); (4) degree of confinement in forms; (5) size of pour; (6) fluid to be displaced--sea water or drilling mud; (7) reinforcement and inserts; (8) total volume; and (9) economical factors. Future direction of underwater concrete construction is discussed. (USBR)
W73-09763

WET-TYPE HYPERBOLIC COOLING TOWERS,
Alfred A. Yee and Paul Rogers, Inc., Los Angeles, Calif.
P. Rogers.
Civil Engineering - ASCE, Vol 42, No 5, p 70-72, May 1972. 6 fig.

Descriptors: *Cooling towers, *Design criteria, *Concrete structures, Electric power production, Thermal powerplants, Water cooling, Design, Reinforced concrete, Structural design, Dead loads, Structural engineering, Wind pressure, Model studies, Graphical analysis, Waste water treatment.

Wet-type hyperbolic reinforced concrete cooling towers operate on principles of convection and evaporation. There are several towers in the design or construction stage of 500 feet height and cooling capacities up to 600,000 gallons per minute. The main parts of a hyperbolic cooling tower are the tower shell, the internal fill, the hot water intake, the collecting pond, the cold water return, and auxiliary installations. The hyperbolic shell and its foundation are discussed. The forces considered are the dead load, wind, and seismic forces. Present stress analysis of shells is usually based on the so-called membrane theory in response to seismic stress. The bottom part of the shell acts as a lintel, transferring loads from the continuous shell to the diagonal columns. A simplified graphical method to obtain the maximum compression and tension forces in columns has been developed. Serious attention has been given to the development of steel towers for future use. (Jerome-Vanderbilt)
W73-09857

8G. Materials

SEISMIC RISK AND DESIGN RESPONSE SPECTRA,
Massachusetts Inst. of Tech., Cambridge.
E. H. Vanmarcke, and C. A. Cornell.
Paper ASCE Special Conference on Safety and Reliability of Metal Structure, Pittsburgh, Pa, p 1-25, Nov 1972. 25 p, 7 fig, 19 ref.

Descriptors: *Earthquake engineering, *Seismic design, *Earthquakes, Dynamics, Statistical methods, Probability, Vibration, Risks, Attenuation, Soil dynamics, Structural design, Mathematical analysis, Bibliographies, Seismic studies, Seismic waves.
Identifiers: Earthquake forecasting, Ground amplitude.

The problem of predicting seismic structural response is enlarged by linking the problems of predicting earthquakes and of specifying the characteristics of ground motions at a given site. A method of seismic risk analysis provides a relationship between the ground motion parameters and the average return period. When combined with factors recommended by Newmark's method, the relationship developed can be used to obtain a design response spectrum corresponding to a specified average return period. The design response spectrum can be used directly to predict the peak response in each mode of a linear elastic structure. Using the response spectrum approach, the individual modal maxima are combined, usually by the square root of the sum of the squares, to provide the design peak response of the complete structure. If a time-history analysis or a random vibration approach are used to estimate structural response, then actual design level ground motions or their principal statistical properties (spectral density funtion, duration) must be specified, often using the design response spectrum as a start. The design response spectrum does not uniquely define the frequency content of potentially damaging ground motions. (USBR)
W73-09761

DYNAMIC DESIGN OF DAMS FOR EARTHQUAKES,
For primary bibliographic entry see Field 08D.
W73-09766

A NEW METHOD OF DETECTING LEAKS IN RESERVOIRS OR CANALS USING LABELLED BITUMEN EMULSIONS,
Centre d'Etude de l'Energie Nucleaire, Grenoble (France).
J. Molinari, J. Guizerix, and R. Chambard.
International Atomic Energy Agency, Ser IAEA-SM-129/47, Isot Hydrol, Vienna, Austria, p 743-760, 1970. Transl from Fr, 1971. 11 fig, 6 ref, disc.

Descriptors: *Reservoir leakage, *Canal seepage, *Emulsions, Leakage, Radioisotopes, Iodine radioisotopes, Hydraulic structures, Bituminous materials.
Identifiers: *Radioactive tracers, *Detection, Field permeability test, France, Detectors, Bitumens.

A method of detecting and localizing leaks in natural or artificial lakes, reservoirs, canals, or wells is described. The method consists of injecting an emulsion labeled with a radioactive tracer into the water. The labeled emulsion is entrained into the leakage areas where emulsion particles separate from the water and accumulate. The distribution of these particles, considered as proportional to the specific infiltration flow for structures with interstitial permeability, is then determined by measuring the radioactivity. The plastic properties of bitumen labeled emulsion promote agglomeration of the particles and adhesion to the materials of the revetment or wall to be studied. Preparation and use of a bitumen emulsion labeled with iodine-131 in detecting leaks in a canal are described. This method of leak detection has a wide range of applications, and can be used to study relative permeabilities of strata in which wells, pits, or boreholes have been drilled, and to distinguish between areas of varying permeabilities. (USBR)
W73-09770

COOLING WATER,
Drew Chemical Corp., Parsippany, N.J.
For primary bibliographic entry see Field 05D.
W73-09861

THE EFFECT OF MARINE FOULING ON CORROSION OF METALS ANTI-CORROSION TECHNIQUES (INFLUENCE DE LA SALISSURE MARINE SUR LA CORROSION DES METAUX ET MOYENS DE PREVENTION),
Electricite de France, (Chatou). Service Etudes et Projets Thermiques et Nucleaires.
M. Bureau, and R. Boyer.
Houille Blanche, Vol 205, Special No 2-3, p 189-198, 1972. 8 fig, 1 tab.

Descriptors: *Coasts, *Electric power plants, Water pollution control, Research, *Fouling, *Corrosion, *Erosion, *Pollutants, Inorganic compounds, Sands, Organic wastes, Bacteria, Algae, Crustaceans, Antifouling materials, Coatings, Chlorination, Cleaning, Equipment, Design.

Electricite de France has stepped up research efforts on marine pollution, resulting metal corrosion, and anti-corrosion techniques due to the rapidly growing number of coastal power stations. The effects of inorganic (sand and various debris) and organic (bacteria, crustaceans, algae, and decomposition products) pollutants on corrosion and on corrosion-erosion are discussed. Protective methods aim at preventing pollution or minimizing pollution effects. Chemical protection methods include antifouling coatings and chlorination of circuit water. Continuous mechanical cleaning of tubes, filtration and seawater heating are examples of the physical protection methods. Chlorine dosing and chlorine production methods are also discussed. General plant design, and equipment are presented. (Ensign-PAI)
W73-10146

WHY CORROSION PRODUCTS MUST BE CONTROLLED IN POWER PLANT WATERS,
Battelle-Pacific Northwest Labs. Richland, Wash.
T. F. Demmitt, and A. B. Johnson, Jr.
Available from NTIS, Springfield, Va., as BNWL-SA-4051; $4 in paper copy, $0.95 in microfiche. Report BNWL-SA-4051, Dec 1971, 8 p, 4 ref.

Descriptors: *Corrosion control, *Cost-benefit analysis, Research and development, *Cooling water, Physicochemical properties, Instrumentation, Remote sensing, Separation techniques, Water pollution control, *Waste water treatment, Powerplants, Programs, Forecasting, Thermal pollution.

The desirability is stressed of establishing corrosion-control programs. Emphasis is placed on the nature of corrosion-product films and mechanisms of their formation. Other important studies are the transport processes of corrosion products in coolant streams, including means of removal (ion exchange; filtration; and electrostatic, electromagnetic, and centrifugal separation methods). Operator-alarm instrumentation has benefitted from developments in methods for predicting deposit formation. Prediction of effects from changes in systems materials will benefit from longer-range studies that are more quantitative. (Bopp-ORNL)
W73-10182

8H. Rapid Excavation

STUDY OF EXPANSION-TYPE ROCK BOLTS IN MINE-GALLERY AND TUNNEL CONSTRUCTION,
M. Ladner.
Zurich Federal Institute of Technology Dissertation, No 4197, p 34-42, 1969. Transl from Ger, Bur Reclam Transl 709, May 1971. 13 p, 7 fig.

Descriptors: *Rock bolts, *Roofs, *Tunnel design, *Underground structures, Tunnel construction, Rock excavation, Mining engineering, Tensile strength, Mechanical properties, Prestressing, Rock mechanics, High strength steels.
Identifiers: *Expansion bolts, Switzerland, Foreign studies, Galleries, Rock pressures.

This translation from the German is the third part or conclusions of the report, Study of Expansion-Type Rock Bolts in Mine-Gallery and Tunnel Construction. (The first and second parts, on rock bolting theory and experimental tests, respectively, have not been translated.) Practical applications of theoretical and experimental research are presented. To ensure safety, the bolts and installation must be dependable. Steel of high tensile strength, yield point, and offset yield point is desirable because thinner bolts may be used and holes drilled quicker since diameter may be smaller. Desirable properties of bolts are: small deformation, high load-bearing capacity, and slow loss of capacity over prolonged usage. To obtain these properties, installation precautions which must be taken are: (1) hole diameter must suit the bolt; (2) drilling method used depends on type of rock, (3) bolt should be screwed into the full depth of the cone, and (4) the tip of the bolt should always be in sound rock. The best rock bolt arrangement in a gallery determined from the theoretical investigation of stress conditions around prestressed bolts is exemplified. Use of inclined bolts to reduce stress concentration in a gallery roof is described. (USBR)
W73-09760

8I. Fisheries Engineering

STREAM CATALOG--PUGET SOUND AND CHEHALIS REGION,
Washington State Dept. of Fisheries, Olympia.
For primary bibliographic entry see Field 07C.
W73-09787

RESERVOIR RELEASE RESEARCH PROJECT.
For primary bibliographic entry see Field 05G.
W73-10261

EVALUATION OF PISCICULTURAL QUALITIES OF ROPSHA CARP, (IN RUSSIAN),
K. V. Kryazheva.
Izv Nauchno-Issled Inst Ozern Rechn Rybn Khoz. 74, p 45-54, 1971. English summary.
Identifiers: *Carp (Ropsha), Growth, Hybrids, Piscicultural studies, *Temperature resistance, USSR.

Ropsha carp hybrids from the 5th breeding generation were evaluated. Underyearlings had a slightly better growth rate. In the second yr of life, hybrids and non-hybrids grew at the same rate. Hybrid vigor was observed for such indices as general viability Hb content of the blood for hybrids of the first generation. Amur wild carp were most resistant to high and low temperatures. Ropshi carp differed from hybrids of the first generation by their decreased resistance to heat and increased resistance to cold.--Copyright 1973, Biological Abstracts, Inc.
W73-10355

CULTIVATION OF CARP IN RICE PADDIES OF WESTERN FERGANA AND THE EFFECT OF FISH ON RICE YIELD, (IN RUSSIAN),
For primary bibliographic entry see Field 03F.
W73-10359

FEEDING OF BABY FISH OF THE GRAND AMUDAR SHOVEL-HEADED STURGEON, (IN RUSSIAN),
N. I. Sagitov.
Uzb Biol Zh. Vol 15, No 5, p 64-65, 1971. Illus.
Identifiers: Baby fish, Chironomid larvae, Feeding (Fish), Larvae, *Sturgeon, USSR.

In May-June of 1964, 34 baby fish were obtained. They were, on the average, 53.4 mm long, and 715 mg in weight. Analysis of the content of esophagus and stomach showed that chironomid larvae was their principal food item.--Copyright 1973, Biological Abstracts, Inc.
W73-10360

TRANSPLANTATION OF FISH AND AQUATIC INVERTEBRATES IN 1968, (IN RUSSIAN),
A. F. Karpevich, and N. K. Lukonina.
Vopr Ikhtiol. Vol 12, No 2, p 364-380, 1972.
Identifiers: Aquatic invertebrates, *Salmon, *Smelt, *Transplantation (Fish), *USSR, Fish.

The introduction of fish and invertebrates into bodies of water in the USSR during 1968 is described, mainly in table form. There was some increase in the number of transplanted salmon from 1967, as well as the number of smelt, which had not been handled during the previous year.--Copyright 1973, Biological Abstracts, Inc.
W73-10366

INFORMATION ON THE REPRODUCTION OF CHINOOK SALMON ONCORHYNCHUS TSCHAWYTSCHA (WALBAUM) IN THE KAMCHATKA RIVER, (IN RUSSIAN),
B. B. Vronskii.
Identifiers: Information, *Kamchatka River, Oncorhynchus-tschawytscha, *Reproduction, Rivers, *Salmon (Chinook), USSR.

The spawning grounds of chinook salmon in the Kamchatka River basin were investigated. Data on the dynamics of runs, spawning duration, variations in the sexual composition during runs and reproduction, and behavior of reproducing fishes at the spawning grounds were collected. Spawning was confined to channel portions of flowing systems. Areas with strong vertical current in the subchannel flow associated with certain microreliefs were selected. The spawning period for the majority was associated with lowered water levels after high flood and coincided in different spawning tributaries. The large loss of eggs during laying is counteracted to a certain extent by the high viability of eggs and larvae. The factors limiting the number of chinook salmon are insufficient numbers of spawning grounds and loss of eggs.--Copyright 1973, Biological Abstracts, Inc.
W73-10367

FEEDING OF MIKIZHA SALMO MYKISS WALB. IN SOME KAMCHATKA WATERS, (IN RUSSIAN),
L. V. Kokhmenko.
Vopr Ikhtiol. Vol 12, No 2, p 319-328, 1972. Illus.
Identifiers: Amphibians, Invertebrates, *Kamchatka River, *Mikizha (Feeding), *Salmo-mykiss, Salmon, Smelt, Stickleback, USSR.

The feeding of underyearlings and yearlings in the Kishimshinaya River and of adult fishes in piedmont tributaries of the middle Kamchatka River was investigated. Mikizha have a wide spectrum of food including aquatic terrestrial invertebrates, fish, including salmon, stickleback, smelt and less often char, kundzha, goldfish, plus amphibians and mammals. Salmon eggs are not a significant factor in their diet, though char eggs are. The role of mikizha in the production of salmon stocks is minor due to its relatively small numbers. By feeding on competitors for food, it improves fattening conditions for salmon.--Copyright 1973, Biological Abstracts, Inc.
W73-10368

THE WATER DIVIDER AND REPRODUCTIVE CONDITIONS OF VOLGA STURGEON, (IN RUSSIAN),
Kaspiiskii Nauchno-Issledovatelskii Institut Rybnogo Khozyaistva, Astrakhan (USSR).
P. N. Khoroshko.
Vopr Ikhtiol. Vol 12, No 2, p 388-391. 1972.
Identifiers: Reproduction, *Sturgeon, USSR, *Volga River, *Water divider, Spawning.

A water divider was constructed to improve spawning conditions for fish after the Volgograd Hydroelectric Power Station was built, but the measure was detrimental to sturgeon. After the power station was constructed, 50% of the natural reserves of Volga-Caspian sturgeon were lost, and the damage increased after construction of the water divider. The deterioration resulted from the nature of flooding, abnormal wintering conditions, silting of spawning grounds, alteration of benthic biocoenoses and loss of more spawning grounds. The construction of artificial beds on the

Available from the National Technical Information Service as COM-72-11116, $3.00 in paper copy, $0.95 in microfiche. University of Southern California Sea Grant Publication No. USC-SG-2-72. January 1972. 146 p, 1 append. SG 2-35227.

Descriptors: *Coastal areas, *Information exchange, Bibliographies, Publications, Environmental effects, California.
Identifiers: *Coastal zone, Los Angeles County, Orange County.

This survey is a preliminary effort at cataloguing governmental sources of information dealing with the coastal zone of Southern California. Governmental agencies are potentially a rich source of information, but the documents and reports are not readily accessible because of the wide dispersion of agencies and a lack of systematic listings. This report, though limited in scope, is an attempt to provide an inventory to aid researchers in identifying and locating many of these source materials. The main part of the survey is a bibliography of government reports arranged according to subject. Other sections include a survey of state agency activities, a listing of selected Los Angeles County agencies, plans and activities of the seventy-seven cities within Los Angeles County, plans and activities in Orange County, major coastal zone plans and studies in Southern California, and libraries and other information sources in Southern California. The appendix includes state legislation affecting the California coast in 1971. (Elfers-North Carolina)
W73-09823

10C. Secondary Publication And Distribution

CLADOPHORA AS RELATED TO POLLUTION AND EUTROPHICATION IN WESTERN LAKE ERIE,
Ohio State Univ., Columbus. Dept. of Botany.
For primary bibliographic entry see Field 05C.
W73-09757

POLLUTION RELATED STRUCTURAL AND FUNCTIONAL CHANGES IN AQUATIC COMMUNITIES WITH EMPHASIS ON FRESHWATER ALGAE AND PROTOZOA,
Virginia Polytechnic Inst. and State Univ., Blacksburg. Dept. of Biology.
For primary bibliographic entry see Field 05C.
W73-09875

NEW METHODS OF SOIL RECLAMATION IN ARID LANDS, (IN RUSSIAN),
For primary bibliographic entry see Field 03F.
W73-09981

GEOMETRY OF SANDSTONE RESERVOIR BODIES,
Shell Oil Co., Houston, Tex.
For primary bibliographic entry see Field 02F.
W73-10026

A REVIEW OF SOME MATHEMATICAL MODELS USED IN HYDROLOGY, WITH OBSERVATIONS ON THEIR CALIBRATION AND USE,
Institute of Hydrology, Wallingford (England).
For primary bibliographic entry see Field 02A.
W73-10061

GEOTHERMAL WATER AND POWER RESOURCE EXPLORATION AND DEVELOPMENT FOR IDAHO,
Boise State Coll., Idaho. Dept. of Geology.
For primary bibliographic entry see Field 06B.
W73-10217

POSSIBLE EFFECTS OF THERMAL EFFLUENTS ON FISH: A REVIEW,
Washington, Univ., Seattle. Coll. of Fisheries.
For primary bibliographic entry see Field 05C.
W73-10238

FISH PROTEIN CONCENTRATE: A REVIEW OF PERTINENT LITERATURE WITH AN EMPHASIS ON PRODUCTION FROM FRESHWATER FISH,
State Univ., Coll., Buffalo, N.Y. Great Lakes Lab.
For primary bibliographic entry see Field 03E.
W73-10313

BENEFIT-COST ANALYSIS FOR WATER RESOURCE PROJECTS,
Tennessee Valley Authority, Muscle Shoals, Ala.
For primary bibliographic entry see Field 06B.
W73-10318

A REVIEW OF METHODS FOR PREDICTING AIR POLLUTION DISPERSION,
National Aeronautics and Space Administration, Hampton, Va. Langley Research Center. National Aeronautics and Space Administration, Langley Station, Va. Langley Research Center.
For primary bibliographic entry see Field 05A.
W73-10324

10F. Preparation of Reviews

COOLING PONDS - A SURVEY OF THE STATE OF THE ART,
Hanford Engineering Development Lab., Richland, Wash.
For primary bibliographic entry see Field 05D.
W73-09849

POLYCHLORINATED BIPHENYLS-ENVIRONMENTAL IMPACT. A REVIEW BY THE PANEL ON HAZARDOUS TRACE SUBSTANCES. MARCH 1972,
Minnesota Univ., St. Paul. Coll. of Veterinary Medicine.
For primary bibliographic entry see Field 05B.
W73-10221

SUBJECT INDEX

AUTHOR INDEX

ORGANIZATIONAL INDEX

CAIRO UNIV., GIZA (EGYPT). DEPT. OF BOTANY.
Ecophysiological Studies on Desert Plants: VII. Water Relations of Leptadenia pyrotechnica (Porsk.) Decne. Growing in the Egyptian Desert,
W73-10364 2I

CALIFORNIA UNIV., BERKELEY.
Hydrodynamics of Fluid Injection,
W73-10032 5B

CALIFORNIA UNIV., BERKELEY. COLL. OF ENGINEERING.
Vertical Turbulent Mixing in Stratified Flow - A Comparison of Previous Experiments,
W73-09852 8B

CALIFORNIA UNIV., DAVIS.
Water-Table Fluctuation in Semipervious Stream-Unconfined Aquifer Systems,
W73-10058 2F

CALIFORNIA UNIV., IRVINE.
Expected Number and Magnitudes of Stream Networks in Random Drainage Patterns,
W73-10335 4A

CALIFORNIA UNIV., LIVERMORE. LAWRENCE LIVERMORE LAB.
Environmental Levels of Radioactivity in the Vicinity of the Lawrence Livermore Laboratory: 1972 Annual Report,
W73-10202 5B

CALIFORNIA UNIV., LOS ANGELES. SCHOOL OF ENGINEERING AND APPLIED SCIENCE.
Hemihydrate Scaling Threshold Enhancement by Magnesium Ion Augmentation,
W73-09789 3A

CALIFORNIA UNIV., SAN DIEGO. INST. OF MARINE RESOURCES.
The Effects of Suspended 'Red Mud' on Mortality, Body Weight, and Growth of the Marine Planktonic Copepod, Calanus helgolandicus,
W73-10127 5C

CALIFORNIA UNIV., SAN DIEGO, LA JOLLA. INST. OF MARINE RESOURCES.
Research on the Marine Food Chain, Progress Report July 1971-June 1972.
W73-10200 5C

CANADIAN WILDLIFE SERVICE, OTTAWA (ONTARIO).
Halogenated Hydrocarbons,
W73-09916 5A

CENTRAL RESEARCH INST. OF ELECTRIC POWER INDUSTRY, TOKYO (JAPAN).
Flow Measurement of Pump-Turbines During Transient Operation,
W73-09771 8B

CENTRE DE'ETUDES ET DE RECHERCHES DE BIOLOGIE ET D'OCEANOGRAPHIE MEDICALE, NICE (FRANCE).
Effects of Chemical Pollution with Respect to Telemediators Occurring in Microbiological and Planktonic Ecology in the Marine Environment, (Effets Des Pollutions Chimiques Vis-A-Vis De Telemediateurs Intervenant Dan s L'Ecologie Microbiologique Et Planctonique En Milieu Marin),
W73-10095 5C

CENTRE D'ETUDE DE L'ENERGIE NUCLEAIRE, GRENOBLE (FRANCE).
A New Method of Detecting Leaks in Reservoirs or Canals Using Labelled Bitumen Emulsions,
W73-09770 8G

CENTRE D'ETUDE DE L'ENERGIE NUCLEAIRE, MOL (BELGIUM). LABORATOIRES.
Behaviour of Ru in an Established Pasture Soil and its Uptake by Grasses,
W73-10192 5B

Distribution of Radiostrontium and Radiocaesium in the Organic and Mineral Fractions of Pasture Soils and Their Subsequent Transfer to Grasses,
W73-10193 5B

CENTRE D'ETUDE ET DE RECHERCHES DE BIOLOGIE ET D'OCEANOGRAPHIE MEDICALE, NICE (FRANCE).
Chemical Pollution and Marine Trophodynamic Chains, (Pollutions Chimiques et Chaines Trophodynamiques Marines),
W73-10091 5C

Study of the Toxicity of Effluents from Paper Mills in the Marine Environment, (Etude de la Toxicite d'Effluents de Papeterie en Milieu Marin),
W73-10093 5C

The Marine Environment and Organic Matter, (Le Milieu Marin Et Les Matieres Organiques),
W73-10097 5C

CENTRE D'ETUDES ET DE RECHERCHES DE BIOLOGIE ET D'OCEANOGRAPHIE MEDICALE, NICE (FRANCE).
Utilization of a Trophodynamic Chain of a Pelagic Type for the Study of Transfer of Metallic Pollution, (Utilisation D'une Chaine Trophodynamique De Type Pelagique Pour L'etude des Pollutions Metalliques),
W73-10092 5C

Restructuring of Shores and Secondary Pollution. Study of the Eutrophication of Harbor Zones. (Restructuration des Rivages et Pollutions Secondaires. Etude de l'Eutrophisation de Zones Portuaires),
W73-10120 5C

Antagonisms and Synergies in the Antibiosis Among Certain Marine Bacteria. Role of Certain Respiratory Enzymes in These Phenomena. (Antagonismes et Synergies Dans L'Antibiose Chez Certaines Bacteries Marines Role de Certains Enzymes Respiratoires Dans Ces Phenomenes),
W73-10121 5C

Influence of Aerosols in the Marine Environment (Influence Des Aerosols Marins Dans L'Environment),
W73-10136 5B

CENTRE TECHNIQUE FORESTIER TROPICAL, NOGENT-SUR-MARNE (FRANCE).
French Speaking Africa South of Sahara and Madagascar, Inland Water Production, (In French),
W73-10372 2I

CENTRE UNIVERSITAIRE DE LUMINY, MARSEILLE (FRANCE). LABORATOIRE D'HYDROBIOLOGIE MARINE.
Ecology of a Eutrophic Environment: Mathematical Analysis of Data, (In French),
W73-10011 5C

CHANCELLOR COLL., LIMBE (MALAWI).
A Note on Lake Chilwa Cyclic Changes,
W73-09844 2H

CHILE UNIV., SANTIAGO.
Reactions of the Monocalcium Phosphate Monohydrate and Anhydrous Diacalcium Phosphate of Volcanic Soils, (In Spanish),
W73-09982 2G

CHRISTIAN BROS. COLL., MEMPHIS, TENN.
Effects of Industrial Effluents on Primary Phytoplankton Indicators,
W73-09873 5C

CLEMSON UNIV., S.C.
Evaluation of Recreational Benefits Accruing to Recreators on Federal Water Projects--A Review Article,
W73-10316 6B

CLEMSON UNIV., S.C. DEPT. OF CIVIL ENGINEERING.
Coastal Zone Pollution Management. A Summary,
W73-10111 5G

COAST GUARD, WASHINGTON, D.C.
Coast Guard Anti-Pollution Engineering,
W73-10122 5G

COMMISSARIAT A L'ENERGIE ATOMIQUE, CADARACHE (FRANCE). CENTRE D'ETUDES NUCLEAIRES.
Investigation of the Organic-Mineral Complexes of Iron in the Biologic Cycle of Plants,
W73-10084 2I

COMMISSARIAT A L'ENERGIE ATOMIQUE, FONTENAY-AUX-ROSES (FRANCE). CENTRE D'ETUDES NUCLEAIRES.
A Comparative Study of Two Methods of Estimating Human Internal Exposure Hazards Resulting from the Ingestion of Sea Foodstuffs,
W73-10208 5C

COMMISSARIAT A L'ENERGIE ATOMIQUE, SACLAY (FRANCE). CENTRE D'ETUDES NUCLEAIRES.
Potentialities and Limits of Multielement Neutron Activation Analysis of Biological Specimens with or without Separation from an Active Matrice, (Possibilities Et Limites De l'Analyse Par Activation Neutronique Multiele mentaire D'Echantillons Biologiques Avec ou Sans Separation de la Matrice Activable.),
W73-10184 5A

COMMONWEALTH EDISON CO., CHICAGO, ILL.
Transmission Interchange Capability--Analysis by Computer,
W73-09764 8C

COPENHAGEN UNIV. (DENMARK). INST. OF PHYSICAL OCEANOGRAPHY.
Suspended Solids and Turbidity,
W73-09921 5A

CORPS OF ENGINEERS, ATLANTA, GA. SOUTH ATLANTIC DIV.
Activities of the Corps of Engineers in Marine Pollution Control,
W73-10103 5G

CORPS OF ENGINEERS, DALLAS, TEX. SOUTHWESTERN DIV.
Restoration of Historic Low Flow Diversion to Arroyo Colorado; Lower Rio Grande Flood Control Project, Texas (Final Environmental Impact Statement).
W73-10264 4A

CORPS OF ENGINEERS, PORTLAND, OREG. NORTH PACIFIC DIV.
Hydro-Power System Simulation Technique,
W73-10019 4A

CORSON (G. AND W. H.), INC., PLYMOUTH MEETING, PA. (ASSIGNEE).
Process for Treating Aqueous Chemical Waste Sludges and Composition Produced Thereby,
W73-09932 5D

COUNCIL ON ENVIRONMENTAL QUALITY, WASHINGTON, D.C.
A Study of the Economic Impact on the Steel Industry of the Costs of Meeting Federal Air and Water Pollution Abatement Requirements, Part 2.
W73-10312 5G

DALHOUSIE UNIV., HALIFAX (NOVA SCOTIA). DEPT. OF BIOLOGY.
Temperature Adaptation of Embryonic Development Rate Among Frogs,
W73-09858 5C

DAMJANICH MUZEUM, SZOLNOK (HUNGARY). LABORATORIUM TISZA-FORSCH.
Contributions to the Knowledge of the Algae of the Alkaline ('Szik') Waters of Hungary. III. The Phytoseston of the Alkaline Ponds Near Kunfeherto,
W73-10397 2H

DELAWARE UNIV., NEWARK. DEPT. OF GEOLOGY.
Trace Metals in Cores From the Great Marsh, Lewes, Delaware,
W73-10141 5A

Trace Metal Environments Near Shell Banks in Delaware Bay,
W73-10250 5B

DENVER WATER BOARD, COLO.
A Quality Control Study for a New Watershed,
W73-09815 5G

DEPARTMENT OF ENERGY, MINES AND RESOURCES, BURLINGTON (ONTARIO). CANADA CENTER FOR INLAND WATERS.
The Environmental Matrix: Input-Output Techniques Applied to Pollution Problems in Ontario,
W73-10020 5G

DEPARTMENT OF PUBLIC HEALTH ENGINEERING, LONDON (ENGLAND).
Current Proposals for the Thames Barrier and the Organization of the Investigations,
W73-10000 8A

DEPARTMENT OF THE ENVIRONMENT (ALBERTA). INLAND WATERS BRANCH.
Groundwater Flow Systems, Past and Present,
W73-09965 2F

DEPARTMENT OF THE ENVIRONMENT, OTTAWA (ONTARIO). INLAND WATERS BRANCH.
Determination of Hydrogeological Properties of Fissured Rocks,
W73-09954 2J

DEPARTMENT OF THE NAVY, WASHINGTON, D.C.
Collection and Disposal of Ship's Sewage,
W73-09931 5G

DEUTSCHE AKADEMIE DER LANDWIRSCHAFTSWISSENSCHAFTEN ZU BERLIN (EAST GERMANY). INSTITUT FUER PFLANZENZUECHTUNG.
Automatic Wick Irrigation of Pot Experiments, (In German),
W73-10073 3F

DEUTSCHE AKADEMIE DER LANDWIRTSCHAFTSWISSENCHAFTEN ZU BERLIN, MUENCHBERG (EAST GERMANY). FORSCHUNGSZENTRUM BODENFRUCHTBARKEIT.
Modes of Action and Methods of Application of Soil Conditioners for Improving Physical Properties of the Soil, (In German),
W73-09985 2G

DEUTSCHE AKADEMIE DER LANDWIRTSCHAFTSWISSENSCHAFTEN ZU BERLIN, LEIPZIG (EAST GERMANY). INSTITUT FUER MINERALDUENG.
The Effectivity of Nitrogen Fertilization Under Different Conditions of Production, (In German),
W73-09986 3F

Considering the Site and the Reached Level of Fertilization when Calculating the Optimal Nitrogen Fertilization of Grain Crops in the Model for Fertilization Systems, (In German),
W73-10074 3F

DIRECTORATE OF LICENSING (AEC), WASHINGTON, D.C.
Final Environmental Statement Related to Operation of Salem Nuclear Generating Station Units 1 and 2.
W73-10177 5B

Final Environmental Statement Related to Operation of Peach Bottom Atomic Power Station Units 2 and 3.
W73-10178 5B

Final Environmental Statement Related to the Proposed San Onofre Nuclear Generating Station - Units 2 and 3.
W73-10179 5B

Final Environmental Statement Related to Continuation of Construction and Operation of Units 1 and 2 and Construction of Units 3 and 4 of North Anna Power Station.
W73-10205 5B

DNEPROPETROVSKII GOSUDARSTVENNYI UNIVERSITET (USSR).
Phenol-and Rhodane Destructing Microflora as a Factor for Natural Purification of the Dnieper Reservoirs, (In Russian),
W73-10394 5C

DREW CHEMICAL CORP., PARSIPPANY, N.J.
Cooling Water,
W73-09861 5D

DSUTSCHE AKADEMIE DER LANDWIRTSCHAFTSWISSENSCHAFTEN ZU BERLIN, MUENCHEBERG (EAST GERMANY). FORSCHUNGSZENTRUM BODENFRUCHTBARKEIT.
On the Hydrophobization of a Soil Layer With the View of Interrupting Capillary Water Movement, (In German),
W73-09984 2D

DUNCAN AND JONES, BERKELEY, CALIF.; AND YODER-TROTTER-ORLOB AND ASSOCIATES, BERKELEY, CALIF.
Initial Water, Sewerage and Flood Control Plan,
W73-09828 6D

DURHAM UNIV. (ENGLAND). DEPT. OF BOTANY.
Changes in the Ecological Balance of Invertebrate Communities in Kelp Holdfast Habitats of Some Polluted North Sea Waters,
W73-10123 5C

EAST AFRICAN FRESHWATER FISHERIES RESEARCH ORGANIZATION, JINJA (UGANDA).
African Journal of Tropical Hydrobiology and Fisheries.
W73-09903 2I

EAST-WEST GATEWAY COORDINATING COUNCIL, ST. LOUIS, MO.
St. Louis Region Water and Sewerage Facilities Supplement Number 1 - Programming.
W73-09826 5E

ELECTRICITE DE FRANCE, (CHATOU). SERVICE ETUDES ET PROJETS THERMIQUES ET NUCLEAIRES.
The Effect of Marine Fouling on Corrosion of Metals Anti-Corrosion Techniques (Influence de la Salissure Marine sur la Corrosion des Metaux et Moyens de Prevention),
W73-10146 8G

ENGINEERING-SCIENCE, INC., BERKELEY, CALIF.
Oceanographic Surveys for Design of Pollution Control Facilities,
W73-10107 5E

ENGINEERING-SCIENCE, INC., PASADENA, CALIF.
Trends in Marine Waste Disposal, Good and Bad,
W73-10105 5E

ENVIRONMENTAL PROTECTION AGENCY, ATHENS, GA. SOUTHEAST ENVIRONMENTAL RESEARCH LAB.
Estuarine Surveys in the Southeast,
W73-10106 5B

ENVIRONMENTAL PROTECTION AGENCY, ATHENS, GA. SOUTHEAST WATER LAB.
Report on Pollution of the St. Marys and Amelia Estuaries, Georgia-Florida,
W73-09788 5B

ENVIRONMENTAL PROTECTION AGENCY, ATLANTA, GA. REGION IV.
Regional Coastal Pollution Control,
W73-10101 5G

ENVIRONMENTAL PROTECTION AGENCY, GULF BREEZE, FLA. GULF BREEZE LAB.
Test, Monitoring and Indicator Organisms,
W73-09923 5A

ENVIRONMENTAL PROTECTION AGENCY, WASHINGTON, D.C. OFFICE OF RESEARCH AND MONITORING.
Upper Potomac Estuary Eutrophication Control Requirements,
W73-10247 5C

ENVIRONMENTAL PROTECTION AGENCY, WASHINGTON, D.C. OFFICE OF WATER PROGRAMS.
National Coastal Pollution Program,
W73-10100 5G

Reaeration Capacity of the Flint, South and Patuxent Rivers,
W73-10343 5B

Relationships between Hydraulic Properties and Reaeration,
W73-10352 5G

GEORGIA UNIV., SAPELO ISLAND. MARINE INST.
Physical and Chemical Properties of the Coastal Waters of Georgia,
W73-09752 2L

GHENT RIJKSUNIVERSITEIT (BELGIUM).
Faunistic Ecological Study of Creeks of East. Flanders and Those Along Scheldt River: Considerations on Their Chemistry Plantonic, Entomological, and Malacological Fauna and a Discussion of Their Current Biological Status, (In French),
W73-10086 2E

GHENT RIJKSUNIVERSITEIT (BELGIUM). DEPT. OF AGRICULTURAL SCIENCE.
Study on the Sensitivity of Soil Moisture Measurements Using Radiation Techniques, Nylon Resistance Units and a Precise Lysimeter System,
W73-10194 2G

GOETEBORG UNIV. (SWEDEN). DEPT. OF ANALYTICAL CHEMISTRY.
Inorganic Chemicals,
W73-09918 5A

GOETTINGEN UNIV. (WEST GERMANY).
Systems Analysis of Mineral Cycling in Forest Ecosystems,
W73-10188 2K

GORKOVSKII INZHENERNO-STROITELNYI INSTITUT (USSR).
Formation of Heat-Cracks in Ice Cover of Rivers and Bodies of Water (Usloviya obrazovaniya termicheskhikh treshchin v ledyanom pokrove rek i vodoyemov),
W73-10049 2C

GOTEBORG UNIV. (SWEDEN). DEPT. OF ANALYTICAL CHEMISTRY.
The Changing Chemistry of the Oceans,
W73-10125 5B

GRENOBLE UNIV. (FRANCE).
Method of Evaluating Water Balance in Situ on the Basis of Water Content Measurements and Suction Effects (Methode d'Evaluation du Bilan Hydrique in Situ a Partir de la Mesure des Teneurs en eau et des Succions),
W73-10197 2A

HAILE SELLASSIE I UNIV., ALEM MAYA (ETHIOPIA). COLL. OF AGRICULTURE.
Movement of Zn65 into Soils from Surface Application of Zn Sulphate in Columns,
W73-10191 2G

HANFORD ENGINEERING DEVELOPMENT LAB., RICHLAND, WASH.
Thermal Effects of Projected Power Growth: South Atlantic and East Gulf Coast River Basins,
W73-09848 5C

Cooling Ponds - A Survey of the State of the Art,
W73-09849 5D

HAWAII INST. OF GEOPHYSICS, HONOLULU.
Kinetics of Carbonate-Seawater Interactions,
W73-09790 2K

HAWAII UNIV., HONOLULU.
Erosion and Accretion of Selected Hawaiian Beaches 1962-1972,
W73-09879 2J

HEIDELBERG UNIV. (WEST GERMANY). SEDIMENT RESEARCH LAB.
Formation and Diagenesis of Inorganic Ca Mg Carbonates in the Lacustrine Environment,
W73-10003 2H

HESSICHES LANDESAMT FUER BODENFORSCHUNG, WIESBADEN (WEST GERMANY).
Hydrogeologic Criteria for the Self-Purification of Polluted Groundwater,
W73-09971 5B

HIROSHIMA UNIV. (JAPAN). DEPT. OF FISHERIES.
Studies on the Coastal Oceanography in the Vicinity of Fukuyama, Hiroshima Pref., I. Distribution Patterns of Temperature, Chlorinity, pH and Inorganic Nutrient (Phosphate-P, Ammonia-N, Nitrite-N, Nitrate-N) Contents of Sea Water in Early February, 1968, (In Japanese),
W73-09899 5B

HORTICULTURAL RESEARCH INST., PRETORIA (SOUTH AFRICA).
Irrigation Trials with Onions,
W73-10139 3F

HUMBOLDT-UNIVERSITAET SUR BERLIN (EAST GERMANY). INSTITUT FUER BODENKUNDE UND PFLANZENERNAEHRUNG.
On the Effect of A Bituminous Mulch Cover on Water Infiltration, Air Composition and Condensation Processes in the Soil, (In German),
W73-09987 2G

HYDRAULICS RESEARCH STATION, WALLINGFORD (ENGLAND).
Siltation Problems in Relation to the Thames Barrier,
W73-09997 8A

IBM WATSON RESEARCH CENTER, YORKTOWN HEIGHTS, N.Y.
An Investigation into the Longterm Persistence of U. S. Hydrologic Records,
W73-09869 4A

Regionalization of Hydrogeologic Parameters for Use in Mathematical Models of Groundwater Flow,
W73-09961 2F

ILLINOIS STATE WATER SURVEY, URBANA.
Aquifer Simulation Model for Use on Disk Supported Small Computer Systems,
W73-10328 2F

ILLINOIS STATE WATER SURVEY, URBANA. ATMOSPHERIC SCIENCE SECTION.
Feasibility Studies of Precipitation Enhancement in Illinois,
W73-09782 3B

ILLINOIS UNIV., CHICAGO. DEPT. OF BIOLOGICAL SCIENCES.
A Critical Test of Methods for Isolation of Viruses for Use in Control of Nuisance Algae,
W73-09753 5A

ILLINOIS UNIV., CHICAGO. DEPT. OF MATERIALS ENGINEERING.
Nutrient Transport by Sediment-Water Interaction,
W73-09754 5C

IMPERIAL COLL. OF SCIENCE AND TECHNOLOGY, LONDON (ENGLAND). DEPT. OF BOTANY.
Growth and Water Use of Lolium perenne: II. Plant Growth,
W73-09974 2I

INSTITUT GIGIENY TRUDA I PROFESSIONALNYKH ZABOLEVANII, MOSCOW (USSR).
Hygienic Substantiation of Permissible Humidity Level in Case of High Temperature, (In Russian),
W73-10390 5G

INSTITUT NATIONAL DE LA RECHERCHE AGRONOMIQUE, VERSAILLES (FRANCE).
Textural Aspect of Soil Porosity, (in French),
W73-10069 2G

INSTITUT PASTEUR, LILLE (FRANCE). LABORATOIRE D'HYDROBIOLOGIE.
Automation of the Bacteriological Analysis of Water, I, Study of a New Specific Test for Fecal Contamination and the Optimal Conditions for its Determination,
W73-09772 5A

INSTITUT PO FURAZHITE, PLEVEN (BULGARIA).
Influence of the Nutrient Solution Concentration Upon the Water Absorption of Pea Plants, (In Bulgarian),
W73-09978 3F

INSTITUT SCIENTIFIQUE ET TECHNIQUE DES PECHES MARITIMES, NANTES (FRANCE).
Choice of Products for Control Against Pollution of the Marine Environment by Hydrocarbons. I. Determining Some Choices, (Choix de Produits Pour Lutter Contre la Pollution Du Milieu Marin Par Les Hydrocarbures. I. Motif Du Choix),
W73-10113 5G

Choice of Products for Control Against Pollution of the Marine Environment by Hydrocarbons. II. Efficacy of Antipetroleum Products, (Choix de Produits Pour Lutter Contre La Pollution du Milieu Marin Les Hydrocarbures. II. Efficacite des Produits Antipetrole),
W73-10114 5G

Choice of Products for Control Against Pollution of the Marine Environment by Hydrocarbons. III. The Relative Toxicity of Antipetroleum Products on Two Marine Organisms, (Choix De Produits Pour Lutter Contre La Pollution Du Milieu Marin Par Les Hydrocarbures. III. Toxicite Relative De Produits Antipetrole Sur Deux Organismes Marins),
W73-10115 5C

Choice of Products for Control Against Pollution of the Marine Environment by Hydrocarbons. IV. The Relative Toxicity of Seven Antipetroleum Emulsion Products, (Choix De Produits Pour Lutter Contre La Pollution Du Milieu Marin Par Les Hydrocarbures. IV. Toxicite Relative De Sept Produits Emulsionnants Antipetrole),
W73-10116 5C

INSTITUT TECHNIQUE DES CEREALES ET FOURRAGES, PARIS (FRANCE).
Optimisation of Sampling for Determining Water Balances in Land under Crops by Means of Neutron Moisture Meters (Optimisation de l'Echantillonnage pour Determiner les Bilans

LOUISIANA STATE UNIV., NEW ORLEANS. URBAN STUDIES INST.
Lakefront New Orleans, Planning and Development, 19261971,
W73-09822 6A

MAGYAR TUDOMANYOS AKADEMIA, TIHANY. BIOLOGICAL RESEARCH INST.
Investigations on Crustacean and Molluscan Remains in the Upper Sedimentary Layer of Lake Balaton,
W73-10005 2H

MAINE UNIV., ORONO. DEPT. OF ELECTRICAL ENGINEERING.
A System for Remote Telemetry of Water Quality Data,
W73-09946 7B

MAINE UNIV., PORTLAND. SCHOOL OF LAW.
Coastal Zone Management: Intergovernmental Relations and Legal Limitations,
W73-10254 6E

MANHATTAN COLL., BRONX, N.Y.
Applications of Estuarine Water Quality Models,
W73-10108 5B

MARIE CURIE-SKLODOWSKA UNIV., LUBLIN (POLAND). NAUK ZIEMI.
Vegetation of the Reserve 'Peat Bog at Czarne Sosnowickie Lake', (In Polish),
W73-10398 2H

MARINE BIOLOGICAL ASSOCIATION OF THE UNITED KINGDOM, PLYMOUTH (ENGLAND). LAB.
Disposal of China Clay Waste in the English Channel,
W73-10145 5B

MARYLAND UNIV., COLLEGE PARK. DEPT. OF CIVIL ENGINEERING.
Efficiency of Partially Penetrating Wells,
W73-10336 4B

MARYLAND UNIV., PRINCE FREDERICK, HALLOWING POINT FIELD STATION.
Current Status of the Knowledge of the Biological Effects of Suspended and Deposited Sediments in Chesapeake Bay,
W73-10234 5C

(Biota of the Chesapeake Bay), Conclusions and Recommendations,
W73-10236 5B

MARYLAND UNIV., SOLOMONS. NATURAL RESOURCES INST.
Current Status of Knowledge Concerning the Cause and Biological Effects of Eutrophication in Chesapeake Bay,
W73-10233 5C

MASSACHUSETTS INST. OF TECH., CAMBRIDGE.
Seismic Risk and Design Response Spectra,
W73-09761 8G

Numerical Model for the Prediction of Transient Water Quality in Estuary Networks,
W73-10112 5B

MASSACHUSETTS INST. OF TECH., CAMBRIDGE. DEPT. OF CIVIL ENGINEERING.
A Marginal Analysis - System Simulation Technique to Formulate Improved Water Resources Configurations to Meet Multiple Objectives,
W73-09832 6A

Estimation Theory Applications to Design of Water Quality Monitoring Systems,
W73-09837 5G

Submerged Diffusers in Shallow Coastal Waters,
W73-10109 5G

MASSACHUSETTS INST. OF TECH., CAMBRIDGE. DEPT. OF MECHANICAL ENGINEERING.
Marine Pollution, Concentrating on the Effects of Hydrocarbons in Seawater,
W73-10144 5B

MASSACHUSETTS INST. OF TECH., CAMBRIDGE. FLUID MECHANICS LAB.
Effects of Oil Under Sea Ice,
W73-10246 5B

MASSACHUSETTS INST. OF TECH., CAMBRIDGE. RALPH M. PARSONS LAB. FOR WATER RESOURCES AND HYDRODYNAMICS.
Time Periods and Parameters in Mathematical Programming for River Basins,
W73-10317 4A

MASSACHUSETTS UNIV., AMHERST. DEPT. OF AGRICULTURAL AND FOOD ECONOMICS.
Investment Sequencing Recognizing Externalities in Water Desalting,
W73-10018 3A

MAX-PLANCK-INSTITUT FUER LIMNOLOGIE, SCHLITZ (WEST GERMANY). LIMNOLOGISCHE FLUSSSTATION.
Methods of Measuring Drift for Production Biology Studies in Running Waters, (in German),
W73-10065 2I

MAX-PLANCK-INSTITUT FUER LIMNOLOGIE, SCHLITZ (WEST GERMANY). LIMNOLOGISCHE FLUSSTATION.
Emergency Control, A New Method for Productivity Studies in Running Waters, (In German),
W73-10078 5G

MCGILL UNIV., MONTREAL (QUEBEC). DEPT. OF METEOROLOGY.
Evaluation of North Water Spring Ice Cover From Satellite Photographs,
W73-09800 2C

MCMASTER UNIV., HAMILTON (ONTARIO). DEPT. OF GEOLOGY.
Cobequid Bay Sedimentology Project: A Progress Report,
W73-09796 2L

MEMORIAL UNIV. OF NEWFOUNDLAND, ST. JOHN'S. MARINE SCIENCES RESEARCH LAB.
Metabolic Rate Independent of Temperature in Mollusc Tissue,
W73-09863 5C

METCALF AND EDDY, INC., BOSTON, MASS.
Preliminary Water and Waste Management Plan.
W73-09813 6A

Cost Apportionment Planning Report.
W73-09827 6C

METEOROLOGY RESEARCH, INC., ALTADENA, CALIF.
Areas of Needed Emphasis to Bring Weather Modification to an Accountable Operational State,
W73-09776 3B

The San Angelo Cumulus Project,
W73-09778 3B

MICHIGAN UNIV., ANN ARBOR. SCHOOL OF NATURAL RESOURCES.
Regional Benefits in Federal Project Evaluations,
W73-09830 6B

MILAN UNIV. (ITALY). LABORATORIO DI ZOOLOGIA.
Study of the Biocenosis of the Bottom of the River Po at Trino Vercellese in the Years 1967-68 (In Italian),
W73-09912 2I

MINISTRY OF AGRICULTURE AND LANDS, KUALA LUMPUR (MALAYSIA). DIV. OF AGRICULTURE.
Studies on the Reclamation of Acid Swamp Soils,
W73-10017 2G

MINISTRY OF AGRICULTURE AND WATER, RIYADH (SAUDI ARABIA).
Observations on Mesozoic Sandstone Aquifers in Saudi Arabia,
W73-09949 2F

MINISTRY OF AGRICULTURE, FISHERIES AND FOOD, LOWESTOFT (ENGLAND). FISHERIES LAB.
Design of a World Monitoring System,
W73-09924 5A

MINISTRY OF AGRICULTURE, FISHERIES AND FOOD, LOWESTOFT (ENGLAND). FISHERIES RADIOBIOLOGICAL LAB.
Artificial Radioactivity in Fresh Water and Estuarine Systems,
W73-09893 5B

Radioactivity,
W73-09922 5A

The Impact of Artificial Radioactivity on the Oceans and on Oceanography,
W73-10226 5B

MINISTRY OF AGRICULTURE, TEL-AVIV (ISRAEL).
The Downy Mildew of Onion (Peronospora Destructor (Berk.) Casp.) in Israel,
W73-10083 3F

MINNESOTA MINING AND MANUFACTURING CO., ST. PAUL. (ASSIGNEE).
Process for Treating Waste Effluent,
W73-09926 5D

MINNESOTA UNIV., MINNEAPOLIS. DEPT. OF GEOLOGY AND GEOPHYSICS.
On the Geometrical Basis of Intrinsic Hydrogeologic Parameters in Porous Media Flow,
W73-09962 2F

MINNESOTA UNIV., ST. PAUL. COLL. OF VETERINARY MEDICINE.
Polychlorinated Biphenyls-Environmental Impact. A Review by the Panel on Hazardous Trace Substances. March 1972,
W73-10221 5B

MISSOURI CLEAN WATER COMMISSION, JEFFERSON CITY.
Storage Facilities (For Hazardous Substances) Regulation Presented at Hearing on March 21, 1973.
W73-10258 5G

ACCESSION NUMBER INDEX

ACCESSION NUMBER INDEX